8A (18)
					Helium 2 **He** 4.0026

3A (13)	4A (14)	5A (15)	6A (16)	7A (17)	
Boron 5 **B** 10.811	Carbon 6 **C** 12.011	Nitrogen 7 **N** 14.0067	Oxygen 8 **O** 15.9994	Fluorine 9 **F** 18.9984	Neon 10 **Ne** 20.1797
Aluminum 13 **Al** 26.9815	Silicon 14 **Si** 28.0855	Phosphorus 15 **P** 30.9738	Sulfur 16 **S** 32.066	Chlorine 17 **Cl** 35.4527	Argon 18 **Ar** 39.948

2B (12)							
Zinc 30 **Zn** 65.39	Gallium 31 **Ga** 69.723	Germanium 32 **Ge** 72.61	Arsenic 33 **As** 74.9216	Selenium 34 **Se** 78.96	Bromine 35 **Br** 79.904	Krypton 36 **Kr** 83.80
Cadmium 48 **Cd** 112.411	Indium 49 **In** 114.818	Tin 50 **Sn** 118.710	Antimony 51 **Sb** 121.760	Tellurium 52 **Te** 127.60	Iodine 53 **I** 126.9045	Xenon 54 **Xe** 131.29
Mercury 80 **Hg** 200.59	Thallium 81 **Tl** 204.3833	Lead 82 **Pb** 207.2	Bismuth 83 **Bi** 208.9804	Polonium 84 **Po** (208.98)	Astatine 85 **At** (209.99)	Radon 86 **Rn** (222.02)
— 112 — Discovered 1996	— 113 — Discovered 2004	— 114 — Discovered 1999	— 115 — Discovered 2004	— 116 — Discovered 1999		

Terbium 65 **Tb** 158.9253	Dysprosium 66 **Dy** 162.50	Holmium 67 **Ho** 164.9303	Erbium 68 **Er** 167.26	Thulium 69 **Tm** 168.9342	Ytterbium 70 **Yb** 173.04	Lutetium 71 **Lu** 174.967
Berkelium 97 **Bk** (247.07)	Californium 98 **Cf** (251.08)	Einsteinium 99 **Es** (252.08)	Fermium 100 **Fm** (257.10)	Mendelevium 101 **Md** (258.10)	Nobelium 102 **No** (259.10)	Lawrencium 103 **Lr** (262.11)

GENERAL Chemistry Now™

Completely integrated with this text!
http://now.brookscole.com/kotz6e

What do you need to learn now?

Take charge of your learning with General ChemistryNow™!

This powerful online learning companion helps you manage and make the most of your study time—and it's included with every new copy of this text! This collection of dynamic technology resources gauges your unique study needs to provide you with a *Personalized Learning Plan* that will enhance your problem-solving skills and understanding of core concepts. Designed to optimize your time investment, **General ChemistryNow**™ helps you succeed by focusing your study time on the concepts you need to master.

GENERAL Chemistry Now™

See the General ChemistryNow CD-ROM or website:
- **Screen 6.17 Product-Favored Systems,** for an exercise on the reaction when a Gummi Bear is placed in molten potassium chlorate

Look for these references in the text, such as this one from page 270. They direct you to the corresponding media-enhanced activities on **General ChemistryNow**. The text's *Chapter Goals Revisited* sections are also reinforced through the *Homework and Goals* found on **General ChemistryNow**.

This precise page-by-page integration enables you to go beyond reading about chemistry—you'll actually experience it in action!

Easy to use

The **General ChemistryNow** system includes two powerful assessment components:

- **WHAT DO I KNOW?**
This diagnostic *Exam-Prep Quiz*, based on the text's *Chapter Goals*, gives you an initial assessment of your understanding of core concepts.

- **WHAT DO I NEED TO LEARN?**
A *Personalized Learning Plan* outlines key elements for review.

With a click of the mouse, **General ChemistryNow's** unique activities allow you to:

- Create a *Personalized Learning Plan* or review for an exam using the *Exam-Prep Quiz* web quizzes

- Explore chemical concepts through tutorials, simulations, and animations

- View *Active Figures* and interact with text illustrations. These *Active Figures* will help you master key concepts from the book. Each figure is paired with corresponding questions to help you focus on chemistry at work to ensure that you understand the concepts played out in the animations.

By providing you with a better understanding of exactly what you need to focus on, **General ChemistryNow** saves you time and brings you a step closer to success!

**Make the most of your study time—
log on to *General ChemistryNow* today!**
http://now.brookscole.com/kotz6e

GENERAL
Chemistry◆Now™

http://chemistry.brookscole.com/kotz6e

The **Media Integration Guide** on the next several pages provides you with a grid that links each chapter to the wealth of interactive media resources you will find at **General ChemistryNow**, a unique web-based, assessment-centered personalized learning system for chemistry students.

Chapter	Exercises	Tutorials	Active Figures	Additional Resources
1 Matter and Measurement	• Screen 1.5: Mixtures and Pure Substances • Screen 1.12: Chemical Changes	• Screen 1.10: Density • Screen 1.15: Temperature • Screen 1.16: The Metric System • Screen 1.17: Using Numerical Information	• 1.1: Classifying Matter • 1.2: States of Matter–Solid, Liquid, and Gas • 1.3: Levels of Matter • 1.15: Comparison of Farenheit, Celsius, and Kelvin Scales	• Screen 1.6: Separation of Mixtures • Screen 1.7: Elements and Atoms • Screen 1.13: Chemical Change on the Molecular Scale
2 Atoms and Elements	• Screen 2.6: Electrons • Screen 2.8: Protons • Screen 2.10: The Nucleus of the Atom • Screen 2.16: The Periodic Table	• Screen 2.11: Summary of Atomic Composition • Screen 2.14: The Mole • Screen 2.15: Moles and Molar Mass of the Elements	• 2.3: Measuring the Electron's Charge to Mass Ratio • 2.6: Rutherford's Experiment to Determine the Structure of the Atom • 2.8: Mass Spectrometer • 2.10: Some of the 113 Known Elements	
3 Molecules, Ions, and Their Compounds	• Screen 3.13: Alkanes • Screen 3.19: Hydrated Compounds	• Screen 3.5: Ions • Screen 3.6: Polyatomic Ions • Screen 3.10: Naming Ionic Compounds • Screen 3.12: Binary Compounds of the Nonmetals • Screen 3.14: Compounds, Molecules, and the Mole • Screen 3.15: Using Molar Mass • Screen 3.16: Percent Composition • Screen 3.17: Determining Empirical Formulas • Screen 3.18: Determining Molecular Formulas • Screen 3.19: Hydrated Compounds	• 3.1: Reaction of the Elements Aluminum and Bromine • 3.4: Ways of Depicting the Methane (CH_4) Molecule • 3.6: Ions • 3.8: Common Ionic Compounds Based on Polyatomic Ions • 3.10: Coulomb's Law and Electrostatic Forces • 3.17: Dehydrating Hydrating Cobalt(II) Chloride, $CoCl_2 \cdot 6H_2O$	• Screen 3.8: Ionic Compounds • Screen 3.13: Alkanes • Screen 3.14: Compounds, Molecules, and the Mole
4 Chemical Equations and Stoichiometry	• Screen 4.3: The Law of Conservation of Mass • Screen 4.5: Weight Relations in Chemical Reactions • Screen 4.8: Limiting Reactants	• Screen 4.4: Balancing Chemical Equations • Screen 4.6: Calculations in Stoichiometry • Screen 4.9: Percent Yield	• 4.2: The Reaction of Iron and Chlorine • 4.4: Oxidation of Ammonia • 4.8: Analysis for the Sulfate Content of a Sample • 4.9: Combustion Analysis of a Hydrocarbon	• Screen 4.5 Weight Relations in Chemical Reactions • Screen 4.7: Reactions Controlled by the Supply of One Reactant • Screen 4.8: Limiting Reactants

Media Integration Guide

Chapter	Exercises	Tutorials	Active Figures	Additional Resources
5 **Reactions in Aqueous Solution**	• Screen 5.13: Oxidation Numbers • Screen 5.14: Recognizing Oxidation–Reduction Reactions • Screen 5.16: Preparing Solutions of Known Concentrations • Screen 5.18: Stoichiometry of Reactions in Solution	• Screen 5.4: Solubility of Ionic Compounds • Screen 5.7: Net Ionic Equations • Screen 5.11: Gas Forming Reactions • Screen 5.13: Oxidation Numbers • Screen 5.15: Solution Concentrations • Screen 5.16: Preparing Solutions of Known Concentrations • Screen 5.17: The pH Scale • Screen 5.19: Titration	• 5.2: Classifying Solutions by Their Ability to Conduct Electricity • 5.3: Guidelines to Predict the Solubility of Ionic Compounds • 5.8: An Acid–Base Reaction, HCl and NaOH • 5.14: The Reaction of Copper with Nitric Acid • 5.18: Making a Solution • 5.20: pH Values of Some Common Substances • 5.23: Titration of an Acid in Aqueous Solution with a Base	• Screen 5.2: Solutions • Screen 5.3: Compounds in Aqueous Solution • Screen 5.4: Solubility of Ionic Compounds • Screen 5.5: Types of Aqueous Solutions • Screen 5.8: Acids • Screen 5.9: Bases • Screen 5.11: Gas Forming Reactions
6 **Principles of Reactivity: Energy and Chemical Reactions**	• Screen 6.3: Forms of Energy • Screen 6.7: Heat Capacity of Pure Substances • Screen 6.10: Calculating Heat Transfer • Screen 6.15: Hess's Law • Screen 6.17: Product-Favored Systems	• Screen 6.5: Energy Units • Screen 6.10: Calculating Heat Transfer • Screen 6.13: Enthalpy Changes for Chemical Reactions • Screen 6.14: Measuring Heats of Reactions • Screen 6.16: Standard Enthalpy of Formation	• 6.3: Energy and its Conversion • 6.8: Exothermic and Endothermic Processes • 6.10: Heat Transfer • 6.11: Heat Transfer and the Temperature Change for Water • 6.12: Changes of State • 6.13: Energy Changes in a Physical Process • 6.15: The Exothermic Combustion of Hydrogen in Air • 6.17: Constant Volume Calorimeter • 6.18: Energy Level Diagrams	• Screen 6.4: Directionality of Heat Transfer • Screen 6.7: Heat Capacity of Pure Substances • Screen 6.10: Calculating Heat Transfer • Screen 6.11: The First Law of Thermodynamics • Screen 6.14: Measuring Heats of Reactions • Screen 6.15: Hess's Law
7 **Atomic Structure**	• Screen 7.5: Planck's Equation • Screen 7.6: Atomic Line Spectrum • Screen 7.13: Shapes of Atomic Orbitals	• Screen 7.3: Electromagnetic Radiation • Screen 7.6: Atomic Line Spectrum • Screen 7.8: Wave Properties of the Electron • Screen 7.12: Quantum Numbers and Orbitals	• 7.1: Electromagnetic Radiation • 7.3: The Electromagnetic Spectrum • 7.8: The Line Emission Spectrum of Hydrogen • 7.10: H Atom in the Bohr Model • 7.11: Absorption of Energy • 7.12: Electronic Transitions That Can Occur in an Excited H Atom • 7.13: Magnesium Oxide • 7.14: Different Views of a 1s ($n = 1$ and $\ell = 0$) Orbital • 7.15: Atomic Orbitals	• Screen 7.4: Electromagnetic Spectrum • Screen 7.5: Planck's Equation • Screen 7.6: Atomic Line Spectrum • Screen 7.9: Heisenberg's Uncertainty Principle

Media Integration Guide

Chapter	Exercises	Tutorials	Active Figures	Additional Resources
8 **Atomic Electron Configurations and Chemical Periodicity**	• Screen 8.6: Effective Nuclear Charge, Z^*	• Screen 8.7: Atomic Electron Configurations • Screen 8.8: Electron Configuration in Ions	• 8.2: Observing and Measuring Paramagnetism • 8.4: Experimentally Determined Order of Subshell Energies • 8.7: Electron Configurations and the Periodic Table • 8.9: Examples of the Periodicity of Group 1A and Group 7A Elements • 8.11: Atomic Radii in Picometers for Main Group Elements • 8.13: First Ionization Energies of the Main Group Elements of the First Four Periods • 8.14: Electron Affinity • 8.15: Relative Sizes of Some Common Ions	• Screen 8.3: Spinning Electrons and Magnetism • Screen 8.6: Effective Nuclear Charge, Z^* • Screen 8.7: Atomic Electron Configurations • Screen 8.8: Electron Configuration in Ions • Screen 8.9: Atomic Properties and Periodic Trends • Screen 8.10: Atomic Sizes • Screen 8.11: Ionization Energy • Screen 8.12: Electron Affinity • Screen 8.14: Ion Size • Screen 8.15: Chemical Reactions and Periodic Properties
9 **Bonding and Molecular Structure: Fundamental Concepts**	• Screen 9.8: Drawing Lewis Structures • Screen 9.14: Determining Molecular Shape	• Screen 9.7: Lewis Electron Dot Structures • Screen 9.8: Drawing Lewis Structures • Screen 9.9: Resonance Structures • Screen 9.10: Exceptions to the Octet Rule • Screen 9.13: Ideal Electron Repulsion Shapes • Screen 9.14: Determining Molecular Shape	• 9.3: Lattice Energy • 9.8: Various Geometries Predicted by VSEPR • 9.14: Electronegativity Values for the Elements According to Pauling • 9.16: Polarity of Triatomic Molecules, AB_2 • 9.17: Polar and Nonpolar Molecules of the Type AB_3	• Screen 9.2: Valence Electrons • Screen 9.4: Lattice Energy • Screen 9.5: Chemical Reactions and Periodic Properties • Screen 9.6: Chemical Bond Formation—Covalent Bonding • Screen 9.13: Ideal Electron Repulsion Shapes • Screen 9.16: Formal Charge • Screen 9.17: Bond Polarity and Electronegativity • Screen 9.18: Molecular Polarity • Screen 9.19: Bond Properties • Screen 9.20: Bond Energy and ΔH_{rxn}
10 **Bonding and Molecular Structure: Orbital Hybridization and Molecular Orbitals**	• Screen 10.8: Molecular Fluxionality • Screen 10.9: Molecular Orbital Theory • Screen 10.11: Homonuclear Diatomic Molecules	• Screen 10.5: Sigma Bonding • Screen 10.6: Determining Hybrid Orbitals • Screen 10.7: Multiple Bonding	• 10.1: Potential Energy Change During H—H Bond Formation • 10.5: Hybrid Orbitals for Two to Six Electron Pairs • 10.6: Bonding in the Methane (CH_4) Molecule • 10.10: The Valence Bond Model of Bonding in Ethylene, C_2H_4 • 10.13: Rotation Around Bonds • 10.22: Molecular Orbital Energy Level Diagram	• Screen 10.3: Valence Bond Theory • Screen 10.4: Hybrid Orbitals • Screen 10.10: Molecular Orbital Configurations

Media Integration Guide

Chapter	Exercises	Tutorials	Active Figures	Additional Resources
11 **Carbon: More Than Just Another Element**	• Screen 11.6: Functional Groups (1): Reactions of Alcohols	• Screen 11.4: Hydrocarbons and Addition Reactions • Screen 11.6: Functional Groups	• 11.2: Optical Isomers • 11.4: Alkanes • 11.7: Bacon Fat and Addition Reactions • 11.13: Polyethylene • 11.18: Nylon-6,6	• Screen 11.3: Hydrocarbons • Screen 11.4: Hydrocarbons and Addition Reactions • Screen 11.6: Functional Groups • Screens 11.9, 11.10: Synthetic Organic Polymers
12 **Gases & Their Properties**	• Screen 12.5: Gas Density • Screen 12.12: Application of the Kinetic-Molecular Theory: Diffusion	• Screen 12.6: Using Gas Laws: Determining Molar Mass • Screen 12.7: Gas Laws and Chemical Reactions: Stoichiometry • Screen 12.8: Gas Mixtures and Partial Pressures	• 12.4: An Experiment to Demonstrate Boyle's Law • 12.6: Charles's Law • 12.18: Gaseous Diffusion	• Screen 12.3: Gas Laws • Screen 12.4: The Ideal Gas Law • Screen 12.5: Gas Density • Screen 12.9: The Kinetic-Molecular Theory of Gases: Gases on the Molecular Scale • Screen 12.10: Gas Laws and Kinetic-Molecular Theory • Screen 12.11: Distribution of Molecular Speeds: Maxwell-Boltzmann Curves • Screen 12.12: Application of the Kinetic-Molecular Theory: Diffusion
13 **Intermolecular Forces, Liquids, and Solids**	• Screen 13.5: Intermolecular Forces (3) • Screen 13.17: Phase Changes	• Screen 13.5: Intermolecular Forces (3) • Screen 13.9: Properties of Liquids	• 13.2: Ion–Dipole Interactions • 13.8: The Boiling Points of Some Simple Hydrogen Compounds • 13.11: The Temperature Dependence of the Densities of Ice and Water • 13.17: Vapor Pressure • 13.18: Vapor Pressure Curves for Diethyl Ether [$(C_2H_5)_2O$], Ethanol (C_2H_5OH), and Water • 13.39: Phase Diagram for Water	• Screen 13.2: Phases of Matter • Screens 13.3, 13.4, 13.5: Intermolecular Forces • Screen 13.6: Hydrogen Bonding • Screen 13.7: The Weird Properties of Water • Screens 13.8, 13.9, 13.10, 13.11: Properties of Liquids • Screens 13.12, 13.13, 13.14, 13.15: Solid Structures • Screens 13.17: Phase Changes
14 **Solutions and Their Behavior**	• Screen 14.2: Solubility • Screen 14.5: Factors Affecting Solubility (1)—Henry's Law and Gas Pressure • Screens 14.7, 14.8: Colligative Properties	• Screens 14.5, 14.6: Factors Affecting Solubility • Screens 14.7, 14.8, 14.9: Colligative Properties	• 14.6: Solubility of Nonpolar Iodine in Polar Water and Nonpolar Carbon Tetrachloride • 14.9: Dissolving an Ionic Solid in Water	• Screen 14.3: The Solution Process: Intermolecular Forces • Screen 14.4: Energetics of Solution Formation—Dissolving Ionic Compounds • Screen 14.9: Colligative Properties
15 **Principles of Reactivity: Chemical Kinetics**	• Screen: 15.4 Concentration Dependence • Screen: 15.5 Determination of the Rate Equation (1) • Screen 15.12: Reaction Mechanisms • Screen 15.13: Reaction Mechanisms and Rate Equations • Screen 15.14: Catalysis and Reaction Rate	• Screen 15.4: Concentration Dependence • Screen 15.5: Determination of the Rate Equation (1) • Screen 15.6: Concentration–Time Relationships • Screen 15.7: Determination of Rate Equation (2) • Screen 15.8: Half-Life • Screen 15.10: Control of Reaction Rates (3)	• 15.2: A Plot of Reactant Concentration Versus Time for the Decomposition of N_2O_5 • 15.7: The Decomposition of H_2O_2 • 15.9: Half-Life of a First-Order Reaction • 15.13: Activation Energy • 15.14: Arrhenius Plot	• Screen 15.2: Rates of Chemical Reactions • Screens 15.3, 15.4, 15.10: Control of Reaction Rates • Screen 15.4: Concentration Dependence • Screen 15.5: Determination of the Rate Equation (1) • Screens 15.9, 15.10: Microscopic View of Reactions • Screen 15.14: Catalysis and Reaction Rate

Media Integration Guide

Chapter	Exercises	Tutorials	Active Figures	Additional Resources
16 **Principles of Reactivity: Chemical Equilibria**		• Screen 16.6: Writing Equilibrium Expressions • Screen 16.8: Determining an Equilibrium Constant • Screen 16.9: Systems at Equilibrium • Screen 16.10: Estimating Equilibrium Concentrations • Screens 16.12, 16.13: Disturbing a Chemical Equilibrium	• 16.3: The Reaction of H_2 and I_2 Reaches Equilibrium • 16.9: Changing Concentrations	• Screen 16.2: The Principle of Microscopic Reversibility • Screen 16.3: Equilibrium State • Screen 16.4: Equilibrium Constant • Screen 16.5: The Meaning of the Equilibrium Constant • Screen 16.6: Writing Equilibrium Expressions • Screen 16.9: Systems at Equilibrium • Screens 16.11, 16.13, 16.14: Disturbing a Chemical Equilibrium
17 **Principles of Reactivity: The Chemistry of Acids and Bases**	• Screen 17.2: Brønsted Acids and Bases	• Screen 17.2: Brønsted Acids and Bases • Screen 17.4: The pH Scale • Screen 17.5: Strong Acids and Bases • Screen 17.8: Determining K_a and K_b Values • Screen 17.9: Estimating the pH of Weak Acid Solutions • Screen 17.11: Estimating the pH Following an Acid-Base Reaction • Screen 17.13: Lewis Acids and Bases • Screen 17.15: Neutral Lewis Acids	• 17.2: pH and pOH	• Screen 17.3: The Acid–Base Properties of Water • Screen 17.4: The pH Scale • Screen 17.6: Weak Acids and Bases • Screen 17.7: Acid–Base Reactions • Screen 17.12: Acid–Base Properties of Salts • Screen 17.14: Cationic Lewis Acids • Screen 17.16: Molecular Interpretation of Acid–Base Behavior
18 **Principles of Reactivity: Other Aspects of Aqueous Equilibria**		• Screen 18.3: Buffer Solutions • Screen 18.4: pH of Buffer Solutions • Screen 18.5: Preparing Buffer Solutions • Screen 18.6: Adding Reagents to a Buffer Solution • Screen 18.7: Titration Curves • Screen 18.12: Solubility Product Constant • Screen 18.13: Determining K_P, Experimentally • Screen 18.14: Estimating Salt Solubility: Using K_P • Screen 18.15: Common Ion Effect • Screen 18.16: Solubility and pH • Screen 18.17: Can a Precipitation Reaction Occur? • Screen 18.19: Complex Ion Formation and Solubility	• 18.2: Buffer Solutions • 18.5: The Change in pH During the Titration of a Weak Acid with a Strong Base	• Screen 18.2: Common Ion Effect • Screen 18.3: Buffer Solutions • Screen 18.4: pH of Buffer Solutions • Screen 18.5: Preparing Buffer Solutions • Screen 18.7: Titration Curves • Screen 18.8: Titration of a Weak Polyprotic Acid • Screen 18.9: Titration of a Weak Base with a Strong Acid • Screen 18.10: Acid-Base Indicators • Screen 18.11: Precipitation Reactions • Screen 18.12: Solubility Product Constant • Screen 18.15: Common Ion Effect • Screen 18.16: Solubility and pH • Screen 18.17: Can a Precipitation Reaction Occur? • Screen 18.18: Simultaneous Equilibria • Screen 18.20: Using Solubility
19 **Principles of Reactivity: Entropy and Free Energy**		• Screen 19.5: Calculating ΔS for a Chemical Reaction • Screen 19.6: The Second Law of Thermodynamics • Screen 19.7: Gibbs Free Energy • Screen 19.8: Free Energy and Temperature • Screen 19.9: Thermodynamics and the Equilibrium Constant	• 19.12: Spontaneity ΔG^o with Temperature • 19.13: Free Energy Changes as a Reaction Approaches Equilibrium	• Screen 19.2: Reaction Spontaneity • Screen 19.3: Directionality of Reactions • Screen 19.4: Entropy: Matter Dispersal and Disorder • Screen 19.6: The Second Law of Thermodynamics • Screen 19.8: Free Energy and Temperature • Screen 19.9: Thermodynamics and the Equilibrium Constant

Media Integration Guide

Chapter	Exercises	Tutorials	Active Figures	Additional Resources
20 **Principles of Reactivity: Electron Transfer Reactions**		• Screen 20.6: Standard Potentials • Screen 20.8: Cells at Nonstandard Conditions • Screen 20.12: Coulometry: Counting Electrons	• 20.13: A Voltaic Cell Using $Zn|Zn^{2+}(aq, 1.0\ M)$ and $H_2|H^+(aq, 1.0\ M)$ Half-Cells	• Screen 20.2: Redox Reactions: Electron Transfer • Screen 20.3: Balancing Equations for Redox Reactions • Screen 20.4: Electrochemical Cells • Screen 20.5: Batteries • Screen 20.5: Electrochemical Cells and Potentials • Screen 20.6: Standard Potentials • Screen 20.11: Electrolysis: Chemical Change from Electrical Energy
21 **The Chemistry of the Main Group Elements**	• Screen 21.4: Boron Hydrides Structures • Screen 21.5: Aluminum Compounds • Screen 21.6: Silicon-Oxygen Compounds: Formulas and Structures • Screen 21.8: Sulfur Allotropes • Screen 21.9: Structures of Sulfur Compounds	• Screen 21.2: Formation of Ionic Compounds by Main Group Elements	• 21.15: Industrial Production of Aluminum • 21.22: Compounds and Oxidation Numbers for Nitrogen • 21.32: A Membrane Cell for the Production of NaOH and Cl_2 Gas from a Saturated, Aqueous Solution of NaCl (Brine)	
22 **The Chemistry of the Transition Elements**	• Screen 22.2: Formulas and Oxidation Numbers in Transition Metal Complexes • Screen 22.5: Geometry of Coordination Compounds • Screen 22.6: Geometric Isomerism in Coordination Compounds		• 22.8: A Blast Furnace	• Screen 21.7: Electronic Structure in Transition Metal Complexes • Screen 21.8: Spectroscopy of Transition Metal Complexes • Screen 22.3: Periodic Trends for Transition Elements
23 **Nuclear Chemistry**	• Screen 23.5: Kinetics of Nuclear Decay	• Screen 23.2: Radioactive Decay • Screen 23.3: Balancing Nuclear Reaction Equations • Screen 23.4: Stability of Atomic Nuclei • Screen 23.5: Kinetics of Nuclear Decay		• Screen 23.4: Stability of Atomic Nuclei • Screen 23.6: Nuclear Fission

Media Integration Guide

Chemistry
& CHEMICAL REACTIVITY

ENHANCED REVIEW EDITION

SIXTH EDITION

John C. Kotz

SUNY Distinguished Teaching Professor
State University of New York
College at Oneonta

Paul M. Treichel

Professor of Chemistry
University of Wisconsin–Madison

Gabriela C. Weaver

Associate Professor of Chemistry
Purdue University

THOMSON ™

BROOKS/COLE

Australia • Canada • Mexico • Singapore • Spain • United Kingdom • United States

THOMSON
™
BROOKS/COLE

Publisher/Executive Editor: DAVID HARRIS
Development Editor: PETER McGAHEY
Assistant Editor: ELLEN BITTER
Editorial Assistant: LAUREN OLIVEIRA
Technology Project Manager: DONNA KELLEY
Executive Marketing Manager: JULIE CONOVER
Senior Marketing Manager: AMEE MOSLEY
Marketing Communications Manager: NATHANIEL BERGSON-MICHELSON
Project Manager, Editorial Production: LISA WEBER
Creative Director: ROB HUGEL
Art Director: LEE FRIEDMAN
Print Buyer: REBECCA CROSS

Permissions Editor: KIELY SEXTON
Production Service: THOMPSON STEELE, INC.
Text Designers: ROB HUGEL AND JOHN WALKER DESIGN
Photo Researcher: JANE SANDERS MILLER
Copy Editor: THOMPSON STEELE, INC.
Developmental Artist: PATRICK A. HARMAN
Illustrators: ROLIN GRAPHICS AND THOMPSON STEELE, INC.
Cover Designer: JOHN WALKER DESIGN
Cover Images: MOTOHIKO MURAKAMI
Cover Printer: TRANSCONTINENTAL PRINTING/INTERGLOBE
Compositor: THOMPSON STEELE, INC.
Printer: TRANSCONTINENTAL PRINTING/INTERGLOBE

Printed in Canada
1 2 3 4 5 6 7 00 09 08 07 06

For more information about our products, contact us at:
THOMSON LEARNING ACADEMIC RESOURCE CENTER
1-800-423-0563

For permission to use material from this text or product, submit a request online at: **http://www.thomsonrights.com**

Any additional questions about permissions can be submitted by email to: **thomsonrights@thomson.com**

Library of Congress Control Number: 2004109955

Student Edition (paper bound): ISBN 0-495-11299-2

Student Edition (case bound): ISBN 0-495-11450-2

Thomson Brooks/Cole
10 Davis Drive
Belmont, CA 94002-3098
USA

Asia
Thomson Learning
5 Shenton Way #01-01
UIC Building
Singapore 068808

Canada
Nelson
1120 Birchmount Road
Toronto, Ontario M1K 5G4
Canada

Australia/New Zealand
Thomson Learning
102 Dodds Street
Southbank, Victoria 3006
Australia

Europe/Middle East/Africa
Thomson Learning
High Holborn House
50/51 Bedford Row
London WC1R 4LR
United Kingdom

About the Cover

What lies beneath the Earth's surface? The mantle of the Earth consists largely of silicon-oxygen based minerals. But about 2900 km below the surface the solid silicate rock of the mantle gives way to the liquid iron alloy core of the planet. To explore the nature of the rocks at the core-mantle boundary, scientists in Japan examined magnesium silicate ($MgSiO_3$) at a high pressure (125 gigapascals) and high temperature (2500 K). The cover image is what they saw. The solid consists of SiO_6 octahedra (blue) and magnesium ions (Mg^{2+}; yellow spheres). Each SiO_6 octahedron shares the four O atoms in opposite edges with two neighboring octahedra, thus forming a chain of octahedra. These chains are interlinked by sharing the O atoms at the "top" and "bottom" of SiO_6 octahedra in neighboring chains. The magnesium ions lie between the layers of interlinked SiO_6 chains. For more information see M. Murakami, K. Hirose, K. Kawamura, N. Sata, and Y. Ohishi, *Science*, Volume 304, page 855, May 7, 2004.

Brief Contents

This text is available in these student versions:
- Complete text ISBN 0-534-99766-X
- Complete text Enhanced Review Edition ISBN 0-495-11299-2
- Volume 1 (Chapters 1–12) ISBN 0-495-01013-8
- Volume 2 (Chapters 12–23) ISBN 0-495-01014-6
- Two-volume set ISBN 0-534-40800-1

Contents

This text is available in these student versions:
- Complete text ISBN 0-534-99766-X
- Complete text Enhanced Review Edition ISBN 0-495-11299-2
- Volume 1 (Chapters 1–12) ISBN 0-495-01013-8
- Volume 2 (Chapters 12–23) ISBN 0-495-01014-6
- Two-volume set ISBN 0-534-40800-1

Charles D. Winters

page 19

Charles D. Winters

page 25

xi

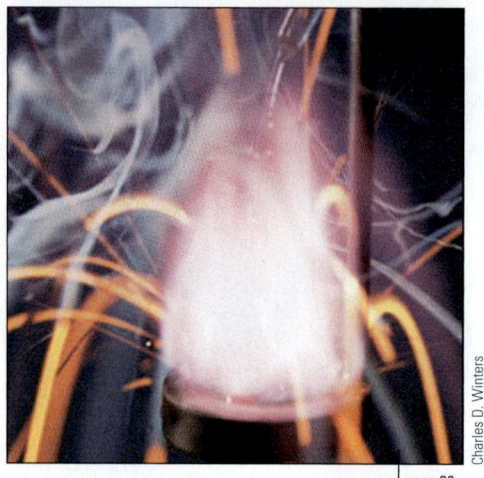

Charles D. Winters

page 82

Charles D. Winters

Charles D. Winters

page 145

page 214

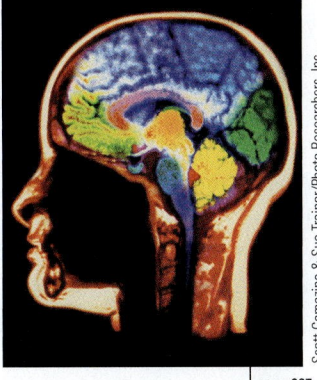

page 337   page 373

Part 3

States of Matter

page 515   page 568

Charles D. Winters

page 646

Simon Fraser/Photo Researchers, Inc.

page 651

Part 4

The Control of Chemical Reactions

page 686   page 763

Charles D. Winters

Charles D. Winters

page 882   page 921

Arthur N. Palmer

© Ludovic Maisant/Corbis

page 1013   page 1052

Preface

We are gratified that *Chemistry & Chemical Reactivity* has been used by more than a million students in its first five editions. Because this is one indication our book has been successful in helping students learn chemistry, we believe the goals we set out in the first edition are still appropriate. Our principal goals have always been to provide a broad overview of the principles of chemistry, the reactivity of the chemical elements and their compounds, and the applications of chemistry. We have organized this approach around the close relation between the observations chemists make of chemical and physical changes in the laboratory and in nature and the way these changes are viewed at the atomic and molecular levels.

Another of our goals has been to convey a sense of chemistry not only as a field that has a lively history but also as one that is highly dynamic, with important new developments occurring every year. Furthermore, we want to provide some insight into the chemical aspects of the world around us. Indeed, a major objective of this book is to provide the tools needed for you to function as a chemically literate citizen. Learning something of the chemical world is just as important as understanding some basic mathematics and biology, and as important as having an appreciation for history, music, and literature. For example, you should know which materials are important to our economy, what some of the reactions in plants and animals and in our environment are, and what role chemists play in protecting the environment.

Among the most exciting and satisfying aspects of our careers as chemists has been our ability to discover new compounds and to find new ways to apply chemical principles and explain what we observe. We hope we have conveyed that sense of enjoyment in this book as well as our awe at what is known about chemistry—and, just as important, what is not known!

W. Keel, U. Alabama/NASA

page 36

Enhanced Review Edition

Why an Enhanced Review Edition?

As authors and publishers we are in constant conversation with instructors and students about their textbooks and courses. To accommodate requests for less costly alternatives to the standard book, we sought a manner in which we could reduce manufacturing costs and pass the savings on to students. Similarly, many users have mentioned a desire for exam preparation materials and study tools in the text itself that better integrate with the book and media. The Enhanced Review Edition is an attempt to satisfy these requests.

What Is an Enhanced Review Edition?

The Enhanced Review Edition of *Chemistry & Chemical Reactivity*, Sixth Edition, is an alternative version that can be used in place of the standard sixth edition. Although

page 52

Charles D. Winters

we believe its changes will provide a better learning tool for those who choose it, the new book's alterations are not substantial enough to warrant it being called a new edition because the core chapters and appendices remain largely unchanged. The standard hardbound sixth edition remains available for purchase and classroom use.

Four new *Let's Review* sections have been added. These cumulative review sections in four of the book parts offer students review aids and questions that bring together material from the several chapters and are similar to those they may see on an exam. The review questions are keyed to text and media material to assist students in preparing for an exam. In order to include these sections, the four supplemental interchapters in the standard sixth edition have been transferred to the book's companion website where they are accessible to users of the Enhanced Review Edition.

The Enhanced Review Edition is offered with a soft paper binding and does not include the General ChemistryNow Interactive CD-ROM. The subsequent reduction in manufacturing costs allows us to offer this version at a reduced price compared to the full standard edition.

> Throughout the text, you will see references to the General ChemistryNow website and the CD-ROM. This alternative edition contains only a password to access the Web-based material. All material available on the CD is also available on the website, so purchasers of this version do not lose material. If you wish to purchase a copy of the CD-ROM, visit **www.brookscole.com**.

Emerging Developments in Content Usage and Delivery

The use of media, presentation tools, and homework management tools has expanded significantly in the last three years. About ten years ago we incorporated electronic media into this text with the first edition of our interactive CD-ROM. It has been used by thousands of students worldwide and has been the most successful attempt to date to encourage students to interact with chemistry.

Multimedia technology has evolved over the past ten years, and so have our students. Students are not only focused on conceptual understanding, but are also keenly aware of the necessity of preparing for examinations. Our challenge as authors and educators is to use students' focus on assessment as a way to help them reach a higher level of conceptual understanding. In light of this goal, we have made major changes in our integrated media program. We have found that few students explore multimedia for its own sake. Therefore, we have redesigned the media so that students now have the opportunity to interact with media based on clearly stated chapter goals that are correlated to end-of-chapter questions. By using new diagnostic tools, students will be directed to specific resources based on their levels of understanding. This new program, called General ChemistryNow, is described in detail later. The closely related OWL homework management system has also been used by tens of thousands of students, and we are pleased to announce that selected end-of-chapter questions are now available for use within the OWL system.

Audience for the Textbook, the General ChemistryNow Website, and OWL

The textbook, CD-ROM and website, and OWL are designed to serve introductory courses in chemistry for students interested in further study in science, whether that science is biology, chemistry, engineering, geology, physics, or related subjects. Our assumption is that students beginning this course have had some preparation in algebra and in general science. Although undeniably helpful, a previous exposure to chemistry is neither assumed nor required.

Philosophy and Approach of the Program

We have had three major, albeit not independent, goals since the first edition of the book. The first goal was to write a book that students would enjoy reading and that would offer, at a reasonable level of rigor, chemistry and chemical principles in a format and organization typical of college and university courses today. Second, we wanted to convey the utility and importance of chemistry by introducing the properties of the elements, their compounds, and their reactions as early as possible and by focusing the discussion as much as possible on these subjects. Finally, with the new, integrated media program, we hope to bring students to a higher level of conceptual understanding.

The American Chemical Society has been urging educators to put "chemistry" back into introductory chemistry courses. We agree with this position wholeheartedly. Therefore, we have tried to describe the elements, their compounds, and their reactions as early and as often as possible in several ways. First, numerous color photographs depict reactions occurring, the elements and common compounds, and common laboratory operations and industrial processes. Second, we have tried to bring material on the properties of elements and compounds as early as possible into the Exercises and Study Questions and to introduce new principles using realistic chemical situations. Finally, relevant highlights are given in Chapters 21 and 22 as a capstone to the principles described earlier.

Organization of the Book

Chemistry & Chemical Reactivity has two overarching themes: *chemical reactivity* and *bonding and molecular structure.* The chapters on *principles of reactivity* introduce the factors that lead chemical reactions to be successful in converting reactants to products. Under this topic you will study common types of reactions, the energy involved in reactions, and the factors that affect the speed of a reaction. One reason for the enormous advances in chemistry and molecular biology in the last several decades has been an understanding of molecular structure. Sections of the book on *principles of bonding and molecular structure* lay the groundwork for understanding these developments. Particular attention is paid to an understanding of the structural aspects of such biologically important molecules as DNA.

A glance at the introductory chemistry texts currently available shows there is a generally common order of topics used by educators. With a few minor variations, we have followed that order as well. That is not to say that the chapters in our book cannot be used in some other order. We have written it to be as flexible as possible. For example, the chapter on the behavior of gases (Chapter 12) is placed with chapters on liquids, solids, and solutions (Chapters 13 and 14) because it logically fits with these topics. It can easily be read and understood, however, after covering only the first four or five chapters of the book. Similarly, chapters on atomic and molecular structure (Chapters 7–10) could be used before the chapters on stoichiometry and common reactions (Chapters 4 and 5). Also, the chapters on chemical equilibria (Chapters 16–18) can be covered before those on solutions and kinetics (Chapters 14 and 15).

Organic chemistry (Chapter 11) is often left to one of the final chapters in chemistry textbooks. We believe that the importance of organic compounds in bio-

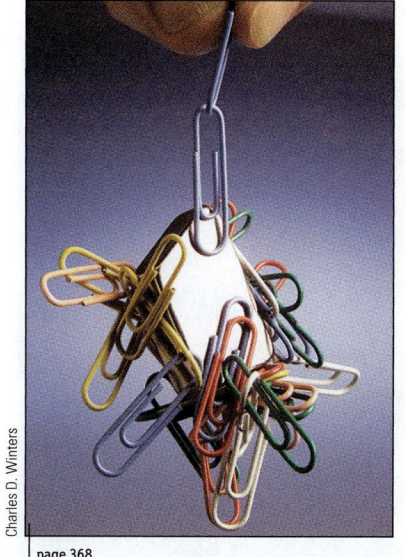

Charles D. Winters

page 368

chemistry and in consumer products means we should present that material earlier in the sequence of chapters. This coverage follows the chapters on structure and bonding, because organic chemistry nicely illustrates the application of models of chemical bonding and molecular structure. However, one can use the remainder of the book without including this chapter.

The order of topics in the text was also devised to introduce as early as possible the background required for the laboratory experiments usually done in General Chemistry courses. For this reason, chapters on chemical and physical properties, common reaction types, and stoichiometry begin the book. In addition, because an understanding of energy is so important in the study of chemistry, thermochemistry is introduced in Chapter 6.

In addition to the regular chapters, uses and applications of chemistry are described in more detail in interchapters on *The Chemistry of Fuels and Energy Sources, The Chemistry of Life: Biochemistry, The Chemistry of Modern Materials,* and *The Chemistry of the Environment.* These chapters, new to this edition, are described in more detail later in this Preface.

Additionally, *Chemical Perspectives* attempt to bring relevance and perspective to a study of chemistry. These features delve into such topics as nanotechnology, using isotopes, what it means to be in the "limelight," the importance of sulfuric acid in the world economy, sunscreens, and the newly recognized importance of the NO molecule. *Historical Perspectives* describe the historical development of chemical principles and the people who made the advances in our understanding of chemistry.

A *Closer Look* boxes describe ideas that form the background to material under discussion or provide another dimension of the subject. For example, in Chapter 11 on organic chemistry, the "A Closer Look" boxes are devoted to a discussion of structural aspects of important molecules, to petroleum, and to fats and oils. In other chapters we delve into molecular modeling, magnetic resonance, and mass spectrometry.

Finally, *Problem-Solving Tips* provide students with important insights into problem solving. They also identify where, from our experience, students often make mistakes and suggest alternative ways to solve problems.

The chapters of *Chemistry & Chemical Reactivity* are organized into five sections, each grouping with a common theme.

Part 1: The Basic Tools of Chemistry

There are fundamental ideas and methods that are the basis of all of chemistry, and these are introduced in Part 1. Chapter 1 defines important terms and reviews units and mathematical methods. Chapters 2 and 3 introduce basic ideas of atoms, molecules, and ions, and Chapter 2 describes the most important organizational device in chemistry, the periodic table. In Chapters 4 and 5, we begin to discuss the principles of chemical reactivity and to introduce the numerical methods used by chemists to extract quantitative information from chemical reactions. Chapter 6 introduces the energy involved in chemical processes. The interchapter *The Chemistry of Fuels and Energy Sources* follows Chapter 6 and uses many of the concepts developed in the preceding chapters.

Part 2: The Structure of Atoms and Molecules

The goal of this section is to outline the current theories of the arrangement of electrons in atoms and some of the historical developments that led to these ideas (Chapters 7 and 8). This discussion is tied closely to the arrangement of elements in the periodic table, so that these properties can be recalled and predictions made. In Chapter 9, we discuss for the first time how the electrons of atoms in a molecule par-

William James Warren/Corbis

page 235

ticipate in chemical bonding and lead to the properties of these bonds. In addition, we show how to derive the three-dimensional structure of simple molecules. Chapter 10 considers the major theories of chemical bonding in more detail.

This part of the book finishes with a discussion of organic chemistry (Chapter 11), primarily from a structural point of view. Organic chemistry is such an enormous area of chemistry that we cannot hope to cover it in detail in this book. Therefore, we have focused on compounds of particular importance, including synthetic polymers and the structures of these materials.

In this section of the book you will find the molecular modeling software on the General ChemistryNow CD-ROM and website to be especially useful.

To cap this section, the interchapter *The Chemistry of Life: Biochemistry* provides an overview of some of the most important aspects of biochemistry.

Part 3: States of Matter

The behavior of the three states of matter—gas, liquid, and solid—is described in that order in Chapters 12 and 13. The discussion of liquids and solids is tied to gases through the description of intermolecular forces, with particular attention given to liquid and solid water. In Chapter 14, we describe the properties of solutions, intimate mixtures of gases, liquids, and solids.

The interchapter *The Chemistry of Modern Materials* is placed after Chapter 13, following coverage of the solid state. Designing and making new materials with useful properties is one of the most exciting areas of modern chemistry.

Part 4: The Control of Chemical Reactions

Part 4 is wholly concerned with the principles of reactivity. Chapter 15 examines the important question of the rates of chemical processes and the factors controlling these rates. With this material on chemical kinetics in mind, we move to Chapters 16–18, which describe chemical reactions at equilibrium. After an introduction to equilibrium in Chapter 16, we highlight reactions involving acids and bases in water (Chapters 17 and 18) and reactions leading to insoluble salts (Chapter 18). To tie together the discussion of chemical equilibria, we again explore thermodynamics in Chapter 19. As a final topic in Part 4, we describe in Chapter 20 a major class of chemical reactions, those involving the transfer of electrons, and the use of these reactions in electrochemical cells.

The *Chemistry of the Environment* interchapter appears at the end of Part 4. This chapter uses ideas from kinetics and chemical equilibria in particular, as well as principles described in earlier chapters in the book.

Part 5: The Chemistry of the Elements and Their Compounds

Although the chemistry of the various elements has been described throughout the book to this point, Part 5 considers this topic in a more systematic way. Chapter 21, which has been expanded for this edition, is devoted to the chemistry of the representative elements, whereas Chapter 22 discusses the transition elements and their compounds. Finally, Chapter 23 offers a brief discussion of nuclear chemistry.

Changes for the Sixth Edition

Colleagues and students often ask why yet another edition of the book has been prepared. We all understand, however, that even the most successful books can be improved. In addition, our experience in the classroom suggests that student interests change and that there are ever more effective ways to help our students learn chemistry. For these reasons, we made a number of changes in this book from the fifth edition. For this new, sixth edition, the material and our approach have been refined further to take students to a higher level of conceptual understanding, and several important ideas have been added.

In summary, while this sixth edition retains the overall structure and goals of the previous five editions, we have done much more than change a few words and illustrations. Significant changes have been made that we believe will aid our students in learning and understanding the important principles of chemistry and in discovering that it is an exciting and dynamic field.

Book Revisions

Readability and Clarity   A hallmark of the first five editions of *Chemistry & Chemical Reactivity* has been the book's readability. Nonetheless, each sentence and paragraph in the book has been examined with an eye toward improving clarity and shortening the material without reducing content coverage or readability. Many of the illustrations have been revised and new ones added.

Expanded Coverage   We have worked to raise the level of the text by introducing new material on, among other things, molecular orbital theory and the solid state and on biochemistry and environmental chemistry. The Clausius-Clapeyron equation has been given greater prominence, and "cumulative" and more challenging Study Questions have been added.

Accuracy   Although previous editions of the book have always been relatively free of errors, even greater effort has been made in this edition, and seven accuracy reviewers—four for the text and three for the supplemental chapters—have been brought into our team.

Supplemental Material on Mathematics

A knowledge of basic mathematics is required to be successful in general chemistry. For students unsure of their abilities, a special section (Section 1.8) has been added that reviews exponential notation, significant figures, dimensional analysis, plotting graphs, and reading graphical information.

Supplemental Interchapters

Applications of chemical principles are pervasive in our lives. Although the sixth edition describes many applications as chemical principles are developed, a number of important and interesting areas are left untouched. Therefore, four areas of chemistry are covered in interchapters in a magazine style.

- *The Chemistry of Fuels and Energy Sources* (page 282). This material explores the energy situation confronting our planet and examines such subjects as alternative energy sources, hybrid cars, fuel cells, and "the hydrogen economy."

- *The Chemistry of Life: Biochemistry* (page 530). Perhaps more chemists work in biochemistry than in any other area. This chapter delves into amino acids and proteins, nucleic acids, and metabolism.

- *The Chemistry of Modern Materials* (page 642). The past few decades have seen the development of new electronic devices (such LEDs in car and traffic lights), nanostructures, superconductors, and new adhesives. This supplemental chapter touches on some of these areas as well as others. In addition, there is a discussion of the molecular orbital approach to bonding in metals and semiconductors, material that was in Chapter 10 in the previous edition.

- ***The Chemistry of the Environment*** (page 998). Environmental issues such as smog, the hole in the earth's ozone layer, global warming, and water quality are regularly encountered in the news. This chapter describes how our water is treated, discusses the effect of particulate pollutants in our atmosphere, and explores the new efforts chemists are making worldwide to produce the products we all rely on in an environmentally safe manner.

For users of *Chemistry & Chemical Reactivity, Sixth Edition: Enhanced Review Edition,* the interchapters are available at **www.brookscole.com/kotz6e**.

Royalty-free/Corbis

page 235

Introducing General ChemistryNow Linked to Chapter Goals

Students have always been concerned about "what's on the exam." Although this is certainly a legitimate concern, our challenge as educators has been to help students come to a conceptual understanding and not have them simply learn patterns of thought and memorize equations. To that end, each chapter in the textbook is introduced by 4–6 Chapter Goals that have a conceptual underpinning and are covered in the chapter. These goals are revisited at the end of the chapter, where each goal is divided into several subtopics with which the student should be familiar. Study Questions relevant to the goals are noted in the Chapter Goals Revisited section and are marked with the ■ icon in the Study Questions.

General ChemistryNow at **http://now.brookscole.com/kotz6e** is a Web-based program (available with each new copy of the book) that incorporates material from our original General Chemistry Interactive CD-ROM and includes *more than 400 new step-by-step tutorial modules keyed to end-of-chapter Study Questions*. The system is completely flexible, so students have access to the material through a variety of methods.

- A **Chapter Outline** screen for each chapter matches the text organization.

- A **Homework and Goals** screen is keyed to the Chapter Goals Revisited section in each chapter. Each goal is linked to Simulations, Exercises, and Tutorials and to selected end-of-chapter Study Questions taken from the book. (These questions are marked in the book with ■.) Students can attempt to answer each of the selected Study Questions any number of times, view feedback on the solution, and submit answers online to the instructor for grading.

- A **Diagnostic Exam-Prep Quiz** ("What Do I Know") provides diagnostic questions that have been carefully crafted to assess student understanding of the Chapter Goals. Upon completing a quiz, students receive feedback and a personalized Learning Plan, and, if applicable, will be directed to the relevant Chapter Goals and accompanying resources.

Study Questions

Several important changes have been made in the end-of-chapter questions:

- As noted earlier, approximately 20 Study Questions in each chapter have been selected as illustrative of the chapter goals, and these questions are available in interactive form in General ChemistryNow. These questions are marked in the book with the ■ icon.

- As in previous editions, a number of Study Questions are provided that refer to a particular section of the book. These questions are paired; that is, there are two similar questions with one question (indicated with a blue number) having an answer in Appendix O and a solution in the Student Solutions Manual. The idea is that you can learn how to solve the question without an answer in the appendix by first doing the question for which an answer is provided. Furthermore, for questions on a given section or subsection of the chapter, we note which Example questions or Exercises are relevant. Also, we refer to a particular screen or screens of General ChemistryNow that may be helpful.

- After the sections containing paired questions on specific topics, General Questions integrate concepts from several parts of the chapter.

- Challenging questions are marked with the ▲ icon. The number of these challenging questions has been increased in this sixth edition.

- Some questions rely more heavily than usual on material in preceding chapters or are more conceptual. These questions, sometimes called "cumulative questions," are set out in a separate section called Summary and Conceptual Questions.

- Some questions have been added that call upon students to understand the chemistry at the molecular level.

Homework Management Options

Thousands of students around the country are successfully using the OWL program (Online Web-based Learning) developed at the University of Massachusetts–Amherst. (OWL is described in detail later.) We have heard from many chemistry instructors that they would like to be able to assign specific, parameterized (algorithmic) questions from the end-of-chapter problem set, so we are pleased to announce that *approximately 20 questions per chapter are available to assign in this new OWL format.* These are the same 20 or so questions marked in the Study Questions section as relevant to the Chapter Goals. In addition, all of the end-of-chapter problems are available in Web CT and Blackboard formats.

Book Design

A major effort was made with the fifth edition to design a book that would aid students by clearly delineating the functions of the various parts of the book. (Although seemingly simple, one of many innovations was to use different typographic fonts for text and chemical equations so that these are clearly separated. Another was to label chemicals or parts of an apparatus in photos so that the reader does not have to move continually between caption and photo to understand the photo's message.) For this new edition, we have continued to put a great deal of thought into book design for functional clarity.

Supporting Materials for the Student

> Visit http://chemistry.brookscole.com to see samples of selected student supplements or to purchase them online from Brooks/Cole. To locate products at your local retailer, provide them with the ISBN.

NEW! General ChemistryNow Website by William Vining, University of Massachusetts–Amherst, and John Kotz, State University of New York–Oneonta.

General ChemistryNow at http://now.brookscole.com/kotz6e is a powerful, assessment-based online learning companion designed to help students master chapter goals by directing them to interactive resources based on their level of conceptual understanding. Incorporating material from the best-selling *General Chemistry Interactive CD-ROM,* this new media resource includes more than 400 new step-by-step tutorial modules keyed to end-of-chapter Study Questions. The system is completely flexible so students have access to the material through a variety of methods:

- A **Chapter Outline** screen matches the text organization.

- A **Homework and Goals** screen is keyed to the Chapter Goals Revisited feature from the sixth edition and provides selected end-of-chapter Study Questions. The goals are linked to simulations, exercises, and tutorials. Students can attempt each question a number of times, and view feedback on the solution. These questions are indicated with the ■ icon.

- An **Exam-Prep Quiz** ("What Do I Know?") provides diagnostic questions that have been carefully crafted to assess students' understanding of the chapter goals. Upon completing a quiz, students will receive feedback and a personalized Learning Plan, and, if applicable, will be directed to the relevant chapter goals screens and accompanying interactive resources.

Access to General ChemistryNow is included with the purchase of a new text.

Enhanced! OWL (Online Web-based Learning system),
University of Massachusetts–Amherst

Learning chemistry takes practice, and that usually means completing homework assignments. With a new, easier-to-use interface, the class-tested, Web-based OWL system at http://owl.thomsonlearning.com presents students with a series of questions—many from the text itself for this new edition—and students respond with numerical answers or with a selection from a menu of choices. Questions are generated from a database of numerical and chemical information, so each student in a course receives a different variant of the question each time he or she accesses an instructional unit. Each question has extensive, question-specific feedback keyed to a student's answer. Instructors can customize the unit by determining when questions are available, how many attempts students may make, and how many questions students must answer successfully before they are considered to have mastered the topic. Gradable reports on each attempt at the unit are provided to the instructor, who has access to course management tools such as a gradebook and report-generating functions. Students find OWL an excellent exam review and studies at the University of Massachusetts–Amherst show a positive correlation between use of the OWL system and course performance. **The end-of-chapter questions in the text that are correlated to the Chapter Goals are now fully assignable within the OWL program.**

Student Solutions Manual by Alton Banks, North Carolina State University

This ancillary contains detailed solutions to selected end-of-chapter Study Questions found in the text. Solutions match the problem-solving strategies used in the text. Sample chapters are available for review at the book's website. ISBN 0-534-99852-6

NEW! *Study Guide* by John R. Townsend, West Chester University of Pennsylvania

This completely new study guide contains learning tools explicitly linked to the goals introduced in each chapter. It includes chapter overviews, key terms and definitions, and sample tests. Emphasis is placed on the chapter goals presented in this text by means of further commentary and study tips, worked-out examples, and direct references back to the text. Sample chapters are available for review at the book's website. ISBN 0-534-99851-8

vMentor included with General ChemistryNow

vMentor is an online live tutoring service from Brooks/Cole in partnership with *Elluminate.* vMentor is included in General ChemistryNow. Whether it's one-to-one tutoring help with daily homework or exam review tutorials, vMentor lets students interact with experienced tutors right from their own computers at school or at home. All tutors have not only specialized degrees in the particular subject area (biology, chemistry, mathematics, physics, or statistics), but also extensive teaching experience. Each tutor also has a copy of the textbook the student is using in class. Students can ask as many questions as they want when they access vMentor—and they don't need to set up appointments in advance! Access is provided with vClass, an Internet-based virtual classroom featuring two-way voice, a shared whiteboard, chat, and more. For proprietary, college, and university adopters only. For additional information, consult your local Thomson representative.

NEW! *Chemistry & Chemical Reactivity, Sixth Edition in Two Hardbound Volumes (Volume 1: Chapters 1–12 and Volume 2: Chapters 12–23)*

We recognize that students are concerned about price and portability of their textbooks, and that some students take only one semester of general chemistry. Therefore, we are pleased to announce that the sixth edition is available in two volumes. Volume 1 covers Chapters 1–12 and Volume 2 covers Chapters 12–23. Note that both volumes contain Chapter 12 so as to serve differing curricula. Both volumes will include full access to all the media resources. Consult your Thomson representative for special pricing options. Volume 1 ISBN 0-495-01013-8; Volume 2 ISBN 0-495-01014-6; Two-volume set ISBN 0-534-40800-1.

Essential Algebra for Chemistry Students, Second Edition by David W. Ball, Cleveland State University

This supplement focuses on the skills needed to survive in General Chemistry, with worked examples showing how these skills translate into successful chemical problem solving. This text is an ideal tool for students lacking in confidence or competency in the essential math skills required for general chemistry. Consult your Thomson representative for special bundling pricing. ISBN 0-495-01327-7.

Survival Guide for General Chemistry with Math Review by Charles H. Atwood, University of Georgia

Designed to help students gain a better understanding of the basic problem-solving skills and concepts of General Chemistry, this guide assists students who lack confidence and/or competency in the essential skills necessary to survive general chemistry. The text can be fully customized so that you can incorporate, if you so wish, your old exams. Consult your Brooks/Cole representative for special bundling pricing. ISBN 0-534-99370-2

Supporting Materials for the Instructor

Supporting instructor materials are available to qualified adopters. Please consult your local Thomson Brooks/Cole representative for details. **Visit http:// chemistry.brookscole.com** to:

- See samples of materials
- Locate your local representative
- Download electronic files of books, PowerPoint slides, and text art
- Request a desk copy
- Purchase a book online

Instructor's Resource Manual by Susan Young, Hartwick College

Contains worked-out solutions to *all* end-of-chapter Study Questions and features ideas for instructors on how to fully utilize resources and technology in their courses. The *Manual* provides questions for electronic response systems, suggests classroom demonstrations, and emphasizes good and innovative teaching practices. Electronic files of the *Instructor's Resource Manual* are available for download on the instructor's website. ISBN 0-534-99856-9

General ChemistryNow Website

A powerful, personalized learning companion that offers your students a variety of tools with which to learn the material, test their knowledge, and identify which tools will best meet their needs. General ChemistryNow is included with every new copy of the book. (Please see the description in the "For the Student" list of ancillary materials.)

Multimedia Manager Instructor CD-ROM

The Multimedia Manager is a dual-platform digital library and presentation tool that provides art, photos, and tables from the main text in a variety of electronic formats that can be used to make transparencies and are easily exported into other software packages. This enhanced CD-ROM also contains simulations, molecular models, and QuickTime movies to supplement lectures as well as electronic files of various print supplements. In addition, instructors can customize presentations by importing personal lecture slides or other selected materials. ISBN 0-534-99855-0

OWL (Online Web-based Learning System)

An online homework, quizzing, and testing tool with course management capability. (Please see the description in the "For the Student" list of ancillary materials.)

PowerPoint Lecture Slides by John Kotz, State University of New York–Oneonta

These class-tested, fully customizable, lecture slides have been used by author John Kotz for many years and are available for instructor download at the text's website at http://chemistry.brookscole.com. Hundreds of slides cover the entire year of general chemistry. Slides use the full power of Microsoft PowerPoint and incorporate videos, animations, and other assets from General ChemistryNow. Instructors can customize their lecture presentations by adding their own slides or by deleting or changing existing slides.

Test Bank by David Treichel, Nebraska Wesleyan University

This printed test bank contains more than 1250 questions, over 90% of which are revised or newly written for this edition. Questions range in difficulty and variety and correlate directly to the chapter sections found in the main text. Numerical, open-ended, or conceptual problems are written in multiple choice, fill-in-the-blank, or short-answer formats. Both single- and multiple-step problems are presented for each chapter. Electronic files of the Test Bank are available for instructor download at the text's website at http://chemistry.brookscole.com. ISBN 0-53-499850-X

Transparencies

A collection of 150 full-color transparencies of key images selected by the authors from the text. Instructors have access on the Multimedia Manager CD-ROM to all text art and many photos to aid in preparing transparencies for material not present in this set. ISBN 0-534-99854-2

iLrn Testing

With a balance of efficiency and high performance, simplicity and versatility, iLrn Testing lets instructors test the way they teach, giving them the power to transform the learning and teaching experience. iLrn Testing is a revolutionary, Internet-ready, cross-platform text-specific testing suite that allows instructors to customize exams and track student progress in an accessible, browser-based format delivered via the Web at www.iLrn.com. Results flow automatically to instructors' gradebooks so that they are better able to assess students' understanding of the material prior to class or an actual test. iLrn offers full algorithmic generation of problems as well as free-response problems using intuitive mathematical notation. **Populated with the questions from the printed Test Bank.** ISBN 0-534-99857-7

JoinIn on TurningPoint for Response Systems

Thomson Brooks/Cole is now pleased to offer book-specific JoinIn content for Response Systems tailored to *Chemistry & Chemical Reactivity*, allowing you to transform your classroom and assess your students' progress with instant in-class quizzes and polls. Our exclusive agreement to offer TurningPoint software lets you pose book-specific questions and display students' answers seamlessly within the Microsoft PowerPoint slides of your own lecture, in conjunction with the "clicker" hardware of your choice. Enhance how your students interact with you, your lecture, and each other. Contact your local Thomson representative to learn more.

WebTutor ToolBox for WebCT and WebTutor ToolBox for Blackboard

Preloaded with content and available via a free access code when packaged with this text, WebTutor ToolBox pairs the content of this text's rich Book Companion website with sophisticated course management functionality. **The end-of-chapter Study Questions in the text are available in WebCT and Blackboard formats.** Instructors can assign materials (including online quizzes) and have the results flow automatically to their gradebooks. ToolBox is ready to use upon logging on—or instructors can customize its preloaded content by uploading images and other resources, adding weblinks, or creating their own practice materials. Students have access only to student resources on the website. Instructors can enter an access code to utilize password-protected Instructor Resources. Contact your Thomson representative for information on packaging WebTutor ToolBox with this text.

For the Laboratory

Chemical Education Resources (CER) at http://www.CERLabs.com

Allows instructors to customize laboratory manuals for their courses from a wide range of more than 300 experiments refereed by the CER board.

Brooks/Cole Laboratory Series for General Chemistry

Brooks/Cole offers a variety of printed manuals to meet all General Chemistry laboratory needs. Instructors can visit the chemistry website at http://chemistry.brookscole.com for a full listing and description of these laboratory manuals and laboratory notebooks. All Brooks/Cole lab manuals can be customized for your specific needs.

Acknowledgments

Because significant changes have been made, preparing this new edition of *Chemistry & Chemical Reactivity* took almost three years of continuous effort. However, as in our work on the first five editions, we have enjoyed the support and encouragement of our families and of some wonderful friends, colleagues, and students.

Brooks/Cole Publishing

The first four editions of this book were published by Saunders College Publishing, a part of Harcourt College Publishing. About a year before the fifth edition was published, however, the company came under new ownership, the Brooks/Cole group of Thomson Higher Education. Throughout the period during which the first five editions were developed, we had the guidance of John Vondeling as our Editor-Publisher and friend. John was responsible for much of the success the book enjoyed, but he passed away in January 2001. Angus McDonald guided us through the final stages of the publication of the fifth edition. We owe Angus a great debt of gratitude for taking over under difficult circumstances and for bringing the project to a successful conclusion.

Following the final acquisition of Harcourt by Thomson Higher Education, we were introduced to our new Editor in Chief, Michelle Julet, and our new Publisher, David Harris. Both have been invaluable in guiding this new edition, and both have become good friends. We look forward to doing future editions with them—and to more sailing with David.

Peter McGahey was the Developmental Editor for the fifth edition and again for this sixth edition. He is blessed with energy, creativity, enthusiasm, intelligence, and good humor. Peter is a trusted friend and confidant. And he cheerfully answered our many questions during almost-daily phone calls.

No book can be successful without proper marketing. Julie Conover is a whiz at marketing and a delight to work with. She is knowledgeable about the market and has worked tirelessly to bring the book to everyone's attention.

Our team at Brooks/Cole is completed with Lisa Weber, Production Manager, and Rob Hugel, Creative Director. Schedules are very demanding in textbook publishing, and Lisa has helped to keep us on schedule. We certainly appreciate her organizational skills. Rob has been involved in product and advertising design for many years, and he has brought his design skills to bear in making this a very attractive book.

People outside of publishing often do not realize the number of people involved in producing a textbook. Karla Maki and Nicole Barone of Thompson Steele, the production company, guided the book through the almost year-long production process. Jane Sanders Miller was the photo researcher for the book and was successful in filling our sometimes off-beat requests for a particular photo. Finally, Jill Hobbs did a very thorough job copyediting the manuscript, and Jay Freedman once again did a masterful job on the index.

Photography, Art, and Design

Most of the color photographs for this edition were again beautifully done by Charles D. Winters. He produced several dozen new images for this book, often under deadline pressure and always with a creative eye. Charlie's work gets better and better with each edition. We have worked with Charlie for almost 20 years and have become close friends. We listen to his jokes, both new and old—and always forget them. When we finish the book, we look forward to a kayaking trip.

When the fifth edition was being planned, we brought in Patrick Harman as a member of the team. Pat designed the first edition of the General ChemistryNow CD-ROM, and we believe its success is in no small way connected to his design skill. For the fifth edition of the book Pat went over almost every figure, and almost every word, to bring a fresh perspective to ways to communicate chemistry. Pat also worked on designing and producing new illustrations for the sixth edition, and his creativity is obvious in their clarity and beauty. As we have worked together so closely for so many years, Pat has become a good friend as well, and we share interests not only in beautiful books but also in interesting music.

Other Collaborators

We have been fortunate to have a number of other colleagues who have played valuable roles in this project.

- Bill Vining (University of Massachusetts–Amherst), the lead author of the General ChemistryNow CD-ROM and website, has been a colleague and friend for many years. Not only has he applied his considerable energy and creativity to preparing a thorough revision of these materials, but he was also a valuable advisor on the book.

- Susan Young (Hartwick College) has been a good friend and collaborator through four editions and has again prepared the *Instructor's Resource Manual*. She has always been helpful in proofreading, in answering questions on content, and in giving us good advice.

- Alton Banks (North Carolina State University) has also been involved for several editions preparing the *Student Solutions Manual*. Both Susan and Alton have been very helpful in ensuring the accuracy of the Study Questions answers in the book as well as in their respective manuals.

- John Townsend (West Chester University) prepared the *Study Guide* for this edition. This book has had a history of excellent study guides, and John's manual follows that tradition. As described later, John also contributed the supplemental chapter on biochemistry.

- Beatrice Botch (University of Massachusetts–Amherst) gave advice on parts of the text and supplied the information for Figure 13.13.

A major task is proofreading the book once it has been set in type. The book is read in its entirety by the authors and accuracy reviewers. After making corrections, the book is read a second time. Any errors remaining at this point are certainly the responsibility of the authors, and students and instructors should contact the authors by email to offer their suggestions. If this is done in a timely manner, corrections can be made when the book is reprinted.

We want to thank the following accuracy reviewers for their invaluable assistance. The book is immeasurably improved by their work.

Rodney Boyer, Ph.D., Hope College
Larry Fishel, Ph.D.
Michael Grady, Ph.D., College of the Redwoods
Frances Houle, Ph.D., IBM Almaden Research Center
Wayne E. Jones, Jr., Ph.D., Binghamton University
Kathy Mitchell, St. Petersburg College
Barbara Mowery, York College of Pennsylvania
David Shinn, Ph.D.

Reviewers for the Sixth Edition

Patricia Amateis, Virginia Tech
Todd L. Austell, University of North Carolina, Chapel Hill
Joseph Bularzik, Purdue University, Calumet
Stephen Carlson, Lansing Community College
Robert L. Carter, University of Massachusetts, Boston
Paul Charlesworth, Michigan Technological University
Paul Gilletti, Mesa Community College
Stan Genda, University of Nevada, Las Vegas
C. Alton Hassell, Baylor University
Margaret Kerr, Worcester State University
Jeffrey A. Mack, California State University, Sacramento
Elizabeth M. Martin, College of Charleston
Shelley D. Minteer, Saint Louis University
Jason R. Telford, University of Iowa
Wayne Tikkanen, California State University, Los Angeles
Mark A. Whitener, Montclair State University
Marcy Whitney, University of Alabama

Reviewers for the Fifth Edition

David W. Ball, Cleveland State University
Roger Barth, West Chester University
John G. Berberian, Saint Joseph's University
Don A. Berkowitz, University of Maryland
Simon Bott, University of Houston
Wendy Clevenger Cory, University of Tennessee, Chattanooga
Richard Cornelius, Lebanon Valley College
James S. Falcone, West Chester University
Martin Fossett, Tabor Academy
Michelle Fossum, Laney College
Sandro Gambarotta, University of Ottawa
Robert Garber, California State University, Long Beach
Michael D. Hampton, University of Central Florida
Paul Hunter, Michigan State University
Michael E. Lipschutz, Purdue University
Shelley D. Minteer, Saint Louis University
Jessica N. Orvis, Georgia Southern University
David Spurgeon, University of Arizona
Stephen P. Tanner, University of West Florida
John Townsend, West Chester University
John A. Weyh, Western Washington University
Marcy Whitney, University of Alabama
Sheila Woodgate, University of Auckland

Contributors

When we designed this edition, we decided to seek chemists outside of our team to author the supplemental interchapters. John Townsend prepared the chapter on The Chemistry of Life: Biochemistry, and Meredith Newman authored the chapter on The Chemistry of the Environment. We thank them for their very valuable contributions.

John R. Townsend, Associate Professor of Chemistry at West Chester University of Pennsylvania, completed his B.A. in Chemistry as well as the Approved Program for Teacher Certification in Chemistry at the University of Delaware. After a career teaching high school science and mathematics, he earned his M.S. and Ph.D. in biophysical chemistry at Cornell University. At Cornell he also performed experiments in the origins of life field and received the DuPont Teaching Award. After teaching at Bloomsburg University, Dr. Townsend joined the faculty at West Chester University, where he coordinates the chemistry education program for prospective high school teachers and the general chemistry program for science majors. He is also the co-leader of his university's local team of the Collaborative for Excellence in Teacher Preparation in Pennsylvania. His research interests lie in the fields of chemical education and biochemistry.

Meredith E. Newman is an associate professor of chemistry and geology at Hartwick College in Oneonta, New York. She received her B.S. in biology and her M.S. and Ph.D. in environmental engineering. After a postdoctoral appointment in the Department of Analytical Chemistry at the University of Geneva, Switzerland, and work at the Idaho National Environmental and Engineering Laboratory, she joined the faculty at Hartwick College. She has been a visiting scientist in the Environmental Engineering Department at Clemson University and the Institute for Alpine and Arctic Research at the University of Colorado in Boulder. Having previously been an affiliate faculty member at the University of Idaho in Idaho Falls, she is currently an affiliate faculty member at Clemson University. Her research on groundwater contaminant transport, subsurface colloid transport, and environmental education has been published in a variety of scientific journals and texts.

Advisory Board

Many decisions on topic placement, level of text, illustrations, and so on must be made when a textbook is being developed. We have benefited from the help of some wonderful colleagues who met with us on several occasions and who carried on email conversations in between. We certainly acknowledge their significant contributions.

Kevin Chambliss, Baylor University
Michael Hampton, University of Central Florida
Andy Jorgensen, University of Toledo
Laura Kibler-Herzog, Georgia State University
Cathy Middlecamp, University of Wisconsin, Madison
Norbert Pienta, University of Iowa
John Townsend, West Chester University

Left to right: Paul Treichel, Gabriela Weaver, and John Kotz

About the Authors

JOHN C. KOTZ, a State University of New York Distinguished Teaching Professor at the College at Oneonta, was educated at Washington and Lee University and Cornell University. He held National Institutes of Health postdoctoral appointments at the University of Manchester Institute for Science and Technology in England and at Indiana University.

He has coauthored three textbooks in several editions (*Inorganic Chemistry, Chemistry & Chemical Reactivity,* and *The Chemical World*) and the General ChemistryNow CD-ROM. He has also published on his research in inorganic chemistry and electrochemistry.

Dr. Kotz was a Fulbright Lecturer and Research Scholar in Portugal in 1979 and a Visiting Professor there in 1992. He was also a Visiting Professor at the Institute for Chemical Education (University of Wisconsin, 1991–1992) and at Auckland University in New Zealand (1999). He has been an invited speaker at a meeting of the South African Chemical Society and at the biennial conference for secondary school chemistry teachers in Christchurch, New Zealand. He was recently named a mentor of the U.S. Chemistry Olympiad Team.

Dr. Kotz has received several awards, among them a State University of New York Chancellor's Award (1979), a National Catalyst Award for Excellence in Teaching (1992), the Estee Lecturership in Chemical Education at the University of South Dakota (1998), the Visiting Scientist Award from the Western Connecticut Section of the American Chemical Society (1999), and the first annual Distinguished Education Award from the Binghamton (New York) Section of the American Chemical Society (2001). He may be contacted by email at kotzjc@oneonta.edu.

PAUL M. TREICHEL received his B.S. degree at the University of Wisconsin in 1958 and a Ph.D. from Harvard University in 1962. After a year of postdoctoral study in London, he assumed a faculty position at the University of Wisconsin–Madison, where he is currently Helfaer Professor of Chemistry. He served as department chair from 1986 through 1995. He has held visiting faculty positions in South Africa (1975) and in Japan (1995). Currently, he teaches courses in general chemistry, inorganic chemistry, and scientific ethics. Dr. Treichel's research in organometallic and metal cluster chemistry and in mass spectrometry, aided by 75 graduate and undergraduate students, has led to publication of more than 170 papers in scientific journals. He may be contacted by email at treichel@chem.wisc.edu.

GABRIELA C. WEAVER received her B.S. in 1989 from the California Institute of Technology and her Ph.D. in 1994 from the University of Colorado at Boulder. She served as Assistant Professor at the University of Colorado at Denver from 1994 to 2001 and as Associate Professor at Purdue University since 2001. She has been an invited speaker at more than 35 national and international meetings, including the 2001 Gordon Conference on Chemical Education Research and the DVD Summit in Dublin, Ireland. She is currently Director of the Center for Authentic Science Practice in Education at Purdue University. Her work in instructional technology development and on active learning has led to numerous publications in addition to her publications on surface physical chemistry. She may be contacted by email at gweaver@purdue.edu.

An Introduction to Chemistry

Dr. Donald Catlin, the director of the Olympic Analytical Laboratory in Los Angeles, California.

Chemical Sleuthing

On June 13, 2003, a colorless liquid arrived at the Olympic Analytical Laboratory in Los Angeles, California. This laboratory, headed by Dr. Donald H. Catlin, annually tests about 25,000 samples for the presence of illegal drugs. Among its clients are the U.S. Olympic Committee, the National Collegiate Athletic Association, and the National Football League.

At about the time of the U.S. Outdoor Track and Field Championships in the summer of 2003, a coach in Colorado tipped off the U.S. Anti-Doping Agency (USADA) that several athletes were using a new steroid. The coach had found a syringe with an unknown substance and sent it to the USADA. The USADA dissolved the contents of the syringe in a few milliliters of an alcohol, and then sent the solution to Catlin's laboratory for analysis. That submission initiated weeks of intense work that led to the identification of a previously unknown steroid that was presumably being used by athletes. The head of the USADA later said that the story behind the discovery suggested a "conspiracy involving chemists, coaches, and certain athletes using . . . undetectable designer steroids to defraud their fellow competitors and the world public."

To identify the unknown substance, chemists at the Olympic Analytical Laboratory used a GC-MS, an instrument widely employed in forensic science work. They first passed the sample through a gas chromatograph (GC), an instrument that can separate different chemical compounds in a mixture of liquids. A GC has a very-small-diameter, coiled tube (a typical inside diameter is 0.025 mm), in which the inside surface has been specially treated so that chemicals are attracted to the surface. This

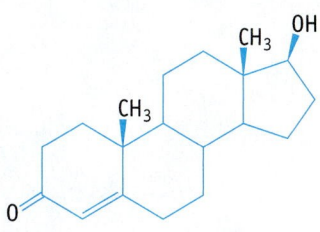

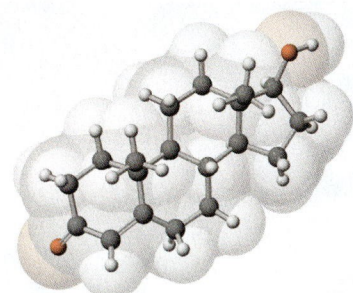

The steroid testosterone.
All steroids, including cholesterol, have the same basic four-ring structure, but they differ in detail.

A molecular model of testosterone.

A photo of crystals of the steroid cholesterol taken with a microscope using polarized light.

tube is placed in an oven and heated to a temperature of 200 °C or higher. Different substances in a mixture are swept along the tube by a stream of helium gas. Because each component in the sample binds differently to the material on the inside surface of the tube, each component moves through the column at a different rate and exits from the end of the column at a different time. Thus, separation of the components in the mixture is achieved.

After exiting the GC, each compound is routed directly into a mass spectrometer (MS). (Scientists would describe the two instruments as being interfaced, or linked together, and operating as a single unit.) In a mass spectrometer, the compounds are bombarded with high-energy electrons, and each compound is turned into an ion, a species with a positive electric charge. These ions are then passed through a strong magnetic field, causing the ions to be deflected. The path an ion takes in the magnetic field (the extent of deflection) is related to its mass. The mass of the particle is a key piece of information that will help to identify the compound.

Such a straightforward process: separate the compounds in a GC and identify them in a MS. What can go wrong? In fact, many things can potentially go awry that require ingenuity to overcome. In this case, the unknown steroid did not survive the high temperatures of the GC. It broke apart into pieces, making it possible to study only the pieces of the original molecule. However, this analysis gave enough evidence to convince scientists that the compound was a steroid. But what steroid? According to Catlin, one hypothesis was that "the new steroid was made by people who knew it was not going to be detectable"—that is, the molecule had been designed in a way that would guarantee that it would not be detected by the standard GC-MS procedure.

Intrigued, Catlin and his colleagues set out to identify the steroid. First, they made the molecule stable during the analysis. This was done by attaching new atoms to the molecule to make what chemists call a *derivative*. A number of approaches were tested, and one gave a molecule that did not break down in the GC. MS data allowed scientists to identify the intact molecule (the derivative). Based on this identification and the chemistry used to prepare the derivative, they now knew the identity of the unknown steroid only a few weeks after they had received the sample.

The final step to solve the mystery was to try to make the compound in the laboratory and then to use the GC-MS on this sample. If the material behaved the same way as the unknown sample, then the scientists could be as certain as possible they knew what they had received from the track coach. These experiments worked, confirming the identity of the compound. It was an entirely new steroid, never seen

A GC-MS (gas chromatograph-mass spectrometer). A GC-MS is one of the major tools used in forensic chemistry. A gas chromatograph (GC) separates chemical compounds in a mixture by using differences in the ability of compounds to bind to a chemically treated surface in a thin, coiled tube. When substances emerge from the chromatograph, they are analyzed and identified by the mass spectrometer (MS). The GC-MS pictured has an automated sample changer (carousel, center). An operator will load dozens of samples into the carousel, and the instrument will then process the samples automatically, with data recorded and stored in a computer.

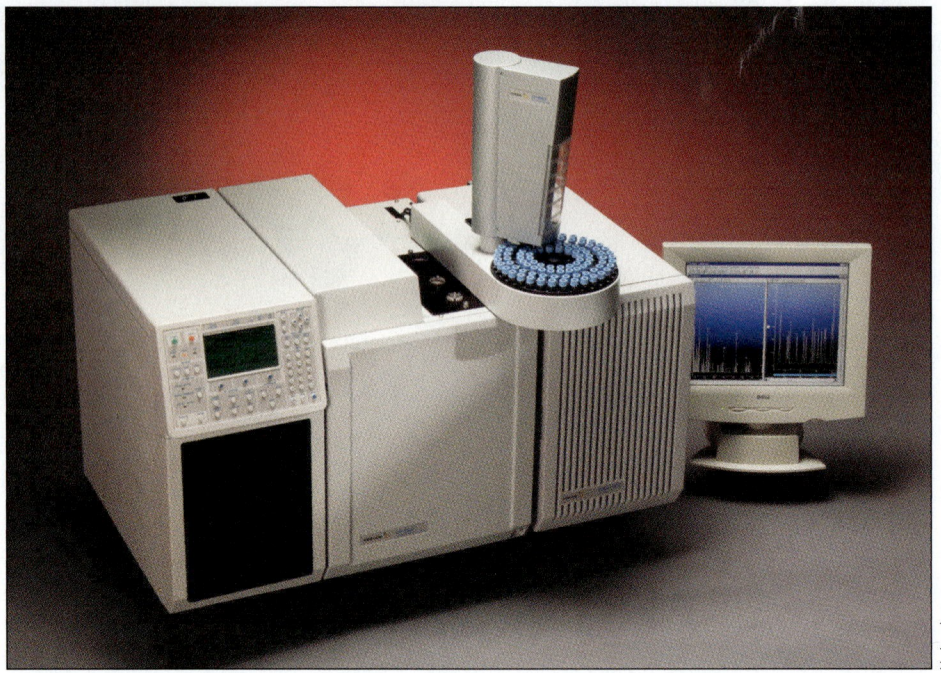

Varian, Inc.

before in nature or in the laboratory. Its formula is $C_{21}H_{28}O_2$, and its name is tetrahydrogestrinone or THG. THG resembled two well-known steroids: gestrinone, used to treat gynecological problems, and trenbolone, a steroid used by ranchers to beef up cattle.

There are two sequels to the story. First, a scientific problem is not solved until its solution has been verified in another laboratory. Not only was this confirmatory analysis done, but a test was soon devised to find THG in urine samples. Second, armed with the new analytical procedures, the USADA asked Catlin's lab to retest 550 urine samples—and THG was found in several.

What is the problem with athletes taking steroids? THG is one of a class of steroids called anabolic steroids. They elevate the body's natural testosterone levels and increase body mass, muscle strength, and muscle definition. They can also improve an athlete's capacity to train and compete at the highest levels. Aside from giving steroid users an illegal competitive advantage, steroids have some damaging potential side effects—liver damage, heart disease, anxiety, and rage.

A check of the Internet shows that there are hundreds of sources of steroids for athletes. The known performance-enhancing drugs can be detected and their users banned from competitive sports. But what about as-yet-unknown steroids? Catlin believes that other steroids are available on the market, made by secret labs without safety standards, a problem he calls horrifying.

Chemistry and Its Methods

Chemistry is about change. It was once only about changing one natural substance into another—wood and oil burn, grape juice turns into wine, and cinnabar (Figure 1), a red mineral from the earth, changes into shiny quicksilver. Today chemistry is still about change, but now chemists focus on the change of one pure substance, whether natural or synthetic, into another (Figure 2).

Charles D. Winters

(a)

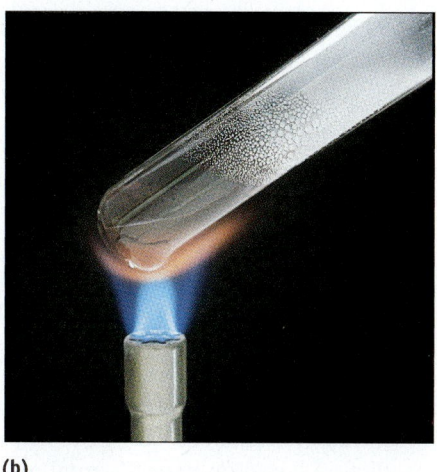

(b)

Figure 1 Cinnabar and mercury.
(a) The red crystals of cinnabar consist of the chemical compound mercury sulfide. It is heated in air to change it into orange mercury oxide (b), which, on further heating, decomposes to the elements oxygen and mercury metal. (The droplets you see on the test tube are mercury.)

Although chemistry is endlessly fascinating—at least to chemists—why should you study it? Each person probably has a different answer, but many of you may be taking this chemistry course because someone else has decided it is an important part of preparing for a particular career. Chemistry is especially useful because it is central to our understanding of disciplines as diverse as biology, geology, materials science, medicine, physics, and many branches of engineering. In addition, chemistry plays a major role in our economy; chemistry and chemicals affect our daily lives in a wide variety of ways. Furthermore, a course in chemistry can help you see how a scientist thinks about the world and how to solve problems. The knowledge and skills developed in such a course will benefit you in many career paths and will

Photos: Charles D. Winters

Solid sodium, Na

Chlorine gas, Cl$_2$

Sodium chloride solid, NaCl

Figure 2 Forming a chemical compound. (Sodium chloride, table salt, can be made by combining sodium metal (Na) and yellow chlorine gas (Cl$_2$). The result is a crystalline solid, common salt. (The spheres show how the atoms are arranged in the substances.)

Figure 3 **The metallic element sodium reacts vigorously with water.** (*See General ChemistryNow Screen 8.15 Chemical Reactions and Periodic Properties, for a video of the reactions of lithium, sodium, and potassium with water.*)

help you become a better-informed citizen in a world that is becoming technologically more complex—and more interesting.

Hypotheses, Laws, and Theories

To begin your study of chemistry, this Preface discusses some fundamental ideas used by scientists of all kinds.

As scientists, we study questions of our own choosing or ones that someone else poses in the hope of finding an answer or of discovering some useful information. In the story of the revelation of the banned steroid, THG, Dr. Catlin and his group of chemists were handed a problem to solve, and they followed the usual methods of science to arrive at the answer.

After some preliminary tests, they recognized that the mystery substance was most likely a steroid. That is, they formed a **hypothesis,** a tentative explanation or prediction based on experimental observations.

After formulating one or more hypotheses, scientists perform experiments that are designed to give results that confirm or invalidate these hypotheses. In chemistry this usually requires that both quantitative and qualitative information be collected. **Quantitative** information is numerical data, such as the temperature at which a chemical substance melts. **Qualitative** information, in contrast, consists of non-numerical observations, such as the color of a substance or its physical appearance.

Catlin and his colleagues assembled a great deal of qualitative and quantitative information. Based on their experience, and on experiments done in the past by chemists around the world, they became more certain they knew the identity of the substance. Their preliminary experiments led them to perform still more experiments, such as looking for a way to stabilize the molecule so that it would not decompose and attempting to make the molecule in the laboratory. Finally, to make certain they had the right molecule and knew how to detect it, their work was confirmed by scientists in other laboratories.

After scientists have performed a number of experiments, and the results have been checked to ensure they are reproducible, a pattern of behavior or results may emerge. At this point it may be possible to summarize the observations in the form of a general rule or conclusion. After making a number of experimental observations, Catlin and his associates could conclude, for example, that the unknown substance was a steroid because it had properties characteristic of many other steroids they had observed.

Finally, after numerous experiments have been conducted by many scientists over an extended period of time, the original hypothesis may become a **law**—a concise verbal or mathematical statement of a behavior or a relation that seems always to be the same under the same conditions. An example might be the law of mass conservation in chemical reactions.

We base much of what we do in science on laws because they help us predict what may occur under a new set of circumstances. For example, we know from experience that if the chemical element sodium comes in contact with water, a violent reaction will occur and new substances will be formed (Figure 3). We also know that the mass of the substances produced in the reaction is exactly the same as the mass of sodium and water used in the reaction. That is, mass is conserved. But the result of an experiment might be different from what is expected based on a general rule. When that happens, chemists get excited because experiments that do not follow the known rules of chemistry are often the most interesting. We know that understanding the exceptions almost invariably gives new insights.

Once enough reproducible experiments have been conducted, and experimental results have been generalized as a law or general rule, it may be possible to conceive a theory to explain the observation. A **theory** is a unifying principle that explains a body of facts and the laws based on them. It is capable of suggesting new hypotheses.

Sometimes nonscientists use the word "theory" to imply that someone has made a guess and that an idea is not yet substantiated. To scientists, a theory is based on carefully determined and reproducible evidence. Theories are the cornerstone of our understanding of the natural world at any given time. Remember, though, that theories are inventions of the human mind. Theories can and do change as new facts are uncovered.

Goals of Science

The sciences, including chemistry, have several goals. Two of these are prediction and control. We do experiments and seek generalities because we want to be able to predict what may occur under a given set of circumstances. We also want to know how we might control the outcome of a chemical reaction or process.

A third goal is explanation and understanding. We know, for example, that certain elements will react vigorously with water (see Figure 3). But why should this be true? And why is this extreme reactivity unique to these elements? To explain and understand this phenomenon, we turn to theories such as those developed in Chapters 9 and 10.

The Importance of Serendipity and Creativity

People who work outside of science usually have the idea that science is an intensely logical field. They picture white-coated chemists moving logically from hypothesis to experiment and then to laws and theories without human emotion or foibles. This picture is a great simplification—and quite wrong!

Often, scientific results and understanding arise quite by accident, otherwise known as **serendipity.** Creativity and insight are needed to transform a fortunate accident into useful and exciting results. The discovery of the cancer drug cisplatin by Barnett Rosenberg in 1965 or of penicillin by Alexander Fleming (1881–1955) in 1928 are wonderful examples of serendipity.

A material familiar to many of you—Teflon®—was found by a combination of serendipity and curiosity. In 1938 Dr. Roy Plunkett was a young scientist working in a DuPont laboratory on the chemistry of fluorine-containing refrigerants (which we now know by their trademark name, Freon). For one experiment, Plunkett and his assistants opened a tank of tetrafluoroethylene gas. The tank supposedly held 1000 g of gas, but only 990 g came out. What happened to the other 10 g? Curiosity is the mark of a good scientist, so they sawed open the tank. A white, waxy substance coated the inside (Figure 4). Following his curiosity further, Plunkett tested the material and found it had remarkable properties. It was more inert than sand! Strong acids and bases did not affect it, nothing could dissolve it, and it was resistant to heat. Unlike sand, it was slippery.

Were it not so expensive, the remarkable properties of this new substance should have led to an immediate search for uses in consumer products. However, Teflon found its first use in the World War II atomic bomb project as a sealant in the equipment used in the separation of uranium. The project was of such national importance that the expense of the material was of no concern. Not until the 1960s did Teflon begin to show up in consumer items. Today one of its most important uses is in medical products (Figure 5). Because it is one of the few substances the body does not reject, it can be used for hip and knee joints, heart valves, and many other body parts.

Hagley Museum and Library

Figure 4 Discovery of Teflon. In a photo taken of a reenactment of the actual event in 1938, Roy Plunkett (*right*) (1910–1994) and his assistants find a white solid coating the inside of a gas cylinder. This solid, now called Teflon, was discovered by accident.

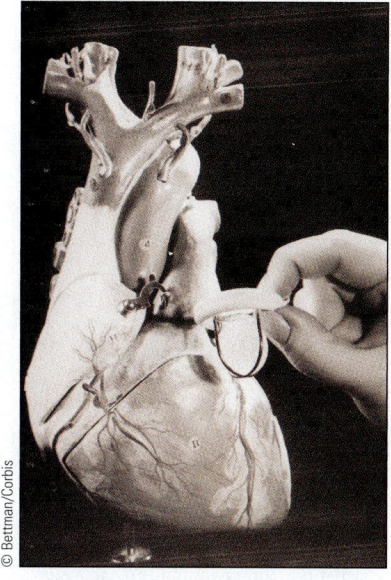

© Bettman/Corbis

Figure 5 Medical products such as heart valves use the polymer Teflon.

Dilemmas and Integrity in Science

You may think research in science is straightforward: Do an experiment, draw a conclusion. In reality, research is seldom that easy. Frustrations and disappointments are common enough, and results can be inconclusive. Complicated experiments often contain some level of uncertainty, and spurious or contradictory data can be collected. For example, suppose you perform an experiment expecting to find a direct relationship between two experimental quantities. You collect six data sets. When plotted, four of the sets lie on a straight line, but two others lie far away from the line. Should you ignore the last two points? Or should you do more experiments when you know the time they take might mean someone else could publish first and thus get the credit for discovering a new scientific principle? What if the two points not on the line indicate that your original hypothesis is wrong, so that you will have to abandon a favorite idea you have worked on for a year? Scientists have a responsibility to remain objective in these situations, but it is sometimes hard to do.

It is important to remember that scientists are human and therefore subject to the same moral pressures and dilemmas as any other person. To help ensure integrity in science, some simple principles have emerged over time that guide scientific practice:

- Experimental results should be reproducible. Furthermore, these results should be reported in sufficient detail that they can be used or reproduced by others.

- Conclusions should be reasonable and unbiased.

- Credit should be given where it is due.

Moral and ethical issues frequently arise in science. Consider the ban on using the pesticide DDT (Figure 6). This is a classic case of the law of "unintended consequences." DDT was developed during World War II and promoted as effective in controlling pests but harmless to people. In fact, it was thought to be so effective that it was used in larger and larger quantities around the world. It was especially effective in controlling mosquitoes carrying malaria. Unfortunately, it soon became evident that there were negative consequences to DDT use. In Borneo, the World Health Organization used large quantities of DDT to kill mosquitoes. The mosquito population did indeed decline, as did malaria incidence. Soon, however, the thatch roofs of people's houses fell down. A parasitic wasp, which ate thatch-eating caterpillars, had also been wiped out by the DDT. Worse still was that geckoes, small lizards, which had eaten DDT-laced caterpillars, were eaten by cats, which then died. The end result was an infestation of rats. Unintended consequences, indeed.

DDT use has been banned in many parts of the world because of its very real, but unforeseen, environmental consequences. The DDT ban occurred in the United States in 1972 because evidence accumulated that the pesticide affected the reproduction of birds such as the bald eagle. DDT is also known to accumulate slowly in human body fat.

The ban on DDT has affected the control of malaria-carrying insects, however. Several million people, primarily children in sub-Saharan Africa, die every year from malaria. The chairman of the Malaria Foundation International has said that "the malaria epidemic is like loading up seven Boeing 747 airliners each day and crashing them into Mt. Kilimanjaro." Consequently, there is a movement to return DDT to the arsenal of weapons in fighting the spread of malaria.

There are many, many moral and ethical issues for chemists. Chemistry has extended and improved the lives of millions of people. But just as certainly, chemicals can cause harm, particularly when misused. It is incumbent on all of us to understand enough science to ask pertinent questions and to evaluate sources of infor-

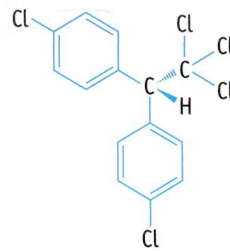

(a) The molecular structure of DDT.

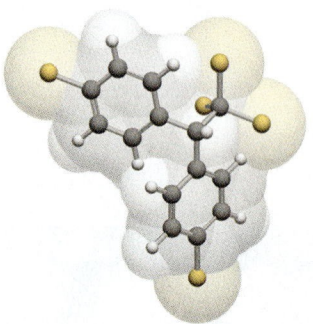

(b) A molecular model of DDT.

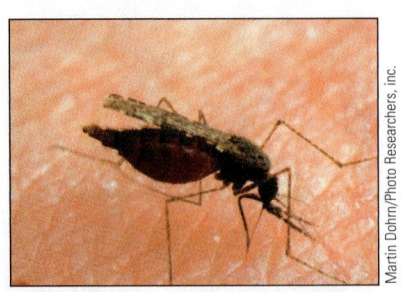

(c) DDT can be used to control malaria-carrying insects such as mosquitos.

Martin Dohrn/Photo Researchers, inc.

Figure 6   The pesticide DDT, an example of the moral and ethical issues in science.

mation sufficiently to reach reasonable conclusions regarding the health and safety of ourselves and our communities.

A Final Word to Students

Why study chemistry? The reasons are clear. Whether you want to become a biologist, a geologist, an engineer, or a physician, or pursue any of dozens of other professions, chemistry will be at the core of your discipline. It will always be useful to you, sometimes when least expected.

In addition, you will be called upon to make many decisions in your life for your own good or for the good of those in your community—whether that be your neighborhood or the world. An understanding of the nature of science in general, and of chemistry in particular, can only serve to help in these decisions.

Because the authors of this book were students once—and still are—we know chemistry can be a challenging area of study. Like anything worthwhile, it takes time and effort to reach genuine understanding. Be sure to give it time, and talk with your professors and your fellow students. We are sure you will find it as exciting, as useful, and as interesting as we do.

Readings About Science

You will find the following books about science both interesting and informative:

- Rachel Carson: *Silent Spring*, New York, Houghton Mifflin, 1962.
- John Emsley: *Molecules at an Exhibition, Portraits of Intriguing Materials in Everyday Life*, New York, Oxford University Press, 1998.
- John Emsley: *The 13th Element: The Sordid Tale of Murder, Fire, and Phosphorus*, New York, John Wiley & Sons, 2000.
- John Emsley: *Nature's Building Blocks, An A–Z Guide to the Elements*, New York, Oxford University Press, 2001.
- Richard Feynman: *What Do You Care What Other People Think?*, New York, W. W. Norton and Company, 1988; and *Surely You're Joking, Mr. Feynman*, New York, W. W. Norton and Company, 1985.
- Arthur Greenberg: *A Chemical History Tour, Picturing Chemistry from Alchemy to Modern Molecular Science*, New York, Wiley-Interscience, 2000.
- Roald Hoffmann and Vivian Torrance: *Chemistry Imagined, Reflections on Science*, Washington, D.C., Smithsonian Institution Press, 1993.
- Primo Levi: *The Periodic Table*, New York, Schocken Books, 1984. An autobiography of a chemist, a resistance fighter in World War II, and a man who survived some years in a concentration camp.
- Sharon D. McGrayne: *Nobel Prize Women in Science*, New York, Birch Lane Press, 1993.
- Royston M. Roberts: *Serendipity: Accidental Discoveries in Science*, New York, John Wiley & Sons, 1989.
- Oliver Sacks: *Uncle Tungsten, Memories of a Chemical Boyhood*, New York, Alfred Knopf, 2001.
- Lewis Thomas: *The Lives of a Cell*, New York, Penguin Books, 1978.
- J. D. Watson: *The Double Helix, A Personal Account of the Discovery of the Structure of DNA*, New York, Atheneum, 1968.

1—Matter and Measurement

Platinum resistance thermometer. This device measures temperatures over a range from about −259 °C to +962 °C.

How Hot Is It?

"It's so hot outside you could fry an egg on the sidewalk!" This is an expression we heard as children. But what does it mean to say that something is hot? We would say it has a high temperature—but what is temperature and how is it measured?

Temperature and heat are related but different concepts. Although we will discuss the difference in more detail in Chapter 6, for the moment it is easy to think of them this way: Temperature determines the direction of heat transfer. That is, heat transfers from something at a higher temperature to something at a lower temperature. If you touch your finger to a hot match, heat is transferred to your finger, and you decide the match is hot.

Early scientists learned that gases, liquids, and solids expand when heated. In his investigations of heat, Galileo Galilei (1564–1642) invented the "thermoscope," a simple device that depended on the expansion of a liquid in a tube with increasing temperature. Others developed instruments based on this principle, using liquids such as alcohol and mercury. Among them was Daniel Gabriel Fahrenheit (1686–1736). To create his scale, Fahrenheit initially assigned the freezing point of water as 7.5 °F and body temperature as 22.5 °F. He multiplied these values by 4, and then later adjusted them so that the freezing point of water was 32 °F and body temperature was 96 °F. After Fahrenheit's death a further revision of the scale established the reference temperatures at their current values, 32 °F for the freezing point of water and 212 °F for the boiling point. On the current scale, normal body temperature is 98.6 °F.

Anders Celsius (1701–1744). Swedish astronomer and geographer.

A significant advance in temperature measurement came from Anders Celsius (1701–1744). Celsius was a Swedish geographer and astronomer who constructed the Celsius thermometer, which used liquid mercury in a glass tube. The Celsius thermometer scale originally used 0 as the boiling point of water, and 100 as the freezing point of water—reference points that were reversed after Celsius's death. His contribution to thermometry was to show experimentally that the freezing point of water is unchanged by atmospheric pressure or the latitude at which the experiment is done. Celsius also showed that, in contrast, the boiling point of water does depend on atmospheric pressure. Both of these observations were important to establishing a standard temperature scale that could be used around the world.

In modern science there is an interest in determining low and high temperatures well outside the ranges where alcohol and mercury are liquids. Scientists have created new temperature measuring devices for this purpose. The platinum resistance thermometer, for example, relies on the fact that the electrical resistance of platinum wire changes with temperature in a predictable manner. Such devices are extremely sensitive and can make measurements to within one thousandth of a degree over temperatures ranging from −259.25 °C to +961.78 °C (the melting point of silver).

How do you measure a very high temperature—say, a temperature high enough to boil mercury or melt glass or platinum? From watching the heater element on a stove or in a toaster, you know that heated objects emit light. It turns out that the wavelength of the emitted light can be correlated with temperature. A pyrometer, an optical device, is commonly used for this purpose.

You might have had your temperature taken with a device that is inserted in your ear. This instrument is essentially a pyrometer. Warm humans emit light, albeit at longer wavelengths than a toaster element. A sensor in the ear thermometer scans the wavelength emitted from the eardrum and reports the temperature. This is a useful measure of body temperature because the eardrum shares blood vessels with the hypothalamus, the area of the brain that regulates body temperature.

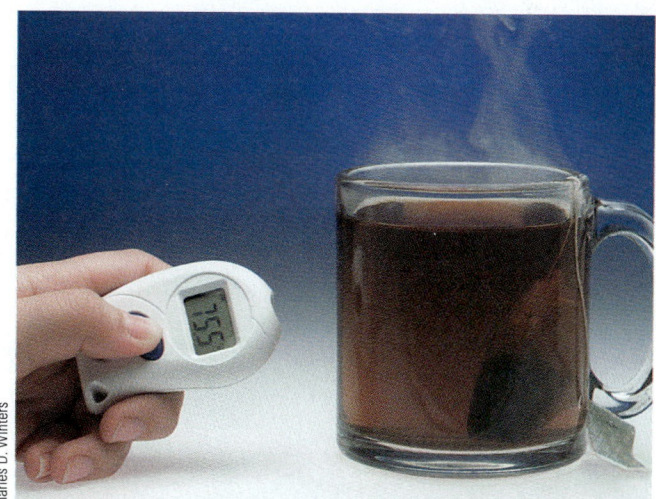

Charles D. Winters

Infrared thermometer. This device depends on the long wavelength radiation emitted by a warm object.

Charles D. Winters

Thinking about matter. Is this a glass of
pure water? How can you prove it is?

Imagine a tall glass filled with a clear liquid. Sunlight from a nearby window causes the liquid to sparkle, and the glass is cool to the touch. A drink of water would certainly taste good, but should you take a sip? If the glass were sitting in your kitchen you might say yes. But what if this scene occurred in a chemical laboratory? How would you know that the glass held pure water? Or, to pose a more "chemical" question, how would you *prove* this liquid is water?

We usually think of the water we drink as being pure, but this is not strictly true. In some instances material may be suspended in it or bubbles of gases such as oxygen may be visible to the eye. Some tap water has a slight color from dissolved iron. In fact, drinking water is almost always a mixture of substances, some dissolved and some not. As with any mixture, we could ask many questions. What are the components of the mixture—dust particles, bubbles of oxygen, dissolved sodium, calcium, or iron salts—and what are their relative amounts? How can these substances be separated from one another, and how are the properties of one substance changed when it is mixed with another?

This chapter begins our discussion of how chemists think about matter. After looking at a way to classify matter, we will turn to some basic ideas about elements, atoms, compounds, and molecules and discover how chemists characterize these building blocks of matter. Finally, we will see how we can use numerical information.

1.1—Classifying Matter

A chemist looks at a glass of drinking water and sees a liquid. This liquid could be the chemical compound water. More likely, the liquid is a homogeneous mixture of water and dissolved substances—that is, a solution. It is also possible the water sample is a heterogeneous mixture, with solids being suspended in the liquid. These descriptions represent some of the ways we can classify matter (Figure 1.1).

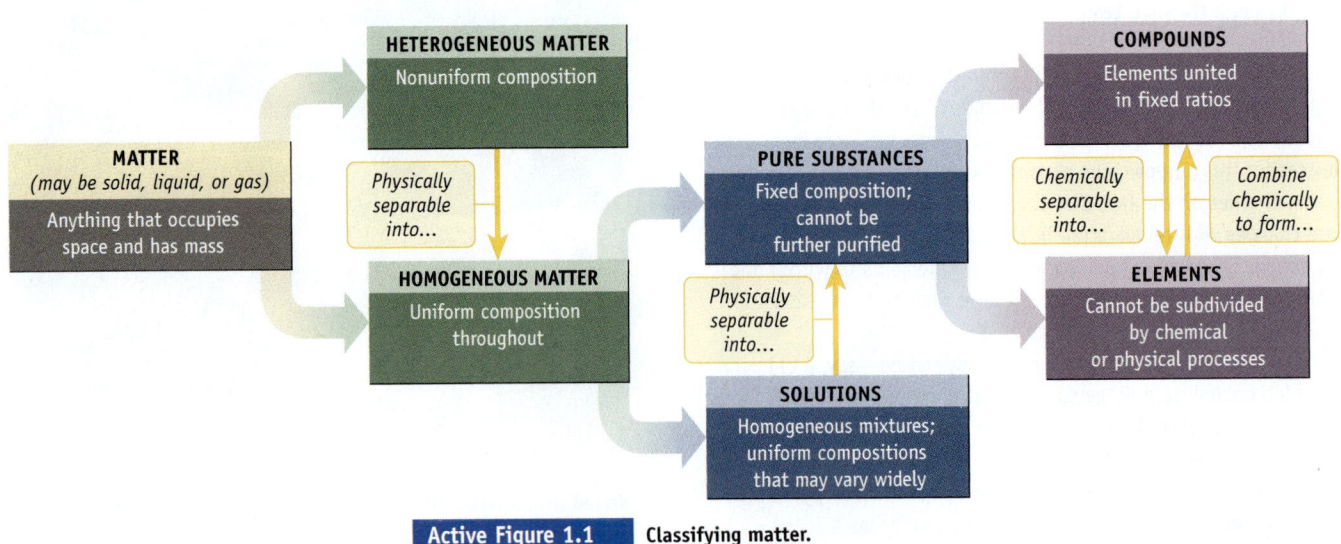

Active Figure 1.1   **Classifying matter.**

GENERAL
Chemistry Now™ See the General ChemistryNow CD-ROM or website to explore an interactive version of this figure accompanied by an exercise.

| Solid | Liquid | Gas |

Bromine solid and liquid   Bromine gas and liquid

Active Figure 1.2   **States of matter—solid, liquid, and gas.** Elemental bromine exists in all three states near room temperature. The tiny spheres represent bromine (Br) atoms. In elemental bromine, two Br atoms join to form a Br_2 molecule. (See Section 1.3 and Chapter 3.)

GENERAL Chemistry Now™ See the General ChemistryNow CD-ROM or website to explore an interactive version of this figure accompanied by an exercise.

States of Matter and Kinetic-Molecular Theory

An easily observed property of matter is its **state**—that is, whether a substance is a solid, liquid, or gas (Figure 1.2). You recognize a solid because it has a rigid shape and a fixed volume that changes little as temperature and pressure change. Like solids, liquids have a fixed volume, but a liquid is fluid—it takes on the shape of its container and has no definite shape of its own. Gases are fluid as well, but the volume of a gas is determined by the size of its container. The volume of a gas varies more than the volume of a liquid with temperature and pressure.

At low enough temperatures, virtually all matter is found in the solid state. As the temperature is raised, solids usually melt to form liquids. Eventually, if the temperature is high enough, liquids evaporate to form gases. Volume changes typically accompany changes in state. For a given mass of material, there is usually a small increase in volume on melting—water being a significant exception—and then a large increase in volume occurs upon evaporation.

The **kinetic-molecular theory** of matter helps us interpret the properties of solids, liquids, and gases. According to this theory, all matter consists of extremely tiny particles (atoms, molecules, or ions), which are in constant motion.

- In solids these particles are packed closely together, usually in a regular array. The particles vibrate back and forth about their average positions, but seldom does a particle in a solid squeeze past its immediate neighbors to come into contact with a new set of particles.

- The atoms or molecules of liquids are arranged randomly rather than in the regular patterns found in solids. Liquids and gases are fluid because the particles are not confined to specific locations and can move past one another.

- Under normal conditions, the particles in a gas are far apart. Gas molecules move extremely rapidly because they are not constrained by their neighbors. The molecules of a gas fly about, colliding with one another and with the container walls. This random motion allows gas molecules to fill their container, so the volume of the gas sample is the volume of the container.

An important aspect of the kinetic-molecular theory is that the higher the temperature, the faster the particles move. The energy of motion of the particles (their **kinetic energy**) acts to overcome the forces of attraction between particles. A solid melts to form a liquid when the temperature of the solid is raised to the point at which the particles vibrate fast enough and far enough to push one another out of the way and move out of their regularly spaced positions. As the temperature increases even more, the particles move even faster until finally they can escape the clutches of their comrades and enter the gaseous state. Increasing temperature corresponds to faster and faster motions of atoms and molecules, a general rule you will find useful in many future discussions.

Matter at the Macroscopic and Particulate Levels

The characteristic properties of gases, liquids, and solids just described are observed by the unaided human senses. They are determined using samples of matter large enough to be seen, measured, and handled. Using such samples, we can also determine, for example, what the color of a substance is, whether it dissolves in water, or whether it conducts electricity or reacts with oxygen. Observations and manipulations generally take place in the **macroscopic** world of chemistry (Figure 1.3). This is the world of experiments and observations.

Now let us move to the level of atoms, molecules, and ions—a world of chemistry we cannot see. Take a macroscopic sample of material and divide it, again and again, past the point where the amount of sample can be seen by the naked eye, past the point where it can be seen using an optical microscope. Eventually you reach the level of individual particles that make up all matter, a level that chemists refer to as the **submicroscopic** or **particulate** world of atoms and molecules (Figures 1.2 and 1.3).

Chemists are interested in the structure of matter at the particulate level. Atoms, molecules, and ions cannot be "seen" in the same way that one views the macroscopic world, but they are no less real to chemists. Chemists imagine what atoms must look like and how they might fit together to form molecules. They create models to represent atoms and molecules (Figures 1.2 and 1.3)—where tiny spheres are used to represent atoms—and then use these models to think about chemistry and to explain the observations they have made about the macroscopic world.

It has been said that chemists carry out experiments at the macroscopic level, but they think about chemistry at the particulate level. They then write down their observations as "symbols," the letters (such as H_2O for water or Br_2 for bromine molecules) and drawings that signify the elements and compounds involved. This is a useful perspective that will help you as you study chemistry. Indeed, one of our goals is to help you make the connections in your own mind among the symbolic, particulate, and macroscopic worlds of chemistry.

Pure Substances

Let us think again about a glass of drinking water. How would you tell whether the water is pure (a single substance) or a mixture of substances? Begin by making a few simple observations. Is solid material floating in the liquid? Does the liquid have an odor or an unexpected taste or color?

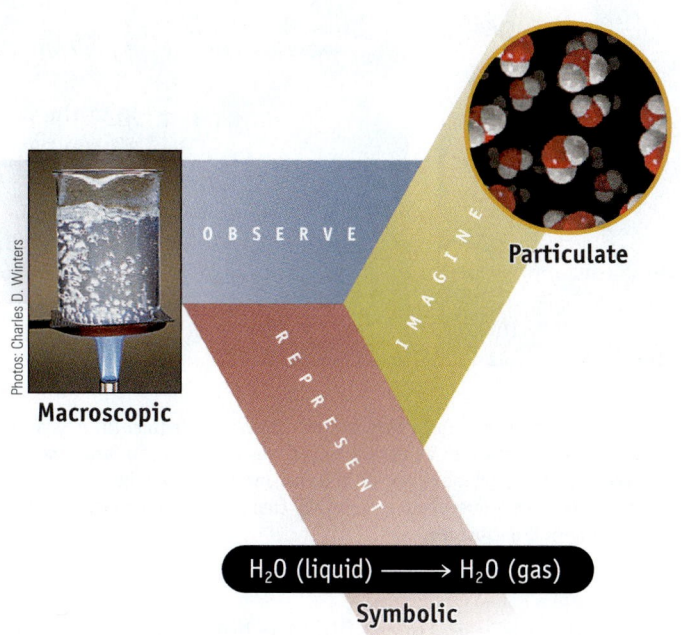

Particulate

OBSERVE

IMAGINE

REPRESENT

Macroscopic

Photos: Charles D. Winters

H₂O (liquid) ⟶ H₂O (gas)

Symbolic

Active Figure 1.3 **Levels of matter.** We observe chemical and physical processes at the macroscopic level. To understand or illustrate these processes, scientists often try to imagine what has occurred at the particulate atomic and molecular levels and write symbols to represent these observations. A beaker of boiling water can be visualized at the particulate level as rapidly moving H_2O molecules. The process is symbolized by indicating that the liquid H_2O molecules are becoming H_2O molecules in the gaseous state.

GENERAL
Chemistry Now™ See the General ChemistryNow CD-ROM or website to explore an interactive version of this figure accompanied by an exercise.

Every substance has a set of unique properties by which it can be recognized. Pure water, for example, is colorless, is odorless, and certainly does not contain suspended solids. If you wanted to identify a substance conclusively as water, you would have to examine its properties carefully and compare them against the known properties of pure water. Melting point and boiling point serve the purpose well here. If you could show that the substance melts at 0 °C and boils at 100 °C at atmospheric pressure, you can be certain it is water. No other known substance melts and boils at precisely these temperatures.

A second feature of a pure substance is that it cannot be separated into two or more different species by any physical technique such as heating in a Bunsen flame. If it could be separated, our sample would be classified as a mixture.

Mixtures: Homogeneous and Heterogeneous

A cup of noodle soup is obviously a mixture of solids and liquids (Figure 1.4a). A **mixture** in which the uneven texture of the material can be detected is called a **heterogeneous** mixture. Heterogeneous mixtures may appear completely uniform but on closer examination are not. Blood, for example, may not look heterogeneous until you examine it under a microscope and red and white blood cells are revealed (Figure 1.4b). Milk appears smooth in texture to the unaided eye, but magnification

(a)

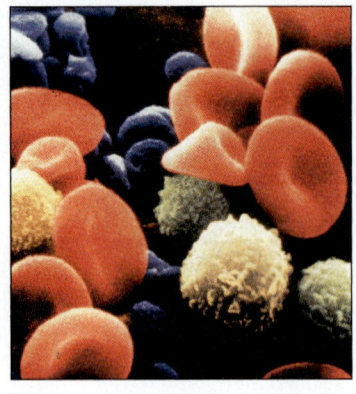

(b)

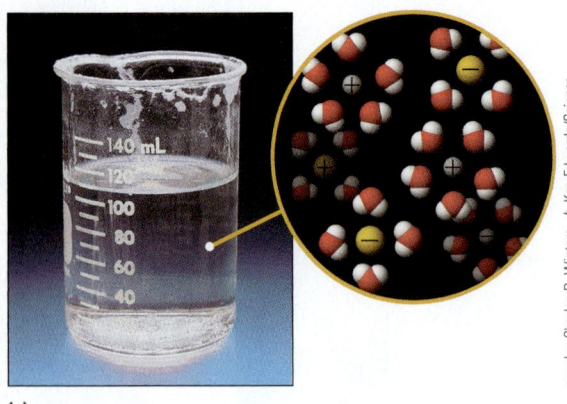

(c)

a and c, Charles D. Winters; b, Ken Edwards/Science Source/Photo Researchers, Inc.

Figure 1.4   Mixtures. (a) A cup of noodle soup is a heterogeneous mixture. (b) A sample of blood may look homogeneous, but examination with an optical microscope shows it is, in fact, a heterogeneous mixture of liquids and suspended particles (blood cells). (c) A homogeneous mixture, here consisting of salt in water. The model shows that salt consists of separate, electrically charged particles (ions) in water, but the particles cannot be seen with an optical microscope.

would reveal fat and protein globules within the liquid. In a heterogeneous mixture the properties in one region are different from those in another region.

A **homogeneous** mixture consists of two or more substances in the same phase (Figure 1.4c). No amount of optical magnification will reveal a homogeneous mixture to have different properties in different regions. Homogeneous mixtures are often called **solutions**. Common examples include air (mostly a mixture of nitrogen and oxygen gases), gasoline (a mixture of carbon- and hydrogen-containing compounds called hydrocarbons), and an unopened soft drink.

When a mixture is separated into its pure components, the components are said to be *purified* (see Figure 1.5). Efforts at separation are often not complete in a sin-

(a)

(b)

a, Charles D. Winters; b, Littleton, Massachusetts, Spectacle Pond Iron and Manganese Treatment Facility

Figure 1.5   Purifying water by filtration. (a) A laboratory setup. A beaker full of muddy water is passed through a paper filter, and the mud and dirt are removed. (b) A water treatment plant uses filtration to remove suspended particles from the water.

gle step, however, and repetition almost always gives an increasingly pure substance. For example, soil particles can be separated from water by filtration (Figure 1.5). When the mixture is passed through a filter, many of the particles are removed. Repeated filtrations will give water a higher and higher state of purity. This purification process uses a property of the mixture, its clarity, to measure the extent of purification. When a perfectly clear sample of water is obtained, all of the soil particles are assumed to have been removed.

Homogeneous and heterogeneous mixtures. Which is homogeneous? See Exercise 1.1.

See the General ChemistryNow CD-ROM or website:

- **Screen 1.5 Mixtures and Pure Substances,** for an exercise on identifying pure substances and types of mixtures
- **Screen 1.6 Separation of Mixtures,** to watch a video on heterogeneous mixtures

Exercise 1.1—Mixtures and Pure Substances

The photo in the margin shows two mixtures. Which is a homogeneous mixture and which is a heterogeneous mixture?

■ **Exercise Answers**
In each chapter of the book you will find a number of Exercises. Their purpose is to help you to check your knowledge of the material in that chapter. Solutions to the Exercises are found in Appendix N.

1.2—Elements and Atoms

Passing an electric current through water can decompose it to gaseous hydrogen and oxygen (Figure 1.6a). Substances like hydrogen and oxygen that are composed of only one type of atom are classified as **elements**. Currently 116 elements are known. Of these, only about 90—some of which are illustrated in Figure 1.6—are found in nature. The remainder have been created by scientists. The name and symbol for each element are listed in the tables at the front and back of this book. Carbon (C), sulfur (S), iron (Fe), copper (Cu), silver (Ag), tin (Sn), gold (Au), mercury (Hg), and lead (Pb) were known to the early Greeks and Romans and to the alchemists of ancient China, the Arab world, and medieval Europe. However, many other elements—such as aluminum (Al), silicon (Si), iodine (I), and helium (He)—were not discovered until the 18th and 19th centuries. Finally, artificial elements—those that do not exist in nature, such as technetium (Tc), plutonium (Pu), and americium (Am)—were made in the 20th century using the techniques of modern physics.

Many elements have names and symbols with Latin or Greek origins. Examples include helium (He), named from the Greek word *helios* meaning "sun," and lead, whose symbol, Pb, comes from the Latin word for "heavy," *plumbum*. More recently discovered elements have been named for their place of discovery or for a person or place of significance. Examples include americium (Am), californium (Cf), and curium (Cm).

The table inside the front cover of this book, in which the symbol and other information for the elements are enclosed in a box, is called the **periodic table**. We will describe this important tool of chemistry in more detail beginning in Chapter 2.

An **atom** is the smallest particle of an element that retains the characteristic chemical properties of that element. Modern chemistry is based on an understanding and exploration of nature at the atomic level. We will have much more to say about atoms and atomic properties in Chapters 2, 7, and 8, in particular.

■ **Writing Element Symbols**
Notice that only the first letter of an element's symbol is capitalized. For example, cobalt is Co, not CO. The notation CO represents the chemical compound carbon monoxide. Also note that the element name is not capitalized, except at the beginning of a sentence.

■ **Periodic Table**
See the periodic table at General ChemistryNow. It can be accessed from Screen 1.5 or from the Toolbox. See also the extensive information on the periodic table and the elements at the American Chemical Society website:
- www.chemistry.org/periodic_table.html
- http://pubs.acs.org/cen/80th/elements.html

Photos: Charles D. Winters

Figure 1.6   Elements. (a) Passing an electric current through water produces the elements hydrogen (test tube on the right) and oxygen (test tube on the left). (b) Chemical elements can often be distinguished by their color and their state at room temperature.

GENERAL
Chemistry ⚛ Now™

See the General ChemistryNow CD-ROM or website:
• **Screen 1.7 Elements and Atoms,** and the Periodic Table tool on this screen or in the Toolbox

---

Exercise 1.2—Elements

Using the periodic table inside the front cover of this book or on the CD-ROM:
(a) Find the names of the elements having the symbols Na, Cl, and Cr.
(b) Find the symbols for the elements zinc, nickel, and potassium.

1.3—Compounds and Molecules

A pure substance like sugar, salt, or water, which is composed of two or more different elements held together by a **chemical bond**, is referred to as a **chemical compound**. Even though only 116 elements are known, there appears to be no limit to the number of compounds that can be made from those elements. More than 20 million compounds are now known, with about a half million added to the list each year.

When elements become part of a compound, their original properties, such as their color, hardness, and melting point, are replaced by the characteristic properties of the compound. Consider common table salt (sodium chloride), which is composed of two elements (Figure 1.7):

• Sodium is a shiny metal that reacts violently with water. It is composed of sodium atoms tightly packed together.

• Chlorine is a light yellow gas that has a distinctive, suffocating odor and is a powerful irritant to lungs and other tissues. The element is composed of Cl_2 units in which two chlorine atoms are tightly bound together.

Photos: Charles D. Winters

Solid sodium, Na

Chlorine gas, Cl₂

Sodium chloride solid, NaCl

Figure 1.7 **Forming a chemical compound.** Sodium chloride, commonly known as table salt, can be made by combining sodium metal (Na) and yellow chlorine gas (Cl₂). The result is a crystalline solid.

- Sodium chloride, or common salt, is a colorless, crystalline solid. Its properties are completely unlike those of the two elements from which it is made (Figure 1.7). Salt is composed of sodium and chlorine bound tightly together. (The meaning of chemical formulas such as NaCl is explored in Sections 3.3 and 3.4.)

It is important to distinguish between a mixture of elements and a chemical compound of two or more elements. Pure metallic iron and yellow, powdered sulfur (Figure 1.8a) can be mixed in varying proportions. In the chemical compound iron pyrite (Figure 1.8b), however, there is no variation in composition. Not only does iron pyrite exhibit properties peculiar to itself and different from those of either iron or sulfur, or a mixture of these two elements, but it also has a definite percentage composition by weight (46.55% Fe and 53.45% S). Thus, two major differences

Charles D. Winters

(a)

(b)

Figure 1.8 **Mixtures and compounds.** (a) The substance in the dish is a mixture of iron chips and sulfur. The iron can be removed easily by using a magnet. (b) Iron pyrite is a chemical compound composed of iron and sulfur. It is often found in nature as perfect, golden cubes.

Figure 1.9   Names, formulas, and models of some common molecules. Models of molecules appear throughout this book. In such models C atoms are gray, H atoms are white, N atoms are blue, and O atoms are red.

NAME	Water	Methane	Ammonia	Carbon dioxide
FORMULA	H_2O	CH_4	NH_3	CO_2
MODEL				

exist between mixtures and pure compounds: Compounds have distinctly different characteristics from their parent elements, and they have a definite percentage composition (by mass) of their combining elements.

Some compounds—such as table salt, NaCl—are composed of **ions**, which are electrically charged atoms or groups of atoms [▶ Chapter 3]. Other compounds—such as water and sugar—consist of **molecules**, the smallest discrete units that retain the composition and chemical characteristics of the compound.

The composition of any compound is represented by its **chemical formula**. In the formula for water, H_2O, for example, the symbol for hydrogen, H, is followed by a subscript "2" indicating that two atoms of hydrogen occur in a single water molecule. The symbol for oxygen appears without a subscript, indicating that one oxygen atom occurs in the molecule.

As you shall see throughout this book, molecules can be represented with models that depict their composition and structure. Figure 1.9 illustrates the names, formulas, and models of the structures of a few common molecules.

1.4—Physical Properties

You recognize your friends by their physical appearance: their height and weight and the color of their eyes and hair. The same is true of chemical substances. You can tell the difference between an ice cube and a cube of lead of the same size not only because of their appearance (one is clear and colorless, and the other is a lustrous metal) (Figure 1.10), but also because one is much heavier (lead) than the other (ice). Properties such as these, which can be observed and measured without changing the composition of a substance, are called **physical properties**. The chemical elements in Figures 1.6 and 1.7, for example, clearly differ in terms of their color, appearance, and state (solid, liquid, or gas). Physical properties allow us to classify and identify substances. Table 1.1 lists a few physical properties of matter that chemists commonly use.

Charles D. Winters

Figure 1.10   Physical properties. An ice cube and a piece of lead can be differentiated easily by their physical properties (such as density, color, and melting point).

Exercise 1.3—Physical Properties

Identify as many physical properties in Table 1.1 as you can for the following common substances: **(a)** iron, **(b)** water, **(c)** table salt (chemical name is sodium chloride), and **(d)** oxygen.

Density

Density, the ratio of the mass of an object to its volume, is a physical property useful for identifying substances.

$$\text{Density} = \frac{\text{mass}}{\text{volume}} \tag{1.1}$$

Table 1.1 Some Physical Properties

Property	Using the Property to Distinguish Substances
Color	Is the substance colored or colorless? What is the color and what is its intensity?
State of matter	Is it a solid, liquid, or gas? If it is a solid, what is the shape of the particles?
Melting point	At what temperature does a solid melt?
Boiling point	At what temperature does a liquid boil?
Density	What is the substance's density (mass per unit volume)?
Solubility	What mass of substance can dissolve in a given volume of water or other solvent?
Electric conductivity	Does the substance conduct electricity?
Malleability	How easily can a solid be deformed?
Ductility	How easily can a solid be drawn into a wire?
Viscosity	How easily will a liquid flow?

Your brain unconsciously uses the density of an object you want to pick up by estimating volume visually and preparing your muscles to lift the expected mass. For example, you can readily tell the difference between an ice cube and a cube of lead of identical size (Figure 1.10). Lead has a high density, 11.35 g/cm³ (11.35 grams per cubic centimeter), whereas the density of ice is slightly less than 0.917 g/cm³. An ice cube with a volume of 16.0 cm³ has a mass of 14.7 g, whereas a cube of lead with the same volume has a mass of 182 g.

Density relates the mass and volume of a substance. If any two of three quantities—mass, volume, and density—are known for a sample of matter, the third can be calculated. For example, the mass of an object is the product of its density and volume.

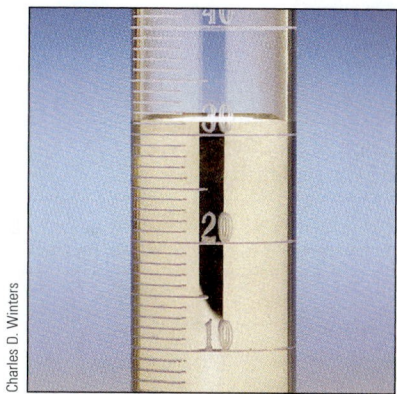
Charles D. Winters

$$\text{Mass (g)} = \text{volume} \times \text{density} = \text{volume (cm}^3) \times \frac{\text{mass (g)}}{\text{volume (cm}^3)}$$

Density, mass, and volume. What is the mass of 32 mL of mercury?

You can use this approach to find the mass of 32 cm³ [or 32 mL (milliliters)] of mercury in the graduated cylinder in the photo. A handbook of information for chemistry lists the density of mercury as 13.534 g/cm³ (at 20 °C).

$$\text{Mass (g)} = 32 \text{ cm}^3 \times \frac{13.534 \text{ g}}{1 \text{ cm}^3} = 430 \text{ g}$$

Be sure to notice that the units of cm³ cancel to leave the answer in units of g as required.

■ **Dimensional Analysis**
The approach to problem solving used in this book is often called *dimensional analysis*. The essence of this approach is to change one number (A) into another (B) using a conversion factor so that the units of A are changed to the desired unit. See Section 1.8.

GENERAL
Chemistry ⚛ Now™

See the General ChemistryNow CD-ROM or website:
• **Screen 1.10 Density,** for two step-by-step tutorials on determining density and volume

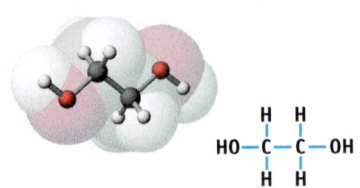

ethylene glycol, $C_2H_6O_2$
density = 1.11 g/cm^3 (or 1.11 g/mL)

$$HO-\overset{\overset{\displaystyle H}{|}}{\underset{\underset{\displaystyle H}{|}}{C}}-\overset{\overset{\displaystyle H}{|}}{\underset{\underset{\displaystyle H}{|}}{C}}-OH$$

■ **Units of Density**
The SI unit of mass is the kilogram and the SI unit of length is the meter. Therefore, the SI unit of density is kg/m^3. In chemistry the more commonly used unit is g/cm^3. To convert from kg/m^3 to g/cm^3, divide by 1000.

Temperature Dependence of Water Density

Temperature (°C)	Density of Water (g/cm^3)
0 (ice)	0.917
0 (liq water)	0.99984
2	0.99994
4	0.99997
10	0.99970
25	0.99707
100	0.95836

Example 1.1—Using Density

Problem Ethylene glycol, $C_2H_6O_2$, is widely used in automobile antifreeze. It has a density of 1.11 g/cm^3 (or 1.11 g/mL). What volume of ethylene glycol will have a mass of 1850 g?

Strategy You know the density and mass of the sample. Because density is the ratio of the mass of a sample to its volume, volume = (mass)(1/density).

Solution

$$\text{Volume (cm}^3\text{)} = 1850 \text{ g} \left(\frac{1 \text{ cm}^3}{1.11 \text{ g}} \right) = 1670 \text{ cm}^3$$

Comment Here we multiply the mass (in grams) by the conversion factor (1 cm^3/1.11 g) so that units of g cancel to leave an answer in the desired unit of cm^3.

Exercise 1.4—Density

The density of dry air is 1.18×10^{-3} g/cm^3 (= 0.00118 g/cm^3; see Section 1.8 on using scientific notation). What volume of air, in cubic centimeters, has a mass of 15.5 g?

Temperature Dependence of Physical Properties

The temperature of a sample of matter often affects the numerical values of its properties. Density is a particularly important example. Although the change in water density with temperature seems small, it affects our environment profoundly. For example, as the water in a lake cools, the density of the water increases, and the denser water sinks (Figure 1.11a). This continues until the water temperature reaches 3.98 °C, the point at which water has its maximum density (0.999973 g/cm^3). If the water temperature drops further, the density decreases slightly, and the colder water floats on top of water at 3.98 °C.

If water is cooled below about 0 °C, solid ice forms. Water is unique among substances in the universe: Ice is much less dense than water, so it floats on water.

Because the density of liquids changes with temperature, it is necessary to report the temperature when you make accurate volume measurements. Laboratory glassware used to make such measurements always specifies the temperature at which it was calibrated (Figure 1.11b).

Problem-Solving Tip 1.1

Finding Data

All the information you need to solve a problem in this book may not be presented in the problem. For example, we could have left out the value of the density in Example 1.1 and assumed you would (a) recognize that you needed density to convert a mass to a volume and (b) know where to find the information. The Appendices of this book contain a wealth of information, and even more is available on the General ChemistryNow CD-ROM and website. Various handbooks of information are available in most libraries; among the best are the *Handbook of Chemistry and Physics* (CRC Press) and *Lange's Handbook of Chemistry* (McGraw-Hill). The most up-to-date source of data is the National Institute for Standards and Technology (www.nist.org). See also the World Wide Web site Webelements (www.webelements.com).

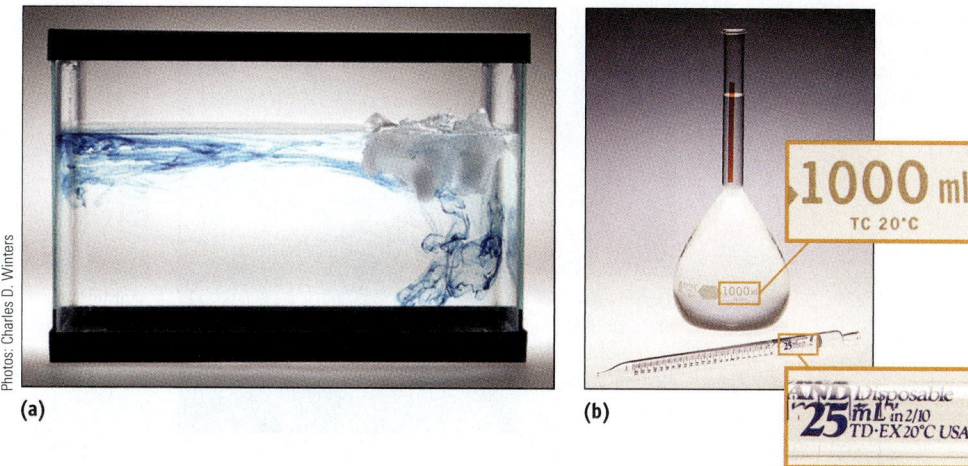

(a)   (b)

**Figure 1.11 Temperature dependence of physical properties. (a) Change in density with tempera-
ture.** Ice cubes were placed in the right side of the tank and blue dye in the left side. The water beneath the
ice is cooler and denser than the surrounding water, so it sinks. The convection current created by this movement of water is traced by the dye movement as the denser, cooler water sinks. **(b) Temperature and
calibration.** Laboratory glassware is calibrated for specific temperatures. This pipet or volumetric flask will
contain the specified volume at the indicated temperature.

Exercise 1.5—Density and Temperature

The density of mercury at 0 °C is 13.595 g/cm^3, at 10 °C it is 13.570 g/cm^3, and at 20 °C it is
13.546 g/cm^3. Estimate the density of mercury at 30 °C.

Extensive and Intensive Properties

Extensive properties depend on the amount of a substance present. The mass and
volume of the samples of elements in Figures 1.2 and 1.6 are extensive properties,
for example. In contrast, **intensive properties** do not depend on the amount of substance. A sample of ice will melt at 0 °C, no matter whether you have an ice cube or
an iceberg. Density is also an intensive property. The density of gold, for example, is
the same (19.3 g/cm^3) whether you have a flake of pure gold or a solid gold ring.

1.5—Physical and Chemical Changes

Changes in physical properties are called **physical changes**. In a physical change the
identity of a substance is preserved even though it may have changed its physical
state or the gross size and shape of its pieces. An example of a physical change is the
melting of a solid. The temperature at which this occurs (the melting point) is often so characteristic that it can be used to identify the solid (Figure 1.12).

A physical property of hydrogen gas (H_2) is its low density, so a balloon filled
with H_2 floats in air (Figure 1.13). Suppose a lighted candle is brought up to the balloon. When the heat causes the skin of the balloon to rupture, the hydrogen combines with the oxygen (O_2) in the air, and the heat of the candle sets off a chemical
reaction (Figure 1.13), producing water, H_2O. This reaction is an example of a
chemical change, in which one or more substances (the **reactants**) are transformed
into one or more different substances (the **products**).

Figure 1.12   A physical property used to distinguish compounds.
Aspirin and naphthalene are both white solids at 25 °C. You can tell them
apart by, among other things, a difference in physical properties. At the
temperature of boiling water, 100 °C, naphthalene is a liquid (*left*),
whereas aspirin is a solid (*right*).

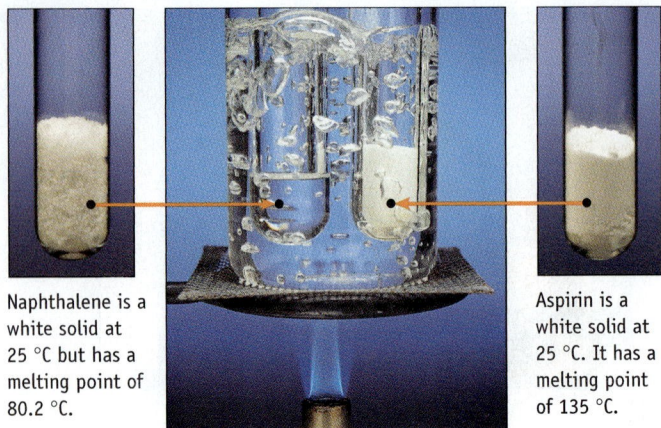

Naphthalene is a
white solid at
25 °C but has a
melting point of
80.2 °C.

Aspirin is a
white solid at
25 °C. It has a
melting point
of 135 °C.

Photos: Charles D. Winters

The reaction of H_2 with O_2 is an example of a chemical property of hydrogen. A chemical property involves a change in the identity of a substance. Here the H atoms of the gaseous H_2 molecules have become incorporated into H_2O. Similarly, a chemical change occurs when gasoline burns in air in an automobile engine or an old car rusts in the air. Burning of gasoline or rusting of iron are characteristic chemical properties of these substances.

A chemical change at the particulate level is illustrated by the reaction of hydrogen and oxygen molecules to form water molecules.

$$2 \ H_2(gas) + O_2(gas) \longrightarrow 2 \ H_2O(gas)$$

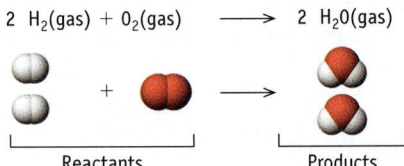

Reactants   Products

The representation of the change with chemical formulas is called a **chemical equation**. It shows that the substances on the left (the **reactants**) produce the substances on the right (the **products**). As this equation shows, there are four atoms of H and two atoms of O before *and* after the reaction, but the molecules before the reaction are different from those after the reaction.

Unlike a chemical change, a physical change does not result in a new chemical substance being produced. The substances (atoms, molecules, or ions) present before and after the change are the same, but they might be farther apart in a gas or closer together in a solid (Figure 1.2).

Finally, as described more fully in Chapter 6, physical changes and chemical changes are often accompanied by transfer of energy. The reaction of hydrogen and oxygen to give water (Figure 1.13), for example, transfers a tremendous amount of energy (in the form of heat and light) to its surroundings.

GENERAL
Chemistry Now™

See the General ChemistryNow CD-ROM or website:
- **Screen 1.12 Chemical Changes,** for an exercise on identifying physical and chemical changes
- **Screen 1.13 Chemical Change on the Molecular Scale,** to watch a video and view an animation of the molecular changes when chlorine gas and solid phosphorus react

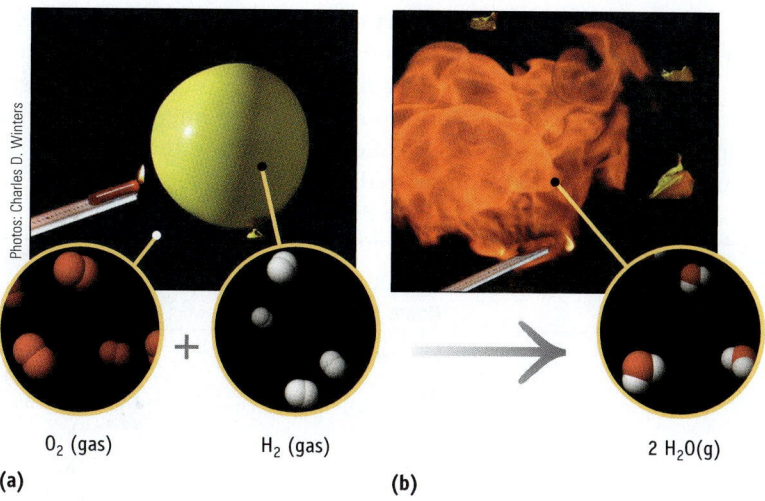

O_2 (gas) H_2 (gas)   2 $H_2O(g)$

(a)   **(b)**

Figure 1.13   A chemical change—the reaction of hydrogen and oxygen. (a) A balloon filled with molecules of hydrogen gas, and surrounded by molecules of oxygen in the air. (The balloon floats in air because gaseous hydrogen is less dense than air.) (b) When ignited with a burning candle, H_2 and O_2 react to form water, H_2O. (*See General ChemistryNow Screen 1.11 Chemical Change, for a video of this reaction.*)

Exercise 1.6—Chemical Reactions and Physical Changes

When camping in the mountains, you boil a pot of water on a campfire. What physical and chemical changes take place in this process?

1.6—Units of Measurement

Doing chemistry requires observing chemical reactions and physical changes. Suppose you mix two solutions in the laboratory and see a golden yellow solid form. Because this new solid is denser than water, it drops to the bottom of the test tube (Figure 1.14). The color and appearance of the substances, and whether heat is involved, are **qualitative** observations. No measurements and numbers were involved.

To understand a chemical reaction more completely, chemists usually make **quantitative** observations. These involve numerical information. For example, if two compounds react with each other, how much product forms? How much heat, if any, is evolved?

In chemistry, quantitative measurements of time, mass, volume, and length, among other things, are common. On page 31 you can read about one of the fastest growing areas of science, nanotechnology, which involves the creation and study of matter on the nanometer scale. A nanometer (nm) is equivalent to 1×10^{-9} m

Chemical and physical changes. A pot of water has been put on a campfire. What chemical and physical changes are occurring here (Exercise 1.6)?

Figure 1.14   Qualitative and quantitative observations. A new substance is formed by mixing two known substances in solution. Of the substance produced we can make several observations. *Qualitative observations:* yellow, fluffy solid. *Quantitative observations:* mass of solid formed.

(meter), a common dimension in chemistry and biology. For example, a typical molecule is only about 1 nm across and a bacterium is about 1000 nm in length.

The scientific community has chosen a modified version of the **metric system** as the standard system for recording and reporting measurements. This decimal system, used internationally in science, is called the Système International d'Unités (International System of Units), abbreviated **SI**.

Table 1.2   Some SI Base Units

Measured Property	Name of Unit	Abbreviation
Mass	kilogram	kg
Length	meter	m
Time	second	s
Temperature	kelvin	K
Amount of substance	mole	mol
Electric current	ampere	A

All SI units are derived from base units, some of which are listed in Table 1.2. Larger and smaller quantities are expressed by using appropriate prefixes with the base unit (Table 1.3).

GENERAL
Chemistry ⋅ Now ™

See the General ChemistryNow CD-ROM or website:
- **Screen 1.16 The Metric System,** for a step-by-step tutorial on converting metric units

Temperature Scales

Three temperature scales are commonly used: the Fahrenheit, Celsius, and Kelvin scales (Figure 1.15). The Fahrenheit scale is used in the United States to report everyday temperatures, but most other countries use the Celsius scale. The Celsius scale is generally used worldwide for measurements in the laboratory. When calculations incorporate temperature data, however, kelvin degrees must be used.

The Celsius Temperature Scale
The size of the Celsius degree is defined by assigning zero as the freezing point of pure water (0 °C) and 100 as its boiling point (100 °C) (page 10). You can readily interconvert Fahrenheit and Celsius temperatures using the equation

$$T(°C) = \frac{5 \ °C}{9 \ °F}[T \ (°F) - 32]$$

but it is best to "calibrate" your senses on the Celsius scale. Pure water freezes at 0 °C, a comfortable room temperature is around 20 °C, your body temperature is 37 °C, and the warmest water you could stand to immerse a finger in is probably about 60 °C.

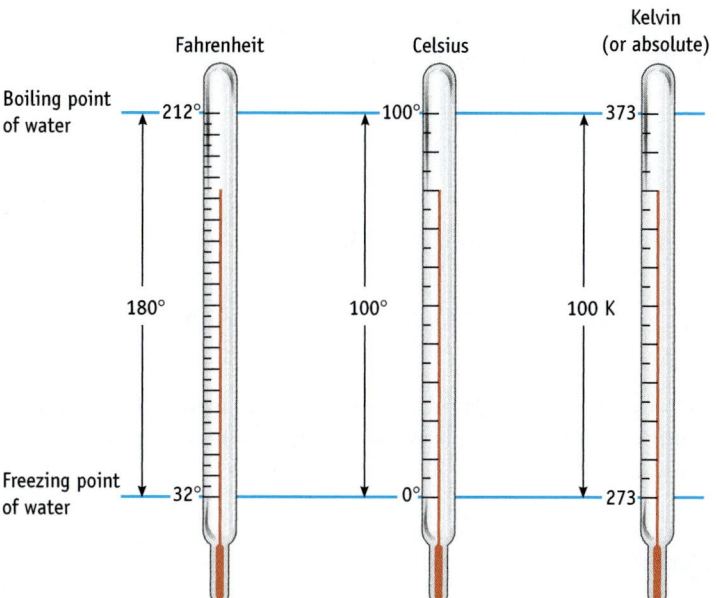

Active Figure 1.15   **A comparison of Fahrenheit, Celsius, and Kelvin scales.** The reference, or starting point, for the Kelvin scale is absolute zero (0 K = −273.15 °C), which has been shown theoretically and experimentally to be the lowest possible temperature.

GENERAL
Chemistry ·⚛· Now™ See General ChemistryNow CD-ROM or website to explore an interactive version of this figure accompanied by an exercise.

Table 1.3   **Selected Prefixes Used in the Metric System**

Prefix	Abbreviation	Meaning	Example
mega-	M	10^6 (million)	1 megaton = 1×10^6 tons
kilo-	k	10^3 (thousand)	1 kilogram (kg) = 1×10^3 g
deci-	d	10^{-1} (tenth)	1 decimeter (dm) = 1×10^{-1} m
centi-	c	10^{-2} (one hundredth)	1 centimeter (cm) = 1×10^{-2} m
milli-	m	10^{-3} (one thousandth)	1 millimeter (mm) = 1×10^{-3} m
micro-	μ	10^{-6} (one millionth)	1 micrometer (μm) = 1×10^{-6} m
nano-	n	10^{-9} (one billionth)	1 nanometer (nm) = 1×10^{-9} m
pico-	p	10^{-12}	1 picometer (pm) = 1×10^{-12} m
femto-	f	10^{-15}	1 femtometer (fm) = 1×10^{-15} m

■ **Common Conversion Factors**
1 kg = 1000 g
1×10^9 nm = 1 m
10 mm = 1 cm
100 cm = 10 dm = 1 m
1000 m = 1 km
Conversion factors for SI units are given in Appendix C and inside the back cover of this book.

■ **Lord Kelvin**
William Thomson (1824–1907), known as Lord Kelvin, was a professor of natural philosophy at the University in Glasgow, Scotland, from 1846 to 1899. He was best known for his work on heat and work, from which came the concept of the absolute temperature scale.

The Kelvin Temperature Scale

William Thomson, known as Lord Kelvin (1824–1907), first suggested the temperature scale that now bears his name. The Kelvin scale uses the same size unit as the Celsius scale, but it assigns zero as the lowest temperature that can be achieved, a point called **absolute zero**. Many experiments have found that this limiting temperature is −273.15 °C (−459.67 °F). Kelvin units and Celsius degrees are the same size. Thus, the freezing point of water is reached at 273.15 K; that is, 0 °C = 273.15 K.

The boiling point of pure water is 373.15 K. Temperatures in Celsius degrees are readily converted to kelvins, and vice versa, using the relation

$$T\ (K) = \frac{1\ K}{1\ °C} [T\ °C + 273.15\ °C] \qquad (1.2)$$

■ **Temperature Conversions**
When converting 23.5 °C to kelvins, adding the two numbers gives 296.65. However, the rules of "significant figures" tell us that the sum or difference of two numbers can have no more decimal places than the number with the fewest decimal places. (See page 40.) Thus, we round the answer to 296.7 K, a number with one decimal place.

Thus, a common room temperature of 23.5 °C is

$$T\ (K) = \frac{1\ K}{1\ °C} (23.5\ °C + 273.15\ °C) = 296.7\ K$$

Finally, notice that the degree symbol (°) is not used with Kelvin temperatures. The name of the unit on this scale is the kelvin (not capitalized), and such temperatures are designated with a capital K.

GENERAL
Chemistry ⚛ Now™

See the General ChemistryNow CD-ROM or website:
• **Screen 1.15 Temperature,** for a step-by-step tutorial on converting temperatures

---

Exercise 1.7—Temperature Scales

Liquid nitrogen boils at 77 K. What is this temperature in Celsius degrees?

Length

The meter is the standard unit of length, but objects observed in chemistry are frequently smaller than 1 meter. Measurements are often reported in units of centimeters or millimeters, and objects on the atomic and molecular scale have dimensions of nanometers (nm; 1 nm = 1.0×10^{-9} m) or picometers (pm; 1 pm = 1×10^{-12} m). Your hand, for example, is about 18 cm from the wrist to the fingertips, and the ant in the photo here is about 1 cm long. Using a special microscope—a scanning electron microscope (SEM)—scientists can zoom in on the face of an ant, then to the ant's eye, and finally to one segment of the eye (Figure 1.16).

If we could continue to zoom in on the ant's eye in Figure 1.16, we would enter the nanoscale molecular world (Figure 1.17). The DNA (deoxyribonucleic acid) in the ant's eye is a helical coil of atoms many nanometers long. The rungs of the DNA ladder are approximately 0.34 nm apart, and the helix repeats itself about every 3.4 nm. Zooming in even more, we might encounter a water molecule. Here the distance between the two hydrogen atoms on either side of the oxygen atom is 0.152 nm or 152 pm (pm; picometer, 1 pm = 1×10^{-12} m).

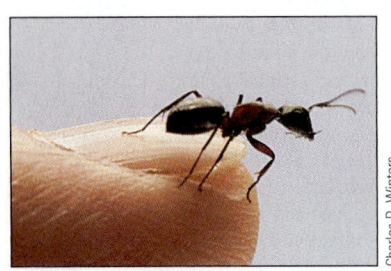

Charles D. Winters

Ant. Your hand is about 18 centimeters long from your wrist to your fingertips. The ant here is about 1 cm in length.

Example 1.2—Distances on the Molecular Scale

Problem The distance between an O atom and an H atom in a water molecule is 95.8 pm. What is this distance in meters (m)? In nanometers (nm)?

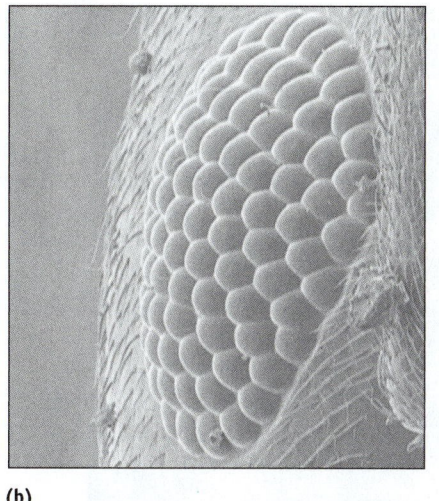

(a)   **(b)**   **(c)**

Figure 1.16 Dimensions in biology. These photos were done at the IBM Laboratories using a scanning electron microscope (SEM). The subject was a dead ant. (a) The head of the ant is about 600 micrometers (microns, μm) wide. (This is equivalent to 6×10^{-4} m or 0.6 mm.) (b) The compound eye of the ant. (c) The scientists at IBM used a special probe to write, on one lens of the ant eye, their advice to science students. The word "homework" is about 1.5 micrometers (microns, μm) long.

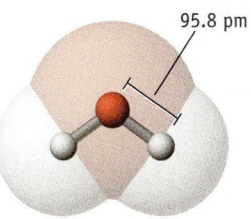

95.8 pm

Strategy You can solve this problem by knowing the conversion factor between the units in the information you are given (picometers) and the desired units (meters or nanometers). (For more about conversion factors and their use in problem solving see Section 1.8.) There is no conversion factor given in Table 1.3 to change nanometers to picometers, but relationships are listed between meters and picometers and between meters and nanometers (Table 1.3). First, we convert picometers to meters, and then we convert meters to nanometers.

$$\times \frac{\text{m}}{\text{pm}} \qquad \times \frac{\text{nm}}{\text{m}}$$
$$\text{Picometers} \longrightarrow \text{Meters} \longrightarrow \text{Nanometers}$$

Solution Using the appropriate conversion factors (1 pm = 1×10^{-12} m and 1 nm = 1×10^{-9} m), we have

$$95.8 \ \cancel{\text{pm}} \times \frac{1 \times 10^{-12} \ \text{m}}{1 \ \cancel{\text{pm}}} = 9.58 \times 10^{-11} \ \text{m}$$

$$9.58 \ \times 10^{-11} \ \cancel{\text{m}} \times \frac{1 \ \text{nm}}{1 \times 10^{-9} \ \cancel{\text{m}}} = 9.58 \times 10^{-2} \ \text{nm} \ \text{ or } \ 0.0958 \ \text{nm}$$

Comment Notice how the units cancel to leave an answer whose unit is that of the numerator of the conversion factor. The process of using units to guide a calculation is called dimensional analysis and is discussed further on pages 41–43.

■ **Powers of Ten**
The book *Powers of Ten* explores the dimensions of our universe (Philip and Phylis Morrison, Scientific American Books, 1982). See also the following website in which the "powers of ten" is elegantly animated. **http://micro.magnet.fsu.edu/ primer/java/scienceopticsu/powersof10/**

Figure 1.17   Dimensions in the molecular world. Objects on the molecular scale are often given in terms of nanometers (1 nm = 1×10^{-9} m) or picometers (1 pm = 1×10^{-12} m). An older non-SI unit is the angstrom unit, where 1 Å = 1.0×10^{-10} m.

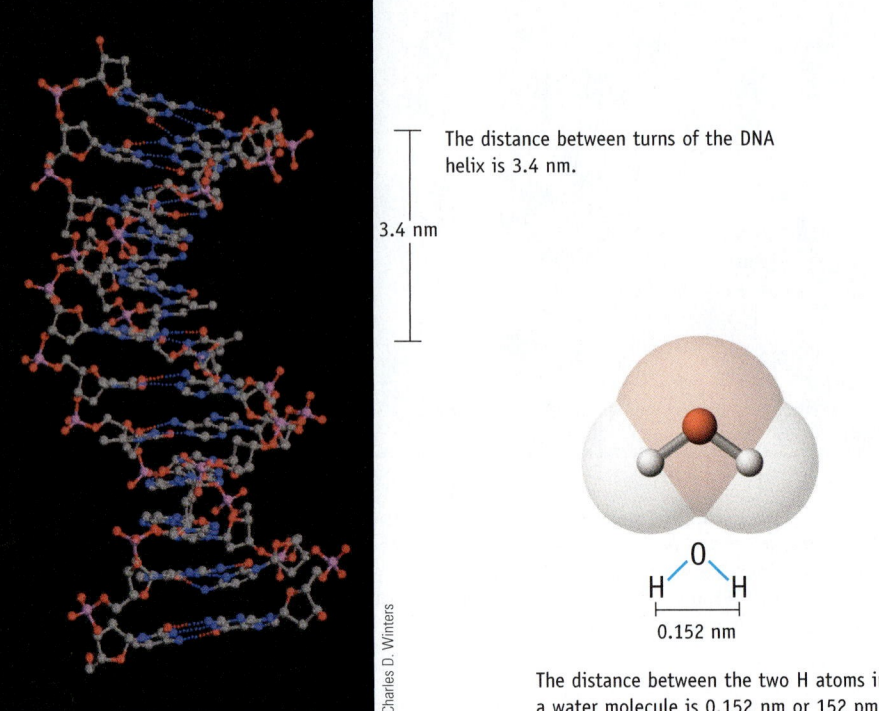

The distance between turns of the DNA helix is 3.4 nm.

3.4 nm

0.152 nm

The distance between the two H atoms in a water molecule is 0.152 nm or 152 pm.

Charles D. Winters

Exercise 1.8—Interconverting Units of Length

The pages of a typical textbook are 25.3 cm long and 21.6 cm wide. What is each dimension in meters? In millimeters? What is the area of a page in square centimeters? In square meters?

Exercise 1.9—Using Units of Length and Density

A platinum sheet is 2.50 cm square and has a mass of 1.656 g. The density of platinum is 21.45 g/cm^3. What is the thickness of the platinum sheet in millimeters?

Volume

Chemists often use glassware such as beakers, flasks, pipets, graduated cylinders, and burets, which are marked in volume units (Figure 1.18). The SI unit of volume is the cubic meter (m^3), which is too large for everyday laboratory use. Therefore, chemists usually use the **liter**, symbolized by **L**. A cube with sides equal to 10 cm (0.1 m) has a volume of 10 cm $\times$ 10 cm $\times$ 10 cm = 1000 cm^3 (or 0.001 m^3). This is defined as 1 liter.

$$1 \text{ liter (L)} = 1000 \text{ mL} = 1000 \text{ cm}^3$$

The liter is a convenient unit to use in the laboratory, as is the milliliter (mL). Because there are exactly 1000 mL (= 1000 cm^3) in a liter, this means that

$$1 \text{ cm}^3 = 0.001 \text{ L} = 1 \text{ mL}$$

Figure 1.18   Some common laboratory glassware. Volumes are marked in units of milliliters (mL). Remember that 1 mL is equivalent to 1 cm^3.

Charles D. Winters

Chemical Perspectives

It's a Nanoworld!

A nanometer is one billionth of a meter, a dimension in the realm of atoms and molecules—eight oxygen atoms in a row span a distance of about 1 nanometer. Nanotechnology is one of the hottest fields in science today because the building blocks of those materials having nanoscale dimension can have unique properties.

Carbon nanotubes are excellent examples of nanomaterials. These lattices of carbon atoms form the walls of tubes

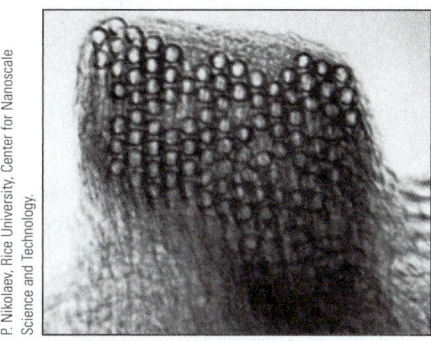

A bundle of carbon nanotubes. Each tube has a diameter of 1.4 nm, and the bundle is 10–20 nm thick.

having diameters of only a few nanometers. Carbon nanotubes are at least 100 times stronger than steel, but only one-sixth as dense. In addition, they conduct heat and electricity far better than copper. As a consequence, carbon nanotubes could be used in tiny, physically strong, conducting devices. Recently, carbon nanotubes have been filled with potassium atoms, making them even better electrical conductors. And even more recently, molecular-sized bearings have been made by sliding one nanotube inside another.

Nanomaterials are by no means new. For the last century tire companies have reinforced tires by adding nanosized particles called carbon black to rubber.

Atomic force microscopy (AFM) is an important tool in chemistry and physics to observe materials at the nanometer level. A tiny probe, often a whisker of a carbon nanotube, moves over the surface of a substance and interacts with individual molecules. Here you see an AFM image of a silicon surface about 460 nm on a side and 5 nm high.

Professor Alex Zettl of the University of California–Berkeley, holding a model of a carbon nanotube.

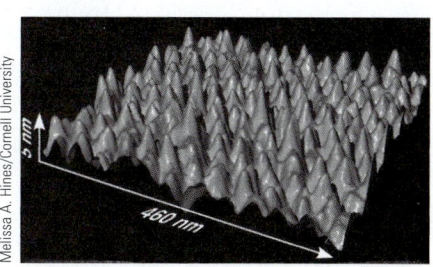

An AFM image of nanobumps on a silicon surface. The average spacing between nanobumps is 38 nm, or about 160 silicon atoms. The average nanobump width is 25 nm or 100 silicon atoms.

The units *milliliter and cubic centimeter* (or "cc") *are interchangeable.* Therefore, a flask that contains exactly 125 mL has a volume of 125 cm^3.

Although not widely used in the United States, the cubic decimeter (dm^3) is a common unit in the rest of the world. A length of 10 cm is called a decimeter (dm). Because a cube 10 cm on a side defines a volume of 1 liter, *a liter is equivalent to a cubic decimeter:* $1 L = 1 dm^3$. Products in Europe and other parts of the world are often sold by the cubic decimeter.

The *deciliter, dL,* which is exactly equivalent to 0.100 L or 100 mL, is widely used in medicine. For example, standards for amounts of environmental contaminants are often set as a certain mass per deciliter. The state of Massachusetts recommends that children with more than 10 micrograms (10×10^{-6} g) of lead per deciliter of blood undergo further testing for lead poisoning.

Example 1.3—Units of Volume

Problem A laboratory beaker has a volume of 0.6 L. What is its volume in cubic centimeters (cm^3), milliliters (mL), and deciliters?

Strategy Use the information in Table 1.3 to interconvert between units, and use dimensional analysis (see *The Mathematics of Chemistry*, pages 41–43) as a guide.

Solution You should multiply 0.6 L by the conversion factor (1000 cm³/L). The units of L cancel to leave an answer with units of cm³.

$$0.6 \, \cancel{L} \cdot \frac{1000 \text{ cm}^3}{1 \, \cancel{L}} = \boxed{600 \text{ cm}^3}$$

Because cubic centimeters and milliliters are equivalent, we can also say that the volume of the beaker is 600 mL. The deciliter is 0.100 L or 100 mL. In deciliters, the volume is

$$600 \, \cancel{mL} \cdot \frac{1 \text{ dL}}{100 \, \cancel{mL}} = \boxed{6 \text{ dL}}$$

Exercise 1.10—Volume

(a) A standard wine bottle has a volume of 750 mL. How many liters does this represent? How many deciliters?

(b) One U.S. gallon is equivalent to 3.7865 L. How many liters are in a 2.0-quart carton of milk? (There are 4 quarts in a gallon.) How many cubic decimeters?

Mass

The mass of a body is the fundamental measure of the quantity of matter, and the SI unit of mass is the kilogram (kg). Smaller masses are expressed in grams (g) or milligrams (mg).

$$1 \text{ kg} = 1000 \text{ g}$$
$$1 \text{ g} = 1000 \text{ mg}$$

■ **Micrograms**
Very small masses are often given in micrograms. A microgram is 1/1000 of a milligram or one millionth of a gram.

Exercise 1.11—Mass

(a) A new U.S. quarter has a mass of 5.59 g. Express this mass in kilograms and milligrams.

(b) An environmental study of a river found a pesticide present to the extent of 0.02 microgram per liter of water. Express this amount in grams per liter.

■ **Accuracy**
The National Institute for Standards and Technology (NIST) is the most important resource for the standards used in science. Comparison with the NIST data is the best test of the accuracy of the measurement. See www.nist.gov.

1.7—Making Measurements: Precision, Accuracy, and Experimental Error

The **precision** of a measurement indicates how well several determinations of the same quantity agree. This is illustrated by the results of throwing darts at a target. In Figure 1.19a, the dart thrower was apparently not skillful, and the precision of the dart's placement on the target is low. In Figures 1.19b and 1.19c, the darts are clustered together, indicating much better consistency on the part of the thrower—that is, greater precision.

Accuracy is the agreement of a measurement with the accepted value of the quantity. Figure 1.19c shows that our thrower was accurate as well as precise—the average of all shots is close to the targeted position, the bull's eye.

(a) Poor precision and poor accuracy

(b) Good precision and poor accuracy

(c) Good precision and good accuracy

Figure 1.19   Precision and accuracy.

Figure 1.19b shows that it is possible to be precise without being accurate—the thrower has consistently missed the bull's eye, although all the darts are clustered precisely around one point on the target. This is analogous to an experiment with some flaw (either in design or in a measuring device) that causes all results to differ from the correct value by the same amount.

The precision of a measurement is often expressed in terms of its **standard deviation**, a technique of data analysis explored in *A Closer Look: Standard Deviation*. For

A Closer Look

Standard Deviation

Laboratory measurements can be in error for two basic reasons. First, there may be "determinate" errors caused by faulty instruments or human errors such as incorrect record keeping. So-called "indeterminate" errors arise from uncertainties in a measurement where the cause is not known and cannot be controlled by the lab worker. One way to judge the indeterminate error in a result is to calculate the standard deviation.

The standard deviation of a series of measurements is equal to the square root of the sum of the squares of the deviations for each measurement divided by the number of measurements. It has a precise statistical significance: 68% of the values collected are expected to be within one standard deviation of the value determined. (This value assumes a large number of measurements is used to calculate the deviation.)

Consider a simple example. Suppose you carefully measured the mass of water delivered by a 10-mL pipet. For five attempts at the measurement (shown in the table, column 2), the standard deviation is found as follows: First, the average of the measurements is calculated (here, 9.984). Next, the deviation of each individual measurement from this value is determined (column 3). These values are squared, giving the values in column 4, and the sum of these values is determined. The standard deviation is then

Determination	Measured Mass, (g)	Difference between Average and Measurement (g)	Square of Difference
1	9.990	0.006	4×10^{-5}
2	9.993	0.009	8×10^{-5}
3	9.973	0.011	12×10^{-5}
4	9.980	0.004	2×10^{-5}
5	9.982	0.002	0.4×10^{-5}

calculated by dividing this number by 5 (the number of determinations) and taking the square root of the result.

$$\text{Average mass} = 9.984 \text{ g}$$
$$\text{Sum of squares of differences} = 26 \times 10^{-5}$$

$$\text{Standard deviation} = \sqrt{\frac{26 \times 10^{-5}}{5}} = \pm 0.007$$

Based on this calculation it would be appropriate to represent the measured mass as 9.984 ± 0.007 g. This would tell a reader that if this experiment were repeated, approximately 68% of the values would fall in the range of 9.977 g to 9.991 g.

Charles D. Winters

example, suppose a series of measurements led to a distance of 2.965 cm, and the standard deviation was 0.006 cm. Because the uncertainty shows up in the thousandths position, the value should be reported to the nearest thousandth—that is, 2.965 cm. A standard deviation of 0.006 cm means that 68% of the random measurements we make will be within 1 standard deviation—that is, within $\pm$ 0.006 cm.

If you are measuring a quantity in the laboratory, you may be required to report the error in the result, the difference between your result and the accepted value,

$$\text{Error} = \text{experimentally determined value} - \text{accepted value}$$

or the **percent error**.

$$\text{Percent error} = \frac{\text{error in measurement}}{\text{accepted value}} \times 100\%$$

Example 1.4—Precision and Accuracy

Problem A coin has an "accepted" diameter of 28.054 mm. In an experiment, two students measure this diameter. Student A makes four measurements of the diameter of a coin using a precision tool called a micrometer. Student B measures the same coin using a simple plastic ruler. The two students report the following results:

Student A	Student B
28.246 mm	27.9 mm
28.244	28.0
28.246	27.8
28.248	28.1

What is the average diameter and percent error obtained in each case? Which student's data are more accurate? Which are more precise?

Strategy For each set of values we calculate the average of the results and then compare this average with 28.054 mm.

Solution The average for each set of data is obtained by summing the four values and dividing by 4.

Student A	Student B
28.246 mm	27.9 mm
28.244	28.0
28.246	27.8
28.248	28.1
Average = 28.246	Average = 28.0

Student A's data are all very close to the average value, so they are quite precise. Student B's data, in contrast, have a wider range and are less precise. However, student A's result is less accurate than that of student B. The average diameter for student A differs from the "accepted" value by 0.192 mm and has a percent error of 0.684%:

$$\text{Percent error} = \frac{28.246 \text{ mm} - 28.054 \text{ mm}}{28.054 \text{ mm}} \times 100\% = 0.684\%$$

Student B's measurement has an error of only about 0.2%.

Comment Possible reasons for the error in Students A's result are incorrect use of the micrometer or a flaw in the instrument.

Exercise 1.12—Error, Precision, and Accuracy

Two students measured the freezing point of an unknown liquid. Student A used an ordinary laboratory thermometer calibrated in 0.1 °C units. Student B used a thermometer certified by NIST and calibrated in 0.01 °C units. Their results were as follows:

Student A: −0.3 °C; 0.2 °C; 0.0 °C; and −0.3 °C

Student B: 273.13 K; 273.17 K; 273.15 K; 273.19 K

Calculate the average value and, knowing that the liquid was water, calculate the percent error for each student. Which student has the more precise values? Which has the smaller error?

1.8—Mathematics of Chemistry

At its core, chemistry is a quantitative science. Chemists make measurements of, among other things, size, mass, volume, time, and temperature. Scientists then manipulate that quantitative numerical information to search for relationships among properties and to provide insight into the molecular basis of matter.

This section reviews some of the mathematical skills you will need in chemical calculations. It also describes ways to perform calculations and ways to handle quantitative information. The background you should have to be successful includes the following skills:

- Ability to express and use numbers in exponential or scientific notation.
- Ability to make unit conversions (such as liters to milliliters).
- Ability to express quantitative information in an algebraic expression and solve that expression. An example would be to solve the equation $a = (b/x)c$ for x.
- Ability to read information from graphs.
- Ability to prepare a graph of numerical information. If the graph produces a straight line, find the slope and equation of the line.

Examples and Exercises using some of these skills follow, and some problems involving unit conversions and solving algebraic expressions are included in the Study Questions at the end of this chapter.

Exponential or Scientific Notation

Lake Otsego in northern New York is also called Glimmerglass, a name suggested by James Fenimore Cooper (1789–1851), the great American author and an early resident of the village now known as Cooperstown. Extensive environmental studies

Charles D. Winters

Figure 1.20 **Lake Otsego.** This lake, with a surface area of 2.33×10^7 m², is located in northern New York. Cooperstown is a village at the base of the lake, where the Susquehanna River originates. To learn more about the environmental biology and chemistry of the lake, go to www.oneonta.edu/academics/biofld/

Figure 1.21   Exponential numbers used in astronomy. The spiral galaxy M-83 is 3.0×10^6 parsecs away and has a diameter of 9.0×10^3 parsecs. The unit used in astronomy, the parsec (pc), is 206265 AU (astronomical units), and 1 AU is 1.496×10^8 km. Therefore, the galaxy is about 9.3×10^{19} km away from Earth.

have been done along this lake (Figure 1.20), and some quantitative information useful to chemists, biologists, and geologists is given in the following table:

Lake Otsego Characteristics	Quantitative Information
Area	2.33×10^7 m^2
Maximum depth	505 m
Dissolved solids in lake water	2×10^2 mg/L
Average rainfall in the lake basin	1.02×10^2 cm/year
Average snowfall in the lake basin	198 cm/year

All of the data collected are in metric units. However, some data are expressed in **fixed notation** (505 m, 198 cm/year), whereas other data are expressed in **exponential**, or **scientific, notation** (2.33×10^7 m^2). Scientific notation is a way of presenting very large or very small numbers in a compact and consistent form that simplifies calculations. Because of its convenience it is widely used in sciences such as chemistry, physics, engineering, and astronomy (Figure 1.21).

In scientific notation the number is expressed as a product of two numbers: $N \times 10^n$. N is the *digit* term and is a number between 1 and 9.9999. . . . The second number, 10^n, the *exponential* term, is some integer power of 10. For example, 1234 is written in scientific notation as 1.234×10^3, or 1.234 multiplied by 10 three times:

$$1.234 = 1.234 \times 10^1 \times 10^1 \times 10^1 = 1.234 \times 10^3$$

Conversely, a number less than 1, such as 0.01234, is written as 1.234×10^{-2}. This notation tells us that 1.234 should be divided twice by 10 to obtain 0.01234:

$$0.01234 = \frac{1.234}{10^1 \times 10^1} = 1.234 \times 10^{-1} \times 10^{-1} = 1.234 \times 10^{-2}$$

Some other examples of scientific notation follow:

$$10000 = 1 \times 10^4 \qquad\qquad 12345 = 1.2345 \times 10^4$$
$$1000 = 1 \times 10^3 \qquad\qquad 1234.5 = 1.2345 \times 10^3$$
$$100 = 1 \times 10^2 \qquad\qquad 123.45 = 1.2345 \times 10^2$$
$$10 = 1 \times 10^1 \qquad\qquad 12.345 = 1.2345 \times 10^1$$
$$1 = 1 \times 10^0 \text{ (any number to the zero power = 1)}$$
$$1/10 = 1 \times 10^{-1} \qquad\qquad 0.12 = 1.2 \times 10^{-1}$$
$$1/100 = 1 \times 10^{-2} \qquad\qquad 0.012 = 1.2 \times 10^{-2}$$
$$1/1000 = 1 \times 10^{-3} \qquad\qquad 0.0012 = 1.2 \times 10^{-3}$$
$$1/10000 = 1 \times 10^{-4} \qquad\qquad 0.00012 = 1.2 \times 10^{-4}$$

When converting a number to scientific notation, notice that the exponent n is positive if the number is greater than 1 and negative if the number is less than 1. The value of n is the number of places by which the decimal is shifted to obtain the number in scientific notation:

$$1\ 2\ 3\ 4\ 5. = 1.2345 \times 10^4$$

(a) Decimal shifted four places to the left. Therefore, n is positive and equal to 4.

■ **Comparing the Earth and a Plant Cell—Powers of Ten**
Earth = 12,760,000 meters wide
  = 12.76 million meters
  = 1.276×10^7 meters
Plant cell = 0.00001276 meter wide
  = 12.76 millionths of a meter
  = 1.276×10^{-5} meters

Problem-Solving Tip 1.2

Using Your Calculator

You will be performing a number of calculations in general chemistry, most of them using a calculator. Many different types of calculators are available, but this problem-solving tip describes several of the kinds of operations you will need to perform on a typical calculator. Be sure to consult your calculator manual for specific instructions to enter scientific notation and to find powers and roots of numbers.

1. Scientific Notation
When entering a number such as 1.23×10^{-4} into your calculator, you first enter 1.23 and then press a key marked EE or EXP (or something similar). This enters the "$\times 10$" portion of the notation for you. You then complete the entry by keying in the exponent of the number, -4. (To change the exponent from $+4$ to -4, press the "$+/-$" key.)

A common error made by students is to enter 1.23, press the multiply key (x), and then key in 10 before finishing by pressing EE or EXP followed by -4. This gives you an entry that is 10 times too large. Try this! Experiment with your calculator so you are sure you are entering data correctly.

2. Powers of Numbers
Electronic calculators usually offer two methods of raising a number to a power. To square a number, enter the number and then press the "x^2" key. To raise a number to any power, use the "y^x" (or similar) key. For example, to raise 1.42×10^2 to the fourth power:

1. Enter 1.42×10^2.
2. Press "y^x".
3. Enter 4 (this should appear on the display).
4. Press "=" and 4.0659×10^8 appears on the display.

3. Roots of Numbers
To take a square root on an electronic calculator, enter the number and then press the "$\sqrt{}x$" key. To find a higher root of a number, such as the fourth root of 5.6×10^{-10}:

1. Enter the number.
2. Press the "$\sqrt[x]{y}$" key. (On many calculators, the sequence you actually use is to press "2ndF" and then "=." Alternatively, you press "INV" and then "y^x".)
3. Enter the desired root, 4 in this case.
4. Press "=". The answer here is 4.8646×10^{-3}.

A general procedure for finding any root is to use the "y^x" key. For a square root, x is 0.5 (or 1/2), whereas it is 0.3333 (or 1/3) for a cube root, 0.25 (or 1/4) for a fourth root, and so on.

$$0.0\ 0\ 1\ 2 = 1.2 \times 10^{-3}$$

(b) Decimal shifted three places to the right. Therefore, *n* is negative and equal to 3.

If you wish to convert a number in scientific notation to one using fixed notation (that is, not using powers of 10), the procedure is reversed:

$$6\ .\ 2\ 7\ 3 \times 10^2 = 627.3$$

(a) Decimal point moved two places to the right because *n* is positive and equal to 2.

$$0\ 0\ 6.273 \times 10^{-3} = 0.006273$$

(b) Decimal point shifted three places to the left because *n* is negative and equal to 3.

Two final points should be made concerning scientific notation. First, be aware that calculators and computers often express a number such as 1.23×10^3 as 1.23E3 or 6.45×10^{-5} as 6.45E-5. Second, some electronic calculators can readily convert numbers in fixed notation to scientific notation. If you have such a calculator, you may be able to do this by pressing the EE or EXP key and then the "=" key (but check your calculator manual to learn how your device operates).

In chemistry you will often have to use numbers in exponential notation in mathematical operations. The following five operations are important:

- *Adding and Subtracting Numbers Expressed in Scientific Notation*

When adding or subtracting two numbers, first convert them to the same powers of 10. The digit terms are then added or subtracted as appropriate:

$$(1.234 \times 10^{-3}) + (5.623 \times 10^{-2}) = (0.1234 \times 10^{-2}) + (5.623 \times 10^{-2})$$
$$= 5.746 \times 10^{-2}$$

- *Multiplication of Numbers Expressed in Scientific Notation*

The digit terms are multiplied in the usual manner, and the exponents are added algebraically. The result is expressed with a digit term with only one nonzero digit to the left of the decimal:

$$(6.0 \times 10^{23})(2.0 \times 10^{-2}) = (6.0)(2.0) \times 10^{23-2} = 12 \times 10^{21} = 1.2 \times 10^{22}$$

- *Division of Numbers Expressed in Scientific Notation*

The digit terms are divided in the usual manner, and the exponents are subtracted algebraically. The quotient is written with one nonzero digit to the left of the decimal in the digit term:

$$\frac{7.60 \times 10^3}{1.23 \times 10^2} = \frac{7.60}{1.23} \times 10^{3-2} = 6.18 \times 10^1$$

- *Powers of Numbers Expressed in Scientific Notation*

When raising a number in exponential notation to a power, treat the digit term in the usual manner. The exponent is then multiplied by the number indicating the power:

$$(5.28 \times 10^3)^2 = (5.28)^2 \times 10^{3 \times 2} = 27.9 \times 10^6 = 2.79 \times 10^7$$

- *Roots of Numbers Expressed in Scientific Notation*

Unless you use an electronic calculator, the number must first be put into a form in which the exponent is exactly divisible by the root. For example, for a square root, the exponent should be divisible by 2. The root of the digit term is found in the usual way, and the exponent is divided by the desired root:

$$\sqrt{3.6 \times 10^7} = \sqrt{36 \times 10^6} = \sqrt{36} \times \sqrt{10^6} = 6.0 \times 10^3$$

Significant Figures

In most experiments several kinds of measurements must be made, and some can be made more precisely than others. It is common sense that a result calculated from experimental data can be no more precise than the least precise piece of information that went into the calculation. This is where the rules for significant figures come in. **Significant figures** are the digits in a measured quantity that reflect the accuracy of the measurement.

When describing standard deviation on page 33, we used the example of a measurement that was known to be 9.984 with an uncertainty of ± 0.007 cm. That is, the last number of our measurement, 0.004 cm, was uncertain to some degree. Our measurement is said to have four significant figures, the last of which is uncertain to some extent.

Suppose we want to calculate the density of a piece of metal (Figure 1.22). The mass and dimensions were determined by standard laboratory techniques. Most of these numbers have two digits to the right of the decimal, but they have different numbers of significant figures.

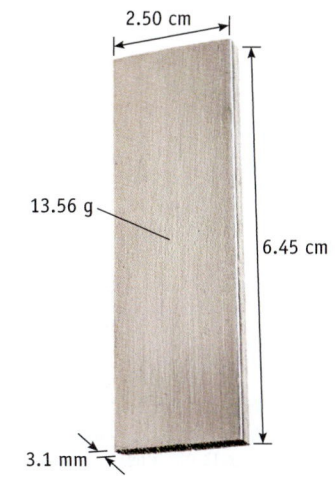

Measurement	Data Collected	Significant Figures
Mass of metal	13.56 g	4
Length	6.45 cm	3
Width	2.50 cm	3
Thickness	3.1 mm	2

The quantity 3.1 mm has two significant figures. That is, the 3 in 3.1 is exactly known, but the 1 is not. In general, *in a number representing a scientific measurement, the last digit to the right is taken to be inexact.* Unless stated otherwise, it is common practice to assign an uncertainty of ±1 to the last significant digit. This means the thickness of the metal piece may have been as small as 3.0 mm or as large as 3.2 mm.

When the data on the piece of metal are combined to calculate the density, the result will be 2.7 g/cm^3, a number with two significant figures. (The complete calculation of the metal density is given on page 41). The reason for this is that a calculated result can be no more precise than the least precise data used, and here the thickness has only two significant figures.

When doing calculations using measured quantities, we follow some basic rules so that the results reflect the precision of all the measurements that go into the calculations. The *rules used for significant figures in this book* are as follows:

Figure 1.22   Data to determine the density of a metal.

Rule 1. To determine the number of significant figures in a measurement, read the number from left to right and count all digits, starting with the first digit that is not zero.

Example	Number of Significant Figures
1.23	3; all nonzero digits are significant.
0.00123 g	3; the zeros to the left of the 1 (the first significant digit) simply locate the decimal point. To avoid confusion, write numbers of this type in scientific notation; thus, 0.00123 = 1.23 × 10^{-3}.
2.040 g	4; when a number is greater than 1, *all zeros to the right of the decimal point are significant.*
0.02040 g	4; for a number less than 1, only zeros to the right of the first nonzero digit are significant.
100 g	1; in numbers that do not contain a decimal point, "trailing" zeros may or may not be significant. *The practice followed in this book is to include a decimal point if the zeros are significant.* Thus, 100. is used to represent three significant digits, whereas 100 has only one significant digit. To avoid confusion, an alternative method is to write numbers in scientific notation because all digits are significant when written in scientific notation. Thus, 1.00 × 10^2 has three significant digits, whereas 1 × 10^2 has only one significant digit.
100 cm/m	Infinite number of significant digits. This is a *defined quantity.* Defined quantities do not limit the number of significant figures in a calculated result.
π = 3.1415926	The value of certain constants such as π is known to a greater number of significant figures than you will ever use in a calculation.

Standard laboratory balance and significant figures. Such balances can determine the mass of an object to the nearest milligram. Thus, an object may have a mass of 13.456 g (13456 mg, five significant figures), 0.123 g (123 mg, three significant figures), or 0.072 g (72 mg, two significant figures).

Rule 2. When adding or subtracting numbers, the number of decimal places in the answer is equal to the number of decimal places in the number with the fewest digits after the decimal.

0.12	2 decimal places	2 significant figures
+ 1.9	1 decimal place	2 significant figures
+10.925	3 decimal places	5 significant figures
12.945	3 decimal places	

The sum should be reported as 12.9, a number with one decimal place, because 1.9 has only one decimal place.

Rule 3. In multiplication or division, the number of significant figures in the answer should be the same as that in the quantity with the fewest significant figures.

$$\frac{0.01208}{0.0236} = 0.512 \text{ or, in scientific notation, } 5.12 \times 10^{-1}$$

Because 0.0236 has only three significant digits and 0.01208 has four, the answer should have three significant digits.

Rule 4. When a number is rounded off, the last digit to be retained is increased by one only if the following digit is 5 or greater.

Full Number	Number Rounded to Three Significant Digits
12.696	12.7
16.349	16.3
18.35	18.4
18.351	18.4

One last word on significant figures and calculations: When working problems, you should do the calculation with all the digits allowed by your calculator and round off only at the end of the calculation. Rounding off in the middle can introduce errors.

GENERAL
Chemistry ⚛ Now™

See the General ChemistryNow CD-ROM or website:
- **Screen 1.17 Using Numerical Information,** for tutorials on multiplying and dividing with significant figures, raising significant figures to a power, and taking square roots of significant figures

Example 1.5—Using Significant Figures

Problem An example of a calculation you will do later in the book (Chapter 12) is

$$\text{Volume of gas (L)} = \frac{(0.120)(0.08206)(273.15 + 23)}{(230/760.0)}$$

■ To Multiply or to Add?
Take the number 4.68.

(a) Take the sum of 4.68 + 4.68 + 4.68. The answer is 14.04, a number with four significant figures.

(b) Multiply 4.68 times 3. The answer can have only three significant figures (14.0).

You should recognize that different outcomes are possible depending on the type of mathematical operation.

■ Who Is Right—You or the Book?
If your answer to a problem in this book does not quite agree with the answers in Appendix N or O, the discrepancy may be the result of rounding the answer after each step and then using that rounded answer in the next step. This book follows these conventions:

(a) Final answers to numerical problems in this book result from retaining full calculator accuracy throughout the calculation and rounding only at the end.

(b) In Example problems, the answer to each step is given to the correct number of significant figures for that step, but the full calculator accuracy is carried to the next step. The number of significant figures in the final answer is dictated by the number of significant figures in the original data.

Calculate the final answer to the correct number of significant figures.

Strategy Let us first decide on the number of significant figures represented by each number (Rule 1), and then apply Rules 2 and 3.

Solution

Number	Number of Significant Figures	Comments
0.120	3	The trailing 0 is significant. See Rule 1.
0.08206	4	The first 0 to the immediate right of the decimal is not significant. See Rule 1.
$273.15 + 23 = 296$	3	23 has no decimal places, so the sum can have none. See Rule 2.
$230/760.0 = 0.30$	2	230 has two significant figures because the last zero is not significant. In contrast, there is a decimal point in 760.0, so there are four significant digits. The quotient may have only two significant digits. See Rules 1 and 3.

Analysis shows that one of the pieces of information is known to only two significant figures. Therefore, the volume of gas is 9.6 L, a number with two significant figures.

Exercise 1.13—Using Significant Figures

(a) How many significant figures are indicated by 2.33×10^7, by 50.5, and by 200?

(b) What are the sum and the product of 10.26 and 0.063?

(c) What is the result of the following calculation?

$$x = \frac{(110.7 - 64)}{(0.056)(0.00216)}$$

Problem Solving by Dimensional Analysis

Suppose you want to find the density of a rectangular piece of metal (Figure 1.22) in units of grams per cubic centimeter (g/cm^3). Because density is the ratio of mass to volume, you need to measure the mass and determine the volume of the piece. To find the volume of the sample in cubic centimeters, you multiply its length by its width and its thickness. First, however, all the measurements must have the same units, meaning that the thickness must be converted to centimeters. Recognizing that there are 10 mm in 1 cm, we use this relationship to get a thickness of 0.31 cm:

$$3.1 \text{ mm} \times \frac{1 \text{ cm}}{10 \text{ mm}} = 0.31 \text{ cm}$$

With all the dimensions in the same unit, the volume and then the density can be calculated:

$$\text{Length} \times \text{width} \times \text{thickness} = \text{volume}$$

$$6.45 \text{ cm} \times 2.50 \text{ cm} \times 0.31 \text{ cm} = 5.0 \text{ cm}^3$$

$$\text{Density} = \frac{13.56 \text{ g}}{5.0 \text{ cm}^3} = 2.7 \text{ g/cm}^3$$

■ **Data to Calculate Metal Density**
(See Figure 1.22)
Mass of metal = 13.56 g
Length = 6.45 cm
Width = 2.50 cm
Thickness = 3.1 mm

Dimensional analysis (sometimes called the *factor-label method*) is a general problem-solving approach that uses the dimensions or units of each value to guide you through calculations. This approach was used above to change 3.1 mm to its equivalent in centimeters. We multiplied the number we wished to convert (3.1 mm) by a **conversion factor** (1 cm/10 mm) to produce the result in the desired unit (0.31 cm). Units are handled like numbers: Because the unit "mm" was in both the numerator and the denominator, dividing one by the other leaves a quotient of 1. The units are said to "cancel out." Here this leaves the answer in centimeters, the desired unit.

A conversion factor expresses the equivalence of a measurement in two different units (1 cm $\equiv$ 10 mm; 1 g $\equiv$ 1000 mg; 12 eggs $\equiv$ 1 dozen; 12 inches $\equiv$ 1 foot). Because the numerator and the denominator describe the same quantity, the conversion factor is equivalent to the number 1. Therefore, multiplication by this factor does not change the measured quantity, only its units. A conversion factor is always written so that it has the form "new units divided by units of original number."

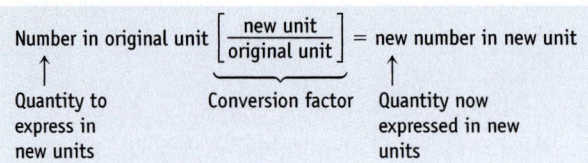

Number in original unit $\left[\dfrac{\text{new unit}}{\text{original unit}}\right]$ = new number in new unit

↑   ↑
Quantity to   Conversion factor   Quantity now
express in   expressed in new
new units   units

Example 1.6—Using Conversion Factors— Density in Different Units

Problem Oceanographers often express the density of sea water in units of kilograms per cubic meter. If the density of sea water is 1.025 g/cm^3 at 15 °C, what is its density in kilograms per cubic meter?

Strategy To simplify this problem, break it into three steps. First, change grams to kilograms. Next, convert cubic centimeters to cubic meters. Finally, calculate the density by dividing the mass in kilograms by the volume in cubic meters.

Solution First convert the mass in grams to kilograms.

$$1.025 \text{ g} \times \frac{1 \text{ kg}}{1000 \text{ g}} = 1.025 \times 10^{-3} \text{ kg}$$

No conversion factor is available in one of our tables to *directly* change units of cubic centimeters to cubic meters. You can find one, however, by cubing (raising to the third power) the relation between the meter and the centimeter:

$$1 \text{ cm}^3 \times \left(\frac{1 \text{ m}}{100 \text{ cm}}\right)^3 = 1 \text{ cm}^3 \times \left(\frac{1 \text{ m}^3}{1 \times 10^6 \text{ cm}^3}\right) = 1 \times 10^{-6} \text{ m}^3$$

Therefore, the density of sea water is

$$\text{Density} = \frac{1.025 \times 10^{-3}\ \text{kg}}{1 \times 10^{-6}\ \text{m}^3} = 1.025 \times 10^3\ \text{kg/m}^3$$

Exercise 1.14—Using Dimensional Analysis

(a) The annual snowfall at Lake Otsego is 198 cm each year. What is this depth in meters? In feet (where 1 foot = 30.48 cm)?

(b) The area of Lake Otsego is $2.33 \times 10^7\ \text{m}^2$. What is this area in square kilometers?

(c) The density of gold is 19,320 kg/m^3. What is this density in g/cm^3?

(d) See Figure 1.21. Show that 9.0×10^3 pc is 2.8×10^{17} km.

Graphing

In a number of instances in this text, graphs are used when analyzing experimental data with a goal of obtaining a mathematical equation. The procedure used will often result in a straight line, which has the equation

$$y = mx + b$$

In this equation, y is usually referred to as the dependent variable; its value is determined from (that is, is dependent on) the values of x, m, and b. In this equation x is called the independent variable and m is the slope of the line. The parameter b is the y-intercept—that is, the value of y when $x = 0$. Let us use an example to investigate two things: (a) how to construct a graph from a set of data points, and (b) how to derive an equation for the line generated by the data.

A set of data points to be graphed is presented in Figure 1.23. We first mark off each axis in increments of the values of x and y. Here our x-data range from -2 to 4, so the x-axis is marked off in increments of 1 unit. The y-data range from 0 to 2.5, so we mark off the y-axis in increments of 0.5. Each data set is marked as a circle on the graph.

After plotting the points on the graph (round circles), we draw a *straight line* that comes as close as possible to representing the trend in the data. (Do not connect the dots!) Because there is always some inaccuracy in experimental data, this line may not pass exactly through every point.

To identify the specific equation corresponding to our data, we must determine the y-intercept (b) and slope (m) for the equation $y = mx + b$. The y-intercept is the point at which $x = 0$. (In Figure 1.23, $y = 1.87$ when $x = 0$). The slope is determined by selecting two points on the line (marked with squares in Figure 1.23) and calculating the difference in values of y ($\Delta y = y_2 - y_1$) and x ($\Delta x = x_2 - x_1$). The slope of the line is then the ratio of these differences, $m = \Delta y/\Delta x$. Here the slope has the value -0.525. With the slope and intercept now known, we can write the equation for the line

$$y = -0.525x + 1.87$$

and we can use this equation to calculate y-values for points that are not part of our original set of x-y data. For example, when $x = 1.50$, $y = 1.08$.

■ **Determining the Slope with a Computer Program—Least-Squares Analysis**
Generally the easiest method of determining the slope and intercept of a straight line (and thus the line's equation) is to use a program such as Microsoft Excel. These programs perform a "least squares" or "linear regression" analysis and give the best straight line based on the data. (This line is referred to in Excel as a *trendline*.) The General ChemistryNow CD-ROM also has a useful plotting program that performs this analysis; see the "Plotting Tool" in the menu on any screen.

Figure 1.23　Plotting Data. Data for the variable x are plotted along the horizontal axis (abscissa), and data for y are plotted along the vertical axis (ordinate). The slope of the line, m in the equation $y = mx + b$, is given by $\Delta y / \Delta x$. The intercept of the line with the y-axis (when $x = 0$) is b in the equation.

Using Microsoft Excel with these data, and doing a linear regression (or least-squares) analysis, we find $y = -0.525x + 1.87$.

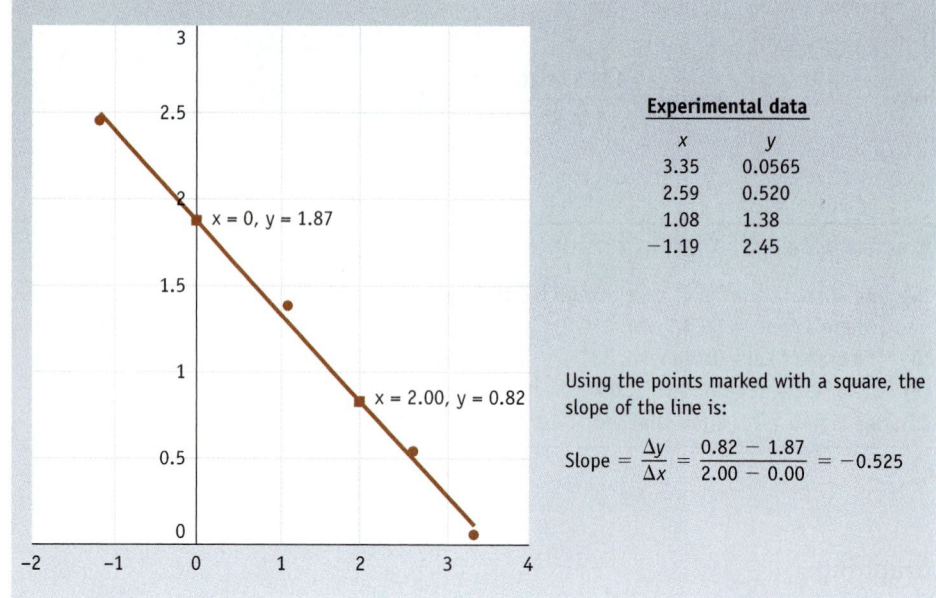

Experimental data

x	y
3.35	0.0565
2.59	0.520
1.08	1.38
−1.19	2.45

x = 0, y = 1.87

x = 2.00, y = 0.82

Using the points marked with a square, the slope of the line is:

$$\text{Slope} = \frac{\Delta y}{\Delta x} = \frac{0.82 - 1.87}{2.00 - 0.00} = -0.525$$

Exercise 1.15—Graphing

To find the mass of 50 jelly beans, we weighed several samples of beans.

Number of Beans	Mass (g)
5	12.82
11	27.14
16	39.30
24	59.04

Plot these data with the number of beans on the horizontal or x-axis, and the mass of beans on the vertical or y-axis. What is the slope of the line? Use your equation of a straight line to calculate the mass of exactly 50 jelly beans.

Problem Solving and Chemical Arithmetic

Problem-Solving Strategy

Some of the calculations in chemistry can be complex. Students frequently find it is helpful to follow a definite plan of attack as illustrated in examples throughout this book.

Step 1: Problem. State the problem. Read it carefully.

Step 2: Strategy. What key principles are involved? What information is known or not known? What information might be there just to place the question in the context of chemistry? Organize the information to see what is required and to discover the relationships among the data given. Try writing the information down in table form. If it is numerical information, be sure to include units.

One of the greatest difficulties for a student in introductory chemistry is picturing what is being asked for. Try sketching a picture of the situation involved. For example, we sketched a picture of the piece of metal whose density we wanted to calculate, and put the dimensions on the drawing (page 39).

Develop a plan. Have you done a problem of this type before? If not, perhaps the problem is really just a combination of several simpler ones you have seen before. Break it down into those simpler components. Try reasoning backward from the units of the answer. What data do you need to find an answer in those units?

Step 3: Solution. Execute the plan. Carefully write down each step of the problem, being sure to keep track of the units on numbers. (Do the units cancel to give you the answer in the desired units?) Don't skip steps. Don't do anything except the simplest steps in your head. Students often say they got a problem wrong because they "made a stupid mistake." Your instructor—and book authors—make them, too, and it is usually because they don't take the time to write down the steps of the problem clearly.

Step 4: Comment and Check Answer. As a final check, ask yourself whether the answer is reasonable.

Example 1.7—Problem Solving

Problem A mineral oil has a density of 0.875 g/cm³. Suppose you spread 0.75 g of this oil over the surface of water in a large dish with an inner diameter of 21.6 cm. How thick is the oil layer? Express the thickness in centimeters.

Strategy It is often useful to begin solving such problems by sketching a picture of the situation.

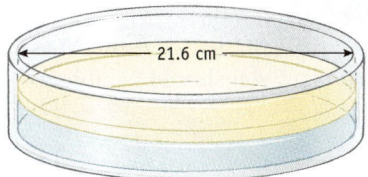

This helps recognize that the solution to the problem is to find the volume of the oil on the water. If we know the volume, then we can find the thickness because

$$\text{Volume of oil layer} = (\text{thickness of layer}) \times (\text{area of oil layer})$$

So, we need two things: (a) the volume of the oil layer and (b) the area of the layer.

Solution First calculate the volume of oil. The mass of the oil layer is known, so combining the mass of oil with its density gives the volume of the oil used:

$$0.75 \text{ g} \times \frac{1 \text{ cm}^3}{0.875 \text{ g}} = 0.86 \text{ cm}^3$$

Next calculate the area of the oil layer. The oil is spread over a circular surface, whose area is given by

$$\text{Area} = \pi \times (\text{radius})^2$$

The radius of the oil layer is one half its diameter (= 21.6 cm) or 10.8 cm, so

$$\text{Area of oil layer} = (3.142)(10.8 \text{ cm})^2 = 366 \text{ cm}^2$$

With the volume and the area of the oil layer known, the thickness can be calculated.

$$\text{Thickness} = \frac{\text{Volume}}{\text{Area}} = \frac{0.86 \text{ cm}^3}{366 \text{ cm}^2} = 0.0023 \text{ cm}$$

Comment In the volume calculation, the calculator shows 0.857143. . . . The quotient should have two significant figures because 0.75 has two significant figures, so the result of this step is 0.86 cm^3. In the area calculation, the calculator shows 366.435. . . . The answer to this step should have three significant figures because 10.8 has three. When these interim results are combined in calculating thickness, however, the final result can have only two significant figures. Premature rounding can lead to errors.

Exercise 1.16—Problem Solving

A particular paint has a density of 0.914 g/cm^3. You need to cover a wall that is 7.6 m long and 2.74 m high with a paint layer 0.13 mm thick. What volume of paint (in liters) is required? What is the mass (in grams) of the paint layer?

For additional preparation for an examination on this chapter see the **Let's Review** section on pages 282–293.

Chapter Goals Revisited

Now that you have studied this chapter, you should ask whether you have met the chapter goals. In particular, you should be able to

Classify matter

  a. Recognize the different states of matter (solids, liquids, and gases) and give their characteristics (Section 1.1).

  b. Appreciate the difference between pure substances and mixtures and the difference between homogeneous and heterogeneous mixtures (Section 1.1).

  c. Recognize the importance of representing matter at the macroscopic level and at the particulate level (Section 1.1).

Apply the kinetic-molecular theory to the properties of matter

  a. Understand the basic ideas of the kinetic-molecular theory (Section 1.1).

Recognize elements, atoms, compounds, and molecules

  a. Identify the name or symbol for an element, given its symbol or name (Section 1.2). General ChemistryNow homework: Study Question(s) 2

  b. Use the terms *atom, element, molecule,* and *compound* correctly (Sections 1.2 and 1.3).

Identify physical and chemical properties and changes

  a. List commonly used physical properties of matter (Section 1.4). General ChemistryNow homework: SQ(s) 8

b. Identify several physical properties of common substances (Section 1.4).

c. Use density to connect the volume and mass of a substance (Sections 1.4, 1.6 and 1.8). General ChemistryNow homework: SQ(s) 11, 13

d. Explain the difference between chemical and physical changes (Sections 1.4 and 1.5).

e. Understand the difference between extensive and intensive properties and give examples of them (Section 1.4).

Use metric units correctly

a. Convert between temperatures on the Celsius and Kelvin scales (Section 1.6). General ChemistryNow homework: SQ(s) 19

b. Recognize and know how to use the prefixes that modify the sizes of metric units (Section 1.6).

Understand and use the mathematics of chemistry

a. Use dimensional analysis to carry out unit conversions. Perform other mathematical operations (Section 1.8). General ChemistryNow homework: SQ(s) 24, 39, 40, 43, 45, 51, 83c, 84a, 89, 94b, 95, 96, 97, 104

b. Know the difference between precision and accuracy and how to calculate percent error (Section 1.7). General ChemistryNow homework: SQ(s) 30

c. Understand the use of significant figures (Section 1.8). General ChemistryNow homework: SQ(s) 78, 80

Key Equations

Equation 1.1 (page 20)
Density is the quotient of the mass of an object divided by its volume. In chemistry the common unit of density is g/cm^3.

$$\text{Density} = \frac{\text{mass}}{\text{volume}}$$

Equation 1.2 (page 28)
The equation allows the conversion between the Kelvin and Celsius temperature scales.

$$T(K) = \frac{1\ K}{1\ °C}\left[T\,(°C) + 273.15\ °C\right]$$

Equation 1.3 (page 34)
The percent error of a measurement is the deviation of the measurement from the accepted value.

$$\text{Percent error} = \frac{\text{error in measurement}}{\text{accepted value}} \times 100\%$$

Study Questions

▲ denotes more challenging questions.

■ denotes questions available in the Homework and Goals section of the General ChemistryNow CD-ROM or website.

Blue numbered questions have answers in Appendix O and fully worked solutions in the *Student Solutions Manual*.

Structures of many of the compounds used in these questions are found on the General ChemistryNow CD-ROM or website in the Models folder.

GENERAL
Chemistry ⚛ Now™ Assess your understanding of this chapter's topics with additional quizzing and conceptual questions at http://now.brookscole.com/kotz6e

Practicing Skills

Matter: Elements and Atoms, Compounds and Molecules
(See Exercise 1.2.)

1. Give the name of each of the following elements:
   (a) C   (c) Cl   (e) Mg
   (b) K   (d) P   (f) Ni

2. ■ Give the name of each of the following elements:
   (a) Mn   (c) Na   (e) Xe
   (b) Cu   (d) Br   (f) Fe

3. Give the symbol for each of the following elements:
   (a) barium   (d) lead
   (b) titanium   (e) arsenic
   (c) chromium   (f) zinc

4. Give the symbol for each of the following elements:
   (a) silver   (d) tin
   (b) aluminum   (e) technetium
   (c) plutonium   (f) krypton

5. In each of the following pairs, decide which is an element and which is a compound.
   (a) Na and NaCl
   (b) sugar and carbon
   (c) gold and gold chloride

6. In each of the following pairs, decide which is an element and which is a compound.
   (a) $Pt(NH_3)_2Cl_2$ and Pt
   (b) copper or copper(II) oxide
   (c) silicon or sand

Physical and Chemical Properties
(See Exercises 1.3 and 1.6.)

7. In each case, decide whether the underlined property is a physical or chemical property.

(a) The normal color of elemental bromine is <u>orange</u>.
(b) Iron <u>turns to rust</u> in the presence of air and water.
(c) Hydrogen can <u>explode</u> when ignited in air.
(d) The <u>density</u> of titanium metal is 4.5 g/cm³.
(e) Tin metal <u>melts</u> at 505 K.
(f) Chlorophyll, a plant pigment, <u>is green</u>.

8. ■ In each case, decide whether the change is a chemical or physical change.
   (a) A cup of household bleach changes the color of your favorite T-shirt from purple to pink.
   (b) Water vapor in your exhaled breath condenses in the air on a cold day.
   (c) Plants use carbon dioxide from the air to make sugar.
   (d) Butter melts when placed in the sun.

9. Which part of the description of a compound or element refers to its physical properties and which to its chemical properties?
   (a) The colorless liquid ethanol burns in air.
   (b) The shiny metal aluminum reacts readily with orange, liquid bromine.

10. Which part of the description of a compound or element refers to its physical properties and which to its chemical properties?
    (a) Calcium carbonate is a white solid with a density of 2.71 g/cm³. It reacts readily with an acid to produce gaseous carbon dioxide.
    (b) Gray, powdered zinc metal reacts with purple iodine to give a white compound.

Using Density
(See Example 1.1. and the General ChemistryNow Screen 1.10.)

11. ■ Ethylene glycol, $C_2H_6O_2$, is an ingredient of automobile antifreeze. Its density is 1.11 g/cm³ at 20 °C. If you need exactly 500. mL of this liquid, what mass of the compound, in grams, is required?

12. A piece of silver metal has a mass of 2.365 g. If the density of silver is 10.5 g/cm³, what is the volume of the silver?

13. ■ A chemist needs 2.00 g of a liquid compound with a density of 0.718 g/cm³. What volume of the compound is required?

14. The *cup* is a volume measure widely used by cooks in the United States. One cup is equivalent to 237 mL. If 1 cup of olive oil has a mass of 205 g, what is the density of the oil (in grams per cubic centimeter)?

15. A sample of unknown metal is placed in a graduated cylinder containing water. The mass of the sample is 37.5 g, and the water levels before and after adding the sample to the cylinder are as shown in the figure. Which metal in the following list is most likely the sample? (*d* is the density of the metal.)

(a) Mg, $d = 1.74$ g/cm^3
(b) Fe, $d = 7.87$ g/cm^3
(c) Ag, $d = 10.5$ g/cm^3
(d) Al, $d = 2.70$ g/cm^3
(e) Cu, $d = 8.96$ g/cm^3
(f) Pb, $d = 11.3$ g/cm^3

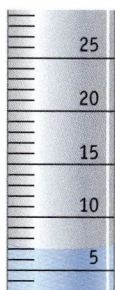 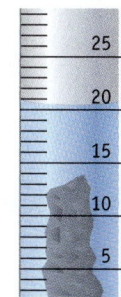

Graduated cylinders with unknown metal (*right*).

16. Iron pyrite is often called "fool's gold" because it looks like gold (see page 19). Suppose you have a solid that looks like gold, but you believe it to be fool's gold. The sample has a mass of 23.5 g. When the sample is lowered into the water in a graduated cylinder (see Study Question 15), the water level rises from 47.5 mL to 52.2 mL. Is the sample fool's gold ($d = 5.00$ g/cm^3) or "real" gold ($d = 19.3$ g/cm^3)?

Temperature Scales

(*See Exercise 1.7. and the General ChemistryNow Screen 1.15.*)

17. Many laboratories use 25 °C as a standard temperature. What is this temperature in kelvins?

18. The temperature on the surface of the sun is 5.5×10^3 °C. What is this temperature in kelvins?

19. ■ Make the following temperature conversions:

°C	K
(a) 16	——
(b) ——	370
(c) 40	——

20. Make the following temperature conversions:

°C	K
(a) ——	77
(b) 63	——
(c) ——	1450

Using Units

(*See Examples 1.2 and 1.3. and the General ChemistryNow Screen 1.14.*)

21. A marathon race covers a distance of 42.195 km. What is this distance in meters? In miles?

22. The average lead pencil, new and unused, is 19 cm long. What is its length in millimeters? In meters?

23. A standard U.S. postage stamp is 2.5 cm long and 2.1 cm wide. What is the area of the stamp in square centimeters? In square meters?

24. ■ A compact disk has a diameter of 11.8 cm. What is the surface area of the disk in square centimeters? In square meters? [Area of a circle $= (\pi)(\text{radius})^2$.]

25. A typical laboratory beaker has a volume of 250. mL. What is its volume in cubic centimeters? In liters? In cubic meters? In cubic decimeters?

26. Some soft drinks are sold in bottles with a volume of 1.5 L. What is this volume in milliliters? In cubic centimeters? In cubic decimeters?

27. A book has a mass of 2.52 kg. What is this mass in grams?

28. A new U. S. dime has a mass of 2.265 g. What is this mass in kilograms? In milligrams?

Accuracy, Precision, and Error

(*See Example 1.4.*)

29. You and your lab partner are asked to determine the density of an aluminum bar. The mass is known accurately (to four significant figures). You use a simple metric ruler to determine its size and calculate the results in A. Your partner uses a precision micrometer and obtains the results in B.

Method A (g/cm^3)	Method B (g/cm^3)
2.2	2.703
2.3	2.701
2.7	2.705
2.4	5.811

The accepted density of aluminum is 2.702 g/cm^3.

(a) Calculate the average density for each method. Should all the experimental results be included in your calculations? If not, justify any omissions.

(b) Calculate the percent error for each method's average value.

(c) Which method's average value is more precise? Which method is more accurate?

30. ■ The accepted value of the melting point of pure aspirin is 135 °C. Trying to verify that value, you obtain the melting points of 134 °C, 136 °C, 133 °C, and 138 °C in four separate trials. Your partner finds melting points of 138 °C, 137 °C, 138 °C, and 138 °C.

(a) Calculate the average value and percent error for you and your partner.

(b) Which of you is more precise? More accurate?

General Questions

These questions are not designated as to type or location in the chapter. They may combine several concepts.

31. A piece of turquoise is a blue-green solid, and has a density of 2.65 g/cm^3 and a mass of 2.5 g.

(a) Which of these observations are qualitative and which are quantitative?

(b) Which of these observations are extensive and which are intensive?

(c) What is the volume of the piece of turquoise?

32. Give a physical property and a chemical property for the elements hydrogen, oxygen, iron, and sodium. (The elements listed are selected from examples given in Chapter 1.)

33. The gemstone called aquamarine is composed of aluminum, silicon, and oxygen.

Aquamarine is the bluish crystal. It is surrounded by aluminum foil and crystalline silicon.

(a) What are the symbols of the three elements that combine to make the gem aquamarine?

(b) Based on the photo, describe some of the physical properties of the elements and the mineral. Are any the same? Are any properties different?

34. Eight observations are listed below. Which of these observations identify chemical properties?

(a) Sugar is soluble in water.

(b) Water boils at 100 °C.

(c) Ultraviolet light converts O_3 (ozone) to O_2 (oxygen).

(d) Ice is less dense than water.

(e) Sodium metal reacts violently with water.

(f) CO_2 does not support combustion.

(g) Chlorine is a yellow gas.

(h) Heat is required to melt ice.

35. Neon, a gaseous element used in neon signs, has a melting point of −248.6 °C and a boiling point of −246.1 °C. Express these temperatures in kelvins.

36. You can identify a metal by carefully determining its density (d). An unknown piece of metal, with a mass of 2.361 g, is 2.35 cm long, 1.34 cm wide, and 1.05 mm thick. Which of the following is this element?

(a) Nickel, $d = 8.90$ g/cm^3

(b) Titanium, $d = 4.50$ g/cm^3

(c) Zinc, $d = 7.13$ g/cm^3

(d) Tin, $d = 7.23$ g/cm^3

37. Molecular distances are usually given in nanometers (1 nm $= 1 \times 10^{-9}$ m) or in picometers (1 pm $= 1 \times 10^{-12}$ m). However, the angstrom (Å) is sometimes

used, where 1 Å $= 1 \times 10^{-10}$ m. (The angstrom is not an SI unit.) If the distance between the Pt atom and the N atom in the cancer chemotherapy drug cisplatin is 1.97 Å, what is this distance in nanometers? In picometers?

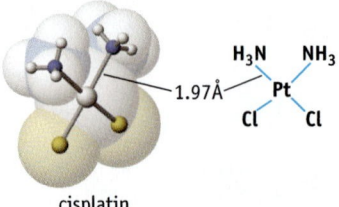

cisplatin

38. The separation between carbon atoms in diamond is 0.154 nm. What is their separation in meters? In picometers?

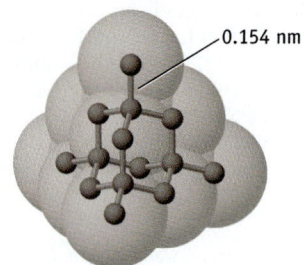

A portion of the diamond structure

39. ■ A red blood cell has a diameter of 7.5 μm (micrometers). What is this dimension in (a) meters, (b) nanometers, and (c) picometers?

40. ■ Which occupies a larger volume, 600. g of water (with a density of 0.995 g/cm^3) or 600. g of lead (with a density of 11.34 g/cm^3)?

41. The platinum-containing cancer drug cisplatin contains 65.0% platinum. If you have 1.53 g of the compound, what mass of platinum (in grams) is contained in this sample?

42. The solder once used by plumbers to fasten copper pipes together consists of 67% lead and 33% tin. What is the mass of lead in a 250-g block of solder?

43. ■ The anesthetic procaine hydrochloride is often used to deaden pain during dental surgery. The compound is packaged as a 10.% solution (by mass; $d = 1.0$ g/mL) in water. If your dentist injects 0.50 mL of the solution, what mass of procaine hydrochloride (in milligrams) is injected?

44. A cube of aluminum (density = 2.70 g/cm^3) has a mass of 7.6 g. What must be the length of the cube's edge (in centimeters)? (*See General ChemistryNow Screen 1.10, Tutorial 2, Density.*)

45. ■ You have a 100.0-mL graduated cylinder containing 50.0 mL of water. You drop a 154-g piece of brass ($d = 8.56$ g/cm^3) into the water. How high does the water rise in the graduated cylinder?

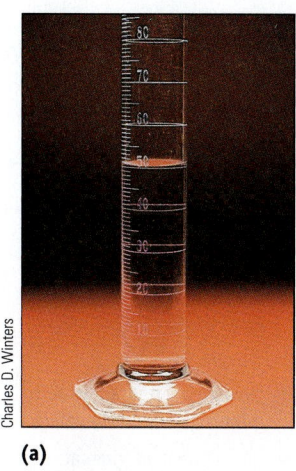

(a)

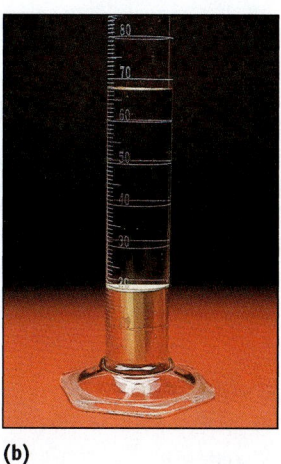

(b)

(a) A graduated cylinder with 50.0 ml of water. (b) A piece of brass is added to the cylinder.

46. You have a white crystalline solid, known to be one of the potassium compounds listed below. To determine which, you measure the solid's density. You measure out 18.82 g and transfer it to a graduated cylinder containing kerosene (in which salts will not dissolve). The level of liquid kerosene rises from 8.5 mL to 15.3 mL. Calculate the density of the solid, and identify the compound from the following list.

   (a) KF, $d = 2.48$ g/cm^3 (c) KBr, $d = 2.75$ g/cm^3

   (b) KCl, $d = 1.98$ g/cm^3 (d) KI, $d = 3.13$ g/cm^3

47. A distant acquaintance has offered to sell you a necklace, said to be pure (24-carat) gold, for $300. You have some doubts, however; perhaps it is gold plated. You decide to run a test. You have a graduated cylinder and a small balance. You partially fill the cylinder with water and immerse the necklace; the height of water rises from 22.5 mL to 26.0 mL. Then you determine the mass to be 67 g. You recall that the density of gold is 19.3 g/cm^3, and that no other element has a density near this value. (Silver has a density of 11.5 g/cm^3.) The price of gold on the open market is $380 per troy ounce (1 troy ounce = 31.1 g). Is the necklace gold? Explain your conclusion. Is $300 a good price?

Conceptual Questions

48. The mineral fluorite contains the elements calcium and fluorine. What are the symbols of these elements? How would you describe the shape of the fluorite crystals in the photo? What can this tell us about the arrangement of the atoms inside the crystal?

The mineral fluorite, calcium fluoride.

49. Small chips of iron are mixed with sand (see the photo). Is this a homogeneous or heterogeneous mixture? Suggest a way to separate the iron from the sand.

Chips of iron mixed with sand.

50. The following photo shows copper balls, immersed in water, floating on top of mercury. What are the liquids and the solids in this photo? Which substance is most dense? Which is least dense?

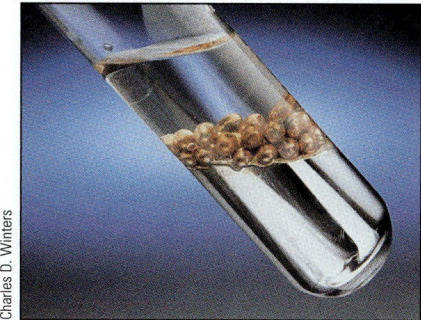

Water, copper, and mercury.

51. ■ Carbon tetrachloride, CCl$_4$, a liquid compound, has a density of 1.58 g/cm^3. If you place a piece of a plastic soda bottle ($d = 1.37$ g/cm^3) and a piece of aluminum ($d = 2.70$ g/cm^3) in liquid CCl$_4$, will the plastic and aluminum float or sink?

52. Figure 1.7 shows a piece of table salt and a representation of its internal structure. Which is the macroscopic view and which is the particulate view? How are the macroscopic and particulate views related?

53. ▲ You have a sample of a white crystalline substance from your kitchen. You know that it is either salt or sugar. Although you could decide by taste, suggest another property that you could use to determine the sample's identity. (*Hint:* You may use the World Wide Web or a handbook of chemistry in the library to find some pertinent information.)

54. Milk in a glass bottle was placed in the freezer compartment of a refrigerator overnight. By morning a column of frozen milk emerged from the bottle. Explain this observation.

Frozen milk in a glass bottle.

55. The element gallium has a melting point of 29.8 °C. If you held a sample of gallium in your hand, should it melt? Explain briefly.

Gallium metal.

56. ▲ The density of pure water is given at various temperatures.

T(°C)	d(g/cm³)
4	0.99997
15	0.99913
25	0.99707
35	0.99406

Suppose your laboratory partner tells you that the density of water at 20 °C is 0.99910 g/cm³. Is this a reasonable number? Why or why not?

57. You can figure out whether a substance floats or sinks if you know its density and the density of the liquid. In which of the liquids listed below will high-density polyethylene (HDPE, a common plastic whose density is 0.97 g/mL) float? (HDPE does not dissolve in these liquids.)

Substance	Density (g/cm³)	Properties, Uses
Ethylene glycol	1.1088	Toxic; the major component of automobile antifreeze
Water	0.9997	
Ethanol	0.7893	The alcohol in alcoholic beverages
Methanol	0.7914	Toxic; gasoline additive to prevent gas line freezing
Acetic acid	1.0492	Component of vinegar
Glycerol	1.2613	Solvent used in home care products.

58. Hexane (C_6H_{14}, $d = 0.766$ g/cm³), perfluorohexane (C_6F_{14}, $d = 1.669$ g/cm³), and water are immiscible liquids; that is, they do not dissolve in one another. You place 10 mL of each liquid in a graduated cylinder, along with pieces of high-density polyethylene (HDPE, $d = 0.97$ g/cm³), polyvinyl chloride (PVC, $d = 1.36$ g/cm³), and Teflon (density = 2.3 g/cm³). None of these common plastics dissolve in these liquids. Describe what you expect to see.

59. Make a drawing, based on the kinetic-molecular theory and the ideas about atoms and molecules presented in this chapter, of the arrangement of particles in each of the cases listed here. For each case draw ten particles of each substance. Your diagram can be two-dimensional. Represent each atom as a circle and distinguish each kind of atom by shading.

   (a) a sample of solid iron (which consists of iron atoms)
   (b) a sample of *liquid* water (which consists of H_2O molecules)
   (c) a sample of water *vapor*
   (d) a homogeneous mixture of water vapor and helium gas (which consists of helium atoms)
   (e) a heterogeneous mixture consisting of liquid water and solid aluminum; show a region of the sample that includes both substances
   (f) a sample of brass (which is a homogeneous mixture of copper and zinc)

60. You are given a sample of a silvery metal. What information would you seek to prove that the metal is silver?

61. Suggest a way to determine whether the colorless liquid in a beaker is water. If it is water, does it contain dissolved salt? How could you discover whether salt is dissolved in the water?

62. Describe an experimental method that can be used to determine the density of an irregularly shaped piece of metal.

63. Three liquids of different densities are mixed. Because they are not miscible (do not form a homogeneous solution with one another), they form discrete layers, one on top of the other. Sketch the result of mixing carbon tetrachloride (CCl_4, $d = 1.58$ g/cm^3), mercury ($d = 13.546$ g/cm^3), and water ($d = 1.00$ g/cm^3).

64. Diabetes can alter the density of urine, so urine density can be used as a diagnostic tool. People with diabetes may excrete too much sugar or too much water. What do you predict will happen to the density of urine under each of these conditions? (*Hint:* Water containing dissolved sugar has a higher density than pure water.)

65. The following photo shows the element potassium reacting with water to form the element hydrogen, a gas, and a solution of the compound potassium hydroxide.

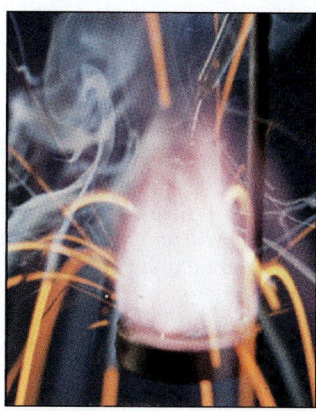

Potassium reacting with water to produce hydrogen gas and potassium hydroxide.

(a) What states of matter are involved in the reaction?

(b) Is the observed change chemical or physical?

(c) What are the reactants in this reaction and what are the products?

(d) What qualitative observations can be made concerning this reaction?

66. A copper-colored metal is found to conduct an electric current. Can you say with certainty that it is copper? Why or why not? Suggest additional information that could provide unequivocal confirmation that the metal is copper.

67. What experiment can you use to:

(a) Separate salt from water?

(b) Separate iron filings from small pieces of lead?

(c) Separate elemental sulfur from sugar?

68. Four balloons (each with a volume of 10 L and a mass of 1.00 g) are filled with a different gas:

Helium, $d = 0.164$ g/L

Neon, $d = 0.825$ g/L

Argon, $d = 1.633$ g/L

Krypton, $d = 4.425$ g/L

If the density of dry air is 1.12 g/L, which balloon or balloons float in air?

69. Many foods are fortified with vitamins and minerals. For example, some breakfast cereals have elemental iron added. Iron chips are used instead of iron compounds because the compounds can be converted by the oxygen in air to a form of iron that is not biochemically useful. Iron chips, in contrast, are converted to useful iron compounds in the gut, and the iron can then be absorbed. Outline a method by which you could remove the iron (as iron chips) from a box of cereal and determine the mass of iron in a given mass of cereal. (*See General ChemistryNow Screens 1.1 and 1.18 Chemical Puzzler.*)

Some breakfast cereals contain iron in the form of elemental iron.

70. Describe what occurs when a hot object comes in contact with a cooler object. (*See General ChemistryNow Screen 1.15 Temperature.*)

71. Study the animation of the conversion of P_4 and Cl_2 molecules to PCl_3 molecules on General ChemistryNow CD-ROM or website Screen 1.13 Chemical Change on the Molecular Scale.

(a) What are the reactants in this chemical change? What are the products?

(b) Describe how the structures of the reactant molecules differ from the structures of the product molecules.

72. The photo below shows elemental iodine dissolving in ethanol to give a solution. Is this a physical or a chemical change?

Elemental iodine dissolving in ethanol.

(*See General ChemistryNow Screen 1.9 Exercise, Physical Properties of Matter.*)

Mathematics of Chemistry

These questions provide an additional review of the mathematical skill used in general chemistry as presented in Section 1.8.

Exponential Notation

73. Express the following numbers in exponential or scientific notation.
    (a) 0.054   (b) 5462   (c) 0.000792

74. Express the following numbers in fixed notation (e.g., $123 \times 10^2 = 123$).
    (a) 1.62×10^3
    (b) 2.57×10^{-4}
    (c) 6.32×10^{-2}

75. Carry out the following operations. Provide the answer with the correct number of significant figures.
    (a) $(1.52)(6.21 \times 10^{-3})$
    (b) $(6.21 \times 10^3) - (5.23 \times 10^2)$
    (c) $(6.21 \times 10^3) \div (5.23 \times 10^2)$

76. Carry out the following operations. Provide the answer with the correct number of significant figures.
    (a) $(6.25 \times 10^2)^3$
    (b) $\sqrt{2.35 \times 10^{-3}}$
    (c) $(2.35 \times 10^{-3})^{1/3}$

Significant Figures

(*See Exercise 1.13.*)

77. Give the number of significant figures in each of the following numbers:
    (a) 0.0123 g   (c) 1.6402 g
    (b) 3.40×10^3 mL   (d) 1.020 L

78. ■ Give the number of significant figures in each of the following numbers:
    (a) 0.00546 g   (c) 2.300×10^{-4} g
    (b) 1600 mL   (d) 2.34×10^9 atoms

79. Carry out the following calculation, and report the answer with the correct number of significant figures.

$$(0.0546)(16.0000)\left[\frac{7.779}{55.85}\right]$$

80. ■ Carry out the following calculation, and report the answer to the correct number of significant figures.

$$(1.68)\left[\frac{23.56 - 2.3}{1.248 \times 10^3}\right]$$

Graphing

(*See Exercise 1.15. Use the plotting program on the General ChemistryNow CD-ROM or website or Microsoft Excel.*)

81. You are asked to calibrate a spectrophotometer in the laboratory and collect the following data. Plot the data with concentration on the *x*-axis and absorbance on the *y*-axis. Draw the best straight line using the points on the graph (or do a least-squares or linear regression analysis using a computer program) and then write the equation for the resulting straight line. What is the slope of the line? What is the concentration when the absorbance is 0.635?

Concentration (M)	Absorbance
0.00	0.00
1.029×10^{-3}	0.257
2.058×10^{-3}	0.518
3.087×10^{-3}	0.771
4.116×10^{-3}	1.021

82. To determine the average mass of a popcorn kernel you collect the following data:

Number of kernels	Mass (g)
5	0.836
12	2.162
35	5.801

Plot the data with number of kernels on the *x*-axis and mass on the *y*-axis. Draw the best straight line using the points on the graph (or do a least-squares or linear regression analysis using a computer program) and then write the equation for the resulting straight line. What is the slope of the line? What does the slope of the line signify about the mass of a popcorn kernel? What is the mass of 50 popcorn kernels? How many kernels are there in a handful of popcorn (20.88 g)?

83. Using the graph below:
    (a) What is the value of *x* when *y* = 4.0?
    (b) What is the value of *y* when *x* = 0.30?
    (c) ■ What are the slope and the *y*-intercept of the line?
    (d) What is the value of *y* when *x* = 1.0?

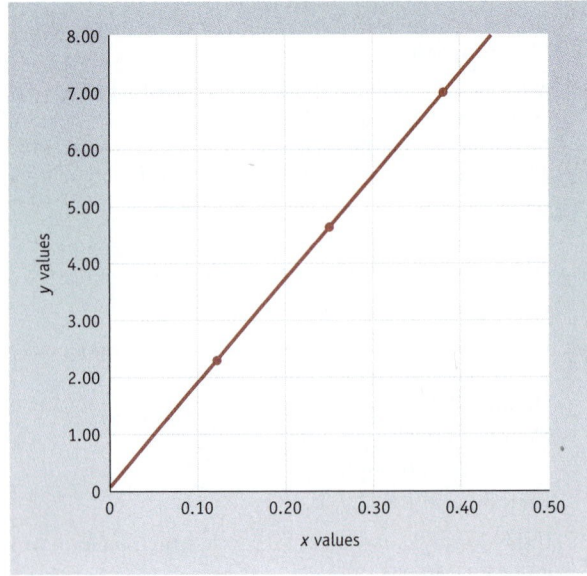

84. ■ Use the graph below to answer the following questions.

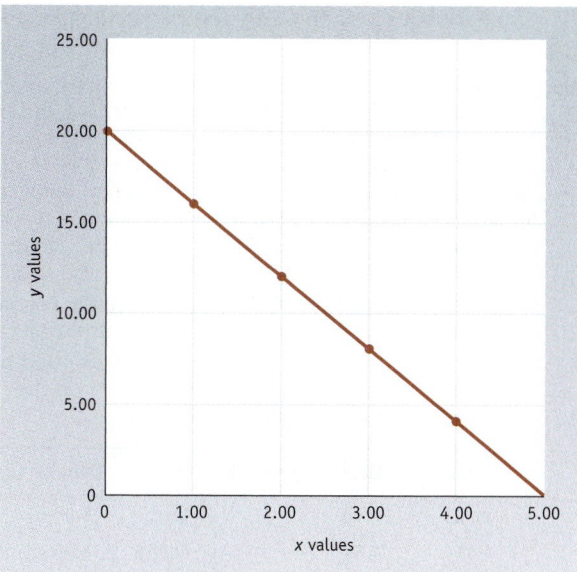

(a) Derive the equation for the straight line, $y = mx + b$.
(b) What is the value of y when $x = 6.0$?

Using Equations

85. Solve the following equation for the unknown value, C.
$$(0.502)(123) = (750.)C$$

86. Solve the following equation for the unknown value, n.
$$(2.34)(15.6) = n(0.0821)(273)$$

87. Solve the following equation for the unknown value, T.
$$(4.184)(244)(T - 292.0) + (0.449)(88.5)(T - 369.0) = 0$$

88. Solve the following equation for the unknown value, n.
$$-246.0 = 1312\left[\frac{1}{2^2} - \frac{1}{n^2}\right]$$

Problem Solving

89. ■ Diamond has a density of 3.513 g/cm³. The mass of diamonds is often measured in "carats," where 1 carat equals 0.200 g. What is the volume (in cubic centimeters) of a 1.50-carat diamond?

90. ▲ The smallest repeating unit of a crystal of common salt is a cube (called a unit cell) with an edge length of 0.563 nm.

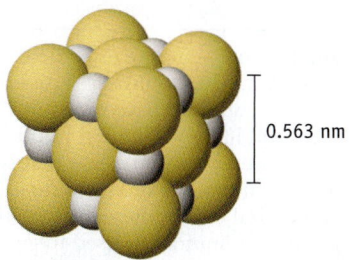

0.563 nm

sodium chloride, NaCl

(a) What is the volume of this cube in cubic nanometers? In cubic centimeters?
(b) The density of NaCl is 2.17 g/cm³. What is the mass of this smallest repeating unit ("unit cell")?
(c) Each repeating unit is composed of four NaCl "molecules." What is the mass of one NaCl molecule?

91. An ancient gold coin is 2.2 cm in diameter and 3.0 mm thick. It is a cylinder for which volume = $(\pi)(\text{radius})^2(\text{thickness})$. If the density of gold is 19.3 g/cm³, what is the mass of the coin in grams?

92. Copper has a density of 8.96 g/cm³. An ingot of copper with a mass of 57 kg (126 lb) is drawn into wire with a diameter of 9.50 mm. What length of wire (in meters) can be produced? [Volume of wire = $(\pi)(\text{radius})^2(\text{length})$].

93. ▲ In July 1983, an Air Canada Boeing 767 ran out of fuel over central Canada on a trip from Montreal to Edmonton. (The plane glided safely to a landing at an abandoned airstrip.) The pilots knew that 22,300 kg of fuel were required for the trip, and they knew that 7682 L of fuel were already in the tank. The ground crew added 4916 L of fuel, which was only about one fifth of what was required. The crew members used a factor of 1.77 for the fuel density—the problem is that 1.77 has units of *pounds* per liter and not *kilograms* per liter! What is the fuel density in units of kg/L? What mass of fuel should have been loaded? (1 lb = 453.6 g.)

94. When you heat popcorn, it pops because it loses water explosively. Assume a kernel of corn, with a mass of 0.125 g, has a mass of only 0.106 g after popping.
(a) What percentage of its mass did the kernel lose on popping?
(b) ■ Popcorn is sold by the pound in the United States. Using 0.125 g as the average mass of a popcorn kernel, how many kernels are there in a pound of popcorn? (1 lb = 453.6 g.)

95. ▲ The aluminum in a package containing 75 ft² of kitchen foil weighs approximately 12 ounces. Aluminum has a density of 2.70 g/cm³. What is the approximate thickness of the aluminum foil in millimeters? (1 oz = 28.4 g.)

96. ▲ The fluoridation of city water supplies has been practiced in the United States for several decades. It is done by continuously adding sodium fluoride to water as it comes from a reservoir. Assume you live in a medium-sized city of 150,000 people and that 660 L (170 gal) of water is consumed per person per day. What mass of sodium fluoride (in kilograms) must be added to the water supply each year (365 days) to have the required fluoride concentration of 1 ppm (part per million)—that is, 1 kilogram of fluoride per 1 million kilograms of water? (Sodium fluoride is 45.0% fluoride, and water has a density of 1.00 g/cm³.)

97. ■ ▲ About two centuries ago, Benjamin Franklin showed that 1 teaspoon of oil would cover about 0.5 acre of still water. If you know that 1.0×10^4 m² = 2.47 acres, and that there is approximately 5 cm³ in a teaspoon, what is the thickness of the layer of oil? How might this thickness be related to the sizes of molecules?

98. ▲ Automobile batteries are filled with an aqueous solution of sulfuric acid. What is the mass of the acid (in grams) in 500. mL of the battery acid solution if the density of the solution is 1.285 g/cm^3 and if the solution is 38.08% sulfuric acid by mass?

99. A piece of copper has a mass of 0.546 g. Show how to set up an expression to find the volume of this piece of copper in units of liters. (Copper density = 8.96 g/cm^3.) (*See General ChemistryNow Screen 1.17 Tutorial 1, Using Numerical Information.*)

100. Evaluate the value of x in the following expressions:

    (a) $x = [(9.345 \times 10^{-4})(6.23 \times 10^6)]^3$
    (b) $x = \sqrt{(1.23 \times 10^{-2})(4.5 \times 10^5)}$
    (c) $x = \sqrt[3]{(1.23 \times 10^{-2})(4.5 \times 10^5)}$

    Show the answers to the correct number of significant figures. (*See General ChemistryNow CD-ROM or website Screen 1.17 Tutorial 4, Using Numerical Information.*)

101. A 26-meter tall statue of Buddha in Tibet is covered with 279 kg of gold. If the gold was applied to a thickness of 0.0015 mm, what surface area is covered (in square meters)? (Gold density = 19.3 g/cm^3.)

102. At 25 °C the density of water is 0.997 g/cm^3, whereas the density of ice at −10 °C is 0.917 g/cm^3.

    (a) If a soft-drink can (volume = 250. mL) is filled completely with pure water at 25 °C and then frozen at −10 °C, what volume does the solid occupy?

    (b) Can the ice be contained within the can?

103. Suppose your bedroom is 18 ft long, 15 ft wide, and the distance from floor to ceiling is 8 ft, 6 in. You need to know the volume of the room in metric units for some scientific calculations.

    (a) What is the room's volume in cubic meters? In liters?

    (b) What is the mass of air in the room in kilograms? In pounds? (Assume the density of air is 1.2 g/L and that the room is empty of furniture.)

104. ■ A spherical steel ball has a mass of 3.475 g and a diameter of 9.40 mm. What is the density of the steel? [The volume of a sphere = $(4/3)\pi r^3$ where r = radius.]

105. ▲ The substances listed below are clear liquids. You are asked to identify an unknown liquid that is known to be one of these liquids. You pipette a 3.50-mL sample into a beaker. The empty beaker had a mass of 12.20 g, and the beaker plus the liquid weighed 16.08 g.

Substance	Known Density at 25 °C (g/cm^3)
Ethylene glycol	1.1088 (the major component of antifreeze)
Water	0.9997
Ethanol	0.7893 (the alcohol in alcoholic beverages)
Acetic acid	1.0492 (the active component of vinegar)
Glycerol	1.2613 (a solvent, used in home care products)

    (a) Calculate the density and identify the unknown.

(b) If you were able to measure the volume to only two significant figures (that is, 3.5 mL, not 3.50 mL), will the results be sufficiently accurate to identify the unknown? Explain.

106. ▲ You have an irregularly shaped chunk of an unknown metal. To identify it, you determine its density and then compare this value with known values that you look up in the chemistry library. The mass of the metal is 74.122 g. Because of the irregular shape, you measure the volume by submerging the metal in water in a graduated cylinder. When you do this, the water level in the cylinder rises from 28.2 mL to 36.7 mL.

    (a) What is the density of the metal? (Use the correct number of significant figures in your answer.)

    (b) The unknown is one of the seven metals listed below. Is it possible to identify the metal based on the density you have calculated? Explain.

Metal	Density (g/cm^3)	Metal	Density (g/cm^3)
zinc	7.13	nickel	8.90
iron	7.87	copper	8.96
cadmium	8.65	silver	10.50
cobalt	8.90		

107. ▲ A 7.50×10^2-mL sample of an unknown gas has a mass of 0.9360 g.

    (a) What is the density of the gas? Express your answer in units of g/L.

    (b) Nine gases and their densities are listed below. Compare the experimentally determined density with these values. Can you determine the identity of the gas based on the experimentally determined density?

    (c) A more accurate measure of volume is made next, and the volume of this sample of gas is found to be 7.496×10^2 mL. Using a more accurate density calculated using this value, can you now determine the identity of the gas?

Gas	Density (g/L)	Gas	Density (g/L)
B_2H_6	1.2345	C_2H_4	1.2516
CH_2O	1.3396	CO	1.2497
Dry air	1.2920	C_2H_6	1.3416
N_2	1.2498	NO	1.2949
O_2	1.4276		

108. ▲ The density of a single, small crystal can be determined by the *flotation method*. This method is based on the idea that if a crystal and a liquid have precisely the same density, the crystal will hang suspended in the liquid. A crystal that is more dense will sink; one that is less dense will float. If the crystal neither sinks nor floats, then the density of the crystal equals the density of the liquid. Generally, mixtures of liquids are used to get the proper density. Chlorocarbons and bromocarbons (see

the list below) are often the liquids of choice. If the two liquids are similar, then volumes are usually additive and the density of the mixture relates directly to composition. (An example: 1.0 mL of $CHCl_3$, d = 1.4832 g/mL, and 1.0 mL of CCl_4, d = 1.5940 g/mL, when mixed, give 2.0 mL of a mixture with a density of 1.5386 g/mL. The density of the mixture is the average of the values of the two individual components.)

The problem: A small crystal of silicon, germanium, tin, or lead (Group 4A in the periodic table) will hang suspended in a mixture made of 61.18% (by volume) $CHBr_3$ and 38.82% (by volume) $CHCl_3$. Calculate the density and identify the element. (You will have to look up the values of the density of the elements in a manual such as the *The Handbook of Chemistry and Physics* in the library or in a World Wide Web site such as WebElements at, **www.webelements.com**.)

Liquid	Density (g/mL)	Liquid	Density (g/mL)
CH_2Cl_2	1.3266	CH_2Br_2	2.4970
$CHCl_3$	1.4832	$CHBr_3$	2.8899
CCl_4	1.5940	CBr_4	2.9609

109. ▲ Suppose you have a cylindrical glass tube with a thin capillary opening, and you wish to determine the diameter of the capillary. You can do this experimentally by weighing a piece of the tubing before and after filling a portion of the capillary with mercury. Using the following information, calculate the diameter of the capillary.

Mass of tube before adding mercury = 3.263 g

Mass of tube after adding mercury = 3.416 g

Length of capillary filled with mercury = 16.75 mm

Density of mercury = 13.546 g/cm^3

Volume of cylindrical capillary filled with mercury = $(\pi)(radius)^2(length)$

 Mentor™

Do you need a live tutor for homework problems?
Access vMentor at General ChemistryNow at
http://now.brookscole.com/kotz6e
for one-on-one tutoring from a chemistry expert

2—Atoms and Elements

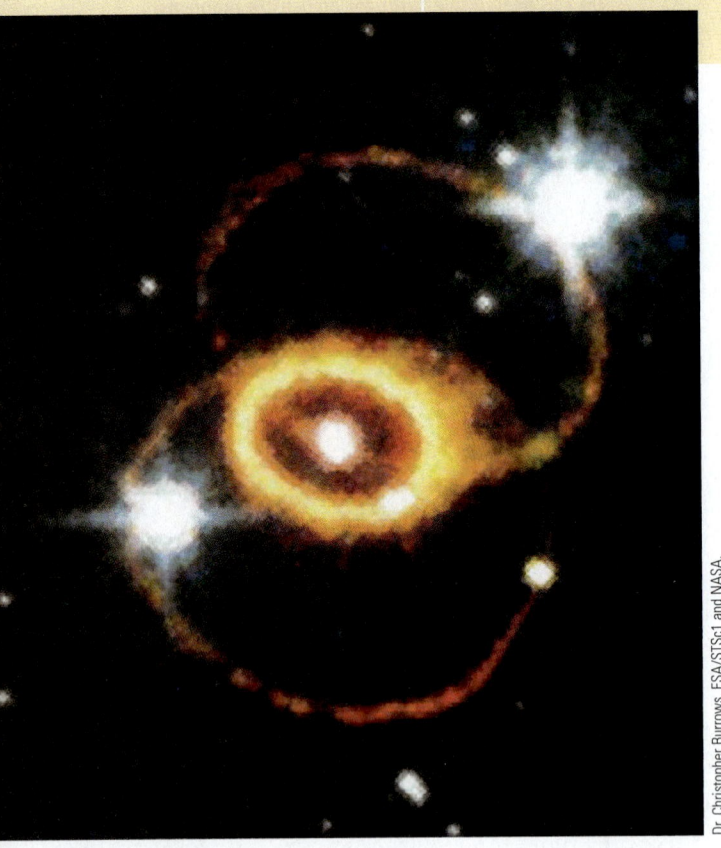

Dr. Christopher Burrows, ESA/STScI and NASA.

The supernova of 1987. When a star becomes more and more dense, and hotter and hotter, it can become a "red giant." The star is unstable and explodes as a "supernova." One such spectacular event occurred in a distant star in 1987. These explosions are the origin of the heavier elements, such as iron, nickel, and cobalt.

Stardust

A wide array of elements make up planet Earth and every living thing on it. What is science's view of the cosmic origin of these elements that we take for granted in our environment and in our lives?

The "big bang" theory is the generally accepted theory for the origin of the universe. This theory holds that an unimaginably dense, grapefruit-sized sphere of matter exploded about 15 billion years ago, spewing the products of that explosion as a rapidly expanding cloud with a temperature in the range of 10^{30} K. Within 1 second, the universe was populated with the particles we explore in this chapter: protons, electrons, and neutrons. Within a few more seconds, the universe had cooled by millions and millions of degrees, and protons and neutrons began to combine to form helium nuclei. After only about 8 minutes, scientists believe, the universe was about one-quarter helium and about three-quarters hydrogen. In fact, this is very close to the composition of the universe today, 15 billion years later. But humans, animals, and plants are built mainly from carbon, oxygen, nitrogen, sulfur, phosphorus, iron, and zinc—heavier elements that have only a trace abundance in the universe as a whole. Where do these heavier elements come from?

The cloud of hydrogen and helium cooled over a period of thousands of years and condensed into stars like our sun. There hydrogen atoms fuse into more helium atoms and energy streams outward. Every second on the sun, 700 million tons of hydrogen is converted to 695 million tons of helium, and 3.9×10^{26} joules of energy is evolved.

Gradually, over millions of years, a hydrogen-burning star becomes more and more dense and hotter and hotter. The helium atoms initially formed in the star begin to fuse into heavier atoms—first carbon, then oxygen, and then neon, magnesium, silicon, phosphorus, and argon. The star becomes even hotter and more dense. Hydrogen is forced to the outer reaches of the star, and the star becomes a red giant. Under certain circumstances, the star will explode, and earth-bound observers see it as a supernova. A supernova can be as much as 10^8 times brighter than the original star. A single supernova is comparable in brightness to the whole of the galaxy in which it is formed!

The supernova that appeared in 1987 gave astronomers an opportunity to study what happens in these element factories. It is here that the heavier atoms such as iron form. In fact, it is with iron that nature reaches its zenith of stability. To make heavier and heavier elements requires energy, rather than having energy as an outcome of element synthesis.

The elements spewing out of an exploding supernova move through space and gradually condense into planets, of which ours is just one.

The mechanism of element formation in stars is reasonably well understood, and much experimental evidence exists to support this theory. However, the way in which these elements are then assembled out of stardust into living organisms on our planet—and perhaps other planets—is not yet understood at all. (*See the General ChemistryNow Screen 2.2 Introduction to Atoms, to watch a video on the "big bang" theory.*)

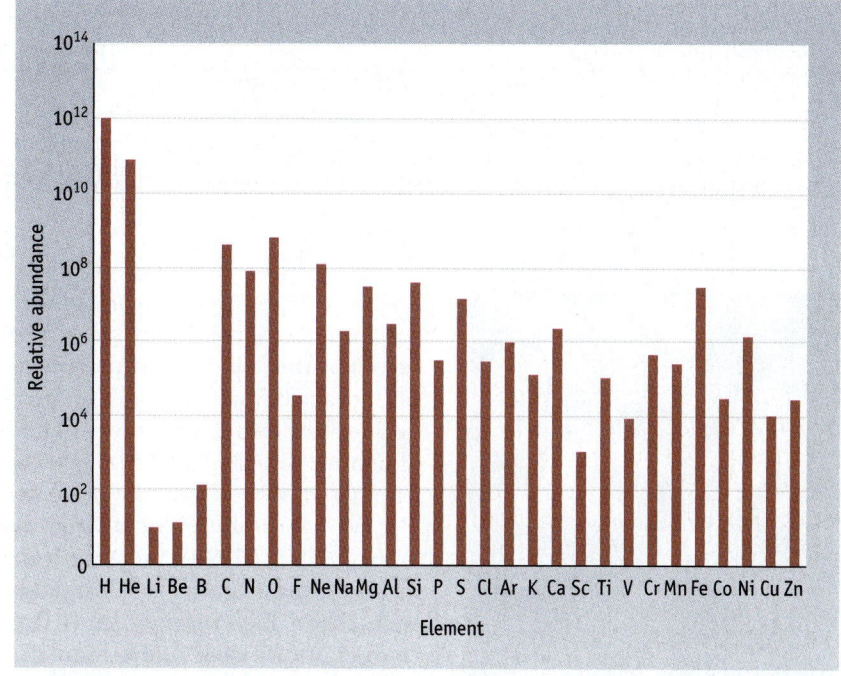

The abundance of the elements in the solar system from H to Zn. The chart shows a general decline in abundance with increasing mass among the first 30 elements. The decline continues above zinc. Notice that the scale on the vertical axis is logarithmic—that is, it progresses in powers of ten. The abundance of nitrogen, for example, is 1/10,000 (1/10^4) of the abundance of hydrogen. (All abundances are plotted as the number of atoms per 10^{12} atoms of H. The fact that the abundances of Li, Be, and B, as well as those of the elements near Fe, do not follow the general decline is a consequence of the way that elements are synthesized in stars.)

GENERAL
Chemistry Now™

Throughout the chapter this icon introduces a list of resources on the General ChemistryNow CD-ROM or website (http://now .brookscole.com/kotz6e) that will:

- help you evaluate your knowledge of the material
- provide homework problems
- allow you to take an exam-prep quiz
- provide a personalized Learning Plan targeting resources that address areas you should study

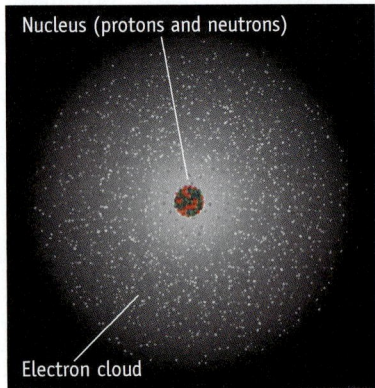

Figure 2.1 The structure of the atom. All atoms contain a nucleus with one or more protons (positive electric charge) and neutrons (no charge). Electrons (negative electric charge) are arranged in space as a "cloud" around the nucleus. In an electrically neutral atom, the number of electrons equals the number of protons. Note that this figure is not drawn to scale. If the nucleus were really the size depicted here, the electron cloud would extend about 800 feet. The atom is mostly empty space!

The chemical elements are forged in stars. What are the similarities among the elements? What are the differences? What are their physical and chemical properties? How can we tell them apart? This chapter begins our exploration of the chemistry of the elements, the building blocks of the science of chemistry.

2.1—Protons, Electrons, and Neutrons: Development of Atomic Structure

Around 1900 a series of experiments done by scientists such as Sir John Joseph Thomson (1856–1940) and Ernest Rutherford (1871–1937) in England established a model of the atom that is still the basis of modern atomic theory. Three **subatomic particles** make up all atoms: electrically positive protons, electrically neutral neutrons, and electrically negative electrons. The model places the more massive protons and neutrons in a very small nucleus, which contains all the positive charge and almost all the mass of an atom. Electrons, with a much smaller mass than protons or neutrons, surround the nucleus and occupy most of the volume (Figure 2.1). Atoms have no net charge; the positive and negative charges balance. *The number of electrons outside the nucleus equals the number of protons within the nucleus.*

What is the experimental basis of atomic structure? How did the work of Thomson, Rutherford, and others lead to this model?

Electricity

Electricity is involved in many of the experiments from which the theory of atomic structure was derived. The fact that objects can bear an electric charge was first observed by the ancient Egyptians, who noted that amber, when rubbed with wool or silk, attracted small objects. You can observe the same thing when you rub a balloon on your hair on a dry day—your hair is attracted to the balloon (Figure 2.2a). A bolt of lightning or the shock you get when touching a doorknob results when an electric charge moves from one place to another.

Two types of electric charge had been discovered by the time of Benjamin Franklin (1706–1790), the American statesman and inventor. He named them positive (+) and negative (−), because they appear as opposites and can neutralize each other. Experiments show that like charges repel each other and unlike charges attract each other. Franklin also concluded that charge is balanced: If a negative charge appears somewhere, a positive charge of the same size must appear somewhere else. The fact that a charge builds up when one substance is rubbed over another implies that the rubbing separates positive and negative charges. By the 19th century it was understood that positive and negative charges are somehow associated with matter—and perhaps with atoms.

Radioactivity

In 1896 the French physicist Henri Becquerel (1852–1908) discovered that a uranium ore emitted rays that could darken a photographic plate, even though the plate was covered by black paper to protect it from being exposed to light. In 1898 Marie

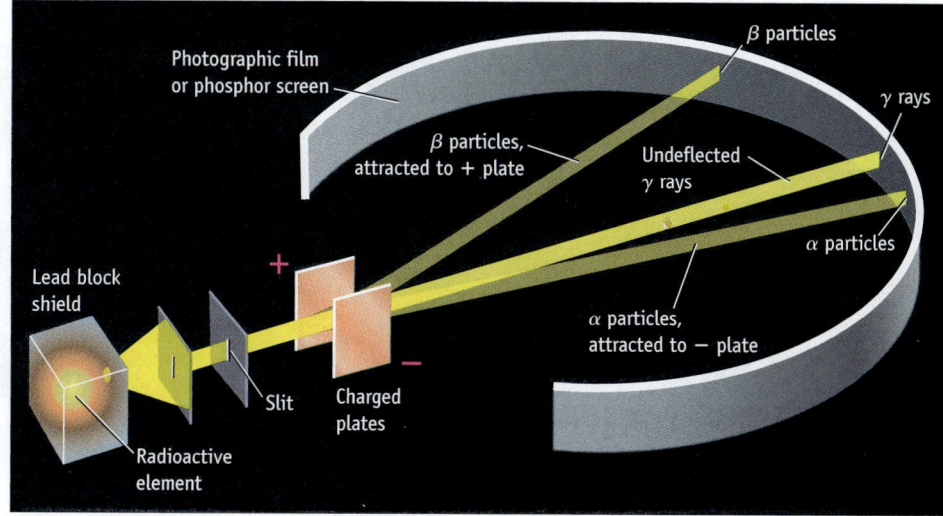

(a) **(b)**

Figure 2.2 **Electricity and radioactivity.** (a) If you brush your hair with a balloon, a static electric charge builds up on the surface of the balloon. Experiments show that objects having opposite electric charges attract each other, whereas objects having the same electric charge repel each other. (*See the General ChemistryNow Screen 2.4 Electricity and Electric Charge, for an exercise on the effects of charge.*) (b) Alpha (α), beta (β), and gamma (γ) rays from a radioactive element are separated by passing them between electrically charged plates. Positively charged α particles are attracted to the negative plate, and negatively charged β particles are attracted to the positive plate. (Note that the heavier α particles are deflected less than the lighter β particles.) Gamma rays have no electric charge and pass undeflected between the charged plates. (*See the General ChemistryNow Screen 2.5 Evidence of Subatomic Particles, for an exercise on this experiment.*)

and Pierre Curie (1867–1934) isolated polonium and radium, which also emitted the same kind of rays, and in 1899 they suggested that atoms of certain substances emit these unusual rays when they disintegrate. They named this phenomenon **radioactivity**, and substances that display this property are said to be **radioactive**.

Early experiments identified three kinds of radiation: alpha (α), beta (β), and gamma (γ) rays. These rays behave differently when passed between electrically charged plates (Figure 2.2b). Alpha and β rays are deflected, but γ rays pass straight through. This implies that α and β rays are electrically charged particles, because they are attracted or repelled by the charged plates. Even though an α particle was found to have an electric charge (+2) twice as large as that of a β particle (−1), α particles are deflected less, which implies that α particles must be heavier than β particles. Gamma rays have no detectable charge or mass; they behave like light rays.

Marie Curie's suggestion that atoms disintegrate contradicted ideas put forward in 1803 by John Dalton (1766–1844) that atoms are indivisible. If atoms can break apart, there must be something smaller than an atom; that is, atoms must be composed of even smaller, subatomic particles.

Cathode-Ray Tubes and the Characterization of Electrons

Further evidence that atoms are composed of smaller particles came from experiments with cathode-ray tubes (Figure 2.3). These are glass tubes from which most of the air has been removed and that contain two metal electrodes. When a sufficiently high voltage is applied to the electrodes, a *cathode ray* flows from the negative electrode (cathode) to the positive electrode (anode). Experiments showed that cathode rays travel in straight lines, cause gases to glow, can heat metal objects red hot, can be deflected by a magnetic field, and are attracted toward positively charged

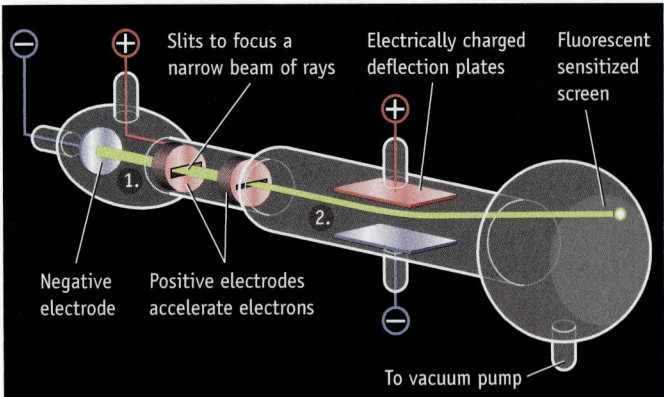

Slits to focus a
narrow beam of rays

Electrically charged
deflection plates

Fluorescent
sensitized
screen

Negative
electrode

Positive electrodes
accelerate electrons

To vacuum pump

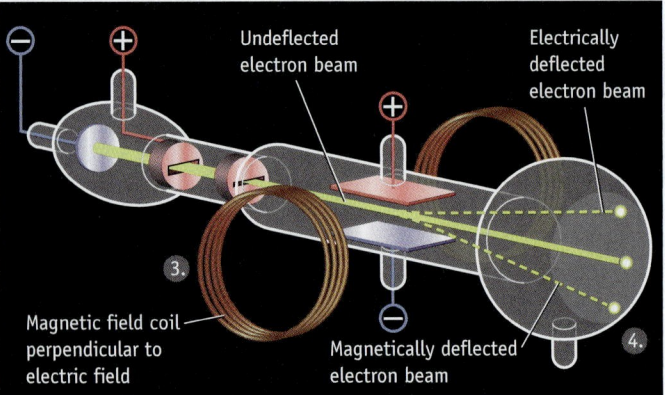

Undeflected
electron beam

Electrically
deflected
electron beam

Magnetic field coil
perpendicular to
electric field

Magnetically deflected
electron beam

1. A beam of electrons (cathode rays) is accelerated through two focusing slits.

2. When passing through an electric field the beam of electrons is deflected.

3. The experiment is arranged so that the electric field causes the beam of electrons to be deflected in one direction. The magnetic field deflects the beam in the opposite direction.

4. By balancing the effects of the electrical and magnetic fields the charge-to-mass ratio of the electron can be determined.

Active Figure 2.3   **Measuring the electron's charge-to-mass ratio.** This experiment was done by J. J. Thomson in 1896–1897.

GENERAL
Chemistry Now™ See the General ChemistryNow CD-ROM or website to explore an interactive version of this figure accompanied by an exercise.

plates. When cathode rays strike a fluorescent screen, light is given off in a series of tiny flashes. We can understand all of these observations if a cathode ray is assumed to be a beam of the negatively charged particles we now know as **electrons**.

You are already familiar with cathode rays. Television pictures and the images on some types of computer monitors are formed by using electrically charged plates to aim cathode rays onto the back of a phosphor screen on which we view the image. Sir Joseph John Thomson (1856–1940) used this principle to prove experimentally the existence of the electron and to study its properties. He applied electric and magnetic fields simultaneously to a beam of cathode rays (Figure 2.3). By balancing the effect of the electric field against that of the magnetic field and using basic laws of electricity and magnetism, Thompson calculated the ratio of the charge to the mass for the particles in the beam. He was not able to determine either charge or mass independently. However, he found the same charge-to-mass ratio in experiments using 20 different metals as cathodes and several different gases. These results suggested that electrons are present in atoms of all elements.

It remained for the American physicist Robert Andrews Millikan (1868–1953) to measure the charge on an electron and thereby enable scientists to calculate its mass (Figure 2.4). In his experiment tiny droplets of oil were sprayed into a chamber. As they settled slowly through the air, the droplets were exposed to x-rays, which caused them to acquire an electric charge. Millikan used a small telescope to observe individual droplets. If the electric charge on the plates above and below the droplets was adjusted, the electrostatic attractive force pulling a droplet upward could be balanced by the force of gravity pulling the droplet downward. From the equations describing these forces, Millikan calculated the charge on various droplets. Different droplets had different charges, but Millikan found that each was a whole-number multiple of the same smaller charge, 1.60×10^{-19} C (where C represents the coulomb, the SI unit of electric charge; Appendix C). Millikan assumed this to be the fundamental unit of charge, the charge on an electron. Because the

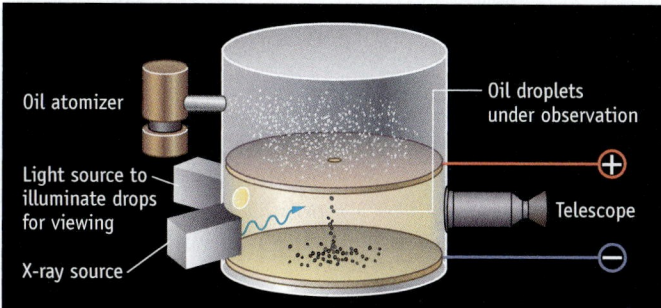

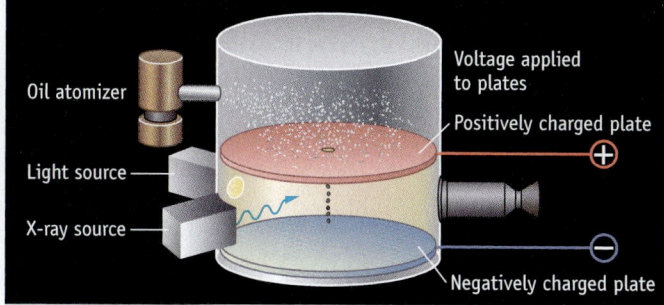

1. A fine mist of oil drops is introduced into one chamber.

2. The droplets fall one by one into the lower chamber under the force of gravity.

3. Gas molecules in the bottom chamber are ionized (split into electrons and a positive fragment) by a beam of x-rays. The electrons adhere to the oil drops, some droplets having one electron, some two, and so on.

These negatively charged droplets continue to fall due to gravity.

4. By carefully adjusting the voltage on the plates, the force of gravity on the droplet is exactly counterbalanced

by the attraction of the negative droplet to the upper, positively charged plate.

Analysis of these forces lead to a value for the charge on the electron.

Figure 2.4 **Electron Charge.** The experiment was done by R. A. Millikan in 1909. (*See the General ChemistryNow Screen 2.7 Charge and Mass of the Electron, for an exercise on this experiment.*)

charge-to-mass ratio of the electron was known, the mass of an electron could be calculated. The currently accepted value for the electron mass is 9.109383×10^{-28} g, and the electron charge is $-1.602176 \times 10^{-19}$ C. When describing the properties of fundamental particles, we always express charge relative to the charge on the electron, which is given the value of -1.

Additional experiments showed that cathode rays have the same properties as the β particles emitted by radioactive elements. This provided further evidence that the electron is a fundamental particle of matter.

Extensive studies with cathode ray tubes in the late nineteenth century provided another dividend. In addition to cathode rays, a second type of radiation was detected. A beam of positively charged particles called **canal rays** was observed using a specially designed cathode-ray tube with a perforated cathode (Figure 2.5).

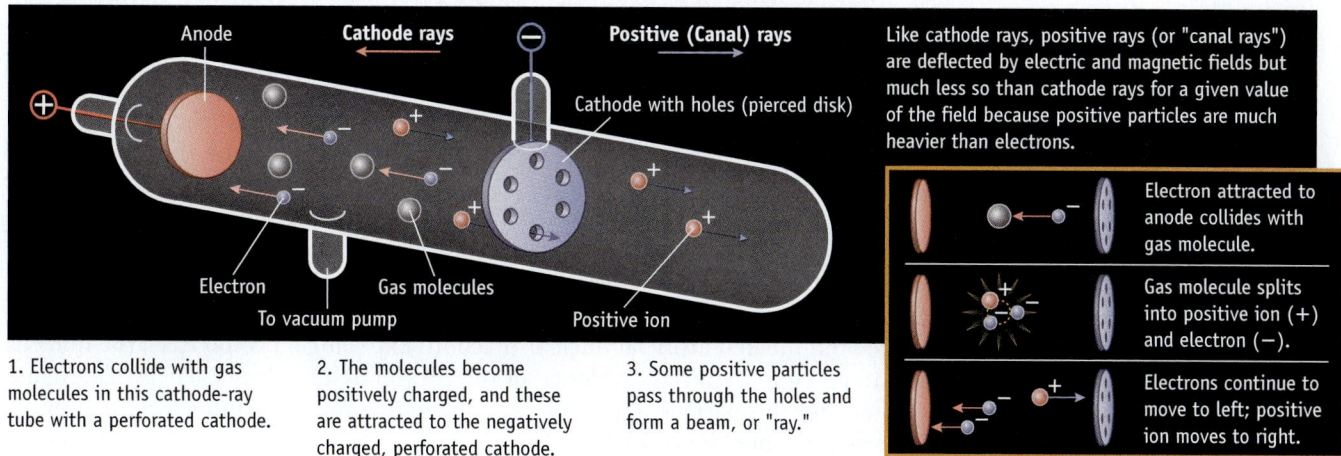

Like cathode rays, positive rays (or "canal rays") are deflected by electric and magnetic fields but much less so than cathode rays for a given value of the field because positive particles are much heavier than electrons.

Electron attracted to anode collides with gas molecule.

Gas molecule splits into positive ion (+) and electron (−).

Electrons continue to move to left; positive ion moves to right.

1. Electrons collide with gas molecules in this cathode-ray tube with a perforated cathode.

2. The molecules become positively charged, and these are attracted to the negatively charged, perforated cathode.

3. Some positive particles pass through the holes and form a beam, or "ray."

Figure 2.5 **Canal rays.** In 1886 E. Goldstein detected a stream of particles traveling in the direction opposite to that of the negatively charged cathode rays. We now know that these particles are positively charged ions, formed by collisions of electrons with gaseous molecules in the cathode-ray tube. (*See the General ChemistryNow Screen 2.8 Protons, to view an animation on this experiment.*)

These particles, which moved in the opposite direction to cathode rays, passed through the holes in the cathode and were detected on the opposite side. Charge-to-mass values for canal rays were much smaller than the corresponding values measured for cathode rays, indicating particles of higher mass. However, the values also varied depending on the nature of the gas in the tube. We now know that canal rays arise through collisions of cathode rays with gaseous atoms within the cathode-ray tube, which cause each atom to fragment into a positive ion and an electron. The positive particles are attracted to the negatively charged cathode.

GENERAL
Chemistry⸱⚛⸱Now™

See the General ChemistryNow CD-ROM or website:
- **Screen 2.6 Electrons,** for an exercise on cathode rays and an animation on cathode-ray deflection
- **Screen 2.8 Protons,** for an exercise on the properties of nuclei in a canal-ray tube

---

Protons

A century after these seminal studies on the structure of the atom, it is easy for us to recognize the proton as the fundamental positively charged particle in an atom. This understanding did not come so easily a hundred years ago, however. This basic fact was not established in one specific experiment or at one specific moment.

With the determination that negatively charged electrons were a component of the atom came recognition that positively charged atomic particles must also exist. One hypothesis suggested that there should be a complementary particle to the electron with a corresponding small mass and a +1 charge, but there was no experimental evidence for such a particle. The positive particles detected and studied in early experiments (α particles from radioactive elements and positive ions making up canal rays) were considerably more massive.

Ernest Rutherford (1871–1937) probably deserves most of the credit for the discovery of the proton. He carried out experiments in the early 1900s in which various elements were irradiated with α particles. One of his better-known experiments involved the irradiation of metals such as gold, which led to the conclusion that atoms contained a small positively charged nucleus with most of the mass of an atom [▶ The Nucleus of the Atom, page 65]. At the same time Rutherford was performing similar experiments using gaseous elements, and, in these experiments, he observed the deflection of α particles as a function of atomic mass. From these observations he concluded, in 1911, that "the hydrogen atom has the simplest possible structure with only one unit charge." However, the formal identification of the proton did not come until almost 10 years later. In experiments in which nitrogen was bombarded with α particles, Rutherford and his collaborators observed highly energetic particles. The values of their charge-to-mass ratio matched those for hydrogen, the positive particle known to have the lowest mass. Unexpectedly they had carried out the first artificial nuclear reaction. Expelling a proton from the nucleus was accepted as definitive evidence of the proton as a nuclear particle. The name "proton" for this particle appears to have been first used by Rutherford in a report in a scientific meeting in 1919.

Neutrons

Because atoms have no net electric charge, the number of positive protons must equal the number of negative electrons in an atom. Most atoms, however, have

Historical Perspectives

Uncovering Atomic Structure

The last few years of the 19th century and the first decades of the 20th century were among the most important in the history of science, in part because the structure of the atom was discovered, setting the stage for the explosion of developments in science in the 20th century.

The notion that matter was built of atoms and that this structure could be used  to explain chemical phenomena was first used by **John Dalton** (1766–1844). Dalton proposed not only that all matter is made of atoms, but also that all atoms of a given element are identical and that atoms are indivisible and indestructible. Dalton's ideas were generally accepted within a few years of his proposal, but we know now the last two postulates are not correct.

Marie Curie (1867–1934) understood the nature of radioactivity and its implications for the nature of the atom. She was born Marya Sklodovska in Poland. When she later lived in France she was known as Marie, but today she is often referred to as Madame Curie. With her husband Pierre she

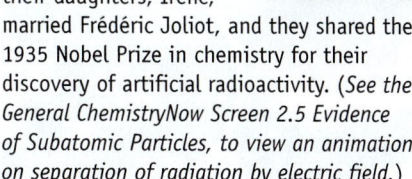

isolated the previously unknown elements polonium and radium from a uranium-bearing ore. They shared the 1911 Nobel Prize in chemistry for this discovery. One of their daughters, Irène, married Frédéric Joliot, and they shared the 1935 Nobel Prize in chemistry for their discovery of artificial radioactivity. (*See the General ChemistryNow Screen 2.5 Evidence of Subatomic Particles, to view an animation on separation of radiation by electric field.*)

Sir Joseph John Thomson (1856–1940) was Cavendish Professor of Experimental Physics at Cambridge University in England. In 1896 he gave a series of lectures at Princeton University in the United States titled the *Discharge of Electricity in Gases*. This work on cathode rays led to his discovery of the electron, which he announced at a lecture on the evening of Friday, April 30, 1897. Thomson later published a number of books on the electron and was awarded the Nobel Prize in physics in 1906. (*See the General ChemistryNow Screen 2.6 Electrons, to view an animation on cathode-ray deflection.*)

Ernest Rutherford (1871–1937) was born in New Zealand in 1871 but went to Cambridge University in England to pursue his Ph.D. in physics in 1895. There he worked with J. J. Thomson, and it was at Cambridge that he discovered α and β radiation. At McGill University in Canada in 1899 Rutherford did  further experiments to prove that α radiation is composed of helium nuclei and that β radiation consists of electrons. He received the Nobel Prize in chemistry for his work in 1908. His research on the structure of the atom was done after he moved to Manchester University in England. In 1919 he returned to Cambridge University, where he took up the position formerly held by Thomson. In his career, Rutherford guided the work of ten future recipients of the Nobel Prize. Element 104 has been named *rutherfordium* in his honor. (*See the General ChemistryNow Screen 2.10 The Nucleus of the Atom, to view an animation on Rutherford's α particle experiment.*)

Photos: (Left and Right) Oesper Collection in the History of Chemistry/University of Cincinnati; (Center Top) E. F. Smith Collection; (Center Bottom) Corbis.

masses greater than would be predicted on the basis of only protons and electrons, which suggested that atoms must also contain relatively massive particles with no electric charge. In 1932, the British physicist James Chadwick (1891–1974), a student of Rutherford, presented experimental evidence for the existence of such particles. Chadwick found very penetrating radiation was released when particles from radioactive polonium struck a beryllium target. This radiation was directed at a paraffin wax target, and Chadwick observed protons coming from that target. He reasoned that only a heavy, noncharged particle emanating from the beryllium could have caused this effect. This particle, now known as the **neutron**, has no electric charge and a mass of 1.674927×10^{-24} g, slightly greater than the mass of a proton.

The Nucleus of the Atom

J. J. Thomson had supposed that an atom was a uniform sphere of positively charged matter within which thousands of electrons were embedded. Thomson and his students thought the only question was the number of electrons

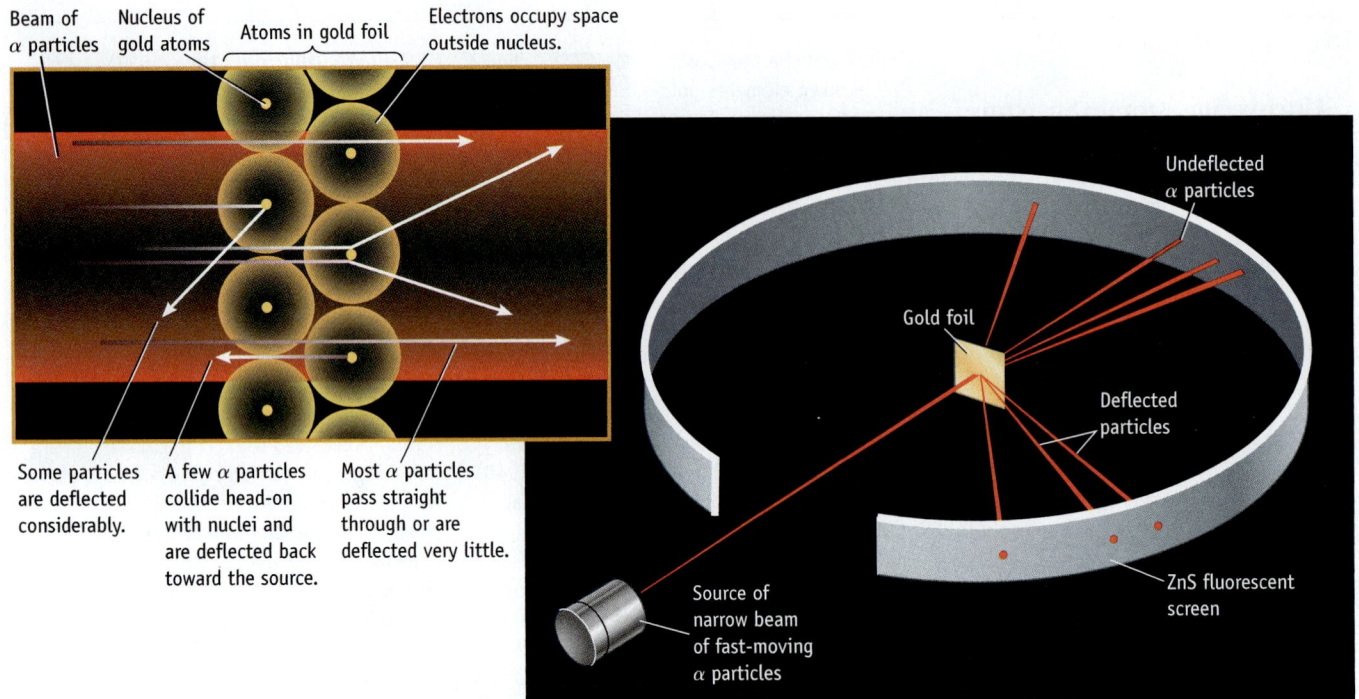

Beam of α particles | Nucleus of gold atoms | Atoms in gold foil | Electrons occupy space outside nucleus.

Some particles are deflected considerably. | A few α particles collide head-on with nuclei and are deflected back toward the source. | Most α particles pass straight through or are deflected very little.

Undeflected α particles

Gold foil

Deflected particles

Source of narrow beam of fast-moving α particles

ZnS fluorescent screen

Active Figure 2.6   **Rutherford's experiment to determine the structure of the atom.** A beam of positively charged α particles was directed at a thin gold foil. A fluorescent screen coated with zinc sulfide (ZnS) was used to detect particles passing through. Most of the particles passed through the foil, but some were deflected from their path. A few were even deflected backward.

GENERAL
Chemistry ⚛ Now™ See the General ChemistryNow CD-ROM or website to explore an interactive version of this figure accompanied by an exercise.

circulating within this sphere. About 1910, Rutherford decided to test Thomson's model. Rutherford had discovered earlier that α rays (see Figure 2.2b) consisted of positively charged particles having the same mass as helium atoms. He reasoned that, if Thomson's atomic model were correct, a beam of such massive particles would be deflected very little as it passed through the atoms in a thin sheet of gold foil. Rutherford, with his associates Hans Geiger (1882–1945) and Ernst Marsden, set up the apparatus diagrammed in Figure 2.6 and observed what happened when α particles hit the foil. Most passed almost straight through, but a few were deflected at large angles, and some came almost straight back! Rutherford later described this unexpected result by saying, "It was about as credible as if you had fired a 15-inch [artillery] shell at a piece of paper and it came back and hit you."

The only way for Rutherford and his colleagues to account for their observations was to propose a new model of the atom, in which all of the positive charge and most of the mass of the atom is concentrated in a very small volume. Rutherford called this tiny core of the atom the **nucleus**. The electrons occupy the rest of the space in the atom. From their results Rutherford, Geiger, and Marsden calculated that the nucleus of a gold atom had a positive charge in the range of 100 ± 20 and a radius of about 10^{-12} cm. The currently accepted values are $+79$ for the charge and about 10^{-13} cm for the radius.

■ **How Small Is an Atom?**
The radius of the typical atom is between 30 and 300 pm (3×10^{-11} m to 3×10^{-10} m). To get a feeling for the incredible smallness of an atom, consider that one teaspoon of water (about 1 cm³) contains about three times as many atoms as the Atlantic Ocean contains teaspoons of water.

GENERAL
Chemistry ·⚛·Now ™

See the General ChemistryNow CD-ROM or website:
- **Screen 2.10 The Nucleus of the Atom,** for an exercise on an experiment investigating the properties of the nuclei

---

Exercise 2.1—Describing Atoms

We know now that the radius of the nucleus is about 0.001 pm, and the radius of an atom is approximately 100 pm. If an atom were a macroscopic object with a radius of 100 m, it would approximately fill a small football stadium. What would be the radius of the nucleus of such an atom? Can you think of an object that is about that size?

2.2—Atomic Number and Atomic Mass

Atomic Number

All atoms of the same element have the same number of protons in the nucleus. Hydrogen is the simplest element, with one nuclear proton. All helium atoms have two protons, all lithium atoms have three protons, and all beryllium atoms have four protons. The number of protons in the nucleus of an element is its **atomic number**, generally given the symbol **Z**.

Currently known elements are listed in the periodic table inside the front cover of this book. The integer number at the top of the box for each element is its atomic number. A sodium atom, for example, has an atomic number of 11, so its nucleus contains 11 protons. A uranium atom has 92 nuclear protons and $Z = 92$.

■ **The Periodic Table Entry for Copper**

Copper
29 ----- Atomic number
Cu ---- Symbol
63.546 --- Atomic weight

Relative Atomic Mass and the Atomic Mass Unit

What is the mass of an atom? Chemists in the 18th and 19th centuries recognized that careful experiments could give *relative* atomic masses. For example, the mass of an oxygen atom was found to be 1.33 times the mass of a carbon atom, and a calcium atom has 2.5 times the mass of an oxygen atom.

Chemistry in the 21st century still uses a system of relative masses. After trying several standards, scientists settled on the current one: A carbon atom having six protons and six neutrons in the nucleus is assigned a mass value of exactly 12.000. An oxygen atom having eight protons and eight neutrons has 1.3333 times the mass of carbon, so it has a relative mass of 16.000. Masses of atoms of other elements have been assigned in a similar manner.

Masses of fundamental atomic particles are often expressed in **atomic mass units (u)**. *One atomic mass unit, 1 u, is one-twelfth of the mass of an atom of carbon with six protons and six neutrons.* Thus, such a carbon atom has a mass of 12.000 u. The atomic mass unit can be related to other units of mass using a conversion factor; that is, $1 \text{ u} = 1.661 \times 10^{-24} \text{ g}$.

Mass Number

Protons and neutrons have masses very close to 1 u (Table 2.1). The electron, in contrast, has a mass only about 1/2000 of this value. Because proton and neutron masses are so close to 1 u, the approximate mass of an atom can be estimated if the

Table 2.1   Properties of Subatomic Particles*

Particle	Mass		Charge	Symbol
	Grams	*Atomic Mass Units*		
Electron	9.109383×10^{-28}	0.0005485799	-1	$_{-1}^{0}e$ or e^-
Proton	1.672622×10^{-24}	1.007276	$+1$	$_{1}^{1}p$ or p^+
Neutron	1.674927×10^{-24}	1.008665	0	$_{0}^{1}n$ or n^0

* These values and others in the book are taken from the National Institute of Standards and Technology website at http://physics.nist.gov/cuu/Constants/index.html

number of neutrons and protons is known. The sum of the number of protons and neutrons for an atom is called its **mass number** and is given the symbol A.

$$A = \text{mass number} = \text{number of protons} + \text{number of neutrons}$$

For example, a sodium atom, which has 11 protons and 12 neutrons in its nucleus, has a mass number of $A = 23$. The most common atom of uranium has 92 protons and 146 neutrons, and a mass number of $A = 238$. Using this information, we often symbolize atoms with the notation

$$\text{Mass number} \rightarrow {}_{Z}^{A}X \leftarrow \text{Element symbol}$$
$$\text{Atomic number} \rightarrow$$

The subscript Z is optional because the element symbol tells us what the atomic number must be. For example, the atoms described previously have the symbols $_{11}^{23}Na$ or $_{92}^{238}U$, or just ^{23}Na or ^{238}U. In words, we say "sodium-23" or "uranium-238."

GENERAL
Chemistry ⚛ Now™

See the General ChemistryNow CD-ROM or website:
- **Screen 2.11 Summary of Atomic Composition,** for a tutorial on the notation for symbolizing atoms

---

Example 2.1—Atomic Composition

Problem What is the composition of an atom of phosphorus with 16 neutrons? What is its mass number? What is the symbol for such an atom? If the atom has an actual mass of 30.9738 u, what is its mass in grams?

Strategy All P atoms have the same number of protons, 15, which is given by the atomic number (see the periodic table inside the front cover of this book). The mass number is the sum of the number of protons and neutrons. The mass of the atom in grams can be obtained from the mass in atomic mass units using the conversion factor $1\,u = 1.661 \times 10^{-24}$ g.

Solution A phosphorus atom has 15 protons and, because it is electrically neutral, also has 15 electrons.

$$\text{Mass number} = \text{number of protons} + \text{number of neutrons} = 15 + 16 = \boxed{31}$$

The atom's complete symbol is $^{31}_{15}P$.

Mass of one ^{31}P atom (g) = (30.9738 u) × (1.661 × 10^{-24} g/u) = $\boxed{5.145 \times 10^{-23}\text{ g}}$

Exercise 2.2—Atomic Composition

(a) What is the mass number of an iron atom with 30 neutrons?

(b) A nickel atom with 32 neutrons has a mass of 59.930788 u. What is its mass in grams?

(c) How many protons, neutrons, and electrons are in a ^{64}Zn atom?

2.3—Isotopes

In only a few instances (for example, aluminum, fluorine, and phosphorus) do all atoms in a naturally occurring sample of a given element have the same mass. Most elements consist of atoms having several different mass numbers. For example, there are two kinds of boron atoms, one with a mass of about 10 u (^{10}B) and a second with a mass of about 11 u (^{11}B). Atoms of tin can have any of 10 different masses. Atoms with the same atomic number but different mass numbers are called **isotopes**.

All atoms of an element have the same number of protons—five in the case of boron. This means that, to have different masses, isotopes must have different numbers of neutrons. The nucleus of a ^{10}B atom ($Z = 5$) contains five protons and five neutrons, whereas the nucleus of a ^{11}B atom contains five protons and six neutrons.

Scientists often refer to a particular isotope by giving its mass number (for example, uranium-238, ^{238}U), but the isotopes of hydrogen are so important that they have special names and symbols. All hydrogen atoms have one proton. When that is the only nuclear particle, the isotope is called *protium*, or just "hydrogen." The isotope of hydrogen with one neutron, $^{2}_{1}H$, is called *deuterium*, or "heavy hydrogen" (symbol = D). The nucleus of radioactive hydrogen-3, $^{3}_{1}H$, or *tritium* (symbol = T), contains one proton and two neutrons.

The substitution of one isotope of an element for another isotope of the same element in a compound sometimes has an interesting effect (Figure 2.7). This is especially true when deuterium is substituted for hydrogen because the mass of deuterium is double that of hydrogen.

Isotope Abundance

A sample of water from a stream or lake will consist almost entirely of H_2O where the H atoms are the ^{1}H isotope. A few molecules, however, will have deuterium (^{2}H) substituted for ^{1}H. We can predict this outcome because we know that 99.985% of all hydrogen atoms on earth are ^{1}H atoms. That is, the **percent abundance** of ^{1}H atoms is 99.985%.

$$\text{Percent abundance} = \frac{\text{number of atoms of a given isotope}}{\text{total number of atoms of all isotopes of that element}} \times 100\%$$

(2.1)

The remainder of naturally occurring hydrogen is deuterium, whose abundance is only 0.015% of the total hydrogen atoms. Tritium, the radioactive ^{3}H isotope, does not occur naturally.

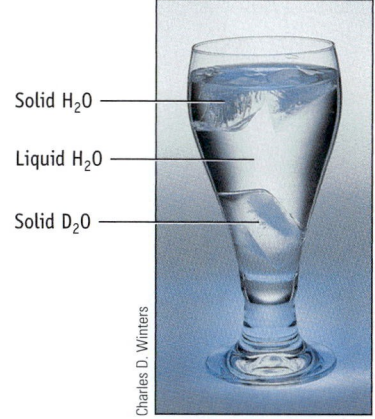

Solid H_2O ——

Liquid H_2O ——

Solid D_2O ——

Charles D. Winters

Figure 2.7 Ice made from "heavy water." Water containing ordinary hydrogen ($^{1}_{1}H$, protium) forms a solid that is less dense ($d = 0.917$ g/cm^3 at 0 °C) than liquid H_2O ($d = 0.997$ g/cm^3 at 25 °C) and so floats in the liquid. (Water is unique in this regard. The solid phase of virtually all other substances sinks in the liquid phase of that substance.) Similarly, "heavy ice" (D_2O, deuterium oxide) floats in "heavy water." D_2O-ice is denser than H_2O, however, so cubes made of D_2O sink in liquid H_2O.

Consider the two isotopes of boron. The boron-10 isotope has an abundance of 19.91%; the abundance of boron-11 is 80.09%. Thus, if you could count out 10,000 boron atoms from an "average" natural sample, 1991 of them would be boron-10 atoms and 8009 of them would be boron-11 atoms.

Example 2.2—Isotopes

Problem Silver has two isotopes, one with 60 neutrons (percent abundance = 51.839%) and the other with 62 neutrons. What are the mass numbers and symbols of these isotopes? What is the percent abundance of the isotope with 62 neutrons?

Strategy Recall that the mass number is the sum of the number of protons and neutrons. The symbol is written as $^A_Z X$, where X is the one or two-letter element symbol. The percent abundances of all isotopes must add up to 100%.

Solution Silver has an atomic number of 47, so each silver atom has 47 protons in its nucleus. The two isotopes, therefore, have mass numbers of 107 and 109.

Isotope 1, with 47 protons and 60 neutrons

$$A = 47 \text{ protons} + 60 \text{ neutrons} = \boxed{107}$$

Isotope 2, with 47 protons and 62 neutrons

$$A = 47 \text{ protons} + 62 \text{ neutrons} = \boxed{109}$$

The first isotope has a symbol $^{107}_{47}\text{Ag}$ and the second is $^{109}_{47}\text{Ag}$.

Silver-107 has a percent abundance of 51.839%. Therefore, the percent abundance of silver-109 is

$$\% \text{ Abundance of } ^{109}\text{Ag} = 100.000\% - 51.839\% = \boxed{48.161\%}$$

■ Atomic Masses of Some Isotopes

Atom	Mass (u)
^{4}He	4.0092603
^{13}C	13.003355
^{16}O	15.994915
^{58}Ni	57.935348
^{60}Ni	59.930791
^{79}Br	78.918338
^{81}Br	80.916291
^{197}Au	196.966552
^{238}U	238.050783

Exercise 2.3—Isotopes

(a) Argon has three isotopes with 18, 20, and 22 neutrons, respectively. What are the mass numbers and symbols of these three isotopes?

(b) Gallium has two isotopes: ^{69}Ga and ^{71}Ga. How many protons and neutrons are in the nuclei of each of these isotopes? If the abundance of ^{69}Ga is 60.1%, what is the abundance of ^{71}Ga?

Determining Atomic Mass and Isotope Abundance

The masses of isotopes and their percent abundances are determined experimentally using a *mass spectrometer* (Figure 2.8). A gaseous sample of an element is introduced into the evacuated chamber of the spectrometer, and the molecules or atoms of the sample are converted to charged particles (ions). A beam of these ions is subjected to a magnetic field, which causes the paths of the ions to be deflected. The extent of deflection depends on particle mass: The less massive ions are deflected more, and the more massive ions are deflected less. The ions, now separated by mass, are detected at the end of the chamber. In early experiments, ions were detected using photographic film, but modern instruments measure the electric current in a detector. The darkness of a spot on photographic film, or the amount of

VAPORIZATION   IONIZATION   ACCELERATION   DEFLECTION   DETECTION

Electron gun   Magnet

Heavy ions
are deflected
too little.

A *mass spectrum* is a plot of the
relative abundance of the charged
particles versus the ratio of
mass/charge (m/z).

$^{20}Ne^+$

$^{22}Ne^+$

To mass
analyzer

$^{21}Ne^+$

Gas inlet Repeller Electron Accelerating
           plate   trap   plates

Magnet   Light ions
         are deflected
         too much.

To vacuum pump   Detector

1. A sample is introduced as a vapor into the ionization chamber.

There it is bombarded with high-energy electrons that strip electrons from the atoms or molecules of the sample.

2. The resulting positive particles are accelerated by a series of negatively charged accelerator plates into an analyzing chamber.

3. This chamber is in a magnetic field, which is perpendicular to the direction of the beam of charged particles.

The magnetic field causes the beam to curve. The radius of curvature depends on the mass and charge of the particles (as well as the accelerating voltage and strength of the magnetic field).

4. Here particles of $^{21}Ne^+$ are focused on the detector, whereas beams of ions of $^{20}Ne^+$ and $^{22}Ne^+$ (of lighter or heavier mass) experience greater and lesser curvature, respectively, and so fail to be detected.

By changing the magnetic field, a beam of charged particles of different mass can be focused on the detector, and a spectrum of masses is observed.

Active Figure 2.8   **Mass spectrometer.**

GENERAL
Chemistry ⚛ Now™ See the General ChemistryNow CD-ROM or website to explore an interactive version of this figure accompanied by an exercise.

current measured, is related to the number of ions of a particular mass and hence to the abundance of the ion.

The mass-to-charge ratio for the ions can also be determined from the extent of curvature in the ion's path to the detector. Knowing that most of the ions within the spectrometer have a +1 charge allows us to derive a value for mass. Chemists using modern instruments can measure isotopic masses with as many as nine significant figures.

A Closer Look

**Atomic Mass
and the Mass Defect**

You might expect that the mass of a deuterium nucleus, ^2H, would be the sum of the masses of its constituent particles, a proton and a neutron.

^1_1p (1.007276 u) + ^1_0n (1.008665 u)
$\longrightarrow$ ^2_1H (2.01355 u)

However, the mass of ^2H is *less* than the sum of its constituents!

Difference in mass, Δm

= mass of product − total mass of reactants
= 2.01355 u − 2.015941 u
= −0.00239 u

This "missing mass" is equated to energy, the binding energy for the nucleus. The binding energy can be calculated from Einstein's equation that relates the mass,

m, to energy, E ($E = mc^2$, where c is the velocity of light).

Although the mass loss on forming an atomic nucleus from its constituent protons and neutrons seems small, the energy equivalent is enormous. In fact, it is the mass loss from fusing protons into helium nuclei on the sun that provides the energy for life on earth. (See the story, *Stardust*, on page 58 and see Chapter 23 for more details.)

Except for carbon-12, whose mass is defined to be exactly 12u, isotopic masses do not have integer values. However, the isotopic masses are always very close to the mass numbers for the isotope. For example, the mass of an atom of boron-11 (^{11}B, 5 protons and 6 neutrons) is 11.0093 u, and the mass of an atom of iron-58 (^{58}Fe, 26 protons and 32 neutrons) is 57.9333 u. Note also that the masses of individual isotopes are always slightly less than the sum of the masses of the protons, neutrons, and electrons making up the atom. This mass difference, called the "mass defect," is related to the energy binding the particles of the nucleus together. (See A Closer Look: Atomic Mass and the Mass Defect.)

2.4—Atomic Weight

Because every sample of boron has some atoms with a mass of 10.0129 u and others with a mass of 11.0093 u, the average atomic mass must be somewhere between these values. The **atomic weight** is the average weight of a representative sample of atoms. For boron, the atomic weight is 10.81. In general, the atomic weight of an element can be calculated using the equation

$$\text{Atomic weight} = \left(\frac{\% \text{ abundance isotope 1}}{100}\right)(\text{mass of isotope 1})$$
$$+ \left(\frac{\% \text{ abundance isotope 2}}{100}\right)(\text{mass of isotope 2}) + \cdots$$

(2.2)

■ **Atomic Weight and Units**
Values of atomic weight are relative to the mass of the carbon-12 isotope and so are unitless numbers.

For boron with two isotopes (^{10}B, 19.91% abundant; ^{11}B, 80.09% abundant), we find

$$\text{Atomic weight} = \left(\frac{19.91}{100}\right) \times 10.0129 + \left(\frac{80.09}{100}\right) \times 11.0093$$
$$= 10.81$$

Equation 2.2 gives an average, weighted in terms of the abundance of each isotope for the element. As illustrated by the data in Table 2.2, *the atomic weight of an element is always closer to the mass of the more abundant isotope or isotopes.*

■ **Fractional Abundance**
The percent abundance of an isotope divided by 100 is called its fractional abundance.

Table 2.2   Isotope Abundance and Atomic Weight

Element	Symbol	Atomic Weight	Mass Number	Isotopic Mass (u)	Natural Abundance (%)
Hydrogen	H	1.00794	1	1.0078	99.985
	D*		2	2.0141	0.015
	T†		3	3.0161	0
Boron	B	10.811	10	10.0129	19.91
			11	11.0093	80.09
Neon	Ne	20.1797	20	19.9924	90.48
			21	20.9938	0.27
			22	21.9914	9.25
Magnesium	Mg	24.305	24	23.9850	78.99
			25	24.9858	10.00
			26	25.9826	11.01

*D = deuterium; †T = tritium, radioactive.

The atomic weight of each stable element has been determined experimentally, and these numbers appear in the periodic table inside the front cover of this book. In the periodic table, each element's box contains the atomic number, the element symbol, and the atomic weight. For unstable (radioactive) elements, the atomic mass or mass number of the most stable isotope is given in parentheses.

■ **The Periodic Table Entry for Copper**

Copper
29 - - - - - Atomic number
Cu - - - Symbol
63.546 - - - Atomic weight

Example 2.3—Calculating Atomic Weight from Isotope Abundance

Problem Bromine (used to make silver bromide, the important component of photographic film) has two naturally occurring isotopes. One has a mass of 78.918338 u and an abundance of 50.69%. The other isotope, of mass 80.916291 u, has an abundance of 49.31%. Calculate the atomic weight of bromine.

Strategy The atomic weight of any element is the weighted average of the masses of the isotopes in a representative sample. To calculate the atomic weight, multiply the mass of each isotope by its percent abundance divided by 100 (Equation 2.2). (*See the General ChemistryNow Screen 2.13 Atomic Mass, for a tutorial on calculating atomic weight from isotope abundance.*)

Solution

Average atomic mass of bromine = (50.69/100)(78.918338) + (49.31/100)(80.916291)

$$= \boxed{79.90}$$

Exercise 2.4—Calculating Atomic Weight

Verify that the atomic weight of chlorine is 35.45, given the following information:

^{35}Cl mass = 34.96885; percent abundance = 75.77%

^{37}Cl mass = 36.96590; percent abundance = 24.23%

2.5—Atoms and the Mole

One of the most exciting aspects of chemical research is the discovery of some new substance, and part of this process of discovery involves quantitative experiments. When two chemicals react with each other, we want to know how many atoms of each are used so that formulas can be established for the reaction's products. To do so, we need some method of counting atoms. That is, we must discover a way of connecting the macroscopic world, the world we can see, with the particulate world of atoms, molecules, and ions. The solution to this problem is to define a convenient amount of matter that contains a known number of particles. That chemical unit is the **mole**.

The mole (abbreviated mol) is the SI base unit for measuring an *amount of a substance* (see Table 1.2) and is defined as follows:

A **mole** is the amount of a substance that contains as many elementary entities (atoms, molecules, or other particles) as there are atoms in exactly 12 g of the carbon-12 isotope.

The key to understanding the concept of the mole is recognizing that *one mole always contains the same number of particles, no matter what the substance.* One mole of sodium

■ **The "Mole"**

The term "mole" was introduced about 1896 by Wilhelm Ostwald (1853–1932), who derived the term from the Latin word *moles*, meaning a "heap" or "pile."

Historical Perspectives

Amedeo Avogadro and His Number

Amedeo Avogadro, Conte di Quaregna (1776–1856) was an Italian nobleman and a lawyer. In about 1800, he turned to science, becoming the first professor of mathematical physics in Italy.

Avogadro did not propose the notion of a fixed number of particles in a chemical unit. Rather, the number was named in his honor because he had performed experiments in the 19th century that laid the groundwork for the concept.

Just how large is Avogadro's number? One mole of unpopped popcorn kernels would cover the continental United States to a depth of about 9 miles. One mole of pennies divided equally among every man, woman, and child in the United States would allow each person to pay off the national debt ($5.7 trillion or 5.7×10^{12} dollars) and there would still be $15 trillion left over!

Is Avogadro's number a unique value like π? No. It is fixed by the definition of the mole as exactly 12 g of carbon-12. If one mole of carbon were defined to have some other mass, then Avogadro's number would have a different value.

Photo: E. F. Smith Collection/Van Pelt Library/ University of Pennsylvania

contains the same number of atoms as one mole of iron. How many particles? Many, many experiments over the years have established that number as

$$1 \text{ mole} = 6.0221415 \times 10^{23} \text{ particles}$$

This value is known as **Avogadro's number** in honor of Amedeo Avogadro, an Italian lawyer and physicist (1776–1856) who conceived the basic idea (but never determined the number).

■ **The Difference Between "Amount" and "Quantity"**
The terms "amount" and "quantity" are used in a specific sense by chemists. The *amount* of a substance is the number of moles of that substance. *Quantity* refers to the mass of the substance. See W. G. Davies and J. W. Moore: *Journal of Chemical Education,* Vol. 57, p. 303, 1980. See also http://physics.nist.gov on the Internet.

Molar Mass

The mass in grams of one mole of any element (6.0221415×10^{23} atoms of that element) is the **molar mass** of that element. Molar mass is conventionally abbreviated with a capital italicized M and has units of grams per mole (g/mol). An element's *molar mass is the amount in grams numerically equal to its atomic weight.* Using sodium and lead as examples,

$$
\begin{aligned}
\text{Molar mass of sodium (Na)} &= \text{mass of 1.000 mol of Na atoms} \\
&= 22.99 \text{ g/mol} \\
&= \text{mass of } 6.022 \times 10^{23} \text{ Na atoms} \\
\text{Molar mass of lead (Pb)} &= \text{mass of 1.000 mol of Pb atoms} \\
&= 207.2 \text{ g/mol} \\
&= \text{mass of } 6.022 \times 10^{23} \text{ Pb atoms}
\end{aligned}
$$

Figure 2.9 shows the relative sizes of a mole of some common elements. Although each of these "piles of atoms" has a different volume and different mass, each contains 6.022×10^{23} atoms.

The mole concept is the cornerstone of quantitative chemistry. It is essential to be able to convert from moles to mass and from mass to moles. Dimensional analysis, which is described in Section 1.8 (pages 41–43), shows that this can be done in the following way:

Figure 2.9 **One-mole of common elements.** (*left to right*) Sulfur powder, magnesium chips, tin, and silicon. (*above*) Copper beads.

Charles D. Winters

MASS ⟷ MOLES CONVERSION

Moles to Mass

$$\text{Moles} \times \frac{\text{grams}}{\text{1 mol}} = \text{grams}$$

↑
molar mass

Mass to Moles

$$\text{Grams} \times \frac{\text{1 mol}}{\text{grams}} = \text{moles}$$

↑
1/molar mass

For example, what mass, in grams, is represented by 0.35 mol of aluminum? Using the molar mass of aluminum (27.0 g/mol), you can determine that 0.35 mol of Al has a mass of 9.5 g.

$$0.35 \text{ mol Al} \times \frac{27.0 \text{ g Al}}{1 \text{ mol Al}} = 9.5 \text{ g Al}$$

Molar masses are generally known to at least four significant figures. The convention followed in calculations in this book is to use a value of the molar mass with one more significant figure than in any other number in the problem. For example, if you weigh out 16.5 g of carbon, you use 12.01 g/mol for the molar mass of C to find the amount of carbon present.

$$16.5 \text{ g C} \times \frac{1 \text{ mol C}}{12.01 \text{ g C}} = 1.37 \text{ mol C}$$

↑
Note that four significant figures are used in the molar mass, but there are three in the sample mass.

Using one more significant figure for the molar mass means the accuracy of this value will not affect the accuracy of the result.

Lead. A 150-mL beaker containing 2.50 mol or 518 g of lead.

Charles D. Winters

Tin. A sample of tin having a mass of 36.5 g (1.85×10^{23} atoms).

Charles D. Winters

Mercury. A graduated cylinder containing 32.0 cm³ of mercury. This is equivalent to 433 g or 2.16 mol of mercury.

Charles D. Winters

GENERAL
Chemistry﹒Now™

See the General ChemistryNow CD-ROM or website:
- **Screen 2.14 the Mole,** for a tutorial on moles and atoms conversion
- **Screen 2.15 Moles and Molar Mass of the Elements,** for two tutorials on molar mass conversion

Example 2.4—Mass, Moles, and Atoms

Problem Consider two elements in the same vertical column of the periodic table, lead and tin.

(a) What mass of lead, in grams, is equivalent to 2.50 mol of lead (Pb, atomic number = 82)?

(b) What amount of tin, in moles, is represented by 36.5 g of tin (Sn, atomic number = 50)? How many atoms of tin are in the sample?

Strategy The molar masses of lead (207.2 g/mol) and tin (118.7 g/mol) are required and can be found in the periodic table inside the front cover of this book. Avogadro's number is needed to convert the amount of each element to number of atoms.

Solution

(a) Convert the amount of lead in moles to mass in grams.

$$2.50 \; \text{mol Pb} \times \frac{207.2 \; \text{g}}{1 \; \text{mol Pb}} = \boxed{518 \; \text{g Pb}}$$

(b) First convert the mass of tin to the amount in moles.

$$36.5 \; \text{g Sn} \times \frac{1 \; \text{mol Sn}}{118.7 \; \text{g Sn}} = \boxed{0.308 \; \text{mol Sn}}$$

Finally, use Avogadro's number to find the number of atoms in the sample.

$$0.308 \; \text{mol Sn} \times \frac{6.022 \times 10^{23} \; \text{atoms Sn}}{1 \; \text{mol Sn}}$$

$$= \boxed{1.85 \times 10^{23} \; \text{atoms Sn}}$$

Example 2.5—Mole Calculation

Problem The graduated cylinder in the photo contains 32.0 cm³ of mercury. If the density of mercury at 25 °C is 13.534 g/cm³, what amount of mercury, in moles, is in the cylinder?

Strategy Volume and moles of mercury are not directly connected. You must first use the density of mercury to find the mass of the metal and then use this value with the molar mass of mercury to calculate the amount in moles.

Volume (cm³) × density (g/cm³) = mass of mercury (g)

Amount of mercury (mol) = mass of mercury (g) × (1/molar mass)(mol/g)

Solution Combining the volume and density gives the mass of the mercury.

$$32.0 \text{ cm}^3 \times \frac{13.534 \text{ g Hg}}{1 \text{ cm}^3} = 433 \text{ g Hg}$$

Finally, the amount of mercury can be calculated from its mass and molar mass.

$$433 \text{ g Hg} \times \frac{1 \text{ mol Hg}}{200.6 \text{ g Hg}} = 2.16 \text{ mol Hg}$$

Example 2.6—Mass of an Atom

Problem What is the average mass of an atom of platinum (Pt)?

Strategy The mass of one mole of platinum is 195.08 g. Each mole contains Avogadro's number of atoms.

Solution Here we divide the mass of a mole by the number of objects in that unit.

$$\frac{195.08 \text{ g Pt}}{1 \text{ mol Pt}} \times \frac{1 \text{ mol Pt}}{6.02214 \times 10^{23} \text{ atoms Pt}} = \frac{3.2394 \times 10^{-22} \text{ g Pt}}{1 \text{ atom Pt}}$$

Comment Notice that the units "mol Pt" cancel and leave an answer with units of g/atom.

Exercise 2.5—Mass/Mole Conversions

(a) What is the mass, in grams, of 1.5 mol of silicon?
(b) What amount (moles) of sulfur is represented by 454 g? How many atoms?
(c) What is the average mass of one sulfur atom?

Exercise 2.6—Atoms

The density of gold, Au, is 19.32 g/cm³. What is the volume (in cubic centimeters) of a piece of gold that contains 2.6 x 10²⁴ atoms? If the piece of metal is a square with a thickness of 0.10 cm, what is the length (in centimeters) of one side of the piece?

2.6—The Periodic Table

The periodic table of elements is one of the most useful tools in chemistry. Not only does it contain a wealth of information, but it can also be used to organize many of the ideas of chemistry. It is important that you become familiar with its main features and terminology.

Features of the Periodic Table

The main organizational features of the periodic table are the following:

- Elements are arranged so that those with similar chemical and physical properties lie in vertical columns called **groups** or **families**. The periodic table commonly used in the United States has groups numbered 1 through 8, with each number followed by a letter: A or B. The A groups are often called the **main group elements** and the B groups are the **transition elements**.

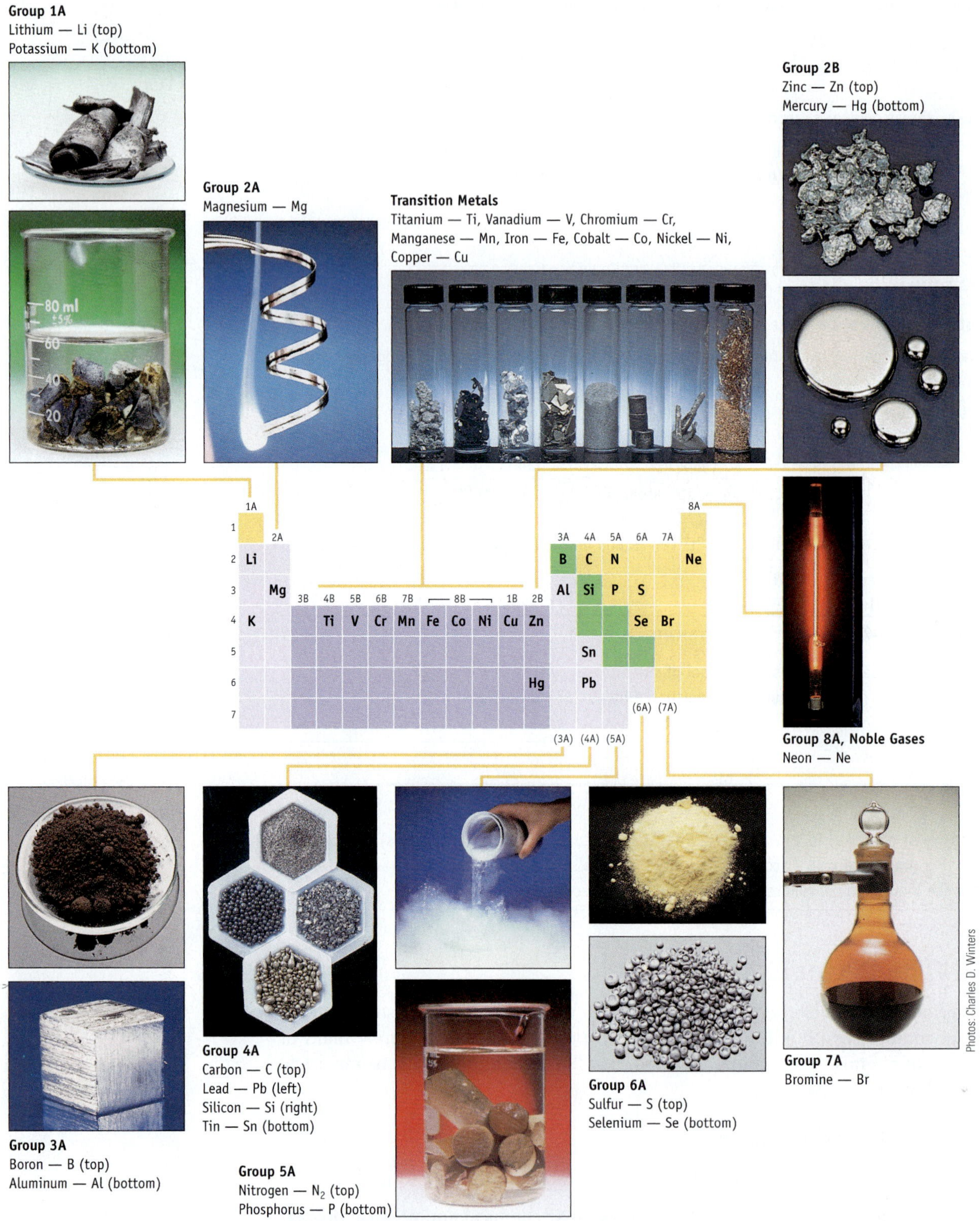
Group 1A
Lithium — Li (top)
Potassium — K (bottom)
Group 2A
Magnesium — Mg
Transition Metals
Titanium — Ti, Vanadium — V, Chromium — Cr,
Manganese — Mn, Iron — Fe, Cobalt — Co, Nickel — Ni,
Copper — Cu
Group 2B
Zinc — Zn (top)
Mercury — Hg (bottom)
80 ml
1A 2A 3B 4B 5B 6B 7B 8B 1B 2B 3A 4A 5A 6A 7A 8A
Li Mg K Ti V Cr Mn Fe Co Ni Cu Zn Hg B C N Al Si P S Se Br Sn Pb Ne
1 2 3 4 5 6 7
(3A) (4A) (5A) (6A) (7A)
Group 8A, Noble Gases
Neon — Ne
Group 3A
Boron — B (top)
Aluminum — Al (bottom)
Group 4A
Carbon — C (top)
Lead — Pb (left)
Silicon — Si (right)
Tin — Sn (bottom)
Group 5A
Nitrogen — N_2 (top)
Phosphorus — P (bottom)
Group 6A
Sulfur — S (top)
Selenium — Se (bottom)
Group 7A
Bromine — Br
Photos: Charles D. Winters

Active Figure 2.10 **Some of the 116 known elements.**

GENERAL Chemistry Now™ See the General ChemistryNow CD-ROM or website to explore an interactive version of this figure accompanied by an exercise.

- The horizontal rows of the table are called **periods**, and they are numbered beginning with 1 for the period containing only H and He. For example, sodium, Na, in Group 1A, is the first element in the third period. Mercury, Hg, in Group 2B, is in the sixth period (or sixth row).

The periodic table can be divided into several regions according to the properties of the elements. On the table inside the front cover of this book, elements that behave as *metals* are indicated in purple, those that are *nonmetals* are indicated in yellow, and elements called *metalloids* appear in green. Elements gradually become less metallic as one moves from left to right across a period, and the metalloids lie along the metal–nonmetal boundary. Some elements are shown in Figure 2.10.

You are probably familiar with many properties of **metals** from everyday experience (Figure 2.11a). Metals are solids (except for mercury), can conduct electricity, are usually ductile (can be drawn into wires) and malleable (can be rolled into sheets), and can form alloys (solutions of one or more metals in another metal). Iron (Fe) and aluminum (Al) are used in automobile parts because of their ductility, malleability, and low cost relative to other metals. Copper (Cu) is used in electric wiring because it conducts electricity better than most other metals. Chromium (Cr) is plated onto automobile parts, not only because its metallic luster makes cars look better but also because chrome-plating protects the underlying metal from reacting with oxygen in the air.

The **nonmetals** lie to the right of a diagonal line that stretches from B to Te in the periodic table and have a wide variety of properties. Some are solids (carbon, sulfur, phosphorus, and iodine). Four elements are gases at room temperature (oxygen, nitrogen, fluorine, and chlorine). One element, bromine, is a liquid at room temperature (Figure 2.11b). With the exception of carbon in the form of graphite, nonmetals do not conduct electricity, which is one of the main features that distinguishes them from metals.

Some of the elements next to the diagonal line from boron (B) to tellurium (Te) have properties that make them difficult to classify as metals or nonmetals. Chemists call them metalloids or, sometimes, semimetals (Figure 2.11c). You should

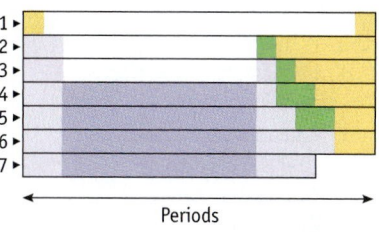

Periods

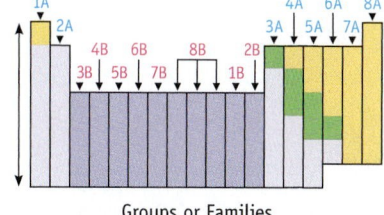

Groups or Families

■ **Two Ways to Designate Groups**
One way to designate periodic groups is to number them 1 through 18 from left to right. This method is generally used outside the United States. The system predominant in the United States labels main group elements as Groups 1A–8A and transition elements as Groups 1B–8B. This book uses the A/B system.

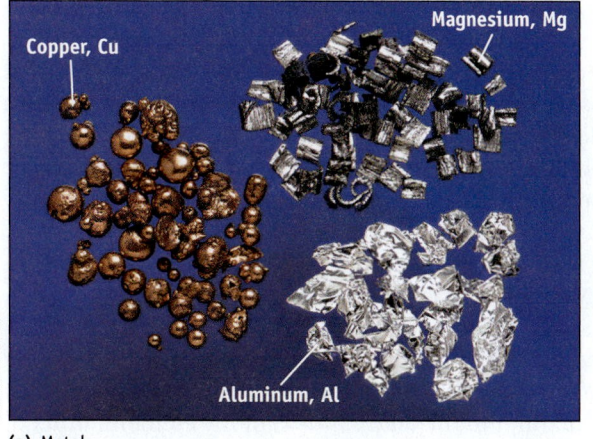

(a) Metals

(b) Nonmetals

(c) Metalloids

Figure 2.11　Representative elements. (a) Magnesium, aluminum, and copper are metals. All can be drawn into wires and conduct electricity. (b) Only 15 or so elements can be classified as nonmetals. Here are orange liquid bromine and purple solid iodine. (c) Only 6 elements are generally classified as metalloids or semimetals. This photograph shows solid silicon in various forms, including a wafer that holds printed electronic circuits.

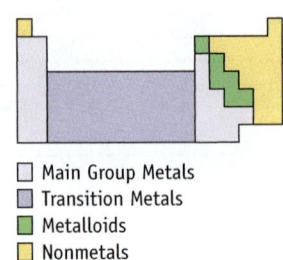

☐ Main Group Metals
☐ Transition Metals
☐ Metalloids
☐ Nonmetals

know, however, that chemists often disagree about what a metalloid is as well as which elements fit into this category. We will define a **metalloid** as an element that has some of the physical characteristics of a metal but some of the chemical characteristics of a nonmetal; we include only B, Si, Ge, As, Sb, and Te in this category. This distinction reflects the ambiguity in the behavior of these elements. Antimony (Sb), for example, conducts electricity as well as many elements that are true metals. Its chemistry, however, resembles that of a nonmetal such as phosphorus.

Developing the Periodic Table

Although the arrangement of elements in the periodic table can now be understood on the basis of atomic structure [▶ Chapter 8], the table was originally developed from many, many experimental observations of the chemical and physical properties of elements and is the result of the ideas of a number of chemists in the 18th and 19th centuries.

In 1869, at the University of St. Petersburg in Russia, Dmitri Ivanovitch Mendeleev (1834–1907) was pondering the properties of the elements as he wrote a textbook on chemistry. On studying the chemical and physical properties of the elements, he realized that, if the elements were arranged in order of increasing atomic mass, elements with similar properties appeared in a regular pattern. That is, he saw a **periodicity** or periodic repetition of the properties of elements. Mendeleev organized the known elements into a table by lining them up in a horizontal row in order of increasing atomic mass. Every time he came to an element with properties similar to one already in the row, he started a new row. For example, the elements Li, Be, B, C, N, O, and F were in a row. Sodium was the next element then known; because its properties closely resembled those of Li, Mendeleev started a new row. The columns, then, contained elements such as Li, Na, and K with similar properties.

An important feature of Mendeleev's table—and a mark of his genius—was that he left an empty space in a column when an element was not known but should exist and have properties similar to the element above it in his table. He deduced that these spaces would be filled by undiscovered elements. For example, a space was left between Si (silicon) and Sn (tin) in what is now Group 4A. Based on the progression of properties in this group, Mendeleev was able to predict the properties of this missing element. With the discovery of germanium in 1886, Mendeleev's prediction was confirmed.

In Mendeleev's table the elements were ordered by increasing mass. A glance at a modern table, however, shows that, on this basis, Ni and Co, Ar and K, and Te and I, should be reversed. Mendeleev assumed the atomic masses known at that time were inaccurate—not a bad assumption based on the analytical methods then in use. In fact, his order was correct and what was wrong was his assumption that element properties were a function of their mass.

In 1913 H. G. J. Moseley (1887–1915), a young English scientist working with Ernest Rutherford, corrected Mendeleev's assumption. Moseley was doing experiments in which he bombarded many different metals with electrons in a cathode-ray tube (Figure 2.3) and examined the x-rays emitted in the process. In seeking some order in his data, he realized that the wavelength of the x-rays emitted by a given element were related in a precise manner to the *atomic number* of the element. Indeed, chemists quickly recognized that organizing the elements in a table by increasing atomic number corrected the inconsistencies in the Mendeleev table. The **law of chemical periodicity** is now stated as "the properties of the elements are periodic functions of atomic number."

■ **About the Periodic Table**
For more information on the periodic table, the central icon of chemistry, we recommend the following:
- The American Chemical Society has a description of every element on its website at **www.cen-online.org.**
- J. Emsley: *Nature's Building Blocks—An A–Z Guide to the Elements,* New York, Oxford University Press, 2001.
- O. Sacks: *Uncle Tungsten—Memories of a Chemical Boyhood,* New York, Alfred A. Knopf, 2001.

■ **Placing H in the Periodic Table**
Where to place H? Tables often show it in Group 1A even though it is clearly not an alkali metal. However, in its reactions it forms a 1+ ion just like the alkali metals. For this reason, H is often placed in Group 1A.

Historical Perspectives

Periodic Table

In his book *Nature's Building Blocks* (p. 527, New York, Oxford University Press, 2001), John Emsley tells us that "As long as chemistry is studied, there will be a periodic table. Even if some day we communicate with another part of the Universe, we can be sure that one thing both cultures will have in common is an ordered system of the elements that will be instantly recognizable by both intelligent life forms."

The person credited with organizing the elements into a periodic table is Dmitri Mendeleev. However, other chemists had long recognized that groups of elements shared similar properties. In 1829 Johann Dobereiner (1780–1849) announced the Law of Triads. He showed that there were groups of three elements (triads), in which the middle element had an atomic weight that was the average of the other two. One such triad consisted of Li, Na, and K; another was made up of Cl, Br, and I.

Perhaps the first revelation of the periodicity of the elements was published by a French geologist, A. E. Béguyer de Chancourtois (1820–1886), in 1862. He listed the elements on a paper tape, and, according to Emsley, "then wound this, spiral like around a cylinder. The cylinder's surface was divided into 16 parts, based on the atomic weight of oxygen. De Chancourtois noted that certain triads came together down the cylinder, such as the alkali metals." He called his model the "telluric screw."

Another attempt at organizing the elements was proposed by John Newlands (1837–1898) in 1864. His "Law of Octaves" proposed that there was a periodic similarity every eight elements, just as the musical scale repeats every eighth note. Unfortunately, his proposal was ridiculed at the time.

Julius Lothar Meyer (1830–1895) came closer than any other to discovering the periodic table. He drew a graph of atomic volumes of elements plotted against their atomic weight. This clearly showed a periodic rise and fall in atomic volume on moving across what we now call the periods of the table. Before publishing the paper, Meyer passed it along to a colleague for comment. His colleague was slow to return the paper, and, unfortunately for Meyer, Mendeleev's paper was published in the interim. Because chemists quickly recognized the importance of Mendeleev's paper, Meyer was not given the recognition he perhaps deserves.

An essay on Mendeleev and his life appears at the beginning of Chapter 8 (pages 332–3).

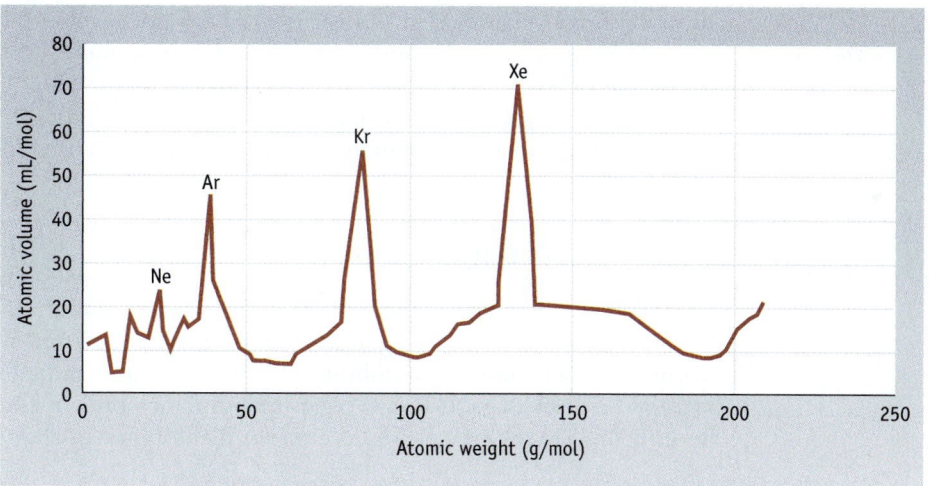

Atomic volume plot. Julius Lothar Meyer (1830–1895) illustrated the periodicity of the elements in 1868 by plotting atomic volume against atomic weight. (This plot uses current data.)

Source: C. N. Singman: *Journal of Chemical Education,* Vol. 61, p. 137, 1984.

GENERAL
Chemistry Now™

See the General ChemistryNow CD-ROM or website:
• **Screen 2.16 The Periodic Table,** for an exercise on the periodic table organization

2.7—An Overview of the Elements, Their Chemistry, and the Periodic Table

The vertical columns, or groups, of the periodic table contain elements having similar chemical and physical properties, and several groups of elements have distinctive names that are useful to know.

 ### Group 1A, Alkali Metals: Li, Na, K, Rb, Cs, Fr

Elements in the leftmost column, Group 1A, are known as the **alkali metals**. All are metals and are solids at room temperature. The metals of Group 1A are all reactive. For example, they react with water to produce hydrogen and alkaline solutions (Figure 2.12). Because of their reactivity, these metals are found in nature only combined in compounds, such as NaCl (Figure 1.7)—never as the free element.

 ### Group 2A, Alkaline Earth Metals: Be, Mg, Ca, Sr, Ba, Ra

The second group in the periodic table, Group 2A, is composed entirely of metals that occur naturally only in compounds (Figure 2.13). Except for beryllium (Be), these elements also react with water to produce alkaline solutions, and most of their oxides (such as lime, CaO) form alkaline solutions; hence, they are known as the **alkaline earth metals**. Magnesium (Mg) and calcium (Ca) are the seventh and fifth most abundant elements in the earth's crust, respectively (Table 2.3). Calcium is one of the important elements in teeth and bones, and it occurs naturally in vast limestone deposits. Calcium carbonate ($CaCO_3$) is the chief constituent of limestone and of corals, sea shells, marble, and chalk (see Figure 2.13b). Radium (Ra), the heaviest alkaline earth element, is radioactive and is used to treat some cancers by radiation.

 ### Group 3A: B, Al, Ga, In, Tl

Group 3A contains one element of great importance, aluminum (Figure 2.14). This element and three others (gallium, indium, and thallium) are metals, whereas boron (B) is a metalloid. Aluminum (Al) is the most abundant metal in the earth's crust at 8.2% by mass. It is exceeded in abundance only by the nonmetals oxygen and silicon. These three elements are found combined in clays and other common

---

■ **Alkali and Alkaline**
The word "alkali" comes from the Arabic language; ancient Arabian chemists discovered that ashes of certain plants, which they called al-qali, gave water solutions that felt slippery and burned the skin. These ashes contain compounds of Group 1A elements that produce alkaline (basic) solutions.

Table 2.3   The Ten Most Abundant Elements in the Earth's Crust

Rank	Element	Abundance (ppm)*
1	Oxygen	474,000
2	Silicon	277,000
3	Aluminum	82,000
4	Iron	41,000
5	Calcium	41,000
6	Sodium	23,000
7	Magnesium	23,000
8	Potassium	21,000
9	Titanium	5,600
10	Hydrogen	1,520

*ppm = g per 1000 kg.

Figure 2.12   Alkali metals. (a) Cutting a bar of sodium with a knife is about like cutting a stick of cold butter. (b) When an alkali metal such as potassium is treated with water, a vigorous reaction occurs, giving an alkaline solution and hydrogen gas, which burns in air. See also Figure 1.7, the reaction of sodium with chlorine.

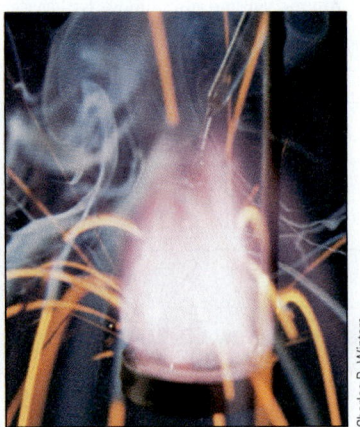

Charles D. Winters

(a) Cutting sodium.

(b) Potassium reacts with water.

(a) Magnesium and strontium in fireworks. (b) Calcium-containing compounds.

Figure 2.13 Alkaline earth metals. (a) When heated in air, magnesium burns to give magnesium oxide. The white sparks you see in burning fireworks are burning magnesium. (b) Some common calcium-containing substances: calcite (the clear crystal); a seashell; limestone; and an over-the-counter remedy for excess stomach acid.

minerals. Boron occurs in the mineral borax, a compound used as a cleaning agent, antiseptic, and flux for metal work.

As a metalloid, boron has a different chemistry than the other elements of Group 3A, all of which are metals. Nonetheless, all form compounds with analogous formulas such as BCl_3 and $AlCl_3$, and this similarity marks them as members of the same periodic group.

Group 4A: C, Si, Ge, Sn, Pb

All of the elements we have described so far, except boron, have been metals. Beginning with Group 4A, however, the groups contain more and more nonmetals. Group 4A includes one nonmetal, carbon (C); two metalloids, silicon (Si) and germanium (Ge), and two metals, tin (Sn) and lead (Pb). Because of the change from nonmetallic to metallic behavior, more variation occurs in the properties of the elements of this group than in most others. Nonetheless, these elements also form

(a) Wagons for hauling borax in Death Valley.

(b) Aluminum-containing minerals.

Figure 2.14 Group 3A elements. (a) Boron is mined as borax, a natural compound used in soap. Borax was mined in Death Valley, California, at the end of the 19th century and was hauled from the mines in wagons drawn by teams of 20 mules. Boron is also a component of borosilicate glass, which is used for laboratory glassware. (b) Aluminum is abundant in the earth's crust; it is found in all clays and in many minerals and gems. It has many commercial applications as the metal as well as in aluminum sulfate, which is used in water purification.

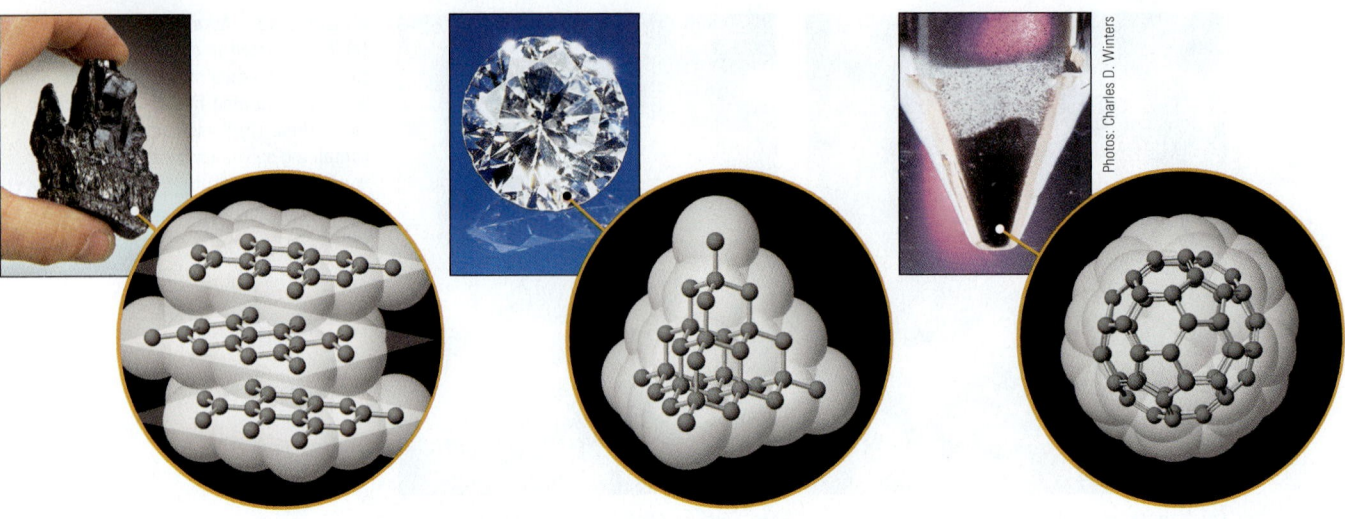

Photos: Charles D. Winters

(a) Graphite   **(b)** Diamond   **(c)** Buckyballs

Figure 2.15   The allotropes of carbon. (a) Graphite consists of layers of carbon atoms. Each carbon atom is linked to three others to form a sheet of six-member, hexagonal rings. (b) In diamond the carbon atoms are also arranged in six-member rings, but the rings are not flat because each C atom is connected tetrahedrally to four other C atoms. (c) A member of the family called buckminsterfullerenes, C_{60} is an allotrope of carbon. Sixty carbon atoms are arranged in a spherical cage that resembles a hollow soccer ball. Notice that each six-member ring shares an edge with three other six-member rings and three five-member rings. Chemists call this molecule a "buckyball." C_{60} is a black powder; it is shown here in the tip of a pointed glass tube.

compounds with analogous formulas (such as CO_2, SiO_2, GeO_2, SnO_2, and PbO_2), so they are assigned to the same group.

Carbon is the basis for the great variety of chemical compounds that make up living things. It is found in the earth's atmosphere as CO_2, on the surface of the earth in carbonates like limestone and coral (see Figure 2.13b), and in coal, petroleum, and natural gas—the fossil fuels.

One of the most interesting aspects of the chemistry of the nonmetals is that a particular element can often exist in several different and distinct forms, called **allotropes**, each having its own properties. Carbon has at least three allotropes, the best known of which are graphite and diamond (Figure 2.15). The flat sheets of carbon atoms in graphite (Figure 2.15a) cling only weakly to one another. One layer can slip easily over another, which explains why graphite is soft, is a good lubricant, and is used in pencil lead. (Pencil "lead" is not the element lead, Pb, but rather a composite of clay and graphite that leaves a trail of graphite on the page as you write.)

In diamond each carbon atom is connected to four others at the corners of a tetrahedron, and this pattern extends throughout the solid (see Figure 2.15b). This structure causes diamonds to be extremely hard, denser than graphite ($d = 3.51$ g/cm^3 for diamond and $d = 2.22$ g/cm^3 for graphite), and chemically less reactive. Because diamonds are not only hard but are also excellent conductors of heat, they are used on the tips of metal- and rock-cutting tools.

In the late 1980s another form of carbon was identified as a component of black soot, the stuff that collects when carbon-containing materials are burned in a deficiency of oxygen. This substance is made up of molecules with 60 carbon atoms arranged as a spherical "cage" (Figure 2.15c). You may recognize that the surface is made up of five- and six-member rings and resembles a hollow soccer ball. The shape reminded its discoverers of an architectural dome invented several decades ago by the American philosopher and engineer, R. Buckminster Fuller. The official name of the allotrope is therefore buckminsterfullerene, and chemists often call these molecules "buckyballs."

Oxides of silicon are the basis of many minerals such as clay, quartz, and beautiful gemstones like amethyst (Figure 2.16). Tin and lead have been known for centuries because they are easily smelted from their ores. Tin alloyed with copper makes bronze, which was used in ancient times in utensils and weapons. Lead has been used in water pipes and paint, even though the element is toxic to humans.

Figure 2.16 Compounds containing silicon. Ordinary clay, sand, and many gemstones are based on compounds of silicon and oxygen. Here clear, colorless quartz and dark purple amethyst lie in a bed of sand. All are made of silicon dioxide, SiO_2. The different colors are due to impurities.

Group 5A: N, P, As, Sb, Bi

Nitrogen, which occurs naturally in the form of N_2 (Figures 2.10 and 2.17), makes up about three-fourths of earth's atmosphere. It is also incorporated in biochemically important substances such as chlorophyll, proteins, and DNA. Scientists have long sought ways to make compounds from atmospheric nitrogen, a process referred to as "nitrogen fixation." Nature accomplishes this transformation easily in plants, but severe conditions (high temperatures, for example) must be used in the laboratory and in industry to cause N_2 to react with other elements (such as H_2 to make ammonia, NH_3, which is widely used as a fertilizer).

Phosphorus is also essential to life. For example, it is an important constituent in bones and teeth. The element glows in the dark if it is in the air, and its name, based on Greek words meaning "light-bearing," reflects this property. This element also has several allotropes, the most important being white (Figure 2.10) and red phosphorus. Both forms of phosphorus are used commercially. White phosphorus ignites spontaneously in air, so it is normally stored under water. When it does react with air, it forms P_4O_{10}, which can react with water to form phosphoric acid (H_3PO_4), a compound used in food products such as soft drinks. Red phosphorus also reacts with oxygen in the air and is used in the striking strips on match books.

As with Group 4A, we again see nonmetals (N and P), metalloids (As and Sb), and a metal (Bi) in Group 5A. In spite of these variations, all of the members of this group form analogous compounds such as the oxides N_2O_5, P_2O_5, and As_2O_5.

Group 6A: O, S, Se, Te, Po

Oxygen, which constitutes about 20% of earth's atmosphere and combines readily with most other elements, is found at the top of Group 6A. Most of the energy that

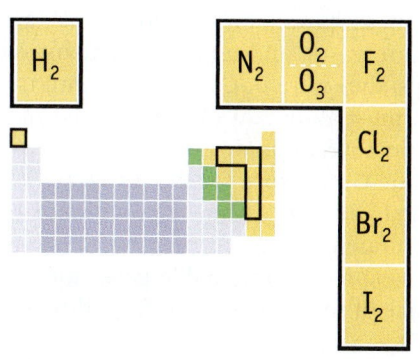

Figure 2.17 Elements that exist as diatomic molecules. Seven of the elements in the periodic table exist as diatomic, or two-atom, molecules. Oxygen has an additional allotrope, ozone, with three O atoms in each molecule.

Figure 2.18 Sulfur. The most common allotrope of sulfur consists of eight-member, crown-shaped rings.

The halogens bromine and iodine. Bromine is a liquid at room temperature and iodine is a solid. However, some of the element exists in the vapor state above the liquid or solid.

powers life on earth is derived from reactions in which oxygen combines with other substances.

Sulfur has been known in elemental form since ancient times as brimstone or "burning stone" (Figure 2.18). Sulfur, selenium, and tellurium are referred to collectively as *chalcogens* (from the Greek word, *khalkos*, for copper) because most copper ores contain these elements. Their compounds can be foul-smelling and poisonous; nevertheless, sulfur and selenium are essential components of the human diet. By far the most important compound of sulfur is sulfuric acid (H_2SO_4), which is manufactured in larger amounts than any other compound.

As in Group 5A, the second- and third-period elements have different structures. Like nitrogen, oxygen is a diatomic molecule (see Figure 2.17). Unlike nitrogen, however, oxygen has an allotrope, the well-known ozone, O_3. Sulfur, which can be found in nature as a yellow solid, has many allotropes. The most common allotrope consists of eight-member, crown-shaped rings of sulfur atoms (see Figure 2.18).

Polonium, a radioactive element, was isolated in 1898 by Marie and Pierre Curie, who separated it from tons of a uranium-containing ore and named it for Madame Curie's native country, Poland.

With Group 6A, once again we observe variations in the properties in a group. Oxygen, sulfur, and selenium are nonmetals, tellurium is a metalloid, and polonium is a metal. Nonetheless, there is a family resemblance in their chemistries. All form oxygen-containing compounds such as SO_2, SeO_2, and TeO_2 and sodium-containing compounds such as Na_2O, Na_2S, Na_2Se, and Na_2Te.

Group 7A, Halogens: F, Cl, Br, I, At

At the far right of the periodic table are two groups composed entirely of nonmetals. The Group 7A elements—fluorine, chlorine, bromine, and iodine—are nonmetals, all of which exist as diatomic molecules (see Figure 2.17). At room temperature, fluorine (F_2) and chlorine (Cl_2) are gases. Bromine (Br_2) is a liquid and iodine (I_2) is a solid, but bromine and iodine vapor are clearly visible over the liquid or solid.

The Group 7A elements are among the most reactive of all elements. All combine violently with alkali metals to form salts such as table salt, NaCl (Figure 1.7). The name for this group, the **halogens**, comes from the Greek words *hals*, meaning "salt," and *genes*, meaning "forming." The halogens also react with other metals and with most nonmetals to form compounds.

Group 8A, Noble Gases: He, Ne, Ar, Kr, Xe, Rn

The Group 8A elements—helium, neon, argon, krypton, xenon, and radon—are the least reactive elements (Figure 2.19). All are gases, and none is abundant on earth or in earth's atmosphere. Because of this, they were not discovered until the end of the 19th century. Helium, the second most abundant element in the universe after hydrogen, was detected in the sun in 1868 by analysis of the solar spectrum. (The name of the element comes from the Greek word for the sun, *helios*.) It was not found on earth until 1895, however. Until 1962, when a compound of xenon was first prepared, it was believed that none of these elements would combine chemically with any other element. The common name **noble gases** for this group, a term meant to denote their general lack of reactivity, derives from this fact.

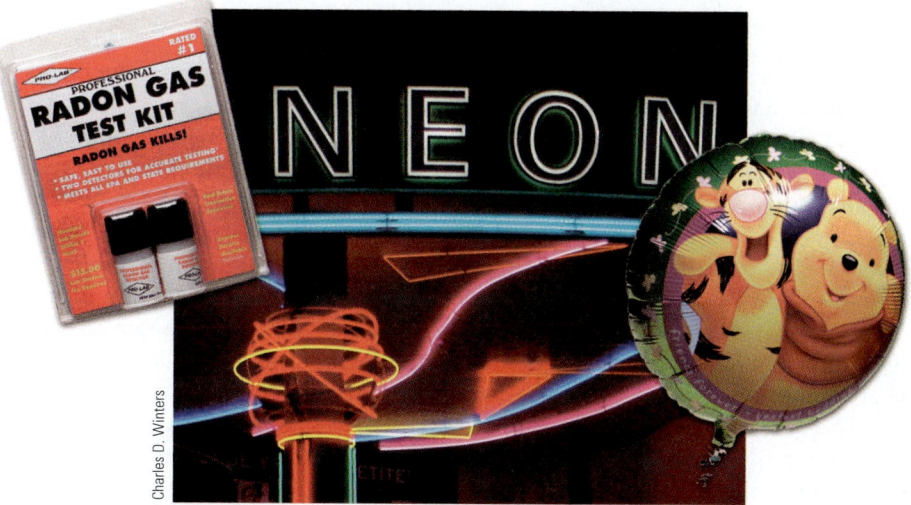

Charles D. Winters

Figure 2.19   The noble gases. This kit is sold for detecting the presence of radon gas in the home. Neon gas is used in advertising signs, and helium-filled balloons are popular.

For the same reason they are sometimes called the *inert gases* or, because of their low abundance, the *rare gases*.

 The Transition Elements

Stretching between Groups 2A and 3A is a series of elements called the **transition elements**. These fill the B-groups (1B through 8B) in the fourth through the seventh periods in the center of the periodic table. All are metals (see Figure 2.10), and 13 of them are in the top 30 elements in terms of abundance in the earth's crust. Some, like iron (Fe), are abundant in nature (Table 2.4). Most occur naturally in combination with other elements, but a few—silver (Ag), gold (Au), and platinum (Pt)—are much less reactive and can be found in nature as pure elements.

Virtually all of the transition elements have commercial uses. They are used as structural materials (iron, titanium, chromium, copper); in paints (titanium, chromium); in the catalytic converters in automobile exhaust systems (platinum, rhodium); in coins (copper, nickel, zinc); and in batteries (manganese, nickel, cadmium, mercury).

A number of the transition elements play important biological roles. For example, iron, a relatively abundant element (see Table 2.3), is the central element in the chemistry of hemoglobin, the oxygen-carrying component of blood.

Two rows at the bottom of the table accommodate the **lanthanides** [the series of elements between the elements lanthanum ($Z = 57$) and hafnium ($Z = 72$)] and the **actinides** [the series of elements between actinium ($Z = 89$) and rutherfordium ($Z = 104$)]. Some lanthanide compounds are used in color television picture tubes, uranium ($Z = 92$) is the fuel for atomic power plants, and americium ($Z = 95$) is used in smoke detectors.

Table 2.4   Abundance of the Ten Most Abundant Transition Elements in the Earth's Crust

Rank	Element	Abundance (ppm)*
4	Iron	41,000
9	Titanium	5,600
12	Manganese	950
18	Zirconium	190
19	Vanadium	160
21	Chromium	100
23	Nickel	80
24	Zinc	75
25	Cerium	68
26	Copper	50

*ppm = g per 1000 kg.

Exercise 2.7—The Periodic Table

How many elements are in the third period of the periodic table? Give the name and symbol of each. Tell whether each element in the period is a metal, metalloid, or nonmetal.

Table 2.5   Relative Amounts of Elements in the Human Body

Element	Percent by Mass
Oxygen	65
Carbon	18
Hydrogen	10
Nitrogen	3
Calcium	1.5
Phosphorus	1.2
Potassium, sulfur, chlorine	0.2
Sodium	0.1
Magnesium	0.05
Iron, cobalt, copper, zinc, iodine	<0.05
Selenium, fluorine	<0.01

2.8—Essential Elements

As our knowledge of biochemistry—the chemistry of living systems—increases, we learn more and more about essential elements. These elements are so important to life that a deficiency in any one will result in either death, severe developmental abnormalities, or chronic ailments. No other element can take the place of an essential element.

Of the 116 known elements, 11 are predominant in many different biological systems and are present in approximately the same relative amounts (Table 2.5). In humans these 11 elements constitute 99.9% of the total number of atoms present, but 4 of these elements—C, H, N, and O—account for 99% of the total. These elements are found in the basic structure of all biochemical molecules. Additionally, H and O are present in water, a major component of all biological systems.

The other 7 elements of the group of 11 elements comprise only 0.9% of the total atoms in the body. These are sodium, potassium, calcium, magnesium, phosphorus, sulfur, and chlorine. These generally occur in the form of ions such as Na^+, K^+, Mg^{2+}, Ca^{2+}, Cl^-, and HPO_4^{2-}.

The 11 essential elements represent 6 of the groups of the periodic table, and all are "light" elements; they have atomic numbers less than 21. Another 17 elements are required by most but not all biological systems. Some may be required by plants, some by animals, and others by only certain plants or animals. With a few exceptions, these elements are generally "heavier" elements, elements having an atomic number greater than 18. They are about evenly divided between metals and nonmetals (or metalloids).

Elements in the Human Body

Major Elements	Trace Elements
99.9% of all atoms (99.5% by mass)	0.1% of all atoms (0.5% by mass)
C, H, N, O	V, Cr, Mo, Mn, Fe, Co, Ni, Cu, Zn
Na, Mg, P, S, Cl	B, Si, Se, F, Br, I, As, Sn
K, Ca	

Sources of Some Biologically Important Elements

Element	Source	mg/100 g
Iron	Brewer's yeast	17.3
	Eggs	2.3
Zinc	Brazil nuts	4.2
	Chicken	2.6
Copper	Oysters	13.7
	Brazil nuts	2.3
Calcium	Swiss cheese	925
	Whole milk	118
	Broccoli	103
Selenium	Butter	0.15
	Cider vinegar	0.09

Although many of the metals in this group are required only in trace amounts, they are often an integral part of specific biological molecules—such as hemoglobin (Fe), myoglobin (Fe), and vitamin B_{12} (Co)—and activate or regulate their functions.

Much of the 3 or 4 g of iron in the body is found in hemoglobin, the substance responsible for carrying oxygen to cells. Iron deficiency is marked by fatigue, infections, and mouth inflammation. The average person also contains about 2 g of zinc. A deficiency of this element will be evidenced as loss of appetite, failure to grow, and changes in the skin.

The human body has about 75 mg of copper, about one third of which is found in the muscles and the remainder in other tissues. Copper is involved in many biological functions, and a deficiency shows up in a variety of ways: anemia, degeneration of the nervous system, impaired immunity, and defects in hair color and structure.

See the General ChemistryNow CD-ROM or website to:

- Assess your understanding with homework questions keyed to each goal
- Check your readiness for an exam by taking the exam-prep quiz and exploring the resources in the personalized Learning Plan it provides

Chapter Goals Revisited

Now that you have studied this chapter, you should ask whether you have met the chapter goals. In particular, you should be able to

Describe atomic structure and define atomic number and mass number

a. Explain the historical development of the atomic theory and identify some of the scientists who made important contributions (Section 2.1). General ChemistryNow homework: Study Question(s) 65

b. Describe electrons, protons, and neutrons, and the general structure of the atom (Section 2.1). General ChemistryNow homework: SQ(s) 6, 12

c. Understand the relative mass scale and the atomic mass unit (Section 2.2). General ChemistryNow homework: SQ(s) 64

Understand the nature of isotopes and calculate atomic weight from isotope abundances and isotopic masses

a. Define isotope and give the mass number and number of neutrons for a specific isotope (Sections 2.2 and 2.3). General ChemistryNow homework: SQ(s) 14

b. Do calculations that relate the atomic weight (atomic mass) of an element and isotopic abundances and masses (Section 2.4). General ChemistryNow homework: SQ(s) 20, 22, 25, 47

Explain the concept of the mole and use molar mass in calculations

a. Understand that the molar mass of an element is the mass in grams of Avogadro's number of atoms of that element (Section 2.5). General ChemistryNow homework: SQ(s) 27, 29, 31

b. Know how to use the molar mass of an element and Avogadro's number in calculations (Section 2.5). General ChemistryNow homework: SQ(s) 33, 57, 67, 77

Know the terminology of the periodic table

a. Identify the periodic table locations of groups, periods, metals, metalloids, nonmetals, alkali metals, alkaline earth metals, halogens, noble gases, and the transition elements (Sections 2.6 and 2.7). General ChemistryNow homework: SQ(s) 38, 39, 41, 49

b. Recognize similarities and differences in properties of some of the common elements of a group. General ChemistryNow homework: SQ(s) 56

Charles D. Winters

Foods rich in essential elements.

For additional preparation for an examination on this chapter see the **Let's Review** section on pages 282–293.

Key Equations

Equation 2.1 (page 69)
Calculate the percent abundance of an isotope.

$$\text{Percent abundance} = \frac{\text{number of atoms of a given isotope}}{\text{total number of atoms of all isotopes of that element}} \times 100\%$$

Equation 2.2 (page 72)
Calculate the atomic mass (atomic weight) from isotope abundances and the exact atomic mass of each isotope of an element.

$$\text{Atomic weight} = \left(\frac{\% \text{ abundance isotope 1}}{100}\right)(\text{mass of isotope 1}) + \left(\frac{\% \text{ abundance isotope 2}}{100}\right)(\text{mass of isotope 2}) + \cdots$$

Study Questions

▲ denotes more challenging questions.

■ denotes questions available in the Homework and Goals section of the General ChemistryNow CD-ROM or website.

Blue numbered questions have answers in Appendix O and fully worked solutions in the *Student Solutions Manual*.

Structures of many of the compounds used in these questions are found on the General ChemistryNow CD-ROM or website in the Models folder.

GENERAL
Chemistry•⚛•Now™ Assess your understanding of this chapter's topics with additional quizzing and conceptual questions at http://now.brookscole.com/kotz6e

Practicing Skills

Atoms: Their Composition and Structure

(See Example 2.1, Exercise 2.2, and the General ChemistryNow Screen 2.11.)

1. What are the three fundamental particles from which atoms are built? What are their electric charges? Which of these particles constitute the nucleus of an atom? Which is the least massive particle of the three?

2. Around 1910 Rutherford carried out his now-famous alpha-particle scattering experiment. What surprising observation did he make in this experiment and what conclusion did he draw from it?

3. What did the discovery of radioactivity reveal about the structure of atoms?

4. What scientific instrument was used to discover that not all atoms of neon have the same mass?

5. If the nucleus of an atom were the size of a medium-sized orange (say, with a diameter of about 6 cm), what would be the diameter of the atom?

6. ■ If a gold atom has a radius of 145 pm, and you could string gold atoms like beads on a thread, how many atoms would you need to have a necklace 36 cm long?

7. The volcanic eruption of Mount St. Helens in the state of Washington in 1980 produced a considerable quantity of a radioactive element in the gaseous state. The element has atomic number 86. What are the symbol and name of this element?

8. Titanium and thallium have symbols that are easily confused with each other. Give the symbol, atomic number, atomic weight, and group and period number of each element. Are they metals, metalloids, or nonmetals?

9. Give the mass number of each of the following atoms: (a) magnesium with 15 neutrons, (b) titanium with 26 neutrons, and (c) zinc with 32 neutrons.

10. Give the mass number of (a) a nickel atom with 31 neutrons, (b) a plutonium atom with 150 neutrons, and (c) a tungsten atom with 110 neutrons.

11. Give the complete symbol ($^A_Z X$) for each of the following atoms: (a) potassium with 20 neutrons, (b) krypton with 48 neutrons, and (c) cobalt with 33 neutrons.

12. ■ Give the complete symbol ($^A_Z X$) for each of the following atoms: (a) fluorine with 10 neutrons, (b) chromium with 28 neutrons, and (c) xenon with 78 neutrons.

13. How many electrons, protons, and neutrons are there in an atom of (a) magnesium-24, ^{24}Mg; (b) tin-119, ^{119}Sn; and (c) thorium-232, ^{232}Th?

14. ■ How many electrons, protons, and neutrons are there in an atom of (a) carbon-13, ^{13}C; (b) copper-63, ^{63}Cu; and (c) bismuth-205, ^{205}Bi?

Isotopes

(See Example 2.2 and the General Chemistry Now Screen 2.12.)

15. The synthetic radioactive element technetium is used in many medical studies. Give the number of electrons, protons, and neutrons in an atom of technetium-99.

16. Radioactive americium-241 is used in household smoke detectors and in bone mineral analysis. Give the number of electrons, protons, and neutrons in an atom of americium-241.

17. Cobalt has three radioactive isotopes used in medical studies. Atoms of these isotopes have 30, 31, and 33 neutrons, respectively. Give the symbol for each of these isotopes.

18. Which of the following are isotopes of element X, the atomic number for which is 9: $^{19}_{9}X$, $^{20}_{9}X$, $^{9}_{18}X$, and $^{21}_{9}X$?

Isotope Abundance and Atomic Mass

(See Exercises 2.3 and 2.4 and the General ChemistryNow Screen 2.13.)

19. Thallium has two stable isotopes, ^{203}Tl and ^{205}Tl. Knowing that the atomic weight of thallium is 204.4, which isotope is the more abundant of the two?

20. ■ Strontium has four stable isotopes. Strontium-84 has a very low natural abundance, but ^{86}Sr, ^{87}Sr, and ^{88}Sr are all reasonably abundant. Knowing that the atomic weight of strontium is 87.62, which of the more abundant isotopes predominates?

21. Verify that the atomic mass of lithium is 6.94, given the following information:
^6Li, mass = 6.015121 u; percent abundance = 7.50%
^7Li, mass = 7.016003 u; percent abundance = 92.50%

22. ■ Verify that the atomic mass of magnesium is 24.31, given the following information:
^{24}Mg, mass = 23.985042 u; percent abundance = 78.99%
^{25}Mg, mass = 24.985837 u; percent abundance = 10.00%
^{26}Mg, mass = 25.982593 u; percent abundance = 11.01%

23. Silver (Ag) has two stable isotopes, ^{107}Ag and ^{109}Ag. The isotopic mass of ^{107}Ag is 106.9051 and the isotopic mass of ^{109}Ag is 108.9047. The atomic weight of Ag, from the periodic table, is 107.868. Estimate the percentage of ^{107}Ag in a sample of the element.

    (a) 0%   (b) 25%   (c) 50%   (d) 75%

24. Copper exists as two isotopes: ^{63}Cu (62.9298 u) and ^{65}Cu (64.9278 u). What is the approximate percentage of ^{63}Cu in samples of this element?

    (a) 10%   (c) 50%   (e) 90%

    (b) 30%   (d) 70%

25. ■ Gallium has two naturally occurring isotopes, ^{69}Ga and ^{71}Ga, with masses of 68.9257 u and 70.9249 u, respectively. Calculate the percent abundances of these isotopes of gallium.

26. Antimony has two stable isotopes, ^{121}Sb and ^{123}Sb, with masses of 120.9038 u and 122.9042 u, respectively. Calculate the percent abundances of these isotopes of antimony.

Atoms and the Mole

(See Examples 2.5–2.7 and the General ChemistryNow Screens 2.14 and 2.15.)

27. ■ Calculate the mass, in grams, of the following:

    (a) 2.5 mol of aluminum

    (b) 1.25×10^{-3} mol of iron

    (c) 0.015 mol of calcium

    (d) 653 mol of neon

28. Calculate the mass, in grams, of

    (a) 4.24 mol of gold

    (b) 15.6 mol of He

    (c) 0.063 mol of platinum

    (d) 3.63×10^{-4} mol of Pu

29. ■ Calculate the amount (moles) represented by each of the following:

    (a) 127.08 g of Cu

    (b) 0.012 g of lithium

    (c) 5.0 mg of americium

    (d) 6.75 g of Al

30. Calculate the amount (moles) represented by each of the following:

    (a) 16.0 g of Na

    (b) 0.876 g of tin

    (c) 0.0034 g of platinum

    (d) 0.983 g of Xe

31. ■ You are given 1.0-g samples of He, Fe, Li, Si, and C. Which sample contains the largest number of atoms? Which contains the smallest?

32. You are given 1.0-mol amounts of He, Fe, Li, Si, and C. Which sample has the largest mass?

33. ■ What is the average mass of one copper atom?

34. What is the average mass of one titanium atom?

The Periodic Table

(See Section 2.6 and Exercise 2.7. See also the Periodic Table Tool on the General ChemistryNow CD-ROM or website.)

35. Give the name and symbol of each of the Group 5A elements. Tell whether each is a metal, nonmetal, or metalloid.

36. Give the name and symbol of each of the fourth-period elements. Tell whether each is a metal, nonmetal, or metalloid.

37. How many periods of the periodic table have 8 elements, how many have 18 elements, and how many have 32 elements?

38. ■ How many elements occur in the seventh period? What is the name given to the majority of these elements and what well-known property characterizes them?

39. ■ Select answers to the questions listed below from the following list elements whose symbols start with the letter C: C, Ca, Cr, Co, Cd, Cl, Cs, Ce, Cm, Cu, and Cf. (You should expect to use some symbols more than once.)

    (a) Which are nonmetals?

    (b) Which are main group elements?

    (c) Which are lanthanides?

    (d) Which are transition elements?

    (e) Which are actinides?

    (f) Which are gases?

40. Give the name and chemical symbol for the following.

    (a) a nonmetal in the second period

    (b) an alkali metal

    (c) the third-period halogen

    (d) an element that is a gas at 20°C and 1 atmosphere pressure

41. ■ Classify the following elements as metals, metalloids, or nonmetals: N, Na, Ni, Ne, and Np.

42. Here are symbols for five of the seven elements whose names begin with the letter B: B, Ba, Bk, Bi, and Br. Match each symbol with one of the descriptions below.

    (a) a radioactive element

    (b) a liquid at room temperature

    (c) a metalloid

    (d) an alkaline earth element

    (e) a Group 5A element

43. Use the elements in the following list to answer the questions: sodium, silicon, sulfur, scandium, selenium, strontium, silver, and samarium. (Some elements will be entered in more than one category.)

    (a) Identify those that are metals.

    (b) Identify those that are main group elements

    (c) Identify those that are transition metals.

44. Compare the elements silicon (Si) and phosphorus (P) using the following criteria:

    (a) metal, metalloid, or nonmetal

    (b) possible conductor of electricity

    (c) physical state at 25 °C (solid, liquid, or gas)

▲ More challenging   ■ In General ChemistryNow   **Blue-numbered questions** answered in Appendix 0

General Questions

These questions are not designed as to type or location in the chapter. They may combine several concepts. More challenging questions are marked with the icon ▲.

45. Fill in the blanks in the table (one column per element).

Symbol	^{58}Ni	^{33}S	____	____
Number of protons	____	____	10	____
Number of neutrons	____	____	10	30
Number of electrons in the neutral atom	____	____	____	25
Name of element	____	____	____	____

46. Fill in the blanks in the table (one column per element).

Symbol	^{65}Cu	^{86}Kr	____	____
Number of protons	____	____	78	____
Number of neutrons	____	____	117	46
Number of electrons in the neutral atom	____	____	____	35
Name of element	____	____	____	____

47. ■ Potassium has three naturally occurring isotopes (^{39}K, ^{40}K, and ^{41}K), but ^{40}K has a very low natural abundance. Which of the other two isotopes is the more abundant? Briefly explain your answer.

48. *Crossword Puzzle:* In the 2×2 box shown here, each answer must be correct four ways: horizontally, vertically, diagonally, and by itself. Instead of words, use symbols of elements. When the puzzle is complete, the four spaces will contain the overlapping symbols of ten elements. There is only one correct solution.

1	2
3	4

Horizontal
1–2: Two-letter symbol for a metal used in ancient times
3–4: Two-letter symbol for a metal that burns in air and is found in Group 5A

Vertical
1–3: Two-letter symbol for a metalloid
2–4: Two-letter symbol for a metal used in U.S. coins

Single squares: all one-letter symbols
1: A colorful nonmetal
2: Colorless gaseous nonmetal
3: An element that makes fireworks green
4: An element that has medicinal uses

Diagonal
1–4: Two-letter symbol for an element used in electronics
2–3: Two-letter symbol for a metal used with Zr to make wires for superconducting magnets

This puzzle first appeared in *Chemical & Engineering News,* p. 86, December 14, 1987 (submitted by S. J. Cyvin) and in *Chem Matters,* October 1988.

49. ■ The chart shown in the *Stardust* story (page 58) plots the logarithm of the abundance of elements 1 through 30 in the solar system on a logarithmic scale.
  (a) What is the most abundant main group metal?
  (b) What is the most abundant nonmetal?
  (c) What is the most abundant metalloid?
  (d) Which of the transition elements is most abundant?
  (e) Which halogens are included on this plot and which is the most abundant?

50. The molecule buckminsterfullerene, commonly called a "buckyball," is one of three common allotropes of a familiar element. Identify two other allotropes of this element.

51. Which of the following is impossible?
  (a) silver foil that is 1.2×10^{-4} m thick
  (b) a sample of potassium that contains 1.784×10^{24} atoms
  (c) a gold coin of mass 1.23×10^{-3} kg
  (d) 3.43×10^{-27} mol of S_8

52. Give the symbol for a metalloid in the third period and then identify a property of this element.

53. Reviewing the periodic table.
  (a) Name an element in Group 2A.
  (b) Name an element in the third period.
  (c) Which element is in the second period in Group 4A?
  (d) Which element is in the third period in Group 6A?
  (e) Which halogen is in the fifth period?
  (f) Which alkaline earth element is in the third period?
  (g) Which noble gas element is in the fourth period?
  (h) Name the nonmetal in Group 6A and the third period.
  (i) Name a metalloid in the fourth period.

54. Reviewing the periodic table:
  (a) Name an element in Group 2B.
  (b) Name an element in the fifth period.
  (c) Which element is in the sixth period in Group 4A?
  (d) Which element is in the third period in Group 6A?
  (e) Which alkali metal is in the third period?
  (f) Which noble gas element is in the fifth period?
  (g) Name the element in Group 6A and the fourth period. Is it a metal, nonmetal, or metalloid?
  (h) Name a metalloid in Group 5A.

55. The plot on the following page shows the variation in density with atomic number for the first 36 elements. Use this plot to answer the following questions:
  (a) Which three elements in this series have the highest density? What is their approximate density? Are these elements metals or nonmetals?

▲ More challenging   ■ In General ChemistryNow   **Blue-numbered questions** answered in Appendix 0

(b) Which element in the second period has the highest density? Which element in the third period has the highest density? What do these two elements have in common?

(c) Some elements have densities so low that they do not show up on the plot. What elements are these? What property do they have in common?

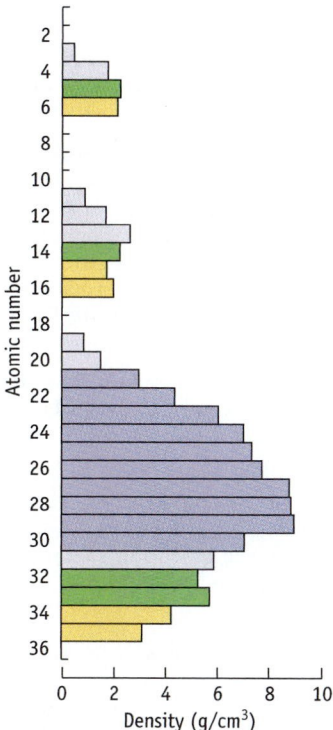

Density (g/cm³)

56. ■ Give two examples of nonmetallic elements that have allotropes. Name those elements and describe the allotropes of each.

57. ■ In each case, decide which represents more mass:
(a) 0.5 mol of Na or 0.5 mol of Si
(b) 9.0 g of Na or 0.50 mol of Na
(c) 10 atoms of Fe or 10 atoms of K

58. A semiconducting material is composed of 52 g of Ga, 9.5 g of Al, and 112 g of As. Which element has the largest number of atoms in the final mixture?

59. You are given 15 g each of yttrium, boron, and copper. Which sample represents the largest number of atoms?

60. Lithium has two stable isotopes: ^6Li and ^7Li. One of them has an abundance of 92.5%, and the other has an abundance of 7.5%. Knowing that the atomic mass of lithium is 6.941, which is the more abundant isotope?

61. Superman comes from the planet Krypton. If you have 0.00789 g of the gaseous element krypton, how many moles does this represent? How many atoms?

62. The recommended daily allowance (RDA) of iron in your diet is 15 mg. How many moles is this? How many atoms?

63. Put the following elements in order from smallest to largest mass:
(a) 3.79×10^{24} atoms Fe   (e) 9.221 mol Na
(b) 19.921 mol H_2 (f) 4.07×10^{24} atoms Al
(c) 8.576 mol C   (g) 9.2 mol Cl_2
(d) 7.4 mol Si

64. ■ ▲ When a sample of phosphorus burns in air, the compound P_4O_{10} forms. One experiment showed that 0.744 g of phosphorus formed 1.704 g of P_4O_{10}. Use this information to determine the ratio of the atomic masses of phosphorus and oxygen (mass P/mass O). If the atomic mass of oxygen is assumed to be 16.000 u, calculate the atomic mass of phosphorus.

65. ■ The data below were collected in a Millikan oil drop experiment.

Oil Drop	Measured Charge on Drop (C)
1	1.59×10^{-19}
2	11.1×10^{-19}
3	9.54×10^{-19}
4	15.9×10^{-19}
5	6.36×10^{-19}

(a) Use these data to calculate the charge on the electron (in coulombs).

(b) How many electrons have accumulated on each oil drop?

(c) The accepted value of the electron charge is 1.60×10^{-19} C. Calculate the percent and error for the value determined by the data in the table.

66. ▲ Although carbon-12 is now used as the standard for atomic masses, this has not always been the case. Early attempts at classification used hydrogen as the standard, with the mass of hydrogen being set equal to 1.0000 u. Later attempts defined atomic masses using oxygen (with a mass of 16.0000 u). In each instance, the atomic masses of the other elements were defined relative to these masses. (To answer this question, you need more precise data on current atomic masses: H, 1.00794 u; O, 15.9994 u.)

(a) If H = 1.0000 u was used as a standard for atomic masses, what would the atomic mass of oxygen be? What would be the value of Avogadro's number under these circumstances?

(b) Assuming the standard is O = 16.0000 u, determine the value for the atomic mass of hydrogen and the value of Avogadro's number.

67. ■ A reagent occasionally used in chemical synthesis is sodium-potassium alloy. (Alloys are mixtures of metals, and Na-K has the interesting property that it is a liquid.) One formulation of the alloy (the one that melts at the lowest temperature) contains 68 atom percent K; that is, out of every 100 atoms, 68 are K and 32 are Na. What is the weight percent of potassium in sodium-potassium alloy?

68. Mass spectrometric analysis showed that there are four isotopes of an unknown element having the following masses and abundances:

Isotope	Mass Number	Isotope Mass	Abundance (%)
1	136	135.9090	0.193
2	138	137.9057	0.250
3	140	139.9053	88.48
4	142	141.9090	11.07

Three elements in the periodic table that have atomic weights near these values are lanthanum (La), atomic number 57, atomic weight 139.9055; cerium (Ce), atomic number 58, atomic weight 140.115; and praeseodymium (Pr), atomic number 59, atomic weight 140.9076. Using the data above, calculate the atomic weight and identify the element if possible.

Summary and Conceptual Questions

The following questions use concepts from the preceding chapter (Chapter 1).

69. Draw a picture showing the approximate positions of all protons, electrons, and neutrons in an atom of helium-4. Make certain that your diagram indicates both the number and position of each type of particle.

70. Draw two boxes, each about 3 cm on a side. In one box, sketch a representation of iron metal. In the other box, sketch a representation of nitrogen gas. How do these drawings differ?

71. ▲ Identify, from the list below, the information needed to calculate the number of atoms in 1 cm³ of iron. Outline the procedure used in this calculation.
    (a) the structure of solid iron
    (b) the molar mass of iron
    (c) Avogadro's number
    (d) the density of iron
    (e) the temperature
    (f) iron's atomic number
    (g) the number of iron isotopes

72. Consider the plot of relative element abundances on page 58. Is there a relationship between abundance and atomic number? Is there any difference between the relative abundance of an element of even atomic number and the relative abundance of an element of odd atomic number?

73. The photo here depicts what happens when a coil of magnesium ribbon and a few calcium chips are placed in water.

Magnesium (*left*) and calcium (*right*) in water.

Charles D. Winters

(a) Based on their relative reactivities, what might you expect to see when barium, another Group 2A element, is placed in water?

(b) Give the period in which each element (Mg, Ca, and Ba) is found. What correlation do you think you might find between the reactivity of these elements and their positions in the periodic table?

74. ▲ In an experiment, you need 0.125 mol of sodium metal. Sodium can be cut easily with a knife (Figure 2.12), so if you cut out a block of sodium, what should the volume of the block be in cubic centimeters? If you cut a perfect cube, what is the length of the edge of the cube? (The density of sodium is 0.97 g/cm³.)

75. ▲ Dilithium is the fuel for the *Starship Enterprise*. Because its density is quite low, however, you need a large space to store a large mass. To estimate the volume required, we shall use the element lithium. If you need 256 mol for an interplanetary trip, what must the volume of the piece of lithium be? If the piece of lithium is a cube, what is the dimension of an edge of the cube? (The density for the element lithium is 0.534 g/cm³ at 20 °C.)

76. An object is coated with a layer of chromium, 0.015 cm thick. The object has a surface area of 15.3 cm². How many atoms of chromium are used in the coating? (The density of chromium = 7.19 g/cm³.)

77. ■ A cylindrical piece of sodium is 12.00 cm long and has a diameter of 4.5 cm. The density of sodium is 0.971 g/cm³. How many atoms does the piece of sodium contain? (The volume of a cylinder is $V = \pi \times r^2 \times length$.)

78. ▲ Consider an atom of ^{64}Zn.
    (a) Calculate the density of the nucleus in grams per cubic centimeter, knowing that the nuclear radius is 4.8×10^{-6} nm and the mass of the ^{64}Zn atom is 1.06×10^{-22} g. Recall that the volume of a sphere is $(4/3)\pi r^3$.
    (b) Calculate the density of the space occupied by the electrons in the zinc atom, given that the atomic radius is 0.125 nm and the electron mass is 9.11×10^{-28} g.
    (c) Having calculated these densities, what statement can you make about the relative densities of the parts of the atom?

79. ▲ Most standard analytical balances can measure accurately to the nearest 0.0001 g. Assume you have weighed out a 2.0000-g sample of carbon. How many atoms are in this sample? Assuming the indicated accuracy of the measurement, what is the largest number of atoms that can be present in the sample?

80. ▲ To estimate the radius of a lead atom:
    (a) You are given a cube of lead that is 1.000 cm on each side. The density of lead is 11.35 g/cm³. How many atoms of lead are in the sample?
    (b) Atoms are spherical; therefore, the lead atoms in this sample cannot fill all the available space. As an approximation, assume that 60% of the space of the cube is filled with spherical lead atoms. Calculate the volume

of one lead atom from this information. From the calculated volume (V), and the formula $V = \frac{4}{3}(\pi r^3)$, estimate the radius (r) of a lead atom.

81. A jar contains some number of jelly beans. To find out precisely how many are in the jar you could dump them out and count them. How could you estimate their number without counting each one? (Chemists need to do just this kind of "bean counting" when we work with atoms and molecules. They are too small to count one by one, so we have worked out other methods to "count atoms.") (*See General ChemistryNow Screen 2.18, Chemical Puzzler.*)

Do you need a live tutor for homework problems?
Access vMentor at General ChemistryNow at
http://now.brookscole.com/kotz6e
for one-on-one tutoring from a chemistry expert

Charles D. Winters

How many jelly beans are in the jar?

3—Molecules, Ions, and Their Compounds

James D. Watson and Francis Crick. In a photo taken in 1953, Watson (*left*) and Crick (*right*) stand by their model of the DNA double helix. Together with Maurice Wilkins, Watson and Crick received the Nobel Prize in medicine and physiology in 1962.

A. Barrington Brown/Science Source/Photo Researchers, Inc.

DNA: The Most Important Molecule

DNA is the substance in every plant and animal that carries the exact blueprint of that plant or animal. The structure of this molecule, the cornerstone of life, was uncovered in 1953, and James D. Watson, Francis Crick, and Maurice Wilkins shared the 1962 Nobel Prize in medicine and physiology for the work. It was one of the most important scientific discoveries of the 20th century, and the story has recently been told by Watson in his book *The Double Helix*.

When Watson was a graduate student at Indiana University, he had an interest in the gene and said he hoped that its biological role might be solved "without my learning any chemistry." Later, however, he and Crick found out just how useful chemistry can be when they began to unravel the structure of DNA.

Solving important problems requires teamwork among scientists of many kinds so Watson went to Cambridge University in England in 1951. There he met Crick, who, Watson said, talked louder and faster than anyone else. Crick shared Watson's belief in the fundamental importance of DNA, and the pair soon learned that Maurice Wilkins and Rosalind Franklin at King's College in London were using a technique called x-ray crystallography to learn more about DNA's structure. Watson and Crick believed that understanding this structure was crucial to understanding genetics. To solve the structural problem, however, they needed experimental data of the type that could come from the experiments at King's College.

The King's College group was initially reluctant to share their data; and, what is more, they did not seem to share Watson and Crick's sense of urgency. There was also an ethical dilemma: Could Watson and Crick work on a problem that others had claimed as theirs? "The English sense of fair play would not allow Francis to move in on Maurice's problem," said Watson.

See Chapter Goals Revisited (page 130). Test you knowledge of these goals by taking the exam-prep quiz on the General ChemistryNow CD-ROM or website.

- Interpret, predict, and write formulas for ionic and molecular compounds.
- Name compounds.
- Understand some properties of ionic compounds.
- Calculate and use molar mass.
- Calculate percent composition for a compound and derive formulas from experimental data.

Watson and Crick approached the problem through a technique chemists now use frequently—model building. They built models of the pieces of the DNA chain, and they tried various chemically reasonable ways of fitting them together. Finally, they discovered that one arrangement was "too pretty not to be true." Ultimately, the experimental evidence of Wilkins and Franklin confirmed the "pretty structure" to be the real DNA structure. As you will see, chemists often use models to help guide them to experimental evidence that is definitive.

The story of how Watson, Crick, Wilkins, and Franklin ultimately came to share information and insight is an interesting human drama and illustrates how scientific progress is often made. For more on this interesting human and scientific drama, read *Rosalind Franklin: The Dark Lady of DNA* by Brenda Maddox and Watson's book *The Double Helix.*

Rosalind Franklin of King's College, London. She died in 1958 at the age of 37. Because Nobel Prizes are never awarded posthumously, she did not share in this honor with Watson, Crick, and Wilkins.

Watson and Crick recognized early on that the overall structure of DNA was a helix; that is, the atomic-level building blocks formed chains that twisted in space like the strands of a grapevine. They also knew which chemical elements it contained and roughly how they were grouped together. What they did not know was the detailed structure of the helix. By the spring of 1953, however, they had the answer. The atomic-level building blocks of DNA form two chains twisted together in a double helix.

Structure of DNA: Sugar, Phosphate, and Bases

DNA is a very large molecule that consists of two chains of atoms (P, C, and O) that twist together. The P, C, and O atoms are parts of phosphate ions (P) and sugar molecules. The chains are joined by four different molecules (adenine, thymine, guanine, and cytosine) belonging to a general class of molecules called bases.

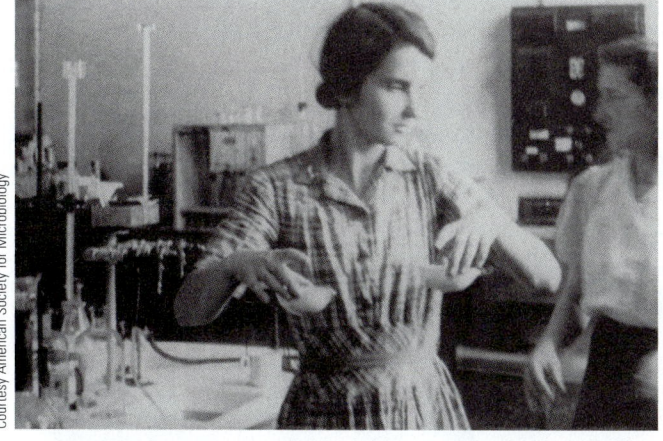

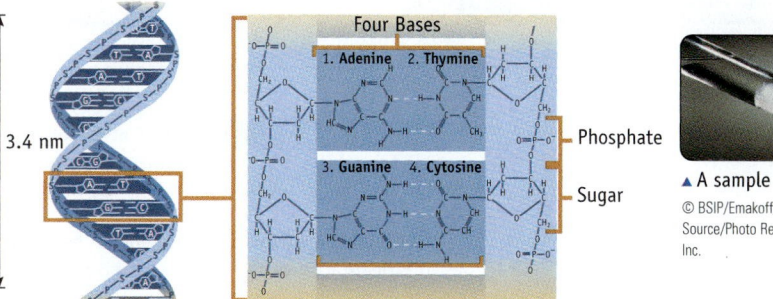

▲ A sample of DNA.
© BSIP/Emakoff/Science Source/Photo Researchers, Inc.

To Review Before You Begin

• Know how to calculate and use molar amounts (Section 2.5)

Throughout the chapter this icon introduces a list of resources on the General ChemistryNow CD-ROM or website (http://now .brookscole.com/kotz6e) that will:

• help you evaluate your knowledge of the material

• provide homework problems

• allow you to take an exam-prep quiz

• provide a personalized Learning Plan targeting resources that address areas you should study

In 1953 the structure of the giant molecule DNA, *deoxyribonucleic acid*, was finally understood (page 96). Chromosomes, which are present in the nuclei of almost all living cells, consist of DNA. Recently discovered knowledge of the human genome, which is the complete structure of the DNA in every one of our 23 chromosomes, is widely expected to revolutionize the practice of medicine. To comprehend modern molecular biology—indeed all of modern chemistry—we have to understand the structures and properties of molecules. This chapter marks the beginning of our attempt to acquaint you with this important subject.

3.1—Molecules, Compounds, and Formulas

A *molecule* is the smallest identifiable unit into which a pure substance like sugar and water can be divided and still retain the composition and chemical properties of the substance. Such substances are composed of identical molecules consisting of atoms of two or more elements bound firmly together. For example, atoms of the element aluminum, Al, combine with molecules of the element bromine, Br_2, to produce the compound aluminum bromide, Al_2Br_6 (Figure 3.1).

$$2 \; Al(s) + 3 \; Br_2(\ell) \longrightarrow Al_2Br_6(s)$$

aluminum + bromine $\longrightarrow$ aluminum bromide

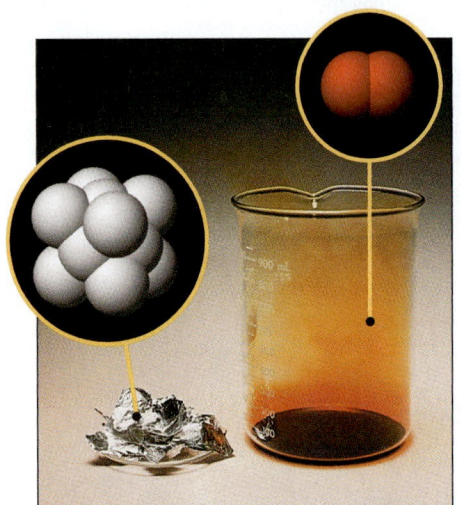

(a)

(b)

(c)

Photos: Charles D. Winters

Active Figure 3.1   **Reaction of the elements aluminum and bromine.** (a) Solid aluminum and (in the beaker) liquid bromine. (b) When the aluminum is added to the bromine, a vigorous chemical reaction produces white, solid aluminum bromide, Al_2Br_6 (c).

Chemistry·Now™ See the General ChemistryNow CD-ROM or website to explore an interactive version of this figure accompanied by an exercise.

NAME	MOLECULAR FORMULA	CONDENSED FORMULA	STRUCTURAL FORMULA	MOLECULAR MODEL
Ethanol	C_2H_6O	CH_3CH_2OH	(structural formula)	(molecular model)
Dimethyl ether	C_2H_6O	CH_3OCH_3	(structural formula)	(molecular model)

Figure 3.2   Four approaches to showing molecular formulas. Here the two molecules have the same molecular formula. However, once they are written as condensed or structural formulas, and illustrated with a molecular model, it is clear that these molecules are different. (*See the General ChemistryNow Screen 3.4 Representing Compounds, for a tutorial on identifying molecular representations.*)

To describe this chemical change (or chemical reaction) on paper, the composition of each element and compound is represented by a symbol or formula. Here one molecule of Al_2Br_6 is composed of two Al atoms and six Br atoms.

How do compounds differ from elements? When a compound is produced from its elements, the characteristics of the constituent elements are lost. Solid, metallic aluminum and red-orange liquid bromine, for example, react to form Al_2Br_6, a white solid (see Figure 3.1).

Formulas

For molecules more complicated than water, there is often more than one way to write the formula. For example, the formula of ethanol (also called ethyl alcohol) can be represented as C_2H_6O (Figure 3.2). This **molecular formula** describes the composition of ethanol molecules—two carbon atoms, six hydrogen atoms, and one atom of oxygen occur per molecule—but it gives us no structural information. Structural information—how the atoms are connected and how the molecule fills space—is important, however, because it helps us understand how a molecule can interact with other molecules, which is the essence of chemistry.

To provide some structural information, it is useful to write a **condensed formula**, which indicates how certain atoms are grouped together. For example, the condensed formula of ethanol, CH_3CH_2OH (see Figure 3.2), informs us that the molecule consists of three "groups": a CH_3 group, a CH_2 group, and an OH group. Writing the formula as CH_3CH_2OH also shows that the compound is not dimethyl ether, CH_3OCH_3, a compound with the same molecular formula but a different structure and distinctly different properties.

That ethanol and dimethyl ether are different molecules is further apparent from their **structural formulas** (see Figure 3.2). This type of formula gives us an even higher level of structural detail, showing how all of the atoms are attached within a molecule. The lines between atoms represent the chemical bonds that hold atoms together in this molecule [▶ Chapters 9 and 10].

■ **Standard Colors for Atoms in Molecular Models**
The colors listed here are used in this book and are generally used by chemists. The colors of some common atoms are:

carbon atoms

hydrogen atoms

oxygen atoms

nitrogen atoms

chlorine atoms

■ **Isomers**
Compounds having the same molecular formula but different structures are called *isomers*. (*See Chapter 11 and General ChemistryNow Screen 3.4 Representing Compounds.*)

Example 3.1—Molecular Formulas

Problem The acrylonitrile molecule is the building block for acrylic plastics (such as Orlon and Acrilan). Its structural formula is shown here. What is the molecular formula for acrylonitrile?

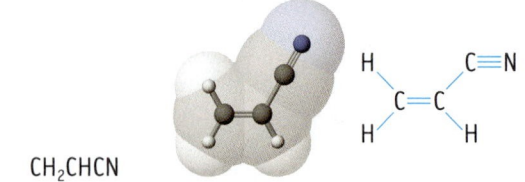

CH$_2$CHCN

Condensed formula   Molecular model   Structural formula

Strategy Count the number of atoms of each type.

Solution Acrylonitrile has three C atoms, three H atoms, and one N atom. Therefore, its molecular formula is C$_3$H$_3$N.

Comment When writing molecular formulas of organic compounds (compounds with C, H, and other elements) the convention is to write C first, then H, and finally other elements in alphabetical order.

Exercise 3.1—Molecular Formulas

The styrene molecule is the building block of polystyrene, a material used for drinking cups and building insulation. What is the molecular formula of styrene?

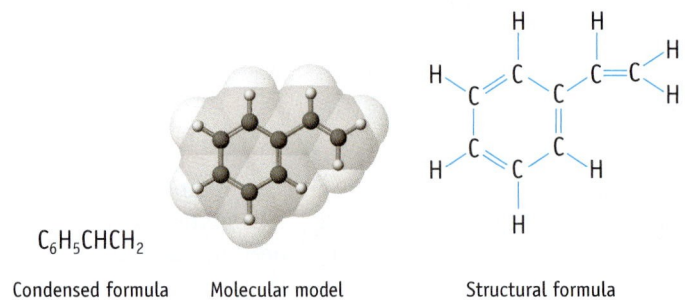

C$_6$H$_5$CHCH$_2$

Condensed formula   Molecular model   Structural formula

3.2—Molecular Models

Molecular structures are often beautiful in the same sense that art is beautiful. For example, there is something intrinsically beautiful about the pattern created by water molecules assembled in ice (Figure 3.3).

More important, however, is the fact that the physical and chemical properties of a molecular compound are often closely related to its structure. For example, two well-known features of ice are easily related to its structure. The first is the shape of ice crystals: The sixfold symmetry of macroscopic ice crystals also appears at the particulate level in the form of six-sided rings of hydrogen and oxygen atoms. The second is water's unique property of being less dense when solid than it is when liquid. The lower density of ice, which has enormous consequences for earth's climate, results from the fact that molecules of water are not packed together tightly.

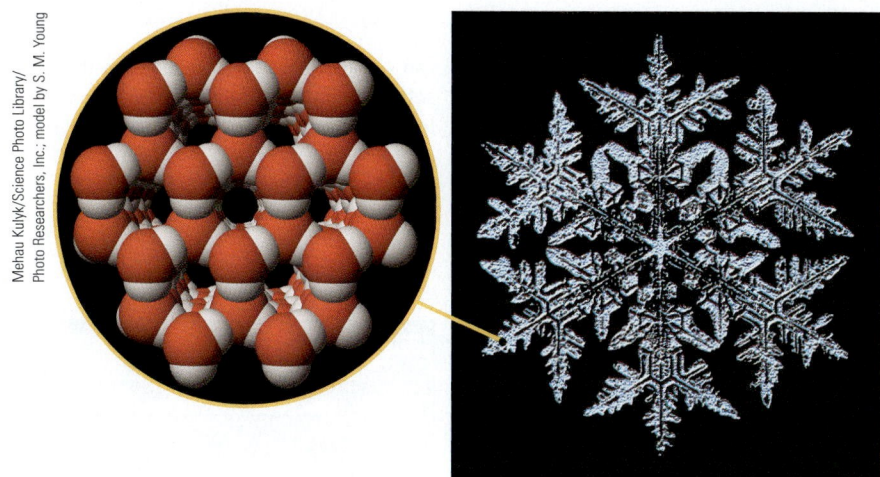

Mehau Kulyk/Science Photo Library/
Photo Researchers, Inc.; model by S. M. Young

Figure 3.3 Ice. Snowflakes are six-sided structures, reflecting the underlying structure of ice. Ice consists of six-sided rings formed by water molecules, in which each side of a ring consists of two O atoms and an H atom.

Because molecules are three-dimensional, it is often difficult to represent their shapes on paper. Certain conventions have been developed, however, that help represent three-dimensional structures on two-dimensional surfaces. Simple perspective drawings are often used (Figure 3.4).

Wood or plastic models are also a useful way of representing molecular structure. These models can be held in the hand and rotated to view all parts of the molecule.

Several kinds of molecular models exist. In the **ball-and-stick model**, spheres, usually in different colors, represent the atoms, and sticks represent the bonds holding them together. These models make it easy to see how atoms are attached to one another. Molecules can also be represented using **space-filling models**. These models are more realistic because they offer a better representation of relative sizes of atoms and their proximity to each other when in a molecule. A disadvantage of pictures of space-filling models is that atoms can often be hidden from view.

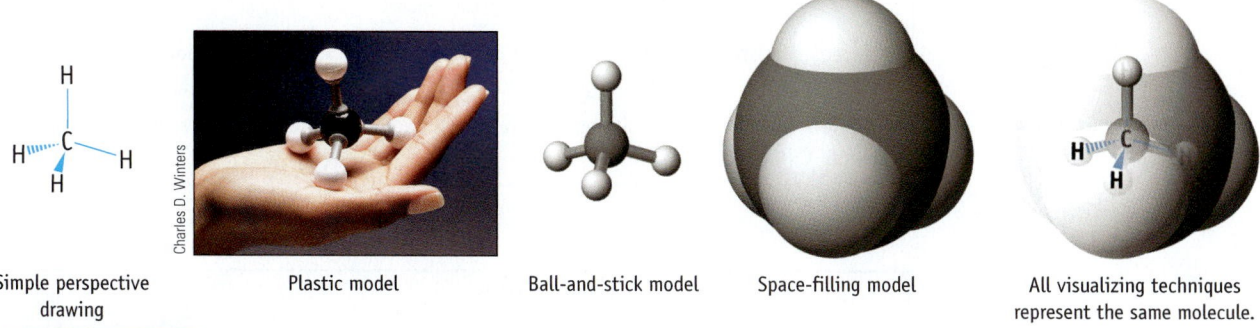

Simple perspective drawing   Plastic model   Ball-and-stick model   Space-filling model   All visualizing techniques represent the same molecule.

Charles D. Winters

Active Figure 3.4   Ways of depicting the methane (CH_4) molecule.

Chemistry Now™ See the General ChemistryNow CD-ROM or website to explore an interactive version of this figure accompanied by an exercise.

A Closer Look

Computer Resources for Molecular Modeling

With the availability of relatively low-cost, high-powered computers, the use of molecular modeling programs has become common. Although the computer screen is two-dimensional, the perspective drawings obtained from molecular-modeling programs are usually quite good. In addition, most programs offer an option to rotate the model on the computer screen to allow the viewer to see the structure from any desired angle. Both ball-and-stick and space-filling representations can be portrayed. Most of the drawings in this book were prepared with the commercial molecular modeling software from CAChe/Fujitsu.

The General ChemistryNow CD-ROM includes a program for visualizing molecules and for measuring atom–atom distances and angles.

The site on the World Wide Web for this textbook (http://www.brookscole.com) contains a link to RasMol and Chime, molecular visualization software. Models of many of the compounds mentioned in this book are available through the General ChemistryNow CD-ROM and website. You can visualize these molecules using the software on the CD-ROM, or, if you download RasMol or Chime and configure your browser properly, you can download files from the Internet will that allow you to visualize these models on your own computer.

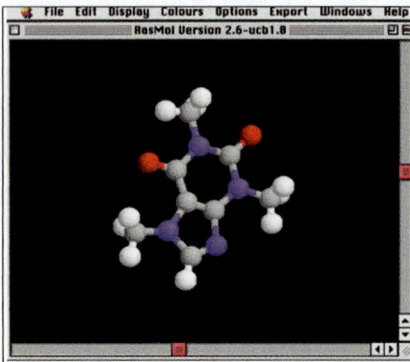 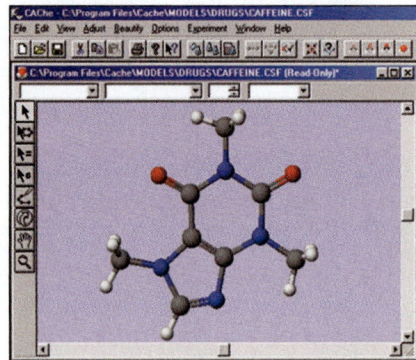

A model of caffeine as viewed with RasMol (*left*) and the CAChe/Fujitsu software (*right*).

Example 3.2—Using Molecular Models

Problem A model of uracil, an important biological molecule, is given here. Write its molecular formula.

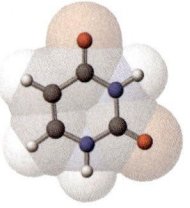

Molecular model

Strategy The standard color codes used for the atoms are as follows: carbon atoms = gray; hydrogen atoms = white; nitrogen atoms = blue; and oxygen atoms = red.

Solution Uracil has four C atoms, four H atoms, two N atoms, and two O atoms, giving a formula of $C_4H_4N_2O_2$.

Exercise 3.2—Formulas of Molecules

Cysteine, whose molecular model and structural formula are illustrated here, is an important amino acid and a constituent of many living things. What is its molecular formula? See Example 3.2 and page 99 for the color coding of the model.

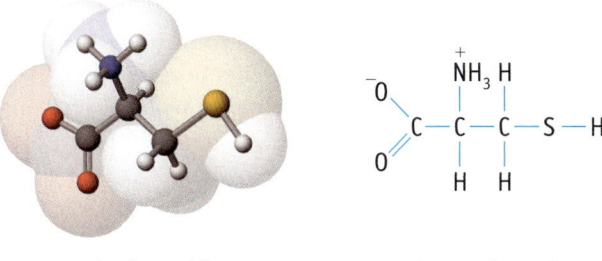

Molecular model   Structural formula

3.3—Ionic Compounds: Formulas, Names, and Properties

The compounds you have encountered so far in this chapter are molecular compounds—that is, compounds that consist of discrete molecules at the particulate level. **Ionic compounds** constitute another major class of compounds. They consist of *ions*, atoms or groups of atoms that bear a positive or negative electric charge. Many familiar compounds are composed of ions (Figure 3.5). Table salt, or sodium chloride ($NaCl$), and lime (CaO) are just two. To recognize ionic compounds, and to be able to write formulas for these compounds, it is important to know the formulas and charges of common ions. You also need to know the names of ions and be able to name the compounds they form.

Hematite, Fe_2O_3

Calcite, $CaCO_3$

Gypsum, $CaSO_4 \cdot 2 H_2O$

Fluorite, CaF_2

Orpiment, As_2S_3

Charles D. Winters

Common Name	Name	Formula	Ions Involved
Calcite	Calcium carbonate	$CaCO_3$	Ca^{2+}, CO_3^{2-}
Fluorite	Calcium fluoride	CaF_2	Ca^{2+}, F^-
Gypsum	Calcium sulfate dihydrate	$CaSO_4 \cdot 2 H_2O$	Ca^{2+}, SO_4^{2-}
Hematite	Iron(III) oxide	Fe_2O_3	Fe^{3+}, O^{2-}
Orpiment	Arsenic sulfide	As_2S_3	As^{3+}, S^{2-}

Figure 3.5 **Some common ionic compounds.**

Ions

Atoms of many elements can lose or gain electrons in the course of a chemical reaction. To be able to predict the outcome of chemical reactions [▶ Section 5.6], you need to know whether an element will likely gain or lose electrons and, if so, how many.

Cations

If an atom loses an electron (which is transferred to an atom of another element in the course of a reaction), the atom now has one fewer negative electrons than it has positive protons in the nucleus. The result is a positively charged ion called a **cation** (see Figure 3.6). (The name is pronounced "cat′-ion.") Because it has an excess of one positive charge, we write the cation's symbol as, for example, Li^+:

$$Li \text{ atom} \longrightarrow e^- + Li^+ \text{ cation}$$
(3 protons and 3 electrons)   (3 protons and 2 electrons)

■ **Writing Ion Formulas**
When writing the formula of an ion, the charge on the ion must be included.

Anions

Conversely, if an atom gains one or more electrons, there is now one or more negatively charged electrons than protons. The result is an **anion** (see Figure 3.6). (The name is pronounced "ann′-ion.")

$$O \text{ atom} + 2 e^- \longrightarrow O^{2-} \text{ anion}$$
(8 protons and 8 electrons)   (8 protons and 10 electrons)

Here the O atom has gained two electrons so we write the anion's symbol as O^{2-}.

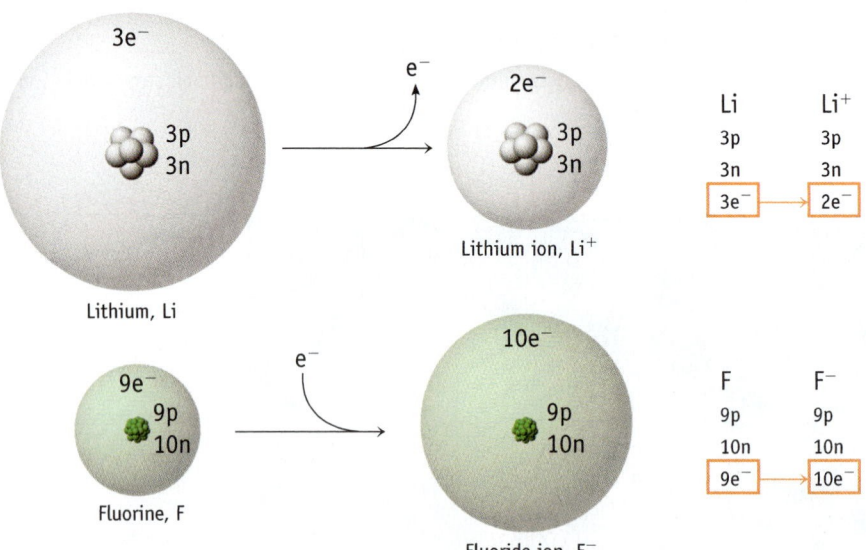

Active Figure 3.6   **Ions.** A lithium-6 atom is electrically neutral because the number of positive charges (three protons) and negative charges (three electrons) are the same. When it loses one electron, it has one more positive charge than negative charge, so it has a net charge of 1+. We symbolize the resulting lithium cation as Li^+. A fluorine atom is also electrically neutral, having nine protons and nine electrons. A fluorine atom can acquire an electron to produce a F^- anion. This anion has one more electron than it has protons, so it has a net charge of 1−.

GENERAL
Chemistry⚛Now™ See the General ChemistryNow CD-ROM or website to explore an interactive version of this figure accompanied by an exercise.

How do you know whether an atom is likely to form a cation or an anion? It depends on whether the element is a metal or a nonmetal.

- Metals generally lose electrons in the course of their reactions to form cations.

- Nonmetals frequently gain one or more electrons to form anions in the course of their reactions.

Monatomic Ions

Monatomic ions are single atoms that have lost or gained electrons. As indicated in Figure 3.7, metals typically lose electrons to form monatomic cations, and nonmetals typically gain electrons to form monatomic anions.

How can you predict the number of electrons gained or lost? Typical charges on such ions are indicated in Figure 3.7. *Metals of Groups 1A–3A form positive ions having a charge equal to the group number of the metal.*

Group	Metal Atom	Electron Change		Resulting Metal Cation
1A	Na (11 protons, 11 electrons)	-1	$\longrightarrow$	Na^+ (11 protons, 10 electrons)
2A	Ca (20 protons, 20 electrons)	-2	$\longrightarrow$	Ca^{2+} (20 protons, 18 electrons)
3A	Al (13 protons, 13 electrons)	-3	$\longrightarrow$	Al^{3+} (13 protons, 10 electrons)

Transition metals (B-group elements) also form cations. Unlike the A-group metals, however, no easily predictable pattern of behavior occurs for transition metal cations. In addition, many transition metals form several different ions. An iron-containing compound, for example, may contain either Fe^{2+} or Fe^{3+} ions. Indeed, 2+ and 3+ ions are typical of many transition metals (see Figure 3.7).

Group	Metal Atom	Electron Change		Resulting Metal Cation
7B	Mn (25 protons, 25 electrons)	-2	$\longrightarrow$	Mn^{2+} (25 protons, 23 electrons)
8B	Fe (26 protons, 26 electrons)	-2	$\longrightarrow$	Fe^{2+} (26 protons, 24 electrons)
8B	Fe (26 protons, 26 electrons)	-3	$\longrightarrow$	Fe^{3+} (26 protons, 23 electrons)

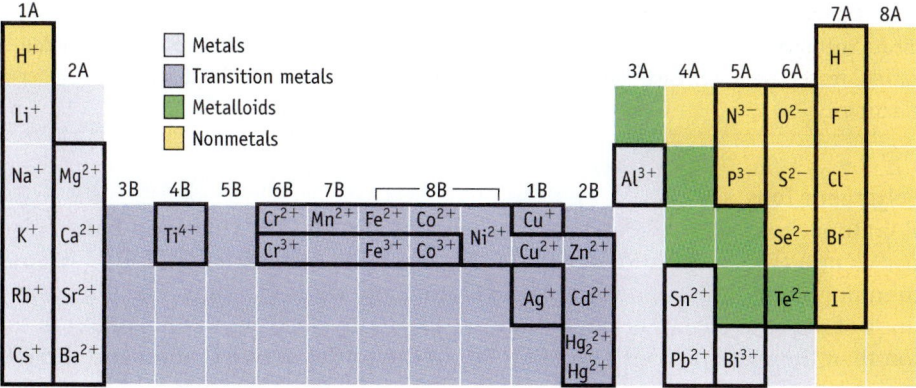

Figure 3.7   Charges on some common monatomic cations and anions. Metals usually form cations and nonmetals usually form anions. (The boxed areas show ions of identical charge.)

Nonmetals often form ions having a negative charge equal to 8 minus the group number of the element. For example, nitrogen is in Group 5A, so it forms an ion having a charge of 3− because a nitrogen atom can gain three electrons.

Group	Nonmetal Atom	Electron Change	Resulting Nonmetal Anion
5A	N (7 protons, 7 electrons)	+3 (= 8 − 5) $\longrightarrow$	N^{3-} (7 protons, 10 electrons)
6A	S (16 protons, 16 electrons)	+2 (= 8 − 6) $\longrightarrow$	S^{2-} (16 protons, 18 electrons)
7A	Br (35 protons, 35 electrons)	+1 (= 8 − 7) $\longrightarrow$	Br^- (35 protons, 36 electrons)

Notice that hydrogen appears at two locations in Figure 3.7. The H atom can either lose or gain electrons, depending on the other atoms it encounters.

Electron lost: H (1 proton, 1 electron) $\longrightarrow$ H^+ (1 proton, 0 electrons) + e^-
Electron gained: H (1 proton, 1 electron) + e^- $\longrightarrow$ H^- (1 proton, 2 electrons)

Finally, the noble gases do not form monatomic cations or anions in chemical reactions.

Ion Charges and the Periodic Table

■ **Cation Charges and the Periodic Table**

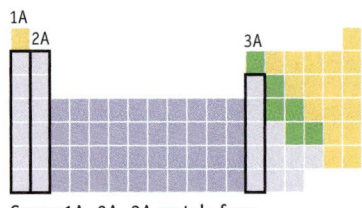

Group 1A, 2A, 3A metals form M^{n+} cations where n = group number.

As illustrated in Figure 3.7, the metals of Groups 1A, 2A, and 3A form ions having 1+, 2+, and 3+ charges; that is, their atoms lose one, two, or three electrons, respectively. *For cations formed from A-group elements, the number of electrons remaining on the ion is the same as the number of electrons in an atom of the noble gas that precedes it in the periodic table.* For example, Mg^{2+} has 10 electrons, the same number as in an atom of the noble gas neon (atomic number 10).

An atom of a nonmetal near the right side of the periodic table would have to lose a great many electrons to achieve the same number as a noble gas atom of lower atomic number. (For instance, Cl, whose atomic number is 17, would have to lose 7 electrons to have the same number of electrons as Ne.) If a nonmetal atom were to gain just a few electrons, however, it would have the same number as a noble gas atom of higher atomic number. For example, an oxygen atom has eight electrons. By gaining two electrons per atom it forms O^{2-}, which has ten electrons, the same number as neon. *Anions having the same number of electrons as the noble gas atom succeeding it in the periodic table are commonly observed in chemical compounds.*

Exercise 3.3—Predicting Ion Charges

Predict formulas for monatomic ions formed from (a) K, (b) Se, (c) Ba, and (d) Cs. In each case indicate the number of electrons gained or lost by an atom of the element in forming the anion or cation, respectively. For each ion, indicate the noble gas atom having the same total number of electrons.

Polyatomic Ions

Polyatomic ions are made up of two or more atoms, and the collection has an electric charge (Figure 3.8 and Table 3.1). For example, carbonate ion, CO_3^{2-}, a common polyatomic anion, consists of one C atom and three O atoms. The ion has two units of negative charge because there are two more electrons (a total of 32) in the ion than there are protons (a total of 30) in the nuclei of one C atom and three O atoms.

A common polyatomic cation is NH_4^+, the ammonium ion. In this case, four H atoms surround an N atom, and the ion has a 1+ electric charge. This ion has ten

Calcite, $CaCO_3$
Calcium carbonate

CO_3^{2-}

Apatite, $Ca_5F(PO_4)_3$
Calcium fluorophosphate

PO_4^{3-}

Celestite, $SrSO_4$
Strontium sulfate

SO_4^{2-}

| **Active Figure 3.8** | **Common ionic compounds based on polyatomic ions.** |

GENERAL
Chemistry⚛Now™ See the General ChemistryNow CD-ROM or website to explore an interactive version of this figure accompanied by an exercise.

electrons, but there are 11 positively charged protons in the nuclei of the N and H atoms (seven and one each, respectively).

Table 3.1   Formulas and Names of Some Common Polyatomic Ions

Formula	Name	Formula	Name
CATION: Positive Ion			
NH_4^+	ammonium ion		
ANIONS: Negative Ions			
Based on a Group 4A element		**Based on a Group 7A element**	
CN^-	cyanide ion	ClO^-	hypochlorite ion
$CH_3CO_2^-$	acetate ion	ClO_2^-	chlorite ion
CO_3^{2-}	carbonate ion	ClO_3^-	chlorate ion
HCO_3^-	hydrogen carbonate ion (or bicarbonate ion)	ClO_4^-	perchlorate ion
Based on a Group 5A element		**Based on a transition metal**	
NO_2^-	nitrite ion	CrO_4^{2-}	chromate ion
NO_3^-	nitrate ion	$Cr_2O_7^{2-}$	dichromate ion
PO_4^{3-}	phosphate ion	MnO_4^-	permanganate ion
HPO_4^{2-}	hydrogen phosphate ion		
$H_2PO_4^-$	dihydrogen phosphate ion		
Based on a Group 6A element			
OH^-	hydroxide ion		
SO_3^{2-}	sulfite ion		
SO_4^{2-}	sulfate ion		
HSO_4^-	hydrogen sulfate ion (or bisulfate ion)		

Formulas of Ionic Compounds

Ionic compounds are composed of ions. For an ionic compound to be electrically neutral—to have no net charge—the numbers of positive and negative ions must be such that the positive and negative charges balance. In sodium chloride, the sodium ion has a 1+ charge (Na^+) and the chloride ion has a 1− charge (Cl^-). These ions must be present in a 1 : 1 ratio, and the formula is NaCl.

Aluminum, a metal in Group 3A, loses three electrons to form the Al^{3+} cation. Oxygen, a nonmetal in Group 6A, gains two electrons to form an O^{2-} anion. Notice that the charge on the cation is the subscript on the anion, and vice versa.

$$2\ Al^{3+} + 3\ O^{2-} \longrightarrow Al_2O_3$$

This often works well, but be careful. The subscripts in $Ti^{4+} + O^{2-}$ are reduced to the simplest ratio (1 Ti to 2 O, rather than, 2 Ti to 4 O).

$$Ti^{4+} + 2\ O^{2-} \longrightarrow TiO_2$$

The gem ruby is largely the compound formed from aluminum ions (Al^{3+}) and oxide ions (O^{2-}). Here the ions have positive and negative charges that are of different absolute value. To have a compound with the same number of positive and negative charges, two Al^{3+} ions [total charge = $2 \times (3+) = 6+$] must combine with three O^{2-} ions [total charge = $3 \times (2-) = 6-$] to give a formula of Al_2O_3.

Calcium is a Group 2A metal, and it forms a cation having a 2+ charge. It can combine with a variety of anions to form ionic compounds such as those in the following table:

Compound	Ion Combination	Overall Charge on Compound
$CaCl_2$	$Ca^{2+} + 2\ Cl^-$	$(2+) + 2 \times (1-) = 0$
$CaCO_3$	$Ca^{2+} + CO_3^{2-}$	$(2+) + (2-) = 0$
$Ca_3(PO_4)_2$	$3\ Ca^{2+} + 2\ PO_4^{3-}$	$3 \times (2+) + 2 \times (3-) = 0$

In writing formulas, the convention is that the symbol of the cation is given first, followed by the anion symbol. Also notice the use of parentheses when more than one polyatomic ion is present.

Example 3.3—Ionic Compound Formulas

Problem For each of the following ionic compounds, write the symbols for the ions present and give the number of each: (a) $MgBr_2$, (b) Li_2CO_3, and (c) $Fe_2(SO_4)_3$.

Strategy Divide the formula of the compound into the cation and the anion. To accomplish this you will have to recognize, and remember, the composition and charges of common ions.

Solution

(a) $MgBr_2$ is composed of one Mg^{2+} ion and two Br^- ions. When a halogen such as bromine is combined only with a metal, you can assume the halogen is an anion with a charge of 1−. Magnesium is a metal in Group 2A and always has a charge of 2+ in its compounds.

(b) Li_2CO_3 is composed of two lithium ions, Li^+, and one carbonate ion, CO_3^{2-}. Li is a Group 1A element and always has a 1+ charge in its compounds. Because the two 1+ charges balance the negative charge of the carbonate ion, the latter must be 2−.

(c) $Fe_2(SO_4)_3$ contains two iron ions, Fe^{3+}, and three sulfate ions, SO_4^{2-}. The way to recognize this is to recall that sulfate has a 2− charge. Because three sulfate ions are present (with a total charge of 6−), the two iron cations must have a total charge of 6+. This is possible only if each iron cation has a charge of 3+.

Comment Remember that the formula for an ion must include its composition and its charge. Formulas for ionic compounds are *always* written with the cation first and then the anion.

Example 3.4—Ionic Compound Formulas

Problem Write formulas for ionic compounds composed of an aluminum cation and each of the following anions: (a) fluoride ion, (b) sulfide ion, and (c) nitrate ion.

Strategy First decide on the formula of the Al cation and the formula of each anion. Combine the Al cation with each type of anion to form an electrically neutral compound.

Solution An aluminum cation is predicted to have a charge of 3+ because Al is a metal in Group 3A.

(a) Fluorine is a Group 7A element. The charge of the fluoride ion is predicted to be $1-$ (from $8 - 7 = 1$). Therefore, we need 3 F^- ions to combine with one Al^{3+}. The formula of the compound is AlF_3.

(b) Sulfur is a nonmetal in Group 6A, so it forms a $2-$ anion. Thus, we need to combine two Al^{3+} ions [total charge is $6+ = 2 \times (3+)$] with three S^{2-} ions [total charge is $6- = 3 \times (2-)$]. The compound has the formula Al_2S_3.

(c) The nitrate ion has the formula NO_3^- (see Table 3.1). The answer here is therefore similar to the AlF_3 case, and the compound has the formula $Al(NO_3)_3$. Here we place parentheses around NO_3 to show that three polyatomic NO_3^- ions are involved.

Comment The most common error students make is not knowing the correct charge on an ion.

Exercise 3.4—Formulas of Ionic Compounds

(a) Give the number and identity of the constituent ions in each of the following ionic compounds: NaF, $Cu(NO_3)_2$, and $NaCH_3CO_2$.

(b) Iron, a transition metal, forms ions having at least two different charges. Write the formulas of the compounds formed between two different iron cations and chloride ions.

(c) Write the formulas of all neutral ionic compounds that can be formed by combining the cations Na^+ and Ba^{2+} with the anions S^{2-} and PO_4^{3-}.

Names of Ions

Naming Positive Ions (Cations)

With a few exceptions (such as NH_4^+), the positive ions described in this text are metal ions. Positive ions are named by the following rules:

1. For a monatomic positive ion (that is, a metal cation) the name is that of the metal plus the word "cation." For example, we have already referred to Al^{3+} as the aluminum cation.

2. Some cases occur, especially in the transition series, in which a metal can form more than one type of positive ion. In these cases the charge of the ion is indicated by a Roman numeral in parentheses immediately following the ion's name. For example, Co^{2+} is the cobalt(II) cation, and Co^{3+} is the cobalt(III) cation.

Finally, you will encounter the ammonium cation, NH_4^+, many times in this book and in the laboratory. Do not confuse the ammonium cation with the ammonia molecule, NH_3, which has no electric charge and one less H atom.

■ **"-ous" and "-ic" Endings**
An older naming system for metal ions uses the ending *-ous* for the ion of lower charge and *-ic* for the ion of higher charge. For example, there are cobaltous (Co^{2+}) and cobaltic (Co^{3+}) ions, and ferrous (Fe^{2+}) and ferric (Fe^{3+}) ions. We do not use this system in this book, but some chemical manufacturers continue to use it.

Problem-Solving Tip 3.1

Formulas for Ions and Ionic Compounds

Writing formulas for ionic compounds takes practice, and it requires that you know the formulas and charges of the most common ions. The charges on monatomic ions are often evident from the position of the element in the periodic table, but you simply have to remember the formula and charges of polyatomic ions; especially the most common ones such as nitrate, sulfate, carbonate, phosphate, and acetate.

If you cannot remember the formula of a polyatomic ion, or if you encounter an ion you have not seen before, you may be able to figure out its formula and the name of one of its compounds. For example, suppose you are told that $NaCHO_2$ is sodium formate. You know that the sodium ion is Na^+, so the formate ion must be the remaining portion of the compound; it must have a charge of $1-$ to balance the $1+$ charge on the sodium ion. Thus, the formate ion must be CHO_2^-.

Finally, when writing the formulas of ions, you must include the charge on the ion (except in an ionic compound formula). Writing Na when you mean sodium ion is incorrect. There is a vast difference in the properties of the element sodium (Na) and those of its ion (Na^+).

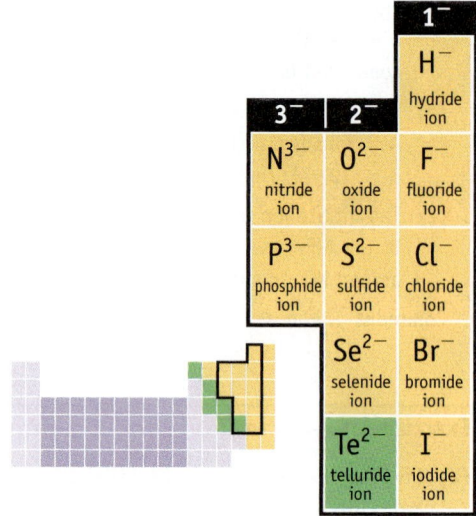

Figure 3.9 **Names and charges of some common monatomic anions.**

Naming Negative Ions (Anions)

There are two types of negative ions: those having only one atom (*monatomic*) and those having several atoms (*polyatomic*).

1. A monatomic negative ion is named by adding *-ide* to the stem of the name of the nonmetal element from which the ion is derived (Figure 3.9). The anions of the Group 7A elements, the halogens, are known as the fluoride, chloride, bromide, and iodide ions and as a group are called **halide ions**.

2. Polyatomic negative ions are common, especially those containing oxygen (called **oxoanions**). The names of some of the most common oxoanions are given in Table 3.1. Although most of these names must simply be learned, some guidelines can help. For example, consider the following pairs of ions:

NO_3^- is the nitrate ion; NO_2^- is the nitrite ion.

SO_4^{2-} is the sulfate ion; SO_3^{2-} is the sulfite ion.

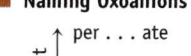

■ **Naming Oxoanions**

per . . . ate

. . . ate

. . . ite

hypo . . . ite

increasing oxygen content

The oxoanion having the *greater number of oxygen atoms* is given the suffix *-ate*, and the oxoanion having the *smaller number of oxygen atoms* has the suffix *-ite*. For a series of oxoanions having more than two members, the ion with the largest number of oxygen atoms has the prefix *per-* and the suffix *-ate*. The ion having the smallest number of oxygen atoms has the prefix *hypo-* and the suffix *-ite*. The oxoanions containing chlorine are good examples.

ClO_4^-	*perchlorate* ion
ClO_3^-	*chlorate* ion
ClO_2^-	*chlorite* ion
ClO^-	*hypochlorite* ion

Oxoanions that contain hydrogen are named by adding the word "hydrogen" before the name of the oxoanion. If two hydrogens are in the compound, we say "dihydrogen." Many hydrogen-containing oxoanions have common names that are used as well. For example, the hydrogen carbonate ion, HCO_3^-, is called the bicarbonate ion.

Ion	Systematic Name	Common Name
HPO_4^{2-}	hydrogen phosphate ion	
$H_2PO_4^-$	dihydrogen phosphate ion	
HCO_3^-	hydrogen carbonate ion	bicarbonate ion
HSO_4^-	hydrogen sulfate ion	bisulfate ion
HSO_3^-	hydrogen sulfite ion	bisulfite ion

Names of Ionic Compounds

The name of an ionic compound is built from the names of the positive and negative ions in the compound. The name of the positive cation is given first, followed by the name of the negative anion. If an element such as titanium can form cations with more than one charge, the charge is indicated by a Roman numeral. Examples of ionic compound names are given below.

Ionic Compound	Ions Involved	Name
$CaBr_2$	Ca^{2+} and $2\ Br^-$	calcium bromide
$NaHSO_4$	Na^+ and HSO_4^-	sodium hydrogen sulfate
$(NH_4)_2CO_3$	$2\ NH_4^+$ and CO_3^{2-}	ammonium carbonate
$Mg(OH)_2$	Mg^{2+} and $2\ OH^-$	magnesium hydroxide
$TiCl_2$	Ti^{2+} and $2\ Cl^-$	titanium(II) chloride
Co_2O_3	$2\ Co^{3+}$ and $3\ O^{2-}$	cobalt(III) oxide

GENERAL
Chemistry Now™

See the General ChemistryNow CD-ROM or website:
- **Screen 3.6 Polyatomic Ions,** for a tutorial on the names of polyatomic ions
- **Screen 3.9 Naming Ionic Compounds,** for a tutorial on naming ionic compounds

Exercise 3.5—Names of Ionic Compounds

1. Give the formula for each of the following ionic compounds. Use Table 3.1 and Figure 3.9.

(a) ammonium nitrate   **(d)** vanadium(III) oxide
(b) cobalt(II) sulfate   **(e)** barium acetate
(c) nickel(II) cyanide   **(f)** calcium hypochlorite

2. Name the following ionic compounds:
(a) $MgBr_2$ **(d)** $KMnO_4$
(b) Li_2CO_3 **(e)** $(NH_4)_2S$
(c) $KHSO_3$ **(f)** $CuCl$ and $CuCl_2$

Properties of Ionic Compounds

What is the "glue" that causes ions of opposite electric charge to be held together and to form an orderly arrangement of ions in an ionic compound? As described in

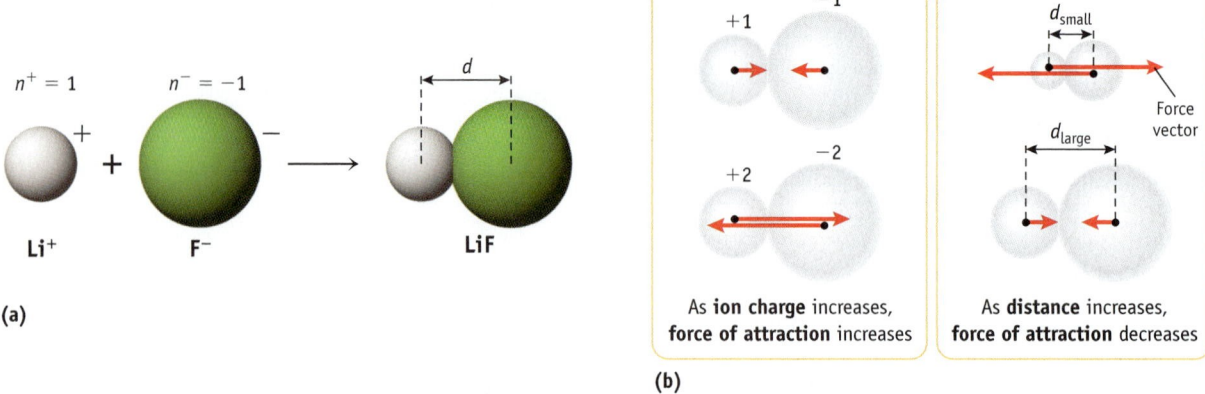

Active Figure 3.10   **Coulomb's law and electrostatic forces.** (a) Ions such as Li⁺ and F⁻ are held together by an electrostatic force. Here a lithium ion is attracted to a fluoride ion, and the distance between the nuclei of the two ions is *d*. (b) Forces of attraction between ions of opposite charge increase with increasing ion charge and decrease with increasing distance (*d*). (*The force of attraction is proportional to the length of the arrow in this figure.*)

GENERAL
Chemistry Now™ See the General ChemistryNow CD-ROM or website to explore an interactive version of this figure accompanied by an exercise.

Section 2.1, when a substance having a negative electric charge is brought near a substance having a positive electric charge, a force of attraction occurs between them (Figure 3.10). In contrast, a force of repulsion occurs when two substances with the same charge—both positive or both negative—are brought together. These forces are called **electrostatic forces**, and the force of attraction or repulsion between ions is given by **Coulomb's law** (Equation 3.1).

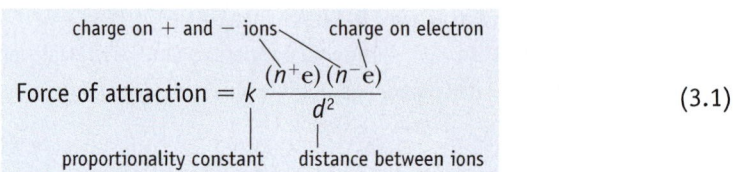

$$\text{Force of attraction} = k\,\frac{(n^+\text{e})\,(n^-\text{e})}{d^2} \tag{3.1}$$

where, for example, n^+ is 3 for Al^{3+} and n^- is 2 for O^{2-}. Based on Coulomb's law, the force of attraction between oppositely charged ions increases

- As the ion charges (n^+ and n^-) increase. Thus, the attraction between ions having charges of 2+ and 2− is greater than that between ions having 1+ and 1− charges (see Figure 3.10).

- As the distance between the ions becomes smaller [Figure 3.10; ▶ Chapter 9].

Ionic compounds do not consist of simple pairs or small groups of positive and negative ions. The simplest ratio of cations to anions in an ionic compound is represented by its formula, but an ionic solid consists of millions upon millions of ions arranged in an extended three-dimensional network called a **crystal lattice**. A portion of the lattice for NaCl, illustrated in Figure 3.11, represents a common way of arranging ions for compounds that have a 1 : 1 ratio of cations to anions.

Ionic compounds have characteristic properties that can be understood in terms of the charges of the ions and their arrangement in the lattice. Because each ion is surrounded by oppositely charged nearest neighbors, it is held tightly in its allotted location. At room temperature each ion can move just a bit around its aver-

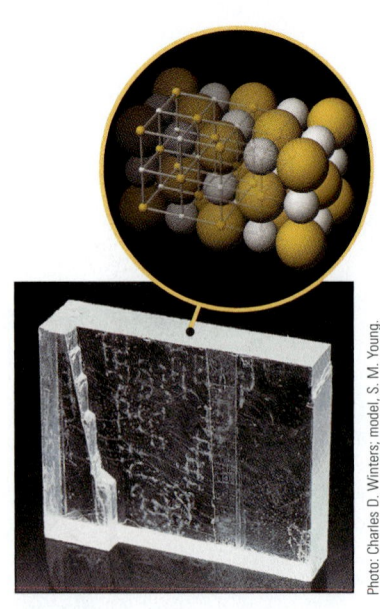

Photo: Charles D. Winters; model: S. M. Young.

Figure 3.11   Sodium chloride. A crystal of NaCl consists of an extended lattice of sodium ions and chloride ions in a 1:1 ratio. (*See General ChemistryNow Screen 3.8 Ionic Compounds, to view an animation on the formation of a sodium chloride crystal lattice.*)

Problem-Solving Tip 3.2

Is a Compound Ionic?

Students often ask how to know whether a compound is ionic. No method works all of the time, but here are some useful guidelines.

1. Most metal-containing compounds are ionic. So, if a metal atom appears in the formula of a compound, a good first guess is that it is ionic. (There are interesting exceptions, but few come up in introductory chemistry.) It is helpful in this regard to recall trends in metallic behavior: All elements to the left of a diagonal line running from boron to tellurium in the periodic table are metallic.

2. If there is no metal in the formula, it is likely that the compound is not ionic. The exceptions here are compounds composed of polyatomic ions based on nonmetals (e.g., NH_4Cl or NH_4NO_3).

3. Learn to recognize the formulas of polyatomic ions (see Table 3.1). Chemists write the formula of ammonium nitrate as NH_4NO_3 (not as $N_2H_4O_3$) to alert others to the fact that it is an ionic compound composed of the common polyatomic ions NH_4^+ and NO_3^-.

As an example of these guidelines, you can be sure that $MgBr_2$ (Mg^{2+} with Br^-) and K_2S (K^+ with S^{2-}) are ionic compounds. On the other hand, the compound CCl_4, formed from two nonmetals, C and Cl, is not ionic.

---

age position. However, considerable energy must be added before an ion can move fast enough and far enough to escape the attraction of its neighboring ions. Only if enough energy is added will the lattice structure collapse and the substance melt. Greater attractive forces mean that ever more energy—higher and higher temperatures—is required to cause melting. Thus, Al_2O_3, a solid composed of Al^{3+} and O^{2-} ions, melts at a much higher temperature (2072 °C) than NaCl (801 °C), a solid composed of Na^+ and Cl^- ions.

Most ionic compounds are "hard" solids. That is, the solids are not pliable or soft. The reason for this characteristic is again related to the lattice of ions. The nearest neighbors of a cation in a lattice are anions, and the force of attraction makes the lattice rigid. However, a blow with a hammer can cause the lattice to cleave cleanly along a sharp boundary. The hammer blow displaces layers of ions just enough to cause ions of like charge to become nearest neighbors. The repulsion between like charges then forces the lattice apart (Figure 3.12).

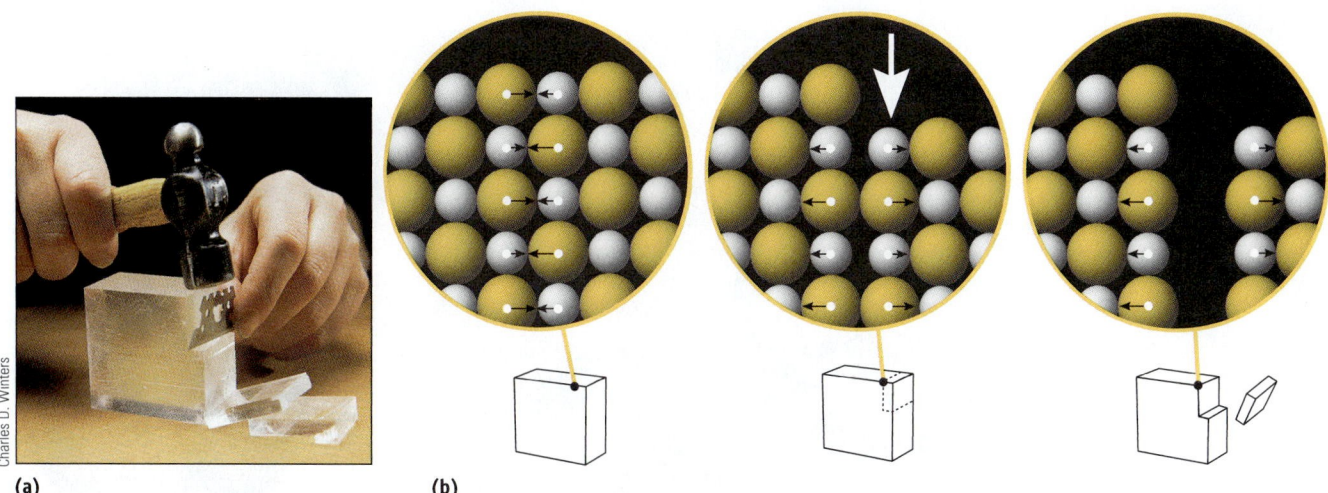

(a) **(b)**

Charles D. Winters

Figure 3.12 **Ionic solids.** (a) An ionic solid is normally rigid owing to the forces of attraction between oppositely charged ions. When struck sharply, however, the crystal can cleave cleanly. (b) When a crystal is struck, layers of ions move slightly, and ions of like charge become nearest neighbors. Repulsions between ions of similar charge cause the crystal to cleave. (*See the General ChemistryNow Screen 3.10 Properties of Ionic Compounds, to watch a video of cleaving a crystal.*)

See the General ChemistryNow CD-ROM or website:

• **Screen 3.8 Ionic Compounds,** to watch a video of the sodium + chlorine reaction and for a simulation on the relationship between cations and anions in ionic compounds

---

Exercise 3.6—Coulomb's Law

Explain why the melting point of MgO (2830 °C), much higher than the melting point of NaCl (801 °C).

3.4—Molecular Compounds: Formulas, Names, and Properties

Many familiar compounds are not ionic, they are molecular: the water you drink, the sugar in your coffee or tea, or the aspirin you take for a headache.

Ionic compounds are generally solids, whereas molecular compounds can range from gases to liquids to solids at ordinary temperatures (see Figure 3.13). As size and molecular complexity increase, compounds generally exist as solids. We will explore some of the underlying causes of these general observations in Chapter 13.

Some molecular compounds have complicated formulas that you cannot, at this stage, predict or even decide if they are correct. However, there are many simple compounds you will encounter often, and you should understand how to name them and, in many cases, know their formulas.

Let us look first at molecules formed from combinations of two nonmetals. These "two-element" compounds of nonmetals, often called **binary compounds,** can be named in a systematic way.

Hydrogen forms binary compounds with all of the nonmetals except the noble gases. For compounds of oxygen, sulfur, and the halogens, the H atom is generally

Photo: Charles D. Winters

Figure 3.13   Molecular compounds. Ionic compounds are generally solids at room temperature. In contrast, molecular compounds can be gases, liquids, or solids. The models are of caffeine (in coffee), water, and citric acid (in lemons).

written first in the formula and is named first. The other nonmetal is named as if it were a negative ion.

Compound	Name
HF	hydrogen fluoride
HCl	hydrogen chloride
H_2S	hydrogen sulfide

Virtually all binary molecular compounds of nonmetals are a combination of elements from Groups 4A–7A with one another or with hydrogen. The formula is generally written by putting the elements in order of increasing group number. When naming the compound, the number of atoms of a given type in the compound is designated with a prefix, such as "di-," "tri-," "tetra-," "penta-," and so on.

Compound	Systematic Name
NF_3	nitrogen trifluoride
NO	nitrogen monoxide
NO_2	nitrogen dioxide
N_2O	dinitrogen monoxide
N_2O_4	dinitrogen tetraoxide
PCl_3	phosphorus trichloride
PCl_5	phosphorus pentachloride
SF_6	sulfur hexafluoride
S_2F_{10}	disulfur decafluoride

Finally, many of the binary compounds of nonmetals were discovered years ago and have common names.

Compound	Common Name
CH_4	methane
C_2H_6	ethane
C_3H_8	propane
C_4H_{10}	butane
NH_3	ammonia
N_2H_4	hydrazine
PH_3	phosphine
NO	nitric oxide
N_2O	nitrous oxide ("laughing gas")
H_2O	water

■ **Formulas of Binary Nonmetal Compounds Containing Hydrogen**
Simple hydrocarbons (compounds of C and H) such as methane and ethane have formulas written with H following C, and the formulas of ammonia and hydrazine have H following N. Water and the hydrogen halides, however, have the H atom preceding O or the halogen atom. Tradition is the only explanation for such irregularities in writing formulas.

■ **Hydrocarbons**
Compounds such as methane, ethane, propane, and butane belong to a class of hydrocarbons called alkanes. (*See Chapter 11 and General ChemistryNow Screen 3.13, Alkanes.*)

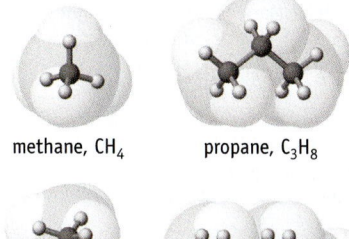

methane, CH_4 propane, C_3H_8

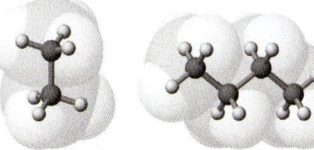

ethane, C_2H_6 butane, C_4H_{10}

GENERAL
Chemistry ⚛ Now™

See the General ChemistryNow CD-ROM or website:

- **Screen 3.12 Binary Compounds of the Nonmetals,** for a tutorial on naming compounds of the nonmetals
- **Screen 3.13 Alkanes,** for a simulation and exercise on naming alkanes

Exercise 3.7—Naming Compounds of the Nonmetals

1. Give the formula for each of the following binary, nonmetal compounds:

   (a) carbon dioxide   (d) boron trifluoride
   (b) phosphorus triiodide   (e) dioxygen difluoride
   (c) sulfur dichloride   (f) xenon trioxide

2. Name the following binary, nonmetal compounds:

   (a) N_2F_4 (c) SF_4 (e) P_4O_{10}
   (b) HBr   (d) BCl_3 (f) ClF_3

3.5—Formulas, Compounds, and the Mole

The formula of a compound tells you the type of atoms or ions in the compound and the relative number of each. For example, one molecule of methane, CH_4, is made up of one atom of C and four atoms of H. But suppose you have Avogadro's number of C atoms (6.022×10^{23}) combined with the proper number of H atoms. The compound's formula tells us that four times as many H atoms are required ($4 \times 6.022 \times 10^{23}$ H atoms) to give Avogadro's number of CH_4 molecules. What masses of atoms are combined, and what is the mass of this many CH_4 molecules?

C	+	4 H	$\longrightarrow$	CH_4
6.022×10^{23} C atoms		$4 \times 6.022 \times 10^{23}$ H atoms		6.022×10^{23} CH_4 molecules
= 1.000 mol of C		= 4.000 mol of H atoms		= 1.000 mol of CH_4 molecules
= 12.01 g of C atoms		= 4.032 g of H atoms		= 16.04 g of CH_4 molecules

Because we know the number of moles of C and H atoms, we know the masses of carbon and hydrogen that combine to form CH_4. It follows that the mass of CH_4 is the sum of these masses. That is, 1 mol of CH_4 has a mass equivalent to the mass of 1 mol of C atoms (12.01 g) plus 4 mol of H atoms (4.032 g). Thus, the *molar mass*, *M*, of CH_4 is 16.04 g/mol [◀ Section 2.5].

■ **Molar Mass or Molecular Weight**
Although chemists often use the term "molecular weight," we should more properly cite a compound's molar mass. The SI unit of molar mass is kg/mol, but chemists worldwide usually express it in units of g/mol. See "NIST Guide to SI Units" at www.NIST.gov

Molar and Molecular Masses

Element or Compound	Molar Mass, *M* (g/mol)	Average Mass of One Molecule* (g/molecule)
O_2	32.00	5.314×10^{-23}
P_4	123.9	2.057×10^{-22}
NH_3	17.03	2.828×10^{-23}
H_2O	18.02	2.992×10^{-23}
CH_2Cl_2	84.93	1.410×10^{-22}

*See text, page 117, for the calculation of the mass of one molecule.

Ionic compounds such as NaCl do not exist as individual molecules. Thus, we write the simplest formula that shows the relative number of each kind of atom in a "formula unit" of the compound, and the molar mass is calculated from this formula. To differentiate substances like NaCl that do not contain molecules, chemists sometimes refer to their *formula weight* instead of their molecular weight.

Figure 3.14 illustrates 1-mol quantities of several common compounds. To find the molar mass of any compound, you need only add up the atomic masses for each element in one formula unit. As an example, let us find the molar mass of aspirin, $C_9H_8O_4$. In one mole of aspirin there are 9 mol of carbon atoms, 8 mol of hydrogen atoms, and 4 mol of oxygen atoms, which add up to 180.2 g/mol of aspirin.

$$\text{Mass of C in 1 mol } C_9H_8O_4 = 9 \text{ mol C} \times \frac{12.01 \text{ g C}}{1 \text{ mol C}} = 108.1 \text{ g C}$$

$$\text{Mass of H in 1 mol } C_9H_8O_4 = 8 \text{ mol H} \times \frac{1.008 \text{ g H}}{1 \text{ mol H}} = 8.064 \text{ g H}$$

$$\text{Mass of O in 1 mol } C_9H_8O_4 = 4 \text{ mol O} \times \frac{16.00 \text{ g O}}{1 \text{ mol O}} = 64.00 \text{ g O}$$

$$\text{Total mass of 1 mol of } C_9H_8O_4 = \text{molar mass of } C_9H_8O_4 = 180.2 \text{ g}$$

As was the case with elements, it is important to be able to convert the mass of a compound to the equivalent number of moles (or moles to mass) [◀ Section 2.5]. For example, if you take 325 mg (0.325 g) of aspirin in one tablet, what amount of the compound have you ingested? Based on a molar mass of 180.2 g/mol, there are 0.00180 mol of aspirin per tablet.

$$0.325 \text{ g aspirin} \times \frac{1 \text{ mol aspirin}}{180.2 \text{ g aspirin}} = 0.00180 \text{ mol aspirin}$$

Using the molar mass of a compound it is possible to determine the number of molecules in any sample from the sample mass and to determine the mass of one molecule. For example, the number of aspirin molecules in one tablet is

$$0.00180 \text{ mol aspirin} \times \frac{6.022 \times 10^{23} \text{ molecules}}{1 \text{ mol aspirin}} = 1.08 \times 10^{21} \text{ molecules}$$

and the mass of one molecule is

$$\frac{180.2 \text{ g aspirin}}{1 \text{ mol aspirin}} \times \frac{1 \text{ mol aspirin}}{6.022 \times 10^{23} \text{ molecules}} = 2.99 \times 10^{-22} \text{ g/molecule}$$

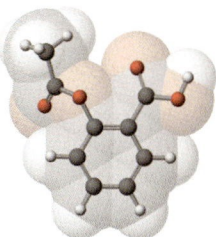

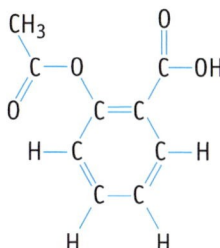

■ **Aspirin Formula**
Aspirin has the molecular formula $C_9H_8O_4$ and a molar mass of 180.2 g/mol. Aspirin is the common name of the compound acetylsalicylic acid.

Figure 3.14 **One-mole quantities of some compounds.**

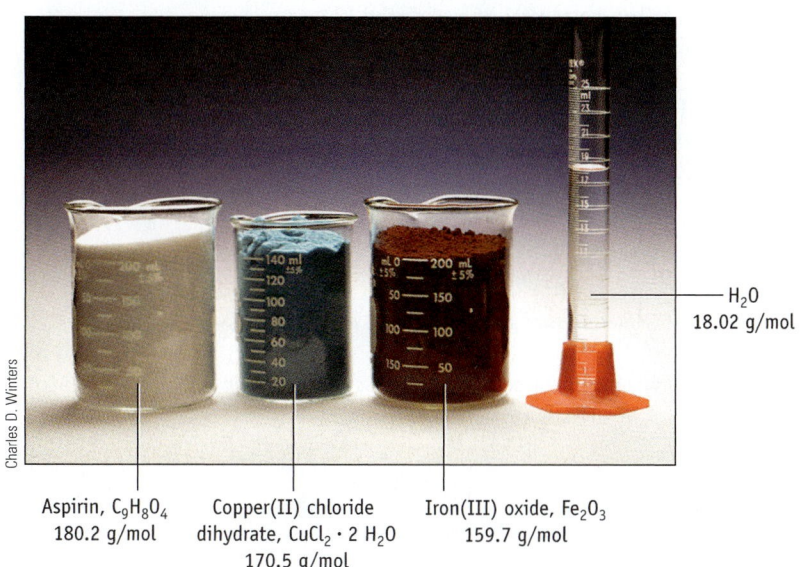

Aspirin, $C_9H_8O_4$
180.2 g/mol

Copper(II) chloride dihydrate, $CuCl_2 \cdot 2 H_2O$
170.5 g/mol

Iron(III) oxide, Fe_2O_3
159.7 g/mol

H_2O
18.02 g/mol

Charles D. Winters

See the General ChemistryNow CD-ROM or website:

- **Screen 3.14 Compounds, Molecules, and the Mole,** for a simulation exploring the relationship between mass, moles, molecules, and atoms, and a tutorial on determining molar mass
- **Screen 3.15 Using Molar Mass,** for a tutorial on determining moles from mass and a second tutorial on determining mass from moles

Example 3.5—Molar Mass and Moles

Problem You have 16.5 g of oxalic acid, $H_2C_2O_4$.

(a) What amount is represented by 16.5 g of oxalic acid?

(b) How many molecules of oxalic acid are in 16.5 g?

(c) How many atoms of carbon are in 16.5 g of oxalic acid?

(d) What is the mass of one molecule of oxalic acid?

Strategy The first step in any problem involving the conversion of mass and moles is to find the molar mass of the compound in question. Then you can perform the other calculations as outlined by the scheme shown here to find the number of molecules from the amount of substance and the number of atoms of a particular kind:

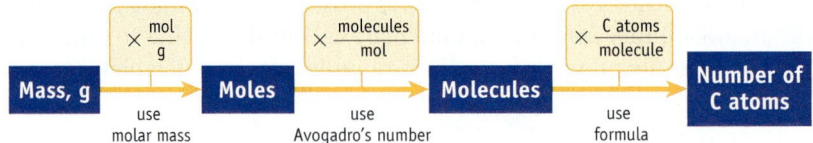

(*See the General ChemistryNow Screen 3.14 Compounds and Moles, and Screen 3.15 Molar Mass.*)

Solution

(a) *Moles represented by 16.5 g*

Let us first calculate the molar mass of oxalic acid:

$$2 \text{ mol C per mol } H_2C_2O_4 \times \frac{12.01 \text{ g C}}{1 \text{ mol C}} = 24.02 \text{ g C per mol } H_2C_2O_4$$

$$2 \text{ mol H per mol } H_2C_2O_4 \times \frac{1.008 \text{ g H}}{1 \text{ mol H}} = 2.016 \text{ g H per mol } H_2C_2O_4$$

$$4 \text{ mol O per mol } H_2C_2O_4 \times \frac{16.00 \text{ g O}}{1 \text{ mol O}} = 64.00 \text{ g O per mol } H_2C_2O_4$$

$$\text{Molar mass of } H_2C_2O_4 = 90.04 \text{ g per mol } H_2C_2O_4$$

Now calculate the amount in moles. The molar mass (expressed in units of 1 mol/90.04 g) is the conversion factor in all mass-to-mole conversions.

$$16.5 \text{ g } H_2C_2O_4 \times \frac{1 \text{ mol}}{90.04 \text{ g } H_2C_2O_4} = \boxed{0.183 \text{ mol } H_2C_2O_4}$$

(b) *Number of molecules*

Use Avogadro's number to find the number of oxalic acid molecules in 0.183 mol of $H_2C_2O_4$.

$$0.183 \text{ mol} \times \frac{6.022 \times 10^{23} \text{ molecules}}{1 \text{ mol}} = 1.10 \times 10^{23} \text{ molecules}$$

(c) *Number of C atoms*

Because each molecule contains two carbon atoms, the number of carbon atoms in 16.5 g of the acid is

$$1.10 \times 10^{23} \text{ molecules} \times \frac{2 \text{ C atoms}}{1 \text{ molecule}} = 2.20 \times 10^{23} \text{ C atoms}$$

(d) *Mass of one molecule*

The units of the desired answer are grams per molecule, which indicates that you should multiply the starting unit of molar mass (grams per mole) by (1/Avogadro's number) (units are mole/molecule), so that the unit "mol" cancels.

$$\frac{90.04 \text{ g}}{1 \text{ mol}} \times \frac{1 \text{ mol}}{6.0221 \times 10^{23} \text{ molecules}} = 1.495 \times 10^{-22} \text{ g/molecule}$$

Exercise 3.8—Molar Mass and Moles-to-Mass Conversions

(a) Calculate the molar mass of citric acid, $C_6H_8O_7$, and $MgCO_3$.

(b) If you have 454 g of citric acid, what amount (moles) does this represent?

(c) To have 0.125 mol of $MgCO_3$, what mass (g) must you have?

3.6—Describing Compound Formulas

Given a sample of an unknown compound, how can its formula be determined? The answer lies in chemical analysis, a major branch of chemistry that deals with the determination of formulas and structures.

Percent Composition

Any sample of a pure compound always consists of the same elements combined in the same proportion by mass. This means *molecular composition* can be expressed in at least three ways:

- In terms of the number of atoms of each type per molecule or per formula unit—that is, by giving the formula of the compound
- In terms of the mass of each element per mole of compound
- In terms of the mass of each element in the compound relative to the total mass of the compound—that is, as a **mass percent**

Suppose you have 1.000 mol of NH_3 or 17.03 g. This mass of NH_3 is composed of 14.01 g of N (1.000 mol) and 3.024 g of H (3.000 mol). If you compare the mass of N to the total mass of compound, 82.27% of the total mass is N (and 17.76% is H).

■ **Molecular Composition**
Molecular composition can be expressed as a percent (mass of an element in a 100-g sample). For example, NH_3 is 82.27% N. Therefore, it has 82.27 g of N in 100.0 g of compound.

82.27% of NH_3 mass is **nitrogen**.

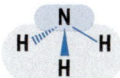

17.76% of NH_3 mass is **hydrogen**.

Note that the %N and %H do not add up to exactly 100%. This is not unusual and does not mean there is an error. The last digit of the answer is limited by the accuracy of the data used.

$$\text{Mass of N per mole of } NH_3 = \frac{1 \text{ mol N}}{1 \text{ mol } NH_3} \times \frac{14.01 \text{ g N}}{1 \text{ mol N}} = 14.01 \text{ g N/1 mol } NH_3$$

$$\begin{aligned}\text{Mass percent N in } NH_3 &= \frac{\text{mass of N in 1 mol } NH_3}{\text{mass of 1 mol } NH_3} \\ &= \frac{14.01 \text{ g N}}{17.03 \text{ g } NH_3} \times 100\% \\ &= 82.27\% \text{ (or 82.27 g N in 100.0 g } NH_3)\end{aligned}$$

$$\text{Mass of H per mole of } NH_3 = \frac{3 \text{ mol H}}{1 \text{ mol } NH_3} \times \frac{1.008 \text{ g H}}{1 \text{ mol H}} = 3.024 \text{ g H/1 mol } NH_3$$

$$\begin{aligned}\text{Mass percent H in } NH_3 &= \frac{\text{mass of H in 1 mol } NH_3}{\text{mass of 1 mol } NH_3} \times 100\% \\ &= \frac{3.024 \text{ g H}}{17.03 \text{ g } NH_3} \times 100\% \\ &= 17.76\% \text{ (or 17.76 g H in 100.0 g } NH_3)\end{aligned}$$

These values represent the mass percent of each element, or percent composition by mass. They tell you that in a 100.0-g sample there are 82.27 g of N and 17.76 g of H.

GENERAL
Chemistry‑❖‑Now™

See the General ChemistryNow CD-ROM or website:
- **Screen 3.16 Percent Composition,** for a tutorial on detemining percent composition

Example 3.6—Using Percent Composition

Problem What is the mass percent of each element in propane, C_3H_8? What mass of carbon is contained in 454 g of propane?

Strategy First find the molar mass of C_3H_8 and then calculate the mass percent of C and H per mole of C_3H_8. Using the knowledge of the mass percent of C, calculate the mass of carbon in 454 g of C_3H_8.

Solution

(a) The molar mass of C_3H_8 is 44.10 g/mol.

(b) Mass percent of C and H in C_3H_8:

$$\frac{3 \text{ mol C}}{1 \text{ mol } C_3H_8} \times \frac{12.01 \text{ g C}}{1 \text{ mol C}} = 36.03 \text{ g C/1 mol } C_3H_8$$

$$\text{Mass percent of C in } C_3H_8 = \frac{36.03 \text{ g C}}{44.10 \text{ g } C_3H_8} \times 100\% = \boxed{81.70\% \text{ C}}$$

$$\frac{8 \text{ mol H}}{1 \text{ mol } C_3H_8} \times \frac{1.008 \text{ g H}}{1 \text{ mol H}} = 8.064 \text{ g H/1 mol } C_3H_8$$

$$\text{Mass percent of H in } C_3H_8 = \frac{8.064 \text{ g H}}{44.10 \text{ g } C_3H_8} \times 100\% = \boxed{18.29\% \text{ H}}$$

(c) Mass of C in 454 g of C_3H_8:

$$454 \text{ g } C_3H_8 \times \frac{81.70 \text{ g C}}{100.0 \text{ g } C_3H_8} = 371 \text{ g C}$$

Exercise 3.9—Percent Composition

(a) Express the composition of ammonium carbonate, $(NH_4)_2CO_3$, in terms of the mass of each element in 1.00 mol of compound and the mass percent of each element.

(b) What is the mass of carbon in 454 g of octane, C_8H_{18}?

Empirical and Molecular Formulas from Percent Composition

Now let us consider the *reverse* of the procedure just described: using relative mass or percent composition data to find a molecular formula. Suppose you know the identity of the elements in a sample and have determined the mass of each element in a given mass of compound by chemical analysis [▶ Section 4.6]. You can then calculate the relative amount (moles) of each element and from this the relative number of atoms of each element in the compound. For example, for a compound composed of atoms of A and B, the steps from percent composition to a formula are

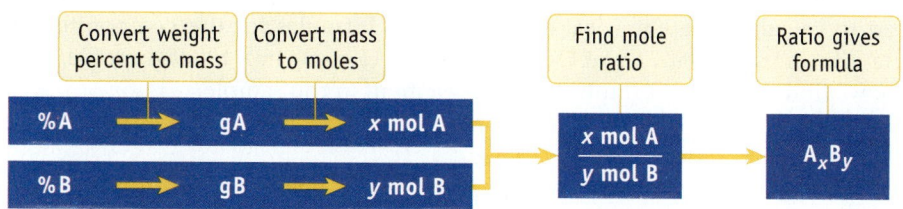

Let us derive the formula for hydrazine, a close relative of ammonia and a compound used to remove oxygen from water used for heating and cooling.

Step 1: *Convert mass percent to mass.* The mass percentages in a sample of hydrazine are 87.42% N and 12.58% H. Thus, in a 100.00-g sample of hydrazine, there are 87.42 g of N and 12.58 g of H.

Step 2: *Convert the mass of each element to moles.* The amount of each element in the 100.00-g sample is

$$87.42 \text{ g N} \times \frac{1 \text{ mol N}}{14.007 \text{ g N}} = 6.241 \text{ mol N}$$

$$12.58 \text{ g H} \times \frac{1 \text{ mol H}}{1.008 \text{ g H}} = 12.48 \text{ mol H}$$

■ **Deriving a Formula**
Percent composition gives the mass of an element in 100 g of sample. However, any amount of sample is appropriate if you know the mass of an element in that sample mass. See Example 3.8.

Step 3: *Find the mole ratio of elements.* Use the amount (moles) of each element in the 100.00-g of sample to find the amount of one element relative to the other. For hydrazine, this ratio is 2 mol of H to 1 mol of N,

$$\frac{12.48 \text{ mol H}}{6.241 \text{ mol N}} = \frac{2.00 \text{ mol H}}{1.00 \text{ mol N}} \longrightarrow NH_2$$

■ **Deriving a Formula—Mole Ratios**
When finding the ratio of moles of one element relative to another, *always* divide the larger number by the smaller one.

showing that there are 2 mol of H atoms for every 1 mol of N atoms in hydrazine. Thus, in one molecule, two atoms of H occur for every atom of N; that is, the formula is NH_2. This simplest whole-number atom ratio is called the **empirical formula**.

Percent composition data allow us to calculate the atom ratios in a compound. A *molecular formula*, however, must convey *two* pieces of information: (1) the relative numbers of atoms of each element in a molecule (the atom ratios) and (2) the total number of atoms in the molecule. For hydrazine there are twice as many H atoms as N atoms, so the molecular formula could be NH_2. Recognize, however, that percent composition data give only the *simplest possible ratio of atoms* in a molecule. The empirical formula of hydrazine is NH_2, but the true *molecular formula* could also be NH_2, N_2H_4, N_3H_6, N_4H_8, or any other formula having a $1:2$ ratio of N to H.

To determine the molecular formula from the empirical formula, the molar mass must be obtained from experiment. For example, experiments show that the molar mass of hydrazine is 32.0 g/mol, twice the formula mass of NH_2, which is 16.0 g/mol. Thus, the molecular formula of hydrazine is two times the empirical formula of NH_2, that is, N_2H_4.

As another example of the usefulness of percent composition data, let us say that you collected the following information in the laboratory for isooctane, the compound used as the standard for determining the octane rating of a fuel: % carbon = 84.12; % hydrogen = 15.88; molar mass = 114.2 g/mol. These data can be used to calculate the empirical and molecular formulas for the compound. The data inform us that 84.12 g of C and 15.88 g of H occur in a 100.0-g sample. From this, we find the amount (moles) of each element in this sample.

$$84.12 \ \text{g C} \times \frac{1 \ \text{mol C}}{12.011 \ \text{g C}} = 7.004 \ \text{mol C}$$

$$15.88 \ \text{g H} \times \frac{1 \ \text{mol H}}{1.0079 \ \text{g H}} = 15.76 \ \text{mol H}$$

This means that, in any sample of isooctane, the ratio of moles of H to C is

$$\text{Mole ratio} \ = \ \frac{15.76 \ \text{mol H}}{7.004 \ \text{mol C}} = \frac{2.250 \ \text{mol H}}{1.000 \ \text{mol C}}$$

Now the task is to turn this decimal fraction into a whole-number ratio of H to C. To do this, recognize that 2.25 is the same as $2\frac{1}{4}$ or $9/4$. Therefore, the ratio of C to H is

$$\text{Mole ratio} = \frac{2.25 \ \text{mol H}}{1.00 \ \text{mol C}} = \frac{2\frac{1}{4} \ \text{mol H}}{1 \ \text{mol C}} = \frac{9/4 \ \text{mol H}}{1 \ \text{mol C}} = \frac{9 \ \text{mol H}}{4 \ \text{mol C}}$$

You now know that nine H atoms occur for every four C atoms in isooctane. Thus, the simplest or *empirical formula* is C_4H_9. If C_4H_9 were the molecular formula, the molar mass would be 57.12 g/mol. However, we know from experiment that the actual molar mass is 114.2 g/mol, twice the value for the empirical formula.

$$\frac{114.2 \ \text{g/mol of isooctane}}{57.12 \ \text{g/mol of } C_4H_9} = 2.00 \ \text{mol } C_4H_9 \ \text{per mol of isooctane}$$

The molecular formula is therefore C_8H_{18}.

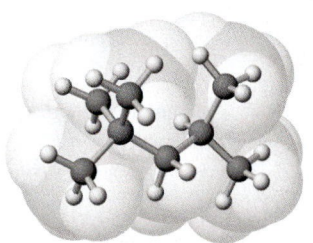

■ Isooctane and the Octane Rating
Isooctane, C_8H_{18}, is the standard against which the octane rating of gasoline is determined. Octane numbers are assigned by comparing the burning performance of gasoline with the burning performance of mixtures of isooctane and heptane. Gasoline with an octane rating of 90 matches the burning characteristics of a mixture of 90% isooctane and 10% heptane.

See the General ChemistryNow CD-ROM or website:

- **Screen 3.17 Determining Empirical Formulas,** for a tutorial on determining empirical formulas
- **Screen 3.18 Determining Molecular Formulas,** for a tutorial on determining molecular formulas

Example 3.7—Calculating a Formula from Percent Composition

Problem Eugenol is the major component in oil of cloves. It has a molar mass of 164.2 g/mol and is 73.14% C and 7.37% H; the remainder is oxygen. What are the empirical and molecular formulas of eugenol?

Strategy To derive a formula we need to know the mass percent of each element. Because the mass percents of all elements must add up to 100.0%, we find the mass percent of O from the difference between 100.0% and the mass percents of C and H. Next, we assume the mass percent of each element is equivalent to its mass in grams, and convert each mass to moles. Finally, the ratio of moles gives the empirical formula. The mass of a mole of compound having the calculated empirical formula is compared with the actual, experimental molar mass to find the true molecular formula.

Solution The mass of O in a 100.0-g sample is

$$100.0 \text{ g} = 73.14 \text{ g C} + 7.37 \text{ g H} + \text{mass of O}$$

$$\text{Mass of O} = 19.49 \text{ g}$$

The amount of each element is

$$73.14 \text{ g C} \times \frac{1 \text{ mol C}}{12.011 \text{ g C}} = 6.089 \text{ mol C}$$

$$7.37 \text{ g H} \times \frac{1 \text{ mol H}}{1.008 \text{ g H}} = 7.31 \text{ mol H}$$

$$19.49 \text{ g O} \times \frac{1 \text{ mol O}}{15.999 \text{ g O}} = 1.218 \text{ mol O}$$

To find the mole ratio, the best approach is to base the ratios on the smallest number of moles present—in this case, oxygen.

$$\frac{\text{mol C}}{\text{mol O}} = \frac{6.089 \text{ mol C}}{1.218 \text{ mol O}} = \frac{4.999 \text{ mol C}}{1.000 \text{ mol O}} = \frac{5 \text{ mol C}}{1 \text{ mol O}}$$

$$\frac{\text{mol H}}{\text{mol O}} = \frac{7.31 \text{ mol H}}{1.218 \text{ mol O}} = \frac{6.00 \text{ mol H}}{1.000 \text{ mol O}} = \frac{6 \text{ mol H}}{1 \text{ mol O}}$$

Now we know there are 5 mol of C and 6 mol of H per 1 mol of O. Thus, the empirical formula is C_5H_6O.

The experimentally determined molar mass of eugenol is 164.2 g/mol. This is twice the mass of C_5H_6O (82.1 g/mol).

$$\frac{164.2 \text{ g/mol of eugenol}}{82.10 \text{ g/mol of } C_5H_6O} = 2.00 \text{ mol } C_5H_6O \text{ per mol of eugenol}$$

The molecular formula is $C_{10}H_{12}O_2$.

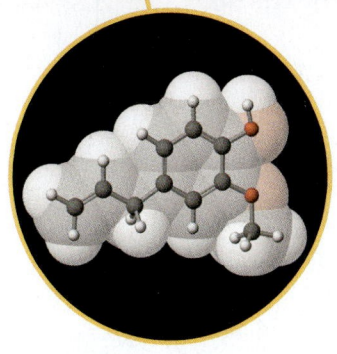

Eugenol, $C_{10}H_{12}O_2$, is an important component in oil of cloves.

Comment There is another approach to finding the molecular formula here. Knowing the percent composition of eugenol and its molar mass, we could calculate that in 164.2 g of eugenol there are 120.1 g of C (10 mol of C), 12.1 g of H (12 mol of H), and 32.00 g of O (2 mol of O). This gives us a molecular formula of $C_{10}H_{12}O_2$. However, *you must recognize that this approach can only be used when you know* **both** *the percent composition and the molar mass.*

Exercise 3.10—Empirical and Molecular Formulas

(a) What is the empirical formula of naphthalene, $C_{10}H_8$?

(b) The empirical formula of acetic acid is CH_2O. If its molar mass is 60.05 g/mol, what is the molecular formula of acetic acid?

Exercise 3.11—Calculating a Formula from Percent Composition

Isoprene is a liquid compound that can be polymerized to form natural rubber. It is composed of 88.17% carbon and 11.83% hydrogen. Its molar mass is 68.11 g/mol. What are its empirical and molecular formulas?

Exercise 3.12—Calculating a Formula from Percent Composition

Camphor is found in "camphor wood," much prized for its wonderful odor. It is composed of 78.90% carbon and 10.59% hydrogen. The remainder is oxygen. What is its empirical formula?

Determining a Formula from Mass Data

The composition of a compound in terms of mass percent gives us the mass of each element in a 100.0-g sample. In the laboratory we often collect information on the composition of compounds slightly differently. We can

1. *Combine known masses of elements to give a sample of the compound of known mass.* Element masses can be converted to moles, and the ratio of moles gives the combining ratio of atoms—that is, the empirical formula. This approach is described in Example 3.8.

2. *Decompose a known mass of an unknown compound into "pieces" of known composition.* If the masses of the "pieces" can be determined, the ratio of moles of the "pieces" gives the formula. An example is a decomposition such as

$$Ni(CO)_4(\ell) \longrightarrow Ni(s) + 4\ CO(g)$$

Problem-Solving Tip 3.3

Finding Empirical and Molecular Formulas

- The experimental data available to find a formula may be in the form of percent composition or the masses of elements combined in some mass of compound. No matter what the starting point, the first step is always to convert masses of elements to moles.

- Be sure to use at least three significant figures when calculating empirical formulas. Using fewer significant figures often gives a misleading result.

- When finding atom ratios, always divide the larger number of moles by the smaller one.

- Empirical and molecular formulas often differ for molecular compounds. In contrast, the formula of an ionic compound is generally the same as its empirical formula.

- Determining the molecular formula of a compound after calculating the empirical formula requires knowing the molar mass.

- When both the percent composition and the molar mass are known for a compound, the alternative method mentioned in the comment to Example 3.7 could be used. However, you must recognize that this approach can only be used when you know both the percent composition and the molar mass.

The masses of Ni and CO can be converted to moles, whose $1:4$ ratio would reveal the formula of the compound. We will describe this approach in Chapter 4 [▶ Section 4.6].

Example 3.8—Formula of a Compound from Combining Masses

Problem Gallium oxide, Ga_xO_y, forms when gallium is combined with oxygen. Suppose you allow 1.25 g of gallium (Ga) to react with oxygen and obtain 1.68 g of Ga_xO_y. What is the formula of the product?

Strategy Calculate the mass of oxygen in 1.68 g of product (which you already know contains 1.25 g of Ga). Next, calculate the amounts of Ga and O (in moles) and find their ratio.

Solution The masses of Ga and O combined in 1.68 g of product are

$$1.68 \text{ g product} - 1.25 \text{ g Ga} = 0.43 \text{ g O}$$

Next, calculate the amount of each reactant:

$$1.25 \text{ g Ga} \times \frac{1 \text{ mol Ga}}{69.72 \text{ g Ga}} = 0.0179 \text{ mol Ga}$$

$$0.43 \text{ g O} \times \frac{1 \text{ mol O}}{16.0 \text{ g O}} = 0.027 \text{ mol O}$$

Find the ratio of moles of O to moles of Ga:

$$\text{Mole ratio} = \frac{0.027 \text{ mol O}}{0.0179 \text{ mol Ga}} = \frac{1.5 \text{ mol O}}{1.0 \text{ mol Ga}}$$

It is 1.5 mol O/1.0 mol Ga, or 3 mol O to 2 mol Ga. Thus, the product is gallium oxide, Ga_2O_3.

Example 3.9—Determining a Formula from Mass Data

Problem Tin metal (Sn) and purple iodine (I_2) combine to form orange, solid tin iodide with an unknown formula.

$$\text{Sn metal} + \text{solid } I_2 \longrightarrow \text{solid } Sn_xI_y$$

Weighed quantities of Sn and I_2 are combined, where the quantity of Sn is more than is needed to react with *all* of the iodine. After Sn_xI_y has been formed, it is isolated by filtration. The mass of excess tin is also determined. The following data were collected:

Mass of tin (Sn) in original mixture	1.056 g
Mass of iodine (I_2) in original mixture	1.947 g
Mass of tin (Sn) recovered after reaction	0.601 g

Strategy The first step is to find the masses of Sn and I that are combined in Sn_xI_y. The masses are then converted to moles, and the ratio of moles reveals the compound's empirical formula.

(a) Weighed samples of tin (left) and iodine (right).

(b) The tin and iodine are heated in a solvent.

(c) The hot reaction mixture is filtered to recover unreacted tin.

(d) When the solvent cools, solid, orange tin iodide forms and is isolated.

Charles D. Winters

The formula of a compound of tin and iodine can be found by determining the mass of iodine that combines with a given mass of tin.

Solution First, let us find the mass of tin that combined with iodine.

Mass of Sn in original mixture	1.056 g
Mass of Sn recovered	−0.601 g
Mass of Sn combined with 1.947 g I_2	0.455 g

Now convert the mass of tin to the amount of tin.

$$0.455 \text{ g Sn} \times \frac{1 \text{ mol Sn}}{118.7 \text{ g Sn}} = 0.00383 \text{ mol Sn}$$

No I_2 was recovered; it all reacted with Sn. Therefore, 0.00383 mol of Sn combined with 1.947 g of I_2. Because we want to know the amount of I that combined with 0.00383 mol of Sn, we calculate the amount of I from the mass of I_2.

$$1.947 \text{ g } I_2 \times \frac{1 \text{ mol } I_2}{253.81 \text{ g } I_2} \times \frac{2 \text{ mol I}}{1 \text{ mol } I_2} = 0.01534 \text{ mol I}$$

Finally, we find the ratio of moles.

$$\frac{\text{mol I}}{\text{mol Sn}} = \frac{0.01534 \text{ mol I}}{0.00383 \text{ mol Sn}} = \frac{4.01 \text{ mol I}}{1.00 \text{ mol Sn}} = \frac{4 \text{ mol I}}{1 \text{ mol Sn}}$$

There are four times as many moles of I as moles of Sn in the sample. Therefore, there are four times as many atoms of I as atoms of Sn per formula unit. The empirical formula is SnI_4.

Exercise 3.13—Determining a Formula from Combining Masses

Analysis shows that 0.586 g of potassium metal combines with 0.480 g of O_2 gas to give a white solid having a formula of K_xO_y. What is the empirical formula of the compound?

Determining a Formula by Mass Spectrometry

We have described chemical methods of determining a molecular formula, but many instrumental methods are available as well. One of them is *mass spectrometry* (Figure 3.15). We introduced this technique in Chapter 2, where it was used to show

A Closer Look

Mass Spectrometry, Molar Mass, and Isotopes

Bromobenzene, C_6H_5Br, has a molecular weight of 157.010. Why, then, are there two prominent lines at 156 and 158 in the mass spectrum of the compound? The answer shows us the influence of isotopes on molecular weight.

Bromine has two naturally occurring isotopes: ^{79}Br and ^{81}Br. They are 50.7% and 49.3% abundant, respectively. What is the mass of C_6H_5Br based on each isotope? If we use the most abundant isotopes of C and H (^{12}C and ^{1}H), the mass of the compound having only the ^{79}Br isotope, $C_6H_5{}^{79}Br$, is 156. The mass of the compound containing only the ^{81}Br isotope, $C_6H_5{}^{81}Br$, is 158. These two lines have the highest mass-to-charge ratio in the spectrum.

The calculated molecular weight of bromobenzene is 157.010, a value calculated from the atomic weights of the elements. These atomic weights reflect the abundances of all isotopes. In contrast, the mass spectrum has a line for each possible combination of isotopes. This explains why there are also small lines at the mass-to-charge ratios of 157 and 159. They arise from various combinations of ^{1}H, ^{12}C, ^{13}C, ^{79}Br, and ^{81}Br atoms. In fact, careful analysis of such patterns can unambiguously identify a molecule.

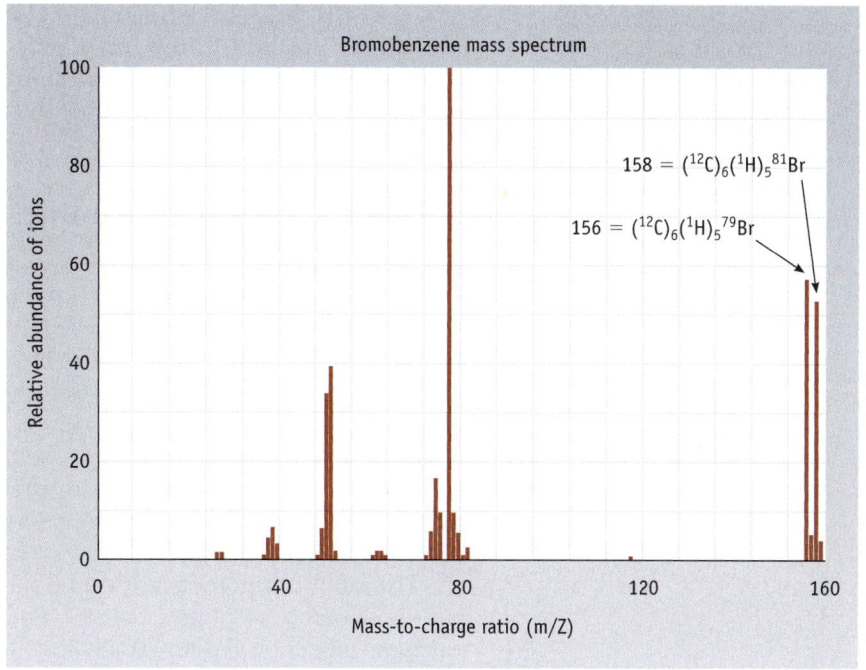

Figure 3.15

Mass spectrum of ethanol, CH_3CH_2OH. A prominent peak or line in the spectrum is the "parent" ion ($CH_3CH_2OH^+$) at mass 46. (The "parent" ion is the heaviest ion observed.) The mass designated by the peak for the "parent" ion confirms the formula of the molecule. Other peaks are for "fragment" ions. This pattern of lines can provide further, unambiguous evidence of the formula of the compound. (The horizontal axis is the mass-to-charge ratio of a given ion. Because almost all observed ions have a charge of $Z = +1$, the value observed is the mass of the ion.) (See A Closer Look: Mass Spectrometry, Molar Mass, and Isotopes.)

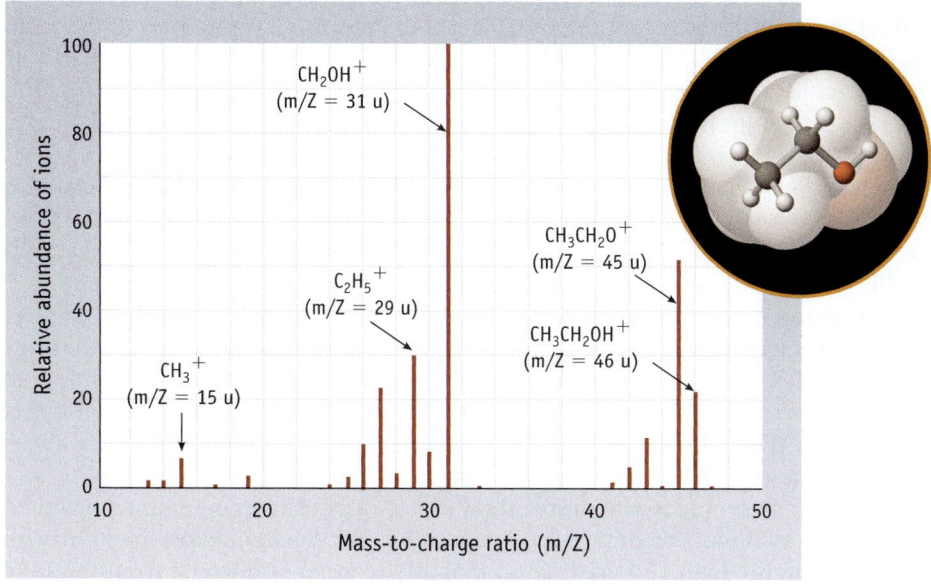

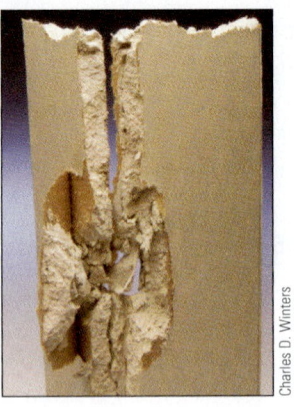

Charles D. Winters

Figure 3.16   Gypsum wallboard. Gypsum is hydrated calcium sulfate, $CaSO_4 \cdot 2\ H_2O$.

the existence of isotopes and to measure their relative abundance [◀ Figure 2.8]. If a compound can be turned into a vapor, the vapor can be passed through an electron beam in a mass spectrometer where high-energy electrons collide with the gas-phase molecules. These high-energy collisions cause the molecule to lose electrons and turn the molecules into positive ions. These ions usually break apart or fragment into smaller pieces. As illustrated in Figure 3.15, the cation created from ethanol ($CH_3CH_2OH^+$) fragments (losing an H atom) to give another cation ($CH_3CH_2O^+$), which further fragments. A mass spectrometer detects and records the masses of the different particles. Analysis of the spectrum can help identify a compound and can give an accurate molar mass.

3.7—Hydrated Compounds

If ionic compounds are prepared in water solution and then isolated as solids, the crystals often have molecules of water trapped within the lattice. Compounds in which molecules of water are associated with the ions of the compound are called **hydrated compounds.** The beautiful blue copper(II) compound in Figure 3.14, for example, has a formula that is conventionally written as $CuCl_2 \cdot 2\ H_2O$. The dot between $CuCl_2$ and $2\ H_2O$ indicates that 2 mol of water is associated with every mole of $CuCl_2$; it is equivalent to writing the formula as $CuCl_2(H_2O)_2$. The name of the compound, copper(II) chloride dihydrate, reflects the presence of 2 mol of water per mole of $CuCl_2$. The molar mass of $CuCl_2 \cdot 2\ H_2O$ is 134.5 g/mol (for $CuCl_2$) plus 36.0 g/mol (for $2\ H_2O$) giving a total mass of 170.5 g/mol.

Hydrated compounds are common. The walls of your home may be covered with wallboard, or "plaster board" (Figure 3.16). These sheets contain hydrated calcium sulfate, or gypsum ($CaSO_4 \cdot 2\ H_2O$), as well as unhydrated $CaSO_4$, sandwiched between paper. Gypsum is a mineral that can be mined. Now, however, it is more commonly a byproduct in the manufacture of superphosphate fertilizer from $Ca_5F(PO_4)_3$ and sulfuric acid.

If gypsum is heated between 120 and 180 °C, the water is partly driven off to give $CaSO_4 \cdot \frac{1}{2}H_2O$, a compound commonly called "plaster of Paris." If you have ever broken an arm or leg and had to have a cast, the cast may have been made of this compound. It is an effective casting material because, when added to water, it forms a thick slurry that can be poured into a mold or spread out over a part of the body. As it takes on more water, the material increases in volume and forms a hard, inflexible solid. These properties also make plaster of Paris a useful material to artists, because the expanding compound fills a mold completely and makes a high-quality reproduction.

Hydrated cobalt(II) chloride is the red solid in Figure 3.17. When heated it turns first purple and then deep blue as it loses water to form anhydrous $CoCl_2$; "**anhydrous**" means a substance without water. On exposure to moist air, anhydrous $CoCl_2$ takes up water and is converted back into the red hydrated compound. It is this property that allows crystals of the blue compound to be used as a humidity indicator. You may have seen them in a small bag packed with a piece of electronic equipment. The compound also makes a good "invisible ink." A solution of cobalt(II) chloride in water is red, but if you write on paper with the solution it cannot be seen. When the paper is warmed, however, the cobalt compound dehydrates to give the deep blue anhydrous compound, and the writing becomes visible.

There is no simple way to predict how much water will be present in a hydrated compound, so it must be determined experimentally. Such an experiment may involve heating the hydrated material so that all the water is released from the solid

Charles D. Winters

Active Figure 3.17   **Dehydrating hydrated cobalt(II) chloride, $CoCl_2 \cdot 6H_2O$.** (*left*) Cobalt chloride hexahydrate, $CoCl_2 \cdot 6H_2O$, is a deep red compound. (*left and center*) When it is heated, the compound loses the water of hydration and forms the deep blue compound $CoCl_2$.

and evaporated. Only the anhydrous compound is left. The formula of hydrated copper(II) sulfate, commonly known as "blue vitriol," is determined in this manner in Example 3.10.

Example 3.10—Determining the Formula of a Hydrated Compound

Problem You want to know the value of x in blue, hydrated copper(II) sulfate, $CuSO_4 \cdot x H_2O$—that is, the number of water molecules for each unit of $CuSO_4$. In the laboratory you weigh out 1.023 g of the solid. After heating the solid thoroughly in a porcelain crucible (see figure), 0.654 g of nearly white, anhydrous copper(II) sulfate, $CuSO_4$, remains.

$$1.023 \text{ g } CuSO_4 \cdot x H_2O + \xrightarrow{\text{heat}} 0.654 \text{ g } CuSO_4 + ? \text{ g } H_2O$$

Strategy To find x we need to know the amount of H_2O per mole of $CuSO_4$. Therefore, first we find the mass of water lost by the sample from the difference between the mass of hydrated compound and the anhydrous form. Finally, we find the ratio of amount of water lost (moles) to the amount of anhydrous $CuSO_4$.

Charles D. Winters

White $CuSO_4$

Blue $CuSO_4 \cdot 5 H_2O$

Heating a Hydrated Compound
The formula of a hydrated compound can be determined by heating a weighed sample enough to cause the compound to release its water of hydration. Knowing the mass of the hydrated compound before heating, and the mass of the anhydrous compound after heating, we can determine the mass of water in the original sample.

Solution Find the mass of water.

Mass of hydrated compound	1.023 g
Mass of anhydrous compound, $CuSO_4$	−0.654 g
Mass of water	0.369 g

Next convert the masses of $CuSO_4$ and H_2O to moles.

$$0.369 \text{ g } H_2O \cdot \frac{1 \text{ mol } H_2O}{18.02 \text{ g } H_2O} = 0.0205 \text{ mol } H_2O$$

$$0.654 \text{ g } CuSO_4 \cdot \frac{1 \text{ mol } CuSO_4}{159.6 \text{ g } CuSO_4} = 0.00410 \text{ mol } CuSO_4$$

The value of x is determined from the mole ratio.

$$\frac{0.0205 \text{ mol } H_2O}{0.00410 \text{ mol } CuSO_4} = \frac{5.00 \text{ mol } H_2O}{1.00 \text{ mol } CuSO_4}$$

The water-to-$CuSO_4$ ratio is 5:1, so the formula of the hydrated compound is $CuSO_4 \cdot 5\ H_2O$. Its name is copper(II) sulfate pentahydrate.

Exercise 3.14—Determining the Formula of a Hydrated Compound

Hydrated nickel(II) chloride is a beautiful green, crystalline compound. When heated strongly, the compound is dehydrated. If 0.235 g of $NiCl_2 \cdot x\ H_2O$ gives 0.128 g of $NiCl_2$ on heating, what is the value of x?

Chapter Goals Revisited

Now that you have studied this chapter, you should ask whether you have met the chapter goals. In particular, you should be able to

Interpret, predict, and write formulas for ionic and molecular compounds

a. Recognize and interpret molecular formulas, condensed formulas, and structural formulas (Section 3.1).

b. Recognize that metal atoms commonly lose one or more electrons to form positive ions, called cations, and nonmetal atoms often gain electrons to form negative ions, called anions (see Figure 3.7).

c. Recognize that the charge on a metal cation in Groups 1A, 2A, and 3A is equal to the group number in which the element is found in the periodic table (M^{n+}, n = Group number) (Section 3.3). Charges on transition metal cations are often 2+ or 3+, but other charges are observed. General ChemistryNow homework: Study Question(s) 11

d. Recognize that the negative charge on a single-atom or monatomic anion, X^{n-}, is given by n = 8 − group number (Section 3.3).

GENERAL Chemistry•Now™

See the General ChemistryNow CD-ROM or website to:

- Assess your understanding with homework questions keyed to each goal
- Check your readiness for an exam by taking the exam-prep quiz and exploring the resources in the personalized Learning Plan it provides

For additional preparation for an examination on this chapter see the **Let's Review** section on pages 282–293.

e. Write formulas for ionic compounds by combining ions in the proper ratio to give no overall charge (Section 3.4).

Name compounds

a. Give the names or formulas of polyatomic ions, knowing their formulas or names, respectively (Table 3.1 and Section 3.3).

b. Name ionic compounds and simple binary compounds of the nonmetals (Sections 3.3 and 3.4). General ChemistryNow homework: SQ(s) 7, 19, 21, 27, 29

Understand some properties of ionic compounds

a. Understand the importance of Coulomb's law (Equation 3.1), which describes the electrostatic forces of attraction and repulsion of ions. Coulomb's law states that the force of attraction between oppositely charged species increases with electric charge and with decreasing distance between the species (Section 3.3). General ChemistryNow homework: SQ(s) 26

Calculate and use molar mass

a. Understand that the molar mass of a compound (often called the molecular weight) is the mass in grams of Avogadro's number of molecules (or formula units) of a compound (Section 3.5). For ionic compounds, which do not consist of individual molecules, the sum of atomic masses is often called the formula mass (or formula weight).

b. Calculate the molar mass of a compound from its formula and a table of atomic weights (Section 3.5). General ChemistryNow homework: SQ(s) 31, 33

c. Calculate the number of moles of a compound that is represented by a given mass, and vice versa (Section 3.5). General ChemistryNow homework: SQ(s) 35

Calculate percent composition for a compound and derive formulas from experimental data

a. Express the composition of a compound in terms of percent composition (Section 3.6). General ChemistryNow homework: SQ(s) 41, 45

b. Use percent composition or other experimental data to determine the empirical formula of a compound (Section 3.6). General ChemistryNow homework: SQ(s) 47, 52, 53, 94

c. Understand how mass spectrometry can be used to find a molar mass (Section 3.6).

d. Use experimental data to find the number of water molecules in a hydrated compound (Section 3.7). General ChemistryNow homework: SQ(s) 55, 57. 59

Key Equations

Equation 3.1 (page 112)

Coulomb's law describes the dependence of the force of attraction between ions of opposite charge (or the force of repulsion between ions of like charge) on ion charge and the distance between ions.

$$\text{Force of attraction} = k \frac{(n^+e)(n^-e)}{d^2}$$

charge on + and − ions — charge on electron

proportionality constant — distance between ions

Study Questions

▲ denotes more challenging questions.

■ denotes questions available in the Homework and Goals section of the General ChemistryNow CD-ROM or website.

Blue numbered questions have answers in Appendix O and fully worked solutions in the *Student Solutions Manual.*

Structures of many of the compounds used in these questions are found on the General ChemistryNow CD-ROM or website in the Models folder.

GENERAL
Chemistry ⚛ Now™ Assess your understanding of this chapter's topics with additional quizzing and conceptual questions at http://now.brookscole.com/kotz6e

Practicing Skills

Molecular Formulas and Models

(See Examples 3.1 and 3.2 and Exercises 3.1 and 3.2.)

1. A ball-and-stick model of sulfuric acid is illustrated here. Write the molecular formula for sulfuric acid and draw the structural formula. Describe the structure of the molecule. Is it flat? That is, are all the atoms in the plane of the paper? (Color code: sulfur atoms are yellow; oxygen atoms are red; and hydrogen atoms are white.)

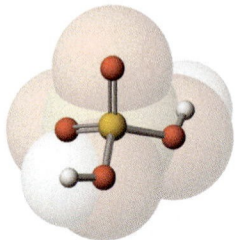

2. A ball-and-stick model of toluene is illustrated here. What is its molecular formula? Describe the structure of the molecule. Is it flat or is only a portion of it flat? (Color code: carbon atoms are gray and hydrogen atoms are white.)

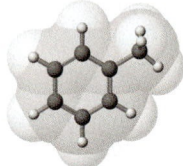

3. A model of the cancer chemotherapy agent cisplatin is given here. Write the molecular formula for the compound and draw its structural formula.

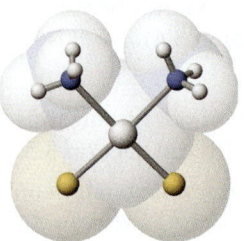

4. The molecule illustrated here is methanol. Using Figure 3.4 as a guide, decide which atoms are in the plane of the paper, which lie above the plane, and which lie below. Sketch a ball-and-stick model. If available to you, go to the General ChemistryNow CD-ROM or website and find the model of methanol.

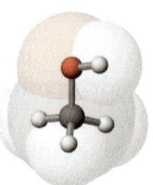

Ions and Ion Charges

(See Exercise 3.3, Figure 3.7, Table 3.1, and the General ChemistryNow Screens 3.5 and 3.6.)

5. What charges are most commonly observed for monatomic ions of the following elements?
   (a) magnesium   (c) nickel
   (b) zinc   (d) gallium

6. What charges are most commonly observed for monatomic ions of the following elements?
   (a) selenium   (c) iron
   (b) fluorine   (d) nitrogen

7. ■ Give the symbol, including the correct charge, for each of the following ions:
   (a) barium ion
   (b) titanium(IV) ion
   (c) phosphate ion
   (d) hydrogen carbonate ion
   (e) sulfide ion
   (f) perchlorate ion
   (g) cobalt(II) ion
   (h) sulfate ion

8. Give the symbol, including the correct charge, for each of the following ions:
   (a) permanganate ion
   (b) nitrite ion
   (c) dihydrogen phosphate ion
   (d) ammonium ion
   (e) phosphate ion
   (f) sulfite ion

9. When a potassium atom becomes a monatomic ion, how many electrons does it lose or gain? What noble gas atom has the same number of electrons as a potassium ion?

10. When oxygen and sulfur atoms become monatomic ions, how many electrons does each lose or gain? Which noble gas atom has the same number of electrons as an oxygen ion? Which noble gas atom has the same number of electrons as a sulfur ion?

Ionic Compounds

(See Examples 3.3 and 3.4 and the General ChemistryNow Screen 3.8.)

11. ■ Predict the charges of the ions in an ionic compound containing the elements barium and bromine. Write the formula for the compound.

12. What are the charges of the ions in an ionic compound containing cobalt(III) and fluoride ions? Write the formula for the compound.

13. For each of the following compounds, give the formula, charge, and the number of each ion that makes up the compound:
    (a) K_2S (d) $(NH_4)_3PO_4$
    (b) $CoSO_4$ (e) $Ca(ClO)_2$
    (c) $KMnO_4$

14. For each of the following compounds, give the formula, charge, and the number of each ion that makes up the compound:
    (a) $Mg(CH_3CO_2)_2$ (d) $Ti(SO_4)_2$
    (b) $Al(OH)_3$ (e) KH_2PO_4
    (c) $CuCO_3$

15. Cobalt forms Co^{2+} and Co^{3+} ions. Write the formulas for the two cobalt oxides formed by these transition metal ions.

16. Platinum is a transition element and forms Pt^{2+} and Pt^{4+} ions. Write the formulas for the compounds of each of these ions with (a) chloride ions and (b) sulfide ions.

17. Which of the following are correct formulas for ionic compounds? For those that are not, give the correct formula.
    (a) $AlCl_2$ (c) Ga_2O_3
    (b) KF_2 (d) MgS

18. Which of the following are correct formulas for ionic compounds? For those that are not, give the correct formula.
    (a) Ca_2O (c) Fe_2O_5
    (b) $SrBr_2$ (d) Li_2O

Naming Ionic Compounds

(See Exercise 3.5 and the General ChemistryNow Screen 3.9.)

19. ■ Name each of the following ionic compounds:
    (a) K_2S (c) $(NH_4)_3PO_4$
    (b) $CoSO_4$ (d) $Ca(ClO)_2$

20. Name each of the following ionic compounds:
    (a) $Ca(CH_3CO_2)_2$ (c) $Al(OH)_3$
    (b) $Ni_3(PO_4)_2$ (d) KH_2PO_4

21. ■ Give the formula for each of the following ionic compounds:
    (a) ammonium carbonate
    (b) calcium iodide
    (c) copper(II) bromide
    (d) aluminum phosphate
    (e) silver(I) acetate

22. Give the formula for each of the following ionic compounds:
    (a) calcium hydrogen carbonate
    (b) potassium permanganate
    (c) magnesium perchlorate
    (d) potassium hydrogen phosphate
    (e) sodium sulfite

23. Write the formulas for the four ionic compounds that can be made by combining each of the cations Na^+ and Ba^{2+} with the anions CO_3^{2-} and I^-. Name each of the compounds.

24. Write the formulas for the four ionic compounds that can be made by combining the cations Mg^{2+} and Fe^{3+} with the anions PO_4^{3-} and NO_3^-. Name each compound formed.

Coulomb's Law

(See Equation 3.1, Figure 3.10, and the General ChemistryNow Screen 3.7.)

25. Sodium ion, Na^+, forms ionic compounds with fluoride, F^-, and iodide, I^-. The radii of these ions are as follows: $Na^+ = 116$ pm; $F^- = 119$ pm; and $I^- = 206$ pm. In which ionic compound, NaF or NaI, are the forces of attraction between cation and anion stronger? Explain your answer.

26. ■ Consider the two ionic compounds NaCl and CaO. In which compound are the cation–anion attractive forces stronger? Explain your answer.

Naming Binary, Nonmetal Compounds

(See Exercise 3.6 and the General ChemistryNow Screen 3.12.)

27. ■ Name each of the following binary, nonionic compounds:
    (a) NF_3 (b) HI (c) BI_3 (d) PF_5

28. Name each of the following binary, nonionic compounds:
    (a) N_2O_5 (b) P_4S_3 (c) OF_2 (d) XeF_4

29. ■ Give the formula for each of the following compounds:
    (a) sulfur dichloride
    (b) dinitrogen pentaoxide
    (c) silicon tetrachloride
    (d) diboron trioxide (commonly called boric oxide)

30. Give the formula for each of the following compounds:
    (a) bromine trifluoride
    (b) xenon difluoride
    (c) hydrazine
    (d) diphosphorus tetrafluoride
    (e) butane

Molecules, Compounds, and the Mole
(See Example 3.5 and the General ChemistryNow Screens 3.14 and 3.15.)

31. ■ Calculate the molar mass of each of the following compounds:
    (a) Fe_2O_3, iron(III) oxide
    (b) BCl_3, boron trichloride
    (c) $C_6H_8O_6$, ascorbic acid (vitamin C)

32. Calculate the molar mass of each of the following compounds:
    (a) $Fe(C_6H_{11}O_7)_2$, iron(II) gluconate, a dietary supplement
    (b) $CH_3CH_2CH_2CH_2SH$, butanethiol, has a skunk-like odor
    (c) $C_{20}H_{24}N_2O_2$, quinine, used as an antimalarial drug

33. ■ Calculate the molar mass of each hydrated compound. Note that the water of hydration is included in the molar mass. (See Section 3.7.)
    (a) $Ni(NO_3)_2 \cdot 6 H_2O$
    (b) $CuSO_4 \cdot 5 H_2O$

34. Calculate the molar mass of each hydrated compound. Note that the water of hydration is included in the molar mass. (See Section 3.7.)
    (a) $H_2C_2O_4 \cdot 2 H_2O$
    (b) $MgSO_4 \cdot 7 H_2O$, Epsom salts

35. ■ What mass is represented by 0.0255 mol of each of the following compounds?
    (a) C_3H_7OH, propanol, rubbing alcohol
    (b) $C_{11}H_{16}O_2$, an antioxidant in foods, also known as BHA (butylated hydroxyanisole)
    (c) $C_9H_8O_4$, aspirin

36. Assume you have 0.123 mol of each of the following compounds. What mass of each is present?
    (a) $C_{14}H_{10}O_4$, benzoyl peroxide, used in acne medications
    (c) $Pt(NH_3)_2Cl_2$, cisplatin, a cancer chemotherapy agent

37. Acetonitrile, CH_3CN, was found in the tail of Comet Hale-Bopp in 1997. What amount (moles) of acetonitrile is represented by 2.50 kg?

38. Acetone, $(CH_3)_2CO$, is an important industrial solvent. If 1260 million kg of this organic compound is produced annually, what amount (moles) is produced?

39. Sulfur trioxide, SO_3, is made industrially in enormous quantities by combining oxygen and sulfur dioxide, SO_2. What amount (moles) of SO_3 is represented by 1.00 kg of sulfur trioxide? How many molecules? How many sulfur atoms? How many oxygen atoms?

40. An Alka-Seltzer tablet contains 324 mg of aspirin ($C_9H_8O_4$), 1904 mg of $NaHCO_3$, and 1000. mg of citric acid ($H_3C_6H_5O_7$). (The last two compounds react with each other to provide the "fizz," bubbles of CO_2, when the tablet is put into water.)
    (a) Calculate the amount (moles) of each substance in the tablet.
    (b) If you take one tablet, how many molecules of aspirin are you consuming?

Percent Composition
(See Exercise 3.6 and the General ChemistryNow Screen 3.16.)

41. ■ Calculate the mass percent of each element in the following compounds.
    (a) PbS, lead(II) sulfide, galena
    (b) C_3H_8, propane
    (c) $C_{10}H_{14}O$, carvone, found in caraway seed oil

42. Calculate the mass percent of each element in the following compounds:
    (a) $C_8H_{10}N_2O_2$, caffeine
    (b) $C_{10}H_{20}O$, menthol
    (c) $CoCl_2 \cdot 6 H_2O$

43. Calculate the weight percent of lead in PbS, lead(II) sulfide. What mass of lead (in grams) is present in 10.0 g of PbS?

44. Calculate the weight percent of iron in Fe_2O_3, iron(III) oxide. What mass of iron (in grams) is present in 25.0 g of Fe_2O_3?

45. ■ Calculate the weight percent of copper in CuS, copper(II) sulfide. If you wish to obtain 10.0 g of copper metal from copper(II) sulfide, what mass of the sulfide (in grams) must you use?

46. Calculate the weight percent of titanium in the mineral ilmenite, $FeTiO_3$. What mass of ilmenite (in grams) is required if you wish to obtain 750 g of titanium?

Empirical and Molecular Formulas
(See Example 3.7 and the General ChemistryNow Screens 3.16–3.18.)

47. ■ Succinic acid occurs in fungi and lichens. Its empirical formula is $C_2H_3O_2$ and its molar mass is 118.1 g/mol. What is its molecular formula?

48. An organic compound has the empirical formula C_2H_4NO. If its molar mass is 116.1 g/mol, what is the molecular formula of the compound?

49. Complete the following table:

	Empirical Formula	Molar Mass (g/mol)	Molecular Formula
(a)	CH	26.0	_____
(b)	CHO	116.1	_____
(c)	_____	_____	C_8H_{16}

▲ More challenging   ■ In General ChemistryNow   **Blue-numbered questions** answered in Appendix O

50. Complete the following table:

Empirical Formula	Molar Mass (g/mol)	Molecular Formula
(a) $C_2H_3O_3$	150.0	_____
(b) C_3H_8	44.1	_____
(c) _____	_____	B_4H_{10}

51. Acetylene is a colorless gas used as a fuel in welding torches, among other things. It is 92.26% C and 7.74% H. Its molar mass is 26.02 g/mol. What are the empirical and molecular formulas of acetylene?

52. ■ A large family of boron-hydrogen compounds has the general formula B_xH_y. One member of this family contains 88.5% B; the remainder is hydrogen. Which of the following is its empirical formula: BH_2, BH_3, B_2H_5, B_5H_7, or B_5H_{11}?

53. ■ Cumene is a hydrocarbon, a compound composed only of C and H. It is 89.94% carbon, and its molar mass is 120.2 g/mol. What are the empirical and molecular formulas of cumene?

54. Nitrogen and oxygen form a series of oxides with the general formula N_xO_y. One of them, a blue solid, contains 36.84% N. What is the empirical formula of this oxide?

55. ■ Mandelic acid is an organic acid composed of carbon (63.15%), hydrogen (5.30%), and oxygen (31.55%). Its molar mass is 152.14 g/mol. Determine the empirical and molecular formulas of the acid.

56. Nicotine, a poisonous compound found in tobacco leaves, is 74.0% C, 8.65% H, and 17.35% N. Its molar mass is 162 g/mol. What are the empirical and molecular formulas of nicotine?

Determining Formulas from Mass Data
(See Examples 3.8–3.10 and the General ChemistryNow Screens 3.17–3.19.)

57. ■ If Epsom salt, $MgSO_4 \cdot x\,H_2O$, is heated to 250 °C, all the water of hydration is lost. On heating a 1.687-g sample of the hydrate, 0.824 g of $MgSO_4$ remains. How many molecules of water occur per formula unit of $MgSO_4$?

58. The "alum" used in cooking is potassium aluminum sulfate hydrate, $KAl(SO_4)_2 \cdot x\,H_2O$. To find the value of x, you can heat a sample of the compound to drive off all of the water and leave only $KAl(SO_4)_2$. Assume you heat 4.74 g of the hydrated compound and that the sample loses 2.16 g of water. What is the value of x?

59. ■ A new compound containing xenon and fluorine was isolated by shining sunlight on a mixture of Xe (0.526 g) and F_2 gas. If you isolate 0.678 g of the new compound, what is its empirical formula?

60. Elemental sulfur (1.256 g) is combined with fluorine, F_2, to give a compound with the formula SF_x, a very stable, colorless gas. If you have isolated 5.722 g of SF_x, what is the value of x?

61. Zinc metal (2.50 g) combines with 9.70 g of iodine to produce zinc iodide, Zn_xI_y. What is the formula of this ionic compound?

62. You combine 1.25 g of germanium, Ge, with excess chlorine, Cl_2. The mass of product, Ge_xCl_y, is 3.69 g. What is the formula of the product, Ge_xCl_y?

General Questions

These questions are not designated as to type or locations in the chapter. They may combine several concepts. More challenging questions are marked with the icon ▲.

63. Write formulas for all of the compounds that can be made by combining the cations NH_4^+ and Ni^{2+} with the anions CO_3^{2-} and SO_4^{2-}.

64. Using the General ChemistryNow CD-ROM or website, find a model for each of the following molecules. Write the molecular formula and draw the structural formula.
    (a) acetic acid
    (b) methylamine
    (c) formaldehyde

65. How many electrons are in a strontium atom (Sr)? Does an atom of Sr gain or lose electrons when forming an ion? How many electrons are gained or lost by the atom? When Sr forms an ion, the ion has the same number of electrons as which one of the noble gases?

66. The compound $(NH_4)_2SO_4$ consists of two polyatomic ions. What are the names and electric charges of these ions? What is the molar mass of this compound?

67. Which of the following compounds has the highest weight percent of chlorine?
    (a) BCl_3 (d) $AlCl_3$
    (b) $AsCl_3$ (e) PCl_3
    (c) $GaCl_3$

68. Which of the following samples has the largest number of ions?
    (a) 1.0 g of $BeCl_2$ (d) 1.0 g of $SrCO_3$
    (b) 1.0 g of $MgCl_2$ (e) 1.0 g of $BaSO_4$
    (c) 1.0 g of CaS

69. Which of the following compounds (NO, CO, MgO, or CaO) has the highest weight percent of oxygen?

70. The chemical compound alum has the formula $KAl(SO_4)_2 \cdot 12\,H_2O$. Give formulas for the ions that make up this ionic compound.

71. Knowing that the formula of sodium borate is Na_3BO_3, give the formula and charge of the borate ion. Is the borate ion a cation or an anion?

72. What is the difference between an empirical formula and a molecular formula? Use the compound ethane, C_2H_6, to illustrate your answer.

73. The structure of one of the bases in DNA, adenine, is shown here. Which represents the greater mass: 40.0 g of adenine or 3.0×10^{23} molecules of the compound?

74. Which has the larger mass, 0.5 mol of $BaCl_2$ or 0.5 mol of $SiCl_4$?

75. ■ A drop of water has a volume of about 0.05 mL. How many molecules of water are in a drop of water? (Assume water has a density of 1.00 g/cm³.)

76. Capsaicin, the compound that gives the hot taste to chili peppers, has the formula $C_{18}H_{27}NO_3$.
    (a) Calculate its molar mass.
    (b) If you eat 55 mg of capsaicin, what amount (moles) have you consumed?
    (c) Calculate the mass percent of each element in the compound.
    (d) What mass of carbon (in milligrams) is there in 55 mg of capsaicin?

77. Calculate the molar mass and the mass percent of each element in the blue solid $Cu(NH_3)_4SO_4 \cdot H_2O$. What are the mass (in grams) of copper and the mass of water in 10.5 g of the compound?

78. Write the molecular formula and calculate the molar mass for each of the molecules shown here. Which has the larger percentage of carbon? Of oxygen?
    (a) Ethylene glycol (used in antifreeze)

    (b) Dihydroxyacetone (used in artificial tanning lotions)

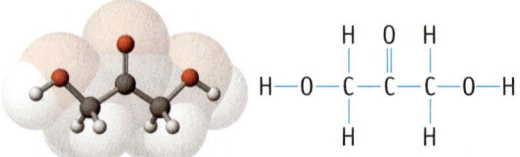

79. Malic acid, an organic acid found in apples, contains C, H, and O in the following ratios: $C_1H_{1.50}O_{1.25}$. What is the empirical formula of malic acid?

80. Your doctor has diagnosed you as being anemic—that is, as having too little iron in your blood. At the drugstore you find two iron-containing dietary supplements: one with

iron(II) sulfate, $FeSO_4$, and the other with iron(II) gluconate, $Fe(C_6H_{11}O_7)_2$. If you take 100. mg of each compound, which will deliver more atoms of iron?

81. ▲ Spinach is high in iron (2 mg per 90-g serving). It is also a source of the oxalate ion, $C_2O_4{}^{2-}$; however, oxalate ion combines with iron ions to form iron oxalate, $Fe_x(C_2O_4)_y$, a substance that prevents your body from absorbing the iron. Analysis of a 0.109-g sample of iron oxalate shows that it contains 38.82% iron. What is the empirical formula of the compound?

82. A compound composed of iron and carbon monoxide, $Fe_x(CO)_y$, is 30.70% iron. What is the empirical formula for the compound?

83. *Ma huang*, an extract from the ephedra species of plants, contains ephedrine. The Chinese have used this herb more than 5000 years to treat asthma. More recently the substance has been used in diet pills that can be purchased over the counter in herbal medicine shops. However, very serious concerns have been raised regarding these pills following reports of serious heart problems with their use.
    (a) Write the molecular formula for ephedrine, draw its structural formula, and calculate its molar mass.
    (b) What is the weight percent of carbon in ephedrine?
    (c) Calculate the amount (moles) of ephedrine in a 0.125-g sample.
    (d) How many molecules of ephedrine are there in 0.125 g? How many C atoms?

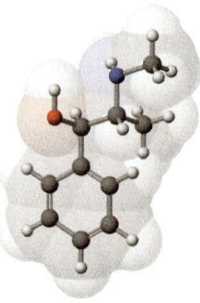

84. Saccharin is more than 300 times sweeter than sugar. It was first made in 1897, a time when it was common practice for chemists to record the taste of any new substances they synthesized.
    (a) Write the molecular formula for the compound and draw its structural formula. (S atoms are yellow.)
    (b) If you ingest 125 mg of saccharin, what amount (moles) of saccharin have you ingested?
    (c) What mass of sulfur is contained in 125 mg of saccharin?

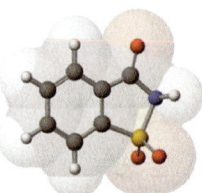

85. Which of the following pairs of elements are likely to form ionic compounds when allowed to react with each other? Write appropriate formulas for the ionic compounds you expect to form, and give the name of each.

  (a) chlorine and bromine
  (b) phosphorus and bromine
  (c) lithium and sulfur
  (d) indium and oxygen
  (e) sodium and argon
  (f) sulfur and bromine
  (g) calcium and fluorine

86. Name each of the following compounds, and tell which ones are best described as ionic:

  (a) ClF_3 (f) OF_2
  (b) NCl_3 (g) KI
  (c) $SrSO_4$ (h) Al_2S_3
  (d) $Ca(NO_3)_2$ (i) PCl_3
  (e) XeF_4 (j) K_3PO_4

87. Write the formula for each of the following compounds, and tell which ones are best described as ionic:

  (a) sodium hypochlorite
  (b) boron triiodide
  (c) aluminum perchlorate
  (d) calcium acetate
  (e) potassium permanganate
  (f) ammonium sulfite
  (g) potassium dihydrogen phosphate
  (h) disulfur dichloride
  (i) chlorine trifluoride
  (j) phosphorus trifluoride

88. Complete the table by placing symbols, formulas, and names in the blanks.

Cation	Anion	Name	Formula
_____	_____	ammonium bromide	_____
Ba^{2+}	_____	_____	BaS
_____	Cl^-	iron(II) chloride	_____
_____	F^-	_____	PbF_2
Al^{3+}	$CO_3{}^{2-}$	_____	_____
_____	_____	iron(III) oxide	_____

89. Complete the table by placing symbols, formulas, and names in the blanks.

Cation	Anion	Name	Formula
_____	_____	_____	$LiClO_4$
_____	_____	aluminum phosphate	_____
_____	Br^-	lithium bromide	_____
_____	_____	_____	$Ba(NO_3)_2$
Al^{3+}	_____	aluminum oxide	_____
_____	_____	iron(III) carbonate	_____

90. Fluorocarbonyl hypofluorite is composed of 14.6% C, 39.0% O, and 46.3% F. If the molar mass of the compound is 82 g/mol, determine the empirical and molecular formulas of the compound.

91. Azulene, a beautiful blue hydrocarbon, is 93.71% C and has a molar mass of 128.16 g/mol. What are the empirical and molecular formulas of azulene?

92. Cacodyl, a compound containing arsenic, was reported in 1842 by the German chemist Robert Wilhelm Bunsen. It has an almost intolerable garlic-like odor. Its molar mass is 210 g/mol, and it is 22.88% C, 5.76% H, and 71.36% As. Determine its empirical and molecular formulas.

93. The action of bacteria on meat and fish produces a compound called cadaverine. As its name and origin imply, it stinks! (It is also present in bad breath and adds to the odor of urine.) It is 58.77% C, 13.81% H, and 27.40% N. Its molar mass is 102.2 g/mol. Determine the molecular formula of cadaverine.

94. ■ ▲ Transition metals can combine with carbon monoxide (CO) to form compounds such as $Fe(CO)_5$ (Study Question 3.82). Assume that you combine 0.125 g of nickel with CO and isolate 0.364 g of $Ni(CO)_x$. What is the value of *x*?

95. ▲ A major oil company has used a gasoline additive called MMT to boost the octane rating of its gasoline. What is the empirical formula of MMT if it is 49.5% C, 3.2% H, 22.0% O, and 25.2% Mn?

96. ▲ Elemental phosphorus is made by heating calcium phosphate with carbon and sand in an electric furnace. What is the weight percent of phosphorus in calcium phosphate? Use this value to calculate the mass of calcium phosphate (in kilograms) that must be used to produce 15.0 kg of phosphorus.

97. ▲ Chromium is obtained by heating chromium(III) oxide with carbon. Calculate the weight percent of chromium in the oxide and then use this value to calculate the quantity of Cr_2O_3 required to produce 850 kg of chromium metal.

98. ▲ Stibnite, Sb_2S_3, is a dark gray mineral from which antimony metal is obtained. What is the weight percent of antimony in the sulfide? If you have 1.00 kg of an ore that contains 10.6% antimony, what mass of Sb_2S_3 (in grams) is in the ore?

99. ▲ Direct reaction of iodine (I_2) and chlorine (Cl_2) produces an iodine chloride, I_xCl_y, a bright yellow solid. If you completely used up 0.678 g of iodine and produced 1.246 g of I_xCl_y, what is the empirical formula of the compound? A later experiment showed that the molar mass of I_xCl_y was 467 g/mol. What is the molecular formula of the compound?

100. ▲ In a reaction 2.04 g of vanadium combined with 1.93 g of sulfur to give a pure compound. What is the empirical formula of the product?

101. ▲ Iron pyrite, often called "fool's gold," has the formula FeS_2. If you could convert 15.8 kg of iron pyrite to iron metal, what mass of the metal would you obtain?

▲ More challenging   ■ In General ChemistryNow   **Blue-numbered questions** answered in Appendix O

102. Which of the following statements about 57.1 g of octane, C_8H_{18}, is (are) not true?
 (a) 57.1 g is 0.500 mol of octane.
 (b) The compound is 84.1% C by weight.
 (c) The empirical formula of the compound is C_4H_9.
 (d) 57.1 g of octane contains 28.0 g of hydrogen atoms.

103. The formula of barium molybdate is $BaMoO_4$. Which of the following is the formula of sodium molybdate?
 (a) Na_4MoO　　(c) Na_2MoO_3　　(e) Na_4MoO_4
 (b) $NaMoO$　　(d) Na_2MoO_4

104. ▲ A metal M forms a compound with the formula MCl_4. If the compound is 74.75% chlorine, what is the identity of M?

105. Pepto-Bismol, which helps provide soothing relief for an upset stomach, contains 300. mg of bismuth subsalicylate, $C_{21}H_{15}Bi_3O_{12}$, per tablet. If you take two tablets for your stomach distress, what amount (in moles) of the "active ingredient" are you taking? What mass of Bi are you consuming in two tablets?

106. ▲ The weight percent of oxygen in an oxide that has the formula MO_2 is 15.2%. What is the molar mass of this compound? What element or elements are possible for M?

107. The mass of 2.50 mol of a compound with the formula ECl_4, in which E is a nonmetallic element, is 385 g. What is the molar mass of ECl_4? What is the identity of E?

108. ▲ The elements A and Z combine to produce two different compounds: A_2Z_3 and AZ_2. If 0.15 mol of A_2Z_3 has a mass of 15.9 g and 0.15 mole of AZ_2 has a mass of 9.3 g, what are the atomic masses of A and Z?

109. ▲ Polystyrene can be prepared by heating styrene with tribromobenzoyl peroxide in the absence of air. A sample prepared by this method has the empirical formula $Br_3C_6H_3(C_8H_8)_n$, where the value of n can vary from sample to sample. If one sample has 10.46% Br, what is the value of n?

110. A sample of hemoglobin is found to be 0.335% iron. If hemoglobin contains one iron atom per molecule, what is the molar mass of hemoglobin? What is the molar mass if there are four iron atoms per molecule?

Summary and Conceptual Questions

The following questions use concepts from the preceding chapters.

111. A piece of nickel foil, 0.550 mm thick and 1.25 cm square, is allowed to react with fluorine, F_2, to give a nickel fluoride.
 (a) How many moles of nickel foil were used? (The density of nickel is 8.902 g/cm^3.)
 (b) If you isolate 1.261 g of the nickel fluoride, what is its formula?
 (c) What is its complete name?

112. An ionic compound can dissolve in water because the cations and anions are attracted to water molecules. The drawing here shows how a cation and a water molecule, which has a negatively charged O atom and positively charged H atoms, can interact. Which of the following cations should be most strongly attracted to water: Na^+, Mg^{2+}, or Al^{3+}? Explain briefly.

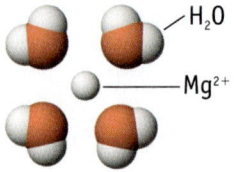

Water molecules interacting with a magnesium ion.

113. ▲ When analyzed, an unknown compound gave these experimental results: C, 54.0%; H, 6.00%; and O, 40.0%. Four different students used these values to calculate the empirical formulas shown here. Which answer is correct? Why did some students not get the correct answer?
 (a) $C_4H_5O_2$　　(c) $C_7H_{10}O_4$
 (b) $C_5H_7O_3$　　(d) $C_9H_{12}O_5$

114. ▲ Two general chemistry students working together in the lab weigh out 0.832 g of $CaCl_2 \cdot 2\ H_2O$ into a crucible. After heating the sample for a short time and allowing the crucible to cool, the students determine that the sample has a mass of 0.739 g. They then do a quick calculation. On the basis of this calculation, what should they do next?
 (a) Congratulate themselves on a job well done.
 (b) Assume the bottle of $CaCl_2 \cdot 2\ H_2O$ was mislabeled; it actually contained something different.
 (c) Heat the crucible again, and then reweigh it.

115. ▲ Uranium is used as a fuel, primarily in the form of uranium(IV) oxide, in nuclear power plants. This question considers some uranium chemistry.
 (a) A small sample of uranium metal (0.169 g) is heated to between 800 and 900 °C in air to give 0.199 g of a dark green oxide, U_xO_y. How many moles of uranium metal were used? What is the empirical formula of the oxide, U_xO_y? What is the name of the oxide? How many moles of U_xO_y must have been obtained?
 (b) The naturally occurring isotopes of uranium are ^{234}U, ^{235}U, and ^{238}U. Knowing that uranium's atomic weight is 238.02 g/mol, which isotope must be the most abundant?
 (c) If the hydrated compound $UO_2(NO_3)_2 \cdot z\ H_2O$ is heated gently, the water of hydration is lost. If you have 0.865 g of the hydrated compound and obtain 0.679 g of $UO_2(NO_3)_2$ on heating, how many molecules of water of hydration are in each formula unit of the original compound? (The oxide U_xO_y is obtained if the hydrate is heated to temperatures over 800 °C in the air.)

---

116. The "simulation" section on General ChemistryNow Screen 3.7 Coulomb's Law, helps you explore Coulomb's law. You can change the charges on the ions and the distance between them. If the ions experience an attractive force, arrows point from one ion to the other. Repulsion is indicated by arrows pointing in opposite directions. Change the ion charges (from $1\pm$ to $2\pm$ to $3\pm$). How does this affect the attractive force? How close can the ions approach before significant repulsive forces set in? How does this distance vary with ion charge?

117. The common chemical compound alum has the formula $KAl(SO_4)_2 \cdot 12\ H_2O$. An interesting characteristic of alum is that it is possible to grow very large crystals of this compound. Suppose you have a crystal of alum in the form of a cube that is 3.00 cm on each side. You want to know how many aluminum atoms are contained in this cube. Outline the steps to determine this value, and indicate the information that you need to carry out each step.

118. Cobalt(II) chloride hexahydrate dissolves readily in water to give a red solution. If we use this solution as an "ink," we can write secret messages on paper. The writing is not visible when the water evaporates from the paper. When the paper is heated, however, the message can be read. Explain the chemistry behind this observation. (*See General ChemistryNow Screen 3.20 Chemical Puzzler.*)

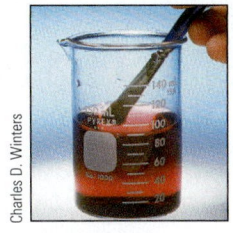  

A solution of $CoCl_2 \cdot 6\ H_2O$.

Using the secret ink to write on paper.

Heating the paper reveals the writing.

Do you need a live tutor for homework problems?
Access vMentor at General ChemistryNow at
http://now.brookscole.com/kotz6e
for one-on-one tutoring from a chemistry expert

4—Chemical Equations and Stoichiometry

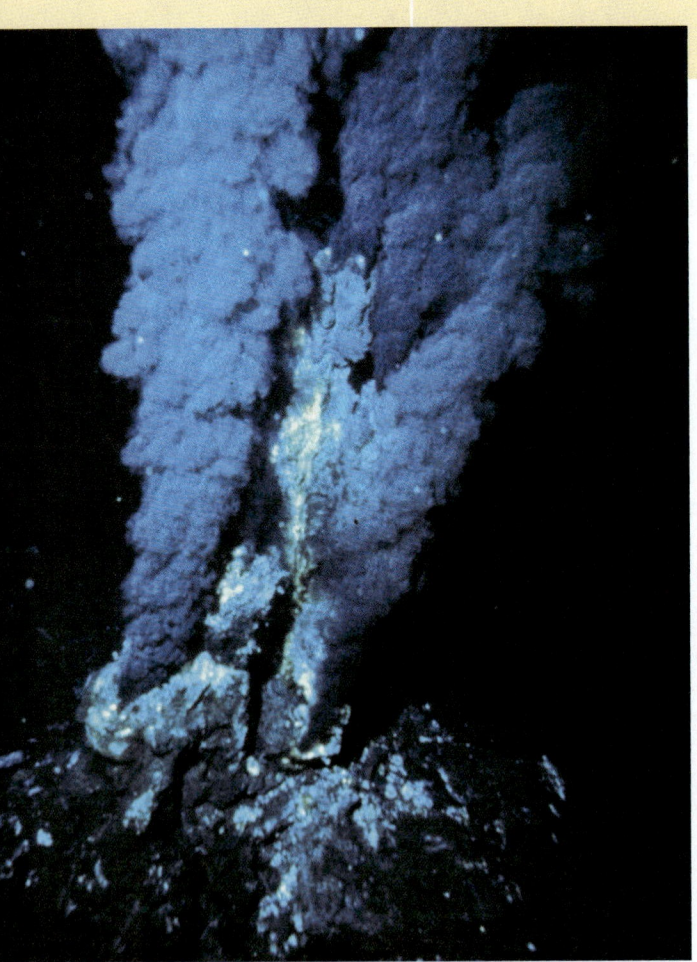

A "black smoker" in the East Pacific Rise.

National Oceanic and Atmospheric Administration/Department of Commerce

Black Smokers and the Origin of Life

"The origin of life appears almost a miracle, so many are the conditions which would have had to be satisfied to get it going."

Francis Crick, quoted by John Horgan, "In the Beginning," *Scientific American*, pp. 116–125, February 1991.

The statement by Francis Crick on the origin of life does not mean that chemists and biologists have not tried to find the conditions under which life might have begun. Charles Darwin thought life might have begun when simple molecules combined to produce molecules of greater and greater complexity. Darwin's idea lives on in experiments such as those done by Stanley Miller in 1953. Attempting to recreate what was thought to be the atmosphere of the primeval earth, Miller filled a flask with the gases methane, ammonia, and hydrogen and added a bit of water. A discharge of electricity acted like lightning in the mixture. The inside of the flask was soon covered with a reddish slime, a mixture found to contain amino acids, the building blocks of proteins. Chemists thought they would soon know in more detail how living organisms began their development—but it was not to be. As Miller said recently, "The problem of the origin of life has turned out to be much more difficult than I, and most other people, envisioned."

Other theories have been advanced to account for the origin of life. The most recent conjecture relates to the discovery of geologically active sites on the ocean floor. Could life have originated in such exotic environments? The evidence is tenuous. As in Miller's experiments, this hypothesis relies on the creation of complex carbon-based molecules from simple ones.

In 1977 scientists were exploring the junction of two of the tectonic plates that form the floor of the Pacific Ocean. There they

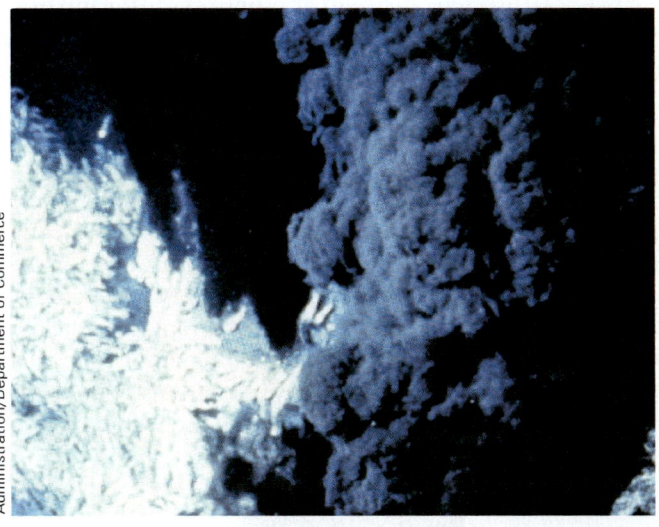

National Oceanic and Atmospheric Administration/Department of Commerce

Black smoker chimney and shrimp on the Mid-Atlantic Ridge.

found thermal springs gushing a hot, black soup of minerals. Water seeping into cracks in the thin surface of the earth is superheated to between 300 and 400 °C by the magma of the earth's core. This superhot water dissolves minerals in the crust and provides conditions for the conversion of sulfate ions in sea water to hydrogen sulfide, H_2S. When this hot water, now laden with dissolved minerals and rich in sulfides, gushes through the surface, it cools. Metal sulfides, such as those of copper, manganese, iron, zinc, and nickel, then precipitate.

Many metal sulfides are black, and the plume of material coming from the sea bottom looks like black "smoke"; for this reason,

the vents have been called "black smokers." The solid sulfides settle around the edges of the vent on the sea floor, eventually forming a "chimney" of precipitated minerals.

Scientists were amazed to find that the black smoker vents were surrounded by primitive animals living in the hot, sulfide-rich environment. Because smokers lie under hundreds of meters of water and sunlight does not penetrate to these depths, the animals have developed a way to live without energy from sunlight. It is currently believed that they derive the energy needed to survive from the reaction of oxygen with hydrogen sulfide, H_2S:

$$H_2S(aq) + 2\ O_2(aq) \longrightarrow H_2SO_4(aq) + energy$$

The hypothesis that life might have originated in this inhospitable location developed out of laboratory experiments by a German lawyer and scientist, G. Wächtershäuser and a colleague, Claudia Huber. They found that metal sulfides such as iron sulfide promote reactions that convert simple carbon-containing molecules to more complex molecules. If this transformation could happen in the laboratory, perhaps similar chemistry might also occur in the exotic environment of black smokers.

To Review Before You Begin

- Review names and formulas of common compounds and ions (Chapter 3)
- Know how to convert mass to moles and moles to mass (Chapters 2 and 3)

Chemistry ❄ Now™

Throughout the chapter this icon introduces a list of resources on the General ChemistryNow CD-ROM or website (http://now .brookscole.com/kotz6e) that will:

- help you evaluate your knowledge of the material
- provide homework problems
- allow you to take an exam-prep quiz
- provide a personalized Learning Plan targeting resources that address areas you should study

■ **Information from Chemical Equations**
Chemical equations show the compounds involved in the chemical reaction and their physical state. Equations usually do *not* show the conditions of the experiment or indicate whether any energy (in the form of heat or light) is involved.

When you think about chemistry, you probably think of chemical reactions. The image of a medieval chemist mixing chemicals in hopes of turning lead into gold lingers in the imagination. Of course, there is much more to chemistry. Just reading this sentence involves an untold number of chemical reactions in your body. Indeed, every activity of living things depends on carefully regulated chemical reactions. Our objective in this chapter is to introduce the quantitative study of chemical reactions. Quantitative studies are needed to determine, for example, how much oxygen is required for the complete combustion of a given quantity of gasoline and what mass of carbon dioxide and water can be obtained. This part of chemistry is fundamental to much of what chemists, chemical engineers, biochemists, molecular biologists, geochemists, and many others do.

4.1—Chemical Equations

When a stream of chlorine gas, Cl_2, is directed onto solid phosphorus, P_4, the mixture bursts into flame, and a chemical reaction produces liquid phosphorus trichloride, PCl_3 (Figure 4.1). We can depict this reaction using a **balanced chemical equation**.

$$P_4(s) + 6\ Cl_2(g) \longrightarrow 4\ PCl_3(\ell)$$
$$\underbrace{}_{\text{Reactants}} \qquad \underbrace{}_{\text{Products}}$$

In a balanced equation, the formulas for the **reactants** (the substances combined in the reaction) are written to the left of the arrow and the formulas for the **products** (the substances produced) are written to the right of the arrow. The physical states of reactants and products can also be indicated. The symbol (s) indicates a solid, (g) a gas, and (ℓ) a liquid. A substance dissolved in water—that is, an *aqueous* solution of a substance—is indicated by (aq). The relative amounts of the reactants and products are shown by numbers, the *coefficients*, before the formulas.

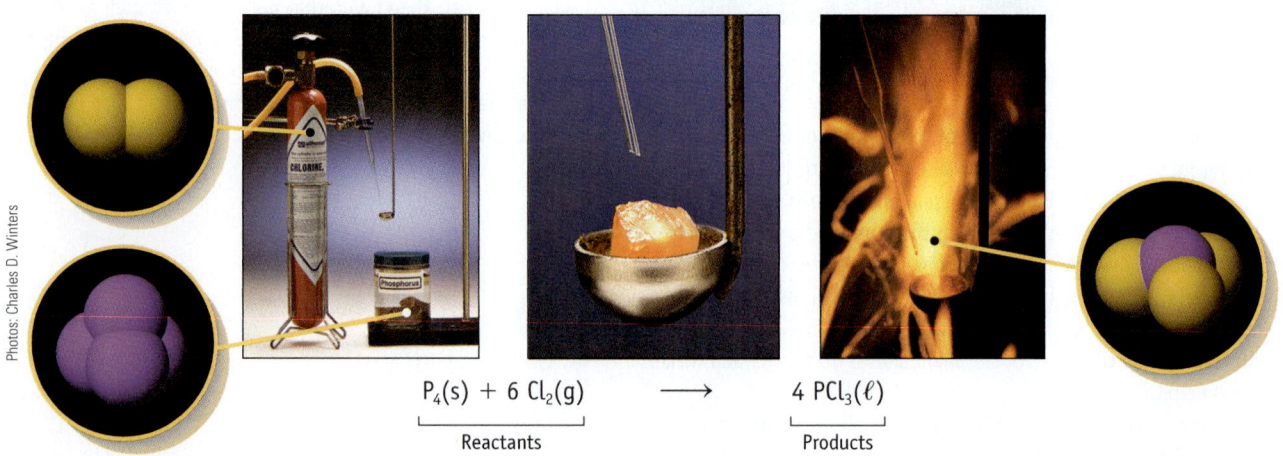

$$P_4(s) + 6\ Cl_2(g)$$
$$\underbrace{}_{\text{Reactants}} \qquad \longrightarrow \qquad \underbrace{4\ PCl_3(\ell)}_{\text{Products}}$$

Figure 4.1   Reaction of solid white phosphorus with chlorine gas. The product is liquid phosphorus trichloride.

Photos: Charles D. Winters

Historical Perspectives

Antoine Laurent Lavoisier (1743–1794)

On Monday, August 7, 1774, the English-man Joseph Priestley (1733–1804) became the first person to isolate oxygen. He heated solid mercury(II) oxide, HgO, causing the oxide to decompose to mercury and oxygen.

$$2 \, HgO(s) \longrightarrow 2 \, Hg(\ell) + O_2(g)$$

Priestley did not immediately understand the significance of his discovery, but he mentioned it to the French chemist Antoine Lavoisier in October 1774. One of Lavoisier's contributions to science was his recognition of the importance of exact scientific measurements and of carefully planned experiments, and he applied these methods to the study of oxygen. From this work he came to believe Priestley's gas was present in all acids, so he named it "oxygen," from the Greek words meaning "to form an acid." In addition, Lavoisier observed that the heat produced by a guinea pig when exhaling a given amount of carbon dioxide is similar to the quantity of heat produced by burning carbon to give the same amount of carbon dioxide. From this and other experiments he concluded that "Respiration is a combustion, slow it is true, but otherwise perfectly similar to that of charcoal." Although he did not understand the details of the process,

Lavoisier and his wife, as painted in 1788 by Jacques-Louis David. Lavoisier was then 45 and his wife, Marie Anne Pierrette Paulze, was 30.

Lavoisier's recognition marked an important step in the development of biochemistry.

Lavoisier was a prodigious scientist and the principles of naming chemical substances that he introduced are still in use today. Further, he wrote a textbook in which he applied for the first time the principles of the conservation of matter to chemistry and used the idea to write early versions of chemical equations.

Because Lavoisier was an aristocrat, he came under suspicion during the Reign of Terror of the French Revolution. He was an investor in the Ferme Générale, the infamous tax-collecting organization in 18th-century France. Tobacco was a monopoly product of the Ferme Générale, and it was a common occurrence to cheat the purchaser by adding water to the tobacco, a practice that Lavoisier opposed. Nonetheless, because of his involvement with the Ferme, his career was cut short by the guillotine on May 8, 1794, on the charge of "adding water to the people's tobacco."

The decomposition of red mercury (II) oxide. The decomposition reaction gives mercury metal and oxygen gas. The mercury is seen as a film on the surface of the test tube.

Photos: (Center) The Metropolitan Museum of Art, Purchase, Mr. and Mrs. Charles Wrightsman gift, in honor of Everett Fahy, 1997. Photograph © 1989 The Metropolitan Museum of Art. (Right) Charles D. Winters.

In the 18th century, the great French scientist Antoine Lavoisier (1743–1794) introduced the **law of conservation of matter**, which states that *matter can be neither created nor destroyed.* This means that if the total mass of reactants is 10 g, and if the reaction completely converts reactants to products, you must end up with 10 g of products. It also means that if 1000 atoms of a particular element are contained in the reactants, then those 1000 atoms must appear in the products in some fashion.

When applied to the reaction of phosphorus and chlorine, the law of conservation of matter tells us that 1 molecule of phosphorus (with 4 phosphorus atoms) and 6 diatomic molecules of Cl_2 (with 12 atoms of Cl) are required to produce 4 molecules of PCl_3. Because each PCl_3 molecule contains 1 P atom and 3 Cl atoms, the 4 PCl_3 molecules are needed to account for 4 P atoms and 12 Cl atoms in the product.

■ **More Information from Chemical Equations**
The same number of atoms must exist after a reaction as before it takes place. However, these atoms are arranged differently. In the phosphorus/chlorine reaction, for example, the P atoms were in the form of P_4 molecules before reaction, but appear as PCl_3 molecules after reaction.

$$6 \times 2 = \qquad 4 \times 3 =$$
$$\text{12 Cl atoms} \qquad \text{12 Cl atoms}$$

$$P_4(s) + 6 \, Cl_2(g) \longrightarrow 4 \, PCl_3(\ell)$$

$$\text{4 P atoms} \qquad\qquad \text{4 P atoms}$$

Photos: Charles D. Winters

$$2\ \text{Fe(s)} + 3\ \text{Cl}_2\text{(g)} \longrightarrow 2\ \text{FeCl}_3\text{(s)}$$

Reactants   Products

Active Figure 4.2   **The reaction of iron and chlorine.** Hot iron gauze is inserted into a flask of chlorine gas. The heat from the reaction causes the iron gauze to glow, and brown iron(III) chloride forms.

GENERAL
Chemistry•⚛•Now™ See the General ChemistryNow CD-ROM or website to explore an interactive version of this figure accompanied by an exercise.

The numbers in front of each formula in a *balanced* chemical equation are required by the law of conservation of matter. Review the equation for the reaction of phosphorus and chlorine, and then consider the balanced equation for the reaction of iron and chlorine (Figure 4.2).

$$2\ \text{Fe(s)} + 3\ \text{Cl}_2\text{(g)} \longrightarrow 2\ \text{FeCl}_3\text{(s)}$$

stoichiometric coefficients

The number in front of each chemical formula can be read as the number of atoms or molecules (2 atoms of Fe and 3 molecules of Cl_2 form 2 formula units of $FeCl_3$). It can refer equally well to amounts of reactants and products: 2 moles of solid iron combine with 3 moles of chlorine gas to produce 2 moles of solid $FeCl_3$. The relationship between the quantities of chemical reactants and products is called **stoichiometry** (pronounced "stoy-key-AHM-uh-tree"), and the coefficients in a balanced equation are the **stoichiometric coefficients**.

Balanced chemical equations are fundamentally important for understanding the quantitative basis of chemistry. *You must always begin with a balanced equation before carrying out a stoichiometry calculation.*

GENERAL
Chemistry•⚛•Now™

See the General ChemistryNow CD-ROM or website:
• **Screen 4.3 The Law of Conservation of Mass,** for two exercises on the conservation of mass in several reactions

Exercise 4.1—Chemical Reactions

The reaction of aluminum with bromine is shown on page 98. The equation for the reaction is

$$2 \text{ Al(s)} + 3 \text{ Br}_2(\ell) \longrightarrow \text{Al}_2\text{Br}_6(s)$$

(a) Name the reactants and products in this reaction and give their states.
(b) What are the stoichiometric coefficients in this equation?
(c) If you were to use 8000 atoms of Al, how many molecules of Br_2 are required to consume the Al completely?

4.2—Balancing Chemical Equations

Balancing an equation ensures that the same number of atoms of each element appear on both sides of the equation. Many chemical equations can be balanced by trial and error, although some will involve more trial than others.

One general class of chemical reactions is the reaction of metals or nonmetals with oxygen to give oxides of the general formula M_xO_y. For example, iron can react with oxygen to give iron(III) oxide (Figure 4.3a),

$$4 \text{ Fe(s)} + 3 \text{ O}_2(g) \longrightarrow 2 \text{ Fe}_2\text{O}_3(s)$$

magnesium and oxygen react to form magnesium oxide (Figure 4.3b),

$$2 \text{ Mg(s)} + \text{O}_2(g) \longrightarrow 2 \text{ MgO(s)}$$

and phosphorus, P_4, reacts vigorously with oxygen to give tetraphosphorus decaoxide, P_4O_{10} (Figure 4.3c),

$$P_4(s) + 5 \text{ O}_2(g) \longrightarrow P_4O_{10}(s)$$

The equations written above are balanced. The same number of metal or phosphorus atoms and oxygen atoms occurs on each side of these equations.

Charles D. Winters

(a) Reaction of iron and oxygen to give iron(III) oxide, Fe_2O_3.

(b) Reaction of magnesium and oxygen to give magnesium oxide, MgO.

(c) Reaction of phosphorus and oxygen to give tetraphosphorus decaoxide, P_4O_{10}.

Figure 4.3 **Reactions of metals and a nonmetal with oxygen.** (*See General ChemistryNow Screen 4.4 Balancing Chemical Equations, for a video of the phosphorus and oxygen reaction.*)

A combustion reaction. Propane, C_3H_8, burns to give CO_2 and H_2O. These simple oxides are always the products of the complete combustion of a hydrocarbon. (*See General ChemistryNow Screen 4.4 Balancing Chemical Equations, for a animation of this reaction.*)

The **combustion**, or burning, of a fuel in oxygen is accompanied by the evolution of heat. You are familiar with combustion reactions such as the burning of octane, C_8H_{18}, a component of gasoline, in an automobile engine:

$$2\ C_8H_{18}(\ell) + 25\ O_2(g) \longrightarrow 16\ CO_2(g) + 18\ H_2O(g)$$

In all combustion reactions, some or all of the elements in the reactants end up as oxides, compounds containing oxygen. When the reactant is a hydrocarbon (a compound containing only C and H), the products of complete combustion are carbon dioxide and water.

When balancing chemical equations, there are two important things to remember:

- Formulas for reactants and products must be correct or the equation is meaningless.

- Subscripts in the formulas of reactants and products cannot be changed to balance equations. Changing the subscripts changes the identity of the substance. For example, you cannot change CO_2 to CO to balance an equation; carbon monoxide, CO, and carbon dioxide, CO_2, are different compounds.

As an example of equation balancing, let us write the balanced equation for the complete combustion of propane, C_3H_8.

Step 1. *Write correct formulas for the reactants and products.*

$$C_3H_8(g) + O_2(g) \xrightarrow{\text{unbalanced equation}} CO_2(g) + H_2O(g)$$

Here propane and oxygen are the reactants, and carbon dioxide and water are the products.

Step 2. *Balance the C atoms.* In combustion reactions such as this it is usually best to balance the carbon atoms first and leave the oxygen atoms until the end (because the oxygen atoms are often found in more than one product). In this case three carbon atoms are in the reactants, so three must occur in the products. Three CO_2 molecules are therefore required on the right side:

$$C_3H_8(g) + O_2(g) \xrightarrow{\text{unbalanced equation}} 3\ CO_2(g) + H_2O(g)$$

Step 3. *Balance the H atoms.* Propane, the reactant, contains 8 H atoms. Each molecule of water has two hydrogen atoms, so four molecules of water account for the required eight hydrogen atoms on the right side:

$$C_3H_8(g) + O_2(g) \xrightarrow{\text{unbalanced equation}} 3\ CO_2(g) + 4\ H_2O(g)$$

Step 4. *Balance the number of O atoms.* Ten oxygen atoms are on the right side ($3 \times 2 = 6$ in CO_2 plus $4 \times 1 = 4$ in water). Therefore, five O_2 molecules are needed to supply the required ten oxygen atoms:

$$C_3H_8(g) + 5\ O_2(g) \longrightarrow 3\ CO_2(g) + 4\ H_2O(g)$$

Step 5. *Verify that the number of atoms of each element is balanced.* The equation shows three carbon atoms, eight hydrogen atoms, and ten oxygen atoms on each side.

GENERAL
Chemistry Now™

See the General ChemistryNow CD-ROM or website:

- **Screen 4.4 Balancing Chemical Equations,** for a tutorial in which you balance a series of combustion reactions.

Example 4.1—Balancing an Equation for a Combustion Reaction

Problem Write the balanced equation for the combustion of ammonia ($NH_3 + O_2$) to give NO and H_2O.

Strategy First write the unbalanced equation. Next balance the N atoms, then balance the H atoms, and finally balance the O atoms.

Solution

Step 1. *Write correct formulas for reactants and products.* The unbalanced equation for the combustion is

$$NH_3(g) + O_2(g) \xrightarrow{\text{unbalanced equation}} NO(g) + H_2O(g)$$

Step 2. *Balance the N atoms.* There is one N atom on each side of the equation. The N atoms are in balance, at least for the moment.

$$NH_3(g) + O_2(g) \xrightarrow{\text{unbalanced equation}} NO(g) + H_2O(g)$$

Step 3. *Balance the H atoms.* There are three H atoms on the left and two on the right. To have the same number on each side, let us use two molecules of NH_3 on the left and three molecules of H_2O on the right (which gives us six H atoms on each side).

$$2\, NH_3(g) + O_2(g) \xrightarrow{\text{unbalanced equation}} NO(g) + 3\, H_2O(g)$$

Notice that when we balance the H atoms, the N atoms are no longer balanced. To bring them into balance, let us use two NO molecules on the right.

$$2\, NH_3(g) + O_2(g) \xrightarrow{\text{unbalanced equation}} 2\, NO(g) + 3\, H_2O(g)$$

Step 4. *Balance the O atoms.* After Step 3, there are two O atoms on the left side and five on the right. That is, there are an even number of O atoms on the left and an odd number on the right. Because there cannot be an odd number of O atoms on the left (O atoms are paired in O_2 molecules), multiply each coefficient on both sides of the equation by 2 so that an even number of oxygen atoms (ten) can now occur on the right side:

$$4\, NH_3(g) + O_2(g) \xrightarrow{\text{unbalanced equation}} 4\, NO(g) + 6\, H_2O(g)$$

Now the oxygen atoms can be balanced by having five O_2 molecules on the left side of the equation:

$$4\, NH_3(g) + 5\, O_2(g) \xrightarrow{\text{balanced equation}} 4\, NO(g) + 6\, H_2O(g)$$

Step 5. *Verify the result.* Four N atoms, 12 H atoms, and 10 O atoms occur on each side of the equation.

Comment An alternative way to write this equation is

$$2\,NH_3(g) + \tfrac{5}{2}\,O_2(g) \longrightarrow 2\,NO(g) + 3\,H_2O(g)$$

where a fractional coefficient has been used. This equation is correctly balanced and will be useful under some circumstances. In general, however, we balance equations with whole-number coefficients.

Exercise 4.2—Balancing the Equation for a Combustion Reaction

(a) Butane gas, C_4H_{10}, can burn completely in air [use $O_2(g)$ as the other reactant] to give carbon dioxide gas and water vapor. Write a balanced equation for this combustion reaction.

(b) Write a balanced chemical equation for the complete combustion of liquid tetraethyllead, $Pb(C_2H_5)_4$ (which was used until the 1970s as a gasoline additive). The products of combustion are $PbO(s)$, $H_2O(g)$, and $CO_2(g)$.

4.3—Mass Relationships in Chemical Reactions: Stoichiometry

A balanced chemical equation shows the quantitative relationship between reactants and products in a chemical reaction. Let us apply this concept to the reaction of phosphorus and chlorine (see Figure 4.1). Suppose you use 1.00 mol of phosphorus (P_4, 124 g/mol) in this reaction. The balanced equation shows that 6.00 mol (= 425 g) of Cl_2 must be used for complete reaction with 1.00 mol of P_4 and that 4.00 mol (= 549 g) of PCl_3 can be produced.

■ **Amounts Tables**
Amounts tables not only are useful here but will also be used extensively when you study chemical equilibria in Chapters 16–18.

Equation	$P_4(s)$	$+$	$6\,Cl_2\,(g)$	$\longrightarrow$	$4\,PCl_3\,(\ell)$
Initial amount (mol)	1.00 mol (124 g)		6.00 mol (425 g)		0 mol (0 g)
Change in amount upon reaction (mol)	−1.00 mol		−6.00 mol		+4.00 mol
Amount after complete reaction (mol)	0 mol (0 g)		0 mol (0 g)		4.00 mol [549 g = 124 g + 425 g]

The mole and mass relationships of reactants and products in a reaction are summarized in an *amounts table*. Such tables identify the amounts of reactants and products and the changes that occur upon reaction.

The balanced equation for a reaction tells us the correct *mole ratios* of reactants and products. Therefore, the equation for the phosphorus and chlorine reaction, for example, applies no matter how much P_4 is used. Suppose 0.0100 mol of P_4 (1.24 g) is used. Now only 0.0600 mol of Cl_2 (4.25 g) is required, and 0.0400 mol of PCl_3 (5.49 g) can form.

Following this line of reasoning, let us decide (a) what mass of Cl_2 is required to react completely with 1.45 g of phosphorus and (b) what mass of PCl_3 is produced.

■ **Mass Balance**
Mass is always conserved in chemical reactions. The total mass before reaction is always the same as that after reaction. This does not mean, however, that the total amount (moles) of reactants is the same as that of the products. Atoms are rearranged into different "units" (molecules) in the course of a reaction. In the $P_4 + Cl_2$ reaction, 7 mol of reactants gives 4 mol of product.

Part (a): Mass of Cl_2 Required
Step 1. *Write the balanced equation* (using correct formulas for reactants and products). This is always the first step when dealing with chemical reactions.

$$P_4(s) + 6\,Cl_2(g) \longrightarrow 4\,PCl_3(\ell)$$

Problem-Solving Tip 4.1

Stoichiometry Calculations

You are asked to determine what mass of product can be formed from a given mass of reactant. Keep in mind that it is not possible to calculate the mass of product in a single step. Instead, you must follow a route such as that illustrated here for the reaction of a reactant A to give the product B according to an equation such as x A $\rightarrow$ y B. Here the mass of reactant A is converted to moles of A. Then, using the stoichiometric factor, you find moles of B. Finally, the mass of B is obtained by multiplying moles of B by its molar mass.

When solving a chemical stoichiometry problem, remember that you will always use a stoichiometric factor at some point.

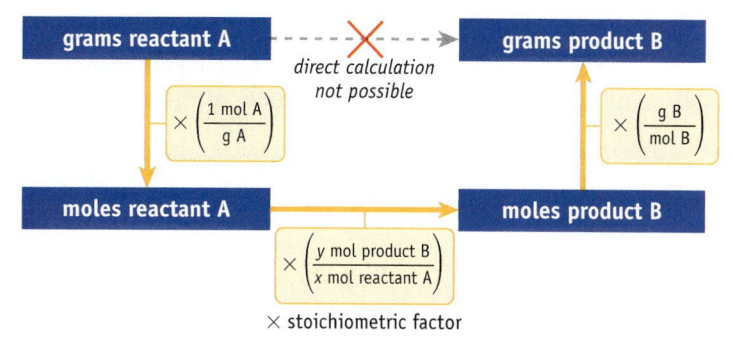

Step 2. *Calculate moles from masses.* From the mass of P_4, calculate the amount of P_4 available.

$$1.45 \text{ g } P_4 \times \frac{1 \text{ mol } P_4}{123.9 \text{ g } P_4} = 0.0117 \text{ mol } P_4$$

Step 3. *Use a stoichiometric factor.* The amount of P_4 available is related to the amount of the other reactant (Cl_2) required by the balanced equation.

$$0.0117 \text{ mol } P_4 \times \frac{6 \text{ mol } Cl_2 \text{ required}}{1 \text{ mol } P_4 \text{ available}} = 0.0702 \text{ mol } Cl_2 \text{ required}$$

⇑ stoichiometric factor (from balanced equation)

To perform this calculation the amount of phosphorus available has been multiplied by a **stoichiometric factor**, a *mole ratio based on the coefficients for the two chemicals in the balanced equation.* This is the reason you must balance chemical equations before proceeding with calculations. Here the balanced equation specifies that 6 mol of Cl_2 is required for each mole of P_4, so the stoichiometric factor is (6 mol Cl_2/ 1 mol P_4). Calculation shows that 0.0702 mol of Cl_2 is required to react with all the available phosphorus (1.45 g, 0.0117 mol).

Step 4. *Calculate mass from moles.* Convert amount (moles) of Cl_2 calculated in Step 3 to quantity (mass in grams) of Cl_2 required.

$$0.0702 \text{ mol } Cl_2 \times \frac{70.91 \text{ g } Cl_2}{1 \text{ mol } Cl_2} = 4.98 \text{ g } Cl_2$$

Part (b) Mass of PCl_3 Produced from P_4 and Cl_2

What mass of PCl_3 can be produced from the reaction of 1.45 g of phosphorus with 4.98 g of Cl_2? Because matter is conserved, the answer can be obtained in this case

■ **Stoichiometric Factor**
The stoichiometric factor is a conversion factor (see page 42). Thus, a stoichiometric factor can also relate moles of a reactant to moles of a product, and vice versa.

■ **Amount and Quantity**
When doing stoichiometry problems, recall from Chapter 2 that the terms "amount" and "quantity" are used in a specific sense by chemists. The *amount* of a substance is the number of moles of that substance. *Quantity* refers to the mass of the substance.

by adding the masses of P_4 and Cl_2 used (giving 1.45 g + 4.98 g = 6.43 g of PCl_3 produced). Alternatively, Steps 3 and 4 can be repeated, but with the appropriate stoichiometric factor and molar mass.

Step 3b. *Use a stoichiometric factor.* Convert the amount of available P_4 to the amount of PCl_3 produced. Here the balanced equation specifies that 4 mol PCl_3 is produced for each mole of P_4 used, so the stoichiometric factor is (4 mol PCl_3/1 mol P_4).

$$0.0117 \; \text{mol } P_4 \times \frac{4 \text{ mol } PCl_3 \text{ produced}}{1 \text{ mol } P_4 \text{ available}} = 0.0468 \text{ mol } PCl_3 \text{ produced}$$

⇑ stoichiometric factor (from balanced equation)

Step 4b. *Calculate mass from moles.* Convert the amount of PCl_3 produced to a mass in grams.

$$0.0468 \; \text{mol } PCl_3 \times \frac{137.3 \text{ g } PCl_3}{1 \text{ mol } PCl_3} = 6.43 \text{ g } PCl_3$$

GENERAL
Chemistry ⚛ Now™

See the General ChemistryNow CD-ROM or website:
- **Screen 4.5 Weight Relations in Chemical Reactions**
  - **(a)** for a video and animation of the phosphorus and chlorine reaction discussed in this section
  - **(b)** for an exercise that examines the reaction between chlorine and elemental phosphorus
- **Screen 4.6 Calculations in Stoichiometry,** for a tutorial on yield

---

Example 4.2—Mass Relations in Chemical Reactions

Problem Glucose reacts with oxygen to give CO_2 and H_2O.

$$C_6H_{12}O_6(s) + 6 \; O_2(g) \longrightarrow 6 \; CO_2(g) + 6 \; H_2O(\ell)$$

What mass of oxygen (in grams) is required for complete reaction of 25.0 g of glucose? What masses of carbon dioxide and water (in grams) are formed?

Strategy After referring to the balanced equation, you can perform the stoichiometric calculations using the scheme in Problem-Solving Tip 4.1.

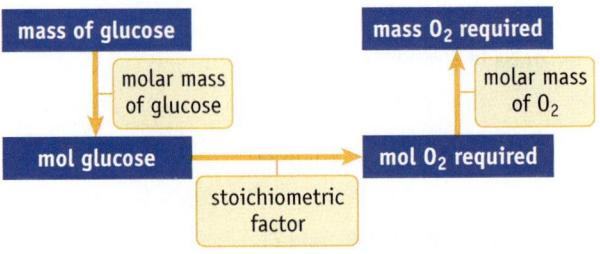

First find the amount of glucose available, then relate it to the amount of O_2 required using the stoichiometric factor based on the coefficients in the balanced equation. Finally, find the mass of O_2 required from the amount of O_2. Follow the same procedure to find the masses of carbon dioxide and water.

Solution

Step 1. *Write a balanced equation.*

$$C_6H_{12}O_6(s) + 6\ O_2(g) \longrightarrow 6\ CO_2(g) + 6\ H_2O(\ell)$$

Step 2. *Convert the mass of glucose to moles.*

$$25.0\ \text{g glucose} \times \frac{1\ \text{mol}}{180.2\ \text{g}} = 0.139\ \text{mol glucose}$$

Step 3. *Use the stoichiometric factor.* Here we calculate the amount of O_2 required.

$$0.139\ \text{mol glucose} \times \frac{6\ \text{mol}\ O_2}{1\ \text{mol glucose}} = 0.832\ \text{mol}\ O_2$$

Step 4. *Calculate mass from moles.* Convert the required amount of O_2 to a mass in grams.

$$0.832\ \text{mol}\ O_2 \times \frac{32.00\ \text{g}}{1\ \text{mol}\ O_2} = 26.6\ \text{g}\ O_2$$

Repeat Steps 3 and 4 to find the mass of CO_2 produced in the combustion. First, relate the amount (moles) of glucose available to the amount of CO_2 produced using a stoichiometric factor. Then convert the amount of CO_2 to the mass in grams.

$$0.139\ \text{mol glucose} \times \frac{6\ \text{mol}\ CO_2}{1\ \text{mol glucose}} \times \frac{44.01\ \text{g}\ CO_2}{1\ \text{mol}\ CO_2} = 36.6\ \text{g}\ CO_2$$

Now, how can you find the mass of H_2O produced? You could go through Steps 3 and 4 again. However, recognize that the total mass of reactants

$$25.0\ \text{g}\ C_6H_{12}O_6 + 26.6\ \text{g}\ O_2 = 51.6\ \text{g of reactants}$$

must be the same as the total mass of products. The mass of water that can be produced is therefore

$$\text{Total mass of products} = 51.6\ \text{g} = 36.6\ \text{g}\ CO_2\ \text{produced} + ?\ \text{g}\ H_2O$$

$$\text{Mass of}\ H_2O\ \text{produced} = 15.0\ \text{g}$$

The amounts table for this problem is

Equation	$C_6H_{12}O_6(s)$	+	$6\ O_2(g)$	$\longrightarrow$	$6\ CO_2(g)$	+	$6\ H_2O(\ell)$
Initial amount (mol)	0.139 mol		6(0.139 mol) = 0.832 mol		0		0
Change (mol)	−0.139 mol		−0.832 mol		+0.832 mol		+0.832 mol
Amount after reaction (mol)	0		0		0.832 mol		0.832 mol

Comment When you know the mass of all but one of the chemicals in a reaction, you can find the unknown mass using the principle of mass conservation (the total mass of reactants must equal the total mass of products).

Exercise 4.3—Mass Relations in Chemical Reactions

What mass of oxygen, O_2, is required to completely combust 454 g of propane, C_3H_8? What masses of CO_2 and H_2O are produced?

$$C_3H_8(g) + 5\ O_2(g) \longrightarrow 3\ CO_2(g) + 4\ H_2O(\ell)$$

4.4—Reactions in Which One Reactant Is Present in Limited Supply

You may have observed in your laboratory experiments that reactions are often carried out with an excess of one reactant over that required by stoichiometry. This is usually done to ensure that one of the reactants in the reaction is consumed completely, even though some of another reactant remains unused.

Suppose you burn a toy "sparkler," a wire coated with magnesium (Figure 4.3b). The magnesium burns in air, consuming oxygen and producing magnesium oxide, MgO.

$$Mg(s) + O_2(g) \longrightarrow 2\ MgO(s)$$

The sparkler burns until the magnesium is consumed completely. What about the oxygen? Two moles of magnesium require one mole of oxygen, but there is much, much more O_2 available in the air than is needed to consume the magnesium in a sparkler. How much MgO is produced? That depends on the quantity of magnesium in the sparkler, not on the quantity of O_2 in the atmosphere. A substance such as the magnesium in this example is called the **limiting reactant** because its amount determines, or limits, the amount of product formed.

Let us look at an example of a limiting reactant situation using the reaction of oxygen and carbon monoxide to give carbon dioxide. The balanced equation for the reaction is

$$2\ CO(g) + O_2(g) \longrightarrow 2\ CO_2(g)$$

Suppose you have a mixture of four CO molecules and three O_2 molecules.

■ **Comparing Reactant Ratios**
For the CO/O_2 reaction, the stoichiometric ratio of reactants should be (2 mol CO/ 1 mol O_2). However, the ratio of amounts of reactants available is (4 mol CO/3 mol O_2) or (1.33 mol CO/1 mol O_2). Clearly, there is not sufficient CO to react with all of the available O_2. Carbon monoxide is the limiting reactant, and some O_2 will be left over when all of the CO is consumed.

Reactants: 4 CO and 3 O_2 Products: 4 CO_2 and 1 O_2

The four CO molecules require only two O_2 molecules (and produce four CO_2 molecules). This means that one O_2 molecule remains after reaction is complete. Because more O_2 molecules are available than are required, the number of CO_2 molecules produced is determined by the number of CO molecules available. Carbon monoxide, CO, is therefore the limiting reactant in this case.

(a)   **(b)**

Active Figure 4.4   **Oxidation of ammonia.** (a) Burning ammonia on the surface of a platinum wire produces so much heat that the wire glows bright red. (b) Billions of kilograms of HNO_3 are made annually starting with the oxidation of ammonia over a wire gauze containing platinum.

GENERAL
Chemistry ✦ Now™ See the General ChemistryNow CD-ROM or website to explore an interactive version of this figure accompanied by an exercise.

A Stoichiometry Calculation with a Limiting Reactant

The first step in the manufacture of nitric acid is the oxidation of ammonia to NO over a platinum-wire gauze (Figure 4.4).

$$4\ NH_3(g) + 5\ O_2(g) \longrightarrow 4\ NO(g) + 6\ H_2O(\ell)$$

Suppose that equal masses of NH_3 and of O_2 are mixed (750. g of each). Are these reactants mixed in the correct stoichiometric ratio or is one of them in short supply? That is, will one of them limit the quantity of NO that can be produced? How much NO can be formed if the reaction using this reactant mixture goes to completion? And how much of the excess reactant is left over when the maximum amount of NO has been formed?

Step 1. Find the amount of each reactant.

$$750.\ g\ NH_3 \times \frac{1\ mol\ NH_3}{17.03\ g\ NH_3} = 44.0\ mol\ NH_3\ available$$

$$750.\ g\ O_2 \times \frac{1\ mol\ O_2}{32.00\ g\ O_2} = 23.4\ mol\ O_2\ available$$

Step 2. What is the limiting reactant? Examine the ratio of amounts of reactants.
Are the reactants present in the correct stoichiometric ratio as given by the balanced equation?

Stoichiometric ratio of reactants **required** by balanced equation
$$= \frac{5\ mol\ O_2}{4\ mol\ NH_3} = \frac{1.25\ mol\ O_2}{1\ mol\ NH_3}$$

Ratio of reactants **actually available** $= \dfrac{23.4\ mol\ O_2}{44.0\ mol\ NH_3} = \dfrac{0.532\ mol\ O_2}{1\ mol\ NH_3}$

Problem-Solving Tip 4.2

More on Reactions with a Limiting Reactant

There is another method of solving limiting reactant problems: Calculate the mass of product expected based on each reactant. The limiting reactant is the reactant that gives the smallest quantity of product. For example, refer to the $NH_3 + O_2$ reaction on page 153. To confirm that O_2 is the limiting reactant, calculate the quantity of NO that can be formed starting with (a) 44.1 mol of NH_3 and unlimited O_2 and (b) with 23.4 mol of O_2 and unlimited NH_3.

1. Quantity of NO produced from 44.1 mol of NH_3 and unlimited O_2

$$44.0 \text{ mol NH}_3 \times \frac{4 \text{ mol NO}}{4 \text{ mol NH}_3} \times \frac{30.01 \text{ g NO}}{1 \text{ mol NO}} = 1320 \text{ g NO}$$

2. Quantity of NO produced from 23.4 mol O_2 and unlimited NH_3

$$23.4 \text{ mol O}_2 \times \frac{4 \text{ mol NO}}{5 \text{ mol O}_2} \times \frac{30.01 \text{ g NO}}{1 \text{ mol NO}} = 562 \text{ g NO}$$

3. Compare the quantities of NO produced. The available O_2 is capable of producing less NO (562 g) than the available NH_3 (1320 g), which confirms that O_2 is the limiting reactant.

As a final note, you may find this approach easier to use when there are more than two reactants, each present initially in some designated quantity.

Dividing moles of O_2 available by moles of NH_3 available shows that the ratio of available reactants is much smaller than the 5 mol O_2/4 mol NH_3 ratio required by the balanced equation. Thus there is not sufficient O_2 available to react with all of the NH_3. In this case, *oxygen, O_2, is the limiting reactant.* That is, 1 mol of NH_3 requires 1.25 mol of O_2, but we have only 0.532 mol of O_2 available for each mole of NH_3.

Step 3. Calculate the mass of product.

We can now calculate the mass of product, NO, expected based on the amount of the limiting reactant, O_2.

$$23.4 \text{ mol O}_2 \times \frac{4 \text{ mol NO}}{5 \text{ mol O}_2} \times \frac{30.01 \text{ g NO}}{1 \text{ mol NO}} = 562 \text{ g NO}$$

Step 4. Calculate the mass of excess reactant.

Ammonia is the "excess reactant" in this NH_3/O_2 reaction because more than enough NH_3 is available to react with 23.4 mol of O_2. Let us calculate the quantity of NH_3 remaining after all the O_2 has been used. To do so, we first need to know the amount of NH_3 required to consume all the limiting reactant, O_2.

$$23.4 \text{ mol O}_2 \text{ available} \times \frac{4 \text{ mol NH}_3 \text{ required}}{5 \text{ mol O}_2} = 18.8 \text{ mol NH}_3 \text{ required}$$

Because 44.0 mol of NH_3 is available, the amount of excess NH_3 can be calculated,

$$\text{Excess NH}_3 = 44.0 \text{ mol NH}_3 \text{ available} - 18.8 \text{ mol NH}_3 \text{ required}$$

$$= 25.2 \text{ mol NH}_3 \text{ remaining}$$

and then converted to a mass,

$$25.2 \text{ mol NH}_3 \times \frac{17.03 \text{ g NH}_3}{1 \text{ mol NH}_3} = 429 \text{ g NH}_3 \text{ in excess of that required}$$

Finally, because 429 g of NH_3 is left over, this means that 321 g of NH_3 has been consumed (= 750. g − 429 g).

It is helpful in limiting reactant problems to summarize your results in an amounts table.

Equation	$4\ NH_3(g)$	+	$5\ O_2(g)$	$\rightarrow$	$4\ NO(g)$	+	$6\ H_2O(g)$
Initial amount (mol)	44.0		23.4		0		0
Change in amount (mol)	−(4/5)(23.4)		−23.4		+(4/5)(23.4)		+(6/5)(23.4)
	= −18.8		−23.4		= +18.8		= +28.1
After complete reaction (mol)	25.2		0		18.8		28.1

All of the limiting reactant, O_2, has been consumed. Of the original 44.0 mol of NH_3, 18.8 mol has been consumed and 25.2 mol remains. The balanced equation indicates that the amount of NO produced is equal to the amount of NH_3 consumed, so 18.8 mol of NO is produced from 18.8 mol of NH_3. In addition, 28.1 mol of H_2O has been produced.

■ **Conservation of Mass**
Mass is conserved in the NH_3 + O_2 reaction. The total mass present before reaction (1500. g) is the same as the total mass produced in the reaction plus the mass of NH_3 remaining. That is, 562 g of NO (18.8 mol) and 506 g of H_2O (28.1 mol) are produced. Because 429 g of NH_3 (25.2 mol) remains, the total mass after reaction (562 g + 506 g + 429 g) is the same as the total mass before reaction.

GENERAL
Chemistry ⚛ Now™

See the General ChemistryNow CD-ROM or website:

- **Screen 4.7 Reactions Controlled by the Supply of One Reactant** for a video and animation of the limiting reactant in the methanol and oxygen reaction
- **Screen 4.8 Limiting Reactants**
  - **(a)** for an exercise on zinc and hydrochloric acid in aqueous solution
  - **(b)** for a simulation using limiting reactants

Example 4.3—A Reaction with a Limiting Reactant

Problem Methanol, CH_3OH, which is used as a fuel, can be made by the reaction of carbon monoxide and hydrogen.

$$CO(g) + 2\ H_2(g) \longrightarrow CH_3OH(\ell)$$
$$\text{methanol}$$

Suppose 356 g of CO and 65.0 g of H_2 are mixed and allowed to react.

(a) Which is the limiting reactant?

(b) What mass of methanol can be produced?

(c) What mass of the excess reactant remains after the limiting reactant has been consumed?

Strategy There are usually two steps to a limiting reactant problem:

(a) After calculating the amount of each reactant, compare the ratio of reactant amounts to the required stoichiometric ratio, 2 mol H_2/1 mol CO.

- If [mol H_2 available/mol CO available] > 2/1, then CO is the limiting reactant.
- If [mol H_2 available/mol CO available] < 2/1, then H_2 is the limiting reactant.

(b) Use the amount of limiting reactant to find the amount of product.

A car that uses methanol as a fuel. In this car methanol is converted to hydrogen, which is then combined with oxygen in a fuel cell. The fuel cell generates electric energy to run the car (see Chapter 20). See Example 4.3.

Solution

(a) *What is the limiting reactant?* The amount of each reactant is

$$\text{Amount of CO} = 356 \text{ g CO} \times \frac{1 \text{ mol CO}}{28.01 \text{ g CO}} = 12.7 \text{ mol CO}$$

$$\text{Amount of H}_2 = 65.0 \text{ g H}_2 \times \frac{1 \text{ mol H}_2}{2.016 \text{ g H}_2} = 32.2 \text{ mol H}_2$$

Are these reactants present in a perfect stoichiometric ratio?

$$\frac{\text{Mol H}_2 \text{ available}}{\text{Mol CO available}} = \frac{32.2 \text{ mol H}_2}{12.7 \text{ mol CO}} = \frac{2.54 \text{ mol H}_2}{1.00 \text{ mol CO}}$$

The required mole ratio is 2 mol of H_2 to 1 mol of CO. Here we see that more hydrogen is available than is required to consume all the CO. It follows that not enough CO is present to use up all of the hydrogen. *CO is the limiting reactant.*

(b) *What is the maximum mass of CH_3OH that can be formed?* This calculation must be based on the amount of limiting reactant.

$$12.7 \text{ mol CO} \times \frac{1 \text{ mol CH}_3\text{OH formed}}{1 \text{ mol CO available}} \times \frac{32.04 \text{ g CH}_3\text{OH}}{1 \text{ mol CH}_3\text{OH}} = 407 \text{ g CH}_3\text{OH}$$

(c) *What amount of H_2 remains when all the CO has been converted to product?* First, we must find the amount of H_2 required to react with all the CO.

$$12.7 \text{ mol CO} \times \frac{2 \text{ mol H}_2}{1 \text{ mol CO}} = 25.4 \text{ mol H}_2 \text{ required}$$

Because 32.2 mol of H_2 is available, but only 25.4 mol is required by the limiting reactant, 32.2 mol − 25.4 mol = 6.8 mol of H_2 is in excess. This is equivalent to 14 g of H_2.

$$6.8 \text{ mol H}_2 \times \frac{2.02 \text{ g H}_2}{1 \text{ mol H}_2} = 14 \text{ g H}_2 \text{ remaining}$$

Comment The amounts table for this reaction is

Equation	$CO(g)$ +	$2 H_2(g)$	$\longrightarrow$ $CH_3OH(\ell)$
Initial amount (mol)	12.7	32.2	0
Change (mol)	−12.7	−2(12.7)	+12.7
After complete reaction (mol)	0	6.8	12.7

The mass of product formed plus the mass of H_2 remaining after reaction (407 g CH_3OH produced + 14 g H_2 remaining = 421 g) is equal to the mass of reactants present before reaction (356 g CO + 65.0 g H_2 = 421 g).

Exercise 4.4—A Reaction With a Limiting Reactant

Titanium is an important structural metal, and a compound of titanium, TiO_2, is the white pigment in paint. In the refining process, titanium ore (impure TiO_2) is first converted to liquid $TiCl_4$ by the following reaction.

$$TiO_2(s) + 2 Cl_2(g) + C(s) \longrightarrow TiCl_4(\ell) + CO_2(g)$$

Using 125 g each of Cl_2 and C, but plenty of TiO_2-containing ore, which is the limiting reactant in this reaction? What mass of $TiCl_4$, in grams, can be produced?

Exercise 4.5—A Reaction with a Limiting Reactant

The thermite reaction produces iron metal and aluminum oxide from a mixture of powdered aluminum metal and iron(III) oxide.

$$Fe_2O_3(s) + 2\ Al(s) \longrightarrow 2\ Fe(s) + Al_2O_3(s)$$

A mixture of 50.0 g each of Fe_2O_3 and Al is used.

(a) Which is the limiting reactant?
(b) What mass of iron metal can be produced?

Thermite reaction Iron(III) oxide reacts with aluminum metal to produce aluminum oxide and iron metal. The reaction produces so much heat that the iron melts and spews out of the reaction vessel. See Exercise 4.5.

4.5—Percent Yield

The maximum quantity of product we calculate can be obtained from a chemical reaction is the **theoretical yield**. Frequently, however, the **actual yield** of a compound—the quantity of material that is actually obtained in the laboratory or a chemical plant—is less than the theoretical yield. Some loss of product often occurs during the isolation and purification steps. In addition, some reactions do not go completely to products, and reactions are sometimes complicated by giving more than one set of products. For all these reasons, the actual yield is likely to be less than the theoretical yield (Figure 4.5).

To provide information to other chemists who might want to carry out a reaction, it is customary to report a **percent yield**. Percent yield, which specifies how much of the theoretical yield was obtained, is defined as

$$\text{Percent yield} = \frac{\text{actual yield}}{\text{theoretical yield}} \times 100\% \qquad (4.1)$$

(a)

Suppose you made aspirin in the laboratory by the following reaction:

$$C_6H_4(OH)CO_2H(s) + (CH_3CO)_2O(\ell) \longrightarrow C_6H_4(OCOCH_3)CO_2H(s) + CH_3CO_2H(\ell)$$

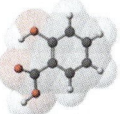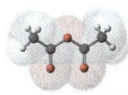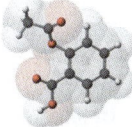

salicylic acid   acetic anhydride   aspirin   acetic acid

and that you began with 14.4 g of salicylic acid and an excess of acetic anhydride. That is, salicylic acid is the limiting reactant. If you obtain 6.26 g of aspirin, what is the percent yield of this product? The first step is to find the amount of the limiting reactant, salicylic acid ($C_6H_4(OH)CO_2H$).

$$14.4\text{ g }C_6H_4(OH)CO_2H \times \frac{1\text{ mol }C_6H_4(OH)CO_2H}{138.1\text{ g }C_6H_4(OH)CO_2H} = 0.104\text{ mol }C_6H_4(OH)CO_2H$$

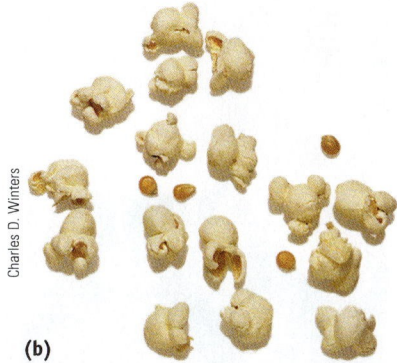

(b)

Figure 4.5   Percent yield. Although not a chemical reaction, popping corn is a good analogy to the difference between a theoretical yield and an actual yield. Here we began with 20 popcorn kernels and found that only 16 of them popped. The percent yield from our "reaction" was (16/20) × 100%, or 80%.

Next, use the stoichiometric factor from the balanced equation to find the amount of aspirin expected based on the limiting reactant, $C_6H_4(OH)CO_2H$.

$$0.104 \text{ mol } C_6H_4(OH)CO_2H \times \frac{1 \text{ mol aspirin}}{1 \text{ mol } C_6H_4(OH)CO_2H} = 0.104 \text{ mol aspirin}$$

The maximum amount of aspirin that can be produced—the theoretical yield—is 0.104 mol. Because the quantity you measure in the laboratory is the mass of the product, it is customary to express the theoretical yield as a mass in grams.

$$0.104 \text{ mol aspirin} \times \frac{180.2 \text{ g aspirin}}{1 \text{ mol aspirin}} = 18.7 \text{ g aspirin}$$

Finally, with the actual yield known to be only 6.26 g, the percent yield of aspirin can be calculated.

$$\text{Percent yield} = \frac{6.26 \text{ g aspirin obtained (actual yield)}}{18.7 \text{ g aspirin expected (theoretical yield)}} \times 100\% = 33.5\% \text{ yield}$$

GENERAL
Chemistry ⚛ Now™

See the General ChemistryNow CD-ROM or website:
- **Screen 4.9 Percent Yield**
  - **(a)** for a tutorial on determining the theoretical yield of a reaction
  - **(b)** for a tutorial on determining the percent yield of a reaction

---

Exercise 4.6—Percent Yield

Methanol, CH_3OH, can be burned in oxygen to provide energy, or it can be decomposed to form hydrogen gas, which can then be used as a fuel (see Example 4.3).

$$CH_3OH(\ell) \longrightarrow 2 \text{ H}_2(g) + CO(g)$$

If 125 g of methanol is decomposed, what is the theoretical yield of hydrogen? If only 13.6 g of hydrogen is obtained, what is the percent yield of this gas?

4.6—Chemical Equations and Chemical Analysis

Analytical chemists use a variety of approaches to identify substances as well as to measure the quantities of components of mixtures. Analytical chemistry is often done now using instrumental methods (Figure 4.6), but classical chemical reactions and stoichiometry play a central role.

Quantitative Analysis of a Mixture

Quantitative chemical analyses generally depend on one or the other of two basic ideas:

- A substance, present in unknown amount, can be allowed to react with a known quantity of another substance. If the stoichiometric ratio for their reaction is known, the unknown amount can be determined.

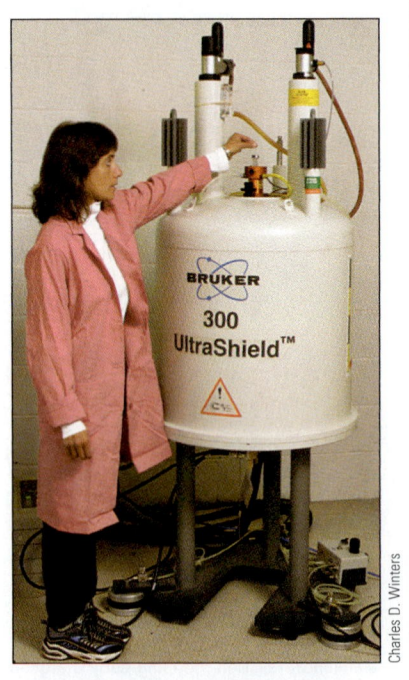

Figure 4.6   A modern analytical instrument. This nuclear magnetic resonance (NMR) spectrometer is closely related to a magnetic resonance imaging (MRI) instrument found in a hospital. The NMR is used to analyze compounds and to decipher their structure.

Charles D. Winters

- A material of unknown composition can be converted to one or more substances of known composition. Those substances can be identified, their amounts determined, and these amounts related to the amount of the original, unknown substance.

An example of the first type of analysis is the analysis of a sample of vinegar containing an unknown amount of acetic acid, the ingredient that makes vinegar acidic. The acid reacts readily and completely with sodium hydroxide.

$$CH_3CO_2H(aq) + NaOH(aq) \longrightarrow CH_3CO_2Na(aq) + H_2O(\ell)$$
$$\text{acetic acid}$$

If the exact amount of sodium hydroxide used in the reaction can be measured, the amount of acetic acid present is also known. This type of analysis is the subject of a major portion of Chapter 5 [▶ Section 5.10].

The second type of analysis is exemplified by the analysis of a sample of a mineral, thenardite, which is largely sodium sulfate, Na_2SO_4, (Figure 4.7). Sodium sulfate is soluble in water. Therefore, to find the quantity of Na_2SO_4 in an impure mineral sample, we would crush the rock and then wash it thoroughly with water to dissolve the sodium sulfate. Next, we would treat this solution with barium chloride to form the water-insoluble compound barium sulfate. The barium sulfate is collected on a filter and weighed (Figure 4.8).

$$Na_2SO_4(aq) + BaCl_2(aq) \longrightarrow BaSO_4(s) + 2\ NaCl(aq)$$

Figure 4.7 Thenardite. The mineral thenardite is sodium sulfate, Na_2SO_4. It is named after the French chemist Louis Thenard (1777–1857), a co-discoverer (with Gay-Lussac and Davy) of boron. Sodium sulfate is used in making detergents, glass, and paper.

■ **Analysis and 100% Yield**
Quantitative analysis requires reactions in which the yield is 100%.

(a)

$Na_2SO_4(aq)$, clear solution

$BaCl_2(aq)$, clear solution

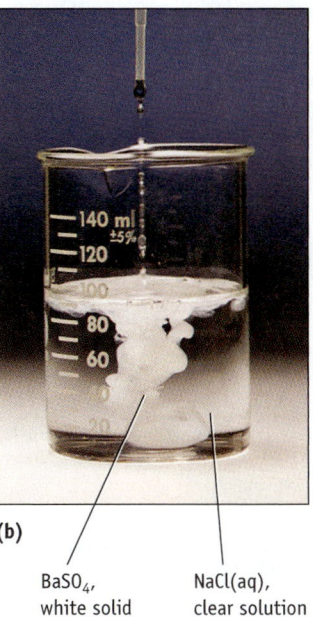

(b)

$BaSO_4$, white solid

$NaCl(aq)$, clear solution

(c)

$NaCl(aq)$, clear solution

$BaSO_4$, white solid caught in filter

(d)

Filter paper weighed

Active Figure 4.8 **Analysis for the sulfate content of a sample.** The sulfate ion in a solution of Na_2SO_4 reacts with barium ion (Ba^{2+}) to form $BaSO_4$. The solid precipitate, barium sulfate ($BaSO_4$), is collected on a filter and weighed. The amount of $BaSO_4$ obtained can be related to the amount of Na_2SO_4 in the sample.

GENERAL
Chemistry⚛Now™ See the General ChemistryNow CD-ROM or website to explore an interactive version of this figure accompanied by an exercise.

We can then find the amount of sulfate in the mineral sample because it is directly related to the amount of $BaSO_4$.

$$1 \text{ mol } Na_2SO_4 \longrightarrow 1 \text{ mol } BaSO_4$$

This approach to the analysis of a mineral is one of many examples of the use of stoichiometry in chemical analysis. Examples 4.4 and 4.5 further illustrate this method.

Example 4.4—Analysis of a Lead-Containing Mineral

Problem The mineral cerussite is mostly lead carbonate, $PbCO_3$, but other substances are present. To analyze for the $PbCO_3$ content, a sample of the mineral is first treated with nitric acid to dissolve the lead carbonate.

$$PbCO_3(s) + 2 \, HNO_3(aq) \longrightarrow Pb(NO_3)_2(aq) + H_2O(\ell) + CO_2(g)$$

On adding sulfuric acid to the resulting solution, lead sulfate precipitates.

$$Pb(NO_3)_2(aq) + H_2SO_4(aq) \longrightarrow PbSO_4(s) + 2 \, HNO_3(aq)$$

Solid lead sulfate is isolated and weighed (as in Figure 4.8). Suppose a 0.583-g sample of mineral produced 0.628 g of $PbSO_4$. What is the mass percent of $PbCO_3$ in the mineral sample?

Strategy The key is to recognize that 1 mol of $PbCO_3$ will ultimately yield 1 mol of $PbSO_4$. Based on the amount of $PbSO_4$ isolated, we can calculate the amount of $PbCO_3$ (in moles), and its mass, in the original sample. When the mass of $PbCO_3$ is known, this is compared with the mass of the mineral sample to give the percent composition.

Solution Let us first calculate the amount of $PbSO_4$.

$$0.628 \text{ g } PbSO_4 \times \frac{1 \text{ mol } PbSO_4}{303.3 \text{ g } PbSO_4} = 0.00207 \text{ mol } PbSO_4$$

From stoichiometry, we can relate the amount of $PbSO_4$ to the amount of $PbCO_3$. (Here the two stoichiometric factors are based on the two balanced equations describing the chemical reactions.)

$$0.00207 \text{ mol } PbSO_4 \times \frac{1 \text{ mol } Pb(NO_3)_2}{1 \text{ mol } PbSO_4} \times \frac{1 \text{ mol } PbCO_3}{1 \text{ mol } Pb(NO_3)_2} = 0.00207 \text{ mol } PbCO_3$$

The mass of $PbCO_3$ is

$$0.00207 \text{ mol } PbCO_3 \times \frac{267.2 \text{ g } PbCO_3}{1 \text{ mol } PbCO_3} = 0.553 \text{ g } PbCO_3$$

Finally, the mass percent of $PbCO_3$ in the mineral sample is

$$\text{Mass percent of } PbCO_3 = \frac{0.553 \text{ g } PbCO_3}{0.583 \text{ g sample}} \times 100\% = \boxed{94.9\%}$$

Example 4.5—Mineral Analysis

Problem Nickel(II) sulfide, NiS, occurs naturally as the relatively rare mineral millerite. One of its occurrences is in meteorites. To analyze a mineral sample for the quantity of NiS, the sample is digested in nitric acid to form a solution of $Ni(NO_3)_2$.

$$NiS(s) + 4\ HNO_3(aq) \longrightarrow Ni(NO_3)_2(aq) + S(s) + 2\ NO_2(g) + 2\ H_2O(\ell)$$

The aqueous solution of $Ni(NO_3)_2$ is then treated with the organic compound dimethylglyoxime ($C_4H_8N_2O_2$, DMG) to give the red solid $Ni(C_4H_7N_2O_2)_2$.

$$Ni(NO_3)_2(aq) + 2\ C_4H_8N_2O_2(aq) \longrightarrow Ni(C_4H_7N_2O_2)_2(s) + 2\ HNO_3(aq)$$

Suppose a 0.468-g sample containing millerite produces 0.206 g of red, solid $Ni(C_4H_7N_2O_2)_2$. What is the mass percent of NiS in the sample?

Strategy The balanced equations show the following "road map":

$$1\ mol\ NiS \longrightarrow 1\ mol\ Ni(NO_3)_2 \longrightarrow 1\ mol\ Ni(C_4H_7N_2O_2)_2$$

Thus, if we know the mass of $Ni(C_4H_7N_2O_2)_2$, we can calculate its amount and thus the amount of NiS. The amount of NiS allows us to calculate the mass and mass percent of NiS.

Solution The molar mass of $Ni(C_4H_7N_2O_2)_2$ is 288.9 g/mol. Thus, the amount of the red solid is

$$0.206\ g\ \cancel{Ni(C_4H_7N_2O_2)_2} \times \frac{1\ mol\ Ni(C_4H_7N_2O_2)_2}{288.9\ g\ \cancel{Ni(C_4H_7N_2O_2)_2}} = 7.13 \times 10^{-4}\ mol\ Ni(C_4H_7N_2O_2)_2$$

Because 1 mol of $Ni(C_4H_7N_2O_2)_2$ is ultimately produced from 1 mol of NiS, the amount of NiS in the sample must have been 7.13×10^{-4} mol.

With the amount of NiS known, we calculate the mass of NiS.

$$7.13 \times 10^{-4}\ \cancel{mol\ NiS} \times \frac{90.76\ g\ NiS}{1\ \cancel{mol\ NiS}} = 0.0647\ g\ NiS$$

Finally, the mass percent of NiS in the 0.468-g sample is

$$Mass\ percent\ NiS = \frac{0.0647\ g\ NiS}{0.468\ g\ sample} \times 100\% = 13.8\%\ NiS$$

Photo: Charles D. Winters

A precipitate of nickel. Red, insoluble $Ni(C_4H_7N_2O_2)_2$ precipitates when dimethylglyoxime ($C_4H_8N_2O_2$) is added to an aqueous solution of nickel(II) ions. (See Example 4.5.)

Exercise 4.7—Analysis of a Mixture

One method for determining the purity of a sample of titanium(IV) oxide, TiO_2, an important industrial chemical, is to combine the sample with bromine trifluoride.

$$3\ TiO_2(s) + 4\ BrF_3(\ell) \longrightarrow 3\ TiF_4(s) + 2\ Br_2(\ell) + 3\ O_2(g)$$

This reaction is known to occur completely and quantitatively. That is, all of the oxygen in TiO_2 is evolved as O_2. Suppose 2.367 g of a TiO_2-containing sample evolves 0.143 g of O_2. What is the mass percent of TiO_2 in the sample?

Determining the Formula of a Compound by Combustion

The empirical formula of a compound can be determined if the percent composition of the compound is known [◀ Section 3.6]. But where do the percent composition data come from? One chemical method that works well for compounds that burn in oxygen is *analysis by combustion*. In this technique, each element in the compound combines with oxygen to produce the appropriate oxide.

Consider an analysis of the hydrocarbon methane, CH_4, as an example of combustion analysis. A balanced equation for the combustion of methane shows that every mole of carbon in the original compound is converted to a mole of CO_2. Every mole of hydrogen in the original compound gives *half* a mole of H_2O. (Here the four moles of H atoms in one mole of CH_4 give two moles of H_2O.)

■ **Finding an Empirical Formula by Chemical Analysis**
Finding the empirical formula of a compound by chemical analysis always uses the following procedure:
1. The unknown but pure compound is decomposed into known products.
2. The reaction products are isolated in pure form and the amount of each is determined.
3. The amount of each product is related to the amount of each element in the original compound to give the empirical formula.

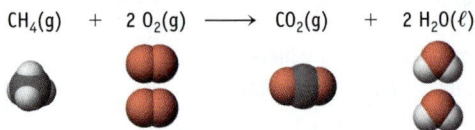

$$CH_4(g) \; + \; 2 \, O_2(g) \longrightarrow CO_2(g) \; + \; 2 \, H_2O(\ell)$$

The gaseous carbon dioxide and water are separated (as illustrated in Figure 4.9) and their masses determined. From these masses it is possible to calculate the amounts of C and H in CO_2 and H_2O, respectively. The ratio of amounts of C and H in a sample of the original compound can then be found. This ratio gives the empirical formula:

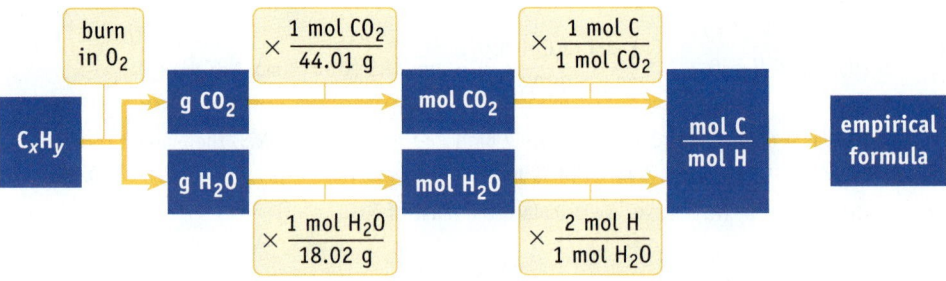

When using this procedure, a key observation is that every atom of C in the original compound appears as CO_2 and every atom of H appears in the form of water. In other words, for every mole of CO_2 observed, there must have been one mole of carbon in the unknown compound. Similarly, for every mole of H_2O observed from combustion, there must have been *two* moles of H atoms in the unknown carbon-hydrogen compound.

Example 4.6—Using Combustion Analysis to Determine the Formula of a Hydrocarbon

Problem When 1.125 g of a liquid hydrocarbon, C_xH_y, was burned in an apparatus like that shown in Figure 4.9, 3.447 g of CO_2 and 1.647 g of H_2O were produced. The molar mass of the compound was found to be 86.2 g/mol in a separate experiment. Determine the empirical and molecular formulas for the unknown hydrocarbon, C_xH_y.

Strategy As outlined in the preceding diagram, we first calculate the amounts of CO_2 and H_2O. These are then converted to amounts of C and H. The ratio (mol H/mol C) gives the empirical formula of the compound.

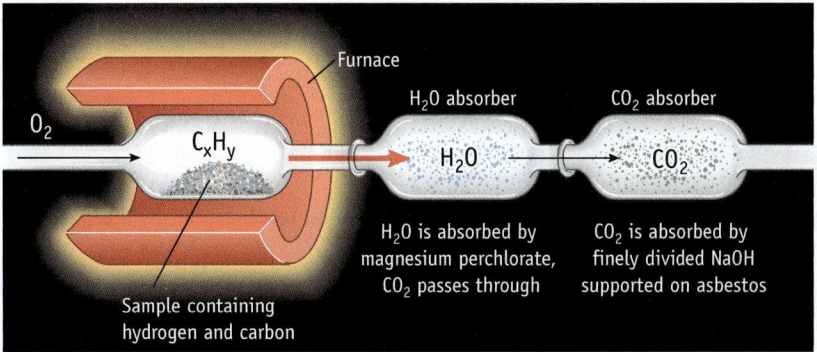

Active Figure 4.9   **Combustion analysis of a hydrocarbon.** If a compound containing C and H is burned in oxygen, CO_2 and H_2O are formed, and the mass of each can be determined. The H_2O is absorbed by magnesium perchlorate, and the CO_2 is absorbed by finely divided NaOH supported on asbestos. The mass of each absorbent before and after combustion gives the masses of CO_2 and H_2O. Only a few milligrams of a combustible compound are needed for analysis.

GENERAL
Chemistry⋅✦⋅**Now**™ **See the General ChemistryNow CD-ROM or website to explore an interactive version of this figure accompanied by an exercise.**

Solution The amounts of CO_2 and H_2O isolated from the combustion are

$$3.447 \text{ g } CO_2 \times \frac{1 \text{ mol } CO_2}{44.010 \text{ g } CO_2} = 0.07832 \text{ mol } CO_2$$

$$1.647 \text{ g } H_2O \times \frac{1 \text{ mol } H_2O}{18.015 \text{ g } H_2O} = 0.09142 \text{ mol } H_2O$$

For every mole of CO_2 isolated, 1 mol of C must have been present in the compound C_xH_y.

$$0.07832 \text{ mol } CO_2 \times \frac{1 \text{ mol C in } C_xH_y}{1 \text{ mol } CO_2} = 0.07832 \text{ mol C}$$

For every mole of H_2O isolated, 2 mol of H must have been present in C_xH_y.

$$0.09142 \text{ mol } H_2O \times \frac{2 \text{ mol H in } C_xH_y}{1 \text{ mol } H_2O} = 0.1828 \text{ mol H in } C_xH_y$$

The original 1.125-g sample of compound therefore contained 0.07832 mol of C and 0.1828 mol of H. To determine the empirical formula of C_xH_y, we find the ratio of moles of H to moles of C [◀ Section 3.6].

$$\frac{0.1828 \text{ mol H}}{0.07832 \text{ mol C}} = \frac{2.335 \text{ mol H}}{1.000 \text{ mol C}}$$

Atoms combine to form molecules in whole-number ratios. The translation of this ratio (2.335/1) to a whole-number ratio can usually be done quickly by trial and error. Multiplying the numerator and denominator by 3 gives 7/3. So, we know the ratio is 7 mol H to 3 mol C, which means the *empirical formula* of the hydrocarbon is C_3H_7.

Comparing the experimental molar mass with the molar mass calculated for the empirical formula,

$$\frac{\text{Experimental molar mass}}{\text{Molar mass of } C_3H_7} = \frac{86.2 \text{ g/mol}}{43.1 \text{ g/mol}} = \frac{2}{1}$$

we find that the molecular formula is twice the empirical formula. That is, the *molecular formula* is $(C_3H_7)_2$, or C_6H_{14}.

Comment As noted in Problem-Solving Tip 3.3 (page 124), for problems of this type be sure to use data with enough significant figures to give accurate atom ratios. Finally, note that the determination of the molecular formula does not end the problem for a chemist. In this case, the formula C_6H_{14} is appropriate for several distinctly different compounds. Two of the five compounds having this formula are shown here:

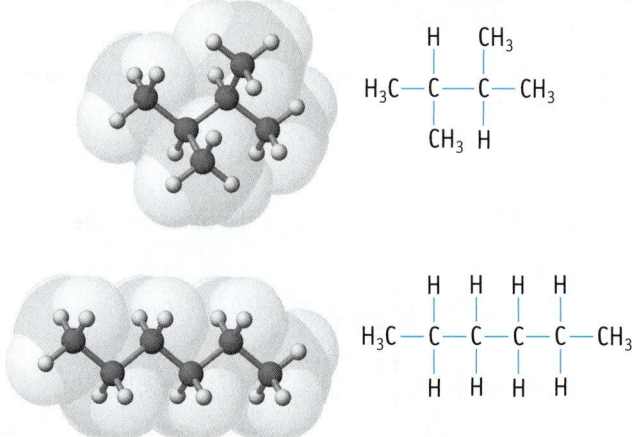

To decide finally the identity of the unknown compound, more laboratory experiments will have to be done.

Exercise 4.8—Determining the Empirical and Molecular Formulas for a Hydrocarbon

A 0.523-g sample of the unknown compound C_xH_y was burned in air to give 1.612 g of CO_2 and 0.7425 g of H_2O. A separate experiment gave a molar mass for C_xH_y of 114 g/mol. Determine the empirical and molecular formulas for the hydrocarbon.

Exercise 4.9—Determining the Empirical and Molecular Formulas for a Compound Containing C, H, and O

A 0.1342-g sample of a compound with C, H, and O ($C_xH_yO_z$) was burned in oxygen, and 0.240 g of CO_2 and 0.0982 g of H_2O were isolated. What is the empirical formula of the compound? If the experimentally determined molar mass was 74.1 g/mol, what is the molecular formula of the compound? (*Hint:* The carbon atoms in the compound are converted to CO_2 and the hydrogen atoms are converted to H_2O. The O atoms are found in both CO_2 and H_2O. To find the mass of O in the original sample, use the masses of CO_2 and H_2O to find the masses of C and H in the 0.1342 g-sample. Whatever of the 0.1342-g sample is not C and H is the mass of O.)

Chapter Goals Revisited

When you have finished studying this chapter, you should ask whether you have met the chapter goals. In particular, you should be able to

Balance equations for simple chemical reactions
a. Understand the information conveyed by a balanced chemical equation (Section 4.1).
b. Balance simple chemical equations (Section 4.2). General ChemistryNow homework: Study Question(s) 2, 12b

Perform stoichiometry calculations using balanced chemical equations
a. Understand the principle of the conservation of matter, which forms the basis of chemical stoichiometry (Section 4.3).
b. Calculate the mass of one reactant or product from the mass of another reactant or product by using the balanced chemical equation (Section 4.3). General ChemistryNow homework: SQ(s) 8, 16, 47, 53, 70, 72
c. Use amounts tables to organize stoichiometric information. General ChemistryNow homework: SQ(s) 16

Understand the impact of a limiting reactant on a chemical reaction
a. Determine which of two reactants is the limiting reactant (Section 4.4). General ChemistryNow homework: SQ(s) 22
b. Determine the yield of a product based on the limiting reactant. General ChemistryNow homework: SQ(s) 20, 24, 26

Calculate the theoretical and percent yields of a chemical reaction
a. Explain the differences among actual yield, theoretical yield, and percent yield, and calculate percent yield (Section 4.5). General ChemistryNow homework: SQ(s) 27

Use stoichiometry to analyze a mixture of compounds or to determine the formula of a compound
a. Use stoichiometry principles to analyze a mixture (Section 4.6). General ChemistryNow homework: SQ(s) 31, 69, 77
b. Find the empirical formula of an unknown compound using chemical stoichiometry (Section 4.6). General ChemistryNow homework: SQ(s) 37, 42, 66

GENERAL
Chemistry Now™

See the General ChemistryNow CD-ROM or website to:
- Assess your understanding with homework questions keyed to each goal
- Check your readiness for an exam by taking the exam-prep quiz and exploring the resources in the personalized Learning Plan it provides

For additional preparation for an examination on this chapter see the **Let's Review** section on pages 282–293.

Key Equation

Equation 4.1 (page 157) Calculating percent yield.

$$\text{Percent yield} = \frac{\text{actual yield (g)}}{\text{theoretical yield (g)}} \times 100\%$$

Study Questions

▲ denotes more challenging questions.

■ denotes questions available in the Homework and Goals section of the General ChemistryNow CD-ROM or website.

Blue numbered questions have answers in Appendix O and fully worked solutions in the *Student Solutions Manual*.

Structures of many of the compounds used in these questions are found on the General ChemistryNow CD-ROM or website in the Models folder.

Chemistry⚛Now™ Assess your understanding of this chapter's topics with additional quizzing and conceptual questions at **http://now.brookscole.com/kotz6e**

Practicing Skills

Balancing Equations

(See Example 4.1 and General ChemistryNow Screen 4.4.)

1. Write a balanced chemical equation for the combustion of liquid pentane.

2. ■ Write a balanced chemical equation for the production of ammonia, $NH_3(g)$, from $N_2(g)$ and $H_2(g)$.

3. Balance the following equations:
   (a) $Cr(s) + O_2(g) \longrightarrow Cr_2O_3(s)$
   (b) $Cu_2S(s) + O_2(g) \longrightarrow Cu(s) + SO_2(g)$
   (c) $C_6H_5CH_3(\ell) + O_2(g) \longrightarrow H_2O(\ell) + CO_2(g)$

4. Balance the following equations:
   (a) $Cr(s) + Cl_2(g) \longrightarrow CrCl_3(s)$
   (b) $SiO_2(s) + C(s) \longrightarrow Si(s) + CO(g)$
   (c) $Fe(s) + H_2O(g) \longrightarrow Fe_3O_4(s) + H_2(g)$

5. Balance the following equations and name each reactant and product:
   (a) $Fe_2O_3(s) + Mg(s) \longrightarrow MgO(s) + Fe(s)$
   (b) $AlCl_3(s) + NaOH(aq) \longrightarrow Al(OH)_3(s) + NaCl(aq)$
   (c) $NaNO_3(s) + H_2SO_4(\ell) \longrightarrow Na_2SO_4(s) + HNO_3(\ell)$
   (d) $NiCO_3(s) + HNO_3(aq) \longrightarrow$
   $Ni(NO_3)_2(aq) + CO_2(g) + H_2O(\ell)$

6. Balance the following equations and name each reactant and product:
   (a) $SF_4(g) + H_2O(\ell) \longrightarrow SO_2(g) + HF(\ell)$
   (b) $NH_3(aq) + O_2(aq) \longrightarrow NO(g) + H_2O(\ell)$
   (c) $BF_3(g) + H_2O(\ell) \longrightarrow HF(aq) + H_3BO_3(aq)$

Mass Relationships in Chemical Reactions: Basic Stoichiometry

(See Example 4.2 and General ChemistryNow Screens 4.5 and 4.6.)

7. Aluminum reacts with oxygen to give aluminum oxide.

$$4\,Al(s) + 3\,O_2(g) \longrightarrow 2\,Al_2O_3(s)$$

What amount of O_2, in moles, is needed for complete reaction with 6.0 mol of Al? What mass of Al_2O_3, in grams, can be produced?

8. ■ What mass of HCl, in grams, is required to react with 0.750 g of $Al(OH)_3$? What mass of water, in grams, is produced?

$$Al(OH)_3(s) + 3\,HCl(aq) \longrightarrow AlCl_3(aq) + 3\,H_2O(\ell)$$

9. Like many metals, aluminum reacts with a halogen to give a metal halide (see Figure 3.1).

$$2\,Al(s) + 3\,Br_2(\ell) \longrightarrow Al_2Br_6(s)$$

What mass of Br_2, in grams, is required for complete reaction with 2.56 g of Al? What mass of white, solid Al_2Br_6 is expected?

10. The balanced equation for a reaction in the process of reducing iron ore to the metal is

$$Fe_2O_3(s) + 3\,CO(g) \longrightarrow 2\,Fe(s) + 3\,CO_2(g)$$

(a) What is the maximum mass of iron, in grams, that can be obtained from 454 g (1.00 lb) of iron(III) oxide?
(b) What mass of CO is required to react with 454 g of Fe_2O_3?

11. Iron metal reacts with oxygen to give iron(III) oxide, Fe_2O_3.
(a) Write a balanced equation for the reaction.
(b) If an ordinary iron nail (assumed to be pure iron) has a mass of 2.68 g, what mass of Fe_2O_3, in grams, is produced if the nail is converted completely to the oxide?
(c) What mass of O_2, in grams, is required for the reaction?

12. Methane, CH_4, burns in oxygen.
(a) What are the products of the reaction?
(b) ■ Write the balanced equation for the reaction.
(c) What mass of O_2, in grams, is required for complete combustion of 25.5 g of methane?
(d) What is the total mass of products expected from the combustion of 25.5 g of methane?

13. Sulfur dioxide, a pollutant produced by burning coal and oil in power plants, can be removed by reaction with calcium carbonate.

$$2\,SO_2(g) + 2\,CaCO_3(s) + O_2(g) \longrightarrow 2\,CaSO_4(s) + 2\,CO_2(g)$$

(a) What mass of $CaCO_3$ is required to remove 155 g of SO_2?
(b) What mass of $CaSO_4$ is formed when 155 g of SO_2 is consumed completely?

14. The formation of water-insoluble silver chloride is useful in the analysis of chloride-containing substances. Consider the following *unbalanced* equation:

$$BaCl_2(aq) + AgNO_3(aq) \longrightarrow AgCl(s) + Ba(NO_3)_2(aq)$$

   (a) Write the balanced equation.
   (b) What mass $AgNO_3$, in grams, is required for complete reaction with 0.156 g of $BaCl_2$? What mass of $AgCl$ is produced?

Amounts Tables and Chemical Stoichiometry

For each question below set up an amounts table that lists the initial amount or amounts of reactants, the changes in amounts of reactants and products, and the amounts of reactants and products after reaction. See page 148 and Example 4.2.

15. A major source of air pollution years ago was the metals industry. One common process involved "roasting" metal sulfides in the air:

$$2\,PbS(s) + 3\,O_2(g) \longrightarrow 2\,PbO(s) + 2\,SO_2(g)$$

   If you heat 2.5 mol of PbS in the air, what amount of O_2 is required for complete reaction? What amounts of PbO and SO_2 are expected?

16. ■ Iron ore is converted to iron metal in a reaction with carbon.

$$2\,Fe_2O_3(s) + 3\,C(s) \longrightarrow 4\,Fe(s) + 3\,CO_2(g)$$

   If 6.2 mol of $Fe_2O_3(s)$ is used, what amount of $C(s)$ is needed and what amounts of Fe and CO_2 are produced?

17. Chromium metal reacts with oxygen to give chromium(III) oxide, Cr_2O_3.
   (a) Write a balanced equation for the reaction.
   (b) If a piece of chromium has a mass of 0.175 g, what mass (in grams) of Cr_2O_3 is produced if the metal is converted completely to the oxide?
   (c) What mass of O_2 (in grams) is required for the reaction?

18. Ethane, C_2H_6, burns in oxygen.
   (a) What are the products of the reaction?
   (b) Write the balanced equation for the reaction.
   (c) What mass of O_2, in grams, is required for complete combustion of 13.6 of ethane?
   (d) What is the total mass of products expected from the combustion of 13.6 g of ethane?

Limiting Reactants and Amounts Tables

(See Example 4.3 and Exercises 4.4 and 4.5. See also the General ChemistryNow Screens 4.7 and 4.8. In each case set up an amounts table.)

19. Sodium sulfide, Na_2S, is used in the leather industry to remove hair from hides. (This is the reason these kinds of plants stink!) The Na_2S is made by the reaction

$$Na_2SO_4(s) + 4\,C(s) \longrightarrow Na_2S(s) + 4\,CO(g)$$

   Suppose you mix 15 g of Na_2SO_4 and 7.5 g of C. Which is the limiting reactant? What mass of Na_2S is produced?

20. ■ Ammonia gas can be prepared by the reaction of a metal oxide such as calcium oxide with ammonium chloride.

$$CaO(s) + 2\,NH_4Cl(s) \longrightarrow \\ 2\,NH_3(g) + H_2O(g) + CaCl_2(s)$$

   If 112 g of CaO and 224 g of NH_4Cl are mixed, what mass of NH_3 can be produced?

21. The compound SF_6 is made by burning sulfur in an atmosphere of fluorine. The balanced equation is

$$S_8(s) + 24\,F_2(g) \longrightarrow 8\,SF_6(g)$$

   If you begin with 1.6 moles of sulfur, S_8, and 35 moles of F_2, which is the limiting reagent?

22. ■ Disulfur dichloride, S_2Cl_2, is used to vulcanize rubber. It can be made by treating molten sulfur with gaseous chlorine:

$$S_8(\ell) + 4\,Cl_2(g) \longrightarrow 4\,S_2Cl_2(\ell)$$

   Starting with a mixture of 32.0 g of sulfur and 71.0 g of Cl_2, which is the limiting reactant?

23. The reaction of methane and water is one way to prepare hydrogen for use as a fuel:

$$CH_4(g) + H_2O(g) \longrightarrow CO(g) + 3\,H_2(g)$$

   If you begin with 995 g of CH_4 and 2510 g of water,
   (a) Which reactant is the limiting reactant?
   (b) What is the maximum mass of H_2 that can be prepared?
   (c) What mass of the excess reactant remains when the reaction is completed?

24. ■ Aluminum chloride, $AlCl_3$, is made by treating scrap aluminum with chlorine.

$$2\,Al(s) + 3\,Cl_2(g) \longrightarrow 2\,AlCl_3(s)$$

   If you begin with 2.70 g of Al and 4.05 g of Cl_2,
   (a) Which reactant is limiting?
   (b) What mass of $AlCl_3$ can be produced?
   (c) What mass of the excess reactant remains when the reaction is completed?

25. Hexane (C_6H_{14}) burns in air (O_2) to give CO_2 and H_2O.
   (a) Write a balanced equation for the reaction.
   (b) If 215 g of C_6H_{14} is mixed with 215 g of O_2, what masses of CO_2 and H_2O are produced in the reaction?
   (c) What mass of the excess reactant remains after the hexane has been burned?

26. ■ Aspirin, $C_6H_4(OCOCH_3)CO_2H$, is produced by the reaction of salicylic acid, $C_6H_4(OH)CO_2H$, and acetic anhydride, $(CH_3CO)_2O$ (page 157).

$$C_6H_4(OH)CO_2H(s) + (CH_3CO)_2O(\ell) \longrightarrow \\ C_6H_4(OCOCH_3)CO_2H(s) + CH_3CO_2H(\ell)$$

   If you mix 100. g of each of the reactants, what is the maximum mass of aspirin that can be obtained?

▲ More challenging   ■ In General ChemistryNow   Blue-numbered questions answered in Appendix O

Percent Yield

(See Exercise 4.6 and General ChemistryNow Screen 4.9)

27. ■ In Example 4.3 you found that a mixture of CO and H_2 produced 407 g CH_3OH.

$$CO(g) + 2 H_2(g) \longrightarrow CH_3OH(\ell)$$

If only 332 g of CH_3OH is actually produced, what is the percent yield of the compound?

28. Ammonia gas can be prepared by the following reaction:

$$CaO(s) + 2 NH_4Cl(s) \longrightarrow 2 NH_3(g) + H_2O(g) + CaCl_2(s)$$

If 112 g of CaO and 224 g of NH_4Cl are mixed, the theoretical yield of NH_3 is 68.0 g (Study Question 20). If only 16.3 g of NH_3 is actually obtained, what is its percent yield?

29. The deep blue compound $Cu(NH_3)_4SO_4$ is made by the reaction of copper(II) sulfate and ammonia.

$$CuSO_4(aq) + 4 NH_3(aq) \longrightarrow Cu(NH_3)_4SO_4(aq)$$

(a) If you use 10.0 g of $CuSO_4$ and excess NH_3, what is the theoretical yield of $Cu(NH_3)_4SO_4$?

(b) If you isolate 12.6 g of $Cu(NH_3)_4SO_4$, what is the percent yield of $Cu(NH_3)_4SO_4$?

30. A reaction studied by Wächtershäuser and Huber (see "Black Smokers and the Origins of Life") is

$$2 CH_3SH + CO \longrightarrow CH_3COSCH_3 + H_2S$$

If you begin with 10.0 g of CH_3SH, and excess CO,

(a) What is the theoretical yield of CH_3COSCH_3?

(b) If 8.65 g of CH_3COSCH_3 is isolated, what is its percent yield?

Analysis of Mixtures

(See Examples 4.4 and 4.5 and General ChemistryNow Screen 4.10.)

31. ■ A mixture of $CuSO_4$ and $CuSO_4 \cdot 5 H_2O$ has a mass of 1.245 g. After heating to drive off all the water, the mass is only 0.832 g. What is the mass percent of $CuSO_4 \cdot 5 H_2O$ in the mixture? (See page 129.)

32. A 2.634-g sample containing $CuCl_2 \cdot 2H_2O$ and other materials was heated. The sample mass after heating to drive off the water was 2.125 g. What was the mass percent of $CuCl_2 \cdot 2H_2O$ in the original sample?

33. A sample of limestone and other soil materials is heated, and the limestone decomposes to give calcium oxide and carbon dioxide.

$$CaCO_3(s) \longrightarrow CaO(s) + CO_2(g)$$

A 1.506-g sample of limestone-containing material gives 0.558 g of CO_2, in addition to CaO, after being heated at a high temperature. What is the mass percent of $CaCO_3$ in the original sample?

34. At higher temperatures $NaHCO_3$ is converted quantitatively to Na_2CO_3.

$$2 NaHCO_3(s) \longrightarrow Na_2CO_3(s) + CO_2(g) + H_2O(g)$$

Heating a 1.7184-g sample of impure $NaHCO_3$ gives 0.196 g of CO_2. What was the mass percent of $NaHCO_3$ in the original 1.7184-g sample?

35. A pesticide contains thallium(I) sulfate, Tl_2SO_4. Dissolving a 10.20-g sample of impure pesticide in water and adding sodium iodide precipitates 0.1964 g of thallium(I) iodide, TlI.

$$Tl_2SO_4(aq) + 2 NaI(aq) \longrightarrow 2 TlI(s) + Na_2SO_4(aq)$$

What is the mass percent of Tl_2SO_4 in the original 10.20-g sample?

36. ▲ The aluminum in a 0.764-g sample of an unknown material was precipitated as aluminum hydroxide, $Al(OH)_3$, which was then converted to Al_2O_3 by heating strongly. If 0.127 g of Al_2O_3 is obtained from the 0.764-g sample, what is the mass percent of aluminum in the sample?

Using Stoichiometry to Determine Empirical and Molecular Formulas

(See Example 4.6, Exercise 4.9, and General ChemistryNow Screen 4.11.)

37. ■ Styrene, the building block of polystyrene, consists of only C and H. If 0.438 g of styrene is burned in oxygen and produces 1.481 g of CO_2 and 0.303 g of H_2O, what is the empirical formula of styrene?

38. Mesitylene is a liquid hydrocarbon. Burning 0.115 g of the compound in oxygen gives 0.379 g of CO_2 and 0.1035 g of H_2O. What is the empirical formula of mesitylene?

39. Cyclopentane is a simple hydrocarbon. If 0.0956 g of the compound is burned in oxygen, 0.300 g of CO_2 and 0.123 g of H_2O are isolated.

(a) What is the empirical formula of cyclopentane?

(b) If a separate experiment gave 70.1 g/mol as the molar mass of the compound, what is its molecular formula?

40. Azulene is a beautiful blue hydrocarbon. If 0.106 g of the compound is burned in oxygen, 0.364 g of CO_2 and 0.0596 g of H_2O are isolated.

(a) What is the empirical formula of azulene?

(b) If a separate experiment gave 128.2 g/mol as the molar mass of the compound, what is its molecular formula?

41. An unknown compound has the formula $C_xH_yO_z$. You burn 0.0956 g of the compound and isolate 0.1356 g of CO_2 and 0.0833 g of H_2O. What is the empirical formula of the compound? If the molar mass is 62.1 g/mol, what is the molecular formula? (See Exercise 4.9.)

42. ■ An unknown compound has the formula $C_xH_yO_z$. You burn 0.1523 g of the compound and isolate 0.3718 g of CO_2 and 0.1522 g of H_2O. What is the empirical formula of the compound? If the molar mass is 72.1 g/mol, what is the molecular formula? (See Exercise 4.9.)

▲ More challenging   ■ In General ChemistryNow   Blue-numbered questions answered in Appendix 0

43. Nickel forms a compound with carbon monoxide, $Ni_x(CO)_y$. To determine its formula, you carefully heat a 0.0973-g sample in air to convert the nickel to 0.0426 g of NiO and the CO to 0.100 g of CO_2. What is the empirical formula of $Ni_x(CO)_y$?

44. To find the formula of a compound composed of iron and carbon monoxide, $Fe_x(CO)_y$, the compound is burned in pure oxygen to give Fe_2O_3 and CO_2. If you burn 1.959 g of $Fe_x(CO)_y$ and obtain 0.799 g of Fe_2O_3 and 2.200 g of CO_2, what is the empirical formula of $Fe_x(CO)_y$?

General Questions on Stoichiometry

These questions are not designated as to type or location in the chapter. They may combine several chapters.

45. Balance the following equations:
(a) The synthesis of urea, a common fertilizer

$$CO_2(g) + NH_3(g) \longrightarrow NH_2CONH_2(s) + H_2O(\ell)$$

(b) Reactions used to make uranium(VI) fluoride for the enrichment of natural uranium

$$UO_2(s) + HF(aq) \longrightarrow UF_4(s) + H_2O(\ell)$$
$$UF_4(s) + F_2(g) \longrightarrow UF_6(s)$$

(c) The reaction to make titanium(IV) chloride, which is then converted to titanium metal

$$TiO_2(s) + Cl_2(g) + C(s) \longrightarrow TiCl_4(\ell) + CO(g)$$
$$TiCl_4(\ell) + Mg(s) \longrightarrow Ti(s) + MgCl_2(s)$$

46. Balance the following equations:
(a) Reaction to produce "superphosphate" fertilizer

$$Ca_3(PO_4)_2(s) + H_2SO_4(aq) \longrightarrow Ca(H_2PO_4)_2(aq) + CaSO_4(s)$$

(b) Reaction to produce diborane, B_2H_6

$$NaBH_4(s) + H_2SO_4(aq) \longrightarrow B_2H_6(g) + H_2(g) + Na_2SO_4(aq)$$

(c) Reaction to produce tungsten metal from tungsten(VI) oxide

$$WO_3(s) + H_2(g) \longrightarrow W(s) + H_2O(\ell)$$

(d) Decomposition of ammonium dichromate

$$(NH_4)_2Cr_2O_7(s) \longrightarrow N_2(g) + H_2O(\ell) + Cr_2O_3(s)$$

47. ■ Suppose 16.04 g of benzene, C_6H_6, is burned in oxygen.
(a) What are the products of the reaction?
(b) What is the balanced equation for the reaction?
(c) What mass of O_2, in grams, is required for complete combustion of benzene?
(d) What is the total mass of products expected from 16.04 g of benzene?

48. If 10.0 g of carbon is combined with an exact, stoichiometric amount of oxygen (26.6 g) to produce carbon dioxide, what is the theoretical yield of CO_2, in grams?

49. The metabolic disorder diabetes causes a buildup of acetone, CH_3COCH_3, in the blood. Acetone, a volatile com-

pound, is exhaled, giving the breath of untreated diabetics a distinctive odor. The acetone is produced by a breakdown of fats in a series of reactions. The equation for the last step is

$$CH_3COCH_2CO_2H \longrightarrow CH_3COCH_3 + CO_2$$

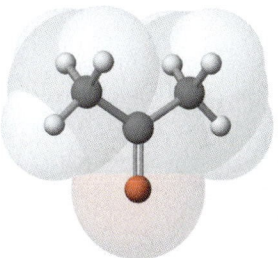

acetone, CH_3COCH_3

What mass of acetone can be produced from 125 mg of acetoacetic acid ($CH_3COCH_2CO_2H$)?

50. Your body deals with excess nitrogen by excreting it in the form of urea, NH_2CONH_2. The reaction producing it is the combination of arginine ($C_6H_{14}N_4O_2$) with water to give urea and ornithine ($C_5H_{12}N_2O_2$).

$$\underset{\text{Arginine}}{C_6H_{14}N_4O_2} + H_2O \longrightarrow \underset{\text{Urea}}{NH_2CONH_2} + \underset{\text{Ornithine}}{C_5H_{12}N_2O_2}$$

If you excrete 95 mg of urea, what mass of arginine must have been used? What mass of ornithine must have been produced?

51. In the Figure 4.2, you see the reaction of iron metal and chlorine gas to give iron(III) chloride.
(a) Write the balanced chemical equation for the reaction.
(b) Beginning with 10.0 g of iron, what mass of Cl_2, in grams, is required for complete reaction? What mass of $FeCl_3$ can be produced?
(c) If only 18.5 g of $FeCl_3$ is obtained from 10.0 g of iron and excess Cl_2, what is the percent yield?
(d) If equal masses of iron and chlorine are combined (10.0 g of each), what is the theoretical yield of iron(III) chloride?

52. Two beakers sit on a balance; the total mass is 161.170 g.

Solutions of KI and $Pb(NO_3)_2$ before reaction.

One beaker contains a solution of KI; the other contains a solution of $Pb(NO_3)_2$. When the solution in one beaker is poured completely into the other, the following reaction occurs:

$$2 KI(aq) + Pb(NO_3)_2(aq) \longrightarrow 2 KNO_3(aq) + PbI_2(s)$$

Solutions after reaction.

What is the total mass of the beakers and solutions after reaction? Explain completely. (*See the General ChemistryNow Screen 4.3, Exercise 1.*)

53. ■ Some metal halides react with water to produce the metal oxide and the appropriate hydrogen halide (see photo). For example,

$$TiCl_4(\ell) + 2 H_2O(\ell) \longrightarrow TiO_2(s) + 4 HCl(g)$$

(a) Name the four compounds involved in this reaction.
(b) If you begin with 14.0 mL of $TiCl_4$ ($d = 1.73$ g/mL), what mass of water, in grams, is required for complete reaction?
(c) What mass of each product is expected?

54. The reaction of 750. g each of NH_3 and O_2 was found to produce 562 g of NO (see pages 153–155).

$$4 NH_3(g) + 5 O_2(g) \longrightarrow 4 NO(g) + 6 H_2O(g)$$

(a) What mass of water is produced by this reaction?
(b) What quantity of O_2 is required to consume 750. g of NH_3?

55. Sodium azide, the explosive chemical used in automobile airbags, is made by the following reaction:

$$NaNO_3 + 3 NaNH_2 \longrightarrow NaN_3 + 3 NaOH + NH_3$$

If you combine 15.0 g of $NaNO_3$ (85.0 g/mol) with 15.0 g of $NaNH_2$, what mass of NaN_3 is produced?

56. Iodine is made by the reaction

$$2 NaIO_3(aq) + 5 NaHSO_3(aq) \longrightarrow$$
$$3 NaHSO_4(aq) + 2 Na_2SO_4(aq) + H_2O(\ell) + I_2(aq)$$

(a) Name the two reactants.
(b) If you wish to prepare 1.00 kg of I_2, what mass of $NaIO_3$ is required? What mass of $NaHSO_3$?

57. Copper(I) sulfide reacts with O_2 upon heating to give copper metal and sulfur dioxide.
(a) Write a balanced equation for the reaction.
(b) What mass of copper metal can be obtained from 500. g of copper(I) sulfide?

58. Saccharin, an artificial sweetener, has the formula $C_7H_5NO_3S$. Suppose you have a sample of a saccharin-containing sweetener with a mass of 0.2140 g. After decomposition to free the sulfur and convert it to the SO_4^{2-} ion, the sulfate ion is trapped as water-insoluble $BaSO_4$ (see Figure 4.8). The quantity of $BaSO_4$ obtained is 0.2070 g. What is the mass percent of saccharin in the sample of sweetener?

59. ▲ Boron forms an extensive series of compounds with hydrogen, all with the general formula B_xH_y.

$$B_xH_y(s) + \text{excess } O_2(g) \longrightarrow \tfrac{x}{2} B_2O_3(s) + \tfrac{y}{2} H_2O(g)$$

If 0.148 g of B_xH_y gives 0.422 g of B_2O_3 when burned in excess O_2, what is the empirical formula of B_xH_y?

60. ▲ Silicon and hydrogen form a series of compounds with the general formula Si_xH_y. To find the formula of one of them, a 6.22-g sample of the compound is burned in oxygen. All of the Si is converted to 11.64 g of SiO_2, and all of the H is converted to 6.980 g of H_2O. What is the empirical formula of the silicon compound?

61. ▲ Menthol, from oil of mint, has a characteristic odor. The compound contains only C, H, and O. If 95.6 mg of menthol burns completely in O_2, and gives 269 mg of CO_2 and 110 mg of H_2O, what is the empirical formula of menthol?

62. ▲ Quinone, a chemical used in the dye industry and in photography, is an organic compound containing only C, H, and O. What is the empirical formula of the compound if 0.105 g of the compound gives 0.257 g of CO_2 and 0.0350 g of H_2O when burned completely in oxygen?

63. ▲ In the Simulation portion of Screen 4.8 of the General ChemistryNow CD-ROM or website, choose the reaction of $FeCl_2$ and Na_2S.
(a) Write the balanced equation for the reaction.
(b) Choose 40 g of Na_2S as one reactant and add 40 g of $FeCl_2$. What is the limiting reactant?

▲ More challenging   ■ In General ChemistryNow   Blue-numbered questions answered in Appendix O

(c) What mass of FeS is produced?

(d) What mass of Na_2S or $FeCl_2$ remains after the reaction?

(e) What mass of $FeCl_2$ is required to react completely with 40 g of Na_2S?

64. Sulfuric acid can be prepared starting with the sulfide ore, cuprite (Cu_2S). If each S atom in Cu_2S leads to one molecule of H_2SO_4, what mass of H_2SO_4 can be produced from 3.00 kg of Cu_2S?

65. ▲ In an experiment 1.056 g of a metal carbonate, containing an unknown metal M, is heated to give the metal oxide and 0.376 g CO_2.

$$MCO_3(s) + heat \longrightarrow MO(s) + CO_2(g)$$

What is the identity of the metal M?

(a) M = Ni   (c) M = Zn

(b) M = Cu   (d) M = Ba

66. ■ ▲ An unknown metal reacts with oxygen to give the metal oxide, MO_2. Identify the metal based on the following information:

Mass of metal = 0.356 g

Mass of sample after converting metal completely to oxide = 0.452 g

67. ▲ Titanium(IV) oxide, TiO_2, is heated in hydrogen gas to give water and a new titanium oxide, Ti_xO_y. If 1.598 g of TiO_2 produces 1.438 g of Ti_xO_y, what is the formula of the new oxide?

68. ▲ Thioridazine, $C_{21}H_{26}N_2S_2$, is a pharmaceutical used to regulate dopamine. (Dopamine, a neurotransmitter, affects brain processes that control movement, emotional response, and ability to experience pleasure and pain.) A chemist can analyze a sample of the pharmaceutical for the thioridazine content by decomposing it to convert the sulfur in the compound to sulfate ion. This is then "trapped" as water-insoluble barium sulfate (see Figure 4.8).

$$SO_4^{2-}(aq, \text{ from thioridazine}) + BaCl_2(aq) \longrightarrow$$
$$BaSO_4(s) + 2\ Cl^-(aq)$$

Suppose a 12-tablet sample of the drug yielded 0.301 g of $BaSO_4$. What is the thioridazine content, in milligrams, of each tablet?

69. ■ ▲ A herbicide contains 2,4-D (2,4-dichlorophenoxy-acetic acid), $C_8H_6Cl_2O_3$. A 1.236-g sample of the herbicide was decomposed to liberate the chlorine as Cl^- ion. This was precipitated as AgCl, with a mass of 0.1840 g. What is the mass percent of 2,4-D in the sample?

70. ■ ▲ Potassium perchlorate is prepared by the following sequence of reactions:

$$Cl_2(g) + 2\ KOH(aq) \longrightarrow KCl(aq) + KClO(aq) + H_2O(\ell)$$
$$3\ KClO(aq) \longrightarrow 2\ KCl(aq) + KClO_3(aq)$$
$$4\ KClO_3(aq) \longrightarrow 3\ KClO_4(aq) + KCl(aq)$$

What mass of $Cl_2(g)$ is required to produce 234 kg of $KClO_4$?

71. ▲ Commercial sodium "hydrosulfite" is 90.1% pure $Na_2S_2O_4$. The sequence of reactions used to prepare the compound is

$$Zn(s) + 2\ SO_2(g) \longrightarrow ZnS_2O_4(s)$$
$$ZnS_2O_4(s) + Na_2CO_3(aq) \longrightarrow ZnCO_3(s) + Na_2S_2O_4(aq)$$

(a) What mass of pure $Na_2S_2O_4$ can be prepared from 125 kg of Zn, 500 g of SO_2, and an excess of Na_2CO_3?

(b) What mass of the commercial product would contain the $Na_2S_2O_4$ produced using the amounts of reactants in part (a)?

72. ■ What mass of lime, CaO, can be obtained by heating 125 kg of limestone that is 95.0% by mass $CaCO_3$?

$$CaCO_3(s) \longrightarrow CaO(s) + CO_2(g)$$

73. Sulfuric acid can be produced from a sulfide ore such as iron pyrite by the following sequence of reactions:

$$4\ FeS_2(s) + 11\ O_2(g) \longrightarrow 2\ Fe_2O_3(s) + 8\ SO_2(g)$$
$$2\ SO_2(g) + O_2(g) \longrightarrow 2\ SO_3(g)$$
$$SO_3(g) + H_2O(\ell) \longrightarrow H_2SO_4(\ell)$$

Starting with 525 kg of FeS_2 (and an excess of other reactants), what mass of pure H_2SO_4 can be prepared?

74. ▲ The elements silver, molybdenum, and sulfur combine to form Ag_2MoS_4. What is the maximum mass of Ag_2MoS_4 that can be obtained if 8.63 g of silver, 3.36 g of molybdenum, and 4.81 g of sulfur are combined?

75. ▲ A mixture of butene, C_4H_8, and butane, C_4H_{10}, is burned in air to give CO_2 and water. Suppose you burn 2.86 g of the mixture and obtain 8.80 g of CO_2 and 4.14 g of H_2O. What is the weight percents of butene and butane in the mixture?

76. ▲ Cloth can be waterproofed by coating it with a silicone layer. This is done by exposing the cloth to $(CH_3)_2SiCl_2$ vapor. The silicon compound reacts with OH groups on the cloth to form a waterproofing film (density = 1.0 g/cm³) of $[(CH_3)_2SiO]_n$, where n is a large integer number.

$$n(CH_3)_2SiCl_2 + 2n\ OH^- \longrightarrow$$
$$2n\ Cl^- + n\ H_2O + [(CH_3)_2SiO]_n$$

The coating is added layer by layer, each layer of $[(CH_3)_2SiO]_n$ being 0.60 nm thick. Suppose you want to waterproof a piece of cloth that is 3.00 m square, and you want 250 layers of waterproofing compound on the cloth. What mass of $(CH_3)_2SiCl_2$ do you need?

77. ▲ Sodium hydrogen carbonate, $NaHCO_3$, can be decomposed quantitatively by heating.

$$2\ NaHCO_3(s) \longrightarrow Na_2CO_3(s) + CO_2(g) + H_2O(g)$$

A 0.682-g sample of impure $NaHCO_3$ yielded a solid residue (consisting of Na_2CO_3 and other solids) with a mass of 0.467 g. What was the mass percent of $NaHCO_3$ in the sample?

78. ▲ Copper metal can be prepared by roasting copper ore, which can contain cuprite (Cu_2S) and copper(II) sulfide.

$$Cu_2S(s) + O_2(g) \longrightarrow 2\ Cu(s) + SO_2(g)$$
$$CuS(s) + O_2(g) \longrightarrow Cu(s) + SO_2(g)$$

Suppose an ore sample contains 11.0% impurity in addition to a mixture of CuS and Cu_2S. Heating 100.0 g of the mixture produces 75.4 g of copper metal with a purity of 89.5%. What is the weight percent of CuS in the ore? The weight percent of Cu_2S?

Summary and Conceptual Questions

The following questions use concepts from the preceding chapters.

79. ▲ A weighed sample of iron (Fe) is added to liquid bromine (Br_2) and allowed to react completely. The reaction produces a single product, which can be isolated and weighed. The experiment was repeated a number of times with different masses of iron but with the same mass of bromine. (See the graph below.)

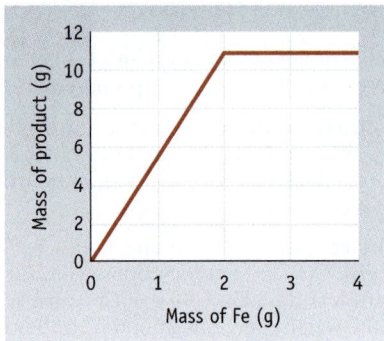

(a) What mass of Br_2 is used when the reaction consumes 2.0 g of Fe?

(b) What is the mole ratio of Br_2 to Fe in the reaction?

(c) What is the empirical formula of the product?

(d) Write the balanced chemical equation for the reaction of iron and bromine.

(e) What is the name of the reaction product?

(f) Which statement or statements best describe the experiments summarized by the graph?

   (i) When 1.00 g of Fe is added to the Br_2, Fe is the limiting reagent.

   (ii) When 3.50 g of Fe is added to the Br_2, there is an excess of Br_2.

   (iii) When 2.50 g of Fe is added to the Br_2, both reactants are used up completely.

   (iv) When 2.00 g of Fe is added to the Br_2, 10.0 g of product is formed. The percent yield must therefore be 20.0%.

80. Chlorine and iodine react according to the balanced equation

$$I_2(g) + 3\ Cl_2(g) \longrightarrow 2\ ICl_3(g)$$

Suppose that you mix I_2 and Cl_2 in a flask and that the mixture is represented by the diagram below.

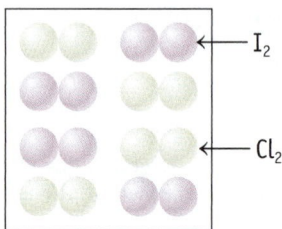

When the reaction between the Cl_2 and I_2 (according to the balanced equation above) is complete, which panel below represents the outcome? Which compound is the limiting reactant?

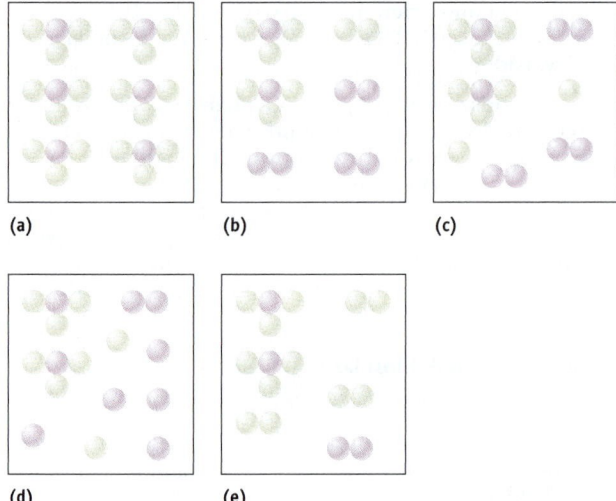

81. Cisplatin [$Pt(NH_3)_2Cl_2$] is a cancer chemotherapy agent. Notice that it contains NH_3 groups attached to platinum.

(a) What is the weight percent of Pt, N, and Cl in the cisplatin?

(b) Cisplatin is made by reacting K_2PtCl_4 with ammonia.

$$K_2PtCl_4(aq) + 2\ NH_3(aq) \longrightarrow Pt(NH_3)_2Cl_2(aq) + 2\ KCl(aq)$$

If you begin with 16.0 g of K_2PtCl_4, what mass of ammonia should be used to completely consume the K_2PtCl_4? What mass of cisplatin will be produced?

---

82. Iron(III) chloride is produced by the reaction of iron and chlorine (Figure 4.2).

    (a) If you place 1.54 g of iron gauze in chlorine gas, what mass of chlorine is required for complete reaction? What mass of iron(III) chloride is produced?

    (b) Iron(III) chloride reacts readily with NaOH to produce iron(III) hydroxide and sodium chloride. If you mix 2.0 g of iron(III) chloride with 4.0 g of NaOH, what mass of iron(III) hydroxide is produced? (*See the General ChemistryNow Screen 4.8 Simulation.*)

83. Let us explore a reaction with a limiting reactant. (*See the General ChemistryNow Screen 4.8.*) Here zinc metal is added to a flask containing aqueous HCl, and H_2 gas is a product.

    $$Zn(s) + 2\,HCl(aq) \longrightarrow ZnCl_2(aq) + H_2(g)$$

    The three flasks each contain 0.100 mol of HCl. Zinc is added to each flask in the following quantities.

    Flask 1: 7.00 g Zn

    Flask 2: 3.27 g Zn

    Flask 3: 1.31 g Zn

Charles D. Winters

When the reactants are combined, the H_2 inflates the balloon attached to the flask. The results are as follows:

Flask 1: Balloon inflates completely but some Zn remains when inflation ceases.

Flask 2: Balloon inflates completely. No Zn remains.

Flask 3: Balloon does not inflate completely. No Zn remains.

Explain these results completely. Perform calculations that support your explanation.

84. The reaction of aluminum and bromine is pictured in Figure 3.1 and below. The white solid on the lip of the beaker at the end of the reaction is Al_2Br_6. In the reaction pictured below, which was the limiting reactant, Al or Br_2? (*See General ChemistryNow Screen 4.2.*)

Charles D. Winters

Before reaction   After reaction

Do you need a live tutor for homework problems?
Access vMentor at General ChemistryNow at
http://now.brookscole.com/kotz6e
for one-on-one tutoring from a chemistry expert

5—Reactions in Aqueous Solution

API/Explorer/Photo Researchers, Inc.

Volcanoes are the chief source of chloride ion in the earth's oceans.

Salt

There is a French legend about a princess who told her father, the king, that she loved him as much as she loved salt. Thinking that this was not a great measure of love, he banished her from the kingdom. Only later did he realize how much he needed, and valued, salt.

Salt has played a key role in history. The earliest written record of salt production dates from around 800 B.C., but the sea has always been a source of salt. Indeed, there is evidence of the Chinese harvesting salt from sea water by 6000 B.C.

Saltiness is one of the basic taste sensations, and a taste of sea water quickly reveals its nature. How did the oceans become salty? What, in addition to salt, is dissolved in sea water?

Sea water contains enormous amounts of dissolved salts. Ions of virtually every element are present as well as dozens of polyatomic ions. What is their origin? And why is chloride ion the most abundant ion?

The carbonate ion and its close relative HCO_3^-, the bicarbonate ion, can come from the interaction of atmospheric CO_2 with water.

$$CO_2(g) + H_2O(\ell) \longrightarrow H_2CO_3(aq)$$
$$H_2CO_3(aq) \longrightarrow H^+(aq) + HCO_3^-(aq)$$

The reaction of CO_2 and H_2O is the reason rain is normally acidic. The slightly acidic rainwater then causes substances such as limestone or corals to dissolve, producing calcium ions and more bicarbonate ions.

$$CaCO_3(s) + CO_2(g) + H_2O(\ell) \longrightarrow Ca^{2+}(aq) + 2\ HCO_3^-(aq)$$

Magnesium ions come from a similar reaction with the mineral dolomite (a mixture of $CaCO_3$ and $MgCO_3$), which is found in terrestrial rocks such as those in Arizona's Grand Canyon and Italy's Dolomite Mountains. ($MgCl_2$ is often found with sea salt and gives the salt a bitter taste.)

Sodium ions arrive in the oceans by a similar reaction with sodium-bearing minerals such as albite, $NaAlSi_3O_8$. Acidic rain falling on the land extracts sodium ions that rivers then carry to the ocean.

The average chloride content of rocks in the earth's crust is only 0.01%, so only a minute proportion of the chloride ions in the oceans can come from the weathering of rocks and minerals. What, then, is the origin of the chloride ions in sea water? Volcanoes. Hydrogen chloride gas, HCl, is an important constituent of volcanic gases. Early in earth's history, the planet was much hotter, and volcanoes were much more widespread. The HCl gas emitted from these volcanoes is very soluble in water and quickly dissolves to give a dilute solution of hydrochloric acid.

$$HCl(g) \longrightarrow H^+(aq) + Cl^-(aq)$$

The chloride ions from dissolved HCl gas and the sodium ions from weathered rocks are the source of the salt in the sea.

The average human body contains about 230 g of salt. Because we continually lose salt in urine, sweat, and other excretions, salt must be a part of our diet. Early humans recognized that salt deficiency causes headaches, cramps, loss of appetite, and, in extreme cases, death. Consuming meat provides salt, but consuming vegetables does not. This is the reason why herbivorous animals seek out "salt licks."

Early humans also learned that salt preserves other materials. Egyptians used salt to make mummies, and fish and meat are often preserved by salting. This ability to protect against decay led to the Jewish tradition of bringing salt to a new home. In medieval France, salt was placed on the tongue of a newborn child and a young child was salted.

The importance of salt in society is reflected in a 16th-century book of table etiquette. It was written that salt could be handled safely only with the middle two fingers. If a person were to use a thumb, his children will die, and using the index finger would cause one to become a murderer.

Salt is so indispensible that it has been, not surprisingly, a source of revenue for governments. One example, which led to an extremely abusive tax, occurred in India in the 20th century. In colonial times the British established a salt tax and outlawed the production of salt from sea water. Salt could only be purchased from British government agents at a price established by the British. What is more, even though the salt tax was eliminated in Great Britain in the 18th century, the tax on salt was doubled in India in 1923. To protest this tax, in March 1930 Mahatma Gandhi led a pilgrimage to the sea, joined by thousands, to collect salt. Thousands were jailed, but strikes and demonstrations continued. A year later the salt tax was relaxed, and Britain's monopoly on salt was broken. This event marked the beginning of the end of British rule in India, and the country became independent in 1947.

For an account of salt in history and society, read *Salt, A World History*, by Mark Kurlansky (New York, Penguin Books, 2003).

The Dead Sea. This sea in the Middle East has the highest salt content of any body of water.

Paul Stephan-Vierow/Photo Researchers, Inc.

GENERAL
Chemistry Now™

Throughout the chapter this icon introduces a list of resources on the General ChemistryNow CD-ROM or website (http://now .brookscole.com/kotz6e) that will:

- help you evaluate your knowledge of the material
- provide homework problems
- allow you to take an exam-prep quiz
- provide a personalized Learning Plan targeting resources that address areas you should study

The human body is two-thirds water. Water is essential because it is involved in every function of the body. It assists in transporting nutrients and waste products in and out of cells and is necessary for all digestive, absorption, circulatory, and excretory functions. We turn now to the study of **aqueous** solutions, chemical systems in which water plays a major role.

5.1—Properties of Compounds in Aqueous Solution

A **solution** is a homogeneous mixture of two or more substances. One substance is generally considered the **solvent**, the medium in which another substance, the **solute**, is dissolved. To understand reactions occurring in aqueous solution, it is important first to understand something about the behavior of compounds in water. The focus here is on compounds that produce ions when dissolved in water.

Ions in Aqueous Solution: Electrolytes

The water you drink every day and the oceans of the world contain many ions, most of which result from dissolving solid materials present in the environment (Table 5.1).

Dissolving an ionic solid requires separating each ion from the oppositely charged ions that surround it in the solid state. Water is especially good at dissolving ionic compounds, because each water molecule has a positively charged end and a negatively charged end. When an ionic compound dissolves in water, each negative ion becomes surrounded by water molecules with their positive ends pointing toward it, and each positive ion becomes surrounded by the negative ends of several water molecules (Figure 5.1).

Table 5.1 Concentrations of Some Cations and Anions in the Environment and in Living Cells

Element	Dissolved Species	Sea Water	Valonia*	Red-Blood Cells	Blood Plasma
Chlorine	Cl^-	550	50	50	100
Sodium	Na^+	460	80	11	160
Magnesium	Mg^{2+}	52	50	2.5	2
Calcium	Ca^{2+}	10	1.5	10^{-4}	2
Potassium	K^+	10	400	92	10
Carbon	HCO_3^-, CO_3^{2-}	30	<10	<10	30
Phosphorus	HPO_4^{2-}	<1	5	3	<3

Data are taken from J. J. R. Fraústo da Silva and R. J. P. Williams: *The Biological Chemistry of the Elements*, Oxford, UK, Clarendon Press, 1991. Concentrations are given in millimoles per liter. (A millimole is 1/1000 of a mole.)

Valonia are single-celled algae that live in sea water.

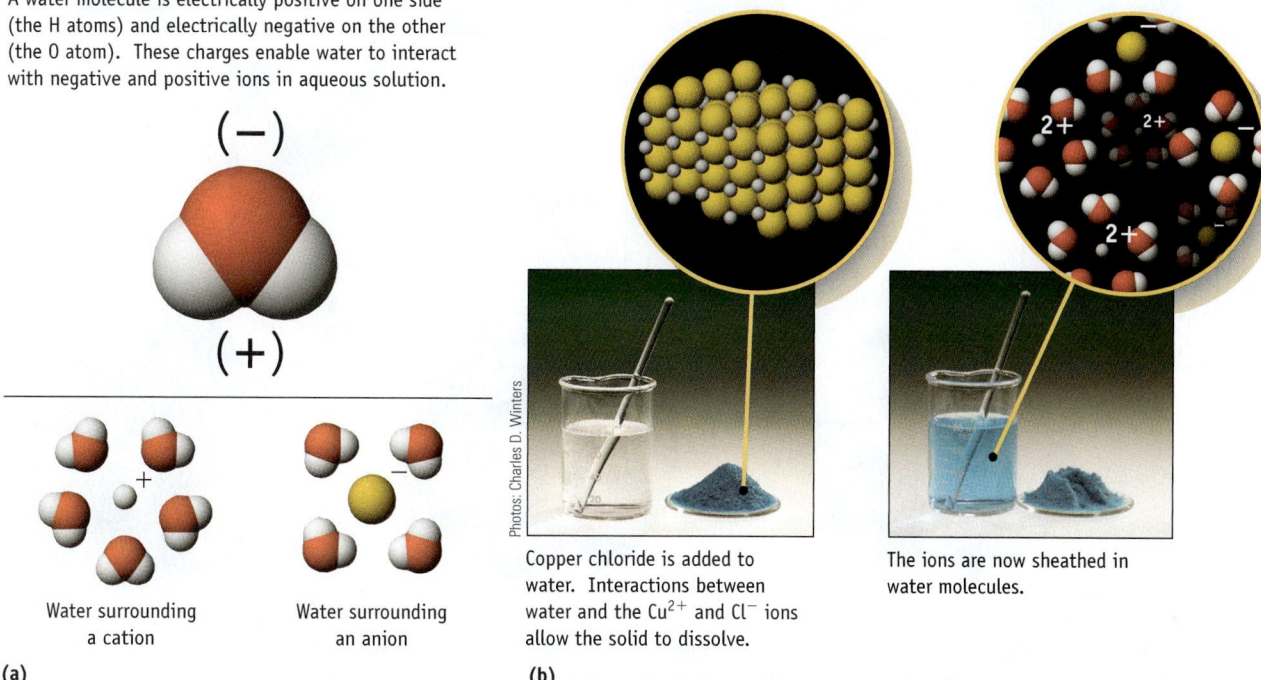

A water molecule is electrically positive on one side (the H atoms) and electrically negative on the other (the O atom). These charges enable water to interact with negative and positive ions in aqueous solution.

(−)

(+)

Water surrounding a cation

Water surrounding an anion

(a)

Photos: Charles D. Winters

Copper chloride is added to water. Interactions between water and the Cu^{2+} and Cl^- ions allow the solid to dissolve.

The ions are now sheathed in water molecules.

(b)

Figure 5.1   Water as a solvent for ionic substances. (a) Water can bind to both positive cations and negative anions in aqueous solution. (b) When an ionic substance dissolves in water, each ion is surrounded by a sheath of water molecules. (The number of H_2O molecules around an ion is often 6.)

The water-encased ions produced by dissolving an ionic compound are free to move about in solution. Under normal conditions, the movement of ions is random, and the cations and anions from a dissolved ionic compound are dispersed uniformly throughout the solution. However, if two **electrodes** (conductors of electricity such as copper wire) are placed in the solution and connected to a battery, ion movement is no longer random. Positive cations move through the solution to the negative electrode, and negative anions move to the positive electrode (Figure 5.2). If a light bulb is inserted into the circuit, the bulb lights, showing that ions are available to conduct charge in the solution, just as electrons conduct charge in the wire part of the circuit. Compounds whose aqueous solutions conduct electricity are called **electrolytes**. *All ionic compounds that are soluble in water are electrolytes.*

Types of Electrolytes

For every mole of NaCl that dissolves, 1 mol of Na^+ ions and 1 mol of Cl^- ions enter the solution.

$$NaCl(s) \longrightarrow Na^+(aq) + Cl^-(aq)$$
$$100\% \text{ Dissociation} \equiv \text{strong electrolyte}$$

Because the solute has dissociated completely into ions, the solution will be a good conductor of electricity. Substances whose solutions are good electrical conductors owing to the presence of ions are called **strong electrolytes** (see Figure 5.2).

Other substances dissociate only partially in solution and so are poor conductors of electricity; they are known as **weak electrolytes** (see Figure 5.2). For example,

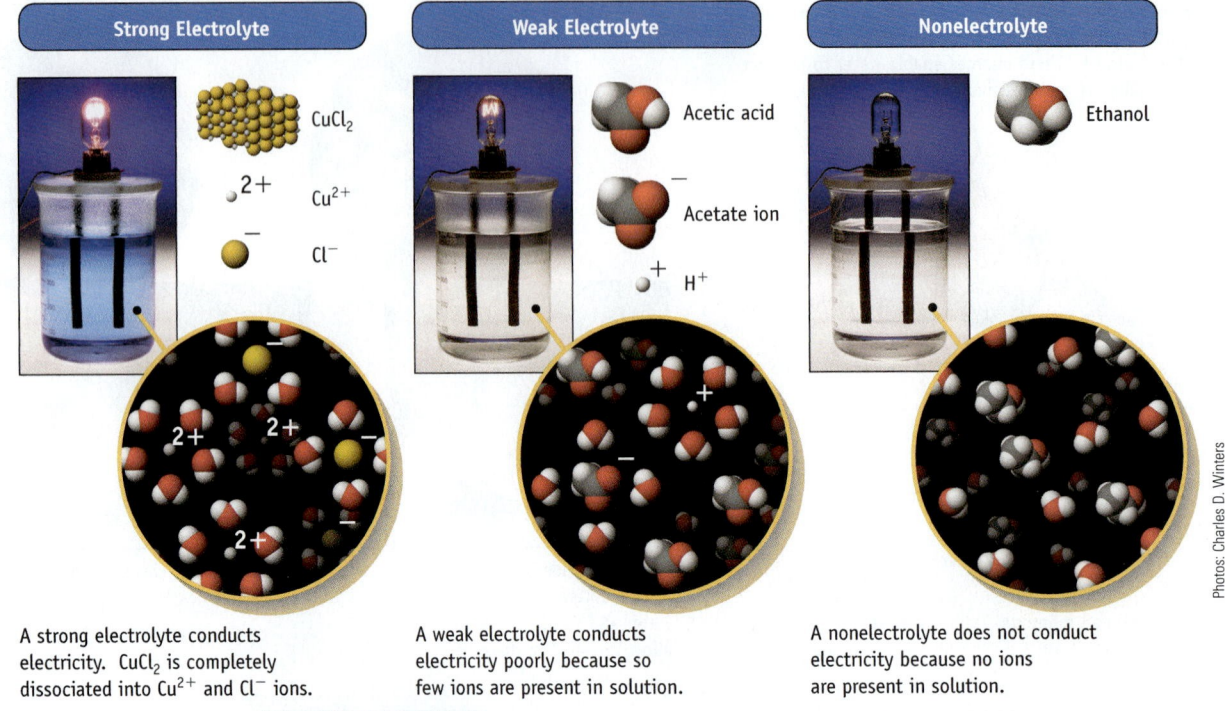

Strong Electrolyte	Weak Electrolyte	Nonelectrolyte

A strong electrolyte conducts electricity. $CuCl_2$ is completely dissociated into Cu^{2+} and Cl^- ions.

A weak electrolyte conducts electricity poorly because so few ions are present in solution.

A nonelectrolyte does not conduct electricity because no ions are present in solution.

Photos: Charles D. Winters

Active Figure 5.2   **Classifying solutions by their ability to conduct electricity.**

GENERAL
Chemistry··Now™ See the General ChemistryNow CD-ROM or website to explore an interactive version of this figure accompanied by an exercise.

when acetic acid—an important ingredient in vinegar—dissolves in water, only a few molecules in every 100 molecules of acetic acid are ionized to form acetate and hydrogen ions.

■ **Double arrows,** $\rightleftharpoons$
The double arrows in the equation for the ionization of acetic acid, and in many other chemical equations, indicate the reactant produces the product, but also that the product ions recombine to produce the original reactant. This is the subject of chemical equilibrium [▶Chapters 16–18].

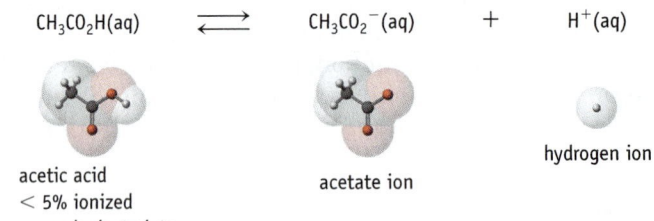

$$CH_3CO_2H(aq) \rightleftharpoons CH_3CO_2^-(aq) + H^+(aq)$$

acetic acid
$< 5\%$ ionized
$=$ weak electrolyte

acetate ion

hydrogen ion

Many other substances dissolve in water but do not ionize. They are called **nonelectrolytes** because their solutions do not conduct electricity (see Figure 5.2). Examples of nonelectrolytes include sucrose ($C_{12}H_{22}O_{11}$), ethanol (CH_3CH_2OH), and antifreeze (ethylene glycol, $HOCH_2CH_2OH$).

GENERAL
Chemistry··Now™

See the General ChemistryNow CD-ROM or website:
- **Screen 5.2 Solutions,** for a video and an animation on the dissolving of an ionic compound
- **Screen 5.3 Compounds in Aqueous Solution,** for an animation on the types of electrolytes

SILVER COMPOUNDS

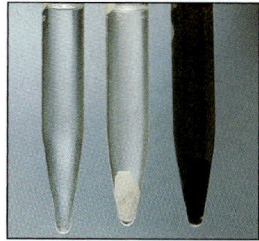

AgNO₃   AgCl   AgOH

(a) Nitrates are generally soluble, as are chlorides (except AgCl). Hydroxides are generally not soluble.

SULFIDES

(NH₄)₂S   CdS   Sb₂S₃   PbS

(b) Sulfides are generally not soluble (exceptions include salts with NH_4^+ and Na^+).

HYDROXIDES

Photos: Charles D. Winters

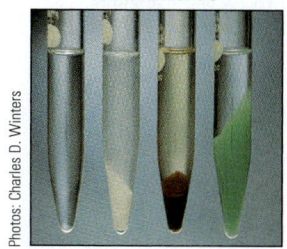

NaOH   Ca(OH)₂   Fe(OH)₃   Ni(OH)₂

(c) Hydroxides are generally not soluble except when the cation is a Group 1A metal.

SOLUBLE COMPOUNDS	EXCEPTIONS
Almost all salts of Na^+, K^+, NH_4^+	

Salts of nitrate, NO_3^-
       chlorate, ClO_3^-
       perchlorate, ClO_4^-
       acetate, $CH_3CO_2^-$	

Almost all salts of Cl^-, Br^-, I^-	Halides of Ag^+, Hg_2^{2+}, Pb^{2+}
Compounds containing F^-	Fluorides of Mg^{2+}, Ca^{2+}, Sr^{2+}, Ba^{2+}, Pb^{2+}
Salts of sulfate, SO_4^{2-}	Sulfates of Ca^{2+}, Sr^{2+}, Ba^{2+}, Pb^{2+}

INSOLUBLE COMPOUNDS	EXCEPTIONS

Most salts of carbonate, CO_3^{2-}
       phosphate, PO_4^{3-}
       oxalate, $C_2O_4^{2-}$
       chromate, CrO_4^{2-}
	Salts of NH_4^+ and the alkali metal cations

Most metal sulfides, S^{2-}	
Most metal hydroxides and oxides	Ba(OH)₂ is soluble

Active Figure 5.3 **Guidelines to predict the solubility of ionic compounds.** If a compound contains one of the ions in the column to the left in the top chart, it is predicted to be at least moderately soluble in water. There are a few exceptions, which are noted at the right. Most ionic compounds formed by the anions listed at the bottom of the chart are poorly soluble (with exceptions such as compounds with NH_4^+ and the alkali metal cations).

GENERAL
Chemistry⚛Now™ See the General ChemistryNow CD-ROM or website to explore an interactive version of this figure accompanied by an exercise.

Exercise 5.1—Electrolytes

Epsom salt, MgSO₄ · 7 H₂O, is sold in drugstores and, as a solution in water, is used for various medical purposes. Methanol, CH₃OH, is dissolved in gasoline in the winter in colder climates to prevent the formation of ice in automobile fuel lines. Which of these compounds is an electrolyte and which is a nonelectrolyte?

Solubility of Ionic Compounds in Water

Not all ionic compounds dissolve completely in water. Many dissolve only to a small extent, and still others are essentially insoluble. Fortunately, we can make some general statements about which ionic compounds are water soluble.

Figure 5.3 lists broad guidelines that help predict whether a particular ionic compound will be soluble in water. For example, sodium nitrate, NaNO₃, contains

■ Solubility Guidelines
Observations such as those shown in Figure 5.3 were used to create the solubility guidelines. Note, however, that these are general guidelines and not rules followed under any circumstance. Some exceptions do exist, but the guidelines are a good place to begin. See B. Blake, *Journal of Chemical Education*, Vol. 80, pp. 1348–1350, 2003.

both an alkali metal cation, Na^+, and the nitrate anion, NO_3^-. The presence of either of these ions ensures that the compound is soluble in water. By contrast, calcium hydroxide is poorly soluble in water (Figure 5.3c). If a spoonful of solid $Ca(OH)_2$ is added to 100 mL of water, only 0.17 g, or 0.0023 mol, will dissolve at 10 °C. Very few Ca^{2+} and OH^- ions are present in solution. Nearly all of the $Ca(OH)_2$ remains as a solid.

$$0.0023 \text{ mol } Ca(OH)_2 \text{ dissolves in 100 mL water at 10 °C} \longrightarrow$$
$$0.0023 \text{ mol } Ca^{2+}(aq) + (2 \times 0.0023) \text{ mol } OH^-(aq)$$

GENERAL
Chemistry⚛Now™

See the General ChemistryNow CD-ROM or website:
- **Screen 5.4 Solubility of Ionic Compounds**

  (a) for a simulation exploring the rules for predicting whether a compound is soluble or insoluble

  (b) for a tutorial on determining whether a compound is soluble in water

■ Soluble Ionic Compounds = Electrolytes
Ionic compounds that dissolve in water are electrolytes. For example, an aqueous solution of $AgNO_3$ (Figure 5.3a) consists of the separated ions $Ag^+(aq)$ and $NO_3^-(aq)$ and is a good conductor of electricity.

Example 5.1—Solubility Guidelines

Problem Predict whether the following ionic compounds are likely to be water-soluble. List the ions present in solution for soluble compounds.

(a) KCl   **(c)** Fe_2O_3

(b) $MgCO_3$ **(d)** $Cu(NO_3)_2$

Strategy You must first recognize the cation and anion involved and then decide the probable water solubility based on the guidelines outlined in Figure 5.3.

Solution

(a) KCl is composed of K^+ and Cl^- ions. The presence of *either* of these ions means that the compound is likely to be soluble in water. The solution consists of K^+ and Cl^- ions.

$$KCl(s) \longrightarrow K^+(aq) + Cl^-(aq)$$

(The solubility of KCl is about 35 g in 100 mL of water at 20 °C.)

(b) Magnesium carbonate is composed of Mg^{2+} and CO_3^{2-} ions. Salts containing the carbonate ion are usually insoluble, unless combined with an ion like Na^+ or NH_4^+. Therefore, $MgCO_3$ is predicted to be insoluble in water. (The experimental solubility of $MgCO_3$ is less than 0.2 g/100 mL of water.)

(c) Iron(III) oxide is composed of Fe^{3+} and O^{2-} ions. Oxides are soluble only when O^{2-} is combined with an alkali metal ion; Fe^{3+} is a transition metal ion, so Fe_2O_3 is insoluble.

(d) Copper(II) nitrate is composed of Cu^{2+} and NO_3^- ions. Almost all nitrates are soluble in water, so $Cu(NO_3)_2$ is water-soluble and produces copper(II) cations and nitrate anions in water.

$$Cu(NO_3)_2(s) \longrightarrow Cu^{2+}(aq) + 2 NO_3^-(aq)$$

Comment Notice that $Cu(NO_3)_2$ gives one Cu^{2+} ion and *two* NO_3^- ions on dissolving in water.

Exercise 5.2—Solubility of Ionic Compounds

Predict whether each of the following ionic compounds is likely to be soluble in water. If it is soluble, write the formulas of the ions present in aqueous solution.

(a) $LiNO_3$ (b) $CaCl_2$ (c) CuO (d) $NaCH_3CO_2$

5.2—Precipitation Reactions

A **precipitation reaction** produces a water-insoluble product, known as a **precipitate**. The reactants in such reactions are generally water-soluble ionic compounds. When these substances dissolve in water, they dissociate to give the appropriate cations and anions. If the cation from one compound can form an insoluble compound with the anion from the other compound in the solution, precipitation occurs. For example, silver nitrate and potassium chloride, both of which are water-soluble ionic compounds, form insoluble silver chloride and soluble potassium nitrate (Figure 5.4).

$$AgNO_3(aq) + KCl(aq) \longrightarrow AgCl(s) + KNO_3(aq)$$

Reactants	Products
$Ag^+(aq) + NO_3^-(aq)$	Insoluble AgCl
$K^+(aq) + Cl^-(aq)$	$K^+(aq) + NO_3^-(aq)$

Many combinations of positive and negative ions give insoluble substances (see Figures 5.4 and 5.5). For example, lead(II) chromate precipitates when a water soluble lead(II) compound is combined with a water-soluble chromate compound (Figure 5.5a).

$$Pb(NO_3)_2(aq) + K_2CrO_4(aq) \longrightarrow PbCrO_4(s) + 2\ KNO_3(aq)$$

Reactants	Products
$Pb^{2+}(aq) + 2\ NO_3^-(aq)$	Insoluble $PbCrO_4$
$2\ K^+(aq) + CrO_4^{2-}(aq)$	$2\ K^+(aq) + 2\ NO_3^-(aq)$

■ **Exchange Reactions**
When two ionic compounds in aqueous solution react to form a solid precipitate, they do so by exchanging ions. For example, silver(I) ions exchange nitrate ions for chloride ions, and potassium ions exchange chloride ions for nitrate ions.

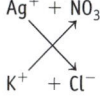

$$Ag^+ + NO_3^-$$
$$K^+ + Cl^-$$

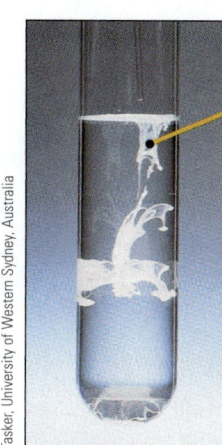

(a)

(b) Initially the Ag^+ ions (silver color) and Cl^- ions (green) are widely separated.

(c) Ag^+ and Cl^- ions approach and form ion pairs.

(d) As more and more Ag^+ and Cl^- ions come together, a precipitate of solid AgCl forms.

Figure 5.4   Precipitation of silver chloride. (a) Mixing aqueous solutions of silver nitrate and potassium chloride produces white, insoluble silver chloride, AgCl. In (b) through (d) you see a model of the process.

Figure 5.5   Precipitation reactions.
Many ionic compounds are insoluble in water. Guidelines for predicting the solubilities of ionic compounds are given in Figure 5.3.

  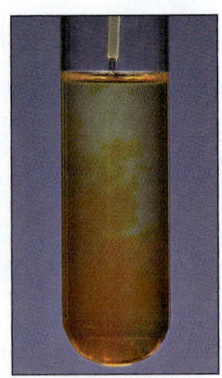 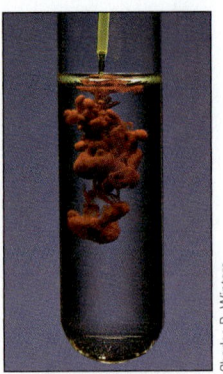

(a) $Pb(NO_3)_2$ and K_2CrO_4 produce yellow, insoluble $PbCrO_4$ and soluble KNO_3.

(b) $Pb(NO_3)_2$ and $(NH_4)_2S$ produce black, insoluble PbS and soluble NH_4NO_3.

(c) $FeCl_3$ and NaOH produce red, insoluble $Fe(OH)_3$ and soluble NaCl.

(d) $AgNO_3$ and K_2CrO_4 produce red, insoluble Ag_2CrO_4 and soluble KNO_3. See Example 5.2.

Charles D. Winters

Almost all metal sulfides are insoluble in water (Figure 5.5b). In nature, if a soluble metal compound comes in contact with a source of sulfide ions, the metal sulfide precipitates.

$$Pb(NO_3)_2(aq) + (NH_4)_2S(aq) \longrightarrow PbS(s) + 2\ NH_4NO_3(aq)$$

Reactants	Products
$Pb^{2+}(aq) + 2\ NO_3^-(aq)$	Insoluble PbS
$2\ NH_4^+(aq) + S^{2-}(aq)$	$2\ NH_4^+(aq) + 2\ NO_3^-(aq)$

In fact, this process is how many sulfur-containing minerals such as iron pyrite (see page 19) are believed to have been formed. (The black "smoke" from undersea volcanoes consists of precipitated metal sulfides arising from sulfide anions and metal cations in the volcanic emissions; see page 140.)

Finally, with the exception of the alkali metal cations (and Ba^{2+}), all metal cations form insoluble hydroxides. Thus, water-soluble iron(III) chloride and sodium hydroxide react to give insoluble iron(III) hydroxide (Figures 5.3c and 5.5c).

$$FeCl_3(aq) + 3\ NaOH(aq) \longrightarrow Fe(OH)_3(s) + 3\ NaCl(aq)$$

Reactants	Products
$Fe^{3+}(aq) + 3\ Cl^-(aq)$	Insoluble $Fe(OH)_3$
$3\ Na^+(aq) + 3\ OH^-(aq)$	$3\ Na^+(aq) + 3\ Cl^-(aq)$

Example 5.2—Writing the Equation for a Precipitation Reaction

Problem Is an insoluble product formed when aqueous solutions of potassium chromate and silver nitrate are mixed? If so, write the balanced equation.

Strategy First decide which ions are formed in solution when the reactants dissolve. Then use information in Figure 5.3 to determine whether a cation from one reactant will combine with an anion from the other reactant to form an insoluble compound.

Solution Both reactants—$AgNO_3$ and K_2CrO_4—are water-soluble. The ions Ag^+, NO_3^-, K^+, and CrO_4^{2-} are released into solution when these compounds are dissolved.

$$AgNO_3(s) \longrightarrow Ag^+(aq) + NO_3^-(aq)$$

$$K_2CrO_4(s) \longrightarrow 2\ K^+(aq) + CrO_4^{2-}(aq)$$

Here Ag^+ could combine with CrO_4^{2-}, and K^+ could combine with NO_3^-. The former combination, Ag_2CrO_4, is an insoluble compound, whereas KNO_3 is soluble in water. Thus, the balanced equation for the reaction of silver nitrate and potassium chromate is

$$2\ AgNO_3(aq) + K_2CrO_4(aq) \longrightarrow Ag_2CrO_4(s) + 2\ KNO_3(aq)$$

Comment This reaction is illustrated in Figure 5.5d.

Exercise 5.3—Precipitation Reactions

In each of the following cases, does a precipitation reaction occur when solutions of two water-soluble reactants are mixed? Give the formula of any precipitate that forms, and write a balanced chemical equation for the precipitation reactions that occur.

(a) Sodium carbonate is mixed with copper(II) chloride.

(b) Potassium carbonate is mixed with sodium nitrate.

(c) Nickel(II) chloride is mixed with potassium hydroxide.

Net Ionic Equations

An aqueous solution of silver nitrate contains Ag^+ and NO_3^- ions, and an aqueous solution of potassium chloride contains K^+ and Cl^- ions. When these solutions are mixed (Figure 5.4), insoluble AgCl precipitates, and the ions K^+ and NO_3^- remain in solution.

$$\underset{\text{before reaction}}{Ag^+(aq) + NO_3^-(aq) + K^+(aq) + Cl^-(aq)} \longrightarrow AgCl(s) + \underset{\text{after reaction}}{K^+(aq) + NO_3^-(aq)}$$

The K^+ and NO_3^- ions are present in solution before and after reaction, so they appear on both the reactant and product sides of the balanced chemical equation. Such ions are often called **spectator ions** because they do not participate in the net reaction; they merely "look on" from the sidelines. Little chemical information is lost if the equation is written without them, and so we can simplify the equation to

$$Ag^+(aq) + Cl^-(aq) \longrightarrow AgCl(s)$$

The balanced equation that results from leaving out the spectator ions is the **net ionic equation** for the reaction. *Only the aqueous ions and nonelectrolytes (which can be insoluble compounds, soluble molecular compounds such as sugar, weak acids or bases (page 177), or gases) that participate in a chemical reaction need to be included in the net ionic equation.*

Leaving out the spectator ions does not imply that K^+ and NO_3^- ions are unimportant in the $AgNO_3 + KCl$ reaction. Indeed, Ag^+ and Cl^- ions cannot exist alone in solution; a negative ion must be present to balance the positive ion charge of Ag^+, for example. Any anion will do, however, as long as it forms water-soluble compounds with Ag^+. Thus, we could have used $AgClO_4$ instead of $AgNO_3$ and NaCl instead of KCl. The net ionic equation would have been the same.

■ **Net ionic equations**
1. All chemical equations must be balanced. The same number of atoms of each kind must appear on both the product and reactant sides. In addition, the sum of positive and negative charges must be the same on both sides of the equation.
2. See Problem-Solving Tip 5.1, page 185.

Finally, notice that there must be a *charge balance* as well as a mass balance in a balanced chemical equation. In the $Ag^+ + Cl^-$ net ionic equation, the cation and anion charges on the left add together to give a net charge of 0, the same as the 0 charge on $AgCl(s)$ on the right.

GENERAL
Chemistry••Now™

See the General ChemistryNow CD-ROM or website:

• **Screen 5.7 Net Ionic Equations,** for a tutorial on writing net ionic equations

---

■ **Dissolving Halides**
When an ionic compound with halide ions dissolves in water, the halide ions are released into aqueous solution. Thus, $BaCl_2$ produces one Ba^{2+} ion and two Cl^- ions for each Ba^{2+} ion (and not Cl_2 or Cl_2^{2-} ions).

Precipitation reaction. The reaction of barium chloride and sodium sulfate produces insoluble barium sulfate and water-soluble sodium chloride. See Example 5.3.

Example 5.3—Writing and Balancing Net Ionic Equations

Problem Write a balanced, net ionic equation for the reaction of aqueous solutions of $BaCl_2$ and Na_2SO_4 to give $BaSO_4$ and $NaCl$.

Strategy First, write a balanced equation for the overall reaction. Next, decide which compounds are soluble in water (Figure 5.3) and determine the ions that these compounds produce in solution. Finally, eliminate ions that appear on both the reactant and product sides of the equation.

Solution

Step 1. Write the balanced equation.

$$BaCl_2 + Na_2SO_4 \longrightarrow BaSO_4 + 2\ NaCl$$

Step 2. Decide on the solubility of each compound. Compounds containing sodium ions are always water-soluble, and those containing chloride ions are almost always soluble. Sulfate salts are also usually soluble, with one important exception being $BaSO_4$. We can therefore write

$$BaCl_2(aq) + Na_2SO_4(aq) \longrightarrow BaSO_4(s) + 2\ NaCl(aq)$$

Step 3. Identify the ions in solution. All soluble ionic compounds dissociate to form ions in aqueous solution. (All are electrolytes.)

$$BaCl_2(s) \longrightarrow Ba^{2+}(aq) + 2\ Cl^-(aq)$$
$$Na_2SO_4(s) \longrightarrow 2\ Na^+(aq) + SO_4^{2-}(aq)$$
$$NaCl(s) \longrightarrow Na^+(aq) + Cl^-(aq)$$

This results in the following ionic equation:

$$Ba^{2+}(aq) + 2\ Cl^-(aq) + 2\ Na^+(aq) + SO_4^{2-}(aq) \longrightarrow BaSO_4(s) + 2\ Na^+(aq) + 2\ Cl^-(aq)$$

Step 4. Identify and eliminate the spectator ions (Na^+ and Cl^-) to give the net ionic equation.

$$Ba^{2+}(aq) + SO_4^{2-}(aq) \longrightarrow BaSO_4(s)$$

Notice that the sum of ion charges is the same on both sides of the equation. On the left, 2+ and 2− give zero; on the right, the charge on $BaSO_4$ is also zero.

Comment The steps followed in this example represent a general approach to writing net ionic equations.

Problem-Solving Tip 5.1

Writing Net Ionic Equations

Net ionic equations are commonly written for chemical reactions in aqueous solution because they describe the actual chemical species involved in a reaction. To write net ionic equations we must know which compounds exist as ions in solution.

1. Strong acids, soluble strong bases, and soluble salts exist as ions in solution. Examples include the acids HCl and HNO_3, a base such as NaOH, and salts such as NaCl and $CuCl_2$ (see Figures 5.1–5.3).

2. All other species should be represented by their complete formulas. Weak acids such as acetic acid (CH_3CO_2H) exist in solutions primarily as molecules. (See Section 5.3.) Insoluble salts such as $CaCO_3$(s) or insoluble bases such as $Mg(OH)_2$(s) should not be written in ionic form, even though they are ionic compounds.

The best way to approach writing net ionic equations is to follow precisely a set of steps.

1. Write a complete, balanced equation. Indicate the state of each substance (aq, s, ℓ, g).

2. Rewrite the equation, writing all strong acids, strong soluble bases, and soluble salts as ions. Look carefully at species labeled with an "(aq)" suffix.

3. Some ions may remain unchanged in the reaction (the ions that appear in the equation as both reactants and products). These spectator ions are not part of the chemistry that is going on. You can cancel them from each side of the equation.

Here are three general net ionic equations it is helpful to remember:

- The net ionic equation for the reaction between any strong acid and any soluble strong base is H^+(aq) + OH^-(aq) $\longrightarrow H_2O(\ell)$.

- The equation for the reaction of any weak acid HX (such as HCN, HF, HOCl, CH_3CO_2H) and a soluble strong base is HX + OH^-(aq) $\longrightarrow H_2O(\ell)$ + X^-(aq). (See Section 5.3.)

- The net ionic equation for the reaction of ammonia with any weak acid HX is NH_3(aq) + HX(aq) $\longrightarrow NH_4^+$(aq) + X^-(aq) and with a strong acid it is NH_3(aq) + H^+(aq) $\longrightarrow NH_4^+$(aq). (See Section 5.3.)

Finally, like molecular equations, net ionic equations must be balanced. The same number of atoms appears on each side of the arrow. But, an additional requirement applies. The sum of the ion charges on the two sides must be equal.

Exercise 5.4—Net Ionic Equations

Write balanced net ionic equations for each of the following reactions:

(a) $AlCl_3$ + Na_3PO_4 $\longrightarrow$ $AlPO_4$ + NaCl (not balanced)

(b) Solutions of iron(III) chloride and potassium hydroxide give iron(III) hydroxide and potassium chloride when combined. See Figure 5.5c.

(c) Solutions of lead(II) nitrate and potassium chloride give lead(II) chloride and potassium nitrate when combined.

5.3—Acids and Bases

Acids and bases, two important classes of compounds, have some related properties. Solutions of acids or bases, for example, can change the colors of vegetable pigments (Figure 5.6). You may have seen acids change the color of litmus, a dye derived from certain lichens, from blue to red. If an acid has made blue litmus paper turn red, then adding a base reverses the effect, making the litmus blue again. Thus, acids and bases seem to be opposites. A base can neutralize the effect of an acid, and an acid can neutralize the effect of a base.

Acids

Acids have characteristic properties. They produce bubbles of CO_2 gas when added to a metal carbonate such as $CaCO_3$, and they react with many metals to produce hydrogen gas, H_2, (Figure 5.6). Although tasting substances is *never* done in a chemistry laboratory, you have probably experienced the sour taste of acids such as acetic

(a) The juice of a red cabbage is normally blue-purple. On adding acid, the juice becomes more red. Adding base produces a yellow color.

(b) A piece of coral (mostly $CaCO_3$) dissolves in acid to give CO_2 gas.

(c) Zinc reacts with hydrochloric acid to produce zinc chloride and hydrogen gas.

Charles D. Winters

Figure 5.6 **Some properties of acids and bases.** (a) The colors of natural dyes, such as the juice from a red cabbage, are affected by acids and bases. (b) Acids react readily with coral ($CaCO_3$) and other metal carbonates to produce gaseous CO_2 (and a salt). (c) Acids react with many metals to produce hydrogen gas (and a metal salt).

acid (in vinegar) or citric acid (commonly found in fruits and added to candies and soft drinks).

The properties of acids can be interpreted in terms of a feature common to all acid molecules:

> An acid is a substance that, when dissolved in water, increases the concentration of hydrogen ions, H^+, in the solution.

Hydrochloric acid, an aqueous solution of gaseous HCl, is a common acid. In water, hydrogen chloride ionizes to form a hydrogen ion, $H^+(aq)$, and a chloride ion, $Cl^-(aq)$.

$$HCl(aq) \longrightarrow H^+(aq) + Cl^-(aq)$$

hydrochloric acid
strong electrolyte
= 100% ionized

Because it is completely converted to ions in aqueous solution, HCl is a **strong acid** (and a strong electrolyte). See Table 5.2 for a list of other common acids.

Many acids, such as sulfuric acid, can provide more than 1 mol of H^+ per mole of acid. This occurs in two steps.

Strong Acid: $\quad H_2SO_4(aq) \longrightarrow H^+(aq) + HSO_4^-(aq)$

$\qquad\qquad\qquad$ sulfuric acid $\qquad\qquad$ hydrogen ion $\qquad$ hydrogen
$\qquad\qquad\qquad$ 100% ionized $\qquad\qquad\qquad\qquad\qquad$ sulfate ion

Weak Acid: $\quad HSO_4^-(aq) \rightleftharpoons H^+(aq) + SO_4^{2-}(aq)$

$\qquad\qquad$ hydrogen sulfate ion $\qquad$ hydrogen ion $\qquad$ sulfate ion
$\qquad\qquad$ <100% ionized

■ **Weak Acids**
Common acids and bases are listed in Table 5.2. There are numerous other weak acids and bases, many of which are natural substances. Oxalic acid and acetic acid are among them. All of these natural acids contain CO_2H groups. (The H of this group is lost as H^+.) This structural feature is characteristic of hundreds of organic acids. (See Chapter 11.)

Oxalic acid
$H_2C_2O_4$

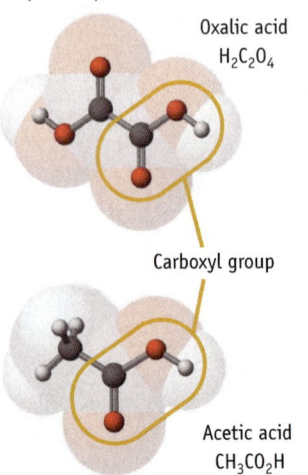

Carboxyl group

Acetic acid
CH_3CO_2H

Chemical Perspectives

Sulfuric Acid

For some years sulfuric acid has been the chemical produced in the largest quantity in the United States (and in many other industrialized countries). Approximately 40–50 billion kilograms (40–50 million metric tons) are made annually in the United States. The acid is so important to the economy of industrialized nations that some economists have said sulfuric acid production is a measure of a nation's industrial strength.

Sulfuric acid is a colorless, syrupy liquid with a density of 1.84 g/mL and a boiling point of 337 °C. It has several desirable properties that have led to its widespread use: It is generally less expensive to produce than other acids, is a strong acid, can be handled in steel containers, reacts readily with many organic compounds to produce useful products, and reacts readily with lime (CaO), the least expensive and most readily available base, to give calcium sulfate.

The first step in the industrial preparation of sulfuric acid is combustion of sulfur in air to give sulfur dioxide.

$$S_8(s) + 8\ O_2(g) \longrightarrow 8\ SO_2(g)$$

This gas is then combined with more oxygen, in the presence of a catalyst, to give sulfur trioxide,

$$2\ SO_2(g) + O_2(g) \longrightarrow 2\ SO_3(g)$$

which can give sulfuric acid when absorbed in water.

$$SO_3(g) + H_2O(\ell) \longrightarrow H_2SO_4(aq)$$

Currently more than two thirds of the production is used in the fertilizer industry, which makes "superphosphate" fertilizer by treating phosphate rock with sulfuric acid.

Sulfur is found in pure form in underground deposits along the coast of the United States in the Gulf of Mexico. It is recovered by pumping superheated steam into the sulfur beds to melt the sulfur. The molten sulfur is brought to the surface by means of compressed air.

$$2\ Ca_5F(PO_4)_3(s) + 7\ H_2SO_4(aq) + 3\ H_2O(\ell)$$
$$\longrightarrow 3\ Ca(H_2PO_4)_2 \cdot H_2O(s) + 7\ CaSO_4(s) + 2\ HF(g)$$

The remainder is used to make pigments, explosives, alcohol, pulp and paper, and detergents, and is employed as a component in storage batteries.

A sulfuric acid plant.

Some products that depend on sulfuric acid for their manufacture or use.

Table 5.2   Common Acids and Bases

Strong Acids (Strong Electrolytes)		Strong Bases (Strong Electrolytes)	
HCl	Hydrochloric acid	LiOH	Lithium hydroxide
HBr	Hydrobromic acid	NaOH	Sodium hydroxide
HI	Hydroiodic acid	KOH	Potassium hydroxide
HNO_3	Nitric acid		
$HClO_4$	Perchloric acid		
H_2SO_4	Sulfuric acid		
Weak Acids (Weak Electrolytes)*		**Weak Base (Weak Electrolyte)**	
H_3PO_4	Phosphoric acid	NH_3	Ammonia
H_2CO_3	Carbonic acid		
CH_3CO_2H	Acetic acid		
$H_2C_2O_4$	Oxalic acid		
$H_2C_4H_4O_6$	Tartaric acid		
$H_3C_6H_5O_7$	Citric acid		
$HC_9H_8O_4$	Aspirin		

* These are representative of hundreds of weak acids.

A Closer Look

The H$^+$ Ion in Water

The H$^+$ ion is a hydrogen atom that has lost its electron. Only the nucleus, a proton, remains. Because a proton is only about 1/100,000 as large as the average atom or ion, water molecules can approach closely, and the proton and the water molecules are strongly attracted. In fact, the H$^+$ ion in water is better represented as H$_3$O$^+$, called the *hydronium ion*. This ion is formed by combining H$^+$ and H$_2$O. Experiments also show that other forms of the ion exist in water, one example being [H$_3$O(H$_2$O)$_3$]$^+$.

For simplicity we will use H$^+$(aq) in this text for the hydronium and similar ions. When discussing the functions of acids in detail, however, we will use H$_3$O$^+$ [▶Chapters 17–18].

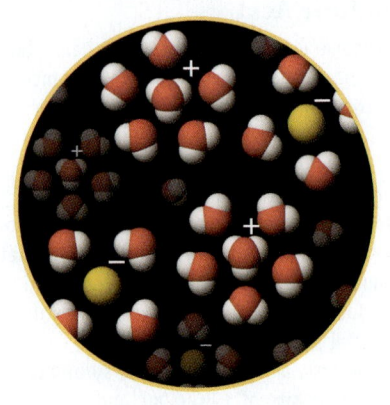

When an acid ionizes in water, it produces a hydronium ion, H$_3$O$^+$, which is surrounded by water molecules.

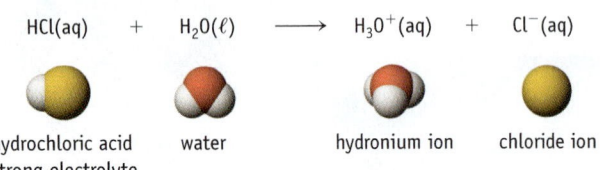

HCl(aq)   +   H$_2$O(ℓ)   ⟶   H$_3$O$^+$(aq)   +   Cl$^-$(aq)

hydrochloric acid   water   hydronium ion   chloride ion
strong electrolyte
= 100% ionized

The first ionization reaction is essentially complete, so sulfuric acid is a strong acid (and, therefore, a strong electrolyte). However, the hydrogen sulfate ion (HSO$_4^-$), like acetic acid (Figure 5.2), is only partially ionized in aqueous solution. Both the hydrogen sulfate ion and acetic acid are therefore classified as **weak acids**.

Bases

The hydroxide ion is characteristic of bases so we can immediately recognize metal hydroxides as bases from their formulas. Although most metal hydroxides are insoluble (see Figure 5.3c), a few dissolve in water, which leads to an increase in the concentration of OH$^-$ ions in solution.

> A base is a substance that, when dissolved in water, increases the concentration of hydroxide ions, OH$^-$, in the solution.

Compounds that contain hydroxide ions, such as sodium hydroxide or potassium hydroxide, are obvious bases. These water-soluble ionic compounds are **strong bases** (and strong electrolytes).

NaOH(s) ⟶ Na$^+$(aq) + OH$^-$(aq)

sodium hydroxide, soluble   hydroxide ion
base, strong electrolyte
= 100% dissociated

Ammonia, NH$_3$, another common base, does not have an OH$^-$ ion as part of its formula. Instead, the OH$^-$ ion is a result of the reaction with water.

NH$_3$(aq) + H$_2$O(ℓ) ⇌ NH$_4^+$(aq) + OH$^-$(aq)

ammonia, base   water   ammonium   hydroxide ion
weak electrolyte   ion
< 100% ionized

Only a small concentration of ammonium and hydroxide ions is present in a solution of NH_3. Therefore, ammonia is a **weak base** (and a weak electrolyte). (See Figure 5.7.)

GENERAL
Chemistry⚛Now™

See the General ChemistryNow CD-ROM or website:
- **Screen 5.8 Acids**

  (a) for a simulation exploring the degree to which different acids ionize to give H^+ ions in aqueous solution

  (b) for animations on weak and strong acids
- **Screen 5.9 Bases,** for animations of weak and strong bases

Exercise 5.5—Acids and Bases

(a) What ions are produced when nitric acid dissolves in water?
(b) Barium hydroxide is moderately soluble in water. What ions are produced when it dissolves in water?

Oxides of Nonmetals and Metals

Each acid shown in Table 5.2 has one or more H atoms in the molecular formula that dissociate in water to form H^+ ions. There are, however, less obvious compounds that form acidic solutions. Oxides of nonmetals, such as carbon dioxide and sulfur trioxide, have no H atoms but react with water to produce H^+ ions. Carbon dioxide, for example, dissolves in water to a small extent, and some of the dissolved molecules react with water to form the weak acid, carbonic acid. This acid then ionizes to a small extent to form the hydrogen ion, H^+, and the hydrogen carbonate (bicarbonate) ion, HCO_3^-.

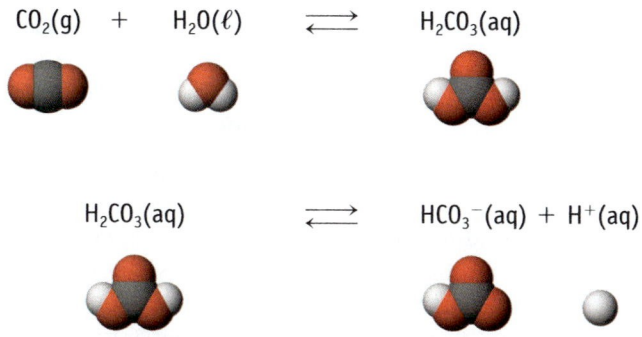

Like the HSO_4^- ion, the HCO_3^- ion can also function as an acid, and it can ionize to produce H^+ and the carbonate ion, CO_3^{2-}.

These reactions are important in our environment and in the human body. Carbon dioxide is normally found in small amounts in the atmosphere, so rainwater is always slightly acidic. In the human body, carbon dioxide is dissolved in body fluids where the HCO_3^- and CO_3^{2-} ions perform an important "buffering" action [▶ Chapter 18].

Figure 5.7 Ammonia, a weak electrolyte. Ammonia, NH_3, interacts with water to produce a very small number of NH_4^+ and OH^- ions per mole of ammonia molecules.

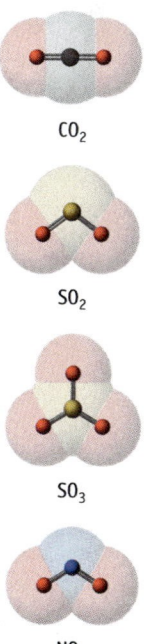

Some common nonmetal oxides that form acids in water.

Chemical Perspectives

Limelight and Metal Oxides

In the 1820s, Lt. Thomas Drummond (1797–1840) of the Royal Engineers was involved in a survey of Great Britain. During the winters he attended the famous public chemistry lectures and demonstrations by the great chemist Michael Faraday at the Royal Institution in London. There he apparently heard about the bright light that is emitted when a piece of lime, CaO, is heated to a high temperature. It occurred to him that this phenomenon could be used to make distant surveying stations visible, especially at night. Soon he developed an apparatus in which a ball of lime was heated by an alcohol flame in a stream of oxygen gas. It was reported at the time that the light from a "ball of lime not larger than a boy's marble" could be seen at a distance of 70 miles! Such lights were adapted to lighthouses and became known as Drummond lights.

Many inventions are soon adapted to warfare, and such was the case with limelights. They were used to illuminate targets in the battle of Charleston, South Carolina, during the U.S. Civil War in the 1860s. The public came to know about limelights when they moved into theaters. Gaslights were used in the early 1800s to illuminate the stage, but they were clearly not adequate. Soon after Drummond's invention, though, actors trod the boards "in the limelight."

Limelight. Metal oxides such as CaO and ThO$_2$ [thorium(IV) oxide] emit a brilliant white light when heated to incandescence.

Oxides like CO$_2$ that can react with water to produce H$^+$ ions are known as **acidic oxides**. Other acidic oxides include those of sulfur and nitrogen, which can be present in significant amounts in polluted air and can ultimately lead to acids and other pollutants. For example, sulfur dioxide, SO$_2$, from human and natural sources can react with oxygen to give sulfur trioxide, SO$_3$, which then forms sulfuric acid with water.

$$2\ SO_2(g) + O_2(g) \longrightarrow 2\ SO_3(g)$$
$$SO_3(g) + H_2O(\ell) \longrightarrow H_2SO_4(aq)$$

Nitrogen dioxide, NO$_2$, reacts with water to give nitric and nitrous acids.

$$2\ NO_2(g) + H_2O(\ell) \longrightarrow HNO_3(aq) + HNO_2(aq)$$
$$\text{nitric acid} \qquad \text{nitrous acid}$$

These reactions are the origin of the acid in so-called acid rain. The acidic oxides arise from the burning of fossil fuels such as coal and gasoline in the United States, Canada, and other industrialized countries. The gaseous oxides mix with water and other chemicals in the troposphere, and the rain that falls is more acidic than if it contained only dissolved CO$_2$. When the rain falls on areas that cannot easily tolerate this greater than normal acidity, such as the northeastern parts of the United States and the eastern provinces of Canada, serious environmental problems can occur.

Oxides of metals are **basic oxides**, so called because they give basic solutions if they dissolve appreciably in water. Perhaps the best example is calcium oxide, CaO, often called *lime*, or *quicklime*. Almost 20 billion kg of lime is produced annually in the United States for use in the metals and construction industries, in sewage and pollution control, in water treatment, and in agriculture. This metal oxide reacts with water to give calcium hydroxide, commonly called *slaked lime*. This compound, although only slightly soluble in water (0.17 g/100 g H$_2$O at 10 °C), is widely used in industry as a base because it is inexpensive.

$$CaO(s) + H_2O(\ell) \longrightarrow Ca(OH)_2(s)$$
$$\text{lime} \qquad\qquad\qquad \text{slaked lime}$$

GENERAL
Chemistry ·.·Now™

See the General ChemistryNow CD-ROM or website:
- **Screen 5.9 Bases,** for a description of strong and weak bases

Exercise 5.6—Acidic and Basic Oxides

For each of the following, indicate whether you expect an acidic or basic solution when the compound dissolves in water. Remember that compounds based on elements in the same group usually behave similarly.

(a) SeO_2 **(b)** MgO **(c)** P_4O_{10}

5.4—Reactions of Acids and Bases

Acids and bases in aqueous solution react to produce a salt and water. For example (Figure 5.8),

$$HCl(aq) + NaOH(aq) \longrightarrow H_2O(\ell) + NaCl(aq)$$

hydrochloric acid   sodium hydroxide   water   sodium chloride

The word "salt" has come into the language of chemistry as a description for any ionic compound whose cation comes from a base (here Na^+ from $NaOH$) and

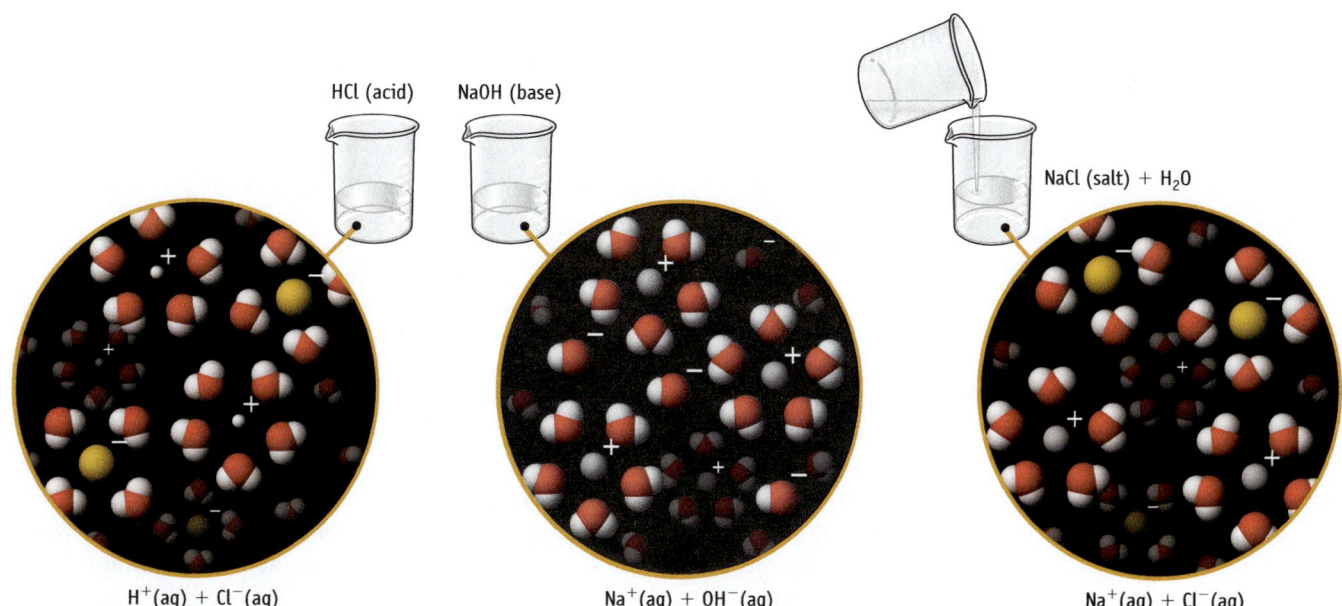

HCl (acid)   NaOH (base)   NaCl (salt) + H_2O

$H^+(aq) + Cl^-(aq)$ $Na^+(aq) + OH^-(aq)$ $Na^+(aq) + Cl^-(aq)$

Active Figure 5.8   **An acid–base reaction, HCl and NaOH.** The acid and base consist of ions in solution. On mixing, the H^+ and OH^- ions combine to produce H_2O, whereas the ions Na^+ and Cl^- remain in solution.

GENERAL
Chemistry ·.·Now™ See the General ChemistryNow CD-ROM or website to explore an interactive version of this figure accompanied by an exercise.

whose anion comes from an acid (here Cl⁻ from HCl). Reaction of any of the acids listed in Table 5.2 with any of the hydroxide-containing bases listed there produces a salt and water. (The reaction of an acid with the weak base NH_3 produces only a salt [▶ Example 5.4].)

Hydrochloric acid and sodium hydroxide are strong electrolytes in water (see Figure 5.8 and Table 5.2), so the complete ionic equation for the reaction of HCl(aq) and NaOH(aq) should be written as

$$\underbrace{H^+(aq) + Cl^-(aq)}_{\text{from HCl(aq)}} + \underbrace{Na^+(aq) + OH^-(aq)}_{\text{from NaOH(aq)}} \longrightarrow \underbrace{H_2O(\ell)}_{\text{water}} + \underbrace{Na^+(aq) + Cl^-(aq)}_{\text{from salt}}$$

Because Na⁺ and Cl⁻ ions appear on both sides of the equation, the *net ionic equation* is just the combination of the ions H⁺ and OH⁻ to give water.

$$H^+(aq) + OH^-(aq) \longrightarrow H_2O(\ell)$$

This is always the net ionic equation when a strong acid reacts with a strong base.

Reactions between *strong acids* and *strong bases* are called **neutralization reactions** because, on completion of the reaction, the solution is neutral; that is, it is neither acidic nor basic. The other ions (the cation of the base and the anion of the acid) remain unchanged. If the water is evaporated, however, the cation and anion form a solid salt. In the preceding example, NaCl can be obtained, whereas nitric acid, HNO_3, and NaOH give the salt sodium nitrate, $NaNO_3$ (and water).

$$HNO_3(aq) + NaOH(aq) \longrightarrow H_2O(\ell) + NaNO_3(aq)$$

One of the major uses of the basic oxide calcium oxide (lime) is in "scrubbing" sulfur oxides from the exhaust gases of power plants fueled by coal and oil. The oxides of sulfur dissolve in water to produce acids (page 190), and these acids can react with a base. Lime produces the base calcium hydroxide when added to water. A water suspension of lime is sprayed into the exhaust stack of the power plant, where it reacts with acids such as H_2SO_4 to produce $CaSO_4 \cdot 2H_2O$.

$$Ca(OH)_2(s) + H_2SO_4(aq) \longrightarrow CaSO_4 \cdot 2\,H_2O(s)$$

Hydrated calcium sulfate, $CaSO_4 \cdot 2\,H_2O$, is also found in the earth as the mineral gypsum. Assuming the gypsum from a coal-burning power plant is not contaminated with compounds that are pollutants, it is environmentally acceptable to put this substance into the earth.

Acetic acid, CH_3CO_2H, is the substance that gives the taste and odor to vinegar. Fermentation of carbohydrates such as sugar produces ethanol, CH_3CH_2OH, and the action of bacteria on the alcohol results in acetic acid. Even a trace of acetic acid will ruin the taste of wine. This characteristic is the source of the name "vinegar," which comes from the French *vin egar* meaning "sour wine." In addition to its use in food products such as salad dressings, mayonnaise, and pickles, acetic acid is used in hair-coloring products and in the manufacture of cellulose acetate, a commonly used synthetic fiber.

Acetic acid is a weak acid. Only a few acetic acid molecules are ionized to form H⁺ and $CH_3CO_2^-$ ions in water (Figure 5.2).

$$CH_3CO_2H(aq) \rightleftharpoons H^+(aq) + CH_3CO_2^-(aq)$$

Nonetheless, like all acids, acetic acid will react with metal carbonates such as calcium carbonate. This carbonate is a common residue from hard water in home heating systems and cooking utensils, so washing with vinegar is a good way to clean the system or utensils because the insoluble calcium carbonate is turned into water-soluble calcium acetate (Figure 5.9).

$$2\ CH_3CO_2H(aq) + CaCO_3(s) \longrightarrow Ca(CH_3CO_2)_2(aq) + H_2O(\ell) + CO_2(g)$$

What is the net ionic equation for this reaction? Acetic acid is a weak acid, so it produces only a trace of ions in solution. Calcium carbonate is insoluble in water. Therefore, the two reactants are simply $CH_3CO_2H(aq)$ and $CaCO_3(s)$. The product, calcium acetate, is water-soluble and forms calcium and acetate ions.

$$2\ CH_3CO_2H(aq) + CaCO_3(s) \longrightarrow Ca^{2+}(aq) + 2\ CH_3CO_2^-(aq) + H_2O(\ell) + CO_2(g)$$

There are no spectator ions in this reaction. (See Problem Solving Tip 5.1, Writing Net Ionic Equations, page 185.)

Charles D. Winters

Figure 5.9 Dissolving limestone (calcium carbonate, CaCO₃) in vinegar. This reaction shows why vinegar can be used as a household cleaning agent. It can be used, for example, to clean the calcium carbonate deposited from hard water in the filter in an electric coffee maker.

Example 5.4—Net Ionic Equation for an Acid–Base Reaction

Problem Ammonia, NH_3, is one of the most important chemicals in industrial economies. Not only is it used directly as a fertilizer but it is also the raw material for the manufacture of nitric acid. As a base, it reacts with acids such as hydrochloric acid. Write a balanced, net ionic equation for this reaction.

Strategy First, write a complete balanced equation for the reaction. Next, indicate whether each reactant and product is a solid, liquid, gas, or soluble in water (aq). Then, write each water-soluble salt or any strong acids and bases as the ions they produce in water. *Insoluble solids and weak acids and bases are not written as ions.* Finally, eliminate any spectator ions to give the net ionic equation.

Solution The complete balanced equation is

$$\underset{\text{ammonia}}{NH_3(aq)} + \underset{\text{hydrochloric acid}}{HCl(aq)} \longrightarrow \underset{\text{ammonium chloride}}{NH_4Cl(aq)}$$

Notice that the reaction produces a salt, NH_4Cl. An H^+ ion from the acid transfers directly to ammonia, a weak base, to give the ammonium ion. To write the net ionic equation, start with the facts that hydrochloric acid is a strong acid and produces H^+ and Cl^- ions and that NH_4Cl is a soluble, ionic compound.

$$NH_3(aq) + H^+(aq) + Cl^-(aq) \longrightarrow NH_4^+(aq) + Cl^-(aq)$$

Eliminating the spectator ion, Cl^-, we have

$$NH_3(aq) + H^+(aq) \longrightarrow NH_4^+(aq)$$

Comment The net ionic equation shows that the important aspect of the reaction between the weak base ammonia and the strong acid HCl is the transfer of an H^+ ion from the acid to the NH_3. Any strong acid could be used here (HBr, HNO_3, $HClO_4$, H_2SO_4) and the net ionic equation would be the same.

Figure 5.10   Muffins rise because of a gas-forming reaction. The acid and sodium bicarbonate in baking powder produce carbon dioxide gas. The acid used in many baking powders is $CaHPO_4$, but $NaAl(SO_4)_2$ is also common. (The aluminum-containing compound forms an acidic solution when placed in water; See Chapter 17.)

Exercise 5.7—Acid–base Reactions

Write the balanced, overall equation and the net ionic equation for the reaction of magnesium hydroxide with hydrochloric acid.

5.5—Gas-Forming Reactions

Have you ever made biscuits or muffins? As you bake the dough, it rises in the oven (Figure 5.10). But what makes it rise? A gas-forming reaction occurs between an acid and baking soda, sodium hydrogen carbonate (bicarbonate of soda, $NaHCO_3$). One acid used for this purpose is tartaric acid, a weak acid found in many foods. The net ionic equation for a typical reaction would be

$$H_2C_4H_4O_6(aq) + HCO_3^-(aq) \longrightarrow HC_4H_4O_6^-(aq) + H_2O(\ell) + CO_2(g)$$

tartaric acid   hydrogen carbonate ion   tartrate ion

In dry baking powder, the acid and $NaHCO_3$ are kept apart by using starch as a filler. When mixed into the moist batter, however, the acid and sodium hydrogen carbonate dissolve and come into contact. Now they can react to produce CO_2, causing the dough to rise.

Several different chemical reactions lead to gas formation (Table 5.3), but the most common are those leading to CO_2 formation. All metal carbonates (and bicarbonates) react with acids to produce a salt and carbonic acid, H_2CO_3, which in turn decomposes rapidly to carbon dioxide and water (Figure 5.6b).

$$CaCO_3(s) + 2\ HCl(aq) \longrightarrow CaCl_2(aq) + H_2CO_3(aq)$$
$$H_2CO_3(aq) \longrightarrow H_2O(\ell) + CO_2(g)$$

Overall reaction: $CaCO_3(s) + 2\ HCl(aq) \longrightarrow CaCl_2(aq) + H_2O(\ell) + CO_2(g)$

If the reaction is done in an open beaker, most of the CO_2 gas bubbles out of the solution.

GENERAL
Chemistry ⚛ Now™

See the General ChemistryNow CD-ROM or website:
- **Screen 5.11 Gas Forming Reactions,** for a tutorial on identifying the type of reaction that will result from the mixing of solutions and to watch videos about four of the most important gases produced in reactions

---

■ **Gas-Forming Reactions**
Metal carbonates such as $CaCO_3$ react with acids to produce a salt and CO_2 gas. See Figure 5.6.

Table 5.3   Gas-Forming Reactions

Metal carbonate or bicarbonate + acid $\longrightarrow$ metal salt + $CO_2(g)$ + $H_2O(\ell)$
$Na_2CO_3(aq) + 2\ HCl(aq) \longrightarrow 2\ NaCl(aq) + CO_2(g) + H_2O(\ell)$
Metal sulfide + acid $\longrightarrow$ metal salt + $H_2S(g)$
$Na_2S(aq) + 2\ HCl(aq) \longrightarrow 2\ NaCl(aq) + H_2S(g)$
Metal sulfite + acid $\longrightarrow$ metal salt + $SO_2(g)$ + $H_2O(\ell)$
$Na_2SO_3(aq) + 2\ HCl(aq) \longrightarrow 2\ NaCl(aq) + SO_2(g) + H_2O(\ell)$
Ammonium salt + strong base $\longrightarrow$ metal salt + $NH_3(g)$ + $H_2O(\ell)$
$NH_4Cl(aq) + NaOH(aq) \longrightarrow NaCl(aq) + NH_3(g) + H_2O(\ell)$

Example 5.5—Gas-Forming Reactions

Problem Write a balanced equation for the reaction that occurs when nickel(II) carbonate is treated with sulfuric acid.

Strategy First, identify the reactants and write their formulas (here $NiCO_3$ and H_2SO_4). Next, recognize this case as a typical gas-forming reaction (Table 5.3) between a metal carbonate (or metal hydrogen carbonate) and an acid. According to Table 5.3, the products are water, CO_2, and a metal salt. The anion of the metal salt is the anion from the acid (SO_4^{2-}), and the cation is from the metal carbonate (Ni^{2+}).

Solution The complete, balanced equation is

$$NiCO_3(s) + H_2SO_4(aq) \longrightarrow NiSO_4(aq) + H_2O(\ell) + CO_2(g)$$

Exercise 5.8—Gas-Forming Reactions

(a) Barium carbonate, $BaCO_3$, is used in the brick, ceramic, glass, and chemical manufacturing industries. Write a balanced equation that shows what happens when barium carbonate is treated with nitric acid. Give the name of each of the reaction products.

(b) Write a balanced equation for the reaction of ammonium sulfate with sodium hydroxide.

5.6—Classifying Reactions in Aqueous Solution

One goal of this chapter is to explore the most common types of reactions that can occur in aqueous solution. This helps you decide, for example, that a gas forming reaction occurs when an Alka-Seltzer tablet (containing citric acid and $NaHCO_3$) is dropped into water (Figure 5.11).

$$\underset{\text{citric acid}}{H_3C_6H_5O_7(aq)} + \underset{\text{hydrogen carbonate ion}}{HCO_3^-(aq)} \longrightarrow \underset{\text{dihydrogen citrate ion}}{H_2C_6H_5O_7^-(aq)} + H_2O(\ell) + CO_2(g)$$

Reactions in aqueous solution are important not only because they provide a way to make useful products, but also because these kinds of reactions occur on the earth and in plants and animals. Therefore, it is useful to look for common reaction patterns to see what their "driving forces" might be and how to predict the products. Most of the reactions described thus far in this chapter are **exchange reactions**, in which *the ions of the reactants changed partners.*

$$A^+B^- + C^+D^- \longrightarrow A^+D^- + C^+B^-$$

Recognizing that cations exchange anions gives us a good way to predict the products of precipitation, acid–base, and gas-forming reactions.

Precipitation Reactions (see Figure 5.5): Ions combine in solution to form an insoluble reaction product.
Overall Equation

$$Pb(NO_3)_2(aq) + 2\ KI(aq) \longrightarrow PbI_2(s) + 2\ KNO_3(aq)$$

Net Ionic Equation

$$Pb^{2+}(aq) + 2\ I^-(aq) \longrightarrow PbI_2(s)$$

Charles D. Winters

Figure 5.11   A Gas-Forming Reaction. An Alka-Seltzer tablet contains an acid (citric acid) and sodium hydrogen carbonate ($NaHCO_3$), the reactants in a gas-forming reaction.

Acid–Base Reactions (see Figure 5.6): Water is a product of an acid–base reaction, and the cation of the base and the anion of the acid form a salt.

Overall Equation for the Reaction of a Strong Acid and a Strong Base

$$HNO_3(aq) + KOH(aq) \longrightarrow HOH(\ell) + KNO_3(aq)$$

Net Ionic Equation for the Reaction of a Strong Acid and a Strong Base

$$H^+(aq) + OH^-(aq) \longrightarrow H_2O(\ell)$$

Overall Equation for the Reaction of a Weak Acid and a Strong Base

$$CH_3CO_2H(aq) + NaOH(aq) \longrightarrow NaCH_3CO_2(aq) + HOH(\ell)$$

Net Ionic Equation for the Reaction of a Weak Acid and a Strong Base

$$CH_3CO_2H(aq) + OH^-(aq) \longrightarrow CH_3CO_2^-(aq) + H_2O(\ell)$$

Gas-Forming Reactions (see Figures 5.9 and 5.11): The most common examples involve metal carbonates and acids but other gas-forming reactions exist (see Table 5.3). One product with a metal carbonate is always carbonic acid, H_2CO_3, most of which decomposes to H_2O and CO_2. Carbon dioxide is the gas in the bubbles you see during these reactions.

$$CuCO_3(s) + 2\ HNO_3(aq) \longrightarrow Cu(NO_3)_2(aq) + H_2CO_3(aq)$$
$$H_2CO_3(aq) \longrightarrow CO_2(g) + H_2O(\ell)$$

Overall Equation

$$CuCO_3(s) + 2\ HNO_3(aq) \longrightarrow Cu(NO_3)_2(aq) + CO_2(g) + H_2O(\ell)$$

Net Ionic Equation

$$CuCO_3(s) + 2\ H^+(aq) \longrightarrow Cu^{2+}(aq) + CO_2(g) + H_2O(\ell)$$

A Summary of Common Reaction Types in Aqueous Solution

Three common "driving forces" responsible for reactions in aqueous solution were outlined above. A fourth, to be discussed in the next section (Section 5.7), is the transfer of electrons from one substance to another. Such reactions are called oxidation–reduction processes.

Reaction Type	Driving Force
Precipitation	Formation of an insoluble compound (Section 5.2)
Acid–base; neutralization	Formation of a salt and water; proton transfer (Section 5.4)
Gas-forming	Evolution of a water-insoluble gas such as CO_2 (Section 5.5)
Oxidation–reduction	Electron transfer (Section 5.7)

These four types of reactions are usually easy to recognize, but keep in mind that a reaction may have more than one driving force. For example, barium hydroxide reacts readily with sulfuric acid to give barium sulfate and water, a reaction that is both a precipitation reaction and an acid–base reaction.

$$Ba(OH)_2(aq) + H_2SO_4(aq) \longrightarrow BaSO_4(s) + 2\ H_2O(\ell)$$

A Closer Look

Product-Favored and Reactant-Favored Reactions

The driving force for a precipitation, acid–base, or gas-forming reaction is, in each case, the formation of a product that removes ions from solution: a solid precipitate, a water molecule, or a gas molecule. These, and all other reactions in which reactants are completely or largely converted to products, are said to be **product-favored.**

The opposite of a product-favored reaction is one that is **reactant-favored.** Such reactions lead to the conversion of little, if any, of the reactants to products. An example would be the formation of hydrochloric acid and sodium hydroxide in a solution of sodium chloride in water.

Reactant-favored:

$$NaCl(aq) + H_2O(\ell)$$

$$-\mathbf{X} \longrightarrow NaOH(aq) + HCl(aq)$$

This reaction, which does not occur to any measurable extent, is the opposite of an acid–base reaction.

The title of this book is *Chemistry and Chemical Reactivity.* One aspect of chemical reactivity, and a goal of this book, is to be able to predict whether a chemical reaction is product- or reactant-favored. Thus far you have learned that certain common reactions are generally product-favored. We will use this idea to organize chemistry many more times in this book, particularly in Chapters 6 and 16–19.

GENERAL
Chemistry ·⚛· Now™

See the General ChemistryNow CD-ROM or website:
- **Screen 5.5 Types of Aqueous Solutions,** to watch videos on the four reaction types

---

Exercise 5.9—Classifying Reactions

Classify each of the following reactions as a precipitation, acid–base, or gas-forming reaction. Predict the products of the reaction, and then balance the completed equation. Write the net ionic equation for each.

(a) $CuCO_3(s) + H_2SO_4(aq) \longrightarrow$

(b) $Ba(OH)_2(s) + HNO_3(aq) \longrightarrow$

(c) $CuCl_2(aq) + (NH_4)_2S(aq) \longrightarrow$

5.7—Oxidation–Reduction Reactions

The terms "oxidation" and "reduction" come from reactions that have been known for centuries. Ancient civilizations learned how to change metal oxides and sulfides to the metal—that is, how to "reduce" ore to the metal. A modern example is the reduction of iron(III) oxide with carbon monoxide to give iron metal (Figure 5.12a).

Fe_2O_3 loses oxygen and is reduced.

$$Fe_2O_3(s) + 3\ CO(g) \longrightarrow 2\ Fe(s) + 3\ CO_2(g)$$

CO is the reducing agent. It
gains oxygen and is oxidized.

In this reaction carbon monoxide is the agent that brings about the reduction of iron ore to iron metal, so it is called the **reducing agent.**

Figure 5.12 Oxidation–reduction.
(a) Iron ore, which is largely Fe_2O_3, is reduced to metallic iron with carbon or carbon monoxide in a blast furnace, a process done on a massive scale. (b) Burning magnesium metal in air produces magnesium oxide.

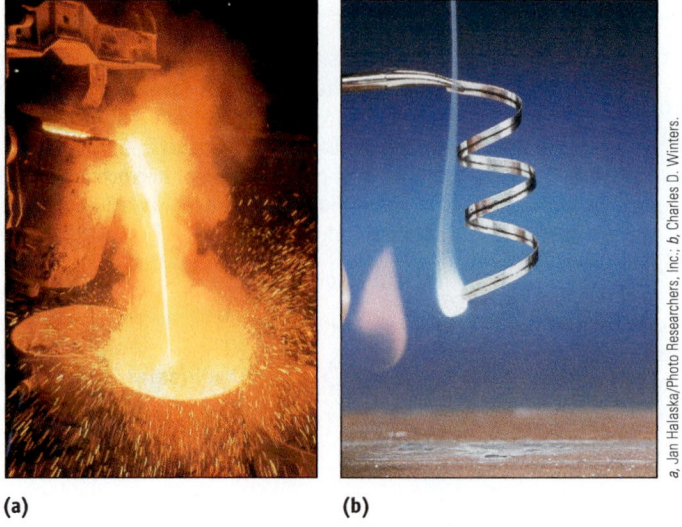

(a)   (b)

a, Jan Halaska/Photo Researchers, Inc.; b, Charles D. Winters.

When Fe_2O_3 is reduced by carbon monoxide, oxygen is removed from the iron ore and added to the carbon monoxide. The carbon monoxide, therefore, is "oxidized" by the addition of oxygen to give carbon dioxide. *Any process in which oxygen is added to another substance is an oxidation.* In the reaction of oxygen with magnesium, for example (see Figure 5.12b), oxygen is the **oxidizing agent** because it is responsible for the oxidation of magnesium.

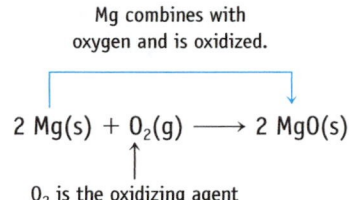

Mg combines with
oxygen and is oxidized.

$$2\ Mg(s) + O_2(g) \longrightarrow 2\ MgO(s)$$

O_2 is the oxidizing agent

The observations outlined here lead to several important conclusions:

- If one substance is oxidized, another substance in the same reaction must be reduced. For this reason, such reactions are often called oxidation–reduction reactions, or **redox reactions** for short.
- The reducing agent is itself oxidized, and the oxidizing agent is reduced.
- Oxidation is the opposite of reduction. For example, the removal of oxygen is reduction and the addition of oxygen is oxidation.

Redox Reactions and Electron Transfer

Not all redox reactions involve oxygen, but *all oxidation and reduction reactions involve transfer of electrons between substances.* When a substance accepts electrons, it is said to be **reduced** because there is a reduction in the positive charge on an atom of the substance. In the net ionic equation for the reaction of a silver salt with copper metal, for example, positively charged Ag^+ ions are reduced to uncharged silver atoms when they accept electrons from copper metal (Figure 5.13).

Ag$^+$ ions accept electrons from Cu and are
reduced to Ag. Ag$^+$ is the oxidizing agent.
$$Ag^+(aq) + e^- \longrightarrow Ag(s)$$

$$2\ Ag^+(aq) + Cu(s) \longrightarrow 2\ Ag(s) + Cu^{2+}(aq)$$

Cu donates electrons to Ag$^+$ and is oxidized to Cu^{2+}.
Cu is the reducing agent.
$$Cu(s) \longrightarrow Cu^{2+}(aq) + 2\ e^-$$

Because copper metal supplies the electrons and causes Ag$^+$ ions to be reduced, Cu is the *reducing agent*.

When a substance *loses electrons*, the positive charge on an atom of the substance increases. The substance is said to have been **oxidized**. In our example, copper metal releases electrons on going to Cu^{2+}, so the metal is oxidized. For this to happen, something must be available to accept the electrons from copper. In this case, Ag$^+$ is the electron acceptor, and its charge is reduced to zero in silver metal. Therefore, Ag$^+$ is the "agent" that causes Cu metal to be oxidized; that is, Ag$^+$ is the *oxidizing agent*.

In every oxidation–reduction reaction, one reactant is reduced (and is therefore the oxidizing agent) and one reactant is oxidized (and is therefore the reducing agent). We can show this by dividing the general redox reaction X + Y → X^{n+} + Y^{n-} into two parts or half-reactions:

Half Reaction	Electron Transfer	Result
X $\longrightarrow$ X^{n+} + ne$^-$	X transfers electrons to Y.	X is **oxidized** to X^{n+}. X is the **reducing agent**.
Y + ne$^-$ $\longrightarrow$ Y^{n-}	Y accepts electron from X.	Y is **reduced** to Y^{n-}. Y is the **oxidizing agent**.

■ **Balancing Equations for Redox Reactions**
The notion that a redox reaction can be divided into an oxidizing portion and a reducing portion will lead us to a method of balancing more complex equations for redox reactions, described in Chapter 20.

Pure copper wire

Copper wire in dilute AgNO$_3$ solution; after several hours

Blue color due to Cu^{2+} ions formed in redox reaction

Silver crystals formed after several weeks

Figure 5.13 **The oxidation of copper metal by silver ions.** A clean piece of copper wire is placed in a solution of silver nitrate, AgNO$_3$. Over time, the copper reduces Ag$^+$ ions, forming silver crystals, and the copper metal is oxidized to copper ions, Cu^{2+}. The blue color of the solution is due to the presence of aqueous copper(II) ions. *(See General ChemistryNow Screen 5.12 Redox Reactions and Electron Transfer, to watch a video of the reaction.)*

Charles D. Winters

In the reaction of magnesium and oxygen (see Figure 5.12b), O_2 is reduced because it gains electrons (four electrons per molecule) on going to two oxide ions. Thus, O_2 is the oxidizing agent.

Mg releases 2 e$^-$ per atom. Mg is oxidized to Mg^{2+} and is the reducing agent.

$$2 \text{ Mg(s)} + O_2(g) \longrightarrow 2 \text{ MgO(s)}$$

O_2 gains 4 e$^-$ per molecule to form 2 O^{2-}. O_2 is reduced and is the oxidizing agent.

In the same reaction, magnesium is the reducing agent because it releases two electrons per atom on being oxidized to the Mg^{2+} ion (and so two Mg atoms are required to supply the four electrons required by one O_2 molecule). All redox reactions can be analyzed in a similar manner.

Oxidation Numbers

How can you tell an oxidation–reduction reaction when you see one? How can you tell which substance has gained (or lost) electrons and so decide which substance is the oxidizing (or reducing) agent? Sometimes it is obvious. For example, if an uncombined element becomes part of a compound (Mg becomes part of MgO, for example), the reaction is definitely a redox process. If it's not obvious, then the answer is to *look for a change in the oxidation number of an element in the course of the reaction.* The **oxidation number** of an atom in a molecule or ion is defined as the charge an atom has, *or appears to have,* as determined by the following guidelines for assigning oxidation numbers.

■ **Why Use Oxidation Numbers?**
The reason for learning about oxidation numbers at this point is to be able to identify which reactions are oxidation–reduction processes and to know which is the oxidizing agent and which is the reducing agent in a reaction. We return to a more detailed discussion of redox reactions in Chapter 20.

1. **Each atom in a pure element has an oxidation number of zero.** The oxidation number of Cu in metallic copper is 0, and it is 0 for each atom in I_2 or S_8.

2. **For monatomic ions, the oxidation number is equal to the charge on the ion.** Elements of Groups 1A–3A form monatomic ions with a positive charge and an oxidation number equal to the group number. Magnesium forms Mg^{2+}, and its oxidation number is therefore +2. (See Section 3.3.)

3. **Fluorine always has an oxidation number of −1 in compounds with all other elements**.

4. **Cl, Br, and I always have oxidation numbers of −1 in compounds, except when combined with oxygen or fluorine.** This means that Cl has an oxidation number of −1 in NaCl (in which Na is +1, as predicted by the fact that it is a member of Group 1A). In the ion ClO^-, however, the Cl atom has an oxidation number of +1 (and O has an oxidation number of −2; see Guideline 5).

5. **The oxidation number of H is +1 and of O is −2 in most compounds.** Although this statement applies to most compounds, a few important exceptions occur.

   • When H forms a binary compound with a metal, the metal forms a positive ion and H becomes a hydride ion, H^-. Thus, in CaH_2 the oxidation number of Ca is +2 (equal to the group number) and that of H is −1.

   • Oxygen can have an oxidation number of −1 in a class of compounds called peroxides. For example, in H_2O_2, hydrogen peroxide, H is assigned its usual oxidation number of +1, so O is −1.

6. **The algebraic sum of the oxidation numbers for the atoms in a neutral compound must be zero; in a polyatomic ion, the sum must be equal to the ion**

A Closer Look

Are Oxidation Numbers "Real"?

Do oxidation numbers reflect the actual electric charge on an atom in a molecule or ion? With the exception of monatomic ions such as Cl^- or Na^+, the answer is no.

Oxidation numbers assume that the atoms in a molecule are positive or negative ions, which is not true. For example, in H_2O, the H atoms are not H^+ ions and the O atoms are not O^{2-} ions. This is not to say, however, that atoms in molecules do not bear an electric charge of any kind. In

Charge on O atom $= -0.4$

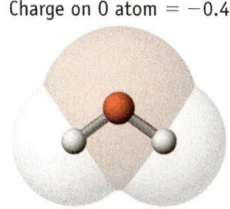

Charge on each H atom $= +0.2$

water, for example, calculations indicate the O atom has a charge of about -0.4 (or 40% of the electron charge) and the

H atoms are each about $+0.2$. (The partial charges on H and O in water are responsible for water molecules' ability to solvate ions in solution. See Figure 5.1.)

So why use oxidation numbers? These numbers provide a way of dividing up the electrons among the atoms in a molecule or polyatomic ion. Because the division of electrons changes in a redox reaction, we use this method as a way to decide whether a redox reaction has occurred, to distinguish the oxidizing and reducing agents, and, as you will see in Chapter 20, to balance equations for redox reactions.

charge. For example, in $HClO_4$ the H atom is assigned $+1$ and the O atom is assigned -2. This means the Cl atom must be $+7$. Additional examples are found in Example 5.6.

Example 5.6—Determining Oxidation Numbers

Problem Determine the oxidation number of the indicated element in each of the following compounds or ions:

(a) aluminum in aluminum oxide, Al_2O_3

(b) phosphorus in phosphoric acid, H_3PO_4

(c) sulfur in the sulfate ion, $SO_4{}^{2-}$

(d) each Cr atom in the dichromate ion, $Cr_2O_7{}^{2-}$

Strategy Follow the guidelines in the text, paying particular attention to Guidelines 5 and 6.

Solution

(a) Al_2O_3 is a neutral compound. Assuming that O has its usual oxidation number of -2, the oxidation number of Al must be $+3$, in agreement with its position in the periodic table.

$$\text{Net charge on } Al_2O_3 = 0$$
$$= \text{sum of oxidation numbers of Al atoms}$$
$$+ \text{sum of oxidation numbers of O atoms}$$
$$= 2(+3) + 3(-2)$$

(b) H_3PO_4 has an overall charge of 0. If each of the oxygen atoms has an oxidation number of -2 and each of the H atoms is $+1$, the oxidation number of P must be $+5$.

$$\text{Net charge on } H_3PO_4 = 0$$
$$= \text{sum of oxidation numbers for H atoms} + \text{oxidation number of P}$$
$$+ \text{sum of oxidation numbers for O atoms}$$
$$= 3(+1) + (+5) + 4(-2)$$

(c) The sulfate ion, $SO_4{}^{2-}$, has an overall charge of $2-$. Because this compound is not a peroxide, O is assigned an oxidation number of -2, which means that S has an oxidation number of $+6$.

$$\text{Net charge on } SO_4{}^{2-} = -2$$
$$= \text{oxidation number of S} + \text{sum of oxidation numbers for O atoms}$$
$$= (+6) + 4(-2)$$

■ **Writing Charges on Ions**
By convention, charges on ions are written as (number, sign), whereas oxidation numbers are written as (sign, number). For example, the oxidation number of the Cu^{2+} ion is $+2$ and it charge is $2+$.

(d) The net charge on $Cr_2O_7^{2-}$ ion is $2-$. Assigning each O atom an oxidation number of -2 means that each Cr atom must have an oxidation number of $+6$.

Net charge on $Cr_2O_7^{2-}$ = -2

= sum of oxidation numbers for Cr atoms + sum of oxidation numbers for O atoms

= $2(+6) + 7(-2)$

Exercise 5.10—Determining Oxidation Numbers

Assign an oxidation number to the underlined atom in each ion or molecule.

(a) $\underline{Fe}_2O_3$ **(b)** $H_2\underline{S}O_4$ **(c)** $\underline{C}O_3^{2-}$ **(d)** $\underline{N}O_2^+$

Recognizing Oxidation–Reduction Reactions

You can tell whether a reaction involves oxidation and reduction by assessing the oxidation number of each element and noting whether any of these numbers change in the course of the reaction. In many cases, however, this analysis will not be necessary. It will be obvious that a redox reaction has occurred if an uncombined element is converted to a compound or involves a well-known oxidizing or reducing agent (Table 5.4).

Like oxygen, O_2, the halogens (F_2, Cl_2, Br_2, and I_2) are always oxidizing agents in their reactions with metals and nonmetals. An example is the reaction of chlorine with sodium metal (see Figure 1.7).

■ **Sodium/Chlorine Reaction**
Sodium metal reduces chlorine gas. See Figure 1.7, page 19.

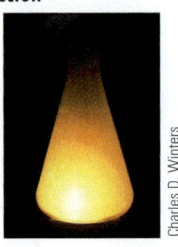

Charles D. Winters

Na releases 1 e^- per atom.
Oxidation number increases.
Na is oxidized to Na^+ and is the reducing agent.

$$2\,Na(s)\ +\ Cl_2(g)\ \longrightarrow\ 2\,NaCl(s)$$

Cl_2 gains 2 e^- per molecule.
Oxidation number decreases by 1 per Cl.
Cl_2 is reduced to Cl^- and is the oxidizing agent.

A chlorine molecule ends up as two Cl^- ions, having acquired two electrons (from two Na atoms). Thus, the oxidation number of each Cl atom has decreased from 0 to -1. This means Cl_2 has been reduced and so is the oxidizing agent.

Figure 5.14 illustrates the chemistry of another excellent oxidizing agent, nitric acid, HNO_3. Here copper metal is oxidized to give copper(II) nitrate, and the nitrate ion is reduced to the brown gas NO_2. The net ionic equation for the reaction is

Oxidation number of Cu changes from 0 to $+2$. Cu is oxidized to Cu^{2+} and is the reducing agent.

$$Cu(s) + 2\,NO_3^-(aq) + 4\,H^+(aq)\ \longrightarrow\ Cu^{2+}(aq) + 2\,NO_2(g) + 2\,H_2O(\ell)$$

N in NO_3^- changes from $+5$ to $+4$ in NO_2. NO_3^- is reduced to NO_2 and is the oxidizing agent.

Table 5.4   Common Oxidizing and Reducing Agents

Oxidizing Agent	Reaction Product	Reducing Agent	Reaction Product
O_2, oxygen	O^{2-}, oxide ion or O combined in H_2O	H_2, hydrogen	$H^+(aq)$, hydrogen ion or H combined in H_2O or other molecule
Halogen, F_2, Cl_2, Br_2, or I_2	Halide ion, F^-, Cl^-, Br^-, or I^-	M, metals such as Na, K, Fe, and Al	M^{n+}, metal ions such as Na^+, K^+, Fe^{2+} or Fe^{3+}, and Al^{3+}
HNO_3, nitric acid	Nitrogen oxides* such as NO and NO_2	C, carbon (used to reduce metal oxides)	CO and CO_2
$Cr_2O_7^{2-}$, dichromate ion	Cr^{3+}, chromium(III) ion (in acid solution)		
MnO_4^-, permanganate ion	Mn^{2+}, manganese(II) ion (in acid solution)		

* NO is produced with dilute HNO_3, whereas NO_2 is a product of concentrated acid.

NO₂ gas

Charles D. Winters

Copper metal oxidized to green $Cu(NO_3)_2$

Active Figure 5.14   **The reaction of copper with nitric acid.** Copper (a reducing agent) reacts vigorously with concentrated nitric acid, an oxidizing agent, to give the brown gas NO_2 and a deep green solution of copper(II) nitrate.

GENERAL Chemistry⬦Now™ See the General ChemistryNow CD-ROM or website to explore an interactive version of this figure accompanied by an exercise.

Nitrogen has been reduced from +5 (in the NO_3^- ion) to +4 (in NO_2); therefore, the nitrate ion in acid solution is an oxidizing agent. Copper metal is the reducing agent; here each metal atom has given up two electrons to produce the Cu^{2+} ion.

In the reactions of sodium with chlorine and copper with nitric acid, the metals are oxidized. This is typical of metals, which are generally good reducing agents. Indeed, the alkali and alkaline earth metals are especially good reducing agents. An example is the reaction of potassium with water. Here potassium reduces the hydrogen in water to H_2 gas (page 82).

$$2\ K(s) + 2\ H_2O(\ell) \longrightarrow 2\ KOH(aq) + H_2(g)$$

reducing agent oxidizing agent

Aluminum metal, a good reducing agent, is capable of reducing iron(III) oxide to iron metal in a reaction called the *thermite reaction* (Figure 5.15).

$$Fe_2O_3(s) + 2\ Al(s) \longrightarrow 2\ Fe(\ell) + Al_2O_3(s)$$

oxidizing agent reducing agent

Such a large quantity of heat is evolved in this reaction that the iron is produced in the molten state.

Tables 5.4 and 5.5 may help you organize your thinking as you look for oxidation–reduction reactions and use their terminology.

Table 5.5   Recognizing Oxidation–Reduction Reactions

	Oxidation	Reduction
In terms of oxidation number	Increase in oxidation number of an atom	Decrease in oxidation number of an atom
In terms of electrons	Loss of electrons by an atom	Gain of electrons by an atom
In terms of oxygen	Gain of one or more O atoms	Loss of one or more O atoms

Charles D. Winters

Figure 5.15   **Thermite reaction.** Here Fe_2O_3 is reduced by aluminum metal to produce iron metal and aluminum oxide.

GENERAL
Chemistry⚛Now™

See the General ChemistryNow CD-ROM or website:

- **Screen 5.13 Oxidation Numbers,** for an exercise that examines the reaction between bromine and elemental aluminum, for an exercise that explores the electron transfer aspects of reactions with hydrogen peroxide, and for a tutorial on assigning oxidation numbers
- **Screen 5.14 Recognizing Oxidation–Reduction Reactions,** for an exercise examining the process of a redox reaction

Example 5.7—Oxidation–Reduction Reaction

Problem For the reaction of iron(II) ion with permanganate ion in aqueous acid,

$$5\ Fe^{2+}(aq) + MnO_4^-(aq) + 8\ H^+(aq) \longrightarrow 5\ Fe^{3+}(aq) + Mn^{2+}(aq) + 4\ H_2O(\ell)$$

decide which atoms are undergoing a change in oxidation number and identify the oxidizing and reducing agents.

Strategy Determine the oxidation numbers of the atoms in each ion or molecule involved in the reaction. Decide which atoms have increased in oxidation number (oxidation) and which have decreased in oxidation number (reduction).

Solution The Mn oxidation number in MnO_4^- is +7, and it decreases to +2 in the product, the Mn^{2+} ion. Thus, the MnO_4^- ion has been reduced and is the oxidizing agent (see Table 5.4).

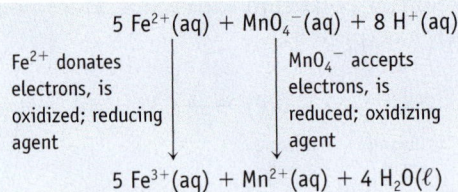

$$5\ Fe^{2+}(aq) + MnO_4^-(aq) + 8\ H^+(aq)$$

Fe^{2+} donates electrons, is oxidized; reducing agent

MnO_4^- accepts electrons, is reduced; oxidizing agent

$$5\ Fe^{3+}(aq) + Mn^{2+}(aq) + 4\ H_2O(\ell)$$

The oxidation number of iron has increased from +2 to +3, so the Fe^{2+} ion has lost electrons upon being oxidized to Fe^{3+} (see Table 5.5). This means the Fe^{2+} ion is the reducing agent.

Comment If one of the reactants in a redox reaction is a simple substance (here Fe^{2+}), it is usually obvious whether its oxidation number has increased or decreased. Once a species has been established as having been reduced (or oxidized), you know another species has been oxidized (or reduced). It is also helpful to recognize common oxidizing and reducing agents (Table 5.4).

Charles D. Winters

MnO_4^- oxidizing agent

$Fe^{2+}(aq)$ reducing agent

The reaction of iron(II) ion and permanganate ion. The reaction of purple permanganate ion (MnO_4^-, the oxidizing agent) with the iron(II) ion (Fe^{2+}, the reducing agent) in acidified aqueous solution gives the nearly colorless manganese(II) ion (Mn^{2+}) and the iron(III) ion (Fe^{3+}).

Example 5.8—Types of Reactions

Problem Classify each of the following reactions as precipitation, acid–base, gas-forming, or oxidation–reduction.

(a) $2\ HNO_3(aq) + Ca(OH)_2(s) \longrightarrow Ca(NO_3)_2(aq) + 2\ H_2O(\ell)$

(b) $SO_4^{2-}(aq) + 2\ CH_2O(aq) + 2\ H^+(aq) \longrightarrow H_2S(aq) + 2\ CO_2(g) + 2\ H_2O(\ell)$

Strategy A good strategy is first to check whether a reaction is one of the three types of exchange reactions. An acid–base reaction is usually easy to distinguish. Next, check the oxidation numbers of each element. If they change, then it is a redox reaction. If there is no change, then it is a simple precipitation or gas-forming process.

Solution Reaction (a) involves a common acid (nitric acid, HNO_3) and a common base [calcium hydroxide, $Ca(OH)_2$]; it produces a salt, calcium nitrate, and water. It is an acid–base reaction. Reaction (b) is a redox reaction because the oxidation numbers of S and C change.

$$SO_4^{2-}(aq) + 2\ CH_2O(aq) + 2\ H^+(aq) \longrightarrow H_2S(aq) + 2\ CO_2(g) + 2\ H_2O(\ell)$$
$$\ \ \ \ \ +6, -2 \qquad\quad 0, +1, -2 \qquad +1 \qquad\qquad +1, -2 \qquad +4, -2 \qquad +1, -2$$

The oxidation number of S changes from +6 to −2, and that of C changes from 0 to +4. Therefore, sulfate, SO_4^{2-}, has been reduced (and is the oxidizing agent), and CH_2O has been oxidized (and is the reducing agent).

No changes occur in the oxidation numbers of the elements in reaction (a).

$$HNO_3(aq) + Ca(OH)_2(s) \longrightarrow Ca(NO_3)_2(aq) + 2\ H_2O(\ell)$$
$$\ +1, +5, -2 \quad +2, -2, +1 \qquad\quad +2, +5, -2 \qquad\ +1, -2$$

Comment If an uncombined element is a reactant or product, the reaction is a redox reaction.

Exercise 5.11—Oxidation–Reduction Reactions

The following reaction occurs in a device for testing the breath for the presence of ethanol. Identify the oxidizing and reducing agents and the substances oxidized and reduced (Figure 5.16).

$$3\ C_2H_5OH(aq) + 2\ Cr_2O_7^{2-}(aq) + 16\ H^+(aq) \longrightarrow 3\ CH_3CO_2H(aq) + 4\ Cr^{3+}(aq) + 11\ H_2O(\ell)$$

ethanol   dichromate ion;   acetic acid   chromium(III)
            orange-red   ion; green

Exercise 5.12—Oxidation–Reduction and Other Reactions

Decide which of the following reactions are oxidation–reduction reactions. In each case explain your choice and identify the oxidizing and reducing agents.

(a) $NaOH(aq) + HNO_3(aq) \longrightarrow NaNO_3(aq) + H_2O(\ell)$
(b) $Cu(s) + Cl_2(g) \longrightarrow CuCl_2(s)$
(c) $Na_2CO_3(aq) + 2\ HClO_4(aq) \longrightarrow CO_2(g) + H_2O(\ell) + 2\ NaClO_4(aq)$
(d) $2\ S_2O_3^{2-}(aq) + I_2(aq) \longrightarrow S_4O_6^{2-}(aq) + 2\ I^-(aq)$

5.8—Measuring Concentrations of Compounds in Solution

Most chemical studies require quantitative measurements, including experiments involving aqueous solutions. When doing such experiments, we continue to use balanced equations and moles, but we measure volumes of solution rather than masses of solids, liquids, or gases. Solution concentration expressed as molarity relates the volume of solution in liters to the amount of substance in moles.

Solution Concentration: Molarity

The concept of concentration is useful in many contexts. For example, about 5,500,000 people live in Wisconsin, and the state has a land area of roughly 56,000 square miles; therefore, the average concentration of people is about $(5.5 \times 10^6$ people/5.6×10^4 square miles) or 96 people per square mile. In chemistry the

■ **Chemical Safety and Redox Reactions**
It is not a good idea to mix a strong oxidizing agent with a strong reducing agent; a violent reaction—even an explosion—may take place. This reason explains why chemicals are not necessarily stored on shelves in alphabetical order. This practice can be unsafe, because such an ordering may place a strong oxidizing agent next to a strong reducing agent.

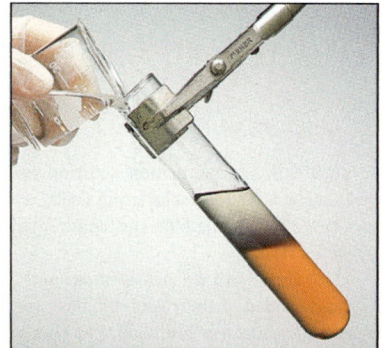

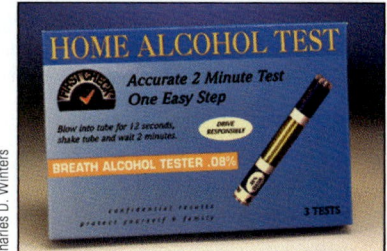

Charles D. Winters

Figure 5.16 **The redox reaction of ethanol and dichromate ion is the basis of the test used in a Breathalyzer.** When ethanol, an alcohol, is poured into a solution of orange-red dichromate ion, it reduces the dichromate ion to green chromium(III) ion. The bottom photo is a breath-tester that can be purchased in grocery or drug stores. See Exercise 5.11.

■ **Molar and Molarity**
Chemists use "molar" as an adjective to describe a solution. We use "molarity" as a noun. For example, we refer to a 0.1 molar solution or say the solution has a molarity of 0.1 mole per liter.

amount of solute dissolved in a given volume of solution, the concentration of the solution, can be found in the same way. A useful unit of solute concentration, c, is **molarity**, which is defined as *amount of solute per liter of solution*.

$$\text{Concentration } (c_{\text{molarity}}) = \frac{\text{amount of solute (mol)}}{\text{volume of solution (L)}} \qquad (5.1)$$

■ **Volumetric Flask**
A volumetric flask is a special flask with a line marked on its neck (see Figures 5.17 and 5.18). If the flask is filled with a solution to this line (at a given temperature), it contains precisely the volume of solution specified.

For example, if 58.4 g, or 1.00 mol, of NaCl is dissolved in enough water to give a total solution volume of 1.00 L, the concentration, c, is 1.00 mol/L, or 1.00 molar. This is often abbreviated as 1.00 M, where the capital "M" stands for "moles per liter." Another common notation is to place the formula of the compound in square brackets; this implies that the concentration of the solute in moles of compound per liter of solution is being specified.

$$c_{\text{molarity}} = 1.00 \text{ M} = [\text{NaCl}]$$

It is important to notice that molarity refers to the amount of solute per liter of *solution* and not per liter of *solvent*. If one liter of water is added to one mole of a solid compound, the final volume probably will not be exactly one liter, and the final concentration will not be exactly one molar (Figure 5.17). When making solutions of a given molarity, it is almost always the case that we dissolve the solute in a volume of solvent smaller than the desired volume of solution, then add solvent until the final solution volume is reached.

Potassium permanganate, $KMnO_4$, which was used at one time as a germicide in the treatment of burns, is a shiny, purple-black solid that dissolves readily in water to give a deep purple solution. Suppose 0.435 g of $KMnO_4$ has been dissolved in enough water to give 250. mL of solution (Figure 5.18). What is the molar concen-

Figure 5.17   Volume of solution versus volume of solvent.
To make a 0.100 M solution of $CuSO_4$, 25.0 g or 0.100 mol of $CuSO_4 \cdot 5 H_2O$ (the blue crystalline solid) was placed in a 1.00-L volumetric flask.

For this photo we measured out exactly 1.00 L of water, which was slowly added to the volumetric flask containing $CuSO_4 \cdot 5 H_2O$. When enough water had been added so that the solution volume was exactly 1.00 L, approximately 8 mL (the quantity in the small graduated cylinder) was left over from the original 1.00 L of water.

This emphasizes that molar concentrations are defined as moles of solute per liter of solution and not per liter of water or other solvent.

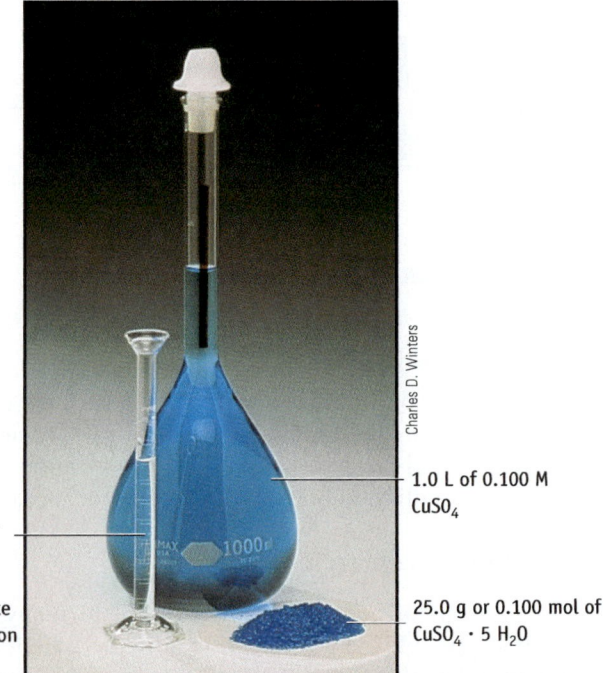

Volume of water remaining when 1.0 L of water was used to make 1.0 L of a solution

1.0 L of 0.100 M $CuSO_4$

25.0 g or 0.100 mol of $CuSO_4 \cdot 5 H_2O$

Charles D. Winters

Distilled water is added to fill the flask with solution just to the mark on the flask.

250 mL volumetric flask

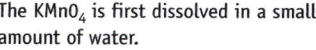
0.435g KMnO₄

The KMnO₄ is first dissolved in a small amount of water.

A mark on the neck of a volumetric flask indicates a volume of exactly 250 mL at 25 °C.

Active Figure 5.18   **Making a solution.** A 0.0110 M solution of KMnO₄ is made by adding enough water to 0.435 g of KMnO₄ to make 0.250 L of solution.

GENERAL
Chemistry Now™ See the General ChemistryNow CD-ROM or website to explore an interactive version of this figure accompanied by an exercise.

tration of $KMnO_4$? The first step is to convert the mass of $KMnO_4$ to an amount (moles) of solute.

$$0.435 \text{ g KMnO}_4 \times \frac{1 \text{ mol KMnO}_4}{158.0 \text{ g KMnO}_4} = 0.00275 \text{ mol KMnO}_4$$

Now that the amount of $KMnO_4$ is known, this information can be combined with the volume of solution—which must be in liters—to give the molarity. Because 250. mL is equivalent to 0.250 L,

$$\text{Concentration of KMnO}_4 = \frac{0.00275 \text{ mol KMnO}_4}{0.250 \text{ L}} = 0.0110 \text{ M}$$

$$[\text{KMnO}_4] = 0.0110 \text{ M}$$

The $KMnO_4$ concentration is 0.0110 mol/L, or 0.0110 M. This is useful information, but it is often equally useful to know the concentration of each type of ion in a solution. Like all soluble ionic compounds, $KMnO_4$ dissociates completely into its ions, K^+ and MnO_4^-, when dissolved in water.

$$\text{KMnO}_4(aq) \xrightarrow{} \text{K}^+(aq) + \text{MnO}_4^-(aq)$$
$$\text{100\% dissociation}$$

One mole of $KMnO_4$ provides 1 mol of K^+ ions and 1 mol of MnO_4^- ions. Accordingly, 0.0110 M $KMnO_4$ gives a concentration of K^+ in the solution of 0.0110 M; similarly, the concentration of MnO_4^- is also 0.0110 M.

Photo: Charles D. Winters

Ion concentrations for a soluble ionic compound. Here 1 mol of $CuCl_2$ dissociates to 1 mol of Cu^{2+} ions and 2 mol of Cl^- ions. Therefore, the Cl^- concentration is twice the stated concentration of $CuCl_2$.

Another example of ion concentrations is provided by the dissociation of an ionic compound such as $CuCl_2$.

$$CuCl_2(aq) \xrightarrow{\text{100\% dissociation}} Cu^{2+}(aq) + 2\ Cl^-(aq)$$

If 0.10 mol of $CuCl_2$ is dissolved in enough water to make 1.0 L of solution, the concentration of the copper(II) ion is $[Cu^{2+}] = 0.10$ M. The concentration of chloride ions, $[Cl^-]$, is 0.20 M because the compound dissociates in water to provide 2 mol of Cl^- ions for each mole of $CuCl_2$.

GENERAL
Chemistry ⚛ Now ™

See the General ChemistryNow CD-ROM or website:
• **Screen 5.15 Solution Concentrations,** for a tutorial on determining solution concentration and for a tutorial on determining ion concentration

Example 5.9—Concentration

Problem If 25.3 g of sodium carbonate, Na_2CO_3, is dissolved in enough water to make 250. mL of solution, what is the molar concentration of Na_2CO_3? What are the concentrations of the Na^+ and CO_3^{2-} ions?

Strategy The molar concentration of Na_2CO_3 is defined as the amount of Na_2CO_3 per liter of solution. We know the volume of solution (0.250 L). We need the amount of Na_2CO_3. To find the concentrations of the individual ions, recognize that the dissolved salt dissociates completely.

$$Na_2CO_3(s) \longrightarrow 2\ Na^+(aq) + CO_3^{2-}(aq)$$

Thus, a 1 M solution of Na_2CO_3 is really a 2 M solution of Na^+ ions and 1 M solution of CO_3^{2-} ions.

Solution Let us first find the amount of Na_2CO_3,

$$25.3\ \text{g } Na_2CO_3 \times \frac{1\ \text{mol } Na_2CO_3}{106.0\ \text{g } Na_2CO_3} = 0.239\ \text{mol } Na_2CO_3$$

and then the molar concentration of Na_2CO_3,

$$\text{Concentration} = \frac{0.239\ \text{mol } Na_2CO_3}{0.250\ \text{L}} = 0.955\ \text{M}$$

$$[Na_2CO_3] = 0.955\ \text{M}$$

The ion concentrations follow from the concentration of Na_2CO_3 and the knowledge that each mole of Na_2CO_3 produces 2 mol of Na^+ ions and 1 mol of CO_3^{2-} ions.

$$0.955\ \text{M } Na_2CO_3(aq) \equiv 2 \times 0.955\ \text{M } Na^+(aq) + 0.955\ \text{M } CO_3^{2-}(aq)$$

That is, $[Na^+] = 1.91$ M and $[CO_3^{2-}] = 0.955$ M.

Exercise 5.13—Concentration

Sodium bicarbonate, $NaHCO_3$, is used in baking powder formulations and in the manufacture of plastics and ceramics, among other things. If 26.3 g of the compound is dissolved in enough water to make 200. mL of solution, what is the molar concentration of $NaHCO_3$? What are the concentrations of the ions in solution?

Preparing Solutions of Known Concentration

A task chemists often must perform is preparing a given volume of solution of known concentration. There are two commonly used ways to do this.

Combining a Weighed Solute with the Solvent

Suppose you wish to prepare 2.00 L of a 1.50 M solution of Na_2CO_3. You have some solid Na_2CO_3 and distilled water. You also have a 2.00-L volumetric flask (see Figures 5.17 and 5.18). To make the solution, you must weigh the necessary quantity of Na_2CO_3 as accurately as possible, carefully place all the solid in the volumetric flask, and then add some water to dissolve the solid. After the solid has dissolved completely, more water is added to bring the solution volume to 2.00 L. The solution then has the desired concentration and the volume specified.

But what mass of Na_2CO_3 is required to make 2.00 L of 1.50 M Na_2CO_3? First, calculate the amount of substance required,

$$2.00 \; \cancel{L} \times \frac{1.50 \; mol \; Na_2CO_3}{1.00 \; \cancel{L} \; solution} = 3.00 \; mol \; Na_2CO_3 \; required$$

and then the mass in grams,

$$3.00 \; \cancel{mol \; Na_2CO_3} \times \frac{106.0 \; g \; Na_2CO_3}{1 \; \cancel{mol \; Na_2CO_3}} = 318 \; g \; Na_2CO_3$$

Thus, to prepare the desired solution, you should dissolve 318 g of Na_2CO_3 in enough water to make 2.00 L of solution.

Exercise 5.14—Preparing Solutions of Known Concentration

An experiment in your laboratory requires 250. mL of a 0.0200 M solution of $AgNO_3$. You are given solid $AgNO_3$, distilled water, and a 250.-mL volumetric flask. Describe how to make up the required solution.

Diluting a More Concentrated Solution

Another method of making a solution of a given concentration is to *begin with a concentrated solution and add water until the desired, lower concentration is reached* (Figure 5.19). Many of the solutions prepared for your laboratory course are probably made by this dilution method. It is more efficient to store a small volume of a concentrated solution and then, when needed, add water to make a much larger volume of a dilute solution.

Suppose you need 500. mL of 0.0010 M potassium dichromate, $K_2Cr_2O_7$, for use in chemical analysis. You have some 0.100 M $K_2Cr_2O_7$ solution available. To make the required 0.0010 M solution, place a measured volume of the more concentrated

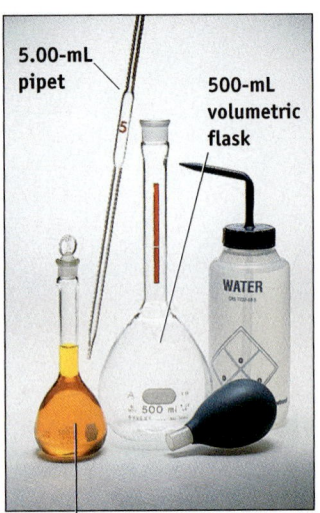

5.00-mL
pipet

500-mL
volumetric
flask

WATER

0.100 M K₂Cr₂O₇

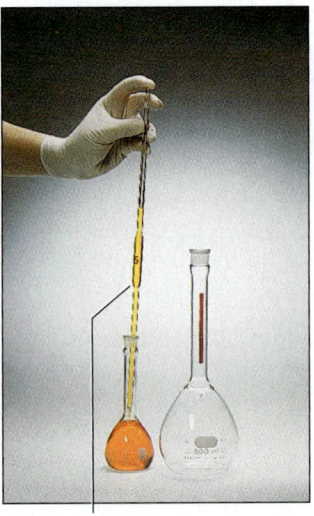

Use a 5.00-mL pipet to withdraw
5.00 mL of 0.100 M
K₂Cr₂O₇ solution.

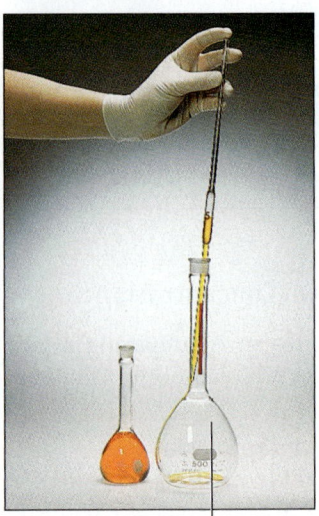

Add the 5.00-mL sample of 0.100 M
K₂Cr₂O₇ solution to a 500-mL
volumetric flask.

Fill the flask to the mark with
distilled water to give 0.00100 M
K₂Cr₂O₇ solution.

Charles D. Winters

Figure 5.19   Making a solution by dilution. Here 5.00 mL of a K₂Cr₂O₇ solution is diluted to 500. mL.
This means the solution is diluted by a factor of 100, from 0.100 M to 0.00100 M *(See General ChemistryNow
Screen 5.16 for a video of this procedure.)*

K₂Cr₂O₇ solution in a flask and then add water until the K₂Cr₂O₇ is contained
in a larger volume of water—that is, until it is less concentrated (or more dilute)
(Figure 5.19).

What volume of a 0.100 M K₂Cr₂O₇ solution must be diluted to make the 0.0010
M solution? In general, if the volume and concentration of a solution are known,
the amount of solute is also known. Therefore, the amount of K₂Cr₂O₇ that must be
in the final dilute solution is

$$\text{Amount of K}_2\text{Cr}_2\text{O}_7 \text{ in dilute solution } = (0.500 \; \cancel{L})\left(\frac{0.0010 \text{ mol}}{\cancel{L}}\right)$$
$$= 0.00050 \text{ mol K}_2\text{Cr}_2\text{O}_7$$

Problem Solving Tips 5.2

Preparing a Solution by Dilution

The preparation of the K₂Cr₂O₇ solution and Example 5.10 suggest a
way to do the calculations for dilutions. The central idea is that the
amount of solute in the final, dilute solution has to be equal to the
amount of solute taken from the more concentrated solution. If c is
the concentration (molarity) and V is the volume (and the subscripts
d and c identify the dilute and concentrated solutions,
respectively), then the amount of solute in either solution (in the
case of the K₂Cr₂O₇ example in the text) can be calculated as follows:

Amount of K₂Cr₂O₇ in the final dilute solution = $c_d V_d$
$$= 0.00050 \text{ mol}$$

Amount of K₂Cr₂O₇ taken from the more concentrated solution =
$$c_c V_c = 0.00050 \text{ mol}$$

Because both cV products are equal to the same amount of solute,
we can use the following equation:

$$c_c V_c = c_d V_d$$

Amount of reagent in concentrated solution =
              Amount of reagent in dilute solution

This equation is valid for all cases in which a more concentrated
solution is used to make a more dilute one. It can be used to find,
for example, the molarity of the dilute solution, c_d, when the
values of c_c, V_c, and V_d are known.

A more concentrated solution containing this amount of $K_2Cr_2O_7$ must be placed in a 500.-mL flask and then diluted to the final volume. The volume of 0.100 M $K_2Cr_2O_7$ that must be transferred and diluted is 5.0 mL.

$$0.00050 \text{ mol } K_2Cr_2O_7 \times \frac{1.00 \text{ L}}{0.100 \text{ mol } K_2Cr_2O_7} = 0.0050 \text{ L or } 5.0 \text{ mL}$$

Thus, to prepare 500. mL of 0.0010 M $K_2Cr_2O_7$, place 5.0 mL of 0.100 M $K_2Cr_2O_7$ in a 500.-mL flask and add water until a volume of 500. mL is reached (Figure 5.19).

■ **Diluting Concentrated Sulfuric Acid** The direction that one can prepare a solution by adding water to a more concentrated solution is correct except for sulfuric acid solutions. When mixing water and sulfuric acid, the resulting solution becomes quite warm. If water is added to concentrated sulfuric acid, so much heat is evolved that the solution may boil over or splash and burn someone nearby. To avoid this problem, chemists always add concentrated sulfuric acid to water to make a dilute solution.

GENERAL
Chemistry ⚛ Now™

See the General ChemistryNow CD-ROM or website:

- **Screen 5.16 Preparing Solutions of Known Concentrations,** for an exercise and a tutorial on the direct addition method of preparing a solution and for an exercise and tutorial on the dilution method of preparing a solution

EXAMPLE 5.10—Preparing a Solution by Dilution

Problem You need a 2.36×10^{-3} M solution of iron(III) ion. A lab procedure suggests this can be done by placing 1.00 mL of 0.236 M iron(III) nitrate in a volumetric flask and diluting to exactly 100.0 mL. Show that this method will work.

Strategy First calculate the amount of iron(III) ion in the 1.00-mL sample. The concentration of the ion in the final, dilute solution is equal to this amount of iron(III) divided by the new volume.

Solution The amount of iron(III) ion in the 1.00 mL sample is

$$c \times V = \frac{0.236 \text{ mol } Fe^{3+}}{\text{L}} \times 1.00 \times 10^{-3} \text{ L}$$

$$= 2.36 \times 10^{-4} \text{ mol } Fe^{3+}$$

This amount of iron(III) ion is distributed in the new volume of 100.0 mL, so the final concentration of the diluted solution is

$$[Fe^{3+}] = \frac{2.36 \times 10^{-4} \text{ mol } Fe^{3+}}{0.100 \text{ L}} = 2.36 \times 10^{-3} \text{ M}$$

Comment The experimental procedure is illustrated in Figure 5.19.

Exercise 5.15—Preparing a Solution by Dilution

In one of your laboratory experiments you are given a solution of $CuSO_4$ that has a concentration of 0.15 M. If you mix 6.0 mL of this solution with enough water to have a total volume of 10.0 mL, what is the concentration of $CuSO_4$ in the new solution?

Exercise 5.16—Preparing a Solution by Dilution

An experiment calls for you to use 250. mL of 1.00 M NaOH, but you are given a large bottle of 2.00 M NaOH. Describe how to make the 1.00 M NaOH in the desired volume.

5.9—pH, a Concentration Scale for Acids and Bases

Vinegar, which contains the weak acid acetic acid, has a hydrogen ion concentration of only 1.6×10^{-3} M and "pure" rainwater has $[H^+] = 2.5 \times 10^{-6}$ M. These extremely small values can be expressed using scientific notation, but this is awkward. A more convenient way to express such numbers is the logarithmic pH scale.

The **pH** of a solution is the negative of the base-10 logarithm of the hydrogen ion concentration.

$$pH = -\log [H^+] \tag{5.2}$$

Taking vinegar, pure water, blood, and ammonia as examples,

pH of vinegar	$= -\log (1.6 \times 10^{-3} \text{ M}) = -(-2.80) = 2.80$
pH of pure water (at 25 °C)	$= -\log (1.0 \times 10^{-7} \text{ M}) = -(7.00) = 7.00$
pH of blood	$= -\log (4.0 \times 10^{-8} \text{ M}) = -(-7.40) = 7.40$
pH of ammonia	$= -\log (1.0 \times 10^{-11} \text{ M}) = -(-11.00) = 11.00$

you see that something you recognize as acidic has a relatively low pH, whereas ammonia, a common base, has a *very* low hydrogen ion concentration and a high pH. Blood, which your common sense tells you is likely to be neither acidic nor basic, has a pH near 7. Indeed, for aqueous solutions at 25 °C, we can say that acids will have pH values less than 7, bases will have values greater than 7, and a pH of 7 represents a neutral solution (Figure 5.20).

Suppose you know the pH of a solution. To find the hydrogen ion concentration you take the antilog of the pH. That is,

$$[H^+] = 10^{-pH} \tag{5.3}$$

■ **Logarithms**
Numbers less than 1 have negative logs. Defining pH as $-\log [H^+]$ produces a positive number. See Appendix A for a discussion of logs.

■ **pH of Pure Water**
Highly purified water, which is said to be "neutral," has a pH of exactly 7 at 25 °C. This is the "dividing line" between acidic substances (pH < 7) and basic substances (pH > 7).

■ **Logs and Your Calculator**
All scientific calculators have a key marked "log." To find an antilog, use the key marked "10^x" or the inverse log. When you enter the value of x for 10^x, make sure it has a negative sign.

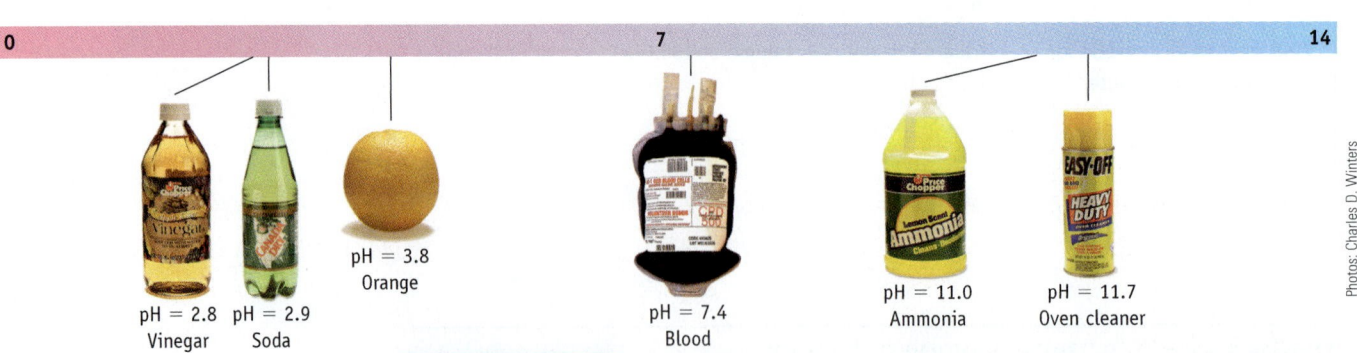

0	7	14

pH = 2.8   pH = 2.9   pH = 3.8   pH = 7.4   pH = 11.0   pH = 11.7
Vinegar   Soda   Orange   Blood   Ammonia   Oven cleaner

Photos: Charles D. Winters

Active Figure 5.20 **pH values of some common substances.** Here the "bar" is colored red at one end and blue at the other. These are the colors of litmus paper, commonly used in the laboratory to decide whether a solution is acidic (litmus is red) or basic (litmus is blue).

GENERAL
Chemistry·⚛·Now™ See the General ChemistryNow CD-ROM or website to explore an interactive version of this figure accompanied by an exercise.

(a)

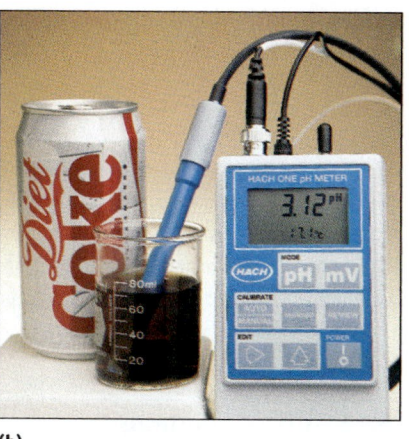

(b)

Figure 5.21 **Determining pH.** (a) Some household products. Each solution contains a few drops of a dye called a pH indicator (in this case a "universal indicator"). A color of yellow or red indicates a pH less than 7. A green to purple color indicates a pH greater than 7. (b) The pH of a soda is measured with a modern pH meter. Soft drinks are often quite acidic owing to the dissolved CO_2 and other ingredients.

For example, the pH of a diet soda is 3.12, and the hydrogen ion concentration of the solution is

$$[H^+] = 10^{-3.12} = 7.6 \times 10^{-4} \text{ M}$$

The approximate pH of a solution may be determined using any of a variety of dyes. The litmus paper you use in the laboratory contains a dye extracted from a variety of lichen, but many other dyes are also available (Figure 5.21a). A more accurate measurement of pH is done with a pH meter such as that shown in Figure 5.21b. Here a pH electrode is immersed in the solution to be tested, and the pH is read from the instrument.

■ **pH Indicating Dyes**
Many natural substances change color in solution as pH changes. See the extract of red cabbage in Figure 5.6a and of red rose petals on page 849. Tea changes color when acidic lemon juice is added.

GENERAL
Chemistry·❄·Now™

See the General ChemistryNow CD-ROM or website:
- **Screen 5.17 The pH Scale,** for a tutorial on determining the pH of a solution

Example 5.11—pH of Solutions

Problem

(a) Lemon juice has $[H^+] = 0.0032$ M. What is its pH?

(b) Sea water has a pH of 8.30. What is the hydrogen ion concentration of this solution?

(c) A solution of nitric acid has $[HNO_3] = 0.0056$ M. What is the pH of this solution?

Strategy Use Equation 5.2 to calculate pH from the H^+ concentration. Use Equation 5.3 to find $[H^+]$ from the pH.

Solution

(a) *Lemon juice:* Because the hydrogen ion concentration is known, the pH is found using Equation 5.2.

$$\text{pH} = -\log[H^+] = -\log(3.2 \times 10^{-3}) = -(-2.49) = \boxed{2.49}$$

(b) *Sea water:* Here pH = 8.30. Therefore,

$$[H^+] = 10^{-\text{pH}} = 10^{-8.30} = \boxed{5.0 \times 10^{-9} \text{ M}}$$

Charles D. Winters

(c) *Nitric acid:* Nitric acid is a strong acid (Table 5.2, page 187) and is completely ionized in aqueous solution. Because $[HNO_3] = 0.0056$ M, the ion concentrations are

$$[H^+] = [NO_3^-] = 0.0056 \text{ M}$$

$$pH = -\log [H^+] = -\log (0.0056 \text{ M}) = \boxed{2.25}$$

Comment A comment on logarithms and significant figures (Appendix A) is useful. The number to the left of the decimal point in a logarithm is called the *characteristic*, and the number to the right is the *mantissa*. The mantissa has as many significant figures as the number whose log was found. For example, the logarithm of 3.2×10^{-3} (two significant figures) is 2.49 (two numbers to the right of the decimal point).

Exercise 5.17—pH of Solutions

(a) What is the pH of a solution of HCl in which $[HCl] = 2.6 \times 10^{-2}$ M?

(b) What is the hydrogen ion concentration in orange juice with a pH of 3.80?

5.10—Stoichiometry of Reactions in Aqueous Solution

General Solution Stoichiometry

Suppose we want to know what mass of $CaCO_3$ is required to react completely with 25 mL of 0.750 M HCl. The first step in finding the answer is to write a balanced equation. In this case, we have an exchange reaction involving a metal carbonate and an aqueous acid (Figure 5.22).

$$CaCO_3(s) + 2 \text{ HCl(aq)} \longrightarrow CaCl_2(aq) + H_2O(\ell) + CO_2(g)$$

$$\text{metal carbonate} + \text{acid} \longrightarrow \text{salt} + \text{water} + \text{carbon dioxide}$$

This problem can be solved in the same way as all the stoichiometry problems you have seen so far, except that the quantity of one reactant is given in volume and concentration units instead of as a mass in grams. The first step is to find the amount of HCl.

$$0.025 \text{ L HCl} \times \frac{0.750 \text{ mol HCl}}{1 \text{ L HCl}} = 0.019 \text{ mol HCl}$$

This is then related to the amount of $CaCO_3$ required.

$$0.019 \text{ mol HCl} \times \frac{1 \text{ mol } CaCO_3}{2 \text{ mol HCl}} = 0.0094 \text{ mol } CaCO_3$$

Finally, the amount of $CaCO_3$ is converted to a mass in grams.

$$0.0094 \text{ mol } CaCO_3 \times \frac{100. \text{ g } CaCO_3}{1 \text{ mol } CaCO_3} = 0.94 \text{ g } CaCO_3$$

Chemists are likely to do such calculations many times in the course of their work. If you follow the general scheme outlined in Problem-Solving Tip 5.3, and pay attention to the units on the numbers, you can successfully carry out any kind of stoichiometry calculations involving concentrations.

Figure 5.22   A commercial remedy for excess stomach acid. The tablet contains calcium carbonate, which reacts with hydrochloric acid, the acid present in the digestive system. The most obvious product is CO_2 gas.

Charles D. Winters

Problem-Solving Tip 5.3

Stoichiometry Calculations Involving Solutions

In Problem-Solving Tip 4.1, you learned about a general approach to stoichiometry problems. We can now modify that scheme for a reaction involving solutions such as $x A(aq) + y B(aq) \longrightarrow$ products.

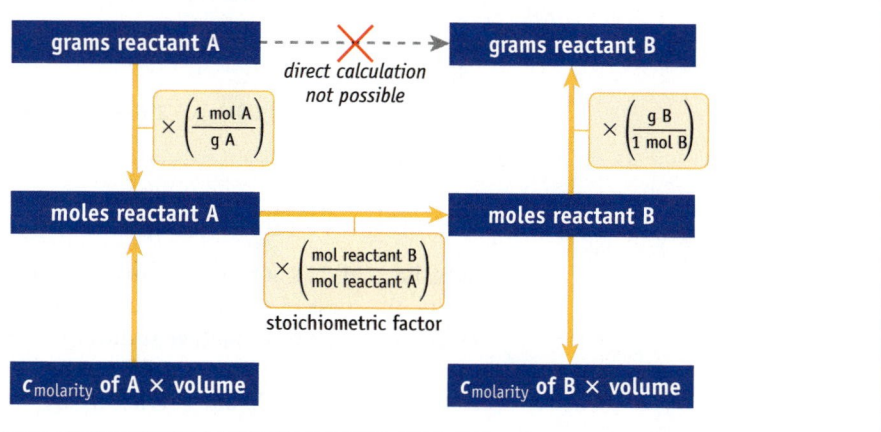

GENERAL
Chemistry · Now™

See the General ChemistryNow CD-ROM or website:

- **Screen 5.18 Stoichiometry of Reactions in Solution,** for an exercise on solution stoichiometry, for a tutorial on determining the mass of a product, and for a tutorial on determining the volume of a reactant

Example 5.12—Stoichiometry of a Reaction in Solution

Problem Metallic zinc reacts with aqueous HCl (see Figure 5.6c).

$$Zn(s) + 2\, HCl(aq) \longrightarrow ZnCl_2(aq) + H_2(g)$$

What volume of 2.50 M HCl, in milliliters, is required to convert 11.8 g of Zn completely to products?

Strategy Here the mass of zinc is known, so you first calculate the amount of zinc. Next, use a stoichiometric factor (= 2 mol HCl/1 mol Zn) to relate amount of HCl required to amount of Zn available. Finally, calculate the volume of HCl from the amount of HCl and its concentration.

Solution Begin by calculating the amount of Zn.

$$11.8\ \text{g Zn} \times \frac{1\ \text{mol Zn}}{65.39\ \text{g Zn}} = 0.180\ \text{mol Zn}$$

Use the stoichiometric factor to calculate the amount of HCl required.

$$0.180\ \text{mol Zn} \times \frac{2\ \text{mol HCl}}{1\ \text{mol Zn}} = 0.360\ \text{mol HCl}$$

Use the amount of HCl and the solution concentration to calculate the volume.

$$0.360\ \text{mol HCl} \times \frac{1.00\ \text{L solution}}{2.50\ \text{mol HCl}} = 0.144\ \text{L HCl}$$

The answer is requested in units of milliliters, so we convert the volume to milliliters and find that 144 mL of 2.50 M HCl is required to convert 11.8 g of Zn completely to products.

Exercise 5.18—Solution Stoichiometry

If you combine 75.0 mL of 0.350 M HCl and an excess of Na_2CO_3, what mass of CO_2, in grams, is produced?

$$Na_2CO_3(s) + 2\ HCl(aq) \longrightarrow 2\ NaCl(aq) + H_2O(\ell) + CO_2(g)$$

Titration: A Method of Chemical Analysis

Oxalic acid, $H_2C_2O_4$, is a naturally occurring acid. Suppose you are asked to determine the mass of this acid in an impure sample. Because the compound is an acid, it reacts with a base such as sodium hydroxide (see Section 5.4).

$$H_2C_2O_4(aq) + 2\ NaOH(aq) \longrightarrow Na_2C_2O_4(aq) + 2\ H_2O(\ell)$$

You can use this reaction to determine the quantity of oxalic acid present in a given mass of sample if the following conditions are met:

- You can determine when the amount of sodium hydroxide added is just enough to react with all the oxalic acid present in solution.

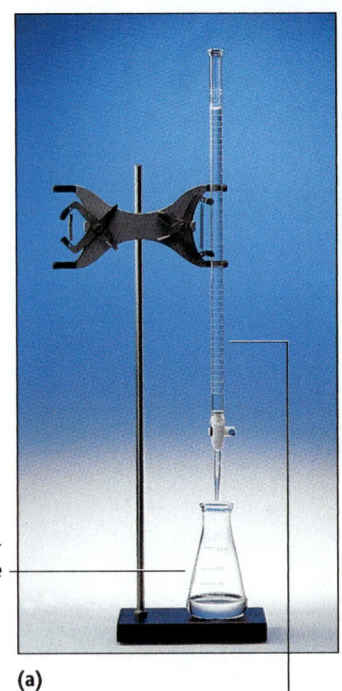

Flask containing aqueous solution of sample being analyzed

(a)
Buret containing aqueous NaOH of accurately known concentration.

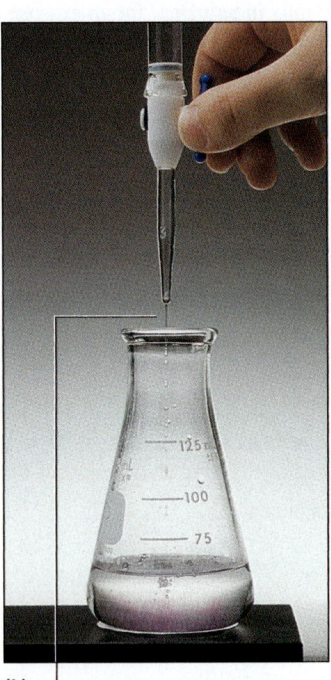

(b)
A solution of NaOH is added slowly to the sample being analyzed.

Charles D. Winters

(c)
When the amount of NaOH added from the buret exactly equals the amount of H^+ supplied by the acid being analyzed, the dye (indicator) changes color.

Active Figure 5.23   **Titration of an acid in aqueous solution with a base.** (a) A buret, a volumetric measuring device calibrated in divisions of 0.1 mL, is filled with an aqueous solution of a base of known concentration. (b) Base is added slowly from the buret to the solution containing the acid being analyzed and an indicator. (c) A change in the color of an indicator signals the equivalence point. (The indicator used here is phenolphthalein.)

GENERAL
Chemistry⚛Now™ See the General ChemistryNow CD-ROM or website to explore an interactive version of this figure accompanied by an exercise.

- You know the concentration of the sodium hydroxide solution and volume that has been added at exactly the point of complete reaction.

These conditions are fulfilled in a **titration**, a procedure illustrated in Figure 5.23. The solution containing oxalic acid is placed in a flask along with an **acid–base indicator**, a dye that changes color when the pH of the reaction solution changes. It is common practice to use a dye that has one color in acid solution and another color in basic solution. Aqueous sodium hydroxide of accurately known concentration is placed in a buret. The sodium hydroxide in the buret is added slowly to the acid solution in the flask. As long as some acid is present in solution, all the base supplied from the buret is consumed, the solution remains acidic, and the indicator color is unchanged. At some point, however, the amount of OH^- added exactly equals the amount of H^+ that can be supplied by the acid. This is called the **equivalence point**. As soon as the slightest excess of base has been added beyond the equivalence point, the solution becomes basic, and the indicator changes color (see Figure 5.23).

When the equivalence point has been reached in a titration, the volume of base added is determined by reading the calibrated buret. From this volume and the concentration of the base, the amount of base used can be found:

Amount of base added (mol) = concentration of base (mol/L) × volume of base (L)

Then, using the stoichiometric factor from the balanced equation, the amount of base added is related to the amount of acid present in the original sample. For the specific problem of finding the mass of oxalic acid in an impure sample, we would convert the amount of acid to a mass. If the mass of oxalic acid is divided by the mass of the impure sample (and the quotient multiplied by 100%), we can express the purity of the sample in terms of a mass percent.

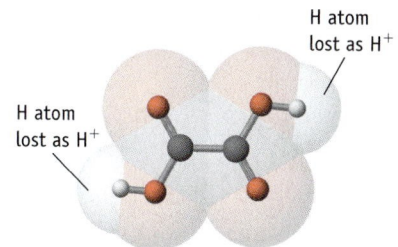

Oxalic acid $H_2C_2O_4$

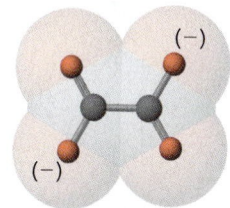

Oxalate anion $C_2O_4{}^{2-}$

Oxalic acid. Oxalic acid has two groups that can supply an H^+ ion to solution. Hence, 1 mol of the acid requires 2 mol of NaOH for complete reaction.

GENERAL
Chemistry Now™

See the General ChemistryNow CD-ROM or website:

- **Screen 5.19 Titration,** for a tutorial on the volume of titrant used, for a tutorial on determining the concentration of acid solution, and for a tutorial on determining the concentration of an unknown acid

Example 5.13—Acid–Base Titration

Problem A 1.034-g sample of impure oxalic acid is dissolved in water and an acid–base indicator added. The sample requires 34.47 mL of 0.485 M NaOH to reach the equivalence point. What is the mass of oxalic acid and what is its mass percent in the sample?

Strategy The balanced equation for the reaction of NaOH and $H_2C_2O_4$ is

$$H_2C_2O_4(aq) + 2\ NaOH(aq) \longrightarrow Na_2C_2O_4(aq) + 2\ H_2O(\ell)$$

The concentration of NaOH and the volume used in the titration are used to determine the amount of NaOH. Use a stoichiometric factor to relate the amount of NaOH to the amount of $H_2C_2O_4$. Finally, the amount of $H_2C_2O_4$ is converted to a mass. The mass percent of acid in the sample is then calculated. See Problem Solving Tip 5.3.

Solution The amount of NaOH is given by

$$c_{NaOH} \times V_{NaOH} = \frac{0.485\ \text{mol NaOH}}{\cancel{L}} \times 0.03447\ \cancel{L} = 0.0167\ \text{mol NaOH}$$

The balanced equation for the reaction shows that 1 mol of oxalic acid requires 2 mol of sodium hydroxide. This is the required stoichiometric factor to obtain the amount of oxalic acid present.

$$0.0167 \; \text{mol NaOH} \times \frac{1 \; \text{mol} \; H_2C_2O_4}{2 \; \text{mol NaOH}} = 0.00836 \; \text{mol} \; H_2C_2O_4$$

The mass of oxalic acid is found from the amount of the acid.

$$0.00836 \; \text{mol} \; H_2C_2O_4 \times \frac{90.04 \; \text{g} \; H_2C_2O_4}{1 \; \text{mol} \; H_2C_2O_4} = 0.753 \; \text{g} \; H_2C_2O_4$$

This mass of oxalic acid represents 72.8% of the total sample mass.

$$\frac{0.753 \; \text{g} \; H_2C_2O_4}{1.034 \; \text{g sample}} \times 100\% = 72.8\% \; H_2C_2O_4$$

Exercise 5.19—Acid–Base Titration

A 25.0-mL sample of vinegar (which contains the weak acid acetic acid, CH_3CO_2H) requires 28.33 mL of a 0.953 M solution of NaOH for titration to the equivalence point. What mass of acetic acid, in grams, is in the vinegar sample, and what is the concentration of acetic acid in the vinegar?

$$CH_3CO_2H(aq) + NaOH(aq) \longrightarrow NaCH_3CO_2(aq) + H_2O(\ell)$$

Standardizing an Acid or Base

In Example 5.13 the concentration of the base used in the titration was given. In actual practice this usually has to be found by a prior measurement. The procedure by which the concentration of an analytical reagent is determined accurately is called **standardization**, and there are two general approaches.

One approach is to weigh accurately a sample of a pure, solid acid or base (known as a **primary standard**) and then titrate this sample with a solution of the base or acid to be standardized (Example 5.14). An alternative approach to standardizing a solution is to titrate it with another solution that is already standardized (Exercise 5.20). This is often done with standard solutions purchased from chemical supply companies.

Example 5.14—Standardizing an Acid by Titration

Problem A sample of sodium carbonate, a base (Na_2CO_3, 0.263 g), requires 28.35 mL of aqueous HCl for titration to the equivalence point. What is the molarity of the HCl?

Strategy The balanced equation for the reaction is written first.

$$Na_2CO_3(aq) + 2 \; HCl(aq) \longrightarrow 2 \; NaCl(aq) + H_2O(\ell) + CO_2(g)$$

The amount of Na_2CO_3 can be calculated from its mass and then, using the stoichiometric factor, the amount of HCl in 28.35 mL can be calculated. The amount of HCl divided by the volume of solution (in liters) gives its molar concentration.

Solution Convert the mass of Na_2CO_3 used as the standard to amount of the base.

$$0.263 \; \text{g} \; Na_2CO_3 \times \frac{1 \; \text{mol} \; Na_2CO_3}{106.0 \; \text{g} \; Na_2CO_3} = 0.00248 \; \text{mol} \; Na_2CO_3$$

Use the stoichiometric factor to calculate amount of HCl in 28.35 mL.

$$0.00248 \text{ mol Na}_2\text{CO}_3 \times \frac{2 \text{ mol HCl required}}{1 \text{ mol Na}_2\text{CO}_3 \text{ available}} = 0.00496 \text{ mol HCl}$$

The 28.35-mL (0.02835-L) sample of aqueous HCl contains 0.00496 mol of HCl, so the concentration of the HCl solution is 0.175 M.

$$[\text{HCl}] = \frac{0.00496 \text{ mol HCl}}{0.02835 \text{ L}} = 0.175 \text{ M}$$

Comment In this example Na_2CO_3 is a primary standard. Sodium carbonate can be obtained in pure form, which can be weighed accurately, and which reacts completely with a strong acid.

Exercise 5.20—Standardization of a Base

Hydrochloric acid, HCl, can be purchased from chemical supply houses with a concentration of 0.100 M, and this solution can be used to standardize the solution of a base. If titrating 25.00 mL of a sodium hydroxide solution to the equivalence point requires 29.67 mL of 0.100 M HCl, what is the concentration of the base?

Determining Molar Mass by Titration

In Chapters 3 and 4 we used analytical data to determine the empirical formula of a compound. The molecular formula could then be derived if the molar mass were known. If the unknown substance is an acid or a base, it is possible to determine the molar mass by titration.

Example 5.15—Determining the Molar Mass of an Acid by Titration

Problem To determine the molar mass of an organic acid, HA, we titrate 1.056 g of HA with standardized NaOH. Calculate the molar mass of HA assuming the acid reacts with 33.78 mL of 0.256 M NaOH according to the equation

$$\text{HA(aq)} + \text{OH}^-(\text{aq}) \longrightarrow \text{A}^-(\text{aq}) + \text{H}_2\text{O}(\ell)$$

Strategy The key to this problem is to recognize that the molar mass of a substance is the ratio of the mass of a sample (g) to the amount of substance (mol) in the sample. Here molar mass of HA = 1.056 g HA/x mol HA. Because 1 mol of HA reacts with 1 mol of NaOH in this case, the amount of acid (x mol) is equal to the amount of NaOH used in the titration, which is given by its concentration and volume.

Solution Let us first calculate the amount of NaOH used in the titration.

$$c_{\text{NaOH}}V_{\text{NaOH}} = (0.256 \text{ mol/L})(0.03378 \text{ L}) = 8.65 \times 10^{-3} \text{ mol NaOH}$$

Next, recognize that the amount of NaOH used in the titration is the same as the amount of acid titrated. That is,

$$8.65 \times 10^{-3} \text{ mol NaOH} \left(\frac{1 \text{ mol HA}}{1 \text{ mol NaOH}} \right) = 8.65 \times 10^{-3} \text{ mol HA}$$

Finally, calculate the molar mass of HA.

$$\text{Molar mass of acid} = \frac{1.056 \text{ g HA}}{8.65 \times 10^{-3} \text{ mol HA}} = 122 \text{ g/mol}$$

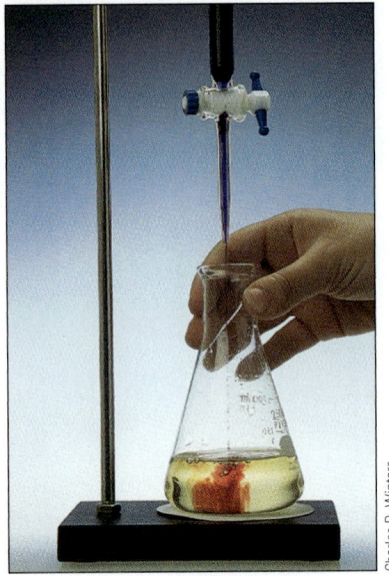

Using an oxidation–reduction reaction for analysis by titration. Purple, aqueous $KMnO_4$ is added to a solution containing Fe^{2+}. As $KMnO_4$ drops into the solution, colorless Mn^{2+} and pale yellow Fe^{3+} form. Here an area of the solution containing unreacted $KMnO_4$ is seen. As the solution is mixed, this disappears until the equivalence point is reached.

Charles D. Winters

Exercise 5.21—Determining the Molar Mass of an Acid by Titration

An acid reacts with NaOH according to the net ionic equation

$$HA(aq) + OH^-(aq) \longrightarrow A^-(aq) + H_2O(\ell)$$

Calculate the molar mass of HA if 0.856 g of the acid requires 30.08 mL of 0.323 M NaOH.

Titrations Using Oxidation–Reduction Reactions

Oxidation–reduction reactions (see Section 5.7) also lend themselves to chemical analysis by titration. Many of these reactions go rapidly to completion in aqueous solution, and methods exist to determine their equivalence point.

Example 5.16—Using an Oxidation–Reduction Reaction in a Titration

Problem We wish to analyze an iron ore for its iron content. The iron in the sample can be converted quantitatively to the iron(II) ion, Fe^{2+}, in aqueous solution, and this solution can then be titrated with aqueous potassium permanganate, $KMnO_4$. The balanced, net ionic equation for the reaction occurring in the course of this titration is

$$MnO_4^-(aq) + 5\ Fe^{2+}(aq) + 8\ H^+(aq) \longrightarrow Mn^{2+}(aq) + 5\ Fe^{3+}(aq) + 4\ H_2O(\ell)$$

purple	colorless		colorless	pale yellow

A 1.026-g sample of iron-containing ore requires 24.35 mL of 0.0195 M $KMnO_4$ to reach the equivalence point. What is the mass percent of iron in the ore?

Strategy Because the volume and molar concentration of the $KMnO_4$ solution are known, the amount of $KMnO_4$ used in the titration can be calculated. Using the stoichiometric factor, the amount of $KMnO_4$ is related to the amount of iron(II) ion. The amount of iron(II) is converted to its mass, and the mass percent of iron in the sample is determined.

Solution First, calculate the amount of $KMnO_4$.

$$c_{KMnO_4} \times V_{KMnO_4} = \frac{0.0195\ mol\ KMnO_4}{\ell} \times 0.02435\ \ell$$

$$= 0.000475\ mol\ KMnO_4$$

Use the stoichiometric factor to calculate the amount of iron(II) ion.

$$0.000475\ mol\ KMnO_4 \times \frac{5\ mol\ Fe^{2+}}{1\ mol\ KMnO_4} = 0.00237\ mol\ Fe^{2+}$$

The mass of iron can now be calculated,

$$0.00237\ mol\ Fe^{2+} \times \frac{55.85\ g\ Fe^{2+}}{1\ mol\ Fe^{2+}} = 0.133\ g\ Fe^{2+}$$

Finally, the mass percent can be determined.

$$\frac{0.133\ g\ Fe^{2+}}{1.026\ g\ sample} \times 100\% = 12.9\%\ iron$$

Comment This is a useful analytical reaction because it is easy to detect when all the iron(II) ion has reacted. The MnO_4^- ion is a deep purple color, but when it reacts with Fe^{2+} the color disappears because the reaction product, Mn^{2+}, is colorless. Thus, as $KMnO_4$ is added from a buret, the purple color disappears as the solutions mix. When all the Fe^{2+} has been converted to Fe^{3+}, any additional $KMnO_4$ will give the solution a permanent purple color. Therefore, $KMnO_4$ solution is added from the buret until the initially colorless, Fe^{2+}-containing solution just turns a faint purple color, the signal that the equivalence point has been reached.

Exercise 5.22—Using an Oxidation–Reduction Reaction in a Titration

Vitamin C, ascorbic acid ($C_6H_8O_6$), is a reducing agent. One way to determine the ascorbic acid content of a sample is to mix the acid with an excess of iodine,

$$C_6H_8O_6(aq) + I_2(aq) \longrightarrow C_6H_6O_6(aq) + 2\,H^+(aq) + 2\,I^-(aq)$$

and then titrate the iodine that did *not* react with the ascorbic acid with sodium thiosulfate. The balanced, net ionic equation for the reaction occurring in this titration is

$$I_2(aq) + 2\,S_2O_3{}^{2-}(aq) \longrightarrow 2\,I^-(aq) + S_4O_6{}^{2-}(aq)$$

Suppose 50.00 mL of 0.0520 M I_2 was added to the sample containing ascorbic acid. After the ascorbic acid/I_2 reaction was complete, the I_2 not used in the reaction required 20.30 mL of 0.196 M $Na_2S_2O_3$ for titration to the equivalence point. Calculate the mass of ascorbic acid in the unknown sample.

Chapter Goals Revisited

Now that you have studied this chapter, you should ask if you have met the chapter goals. In particular, you should be able to

Understand the nature of ionic substances dissolved in water
a. Explain the difference between electrolytes and nonelectrolytes and recognize examples of each (Section 5.1 and Figure 5.2).
b. Predict the solubility of ionic compounds in water (Section 5.1 and Figure 5.3). General ChemistryNow homework: Study Question(s) 7
c. Recognize which ions are formed when an ionic compound or acid or base dissolves in water (Sections 5.1–5.3). General ChemistryNow homework: SQ(s) 13

Recognize common acids and bases and understand their behavior in aqueous solution (Section 5.3 and Table 5.2)
a. Know the names and formulas of common acids and bases. General ChemistryNow homework: SQ(s) 13, 18
b. Categorize acids and bases as strong or weak.

Recognize and write equations for the common types of reactions in aqueous solution
a. Predict the products of precipitation reactions (Section 5.2), which involve the formation of an insoluble reaction product by the exchange of anions between the cations of the reactants. General ChemistryNow homework: SQ(s) 11
b. Write net ionic equations and show how to arrive at such an equation for a given reaction (Sections 5.2 and 5.6). General ChemistryNow homework: SQ(s) 11

GENERAL
Chemistry Now™

See the General ChemistryNow CD-ROM or website to:
- Assess your understanding with homework questions keyed to each goal
- Check your readiness for an exam by taking the exam-prep quiz and exploring the resources in the personalized Learning Plan it provides

For additional preparation for an examination on this chapter see the **Let's Review** section on pages 282–293.

c. Predict the products of acid–base reactions involving common acids and strong bases (Section 5.4). General ChemistryNow homework: SQ(s) 19

d. Understand that the net ionic equation for the reaction of a strong acid with a strong base is $H^+(aq) + OH^-(aq) \longrightarrow H_2O(\ell)$ (Section 5.4).

e. Predict the products of gas-forming reactions (Section 5.5), the most common of which are those between a metal carbonate and an acid.

$$NiCO_3(s) + 2\ HNO_3(aq) \longrightarrow Ni(NO_3)_2(aq) + CO_2(g) + H_2O(\ell)$$

f. Use the ideas developed in Sections 5.2–5.7 as an aid in recognizing four of the common types of reactions that occur in aqueous solution, and write balanced equations for such reactions (Section 5.6).

Reaction Type	Driving Force
Precipitation	Formation of an insoluble compound
Acid–strong base	Formation of a salt and water
Gas-forming	Evolution of a water-insoluble gas such as CO_2
Oxidation–reduction	Transfer of electrons

The first three of these reaction types involve the exchange of anions between the cations involved, and so are called exchange reactions. The fourth type (redox reactions) involves the transfer of electrons. General ChemistryNow homework: SQ(s) 29

g. Identify reactant- and product-favored reactions. General ChemistryNow homework: SQ(s) 33

Recognize common oxidizing and reducing agents and identify oxidation–reduction reactions

a. Determine oxidation numbers of elements in a compound and understand that these numbers represent the charge an atom has, or appears to have, when the electrons of the compound are counted according to a set of guidelines (Section 5.7). General ChemistryNow homework: SQ(s) 35

b. Identify oxidation–reduction reactions (redox reactions) and identify the oxidizing and reducing agents and substances oxidized and reduced in the reaction (Section 5.7 and Tables 5.4 and 5.5). General ChemistryNow homework: SQ(s) 39

Define and use molarity in solution stoichiometry

a. Calculate the concentration of a solute in a solution in units of moles per liter (molarity), and use concentrations in calculations (Section 5.8). General ChemistryNow homework: SQ(s) 41, 43, 45

b. Describe how to prepare a solution of a given molarity from the solute and a solvent or by dilution from a more concentrated solution (Section 5.8). General ChemistryNow homework: SQ(s) 50, 51

c. Calculate the pH of a solution containing an acid or a base and know what this means in terms of the relative amount of hydrogen ion in the solution. Calculate the hydrogen ion concentration of a solution from the pH (Section 5.9). General ChemistryNow homework: SQ(s) 56, 57

d. Solve stoichiometry problems using solution concentrations (Section 5.10). General ChemistryNow homework: SQ(s) 61, 64

e. Explain how a titration is carried out, explain the procedure of standardization, and calculate concentrations or amounts of reactants from titration data (Section 5.10). General ChemistryNow homework: SQ(s) 69, 73

Key Equations

Equation 5.1 (page 206)
Definition of molarity, a measure of the concentration of a solute in a solution.

$$\text{Concentration } (c_{molarity}) = \frac{\text{amount of solute (mol)}}{\text{volume of solution (L)}}$$

A useful form of this equation is

$$\text{Amount of solute (moles)} = c_{molarity} \times \text{volume of solution (L)}$$

Related to this equation is the "shortcut" used when diluting a concentrated solution to obtain a more dilute solution. The product of the concentration and volume of a more concentrated solution (c) must be the same as that for the diluted solution (d).

$$c_c \times V_c = c_d \times V_d$$

If any three of these parameters is known (say c_c, V_c, and c_d), the fourth may be calculated (say V_d).

Equation 5.2 (page 212)
The pH of a solution is the negative logarithm of the hydrogen ion concentration.

$$pH = -\log [H^+]$$

Equation 5.3 (page 212)
The equation for calculating the hydrogen ion concentration of a solution from the pH of the solution.

$$[H^+] = 10^{-pH}$$

Study Questions

▲ denotes more challenging questions.

■ denotes questions available in the Homework and Goals section of the General ChemistryNow CD-ROM or website.

Blue numbered questions have answers in Appendix O and fully worked solutions in the *Student Solutions Manual.*

Structures of many of the compounds used in these questions are found on the General ChemistryNow CD-ROM or website in the Models folder.

GENERAL
Chemistry · Now™ Assess your understanding of this chapter's topics with additional quizzing and conceptual questions at http://now.brookscole.com/kotz6e

Practicing Skills

Electrolytes and Solubility of Compounds
(See Exercise 5.1, Example 5.1, and General ChemistryNow Screens 5.3 and 5.4.)

1. What is an electrolyte? How can you differentiate experimentally between a weak electrolyte and a strong electrolyte? Give an example of each.

2. Name two acids that are strong electrolytes and one acid that is a weak electrolyte. Name two bases that are strong electrolytes and one base that is a weak electrolyte.

3. Which compound or compounds in each of the following groups is (are) expected to be soluble in water?
   (a) CuO, $CuCl_2$, $FeCO_3$
   (b) AgI, Ag_3PO_4, $AgNO_3$
   (c) K_2CO_3, KI, $KMnO_4$

4. Which compound or compounds in each of the following groups is (are) expected to be soluble in water?
   (a) $BaSO_4$, $Ba(NO_3)_2$, $BaCO_3$
   (b) Na_2SO_4, $NaClO_4$, $NaCH_3CO_2$
   (c) $AgBr$, KBr, Al_2Br_6

5. The following compounds are water-soluble. What ions are produced by each compound in aqueous solution?
   (a) KOH (c) $LiNO_3$
   (b) K_2SO_4 (d) $(NH_4)_2SO_4$

6. The following compounds are water-soluble. What ions are produced by each compound in aqueous solution?
   (a) KI (c) K_2HPO_4
   (b) $Mg(CH_3CO_2)_2$ (d) $NaCN$

7. ■ Decide whether each of the following is water-soluble. If soluble, tell what ions are produced.
   (a) Na_2CO_3 (c) NiS
   (b) $CuSO_4$ (d) $BaBr_2$

8. Decide whether each of the following is water-soluble. If soluble, tell what ions are produced.
   (a) $NiCl_2$ (c) $Pb(NO_3)_2$
   (b) $Cr(NO_3)_3$ (d) $BaSO_4$

Precipitation Reactions and Net Ionic Equations

(See Examples 5.2 and 5.3 and General ChemistryNow Screens 5.5–5.7.)

9. Balance the equation for the following precipitation reaction, and then write the net ionic equation. Indicate the state of each species (s, ℓ, aq, or g).

$$CdCl_2 + NaOH \longrightarrow Cd(OH)_2 + NaCl$$

10. Balance the equation for the following precipitation reaction, and then write the net ionic equation. Indicate the state of each species (s, ℓ, aq, or g).

$$Ni(NO_3)_2 + Na_2CO_3 \longrightarrow NiCO_3 + NaNO_3$$

11. ■ Predict the products of each precipitation reaction. Balance the completed equation, and then write the net ionic equation.
    (a) $NiCl_2(aq) + (NH_4)_2S(aq) \longrightarrow$?
    (b) $Mn(NO_3)_2(aq) + Na_3PO_4(aq) \longrightarrow$?

12. Predict the products of each precipitation reaction. Balance the completed equation, and then write the net ionic equation.
    (a) $Pb(NO_3)_2(aq) + KBr(aq) \longrightarrow$?
    (b) $Ca(NO_3)_2(aq) + KF(aq) \longrightarrow$?
    (c) $Ca(NO_3)_2(aq) + Na_2C_2O_4(aq) \longrightarrow$?

Acids and Bases

(See Exercises 5.5 and 5.6 and General ChemistryNow Screens 5.8 and 5.9.)

13. ■ Write a balanced equation for the ionization of nitric acid in water.

14. Write a balanced equation for the ionization of perchloric acid in water.

15. Oxalic acid, $H_2C_2O_4$, which is found in certain plants, can provide two hydrogen ions in water. Write balanced equations (like those for sulfuric acid on page 186) to show how oxalic acid can supply one and then a second H^+ ion.

16. Phosphoric acid can supply one, two, or three H^+ ions in aqueous solution. Write balanced equations (like those for sulfuric acid on page 186) to show this successive loss of hydrogen ions.

17. Write a balanced equation for reaction of the basic oxide, magnesium oxide, with water.

18. ■ Write a balanced equation for the reaction of sulfur trioxide with water.

Reactions of Acids and Bases

(See Example 5.4, Exercise 5.7, and General ChemistryNow Screens 5.5 and 5.10.)

19. ■ Complete and balance the following acid–base reactions. Name the reactants and products.
    (a) $CH_3CO_2H(aq) + Mg(OH)_2(s) \longrightarrow$
    (b) $HClO_4(aq) + NH_3(aq) \longrightarrow$

20. Complete and balance the following acid–base reactions. Name the reactants and products.
    (a) $H_3PO_4(aq) + KOH(aq) \longrightarrow$
    (b) $H_2C_2O_4(aq) + Ca(OH)_2(s) \longrightarrow$

    ($H_2C_2O_4$ is oxalic acid, an acid capable of donating two H^+ ions.)

21. Write a balanced equation for the reaction of barium hydroxide with nitric acid.

22. Write a balanced equation for the reaction of aluminum hydroxide with sulfuric acid.

Writing Net Ionic Equations

(See Example 5.3 and General ChemistryNow Screen 5.7.)

23. Balance the following equations, and then write the net ionic equation.
    (a) $(NH_4)_2CO_3(aq) + Cu(NO_3)_2(aq) \longrightarrow$
    $$CuCO_3(s) + NH_4NO_3(aq)$$
    (b) $Pb(OH)_2(s) + HCl(aq) \longrightarrow PbCl_2(s) + H_2O(\ell)$
    (c) $BaCO_3(s) + HCl(aq) \longrightarrow$
    $$BaCl_2(aq) + H_2O(\ell) + CO_2(g)$$

24. Balance the following equations, and then write the net ionic equation:
    (a) $Zn(s) + HCl(aq) \longrightarrow H_2(g) + ZnCl_2(aq)$
    (b) $Mg(OH)_2(s) + HCl(aq) \longrightarrow MgCl_2(aq) + H_2O(\ell)$
    (c) $HNO_3(aq) + CaCO_3(s) \longrightarrow$
    $$Ca(NO_3)_2(aq) + H_2O(\ell) + CO_2(g)$$

25. Balance the following equations, and then write the net ionic equation. Show states for all reactants and products (s, ℓ, g, aq).
    (a) the reaction of silver nitrate and potassium iodide to give silver iodide and potassium nitrate
    (b) the reaction of barium hydroxide and nitric acid to give barium nitrate and water

---

▲ More challenging   ■ In General ChemistryNow   Blue-numbered questions answered in Appendix O

(c) the reaction of sodium phosphate and nickel(II) nitrate to give nickel(II) phosphate and sodium nitrate

26. Balance each of the following equations, and then write the net ionic equation. Show states for all reactants and products (s, ℓ, g, aq).
    (a) the reaction of sodium hydroxide and iron(II) chloride to give iron(II) hydroxide and sodium chloride
    (b) the reaction of barium chloride with sodium carbonate to give barium carbonate and sodium chloride

Gas-Forming Reactions

(See Example 5.5 and General ChemistryNow Screens 5.5 and 5.11.)

27. Siderite is a mineral consisting largely of iron(II) carbonate. Write an overall, balanced equation for its reaction with nitric acid, and name each reactant and product.

28. The beautiful red mineral rhodochrosite is manganese(II) carbonate. Write an overall, balanced equation for the reaction of the mineral with hydrochloric acid. Name each reactant and product.

Rhodochrosite, a mineral consisting largely of MnCO$_3$

Types of Reactions in Aqueous Solution

(See Exercise 5.9, Example 5.8, and General ChemistryNow Screen 5.5.)

29. ■ Balance the following reactions and then classify each as a precipitation, acid–base, or gas-forming reaction.
    (a) $Ba(OH)_2(aq) + HCl(aq) \longrightarrow BaCl_2(aq) + H_2O(\ell)$
    (b) $HNO_3(aq) + CoCO_3(s) \longrightarrow$
      $Co(NO_3)_2(aq) + H_2O(\ell) + CO_2(g)$
    (c) $Na_3PO_4(aq) + Cu(NO_3)_2(aq) \longrightarrow$
      $Cu_3(PO_4)_2(s) + NaNO_3(aq)$

30. Balance the following reactions and then classify each as a precipitation, acid–base reaction, or a gas-forming reaction.
    (a) $K_2CO_3(aq) + Cu(NO_3)_2(aq) \longrightarrow$
      $CuCO_3(s) + KNO_3(aq)$
    (b) $Pb(NO_3)_2(aq) + HCl(aq) \longrightarrow PbCl_2(s) + HNO_3(aq)$
    (c) $MgCO_3(s) + HCl(aq) \longrightarrow$
      $MgCl_2(aq) + H_2O(\ell) + CO_2(g)$

31. Balance the following reactions and then classify each as a precipitation, acid–base reaction, or gas-forming reaction. Show states for the products (s, ℓ, g, aq) and then balance the completed equation. Write the net ionic equation.

(a) $MnCl_2(aq) + Na_2S(aq) \longrightarrow MnS + NaCl$
(b) $K_2CO_3(aq) + ZnCl_2(aq) \longrightarrow ZnCO_3 + KCl$

32. Balance the following reactions and then classify each as a precipitation, acid–base, or gas-forming reaction. Write the net ionic equation.
    (a) $Fe(OH)_3(s) + HNO_3(aq) \longrightarrow Fe(NO_3)_3 + H_2O$
    (b) $FeCO_3(s) + HNO_3(aq) \longrightarrow Fe(NO_3)_2 + CO_2 + H_2O$

Product- or Reactant-Favored Reactions

33. ■ What feature causes the following reactions to be product-favored?
    (a) $CuCl_2(aq) + H_2S(aq) \longrightarrow CuS(s) + 2\,HCl(aq)$
    (b) $H_3PO_4(aq) + 3\,KOH(aq) \longrightarrow 3\,H_2O(\ell) + K_3PO_4(aq)$

34. Which of the following reactions is predicted to be product-favored?
    (a) $Zn(s) + 2\,HCl(aq) \longrightarrow H_2(g) + ZnCl_2(aq)$
    (b) $MgCl_2(aq) + 2\,H_2O(\ell) \longrightarrow Mg(OH)_2(s) + 2\,HCl(aq)$

Oxidation Numbers

(See Example 5.6 and General ChemistryNow Screen 5.13.)

35. ■ Determine the oxidation number of each element in the following ions or compounds.
    (a) BrO_3^- (d) CaH_2
    (b) $C_2O_4^{2-}$ (e) H_4SiO_4
    (c) F^- (f) HSO_4^-

36. Determine the oxidation number of each element in the following ions or compounds.
    (a) PF_6^- (d) N_2O_5
    (b) $H_2AsO_4^-$ (e) $POCl_3$
    (c) UO^{2+} (f) XeO_4^{2-}

Oxidation–Reduction Reactions

(See Example 5.7 and General ChemistryNow Screens 5.12–5.14.)

37. Which two of the following reactions are oxidation–reduction reactions? Explain your answer in each case. Classify the remaining reaction.
    (a) $Zn(s) + 2\,NO_3^-(aq) + 4\,H^+(aq) \longrightarrow$
      $Zn^{2+}(aq) + 2\,NO_2(g) + 2\,H_2O(\ell)$
    (b) $Zn(OH)_2(s) + H_2SO_4(aq) \longrightarrow ZnSO_4(aq) + 2\,H_2O(\ell)$
    (c) $Ca(s) + 2\,H_2O(\ell) \longrightarrow Ca(OH)_2(s) + H_2(g)$

38. Which two of the following reactions are oxidation–reduction reactions? Explain your answer briefly. Classify the remaining reaction.
    (a) $CdCl_2(aq) + Na_2S(aq) \longrightarrow CdS(s) + 2\,NaCl(aq)$
    (b) $2\,Ca(s) + O_2(g) \longrightarrow 2\,CaO(s)$
    (c) $4\,Fe(OH)_2(s) + 2\,H_2O(\ell) + O_2(g) \longrightarrow 4\,Fe(OH)_3(aq)$

39. ■ In the following reactions, decide which reactant is oxidized and which is reduced. Designate the oxidizing agent and the reducing agent.
    (a) $C_2H_4(g) + 3\,O_2(g) \longrightarrow 2\,CO_2(g) + 2\,H_2O(g)$
    (b) $Si(s) + 2\,Cl_2(g) \longrightarrow SiCl_4(\ell)$

40. In the following reactions, decide which reactant is oxidized and which is reduced. Designate the oxidizing agent and the reducing agent.

(a) $Cr_2O_7^{2-}(aq) + 3\ Sn^{2+}(aq) + 14\ H^+(aq) \longrightarrow$
$2\ Cr^{3+}(aq) + 3\ Sn^{4+}(aq) + 7\ H_2O(\ell)$

(b) $FeS(s) + 3\ NO_3^-(aq) + 4\ H^+(aq) \longrightarrow$
$3\ NO(g) + SO_4^{2-}(aq) + Fe^{3+}(aq) + 2\ H_2O(\ell)$

Solution Concentration

(See Example 5.9 and General ChemistryNow Screen 5.15.)

41. ■ If 6.73 g of Na_2CO_3 is dissolved in enough water to make 250. mL of solution, what is the molar concentration of the sodium carbonate? What are the molar concentrations of the Na^+ and CO_3^{2-} ions?

42. Some potassium dichromate ($K_2Cr_2O_7$), 2.335 g, is dissolved in enough water to make exactly 500. mL of solution. What is the molar concentration of the potassium dichromate? What are the molar concentrations of the K^+ and $Cr_2O_7^{2-}$ ions?

43. ■ What is the mass of solute, in grams, in 250. mL of a 0.0125 M solution of $KMnO_4$?

44. What is the mass of solute, in grams, in 125 mL of a 1.023×10^{-3} M solution of Na_3PO_4? What are the molar concentrations of the Na^+ and PO_4^{3-} ions?

45. ■ What volume of 0.123 M NaOH, in milliliters, contains 25.0 g of NaOH?

46. What volume of 2.06 M $KMnO_4$, in liters, contains 322 g of solute?

47. For each solution, identify the ions that exist in aqueous solution, and specify the concentration of each ion.

(a) 0.25 M $(NH_4)_2SO_4$

(b) 0.123 M Na_2CO_3

(c) 0.056 M HNO_3

48. For each solution, identify the ions that exist in aqueous solution, and specify the concentration of each ion.

(a) 0.12 M $BaCl_2$

(b) 0.0125 M $CuSO_4$

(c) 0.500 M $K_2Cr_2O_7$

Preparing Solutions

(See Exercise 5.14, Example 5.10, and General ChemistryNow Screen 5.16.)

49. An experiment in your laboratory requires exactly 500. mL of a 0.0200 M solution of Na_2CO_3. You are given solid Na_2CO_3, distilled water, and a 500.-mL volumetric flask. Describe how to prepare the required solution.

50. ■ What mass of oxalic acid, $H_2C_2O_4$, is required to prepare 250. mL of a solution that has a concentration of 0.15 M $H_2C_2O_4$?

51. ■ If you dilute 25.0 mL of 1.50 M hydrochloric acid to 500. mL, what is the molar concentration of the dilute acid?

52. If 4.00 mL of 0.0250 M $CuSO_4$ is diluted to 10.0 mL with pure water, what is the molar concentration of copper(II) sulfate in the diluted solution?

53. Which of the following methods would you use to prepare 1.00 L of 0.125 M H_2SO_4?

(a) Dilute 20.8 mL of 6.00 M H_2SO_4 to a volume of 1.00 L.

(b) Add 950. mL of water to 50.0 mL of 3.00 M H_2SO_4.

54. Which of the following methods would you use to prepare 300. mL of 0.500 M $K_2Cr_2O_7$?

(a) Add 30.0 mL of 1.50 M $K_2Cr_2O_7$ to 270. mL of water.

(b) Dilute 250. mL of 0.600 M $K_2Cr_2O_7$ to a volume of 300. mL.

pH

(See Example 5.11 and General ChemistryNow Screen 5.17.)

55. A table wine has a pH of 3.40. What is the hydrogen ion concentration of the wine? Is it acidic or basic?

56. ■ A saturated solution of milk of magnesia, $Mg(OH)_2$, has a pH of 10.5. What is the hydrogen ion concentration of the solution? Is the solution acidic or basic?

57. ■ What is the hydrogen ion concentration of a 0.0013 M solution of HNO_3? What is its pH?

58. What is the hydrogen ion concentration of a 1.2×10^{-4} M solution of $HClO_4$? What is its pH?

59. Make the following conversions. In each case, tell whether the solution is acidic or basic.

	pH	$[H^+]$
(a)	1.00	_____
(b)	10.50	_____
(c)	_____	1.3×10^{-5} M
(d)	_____	2.3×10^{-8} M

60. Make the following conversions. In each case, tell whether the solution is acidic or basic.

	pH	$[H^+]$
(a)	_____	6.7×10^{-10} M
(b)	_____	2.2×10^{-6} M
(c)	5.25	_____
(d)	_____	2.5×10^{-2} M

Stoichiometry of Reactions in Solution

(See Example 5.12 and General ChemistryNow Screen 5.18.)

61. ■ What volume of 0.109 M HNO_3, in milliliters, is required to react completely with 2.50 g of $Ba(OH)_2$?

$2\ HNO_3(aq) + Ba(OH)_2(s) \longrightarrow 2\ H_2O(\ell) + Ba(NO_3)_2(aq)$

62. What mass of Na_2CO_3, in grams, is required for complete reaction with 50.0 mL of 0.125 M HNO_3?

$Na_2CO_3(aq) + 2\ HNO_3(aq) \longrightarrow$
$2\ NaNO_3(aq) + CO_2(g) + H_2O(\ell)$

63. When an electric current is passed through an aqueous solution of NaCl, the valuable industrial chemicals $H_2(g)$, $Cl_2(g)$, and NaOH are produced.

$2\ NaCl(aq) + 2\ H_2O(\ell) \longrightarrow H_2(g) + Cl_2(g) + 2\ NaOH(aq)$

▲ More challenging   ■ In General ChemistryNow   Blue-numbered questions answered in Appendix O

What mass of NaOH can be formed from 15.0 L of 0.35 M NaCl? What mass of chlorine is obtained?

64. ■ Hydrazine, N_2H_4, a base like ammonia, can react with an acid such as sulfuric acid.

$$2\ N_2H_4(aq) + H_2SO_4(aq) \longrightarrow 2\ N_2H_5^+(aq) + SO_4^{2-}(aq)$$

What mass of hydrazine reacts with 250. mL of 0.146 M H_2SO_4?

65. In the photographic developing process, silver bromide is dissolved by adding sodium thiosulfate:

$$AgBr(s) + 2\ Na_2S_2O_3(aq) \longrightarrow Na_3Ag(S_2O_3)_2(aq) + NaBr(aq)$$

If you want to dissolve 0.225 g of AgBr, what volume of 0.0138 M $Na_2S_2O_3$, in milliliters, should be used?

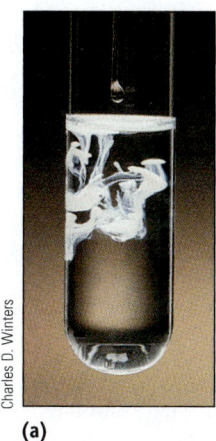 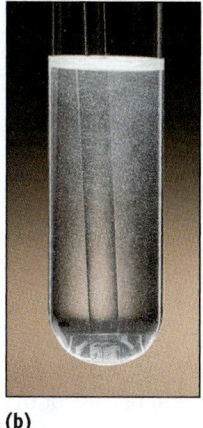

(a)　　　　**(b)**

Silver Chemistry. (a) A precipitate of AgBr formed by adding $AgNO_3(aq)$ to KBr(aq). (b) On adding $Na_2S_2O_3(aq)$, sodium thiosulfate, the solid AgBr dissolves.

66. You can dissolve an aluminum soft-drink can in an aqueous base such as potassium hydroxide.

$$2\ Al(s) + 2\ KOH(aq) + 6\ H_2O(\ell) \longrightarrow 2\ KAl(OH)_4(aq) + 3\ H_2(g)$$

If you place 2.05 g of aluminum in a beaker with 185 mL of 1.35 M KOH, will any aluminum remain? What mass of $KAl(OH)_4$ is produced?

67. What volume of 0.750 M $Pb(NO_3)_2$, in milliliters, is required to react completely with 1.00 L of 2.25 M NaCl solution? The balanced equation is

$$Pb(NO_3)_2(aq) + 2\ NaCl(aq) \longrightarrow PbCl_2(s) + 2\ NaNO_3(aq)$$

68. What volume of 0.125 M oxalic acid, $H_2C_2O_4$ is required to react with 35.2 mL of 0.546 M NaOH?

$$H_2C_2O_4(aq) + 2\ NaOH(aq) \longrightarrow Na_2C_2O_4(aq) + 2\ H_2O(aq)$$

Titrations

(See Examples 5.13–5.16 and General ChemistryNow Screen 5.19.)

69. ■ What volume of 0.812 M HCl, in milliliters, is required to titrate 1.45 g of NaOH to the equivalence point?

$$NaOH(aq) + HCl(aq) \longrightarrow H_2O(\ell) + NaCl(aq)$$

70. What volume of 0.955 M HCl, in milliliters, is required to titrate 2.152 g of Na_2CO_3 to the equivalence point?

$$Na_2CO_3(aq) + 2\ HCl(aq) \longrightarrow H_2O(\ell) + CO_2(g) + 2\ NaCl(aq)$$

71. If 38.55 mL of HCl is required to titrate 2.150 g of Na_2CO_3 according to the following equation, what is the molarity of the HCl solution?

$$Na_2CO_3(aq) + 2\ HCl(aq) \longrightarrow 2\ NaCl(aq) + CO_2(g) + H_2O(\ell)$$

72. Potassium hydrogen phthalate, $KHC_8H_4O_4$, is used to standardize solutions of bases. The acidic anion reacts with strong bases according to the following net ionic equation:

$$HC_8H_4O_4^-(aq) + OH^-(aq) \longrightarrow C_8H_4O_4^{2-}(aq) + H_2O(\ell)$$

If a 0.902-g sample of potassium hydrogen phthalate is dissolved in water and titrated to the equivalence point with 26.45 mL of NaOH, what is the molar concentration of the NaOH?

73. ■ You have 0.954 g of an unknown acid, H_2A, which reacts with NaOH according to the balanced equation

$$H_2A(aq) + 2\ NaOH(aq) \longrightarrow Na_2A(aq) + 2\ H_2O(\ell)$$

If 36.04 mL of 0.509 M NaOH is required to titrate the acid to the equivalence point, what is the molar mass of the acid?

74. ▲ An unknown solid acid is either citric acid or tartaric acid. To determine which acid you have, you titrate a sample of the solid with NaOH. The appropriate reactions are as follows:

Citric acid:
$$H_3C_6H_5O_7(aq) + 3\ NaOH(aq) \longrightarrow 3\ H_2O(\ell) + Na_3C_6H_5O_7(aq)$$

Tartaric acid:
$$H_2C_4H_4O_6(aq) + 2\ NaOH(aq) \longrightarrow 2\ H_2O(\ell) + Na_2C_4H_4O_6(aq)$$

A 0.956-g sample requires 29.1 mL of 0.513 M NaOH for titration to the equivalence point. What is the unknown acid?

75. To analyze an iron-containing compound, you convert all the iron to Fe^{2+} in aqueous solution and then titrate the solution with standardized $KMnO_4$. The balanced, net ionic equation is

$$MnO_4^-(aq) + 5\ Fe^{2+}(aq) + 8\ H^+(aq) \longrightarrow Mn^{2+}(aq) + 5\ Fe^{3+}(aq) + 4\ H_2O(\ell)$$

A 0.598-g sample of the iron-containing compound requires 22.25 mL of 0.0123 M $KMnO_4$ for titration to the equivalence point. What is the mass percent of iron in the sample?

76. Vitamin C is the simple compound $C_6H_8O_6$. Besides being an acid, it is a reducing agent. One method for determining the amount of vitamin C in a sample is therefore to titrate it with a solution of bromine, Br_2, an oxidizing agent.

$$C_6H_8O_6(aq) + Br_2(aq) \longrightarrow 2\ HBr(aq) + C_6H_6O_6(aq)$$

A 1.00-g "chewable" vitamin C tablet requires 27.85 mL of 0.102 M Br_2 for titration to the equivalence point. What is the mass of vitamin C in the tablet?

General Questions

These questions are not designated as to type or location in the chapter. They may combine several concepts.

77. Give the formula for the following:
    (a) a soluble compound containing the bromide ion
    (b) an insoluble hydroxide
    (c) an insoluble carbonate
    (d) a soluble nitrate-containing compound

78. Give the formula for the following:
    (a) a soluble compound containing the acetate ion
    (b) an insoluble sulfide
    (c) a soluble hydroxide
    (d) an insoluble chloride

79. Which of the following copper(II) salts are soluble in water and which are insoluble: $Cu(NO_3)_2$, $CuCO_3$, $Cu_3(PO_4)_2$, $CuCl_2$?

80. Name two anions that combine with Al^{3+} ion to produce water-soluble compounds.

81. Identify the spectator ion or ions in the reaction of nitric acid and magnesium hydroxide, and write the net ionic equation. What type of exchange reaction is this?

$$2\ H^+(aq) + 2\ NO_3^-(aq) + Mg(OH)_2(s) \longrightarrow \\ 2\ H_2O(\ell) + Mg^{2+}(aq) + 2\ NO_3^-(aq)$$

82. Identify the water-insoluble product in each reaction and write the net ionic equation:
    (a) $CuCl_2(aq) + H_2S(aq) \longrightarrow CuS + 2\ HCl$
    (b) $CaCl_2(aq) + K_2CO_3(aq) \longrightarrow 2\ KCl + CaCO_3$
    (c) $AgNO_3(aq) + NaI(aq) \longrightarrow AgI + NaNO_3$

83. Bromine is obtained from sea water by the following reaction:

$$Cl_2(g) + 2\ NaBr(aq) \longrightarrow 2\ NaCl(aq) + Br_2(\ell)$$

    (a) What has been oxidized? What has been reduced?
    (b) Identify the oxidizing and reducing agents.
    (c) What mass of Cl_2 is required to react completely with 125 mL of 0.153 M NaBr?

84. Identify each of the following substances as an oxidizing or reducing agent: HNO_3, Na, Cl_2, O_2, $KMnO_4$.

85. Which contains the greater mass of solute: 1 L of 0.1 M NaCl or 1 L of 0.06 M Na_2CO_3?

86. Describe each of the following as product- or reactant-favored.
    (a) $BaBr_2(aq) + 2\ H_2O(\ell) \longrightarrow \\ Ba(OH)_2(aq) + 2\ HBr(aq)$
    (b) $NaOH(aq) + FeCl_3(aq) \longrightarrow NaCl(aq) + Fe(OH)_3(s)$

87. You have a bottle of solid Na_2CO_3 and a 500.0-mL volumetric flask. Explain how you would make a 0.20 M solution of sodium carbonate.

88. You have 0.500 mol of KCl, some distilled water, and a 250.-mL volumetric flask. Describe how you would make a 0.500 M solution of KCl.

89. Which has the larger concentration of hydrogen ions, 0.015 M HCl or a hydrochloric acid solution with a pH of 1.2?

90. What volume of 0.054 M H_2SO_4 is required to react completely with 1.56 g of KOH?

91. The mineral dolomite contains magnesium carbonate.

$$MgCO_3(s) + 2\ HCl(aq) \longrightarrow CO_2(g) + MgCl_2(aq) + H_2O(\ell)$$

    (a) Write the net ionic equation for the reaction of magnesium carbonate and hydrochloric acid, and name the spectator ions.
    (b) What type of reaction is this?
    (c) What mass of $MgCO_3$ will react with 125 mL of HCl(aq) with a pH of 1.56?

92. Ammonium sulfide, $(NH_4)_2S$, reacts with $Hg(NO_3)_2$ to produce HgS and NH_4NO_3.
    (a) Write the overall balanced equation for the reaction. Indicate the state (s, aq) for each compound.
    (b) Name each compound.
    (c) What type of reaction is this?

93. What species (atoms, molecules, or ions) are present in an aqueous solution of each of the following compounds?
    (a) NH_3 (c) NaOH
    (b) CH_3CO_2H (d) HBr

94. Suppose an Alka-Seltzer tablet contains exactly 100 mg of citric acid, $H_3C_6H_5O_7$, plus some sodium bicarbonate. If the following reaction occurs, what mass of sodium bicarbonate must the tablet also contain?

$$H_3C_6H_5O_7(aq) + 3\ NaHCO_3(aq) \longrightarrow \\ 3\ H_2O(\ell) + 3\ CO_2(g) + Na_3C_6H_5O_7(aq)$$

95. ▲ Sodium bicarbonate and acetic acid react according to the equation

$$NaHCO_3(aq) + CH_3CO_2H(aq) \longrightarrow \\ NaCH_3CO_2(aq) + CO_2(g) + H_2O(\ell)$$

    What mass of sodium acetate can be obtained from mixing 15.0 g of $NaHCO_3$ with 125 mL of 0.15 M acetic acid?

96. A noncarbonated soft drink contains an unknown amount of citric acid, $H_3C_6H_5O_7$. If 100. mL of the soft drink requires 33.51 mL of 0.0102 M NaOH to neutralize the citric acid completely, what mass of citric acid does the soft drink contain per 100. mL? The reaction of citric acid and NaOH is

$$H_3C_6H_5O_7(aq) + 3\ NaOH(aq) \longrightarrow Na_3C_6H_5O_7(aq) + 3\ H_2O(\ell)$$

97. Sodium thiosulfate, $Na_2S_2O_3$, is used as a "fixer" in black-and-white photography. Suppose you have a bottle of sodium thiosulfate and want to determine its purity. The

thiosulfate ion can be oxidized with I_2 according to the balanced, net ionic equation

$$I_2(aq) + 2\,S_2O_3^{2-}(aq) \longrightarrow 2\,I^-(aq) + S_4O_6^{2-}(aq)$$

If you use 40.21 mL of 0.246 M I_2 in a titration, what is the weight percent of $Na_2S_2O_3$ in a 3.232-g sample of impure material?

98. You have a 4.554-g sample that is a mixture of oxalic acid, $H_2C_2O_4$, and another solid that does not react with sodium hydroxide. If 29.58 mL of 0.550 M NaOH is required to titrate the oxalic acid in the 4.554-g sample to the equivalence point, what is the weight percent of oxalic acid in the mixture? Oxalic acid and NaOH react according to the equation

$$H_2C_2O_4(aq) + 2\,NaOH(aq) \longrightarrow Na_2C_2O_4(aq) + 2\,H_2O(\ell)$$

99. (a) Name two water-soluble compounds containing the Cu^{2+} ion. Name two water-insoluble compounds based on the Cu^{2+} ion.

    (b) Name two water-soluble compounds containing the Ba^{2+} ion. Name two water-insoluble compounds based on the Ba^{2+} ion.

100. Balance these reactions and then classify each one as a precipitation, acid–base, or gas-forming reaction. Show states for the products (s, ℓ, g, aq), and write the net ionic equation.

    (a) $K_2CO_3(aq) + HClO_4(aq) \longrightarrow KClO_4 + CO_2 + H_2O$

    (b) $FeCl_2(aq) + (NH_4)_2S(aq) \longrightarrow FeS + NH_4Cl$

    (c) $Fe(NO_3)_2(aq) + Na_2CO_3(aq) \longrightarrow FeCO_3 + NaNO_3$

101. For each reaction, write an overall, balanced equation and the net ionic equation.

    (a) the reaction of aqueous lead(II) nitrate and aqueous potassium hydroxide

    (b) the reaction of aqueous copper(II) nitrate and aqueous sodium carbonate

102. (a) What is the pH of a 0.105 M HCl solution?

    (b) What is the hydrogen ion concentration in a solution with a pH of 2.56? Is the solution acidic or basic?

    (c) A solution has a pH of 9.67. What is the hydrogen ion concentration in the solution? Is the solution acidic or basic?

    (d) A 10.0-mL sample of 2.56 M HCl is diluted with water to 250. mL. What is the pH of the dilute solution?

103. A solution of hydrochloric acid has a volume of 125 mL and a pH of 2.56. What mass of $NaHCO_3$ must be added to completely consume the HCl?

104. ▲ One-half liter (500. mL) of 2.50 M HCl is mixed with 250. mL of 3.75 M HCl. Assuming the total solution volume after mixing is 750. mL, what is the concentration of hydrochloric acid in the resulting solution? What is its pH?

105. A solution of hydrochloric acid has a volume of 250. mL and a pH of 1.92. Exactly 250. mL of 0.0105 M NaOH is added. What is the pH of the resulting solution?

106. Suppose you dilute 25.0 mL of a 0.110 M solution of Na_2CO_3 to exactly 100.0 mL. You then take exactly 10.0 mL of this diluted solution and add it to a 250-mL volumetric flask. After filling the volumetric flask to the mark with distilled water (indicating the volume of the new solution is exactly 250 mL), what is the concentration of the diluted Na_2CO_3 solution?

107. On General ChemistryNow CD-ROM or website Screen 4.12, Chemical Puzzler, you can explore the reaction of baking soda ($NaHCO_3$) with the acetic acid in vinegar. Suppose you place exactly 200 mL of vinegar in the beaker and add baking soda. The reaction occurring is

$$CH_3CO_2H(aq) + NaHCO_3(aq) \longrightarrow$$
$$NaCH_3CO_2(aq) + CO_2(g) + H_2O(\ell)$$

How many spoonfuls of baking soda is required to consume the acetic acid in the 200-mL sample? (Assume there is 50.0 g of acetic acid per liter of vinegar and a spoonful of baking soda has a mass of 3.8 g.) Are three spoonfuls sufficient? Are four spoonfuls enough?

108. The following reaction can be used to prepare iodine in the laboratory. (See photos.)

$$2\,NaI(s) + 2\,H_2SO_4(aq) + MnO_2(s) \longrightarrow$$
$$Na_2SO_4(aq) + MnSO_4(aq) + I_2(g) + 2\,H_2O(\ell)$$

    (a) Determine the oxidation number of each atom in the equation.

    (b) What is the oxidizing agent and what has been oxidized? What is the reducing agent and what has been reduced?

    (c) What quantity of iodine can be obtained if 20.0 g of NaI is mixed with 10.0 g of MnO_2 (and a stoichiometric excess of sulfuric acid)?

Charles D. Winters

Preparation of iodine. A mixture of sodium iodide and manganese(IV) oxide was placed in a flask in a hood (*left*). On adding concentrated sulfuric acid (*right*), brown gaseous I_2 was involved.

109. ▲ You place 2.56 g of $CaCO_3$ in a beaker containing 250. mL of 0.125 M HCl (Figure 5.5). When the reaction has ceased, does any calcium carbonate remain? What mass of $CaCl_2$ can be produced?

$$CaCO_3(s) + 2\,HCl(aq) \longrightarrow CaCl_2(aq) + CO_2(g) + H_2O(\ell)$$

110. ▲ A compound has been isolated that can have either of two possible formulas: (a) $K[Fe(C_2O_4)_2(H_2O)_2]$ or (b) $K_3[Fe(C_2O_4)_3]$. To find which is correct, you dissolve a weighed sample of the compound in acid and then titrate the oxalate ion $(C_2O_4^{2-})$ that comes from the compound with potassium permanganate, $KMnO_4$ (the source of the MnO_4^- ion). The balanced, net ionic equation for the titration is

$$5\ C_2O_4^{2-}(aq) + 2\ MnO_4^-(aq) + 16\ H^+(aq) \longrightarrow$$
$$2\ Mn^{2+}(aq) + 10\ CO_2(g) + 8\ H_2O(\ell)$$

Titration of 1.356 g of the compound requires 34.50 mL of 0.108 M $KMnO_4$. Which is the correct formula of the iron-containing compound: (a) or (b)?

111. ▲ Chromium(III) ion forms many compounds with ammonia. To find the formula of one of these compounds, you titrate the NH_3 in the compound with standardized acid.

$$Cr(NH_3)_xCl_3(aq) + x\ HCl(aq) \longrightarrow$$
$$x\ NH_4^+(aq) + Cr^{3+}(aq) + (x+3)\ Cl^-(aq)$$

Assume that 24.26 mL of 1.500 M HCl is used to titrate 1.580 g of $Co(NH_3)_xCl_3$. What is the value of x?

112. ▲ The cancer chemotherapy drug cisplatin, $Pt(NH_3)_2Cl_2$, can be made by reacting $(NH_4)_2PtCl_4$ with ammonia in aqueous solution. Besides cisplatin, the other product is NH_4Cl.

(a) Write a balanced equation for this reaction.

(b) To obtain 12.50 g of cisplatin, what mass of $(NH_4)_2PtCl_4$ is required? What volume of 0.125 M NH_3 is required?

(c) Cisplatin can react with the organic compound pyridine, C_5H_5N, to form a new compound.

$$Pt(NH_3)_2Cl_2(aq) + x\ C_5H_5N(aq) \longrightarrow Pt(NH_3)_2Cl_2(C_5H_5N)_x(s)$$

Suppose you treat 0.150 g of cisplatin with what you believe is an excess of liquid pyridine (1.50 mL; $d = 0.979$ g/mL). When the reaction is complete, you can find out how much pyridine was not used by titrating the solution with standardized HCl. If 37.0 mL of 0.475 M HCl is required to titrate the excess pyridine,

$$C_5H_5N(aq) + HCl(aq) \longrightarrow C_5H_5NH^+(aq) + Cl^-(aq)$$

what is the formula of the unknown compound $Pt(NH_3)_2Cl_2(C_5H_5N)_x$?

113. You need to know the volume of water in a small swimming pool, but, owing to the pool's irregular shape, it is not a simple matter to determine its dimensions and calculate the volume. To solve the problem you stir in a solution of a dye (1.0 g of methylene blue, $C_{16}H_{18}ClN_3S$, in 50.0 mL of water). After the dye has mixed with the water in the pool, you take a sample of the water. Using an instrument such as a spectrophotometer, you determine that the concentration of the dye in the pool is 4.1×10^{-8} M. What is the volume of water in the pool?

114. ▲ In some laboratory analyses the preferred technique is to dissolve a sample in an excess of acid or base and then "back-titrate" the excess with a standard base or acid. This technique is used to assess the purity of a sample of $(NH_4)_2SO_4$. Suppose you dissolve a 0.475-g sample of impure $(NH_4)_2SO_4$ in aqueous KOH.

$$(NH_4)_2SO_4(aq) + KOH(aq) \longrightarrow$$
$$NH_3(aq) + K_2SO_4(aq) + 2\ H_2O(\ell)$$

The NH_3 liberated in the reaction is distilled from the solution into a flask containing 50.0 mL of 0.100 M HCl. The ammonia reacts with the acid to produce NH_4Cl, but not all of the HCl is used in this reaction. The amount of excess acid is determined by titrating the solution with standardized NaOH. This titration consumes 11.1 mL of 0.121 M NaOH. What is the weight percent of $(NH_4)_2SO_4$ in the 0.475-g sample?

115. You wish to determine the weight percent of copper in a copper-containing alloy. After dissolving a 0.251-g sample of the alloy in acid, an excess of KI is added, and the Cu^{2+} and I^- ions undergo the reaction

$$2\ Cu^{2+}(aq) + 5\ I^-(aq) \longrightarrow 2\ CuI(s) + I_3^-(aq)$$

The liberated I_3^- is titrated with sodium thiosulfate according to the equation

$$I_3^-(aq) + 2\ S_2O_3^{2-}(aq) \longrightarrow S_4O_6^{2-}(aq) + 3\ I^-(aq)$$

(a) Designate the oxidizing and reducing agents in the two reactions above.

(b) If 26.32 mL of 0.101 M $Na_2S_2O_3$ is required for titration to the equivalence point, what is the weight percent of Cu in 0.251-g sample of the alloy?

116. ▲ Calcium and magnesium carbonates occur together in the mineral dolomite. Suppose you heat a sample of the mineral to obtain the oxides, CaO and MgO, and then treat the oxide sample with hydrochloric acid. If 7.695 g of the oxide sample requires 125 mL of 2.55 M HCl,

$$CaO(s) + 2\ HCl(aq) \longrightarrow CaCl_2(aq) + H_2O(\ell)$$
$$MgO(s) + 2\ HCl(aq) \longrightarrow MgCl_2(aq) + H_2O(\ell)$$

What is the weight percent of each oxide (CaO and MgO) in the sample?

117. Gold can be dissolved from gold-bearing rock by treating the rock with sodium cyanide in the presence of oxygen.

$$4\ Au(s) + 8\ NaCN(aq) + O_2(g) + 2\ H_2O(\ell) \longrightarrow$$
$$4\ NaAu(CN)_2(aq) + 4\ NaOH(aq)$$

(a) Name the oxidizing and reducing agents in this reaction. What has been oxidized and what has been reduced?

(b) If you have exactly one metric ton (1 metric ton = 1000 kg) of gold-bearing rock, what volume of 0.075 M NaCN, in liters, do you need to extract the gold if the rock is 0.019% gold?

118. ▲ You mix 25.0 mL of 0.234 M FeCl₃ with 42.5 mL of 0.453 M NaOH.

    (a) What mass of $Fe(OH)_3$ (in grams) will precipitate from this reaction mixture?

    (b) On of the reactants (FeCl₃ or NaOH) is present in a stoichiometric excess. What is the molar concentration of the excess reactant remaining in solution after $Fe(OH)_3$ has been precipitated?

Summary and Conceptual Questions

The following questions use concepts from the preceding chapters.

119. ▲ Two students titrate different samples of the same solution of HCl using 0.100 M NaOH solution and phenolphthalein indicator (see Figure 5.23). The first student pipets 20.0 mL of the HCl solution into a flask, adds 20 mL of distilled water and a few drops of phenolphthalein solution, and titrates until a lasting pink color appears. The second student pipets 20.0 mL of the HCl solution into a flask, adds 60 mL of distilled water and a few drops of phenolphthalein solution, and titrates to the first lasting pink color. Each student correctly calculates the molarity of a HCl solution. What will the second student's result be?

    (a) four times less than the first student's result

    (b) four times greater than the first student's result

    (c) two times less than the first student's result

    (d) two times greater than the first student's result

    (e) the same as the first student's result

120. On General ChemistryNow CD-ROM or website Screen 5.18, Exercise, Stoichiometry of Reactions in Solution, the video shows the reaction of Fe^{2+} with MnO_4^- in aqueous solution.

    (a) What is the balanced equation for the reaction that occurred?

    (b) What is the oxidizing agent and what is the reducing agent?

    (c) Equal volumes of Fe^{2+}-containing solution and MnO_4^--containing solution were mixed. The amount of Fe^{2+} was just sufficient to consume all of the MnO_4^-. Which ion (Fe^{2+} or MnO_4^-) was initially present in larger concentration?

121. ▲ General ChemistryNow CD-ROM or website Screen 4.8 Limiting Reactants, explores the reaction of zinc and hydrochloric acid.

$$Zn(s) + 2\,HCl(aq) \longrightarrow ZnCl_2(aq) + H_2(g)$$

Different quantities of zinc are added to three flasks, each containing exactly 100 mL of 0.10 M HCl.

Flask 1: 7.00 g Zn

Flask 2: 3.27 g Zn

Flask 3: 1.31 g Zn

The same amount of H₂ gas was generated in Flasks 1 and 2, but a smaller amount was generated in Flask 3. The zinc was completely consumed in Flasks 2 and 3, but some remained in Flask 1. Explain these observations.

122. ▲ You want to prepare barium chloride, $BaCl_2$, using an exchange reaction of some type. To do so, you have the following reagents from which to select the reactants: $BaSO_4$, $BaBr_2$, $BaCO_3$, $Ba(OH)_2$, HCl, $HgSO_4$, $AgNO_3$, and HNO_3. Write a complete, balanced equation for the reaction chosen. *(Note: There are several possibilities.)*

123. Describe how to prepare $BaSO_4$, barium sulfate, by (a) a precipitation reaction and (b) a gas-forming reaction. To do so, you have the following reagents from which to select the reactants: $BaCl_2$, $BaCO_3$, $Ba(OH)_2$, H_2SO_4, and Na_2SO_4. Write complete, balanced equations for the reactions chosen. (See Figure 4.8 for an illustration of the preparation of a compound.)

124. Describe how to prepare zinc chloride by (a) an acid–base reaction, (b) a gas-forming reaction, and (c) an oxidation–reduction reaction. The available starting materials are $ZnCO_3$, HCl, Cl_2, HNO_3, $Zn(OH)_2$, NaCl, $Zn(NO_3)_2$, and Zn. Write complete, balanced equations for the reactions chosen.

125. In some states a person will receive a "driving while intoxicated" (DWI) ticket if the blood alcohol level (BAL) is 100 mg per deciliter (dL) of blood or higher. Suppose a person is found to have a BAL of 0.033 mol of ethanol (C_2H_5OH) per liter of blood. Will the person receive a DWI ticket?

Do you need a live tutor for homework problems?
Access vMentor at General ChemistryNow at
http://now.brookscole.com/kotz6e
for one-on-one tutoring from a chemistry expert

6—Principles of Reactivity: Energy and Chemical Reactions

The Rolex Awards for Enterprise/Tomas Bertelsen/Scientific American, Nov. 2000, p. 26

Mohammed Bah Abba. Abba's family were potmakers. As a boy Abba was fascinated by earthenware objects and their ability to absorb water yet remain structurally intact. Abba earned a college degree in business and, while still in his twenties, became an instructor in a college of business in Jigwa, Nigeria, and a consultant to the United Nations Development Program. That brought him back into close contact with rural communities in northern Nigeria and made him aware of the hardships of the families there. Making and distributing the pot-in-pot device for safe food storage closed this interesting circle.

Abba's Refrigerator

If you put a pot of water on a kitchen stove or a campfire, or if you put the pot in the sun, the water will evaporate. You must supply energy in some form because evaporation requires the input of energy. This well-known principle was applied in a novel way by a young African teacher, Mohammed Bah Abba, to improve the life of his people in Nigeria.

Life is hard in northern Nigerian communities. In this rural semi-desert area, most people eke out a living through subsistence

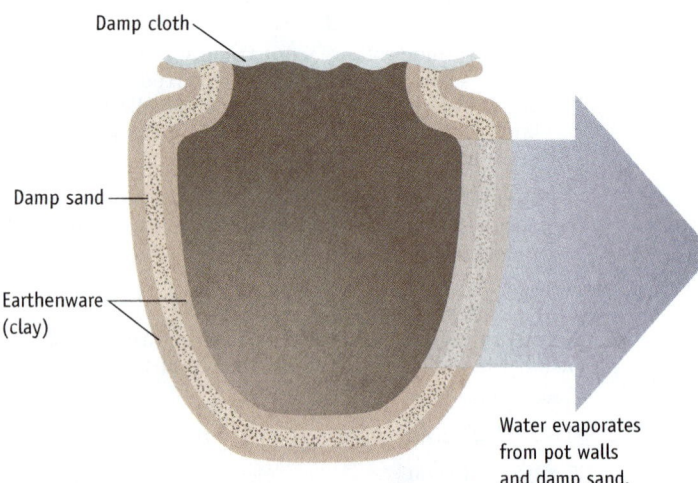

Damp cloth

Damp sand

Earthenware (clay)

Water evaporates from pot walls and damp sand.

The pot-in-pot refrigerator. Water seeps through the outer pot from the damp sand layer separating the pots, or from food stored in the inner pot. As the water evaporates from the surface of the outer pot, the food is cooled.

See Chapter Goals Revisited (page 270). Test your knowledge of these goals by taking the exam-prep quiz on the General ChemistryNow CD-ROM or website.

- Assess heat transfer associated with changes in temperature and changes of state.

- Apply the first law of thermodynamics.

- Define and understand the state functions enthalpy and internal energy.

- Calculate the energy changes occurring in chemical reactions and learn how these changes are measured.

James Cowlin/Image Enterprises, Phoenix, AZ.

Swamp coolers. These inexpensive air-conditioners work on the same principle as Abba's pot. A trickle of water washes over a bed of straw or other porous material. As air is drawn over the moist material, the air is cooled as the water takes energy from the air to evaporate.

farming. Because of the dearth of modern refrigeration, food spoilage is a major problem. Using a simple thermodynamic principle, Abba developed a refrigerator that cost about 30 cents to make and does not use electricity.

Abba's refrigerator consists of two earthen pots, one inside the other, separated by a layer of sand. The pots are covered with a damp cloth and placed in a well ventilated area. Water seeps

through the pot's outer wall and rapidly evaporates in the dry desert air. The water remaining in the pot and its contents drop in temperature. Food in the inner pot can stay cool for days and not spoil.

In the 1990s, at his own expense, Abba made and distributed almost 10,000 pots in the villages of northern Nigeria. He estimates that about 75% of the families in this area are now using his refrigerator. The impact of this simple device has implications not only for the health of his people but also for their economy and their social structure. Prior to the development of the pot-in-pot device for food storage, it was necessary to sell produce immediately upon harvesting it. The young girls in the family who sold food on the street daily could now be released from this chore to attend school and improve their lives.

Every two years, the Rolex Company, the Swiss maker of timepieces, gives a series of awards for enterprise. For his pot-in-pot refrigerator, Abba was one of the five recipients of a Rolex Award in 2000.

Charles D. Winters

Evaporative cooling. The same principle that cools Abba's refrigerator cools you down if you wear a strip of damp cloth, a "neck cooler," around your neck on a hot day.

To Review Before You Begin

- Know how to write balanced chemical equations (Chapter 4)
- Review product-favored and reactant-favored reactions (page 197)
- Know how to use Kelvin and Celsius temperature scales (Section 1.6)
- Review states of matter and changes of state (Section 1.5)

Energy transfer accompanies both chemical and physical changes. Our bodies are cooled when we perspire—the evaporation of water in sweat, a physical change, draws energy from our body and causes us to feel cooler. When water vapor condenses, heat is given off, a process that has a significant impact on the weather (Figure 6.1). The sun's energy can be stored as chemical energy by the formation of carbohydrates and oxygen from carbon dioxide and water in the process of photosynthesis, a chemical change.

$$6\ CO_2(g) + 6\ H_2O(g) + energy \longrightarrow C_6H_{12}O_6(s) + 6\ O_2(g)$$

This chemical energy can be released in a chemical reaction of carbohydrate and oxygen, whether in the laboratory (Figure 6.2), in living tissue, or in a forest fire.

$$C_6H_{12}O_6(s) + 6\ O_2(g) \longrightarrow 6\ CO_2(g) + 6\ H_2O(g) + energy$$

When studying chemistry, it is important to know something about energy. The most common form of energy we see in chemical processes is heat. Changes of heat

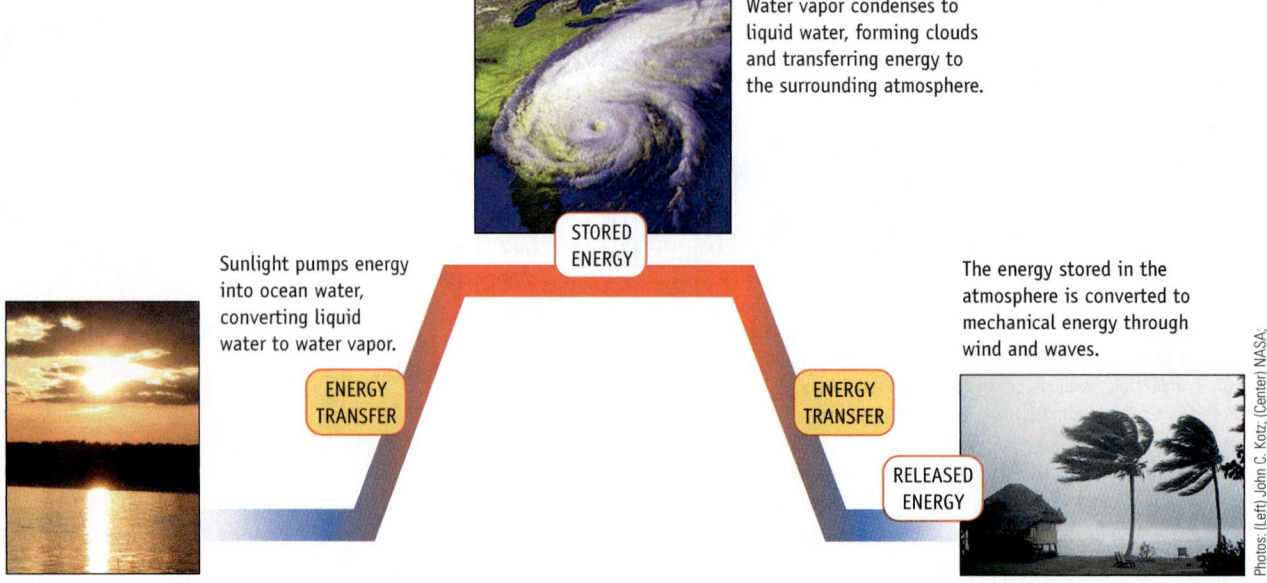

Figure 6.1   Energy transfer in nature. Hurricanes and other forms of violent weather involve the storage and release of energy. The average hurricane releases energy equivalent to the annual U.S. production of electricity.

Water vapor condenses to liquid water, forming clouds and transferring energy to the surrounding atmosphere.

STORED ENERGY

Sunlight pumps energy into ocean water, converting liquid water to water vapor.

ENERGY TRANSFER

The energy stored in the atmosphere is converted to mechanical energy through wind and waves.

ENERGY TRANSFER

RELEASED ENERGY

Photos: (Left) John C. Kotz; (Center) NASA; (Right) Frederick Ayer/Photo Researchers, Inc.

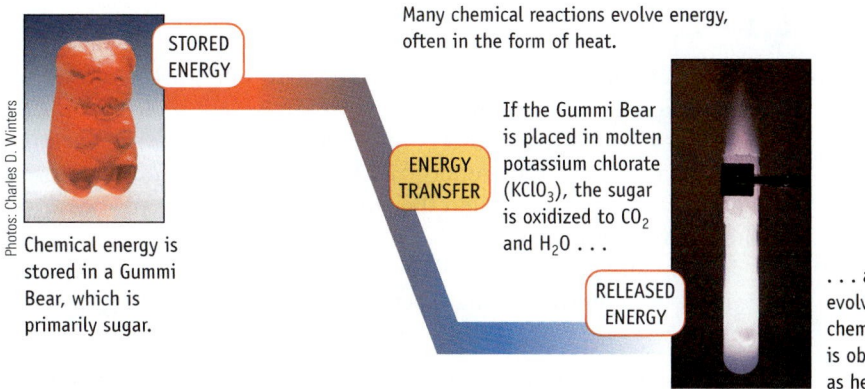

Photos: Charles D. Winters

STORED ENERGY

Chemical energy is stored in a Gummi Bear, which is primarily sugar.

Many chemical reactions evolve energy, often in the form of heat.

ENERGY TRANSFER

If the Gummi Bear is placed in molten potassium chlorate ($KClO_3$), the sugar is oxidized to CO_2 and H_2O . . .

RELEASED ENERGY

. . . and the energy evolved in the chemical reaction is observed as heat and light.

Figure 6.2 **Energy transfer in a chemical reaction.** (*See General ChemistryNow Screen 6.17 Product-Favored Systems, to watch a video of this reaction.*)

content and the transfer of heat between objects are major themes of **thermodynamics**, the science of heat and work—and the subject of this chapter and a later one (Chapter 19). As described in the "The Chemistry of Fuels and Energy Sources" (pages 282–293), the principles of thermodynamics apply to energy use in your home, to ways of conserving energy, to recycling of materials, and to problems of current and future energy availability and use in our economy.

a, Bruce Roberts/Photo Researchers, Inc.; b, Royalty-Free/Corbis; c, William James Warren/Corbis

(a) Gravitational energy

(b) Chemical potential energy

(c) Electrostatic energy

Active Figure 6.3 **Energy and its conversion** (a) Water at the top of a water wheel represents stored, or potential, energy. As water flows over the wheel, its potential energy is converted to mechanical energy. (b) Chemical potential energy is converted to heat and then to work. (c) Lightning converts electrostatic energy into radiant and thermal energy.

GENERAL
Chemistry ·♦· Now™ See the General ChemistryNow CD-ROM or website to explore an interactive version of this figure accompanied by an exercise.

6.1—Energy: Some Basic Principles

Energy is defined as the capacity to do work. You do work against the force of gravity when carrying yourself and hiking equipment up a mountain. You can do this work because you have the energy to do so, the energy having been provided by the food you have eaten. Food energy is chemical energy—energy stored in chemical compounds and released when the compounds undergo the chemical reactions of metabolism in your body.

Energy can be classified as kinetic or potential. Kinetic energy, as noted in the discussion of kinetic-molecular theory (Section 1.5), is energy associated with motion, such as

- *Thermal energy* of atoms, molecules, or ions in motion at the submicroscopic level. All matter has thermal energy.
- *Mechanical energy* of a macroscopic object like a moving tennis ball or automobile.
- *Electrical energy* of electrons moving through a conductor.
- *Sound,* which corresponds to compression and expansion of the spaces between molecules.

Potential energy, energy that results from an object's position (Figure 6.3), includes:

- *Gravitational energy*, such as that possessed by a ball held above the floor and by water at the top of a waterfall (Figure 6.3a).
- *Chemical potential energy.* The energy stored in coal, for example, is converted to heat when burned, and the heat is converted to work (Figure 6.3b). All chemical reactions involve a change in chemical potential energy.
- *Electrostatic energy*, potential energy associated with the separation of two dissimilar electrical charges. The energy is released (as light, heat, and sound) when the opposite charges are neutralized, as happens when a bolt of lightning darts between clouds and the ground (Figure 6.3c).

Potential energy is stored energy and can be converted into kinetic energy. For example, as water falls over a waterfall, its potential energy is converted into kinetic energy. Similarly, kinetic energy can be converted into potential energy: The kinetic energy of falling water can turn a turbine to produce electricity, which can then be used to convert water into H_2 and O_2 (Figure 1.6, page 18). The H_2 gas represents stored chemical potential energy because it can be burned to produce heat and light (Figure 1.13, page 25) or used in a fuel cell (as in the Space Shuttle) to produce electrical energy.

Conservation of Energy

Standing on a diving board, you have considerable potential energy because of your position above the water. Once you jump off the board, some of that potential energy is converted into kinetic energy (Figure 6.4). During the dive, the force of gravity accelerates your body so that it moves faster and faster. Your kinetic energy increases and your potential energy decreases. At the moment you hit the water, your velocity is abruptly reduced, and much of your kinetic energy is converted to mechanical energy; the water splashes as your body moves it aside by doing work on it. Eventually you float on the surface, and the water becomes still again. If you could see them, however, you would find that the water molecules are moving a lit-

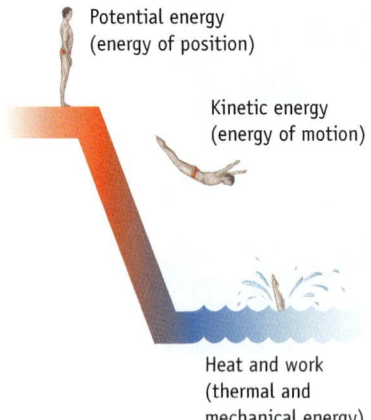

Potential energy
(energy of position)

Kinetic energy
(energy of motion)

Heat and work
(thermal and
mechanical energy)

Figure 6.4 The law of energy conservation. The diver's potential energy is converted to kinetic energy and then to thermal energy, illustrating the law of conservation of energy (*See General ChemistryNow Screen 6.2 Energy to view an animation based on this figure.*)

tle faster in the vicinity of your dive; that is, the kinetic energy of the water is slightly higher.

This series of energy conversions illustrates the **law of conservation of energy**, otherwise known as the **first law of thermodynamics**. These terms are synonymous; both state that energy can neither be created nor destroyed. Or, to state this law differently, *the total energy of the universe is constant.* These statements summarize the results of a great many experiments in which heat, work, and other forms of energy transfer have been measured and the total energy content found to be the same before and after an event.

Another example of the law of energy conservation is burning oil or coal to heat your house or to drive a locomotive (Figure 6.3b). These fuels are an energy resource. When burned, the chemical energy present in the oil or gas is converted to an equal quantity of energy, now in the form of heat for your home and the thermal energy of the gases going up the chimney.

GENERAL

See the General ChemistryNow CD-ROM or website:
- **Screen 6.3 Forms of Energy,** for an exercise that examines the energy conversions in several situations

---

Exercise 6.1—Energy

A battery stores chemical potential energy. Into what types of energy can this potential energy be converted?

Temperature and Heat

The temperature of an object is a measure of its heat energy content and of its ability to transfer heat. One way to measure temperature is with a thermometer containing mercury or some other liquid (Figure 6.5). When the thermometer is placed in hot water, heat is transferred from the water to the thermometer. The increased energy causes the mercury atoms, for example, to move about more rapidly and the space between atoms to increase slightly. You observe this effect as an expansion in the volume of the mercury, such that the column of mercury rises higher in the thermometer tube.

Three important aspects of thermal energy and temperature should be understood:

- Heat is not the same as temperature.
- The more thermal energy a substance has, the greater the motion of its atoms and molecules.
- The total thermal energy in an object is the sum of the individual energies of all the atoms, molecules, or ions in that object.

The thermal energy of a given substance depends not only on temperature but also on the amount of substance. Thus, a cup of hot coffee may contain less thermal energy than a bathtub full of warm water, even though the coffee is at a higher temperature.

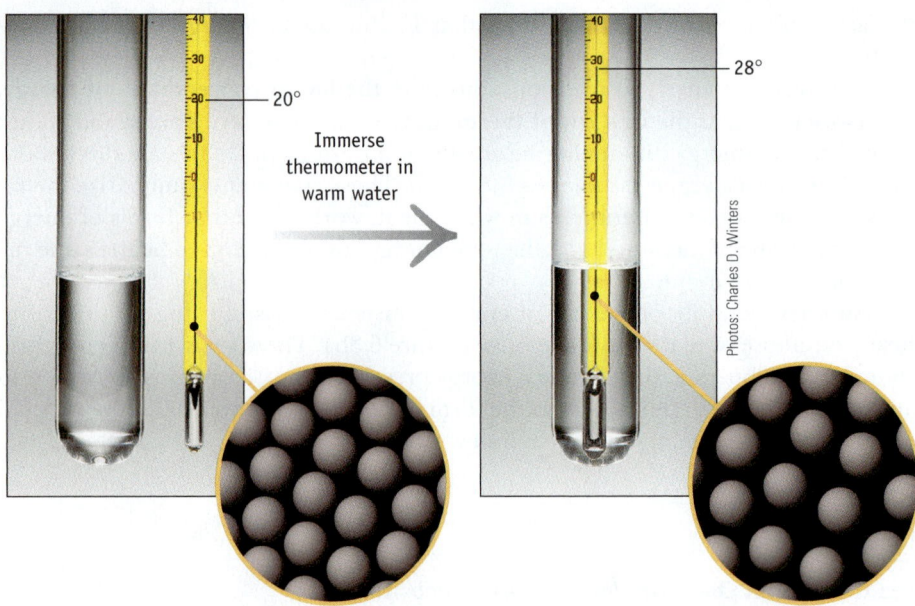

Photos: Charles D. Winters

Figure 6.5 Measuring temperature. The volume of liquid mercury in a thermometer increases slightly when immersed in warm water. The volume increase causes the mercury to rise in the thermometer, which is calibrated to indicate the temperature.

Systems and Surroundings

In thermodynamics, the terms "system" and "surroundings" have precise and important scientific meanings. A **system** is defined as the object, or collection of objects, being studied (Figure 6.6). The **surroundings** include everything outside the system that can exchange energy with the system. In the discussions that follow, we will need to identify systems precisely. If we are studying the heat evolved in a chemical reaction, for example, the system might be defined as a reaction vessel and its contents. The surroundings would be the air in the room and anything else in contact with the vessel. At the atomic level, the system could be a single atom or molecule and the surroundings would be the atoms or molecules in its vicinity. In general, we choose how we define the system and its surroundings for each situation, depending on the information we are trying to obtain.

This concept of a system and its surroundings applies to nonchemical situations as well. If we want to study the energy balance on this planet, we might choose to define the earth as the system and outer space as the surroundings. On a cosmic level, the solar system might be defined as the system being studied, and the rest of the galaxy would be the surroundings.

Directionality of Heat Transfer: Thermal Equilibrium

Heat transfer occurs when two objects at different temperatures are brought into contact. In Figure 6.7, for example, the beaker of water and the piece of metal being heated in a Bunsen burner flame have different temperatures. When the hot metal is plunged into the cold water, heat is transferred from the metal to the water. Eventually, the two objects reach the same temperature. At that point, the system has reached **thermal equilibrium**. The distinguishing feature of thermal equilibrium is that, on the macroscopic scale, no further temperature change occurs and the temperature throughout the entire system (metal plus water) is the same.

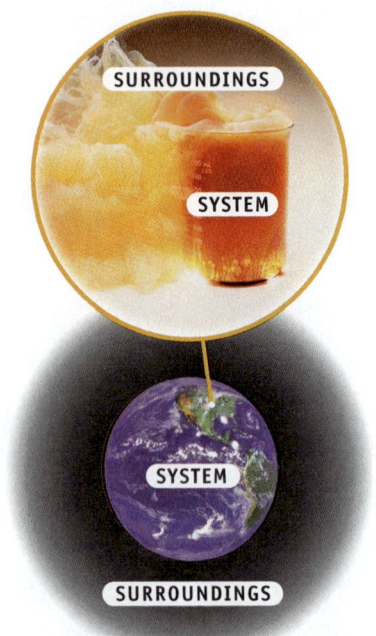

Photos: (Top) Charles D. Winters; (Bottom) NASA

Figure 6.6 Systems and their surroundings. Earth can be considered a thermodynamic system, with the rest of the universe as its surroundings. A chemical reaction occurring in a laboratory is also a system, with the laboratory as its surroundings.

Figure 6.7 **Energy transfer.** Heat is transferred from the hotter metal bar to the cooler water. Eventually the water and metal reach the same temperature and are said to be in thermal equilibrium. *(See General ChemistryNow Screen 6.9 Heat Transfer Between Substances, for a simulation and tutorial.)*

The manipulation of the hot metal bar and the beaker of water may seem like a rather simple experiment. Embedded in the experiment, however, are some principles that will be very important in our further discussion:

- Heat transfer always occurs from an object at a higher temperature to an object at a lower temperature. The directionality of heat transfer is an important principle of thermodynamics. (See "A Closer Look: Why Doesn't the Heat in a Room Cause Your Cup of Coffee to Boil?")

- Transfer of heat continues until both objects are at the same temperature (thermal equilibrium).

For the specific case where heat transfer occurs within a system, we can also say that the quantity of heat lost by a hotter object and the quantity of heat gained by a cooler object when they are in contact are numerically equal. (This is required by the law of conservation of energy.)

When heat transfer occurs across the boundary between system and surroundings, we can describe the directionality of heat transfer as exothermic or endothermic (Figure 6.8).

- In an **exothermic** process, heat is transferred from a system to the surroundings.

- An **endothermic** process is the opposite of an exothermic process: Heat is transferred from surroundings to the system.

■ **Thermal Equilibrium**
Although no change is evident at the macroscopic level when thermal equilibrium is reached, on the molecular level transfer of energy between individual molecules will continue to occur. This feature—no change visible on a macroscopic level, but processes still occurring at the particulate level—is a general feature of equilibria that we will encounter again (Chapters 16–18).

A Closer Look

Why Doesn't the Heat in a Room Cause Your Cup of Coffee to Boil?

If a cup of coffee or tea is hotter than its surroundings, heat is transferred to the surroundings until the hot coffee cools off and the surroundings warm up a bit. It is interesting and useful to think about why the opposite process doesn't occur. Why doesn't the heat in a room cause a cup of cold coffee to boil? The law of energy conservation would not be violated

if the coffee got hotter and hotter and the surroundings in the room got cooler and cooler. However, we know from experience that this will never happen. The directionality in heat transfer—heat energy always transfers from a hotter object to a cooler one, never the reverse—corresponds to a spreading out of energy over the greatest possible number of atoms, ions, or molecules. Energy transfers from a relatively small number of molecules in a hot cup of coffee to a large number of atoms and molecules surrounding the cup.

Similarly, the large number of particles in the surrounding environment will heat a glass of ice water by transferring some of their energy to the glass, ice, and water molecules. As in the previous example, the end result is to spread thermal energy more evenly over the maximum number of particles. The opposite process, concentrating energy in only a few particles at the expense of many, is never observed.

The directionality of energy transfer, which plays an important role in thermodynamics, will be discussed further in Chapter 19.

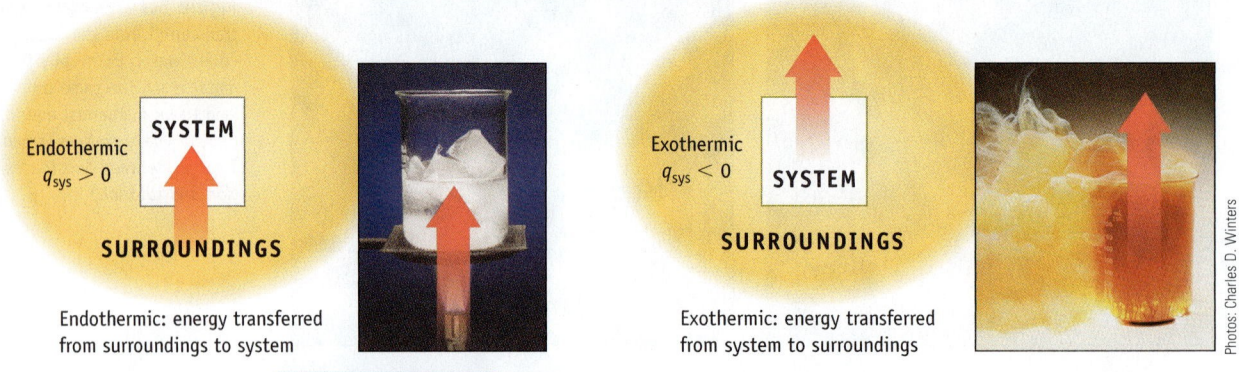

Endothermic: energy transferred
from surroundings to system

Exothermic: energy transferred
from system to surroundings

Photos: Charles D. Winters

Active Figure 6.8   **Exothermic and endothermic processes.** The symbol q represents heat transferred, and the subscript "sys" refers to the system.

GENERAL
Chemistry ⚛ Now™ See the General ChemistryNow CD-ROM or website to explore an interactive version of this figure accompanied by an exercise.

GENERAL
Chemistry ⚛ Now™

See the General ChemistryNow CD-ROM or website:
- **Screen 6.4 Directionality of Heat Transfer,** to view an animation on endothermic and exothermic systems

---

■ **Kinetic Energy**
Kinetic energy is calculated by the equation $KE = \frac{1}{2} mv^2$. One joule is the kinetic energy of a 2.0 kg mass (m) moving at 1.0 m/s (v).

$$KE = \frac{1}{2}(2.0 \text{ kg})(1.0 \text{ m/s})^2$$
$$= 1.0 \text{ kg} \cdot \text{m}^2/\text{s}^2 = 1.0 \text{ J}$$

Energy Units

When expressing energy quantities, most chemists (and much of the world outside the United States) use the **joule** (J), the SI unit of thermal energy. The joule is preferred in scientific study because it is related directly to the units used for mechanical energy: 1 J equals $1 \text{ kg} \cdot \text{m}^2/\text{s}^2$. However, the joule can be inconveniently small as a unit for use in chemistry, so the kilojoule (kJ), equivalent to 1000 joules, is often used.

To give you some feeling for joules, suppose you drop a six-pack of soft-drink cans, each full of liquid, on your foot. Although you probably will not take time to calculate the kinetic energy at the moment of impact, it is between 4 J and 10 J.

An older unit for measuring heat is the calorie (cal). It is defined as the heat required to raise the temperature of 1.00 g of pure liquid water from 14.5 °C to 15.5 °C. A kilocalorie (kcal) is equivalent to 1000 calories. The conversion factor relating joules and calories is

$$1 \text{ calorie (cal)} = 4.184 \text{ joules (J)}$$

The dietary Calorie (with a capital C) is often used in the United States to represent the energy content of foods. This unit is encountered when reading the nutritional information on a food label. The dietary Calorie (Cal) is equivalent to the kilocalorie or 1000 calories. Thus, a breakfast cereal that gives you 100.0 Calories of nutritional energy per serving provides 100.0 kcal or 418.4 kJ.

Chemical Perspectives

Food and Calories

The U.S. Food and Drug Administration (FDA) mandates that nutritional data, including energy content, be included on almost all packaged food. The Nutrition Labeling and Education Act of 1990 requires that the total energy from protein, carbohydrates, fat, and alcohol be specified. How is this determined? Initially the method used was calorimetry. In this method, which is described in Section 6.6, a food product is burned and the heat evolved in the combustion is measured. Now, however, all energy content is estimated using the Atwater system. This method specifies the following average values for energy sources in foods:

1 g protein = 4 kcal (17 kJ)

1 g carbohydrate = 4 kcal (17 kJ)

1 g fat = 9 kcal (38 kJ)

1 g alcohol = 7 kcal (29 kJ)

Because carbohydrates contain some indigestible fiber, the mass of fiber is subtracted from the mass of carbohydrate when calculating the energy from carbohydrates.

As an example, one serving of cashew nuts (about 28 g) has

14 g fat = 126 kcal

6 g protein = 24 kcal

7 g carbohydrates − 1 g fiber = 24 kcal

Total = 174 kcal (728 kJ)

A value of 170 kcal is reported on the package.

You can find data on more than 6000 foods at the Nutrient Data Laboratory Website (**www.nal.usda.gov/fnic/foodcomp/**). See also **nat.crgq.com** for an online tool that allows you to find the energy content of foods.

Nutrition Facts
Serving Size 1 cup (30g)
Children Under 4 - ¾ cup (20g)
Servings Per Container About 19
Children Under 4 - About 28

Amount Per Serving	Cheerios	with ½ cup skim milk	Cereal for Children Under 4
Calories	110	150	70
Calories from Fat	15	20	10
		% Daily Value**	
Total Fat 2g*	3%	3%	1g
Saturated Fat 0g	0%	3%	0g
Polyunsaturated Fat 0.5g			0g
Monounsaturated Fat 0.5g			0g
Cholesterol 0mg	0%	1%	0mg
Sodium 210mg	9%	12%	140mg
Potassium 200mg	6%	12%	130mg

Energy and food labels. All packaged foods must have labels specifying nutritional values, with energy given in Calories (where 1 Cal = 1 kilocalorie).

Charles D. Winters

GENERAL
Chemistry Now™

See the General ChemistryNow CD-ROM or website:
- **Screen 6.5 Energy Units,** for a tutorial on converting energy units

---

Exercise 6.2—Energy Units

(a) In an old textbook you read that the oxidation of 1.00 g of hydrogen to form liquid water produces 3800 calories. What is this energy in units of joules?

(b) The label on a cereal box indicates that one serving (with skim milk) provides 250 Cal. What is this energy in kilojoules (kJ)?

6.2—Specific Heat Capacity and Heat Transfer

The quantity of heat transferred to or from an object depends on three things:

- The quantity of material
- The size of the temperature change
- The identity of the material gaining or losing heat

The **specific heat capacity (C)** is related to these three parameters. The specific heat capacity is the quantity of heat required to raise the temperature of 1 gram of a substance by one kelvin. It has units of joules per gram per kelvin ($J/g \cdot K$).

Energy content of foods. In many countries that use standardized SI units, food energy is also measured in joules. The diet soda in this can (from Australia) is said to have an energy content of only 1 joule.

The quantity of heat gained or lost when a given mass of a substance is warmed or cooled is calculated using Equation 6.1.

Specific heat capacity (J/g · K)   Change in temperature (K)

$$q = C \times m \times \Delta T \tag{6.1}$$

Heat transferred (J)   Mass of substance (g)

Here, q is the quantity of heat transferred to or from a given mass of substance (m), C is the specific heat capacity, and ΔT is the change in temperature. The capital Greek letter delta, Δ, means "change in." The change in temperature, ΔT, is calculated as the final temperature minus the initial temperature.

$$\Delta T = T_{final} - T_{initial} \tag{6.2}$$

Calculating a change in temperature as in Equation 6.2 will give a result with an algebraic sign that indicates the direction of heat transfer ("A Closer Look, Sign Conventions"). For example, we can use the specific heat capacity of copper, $0.385\,J/g \cdot K$, to calculate the change in heat content of a 10.0-g sample of copper if its temperature is raised from 298 K (25 °C) to 598 K (325 °C).

$$q = \left(0.385\,\frac{J}{g \cdot K}\right)(10.0\ g)(598\ K - 298\ K) = +1160\ J$$

T_{final} $T_{initial}$
Final temp.   Initial temp.

Notice that the answer has a positive sign. This indicates that the heat content of the sample of copper has *increased* by 1160 J because heat has transferred *to* the copper (the system) *from* the surroundings.

Specific heat capacities of some metals, compounds, and common substances are listed in Table 6.1. Notice that water has one of the highest values, $4.184\,J/g \cdot K$. In contrast, the specific heat capacities of most common materials are considerably smaller. For example, the specific heat capacity of iron is $0.449\,J/g \cdot K$; to raise the temperature of a gram of water by 1 K requires about nine times as much heat as is required to cause a 1 K change in temperature for a gram of iron.

The high specific heat capacity of water has major significance. A great deal of energy must be absorbed by a large body of water to raise its temperature by just a degree or so. Thus, large bodies of water have a profound influence on our weather. In spring, lakes tend to warm up more slowly than the air. In autumn, the heat given off by a large lake moderates the drop in air temperature.

The greater the specific heat and the larger the mass, the more thermal energy a substance can store. This relationship has numerous implications. For example, you might wrap some bread in aluminum foil and heat it in an oven. You can remove the foil with your fingers after taking the bread from the oven, even though the bread is very hot. A small quantity of aluminum foil is used and the metal has a low specific heat capacity; thus, when you touch the hot foil, only a small quantity of heat will be transferred to your fingers (which have a larger mass and a higher specific heat capacity). This is also the reason why a chain of fast-food restaurants warns you that the filling of an apple pie can be much warmer than the paper wrapper or the pie crust (Figure 6.9). Although the wrapper, pie crust, and filling are at the same temperature, the heat content of the filling is greater than that of the wrapper and crust.

■ Change In Temperature, ΔT

Sign of ΔT	Meaning
Positive	$T_{final} > T_{initial}$, so T has increased, and q will be positive. Heat has been transferred to the object under study.
Negative	$T_{final} < T_{initial}$, so T has decreased, and q will be negative. Heat has been transferred out of the object under study.

■ Molar Heat Capacity

Heat capacities can be expressed on a per mole basis. The molar heat capacity is the amount of heat required to raise the temperature of one mole of a substance by one degree kelvin. The molar heat capacity of metals at room temperature is always near 25 kJ/mol · K.

Figure 6.9   A practical example of specific heat capacity. The filling of the apple pie has a higher specific heat capacity (and larger mass) than the pie crust and wrapper. Notice the warning on the wrapper.

A Closer Look

Sign Conventions

Whenever you take the difference between two quantities in chemistry, you should always subtract the initial quantity from the final quantity. A consequence of this convention is that the algebraic sign of the result indicates an increase (+) or a decrease (−) in the quantity being studied. This is an important point, as you will see not only in this chapter but also in other chapters of this book.

Thus far, we have described temperature changes and the direction of heat transfer. The table below summarizes the conventions used.

When discussing the quantity of heat, we use an *unsigned number*. If we want to indicate the *direction of transfer* in a process, however, we attach a sign, either negative (heat transferred from the substance) or positive (heat transferred to the substance), to q. The sign of q "signals" the direction of heat transfer. Heat, a quantity of energy, cannot be negative but the heat content of an object can increase or decrease, depending on the direction of heat transfer.

An analogy might make this point clearer. Consider your bank account. Assume you have $260 in your account ($A_{initial}$) and after a withdrawal you have $200 ($A_{final}$). The cash flow is thus

$$\text{Cash flow} = A_{final} - A_{initial}$$
$$= \$200 - \$260$$
$$= -\$60$$

The negative sign on the $60 indicates that a withdrawal has been made; the cash itself is not a negative quantity.

ΔT of System	Sign of ΔT	Sign of q	Direction of Heat Transfer
Increase	+	+	Heat transferred from surroundings to system (an endothermic process)
Decrease	−	−	Heat transferred from system to surroundings (an exothermic process)

GENERAL
Chemistry Now™

See the General ChemistryNow CD-ROM or website:

- **Screen 6.7 Heat Capacity of Pure Substances,** for a simulation and exercise on the change in thermal energy when various material are heated

Table 6.1 Specific Heat Capacity Values for Some Elements, Compounds, and Common Solids

Substance	Specific Heat Capacity (J/g · K)	Molar Heat Capacity (J/mol · K)
Elements		
Al, aluminum	0.897	24.2
C, graphite	0.685	8.23
Fe, iron	0.449	25.1
Cu, copper	0.385	24.5
Au, gold	0.129	25.4
Compounds		
$NH_3(\ell)$, ammonia	4.70	80.0
$H_2O(\ell)$, water (liquid)	4.184	75.4
$C_2H_5OH(\ell)$, ethanol	2.44	11.2
$HOCH_2CH_2OH(\ell)$, ethylene glycol (antifreeze)	2.39	14.8
$H_2O(s)$, water (ice)	2.06	37.1
Common Solids		
wood	1.8	
cement	0.9	
glass	0.8	
granite	0.8	

Charles D. Winters

Specific heat capacity. Metals have different values of specific heat capacity on a per-gram basis. However, their molar heats capacities are all in the range of 25 J/mol · K. Among common substances, liquid water has the highest specific heat capacity on a per gram basis, a fact that plays a significant role in the earth's weather and climate.

Example 6.1—Specific Heat Capacity

Problem Determine the quantity of heat that must be added to raise the temperature of a cup of coffee (250 mL) from 20.5 °C (293.7 K) to 95.6 °C (368.8 K). Assume that water and coffee have the same density (1.00 g/mL) and the same specific heat capacity.

Strategy Use Equation 6.1. For this calculation, you will need the specific heat capacity for H_2O from Table 6.1 (4.184 J/g · K), the mass of the coffee (calculated from its density and volume), and the change in temperature ($T_{final} - T_{initial}$).

Solution

$$\text{Mass of coffee} = (250 \text{ mL})(1.00 \text{ g/mL}) = 250 \text{ g}$$

$$\Delta T = T_{final} - T_{initial} = 368.8 \text{ K} - 293.7 \text{ K} = 75.1 \text{ K}$$

$$q = C \times m \times \Delta T$$

$$q = (4.184 \text{ J/g} \cdot \text{K})(250 \text{ g})(75.1 \text{ K})$$

$$\Delta T = \boxed{79{,}000 \text{ J (or 79 kJ)}}$$

Comment Notice that heat has been transferred to the coffee from the surroundings. The heat content of the coffee has increased.

Exercise 6.3—Specific Heat Capacity

In an experiment it was determined that 59.8 J was required to change the temperature of 25.0 g of ethylene glycol (a compound used as antifreeze in automobile engines) by 1.00 K. Calculate the specific heat capacity of ethylene glycol from these data.

Quantitative Aspects of Heat Transfer

Like melting point, boiling point, and density, the specific heat capacity is a characteristic property of a pure substance. The specific heat capacity of a substance can be determined experimentally by accurately measuring temperature changes that occur when heat is transferred from the substance to a known quantity of water (whose specific heat capacity is known).

Suppose a 55.0-g piece of metal is heated in boiling water to 99.8 °C and then dropped into cool water in an insulated beaker (Figure 6.10). Assume the beaker contains 225 g of water and its initial temperature (before the metal was dropped in) was 21.0 °C. The final temperature of the metal and water is 23.1 °C. What is the specific heat capacity of the metal? Here are the most important aspects of this experiment:

- The metal and the water are the system, and the beaker and environment are the surroundings. We assume that heat is transferred only within the system and not between the system and the surroundings. (For an accurate calculation, we would want to include heat transfer to the surroundings.)

- The water and the metal bar end up at the same temperature. (T_{final} is the same for both.)

- The heat transferred from the metal to the water, q_{metal}, has a negative value because the temperature of the metal dropped as heat was transferred out of it to the water. Conversely, q_{water} has a positive value because its temperature increased as heat was transferred into the water from the metal.

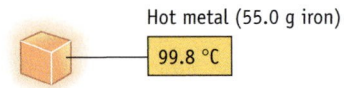

Hot metal (55.0 g iron)

99.8 °C

Cool water (225 g)

21.0 °C

Immerse hot metal in water

Metal cools in exothermic process.

ΔT of metal is negative.

q_{metal} is negative.

23.1 °C

Water is warmed in endothermic process.

ΔT of water is positive.

q_{water} is positive.

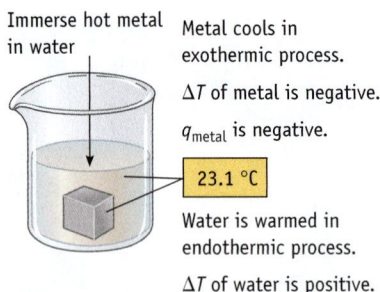

Active Figure 6.10   Heat transfer.
When heat transfers from a hot metal to cool water, the heat transferred from the metal, q_{metal}, has a negative value. The heat transferred to the water, q_{water}, is positive. (See also Figure 6.7.)

GENERAL
Chemistry Now™ See the General ChemistryNow CD-ROM or website to explore an interactive version of this figure accompanied by an exercise.

- The values of q_{water} and q_{metal} are numerically equal but of opposite sign; that is, $q_{water} = -q_{metal}$. Expressed another way, $q_{water} + q_{metal} = 0$. To paraphrase this equation: *The sum of thermal energy changes in this system is zero.*

Problems involving heat transfer can be approached by assuming that the sum of the heat content changes within a given system is zero (Equation 6.3).

$$q_1 + q_2 + q_3 \cdots = 0 \qquad (6.3)$$

The quantities q_1, q_2, and so on represent the changes in thermal energy for the individual parts of the system. For this specific problem, there are two heat content changes, q_{water} and q_{metal}, and

$$q_{water} + q_{metal} = 0$$

Each of these quantities is related to the specific heat capacities, mass, and change of temperature of the water and metal, as defined by Equation 6.1. Thus

$$[C_{water} \times m_{water} \times (T_{final} - T_{initial, water})] + [C_{metal} \times m_{metal} \times (T_{final} - T_{initial, metal})] = 0$$

The specific heat capacity of the metal is the unknown in this problem. Using the specific heat capacity of water from Table 6.1 and converting Celsius temperatures to kelvin gives

$$[(4.184 \text{ J/g} \cdot \text{K})(225 \text{ g})(296.3 \text{ K} - 294.2 \text{ K})]$$
$$+ [(C_{metal})(55.0 \text{ g})(296.3 \text{ K} - 373.0 \text{ K})] = 0$$
$$C_{metal} = 0.469 \text{ J/g} \cdot \text{K}$$

■ **Heat Transfer**
Remember that $T_{initial}$ for the metal and $T_{initial}$ for the water in this problem have different values.

GENERAL
Chemistry ⚛ Now™

See the General ChemistryNow CD-ROM or website:

- **Screen 6.8 Calculating Heat Transfer**

   (a) for a tutorial on calculations using specific heat capacity

   (b) for a simulation and exercise on determining the temperature of thermal equilibrium when two objects are in contact

   (c) for a tutorial on calculating the final temperature of thermal equilibrium

Example 6.2—Using Specific Heat Capacity

Problem A 88.5-g piece of iron whose temperature is 78.8 °C (352.0 K) is placed in a beaker containing 244 g of water at 18.8 °C (292.0 K). When thermal equilibrium is reached, what is the final temperature? (Assume no heat is lost to warm the beaker and surroundings.)

Strategy First, define the system as the iron and water. Two changes occur within the system: Iron gives up heat and water gains heat. The sum of the heat quantities of these changes must equal zero. Each quantity of heat is related to the specific heat capacity, mass, and temperature change of the substance using Equation 6.1 [$q = C \times m \times (T_{final} - T_{initial})$]. The final temperature is unknown. The specific heat capacities of iron and water are given in Table 6.1. The change in temperature, ΔT, may be in °C or K. See Problem-Solving Tip 6.1.

Problem Solving Tip 6.1

Units for *T* and Specific Heat Capacity

(a) *Calculating ΔT.* Notice that specific heat values are given in units of joules per-gram per kelvin (J/g · K). Virtually all calculations that involve temperature in chemistry are expressed in kelvins. In calculating ΔT, however, we could use Celsius temperatures because a kelvin and a Celsius degree are the

same size, so that the difference between two temperatures is the same on both scales. For example, the difference between the boiling and freezing points of water is

ΔT, Celsius = 100 °C − 0 °C = 100 °C

ΔT, kelvin = 373 K − 273 K = 100 K

(b) *Units of Specific Heat Capacity.* Specific heat capacities are given in this book in units of joules per gram per kelvin (J/g · K). Often, however, specific heat

capacity values found in handbooks (such as the *CRC Handbook of Chemistry and Physics*) or the NIST Webbook (**webbook.nist.gov**) will have units of J/mol · K; that is, they are molar heat capacities. For example, liquid water has a specific heat capacity of 4.184 J/g · K or 75.40 J/mol · K. The values are related as follows:

$$(4.184 \text{ J/g} \cdot \text{K})(18.02 \text{ g/mol}) =$$
$$75.40 \text{ J/mol} \cdot \text{K}$$

Solution

$$q_{water} + q_{metal} = 0$$

$$[C_{water} \times m_{water} \times (T_{final} - T_{initial, \, water})] + [C_{Fe} \times m_{Fe} \times (T_{final} - T_{initial, \, Fe})] = 0$$

$$[(4.184 \text{ J/g} \cdot \text{K})(244 \text{ g})(T_{final} - 292.0 \text{ K})] + [(0.449 \text{ J/g} \cdot \text{K})(88.5 \text{ g})(T_{final} - 352.0 \text{ K})] = 0$$

$$T_{final} = 295 \text{ K} \, (22 \text{ °C})$$

Comment The low specific heat capacity of iron and the small quantity of iron result in the temperature of iron being reduced by about 60 degrees whereas the temperature of water has been raised by only a few degrees.

Exercise 6.4—Using Specific Heat Capacity

A 15.5-g piece of chromium, heated to 100.0 °C, is dropped into 55.5 g of water at 16.5 °C. The final temperature of the metal and the water is 18.9 °C. What is the specific heat capacity of chromium? (Assume no heat is lost to the container or to the surrounding air.)

Exercise 6.5—Heat Transfer Between Substances

A piece of iron (400. g) is heated in a flame and then dropped into a beaker containing 1000. g of water. The original temperature of the water was 20.0 °C, and the final temperature of the water and iron is 32.8 °C after thermal equilibrium has been attained. What was the original temperature of the hot iron bar? (Assume no heat is lost to the beaker or to the surrounding air.)

6.3—Energy and Changes of State

When a solid melts, its atoms, molecules, or ions move about vigorously enough to break free of the constraints imposed by their neighbors in the solid. When a liquid boils, particles move much farther apart from one another. A change between solid and liquid or between liquid and gas is called a **change of state**. In both cases, energy must be furnished to overcome attractive forces among the particles. The heat required to convert a substance from a solid at its melting point to a liquid is called the **heat of fusion**. The heat required to convert liquid at its boiling point to gas is called the **heat of vaporization**. Heats of fusion and vaporization for many pure substances are provided along with other physical properties in reference books.

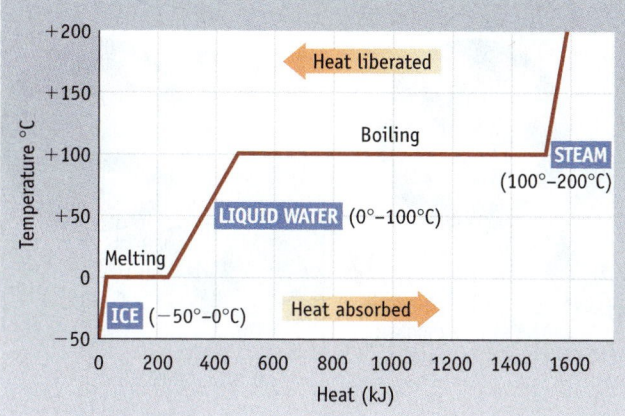

Active Figure 6.11 **Heat transfer and the temperature change for water.** This graph shows the quantity of heat absorbed and the consequent temperature change as 500. g of water warms from −50 °C to 200 °C.

GENERAL
Chemistry ᐧ◈ᐧ **Now**™ See the General ChemistryNow CD-ROM or website to explore an interactive version of this figure accompanied by an exercise.

For water, the heat of fusion at 0 °C is 333 J/g and the heat of vaporization at 100 °C is 2256 J/g. These values are used when calculating the quantity of heat required or evolved when water boils or freezes. For example, the heat required to convert 500. g of water from the liquid to gaseous state at 100 °C is

$$(2256 \text{ J/g})(500. \text{ g}) = 1.13 \times 10^6 \text{ J (or 1130 kJ)}$$

If the same quantity of liquid water at 0 °C freezes to ice, the quantity of heat evolved is

$$(333 \text{ J/g})(500. \text{ g}) = 1.67 \times 10^5 \text{ J (or 167 kJ)}$$

Figure 6.11 illustrates the quantity of heat absorbed and the consequent temperature change as 500. g of water is warmed from −50 °C to 200 °C. First, the temperature of the ice increases as heat is added. On reaching 0 °C, however, the temperature remains constant as sufficient heat (167 kJ) is absorbed to melt the ice to liquid water. When all the ice has melted, the liquid absorbs heat and is warmed to 100 °C, the boiling point of water. The temperature again remains constant as enough heat is absorbed (1130 kJ) to convert the liquid completely to vapor. Any further heat added raises the temperature of the water vapor. The heat absorbed at other steps in Figure 6.11 and the total heat absorbed are calculated in Example 6.3.

It is important to notice that *temperature is constant throughout a change of state* (see Figure 6.11). During a change of state, the added energy is used to overcome the forces holding one molecule to another, not to increase the temperature of the substance (see Figures 6.11 and 6.12).

■ **Heats of Fusion and Vaporization for H_2O**

Heat of fusion = 333 J/g
= 6.00 kJ/mol

Heat of vaporization = 2256 J/g
= 40.65 kJ/mol

Example 6.3—Energy and Changes of State

Problem Calculate the quantity of heat involved in each step shown in Figure 6.11 and the total quantity of heat required to convert 500. g of ice at −50.0 °C to steam at 200.0 °C. The heat of fusion of water is 333 J/g and the heat of vaporization is 2256 J/g. The specific heat capacity of steam at 200 °C is 1.92 J/g · K. See also Table 6.1.

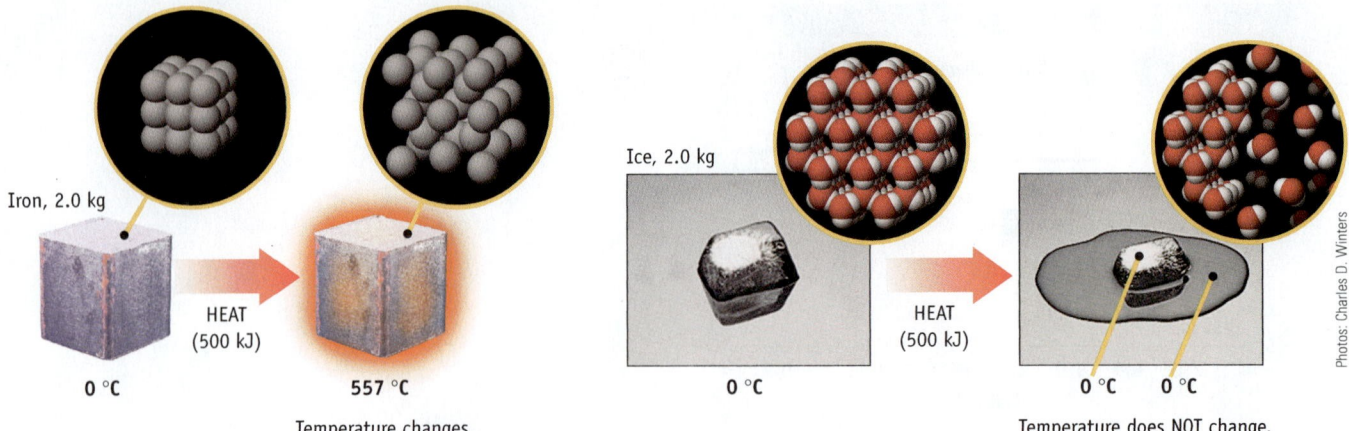

Iron, 2.0 kg

HEAT
(500 kJ)

0 °C   557 °C

Temperature changes.
State does NOT change.

Ice, 2.0 kg

HEAT
(500 kJ)

0 °C   0 °C   0 °C

Temperature does NOT change.
State changes.

Photos: Charles D. Winters

Active Figure 6.12   **Changes of state.** Adding 500 kJ of heat to 2.0 kg of iron at 0 °C will cause the iron's temperature to increase to 557 °C (and the metal expands slightly). In contrast, adding 500 kJ of heat to 2.0 kg of ice will cause 1.5 kg of ice to melt to water at 0 °C (and 0.5 kg of ice will remain). No temperature change occurs.

GENERAL
Chemistry⚛Now™ See the General ChemistryNow CD-ROM or website to explore an interactive version of this figure accompanied by an exercise.

Strategy The problem is broken down into a series of steps: (1) warm the ice from −50 °C to 0 °C; (2) melt the ice at 0 °C; (3) raise the temperature of the liquid water from 0 °C to 100 °C; (4) evaporate the water at 100 °C; (5) raise the temperature of the steam from 100 °C to 200 °C. Use Equation 6.1 to calculate the heats associated with temperature changes. Use the heat of fusion and the heat of vaporization for heats associated with changes of state. The total heat required is the sum of the heats of the individual steps.

Solution

Step 1.

q (to warm ice from −50 °C to 0 °C) = (2.06 J/g · K)(500. g)(273.2 K − 223.2 K)

$$= 5.15 \times 10^4 \text{ J}$$

Step 2.

q (to melt ice at 0 °C) = (500. g)(333 J/g) = 1.67×10^5 J

Step 3.

q (to raise temperature of water from 0 °C to 100 °C)

$$= (4.184 \text{ J/g} \cdot \text{K})(500. \text{ g})(373.2 \text{ K} - 273.2 \text{ K})$$
$$= 2.09 \times 10^5 \text{ J}$$

Step 4.

q (to evaporate water at 100 °C) = (2256 J/g)(500. g) = 1.13×10^6 J

Step 5.

q (to raise temperature of steam from 100 °C to 200 °C)

$$= (1.92 \text{ J/g} \cdot \text{K})(500. \text{ g})(473.2 \text{ K} - 373.2 \text{ K})$$
$$= 9.60 \times 10^4 \text{ J}$$

The total thermal energy required is the sum of the thermal energy required in each step.

$$q_{total} = q_1 + q_2 + q_3 + q_4 + q_5$$
$$q_{total} = 1.60 \times 10^6 \text{ J (or 1600 kJ)}$$

Comment The conversion of liquid water to steam is the largest increment of energy added by a considerable margin. (You may have noticed that it does not take much time to heat water to boiling on a stove, but to boil off the water takes a much greater time.)

Example 6.4—Change of State

Problem What is the minimum amount of ice at 0 °C that must be added to the contents of a can of diet cola (340. mL) to cool it from 20.5 °C to 0 °C? Assume that the specific heat capacity and density of diet cola are the same as for water, and that no heat is gained or lost to the surroundings.

Strategy The system here is defined as the ice and the cola, and heat is transferred between these substances. Two energy quantities, the heat change in cooling the soda and the heat change in melting the ice, are needed. The first is calculated using the specific heat capacity and Equation 6.1 ($q_{cola} = C_{cola} \times m_{cola} \times \Delta T$); the second uses the heat of fusion of water [$q_{ice} =$ (heat of fusion)(mass of ice)]. The law of conservation of energy requires that the sum of these two quantities of energy is zero (Equation 6.3).

Solution The mass of cola is

$$(340. \text{ mL})(1.00 \text{ g/mL}) = 340. \text{ g}$$

and its temperature changes from 293.7 K to 273.2 K. The heat of fusion of water is 333 J/g, and the mass of ice is the unknown.

$$q_{cola} + q_{ice} = 0$$
$$C_{cola} \times m_{cola} \times (T_{final} - T_{initial}) + q_{ice} = 0$$
$$[(4.184 \text{ J/g} \cdot \text{K})(340. \text{ g})(273.2 \text{ K} - 293.7 \text{ K})] + [(333 \text{ J/g})(m_{ice})] = 0$$
$$m_{ice} = 87.6 \text{ g}$$

Comment This quantity of ice is just sufficient to cool the cola to 0 °C. If more than 87.6 g of ice is added then, when thermal equilibrium is reached, the temperature will be 0 °C and some ice will remain (see Exercise 6.7). If less than 87.6 g of ice is added, the final temperature will be greater than 0 °C. In this case, all the ice will melt and the liquid water formed by melting the ice will absorb additional heat to warm up to the final temperature (an example is given in Study Question 77, page 277).

Exercise 6.6—Changes of State

How much heat must be absorbed to warm 25.0 g of liquid methanol, CH_3OH, from 25.0 °C to its boiling point (64.6 °C) and then to evaporate the methanol completely at that temperature? The specific heat capacity of liquid methanol is 2.53 J/g · K. The heat of vaporization of methanol is 2.00×10^3 J/g.

Exercise 6.7—Changes of State

To make a glass of ice tea, you pour 250 mL of tea, whose temperature is 18.2 °C, into a glass containing 5 ice cubes. Each cube has a mass of about 15 g. What quantity of ice will melt, and how much ice will remain to float at the surface in this beverage? Ice tea has a density of 1.0 g/cm³ and a specific heat capacity of 4.2 J/g · K. Assume that no heat is lost in cooling the glass or the surroundings.

6.4—The First Law of Thermodynamics

To this point, we have considered only energy in the form of heat. Now we need to broaden the discussion. Recall the definition given on page 235: *Thermodynamics is the science of heat and work.* Let us first note that *work* is done whenever something is moved against an opposing force. If a system does work on its surroundings, energy must be expended and the energy content of the system will decrease. Conversely, if work is done by the surroundings on a system, the energy content of the system increases. As with heat gained or lost, work done by a system or on a system will change its energy content (see "Historical Perspectives: Heat, Cannons, Soup, and Beer"). Therefore, we next want to introduce work into the equations for energy transfer.

An example of a system doing work on its surroundings is illustrated by the experiment shown in Figure 6.13. A small quantity of dry ice [$CO_2(s)$] is sealed inside a plastic bag, and a weight (a book in Figure 6.13) is placed on top of the bag. Dry ice has the interesting property that when it absorbs heat from its surroundings it changes directly from solid to gas at $-78\ °C$, in a process called **sublimation**:

$$CO_2(s,\ -78\ °C) \xrightarrow[+\ heat]{} CO_2(gas,\ -78\ °C)$$

As the experiment proceeds, the gaseous CO_2 expands within the plastic bag, lifting the book. To lift the book against the force of gravity requires that work be done. The system (the CO_2 inside the bag) is expending energy to do this work.

Even if the book had not been on top of the plastic bag, work would have been done by the expanding gas. This is because a gas must push back the atmosphere when it expands. Instead of raising a book, the expanding gas moves a part of the atmosphere.

Now let us recast this example as an experiment in thermodynamics. First, we must precisely identify the system and the surroundings. The system is the CO_2, a solid initially, and later a mixture of solid and gas. The surroundings consist of the objects that exchange energy with the system—that is, those objects in contact with the CO_2. They include the plastic bag, the book, the table-top, and the surrounding air. Thermodynamics focuses on the energy transfer that is occurring in the experi-

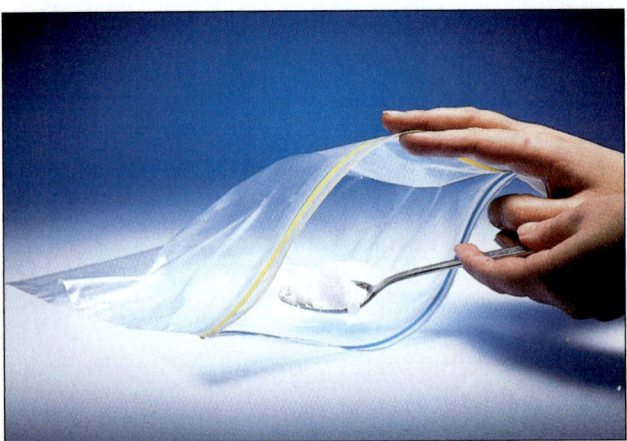

(a) Pieces of dry ice [$CO_2(s)$, $-78°C$] are placed in a plastic bag. The dry ice will sublime (change directly from a solid to a gas) upon the input of heat.

(b) Heat is absorbed by $CO_2(s)$ when it sublimes and the system (the contents of the bag) does work on its surroundings by lifting the book against the force of gravity.

Charles D. Winters

Active Figure 6.13   **Energy changes in a physical process.**

GENERAL
Chemistry ⚛ Now™ See the General ChemistryNow CD-ROM or website to explore an interactive version of this figure accompanied by an exercise.

Historical Perspectives

Work, Heat, Cannons, Soup, and Beer

Benjamin Thompson (1753–1814) is one of the more colorful characters in the history of science. He was born in the state of Massachusetts, but fled to London, England, before the American Revolution

because of his sympathy with the royalists. Thompson later moved to Munich, Germany, where he contributed so greatly to society that he was given the title of Count Rumford by the King of Bavaria in 1792. Among his contributions in Munich were the famous English Gardens and a unique system to care for the poor. He also created a candle so consistent in its light level that it became the international standard for measuring "candle power." Thompson became a nutritional expert, stressing the potato, and concocted a soup still known as *Rumfordsuppe*. He invented the modern kitchen range and convection oven, a double boiler, and a pressure cooker. And the efficient fireplace he designed is still known as a "Rumford fireplace."

Count Rumford is best known today for the experiments he did on heat. When visiting a cannon-boring factory, he

Work and Heat. A classic experiment that showed the relationship between work and heat was performed by Benjamin Thompson, Count Rumford, using the apparatus shown here. Thompson measured the rise in temperature of water (in the vessel mostly hidden at the back of the apparatus) that resulted from the energy expended to turn the crank.

noticed that the cannon barrels were hot, and the bore-hole shavings were even hotter. This had been observed for centuries, but Thompson was interested in what caused the heat and how it was passed along. Convinced that heat could not be a substance, as some then believed, he set up experiments to answer these questions.

Thompson eventually returned to London, and settled finally in Paris where he was acclaimed by Napoleon and elected to

the French Academy. There he also met and married Madame Lavoisier, the widow of Antoine Lavoisier (page 143). He first described her as an "incarnation of goodness," but they divorced in 1809. Thompson died in France in 1814.

The unit of heat, the joule, is named for **James P. Joule** (1818–1889), the son of a wealthy brewer in Manchester, England, and a student of John Dalton. The family wealth, and a

workshop in the brewery, gave Joule the opportunity to pursue scientific studies. Among the topics Joule studied was the issue of whether heat was a massless fluid, which some scientists called the caloric hypothesis. This had been a source of controversy for several decades, and it had not been resolved by the early experiments and advocacy of Rumford. Joule's careful experiments convincingly showed that heat and mechanical work can be interconverted and that heat is not a fluid. The caloric hypothesis was finally abandoned.

See G. I. Brown, *Count Rumford, The Extraordinary Life of a Scientific Genius*, Trowbridge, England, Sutton Publishing, 1999.

ment. Sublimation of CO_2 requires heat, which is transferred to the CO_2 from the surroundings. At the same time, the system does work on the surroundings by lifting the book. An energy balance for the system will include both quantities; that is, the change in energy content for the system (ΔE) will equal the sum of heat transferred (q) to or from the system and the work done by or to the system (w). We can express this explicitly as an equation:

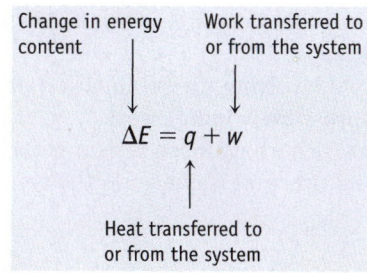

$$\Delta E = q + w \tag{6.4}$$

Change in energy content

Work transferred to or from the system

Heat transferred to or from the system

A Closer Look

P-V Work

Work is done when an object of some mass is moved against an external resisting force. We know this well from common experience, such as when we use a pump to blow up a bicycle tire.

To evaluate the work done when a gas is compressed we can use, for example, a cylinder with a movable piston, as would occur in a bicycle pump (see figure). The drawing on the left shows the initial position of the piston, and the one on the right shows its final position. To depress the piston, we would have to expend some energy (the energy of this process coming from the energy obtained by food metabolism in our body.) The work required to depress the piston is calculated from a law of physics, $w = F \times d$, or work equals the magnitude of the force applied times the distance (d) over which the force is applied.

Pressure is defined as a force divided by the area over which the force is applied: $P = F/A$. In this example, the force is being applied to a piston with an area A. Substituting $P \times A$ for F in the equation gives $w = (P \times A) \times d$. However, since the product of $A \times d$ is the change of volume, ΔV, we can rewrite our equation for work as $w = -P\Delta V$.

Pushing down on the piston means we have done work on the system, the gas contained within the cylinder. The gas is now compressed to a smaller volume and has attained a higher energy as a consequence. The additional energy is equal to $-P\Delta V$.

Notice how we have allowed energy to be converted from one form to another— from chemical energy in food to mechanical energy used to depress the piston, to potential energy stored in a system of a gas at a higher pressure. In each step, energy was conserved, not lost, and the total energy of the universe remained constant.

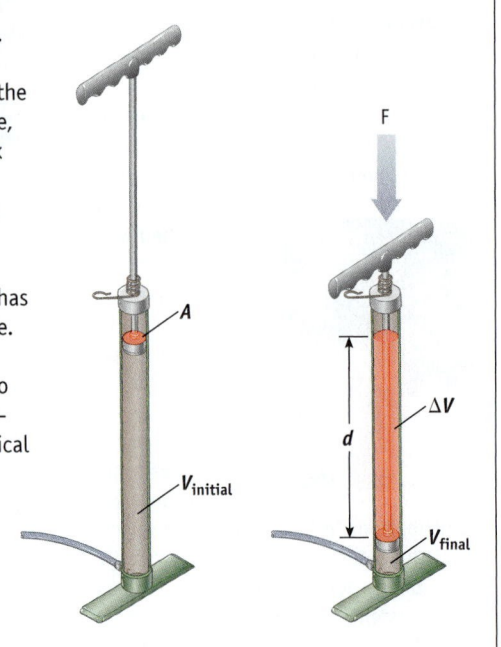

Equation 6.4 is a mathematical statement of the *first law of thermodynamics:* The energy change for a system is the sum of heat transferred between the system and its surroundings and the work done on the system by the surroundings or on the surroundings by the system. You will notice that this equation is a version of the general principle of conservation of energy applied specifically to the system.

The quantity E in Equation 6.4 has a formal name and a precise meaning in thermodynamics: **internal energy**. The internal energy in a chemical system is the sum of the potential and kinetic energies of the atoms, molecules, or ions in the system. Potential energy is the energy associated with the attractive and repulsive forces between all the nuclei and electrons in the system. It includes the energy associated with bonds in molecules, forces between ions, and forces between molecules in the liquid and solid state. Kinetic energy is the energy of motion of the atoms, ions, and molecules in the system. A value of internal energy is extremely difficult to determine but fortunately this step is not necessary. As the equation indicates, we are evaluating the *change* of internal energy, ΔE, which is a measurable quantity. In fact, the equation tells us how *to determine ΔE: Measure the heat transferred and the work done to or by the system.*

The work in the example involving the sublimation of CO_2 (Figure 6.13) is of a specific type, called ***P-V* (pressure-volume) work**. It is the work associated with a change in volume (ΔV) that occurs against a resisting external pressure (P). For a system in which the external pressure is constant, the value of *P-V* work can be calculated by Equation 6.5:

■ **Work**
Electrical work is another type of work commonly encountered in chemistry.

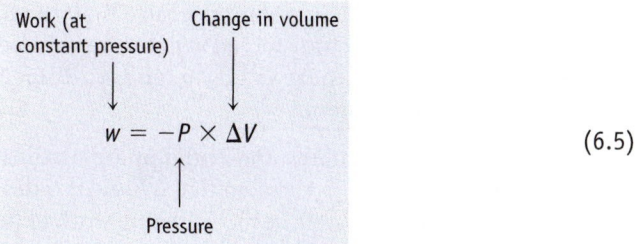

$$w = -P \times \Delta V \qquad (6.5)$$

The origin of this relationship is explained in "A Closer Look: P-V Work".

The sign convention for Equation 6.4 is important. The following table summarizes how the internal energy of a system is affected by heat and work.

Sign Conventions for q and w of the System

Change	Sign Convention	Effect on E_{system}
Heat transferred to system from surroundings	$q > 0$ (+)	E increases
Heat transferred from system to surroundings	$q < 0$ (−)	E decreases
Work done on system by surroundings	$w > 0$ (+)	E increases
Work done by system on surroundings	$w < 0$ (−)	E decreases

GENERAL
Chemistry ⚛ Now™

See the General ChemistryNow CD-ROM or website:
- **Screen 6.11 The First Law of Thermodynamics,** for a video of energy change in a physical change

Enthalpy

Most experiments in a chemical laboratory are carried out in beakers or flasks open to the atmosphere. Similarly, chemical processes that occur in living systems are open to the atmosphere. The fact that pressure is constant under these conditions is an important consideration when applying the first law of thermodynamics to heat measurements.

Because heat at constant pressure is so frequently the focus of attention in chemistry and biology, it is useful to have a specific measure of heat transfer under these conditions. The heat content of a substance at constant pressure is called **enthalpy** and is given the symbol H. In experiments at constant pressure, the **enthalpy change, ΔH**, is the difference between the final and initial enthalpy content. With enthalpy, as with internal energy, attention is focused on changes (that is, ΔH) rather than on the value of H itself. It is the value of the enthalpy change, ΔH, that is measured in chemical and physical processes.

Similar sign and symbol conventions apply to both ΔE and ΔH.

- Negative values of ΔE and ΔH specify that energy is transferred from the system to the surroundings.

- Positive values of ΔE and ΔH refer to energy transferred from the surroundings to the system.

- The heat transferred at constant pressure is often symbolized by q_p and is equivalent to ΔH. The heat transferred at constant volume, symbolized by q_v, is equivalent to ΔE. The two heat values, q_p and q_v, differ by the amount of work, w, done on or by the system.

Changes in internal energy and enthalpy are mathematically related by the general equation $\Delta E = \Delta H + w$, showing that ΔE and ΔH differ by the quantity of energy transferred to or from a system as work. Taking work to be $-P\,\Delta V$, we observe that in many processes—such as the melting of ice—ΔV is small and hence the amount of work is small. Under these circumstances, ΔE and ΔH are of similar magnitude. The amount of work can be significant, however, in processes in which the volume change is large. This usually occurs when gases are formed or consumed. In the evaporation or condensation of water, the sublimation of CO_2 (see Figure 6.13), and chemical reactions in which gas volumes change, for example, ΔE and ΔH are significantly different.

State Functions

Both internal energy and enthalpy share a significant characteristic—namely, changes in these quantities that accompany chemical or physical changes do not depend on the path chosen to go from the initial state to the final state. No matter how you go from reactants to products in a reaction, for example, the value of ΔH or ΔE for the reaction is always the same. A quantity that has this characteristic property is called a **state function**.

Many commonly measured quantities, such as the pressure of a gas, the volume of a gas or liquid, the temperature of a substance, and the size of your bank account, are state functions. For example, you could have arrived at a current bank balance of $25 by having deposited $25, or you could have deposited $100 and then withdrawn $75.

The volume of a balloon is also a state function. You can blow up a balloon to a large volume and then let some air out to arrive at the desired volume. Alternatively, you can blow up the balloon in stages, adding tiny amounts of air at each stage. The final volume does not depend on how you got there. For bank balances and balloons, there are an infinite number of ways to arrive at the final state, but the final value depends only on the size of the bank balance or the balloon, not on the path taken from the initial state to the final state.

Not all quantities are state functions. For instance, distance traveled is not a state function (Figure 6.14). The travel distance from Oneonta, New York, to Madison, Wisconsin, depends on the route taken. Nor is the elapsed time of travel between these two locations a state function. In contrast, the altitude above sea level is a state function; in going from Oneonta (538 m above sea level) to Madison (280 m above sea level), there is an altitude change of 258 m, regardless of the route followed. Interestingly, neither heat nor work individually is a state functions but their sum, the change in internal energy, ΔE, is. The value of ΔE is fixed by $E_{initial}$ and E_{final}, but a transition between the initial and final states can be accomplished by different routes having different values of q and w. Enthalpy is also a state function. The enthalpy change occurring when 1.0 g of water is heated from 20 °C to 50 °C, or when 1.0 g of water is evaporated at 100 °C, is independent of the way in which the process is carried out.

Taxi/Getty Images

Figure 6.14 State functions. There are many ways to climb a mountain, but the change in altitude from the base of the mountain to its summit is the same.
The change in altitude is a state function. The distance traveled to reach the summit is not.

6.5—Enthalpy Changes for Chemical Reactions

Enthalpy changes accompany chemical reactions. For example, for the decomposition of 1 mol of water vapor to its elements, 1 mol of H_2 and $\frac{1}{2}$ mol of O_2, the enthalpy change $\Delta H = +241.8$ kJ at 25 °C.

$$H_2O(g) \longrightarrow H_2(g) + \tfrac{1}{2} O_2(g) \qquad \Delta H = +241.8 \text{ kJ}$$

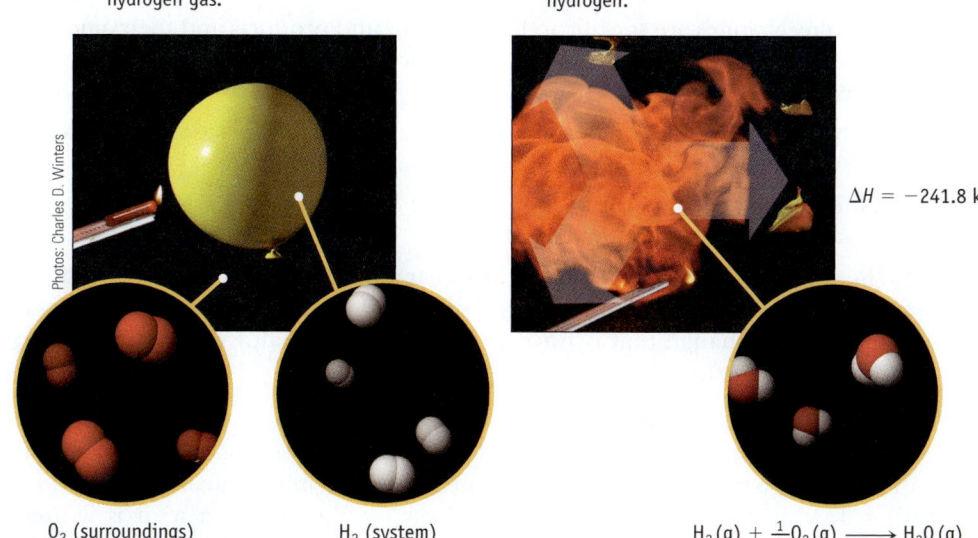

(a) A lighted candle is brought up to a balloon filled with hydrogen gas.

(b) When the balloon breaks, the candle flame ignites the hydrogen.

$\Delta H = -241.8$ kJ

Photos: Charles D. Winters

O_2 (surroundings) H_2 (system) $H_2(g) + \frac{1}{2}O_2(g) \longrightarrow H_2O(g)$

Active Figure 6.15 **The exothermic combustion of hydrogen in air.** The reaction transfers energy to the surroundings in the form of heat, work, and light.

GENERAL
Chemistry Now™ See the General ChemistryNow CD-ROM or website to explore an interactive version of this figure accompanied by an exercise.

The positive sign of ΔH indicates that the decomposition is an endothermic process. That is, the reaction requires that 241.8 kJ be transferred to the system, $H_2O(g)$, from the surroundings.

Now consider the opposite reaction, the combination of hydrogen and oxygen to form water. The quantity of heat energy evolved in this reaction is the same as is required for the decomposition reaction, except that the sign of ΔH is reversed. The exothermic formation of 1 mol of water vapor from H_2 and $\frac{1}{2}$ mol of O_2 transfers 241.8 kJ to the surroundings (Figure 6.15).

$$H_2(g) + \tfrac{1}{2}O_2(g) \longrightarrow H_2O(g) \qquad \Delta H = -241.8 \text{ kJ}$$

The quantity of heat transferred during a chemical change depends on the amounts of reactants used or products formed. Thus, the formation of 2 mol of water vapor from the elements produces twice as much heat as the formation of 1 mol of water.

$$2\,H_2(g) + O_2(g) \longrightarrow 2\,H_2O(g) \qquad \Delta H = -483.6 \text{ kJ} \; (= 2 \times -241.8 \text{ kJ})$$

It is important to identify the states of reactants and products in a reaction because the magnitude of ΔH also depends on whether they are solids, liquids, or gases. Formation of 1 mol of *liquid* water from the elements is accompanied by the evolution of 285.8 kJ of energy.

$$H_2(g) + \tfrac{1}{2}O_2(g) \longrightarrow H_2O(\ell) \qquad \Delta H = -285.8 \text{ kJ}$$

The additional energy evolved relative to the formation of water vapor arises from the energy released when 1 mol of water vapor condenses to 1 mol of liquid water.

■ **Fractional Stoichiometric Coefficients** When writing balanced equations to define thermodynamic quantities, chemists often use fractional stoichiometric coefficients. For example, when we wish to define ΔH for the decomposition or formation of 1 mol of H_2O, the coefficient for O_2 must be $\frac{1}{2}$.

These examples illustrate several features of the enthalpy changes for chemical reactions.

- Enthalpy changes are specific to the reactants and products and their amounts. Both the identities of reactants and products and their states (s, ℓ, g) are important.

- ΔH has a negative value if heat is evolved (an exothermic reaction). It has a positive value if heat is required (an endothermic reaction.)

- Values of ΔH are numerically the same, but opposite in sign, for chemical reactions that are the reverse of each other.

- The enthalpy change depends on the molar amounts of reactants and products. The formation of 2 mol of $H_2O(g)$ from the elements, for example, results in an enthalpy change that is twice as large as the enthalpy change in forming 1 mol of $H_2O(g)$.

Enthalpies of reactions are usually provided in one of two ways. They may be expressed as energy per mole of a reactant or per mole of a product. Alternatively, the enthalpy change may be given along with a balanced chemical equation, as was done earlier. In this case the value of ΔH is given for the equation as it is written. Whichever way the enthalpy change is presented, the value can be used to calculate the quantity of heat transferred by any given mass of a reactant or product. Suppose, for example, you want to know the enthalpy change if 454 g of propane, C_3H_8, is burned, given the equation for the exothermic combustion and the enthalpy change for the reaction.

$$C_3H_8(g) + 5\ O_2(g) \longrightarrow 3\ CO_2(g) + 4\ H_2O(\ell) \qquad \Delta H = -2220 \text{ kJ}$$

Two steps are needed. First, find the amount of propane present in the sample:

$$454 \text{ g } C_3H_8 \left(\frac{1 \text{ mol } C_3H_8}{44.10 \text{ g } C_3H_8} \right) = 10.3 \text{ mol } C_3H_8$$

Second, multiply the quantity of heat transferred per mole of propane by the amount of propane:

$$\Delta H = 10.3 \text{ mol } C_3H_8 \left(\frac{-2220 \text{ kJ}}{1 \text{ mol } C_3H_8} \right) = -22,900 \text{ kJ}$$

GENERAL
Chemistry · Now™

See the General ChemistryNow CD-ROM or website:
- **Screen 6.13 Enthalpy Changes for Chemical Reactions,** for a tutorial on calculating the enthalpy change for a reaction

■ **Chemical Potential Energy** Gummi Bears are mostly sugar, and you can see in Figure 6.2 that their oxidation is highly exothermic. The enthalpy change for the oxidation of 1 teaspoonful of sugar, such as you might have in a large Gummi Bear, is about 100 kJ. See Example 6.5.

Charles D. Winters

Example 6.5—Enthalpy Calculation

Problem Sucrose (sugar, $C_{12}H_{22}O_{11}$) is oxidized to CO_2 and H_2O. The enthalpy change for the reaction can be measured in the laboratory

$$C_{12}H_{22}O_{11}(s) + 12\ O_2(g) \longrightarrow 12\ CO_2(g) + 11\ H_2O(\ell) \qquad \Delta H = -5645 \text{ kJ}$$

What is the enthalpy change for the oxidation of 5.00 g (1 teaspoonful) of sugar?

Strategy We will first determine the amount of sucrose in 5.00 g, then use this with the value given for the enthalpy change for the oxidation of 1 mol of sucrose.

Solution

$$5.00 \text{ g sucrose} \times \frac{1 \text{ mol sucrose}}{342.3 \text{ g sucrose}} = 1.46 \times 10^{-2} \text{ mol sucrose}$$

$$q = 1.46 \times 10^{-2} \text{ mol sucrose} \left(\frac{-5645 \text{ kJ}}{1 \text{ mol sucrose}} \right)$$

$$q = -82.5 \text{ kJ}$$

Comment Persons concerned about their diets might be interested to note that a (level) tea-spoonful of sugar supplies about 25 Calories (dietary Calories; the conversion is 4.184 kJ = 1 Cal). As diets go, a single spoonful of sugar doesn't have a large caloric content. But will you use a level teaspoonful? And will you stop with just one?

Exercise 6.8—Enthalpy Calculation

(a) What quantity of heat energy is required to decompose 12.6 g of liquid water to the elements?

(b) The combustion of ethane, C_2H_6, has an enthalpy change of -2857.3 kJ for the reaction as written below. Calculate the value of ΔH when 15.0 g of C_2H_6 is burned.

$$2 \text{ } C_2H_6(g) + 7 \text{ } O_2(g) \longrightarrow 4 \text{ } CO_2(g) + 6 \text{ } H_2O(g) \qquad \Delta H = -2857.3 \text{ kJ}$$

6.6—Calorimetry

The heat transferred in a chemical or physical process is measured by an experimental technique called **calorimetry.** The apparatus used in this kind of experiment is a calorimeter, of which there are two basic types. A **constant pressure calorimeter** allows measurement of heats evolved or required under constant pressure conditions. In a **constant volume calorimeter**, the volume cannot change. The two types of calorimetry highlight the differences between enthalpy and internal energy. Heat transferred at constant pressure, q_p, is, by definition, ΔH, whereas the heat transferred at constant volume, q_v, is ΔE.

Constant Pressure Calorimetry: Measuring ΔH

Heat changes at constant pressure are often measured in the general chemistry laboratory by using a "coffee-cup calorimeter." This inexpensive device consists of two nested Styrofoam coffee cups with a loose-fitting lid and a temperature-measuring device such as a thermometer (Figure 6.16). The cup contains a solution of the reactants. The mass and specific heat capacity of the solution, and the amount of reactants, must be known. If heat is evolved in the process under study, the temperature of the solution rises. If heat is required, it is furnished by the solution and a decrease in temperature will be seen. In each case the change in temperature is measured. From mass, specific heat capacity, and temperature change, the heat change for the contents of the calorimeter can be calculated.

In the terminology of thermodynamics, the contents of the coffee-cup calorimeter are the system, and the cup and the immediate environment around the apparatus are the surroundings. Two heat changes occur within the system. One is the change that takes place as the chemical (potential) energy stored in the reactants is released as heat during the reaction. We label this heat quantity q_{rxn} (where

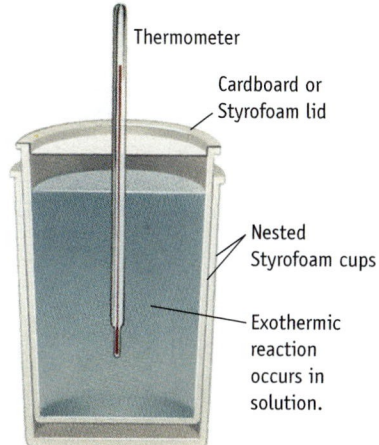

Figure 6.16 A coffee-cup calorimeter. A chemical reaction produces a change in the temperature of the solution in the calorimeter. The Styrofoam container is fairly effective in preventing heat transfer between the solution and its surroundings. Because the cup is open to the atmosphere, this is a constant pressure measurement.

Thermometer

Cardboard or Styrofoam lid

Nested Styrofoam cups

Exothermic reaction occurs in solution.

"rxn" is an abbreviation for "reaction"). The other is the heat gained or lost by the solution ($q_{solution}$). Assuming no heat transfer between the system and the surroundings, the sum of the heat changes within the system is zero.

$$q_{rxn} + q_{solution} = 0$$

The change in heat content of the solution ($q_{solution}$) can be calculated from its heat capacity, mass, and change in temperature. The quantity of heat evolved or required for the reaction (q_{rxn}) is the unknown in the equation. Because the reaction is carried out at constant pressure, the heat being measured is an enthalpy change, ΔH.

The accuracy of a calorimeter experiment depends on the accuracy of the measured quantities (temperature, mass, specific heat capacity). In addition, it depends on how closely the assumption of no heat transfer between system and surroundings is followed. A coffee-cup calorimeter is an unsophisticated apparatus and the results obtained with it are not highly accurate, largely because the latter assumption is poorly met. In research laboratories, scientists utilize calorimeters that more effectively limit the heat transfer between system and surroundings, and they may also estimate and correct for any minimal heat transfer that does occur between the system and the surroundings.

Example 6.6—Using a Coffee-Cup Calorimeter

Problem Suppose you place 0.500 g of magnesium chips in a coffee-cup calorimeter and then add 100.0 mL of 1.00 M HCl. The reaction that occurs is

$$Mg(s) + 2 HCl(aq) \longrightarrow H_2(g) + MgCl_2(aq)$$

The temperature of the solution increases from 22.2 °C (295.4 K) to 44.8 °C (318.0 K). What is the enthalpy change for the reaction per mole of Mg? (Assume that the specific heat capacity of the solution is 4.20 J/g · K and the density of the HCl solution is 1.00 g/mL.)

Strategy Two changes in heat content take place within the system: the heat evolved in the reaction (q_{rxn}) and the heat gained by the solution to increase its temperature ($q_{solution}$). The problem solution has three steps. First, calculate $q_{solution}$ from the values of the mass, specific heat capacity, and ΔT using Equation 6.1. Second, calculate q_{rxn}, assuming no energy transfer occurs between the system and the surroundings (so the sum of heat changes in the system $q_{rxn} + q_{solution} = 0$). Third, use the value of q_{rxn} and the amount of Mg to calculate the enthalpy change per mole.

Solution

Step 1. *Calculate $q_{solution}$.* The mass of the solution is approximately the mass of the 100.0 mL of HCl plus the mass of magnesium, or 100.5 g.

$$q_{solution} = (100.5 \text{ g})(4.20 \text{ J/g} \cdot \text{K})(318.0 \text{ K} - 295.4 \text{ K})$$
$$= 9.54 \times 10^3 \text{ J}$$

Step 2. *Calculate q_{rxn}.*

$$q_{rxn} + q_{solution} = 0$$
$$q_{rxn} + 9.54 \times 10^3 \text{ J} = 0$$
$$q_{rxn} = -9.54 \times 10^3 \text{ J}$$

Step 3. *Calculate the value of ΔH per mole.* The quantity of heat found in Step 2 is produced by the reaction of 0.500 g of Mg. The heat produced by the reaction of 1.00 mol of Mg is therefore

$$\Delta H = \left(\frac{-9.54 \times 10^3 \text{ J}}{0.500 \text{ g Mg}}\right)\left(\frac{24.31 \text{ g Mg}}{1 \text{ mol Mg}}\right)$$

$$\Delta H = -4.64 \times 10^5 \text{ J/mol Mg}$$

Comment The calculation will give the correct sign of q_{rxn} and ΔH. The negative sign indicates that this is an exothermic reaction.

Exercise 6.9—Using a Coffee-Cup Calorimeter

Assume you mix 200. mL of 0.400 M HCl with 200. mL of 0.400 M NaOH in a coffee-cup calorimeter. The temperature of the solutions before mixing was 25.10 °C; after mixing and allowing the reaction to occur, the temperature is 27.78 °C. What is the molar enthalpy of neutralization of the acid? (Assume that the densities of all solutions are 1.00 g/mL and their specific heat capacities are 4.20 J/g · K.)

Constant Volume Calorimetry: Measuring ΔE

Constant volume calorimetry is often used to evaluate heats of combustion of fuels and the caloric value of foods. A weighed sample of a combustible solid or liquid is placed inside a "bomb," often a cylinder about the size of a large fruit-juice can with thick steel walls and ends (Figure 6.17). The bomb is placed in a water-filled container

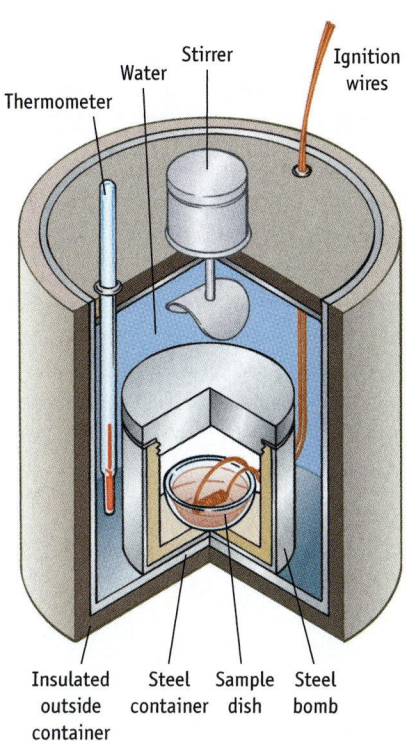
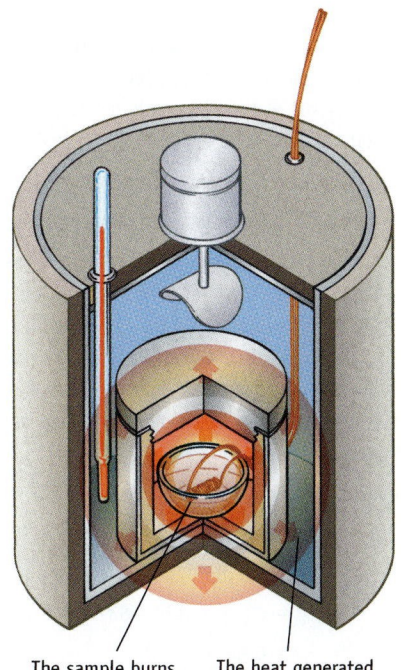

Thermometer · Water · Stirrer · Ignition wires

Insulated outside container · Steel container · Sample dish · Steel bomb

The sample burns in pure oxygen, warming the bomb

The heat generated warms the water and ΔT is measured by the thermometer

Active Figure 6.17 **Constant volume calorimeter.** A combustible sample is burned in pure oxygen in a sealed metal container or "bomb." The heat generated warms the bomb and the water surrounding it. By measuring the increase in temperature, the heat evolved in the reaction can be determined.

GENERAL Chemistry⚛Now™ See the General ChemistryNow CD-ROM or website to explore an interactive version of this figure accompanied by an exercise.

with well-insulated walls. After filling the bomb with pure oxygen, the sample is ignited, usually by an electric spark. The heat generated by the combustion reaction warms the bomb and the water around it. The bomb, its contents, and the water are defined as the system. Assessment of heat transfer within the system shows that

$$q_{rxn} + q_{bomb} + q_{water} = 0$$

Because the volume does not change in a constant volume calorimeter, energy transfer as work cannot occur. Therefore, the heat measured at constant volume (q_v) is the change in internal energy, ΔE.

GENERAL
Chemistry ⚛ Now™

See the General ChemistryNow CD-ROM or website:
- **Screen 6.14 Measuring Heats of Reactions**
    - **(a)** for a simulation and exercise exploring reactions in a bomb calorimeter
    - **(b)** for a tutorial on calculating the heat of a reaction using a calorimeter

---

Example 6.7—Constant Volume Calorimetry

Problem Octane, C_8H_{18}, a primary constituent of gasoline, burns in air:

$$C_8H_{18}(\ell) + 25/2\ O_2(g) \longrightarrow 8\ CO_2(g) + 9\ H_2O(\ell)$$

A 1.00-g sample of octane is burned in a constant volume calorimeter similar to that shown in Figure 6.17. The calorimeter is in an insulated container with 1.20 kg of water. The temperature of the water and the bomb rises from 25.00 °C (298.15 K) to 33.20 °C (306.35 K). The heat capacity of the bomb, C_{bomb}, is 837 J/K.

(a) What is the heat of combustion per gram of octane?

(b) What is the heat of combustion per mole of octane?

Strategy (a) The sum of all heat changes in the system will be zero; that is, $q_{rxn} + q_{bomb} + q_{water} = 0$. The first term, q_{rxn}, is the unknown. The second and third terms in the equation can be calculated from the data given: q_{bomb} is calculated from the bomb's heat capacity and ΔT, and q_{water} is determined from the specific heat capacity, mass, and ΔT for water. (b) The value of q_{rxn} calculated in part (a) is the heat evolved in the combustion of 1.00 g of octane. Use this value and the molar mass of octane (114.2 g/mol) to calculate the heat evolved per mole of octane.

Solution

(a) $q_{water} = C_{water} \times m_{water} \times \Delta T$

$= (4.184\ \text{J/g} \cdot \text{K})(1.20 \times 10^3\ \text{g})(306.35\ \text{K} - 298.15\ \text{K}) = +41.2 \times 10^3\ \text{J}$

$q_{bomb} = C_{bomb} \times \Delta T = 837\ \text{J/K}\ (306.35\ \text{K} - 298.15\ \text{K})$

$= 6.86 \times 10^3\ \text{J}$

$$q_{rxn} + q_{water} + q_{bomb} = 0$$

$q_{rxn} + 41.2 \times 10^3\ \text{J} + 6.86 \times 10^3\ \text{J} = 0$

$$q_{rxn} = -48.1 \times 10^3\ \text{J}$$

$$(\text{or } -48.1\ \text{kJ})$$

Heat of combustion per gram $= -48.1$ kJ

(b) Heat of combustion per mol $= (-48.1\ \text{kJ/g})(114.2\ \text{g/mol}) = -5.49 \times 10^3\ \text{kJ/mol}$

Comment Because the volume does not change, no energy transfer in the form of work occurs. The change of internal energy, ΔE, for the combustion of $C_8H_{18}(\ell)$ is -5.49×10^3 kJ/mol. Also note that C_{bomb} has no mass units. It is the heat required to warm the whole object by 1 kelvin.

Exercise 6.10—Constant Volume Calorimetry

A 1.00-g sample of ordinary table sugar (sucrose, $C_{12}H_{22}O_{11}$) is burned in a bomb calorimeter. The temperature of 1.50×10^3 g of water in the calorimeter rises from 25.00 °C to 27.32 °C. The heat capacity of the bomb is 837 J/K, and the specific heat capacity of the water is 4.20 J/g · K.

(a) Calculate the heat evolved per gram of sucrose.
(b) Calculate the heat evolved per mole of sucrose.

6.7—Hess's Law

Measuring a heat of reaction using a calorimeter is not possible for many chemical reactions. Consider, for example, the oxidation of carbon to carbon monoxide.

$$C(s) + \tfrac{1}{2} O_2(g) \longrightarrow CO(g)$$

Some CO_2 will always form in reactions of carbon and oxygen, even if there is a deficiency of oxygen. The reaction of CO and O_2 is very favorable; thus, as soon as CO is formed, it will react with O_2 to form CO_2. Therefore, using calorimetry to measure the heat evolved in the formation of CO is not possible.

Fortunately, the heat evolved in the reaction forming $CO(g)$ from $C(s)$ and $O_2(g)$ can be calculated from heats measured for other reactions. The calculation is based on **Hess's law**, which states that *if a reaction is the sum of two or more other reactions, ΔH for the overall process is the sum of the ΔH values of those reactions.*

The oxidation of $C(s)$ to $CO_2(g)$ can be viewed as occurring in two steps: first the oxidation of $C(s)$ to $CO(g)$ (Equation 1), and then the oxidation of $CO(g)$ to $CO_2(g)$ (Equation 2). Adding these two equations gives the equation for the oxidation of $C(s)$ to $CO_2(g)$ (Equation 3).

Equation 1: $C(s) + \tfrac{1}{2} O_2(g) \longrightarrow CO(g)$ $\Delta H_1 = ?$

Equation 2: $CO(g) + \tfrac{1}{2} O_2(g) \longrightarrow CO_2(g)$ $\Delta H_2 = -283.0$ kJ

Equation 3: $C(s) + O_2(g) \longrightarrow CO_2(g)$ $\Delta H_3 = -393.5$ kJ

Hess's law tells us that the enthalpy change for overall reaction (ΔH_3) will equal the sum of the enthalpy changes for reactions 1 and 2 ($\Delta H_1 + \Delta H_2$). Both ΔH_2 and ΔH_3 can be measured, and these values are then used to determine the enthalpy change for reaction 1.

$$\Delta H_3 = \Delta H_1 + \Delta H_2$$
$$-393.5 \text{ kJ} = \Delta H_1 + (-283.0 \text{ kJ})$$
$$\Delta H_1 = -110.5 \text{ kJ}$$

Hess's law applies to physical processes, too. The enthalpy change for the reaction of $H_2(g)$ and $O_2(g)$ to form 1 mol of liquid H_2O is different from the enthalpy change to form 1 mol of H_2O vapor (page 255). The difference is the heat of vaporization of water, ΔH_2.

Equation 1:	$H_2(g) + \frac{1}{2} O_2(g) \longrightarrow H_2O(\ell)$	$\Delta H_1 = -285.8$ kJ
Equation 2:	$H_2O(\ell) \longrightarrow H_2O(g)$	$\Delta H_2 = ?$
Equation 3:	$H_2(g) + \frac{1}{2} O_2(g) \longrightarrow H_2O(g)$	$\Delta H_3 = -241.8$ kJ

The relationship $\Delta H_3 = \Delta H_1 + \Delta H_2$ makes it possible to calculate the value of ΔH_2, the heat of vaporization of water (44.0 kJ, with all substances at 25 °C).

Energy Level Diagrams

When using Hess's law, it is often helpful to represent enthalpy data schematically in an **energy level diagram**. In such a drawing, various substances—for example, the reactants and products in a chemical reaction—are placed on an arbitrary (potential) energy scale. The relative energy of each substance is given by its position on the vertical axis, and numerical differences in energy between them are shown by the vertical arrows. Such diagrams provide an easy-to-read perspective on the magnitude and direction of energy changes and show how energy of the substances are related.

Energy level diagrams that summarize the two examples of Hess's law discussed earlier appear in Figure 6.18. In Figure 6.18a, the elements, C(s) and $O_2(g)$ are at

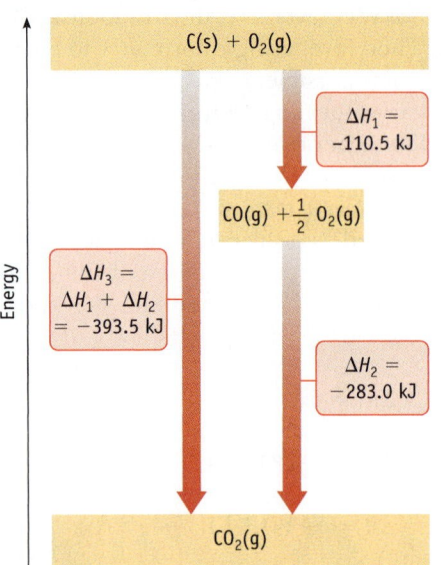

(a) The formation of CO_2 can occur in a single step or in a succession of steps. ΔH for the overall process is -393.5 kJ, no matter which path is followed.

(b) The formation of $H_2O(\ell)$ can occur in a single step or in a succession of steps. ΔH for the overall process is -285.8 kJ, no matter which path is followed.

Active Figure 6.18   Energy level diagrams. (a) Relating enthalpy changes in the formation of $CO_2(g)$. (b) Relating enthalpy changes in the formation of $H_2O(\ell)$. Enthalpy changes associated with changes between energy levels are given alongside the vertical arrows.

GENERAL
Chemistry Now™ See the General ChemistryNow CD-ROM or website to explore an interactive version of this figure accompanied by an exercise.

the highest potential energy. Converting carbon and oxygen to CO_2 lowers the potential energy by 393.5 kJ. This can occur either in a single step, shown on the left in Figure 6.18a, or in two steps, shown on the right. Similarly, in Figure 6.18b, the potential energy of the elements is at the highest potential energy. The product, liquid or gaseous water, has a lower potential energy, with the difference between the two being the heat of vaporization.

GENERAL
Chemistry·⚛·Now™

See the General ChemistryNow CD-ROM or website:
• **Screen 6.15 Hess's Law,** for a simulation and exercise on "adding" reactions

Example 6.8—Using Hess's Law

Problem Suppose you want to know the enthalpy change for the formation of methane, CH_4, from solid carbon (as graphite) and hydrogen gas:

$$C(s) + 2 H_2(g) \longrightarrow CH_4(g) \qquad \Delta H = ?$$

The enthalpy change for this reaction cannot be measured in the laboratory because the reaction is very slow. We can, however, measure enthalpy changes for the combustion of carbon, hydrogen, and methane.

Equation 1: $\quad C(s) + O_2(g) \longrightarrow CO_2(g)$ $\qquad\qquad\qquad \Delta H_1 = -393.5$ kJ

Equation 2: $\quad H_2(g) + \frac{1}{2} O_2(g) \longrightarrow H_2O(\ell)$ $\qquad\qquad \Delta H_2 = -285.8$ kJ

Equation 3: $\quad CH_4(g) + 2 O_2(g) \longrightarrow CO_2(g) + 2 H_2O(\ell)$ $\quad \Delta H_3 = -890.3$ kJ

Use these energies to obtain ΔH for the formation of methane from its elements.

Strategy The three reactions (1, 2, and 3), as they are written, cannot be added together to obtain the equation for the formation of CH_4 from its elements. Methane, CH_4, is a product in a reaction whose enthalpy is sought, but it is a reactant in Equation 3. Water appears in two of these equations although it is not a component of the reaction forming CH_4 from carbon and hydrogen. To use Hess's law to solve this problem, we will have to manipulate the equations and adjust the heats accordingly. Recall, from Section 6.5, that writing an equation in the reverse direction changes the sign of ΔH, and that doubling the amount of reactants and products doubles the value of ΔH. Adjustments to Equations 2 and 3 will produce new equations that, along with Equation 1, can be combined to give the desired net reaction.

Solution To make CH_4 a product in the overall reaction, we reverse Equation 3 while changing the sign of ΔH. (If a reaction is exothermic in one direction, its reverse must be endothermic):

Equation 3′: $\quad CO_2(g) + 2 H_2O(\ell) \longrightarrow CH_4(g) + 2 O_2(g) \qquad \Delta H_3' = -\Delta H_3 = +890.3$ kJ

Next, we see that 2 mol of $H_2(g)$ is on the reactant side in our desired equation. Equation 2 is written for only 1 mol of $H_2(g)$ as a reactant, however. We therefore multiply the stoichiometric coefficients in Equation 2 by 2 and multiply the value of ΔH by 2.

Equation 2: $\quad 2 H_2(g) + O_2(g) \longrightarrow 2 H_2O(\ell) \qquad 2 \Delta H_2 = 2(-285.8 \text{ kJ}) = -571.6$ kJ

With these modifications, we rewrite the three equations. When added together, $O_2(g)$, $H_2O(\ell)$, and $CO_2(g)$ all cancel to give the equation for the formation of methane from its elements.

Equation 1: $C(s) + O_2(g) \longrightarrow CO_2(g)$ $\Delta H_1 = -393.5$ kJ

Equation 2: $2 H_2(g) + O_2(g) \longrightarrow 2 H_2O(\ell)$ $2 \Delta H_2 = 2(-285.8 \text{ kJ}) = -571.6$ kJ

Equation 3: $CO_2(g) + 2 H_2O(\ell) \longrightarrow CH_4(g) + 2 O_2(g)$ $\Delta H_3' = -\Delta H_3 = +890.3$ kJ

Net Equation: $C(s) + 2 H_2(g) \longrightarrow CH_4(g)$ $\Delta H_{net} = \boxed{-74.8 \text{ kJ}}$

$\Delta H_{net} = \Delta H_1 + 2 \Delta H_2 + (-\Delta H_3)$

Comment You can construct an energy level diagram summarizing the energies of this process.

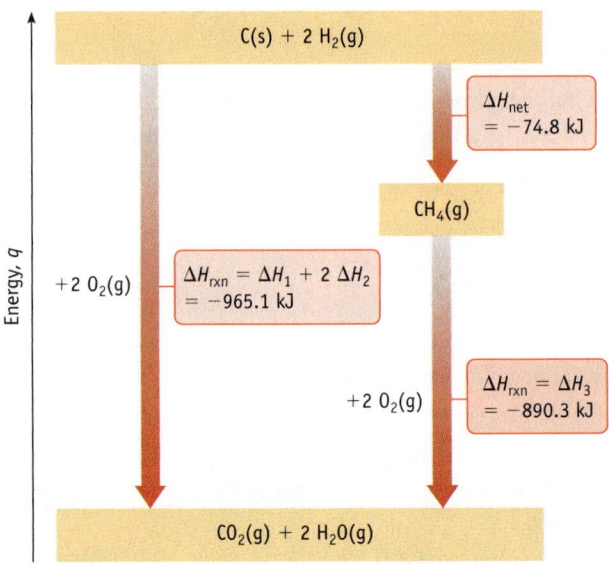

This diagram shows there are two ways to go from $C(g) + 2 H_2(g)$ to $CO_2(g) + 2 H_2O(g)$. The enthalpy changes along these two paths were $\Delta H_1 + 2 \Delta H_2$ and $\Delta H_{net} + \Delta H_3$. According to Hess's law,

$$\Delta H_1 + 2 \Delta H_2 = \Delta H_{net} + \Delta H_3$$

so $$\Delta H_{net} = \Delta H_1 + 2 \Delta H_2 + (-\Delta H_3).$$

Exercise 6.11—Using Hess's Law

Graphite and diamond are two allotropes of carbon. The enthalpy change for the process

$$C(\text{graphite}) \longrightarrow C(\text{diamond})$$

cannot be measured directly, but it can be evaluated using Hess's law.

(a) Determine this enthalpy change, using experimentally measured heats of combustion of graphite (-393.5 kJ/mol) and diamond (-395.4 kJ/mol).

(b) Draw an energy level diagram for this system.

Exercise 6.12—Using Hess's Law

Use Hess's law to calculate the enthalpy change for the formation of $CS_2(\ell)$ from $C(s)$ and $S(s)$ from the following enthalpy values.

$C(s) + O_2(g) \longrightarrow CO_2(g)$ $\Delta H = -393.5$ kJ

$S(s) + O_2(g) \longrightarrow SO_2(g)$ $\Delta H = -296.8$ kJ

$CS_2(\ell) + 3 O_2(g) \longrightarrow CO_2(g) + 2 SO_2(g)$ $\Delta H = -1103.9$ kJ

$C(s) + 2 S(s) \longrightarrow CS_2(\ell)$ $\Delta H = ?$

Problem-Solving Tip 6.2

Using Hess's Law

How did we know how the three equations should be adjusted in Example 6.8? Here is a general strategy for solving this type of problem.

Step 1. Inspect the equation whose ΔH you wish to calculate, identifying reactants and products, and locate those substances in the equations available to be added. In Example 6.8 the reactants, C(s)

and $H_2(g)$, are reactants in Equations 1 and 2, and the product, $CH_4(g)$, is a reactant in Equation 3. Equation 3 was reversed to get CH_4 on the product side where it is located in the target equation.

Step 2. Get the correct amount of the reagents on each side. In Example 6.8 only one adjustment was needed. There was 1 mol of H_2 on the left (reactant side) in Equation 2. We needed 2 mol of H_2 in the overall equation; this required doubling the quantities in Equation 2.

Step 3. Make sure other reagents in the equations will cancel when the equations are added. In Example 6.8, equal amounts of O_2 and H_2O appeared on the left and right sides in the three equations, so they cancelled when the equations were added together.

Each manipulation requires adjustment of the energy quantities. Summing the equations and the adjusted enthalpies gives the overall equation and its enthalpy change.

6.8—Standard Enthalpies of Formation

Calorimetry and the application of Hess's law have made available a great many ΔH values for chemical reactions. Often, these values are assembled into tables to make it easy to retrieve and use the data (see Table 6.2 or Appendix L). A very useful table contains **standard molar enthalpies of formation, ΔH_f°**. *The standard molar enthalpy of formation is the enthalpy change for the formation of 1 mol of a compound directly from its component elements in their standard states.* The **standard state** of an element or a compound is defined as the most stable form of the substance in the physical state that exists at a pressure of 1 bar and at a specified temperature. Most tables report standard molar enthalpies of formation at 25 °C (298 K).

Several examples of standard molar enthalpies of formation will be helpful to illustrate the meaning of these definitions.

ΔH_f° for CO_2(g): At 25 °C and 1 bar, the standard states of carbon and oxygen are solid graphite and $O_2(g)$, respectively. The standard enthalpy of formation of $CO_2(g)$ is defined as the enthalpy change that occurs in the formation of 1 mol of $CO_2(g)$ from 1 mol of C(s, graphite) and 1 mol of $O_2(g)$; that is, it is the enthalpy change for the process

$$C(s) + O_2(g) \longrightarrow CO_2(g) \qquad \Delta H_f^\circ = -393.5 \text{ kJ}$$

ΔH_f° for NaCl(s): At 25 °C and 1 bar, Na is a solid and Cl_2 is a gas. The standard enthalpy of formation of NaCl(s) is defined as the enthalpy change that occurs if 1 mol of NaCl(s) is formed from 1 mol of Na(s) and $\frac{1}{2}$ mol of $Cl_2(g)$.

$$Na(s) + \tfrac{1}{2} Cl_2(g) \longrightarrow NaCl(s) \qquad \Delta H_f^\circ = -411.12 \text{ kJ}$$

ΔH_f° for $C_2H_5OH(\ell)$: At 25 °C and 1 bar, the standard states of the elements are C(s, graphite), $H_2(g)$, and $O_2(g)$. The standard enthalpy of formation of $C_2H_5OH(\ell)$ is defined as the enthalpy change that occurs if 1 mol of $C_2H_5OH(\ell)$ is formed from 2 mol of C(s), 3 mol of $H_2(g)$, and $\frac{1}{2}$ mol of $O_2(g)$.

$$2 \text{ C(s)} + 3 \text{ H}_2(g) + \tfrac{1}{2} O_2(g) \longrightarrow C_2H_5OH(\ell) \qquad \Delta H_f^\circ = -277.0 \text{ kJ}$$

Notice that the reaction defining the heat of formation need not be (and most often is not) a reaction that a chemist is likely to carry out in the laboratory. Ethanol, for example, is not made by a reaction of the elements.

■ **ΔH Under Standard Conditions** The superscript ° indicating standard conditions is applied to other types of thermodynamic data, such as the heat of fusion and vaporization (ΔH_{fus}° and ΔH_{vap}°) and the heat of a reaction (ΔH_{rxn}°).

■ **ΔH_f° Pressure and Standard Conditions** The bar is the unit of pressure for thermodynamic quantities. One bar is approximately one atmosphere. (1 atm = 1.013 bar; see Appendix B).

Table 6.2 (and Appendix L) list values of ΔH_f° for some common substances. These values are for the formation of one mole of the compound in its standard state from its elements in their standard states. A review of these values leads to some important observations.

- The standard enthalpy of formation for an element in its standard state is zero.
- Values for compounds in solution refer to the enthalpy change for the formation of a 1 M solution of the compound from the elements making up the compound plus the enthalpy change occurring when the substance dissolves in water.
- Most ΔH_f° values are negative, indicating that formation of most compounds from the elements is exothermic. Heat evolution generally indicates that forming compounds from their elements (under standard conditions) is product-favored (see Section 6.9, page 269).
- Values of ΔH_f° can be used to compare thermal stabilities of related compounds. Consider the values of ΔH_f° for the hydrogen halides in Table 6.3. Hydrogen fluoride is the most stable of these compounds, whereas HI is the least stable.

■ ΔH_f° **Values**
Enthalpy of formation values are found in this book in Table 6.2 or Appendix L. Consult the National Institute for Standards and Technology website (**webbook.nist.gov**) for an extensive compilation of data.

Table 6.2   Selected Standard Molar Enthalpies of Formation at 298 K

Substance	Name	Standard Molar Enthalpy of Formation (kJ/mol)
C(graphite)	graphite	0
C(diamond)	diamond	+1.8
$CH_4(g)$	methane	−74.87
$C_2H_6(g)$	ethane	−83.85
$C_3H_8(g)$	propane	−104.7
$C_2H_4(g)$	ethene (ethylene)	+52.47
$CH_3OH(\ell)$	methanol	−238.4
$C_2H_5OH(\ell)$	ethanol	−277.0
$C_{12}H_{22}O_{11}(s)$	sucrose	−2221.2
$CO(g)$	carbon monoxide	−110.53
$CO_2(g)$	carbon dioxide	−393.51
$CaCO_3(s)$*	calcium carbonate	−1207.6
$CaO(s)$	calcium oxide	−635.1
$H_2(g)$	hydrogen	0
$H_2O(\ell)$	liquid water	−285.83
$H_2O(g)$	water vapor	−241.83
$N_2(g)$	nitrogen	0
$NH_3(g)$	ammonia	−45.90
$NH_4Cl(s)$	ammonium chloride	−314.55
$NO(g)$	nitrogen monoxide	+90.29
$NO_2(g)$	nitrogen dioxide	+33.10
$NaCl(s)$	sodium chloride	−411.12
$S_8(s)$	sulfur	0
$SO_2(g)$	sulfur dioxide	−296.81
$SO_3(g)$	sulfur trioxide	−395.77

Data from the NIST Webbook (**http://webbook.nist.gov**).
* Data not in NIST database. Value is from J. Dean (editor): *Lange's Handbook of Chemistry*, 14th edition, New York, McGraw-Hill, 1992.

Exercise 6.13—Standard States

What are the standard states of the following elements or compounds (at 25 °C): bromine, mercury, sodium sulfate, ethanol?

Exercise 6.14—Standard Heats of Formation

Write equations for the reactions that define the standard enthalpy of formation of $FeCl_3(s)$ and sucrose (sugar, $C_{12}H_{22}O_{11}$). What are the standard states of the reactants in each equation?

Table 6.3 Standard Molar Enthalpies of Formation of the Hydrogen Halides (at 298 K)

Compound	ΔH_f° (kJ/mol)
HF(g)	−273.3
HCl(g)	−92.3
HBr(g)	−36.3
HI(g)	+26.5

Enthalpy Change for a Reaction

The enthalpy change for a reaction under standard conditions can be calculated using Equation 6.6 *if* the standard molar enthalpies of formation are known for *all* reactants and products.

$$\Delta H_{rxn}^\circ = \sum \left[\Delta H_f^\circ (\text{products})\right] - \sum \left[\Delta H_f^\circ (\text{reactants})\right] \qquad (6.6)$$

In this equation, the symbol Σ (the Greek capital letter sigma) means "take the sum." To find ΔH_{rxn}°, add up the molar enthalpies of formation of the products and subtract from this the sum of the molar enthalpies of formation of the reactants. This equation is a logical consequence of the definition of ΔH_f° and Hess's law (see "A Closer Look: Hess's Law and Equation 6.6").

Suppose you want to know how much heat is required to decompose one mole of calcium carbonate (limestone) to calcium oxide (lime) and carbon dioxide under standard conditions:

$$CaCO_3(s) \longrightarrow CaO(s) + CO_2(g) \qquad \Delta H_{rxn}^\circ = ?$$

To do so, you would use the following enthalpies of formation from Table 6.2 (or Appendix L):

Compound	ΔH_f° (kJ/mol)
$CaCO_3(s)$	−1207.6
$CaO(s)$	−635.1
$CO_2(g)$	−393.5

■ Δ = **Final − Initial**
Equation 6.6 is another example of the principle that a change (Δ) is always calculated by subtracting the initial state (the reactants) from the final state (the products).

and then use Equation 6.6 to find the standard enthalpy change for the reaction, ΔH_{rxn}°.

$$\begin{aligned}
\Delta H_{rxn}^\circ &= \Delta H_f^\circ \left[CaO(s)\right] + \Delta H_f^\circ \left[CO_2(g)\right] - \Delta H_f^\circ \left[CaCO_3(s)\right] \\
&= \left[1 \text{ mol} (-635.1 \text{ kJ/mol}) + 1 \text{ mol} (-393.5 \text{ kJ/mol})\right] \\
&\qquad\qquad\qquad\qquad\qquad\qquad\qquad - \left[1 \text{ mol} (-1207.6 \text{ kJ/mol})\right] \\
&= +179.0 \text{ kJ}
\end{aligned}$$

The decomposition of limestone to lime and CO_2 is endothermic. That is, energy (179.0 kJ/mol of $CaCO_3$) must be supplied to decompose $CaCO_3(s)$ to $CaO(s)$ and $CO_2(g)$.

A Closer Look

Hess's Law and Equation 6.6

Equation 6.6 is an application of Hess's law. To illustrate this, let us look again at the decomposition of calcium carbonate.

$$CaCO_3(s) \longrightarrow CaO(s) + CO_2(g) \qquad \Delta H^\circ_{rxn} = ?$$

We know the enthalpy changes for the decomposition of both the reactant and the products to the elements. These correspond to ΔH°_1, ΔH°_2, and ΔH°_3 on the diagram, and each of these is the negative of the enthalpy of formation of the respective compound. We do not know ΔH°_{rxn} for $CaCO_3$'s decomposition to CaO and CO_2. However, we do know from Hess's law that

$$\Delta H^\circ_1 = \Delta H^\circ_{rxn} + \Delta H^\circ_2 + \Delta H^\circ_3$$

$$-\Delta H^\circ_f[CaCO_3(s)] = \Delta H^\circ_{rxn} - \Delta H^\circ_f\,[CaO(s)] - \Delta H^\circ_f\,[CO_2(g)]$$

$$\Delta H^\circ_{rxn} = \Delta H^\circ_f\,[CaO(s)] + \Delta H^\circ_f\,[CO_2(g)]$$
$$- \Delta H^\circ_f[CaCO_3(s)]$$
$$= (-635.1 \text{ kJ}) + (-393.5 \text{ kJ}) - (-1207.6 \text{ kJ})$$

$$\Delta H^\circ_{rxn} = +179.0 \text{ kJ}$$

This is exactly the result we obtain by applying Equation 6.6. The enthalpy change for the reaction is indeed the sum of the enthalpies of formation of the products minus that of the reactant.

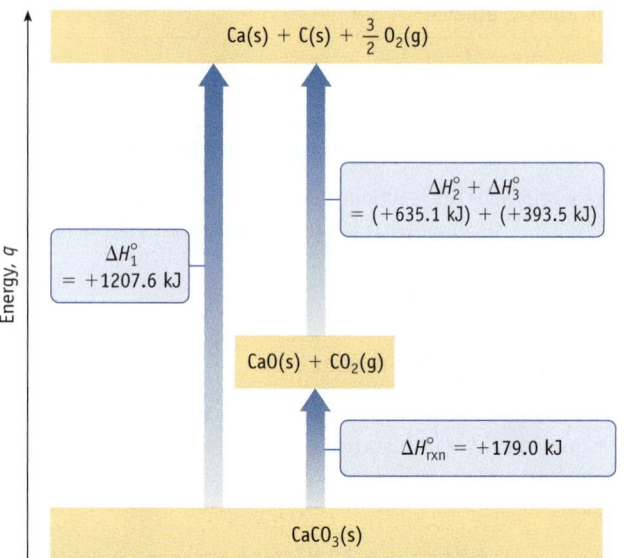

Energy level diagram for the decomposition of $CaCO_3(s)$

$Ca(s) + C(s) + \frac{3}{2} O_2(g)$

$\Delta H^\circ_2 + \Delta H^\circ_3$
$= (+635.1 \text{ kJ}) + (+393.5 \text{ kJ})$

$\Delta H^\circ_1 = +1207.6 \text{ kJ}$

Energy, q

$CaO(s) + CO_2(g)$

$\Delta H^\circ_{rxn} = +179.0 \text{ kJ}$

$CaCO_3(s)$

GENERAL
Chemistry Now™

See the General ChemistryNow CD-ROM or website:

- **Screen 6.16 Standard Enthalpy of Formation,** for a tutorial on calculating the standard enthalpy change for a reaction

---

Example 6.9—Using Enthalpies of Formation

Problem Nitroglycerin is a powerful explosive that forms four different gases when detonated:

$$2\ C_3H_5(NO_3)_3(\ell) \longrightarrow 3\ N_2(g) + \tfrac{1}{2}\ O_2(g) + 6\ CO_2(g) + 5\ H_2O(g)$$

Calculate the enthalpy change when 10.0 g of nitroglycerin is detonated. The enthalpy of formation of nitroglycerin, ΔH°_f, is −364 kJ/mol. Use Table 6.2 or Appendix L to find other ΔH°_f values that are needed.

Strategy Use values of ΔH°_f for the reactants and products in Equation 6.6 to calculate the enthalpy change produced by the detonation of 2 moles of nitroglycerin (ΔH°_{rxn}). From Table 6.2, $\Delta H^\circ_f\,[CO_2(g)] = -393.5$ kJ/mol, $\Delta H^\circ_f\,[H_2O(g)] = -241.8$ kJ/mol, and $\Delta H^\circ_f = 0$ for $N_2(g)$ and $O_2(g)$. Determine the amount represented by 10.0 g of nitroglycerin, then use this value with ΔH°_{rxn} to obtain the answer.

Solution Using Equation 6.6, we find the enthalpy change for the explosion of 2 mol of nitroglycerin is

$$\Delta H^\circ_{rxn} = 6 \text{ mol} \times \Delta H^\circ_f\,[CO_2(g)] + 5 \text{ mol} \times \Delta H^\circ_f\,[H_2O(g)] - 2 \text{ mol} \times \Delta H^\circ_f\,[C_3H_5(NO_3)_3(\ell)]$$

$$= 6 \text{ mol}\,(-393.5 \text{ kJ/mol}) + 5 \text{ mol}\,(-241.8 \text{ kJ/mol}) - 2 \text{ mol}\,(-364 \text{ kJ/mol})$$

$$= -2842 \text{ kJ for 2 mol nitroglycerin}$$

The problem asks for the enthalpy change using 10.0 g of nitroglycerin. We next need to determine the amount of nitroglycerin in 10.0 g.

$$10.0 \text{ g nitroglycerin} \times \frac{1 \text{ mol nitroglycerin}}{227.1 \text{ g nitroglycerin}} = 0.0440 \text{ mol nitroglycerin}$$

The enthalpy change for the detonation of 0.0440 mol is

$$\Delta H^\circ_{rxn} = 0.0440 \text{ mol nitroglycerin}\left(\frac{-2842 \text{ kJ}}{2 \text{ mol nitroglycerin}}\right)$$

$$= -62.6 \text{ kJ}$$

Comment The large exothermic value of ΔH°_{rxn} is in accord with the fact that this reaction is highly energetic.

Exercise 6.15—Using Enthalpies of Formation

Calculate the standard enthalpy of combustion for benzene, C_6H_6.

$$C_6H_6(\ell) + 7.5 \text{ } O_2(g) \longrightarrow 6 \text{ } CO_2(g) + 3 \text{ } H_2O(\ell) \qquad \Delta H^\circ_{rxn} = ?$$

$\Delta H^\circ_f[C_6H_6(\ell)] = +49.0$ kJ/mol. Other values needed can be found in Table 6.2 and Appendix L.

6.9—Product- or Reactant-Favored Reactions and Thermochemistry

Reactions in which reactants are largely converted to products are said to be product-favored [◄ page 197]. One aspect of chemical reactivity, and a goal of this book, is to be able to predict whether a chemical reaction will be product- or reactant-favored. In Chapter 5 you learned that certain common reactions occurring in aqueous solution—precipitation, acid–base, and gas-forming reactions—and reactions such as combustions are generally product-favored. Our discussion of the energy changes in chemical reactions allows us to begin to understand more about predicting which reactions may be product-favored.

The oxidation reactions of hydrogen and carbon (see Figures 6.15 and 6.18), Gummi Bears (Figure 6.2), and iron (Figure 6.19)

$$4 \text{ Fe}(s) + 3 \text{ } O_2(g) \longrightarrow 2 \text{ } Fe_2O_3(s)$$
$$\Delta H^\circ_{rxn} = 2 \text{ } \Delta H^\circ_f[Fe_2O_3(s)] = 2(-825.5 \text{ kJ}) = -1651.0 \text{ kJ}$$

are exothermic. All have negative values for ΔH°_{rxn}, and transfer energy to their surroundings. They are also all product-favored reactions.

Conversely, the reactant-favored decomposition of calcium carbonate is endothermic. Heat is required for the reaction to occur, and ΔH°_{rxn} is positive.

$$CaCO_3(s) \longrightarrow CaO(s) + CO_2(g) \qquad \Delta H^\circ_{rxn} = +179.0 \text{ kJ}$$

Are all exothermic reactions product-favored and all endothermic reactions reactant-favored? From these examples, we might formulate this idea as a hypothesis that can be tested by experiment and by examination of many other examples. We would find that *in most cases product-favored reactions have negative values of ΔH°_{rxn} and reactant-favored reactions have positive values of ΔH°_{rxn}.* But this is not *always* true; there are exceptions, and we shall return to the issue in Chapter 19.

■ **Reactant- or Product-Favored?** In most cases exothermic reactions are product-favored and endothermic reactions are reactant-favored.

Charles D. Winters

Figure 6.19 The product-favored oxidation of iron. Iron powder, sprayed into a bunsen burner flame, is rapidly oxidized. The reaction is exothermic and is product-favored.

■ **More About Energy**
See the interchapter "The Chemistry of Fuels" that follows on pages 282–293. It explores the types of fuels used now and new energy sources.

GENERAL
Chemistry✦Now™

See the General ChemistryNow CD-ROM or website:
- **Screen 6.17 Product-Favored Systems,** for an exercise on the reaction when a Gummi Bear is placed in molten potassium chlorate

Exercise 6.16—Product- or Reactant-Favored?

Calculate ΔH°_{rxn} for each of the following reactions and decide whether the reaction may be product- or reactant-favored.

(a) $2 \, HBr(g) \longrightarrow H_2(g) + Br_2(g)$
(b) $C(diamond) \longrightarrow C(graphite)$

GENERAL
Chemistry✦Now™

See the General ChemistryNow CD-ROM or website to:
- Assess your understanding with homework questions keyed to each goal
- Check your readiness for an exam by taking the exam-prep quiz and exploring the resources in the personalized Learning Plan it provides

For additional preparation for an examination on this chapter see the **Let's Review** section on pages 282–293.

Chapter Goals Revisited

When you have studied this chapter, you should ask whether you have met the chapter goals. In particular, you should be able to

Assess heat transfer associated with changes in temperature and changes of state
a. Describe various forms of energy and the nature of heat and thermal energy transfer (Section 6.1).
b. Use the most common energy unit, the joule, and convert between other energy units and joules (Section 6.1). General ChemistryNow homework: Study Question(s) 5
c. Recognize and use the language of thermodynamics: the system and its surroundings; exothermic and endothermic reactions (Section 6.1).
d. Use specific heat capacity in calculations of heat transfer and temperature changes (Section 6.2). General ChemistryNow homework: SQ(s) 8, 10, 12, 16, 18, 34, 42
e. Understand the sign conventions in thermodynamics (Section 6.2).
f. Use heat of fusion and heat of vaporization to find the quantity of thermal energy involved in changes of state (Section 6.3). General ChemistryNow homework: SQ(s) 22, 26, 75, 77

Apply the first law of thermodynamics
a. Understand the basis of the first law of thermodynamics (Section 6.4).

Define and understand the state functions enthalpy and internal energy
a. Recognize state functions whose values are determined only by the state of the system and not by the pathway by which that state was achieved (Section 6.4).

Calculate the energy changes occurring in chemical reactions and learn how changes are measured
a. Recognize that when a process is carried out under constant pressure conditions, the heat transferred is the enthalpy change, ΔH (Section 6.5). General ChemistryNow homework: SQ(s) 30
b. Describe how to measure the quantity of heat energy transferred in a reaction by using calorimetry (Section 6.6). General ChemistryNow homework: SQ(s) 32, 38

c. Apply Hess's law to find the enthalpy change for a reaction (Section 6.7). General ChemistryNow homework: SQ(s) 44a

d. Know how to draw and interpret energy level diagrams (Section 6.7).

e. Use standard molar enthalpy of formation, ΔH_f°, to calculate the enthalpy change for a reaction, ΔH_{rxn}° (Section 6.8). General ChemistryNow homework: SQ(s) 49b, 53b, 58, 84a

Key Equations

Equation 6.1 (page 242)

The heat transferred when the temperature of a substance changes (q). Calculated from the specific heat capacity (C), mass (m), and change in temperature (ΔT).

$$q(J) = C(J/g \cdot K) \times m(g) \times \Delta T \, (K)$$

Equation 6.2 (page 242)

Temperature changes are always calculated as final temperature minus initial temperature.

$$\Delta T = T_{final} - T_{initial}$$

Equation 6.3 (page 245)

If no heat is transferred between a system and its surroundings, the sum of heat changes within the system equals zero

$$q_1 + q_2 + q_3 \cdots = 0$$

Equation 6.4 (page 251)

The first law of thermodynamics: the change in internal energy (ΔE) in a system is the sum of the heat transferred (q) and work done (w).

$$\Delta E = q + w$$

Equation 6.5 (page 253)

Work (w) at constant pressure is the product of pressure (P) and change in volume (ΔV)

$$w = -P \times \Delta V$$

Equation 6.6 (page 267)

This equation is used to calculate the standard enthalpy change of a reaction (ΔH_{rxn}°) when the heats of formation of all of the reactants and products are known.

$$\Delta H_{rxn}^\circ = \sum \left[\Delta H_f^\circ (\text{products}) \right] - \sum \left[\Delta H_f^\circ (\text{reactants}) \right]$$

Study Questions

▲ denotes more challenging questions.

■ denotes questions available in the Homework and Goals section of the General ChemistryNow CD-ROM or website.

Blue numbered questions have answers in Appendix O and fully worked solutions in the *Student Solutions Manual*.

Structures of many of the compounds used in these questions are found on the General ChemistryNow CD-ROM or website in the Models folder.

GENERAL ChemistryNow™ Assess your understanding of this chapter's topics with additional quizzing and conceptual questions at http://now.brookscole.com/kotz6e

Practicing Skills

Energy

(*See Exercise 6.1 and General ChemistryNow Screen 6.3.*)

1. The flashlight in the photo does not use batteries. Instead you move a lever, which turns a geared mechanism and results finally in light from the bulb. What type of energy is used to move the lever? What type or types of energy are produced?

A hand-operated flashlight.

2. A solar panel is pictured in the photo. When light shines on the panel, a small electric motor propels the car. What types of energy are involved in this setup?

A solar panel operates a toy car.

Energy Units

(*See Exercise 6.2 and General ChemistryNow Screen 6.5.*)

3. You are on a diet that calls for eating no more than 1200 Cal/day. How many joules would this be?

4. A 2-in. piece of chocolate cake with frosting provides 1670 kJ of energy. What is this in dietary Calories (Cal)?

5. ■ One food product has an energy content of 170 kcal per serving and another has 280 kJ per serving. Which food has a greater energy content per serving?

6. Which has a greater energy content, a raw apple or a raw apricot? Go to the USDA Nutrient Database on the World Wide Web for the information (http://www.nal.usda.gov/fnic/foodcomp/). Report the energy content of the fruit in kcal and kJ.

Specific Heat Capacity

(*See Examples 6.1 and 6.2 and General ChemistryNow Screens 6.7–6.9.*)

7. The molar heat capacity of mercury is 28.1 J/mol · K. What is the specific heat capacity of this metal in J/g · K?

8. ■ The specific heat capacity of benzene (C_6H_6) is 1.74 J/g · K. What is its molar heat capacity (in J/mol · K)?

9. The specific heat capacity of copper is 0.385 J/g · K. What quantity of heat is required to heat 168 g of copper from $-12.2\,°C$ to $+25.6\,°C$?

10. ■ What quantity of heat is required to raise the temperature of 50.00 mL of water from 25.52 °C to 28.75 °C? The density of water at this temperature is 0.997 g/mL.

11. The initial temperature of a 344-g sample of iron is 18.2 °C. If the sample absorbs 2.25 kJ of heat, what is its final temperature?

12. ■ After absorbing 1.850 kJ of heat, the temperature of a 0.500-kg block of copper is 37 °C. What was its initial temperature?

13. A 45.5-g sample of copper at 99.8 °C is dropped into a beaker containing 152 g of water at 18.5 °C. What is the final temperature when thermal equilibrium is reached?

14. A 182-g sample of gold at some temperature is added to 22.1 g of water. The initial water temperature is 25.0 °C, and the final temperature is 27.5 °C. If the specific heat capacity of gold is 0.128 J/g · K, what was the initial temperature of the gold?

15. One beaker contains 156 g of water at 22 °C and a second beaker contains 85.2 g of water at 95 °C. The water in the two beakers is mixed. What is the final water temperature?

16. ■ When 108 g of water at a temperature of 22.5 °C is mixed with 65.1 g of water at an unknown temperature, the final temperature of the resulting mixture is 47.9 °C. What was the initial temperature of the second sample of water?

17. A 13.8-g piece of zinc was heated to 98.8 °C in boiling water and then dropped into a beaker containing 45.0 g of water at 25.0 °C. When the water and metal come to ther-

mal equilibrium, the temperature is 27.1 °C. What is the specific heat capacity of zinc?

18. ■ A 237-g piece of molybdenum, initially at 100.0 °C, is dropped into 244 g of water at 10.0 °C. When the system comes to thermal equilibrium, the temperature is 15.3 °C. What is the specific heat capacity of molybdenum?

Changes of State

(*See Examples 6.3 and 6.4 and General ChemistryNow Screen 6.10.*)

19. What quantity of heat is evolved when 1.0 L of water at 0 °C solidifies to ice? The heat of fusion of water is 333 J/g.

20. The heat energy required to melt 1.00 g of ice at 0 °C is 333 J. If one ice cube has a mass of 62.0 g, and a tray contains 16 ice cubes, what quantity of energy is required to melt a tray of ice cubes to form liquid water at 0 °C?

21. What quantity of heat is required to vaporize 125 g of benzene, C_6H_6, at its boiling point, 80.1 °C? The heat of vaporization of benzene is 30.8 kJ/mol.

22. ■ Chloromethane, CH_3Cl, arises from the oceans and from microbial fermentation and is found throughout the environment. It is used in the manufacture of various chemicals and has been used as a topical anesthetic. What quantity of heat must be absorbed to convert 92.5 g of liquid to a vapor at its boiling point, −24.09 °C? The heat of vaporization of CH_3Cl is 21.40 kJ/mol.

23. The freezing point of mercury is −38.8 °C. What quantity of heat energy, in joules, is released to the surroundings if 1.00 mL of mercury is cooled from 23.0 °C to −38.8 °C and then frozen to a solid? (The density of liquid mercury is 13.6 g/cm³. Its specific heat capacity is 0.140 J/g · K and its heat of fusion is 11.4 J/g.)

24. What quantity of heat energy, in joules, is required to raise the temperature of 454 g of tin from room temperature, 25.0 °C, to its melting point, 231.9 °C, and then melt the tin at that temperature? The specific heat capacity of tin is 0.227 J/g · K, and the heat of fusion of this metal is 59.2 J/g.

25. Ethanol, C_2H_5OH, boils at 78.29 °C. What quantity of heat energy, in joules, is required to raise the temperature of 1.00 kg of ethanol from 20.0 °C to the boiling point and then to change the liquid to vapor at that temperature? (The specific heat capacity of liquid ethanol is 2.44 J/g · K and its enthalpy of vaporization is 855 J/g.)

26. ■ A 25.0-mL sample of benzene at 19.9 °C was cooled to its melting point, 5.5 °C, and then frozen. How much heat was given off in this process? The density of benzene is 0.80 g/mL, its specific heat capacity is 1.74 J/g · K, and its heat of fusion is 127 J/g.

Enthalpy

(*See Example 6.5 and General ChemistryNow Screens 6.12 and 6.13.*)

27. Nitrogen monoxide, a gas recently found to be involved in a wide range of biological processes, reacts with oxygen to give brown NO_2 gas.

$$2\ NO(g) + O_2(g) \longrightarrow 2\ NO_2(g) \qquad \Delta H°_{rxn} = -114.1\ kJ$$

Is this reaction endothermic or exothermic? If 1.25 g of NO is converted completely to NO_2, what quantity of heat is absorbed or evolved?

28. Calcium carbide, CaC_2, is manufactured by the reaction of CaO with carbon at a high temperature. (Calcium carbide is then used to make acetylene.)

$$CaO(s) + 3\ C(s) \longrightarrow CaC_2(s) + CO(g)$$
$$\Delta H°_{rxn} = +464.8\ kJ$$

Is this reaction endothermic or exothermic? If 10.0 g of CaO is allowed to react with an excess of carbon, what quantity of heat is absorbed or evolved by the reaction?

29. Isooctane (2,2,4-trimethylpentane), one of the many hydrocarbons that make up gasoline, burns in air to give water and carbon dioxide.

$$2\ C_8H_{18}(\ell) + 25\ O_2(g) \longrightarrow 16\ CO_2(g) + 18\ H_2O(\ell)$$
$$\Delta H°_{rxn} = -10{,}922\ kJ$$

If you burn 1.00 L of isooctane (density = 0.69 g/mL), what quantity of heat is evolved?

30. ■ Acetic acid, CH_3CO_2H, is made industrially by the reaction of methanol and carbon monoxide.

$$CH_3OH(\ell) + CO(g) \longrightarrow CH_3CO_2H(\ell)$$
$$\Delta H°_{rxn} = -355.9\ kJ$$

If you produce 1.00 L of acetic acid (density = 1.044 g/mL) by this reaction, what quantity of heat is evolved?

Calorimetry

(*See Examples 6.6 and 6.7 and General ChemistryNow Screens 6.8, 6.9, and 6.14.*)

31. Assume you mix 100.0 mL of 0.200 M CsOH with 50.0 mL of 0.400 M HCl in a coffee-cup calorimeter. The following reaction occurs:

$$CsOH(aq) + HCl(aq) \longrightarrow CsCl(aq) + H_2O(\ell)$$

The temperature of both solutions before mixing was 22.50 °C, and it rises to 24.28 °C after the acid–base reaction. What is the enthalpy change for the reaction per mole of CsOH? Assume the densities of the solutions are all 1.00 g/mL and the specific heat capacities of the solutions are 4.2 J/g · K.

32. ■ You mix 125 mL of 0.250 M CsOH with 50.0 mL of 0.625 M HF in a coffee-cup calorimeter, and the temperature of both solutions rises from 21.50 °C before mixing to 24.40 °C after the reaction.

$$CsOH(aq) + HF(aq) \longrightarrow CsF(aq) + H_2O(\ell)$$

What is the enthalpy of reaction per mole of CsOH? Assume the densities of the solutions are all 1.00 g/mL and the specific heats of the solutions are 4.2 J/g · K.

33. A piece of titanium metal with a mass of 20.8 g is heated in boiling water to 99.5 °C and then dropped into a coffee-cup calorimeter containing 75.0 g of water at 21.7 °C. When thermal equilibrium is reached, the final temperature is 24.3 °C. Calculate the specific heat capacity of titanium.

34. ■ A piece of chromium metal with a mass of 24.26 g is heated in boiling water to 98.3 °C and then dropped into a coffee-cup calorimeter containing 82.3 g of water at 23.3 °C. When thermal equilibrium is reached, the final temperature is 25.6 °C. Calculate the specific heat capacity of chromium.

35. Adding 5.44 g of $NH_4NO_3(s)$ to 150.0 g of water in a coffee-cup calorimeter (with stirring to dissolve the salt) resulted in a decrease in temperature from 18.6 °C to 16.2 °C. Calculate the enthalpy change for dissolving $NH_4NO_3(s)$ in water, in kJ/mol. Assume that the solution (whose mass is 155.4 g) has a specific heat capacity of 4.2 J/g · K. (Cold packs take advantage of the fact that dissolving ammonium nitrate in water is an endothermic process.)

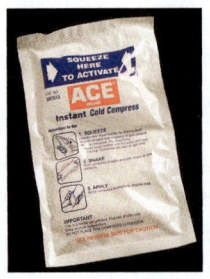

A cold pack uses the endothermic heat of solution of ammonium nitrate.

36. You should use care when dissolving H_2SO_4 in water because the process is highly exothermic. To measure the enthalpy change, 5.2 g $H_2SO_4(\ell)$ was added (with stirring) to 135 g of water in a coffee-cup calorimeter. This resulted in an increase in temperature from 20.2 °C to 28.8 °C. Calculate the enthalpy change for the process $H_2SO_4(\ell) \longrightarrow H_2SO_4(aq)$, in kJ/mol.

37. Sulfur (2.56 g) is burned in a constant volume calorimeter with excess $O_2(g)$. The temperature increases from 21.25 °C to 26.72 °C. The bomb has a heat capacity of 923 J/K, and the calorimeter contains 815 g of water. Calculate the heat evolved, per mole of SO_2 formed, for the reaction

$$S_8(s) + 8 O_2(g) \longrightarrow 8 SO_2(g)$$

Sulfur burns in oxygen with a bright blue flame to give $SO_2(g)$.

38. ■ Suppose you burn 0.300 g of C(graphite) in an excess of $O_2(g)$ in a constant volume calorimeter to give $CO_2(g)$.

$$C(graphite) + O_2(g) \longrightarrow CO_2(g)$$

The temperature of the calorimeter, which contains 775 g of water, increases from 25.00 °C to 27.38 °C. The heat capacity of the bomb is 893 J/K. What quantity of heat is evolved per mole of carbon?

39. Suppose you burn 1.500 g of benzoic acid, $C_6H_5CO_2H$, in a constant volume calorimeter and find that the temperature increases from 22.50 °C to 31.69 °C. The calorimeter contains 775 g of water, and the bomb has a heat capacity of 893 J/K. What quantity of heat is evolved in this combustion reaction, per mole of benzoic acid?

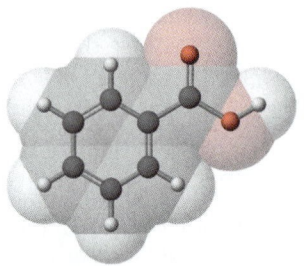

Benzoic acid, $C_6H_5CO_2H$, occurs naturally in many berries. Its heat of combustion is well known so it is used as a standard to calibrate calorimeters.

40. A 0.692-g sample of glucose, $C_6H_{12}O_6$, is burned in a constant volume calorimeter. The temperature rises from 21.70 °C to 25.22 °C. The calorimeter contains 575 g of water and the bomb has a heat capacity of 650 J/K. What quantity of heat is evolved per mole of glucose?

41. An "ice calorimeter" can be used to determine the specific heat capacity of a metal. A piece of hot metal is dropped onto a weighed quantity of ice. The quantity of heat transferred from the metal to the ice can be determined from the amount of ice melted. Suppose you heat a 50.0-g piece of silver to 99.8 °C and then drop it onto ice. When the metal's temperature has dropped to 0.0 °C, it is found that 3.54 g of ice has melted. What is the specific heat capacity of silver?

42. ■ A 9.36-g piece of platinum is heated to 98.6 °C in a boiling water bath and then dropped onto ice. (See Study Question 41.) When the metal's temperature has dropped to 0.0 °C, it is found that 0.37 g of ice has melted. What is the specific heat capacity of platinum?

Hess's Law

(*See Example 6.8 and General ChemistryNow Screen 6.15.*)

43. The enthalpy changes for the following reactions can be measured:

$$CH_4(g) + 2 O_2(g) \longrightarrow CO_2(g) + 2 H_2O(g)$$
$$\Delta H° = -802.4 \text{ kJ}$$

$$CH_3OH(g) + \tfrac{3}{2} O_2(g) \longrightarrow CO_2(g) + 2 H_2O(g)$$
$$\Delta H° = -676 \text{ kJ}$$

(a) Use these values and Hess's law to determine the enthalpy change for the reaction

$$CH_4(g) + \tfrac{1}{2} O_2(g) \longrightarrow CH_3OH(g)$$

(b) Draw an energy level diagram that shows the relationship between the energy quantities involved in this problem.

44. The enthalpy changes of the following reactions can be measured:

$$C_2H_4(g) + 3 O_2(g) \longrightarrow 2 CO_2(g) + 2 H_2O(\ell)$$
$$\Delta H^\circ = -1411.1 \text{ kJ}$$
$$C_2H_5OH(\ell) + 3 O_2(g) \longrightarrow 2 CO_2(g) + 3 H_2O(\ell)$$
$$\Delta H^\circ = -1367.5 \text{ kJ}$$

(a) ■ Use these values and Hess's law to determine the enthalpy change for the reaction

$$C_2H_4(g) + H_2O(\ell) \longrightarrow C_2H_5OH(\ell)$$

(b) Draw an energy level diagram that shows the relationship between the energy quantities involved in this problem.

45. Enthalpy changes for the following reactions can be determined experimentally:

$$N_2(g) + 3 H_2(g) \longrightarrow 2 NH_3(g) \qquad \Delta H^\circ = -91.8 \text{ kJ}$$
$$4 NH_3(g) + 5 O_2(g) \longrightarrow 4 NO(g) + 6 H_2O(g)$$
$$\Delta H^\circ = -906.2 \text{ kJ}$$
$$H_2(g) + \tfrac{1}{2} O_2(g) \longrightarrow H_2O(g) \qquad \Delta H^\circ = -241.8 \text{ kJ}$$

Use these values to determine the enthalpy change for the formation of $NO(g)$ from the elements (an enthalpy change that cannot be measured directly because the reaction is reactant-favored).

$$\tfrac{1}{2} N_2(g) + \tfrac{1}{2} O_2(g) \longrightarrow NO(g) \qquad \Delta H^\circ = ?$$

46. You wish to know the enthalpy change for the formation of liquid PCl_3 from the elements.

$$P_4(s) + 6 Cl_2(g) \longrightarrow 4 PCl_3(\ell) \qquad \Delta H^\circ = ?$$

The enthalpy change for the formation of PCl_5 from the elements can be determined experimentally, as can the enthalpy change for the reaction of $PCl_3(\ell)$ with more chlorine to give $PCl_5(s)$:

$$P_4(s) + 10 Cl_2(g) \longrightarrow 4 PCl_5(s) \qquad \Delta H^\circ = -1774.0 \text{ kJ}$$
$$PCl_3(\ell) + Cl_2(g) \longrightarrow PCl_5(s) \qquad \Delta H^\circ = -123.8 \text{ kJ}$$

Use these data to calculate the enthalpy change for the formation of 1.00 mol of $PCl_3(\ell)$ from phosphorus and chlorine.

Standard Enthalpies of Formation

(*See Example 6.9 and General ChemistryNow Screen 6.16.*)

47. Write a balanced chemical equation for the formation of $CH_3OH(\ell)$ from the elements in their standard states. Find the value for ΔH_f° for $CH_3OH(\ell)$ in Appendix L.

48. Write a balanced chemical equation for the formation of $CaCO_3(s)$ from the elements in their standard states. Find the value for ΔH_f° for $CaCO_3(s)$ in Appendix L.

49. (a) Write a balanced chemical equation for the formation of 1 mol of $Cr_2O_3(s)$ from Cr and O_2 in their standard states. Find the value for ΔH_f° for $Cr_2O_3(s)$ in Appendix L.

(b) ■ What is the standard enthalpy change if 2.4 g of chromium is oxidized to $Cr_2O_3(s)$?

50. (a) Write a balanced chemical equation for the formation of 1 mol of $MgO(s)$ from the elements in their standard states. Find the value for ΔH_f° for $MgO(s)$ in Appendix L.

(b) What is the standard enthalpy change for the reaction of 2.5 mol of Mg with oxygen?

51. Use standard heats of formation in Appendix L to calculate standard enthalpy changes for the following:

(a) 1.0 g of white phosphorus burns, forming $P_4O_{10}(s)$
(b) 0.20 mol of $NO(g)$ decomposes to $N_2(g)$ and $O_2(g)$
(c) 2.40 g of NaCl is formed from Na(s) and excess $Cl_2(g)$
(d) 250 g of iron is oxidized with oxygen to $Fe_2O_3(s)$

52. Use standard heats of formation in Appendix L to calculate standard enthalpy changes for the following:

(a) 0.054 g of sulfur burns, forming $SO_2(g)$
(b) 0.20 mol of $HgO(s)$ decomposes to $Hg(\ell)$ and $O_2(g)$
(c) 2.40 g of $NH_3(g)$ is formed from $N_2(g)$ and excess $H_2(g)$
(d) 1.05×10^{-2} mol of carbon is oxidized to $CO_2(g)$

53. The first step in the production of nitric acid from ammonia involves the oxidation of NH_3.

$$4 NH_3(g) + 5 O_2(g) \longrightarrow 4 NO(g) + 6 H_2O(g)$$

(a) Use standard enthalpies of formation to calculate the standard enthalpy change for this reaction.

(b) ■ What quantity of heat is evolved or absorbed in the oxidation of 10.0 g of NH_3?

54. The Romans used calcium oxide, CaO, to produce a strong mortar to build stone structures. The CaO was mixed with water to give $Ca(OH)_2$, which reacted slowly with CO_2 in the air to give $CaCO_3$.

$$Ca(OH)_2(s) + CO_2(g) \longrightarrow CaCO_3(s) + H_2O(g)$$

(a) Calculate the standard enthalpy change for this reaction.

(b) What quantity of heat is evolved or absorbed if 1.00 kg of $Ca(OH)_2$ reacts with a stoichiometric amount of CO_2?

55. The standard enthalpy of formation of solid barium oxide, BaO, is -553.5 kJ/mol, and the enthalpy of formation of barium peroxide, BaO_2, is -634.3 kJ/mol.

(a) Calculate the standard enthalpy change for the following reaction. Is the reaction exothermic or endothermic?

$$BaO_2(s) \longrightarrow BaO(s) + \tfrac{1}{2} O_2(g)$$

(b) Draw an energy level diagram that shows the relationship between the enthalpy change of the decomposition of BaO_2 to BaO and O_2 and the enthalpies of formation of BaO(s) and $BaO_2(s)$.

56. An important step in the production of sulfuric acid is the oxidation of SO_2 to SO_3.

$$SO_2(g) + \tfrac{1}{2} O_2(g) \longrightarrow SO_3(g)$$

Formation of SO_3 from the air pollutant SO_2 is also a key step in the formation of acid rain.

(a) Use standard enthalpies of formation to calculate the enthalpy change for the reaction. Is the reaction exothermic or endothermic?

(b) Draw an energy level diagram that shows the relationship between the enthalpy change for the oxidation of SO_2 to SO_3 and the enthalpies of formation of $SO_2(g)$ and $SO_3(g)$.

57. The enthalpy change for the oxidation of naphthalene, $C_{10}H_8$, is measured by calorimetry.

$$C_{10}H_8(s) + 12\,O_2(g) \longrightarrow 10\,CO_2(g) + 4\,H_2O(\ell)$$
$$\Delta H°_{rxn} = -5156.1 \text{ kJ}$$

Use this value, along with the standard heats of formation of $CO_2(g)$ and $H_2O(\ell)$, to calculate the enthalpy of formation of naphthalene, in kJ/mol.

58. ■ The enthalpy change for the oxidation of styrene, C_8H_8, is measured by calorimetry.

$$C_8H_8(\ell) + 10\,O_2(g) \longrightarrow 8\,CO_2(g) + 4\,H_2O(\ell)$$
$$\Delta H°_{rxn} = -4395.0 \text{ kJ}$$

Use this value, along with the standard heats of formation of $CO_2(g)$ and $H_2O(\ell)$, to calculate the enthalpy of formation of styrene, in kJ/mol.

Product- and Reactant-Favored Reactions

59. Use your "chemical sense" to decide whether each of the following reactions is product- or reactant-favored. Calculate $\Delta H°_{rxn}$ in each case, and draw an energy level diagram like those in Figure 6.18.

(a) the reaction of aluminum and chlorine to produce $AlCl_3(s)$

(b) the decomposition of mercury(II) oxide to produce liquid mercury and oxygen gas

60. Use your "chemical sense" to decide whether each of the following reactions is product- or reactant-favored. Calculate $\Delta H°_{rxn}$ in each case, and draw an energy level diagram like those in Figure 6.18.

(a) the decomposition of ozone, O_3, to oxygen molecules

(b) the decomposition of $MgCO_3(s)$ to give $MgO(s)$ and $CO_2(g)$

General Questions on Thermochemistry

These questions are not designated as to type or location in the chapter. They may combine several concepts.

61. The following terms are used extensively in thermodynamics. Define each and give an example.

(a) exothermic and endothermic

(b) system and surroundings

(c) specific heat capacity

(d) state function

(e) standard state

(f) enthalpy change, ΔH

(g) standard enthalpy of formation

62. For each of the following, tell whether the process is exothermic or endothermic. (No calculations are required.)

(a) $H_2O(\ell) \longrightarrow H_2O(s)$

(b) $2\,H_2(g) + O_2(g) \longrightarrow 2\,H_2O(g)$

(c) $H_2O(\ell, 25\ °C) \longrightarrow H_2O(\ell, 15\ °C)$

(d) $H_2O(\ell) \longrightarrow H_2O(g)$

63. For each of the following, define a system and its surroundings and give the direction of heat transfer between system and surroundings.

(a) Methane is burning in a gas furnace in your home.

(b) Water drops, sitting on your skin after a dip in a swimming pool, evaporate.

(c) Water, at 25 °C, is placed in the freezing compartment of a refrigerator, where it cools and eventually solidifies.

(d) Aluminum and $Fe_2O_3(s)$ are mixed in a flask sitting on a laboratory bench. A reaction occurs, and a large quantity of heat is evolved.

64. Which of the following are state functions?

(a) the volume of a balloon

(b) the time it takes to drive from your home to your college or university

(c) the temperature of the water in a coffee cup

(d) the potential energy of a ball held in your hand

65. Define the first law of thermodynamics using a mathematical equation and explain the meaning of each term in the equation.

66. What does the term "standard state" mean? What are the standard states of the following substances at 298 K: H_2O, NaCl, Hg, CH_4?

67. Use Appendix L to find the standard enthalpies of formation of oxygen atoms, oxygen molecules (O_2), and ozone (O_3). What is the standard state of oxygen? Is the formation of oxygen atoms from O_2 exothermic? What is the enthalpy change for the formation of 1 mol of O_3 from O_2?

68. See General ChemistryNow CD-ROM or website Screen 6.9 Heat Transfer Between Substances. Use the Simulation section of this screen to do the following experiment: Add 10.0 g of Al at 80 °C to 10.0 g of water at 20 °C. What is the final temperature when equilibrium is achieved? Use this value to estimate the specific heat capacity of aluminum.

69. See General ChemistryNow CD-ROM or website Screen 6.15 Hess's Law. Use the Simulation section of this screen to find the value of $\Delta H°_{rxn}$ for

$$SnBr_2(s) + TiCl_4(\ell) \longrightarrow SnCl_4(\ell) + TiBr_2(s)$$

▲ More challenging   ■ In General ChemistryNow   Blue-numbered questions answered in Appendix O

70. A piece of lead with a mass of 27.3 g was heated to 98.90 °C and then dropped into 15.0 g of water at 22.50 °C. The final temperature was 26.32 °C. Calculate the specific heat capacity of lead from these data.

71. Which gives up more heat on cooling from 50 °C to 10 °C, 50.0 g of water or 100. g of ethanol (specific heat capacity of ethanol = 2.46 J/g · K)?

72. A 192-g piece of copper is heated to 100.0 °C in a boiling water bath and then dropped into a beaker containing 751 g of water (density = 1.00 g/cm^3) at 4.0 °C. What is the final temperature of the copper and water after thermal equilibrium is reached? (The specific heat capacity of copper is 0.385 J/g · K).

73. You determine that 187 J of heat is required to raise the temperature of 93.45 g of silver from 18.5 °C to 27.0 °C. What is the specific heat capacity of silver?

74. Calculate the quantity of heat required to convert 60.1 g of $H_2O(s)$ at 0.0 °C to $H_2O(g)$ at 100.0 °C. The heat of fusion of ice at 0 °C is 333 J/g; the heat of vaporization of liquid water at 100 °C is 2260 J/g.

75. ■ You add 100.0 g of water at 60.0 °C to 100.0 g of ice at 0.00 °C. Some of the ice melts and cools the water to 0.00 °C. When the ice and water mixture has come to a uniform temperature of 0 °C, how much ice has melted?

76. ▲ Three 45-g ice cubes at 0 °C are dropped into 5.00×10^2 mL of tea to make ice tea. The tea was initially at 20.0 °C; when thermal equilibrium was reached, the final temperature was 0 °C. How much of the ice melted and how much remained floating in the beverage? Assume the specific heat capacity of tea is the same as that of pure water.

77. ▲ ■ Suppose that only two 45-g ice cubes had been added to your glass containing 5.00×10^2 mL of tea (See Study Question 76). When thermal equilibrium is reached, all of the ice will have melted and the temperature of the mixture will be somewhere between 20.0 °C and 0 °C. Calculate the final temperature of the beverage. (Note: The 90 g of water formed when the ice melts must be warmed from 0 °C to the final temperature.)

78. You take a diet cola from the refrigerator, and pour 240 mL of it into a glass. The temperature of the beverage is 10.5 °C. You then add one ice cube (45 g). Which of the following describes the system when thermal equilibrium is reached?
   (a) The temperature is 0 °C and some ice remains.
   (b) The temperature is 0 °C and no ice remains.
   (c) The temperature is higher than 0 °C and no ice remains.
   Determine the final temperature and the amount of ice remaining, if any.

79. Insoluble AgCl(s) precipitates when solutions of $AgNO_3(aq)$ and NaCl(aq) are mixed.

$$AgNO_3(aq) + NaCl(aq) \longrightarrow AgCl(s) + NaNO_3(aq)$$
$$\Delta H°_{rxn} = ?$$

To measure the heat evolved in this reaction, 250. mL of 0.16 M $AgNO_3(aq)$ and 125 mL of 0.32 M NaCl(aq) are mixed in a coffee-cup calorimeter. The temperature of the mixture rises from 21.15 °C to 22.90 °C. Calculate the enthalpy change for the precipitation of AgCl(s), in kJ/mol. (Assume the density of the solution is 1.0 g/mL and its specific heat capacity is 4.2 J/g · K.)

80. Insoluble $PbBr_2(s)$ precipitates when solutions of $Pb(NO_3)_2(aq)$ and NaBr(aq) are mixed.

$$Pb(NO_3)_2(aq) + 2\,NaBr(aq) \longrightarrow PbBr_2(s) + 2\,NaNO_3(aq)$$
$$\Delta H°_{rxn} = ?$$

To measure the heat evolved, 200. mL of 0.75 M $Pb(NO_3)_2(aq)$ and 200 mL of 1.5 M NaBr(aq) are mixed in a coffee-cup calorimeter. The temperature of the mixture rises by 2.44 °C. Calculate the enthalpy change for the precipitation of $PbBr_2(s)$, in kJ/mol. (Assume the density of the solution is 1.0 g/mL and its specific heat capacity is 4.2 J/g · K.)

81. The heat evolved in the decomposition of 7.647 g of ammonium nitrate can be measured in a bomb calorimeter. The reaction that occurs is

$$NH_4NO_3(s) \longrightarrow N_2O(g) + 2\,H_2O(g)$$

The temperature of the calorimeter, which contains 415 g of water, increases from 18.90 °C to 20.72 °C. The heat capacity of the bomb is 155 J/K. What quantity of heat is evolved in this reaction, in kJ/mol?

82. A bomb calorimetric experiment was run to determine the heat of combustion of ethanol (a common fuel additive). The reaction is

$$C_2H_5OH(\ell) + 3\,O_2(g) \longrightarrow 2CO_2(g) + 3\,H_2O(\ell)$$

The bomb had a heat capacity of 550 J/K, and the calorimeter contained 650 g of water. Burning 4.20 g of ethanol, $C_2H_5OH(\ell)$ resulted in a rise in temperature from 18.5 °C to 22.3 °C. Calculate the heat of combustion of ethanol, in kJ/mol.

83. ▲ The standard molar enthalpy of formation of diborane, $B_2H_6(g)$, cannot be determined directly because the compound cannot be prepared by the reaction of boron and hydrogen. It can be calculated from other enthalpy changes, however. The following enthalpy changes can be measured.

$$4\,B(s) + 3\,O_2(g) \longrightarrow 2\,B_2O_3(s) \qquad \Delta H°_{rxn} = -2543.8 \text{ kJ}$$
$$H_2(g) + \tfrac{1}{2}\,O_2(g) \longrightarrow H_2O(g) \qquad \Delta H°_{rxn} = -241.8 \text{ kJ}$$
$$H_6(g) + 3\,O_2(g) \longrightarrow B_2H_6(s) + 3\,H_2O(g)$$
$$\Delta H°_{rxn} = -2032.9 \text{ kJ}$$

   (a) Show how these equations can be added together to give the equation for the formation of $B_2H_6(g)$ from B(s) and $H_2(g)$ in their standard states. Assign enthalpy changes to each reaction.
   (b) Calculate $\Delta H°_f$ for $B_2H_6(g)$.
   (c) Draw an energy level diagram that shows how the various enthalpies in this problem are related.

(d) Is the formation of $B_2H_6(g)$ from its elements product- or reactant-favored?

84. Chloromethane, CH_3Cl, a compound found ubiquitously in the environment, is formed in the reaction of chlorine atoms with methane.

$$CH_4(g) + 2\,Cl(g) \longrightarrow CH_3Cl(g) + HCl(g)$$

(a) ■ Calculate the enthalpy change for the reaction of $CH_4(g)$ and Cl atoms to give $CH_3Cl(g)$ and $HCl(g)$. Is the reaction product- or reactant-favored?

(b) Draw an energy level diagram that shows how the various enthalpies in this problem are related.

85. The meals-ready-to-eat (MREs) in the military can be heated on a flameless heater. The source of energy in the heater is

$$Mg(s) + 2\,H_2O(\ell) \longrightarrow Mg(OH)_2(s) + H_2(g)$$

Calculate the enthalpy change under standard conditions, in joules, for this reaction. What quantity of magnesium is needed to supply the heat required to warm 25 mL of water ($d = 1.00$ g/mL) from 25 °C to 85 °C? (See W. Jensen: *Journal of Chemical Education*, Vol. 77, pp. 713–717, 2000.)

86. Hydrazine, $N_2H_4(\ell)$, is an efficient oxygen scavenger. It is sometimes added to steam boilers to remove traces of oxygen that can cause corrosion in these systems. Combustion of hydrazine gives the following information:

$$N_2H_4(\ell) + O_2(g) \longrightarrow N_2(g) + 2\,H_2O(g)$$
$$\Delta H^\circ_{rxn} = -534.3\ kJ$$

(a) Is the reaction product- or reactant-favored?

(b) Use the value for ΔH°_{rxn} with the enthalpy of formation of $H_2O(g)$ to calculate the molar enthalpy of formation of $N_2H_4(\ell)$.

87. When heated to a high temperature, coke (mainly carbon, obtained by heating coal in the absence of air) and steam produce a mixture called water gas, which can be used as a fuel or as a chemical feedstock for other reactions. The equation for the production of water gas is

$$C(s) + H_2O(g) \longrightarrow CO(g) + H_2(g)$$

(a) Use standard heats of formation to determine the enthalpy change for this reaction.

(b) Is the reaction product- or reactant-favored?

(c) What quantity of heat is involved if 1.0 metric ton (1000.0 kg) of carbon is converted to water gas?

88. Camping stoves are fueled by propane (C_3H_8), butane [$C_4H_{10}(g)$, $\Delta H^\circ_f = -127.1$ kJ/mol], gasoline, or ethanol (C_2H_5OH). Calculate the heat of combustion per gram of each of these fuels. [Assume that gasoline is represented by isooctane, $C_8H_{18}(\ell)$, with $\Delta H^\circ_f = -259.2$ kJ/mol.] Do you notice any great differences among these fuels? Are these differences related to their composition?

89. Methanol, CH_3OH, a compound that can be made relatively inexpensively from coal, is a promising substitute for gasoline. The alcohol has a smaller energy content than gasoline, but, with its higher octane rating, it burns more efficiently than gasoline in combustion engines. (It has the added advantage of contributing to a lesser degree to some air pollutants.) Compare the heat of combustion per gram of CH_3OH and C_8H_{18} (isooctane), the latter being representative of the compounds in gasoline. ($\Delta H^\circ_f = -259.2$ kJ/mol for isooctane.)

90. Hydrazine and 1,1-dimethylhydrazine both react spontaneously with O_2 and can be used as rocket fuels.

$$\underset{\text{hydrazine}}{N_2H_4(\ell)} + O_2(g) \longrightarrow N_2(g) + 2\,H_2O(g)$$

$$\underset{\text{1,1-dimethylhydrazine}}{N_2H_2(CH_3)_2(\ell)} + 4\,O_2(g) \longrightarrow$$
$$2\,CO_2(g) + 4\,H_2O(g) + N_2(g)$$

The molar enthalpy of formation of $N_2H_4(\ell)$ is +50.6 kJ/mol, and that of $N_2H_2(CH_3)_2(\ell)$ is +48.9 kJ/mol. Use these values, with other ΔH°_f values, to decide whether the reaction of hydrazine or, -dimethylhydrazine with oxygen gives more heat per gram.

A control rocket in the Space Shuttle uses hydrazine as the fuel.

91. (a) Calculate the enthalpy change, ΔH°, for the formation of 1.00 mol of strontium carbonate (the material that gives the red color in fireworks) from its elements.

$$Sr(s) + C(graphite) + \tfrac{3}{2}O_2(g) \longrightarrow SrCO_3(s)$$

The experimental information available is

$Sr(s) + \tfrac{1}{2}O_2(g) \longrightarrow SrO(s)$	$\Delta H^\circ_f = -592\ kJ$
$SrO(s) + CO_2(g) \longrightarrow SrCO_3(s)$	$\Delta H^\circ_{rxn} = -234\ kJ$
$C(graphite) + O_2(g) \longrightarrow CO_2(g)$	$\Delta H^\circ_f = -394\ kJ$

(b) Draw an energy level diagram relating the energy quantities in this problem.

92. You drink 350 mL of diet soda that is at a temperature of 5 °C.

(a) How much energy will your body expend to raise the temperature of this liquid to body temperature (37 °C)? Assume that the density and specific heat capacity of diet soda are the same as for water.

(b) Compare the value in part (a) with the caloric content of the beverage. (The label says that it has a caloric content of 1 Calorie.) What is the net energy change in your body resulting from drinking this beverage?

(c) Carry out a comparison similar to that in part (b) for a nondiet beverage whose label indicates a caloric content of 240 Calories.

93. Chloroform, $CHCl_3$, is formed from methane and chlorine in the following reaction.

$$CH_4(g) + 3 Cl_2(g) \longrightarrow 3 HCl(g) + CHCl_3(g)$$

Calculate ΔH°_{rxn}, the enthalpy change for this reaction, using the enthalpy of formation of $CHCl_3(g)$, $\Delta H^\circ_f = -103.1$ kJ/mol), and the enthalpy changes for the following reactions:

$$CH_4(g) + 2 O_2(g) \longrightarrow 2 H_2O(\ell) + CO_2(g)$$

$$\Delta H^\circ_{rxn} = -890.4 \text{ kJ}$$

$$2 HCl(g) \longrightarrow H_2(g) + Cl_2(g) \qquad \Delta H^\circ_{rxn} = +184.6 \text{ kJ}$$

$$C(graphite) + O_2(g) \longrightarrow CO_2(g) \qquad \Delta H^\circ_f = -393.5 \text{ kJ}$$

$$H_2(g) + \tfrac{1}{2} O_2(g) \longrightarrow H_2O(\ell) \qquad \Delta H^\circ_f = -285.8 \text{ kJ}$$

94. Water gas, a mixture of carbon monoxide and hydrogen, is produced by treating carbon (in the form of coke or coal) with steam at high temperatures. (See Question 87.)

$$C(s) + H_2O(g) \longrightarrow CO(g) + H_2(g)$$

Not all of the carbon available is converted to water gas as some is burned to provide the heat for the endothermic reaction of carbon and water. What mass of carbon must be burned (to CO_2 gas) to provide the heat to convert 1.00 kg of carbon to water gas?

95. Compare the heat evolved by burning 1.00 kg of carbon (to CO_2 gas) with the heat evolved by the water gas $[CO(g) + H_2(g)]$ obtained from 1.00 kg of carbon (assuming a 100% yield). (See Question 94.) Which provides more energy?

96. ▲ Isomers are molecules with the same elemental composition but a different atomic arrangement. Three isomers with the formula C_4H_8 are shown in the models below. The enthalpy of combustion of each isomer, determined using a calorimeter, is:

Compound	$\Delta H_{combustion}$ (kJ/mol)
cis-2 butene	−2687.5
trans-2-butene	−2684.2
1-butene	−2696.7

(a) Draw an energy level diagram relating the energy content of the three isomers to the energy content of the combustion products, $CO_2(g)$ and $H_2O(g)$.

(b) Use the $\Delta H_{combustion}$ data in part (a), along with the enthalpies of formation of $CO_2(g)$ and $H_2O(g)$ from Appendix L, to calculate the enthalpy of formation for each of the isomers.

(c) Draw an energy level diagram that relates the heats of formation of the three isomers to the energy of the elements in their standard states.

(d) What is the enthalpy change for the conversion of *cis*-2-butene to *trans*-2-butene?

Summary and Conceptual Questions

The following questions may use concepts from preceding chapters.

97. The first law of thermodynamics is often described as another way of stating the law of conservation of energy. Discuss whether this is an accurate portrayal.

98. Many people have tried to make a perpetual motion machine, but none have been successful although some have claimed success. Use the law of conservation of energy to explain why such a device is impossible.

99. Without doing calculations, decide whether each of the following is product- or reactant-favored.

(a) the combustion of natural gas

(b) the decomposition of glucose, $C_6H_{12}O_6$, to carbon and water

100. See General ChemistryNow CD-ROM or website Screen 6.18 Control of Chemical Reactions. What is the difference between thermodynamics and kinetics?

101. See General ChemistryNow CD-ROM or website Screen 6.9 Heat Transfer Between Substances.

(a) Explain what happens in terms of molecular motions when a hotter object comes in contact with a cooler one.

(b) What does it mean when two objects have come to thermal equilibrium?

$\Delta H_{combustion} = -2687.5$ kJ/mol

$\Delta H_{combustion} = -2684.2$ kJ/mol

$\Delta H_{combustion} = -2696.7$ kJ/mol

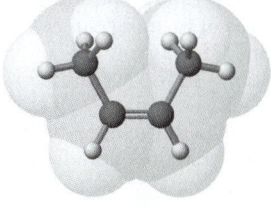

cis-2-butene

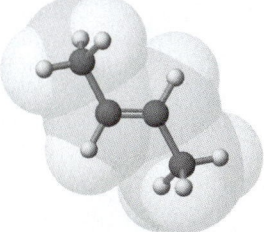

trans-2-butene

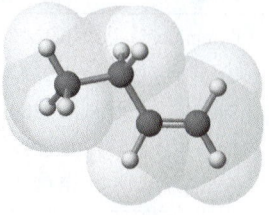

1-butene

102. The photograph here shows a toy car. A solar panel collects light, which generates electricity. This energy is used to electrolyze water to H_2 and O_2 gas, and these gases are recombined in a fuel cell (a special type of battery) to drive the car.

A toy car that uses a solar panel to collect light.
The electricity generated by the panel generates hydrogen and oxygen gases, which are used in a fuel cell.

Describe the form of energy involved in the various processes in the toy car.

103. ▲ You want to determine the value for the enthalpy of formation of $CaSO_4(s)$.

$$Ca(s) + \tfrac{1}{8} S_8(s) + 2\, O_2(g) \longrightarrow CaSO_4(s)$$

This reaction cannot be done directly. You know, however, that both calcium and sulfur react with oxygen to produce oxides in reactions that can be studied calorimetrically. You also know that the basic oxide CaO reacts with the acidic oxide SO_3 (g) to produce $CaSO_4(s)$ with $\Delta H^\circ_{rxn} = -402.7$ kJ. Outline a method for determining ΔH°_f for $CaSO_4(s)$ and identify the information that must be collected by experiment. Using information in Table 6.2, confirm that ΔH°_f for $CaSO_4(s) = -1433.5$ kJ/mol.

104. Prepare a graph of heat capacities for metals versus their atomic weights. Combine the data in Table 6.1 and the values in the following table. What is the relationship between specific heat capacity and atomic weight? Use this relationship to predict the specific heat capacity of platinum. The specific heat capacity for platinum is given in the literature as 0.133 J/g · K. How good is the agreement between the predicted and actual values?

Metal	Specific Heat Capacity (J/g · K)
Chromium	0.450
Lead	0.127
Silver	0.236
Tin	0.227
Titanium	0.522

105. Observe the molar heat capacity values for the metals in Table 6.1. What observation can you make about these values—specifically, are they widely different or very similar? Using this information, estimate the specific heat capacity for silver. Compare this estimate with the correct value for silver, 0.236 J/g · K.

106. ▲ Suppose you are attending summer school and are living in a very old dormitory. The day is oppressively hot. There is no air-conditioner, and you can't open the windows of your room because they are stuck shut from layers of paint. There is a refrigerator in the room, however. In a stroke of genius you open the door of the refrigerator, and cool air cascades out. The relief does not last long, though. Soon the refrigerator motor and condenser begin to run, and not long thereafter the room is hotter than it was before. Why did the room warm up?

107. You want to heat the air in your house with natural gas (CH_4). Assume your house has 275 m^2 (about 2800 ft^2) of floor area and that the ceilings are 2.50 m from the floors. The air in the house has a molar heat capacity of 29.1 J/mol · K. (The number of moles of air in the house can be found by assuming that the average molar mass of air is 28.9 g/mol and that the density of air at these temperatures is 1.22 g/L.) What mass of methane do you have to burn to heat the air from 15.0 °C to 22.0 °C?

108. Water can be decomposed to its elements, H_2 and O_2, using electrical energy or in a series of chemical reactions. The following sequence of reactions is one possibility:

$$CaBr_2(s) + H_2O(g) \longrightarrow CaO(s) + 2\, HBr(g)$$
$$Hg(\ell) + 2\, HBr(g) \longrightarrow HgBr_2(s) + H_2(g)$$
$$HgBr_2(s) + CaO(s) \longrightarrow HgO(s) + CaBr_2(s)$$
$$HgO(s) \longrightarrow Hg(\ell) + \tfrac{1}{2} O_2(g)$$

(a) Show that the net result of this series of reactions is the decomposition of water to its elements.

(b) If you use 1000. kg of water, what mass of H_2 can be produced?

(c) Calculate the value of ΔH°_{rxn} for each step in the series. Are the reactions predicted to be product- or reactant favored?

$$\Delta H^\circ_f\ [CaBr_2(s)] = -683.2\ \text{kJ/mol}$$
$$\Delta H^\circ_f\ [HgBr_2(s)] = -169.5\ \text{kJ/mol}$$

(e) Comment on the commercial feasibility of using this series of reactions to produce $H_2(g)$ from water.

109. Suppose that an inch of rain falls over a square mile of ground. (A density of 1.0 g/cm^3 is assumed.) The heat of vaporization of water at 25 °C is 44.0 kJ/mol. Calculate the quantity of heat transferred to the surroundings from the condensation of water vapor in forming this quantity of liquid water. (The huge number tells you how much energy is "stored" in water vapor and why we think of storms as such great forces of energy in nature. It is interesting to compare this result with the energy given off, 4.2×10^6 kJ, when a ton of dynamite explodes.)

110. ▲ Peanuts and peanut oil are organic materials and burn in air. How many burning peanuts does it take to provide the energy to boil a cup of water (250 mL of water)? To solve this problem we assume each peanut, with an average mass of 0.73 g, is 49% peanut oil and 21% starch; the remainder is noncombustible. We further assume peanut oil is palmitic acid, $C_{16}H_{32}O_2$, with an enthalpy of formation of -848.4 kJ/mol. Starch is a long chain of $C_6H_{10}O_5$ units, each unit having an enthalpy of formation of -960 kJ. (*See General ChemistryNow Screens 6.1 and 6.19: Chemical Puzzler.*)

Charles D Winters

How many burning peanuts are required to provide the heat to boil 250 mL of water?

Do you need a live tutor for homework problems?
Access vMentor at General ChemistryNow at
http://now.brookscole.com/kotz6e
for one-on-one tutoring from a chemistry expert

Let's Review Chapters 1–6

Photo source: Charles D. Winters

When solid NH_4Cl is heated it is converted to gaseous NH_3 and HCl. The white "smoke" you see consists of tiny particles of solid NH_4Cl formed by the recombination of gaseous NH_3 and HCl. At this point in your study of chemistry you have acquired skills to do the following:

(a) write a balanced chemical equation for the decomposition of NH_4Cl,

(b) name each of the compounds involved,

(c) calculate the amount of heat required to decompose the NH_4Cl,

(d) and determine the mass of product expected if you know the masses of the reactants.

This section, called *Let's Review*, will explore questions about this and other reactions.

The Purpose of *Let's Review*

- *Let's Review* provides additional homework questions for Chapters 1 through 6. Routine questions covering each of the concepts in a chapter are found in that chapter and are those usually identified by topic. In contrast, many of the *Let's Review* questions combine several concepts from one or more chapters. Many come from the examinations given by the authors and others are based on actual experiments or processes in chemical research or in the chemical industry.

- *Let's Review* provides guidance for Chapters 1 through 6 as you prepare for an exam covering material in these chapters. Although this is designated for Chapters 1 through 6 you may choose only material appropriate to the exam in your specific course.

- Each Comprehensive Question is correlated with tutorials and exercises on the General ChemistryNow website and with the OWL online homework system to which you may have purchased access. Some questions include a screen shot of a tutorial so you see what resources are available to help you review.

Preparing for an Examination on Chapters 1-6

1. **GENERAL Chemistry ⚛ Now™** Use General ChemistryNow online to take a chapter **Pre-Test.** A **Personalized Learning Plan** will indicate sections of the textbook that should be reviewed.

2. Use General ChemistryNow online to work though media-based **Exercises, Guided Simulations,** and **Intelligent Tutors.**

3. If you subscribe to OWL, use the **Tutorials** in that system.

4. Work on the questions below that are relevant to a particular chapter or chapters. See the solutions to those questions at the end of this section.

5. For background and help with a question, use the General ChemistryNow information or OWL questions that are correlated with the question.

Key Points to Know for Chapters 1–6

Here are some of the key points you must know to be successful in Chapters 1 through 6.

- The names and symbols of the chemical elements.
- Mass to moles and moles to mass conversions.
- Calculating the molar mass of a chemical compound.
- Names and formulas of chemical compounds.
- Balancing chemical equations.
- Basic chemical stoichiometry.
- The names and formulas of common cations and anions and of acids and bases.
- Writing chemical equations for precipitation, acid-base, and gas-forming reactions.
- Calculating and using solution concentrations.
- The terminology of thermochemistry.
- The relation of heat, mass, temperature, and specific heat capacity.
- Calculating the amount of heat involved in a chemical reaction.

Examination Preparation Questions

▲ denotes more challenging questions.

Important information about the questions that follow:

- See the Study Questions in each chapter for questions on basic concepts.

- Some questions arise from recent research in chemistry and the other sciences. They often involve concepts from more than one chapter and may be more challenging than those in earlier chapters. Not all chapter goals or concepts are necessarily addressed in these questions.

- Assessing Key Points questions are short-answer questions covering the key points listed on this page.

- Each Comprehensive Question is correlated with the text section covering that topic, with General ChemistryNow material, and with questions in OWL that may provide additional background.

- The computer screens are largely taken from General ChemistryNow but are also contained in the OWL system if you subscribe to it. They illustrate **Tutorials** or other resources that may help in answering the questions.

Assessing Key Points

1. Sulfur is a *(metal)(nonmetal)(metalloid)* ___ and its symbol is ___. The element has ___ protons in the nucleus. Sulfur-32 and sulfur-34 account for 99.2% of all S atoms; ___ is the more abundant of the two. In the sulfur-34 isotope there are ___ neutrons in the nucleus. Sulfur forms a common monatomic ion whose symbol is ___.

2. A flask having 0.0123 g of Ar contains ____ mol and ___ atoms.

3. Fill in the table.

Cation	Anion	Name	Formula
NH_4^+		ammonium bromide	
		iron(II) sulfate	
Mg^{2+}			$Mg(CH_3CO_2)_2$
Al^{3+}	NO_3^-		

4. Which contains more mass, 0.50 mol of silicon dioxide or 0.50 mol of iron?

5. What is the molar mass of Epsom salt, $MgSO_4 \cdot 7H_2O$?

6. All of the formulas below are correct EXCEPT
   - (a) $Ba(NO_3)_2$
   - (b) $KClO_4$
   - (c) Na_3N
   - (d) $Al_2(SO_4)_3$
   - (e) Ca_2HPO_4

7. Sodium oxalate has the formula $Na_2C_2O_4$. Based on this information, the formula for iron(III) oxalate is
   (a) FeC_2O_4
   (d) $Fe_2(C_2O_4)_3$
   (b) $Fe(C_2O_4)_2$
   (e) $Fe_3(C_2O_4)_2$
   (c) $Fe(C_2O_4)_3$

8. Which of the following series contains only *known nonmetal anions*?
   (a) S^{2-}, Br^-, Al^{3+}
   (d) Cl^-, Fe^{3+}, S^{2-}
   (b) N^{2-}, I^-, O^{2-}
   (e) In^+, Br^{2-}, Te^{2-}
   (c) P^{3-}, F^-, Se^{2-}

9. Balance the following chemical equations.
   (a) $Fe_2O_3(s) + Mg(s) \longrightarrow MgO(s) + Fe(s)$
   (b) $C_6H_5CH_3(\ell) + O_2(g) \longrightarrow H_2O(\ell) + CO_2(g)$

10. The reaction of iron with oxygen produces iron(III) oxide.

    $$4\,Fe(s) + 3\,O_2(g) \longrightarrow 2\,Fe_2O_3(s)$$

    If you have 1.6 mol of Fe, what amount of O_2 is needed for complete reaction and what amount of Fe_2O_3 is produced? What are the masses of the reactants and products involved?

11. Give the oxidation number of each underlined atom:
    (a) $\underline{Br}O_3^-$;   (b) $H_2\underline{C}_2O_4$;   (c) $\underline{S}O_4^{2-}$

12. Zinc reacts readily with nitric acid to give the metal ion and such products as NO_2.

    $Zn(s) + 2\,NO_3^-(aq) + 4\,H^+(aq)$
    $\qquad \longrightarrow Zn^{2+}(aq) + 2\,NO_2(g) + 2\,H_2O(\ell)$

    The oxidation number of N in NO_3^- is ___. The substance oxidized is _____ and the oxidizing agent is ___.

13. What mass of Na_2CO_3 (molar mass = 106 g/mol) must be used to make 250. mL of a 0.100 M solution of sodium carbonate? What is the concentration of Na^+ ions in this solution?

14. In the laboratory you added water to 125 mL of 0.160 M H_2SO_4. If the final volume of the diluted solution is 1.00 L, what is the concentration of the acid in the diluted solution?

15. Which of the following are water-soluble?
    $BaSO_4$ $Ba(NO_3)_2$ $BaCO_3$ Na_2SO_4
    $AgBr$ KCl $Mg(CH_3CO_2)_2$

16. Complete and balance the following equation. Describe it as an acid-base reaction, a precipitation, a gas-forming reaction, or a redox reaction:

    $$Na_2CO_3(aq) + HNO_3(aq) \longrightarrow$$

17. Vinegar has a pH of 4.52. What is the hydrogen ion concentration in the vinegar?

18. A sample of Na_2CO_3 (0.412 g) is titrated to the equivalence point with 35.63 mL of HNO_3. What is the concentration of the HNO_3 solution?

19. Write the net ionic equation for the reaction of aqueous solutions of sodium hydroxide and iron(II) chloride.

20. What quantity of heat is required to warm 225 g water from room temperature (25 °C) to 93 °C?

21. What is the enthalpy change for the combustion of exactly one mole of gaseous propane? Is the reaction exo- or endothermic?

    $$C_3H_8(g) + 5\,O_2(g) \longrightarrow 3\,CO_2(g) + 4\,H_2O(g)$$

Comprehension Questions

22. (Chapters 1 and 2) Gold and nanochemistry.
    (a) Gold has only one naturally occurring isotope. How many neutrons are there in an atom of gold?
    (b) Gold's density is 19.3 g/cm^3. If you have 0.0125 mol of gold, what is the volume of the piece? How many atoms are contained in the piece?
    (c) ▲ A spherical gold nanoparticle has a diameter of 3.00 nm. If the radius of a gold atom is 0.144 nm, estimate the number of gold atoms in the nanoparticle. [Assume the gold atoms are tiny spheres and that they occupy 74.0% of the available space in the spherical nanoparticle. Volume of a sphere = $(4/3)\pi r^3$]

    Text Sections: 1.2, 1.4, 1.6, 1.8, 2.5

    General ChemistryNow Screens: 1.10, 1.14–1.17, 2.12, 2.14

    OWL Questions: 1-2, 1-5-1-7, 2-2a, 2-3

23. (Chapters 1-3) Copper is commonly used, in spite of the fact that its abundance on Earth is only 50 parts per million.
    (a) Copper has two isotopes: ^{63}Cu (62.930 u) and ^{65}Cu (64.928 u). Which is more abundant?
    (b) ▲What are the relative abundances of the two isotopes?
    (c) Two copper-containing compounds are $CuCO_3$ and Cu_2S. Name each compound and give the charge on the copper ion in each.

    Text Sections: 1.4, 1.6, 1.8, 2.5, 3.3

    General ChemistryNow Screens: 1.10, 1.14–1.17, 2.14, 3.8, 3.9

    OWL Questions: 1-2, 1-5-1-7, 2-3, 3-6

3.8 Ionic Compound Formulas

SIMULATION	Cation:	Anion:	
	○ NH_4^+	○ F^-	
	○ Li^+	◉ Cl^-	Name:
	○ Na^+	○ Br^-	calcium chloride
	◉ Ca^{2+}	○ OH^-	
	○ Ba^{2+}	○ S^{2-}	
	○ Ag^+	○ CO_3^{2-}	Formula:
	○ Fe^{2+}	○ SO_4^{2-}	$CaCl_2$
	○ Fe^{3+}	○ PO_4^{3-}	
	○ Al^{3+}	○ NO_3^-	
	○ Pb^{2+}	○ ClO_4^-	

General ChemistryNow Screen 3.8 Tutorial on naming and deriving formulas for ionic compounds.

24. (Chapters 1-3) Nuclear power plants are relatively common in much of the world and most use uranium.
    (a) Natural uranium exists mostly as the ^{238}U isotope, but fissionable uranium is the ^{235}U isotope. Give the number of protons and neutrons is each of these isotopes.

(b) Uranium is found or processed as (i) UO_2, (ii) $U(SO_4)_2$ and, (iii) UF_6. Name each compound, calculate its molar mass, and give the charge on the uranium ion.

Text Sections: 1.8, 2.3, 2.5, 3.3

General ChemistryNow Screens: 2.12, 2.15, and 3.9

OWL Questions: 2-2a, 2-3, 3-5

25. **(Chapter 2)** Silver, a Group 1B element, has two stable isotopes. If one of the silver isotopes has a mass of 106.905 and an abundance of 51.839%, what is the mass of the other silver isotope. What is its mass number?

Text Section: 2.3

General ChemistryNow Screens: 2.12 and 2.13 and Active Figure 2.10

OWL Question: 2-2

26. **(Chapters 3 and 4)** The common mineral fluorite has the formula CaF_2. (See photo on page 51)

(a) What is the proper chemical name of the mineral?

(b) How many electrons do the ions Ca^{2+} and F^- have? Do these have the same number of electrons as a noble gas? Which of the gases?

(c) Fluorite is the commercial source of hydrogen fluoride. Balance the equation for the reaction of sulfuric acid, H_2SO_4, with CaF_2 to give calcium sulfate and hydrogen fluoride.

(d) If you combine 1.0 kg of CaF_2 with excess sulfuric acid, what mass of hydrogen fluoride can be produced?

Text Sections: 3.3, 3.5, 4.2, and 4.3

General ChemistryNow Screens: 3.5, 3.9, 3.14, 3.15, 4.4, and 4.6

OWL Questions: 3-3, 3-10, 3-11, 4-1, and 4-3.

27. **(Chapter 3)** "Green chemistry" is a movement within the chemistry community to produce chemicals in a way that does not lead to environmental problems. Caprolactam, used to make nylon, is made in large quantities. One problem is that production of each kilogram of caprolactam results in 4 kilograms of ammonium sulfate. Ammonium sulfate has few commercial uses, so it is often deposited in landfills.

(a) Caprolactam has the structure shown here (where C atoms are gray, H atoms are white, N atoms are blue, and O atoms are red). What is the formula of caprolactam?

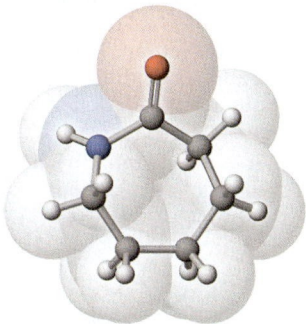

Structure of a molecule of caprolactam

(b) What is the molar mass of caprolactam?

(c) What is the weight percent of each element in caprolactam?

Text Sections: 3.2–3.6

General ChemistryNow Screens: 3.14, 3.15, 3.16

OWL Questions: 3-10, 3-11, 3-12.

28. **(Chapters 2, 3, and 5)** A recent article (C. M. Johnson and B. L. Beard, *Science*, Vol. 309, page 1025, 2005) described the biogeochemical cycling of iron isotopes. The authors stated that "In terms of isotopic studies of the transition elements, iron has received the most attention because of its high abundance on Earth and its prominent role in biogeochemical processes."

(a) Give the number of electrons and protons in an atom of Fe and in ions of Fe^{2+} and Fe^{3+}. Which is the more oxidized form?

(b) Name the following iron compounds: $Fe(CH_3CO_2)_2$, $FePO_4$, $Fe(ClO_4)_3$.

Text Sections: 2.11, 3.5 and 3.9

General ChemistryNow Screens: 2.11, 3.5, and 3.9

OWL Questions: 2-1d, 3-3, 3-4, 3-5, and 3-6

29. **(Chapter 3)** Inorganic nanotubes have been prepared based on tungsten and sulfur. The tungsten content of one such nanotube is 74.14%. What is the empirical formula of these nanotubes?

Text Section: 3.6

General ChemistryNow Screen: 3.17 and 3.18

OWL Question: 3-12

30. **(Chapter 3)** Nepetalactone is the chemical name for catnip. It is one of a family of compounds that sets many cats into a frenzy. It is 72.26% carbon, 8.49% hydrogen, and the remainder is oxygen. It has a molar mass of 166.21 g/mol. What are the empirical and molecular formulas of catnip?

Text Section: 3.6

General ChemistryNow Screen: 3.17 and 3.18

OWL Question: 3-12

3.17 Empirical Formulas

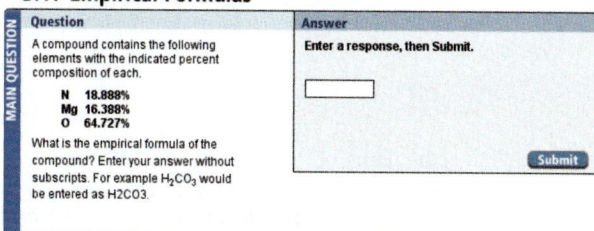

General ChemistryNow Screen 3.17 Tutorial on deriving empirical formulas.

31. **(Chapter 4)** There is great interest in finding inexpensive ways to produce hydrogen gas, which can be used as a fuel. A recently discovered method uses a catalyst to promote the reaction of ethanol, water, and oxygen.

$$2\ C_2H_5OH(g) + 4\ H_2O(g) + O_2(g)$$
$$\longrightarrow 4\ CO_2(g) + 10\ H_2(g)$$

   (a) What mass of H_2 can be produced from 1.00 kg of ethanol and unlimited amounts of water and oxygen?

   (b) What masses of H_2O and O_2 are required to react with 1.00 kg of ethanol?

   Text Sections: 4.2–4.4

   General ChemistryNow Screens: 4.4–4.8

   OWL Questions: 4-1, 4-2, 4-3, and 4-4

32. **(Chapters 4 and 6)** Methane, CH_4, can be converted to gaseous CH_3OH under special circumstances.

   (a) Write a balanced equation for the conversion of CH_4 to gaseous CH_3OH with O_2.

   (b) If you combine 125 g of CH_4 with 145 g of O_2 what mass of CH_3OH would be obtained?

   (c) Is the conversion of CH_4 to CH_3OH with O_2 exothermic or endothermic? What is the enthalpy change for the conversion of 1.00 kg of CH_4 to $CH_3OH(g)$?

   Text Sections: 4.2, 4.3, 6.1, and 6.8

   General ChemistryNow Screens: 4.4–4.8, 6.4, and 6.16

   OWL Questions: 4-1, 4-2, 4-3, and 6-6

6.16 Enthalpy of Formation and Enthalpy Change

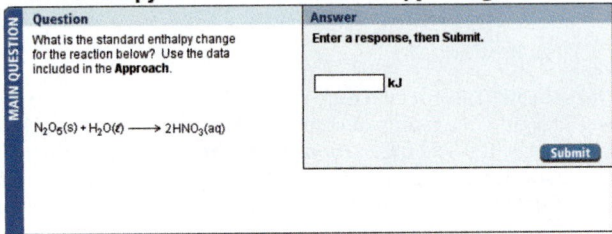

General ChemistryNow Screen 6.16 Tutorial on determining the enthalpy change for a reaction.

33. **(Chapters 4 and 5)** Gold is an important commodity in our economy. Gold (in rocks) dissolves in the presence of CN^- ion (from KCN) and oxygen to form the stable anion $[Au(CN)_2]^-$.

$$4\ Au(s) + 8\ CN^-(aq) + O_2(g) + 2\ H_2O(\ell)$$
$$\longrightarrow 4\ [Au(CN)_2]^-(aq) + 4\ OH^-(aq)$$

   (a) What is the name of the CN^- ion and of KCN?

   (b) This is an oxidation–reduction reaction. Describe how you would reach that conclusion and then identify the substance oxidized, the substance reduced, and the oxidizing and reducing agents.

   (c) What mass of KCN would be needed to dissolve 1.00 kg of gold?

   (d) If the KCN is used as a 0.15 M solution, what volume of solution is used to supply the KCN required to dissolve 1.00 kg of gold?

   Text Sections: 4.3, 5.7, 5.8, and 5.11

   General ChemistryNow Screens: 4.5, 4.6, 5.14, 5.15, and 5.18

   OWL Questions: 4-2, 4-3, 5-8, 5-9, 5-11

34. ▲ **(Chapters 4 and 5)** Linoleic acid is an essential fatty acid and a component of peanut, corn, safflower, and soybean oils.

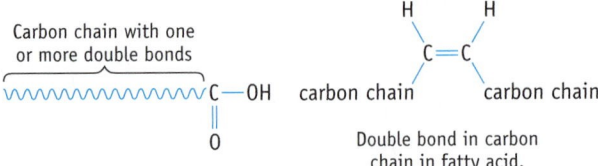

   All fatty acids have a long carbon chain, which consists of a CH_3 group at the end, an acid $—CO_2H$ group at the other end, and some number of CH_2 groups along the chain, and, often, one or more carbon-carbon double bonds.

   (a) Analysis of linoleic acid indicates it is 77.09% carbon and 11.50% H; the remainder is oxygen. What is the empirical formula of linoleic acid?

   (b) A 1.234-g sample is titrated with 0.113 M NaOH. The volume of base required is 38.94 mL. What is the molar mass of the acid?

   (c) One can determine the number of double bonds in a fatty acid by reacting it with iodine, I_2. Each double bond reacts with one I_2 molecule. If 0.351 g of linoleic acid requires 0.636 g of I_2, how many double bonds does each molecule of the fatty acid contain?

   Text Sections: 4.6, 5.10

   General ChemistryNow Screens: 3.16, 4.5, 4.6, 4.11, 5.19

   OWL Questions: 3-12, 4-6, 5-12

35. **(Chapter 5)** Oyster beds in the oceans require chloride ions for growth. The minimum concentration is 8 mg/L (8 parts per million). To analyze for the amount of chloride ion in a 50.0-mL sample of water, you add a few drops of aqueous potassium chromate and then titrate the sample with 25.60 mL of 0.001036 M silver nitrate. The silver nitrate reacts with chloride ion, and, when the ion is completely removed, the silver nitrate reacts with potassium chromate to give a red precipitate.

   (a) Write a balanced net ionic equation for the reaction of silver nitrate with chloride ion in aqueous solution.

(b) Write a complete balanced equation and a net ionic equation for the reaction of silver nitrate with potassium chromate, indicating whether each compound is water-soluble or not.

(c) What is the concentration of chloride ions in the sample? Is it sufficient to promote oyster growth?

Text Sections: 5.1, 5.2, 5.8, 5.10

General ChemistryNow Screens: 5.4, 5.6, 5.15, 5.18, and 5.19

OWL Questions: 5-1, 5-2, 5-3, 5-9, 5-11, and 5-12.

5.4 Solubility of Ionic Compounds

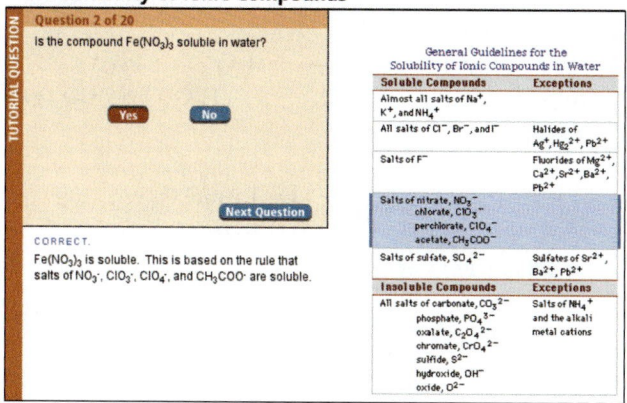

General ChemistryNow Screen 5.4 Exercise on the solubility of ionic compounds

5.7 Net Ionic Equations

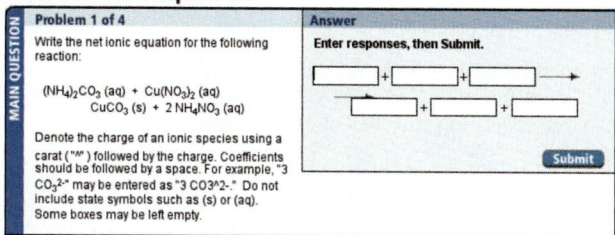

General ChemistryNow Screen 5.7 Tutorial on writing net ionic equations

36. ▲ **(Chapter 3 and 5)** A 0.463-g sample of a compound composed of Ti, Cl, C and H was burned in air to produce 0.818 g of CO_2 and 0.168 g of H_2O. Aqueous silver nitrate was added to the solid residue from the combustion and produced 0.533 g of AgCl. What is the empirical formula of the compound?

Text Sections: 3.6, 5.2, and 5.10

General ChemistryNow Screens: 3.17 and 3.18

OWL Questions: 3-12

37. ▲ **(Chapter 5)** You have a bottle of solid barium hydroxide and some dilute sulfuric acid. You place some of the barium hydroxide in water and slowly add sulfuric acid to the mixture. While adding the sulfuric acid, you measure the conductivity of the mixture (as in Active Figure 5.2, page 178)

(a) Write the complete, balanced equation and the net ionic equation for the reaction occurring when barium hydroxide and sulfuric acid are mixed.

(b) Which diagram below represents the change in conductivity as the acid is added to the aqueous barium hydroxide? Explain briefly.

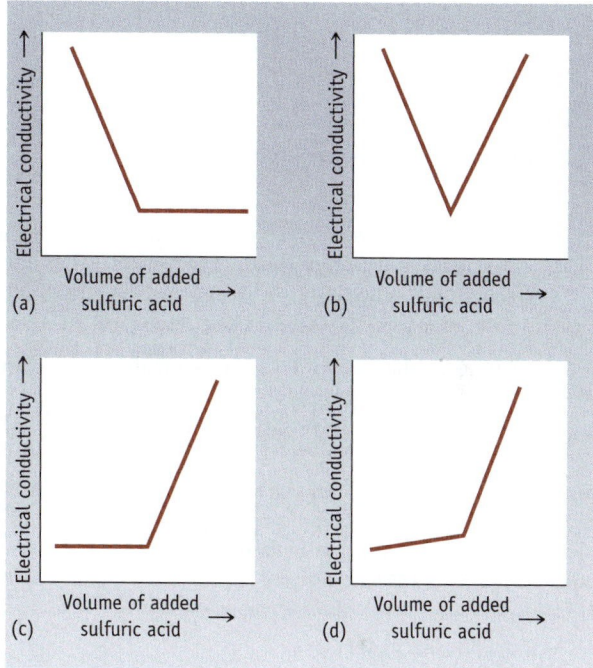

Text Sections: 5.2, 5.7, 5.8, and 5.10

General ChemistryNow Screens: 5.4, 5.5–5.7, and 5.10.

OWL Questions: 5-1, 5-2, 5-3

38. **(Chapter 6)** A 9.40-g sample of KBr was dissolved in 105 g of water at 23.6 °C in a coffee cup calorimeter. After mixing, the temperature of the solution was 20.3 °C. Assume no heat is transferred to the cup or the surroundings. Find the heat of solution, ΔH_{soln}, of KBr. (The specific heat capacity of the solution is assumed to be 4.184 J/g · °C.)

Text Section: 6.6

General ChemistryNow Screen: 6.14

OWL Question: 6-5

39. **(Chapter 6)** The heat of combustion of isooctane (C_8H_{18}) is 5.45×10^3 kJ/mol. Calculate the heat evolved per gram of isooctane and per liter of isooctane (d = 0.688 g/mL). (Isooctane is one of many hydrocarbons in gasoline, and its heat of combustion will approximate the energy obtained when gasoline burns.)

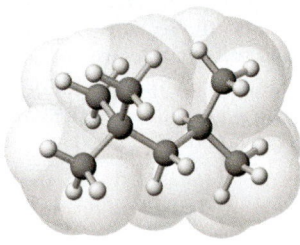

Isooctane
C_8H_{18}

Text Section: 6.6

General ChemistryNow Screen: 6.14

OWL Question: 6-5

40. (Chapters 4 and 6) Hydrogen can be produced (along with CO) using the reaction of steam (H_2O) with carbon (as coal), methane, and other hydrocarbons. For example,

$$C(s) + H_2O(g) \longrightarrow H_2(g) + CO(g)$$

(a) Compare the mass of H_2 expected from the reaction of steam with 100. g each of methane, petroleum, and coal. (Assume complete reaction in each case. Use CH_2 and C as representative formula for petroleum and coal, respectively.)

(b) Compare the heat involved in the reactions of carbon and methane with steam. Which involves more heat?

Text Sections: 4.3 and 6.8

General ChemistryNow Screens: 4.6 and 6.16

OWL Questions: 4-3 and 6-6

41. (All chapters) On page 282 you see the decomposition of solid NH_4Cl to form gaseous NH_3 and HCl.

(a) Write a balanced chemical equation for the decomposition of NH_4Cl.

(b) When NH_3 and HCl recombine to form NH_4Cl, is this an oxidation-reduction reaction?

(c) Name each of the compounds involved.

(d) Calculate the amount of heat required to decompose 1.00 mol of NH_4Cl.

(e) Determine the mass of NH_3 expected if you decompose 1.00 g of NH_4Cl.

(f) If the HCl from the decomposition of 1.00 g of NH_4Cl is absorbed by 126 mL of water, what is the pH of the solution?

Answers to Assessing Key Point Questions

1. Sulfur (S) is a nonmetal and an atom has 16 protons. The ^{32}S isotope, with 16 neutrons, is more abundant because the atomic mass is very close to 32. ^{34}S has 18 neutrons. A common ion is S^{2-}.

2. 3.08×10^{-4} mol Ar and 1.85×10^{20} atoms

3. NH_4^+, Br^-, ammonium bromide, NH_4Br
   Fe^{2+}, SO_4^{2-}, iron(II) sulfate, $FeSO_4$

Mg^{2+}, $CH_3CO_2^-$, magnesium acetate, $Mg(CH_3CO_2)_2$
Al^{3+}, NO_3^-, aluminum nitrate, $Al(NO_3)_3$

4. 0.50 mol of SiO_2 (with a molar mass of 60.1 g/mol) has more mass than 0.50 mol of Fe (55.85 g/mol).

5. 246.5 g/mol. **Comment:** Make sure to add in 7 mol of H_2O.

6. (e) The calcium ion is always Ca^{2+}, and the hydrogen phosphate ion is HPO_4^{2-}. The correct formula is $CaHPO_4$.

7. Based on the formula of $Na_2C_2O_4$ you know the oxalate ion has a 2− charge, $C_2O_4^{2-}$. Thus, the correct formula is (d).

8. (c) Answers (a) and (d) have correct ions, but two of the ions are based on metals.

9. (a) $Fe_2O_3(s) + 3\,Mg(s) \longrightarrow 3\,MgO(s) + 2\,Fe(s)$
   (b) $C_6H_5CH_3(\ell) + 9\,O_2(g) \longrightarrow 4\,H_2O(\ell) + 7\,CO_2(g)$

10. Stoichiometry.

$$1.6\ \text{mol Fe} \left(\frac{3\ \text{mol O}_2}{4\ \text{mol Fe}} \right) = 1.2\ \text{mol O}_2\ \text{required}$$

$$1.6\ \text{mol Fe} \left(\frac{2\ \text{mol Fe}_2O_3}{4\ \text{mol Fe}} \right) = 0.80\ \text{mol Fe}_2O_3\ \text{produced}$$

$$1.6\ \text{mol Fe} \left(\frac{55.85\ \text{g Fe}}{\text{mol Fe}} \right) = 89\ \text{g Fe}$$

$$1.2\ \text{mol O}_2 \left(\frac{32.0\ \text{g O}_2}{\text{mol O}_2} \right) = 38\ \text{g O}_2$$

$$0.80\ \text{mol Fe}_2O_3 \left(\frac{159.7\ \text{g Fe}_2O_3}{\text{mol Fe}_2O_3} \right) = 130\ \text{g Fe}_2O_3$$

11. $Br = +5$; $C = +3$; $S = +6$

12. N is NO_3^- is +5. Zn is oxidized (oxidation number goes from 0 to +2) and the oxidizing agent is NO_3^- (because the oxidation number of N goes from +5 to +4).

13. Solutions

$$0.250\ \text{L} \left(\frac{0.100\ \text{mol}}{\text{L}} \right) \left(\frac{106.0\ \text{g Na}_2CO_3}{1\ \text{mol Na}_2CO_3} \right) = 2.65\ \text{g Na}_2CO_3$$

The Na^+ concentration is $2 \times 0.100\ \text{M} = 0.200\ \text{M}$

14. Here we can use the shortcut equation $C_{conc} \cdot V_{conc} = C_{dilute} \cdot V_{dilute}$. See Problem-Solving Tip 5.2, page 210.

$$\left(\frac{0.16\ \text{mol H}_2SO_4}{\text{L}} \right)(0.125\ \text{L}) = (C_{dilute})(1.00\ \text{L})$$

$$C_{dilute} = \frac{0.020\ \text{mol H}_2SO_4}{\text{L}} = 0.020\ \text{M}$$

15. $Ba(NO_3)_2$, Na_2SO_4, KCl, $Mg(CH_3CO_2)_2$. Nitrates, sodium salts, potassium salts, and acetates are almost always soluble in water.

16. $Na_2CO_3(aq) + 2\,HNO_3(aq)$
$$\longrightarrow 2\,NaNO_3(aq) + CO_2(g) + H_2O(\ell)$$
This is a gas-forming reaction.

17. $pH = -\log[H^+]$ or $[H^+] = 10^{-pH}$. See page 212, Equations 5.2 and 5.3.

$$[H^+] = 1 \times 10^{-4.52} = 3.0 \times 10^{-5}\ M$$

Comment: Don't forget that the power of 10 is a negative number when using the equation $[H^+] = 10^{-pH}$.

18. The first step is to write a balanced equation for the titration. See the answer to 16 above.

$$0.412\ g\ Na_2CO_3\left(\frac{1\ mol\ Na_2CO_3}{106.0\ g\ Na_2CO_3}\right)$$
$$= 3.89 \times 10^{-3}\ mol\ Na_2CO_3$$

$$3.89 \times 10^{-3}\ mol\ Na_2CO_3\left(\frac{2\ mol\ HNO_3}{1\ mol\ Na_2CO_3}\right)$$
$$= 7.77 \times 10^{-3}\ mol\ HNO_3$$

$$\frac{7.77 \times 10^{-3}\ mol\ HNO_3}{0.03563\ L} = 0.218\ M\ HNO_3$$

19. Complete balanced equation:

$$2\ NaOH(aq) + FeCl_2(aq) \longrightarrow 2\ NaCl(aq) + Fe(OH)_2(s)$$

Net ionic equation:

$$2\ OH^-(aq) + Fe^{2+}(aq) \longrightarrow Fe(OH)_2(s)$$

20. See Equation 6.1 on page 242.
$$q = (225\ g)(4.184\ J/g \cdot K)(366\ K - 298\ K) = 6.40 \times 10^4\ J$$

21. See Equation 6.6 on page 267.
$$\Delta H^\circ_{rxn} = 3\,\Delta H^\circ_f[CO_2(g)] + 4\,\Delta H^\circ_f[H_2O(g)]$$
$$-\{\Delta H^\circ_f[C_3H_8(g)] + 5\,\Delta H^\circ_f[O_2(g)]\}$$
$$= 3\ mol\ (-393.5\ kJ/mol) + 4\ mol\ (-241.8\ kJ/mol)$$
$$- [1\ mol\ (-104.7\ kJ/mol) + 5\ mol\ (0\ kJ/mol)]$$
$$= -2043.0\ kJ$$

Reaction is exothermic.

Solutions to Comprehensive Questions

22. (a) The atomic mass of gold is 196.967 u, so the mass number of the single gold isotope is 197 ($^{197}_{79}Au$).

Mass number = 197 = No. of protons + no. of neutrons

No. of neutrons = 118

(b) Calculate the volume of the piece of gold. Let us first convert amount of gold (mol) to mass (g) using the molar mass and then convert the mass to a volume using the density.

$$0.0125\ mol\left(\frac{196.967\ g}{1\ mol}\right) = 2.46\ g$$

$$2.46\ g\left(\frac{1\ cm^3}{19.3\ g}\right) = 0.128\ cm^3$$

Calculate the number of atoms of gold from the amount of gold using Avogadro's number.

$$0.0125\ mol\ Au\left(\frac{6.022 \times 10^{23}\ atoms\ Au}{1\ mol\ Au}\right)$$
$$= 7.53 \times 10^{21}\ atoms\ Au$$

(c) ▲ The strategy here is to calculate the volume of the nanoparticle. Next find the volume of the particle occupied by gold atoms. To find the number of atoms in the particle we need to know the volume of one atom.

$$\text{Volume of gold nanoparticle} = \frac{4}{3}\pi r^3$$
$$= \frac{4}{3}\pi(1.50\ nm)^3 = 14.1\ nm^3$$

Volume of particle occupied by gold atoms
$$= 14.1\ nm^3\ (0.740) = 10.5\ nm^3$$

$$\text{Volume of one gold atom} = \frac{4}{3}\pi r^3$$
$$= \frac{4}{3}\pi(0.144\ nm)^3 = 0.0125\ nm^3\ per\ atom$$

Number of atoms in the nanoparticle
$$= 10.5\ nm^3\left(\frac{1\ atom\ Au}{0.0125\ nm^3}\right) = 839\ Au\ atoms$$

23. (a) The atomic mass of an element is the weighted average of the isotope masses and their abundances. The atomic mass of copper is closer to 63 than to 65 so ^{63}Cu is more abundant than ^{65}Cu.

(b) To calculate the isotopes abundances begin by assuming the percent abundance of ^{63}Cu is X and that of ^{65}Cu is Y. Therefore,

63.546 u = (X/100)(62.930 u) + (Y/100)(64.928 u)

Because X + Y must equal 100, this means that Y = 100 − X, and so

63.546 u = (X/100)(62.930 u)
$$+ [(100 - X)/100](64.928\ u)$$

Solving for X, we find a percent abundance of 69.17%. This means Y is 30.83%.

(c) $CuCO_3$, copper(II) carbonate, Cu^{2+}

Cu_2S, copper(I) sulfide, Cu^+

24. (a) ^{238}U has 92 proton and 146 neutrons. ^{235}U has 92 protons and 143 neutrons.

(b) Formulas

Formula	Name	Molar Mass (g/mol)	Charge on U
UO_2	uranium(IV) oxide	270.0	+4
$U(SO_4)_2$	uranium(IV) sulfate	430.2	+4
UF_6	uranium(VI) fluoride	352.0	+6

25. This question is similar to question 23b above. Here we want to find the mass of one isotope knowing the mass of the other isotope and the percent abundance of each. Here we have

Atomic mass of silver $= 107.868$ u
$\qquad = (51.839/100)(106.905 \text{ u}) + X\,[(100 - 51.839)/100]$

where $X =$ mass of other silver isotope $= 108.905$ u. The mass number of the isotope is 109 (^{109}Ag).

26. (a) The name of the mineral is calcium fluoride.
   (b) Ca^{2+} ions have 18 electrons, the same as the number of electrons in argon, Ar. F^- ions have 10 electrons, the same as the number of electrons in Ne.
   (c) $CaF_2 + H_2SO_4 \longrightarrow CaSO_4 + 2\,HF$
   (d) See Problem-Solving Tip 4.1 on page 149. First convert the mass of CaF_2 to amount (mol). Next, use the stoichiometric factor to calculate the amount of HF expected (the theoretical yield). Finally, convert this amount to mass.

$$1.0 \times 10^3 \text{ g CaF}_2 \left(\frac{1 \text{ mol}}{78.07 \text{ g CaF}_2} \right) = 13 \text{ mol CaF}_2$$

$$13 \text{ mol CaF}_2 \left(\frac{2 \text{ mol HF}}{1 \text{ mol CaF}_2} \right) = 26 \text{ mol HF}$$

$$26 \text{ mol HF} \left(\frac{20.0 \text{ g}}{1 \text{ mol HF}} \right) = 5.1 \times 10^2 \text{ g HF}$$

27. (a) Molecular formula: $C_6H_{11}NO$
   Comment: Note that it is usual to write formulas of compounds having C and H with those elements first. Any other elements follow in alphabetical order.
   (b) Molar mass: 113.2 g/mol
   (c) The mass percent is calculated as on page 119.

   Mass percent of carbon

$$= \frac{6 \text{ mol C}(12.0 \text{ g/mol C})}{113.2 \text{ g/mol caprolactam}} \times 100\% = 63.6\%$$

   In the same manner, we find 9.80% for H, 12.4% for N, and 14.1% for O.

28. **Iron chemistry**
   (a)

Atom/Ion	Protons	Electrons
Fe	26	26
Fe^{2+}	26	24
Fe^{3+}	26	23

   Comment: Students occasionally add electrons to make a positive ion and subtract them to make a negative, the opposite of the correct procedure. Positive ions occur when there are fewer electrons than protons, and negative ions result when there are more electrons than protons.
   (b)

Formula	Name
$Fe(CH_3CO_2)_2$	Iron(II) acetate
$FePO_4$	Iron(III) phosphate
$Fe(ClO_4)_3$	Iron(III) perchlorate

29. We assume that the weight percent of an element is the mass of the element in a 100-g sample. To find the empirical formula, we first find the amount (mol) of the element in the 100-g sample, then compare the amounts of all elements in the 100-g sample. These amounts are in the same ratio as the number of atoms in the molecule.

$$74.14 \text{ g W}\left(\frac{1 \text{ mol W}}{183.84 \text{ g}} \right) = 0.4033 \text{ mol W}$$

$$25.86 \text{ g S}\left(\frac{1 \text{ mol S}}{32.066 \text{ g}} \right) = 0.8065 \text{ mol S}$$

$$\frac{0.8065 \text{ mol S}}{0.4033 \text{ mol W}} = \frac{2 \text{ mol S}}{1 \text{ mol W}}$$

The empirical formula of the compound is WS_2.

Comment: When finding a formula from weight percent data, always use as many significant figures as possible (usually at least 3). See Problem-Solving Tip 3.3 on page 124.

30. This follows the same procedure as in question 29.

$$72.26 \text{ g C}\left(\frac{1 \text{ mol C}}{12.011 \text{ g}} \right) = 6.016 \text{ mol C}$$

$$8.49 \text{ g H}\left(\frac{1 \text{ mol H}}{1.008 \text{ g}} \right) = 8.423 \text{ mol H}$$

$$19.25 \text{ g O}\left(\frac{1 \text{ mol O}}{15.999 \text{ g}} \right) = 1.203 \text{ mol O}$$

With the amount of each element in a 100-g sample known, we divide each amount by the amount of the element present in the smallest amount (here oxygen).

$$\frac{6.016 \text{ mol C}}{1.203 \text{ mol O}} = 5 \text{ C to 1 O} \qquad \frac{8.423 \text{ mol H}}{1.203 \text{ mol O}} = 7 \text{ H to 1 O}$$

This leads to an empirical formula of C_5H_7O. Because the experimental molar mass (166.21 g/mol) is twice the empirical formula mass, the molecular formula is $C_{10}H_{14}O_2$. The structure of the active ingredient in catnip is illustrated here.

Nepetalactone, catnip

31. (a) After calculating the amount of C_2H_5OH, we use a stoichiometric factor to determine the amount of H_2 produced.

$$1.00 \times 10^3 \text{ g } C_2H_5OH \left(\frac{1 \text{ mol } C_2H_5OH}{46.07 \text{ g } C_2H_5OH} \right) = 21.7 \text{ mol } C_2H_5OH$$

$$21.7 \text{ mol } C_2H_5OH \left(\frac{10 \text{ mol } H_2}{2 \text{ mol } C_2H_5OH} \right) = 109 \text{ mol } H_2$$

$$109 \text{ mol } H_2 \left(\frac{2.016 \text{ g}}{1 \text{ mol } H_2} \right) = 219 \text{ g } H_2$$

(b) The amount of C_2H_5OH is known from part (a), and this can be related to the amount and mass of H_2O and O_2 required, again using the appropriate stoichiometric factors.

$$21.7 \text{ mol } C_2H_5OH \left(\frac{1 \text{ mol } O_2 \text{ required}}{2 \text{ mol } C_2H_5OH \text{ available}} \right)$$
$$= 10.9 \text{ mol } O_2 \text{ required}$$

$$10.9 \text{ mol } O_2 \left(\frac{32.0 \text{ g } O_2}{1 \text{ mol } O_2} \right) = 347 \text{ g } O_2 \text{ required}$$

$$21.7 \text{ mol } C_2H_5OH \left(\frac{4 \text{ mol } H_2O \text{ required}}{2 \text{ mol } C_2H_5OH \text{ available}} \right)$$
$$= 43.4 \text{ mol } H_2O \text{ required}$$

$$43.4 \text{ mol } H_2O \left(\frac{18.02 \text{ g } H_2O}{1 \text{ mol } H_2O} \right) = 782 \text{ g } H_2O \text{ required}$$

32. Stoichiometry and thermochemistry of methanol production

(a) $2 CH_4(g) + O_2(g) \longrightarrow 2 CH_3OH(g)$

(b) This is a limiting reactant problem, so we first calculate the amounts of each reactant and then determine if they are in the correct stoichiometric ratio or if one is in limited supply.

$$125 \text{ g } CH_4 \left(\frac{1 \text{ mol } CH_4}{16.04 \text{ g } CH_4} \right) = 7.79 \text{ mol } CH_4$$

$$145 \text{ g } O_2 \left(\frac{1 \text{ mol } O_2}{31.999 \text{ g } O_2} \right) = 4.53 \text{ mol } O_2$$

$$\text{Ratio of amounts} = \frac{7.79 \text{ mol } CH_4}{4.53 \text{ mol } O_2} = \frac{1.72 \text{ mol } CH_4}{1.00 \text{ mol } O_2}$$

The balanced equation specifies there should be twice as much CH_4 as O_2. The ratio of amounts available is not 2 to 1, so CH_4 is the limiting reactant.

$$7.79 \text{ mol } CH_4 \left(\frac{2 \text{ mol } CH_3OH}{2 \text{ mol } CH_4} \right) = 7.79 \text{ mol } CH_3OH$$

$$7.79 \text{ mol } CH_3OH \left(\frac{32.04 \text{ g } CH_3OH}{1 \text{ mol } CH_3OH} \right) = 250. \text{ g } CH_3OH$$

(c) See Equation 6.6 on page 267 for the way to calculate the enthalpy change for a reaction from heats of formation.

$$\Delta H^\circ_{rxn} = 2 \, \Delta H^\circ_f [CH_3OH(g)]$$
$$\quad -\{2\Delta H^\circ_f [CH_4(g)] + \Delta H^\circ_f [O_2(g)]\}$$
$$\Delta H^\circ_{rxn} = 2 \text{ mol } CH_3OH \, (-201.0 \text{ kJ/mol})$$
$$\quad -2 \text{ mol } CH_4 \, (-74.87 \text{ kJ/mol})$$
$$\Delta H^\circ_{rxn} = -252.3 \text{ kJ for 2 mol } CH_4$$

The reaction is exothermic.

$$1.00 \times 10^3 \text{ g } CH_4 \left(\frac{1 \text{ mol } CH_4}{16.04 \text{ g } CH_4} \right) \left(\frac{-252.3 \text{ kJ}}{2 \text{ mol } CH_4} \right)$$
$$= -7860 \text{ kJ for 1.00 kg of } CH_4$$

33. (a) CN^- is the cyanide ion and KCN is potassium cyanide.

(b) Oxygen begins with an oxidation number of 0 in O_2 and appears as an O atom with an oxidation number of -2 in H_2O. Thus, the oxygen is reduced and is the oxidizing agent. Gold begins as gold(0), whereas it is a Au^+ ion in the anion $[Au(CN)_2]^-$. This means the gold has been oxidized and is the reducing agent.

(c) Mass of KCN required by 1.00 kg of gold

$$1.00 \times 10^3 \text{ g } Au \left(\frac{1 \text{ mol } Au}{196.97 \text{ g } Au} \right) = 5.08 \text{ mol } Au$$

$$5.08 \text{ mol } Au \left(\frac{8 \text{ mol } KCN \text{ required}}{4 \text{ mol } Au \text{ available}} \right) = 10.2 \text{ mol } KCN \text{ required}$$

$$10.2 \text{ mol } KCN \left(\frac{65.12 \text{ g } KCN}{1 \text{ mol } KCN} \right) = 661 \text{ g } KCN$$

(d) Volume of 0.15 M KCN solution required by 1.00 kg of gold

$$10.2 \text{ mol } KCN \left(\frac{1 \text{ L}}{0.15 \text{ mol } KCN} \right) = 68 \text{ L KCN solution}$$

34. (a) Use the percent composition to determine the empirical formula.

$$77.09 \text{ g } C \left(\frac{1 \text{ mol } C}{12.011 \text{ g } C} \right) = 6.418 \text{ mol } C$$

$$11.50 \text{ g } H \left(\frac{1 \text{ mol } H}{1.008 \text{ g } H} \right) = 11.41 \text{ mol } H$$

$$11.41 \text{ g } O \left(\frac{1 \text{ mol } O}{15.999 \text{ g } O} \right) = 0.7132 \text{ mol } O$$

To find the empirical formula, find the ratio of amounts of the elements.

$$\frac{6.418 \text{ mol } C}{0.7132 \text{ mol } O} = \frac{9 \text{ mol } C}{1 \text{ mol } O} \qquad \frac{11.41 \text{ mol } H}{0.7132 \text{ mol } O} = \frac{16 \text{ mol } H}{1 \text{ mol } O}$$

The empirical formula is $C_9H_{16}O$.

(b) Use titration data to determine the molar mass and thus the molecular formula.

$$0.03894 \text{ L NaOH} \left(\frac{0.113 \text{ mol NaOH}}{1 \text{ L NaOH}} \right)$$
$$= 4.40 \times 10^{-3} \text{ mol NaOH}$$

Because each molecule of fatty acid has only one acid group to react with a base, the amount of acid in the sample is 4.40×10^{-3} mol. The molar mass is

$$\text{Molar mass} = \frac{\text{sample mass}}{\text{amount of acid in sample}}$$
$$= \frac{1.234 \text{ g}}{4.40 \times 10^{-3} \text{ mol}} = 2.80 \times 10^2 \text{ g/mol}$$

The molecular formula is therefore $C_{18}H_{32}O_2$.

(c) Number of double bonds in the carbon chain.

$$0.636 \text{ g } I_2 \left(\frac{1 \text{ mol } I_2}{253.8 \text{ g } I_2} \right)$$

$$= 2.51 \times 10^{-3} \text{ mol } I_2 \text{ and mol of}$$

double bonds in carbon chain

$$0.351 \text{ g fatty acid} \left(\frac{1 \text{ mol fatty acid}}{280.4 \text{ g fatty acid}} \right)$$

$$= 1.25 \times 10^{-3} \text{ mol fatty acid}$$

$$\frac{2.51 \times 10^{-3} \text{ mol double bonds}}{1.25 \times 10^{-3} \text{ mol fatty acid}}$$

$$= 2 \text{ mol double bonds per mol of fatty acid}$$

35. (a) $Ag^+(aq) + Cl^-(aq) \longrightarrow AgCl(s)$

(b) Complete equation:

$$2 \text{ AgNO}_3(aq) + K_2CrO_4(aq) \longrightarrow Ag_2CrO_4(s) + 2 \text{ KNO}_3(aq)$$

Net ionic equation:

$$2 \text{ Ag}^+(aq) + CrO_4^{2-}(aq) \longrightarrow Ag_2CrO_4(s)$$

(c) Concentration of chloride ions: From the volume and concentration of silver ions used in the titration, we know the amount of silver used.

$$0.02560 \text{ L AgNO}_3 \left(\frac{1.036 \times 10^{-3} \text{ mol Ag}^+}{1 \text{ L AgNO}_3} \right)$$

$$= 2.652 \times 10^{-5} \text{ mol Ag}^+$$

Because each silver ion reacts with one Cl^- ion, the 50.0-mL sample contained 2.652×10^{-5} mol Cl^-.

$$2.652 \times 10^{-5} \text{ mol Cl}^- \left(\frac{35.453 \text{ g Cl}}{1 \text{ mol Cl}} \right) \left(\frac{1000 \text{ mg}}{1 \text{ g}} \right)$$

$$= 0.9403 \text{ mg Cl}^-$$

$$Cl^- \text{ concentration in the water} = \frac{\text{mass of Cl}^-}{\text{sample volume}}$$

$$= \left(\frac{0.9403 \text{ mg Cl}^-}{50.0 \text{ mL}} \right) \left(\frac{1000 \text{ mL}}{\text{L}} \right) = 18.8 \text{ mg/L}$$

The Cl^- concentration is sufficient.

36. From the combustion of the compound we can find the amount and mass of C and H in the 0.463-g sample.

$$0.818 \text{ g CO}_2 \left(\frac{1 \text{ mol CO}_2}{44.01 \text{ g CO}_2} \right) \left(\frac{1 \text{ mol C}}{1 \text{ mol CO}_2} \right) = 0.0186 \text{ mol C}$$

$$0.0186 \text{ mol C} \left(\frac{12.011 \text{ g C}}{1 \text{ mol C}} \right) = 0.223 \text{ g C}$$

$$0.168 \text{ g H}_2O \left(\frac{1 \text{ mol H}_2O}{18.02 \text{ g H}_2O} \right) \left(\frac{2 \text{ mol H}}{1 \text{ mol H}_2O} \right) = 0.0186 \text{ mol H}$$

$$0.0186 \text{ mol H} \left(\frac{1.008 \text{ g H}}{1 \text{ mol H}} \right) = 0.0188 \text{ g H}$$

By precipitating the Cl as AgCl, we find the amount and mass of Cl in the sample.

$$0.533 \text{ g AgCl} \left(\frac{1 \text{ mol AgCl}}{143.3 \text{ g AgCl}} \right) \left(\frac{1 \text{ mol Cl}}{1 \text{ mol AgCl}} \right) = 0.00372 \text{ mol Cl}$$

$$0.00372 \text{ mol Cl} \left(\frac{35.45 \text{ g Cl}}{1 \text{ mol Cl}} \right) = 0.132 \text{ g Cl}$$

The total mass of C, H, and Cl in the 0.463-g sample is 0.374 g. Thus, the mass of Ti is 0.089 g or 1.9×10^{-3} mol.

$$0.089 \text{ g Ti} \left(\frac{1 \text{ mol Ti}}{47.9 \text{ g Ti}} \right) = 0.0019 \text{ mol Ti}$$

Ratio of amounts:

$$\frac{0.0186 \text{ mol C}}{0.0019 \text{ mol Ti}} = 10 \text{ C to 1 Ti} \qquad \frac{0.0186 \text{ mol H}}{0.0019 \text{ mol Ti}} = 10 \text{ H to 1 Ti}$$

$$\frac{0.00372 \text{ mol Cl}}{0.0019 \text{ mol Ti}} = 2 \text{ Cl to 1 Ti}$$

The empirical formula of the compound is $C_{10}H_{10}TiCl_2$.

37. Solubility, net ionic equations, and conductivity

(a) $Ba(OH)_2(aq) + H_2SO_4(aq) \longrightarrow BaSO_4(s) + 2 H_2O(\ell)$
$Ba^{2+}(aq) + 2 OH^-(aq) + 2 H^+(aq) + SO_4^{2-}(aq)$
$\longrightarrow BaSO_4(s) + 2 H_2O(\ell)$

(b) Plot (b). As H_2SO_4 is added to the $Ba(OH)_2$ solution, both are consumed, and the products are both nonconducting. The number of ions in solution declines and so does the conductivity. At some point both of the reactants are completely consumed, so the conductivity drops to a minimum. If addition of H_2SO_4 continues, the conductivity increases.

38. Calculate the heat involved in the solution process.

$$q_{\text{solution}} = (105 \text{ g water} + 9.4 \text{ g KBr})(4.184 \text{ J/g} \cdot \text{K})(-3.3 \text{ K})$$

$$= -1600 \text{ J}$$

The sign of the heat involved in the solution process is negative, indicating that the water has given heat up to the KBr.

Now we recognize that

$$q_{\text{solution}} + q_{\text{KBr}} = 0 \quad \text{and} \quad \text{so } q_{\text{KBr}} = +1600 \text{ J}$$

Calculate the amount of KBr.

$$9.40 \text{ g KBr} \left(\frac{1 \text{ mol KBr}}{119.0 \text{ g KBr}} \right) = 0.0790 \text{ mol KBr}$$

Calculate the heat involved per mole.

$$\text{Heat of solution per mol} = \frac{+ 1600 \text{ J}}{0.0790 \text{ mol KBr}}$$

$$= 2.0 \times 10^4 \text{ J/mol or 20. kJ/mol}$$

Comment: The solution process is endothermic, so the temperature drops when KBr dissolves, and q for the solution process is a negative quantity. This means the heat of solution of KBr is a positive quantity. This can be verified using the heats of formation.

$$\Delta H^\circ_{\text{solution}} = \Delta H^\circ_f [\text{KBr}(aq)] - \Delta H^\circ_f [\text{KBr}(s)]$$

$$= -373.9 \text{ kJ/mol} - (-393.8 \text{ kJ/mol}) = +19.9 \text{ kJ/mol}$$

39. Burning isooctane.

$$\left(\frac{5.45 \times 10^3 \text{ kJ}}{1 \text{ mol isooctane}}\right)\left(\frac{1 \text{ mol isooctane}}{114.2 \text{ g}}\right) = 47.7 \text{ kJ/g}$$

$$\left(\frac{47.7 \text{ kJ}}{1 \text{ g isooctane}}\right)\left(\frac{0.688 \text{ g isooctane}}{\text{mL}}\right)\left(\frac{1000 \text{ mL}}{\text{L}}\right)$$
$$= 3.28 \times 10^4 \text{ kJ/L}$$

40. (a) Compare the mass of hydrogen obtained from methane, petroleum, and coal.

Methane: $CH_4(g) + H_2O(g) \longrightarrow 3 H_2(g) + CO(g)$

$$100. \text{ g}\left(\frac{1 \text{ mol CH}_4}{16.04 \text{ g}}\right)\left(\frac{3 \text{ mol H}_2}{1 \text{ mol CH}_4}\right)\left(\frac{2.016 \text{ g H}_2}{1 \text{ mol H}_2}\right) = 37.7 \text{ g H}_2$$

Petroleum: $CH_2(g) + H_2O(g) \longrightarrow 2 H_2(g) + CO(g)$

$$100. \text{ g}\left(\frac{1 \text{ mol CH}_2}{14.03 \text{ g}}\right)\left(\frac{2 \text{ mol H}_2}{1 \text{ mol CH}_2}\right)\left(\frac{2.016 \text{ g H}_2}{1 \text{ mol H}_2}\right) = 28.7 \text{ g H}_2$$

Coal: $C(s) + H_2O(g) \longrightarrow H_2(g) + CO(g)$

$$100. \text{ g}\left(\frac{1 \text{ mol C}}{12.01 \text{ g}}\right)\left(\frac{1 \text{ mol H}_2}{1 \text{ mol C}}\right)\left(\frac{2.016 \text{ g H}_2}{1 \text{ mol H}_2}\right) = 16.8 \text{ g H}_2$$

(b) Heat involved in reaction of steam with methane and coal.

Methane:

$$\Delta H^\circ_{rxn} = \Delta H^\circ_f[CO(g)] + 3 \Delta H^\circ_f[H_2(g)]$$
$$- \{\Delta H^\circ_f[CH_4(g)] + \Delta H^\circ_f[H_2O(g)]\}$$
$$= 1 \text{ mol } (-110.525 \text{ kJ/mol}) + 3 \text{ mol } (0 \text{ kJ/mol})$$
$$- [1 \text{ mol } (-74.87 \text{ kJ/mol}) + 1 \text{ mol } (-241.83 \text{ kJ/mol})]$$
$$= +206.18 \text{ kJ}$$

Coal:

$$\Delta H^\circ_{rxn} = \Delta H^\circ_f[CO(g)] + \Delta H^\circ_f[H_2(g)]$$
$$- \{\Delta H^\circ_f[C(s)] + \Delta H^\circ_f[H_2O(g)]\}$$
$$= 1 \text{ mol } (-110.525 \text{ kJ/mol}) + 1 \text{ mol } (0 \text{ kJ/mol})$$
$$- [1 \text{ mol } (0 \text{ kJ/mol}) + 1 \text{ mol}(-241.83 \text{ kJ/mol})]$$
$$= +131.31 \text{ kJ}$$

Methane has a higher heat requirement than coal.

41. (a) $NH_4Cl(s) \longrightarrow NH_3(g) + HCl(g)$

(b) The combination of NH_3 and HCl is an acid-base reaction, not a redox reaction. (None of the atoms involved changes in oxidation number.)

(c) Ammonium chloride, ammonia, and hydrogen chloride

(d) $\Delta H^\circ_{rxn} = \Delta H^\circ_f[NH_3(g)] + \Delta H^\circ_f[HCl(g)]$
$$- \Delta H^\circ_f[NH_4Cl(s)]$$
$$= 1 \text{ mol } (-45.90 \text{ kJ/mol}) + 1 \text{ mol } (-92.31 \text{ kJ/mol})$$
$$- 1 \text{ mol } (-314.55 \text{ kJ/mol})$$
$$= +176.34 \text{ kJ}$$

(e) Mass of ammonia expected from 1.0 g of NH_4Cl

$$1.0 \text{ g NH}_4Cl\left(\frac{1 \text{ mol NH}_4Cl}{53.49 \text{ g}}\right)\left(\frac{1 \text{ mol NH}_3}{1 \text{ mol NH}_4Cl}\right)\left(\frac{17.03 \text{ g NH}_3}{1 \text{ mol NH}_3}\right)$$
$$= 0.32 \text{ g NH}_3$$

(f) pH of solution

$$1.00 \text{ g NH}_4Cl\left(\frac{1 \text{ mol NH}_4Cl}{53.49 \text{ g}}\right)\left(\frac{1 \text{ mol HCl}}{1 \text{ mol NH}_4Cl}\right)$$
$$= 0.0187 \text{ mol HCl}$$

$$\text{Concentration of HCl solution} = \frac{0.0187 \text{ mol HCl}}{0.126 \text{ L}} = 0.148 \text{ M}$$

Hydrochloric acid is a strong acid, so the concentration of H^+ ion in solution is also 0.148 M. This gives a pH of 0.830.

$$pH = -\log[H^+] = -\log(0.148) = -(-0.830) = 0.830$$

7—Atomic Structure

A fireworks display by the famous Grucci Company of New York.

Colors in the Sky

The discovery of black powder, the predecessor of gunpowder, occurred well before 1000 A.D., most likely in China. It was not until the Middle Ages, however, that black powder was known in the Western world. In 1252, Roger Bacon in England described the preparation of black powder from "saltpetre [potassium nitrate], young willow, and sulfur," and its use by the military and for fireworks spread to the European continent. By the time of the American Revolution, fireworks formulations and manufacturing methods had been worked out that are still in use today.

Typical fireworks have several important chemical components. For example, there must be an oxidizer. Today this is usually potassium perchlorate ($KClO_4$), potassium chlorate ($KClO_3$), or potassium nitrate (KNO_3). Potassium salts are used instead of sodium salts because the latter have two important drawbacks. They are hygroscopic—they absorb water from the air—and so do not remain dry on storage. Also, when heated, sodium salts give off an intense, yellow light that is so bright it can mask other colors.

The parts of any fireworks display we remember best are the vivid colors and brilliant flashes. White light can be produced by oxidizing magnesium or aluminum metal at high temperatures. The flashes you see at rock concerts or similar events, for example, are typically $Mg/KClO_4$ mixtures.

Yellow light is easiest to produce because sodium salts give an intense light with a wavelength of 589 nm. Fireworks mixtures usually contain sodium in the form of nonhygroscopic compounds such as cryolite, Na_3AlF_6. Strontium salts are most often used to produce a red light, and green is produced by barium salts such as $Ba(NO_3)_2$.

294

See Chapter Goals Revisited (page 324). Test your knowledge of these goals by taking the exam-prep quiz on the General ChemistryNow CD-ROM or website.

- Describe the properties of electromagnetic radiation.
- Understand the origin of light from excited atoms and its relationship to atomic structure.
- Describe the experimental evidence for wave-particle duality.
- Describe the basic ideas of quantum mechanics.
- Define the three quantum numbers (n, ℓ, and m_ℓ) and their relationship to atomic structure.

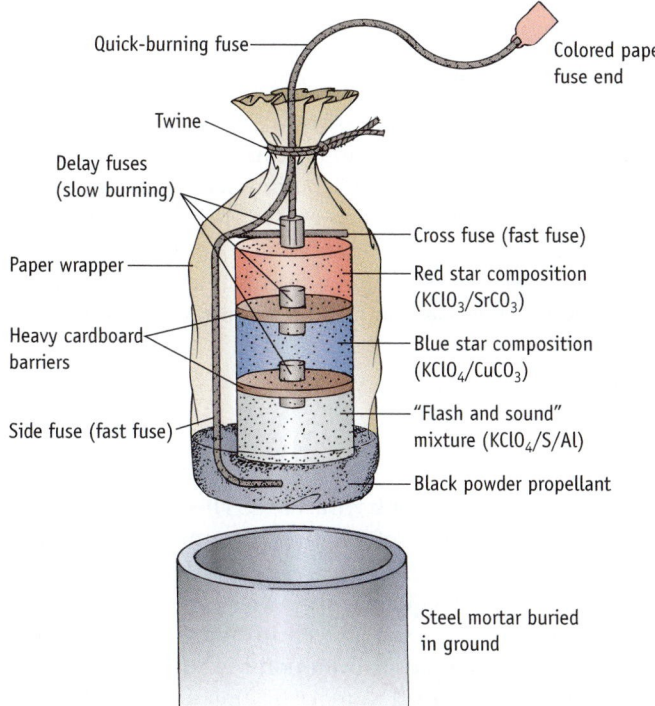

The design of an aerial rocket for a fireworks display. When the fuse is ignited, it burns quickly to the delay fuse at the top of the red star mixture as well as to the black powder propellant at the bottom. The propellant ignites, sending the shell into the air. Meanwhile, the delay fuses burn. If the timing is correct, the shell bursts high in the sky into a red star. This is followed by a blue burst and then a flash and sound.

The next time you see a fireworks display, watch for the ones that are blue. Blue has always been the most difficult color to produce. Recently, however, fireworks designers have learned that the best way to get a really good "blue" is to decompose copper(I) chloride at low temperatures. To achieve this effect, CuCl is mixed with $KClO_4$, copper powder, and the organic chlorine-containing compound hexachloroethane, C_2Cl_6.

Why are chemists—and many others—interested in fireworks? Because their colors arise from energetically excited atoms and molecules. The way that atoms can produce colored light provides insight into the structure of the atom, the subject of this chapter.

Charles D. Winters

Colors in fireworks. Fireworks displays are usually colored. Here the white, solid salts sodium chloride (*left*), strontium chloride (*center*), and boric acid (*right*) were soaked in methanol and then the methanol was ignited. The compounds were entrained in the burning fuel, and the energy from the combustion excited the atoms. The colors you observe are characteristic of sodium, strontium, and boron. (See *General ChemistryNow Screen 7.1 Chemical Puzzler*, for a description of colors in fireworks.)

To Review Before You Begin

- Review metric units of measurements (Chapter 1)
- Review the structure of the atom (Section 2.1)

Chemical elements that exhibit similar properties are found in the same column of the periodic table. But why should there be similarities among elements? The discovery of the electron, proton, and neutron [◄ Section 2.1] prompted scientists to look for relationships between atomic structure and chemical behavior. As early as 1902 Gilbert N. Lewis (1875–1946) suggested the idea that electrons in atoms might be arranged in shells, starting close to the nucleus and building outward. Lewis explained the similarity of chemical properties for elements in a given group by assuming that all the elements of that group have the same number of electrons in the outer shell.

Lewis's model of the atom raises a number of questions. Where are the electrons located? Do they have different energies? What experimental evidence supports this model? These questions were the reason for many of the experimental and theoretical studies that began around 1900 and continue to this day. This chapter and the next one outline the current theories of electronic structure.

7.1—Electromagnetic Radiation

You are familiar with water waves, and you may also know that some properties of radiation such as visible light and radio waves can be explained by wave motion. Our understanding of light as waves came from the experiments of physicists in the 19th century, among them a Scot, James Clerk Maxwell (1831–1879). In 1864, he developed an elegant mathematical theory to describe all forms of radiation in terms of oscillating, or wave-like, electric and magnetic fields (Figure 7.1). Hence, radiation, such as light, microwaves, television and radio signals, and x-rays, is collectively called **electromagnetic radiation**.

Wave Properties

The distance between successive crests or high points of a wave (or between successive troughs or low points) is the **wavelength** of a wave. This distance can be given in meters, nanometers, or whatever unit of length is convenient. The symbol for wavelength is the Greek letter λ (lambda).

Waves are also characterized by their **frequency**, symbolized by the Greek letter ν (nu). For any wave motion—whether water waves or electromagnetic radiation—the frequency is the number of waves that pass a given point in some unit of time, usually per second (Figure 7.1). The unit for frequency is often written s^{-1}, which stands for 1 per second, 1/s, and is now called the **hertz**.

If you enjoy water sports, you are familiar with the height of waves. In more scientific terms, the maximum height of a wave is its **amplitude**. In Figure 7.1, notice that the wave has zero amplitude at certain intervals along the wave. Points of zero amplitude, called **nodes**, occur at intervals of $\lambda/2$.

Finally, the speed of a moving wave is an important factor. As an analogy, consider cars in a traffic jam traveling bumper to bumper. If each car is 5 m long, and if a car passes you every 4 s (that is, the frequency is 1 per 4 seconds, or $\frac{1}{4}\,s^{-1}$), then

■ **Hertz**
Heinrich Hertz (1957–1894) was the first to send and receive radio waves. He showed that they could be reflected and refracted the same as light, confirming Maxwell's prediction that light waves were electromagnetic radiation. In his honor scientists use "hertz" as the unit of frequency (number of cycles per second, s^{-1}) of radiation.

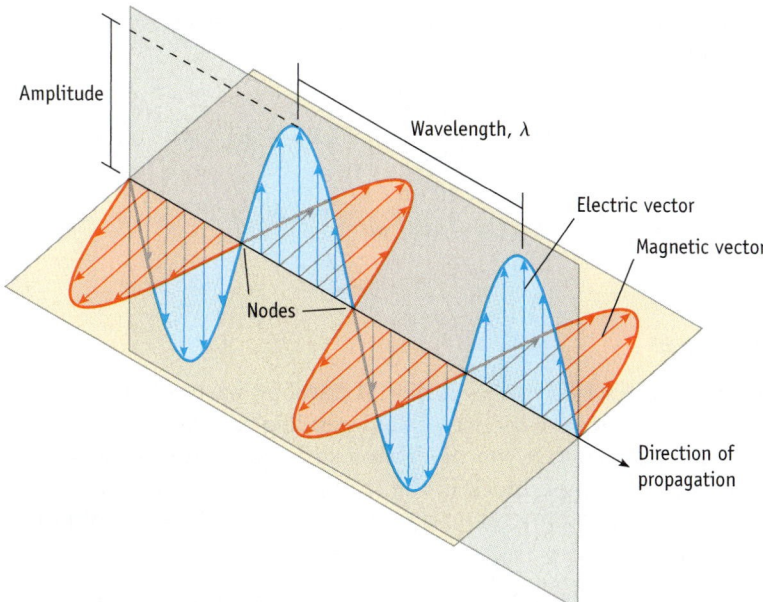

Figure 7.1 Electromagnetic radiation. In the 1860s James Clerk Maxwell developed the currently accepted theory that all forms of radiation are propagated through space as vibrating electric and magnetic fields at right angles to one another. Each of the fields is described by a sine wave (the mathematical function describing the wave). Such oscillating fields emanate from vibrating charges in a source such as a light bulb or radio antenna.

the traffic is "moving" at the speed of $(5 \text{ m}) \times (\frac{1}{4}\,\text{s}^{-1})$, or $1.25 \text{ m} \cdot \text{s}^{-1}$. The speed for any periodic motion, including a wave, is the product of the wavelength and the frequency of the wave:

$$\text{Speed } (\text{m} \cdot \text{s}^{-1}) = \text{wavelength } (\text{m}) \times \text{frequency } (\text{s}^{-1})$$

This equation also applies to electromagnetic radiation, where the speed of light, c, is the product of the wavelength and frequency of a light wave.

Speed of light $(\text{m} \cdot \text{s}^{-1})$

$$c = \lambda \times \nu \tag{7.1}$$

Wavelength (m)   Frequency (s^{-1})

■ **Speed of Light**
The speed of light passing through a substance (air, glass, water, and so on) depends on the chemical constitution of the substance and the wavelength of the light. This is the basis for using a glass prism to disperse light and is the explanation for rainbows. The speed of sound also depends on the material through which it passes.

The speed of visible light and all other forms of electromagnetic radiation in a vacuum is a constant, c (= $2.99792458 \times 10^8 \text{ m} \cdot \text{s}^{-1}$; approximately 186,000 miles · s^{-1}). Given this value, and knowing the wavelength of a light wave, you can calculate the frequency, and vice versa. For example, what is the frequency of orange light, which has a wavelength of 625 nm? Because the speed of light is expressed in meters

■ **Speed of Light and Significant Figures**
The speed of light is known to nine significant figures. For calculations we will generally use four or fewer significant figures.

per second, the wavelength in nanometers must be changed to meters before substituting into Equation 7.1:

$$625 \text{ nm} \times \frac{1 \times 10^{-9} \text{ m}}{1 \text{ nm}} = 6.25 \times 10^{-7} \text{ m}$$

$$\nu = \frac{c}{\lambda} = \frac{2.998 \times 10^8 \text{ m} \cdot \text{s}^{-1}}{6.25 \times 10^{-7} \text{ m}} = 4.80 \times 10^{14} \text{ s}^{-1}$$

Standing Waves

The wave motion described so far is that of traveling waves such as sound or water waves. Another type of wave motion, called **standing** or stationary waves, is relevant to modern atomic theory. If you tie down a string at both ends, as you would the string of a guitar, and then pluck it, the string vibrates as a standing wave (Figure 7.2). Several important points about standing waves are relevant to our discussion of electrons in atoms:

- A standing wave is characterized by having two or more points of no amplitude; that is, the wave amplitude is zero at the nodes.

- As with traveling waves, the distance between consecutive nodes is always $\lambda/2$.

- Only certain wavelengths are possible for standing waves. The only allowed vibrations have wavelengths of $n(\lambda/2)$, where n is an integer.

Figure 7.2 illustrates the third point. In the first of the vibrations illustrated, the distance between the ends of the string is half a wavelength, or $\lambda/2$. In the second vibration, the string length equals one complete wavelength, or $2(\lambda/2)$. In the third vibration, the string length is $3(\lambda/2)$, or $(3/2)\lambda$. Could the distance between the ends of a standing wave vibration ever be $(3/4)\lambda$? The answer is no. For standing waves, only certain wavelengths are possible. Because the ends of a standing wave must be nodes, the only allowed vibrations are those in which the distance from one end, or "boundary," to the other is $n(\lambda/2)$, where n is an integer $(1, 2, 3, \ldots)$.

■ **Standing Waves**
Only certain wavelengths are allowed for standing waves. This is an example of *quantization*, a concept we describe in the sections that follow.

Figure 7.2   Standing waves. In the first wave, the end-to-end distance is $(1/2)\lambda$, in the second wave it is λ, and in the third wave it is $(3/2)\lambda$.

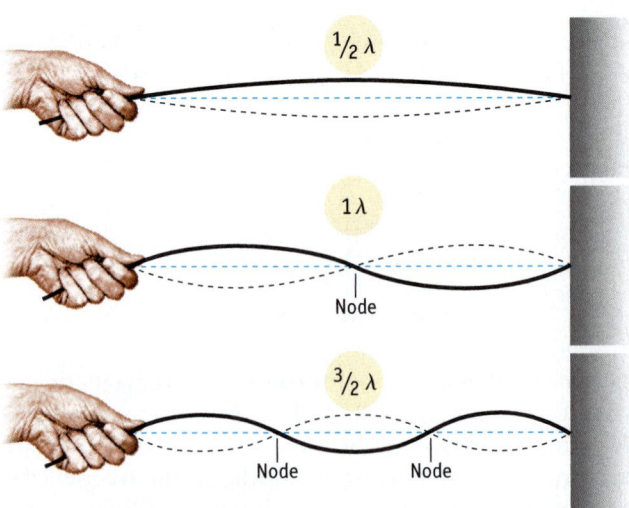

Exercise 7.1—Standing Waves

The line shown here is 10 cm long.

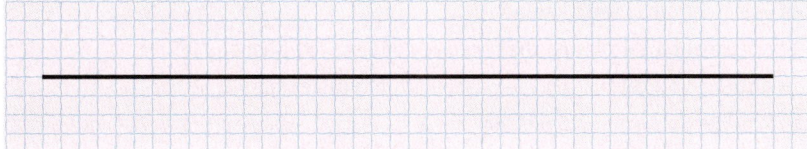

Using this line,

(a) Draw a standing wave with one node between the ends. What is the wavelength of this wave?

(b) Draw a standing wave with three evenly spaced nodes between the ends. What is its wavelength?

(c) If the wavelength of the standing wave is 2.5 cm, how many waves fit within the boundaries? How many nodes are there between the ends?

The Visible Spectrum of Light

Visible light consists of a spectrum of colors, ranging from red light at the long-wavelength end of the spectrum to violet light at the short-wavelength end (Figure 7.3). Visible light is, however, merely a small portion of the total electromagnetic spectrum. Ultraviolet (UV) radiation, the radiation that can lead to sunburn, has wavelengths shorter than those of visible light; x-rays and γ rays, the latter emitted in the process of radioactive disintegration of some atoms, have even shorter wavelengths. At longer wavelengths than those of visible light, we first encounter infrared radiation (IR), the type that is sensed as heat. Longer still is the wavelength of the radiation used in microwave ovens and in television and radio transmissions.

■ **ROY G BIV**
You can remember the colors of visible light, in order of decreasing wavelength, by the well-known mnemonic phrase ROY G BIV: red, orange, yellow, green, blue, indigo, and violet.

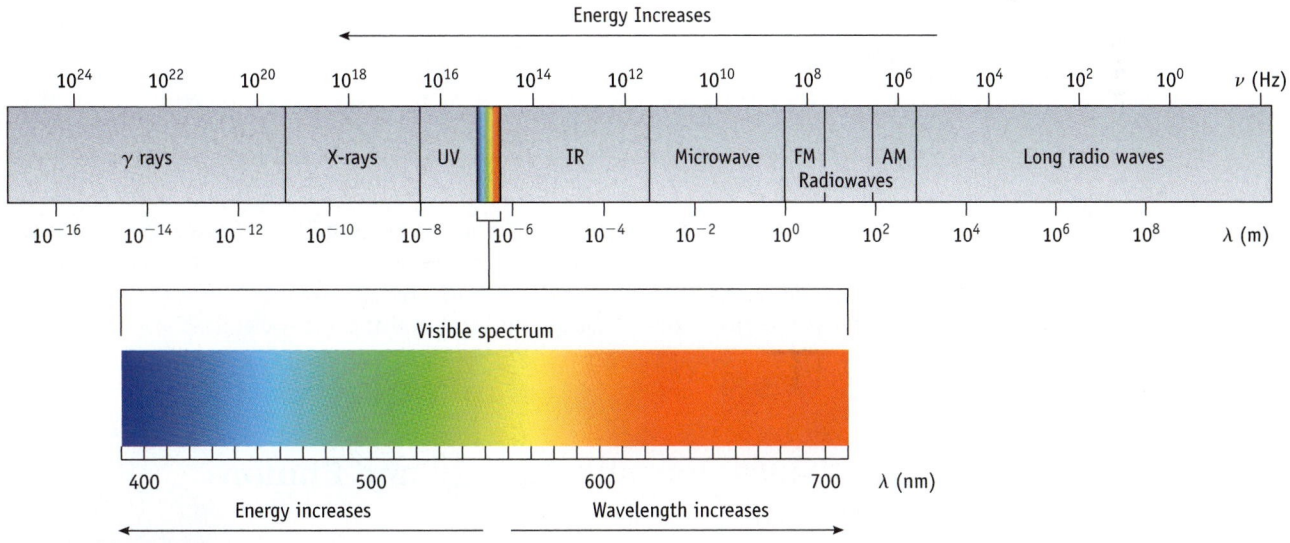

Active Figure 7.3 **The electromagnetic spectrum.** Visible light (enlarged portion) is a very small part of the entire spectrum. The radiation's energy increases from the radiowave end of the spectrum (low frequency, ν, and long wavelength, λ) to the γ-ray end (high frequency and short wavelength).

GENERAL
Chemistry⚛Now™ See the General ChemistryNow CD-ROM or website to explore an interactive version of this figure accompanied by an exercise.

See the General ChemistryNow CD-ROM or website:

- **Screen 7.3 Electromagnetic Radiation**

  (a) for a tutorial on calculating the frequency of ultraviolet light

  (b) for a tutorial on calculating the wavelength of visible light

- **Screen 7.4 Electromagnetic Spectrum,** for a simulation exploring the wavelength and frequency of the visible portion of the electromagnetic spectrum

Example 7.1—Wavelength–Frequency Conversions

Problem The frequency of the radiation used in all microwave ovens sold in the United States is 2.45 GHz. (The unit GHz stands for "gigahertz"; 1 GHz is 1 billion cycles per second, or 10^9 s^{-1}.) What is the wavelength, in meters, of this radiation? Compare the wavelength of microwave radiation with the wavelength of light in the visible region—say, orange light with $\lambda = 625$ nm.

Strategy The wavelength of microwave radiation in meters can be calculated directly from Equation 7.1. Convert 625 nm to a wavelength in meters so that units are compatible.

Solution

$$\lambda = \frac{c}{\nu} = \frac{2.998 \times 10^8 \text{ m} \cdot \text{s}^{-1}}{2.45 \times 10^9 \text{ s}^{-1}} = \boxed{0.122 \text{ m}}$$

Orange light has a wavelength, in meters, of

$$625 \text{ nm} \times \frac{1 \times 10^{-9} \text{ m}}{1 \text{ nm}} = \boxed{6.25 \times 10^{-7} \text{ m}}$$

The wavelength of microwave radiation is about 200,000 times *longer* than that of orange light.

Exercise 7.2—Radiation, Wavelength, and Frequency

(a) Which color in the visible spectrum has the highest frequency? Which has the lowest frequency?

(b) Is the frequency of the radiation used in a microwave oven higher or lower than that from your favorite FM radio station (91.7 MHz), where MHz (megahertz) = 10^6 s^{-1}?

(c) Is the wavelength of x-rays longer or shorter than that of ultraviolet light?

7.2—Planck, Einstein, Energy, and Photons

Planck's Equation

If you heat a piece of metal, it emits electromagnetic radiation with wavelengths that depend on temperature. At first its color is a dull red. At higher temperatures the red color brightens (Figure 7.4a), and at still higher temperatures the redness turns to a brilliant white light. For example, the heating element of a toaster becomes "red hot," and the filament of an incandescent light bulb glows "white hot."

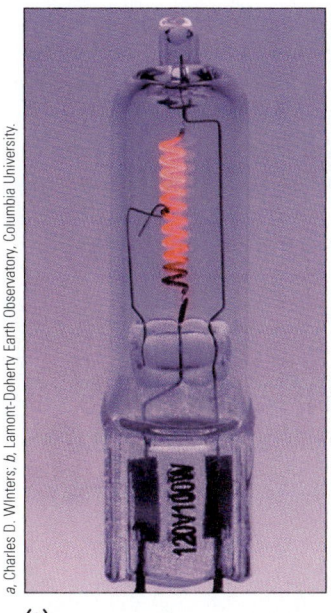

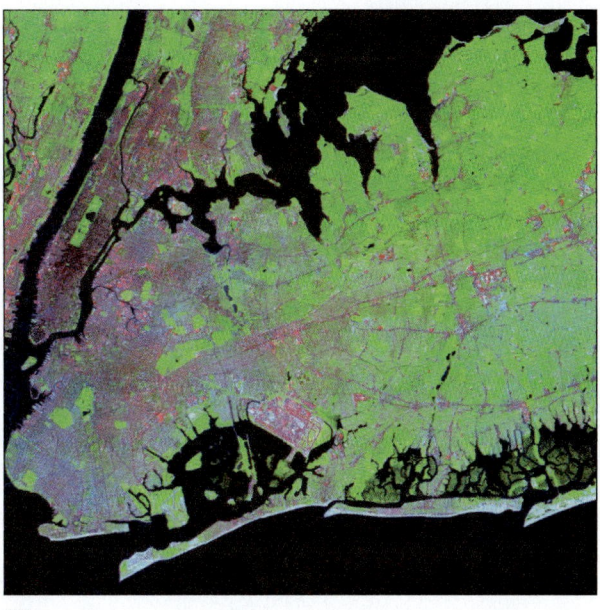

(a)   **(b)**

a, Charles D. Winters; *b*, Lamont-Doherty Earth Observatory, Columbia University.

Figure 7.4 **Infrared radiation.** IR radiation has longer wavelengths than visible light. (a) The filament of an incandescent light bulb emitting radiation at the long-wavelength or red end of the visible spectrum. (b) A photo of the New York City area taken from a satellite using film sensitive to infrared light. Water is dark blue, pavement is light blue, and vegetation is green.

Your eyes detect the radiation from a piece of heated metal that occurs in the visible region of the electromagnetic spectrum. Of course, these are not the only wavelengths of the light emitted by the metal. Radiation is also emitted with wavelengths both shorter (in the ultraviolet region) and longer (in the infrared region; Figure 7.4b) than those of visible light. That is, a spectrum of electromagnetic radiation is emitted (Figure 7.5), with some wavelengths being more intense than others. As the metal is heated, the maximum in the curve of light intensity versus wavelength is shifted more and more to the ultraviolet region. The color of the glowing object shifts from red to yellow, and, if it does not melt, it will finally glow white hot.

At the end of the 19th century, scientists were trying to explain the relationship between the intensity and the wavelength for radiation given off by heated objects.

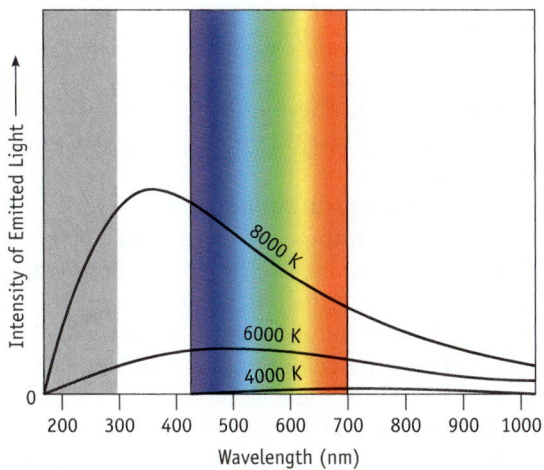

Figure 7.5 **The spectrum of the radiation given off by a heated body.** When an object is heated, it emits radiation covering a spectrum of wavelengths. For a given temperature, some of the radiation is emitted at long wavelengths and some at short wavelengths. Most, however, is emitted at some intermediate wavelength, the maximum in the curve. As the temperature of the object increases, the maximum moves from the red end of the spectrum to the violet end. At still higher temperatures intense light is emitted at all wavelengths in the visible region, and the maximum in the curve is in the ultraviolet region. Such an object is described as "white hot." (Stars are often referred to as "red giants" or "white dwarfs," a reference to their temperatures and relative sizes.) (*See the General ChemistryNow Screen 7.5 Planck's Equation, to watch a video on the light emitted by a heated metal bar.*)

All of their attempts were unsuccessful, however. Theories available at the time predicted that the intensity of radiation should increase continuously with decreasing wavelength (instead of declining with decreasing wavelength as is observed experimentally; Figure 7.5). This perplexing situation became known as the "ultraviolet catastrophe" because predictions failed in the ultraviolet region. Classical physics did not provide a satisfactory explanation, so a new way to look at matter and energy was needed.

In 1900, a German physicist, Max Planck (1858–1947), offered an explanation. Following classical theory, he assumed that vibrating atoms in a heated object give rise to the emitted electromagnetic radiation. He also introduced an important new assumption: These vibrations are **quantized**. In Planck's model, quantization means that only certain vibrations, with specific frequencies, are allowed.

Planck introduced an important equation, now called Planck's equation, that states that the energy of a vibrating system is proportional to the frequency of vibration. The proportionality constant h is called **Planck's constant** in his honor. It has the value $6.6260693 \times 10^{-34}$ J · s.

$$
\begin{array}{cc}
\text{Energy (J)} & \text{Planck's constant (J · s)} \\
\downarrow & \downarrow \\
E & = \quad h\nu \\
& \uparrow \\
& \text{Frequency (s}^{-1}\text{)}
\end{array}
\tag{7.2}
$$

Now, assume as Planck did that there must be a *distribution* of vibrations of atoms in an object—some atoms are vibrating at a high frequency, some are vibrating at a low frequency, but most have some intermediate frequency. The few atoms with high-frequency vibrations are responsible for some of the light, as are those few with low-frequency vibrations. Nevertheless, most of the light must come from the majority of the atoms that have intermediate vibrational frequencies. That is, a spectrum of light is emitted with a maximum intensity at some wavelength, in accord with experiment. The intensity should not become greater and greater on approaching the ultraviolet region. With this realization, the ultraviolet catastrophe was solved.

Einstein and the Photoelectric Effect

As almost always occurs, the explanation of a fundamental phenomenon—such as the spectrum of light from a hot object—leads to another fundamental discovery. A few years after Planck's work, Albert Einstein (1879–1955) incorporated Planck's ideas into an explanation of the photoelectric effect.

Photoelectric cells are commonly used in automatic door openers in stores and elevators. They depend on the **photoelectric effect**, the ejection of electrons when light strikes the surface of a metal. In the cell in Figure 7.6, an electric potential is applied to the cell. When light strikes the cathode of the cell, electrons are ejected from the cathode surface and move to a positively charged anode. A stream of electrons—a current—flows through the cell. Thus, the cell can act as a light-activated switch in an electric circuit.

Experiments with photoelectric cells show that electrons are ejected from the surface *only* if the frequency of the light is high enough. If lower-frequency light is

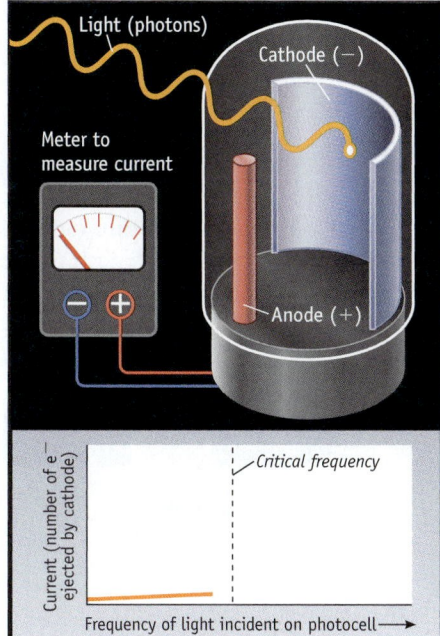

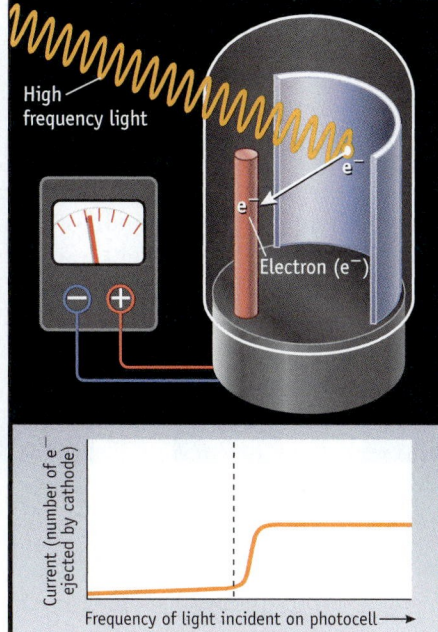

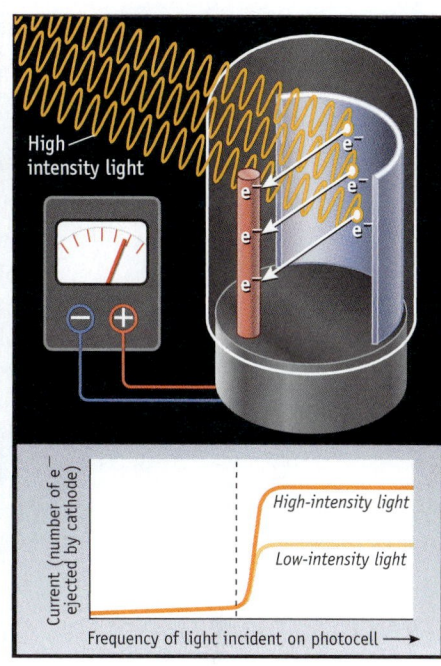

(a) A photocell operates by the photoelectric effect. The main part of the cell is a light-sensitive cathode. This is a material, usually a metal, that ejects electrons if struck by photons of light of sufficient energy. No current is observed until the critical frequency is reached.

(b) When light of higher frequency than the minimum is used, the excess energy of the photon allows the electron to escape the atom with greater velocity. The ejected electrons move to the anode and a current flows in the cell. Such a device can be used as a switch in electric circuits.

(c) If higher intensity light is used, the only effect is to cause more electrons to be released from the surface. The onset of current is observed at the same frequency as with lower intensity light, but more current flows.

Figure 7.6 **A photoelectric cell.**

used, no effect is observed, regardless of the light's intensity (brightness). In contrast, if the frequency is above the minimum, increasing the light intensity causes a higher current to flow because more and more electrons are ejected.

Einstein decided the experimental observations could be explained by combining Planck's equation ($E = h\nu$) with a new idea: Light has particle-like properties. Einstein assumed these massless "particles," now called **photons**, are packets of energy. The energy of each photon is proportional to the frequency of the radiation, as given by Planck's relation.

Einstein's proposal helps us understand the photoelectric effect. It is reasonable to suppose that a high-energy particle would have to bump into an atom to cause the atom to lose an electron. It is also reasonable to suppose that an electron can be torn away from the atom only if the photon has enough energy. If electromagnetic radiation is described as a stream of photons, as Einstein said, then the greater the intensity of light, the more photons are available to strike a surface per unit of time. However, the atoms of a metal surface will not lose electrons when the metal is bombarded by millions of photons if no individual photon has enough energy to remove an electron from an atom. Only if the critical minimum energy (that is, minimum light frequency) is exceeded will the energy content be sufficient to displace an electron from a metal atom. The greater the number of photons with this energy that strike the surface, the greater the number of electrons dislodged. Thus, the connection is made between light intensity and the number of electrons ejected.

Energy and Chemistry: Using Planck's Equation

Compact disc players use lasers that emit red light with a wavelength of 685 nm. What is the energy of one photon of this light? What is the energy of one mole of photons of red light? To answer these questions, first convert the wavelength to the frequency of the radiation and then use the frequency to calculate the energy per photon. Finally, calculate the energy of a mole of photons by multiplying the energy per photon by Avogadro's number:

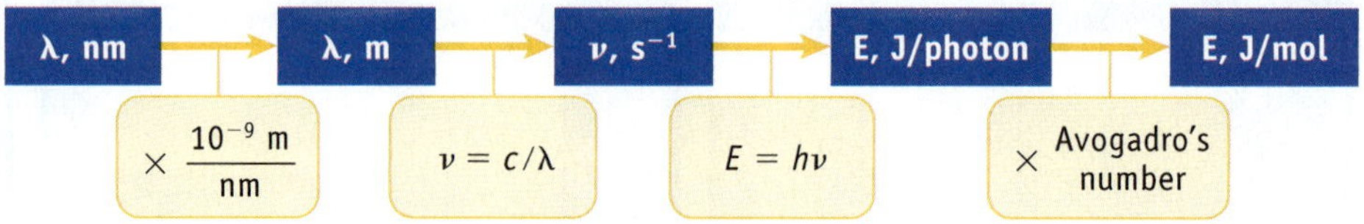

λ, nm	λ, m	ν, s^{-1}	E, J/photon	E, J/mol
$\times \dfrac{10^{-9}\ m}{nm}$	$\nu = c/\lambda$	$E = h\nu$	$\times$ Avogadro's number	

$$685\ nm\ (10^{-9}\ m/nm) = 6.85 \times 10^{-7}\ m$$

$$\nu = \frac{c}{\lambda} = \frac{2.998 \times 10^{8}\ m \cdot s^{-1}}{6.85 \times 10^{-7}\ m} = 4.38 \times 10^{14}\ s^{-1}$$

$E = h\nu$

$$= (6.626 \times 10^{-34}\ J \cdot s/photon)(4.38 \times 10^{14}\ s^{-1}) = 2.90 \times 10^{-19}\ J/photon$$

$$= (2.90 \times 10^{-19}\ J/photon)(6.022 \times 10^{23}\ photons/mol) = 1.75 \times 10^{5}\ J/mol$$

The energy of a mole of photons of red light is equivalent to 175 kJ. A mole of photons of blue light ($\lambda = 400$ nm) has an energy of about 300 kJ. These energies are in a range that can affect the bonds between atoms in molecules. It should not be surprising, therefore, that light can cause chemical reactions to occur. For example, you may have seen paint or dye that has faded or even decomposed from exposure to light.

The previous calculation shows that, as the frequency of radiation increases, the energy of the radiation also increases (see Figure 7.3). Similarly, the energy increases as the wavelength of radiation decreases:

As frequency (ν) increases, energy (E) increases

$$E = h\nu = \frac{hc}{\lambda}$$

As wavelength (λ) decreases, energy (E) increases

Photons of ultraviolet radiation—with wavelengths shorter than those of visible light—have higher energy than visible light. Because visible light has enough energy to affect the bonds between atoms, obviously ultraviolet light does as well. That is the reason ultraviolet radiation can cause a sunburn. In contrast, photons of infrared radiation—with wavelengths longer than those of visible light—have lower

Chemical Perspectives

UV Radiation, Skin Damage, and Sunscreens

Most of us are well aware of the effects of exposure to the sun. A sunburn results, and over the long term permanent skin damage can occur. Most of this problem results from the damage to organic molecules caused by ultraviolet (UV) radiation.

UV radiation is often divided into three categories: UVA (315–400 nm), UVB (290–315 nm), and UVC (100–290 nm). UVC radiation has a high energy, but it is absorbed by the earth's ozone layer. UVB light is responsible for your sunburn. Tanning occurs when the light strikes your skin and activates the melanocytes in the skin so that they produce melanin. UVA light also produces damage such as the alteration of connective tissue in the dermis.

We can calculate that the energy of a mole of photons in the ultraviolet region (at 300 nm) is about 400 kJ.

For $\lambda = 300.$ nm, $\nu = 9.99 \times 10^{14}$ s^{-1}
$E = h\nu$
$\quad = (6.626 \times 10^{-34}$ J $\cdot$ s/photon$)$
$\quad\quad \times (9.99 \times 10^{14}$ s$^{-1})$
$\quad = 6.62 \times 10^{-19}$ J/photon
$E = 3.99 \times 10^{5}$ J/mol

This energy is significantly greater than the energy of light in the visible region. Indeed, the energy of UV light is in the range of the energies necessary to break the chemical bonds in proteins.

Various manufacturers have developed mixtures of compounds that protect skin from UVA and UVB radiation. These sunscreens are given "sun protection factor" (SPF) labels that indicate how long the user can stay in the sun without burning.

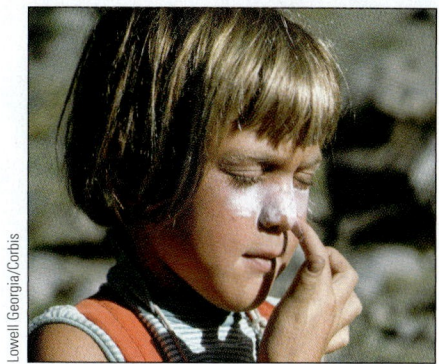

Sunscreens produced by Coppertone, for example, contain the organic compounds 2-ethylhexyl-p-methoxycinnamate and oxybenzone. These molecules absorb UV radiation, preventing it from reaching your skin.

---

energy than visible light. They are generally not energetic enough to cause chemical reactions, but they can affect the vibrations of molecules. We sense infrared radiation as heat, such as the heat given off by a glowing burner on an electric stove.

GENERAL
Chemistry ⚛ Now™

See the General ChemistryNow CD-ROM or website:

- **Screen 7.5 Planck's Equation**

    (a) for a simulation exploring the relationship between wavelength, frequency, and photon energy

    (b) for an exercise on using Planck's equation to calculate wavelength

---

Exercise 7.3—Photon Energies

Compare the energy of a mole of photons of orange light (625 nm) with the energy of a mole of photons of microwave radiation having a frequency of 2.45 GHz (1 GHz = 10^9 s^{-1}). Which has the greater energy? By what factor is one greater than the other? (See Example 7.1.)

7.3—Atomic Line Spectra and Niels Bohr

Atomic Line Spectra

If a high voltage is applied to atoms of an element in the gas phase at low pressure, the atoms absorb energy and are said to be "excited." The excited atoms emit light. This light is different, however, from the *continuous spectrum* of wavelengths from

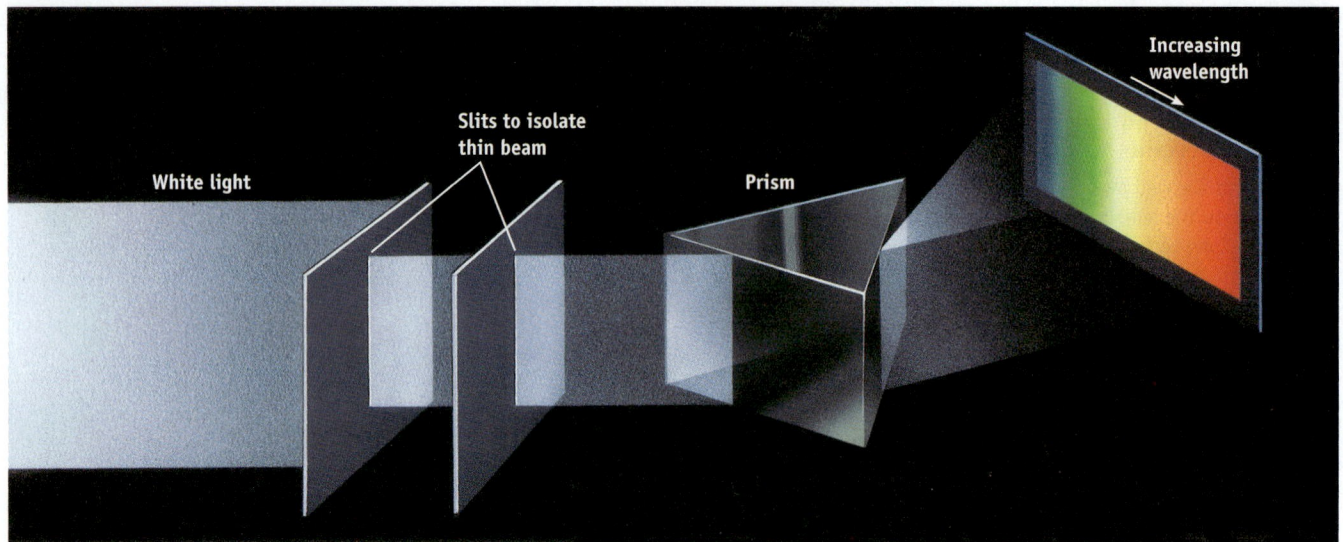

Figure 7.7   A spectrum of white light, produced by refraction in a prism. The light is first passed through a narrow slit to isolate a thin beam, or line, of light. The beam is then passed through a prism (or, in modern instruments, a diffraction grating is used). See the spectrum of visible light in Figure 7.3.

white light (Figure 7.7). Excited atoms in the gas phase emit only certain wavelengths of light. We know this because when this light is passed through a prism, only a few colored lines are seen. This phenomenon is called a **line emission spectrum** (Figure 7.8). A familiar example is the light from a neon advertising sign, in which excited neon atoms emit orange-red light.

The line emission spectra of hydrogen, mercury, and neon are shown in Figure 7.9. Every element has a unique line spectrum. Indeed, the characteristic lines in

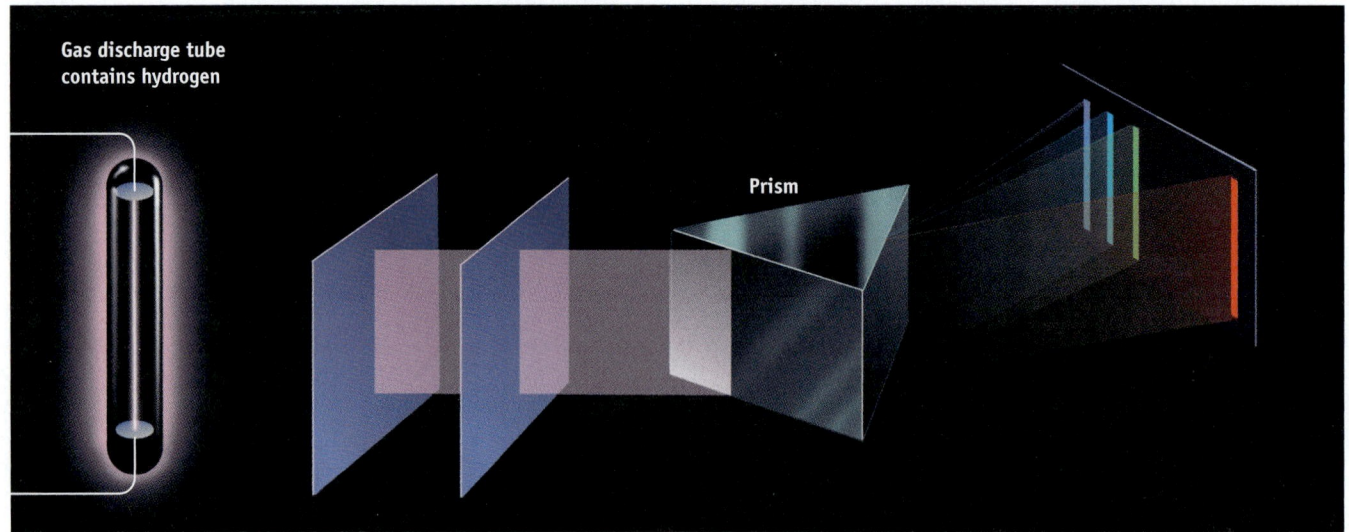

Active Figure 7.8   **The line emission spectrum of hydrogen.** The emitted light is passed through a series of slits to create a narrow beam of light, which is then separated into its component wavelengths by a prism. A photographic plate or photocell can be used to detect the separate wavelengths as individual lines. Hence, the name "line spectrum" is given to the light emitted by a glowing gas.

GENERAL
Chemistry ⚛ Now™ See the General ChemistryNow CD-ROM or website to explore an interactive version of this figure accompanied by an exercise.

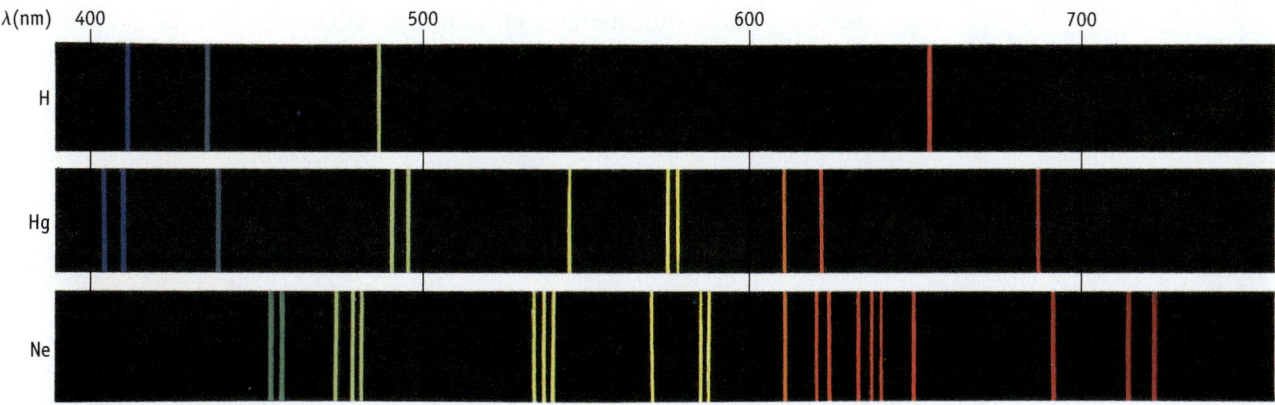

Figure 7.9 Line emission spectra of hydrogen, mercury, and neon. Excited gaseous elements produce characteristic spectra that can be used to identify the elements as well as to determine how much of each element is present in a sample.

the emission spectrum of an element can be used in chemical analysis, both to identify the element and to determine how much of it is present.

A goal of scientists in the late 19th century was to explain why gaseous atoms emitted light of only certain frequencies and to find a mathematical relationship among the observed frequencies. (It is always useful if experimental data can be related by a mathematical equation, because a regular pattern of information implies a logical explanation.) The first steps in this direction were made by Johann Balmer (1825–1898) and later Johannes Rydberg (1854–1919). They developed an equation—now called the **Rydberg equation**—from which it was possible to calculate the wavelength of the red, green, and blue lines in the visible emission spectrum of hydrogen atoms (Figure 7.9).

$$\frac{1}{\lambda} = R\left(\frac{1}{2^2} - \frac{1}{n^2}\right) \qquad \text{when } n > 2 \qquad (7.3)$$

In this equation n is an integer, and R, now called the **Rydberg constant**, has the value $1.0974 \times 10^7 \ \text{m}^{-1}$. If $n = 3$, the wavelength of the red line in the hydrogen spectrum is obtained (6.563×10^{-7} m, or 656.3 nm). If $n = 4$, the wavelength for the green line is obtained. The value $n = 5$ gives the wavelength of the blue line. This group of visible lines in the spectrum of hydrogen atoms (and others for which $n = 6, 7, 8, \ldots$) is now called the **Balmer series**.

The Bohr Model of the Hydrogen Atom

Niels Bohr (1885–1962), a Danish physicist, provided the first connection between the spectra of excited atoms and the quantum ideas of Planck and Einstein. From Rutherford's work [◄ Section 2.1], it was known that electrons are arranged in space outside the atom's nucleus. For Bohr the simplest model of a hydrogen atom was one in which the electron moved in a circular orbit around the nucleus, just as the planets revolve about the sun. In proposing this hypothesis, however, he had to contradict the laws of classical physics. According to the theories at the time, a charged electron moving in the positive electric field of the nucleus should lose energy. Eventually the electron should crash into the nucleus. This is clearly not the case; if it were so, all matter would eventually self-destruct.

To solve the contradiction with the laws of classical physics, Bohr postulated that an electron could occupy only *certain* orbits or energy levels in which it is stable. That is, the energy of the electron in the atom is quantized. By combining this

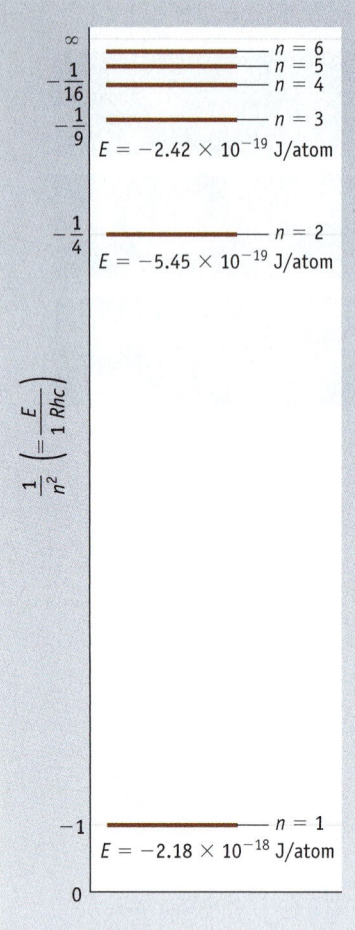

quantization postulate with the laws of motion from classical physics, Bohr showed that the energy possessed by the single electron in the *n*th orbit or energy level of the H atom is given by the equation

Planck's constant

Rydberg constant   Speed of light

$$\text{Potential energy of electron in the } n\text{th level} = E_n = -\frac{Rhc}{n^2} \qquad (7.4)$$

Principal quantum number

which gives the energy in units of J/atom. Each allowed orbit was assigned a value of *n*, a unitless integer having values of 1, 2, 3, and so on. This integer is now known as the **principal quantum number**.

Equation 7.4 has several important features. First, the energy of the electron has a negative value. This follows from Coulomb's law [◄ Section 3.3]. The energy of attraction between oppositely charged bodies (a negative electron and the positive nuclear proton) has a negative value, and that value becomes more negative (the attraction increases) as the bodies move closer together. Bohr's equation shows that as the value of *n* increases, the value of the energy becomes *less* negative (Figure 7.10). Bohr also showed that as *n* increases (and the energy becomes less negative), the distance of the electron from the nucleus increases. An electron in the *n* = 1 orbit is closest to the nucleus and has the lowest or most negative energy. The electron of the hydrogen atom is normally in this energy level.

An atom with its electrons in the lowest possible energy levels is said to be in its **ground state**. When the electron of a hydrogen atom occupies an orbit with *n* greater than 1, the electron is farther from the nucleus, the value of its energy is less negative, and it is said to be in an **excited state**. The energies of the ground state and an excited state are calculated in Example 7.2.

Example 7.2—Energies of the Ground and Excited States of the H Atom

Problem Calculate the energies of the $n = 1$ and $n = 2$ states of the hydrogen atom in joules per atom and in kilojoules per mole. What is the difference in energy of these two states?

Strategy Here we use Equation 7.4 with the following constants: $R = 1.097 \times 10^7$ m^{-1}, $h = 6.626 \times 10^{-34}$ J · s, and $c = 2.998 \times 10^8$ m · s^{-1}.

Solution When $n = 1$, the energy of an electron in a single H atom is

$$E_1 = -\frac{Rhc}{n^2} = -\frac{Rhc}{1^2} = -Rhc$$

$$= -(1.097 \times 10^7 \text{ m}^{-1})(6.626 \times 10^{-34} \text{ J · s})(2.998 \times 10^8 \text{ m · s}^{-1})$$

$$= -2.179 \times 10^{-18} \text{ J/atom}$$

In units of kJ/mol, we have

$$E_1 = \frac{-2.179 \times 10^{-18} \text{ J}}{\text{atom}} \times \frac{6.022 \times 10^{23} \text{ atoms}}{\text{mol}} \times \frac{1 \text{ kJ}}{1000 \text{ J}}$$

$$= -1312 \text{ kJ/mol}$$

When n = 2, the energy is

$$E_2 = -\frac{Rhc}{2^2} = -\frac{E_1}{4} = -\frac{2.179 \times 10^{-18} \text{ J/atom}}{4}$$

$$= -5.448 \times 10^{-19} \text{ J/atom}$$

Finally, because $E_2 = E_1/4$, we calculate E_2 to be -328.1 kJ/mol.

The difference in energy, ΔE, between the first two energy states of the H atom is

$$\Delta E = E_2 - E_1 = (-328.1 \text{ kJ/mol}) - (-1312 \text{ kJ/mol}) = 984 \text{ kJ/mol}$$

Comment Notice that the calculated energies are negative for an electron in the $n = 1$ or $n = 2$ state, with E_1 more negative than E_2. Notice also that the $n = 2$ state is higher in energy than the $n = 1$ state by 984 kJ/mol. For more on this point, see Figure 7.11 and the discussion below.

Exercise 7.4—Electron Energies

Calculate the energy of the $n = 3$ state of the H atom in (a) joules per atom and (b) kilojoules per mole.

You can think of the energy levels in the Bohr model as the rungs of a ladder climbing out of the basement of an "atomic building," where the energy of the H atom is -2.18×10^{-18} J/atom (Example 7.2), to the ground level, where the energy is 0 (Figure 7.10). Each step represents a quantized energy level; as you climb the ladder, you can stop on any rung but not between them. Unlike the rungs of a real ladder, however, Bohr's energy levels get closer and closer together as n increases.

The Bohr Theory and the Spectra of Excited Atoms

A major assumption of Bohr's theory was that an electron in an atom would remain in its lowest energy level unless disturbed. Energy must be absorbed or evolved if the electron changes from one energy level to another, in agreement with the first law of thermodynamics [◄ Section 6.4]. This idea allowed Bohr to explain the spectra of excited gases.

When the H atom electron has $n = 1$ and so is in its ground state, the energy is a large negative value. As we climb the ladder (see Figure 7.10) to the $n = 2$ level, the electron is less strongly attracted to the nucleus, and the energy of an $n = 2$ electron is less negative. Therefore, to move an electron in the $n = 1$ state to the $n = 2$ state, the atom must absorb energy, just as energy must be expended in climbing a ladder. The electron must be excited (Figure 7.11).

Using Bohr's equation we can calculate the energy required to carry the H atom from the ground state ($n = 1$) to its first excited state ($n = 2$). As you learned in Chapter 6, the difference in energy between two states is always

$$\Delta E = E_{\text{final state}} - E_{\text{initial state}}$$

When E_{final} has $n = 2$, and E_{initial} has $n = 1$, we can calculate ΔE from the equation

$$\Delta E = E_{\text{final}} - E_{\text{initial}} = \left(-\frac{Rhc}{2^2}\right) - \left(-\frac{Rhc}{1^2}\right)$$

where Rhc has the value 1312 kJ/mol (as calculated in Example 7.2).

$$\Delta E = E_{\text{final}} - E_{\text{initial}} = (-Rhc/2^2) - (-Rhc/1^2) = (\tfrac{3}{4})Rhc = 984 \text{ kJ/mol}$$

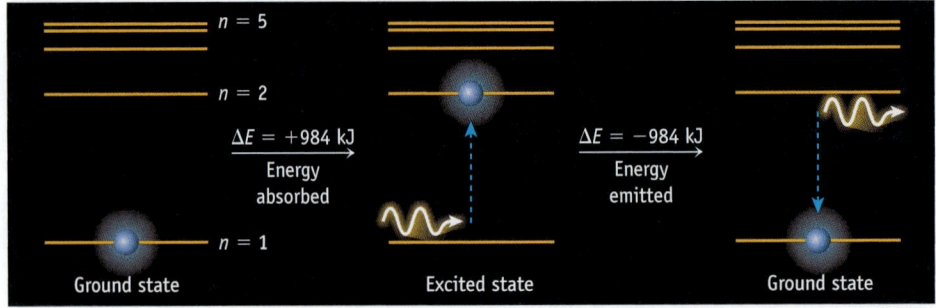

Active Figure 7.11   **Absorption of energy by the atom as the electron moves to an excited state.**
Energy is absorbed when an electron moves from the $n = 1$ state to the $n = 2$ state ($\Delta E > 0$). When the
electron returns to the $n = 1$ state from $n = 2$, energy is evolved ($\Delta E < 0$). The change in energy is
984 kJ/mol, as calculated in Example 7.2.

GENERAL
Chemistry Now™ See the General ChemistryNow CD-ROM or website to explore an interactive version of
this figure accompanied by an exercise.

The amount of energy that must be absorbed by the atom so that an electron can
move from the first to the second energy state is $0.75Rhc$ or 984 kJ/mol of atoms—
no more and no less. If $0.7Rhc$ or $0.8Rhc$ is provided, no transition between states is
possible. *Energy levels in the H atom are quantized*, with the consequence that only certain amounts of energy may be absorbed or emitted.

Moving an electron from a state of low n to one of higher n requires that energy
is absorbed, and the sign of the value of ΔE is positive. The opposite process, in
which an electron "falls" from a level of higher n to one of lower n, emits energy
(Figure 7.11). For example, for a transition from $n = 2$ to $n = 1$,

$$\Delta E = E_{\text{final state}} - E_{\text{initial state}}$$

$$\Delta E = E_{\text{final}} - E_{\text{initial}} = \left(-\frac{Rhc}{1^2}\right) - \left(-\frac{Rhc}{2^2}\right) = -\left(\frac{3}{4}\right)Rhc = -984 \text{ kJ/mol}$$

The negative sign indicates energy is evolved; that is, 984 kJ must be *emitted* per mole
of H atoms.

Depending on how much energy is added to a collection of H atoms, some
atoms have their electrons excited from the $n = 1$ to the $n = 2$ or 3 or higher states.
After absorbing energy, these electrons naturally move back down to lower levels
(either directly or in a series of steps to $n = 1$) and release the energy the atom originally absorbed. The energy emitted is observed as light. *This is the source of the lines
observed in the emission spectrum of H atoms*, and the same basic explanation holds for
the spectra of other elements and for the colors of fireworks.

For hydrogen, a series of emission lines having energies in the ultraviolet region
(called the **Lyman series**; Figure 7.12) arises from electrons moving from states with
$n > 1$ to the $n = 1$ state. The series of lines that have energies in the visible region—the **Balmer series**—arises from electrons moving from states with $n > 2$ to
the lower state with $n = 2$.

In summary, we now recognize that *the origin of atomic spectra is the movement of
electrons between quantized energy states*. If an electron is excited from a lower energy
state to a higher one, energy is absorbed. Conversely, if an electron moves from a
higher energy state to a lower one, energy is emitted. If the energy is emitted as electromagnetic radiation, an emission line is observed. The energy of a specific emission line for excited hydrogen atoms is

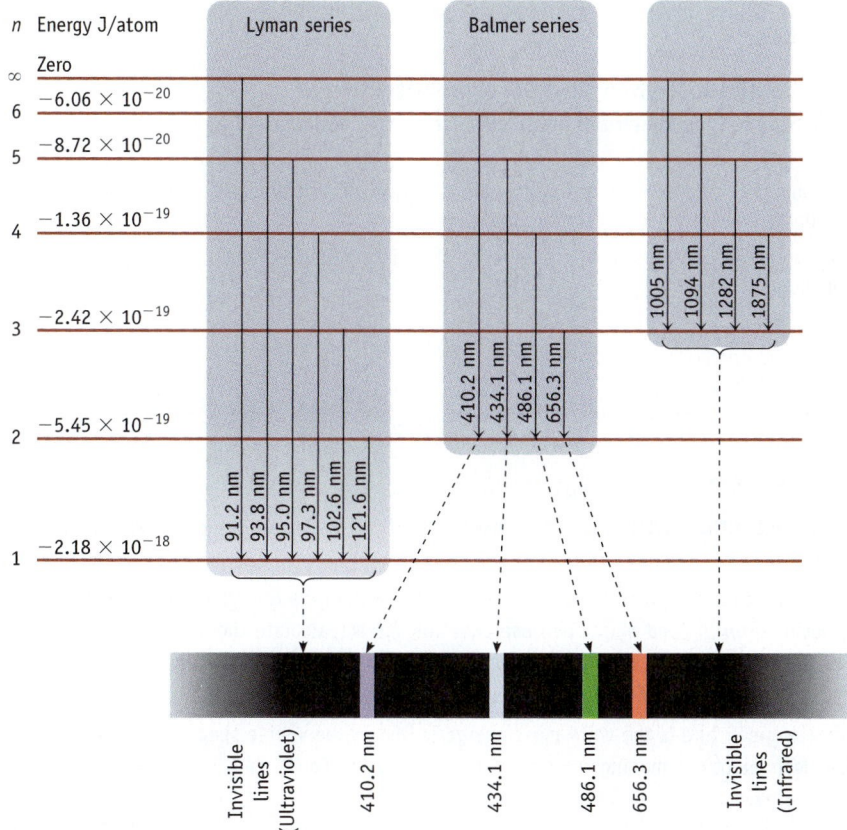

Active Figure 7.12 **Some of the electronic transitions that can occur in an excited H atom.** The Lyman series of lines in the ultraviolet region results from transitions to the $n = 1$ level. Transitions from levels with values of $n > 2$ to $n = 2$ occur in the visible region (Balmer series; see Figure 7.8). Lines in the infrared region result from transitions from levels with $n > 3$ or 4 to the $n = 3$ or 4 levels. (Only the series ending at $n = 3$ is illustrated.)

GENERAL
Chemistry⚛Now™ See the General ChemistryNow CD-ROM or website to explore an interactive version of this figure accompanied by an exercise.

$$\Delta E = E_{final} - E_{initial} = -Rhc \left(\frac{1}{n_{final}^2} - \frac{1}{n_{initial}^2} \right) \tag{7.5}$$

where Rhc is 2.179×10^{-18} J/atom, or 1312 kJ/mol.

Bohr was able to use his model of the atom to calculate the wavelengths of the lines in the hydrogen spectrum. He had tied the unseen (the interior of the atom) to the seen (the observable lines in the hydrogen spectrum)—a fantastic achievement! In addition, he introduced the concept of energy quantization in describing atomic structure, a concept that remains an important part of modern science.

As mentioned previously, agreement between theory and experiment is taken as evidence that the theoretical model is valid. It soon became apparent, however, that a flaw existed in Bohr's theory. Bohr's model of the atom explained only the spectrum of H atoms and of other systems having one electron (such as He⁺). Furthermore, *the idea that electrons are particles moving about the nucleus with a path of fixed radius, like that of the planets about the sun, is no longer the accepted model for the atom.*

See the General ChemistryNow CD-ROM or website:

- **Screen 7.6 Atomic Line Spectrum,** for an exercise on the light emitted by excited hydrogen atoms

  (a) for a simulation and exercise examining the radiation emitted when the electrons of excited hydrogen atoms return to the ground state

  (b) for a tutorial on calculating the wavelength of radiation emitted when electrons change energy levels

Example 7.3—Energies of Emission Lines for Excited Atoms

Problem Calculate the wavelength of the green line in the visible spectrum of excited H atoms using the Bohr theory.

Strategy First locate the green line in Figure 7.12 and determine the quantum states involved. That is, decide on $n_{initial}$ and n_{final}. Then use Equation 7.5 to calculate the difference in energy, ΔE, between these states, a difference that appears as visible light in the green region of the spectrum. Finally, express ΔE in terms of a wavelength.

Solution The green line is the third most energetic line in the visible spectrum of hydrogen and arises from electrons moving from $n = 4$ to $n = 2$. Using Equation 7.5 where $n_{final} = 2$ and $n_{initial} = 4$, we have

$$\Delta E = E_{final} - E_{initial} = \left(-\frac{Rhc}{2^2}\right) - \left(-\frac{Rhc}{4^2}\right)$$

$$\Delta E = -Rhc\left(\frac{1}{4} - \frac{1}{16}\right) = -Rhc(0.1875)$$

Earlier, we found that Rhc is 1312 kJ/mol, so the $n = 4$ to $n = 2$ transition involves an energy change of

$$\Delta E = -(1312 \text{ kJ/mol})(0.1875) = -246.0 \text{ kJ/mol}$$

The wavelength can now be calculated. First the photon energy, E_{photon}, is expressed as J/photon.

$$E_{photon} = \frac{\left(246.0 \, \frac{\text{kJ}}{\text{mol}}\right)\left(1 \times 10^3 \, \frac{\text{J}}{\text{kJ}}\right)}{6.022 \times 10^{23} \, \frac{\text{photons}}{\text{mol}}} = 4.085 \times 10^{-19} \, \frac{\text{J}}{\text{photon}}$$

Now apply Planck's equation where $E_{photon} = h\nu = hc/\lambda$, and so $\lambda = hc/E_{photon}$.

$$\lambda = \frac{hc}{E_{photon}} = \frac{\left(6.626 \times 10^{-34} \, \frac{\text{J} \cdot \text{s}}{\text{photon}}\right)(2.998 \times 10^8 \text{ m} \cdot \text{s}^{-1})}{4.085 \times 10^{-19} \text{ J/photon}}$$

$$= 4.863 \times 10^{-7} \text{ m}$$

$$= (4.863 \times 10^{-7} \text{ m})(1 \times 10^9 \text{ nm/m})$$

$$= 486.3 \text{ nm}$$

The experimental value is 486.1 nm (see Figure 7.12). This represents excellent agreement between experiment and theory.

A Closer Look

Experimental Evidence for Bohr's Theory

Niels Bohr's model of the hydrogen atom was powerful because it could reproduce experimentally observed line spectra. But additional experimental confirmation of the model soon arose.

If the electron in the hydrogen atom is moved from the ground state, where $n = 1$, to the energy level where $n =$ infinity, the electron is considered to have been removed from the atom. That is, the atom has been ionized.

$$H(g) \longrightarrow H^+(g) + e^-$$

We can calculate the energy for this process from Equation 7.5 where $n_{final} = \infty$ and $n_{initial} = 1$.

$$\Delta E = -Rhc\left(\frac{1}{n_{final}^2} - \frac{1}{n_{initial}^2}\right)$$

$$\Delta E = -Rhc\left(\frac{1}{\infty^2} - \frac{1}{1^2}\right) = Rhc$$

Because $Rhc = 1312$ kJ/mol, the energy to move an electron from $n = 1$ to $n = \infty$ is 1312 kJ/mol of H atoms. We now call this the *ionization energy* of the atom [▶ Section 8.6] and can measure it in the laboratory. The experimental value is 1312 kJ/mol, in exact agreement with the result calculated from Bohr's theory!

$H^+(g)$ ─┬─ $n = \infty$ $E = 0$ kJ/mol

$\Delta E = +Rhc = +1312$ kJ/mol

$H(g)$ ─┴─ $n = 1$ $E = -1312$ kJ/mol

Exercise 7.5—Energy of an Atomic Spectral Line

The Lyman series of spectral lines for the H atom occurs in the ultraviolet region. They arise from transitions from higher levels to $n = 1$. Calculate the frequency and wavelength of the least energetic line in this series.

7.4—The Wave Properties of the Electron

Einstein used the photoelectric effect to demonstrate that light, usually thought of as having wave properties, can also have the properties of particles, albeit without mass [◀ page 303]. This fact was pondered by Louis Victor de Broglie (1892–1987). If light can be considered as having both wave and particle properties, would matter behave similarly? That is, could a tiny object such as an electron, normally considered a particle, also exhibit wave properties in some circumstances? In 1925, de Broglie proposed that a free electron of mass m moving with a velocity v should have an associated wavelength given by the equation

$$\lambda = \frac{h}{mv} \tag{7.6}$$

This idea was revolutionary because it linked the particle properties of the electron (m and v) with a wave property (λ). Experimental proof was soon produced. In 1927, C. J. Davisson (1881–1958) and L. H. Germer (1896–1971), working at Bell Telephone Laboratories in New Jersey, found that a beam of electrons was diffracted like light waves by the atoms of a thin sheet of metal foil (Figure 7.13) and that de Broglie's relation was followed quantitatively. Because diffraction is an effect best explained based on the wave properties of radiation, it follows that *electrons can be described as having wave properties under some circumstances.*

De Broglie's equation suggests that any moving particle has an associated wavelength. For λ to be measurable, however, the product of m and v must be very small

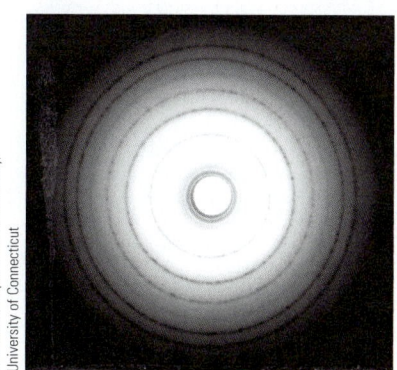

Figure 7.13   **Electron diffraction pattern obtained for magnesium oxide (MgO).**

because h is so small. For example, a 114-g baseball traveling at 110 mph has a large mv product (5.6 kg · m/s) and therefore the incredibly small wavelength of 1.2×10^{-34} m! This tiny value cannot be measured with any instrument now available. As consequence, we will never assign wave properties to a baseball or any other massive object. It is possible to observe wave-like properties only for particles of extremely small mass, such as protons, electrons, and neutrons.

GENERAL
Chemistry• •Now™

See the General ChemistryNow CD-ROM or website:
• **Screen 7.8 Wave Properties of the Electron,** for a tutorial on calculating the wavelength of a moving electron

---

Example 7.4—Using de Broglie's Equation

Problem Calculate the wavelength associated with an electron of mass $m = 9.109 \times 10^{-28}$ g that travels at 40.0% of the speed of light.

Strategy First, consider the units involved. Wavelength is calculated from h/mv, where h is Planck's constant expressed in units of joule seconds (J · s). As discussed in Chapter 6, 1 J = 1 kg · m²/s². Therefore, the mass must be in kilograms and speed in meters per second.

Solution

Electron mass = 9.109×10^{-31} kg

Electron speed (40.0% of light speed) = $(0.400)(2.998 \times 10^8$ m · s$^{-1}) = 1.20 \times 10^8$ m · s^{-1}

Substituting these values into de Broglie's equation, we have

$$\lambda = \frac{h}{mv}$$

$$= \frac{6.626 \times 10^{-34}(\text{kg} \cdot \text{m}^2/\text{s}^2)(\text{s})}{(9.109 \times 10^{-31} \text{ kg})(1.20 \times 10^8 \text{ m/s})}$$

$$= 6.07 \times 10^{-12} \text{ m}$$

In nanometers, the wavelength is

$$\lambda = (6.07 \times 10^{-12} \text{ m})(1.00 \times 10^9 \text{ nm/m}) = 6.07 \times 10^{-3} \text{ nm}$$

Comment The calculated wavelength is about $\frac{1}{12}$ of the diameter of the H atom.

Exercise 7.6—De Broglie's Equation

Calculate the wavelength associated with a neutron having a mass of 1.675×10^{-24} g and a kinetic energy of 6.21×10^{-21} J. (Recall that the kinetic energy of a moving particle is $E = \frac{1}{2} mv^2$.)

7.5—Quantum Mechanical View of the Atom

After World War I, Erwin Schrödinger (1887–1961), an Austrian, worked toward a comprehensive theory of the behavior of electrons in atoms. Starting with de Broglie's hypothesis that an electron in an atom could be described by equations for

Historical Perspectives

20th-Century Giants of Science

Many of the advances in science described in this chapter occurred during the early part of the 20th century, as the result of theoretical studies by some of the greatest minds in the history of science.

Max Karl Ernst Ludwig Planck

(1858–1947) was raised in Germany, where

Max Planck

his father was a professor at a university. While still in his teens Planck decided to become a physicist, against the advice of the head of the physics department at Munich, who told him, "the important discoveries [in physics] have been made. It is hardly worth entering physics anymore." Fortunately, Planck did not take this advice and went on to study thermodynamics. This interest led him eventually to consider the ultraviolet catastrophe and to develop his revolutionary hypothesis, which was announced two weeks before Christmas in

1900. He was awarded the Nobel Prize in physics in 1918 for this work. Einstein later said it was a longing to find harmony and order in nature, a "hunger in his soul," that spurred Planck on.

Erwin Schrödinger

(1887–1961) was born in Vienna, Austria. Following his service as an artillery officer in World War I, he became a professor of physics. In 1928, he succeeded Planck as professor of physics at the University of Berlin. He shared the Nobel Prize in physics in 1933.

Erwin Schrödinger

Niels Bohr

(1885–1962) was born in Copenhagen, Denmark. He earned a Ph.D. in physics in Copenhagen in 1911 and then went to work first with J. J. Thomson and later with Ernest Rutherford in England. It was there that he began to develop his theory of atomic structure and

Niels Bohr

his explanation of atomic spectra. (He received the Nobel Prize in physics in 1922 for this work.) Bohr returned to Copenhagen, where he eventually became director of the Institute for Theoretical Physics. Many young physicists worked with him at the Institute, seven of whom eventually received Nobel Prizes in chemistry and physics. Among these scientists were Werner Heisenberg, Wolfgang Pauli, and Linus Pauling. Element 107 was recently named bohrium in Bohr's honor.

Werner Heisenberg

(1901–1976) studied with Max Born and later with Bohr. He received the Nobel Prize in physics in 1932. The recent play *Copenhagen*, which has been staged in London and New York,

Werner Heisenberg

centers on the relationship between Bohr and Heisenberg and their involvement in the development of atomic weapons in World War II.

wave motion, Schrödinger developed the concept that has come to be called **quantum mechanics** or **wave mechanics**.

The Uncertainty Principle

De Broglie's suggestion that an electron can be described as having wave properties was confirmed by experiment [◀ Section 7.4]. J. J. Thomson's experiments were interpreted on the basis of the particle-like nature of the electron (see page 62). But how can an electron be both a particle and a wave? No single experiment can be done to show the electron behaves *simultaneously* as a wave *and* a particle. Scientists now accept **wave-particle duality**—that is, the idea that the electron indeed has the properties of both.

What does wave-particle duality have to do with electrons in atoms? Werner Heisenberg (1901–1976) and Max Born (1882–1970) provided the answer. Heisenberg concluded, in what is now known as the **uncertainty principle**, that it is impossible to fix both the position of an electron in an atom and its energy with any degree of certainty. Attempting to determine accurately either the location or the energy leaves the other uncertain. (Contrast this principle with the world around you: For objects larger than those on the atomic level—say, an automobile—you can determine, with considerable accuracy, both their energy and location at a given time.)

Based on Heisenberg's idea, Born proposed that the results of quantum mechanics should be interpreted as follows: If we choose to know the energy of an electron in an atom with only a small uncertainty, then we must accept a correspondingly large uncertainty in its position in the space about the atom's nucleus. In practical terms, this means we can assess only the likelihood, or **probability**, of finding an electron with a given energy within a given region of space. In the next section you will see that the result of this viewpoint is the definition of the regions around an atom's nucleus in which there is the highest probability of finding a given electron.

Schrödinger's Model of the Hydrogen Atom and Wave Functions

■ **Wave Functions and Energy**
In Bohr's theory, the electron energy for the H atom is given by $E_n = -Rhc/n^2$. This same result came from Schrödinger's electron wave model.

Schrödinger's model of the hydrogen atom is based on the premise that the electron can be described as a wave and not as a particle. Unlike Bohr's model, Schrödinger's approach resulted in mathematical equations that are complex and difficult to solve except in simple cases. We need not be concerned here with the mathematics, but the solutions to these equations—called **wave functions** and symbolized by the Greek letter ψ (psi)—are important. Understanding the implications of these wave functions is essential to understanding the modern view of the atom. The following important points can be made concerning wave functions:

1. The behavior of the electron in the atom is best described as a standing wave. In a vibrating string, only certain vibrations or standing waves (see Figure 7.2) can be observed. Similarly, *only certain wave functions are allowed for the electron in the atom*.

2. Each wave function ψ is associated with an allowed energy value, E_n, for the electron.

3. Taken together, points 1 and 2 say that *the energy of the electron is quantized*; that is, the electron can have only certain values of energy.

4. The concept of energy quantization enters Schrödinger's theory naturally with the basic assumption that an electron is a standing wave. This is in contrast with Bohr's theory, in which quantization was imposed as a postulate at the start.

5. The square of the wave function (ψ^2) is related to the probability of finding the electron within a given region of space. Scientists refer to this probability as the **electron density**.

6. Schrödinger's theory defines the energy of the electron precisely. The uncertainty principle, however, tells us there must be a large uncertainty in the electron's position. Thus, we can describe only the *probability* of the electron being within a certain region in space when in a given energy state. The region of space in which an electron of a given energy is most probably located is called its **orbital**.

7. To solve Schrödinger's equation for an electron in three-dimensional space, three integer numbers—the **quantum numbers n, ℓ, and m_ℓ**—are an integral part of the mathematical solution. These quantum numbers may have only certain combinations of values, as outlined below.

Quantum numbers are used to define the energy states and orbitals available to the electron. Let us first describe the quantum numbers and the information they provide. We will then turn to the connection between quantum numbers and the energies and shapes of atomic orbitals.

Quantum Numbers

Before looking into the meanings of the three quantum numbers n, ℓ, and m_ℓ, it is important to note two points:

- The quantum numbers are all integers, but their values cannot be selected randomly.

- The three quantum numbers (and their values) are not parameters that scientists dreamed up. Instead, when the behavior of the electron in the hydrogen atom is described mathematically as a wave, the quantum numbers are a natural consequence.

n, the Principal Quantum Number = 1, 2, 3, . . .

The principal quantum number n can have any integer value from 1 to infinity. The value of n is the primary factor in determining the energy of an electron. Indeed, for the hydrogen atom (with its single electron), the energy of the electron varies *only* with the value of n and is given by the same equation derived by Bohr for the H atom: $E_n = -Rhc/n^2$.

The value of n also defines the size of an orbital: The greater the value of n, the greater the electron's average distance from the nucleus.

Each electron is labeled according to its value of n. In atoms having more than one electron, two or more electrons may have the same n value. These electrons are then said to be in the same **electron shell** or same **electron level**.

ℓ, the Angular Momentum Quantum Number = 0, 1, 2, 3, . . . , $n - 1$

The electrons of a given shell can be grouped into **subshells**, where each subshell is characterized by a different value of the quantum number ℓ and by a characteristic shape. The quantum number ℓ can have any integer value from 0 to $n - 1$. *Each value of ℓ corresponds to a different orbital shape or orbital type.*

Because ℓ can be no larger than $n - 1$, the value of n limits the number of subshells possible for the nth shell. Thus, for $n = 1$, ℓ must equal 0. Because ℓ has only one value when $n = 1$, only one subshell is possible for an electron assigned to $n = 1$. When $n = 2$, ℓ can be either 0 or 1. Because two values of ℓ are now possible, there are two subshells in the $n = 2$ electron shell.

The values of ℓ are usually coded by letters according to the following scheme:

Value of ℓ	Corresponding Subshell Label
0	*s*
1	*p*
2	*d*
3	*f*

For example, an $\ell = 1$ subshell is called a "*p* subshell," and an orbital found in that subshell is called a "*p* orbital." Conversely, an electron assigned to a *p* subshell has an ℓ value of 1.

m_ℓ, the Magnetic Quantum Number = 0, ±1, ±2, ±3, . . . , ±ℓ

The magnetic quantum number, m_ℓ, is related to the orientation in space of the orbitals within a subshell. *Orbitals in a given subshell differ only in their orientation in space, not in their energy.*

The value of ℓ limits the integer values assigned to m_ℓ: m_ℓ can range from $+\ell$ to $-\ell$ with 0 included. For example, when $\ell = 2$, m_ℓ has five values: -2, -1, 0, $+1$, and $+2$. The number of values of m_ℓ for a given subshell ($= 2\ell + 1$) specifies the number of orientations that exist for the orbitals of that subshell and thus the number of orbitals in the subshell.

■ **Electron Energy and Quantum Numbers**
The electron energy in the H atom depends *only* on the value of n. In atoms with more electrons, the energy depends on *both* n and ℓ. This is discussed in more detail in Section 8.3.

■ **Orbital Symbols**
Early studies of the emission spectra of elements classified lines into four groups on the basis of their appearance. These groups were labeled *sharp, principal, diffuse,* and *fundamental*. From these names came the labels we now apply to orbitals: *s, p, d,* and *f*.

Useful Information from Quantum Numbers

The three quantum numbers introduced thus far are a kind of ZIP code for electrons. For example, suppose you live in an apartment building. You could specify your location as being on a particular floor (n), in a particular apartment on that floor (ℓ), and in a particular room in the apartment (m_ℓ). Analogously, n describes the shell to which an electron is assigned in an atom, ℓ describes the subshell within that shell, and m_ℓ is related to the orientation of the orbital within that subshell.

Allowed values of the three quantum numbers are summarized in Table 7.1. Before describing the composition of the first four electron shells ($n = 1, 2, 3,$ and 4), let us summarize some useful points:

- Electrons in atoms are assigned to orbitals, which are grouped into subshells. One or more subshells with the same value of n constitute an electron shell.
- Electron subshells are labeled by first giving the value of n and then the value of ℓ in the form of its letter code. For $n = 1$ and $\ell = 0$, for example, the label is $1s$.

If you describe sets of quantum numbers, starting with a given value of n and then deciding the values of ℓ and then m_ℓ that follow (see Table 7.1), you would discover the following:

- n = the number of subshells in a shell
- $2\ell + 1$ = the number of orbitals in a subshell = the number of values of m_ℓ
- n^2 = the number of orbitals in a shell

First Electron Shell, $n = 1$

When $n = 1$ the value of ℓ can only be 0, and so m_ℓ must also have a value of 0. This means that, in the electron shell closest to the nucleus, only one subshell exists, and that subshell consists of only a single orbital, the $1s$ orbital.

For the Second Shell, $n = 2$

When $n = 2$, ℓ can have two values (0 and 1), so two subshells or two types of orbitals occur in the second shell. One of these is the $2s$ subshell ($n = 2$ and $\ell = 0$), and the other is the $2p$ subshell ($n = 2$ and $\ell = 1$). Because the values of m_ℓ can be $-1, 0,$ and $+1$ when $\ell = 1$, three p orbitals exist. All three orbitals have $\ell = 1$ so they all have the same shape. However, because each has a different m_ℓ value, the three orbitals differ in their orientation in space.

For the Third Shell, $n = 3$

When $n = 3$, three subshells, or orbital types, are possible for an electron because ℓ has the values 0, 1, and 2. Because you see ℓ values of 0 and 1 again, you know that two of the subshells within the $n = 3$ shell are $3s$ ($\ell = 0$, one orbital) and $3p$ ($\ell = 1$, three orbitals). The third subshell is d, indicated by $\ell = 2$. Because m_ℓ has five values ($-2, -1, 0, +1,$ and $+2$) when $\ell = 2$, five d orbitals (no more and no less) occur in the $\ell = 2$ subshell.

For the Fourth Shell, $n = 4$, and Beyond

Table 7.1 shows that there are four subshells in the $n = 4$ shell. In addition to s, p, and d orbitals, there is an f subshell; that is, there are orbitals for which $\ell = 3$. Seven such orbitals exist because there are seven values of m_ℓ when $\ell = 3$ ($-3, -2, -1, 0, +1, +2,$ and $+3$).

■ **Subshells and Orbitals**

Subshell	Number of Orbitals in Subshell
s	1
p	3
d	5
f	7

Table 7.1 Summary of the Quantum Numbers, Their Interrelationships, and the Orbital Information Conveyed

Principal Quantum Number	Angular Momentum Quantum Number	Magnetic Quantum Number	Number and Type of Orbitals in the Subshell
Symbol $= n$ Values $= 1, 2, 3, \ldots$ $n =$ number of subshells	Symbol $= \ell$ Values $= 0 \ldots n - 1$	Symbol $= m_\ell$ Values $= -\ell \ldots 0 \ldots +\ell$	Number of orbitals in shell $= n^2$ and number of orbitals in subshell $= 2\ell + 1$
1	0	0	one 1s orbital (one orbital of one type in the $n = 1$ shell)
2	0 1	0 $+1, 0, -1$	one 2s orbital three 2p orbitals (four orbitals of two types in the $n = 2$ shell)
3	0 1 2	0 $+1, 0, -1$ $+2, +1, 0, -1, -2$	one 3s orbital three 3p orbitals five 3d orbitals (nine orbitals of three types in the $n = 3$ shell)
4	0 1 2 3	0 $+1, 0, -1$ $+2, +1, 0, -1, -2$ $+3, +2, +1, 0, -1, -2, -3$	one 4s orbital three 4p orbitals five 4d orbitals seven 4f orbitals (16 orbitals of four types in the $n = 4$ shell)

GENERAL
Chemistry ⚛ Now™

See the General ChemistryNow CD-ROM or website:
- **Screen 7.9 Heisenberg's Uncertainty Principle,** to view an animation on the quantum mechanical view of the atom
- **Screen 7.12 Quantum Numbers and Orbitals,** for a tutorial on determining values for the quantum numbers for an orbital

Exercise 7.7—Using Quantum Numbers

Complete the following statements:

(a) When $n = 2$, the values of ℓ can be _____ and _____.

(b) When $\ell = 1$, the values of m_ℓ can be _____ , _____ , and _____ , and the subshell has the letter label _____.

(c) When $\ell = 2$, the subshell is called a _____ subshell.

(d) When a subshell is labeled s, the value of ℓ is _____ and m_ℓ has the value _____.

(e) When a subshell is labeled p, _____ orbitals occur within the subshell.

(f) When a subshell is labeled f, there are _____ values of m_ℓ, and _____ orbitals occur within the subshell.

7.6—The Shapes of Atomic Orbitals

The chemistry of an element and its compounds is determined by the atom's electrons, and particularly by the electrons with the highest value of n, which are often called *valence electrons* [▶ Section 9.1]. The types of orbitals to which these electrons are assigned are also important, so we turn now to the question of orbital shape and orientation.

s Orbitals

When an electron has $\ell = 0$, we often say the electron is assigned to, or "occupies," an s orbital. But what does this mean? What is an s orbital? What does it look like? To answer these questions, we begin with the wave function for an electron with $n = 1$ and $\ell = 0$, that is, with a $1s$ orbital. If we assume the electron is a tiny particle and not a wave, and if we could photograph the $1s$ electron at 1-second intervals for a few thousand seconds, the composite picture would look like the drawing in Figure 7.14a. It resembles a cloud of dots, so chemists refer to such representations of electron orbitals as **electron cloud pictures**.

The fact that the density of dots is greater close to the nucleus (the electron cloud is denser close to the nucleus) indicates that the electron is most often found near the nucleus (or, conversely, it is less likely to be found farther away). Putting this statement in the language of quantum mechanics, we say the greatest probability of finding the electron is in a tiny volume of space close to the nucleus. Conversely, the probability of finding the electron declines upon moving away from the nucleus; it is less probable that the electron is farther away. The thinning of the electron cloud at increasing distance, shown by the decreasing density of dots in Figure 7.14a, is illustrated in a different way in Figure 7.14b. Here we plotted the square of the wave function for the electron in a $1s$ orbital (ψ^2), times 4π and the distance squared ($4\pi r^2$), as a function of the distance of the electron from the nucleus. The units of $4\pi r^2\psi^2$ at each point are $1/\text{distance}$, so the vertical axis of this plot represents the probability of finding the electron in a thin spherical shell a distance r from the nucleus. For this reason, $4\pi r^2\psi^2$ is sometimes called a **surface density plot** or a **radial distribution plot**. For the $1s$ orbital, $4\pi r^2\psi^2$ is zero at the nucleus—there is no probability the electron will be *at* the nucleus—but the probability is very high a short distance from the nucleus and decreases rapidly as the distance from the nucleus increases. Notice that the probability of finding the electron approaches but never quite reaches zero, even at very large distances.

For the $1s$ orbital, Figure 7.14a shows that the electron is most likely found within a sphere with the nucleus at the center. No matter in which direction you proceed from the nucleus, the probability of finding an electron is the same at the same distance from the nucleus (Figure 7.14b). *The 1s orbital is spherical in shape.*

The visual image in Figure 7.14a is that of a cloud whose density is small at large distances from the center; there is no sharp boundary beyond which the electron is never found. The s and other orbitals, however, are often depicted as having a sharp boundary surface (Figure 7.14c), largely because it is easier to draw such pictures. To arrive at the diagram in Figure 7.14c, we drew a sphere about the nucleus in such a way that the chance of finding the electron somewhere inside is 90%.

Misconceptions exist about pictures such as Figure 7.14c. First, there is not an impenetrable surface within which the electron is "contained." Second, the proba-

■ **Surface Density Plot for 1s**
The wave nature of the electron is evident from Figure 7.14b. The maximum amplitude of the electron wave occurs at 0.0529 nm. It is interesting to note that this maximum is at exactly the same distance from the nucleus as Niels Bohr calculated for the radius of the orbit occupied by the $n = 1$ electron.

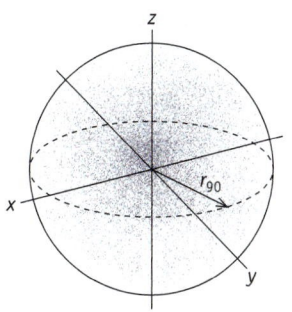

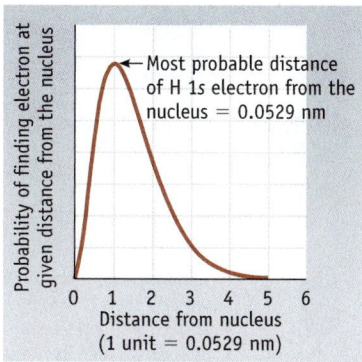

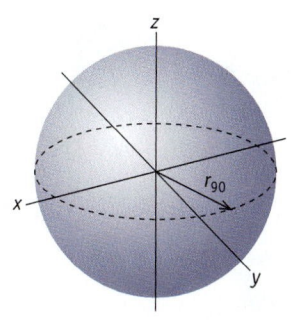

(a) Dot picture of an electron in a 1s orbital. Each dot represents the position of the electron at a different instant in time. Note that the dots cluster closest to the nucleus. r_{90} is the radius of a sphere within which the electron is found 90% of the time.

(b) A plot of the surface density ($4\pi r^2\psi^2$) as a function of distance for a hydrogen atom 1s orbital. This gives the probability of finding the electron at a given distance from the nucleus.

(c) The surface of the sphere within which the electron is found 90% of the time for a 1s orbital. This surface is often called a "boundary surface." (A 90% surface was chosen arbitrarily. If the choice was the surface within which the electron is found 50% of the time, the sphere would be considerably smaller.)

Active Figure 7.14 Different views of a 1s ($n = 1$ and $\ell = 0$) orbital.

GENERAL
Chemistry Now™ See the General ChemistryNow CD-ROM or website to explore an interactive version of this figure accompanied by an exercise.

bility of finding the electron is not the same throughout the volume enclosed by the surface. For example, the electron in the H atom 1s orbital has a greater probability of being at 0.0529 nm from the nucleus than closer or farther away. Third, the terms "electron cloud" and "electron distribution" seem to imply that the electron is a particle, but quantum mechanics treats the electron as having wave properties.

Finally, an important feature of all *s* orbitals (1s, 2s, 3s, and so on) is that they are *spherical in shape*. One important difference between *s* orbitals with different *n* values, however, is that *the size of s orbitals increases as n increases* (Figure 7.15). Thus, the 1s orbital is more compact than the 2s orbital, which is in turn more compact than the 3s orbital.

p Orbitals

Atomic orbitals for which $\ell = 1$, *p* orbitals, all have the same basic shape. *All p orbitals have one imaginary plane that slices through the nucleus and that divides the region of electron density in half* (Figures 7.15 and 7.16). This imaginary plane is called a **nodal surface**, a planar surface on which there is zero probability of finding the electron. The electron can never be found on the nodal surface; the regions of electron density lie on either side of the nucleus. A plot of electron probability ($4\pi r^2\psi^2$) versus distance would start at zero at the nucleus, rise to a maximum, and then drop off at still greater distances.

If you enclose 90% of the electron density within a surface, the views in Figure 7.16 are appropriate. The electron cloud has a shape that resembles a weight lifter's "dumbbell," so chemists often describe *p* orbitals as having dumbbell shapes.

According to Table 7.1, when $\ell = 1$, then m_ℓ can only be -1, 0, or $+1$. That is, three orientations are possible for $\ell = 1$ or *p* orbitals. There are three mutually perpendicular directions in space (*x, y,* and *z*), and the *p* orbitals are commonly visualized as lying along those directions (with the nodal surface perpendicular to the axis). Each orbital is labeled according to the axis along which it lies (p_x, p_y, or p_z).

■ **Standing Waves and Nodal Surfaces**
Recall that standing waves have nodes (Figure 7.2). Similarly, the electron waves in an atom have nodes.

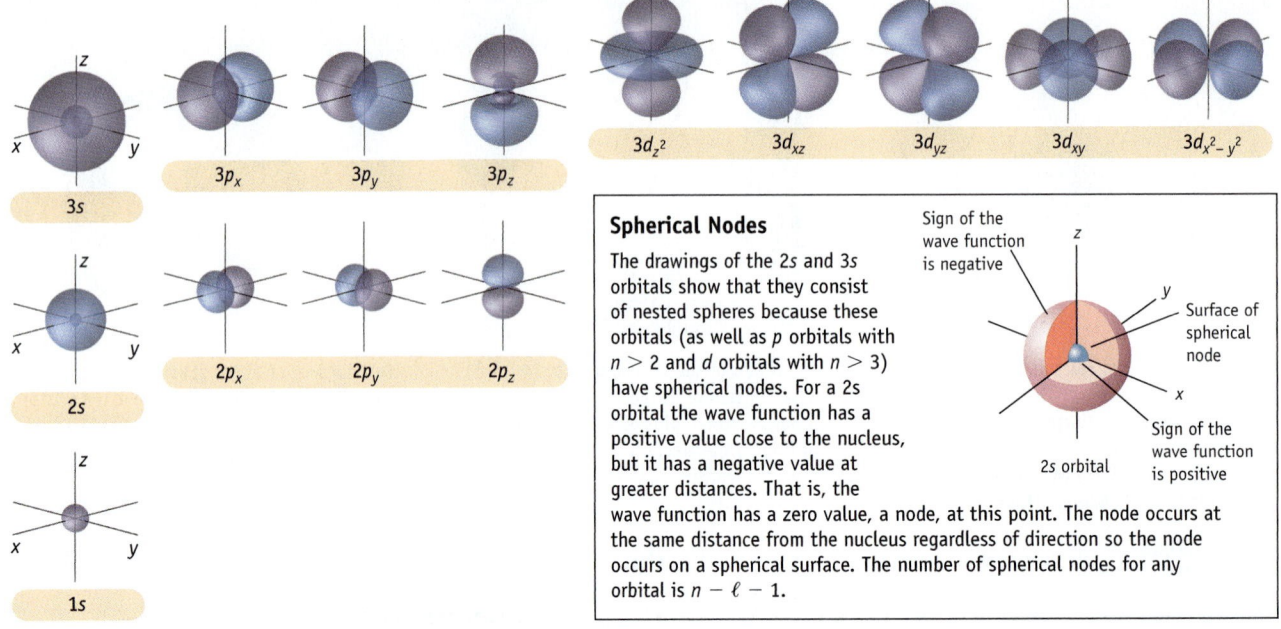

Spherical Nodes

The drawings of the 2s and 3s orbitals show that they consist of nested spheres because these orbitals (as well as p orbitals with $n > 2$ and d orbitals with $n > 3$) have spherical nodes. For a 2s orbital the wave function has a positive value close to the nucleus, but it has a negative value at greater distances. That is, the wave function has a zero value, a node, at this point. The node occurs at the same distance from the nucleus regardless of direction so the node occurs on a spherical surface. The number of spherical nodes for any orbital is $n - \ell - 1$.

Active Figure 7.15 **Atomic Orbitals.** Boundary surface diagrams for electron densities of 1s, 2s, 2p, 3s, 3p, and 3d orbitals for a hydrogen atom. For the p orbitals, the subscript letter on the orbital notation indicates the cartesian axis along which the orbital lies.

GENERAL
Chemistry ⚛ Now™ See the General ChemistryNow CD-ROM or website to explore an interactive version of this figure accompanied by an exercise.

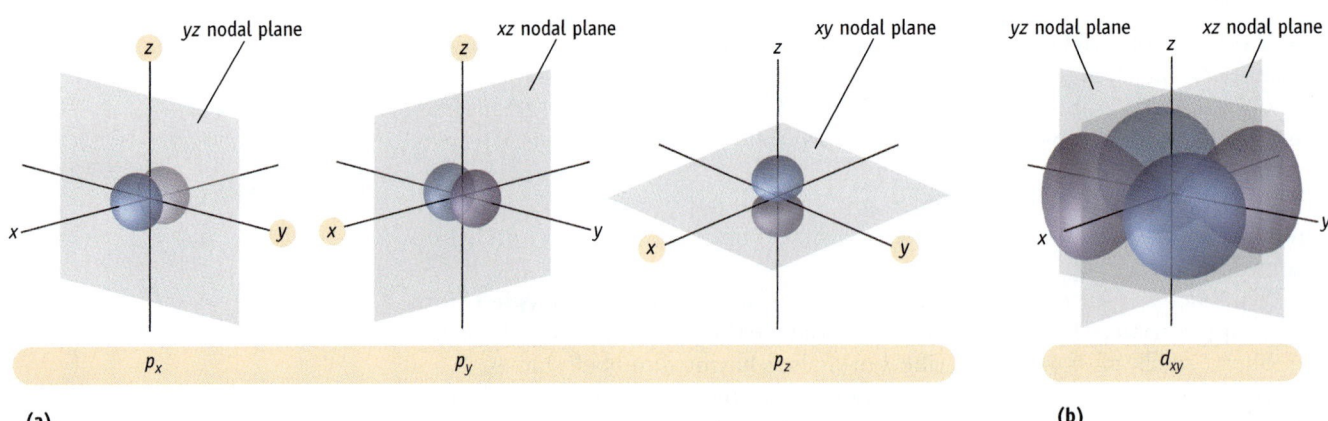

(a)   (b)

Figure 7.16 Nodal surfaces in p and d orbitals. A plane passing through the nucleus (perpendicular to this axis) is called a nodal surface. (a) The three p orbitals each have one nodal surface ($\ell = 1$). (b) The d_{xy} orbital. All five d orbitals have two nodal surfaces ($\ell = 2$). Here the nodal surfaces are the xz- and yz-planes, so the regions of electron density lie in the xy-plane and between the x- and y-axes.

d Orbitals

The value of ℓ is equal to the number of nodal surfaces that slice through the nucleus. Thus, s orbitals, for which $\ell = 0$, have no nodal surfaces, and p orbitals, for which $\ell = 1$, have one planar nodal surface. It follows that the five d orbitals, for which $\ell = 2$, have two nodal surfaces, which results in four regions of electron density. The d_{xy} orbital, for example, lies in the xy-plane and the two nodal surfaces are the xz- and yz-planes (see Figure 7.16). Two other orbitals, d_{xz} and d_{yz}, lie in planes defined by the xz- and yz-axes, respectively; they also have two, mutually perpendicular nodal surfaces (Figure 7.15).

Of the two remaining d orbitals, the $d_{x^2-y^2}$ orbital is easier to visualize. Like the d_{xy} orbital, the $d_{x^2-y^2}$ orbital results from two vertical planes slicing the electron density into quarters. Now, however, the planes bisect the x- and y-axes, so the regions of electron density lie along the x- and y-axes.

The final d orbital, d_{z^2} (Figure 7.15), has two main regions of electron density along the z-axis, but a "doughnut" of electron density also occurs in the xy-plane. This orbital has two nodal surfaces, but the surfaces are not flat.

f Orbitals

The seven f orbitals all have $\ell = 3$. The three nodal surfaces cause the electron density to lie in eight regions of space. These orbitals are less easily visualized, but one f orbital is illustrated in Figure 7.17.

■ **ℓ and Nodal Surfaces**

Orbital	ℓ	Number of Nodal Surfaces
s	0	0
p	1	1
d	2	2
f	3	3

■ **Nodal surfaces**
Nodal surfaces occur for all p, d, and f orbitals. These surfaces are usually flat, so they are referred to as nodal planes. In some cases (for example, d_{z^2}), however, the "plane" is not flat and so is better referred to as a "surface."

GENERAL
Chemistry Now™

See the General ChemistryNow CD-ROM or website:
- **Screen 7.13 Shapes of Atomic Orbitals,** for exercises on orbital shapes, quantum numbers, and nodes

Exercise 7.8—Orbital Shapes

(a) What are the n and ℓ values for each of the following orbitals: 6s, 4p, 5d, and 4f?

(b) How many nodal planes exist for a 4p orbital? For a 6d orbital?

7.7—Atomic Orbitals and Chemistry

We close with some questions to ponder: When an element is part of a molecule, are the orbitals the same? Do they have the same shapes? What do the shapes of orbitals have to do with the chemistry of an element? We will take up these questions in the rest of the book, but a few answers are in order here.

Schrödinger's wave equation can be solved exactly for the hydrogen atom but not for heavier atoms or their ions. Nonetheless, chemists make the assumption that orbitals in other atoms are hydrogen-like, even when those atoms are part of a molecule. This approach has allowed chemists to make predictions, using computer-based simulations, about the behavior of molecules. Such predictions are often remarkably accurate, as confirmed by experimental observations.

Charles D. Winters

Figure 7.17 **One of the seven possible f orbitals.** Notice the presence of three nodal planes as required by an orbital with $\ell = 3$.

Chemistry is the study of molecules and their transformations. By thinking about the orbitals of the atoms in molecules, and by making the simple assumption that they resemble those of the hydrogen atom, we can understand much of the chemistry of even complex systems such as those in plants and animals.

Chapter Goals Revisited

When you have finished studying this chapter, you should ask whether you have met the chapter goals. In particular, you should be able to

Describe the properties of electromagnetic radiation

a. Use the terms *wavelength, frequency, amplitude,* and *node* (Section 7.1). General ChemistryNow homework: Study Question(s) 3

b. Use Equation 7.1 ($c = \lambda\nu$), the relationship between the wavelength (λ) and frequency (ν) of electromagnetic radiation and the speed of light (c).

c. Recognize the relative wavelength (or frequency) of the various types of electromagnetic radiation (Figure 7.3). General ChemistryNow homework: SQ(s) 1

d. Understand that the energy of a photon, a massless particle of radiation, is proportional to its frequency (Planck's equation, Equation 7.2). This is an extension of Planck's idea that energy at the atomic level is quantized (Section 7.2). General ChemistryNow homework: SQ(s) 5, 12, 14, 54, 56, 61, 62, 76c

Understand the origin of light from excited atoms and its relationship to atomic structure

a. Describe the Bohr model of the atom, its ability to account for the emission line spectra of excited hydrogen atoms, and the limitations of the model (Section 7.3).

b. Understand that, in the Bohr model of the H atom, the electron can occupy only certain energy levels, each with an energy proportional to $1/n^2$ ($E = -Rhc/n^2$), where n is the principal quantum number (Equation 7.4, Section 7.3). If an electron moves from one energy state to another, the amount of energy absorbed or emitted in the process is equal to the difference in energy between the two states (Equation 7.5, Section 7.3). General ChemistryNow homework: SQ(s) 18, 22, 58

Describe the experimental evidence for wave-particle duality

a. Understand that in the modern view of the atom, electrons are described by the physics of waves (Section 7.4). The wavelength of an electron or any subatomic particle is given by de Broglie's equation (Equation 7.6). General ChemistryNow homework: SQ(s) 24

Describe the basic ideas of quantum mechanics

a. Recognize the significance of quantum mechanics in describing the modern view of atomic structure (Section 7.5).

b. Understand that an orbital for an electron in an atom corresponds to the allowed energy of that electron.

c. Understand that the position of the electron is not known with certainty; only the probability of the electron being at a given point of space can be calculated. This is the interpretation of the quantum mechanical model and embodies the postulate called the Heisenberg uncertainty principle.

Define the three quantum numbers (n, ℓ, and m_ℓ) and their relationship to atomic structure

a. Describe the allowed energy states of the electron in an atom using three quantum numbers n, ℓ, and m_ℓ (Section 7.5). General ChemistryNow homework: SQ(s) 28, 30, 36, 38, 40, 74

b. Describe the shapes of the orbitals (Section 7.6). General ChemistryNow homework: SQ(s) 44, 51, 65f

Key Equations

Equation 7.1 (page 297)
This equation states that the product of the wavelength (λ) and frequency (ν) of electromagnetic radiation is equal to the speed of light (c).

$$c = \lambda \times \nu$$

Equation 7.2 (page 302)
Planck's equation states that the energy of a photon, a massless particle of radiation, is proportional to its frequency (ν).

$$E = h\nu$$

where h is Planck's constant ($6.626 \times 10^{-34}\,\text{J} \cdot \text{s}$).

Equation 7.4 (page 308)
In Bohr's theory, the potential energy of the electron, E_n, in the nth quantum level of the H atom is proportional to $1/n^2$.

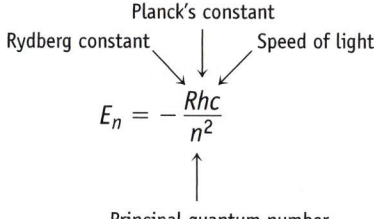

where n is an integer equal to or greater than 1 and $Rhc = 2.179 \times 10^{-18}\,\text{J/atom}$ or 1312 kJ/mol.

Equation 7.5 (page 311)
The change in energy for an electron moving between two quantum levels (n_{final} and n_{initial}) in the H atom.

$$\Delta E = E_{\text{final}} - E_{\text{initial}} = -Rhc \left(\frac{1}{n_{\text{final}}^2} - \frac{1}{n_{\text{initial}}^2} \right)$$

Equation 7.6 (page 313)
De Broglie's equation relates the wavelength of the electron (λ) to its mass (m) and speed (v). h is Planck's constant.

$$\lambda = \frac{h}{mv}$$

Study Questions

▲ denotes more challenging questions.

■ denotes questions available in the Homework and Goals section of the General ChemistryNow CD-ROM or website.

Blue numbered questions have answers in Appendix O and fully worked solutions in the *Student Solutions Manual*.

Structures of many of the compounds used in these questions are found on the General ChemistryNow CD-ROM or website in the Models folder.

GENERAL Chemistry·⚛·Now™ Assess your understanding of this chapter's topics with additional quizzing and conceptual questions at http://now.brookscole.com/kotz6e

Practicing Skills

Electromagnetic Radiation

(See Example 7.1, Exercise 7.1, Figure 7.3, and General ChemistryNow Screen 7.3.)

1. ■ Answer the following questions based on Figure 7.3:
   (a) Which type of radiation involves less energy, x-rays or microwaves?
   (b) Which radiation has the higher frequency, radar or red light?
   (c) Which radiation has the longer wavelength, ultraviolet or infrared light?

2. Consider the colors of the visible spectrum.
   (a) Which colors of light involve less energy than green light?
   (b) Which color of light has photons of greater energy, yellow or blue?
   (c) Which color of light has the greater frequency, blue or green?

3. ■ Traffic signals are often now made of LEDs (light-emitting diodes). Amber and green ones are pictured here.
   (a) The light from an amber signal has a wavelength of 595 nm, and that from a green signal has wavelength of 500 nm. Which has the higher frequency?
   (b) Calculate the frequency of amber light.

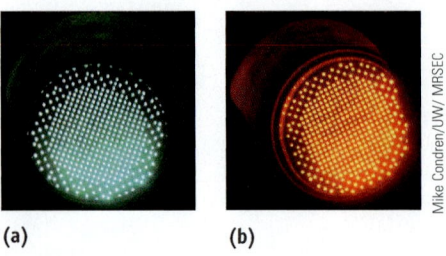

(a)   (b)

Mike Condren/UW/ MRSEC

4. Suppose you are standing 225 m from a radio transmitter. What is your distance from the transmitter in terms of the number of wavelengths if
   (a) The station is broadcasting at 1150 kHz (on the AM radio band)? (1 kHZ = 1×10^3 Hz or 1000 cycles per second.)
   (b) The station is broadcasting at 98.1 MHz (on the FM radio band)? (1 MHz = 10^6 Hz, or cycles per second.)

Electromagnetic Radiation and Planck's Equation

(See page 304, Exercise 7.2, and General ChemistryNow Screens 7.4 and 7.5.)

5. ■ Green light has a wavelength of 5.0×10^2 nm. What is the energy, in joules, of one photon of green light? What is the energy, in joules, of 1.0 mol of photons of green light?

6. Violet light has a wavelength of about 410 nm. What is its frequency? Calculate the energy of one photon of violet light. What is the energy of 1.0 mol of violet photons? Compare the energy of photons of violet light with those of red light. Which is more energetic?

7. The most prominent line in the spectrum of aluminum is at 396.15 nm. What is the frequency of this line? What is the energy of one photon with this wavelength? Of 1.00 mol of these photons?

8. The most prominent line in the spectrum of magnesium is 285.2 nm. Other lines are found at 383.8 and 518.4 nm. In what region of the electromagnetic spectrum are these lines found? Which is the most energetic line? What is the energy of 1 mol of photons with the wavelength of the most energetic line?

9. Place the following types of radiation in order of increasing energy per photon:
   (a) yellow light from a sodium lamp
   (b) x-rays from an instrument in a dentist's office
   (c) microwaves in a microwave oven
   (d) your favorite FM music station at 91.7 MHz

10. Place the following types of radiation in order of increasing energy per photon.
    (a) radar signals
    (b) radiation within a microwave oven
    (c) gamma rays from a nuclear reaction
    (d) red light from a neon sign
    (e) ultraviolet radiation from a sun lamp

Photoelectric Effect

(See page 303 and Figure 7.6.)

11. An energy of 2.0×10^2 kJ/mol is required to cause a cesium atom on a metal surface to lose an electron. Calculate the longest possible wavelength of light that can ionize a cesium atom. In what region of the electromagnetic spectrum is this radiation found?

12. ■ You are an engineer designing a switch that works by the photoelectric effect. The metal you wish to use in your device requires 6.7×10^{-19} J/atom to remove an electron. Will the switch work if the light falling on the metal has a wavelength of 540 nm or greater? Why or why not?

Atomic Spectra and the Bohr Atom

(See Examples 7.2 and 7.3, Figures 7.9 –7.12, and General ChemistryNow Screens 7.6 and 7.7.)

13. The most prominent line in the spectrum of mercury is at 253.652 nm. Other lines are located at 365.015 nm, 404.656 nm, 435.833 nm, and 1013.975 nm.
    (a) Which of these lines represents the most energetic light?
    (b) What is the frequency of the most prominent line? What is the energy of one photon with this wavelength?
    (c) Are any of these lines found in the spectrum of mercury shown in Figure 7.9? What color or colors are these lines?

14. ■ The most prominent line in the spectrum of neon is found at 865.438 nm. Other lines are located at 837.761 nm, 878.062 nm, 878.375 nm, and 1885.387 nm.
    (a) In what region of the electromagnetic spectrum are these lines found?
    (b) Are any of these lines found in the spectrum of neon shown in Figure 7.9?
    (c) Which of these lines represents the most energetic light?
    (d) What is the frequency of the most prominent line? What is the energy of one photon with this wavelength?

15. A line in the Balmer series of emission lines of excited H atoms has a wavelength of 410.2 nm (Figure 7.12). What color is the light emitted in this transition? What quantum levels are involved in this emission line? What are the values of $n_{initial}$ and n_{final}?

16. What are the wavelength and frequency of the radiation involved in the least energetic emission line in the Lyman series? What quantum levels are involved in this emission line? What are the values of $n_{initial}$ and n_{final}?

17. Consider only transitions involving the $n = 1$ through $n = 5$ energy levels for the H atom (where the energy level spacings below are not to scale).

    _____ $n = 5$

    _____ $n = 4$

    _____ $n = 3$

    _____ $n = 2$

    _____ $n = 1$

    (a) How many emission lines are possible, considering only the five quantum levels?

    (b) Photons of the highest frequency are emitted in a transition from the level with $n =$ ____ to a level with $n =$ ____.
    (c) The emission line having the longest wavelength corresponds to a transition from the level with $n =$ ____ to the level with $n =$ ____.

18. ■ Consider only transitions involving the $n = 1$ through $n = 4$ energy levels for the hydrogen atom (using the diagram in Study Question 17).
    (a) How many emission lines are possible, considering only the four quantum levels?
    (b) Photons of the lowest energy are emitted in a transition from the level with $n =$ ____ to a level with $n =$ ____.
    (c) The emission line having the shortest wavelength corresponds to a transition from the level with $n =$ ____ to the level with $n =$ ____.

19. The energy emitted when an electron moves from a higher energy state to a lower energy state in any atom can be observed as electromagnetic radiation.
    (a) Which involves the emission of less energy in the H atom, an electron moving from $n = 4$ to $n = 2$ or an electron moving from $n = 3$ to $n = 2$?
    (b) Which involves the emission of more energy in the H atom, an electron moving from $n = 4$ to $n = 1$ or an electron moving from $n = 5$ to $n = 2$? Explain fully.

20. If energy is absorbed by a hydrogen atom in its ground state, the atom is excited to a higher energy state. For example, the excitation of an electron from the level with $n = 1$ to the level with $n = 3$ requires radiation with a wavelength of 102.6 nm. Which of the following transitions would require radiation of *longer wavelength* than this?
    (a) $n = 2$ to $n = 4$ (c) $n = 1$ to $n = 5$
    (b) $n = 1$ to $n = 4$ (d) $n = 3$ to $n = 5$

21. Calculate the wavelength and frequency of light emitted when an electron changes from $n = 3$ to $n = 1$ in the H atom. In what region of the spectrum is this radiation found?

22. ■ Calculate the wavelength and frequency of light emitted when an electron changes from $n = 4$ to $n = 3$ in the H atom. In what region of the spectrum is this radiation found?

DeBroglie and Matter Waves

(See Example 7.4 and General ChemistryNow Screen 7.8.)

23. An electron moves with a velocity of 2.5×10^8 cm $\cdot$ s^{-1}. What is its wavelength?

24. ■ A beam of electrons ($m = 9.11 \times 10^{-31}$ kg/electron) has an average speed of 1.3×10^8 m $\cdot$ s^{-1}. What is the wavelength of electrons having this average speed?

25. Calculate the wavelength, in nanometers, associated with a 1.0×10^2-g golf ball moving at 30. m $\cdot$ s^{-1} (about 67 mph). How fast must the ball travel to have a wavelength of 5.6×10^{-3} nm?

26. A rifle bullet (mass = 1.50 g) has a velocity of 7.00×10^2 mph. What is the wavelength associated with this bullet?

Quantum Mechanics

(See Sections 7.5 and 7.6 and General ChemistryNow Screens 7.9–7.14.)

27. (a) When $n = 4$, what are the possible values of ℓ?
    (b) When ℓ is 2, what are the possible values of m_ℓ?
    (c) For a 4s orbital, what are the possible values of n, ℓ, and m_ℓ?
    (d) For a 4f orbital, what are the possible values of n, ℓ, and m_ℓ?

28. ■ (a) When $n = 4$, $\ell = 2$, and $m_\ell = -1$, to what orbital type does this refer? (Give the orbital label, such as 1s.)
    (b) How many orbitals occur in the $n = 5$ electron shell? How many subshells? What are the letter labels of the subshells?
    (c) If a subshell is labeled f, how many orbitals occur in the subshell? What are the values of m_ℓ?

29. A possible excited state of the H atom has the electron in a 4p orbital. List all possible sets of quantum numbers n, ℓ, and m_ℓ for this electron.

30. ■ A possible excited state for the H atom has an electron in a 5d orbital. List all possible sets of quantum numbers n, ℓ, and m_ℓ for this electron.

31. How many subshells occur in the electron shell with the principal quantum number $n = 4$?

32. How many subshells occur in the electron shell with the principal quantum number $n = 5$?

33. Explain briefly why each of the following is not a possible set of quantum numbers for an electron in an atom.
    (a) $n = 2$, $\ell = 2$, $m_\ell = 0$
    (b) $n = 3$, $\ell = 0$, $m_\ell = -2$
    (c) $n = 6$, $\ell = 0$, $m_\ell = 1$

34. Which of the following represent valid sets of quantum numbers? For a set that is invalid, explain briefly why it is not correct.
    (a) $n = 3$, $\ell = 3$, $m_\ell = 0$ (c) $n = 6$, $\ell = 5$, $m_\ell = -1$
    (b) $n = 2$, $\ell = 1$, $m_\ell = 0$ (d) $n = 4$, $\ell = 3$, $m_\ell = -4$

35. What is the maximum number of orbitals that can be identified by each of the following sets of quantum numbers? When "none" is the correct answer, explain your reasoning.
    (a) $n = 3$, $\ell = 0$, $m_\ell = +1$ (c) $n = 7$, $\ell = 5$
    (b) $n = 5$, $\ell = 1$ (d) $n = 4$, $\ell = 2$, $m_\ell = -2$

36. ■ What is the maximum number of orbitals that can be identified by each of the following sets of quantum numbers? When "none" is the correct answer, explain your reasoning.
    (a) $n = 4$, $\ell = 3$ (c) $n = 2$, $\ell = 2$
    (b) $n = 5$ (d) $n = 3$, $\ell = 1$, $m_\ell = -1$

37. State which of the following orbitals cannot exist according to the quantum theory: 2s, 2d, 3p, 3f, 4f, and 5s. Briefly explain your answers.

38. ■ State which of the following are incorrect designations for orbitals according to the quantum theory: 3p, 4s, 2f, and 1p. Briefly explain your answers.

39. Write a complete set of quantum numbers (n, ℓ, and m_ℓ) that quantum theory allows for each of the following orbitals: (a) 2p, (b) 3d, and (c) 4f.

40. ■ Write a complete set of quantum numbers (n, ℓ, and m_ℓ) for each of the following orbitals: (a) 5f, (b) 4d, and (c) 2s.

41. A particular orbital has $n = 4$ and $\ell = 2$. What must this orbital be: (a) 3p, (b) 4p, (c) 5d, or (d) 4d?

42. A given orbital has a magnetic quantum number of $m_\ell = -1$. This could *not* be a (an)
    (a) f orbital   (c) p orbital
    (b) d orbital   (d) s orbital

43. How many nodal surfaces are associated with each of the following orbitals?
    (a) 2s   (b) 5d   (c) 5f

44. ■ How many nodal surfaces are associated with each of the following atomic orbitals?
    (a) 4f   (b) 2p   (c) 6s

General Questions on Atomic Structure

These questions are not designated as to type or location in the chapter. They may combine several concepts. More challenging questions are indicated by ▲.

45. Which of the following are applicable when explaining the photoelectric effect? Correct any statements that are wrong.
    (a) Light is electromagnetic radiation.
    (b) The intensity of a light beam is related to its frequency.
    (c) Light can be thought of as consisting of massless particles whose energy is given by Planck's equation, $E = h\nu$.

46. In what region of the electromagnetic spectrum for hydrogen is the Lyman series of lines found? The Balmer series?

47. Give the number of nodal surfaces for each orbital type: s, p, d, and f.

48. What is the maximum number of s orbitals found in a given electron shell? The maximum number of p orbitals? Of d orbitals? Of f orbitals?

49. Match the values of ℓ shown in the table with orbital type (s, p, d, or f).

ℓ Value	Orbital Type
3	_____
0	_____
1	_____
2	_____

50. Sketch a picture of the 90% boundary surface of an *s* orbital and the p_x orbital. Be sure the latter drawing shows why the *p* orbital is labeled p_x and not p_y, for example.

51. ■ Complete the following table.

Orbital Type	Number of Orbitals in a Given Subshell	Number of Nodal Surfaces
s	_____	_____
p	_____	_____
d	_____	_____
f	_____	_____

52. Excited H atoms have many emission lines. One series of lines, called the Pfund series, occurs in the infrared region. It results when an electron changes from higher energy levels to a level with $n = 5$. Calculate the wavelength and frequency of the lowest energy line of this series.

53. An advertising sign gives off red light and green light.
    (a) Which light has the higher-energy photons?
    (b) One of the colors has a wavelength of 680 nm and the other has a wavelength of 500 nm. Which color has which wavelength?
    (c) Which light has the higher frequency?

54. ■ Radiation in the ultraviolet region of the electromagnetic spectrum is quite energetic. It is this radiation that causes dyes to fade and your skin to develop a sunburn. If you are bombarded with 1.00 mol of photons with a wavelength of 375 nm, what amount of energy, in kilojoules per mole of photons, are you being subjected to?

55. A cell phone sends signals at about 850 MHz (1 MHz = 1×10^6 Hz or cycles per second).
    (a) What is the wavelength of this radiation?
    (b) What is the energy of 1.0 mol of photons with a frequency of 850 MHz?
    (c) Compare the energy in part (b) with the energy of a mole of photons of blue light (420 nm).
    (d) Comment on the difference in energy between 850 MHz radiation and blue light.

56. ■ Assume your eyes receive a signal consisting of blue light, $\lambda = 470$ nm. The energy of the signal is 2.50×10^{-14} J. How many photons reach your eyes?

57. If sufficient energy is absorbed by an atom, an electron can be lost by the atom and a positive ion formed. The amount of energy required is called the ionization energy. In the H atom, the ionization energy is that required to change the electron from $n = 1$ to $n =$ infinity. (See "A Closer Look: Experimental Evidence for Bohr's Theory," page 313.) Calculate the ionization energy for He$^+$ ion. Is the ionization energy of the He$^+$ more or less than that of H? (Bohr's theory applies to He$^+$ because it, like the H atom, has a single electron. The electron energy, however, is now given by $E = -Z^2 Rhc/n^2$, where Z is the atomic number of helium.)

58. ■ Suppose hydrogen atoms absorb energy so that electrons are excited to the $n = 7$ energy level. Electrons then undergo these transitions, among others: (a) $n = 7 \longrightarrow n = 1$; (b) $n = 7 \longrightarrow n = 6$; and (c) $n = 2 \longrightarrow n = 1$. Which transition produces a photon with (i) the smallest energy, (ii) the highest frequency, and (iii) the shortest wavelength?

59. Rank the following orbitals in the H atom in order of increasing energy: 3*s*, 2*s*, 2*p*, 4*s*, 3*p*, 1*s*, and 3*d*.

60. How many orbitals correspond to each of the following designations?
    (a) 3*p*   (d) 6*d*   (g) $n = 5$
    (b) 4*p*   (e) 5*d*   (h) 7*s*
    (c) $4p_x$ (f) 5*f*

61. ■ Cobalt-60 is a radioactive isotope used in medicine for the treatment of certain cancers. It produces β particles and γ rays, the latter having energies of 1.173 and 1.332 MeV. (1 MeV = 1 million electron-volts and 1 eV = 9.6485×10^4 J/mol.) What are the wavelength and frequency of a γ-ray photon with an energy of 1.173 MeV?

62. ▲ ■ Exposure to high doses of microwaves can cause damage. Estimate how many photons, with $\lambda = 12$ cm, must be absorbed to raise the temperature of your eye by 3.0 °C. Assume the mass of an eye is 11 g and its specific heat capacity is 4.0 J/g · K.

63. When the *Sojourner* spacecraft landed on Mars in 1997, the planet was approximately 7.8×10^7 km from the earth. How long did it take for the television picture signal to reach earth from Mars?

64. The most prominent line in the emission spectrum of chromium is found at 425.4 nm. Other lines in the chromium spectrum are found at 357.9 nm, 359.3 nm, 360.5 nm, 427.5 nm, 429.0 nm, and 520.8 nm.
    (a) Which of these lines represents the most energetic light?
    (b) What color is light of wavelength 425.4 nm?

65. Answer the following questions as a summary quiz on the chapter.
    (a) The quantum number *n* describes the _____ of an atomic orbital.
    (b) The shape of an atomic orbital is given by the quantum number _____.
    (c) A photon of green light has _____ (less or more) energy than a photon of orange light.
    (d) The maximum number of orbitals that may be associated with the set of quantum numbers $n = 4$ and $\ell = 3$ is _____.
    (e) The maximum number of orbitals that may be associated with the quantum number set $n = 3$, $\ell = 2$, and $m_\ell = -2$ is _____.

▲ More challenging   ■ In General ChemistryNow   **Blue-numbered questions** answered in Appendix O

(f) ■ Label each of the following orbital pictures with the appropriate letter:

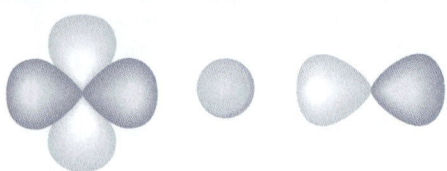

(g) When $n = 5$, the possible values of ℓ are _____.

(h) The number of orbitals in the $n = 4$ shell is _____.

66. Answer the following questions as a summary quiz on this chapter.

(a) The quantum number n describes the _____ of an atomic orbital and the quantum number ℓ describes its _____.

(b) When $n = 3$, the possible values of ℓ are _____.

(c) What type of orbital corresponds to $\ell = 3$? _____

(d) For a 4d orbital, the value of n is _____, the value of ℓ is _____, and a possible value of m_ℓ is _____.

(e) Each of the following drawings represents a type of atomic orbital. Give the letter designation for the orbital, give its value of ℓ, and specify the number of nodal surfaces.

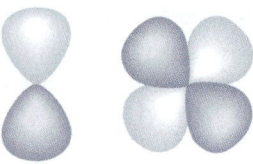

Letter =   _____   _____
ℓ value =   _____   _____
Nodal surfaces =   _____   _____

(f) An atomic orbital with three nodal surfaces is _____.

(g) Which of the following orbitals *cannot* exist according to modern quantum theory: 2s, 3p, 2d, 3f, 5p, 6p?

(h) Which of the following is *not* a valid set of quantum numbers?

n	ℓ	m_ℓ
3	2	1
2	1	2
4	3	0

(i) What is the maximum number of orbitals that can be associated with each of the following sets of quantum numbers? (One possible answer is "none.")

(i) $n = 2$ and $\ell = 1$

(ii) $n = 3$

(iii) $n = 3$ and $\ell = 3$

(iv) $n = 2$, $\ell = 1$, and $m_\ell = 0$

Summary and Conceptual Questions

The following questions use concepts from the previous chapters.

67. What are two major assumptions of Bohr's theory of atomic structure?

68. Bohr pictured the electrons of the atom as being located in definite orbits about the nucleus, just as the planets orbit the sun. Criticize this model.

69. Light is given off by a sodium- or mercury-containing streetlight when the atoms are excited. The light you see arises for which of the following reasons?

(a) Electrons are moving from a given energy level to one of higher n.

(b) Electrons are being removed from the atom, thereby creating a metal cation.

(c) Electrons are moving from a given energy level to one of lower n.

70. How do we interpret the physical meaning of the square of the wave function? What are the units of $4\pi r^2 \psi^2$?

71. What does "wave-particle duality" mean? What are its implications in our modern view of atomic structure?

72. Which of these are observable?

(a) position of an electron in an H atom

(b) frequency of radiation emitted by H atoms

(c) path of an electron in an H atom

(d) wave motion of electrons

(e) diffraction patterns produced by electrons

(f) diffraction patterns produced by light

(g) energy required to remove electrons from H atoms

(h) an atom

(i) a molecule

(j) a water wave

73. In principle, which of the following can be determined?

(a) the energy of an electron in the H atom with high precision and accuracy

(b) the position of a high-speed electron with high precision and accuracy

(c) at the same time, both the position and the energy of a high-speed electron with high precision and accuracy

74. ▲ ■ Suppose you live in a different universe where a different set of quantum numbers is required to describe the atoms of that universe. These quantum numbers have the following rules:

N, principal   1, 2, 3, . . . , ∞
L, orbital $= N$
M, magnetic $-1, 0, +1$

How many orbitals are there altogether in the first three electron shells?

75. A photon with a wavelength of 93.8 nm strikes a hydrogen atom, and light is emitted by the atom. How many emission lines would be observed? At what wavelengths? Explain briefly. (See Figure 7.12.)

76. ▲ Technetium is not found naturally on earth; it must be synthesized in the laboratory. Nonetheless, because it is radioactive it has valuable medical uses. For example, the element in the form of sodium pertechnetate ($NaTcO_4$) is used in imaging studies of the brain, thyroid, and salivary glands and in renal blood flow studies, among other things.

(a) In what group and period of the periodic table is the element found?

(b) The valence electrons of technetium are found in the $5s$ and $4d$ subshells. What is a set of quantum numbers (n, ℓ, and m_ℓ) for one of the electrons of the $5s$ subshell?

(c) ■ Technetium emits a γ-ray with an energy of 0.141 MeV. (1 MeV = 1 million electron-volts, where 1 eV = 9.6485×10^4 J/mol.) What are the wavelength and frequency of a γ-ray photon with an energy of 0.141 MeV?

(d) To make $NaTcO_4$, the metal is dissolved in nitric acid.

$$7\ HNO_3(aq)\ +\ Tc(s) \longrightarrow$$
$$HTcO_4(aq)\ +\ 7\ NO_2(g) +\ 3\ H_2O(\ell)$$

and the product, $HTcO_4$, is treated with NaOH to make $NaTcO_4$.

(i) Write a balanced equation for the reaction of $HTcO_4$ with NaOH.

(ii) If you begin with 4.5 mg of Tc metal, how much $NaTcO_4$ can be made? What mass of NaOH, in grams, is required to convert all of the $HTcO_4$ into $NaTcO_4$?

77. Explain why you could or could not measure the wavelength of a golf ball in flight.

78. See the General ChemistryNow CD-ROM or website, Screen 7.1 Chemical Puzzler. This screen shows that light of different colors can come from a "neon" sign or from certain salts when they are placed in a burning organic liquid. ("Neon" signs are glass tubes filled with neon, argon, and other gases, and the gases are excited by an electric current. They are very similar in this regard to common fluorescent lights, although the light in fluorescent tubes comes from the phosphor that coats the inside of the tube.) What do these two sources of light have in common? How is the light generated in each case?

79. A large pickle is attached to two electrodes, which are then attached to a 110-V power supply (see the problem on Screen 7.7 of the General ChemistryNow CD-ROM or website). As the voltage is increased across the pickle, it begins to glow with a yellow color. Knowing that pickles are made by soaking the vegetable in a concentrated salt solution, describe why the pickle might emit light when electrical energy is added.

Charles D. Winters

The "electric pickle."

80. See the General ChemistryNow CD-ROM or website, Screen 7.7 Bohr's Model of the Hydrogen Atom, Simulation. A photon with a wavelength of 97.3 nm is fired at a hydrogen atom and leads to the emission of light. How many emission lines are emitted? Explain why more than one line is emitted.

○Mentor

Do you need a live tutor for homework problems?
Access vMentor at General ChemistryNow at
http://now.brookscole.com/kotz6e
for one-on-one tutoring from a chemistry expert

▲ More challenging ■ In General ChemistryNow Blue-numbered questions answered in Appendix O

8—Atomic Electron Configurations and Chemical Periodicity

Courtesy of Chemical and Engineering News

Oliver Sacks. Born in London in 1933 to two physicians, Sacks now lives in New York City, where he is a practicing neurologist. He is a member of the American Academy of Arts and Letters and is the author of books such as *The Man Who Mistook His Wife for a Hat* and *Awakenings*. His most recent book, *Uncle Tungsten* (New York, Alfred Knopf, 2001), describes his lifelong fascination with chemistry.

Everything in Its Place

The periodic table of elements has put "everything in its place," according to Oliver Sacks. Sacks is a well-known neurologist, but he is also a writer of books such as *The Man Who Mistook His Wife for a Hat* and *Awakenings*. Less well known is the fact that he has had a love affair with chemistry since he was a boy growing up in London during World War II. On a trip to the London Science Museum, he saw a wall-sized periodic table that displayed samples of many of the 92 chemical elements known at that time. Said Sacks, "Seeing the table, with its actual samples of the elements, was one of the formative experiences of my boyhood and showed me, with the force of revelation, the beauty of science. The periodic table seemed so economical and simple: everything, the whole 92-ishness, reduced to two axes, and yet along each axis an ordered progression of different properties."

Dmitri Mendeleev, one of two people responsible for the creation of the periodic table, was born in Tobolsk in western Siberia on February 8, 1834. He was the youngest of 14 or 17 children (the number is not certain). His father became incapacitated shortly after Dmitri's birth, so, to support the large family, his mother took over a glass manufacturing business begun by her father.

Catastrophe struck the family in 1848 and 1849, when Mendeleev's father died and the glass factory burned. Young Mendeleev's mother was determined to ensure that he be schooled properly, so they journeyed 1300 miles to Moscow so that the boy could enroll in the university. Once in Moscow they found that students from Siberia were not permitted at the university, so they went another 400 miles to St. Petersburg. There Mendeleev's mother was able to secure a place for him at the Central Pedagogical Institute. She died shortly thereafter.

Mendeleev was an extraordinary student of science and published original work before he was 20, even though he was afflicted with tuberculosis and had to do much of his writing in bed. He took the gold medal as the top student at the Institute in 1855 and shortly thereafter was sent to the Crimea as a teacher. The climate in the Crimea was hospitable, much suited to recovering from his illness. However, one reason he was sent far away from St. Petersburg was because he had a terrible temper and was less than beloved by his former teachers and colleagues.

Within a few years Mendeleev returned to St. Petersburg as a lecturer at the university. Soon thereafter, he went to study and do research in Paris, France, and Heidelberg, Germany. In Heidelberg he worked briefly with Robert Bunsen, the inventor of a burner used for spectroscopic studies. There also Mendeleev's temper got the better of him, and he was forced to retreat to a small room where he worked in isolation.

A defining moment for Mendeleev came in 1860 at a conference in Karlsruhe, Germany, where leading chemists from all over Europe came to settle on a system for determining atomic weights. This system, once in place, was crucial to Mendeleev's discovery a few years later of the periodic law and his publication of the first periodic table.

Novosti/Science Photo Library/Photo Researchers, Inc.

Dmitri Mendeleev seated at his desk. Every picture of him shows his long hair and beard. He was in the habit of having it cut only once a year. For more on the story of Mendeleev and the periodic table, see *Mendeleyev's Dream* by P. Strathern: New York, St. Martin's Press, 2001.

In 1861 Mendeleev returned to St. Petersburg and joined the faculty of the Technical Institute. His love of chemistry, as well as his intense blue eyes and flowing beard and hair, made him a popular teacher. He also realized that the teaching of chemistry in Russia was in a sorry state. To remedy this situation, he wrote a 500-page textbook of organic chemistry in only 60 days!

At the age of 32 Mendeleev was appointed professor of general chemistry at the University of St. Petersburg. By 1869 he had completed the first volume of a new textbook, *The Principles of Chemistry*, which was subsequently translated into all the major languages of the world. As he began the second volume, Mendeleev was searching for an organizing principle underlying chemistry. To look for patterns in the chemical and physical behaviors of the elements, he wrote lists of those properties on small cards, one for each element. After four days of pondering the problem for hours on end, he was so exhausted he fell asleep at his desk. In his words, "I saw in a dream a table where all the elements fell into place as required. Awakening, I immediately wrote it down on a piece of paper." This was the beginning of the periodic table chemists use today.

To Review Before You Begin

- Review Chapter 2 on atoms and atomic structure
- Review Chapter 7 on quantum numbers

GENERAL
Chemistry ⊶Now™

Throughout the chapter this icon introduces a list of resources on the General ChemistryNow CD-ROM or website (http://now .brookscole.com/kotz6e) that will:

- help you evaluate your knowledge of the material
- provide homework problems
- allow you to take an exam-prep quiz
- provide a personalized Learning Plan targeting resources that address areas you should study

The wave mechanical model of the atom accurately describes atoms or ions that have a single electron, such as H and He$^+$. It is obvious, however, that a truly useful model must be applicable to atoms with more than one electron—that is, to all the other known elements. One objective of this chapter, therefore, is to develop a workable model for the electronic structure of elements other than hydrogen.

A second objective is to explore some of the physical properties of elements, among them the ease with which atoms lose or gain electrons to form ions and the sizes of atoms and ions. These properties are directly related to the arrangement of electrons in atoms and thus to the chemistry of the elements and their compounds.

8.1—Electron Spin

Around 1920 it was demonstrated experimentally that the electron behaves as though it has a spin, just as the earth has a spin. To understand this property and its relationship to atomic structure requires understanding some aspects of the general phenomenon of magnetism. You will see that electron spin must be represented by a fourth quantum number, the **electron spin magnetic quantum number, m_s.** That is, the complete description of an electron in an atom requires *four* quantum numbers (n, ℓ, m_ℓ, and m_s).

Magnetism

In 1600, William Gilbert (1544–1603) concluded that the earth is a large spherical magnet giving rise to a magnetic field that surrounds the planet (Figure 8.1). The needle of a compass, itself a small magnet, lines up with earth's magnetic field, with

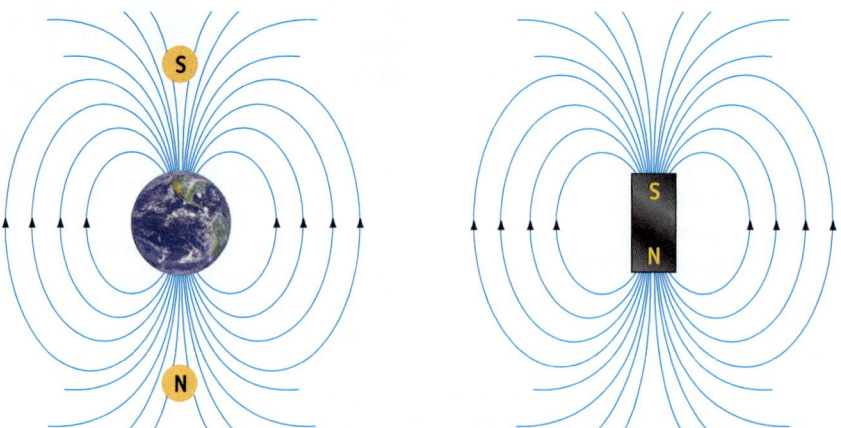

Figure 8.1   **The magnetic fields of the earth and of a bar magnet.** The lines of magnetic force of the earth come from one pole, arbitrarily called the "north magnetic pole" (N) and loop toward the "south magnetic pole" (S). (The geographic North Pole of the earth, named before the introduction of the term "magnetic pole," is actually the magnetic south pole.)

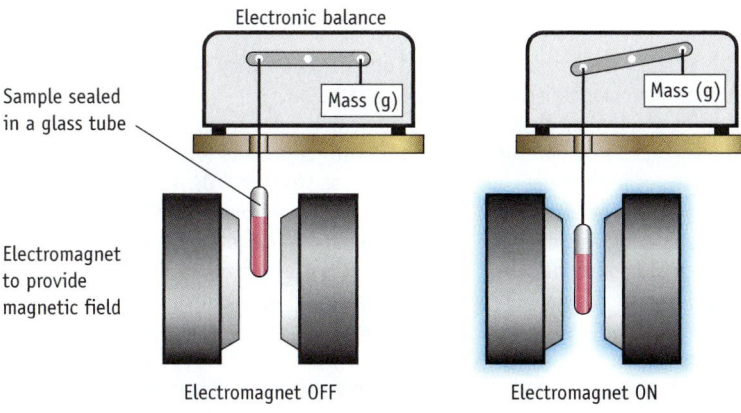

(a)

(b)

Charles D. Winters

Active Figure 8.2 **Observing and measuring paramagnetism.** (a) A magnetic balance is used to measure the magnetism of a sample. The sample is first weighed with the electromagnet turned off. The magnet is then turned on and the sample reweighed. If the substance is paramagnetic, the sample is drawn into the magnetic field and the apparent weight increases. (b) Liquid oxygen (boiling point 90.2 K) clings to the poles of a strong magnet. Elemental oxygen is paramagnetic because it has unpaired electrons. (See Chapter 10.)

GENERAL
Chemistry Now™ See the General ChemistryNow CD-ROM or website to explore an interactive version of this figure accompanied by an exercise.

one end of the needle pointing approximately to the earth's geographic North Pole. We say the end of the compass needle pointing north is the magnet's "magnetic north pole" or simply its "north pole" (N). The other end of the needle is its "south pole" (S). Because opposite poles (N–S) attract, this means that earth's geographic north pole is its magnetic south pole.

Paramagnetism and Unpaired Electrons

Most substances are slightly repelled by a strong magnet; that is, they are **diamagnetic**. In contrast, some metals and compounds are attracted to a magnetic field. Such substances are called **paramagnetic**, and the magnitude of the effect can be determined with an apparatus such as that illustrated in Figure 8.2a.

The magnetism of most paramagnetic materials is so weak that you can observe the effect only in the presence of a strong magnetic field. For example, the oxygen we breathe is paramagnetic; it sticks to the poles of a strong magnet (Figure 8.2b).

Paramagnetism arises from electron spins. An electron in an atom has the magnetic properties expected for a spinning, charged particle (Figure 8.3). Experiments have shown that, if an atom with a single unpaired electron is placed in a magnetic field, only two orientations are possible for the electron spin: aligned with the field or opposed to the field. One orientation is associated with an electron spin quantum number value of $m_s = +\frac{1}{2}$ and the other with an m_s value of $-\frac{1}{2}$. *Electron spin is quantized.*

When one electron is assigned to an orbital in an atom, the electron's spin orientation can take either value of m_s. We observe experimentally that hydrogen atoms, each of which has a single electron, are paramagnetic; when an external magnetic field is applied, the electron magnets align with the field—like the needle of a compass—and experience an attractive force. Helium, with two electrons, is diamagnetic. To account for this observation, we assume that *the two electrons have*

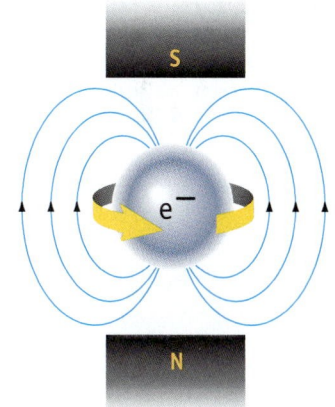

Figure 8.3 **Electron spin and magnetism.** The electron, with its spin and negative electric charge, acts as a "micromagnet." Relative to a magnetic field, only two spin directions are possible: clockwise or counterclockwise. The north pole of the spinning electron can therefore be either aligned with an external magnetic field or opposed to that field. (*See General ChemistryNow Screen 8.3 Spinning Electrons and Magnetism,* to view an animation of concepts in this figure.)

A Closer Look

Paramagnetism and Ferromagnetism

Magnetic materials are relatively common and many are important in our economy. For example, a large magnet is at the heart of the magnetic resonance imaging (MRI) used in medicine, and tiny magnets are found in stereo speakers and in telephone handsets. Magnetic oxides are used in recording tapes and computer disks.

The magnetic materials we use are *ferromagnetic*. The magnetic effect of ferro-

magnetic materials is much larger than that of paramagnetic ones. Ferromagnetism occurs when the spins of unpaired electrons in a cluster of atoms (called a *domain*) in the solid align themselves in the same direction. Only the metals of the iron, cobalt, and nickel subgroups, as well as a few other metals such as neodymium, exhibit this property. They are also unique in that, once the domains are aligned in a magnetic field, the metal is permanently magnetized.

Many alloys exhibit greater ferromagnetism than do the pure metals themselves. One example of such a material is Alnico,

and another is an alloy of neodymium, iron, and boron.

Audio and video tapes are plastics coated with crystals of ferromagnetic oxides such as Fe_2O_3 or CrO_2. The recording head uses an electromagnetic field to create a varying magnetic field based on signals from a microphone. This magnetizes the tape as it passes through the head, with the strength and direction of magnetization varying with the frequency of the sound to be recorded. When the tape is played back, the magnetic field of the moving tape induces a current, which is amplified and sent to the speakers.

Magnets. Many common consumer products contain magnetic materials.

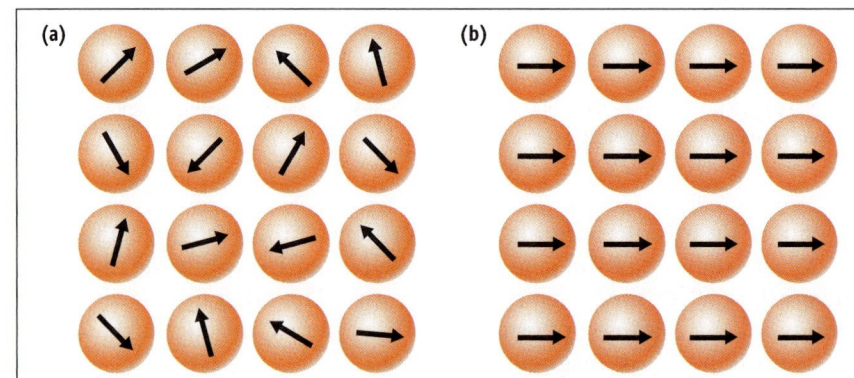

Magnetism. (a) Paramagnetism: In the absence of an external magnetic field, the unpaired electrons in the atoms or ions of the substance are randomly oriented. If a magnetic field is imposed, however, these spins will tend to become aligned with the field. (b) Ferromagnetism: The spins of the unpaired electrons in a cluster of atoms or ions align themselves in the same direction.

opposite spin orientations. We say their *spins are paired*, which means that the magnetic field of one electron is canceled out by the magnetic field of the second electron with opposite spin.

In summary, *paramagnetism is the attraction to a magnetic field of substances in which the constituent ions or atoms contain unpaired electrons*. Substances in which all electrons are paired with partners of opposite spin are diamagnetic. This explanation opens the way to understanding the arrangement of electrons in atoms with more than one electron.

GENERAL
Chemistry⚛Now™

See the General ChemistryNow CD-ROM or website:

- **Screen 8.3 Spinning Electrons and Magnetism,** to watch a video demonstrating the paramagnetism of liquid oxygen

Chemical Perspectives

Quantized Spins and MRI

Just as electrons have a spin, so do atomic nuclei. In the hydrogen atom, the single proton of the nucleus spins on its axis. For most heavier atoms, such as carbon, the atomic nucleus includes both protons and neutrons, and the entire entity has a spin. This property is important, because nuclear spin allows scientists to detect these atoms in molecules and to learn something about their chemical environments.

The technique used to detect the spins of atomic nuclei is **nuclear magnetic resonance (NMR)**. It is one of the most powerful methods currently available to determine molecular structures. About 20 years ago it was adapted as a diagnostic technique in medicine, where it is known as **magnetic resonance imaging (MRI)**.

Just as electron spin is quantized, so too is nuclear spin. The H atom nucleus can spin in either of two directions. If the H atom is placed in a strong, external magnetic field, however, the spinning nuclear magnet can align itself with the external field or against. If a sample of ethanol (CH_3CH_2OH), for example, is placed in a strong magnetic field, a slight excess of the H atom nuclei (and C atom nuclei) is aligned with the lines of force of the field.

The nuclei aligned with the field have a slightly lower energy than those not aligned. The NMR and MRI technologies depend on the fact that energy in the

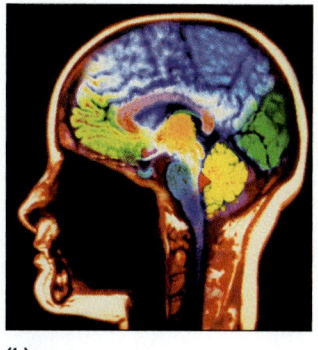

(a)

(b)

Magnetic resonance imaging. (a) MRI instrument. The patient is placed inside a large magnet, and the tissues to be examined are irradiated with radio-frequency radiation. (b) An MRI image of the human brain.

radio-frequency region can be absorbed by the sample and can cause the nuclear spins to go out of alignment—that is, to move to a higher energy state. This absorption of energy is detected by the instrument.

The most important aspect of the magnetic resonance technique is that the difference in energy between two different spin states depends on the locations of atoms in the molecule. In the case of ethanol, the three CH_3 protons are different from the two CH_2 protons, and both sets are different from the OH proton. These three different sets of H atoms absorb radiation of slightly different energies. The instrument measures the frequencies absorbed, and a scientist familiar with the technique can quickly distinguish the three different environments in the molecule.

The MRI technique closely resembles the NMR method. Hydrogen is abundant in the human body as water and in numerous organic molecules. In the MRI device, the patient is placed in a strong magnetic field, and the tissues being examined are irradiated with pulses of radio-frequency radiation.

The MRI image is produced by detecting how fast the excited nuclei "relax" from the higher energy state to the lower energy state. The "relaxation time" depends on the type of tissue. When the tissue is scanned, the H atoms in different regions of the body show different relaxation times, and an accurate "image" is built up.

MRI gives information on soft tissue—muscle, cartilage, and internal organs—which is unavailable from x-ray scans. This technology is also noninvasive, and the magnetic fields and radio-frequency radiation used are not harmful to the body.

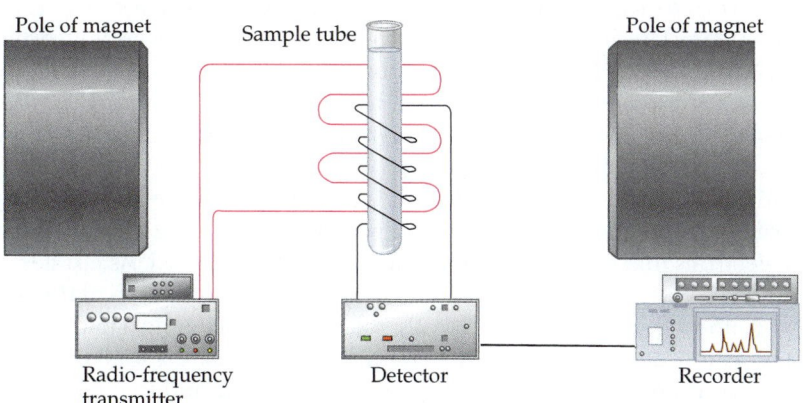

(a)

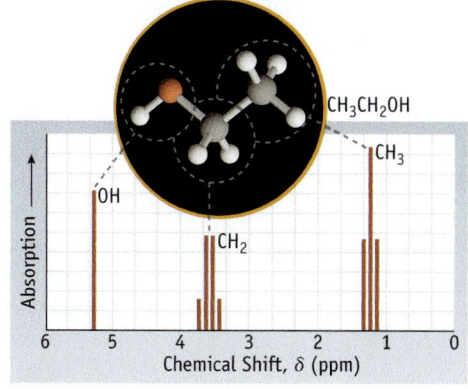

(b)

Nuclear magnetic resonance. (a) A schematic diagram of an NMR spectrometer. (b) The NMR spectrum of ethanol, showing that the three different types of protons appear in distinctly different regions of the spectrum. The pattern observed for the CH_2 and CH_3 protons, for example, is characteristic of these groups of atoms and signals the chemist that they are present in the molecule.

8.2—The Pauli Exclusion Principle

To make the quantum theory consistent with experiment, the Austrian physicist Wolfgang Pauli (1900–1958) stated in 1925 his **exclusion principle**:

> **Pauli Exclusion Principle**
> No two electrons in an atom can have the same set of four quantum numbers (n, ℓ, m_ℓ, and m_s).
>
> ⇓ which leads to
> No atomic orbital can contain
> more than two electrons.

The $1s$ orbital of the H atom has the set of quantum numbers $n = 1$, $\ell = 0$, and $m_\ell = 0$. If an electron is in this orbital, the electron spin direction must also be specified. Let us represent an orbital by a box and the electron by an arrow ($\uparrow$ or $\downarrow$). A representation of the *hydrogen* atom is then as follows:

Electron in 1s orbital: $\boxed{\uparrow}$ Quantum number set
                           1s $n = 1$, $\ell = 0$, $m_\ell = 0$, $m_s = +\frac{1}{2}$

[The direction of the electron spin arrow is arbitrary; that is, it may point in either direction. Here we associate $m_s = +\frac{1}{2}$ with an arrow pointing up ($\uparrow$), but the electron could equally well be depicted as $\downarrow$.] Diagrams such as these are called **orbital box diagrams**.

For a *helium* atom, which has two electrons, both electrons are assigned to the $1s$ orbital. From the Pauli exclusion principle, you know that each electron must have a different set of quantum numbers, so the orbital box diagram now is:

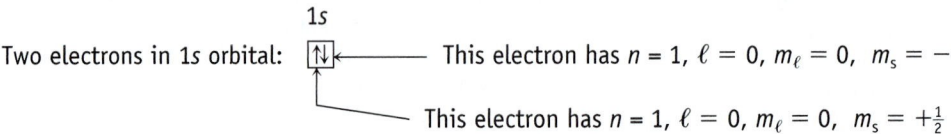

Two electrons in 1s orbital: $\boxed{\uparrow\downarrow}$ ← This electron has $n = 1$, $\ell = 0$, $m_\ell = 0$, $m_s = -\frac{1}{2}$

This electron has $n = 1$, $\ell = 0$, $m_\ell = 0$, $m_s = +\frac{1}{2}$

Each of the two electrons in the $1s$ orbital of a He atom has a different set of the four quantum numbers. The first three numbers of a set describe this as a $1s$ orbital. There are only two choices for the fourth number, $m_s = +\frac{1}{2}$ or $-\frac{1}{2}$. Thus, *the $1s$ orbital, and any other atomic orbital, can be occupied by no more than two electrons, and these two electrons must have opposite spin directions.* The consequence is that the helium atom is diamagnetic, as experimentally observed.

Our understanding of orbitals [◄ Table 7.1, page 319], and the knowledge that an orbital can accommodate no more than two electrons, tells us the maximum number of electrons that can occupy each electron shell or subshell. As just demonstrated, only two electrons can be assigned to an s orbital. Because each of the three orbitals in a p subshell can hold two electrons, that subshell can hold a maximum of six electrons. The five orbitals of a d subshell can accommodate a total of 10 electrons. Recall that there are always n subshells in the nth shell, and that there are n^2 orbitals in that shell [◄ Table 7.1, page 319]. Thus, the maximum number of electrons in any shell is $2n^2$. The relationships among the quantum numbers and the numbers of electrons are shown in Table 8.1.

■ **Orbitals Are Not Boxes**
Orbitals are not literally things or boxes in which electrons are placed. Orbitals are electron waves. Thus, it is not conceptually correct to talk about electrons being *in* orbitals or *occupying* orbitals, although this is commonly done for the sake of simplicity.

■ **Spin Quantum Number and** ↑
We arbitrarily use $m_s = +\frac{1}{2}$ for an arrow pointing up (↑), and $m_s = -\frac{1}{2}$ for an arrow pointing down (↓).

Table 8.1   Number of Electrons Accommodated in Electron Shells and Subshells with $n = 1$ to 6

Electron Shell (n)	Subshells Available	Orbitals Available ($2\ell + 1$)	Number of Electrons Possible in Subshell $[2(2\ell + 1)]$	Maximum Electrons Possible for nth Shell ($2n^2$)
1	s	1	2	2
2	s	1	2	8
	p	3	6	
3	s	1	2	18
	p	3	6	
	d	5	10	
4	s	1	2	32
	p	3	6	
	d	5	10	
	f	7	14	
5	s	1	2	50
	p	3	6	
	d	5	10	
	f	7	14	
	g^*	9	18	
6	s	1	2	72
	p	3	6	
	d	5	10	
	f	7	14	
	g^*	9	18	
	h^*	11	22	

*These orbitals are not occupied in the ground state of any known element.

8.3—Atomic Subshell Energies and Electron Assignments

Our goal is to understand and predict the distribution of electrons in atoms with many electrons. The basic principle involved is the *aufbau,* or "building up," princi-ple in which electrons are assigned to shells (defined by the quantum number n) of increasingly higher energy. Within a given shell, electrons are assigned to subshells (defined by the quantum number ℓ) of successively higher energy. Electrons are as-signed in such a way that the total energy of the atom is as low as possible. Now the relevant question becomes the order of energy of shells and subshells.

Order of Subshell Energies and Assignments

Quantum theory and the Bohr model of the atom state that the energy of the H atom, with a single electron, depends only on the value of n ($E = -Rhc/n^2$, Equation 7.4). For atoms with more than one electron, however, the situation is more complex.

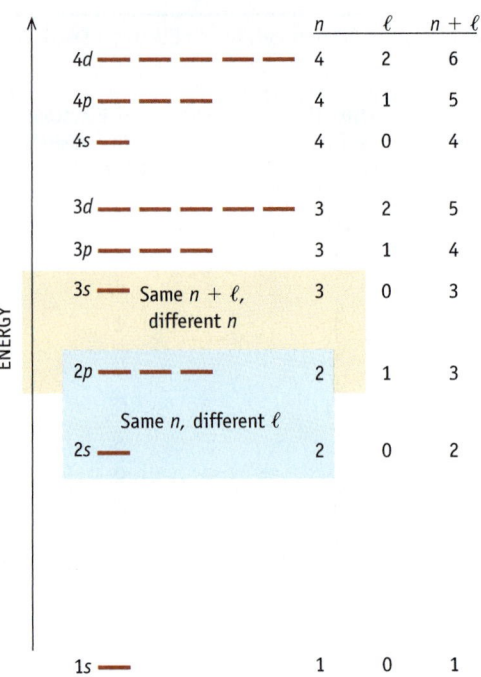

	n	ℓ	$n + \ell$
4d	4	2	6
4p	4	1	5
4s	4	0	4
3d	3	2	5
3p	3	1	4
3s — Same $n + \ell$, different n	3	0	3
2p	2	1	3
Same n, different ℓ			
2s	2	0	2
1s	1	0	1

Active Figure 8.4   Experimentally determined order of subshell energies. Energies of electron shells increase with increasing n and, within a shell, subshell energies increase with increasing ℓ. (The energy axis is not to scale.) The energy gaps between subshells of a given shell become smaller as n increases. Note that the order of orbital energies does not correspond to the order of orbital filling for the heavier elements. For the filling order, see Figure 8.5.

GENERAL
Chemistry ⚛ Now™ See the General ChemistryNow CD-ROM or website to explore an interactive version of this figure accompanied by an exercise.

The experimentally determined order of subshell energies in Figure 8.4 shows that the *subshell energies of multielectron atoms depend on both n and ℓ.* The subshells with $n = 3$, for example, have different energies; for a given atom they are in the order $3s < 3p < 3d$.

The subshell energy order in Figure 8.4 and the actual electron arrangements of the elements lead to two general rules that help us predict these arrangements:

- Electrons are assigned to subshells in order of increasing "$n + \ell$" value.
- For two subshells with the same value of "$n + \ell$," electrons are assigned first to the subshell of lower n.

The following are examples of these rules.

- Electrons are assigned to the $2s$ subshell ($n + \ell = 2 + 0 = 2$) before the $2p$ subshell ($n + \ell = 2 + 1 = 3$).
- Electrons are assigned in the order $3s$ ($n + \ell = 3 + 0 = 3$) before 3p ($n + \ell = 3 + 1 = 4$) before $3d$ ($n + \ell = 3 + 2 = 5$).
- Electrons fill the $4s$ subshell ($n + \ell = 4$) before filling the $3d$ subshell ($n + \ell = 5$).

These filling orders, summarized in Figure 8.5, have been amply verified by experiment.

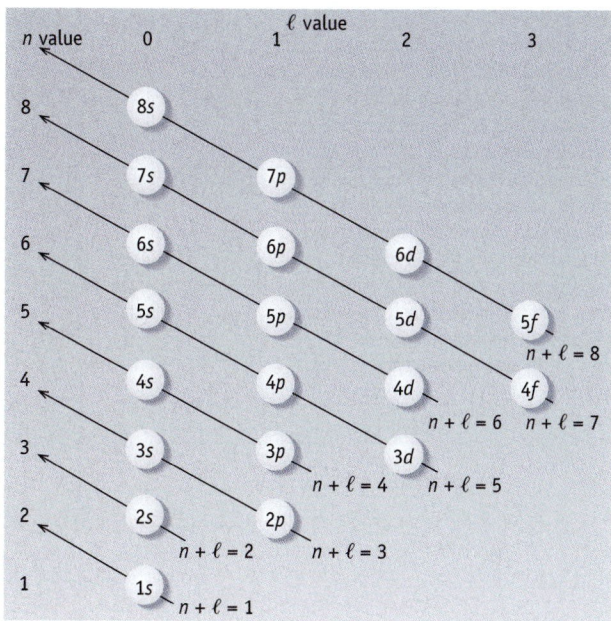

Figure 8.5 Subshell filling order.
Subshells in atoms are filled in order of increasing $n + \ell$. When two subshells have the same $n + \ell$ value, the subshell of lower n is filled first. To use the diagram, begin at $1s$ and follow the arrows of increasing $n + \ell$. (Thus, the order of filling is $1s \Rightarrow 2s \Rightarrow 2p \Rightarrow 3s \Rightarrow 3p \Rightarrow 4s \Rightarrow 3d$ and so on.)

Exercise 8.1—Order of Subshell Assignments

Using the "$n + \ell$" rules, you can generally predict the order of subshell assignments (the electron filling order) for a multielectron atom. To which of the following subshells should an electron be assigned first?

(a) $4s$ or $4p$ **(b)** $5d$ or $6s$ **(c)** $4f$ or $5s$

Effective Nuclear Charge, Z*

The order in which electrons are assigned to subshells in an atom, and many atomic properties, can be rationalized by the concept of **effective nuclear charge (Z*)**. This is the nuclear charge experienced by a particular electron in a multielectron atom, as modified by the presence of the other electrons.

In the hydrogen atom, with only one electron, the $2s$ and $2p$ subshells have the same energy. However, for lithium, an atom with three electrons, the presence of the $1s$ electrons alters the energy of the $2s$ and $2p$ subshells. Why should this be true? This question can be answered in part by referring to Figure 8.6.

Figure 8.6 plots, qualitatively, the surface density function $(4\pi r^2\psi^2)$ for a $2s$ electron [◀ Figure 7.14]. The probability of finding the electron (vertical axis) changes as one moves away from the nucleus (horizontal axis). Lightly shaded on this figure is the region occupied by the $1s$ electrons of lithium. Observe that the $2s$ electron wave occurs partly within the region occupied by $1s$ electrons. Chemists say that the $2s$ electron density region *penetrates* the $1s$ electron density region. This alters the energy of the $2s$ electron relative to what it would be in the H atom where there are no other electrons. As more electrons are added to an atom, the outermost electrons will penetrate the region occupied by the inner electrons, but the penetration is different for ns, np, and nd orbitals, and their energies are altered by differing amounts.

■ **More About Z***

For a more complete discussion of effective nuclear charge, see D. M. P. Mingos: *Essential Trends in Inorganic Chemistry*, New York, Oxford University Press, 1998.

Figure 8.6   Effective nuclear charge, Z*. The two 1s electrons of lithium approximately occupy the shaded region, but this region is penetrated by the 2s electron (whose approximate probability distribution curve is shown here). When the 2s electron is at some distance from the nucleus, it experiences a charge of +1 because the +3 charge of the lithium nucleus is screened by the two 1s electrons. As the 2s electron penetrates the 1s region, however, the 2s electron experiences an increasingly larger charge, to a maximum of +3. On average, the 2s electron experiences a charge, called the effective nuclear charge (Z*), that is much smaller than +3 but greater than +1. (See *General ChemistryNow Screen 8.6 Effective Nuclear Charge, Z*, to view an animation of the concepts in this figure.*)

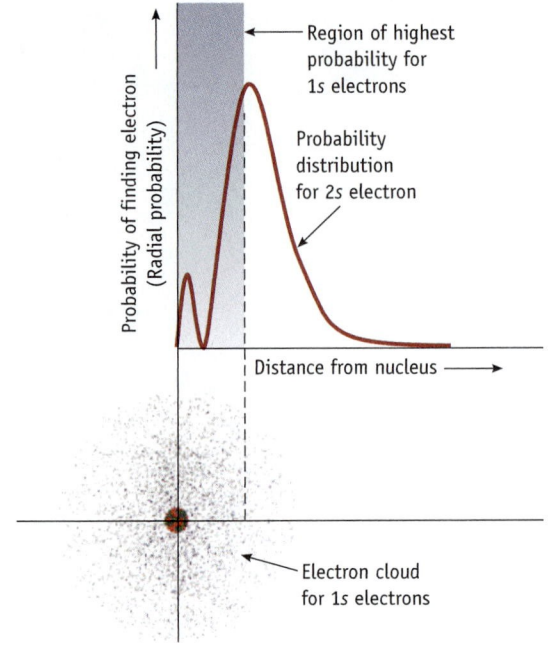

■ **Z* for s and p Subshells**

Z* is greater for s electrons than for p electrons in the same shell. This difference becomes larger as n becomes larger. For example, compare the Group 4A elements.

Atom	Z* (ns)	Z* (np)	Value of n
C	3.22	3.14	2
Si	4.90	4.29	3
Ge	8.04	6.78	4

Lithium has three protons in the nucleus. Suppose the two 1s electrons have been added to the atom. If a third electron (a 2s electron) is at a large distance from the nucleus (Figure 8.6), it would experience a +1 charge because there are two electrons (total charge = −2) between the 2s electron and the +3 charge on the nucleus. Chemists say that the 1s electrons *screen* the effect of the nuclear charge from the 2s electron.

The screening of the nuclear charge varies with the distance of the 2s electron from the nucleus, however. If a 2s electron were to penetrate the 1s electron region, it would experience an increasingly higher positive charge, eventually seeing a charge of +3 if it comes very close to the nucleus. Figure 8.6 shows that a 2s electron has some probability of being both inside and outside the region occupied by the 1s electrons. Thus, on average, a 2s electron experiences a positive charge greater than +1 but smaller than +3. Because of the penetration of the inner electron region by an outer electron, and the screening of the nuclear charge by the inner electrons, an outer electron experiences an *average* nuclear charge, the *effective nuclear charge, Z*.*

Values of Z* for s and p electrons for most second-period elements are listed in Table 8.2. In each case Z* is greater for s electrons than for p electrons and explains why s electrons always have a lower energy than p electrons in the same quantum shell (Figure 8.4).

Another observation regarding Z* for the second-period elements in Table 8.2 is that the value of Z* increases across the period. As you will see in Section 8.6, this effect is important in understanding the change in properties of elements across a period.

Extending these arguments to other subshells, it is observed that the relative penetrating power of subshells is s > p > d > f, so the effective nuclear charge experienced by orbitals is in the order ns > np > nd > nf. One consequence of the differences in orbital penetration and electron shielding is that subshells within an electron shell are filled in the order ns before np before nd before nf.

What emerges from this analysis is the order of shell and subshell energies depicted in Figure 8.4 and the filling order in Figure 8.5. With this understanding, we turn to the periodic table and its use as a guide to electron arrangements in atoms.

GENERAL
Chemistry⚛Now™

See the General ChemistryNow CD-ROM or website:
• **Screen 8.6 Effective Nuclear Charge, Z*,** for a simulation and exercise exploring effective nuclear charge and shielding value

Table 8.2 Effective Nuclear Charges, Z*, for $n = 2$ Elements

Atom	Z*(2s)	Z*(2p)
Li	1.28	
B	2.58	2.42
C	3.22	3.14
N	3.85	3.83
O	4.49	4.45
F	5.13	5.10

8.4—Atomic Electron Configurations

The arrangements of electrons in the elements up to 109—the **electron configurations** of the elements—are given in Table 8.3. These are the ground state electron configurations, where electrons are found in the shells, subshells, and orbitals that result in the lowest energy for the atom. In general, electrons are assigned to orbitals in order of increasing $n + \ell$ (see Figure 8.4). The emphasis here, however, will be to connect the configurations of the elements with their positions in the periodic table, which will allow us ultimately to relate electron configurations to a large number of chemical facts.

Electron Configurations of the Main Group Elements

Hydrogen, the first element in the periodic table, has one electron in a $1s$ orbital. One way to depict its electron configuration is with the orbital box diagram used earlier, but an alternative and more frequently used method is the **spdf notation.** Using this method, the electron configuration of H is $1s^1$, or "one s one."

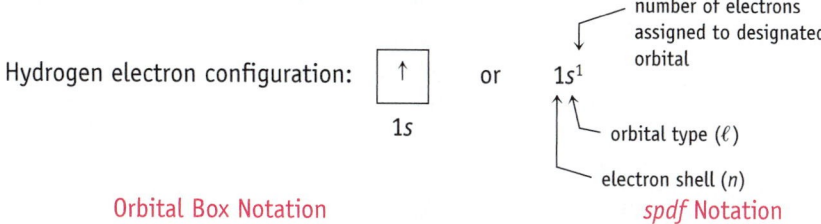

Hydrogen electron configuration: ↑ (1s)   or $1s^1$

number of electrons assigned to designated orbital

orbital type (ℓ)

electron shell (n)

Orbital Box Notation   spdf Notation

Lithium (Li) and Other Elements of Group 1A

Lithium, with three electrons, is the first element in the second period of the periodic table. The first two electrons are in the $1s$ subshell, and the third electron must be in the $n = 2$ shell. According to the energy level diagram in Figure 8.4, that electron must be in the $2s$ subshell. The spdf notation, $1s^2 2s^1$, is read "one s two, two s one."

Lithium: spdf notation $1s^2 2s^1$

Box notation   ↑↓ ↑ ☐☐☐
               1s 2s   2p

Table 8.3   Electron Configurations of Atoms in the Ground State

Z	Element	Configuration	Z	Element	Configuration	Z	Element	Configuration
1	H	$1s^1$	37	Rb	$[Kr]5s^1$	74	W	$[Xe]4f^{14}5d^46s^2$
2	He	$1s^2$	38	Sr	$[Kr]5s^2$	75	Re	$[Xe]4f^{14}5d^56s^2$
3	Li	$[He]2s^1$	39	Y	$[Kr]4d^15s^2$	76	Os	$[Xe]4f^{14}5d^66s^2$
4	Be	$[He]2s^2$	40	Zr	$[Kr]4d^25s^2$	77	Ir	$[Xe]4f^{14}5d^76s^2$
5	B	$[He]2s^22p^1$	41	Nb	$[Kr]4d^45s^1$	78	Pt	$[Xe]4f^{14}5d^96s^1$
6	C	$[He]2s^22p^2$	42	Mo	$[Kr]4d^55s^1$	79	Au	$[Xe]4f^{14}5d^{10}6s^1$
7	N	$[He]2s^22p^3$	43	Tc	$[Kr]4d^55s^2$	80	Hg	$[Xe]4f^{14}5d^{10}6s^2$
8	O	$[He]2s^22p^4$	44	Ru	$[Kr]4d^75s^1$	81	Tl	$[Xe]4f^{14}5d^{10}6s^26p^1$
9	F	$[He]2s^22p^5$	45	Rh	$[Kr]4d^85s^1$	82	Pb	$[Xe]4f^{14}5d^{10}6s^26p^2$
10	Ne	$[He]2s^22p^6$	46	Pd	$[Kr]4d^{10}$	83	Bi	$[Xe]4f^{14}5d^{10}6s^26p^3$
11	Na	$[Ne]3s^1$	47	Ag	$[Kr]4d^{10}5s^1$	84	Po	$[Xe]4f^{14}5d^{10}6s^26p^4$
12	Mg	$[Ne]3s^2$	48	Cd	$[Kr]4d^{10}5s^2$	85	At	$[Xe]4f^{14}5d^{10}6s^26p^5$
13	Al	$[Ne]3s^23p^1$	49	In	$[Kr]4d^{10}5s^25p^1$	86	Rn	$[Xe]4f^{14}5d^{10}6s^26p^6$
14	Si	$[Ne]3s^23p^2$	50	Sn	$[Kr]4d^{10}5s^25p^2$	87	Fr	$[Rn]7s^1$
15	P	$[Ne]3s^23p^3$	51	Sb	$[Kr]4d^{10}5s^25p^3$	88	Ra	$[Rn]7s^2$
16	S	$[Ne]3s^23p^4$	52	Te	$[Kr]4d^{10}5s^25p^4$	89	Ac	$[Rn]6d^17s^2$
17	Cl	$[Ne]3s^23p^5$	53	I	$[Kr]4d^{10}5s^25p^5$	90	Th	$[Rn]6d^27s^2$
18	Ar	$[Ne]3s^23p^6$	54	Xe	$[Kr]4d^{10}5s^25p^6$	91	Pa	$[Rn]5f^26d^17s^2$
19	K	$[Ar]4s^1$	55	Cs	$[Xe]6s^1$	92	U	$[Rn]5f^36d^17s^2$
20	Ca	$[Ar]4s^2$	56	Ba	$[Xe]6s^2$	93	Np	$[Rn]5f^46d^17s^2$
21	Sc	$[Ar]3d^14s^2$	57	La	$[Xe]5d^16s^2$	94	Pu	$[Rn]5f^67s^2$
22	Ti	$[Ar]3d^24s^2$	58	Ce	$[Xe]4f^15d^16s^2$	95	Am	$[Rn]5f^77s^2$
23	V	$[Ar]3d^34s^2$	59	Pr	$[Xe]4f^36s^2$	96	Cm	$[Rn]5f^76d^17s^2$
24	Cr	$[Ar]3d^54s^1$	60	Nd	$[Xe]4f^46s^2$	97	Bk	$[Rn]5f^97s^2$
25	Mn	$[Ar]3d^54s^2$	61	Pm	$[Xe]4f^56s^2$	98	Cf	$[Rn]5f^{10}7s^2$
26	Fe	$[Ar]3d^64s^2$	62	Sm	$[Xe]4f^66s^2$	99	Es	$[Rn]5f^{11}7s^2$
27	Co	$[Ar]3d^74s^2$	63	Eu	$[Xe]4f^76s^2$	100	Fm	$[Rn]5f^{12}7s^2$
28	Ni	$[Ar]3d^84s^2$	64	Gd	$[Xe]4f^75d^16s^2$	101	Md	$[Rn]5f^{13}7s^2$
29	Cu	$[Ar]3d^{10}4s^1$	65	Tb	$[Xe]4f^96s^2$	102	No	$[Rn]5f^{14}7s^2$
30	Zn	$[Ar]3d^{10}4s^2$	66	Dy	$[Xe]4f^{10}6s^2$	103	Lr	$[Rn]5f^{14}6d^17s^2$
31	Ga	$[Ar]3d^{10}4s^24p^1$	67	Ho	$[Xe]4f^{11}6s^2$	104	Rf	$[Rn]5f^{14}6d^27s^2$
32	Ge	$[Ar]3d^{10}4s^24p^2$	68	Er	$[Xe]4f^{12}6s^2$	105	Db	$[Rn]5f^{14}6d^37s^2$
33	As	$[Ar]3d^{10}4s^24p^3$	69	Tm	$[Xe]4f^{13}6s^2$	106	Sg	$[Rn]5f^{14}6d^47s^2$
34	Se	$[Ar]3d^{10}4s^24p^4$	70	Yb	$[Xe]4f^{14}6s^2$	107	Bh	$[Rn]5f^{14}6d^57s^2$
35	Br	$[Ar]3d^{10}4s^24p^5$	71	Lu	$[Xe]4f^{14}5d^16s^2$	108	Hs	$[Rn]5f^{14}6d^67s^2$
36	Kr	$[Ar]3d^{10}4s^24p^6$	72	Hf	$[Xe]4f^{14}5d^26s^2$	109	Mt	$[Rn]5f^{14}6d^77s^2$
			73	Ta	$[Xe]4f^{14}5d^36s^2$			

Electron configurations are often written in abbreviated form by combining the **noble gas notation** with the *spdf* or orbital box notation. The arrangement preceding the $2s$ electron is that of the noble gas helium so, instead of writing out $1s^22s^1$, the completed electron shells are represented by placing the symbol of the corresponding noble gas in brackets. Thus, lithium's configuration would be written as $[He]2s^1$.

The electrons included in the noble gas notation are often referred to as the **core electrons** of the atom. Not only is it a time-saving way to write electron config-

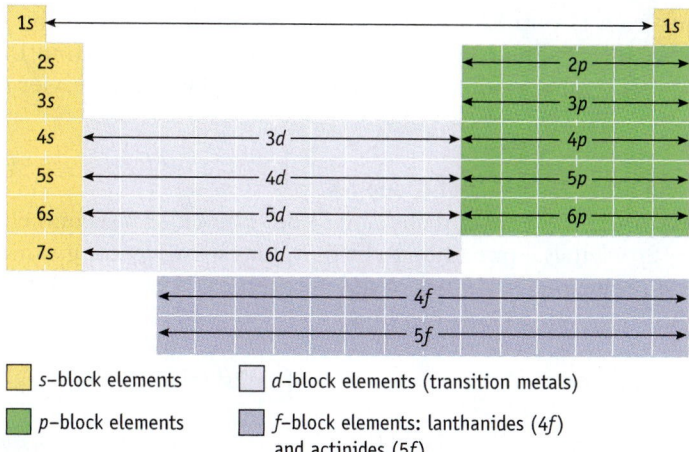

▨ s–block elements	▢ d–block elements (transition metals)
▩ p–block elements	▨ f–block elements: lanthanides (4f) and actinides (5f)

Active Figure 8.7 **Electron configurations and the periodic table.** The outermost electrons of an element are assigned to the indicated orbitals. See Table 8.3.

GENERAL
Chemistry ⚛ Now™ See the General ChemistryNow CD-ROM or website to explore an interactive version of this figure accompanied by an exercise.

urations, but the noble gas notation also conveys the idea that the core electrons can generally be ignored when considering the chemistry of an element. The electrons beyond the core electrons—the $2s^1$ electron in the case of lithium—are the **valence electrons**, the electrons that determine the chemical properties of an element.

The position of lithium in the periodic table tells you its configuration immediately. All the elements of Group 1A have one electron assigned to an s orbital of the nth shell, for which n is the number of the period in which the element is found (Figure 8.7). For example, potassium is the first element in the $n = 4$ row (the fourth period), so potassium has the electron configuration of the element preceding it in the table (Ar) plus a final electron assigned to the $4s$ orbital: $[Ar]4s^1$.

Beryllium (Be) and Other Elements of Group 2A
Beryllium, in Group 2A, has two electrons in the 1s orbital plus two additional electrons.

Beryllium: *spdf* notation $1s^2 2s^2$ or $[He]2s^2$

Box notation   ⇅ ⇅ ▢▢▢
            $1s$ $2s$ $2p$

All elements of Group 2A have electron configurations of [*electrons of preceding noble gas*] ns^2, where n is the period in which the element is found in the periodic table. Because all the elements of Group 1A have the valence electron configuration ns^1, and those in Group 2A have ns^2, these elements are called **s-block elements.**

Boron (B) and Other Elements of Group 3A
Boron (Group 3A) is the first element in the block of elements on the right side of the periodic table. Because the $1s$ and $2s$ orbitals are filled in a boron atom, the fifth electron must be assigned to a $2p$ orbital.

Boron: *spdf* notation $1s^2 2s^2 2p^1$ or $[He]2s^2 2p^1$

Box notation   ⇅ ⇅ ↑▢▢
            $1s$ $2s$ $2p$

Elements from Group 3A through Group 8A are often called the **p-block elements**. All have the general configuration ns^2np^x, where x varies from 1 to 6 (and is equal to the group number minus 2) (plus filled d orbitals for heavier elements as outlined below).

Carbon (C) and Other Elements of Group 4A

Carbon (Group 4A) is the second element in the p block, so a second electron is assigned to the $2p$ orbitals. For carbon to be in its lowest energy (or ground state) this electron must be assigned to either of the remaining p orbitals, and it will have the same spin direction as the first p electron.

Carbon: *spdf* notation $1s^22s^22p^2$ or $[He]2s^22p^2$

Box notation
| ⇅ | ⇅ | ↑ | ↑ | |
| 1s | 2s | 2p | | |

In general, when electrons are assigned to p, d, or f orbitals, each successive electron is assigned to a different orbital of the subshell, and each electron has the same spin as the previous one; this pattern continues until the subshell is half full. Additional electrons must then be assigned to half-filled orbitals. This procedure follows **Hund's rule**, which states that the most stable arrangement of electrons is that with the maximum number of unpaired electrons, all with the same spin direction. This arrangement makes the total energy of an atom as low as possible.

Carbon is the second element in the p block of elements, so it has two electrons in p orbitals. Because carbon is a second-period element, the p orbitals involved are $2p$. Thus, you can immediately write the carbon electron configuration by referring to the periodic table: Starting at H and moving from left to right across the successive periods, you write $1s^2$ to reach the end of period 1, and then $2s^2$ and finally $2p^2$ to bring the electron count to six. Carbon, the lightest element of Group 4A, has four electrons in the $n = 2$ shell.

Nitrogen (N) and Oxygen (O) and Elements of Groups 5A and 6A

Nitrogen (Group 5A) has five valence electrons. Besides the two $2s$ electrons, it has three electrons, all with the same spin, in three different $2p$ orbitals.

Nitrogen: *spdf* notation $1s^22s^22p^3$ or $[He]2s^22p^3$

Box notation
| ⇅ | ⇅ | ↑ | ↑ | ↑ |
| 1s | 2s | 2p | | |

Oxygen (Group 6A) has six valence electrons. Two of these six electrons are assigned to the $2s$ orbital, and, as oxygen is the fourth element in the p block, the other four electrons are assigned to $2p$ orbitals.

Oxygen: *spdf* notation $1s^22s^22p^4$ or $[He]2s^22p^4$

Box notation
| ⇅ | ⇅ | ⇅ | ↑ | ↑ |
| 1s | 2s | 2p | | |

This means the fourth $2p$ electron must pair up with one already present. It makes no difference to which orbital this electron is assigned (the $2p$ orbitals all have the same energy), but it must have a spin opposite to the other electron already assigned to that orbital so that each electron has a different set of quantum numbers (the Pauli exclusion principle).

Fluorine (F) and Neon (Ne) and Elements of Groups 7A and 8A

Fluorine (Group 7A) has seven electrons in the $n = 2$ shell. Two of these electrons occupy the 2s subshell, and the remaining five electrons occupy the 2p subshell.

Fluorine: *spdf* notation $1s^2 2s^2 2p^5$ or $[He]2s^2 2p^5$

           Box notation
| ↑↓ | | ↑↓ | | ↑↓ | ↑↓ | ↑ |
           1s   2s   2p

All halogen atoms have a similar configuration, $ns^2 np^5$, where n is the period in which the element is located.

Like the other elements in Group 8A, neon is a noble gas. All Group 8A elements (except helium) have eight electrons in the shell of highest n value, so all have the configuration $ns^2 np^6$, where n is the period in which the element is found. That is, all the noble gases have filled ns and np subshells. As you will see, the nearly complete chemical inertness of the noble gases correlates with this electron configuration.

Neon: *spdf* notation $1s^2 2s^2 2p^6$ or $[He]2s^2 2p^6$

           Box notation
| ↑↓ | | ↑↓ | | ↑↓ | ↑↓ | ↑↓ |
           1s   2s   2p

Elements of Period 3

The first element of the third period, sodium, is in Group 1A. The electron configuration of the element is that of a neon core plus one 3s electron.

Sodium: *spdf* notation $1s^2 2s^2 2p^6 3s^1$ or $[Ne]3s^1$

           Box notation
| ↑↓ | | ↑↓ | | ↑↓ | ↑↓ | ↑↓ | | ↑ |
           1s   2s   2p   3s

Moving across the third period, we come to silicon. This element is in Group 4A and so has four electrons beyond the neon core. Because it is the second element in the p block, it has two electrons in $3p$ orbitals. Thus, its electron configuration is

Silicon: *spdf* notation $1s^2 2s^2 2p^6 3s^2 3p^2$ or $[Ne]3s^2 3p^2$

           Box notation
| ↑↓ | | ↑↓ | | ↑↓ | ↑↓ | ↑↓ | | ↑↓ | | ↑ | ↑ | |
           1s   2s   2p   3s   3p

From silicon to the end of the third period, electrons are added to the $3p$ orbitals in the same manner as the elements in the second period. Finally, at argon the $3p$ subshell is completed with six electrons.

GENERAL
Chemistry⚛Now™

See the General ChemistryNow CD-ROM or website:

• **Screen 8.7 Atomic Electron Configurations**

(a) for a simulation exploring the relationship between an element's electron configuration and its position in the periodic table

(b) for a tutorial on determining an element's box notation

(c) for a tutorial on detemining an element's *spdf* notation

(d) for a tutorial on determining whether an element is diamagnetic or paramagnetic

---

Example 8.1—Electron Configurations

Problem Give the electron configuration of sulfur, using the *spdf*, noble gas, and orbital box notations.

Strategy Sulfur, atomic number 16, is the sixth element in the third period ($n = 3$), and is in the p block. The last six electrons assigned to the atom, therefore, have the configuration $3s^2 3p^4$. These are preceded by the completed shells $n = 1$ and $n = 2$, the electron arrangement for Ne.

Solution The electron configuration of sulfur is

Complete *spdf* notation:	$1s^2 2s^2 2p^6 3s^2 3p^4$
spdf with noble gas notation:	$[Ne]3s^2 3p^4$
Orbital box notation:	[Ne] ⇅ ⇅ ↑ ↑
	3s 3p

Example 8.2—Electron Configurations and Quantum Numbers

Problem Write the electron configuration for Al using the noble gas notation, and give a set of quantum numbers for each of the electrons with $n = 3$ (the valence electrons).

Strategy Aluminum is the third element in the third period. It therefore has three electrons with $n = 3$. Because Al is in the p block of elements, two of the electrons are assigned to $3s$ and the remaining electron is assigned to $3p$.

Solution The element is preceded by the noble gas neon, so the electron configuration is $[Ne]3s^2 3p^1$. Using box notation, the configuration is

Aluminum configuration:	[Ne] ⇅ ↑
	3s 3p

The possible sets of quantum numbers for the two 3s electrons are

	n	ℓ	m_ℓ	m_s
For ↑	3	0	0	$+\frac{1}{2}$
For ↓	3	0	0	$-\frac{1}{2}$

For the single 3p electron, one of six possible sets is $n = 3$, $\ell = 1$, $m_\ell = +1$, and $m_s = +\frac{1}{2}$.

Exercise 8.2—*spdf* Notation, Orbital Box Diagrams, and Quantum Numbers

(a) Which element has the configuration $1s^2 2s^2 2p^6 3s^2 3p^5$?

(b) Using *spdf* notation and a box diagram, show the electron configuration of phosphorus.

(c) Write one possible set of quantum numbers for the valence electrons of calcium.

Electron Configurations of the Transition Elements

The elements of the fourth through the seventh periods use d or f subshells, in addition to s and p subshells, to accommodate electrons (see Figure 8.7 and Table 8.4). Elements whose atoms are filling d subshells are described as *transition elements*. Those for which f subshells are filling are sometimes called the inner transition elements or, more usually, the **lanthanides** (filling $4f$ orbitals) and **actinides** (filling $5f$ orbitals).

The transition elements are always preceded in the periodic table by two s-block elements (Figure 8.7). Accordingly, scandium, the first transition element, has the configuration $[\text{Ar}]3d^1 4s^2$, and titanium follows with $[\text{Ar}]3d^2 4s^2$ (Table 8.4).

The general procedure for assigning electrons would suggest that the configuration of the chromium atom is $[\text{Ar}]3d^4 4s^2$. The actual configuration, however, has one electron assigned to each of the six available $3d$ and $4s$ orbitals: $[\text{Ar}]3d^5 4s^1$. This phenomenon is explained by assuming that the $4s$ and $3d$ orbitals have approximately the same energy in Cr, and each of the six valence electrons of chromium is assigned to one of these orbitals. This element illustrates the fact that occasionally minor differences crop up between the predicted and actual configurations. These discrepancies have little or no effect on the chemistry of the element, however.

Following chromium, atoms of manganese, iron, and nickel have the configurations that would be expected from the order of orbital filling in Figure 8.5. The Group 1B element copper, however, has a single electron in the $4s$ orbital, and the remaining ten electrons beyond the argon core are assigned to the $3d$ orbitals. Zinc ends the first transition series. This Group 2B element has two electrons assigned to the $4s$ orbital, and the $3d$ orbitals are completely filled with ten electrons.

Lanthanides and Actinides

The fifth period ($n = 5$) follows the pattern of the fourth period with minor variations. The sixth period, however, includes the lanthanide series beginning with lanthanum, La. As the first element in the d block, lanthanum has the configuration $[\text{Xe}]5d^1 6s^2$. The next element, cerium (Ce), is set out in a separate row at the bottom of the periodic table, and it is with the elements in this row (Ce through Lu) that electrons are first assigned to f orbitals. Thus, the configuration of cerium is $[\text{Xe}]4f^1 5d^1 6s^2$. Moving across the lanthanide series, the pattern continues with

■ **Writing Electron Configurations**
Although it does not necessarily reflect the filling order, we follow the convention of writing the orbitals in order of increasing n when writing electron configurations. For a given n, the subshells are listed in order of increasing ℓ.

Table 8.4   Orbital Box Diagrams for the Elements Ca Through Zn

		3d	4s
Ca	$[Ar]4s^2$	☐ ☐ ☐ ☐ ☐	↑↓
Sc	$[Ar]3d^14s^2$	↑ ☐ ☐ ☐ ☐	↑↓
Ti	$[Ar]3d^24s^2$	↑ ↑ ☐ ☐ ☐	↑↓
V	$[Ar]3d^34s^2$	↑ ↑ ↑ ☐ ☐	↑↓
Cr*	$[Ar]3d^54s^1$	↑ ↑ ↑ ↑ ↑	↑
Mn	$[Ar]3d^54s^2$	↑ ↑ ↑ ↑ ↑	↑↓
Fe	$[Ar]3d^64s^2$	↑↓ ↑ ↑ ↑ ↑	↑↓
Co	$[Ar]3d^74s^2$	↑↓ ↑↓ ↑ ↑ ↑	↑↓
Ni	$[Ar]3d^84s^2$	↑↓ ↑↓ ↑↓ ↑ ↑	↑↓
Cu*	$[Ar]3d^{10}4s^1$	↑↓ ↑↓ ↑↓ ↑↓ ↑↓	↑
Zn	$[Ar]3d^{10}4s^2$	↑↓ ↑↓ ↑↓ ↑↓ ↑↓	↑↓

*These configurations do not follow the "$n + \ell$" rule.

some variation, with 14 electrons being assigned to the seven $5f$ orbitals in the last element, lutetium (Lu, $[Xe]4f^{14}5d^16s^2$) (see Table 8.3).

The seventh period also includes an extended series of elements utilizing f orbitals, the actinides, which begins with actinium (Ac, $[Rn]6d^17s^2$). The next element is thorium (Th), which is followed by protactinium (Pa) and uranium (U). The electron configuration of uranium is $[Rn]5f^36d^17s^2$. The third element in the actinide series, it has three $5f$ electrons.

When you have completed this section, you should be able to depict accurately the electron configuration of any element in the s and p blocks using the periodic table as a guide. Prediction of the electron configurations for atoms of elements in the d and f blocks (Table 8.3) is somewhat less precise, but you are reminded that these small "anomalies" have little effect on the chemical behavior of the elements.

Example 8.3—Electron Configurations of the Transition Elements

Problem Using the *spdf* and noble gas notations, give electron configurations for (a) technetium, Tc, and (b) osmium, Os.

Strategy Base your answer on the positions of the elements in the periodic table. That is, for each element, find the preceding noble gas and then note the number of s, p, d, and f electrons that lead from the noble gas to the element.

Solution

(a) *Technetium, Tc:* The noble gas that precedes Tc is krypton, Kr, at the end of the $n = 4$ row. After the 36 electrons of Kr are assigned, seven electrons remain. Two of these electrons are in the 5s orbital, and the remaining five are in 4d orbitals. Therefore, the technetium configuration is $[\mathrm{Kr}]4d^55s^2$.

(b) *Osmium, Os:* Osmium is a sixth-period element and the twenty-second element following the noble gas xenon. Of the 22 electrons to be added after the Xe core, 2 are assigned to the 6s orbital and 14 to 4f orbitals. The remaining 6 are assigned to 5d orbitals. Thus, the osmium configuration is $[\mathrm{Xe}]4f^{14}5d^66s^2$.

Exercise 8.3—Electron Configurations

Using the periodic table and without looking at Table 8.3, write electron configurations for the following elements:

(a) P   **(c)** Zr   **(e)** Pb
(b) Zn   **(d)** In   **(f)** U

Use the *spdf* and noble gas notations. When you have finished, check your answers with Table 8.3.

8.5—Electron Configurations of Ions

Much of the chemistry of the elements involves the formation of ions, and we can write their electron configurations as well as those of the elements. To form a cation from a neutral atom, one or more of the valence electrons is removed; that is, electrons are removed from the electron shell of highest n. If several subshells are present within the nth shell, the electron or electrons of maximum ℓ are removed. Thus, a sodium ion is formed by removing the $3s^1$ electron from the Na atom,

$$\mathrm{Na}: [1s^22s^22p^63s^1] \longrightarrow \mathrm{Na}^+: [1s^22s^22p^6] + e^-$$

and Ge^{2+} is formed by removing two $4p$ electrons from a germanium atom,

$$\mathrm{Ge}: [\mathrm{Ar}]3d^{10}4s^24p^2 \longrightarrow \mathrm{Ge}^{2+}: [\mathrm{Ar}]3d^{10}4s^2 + 2\ e^-$$

The same general rule applies to transition metal atoms. This means the titanium(II) cation has the configuration $[\mathrm{Ar}]3d^2$, for example:

$$\mathrm{Ti}: [\mathrm{Ar}]3d^24s^2 \longrightarrow \mathrm{Ti}^{2+}: [\mathrm{Ar}]3d^2 + 2e^-$$

The iron(II) and iron(III) cations have the configurations $[\mathrm{Ar}]3d^6$ and $[\mathrm{Ar}]3d^5$, respectively:

$$\mathrm{Fe}: [\mathrm{Ar}]3d^64s^2 \longrightarrow \mathrm{Fe}^{2+}: [\mathrm{Ar}]3d^6 + 2\ e^-$$
$$\mathrm{Fe}^{2+}: [\mathrm{Ar}]3d^6 \longrightarrow \mathrm{Fe}^{3+}: [\mathrm{Ar}]3d^5 + e^-$$

All common transition metal cations have electron configurations of the general type [noble gas core]$(n-1)d^x$. That is, in the process of ionization the ns electrons are lost before $(n-1)d$ electrons. It is important to remember this point because the chemical and physical properties of transition metal cations are determined by the presence of electrons in d orbitals.

Atoms and ions with unpaired electrons are paramagnetic; that is, they are capable of being attracted to a magnetic field [◀ Section 8.1]. Paramagnetism is

Charles D. Winters

Formation of iron(III) chloride. When iron reacts with chlorine (Cl_2) to produce $FeCl_3$, each iron atom loses three electrons to give a paramagnetic Fe^{3+} ion with the configuration $[\mathrm{Ar}]3d^5$.

Figure 8.8 Paramagnetism. (a) A sample of iron(III) oxide is packed into a plastic tube and suspended from a thin nylon filament. (b) When a powerful magnet is brought near it, the paramagnetic iron(III) ions in Fe_2O_3 cause the sample to be attracted to the magnet. [The magnet is made of neodymium, iron, and boron ($Nd_2Fe_{14}B$). These magnets are powerful enough to attract a U.S. $1 bill, which is printed with ink containing a small quantity of an iron-based compound.]

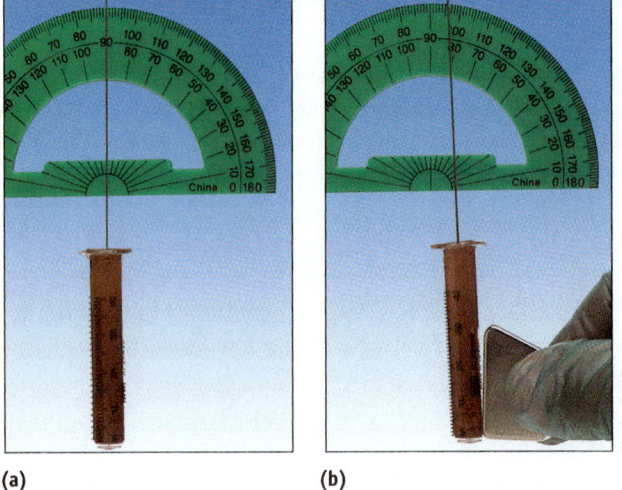

(a)   (b)

Charles D. Winters

important here because it provides experimental evidence that transition metal ions with charges of 2+ or greater have no *ns* electrons. For example, the Fe^{3+} ion is paramagnetic to the extent of five unpaired electrons (Figure 8.8). If three $3d$ electrons had been removed instead to form Fe^{3+}, the ion would still be paramagnetic but only to the extent of three unpaired electrons.

GENERAL
Chemistry ⚛ Now™

See the General ChemistryNow CD-ROM or website:
- **Screen 8.8 Electron Configuration in Ions**

   (a) for a simulation exploring the changes to an element's electron configuration when it ionizes

   (b) for a tutorial on determining an ion's box notation

Example 8.4—Configurations of Transition Metal Ions

Problem Give the electron configurations for copper, Cu, and for its 1+ and 2+ ions. Are either of these ions paramagnetic? How many unpaired electrons does each have?

Strategy Observe the configuration of copper in Table 8.4. Recall that *s* and then *d* electrons are removed to form a transition metal ion.

Solution Copper has only one electron in the $4s$ orbital and ten electrons in $3d$ orbitals:

Cu: [Ar]$3d^{10}4s^1$ ⇅|⇅|⇅|⇅|⇅   ↑
                          $3d$ $4s$

When copper is oxidized to Cu^+, the $4s$ electron is lost.

Cu⁺:   [Ar]$3d^{10}$ ⇅|⇅|⇅|⇅|⇅   ☐
                        $3d$ $4s$

The copper(II) ion is formed from copper(I) by removal of one of the $3d$ electrons.

$$\text{Cu}^{2+}: \quad [\text{Ar}]3d^9$$

3d　　　4s

Copper(II) ions (Cu^{2+}) have one unpaired electron, so they should be paramagnetic. In contrast, Cu^+ has no unpaired electrons, so the ion and its compounds are diamagnetic.

Exercise 8.4—Metal Ion Configurations

Depict the electron configurations for V^{2+}, V^{3+}, and Co^{3+}. Use orbital box diagrams and noble gas notation. Are any of the ions paramagnetic? If so, give the number of unpaired electrons.

8.6—Atomic Properties and Periodic Trends

Once electron configurations were understood, chemists realized that *similarities in properties of the elements are the result of similar valence shell electron configurations.* An objective of this section is to describe how atomic electron configurations are related to some of the physical and chemical properties of the elements and why those properties change in a reasonably predictable manner when moving down groups and across periods (Figure 8.9). This background should make the periodic table an even more useful tool in your study of chemistry. With an understanding of electron configurations and their relation to properties, you should be able to organize and predict chemical and physical properties of the elements and their compounds. We will concentrate on physical properties in this section and then look briefly at chemical behavior in Section 8.7.

Atomic Size

An orbital has no sharp boundary [◀ Figure 7.14, page 321], so how can we define the size of an atom? There are actually several ways, and they can give slightly different results.

　　One of the simplest and most useful ways to define atomic size is to say that it is the distance between atoms in a sample of the element. Let us take a diatomic molecule such as Cl_2 (Figure 8.10a). The radius of a Cl atom is assumed to be one half the experimentally determined distance between the centers of the two atoms. This distance is 198 pm, so the radius of one Cl atom is 99 pm. Similarly, the C — C distance in diamond is 154 pm, so a radius of 77 pm can be assigned to carbon. To test these estimates, we can add them together to estimate the distance between Cl and C in CCl_4. The predicted distance of 176 pm agrees with the experimentally measured C — Cl distance of 176 pm.

　　This approach to determining atomic radii will apply only if molecular compounds of the element exist. For metals, atomic radius can be estimated from measurements of the atom-to-atom distance in a crystal of the element (Figure 8.10b).

　　A set of atomic radii has been assembled (Figure 8.11), and some interesting periodic trends are seen immediately. *For the main group elements, atomic radii generally increase going down a group in the periodic table and decrease going across a period.* These trends reflect two important effects:

■ **Atomic Radii—Caution**
Numerous tabulations of atomic and covalent radii exist, and the values quoted in them may differ. The variation comes about because several methods are used to determine the radii of atoms, and the different methods can give slightly different values.

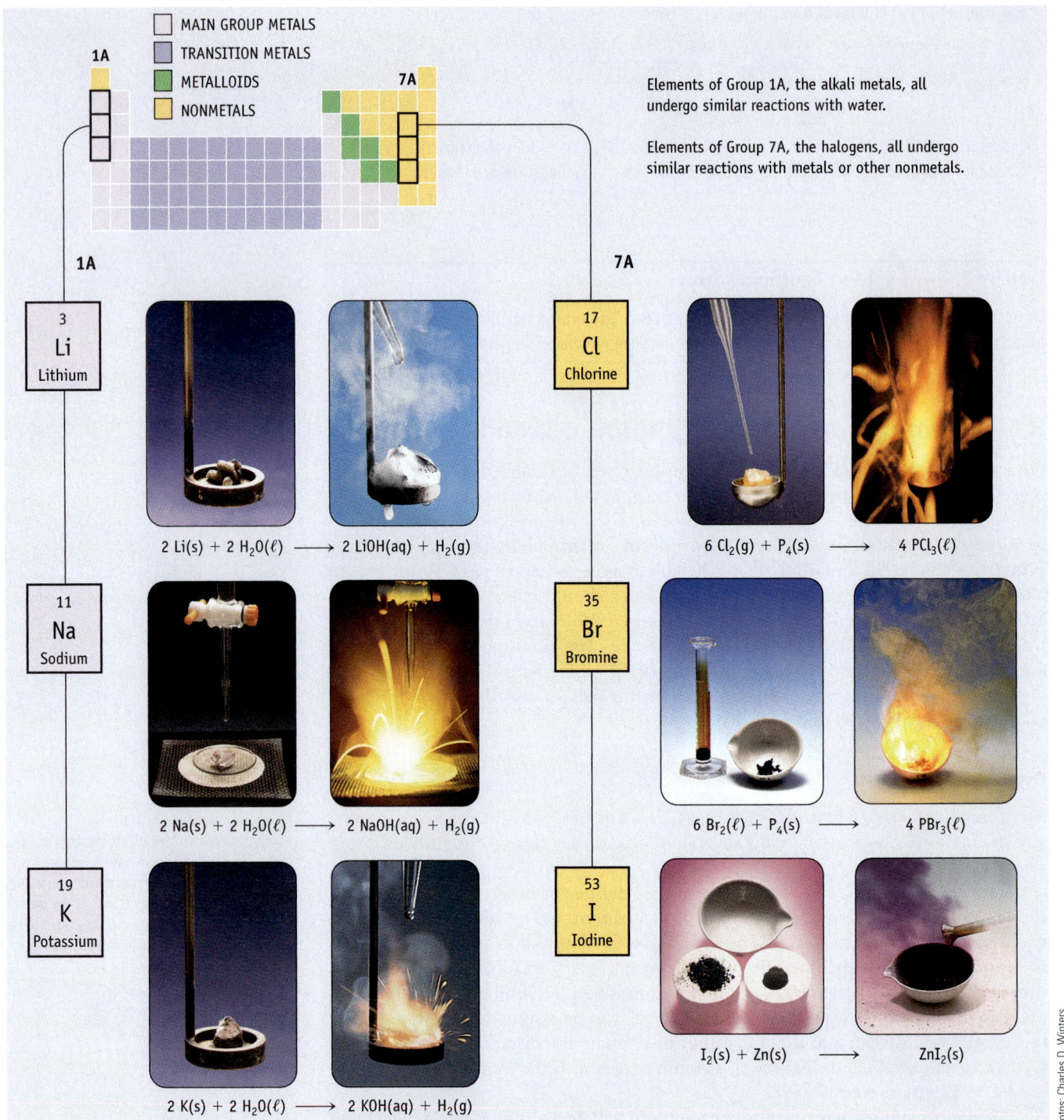

MAIN GROUP METALS
TRANSITION METALS
METALLOIDS
NONMETALS

1A

7A

Elements of Group 1A, the alkali metals, all undergo similar reactions with water.

Elements of Group 7A, the halogens, all undergo similar reactions with metals or other nonmetals.

1A

7A

| 3 |
| Li |
| Lithium |

$2\ Li(s) + 2\ H_2O(\ell) \longrightarrow 2\ LiOH(aq) + H_2(g)$

| 17 |
| Cl |
| Chlorine |

$6\ Cl_2(g) + P_4(s) \longrightarrow 4\ PCl_3(\ell)$

| 11 |
| Na |
| Sodium |

$2\ Na(s) + 2\ H_2O(\ell) \longrightarrow 2\ NaOH(aq) + H_2(g)$

| 35 |
| Br |
| Bromine |

$6\ Br_2(\ell) + P_4(s) \longrightarrow 4\ PBr_3(\ell)$

| 19 |
| K |
| Potassium |

$2\ K(s) + 2\ H_2O(\ell) \longrightarrow 2\ KOH(aq) + H_2(g)$

| 53 |
| I |
| Iodine |

$I_2(s) + Zn(s) \longrightarrow ZnI_2(s)$

Photos: Charles D. Winters

Active Figure 8.9   **Examples of the Periodicity of Group 1A and Group 7A Elements.** Dimitri Mendeleev developed the first periodic table by listing elements in order of increasing atomic weight. Every so often an element had properties similar to those of a lighter element, and these elements were placed in vertical columns or groups. We now recognize that the elements should be listed in order of increasing atomic number and that the periodic occurrence of similar properties is related to the electron configurations of the elements.

GENERAL
Chemistry⚛Now™ See the General ChemistryNow CD-ROM or website to explore an interactive version of this figure accompanied by an exercise.

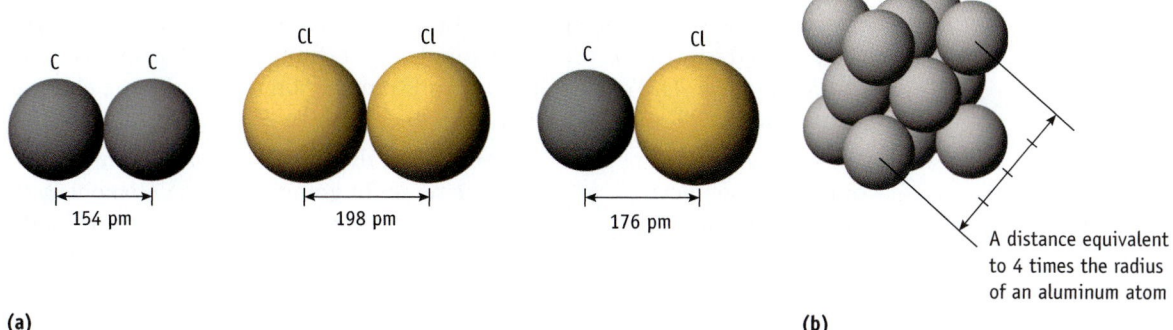

C
C
154 pm
Cl
Cl
198 pm
C
Cl
176 pm
A distance equivalent to 4 times the radius of an aluminum atom
(a)
(b)

Figure 8.10 Determining atomic radii. (a) The sum of the atomic radii of C and Cl provides a good estimate of the C—Cl distance in a molecule having such a bond. (b) Each sphere in this tiny piece of an aluminum crystal represents an aluminum atom. Measuring the distance shown, for example, allows a scientist to estimate the radius of an aluminum atom.

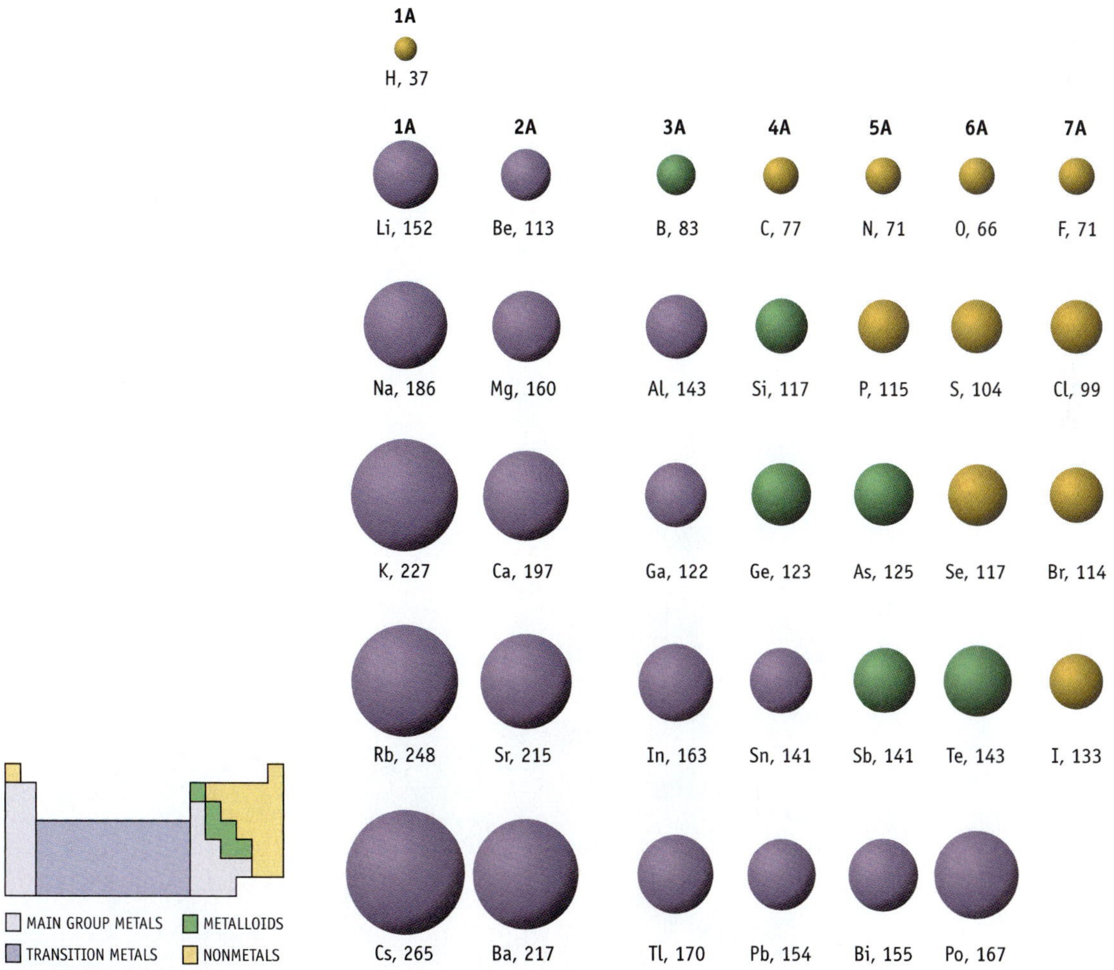

1A
H, 37
1A
2A
3A
4A
5A
6A
7A
Li, 152
Be, 113
B, 83
C, 77
N, 71
O, 66
F, 71
Na, 186
Mg, 160
Al, 143
Si, 117
P, 115
S, 104
Cl, 99
K, 227
Ca, 197
Ga, 122
Ge, 123
As, 125
Se, 117
Br, 114
Rb, 248
Sr, 215
In, 163
Sn, 141
Sb, 141
Te, 143
I, 133
Cs, 265
Ba, 217
Tl, 170
Pb, 154
Bi, 155
Po, 167
MAIN GROUP METALS
METALLOIDS
TRANSITION METALS
NONMETALS

Active Figure 8.11 Atomic radii in picometers for main group elements. 1 pm = 1×10^{-12} m. Data taken from J. Emsley: *The Elements,* 3rd ed., Oxford, Clarendon Press, 1998.

GENERAL **ChemistryNow**™ See the General ChemistryNow CD-ROM or website to explore an interactive version of this figure accompanied by an exercise.

- The size of an atom is determined by the outermost electrons. In going from the top to the bottom of a group in the periodic table, the outermost electrons are assigned to orbitals with increasingly higher values of the principal quantum number, *n*. The underlying electrons require some space, so the electrons of the outer shell must be farther from the nucleus.

- For main group elements of a given period, the principal quantum number, *n*, of the valence electron orbitals is the same. In going from one element to the next across a period, a proton is added to each nucleus and an electron is added to each outer shell. In each step, the effective nuclear charge, Z* [◄ Table 8.2] increases slightly because the effect of each additional proton is more important than the effect of an additional electron. The result is that attraction between the nucleus and electrons increases, and atomic radius decreases.

The periodic trend in the atomic radii of transition metal atoms (Figure 8.12) is somewhat different from that for main group elements. Going from left to right across a given period, the radii initially decrease across the first few elements. The sizes of the elements in the middle of a transition series then change very little until a small increase in size occurs at the end of the series. The size of the atom is determined largely by electrons in the outermost shell—that is, by the electrons of the *ns* subshell. In the first transition series, for example, the outer shell contains the 4*s* electrons, but electrons are being added to the 3*d* orbitals across the series. The increased nuclear charge on the atoms as one moves from left to right should cause the radius to decrease. This effect, however, is mostly cancelled out by increased electron–electron repulsion among the electrons. On reaching the Groups 1B and 2B elements at the end of the series, the size increases slightly because the *d* subshell is filled, and electron–electron repulsions cause the size to increase.

■ **Trends in Atomic Radii**
General trends in atomic radii of *s*- and *p*-block elements with position in the periodic table.

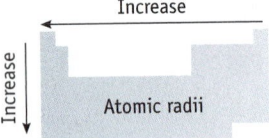

Figure 8.12   Trends in atomic radii for the transition elements. Atomic radii of the Group 1A and 2A metals and the transition metals of the fourth, fifth, and sixth periods.

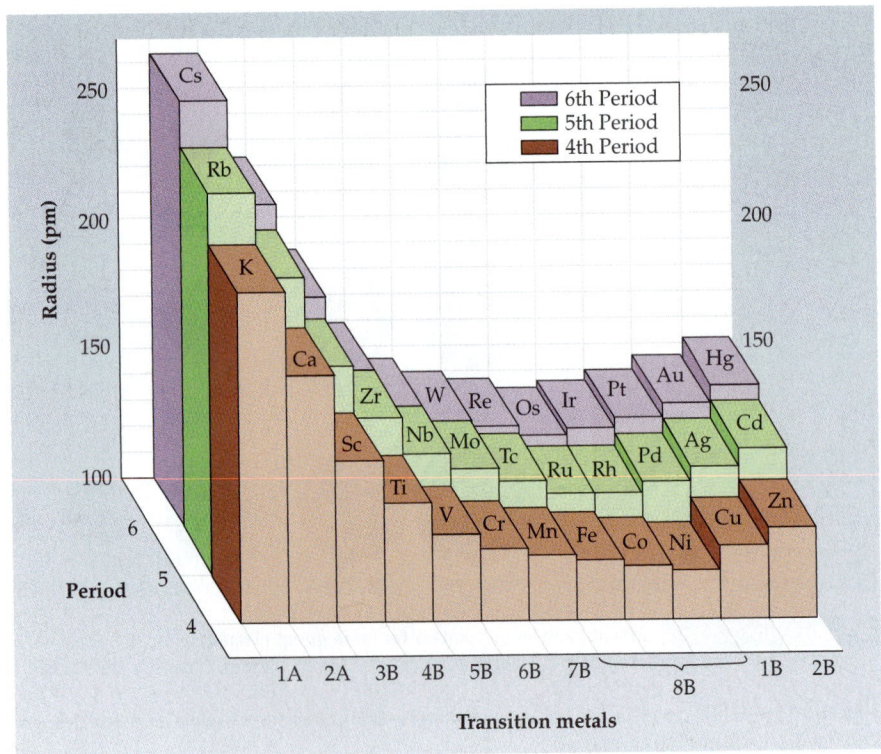

GENERAL
Chemistry Now™

See the General ChemistryNow CD-ROM or website:
- **Screen 8.9 Atomic Properties and Periodic Trends,** for a simulation exploring energy levels of orbitals and the ability to retain electrons
- **Screen 8.10 Atomic Sizes,** for a simulation exploring the trends in atomic size moving across and down the periodic table

---

Exercise 8.5—Periodic Trends in Atomic Radii

Place the three elements Al, C, and Si in order of increasing atomic radius.

Exercise 8.6—Estimating Atom–Atom Distances

(a) Using Figure 8.11, estimate the H — O and H — S distances in H_2O and H_2S, respectively.

(b) If the interatomic distance in Br_2 is 228 pm, what is the radius of Br? Using this value, and that for Cl (99 pm), estimate the distance between atoms in BrCl.

Ionization Energy

Ionization energy is the energy required to remove an electron from an atom in the gas phase.

$$\text{Atom in ground state(g)} \longrightarrow \text{Atom}^+ (g) + e^-$$
$$\Delta E \equiv \text{ionization energy, } IE$$

To separate an electron from an atom, energy must be supplied to overcome the attraction of the nuclear charge. Because energy must be supplied (an endothermic process), ionization energies always have positive values.

Atoms other than hydrogen have a series of ionization energies, because more than one electron can always be removed [◀ page 351]. For example, the first three ionization energies of magnesium are

First ionization energy, $IE_1 = 738$ kJ/mol
$$\underset{1s^2 2s^2 2p^6 3s^2}{Mg(g)} \longrightarrow \underset{1s^2 2s^2 2p^6 3s^1}{Mg^+(g)} + e^-$$

Second ionization energy, $IE_2 = 1451$ kJ/mol
$$\underset{1s^2 2s^2 2p^6 3s^1}{Mg^+(g)} \longrightarrow \underset{1s^2 2s^2 2p^6 3s^0}{Mg^{2+}(g)} + e^-$$

Third ionization energy, $IE_3 = 7733$ kJ/mol
$$\underset{1s^2 2s^2 2p^6}{Mg^{2+}(g)} \longrightarrow \underset{1s^2 2s^2 2p^5}{Mg^{3+}(g)} + e^-$$

■ **Valence and Core Electrons**
Removal of core electrons requires much more energy than removal of a valence electron. Core electrons are not lost in chemical reactions.

Notice that removing each subsequent electron requires more energy because the electron is being removed from an increasingly positive ion. Most importantly, notice the large increase in ionization energy for removing the third electron to give Mg^{3+}. *This large increase is experimental evidence for the electron shell structure of atoms.* The first two ionization steps are for the removal of electrons from the outermost or valence shell of electrons. The third electron, however, must come from the $2p$ subshell. This subshell is significantly lower in energy than the 3s subshell (page 340), and considerably more energy is required to remove the $n = 2$ electron than the $n = 3$ electrons.

■ **Trends in Ionization Energy**
General trends in first ionization energies of s- and p-block elements with position in the periodic table.

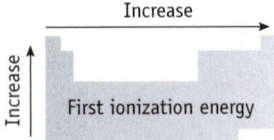

As another example, consider the first two ionization energies for lithium.

First ionization energy, $IE_1 = 513.3$ kJ/mol

$$\underset{1s^2 2s^1}{Li(g)} \longrightarrow \underset{1s^2}{Li^+(g)} + e^-$$

Second ionization energy, $IE_2 = 7298$ kJ/mol

$$\underset{1s^2}{Li^+(g)} \longrightarrow \underset{1s^1}{Li^{2+}(g)} + e^-$$

The ionization energy for the removal of the second electron is large because the second electron is removed from a much lower energy (inner) subshell.

For main group (s- and p-block) elements, *first ionization energies generally increase across a period and decrease down a group* (Figure 8.13 and Appendix F). The trend *across a period* is rationalized by the increase in effective nuclear charge, Z*, with increasing atomic number. Not only does this mean that the atomic radius decreases, but the energy required to remove an electron also increases. The general decrease in ionization energy *down a group* occurs because the electron removed is increasingly farther from the nucleus, thus reducing the nucleus-electron attractive force.

A closer look at ionization energies reveals that the trend across a given period is not smooth, particularly in the second period. Variations are seen on going from s-block to p-block elements—from beryllium to boron, for example. This occurs because the 2p electrons are slightly higher in energy than the 2s electrons (see Figure 8.4, page 340), so the ionization energy for boron is lower than that for beryllium.

Moving from boron to carbon and then to nitrogen, the effective nuclear charge increases (see Table 8.2), which again means an increase in ionization energy. Another dip to lower ionization energy occurs on passing from Group 5A to Group 6A. This is especially noticeable in the second period (N and O). No change occurs in either n or ℓ, but electron–electron repulsions increase for the following

■ **Factors Controlling Trends in Ionization Energies**
The ionization energy of an atom is always a balance between electron–nuclear attraction (which depends on Z) and electron–electron repulsion.

Active Figure 8.13 **First ionization energies of the main group elements of the first four periods.** (For data on all the elements see Appendix F.)

GENERAL
Chemistry⚛Now™ See the General ChemistryNow CD-ROM or website to explore an interactive version of this figure accompanied by an exercise.

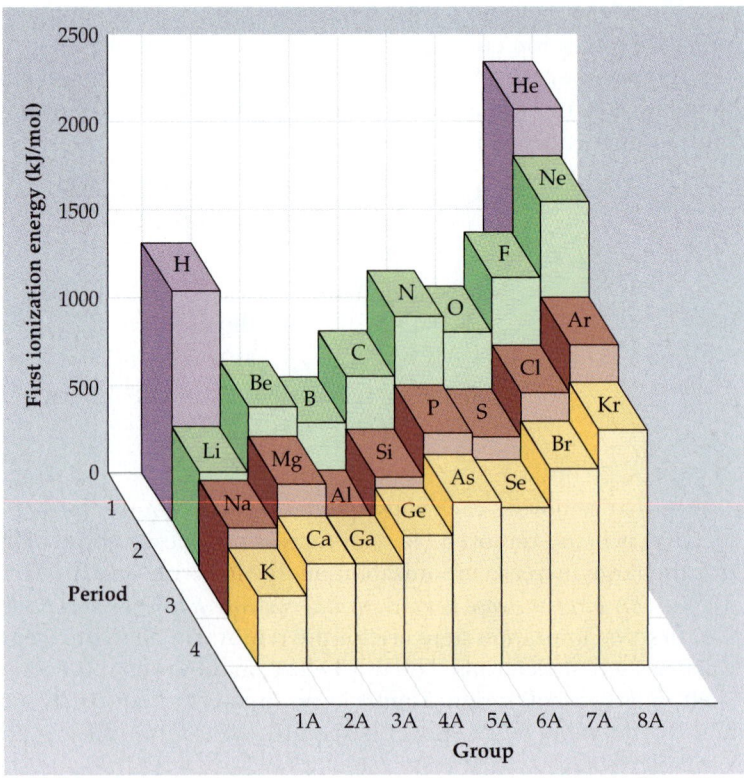

reason. In Groups 3A–5A, electrons are assigned to separate p orbitals (p_x, p_y, and p_z). Beginning in Group 6A, however, two electrons are assigned to the same p orbital. The fourth p electron shares an orbital with another electron and thus experiences greater repulsion than it would if it had been assigned to an orbital of its own:

$$\text{O (oxygen atom)} \xrightarrow{\;+1314 \text{ kJ/mol}\;} \text{O}^+ \text{ (oxygen cation)} + \text{e}^-$$

$$[\text{Ne}]\,\boxed{\uparrow\downarrow}\quad\boxed{\uparrow\downarrow\;\uparrow\;\uparrow}\qquad\qquad [\text{Ne}]\,\boxed{\uparrow\downarrow}\quad\boxed{\uparrow\;\uparrow\;\uparrow}$$
$$\quad\;\; 2s \qquad 2p \qquad\qquad\qquad\quad 2s \qquad 2p$$

The greater repulsion experienced by the fourth $2p$ electron makes it easier to remove, and each of the remaining p electrons has an orbital of its own. The usual trend resumes on going from oxygen to fluorine to neon, however, reflecting the increase in Z*.

Electron Affinity

Some atoms have an affinity, or "liking," for electrons and can acquire one or more electrons to form a negative ion. The **electron affinity**, *EA*, of an atom is defined as the energy of a process in which an electron is acquired by the atom in the gas phase (Figure 8.14 and Appendix F).

$$A(g) + e^-(g) \rightarrow A^-(g) \qquad \Delta E \equiv \text{electron affinity, } EA$$

The greater the affinity an atom has for an electron, the more negative the value of *EA* will be. For example, the electron affinity of fluorine is −328 kJ/mol, a large value indicating an exothermic, product-favored reaction to form the anion, F⁻. Boron has a much lower electron affinity for an electron, as indicated by a less negative *EA* value of −26.7 kJ/mol.

■ **Electron Affinity and Sign Conventions** For a useful discussion of electron affinity, see J. C. Wheeler: "Electron affinities of the alkaline earth metals and the sign convention for electron affinity." *Journal of Chemical Education*, Vol. 74, pp. 123–127, 1997. Numerical values for *EA* are given in Appendix F.

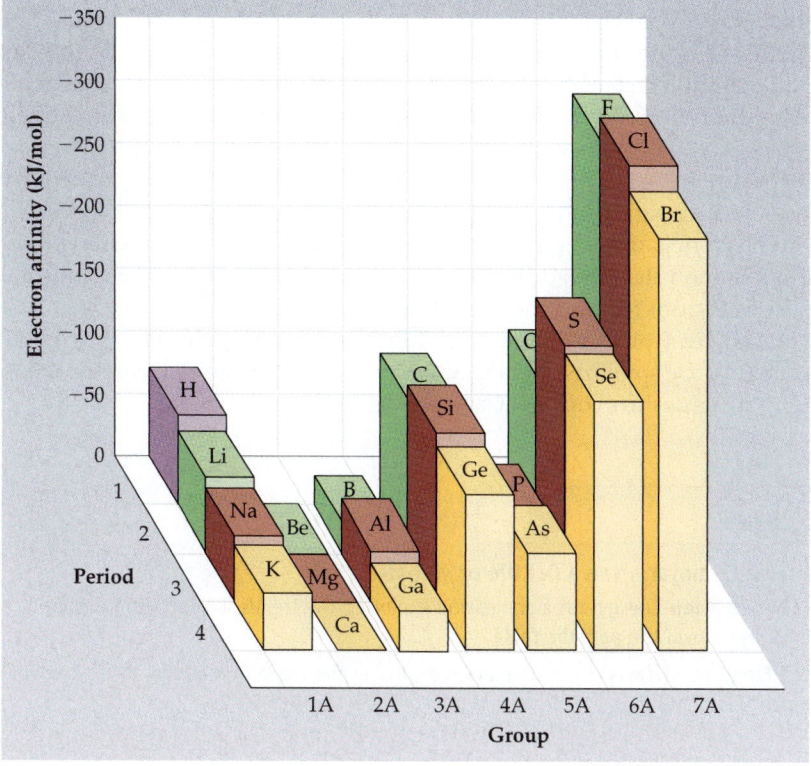

Active Figure 8.14 **Electron affinity.** The larger the affinity (*EA*) of an atom for an electron, the more negative the value. For numerical data, see Appendix F. (Data were taken from H. Hotop and W. C. Lineberger: "Binding energies of atomic negative ions," *Journal of Physical Chemistry*, Reference Data, Vol. 14, p. 731, 1985.)

GENERAL **Chemistry⚛Now**™ See the General ChemistryNow CD-ROM or website to explore an interactive version of this figure accompanied by an exercise.

■ *EA* **Values of Zero**
The value of *EA* for Be is not measurable because the Be⁻ anion does not exist. Most tables assign a value of 0 to the *EA* for this element and similar cases (in particular the Group 2A elements).

■ **Trends In *EA***
General trends in electron affinities of A-group elements. Exceptions occur at Groups 2A and 5A.

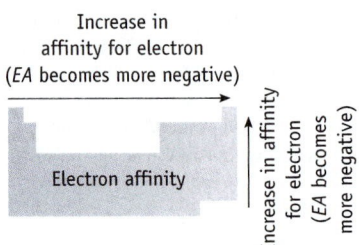

Electron affinity and ionization energy represent the energy involved in the gain or loss of an electron by an atom, respectively. It is therefore not surprising that periodic trends in electron affinity are related to the periodic trends in ionization energy. The effective nuclear charge of atoms increases across a period (Table 8.2), not only making it more difficult to ionize the atom but also increasing the attraction of the atom for an additional electron. Thus, *an element with a high ionization energy generally has a high affinity for an electron.* As seen in Figure 8.14, the values of *EA* generally become more negative on moving across a period as the affinity for electrons increases.

One result of increasing Z* across a period is that nonmetals generally have much more negative values of *EA* than do metals, reflecting the greater affinity of nonmetals for electrons. This prediction agrees with chemical experience, which tells us that metals generally do not form negative ions and that nonmetals have an increasing tendency to form anions as we proceed across a period.

The trend to more negative electron affinities across a period is not smooth, however. For example, beryllium has no affinity for an electron. A beryllium anion, Be⁻, is not stable because the added electron must be assigned to a higher energy subshell ($2p$) than the valence electrons ($2s$) (see page 340). Nitrogen atoms also have no affinity for electrons. Here an electron pair must be formed when an N atom acquires an electron. Significant electron–electron repulsions occur in an N⁻ ion, making the ion much less stable. The increase in Z* on going from carbon to nitrogen cannot overcome the effect of these electron–electron repulsions.

The noble gases are not included in a discussion of electron affinity. They have no affinity for electrons, because any additional electron must be added to the next higher electron shell. The higher Z* of the noble gases is not sufficient to overcome this effect.

The affinity for an electron generally declines on descending a group of the periodic table (Figure 8.14). Electrons are added increasingly farther from the nucleus, so the attractive force between the nucleus and electrons decreases. However, this general trend does not extend to the elements in period 2. For example, the affinity of the fluorine atom for an electron is lower than that of chlorine (*EA* for F is less negative than *EA* for Cl), and the same phenomenon is observed in Groups 3A through 6A. One explanation is that significant electron–electron repulsions occur in small anions such as the F⁻ ion. That is, adding an electron to the seven electrons already present in the $n = 2$ shell of the small F atom leads to considerable repulsion between electrons. Chlorine has a larger atomic volume than fluorine, so adding an electron does not result in such significant electron–electron repulsions in the Cl⁻ anion.

No atom has a negative electron affinity for a *second* electron. Attaching a second electron to an ion that already has a negative charge leads to severe repulsions. So how can you account for ions such as O^{2-}, which is present in so many naturally occurring substances (for example, CaO)? The answer is that doubly charged anions can sometimes be stabilized in crystalline environments by electrostatic attraction to neighboring positive ions [▶ Chapters 9 and 13].

GENERAL
Chemistry✦Now™

See the General ChemistryNow CD-ROM or website:
- **Screen 8.11 Ionization Energy,** for a simulation exploring the trends in ionization energy moving across and down the periodic table
- **Screen 8.12 Electron Affinity,** for a simulation exploring the trend in electron affinity moving across the periodic table

Example 8.5—Periodic Trends

Problem Compare the three elements C, O, and Si.

(a) Place them in order of increasing atomic radius.

(b) Which has the largest ionization energy?

(c) Which has the more negative electron affinity, O or C?

Strategy Review the trends in atomic properties in Figures 8.11–8.14 and Appendix F.

Solution

(a) *Atomic size:* Atomic radius declines on moving across a period, so oxygen must have a smaller radius than carbon. However, radius increases on moving down a periodic group. Because C and Si are in the same group (Group 4A), Si must be larger than C. In order of increasing size, the trend is $O < C < Si$.

(b) *Ionization energy:* Ionization energy generally increases across a period and decreases down a group; a large decrease in *IE* occurs from the second- to the third-period elements. Thus, the trend in ionization energies should be $Si < C < O$.

(c) *Electron affinity:* Electron affinity values generally become more negative across a period and less negative down a group. Therefore, the *EA* for O should be more negative than the *EA* for C. That is, O ($EA = -141.0$ kJ/mol) has a greater affinity for an electron than does C ($EA = -121.9$ kJ/mol).

Comment *EA* for third-period elements (Si, $EA = -133.6$ kJ/mol) is generally slightly more negative than *EA* for second-period elements (C, $EA = -121.85$ kJ/mol). This trend occurs because of electron–electron repulsions; such repulsions are larger in the small C^- ion than in the larger Si^- ion.

Exercise 8.7—Periodic Trends

Compare the three elements B, Al, and C.

(a) Place the three elements in order of increasing atomic radius.
(b) Rank the elements in order of increasing ionization energy. (Try to do this without looking at Figure 8.13; then compare your estimates with the graph.)
(c) Which element is expected to have the most negative electron affinity value?

Ion Sizes

Having considered the energies involved in forming positive and negative ions, let us now look at periodic trends in ion radii.

Periodic trends in the sizes of ions in the same group are the same as those for neutral atoms: Positive or negative ions increase in size when descending the group (Figure 8.15). Pause for a moment, however, and compare the ionic radii in Figure 8.15 with the atomic radii in Figure 8.11. When an electron is removed from an atom to form a cation, the size shrinks considerably; *the radius of a cation is always smaller than that of the atom from which it is derived.* For example, the radius of Li is 152 pm, whereas the radius of Li^+ is only 78 pm. When an electron is removed from a Li atom, the attractive force of three protons is now exerted on only two electrons, so the remaining electrons contract toward the nucleus. The decrease in ion size is especially great when the last electron of a particular shell is removed, as is the case

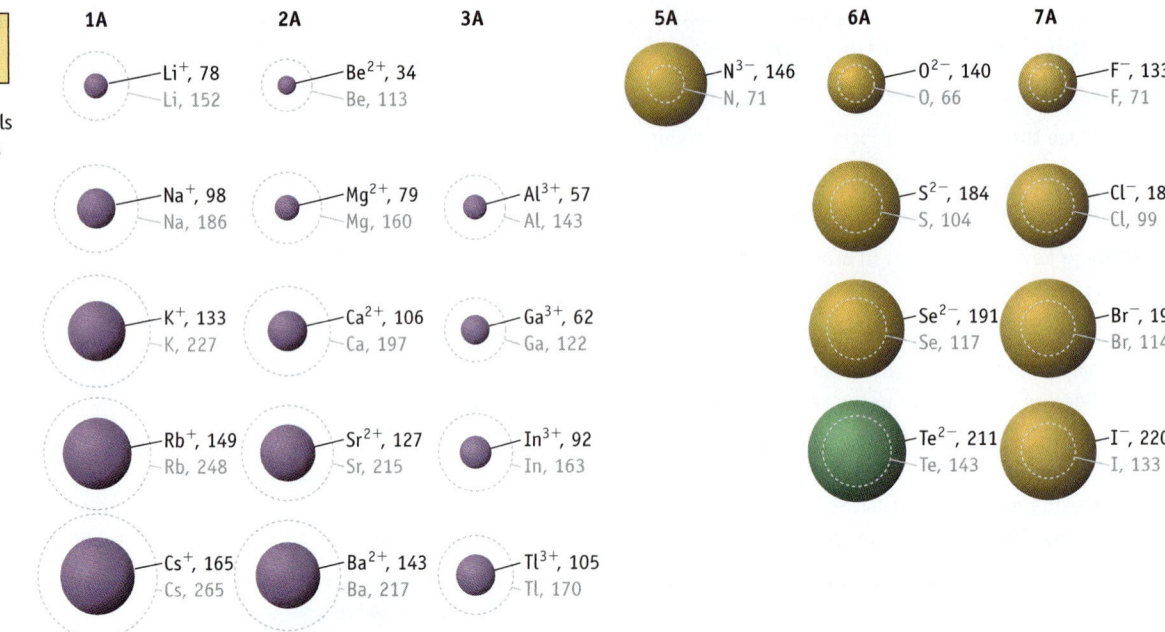

- ☐ Main Group Metals
- ☐ Transition Metals
- ☐ Metalloids
- ☐ Nonmetals

	1A	2A	3A	5A	6A	7A
	Li⁺, 78	Be²⁺, 34		N³⁻, 146	O²⁻, 140	F⁻, 133
	Li, 152	Be, 113		N, 71	O, 66	F, 71
	Na⁺, 98	Mg²⁺, 79	Al³⁺, 57		S²⁻, 184	Cl⁻, 181
	Na, 186	Mg, 160	Al, 143		S, 104	Cl, 99
	K⁺, 133	Ca²⁺, 106	Ga³⁺, 62		Se²⁻, 191	Br⁻, 196
	K, 227	Ca, 197	Ga, 122		Se, 117	Br, 114
	Rb⁺, 149	Sr²⁺, 127	In³⁺, 92		Te²⁻, 211	I⁻, 220
	Rb, 248	Sr, 215	In, 163		Te, 143	I, 133
	Cs⁺, 165	Ba²⁺, 143	Tl³⁺, 105			
	Cs, 265	Ba, 217	Tl, 170			

Active Figure 8.15   **Relative sizes of some common ions compared with neutral atom size.** Radii are given in picometers (1 pm = 1×10^{-12} m). (Data taken from J. Emsley: *The Elements*, 3rd ed., Oxford, Clarendon Press, 1998.)

GENERAL
Chemistry⚛Now™ See the General ChemistryNow CD-ROM or website to explore an interactive version of this figure accompanied by an exercise.

for Li. The loss of the 2s electron from Li leaves Li⁺ with no electrons in the $n = 2$ shell.

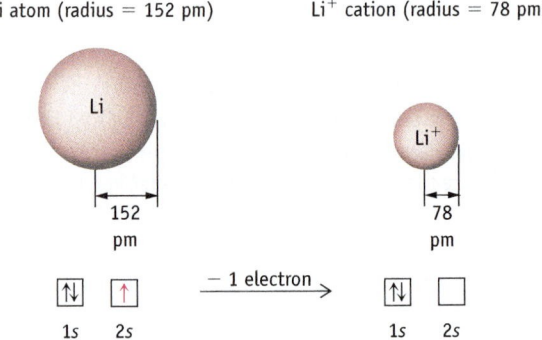

Li atom (radius = 152 pm)   Li⁺ cation (radius = 78 pm)

The shrinkage will also be great when two or more electrons are removed, as for Al³⁺ in which it exceeds 50%:

Al atom (radius = 143 pm)   Al³⁺ cation (radius = 57 pm)

[Ne]↑↓ ↑☐☐☐   —3 electrons→   [Ne]☐ ☐☐☐
    3s 3p   3s 3p

You can also see by comparing Figures 8.11 and 8.15 that *anions are always larger than the atoms from which they are derived.* Here the argument is the opposite of that used to explain positive ion radii. The F atom, for example, has nine protons and nine electrons. On forming the anion, the nuclear charge is still +9, but now ten electrons are in the anion. The F^- ion is much larger than the F atom because of increased electron–electron repulsions.

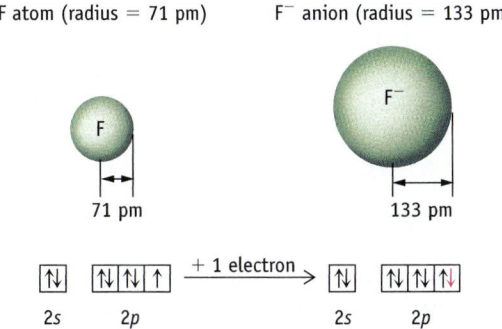

F atom (radius = 71 pm) F^- anion (radius = 133 pm)

71 pm   133 pm

Finally, it is useful to compare the sizes of isoelectronic ions across the periodic table. **Isoelectronic** ions have the same number of electrons (but a different number of protons). One such series of commonly occurring ions is O^{2-}, F^-, Na^+, and Mg^{2+}:

Ion	O^{2-}	F^-	Na^+	Mg^{2+}
Number of electrons	10	10	10	10
Number of nuclear protons	8	9	11	12
Ionic radius (pm)	140	133	98	79

All these ions have a total of ten electrons. The O^{2-} ion, however, has only 8 protons in its nucleus to attract these electrons, whereas F^- has 9, Na^+ has 11, and Mg^{2+} has 12. As the number of protons increases in a series of isoelectronic ions, the balance in electron–proton attraction and electron–electron repulsion shifts in favor of attraction, and the radius decreases. As you can see in Figure 8.15, this is true for all isoelectronic series of ions.

GENERAL
Chemistry Now™

See the General ChemistryNow CD-ROM or website:

- **Screen 8.14 Ion Sizes**

    (a) for a simulation exploring the relationship between ion formation and orbital energies in main group elements

    (b) for a simulation exploring the relationship between orbital energies and electron configurations on the size of the main group element ions

Exercise 8.8—Ion Sizes

What is the trend in sizes of the ions N^{3-}, O^{2-}, and F^-? Briefly explain why this trend exists.

8.7—Periodic Trends and Chemical Properties

Atomic and ionic radii, ionization energies, and electron affinities are properties associated with atoms and their ions. It is reasonable to expect that knowledge of these properties will be useful as we explore the chemistry of the elements. Let us consider just one example here, the formation of ionic compounds.

Charles D. Winters

Reaction of sodium metal and chlorine gas. This reaction produces the ionic compound NaCl, which consists of Na^+ and Cl^- ions. Because of their electron configurations, and their values of ionization energy and electron affinity, the reaction does not produce ions such as Na^{2+}, Cl^+, or Cl^{2-}. (See also Figure 1.7 on page 19.)

As described in Section 2.6, the periodic table was created by grouping together elements having similar chemical properties. Alkali metals, for example, characteristically form compounds in which the metal is in the form of a $1+$ ion, such as Li^+, Na^+, or K^+. Thus, the reaction between sodium and chlorine gives the ionic compound, NaCl (composed of Na^+ and Cl^- ions) [◀ Figure 1.7, page 19], and potassium and water react to form an aqueous solution of KOH, a solution containing the hydrated ions $K^+(aq)$ and $OH^-(aq)$ (Figure 8.9).

$$2\ Na(s) + Cl_2(g) \longrightarrow 2\ NaCl(s)$$
$$2\ K(s) + 2\ H_2O(\ell) \longrightarrow 2\ K^+(aq) + 2\ OH^-(aq) + H_2(g)$$

Both of these observations agree with the fact that alkali metals have electron configurations of the type [noble gas core]ns^1 and have low ionization energies.

Ionization energies also account in part for the fact that these reactions of sodium and potassium do not produce compounds such as $NaCl_2$ or $K(OH)_2$. The formation of a Na^{2+} or K^{2+} ion is clearly a very unfavorable process. Removing a second electron from these metals requires a great deal of energy because this electron must come from the atom's core electrons. Indeed, removal of core electrons from any atom is exceedingly unfavorable. This is the underlying reason that *main group metals generally form cations with an electron configuration equivalent to that of the nearest noble gas.*

Why isn't Na_2Cl another possible product from the sodium and chlorine reaction? This formula would imply that the compound contains Na^+ and Cl^{2-} ions. Chlorine atoms have a relatively high electron affinity, but only for the addition of one electron. Adding two electrons per atom means that the second electron must enter the next higher shell at much higher energy. An anion such as Cl^{2-} is simply not stable. This example leads us to a general statement: *Nonmetals generally acquire enough electrons to form an anion with the electron configuration of the next, higher noble gas.*

We can use similar logic to rationalize results of other reactions. Ionization energies increase on going from left to right across a period. We have seen that elements from Groups 1A and 2A form ionic compounds, an observation directly related to the low ionization energies for these elements. Ionization energies for elements toward the middle and right side of a period, however, are sufficiently large that cation formation is unfavorable. Thus, we generally do not expect to encounter ionic compounds containing carbon; instead, we find carbon *sharing* electrons with other elements in compounds such as CO_2 and CCl_4. On the right side of the second period, oxygen and fluorine much prefer taking on electrons to giving them up; these elements have high ionization energies and relatively large, negative electron affinities. Thus, oxygen and fluorine form anions and not cations when they react.

GENERAL
Chemistry ⚛ Now™

See the General ChemistryNow CD-ROM or website:
- **Screen 8.15 Chemical Reactions and Periodic Properties,** to watch videos on the relationship of atomic electron configurations and orbital energies on periodic trends

---

Exercise 8.9—Energies and Compound Formation

Give a plausible explanation for the observation that magnesium and chlorine react to form $MgCl_2$ and not $MgCl_3$.

Chapter Goals Revisited

Now that you have studied this chapter, you should ask whether you have met the chapter goals. In particular, you should be able to

Understand the role magnetism plays in determining and revealing atomic structure

a. Classify substances as paramagnetic (attracted to a magnetic field; characterized by unpaired electron spins) or diamagnetic (repelled by a magnetic field) (Section 8.1).

b. Recognize that each electron in an atom has a different set of the four quantum numbers, n, ℓ, m_ℓ, and m_s, where m_s, the spin quantum number, has values of $+\frac{1}{2}$ or $-\frac{1}{2}$ (Section 8.2). General ChemistryNow homework: Study Question(s) 18, 19

c. Understand that the Pauli exclusion principle leads to the conclusion that no atomic orbital can be assigned more than two electrons and that the two electrons in an orbital must have opposite spins (different values of m_s) (Section 8.2).

Understand effective nuclear charge and its role in determining atomic properties

a. Understand effective nuclear charge, Z^*, and its ability to explain why different subshells in the same shell have different energies. Also, understand the role of Z^* in determining the properties of atoms (Sections 8.3 and 8.6).

Write the electron configuration for elements and monatomic ions

a. Using the periodic table as a guide, depict electron configurations of the elements and monatomic ions using orbital box or the *spdf* notation. In both cases, configurations can be abbreviated with the noble gas notation (Sections 8.3 and 8.4). General ChemistryNow homework: SQ(s) 2, 3, 6, 12, 14, 21, 37, 38, 39, 48

b. Recognize that electrons are assigned to the subshells of an atom in order of increasing subshell energy. In the H atom the subshell energies increase with increasing n, but, in a many-electron atom, the energies depend on both n and ℓ (see Figure 8.3).

c. When assigning electrons to atomic orbitals, apply the Pauli exclusion principle and Hund's rule (Sections 8.3 and 8.4).

Understand the fundamental physical properties of the elements and their periodic trends

a. Predict how properties of atoms—size, ionization energy (*IE*), and electron affinity (*EA*)—change on moving down a group or across a period of the periodic table (Section 8.6). The general periodic trends for these properties are as follows:

(i) Atomic size decreases across a period and increases down a group.

(ii) *IE* increases across a period and decreases down a group.

(iii) The affinity for an electron generally increases across a period (the value of *EA* becomes more negative) and decreases down a group. General ChemistryNow homework: SQ(s) 26, 28, 30, 45, 46, 50, 53

b. Recognize the role that ionization energy and electron affinity play in the chemistry of the elements (Section 8.7). General ChemistryNow homework: SQ(s) 62

Study Questions

▲ denotes more challenging questions.

■ denotes questions available in the Homework and Goals section of the General ChemistryNow CD-ROM or website.

Blue numbered questions have answers in Appendix O and fully worked solutions in the *Student Solutions Manual*.

Structures of many of the compounds used in these questions are found on the General ChemistryNow CD-ROM or website in the Models folder.

GENERAL Chemistry ⚛ Now™ Assess your understanding of this chapter's topics with additional quizzing and conceptual questions at http://now.brookscole.com/kotz6e

Practicing Skills

Writing Electron Configurations of Atoms
(See Examples 8.1–8.3; Tables 8.1, 8.3, and 8.4; and the Toolbox on the General ChemistryNow.)

1. Write the electron configurations for P and Cl using both *spdf* notation and orbital box diagrams. Describe the relationship between each atom's electron configuration and its position in the periodic table.

2. ■ Write the electron configurations for Mg and Ar using both *spdf* notation and orbital box diagrams. Describe the relation of the atom's electron configuration to its position in the periodic table.

3. ■ Using *spdf* notation, write the electron configurations for atoms of chromium and iron, two of the major components of stainless steel.

4. Using *spdf* notation, give the electron configuration of vanadium, V, an element found in some brown and red algae and some toadstools.

5. Depict the electron configuration for each of the following atoms using *spdf* and noble gas notations.
   (a) Arsenic, As. A deficiency of As can impair growth in animals even though larger amounts are poisonous.
   (b) Krypton, Kr. It ranks seventh in abundance of the gases in the earth's atmosphere.

6. ■ Using *spdf* and noble gas notations, write electron configurations for atoms of the following elements and then check your answers with Table 8.3.
   (a) Strontium, Sr. This element is named for a town in Scotland.
   (b) Zirconium, Zr. The metal is exceptionally resistant to corrosion and so has important industrial applications. Moon rocks show a surprisingly high zirconium content compared with rocks on earth.
   (c) Rhodium, Rh. This metal is used in jewelry and in catalysts in industry.

(d) Tin, Sn. The metal was used in the ancient world. Alloys of tin (solder, bronze, and pewter) are important.

7. Use noble gas and *spdf* notations to depict electron configurations for the following metals of the third transition series.
   (a) Tantalum, Ta. The metal and its alloys resist corrosion and are often used in surgical and dental tools.
   (b) Platinum, Pt. This metal was used by pre-Columbian Indians in jewelry. It is used now in jewelry and for anticancer drugs and industrial catalysts.

8. The lanthanides, once called the rare earth elements, are really only "medium rare." Using noble gas and *spdf* notations, depict reasonable electron configurations for the following elements.
   (a) Samarium, Sm. This lanthanide is used in magnetic materials.
   (b) Ytterbium, Yb. This element was named for the village of Ytterby in Sweden, where a mineral source of the element was found.

9. The actinide americium, Am, is a radioactive element that has found use in home smoke detectors. Depict its electron configuration using noble gas and *spdf* notations.

10. Predict reasonable electron configurations for the following elements of the actinide series of elements. Use noble gas and *spdf* notations.
    (a) Plutonium, Pu. The element is best known as a byproduct of nuclear power plant operations.
    (b) Curium, Cm. This actinide was named for Madame Curie (page 57).

Electron Configurations of Atoms and Ions and Magnetic Behavior
(See Example 8.4 and General ChemistryNow Screens 8.3, 8.7, and 8.8.)

11. Using orbital box diagrams, depict an electron configuration for each of the following ions: (a) Mg^{2+}, (b) K^+, (c) Cl^-, and (d) O^{2-}.

12. ■ Using orbital box diagrams, depict an electron configuration for each of the following ions: (a) Na^+, (b) Al^{3+}, (c) Ge^{2+}, and (d) F^-.

13. Using orbital box diagrams and noble gas notation, depict the electron configurations of (a) V, (b) V^{2+}, and (c) V^{5+}. Are any of the ions paramagnetic?

14. ■ Using orbital box diagrams and noble gas notation, depict the electron configurations of (a) Ti, (b) Ti^{2+}, and (c) Ti^{4+}. Are any of the ions paramagnetic?

15. Manganese is found as MnO_2 in deep ocean deposits.
    (a) Depict the electron configuration of this element using the noble gas notation and an orbital box diagram.
    (b) Using an orbital box diagram, show the electrons beyond those of the preceding noble gas for the 2+ ion.
    (c) Is the 2+ ion paramagnetic?
    (d) How many unpaired electrons does the Mn^{2+} ion have?

16. Nickel generally forms 2+ ions but alkaline batteries have Ni^{3+} ions in NiOOH. Using orbital box diagrams and the noble gas notation, show electron configurations of these ions. Are either of these ions paramagnetic?

Quantum Numbers and Electron Configurations

(See Example 8.2 and General ChemistryNow Screens 7.12, 8.4, and 8.7.)

17. Explain briefly why each of the following is not a possible set of quantum numbers for an electron in an atom. In each case, change the incorrect value (or values) to make the set valid.
    (a) $n = 4, \ell = 2, m_\ell = 0, m_s = 0$
    (b) $n = 3, \ell = 1, m_\ell = -3, m_s = -\frac{1}{2}$
    (c) $n = 3, \ell = 3, m_\ell = -1, m_s = +\frac{1}{2}$

18. ■ Explain briefly why each of the following is not a possible set of quantum numbers for an electron in an atom. In each case, change the incorrect value (or values) to make the set valid.
    (a) $n = 2, \ell = 2, m_\ell = 0, m_s = +\frac{1}{2}$
    (b) $n = 2, \ell = 1, m_\ell = -1, m_s = 0$
    (c) $n = 3, \ell = 1, m_\ell = +2, m_s = +\frac{1}{2}$

19. ■ What is the maximum number of electrons that can be identified with each of the following sets of quantum numbers? In one case, the answer is "none." Explain why this is true.
    (a) $n = 4, \ell = 3$
    (b) $n = 6, \ell = 1, m_\ell = -1$
    (c) $n = 3, \ell = 3, m_\ell = -3$

20. What is the maximum number of electrons that can be identified with each of the following sets of quantum numbers? In some cases, the answer may be "none." In such cases, explain why "none" is the correct answer.
    (a) $n = 3$
    (b) $n = 3$ and $\ell = 2$
    (c) $n = 4, \ell = 1, m_\ell = -1$, and $m_s = -\frac{1}{2}$
    (d) $n = 5, \ell = 0, m_\ell = +1$

21. ■ Depict the electron configuration for magnesium using an orbital box diagram and noble gas notation. Give a complete set of four quantum numbers for each of the electrons beyond those of the preceding noble gas.

22. Depict the electron configuration for phosphorus using an orbital box diagram and noble gas notation. Give one possible set of four quantum numbers for each of the electrons beyond those of the preceding noble gas.

23. Using an orbital box diagram and noble gas notation, show the electron configuration of gallium, Ga. Give a set of quantum numbers for the highest-energy electron.

24. Using an orbital box diagram and noble gas notation, show the electron configuration of titanium. Give one possible set of four quantum numbers for each of the electrons beyond those of the preceding noble gas.

Periodic Properties

(See Section 8.6, Example 8.5, and General ChemistryNow Screens 8.9–8.12.)

25. Arrange the following elements in order of increasing size: Al, B, C, K, and Na. (Try doing it without looking at Figure 8.11, and then check yourself by looking up the necessary atomic radii.)

26. ■ Arrange the following elements in order of increasing size: Ca, Rb, P, Ge, and Sr. (Try doing it without looking at Figure 8.11, then check yourself by looking up the necessary atomic radii.)

27. Select the atom or ion in each pair that has the larger radius.
    (a) Cl or Cl^- (c) In or I
    (b) Al or O

28. ■ Select the atom or ion in each pair that has the larger radius.
    (a) Cs or Rb   (c) Br or As
    (b) O^{2-} or O

29. Which of the following groups of elements is arranged correctly in order of increasing ionization energy?
    (a) C < Si < Li < Ne   (c) Li < Si < C < Ne
    (b) Ne < Si < C < Li   (d) Ne < C < Si < Li

30. ■ Arrange the following atoms in order of increasing ionization energy: Li, K, C, and N.

31. Compare the elements Na, Mg, O, and P.
    (a) Which has the largest atomic radius?
    (b) Which has the most negative electron affinity?
    (c) Place the elements in order of increasing ionization energy.

32. Compare the elements B, Al, C, and Si.
    (a) Which has the most metallic character?
    (b) Which has the largest atomic radius?
    (c) Which has the most negative electron affinity?
    (d) Place the three elements B, Al, and C in order of increasing first ionization energy.

33. Explain each answer briefly.
    (a) Place the following elements in order of increasing ionization energy: F, O, and S.
    (b) Which has the largest ionization energy: O, S, or Se?
    (c) Which has the most negative electron affinity: Se, Cl, or Br?
    (d) Which has the largest radius: O^{2-}, F^-, or F?

34. Explain each answer briefly.
    (a) Rank the following in order of increasing atomic radius: O, S, and F.
    (b) Which has the largest ionization energy: P, Si, S, or Se?
    (c) Place the following in order of increasing radius: O^{2-}, N^{3-}, and F^-.
    (d) Place the following in order of increasing ionization energy: Cs, Sr, and Ba.

▲ More challenging   ■ In General ChemistryNow   **Blue-numbered questions** answered in Appendix O

General Questions

These questions are not designated as to type or location in the chapter. They may combine several concepts. More challenging questions are indicated by ▲.

35. The diagrams below represent a small section of a solid. Each circle represents an atom and an arrow represents an electron.

   (a)   **(b)**   **(c)**

   (a) Which represents a diamagnetic solid, which a paramagnetic solid, and which a ferromagnetic solid?

   (b) Which is most strongly attracted to a magnetic field? Which is least strongly attracted?

36. The name rutherfordium, Rf, has been given to element 104 to honor the physicist Ernest Rutherford (page 65). Depict its electron configuration using *spdf* and noble gas notations.

37. ■ Using an orbital box diagram and noble gas notation, show the electron configurations of uranium and of the uranium(IV) ion. Is either of these paramagnetic?

38. ■ The rare earth elements, or lanthanides, commonly exist as 3+ ions. Using an orbital box diagram and noble gas notation, show the electron configurations of the following elements and ions.

   (a) Ce and Ce^{3+} (cerium)   (b) Ho and Ho^{3+} (holmium)

39. ■ A neutral atom has two electrons with $n = 1$, eight electrons with $n = 2$, eight electrons with $n = 3$, and two electrons with $n = 4$. Assuming this element is in its ground state, supply the following information:

   (a) atomic number

   (b) total number of s electrons

   (c) total number of p electrons

   (d) total number of d electrons

   (e) is the element a metal, metalloid, or nonmetal?

40. Element 109, now named meitnerium (in honor of the Austrian–Swedish physicist, Lise Meitner [1878–1968]), was produced in August 1982 by a team at Germany's Institute for Heavy Ion Research. Depict its electron configuration using *spdf* and noble gas notations. Name another element found in the same group as meitnerium.

41. Which of the following is *not* an allowable set of quantum numbers? Explain your answer briefly.

	n	ℓ	m_ℓ	m_s
(a)	2	0	0	$-\frac{1}{2}$
(b)	1	1	0	$+\frac{1}{2}$
(c)	2	1	-1	$-\frac{1}{2}$
(d)	4	3	$+2$	$-\frac{1}{2}$

42. A possible excited state for the H atom has an electron in a $4p$ orbital. List all possible sets of quantum numbers (n, ℓ, m_ℓ, and m_s) for this electron.

43. The magnet in the photo is made from neodymium, iron, and boron.

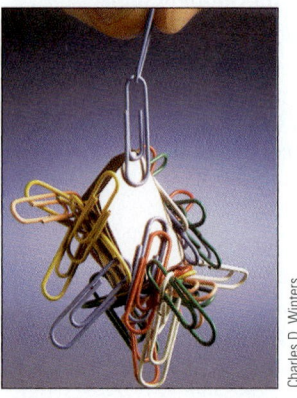

Charles D. Winters

A magnet made of an alloy containing the elements Nd, Fe, and B.

   (a) Write the electron configuration of each of these elements using an orbital box diagram and noble gas notation.

   (b) Are these elements paramagnetic or diamagnetic?

   (c) Write the electron configurations of Nd^{3+} and Fe^{3+} using orbital box diagrams and noble gas notation. Are these ions paramagnetic or diamagnetic?

44. Name the element corresponding to each characteristic below.

   (a) the element with the electron configuration $1s^2 2s^2 2p^6 3s^2 3p^3$

   (b) the alkaline earth element with the smallest atomic radius

   (c) the element with the largest ionization energy in Group 5A

   (d) the element whose 2+ ion has the configuration $[Kr]4d^5$

   (e) the element with the most negative electron affinity in Group 7A

   (f) the element whose electron configuration is $[Ar]3d^{10}4s^2$

45. ■ Arrange the following atoms in the order of increasing ionization energy: Si, K, P, and Ca.

46. ■ Rank the following in order of increasing ionization energy: Cl, Ca^{2+}, and Cl^-. Briefly explain your answer.

47. Answer the questions below about the elements A and B, which have the electron configurations shown.

$$A = [Kr]5s^1 \qquad B = [Ar]3d^{10}4s^24p^4$$

   (a) Is element A a metal, nonmetal, or metalloid?

   (b) Which element has the greater ionization energy?

   (c) Which element has the less negative electron affinity?

   (d) Which element has the larger atomic radius?

48. ■ Answer the following questions about the elements with the electron configurations shown here:

$$A = [Ar]4s^2 \qquad B = [Ar]3d^{10}4s^24p^5$$

(a) Is element A a metal, metalloid, or nonmetal?

(b) Is element B a metal, metalloid, or nonmetal?

(c) Which element is expected to have the larger ionization energy?

(d) Which element has the smaller atomic radius?

49. Which of the following ions are unlikely to be found in a chemical compound: Cs^+, In^{4+}, Fe^{6+}, Te^{2-}, Sn^{5+}, and I^-? Explain briefly.

50. ■ Place the following elements and ions in order of decreasing size: K^+, Cl^-, S^{2-}, and Ca^{2+}.

51. Answer each of the following questions:

(a) Of the elements S, Se, and Cl, which has the largest atomic radius?

(b) Which has the larger radius, Br or Br^-?

(c) Which should have the largest difference between the first and second ionization energy: Si, Na, P, or Mg?

(d) Which has the largest ionization energy: N, P, or As?

(e) Which of the following has the largest radius: O^{2-}, N^{3-}, or F^-?

52. The following are isoelectronic species: Cl^-, K^+, and Ca^{2+}. Rank them in order of increasing (a) size, (b) ionization energy, and (c) electron affinity.

53. ■ Compare the elements Na, B, Al, and C with regard to the following properties:

(a) Which has the largest atomic radius?

(b) Which has the most negative electron affinity?

(c) Place the elements in order of increasing ionization energy.

54. ▲ Two elements in the second transition series (Y through Cd) have four unpaired electrons in their 3+ ions. What elements fit this description?

55. The configuration for an element is given here.

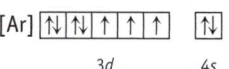

(a) What is the identity of the element with this configuration?

(b) Is a sample of the element paramagnetic or diamagnetic?

(c) How many unpaired electrons does a 3+ ion of this element have?

56. The configuration of an element is given here.

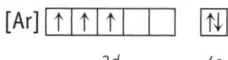

(a) What is the identity of the element?

(b) In what group and period is the element found?

(c) Is the element a nonmetal, a main group element, a transition metal, a lanthanide, or an actinide?

(d) Is the element diamagnetic or paramagnetic? If paramagnetic, how many unpaired electrons are there?

(e) Write a complete set of quantum numbers (n, ℓ, m_ℓ, m_s) for each of the valence electrons.

(f) What is the configuration of the 2+ ion formed from this element? Is the ion diamagnetic or paramagnetic?

Summary and Conceptual Questions

The following questions use concepts from the previous chapters.

57. Why is the radius of Li^+ so much smaller than the radius of Li? Why is the radius of F^- so much larger than the radius of F?

58. Which ions in the following list are not likely to be found in chemical compounds: K^{2+}, Cs^+, Al^{4+}, F^{2-}, and Se^{2-}? Explain briefly.

59. Write electron configurations to show the first two ionization processes for potassium. Explain why the second ionization energy is much greater than the first.

60. Explain how the ionization energy of atoms changes and why the change occurs when proceeding down a group of the periodic table.

61. (a) Explain why the sizes of atoms change when proceeding across a period of the periodic table.

(b) Explain why the sizes of transition metal atoms change very little across a period.

62. ■ Which of the following elements has the greatest difference between its first and second ionization energies: C, Li, N, Be? Explain your answer.

63. What arguments would you use to convince another student in general chemistry that MgO consists of the ions Mg^{2+} and O^{2-} and not the ions Mg^+ and O^-? What experiments could be done to provide some evidence that the correct formulation of magnesium oxide is $Mg^{2+}O^{2-}$?

64. Explain why the first ionization energy of Ca is greater than that of K, whereas the second ionization energy of Ca is lower than the second ionization energy of K.

65. The energies of the orbitals in many elements have been determined. For the first two periods they have the following values:

Element	1s (kJ/mol)	2s (kJ/mol)	2p (kJ/mol)
H	−1313		
He	−2373		
Li		−520.0	
Be		−899.3	
B		−1356	−800.8
C		−1875	−1029
N		−2466	−1272
O		−3124	−1526
F		−3876	−1799
Ne		−4677	−2083

(a) Why does the energy generally become more negative on proceeding across the second period?

(b) How are these values related to the ionization energy and electron affinity of the elements?

(c) Use these energy values to explain the observation that the ionization energies of the first four second-period elements are in the order Li < Be > B < C.

Note that these energy values are the basis for the discussion in the Simulation on the General ChemistryNow CD-ROM or website Screen 8.9. Data from J. B. Mann, T. L. Meek, and L. C. Allen: *Journal of the American Chemical Society*, Vol. 122, p. 2780, 2000.

66. ▲ The ionization energies for the removal of the first electron in Si, P, S, and Cl are as listed in the table below. Briefly rationalize this trend.

Element	First Ionization Energy (kJ/mol)
Si	780
P	1060
S	1005
Cl	1255

67. Using your knowledge of the trends in element sizes on going across the periodic table, explain briefly why the density of the elements increases from K through V.

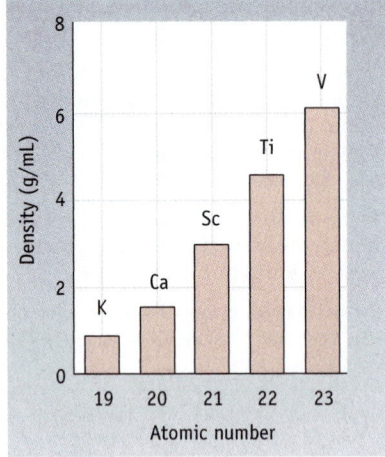

68. The densities (in g/cm³) of elements in Groups 6B, 8B, and 1B are given in the table below.

Period 4	Cr, 7.19	Co, 8.90	Cu, 8.96
Period 5	Mo, 10.22	Rh, 12.41	Ag, 10.50
Period 6	W, 19.30	Ir, 22.56	Au, 19.32

Transition metals in the sixth period all have much greater densities than the elements in the same groups in the fourth and fifth periods. Refer to Figure 8.12 and explain this observation.

69. The discovery of two new elements (atomic numbers 113 and 115) was announced in February 2004.

Some members of the team that discovered elements 113 and 115 at the Lawrence Livermore National Laboratory (left to right): Jerry Landrum, Dawn Shaughnessy, Joshua Patin, Philip Wilk, and Kenton Moody.

(a) Use *spdf* and noble gas notations to give the electron configurations of these two elements.

(b) Name an element in the same periodic group as the two elements.

(c) Element 113 was made by firing a light atom at a heavy americium atom. The two combine to give a nucleus with 113 protons. What light atom was used as a projectile?

70. Explain why the reaction of calcium and fluorine does *not* form CaF_3.

71. ▲ Thionyl chloride, $SOCl_2$, is an important chlorinating and oxidizing agent in organic chemistry. It is prepared industrially by oxygen atom transfer from SO_3 to SCl_2.

$$SO_3(g) + SCl_2(g) \longrightarrow SO_2(g) + SOCl_2(g)$$

(a) Give the electron configuration for an atom of sulfur using an orbital box diagram. Do not use the noble gas notation.

(b) Using the configuration given in part (a), write a set of quantum numbers for the highest-energy electron in a sulfur atom.

(c) What element involved in this reaction (O, S, Cl) should have the smallest ionization energy? The smallest radius?

(d) Which should be smaller: the sulfide ion, S^{2-}, or a sulfur atom, S?

(e) If you want to make 675 g of $SOCl_2$, what mass of SCl_2 is required?

(f) If you use 10.0 g of SO_3 and 10.0 g of SCl_2, what is the theoretical yield of $SOCl_2$?

(g) ΔH°_{rxn} for the reaction of SO_3 and SCl_2 is −96.0 kJ/mol $SOCl_2$ produced. Using data in Appendix L, calculate the standard molar enthalpy of formation of SCl_2.

▲ More challenging   ■ In General ChemistryNow   Blue-numbered questions answered in Appendix O

72. Sodium metal reacts readily with chlorine gas to give sodium chloride. (*See General ChemistryNow CD-ROM or website Screen 8.16 Chemical Puzzler.*)

$$Na(s) + \tfrac{1}{2}Cl_2(g) \longrightarrow NaCl(s)$$

(a) What is the reducing agent in this reaction? What property of the element contributes to its ability as a reducing agent?

(b) What is the oxidizing agent in this reaction? What property of the element contributes to its ability as an oxidizing agent?

(c) Why does the reaction produce NaCl and not a compound such as Na_2Cl or $NaCl_2$?

73. If a C atom is attached or "bonded" to a Cl atom, the calculated distance between the atoms is the sum of their radii. Calculate the expected distance between the pairs of atoms in the following table. Then use model molecules in the Molecular Models folder on the General ChemistryNow CD-ROM or website to examine the appropriate distance in the designated molecules. Is there reasonably good agreement between the calculated and measured distances?

Molecule	Atom Distance	Calculated (pm)	Measured (pm)
BF_3	B — F	_____	_____
PF_3	P — F	_____	_____
CH_4	C — H	_____	_____
H_3COH	C — O	_____	_____

(Note that BF_3 and PF_3 are in the Inorganic folder. CH_4 and CH_3OH are in the Organic folder. The latter (CH_3OH) is called methanol and is in the Alcohol folder. The distances given on these models are in angstrom units, where 1 Å = 100 pm.)

Do you need a live tutor for homework problems?
Access vMentor at General ChemistryNow at
http://now.brookscole.com/kotz6e
for one-on-one tutoring from a chemistry expert

9—Bonding and Molecular Structure: Fundamental Concepts

Image courtesy of NRAO/AUI

The 12-meter radio telescope at the National Radio Astronomy Observatory. It is used to search for molecules in deep space.

© 1997, Fred Espenak, www.mreclipse.com

Comets deliver many complex molecules to earth. This is the Hale-Bopp comet in 1997.

Molecules in Space

Life is based on simple molecules like water and ammonia, slightly more complex ones like sugars, and very complex ones like DNA and hemoglobin. Where do they come from? How are they formed? What do they look like? Are their properties connected to how they look— that is, to their structures?

The origins of molecules is a topic eagerly studied by astronomers. Since the 1960s some space scientists have surmised that comets bring water, ammonia, and even more complex molecules of all kinds to earth from outer space. Every day an average of about 30 tons of organic material arrive on earth from space.

More than 120 molecules have been identified by radio astronomers in the far reaches of our galaxy. These range from hydrogen molecules to other simple molecules such as CO, H_2O, NH_3, and HCl.

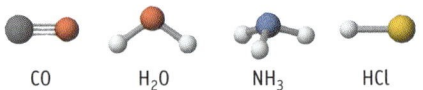

Some molecules from deep space.

Recently, more complex molecules have been observed, including a simple sugar, glycolaldehyde, $C_2H_4O_2$, discovered in 2001 in a cloud of gas and dust about 26,000 light-years from earth. According to a researcher at the NASA Goddard Space Center, "The discovery of this sugar molecule in a cloud from which new stars are forming means it is increasingly likely that the chemical precursors to life are formed in such clouds long before planets develop around the stars."

Glycolaldehyde is a member of the carbohydrate family, all of which have the general formula $C_a(H_2O)_b$. The molecule has two C atoms in its "backbone." One C atom is attached to an H atom and an O atom. The other C atom has two H atoms and one OH group attached. Indeed, its structure is quite predictable, and recognizing that pattern is one objective of this chapter. We want to know, for example, why the angles made by the atom attachments are not 90°, and why there is a difference in the C — O links. We would also like to know how this structure influences its chemical and physical properties, so that we might predict how it would interact with other molecules.

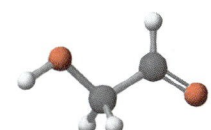

HOCH$_2$CHO

Glycolaldehyde.

How are some of these complex molecules formed? Temperatures in deep space hover near absolute zero, and astronomers believe that simple molecules such as water, CO, CO$_2$, and CH$_3$OH (methanol) freeze onto the surface of minute pieces of interstellar dust. These dust particles are subjected to intense radiation from nearby stars, causing the molecules to fragment (much as you saw in the mass spectrum in Figure 3.15). The fragments rearrange and combine in new ways, forming larger molecules such as glycolaldehyde.

Other molecules found recently in space include hydrocarbons (compounds composed only of C and H) such as anthracene. Anthracene is a member of a large class of compounds called polycyclic aromatic hydrocarbons. You may be aware of them because they are carcinogenic pollutants on earth, and you produce minute quantities when you cook a hamburger on a charcoal grill.

J. Hester and P. Scowan, of Arizona State University, and NASA.

The Eagle Nebula. These pillar-like structures are vast columns of gas and dust, within which new stars have recently formed. It is here that many molecules are created. The tallest of the pillars (at left) is about one light-year in length from base to tip. The Eagle Nebula is a star-forming region 7000 light-years away in the constellation Serpens.

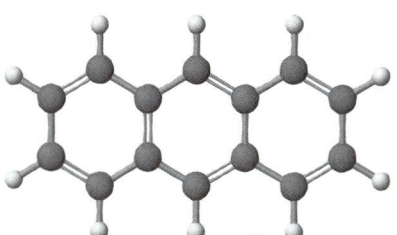

Anthracene, a polycyclic aromatic hydrocarbon.

Why are these compounds flat? What relation do they have to other carbon based compounds? These are just a few of the subjects we begin to explore in this and subsequent chapters.

To Review Before You Begin

- Know the names of common compounds and ions (Chapter 3)
- Review Coulomb's law (page 112)
- Understand energy changes in chemical reactions (Chapter 6)

Scientists have long known that the key to interpreting the properties of a chemical substance is first to recognize and understand its structure and bonding. **Structure** refers to the way atoms are arranged in space, and **bonding** describes the forces that hold adjacent atoms together. In Chapter 3, we told the story of how the basic structure of DNA was uncovered. This structure raises many interesting questions, such as why DNA chains have a helical shape. The answer is related to the geometry of the chemical bonds around each of the carbon, phosphorus, and oxygen atoms of the chain. Just how this relationship works will become more evident as you learn more about the topics of structure and bonding.

The goal of this and the next two chapters is to explain how atoms are arranged in chemical compounds and what holds them together. At the same time, we want to begin to relate the structure and bonding in a molecule to its chemical and physical properties.

Our discussion of structure and bonding begins with small molecules and ions, and then progresses to larger molecules. From compound to compound, atoms of the same element participate in bonding and structure in a predictable manner. This consistency allows us to develop a group of principles that apply to many different chemical compounds, including such complex structures as DNA.

9.1—Valence Electrons

The electrons in an atom can be divided into two groups: *valence electrons* and *core electrons*. Valence electrons determine the chemical properties of the atom because chemical reactions result in the loss, gain, or rearrangement of these electrons [◀ page 345]. The remaining electrons, the core electrons, are not involved in chemical behavior.

For main group elements (elements of the A groups in the periodic table), the valence electrons are the *s* and *p* electrons in the outermost shell (Table 9.1). All electrons in inner shells (such as those in filled *d* subshells) are core electrons. A useful guideline for *main group elements* is that *the number of valence electrons is equal to the group number*. The fact that all elements in a periodic group have the same number of valence electrons accounts for the similarity of chemical properties among members of the group.

Valence electrons for *transition elements* include the electrons in the *ns* and $(n-1)d$ orbitals (see Table 9.1). The remaining electrons are core electrons. As with main group elements, the valence electrons for transition metals determine the chemical properties of these elements.

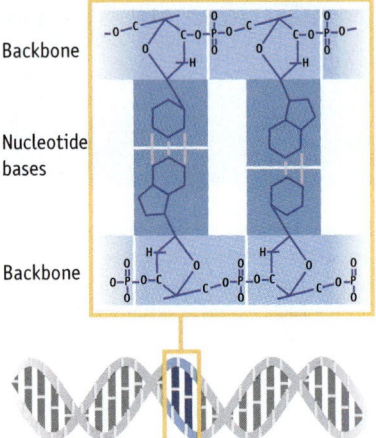

Backbone

Nucleotide bases

Backbone

The arrangement of atoms in DNA.

See General ChemistryNow CD-ROM or website:

- **Screen 9.2 Valence Electrons,** for the correlation of the periodic table and valence electrons

Table 9.1 Core and Valence Electrons for Several Common Elements

Element	Periodic Group	Core Electrons	Valence Electrons	Total Configuration
Main Group Elements				
Na	1A	$1s^2 2s^2 2p^6 = [Ne]$	$3s^1$	$[Ne]3s^1$
Si	4A	$1s^2 2s^2 2p^6 = [Ne]$	$3s^2 3p^2$	$[Ne]3s^2 3p^2$
As	5A	$1s^2 2s^2 2p^6 3s^2 3p^6 3d^{10} = [Ar]3d^{10}$	$4s^2 4p^3$	$[Ar]3d^{10}4s^2 4p^3$
Transition Elements				
Ti	4B	$1s^2 2s^2 2p^6 3s^2 3p^6 = [Ar]$	$3d^2 4s^2$	$[Ar]3d^2 4s^2$
Co	8B	$[Ar]$	$3d^7 4s^2$	$[Ar]3d^7 4s^2$
Mo	6B	$[Kr]$	$4d^5 5s^1$	$[Kr]4d^5 5s^1$

Lewis Symbols for Atoms

G. N. Lewis (1875–1946) introduced a useful way to describe electrons in the valence shell of an atom. In the system he developed, the element's symbol represents the atomic nucleus together with the core electrons. Up to four valence electrons, represented by dots, are placed one at a time around the symbol; then, if any valence electrons remain, they are placed next to ones already there. Chemists now refer to these pictures as **Lewis electron dot symbols**. Lewis dot symbols for the main group elements of the second and third periods are shown in Table 9.2.

Arranging the valence electrons of a main group element around an atom in four groups suggests that the valence shell can accommodate a maximum of four pairs of electrons. Because this arrangement represents eight electrons, it is referred to as an *octet* of electrons. An octet of electrons surrounding an atom is regarded as a stable configuration. The noble gases, with the exception of helium, have eight valence electrons and demonstrate a notable lack of reactivity. (Helium, neon, and argon do not undergo any chemical reactions, and the other noble gases have very limited chemical reactivity.) Because chemical reactions involve changes in the valence electron shell, the limited reactivity of the noble gases is taken as evidence of the stability of their noble gas ($ns^2 np^6$) electron configuration. Hydrogen, which in its compounds has two electrons in its valence shell, obeys the spirit of this rule by matching the electron configuration of He.

■ **H Atoms and Electron Octets**
Hydrogen cannot be surrounded by an octet of electrons. An atom of H, which has only a $1s$ valence electron orbital, can accommodate only a pair of electrons.

Table 9.2 Lewis Dot Symbols for Main Group Atoms

1A ns^1	2A ns^2	3A $ns^2 np^1$	4A $ns^2 np^2$	5A $ns^2 np^3$	6A $ns^2 np^4$	7A $ns^2 np^5$	8A $ns^2 np^6$
Li·	·Be·	·B·	·C·	·N·	:O·	:F·	:Ne:
Na·	·Mg·	·Al·	·Si·	·P·	:S·	:Cl·	:Ar:

Example 9.1—Valence Electrons

Problem Give the number of valence electrons for Ca and Se. Draw the Lewis electron dot symbol for each element.

Strategy Locate the elements in the periodic table. Note that, for main group elements, the number of valence electrons equals the group number.

Solution Calcium, in Group 2A, has two valence electrons, and selenium, in Group 6A, has six. Dots representing electrons are placed around the element symbol one at a time until there are four electrons. Subsequent electrons are paired with those already present:

$$\cdot \text{Ca} \cdot \qquad\qquad \cdot \overset{\displaystyle ..}{\underset{\displaystyle ..}{\text{Se}}} \cdot$$

calcium   selenium

Exercise 9.1—Electrons

Give the number of valence electrons for Ba, As, and Br. Draw the Lewis dot symbol for each of these elements.

9.2—Chemical Bond Formation

When a chemical reaction occurs between two atoms, their valence electrons are reorganized so that a net attractive force—a **chemical bond**—occurs between atoms. There are two general types of bonds: ionic and covalent. Their formation can be depicted using Lewis symbols.

An **ionic bond** forms when *one or more valence electrons are transferred from one atom to another*, creating positive and negative ions. When sodium and chlorine react (Figure 9.1a), an electron is transferred from a sodium atom to a chlorine atom to form Na^+ and Cl^-.

Figure 9.1 **Formation of ionic compounds.** Both reactions shown here are quite exothermic, as reflected by the very negative molar enthalpies of formation for the reaction products. (*See General ChemistryNow Screen 9.5 Chemical Reactions and Periodic Properties, to watch a video of the sodium–chlorine reaction.*)

(a) The reaction of elemental sodium and chlorine.
ΔH_f° [NaCl(s)] = −411.12 kJ/mol

(b) The reaction of elemental calcium and oxygen to give calcium oxide.
ΔH_f° [CaO(s)] = −635.09 kJ/mol

Charles D. Winters

$$Na\cdot + \cdot \overset{..}{\underset{..}{Cl}}: \longrightarrow \left[Na \cdot \curvearrowright \cdot \overset{..}{\underset{..}{Cl}}: \right] \longrightarrow \left[Na^+ \quad : \overset{..}{\underset{..}{Cl}}:^- \right]$$

| Metal atom | Nonmetal atom | Electron transfer from reducing agent to oxidizing agent. | Ionic compound. Ions have noble gas electron configurations. |

The "bond" is the attractive force between the positive and negative ions.

Covalent bonding, in contrast, *involves sharing of valence electrons between atoms.* Two chlorine atoms, for example, share a pair of electrons, one electron from each atom, to form a covalent bond.

$$: \overset{..}{\underset{..}{Cl}} \cdot + \cdot \overset{..}{\underset{..}{Cl}}: \longrightarrow : \overset{..}{\underset{..}{Cl}} : \overset{..}{\underset{..}{Cl}}:$$

It is useful to reflect on the differences in the Lewis electron dot structure representations of ionic and covalent bonding. In both processes, unpaired electrons in the reactants are paired up. Both processes give products in which each atom is surrounded by eight electrons (an octet). The position of the electron pair between the two bonded atoms differs significantly, however. In a chlorine molecule (Cl_2), the electron pair is shared equally by the two atoms. In contrast, the electron pair in sodium chloride has become part of the valence shell of chlorine.

As bonding is described in greater detail, you will discover that the two types of bonding—complete electron transfer and the equal sharing of electrons—are extreme cases. In most chemical compounds electrons are shared unequally, with the extent of sharing varying widely from very little sharing (largely ionic) to considerable sharing (largely covalent).

GENERAL
Chemistry ❄️ Now ™

See General ChemistryNOW CD-ROM or website:
- **Screen 9.5 Chemical Reactions and Periodic Properties,** to watch a video of the formation of sodium chloride from the elements

9.3—Bonding in Ionic Compounds

Metallic sodium reacts vigorously with gaseous chlorine to give sodium chloride, (Figure 9.1a), and calcium metal and oxygen react to give calcium oxide (Figure 9.1b). In each case, the product is an ionic compound: NaCl contains Na^+ and Cl^- ions, whereas CaO is composed of Ca^{2+} and O^{2-} ions.

$$Na(s) + \tfrac{1}{2} Cl_2(g) \longrightarrow NaCl(s) \qquad \Delta H° = -411.12 \text{ kJ}$$
$$Ca(s) + \tfrac{1}{2} O_2(g) \longrightarrow CaO(s) \qquad \Delta H° = -635.09 \text{ kJ}$$

These exothermic reactions, which are examples of the chemical behavior demonstrated by elements in these periodic groups, can be understood based on the atomic properties described in Chapter 8. The alkali and alkaline earth metals have relatively low ionization energies. Loss of the *ns* valence electrons from these elements leads to cations with a noble gas configuration $(n-1)s^2(n-1)p^6$. In

■ **Valence Electron Configurations and Ionic Compound Formation**
For the formation of NaCl:

Na changes from $1s^2 2s^2 2p^6 3s^1$ to Na^+ with $1s^2 2s^2 2p^6$, equivalent to the Ne configuration.

Cl changes from $[Ne]3s^2 3p^5$ to Cl^- with $[Ne]3s^2 3p^6$, equivalent to the Ar configuration.

contrast, elements immediately preceding Group 8A (the halogens and the Group 6A elements) have high affinities for electrons. These elements typically form anions by adding electrons, giving an ion with an electron configuration equivalent to that of the next noble gas. The tendency to achieve a noble gas configuration by gain or loss of electrons is an important aspect in the chemistry of main group elements.

Ion Attraction and Lattice Energy

To understand bonding in ionic compounds, it is useful to think about the energy involved in their formation. We begin by asking what the energy change is for the formation of the ion pair $[Na^+,Cl^-]$ in the gas phase starting with sodium and chlorine atoms, also in the gas phase. The overall energy of this reaction can be thought of as the sum of three individual steps: (1) the ionization of a sodium atom to form Na^+ (the energy of this process is the *ionization energy* of the element); (2) the addition of an electron to a chlorine atom to form the Cl^- ion (the energy here is the *electron affinity* of the element); and (3) the formation of the $[Na^+,Cl^-](g)$ ion pair from $Na^+(g)$ and $Cl^-(g)$.

1. Formation of $Na^+(g)$ and an electron

    $Na(g) \longrightarrow Na^+(g) + e^-$ ΔE_{ion} = ionization energy of Na = +496 kJ/mol

2. Formation of $Cl^-(g)$ from a Cl atom and an electron

    $Cl(g) + e^- \longrightarrow Cl^-(g)$ ΔE_{EA} = electron affinity of Cl = −349 kJ/mol

3. Formation of the ion pair

    $Na^+(g) + Cl^-(g) \longrightarrow [Na^+,Cl^-](g)$ $\Delta E_{ion\ pair}$ = −498 kJ/mol

The energy for the last step in this process can be calculated from an equation related to Coulomb's law [◀ Equation 3.1, page 112].

$$E_{ion\ pair} = C(N)\left(\frac{(n^+e)(n^-e)}{d}\right)$$

The symbol C represents a constant, d is the distance between the ion centers, n is the number of positive (n^+) and negative (n^-) charges on an ion, and e is the charge on an electron. Including Avogadro's number, N, allows us to calculate the energy change for 1 mol of ion pairs. Because the charges are opposite in sign, the energy value is negative. Inspection of this equation reveals that the energy of attraction between ions of opposite charge depends on two factors:

- *The magnitude of the ion charges.* The higher the ion charges, the greater the attraction, so ΔE for ion-pair formation has a larger negative value. For example, the attraction between Ca^{2+} and O^{2-} ions will be about four times larger $[(2+) \times (2-)]$ than the attraction between Na^+ and Cl^- ions, and the energy will be more negative by a factor of about 4.

- *The distance between the ions.* This is an inverse relationship because, as the distance between ions becomes greater (d becomes larger), the attractive force between the ions declines and the energy is less negative. The distance is determined by the sizes of the ions [◀ Figure 8.15].

Ionic compounds exist as solids under normal conditions. Their structures contain not ion pairs but rather positive and negative ions arranged in a three-dimensional lattice [▶ Chapter 13]. Models of a small segment of a NaCl lattice are pictured in Figure 9.2. In crystalline NaCl, each Na^+ cation is surrounded by six Cl^-

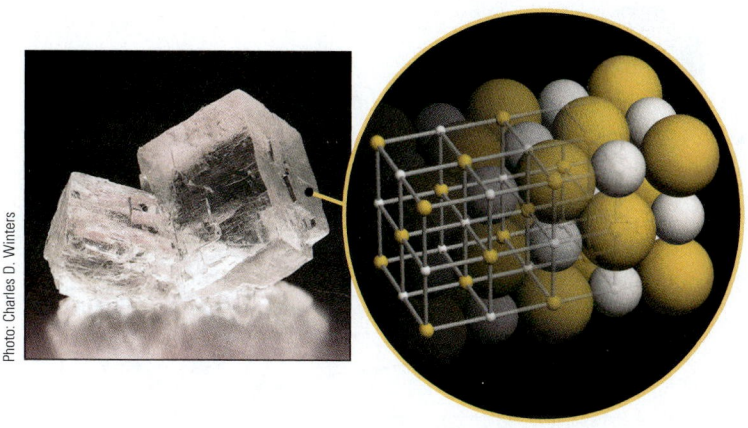

Figure 9.2 **Crystalline NaCl and models of the sodium chloride crystal lattice.** (*left*) A ball-and-stick model. The "sticks" in the model are there to help identify the locations of the atoms. (*right*) A space-filling model. These models represent only a small portion of the lattice. Ideally, it extends infinitely in all directions. Sodium ions are colored silver and chloride ions are colored yellow to distinguish them in this illustration.

Photo: Charles D. Winters

anions, and six Na^+ ions are nearest neighbors to each Cl^-. The equation for $E_{\text{ion pair}}$ can be modified to take into account the extensive attractions between ions of opposite charge and repulsions between ions of like charge in a crystal lattice. That is, we can calculate the **lattice energy**, $\Delta E_{\text{lattice}}$, which is the energy of formation of one mole of a solid crystalline ionic compound when ions in the gas phase combine (see Table 9.3).

$$Na^+(g) + Cl^-(g) \longrightarrow NaCl(s) \qquad \Delta E_{\text{lattice}} = -786 \text{ kJ/mol}$$

Lattice energy is a measure of the strength of ionic bonding in solid compounds. As we shall see, these energy values are closely related to the temperatures required to melt ionic compounds [▶ Section 13.6].

What is important about lattice energy here is the dependence of $\Delta E_{\text{lattice}}$ on ion charges and sizes. The effect of ion charge is illustrated by the lattice energies of MgO and NaF. The value of $\Delta E_{\text{lattice}}$ for MgO (-4050 kJ/mol) is about four times more negative than the value for NaF (-926 kJ/mol) because the charges on the Mg^{2+} and O^{2-} ions are each twice as large as those on Na^+ and F^- ions.

The effect of ion size on lattice energy is also predictable: A lattice built from smaller ions generally leads to a more negative value for the lattice energy (Table 9.3 and Figure 9.3). For alkali metal halides, for example, the lattice energy for lithium compounds is generally more negative than that for potassium compounds because the Li^+ ion is much smaller than the K^+ cation. Similarly, fluorides are more strongly bonded than are iodides with the same cation.

Table 9.3 Lattice Energies of Some Ionic Compounds

Compound	$\Delta E_{\text{lattice}}$ (kJ/mol)
LiF	-1037
LiCl	-852
LiBr	-815
LiI	-761
NaF	-926
NaCl	-786
NaBr	-752
NaI	-702
KF	-821
KCl	-717
KBr	-689
KI	-649

Source: D. Cubicciotti: "Lattice energies of the alkali halides and electron affinities of the halogens." *Journal of Chemical Physics,* Vol. 31, p. 1646, 1959.

Calculating a Lattice Energy

The lattice energies in Table 9.3 were calculated using a thermodynamic energy level diagram known as a Born-Haber cycle. This calculation is an application of Hess's law, which says that the energy involved in one pathway from reactants to products (the heat of formation of the compound, $\Delta H_f°$) is the sum of the energies involved in another pathway (Steps 1–3). Such a cycle is illustrated in Figure 9.4 for solid sodium chloride.

Steps 1 and 2 in Figure 9.4 involve formation of $Na^+(g)$ and $Cl^-(g)$ ions from the elements, and the enthalpy change of each step is available from experiments (Appendices F and L). Step 3 in Figure 9.4 gives the lattice enthalpy, $\Delta H_{\text{lattice}}$, and

■ **Born-Haber Cycles**
These energy level diagrams are named for Max Born (1882–1970) and Fritz Haber (1868–1934), German scientists who played prominent roles in thermodynamic research.

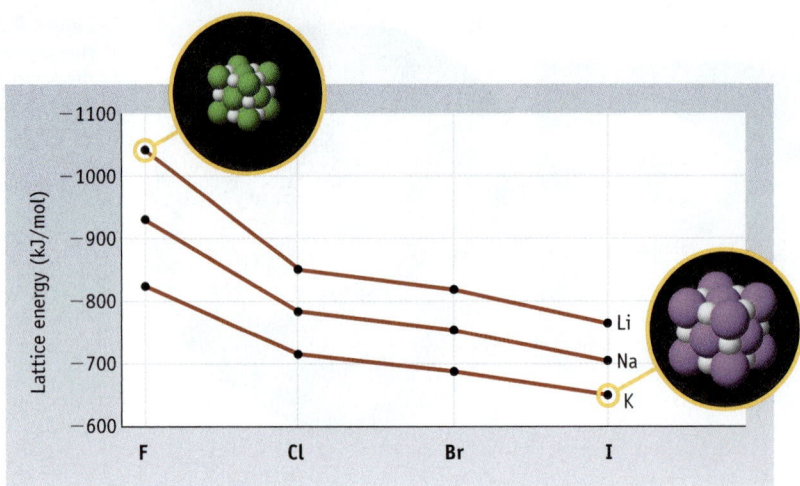

Figure 9.3   Lattice energy. $\Delta E_{\text{lattice}}$ is illustrated for the formation of the alkali metal halides, MX(s), from the ions $M^+(g) + X^-(g)$.

ΔH_f° is the standard molar enthalpy of formation of NaCl(s) (Appendix L). The enthalpy values for each step are related by the following equation:

$$\Delta H_f^\circ \, [\text{NaCl(s)}] = \Delta H_{\text{Step 1a}} + \Delta H_{\text{Step 1b}} + \Delta H_{\text{Step 2a}} + \Delta H_{\text{Step 2b}} + \Delta H_{\text{Step 3}}$$

Because the values for all of these quantities are known except for $\Delta H_{\text{Step 3}}$ ($\Delta H_{\text{lattice}}$), the value for this step can be calculated.

Step 1a. Enthalpy of formation of Cl(g) = +121.3 kJ/mol (Appendix L)

Step 1b. ΔH for Cl(g) + e$^-$ $\longrightarrow$ Cl$^-$(g) = -349 kJ/mol (Appendix F)

Step 2a. Enthalpy of formation of Na(g) = +107.3 kJ/mol (Appendix L)

Figure 9.4   Born-Haber cycle for the formation of NaCl(s) from the elements. The calculation here uses enthalpy values, and the value obtained is the lattice enthalpy, $\Delta H_{\text{lattice}}$. The difference between $\Delta H_{\text{lattice}}$ and $\Delta E_{\text{lattice}}$ is generally not significant and can be corrected for, if desired.

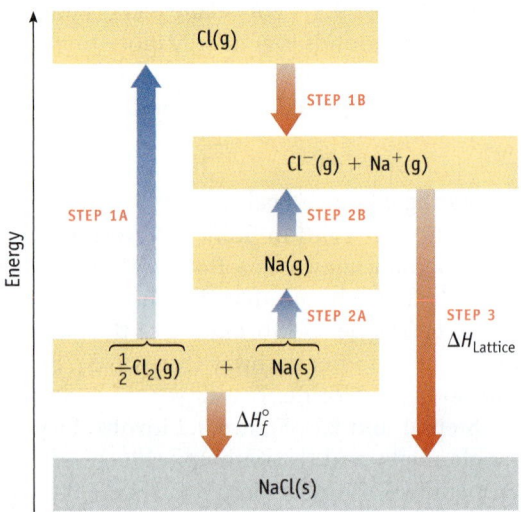

Step 2b. ΔH for $\mathrm{Na(g)} \longrightarrow \mathrm{Na^+(g)} + \mathrm{e^-} = +496$ kJ/mol (Appendix F)

Given that ΔH_f°, the standard heat of formation of NaCl(s) is -411.12 kJ/mol, we can calculate ΔH_3, which is the lattice enthalpy, $\Delta H_{\mathrm{lattice}}$.

Step 3.   Formation of NaCl(s) from the ions in the gas phase $= \Delta H_3 = \Delta H_{\mathrm{lattice}} = -787$ kJ/mol

---

GENERAL
Chemistry⚛Now™

See General ChemistryNOW CD-ROM or website:
- **Screen 9.4 Lattice Energy,** for an illustration of lattice and lattice energy

---

Exercise 9.2—Using Lattice Energies

Calculate the molar enthalpy of formation, ΔH_f°, of solid sodium iodide using the approach outlined in Figure 9.4. The required data can be found in Appendices F and L and in Table 9.3.

Why Don't Compounds Such as NaCl₂ and NaNe Exist?

The Born-Haber energy cycle in Figure 9.4 can help us answer some interesting questions. For example, why is it unlikely that a compound such as $NaCl_2$ will exist? Here the sodium is present as the Na^{2+} ion, and the formation of this ion (with the electron configuration $1s^2 2s^2 2p^5$) would require the loss of *two electrons* from a sodium atom. Because the second electron must be removed from the $n = 2$ shell, formation of Na^{2+} requires a substantial amount of energy. That is, the total energy for Step 2b in the Born-Haber cycle in Figure 9.4 would be the sum of the first *and* second ionization energies for Na (496 kJ + 4562 kJ). We can also make the reasonable assumption that the lattice energy for $NaCl_2$ (Step 3) is negative and has a value at least double that of NaCl (the increase reflects the facts that the cation charge has doubled and the size of Na^{2+} is less than that of Na^+). If we add up the energies for Steps 1 and 2, and then use a lattice energy in Step 3 of about 1500 kJ, we would estimate a very positive value for ΔH_f° of $NaCl_2$ (about +3400 kJ/mol). The formation of $NaCl_2$ from Na(s) and Cl_2(g) is quite unfavorable because the lattice energy is not enough to offset the high energy of formation of Na^{2+}.

Because sodium is a good reducing agent, why doesn't it reduce neon to Ne^- and form NaNe? Again, we can think about this question in terms of the Born-Haber cycle for NaCl. Put Ne in place of Cl_2. Because neon exists in the form of atoms, $\Delta H_{\mathrm{Step\ 1a}}$ is not required. More important to the outcome is the fact that neon's affinity for an electron should be extremely low, and $\Delta H_{\mathrm{Step\ 1b}}$ will be quite positive. (The additional electron has to be placed in the next higher electron shell, and the $n = 3$ shell is much higher in energy than the $n = 2$ shell.)

$$\mathrm{Ne}\ (1s^2 2s^2 2p^6) + \mathrm{e^-} \longrightarrow \mathrm{Ne^-}\ (1s^2 2s^2 2p^6 3s^1) \qquad \Delta H \gg 0$$

The lattice energy of NaNe is not expected to be exothermic enough to overcome this and other endothermic steps, so an overall positive enthalpy change is again expected. The formation of NaNe is energetically unfavorable.

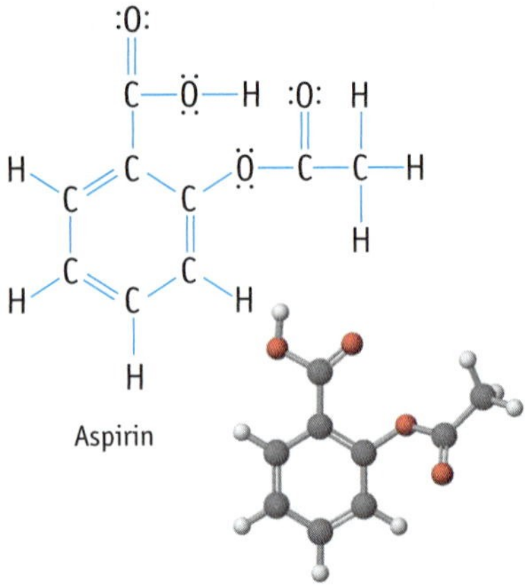

Aspirin

One goal in this chapter is to understand why a molecule such as aspirin has the shape that it exhibits.

9.4—Covalent Bonding and Lewis Structures

The remainder of this chapter is concerned with covalent bonding, in which electron pairs are shared between bonded atoms. Examples of compounds having covalent bonds include gases in our atmosphere (O_2, N_2, H_2O, and CO_2), common fuels (CH_4), and most of the compounds in your body. Covalent bonding is also responsible for the atom-to-atom connections in common ions such as CO_3^{2-}, CN^-, NH_4^+, NO_3^-, and PO_4^{3-}. We will develop the basic principles of structure and bonding using as examples molecules and ions made up of only a few atoms, but the same principles apply to larger molecules from aspirin to proteins and DNA with thousands of atoms.

The molecules and ions just mentioned are composed entirely of *nonmetal* atoms. A point that needs special emphasis is that, in molecules or ions made up *only* of nonmetal atoms, the atoms are attached by covalent bonds. Conversely, the presence of a metal in a formula is usually a signal that the compound is likely to be ionic.

Lewis Electron Dot Structures

In a simple description of covalent bonding, a bond results when one or more electron pairs are shared between two atoms. The electron-pair bond between the two atoms of an H_2 molecule is represented by a pair of dots or, alternatively, a line.

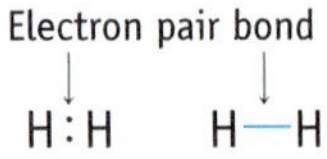

Electron pair bond

H:H   H—H

The representation of molecules such as H_2 in this fashion is called a **Lewis electron dot structure** or just a **Lewis structure** in honor of the American chemist Gilbert Newton Lewis (1875–1946) (page 415).

Simple Lewis structures can be drawn by starting with Lewis dot symbols for atoms and arranging the valence electrons to form bonds. To create the Lewis structure for F_2, for example, we could start with the Lewis dot symbol for a fluorine atom. Fluorine, an element in Group 7A, has seven valence electrons. The Lewis symbol shows that an F atom has a single unpaired electron along with three electron pairs. In F_2, the single electrons, one on each F atom, pair up in the covalent bond.

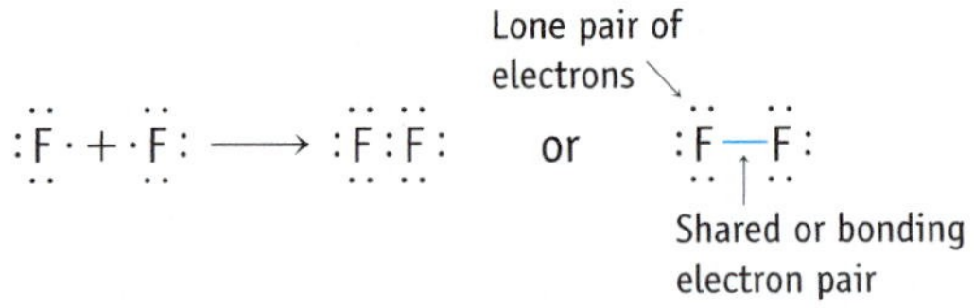

Lone pair of electrons

:F· + ·F: ⟶ :F:F: or   :F—F:

Shared or bonding electron pair

In the Lewis dot structure for F_2 the pair of electrons in the F—F bond is the bonding pair, or **bond pair**. The other six pairs reside on single atoms and are called **lone pairs**. Because they are not involved in bonding; they are also called **nonbonding electrons**.

Carbon dioxide, CO_2, and dinitrogen, N_2, are examples of molecules in which two atoms are multiply bonded, that is, they share more than one electron pair.

O=C=O   :N≡N:

In carbon dioxide, the carbon atom shares two pairs of electrons with each oxygen and so is linked to each O atom by a **double bond**. The valence shell of each oxygen atom in CO_2 has two bonding pairs and two lone pairs. In dinitrogen, the two nitrogen atoms share three pairs of electrons, so they are linked by a **triple bond**. In addition, each N atom has a single lone pair.

■ Importance of Lone Pairs
Lone pairs can be important in a structure. Being in the same valence electron shell as the bonding electrons, they can influence molecular shape. See Section 9.7.

The Octet Rule

An important observation can be made about the molecules you have seen so far: Each atom (except H) has a share in four pairs of electrons, so each has achieved a *noble gas configuration. Each atom is surrounded by an octet of eight electrons.* (Hydrogen typically forms a bond to only one other atom, resulting in two electrons in its valence shell.) *The tendency of molecules and polyatomic ions to have structures in which eight electrons surround each atom* is known as the **octet rule**. As an example, a triple bond is necessary in dinitrogen to have an octet around each nitrogen atom. The carbon atom and both oxygen atoms in CO_2 achieve the octet configuration by forming double bonds.

Octet of electrons around each O atom (four in double bond and four in lone pairs)

Octet of electrons around each N atom (six in triple bond and two in lone pair)

Octet of electrons around the C atom (four in each of two double bonds)

The octet rule is extremely useful, but keep in mind that it is more a *guideline* than a rule. It directs you to seek a Lewis structure in which each atom has eight electrons in its valence shell (or two in the case of hydrogen). Particularly for the second-period elements C, N, O, and F, a Lewis structure in which each atom achieves an octet is likely to be correct. Although some exceptions exist, if an atom such as C, N, O, or F in a Lewis structure does not follow the octet rule, you should probably doubt the structure's validity. If a structure obeying the octet rule cannot be written, then it is possible an incorrect formula has been assigned to the compound.

There is a systematic approach to constructing Lewis structures of molecules and ions. Let us take formaldehyde, CH_2O, as an example.

1. *Determine the arrangement of atoms within a molecule.* The central atom is *usually* the one with the least negative electron affinity. In CH_2O the central atom is C. You will come to recognize that certain elements often appear as the center atom, among them C, N, P, and S. Halogens are often terminal atoms forming a single bond to one other atom, but they can be the central atom when combined

■ Exceptions to the Octet Rule
Although the octet rule is widely applicable, exceptions to it do exist. Fortunately, many will be obvious, such as when there are more than four bonds to an element or when an odd number of electrons occurs.

■ Choosing the Central Atom
1. The relative electronegativities of atoms can also be used to choose the central atom. Electronegativity is discussed in Section 9.8.
2. For simple compounds, the first atom in a formula is often the central atom (e.g., SO_2, NH_4^+, NO_3^-). This is not always a reliable predictor, however. Notable exceptions include water (H_2O) and most common acids (HNO_3, H_2SO_4), in which the acidic hydrogen is usually written first but where N or S is the central atom.

with O in oxoacids (such as $HClO_4$). Oxygen is the central atom in water, but in conjunction with carbon, nitrogen, phosphorus, and the halogens it is usually a terminal atom. Hydrogen is a terminal atom because it typically bonds to only one other atom.

2. *Determine the total number of valence electrons in the molecule or ion.* In a neutral molecule this number will be the sum of the valence electrons for each atom. For an anion, *add* a number of electrons equal to the negative charge; for a cation, *subtract* the number of electrons equal to the positive charge. The number of valence electron pairs will be half the total number of valence electrons. For CH_2O,

$$\text{Valence electrons} = 12 \text{ electrons (or 6 electron pairs)}$$
$$= 4 \text{ for C} + (2 \times 1) \text{ for two H atoms} + 6 \text{ for O}$$

3. *Place one pair of electrons between each pair of bonded atoms to form a single bond.*

Here three electron pairs are used to make three single bonds (which are represented by single lines). Three pairs of electrons remain to be used.

4. *Use any remaining pairs as lone pairs around each terminal atom (except H) so that each terminal atom is surrounded by eight electrons.* If there are electrons left over after this step, assign them to the central atom. If the central atom is an element in the third or higher period, it can have more than eight electrons.

Here all six pairs have been assigned, but notice that the C atom has a share in only three pairs.

5. *If the central atom has fewer than eight electrons at this point, move one or more of the lone pairs on the terminal atoms into a position intermediate between the center and the terminal atom to form multiple bonds.*

As a general rule, double or triple bonds are formed *only* when both atoms are from the following list: C, N, O, or S. That is, bonds such as C=C, C=N, and C=O, and C=S will be encountered.

GENERAL
Chemistry ⚛ Now™

See General ChemistryNOW CD-ROM or website:
• **Screen 9.8 Drawing Lewis Structures,** for a tutorial and exercise on drawing Lewis structures

Example 9.2—Drawing Lewis Structures

Problem Draw Lewis structures for ammonia (NH_3), the hypochlorite ion (ClO^-), and the nitronium ion (NO_2^+).

Strategy Follow the five steps outlined for CH_2O in the text.

Solution for NH_3

1. *Decide on the central atom.* Hydrogen atoms are always terminal atoms, so nitrogen must be the central atom in the molecule.

2. *Count the number of valence electrons.* The total is eight (four valence pairs).

$$\text{Valence electrons} = 5 \text{ (for N)} + 3 \text{ (1 for each H)}$$

3. *Form single covalent bonds between each pair of atoms.* This uses three of the four pairs available.

$$H-N-H \atop \qquad | \atop \qquad H$$

4. *Place the remaining pair of electrons on the central atom.*

$$H-\overset{\cdot\cdot}{N}-H \atop \qquad | \atop \qquad H$$

Each H atom has a share in one pair of electrons as required, and the central N atom has achieved an octet configuration with four electron pairs. No additional steps are required; this is the correct Lewis structure.

Solution for ClO^- Ion

1. With two atoms, there is no "central" atom.

2. Valence electrons = 14 (7 valence pairs)

$$= 7 \text{ (for Cl)} + 6 \text{ (for O)} + 1 \text{ (for the negative charge on the ion)}$$

3. One electron pair is used in the Cl — O bond: Cl — O.

4. Distribute the six remaining electron pairs around the "terminal" atoms.

$$\left[:\overset{\cdot\cdot}{\underset{\cdot\cdot}{Cl}} - \overset{\cdot\cdot}{\underset{\cdot\cdot}{O}}: \right]^-$$

5. As no electrons remain to be assigned and both atoms have an octet of electrons, this is the correct Lewis structure.

Solution for NO_2^+ Ion

1. Nitrogen is the center atom, because its electron affinity is less negative than that of oxygen.

2. Valence electrons = 16 (8 valence pairs)

$$= 5 \text{ (for N)} + 12 \text{ (six for each O)} - 1 \text{ (for the positive charge)}$$

3. Two electron pairs form the single bonds from the nitrogen to each oxygen:

$$O-N-O$$

4. Distribute the remaining six pairs of electrons on the terminal O atoms:

$$\left[:\overset{\cdot\cdot}{\underset{\cdot\cdot}{O}} - N - \overset{\cdot\cdot}{\underset{\cdot\cdot}{O}}: \right]^+$$

5. The central nitrogen atom is two electron pairs short of an octet. Thus, a lone pair of electrons on each oxygen atom is converted to a bonding electron pair to give two $N = O$ double bonds. Each atom in the ion now has four electron pairs. Nitrogen has four bonding pairs, and each oxygen atom has two lone pairs and shares two bond pairs.

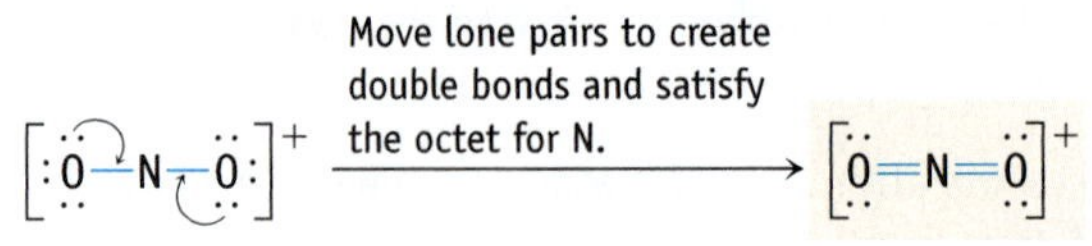

Move lone pairs to create double bonds and satisfy the octet for N.

Exercise 9.3—Drawing Lewis Structures

Draw Lewis structures for NH_4^+, CO, NO^+, and SO_4^{2-}.

Predicting Lewis Structures

Lewis structures are useful in gaining a perspective on the structure and chemistry of a molecule or ion. The guidelines for drawing Lewis structures are helpful, but chemists also rely on patterns of bonding in related molecules.

Hydrogen Compounds

Some common compounds and ions formed from second-period nonmetal elements and hydrogen are shown in Table 9.4. Their Lewis structures illustrate the fact that the Lewis symbol for an element is a useful guide in determining the number of bonds formed by that element. For example, if there is no charge, nitrogen has five valence electrons. Two electrons occur as a lone pair; the other three occur as unpaired electrons. To achieve an octet, it is necessary to pair each of the unpaired electrons with an electron from another atom. Thus, N is predicted to form three bonds in uncharged molecules, which is indeed the case. Similarly, carbon is expected to form four bonds, oxygen two, and fluorine one.

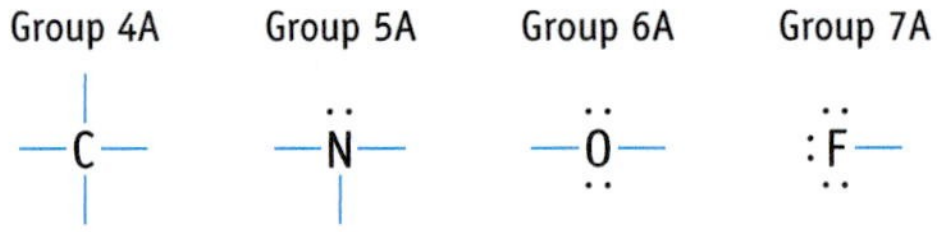

Hydrocarbons are compounds formed from carbon and hydrogen, and the first two members of the series called the *alkanes* are CH_4 and C_2H_6 (see Table 9.4). What is the Lewis structure of the third member of the series, propane (C_3H_8)? We can rely on the idea that the atoms in this species all bond in predictable ways. Carbon is expected to form four bonds, and hydrogen can bond to only one other atom. The only arrangement of atoms that meets these criteria has three atoms of carbon linked together by carbon–carbon single bonds. The remaining positions around the carbon atoms are filled in with hydrogen—three hydrogen atoms on the end carbons and two on the middle carbon:

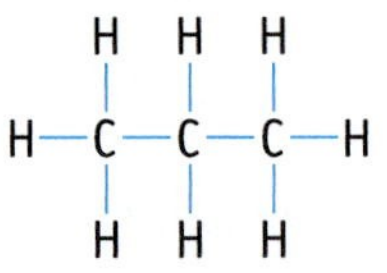

propane, C_3H_8

Table 9.4 Common Hydrogen-Containing Compounds and Ions of the Second-Period Elements

Group 4A		Group 5A		Group 6A		Group 7A	
CH_4 methane	H | H—C—H | H	NH_3 ammonia	H—N̈—H | H	H_2O water	H—Ö—H	HF hydrogen fluoride	H—F̈:
C_2H_6 ethane	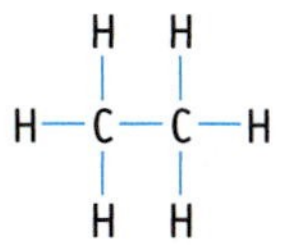 H H | | H—C—C—H | | H H	N_2H_4 hydrazine	H—N̈—N̈—H | | H H	H_2O_2 hydrogen peroxide	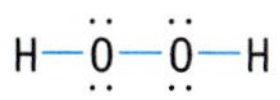 H—Ö—Ö—H		
C_2H_4 ethylene	H—C=C—H | | H H	NH_4^+ ammonium ion	$\left[\begin{array}{c} H \\ \| \\ H—N—H \\ \| \\ H \end{array}\right]^+$	H_3O^+ hydronium ion	$\left[\begin{array}{c} \ddot{} \\ H—O—H \\ \| \\ H \end{array}\right]^+$		
C_2H_2 acetylene	 H—C≡C—H	NH_2^- amide ion	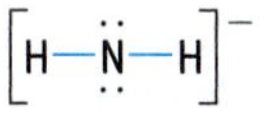 $\left[H—\ddot{N}—H \right]^-$	OH^- hydroxide ion	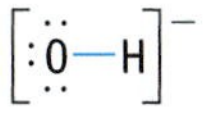 $\left[:\ddot{O}—H \right]^-$		

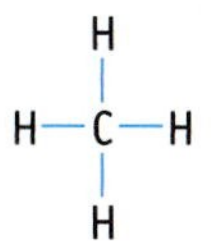

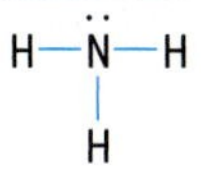

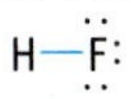

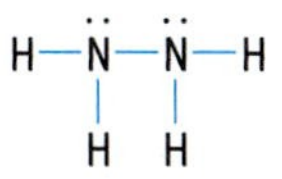

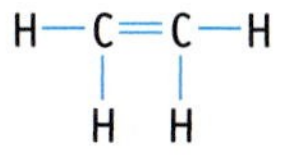

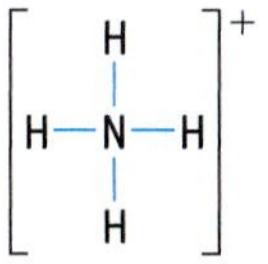

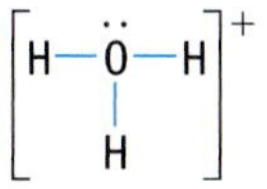

Example 9.3—Predicting Lewis Structures

Problem Draw Lewis electron dot structures for CCl_4 and NF_3.

Strategy One way to solve this problem is to recognize that CCl_4 and NF_3 are similar to CH_4 and NH_3, respectively, except that H atoms have been replaced by halogen atoms.

Solution Recall that carbon is expected to form four bonds and nitrogen three bonds to give an octet of electrons. In addition, halogen atoms have seven valence electrons, so both Cl and F can attain an octet by forming one covalent bond, just as hydrogen does.

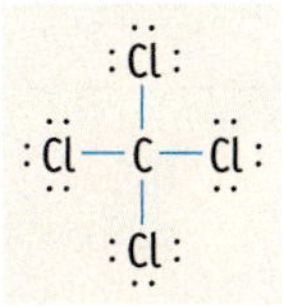

$$
\begin{array}{cc}
:\ddot{C}l: & \\
\| & \\
:\ddot{C}l—C—\ddot{C}l: & :\ddot{F}—\ddot{N}—\ddot{F}: \\
\| & \| \\
:\ddot{C}l: & :\ddot{F}:
\end{array}
$$

carbon tetrachloride 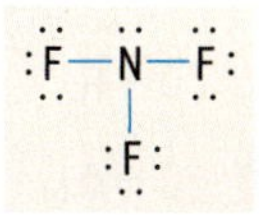   nitrogen trifluoride

As a check, count the number of valence electrons for each molecule and verify that all are present.

CCl_4: Valence electrons = 4 for C + 4 × 7 (for Cl) = 32 electrons (16 pairs)

The structure shows 8 electrons in single bonds and 24 electrons as lone-pair electrons, for a total of 32 electrons. The structure is correct.

NF_3: Valence electrons = 5 for N + 3 × 7 (for F) = 26 electrons (13 pairs)

The structure shows 6 electrons in single bonds and 20 electrons as lone-pair electrons, for a total of 26 electrons. The structure is correct.

Exercise 9.4—Predicting Lewis Structures

Predict Lewis structures for methanol, CH_3OH and hydroxylamine, NH_2OH. (*Hint:* The formulas of these compounds are written to guide you in choosing the correct arrangement of atoms.)

Problem-Solving Tip 9.1

Useful Ideas to Consider When Drawing Lewis Electron Dot Structures

- The octet rule is a useful guideline when drawing Lewis structures.
- Carbon can form four bonds (4 single bonds; 2 double bonds; 2 single bonds and 1 double bond; or one single bond

and 1 triple bond). In uncharged species, nitrogen forms three bonds and oxygen forms two bonds. Hydrogen typically forms only one bond to another atom.

- When multiple bonds are formed, both of the atoms involved are usually one of the following: C, N, O and S. Oxygen has the ability to form multiple bonds with a variety of elements. Carbon forms many compounds having multiple bonds to another carbon or to N or O.

- Nonmetals may form single, double, and triple bonds but never quadruple bonds.
- Always account for single bonds and lone pairs before determining whether multiple bonds are present.
- Be alert for the possibility that the molecule or ion you are considering is isoelectronic (page 389) with a species you have seen before.

Oxoacids and Their Anions

Lewis structures of common acids and their anions are illustrated in Table 9.5. In the absence of water, these acids are covalently bonded molecular compounds, a conclusion that we should draw because all elements in the formula are nonmetals. (Nitric acid, for example, has properties that we associate with a covalent molecule: It is a colorless liquid with a boiling point of 83 °C.) In aqueous solution, however, HNO_3, H_2SO_4, and $HClO_4$ are ionized to give a hydrogen ion and the appropriate anion. A Lewis structure for the nitrate ion, for example, can be created using the guidelines on page 383, and the result is a structure with two N—O single bonds and one N=O double bond. To form nitric acid from the nitrate ion, a hydrogen ion is attached to one of the O atoms that has a single bond to the central N.

Table 9.5　Lewis Structures of Common Oxoacids and Their Anions

$$\left[\ddot{O}-N\!\!=\!\!\ddot{O}\,\Big|\; \begin{array}{c}\\ \ddot{O}\\ \end{array}\right]^{-} \quad \xrightarrow[-H^{+}]{+H^{+}} \quad H-\ddot{O}-N\!\!=\!\!\ddot{O}\;\begin{array}{c}\\ \ddot{O}\\ \end{array}$$

nitrate ion   nitric acid

A characteristic property of acids in aqueous solution is their ability to donate a hydrogen ion (H^+). The NO_3^- anion is formed when the acid, HNO_3, loses a hydrogen ion. The H^+ ion separates from the acid by breaking the $H—O$ bond, the electrons of the bond staying with the O atom. As a result, HNO_3 and NO_3^- have the same number of electrons, 24, and their structures are closely related.

Exercise 9.5—Lewis Structures of Acids and Their Anions

Draw a Lewis structure for the anion $H_2PO_4^-$, which is derived from phosphoric acid.

Isoelectronic Species

In what way are NO^+, N_2, CO, and CN^- similar? Most important, all of them have two atoms and the same total number of valence electrons, 10, which leads to the same Lewis structure for each molecule or ion. The two atoms in each are linked with a triple bond. With three bonding pairs and one lone pair, each atom thus has an octet of electrons.

$$\left[:\!N\!\equiv\!O:\right]^{+} \qquad :N\!\equiv\!N: \qquad :C\!\equiv\!O: \qquad \left[:\!C\!\equiv\!N:\right]^{-}$$

Molecules and ions having the *same number of valence electrons and the same Lewis structures* are said to be **isoelectronic** (Table 9.6). You will find it helpful to think in terms of isoelectronic molecules and ions because this perspective offers another way to see relationships in bonding among common chemical substances.

There are both similarities and important differences in chemical properties of isoelectronic species. For example, both carbon monoxide, CO, and cyanide ion, CN^-, are very toxic, which results from the fact that they can bind to the iron of hemoglobin in blood and block the uptake of oxygen. They differ, though, in terms of their acid–base chemistry. In aqueous solution, cyanide ion readily adds H^+ to form hydrogen cyanide, whereas CO does not. The isoelectronic species Cl_2 and OCl^- provide a similar example. Attachment of H^+ to OCl^- forms hypochlorous acid, HOCl. In contrast, Cl_2 does not add a proton.

■ **Isoelectronic and Isostructural**
The term **isostructural** is often used in conjunction with isoelectronic species. Species that are isostructural have the same structure. For example, the PO_4^{3-}, SO_4^{2-}, and ClO_4^- ions in Table 9.6 all have four oxygens bonded to the central atom. In addition, they are isoelectronic in that all have 32 valence electrons.

Table 9.6   Some Common Isoelectronic Molecules and Ions

Formulas	Representative Lewis Structure	Formulas	Representative Lewis Structure
BH_4^-, CH_4, NH_4^+	$\left[\begin{array}{c} H \\ H-N-H \\ H \end{array}\right]^{+}$	CO_3^{2-}, NO_3^-	$\left[\ddot{O}-N\!\!=\!\!\ddot{O}\;\ddot{O}\right]^{-}$
NH_3, H_3O^+	$H-\ddot{N}-H$ over H	PO_4^{3-}, SO_4^{2-}, ClO_4^-	$\left[\ddot{O}-\overset{\ddot{O}}{P}-\ddot{O}\;\ddot{O}\right]^{3-}$
CO_2, OCN^-, SCN^-, N_2O NO_2^+, OCS, CS_2	$\ddot{O}\!\!=\!\!C\!\!=\!\!\ddot{O}$		

Exercise 9.6—Identifying Isoelectronic Species

(a) Is the acetylide ion, C_2^{2-}, isoelectronic with N_2?

(b) Identify a common molecular (uncharged) species that is isoelectronic with nitrite ion, NO_2^-. Identify a common ion that is isoelectronic with HF.

9.5—Resonance

Ozone, O_3, an unstable, blue, diamagnetic gas with a characteristic pungent odor, protects the earth and its inhabitants from intense ultraviolet radiation from the sun. An important feature of its structure is that the two oxygen–oxygen bonds are the same length. This suggests that the two oxygen–oxygen bonds are equivalent. That is, equal O—O bond lengths imply an equal number of bond pairs in each O—O bond. Using the guidelines for drawing Lewis structures, however, you might come to a different conclusion. There are two possible ways of writing the Lewis structure for the molecule:

Alternative ways of creating the Lewis structure of ozone

Double bond on the right: :O—O—O: ⟶ :O—O=O:

Double bond on the left: :O—O—O: ⟶ :O=O—O:

These structures are equivalent in that each has a double bond on one side of the central oxygen atom and a single bond on the other side. If either were the actual structure of ozone, one bond (O=O) should be shorter than the other (O—O). The actual structure of ozone shows this is not the case. The inescapable conclusion is that these Lewis structures do not correctly represent the bonding in ozone.

Linus Pauling proposed the theory of **resonance** to reconcile this discrepancy. *Resonance structures are used to represent bonding in a molecule or ion when a single Lewis structure fails to describe accurately the actual electronic structure.* The alternative structures shown for ozone are called **resonance structures**. They have identical patterns of bonding and equal energy. The actual structure of this molecule is a *composite*, or **resonance hybrid**, of the equivalent resonance structures. This conclusion is reasonable because we see that the O—O bonds both have a length of 127.8 pm, intermediate between the average length of an O=O double bond (121 pm) and an O—O single bond (132 pm).

Benzene is the classic example of the use of resonance to represent a structure. The benzene molecule is a six-member ring of carbon atoms with six equivalent carbon–carbon bonds (and a hydrogen atom attached to each carbon atom). The carbon–carbon bonds are 139 pm long, intermediate between the average length of a C=C double bond (134 pm) and a C—C single bond (154 pm).

■ **Linus Pauling (1901–1994)**

Linus Pauling was born in Portland, Oregon, earned a B.S. degree in chemical engineering from Oregon State College in 1922, and completed his Ph.D. in chemistry at the California Institute of Technology in 1925. In chemistry he is widely known for his book *The Nature of the Chemical Bond*. For more on Pauling, see page 436.

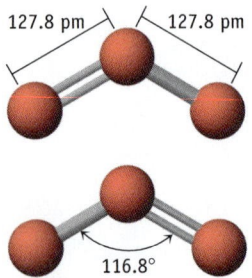

Ozone, O_3. This bent molecule has oxygen–oxygen bonds of the same length.

Resonance structures of benzene, C_6H_6 Abbreviated representation of resonance structures

Problem-Solving Tip 9.2

Resonance Structures

- Resonance is a means of representing the bonding when a single Lewis structure fails to give an accurate picture.

- The atoms must have the same structural arrangement in each resonance structure. Attaching the atoms in a different fashion creates a different compound.

- Resonance structures differ only in the assignment of electron-pair positions, never in their atom positions.

- Resonance structures differ in the number of bond pairs between a given pair of atoms.

- Even though the formal process of converting one resonance structure to another seems to move electrons about, resonance is not meant to indicate the motion of electrons.

- The actual structure of a molecule is a composite or hybrid of the resonance structures.

- There will always be at least one multiple bond (double or triple) in each resonance structure.

Two resonance structures can be written for the molecule that differ only in double bond placement. A hybrid of these two structures, however, will lead to a molecule with six equivalent carbon–carbon bonds.

Let us apply the concept of resonance to describe bonding in the carbonate ion, CO_3^{2-}. This anion has 24 valence electrons (12 pairs).

Three equivalent structures can be drawn for this ion, differing only in the location of the $C\!=\!O$ double bond. This fits the classical situation for resonance, so it is appropriate to conclude that no single structure correctly describes this ion. Instead, the actual structure is a hybrid of the three structures, in good agreement with experimental results. In the CO_3^{2-} ion, all three carbon–oxygen bond distances are 129 pm, intermediate between the $C\!-\!O$ single bond (143 pm) and $C\!=\!O$ double bond (122 pm) distances.

In aqueous solution, a hydrogen ion can be attached to the carbonate ion to give the hydrogen carbonate, or bicarbonate, ion. This ion can be described as a resonance hybrid of two Lewis structures.

■ **Depicting Resonance Structures**
The use of an arrow (↔) as a symbol to link resonance structures and the name "resonance" are somewhat unfortunate. An arrow might seem to imply that a change is occurring, and the term *resonance* has the connotation of vibrating or alternating back and forth between different forms. Neither view is correct. Resonance is simply a way of representing a structure. Electron pairs are not actually moving from one place to another.

A Closer Look

Resonance Structures, Lewis Structures, and Molecular Models

When drawing structures of molecules or ions that have resonance structures, or when illustrating their structures with computer-based molecular models, we generally show only one resonance structure. Thus, a model of benzene, C_6H_6, would have alternating double bonds. This is one of the two possible resonance structures. A model of the nitrate ion would have one double bond and two single bonds in each of the three possible resonance structures.

$$\begin{bmatrix} \ddot{O}=C-\ddot{O}: \\ | \\ :\ddot{O}-H \end{bmatrix}^{-} \longleftrightarrow \begin{bmatrix} :\ddot{O}-C=\ddot{O}: \\ | \\ :\ddot{O}-H \end{bmatrix}^{-}$$

Finally, notice that in each of the examples of resonance structures, the structures have been "linked" by a double-headed arrow ($\longleftrightarrow$). This convention is followed throughout chemistry.

GENERAL
Chemistry·Now™

See General ChemistryNOW CD-ROM or website:
• **Screen 9.9 Resonance Structures**, for a tutorial on drawing resonance structures

Example 9.4—Drawing Resonance Structures

Problem Draw resonance structures for the nitrite ion, NO_2^-. Are the N—O bonds single, double, or intermediate in value?

Strategy Draw the dot structure in the usual manner. If multiple bonds are required, resonance structures may exist. This will be the case if the octet of an atom can be completed by using an electron pair from more than one terminal atom to form a multiple bond. Bonds to the central atom cannot then be "pure" single or double bonds but rather are somewhere between the two.

Solution Nitrogen is the center atom in the nitrite ion, which has a total of 18 valence electrons (9 pairs).

Valence electrons = 5 (for the N atom) + 12 (6 for each O atom) + 1 (for negative charge)

After forming N—O single bonds, and distributing lone pairs on the terminal O atoms, a pair remains, which is placed on the central N atom.

$$\begin{bmatrix} :\ddot{O}-\ddot{N}-\ddot{O}: \end{bmatrix}^{-}$$

To complete the octet of electrons about the N atom, form an N═O double bond.

$$\begin{bmatrix} :O=\ddot{N}-\ddot{O}: \end{bmatrix}^{-} \longleftrightarrow \begin{bmatrix} :\ddot{O}-\ddot{N}=O: \end{bmatrix}^{-}$$

Because there are two ways to do this, two equivalent structures can be drawn, and the actual structure must be a resonance hybrid of these two structures. The nitrogen–oxygen bonds are neither single nor double bonds, but rather have an intermediate value.

Exercise 9.7—Drawing Resonance Structures

Draw resonance structures for the nitrate ion, NO_3^-. Sketch a plausible Lewis dot structure for nitric acid, HNO_3.

9.6—Exceptions to the Octet Rule

Although the vast majority of molecular compounds and ions obey the octet rule, there are exceptions. These include molecules and ions that have fewer than four pairs of electrons on a central atom, those that have more than four pairs, and those that have an odd number of electrons.

Compounds in Which an Atom Has Fewer Than Eight Valence Electrons

Boron, a nonmetal in Group 3A, has three valence electrons and so is expected to form three covalent bonds with other nonmetallic elements. This behavior results in a valence shell for boron in its compounds with only six electrons, two short of an octet. Many boron compounds of this type are known, including such common compounds as boric acid [$B(OH)_3$], borax [$Na_2B_4O_5(OH)_4 \cdot 8 H_2O$] (Figure 9.5), and the boron trihalides (BF_3, BCl_3, BBr_3, and BI_3).

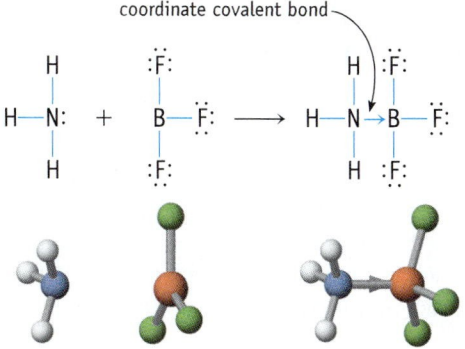

boron trifluoride   boric acid

Boron compounds such as BF_3 that are two electrons short of an octet can be quite reactive. The boron atom can accommodate a fourth electron pair when that pair is provided by another atom. In general, molecules or ions with lone pairs can fulfill this role. Ammonia, for example, reacts with BF_3 to form $H_3N \rightarrow BF_3$. The bond between the B and N atoms in this compound uses an electron pair that originated on the N atom. The reaction of an F^- ion with BF_3 to form BF_4^- is another example.

coordinate covalent bond

If a bonding pair of electrons originates on one of the bonded atoms, the bond is called a **coordinate covalent bond**. In Lewis structures, a coordinate covalent bond is often designated by an arrow that points away from the atom donating the electron pair.

Compounds in Which an Atom Has More Than Eight Valence Electrons

Elements in the third and higher periods often form compounds and ions in which the central element is surrounded by more than four valence electron pairs (Table 9.7). With most compounds and ions in this category, the central atom is bonded to fluorine, chlorine, or oxygen.

It is often obvious from the formula of a compound that an octet around an atom has been exceeded. As an example, consider sulfur hexafluoride, SF_6, a gas formed by the reaction of sulfur and excess fluorine. Sulfur is the central atom in this compound, and fluorine typically bonds to only one other atom with a single electron-pair bond (as in HF and CF_4). Six S — F bonds are required in SF_6, meaning there will be six electron pairs in the valence shell of the sulfur atom.

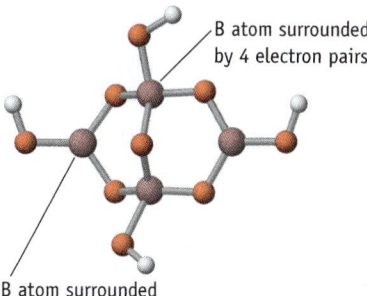

B atom surrounded by 4 electron pairs

B atom surrounded by 3 electron pairs

Figure 9.5 **Borax.** This common material, which is used in soaps, contains an interesting anion, $B_4O_5(OH)_4^{2-}$. This boron–oxygen ion has two B atoms surrounded by four electron pairs, and two B atoms surrounded by only three electron pairs.

Table 9.7   Lewis Structures in Which the Central Atom Exceeds an Octet

Group 4A	Group 5A	Group 6A	Group 7A	Group 8
SiF_5^-	PF_5	SF_4	ClF_3	XeF_2
SiF_6^{2-}	PF_6^-	SF_6	BrF_5	XeF_4

More than four groups bonded to a central atom is a reliable signal that there are more than eight electrons around a central atom. But be careful—the central atom octet can also be exceeded with four or fewer atoms bonded to the central atom. Consider three examples from Table 9.7: The central atom in SF_4, ClF_3, and XeF_2 has five electron pairs in its valence shell.

A useful observation is that *only elements of the third and higher periods in the periodic table form compounds and ions in which an octet is exceeded.* Second-period elements (B, C, N, O, and F) are restricted to a maximum of eight electrons in their compounds. For example, nitrogen forms compounds and ions such as NH_3, NH_4^+, and NF_3, but NF_5 is unknown. Phosphorus, the third-period element just below nitrogen in the periodic table, forms many compounds similar to nitrogen (PH_3, PH_4^+, PF_3), but it also readily accommodates five or six valence electron pairs in compounds such as PF_5 or in ions such as PF_6^-. Arsenic, antimony, and bismuth—the elements below phosphorus in Group 5A—resemble phosphorus in their behavior.

The usual explanation for the contrasting behavior of second- and third-period elements centers on the number of orbitals in the valence shell of an atom. Second-period elements have four valence orbitals (one $2s$ and three $2p$ orbitals). Two electrons per orbital result in a total of eight electrons being accommodated around an atom. For elements in the third and higher periods, the d orbitals in the outer shell are traditionally included among valence orbitals for the elements. Thus, for phosphorus, the $3d$ orbitals are included with the $3s$ and $3p$ orbitals as valence orbitals. The extra orbitals provide the element with an opportunity to accommodate up to 12 electrons.

See General ChemistryNow CD-ROM or website:
• **Screen 9.10 Exceptions to the Octet Rule,** for a tutorial on identifying compounds that do not follow the octet rule

Example 9.5—Lewis Structures in Which the Central Atom Has More Than Eight Electrons

Problem Sketch the Lewis structure of the $[ClF_4]^-$ ion.

Strategy Use the guidelines on page 384.

Solution

1. The Cl atom is the central atom.

2. This ion has 36 valence electrons [= 7 for Cl + 4 × (7 for F) + 1 for ion charge] or 18 pairs.

3. Draw the ion with four single covalent Cl—F bonds.

$$\left[\begin{array}{c} F \\ | \\ F-Cl-F \\ | \\ F \end{array} \right]^-$$

4. Place lone pairs on the terminal atoms. Because two electron pairs remain after placing lone pairs on the four F atoms, and because we know that Cl can accommodate more than four pairs, these two pairs are placed on the central Cl atom.

The last two electron pairs are added to the central Cl atom.

Exercise 9.8—Lewis Structures in Which the Central Atom Has More Than Eight Electrons

Sketch the Lewis structures for $[ClF_2]^+$ and $[ClF_2]^-$. How many lone pairs and bond pairs surround the Cl atom in each ion?

Molecules with an Odd Number of Electrons

Two nitrogen oxides—NO, with 11 valence electrons, and NO_2, with 17 valence electrons—are among a very small group of stable molecules with an odd number of electrons. Because these compounds have an odd number of electrons, it is impossible to draw a structure obeying the octet rule; at least one electron must be unpaired.

Even though NO_2 does not obey the octet rule, an electron dot structure can be written that approximates the bonding in the molecule. This Lewis structure places the unpaired electron on nitrogen. Two resonance structures show that the nitrogen–oxygen bonds are equivalent, as observed experimentally.

Experimental evidence for NO indicates that the bonding between N and O is intermediate between a double bond and a triple bond. It is not possible to write a

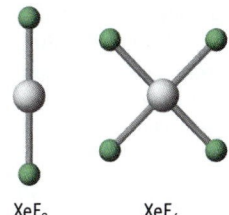

■ **Xenon Compounds**
Compounds of xenon are among the more interesting entries in Table 9.7. Noble gas compounds were not discovered until the early 1960s. One of the more intriguing compounds is XeF_2, in part because of the simplicity of its synthesis. Xenon difluoride can be made by placing a flask containing xenon gas and fluorine gas in sunlight. After several weeks, crystals of colorless XeF_2 are found in the flask.

XeF_2 XeF_4

Chemical Perspectives

The Importance of Odd-Electron Molecules

Small molecules such as H_2, O_2, H_2O, CO, and CO_2 are among the most important molecules commercially, environmentally, and biologically. Imagine the surprise of chemists and biologists when it was discovered a few years ago that nitrogen monoxide (nitric oxide, NO), which was widely considered to be toxic, plays an important biological role.

Nitric oxide is a colorless, paramagnetic gas that is moderately soluble in water. In the laboratory, it can be synthesized by the reduction of nitrite ion with iodide ion:

$$KNO_2(aq) + KI(aq) + H_2SO_4(aq) \longrightarrow$$
$$NO(g) + K_2SO_4(aq) + H_2O(\ell) + \tfrac{1}{2} I_2(aq)$$

The formation of NO from the elements was an unfavorable, energetically uphill reaction ($\Delta H_f^\circ = 90.2$ kJ/mol). Nevertheless, small quantities of this compound form from nitrogen and oxygen at high temperatures. For example, conditions in an internal combustion engine favor this reaction.

Nitric oxide reacts rapidly with O_2 to form the reddish brown gas NO_2.

$$2\ NO(\text{colorless, g}) + O_2(g) \longrightarrow$$
$$2\ NO_2(\text{brown, g})$$

The result is that NO (and compounds such as NO_2 and HNO_3 arising from reactions of NO with O_2 and H_2O) are among the air pollutants produced by automobiles.

A few years ago chemists learned that NO is synthesized in a biological process by animals as diverse as barnacles, fruit flies, horseshoe crabs, chickens, trout, and humans. Even more recently they have found that NO is important in an astonishing range of physiological processes in humans and other animals. For example, it has a role in neurotransmission, blood clotting, and blood pressure control as well as in the immune system's ability to kill tumor cells and intracellular parasites.

On September 11, 2001, the United States was struck by terrorists. These attacks were followed in October when an unknown person or persons mailed letters containing anthrax to various people, including the leaders of the U.S. Senate. No one in the Senate was taken ill, but the building had to be decontaminated.

A hazardous materials worker is sprayed down on Capitol Hill on October 24, 2001, as buildings are checked for anthrax contamination.

One way to kill anthrax spores is to fumigate the contaminated area with chlorine dioxide, ClO_2. Chlorine dioxide, an odd-electron molecule with 19 valence electrons, was the first oxide of chlorine discovered. It was prepared by Humphry Davy in 1811. It is now made in several ways, all involving the reduction of sodium chlorate, $NaClO_3$. The compound is very reactive. Despite this tendency, thousands of tons are made annually, primarily for water-treatment and bleaching wood pulp to make paper.

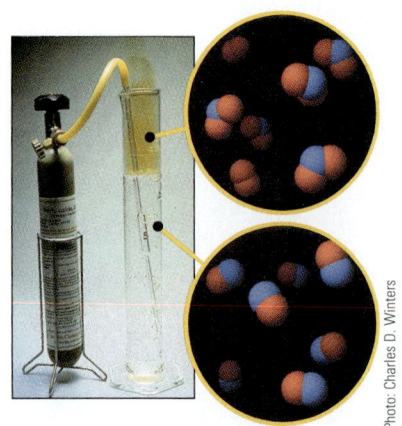

The colorless gas NO is bubbled into water from a high-pressure tank. When the gas emerges into the air, the NO reacts rapidly with O_2 to give brown NO_2 gas.

Lewis structure for NO that is in accord with the properties of this substance, so a different theory is needed to understand bonding in this molecule. We shall return to compounds of this type when molecular orbital theory is introduced in Chapter 10.

The two nitrogen oxides, NO and NO_2, are members of a class of chemical substances called free radicals. **Free radicals** are chemical species—both atomic and molecular—with an unpaired electron. Free radicals are generally quite reactive. Free atoms such as H and Cl, for example, readily combine with other atoms to give molecules such as H_2, Cl_2, and HCl.

Free radicals are involved in many reactions in the environment. For example, small amounts of NO are released from vehicle exhausts. The NO rapidly forms NO_2, which decomposes in the presence of sunlight and oxygen to give more NO as well as ozone, O_3, an air pollutant that affects the respiratory system.

$$NO_2(g) + O_2(g) \longrightarrow NO(g) + O_3(g)$$

The two nitrogen oxides, NO and NO_2, are unique in that they can be isolated, and neither has the extreme reactivity of most free radicals. When cooled, however, two NO_2 molecules join or "dimerize" to form colorless N_2O_4; the unpaired electrons combine to form an N—N bond in N_2O_4. Even though this bond is weak, the reaction is easily observed in the laboratory (Figure 9.6).

A flask of brown NO_2 gas in warm water

When cooled, NO_2 free radicals couple to form N_2O_4 molecules.

N_2O_4 gas is colorless.

→

A flask of NO_2 gas in ice water

Figure 9.6 **Free radical chemistry.** When cooled, the brown gas NO_2, a free radical, forms colorless N_2O_4, a molecule with an N — N single bond. The coupling of two free radicals is a common type of chemical reactivity. Because two identical free radicals come together, the product is called a dimer, and the process is called a dimerization. (*See the General ChemistryNow Screen 9.11 Free Radicals, to watch a video of this process.*)

9.7—Molecular Shapes

One reason for drawing Lewis electron dot structures is to be able to predict the three-dimensional geometry of molecules and ions. Because the physical and chemical properties of compounds are tied to their structures, the importance of this subject cannot be overstated.

The **valence shell electron-pair repulsion (VSEPR)** model provides a reliable method for predicting the shapes of covalent molecules and polyatomic ions. The VSEPR model is based on the idea that *bond and lone electron pairs in the valence shell of an element repel each other and seek to be as far apart as possible.* The positions assumed by the valence electrons of an atom thus define the angles between bonds to surrounding atoms. The VSEPR theory is remarkably successful in predicting structures of molecules and ions of main group elements. However, it is less effective (and seldom used) to predict structures of compounds containing transition metals.

To get a sense of how valence shell electron pairs repel one another and determine structure, blow up several balloons to a similar size. Imagine that each balloon represents an electron cloud. A repulsive force prevents other balloons from occupying the same space. When two, three, four, five, or six balloons are tied together at a central point (representing the nucleus and core electrons of a central atom), the balloons naturally form the shapes shown in Figure 9.7. These geometric arrangements minimize interactions between the balloons.

■ **VSEPR Theory**
The VSEPR theory was devised by Ronald J. Gillespie (1924–) and Ronald S. Nyholm (1917–1971).

GENERAL
Chemistry Now™

See General ChemistryNow CD-ROM or website:

- **Screen 9.13 Ideal Electron Repulsion Shapes,** to view an animation of the electron-pair geometries and a tutorial on identifying geometries
- **Screen 9.14 Determining Molecular Shape,** for an exercise and a tutorial on predicting molecular geometry

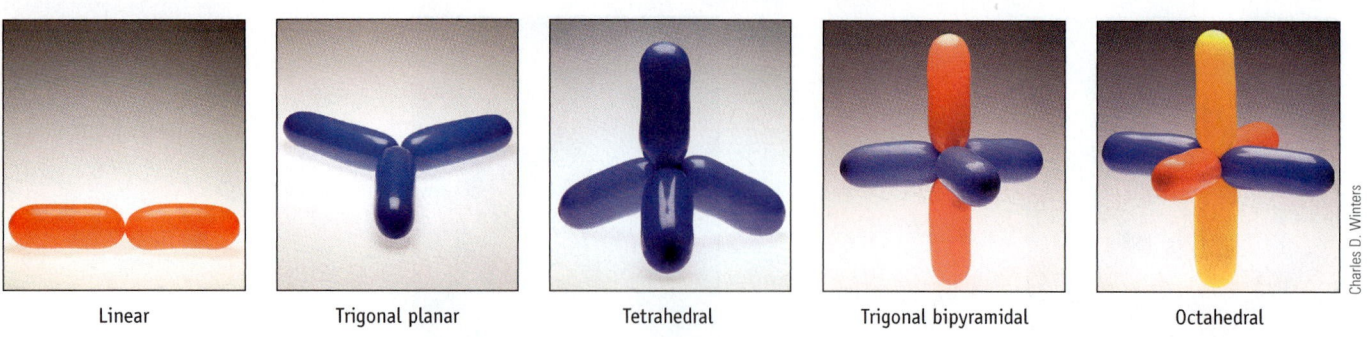

| Linear | Trigonal planar | Tetrahedral | Trigonal bipyramidal | Octahedral |

Charles D. Winters

Figure 9.7 **Balloon models of electron-pair geometries for two to six electron-pairs.** If two to six balloons of similar size and shape are tied together, they will naturally assume the arrangements shown. These pictures illustrate the predictions of the VSEPR.

Central Atoms Surrounded Only by Single-Bond Pairs

The simplest application of VSEPR theory is to molecules and ions in which all of the electron pairs around the central atom are involved in single covalent bonds. Figure 9.8 illustrates the geometries predicted for molecules or ions with the general formula AX_n, where A is the central atom and n is the number of X groups bonded to it.

The **linear** geometry for two bond pairs and the **trigonal-planar** geometry for three bond pairs involve a central atom that does not have an octet of electrons (see Section 9.6). The central atom in a **tetrahedral** molecule obeys the octet rule with four bond pairs. The central atoms in **trigonal-bipyramidal** and **octahedral** molecules have five and six bonding pairs, respectively, and are expected only when the central atom is an element in Period 3 or higher of the periodic table [▶ page 401].

Example 9.6—Predicting Molecular Shapes

Problem Predict the shape of silicon tetrachloride, $SiCl_4$.

Strategy The first step is to draw the Lewis structure. The Lewis structure does not need to be drawn in any particular way because its purpose is merely to describe the number of bonds around an atom and to indicate whether there are any lone pairs. The number of bond and lone pairs of electrons around the central atom determines the molecular shape (Figure 9.8).

Solution The Lewis structure of $SiCl_4$ has four electron pairs, all bond pairs, around the central Si atom. Therefore, a tetrahedral structure is predicted for the $SiCl_4$ molecule, with Cl—Si—Cl bond angles of 109.5°. This agrees with the actual structure for $SiCl_4$.

Lewis structure   Molecular geometry

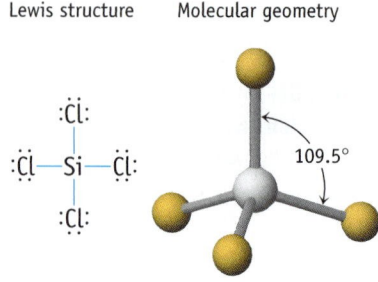

109.5°

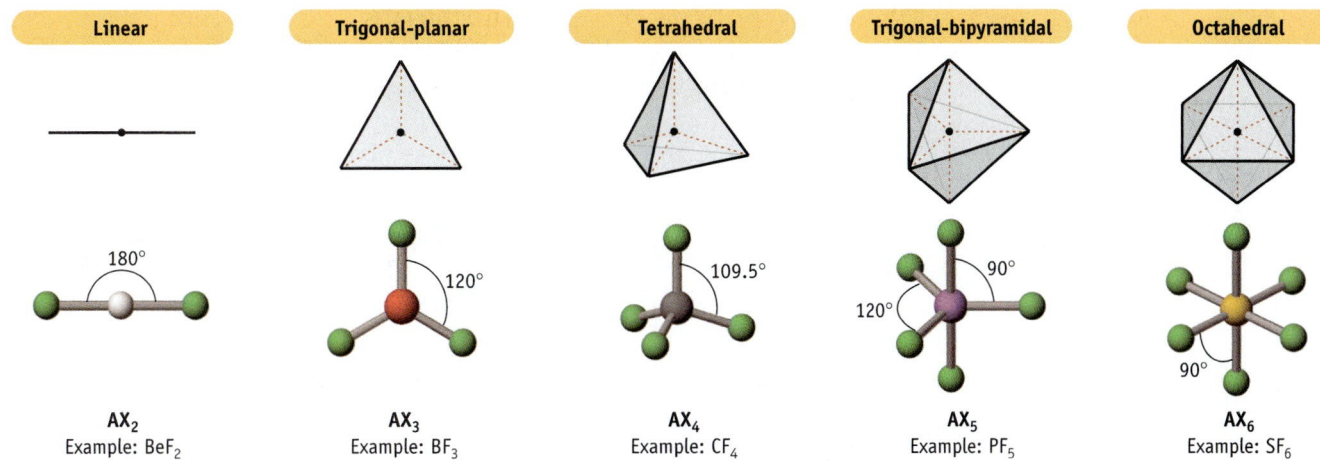

Linear	Trigonal-planar	Tetrahedral	Trigonal-bipyramidal	Octahedral
$180°$	$120°$	$109.5°$	$90°$ $120°$	$90°$
AX_2	AX_3	AX_4	AX_5	AX_6
Example: BeF_2	Example: BF_3	Example: CF_4	Example: PF_5	Example: SF_6

Active Figure 9.8   **Various geometries predicted by VSEPR.** Geometries predicted by VSEPR for molecules that contain only single covalent bonds around the central atom.

GENERAL
Chemistry Now™ See the General ChemistryNow CD-ROM or website to explore an interactive version of this figure accompanied by an exercise.

Exercise 9.9—Predicting Molecular Shapes

What is the shape of the dichloromethane (CH_2Cl_2) molecule? Predict the Cl—C—Cl bond angle.

Central Atoms with Single-Bond Pairs and Lone Pairs

To see how *lone pairs* affect the geometry of a molecule or polyatomic ion, return to the balloon models in Figure 9.7. Recall that the balloons represented *all* of the electron pairs in the valence shell. The balloon model therefore predicts the "electron-pair geometry" rather than the "molecular geometry." The **electron-pair geometry** is the geometry taken up by *all* valence electron pairs around a central atom, whereas the **molecular geometry** describes the arrangement in space of the central atom and the atoms directly attached to it. It is important to recognize that *lone pairs of electrons on the central atom occupy spatial positions even though their locations are not included in the verbal description of the shape of the molecule or ion.*

Let us use the VSEPR model to predict the molecular geometry and bond angles in the NH_3 molecule, which has a lone pair on the central atom. First, draw the Lewis structure and count the total number of electron pairs around the central nitrogen atom. There are four pairs of electrons in the nitrogen valence shell, so the *electron-pair geometry* is predicted to be *tetrahedral.* We have drawn a tetrahedron with nitrogen as the central atom and the three bond pairs represented by lines. The lone pair is included here to indicate its spatial position in the tetrahedron. The *molecular geometry* is described as a *trigonal pyramid.* The nitrogen atom is at the apex of the pyramid, and the three hydrogen atoms form the trigonal base.

H—N̈—H
   |
   H

Lewis structure

→

N̈
(H, H, H)

Electron-pair
geometry, tetrahedral

→

Actual H–N–H
angle = 107.5°

Molecular geometry,
trigonal pyramidal

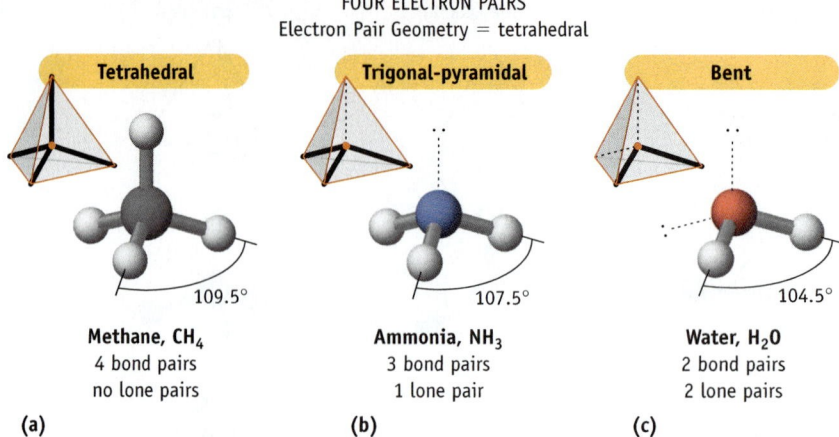

FOUR ELECTRON PAIRS
Electron Pair Geometry = tetrahedral

| Tetrahedral | Trigonal-pyramidal | Bent |

109.5°   107.5°   104.5°

Methane, CH$_4$
4 bond pairs
no lone pairs

Ammonia, NH$_3$
3 bond pairs
1 lone pair

Water, H$_2$O
2 bond pairs
2 lone pairs

(a)   (b)   (c)

Figure 9.9   The molecular geometries of methane, ammonia, and water. All have four electron pairs around the central atom, so all have a tetrahedral electron-pair geometry. (a) Methane has four bond pairs, so it has a tetrahedral molecular shape. (b) Ammonia has three bond pairs and one lone pair, so it has a trigonal-pyramidal molecular shape. (c) Water has two bond pairs and two lone pairs, so it has a bent, or angular, molecular shape. The decrease in bond angles in the series can be explained by the fact that the lone pairs have a larger spatial requirement than the bond pairs.

Effect of Lone Pairs on Bond Angles

Because the electron-pair geometry in NH$_3$ is tetrahedral, we would expect the H—N—H bond angle to be 109.5°. In fact, the experimentally determined bond angles in NH$_3$ are 107.5°, and the H—O—H angle in water is smaller still (104.5°) (Figure 9.9). These angles are close to the tetrahedral angle but not exactly that value. This discrepancy highlights the fact that VSEPR is not a precise model; it can only predict the approximate geometry. Small variations in geometry (e.g., bond angles that are a few degrees different from those predicted) are quite common and often arise because of differences between the spatial requirements of lone pairs and bond pairs. Lone pairs of electrons seem to occupy a larger volume than bonding pairs, and the increased volume of lone pairs causes bond pairs to squeeze closer together. In general, the relative strengths of repulsions are in the order

Lone pair–lone pair > lone pair–bond pair > bond pair–bond pair

The different spatial requirements of lone pairs and bond pairs are important and are included as part of the VSEPR model. For example, the VSEPR model can be used to predict variations in the bond angles in the series of molecules CH$_4$, NH$_3$, and H$_2$O. The bond angles decrease in this series as the number of lone pairs on the central atom increases (Figure 9.9).

Example 9.7—Finding the Shapes of Molecules and Ions

Problem What are the shapes of the ions (a) H$_3$O$^+$ and (b) ClF$_2^+$?

Strategy Draw the Lewis structures for each ion. Count the number of lone and bond pairs around each central atom. Use Figure 9.8 to decide on the electron-pair geometry. Finally, the location of the atoms in the ion—which are determined by the bond and lone pairs—gives the geometry of the ion.

Solution

(a) The Lewis structure of the hydronium ion, H_3O^+, shows that the oxygen atom is surrounded by four electron pairs, so the electron-pair geometry is tetrahedral.

Lewis structure	Electron-pair geometry, tetrahedral	Molecular geometry, trigonal pyramid

Because three of the four pairs are used to bond terminal atoms, the central O atom and the three H atoms form a trigonal-pyramidal molecular shape like that of NH_3.

(b) Chlorine is the central atom in ClF_2^+. It is surrounded by four electron pairs, so the electron-pair geometry around chlorine is tetrahedral. Because only two of the four pairs are bonding pairs, the ion has a bent geometry .

Lewis structure	Electron-pair geometry, tetrahedral	Molecular geometry, bent or angular

---

Exercise 9.10—VSEPR and Molecular Geometry

Give the electron-pair geometry and molecular geometry for BF_3 and BF_4^-. What is the effect on the molecular geometry of adding an F^- ion to BF_3 to give BF_4^-?

Central Atoms with More Than Four Valence Electron Pairs

The situation becomes more complicated if the central atom has five or six electron pairs, some of which are lone pairs. A trigonal-bipyramidal structure (Figures 9.8 and 9.10) has two sets of positions that are not equivalent. The positions in the trigonal plane lie in the equator of an imaginary sphere around the central atom and are called the *equatorial* positions. The north and south poles in this representation are called the *axial* positions. Each equatorial atom has two neighboring groups (the axial atoms) at 90°, and each axial atom has three groups (the equatorial atoms) at 90°. The result is that the lone pairs, which require more space than bonding pairs, prefer to occupy equatorial positions rather than axial positions.

The entries in the top line of Figure 9.11 show species having a total of five valence electron pairs, with zero, one, two, and three lone pairs. In SF_4, with one lone pair, the molecule assumes a seesaw shape with the lone pair in one of the equatorial positions. The ClF_3 molecule has three bond pairs and two lone pairs. The two lone pairs in ClF_3 are in equatorial positions; two bond pairs are axial and the third is in the equatorial plane, so the molecular geometry is T-shaped. The third molecule shown is XeF_2. Here, all three equatorial positions are occupied by lone pairs, so the molecular geometry is linear.

The geometry assumed by six electron pairs is octahedral (see Figure 9.11), and all the angles at adjacent positions are 90°. Unlike the trigonal bipyramid, the

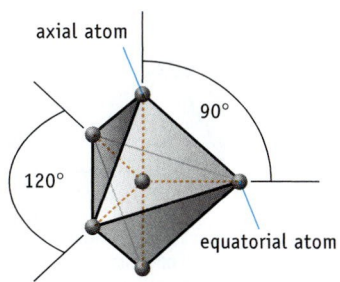

Figure 9.10 The trigonal bipyramid showing the axial and equatorial atoms. The angles between atoms in the equatorial position are 120°. The angles between equatorial and axial atoms are 90°.

octahedron has no distinct axial and equatorial positions; instead, all positions are the same. Therefore, if the molecule has one lone pair, as in BrF_5, it makes no difference which position it occupies. The lone pair is often drawn in the top or bottom position to make it easier to visualize the molecular geometry, which in this case is square-pyramidal. If two pairs of the electrons in an octahedral arrangement are lone pairs, they seek to be as far apart as possible. The result is a square-planar molecule, as illustrated by XeF_4.

Example 9.8—Predicting Molecular Shapes

Problem What is the shape of the ICl_4^- ion?

Strategy Draw the Lewis structure and then decide on the electron-pair geometry. The position of the atoms gives the geometry of the ion. See Example 9.7.

Solution A Lewis structure for the ICl_4^- ion shows that the central iodine atom has six electron pairs in its valence shell. Two of these are lone pairs. Placing the lone pairs on opposite sides leaves the four chlorine atoms in a square-planar geometry .

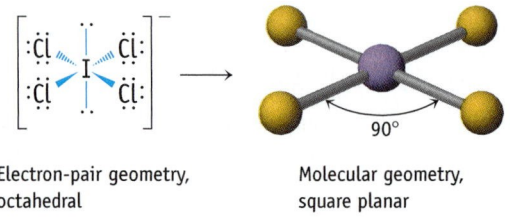

Electron-pair geometry,
octahedral

Molecular geometry,
square planar

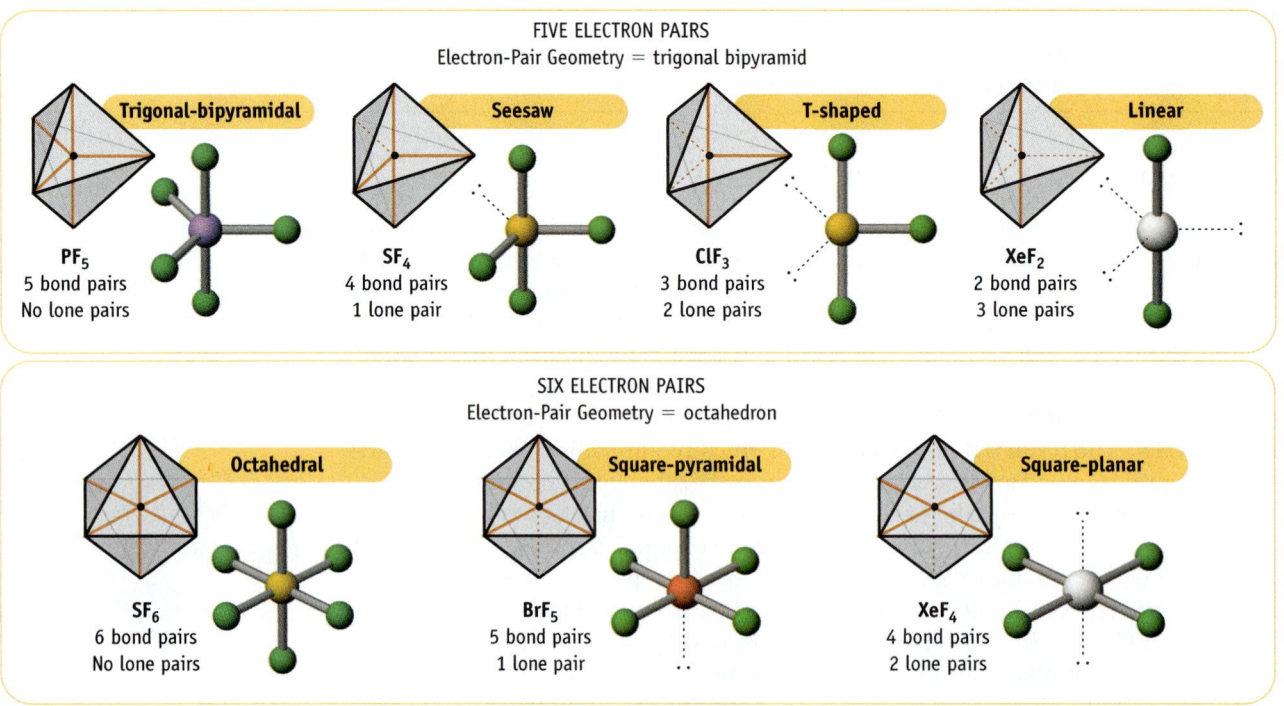

FIVE ELECTRON PAIRS
Electron-Pair Geometry = trigonal bipyramid

Trigonal-bipyramidal

PF₅
5 bond pairs
No lone pairs

Seesaw

SF₄
4 bond pairs
1 lone pair

T-shaped

ClF₃
3 bond pairs
2 lone pairs

Linear

XeF₂
2 bond pairs
3 lone pairs

SIX ELECTRON PAIRS
Electron-Pair Geometry = octahedron

Octahedral

SF₆
6 bond pairs
No lone pairs

Square-pyramidal

BrF₅
5 bond pairs
1 lone pair

Square-planar

XeF₄
4 bond pairs
2 lone pairs

Figure 9.11 Electron-pair geometries and molecular shapes for molecules and ions with five (*top*) or six (*bottom*) electron pairs around the central atom.

Exercise 9.11—Predicting Molecular Shapes

Draw the Lewis structure for ICl_2^- and then decide on the geometry of the ion.

Multiple Bonds and Molecular Geometry

Double and triple bonds involve more electron pairs than single bonds, but this characteristic does not affect the overall molecular shape. All of the electron pairs in a multiple bond are shared between the same two nuclei and therefore occupy the same region of space. Because they must remain in that region, two electron pairs in a double bond (or three electron pairs in a triple bond) behave like a single balloon in Figure 9.7 rather than like two or three balloons. All electron pairs in a multiple bond count as one bond and contribute to molecular geometry the same as a single bond does. For example, the carbon atom in CO_2 has no lone pairs and participates in two double bonds. Each double bond counts as one for the purpose of predicting geometry, so the structure of CO_2 is linear.

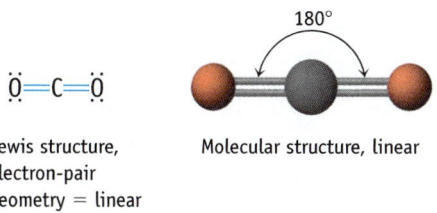

Lewis structure,
electron-pair
geometry = linear

Molecular structure, linear

When resonance structures are possible, the geometry can be predicted from any of the Lewis resonance structures or from the resonance hybrid structure. For example, the geometry of the CO_3^{2-} ion is predicted to be trigonal-planar because the carbon atom has three sets of bonds and no lone pairs.

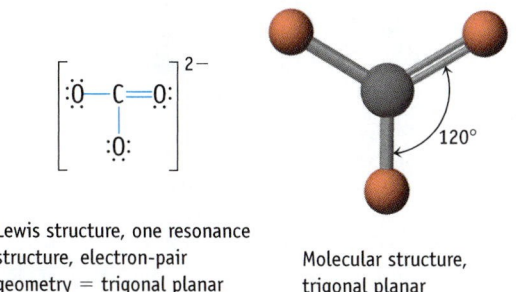

Lewis structure, one resonance
structure, electron-pair
geometry = trigonal planar

Molecular structure,
trigonal planar

The NO_2^- ion also has a trigonal-planar electron-pair geometry. Because there is a lone pair on the central nitrogen atom, and two bonds in the other two positions, the geometry of the ion is angular or bent.

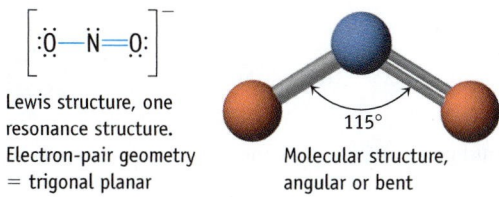

Lewis structure, one
resonance structure.
Electron-pair geometry
= trigonal planar

Molecular structure,
angular or bent

The techniques just outlined can be used to find the geometries of the atoms in more complicated molecules. Consider, for example, cysteine, one of the natural amino acids.

Cysteine, $HSCH_2CH(NH_2)CO_2H$

Four pairs of electrons occur around the S, N, C_2, and C_3 atoms, so the electron-pair geometry around each is tetrahedral. Thus, the H—S—C and H—N—H angles are predicted to be approximately 109°. The O atom in the grouping C_1—O—H also is surrounded by four pairs, so this angle is likewise approximately 109°. Finally, the angle made by O—C_1—O is 120° because the electron-pair geometry around C_1 is planar and trigonal.

Example 9.9—Finding the Shapes of Molecules and Ions

Problem What are the shapes of the (a) nitrate ion, NO_3^-, and (b) $XeOF_4$?

Strategy Draw the Lewis structure and then decide on the electron-pair geometry. The positions of the atoms give the molecular geometry of the ion. Follow the procedure used in Examples 9.7 and 9.8.

Solution

(a) The NO_3^- and CO_3^{2-} ions are isoelectronic. Thus, like the carbonate ion, the electron-pair geometry and molecular shape of NO_3^- are trigonal-planar .

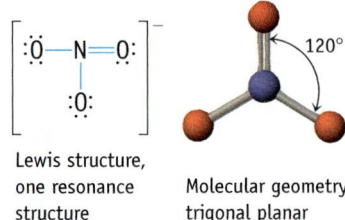

Lewis structure, one resonance structure

Molecular geometry, trigonal planar

(b) The $XeOF_4$ molecule has a Lewis structure with a total of six electron pairs about the central Xe atom, one of which is a lone pair. It has a square-pyramidal molecular structure . Two structures are possible based on the position occupied by the oxygen atom, but there is no way to predict which one is correct. The actual structure is the one shown, with the oxygen in the apex of the square pyramid.

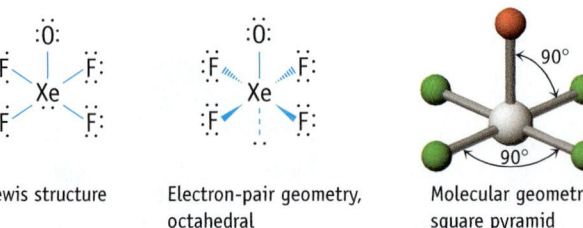

Lewis structure   Electron-pair geometry, octahedral   Molecular geometry, square pyramid

---

Exercise 9.12—Determining Molecular Shapes

Use Lewis structures and the VSEPR model to determine the electron-pair and molecular geometries for the following:

(a) phosphate ion, PO_4^{3-}

(b) sulfite ion, SO_3^{2-}

(c) IF_5

9.8—Charge Distribution In Covalent Bonds and Molecules

Lewis structures generally provide a fairly good picture of bonding in a covalently bonded molecule or ion. It is possible to "fine-tune" this picture, however, to get a more precise description of the distribution of the electrons. This effort will provide further insight into the chemical and physical properties of covalent molecules.

Closer analysis of covalently bonded molecules reveals that the valence electrons are not distributed among the atoms as evenly as Lewis structures might suggest. Some atoms may have a slight negative charge; others may have a slight positive charge. This situation occurs because the electron pair or pairs in a given bond may be drawn more strongly toward one atom than the other. The way the electrons are distributed in the molecule is called its *charge distribution*.

Charge distribution affects the properties of the molecule. Consider a diatomic (two-atom) molecule in which one atom is partially positive and the other is partially negative. In the solid or liquid state, for example, the molecules could be expected to line up with the positive end of one molecule near the negative end of another. The intermolecular or "between molecule" force of attraction would be enhanced by the attraction of opposite charges, affecting properties of the substance that are related to intermolecular forces, such as boiling point.

Positive or negative charges in a molecule or ion will influence, among other things, the site at which reactions occur. For example, does a positive H^+ ion attach itself to the O or the Cl of OCl^-? Is the product HOCl or HClO? It is reasonable to expect H^+ to attach to the more negatively charged atom. We can predict this outcome by evaluating atom formal charges in molecules and ions.

Formal Charges on Atoms

The **formal charge** for an atom in a molecule or ion is the charge calculated for that atom based on the Lewis structure of the molecule or ion, using Equation 9.1.

$$\text{Formal charge of an atom in a molecule or ion} = \text{group number of the atom} - \left[\text{LPE} + \tfrac{1}{2}(\text{BE})\right] \quad (9.1)$$

In this equation,

- The group number gives the number of valence electrons brought by a particular atom to the molecule or ion.
- LPE = number of lone-pair electrons on an atom.
- BE = number of bonding electrons around an atom.

The term in square brackets is the number of electrons assigned by the Lewis structure to an atom in a molecule or ion. The difference between this term and the group number is the formal charge. An atom in a molecule or ion will be positive if it "contributes" more electrons to bonding than it "gets back." The atom's formal charge will be negative if the reverse is true.

Two important assumptions are inherent in Equation 9.1. First, lone pairs are assumed to belong to the atom on which they reside in the Lewis structure. Second, bond pairs are assumed to be divided equally between the bonded atoms. (The factor of $\frac{1}{2}$ divides the bonding electrons equally between the atoms linked by the bond.)

The sum of the formal charges on the atoms in a molecule or ion always equals the net charge on the molecule or ion. Consider the hydroxide ion. Oxygen is in Group 6A and so has six valence electrons. In the hydroxide ion, however, oxygen can lay claim to seven electrons (six lone-pair electrons and one bonding electron), so the atom has a formal charge of -1. The O atom has "formally" gained an electron as part of the hydroxide ion.

■ **Calculating Formal Charge**
We shall apply formal charge calculations only to atoms of main group elements.

$$\text{Formal charge} = -1 = 6 - [6 + \tfrac{1}{2}(2)]$$

$$\left[:\!\ddot{O}\!-\!H\right]^{-}$$

$$\text{Formal charge} = 0 = 1 - [0 + \tfrac{1}{2}(2)]$$

Sum of formal charges = -1

Assume a covalent bond, so bonding electrons are divided equally between O and H

The formal charge on the hydrogen atom in OH^- is zero. We have -1 for oxygen and 0 for hydrogen, which equals the net charge of -1 for the ion. An important conclusion we can draw from the charge distribution in OH^- is that, if an H^+ ion approaches an OH^- ion, it should attach itself to the negatively charged O atom. This, of course, leads to water, as is indeed observed.

Formal charges can also be calculated for more complicated species such as the nitrate ion. Using one of the resonance structures for the ion, we find that the central N atom has a formal charge of $+1$, and the singly bonded O atoms *both* have formal charges of -1. The doubly bonded O atom has no charge. The net charge for the ion is thus -1.

$$\text{Formal charge} = 0 = 6 - [4 + \tfrac{1}{2}(4)]$$

$$\left[\begin{array}{c} :\!\ddot{O}\!: \\ \| \\ :\!\ddot{O}\!-\!N\!-\!\ddot{O}\!: \end{array}\right]^{-}$$

Sum of formal charges = -1

$$\text{Formal charge} = +1 = 5 - [0 + \tfrac{1}{2}(8)]$$
$$\text{Formal charge} = -1 = 6 - [6 + \tfrac{1}{2}(2)]$$

Is this a reasonable charge distribution for the nitrate ion? The answer is no. The problem is that the actual structure of the nitrate ion is a resonance hybrid of three equivalent resonance structures. Because the three oxygen atoms in NO_3^- are equivalent, the charge on one oxygen atom should not be different from the charge on the other two. This problem can be resolved by averaging the formal charges to give a formal charge of $-(\frac{2}{3})$ on the oxygen atoms. Summing the charges on the three oxygen atoms and the $+1$ charge on the nitrogen atom then gives -1, the charge on the ion.

In the resonance structures for O_3, CO_3^{2-}, and NO_3^-, for example, all the possible resonance structures are equally likely; they are "equivalent" structures. The molecule or ion therefore has a symmetrical distribution of electrons over all the

A Closer Look

Formal Charge and Oxidation Number

In Chapter 5 you learned to calculate the oxidation number of an atom as a way to tell whether a reaction is an oxidation–reduction reaction. There is an important difference between an oxidation number and an atom's formal charge.

To illustrate, look again at the hydroxide ion, OH^-. The formal charges are

$$\text{Formal charge} = -1 = 6 - [6 + \tfrac{1}{2}(2)]$$

$$\left[:\!\ddot{O}\!-\!H \right]^- \quad \textit{Sum of formal charges} = -1$$

$$\text{Formal charge} = 0 = 1 - [0 + \tfrac{1}{2}(2)]$$

Recall that these formal charges are calculated assuming the O—H bond electrons are shared equally; the O—H bond is covalent.

In contrast, in Chapter 5 (page 200), you learned that O has an oxidation number of -2 and H has an oxidation number of $+1$. Oxidation numbers are determined by assuming that the bond between a pair of atoms is ionic, not covalent. For OH^- this means that the pair of electrons between O and H is located fully on the O atom. Thus, the O atom now has eight valence electrons instead of six and a charge of -2. The H atom now has no valence electrons and a charge of $+1$.

$$\text{Oxidation number} = -2$$

$$\left[:\!\ddot{O}:\ H \right]^- \quad \begin{array}{l}\textit{Sum of oxidation}\\ \textit{numbers} = -1\end{array}$$

Assume an   Oxidation number $= +1$
ionic bond

Formal charges and oxidation numbers are calculated using different assumptions. Both are useful, but for different purposes. Oxidation numbers allow us to follow changes in redox reactions. Formal charges more nearly resemble atom charges in molecules and polyatomic ions.

---

atoms involved—that is, its electronic structure consists of an equal "mixture," or "hybrid," of the resonance structures.

GENERAL
Chemistry⚛Now™

See the General ChemistryNow CD-ROM or website:
• **Screen 9.16 Formal Charge,** for practice in determining formal charge

Example 9.10—Calculating Formal Charges

Problem Calculate formal charges for the atoms (a) in NH_4^+ and (b) in one resonance structure of CO_3^{2-}.

Strategy The first step is always to write the Lewis structure for the molecule or ion. Then Equation 9.1 can be used to calculate the formal charges.

Solution

(a) *Formal charge for the NH_4^+ ion*

$$\text{Formal charge} = 0$$
$$= 1 - [0 + \tfrac{1}{2}(2)]$$

$$\left[\begin{array}{c} H \\ | \\ H\!-\!N\!-\!H \\ | \\ H \end{array} \right]^+$$

$$\text{Formal charge} = +1$$
$$= 5 - [0 + \tfrac{1}{2}(8)]$$

(b) *Formal charges for the CO_3^{2-} ion*

$$\text{Formal charge} = 0$$
$$= 6 - [4 + \tfrac{1}{2}(4)]$$

$$\begin{bmatrix} & & :\!O\!: & \\ & & \| & \\ :\ddot{O} & \!-\!C\!-\! & \ddot{O}\!: \\ & & \end{bmatrix}^{2-}$$

Formal charge $= -1$
$= 6 - [6 + \tfrac{1}{2}(2)]$

Formal charge $= 0$
$= 4 - [0 + \tfrac{1}{2}(8)]$

Formal charge $= -1$
$= 6 - [6 + \tfrac{1}{2}(2)]$

In each case notice that the sum of the atom's formal charges is the charge on the ion. In the carbonate ion, which has three resonance structures, the average charge on the O atoms is $-\left(\tfrac{2}{3}\right)$.

Exercise 9.13—Calculating Formal Charges

Calculate the formal charge on each atom in the following:

(a) CN^- **(b)** SO_3

Bond Polarity and Electronegativity

The models used to represent covalent and ionic bonding are the extreme situations in bonding. Pure covalent bonding, in which atoms share an electron pair equally, occurs *only* when two identical atoms are bonded. When two dissimilar atoms form a covalent bond, the electron pair will be unequally shared. The result is a **polar covalent bond,** a bond in which the two atoms have residual or partial charges (Figure 9.12).

Bonds are polar because not all atoms hold onto their valence electrons with the same force, nor do atoms take on additional electrons with equal ease. Recall from the discussion of atomic properties that different elements have different values of ionization energy and electron affinity (Section 8.6). These differences in behavior for free atoms carry over to atoms in molecules.

If a bond pair is not equally shared between atoms, the bonding electrons are nearer to one of the atoms. The atom toward which the pair is displaced has a larger share of the electron pair and thus acquires a partial negative charge. At the same time, the atom at the other end of the bond is depleted in electrons and acquires a partial positive charge. The bond between the two atoms has a positive end and a negative end; that is, it has negative and positive poles. The bond is called a **polar bond**, and the molecule is said to be **dipolar** (having two poles).

In ionic compounds, displacement of the bonding pair to one of the two atoms is essentially complete, and $+$ and $-$ symbols are written alongside the atom symbols in the Lewis drawings. For a polar covalent bond, the polarity is indicated by writing the symbols δ^+ and δ^- alongside the atom symbols, where δ (the Greek letter "delta") stands for a *partial* charge. Hydrogen fluoride, water, and ammonia are three simple molecules having polar, covalent bonds (Figure 9.13).

With so many atoms to use in covalent bond formation, it is not surprising that bonds between atoms can fall along in a continuum from pure ionic to pure covalent. There is no sharp dividing line between an ionic bond and a covalent bond.

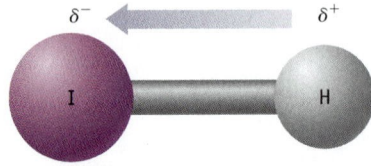

Figure 9.12 A polar covalent bond in HI. Iodine has a larger share of the bonding electrons and hydrogen has the smaller share. The result is that I has a partial negative charge (δ^-), and H has a partial positive charge (δ^+).

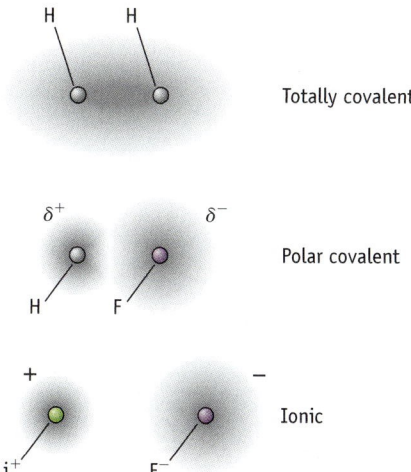

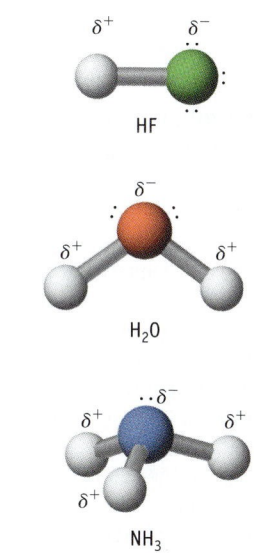

Figure 9.13 **Three simple molecules with polar covalent bonds.** In each case F, O, and N are more electronegative than H. See Figure 9.14.

In the 1930s, Linus Pauling proposed a parameter called atom electronegativity that allows us to decide whether a bond is polar, which atom of the bond is negative and which is positive, and whether one bond is more polar than another. The **electronegativity**, χ, of an atom is defined as a measure of *the ability of an atom in a molecule to attract electrons to itself.*

Values of electronegativity are given in Figure 9.14. Several features and periodic trends are apparent. The element with the largest electronegativity is fluorine; it is assigned a value of $\chi = 4.0$. The element with the smallest value is the alkali metal cesium. Electronegativities generally increase from left to right across a period and decrease down a group—the opposite of the trend observed for metallic character. Metals typically have low values of electronegativity, ranging from slightly less than 1 to about 2. Electronegativity values for the metalloids are around 2,

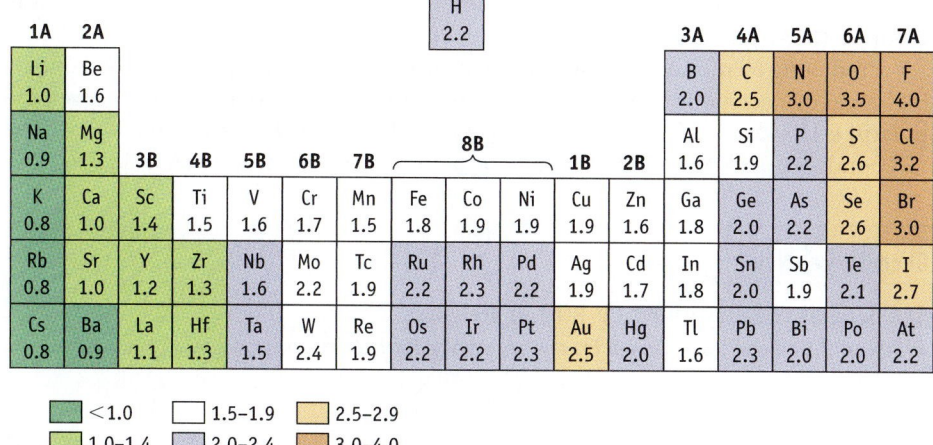

Figure 9.14 **Electronegativity values for the elements according to Pauling.** Trends for electronegativities are the opposite of the trends defining metallic character. Nonmetals have high values of electronegativity, the metalloids have intermediate values, and the metals have low values. (Values to one decimal place. J. Emsley: *The Elements*, 3rd ed., Clarendon Press, Oxford, 1998.)

A Closer Look

Electronegativity

Electronegativity is a useful, if somewhat vague, concept. It is, however, related to the ionic character of bonds. Chemists have found, as illustrated in the figure, that a correlation exists between the difference in electronegativity of bonded atoms and the degree of ionicity expressed as "% ionic character."

As the difference in electronegativity increases, ionic character increases. Does this trend allow us to say that one compound is ionic and another is covalent? No, we can say only that one bond is more ionic or more covalent than another.

Electron affinity was introduced in Section 8.6. At first glance it may appear that electronegativity and electron affinity measure the same property, but they do not. Electronegativity is a parameter that applies only to atoms in molecules, whereas electron affinity is a measurable energy quantity that refers to isolated atoms.

Although electron affinity was introduced earlier as a criterion with which to predict the central atom in a molecule, experience indicates that electronegativity is a better choice. That is, *the central atom is generally the atom of lowest electronegativity.*

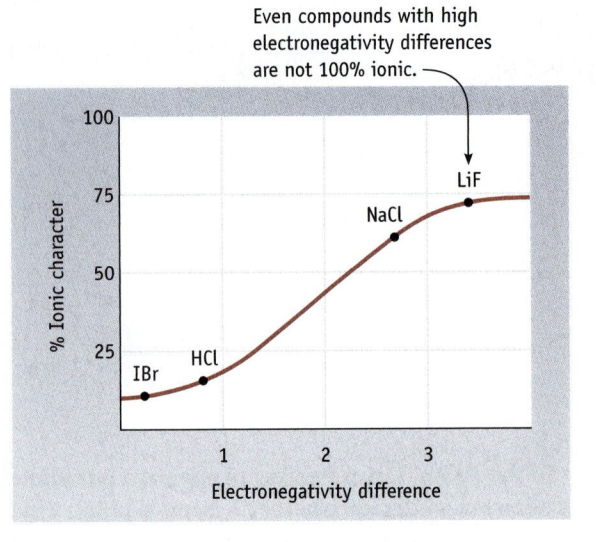

whereas nonmetals have values greater than 2. No values are given for He, Ne, and Ar because these elements do not form chemical compounds.

A large *difference* in electronegativities for atoms is observed when moving from the left- and right-hand side of the periodic table. For cesium fluoride, for example, the difference in electronegativity values, $\Delta\chi$, is 3.2 [= 4.0 (for F) − 0.8 (for Cs)]. The bond is ionic in CsF, therefore, with Cs as the cation (Cs^+) and F as the anion (F^-). In contrast, the electronegativity difference between H and Cl in HCl is only 1.0 [= 3.2 (for Cl) − 2.2 (for H)]. We conclude that bonding in HCl must be more covalent, as expected for a compound formed from two nonmetals. The H—Cl bond is polar, however, with hydrogen being the positive end of the molecule and chlorine the negative end ($H^{\delta+}$—$Cl^{\delta-}$).

Predicting trends in bond polarity in groups of related compounds is possible using values of electronegativity. Among the hydrogen halides, for example, the trend in polarity is HF ($\Delta\chi = 1.8$) > HCl ($\Delta\chi = 1.0$) > HBr ($\Delta\chi = 0.8$) > HI ($\Delta\chi = 0.5$). The HX bonds become less polar on going down the elements of Group 7A from F to I.

GENERAL
Chemistry ⚛ Now™

See General ChemistryNOW CD-ROM or Website:
- **Screen 9.17 Bond Polarity and Electronegativity,** for relative electronegativity values

Example 9.11—Estimating Bond Polarities

Problem For each of the following bond pairs, decide which is the more polar and indicate the negative and positive poles.

(a) B—F and B—Cl   **(b)** Si—O and P—P   **(c)** C═O and C═S

Strategy Locate the elements in the periodic table. Recall that electronegativity generally increases across a period and up a group.

Solution

(a) B and F lie relatively far apart in the periodic table. B is a metalloid and F is a nonmetal. Here χ for B = 2.0 and χ for F = 4.0. Similarly, B and Cl are relatively far apart in the periodic table, but Cl is below F in the periodic table (χ for Cl = 3.2) and is therefore less electronegative than F. The difference in electronegativity for B—F is 2.0: for B—Cl it is 1.2. Both bonds are expected to be polar, with B positive and the halide atom negative, but a B—F bond is more polar than a B—Cl bond .

(b) Because the bond is between two atoms of the same kind, the P—P bond is nonpolar. Silicon is in Group 4A and the third period, whereas O is in Group 6A and the second period. Consequently, O has a greater electronegativity (3.5) than Si (1.9), so the Si—O bond is highly polar ($\Delta\chi$ = 1.6), with O the more negative atom.

(c) Oxygen lies above sulfur in the periodic table, so oxygen is more electronegative than S. This means the C—O bond is more polar than the C—S bond . For the C—O bond, O is the more negative atom. The value of $\Delta\chi$ (1.0) for CO indicates a moderately polar bond.

Exercise 9.14—Bond Polarity

For each of the following pairs of bonds, decide which is the more polar. For each polar bond, indicate the positive and negative poles. First make your prediction from the relative positions of the atoms in the periodic table; then check your prediction by calculating $\Delta\chi$.

(a) H—F and H—I

(b) B—C and B—F

(c) C—Si and C—S

Combining Formal Charge and Bond Polarity

Using formal charge calculations alone to locate the site of a charge in an ion can sometimes lead to results that seem incorrect. The ion BF_4^- illustrates this point. Boron has a formal charge of −1 in this ion, whereas the formal charge calculated for the fluorine atoms is 0. This is not logical: Fluorine is the more electronegative atom, so the negative charge should reside on F and not on B.

To resolve this dilemma, we must consider electronegativity in conjunction with formal charge. Based on the electronegativity difference between fluorine and boron ($\Delta\chi$ = 2.0) the B—F bonds are expected to be polar, with fluorine being the negative end of the bond, $B^{\delta+}$—$F^{\delta-}$. So, in this instance, predictions based on electronegativity and formal charge work in opposite directions. The formal charge calculation places the negative charge on boron, but the electronegativity difference says that the negative charge on boron is distributed onto the fluorine atoms. In effect, the charge is "spread out" over the molecule.

Linus Pauling pointed out two basic guidelines to use when describing charge distributions in molecules and ions. First, the **electroneutrality principle** declares that electrons will be distributed in such a way that the charges on all atoms are as close to zero as possible. Second, if a negative charge is present, it should reside on the most electronegative atoms. Similarly, positive charges would be expected on the least electronegative atoms. The effect of these principles is clearly seen in the case of BF_4^-, where the negative charge is distributed over the four fluorine atoms rather than residing on the boron atom.

Considering the concepts of electronegativity and formal charge together can also help to decide which of several resonance structures is the more important. For example, Lewis structure A for CO_2 is the logical one to draw. But what is wrong with B, in which each atom also has an octet of electrons?

Formal charges   0   0   0   +1 0   −1

Resonance structures   Ö=C=Ö   :Ö≡C—Ö:

                              A   B

In structure A, each atom has a formal charge of 0, a favorable situation. In structure B, one oxygen atom has a formal charge of +1 and the other has −1. This is contrary to the principle of electroneutrality. In addition, B places a positive charge on the very electronegative O atom. Thus, we can conclude that structure B is a less satisfactory structure than A.

Now use what you have learned with CO_2 to decide which of the three possible resonance structures for the OCN^- ion is the most reasonable. Formal charges for each atom are given above the element's symbol.

Formal charges   −1 0   0   0   0   −1   +1 0   −2

Resonance structures $\left[:\ddot{O}—C≡N:\right]^-$ ⟷ $\left[:\ddot{O}=C=\ddot{N}:\right]^-$ ⟷ $\left[:O≡C—\ddot{N}:\right]^-$

                              A   B   C

■ Formal Charges in OCN⁻
Example of formal charge calculation: For resonance form C for OCN^-, we have

O: $6 − [2 + \frac{1}{2}(6)] = +1$

C: $4 − [0 + \frac{1}{2}(8)] = 0$

N: $5 − [6 + \frac{1}{2}(2)] = −2$

Sum of formal charges = −1 = charge on the ion

Structure C will not contribute significantly to the overall electronic structure of the ion. It has a −2 formal charge on the N atom and a +1 formal charge on the O atom, whereas the formal charges in the other structures are 0 or −1. Structure A is more significant than structure B because the negative charge in A is placed on the most electronegative atom (O). We predict, therefore, that structure A is the best representation for this ion and that the carbon–nitrogen bond will resemble a triple bond.

The result for OCN^- also allows us to predict that protonation of the ion will lead to HOCN and not HNCO. That is, an H^+ ion will add to the more negative oxygen atom.

Example 9.12—Calculating Formal Charges

Problem Boron-containing compounds often have a boron atom with only three bonds (and no lone pairs). Why not form a double bond with a terminal atom to complete the boron octet? To answer this question, consider possible resonance structures of BF_3 and calculate the atom formal charges. Are the bonds polar in BF_3? If so, which is the more negative atom?

Strategy Calculate the formal charges on each atom in the resonance structures. The preferred structure will have atoms with low formal charges. Negative formal charges should be on the most electronegative atoms.

Solution The two possible structures for BF_3 are illustrated here with the calculated formal charges on the B and F atoms.

Formal charge = 0   Formal charge = +1
        $= 7 − [6 + \frac{1}{2}(2)]$ $= 7 − [4 + \frac{1}{2}(4)]$

        :F: :F:
         | ‖
   :F—B—F: :F—B—F:

Formal charge = 0   Formal charge = −1
        $= 3 − [0 + \frac{1}{2}(6)]$ $= 3 − [0 + \frac{1}{2}(8)]$

The structure on the left is preferred because all atoms have a zero formal charge and the very electronegative F atom does not have a charge of 1+.

F ($\chi = 4.0$) is more electronegative than B ($\chi = 2.0$), so the B—F bond is polar, with the F atom being partially negative and the B atom being partially positive.

Exercise 9.15—Formal Charge, Bond Polarity, and Electronegativity

Consider all possible resonance structures for SO_2. What is the formal charge on each atom in each resonance structure? What are the bond polarities? Do they agree with the formal charges?

9.9—Molecular Polarity

The term "polar" was used in Section 9.8 to describe a bond in which one atom has a partial positive charge and the other has a partial negative charge. Because most molecules have polar bonds, molecules as a whole can also be polar. In a polar molecule, electron density accumulates toward one side of the molecule, giving that side a negative charge, $-\delta$, and leaving the other side with a positive charge of equal value, $+\delta$ (Figure 9.15).

Before describing the factors that determine whether a molecule is polar, let us look at the experimental measurement of the polarity of a molecule. When placed in an electric field, polar molecules experience a force that tends to align them with the field (Figure 9.15). When the electric field is created by a pair of oppositely charged plates, the positive end of each molecule is attracted to the negative plate, and the negative end is attracted to the positive plate. The extent to which the molecules line up with the field depends on their **dipole moment**, μ, which is defined as the product of the magnitude of the partial charges ($+\delta$ and $-\delta$) on the molecule and the distance by which they are separated. The SI unit of the dipole moment is the coulomb-meter, but dipole moments have traditionally been given using a derived unit called the debye (D; $1\ D = 3.34 \times 10^{-30}\ C \cdot m$). Experimental values of some dipole moments are listed in Table 9.8.

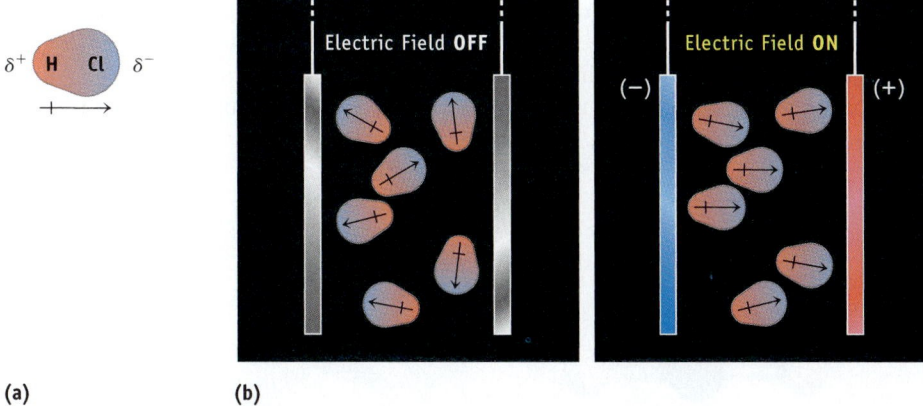

(a)   **(b)**

Figure 9.15 **Polar molecules in an electric field.** (a) A representation of a polar molecule. To indicate the direction of molecular polarity, an arrow is drawn with the head pointing to the negative side and a plus sign placed at the positive end. (b) When placed in an electric field (between charged plates), polar molecules experience a force that tends to align them with the field. The negative ends of the molecules are drawn to the positive plate, and vice versa. The orientation of the polar molecules affects the electrical capacitance of the plates (their ability to hold a charge), which provides a way to measure experimentally the magnitude of the dipole.

Table 9.8   Dipole Moments of Selected Molecules

Molecule (AB)	Moment (μ, D)	Geometry	Molecule (AB_2)	Moment (μ, D)	Geometry
HF	1.78	linear	H_2O	1.85	bent
HCl	1.07	linear	H_2S	0.95	bent
HBr	0.79	linear	SO_2	1.62	bent
HI	0.38	linear	CO_2	0	linear
H_2	0	linear			

Molecule (AB_3)	Moment (μ, D)	Geometry	Molecule (AB_4)	Moment (μ, D)	Geometry
NH_3	1.47	trigonal-pyramidal	CH_4	0	tetrahedral
NF_3	0.23	trigonal-pyramidal	CH_3Cl	1.92	tetrahedral
BF_3	0	trigonal-planar	CH_2Cl_2	1.60	tetrahedral
			$CHCl_3$	1.04	tetrahedral
			CCl_4	0	tetrahedral

The force of attraction between the negative end of one polar molecule and the positive end of another (called a *dipole–dipole force* and discussed in Section 13.2) affects the properties of polar compounds. Intermolecular forces (forces between molecules) influence the temperature at which a liquid freezes or boils, for example. These forces will also help determine whether a liquid dissolves certain gases or solids or whether it mixes with other liquids, and whether it adheres to glass or other solids.

To predict whether a molecule is polar, we need to consider whether the molecule has polar bonds and how these bonds are positioned relative to one another. Diatomic molecules composed of two atoms with different electronegativities are always polar (see Table 9.8); there is one bond, and the molecule has a positive and a negative end. But what happens with a molecule composed of three or more atoms, in which there are two or more polar bonds? Let us look at a series of molecules with stoichiometry AX_2, AX_3, and AX_4, evaluating how the choice of substituent or "terminal" groups (X) and molecular geometry influence the molecular polarity.

Consider first a linear triatomic molecule such as carbon dioxide, CO_2 (Figure 9.16). Here each C—O bond is polar, with the oxygen atom being the negative end of the bond dipole. The terminal atoms are at the same distance from the C atom, both have the same δ^- charge, and they are symmetrically arranged around the cen-

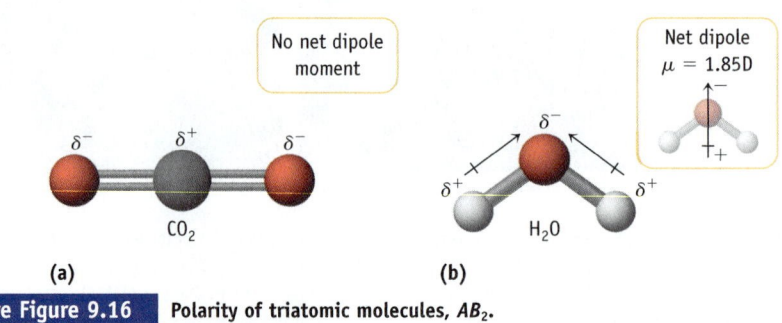

(a)   (b)

Active Figure 9.16   Polarity of triatomic molecules, AB_2.

GENERAL
Chemistry ⚛ Now™ See the General ChemistryNow CD-ROM or website to explore an interactive version of this figure accompanied by an exercise.

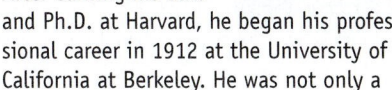

tral C atom. Therefore, CO_2 has no molecular dipole, even though each bond is polar. This is analogous to a tug-of-war in which the people at opposite ends of the rope are pulling with equal force.

In contrast, water is a bent triatomic molecule. Because O has a larger electronegativity ($\chi = 3.5$) than H ($\chi = 2.2$), each of the O—H bonds is polar, with the H atoms having the same δ^+ charge and oxygen having a negative charge (δ^-) (Figure 9.16). Electron density accumulates on the O side of the molecule, making the molecule electrically "lopsided" and therefore polar ($\mu = 1.85$ D).

In trigonal-planar BF_3, the B—F bonds are highly polar because F is much more electronegative than B (χ of B = 2.0 and χ of F = 4.0) (Figure 9.17). The molecule is nonpolar, however, because the three terminal F atoms have the same δ^- charge, are located the same distance from the boron atom, and are arranged symmetrically and in the same plane as the central boron atom. In contrast, the planar-trigonal molecule phosgene, Cl_2CO, is polar ($\mu = 1.17$ D) (Figure 9.17). Here the angles are all approximately 120°, so the O and Cl atoms are symmetrically arranged around the C atom. The electronegativities of the three atoms in the molecule differ, however: $\chi(O) > \chi(Cl) > \chi(C)$. As a consequence, there is a net displacement of electron density away from the center of the molecule, mostly toward the O atom.

Ammonia, like BF_3, has AX_3 stoichiometry and polar bonds. In contrast to BF_3, however, NH_3 is a trigonal-pyramidal molecule. The slightly positive H atoms are located in the base of the pyramid, and the slightly negative N atom is at the apex of the pyramid. As a consequence, NH_3 is polar (Figure 9.17). Indeed, trigonal-pyramidal molecules are generally polar.

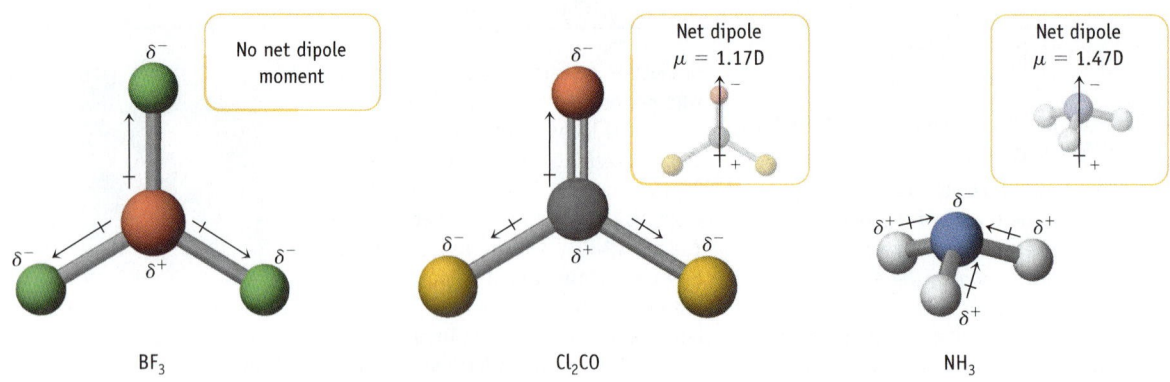

Active Figure 9.17   Polar and nonpolar molecules of the type AB_3.

GENERAL
Chemistry ⚛ Now™ See the General ChemistryNow CD-ROM or website to explore an interactive version of this figure accompanied by an exercise.

Molecules like carbon tetrachloride, CCl_4, and methane, CH_4, are nonpolar owing to their symmetrical, tetrahedral structures. The four atoms bonded to C have the same partial charge and are located the same distance from the C atom. In contrast, tetrahedral molecules with both Cl and H atoms ($CHCl_3$, CH_2Cl_2, and CH_3Cl) are polar (Table 9.8 and Figure 9.8). The electronegativity for H atoms (2.2) is less than that of Cl atoms (3.2), and the carbon–hydrogen distance is different from the carbon–chlorine distances. Because Cl is more electronegative than H, the Cl atoms are on the more negative side of the molecule. Thus, the positive end of the molecular dipole is toward the H atom.

To summarize this discussion of molecular polarity, look again at Figure 9.8 (page 399). These sketches show molecules of the type AX_n where A is the central atom and X is a terminal atom. You can predict that a molecule AX_n will *not* be polar, regardless of whether the A—X bonds are polar, if

- All of the terminal atoms (or groups), X, are identical, and
- All of the X atoms (or groups) are arranged symmetrically around the central atom, A, in the geometries shown.

Chemical Perspectives

Cooking with Microwaves

Microwave ovens are common appliances in homes, dorm rooms, and offices. They work by capitalizing on the polarity of water.

Microwaves are generated in a magnetron, a device invented during World War II for antiaircraft radar. The microwaves (frequency = $2.45 \times 10^9 \ s^{-1}$) bounce off the metal walls of the oven and strike the food from many angles. They pass through glass or plastic dishes with little effect.

Because electromagnetic radiation consists of oscillating electric (and magnetic) fields (Figure 7.1), however, microwaves can affect mobile, charged particles such as dissolved ions or the polar water molecules commonly found in food. As each wave crest approaches a water molecule, the molecule turns to align itself with the wave and continues turning over or rotating as the trough of the wave passes. Thus, the molecules absorb energy through increased molecular motions, which translates to a higher temperature.

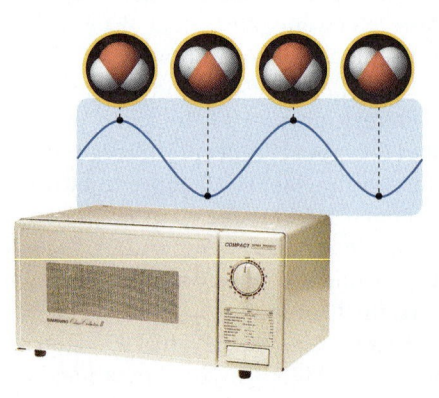

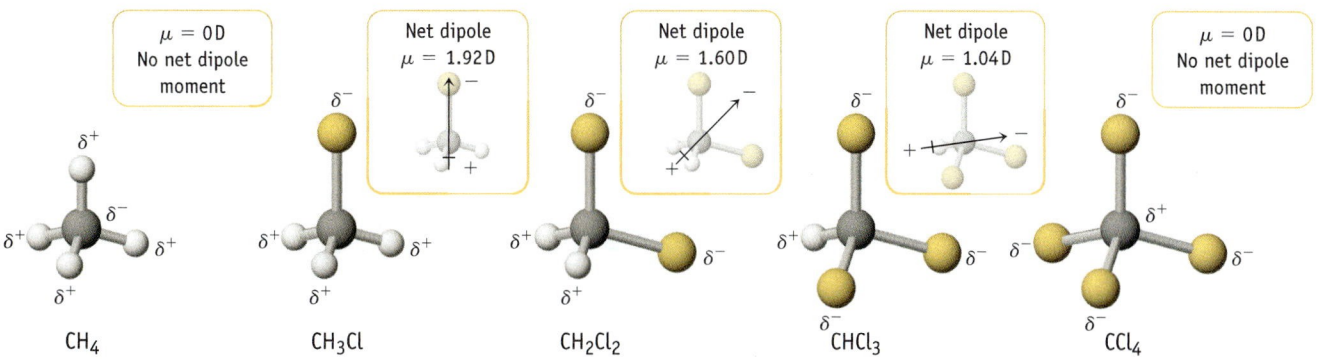

Figure 9.18   Polarity of tetrahedral molecules. The electronegativities of the atoms involved are in the order Cl (3.2) > C (2.5) > H (2.2). This means the C—H and C—Cl bonds are polar with a net displacement of electron density away from the H atoms and toward the Cl atoms [H(δ^+)—C(δ^-) and C(δ^+)—Cl(δ^-)]. Although the electron-pair geometry around the C atom in each molecule is tetrahedral, only in CH_4 and CCl_4 are the polar bonds totally symmetrical in their arrangement. Therefore, CH_3Cl, CH_2Cl_2, and $CHCl_3$ are polar molecules, with the negative end being toward the Cl atoms and the positive end being toward the H atoms.

On the other hand, if one of the *X* atoms (or groups) is different in the structures in Figure 9.8 (as in Figures 9.17 and 9.18), or if one of the *X* positions is occupied by a lone pair, the molecule will be polar.

GENERAL
Chemistry⚛Now™

See the General ChemistryNOW CD-ROM or website:
• **Screen 9.18 Molecular Polarity,** for practice in determining polarity

Example 9.13—Molecular Polarity

Problem Are (a) nitrogen trifluoride, NF_3, and (b) sulfur tetrafluoride, SF_4, polar or nonpolar? If polar, indicate the negative and positive sides of the molecule.

Strategy You cannot decide whether a molecule is polar without determining its structure. Therefore, start with the Lewis structure, decide on the electron-pair geometry, and then decide on the molecular geometry. If the molecular geometry is one of the highly symmetrical geometries in Figure 9.8 on page 399, the molecule is *not* polar. If it does not fit one of these categories, it will be polar.

Solution

(a) NF_3 has the same pyramidal structure as NH_3. Because F is more electronegative than N, each bond is polar, with the more negative end being the F atom. This means that the NF_3 molecule as a whole is polar.

(b) Sulfur tetrafluoride, SF_4, has an electron-pair geometry of a trigonal bipyramid (see Figure 9.11). Because the lone pair occupies one of the positions, the S—F bonds are not arranged symmetrically. Furthermore, the S—F bonds are highly polar, with the bond dipole having F as the negative end (χ for S is 2.6 and χ for F is 4.0). SF_4 is therefore a polar molecule. The axial S—F bond dipoles cancel each other because they point in opposite directions. The equatorial S—F bonds, however, both point to one side of the molecule.

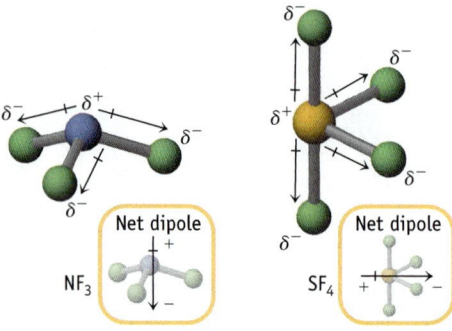

NF₃   SF₄

Example 9.14—Molecular Polarity

Problem 1,2-Dichloroethylene can exist in two forms. Is either of these molecules polar?

$$
\begin{array}{cc}
\text{H} \quad \text{H} & \text{Cl} \quad \text{H} \\
\text{C}=\text{C} & \text{C}=\text{C} \\
\text{Cl} \quad \text{Cl} & \text{H} \quad \text{Cl} \\
\text{A} & \text{B}
\end{array}
$$

Strategy To decide whether a molecule is polar we first sketch the structure. Then, using electronegativity values, we decide on the bond polarity. Finally, we decide whether the electron density in the bonds is distributed symmetrically or is shifted to one side of the molecule.

Solution Here the H and Cl atoms are arranged around the C=C double bonds with all bond angles being 120°. The electronegativities of the atoms involved are in the order Cl (3.2) > C (2.5) > H (2.2) (Figures 9.14 and 9.18). This means the C—H and C—Cl bonds are polar with a net displacement of electron density away from the H atoms and toward the Cl atoms [$H^{\delta+}$—$C^{\delta-}$ and $C^{\delta+}$—$Cl^{\delta-}$]. In structure A, the Cl atoms are located on one side of the molecule, so electrons in the H—C and C—Cl bonds move toward the side of the molecule with Cl atoms and away from the side with the H atoms. Molecule A is polar .

In molecule B, the movement of electron density toward the Cl atom on one end of the molecule is counterbalanced by an opposing movement on the other end. Molecule B is not polar .

Overall displacement of bonding electrons

$$
\begin{array}{cc}
\overset{\delta+}{\text{H}} \quad \overset{\delta+}{\text{H}} \\
\text{C}=\text{C} \\
\underset{\delta-}{\text{Cl}} \quad \underset{\delta-}{\text{Cl}}
\end{array}
$$

A, polar, diplacement of bonding electrons to one side of the molecule.

Displacement of bonding electrons

$$
\begin{array}{cc}
\overset{\delta-}{\text{Cl}} \quad \overset{\delta+}{\text{H}} \\
\text{C}=\text{C} \\
\underset{\delta+}{\text{H}} \quad \underset{\delta-}{\text{Cl}}
\end{array}
$$

Displacement of bonding electrons

B, not polar, no net displacement of bonding electrons to one side of the molecule.

Exercise 9.16—Molecular Polarity

For each of the following molecules, decide whether the molecule is polar and which side is positive and which negative: $BFCl_2$, NH_2Cl, and SCl_2.

9.10—Bond Properties: Order, Length, and Energy

Bond Order

The **order of a bond** is the number of bonding electron pairs shared by two atoms in a molecule (Figure 9.19). You will encounter bond orders of 1, 2, and 3, as well as fractional bond orders.

When the bond order is 1, only a single covalent bond exists between a pair of atoms. Examples are the bonds in molecules such as H_2, NH_3, and CH_4. The bond order is 2 when two electron pairs are shared between atoms, such as the $C{=}O$ bonds in CO_2 and the $C{=}C$ bond in ethylene, $H_2C{=}CH_2$. The bond order is 3 when two atoms are connected by three bonds. Examples include the carbon–oxygen bond in carbon monoxide, CO, and the nitrogen–nitrogen bond in N_2.

Fractional bond orders occur in molecules and ions having resonance structures. For example, what is the bond order for each oxygen–oxygen bond in O_3? Each resonance structure of O_3 has one $O{-}O$ single bond and one $O{=}O$ double bond, for a total of three shared bonding pairs accounting for two oxygen–oxygen links.

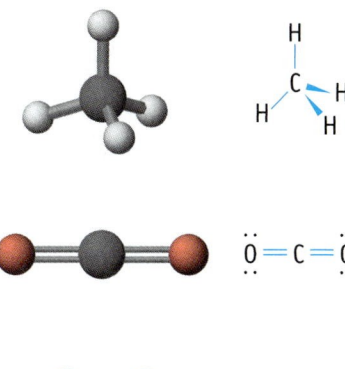

Figure 9.19   Bond order. The four $C{-}H$ bonds in methane each have a bond order of 1. The two $C{=}O$ bonds of CO_2 each have a bond order of two, whereas the nitrogen–nitrogen bond in N_2 has a bond order of 3.

Bond order = 1

O Bond order = 2   Bond order for each oxygen–oxygen bond $= \frac{3}{2}$, or 1.5

One resonance structure

The bond order between any bonded pair of atoms X and Y is defined as

$$\text{Bond order} = \frac{\text{number of shared pairs linking X and Y}}{\text{number of X}-\text{Y links in the molecule or ion}} \qquad (9.2)$$

For ozone there are three bond pairs involved in two oxygen–oxygen bonds, so the bond order is $\frac{3}{2}$, or 1.5.

Bond Length

Bond length is the distance between the nuclei of two bonded atoms. Bond lengths are therefore related to the sizes of the atoms (Section 8.6), but, for a given pair of atoms, the order of the bond determines the final value of the distance.

Table 9.9 lists average bond lengths for a number of common chemical bonds. It is important to recognize that these are *average* values. Neighboring parts of a molecule can affect the length of a particular bond. For example, Table 9.9 specifies that the average $C{-}H$ bond has a length of 110 pm. In methane, CH_4, the measured bond length is 109.4 pm, whereas the $C{-}H$ bond is only 105.9 pm long in acetylene, $H{-}C{\equiv}C{-}H$. Variations as great as 10% from the average values listed in Table 9.9 are possible.

Because atom sizes vary in a regular fashion with the position of the element in the periodic table (Figure 8.11), predictions of trends in bond length can be made quickly. For example, the $H{-}X$ distance in the hydrogen halides increases in the order predicted by the relative sizes of the halogens: $H{-}F < H{-}Cl < H{-}Br < H{-}I$. Likewise, bonds between carbon and another element in a given period decrease going from left to right, in a predictable fashion; for example, $C{-}C > C{-}N > C{-}O > C{-}F$. Trends for multiple bonds are similar. A $C{=}O$ bond is shorter than a $C{=}S$ bond, and a $C{=}N$ bond is shorter than a $C{=}C$ bond.

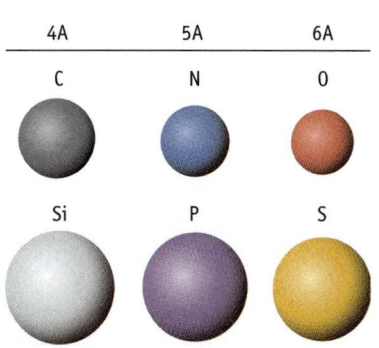

Relative sizes of some atoms of Groups 4A, 5A, and 6A.

Table 9.9   Some Average Single- and Multiple-Bond Lengths in Picometers (pm)*

Single Bond Lengths

Group

	1A	4A	5A	6A	7A	4A	5A	6A	7A	7A	7A
	H	C	N	O	F	Si	P	S	Cl	Br	I
H	74	110	98	94	92	145	138	132	127	142	161
C		154	147	143	141	194	187	181	176	191	210
N			140	136	134	187	180	174	169	184	203
O				132	130	183	176	170	165	180	199
F					128	181	174	168	163	178	197
Si						234	227	221	216	231	250
P							220	214	209	224	243
S								208	203	218	237
Cl									200	213	232
Br										228	247
I											266

Multiple Bond Lengths

$C=C$	134	$C\equiv C$	121
$C=N$	127	$C\equiv N$	115
$C=O$	122	$C\equiv O$	113
$N=O$	115	$N\equiv O$	108

*1 pm = 10^{-12} m.

The effect of bond order is evident when bonds between the same two atoms are compared. For example, the bonds become shorter as the bond order increases in the series $C-O$, $C=O$, and $C\equiv O$:

Bond	$C-O$	$C=O$	$C\equiv O$
Bond order	1	2	3
Bond length (pm)	143	122	113

Double bonds are shorter than single bonds between the same set of atoms, and triple bonds between those same atoms are shorter still.

The carbonate ion, CO_3^{2-}, has three equivalent resonance structures. It has a CO bond order of 1.33 (or $\frac{4}{3}$) because four electron pairs are used to form three carbon–oxygen links. The CO bond distance (129 pm) is intermediate between a $C-O$ single bond (143 pm) and a $C=O$ double bond (122 pm).

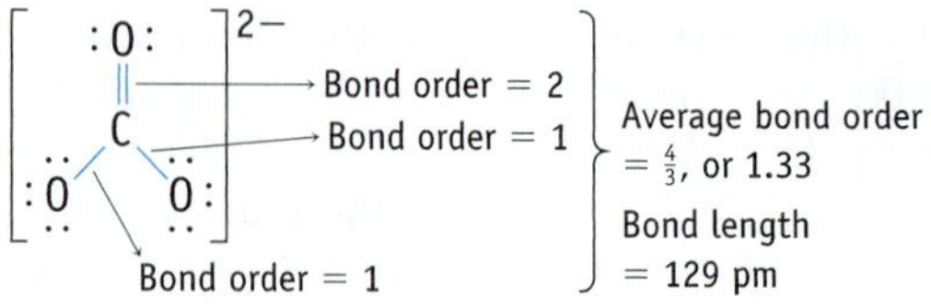

Exercise 9.17—Bond Order and Bond Length

(a) Give the bond order of each of the following bonds and arrange them in order of decreasing bond distance: $C=N$, $C\equiv N$, and $C-N$.

(b) Draw resonance structures for NO_2^-. What is the NO bond order in this ion? Consult Table 9.9 for $N-O$ and $N=O$ bond lengths. Compare these with the NO bond length in NO_2^- (124 pm). Account for any differences you observe.

Bond Energy

The **bond dissociation energy,** symbolized by D, is the enthalpy change for breaking a bond in a molecule with the reactants and products in the gas phase.

$$\text{Molecule (g)} \underset{\text{Energy released} = -D}{\overset{\text{Energy supplied} = D}{\rightleftharpoons}} \text{Molecular fragments (g)}$$

Suppose you wish to break the carbon–carbon bonds in ethane (H_3C-CH_3), ethylene ($H_2C=CH_2$), and acetylene ($HC\equiv CH$), for which the bond orders are 1, 2, and 3, respectively. For the same reason that the ethane $C-C$ bond is the longest of the series, and the acetylene $C\equiv C$ bond is the shortest, bond breaking requires the least energy for ethane and the most energy for acetylene.

$$\begin{array}{lll}
H_3C-CH_3(g) & \longrightarrow H_3C(g) + CH_3(g) & D = +346 \text{ kJ} \\
H_2C=CH_2(g) & \longrightarrow H_2C(g) + CH_2(g) & D = +610 \text{ kJ} \\
HC\equiv CH(g) & \longrightarrow HC(g) + CH(g) & D = +835 \text{ kJ}
\end{array}$$

Because D represents the energy transferred to the molecule from its surroundings, D has a positive value; that is, *the process of breaking bonds in a molecule is always endothermic.*

The energy supplied to break carbon–carbon bonds must be the same as the energy released when the same bonds form. *The formation of bonds from atoms or radicals in the gas phase is always exothermic.* This means, for example, that ΔH for the formation of H_3C-CH_3 from two $CH_3(g)$ radicals is -346 kJ/mol.

$$H_3C\cdot(g) + \cdot CH_3(g) \rightarrow H_3C-CH_3(g) \qquad \Delta H = -D = -346 \text{ kJ}$$

Generally, the bond energy for a given type of bond (a $C-C$ bond, for example) varies somewhat depending on the compound, just as bond lengths vary from one molecule to another. They are sufficiently similar, however, that it is possible to create a table of *average bond energies* (Table 9.10). The values in such tables may be used to *estimate* enthalpies of reactions, as described below.

In reactions between molecules, bonds in reactants are broken and new bonds are formed as products form. If the total energy released when new bonds form exceeds the energy required to break the original bonds, the overall reaction is

■ **Bond Energy and Electronegativity** Linus Pauling derived electronegativity values from a consideration of bond energies. He recognized that the energy of a bond between two different atoms is often greater than expected if the bond electrons are shared equally. He postulated that the "extra energy" arises from the fact that the atoms do not share electrons equally. One atom is slightly positive and the other is slightly negative. Thus, there is a small coulombic force of attraction involving oppositely charged ions in addition to the force of attraction arising from the sharing of electrons. This coulombic force enhances the overall force of attraction.

Table 9.10   Some Average Single- and Multiple-Bond Energies (kJ/mol)*

	Single Bonds										
	H	C	N	O	F	Si	P	S	Cl	Br	I
H	436	413	391	463	565	328	322	347	432	366	299
C		346	305	358	485	—	—	272	339	285	213
N			163	201	283	—	—	—	192	—	—
O				146	—	452	335	—	218	201	201
F					155	565	490	284	253	249	278
Si						222	—	293	381	310	234
P							201	—	326	—	184
S								226	255	—	—
Cl									242	216	208
Br										193	175
I											151

Multiple Bonds			
N$=$N	418	C$=$C	610
N$\equiv$N	945	C$\equiv$C	835
C$=$N	615	C$=$O	745
C$\equiv$N	887	C$\equiv$O	1046
O$=$O (in O_2)	498		

*Sources: I. Klotz and R. M. Rosenberg: Chemical Thermodynamics, 4th ed., p. 55, New York, John Wiley, 1994, and J. E. Huheey, E. A. Keiter, and R. L. Keiter: Inorganic Chemistry, 4th ed., Table E.1, New York, Harper-Collins, 1993.

exothermic. If the opposite is true, then the overall reaction is endothermic. Let us see how this works in practice.

Let us use bond energies to estimate the enthalpy change for the hydrogenation of propene to propane:

■ **Hydrogenation Reactions**
Adding hydrogen to a double (or triple) bond is called a hydrogenation reaction. It is commonly done to convert vegetable oils, whose molecules contain C$=$C double bonds, to solid fats.

The first step is to examine the reactants and product to see what bonds are broken and what bonds are formed. In this case, the C$=$C bond in propene and the H—H bond in hydrogen are broken. A C—C bond and two C—H bonds are formed.

Bonds broken: 1 mol of C$=$C bonds and 1 mol of H—H bonds

Energy required per mole = 610 kJ for C$=$C bonds + 436 kJ for H—H bonds = 1046 kJ/mol

Bonds formed: 1 mol of C — C bonds and 2 mol of C — H bonds

$$
\begin{array}{ccccccc}
& \text{H} & & \text{H} & & \text{H} & \\
& | & & | & & | & \\
\text{H} & - & \text{C} & - & \text{C} & - & \text{C} & - \text{H(g)} \\
& | & & | & & | & \\
& \text{H} & & \text{H} & & \text{H} &
\end{array}
$$

Energy evolved = 346 kJ for C — C bonds + 2 mol $\times$ 413 kJ/mol for C — H bonds = 1172 kJ

By combining the energy required to break bonds and the energy evolved in making bonds, we can estimate ΔH°_{rxn} and see that the reaction is exothermic.

$$\Delta H^{\circ}_{rxn} = 1046 \text{ kJ} - 1172 \text{ kJ} = -126 \text{ kJ}$$

The example of the propene–hydrogen reaction illustrates the fact that the enthalpy change for any reaction can be estimated using the equation

$$\Delta H^{\circ}_{rxn} = \sum D(\text{bonds broken}) - \sum D(\text{bonds formed}) \qquad (9.3)$$

To use this equation, first identify all the bonds in the reactants that are broken and add up their bond energies. Then, identify all the new bonds formed in the products and add up their bond energies. The difference between energy required to break bonds [$= \Sigma \, D$(bonds broken)] and the energy evolved when bonds are made [$= \Sigma \, D$(bonds formed)] gives the estimated enthalpy change for the reaction. Bond energy calculations can give acceptable results in many cases.

GENERAL
Chemistry ⚛ Now™

See the General ChemistryNOW CD-ROM or website:
- **Screen 9.20 Bond Energy and ΔH_{rxn},** to explore how reactant and product bond energies influence the energy of reaction, and to work a simulation on bond energies of simple molecules

Example 9.15—Using Bond Energies

Problem Acetone, a common industrial solvent, can be converted to isopropanol, rubbing alcohol, by hydrogenation. Calculate the enthalpy change for this reaction using bond energies.

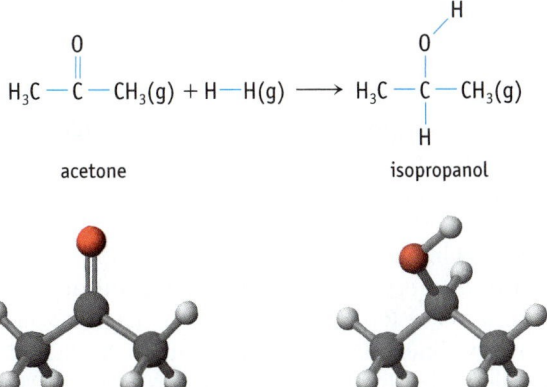

acetone   isopropanol

Strategy Examine the reactants and products to determine which bonds are broken and which are formed. Add up the energies required to break bonds in the reactants and the energy evolved to form bonds in the product. The difference in these energies is an estimate of the enthalpy change of the reaction (Equation 9.3).

Solution

Bonds broken: 1 mol of C=O bonds and 1 mol of H—H bonds

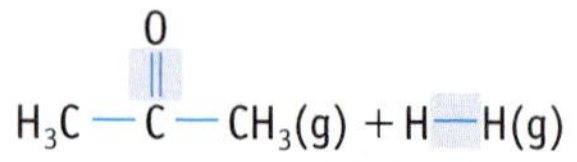

$$\underset{\substack{\| \\ H_3C - C - CH_3(g) + H-H(g)}}{O}$$

$\sum D\text{(bonds broken)} = 745$ kJ for C=O bonds + 436 kJ for H—H bonds = 1181 kJ

Bonds formed: 1 mol of C—H bonds, 1 mol of C—O bonds, and 1 mol of O—H bonds

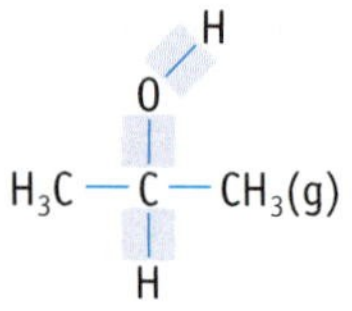

$$\underset{\substack{| \\ H_3C - C - CH_3(g) \\ |\\ H}}{\overset{\substack{H \\ / \\ O \\ |}}{}}$$

$\sum D\text{(bonds formed)} = 413$ kJ for C—H + 358 kJ for C—O + 463 kJ for O—H
$$= 1234 \text{ kJ}$$

$\Delta H^\circ_{rxn} = \sum D\text{(bonds broken)} - \sum D\text{(bonds formed)}$

$\Delta H^\circ_{rxn} = 1181 \text{ kJ} - 1234 \text{ kJ} = -53 \text{ kJ}$

The overall reaction is predicted to be exothermic by 53 kJ per mol of product formed. This is in good agreement with the value calculated from ΔH°_f values ($= -55.8$ kJ).

Exercise 9.18—Using Bond Energies

Using the bond energies in Table 9.10, estimate the heat of combustion of gaseous methane, CH_4. That is, estimate ΔH°_{rxn} for the reaction of methane with O_2 to give water vapor and carbon dioxide gas.

9.11—The DNA Story—Revisited

Chapter 3 opened with the story about the discovery of the structure of DNA, one of the key molecules in all biological systems. The tools are now in place to say more about the structure of this important molecule and why it looks the way it does.

As shown in Figure 9.20, each strand of the double-stranded DNA molecule consists of three units: a phosphate, a deoxyribose molecule (a sugar molecule with a five-member ring), and a nitrogen-containing base. (The bases in DNA can be one of four molecules: adenine, guanine, cytosine, and thymine; in Figure 9.20 the base is adenine.) Two units of the backbone (without the adenine on the deoxyribose ring) are also illustrated in Figure 9.20.

The important point here is that the repeating unit in the backbone of DNA consists of the atoms O—P—O—C—C—C. Each atom has a tetrahedral electron-pair geometry. Therefore, the chain cannot be linear. In fact, the chain twists as one moves along the the backbone. This twisting gives DNA its helical shape.

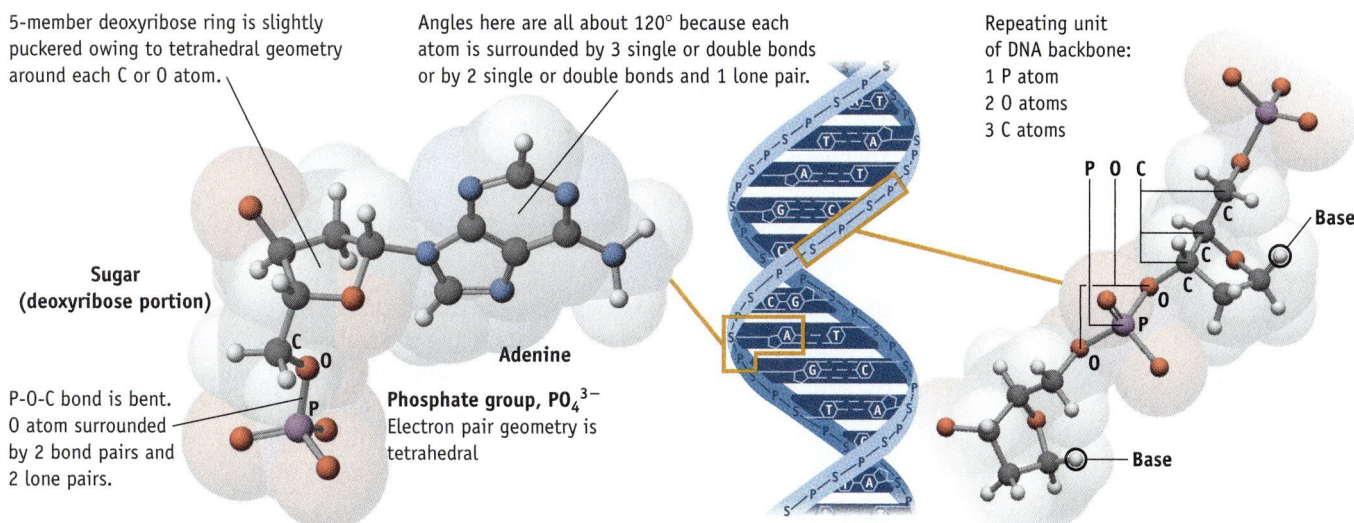

5-member deoxyribose ring is slightly puckered owing to tetrahedral geometry around each C or O atom.

Angles here are all about 120° because each atom is surrounded by 3 single or double bonds or by 2 single or double bonds and 1 lone pair.

Repeating unit of DNA backbone:
1 P atom
2 O atoms
3 C atoms

Sugar (deoxyribose portion)

Adenine

P-O-C bond is bent. O atom surrounded by 2 bond pairs and 2 lone pairs.

Phosphate group, PO₄³⁻ Electron pair geometry is tetrahedral

Base

Base

Figure 9.20 **A portion of the DNA molecule.** A repeating unit consists of a phosphate portion, a deoxyribose portion (a sugar molecule with a five-member ring), and a nitrogen-containing base (here adenine) attached to the deoxyribose ring.

Why are there two strands in DNA with the O—P—O—C—C—C backbone on the outside and the nitrogen-containing bases on the inside? This structure arises from the polarity of the bonds in the base molecules attached to the backbone. For example, the H atoms attached to N in the adenine molecule are very positively charged, which leads to a special form of bonding—hydrogen bonding— to the base molecule in the neighboring chain. More about this in Chapter 13 when we explore intermolecular bonding and again in "*The Chemistry of Life: Biochemistry*" (pages 530–545).

Chapter Goals Revisited

Now that you have studied this chapter, you should ask whether you have met the chapter goals. In particular you should be able to

Understand the difference between ionic and covalent bonds

a. Describe the basic forms of chemical bonding—ionic and covalent—and the differences between them (Section 9.2).

b. Predict from the formula whether a compound has ionic or covalent bonding, based on whether a metal is part of the formula (Section 9.2).

c. Write Lewis symbols for atoms (Section 9.1).

d. Describe the basic ideas underlying ionic bonding and explain how such bonds are affected by the sizes and charges of the ions (Section 9.3). General ChemistryNow homework: Study Question(s) 10

e. Understand lattice energy and know how lattice energies are calculated (Born-Haber cycle); recognize trends in lattice energy and how melting points of ionic compounds are correlated with lattice energy (Section 9.3).

For additional preparation for an examination on this chapter see the **Let's Review** section on pages 530–545.

Draw Lewis electron dot structures for small molecules and ions

    a. Draw Lewis structures for molecular compounds and ions (Section 9.4). General ChemistryNow homework: SQ(s) 12, 14

    b. Understand and apply the octet rule; recognize exceptions to the octet rule (Sections 9.4 and 9.6). General ChemistryNow homework: SQ(s) 18

    c. Write resonance structures, understand what resonance means, and know how and when to use this means of representing bonding (Section 9.5). General ChemistryNow homework: SQ(s) 16

Use the valence shell electron-pair repulsion theory (VSEPR) to predict the shapes of simple molecules and ions and to understand the structures of more complex molecules

    a. Predict the shape or geometry of molecules and ions of main group elements using VSEPR theory (Section 9.7). Table 9.11 summarizes the relation between valence electron pairs, electron-pair and molecular geometry, and molecular polarity. General ChemistryNow homework: SQ(s) 20, 24, 28, 87a, 90b, 97a

Use electronegativity and formal charge to predict the charge distribution in molecules and ions and to define the polarity of bonds

    a. Calculate formal charges for atoms in a molecule based on the Lewis structure (Section 9.8). General ChemistryNow homework: SQ(s) 30

    b. Define electronegativity and understand how it is used to describe the unequal sharing of electrons between atoms in a bond (Section 9.8).

    c. Combine formal charge and electronegativity to gain a perspective on the charge distribution in covalent molecules and ions (Section 9.8). General ChemistryNow homework: SQ(s) 38, 40

Predict the polarity of molecules

    a. Understand why some molecules are polar whereas others are nonpolar (Section 9.9). See Table 9.8. General ChemistryNow homework: SQ(s) 34, 44

    b. Predict the polarity of a molecule (Section 9.9 and Examples 9.13 and 9.14). General ChemistryNow homework: SQ(s) 35

Table 9.11   Summary of Molecular Shapes and Molecular Polarity

Valence Electron Pairs	Electron-Pair Geometry	Number of Bond Pairs	Number of Lone Pairs	Molecular Geometry	Molecular Dipole?*	Examples
2	linear	2	0	linear	no	$BeCl_2$
3	trigonal planar	3	0	trigonal planar	no	BF_3, BCl_3
		2	1	bent (V-shaped)	yes	$SnCl_2(g)$
4	tetrahedral	4	0	tetrahedral	no	CH_4, BF_4^-
		3	1	trigonal-pyramidal	yes	NH_3, PF_3
		2	2	bent (V-shaped)	yes	H_2O, SCl_2
5	trigonal bipyramid	5	0	trigonal bipyramidal	no	PF_5
		4	1	seesaw	yes	SF_4
		3	2	T-shaped	yes	ClF_3
		2	3	linear	no	XeF_2, I_3^-
6	octahedral	6	0	octahedral	no	SF_6, PF_6^-
		5	1	square-pyramidal	yes	ClF_5
		4	2	square-planar	no	XeF_4

*For molecules form of the form AX_n, where the X atoms are identical.

Understand the properties of covalent bonds and their influence on molecular structure

a. Define and predict trends in bond order, bond length, and bond dissociation energy (Section 9.10). General ChemistryNow homework: SQ(s) 48, 50, 54

b. Use bond dissociation energies, *D*, in calculations (Section 9.10 and Example 9.15). General ChemistryNow homework: SQ(s) 56, 57, 58

Key Equations

Equation 9.1 (page 405)

Calculating the formal charge on an atom in a molecule

$$\text{Formal charge of an atom in a molecule or ion} = \text{group number} - \left[\text{LPE} + \tfrac{1}{2}(\text{BE})\right]$$

Equation 9.2 (page 419)

Calculating bond order

$$\text{Bond order} = \frac{\text{number of shared pairs linking X and Y}}{\text{number of X}-\text{Y links in the molecule or ion}}$$

Equation 9.3 (page 423)

Calculating the enthalpy change for a reaction using bond dissociation energies (*D*)

$$\Delta H^{\circ}_{\text{rxn}} = \sum D(\text{bonds broken}) - \sum D(\text{bonds formed})$$

Study Questions

▲ denotes more challenging questions.

■ denotes questions available in the Homework and Goals section of the General ChemistryNow CD-ROM or website.

Blue numbered questions have answers in Appendix O and fully worked solutions in the *Student Solutions Manual.*

Structures of many of the compounds used in these questions are found on the General ChemistryNow CD-ROM or website in the Models folder.

GENERAL
Chemistry⚬∴⚬Now™ Assess your understanding of this chapter's topics with additional quizzing and conceptual questions at http://now.brookscole.com/kotz6e

Practicing Skills

Valence Electrons and the Octet Rule

(See Example 9.1 and General ChemistryNow Screen 9.2.)

1. Give the periodic group number and number of valence electrons for each of the following atoms.

  (a) O   (d) Mg
  (b) B   (e) F
  (c) Na   (f) S

2. Give the periodic group number and number of valence electrons for each of the following atoms.

  (a) C   (d) Si
  (b) Cl   (e) Se
  (c) Ne   (f) Al

3. For elements in Groups 3A–7A of the periodic table, give the number of bonds an element is expected to form if it obeys the octet rule.

4. Which of the following elements are capable of forming compounds in which the indicated atom has more than four valence electron pairs?

  (a) C   (d) F   (g) Se
  (b) P   (e) Cl   (h) Sn
  (c) O   (f) B

Ionic Compounds

(See Section 9.3 and General ChemistryNow Screens 9.3 and 9.4.)

5. Which compound has the most negative energy of ion pair formation? Which has the least negative value?

  (a) NaCl   (b) MgS   (c) KI

6. Which of the following ionic compounds are *not* likely to exist: $MgCl$, $ScCl_3$, BaF_3, $CsKr$, Na_2O? Explain your choices.

7. List the following compounds in order of increasing lattice energy (from least negative to most negative): LiI, LiF, CaO, RbI.

8. Calculate the molar enthalpy of formation, ΔH_f°, of solid lithium fluoride using the approach outlined on pages 378–381. $\Delta H_f^\circ[Li(g)] = 159.37$ kJ/mol, and other required data can be found in Appendices F and L. (See also Exercise 9.2.)

9. To melt an ionic solid, energy must be supplied to disrupt the forces between ions so the regular array of ions collapses. If the distance between the anion and the cation in a crystalline solid decreases (but ion charges remain the same), should the melting point decrease or increase? Explain.

10. ■ Which compound in each of the following pairs should require the higher temperature to melt? (See Study Question 9.)
    (a) $NaCl$ or $RbCl$
    (b) BaO or MgO
    (c) $NaCl$ or MgS

Lewis Electron Dot Structures

(See Examples 9.2–9.5 and General ChemistryNow Screens 9.7–9.8.)

11. Draw a Lewis structure for each of the following molecules or ions.
    (a) NF_3
    (b) ClO_3^-
    (c) $HOBr$
    (d) SO_3^{2-}

12. ■ Draw a Lewis structure for each of the following molecules or ions:
    (a) CS_2
    (b) BF_4^-
    (c) NO_2^-
    (d) $SOCl_2$

13. Draw a Lewis structure for each of the following molecules:
    (a) Chlorodifluoromethane, $CHClF_2$ (C is the central atom)
    (b) Acetic acid, CH_3CO_2H. Its basic structure is pictured.

$$H-\underset{\underset{H}{|}}{\overset{\overset{H}{|}}{C}}-\overset{\overset{O}{\parallel}}{C}-O-H$$

(c) Acetonitrile, CH_3CN (the framework is H_3C-C-N)
(d) Allene, H_2CCCH_2

14. ■ Draw a Lewis structure for each of the following molecules.
    (a) Methanol, CH_3OH (C is the central atom)
    (b) Vinyl chloride, $H_2C=CHCl$, the molecule from which PVC plastics are made.
    (c) Acrylonitrile, $H_2C=CHCN$, the molecule from which materials such as Orlon are made

$$H-\underset{}{\overset{\overset{H}{|}}{C}}-\underset{}{\overset{\overset{H}{|}}{C}}-C-N$$

15. Show all possible resonance structures for each of the following molecules or ions.
    (a) SO_2
    (b) NO_2^-
    (c) SCN^-

16. ■ Show all possible resonance structures for each of the following molecules or ions:
    (a) Nitrate ion, NO_3^-
    (b) Nitric acid, HNO_3
    (c) Nitrous oxide (laughing gas), N_2O

17. Draw a Lewis structure for each of the following molecules or ions.
    (a) BrF_3
    (b) I_3^-
    (c) XeO_2F_2
    (d) XeF_3^+

18. ■ Draw a Lewis structure for each of the following molecules or ions:
    (a) BrF_5
    (b) IF_3
    (c) IBr_2^-
    (d) BrF_2^+

Molecular Geometry

(See Examples 9.6–9.9 and General ChemistryNow Screens 9.12–9.14. Note that many of these molecular structures are available on the General ChemistryNow CD-ROM or website.)

19. Draw a Lewis structure for each of the following molecules or ions. Describe the electron-pair geometry and the molecular geometry around the central atom.
    (a) NH_2Cl
    (b) Cl_2O (O is the central atom)
    (c) SCN^-
    (d) HOF

20. ■ Draw a Lewis structure for each of the following molecules or ions. Describe the electron-pair geometry and the molecular geometry around the central atom.
    (a) ClF_2^+
    (b) $SnCl_3^-$
    (c) PO_4^{3-}
    (d) CS_2

21. The following molecules or ions all have two oxygen atoms attached to a central atom. Draw a Lewis structure for each one and then describe the electron-pair geometry and the molecular geometry around the central atom. Comment on similarities and differences in the series.

(a) CO_2

(b) NO_2^-

(c) O_3

(d) ClO_2^-

22. The following molecules or ions all have three oxygen atoms attached to a central atom. Draw a Lewis structure for each one and then describe the electron-pair geometry and the molecular geometry around the central atom. Comment on similarities and differences in the series.

(a) CO_3^{2-}

(b) NO_3^-

(c) SO_3^{2-}

(d) ClO_3^-

23. Draw a Lewis structure for each of the following molecules or ions. Describe the electron-pair geometry and the molecular geometry around the central atom.

(a) ClF_2^-

(b) ClF_3

(c) ClF_4^-

(d) ClF_5

24. ■ Draw a Lewis structure of each of the following molecules or ions. Describe the electron-pair geometry and the molecular geometry around the central atom.

(a) SiF_6^{2-}

(b) PF_5

(c) SF_4

(d) XeF_4

25. Give approximate values for the indicated bond angles.

(a) $O-S-O$ in SO_2

(b) $F-B-F$ angle in BF_3

(c) $Cl-C-Cl$ angle in Cl_2CO

(c) $H-C-H$ (angle 1) and $C-C\equiv N$ (angle 2) in acetonitrile

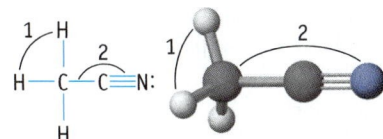

26. Give approximate values for the indicated bond angles.

(a) $Cl-S-Cl$ in SCl_2

(b) $N-N-O$ in N_2O

(c) Bond angles in vinyl alcohol (a component of polymers and another molecule found in space).

27. Phenylalanine is one of the natural amino acids and is a "breakdown" product of aspartame. Estimate the values of the indicated angles in the amino acid. Explain why the $-CH_2-CH(NH_2)-CO_2H$ chain is not linear.

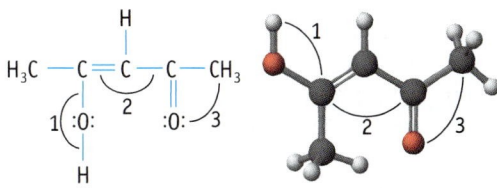

28. ■ Acetylacetone has the structure shown here. Estimate the values of the indicated angles.

Formal Charge

(See Example 9.10 and General ChemistryNow Screen 9.16.)

29. Determine the formal charge on each atom in the following molecules or ions:

(a) N_2H_4 (c) BH_4^-

(b) PO_4^{3-} (d) NH_2OH

30. ■ Determine the formal charge on each atom in the following molecules or ions.

(a) SCO

(b) HCO_2^- (formate ion)

(c) O_3

(d) HCO_2H (formic acid)

31. Determine the formal charge on each atom in the following molecules and ions.

(a) NO_2^+ (c) NF_3

(b) NO_2^- (d) HNO_3

32. Determine the formal charge on each atom in the following molecules and ions.

(a) SO_2 (c) SO_2Cl_2

(b) $SOCl_2$ (d) FSO_3^-

Bond Polarity and Electronegativity

(See Example 9.11 and General ChemistryNow Screen 9.17.)

33. For each pair of bonds, indicate the more polar bond and use an arrow to show the direction of polarity in each bond.

(a) $C-O$ and $C-N$ (c) $B-O$ and $B-S$

(b) $P-Br$ and $P-Cl$ (d) $B-F$ and $B-I$

34. ■ For each of the bonds listed below, tell which atom is the more negatively charged.
    (a) C—N   (c) C—Br
    (b) C—H   (d) S—O

35. ■ Acrolein, C_3H_4O, is the starting material for certain plastics.

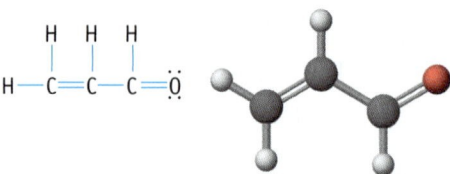

    (a) Which bonds in the molecule are polar and which are nonpolar?
    (b) Which is the most polar bond in the molecule? Which is the more negative atom of this bond?

36. Urea, $(NH_2)_2CO$, is used in plastics and fertilizers. It is also the primary nitrogen-containing substance excreted by humans.
    (a) Which bonds in the molecule are polar and which are nonpolar?
    (b) Which is the most polar bond in the molecule? Which atom is the negative end of the bond dipole?

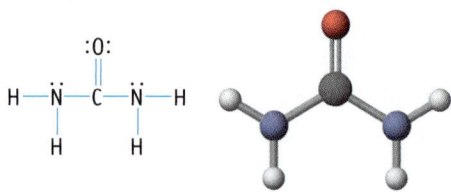

Bond Polarity and Formal Charge
(See Example 9.12 and General ChemistryNow Screens 9.16 and 9.17.)

37. Considering both formal charges and bond polarities, predict on which atom or atoms the negative charge resides in the following anions:
    (a) OH^- (b) BH_4^- (c) $CH_3CO_2^-$

38. ■ Considering both formal charge and bond polarities, predict on which atom or atoms the positive charge resides in the following cations.
    (a) H_3O^+ (c) NO_2^+
    (b) NH_4^+ (d) NF_4^+

39. Three resonance structures are possible for dinitrogen oxide, N_2O.
    (a) Draw the three resonance structures.
    (b) Calculate the formal charge on each atom in each resonance structure.
    (c) Based on formal charges and electronegativity, predict which resonance structure is the most reasonable.

40. ■ Compare the electron dot structures of the carbonate (CO_3^{2-}) and borate (BO_3^{3-}) ions.
    (a) Are these ions isoelectronic?
    (b) How many resonance structures does each ion have?
    (c) What are the formal charges of each atom in these ions?
    (d) If an H^+ ion attaches to CO_3^{2-} to form the bicarbonate ion, HCO_3^-, does it attach to an O atom or to the C atom?

41. Two resonance structures are possible for NO_2^-. Draw these structures and then find the formal charge on each atom in each resonance structure. If an H^+ ion is attached to NO_2^- (to form the acid HNO_2), does it attach to O or N?

42. Draw the resonance structures for the formate ion, HCO_2^- and find the formal charge on each atom. If an H^+ ion is attached to HCO_2^- (to form formic acid), does it attach to C or O?

Molecular Polarity
(See Examples 9.13 and 9.14 and General ChemistryNow Screen 9.18.)

43. Consider the following molecules:
    (a) H_2O (c) CO_2 (e) CCl_4
    (b) NH_3 (d) ClF
        (i)   In which compound are the bonds most polar?
        (ii) Which compounds in the list are *not* polar?
        (iii) Which atom in ClF is more negatively charged?

44. ■ Consider the following molecules:
    (a) CH_4 (c) BF_3
    (b) NH_2Cl (d) CS_2
        (i)   Which compound has the most polar bonds?
        (ii) Which compounds in the list are *not* polar?

45. Which of the following molecules is (are) polar? For each polar molecule, indicate the direction of polarity—that is, which is the negative end and which is the positive end of the molecule.
    (a) $BeCl_2$ (c) CH_3Cl
    (b) HBF_2 (d) SO_3

46. Which of the following molecules is (are) not polar? Which molecule has the most polar bonds?
    (a) CO (d) PCl_3
    (b) BCl_3 (e) GeH_4
    (c) CF_4

Bond Order and Bond Length
(See Exercise 9.17 and General ChemistryNow Screen 9.19.)

47. Give the bond order for each bond in the following molecules or ions.
    (a) CH_2O (c) NO_2^+
    (b) SO_3^{2-} (d) $NOCl$

48. ■ Give the bond order for each bond in the following molecules or ions.
    (a) CN^-
    (b) CH_3CN
    (c) SO_3
    (d) $CH_3CH{=}CH_2$

49. In each pair of bonds, predict which is shorter.
    (a) B—Cl or Ga—Cl
    (b) Sn—O or C—O
    (c) P—S or P—O
    (d) C=O or C=N

50. ■ In each pair of bonds, predict which is shorter.
    (a) Si—N or Si—O
    (b) Si—O or C—O
    (c) C—F or C—Br
    (d) The C—N bond or the C≡N bond in $H_2NCH_2C{\equiv}N$

51. Consider the nitrogen–oxygen bond lengths in NO_2^+, NO_2^-, and NO_3^-. In which ion is the bond predicted to be longest? In which is it predicted to be the shortest? Explain briefly.

52. Compare the carbon–oxygen bond lengths in the formate ion (HCO_2^-), in methanol (CH_3OH), and in the carbonate ion (CO_3^{2-}). In which species is the carbon–oxygen bond predicted to be longest? In which is it predicted to be shortest? Explain briefly.

Bond Energy

(See Table 9.10, Example 9.9, and General ChemistryNow Screen 9.20.)

53. Consider the carbon–oxygen bond in formaldehyde (CH_2O) and carbon monoxide (CO). In which molecule is the CO bond shorter? In which molecule is the CO bond stronger?

54. ■ Compare the nitrogen–nitrogen bond in hydrazine, H_2NNH_2, with that in "laughing gas," N_2O. In which molecule is the nitrogen–nitrogen bond shorter? In which is the bond stronger?

55. Hydrogenation reactions, which involve the addition of H_2 to a molecule, are widely used in industry to transform one compound into another. For example, 1-butene (C_4H_8) is converted to butane (C_4H_{10}) by addition of H_2.

$$
\begin{array}{cccc}
H & H & H & H \\
| & | & | & | \\
H-C-C-C{=}C-H(g) & + & H_2(g) & \longrightarrow \\
| & | & & \\
H & H & &
\end{array}
$$

$$
\begin{array}{cccc}
H & H & H & H \\
| & | & | & | \\
H-C-C-C-C-H(g) & & & \\
| & | & | & | \\
H & H & H & H
\end{array}
$$

Use the bond energies of Table 9.10 to estimate the enthalpy change for this hydrogenation reaction.

56. ■ Phosgene, Cl_2CO, is a highly toxic gas that was used as a weapon in World War I. Using the bond energies of Table 9.10, estimate the enthalpy change for the reaction of carbon monoxide and chlorine to produce phosgene. (*Hint:* First draw the electron dot structures of the reactants and products so you know the types of bonds involved.)

$$CO(g) + Cl_2(g) \longrightarrow Cl_2CO(g)$$

57. ■ The compound oxygen difluoride is quite reactive, giving oxygen and HF when treated with water:

$$OF_2(g) + H_2O(g) \longrightarrow O_2(g) + 2\,HF(g)$$
$$\Delta H^{\circ}_{rxn} = -318 \text{ kJ}$$

Using bond energies, calculate the bond dissociation energy of the O—F bond in OF_2.

58. ■ Oxygen atoms can combine with ozone to form oxygen:

$$O_3(g) + O(g) \longrightarrow 2\,O_2(g) \qquad \Delta H^{\circ}_{rxn} = -394 \text{ kJ}$$

Using ΔH°_{rxn} and the bond energy data in Table 9.10, estimate the bond energy for the oxygen–oxygen bond in ozone, O_3. How does your estimate compare with the energies of an O—O single bond and an O=O double bond? Does the oxygen–oxygen bond energy in ozone correlate with its bond order?

General Questions on Bonding and Molecular Structure

These questions are not designated as to type or location in the chapter. They may combine several concepts. More challenging questions are indicated by ▲.

59. Specify the number of valence electrons for Li, Ti, Zn, Si, and Cl.

60. Describe the formation of KF from K and F atoms using Lewis symbols. Is bonding in KF ionic or covalent?

61. Predict whether the following compounds are ionic or covalent: KI, MgS, CS_2, P_4O_{10}.

62. Define lattice energy. Which should have the more negative lattice energy, LiF or CsF? Explain.

63. Which compound is not likely to exist: $CaCl_2$ or $CaCl_4$? Explain.

64. In boron compounds the B atom often is not surrounded by four valence electron pairs. Illustrate this with BCl_3. Show how the molecule can achieve an octet configuration by forming a coordinate covalent bond with ammonia (NH_3).

65. Which of the following compounds or ions do *not* have an octet of electrons surrounding the central atom: BF_4^-, SiF_4, SeF_4, BrF_4^-, XeF_4?

66. In which of the following does the central atom obey the octet rule: NO_2, SF_4, NH_3, SO_3, O_2^-? Are any of these species odd-electron molecules or ions?

67. Give the bond order of each bond in acetylene, H—C≡C—H, and phosgene, Cl_2CO.

68. Draw resonance structures for the formate ion, HCO_2^- and then determine the C—O bond order in the ion.

69. Determine the N—O bond order in the nitrate ion, NO_3^-.

70. Consider a series of molecules in which carbon is bonded by single bonds to atoms of second-period elements: C—O, C—F, C—N, C—C, and C—B. Place these bonds in order of increasing bond length.

71. To estimate the enthalpy change for the reaction

$$O_2(g) + 2\,H_2(g) \longrightarrow 2\,H_2O(g)$$

what bond energies do you need? Outline the calculation, being careful to show correct algebraic signs.

72. What is the principle of electroneutrality? Use this rule to exclude a possible resonance structure of CO_2.

73. Draw Lewis structures (and resonance structures where appropriate) for the following molecules and ions. What similarities and differences are there in this series?

    (a) CO_2 (b) N_3^- (c) OCN^-

74. Does SO_2 have a dipole moment? If so, what is the direction of the net dipole in SO_2?

75. What are the orders of the N—O bonds in NO_2^- and NO_2^+? The nitrogen–oxygen bond length in one of these ions is 110 pm and 124 pm in the other. Which bond length corresponds to which ion? Explain briefly.

76. Which has the greater O—N—O bond angle, NO_2^- or NO_2^+? Explain briefly.

77. Compare the F—Cl—F angles in ClF_2^+ and ClF_2^-. Using Lewis structures, determine the approximate bond angle in each ion. Decide which ion has the greater bond angle and explain your reasoning.

78. Draw an electron dot structure for the cyanide ion, CN^-. In aqueous solution this ion interacts with H^+ to form the acid. Should the acid formula be written as HCN or CNH?

79. Draw the electron dot structure for the sulfite ion, SO_3^{2-}. In aqueous solution the ion interacts with H^+. Does H^+ attach itself to the S atom or the O atom of SO_3^{2-}?

80. Dinitrogen monoxide, N_2O, can decompose to nitrogen and oxygen gas:

$$2\,N_2O(g) \longrightarrow 2\,N_2(g) + O_2(g)$$

Use bond energies to estimate the enthalpy change for this reaction.

81. ▲ The equation for the combustion of gaseous methanol is

$$2\,CH_3OH(g) + 3\,O_2(g) \longrightarrow 2\,CO_2(g) + 4\,H_2O(g)$$

    (a) Using the bond energies in Table 9.10, estimate the enthalpy change for this reaction. What is the heat of combustion of one mole of gaseous methanol?

    (b) Compare your answer in part (a) with a calculation of ΔH°_{rxn} using thermochemical data and the methods of Chapter 6 (see Equation 6.6).

82. Acrylonitrile, C_3H_3N, is the building block of the synthetic fiber Orlon.

    (a) Give the approximate values of angles 1, 2, and 3.
    (b) Which is the shorter carbon–carbon bond?
    (c) Which is the stronger carbon–carbon bond?
    (d) Which is the most polar bond?

83. ▲ The cyanate ion, NCO^-, has the least electronegative atom, C, in the center. The very unstable fulminate ion, CNO^-, has the same formula, but the N atom is in the center.

    (a) Draw the three possible resonance structures of CNO^-.
    (b) On the basis of formal charges, decide on the resonance structure with the most reasonable distribution of charge.
    (c) Mercury fulminate is so unstable it is used in blasting caps. Can you offer an explanation for this instability? (*Hint:* Are the formal charges in any resonance structure reasonable in view of the relative electronegativities of the atoms?)

84. Vanillin is the flavoring agent in vanilla extract and in vanilla ice cream. Its structure is shown here:

    (a) Give values for the three bond angles indicated.
    (b) Indicate the shortest carbon–oxygen bond in the molecule.
    (c) Indicate the most polar bond in the molecule.

85. ▲ Given that the spatial requirement of a lone pair is much greater than that of a bond pair, explain why
    (a) XeF_2 has a linear molecular structure and not a bent one.
    (b) ClF_3 has a T-shaped structure and not a trigonal-planar one.

86. The formula for nitryl chloride is $ClNO_2$. Draw the Lewis structure for the molecule, including all resonance structures. Describe the electron-pair and molecular geometries, and give values for all bond angles.

---

87. Hydroxyproline is a less common amino acid.

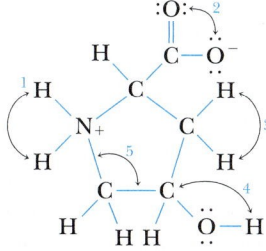

(a) ■ Give approximate values for the indicated bond angles.

(b) Which are the most polar bonds in the molecule?

88. Amides are an important class of organic molecules. They are usually drawn as sketched here, but another resonance structure is possible.

$$H-\overset{\overset{\displaystyle H}{|}}{\underset{\underset{\displaystyle H}{|}}{C}}-\overset{\displaystyle :O:}{\overset{\|}{C}}-\overset{\displaystyle ..}{N}-H$$

(a) Draw that structure, and then suggest why it is usually not pictured.

(b) Suggest a reason for the fact that the H — N — H angle is close to 120°.

89. Use the bond energies in Table 9.10 to calculate the enthalpy change for the decomposition of urea (Study Question 36) to hydrazine, $H_2N — NH_2$, and carbon monoxide. (Assume all compounds are in the gas phase.)

90. The molecule shown here, 2-furylmethanethiol, is responsible for the aroma of coffee:

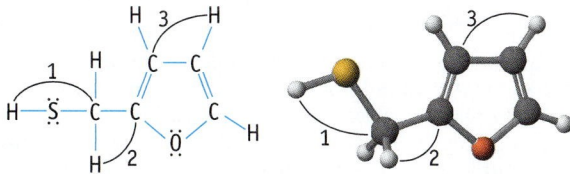

(a) What are the formal charges on the S and O atoms?

(b) ■ Give approximate values of angles 1, 2, and 3.

(c) Which are the shorter carbon–carbon bonds in the molecule?

(d) Which bond in this molecule is the most polar?

(e) Is the molecule as a whole polar or nonpolar?

(f) The molecular model makes it clear that the four C atoms of the ring are all in a plane. Is the O atom in that same plane (making the five-member ring planar), or is the O atom bent above or below the plane?

91. ▲ Dihydroxyacetone is a component of quick-tanning lotions. (It reacts with the amino acids in the upper layer of skin and colors them brown in a reaction similar to that occurring when food is browned as it cooks.)

(a) Supposing you can make this compound by treating acetone with oxygen, use bond energies to estimate the enthalpy change for the following reaction (which is assumed to occur in the gas phase). Is the reaction exothermic or endothermic?

$$H-\overset{\overset{\displaystyle H}{|}}{\underset{\underset{\displaystyle H}{|}}{C}}-\overset{\displaystyle :O:}{\overset{\|}{C}}-\overset{\overset{\displaystyle H}{|}}{\underset{\underset{\displaystyle H}{|}}{C}}-H + O_2 \longrightarrow H-\overset{\displaystyle ..}{\underset{\displaystyle ..}{O}}-\overset{\overset{\displaystyle H}{|}}{\underset{\underset{\displaystyle H}{|}}{C}}-\overset{\displaystyle :O:}{\overset{\|}{C}}-\overset{\overset{\displaystyle H}{|}}{\underset{\underset{\displaystyle H}{|}}{C}}-\overset{\displaystyle ..}{\underset{\displaystyle ..}{O}}-H$$

acetone   dihydroxyacetone

(b) Is acetone polar?

(c) Positive H atoms can sometimes be removed (as H^+) from molecules with strong bases (which is in part what happens in the tanning reaction). Which H atoms are the most positive in dihydroxyacetone?

92. Nitric acid, HNO_3, has three resonance structures. One of them, however, contributes much less to the resonance hybrid than the other two. Sketch the three resonance structures and assign a formal charge to each atom. Which one of your structures is the least important?

93. ▲ Acrolein is used to make plastics. Suppose this compound can be prepared by inserting a carbon monoxide molecule into the C — H bond of ethylene.

$$\underset{\substack{| \\ H}}{\overset{\substack{H \\ |}}{C}}=\underset{\substack{| \\ H}}{\overset{\substack{H \\ |}}{C}} + :C\equiv O: \longrightarrow \underset{\substack{| \\ H}}{\overset{\substack{H \\ |}}{C}}=\underset{\substack{| \\ H}}{\overset{\substack{H \\ |}}{C}}\overset{\substack{H \\ | \\ C=\overset{..}{\underset{..}{O}}}}{}$$

ethylene   acrolein

(a) Which is the stronger carbon–carbon bond in acrolein?

(b) Which is the longer carbon–carbon bond in acrolein?

(c) Is ethylene or acrolein polar?

(d) Is the reaction of CO with C_2H_4 to give acrolein endothermic or exothermic?

94. (a) Glycolaldehyde was featured in the story "Molecules in Space" (page 372). Indicate the unique bond angles in this molecule.

(b) One molecule found in the 1995 Hale-Bopp comet is HC_3N. Suggest a structure for this molecule. (*Hint:* it is based on a chain of atoms.)

95. 1,2-Dichloroethylene can be synthesized by adding Cl_2 to the carbon–carbon triple bond of acetylene.

$$H-C\equiv C-H + Cl_2 \longrightarrow \overset{\substack{H \qquad Cl \\ \diagdown \quad \diagup}}{\underset{\substack{\diagup \quad \diagdown \\ Cl \qquad H}}{C=C}}$$

Using bond energies, estimate the enthalpy change for this reaction in the gas phase.

96. The following molecules or ions have fluorine atoms attached to a central atom from Groups 3A through 7A. Draw the Lewis structure for each one and then describe the electron-pair geometry and the molecular geometry. Comment on similarities and differences in the series.
   (a) BF_3
   (b) CF_4
   (c) PF_3
   (d) OF_2
   (e) HF

97. The molecule pictured below is epinephrine, a compound used as a bronchodilator and antiglaucoma agent.

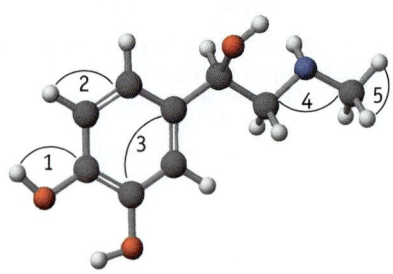

   (a) ■ Give a value for each of the indicated bond angles.
   (b) What are the most polar bonds in the molecule?

Summary and Conceptual Questions

The following questions use concepts from the previous chapters.

98. Define "bond dissociation energy." Does the enthalpy change for a bond-breaking reaction [e.g., $C-H(g) \longrightarrow C(g) + H(g)$] always have a positive sign, always have a negative sign, or vary? Explain briefly.

99. A molecule has four electron pairs around a central atom. Explain how the molecule can have a pyramidal structure. How can the molecule have a bent structure? What bond angles are predicted in each case?

100. What is the difference between the electron-pair geometry and the molecular geometry of a molecule? Use the water molecule as an example in your discussion.

101. Bromine plays a role in environmental chemistry. It is, for example, evolved in volcanic eruptions.

   (a) The following molecules are important in bromine environmental chemistry: HBr, BrO, HOBr, and OH. Which are odd-electron molecules?
   (b) Use bond energies to estimate the energies of three reactions of bromine:
   $Br_2(g) \longrightarrow 2\ Br(g)$
   $2\ Br(g) + O_2(g) \longrightarrow 2\ BrO(g)$
   $BrO(g) + H_2O(g) \longrightarrow HOBr(g) + OH(g)$
   (c) Using bond energies, estimate the standard heat of formation of HOBr(g) from $H_2(g)$, $O_2(g)$, and $Br_2(g)$.
   (d) Are the reactions in parts (b) and (c) exothermic or endothermic?

102. The simple molecule acrylamide, $H_2C=CHC(=O)NH_2$, is a known neurotoxin and possible carcinogen. It was a shock to all consumers of potato chips and french fries a few years ago when was found to occur in those products. (Acrylamide arises during the cooking process from a reaction of the sugar glucose and the amino acid asparagine, both naturally found in many foods.)
   (a) Draw an electron dot structure for acrylamide, showing any possible resonance structures.
   (b) Sketch the molecular structure of acrylamide, showing all unique bond angles.
   (c) Indicate which carbon–carbon bond is the stronger of the two.
   (d) Is the molecule polar or nonpolar?
   (e) The amount of acrylamide found in potato chips is 1.7 mg/kg. If a serving of potato chips is 28 g, how many moles of acrylamide are you consuming?

103. Examine the trends in lattice energy in Table 9.3. The value of the lattice energy becomes somewhat more negative on going from NaI to NaBr to NaCl, and all are in the range of -700 to -800 kJ/mol. Suggest a reason for the observation that the lattice energy of NaF ($\Delta E_{lattice} = -926$ kJ/mol) is much more negative than those of the other sodium halides.

104. Locate the molecules in the table shown here in the Molecular Models folder on the General ChemistryNow CD-ROM or website. Measure the carbon–carbon bond length in each and complete the table. (Note that the bond lengths are given in angstrom units, where 1 Å = 0.1 nm.)

Measured Bond Formula	Bond Distance (Å)	Order
ethane, C_2H_6	_____	_____
butane, C_4H_{10}	_____	_____
ethylene, C_2H_4	_____	_____
acetylene, C_2H_2	_____	_____
benzene, C_6H_6	_____	_____

What relationship between bond order and carbon–carbon bond length do you observe?

105. See General ChemistryNow CD-ROM or website Screen 9.18 Molecular Polarity. Use the Molecular Polarity tool on this screen to explore the polarity of molecules.

    (a) Is BF_3 a polar molecule? Does the molecular polarity change as F is replaced by H on BF_3? Does the polarity change as F is replaced by H? What happens when two F atoms are replaced by H?

    (b) Is $BeCl_2$ a polar molecule? Does the polarity change when Cl is replaced by Br?

106. Locate the following molecules in the Molecular Models folder on the General ChemistryNow CD-ROM or website. In each case, measure unique bond angles and bond lengths and use them to label a sketch of the molecule.

    (a) Tylenol (Drugs folder)

    (b) ClF_3 (Inorganic folder)

    (c) Ethylene glycol (Organic Alcohols folder)

Do you need a live tutor for homework problems?
Access vMentor at General ChemistryNow at
http://now.brookscole.com/kotz6e
for one-on-one tutoring from a chemistry expert

10—Bonding and Molecular Structure: Orbital Hybridization and Molecular Orbitals

Linus Pauling (1901–1994).

Thomas Hollyman/Photo Researchers, Inc.

Linus Pauling: A Life of Chemical Thought

Linus Pauling received the Nobel Prize for chemistry in 1954, an honorary high school diploma in 1962, and the Nobel Peace Prize in 1962. That is an extraordinary sequence—but then Pauling was an extraordinary man. He was born on February 28, 1901, in Portland, Oregon. His father, an itinerant pharmaceutical salesman, died when he was nine, and Pauling soon became a partial provider for his mother and two younger sisters. Although an excellent high school student, he refused to wait around to complete a civics requirement and so did not graduate. It was merely the first of many civil disobediences.

Against the wishes of his mother in 1917, Pauling enrolled as a chemical engineering major at Oregon Agricultural College. His interests soon turned to chemistry and, after taking a year off to help support his family, he graduated in 1922. He decided to embark on graduate work at the California Institute of Technology, then a fledgling institution, unlike today's research powerhouse. Pauling's research involved the use of the relatively new technique of x-ray crystallography to determine the atomic-level structure of crystals. Experiment alone, however, was not sufficient; he also needed to master the relevant theory. Consequently, Pauling followed his Ph.D. studies with a tour of European centers of the emerging discipline of quantum mechanics. He was superbly—perhaps uniquely—equipped for his life's work.

Returning to Caltech, Pauling began an intensive program of structural determination, using x-ray crystallography for solids and electron diffraction for vapors. Interatomic distances and angles were digested and analyzed, and quantum mechanical calculations were made. Prediction became possible. This endeavor was superbly summarized in his 1939 book, *The Nature of the Chemical Bond, and the Structure of Molecules and Crystals,* which was to prove the most influential chemistry text of the 20th century.

Having largely solved the structural chemistry of inorganic and simple organic substances, Pauling then turned his attention to biochemical materials. He was to become, in Francis Crick's words, "one of the founders of molecular biology." Pauling, along with coworker Robert Corey, systematically tackled the basic structural chemistry of proteins. On his fiftieth birthday, he communicated his landmark paper on the α-helix to the *Proceedings of the National Academy of Sciences.* It was his work on proteins, together with his studies of the nature of the chemical bond, that was cited in the award of the 1954 Nobel Prize for chemistry.

But Pauling's scientific career was not yet half over. He met with both disappointments (his failure to solve the structure of DNA and the nonacceptance of his spheron model of nuclear stability) and successes (a diagnosis of sickle cell anemia as a "molecular disease" and the introduction of the molecular evolutionary clock). However, from about 1950, science was to be merely one part, and at times even a minor part, of his active life.

Born into a relatively conservative family, Pauling became, under the urgent prompting of his wife, Ava Helen, an active political propagandist and agitator.

In particular, he played a major role in bringing about the nuclear test ban treaty of 1962. For this effort, he received the 1962 Nobel Peace prize. While his chemistry prize was universally praised, his peace prize was widely denounced in the conservative press. Only much later would his contributions be given their due.

The last years of Pauling's life were mainly spent in the advocacy of what he called "ortho-molecular medicine"—the optimization of levels of various minerals and vitamins in the human body. The most familiar was the prescription of megadoses of vitamin C to treat various ailments, especially the common cold. The nutritional establishment was outraged. The RDA value was vastly smaller than the 1 to 10 grams per day recommended by Pauling, who argued that the amount of vitamin C necessary to prevent scurvy was not necessarily sufficient to contribute maximally to bodily health. While the jury is still out on some of Pauling's more extreme claims, the medical establishment has suddenly developed a fondness for antioxidants such as vitamin C.

Active and optimistic to the end, Linus Pauling died at his ranch on the Big Sur coast of California on August 19, 1994.

Linus and Ava Helen Pauling. Dr. Pauling received the Nobel Peace prize in 1962, but his wife also played a major role in the effort to bring about a nuclear test ban treaty.

Essay by Derek Davenport, Professor Emeritus of Chemistry, Purdue University

Just how are molecules held together? How can two distinctly different molecules have the same formula? Why is oxygen paramagnetic, and how is this property connected with bonding in the molecule? These are just a few of the fundamental and interesting questions that are raised in this chapter and that require us to take a more advanced look at bonding.

10.1—Orbitals and Bonding Theories

Orbitals, both atomic and molecular, are the focus of this chapter. The quantum mechanical model for the atom, which is the most successful way to explain the properties of atoms that scientists have devised, describes electrons in atoms as waves. An atomic orbital has a specific energy related to electrostatic forces: an attractive force due to the positively charged atomic nucleus acting on an electron in that orbital, and a repulsive force acting on the electron due to the other electrons in the atom. If the energy of the orbital is known accurately, an electron's position is known less well (the Heisenberg uncertainty principle). For this reason, we think of orbitals as regions in space in which there is a high probability of finding the electron (Figures 7.14 and 7.15).

From Chapters 7 and 8 you know that the locations of the valence electrons in atoms are described by an orbital model. It seems reasonable that an orbital model could also be used to describe electrons in molecules.

Two common approaches to rationalizing chemical bonding based on orbitals are the **valence bond (VB) theory** and the **molecular orbital (MO) theory**. The former was developed largely by Linus Pauling (page 436) and the latter by Robert S. Mulliken, another American physicist. The valence bond approach is closely tied to Lewis's idea of bonding electron pairs between atoms and lone pairs of electrons localized on a particular atom. In contrast, Mulliken's approach was to derive molecular orbitals that are "spread out," or *delocalized*, over the molecule. One way to do so is to combine atomic orbitals to form a set of orbitals that are the property of the molecule, and then distribute the electrons of the molecule within these orbitals.

Why are two theories used? Isn't one more correct than the other? Actually, both give good descriptions of the bonding in molecules and polyatomic ions, but they are used for different purposes. Valence bond theory is generally the method of choice to provide a qualitative, visual picture of molecular structure and bonding. This theory is particularly useful for molecules made up of many atoms. In contrast, molecular orbital theory is used when a more quantitative picture of bonding is needed. Furthermore, VB theory provides a good description of bonding for molecules in their ground, or lowest, energy state. In contrast, MO theory is essential if we want to describe molecules in higher-energy excited states. Among other things, it is important in explaining the colors of compounds. Finally, for a few molecules such as NO and O_2, MO theory is the only theory to describe their bonding accurately.

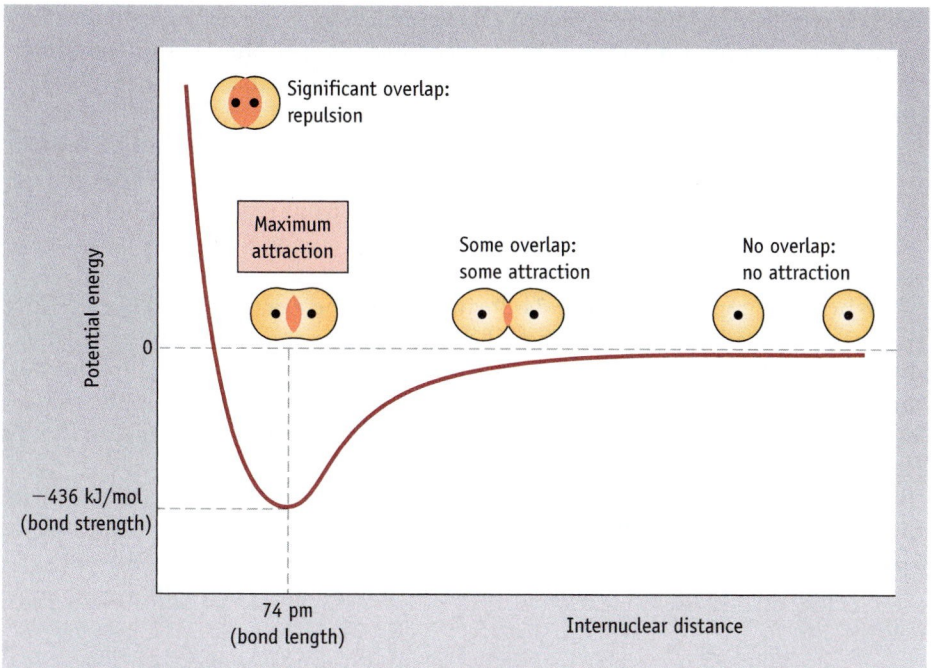

Active Figure 10.1 **Potential energy change during H — H bond formation from isolated hydrogen atoms.** The lowest energy is reached at an H — H separation of 74 pm, where there is overlap of 1s orbitals. At greater distances the overlap is less, and the bond is weaker. At H — H distances less than 74 pm, repulsions between the nuclei and between the electrons of the two atoms increase rapidly, and the potential energy curve rises steeply. Thus, an H_2 molecule is expected to be less stable when the distance between the atoms is very small.

GENERAL
Chemistry Now™ See the General ChemistryNow CD-ROM or website to explore an interactive version of this figure accompanied by an exercise.

10.2—Valence Bond Theory

Orbital Overlap Model of Bonding

What happens if two atoms at an infinite distance apart are brought together to form a bond? This process is often illustrated with H_2 because, with just two electrons and two nuclei, it is the simplest molecular compound known (Figure 10.1). Initially, when two hydrogen atoms are widely separated, they do not interact. If the atoms move closer together, however, the electron on one atom begins to experience an attraction to the positive charge of the nucleus of the other atom. Because of the attractive forces, the electron clouds on the atoms become distorted as the electron of one atom is drawn toward the nucleus of the second atom, and the potential energy of the system is lowered. Calculations show that when the distance between the H atoms is 74 pm, the potential energy reaches a minimum and the H_2 molecule is most stable. Significantly, 74 pm corresponds to the experimentally measured bond distance in the H_2 molecule.

Each individual hydrogen atom has a single electron. In H_2 the two electrons pair up to form the bond. There is a net stabilization, representing the extent to which the energies of the two electrons are lowered from their values in the free atoms. The net stabilization (the extent by which the potential energy is lowered)

■ **Bonds in Valence Bond Theory**
In the language of valence bond theory, a pair of electrons of opposite spin located between a pair of atoms constitutes a bond.

can be calculated, and the calculated value approximates the experimentally determined bond energy [◀ Section 9.10]. Agreement between theory and experiment on both bond distance and energy constitutes evidence that this theoretical approach has merit.

Bond formation is depicted in Figures 10.1 and 10.2 as occurring when the electron clouds on the two atoms interpenetrate, or overlap. This **orbital overlap** increases the probability of finding the bonding electrons in the region of space between the two nuclei. *The idea that bonds are formed by overlap of atomic orbitals is the basis for valence bond theory.*

When the single covalent bond is formed in H_2, the electron cloud of each atom becomes distorted in a way that gives the electrons a higher probability of being in the region between the two hydrogen atoms. This outcome makes sense, because the distortion results in the electrons being situated so that they can be attracted equally to the two positively charged nuclei. Placing the electrons between the nuclei also matches the Lewis electron dot model.

The covalent bond that arises from the overlap of two s orbitals, one from each of two atoms as in H_2, is called a **sigma (σ) bond**. *The electron density of a sigma bond is greatest along the axis of the bond.*

In summary, the main points of the valence bond approach to bonding are as follows:

- Orbitals overlap to form a bond between two atoms (see Figure 10.2).

- Two electrons, of opposite spin, can be accommodated in the overlapping orbitals. Usually one electron is supplied by each of the two bonded atoms.

- Because of orbital overlap, the bonding electrons have a higher probability of being found within a region of space influenced by both nuclei. Both electrons are simultaneously attracted to both nuclei.

What happens with elements beyond hydrogen? In the Lewis structure of HF, for example, a bonding electron pair is placed between H and F, and three lone pairs of electrons are depicted as localized on the F atom (Figure 10.2b). To use an orbital approach, look at the valence shell electrons and orbitals for each atom that will overlap. The hydrogen atom will use its $1s$ orbital in bond formation. The electron configuration of fluorine is $1s^2 2s^2 2p^5$, and the unpaired electron for this atom is assigned to one of the $2p$ orbitals. A sigma bond results from overlap of the hydrogen $1s$ and the fluorine $2p$ orbital.

Formation of the H—F bond is similar to formation of an H—H bond. A hydrogen atom approaches a fluorine atom along the axis containing the $2p$ orbital with a single electron. The orbitals ($1s$ on H and $2p$ on F) become distorted as each atomic nucleus influences the electron and orbital of the other atom. Still closer together, the $1s$ and $2p$ orbitals overlap, and the two electrons pair up to give a σ bond (see Figure 10.2b). There is an optimal distance (92 pm) at which the energy is lowest, which corresponds to the bond distance in HF. The net stabilization achieved in this process is the energy for the H—F bond.

The remaining electrons on the fluorine atom (two electrons in the $2s$ orbital and four electrons in the other two $2p$ orbitals) are not involved in bonding. The lone pairs associated with this element in the Lewis structure are nonbonding electrons.

Extension of this model gives a description of bonding in F_2. The $2p$ orbitals on the two atoms overlap, and the single electron from each atom is paired in the resulting σ bond (Figure 10.2c). The $2s$ and $2p$ electrons not involved in the bond are the lone pairs on each atom.

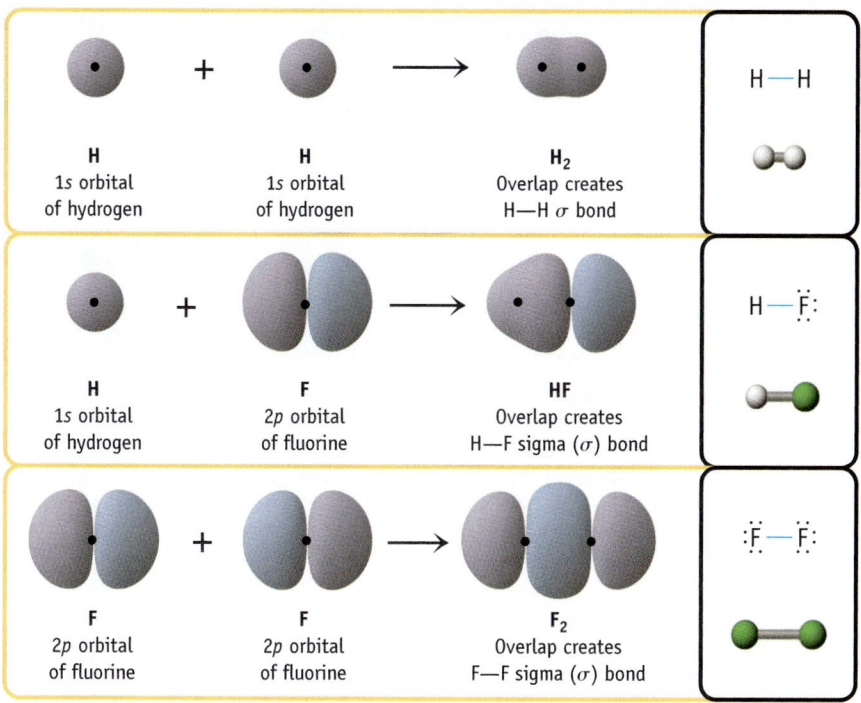

Figure 10.2 **Covalent bond formation in H₂, HF, and F₂.**

(a) Overlap of hydrogen 1s orbitals to form the H—H sigma bond.

(b) Overlap of hydrogen 1s and fluorine 2p orbitals to form the sigma (σ) bond in HF.

(c) Overlap of 2p orbitals on two fluorine atoms to form the sigma (σ) bond in F₂.

GENERAL
Chemistry·⚛·Now™

See the General ChemistryNow CD-ROM or website:
• **Screen 10.3 Valence Bond Theory,** for an animation of bond formation

Hybridization of Atomic Orbitals

The simple picture using orbital overlap to describe bonding in H_2, HF, and F_2 works well, but we run into difficulty when molecules with more atoms are considered. For example, a Lewis dot structure of methane, CH_4, shows four C—H covalent bonds. VSEPR theory predicts, and experiments confirm, that the electron-pair geometry of the C atom in CH_4 is tetrahedral, with an angle of 109.5° between the bond pairs. The hydrogens are identical in this structure. Thus four equivalent bonding electron pairs occur around the C atom. An orbital picture of the bonds should convey both the geometry and the fact that all C—H bonds are the same.

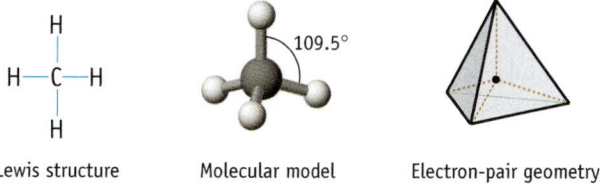

Lewis structure   Molecular model   Electron-pair geometry

If we apply the orbital overlap model used for H_2 and F_2 without modification to describe the bonding in CH_4, a problem arises. The three orbitals for the $2p$ valence electrons of carbon are at right angles, 90° (Figure 10.3), and do not match the tetrahedral angle of 109.5°. The spherical $2s$ orbital could bond in any direction.

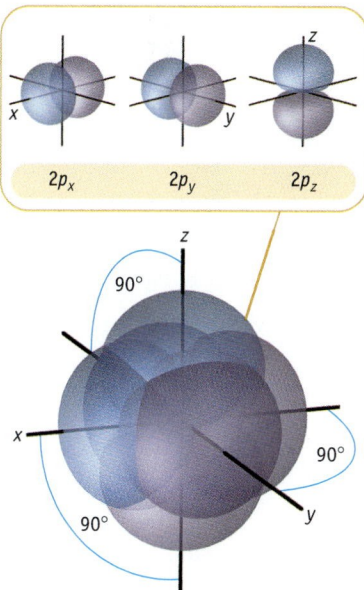

Figure 10.3 **The 2p orbitals of an atom.** The $2p_x$, $2p_y$, and $2p_z$ orbitals lie long the x-, y-, and z-axes, 90° to each other.

Figure 10.4   **Hybridization: an analogy.** Atomic orbitals can mix, or hybridize, to form hybrid orbitals. When two atomic orbitals on an atom combine, two new orbitals are produced on that atom. The new orbitals have a different direction in space than the original orbitals. An analogy is mixing two different colors (*left*) to produce a third color, which is a "hybrid" of the original colors (*center*). After mixing there are still two beakers (*right*), each containing the same volume of solution as before, but the color is a "hybrid" color. (*See General ChemistryNow Screens 10.4 Hybrid Orbitals, and 10.6 Determining Hybrid Orbitals.*)

Furthermore, a carbon atom in its ground state ($1s^2 2s^2 2p^2$) has only two unpaired electrons (in the $2p$ orbitals), not the four that are needed to allow formation of four bonds.

To describe the bonding in methane and other molecules, Linus Pauling proposed the theory of **orbital hybridization** (Figure 10.4). He suggested that a new set of orbitals, called **hybrid orbitals**, could be created by mixing the *s*, *p*, and (when required) *d* atomic orbitals on an atom. Two important principles govern the outcome. First, *the number of hybrid orbitals is always the same as the number of atomic orbitals that are mixed to create the hybrid orbital set.* Second, *the hybrid orbitals are more directed from the central atom toward the terminal atoms than are the unhybridized atomic orbitals, leading to better orbital overlap and a stronger bond between the central and terminal atoms.*

The sets of hybrid orbitals that arise from mixing *s*, *p*, and *d* atomic orbitals are illustrated in Figure 10.5. The following features are important:

- The hybrid orbitals required by an atom in a molecule or ion are determined by the electron-pair geometry around that atom. A hybrid orbital is required for each sigma bond or lone electron pair on a central atom.

- If the valence shell *s* orbital on the central atom in a molecule or ion is mixed with a valence shell *p* orbital on that same atom, two hybrid orbitals are created. They are separated by 180°. The set of two orbitals is labeled *sp*.

- If an *s* orbital is combined with two *p* orbitals, all in the same valence shell, three hybrid orbitals are created. They are separated by 120°, and the set of three orbitals is labeled sp^2.

- When the *s* orbital in a valence shell is combined with three *p* orbitals, the result is four hybrid orbitals, each labeled sp^3. The hybrid orbitals are separated by 109.5°, the tetrahedral angle.

- If one or two *d* orbitals are added to the sp^3 set, then two other hybrid orbital sets are created. They are utilized by the central atom of a molecule or ion with a trigonal-bipyramidal or octahedral electron-pair geometry.

Let us examine a case of each type of hybridization in simple molecules, returning first to the case of methane. Keep in mind, however, that these principles apply to atoms in even the most complex molecules, such as DNA.

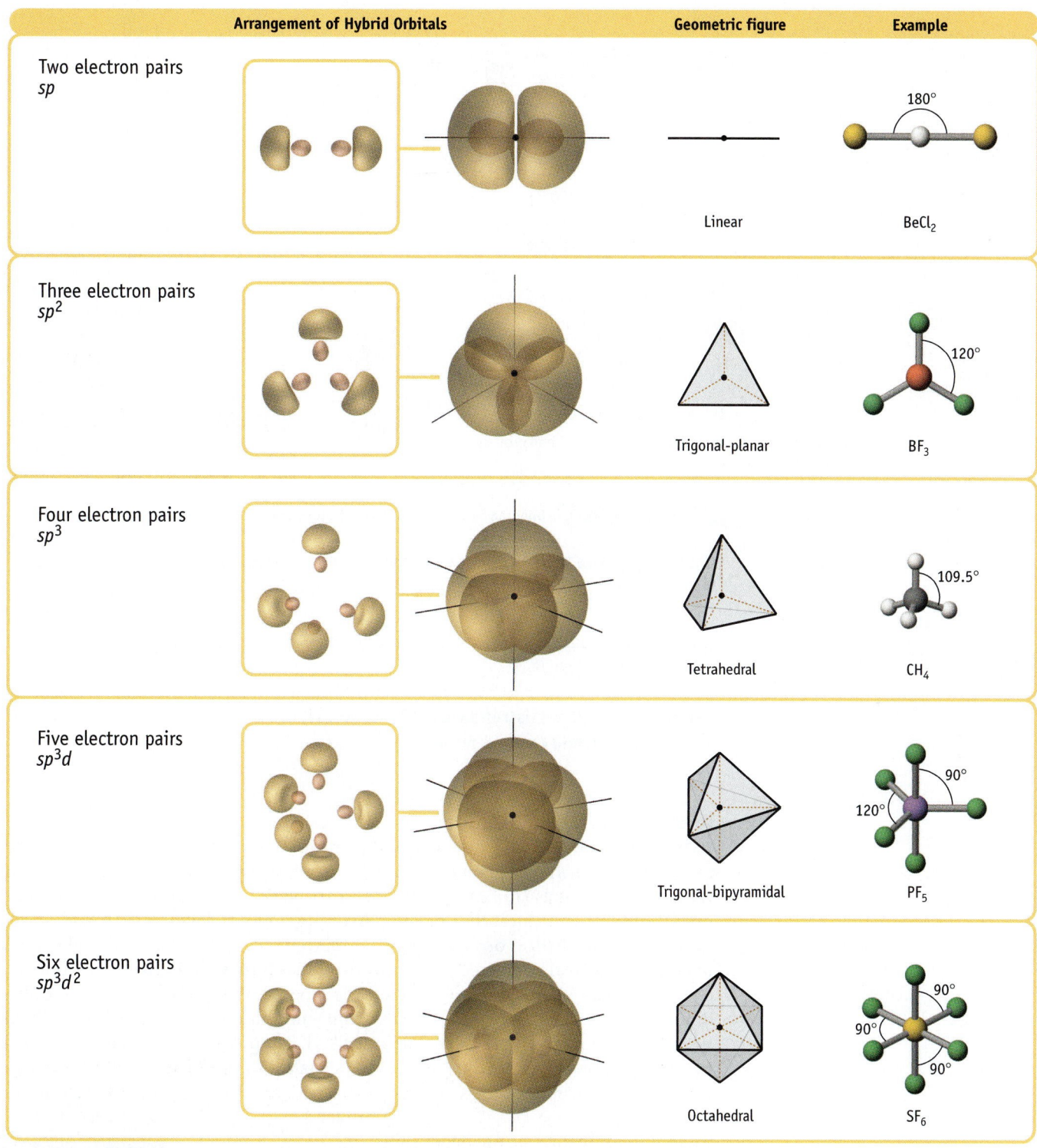

Arrangement of Hybrid Orbitals		Geometric figure	Example
Two electron pairs *sp*		Linear	180° BeCl$_2$
Three electron pairs *sp^2*		Trigonal-planar	120° BF$_3$
Four electron pairs *sp^3*		Tetrahedral	109.5° CH$_4$
Five electron pairs *sp^{3}d*		Trigonal-bipyramidal	90° 120° PF$_5$
Six electron pairs *sp^{3}d^2*		Octahedral	90° 90° 90° SF$_6$

Active Figure 10.5 **Hybrid orbitals for two to six electron pairs.** The geometry of the hybrid orbital sets for two to six valence shell electron pairs is given in the right column. In forming a hybrid orbital set, the *s* orbital is always used, plus as many *p* orbitals (and *d* orbitals) as are required to give the necessary number of σ-bonding and lone-pair orbitals.

GENERAL
Chemistry ⚛ Now™ See the General ChemistryNow CD-ROM or website to explore an interactive version of this figure accompanied by an exercise.

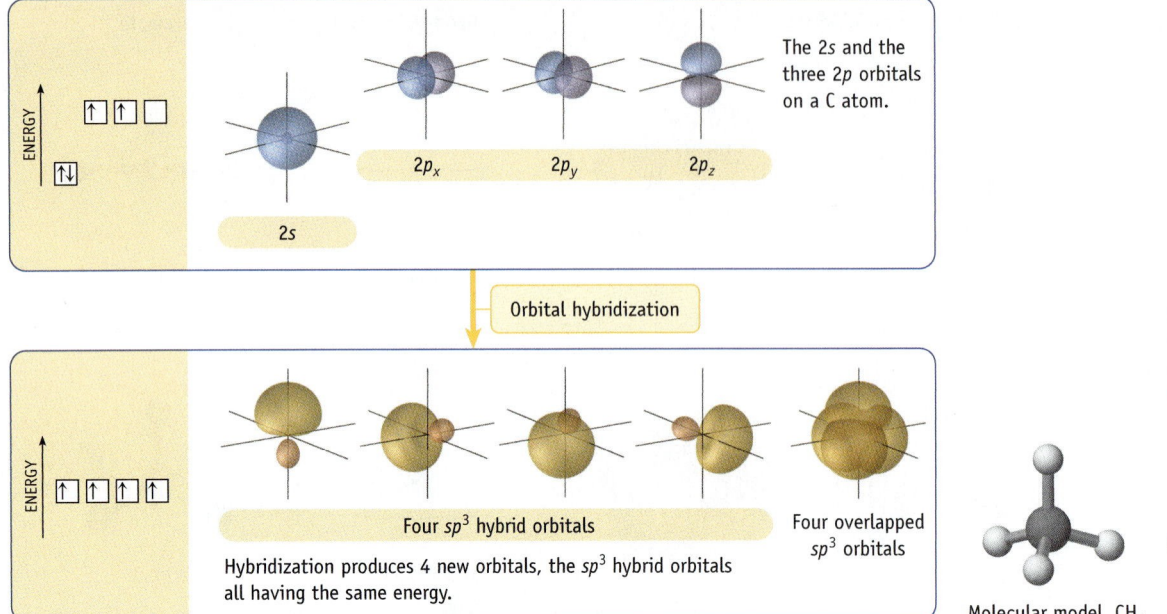

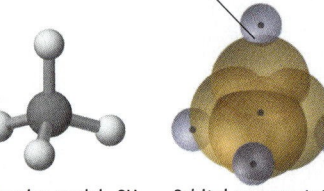

The 2s and the three 2p orbitals on a C atom.

2p_x   2p_y   2p_z

2s

Orbital hybridization

ENERGY

Four sp³ hybrid orbitals

Four overlapped sp³ orbitals

Hybridization produces 4 new orbitals, the sp³ hybrid orbitals all having the same energy.

Each C—H bond uses one C atom sp³ hybrid orbital and a H atom 1s orbital

Molecular model, CH₄   Orbital representation

Active Figure 10.6   Bonding in the methane (CH₄) molecule.

GENERAL
Chemistry ⚛ Now™ See the General ChemistryNow CD-ROM or website to explore an interactive version of this figure accompanied by an exercise.

GENERAL
Chemistry ⚛ Now™

See the General ChemistryNow CD-ROM or website:
- **Screen 10.4 Hybrid Orbitals,** for an animation of the formation of sp³ hybrid orbitals

Valence Bond Theory for Methane, CH₄

In methane, four orbitals directed to the corners of a tetrahedron are needed to match the electron-pair geometry on the central carbon atom. By mixing the four valence shell orbitals (the 2s and all three 2p orbitals on carbon), a new set of four hybrid orbitals is created that has tetrahedral geometry (Figures 10.5 and 10.6). Each of the four hybrid orbitals is labeled sp³ to indicate the atomic orbital combination (an s orbital and three p orbitals) from which they are derived. The four sp³ orbitals have an identical shape, and the angle between them is 109.5°, the tetrahedral angle. Because the orbitals have the same energy, one electron can be assigned to each according to Hund's rule [◀ Section 8.4]. Then, each C—H bond is formed by overlap of one of the carbon sp³ hybrid orbitals with the 1s orbital of a hydrogen atom; one electron from the C atom is paired with an electron from an H atom.

Valence Bond Theory for Ammonia, NH₃

The Lewis structure for ammonia shows four electron pairs in the valence shell of nitrogen: three bond pairs and a lone pair (Figure 10.7). VSEPR theory predicts a tetrahedral electron-pair geometry and a trigonal-pyramidal molecular geometry. Structure evidence is a close match to prediction; the H — N — H bond angles are 107.5° in this molecule.

■ **Hybrid Orbitals & Atomic Orbitals**
Note that *four* atomic orbitals produce *four* hybrid orbitals. The number of hybrid orbitals produced is always the same as the number of atomic orbitals used.

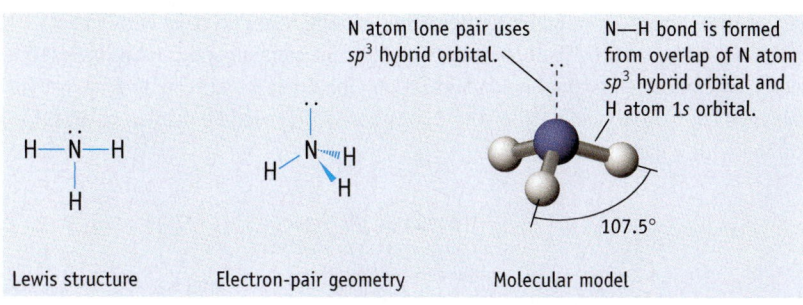

Figure 10.7 **Bonding in ammonia, NH₃, and water, H₂O.**

N atom lone pair uses sp^3 hybrid orbital.

N—H bond is formed from overlap of N atom sp^3 hybrid orbital and H atom 1s orbital.

107.5°

Lewis structure   Electron-pair geometry   Molecular model

O atom lone pairs use sp^3 hybrid orbitals.

O—H bond is formed from overlap of O atom sp^3 hybrid orbital and H atom 1s orbital.

104.5°

Lewis structure   Electron-pair geometry   Molecular model

Based on the electron-pair geometry of NH_3, we predict sp^3 hybridization to accommodate the four electron pairs on the N atom. The lone pair is assigned to one of the hybrid orbitals, and each of the other three hybrid orbitals is occupied by a single electron. Overlap of each of the singly occupied, sp^3 hybrid orbitals with a 1s orbital for hydrogen, and pairing of the electrons in these orbitals, creates the N—H bonds about 109° apart.

Valence Bond Theory for Water, H₂O

The oxygen atom of water has two bonding pairs and two lone pairs in its valence shell, and the H—O—H angle is 104.5° (Figure 10.7). Four sp^3 hybrid orbitals are created from the 2s and 2p atomic orbitals of oxygen. Two of these sp^3 orbitals are occupied by unpaired electrons and are used to form O—H bonds. Lone pairs occupy the other two hybrid orbitals.

GENERAL
Chemistry ⚛ Now™

See the General ChemistryNow CD-ROM or website:
- **Screen 10.5 Sigma Bonding,** for a tutorial on sigma bond formation
- **Screen 10.6 Determining Hybrid Orbitals,** for a tutorial on determining hybrid orbitals

■ **Hybridization and Geometry**
Hybridization reconciles the electron-pair geometry with the orbital overlap criterion of bonding. A statement such as "the atom is tetrahedral because it is sp^3 hybridized" is backward. That the electron-pair geometry around the atom is tetrahedral is a fact. Hybridization is one way to rationalize that fact.

Example 10.1—Valence Bond Description of Bonding in Ethane

Problem Describe the bonding in ethane, C_2H_6, using valence bond theory.

Strategy First, draw the Lewis structure and predict the electron-pair geometry at both carbon atoms. Next, assign a hybridization to these atoms. Finally, describe covalent bonds that arise based on orbital overlap, and place electron pairs in their proper locations.

Solution Each carbon atom has an octet configuration, sharing electron pairs with three hydrogen atoms and with the other carbon atom. The electron pairs around carbon have tetrahedral geometry, so carbon is assigned sp^3 hybridization. The C—C bond is formed by overlap of sp^3 orbitals on each C atom, and each of the C—H bonds is formed by overlap of an sp^3 orbital on carbon with a hydrogen 1s orbital.

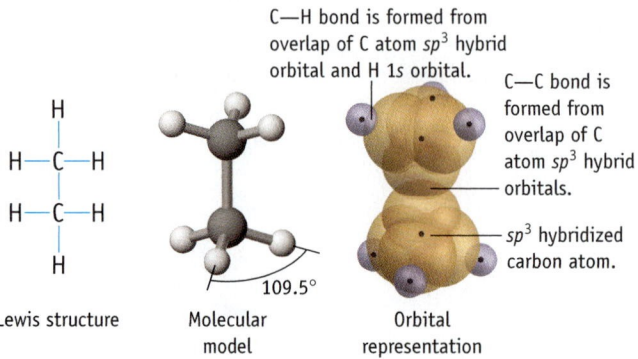

C—H bond is formed from overlap of C atom sp^3 hybrid orbital and H 1s orbital.

C—C bond is formed from overlap of C atom sp^3 hybrid orbitals.

sp^3 hybridized carbon atom.

| Lewis structure | Molecular model | Orbital representation |

Example 10.2—Valence Bond Description of Bonding in Methanol

Problem Describe the bonding in the methanol molecule, CH_3OH, using valence bond theory.

Strategy Construct the Lewis structure for the molecule. The electron-pair geometry around each atom determines the hybrid orbital set used by that atom.

Solution The electron-pair geometry around both the C and O atoms in CH_3OH is tetrahedral. Thus, we may assign sp^3 hybridization to each atom, and the C—O bond is formed by overlap of sp^3 orbitals on these atoms. Each C—H bond is formed by overlap of a carbon sp^3 orbital with a hydrogen 1s orbital, and the O—H bond is formed by overlap of an oxygen sp^3 orbital with the hydrogen 1s orbital. Two lone pairs on oxygen occupy the remaining sp^3 orbitals.

Comment Notice that one end of the CH_3OH molecule (the CH_3 or methyl group) is just like the CH_3 group in the ethane molecule (Example 10.1), and the OH group resembles the OH group in water. It is helpful to recognize pieces of molecules and their bonding descriptions.

This example also shows how to predict the structure and bonding in a complicated molecule by looking at each atom separately. This important principle is essential when dealing with molecules made up of many atoms.

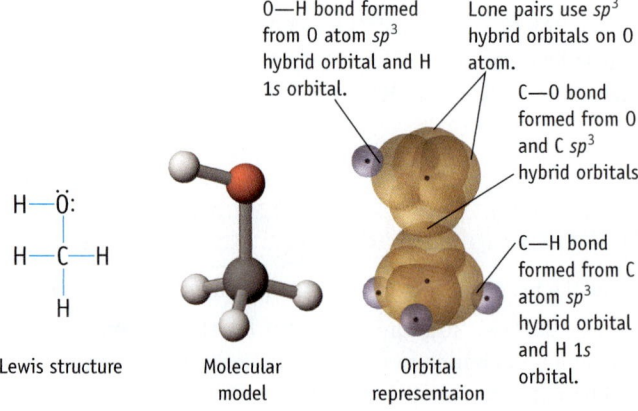

O—H bond formed from O atom sp^3 hybrid orbital and H 1s orbital.

Lone pairs use sp^3 hybrid orbitals on O atom.

C—O bond formed from O and C sp^3 hybrid orbitals.

C—H bond formed from C atom sp^3 hybrid orbital and H 1s orbital.

| Lewis structure | Molecular model | Orbital representaion |

Exercise 10.1—Valence Bond Description of Bonding

Use valence bond theory to describe the bonding in the hydronium ion, H_3O^+, and methylamine, CH_3NH_2.

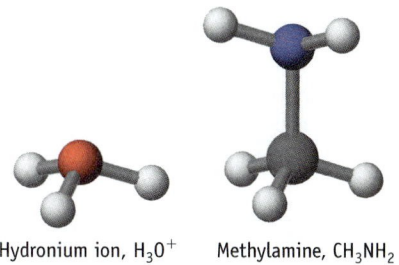

Hydronium ion, H_3O^+ Methylamine, CH_3NH_2

Hybrid Orbitals for Molecules and Ions with Trigonal-Planar Electron-Pair Geometries

Atoms having trigonal-planar geometries are commonly encountered in molecules and ions. For example, BF_3 and other boron halides are trigonal-planar, as are a number of other species, such as NO_3^- and CO_3^{2-}. The carbon atoms in ethylene, $CH_2 = CH_2$, are also trigonal-planar, and the electron-pair geometry of O_3 and NO_2^- is trigonal-planar.

A trigonal-planar electron-pair geometry requires a central atom with three hybrid orbitals in a plane, 120° apart. Three hybrid orbitals mean three atomic orbitals must be combined, and the combination of an s orbital with two p orbitals is appropriate (Figure 10.5). If p_x and p_y orbitals are used in hybrid orbital formation, the three hybrid sp^2 orbitals will lie in the xy-plane. The p_z orbital not used to form these hybrid orbitals is perpendicular to the plane containing the three sp^2 orbitals (Figure 10.8).

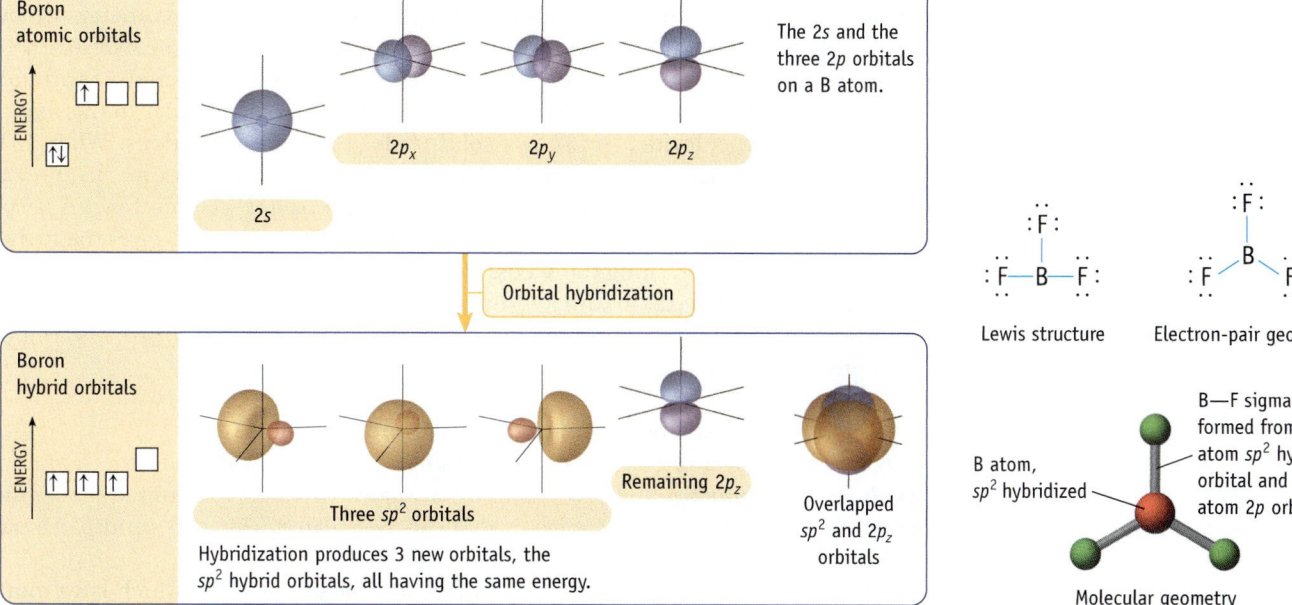

Figure 10.8 **Bonding in a trigonal-planar molecule.**

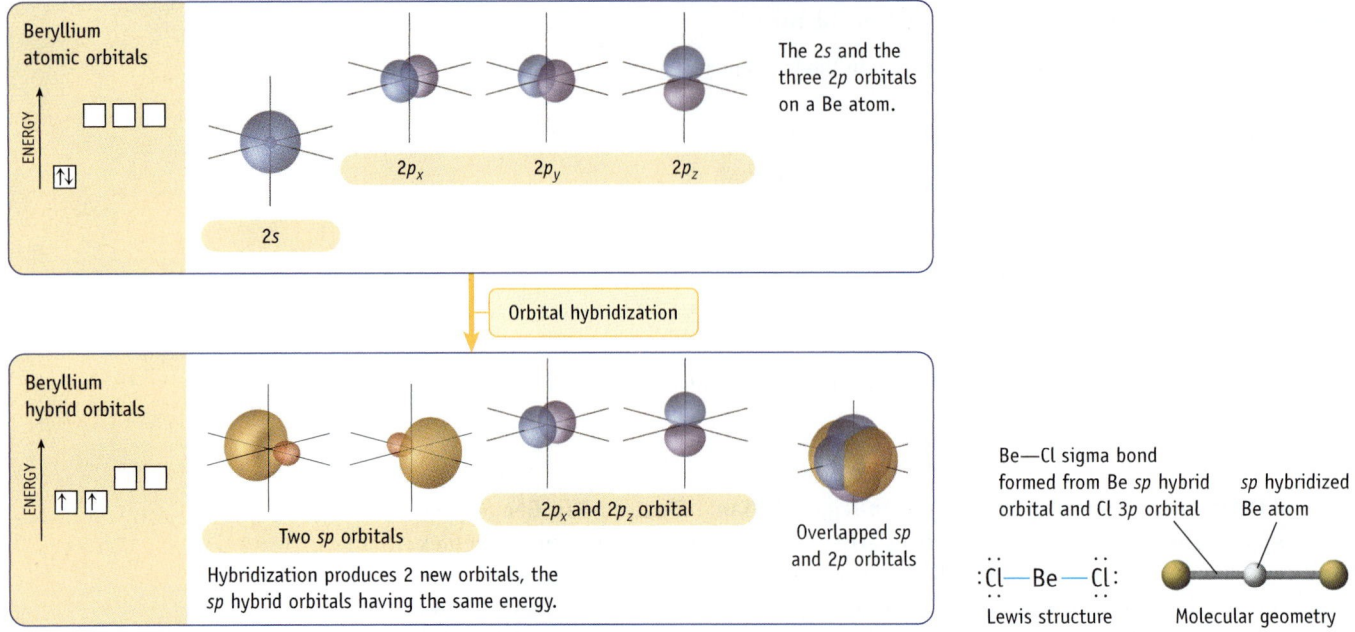

Figure 10.9   Bonding in a linear molecule. Because only one *p* orbital is incorporated in the hybrid orbital, two *p* orbitals remain. These orbitals are perpendicular to each other and to the axis along which the two *sp* hybrid orbitals lie.

Boron trifluoride has trigonal-planar electron-pair and molecular geometries. Each boron–fluorine bond in this compound results from overlap of an sp^2 orbital on boron with a *p* orbital on fluorine. Notice that the p_z orbital on boron, which is not used to form the sp^2 hybrid orbitals, is not occupied by electrons.

Hybrid Orbitals for Molecules and Ions with Linear Electron-Pair Geometries

For molecules in which the central atom has a linear electron-pair geometry, two hybrid orbitals, 180° apart, are required. One *s* and one *p* orbital can be hybridized to form two *sp* hybrid orbitals (Figure 10.9). If the p_x orbital is used, then the *sp* orbitals are oriented along the *x*-axis. The p_y and p_z orbitals are perpendicular to this axis.

Beryllium dichloride, $BeCl_2$, is a solid under ordinary conditions. When it is heated to more than 520 °C, however, it vaporizes to give $BeCl_2$ vapor. In the gas phase, $BeCl_2$ is a linear molecule, so *sp* hybridization is appropriate for the beryllium atom in this species. Combining beryllium's 2*s* and $2p_x$ orbitals gives the two *sp* hybrid orbitals that lie along the *x*-axis. Each Be — Cl bond arises by overlap of an *sp* hybrid orbital on beryllium with a 3*p* orbital on chlorine. In this molecule, there are only two electron pairs around the beryllium atom, so the p_y and p_z orbitals are not occupied (Figure 10.9).

Hybrid Orbitals Involving *s*, *p*, and *d* Atomic Orbitals

A basic assumption of Pauling's valence bond theory is that *the number of hybrid orbitals equals the number of valence orbitals used in their creation.* As a consequence, the maximum number of hybrid orbitals that can be created from the *s* and *p* orbitals for an atom is four.

How, then, should we deal with compounds like PF_5 and SF_6, which have more than four electron pairs in their valence shells? To describe the bonding in compounds having five or six electron pairs on a central atom requires the atom to have five or six hybrid orbitals, which must be created from five or six atomic orbitals. This is possible if additional atomic orbitals from the d subshell are used in hybrid orbital formation. The d orbitals are considered to be valence shell orbitals for main group elements of the third and higher periods.

To accommodate six electron pairs in the valence shell of an element, six sp^3d^2 hybrid orbitals can be created from the one s, three p, and two d orbitals. The six sp^3d^2 hybrid orbitals are directed to the corners of an octahedron (Figure 10.5). Thus, they are oriented to accommodate the valence electron pairs for a compound that has an octahedral electron-pair geometry. Five coordination and trigonal-bipyramidal geometry are matched to sp^3d hybridization. One s, three p, and one d orbital combine to produce five sp^3d hybrid orbitals.

Example 10.3—Hybridization Involving d Orbitals

Problem Describe the bonding in PF_5 using valence bond theory.

Strategy The first step is to establish the electron pair and molecular geometries of PF_5. The electron-pair geometry around the P atom gives the number of hybrid orbitals required. If five hybrid orbitals are required, the combination of atomic orbitals is sp^3d.

Solution Here the P atom is surrounded by five electron pairs, so PF_5 has trigonal-bipyramidal electron-pair and molecular geometries. The hybridization scheme is therefore sp^3d .

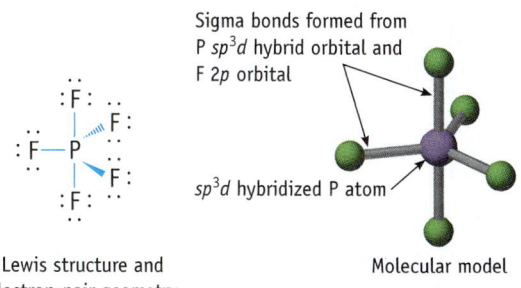

Sigma bonds formed from
P sp^3d hybrid orbital and
F $2p$ orbital

sp^3d hybridized P atom

Lewis structure and
electron-pair geometry

Molecular model

Example 10.4—Recognizing Hybridization

Problem Identify the hybridization of the central atom in the following compounds and ions:

(a) SF_3^+ **(b)** SO_4^{2-} **(c)** SF_4 **(d)** I_3^-

Strategy The hybrid orbitals used by a central atom are determined by the electron-pair geometry (see Figure 10.5). To answer this question, first write the Lewis structure and then predict the electron-pair geometry.

Solution Following the procedures in Chapter 9, the Lewis structures for SF_3^+ and SO_4^{2-} can be written as follows:

Four electron-pairs surround the center atom in each of these ions, and the electron-pair geometry for these atoms is tetrahedral. Thus, sp^3 hybridization for the central atom is used to describe the bonding.

For SF_4 and I_3^-, five pairs of electrons are in the valence shell of the center atom. For these, sp^3d hybridization is appropriate for the central S or I atom.

Exercise 10.2—Hybridization Involving d Orbitals

Describe the bonding in XeF_4 using hybrid orbitals. Remember to consider first the Lewis structure, then the electron-pair geometry (based on VSEPR theory), and finally the molecular shape.

Exercise 10.3—Recognizing Hybridization

Identify the hybridization of the central atom in the following compounds and ions:

(a) BH_4^- (c) OSF_4 (e) BCl_3

(b) SF_5^- (d) ClF_3 (f) XeO_6^{4-}

Multiple Bonds

According to valence bond theory, bond formation requires that two orbitals on adjacent atoms overlap. Many molecules have two or three bonds between pairs of atoms. Therefore, according to valence bond theory, a double bond requires *two* sets of overlapping orbitals and *two* electron pairs. For a triple bond, *three* sets of atomic orbitals are required, with each set accommodating a pair of electrons.

■ Multiple Bonds

C=C

Double bond requires two sets of overlapping orbitals and two pairs of electrons.

C≡C

Triple bond requires three sets of overlapping orbitals and three pairs of electrons.

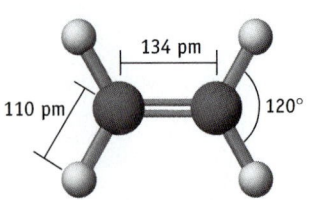

Ethylene, C_2H_4

134 pm

110 pm

120°

Double Bonds

Consider ethylene, $H_2C=CH_2$, one of the more common molecules with a double bond. The molecular structure of ethylene places all six atoms in a plane, with $H-C-H$ and $H-C-C$ angles of approximately 120°. Each carbon atom has trigonal-planar geometry, so sp^2 hybridization is assumed for these atoms. Thus, a description of bonding in ethylene starts with each carbon atom having three sp^2 hybrid orbitals in the molecular plane and an unhybridized p orbital perpendicular to that plane (see Figure 10.8). Because each carbon atom is involved in four bonds, a single unpaired electron is placed in each of these orbitals.

[↑] Unhybridized p orbital. Used for π bonding in C_2H_4.

[↑↑↑↑] Three sp^2 hybrid orbitals. Used for C—H and C—C σ bonding in C_2H_4.

Now we can visualize the C—H bonds, which arise from overlap of sp^2 orbitals on carbon with hydrogen $1s$ orbitals. After accounting for the C—H bonds, one sp^2 orbital on each carbon atom remains. These orbitals point toward each other and over-

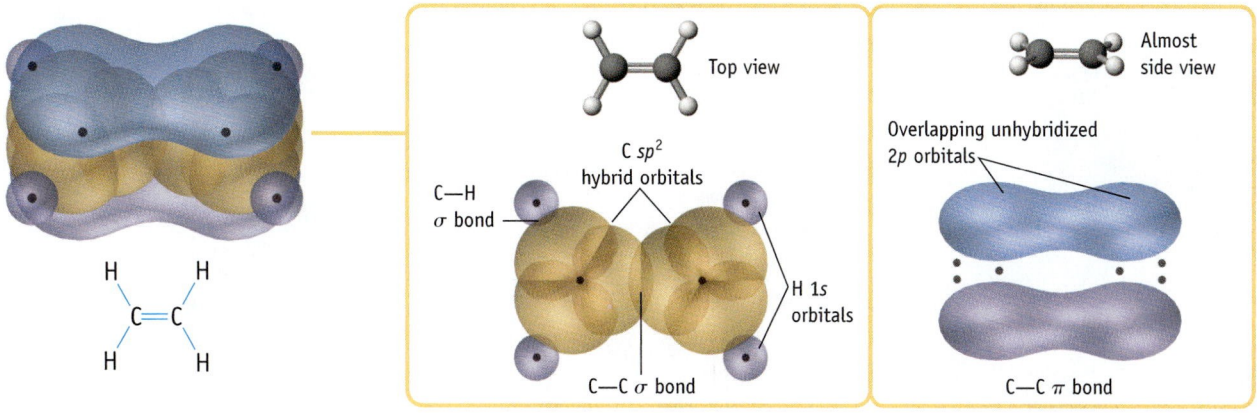

(a) Lewis structure and bonding of ethylene, C_2H_4.

(b) The C—H σ bonds are formed by overlap of C atom sp^2 hybrid orbitals with H atom $1s$ orbitals. The σ bond between C atoms arises from overlap of sp^2 orbitals.

(c) The carbon-carbon π bond is formed by overlap of an unhybridized $2p$ orbital on each atom. Note the lack of electron density along the C—C bond axis.

Active Figure 10.10   **The valence bond model of bonding in ethylene, C_2H_4.** Each C atom is assumed to be sp^2 hybridized.

GENERAL
Chemistry ⚛ Now™ See the General ChemistryNow CD-ROM or website to explore an interactive version of this figure accompanied by an exercise.

lap to form one of the bonds linking the carbon atoms (Figure 10.10a). This leaves only one other orbital unaccounted for on each carbon, an unhybridized p orbital (see Figure 10.8). These orbitals can be used to create the second bond between carbon atoms in C_2H_4. If they are aligned correctly, the unhybridized p orbitals on the two carbons can overlap, allowing the electrons in these orbitals to be paired. The overlap does not occur directly along the C — C axis, however. Instead, the arrangement compels these orbitals to overlap sideways, and the electron pair occupies an orbital with electron density above and below the plane containing the six atoms (Figure 10.10c).

This description results in two types of bonds in C_2H_4. One type is the C — H and C — C bonds that arise from the overlap of atomic orbitals so that the bonding electrons that lie along the bond axis form sigma (σ) bonds. The other is the bond formed by sideways overlap of p atomic orbitals, called a **pi (π) bond**. In a π bond, the overlap region is above and below the internuclear axis, and the electron density of the π bond is above and below the σ bond axis (Figures 10.10b and 10c).

Notice that a π bond can form *only* if (1) there are unhybridized p orbitals on adjacent atoms and (2) the p orbitals are perpendicular to the plane of the molecule and parallel to one another. This happens only if the sp^2 orbitals of both carbon atoms are in the same plane. Thus, *the π bond requires that all six atoms of the molecule lie in one plane.*

Double bonds between carbon and oxygen, sulfur, or nitrogen are quite common. Consider formaldehyde, CH_2O, in which a carbon–oxygen π bond occurs (Figure 10.11). A trigonal-planar electron-pair geometry indicates sp^2 hybridization for the C atom. The σ bonds from the C atom to the O atom and the two H atoms form by overlap of sp^2 hybrid orbitals with half-filled orbitals from the oxygen and two hydrogen atoms. An unhybridized p orbital on carbon is oriented perpendicular to the molecular plane (just as for the carbon atoms of C_2H_4). This p orbital is available for π bonding, this time with an oxygen orbital.

What orbitals on oxygen are used in this model? The approach in Figure 10.11 assumes sp^2 hybridization for oxygen. This uses one O atom sp^2 orbital in σ bond

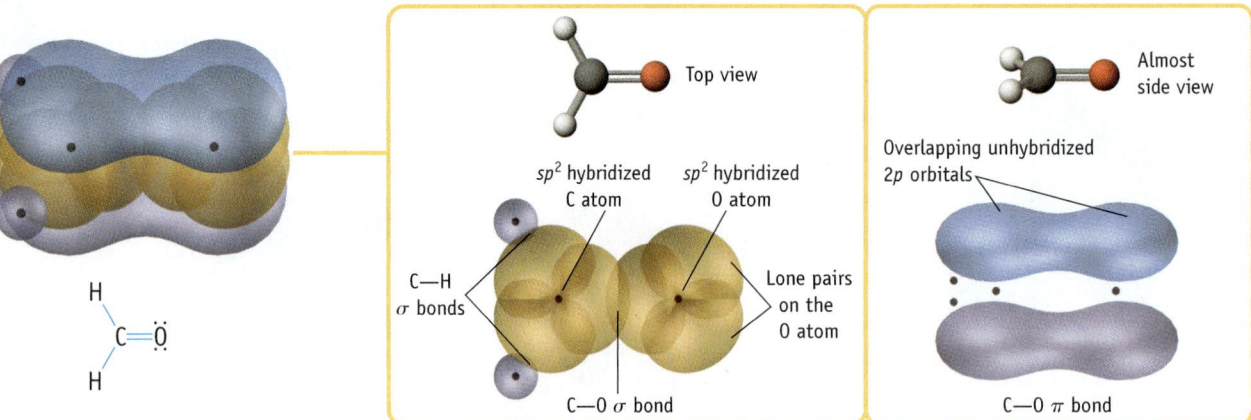

(a) Lewis structure and bonding of formaldehyde, CH_2O.

(b) The C—H σ bonds are formed by overlap of C atom sp^2 hybrid orbitals with H atom $1s$ orbitals. The σ bond between C and O atoms arises from overlap of sp^2 orbitals.

(c) The C—O π bond comes from the side-by-side overlap of p orbitals on the two atoms.

Figure 10.11   Valence bond description of bonding in formaldehyde, CH_2O.

formation, leaving two sp^2 orbitals to accommodate lone pairs. The remaining p orbital on the O atom participates in the π bond.*

GENERAL
Chemistry Now™

See the General ChemistryNow CD-ROM or website:
• **Screen 10.7 Multiple Bonding,** for a tutorial on hybrid orbitals and σ and π bonding

Example 10.5—Bonding in Acetic Acid

Problem Using valence bond theory, describe the bonding in acetic acid, CH_3CO_2H, the important ingredient in vinegar.

Strategy Write a Lewis electron dot structure and determine the electron-pair geometry around each atom using VSEPR. Use this geometry to decide on the hybrid orbitals used in σ bonding. If unhybridized p orbitals are available, then C=O π bonding can occur.

Solution The carbon atom of the CH_3 group has tetrahedral electron-pair geometry, which means that it is sp^3 hybridized. Three sp^3 orbitals are used to form the C—H bonds. The fourth sp^3 orbital is used to bond to the adjacent carbon atom. This carbon atom has a trigonal-planar electron-pair geometry, so it must be sp^2 hybridized. The C—C bond is formed using one of these orbitals, and the other two sp^2 orbitals are used to form the σ bonds to the two oxygens. The oxygen of the O—H group has four electron pairs, so it must be tetrahedral and sp^3 hybridized. Thus, this O atom uses two sp^3 orbitals to bond to the adjacent carbon and the hydrogen, and two sp^3 orbitals accommodate the two lone pairs.

---

* A second approach is to use unhybridized orbitals on oxygen in bonding. If unhybridized oxygen is assumed, the two p orbitals that are oriented at right angles, and that each contain a single electron, are used to create σ and π bonds. The argument favoring hybridization for oxygen is that it adds consistency to the valence bond approach; because hybridization is required for some atoms, it makes sense to use it for all of them. The objection is that hybridization was introduced simply to explain molecular geometry. The O atom is bonded to only one other atom, so there is no geometry to explain; that is, hybridization does not add anything to the explanation, and it could be regarded as an additional complication.

Finally, the carbon–oxygen double bond can be described exactly as in the CH_2O molecule (Figure 10.11). Both the C and O atoms are assumed to be sp^2 hybridized, and the unhybridized p orbital remaining on each atom is used to form the carbon–oxygen π bond.

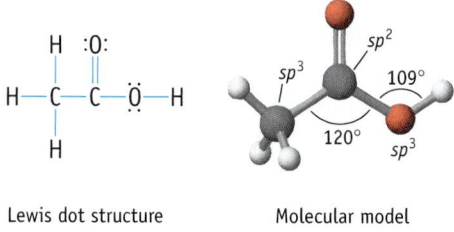

Lewis dot structure   Molecular model

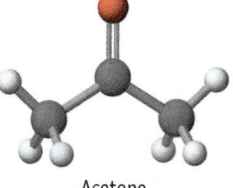

Acetone

Exercise 10.4—Bonding in Acetone

Use valence bond theory to describe the bonding in acetone, CH_3COCH_3.

Triple Bonds

Acetylene, $H—C\equiv C—H$, is an example of a molecule with a triple bond. VSEPR allows us to predict that the four atoms lie in a straight line with $H—C—C$ angles of $180°$. This arrangement implies that the carbon atom is sp hybridized (Figure 10.9). For each carbon atom, there are two sp orbitals: one directed toward hydrogen and used to create the $C—H$ σ bond, and one directed toward the other carbon and used to create a σ bond between the two carbon atoms. Two unhybridized p orbitals remain on each carbon, and they are oriented so that it is possible to form *two* π bonds in $H—C\equiv C—H$ (Figures 10.9 and 10.12).

⊞⊞ Two unhybridized p orbitals. Used for π bonding in C_2H_2.

⊞⊞ Two sp hybrid orbitals. Used for $C—H$ and $C—C$ σ bonding in C_2H_2.

These π bonds are perpendicular to the molecular axis and perpendicular to each other. Three electrons on each carbon atom are paired to form the triple bond consisting of a σ bond and two π bonds (Figure 10.12).

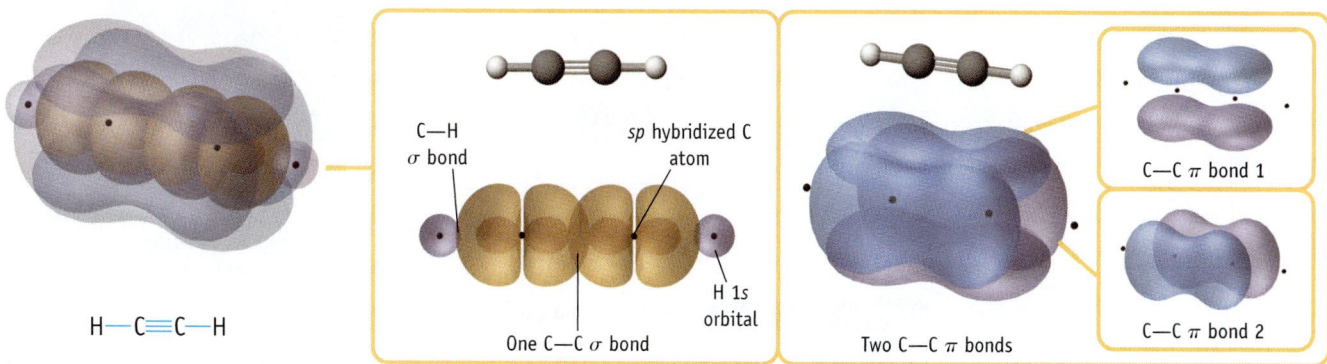

Figure 10.12 **Bonding in acetylene.**

Now that we have examined two cases of multiple bonds, let us summarize several important points:

- A double bond always consists of a σ bond and a π bond. Similarly, a triple bond always consists of a σ bond and *two* π bonds.

- A π bond may form only if unhybridized p orbitals remain on the bonded atoms.

- If a Lewis structure shows multiple bonds, the atoms involved must be either sp^2 or sp hybridized. Only in this manner will unhybridized p orbitals be available to form a π bond.

Exercise 10.5—Triple Bonds Between Atoms

Describe the bonding in a nitrogen molecule, N_2.

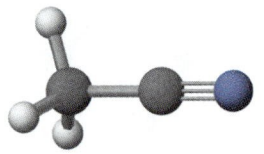
Acetonitrile, CH_3CN

Exercise 10.6—Bonding and Hybridization

Estimate values for the H—C—H, H—C—C, and C—C—N angles in acetonitrile, $CH_3C\equiv N$. Indicate the hybridization of both carbon atoms and the nitrogen atom, and analyze the bonding using valence bond theory.

Cis-Trans Isomerism: A Consequence of π Bonding

Ethylene, C_2H_4, is a planar molecule. This geometry allows the unhybridized p orbitals on the two carbon atoms to line up and form a π bond (see Figure 10.10). Let us speculate on what would happen if one end of the ethylene molecule is twisted relative to the other end (Figure 10.13). This action would distort the molecule away from planarity, and the p orbitals would rotate out of alignment. Rotation

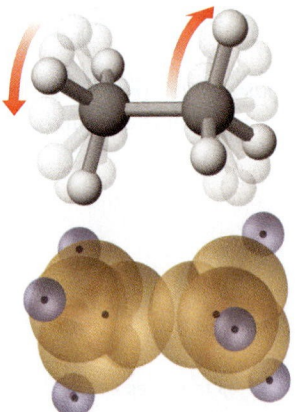

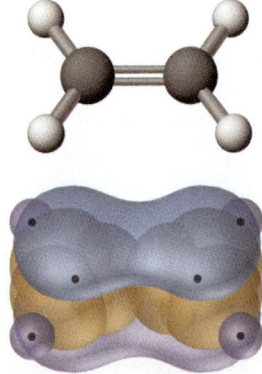

(a) Free rotation can occur around the axis of a single (σ) bond.

(b) Rotation is severely restricted around double bonds because doing so would break the π bond, a process generally requiring a great deal of energy.

Active Figure 10.13   **Rotation around bonds.**

GENERAL
Chemistry Now™ See the General ChemistryNow CD-ROM or website to explore an interactive version of this figure accompanied by an exercise.

would diminish the extent of overlap of these orbitals. If a twist of 90° was achieved, the two p orbitals would no longer overlap; the π bond would be broken. However, so much energy is required to break this bond (about 260 kJ/mol) that rotation around a C=C bond is not expected to occur at room temperature.

A consequence of restricted rotation is that isomers occur for many compounds containing a C=C bond. **Isomers** are compounds that have the same formula but different structures. In this case, the two isomeric compounds differ with respect to the orientation of the groups attached to the carbons of the double bond. Two isomeric compounds with the formula $C_2H_2Cl_2$ are *cis-* and *trans*-1,2-dichloroethylene. Their structures resemble ethylene, except that two hydrogen atoms have been replaced by chlorine atoms. Because a large amount of energy is required to break the π bond, the *cis* compound cannot rearrange to form the *trans* compound under ordinary conditions. Each compound can be obtained separately, and each has its own identity. *Cis*-1,2-dichloroethylene boils at 60.3 °C, whereas *trans*-1,2-dichloroethylene boils at 47.5 °C.

■ **Cis and Trans Isomers**
Compounds having the same formula, but different structures, are isomers. *Trans* isomers have distinguishing groups on opposite sides of a double bond. *Cis* isomers have these groups on the same side of the double bond.

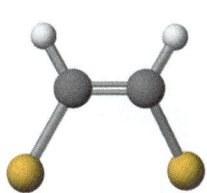
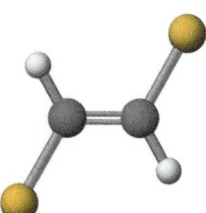

cis-1,2-dichloroethylene   *trans*-1,2-dichloroethylene

Although *cis* and *trans* isomers do not interconvert at ordinary temperatures, they will do so at higher temperatures. According to the kinetic theory of matter [◀ Section 1.1], molecules in the gas and liquid phases move rapidly and often collide with one another. Molecules also constantly flex or vibrate along or around the bonds holding them together. If the temperature is sufficiently high, the molecular motions can become sufficiently energetic that rotation around the C=C bond can occur. It may also occur under other special conditions, such as when the molecule absorbs light energy. Indeed, this specific situation is found to occur in the physiological process that allows us to see (Figure 10.14).

GENERAL
Chemistry · · Now™

See the General ChemistryNow CD-ROM or website:
• **Screen 10.8 Molecular Fluxionality,** for an exercise on isomers and multiple bonds

Benzene: A Special Case of π Bonding

Benzene, C_6H_6, is the simplest member of a large group of substances known as *aromatic* compounds, a historical reference to their odor. It occupies a pivotal place in the history and practice of chemistry.

To 19th-century chemists, benzene was a perplexing substance with an unknown structure. Based on its chemical reactions, however, August Kekulé (1829–1896) suggested that the molecule has a planar, symmetrical ring structure. We know now that he was correct. The ring is flat, and all of the carbon–carbon bonds are the same length (139 pm) a distance intermediate between the average single bond (154 pm) and double bond (134 pm) lengths. Assuming the molecule

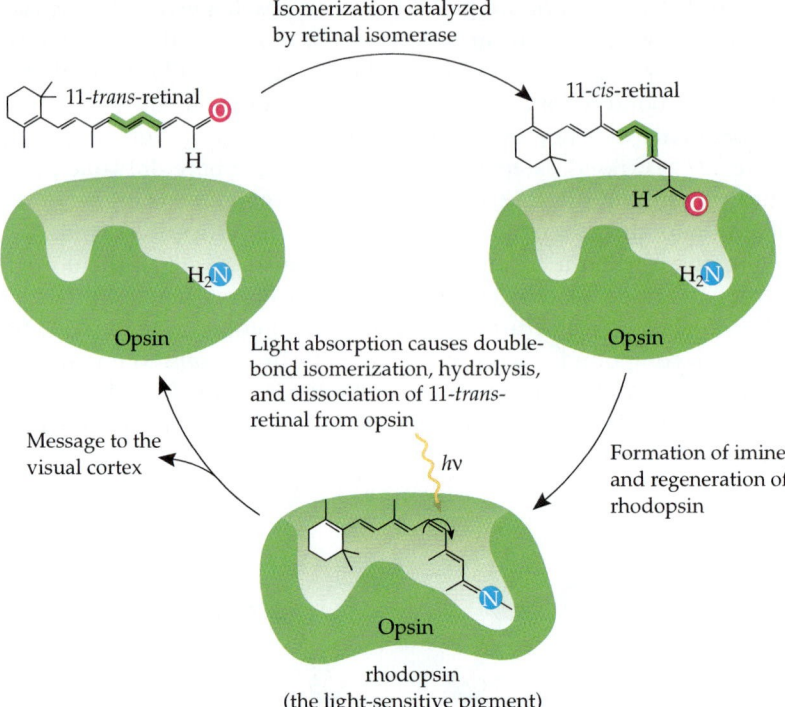

The primary chemical reaction of vision, occurring in the photoreceptor cells of the eyes, is absorption of light by rhodopsin, followed by isomerization of a carbon–carbon double bond from a *cis* configuration to a *trans* configuration.

Figure 10.14 The chemistry of vision.   Rotation around a double bond occurs in the reactions that allow you to see. A yellow-orange compound, *β*-carotene, which is the natural coloring agent in carrots, breaks down in your liver to produce vitamin A, also called retinol. Retinol is oxidized to 11-*trans*-retinal, which isomerizes to 11-*cis*-retinal. The *cis* isomer reacts with the protein opsin in the eye to give the pigment rhodopsin. This light-sensitive combination absorbs light in the blue-green region of the visible spectrum. Light striking the pigment triggers rotation around a carbon–carbon double bond, transforming rhodopsin into meta-rhodopsin. This change in molecular shape causes a nerve impulse to be sent to your brain, and you perceive a visual image.

Eventually meta-rhodopsin reacts chemically to produce 11-*trans*-retinal, and the cycle of chemical changes begins again. Conversion of meta-rhodopsin back to 11-*trans*-retinal is not as rapid as its formation, however, and an image formed on the retina persists for a tenth of a second or so. This persistence of vision allows you to perceive movies and videos as continuously moving images, even though they actually consist of separate pictures, each captured on a piece of film or tape for a thirtieth of a second. (*See General ChemistryNow Screen 10.13 Molecular Orbitals and Vision.*)

has two resonance structures with alternating double bonds, the observed structure is rationalized [◀ Section 9.5]. The C — C bond order in C_6H_6 (1.5) is the average of a single bond and a double bond.

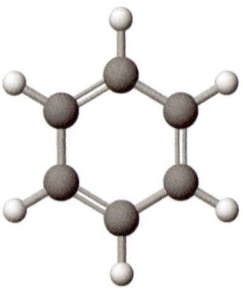

Benzene, C_6H_6

resonance structures   or   resonance hybrid

Understanding the bonding in benzene (Figure 10.15) is important because the benzene ring structure occurs in an enormous number of chemical compounds.

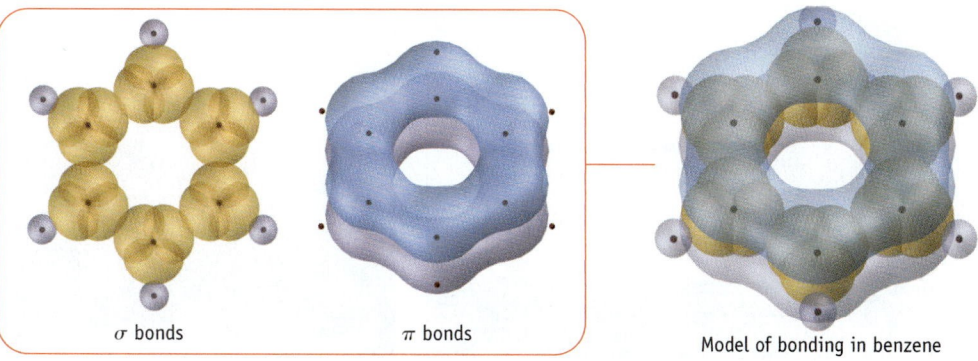

σ bonds   π bonds   Model of bonding in benzene

σ and π bonding in benzene

Figure 10.15 **Bonding in benzene, C_6H_6.** (*left*) The C atoms of the ring are bonded to each other through σ bonds using sp^2 hybrid orbitals of the C atom. The C — H bonds also use C atom sp^2 hybrid orbitals. The π bonding framework of the molecule arises from overlap of C atom p orbitals not used in hybrid orbital formation. As these orbitals are perpendicular to the ring, π electron density is above and below the plane of the ring. (*right*) A composite of σ and π bonding in benzene.

We assume that the trigonal-planar carbon atoms have sp^2 hybridization. Each C — H bond is formed by overlap of an sp^2 orbital of a carbon atom with a $1s$ orbital of hydrogen, and the C — C σ bonds arise by overlap of sp^2 orbitals on adjacent carbon atoms. After accounting for the σ bonding, an unhybridized p orbital remains on each C atom, and each is occupied by a single electron. These six orbitals and six electrons form three π bonds. Because all carbon–carbon bond lengths are the same, each p orbital overlaps equally well with the p orbitals of both adjacent carbons, and the π interaction is unbroken around the six-member ring.

The orbital picture of benzene underscores an important point. The basis of valence bond theory, which states that a bond is described as a pair of electrons between two atoms, does not work well for the π electrons in benzene—nor does it work whenever resonance is needed to describe a structure. However, molecular orbital theory does give us a better view, and that is the subject of the next section.

10.3—Molecular Orbital Theory

Molecular orbital (MO) theory is an alternative way to view orbitals in molecules. In contrast to the localized bond and lone-pair electrons of valence bond theory, MO theory assumes that pure s and p atomic orbitals of the atoms in the molecule combine to produce orbitals that are spread out, or delocalized, over several atoms or even over an entire molecule. These orbitals are called **molecular orbitals**.

One reason for learning about the MO concept is that it correctly predicts the electronic structures of molecules such as O_2 that do not follow the electron-pairing assumptions of the Lewis approach. The rules of Chapter 9 would guide you to draw the electron dot structure of O_2 with all the electrons paired, which fails to explain its paramagnetism (Figure 10.16). The molecular orbital approach can account for this property, but valence bond theory cannot. To see how MO theory can be used to describe the bonding in O_2 and other diatomic molecules, we shall first describe four principles used to develop the theory.

Principles of Molecular Orbital Theory

In MO theory we begin with a given arrangement of atoms in the molecule at the known bond distances. We then determine the *sets* of molecular orbitals. One way to do so is to combine available valence orbitals on all the constituent atoms. These

■ **A Failure of the Valence Bond Theory** Lewis electron dot structures fail to describe the bonding correctly in a well-known diatomic molecule, O_2. The O_2 molecule is paramagnetic, which requires the presence of unpaired electrons. The obvious Lewis structure, however, has all electrons paired. The molecular orbital approach shows that the molecule has two unpaired electrons.

■ **Diatomic Molecules** Molecules such as H_2, Li_2, and N_2, in which two identical atoms are bonded, are often called *homonuclear* diatomic molecules.

Charles D. Winters

Figure 10.16   Liquid oxygen. Oxygen gas condenses (*left*) to a pale blue liquid at −183 °C (*middle*). Oxygen in the liquid state is paramagnetic and clings to the poles of a magnet (*right*). (*See General ChemistryNow Screen 10.12 Paramagnetism, to watch a video of this figure.*)

molecular orbitals more or less encompass all the atoms of the molecule, and the valence electrons for all the atoms in the molecule are assigned to the molecular orbitals. Just as with orbitals in atoms, electrons are assigned according to the Pauli exclusion principle and Hund's rule [◀ Sections 8.2 and 8.4].

The **first principle of molecular orbital theory** is that *the total number of molecular orbitals is always equal to the total number of atomic orbitals contributed by the atoms that have combined.* To illustrate this orbital conservation principle, let us consider the H_2 molecule.

Molecular Orbitals for H_2

Molecular orbital theory specifies that when the $1s$ orbitals of two hydrogen atoms overlap, *two* molecular orbitals result. In the molecular orbital resulting from *addition* of the atomic orbitals, the $1s$ regions of electron density add together, leading to an increased probability that electrons will reside in the bond region between the two nuclei (Figure 10.17). This **bonding molecular orbital** is the same as the chemical bond described by valence bond theory. It is also a σ orbital because the region of electron probability lies directly along the bond axis. This molecular orbital is labeled σ_{1s}, where the subscript $1s$ indicates that $1s$ atomic orbitals were used to create the molecular orbital.

The other molecular orbital is constructed by *subtracting* one atomic orbital from the other (see Figure 10.17). When this happens, the probability of finding an electron between the nuclei in the molecular orbital is reduced, and the probability of finding the electron in other regions is higher. Without significant electron density between them, the nuclei repel one another. This type of orbital is called an **antibonding molecular orbital**. Because it is also a σ orbital, it is labeled σ_{1s}^{*}. The asterisk signifies that it is antibonding. *Antibonding orbitals have no counterpart in valence bond theory.*

■ **Orbitals and Electron Waves**
Orbitals are characterized as electron waves; therefore, a way to view molecular orbital formation is to assume that two electron waves, one from each atom, interfere with each other. The interference can be constructive, giving a bonding MO, or destructive, giving an antibonding MO.

The **second principle of molecular orbital theory** is that *the bonding molecular orbital is lower in energy than the parent orbitals, and the antibonding orbital is higher in energy* (Figure 10.17). As a result, the energy of a group of atoms is lower than the energy of the separated atoms when electrons are assigned to bonding molecular orbitals. Chemists say the system is "stabilized" by chemical bond formation. Conversely, the system is "destabilized" when electrons are assigned to antibonding orbitals because the energy of the system is higher than that of the atoms themselves.

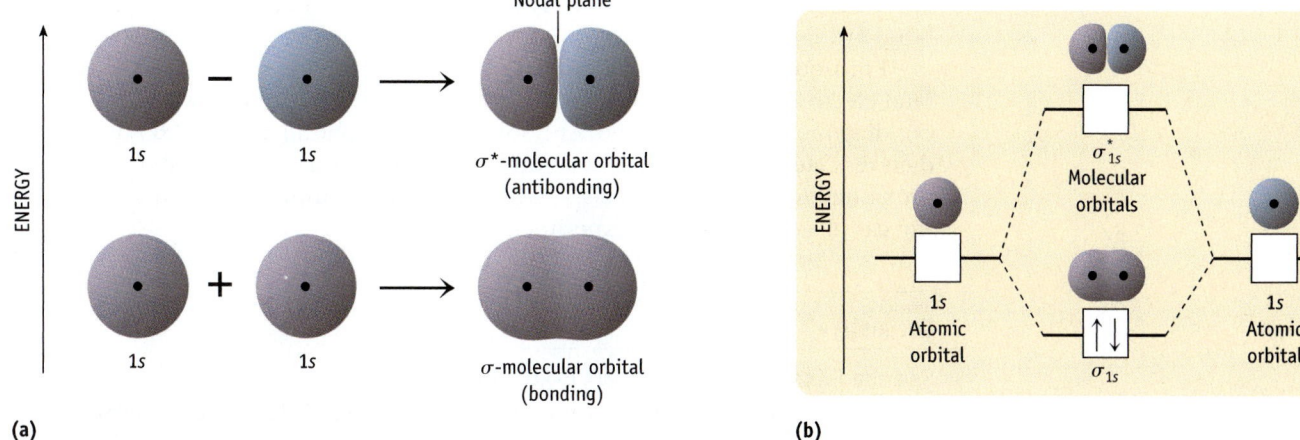

(a)

(b)

Figure 10.17 **Molecular orbitals.** (a) Bonding and antibonding σ molecular orbitals are formed from two $1s$ atomic orbitals on adjacent atoms. Notice the presence of a node in the antibonding orbital. (The node is a plane on which there is zero probability of finding an electron.) (b) A molecular orbital diagram for H_2. The two electrons are placed in the σ_{1s} orbital, the molecular orbital lower in energy. (*See General ChemistryNow Screen 10.9 Molecular Orbital Theory, to view animations based on this figure.*)

The **third principle of molecular orbital theory** is that the *electrons of the molecule are assigned to orbitals of successively higher energy* according to the Pauli exclusion principle and Hund's rule. This is analogous to the procedure for building up electronic structures of atoms. Thus, electrons occupy the lowest energy orbitals available: when two electrons are assigned to an orbital, their spins must be paired. Because the energy of the electrons in the bonding orbital of H_2 is lower than that of either parent $1s$ electron (see Figure 10.17b), the H_2 molecule is stable. We write the electron configuration of H_2 as $(\sigma_{1s})^2$.

What would happen if we try to combine two helium atoms to form dihelium, He_2? Both He atoms have a $1s$ valence orbital that can be added and subtracted to produce the same kind of molecular orbitals as in H_2. Unlike in H_2, however, four electrons need to be assigned to these orbitals (Figure 10.18). The pair of electrons in the σ_{1s} orbital stabilizes He_2. The two electrons in the σ_{1s}^* orbital, however, destabilize the He_2 molecule. The energy decrease from the electrons in the σ_{1s} bonding molecular orbital is offset by the energy increase due to the electrons in the σ_{1s}^* antibonding molecular orbital. Thus, molecular orbital theory predicts that He_2 has no net stability; two He atoms have no tendency to combine. This confirms what we already know—elemental helium exists in the form of single atoms and not as a diatomic molecule.

Bond Order

Bond order was defined in Chapter 9 as the net number of bonding electron pairs linking a pair of atoms. This same concept can be applied directly to molecular orbital theory, but now bond order is defined as

$$\text{Bond order} = \tfrac{1}{2}(\text{number of electrons in bonding MOs} - \text{number of electrons in antibonding MOs}) \quad (10.1)$$

In the H_2 molecule, there are two electrons in a bonding orbital and none in an antibonding orbital, so H_2 has a bond order of 1. In contrast, in He_2 the stabilizing

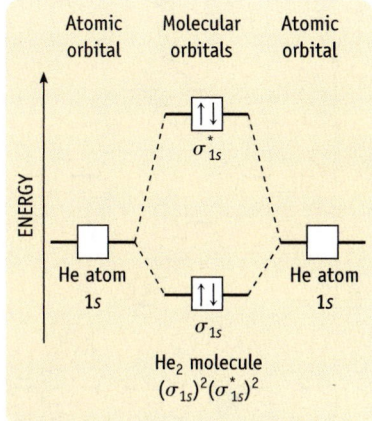

Figure 10.18 **A molecular orbital energy level diagram for the dihelium molecule, He_2.** This diagram provides a rationalization for the nonexistence of the molecule. In He_2 both the bonding (σ_{1s}) and antibonding orbitals (σ_{1s}^*) would be fully occupied. Note that occupation of antibonding orbitals leads to a greater destabilization than occupation of bonding orbitals leads to stabilization.

effect of the σ_{1s} pair is canceled by the destabilizing effect of the σ_{1s}^* pair, so the bond order is 0.

Fractional bond orders are possible. Consider the ion He_2^+. Its molecular orbital electron configuration is $(\sigma_{1s})^2(\sigma_{1s}^*)^1$. In this ion, there are two electrons in a bonding molecular orbital, but only one in an antibonding orbital. MO theory predicts that He_2^+ should have a bond order of 0.5; that is, a weak bond should exist between helium atoms in such a species. Interestingly, this ion has been identified in the gas phase using special experimental techniques.

Example 10.6—Molecular Orbitals and Bond Order

Problem Write the electron configuration of the H_2^- ion in molecular orbital terms. What is the bond order of the ion?

Strategy Count the number of valence electrons in the ion and then place those electrons in the MO diagram for the H_2 molecule. Find the bond order from Equation 10.1.

Solution This ion has three electrons (one each from the H atoms plus one for the negative charge). Therefore, its electronic configuration is $(\sigma_{1s})^2(\sigma_{1s}^*)^1$, identical with the configuration for He_2^+. This means H_2^- also has a net bond order of 0.5. The H_2^- ion is thus predicted to exist under special circumstances.

Exercise 10.7—Molecular Orbitals and Bond Order

What is the electron configuration of the H_2^+ ion? Compare the bond order of this ion with those of He_2^+ and H_2^-. Do you expect H_2^+ to exist?

Molecular Orbitals of Li$_2$ and Be$_2$

The **fourth principle of molecular orbital theory** is that *atomic orbitals combine to form molecular orbitals most effectively when the atomic orbitals are of similar energy.* This principle becomes important when we move past He_2 to Li_2 (dilithium) and to even heavier molecules.

A lithium atom has electrons in two orbitals of the s type ($1s$ and $2s$), so a $1s \pm 2s$ combination is theoretically possible. Because the $1s$ and $2s$ orbitals are quite different in energy, however, this interaction can be disregarded. Thus, the molecular orbitals come only from $1s \pm 1s$ and $2s \pm 2s$ combinations (Figure 10.19). This means the molecular orbital electron configuration of dilithium, Li_2, is

$$Li_2 \text{ MO configuration: } (\sigma_{1s})^2(\sigma_{1s}^*)^2(\sigma_{2s})^2$$

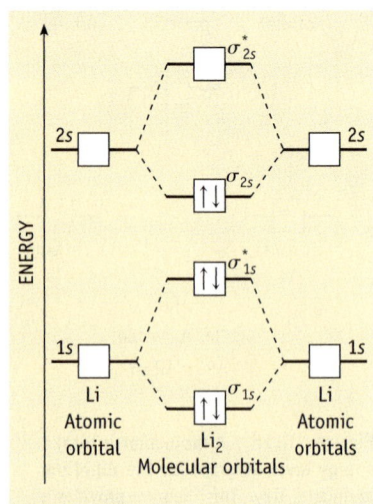

Figure 10.19 **Energy level diagram for the combination of two Li atoms.** Notice that the molecular orbitals are created by combining orbitals of similar energies. The electron configuration is shown for Li$_2$.

The bonding effect of the σ_{1s} electrons is canceled by the antibonding effect of the σ_{1s}^* electrons, so these pairs make no net contribution to bonding in Li_2. Bonding in Li_2 is due to the electron pair assigned to the σ_{2s} orbital, and the bond order is 1.

The fact that the σ_{1s} and σ_{1s}^* electron pairs of Li_2 make no net contribution to bonding is exactly what you observed in drawing electron dot structures in Chapter 9: *Core electrons are ignored.* In molecular orbital terms, core electrons are assigned to bonding and antibonding molecular orbitals that offset one another.

A diberyllium molecule, Be_2, is not expected to exist. Its electron configuration is

$$Be_2 \text{ MO configuration: [core electrons]}(\sigma_{2s})^2(\sigma_{2s}^*)^2$$

The effects of σ_{2s} and σ_{2s}^* electrons cancel, and there is no net bonding. The bond order is 0, so the molecule does not exist.

Example 10.7—Molecular Orbitals in Diatomic Molecules

Problem Be_2 does not exist. But what about the Be_2^+ ion? Describe its electron configuration in molecular orbital terms and give the net bond order. Do you expect the ion to exist?

Strategy Count the number of electrons in the ion and place them in the MO diagram in Figure 10.19. Write the electron configuration and calculate the bond order from Equation 10.1.

Solution The Be_2^+ ion has seven electrons (Be_2 has eight), of which four are core electrons. (The core electrons are assigned to σ_{1s} and σ_{1s}^* molecular orbitals.) The remaining three electrons are assigned to the σ_{2s} and σ_{2s}^* molecular orbitals, so the MO electron configuration is $[\text{core electrons}](\sigma_{2s})^2(\sigma_{2s}^*)^1$. This means the net bond order is 0.5, and so Be_2^+ is predicted to exist under special circumstances.

Exercise 10.8—Molecular Orbitals in Diatomic Molecules

Could the anion Li_2^- exist? What is the ion's bond order?

Molecular Orbitals from Atomic *p* Orbitals

With the principles of molecular orbital theory in place, we are ready to account for bonding in such important homonuclear diatomic molecules as N_2, O_2, and F_2. To describe the bonding in these molecules we will use both *s* and *p* valence orbitals in forming molecular orbitals.

Sigma-bonding and antibonding molecular orbitals are formed by *s* orbitals interacting as illustrated in Figure 10.19. Similarly, it is possible for a *p* orbital on one atom to interact with a *p* orbital on the other atom to produce a pair of σ-bonding and σ^*-antibonding molecular orbitals (Figure 10.20).

In addition, each atom has *two p* orbitals in planes perpendicular to the σ bond connecting the two atoms. These *p* orbitals can interact sideways to give π-bonding and π-antibonding molecular orbitals (Figure 10.21). Combining these two *p* orbitals on each atom produces *two π-bonding* molecular orbitals (π_p) and *two pi-antibonding* molecular orbitals (π_p^*).

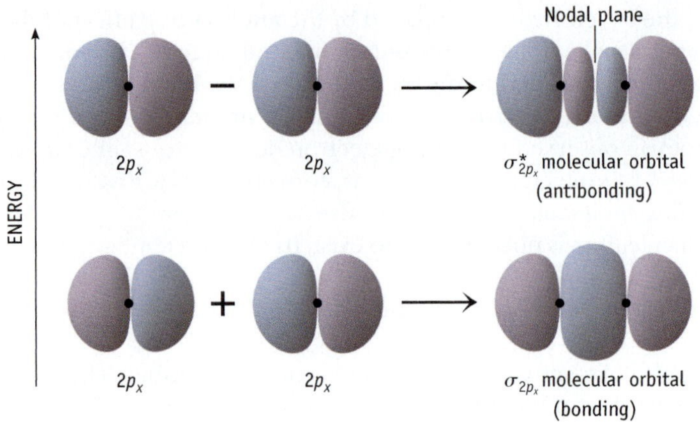

Figure 10.20   Sigma molecular orbitals from *p* atomic orbitals. Sigma-bonding (σ_{2p}) and antibonding (σ_{2p}^*) molecular orbitals arise from overlap of 2*p* orbitals. Each orbital can accommodate two electrons. The *p* orbitals in electron shells of higher *n* give molecular orbitals of the same basic shape.

Figure 10.21   Formation of π molecular orbitals. Sideways overlap of atomic 2*p* orbitals that lie in the same direction in space gives rise to pi-bonding (π_{2p}) and pi-antibonding (π_{2p}^*) molecular orbitals. The *p* orbitals in shells of higher *n* give molecular orbitals of the same basic shape.

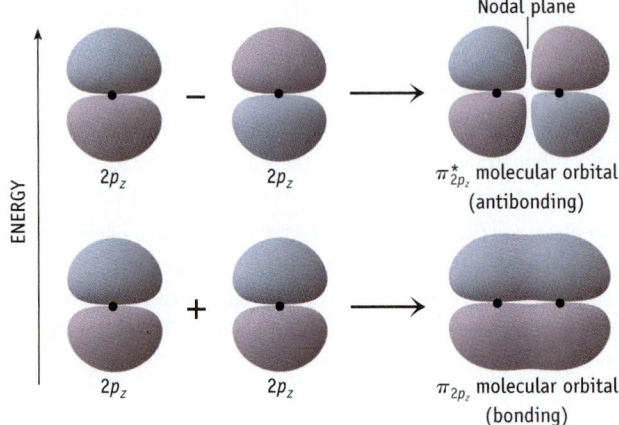

Electron Configurations for Homonuclear Molecules for Boron Through Fluorine

Orbital interactions in a second-period, homonuclear, diatomic molecule lead to the energy level diagram shown in Figure 10.22. Electron assignments can be made using this diagram, and the results for the diatomic molecules B_2 through F_2 are tabulated in Table 10.1, which has two noteworthy features.

First, notice the correlation between the electron configurations and the bond orders, bond lengths, and bond energies at the bottom of Table 10.1. As the bond order between a pair of atoms increases, the energy required to break the bond increases, and the bond distance decreases. Dinitrogen, N_2, with a bond order of 3, has the largest bond energy and the shortest bond distance.

Second, notice the configuration for dioxygen, O_2. Dioxygen has 12 valence electrons (6 from each atom), so it has the molecular orbital configuration

$$O_2 \text{ MO configuration: [core electrons]}(\sigma_{2s})^2(\sigma_{2s}^*)^2(\pi_{2p})^4(\sigma_{2p})^2(\pi_{2p}^*)^2$$

This configuration leads to a bond order of 2 in agreement with experiment, and it specifies two unpaired electrons (in π_{2p}^* molecular orbitals). Thus, molecular orbital theory succeeds where valence bond theory fails. MO theory explains both the observed bond order and the paramagnetic behavior of O_2.

■ **Highest Occupied Molecular Orbital (HOMO)**
Chemists often refer to the highest energy MO that contains electrons as the HOMO. For O_2 this is the π_{2p}^* orbital. Chemists also use the term LUMO, for the lowest unoccupied molecular orbital. For O_2, it would be σ_{2p}^*.

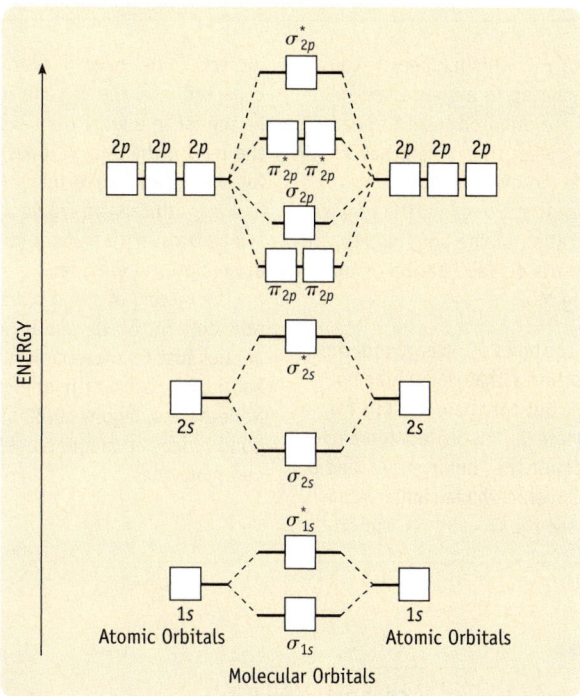

Active Figure 10.22 **Molecular orbital energy level diagram for homonuclear diatomic molecules of second period elements.** Although the diagram leads to the correct conclusions regarding bond order and magnetic behavior for O_2, N_2, and F_2, the energy ordering of the MOs is correct only for B_2, C_2, and N_2. For O_2 and F_2, the σ_{2p} MO is lower in energy than the π_{2p} MOs. See "A Closer Look," page 464.

GENERAL
Chemistry ⚛ Now™ See the General ChemistryNow CD-ROM or website to explore an interactive version of this figure accompanied by an exercise.

Table 10.1 Molecular Orbital Occupations and Physical Data for Homonuclear Diatomic Molecules of Second-Period Elements

	B_2	C_2	N_2	O_2	F_2
σ_{2p}^*	☐	☐	☐	☐	☐
π_{2p}^*	☐☐	☐☐	☐☐	↑ ↑	↑↓ ↑↓
σ_{2p}	☐	☐	↑↓	↑↓	↑↓
π_{2p}	↑ ↑	↑↓ ↑↓	↑↓ ↑↓	↑↓ ↑↓	↑↓ ↑↓
σ_{2s}^*	↑↓	↑↓	↑↓	↑↓	↑↓
σ_{2s}	↑↓	↑↓	↑↓	↑↓	↑↓
Bond order	One	Two	Three	Two	One
Bond-dissociation energy (kJ/mol)	290	620	945	498	155
Bond distance (pm)	159	131	110	121	143
Observed magnetic behavior (paramagnetic or diamagnetic)	Para	Dia	Dia	Para	Dia

A Closer Look

Molecular Orbitals for Compounds Formed from *p*-Block Elements

Several features of the molecular orbital energy level diagram in Figure 10.22 might be described in more detail.

- The bonding and antibonding σ orbitals from 2*s* interactions are lower in energy than the σ and π MOs from 2*p* interactions. The reason is that 2*s* orbitals have a lower energy than 2*p* orbitals in the separated atoms.

- The energy separation of the bonding and antibonding orbitals is greater for σ_{2p} than for π_{2p}. This happens because *p* orbitals overlap to a greater extent when they are oriented head to head (to give σ_{2p} MOs) than when they are side by side (to give π_{2p} MOs). The greater the orbital overlap, the greater the stabilization of the bonding MO and the greater the destabilization of the antibonding MO.

Figure 10.22 shows an energy ordering of molecular orbitals that you might not have expected, but there are reasons for this order. A more sophisticated approach takes into account the "mixing" of *s* and *p* atomic orbitals, which have similar energies. This causes the σ_{2s} and σ_{2s}^{*} molecular

orbitals to be lower in energy than expected, and the σ_{2p} and σ_{2p}^{*} orbitals to be higher in energy than expected. This is the reason the energy lowering and raising for the σ_{2s} and σ_{2s}^{*} orbitals (and for the σ_{2p} and σ_{2p}^{*} orbitals) in Figure 10.22 is not symmetrical with respect to the 2*s* and 2*p* atomic orbital energies.

The mixing of *s* and *p* orbitals is important only for B_2, C_2, and N_2, so the figure applies just to these molecules. For O_2 and F_2, σ_{2p} is lower in energy than π_{2p}. Nonetheless, Figure 10.22 gives the correct bond order and magnetic behavior for these two molecules.

GENERAL
Chemistry ⚛ Now™

See the General ChemistryNow CD-ROM or website:
- **Screen 10.11 Homonuclear Diatomic Molecules,** for an exercise on molecular orbital configurations

Example 10.8—Electron Configuration for a Homonuclear Diatomic Ion

Problem When potassium reacts with O_2, potassium superoxide, KO_2, is one of the products. This is an ionic compound, in which the anion is the superoxide ion, O_2^{-}. Write the molecular orbital electron configuration for the ion. Predict its bond order and magnetic behavior.

Strategy Use the energy level diagram of Figure 10.22 to generate the configuration of this ion. Use Equation 10.1 to determine the bond order.

Solution The MO configuration for O_2^{-} is

$$O_2^{-} \text{ MO configuration: } [\text{core electrons}](\sigma_{2s})^2(\sigma_{2s}^{*})^2(\pi_{2p})^4(\sigma_{2p})^2(\pi_{2p}^{*})^3$$

The ion is predicted to be paramagnetic to the extent of one unpaired electron, a prediction confirmed by experiment. The bond order is 1.5, because there are eight bonding electrons and five antibonding electrons. The bond order for O_2^{-} is lower than that for O_2, so we predict that the O—O bond length in O_2^{-} will be longer than the oxygen–oxygen bond length in O_2. In fact, the superoxide ion has an O—O bond length of 134 pm, whereas the bond length in O_2 is 121 pm.

Comment You should quickly spot the fact that the superoxide ion (O_2^{-}) contains an odd number of electrons. It is another diatomic species (in addition to NO and O_2) for which it is not possible to write a Lewis structure that accurately represents the bonding.

Exercise 10.9—Molecular Electron Configurations

The cations O_2^{+} and N_2^{+} are important components of the earth's upper atmosphere. Write the electron configuration of O_2^{+}. Predict its bond order and magnetic behavior.

Electron Configurations for Heteronuclear Diatomic Molecules

The compounds NO, CO, and ClF—all molecules containing two different elements—are examples of **heteronuclear diatomic molecules**. MO descriptions for heteronuclear diatomic molecules generally resemble those for homonuclear diatomic molecules. As a consequence, an energy level diagram like Figure 10.22 can be used to judge the bond order and magnetic behavior for heteronuclear diatomics.

Consider nitrogen monoxide, NO. Nitrogen monoxide has 11 molecular valence electrons. If they are assigned to the MOs for a homonuclear diatomic molecule, the molecular electron configuration is

NO MO configuration: $[\text{core electrons}](\sigma_{2s})^2(\sigma_{2s}^{\star})^2(\pi_{2p})^4(\sigma_{2p})^2(\pi_{2p}^{\star})^1$

The net bond order is 2.5, in accordance with the bond length information. The single unpaired electron is assigned to the $\pi_{2p}^{\star}$ molecular orbital. The molecule is paramagnetic, as predicted for a molecule with an odd number of electrons.

Resonance and MO Theory

Ozone, O_3, is a simple triatomic molecule with equal oxygen–oxygen bond lengths. Equal $X — O$ bond lengths are also observed in other triatomic molecules and ions, such as SO_2, NO_2^-, and HCO_2^-. Valence bond theory introduced resonance to rationalize the equivalent bonding to the oxygen atoms in these structures. MO theory provides another view of this problem.

O_3 SO_2 NO_2^- HCO_2^-

To visualize the bonding in ozone, begin by assuming that all three O atoms are sp^2 hybridized. The central atom uses its sp^2 hybrid orbitals to form two σ bonds and to accommodate a lone pair. The terminal atoms use their sp^2 hybrid orbitals to form one σ bond and to accommodate two lone pairs. In total, the lone pairs and bonding pairs in the σ framework of O_3 account for seven of the nine valence electron pairs in O_3.

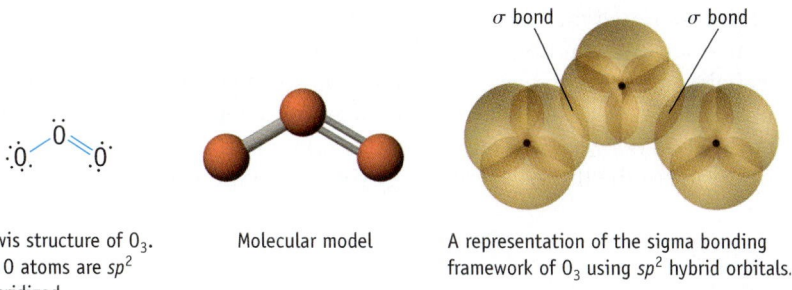

σ bond σ bond

Lewis structure of O_3. All O atoms are sp^2 hybridized.

Molecular model

A representation of the sigma bonding framework of O_3 using sp^2 hybrid orbitals.

The π bond in ozone arises from the two remaining pairs (Figure 10.23). Because we have assumed that each oxygen atom in O_3 is sp^2 hybridized, an unhybridized p orbital perpendicular to the O_3 plane remains on each of the three oxygen atoms. The orbitals are in the correct orientation to form π bonds. *A principle of MO theory is that the number of molecular orbitals must equal the number of atomic orbitals.* Thus, the three $2p$ atomic orbitals must be combined in a way that forms three molecular orbitals.

One π_p MO for ozone is a bonding orbital because the three p orbitals are "in phase" across the molecule. Another π_p MO is an antibonding orbital because the

Figure 10.23 Pi-bonding in ozone, O_3. Each O atom in O_3 is sp^2 hybridized. The three $2p$ orbitals, one on each atom, are used to create the three π molecular orbitals. Two pairs of electrons are assigned to the orbitals: one pair in the bonding orbital and one pair in the nonbonding orbital. The π bond order is 0.5, as one bonding pair is spread across two bonds.

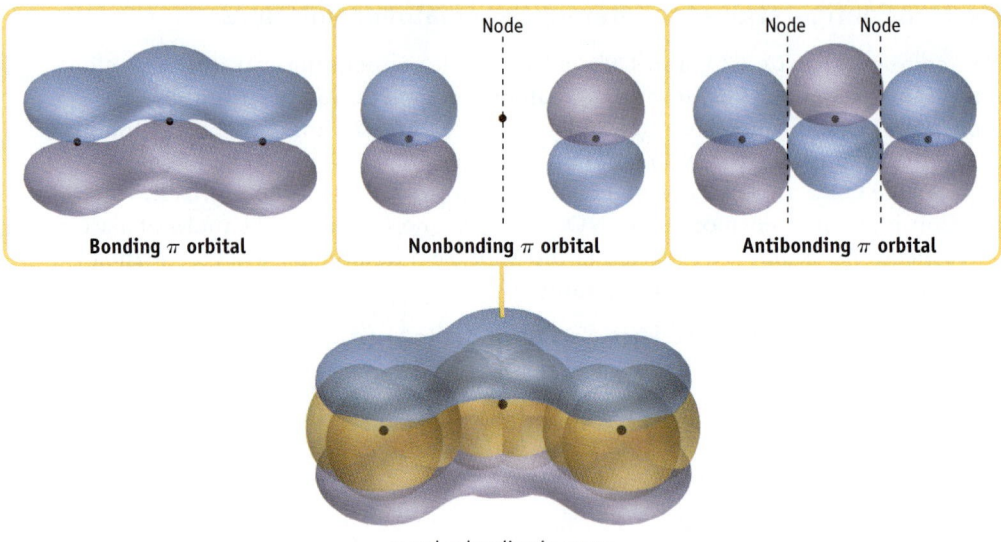

σ and π bonding in ozone

atomic orbital on the central atom is "out of phase" with the terminal atom p orbitals. The third π_p MO is a nonbonding orbital because the middle p orbital does not participate in the MO. The bonding π_p MO is filled by a pair of electrons that is delocalized, or "spread over," the molecule, just as the resonance hybrid implies. The nonbonding orbital is also occupied, but the electrons in this orbital are concentrated near the two terminal oxygens. As the name implies, electrons in this molecular orbital neither help nor hinder the bonding in the molecule. The π bond order of O_3 is 0.5, since one bond pair is spread over two O—O linkages. Because the σ bond order is 1.0 and the π bond order is 0.5, the net oxygen–oxygen bond order is 1.5—the same value given by valence bond theory.

The observation that two of the π molecular orbitals for ozone extend over three atoms illustrates an important point regarding molecular orbital theory: Orbitals can extend beyond two atoms. In valence bond theory, in contrast, all representations for bonding were based on being able to localize pairs of electrons in bonds between two atoms. To further illustrate the MO approach, look again at benzene (Figure 10.24). On page 457 we noted that the π electrons in this molecule were spread out over all six carbon atoms. We can now see how the same case can be made with MO theory. Six p orbitals contribute to the π system. Based on the premise that the number of molecular orbitals must equal the number of atomic orbitals, there must be six π molecular orbitals in benzene. An energy level diagram for benzene shows that the six p electrons reside in the three lowest-energy (bonding) molecular orbitals.

■ **Metals and Molecular Orbitals**
The bonding in metals can be described best using molecular orbital theory. See "The Chemistry of Materials," page 642.

Figure 10.24 Molecular orbital energy level diagram for benzene. Because there are six unhybridized p orbitals, six π molecular orbitals can be formed—three bonding and three antibonding. The three bonding molecular orbitals accommodate the six π electrons.

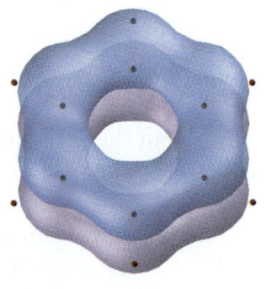

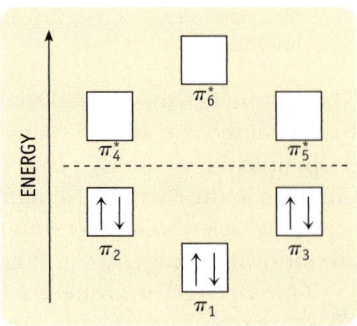

GENERAL
Chemistry‑Now™

See the General ChemistryNow
CD-ROM or website to:

- Assess your understanding with homework questions keyed to each goal
- Check your readiness for an exam by taking the exam-prep quiz and exploring the resources in the personalized Learning Plan it provides

For additional preparation for an examination on this chapter see the **Let's Review** section on pages 530–545.

Chapter Goals Revisited

When you have finished studying this chapter, you should ask whether you have met the chapter goals. In particular, you should be able to

Understand the differences between valence bond theory and molecular orbital theory

a. Describe the main features of valence bond theory and molecular orbital theory, the two commonly used theories for covalent bonding (Section 10.1).

b. Recognize that the premise for valence bond theory is that bonding results from the overlap of atomic orbitals. By virtue of the overlap of orbitals, electrons are concentrated (or localized) between two atoms (Section 10.2).

c. Distinguish how sigma (σ) and pi (π) bonds arise. For σ bonding, orbitals overlap in a head-to-head fashion, concentrating electrons along the bond axis. Sideways overlap of p atomic orbitals results in π bond formation, with electrons above and below the molecular plane (Section 10.2).

d. Understand how molecules having double bonds can have isomeric forms.
General ChemistryNow homework: Study Question(s) 14

Identify the hybridization of an atom in a molecule or ion

a. Use the concept of hybridization to rationalize molecular structure (Section 10.2). General ChemistryNow homework: SQ(s) 3, 4, 6, 8, 11, 22, 27, 32, 34, 38, 51

Hybrid Orbitals	Atomic Orbitals Used	Number of Hybrid Orbitals	Electron-Pair Geometry
sp	$s + p$	2	Linear
sp^2	$s + p + p$	3	Trigonal-planar
sp^3	$s + p + p + p$	4	Tetrahedral
sp^3d	$s + p + p + p + d$	5	Trigonal-bipyramidal
sp^3d^2	$s + p + p + p + d + d$	6	Octahedral

Understand the differences between bonding and antibonding molecular orbitals

a. Understand molecular orbital theory (Section 10.3), in which atomic orbitals are combined to form bonding orbitals, nonbonding orbitals, or antibonding orbitals that are delocalized over several atoms. In this description, the electrons of the molecule or ion are assigned to the orbitals beginning with the one at lowest energy, according to the Pauli exclusion principle and Hund's rule.

b. Use molecular orbital theory to explain the properties of O_2 and other diatomic molecules. General ChemistryNow homework: SQ(s) 15, 16, 18, 20, 42, 44

Key Equations

Equation 10.1 (page 459)

Calculating the order of a bond from the molecular orbital electron configuration

$$\text{Bond order} = \tfrac{1}{2}(\text{number of electrons in bonding MOs} - \text{number of electrons in antibonding MOs})$$

Study Questions

▲ denotes more challenging questions.

■ denotes questions available in the Homework and Goals section of the General ChemistryNow CD-ROM or website.

Blue numbered questions have answers in Appendix O and fully worked solutions in the *Student Solutions Manual*.

Structures of many of the compounds used in these questions are found on the General ChemistryNow CD-ROM or website in the Models folder.

GENERAL Chemistry⚛Now™ Assess your understanding of this chapter's topics with additional quizzing and conceptual questions at http://now.brookscole.com/kotz6e

Practicing Skills

Valence Bond Theory

(See Examples 10.1–10.5 and General ChemistryNow Screens 10.2–10.7)

1. Draw the Lewis structure for chloroform, $CHCl_3$. What are its electron-pair and molecular geometries? What orbitals on C, H, and Cl overlap to form bonds involving these elements?

2. ■ Draw the Lewis structure for NF_3. What are its electron-pair and molecular geometries? What is the hybridization of the nitrogen atom? What orbitals on N and F overlap to form bonds between these elements?

3. Specify the electron-pair and molecular geometry for each of the following. Describe the hybrid orbital set used by the underlined atom in each molecule or ion.
   (a) $\underline{B}Br_3$ (b) $\underline{C}O_2$ (c) $\underline{C}H_2Cl_2$ (d) $\underline{C}O_3{}^{2-}$

4. ■ Specify the electron-pair and molecular geometry for each of the following. Describe the hybrid orbital set used by the underlined atom in each molecule or ion.
   (a) $\underline{C}Se_2$ (b) $\underline{S}O_2$ (c) $\underline{C}H_2O$ (d) $\underline{N}H_4{}^+$

5. ■ Describe the hybrid orbital set used by each of the indicated atoms in the molecules below:
   (a) the carbon atoms and the oxygen atom in dimethyl ether, H_3COCH_3
   (b) each carbon atom in propene

$$H_3C-\underset{\underset{H}{|}}{C}=CH_2$$

   (c) the two carbon atoms and the nitrogen atom in the amino acid glycine

$$H-\underset{\underset{\ddot{\,}}{N}}{\overset{H}{\underset{|}{|}}}-\underset{\underset{H}{|}}{\overset{H}{C}}-\overset{:\!O\!:}{\underset{}{C}}-\ddot{O}-H$$

6. ■ Give the hybrid orbital set used by each of the underlined atoms in the following molecules.

   (a) $H-\underset{\ddot{\,}}{\overset{H}{\underset{|}{N}}}-\overset{:O:}{\underset{\parallel}{\underline{C}}}-\underset{\ddot{\,}}{\overset{H}{\underset{|}{N}}}-H$

   (b) $H_3\underline{C}-\underline{C}=\underset{\underset{H}{|}}{\overset{H}{C}}-\underline{C}=\ddot{O}$

   (c) $H-\underset{\underset{H}{|}}{\overset{H}{\underline{C}}}=\underset{\underset{H}{|}}{\overset{H}{\underline{C}}}-\underline{C}\equiv N:$

7. Draw the Lewis structure and then specify the electron-pair and molecular geometries for each of the following molecules or ions. Identify the hybridization of the central atom.
   (a) $SiF_6{}^{2-}$ (b) SeF_4 (c) $ICl_2{}^-$ (d) XeF_4

8. ■ Draw the Lewis structure and then specify the electron-pair and molecular geometries for each of the following molecules or ions. Identify the hybridization of the central atom.
   (a) $XeOF_4$ (c) OSF_4
   (b) BrF_5 (d) central Br in $Br_3{}^-$

9. Draw the Lewis structures of the acid HPO_2F_2 and its anion $PO_2F_2{}^-$. What is the molecular geometry and hybridization for the phosphorus atom in each species? (H is bonded to the O atom in the acid.)

10. Draw the Lewis structures of HSO_3F and SO_3F^-. What is the molecular geometry and hybridization for the sulfur atom in each species? (H is bonded to the O atom in the acid.)

11. ■ What is the hybridization of the carbon atom in phosgene, Cl_2CO? Give a complete description of the σ and π bonding in this molecule.

12. What is the hybridization of the sulfur atom in sulfuryl fluoride, SO_2F_2?

13. The arrangement of groups attached to the C atoms involved in a C=C double bond leads to *cis* and *trans* isomers. For each compound below, draw the other isomer.

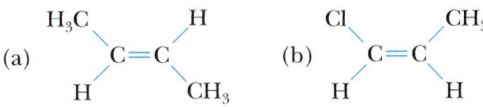

14. ■ For each compound below decide whether *cis* and *trans* isomers are possible. If isomerism is possible, draw the other isomer.

   (a) $\overset{H_3C}{}\underset{H}{}C=C\overset{H}{\underset{CH_2CH_3}{}}$ (c) $\overset{Cl}{}\underset{H}{}C=C\overset{CH_2OH}{\underset{H}{}}$

   (b) $\overset{H}{}\underset{H}{}C=C\overset{CH_3}{\underset{H}{}}$

Molecular Orbital Theory

(See Examples 10.6–10.8 and General ChemistryNow Screens 10.9–10.12.)

15. ■ The hydrogen molecular ion, H_2^+, can be detected spectroscopically. Write the electron configuration of the ion in molecular orbital terms. What is the bond order of the ion? Is the hydrogen–hydrogen bond stronger or weaker in H_2^+ than in H_2?

16. ■ Give the electron configurations for the ions Li_2^+ and Li_2^- in molecular orbital terms. Compare the Li — Li bond order in these ions with the bond order in Li_2.

17. Calcium carbide, CaC_2, contains the acetylide ion, C_2^{2-}. Sketch the molecular orbital energy level diagram for the ion. How many net σ and π bonds does the ion have? What is the carbon–carbon bond order? How has the bond order changed on adding electrons to C_2 to obtain C_2^{2-}? Is the C_2^{2-} ion paramagnetic?

18. ■ Oxygen, O_2, can acquire one or two electrons to give O_2^- (superoxide ion) or O_2^{2-} (peroxide ion). Write the electron configuration for the ions in molecular orbital terms, and then compare them with the O_2 molecule on the following bases.
    (a) magnetic character
    (b) net number of σ and π bonds
    (c) bond order
    (d) oxygen–oxygen bond length

19. Assume the energy level diagram for homonuclear diatomic molecules (Figure 10.22) can be applied to heteronuclear diatomics such as CO.
    (a) Write the electron configuration for carbon monoxide, CO.
    (b) What is the highest-energy, occupied molecular orbital? (Chemists call this the HOMO.)
    (c) Is the molecule diamagnetic or paramagnetic?
    (d) What is the net number of σ and π bonds? What is the CO bond order?

20. ■ The nitrosyl ion, NO^+, has an interesting chemistry.
    (a) Is NO^+ diamagnetic or paramagnetic? If paramagnetic, how many unpaired electrons does it have?
    (b) Assume the molecular orbital diagram for a homonuclear diatomic molecule (Figure 10.22) applies to NO^+. What is the highest-energy molecular orbital occupied by electrons?
    (c) What is the nitrogen–oxygen bond order?
    (d) Is the N — O bond in NO^+ stronger or weaker than the bond in NO?

General Questions on Valence Bond and Molecular Orbital Theory

These questions are not designated as to type or location in the chapter. They may combine several concepts.

21. Draw the Lewis structure for AlF_4^-. What are its electron-pair and molecular geometries? What orbitals on Al and F overlap to form bonds between these elements?

22. ■ Draw the Lewis structure for ClF_3. What are its electron-pair and molecular geometries? What is the hybridization of the chlorine atom? What orbitals on Cl and F overlap to form bonds between these elements?

23. Describe the O — S — O angle and the hybrid orbital set used by sulfur in each of the following molecules or ions:
    (a) SO_2 (c) SO_3^{2-}
    (b) SO_3 (d) SO_4^{2-}

    Do all have the same value for the O — S — O angle? Does the S atom in all these species use the same hybrid orbitals?

24. Sketch the Lewis structures of ClF_2^+ and ClF_2^-. What are the electron-pair and molecular geometries of each ion? Do both have the same F — Cl — F angle? What hybrid orbital set is used by Cl in each ion?

25. Sketch the resonance structures for the nitrite ion, NO_2^-. Describe the electron-pair and molecular geometries of the ion. From these geometries, decide on the O — N — O bond angle, the average NO bond order, and the N atom hybridization.

26. Sketch the resonance structures for the nitrate ion, NO_3^-. Is the hybridization of the N atom the same or different in each structure? Describe the orbitals involved in bond formation by the central N atom.

27. ■ Sketch the resonance structures for the N_2O molecule. Is the hybridization of the N atoms the same or different in each structure? Describe the orbitals involved in bond formation by the central N atom.

28. Compare the structure and bonding in CO_2 and CO_3^{2-} with regard to the O — C — O bond angles, the CO bond order, and the C atom hybridization.

29. Numerous molecules are detected in deep space (page 372). Three of them are illustrated here.

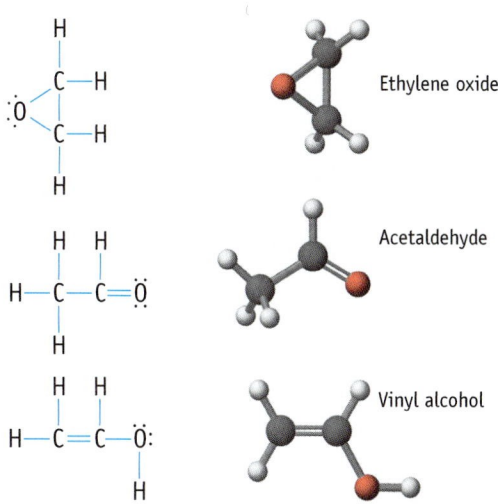

Ethylene oxide

Acetaldehyde

Vinyl alcohol

    (a) Comment on the similarities or differences in the formulas of these compounds. Are they isomers?
    (b) Indicate the hybridization of each C atom in each molecule.

(c) Indicate the value of the H—C—H angle in each of the three molecules.

(d) Are any of these molecules polar?

(e) Which molecule should have the strongest carbon–carbon bond? The strongest carbon–oxygen bond?

30. Acrolein, a component of photochemical smog, has a pungent odor and irritates eyes and mucous membranes.

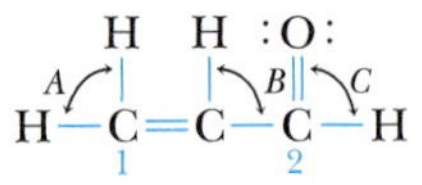

(a) What are the hybridizations of carbon atoms 1 and 2?

(b) What are the approximate values of angles A, B, and C?

(c) Is *cis-trans* isomerism possible here?

31. The organic compound below is a member of a class known as oximes.

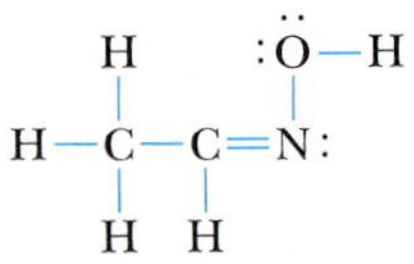

(a) What are the hybridizations of the two C atoms and of the N atom?

(b) What is the approximate C—N—O angle?

32. ■ The compound sketched below is acetylsalicylic acid, commonly known as aspirin:

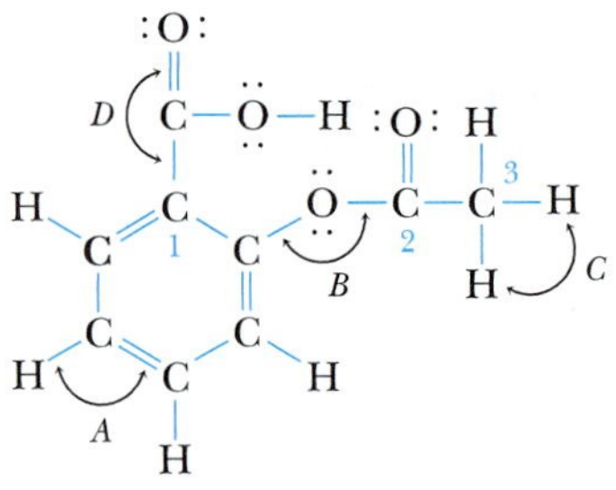

(a) What are the approximate values of the angles marked A, B, C, and D?

(b) What hybrid orbitals are used by carbon atoms 1, 2, and 3?

33. Phosphoserine is a less common amino acid.

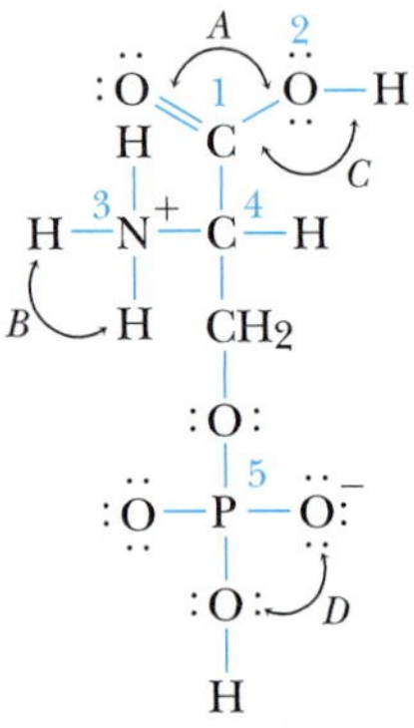

(a) Describe the hybridizations of atoms 1 through 5.

(b) What are the approximate values of the bond angles A, B, C, and D?

(c) What are the most polar bonds in the molecule?

34. ■ Lactic acid is a natural compound found in sour milk.

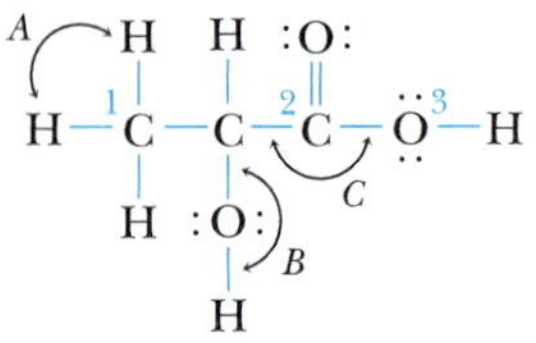

(a) How many π bonds occur in lactic acid? How many σ bonds?

(b) Describe the hybridization of atoms 1, 2, and 3.

(c) Which CO bond is the shortest in the molecule? Which CO bond is the strongest?

(d) What are the approximate values of the bond angles A, B, and C?

35. ■ Boron trifluoride, BF₃, can accept a pair of electrons from another molecule to form a coordinate covalent bond, as in the following reaction with ammonia:

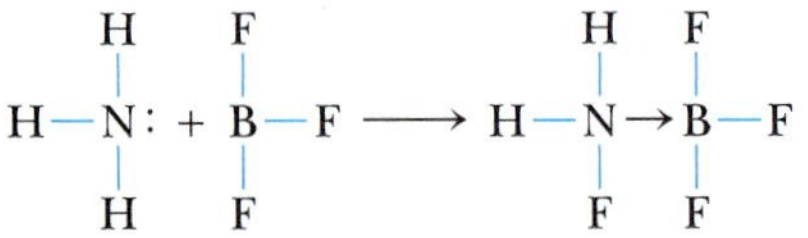

(a) What is the geometry about the boron atom in BF_3? In $H_3N \longrightarrow BF_3$?

(b) What is the hybridization of the boron atom in the two compounds?

(c) Does the boron atom's hybridization change on formation of the coordinate covalent bond?

36. The sulfamate ion, $H_2NSO_3^-$, can be thought of as having been formed from the amide ion, NH_2^-, and sulfur trioxide, SO_3.

(a) Sketch a structure for the sulfamate ion and estimate the bond angles.

(b) What changes in hybridization do you expect for N and S in the course of the reaction $NH_2^- + SO_3 \longrightarrow H_2N—SO_3^-$?

37. Cinnamaldehyde occurs naturally in cinnamon oil.

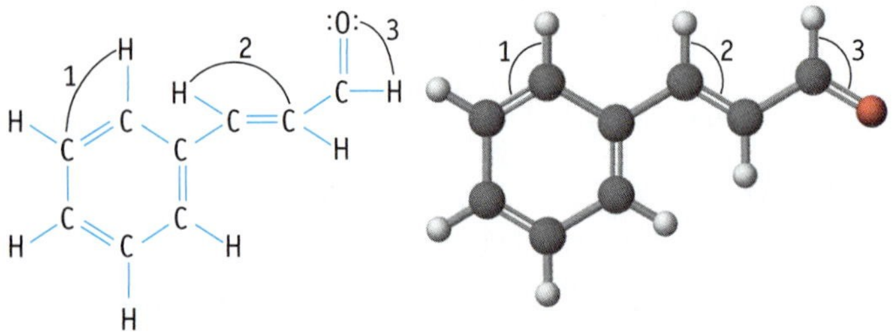

Cinnamaldehyde

(a) What is the most polar bond in the molecule?

(b) How many sigma (σ) bonds and how many pi (π) bonds are there?

(c) Is *cis-trans* isomerism possible? If so, draw the isomers of the molecule.

(d) Give the hybridization of the C atoms in the molecule.

(e) What are the values of the bond angles 1, 2, and 3?

38. ■ Iodine and oxygen form a complex series of ions, among them IO_4^- and IO_5^{3-}. Draw the Lewis structures for these ions, and specify their electron-pair geometries and the shapes of the ions. What is the hybridization of the I atom in these ions?

39. Antimony pentafluoride reacts with HF according to the equation

$$2\ HF + SbF_5 \longrightarrow [H_2F]^+[SbF_6]^-$$

(a) What is the hybridization of the Sb atom in the reactant and product?

(b) Draw a Lewis structure for H_2F^+. What is the geometry of H_2F^+? What is the hybridization of F in H_2F^+?

40. Xenon forms well-characterized compounds. Two xenon–oxygen compounds are XeO_3 and XeO_4. Draw the Lewis structures of these compounds, and give their electron-pair and molecular geometries. What are the hybrid orbital sets used by xenon in these two oxides?

41. The simple valence bond picture of O_2 does not agree with the molecular orbital view. Compare these two theories with regard to the peroxide ion, O_2^{2-}.

(a) Draw an electron dot structure for O_2^{2-}. What is the bond order of the ion?

(b) Write the molecular orbital electron configuration for O_2^{2-}. What is the bond order based on this approach?

(c) Do the two theories of bonding lead to the same magnetic character and bond order for O_2^{2-}?

42. ■ Nitrogen, N_2, can ionize to form N_2^+ or add an electron to give N_2^-. Using molecular orbital theory, compare these species with regard to (a) their magnetic character, (b) net number of π bonds, (c) bond order, (d) bond length, and (e) bond strength.

43. Which of the homonuclear, diatomic molecules of the second-period elements (from Li_2 to Ne_2) are paramagnetic? Which have a bond order of 1? Which have a bond order of 2? Which diatomic molecule has the highest bond order?

44. ■ Which of the following molecules or molecule ions should be paramagnetic? What is the highest occupied molecular orbital (HOMO) in each one? Assume the molecular orbital diagram in Figure 10.22 applies to all of them.

(a) NO   (c) O_2^{2-} (e) CN

(b) OF^- (d) Ne_2^+

45. The CN molecule has been found in interstellar space. Assuming the electronic structure of the molecule can be described using the molecular orbital energy level diagram in Figure 10.22, answer the following questions.

(a) What is the highest energy occupied molecular orbital (HOMO) to which an electron (or electrons) is (are) assigned?

(b) What is the bond order of the molecule?

(c) How many net σ bonds are there? How many net π bonds?

(d) Is the molecule paramagnetic or diamagnetic?

46. Amphetamine is a stimulant. Replacing one H atom on the NH_2, or amino, group with CH_3 gives methamphetamine, a particularly dangerous drug commonly known as "speed."

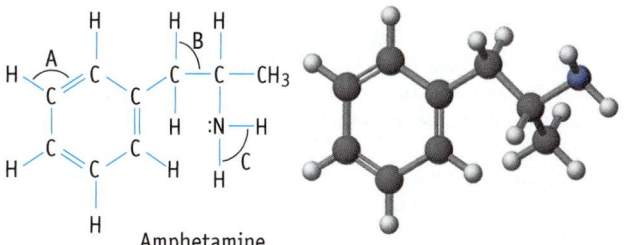

Amphetamine

(a) What are the hybrid orbitals used by the C atoms of the C_6 ring, by the C atoms of the side chain, and by the N atom?

(b) Give approximate values for the bond angles *A*, *B*, and *C*.

(c) How many σ bonds and π bonds are in the molecule?

(d) Is the molecule polar or nonpolar?

(e) Amphetamine reacts readily with a proton (H^+) in aqueous solution. Where does this proton attach to the molecule?

47. Menthol is used in soaps, perfumes, and foods. It is present in the common herb mint, and it can be prepared from turpentine.

(a) What are the hybridizations used by the C atoms in the molecule?

(b) What is the approximate C—O—H bond angle?

(c) Is the molecule polar or nonpolar?

(d) Is the six-member carbon ring planar or nonplanar? Explain why or why not.

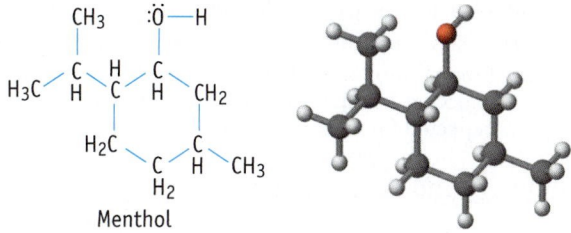

Menthol

48. The elements of the second period from boron to oxygen form compounds of the type X_nE—EX_n, where X can be H or a halogen. Sketch possible molecular structures for B_2F_4, C_2H_4, N_2H_4, and O_2H_2. Give the hybridizations of E in each molecule and specify approximate X—E—E bond angles.

49. ▲ The compound whose structure is shown here is acetylacetone. It exists in two forms: the *enol* form and the *keto* form.

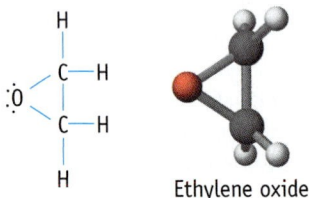

enol form   *keto* form

The molecule reacts with OH^- to form an anion, $[CH_3COCHCOCH_3]^-$ (often abbreviated acac$^-$ for acetylacetonate ion). One of the most interesting aspects of this anion is that one or more of them can react with a transition metal cations to give very stable, highly colored compounds.

(a) Are the *keto* and *enol* forms of acetylacetone contributing resonance forms? Explain your answer.

(b) What is the hybridization of each atom (except H) in the *enol* form? What changes in hybridization occur when it is transformed into the *keto* form?

(c) What is the electron-pair geometry and molecular geometry around each C atom in the *keto* and *enol* forms? What changes in geometry occur when the *keto* form changes to the *enol* form?

(d) Draw two possible resonance structures for the acac$^-$ ion.

(e) Is *cis-trans* isomerism possible in either the *enol* or the *keto* form?

50. ▲ Ethylene oxide has a three-member ring of two C atoms and an O atom.

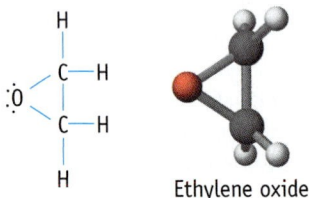

Ethylene oxide

(a) What are the expected bond angles in the ring?

(b) What is the hybridization of each atom in the ring?

(c) Comment on the relation between the bond angles expected based on hybridization and the bond angles expected for a three-member ring.

Summary and Conceptual Questions

The following questions may use concepts from the previous chapters.

51. ■ What is the maximum number of hybrid orbitals that a carbon atom may form? What is the minimum number? Explain briefly.

52. Consider the three fluorides BF_4^-, SiF_4, and SF_4.

(a) Identify a molecule that is isoelectronic with BF_4^-.

(b) Are SiF_4 and SF_4 isoelectronic?

(c) What is the hybridization of the central atom in each of these species?

53. ▲ When two amino acids react with each other, they form a linkage called an amide group, or a peptide link. (If more linkages are added, a protein or polypeptide is formed.)

(a) What are the hybridizations of the C and N atoms in the peptide linkage?

(b) Is the structure illustrated the only resonance structure possible for the peptide linkage? If another resonance structure is possible, compare it with the one shown. Decide which is the more important structure.

(c) The computer-generated structure shown here, which contains a peptide linkage, shows that the linkage is flat. This is an important feature of proteins. Speculate on reasons that the CO—NH linkage is planar.

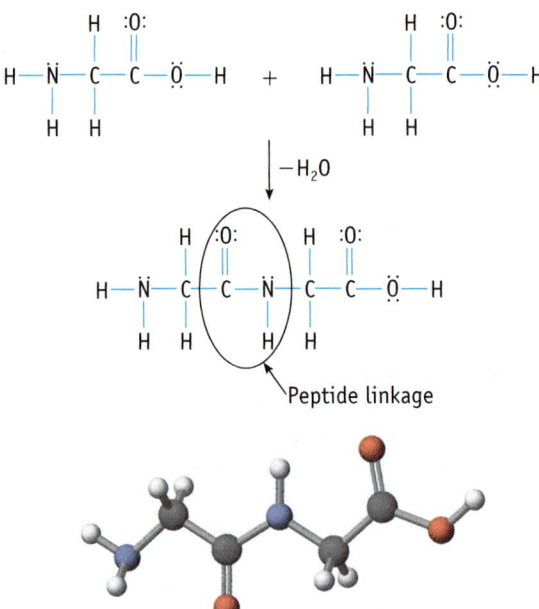

Peptide linkage

54. What is the connection between bond order, bond length, and bond energy? Use ethane (C_2H_6), ethylene (C_2H_4), and acetylene (C_2H_2) as examples.

55. When is it desirable to use MO theory rather than valence bond theory?

56. How do valence bond theory and molecular orbital theory differ in their explanation of the bond order of 1.5 for ozone?

57. Examine the Hybrid Orbitals tool on Screen 10.6 of the General ChemistryNow CD-ROM or website. Use this tool to systematically combine atomic orbitals to form hybrid atomic orbitals.

(a) What is the relationship between the number of hybrid orbitals produced and the number of atomic orbitals used to create them?

(b) Do hybrid atomic orbitals form between different *p* orbitals without involving *s* orbitals?

(c) What is the relationship between the energy of hybrid atomic orbitals and the atomic orbitals from which they are formed?

(d) Compare the shapes of the hybrid orbitals formed from an *s* orbital and a p_x orbital with the hybrid atomic orbitals formed from an *s* orbital and a p_z orbital.

(e) Compare the shape of the hybrid orbitals formed from *s*, p_x, and p_y orbitals with the hybrid atomic orbitals formed from *s*, p_x, and p_z orbitals.

58. Screen 10.2 of the General ChemistryNow CD-ROM or website shows the change in energy as a function of the H—H distance when H_2 forms from separated H atoms.

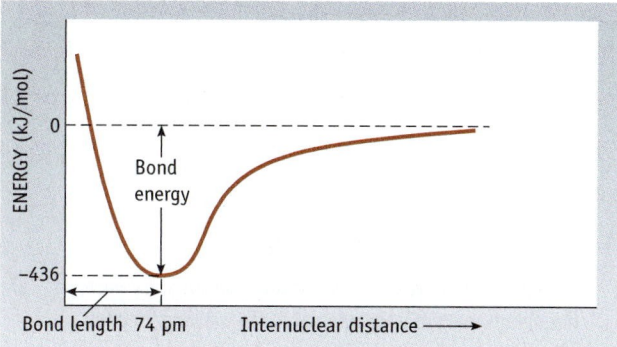

(a) Screen 10.3 describes the attractive and repulsive forces that occur when two atoms approach each other. What must be true about the relative strengths of those attractive and repulsive forces if a covalent bond is to form?

(b) When two atoms are widely separated, the energy of the system is defined as zero. As the atoms approach each other, the energy drops, reaches a minimum, and then increases as they approach still more closely. Explain these observations.

(c) For a bond to form, orbitals on adjacent atoms must overlap, and each pair of overlapping orbitals will contain two electrons. Explain why neon does not form a diatomic molecule, Ne_2, whereas fluorine forms F_2.

59. Examine the bonding in ethylene, C_2H_4, on Screen 10.7 of the General ChemistryNow CD-ROM or website and then go to the "A Closer Look" Auxiliary screen.

(a) Explain why the allene molecule is not flat. That is, explain why the CH_2 groups at opposite ends do not lie in the same plane.

(b) Based on the theory of orbital hybridization, explain why benzene is a planar, symmetrical molecule.

(c) What are the hybrid orbitals used by the three C atoms of allyl alcohol?

$$
\overset{\text{H}}{\underset{\text{H}}{\overset{|}{\underset{|}{\text{C}}}}} \overset{3}{=} \overset{\text{H}}{\underset{\text{H}}{\overset{|}{\underset{|}{\text{C}}}}} \overset{2}{-} \overset{\text{H}}{\overset{|}{\underset{1}{\text{C}}}} \overset{1}{-} \overset{\cdot\cdot}{\underset{\cdot\cdot}{\text{O}}} - \text{H}
$$

60. Screen 10.8 of the General ChemistryNow CD-ROM or website describes the motions of molecules.

(a) Observe the animations of the rotations of *trans*-2-butene and butane about their carbon–carbon bonds.

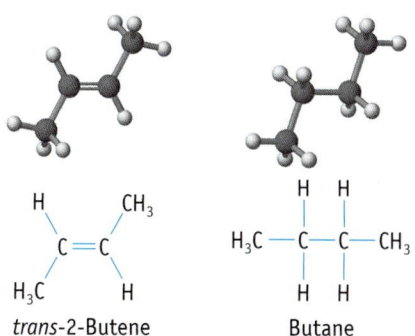

trans-2-Butene   Butane

As one end of *trans*-2-butene rotates relative to the other end, the energy increases about 200 kJ/mol and then drops as the rotation produces *cis*-2-butene. In contrast, the rotation of the butane molecule requires much less energy (only 60 kJ/mol). When butane has reached the halfway point in its rotation, the energy has reached a maximum. Why does *trans*-2-butene require so much more energy to rotate about the central carbon–carbon bond than does butane?

(b) The structure of propene, C_3H_6, is pictured here. Which carbon–hydrogen group (CH_3 or CH_2) can rotate freely with respect to the rest of the molecule?

$$\overset{\text{H}}{\overset{|}{\text{H}_3\text{C}-\text{C}}}=\text{CH}_2$$

(c) Can the two CH_2 fragments of allene (see Screen 10.7SB) rotate with respect to each other? Briefly explain why or why not.

Mentor™

Do you need a live tutor for homework problems?
Access vMentor at General ChemistryNow at
http://now.brookscole.com/kotz6e
for one-on-one tutoring from a chemistry expert

11—Carbon: More Than Just Another Element

A Colorful Beginning

The color purple was once associated with royalty because the dyes were rare and expensive. William Henry Perkin changed everything.

Among the roots of modern organic chemistry one finds the discovery, in 1856, of the compound mauveine (or mauve) by William Henry Perkin (1838–1907). This discovery dates from before the creation of the first periodic table; before the discovery of electrons, protons, and neutrons; before chemists knew anything about bonding; and even before the tetrahedral geometry of carbon was recognized. Perkin's work led to a flourishing dye industry in the latter part of the 19th century, which represented one of the first chemical industries to gain major importance. By 1900, more than 1000 synthetic dyes were known and in use.

Before the discovery of mauve, almost all dyes came from natural sources. Because the dye used for the color purple, Tyrian purple, was the rarest and most expensive, it became the exclusive color of royalty. Tyrian purple was the origin of both fame and fortune for the ancient empire of Tyre, because the dye was obtained only from a small mollusk found in the Mediterranean Sea in that region. More than 9000 mollusks were needed to obtain 1 gram of dye!

Original, stoppered bottle of mauveine prepared by Perkin. The structure of the mauveine cation is shown here.

William Henry Perkin (1838–1907). See the book on Perkin's life, *Mauve*, S. Garfield: New York, W. W. Norton, 2001.

The discovery of mauve by Perkin is an interesting tale of
serendipity. At the age of 13, Perkin enrolled at the City of
London School. His father paid an extra fee for him to attend a lunchtime chemistry course and set up a lab at home
for him to do experiments. Hooked on chemistry, Perkin
attended the public lectures that Michael Faraday gave on
Saturdays at the Royal Institution. At 15, he enrolled in
the Royal College of Science in London to study chemistry under the famous chemist August Wilhelm von
Hofmann. Perkin completed his studies at age 17 (the
field of chemistry was a lot smaller then than it is
today) and took a position at the college as Hofmann's assistant, rather a great honor.

Perkin's first chemistry project was to try to
synthesize quinine ($C_{20}H_{24}N_2O_2$), an antimalarial
drug. The route he proposed involved oxidizing
anilinium sulfate [$(C_6H_5NH_3)_2SO_4$]. Instead of
quinine, he obtained a black solid that dissolved in a water–ethanol mixture to give a
purple solution. Using a cloth to mop up a spill
on the lab bench, he noticed that the substance
stained the cloth a beautiful purple color. Furthermore, the color didn't wash out, an essential feature for
a useful dye. Later it was learned that the anilinium
sulfate used in the original reaction was impure and
that the impurity was essential in the synthesis. Had Perkin used a
pure sample as his starting reagent, the discovery of mauve would
not have happened, at least not in this way. A study in 1994 on

Science & Society Picture Library/Science Museum/London

**A silk dress dyed with Perkin's orig-
inal sample of mauve in 1862, at
the dawning of the synthetic dye
industry.**

samples of mauve preserved in museums
determined that Perkin's mauve was a mixture
of primarily two compounds, which have
closely related structures, along with traces
of several others.

At the age of 18, Perkin quit his assis-
tantship to exploit this new discovery. It
was not an easy decision because it in-
curred the great displeasure of his mentor,
Professor Hofmann. With financial help
from his family, Perkin set up a factory
outside of London. Although the road to
success was not smooth, Perkin perse-
vered and by the age of 36 he was a very
wealthy man. He then retired from the
dye business and devoted the rest of his
life to chemical research on various
topics including the synthesis of fragrances. He also studied optical activity, the ability of certain compounds to
rotate polarized light. During his life he received numerous honors for his research. One
honor, however, came many years after his
death. In 1972, when the Chemical Society of
London renamed its research journals after
famous society members, it chose Perkin's name for the journals in
which organic chemists publish their research.

To Review Before You Begin

- Review writing Lewis structures and predicting molecular structures (Section 9.4)
- Recall how to draw structures of molecules (Section 9.7)
- Review covalent bonding: valence bond and molecular orbital theory (Chapter 10)

The vast majority of the 20 million chemical compounds currently known are organic; that is, they are compounds built on a carbon framework. Organic compounds vary greatly in size and complexity, from the simplest hydrocarbon, methane, to molecules made up of many thousands of atoms. As you read this chapter, you will see that the range of possible materials is huge.

11.1—Why Carbon?

We begin this discussion of organic chemistry with a question: What features of carbon lead to both the abundance and the complexity of organic compounds? Answers fall into two categories: structural diversity and stability.

Structural Diversity

With four electrons in its outer shell, carbon will form four bonds to reach an octet configuration. In contrast, the elements boron and nitrogen form three bonds in molecular compounds, oxygen forms two bonds, and hydrogen and the halogens form one bond. With a larger number of bonds comes the opportunity to create more complex structures. This will become increasingly evident in this brief tour of organic chemistry.

A carbon atom can reach an octet of electrons in various ways (Figure 11.1):

- *By forming four single bonds.* A carbon atom can bond to four other atoms, which can be either atoms of other elements (often H, N, or O) or other carbon atoms.
- *By forming a double bond and two single bonds.* The carbon atoms in ethylene, $H_2C\!=\!CH_2$, are linked to other atoms in this way.
- *By forming two double bonds,* as in carbon dioxide ($O\!=\!C\!=\!O$).
- *By forming a triple bond and a single bond,* an arrangement seen in acetylene, $HC\!\equiv\!CH$.

Recognize, with each of these arrangements, the various possible geometries around carbon: tetrahedral, trigonal planar, and linear. Carbon's tetrahedral geometry is of special significance because it leads to three-dimensional chains and rings of carbon atoms, as in propane and cyclopentane. The ability to form multiple bonds leads to whole families of compounds based on structures such as ethylene, acetylene, and benzene.

GENERAL
Chemistry ⚛ Now™

Throughout the chapter this icon introduces a list of resources on the General ChemistryNow CD-ROM or website (http://now.brookscole.com/kotz6e) that will:

- help you evaluate your knowledge of the material
- provide homework problems
- allow you to take an exam-prep quiz
- provide a personalized Learning Plan targeting resources that address areas you should study

propane, C_3H_8 cyclopentane, C_5H_{10} benzene, C_6H_6

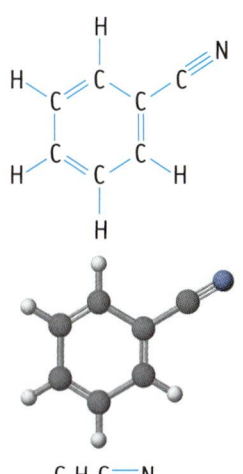

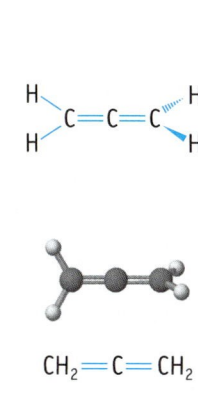

(a) **Acetic acid.** One carbon atom in this compound is attached to 4 other atoms by single bonds and has tetrahedral geometry. The second carbon atom, connected by a double bond to one oxygen, and by single bonds to the other oxygen and to carbon, has trigonal planar geometry.

(b) **Benzonitrile.** Six trigonal planar carbon atoms make up the benzene ring. The seventh C atom, bonded by a single bond to carbon and a triple bond to nitrogen, has a linear geometry.

(c) Carbon is linked by double bonds to two other carbon atoms in C_3H_4, a linear molecule commonly called allene.

Figure 11.1   Ways that carbon atoms bond.

Isomers

A hallmark of carbon chemistry is the remarkable array of isomers that can exist. **Isomers** are compounds that have identical composition but different structures. Two broad categories of isomers exist: structural isomers and stereoisomers.

   Structural isomers are compounds having the same elemental composition, but in which the atoms are linked together in different ways. Ethanol and dimethyl ether are structural isomers, as are 1-butene and 2-methylpropene.

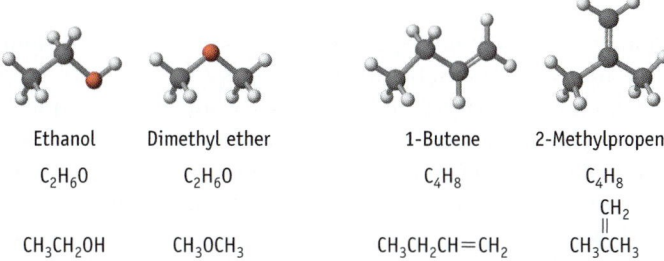

Ethanol	Dimethyl ether	1-Butene	2-Methylpropene
C_2H_6O	C_2H_6O	C_4H_8	C_4H_8
			CH_2
			$\|$
CH_3CH_2OH	CH_3OCH_3	$CH_3CH_2CH{=}CH_2$	CH_3CCH_3

   Stereoisomers are compounds with the same formula and in which there is a similar attachment of atoms. However, the atoms have different orientations in space. Two types of stereoisomers exist: geometric isomers and optical isomers.

   Cis- and *trans*-2-butene are **geometric isomers.** Geometric isomerism in these compounds occurs as a result of the C=C double bond. Recall that the carbon atom and the attached groups cannot rotate around a double bond (page 455). Thus, the geometry around the C=C double bond is fixed in space. If two groups occur on the adjacent carbon atoms and on the same side of the double bond, a *cis* isomer is produced. If groups appear on opposite sides, a *trans* isomer is produced.

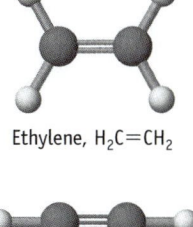

Ethylene, $H_2C{=}CH_2$

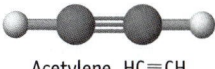

Acetylene, HC≡CH

Ethylene and acetylene. These two-carbon hydrocarbons can be the building blocks of more complex molecules. These are their common names, but their systematic names are ethene and ethyne.

A Closer Look

Writing Formulas and Drawing Structures

In Chapter 3 you learned that there are various ways of presenting structures (page 101). It is appropriate to return to that point as we look at organic compounds. Consider methane and ethane, for example. We can represent these molecules in several ways:

1. *Molecular formula:* CH_4 or C_2H_6. This type of formula gives information only on composition.

2. *Condensed formula:* For ethane this would be written CH_3CH_3 (or as H_3CCH_3). This method of writing the formula gives some information on the way atoms are connected.

3. *Structural formula:* You will recognize this formula as the Lewis structure. An elaboration on the condensed formula in (2), this representation defines more clearly how each atom is connected, but it fails to describe the shapes of molecules.

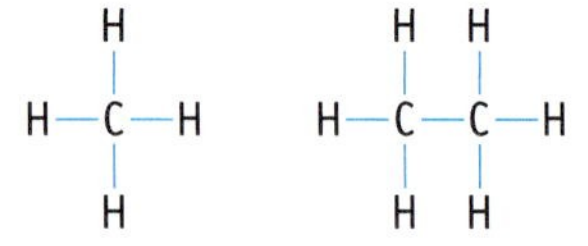

Methane, CH_4 Ethane, C_2H_6

4. *Perspective drawings:* These drawings are used to convey the three-dimensional nature of molecules. Bonds extending out of the plane of the paper are drawn with wedges, and bonds behind the plane of the paper are represented as dashed wedges (page 101). Using these guidelines, the structures of methane and ethane could be drawn as follows:

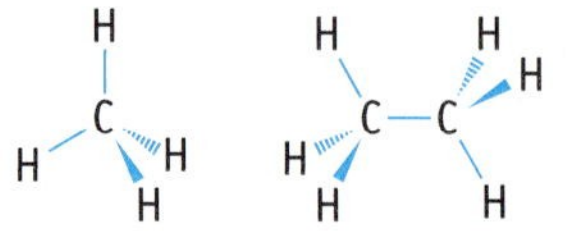

5. *Computer-drawn ball-and-stick and space-filling models.*

Ball-and stick

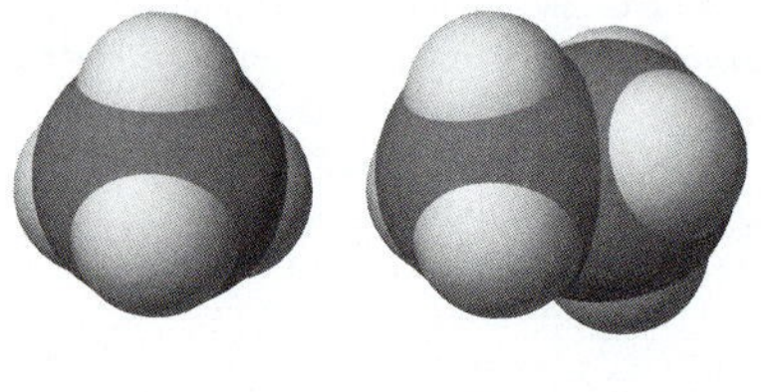

Space-filling

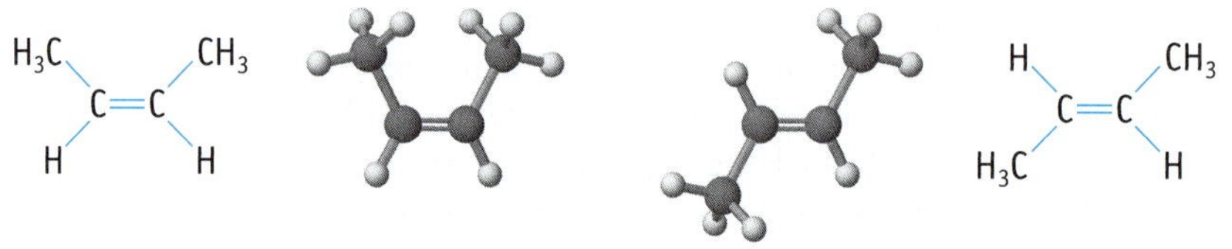

Cis-2-butene, C_4H_8 Trans-2-butene, C_4H_8

Optical isomerism is a second type of stereoisomerism. Optical isomers are molecules that have nonsuperimposable mirror images (Figure 11.2). Molecules (and other objects) that have nonsuperimposable mirror images are termed **chiral**. Pairs of non-superimposable molecules are called **enantiomers**.

Pure samples of enantiomers have the same physical properties, such as melting point, boiling point, density, and solubility in common solvents. They differ in one significant way, however: When a beam of plane-polarized light passes through a solution of a pure enantiomer, the plane of polarization rotates. The two enantiomers rotate polarized light to an equal extent, but in opposite directions (Figure 11.3). The term "optical isomerism" is used because this effect involves light (see "A Closer Look: Optical Isomers").

The most common examples of chiral compounds are those in which four different atoms (or groups of atoms) are attached to a tetrahedral carbon atom. Lactic acid, found in milk and a product of normal human metabolism, is an example of one such chiral compound (Figure 11.2). Optical isomerism is particularly important in the amino acids and other biologically important molecules.

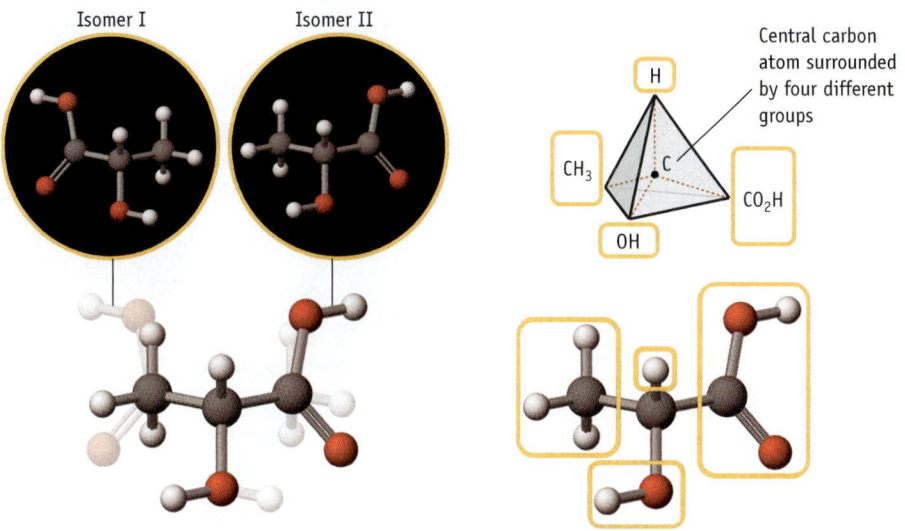

(a) Lactic acid isomers are nonsuperimposable

(b) Lactic acid, $CH_3CH(OH)CO_2H$

Active Figure 11.2 **Optical isomers.** (a) Optical isomerism occurs if a molecule and its mirror image cannot be superimposed. The situation is seen if four different groups are attached to carbon. (b) Lactic acid, a chiral molecule. Four different groups (H, OH, CH_3, and CO_2H) are attached to the central carbon atom.

Lactic acid is produced from milk when milk is fermented to make cheese. It is also found in other sour foods such as sauerkraut and is a preservative in pickled foods such as onions and olives. In our bodies it is produced by muscle activity and normal metabolism.

GENERAL Chemistry Now™ See the General ChemistryNow CD-ROM or website to explore an interactive version of this figure accompanied by an exercise.

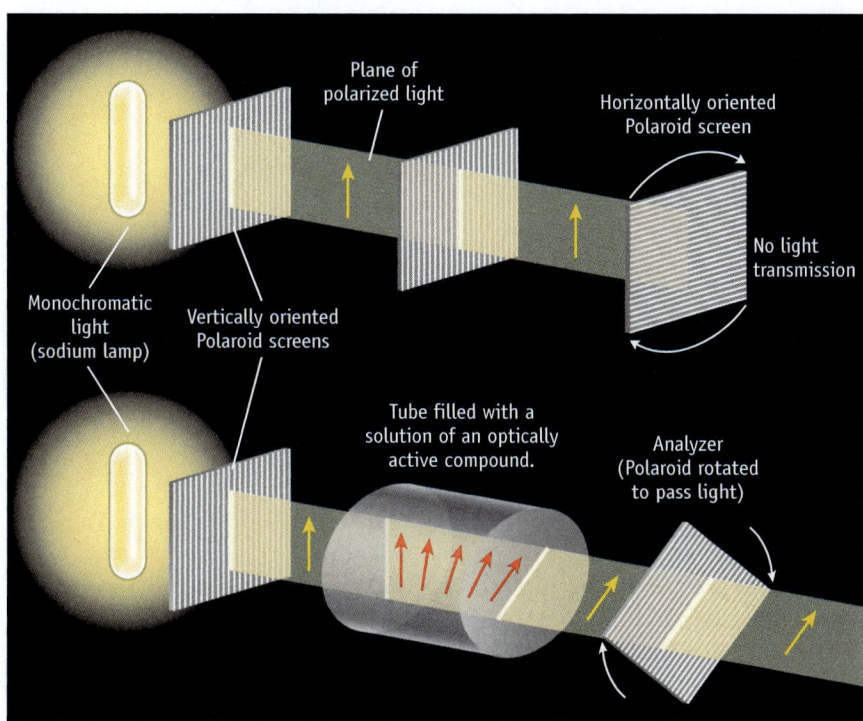

Figure 11.3 **Rotation of plane-polarized light by an optical isomer.** (*Top*) Monochromatic light (light of only one wavelength) is produced by a sodium lamp. After it passes through a polarizing filter, the light vibrates in only one direction—it is polarized. Polarized light will pass through a second polarizing filter if this filter is positioned parallel to the first filter, but not if the second filter is perpendicular.

(*Bottom*) A solution of an optical isomer placed between the first and second polarizing filters causes rotation of the plane of polarized light. The angle of rotation can be determined by rotating the second filter until maximum light transmission occurs. The magnitude and direction of rotation are unique physical properties of the optical isomer being tested.

A Closer Look

Optical Isomers and Chirality

Everyone has accidentally put a left shoe on a right foot, or a left-handed glove on a right hand. It doesn't work very well. Even though our two hands and two feet appear generally similar, a very important distinction separates them. Left hands and feet are mirror images of right hands and feet. Most importantly, these mirror images cannot be superimposed. We describe them as *chiral*.

Many common objects have this property. Some seashells are chiral, for example. Wood screws and machine bolts are also chiral, being distinguished by left-handed or right-handed threads.

Certain molecules have the same characteristic as gloves and hands: A given structure and its mirror image—its enantiomer—cannot be superimposed. There are various ways to visualize that two enantiomers are different. Imagine a tetrahedral carbon atom attached to four other atoms or groups, all different. For simplicity the atoms bonded to the central C atom in the amino acid alanine are shown as 1, 2, 3, and 4 in the drawing. Sight down one of the bonds to carbon (say, the bond from atom 1 to C) in one enantiomer. The other three atoms (2, 3, and 4) will then appear in a clockwise order. In the second enantiomer, atoms 2, 3, and 4 will appear in counterclockwise order.

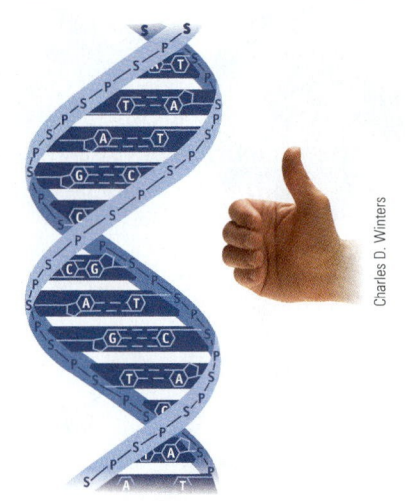

The helical chain of DNA is like the threads of a screw. It twists to the left or it twists to the right. Here it twists to the right. If you curl your right hand around the chain, with your thumb extended, your fingers will show the direction of the twist and your thumb will point along the chain.

The handedness of seashells. Seashells are almost all right-handed. This photo shows the egg cases for whelk shells. Each egg case is about 3 cm in diameter and about 2–3 mm thick. Each egg case is attached to a spine, and the arrangement of egg cases around the spine is right-handed.

Enantiomers of alanine.

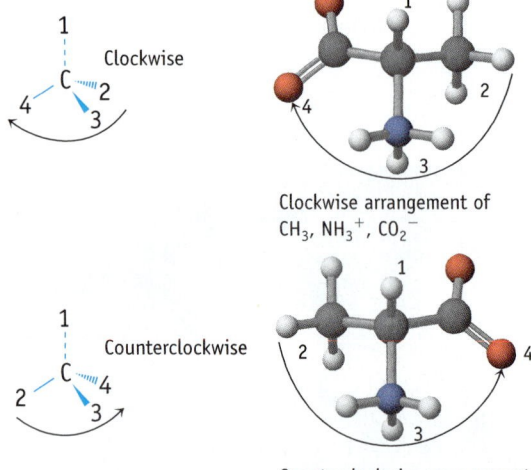

Clockwise arrangement of CH_3, NH_3^+, CO_2^-

Counterclockwise arrangement of CH_3, NH_3^+, CO_2^-

Stability of Carbon Compounds

Carbon compounds are notable for their resistance to chemical change. Were this not so, far fewer compounds of carbon would be known.

Strong bonds are needed for molecules to survive in their environment. Molecular collisions in gases, liquids, and solutions often provide enough energy to break some chemical bonds, and bonds can be broken if the energy associated with photons of visible and ultraviolet light exceeds the bond energy. Carbon–carbon bonds are relatively strong, however, as are the bonds between carbon and most other atoms. The average C—C bond energy is 346 kJ/mol, the C—H bond energy is

413 kJ/mol, and carbon–carbon double and triple bond energies are even higher [◀ Section 9.10]. Contrast these values with bond energies for the Si—H bond (328 kJ/mol) and the Si—Si bond (222 kJ/mol). The consequence of high bond energies for bonds to carbon is that, for the most part, organic compounds do not degrade under normal conditions.

Oxidation of most organic compounds is strongly product-favored, but most organic compounds can survive lengthy contact with O_2. The reason is that these reactions occur slowly. Most organic compounds burn only if their combustion is initiated by heat or by a spark. As a consequence, oxidative degradation is not a barrier to the existence of organic compounds.

11.2—Hydrocarbons

Hydrocarbons, compounds made of carbon and hydrogen only, are classified into several subgroups: alkanes, cycloalkanes, alkenes, alkynes, and aromatic compounds (Table 11.1). We begin our discussion by considering compounds that have carbon atoms with four single bonds, the alkanes and cycloalkanes.

GENERAL
Chemistry ᐧ᛫Now™

See the General ChemistryNow CD-ROM or website:
- **Screen 11.3 Hydrocarbons,** for a description of the classes of hydrocarbons

---

Table 11.1   Some Types of Hydrocarbons

Type of Hydrocarbon	Characteristic Features	General Formula	Example
alkanes	C—C single bonds and all C atoms have four single bonds	C_nH_{2n+2}	CH_4, methane C_2H_6, ethane
cyclic alkanes	C—C single bonds and all C atoms have four single bonds	C_nH_{2n}	C_6H_{12}, cyclohexane
alkenes	C=C double bond	C_nH_{2n}	$H_2C=CH_2$, ethylene
alkynes	C≡C triple bond	C_nH_{2n-2}	HC≡CH, acetylene
aromatics	rings with π bonding extending over several C atoms	—	benzene, C_6H_6

Alkanes

Alkanes have the general formula C_nH_{2n+2}, with n having integer values (Table 11.2). Formulas of specific compounds can be generated from this general formula, the first four of which are CH_4 (methane), C_2H_6 (ethane), C_3H_8 (propane), and C_4H_{10} (butane) (Figure 11.4). Methane has four hydrogen atoms arranged tetrahedrally around a single carbon atom. Replacing a hydrogen atom in methane by a —CH_3 group gives ethane. If an H atom of ethane is replaced by yet another —CH_3 group, propane results. Butane is derived from propane by replacing an H atom of one of the chain-ending carbon atoms with a —CH_3 group. In all of these compounds each C atom is attached to four other atoms, either C or H, so alkanes are often called **saturated compounds**.

**Table 11.2   Selected Hydrocarbons of the Alkane Family, C_nH_{2n+2}*

Name	Molecular Formula	State at Room Temperature
methane	CH_4	
ethane	C_2H_6	gas
propane	C_3H_8	
butane	C_4H_{10}	
pentane	C_5H_{12} (pent- = 5)	
hexane	C_6H_{14} (hex- = 6)	
heptane	C_7H_{16} (hept- = 7)	liquid
octane	C_8H_{18} (oct- = 8)	
nonane	C_9H_{20} (non- = 9)	
decane	$C_{10}H_{22}$ (dec- = 10)	
octadecane	$C_{18}H_{38}$ (octadec- = 18)	solid
eicosane	$C_{20}H_{42}$ (eicos- = 20)	

* This table lists only selected alkanes. Liquid compounds with 11 to 16 carbon atoms are also known. Many solid alkanes with more than 20 carbon atoms also exist.

Structural Isomers

The formulas for alkanes do not hint at their structural diversity. Structural isomers are possible for all alkanes larger than propane. For example, there are two structural isomers for C_4H_{10} and three for C_5H_{12}. As the number of carbon atoms in an alkane increases, the number of possible structural isomers greatly increases; there are 5 isomers possible for C_6H_{14}, 9 isomers for C_7H_{16}, 18 for C_8H_{18}, 75 for $C_{10}H_{22}$, and 366,319 for $C_{20}H_{42}$.

To recognize the isomers corresponding to a given formula, keep in mind the following points:

- Each alkane is built upon a framework of tetrahedral carbon atoms, and each carbon must have four single bonds.

- An effective approach is to create a framework of carbon atoms and then fill the remaining positions around carbon with H atoms so that each C atom has four bonds.

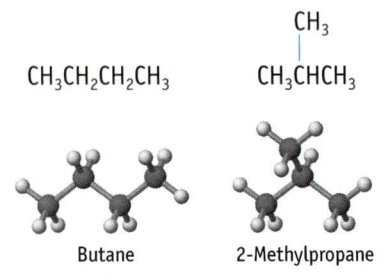

$CH_3CH_2CH_2CH_3$ CH_3CHCH_3
                                          |
                                        CH_3

Butane   2-Methylpropane

Structural isomers of butane, C_4H_{10}.

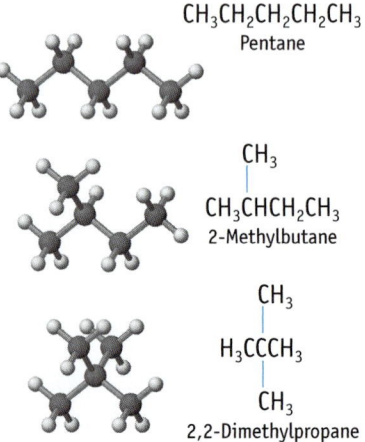

$CH_3CH_2CH_2CH_2CH_3$
Pentane

$CH_3CHCH_2CH_3$
        |
      CH_3
2-Methylbutane

H_3CCCH_3
      |
    CH_3
2,2-Dimethylpropane

Structural isomers of pentane, C_5H_{12}.

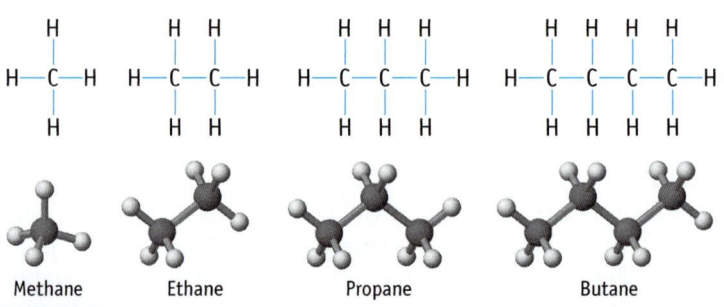

Methane   Ethane   Propane   Butane

Active Figure 11.4   Alkanes. The lowest-molecular-weight alkanes, all gases under normal conditions, are methane, ethane, propane, and butane.

GENERAL
Chemistry Now™ See the General ChemistryNow CD-ROM or website to explore an interactive version of this figure accompanied by an exercise.

• Free rotation occurs around carbon–carbon single bonds. Therefore, when atoms are assembled to form the skeleton of an alkane, the emphasis is on how carbon atoms are attached to one another and not on how they might lie relative to one another in the plane of the paper.

Example 11.1—Drawing Structural Isomers of Alkanes

Problem Draw structures of the five isomers of C_6H_{14}. Are any of these isomers chiral?

Strategy Focus first on the different frameworks that can be built from six carbon atoms. Having created a carbon framework, fill hydrogen atoms into the structure so that each carbon has four bonds.

Solution

Step 1. Placing six carbon atoms in a chain gives the framework for the first isomer. Now fill in hydrogen atoms: three on the carbons on the ends of the chain, two on each of the carbons in the middle. You have created the first isomer, hexane.

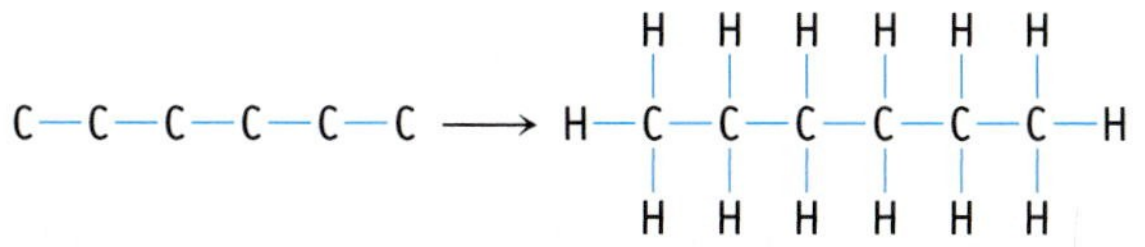

carbon framework of hexane   hexane

Step 2. Draw a chain of five carbon atoms, then add the sixth carbon atom to one of the carbons in the middle of this chain. (Adding it to a carbon at the end of the chain gives a six-carbon chain, the same framework drawn in Step 1.) Two different carbon frameworks can be built from the five-carbon chain, depending on whether the sixth carbon is linked to the 2 or 3 position. For each of these frameworks, fill in the hydrogens.

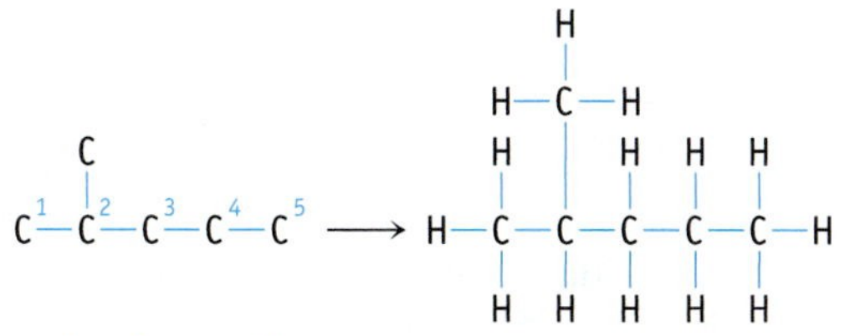

carbon framework
of methylpentane isomers   2-methylpentane

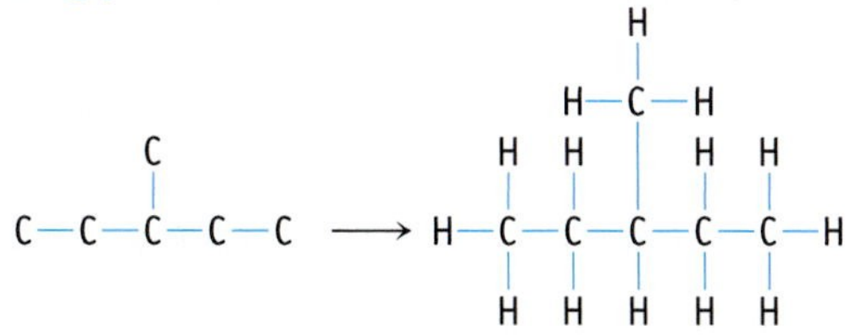

3-methylpentane

Step 3. Draw a chain of four carbon atoms. Add in the two remaining carbons, again being careful not to extend the chain length. Two different structures are possible: one with the remaining carbon atoms each in the 2 and 3 positions, and another with both extra carbon atoms attached at the 2 position. Fill in the 14 hydrogens. You have now drawn the fourth and fifth isomers.

■ **Chirality in Alkanes**
To be chiral, a compound must have at least one C atom attached to four different groups. Thus, the C_7H_{16} isomer here is chiral.

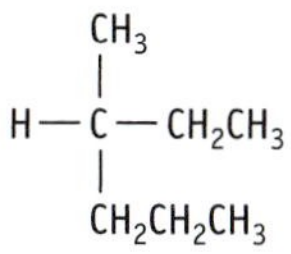

carbon atom frameworks
for dimethylbutane isomers

2,3-dimethylbutane

2,2-dimethylbutane

None of the isomers of C_6H_{14} is chiral. To be chiral, a compound must have at least one C atom with four different groups attached. This condition is not met in any of these isomers.

Comment Should we look for structures in which the longest chain is three carbon atoms? Try it, but you will see that it is not possible to add the three remaining carbons to a three-carbon chain without creating one of the carbon chains already drawn in a previous step. Thus, we have completed the analysis, with five isomers of this compound being identified.

Names have been given to each of these compounds. See the text that follows this Example and see Appendix E for guidelines on nomenclature.

Exercise 11.1—Drawing Structural Isomers of Alkanes

(a) Draw the nine isomers having the formula C_7H_{16}. [*Hint*: There is one structure with a seven-carbon chain, two structures with six-carbon chains, five structures in which the longest chain has five carbons (one is illustrated in the margin), and one structure with a four-carbon chain.]

(b) Identify the isomers of C_7H_{16} that are chiral.

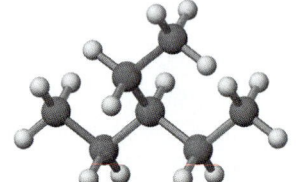

One possible isomer of an alkane with the formula C_7H_{16}.

■ **Naming Guidelines**
For more details on naming organic compounds, see Appendix E.

Naming Alkanes

With so many possible isomers for a given alkane, chemists need a systematic way of naming them. The rules for naming alkanes and their derivatives follow:

- The names of alkanes end in "-ane."
- The names of alkanes with chains of one to ten carbon atoms are given in Table 11.2. After the first four compounds, the names are derived from Latin numbers—pentane, hexane, heptane, octane, nonane, decane—and this regular naming continues for higher alkanes.
- When naming a specific alkane, the root of the name corresponds to the longest carbon chain in the compound. One isomer of C_5H_{12} has a three-carbon chain with two —CH_3 groups on the second C atom of the chain. Thus, its name is based on propane.

Problem Solving Tip 11.1

Drawing Structural Formulas

An error students sometimes make is to suggest that the three carbon skeletons drawn here are different. They are, in fact, the same. All are five-carbon chains with another C atom in the 2 position.

$$
\begin{array}{c}
\text{C}\\
|\\
\text{C}-\text{C}-\text{C}-\text{C}-\text{C}\\
1\ \ \ 2\ \ \ 3\ \ \ 4\ \ \ 5
\end{array}
\qquad
\begin{array}{c}
\text{C}\\
|\\
\text{C}-\text{C}-\text{C}-\text{C}\\
2\ \ \ 3\ \ \ 4\ \ \ 5\\
|\\
\text{C}\\
1
\end{array}
\qquad
\begin{array}{c}
\text{C}\\
|\\
\text{C}-\text{C}-\text{C}-\text{C}\\
5\ \ \ 4\ \ \ 3\ \ \ 2\ \ \ 1
\end{array}
$$

Remember that Lewis structures do not indicate the geometry of molecules.

$$
\begin{array}{c}
\text{CH}_3\\
|\\
\text{H}_3\text{C}-\text{C}-\text{CH}_3\\
|\\
\text{CH}_3
\end{array}
$$

2,2-dimethylpropane

- Substituent groups on a hydrocarbon chain are identified by a name and the position of substitution in the carbon chain; this information precedes the root of the name. The position is indicated by a number that refers to the carbon atom to which the substituent is attached. (Numbering of the carbon atoms in a chain should begin at the end of the carbon chain that allows the substituent groups to have the lowest numbers.) Both —CH$_3$ groups in 2,2-dimethylpropane are located at the 2 position.

- Names of hydrocarbon substituents, called **alkyl groups**, are derived from the name of the hydrocarbon. The group —CH$_3$, derived by taking a hydrogen from methane, is called the methyl group; the C$_2$H$_5$ group is the ethyl group.

- If two or more of the same substituent groups occur, the prefixes di-, tri-, and tetra- are added. When different substituent groups are present, they are generally listed in alphabetical order.

■ **Systematic and Common Names**
Many organic compounds are known by common names. For example, 2,2-dimethylpropane is also called neopentane. However, the IUPAC (International Union of Pure and Applied Chemistry) has formulated rules for systematic names, which are generally used in this book. See Appendix E.

Example 11.2—Naming Alkanes

Problem Give the systematic name for

$$
\begin{array}{ccc}
\text{CH}_3 & & \text{C}_2\text{H}_5\\
| & & |\\
\text{CH}_3\text{CHCH}_2\text{CH}_2\text{CHCH}_2\text{CH}_3
\end{array}
$$

Strategy Identify the longest carbon chain and base the name of the compound on that alkane. Identify the substituent groups on the chain and their locations. When there are two or more substituents (the groups attached to the chain), number the parent chain from the end that gives the lower number to the substituent encountered first. If the substituents are different, list them in alphabetical order. (For more on naming compounds, see Appendix E.)

Solution Here the longest chain has seven C atoms, so the root of the name is *heptane*. There is a methyl group on C-2 and an ethyl group on C-5. Giving the substituents in alphabetic order, and numbering the chain from the end having the methyl group, the systematic name is 5-ethyl-2-methylheptane.

Figure 11.5   Paraffin wax and mineral oil. These common consumer products are mixtures of alkanes.

Exercise 11.2—Naming Alkanes

Name the nine isomers of C_7H_{16} in Exercise 11.1.

Properties of Alkanes

Methane, ethane, propane, and butane are gases at room temperature and pressure, whereas the higher-molecular-weight compounds are liquids or solids (Table 11.2). An increase in melting point and boiling point with molecular weight is a general phenomenon that reflects the increased forces of attraction between molecules [▶ Section 13.2].

You already know about alkanes in a nonscientific context because several are common fuels. Natural gas, gasoline, kerosene, fuel oils, and lubricating oils are all mixtures of various alkanes. White mineral oil is also a mixture of alkanes, as is paraffin wax (Figure 11.5).

Pure alkanes are colorless. (The colors seen in gasoline and other petroleum products are due to additives.) The gases and liquids have noticeable but not unpleasant odors. All of these substances are insoluble in water, which is typical of compounds that are nonpolar or nearly so. Low polarity is expected for alkanes because the electronegativity of carbon ($\chi = 2.5$) and hydrogen ($\chi = 2.2$) are not greatly different [◀ Section 9.9].

All alkanes burn readily in air to give CO_2 and H_2O in very exothermic reactions. This is, of course, the reason they are widely used as fuels.

$$CH_4(g) + 2\ O_2(g) \longrightarrow CO_2(g) + 2\ H_2O(\ell) \qquad \Delta H^\circ_{rxn} = -890.3\ kJ$$

Other than in combustion reactions, alkanes exhibit relatively low chemical reactivity. One reaction that does occur, however, is the replacement of the hydrogen atoms of an alkane by chlorine atoms on reaction with Cl_2. It is formally an oxidation because Cl_2, like O_2, is a strong oxidizing agent. These reactions, which can be initiated by ultraviolet radiation, are free radical reactions. Highly reactive Cl atoms are formed from Cl_2 under UV radiation. Reaction of methane with Cl_2 under these conditions proceeds in a series of steps, eventually yielding CCl_4, commonly known as carbon tetrachloride. (HCl is the other product of these reactions.)

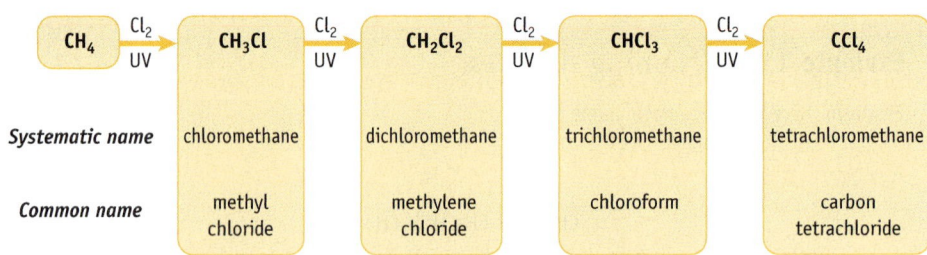

	CH_4	$\xrightarrow[\text{UV}]{Cl_2}$	CH_3Cl	$\xrightarrow[\text{UV}]{Cl_2}$	CH_2Cl_2	$\xrightarrow[\text{UV}]{Cl_2}$	$CHCl_3$	$\xrightarrow[\text{UV}]{Cl_2}$	CCl_4
Systematic name			chloromethane		dichloromethane		trichloromethane		tetrachloromethane
Common name			methyl chloride		methylene chloride		chloroform		carbon tetrachloride

The last three compounds are used as solvents, albeit less frequently today because of their toxicity. Carbon tetrachloride was also once widely used as a dry cleaning fluid and, because it does not burn, in fire extinguishers.

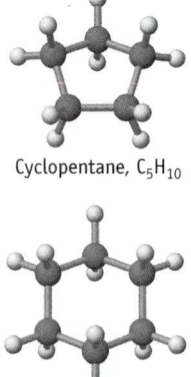

Cyclopentane, C_5H_{10}

Cyclohexane, C_6H_{12}

The structures of cyclopentane, C_5H_{10}, and cyclohexane, C_6H_{12}. The C_5 ring is nearly planar. In contrast, the tetrahedral geometry around carbon means that the C_6 ring is decidedly puckered.

Cycloalkanes, C_nH_{2n}

Cycloalkanes are constructed with tetrahedral carbon atoms joined together to form a ring. For example, cyclopentane, C_5H_{10}, consists of a ring of five carbon atoms. Each carbon atom is bonded to two adjacent carbon atoms and to two

A Closer Look

Flexible Molecules

Most organic molecules are flexible; that is, they can twist and bend in various ways. Few molecules better illustrate this behavior than cyclohexane. Two structures are possible, "chair" and "boat" forms.

These forms can interconvert by partial rotation of several bonds.

The more stable structure is the chair form which allows the hydrogen atoms to remain as far apart as possible. A side view of this form of cyclohexane reveals two sets of hydrogen atoms in this molecule. Six hydrogen atoms, called the equatorial

hydrogens, lie in a plane around the carbon ring. The other six hydrogens are positioned above and below the plane and are called axial hydrogens. Flexing the ring (a rotation around the C—C single bonds) moves the hydrogen atoms between axial and equatorial environments.

chair form ⇌ boat form ⇌ chair form

hydrogen atoms. Notice that the five carbon atoms fall very nearly in a plane. This is because the internal angles of a pentagon, 110°, closely match the tetrahedral angle of 109.5°. The small distortion from planarity allows hydrogen atoms on adjacent carbon atoms to be a little farther apart.

Cyclohexane has a nonpolar ring with six —CH_2 groups. If the carbon atoms were in the form of a regular hexagon with all carbon atoms in one plane, the C—C—C bond angles would be 120°. To have tetrahedral bond angles of 109.5° around each C atom, the ring has to pucker. The C_6 ring is flexible, however, and exists in two interconverting forms (see "A Closer Look: Flexible Molecules").

Interestingly, cyclobutane and cyclopropane are also known, although the bond angles in these species are much less than 109.5°. These compounds are examples of **strained hydrocarbons**, so named because an unfavorable geometry is imposed around carbon. One of the features of strained hydrocarbons is that the C—C bonds are weaker and the molecules readily undergo ring-opening reactions that relieve the bond angle strain.

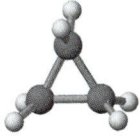 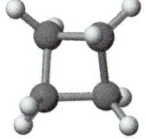

Cyclopropane, C_3H_6 Cyclobutane, C_4H_8

Cyclopropane and cyclobutane.
Cyclopropane was at one time used as a general anesthetic in surgery. However, its explosive nature when mixed with oxygen soon eliminated this application. The *Columbia Encyclopedia* states that "cyclopropane allowed the transport of more oxygen to the tissues than did other common anesthetics and also produced greater skeletal muscle relaxation. It is not irritating to the respiratory tract. Because of the low solubility of cyclopropane in the blood, postoperative recovery was usually rapid but nausea and vomiting were common."

Alkenes and Alkynes

The abundance and diversity of alkanes are repeated with **alkenes**, hydrocarbons with one or more C=C double bonds. The presence of the double bond adds two features missing in alkanes: the possibility of geometric isomerism and increased reactivity.

The general formula for alkenes is C_nH_{2n}. The first two members of the series of alkenes are ethene, C_2H_4 (common name, ethylene), and propene, C_3H_6 (common name, propylene). Only a single structure can be drawn for these compounds. As with alkanes, the occurrence of isomers begins with species containing four carbon atoms. Four alkene isomers have the formula C_4H_8, and each has distinct chemical and physical properties (Table 11.3).

C_2H_4
Systematic name:
Ethene
Common name:
Ethylene

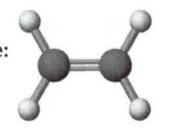

C_3H_6
Systematic name:
Propene
Common name:
Propylene

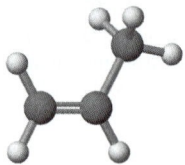

Table 11.3   Properties of Butene Isomers

Name	Boiling Point	Melting Point	Dipole Moment (D)	ΔH°_f (gas) (kJ/mol)
1-butene	−6.26 °C	−185.4 °C	—	−20.5
2-methylpropene	−6.95 °C	−140.4 °C	0.503	−37.5
cis-2-butene	3.71 °C	−138.9 °C	0.253	−29.7
trans-2-butene	0.88 °C	−105.5 °C	0	−33.0

1-butene   2-methylpropene   cis-2-butene   trans-2-butene

Alkene names end in "-ene." As with alkanes, the root name for alkenes is that of the longest carbon chain. The position of the double bond is indicated with a number, and, when appropriate, the prefix *cis* or *trans* is added. Three of the C_4H_8 isomers have four-carbon chains and so are butenes. One has a three-carbon chain and is a propene. Notice that the carbon chain is numbered from the end that gives the double bond the lowest number. In the first isomer at the left, the double bond is between C atoms 1 and 2, so the name is 1-butene and not 3-butene.

Example 11.3—Determining Isomers of Alkenes from a Formula

Problem Draw structures for the six possible alkene isomers with the formula C_5H_{10}. Give the systematic name of each.

Strategy A procedure that involved drawing the carbon skeleton and then adding hydrogen atoms served well when drawing structures of alkanes (Example 11.1), and a similar approach can be used here. It will be necessary to put one double bond into the framework and to be alert for *cis-trans* isomerism.

Solution

1.   A five-carbon chain with one double bond can be constructed in two ways. One gives rise to *cis–trans* isomers.

C=C—C—C—C ⟶ 1-pentene

C—C=C—C—C cis-2-pentene / trans-2-pentene

2. Draw the possible four-carbon chains containing a double bond. Add the fifth carbon atom to either the 2 or 3 position. When all three possible combinations are found, fill in the hydrogen atoms. This results in three more structures:

2-methyl-1-butene

3-methyl-1-butene

2-methyl-2-butene

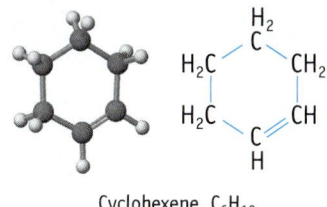

Cyclohexene, C_6H_{10}

Exercise 11.3—Determining Structural Isomers of Alkenes from a Formula

There are 17 possible alkene isomers with the formula C_6H_{12}. Draw structures of the five isomers in which the longest chain has six carbon atoms and give the name of each. Which of these isomers is chiral? (There are also eight isomers in which the longest chain has five carbon atoms, and four isomers in which the longest chain has four carbon atoms. How many can you find?)

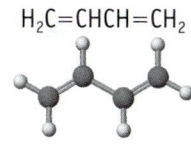

1,3-Butadiene, C_4H_6

Cycloalkenes and dienes. Cyclohexene, C_6H_{10} (*top*), and 1,3-butadiene (C_4H_6) (*bottom*).

More than one double bond can be present in a hydrocarbon. Butadiene, for example, has two double bonds and is known as a *diene*. Many natural products have numerous double bonds (Figure 11.6). There are also cyclic hydrocarbons, such as cyclohexene, with double bonds.

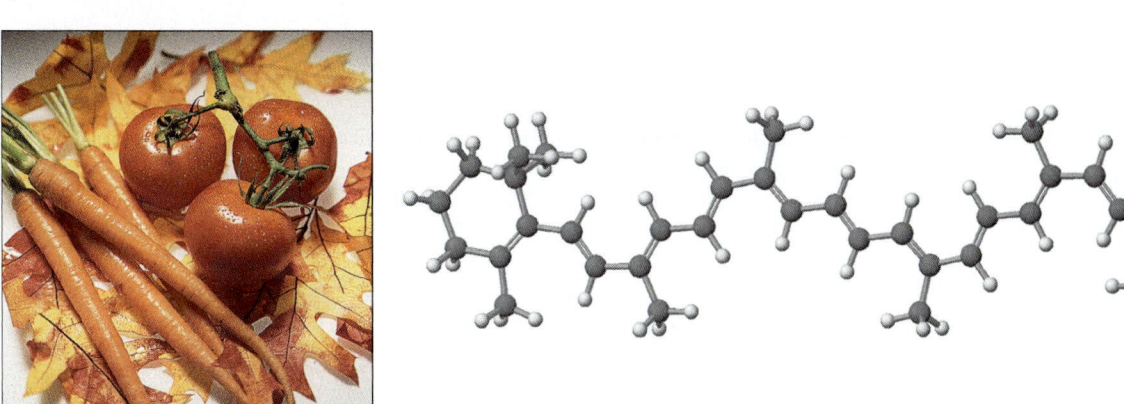

Figure 11.6 Carotene, a naturally occurring compound with 11 C=C bonds. The π electrons can be excited by visible light in the blue-violet region of the spectrum. As a result, carotene appears orange-yellow to the observer. Carotene or carotene-like molecules are partnered with chlorophyll in nature in the role of assisting in the harvesting of sunlight. Green leaves have a high concentration of carotene. In autumn, green chlorophyll molecules are destroyed and the yellows and reds of carotene and related molecules are seen. The red color of tomatoes, for example, comes from a molecule very closely related to carotene. As a tomato ripens, its chlorophyll disintegrates and the green color is replaced by the red of the carotene-like molecule.

An oxy-acetylene torch. The reaction of ethyne (acetylene) with oxygen produces a very high temperature. Oxy-acetylene torches, used in welding, take advantage of this fact.

Alkynes, compounds with a carbon–carbon triple bond, have the general formula (C_nH_{2n-2}). Table 11.4 lists alkynes that have four or fewer carbon atoms. The first member of this family is ethyne (common name, acetylene), a gas used as a fuel in metal cutting torches.

Table 11.4   Some Simple Alkynes C_nH_{2n-2}

Structure	Systematic Name	Common Name	BP (°C)
$HC \equiv CH$	ethyne	acetylene	−85
$CH_3C \equiv CH$	propyne	methylacetylene	−23
$CH_3CH_2C \equiv CH$	1-butyne	ethylacetylene	9
$CH_3C \equiv CCH_3$	2-butyne	dimethylacetylene	27

Properties of Alkenes and Alkynes

Like alkanes, alkenes and alkynes are colorless. Low-molecular-weight compounds are gases, whereas compounds with higher molecular weights are liquids or solids. Alkanes, alkenes, and alkynes are also oxidized by O_2 to give CO_2 and H_2O.

In contrast to alkanes, alkenes and alkynes have an elaborate chemistry. We gain an insight into their chemical behavior by noting that they are called **unsaturated compounds**. Carbon atoms are capable of bonding to a maximum of four other atoms, and they do so in alkanes and cycloalkanes. In alkenes, however, the carbon atoms linked by a double bond are bonded to only three atoms; in alkynes, they bond to two atoms. It is possible to increase the number of bonds to carbon by **addition reactions** in which molecules with the general formula X—Y (such as hydrogen, halogens, hydrogen halides, and water) add across the carbon–carbon double bond (Figure 11.7). The result is a compound with four atoms bonded to each carbon.

$$X—Y = H_2, Cl_2, Br_2;\ H—Cl,\ H—Br,\ H—OH,\ HO—Cl$$

The products of addition reactions are substituted alkanes. For example, the addition of bromine to ethylene forms 1,2-dibromoethane.

1,2-dibromoethane

The addition of 2 mol of chlorine to acetylene gives 1,1,2,2-tetrachloroethane.

$$HC \equiv CH + 2\ Cl_2 \longrightarrow$$

1,1,2,2-tetrachloroethane

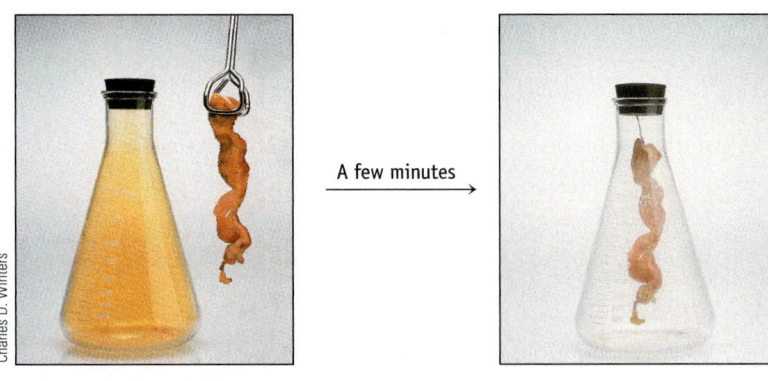

Figure 11.7 Bacon fat and addition reactions. The fat in bacon is partially unsaturated. Like other unsaturated compounds, bacon fat reacts with Br_2 in an addition reaction. Here you see the color of Br_2 vapor fade when a strip of bacon is introduced.

If the reagent added to a double bond is hydrogen ($X—Y = H_2$), the reaction is called **hydrogenation** and the product is an alkane. Hydrogenation is usually a very slow reaction, but it can be speeded up in the presence of a catalyst, often a specially prepared form of a metal, such as platinum, palladium, and rhodium. You may have heard the term hydrogenation because certain foods contain "hydrogenated" or "partially hydrogenated" ingredients. One brand of crackers has a label that says, "Made with 100% pure vegetable shortening . . . (partially hydrogenated soybean oil with hydrogenated cottonseed oil)." One reason for hydrogenating an oil is to make it less susceptible to spoilage; another is to convert it from a liquid to a solid.

■ **Catalysts**
A substance that causes a reaction to occur at a faster rate is called a catalyst. We will describe catalysts in more detail in Chapter 15.

GENERAL
Chemistry ·✦· Now™

See the General ChemistryNow CD-ROM or website:
- **Screen 11.4 Hydrocarbons and Addition Reactions,** for a simulation and tutorial on alkene addition reactions

---

Example 11.4—Reaction of an Alkene

Problem Draw the structure of the compound obtained from the reaction of Br_2 with propene and name the compound.

Strategy Bromine will add across the $C\!=\!C$ double bond. The name will include the name of the carbon chain and indicate the positions of the Br atoms.

Solution

$$
\begin{array}{c}
\underset{\text{propene}}{
\begin{array}{c}
\text{H} \\
 \\
\text{C} \\
 \\
\text{H}
\end{array}
=
\begin{array}{c}
\text{H} \\
 \\
\text{C} \\
 \\
\text{CH}_3
\end{array}}
+ Br_2 \longrightarrow
\underset{\text{1,2-dibromopropane}}{
H-\overset{\overset{\displaystyle Br}{|}}{\underset{\underset{\displaystyle H}{|}}{C}}-\overset{\overset{\displaystyle Br}{|}}{\underset{\underset{\displaystyle H}{|}}{C}}-CH_3}
$$

Exercise 11.4—Reactions of Alkenes

(a) Draw the structure of the compound obtained from the reaction of HBr with ethylene and name the compound.

(b) Draw the structure of the product of the reaction of Br_2 with *cis*-2-butene and name this compound.

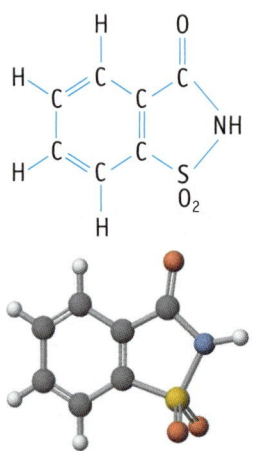

Saccharin ($C_7H_5NO_3S$). This compound, an artificial sweetener, is a benzene derivative.

Aromatic Compounds

Benzene, C_6H_6, is a key molecule in chemistry. It is the simplest **aromatic compound**, one of a class of compounds so named because they have significant, and usually not unpleasant, odors. Other members of this class, which are all based on benzene, include toluene and naphthalene. A source of many aromatic compounds is coal and the volatile substances that are released when coal is heated to a high temperature in the absence of air (Table 11.5).

benzene   toluene   naphthalene

Benzene occupies a pivotal place in the history and practice of chemistry. Michael Faraday discovered this compound in 1825 as a byproduct of illuminating gas, itself produced by heating coal. Today, benzene is an important industrial chemical, usually ranking among the top 25 chemicals in production annually in the United States. It is used as a solvent and is also the starting point for making thousands of different compounds by replacing the H atoms of the ring.

Toluene was originally obtained from Tolu balsam, the pleasant-smelling gum of a South American tree, *Toluifera balsamum*. This balsam has been used in cough syrups and perfumes. Naphthalene is an ingredient in "moth balls," although 1,4-dichlorobenzene is now more commonly used. Aspartame and another artificial sweetener, saccharin, are also benzene derivatives.

Table 11.5   Some Aromatic Compounds from Coal Tar

Common Name	Formula	Boiling Point (°C)	Melting Point (°C)
benzene	C_6H_6	80	+6
toluene	$C_6H_5CH_3$	111	−95
o-xylene	$1,2\text{-}C_6H_4(CH_3)_2$	144	−25
m-xylene	$1,3\text{-}C_6H_4(CH_3)_2$	139	−48
p-xylene	$1,4\text{-}C_6H_4(CH_3)_2$	138	+13
naphthalene	$C_{10}H_8$	218	+80

The Structure of Benzene

The formula of benzene suggested to 19th-century chemists that this compound should be unsaturated, but, if viewed this way, its chemistry was perplexing. Whereas alkenes readily add Br_2, for example, benzene does not do so under similar conditions. The benzene structural question was finally solved by August Kekulé (1829–1896). We now recognize that benzene's different reactivity relates to its structure and bonding, both of which are quite different from the structure and bonding in alkenes. Benzene has equivalent carbon–carbon bonds, 139 pm in length, intermediate between a C—C single bond (154 pm) and a C=C double bond (134 pm). The π bonds are formed by the continuous overlap of the p orbitals on the six carbon atoms (page 455). Using valence bond terminology, the structure is a hybrid of two resonance structures.

Some products containing compounds based on benzene. Examples include sodium benzoate in soft drinks, ibuprofen in Advil, and benzoyl peroxide in Oxy-10.

Representations of benzene, C_6H_6

Benzene Derivatives

Toluene, chlorobenzene, styrene, benzoic acid, aniline, and phenol are common examples of benzene derivatives.

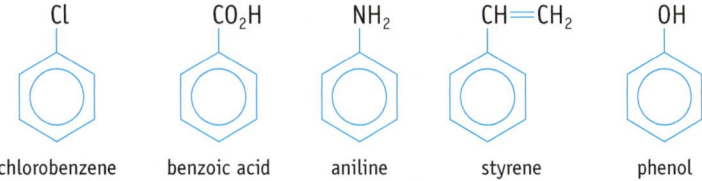

chlorobenzene   benzoic acid   aniline   styrene   phenol

If more than one H atom of benzene is replaced, isomers can arise. Thus, the systematic nomenclature for benzene derivatives involves naming substituent groups and identifying their positions on the ring by numbering the six carbon atoms [▶ Appendix E]. Some common names, which are based on an older naming scheme, are also regularly used. This scheme identified isomers of disubstituted benzenes with the prefixes ***ortho*** (*o-*, substituent groups on adjacent carbons in the benzene ring), ***meta*** (*m-*, substituents separated by one carbon atom), and ***para*** (*p-*, substituent groups on carbons on opposite sides of the ring).

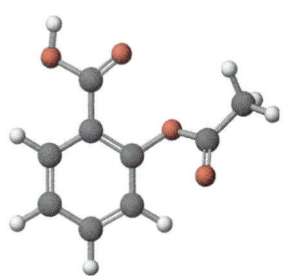

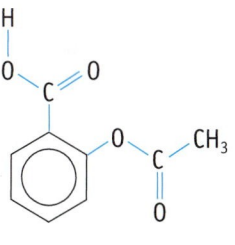

Aspirin, a commonly used analgesic. It is based on benzoic acid with an acetate group, —O_2CCH_3, in the *ortho* position.

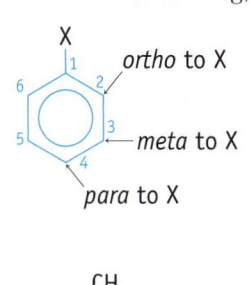

ortho to X
meta to X
para to X

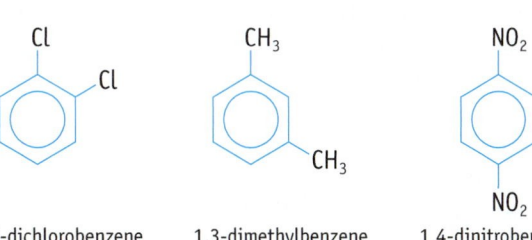

Systematic name: 1,2-dichlorobenzene   1,3-dimethylbenzene   1,4-dinitrobenzene
Common name: *o*-dichlorobenzene   *m*-xylene   *p*-dinitrobenzene

Example 11.5—Isomers of Substituted Benzenes

Problem Draw and name the isomers of $C_6H_3Cl_3$.

Strategy Begin by drawing the structure of C_6H_5Cl. Place a second Cl atom on the ring in the *ortho*, *meta*, and *para* positions. Add the third Cl in one of the remaining positions.

Solution The three isomers of $C_6H_3Cl_3$ are shown here. They are named as derivatives of benzene by specifying the number of substituent groups with the prefix "tri-," the name of the substituent, and the positions of the three groups around the six-member ring.

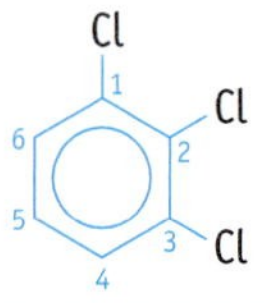

1,2,3-trichlorobenzene

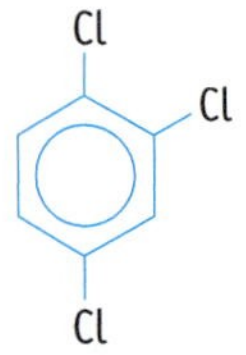

1,2,4-trichlorobenzene 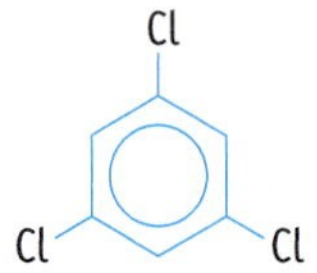   1,3,5-trichlorobenzene

Exercise 11.5—Isomers of Substituted Benzenes

Aniline, $C_6H_5NH_2$, is the common name for aminobenzene. Draw a structure for *p*-diaminobenzene, a compound used in dye manufacture. What is the systematic name for *p*-diaminobenzene?

Properties of Aromatic Compounds

Benzene is a colorless liquid, and simple substituted benzenes are liquids or solids under normal conditions. The properties of aromatic compounds are typical of hydrocarbons in general: They are insoluble in water, soluble in nonpolar solvents, and oxidized by O_2 to form CO_2 and H_2O.

One of the most important properties of benzene and other aromatic compounds is an unusual stability that is associated with the unique π bonding in this molecule. Because the π bonding in benzene is typically described using resonance structures, the extra stability is termed **resonance stabilization**. The extent of resonance stabilization in benzene is evaluated by comparing the energy evolved in the hydrogenation of benzene to form cyclohexane

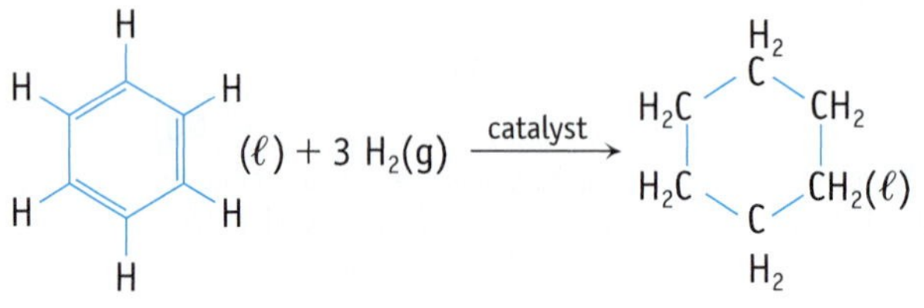

$$\Delta H^{\circ}_{rxn} = -206.7 \text{ kJ}$$

A Closer Look

Petroleum Chemistry

Much of the world's current technology relies on petroleum. Burning fuels derived from petroleum provides by far the largest amount of energy in the industrial world (see "The Chemistry of Fuels and Energy Sources", page 282). Petroleum and natural gas are also the chemical raw materials used in the manufacture of plastics, rubber, pharmaceuticals, and a vast array of other compounds.

The petroleum that is pumped out of the ground is a complex mixture whose composition varies greatly depending on its source. The primary components of petroleum are always alkanes, but, to varying degrees, nitrogen and sulfur-containing compounds are also present. Aromatic compounds are present as well, but alkenes and alkynes are not.

A modern petrochemical plant.

An early step in the petroleum refining process is distillation (Chapter 14), in which the crude mixture is separated into a series of fractions based on boiling point: first a gaseous fraction (mostly alkanes with one to four carbon atoms; this fraction is often burned off), and then gasoline, kerosene, and fuel oils. After distillation, considerable material, in the form of a semisolid, tar-like residue, remains.

The petrochemical industry seeks to maximize the amounts of the higher-valued fractions of petroleum produced and to make specific compounds for which a particular need exists. This means carrying out chemical reactions involving the raw materials on a huge scale. One process to which petroleum is subjected is known as *cracking*. At very high temperatures, bond breaking or "cracking" can occur, and longer-chain hydrocarbons will fragment into smaller molecular units. These reactions are carried out in the presence of a wide array of catalysts, materials that speed up reactions and direct them toward specific products. Among the important products of cracking are ethylene and other alkenes, which serve as the raw materials for the formation of materials such as polyethylene. Cracking also produces gaseous hydrogen, a widely used raw material in the chemical industry.

Other important reactions involving petroleum are run at elevated temperatures and in the presence of specific catalysts.

Such reactions include *isomerization* reactions, in which the carbon skeleton of an alkane rearranges to form a new isomeric species, and *reformation* processes, in which smaller molecules combine to form new molecules. Each process is directed toward achieving a specific goal, such as increasing the proportion of branched-chain hydrocarbons in gasoline to obtain higher octane ratings. A great amount of chemical research has gone into developing and understanding these highly specialized processes.

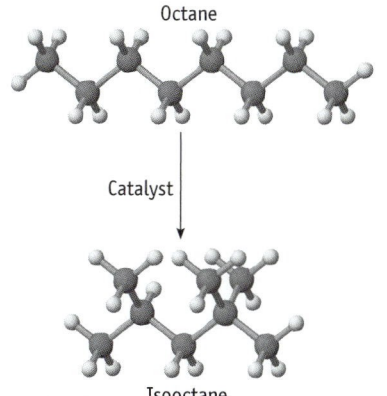

Octane

Catalyst

Isooctane

Producing gasoline. Branched hydrocarbons have a higher octane rating in gasoline. Therefore, an important process in producing gasoline is the isomerization of octane to a branched hydrocarbon such as isooctane, 2,2,4-trimethylpentane.

---

with the energy evolved in hydrogenation of three isolated double bonds.

$$3\ H_2C{=}CH_2(g) + 3\ H_2(g) \longrightarrow 3\ C_2H_6(g) \qquad \Delta H° = -410.8\ kJ$$

The hydrogenation of benzene is about 200 kJ less exothermic than the hydrogenation of three moles of ethylene. The difference is attributable to the added stability associated with π bonding in benzene.

Although aromatic compounds are unsaturated hydrocarbons, they do not undergo the addition reactions typical of alkenes and alkynes. Instead, *substitution reactions* occur, in which one or more hydrogen atoms are replaced by other groups. Such reactions require a second reagent, such as H_2SO_4, $AlCl_3$, or $FeBr_3$.

Nitration: $C_6H_6(\ell) + HNO_3(\ell) \xrightarrow{H_2SO_4} C_6H_5NO_2(\ell) + H_2O(\ell)$

Alkylation: $C_6H_6(\ell) + CH_3Cl(\ell) \xrightarrow{AlCl_3} C_6H_5CH_3(\ell) + HCl(g)$

Halogenation: $C_6H_6(\ell) + Br_2(\ell) \xrightarrow{FeBr_3} C_6H_5Br(\ell) + HBr(g)$

Table 11.6   Common Functional Groups and Derivatives of Alkanes

Functional Group*	General Formula*	Class of Compound	Examples
F, Cl, Br, I	RF, RCl, RBr, RI	haloalkane	CH_3CH_2Cl, chloroethane
OH	ROH	alcohol	CH_3CH_2OH, ethanol
OR'	ROR'	ether	$(CH_3CH_2)_2O$, diethyl ether
NH_2[†]	RNH_2	(primary) amine	$CH_3CH_2NH_2$, ethylamine
$\overset{O}{\overset{\|}{-}}CH$	RCHO	aldehyde	CH_3CHO, ethanal (acetaldehyde)
$\overset{O}{\overset{\|}{-}}C-R'$	RCOR'	ketone	CH_3COCH_3, propanone (acetone)
$\overset{O}{\overset{\|}{-}}C-OH$	RCO_2H	carboxylic acid	CH_3CO_2H, ethanoic acid (acetic acid)
$\overset{O}{\overset{\|}{-}}C-OR'$	RCO_2R'	ester	$CH_3CO_2CH_3$, methyl acetate
$\overset{O}{\overset{\|}{-}}C-NH_2$	$RCONH_2$	amide	CH_3CONH_2, acetamide

* R and R' can be the same or different hydrocarbon groups.
† Secondary amines (R_2NH) and tertiary amines (R_3N) are also possible, see discussion in the text.

David Young/Tom Stack & Associates

Alcohol racing fuel. Methanol, CH_3OH, is used as the fuel in cars of the type that race in Indianapolis.

11.3—Alcohols, Ethers, and Amines

Other types of organic compounds arise as elements other than carbon and hydrogen are included in the compound. Two elements in particular, oxygen and nitrogen, add a rich dimension to carbon chemistry.

Organic chemistry organizes compounds containing elements other than carbon and hydrogen as derivatives of hydrocarbons. Formulas (and structures) are represented by substituting one or more hydrogens in a hydrocarbon molecule by a **functional group**. A functional group is an atom or group of atoms attached to a carbon atom in the hydrocarbon. Formulas of hydrocarbon derivatives are then written as R—X, in which R is a hydrocarbon lacking a hydrogen atom, and X is the functional group that has replaced the hydrogen in the structure. The chemical and physical properties of the hydrocarbon derivatives are a blend of the properties associated with hydrocarbons and the group that has been substituted for hydrogen.

Table 11.6 identifies some common functional groups and the families of organic compounds resulting from their attachment to a hydrocarbon.

GENERAL
Chemistry ⚛ Now™

See the General ChemistryNow CD-ROM or website:
• **Screen 11.5 Functional Groups,** for a description of the types of organic functional groups and for tutorials on their structures, bonding, and chemistry

Alcohols and Ethers

If one of the hydrogen atoms of an alkane is replaced by a hydroxyl (—OH) group, the result is an **alcohol**, ROH. Methanol, CH_3OH, and ethanol, CH_3CH_2OH, are the most important alcohols, but others are also commercially important (Table 11.7). Notice that several have more than one OH functional group.

Table 11.7 Some Important Alcohols

Condensed Formula	BP (°C)	Systematic Name	Common Name	Use
CH_3OH	65.0	methanol	methyl alcohol	fuel, gasoline additive, making formaldehyde
CH_3CH_2OH	78.5	ethanol	ethyl alcohol	beverages, gasoline additive, solvent
$CH_3CH_2CH_2OH$	97.4	1-propanol	propyl alcohol	industrial solvent
$CH_3CH(OH)CH_3$	82.4	2-propanol	isopropyl alcohol	rubbing alcohol
$HOCH_2CH_2OH$	198	1,2-ethanediol	ethylene glycol	antifreeze
$HOCH_2CH(OH)CH_2OH$	290	1,2,3-propanetriol	glycerol (glycerin)	moisturizer in consumer products

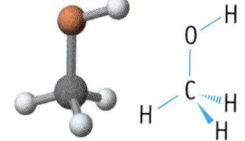

Methanol, CH_3OH, is the simplest alcohol. Methanol is often called "wood alcohol" because it was originally produced by heating wood in the absence of air.

More than 5×10^8 kg of methanol is produced in the United States annually. Most of this production is used to make formaldehyde (CH_2O) and acetic acid (CH_3CO_2H), both important chemicals in their own right. Methanol is also used as a solvent, as a de-icer in gasoline, and as a fuel in high-powered racing cars. It is found in low concentration in new wine, where it contributes to the odor, or "bouquet." Like ethanol, methanol causes intoxication, but methanol differs in being more poisonous, largely because the human body converts it to formic acid (HCO_2H) and formaldehyde (CH_2O). These compounds attack the cells of the retina in the eye, leading to permanent blindness.

■ **Aerobic Fermentation**
Aerobic fermentation (in the presence of O_2) of ethanol leads to the formation of acetic acid. This is how wine vinegar is made.

Ethanol is the "alcohol" of alcoholic beverages, in which it is formed by the anaerobic (without air) fermentation of sugar. For many years, industrial alcohol, which is used as a solvent and as a starting material for the synthesis of other compounds, was made by fermentation. In the last several decades, however, it has become cheaper to make ethanol from petroleum byproducts—specifically, by the addition of water to ethylene.

$$H_2C{=}CH_2 \,(g) + H_2O(g) \xrightarrow{\text{catalyst}} H{-}CH_2{-}CH_2{-}OH(\ell)$$

ethylene → ethanol

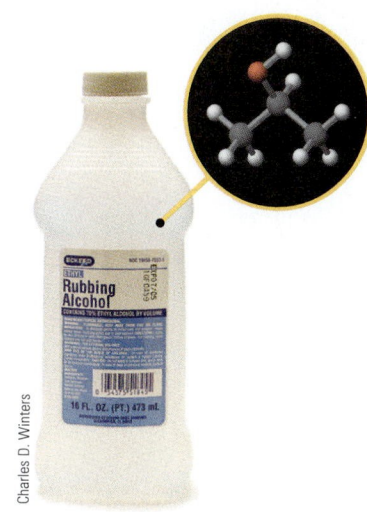

Rubbing alcohol. Common rubbing alcohol is 2-propanol, also called isopropyl alcohol.

Beginning with three-carbon alcohols, structural isomers are possible. For example, 1-propanol and 2-propanol (common name, isopropyl alcohol) are different compounds (Table 11.7).

Ethylene glycol and glycerol are common alcohols having two and three —OH groups, respectively. Ethylene glycol is used as antifreeze in automobiles. Glycerol's most common use is as a softener in soaps and lotions. It is also a raw material for the preparation of nitroglycerin (Figure 11.8).

Charles D. Winters

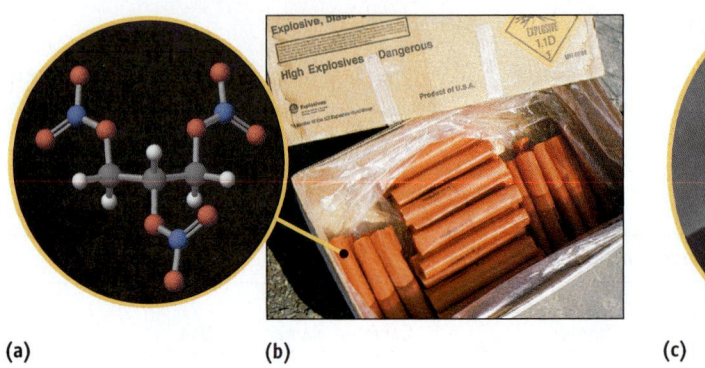

Figure 11.8   Nitroglycerin. (a) Concentrated nitric acid and glycerin react to form an oily, highly unstable compound called nitroglycerin, $C_3H_5(ONO_2)_3$. (b) Nitroglycerin is more stable if absorbed onto an inert solid, a combination called dynamite. (c) The fortune of Alfred Nobel (1833–1896), built on the manufacture of dynamite, now funds the Nobel Prizes.

	H H	H H H
	│ │	│ │ │
	H—C—C—H	H—C—C—C—H
	│ │	│ │ │
	OH OH	OH OH OH
Systematic name:	1,2-ethanediol	1,2,3-propanetriol
Common name:	ethylene glycol	glycerol or glycerin

Example 11.6—Structural Isomers of Alcohols

Problem How many different alcohols are derivatives of pentane? Draw structures and name each alcohol.

Strategy Pentane, C_5H_{12}, has a five-carbon chain. An —OH group can replace a hydrogen atom on one of the carbon atoms. Alcohols are named as derivatives of the alkane (pentane) by replacing the "-e" at the end with "-ol" and indicating the position of the —OH group by a numerical prefix (Appendix E.).

Solution Three different alcohols are possible, depending on whether the —OH group is placed on the first, second, or third carbon atom in the chain. (The fourth and fifth positions are identical to the second and first positions in the chain, respectively.)

1-pentanol	2-pentanol
H H H H H │¹ │² │³ │⁴ │⁵ HO—C—C—C—C—C—H │ │ │ │ │ H H H H H	H OH H H H │ │ │ │ │ H—C—C—C—C—C—H │ │ │ │ │ H H H H H

3-pentanol

H H OH H H
│ │ │ │ │
H—C—C—C—C—C—H
│ │ │ │ │
H H H H H

Comment Additional structural isomers with the formula $C_5H_{11}OH$ are possible in which the longest carbon chain has three C atoms (one isomer) or four C atoms (four isomers).

Exercise 11.6—Structures of Alcohols

Draw the structure of 1-butanol and alcohols that are structural isomers of the compound.

Properties of Alcohols and Ethers

Methane, CH_4, is a gas (boiling point, -161 °C) with low solubility in water. Methanol, CH_3OH, by contrast, is a liquid that is *miscible* with water in all proportions. The boiling point of methanol, 65 °C, is 226 °C higher than the boiling point of methane. What a difference the addition of a single atom into the structure can make in the properties of simple molecules!

Alcohols are related to water, with one of the H atoms of H_2O being replaced by an organic group. If a methyl group is substituted for one of the hydrogens of water, methanol results. Ethanol has a $—C_2H_5$ (ethyl) group, and propanol has a $—C_3H_7$ (propyl) group in place of one of the hydrogens of water. Viewing alcohols as related to water also helps in understanding the properties of alcohols.

The two parts of methanol, the $—CH_3$ group and the $—OH$ group, contribute to its properties. For example, methanol will burn, a property associated with hydrocarbons. On the other hand, its boiling point is more like that of water. The temperature at which a substance boils is related to the forces of attraction between molecules, called *intermolecular forces:* The stronger the attractive, intermolecular forces in a sample, the higher the boiling point [▶ Section 13.5]. These forces are particularly strong in water, a result of the polarity of the $—OH$ group in this molecule [◀ Section 9.9]. Methanol is also a polar molecule, and it is the polar $—OH$ group that leads to methanol's high boiling point. In contrast, methane is nonpolar and its low boiling point is the result of weak intermolecular forces.

It is also possible to explain the differences in the solubility of methane and methanol in water. The solubility of methanol is conferred by the polar $—OH$ portion of the molecule. Methane, which is nonpolar, has low water solubility.

■ **Hydrogen Bonding**
The intermolecular forces of attraction of compounds with hydrogen attached to a highly electronegative atom, like O, N, or F, are so exceptional that they are accorded a special name: hydrogen bonding. We will discuss hydrogen bonding in Section 13.3.

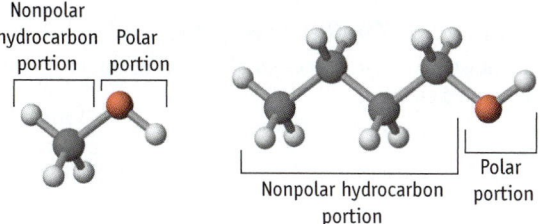

Nonpolar hydrocarbon portion · Polar portion

Nonpolar hydrocarbon portion · Polar portion

As the size of the alkyl group in an alcohol increases, the alcohol's boiling point rises, a general trend seen in families of similar compounds (see Table 11.7). The solubility in water in this series decreases. Methanol and ethanol are completely miscible in water, whereas 1-propanol is moderately water-soluble, and 1-butanol is less soluble than 1-propanol. With an increase in the size of the hydrocarbon group, the organic group (the nonpolar part of the molecule) has become a larger fraction of the molecule, and properties associated with nonpolarity begin to dominate. Space-filling models show that in methanol, the polar and nonpolar parts of the molecule are approximately similar in size, but in 1-butanol the $—OH$ group is less than 20% of the molecule. The molecule is less like water and more "organic."

Attaching more than one $—OH$ group to a hydrocarbon framework has an effect that is opposite to the effect of increased hydrocarbon size. Two $—OH$ groups on a three-carbon framework, as found in propylene glycol, convey complete miscibility with water, in contrast to the limited solubility of 1-propanol and 2-propanol (Figure 11.9).

Methanol is often added to automobile gasoline tanks in the winter to prevent fuel lines from freezing. It is soluble in water and lowers the water's freezing point.

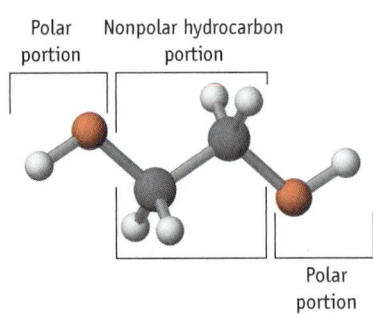

Ethylene glycol is used in automobile radiators. It is soluble in water, and lowers the freezing point and raises the boiling point of the water in the cooling system. (See Section 14.4.)

Ethylene glycol, a major component of automobile antifreeze, is completely miscible with water.

Photos: Charles D. Winters

Figure 11.9   Properties and uses of methanol and ethylene glycol.

Charles D. Winters

Safe antifreeze—propylene glycol, $CH_3CHOHCH_2OH$. Most antifreeze sold today consists of about 95% ethylene glycol. Cats and dogs are attracted by the smell and taste of the compound, but it is toxic. In fact, only a few milliliters can prove fatal to a small dog or cat. In the first stage of poisoning, an animal may appear drunk, but within 12–36 hours the kidneys stop functioning and the animal slips into a coma. To avoid accidental poisoning of domestic and wild animals, you can use propylene glycol antifreeze. This compound affords the same antifreeze protection but is much less toxic.

Ethers have the general formula ROR′. The best known ether is diethyl ether, $CH_3CH_2OCH_2CH_3$. Lacking an —OH group, the properties of ethers are in sharp contrast to those of alcohols. Diethyl ether, for example, has a lower boiling point (34.5 °C) than ethanol, CH_3CH_2OH (78.3 °C), and is only slightly soluble in water.

Amines

It is often convenient to think about water and ammonia as being similar molecules: They are the simplest hydrogen compounds of adjacent second-period elements. Both are polar, and they exhibit some similar chemistry, such as protonation (to give H_3O^+ and NH_4^+) and deprotonation (to give OH^- and NH_2^-).

This comparison of water and ammonia can be extended to alcohols and amines. Alcohols have formulas related to water in which one hydrogen in H_2O is replaced with an organic group (R—OH). In organic **amines**, one or more hydrogen atoms of NH_3 are replaced with an organic group. Amine structures are similar to ammonia's structure; that is, the geometry about the N atom is trigonal-pyramidal.

Amines are categorized based on the number of organic substituents as primary (one organic group), secondary (two organic groups), or tertiary (three organic groups). As examples, consider the three amines with methyl groups: CH_3NH_2, $(CH_3)_2NH$, and $(CH_3)_3N$.

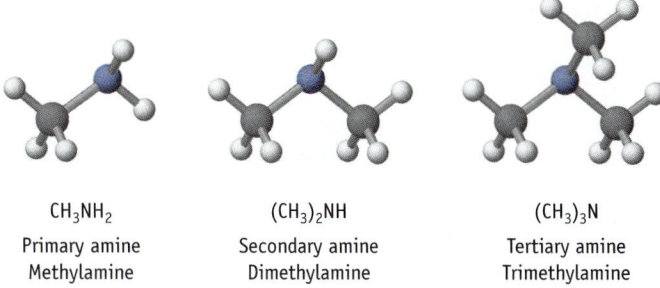

CH₃NH₂	(CH₃)₂NH	(CH₃)₃N
Primary amine	Secondary amine	Tertiary amine
Methylamine	Dimethylamine	Trimethylamine

Properties of Amines

Amines usually have offensive odors. You know what the odor is if you have ever smelled decaying fish. Two appropriately named amines, putrescine and cadaverine, add to the odor of urine, rotten meat, and bad breath.

$$H_2NCH_2CH_2CH_2CH_2NH_2$$
putrescine
1,4-butanediamine

$$H_2NCH_2CH_2CH_2CH_2CH_2NH_2$$
cadaverine
1,5-pentanediamine

The smallest amines are water-soluble, but most amines are not. All amines are bases, however, and they react with acids to give salts, many of which are water-soluble. As with ammonia, the reactions involve adding H^+ to the lone pair of electrons on the N atom. This is illustrated by the reaction of aniline (aminobenzene) with H_2SO_4 to give anilinium sulfate.

$$2\ C_6H_5NH_2(aq) + H_2SO_4(aq) \longrightarrow 2\ C_6H_5NH_3^+(aq) + SO_4^{2-}(aq)$$

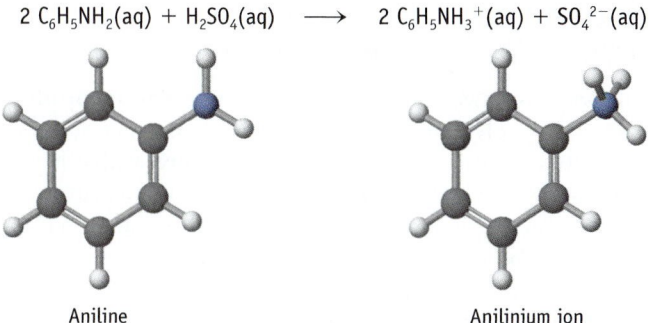

Aniline   Anilinium ion

Recall that Perkin started with this salt in his serendipitous discovery of the dye mauve [◄ page 474].

The facts that an amine can be protonated, and that the proton can be removed again by treating the compound with a base have practical and physiological importance. Nicotine in cigarettes is normally found in the protonated form. (This water-soluble form is often used in insecticides.) Adding a base such as ammonia removes the H^+ ion to leave nicotine in its "free-base" form.

$$NicH_2^{2+}\ (aq) + 2\ NH_3(aq) \longrightarrow Nic(aq) + 2\ NH_4^+(aq)$$

In this form, nicotine is much more readily absorbed by the skin and mucous membranes, so the compound is a much more potent poison.

Nicotine

Nicotine Two nitrogen atoms in the nicotine molecule can be protonated, which is the form in which nicotine is normally found. The protons can be removed, however, by treating it with a base. This "free-base" form is much more poisonous and addictive. See J. F. Pankow: *Environmental Science & Technology,* Vol 31, p. 2428, August 1997.

11.4—Compounds with a Carbonyl Group

Formaldehyde, acetic acid, and acetone are among the organic compounds referred to in previous examples. These compounds have a common structural feature: Each contains a trigonal-planar carbon atom doubly bonded to an oxygen. The C=O group is called the **carbonyl group**, and all of these compounds are members of a large class of compounds called **carbonyl compounds**.

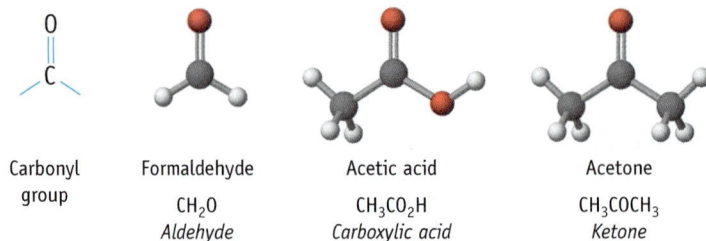

Carbonyl group	Formaldehyde	Acetic acid	Acetone
	CH_2O	CH_3CO_2H	CH_3COCH_3
	Aldehyde	*Carboxylic acid*	*Ketone*

In this section, we will examine five groups of carbonyl compounds (Table 11.6, page 496):

- *Aldehydes* (RCHO) have an organic group (—R) and an H atom attached to a carbonyl group.
- *Ketones* (RCOR′) have two —R groups attached to the carbonyl carbon; they may be the same groups, as in acetone, or different groups.
- *Carboxylic acids* (RCO₂H) have an —R group and an —OH group attached to the carbonyl carbon.
- *Esters* (RCO₂R′) have —R and —OR′ groups attached to the carbonyl carbon.
- *Amides* (RCONR₂′, RCONHR′, and RCONH₂) have an —R group and an amino group (—NH₂, —NHR, —NR₂) bonded to the carbonyl carbon.

Aldehydes, ketones, and carboxylic acids are oxidation products of alcohols and, indeed, are commonly made by this route. The product obtained through oxidation of an alcohol depends on the alcohol's structure, which is classified according to the number of carbon atoms bonded to the C atom bearing the —OH group. *Primary alcohols* have one carbon and two hydrogen atoms attached, whereas *secondary alcohols* have two carbon atoms and one hydrogen atom attached. *Tertiary alcohols* have three carbon atoms attached to the C atom bearing the —OH group.

A primary alcohol is oxidized in two steps. It is first oxidized to an aldehyde and then in a second step to a carboxylic acid:

$$\text{R}-\text{CH}_2-\text{OH} \xrightarrow[\text{agent}]{\text{oxidizing}} \underset{\text{aldehyde}}{\text{R}-\overset{\overset{\displaystyle O}{\|}}{\text{C}}-\text{H}} \xrightarrow[\text{agent}]{\text{oxidizing}} \underset{\text{carboxylic acid}}{\text{R}-\overset{\overset{\displaystyle O}{\|}}{\text{C}}-\text{OH}}$$

primary alcohol

For example, the air oxidation of ethanol in wine produces wine vinegar, the most important ingredient of which is acetic acid.

$$\underset{\text{ethanol}}{\text{H}-\overset{\overset{\displaystyle H}{|}}{\underset{\underset{\displaystyle H}{|}}{\text{C}}}-\overset{\overset{\displaystyle H}{|}}{\underset{\underset{\displaystyle H}{|}}{\text{C}}}-\text{OH}(\ell)} \xrightarrow{\text{oxidizing agent}} \underset{\text{acetic acid}}{\text{H}-\overset{\overset{\displaystyle H}{|}}{\underset{\underset{\displaystyle H}{|}}{\text{C}}}-\overset{\overset{\displaystyle O}{\|}}{\text{C}}-\text{OH}(\ell)}$$

Primary alcohol: ethanol

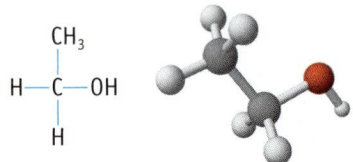

Secondary alcohol: 2-propanol

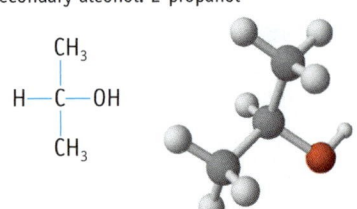

Tertiary alcohol: 2-methyl-2-propanol

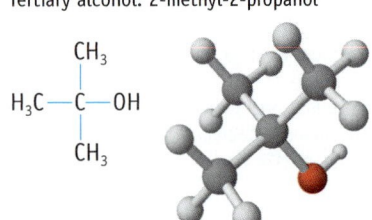

Acids have a sour taste. The word "vinegar" (from the French *vin aigre*) means sour wine. A device to test one's breath for alcohol relies on a similar oxidation of ethanol (Figures 5.16 and 11.10).

In contrast to primary alcohols, oxidation of a secondary alcohol produces a ketone:

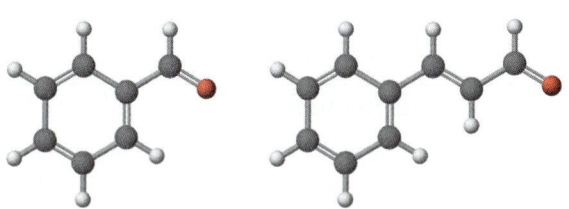

secondary alcohol   ketone

(—R and —R′ are organic groups. They may be the same or different.)

Common oxidizing agents used for these reactions are reagents such as $KMnO_4$ and $K_2Cr_2O_7$ (Table 5.4).

Finally, tertiary alcohols do *not* react with the usual oxidizing agents.

$$(CH_3)_3COH \xrightarrow{\text{oxidizing agent}} \text{no reaction}$$

Aldehydes and Ketones

Aldehydes and **ketones** have pleasant odors and are often used in fragrances. Benzaldehyde is responsible for the odor of almonds and cherries, cinnamaldehyde is found in the bark of the cinnamon tree, and the ketone *p*-hydroxyphenyl-2-butanone is responsible for the odor of ripe raspberries (a favorite of the authors of this book). Table 11.8 lists several simple aldehydes and ketones.

Benzaldehyde, C_6H_5CHO *trans*-Cinnamaldehyde, $C_6H_5CH=CHCHO$

Figure 11.10 Alcohol tester. This device for testing a person's breath for the presence of ethanol relies on the oxidation of the alcohol. If present, ethanol is oxidized by potassium dichromate, $K_2Cr_2O_7$, to acetaldehyde, and then to acetic acid. The yellow-orange dichromate ion is reduced to green $Cr^{3+}(aq)$, the color change indicating that ethanol was present.

Table 11.8 Simple Aldehydes and Ketones

Structure	Common Name	Systematic Name	BP (°C)
HCH (O)	formaldehyde	methanal	−19
CH_3CH (O)	acetaldehyde	ethanal	20
CH_3CCH_3 (O)	acetone	propanone	56
$CH_3CCH_2CH_3$ (O)	methyl ethyl ketone	butanone	80
$CH_3CH_2CCH_2CH_3$ (O)	diethyl ketone	3-pentanone	102

Aldehydes and odors. The odors of almonds and cinnamon are due to aldehydes, but the odor of fresh raspberries comes from a ketone.

Aldehydes and ketones are the oxidation products of primary and secondary alcohols, respectively. The reverse reactions—reduction of aldehydes to primary alcohols, and reduction of ketones to secondary alcohols—are also known. Commonly used reagents for such reductions are NaBH$_4$ and LiBH$_4$, although H$_2$ is used on an industrial scale.

$$
\underset{\text{aldehyde}}{R-\overset{\overset{\textstyle O}{\|}}{C}-H} \xrightarrow{\text{NaBH}_4 \text{ or LiAlH}_4} \underset{\text{primary alcohol}}{R-\overset{\overset{\textstyle OH}{|}}{\underset{\underset{\textstyle H}{|}}{C}}-H}
$$

$$
\underset{\text{ketone}}{R-\overset{\overset{\textstyle O}{\|}}{C}-R} \xrightarrow{\text{NaBH}_4 \text{ or LiAlH}_4} \underset{\text{secondary alcohol}}{R-\overset{\overset{\textstyle OH}{|}}{\underset{\underset{\textstyle H}{|}}{C}}-R}
$$

Exercise 11.7—Aldehydes and Ketones

(a) Draw the structural formula for 2-pentanone. Draw structures for a ketone and two aldehydes that are isomers of 2-pentanone, and name each of these compounds.

(b) What is the product of the reduction of 2-pentanone with LiBH$_4$?

Exercise 11.8—Aldehydes and Ketones

Draw the structures and name the aldehyde or ketone formed upon oxidation of the following alcohols: (a) 1-butanol, (b) 2-butanol, (c) 2-methyl-1-propanol. Are these three alcohols structural isomers?

Carboxylic Acids

Acetic acid is the most common and most important **carboxylic acid**. For many years, acetic acid was made by oxidizing ethanol produced by fermentation. Now, however, acetic acid is generally made by combining carbon monoxide and methanol in the presence of a catalyst:

$$
\underset{\text{methanol}}{CH_3OH(\ell)} + CO(g) \xrightarrow{\text{catalyst}} \underset{\text{acetic acid}}{CH_3CO_2H(\ell)}
$$

Charles D. Winters

Figure 11.11 Acetic acid in bread. Acetic acid is produced in bread when leavened with the yeast *Saccharomyces exigus*. Another group of bacteria, *Lacto-bacillus sanfrancisco*, contribute to the flavor of sourdough bread. These bacteria metabolize the sugar maltose, excreting acetic acid and lactic acid, CH$_3$CH(OH)CO$_2$H, thereby giving the bread its unique sour taste.

About 1 billion kilograms of acetic acid is produced annually in the United States for use in plastics, synthetic fibers, and fungicides.

Many organic acids are found naturally (Table 11.9). Acids are recognizable by their sour taste (Figure 11.11) and are found in common foods: Citric acid in fruits, acetic acid in vinegar, and tartaric acid in grapes are just three examples.

Some carboxylic acids have common names derived from the source of the acid (Table 11.9). Because formic acid is found in ants, its name comes from the Latin

Table 11.9 Some Naturally Occurring Carboxylic Acids

Name	Structure	Natural Source
benzoic acid	⬡—CO₂H	berries
citric acid	HO₂C—CH₂—C(OH)(CO₂H)—CH₂—CO₂H	citrus fruits
lactic acid	H₃C—CH(OH)—CO₂H	sour milk
malic acid	HO₂C—CH₂—CH(OH)—CO₂H	apples
oleic acid	CH₃(CH₂)₇—CH=CH—(CH₂)₇—CO₂H	vegetable oils
oxalic acid	HO₂C—CO₂H	rhubarb, spinach cabbage, tomatoes
stearic acid	CH₃(CH₂)₁₆—CO₂H	animal fats
tartaric acid	HO₂C—CH(OH)—CH(OH)—CO₂H	grape juice, wine

Charles D. Winters

Formic acid, HCO₂H. This acid puts the sting in ant bites.

word for ant (*formica*). Butyric acid gives rancid butter its unpleasant odor, and the name is related to the Latin word for butter (*butyrum*). The systematic names of acids (Table 11.10) are formed by dropping the "-e" on the name of the corresponding alkane and adding "-oic" (and the word "acid").

Because of the substantial electronegativity of oxygen, we expect the two O atoms of the carboxylic acid group to be slightly negatively charged, and the H atom of the —OH group to be positively charged. This distribution of charges has several important implications:

Table 11.10 Some Simple Carboxylic Acids

Structure	Common Name	Systematic Name	BP (°C)
HCOH (with C=O)	formic acid	methanoic acid	101
CH₃COH (with C=O)	acetic acid	ethanoic acid	118
CH₃CH₂COH (with C=O)	propionic acid	propanoic acid	141
CH₃(CH₂)₂COH (with C=O)	butyric acid	butanoic acid	163
CH₃(CH₂)₃COH (with C=O)	valeric acid	pentanoic acid	187

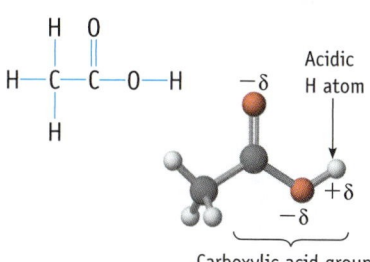

Acetic Acid. The H atom of the carboxylic acid group (—CO₂H) is the acidic proton of this and other carboxylic acids.

A Closer Look

Glucose and Sugars

Having described alcohols and carbonyl compounds, we now pause to look at glucose, the most common, naturally occurring carbohydrate.

As their name implies, formulas of carbohydrates can be written as though they are a combination of carbon and water, $C_x(H_2O)_y$. Thus, the formula of glucose, $C_6H_{12}O_6$, is equivalent to $C_6(H_2O)_6$. This compound is a sugar, or, more accurately, a **monosaccharide**.

Carbohydrates are polyhydroxy aldehydes or ketones. Glucose is an interesting molecule that exists in three different isomeric forms. Two of the isomers contain six-member rings; the third isomer features a chain structure. In solution, the three forms rapidly interconvert.

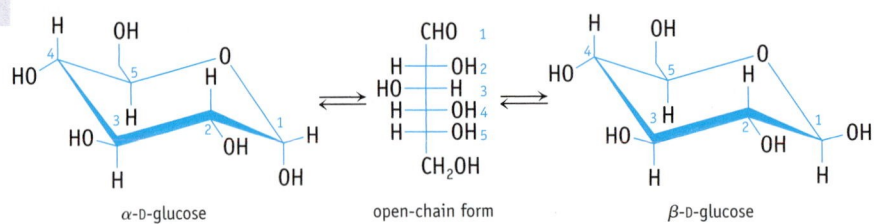

α-D-glucose   open-chain form   β-D-glucose

Notice that glucose is a chiral molecule. In the chain structure, four of the carbon atoms are bonded to four different groups. In nature, glucose occurs in just one of its enantiomeric forms; thus, a solution of glucose rotates polarized light.

Knowing glucose's structure allows one to predict some of its properties. With five polar —OH groups in the molecule, glucose is, not surprisingly, soluble in water.

The aldehyde group is susceptible to chemical oxidation to form a carboxylic acid. Detection of glucose (in urine or blood) takes advantage of this fact; diagnostic tests for glucose involve oxidation with subsequent detection of the products.

Glucose is in a class of sugar molecules called hexoses, molecules having six carbon atoms. 2-Deoxyribose, the sugar in the backbone of the DNA molecule, is a pentose, a molecule with five carbon atoms.

Glucose and other monosaccharides serve as the building blocks for larger carbohydrates. Sucrose, a disaccharide, is formed from a molecule of glucose and a molecule of fructose, another monosaccharide. Starch is a polymer composed of many monosaccharide units.

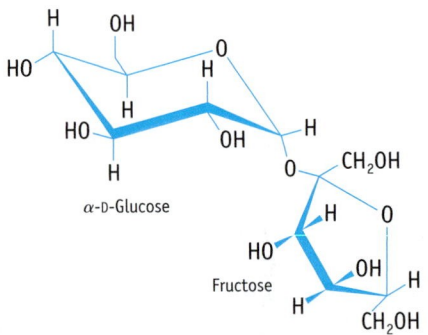

α-D-Glucose

Fructose

The structure of sucrose. Sucrose is formed from the hexoses α-D-glucose and fructose. An ether linkage is formed by loss of H_2O from two —OH groups.

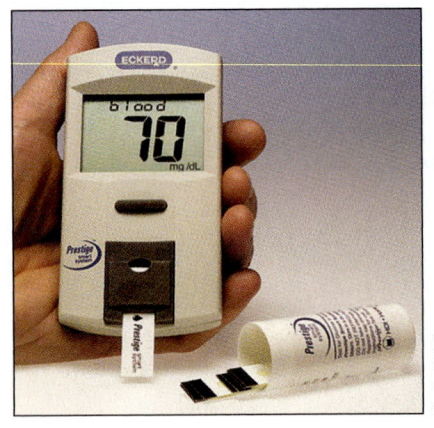

Laboratory test for glucose.

Charles D. Winters

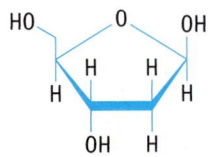

deoxyribose, a pentose in the DNA backbone

- The polar acetic acid molecule dissolves readily in water, which you already know because vinegar is an aqueous solution of acetic acid. (Acids with larger organic groups are less soluble, however.)

- The hydrogen of the —OH group is the acidic hydrogen. As noted in Chapter 5, acetic acid is a weak acid in water, as are all other organic acids.

Carboxylic acids undergo a number of reactions. Among these is the reduction of the acid (with reagents such as $LiAlH_4$ or $NaBH_4$) first to an aldehyde and then to an alcohol. For example, acetic acid is reduced first to acetaldehyde and then to ethanol.

$$CH_3CO_2H \xrightarrow{LiAlH_4} CH_3CHO \xrightarrow{LiAlH_4} CH_3CH_2OH$$

acetic acid   acetaldehyde   ethanol

Chemical Perspectives

Aspirin Is More Than 100 Years Old!

Aspirin is one of the most successful non-prescription drugs ever made. Americans swallow more than 50 million aspirin tablets a day, mostly for the pain-relieving (analgesic) effects of the drug. Aspirin also wards off heart disease and thrombosis (blood clots), and it has even been suggested as a possible treatment for certain cancers and for senile dementia.

Hippocrates (460–370 BC), the ancient Greek physician, recommended an infusion of willow bark to ease the pain of childbirth. It was not until the 19th century that an Italian chemist, Raffaele Piria, isolated salicylic acid, the active compound in the bark. Soon thereafter, it was found that the acid could be extracted from a wildflower, *Spiraea ulmaria*. It is from the name of this plant that the name "aspirin" (a + spiraea) is derived.

Hippocrates's willow bark extract, salicylic acid, is an analgesic, but it is also very irritating to the stomach lining. It was therefore an important advance when Felix Hoffmann and Henrich Dreser of Bayer Chemicals in Germany found, in 1897, that a derivative of salicylic acid, acetylsalicylic acid, was also a useful drug and had fewer side effects. This derivative is the compound we now call "aspirin."

Acetylsalicylic acid slowly reverts to salicylic acid and acetic acid in the presence of moisture. Indeed, if you smell the characteristic odor of acetic acid in a bottle of aspirin tablets, they are too old and should be discarded.

Aspirin is a component of various over-the-counter medicines, such as Anacin, Ecotrin, Excedrin, and Alka-Seltzer. The latter is a combination of aspirin with citric acid and sodium bicarbonate. Sodium bicarbonate is a base, and it reacts with the acid to produce the sodium salt of acetylsalicylic acid, a form of aspirin that is water-soluble and quicker-acting.

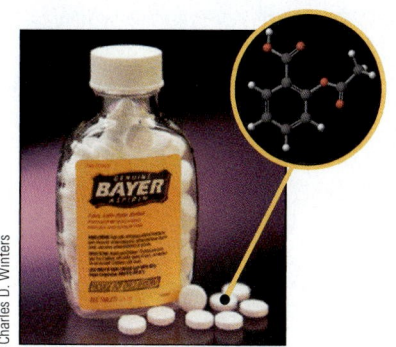

Charles D. Winters

Acetylsalicylic acid, aspirin.

Yet another important aspect of carboxylic acid chemistry is these acids' reaction with bases to give carboxylate anions. For example, acetic acid reacts with sodium hydroxide to give sodium acetate (sodium ethanoate).

$$CH_3CO_2H(aq) + OH^-(aq) \longrightarrow CH_3CO_2^-(aq) + H_2O(\ell)$$

Esters

Carboxylic acids (RCO_2H) react with alcohols ($R'OH$) to form esters (RCO_2R') in an **esterification** reaction. (These reactions are generally run in the presence of strong acids because acids accelerate the reaction.)

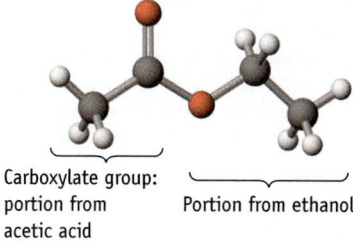

Carboxylate group: portion from acetic acid — Portion from ethanol

Ethyl acetate, an ester
$CH_3CO_2CH_2CH_3$

Table 11.11 lists a few common esters and the acid and alcohol from which they are formed. The two-part name of an ester is given by (1) the name of the hydrocarbon group from the alcohol and (2) the name of the carboxylate group derived from the acid name by replacing "-ic" with "-ate." For example, ethanol (commonly called ethyl alcohol) and acetic acid combine to give the ester ethyl acetate.

Esters. Many fruits such as bananas and strawberries as well as consumer products (here perfume and oil of wintergreen) contain esters.

Table 11.11   Some Acids, Alcohols, and Their Esters

Acid	Alcohol	Ester	Odor of Ester
CH_3CO_2H acetic acid	CH_3 | $CH_3CHCH_2CH_2OH$ 3-methyl-1-butanol	O CH_3 || | $CH_3COCH_2CH_2CHCH_3$ 3-methylbutyl acetate	banana
$CH_3CH_2CH_2CO_2H$ butanoic acid	$CH_3CH_2CH_2CH_2OH$ 1-butanol	O || $CH_3CH_2CH_2COCH_2CH_2CH_3$ butyl butanoate	pineapple
$CH_3CH_2CH_2CO_2H$ butanoic acid	⬡—CH_2OH benzyl alcohol	O || $CH_3CH_2CH_2COCH_2$—⬡ benzyl butanoate	rose

An important reaction of esters is their **hydrolysis** (literally, reaction with water), a reaction that is the reverse of the formation of the ester. The reaction, generally done in the presence of a base such as NaOH, produces the alcohol and a sodium salt of the carboxylic acid:

$$\underset{\text{ester}}{\overset{\displaystyle O}{\overset{\displaystyle \|}{RCOR'}}} + NaOH \xrightarrow[\text{in water}]{\text{heat}} \underset{\text{carboxylate salt}}{\overset{\displaystyle O}{\overset{\displaystyle \|}{RCO^-Na^+}}} + \underset{\text{alcohol}}{R'OH}$$

$$\underset{\text{ethyl acetate}}{\overset{\displaystyle O}{\overset{\displaystyle \|}{CH_3COCH_2CH_3}}} + NaOH \xrightarrow[\text{in water}]{\text{heat}} \underset{\text{sodium acetate}}{\overset{\displaystyle O}{\overset{\displaystyle \|}{CH_3CO^-Na^+}}} + \underset{\text{ethanol}}{CH_3CH_2OH}$$

The carboxylic acid can be recovered if the sodium salt is treated with a strong acid such as HCl:

$$\underset{\text{sodium acetate}}{\overset{\displaystyle O}{\overset{\displaystyle \|}{CH_3CO^-Na^+(aq)}}} + HCl(aq) \longrightarrow \underset{\text{acetic acid}}{\overset{\displaystyle O}{\overset{\displaystyle \|}{CH_3COH(aq)}}} + NaCl(aq)$$

■ **Saponification**
Fats and oils are esters of glycerol and long-chain acids. When reacted with a strong base (NaOH or KOH), they produce glycerol and a salt of the long-chain acid. Because this product is used as soap, the reaction is called *saponification*. See "A Closer Look: Fats and Oils", page 510.

Unlike the acids from which they are derived, esters often have pleasant odors (see Table 11.11). Typical examples are methyl salicylate, or "oil of wintergreen," and benzyl acetate. Methyl salicylate is derived from salicylic acid, the parent compound of aspirin.

$$\underset{\substack{\text{salicylic acid} \\ \text{}}}{\underset{OH}{\overset{\displaystyle O}{⬡\overset{\|}{-}COH}}} + \underset{\text{methanol}}{CH_3OH} \longrightarrow \underset{\substack{\text{methyl salicylate,}\\\text{oil of wintergreen}}}{\underset{OH}{\overset{\displaystyle O}{⬡\overset{\|}{-}COCH_3}}} + H_2O$$

Benzyl acetate, the active component of "oil of jasmine," is formed from benzyl alcohol ($C_6H_5CH_2OH$) and acetic acid. The chemicals are inexpensive, so synthetic jasmine is a common fragrance in less expensive perfumes and toiletries.

$$\text{CH}_3\overset{\displaystyle O}{\overset{\|}{\text{C}}}\text{OH} + \left\langle\bigcirc\right\rangle\!\!-\!\text{CH}_2\text{OH} \longrightarrow \text{CH}_3\overset{\displaystyle O}{\overset{\|}{\text{C}}}\text{OCH}_2\!\!-\!\left\langle\bigcirc\right\rangle + \text{H}_2\text{O}$$

acetic acid   benzyl alcohol   benzyl acetate
oil of jasmine

Exercise 11.9—Esters

Draw the structure and name the ester formed from each of the following reactions:

(a) propanoic acid and methanol
(b) butanoic acid and 1-butanol
(c) hexanoic acid and ethanol

Exercise 11.10—Esters

Draw the structure and name the acid and alcohol from which the following esters are derived:

(a) propyl acetate
(b) 3-methylpentyl benzoate
(c) ethyl salicylate

Amides

An acid and an alcohol react by loss of water to form an ester. In a similar manner, another class of organic compounds—amides—form when an acid reacts with an amine, again with loss of water.

$$\underset{\text{ester}}{R-\overset{\displaystyle O}{\overset{\|}{C}}-OR'} \xleftarrow[\;-H_2O\;]{+\text{ alcohol, } R'OH} \underset{\text{acid}}{R-\overset{\displaystyle O}{\overset{\|}{C}}-OH} \xrightarrow[\;-H_2O\;]{+\text{ amine, } NHR'_2} \underset{\text{amide}}{R-\overset{\displaystyle O}{\overset{\|}{C}}-NR'_2}$$

Amides have an organic group and an amino group ($-NH_2$, $-NHR'$, or $-NR'R$) attached to the carbonyl group.

The structure of the amide group offers a surprise. The C atom involved in the amide bond has three bonded groups and no lone pairs around it. We would predict it should be sp^2 hybridized with trigonal-planar geometry and bond angles of approximately 120°—and this is what is found. However, the N atom is also observed to have trigonal-planar geometry with bonds to three attached atoms at 120°. Because the amide nitrogen is apparently surrounded by four pairs of electrons, we would have predicted the N atom would have sp^3 hybridization and bond angles of about 109°.

Based on the observed geometry of the amide N atom, the atom is assigned sp^2 hybridization. To explain the observed angle and to rationalize sp^2 hybridization, we can introduce a second resonance form of the amide.

$$\underset{\textbf{A}}{R-\overset{\displaystyle :\!\overset{\ddot{}}{O}\!:}{\overset{\|}{C}}\underset{\underset{R}{|}}{\overset{H}{\underset{\ddot{}}{N}}}} \longleftrightarrow \underset{\textbf{B}}{R-\overset{\displaystyle :\!\ddot{O}\!:^-}{\overset{\|}{C}}\underset{\underset{R}{|}}{\overset{H}{\overset{+}{N}}}}$$

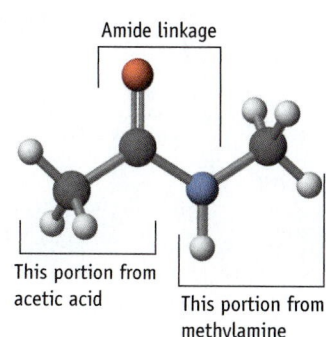

Amide linkage

This portion from acetic acid

This portion from methylamine

An amide, *N*-methylacetamide. The *N*-methyl portion of the name derives from the amine portion of the molecule, where the *N* indicates that the methyl group is attached to the nitrogen atom. The "-aceta" portion of the name indicates the acid on which the amide is based.

A Closer Look

Fats and Oils

Fats and oils are among the many compounds found in plants and animal tissues. In the body, these substances serve several functions, a primary one being the storage of energy.

Fats (solids) and oils (liquids) are triesters formed from glycerol (1,2,3-propanetriol) and three carboxylic acids that can be the same or different.

$$
\begin{array}{c}
H_2C-O-\overset{\displaystyle O}{\overset{\|}{C}}R \\[6pt]
HC-O-\overset{\displaystyle O}{\overset{\|}{C}}R \\[6pt]
H_2C-O-\overset{\displaystyle O}{\overset{\|}{C}}R
\end{array}
$$

The carboxylic acids in fats and oils, known as *fatty acids*, have a lengthy carbon chain, usually containing between 12 and 18 carbon atoms. The hydrocarbon chains can be saturated or may include one or more double bonds. The latter are referred to as monounsaturated or polyunsat-

About 94% of the fatty acids in olive oil are monounsaturated. The major fatty acid is oleic acid.

Common Fatty Acids

Name	Number of C Atoms	Formula
Saturated Acids		
lauric	C_{12}	$CH_3(CH_2)_{10}CO_2H$
myristic	C_{14}	$CH_3(CH_2)_{12}CO_2H$
palmitic	C_{16}	$CH_3(CH_2)_{14}CO_2H$
stearic	C_{18}	$CH_3(CH_2)_{16}CO_2H$
Unsaturated Acid		
oleic	C_{18}	$CH_3(CH_2)_7CH=CH(CH_2)_7CO_2H$

urated, depending on the number of double bonds. Saturated compounds are more common in animal products, while unsaturated fats and oils are more common in plants.

In general, fats containing saturated fatty acids are solids and those containing unsaturated fatty acids are liquids at room temperature. The difference in melting point relates to the molecular structure. With only single bonds linking carbon atoms in saturated fatty acids, the hydrocarbon group is flexible, allowing the molecules to pack more closely together. The double bonds in unsaturated fats introduce kinks that make the hydrocarbon group less flexible; consequently, the molecules pack less tightly together.

Food companies often hydrogenate vegetable oils to reduce unsaturation. The chemical rationale is that double bonds are reactive and unsaturated compounds are more susceptible to oxidation, which results in unpleasant odors. There are also aesthetic reasons for this practice. Food

Polar bear fat. Polar bears feed primarily on seal blubber and build up a huge fat reserve during winter. During summer, they maintain normal activity but eat nothing, relying entirely on body fat for sustenance. A polar bear will burn about 1 to 1.5 kg of fat per day.

processors often want solid fats to improve the quality and appearance of the food. If liquid vegetable oil is used in a cake icing, for example, the icing may slide off the cake.

Like other esters, fats and oils can undergo hydrolysis. This process is catalyzed by enzymes in the body. In industry, hydrolysis is carried out using aqueous NaOH or KOH to produce a mixture of glycerol and the sodium salts of the fatty acids. This reaction is called *saponification*, a term meaning "soap making."

Glyceryl stearate, a fat
$$R = -(CH_2)_{16}CH_3$$

$$
\begin{array}{c}
H_2C-O-\overset{\displaystyle O}{\overset{\|}{C}}R \\[6pt]
HC-O-\overset{\displaystyle O}{\overset{\|}{C}}R + 3\ NaOH \\[6pt]
H_2C-O-\overset{\displaystyle O}{\overset{\|}{C}}R
\end{array}
$$

$$\downarrow$$

$$
\begin{array}{c}
H_2C-O-H \\[6pt]
HC-O-H \ + \quad 3\ R\overset{\displaystyle O}{\overset{\|}{C}}-O^-\,Na^+ \\[6pt]
H_2C-O-H
\end{array}
$$

glycerol   sodium stearate, a soap

Simple soaps are sodium salts of fatty acids. The anion in these compounds has an ionic end (the carboxylate group) and a nonpolar end (the large hydrocarbon tail). The ionic end allows these molecules to interact with water, and the nonpolar end enables them to mix with oily and greasy substances to form an emulsion that can be washed away with water.

Charles D. Winters

Taxi/Getty Images

Form B contains a C=N double bond, and the O and N atoms have negative and positive charges, respectively. The N atom can be assigned sp^2 hybridization, and the π bond in B arises from overlap of p orbitals on C and N.

The existence of a second resonance structure for an amide link explains why the carbon–nitrogen bond is relatively short, about 132 pm, a value between that of a C—N single bond (149 pm) and a C=N double bond (127 pm). In addition, restricted rotation occurs around the C=N bond, making it possible for isomeric species to exist if the two groups bonded to N are different.

The amide grouping is particularly important in some synthetic polymers (Section 11.5) and in many naturally occurring compounds, especially proteins (page 531), where it is referred to as a *peptide* link. The compound *N*-acetyl-*p*-aminophenol, an analgesic known by the generic name acetaminophen and sold under the brand names Tylenol, Datril, and Momentum, among others, is another amide. Use of this compound as an analgesic was apparently discovered by accident when a common organic compound called acetanilide (like acetaminophen but without the —OH group) was mistakenly put into a prescription for a patient. Acetanilide acts as an analgesic, but it can be toxic. An —OH group *para* to the amide group makes the compound nontoxic, an interesting example of the relationship between molecular structure and chemical function.

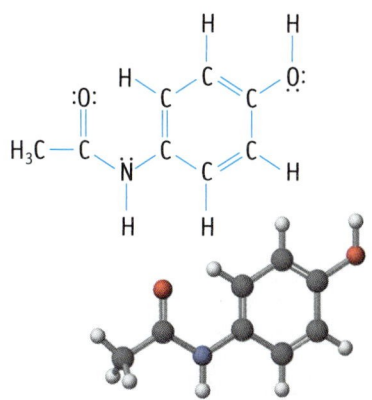

Acetaminophen, *N*-acetyl-*p*-aminophenol. This analgesic is an amide. It is used in over-the-counter painkillers such as Tylenol.

■ **Amides, Peptides, and Proteins** When amino acids combine, they form amide or peptide links. Polymers of amino acids are proteins. For more on amino acids and proteins, see "The Chemistry of Life: Biochemistry," pages 530–545.

GENERAL
Chemistry Now™

See the General ChemistryNow CD-ROM or website:

• **Screen 11.5 Functional Groups,** for a description of the types of organic functional groups and for tutorials on their structures, bonding, and chemistry

Example 11.7—Functional Group Chemistry

Problem

(a) Name the product of the reaction between ethylene and HCl.

(b) Draw the structure of the product of the reaction between propanoic acid and 1-propanol. What is the systematic name of the reaction product, and what functional group does it contain?

(c) What is the result of reacting 2-butanol with an oxidizing agent? Give the name and draw the structure of the reaction product.

Strategy Ethylene is an alkene (page 487), propanoic acid is a carboxylic acid (page 505), and 2-butanol is an alcohol (page 497). Consult the discussion regarding their chemistry.

Solution

(a) HCl will add to the double bond of ethylene to produce chloroethane.

$$H_2C=CH_2 + HCl \longrightarrow H-\underset{\underset{H}{|}}{\overset{\overset{H}{|}}{C}}-\underset{\underset{H}{|}}{\overset{\overset{H}{|}}{C}}-Cl$$

ethylene   chloroethane

(b) Carboxylic acids such as propanoic acid react with alcohols to give esters.

$$CH_3CH_2\overset{\displaystyle O}{\overset{\|}{C}}OH + CH_3CH_2CH_2OH \longrightarrow CH_3CH_2\overset{\displaystyle O}{\overset{\|}{C}}OCH_2CH_2CH_3 + H_2O$$

propanoic acid   1-propanol   propyl propanoate, an ester

(c) 2-Butanol is a secondary alcohol. Such alcohols are oxidized to ketones.

$$CH_3\overset{\displaystyle OH}{\overset{|}{C}}HCH_2CH_3 \xrightarrow{\text{oxidizing agent}} CH_3\overset{\displaystyle O}{\overset{\|}{C}}CH_2CH_3$$

2-butanol   butanone, a ketone

Exercise 11.11—Functional Groups

(a) Name each of the following compounds and its functional group.

$$1.\ CH_3CH_2CH_2OH \qquad 2.\ CH_3\overset{\displaystyle O}{\overset{\|}{C}}OH \qquad 3.\ CH_3CH_2NH_2$$

(b) Name the product from the reaction of compounds 1 and 2.
(c) What is the name and structure of the product from the oxidation of 1?
(d) What compound could result from combining compounds 2 and 3?
(e) What is the result of adding an acid (say HCl) to compound 3?

11.5—Polymers

We now turn to the very large molecules known as polymers. These can be either synthetic materials or naturally occurring substances such as proteins or nucleic acids. Although these materials have widely varying compositions, their structures and properties are understandable based on the principles developed for small molecules.

Classifying Polymers

The word *polymer* means "many parts" (from the Greek, *poly* and *meros*). **Polymers** are giant molecules made by chemically joining many small molecules called **monomers**. Polymer molecular weights range from thousands to millions.

Extensive use of synthetic polymers is a fairly recent development. A few synthetic polymers (Bakelite, rayon, and celluloid) were made early in the 20th century, but most of the products with which you are familiar originated in the last 50 years. By 1976, synthetic polymers outstripped steel as the most widely used material in the United States. The average production of synthetic polymers in the United States is approximately 150 kg per person annually.

The polymer industry classifies polymers in several different ways. One is their response to heating. **Thermoplastics** (such as polyethylene) soften and flow when they are heated and harden when they are cooled. **Thermosetting plastics** (such as Formica) are initially soft but set to a solid when heated and cannot be resoftened.

■ **Biochemical Polymers**
Polymer chemistry extends to biochemistry where chemists study proteins and other large molecules. See "The Chemistry of Life: Biochemistry," pages 530–545.

(a)   (b)   (c)

Figure 11.12   Common polymer-based consumer products. (a) Packaging materials from high-density polyethylene; (b) from polystyrene; and (c) from polyvinyl chloride. Recycling information is provided on most plastics (often molded into the bottom of bottles). High-density polyethylene is designated with a "2" inside a triangular symbol and the letters "HDPE." PVC is designated with a "3" inside a triangular symbol with the letter "V" below.

Another classification scheme depends on the end use of the polymer—for example, plastics, fibers, elastomers, coatings, and adhesives.

A more chemically oriented approach to polymer classification is based on their method of synthesis. **Addition polymers** are made by directly adding monomer units together. **Condensation polymers** are made by combining monomer units and splitting out a small molecule, often water.

Addition Polymers

Polyethylene, polystyrene, and polyvinyl chloride (PVC) are common addition polymers (Figure 11.12). They are built by "adding together" simple alkenes such as ethylene ($CH_2 = CH_2$), styrene ($C_6H_5CH = CH_2$), and vinyl chloride ($CH_2 = CHCl$). These and other addition polymers (Table 11.12), all derived from alkenes, have widely varying properties and uses.

GENERAL
Chemistry ⚛ Now™

See the General ChemistryNow CD-ROM or website:
• **Screen 11.9 Synthetic Organic Polymers (1),** for an animation of addition polymerization

Polyethylene and Other Polyolefins

Polyethylene is by far the leader in terms of addition polymer production. Ethylene (C_2H_4), the monomer from which polyethylene is made, is a product of petroleum refining and one of the top five chemicals produced in the United States. When ethylene is heated to between 100 and 250 °C at a pressure of 1000 to 3000 atm in the presence of a catalyst, polymers with molecular weights up to

Charles D. Winters

Table 11.12   Ethylene Derivatives That Undergo Addition Polymerization

Formula	Monomer Common Name	Polymer Name (Trade Names)	Uses	U.S. Polymer Production (Metric tons/year)*
H₂C=CH₂ (ethylene structure)	ethylene	polyethylene (polythene)	squeeze bottles, bags, films, toys and molded objects, electric insulation	7 million
(propylene structure)	propylene	polypropylene (Vectra, Herculon)	bottles, films, indoor-outdoor carpets	1.2 million
(vinyl chloride structure)	vinyl chloride	polyvinyl chloride (PVC)	floor tile, raincoats, pipe	1.6 million
(acrylonitrile structure)	acrylonitrile	polyacrylonitrile (Orlan, Acrilan)	rugs, fabrics	0.5 million
(styrene structure)	styrene	polystyrene (Styrofoam, Styron)	food and drink coolers, building material insulation	0.9 million
(vinyl acetate structure)	vinyl acetate	polyvinyl acetate (PVA)	latex paint, adhesives, textile coatings	200,000
(methyl methacrylate structure)	methyl methacrylate	polymethyl methacrylate (Plexiglass, Lucite)	high-quality transparent objects, latex paints, contact lenses	200,000
(tetrafluoroethylene structure)	tetrafluoroethylene	polytetrafluoroethylene (Teflon)	gaskets, insulation, bearings, pan coatings	6,000

* One metric ton = 1000 kg

several million are formed. The reaction can be expressed as a balanced chemical equation:

$$n\ H_2C=CH_2 \longrightarrow \left(\begin{array}{cc} H & H \\ | & | \\ -C-C- \\ | & | \\ H & H \end{array}\right)_n$$

ethylene   polyethylene

The abbreviated formula of the reaction product, $(-CH_2CH_2-)_n$, shows that polyethylene is a chain of carbon atoms, each bearing two hydrogens. The chain length for polyethylene can be very long. A polymer with a molecular weight of 1 million would contain almost 36,000 ethylene molecules linked together.

Polyethylene formed under various pressures and catalytic conditions has different properties, as a result of their different molecular structures. For ex-

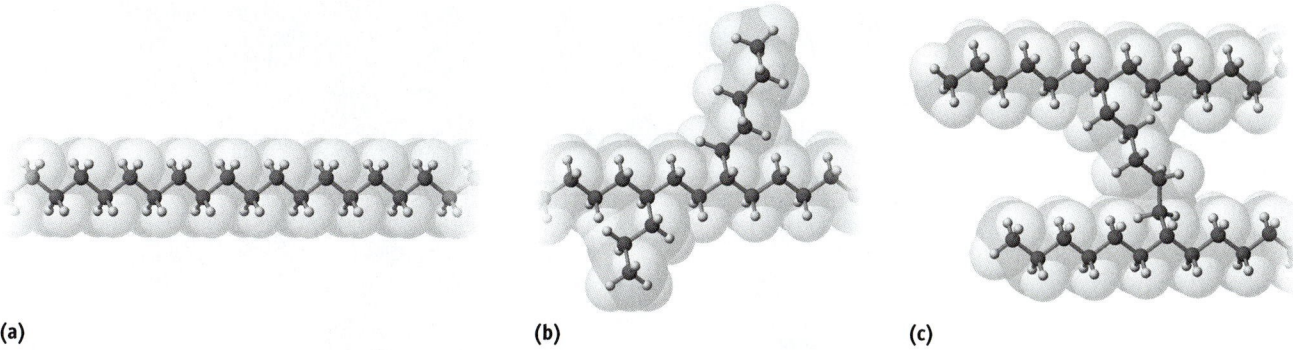

(a)   **(b)**   **(c)**

Figure 11.13 Polyethylene. (a) The linear form, high-density polyethylene (HDPE). (b) Branched chains occur in low-density polyethylene (LDPE). (c) Cross-linked polyethylene (CLPE).

ample, when chromium oxide is used as a catalyst, the product is almost exclusively a linear chain (Figure 11.13a). If ethylene is heated to 230 °C at high pressure, however, irregular branching occurs. Still other conditions lead to cross-linked polyethylene, in which different chains are linked together (Figures 11.13b and c).

The high-molecular-weight chains of linear polyethylene pack closely together and result in a material with a density of 0.97 g/cm^3. This material, referred to as high-density polyethylene (HDPE), is hard and tough, which makes it suitable for items such as milk bottles. If the polyethylene chain contains branches, however, the chains cannot pack as closely together, and a lower-density material (0.92 g/cm^3) known as low-density polyethylene (LDPE) results. This material is softer and more flexible than HDPE. It is used in plastic wrap and sandwich bags, among other things. Linking up the polymer chains in cross-linked polyethylene (CLPE) causes the material to be even more rigid and inflexible. Plastic bottle caps are often made of CLPE.

Polymers formed from substituted ethylenes ($CH_2{=}CHX$) have a range of properties and uses (see Table 11.12). Sometimes the properties are predictable based on the molecule's structure. Polymers without polar substituent groups, such as polystyrene, often dissolve in organic solvents, a property useful for some types of fabrication (Figure 11.14).

Polymers based on substituted ethylenes, $H_2C{=}CHX$

$$\left(\!-CH_2CH-\!\right)_n \qquad \left(-CH_2CH-\right)_n \qquad \left(-CH_2CH-\right)_n$$
$$\quad\;\;|\qquad\qquad\qquad\;\;| \qquad\qquad\qquad\;\;|$$
$$\quad\;\;OH\qquad\qquad\quad OCCH_3\qquad\qquad$$
$$\qquad\qquad\qquad\qquad\qquad\;\; \|\qquad\qquad$$
$$\qquad\qquad\qquad\qquad\qquad\;\; O\qquad\qquad$$

polyvinyl alcohol   polyvinyl acetate   polystyrene

Polyvinyl alcohol is a polymer with little affinity for nonpolar solvents but an affinity for water, which is not surprising based on the large number of polar —OH groups (Figure 11.15). Vinyl alcohol itself is not a stable compound (it isomerizes to acetaldehyde CH_3CHO), so polyvinyl alcohol cannot be made from this compound. Instead, it is made by hydrolyzing the ester groups in polyvinyl acetate.

Solubility in water or organic solvents can be a liability for polymers. The many uses of polytetrafluoroethylene [Teflon, $(-CF_2CF_2-)_n$] stem from the fact that it does not interact with water or organic solvents [◀ page 7].

Christopher Springmann/Corbisstockmarket.com

Polyethylene film. The polymer film is produced by extruding the molten plastic through a ringlike gap and inflating the film like a balloon.

(a)   **(b)**

Figure 11.14   **Polystyrene.** (a) The polymer is a clear, hard, colorless solid, but it may be more familiar as a light, foamlike material called Styrofoam. (b) Styrofoam has no polar groups and thus dissolves well in organic solvents such as acetone. See also Figure 11.12b.

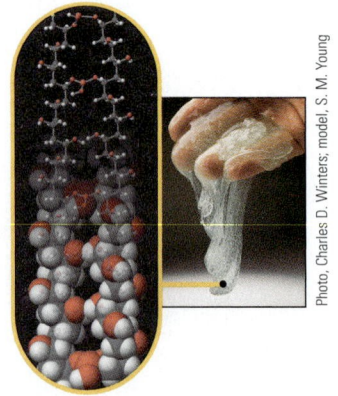

Figure 11.15   **Slime.** When boric acid, $B(OH)_3$, is added to an aqueous suspension of polyvinyl alcohol, $(CH_2CHOH)_n$, the mixture becomes very viscous. This is because boric acid reacts with the —OH groups on the polymer chain, causing cross-linking to occur. (The model shows an idealized structure of a portion of the polymer.)

Isoprene, 2-methyl-1,3-butadiene.

Polystyrene, with $n = 5700$, is a clear, hard, colorless solid that can be molded easily at 250 °C. You are probably more familiar with the very light, foamlike material known as Styrofoam that is used widely for food and beverage containers and for home insulation (Figure 11.14). Styrofoam is produced by a process called "expansion molding." Polystyrene beads containing 4% to 7% of a low-boiling liquid like pentane are placed in a mold and heated with steam or hot air. Heat causes the solvent to vaporize, creating a foam in the molten polymer that expands to fill the shape of the mold.

Natural and Synthetic Rubber

Natural rubber was first introduced in Europe in 1740, but it remained a curiosity until 1823, when Charles Macintosh invented a way of using it to waterproof cotton cloth. The mackintosh, as rain coats are still sometimes called, became popular despite major problems: Natural rubber is notably weak and is soft and tacky when warm but brittle at low temperatures. In 1839, after five years of research on natural rubber, the American inventor Charles Goodyear (1800–1860) discovered that heating gum rubber with sulfur produces a material that is elastic, water-repellent, resilient, and no longer sticky.

Rubber is a naturally occurring polymer, the monomers of which are molecules of 2-methyl-1,3-butadiene, commonly called *isoprene*. In natural rubber, isoprene monomers are linked together through carbon atoms 1 and 4—that is, through the end carbon atoms of the C_4 chain (Figure 11.16). This leaves a double bond between carbon atoms 2 and 3. In natural rubber, these double bonds have a *cis* configuration.

In vulcanized rubber, the material that Goodyear discovered, the polymer chains of natural rubber are cross-linked by short chains of sulfur atoms. Cross-linking helps to align the polymer chains so the material does not undergo a permanent change when stretched. As a result, it springs back when the stress is removed. Substances that behave this way are called **elastomers**.

With a knowledge of the composition and structure of natural rubber, chemists began searching for ways to make synthetic rubber. When they first tried to make the polymer by linking isoprene monomers together, however, what they made was sticky and useless. The problem was that synthesis procedures gave a mixture of *cis* and *trans* polyisoprene. In 1955, however, chemists at the Goodyear and Firestone companies discovered special catalysts to prepare the all-*cis* polymer. This synthetic material, which was structurally identical to natural rubber, is now manufactured cheaply. In fact, more than 8.0×10^8 kg of synthetic polyisoprene is produced annually in the United States.

Other kinds of polymers have further expanded the repertoire of elastomeric materials now available. Polybutadiene, for example, is currently used in the production of tires, hoses, and belts. Some elastomers, called **copolymers**, are formed by polymerization of two (or more) different monomers. A copolymer of styrene and butadiene, made with a 1:3 ratio of these raw materials, is the most important synthetic rubber now made; more than about 1 billion kg of styrene-butadiene rubber (SBR) is produced each year in the United States for making tires.

Photo, Kevin Schafer/Tom Stack & Associates; model, S. M. Young

Figure 11.16 Natural rubber. The sap that comes from the rubber tree is a natural polymer of isoprene. All the linkages in the carbon chain are *cis*. When natural rubber is heated strongly in the absence of air, it smells of isoprene. This observation provided a clue that rubber is composed of this building block.

And a little is left over each year to make bubble gum. The stretchiness of bubble gum once came from natural rubber, but SBR is now used to help you blow bubbles.

Condensation Polymers

A chemical reaction in which two molecules react by splitting out, or eliminating, a small molecule is called a **condensation reaction**. The reaction of an alcohol with a carboxylic acid to give an ester is an example. A polymer can be formed in a condensation reaction if two different reactant molecules, each containing *two* functional groups, are used. This is one route used to make polyesters and polyamides, two important types of condensation polymers.

GENERAL
Chemistry·⚛·Now™

See the General ChemistryNow CD-ROM or website:

- **Screen 11.10 Synthetic Organic Polymers,** to view an animation of condensation polymerization and to watch a video of the synthesis of nylon.

Polyesters

Terephthalic acid contains two carboxylic acid groups, and ethylene glycol contains two alcohol groups. When mixed, the acid and alcohol functional groups at both ends of these molecules can react to form ester linkages, splitting out water.

Charles D. Winters

Copolymer of styrene and butadiene, SBR rubber. The elasticity of bubble gum comes from SBR rubber. (*See General Chemistry-Now Screen 11.11 Puzzler, for a description of the polymer used in bubble gum.*)

Figure 11.17 Polyesters. Polyethylene terephthalate is used to make clothing and soda bottles. The two students are wearing jackets made from recycled PET soda bottles. Mylar film, another polyester, is used to make recording tape as well as balloons. Because the film has very tiny pores, Mylar can be used for helium-filled balloons; the atoms of gaseous helium move through the pores in the film very slowly.

The result is a polymer called polyethylene terephthalate (PET). The multiple ester linkages make this substance a **polyester.**

$$n \; HOC{-}\langle\bigcirc\rangle{-}COH \;+\; n \; HOCH_2CH_2OH \;\longrightarrow\; \left(\!{-}C{-}\langle\bigcirc\rangle{-}COCH_2CH_2O{-}\!\right)_{\!n} \;+\; 2n \; H_2O$$

terephthalic acid   ethylene glycol   polyethylene terephthalate (PET), a polyester

Polyester textile fibers made from PET are marketed as Dacron and Terylene. The inert, nontoxic, noninflammatory, and non–blood-clotting properties of Dacron polymers make Dacron tubing an excellent substitute for human blood vessels in heart bypass operations, and Dacron sheets are sometimes used as temporary skin for burn victims. A polyester film, Mylar, has unusual strength and can be rolled into sheets one-thirtieth the thickness of a human hair. Magnetically coated Mylar films are used to make audio and video tapes (Figure 11.17).

Polyamides

In 1928, the DuPont Company embarked on a basic research program headed by Wallace Carothers (1896–1937). Carothers was interested in high molecular weight compounds, such as rubbers, proteins, and resins. In 1935, his research yielded nylon-6,6 (Figure 11.18), a **polyamide** prepared from adipoyl chloride, a derivative of adipic acid, a diacid, and hexamethylenediamine, a diamine:

$$n \; ClC(CH_2)_4CCl \;+\; n \; H_2N(CH_2)_6NH_2 \;\longrightarrow\; \left(\!{-}C(CH_2)_4C{-}N(CH_2)_6N{-}\!\right)_{\!n} \;+\; n \; HCl$$

adipoyl chloride   hexamethylenediamine   amide link in nylon-6,6 a polyamide

Nylon can be extruded easily into fibers that are stronger than natural fibers and chemically more inert. The discovery of nylon jolted the American textile industry at a critical time. Natural fibers were not meeting 20th-century needs. Silk was expensive and not durable, wool was scratchy, linen crushed easily, and cotton did not have a high-fashion image. Perhaps the most identifiable use for the new fiber was in nylon stockings. The first public sale of nylon hosiery took place on October 24, 1939, in Wilmington, Delaware (the site of DuPont's main office). This use of nylon in commercial products ended shortly thereafter, however, with the start of World War II. All nylon was diverted to making parachutes and other military gear. It was not until about 1952 that nylon reappeared in the consumer marketplace.

Figure 11.19 illustrates why nylon makes such a good fiber. To have good tensile strength (the ability to resist tearing), the polymer chains should be able to attract one another, albeit not so strongly that the plastic cannot be initially extended to form fibers. Ordinary covalent bonds between the chains (cross-linking) would be too strong. Instead, cross-linking occurs by a somewhat weaker intermolecular force called *hydrogen bonding* [▶ Section 13.3] between the hydrogens of N—H groups on one chain and the carbonyl oxygens on another chain. The polarities of the $N^{\delta-}$—$H^{\delta+}$ group and the $C^{\delta+}$=$O^{\delta-}$ group lead to attractive forces between the polymer chains of the desired magnitude.

Charles D. Winters

Active Figure 11.18   **Nylon-6,6.**
Hexamethylenediamine is dissolved in water (bottom layer), and adipoyl chloride (a derivative of adipic acid) is dissolved in hexane (top layer). The two compounds react at the interface between the layers to form nylon, which is being wound onto a stirring rod.

GENERAL
Chemistry ⚛ Now™ See the General ChemistryNow CD-ROM or website to explore an interactive version of this figure accompanied by an exercise.

Example 11.8—Condensation Polymers

Problem What is the repeating unit of the condensation polymer obtained by combining $HO_2CCH_2CH_2CO_2H$ (succinic acid) and $H_2NCH_2CH_2NH_2$ (1,2-ethylenediamine)?

Strategy Recognize that the polymer will link the two monomer units through the amide linkage. The smallest repeating unit of the chain will contain two parts, one from the diacid and the other from the diamine.

Solution The repeating unit of this polyamide is

$$\left(\begin{array}{c} \overset{\displaystyle O}{\overset{\|}{C}}CH_2CH_2\overset{\displaystyle O}{\overset{\|}{C}} - \underset{\underset{H}{|}}{N}CH_2CH_2\underset{\underset{H}{|}}{N} \end{array} \right)_n$$

amide linkage

Exercise 11.12—Condensation Polymers

Draw the structure of the repeating unit in the condensation polymer obtained from the reaction of propylene glycol with maleic acid:

$$HO - \underset{\underset{H}{|}}{\overset{\overset{H_3C}{|}}{C}} - \underset{\underset{H}{|}}{\overset{\overset{H}{|}}{C}} - OH \quad + \quad HO - \overset{\overset{O}{\|}}{C} - \underset{\underset{H}{|}}{\overset{\overset{H}{|}}{C}} = \underset{\underset{H}{|}}{\overset{\overset{H}{|}}{C}} - \overset{\overset{O}{\|}}{C} - OH$$

propylene glycol   maleic acid

(A closely related material is combined with glass fibers [fiberglass] to make hulls for small boats and automobile panels and parts.) Is this a polyamide or a polyester?

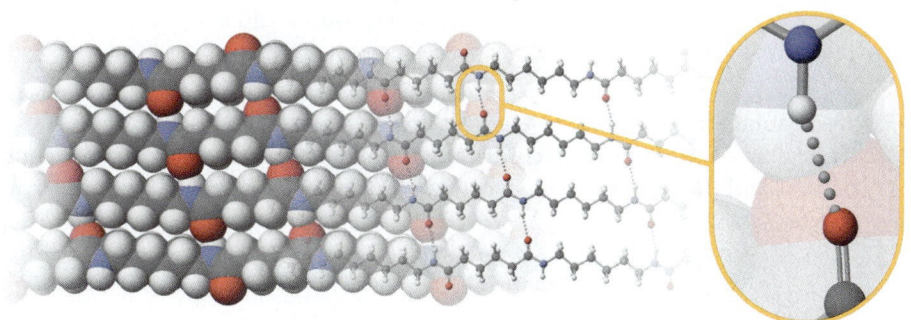

Figure 11.19 Hydrogen bonding between polyamide chains. Carbonyl oxygen atoms with a partial negative charge on one chain interact with an amine hydrogen with a partial positive charge on a neighboring chain. (This form of bonding is described in more detail in Section 13.3.)

Chemical Perspectives

Super Diapers

Disposable diapers are a miracle of modern chemistry: Most of the materials used are synthetic polymers. The outer layer is mostly microporous polyethylene; it keeps the urine in but remains breathable. The inside layer is polypropylene, a material prized by winter-camping enthusiasts. It stays soft and dry while wicking moisture away from the skin. Sandwiched between these layers is powdered sodium polyacrylate combined with cellulose, the latter the only natural part of the materials used. The package is completed with elasticized hydrophobic polypropylene cuffs around the baby's thighs, and Velcro tabs hold the diaper on the baby.

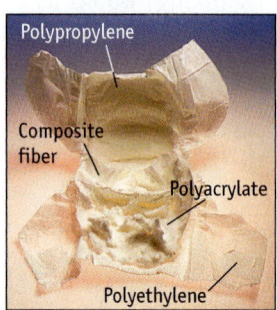

Polypropylene

Composite fiber

Polyacrylate

Polyethylene

Charles D. Winters

The secret ingredient in the diaper is the polyacrylate polymer filling. This substance can absorb up to 800 times its weight in water. When dry, the polymer has a carboxylate group associated with sodium ions. When placed in water, osmotic pressure causes water molecules to enter the polymer (because the ion concentration in the polymer is higher than in water; see Chapter 14). As water enters, the sodium ions dissociate from the polymer, and the polar water molecules are attracted to these positive ions and to the negative carboxylate groups of the polymer. At the same time, the negative carboxylate groups repel one another, forcing them apart and causing the polymer to unwind. Evidence for the unwinding of the polymer is seen as swelling of the diaper. In addition, because it contains so much water, the polymer becomes gel-like.

If the gelled polymer is put into a salt solution, water is attracted to the Na^+ and Cl^- ions and is drawn from the polymer. Thus, the polymer becomes solid once again. The diminished ability of sodium polyacrylate to absorb water in a salt solution is the reason that disposable diapers do not absorb urine as well as pure water.

These kinds of superabsorbent materials—sodium polyacrylate and a related material, polyacrylamide—are useful not only in diapers but also for cleaning up spills in hospitals, for protecting power and optical cables from moisture, for filtering water out of aviation gasoline, and for conditioning garden soil to retain water. You will also find them in the toy store as "gro-creatures."

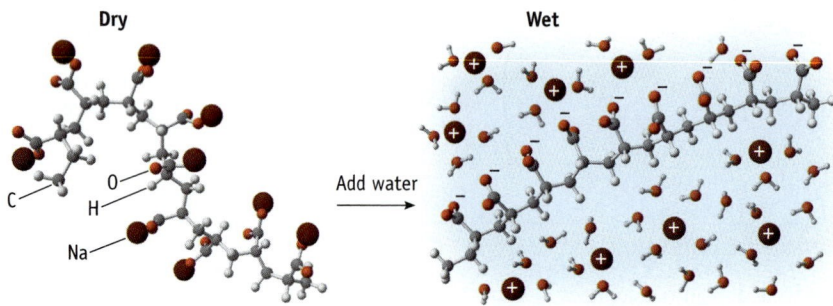

Dry

Wet

Add water

C

O

H

Na

Chapter Goals Revisited

When you have finished studying this chapter, you should ask whether you have met the chapter goals. In particular, you should be able to

Classify organic compounds based on formula and structure

a. Understand the factors that contribute to the large numbers of organic compounds and the wide array of structures (Section 11.1). General ChemistryNow homework: Study Question(s) 3

Recognize and draw structures of structural isomers and stereoisomers for carbon compounds

a. Recognize and draw structures of geometric isomers and optical isomers (Section 11.1). General ChemistryNow homework: SQ(s) 11, 15

Name and draw structures of common organic compounds

a. Draw structural formulas and name simple hydrocarbons, including alkanes, alkenes, alkynes, and aromatic compounds (Section 11.2). General ChemistryNow homework: SQ(s) 1, 5, 7

b. Identify possible isomers for a given formula (Section 11.2).

c. Name and draw structures of alcohols and amines (Section 11.3). General ChemistryNow homework: SQ(s) 31, 32

d. Name and draw structures of carbonyl compounds—aldehydes, ketones, acids, esters, and amides (Section 11.4). General ChemistryNow homework: SQ(s) 39, 40, 41, 51

Know the common reactions of organic functional groups

a. This goal applies specifically to the reactions of alkenes, alcohols, amines, aldehydes and ketones, and carboxylic acids. General ChemistryNow homework: SQ(s) 19, 21, 46

Relate properties to molecular structure

a. Describe the physical and chemical properties of the various classes of hydrocarbon compounds (Section 11.2).

b. Recognize the connection between the structures and the properties of alcohols (Section 11.3).

c. Know the structures and properties of several natural products, including carbohydrates (Section 11.4) and fats and oils (Section 11.4). General ChemistryNow homework: SQ(s) 49, 50

Identify common polymers

a. Write equations for the formation of addition polymers and condensation polymers, and describe their structures (Section 11.5).

b. Relate properties of polymers to their structures (Section 11.5). General ChemistryNow homework: SQ(s) 93

For additional preparation for an examination on this chapter see the **Let's Review** section on pages 530–545.

Study Questions

▲ denotes more challenging questions.

■ denotes questions available in the Homework and Goals section of the General ChemistryNow CD-ROM or website.

Blue numbered questions have answers in Appendix O and fully worked solutions in the *Student Solutions Manual*.

Structures of many of the compounds used in these questions are found on the General ChemistryNow CD-ROM or website in the Models folder.

GENERAL Chemistry ⚛ Now™ Assess your understanding of this chapter's topics with additional quizzing and conceptual questions at http://now.brookscole.com/kotz6e

Practicing Skills

Alkanes and Cycloalkanes

(See Examples 11.1 and 11.2, and General ChemistryNow Screen 11.3.)

1. ■ What is the name of the straight (unbranched) chain alkane with the formula C_7H_{16}?

2. What is the molecular formula for an alkane with 12 carbon atoms?

3. ■ Which of the following is an alkane? Which could be a cycloalkane?
   (a) C_2H_4
   (b) C_5H_{10}
   (c) $C_{14}H_{30}$
   (d) C_7H_8

4. Isooctane, 2,2,4-trimethylpentane, is one of the possible structural isomers with the formula C_8H_{18}. Draw the structure of this isomer, and draw and name structures of two other isomers of C_8H_{18} in which the longest carbon chain is five atoms.

5. ■ Give the systematic name for the following alkane:

$$\begin{array}{c} CH_3 \\ | \\ CH_3CHCHCH_3 \\ | \\ CH_3 \end{array}$$

6. Give the systematic name for the following alkane. Draw a structural isomer of the compound and give its name.

$$\begin{array}{c} CH_3 \\ | \\ CH_3CHCH_2CH_2CHCH_3 \\ | \\ CH_2CH_3 \end{array}$$

7. ■ Draw the structure of each of the following compounds:
   (a) 2,3-dimethylhexane
   (b) 2,3-dimethyloctane
   (c) 3-ethylheptane
   (d) 3-ethyl-2-methylhexane

8. Draw structures for 3-ethylpentane and 2,3-dimethylpentane.

9. Draw Lewis structures and name all possible compounds that have a seven-carbon chain with one methyl substituent group. Which of these isomers has a chiral carbon center?

10. Draw a structure for cycloheptane. Is the seven-member ring planar? Explain your answer.

11. ■ There are two ethylheptanes (compounds with a seven-carbon chain and one ethyl substituent). Draw the structures and name these compounds. Is either isomer chiral?

12. Among the 18 structural isomers with the formula C_8H_{18} are two with a five-carbon chain having one ethyl and one methyl substituent group. Draw the structures and name these two isomers.

13. List several typical physical properties of C_4H_{10}. Predict the following physical properties of dodecane, $C_{12}H_{26}$: color, state (s, ℓ, g), solubility in water, solubility in a nonpolar solvent.

14. Write balanced equations for the following reactions of alkanes.
   (a) the reaction of methane with excess chlorine
   (b) complete combustion of cyclohexane, C_6H_{12}, with excess oxygen

Alkenes and Alkynes

(See Examples 11.3 and 11.4 and General ChemistryNow Screens 11.3 and 11.4.)

15. ■ Draw structures for the *cis* and *trans* isomers of 4-methyl-2-hexene.

16. What structural requirement is necessary for an alkene to have *cis* and *trans* isomers? Can *cis* and *trans* isomers exist for an alkane? For an alkyne?

17. A hydrocarbon with the formula C_5H_{10} can be either an alkene or a cycloalkane.
   (a) Draw a structure for each of the possible isomers for C_5H_{10}, assuming it is an alkene. Six isomers are possible. Give the systematic name of each isomer you have drawn.
   (b) Draw a structure for a cycloalkane having the formula C_5H_{10}.

18. Five alkenes have the formula C_7H_{14} and a seven-carbon chain. Draw their structures and name them.

19. ■ Draw the structure and give the systematic name for the products of the following reactions:
   (a) $CH_3CH{=}CH_2 + Br_2 \longrightarrow$
   (b) $CH_3CH_2CH{=}CHCH_3 + H_2 \longrightarrow$

20. Draw the structure and give the systematic name for the products of the following reactions:

(a) 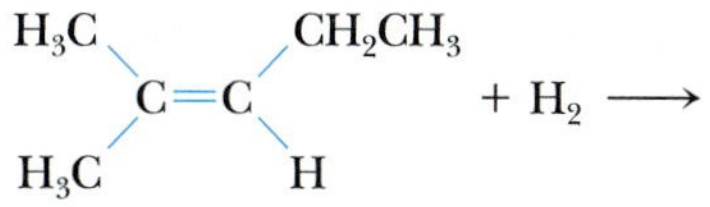

(b) $CH_3C{\equiv}CCH_2CH_3 + 2\ Br_2 \longrightarrow$

21. ■ The compound 2-bromobutane is a product of addition of HBr to an alkene. Identify the alkene and give its name.

22. The compound 2,3-dibromo-2-methylhexane is formed by addition of Br_2 to an alkene. Identify the alkene, and write an equation for this reaction.

23. Draw structures for alkenes that have the formula C_3H_5Cl and name each compound. (In these derivatives of propene, a chlorine atom replaces one hydrogen atom.)

24. Elemental analysis of a colorless liquid has given its formula as C_5H_{10}. You recognize that this compound could be either a cycloalkane or an alkene. A chemical test to determine the class to which it belongs involves adding bromine. Explain how this reaction would allow you to distinguish between the two classes.

Aromatic Compounds

(See Example 11.5, Exercise 11.5, and the General Chemistry Screen 11.3.)

25. Draw structural formulas for the following compounds:
   (a) 1,3-dichlorobenzene (alternatively called *m*-dichlorobenzene)
   (b) 1-bromo-4-methylbenzene (alternatively called *p*-bromotoluene)

26. Give the systematic name for each of the following compounds:

(a) 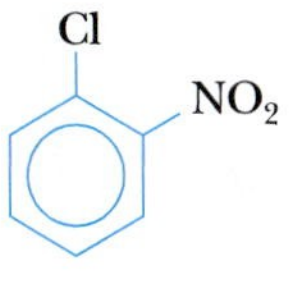

(c) 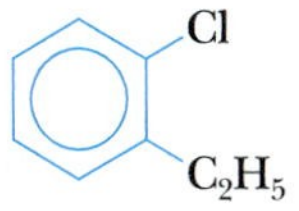

(b) 

27. Write an equation for the preparation of ethylbenzene from benzene and an appropriate compound containing an ethyl group.

28. Write an equation for the preparation of hexylbenzene from benzene and other appropriate reagents.

29. A single compound is formed by alkylation of 1,4-dimethylbenzene. Write the equation for the reaction of this compound with CH_3Cl and $AlCl_3$. What is the structure and name of the product?

30. Nitration of toluene gives a mixture of two products, one with the nitro group (—NO_2) in the *ortho* position and one with the nitro group in a *para* position. Draw the structures of the two products.

Alcohols, Ethers, and Amines

(See Example 11.6 and General ChemistryNow Screen 11.5.)

31. ■ Give the systematic name for each of the following alcohols, and tell whether each is a primary, secondary, or tertiary alcohol:
   (a) $CH_3CH_2CH_2OH$
   (b) $CH_3CH_2CH_2CH_2OH$
   (c) 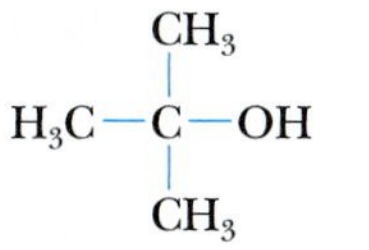
   (d) 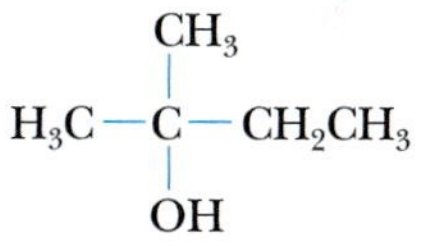

32. ■ Draw structural formulas for the following alcohols, and tell whether each is primary, secondary, or tertiary:
   (a) 1-butanol
   (b) 2-butanol
   (c) 3,3-dimethyl-2-butanol
   (d) 3,3-dimethyl-1-butanol

33. Write the formula and draw the structure for each of the following amines:
   (a) ethylamine
   (b) dipropylamine
   (c) butyldimethylamine
   (d) triethylamine

34. Name the following amines
   (a) $CH_3CH_2CH_2NH_2$
   (b) $(CH_3)_3N$
   (c) $(CH_3)(C_2H_5)NH$
   (d) $C_6H_{13}NH_2$

35. Draw structural formulas for all the alcohols with the formula $C_4H_{10}O$. Give the systematic name of each.

36. Draw structural formulas for all the primary amines with the formula $C_4H_9NH_2$.

37. Complete and balance the following equations:
(a) $C_6H_5NH_2(\ell) + HCl(aq) \longrightarrow$
(b) $(CH_3)_3N(aq) + H_2SO_4(aq) \longrightarrow$

38. ■ Aldehydes and carboxylic acids are formed by oxidation of primary alcohols, and ketones are formed when secondary alcohols are oxidized. Give the name and formula for the alcohol that, when oxidized, gives the following products:
(a) $CH_3CH_2CH_2CHO$
(b) 2-hexanone

Compounds with a Carbonyl Group

(See Exercises 11.7–11.10 and the General ChemistryNow Screen 11.5.)

39. ■ Draw structural formulas for (a) 2-pentanone, (b) hexanal, and (c) pentanoic acid.

40. ■ Identify the class of each the following compounds and give the systematic name for each:

(a) CH_3CCH_3

(c) $CH_3CCH_2CH_2CH_3$

(b) $CH_3CH_2CH_2CH$

41. ■ Identify the class of each the following compounds and give the systematic name for each:

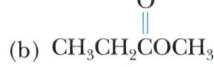
(a) $CH_3CH_2CHCH_2CO_2H$

(b) $CH_3CH_2COCH_3$

(c)
$$CH_3\overset{\overset{\displaystyle O}{\|}}{C}OCH_2CH_2CH_2CH_3$$

(d) Br—⟨○⟩—$\overset{\overset{\displaystyle O}{\|}}{C}OH$

42. Draw structural formulas for the following acids and esters:
(a) 2-methylhexanoic acid
(b) pentyl butanoate (which has the odor of apricots)
(c) octyl acetate (which has the odor of oranges)

43. Give the structural formula and systematic name for the product, if any, from the reactions of each of the following pairs of compounds:
(a) pentanal and $KMnO_4$
(b) pentanal and $LiAlH_4$
(c) 2-octanone and $LiAlH_4$
(d) 2-octanone and $KMnO_4$

44. Describe how to prepare 2-pentanol beginning with the appropriate ketone.

45. Describe how to prepare propyl propanoate beginning with 1-propanol as the only carbon-containing reagent.

46. ■ Give the name and structure of the product of the reaction of benzoic acid and 2-propanol.

47. Draw structural formulas and give the names for the products of the following reaction:

$$CH_3\overset{\overset{\displaystyle O}{\|}}{C}OCH_2CH_2CH_2CH_3 + NaOH$$

48. Draw structural formulas and give the names for the products of the following reaction:

⟨○⟩—$\overset{\overset{\displaystyle O}{\|}}{C}$—O—$\overset{\overset{\displaystyle CH_3}{|}}{\underset{\underset{\displaystyle CH_3}{|}}{CH}}$ + NaOH

49. ■ The Lewis structure of phenylalanine, one of the 20 amino acids that make up proteins, is drawn below (without lone pairs of electrons). The carbon atoms are numbered for the purpose of this question.
(a) What is the geometry around C_3?
(b) What is the O—C—O bond angle?
(c) Is this molecule chiral? If so, which carbon atom is chiral?
(d) Which hydrogen atom in this compound is acidic?

(structure of phenylalanine)

50. ■ The Lewis structure of vitamin C, whose chemical name is ascorbic acid, is drawn below (without lone pairs of electrons).

(structure of ascorbic acid)

(a) What is the approximate value for the C—O—C bond angle in the five-member ring?
(b) There are four OH groups in this structure. Estimate the C—O—H bond angles for these groups. Will they be the same value (more or less), or should there be significant differences in these bond angles?

(c) Is the molecule chiral? How many chiral carbon atoms can be identified in this structure?

(d) Identify the shortest bond in this molecule.

(e) What are the functional groups of the molecule?

Functional Groups

(See Example 11.7 and the General ChemistryNow Screen 11.5.)

51. Identify the functional groups in the following molecules.

(a) $CH_3CH_2CH_2OH$

(b)
$H_3CCNHCH_3$

(c) CH_3CH_2COH

(d) $CH_3CH_2COCH_3$

52. Consider the following molecules:

1.
$CH_3CH_2CCH_3$

2. CH_3CH_2COH

3. $H_2C=CHCH_2OH$

4.
$CH_3CH_2CHCH_3$

(a) What is the result of treating compound 1 with $NaBH_4$? What is the functional group in the product? Name the product.

(b) Draw the structure of the reaction product from compounds 2 and 4. What is the functional group in the product?

(c) What compound results from adding H_2 to compound 3? Name the reaction product.

(d) What compound results from adding NaOH to compound 2?

Polymers

(See Example 11.8, Exercise 11.12, and the General ChemistryNow Screens 11.9 and 11.10.)

53. Polyvinyl acetate is the binder in water-based paints.

(a) Write an equation for its formation from vinyl acetate.

(b) Show a portion of this polymer with three monomer units.

(c) Describe how to make polyvinyl alcohol from polyvinyl acetate.

54. Neoprene (polychloroprene, a kind of rubber) is a polymer formed from the chlorinated butadiene $H_2C=CHCCl=CH_2$.

(a) Write an equation showing the formation of polychloroprene from the monomer.

(b) Show a portion of this polymer with three monomer units.

55. Saran is a copolymer of 1,1-dichloroethene and chloroethene (vinyl chloride). Draw a possible structure for this polymer.

56. The structure of methyl methacrylate is given in Table 11.12. Draw the structure of a polymethyl methacrylate (PMMA) polymer that has four monomer units. (PMMA has excellent optical properties and is used to make hard contact lenses.)

General Questions on Organic Chemistry

These questions are not designated as to type or location in the chapter. They may combine several concepts.

57. Three different compounds with the formula $C_2H_2Cl_2$ are known.

(a) Two of these compounds are geometric isomers. Draw their structures.

(b) The third compound is a structural isomer of the other two. Draw its structure.

58. Draw the structure of 2-butanol. Identify the chiral carbon atom in this compound. Draw the mirror image of the structure you first drew. Are the two molecules superimposable?

59. Draw Lewis structures and name three structural isomers with the formula C_6H_{12}. Are any of these isomers chiral?

60. Draw structures and name the four alkenes that have the formula C_4H_8.

61. Write equations for the reactions of *cis*-2-butene with the following reagents, representing the reactants and products using structural formulas.

(a) H_2O

(b) HBr

(c) Cl_2

62. Draw the structure and name the product formed if the following alcohols are oxidized. Assume that an excess of the oxidizing agent is used. If the alcohol is not expected to react with a chemical oxidizing agent, write NR (no reaction).

(a) $CH_3CH_2CH_2CH_2OH$

(b) 2-butanol

(c) 2-methyl-2-propanol

(d) 2-methyl-1-propanol

63. Write equations for the following reactions, representing the reactants and products using structural formulas.

(a) the reaction of acetic acid and sodium hydroxide

(b) the reaction of methylamine with HCl

64. Write equations for the following reactions, representing the reactants and products using structural formulas.

(a) the formation of ethyl acetate from acetic acid and ethanol

(b) the hydrolysis of glyceryl tristearate (the triester of glycerol with stearic acid, a fatty acid)

65. Write an equation for the formation of the following polymers.

(a) polystyrene, from styrene ($C_6H_5CH{=}CH_2$)

(b) PET (polyethylene terephthalate), from ethylene glycol and terephthalic acid

66. Write equations for the following reactions, representing the reactants and products using structural formulas.

(a) the hydrolysis of the amide, $C_6H_5CONHCH_3$ to form benzoic acid and methylamine

(b) the hydrolysis of nylon-66,

$[-CO(CH_2)_4CONH(CH_2)_6NH-]_x$

a polyamide, to give a carboxylic acid and an amine

67. Draw the structure of each of the following compounds:

(a) 2,2-dimethylpentane

(b) 3,3-diethylpentane

(c) 3-ethyl-2-methylpentane

(d) 3-ethylhexane

68. ▲ Structural Isomers

(a) Draw all of the isomers possible for C_3H_8O. Give the systematic name of each and tell into which class of compounds it fits.

(b) Draw the structural formula for an aldehyde and a ketone with the molecular formula C_4H_8O. Give the systematic name of each.

69. ▲ Draw structural formulas for possible isomers of dichlorinated propane, $C_3H_6Cl_2$. Name each compound.

70. ■ Draw structural formulas for possible isomers with the formula C_3H_6ClBr. Name each isomer.

71. Give structural formulas and systematic names for the three structural isomers of trimethylbenzene, $C_6H_3(CH_3)_3$.

72. Give structural formulas and systematic names for possible isomers of dichlorobenzene, $C_6H_4Cl_2$.

73. Voodoo lilies depend on carrion beetles for pollination. Carrion beetles are attracted to dead animals, and because dead and putrefying animals give off the horrible-smelling amine cadaverine, the lily likewise releases cadaverine (and the closely related compound putrescine, page 501). A biological catalyst, an enzyme, converts the naturally occurring amino acid lysine to cadaverine.

$$H_2NCH_2CH_2CH_2CH_2-\overset{\displaystyle H}{\underset{\displaystyle \underset{O}{\overset{\|}{C}}-OH}{C}}-NH_2$$

lysine

What group of atoms must be replaced in lysine to make cadaverine? (Lysine is essential to human nutrition but is not synthesized in the human body.)

74. Benzoic acid occurs in many berries. When humans eat berries, benzoic acid is converted to hippuric acid in the body by reaction with the amino acid glycine, $H_2NCH_2CO_2H$. Draw the structure of hippuric acid, recognizing that it is an amide formed by reaction of the carboxylic acid group of benzoic acid and the amino group of glycine. Why is hippuric acid referred to as an acid?

75. Consider the reaction of *cis*-2-butene with H_2 (in the presence of a catalyst).

(a) Draw the structure and give the name of the reaction product. Is this reaction product chiral?

(b) Draw an isomer of the reaction product.

76. ■ Give the name of each compound below and name the functional group involved.

(a) $H_3C-\overset{\displaystyle OH}{\underset{\displaystyle H}{C}}-CH_2CH_2CH_3$

(b) $H_3C-\overset{\displaystyle O}{\overset{\|}{C}}-CH_2CH_2CH_3$

(c) $H_3C-\overset{\displaystyle H}{\underset{\displaystyle CH_3}{C}}-\overset{\displaystyle O}{\overset{\|}{C}}-H$

(d) $H_3CCH_2CH_2-\overset{\displaystyle O}{\overset{\|}{C}}-OH$

77. Which of the following compounds produce acetic acid when treated with an oxidizing agent such as $KMnO_4$?

(a) H_3C-CH_3

(b) $H_3C-\overset{\displaystyle O}{\overset{\|}{C}}-H$

(c) $H_3C-\overset{\displaystyle OH}{\underset{\displaystyle H}{C}}-H$

(d) $H_3C-\overset{\displaystyle O}{\overset{\|}{C}}-CH_3$

78. ■ Consider the reactions of C_3H_7OH.

(a) Name the reactant C_3H_7OH.

(b) Draw a structural isomer of the reactant and give its name.

(c) Name the product of reaction A.

(d) Name the product of reaction B.

79. Kevlar is a polyamide made from *p*-phenylenediamine and terephthalic acid. (It is used to make bullet-proof vests, among other things.) Draw the repeating unit of the Kevlar polymer.

p-phenylenediamine   terephthalic acid

80. ▲ A well-known company selling outdoor clothing has recently introduced jackets made of recycled polyethylene terephthalate (PET), the principal material in many soft-drink bottles. Another company makes PET fibers by treating recycled bottles with methanol to give the diester dimethylterephthalate and ethylene glycol and then re-polymerizing these compounds to give new PET. Write a chemical equation to show how the reaction of PET with methanol can give dimethylterephthalate and ethylene glycol.

81. Draw the structure of glyceryl trilaurate. When this triester is saponified, what are the products?

82. Write a chemical equation describing the reaction between glycerol and stearic acid to give glyceryl tristearate.

83. You have a liquid that is either cyclohexene or benzene. When the liquid is exposed to dark red bromine vapor, the vapor is immediately decolorized. What is the identity of the liquid? Write an equation for the chemical reaction that has occurred.

84. ▲ Hydrolysis of an unknown ester of butyric acid, $CH_3CH_2CH_2CO_2R$, produces an alcohol A and butanoic acid. Oxidation of this alcohol forms an acid B that is a structural isomer of butanoic acid. Give the names and structures for alcohol A and acid B.

85. ▲ You are asked to identify an unknown colorless, liquid carbonyl compound. Analysis has determined that the formula for this unknown is C_3H_6O. Only two compounds match this formula.

(a) Draw structures for the two possible compounds.

(b) To decide which of the two structures is correct, you react the compound with an oxidizing agent, and isolate from that reaction a compound that is found to give an acidic solution in water. Use this result to identify the structure of the unknown.

(c) Name the acid formed by oxidation of the unknown.

86. An unknown colorless liquid has the formula C_3H_8O. Draw the structures for the three compounds that have this formula.

87. ▲ Addition of water to alkene X gives an alcohol Y. Oxidation of Y produces 3,3-dimethyl-2-pentanone. Identify X and Y, and write equations for the two reactions.

88. ▲ An unknown ester has the formula $C_4H_8O_2$. Hydrolysis gives methanol as one product. Identify the ester and write an equation for the hydrolysis reaction.

89. Identify the reaction products and write an equation for the following reactions of $CH_2 = CHCH_2OH$.

(a) H_2 (hydrogenation, in the presence of a catalyst)

(b) oxidation (excess oxidizing agent)

(c) addition polymerization

(d) ester formation, using acetic acid

90. Recently, the commercialization of a new polyester was reported in the news media. It was prepared from lactic acid ($CH_3CH(OH)CO_2H$), whose structure is shown on page 479. Write a balanced chemical equation for the formation of this polymer.

Summary and Conceptual Questions

The following questions use concepts from the previous chapters.

91. Carbon atoms appear in organic compounds in several different ways with single, double, and triple bonds combining to give an octet configuration. Describe the various ways that carbon can bond to reach an octet. Give the name and draw the structure of a compound that illustrates that mode of bonding.

92. There is a high barrier to rotation around a carbon–carbon double bond, whereas the barrier to rotation around a carbon–carbon single bond is very small. Use the orbital overlap model of bonding (Chapter 10) to explain why restricted rotation occurs around a double bond.

93. ■ What important properties do the following characteristics impart to an polymer?

(a) cross-linking in polyethylene

(b) the OH groups in polyvinyl alcohol

(c) hydrogen bonding in a polyamide-like nylon

▲ More challenging   ■ In General ChemistryNow   Blue-numbered questions answered in Appendix O

94. One of the resonance structures for pyridine is illustrated here. Draw another resonance structure for the molecule. Comment on the similarity between this compound and benzene.

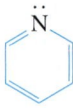

pyridine

95. Write balanced equations for the combustion of ethane and ethanol.

(a) Calculate the heat of combustion for each compound. Which has the more negative enthalpy change for combustion per gram?

(b) If ethanol is assumed to be partially oxidized ethane, what effect does this have on the heat of combustion?

96. Describe a simple chemical test to tell the difference between $CH_3CH_2CH_2CH=CH_2$ and its isomer cyclopentane.

97. Describe a simple chemical test to tell the difference between 2-propanol and its isomer methyl ethyl ether.

98. Plastics make up about 20% of the volume of landfills. There is, therefore, considerable interest in reusing or recycling these materials. To identify common plastics, a set of universal symbols is now used, five of which are illustrated here. They symbolize low- and high-density polyethylene, poly(vinyl chloride), polypropylene, and polyethylene terephthalate.

PETE

HDPE

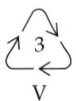
V

LDPE

PP

(a) Tell which symbol belongs to which type of plastic.

(b) Find an item in the grocery or drug store made from each of these plastics.

(c) Properties of several plastics are listed in the table. Based on this information, describe how to separate samples of these plastics from one another.

Plastic	Density (g/cm^3)	Melting Point (°C)
Polypropylene	0.92	170
High-density polyethylene	0.97	135
Polyethylene terephthalate	1.34–1.39	245

99. ▲ Maleic acid is prepared by the catalytic oxidation of benzene. It is a dicarboxylic acid; that is, it has two carboxylic acid groups.

(a) Combustion of 0.125 g of the acid gives 0.190 g of CO_2 and 0.0388 g of H_2O. What is the empirical formula of the acid?

(b) A 0.261-g sample of the acid requires 34.60 mL of 0.130 M NaOH for complete titration (so that the H^+ ions from both carboxylic acid groups are used). What is the molecular formula of the acid?

(c) Draw a Lewis structure for the acid.

(d) Describe the hybridization used by the C atoms.

(e) What are the bond angles around each C atom?

100. Benzene, C_6H_6, is a planar molecule. As General ChemistryNow CD-ROM or website Screen 11.2 shows, another six-carbon cyclic molecule, cyclohexane (C_6H_{12}), is not planar.

(a) Contrast the carbon atom hybridization in these two molecules.

(b) Why is π electron delocalization possible in benzene?

(c) Why is cyclohexane not planar?

101. ▲ Addition reactions of hydrocarbons are described on the General ChemistryNow CD-ROM or website Screen 11.4. In the Simulation you learn that the product of an addition reaction of an alkene is controlled by Markovnikov's rule.

(a) Draw the structure of the product obtained by adding HBr to propene, and give its name.

(b) Draw the structure and give the name of the compound that results from adding H_2O to 2-methyl-1-butene.

(c) If you add H_2O to 2-methyl-2-butene, is the product the same or different than the product from the reaction in part (b)?

102. Refer to the General ChemistryNow CD-ROM or website Screen 11.5, and then describe the hybrid orbitals used by the indicated atoms:

(a) the O atom in an alcohol

(b) the $C=O$ carbon in an aldehyde

(c) the $C=O$ carbon in a carboxylic acid

(d) the $C-O-C$ oxygen atom in an ester

(e) the N atom in an amine

103. Addition and substitution reactions are described in Chapter 11. Another type of reaction of organic compounds, elimination, is described on the General ChemistryNow CD-ROM or website Screen 11.6.

(a) What is the difference between a substitution reaction and an elimination reaction?

(b) Compare the elimination reaction shown on this screen with the hydrogenation reaction shown on Screen 11.4. In what ways are they similar or dissimilar?

▲ More challenging   ■ In General ChemistryNow   **Blue-numbered questions** answered in Appendix O

104. Properties of fats and oils are described on the General ChemistryNow CD-ROM or website Screen 11.7, and on page 510.

    (a) What type of reaction is used to make a fat or oil from glycerol and a fatty acid: addition, substitution, or elimination?

    (b) What is the primary structural difference between fats and oils? What types of functional groups do each contain?

    (c) What structural feature of oil molecules prevents them from coiling up on themselves as fat molecules do?

105. Addition polymerization is described on Screen 11.9 of the General ChemistryNow CD-ROM and website.

    (a) What is the primary structural feature of the molecules used to form addition polymers?

    (b) Consider the animation of a polymerization reaction shown on this screen. The polymer made here has a chain of 14 carbon atoms. Could the chain have been shorter or longer? Explain briefly.

    (c) What controls the length of the polymer chains formed?

    (d) Can the addition polymerization reaction be classified as one of the reaction types studied earlier: addition, substitution, or elimination?

106. Condensation polymerization is described on Screen 11.10 of the General ChemistryNow CD-ROM or website.

    (a) What is the primary structural feature necessary for a molecule to be useful in a condensation polymerization reaction?

    (b) Describe the appearance of the nylon being made in this video.

    (c) What does the designation "6,6" mean in nylon-6,6?

Do you need a live tutor for homework problems?
Access vMentor at General ChemistryNow at
http://now.brookscole.com/kotz6e
for one-on-one tutoring from a chemistry expert

Let's Review Chapters 7–11

Charles D. Winters

Charles D. Winters

Everyone enjoys a good fireworks display, and neon signs can be intriguing and even beautiful. Both owe their colorful display to the emission of light, and these chapters have described some of the principles involved in this effect. When you have finished these chapters you will know more about the colors of a fireworks display and the light emitted by neon signs and what they have in common.

The Purpose of *Let's Review*

- *Let's Review* provides additional homework questions for Chapters 7 through 11. Routine questions covering each of the concepts in a chapter are found in that chapter and are those usually identified by topic. In contrast, many of the *Let's Review* questions combine several concepts from one or more chapters. Many come from the examinations given by the authors and others are based on actual experiments or processes in chemical research or in the chemical industry.

- *Let's Review* provides guidance for Chapters 7 through 11 as you prepare for an exam covering material in these chapters. Although this is designated for Chapters 7 through 11 you may choose only material appropriate to the exam in your specific course.

- Each Comprehensive Question is correlated with tutorials and exercises on the General ChemistryNow website and with the OWL online homework system to which you may have purchased access. Some questions may include a screen shot of a tutorial so you see what resources are available to help you review.

Preparing for an Examination on Chapters 7–11

1. **GENERAL** **Chemistry⚛Now™** Use General ChemistryNow online to take a chapter **Pre-Test.** A **Personalized Learning Plan** will indicate sections of the textbook that should be reviewed.

2. Use General ChemistryNow online to work though media-based **Exercises, Guided Simulations,** and **Intelligent Tutors.**

3. If you subscribe to OWL, use the **Tutorials** in that system.

4. Work on the questions below that are relevant to a particular chapter or chapters. See the solutions to those questions at the end of this section.

5. For background and help with a question, use the General ChemistryNow information or OWL questions that are correlated with the question.

Key Points to Know for Chapters 7–11

Here are some of the key points you must know to be successful in Chapters 7 through 11.

- The properties of electromagnetic radiation.
- The origin of light from excited atoms.
- The quantum numbers and their relation to atomic structure.
- Electron configurations for the elements and common monatomic ions.
- Periodic trends in properties of the elements.
- Lewis structures for small molecules and ions.
- VSEPR theory to predict the shapes of simple molecules and ions and to understand the structures of more complex molecules.
- Electronegativity and its relation to bond and molecular polarity.
- Atom hybridization in molecules or ions.
- The principles of molecular orbital theory.
- Classification of organic molecules by functional group.
- Isomerism in organic molecules.
- Structures and names of common organic compounds.
- Reactions of organic functional groups.
- Common polymers.

Examination Preparation Questions

▲ denotes more challenging questions.

Important information about the questions that follow:

- See the Study Questions in each chapter for questions on basic concepts.

- Some questions arise from recent research in chemistry and the other sciences. They often involve concepts from more than one chapter and may be more challenging than those in earlier chapters. Not all chapter goals or concepts are necessarily addressed in these questions.

- Assessing Key Points questions are short-answer questions covering the key points listed on this page.

- Each Comprehensive Question is correlated with the text section covering that topic, with General ChemistryNow material, and with questions in OWL that may provide additional background.

- The computer screens are largely taken from General ChemistryNow but are also contained in the OWL system if you subscribe to it. They illustrate **Tutorials** or other resources that may help in answering the questions.

Assessing the Key Points

1. Place the following types of radiation in order of increasing energy: (a) x-rays, (b) radio waves, (c) blue light, and (d) light with $\lambda = 520$ nm

2. Radiation called UVB (290–315 nm) is responsible for your sunburn at the beach. Another type of radiation [UVA (315–400 nm)] is also responsible for tissue damage (page 305). Which has the greater energy, UVA or UVB?

3. The quantum number n describes the _____ of an atomic orbital and the quantum number ℓ describes its _____.

4. For a $4d$ orbital, what are the values of n and ℓ? What is one possible value of m_ℓ?

5. What elements have these ground state electron configurations? (a) $[Kr]5s^1$ (b) $[Ar]3d^{10}4s^24p^4$

6. Which has the largest first ionization energy: N, P, or As?

7. Which of the following is NOT a correct Lewis resonance structure for the N_2O molecule?

(a) $\ddot{N}=N=\ddot{O}$ (c) $:N\equiv N-\ddot{O}:$

(b) $:\ddot{N}-N\equiv O:$ (d) $:N\equiv N=\ddot{O}$

8. Consider the N_2O molecule in question 7.

(a) Is the structure of the molecule bent or linear?

(b) Is the N—O bond polar? Is the molecule polar?

(c) What is the hybridization of the central N atom and what is its formal charge?

9. If, in a molecule, three atomic orbitals combine, how many molecular orbitals will result?

10. To which class of organic molecules does each of these belong?

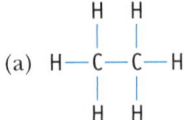 (a)

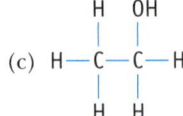 (c)

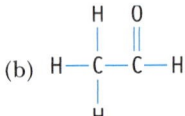 (b)

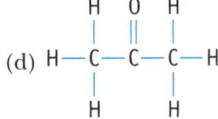 (d)

11. Name compound (d) in question 10 and draw the structure of an isomer of this compound.

12. Name the product of the oxidation with $KMnO_4$ of compound (c) in question 10.

13. Which compound or compounds could be used to synthesize a polymer?

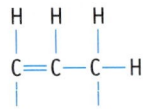   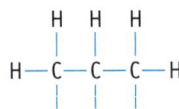   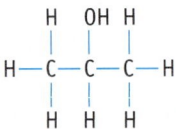

Comprehensive Questions

14. (Chapter 7) Blue light has a wavelength, λ, of 420. nm.
    (a) What is the frequency of the light?
    (b) If the total energy of a signal is 2.50×10^{-14} J, how many photons reach your eyes?

Text Sections: 7.1, 7.2

General ChemistryNow Screens: 7.3–7.5

OWL Questions: 7.1, 7.2

15. (Chapters 7 and 8) Quantum Numbers and Electron Configuration
    (a) When $n = 3$, $\ell = 1$, and $m_\ell = -1$, to what orbital does this refer? (Give the orbital label, such as 1s.)
    (b) For the $n = 5$ shell, there are _____ subshells and _____ orbitals. How many electrons can be accommodated in the $n = 5$ shell? _____
    (c) What type of orbital is not possible based on quantum theory: $2p$, $3s$, $5g$, $3f$, $7d$?
    (d) What is the maximum number of orbitals that can be associated with each of the following sets of quantum numbers? (One possible answer is "none.")
        (i) $n = 3$
        (ii) $n = 3$ and $\ell = 3$
        (iii) $n = 2$, $\ell = 1$, and $m_\ell = 0$

(e) Which ion or ions are unlikely based on your knowledge of electron configurations and ionization energies? Ba^{3+}, Cr^{3+}, Al^{3+}, S^{3-}, Cu^+

(f) At what element is the $n = 3$ shell just completed?

Text Sections: 7.5, 7.11, 7.12, 8.7, 8.8

General ChemistryNow Screens: 7.11-7.13

OWL Questions: 7.7, 7.8

16. ▲ (Chapters 6 and 7) To prepare some tea, you want to heat 254 g of water from 12 °C to 90 °C in a microwave oven, which operates at a frequency of 2.45 Gigahertz ($\nu = 2.45 \times 10^9$ sec^{-1}).
    (a) What is the wavelength of microwave radiation?
    (b) Suppose a microwave oven is 49 cm wide. How many wavelengths of the given radiation are equivalent to 49 cm?
    (c) How many moles of microwave photons must be absorbed to heat your mug of water to 90. °C?

Text Sections: 6.2, 7.2

General ChemistryNow Screens: 7.3, 7.4

OWL Questions: 6.1, 7.1, 7.2

17. ▲ (Chapters 6 and 7) Photosynthesis
    (a) Calculate the enthalpy change for the production of one mole of glucose by the process of photosynthesis at 25 °C. ΔH_f° [glucose(s)] = −1273.3 kJ/mol.

$$6\,CO_2(g) + 6\,H_2O(g) \longrightarrow C_6H_{12}O_6(s) + 6\,O_2(g)$$

    (b) What is the enthalpy change involved in producing one molecule of glucose by this process?
    (c) Chlorophyll molecules absorb light of various wavelengths. One wavelength absorbed is 650. nm. Calculate the energy of a photon of light having this wavelength.
    (d) How many photons with a wavelength of 650. nm are required to produce one glucose molecule, assuming all of the light energy is converted to chemical energy?

Text Sections: 6.5, 6.8, 7.2

General ChemistryNow Screens: 6.13, 6.16, 7.5

OWL Questions: 6.6, 7.3

18. (Chapter 7) The energy level diagram in Figure 7.10 describes a hydrogen atom using the Bohr model.
    (a) What quantum levels are involved in the emission of UV, visible, and infrared light in the hydrogen spectrum?
    (b) What is the energy of the hydrogen atom's electron in its ground state?
    (c) What is the energy of an electron in the $n = 4$ state?
    (d) Calculate the energy of the transition from the $n = 2$ level to the $n = 1$ level.
    (e) Calculate the energy (in kJ/mol) required to ionize 1.0 mole of hydrogen atoms.

(f) Calculate the wavelength of light required to cause ionization of an H atom. In what region of the electromagnetic spectrum is this radiation found?

Text Sections: 7.2, 7.3

General ChemistryNow Screens: 7.5–7.7

OWL Questions: 7.3, 7.4

19. (Chapter 8) Atom A has the ground state electron configuration [Ne] $3s^2\, 3p^5$. Atom B has the ground state electron configuration [Ar] $4s^2$. Identify A and B from their electron configurations, then identify the compound formed when these two elements react. Is this compound diamagnetic or paramagnetic? Write an equation for this reaction.

Text Section: 8.4

General ChemistryNow Screen: 8.7

OWL Question: 8.4

20. (Chapter 8) Electron Configurations and Periodic Trends
  (a) What element has the ground state electron configuration [Ar] $3d^6 4s^2$?
  (b) What element has a 2+ ion with the ground state configuration [Ar] $3d^5$? Is the ion paramagnetic or diamagnetic?
  (c) How many unpaired electrons are there in a ground state Ni^{2+} ion?
  (d) The ground state configuration for an element is given here.

[Ar] $\uparrow\downarrow$ $\uparrow\downarrow$ $\uparrow\downarrow$ $\uparrow\downarrow$ $\uparrow\downarrow$ $\uparrow\downarrow$ $\uparrow$ $\uparrow$ $\uparrow$
         3d orbitals   4s orbital   4p orbitals

What is the identity of the element? Is an atom of the element paramagnetic or diamagnetic? How many unpaired electrons does a 3− ion of this element have?
  (e) What element has the following ground state electron configuration?

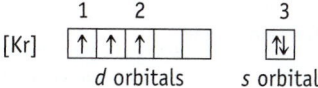

          1   2   3
[Kr] $\uparrow$ $\uparrow$ $\uparrow$ $\uparrow\downarrow$
        d orbitals   s orbital

Is the element paramagnetic or diamagnetic?

Write a complete set of quantum numbers for electrons 1–3.

Electron	n	ℓ	m_ℓ	m_s
1	_____	_____	_____	_____
2	_____	_____	_____	_____
3	_____	_____	_____	_____

(f) Answer the questions below about the elements A and B, which have the ground state electron configurations shown.

A = [Kr]$5s^2$ B = [Kr]$4d^{10}5s^2 5p^5$

Is element A a metal, nonmetal, or metalloid?

Which element has the greater ionization energy?

Which element has larger atoms?

Which is more likely to form a cation?

What is a likely formula for a compound formed between A and B?

Text Sections: 8.4–8.7

General ChemistryNow Screens: 8.7–8.14

OWL Questions: 8.4–8.12

21. (Chapter 8) Periodic Trends

Part 1: General Periodic Trends
  (a) Of the elements S, Se, and Cl, which has the largest atomic radius?
  (b) Which has the larger radius, Br or Br^-?
  (c) Which should have the most negative electron affinity: N, O, S, or Cl?
  (d) Which has the largest first ionization energy: B, Al, or C?
  (e) Which of the following has the largest radius: O^{2-}, N^{3-}, or F^-?

Part 2: Consider the elements Al, C, Ca, Mg, K:

Of these elements, _____ has the lowest ionization energy whereas _____ has the most negative electron affinity. The element with the largest radius is _____, and the element with the smallest radius is_____. The element with the largest difference between the first and second ionization energies is_____.

Text Section: 8.6

General ChemistryNow Screens: 8.9–8.14

OWL Questions: 8.8–8.12

22. (Chapter 8) Electron configurations.
  (a) Using the *spdf* notation [e.g., $1s^2 2s^2$], write ground state electron configurations for each of the following: arsenic, manganese, and plutonium (use the noble gas notation for this case)
  (b) Using the orbital box notation, write electron configurations for the following atoms or ions: tin(II) ion, cobalt(III) ion, and oxide ion. Use the noble gas notation in all cases.

Text Sections: 8.4–8.5

General ChemistryNow Screens: 8.7, 8.8

OWL Questions: 8.4–8.6

23. (Chapter 8) Ionization Energies

   (a) Generally ionization energies increase on proceeding across a period but this is not true for magnesium (738 kJ/mol) and aluminum (578 kJ/mol). Explain this observation.

   (b) Explain why the ionization energy of phosphorus (1012 kJ/mol) is greater than that of sulfur (1000 kJ/mol) when the general trend in ionization energies in a period would predict the opposite.

Text Section: 8.6

General ChemistryNow Screen: 8.11

OWL Question: 8.9

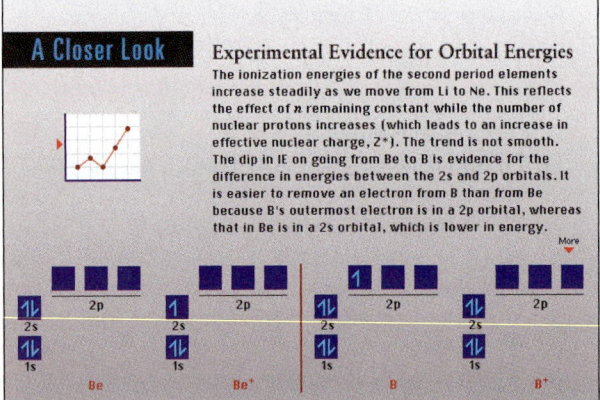

General ChemistryNow Screen 8.11 This "Closer Look" box further discusses trends in ionization energy, electron configurations, and effective nuclear charge.

24. (Chapters 9 and 10) Sketch the electron dot structure for each of the following, give the hybridization of the N atom, and arrange them in order of increasing N — O bond length: H_2NOH, NO_3^-, NO_2^-, NO^+.

Text Sections: 9.4, 9.5, 10.2

General ChemistryNow Screens: 9.7–9.9, 10.4

OWL Questions: 9.1, 9.2, 10.2

25. (Chapters 6, 9, and 10) Brown NO_2 gas is a product of the reaction of copper with nitric acid. (See Figure 5.14, page 203, and Active Figure 5.14 on General ChemistryNow.) Some NO_2 molecules form N_2O_4 where two NO_2 molecules are bonded through an N — N bond.

   (a) Draw the Lewis electron dot structure of N_2O_4, specify the formal charges on each atom and the hybridization of the N atoms, indicate the N — O bond order, and give the bond angles.

   (b) Is the reaction of NO_2 to form N_2O_4 endothermic or exothermic? (You can answer this by considering bond energies and could confirm it using enthalpies of formation.)

Text Sections: 6.8, 9.4, 9.5, 10.2

General ChemistryNow Screens: 6.16, 9.8, 9.10, 10.4

OWL Questions: 6.6, 9.1, 9.2, 9.10, 10.2

9.8 Drawing Lewis Structures

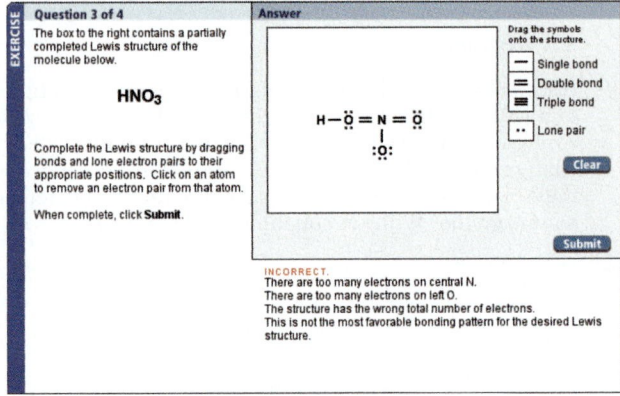

General ChemistryNow Screen 9.8 Exercise on building Lewis structures.

26. (Chapters 9 and 10) Sulfur can react with the sulfite ion to give the thiosulfate ion, $S_2O_3^{2-}$.

   (a) Draw the Lewis electron dot structure for the sulfite ion, specify the formal charge on each atom and the S atom hybridization, and give the S — O bond order.

   (b) Knowing that the thiosulfate ion is analogous with the sulfate ion, where a S atom has replaced an O atom, draw the electron dot structure for thiosulfate ion.

Text Sections: 9.4, 9.5, 10.2

General ChemistryNow Screens: 9.8–9.10, 10.4

OWL Questions: 9.1, 9.2, 10.2

27. (Chapters 9 and 10) Chemistry of chlorine trifluoride.

   (a) Sketch the Lewis electron dot structure for ClF_3 and indicate the hybridization of the central Cl atom.

   (b) There are several geometries possible for ClF_3. Which is the most reasonable and why?

9.13 Ideal Repulsion Shapes

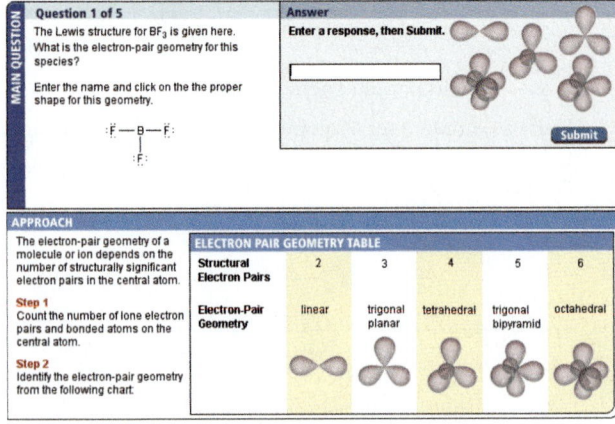

General ChemistryNow Screen 9.13 A tutorial on VSEPR

(c) Liquid ClF_3 is weakly conducting, a behavior attributed to the presence of ClF_2^+ and ClF_4^- ions. Sketch the structures of these two ions and compare their geometries.

Text Sections: 9.4–9.6, 10.2

General ChemistryNow Screens: 9.8–9.10, 10.4

OWL Questions: 9.1–9.3, 10.2

28. (Chapter 9) Lattice energy
  (a) The lattice energy for Na_2O is -2488 kJ/mol, whereas the lattice energy for MgO is -3800 kJ/mol. Explain why that for MgO is so much more negative than for Na_2O.
  (b) The lattice energies for the Group 2A oxides are:

Compound	Lattice Energy (kJ/mol)
MgO	-3800
CaO	-3419
SrO	-3222
BaO	-3034

Rationalize the trend observed in these values.

Text Section: 9.3

General ChemistryNow Screen: 9.4

OWL Questions: —

29. (Chapters 9–11) The Lewis structure of asparagine, one of the naturally occurring amino acids, is drawn below. The questions that follow are about this compound. The carbons are labeled with subscripts for the purpose of this question. The first five questions refer to the letters accompanying the arrows in the figure.

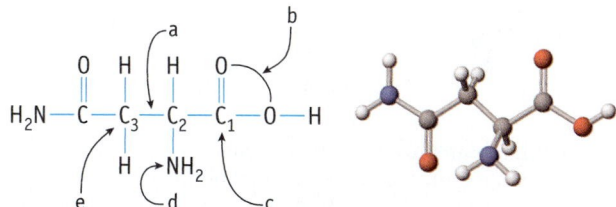

(a) What orbitals overlap to form bond (a)?
(b) What is the O—C—O bond angle labeled (b)?
(c) What hybridization is assigned to C_1?
(d) What is the molecular geometry around this N?
(e) What is the H—C_3—H bond angle?
(f) Identify any chiral carbon atoms in this compound.
(g) How many lone pairs are missing in this drawing? Where should they be located?
(h) Name the three functional groups in the molecule.

Text Sections: 9.7, 10.2, 11.3, 11.4

General ChemistryNow Screens: 9.12, 9.13, 10.4, 11.5

OWL Questions: 9.4, 10.2, 11.7

30. (Chapters 10 and 11) Organic chemistry and hybridization
  (a) Naphthalene is an aromatic hydrocarbon. What is the hybridization of the C atoms in naphthalene? How many resonance structures does naphthalene have?

Naphthalene, $C_{10}H_8$

  (b) What is the hybridization of the C atoms in diamond and in graphite? (See Figure 2.15, page 84, for the structures of diamond and graphite.)

Text Sections: 10.2, 111.2

General ChemistryNow Screens: 10.4, 11.3

OWL Questions: 10.2, 11.4

31. (Chapter 10) The ion Si_2^- was reported in a laboratory experiment in 1996.
  (a) Using molecular orbital theory, predict the bond order for the ion.
  (b) What is the predicted bond order?
  (c) Is the ion paramagnetic or diamagnetic?
  (d) What is the highest energy molecular orbital that contains one or more electrons?

Text Section: 10.3

General ChemistryNow Screens: 10.9–10.11

OWL Question: 10.5

32. (Chapters 9–11) The compound pictured below (anethole) is the substance that gives licorice its odor.

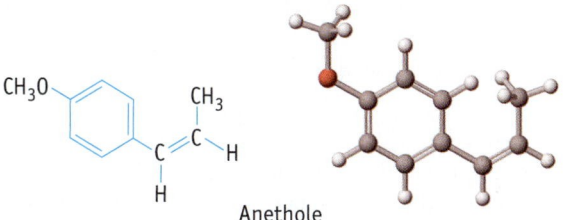

Anethole

(a) Draw Lewis structures for the following: 1) a second resonance structure; 2) any structural isomer of this compound; 3) a geometric isomer of this compound.
(b) What orbitals overlap to form a C=C double bond in this compound?

(c) Describe the bonding in the C_6H_4 ring using molecular orbital theory.

Text Sections: 9.4, 10.2, 11.2

General ChemistryNow Screens: 9.8, 10.9, 11.3

OWL Questions: 9.1, 10.2, 10.3, 10.4, 11.2

33. (Chapters 4 and 9) Acetaminophen, which is sold under the tradenames Tylenol and Excedrin, among others, has the structure shown here.

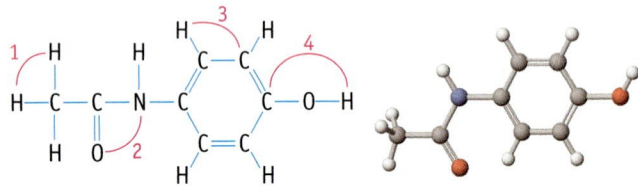

Acetaminophen

(a) Which is the most polar bond?
(b) Which is the strongest carbon-oxygen bond?
(c) Give approximate values for the indicated bond angles in the molecule.
(d) One Excedrin tablet contains 250 mg of acetaminophen. What amount (moles) of acetaminophen are you consuming in one tablet?
(e) What is the weight percent of carbon in acetaminophen?

Text Sections: 4.3, 9.7, 9.9

General ChemistryNow Screens: 4.6, 9.13, 9.18

OWL Questions: 4.3, 9.4, 9.8

34. (Chapter 9) Hydrazine, N_2H_4, is a useful commercial reducing agent

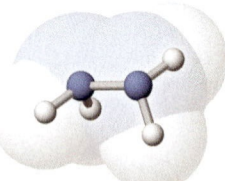

Hydrazine, N_2H_4

(a) Draw a Lewis electron dot structure for the molecule and specify the bond angles.
(b) Specify the electron pair geometry around the N atoms and their hybridization.
(c) Suppose hydrazine, N_2H_4, can be made from ammonia by the reaction

$$2\,NH_3(g) \longrightarrow H_2N-NH_2(g) + H_2(g)$$

Does the geometry around the N atom change in the course of the reaction of ammonia to produce hydrazine?

(d) Use bond energies to calculate the enthalpy change for this reaction. Is it predicted to be endo- or exothermic?

Text Sections: 9.4, 9.7, 9.10

General ChemistryNow Screens: 9.7, 9.14, 9.20

OWL Questions: 9.1, 9.4, 9.10

35. ▲ (Chapters 9, 10, and 11) Urea reacts with malonic acid to produce barbituric acid, a member of the class of compounds called phenobarbitals, which are widely prescribed as sedatives.

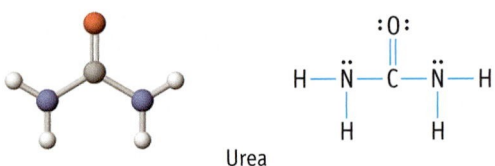

Urea

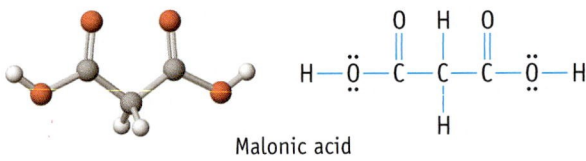

Malonic acid

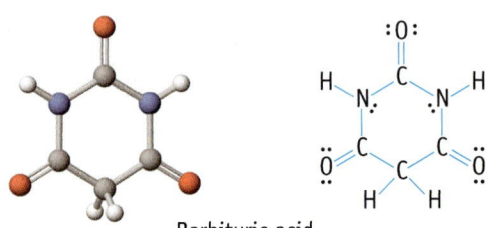

Barbituric acid

(a) What bonds are broken and what bonds are made when malonic acid and urea combine to make barbituric acid? Is the reaction predicted to be exo- or endothermic?
(b) Write a balanced equation for the reaction.
(c) Specify the bond angles in barbituric acid.
(d) Give the hybridization of the C atoms in barbituric acid.
(e) What is (are) the most polar bond(s) in barbituric acid?
(f) Is the molecule polar?

36. (Chapter 11) The alkene *cis*-2-pentene reacts with hydrogen to produce another hydrocarbon.
    (a) Draw the structures of the reactants and products.
    (b) What type of reaction is illustrated here (addition, elimination, condensation, or esterification)?
    (c) Name the product of the reaction.
    (d) Sketch a geometric isomer of the reactant.
    (e) Draw the structure and give the systematic name of each of the isomers of C_5H_{12}.

37. (Chapter 9-11) In order to convert *cis*-2-butene to *trans*-2-butene, it is necessary to heat this compound to around 450 °C. Explain, in a short phrase or sentence, why it is necessary to supply so much energy.

38. (Chapter 11) The structure drawn below is for the molecule aspartame, one of several common artificial sweeteners.

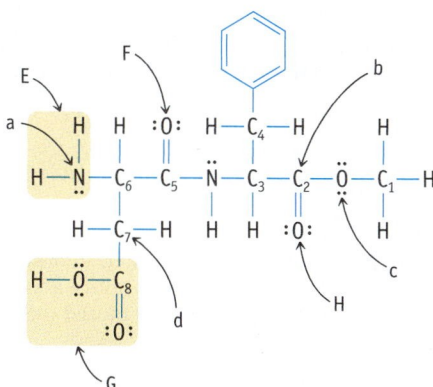

(a) What is the approximate H—N—H bond angle (a)?
(b) What is the geometry of the indicated carbon atom (C_2) (b)?
(c) What is the hybridization of oxygen (c)?
(d) What is the electron pair geometry of carbon (C_7) (d)?
(e) Name the functional group enclosed in box (E).
(f) Name the functional group that contains this carbonyl group (F).
(g) Name the functional group enclosed in this box (G).
(h) Name the functional group that contains this carbonyl group (H).

(i) Carbon atoms in this drawing (except those in the C_6H_5— group) are numbered 1–8. Which of these carbons is/are a chiral center?
(j) Hydrolysis of aspartame (reaction with water) will cleave this large molecule into three different compounds. Draw the structures of the products formed by this hydrolysis.

39. ▲ (Chapter 11) You have two unlabeled flasks containing colorless liquids. One liquid is pentene, the other cyclopentane, but you don't know which is which. To determine their identities you add bromine to each flask.
    (a) What are the empirical and molecular formulas of *cis*-2-pentene and cyclopentane?
    (b) Draw the structure of each compound.
    (c) Describe what you observe when you add bromine to these compounds and indicate how the observation will allow you to tell the identity of the liquids in the flask.
    (d) Draw the structure of the product of the reaction with bromine, if any, for the two molecules.

40. ▲ (Chapter 11) You have two beakers, one containing propanal and the other propanone. To tell which is which, you add a reagent, aqueous acidic $Na_2Cr_2O_7$.
    (a) Draw the structures of propanal and propanone.
    (b) Describe what you observe when you treat these compounds with aqueous acidic sodium dichromate and indicate how the observation will allow you to tell the identity of the liquids in flask.
    (c) Draw the structure of the product of the reaction, if any, in each case.

41. (Chapter 11) Consider the molecule illustrated below.

$$
\begin{array}{ccccccc}
& H & H & O & & CH_3 & \\
& | & | & || & & | & \\
H- & C- & C- & C- & O- & C- & H \\
& | & | & & & | & \\
& H & H & & & CH_3 &
\end{array}
$$

(a) To what class of organic compounds does this belong?
(b) Name the compound.
(c) Draw structures for and name the products that result when the compound reacts with aqueous sodium hydroxide followed by hydrochloric acid.

Text Section: 11.4

General ChemistryNow Screens: 11.5–11.7

OWL Questions: 11.6, 11.7

42. (Chapter 11) Organic reactions.

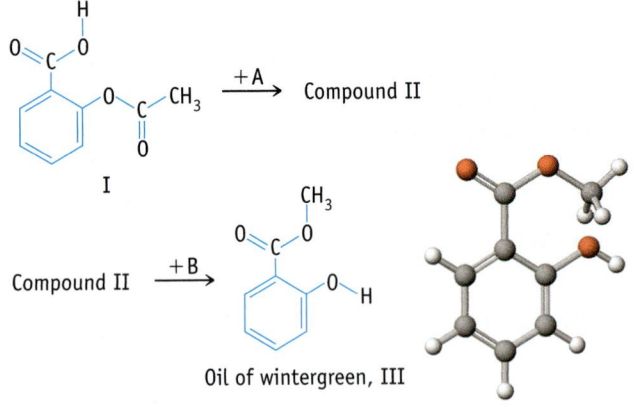

Compound 1

Reaction A
+ H_2O

Compound 2

(a) Name compounds 1 and 2.
(b) Draw structures for the reaction products, compounds 3 and 4, and name each one.

Text Section: 11.3

General ChemistryNow Screens: 11.3, 11.4, 11.6

OWL Questions: 11.3, 11.5

43. (Chapter 11) Compound I, aspirin, can be converted to compound III (oil of wintergreen) in two steps. Identify reactants A and B and draw the structure of compound II.

I

Compound II $\xrightarrow{+B}$

Oil of wintergreen, III

Text Sections: 11.3, 11.4

General ChemistryNow Screens: 11.5, 11.6

OWL Questions: 11.5–11.7

44. (Chapter 11) Give the structural formula for the product formed by polymerization of $H_2NCH_2CH_2CH_2CH_2CH_2CO_2H$. Is the product an addition polymer or a condensation polymer? Is it a polyester or a polyamide?

Text Section: 11.5

General ChemistryNow Screens: 11.9, 11.10

OWL Question: 11.8

45. ▲ (Chapter 11) The reaction of 1-octene, $C_6H_{13}CH=CH_2$, and water could, potentially, form two different isomeric products.
(a) Draw their structures and name the two possible isomeric products.
(b) In fact, only one product is formed in the water reaction. To identify the product, you first add a solution of $Na_2Cr_2O_7$ to this product and observe that a reaction occurs. Next you add aqueous NaOH to the product obtained from the dichromate reaction. There is no evidence of a reaction with NaOH (no heat was evolved and the organic product didn't dissolve in the aqueous solution.) Based on these observations which isomer was formed?

Text Section: 11.2

General ChemistryNow Screens: 11.4, 11.6

OWL Questions: 11.2, 11.3, 11.5

46. (Chapter 11) Draw structures and name the following:
(a) A chiral compound with the formula C_7H_{16}, in which the longest carbon chain contains 5 carbon atoms.
(b) An ester that is a structural isomer of butanoic acid.
(c) An aldehyde and a ketone with the formula C_4H_8O.
(d) An alkene, C_7H_{14}, in which the longest carbon chain is 5 carbons, and there is an ethyl group as a substituent.

Text Sections: 11.2–11.4

General ChemistryNow Screens: 11.3, 11.5

OWL Questions: 11.2, 11.6, 11.7

47. ▲ (Chapter 11) In the lab, you have inadvertently spilled some 1.0 M NaOH (aq) on the sleeve of your polyester shirt. About a half an hour later, you notice a hole where the spill was. Explain why the NaOH caused this to happen.

Text Section: 11.4

General ChemistryNow Screens: 11.7, 11.10

OWL Questions: 11.6, 11.8

48. (Chapter 11) Analysis of an unknown ester has determined that its formula is $C_4H_8O_2$. Hydrolysis of the ester, under acidic conditions, yields methanol. Draw the structure and give the name of the second product of hydrolysis and write a chemical equation for the hydrolysis reaction.

Text Section: 11.4

General ChemistryNow Screen: 11.5

OWL Question: 11.6

49. (Chapter 11) When an acid reacts with an amine, an amide linkage is formed. (See Example 11.8.) Amino acids are characterized by having both a carboxylic acid and an amine functional group, so if an amino acid reacts with another amino acid of the same or different kind, an amide link is formed (and the product is called a peptide). (If many amino acids react a polymer called a protein is formed.)

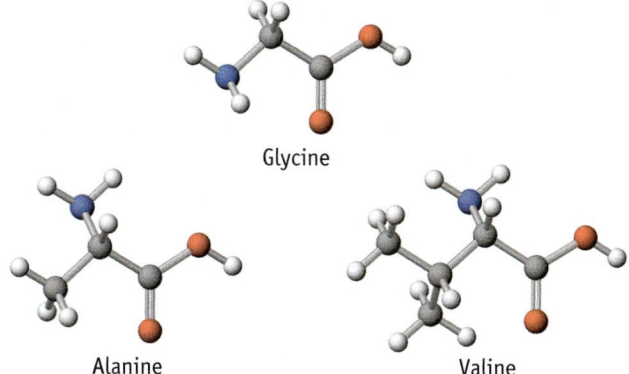

Glycine

Alanine   Valine

(a) Draw structures for the dipeptides formed by reacting glycine and alanine.

(b) How many tripeptides are possible when glycine, alanine, and valine react?

Text Section: 11.5

General ChemistryNow Screen: 11.8

OWL Question: 11.9

50. Fireworks and neon signs.

(a) As described on pages 294–295 fireworks commonly use salts of sodium, strontium, barium, and copper, among other metals. Neon signs are filled with gases such as neon or a mixture of argon with minute particles of mercury. Based on the discussion of atomic spectra in Chapter 7, what do fireworks and neon signs have in common? How do they generate light?

(b) Chlorate and perchlorate salts are common ingredients in fireworks. After drawing the Lewis structures for these ions, describe their geometry.

(c) What are the formal charges on the atoms in the perchlorate ion?

Answers to Assessing Key Points Questions

1. See Figure 7.3: radio < 520 nm < blue light < x-rays

2. Energy UVB > energy UVA

3. n describes orbital size and energy and ℓ describes orbital shape. See page 317.

4. For $n = 4$ and $\ell = 2$, m_ℓ could range from -2 to $+2$ including 0.

5. (a) Rb and (b) Se. See Table 8.3.

6. Nitrogen. See Active Figure 8.13.

7. (d) Too many electron pairs around the central N atom. The other structures are resonance structures of N_2O.

8. (a) Linear. (b) The N—O bond is polar, and the molecule is polar. (c) The central N atom is sp hybridized and has a +1 formal charge.

9. The number of MOs formed is always equal to the number of combining atomic orbitals. See page 458.

10. (a) alkane. (b) aldehyde (c) alcohol (d) ketone

11. (d) is 2-propanone, commonly called acetone. An isomer would be propanal, an aldehyde.

$$
\begin{array}{ccc}
\text{H} & \text{H} & \text{O} \\
| & | & || \\
\text{H}-\text{C}-&\text{C}-&\text{CH} \\
| & | & \\
\text{H} & \text{H} &
\end{array}
$$

12. Oxidizing (c) (ethanol) first produces (b) in question 10 (the aldehyde ethanal) and then the acid acetic acid, CH_3CO_2H.

13. Only (a), an alkene, can be used to make a polymer, in this case polypropylene. See page 514.

Solutions to Comprehensive Questions

14. (a) Convert the wavelength in nm to meters (1 nm = 1×10^{-9} m) then use the relation $\lambda \cdot \nu$ = velocity of light.

$$\text{Frequency} = \frac{\text{Velocity of light}}{\text{Wavelength}} = \frac{2.9979 \times 10^8 \text{ m} \cdot \text{s}^{-1}}{4.20 \times 10^{-7} \text{ m}}$$

$$= 7.14 \times 10^{14} \text{ s}^{-1}$$

(b) Use Planck's equation to calculate the energy per photon, then find the number of photons.

$$E \text{ per photon} = h\nu = (6.626 \times 10^{-34} \text{ J} \cdot \text{s})(7.14 \times 10^{14} \text{ s}^{-1})$$

$$E = 4.73 \times 10^{-19} \text{ J/photon}$$

Number of photons

$$= (2.50 \times 10^{-14} \text{ J})(1 \text{ photon}/4.73 \times 10^{-19} \text{ J})$$

$$= 5.29 \times 10^4 \text{ photons}$$

15. (a) One of the $3p$ orbitals

(b) There are 5 subshells and $n^2 = 25$ orbitals (1 s, 3 p, 5 d, 7 f, 9 g). Number of electrons = $2 \times 25 = 50$.

(c) $3f$ orbitals are not possible as this means ℓ would be 3, and n and ℓ cannot be the same.

(d) (i) When $n = 3$ there can be 9 orbitals (1 s, 3 p, 5 d). (ii) There are no orbitals for $n = 3$ and $\ell = 3$ (as explained in c above). (iii) This refers to one of the $2p$ orbitals.

(e) Barium has the configuration $[Xe]6s^2$, so only two electrons can be removed easily. A 3+ ion is not feasible. The S^{3-} ion is not likely as the S atom would need to acquire three electrons, whereas only two additional electrons can be accommodated in its $3s$ and $3p$ valence orbitals. (S has the ground state configuration $[Ne]3s^23p^4$.)

(f) The $n = 3$ shell is completed with copper.

16. (a) Use the relation $c = \lambda \cdot \nu$ to calculate wavelength from frequency.

$$\lambda = c/\nu = (2.998 \times 10^8 \text{ m} \cdot \text{s}^{-1})/(2.45 \times 10^9 \text{ s}^{-1}) = 0.122 \text{ m}$$

(b) Number of wavelengths = 0.49 m/0.122 m = 4.0

(c) We first need to know the energy per mole of photons and the heat needed to warm the water.

$$E \text{ per photon} = h\nu = (6.626 \times 10^{-34} \text{ J} \cdot \text{s})(2.45 \times 10^9 \text{ s}^{-1})$$
$$= 1.62 \times 10^{-24} \text{ J per photon}$$

E per mol of photons =
$$= (1.62 \times 10^{-24} \text{ J/photon})(6.022 \times 10^{23} \text{ photons/mol})$$
$$= 0.978 \text{ J/mol of photons}$$

$$\text{Heat required} = q = (254 \text{ g})(4.184 \text{ J/g} \cdot \text{K})(363 \text{ K} - 285 \text{ K})$$
$$= 8.29 \times 10^4 \text{ J}$$

Now we can calculate the amount (mol) of photons required.

$$\text{Amount of photons} = (8.29 \times 10^4 \text{ J})(1 \text{ mol photons}/0.978 \text{ J})$$
$$= 8.48 \times 10^4 \text{ mol photons}$$

17. (a) Use Hess's law, Equation 6.6 (page 267).

$$\Delta H^\circ_{\text{rxn}} = \Delta H^\circ_f [\text{glucose}] + 6 \Delta H^\circ_f [\text{O}_2]$$
$$- \{6 \Delta H^\circ_f [\text{CO}_2(\text{g})] + 6 \Delta H^\circ_f [\text{H}_2\text{O}(\text{g})]\}$$

$$\Delta H^\circ_{\text{rxn}} = 1 \text{ mol} (-1273.3 \text{ kJ/mol}) + 6 \text{ mol} [0]$$
$$- [6 \text{ mol} (-393.5 \text{ kJ/mol})$$
$$+ 6 (-241.8 \text{ kJ/mol})]$$
$$= +2538.5 \text{ kJ per mol of glucose}$$

(b) Energy per molecule =
$$= 2538.5 \text{ kJ/mol}(1 \text{ mol}/6.022 \times 10^{23} \text{ molecules})$$
$$= 4.215 \times 10^{-21} \text{ kJ/molecule or } 4.215 \times 10^{-18} \text{ J/molecule}$$

(c) Using the approach in questions 14 and 16, we find light with a wavelength of 650. nm has an energy of 3.06×10^{-19} J/photon.

(d) About 14 photons with $\lambda = 650.$ nm are required to make one molecule of glucose.

$$(4.215 \times 10^{-18} \text{ J/molecule})(1 \text{ photon}/3.06 \times 10^{-19} \text{ J})$$
$$= 13.8 \text{ photons}$$

18. (a) Emitting light in the UV region would require a transition from a higher level to $n = 1$. Visible light is emitted by transitions to $n = 2$ from higher energy levels. Infrared radiation arises from transitions to $n = 3$ from levels of higher energy (>3). See Figure 7.12.

(b) -2.179×10^{-18} J/atom (See Example 7.2)

(c) E (for $n = 4$) $= -Rhc/n^2 = -(2.179 \times 10^{-18} \text{ J/atom})/4^2$
$$E = -1.362 \times 10^{-19} \text{ J/atom}$$

(d) $\Delta E = E_2 - E_1 = (-5.45 \times 10^{-19} \text{ J}) - (-2.18 \times 10^{-18} \text{ J})$
$$= -1.64 \times 10^{-18} \text{ J}$$
or $\Delta E = -(3/4)Rhc$

(e) The energy to ionize the H atom is $+2.18 \times 10^{-18}$ J/atom. Therefore, multiply by Avogadro's number to find the energy per mole ($= 1312$ kJ/mol).

(f) To ionize an H atom requires 2.18×10^{-18} J/atom. Use Planck's equation to calculate the frequency of radiation equivalent to this energy.

$$2.18 \times 10^{-18} \text{ J} = (6.626 \times 10^{-34} \text{ J} \cdot \text{s})(\nu)$$
$$\nu = 3.29 \times 10^{15} \text{ s}^{-1}$$
$$\lambda = c/\nu = (2.998 \times 10^8 \text{ m} \cdot \text{s}^{-1})/(3.29 \times 10^{15} \text{ s}^{-1})$$
$$= 9.11 \times 10^{-8} \text{ m (or 91.1 nm)}$$

Radiation of this wavelength is found in the high energy UV region. See Active Figure 7.3.

19. Atom A is Cl and atom B is Ca. The two elements react to give $CaCl_2$, with Ca^{2+} and Cl^- ions. Neither of these ions has an unpaired electron and so both are diamagnetic.

$$\text{Ca(s)} + \text{Cl}_2(\text{g}) \longrightarrow \text{CaCl}_2(\text{s})$$

20. (a) Iron, Fe

(b) Manganese(II), Mn^{2+}. This ion has 5 unpaired electrons and so is paramagnetic.

(c) Ni^{2+} has the electron configuration $[Ar]3d^8$ and has two unpaired electrons.

(d) The element is arsenic, As. It is paramagnetic but the 3− ion has no unpaired electrons $\{[Ar]3d^{10}4s^24p^6\}$.

(e) Nb, niobium. The element is paramagnetic.

Electron	n	ℓ	m_ℓ	m_s
1	4	2	−2	+1/2
2	4	2	0	+1/2
3	5	0	0	−1/2

Comment: The m_ℓ values for electrons 1 and 2 can have any value from −2 to +2 including 0. However, they cannot both have the same value. The m_ℓ value for electron 3, however, is fixed by the fact that $\ell = 0$. Also, m_s can be either ±1/2 for electrons 1 and 2, but they must have the same value. Finally, whatever the value of m_s for electrons 1 and 2, that for electron 3 must be the opposite because its spin is in the opposite direction.

(f) A is Sr, a metal, and B is I, a nonmetal. Iodine (B) has a higher ionization energy, and its atoms are smaller than those of Sr. (See App. F, page A-19 and Figure 8.11.) Strontium, Sr, is a metal and so is likely to form cations (here Sr^{2+}), and so A and B react to form SrI_2.

21. **Part 1**

(a) Se (Figure 8.11)

(b) Anions are always larger than the atoms from which they come. See Figure 8.15

(c) Cl (App. F, page A-19)

(d) C (Figure 8.13)

(e) These ions are isoelectronic, that is, they have the same number of electrons (here 10 electrons). For such a series, the ion with the largest negative charge and smallest nuclear charge will be the largest ion, here N^{3-}.

Part 2: K has the lowest ionization energy and C the most negative electron affinity. K has the largest radius and C has the smallest radius. K has the largest difference in 1st and 2nd ionization energy (because the 2nd electron comes from an inner shell). See page 358.

22. (a) See Table 8.3 on page 344.

(b) Tin(II), Sn^{2+}

[Kr] ⇅ ⇅ ⇅ ⇅ ⇅   ⇅   ☐ ☐ ☐
     4*d* orbitals   5*s* orbital 5*p* orbitals

Cobalt(III), Co^{3+}

[Ar] ⇅ ↑ ↑ ↑ ↑   ☐
     3*d* orbitals   4*s* orbital

Oxide ion, O^{2-}. Its ground state electron configuration is isoelectronic with neon.

[He] ⇅   ⇅ ⇅ ⇅
    2*s* orbital 2*p* orbitals

23. (a) For Mg one electron of a pair is removed from a 3*s* orbital,

Mg, $1s^2 2s^2 2p^6 3s^2 \longrightarrow Mg^+, 1s^2 2s^2 2p^6 3s^1$

whereas an Al atom loses an electron from a 3*p* orbital of slightly *higher* energy.

Al, $1s^2 2s^2 2p^6 3s^2 3p^1 \longrightarrow Al^+, 1s^2 2s^2 2p^6 3s^2$

Comment: This explanation is similar to that for the difference in the ionization energies of Be and B as discussed in the "Closer Look" box on the ChemistryNow Screen 8.11.

(b) This illustrates the effect on ionization energy of electron-electron repulsion in an electron pair. For S we have

[Ne] ⇅   ⇅ ↑ ↑
    3*s* orbital 3*p* orbitals

The electron lost is one of the pair in the 3*p* orbitals. Removal of an electron is "assisted" by the loss of electron-electron repulsion when the electron is removed. Such pairing is not present in P.

24. Nitrogen-oxygen bond lengths.

Increasing N—O bond length = decreasing bond order →

Molecule	N-O Bond order	N Atom Hybridization
NO^+	3	sp
NO_2^-	1.5	sp^2
NO_3^-	1.3	sp^2
H_2NOH	1	sp^3

25. (a) Dot structure of N_2O_4

The N hybridization is sp^2 and the N—O bond order is 1.5.

(b) $\Delta H^\circ_{rxn} = \Delta H^\circ_f [N_2O_4(g)] - 2 \Delta H^\circ_f [NO_2(g)]$
     $= 1 \text{ mol } (9.08 \text{ kJ/mol}) - 2 \text{ mol } (33.1 \text{ kJ/mol})$
     $= -57.1 \text{ kJ/mol}$

The reaction involves only the formation of an N—N single bond, and, like the formation of any chemical bond, it is an exothermic process. N—N single bonds are often notably weak, and this is reflected in this value calculated from heats of formation.

26. (a) Sulfite ion, SO_3^{2-}

The formal charge on each O atom is -1, whereas it is $+1$ on the central S. The SO bond order is 1, and S atom hybridization is sp^3.

(b) Thiosulfate ion, $S_2O_3^{2-}$

Comment: You can think of this structure as an S atom attached to the central S atom lone pair in SO_3^{2-}

27. (a) and (b) ClF_3 has an electron pair geometry of trigonal bipyramidal with a sp^3d hybrid Cl atom. The most reasonable structure is T-shaped. The structure adopted is the one that minimizes 90° interactions between lone pairs and between lone pairs and atoms. In the T-shape there are four lone pair-atom interactions at 90°. In an alternate structure, trigonal planar, there would be six such interactions.

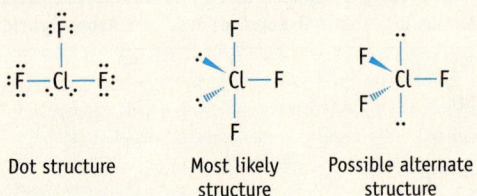

Dot structure   Most likely structure   Possible alternate structure

(c) The ClF_2^+ ion has a tetrahedral electron pair geometry and a bent molecular geometry. The ClF_4^- ion has an octahedral electron pair geometry and a square planar molecular geometry.

28. (a) Lattice energy depends in part on the attraction between anion and cation (the ion pair energy, page 378). The larger the charge, and the smaller the distance of ion separation, the stronger the attraction. In Na_2O the Na cation has a 1+ charge, whereas Mg in MgO has a 2+ charge.

(b) Here the Group 2A cation is becoming larger on descending the group. Ion pair energy declines and the lattice energy becomes less negative.

29. (a) sp^3 orbitals on each C atom overlap to form the C—C bond.

(b) 120°

(c) C_1 carbon is sp^2 hybridized.

(d) Trigonal pyramidal. The N atom is surrounded by two H atoms, a bond to C_2, and a lone pair.

(e) 109°, a tetrahedral angle

(f) C_2 is a chiral carbon. It is surrounded by four *different* groups.

(g) There are eight lone pairs missing. Each O atom needs two and each N atom needs one.

(h) At the left end of the molecule is an amide group, at the right end there is a carboxylic acid group, and the NH_2 on C_2 is an amine.

30. (a) Naphthalene has sp^2 hybridized C atoms and three resonance structures.

(b) The C atoms of graphite are sp^2 hybridized (to account for the planar trigonal C atoms) and the tetrahedral C atoms of diamond are sp^3 hybridized.

31. (a) There are nine valence electrons (four for each Si plus one for the charge). Assuming that the MOs formed by

the valence orbitals ($3s$ and $3p$) are identical to those formed by $2s$ and $2p$ in Figure 10.22, the configuration would be [core] $(\sigma_{3s})^2(\sigma_{3s}^*)^2(\pi_{3p})^4(\sigma_{3p})^1$ orbital.

(b) The bond order is 2.5.

(c) Paramagnetic

(d) σ_{3p}

32. (a) Anethole alternate resonance structure and isomers

Resonance structure   Structural isomer

Geometric isomer

(b) See Figure 10.10. C=C bonds consist of one σ bond formed by direct overlap of hybridized sp^2 orbitals and one π bond formed by sideways overlap of unhybridized $2p$ orbitals.

(c) Six p orbitals, one from each of six C atoms, contribute to form six molecular orbitals (three bonding and three antibonding). Six electrons, one from each of the C atoms, fill the bonding MOs to give three π MOs. See Figure 10.15.

33. (a) Most polar bond, O—H.

(b) C=O bond is strongest bond.

(c) 1 = 109.5°; 2 = 120°; 3 = 120°; 4 = 109.5°

(d) Molar mass of $C_8H_9NO_2$ is 151.165 g/mol. 250 mg is equivalent to 0.0017 mol.

(e) Weight percent C = 63.57%

34. (a) and (b) Electron dot structure of hydrazine.

109° :N—N: ← sp^3 hybrid N with tetrahedral electron pair geometry

(c) N atom geometry does not change. It has a tetrahedral electron pair geometry (and trigonal pyramidal molecular geometry) in ammonia and hydrazine.

(d) Bonds broken on reaction = 2 NH

Energy = 2 mol (391 kJ/mol) = 782 kJ

Bonds made = 1 NN + 1 HH

Energy = 1 mol (163 kJ/mol) + 1 mol (436 kJ/mol)
    = 599 kJ

Net energy = +782 kJ − 599 kJ = +183 kJ

The reaction is endothermic.

35. (a) Two N—H bonds in urea and two C—O bonds and two O—H bonds in malonic acid are broken. Two N—C bonds are made in barbituric acid. Two molecules of water form, involving the formation of four O—H bonds.

Energy required to break bonds =
= 2 NH bonds + 2 CO bonds + 2 OH bonds
= 2 mol (391 kJ/mol) + 2 mol (358 kJ/mol)
    + 2 mol (463 kJ/mol)
= 2424 kJ

Energy from making bonds =
= 2 NC bonds + 4 OH bonds
= 2 mol (305 kJ/mol) + 4 mol (463 kJ/mol) = 2462 kJ

Net energy for reaction = +2424 kJ − 2462 kJ = −38 kJ

The reaction is predicted to be exothermic.

(b) $(NH_2)_2CO + CH_2(CO_2H)_2 \longrightarrow C_4H_4N_2O_3 + 2\,H_2O$

(c) All angles are 120° except for the C—C—C, C—C—H, and H—C—H. (See page 509 for the C—N—C and C—N—H angles.)

(d) The CH_2 C atom is sp^3 hybridized, and the C atoms bonded to O are sp^2 hybridized.

(e) The C—O bonds are the most polar (with a difference in electronegativity of 1.0).

(f) The molecule is polar.

36. (a) Hydrogenation reaction

(b) Addition reaction (of H_2).

(c) pentane

(d) See Example 11.3 on page 488.

cis-2-pentene   trans-2-pentene

(e) See the pentane isomers in the margin of page 482.

37. Energy must be supplied to break the C≡C π bond to allow free rotation of one end of the molecule relative to the other. See page 455.

38. (a) 109°
    (b) trigonal planar
    (c) sp^3
    (d) tetrahedral
    (e) Amine
    (f) Amide
    (g) Carboxylic acid
    (h) Ester
    (i) C_3 and C_6 are chiral centers

(j) Reaction with water breaks the ester and amide links to give three products.

39. (a) Both have the same empirical (CH_2) and molecular formulas (C_5H_{10}).

(b) C_5H_{10} structures

cis-2-pentene   cyclopentane

(c) The alkene, 2-pentene will add bromine across the double bond to give 2,3-dibromopentane, and the color of Br_2 will disappear rapidly. (See Figure 11.7.) Cyclopentane will not react with bromine.

(d) Adding Br_2 to an alkene.

40. (a) and (c) Structures of propanone and propanal and reaction products.

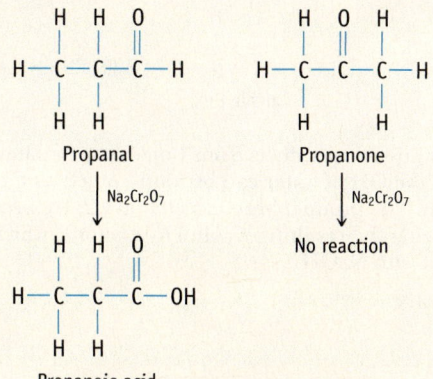

Propanal   Propanone

$Na_2Cr_2O_7$ $Na_2Cr_2O_7$

Propanoic acid   No reaction

Of the isomers, only propanal reacts with the oxidizing agent in acid. The yellow color of $Na_2Cr_2O_7$ fades and is replaced by the green color of Cr^{3+}.

41. (a) Ester

(b) 2-propyl propanoate

(c) Products of reaction with NaOH followed by HCl.

Propanoic acid   2-Propanol

42. (a) 1 = 1-Propanol and 2 = propene

(b) Compounds 3 and 4.

Propanoic acid   2-Bromopropane

Comment: When propene reacts with HBr, the HBr can add in two ways. The H atom could attach to the CH_2 carbon or to the CH carbon. The former is generally observed and is called a Markovnikov addition. See General ChemistryNow, Screen 11.4.

43. The group *ortho* to the carboxylic acid group is an ester. Therefore, first react aspirin with NaOH (=A) (followed by acid) to give compound II, 2-hydroxybenzoic acid (but better known as salicylic acid). Second, react II with methanol (=B) to give the ester (oil of wintergreen).

2-hydroxybenzoic acid
or
salicylic acid

44. The compound forms a polyamide, a condensation polymer. See Example 11.8 on page 519.

amide link

45. The two possible products from 1-octene with water arise because the OH of water can be found on C1 or C2 of the C8 chain. The product here has OH on C2. We prove this by oxidizing the resulting alcohol to a ketone, which does not react with NaOH

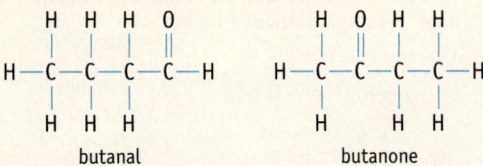

2-octanol   1-octanol

Ketone   Carboxylic acid

If the OH group was on C1, the product of oxidation would be a carboxylic acid, which would react with NaOH.

46. (a) Chiral carbon is marked with *.

(b) Isomer of butanoic acid.

methylpropanoate

(c) Aldehyde and ketone

butanal   butanone

(d) Isomer of C_7H_{14} (One of several.)

3-ethyl-1-pentene

47. An ester can react with NaOH to produce an alcohol and a carboxylic acid, so reaction with NaOH can destroy the polyester in the material. See page 508.

48. Ester hydrolysis. The reaction is, in effect, the addition of water to the reactant, with the ester forming a carboxylic acid and an alcohol. It is done under basic conditions followed by the addition of an acid such as HCl.

Methyl propanoate

Propanoic acid   Methanol

49. (a) Four peptides are possible on reacting a mixture of two amino acids.

glycine   alanine

alanine   glycine

glycine   glycine

alanine   alanine

(b) 3^3 or 27 peptides are possible. One is illustrated here.

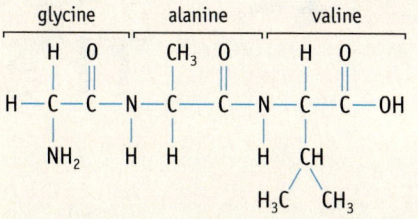

glycine   alanine   valine

50. (a) The colors of fireworks and neon signs arise in the same way: from excited atoms, whether these are strontium atoms or neon atoms. As the atoms are excited by the input of energy, electrons move to higher quantum levels. Light is emitted as the electrons move back to lower quantum levels.

(b) Perchlorate and chlorate both have a tetrahedral electron pair geometry. The former has a tetrahedral shape, whereas the latter is trigonal pyramidal.

Perchlorate ion   Chlorate ion

(c) The formal charge on each O atom in ClO_4^- is -1, whereas it is $+3$ for Cl.

12—Gases & Their Properties

Modern hot air balloons. A hot-air balloon rises because the heated air in the balloon has a lower density than the surrounding atmosphere.

Royalty-Free/Corbis

Up, Up, and Away!

The invention of the balloon appears...to be a discovery of great importance.

Benjamin Franklin, about 1784

The first flight of a manned hot-air balloon took place in Paris, France, on November 21, 1783. The balloon was designed by Joseph and Étienne Montgolfier and piloted by Pilatre de Rozier and the Marquis d'Arlandes. The two pilots traveled 12 kilometers in less than half an hour at about 900 meters in altitude.

How does a hot-air balloon fly? As you will learn in this chapter, it can ascend because heating affects the density of the air in the balloon. When the air inside a hot-air balloon is heated (usually with a propane heater in a modern balloon), the gas expands. Initially, this expansion is used to inflate the balloon. At a certain point, however, the balloon no longer increases in volume. Air is then forced from the inside of the balloon as the air continues to expand on heating. As a result, less air remains inside. The smaller mass of air in the same volume means that the gas inside the balloon has a lower density than the surrounding atmosphere, so the balloon ascends.

Jacques Charles and M. S. Roberts ascended over Paris on December 1, 1783, in a hydrogen-filled balloon.

Smithsonian National Air and Space Museum

Chapter Goals

See Chapter Goals Revisited (page 578). Test your knowledge of these goals by taking the exam-prep quiz on the General ChemistryNow CD-ROM or website.

- Understand the basis of the gas laws and know how to use those laws (Boyle's law, Charles's law, Avogadro's hypothesis, Dalton's law).
- Use the ideal gas law.
- Apply the gas laws to stoichiometric calculations.
- Understand kinetic-molecular theory as it is applied to gases, especially the distribution of molecular speeds (energies).
- Recognize why real gases do not behave like ideal gases.

The typical modern hot-air balloon is about 18 meters tall (60 feet) and has a volume of about 2250 cubic meters of air heated with a propane burner. These balloons can carry enough propane fuel to fly for about 2 hours.

Historians have speculated that the Montgolfier brothers got their idea for a hot-air balloon after reading about the experiments on gases made by Joseph Black (1728–1799) of Scotland. Indeed, the 18th century was a time of great discoveries about the nature of chemistry. Experiments on gases by scientists such as Black, Henry Cavendish (1731–1810), Joseph Priestley (1733–1804), and Antoine Lavoisier (1743–1794) gave birth to modern chemistry.

Among the chemists working on gases was Jacques Charles (1746–1823). In August 1783, Charles exploited his recent studies on hydrogen gas by inflating a balloon with this gas. Because hydrogen would escape easily from a paper bag, Charles made a silk bag coated with rubber. Inflating the bag took several days and required nearly 225 kg of sulfuric acid and 450 kg of iron to produce the H_2 gas. The balloon stayed aloft for almost 45 minutes and traveled about 15 miles. When it landed in a village, however, the people were so terrified that they tore it to shreds. Several months later, Charles and a passenger flew a new hydrogen-filled balloon some distance across the French countryside and ascended to the then-incredible altitude of 2 miles.

Balloons that remain aloft for long periods of time typically need to use a lighter-than-air gas such as helium or hydrogen to produce lift. This approach was first tried in June 1785 by Montgolfier's first pilot, de Rozier. He and a friend, Pierre Romain, tried to fly from Paris to London in a balloon containing a hydrogen-filled cell and an air-filled cell heated by a flame. Unfortunately, at an altitude of about 900 feet, the hydrogen gas exploded, killing the two pilots. They were the first people to die in manned flight.

The latest balloons designed for long-distance flight are now called Rozier balloons. They have one or more cells filled with nonflammable helium as well as a cell in which the air can be heated. The helium provides much of the lift for the balloon, while the cell containing the hot air allows for adjustments to be made in amount of lift—that is, to move to a different altitude or to compensate for the cooling of the helium at night (and the consequent contraction in the volume of the helium). For example, at night when the air is colder, the pilot heats the air in an unsealed cell with a propane burner, which in turn transfers heat to upper helium cells. Such a device was used in the first successful circumnavigation of the globe by a balloon in March 1999.

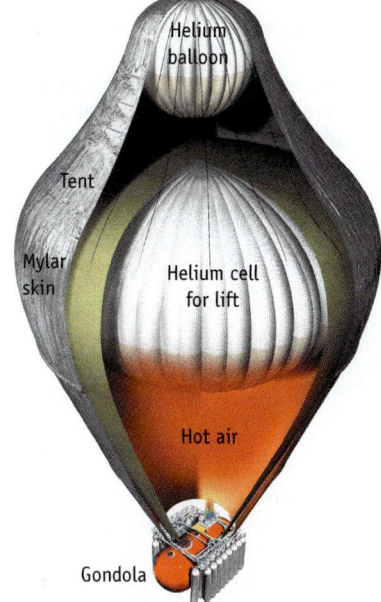

The Rozier balloon. A balloon of this design set a ballooning distance record by flying around the globe in 1999. The upper helium cell "stakes out the tent" around the larger helium cell, which helps to insulate the latter cell. Lift is provided by both the helium cells and heated air. The gondola below the balloon is insulated and sealed with life-support equipment for the crew. The balloon flies at an altitude of 6000 to 15,000 meters, a height at which the outside temperature can be about −60 °C.

To Review Before You Begin

- Review chemical stoichiometry in Chapters 4 and 5

Throughout the chapter this icon introduces a list of resources on the General ChemistryNow CD-ROM or website (http://now .brookscole.com/kotz6e) that will:

- help you evaluate your knowledge of the material
- provide homework problems
- allow you to take an exam-prep quiz
- provide a personalized Learning Plan targeting resources that address areas you should study

Hot-air balloons, SCUBA diving equipment, and automobile air bags (Figure 12.1) depend on the properties of gases. Aside from understanding how these devices work, there are at least three reasons for studying gases. First, some common elements and compounds (such as helium, hydrogen, oxygen, nitrogen, and methane) exist in the gaseous state under normal conditions of pressure and temperature. Furthermore, many common liquids such as water can be vaporized, and the physical properties of these vapors are important. Second, our gaseous atmosphere provides one means of transferring energy and material throughout the globe, and it is the source of life-giving chemicals.

The third reason for studying gases is also compelling. Of the three states of matter, gases are reasonably simple when viewed at the molecular level and, as a result, gas behavior is well understood. It is possible to describe the properties of gases *qualitatively* in terms of the behavior of the molecules that make up the gas. Even more impressive, it is possible to describe the properties of gases *quantitatively* using simple mathematical models. One objective of scientists is to develop precise mathematical and conceptual models of natural phenomena, and a study of gas behavior will introduce you to this approach.

12.1—The Properties of Gases

To describe gases, chemists have learned that only four quantities are needed: the pressure (P), volume (V), and temperature (T, kelvins) of the gas, and its amount (n, mol). Let us examine the first of these parameters, gas pressure, and its units.

Gas Pressure

You are already familiar with pressure. Meteorologists tell us that the pressure of the atmosphere is rising when nice weather approaches and that it is falling when a storm approaches. They also often speak of rising or falling barometric pressure— a barometer is the device used to measure atmospheric pressure.

Figure 12.1 Automobile air bags. Most automobiles are now equipped with air bags to protect the driver and the front-seat passenger in the event of a head-on or side crash. Such bags are inflated with nitrogen gas, which is generated by the explosive decomposition of sodium azide in the event of a crash.

$$2\ NaN_3(s) \longrightarrow 2\ Na(s) + 3\ N_2(g)$$

The air bag is fully inflated in about 0.050 second. This is important because the typical automobile collision lasts about 0.125 second. (*See General ChemistryNow Screen 12.1 Puzzler: Air Bags, for questions about automobile air bags.*)

SAAB Car USA, Inc.

A barometer can be made by filling a tube with a liquid, often mercury, and inverting the tube in a dish containing the same liquid (Figure 12.2). If the air has been removed completely from the vertical tube, the liquid in the tube assumes a level such that the pressure exerted by the mass of the column of liquid in the tube is balanced by the pressure of the atmosphere pressing down on the surface of the liquid in the dish.

At sea level, the mercury in a mercury-filled barometer will rise about 760 mm above the surface of the mercury in the dish. Thus, pressure is often reported in units of **millimeters of mercury (mm Hg).** Pressures are also reported as **standard atmospheres (atm),** a unit defined as follows:

$$1 \text{ standard atmosphere (1 atm)} = 760 \text{ mm Hg (exactly)}$$

Though it is not the SI unit, *the atmosphere is the pressure unit used in this book.*

The SI unit of pressure is the **pascal (Pa),** named for the French mathematician and philosopher Blaise Pascal (1623–1662). **Pressure** is defined as the force exerted on an object divided by the area over which it is exerted, and the pascal is the only pressure unit that is expressed in these terms.

$$1 \text{ pascal (Pa)} = 1 \text{ newton/meter}^2$$

(The newton is the SI unit of force.) Because the pascal is a very small unit compared with ordinary pressures, the unit kilopascal (kPa) is used more frequently.

The pascal has a simple relationship to another unit of pressure called the **bar,** where 1 bar = 100,000 Pa. The thermodynamic data in Chapters 6 and 19 and in Appendix L are given for gas pressures of 1 bar. To summarize, the units used in science for pressure are

$$1 \text{ atm} = 760 \text{ mm Hg (exactly)} = 101.3 \text{ kilopascals (kPa)} = 1.013 \text{ bar}$$

or

$$1 \text{ bar} = 1 \times 10^5 \text{ Pa (exactly)} = 1 \times 10^2 \text{ kPa} = 0.9872 \text{ atm}$$

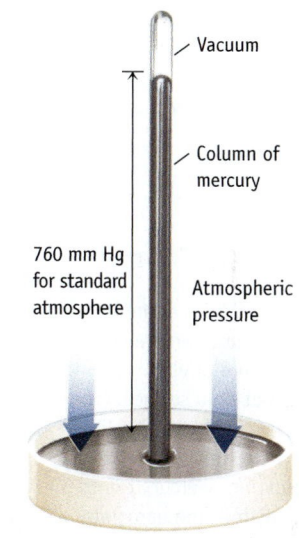

Figure 12.2 A barometer. The pressure of the atmosphere on the surface of the mercury in the dish is balanced by the downward pressure exerted by the column of mercury. The barometer was invented in 1643 by Evangelista Torricelli (1608–1647). A unit of pressure called the *torr* in his honor is equivalent to 1 mm Hg.

Example 12.1—Pressure Unit Conversions

Problem Convert a pressure of 635 mm Hg into its corresponding value in units of atmospheres (atm), bars, and kilopascals (kPa).

Strategy Use the relationships between millimeters of mercury, atmospheres, bars, and pascals described earlier in the text.

Solution The relationship between millimeters of mercury and atmospheres is 1 atm = 760 mm Hg. Notice that the given pressure is less than 760 mm Hg—that is, less than 1 atm:

$$635 \text{ mm Hg} \left(\frac{1 \text{ atm}}{760. \text{ mm Hg}} \right) = \boxed{0.836 \text{ atm}}$$

The relationship between atmospheres and bars is 1 atm = 1.013 bar. We have

$$0.836 \text{ atm} \left(\frac{1.013 \text{ bar}}{1 \text{ atm}} \right) = \boxed{0.847 \text{ bar}}$$

The factor relating units of millimeters of mercury and kilopascals is 101.325 kPa = 760 mm Hg. Therefore,

$$635 \text{ mm Hg} \left(\frac{101.3 \text{ kPa}}{760. \text{ mm Hg}} \right) = \boxed{84.6 \text{ kPa}}$$

A Closer Look

Measuring Gas Pressure

Pressure is the force exerted on an object divided by the area over which the force is exerted:

$$\text{Pressure} = \frac{\text{force}}{\text{area}}$$

This book, for example, weighs more than 4 lb and has an area of 82 in.^{2}, so it exerts a pressure of about 0.05 lb/in.^{2} when it lies flat on a surface. (In metric units, the pressure is about 3 g/cm^2.)

Now consider the pressure that the column of mercury exerts on the mercury in the dish in the barometer shown in Figure 12.2. This pressure exactly balances the pressure of the atmosphere. Thus the pressure of the atmosphere (or of any other gas) can be measured by relating it to the height of the column of mercury (or any other liquid) that the gas can support.

Mercury is the liquid of choice for barometers because of its high density. The height of a barometer filled with water would exceed 10 m. [A column of water is about 13.6 times as high as a column of mercury because mercury's density (13.53 g/cm^3) is about 13.6 times that of water (density = 0.997 g/cm^3, at 25 °C).]

In the laboratory we often use a U-tube manometer, which is a mercury-filled, U-shaped glass tube. The closed side of the tube is evacuated so that no gas remains to exert pressure on the mercury on that side. The other side is open to the gas whose pressure we want to measure. When the gas presses on the mercury in the open side, the gas pressure is read directly (in mm Hg) as the difference in mercury levels on the closed and open sides.

You may have used a tire gauge to check the pressure in your car or bike tires. In the United States, such gauges usually indicate the pressure in pounds per square inch (psi) where 1 atm = 14.7 psi. Some newer gauges give the pressure in kilopascals as well. The reading on the scale refers to the pressure *in excess of atmospheric pressure*. (A flat tire is not a vacuum; it contains air at atmospheric pressure.) For example, if the gauge reads 35 psi (2.4 atm), the pressure in the tire is actually about 50 psi (3.4 atm). (*See General ChemistryNow Screen 12.2.*)

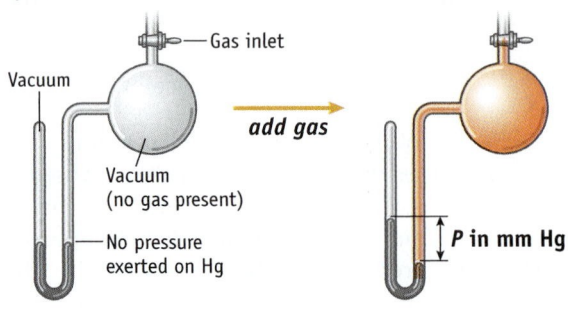

Exercise 12.1—Pressure Unit Conversions

Rank the following pressures in decreasing order of magnitude (from largest to smallest): 75 kPa, 250 mm Hg, 0.83 bar, 0.63 atm.

12.2—Gas Laws: The Experimental Basis

Experimentation in the 17th and 18th centuries led to three gas laws that provide the basis for understanding gas behavior.

The Compressibility of Gases: Boyle's Law

When you pump up the tires of your bicycle, the pump squeezes the air into a smaller volume (Figure 12.3). This property of a gas is called its **compressibility.** While studying the compressibility of gases, Robert Boyle (1627–1691) observed that the volume of a fixed amount of gas at a given temperature is inversely proportional to the pressure exerted by the gas. All gases behave in this manner, and we now refer to this relationship as **Boyle's law.**

Boyle's law can be demonstrated in many ways. In Figure 12.4 a hypodermic syringe is filled with air and sealed. When pressure is applied to the movable plunger of the syringe, the air inside is compressed. As the pressure (P) increases on the syringe, the gas volume in the syringe (V) decreases. When the pressure of the gas in the syringe is plotted as a function of $1/V$, a straight line results. This type of plot

Figure 12.3 A bicycle pump—Boyle's law in action. The pump compresses air into a smaller volume. You experience Boyle's law because you can feel the increasing pressure of the gas as you press down on the plunger.

Charles D. Winters

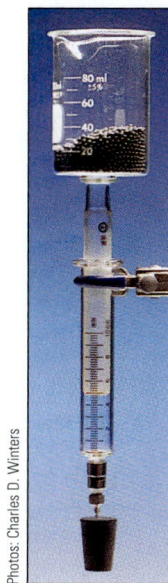 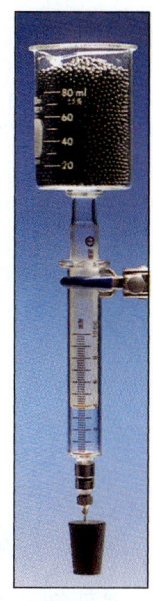

Photos: Charles D. Winters

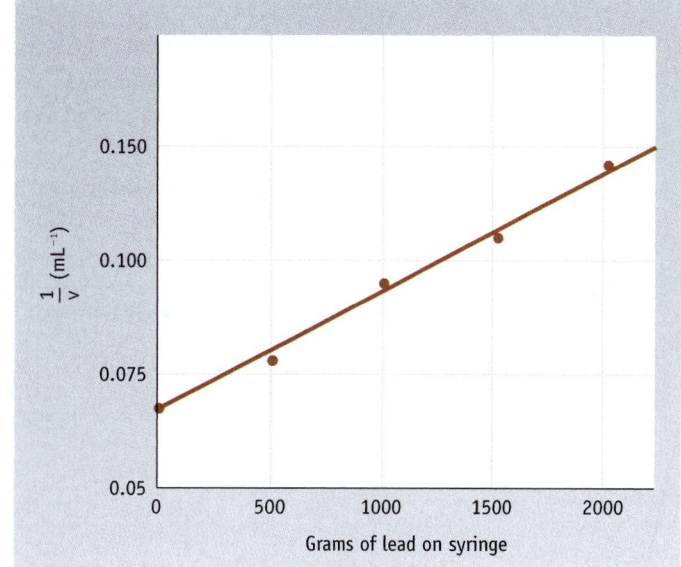

Active Figure 12.4 **An experiment to demonstrate Boyle's law.** A syringe filled with air is sealed. Pressure is applied by adding lead shot to the beaker on top of the syringe. As the mass of lead increases, the pressure on the air in the sealed syringe increases and the gas is compressed. A plot of (1/volume of air in the syringe) versus P (as measured by the mass of lead) is a straight line.

GENERAL
Chemistry·⚛·Now™ See the General ChemistryNow CD-ROM or website to explore an interactive version of this figure accompanied by an exercise.

demonstrates that the pressure and volume of the gas are inversely proportional; that is, they change in opposite directions. Mathematically, we can write this as:

$$P \propto \frac{1}{V} \text{ when } n \text{ and } T \text{ are constant}$$

Where the symbol $\propto$ means "proportional to." For a given amount of gas (n) at a fixed temperature (T), the gas volume decreases if the pressure increases. Conversely, if the pressure is lowered, then the gas volume increases.

Boyle's experimentally determined relationship can be put into a useful mathematical form. When two quantities are proportional to each other, they can be equated if a *proportionality constant*, here called C_B, is introduced. Thus,

$$P = C_B \times \frac{1}{V} \quad \text{or} \quad PV = C_B \text{ when } n \text{ and } T \text{ are constant}$$

This form of Boyle's law expresses the fact that *the product of the pressure and volume of a gas sample is a constant at a given temperature*, where the constant C_B is determined by the amount of gas (in moles) and its temperature (in kelvins). It follows from this that, if the pressure–volume product is known for a gas sample under one set of conditions (P_1 and V_1), then it is known for another set of conditions (P_2 and V_2). Under either set of conditions, the PV product is equal to C_B, so

$$P_1V_1 = P_2V_2 \quad \text{at constant } n \text{ and } T \tag{12.1}$$

This form of Boyle's law is useful when we want to know, for example, what happens to the volume of a given quantity of gas when the pressure changes at a constant temperature.

Example 12.2—Boyle's Law

Problem A sample of gaseous nitrogen in a 65.0-L automobile air bag has a pressure of 745 mm Hg. If this sample is transferred to a 25.0-L bag at the same temperature, what is the pressure of the gas in the 25.0-L bag?

Strategy Here we use Boyle's law, Equation 12.1. The original pressure and volume are known (P_1 and V_1) as well as the new volume (V_2).

Solution It is often useful to make a table of the information provided.

Original Conditions	Final Conditions
$P_1 = 745$ mm Hg	$P_2 = ?$
$V_1 = 65.0$ L	$V_2 = 25.0$ L

You know that $P_1V_1 = P_2V_2$. Therefore,

$$P_2 = \frac{P_1V_1}{V_2} = \frac{(745 \text{ mm Hg})(65.0 \text{ L})}{25.0 \text{ L}} = \boxed{1940 \text{ mm Hg}}$$

Comment The essence of Boyle's law is that P and V *change in opposite directions*. Because the volume has decreased, you know that the new pressure (P_2) must be greater than the original pressure (P_1); thus, P_1 must be multiplied by a volume factor that has a value *greater than 1* to reflect the fact that P_2 must be greater than P_1.

$$P_2 = P_1 \times \frac{65.0 \text{ L}}{25.0 \text{ L}} = 1940 \text{ mm Hg}$$

■ **Using pressure units.**
Pressure can be expressed in mm Hg, atm, or other convenient unit when comparing pressures as in Boyle's law and general gas law.

Exercise 12.2—Boyle's Law

A sample of CO_2 with a pressure of 55 mm Hg in a volume of 125 mL is compressed so that the new pressure of the gas is 78 mm Hg. What is the new volume of the gas? (Assume the temperature is constant.)

The Effect of Temperature on Gas Volume: Charles's Law

In 1787, the French scientist Jacques Charles (1746–1823) discovered that the volume of a fixed quantity of gas at constant pressure decreases with decreasing temperature (Figure 12.5).

Figure 12.6 illustrates how the volumes of two different gas samples change with temperature (at a constant pressure). When the plots of volume versus temperature are extended to lower temperatures, they all reach zero volume at the same temperature, $-273.15\,°C$. (Of course, gases will not actually reach zero volume; they liquefy above that temperature.) This temperature is quite significant. William Thomson (1824–1907), also known as Lord Kelvin, proposed a temperature scale—now known as the Kelvin scale—for which the zero point is $-273.15\,°C$ [◀ page 27].

When Kelvin temperatures are used with volume measurements, the volume–temperature relationship is

$$V = C_c \times T$$

■ **Boyle's and Charles's Laws**
Neither Boyle's law nor Charles's law depends on the identity of the gas being studied. These laws describe the behavior of any gaseous substance, regardless of its identity.

where C_c is a proportionality constant (which depends on the amount of gas and its pressure). This is **Charles's law,** which states that if a given quantity of gas is held at a constant pressure, its volume is directly proportional to the Kelvin temperature.

(a)   (b)   (c)

Figure 12.5 **A dramatic illustration of Charles's law.** (a) Air-filled balloons are placed in liquid nitrogen (77 K). The volume of the gas in the balloons is dramatically reduced at this temperature. (b) All of the balloons have been placed in the liquid nitrogen. (c) The balloons are removed; as they warm to room temperature they reinflate to their original volume.

Writing Charles's law another way, we have $V/T = C_c$; that is, the volume of a gas divided by the temperature of the gas (in kelvins) is constant for a given sample of gas at a specified pressure. Therefore, if we know the volume and temperature of a given quantity of gas (V_1 and T_1), we can find the volume, V_2, at some other temperature, T_2, using the equation

$$\frac{V_1}{T_1} = \frac{V_2}{T_2} \quad \text{at constant } n \text{ and } P \qquad (12.2)$$

Calculations using Charles's law are illustrated by the following example and exercise. Be sure to notice that the temperature T *must always be expressed in kelvins.*

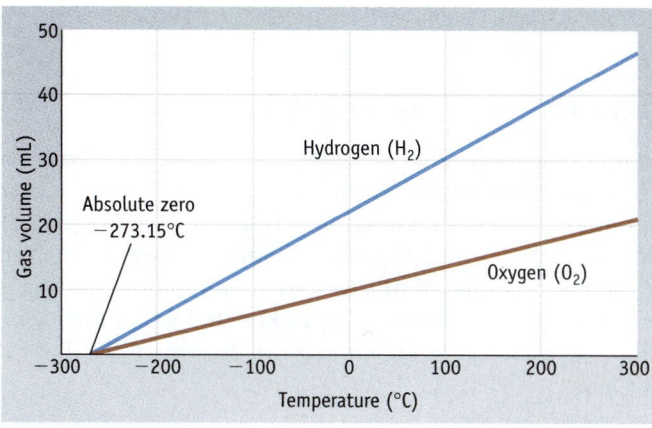

T (°C)	T (K)	Vol. H$_2$ (mL)	Vol. O$_2$ (mL)
300	573	47.0	21.1
200	473	38.8	17.5
100	373	30.6	13.8
0	273	22.4	10.1
−100	173	14.2	6.39
−200	73	6.00	—

Active Figure 12.6 **Charles's law.** The solid lines represent the volumes of samples of H$_2$ gas (0.00200 g) and O$_2$ gas (0.0144 g) at a pressure of 1.00 atm and different temperatures. The volumes decrease as the temperature is lowered at constant pressure. These lines, if extended, intersect the temperature axis at approximately −273 °C.

GENERAL
Chemistry ⚛ Now™ See the General ChemistryNow CD-ROM or website to explore an interactive version of this figure accompanied by an exercise.

Example 12.3—Charles's Law

Problem Suppose you have a sample of CO_2 in a gas-tight syringe (as in Figure 12.4). The gas volume is 25.0 mL at room temperature (20.0 °C). What is the final volume of the gas if you hold the syringe in your hand to raise its temperature to 37 °C?

Strategy Because a given quantity of gas is heated (at a constant pressure), Charles's law applies. Because we know the original V and T, and we want to calculate a new volume at a new, but known, temperature, we can use Equation 12.2.

Solution Organize the information in a table. Notice that the temperature *must* be converted to kelvins.

Original Conditions	Final Conditions
$V_1 = 25.0$ mL	$V_2 = ?$
$T_1 = 20.0 + 273 = 293$ K	$T_2 = 37 + 273 = 310.$ K

Substitute the known quantities into Equation 12.2 and solve for V_2:

$$V_2 = T_2 \times \frac{V_1}{T_1} = (310.\ \text{K}) \times \frac{25.0\ \text{mL}}{293\ \text{K}} = \boxed{26.5\ \text{mL}}$$

Comment As expected, the volume of the gas increased with a temperature increase. The new volume (V_2) must equal the original volume (V_1) multiplied by a temperature fraction that is greater than 1 to reflect the effect of the temperature increase. That is,

$$V_2 = V_1 \times \frac{310.\ \text{K}}{293\ \text{K}} = 26.5\ \text{mL}$$

Exercise 12.3—Charles's Law

A balloon is inflated with helium to a volume of 45 L at room temperature (25 °C). If the balloon is cooled to −10 °C, what is the new volume of the balloon? Assume that the pressure does not change.

Combining Boyle's and Charles's Laws: The General Gas Law

The volume of a given amount of gas is inversely proportional to its pressure at constant temperature (Boyle's law) and directly proportional to the Kelvin temperature at constant pressure (Charles's law). But what if we need to know what happens to the gas when two of the three parameters (P, V, and T) change? For example, what would happen to the pressure of a sample of nitrogen in an automobile air bag if the same amount of gas were placed in a smaller bag and heated to a higher temperature? You can deal with this situation by combining the two equations that express Boyle's and Charles's laws:

$$\frac{P_1V_1}{T_1} = \frac{P_2V_2}{T_2} \quad \text{for a given amount of gas, } n \tag{12.3}$$

This equation is sometimes called the **general gas law** or **combined gas law**. It applies specifically to situations in which the *amount of gas does not change.*

Example 12.4—General Gas Law

Problem Helium-filled balloons are used to carry scientific instruments high into the atmosphere. Suppose a balloon is launched when the temperature is 22.5 °C and the barometric pressure is 754 mm Hg. If the balloon's volume is 4.19×10^3 L (and no helium escapes from the balloon), what will the volume be at a height of 20 miles, where the pressure is 76.0 mm Hg and the temperature is −33.0 °C?

Strategy Here we know the initial volume, temperature, and pressure of the gas. We want to know the volume of the same amount of gas at a new pressure and temperature. It is most convenient to use Equation 12.3, the general gas law.

Solution Begin by setting out the information given in a table.

Initial Conditions	Final Conditions
$V_1 = 4.19 \times 10^3$ L	$V_2 = ?$ L
$P_1 = 754$ mm Hg	$P_2 = 76.0$ mm Hg
$T_1 = 22.5$ °C (295.7 K)	$T_2 = -33.0$ °C (240.2 K)

We can rearrange the general gas law to calculate the new volume, V_2:

$$V_2 = \left(\frac{T_2}{P_2}\right) \times \left(\frac{P_1 V_1}{T_1}\right) = V_1 \times \frac{P_1}{P_2} \times \frac{T_2}{T_1}$$

$$= 4.19 \times 10^3 \text{ L} \times \frac{754 \text{ mm Hg}}{76.0 \text{ mm Hg}} \times \frac{240.2 \text{ K}}{295.7 \text{ K}}$$

$$= \boxed{3.38 \times 10^4 \text{ L}}$$

Comment The pressure decreased by almost a factor of 10, which should lead to about a 10-fold volume increase. This increase is partly offset by a drop in temperature, which leads to a volume decrease. On balance, the volume increases because the pressure has dropped so substantially.

Notice that the solution was to multiply the original volume (V_1) by a pressure factor (larger than 1 because the volume increases with a lower pressure) and a temperature factor (smaller than 1 because volume decreases with a decrease in temperature).

A weather balloon is filled with helium. As it ascends into the troposphere, does the volume increase or decrease?

Exercise 12.4—The General Gas Law

You have a 22.-L cylinder of helium at a pressure of 150 atm and a temperature of 31 °C. How many balloons can you fill, each with a volume of 5.0 L, on a day when the atmospheric pressure is 755 mm Hg and the temperature is 22 °C?

The general gas law leads to other useful predictions of gas behavior. For example, if a given amount of gas is held in a closed container, the pressure of the gas will increase with increasing temperature.

$$\frac{P_1}{T_1} = \frac{P_2}{T_2} \quad \text{when } V_1 = V_2 \text{ and so } P_2 = P_1 \times \frac{T_2}{T_1}$$

That is, when T_2 is greater than T_1, P_2 will be greater than P_1. In fact, this is the reason tire manufacturers recommend checking tire pressures when the tires are cold. After driving for some distance, friction warms a tire and increases the internal pressure. Filling a warm tire to the recommended pressure may lead to a dangerously underinflated tire.

■ **Air bag technology**
Many automobile air bags use the sodium azide system, but new technologies are also being developed.

When a car decelerates in a collision, an electric current initiates a chemical reaction, which releases a gas such as nitrogen, and the folded nylon bag explodes out of the plastic housing. Driver side air bags inflate with about 35–70 L of gas.

The bag deflates within 0.2 s, the gas escaping through holes in the bottom of the bag.

Figure 12.7   Automobile air bags. (*See Figure 12.1 and General ChemistryNow Screen 12.13 Puzzler: Air Bags.*)

Avogadro's Hypothesis

The air bag is a safety device found in most of today's automobiles. In the event of an accident, it is rapidly inflated with nitrogen gas generated by a chemical reaction. The air bag unit has a sensor that is sensitive to sudden deceleration of the vehicle and will send an electrical signal that will trigger the reaction (Figures 12.1 and 12.7). The explosion of sodium azide generates nitrogen gas.

$$2\ NaN_3(s) \longrightarrow 2\ Na(s) + 3\ N_2(g)$$

Driver-side air bags inflate to a volume of about 35–70 L, and passenger-side air bags inflate to about 60–160 L. The final volume of the bag will depend on the amount of nitrogen gas generated.

The relationship between volume and amount of gas was first noted by Amedeo Avogadro. In 1811 he used work on gases by the chemist (and early experimenter with hot-air balloons) Joseph Guy-Lussac (1778–1850) to propose that *equal volumes of gases under the same conditions of temperature and pressure have equal numbers of particles* (either molecules or atoms, depending on the composition of the gas). This idea came to be known as **Avogadro's hypothesis**. Stated another way, the volume of a gas at a given temperature and pressure is directly proportional to the amount of gas in moles:

$$V \propto n \text{ at constant } T \text{ and } P$$

GENERAL
Chemistry•Now™

See the General ChemistryNow CD-ROM or website:
• **Screen 12.3 Gas Laws,** for interactive versions the the three gas laws

Example 12.5—Avogadro's Hypothesis

Problem Ammonia can be made directly from the elements:

$$N_2(g) + 3\ H_2(g) \longrightarrow 2\ NH_3(g)$$

If you begin with 15.0 L of $H_2(g)$, what volume of $N_2(g)$ is required for complete reaction (both gases being at the same T and P)? What is the theoretical yield of NH_3, in liters, under the same conditions?

Strategy From Avogadro's hypothesis we know that gas volume is proportional to the amount of gas. Therefore, we can substitute gas volumes for moles in this stoichiometry problem.

Solution Calculate the volumes of N_2 required and of NH_3 produced (in liters) by multiplying the volume of H_2 available by a stoichiometric factor (also in liters) obtained from the chemical equation:

$$V(N_2 \text{ required}) = 15.0 \text{ L } H_2 \text{ available} \times \frac{1 \text{ L } N_2}{3 \text{ L } H_2} = 5.00 \text{ L } N_2$$

$$V(NH_3 \text{ produced}) = 15.0 \text{ L } H_2 \text{ available} \times \frac{2 \text{ L } NH_3}{3 \text{ L } H_2} = 10.0 \text{ L } NH_3$$

Exercise 12.5—Avogadro's Hypothesis

Methane burns in oxygen to give CO_2 and H_2O, according to the equation

$$CH_4(g) + 2 O_2(g) \longrightarrow CO_2(g) + 2 H_2O(g)$$

If 22.4 L of gaseous CH_4 is burned, what volume of O_2 is required for complete combustion? What volumes of CO_2 and H_2O are produced? Assume all gases are at the same temperature and pressure.

12.3—The Ideal Gas Law

Four interrelated quantities can be used to describe a gas: pressure, volume, temperature, and amount (moles). We know from experiments that three gas laws can be used to describe the relationship of these properties (Section 12.2).

Boyle's Law	Charles's Law	Avogadro's Hypothesis
$V \propto (1/P)$	$V \propto T$	$V \propto n$
(constant T, n)	(constant P, n)	(constant T, P)

If all three relationships are combined, the result is

$$V \propto \frac{nT}{P}$$

This can be made into a mathematical equation by introducing a proportionality constant, R, called the **gas constant**. It is a *universal constant*—a number that you can use to interrelate the properties of any gas:

$$V = R\left(\frac{nT}{P}\right)$$

or

$$PV = nRT \qquad\qquad (12.4)$$

The equation $PV = nRT$ is called the **ideal gas law**. It describes the behavior of an "ideal" gas. As you will learn in Section 12.9, there is no such thing as an "ideal" gas.

■ **Properties of an ideal gas**
For ideal gases it is assumed that there are no forces of attraction between molecules and the molecules themselves occupy no volume.

However, real gases at pressures around one atmosphere or less and temperatures around room temperature usually behave close enough to ideality that $PV = nRT$ adequately describes their behavior.

To use the equation $PV = nRT$, we need a value for R. It is readily determined experimentally. By carefully measuring P, V, n, and T for a sample of gas, we can calculate the value of R from these values using the ideal gas law equation. For example, under conditions of **standard temperature and pressure (STP)**, a gas temperature of 0 °C or 273.15 K and a pressure of 1 atm, 1 mol of gas occupies 22.414 L, a quantity called the **standard molar volume**. Substituting these values into the ideal gas law equation gives a value for R:

$$R = \frac{PV}{nT} = \frac{(1.0000 \text{ atm})(22.414 \text{ L})}{(1.0000 \text{ mol})(273.15 \text{ K})} = 0.082057 \frac{\text{L} \cdot \text{atm}}{\text{K} \cdot \text{mol}}$$

With a value for R, we can now use the ideal gas law in calculations.

■ **STP—What Is It?**
A gas is at STP, or standard temperature and pressure, when its temperature is 0 °C or 273.15 K and its pressure is 1 atm. Under these conditions, exactly 1 mol of a gas occupies 22.414 L.

GENERAL
Chemistry Now™

See the General ChemistryNow CD-ROM or website:
- **Screen 12.4 The Ideal Gas Law,** for a simulation of the ideal gas law and a tutorial

Example 12.6—Ideal Gas Law

Problem The nitrogen gas in an automobile air bag, with a volume of 65 L, exerts a pressure of 829 mm Hg at 25 °C. What amount of N_2 gas (in moles) is in the air bag?

Strategy You are given P, V, and T and want to calculate the amount of gas (n). Use the ideal gas law, Equation 12.4.

Solution First list the information provided.

$$P = 829 \text{ mm Hg} \quad V = 65 \text{ L} \quad T = 25 \text{ °C} \quad n = ?$$

To use the ideal gas law with R having units of (L · atm/K · mol), the pressure must be expressed in atmospheres and the temperature in kelvins. Therefore,

$$P = 829 \text{ mm Hg} \times \frac{1 \text{ atm}}{760 \text{ mm Hg}} = 1.09 \text{ atm}$$

$$T = 25 + 273 = 298 \text{ K}$$

Now substitute the values of P, V, T, and R into the ideal gas law and solve for the amount of gas, n:

$$n = \frac{PV}{RT} = \frac{(1.09 \text{ atm})(65 \text{ L})}{(0.082057 \text{ L} \cdot \text{atm/K} \cdot \text{mol})(298 \text{ K})} = 2.9 \text{ mol } N_2$$

Notice that the units of atmospheres, liters, and kelvins cancel to leave the answer in units of moles.

Exercise 12.6—Ideal Gas Law

The balloon used by Jacques Charles in his historic flight in 1783 was filled with about 1300 mol of H_2. If the temperature of the gas was 23 °C and its pressure was 750 mm Hg, what was the volume of the balloon?

(a)   (b)

Figure 12.8 Gas density. (a) The balloons are filled with nearly equal amounts of gas at the same temperature and pressure. One yellow balloon contains helium, a low-density gas ($d = 0.179$ g/L at STP). The other balloons contain air, a higher-density gas ($d = 1.2$ g/L). (b) A hot-air balloon rises because the heated air has a lower density.

The Density of Gases

The density of a gas at a given temperature and pressure (Figure 12.8) is a useful quantity. Let us see how density is related to the ideal gas law. Because the amount (n, mol) of any compound is given by its mass (m) divided by its molar mass (M), we can substitute m/M for n in the ideal gas equation.

$$PV = \left(\frac{m}{M}\right)RT$$

Density (d) is defined as mass divided by volume (m/V). We can rearrange this form of the gas law to give the following equation, which has the term (m/V) on the left. This is the density of the gas.

$$d = \frac{m}{V} = \frac{PM}{RT} \tag{12.5}$$

Gas density is directly proportional to the pressure and molar mass and inversely proportional to the temperature. Equation 12.5 is useful because gas density can be calculated from the molar mass, or the molar mass can be found from a measurement of gas density of a gas at a given pressure and temperature.

Example 12.7—Density and Molar Mass

Problem Calculate the density of CO_2 at STP. Is CO_2 more or less dense than air (1.2 g/L)?

Strategy Use Equation 12.5, the equation relating gas density and molar mass. Here we know the molar mass (44.0 g/mol), pressure ($P = 1.00$ atm), temperature ($T = 273.15$ K), and the gas constant (R). Only the density (d) is unknown.

Solution The known values are substituted into Equation 12.5, which is then solved for density:

$$d = \frac{PM}{RT} = \frac{(1.00 \text{ atm})(44.0 \text{ g/mol})}{(0.082057 \text{ L} \cdot \text{atm/K} \cdot \text{mol})(273 \text{ K})} = 1.96 \text{ g/L}$$

The density of CO_2 is considerably greater than that of dry air at STP (1.2 g/L).

Charles D. Winters

Figure 12.9   Gas density. Because carbon dioxide from fire extinguishers is denser than air, it settles on top of a fire and smothers it. (When CO_2 gas is released from the tank, it expands and cools significantly. The white cloud is solid CO_2 and condensed moisture from the air.)

Exercise 12.7—Gas Density Calculation

The density of an unknown gas is 5.02 g/L at 15.0 °C and 745 mm Hg. Calculate its molar mass.

Gas density has practical implications. From the equation $d = PM/RT$ we recognize that the density of a gas is directly proportional to its molar mass. Dry air, which has an average molar mass of about 29 g/mol, has a density of about 1.2 g/L at 1 atm and 25 °C. Gases or vapors with molar masses greater than 29 g/mol have densities larger than 1.2 g/L under these same conditions (1 atm and 25 °C). As a consequence, gases such as CO_2, SO_2, and gasoline vapor settle along the ground if released into the atmosphere (Figure 12.9). Conversely, gases such as H_2, He, CO, CH_4 (methane), and NH_3 rise if released into the atmosphere.

The significance of gas density was tragically revealed in several recent events. One occurred in the African country of Cameroon in 1986, when Lake Nyos expelled a huge bubble of CO_2 into the atmosphere. Because CO_2 is denser than air, the CO_2 cloud hugged the ground, killing 1700 people living in a nearby village.

Calculating the Molar Mass of a Gas from *P*, *V*, and *T* Data

When a new compound is isolated in the laboratory, one of the first things to be done is to determine its molar mass. If the compound is in the gas phase, a classical method of determining the molar mass is to measure the pressure and volume exerted by a given mass of the gas at a given temperature.

GENERAL
Chemistry⚛Now™

See the General ChemistryNow CD-ROM or website:

- **Screen 12.5 Gas Density,** to watch a video of a hot-air balloon and to work an exercise on gas density and molar mass
- **Screen 12.6 Using Gas Laws: Determining Molar Mass,** for a tutorial on gas density

Example 12.8—Calculating the Molar Mass of a Gas from *P*, *V*, and *T* Data

Problem You are trying to determine, by experiment, the empirical formula of a gaseous compound to replace chlorofluorocarbons in air conditioners. Your results give an empirical formula of CHF_2. Now you need the molar mass of the compound to find the molecular formula. You conduct another experiment and find that a 0.100-g sample of the compound exerts a pressure of 70.5 mm Hg in a 256-mL container at 22.3 °C. What is the molar mass of the compound? What is its molecular formula?

Strategy Here you know the mass of a gas in a given volume (*V*), so you can calculate its density (*d*). Then, knowing the gas pressure and temperature, you can use Equation 12.5 to calculate the molar mass.

Solution Begin by organizing the data:

$$m = \text{mass of gas} = 0.100 \text{ g}$$
$$P = 70.5 \text{ mm Hg, or } 0.0928 \text{ atm}$$
$$V = 256 \text{ mL, or } 0.256 \text{ L}$$
$$T = 22.3 \text{ °C, or } 295.5 \text{ K}$$

The density of the gas is the mass of the gas divided by the volume:

$$d = \frac{0.100 \text{ g}}{0.256 \text{ L}} = 0.391 \text{ g/L}$$

Use this value of density, along with the values of pressure and temperature in Equation 12.5 ($d = PM/RT$), and solve for the molar mass (M).

$$M = \frac{dRT}{P} = \frac{(0.391 \text{ g/L})(0.082057 \text{ L} \cdot \text{atm/K} \cdot \text{mol})(295.5 \text{ K})}{0.0928 \text{ atm}} = \boxed{102 \text{ g/mol}}$$

With this result, you can compare the experimentally determined molar mass with the mass of a mole of gas having the empirical formula CHF_2.

$$\frac{\text{Experimental molar mass}}{\text{Mass of 1 mol } CHF_2} = \frac{102 \text{ g/mol}}{51.0 \text{ g/formula unit}} = 2 \text{ formula units of } CHF_2 \text{ per mol}$$

Therefore, the formula of the compound is $\boxed{C_2H_2F_4}$.

Comment Alternatively, you can use the ideal gas law. Here you know P and T for a gas in a given volume (V), so you can calculate the amount of gas (n).

$$n = \frac{PV}{RT} = \frac{(0.0928 \text{ atm})(0.256 \text{ L})}{(0.082057 \text{ L} \cdot \text{atm/K} \cdot \text{mol})(295.5 \text{ K})} = 9.80 \times 10^{-4} \text{ mol}$$

You now know that 0.100 g of gas is equivalent to 9.80×10^{-4} mol. Therefore,

$$\text{Molar mass} = \frac{0.100 \text{ g}}{9.80 \times 10^{-4} \text{ mol}} = 102 \text{ g/mol}$$

Exercise 12.8—Molar Mass from P, V, and T Data

A 0.105-g sample of a gaseous compound has a pressure of 561 mm Hg in a volume of 125 mL at 23.0 °C. What is its molar mass?

12.4—Gas Laws and Chemical Reactions

Many industrially important reactions involve gases. Two examples are the combination of nitrogen and hydrogen to produce ammonia,

$$N_2(g) + 3 \; H_2(g) \longrightarrow 2 \; NH_3(g)$$

and the electrolysis of aqueous NaCl to produce hydrogen and chlorine,

$$2 \; NaCl(aq) + 2 \; H_2O(\ell) \longrightarrow 2 \; NaOH(aq) + H_2(g) + Cl_2(g)$$

If we want to understand the quantitative aspects of such reactions, we need to carry out stoichiometry calculations. The scheme in Figure 12.10 connects these calculations for gas reactions with the stoichiometry calculations done in Chapters 4 and 5.

GENERAL
Chemistry❖Now™

See the General ChemistryNow CD-ROM or website:
- **Screen 12.7 Gas Laws and Chemical Reactions: Stoichiometry,** for a tutorial on gas laws and chemical reactions

Figure 12.10 **A scheme for performing stoichiometry calculations.** Here A and B may be either reactants or products. The amount of A (mol) can be calculated from its mass in grams, from the concentration and volume of a solution, or from P, V, and T data by using the ideal gas law. Once the amount of B is determined, this value can be converted to a mass or solution concentration or volume, or to a volume of gas at a given pressure and temperature.

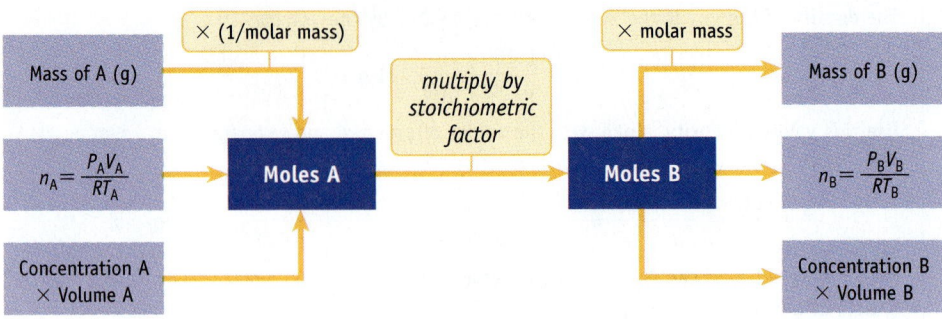

Example 12.9—Gas Laws and Stoichiometry

Problem You are asked to design an air bag for a car. You know that the bag should be filled with gas having a pressure higher than atmospheric pressure, say 829 mm Hg, at a temperature of 22.0 °C. The bag has a volume of 45.5 L. What quantity of sodium azide, NaN_3, should be used to generate the required quantity of gas? The gas-producing reaction is

$$2\ NaN_3(s) \longrightarrow 2\ Na(s) + 3\ N_2(g)$$

Strategy The general logic to be used here follows one of the pathways in Figure 12.10 (middle left to upper right).

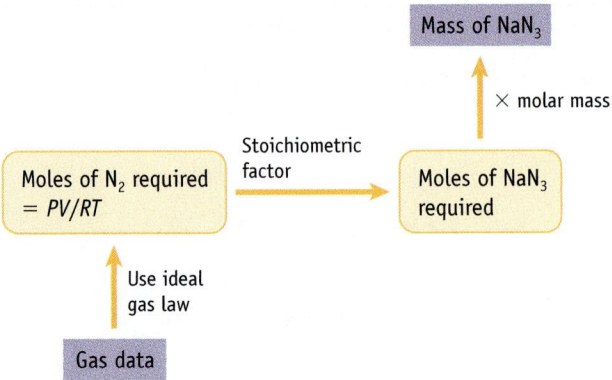

Solution The first step is to find the amount (mol) of gas required so that it can be related in the next step to the amount of sodium azide required:

$$P = 829\ \text{mm Hg}\ (1\ \text{atm}/760\ \text{mm Hg}) = 1.09\ \text{atm}$$

$$V = 45.5\ \text{L}$$

$$T = 22.0\ °C,\ \text{or}\ 295.2\ \text{K}$$

$$n = N_2\ \text{required (mol)} = \frac{PV}{RT}$$

$$n = \frac{(1.09\ \text{atm})(45.5\ \text{L})}{(0.082057\ \text{L} \cdot \text{atm}/\text{K} \cdot \text{mol})(295.2\ \text{K})} = 2.05\ \text{mol}\ N_2$$

Now that the required amount of nitrogen has been calculated, we can calculate the quantity of sodium azide that will produce 2.05 mol of N_2 gas.

$$\text{Mass of } NaN_3 = 2.05\ \text{mol } N_2 \left(\frac{2\ \text{mol } NaN_3}{3\ \text{mol } N_2}\right)\left(\frac{65.01\ \text{g}}{1\ \text{mol } NaN_3}\right) = 88.8\ \text{g } NaN_3$$

Example 12.10—Gas Laws and Stoichiometry

Problem You wish to prepare some deuterium gas, D_2, for use in an experiment. One technique is to react heavy water, D_2O, with an active metal such as lithium.

$$2 \text{ Li}(s) + 2 \text{ D}_2O(\ell) \longrightarrow 2 \text{ LiOD}(aq) + D_2(g)$$

Suppose you place 0.125 g of Li metal in 15.0 mL of D_2O ($d = 1.11$ g/mL). What amount of D_2 (in moles) can be prepared? If dry D_2 gas is captured in a 1450-mL flask at 22.0 °C, what is the pressure of the gas (in atm)? (Deuterium has an atomic weight of 2.0147 g/mol.)

Strategy You are combining two reactants with no guarantee that they are in the correct stoichiometric ratio. This reaction must therefore be approached as a *limiting reactant problem*. You have to find the amount of each substance and then see if one of them is present in a limited amount. Once the limiting reactant is known, the amount of D_2 produced and its pressure under the conditions given can be calculated.

Lithium metal (in the spoon) reacts with drops of water, H_2O, to produce LiOH and hydrogen gas, H_2. If heavy water, D_2O, is used, deuterium gas, D_2, can be produced.

Charles D. Winters

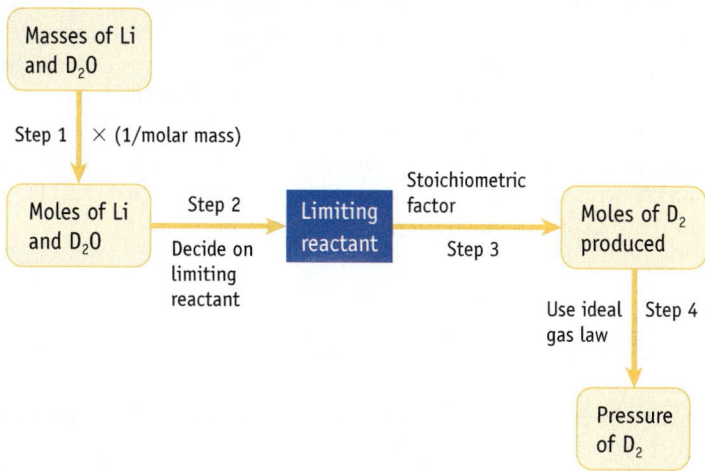

Solution

Step 1. *Calculate the amount (mol) of Li and of D_2O:*

$$0.125 \text{ g Li} \times \frac{1 \text{ mol Li}}{6.941 \text{ g Li}} = 0.0180 \text{ mol Li}$$

$$15.0 \text{ mL D}_2O \times \frac{1.11 \text{ g D}_2O}{1 \text{ mL D}_2O} \times \frac{1 \text{ mol D}_2O}{20.03 \text{ g D}_2O} = 0.831 \text{ mol D}_2O$$

Step 2. *Decide which reactant is the limiting reactant:*

$$\text{Ratio of moles of reactants available} = \frac{0.831 \text{ mol D}_2O}{0.0180 \text{ mol Li}} = \frac{46.2 \text{ mol D}_2O}{1 \text{ mol Li}}$$

The balanced equation shows that the ratio should be 1 mol of D_2O to 1 mol of Li. From the calculated values, we can see that D_2O is in large excess and Li is the limiting reactant. Therefore, further calculations are based on the amount of Li available.

Step 3. *Use the limiting reactant to calculate the amount of D_2 produced:*

$$0.0180 \text{ mol Li} \left(\frac{1 \text{ mol D}_2 \text{ produced}}{2 \text{ mol Li}} \right) = 0.00900 \text{ mol D}_2 \text{ produced}$$

Step 4. *Calculate the pressure of D_2:*

$P = ?$ $\qquad\qquad\qquad$ $T = 22.0$ °C, or 295.2 K

$V = 1450$ mL, or 1.45 L $\qquad$ $n = 0.00900$ mol D_2

$$P = \frac{nRT}{V} = \frac{(0.00900 \text{ mol})(0.082057 \text{ L} \cdot \text{atm/K} \cdot \text{mol})(295.2 \text{ K})}{1.45 \text{ L}} = \boxed{0.150 \text{ atm}}$$

Exercise 12.9—Gas Laws and Stoichiometry

Gaseous ammonia is synthesized by the reaction

$$N_2(g) + 3\ H_2(g) \xrightarrow[\text{500 °C}]{\text{iron catalyst}} 2\ NH_3(g)$$

Assume that 355 L of H_2 gas at 25.0 °C and 542 mm Hg is combined with excess N_2 gas. What amount of NH_3 gas, in moles, can be produced? If this amount of NH_3 gas is stored in a 125-L tank at 25.0 °C, what is the pressure of the gas?

12.5—Gas Mixtures and Partial Pressures

The air you breathe is a mixture of nitrogen, oxygen, carbon dioxide, water vapor, and small amounts of other gases (Table 12.1). Each of these gases exerts its own pressure, and atmospheric pressure is the sum of the pressures exerted by each gas. The pressure of each gas in the mixture is called its **partial pressure**.

John Dalton (1766–1844) was the first to observe that the pressure of a mixture of gases is the sum of the pressures of the various gases in the mixture. This observation is now known as **Dalton's law of partial pressures** (Figure 12.11). Mathematically, we can write Dalton's law of partial pressures as

$$P_{\text{total}} = P_1 + P_2 + P_3 + \cdots \tag{12.6}$$

where P_1, P_2, and P_3 are the pressures of the different gases in a mixture and P_{total} is the total pressure.

In a mixture of gases, each gas behaves independently of all others in the mixture. Therefore, we can consider the behavior of each gas in a mixture separately. As an example let us take a mixture of three ideal gases, labeled A, B, and C. There are n_A moles of A, n_B moles of B, and n_C moles of C. Assume that the mixture ($n_{\text{total}} = n_A + n_B + n_C$) is contained in a given volume (V) at a given temperature (T). We can calculate the pressure exerted by each gas from the ideal gas law equation:

$$P_A V = n_A RT \quad P_B V = n_B RT \quad P_C V = n_C RT$$

■ **John Dalton (1766–1844)**
For a short biography of John Dalton, see page 65.

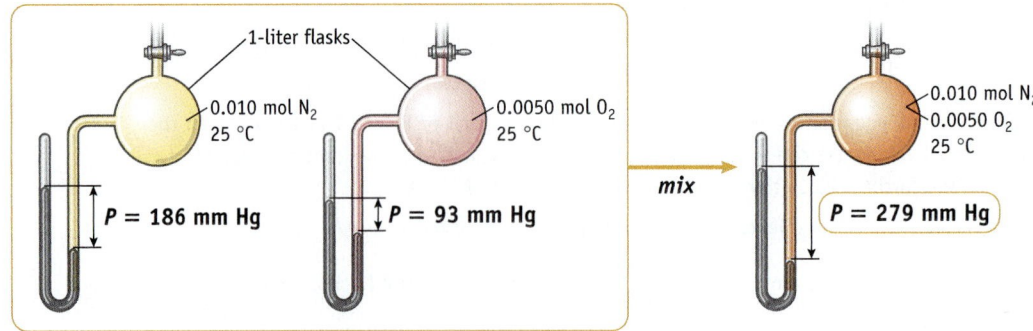

Figure 12.11   Dalton's law. In a 1.0-L flask at 25 °C, 0.010 mol of N_2 exerts a pressure of 186 mm Hg, and 0.0050 mol of O_2 in a 1.0-L flask at 25 °C exerts a pressure of 93 mm Hg (*left and middle*). The N_2 and O_2 samples are mixed in a 1.0-L flask at 25 °C (*right*). The total pressure, 279 mm Hg, is the sum of the partial pressures that each gas alone exerts in the flask.

Table 12.1   **Components of Atmospheric Dry Air**

Constituent	Molar Mass*	Mole Percent	Partial Pressure at STP (atm)
N_2	28.01	78.08	0.7808
O_2	32.00	20.95	0.2095
CO_2	44.01	0.033	0.00033
Ar	39.95	0.934	0.00934

*The average molar mass of dry air = 28.960 g/mol.

where each gas (A, B, and C) is in the same volume V and is at the same temperature T. According to Dalton's law, the total pressure exerted by the mixture is the sum of the pressures exerted by each component:

$$P_{total} = P_A + P_B + P_C = n_A\left(\frac{RT}{V}\right) + n_B\left(\frac{RT}{V}\right) + n_C\left(\frac{RT}{V}\right)$$

$$P_{total} = (n_A + n_B + n_C)\left(\frac{RT}{V}\right)$$

$$P_{total} = n_{total}\left(\frac{RT}{V}\right) \qquad (12.7)$$

For mixtures of gases, it is convenient to introduce a quantity called the **mole fraction, X**, which is defined as the number of moles of a particular substance in a mixture divided by the total number of moles of all substances present. Mathematically, the mole fraction of a substance A in a mixture with B and C is expressed as

$$X_A = \frac{n_A}{n_A + n_B + n_C} = \frac{n_A}{n_{total}}$$

Now we can combine this equation (written as $n_{total} = n_A/X_A$) with the equations for P_A and P_{total}, and derive the equation

$$P_A = X_A P_{total} \qquad (12.8)$$

This equation is useful because it tells us that the pressure of a gas in a mixture of gases is the product of its mole fraction and the total pressure of the mixture. For example, the mole fraction of N_2 in air is 0.78, so, at STP, its partial pressure is 0.78 atm or 590 mm Hg.

GENERAL
Chemistry ⚛ Now™

See the General ChemistryNow CD-ROM or website:
- **Screen 12.8 Gas Mixtures and Partial Pressures,** for two tutorials on Dalton's Law

Example 12.11—Partial Pressures of Gases

Problem Halothane, $C_2HBrClF_3$, is a nonflammable, nonexplosive, and nonirritating gas that is commonly used as an inhalation anesthetic.

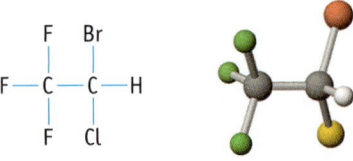

1,1,1-trifluorobromochloroethane, halothane

Suppose you mix 15.0 g of halothane vapor with 23.5 g of oxygen gas. If the total pressure of the mixture is 855 mm Hg, what is the partial pressure of each gas?

Strategy One way to solve this problem is to recognize that the partial pressure of a gas is given by the total pressure of the mixture multiplied by the mole fraction of the gas.

Solution Let us first calculate the mole fractions of halothane and of O_2.

Step 1. *Calculate mole fractions:*

$$\text{Amount } C_2HBrClF_3 = 15.0 \text{ g} \left(\frac{1 \text{ mol}}{197.4 \text{ g}} \right) = 0.0760 \text{ mol}$$

$$\text{Amount } O_2 = 23.5 \text{ g} \left(\frac{1 \text{ mol}}{32.00 \text{ g}} \right) = 0.734 \text{ mol}$$

$$\text{Mole fraction } C_2HBrClF_3 = \frac{0.0760 \text{ mol } C_2HBrClF_3}{0.810 \text{ total moles}} = 0.0938$$

Because the sum of the mole fractions of halothane and of O_2 must equal 1.000, this means that the mole fraction of oxygen is 0.906.

$$X_{halothane} + X_{oxygen} = 1.000$$
$$0.0938 + X_{oxygen} = 1.000$$
$$X_{oxygen} = 0.906$$

Step 2. *Calculate partial pressures:*

$$\text{Partial pressure of halothane} = P_{halothane} = X_{halothane} \times P_{total}$$
$$P_{halothane} = 0.0938 \times P_{total} = 0.0938 \, (855 \text{ mm Hg})$$
$$P_{halothane} = 80.2 \text{ mm Hg}$$

The total pressure of the mixture is the sum of the partial pressures of the gases in the mixture.

$$P_{halothane} + P_{oxygen} = 855 \text{ mm Hg}$$

and so

$$P_{oxygen} = 855 \text{ mm Hg} - P_{halothane}$$
$$P_{oxygen} = 855 \text{ mm Hg} - 80.2 \text{ mm Hg} = 775 \text{ mm Hg}$$

Exercise 12.10—Partial Pressures of Gases

The halothane–oxygen mixture described in Example 12.11 is placed in a 5.00-L tank at 25.0 °C. What is the total pressure (in mm Hg) of the gas mixture in the tank? What are the partial pressures (in mm Hg) of the gases?

Historical Perspectives

Studies on Gases

As described in "Up, Up, and Away" on page 546, the sport of ballooning grew out of the fascination of early chemists with gases and their properties.

Robert Boyle

(1627–1691) was born in Ireland as the 14th and last child of the first Earl of Cork. In his book *Uncle Tungsten*, Oliver Sacks tells us that, "Chemistry as a true science made its first emergence with the work of Robert Boyle in the middle of the seventeenth century. Twenty years [Isaac] Newton's senior, Boyle was born at a time when the practice of alchemy still held sway, and he still maintained a variety of alchemical beliefs and practices, side by side with his scientific ones. He believed gold could be created, and that he had succeeded in creating it (Newton, also an alchemist, advised him to keep silent about this)."

Boyle examined crystals, explored color, devised an acid–base indicator from the syrup of violets, and provided the first modern definition of an element. He was also a physiologist, and was the first to show that the healthy human body has a constant temperature. Today Boyle is best known for his studies of gases, which were described in his book, *The Sceptical Chymist*, published in 1680.

The French chemist and inventor **Jacques Alexandre César Charles** was born on November 12, 1746. He began his career as a clerk in the finance ministry, but his real interest was science. Charles developed several inventions and was best known in his lifetime for inventing the hydrogen balloon. Today we remember him for his work on the properties of gases.

See Oliver Sacks: *Uncle Tungsten*, p. 102, New York, Alfred Knopf, 2001. See also this book, page 546.

Photos: (Left) Oesper Collection in the History of Chemistry, University of Cincinnati; (Right) The Bettman Archive/Corbis.

12.6—The Kinetic-Molecular Theory of Gases

So far we have discussed the macroscopic properties of gases, properties such as pressure and volume that result from the behavior of a system with a large number of particles. Now we turn to the kinetic-molecular theory [◀ Section 1.5] for a description of the behavior of matter at the molecular or atomic level. Hundreds of experimental observations have led to the following postulates regarding the behavior of gases:

- Gases consist of particles (molecules or atoms), whose separation is much greater than the size of the particles themselves (see Figure 12.12).

- The particles of a gas are in continual, random, and rapid motion. As they move, they collide with one another and with the walls of their container, but they do so without loss of energy.

- The average kinetic energy of gas particles is proportional to the gas temperature. *All gases, regardless of their molecular mass, have the same average kinetic energy at the same temperature.*

Let us discuss the behavior of gases from this point of view.

Molecular Speed and Kinetic Energy

If your friend walks into your room carrying a pizza, how do you know it? In scientific terms, we know that the odor-causing molecules of food enter the gas phase and drift through space until they reach the cells of your body that react to odors. The same thing happens in the laboratory when bottles of aqueous ammonia (NH_3) and hydrochloric acid (HCl) sit side by side (Figure 12.13). Molecules of the two compounds enter the gas phase and drift along until they encounter one another, at which time they react and form a cloud of tiny particles of solid ammonium chloride (NH_4Cl).

Photo: Charles D. Winters

Figure 12.12 A molecular view of gases and liquids. The fact that a large volume of N_2 gas can be condensed to a small volume of liquid indicates that the distances between molecules in the gas phase are very large as compared with the distances between molecules in liquids. (Liquid N_2 boils at −196 °C.)

Charles D. Winters

Figure 12.13 **The movement of gas molecules.** Open dishes of aqueous ammonia and hydrochloric acid were placed side by side. When molecules of NH_3 and HCl escape from solution to the atmosphere and encounter one another, we observe a cloud of solid ammonium chloride, NH_4Cl.

If you change the temperature of the environment of the containers in Figure 12.13 and measure the time needed for the cloud of ammonium chloride to form, you would find that this time is longer at lower temperatures. The reason is that the speed at which molecules move depends on the temperature. Let us expand on this idea.

The molecules in a gas sample do not all move at the same speed. Rather, as illustrated in Figure 12.14 for O_2 molecules, there is a distribution of speeds. Figure 12.14 shows the number of particles in a gas sample that are moving at certain speeds at a given temperature. We can make two important observations. First, at a given temperature, some molecules have high speeds and others have low speeds. Most of the molecules, however, have some intermediate speed, and their most probable speed corresponds to the maximum in the curve. For oxygen gas at 25 °C, for example, most molecules have speeds in the range of 200 m/s to 700 m/s, and their most probable speed is about 400 m/s. (These are very high speeds, indeed. A speed of 400 m/s corresponds to about 900 miles per hour!)

A second observation regarding the distribution of speeds is that as the temperature increases, the most probable speed increases, and the number of molecules traveling at very high speeds increases greatly.

The kinetic energy of a single molecule of mass m in a gas sample is given by the equation

$$KE = \frac{1}{2}(\text{mass})(\text{speed})^2 = \frac{1}{2}mu^2$$

where u is the speed of that molecule. We can calculate the kinetic energy of a single gas molecule from this equation but not the kinetic energy of a collection of molecules, because not all of the molecules in a gas sample are moving at the same speed. However, we can calculate the *average* kinetic energy of a collection of molecules by relating it to other averaged quantities of the system. In particular, the average kinetic energy is related to the average speed.

$$\overline{KE} = \frac{1}{2}m\overline{u^2}$$

Figure 12.14 The distribution of molecular speeds. A graph of the number of molecules with a given speed versus that speed shows the distribution of molecular speeds. The red curve shows the effect of increased temperature. Even though the curve for the higher temperature is "flatter" and broader than the curve for the lower temperature, the areas under the curves are the same because the number of molecules in the sample is fixed.

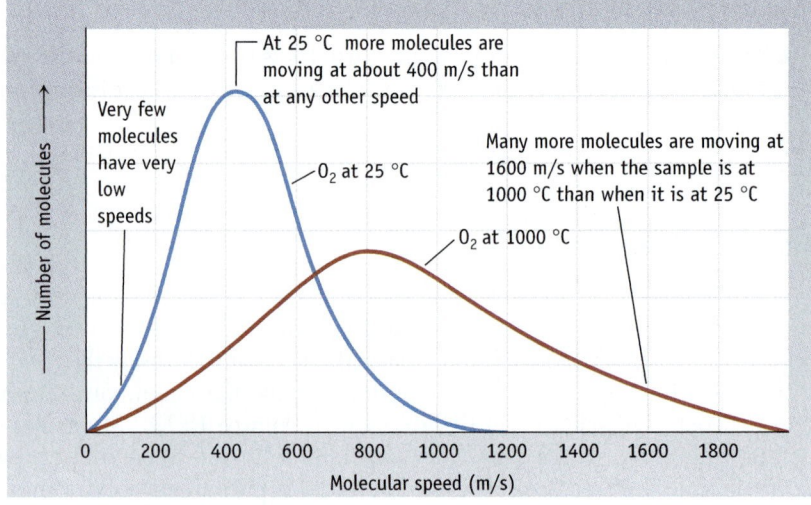

(The horizontal bar over the symbols KE and u indicate an average value.) This equation states that the average kinetic energy of the molecules in a gas sample, $\overline{KE}$, is related to $\overline{u^2}$, the average of the squares of their speeds (called the "mean square speed").

Experiments also show that the average kinetic energy, $\overline{KE}$, of a sample of gas molecules is directly proportional to temperature with a proportionality constant of $\frac{3}{2}R$:

$$\overline{KE} = \frac{3}{2}RT$$

where R is the gas constant expressed in SI units ($8.314472\ \text{J/K} \cdot \text{mol}$).

Because $\overline{KE}$ is proportional to both $\frac{1}{2}m\overline{u^2}$ and T, temperature and $\frac{1}{2}m\overline{u^2}$ must also be proportional; that is, $\frac{1}{2}m\overline{u^2} \propto T$. This relationship among mass, average speed, and temperature is expressed in Equation 12.9. Here the square root of the mean square speed ($\sqrt{\overline{u^2}}$, called the **root-mean-square** or **rms speed**), the temperature (T, in kelvins), and the molar mass (M) are related.

$$\sqrt{\overline{u^2}} = \sqrt{\frac{3RT}{M}} \tag{12.9}$$

This equation, sometimes called *Maxwell's equation* after James Clerk Maxwell [◀ page 296], shows that the speeds of gas molecules are indeed related directly to the temperature (Figure 12.14). The rms speed is a useful quantity because of its direct relationship to the average kinetic energy and because it is very close to the true average speed for a sample. (The average speed is 92% of the rms speed.)

All gases have the same average kinetic energy at the same temperature. However, if you compare a sample of one gas with another—say, compare O_2 and N_2—the molecules do not necessarily have the same average speed (Figure 12.15). Instead, Maxwell's equation shows that the smaller the molar mass of the gas, the greater the rms speed.

■ **Maxwell-Boltzmann Curves**
Plots showing the relationship between the number of molecules and their speed or energy (Figure 12.14) are often called Maxwell–Boltzmann distribution curves. They are named after James Clerk Maxwell (1831–1879) and Ludwig Boltzmann (1844–1906). The distribution of speeds (or kinetic energies) of molecules (as illustrated in Figures 12.14 and 12.15) is often used when explaining chemical phenomena.

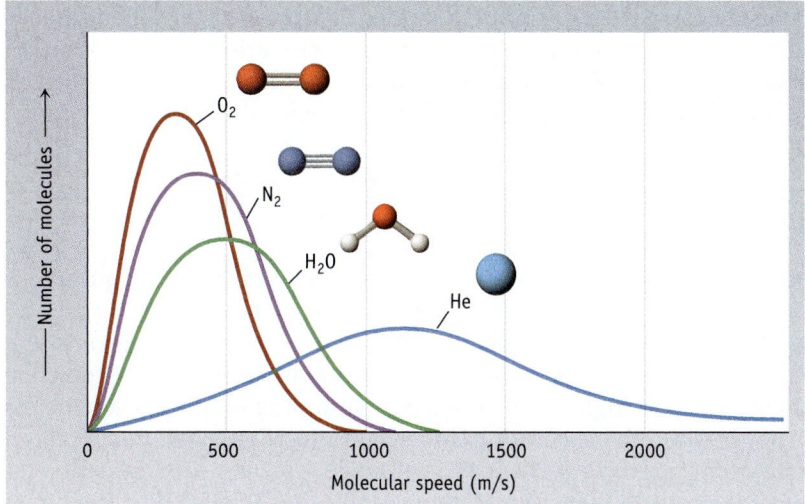

Figure 12.15 The effect of molecular mass on the distribution of speeds. At a given temperature, molecules with higher masses have lower speeds.

GENERAL
Chemistry⚛️**Now**™

See the General ChemistryNow CD-ROM or website:

- **Screen 12.9 The Kinetic-Molecular Theory of Gases: Gases on the Molecular Scale,** to view an animation of gases at different temperatures
- **Screen 12.11 Distribution of Molecular Speeds: Maxwell-Boltzmann Curves,** to view an animation of a Boltzmann distribution and to see a simulation in which distribution curves are calculated

Example 12.12—Molecular Speed

Problem Calculate the rms speed of oxygen molecules at 25 °C.

Strategy We must use Equation 12.9 with M in units of kg/mol. The reason is that R is in units of J/K · mol, and $1\ J = 1\ kg \cdot m^2/s^2$.

Solution The molar mass of O_2 is 32.0×10^{-3} kg/mol.

$$\sqrt{\overline{u^2}} = \sqrt{\frac{3(8.3145\ \text{J/K} \cdot \text{mol})(298\ \text{K})}{32.0 \times 10^{-3}\ \text{kg/mol}}} = \sqrt{2.32 \times 10^5\ \text{J/kg}}$$

To obtain the answer in meters per second, we use the relation $1\ J = 1\ kg \cdot m^2/s^2$. This means we have

$$\sqrt{\overline{u^2}} = \sqrt{2.32 \times 10^5\ \text{kg} \cdot \text{m}^2/(\text{kg} \cdot \text{s}^2)} = \sqrt{2.32 \times 10^5\ \text{m}^2/\text{s}^2} = \boxed{482\ \text{m/s}}$$

This speed is equivalent to about 1100 miles per hour!

Exercise 12.11—Molecular Speed

Calculate the rms speeds of helium atoms and N_2 molecules at 25 °C.

Kinetic-Molecular Theory and the Gas Laws

The gas laws, which come from experiment, can be explained by the kinetic-molecular theory. The starting place is to describe how pressure arises from collisions of gas molecules with the walls of the container holding the gas (Figure 12.16). Recall that pressure is related to the force of the collisions (see Section 12.1).

$$\text{Gas pressure from collisions} = \frac{\text{force of collisions}}{\text{area}}$$

The force exerted by the collisions depends on the number of collisions and the average force per collision. When the temperature of a gas increases, the average kinetic energy of the molecules increases as well. In turn, the average force of the collisions with the walls increases. (This is akin to the difference in the force exerted by a car traveling at high speed versus one moving at only a few kilometers per hour.) Also, because the speed of gas molecules increases with temperature, more collisions occur per second. Thus, the collective force per square centimeter is greater, and the pressure increases. Mathematically, this is related to the direct proportionality between P and T when n and V are fixed; that is, $P = (nR/V)T$.

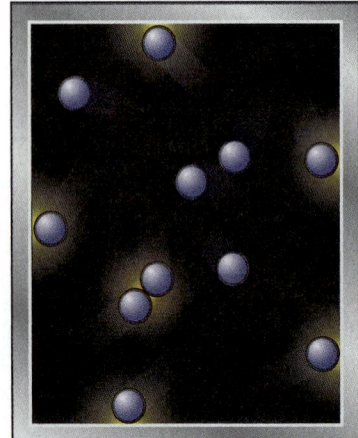

Figure 12.16 Gas pressure. According to the kinetic-molecular theory, gas pressure is caused by gas molecules bombarding the container walls.

Increasing the number of molecules of a gas at a fixed temperature and volume does not change the average collision force, but it does increase the number of collisions occurring per second. Thus, the pressure increases, and we can say that P is proportional to n when V and T are constant; that is, $P = n(RT/V)$.

If the pressure is to remain constant when either the number of molecules of gas or the temperature is increased, then the volume of the container (and the area over which the collisions can take place) must increase. This is expressed by stating that V is proportional to nT when P is constant [$V = nT(R/P)$], a statement that is a *combination of Avogadro's hypothesis and Charles's law.*

Finally, if the temperature is constant, the average impact force of molecules of a given mass with the container walls must be constant. If n is kept constant while the volume of the container becomes smaller, the number of collisions with the container walls per second must increase. This means the pressure increases, so P is proportional to $1/V$ when n and T are constant, as stated by *Boyle's law;* that is, $P = (1/V)(nRT)$.

GENERAL
Chemistry-⚛-Now™

See the General ChemistryNow CD-ROM or website:
- **Screen 12.10 Gas Laws and Kinetic-Molecular Theory,** to view an animation and to see a simulation of the gas laws at the molecular level.

12.7—Diffusion and Effusion

When a pizza is brought into a room, the volatile aroma-causing molecules vaporize into the atmosphere, where they mix with the oxygen, nitrogen, carbon dioxide, water vapor, and other gases present. Even if there were no movement of the air in the room caused by fans or people moving about, the odor would eventually reach everywhere in the room. This mixing of molecules of two or more gases due to their random molecular motions is called **diffusion**. Given time, the molecules of one component in a gas mixture will thoroughly and completely mix with all other components of the mixture (Figure 12.17).

(a)   **(b)**

Charles D. Winters

Figure 12.17 Diffusion. (a) Liquid bromine, Br_2, was placed in a small flask inside a larger container. (b) The cork was removed from the flask and, with time, bromine vapor diffused into the larger container. Bromine vapor is now distributed evenly in the containers.

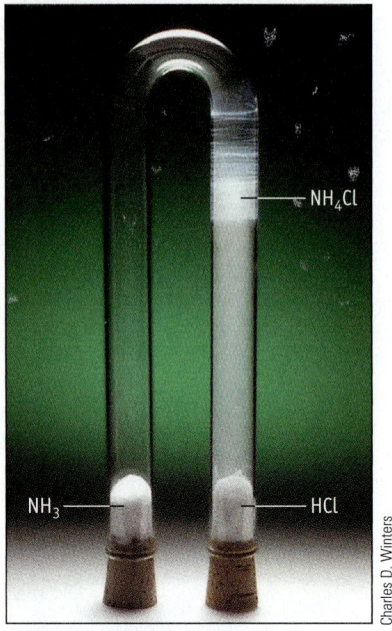

Active Figure 12.18 **Gaseous diffusion.** HCl gas (from hydrochloric acid) and NH₃ gas (from aqueous ammonia) diffuse from opposite ends of a glass U-tube. When they meet, they react to form white, solid NH₄Cl. The NH₄Cl is formed closer to the end from which the HCl gas begins because HCl molecules are heavier than NH₃ molecules and diffuse slower. See also Figure 12.13.

GENERAL
Chemistry⚛Now™ See the General ChemistryNow CD-ROM or website to explore an interactive version of this figure accompanied by an exercise.

Diffusion is also illustrated by the experiment shown in Figure 12.18. Here cotton moistened with hydrochloric acid is placed at one end of a U-tube, and cotton moistened with aqueous ammonia is placed at the other end. Molecules of HCl and NH₃ diffuse into the tube. When they meet, they produce white, solid NH₄Cl (just as in Figure 12.13).

$$HCl(g) + NH_3(g) \longrightarrow NH_4Cl(s)$$

Notice that the gases do not meet in the middle. Rather, because the heavier HCl molecules diffuse less rapidly than the lighter NH₃ molecules, the molecules meet closer to the HCl end of the U-tube.

Closely related to diffusion is **effusion**, which is the movement of gas through a tiny opening in a container into another container where the pressure is very low (Figure 12.19). Thomas Graham (1805–1869), a Scottish chemist, studied the effusion of gases and found that the rate of effusion of a gas—the amount of gas moving from one place to another in a given amount of time—is inversely proportional to the square root of its molar mass. Based on these experimental results, the rates of effusion of two gases can be compared:

$$\frac{\text{Rate of effusion of gas 1}}{\text{Rate of effusion of gas 2}} = \sqrt{\frac{\text{molar mass of gas 2}}{\text{molar mass of gas 1}}} \qquad (12.10)$$

The relationship in Equation 12.10—now known as **Graham's law**—is readily derived from Maxwell's equation by recognizing that the rate of effusion depends on the speed of the molecules. The ratio of the rms speeds is the same as the ratio of the effusion rates:

$$\frac{\text{Rate of effusion of gas 1}}{\text{Rate of effusion of gas 2}} = \frac{\sqrt{u^2 \text{ of gas 1}}}{\sqrt{u^2 \text{ of gas 2}}} = \frac{\sqrt{3RT/(M \text{ of gas 1})}}{\sqrt{3RT/(M \text{ of gas 2})}}$$

Canceling out like terms gives the expression in Equation 12.10.

GENERAL
Chemistry⚛Now™

See the General ChemistryNow CD-ROM or website:
- **Screen 12.12 Application of the Kinetic-Molecular Theory: Diffusion,** to watch a video of diffusion and for an interactive exercise

Figure 12.19 **Effusion.** H₂ and N₂ gas molecules effuse through the pores of a porous barrier. Lighter molecules (H₂) with higher average speeds strike the barrier more often and pass more often through it than heavier, slower molecules (N₂) at the same temperature. According to Graham's law, H₂ molecules effuse 3.73 times faster than N₂ molecules.

Before effusion

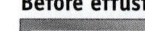

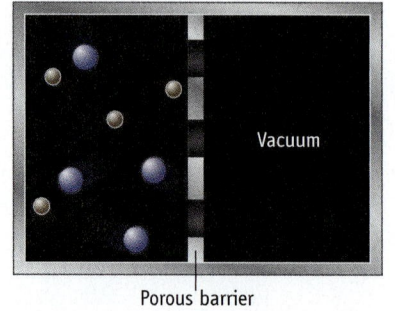

Vacuum

Porous barrier

During effusion

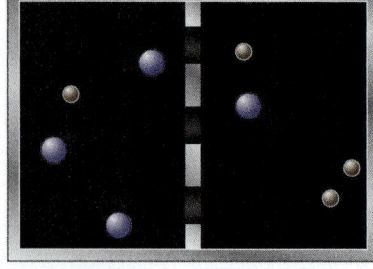

**Example 12.13—Using Graham's Law
of Effusion to Calculate Molar Mass**

Problem Tetrafluoroethylene, C_2F_4, effuses through a barrier at a rate of 4.6×10^{-6} mol/h. An unknown gas, consisting of only boron and hydrogen, effuses at a rate of 5.8×10^{-6} mol/h under the same conditions. What is the molar mass of the unknown gas?

Strategy From Graham's law we know that a light molecule will effuse more rapidly than a heavier one. Because the unknown gas effuses more rapidly than C_2F_4 ($M = 100.0$ g/mol), the unknown must have a molar mass less than 100 g/mol. Substitute the experimental data into Graham's law (Equation 12.10).

Solution

$$\frac{5.8 \times 10^{-6} \text{ mol/h}}{4.6 \times 10^{-6} \text{ mol/h}} = 1.3 = \sqrt{\frac{100.0 \text{ g/mol}}{M \text{ of unknown}}}$$

To solve for the unknown molar mass, square both sides of the equation and rearrange to find M for the unknown.

$$1.6 = \frac{100.0 \text{ g/mol}}{M \text{ of unknown}}$$

$$M = 63 \text{ g/mol}$$

Comment A boron–hydrogen compound corresponding to this molar mass is B_5H_9, called pentaborane.

Exercise 12.12—Graham's Law

A sample of pure methane, CH_4, is found to effuse through a porous barrier in 1.50 min. Under the same conditions, an equal number of molecules of an unknown gas effuse through the barrier in 4.73 min. What is the molar mass of the unknown gas?

12.8—Some Applications of the Gas Laws and Kinetic-Molecular Theory

Separating Isotopes

The effusion process played a central role in the development of the atomic bomb in World War II and is still used today to prepare fissionable uranium for nuclear power plants. Naturally occurring uranium exists primarily as two isotopes: ^{235}U (0.720% abundant) and ^{238}U (99.275% abundant). Because only the lighter isotope, ^{235}U, is suitable as a fuel in reactors, uranium ore must be enriched in this isotope.

Gas effusion is one way to separate the ^{235}U and ^{238}U isotopes. A uranium oxide sample is first converted to uranium hexafluoride, UF_6. This solid fluoride sublimes readily; it has a vapor pressure of 760 mm Hg at 55.6 °C. When UF_6 vapor is placed in a chamber with porous walls, the lighter, more rapidly moving $^{235}UF_6$ molecules effuse through the walls to a greater extent than the heavier $^{238}UF_6$ molecules.

To assess the separation of uranium isotopes, let us compare the rates of effusion of $^{235}UF_6$ and $^{238}UF_6$. Using Graham's law,

$$\frac{\text{Rate of } ^{235}UF_6}{\text{Rate of } ^{238}UF_6} = \sqrt{\frac{238.051 + 6(18.998)}{235.044 + 6(18.998)}} = 1.0043$$

■ **Vapor pressure**
The vapor pressure of volatile liquids and solids is described in detail in Section 13.5.

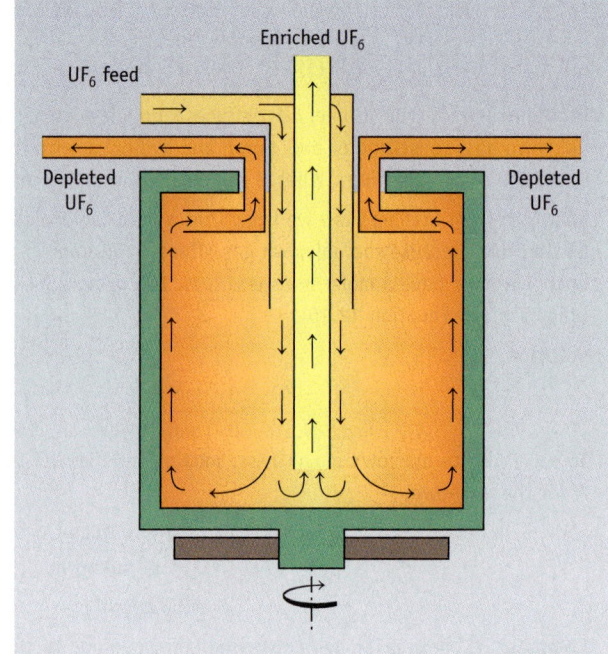

Figure 12.20   Isotope separation. Separation of uranium isotopes for use in atomic weaponry or in nuclear power plants was originally done by gas effusion. (These types of plants are still in use in the United States at Piketon, Ohio, and Paducah, Kentucky.) The more modern approach is to use a gas centrifuge (*left*). (*right*) UF_6 gas is injected into the centrifuge from a tube passing down through the center of a tall, spinning cylinder. The heavier $^{238}UF_6$ molecules experience more centrifugal force and move to the outer wall of the cylinder; the lighter $^{235}UF_6$ molecules stay closer to the center. A temperature difference inside the rotor causes the $^{235}UF_6$ molecules to move to the top of the cylinder. (See *The New York Times*, p. F1, March 23, 2004.)

we find that $^{235}UF_6$ will pass through a porous barrier 1.0043 times faster than $^{238}UF_6$. In other words, if we sample the gas that passes through the barrier, the fraction of $^{235}UF_6$ molecules will be larger. If the process is carried out again on the sample now higher in $^{235}UF_6$ concentration, the fraction of $^{235}UF_6$ would again increase in the effused sample, and the separation factor is now 1.0043 × 1.0043. If the cycle is repeated over and over again, the separation factor is 1.0043^n, where *n* is the number of enrichment cycles. To achieve a separation of about 99%, several hundred cycles are required (Figure 12.20)!

Deep Sea Diving

Diving with a self-contained underwater breathing apparatus (SCUBA) is exciting. If you want to dive much beyond about 60 ft (18 m), however, you need to take special precautions.

When you breathe air from a SCUBA tank (Figure 12.21), the pressure of the gas in your lungs is equal to the pressure exerted on your body. When you are at the surface, atmospheric pressure is about 1 atm and, because air has an oxygen concentration of 21%, the partial pressure of O_2 is about 0.21 atm. If you are at a depth of about 33 ft, the water pressure is 2 atm. Thus the oxygen partial pressure at this depth is double the surface partial pressure, or about 0.4 atm. Similarly, the partial pressure of N_2, which is about 0.8 atm at the surface, doubles to about 1.6 atm at a depth of 33 ft. The solubility of gases in water (and in blood) is directly proportional

Figure 12.21 **SCUBA diving.** Ordinary recreational dives can be made with compressed air to depths of about 60 feet or so. With a gas mixture called Nitrox (which contains a maximum of 64% N_2), a person can stay at such depths for a longer period. To go even deeper, however, divers must breathe special gas mixtures such as Trimix. This breathing mixture consists of oxygen, helium, and nitrogen.

to pressure. Therefore, more oxygen and nitrogen dissolve in blood under these conditions, which creates a problem called "nitrogen narcosis."

Nitrogen narcosis, also called "rapture of the deep" or the "martini effect," results from the toxic effect on nerve conduction of N_2 dissolved in blood. Its effect is comparable to drinking a martini on an empty stomach or taking laughing gas (nitrous oxide, N_2O) at the dentist; it makes you slightly giddy. In severe cases, it can impair a diver's judgment and even cause a diver to take the regulator out of his or her mouth and hand it to a fish! Some people can go as deep as 130 ft with no problem, but others experience nitrogen narcosis at 80 ft.

Another problem with breathing air at depths beyond 100 ft or so is oxygen toxicity. Our bodies are regulated for a partial pressure of O_2 of 0.21 atm. At a depth of 130 ft, the partial pressure of O_2 is comparable to breathing 100% oxygen at sea level. These higher partial pressures can harm the lungs and cause central nervous system damage. Oxygen toxicity is the reason deep dives are done not with compressed air but rather with gas mixtures containing a much lower percentage of O_2—say, about 10%.

Because of the risk of nitrogen narcosis, divers going beyond about 130 ft, such as those who work for offshore oil drilling companies, use a mixture of oxygen and helium. This solves the nitrogen narcosis problem, but it introduces another side effect. If the diver has a voice link to the surface, the diver's speech sounds like Donald Duck! Speech is altered because the velocity of sound in helium is different from that in air, and the density of gas at several hundred feet is much higher than at the surface.

12.9—Nonideal Behavior: Real Gases

If you are working with a gas at approximately room temperature and a pressure of 1 atm or less, the ideal gas law is remarkably successful in relating the amount of gas and its pressure, volume, and temperature. At higher pressures or lower temperatures, however, deviations from the ideal gas law occur. The origin of these deviations is explained by the breakdown of the assumptions used when describing ideal

■ **Assumptions of the KMT—Revisited**
The assumptions of the kinetic-molecular theory were given on page 567.

1. Gases consist of particles (molecules or atoms), whose separation is much greater than the size of the particles themselves.
2. The particles of a gas are in continual, random, and rapid motion. As they move, they collide with one another and with the walls of their container, but they do so without loss of energy.
3. The average kinetic energy of gas particles is proportional to the gas temperature. All gases, regardless of their molecular mass, have the same average kinetic energy at the same temperature.

Table 12.2 Van der Waals Constants

Gas	a Values (atm · L²/mol²)	b Values (L/mol)
He	0.034	0.0237
Ar	1.34	0.0322
H_2	0.244	0.0266
N_2	1.39	0.0391
O_2	1.36	0.0318
CO_2	3.59	0.0427
Cl_2	6.49	0.0562
H_2O	5.46	0.0305

gases. Specifically, ideality assumes gas molecules have no volume and that no forces act between them.

At standard temperature and pressure (STP), the volume occupied by a single molecule is *very* small relative to its share of the total gas volume. A helium atom with a radius of 31 pm, for example, has roughly the same space to move about as a pea has inside a basketball. Now suppose the pressure is increased significantly, to 1000 atm. The volume available to each molecule is a sphere with a radius of only about 200 pm, which means the situation is now like that of a pea inside a sphere a bit larger than a ping-pong ball.

The kinetic-molecular theory and the ideal gas law are concerned with the volume available to the molecules to move about, not the volume of the molecules themselves. It is clear that the volume occupied by gas molecules is not negligible at higher pressures. For example, suppose you have a flask marked with a volume of 500 mL. This does not mean the space available to molecules is 500 mL. In reality, the available volume is less than 500 mL, especially at high gas pressures, because the molecules themselves occupy some of the volume.

Another assumption of the kinetic-molecular theory is that collisions between molecules are elastic—that is, that the atoms or molecules of the gas never stick to one another by some type of intermolecular force. This is also clearly not true. All gases can be liquefied, although some gases require a very low temperature to do so (see Figure 12.12). The only way that this phase change can happen is if there are forces between the molecules. When a molecule is about to strike the wall of its container, other molecules in its vicinity exert a slight attraction for the molecule and pull it away from the wall. As a result of the intermolecular forces, molecules strike the wall with less force than they would in the absence of intermolecular attractive forces. Thus, because collisions between molecules in a real gas and the wall are softer, the observed gas pressure is less than that predicted by the ideal gas law. This effect can be particularly pronounced when the temperature is low.

The Dutch physicist Johannes van der Waals (1837–1923) studied the breakdown of the ideal gas law equation and developed an equation to correct for the errors arising from nonideality. This equation is known as the **van der Waals equation**:

Observed pressure   Container V

$$\left(P + a\left[\frac{n}{V}\right]^2\right)(V - bn) = nRT \qquad (12.11)$$

Correction for intermolecular forces   Correction for molecular volume

where a and b are experimentally determined constants (Table 12.2). Although Equation 12.11 might seem complicated at first glance, the terms in parentheses are those of the ideal gas law, each corrected for the effects discussed previously. The pressure correction term, $a(n/V)^2$, accounts for intermolecular forces. Owing to intermolecular forces the observed gas pressure is lower than the ideal pressure ($P_{observed} < P_{ideal}$, where P_{ideal} is calculated using the equation $PV = nRT$). Therefore, the term $a(n/V)^2$ is added to the observed pressure. The constant a typically has values in the range 0.01 to 10 atm · L²/mol². The actual volume available to the molecules is smaller than the volume of the container because the molecules themselves take up space. Therefore, we subtract an amount from the container volume (= bn) to take this factor into account. Here n is the number of moles of gas, and b is an experimental quantity that corrects for the molecular volume. Typical values of b range from 0.01 to 0.1 L/mol, roughly increasing with increasing molecular size.

Chemical Perspectives

The Earth's Atmosphere

Earth's atmosphere is a fascinating mixture of gases in more or less distinct layers with widely differing temperatures.

Up to the tropopause, there is a gradual decline in temperature (and pressure) with altitude. The temperature climbs again in the stratosphere due to the absorption of energy from the sun by stratospheric ozone, O_3.

Above the stratosphere, the pressure declines because fewer molecules are present. At still higher altitudes, we observe a dramatic increase in temperature in the thermosphere. This trend illustrates the difference between *temperature* and *heat*. The temperature of a gas reflects the *average* kinetic energy of the molecules of the gas, whereas the heat present in an object is the *total* kinetic energy of the molecules. In the thermosphere, the few molecules present have a very high temperature, but the heat content is exceedingly small because there are so few molecules.

Gases within the troposphere are well mixed by convection. Pollutants that are evolved on the earth's surface can rise into the stratosphere, but the stratosphere then acts as a "thermal lid" on the troposphere, preventing significant mixing of polluting gases into the stratosphere and beyond.

The pressure of the atmosphere declines with altitude; in conjunction with this trend, the partial pressure of O_2 declines. The figure shows why climbers have a hard time breathing on Mount Everest. At the mountain's peak, the altitude is 29,028 ft (8848 m), and the O_2 partial pressure is only 30% of the sea-level partial pressure. With proper training, a climber can reach the summit without supplemental oxygen. However, this same feat would not be possible if Mount Everest were located farther north. The earth's atmosphere thins toward the poles, so the O_2 partial pressure would be even lower if Mount Everest's summit were in North America, for example.

See G. N. Eby: *Environmental Geochemistry*, Belmont, CA, Thomson/Brooks/Cole, 2004.

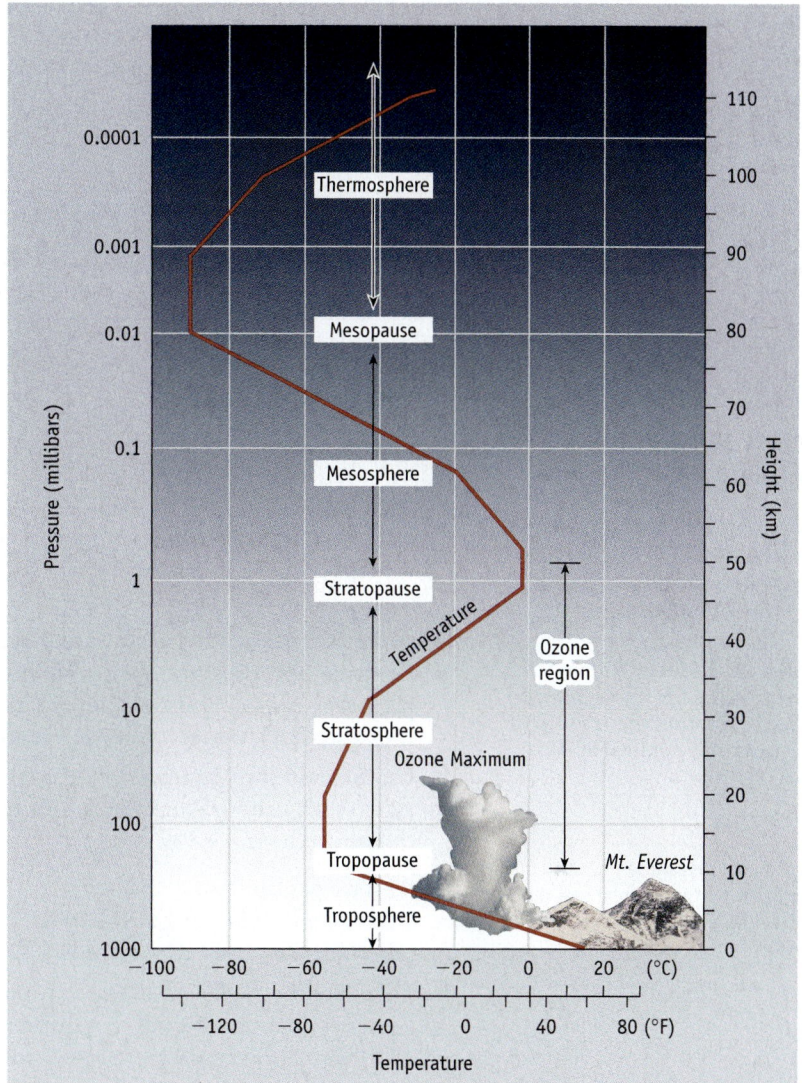

Average Composition of the Earth's Atmosphere to a Height of 25 km

Gas	Volume %	Source
N_2	78.08	biologic
O_2	20.95	biologic
Ar	0.93	radioactivity
Ne	0.0018	Earth's interior
He	0.0005	radioactivity
H_2O	0 to 4	evaporation
CO_2	0.036	biologic, industrial
CH_4	0.00017	biologic
N_2O	0.00003	biologic, industrial
O_3	0.000004	photochemical

As an example of the importance of these corrections, consider a sample of 8.00 mol of chlorine gas, Cl_2, in a 4.00-L tank at 27.0 °C. The ideal gas law would lead you to expect a pressure of 49.2 atm. A better estimate of the pressure, obtained from the van der Waals equation, is 29.5 atm, about 20 atm less than the ideal pressure!

Exercise 12.13—The van der Waals Equation

Using both the ideal gas law and the van der Waals equation, calculate the pressure expected for 10.0 mol of helium gas in a 1.00-L container at 25 °C.

GENERAL **ChemistryNow™**

See the General ChemistryNow CD-ROM or website to:

- Assess your understanding with homework questions keyed to each goal
- Check your readiness for an exam by taking the exam-prep quiz and exploring the resources in the personalized Learning Plan it provides

For additional preparation for an examination on this chapter see the **Let's Review** section on pages 642–655.

Chapter Goals Revisited

When you have finished studying this chapter, you should ask if you have met the chapter goals. In particular you should be able to

Understand the basis of the gas laws and know how to use those laws

a. Describe how pressure measurements are made and work with the units of pressure, especially atmospheres (atm) and millimeters of mercury (mm Hg) (Section 12.1). General ChemistryNow homework: Study Question(s) 1

b. Understand the origins of the gas laws (Boyle's law, Charles's law, and Avogadro's hypothesis) and know how to apply them (Section 12.2). General ChemistryNow homework: SQ(s) 6, 8, 10, 12

Use the ideal gas law

a. Understand the origin of the ideal gas law and know how to use the equation (Section 12.3). General ChemistryNow homework: SQ(s) 18, 22, 24, 65

b. Calculate the molar mass of a compound from a knowledge of the pressure of a known quantity of a gas in a given volume at a known temperature (Section 12.3). General ChemistryNow homework: SQ(s) 26, 30

Apply the gas laws to stoichiometric calculations (Section 12.4)

a. Stoichiometric calculations involving gases. General ChemistryNow homework: SQ(s) 32, 34, 71, 93

b. Use Dalton's law of partial pressures (Section 12.5). General ChemistryNow homework: SQ(s) 39, 40, 74

Understand kinetic-molecular theory as it is applied to gases, especially the distribution of molecular speeds (energies) (Section 12.6)

a. Apply the kinetic-molecular theory of gas behavior at the molecular level (Section 12.6). General ChemistryNow homework: SQ(s) 41, 45

b. Understand the phenomena of diffusion and effusion and know how to use Graham's law (Section 12.7). General ChemistryNow homework: SQ(s) 47

Recognize why real-world gases do not behave like ideal gases

a. Appreciate the fact that gases usually do not behave as ideal gases (Section 12.9). Deviations from ideal behavior are largest at high pressure and low temperature.

Key Equations

Equation 12.1 (page 551)
Boyle's law (where P is the gas pressure and V is its volume)
$$P_1V_1 = P_2V_2$$

Equation 12.2 (page 553)
Charles's law (where T is the temperature in kelvins)
$$\frac{V_1}{T_1} = \frac{V_2}{T_2}$$

Equation 12.3 (page 554)
General gas law (combined gas law)
$$\frac{P_1V_1}{T_1} = \frac{P_2V_2}{T_2}$$

Equation 12.4 (page 557)
Ideal gas law (where n is the amount of gas in moles and R is the universal gas constant, $0.082057\ \text{L} \cdot \text{atm}/\text{K} \cdot \text{mol}$)
$$PV = nRT$$

Equation 12.5 (page 559)
Density of gases (where d is the gas density in grams per liter)
$$d = \frac{m}{V} = \frac{PM}{RT}$$

Equation 12.6 (page 564)
Dalton's law of partial pressures: The total pressure of a gas mixture is the sum of the partial pressures of the component gases (P_n)
$$P_{\text{total}} = P_1 + P_2 + P_3 + \cdots$$

Equation 12.7 (page 565)
The total pressure of a gas mixture is equal to the total number of moles of gases multiplied by (RT/V)
$$P_{\text{total}} = n_{\text{total}}\left(\frac{RT}{V}\right)$$

Equation 12.8 (page 565)
The pressure of a gas (A) in a mixture is the product of its mole fraction (X_A) and the total pressure of the mixture
$$P_A = X_A P_{\text{total}}$$

Equation 12.9 (page 569)
Maxwell's equation, which relates the rms speed $\sqrt{\overline{u^2}}$ to the molar mass of a gas (M) and its temperature (T) (where R = $8.314472\ \text{J}/\text{K} \cdot \text{mol}$)
$$\sqrt{\overline{u^2}} = \sqrt{\frac{3RT}{M}}$$

Equation 12.10 (page 572)
Graham's law: The rate of effusion of a gas—the amount of material moving from one place to another in a given time—is inversely proportional to the square root of its molar mass
$$\frac{\text{Rate of effusion of gas 1}}{\text{Rate of effusion of gas 2}} = \sqrt{\frac{\text{molar mass of gas 2}}{\text{molar mass of gas 1}}}$$

Equation 12.11 (page 576)

The van der Waals equation: Relates pressure, volume, temperature, and amount of gas for a nonideal gas

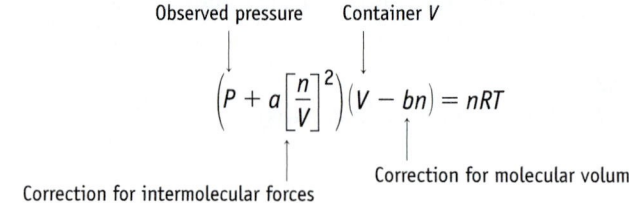

Observed pressure   Container V

$$\left(P + a\left[\frac{n}{V}\right]^2\right)(V - bn) = nRT$$

Correction for intermolecular forces

Correction for molecular volume

Study Questions

▲ denotes more challenging questions.

■ denotes questions available in the Homework and Goals section of the General ChemistryNow CD-ROM or website.

Blue numbered questions have answers in Appendix O and fully worked solutions in the *Student Solutions Manual.*

Structures of many of the compounds used in these questions are found on the General ChemistryNow CD-ROM or website in the Models folder.

GENERAL Chemistry··Now™ Assess your understanding of this chapter's topics with additional quizzing and conceptual questions at **http://now.brookscole.com/kotz6e**

Practicing Skills

Pressure
(See Example 12.1 and the General ChemistryNow Screen 12.2.)

1. ■ The pressure of a gas is 440 mm Hg. Express this pressure in units of (a) atmospheres, (b) bars, and (c) kilopascals.

2. The average barometric pressure at an altitude of 10 km is 210 mm Hg. Express this pressure in atmospheres, bars, and kilopascals.

3. Indicate which represents the higher pressure in each of the following pairs:
    (a) 534 mm Hg or 0.754 bar
    (b) 534 mm Hg or 650 kPa
    (c) 1.34 bar or 934 kPa

4. Put the following in order of increasing pressure: 363 mm Hg, 363 kPa, 0.256 atm, and 0.523 bar.

Boyle's Law and Charles's Law
(See Examples 12.2 and 12.3 and the General ChemistryNow Screen 12.3.)

5. A sample of nitrogen gas has a pressure of 67.5 mm Hg in a 500.-mL flask. What is the pressure of this gas sample when it is transferred to a 125-mL flask at the same temperature?

6. ■ A sample of CO_2 gas has a pressure of 56.5 mm Hg in a 125-mL flask. The sample is transferred to a new flask, where it has a pressure of 62.3 mm Hg at the same temperature. What is the volume of the new flask?

7. You have 3.5 L of NO at a temperature of 22.0 °C. What volume would the NO occupy at 37 °C? (Assume the pressure is constant.)

8. ■ A 5.0-mL sample of CO_2 gas is enclosed in a gas-tight syringe (see Figure 12.4) at 22 °C. If the syringe is immersed in an ice bath (0 °C), what is the new gas volume, assuming that the pressure is held constant?

The General Gas Law
(See Example 12.4.)

9. You have 3.6 L of H_2 gas at 380 mm Hg and 25 °C. What is the pressure of this gas if it is transferred to a 5.0-L flask at 0.0 °C?

10. ■ You have a sample of CO_2 in a flask A with a volume of 25.0 mL. At 20.5 °C, the pressure of the gas is 436.5 mm Hg. To find the volume of another flask B, you move the CO_2 to that flask and find that its pressure is now 94.3 mm Hg at 24.5 °C. What is the volume of flask B?

11. You have a sample of gas in a flask with a volume of 250 mL. At 25.5 °C the pressure of the gas is 360 mm Hg. If you decrease the temperature to −5.0 °C, what is the gas pressure at the lower temperature?

12. ■ A sample of gas occupies 135 mL at 22.5 °C; the pressure is 165 mm Hg. What is the pressure of the gas sample when it is placed in a 252-mL flask at a temperature of 0.0 °C?

13. One of the cylinders of an automobile engine has a volume of 400. cm^3. The engine takes in air at a pressure of 1.00 atm and a temperature of 15 °C and compresses the air to a volume of 50.0 cm^3 at 77 °C. What is the final pressure of the gas in the cylinder? (The ratio of before and after volumes—in this case, 400:50 or 8:1—is called the compression ratio.)

14. A helium-filled balloon of the type used in long-distance flying contains 420,000 ft^3 (1.2×10^7 L) of helium. Suppose you fill the balloon with helium on the ground, where the pressure is 737 mm Hg and the temperature is 16.0 °C. When the balloon ascends to a height of 2 miles, where the pressure is only 600. mm Hg and the temperature is −33 °C, what volume is occupied by the helium gas? Assume the pressure inside the balloon matches the external pressure. Comment on the result.

Avogadro's Hypothesis

(See Example 12.5 and the General ChemistryNow Screen 12.3.)

15. Nitrogen monoxide reacts with oxygen to give nitrogen dioxide.

$$2\,NO(g) + O_2(g) \longrightarrow 2\,NO_2(g)$$

(a) If you mix NO and O_2 in the correct stoichiometric ratio, and NO has a volume of 150 mL, what volume of O_2 is required (at the same pressure and temperature)?

(b) After reaction is complete between 150 mL of NO and the stoichiometric volume of O_2, what is the volume of NO_2 (at the same pressure and temperature)?

16. Ethane, C_2H_6, burns in air according to the equation

$$2\,C_2H_6(g) + 7\,O_2(g) \longrightarrow 4\,CO_2(g) + 6\,H_2O(g)$$

What volume of O_2 (L) is required for complete reaction with 5.2 L of C_2H_6? What volume of H_2O vapor (L) is produced? Assume all gases are measured at the same temperature and pressure.

Ideal Gaw Law

(See Example 12.6 and the General ChemistryNow Screen 12.4.)

17. A 1.25-g sample of CO_2 is contained in a 750.-mL flask at 22.5 °C. What is the pressure of the gas?

18. ■ A balloon holds 30.0 kg of helium. What is the volume of the balloon if the final pressure is 1.20 atm and the temperature is 22 °C?

19. A flask is first evacuated so that it contains no gas at all. Then, 2.2 g of CO_2 is introduced into the flask. On warming to 22 °C, the gas exerts a pressure of 318 mm Hg. What is the volume of the flask?

20. A steel cylinder holds 1.50 g of ethanol, C_2H_5OH. What is the pressure of the ethanol vapor if the cylinder has a volume of 251 cm^3 and the temperature is 250 °C? (Assume all of the ethanol is in the vapor phase at this temperature.)

21. A balloon for long-distance flying contains 1.2×10^7 L of helium. If the helium pressure is 737 mm Hg at 25 °C, what mass of helium (in grams) does the balloon contain? (See Study Question 14 and page 546.)

22. ■ What mass of helium, in grams, is required to fill a 5.0-L balloon to a pressure of 1.1 atm at 25 °C?

Gas Density

(See Examples 12.7 and 12.8 and the General ChemistryNow Screen 12.5.)

23. Forty miles above the earth's surface the temperature is 250 K and the pressure is only 0.20 mm Hg. What is the density of air (in grams per liter) at this altitude? (Assume the molar mass of air is 28.96 g/mol.)

24. ■ Diethyl ether, $(C_2H_5)_2O$, vaporizes easily at room temperature. If the vapor exerts a pressure of 233 mm Hg in a flask at 25 °C, what is the density of the vapor?

25. A gaseous organofluorine compound has a density of 0.355 g/L at 17 °C and 189 mm Hg. What is the molar mass of the compound?

26. ■ Chloroform is a common liquid used in the laboratory. It vaporizes readily. If the pressure of chloroform vapor in a flask is 195 mm Hg at 25.0 °C, and the density of the vapor is 1.25 g/L, what is the molar mass of chloroform?

Ideal Gas Laws and Determining Molar Mass

(See Examples 12.7 and 12.8 and the General ChemistryNow Screen 12.6.)

27. A 1.007-g sample of an unknown gas exerts a pressure of 715 mm Hg in a 452-mL container at 23 °C. What is the molar mass of the gas?

28. A 0.0125-g sample of a gas with an empirical formula of CHF_2 is placed in a 165-mL flask. It has a pressure of 13.7 mm Hg at 22.5 °C. What is the molecular formula of the compound?

29. A new boron hydride, B_xH_y, has been isolated. To find its molar mass, you measure the pressure of the gas in a known volume at a known temperature. The following experimental data are collected:

Mass of gas = 12.5 mg

Pressure of gas = 24.8 mm Hg

Temperature = 25 °C

Volume of flask = 125 mL

Which formula corresponds to the calculated molar mass?

(a) B_2H_6

(b) B_4H_{10}

(c) B_5H_9

(d) B_6H_{10}

(e) $B_{10}H_{14}$

30. ■ Acetaldehyde is a common liquid compound that vaporizes readily. Determine the molar mass of acetaldehyde from the following data:

Sample mass = 0.107 g   Volume of gas = 125 mL

Temperature = 0.0 °C   Pressure = 331 mm Hg

Gas Laws and Stoichiometry

(See Examples 12.9 and 12.10 and the General ChemistryNow Screen 12.7.)

31. Iron reacts with hydrochloric acid to produce iron(II) chloride and hydrogen gas:

$$Fe(s) + 2\,HCl(aq) \longrightarrow FeCl_2(aq) + H_2(g)$$

The H_2 gas from the reaction of 2.2 g of iron with excess acid is collected in a 10.0-L flask at 25 °C. What is the pressure of the H_2 gas in this flask?

32. ■ Silane, SiH_4, reacts with O_2 to give silicon dioxide and water:

$$SiH_4(g) + 2\,O_2(g) \longrightarrow SiO_2(s) + 2\,H_2O(\ell)$$

A 5.20-L sample of SiH_4 gas at 356 mm Hg pressure and 25 °C is allowed to react with O_2 gas. What volume of O_2 gas, in liters, is required for complete reaction if the oxygen has a pressure of 425 mm Hg at 25 °C?

33. Sodium azide, the explosive compound in automobile air bags, decomposes according to the following equation:

$$2\,NaN_3(s) \longrightarrow 2\,Na(s) + 3\,N_2(g)$$

What mass of sodium azide is required to provide the nitrogen needed to inflate a 75.0-L bag to a pressure of 1.3 atm at 25 °C?

34. ■ The hydrocarbon octane (C_8H_{18}) burns to give CO_2 and water vapor:

$$2\,C_8H_{18}(g) + 25\,O_2(g) \longrightarrow 16\,CO_2(g) + 18\,H_2O(g)$$

If a 0.095-g sample of octane burns completely in O_2, what will be the pressure of water vapor in a 4.75-L flask at 30.0 °C? If the O_2 gas needed for complete combustion was contained in a 4.75-L flask at 22 °C, what would its pressure be?

35. Hydrazine reacts with O_2 according to the following equation:

$$N_2H_4(g) + O_2(g) \longrightarrow N_2(g) + 2\,H_2O(\ell)$$

Assume the O_2 needed for the reaction is in a 450-L tank at 23 °C. What must the oxygen pressure be in the tank to have enough oxygen to consume 1.00 kg of hydrazine completely?

36. A self-contained breathing apparatus uses canisters containing potassium superoxide. The superoxide consumes the CO_2 exhaled by a person and replaces it with oxygen.

$$4\,KO_2(s) + 2\,CO_2(g) \longrightarrow 2\,K_2CO_3(s) + 3\,O_2(g)$$

What mass of KO_2, in grams, is required to react with 8.90 L of CO_2 at 22.0 °C and 767 mm Hg?

Gas Mixtures and Dalton's Law

(See Example 12.11 and the General ChemistryNow Screen 12.8.)

37. What is the total pressure in atmospheres of a gas mixture that contains 1.0 g of H_2 and 8.0 g of Ar in a 3.0-L container at 27 °C? What are the partial pressures of the two gases?

38. A cylinder of compressed gas is labeled "Composition (mole %): 4.5% H_2S, 3.0% CO_2, balance N_2." The pressure gauge attached to the cylinder reads 46 atm. Calculate the partial pressure of each gas, in atmospheres, in the cylinder.

39. ■ A halothane–oxygen mixture ($C_2HBrClF_3 + O_2$) can be used as an anesthetic. A tank containing such a mixture has the following partial pressures: P (halothane) = 170 mm Hg and P (O_2) = 570 mm Hg.

   (a) What is the ratio of the number of moles of halothane to the number of moles of O_2?

   (b) If the tank contains 160 g of O_2, what mass of $C_2HBrClF_3$ is present?

40. ■ A collapsed balloon is filled with He to a volume of 12.5 L at a pressure of 1.00 atm. Oxygen, O_2 is then added so that the final volume of the balloon is 26 L with a total pressure of 1.00 atm. The temperature, which remains constant throughout, is 21.5 °C.

   (a) What mass of He does the balloon contain?

   (b) What is the final partial pressure of He in the balloon?

   (c) What is the partial pressure of O_2 in the balloon?

   (d) What is the mole fraction of each gas?

Kinetic-Molecular Theory

(See Section 12.6, Example 12.12, and the General ChemistryNow Screens 12.9–12.12.)

41. ■ You have two flasks of equal volume. Flask A contains H_2 at 0 °C and 1 atm pressure. Flask B contains CO_2 gas at 25 °C and 2 atm pressure. Compare these two gases with respect to each of the following:

   (a) average kinetic energy per molecule

   (b) average molecular velocity

   (c) number of molecules

   (d) mass of gas

42. Equal masses of gaseous N_2 and Ar are placed in separate flasks of equal volume at the same temperature. Tell whether each of the following statements is true or false. Briefly explain your answer in each case.

   (a) There are more molecules of N_2 present than atoms of Ar.

   (b) The pressure is greater in the Ar flask.

   (c) The Ar atoms have a greater average speed than the N_2 molecules.

   (d) The N_2 molecules collide more frequently with the walls of the flask than do the Ar atoms.

43. If the speed of an oxygen molecule is 4.28×10^4 cm/s at 25 °C, what is the speed of a CO_2 molecule at the same temperature?

44. Calculate the rms speed for CO molecules at 25 °C. What is the ratio of this speed to that of Ar atoms at the same temperature?

45. ■ Place the following gases in order of increasing average molecular speed at 25 °C: Ar, CH_4, N_2, CH_2F_2.

---

46. The reaction of SO_2 with Cl_2 gives dichlorine oxide, which is used to bleach wood pulp and to treat wastewater:

$$SO_2(g) + 2\ Cl_2(g) \longrightarrow OSCl_2(g) + Cl_2O(g)$$

All of the compounds involved in the reaction are gases. List them in order of increasing average speed.

Diffusion and Effusion

(See Example 12.13 and the General ChemistryNow Screen 12.12.)

47. ■ In each pair of gases below, tell which will effuse faster:
    (a) CO_2 or F_2
    (b) O_2 or N_2
    (c) C_2H_4 or C_2H_6
    (d) two chlorofluorocarbons: $CFCl_3$ or $C_2Cl_2F_4$

48. Argon gas is ten times denser than helium gas at the same temperature and pressure. Which gas is predicted to effuse faster? How much faster?

49. A gas whose molar mass you wish to know effuses through an opening at a rate one-third as fast as that of helium gas. What is the molar mass of the unknown gas?

50. ▲ A sample of uranium fluoride is found to effuse at the rate of 17.7 mg/h. Under comparable conditions, gaseous I_2 effuses at the rate of 15.0 mg/h. What is the molar mass of the uranium fluoride? (*Hint:* Rates must be converted to units of moles per time.)

Nonideal Gases

(See Section 12.9.)

51. In the text it is stated that the pressure of 8.00 mol of Cl_2 in a 4.00-L tank at 27.0 °C should be 29.5 atm if calculated using the van der Waals's equation. Verify this result and compare it with the pressure predicted by the ideal gas law.

52. You want to store 165 g of CO_2 gas in a 12.5-L tank at room temperature (25 °C). Calculate the pressure the gas would have using (a) the ideal gas law and (b) the van der Waals equation. (For CO_2, $a = 3.59$ atm · L^2/mol^2 and $b = 0.0427$ L/mol.)

General Questions

These questions are not designated as to type or location in the chapter. They may combine several concepts.

53. Complete the following table:

	atm	mm Hg	kPa	bar
Standard atmosphere	___	___	___	___
Partial pressure of N_2 in the atmosphere	___	593	___	___
Tank of compressed H_2	___	___	___	133
Atmospheric pressure at the top of Mount Everest	___	___	33.7	___

54. You want to fill a cylindrical tank with CO_2 gas at 865 mm Hg and 25 °C. The tank is 20.0 m long with a 10.0-cm radius. What mass of CO_2 (in grams) is required?

55. On combustion, 1.0 L of a gaseous compound of hydrogen, carbon, and nitrogen gives 2.0 L of CO_2, 3.5 L of H_2O vapor, and 0.50 L of N_2 at STP. What is the empirical formula of the compound?

56. To what temperature, in degrees Celsius, must a 25.5-mL sample of oxygen at 90 °C be cooled for its volume to decrease to 21.5 mL? Assume the pressure and mass of the gas are constant.

57. ▲ You have a sample of helium gas at −33 °C, and you want to increase the average speed of helium atoms by 10.0%. To what temperature should the gas be heated to accomplish this?

58. If 12.0 g of O_2 is required to inflate a balloon to a certain size at 27 °C, what mass of O_2 is required to inflate it to the same size (and pressure) at 5.0 °C?

59. You have two gas-filled balloons, one containing He and the other containing H_2. The H_2 balloon is twice the size of the He balloon. The pressure of gas in the H_2 balloon is 1 atm, and that in the He balloon is 2 atm. The H_2 balloon is outside in the snow (−5 °C), and the He balloon is inside a warm building (23 °C).
    (a) Which balloon contains the greater number of molecules?
    (b) Which balloon contains the greater mass of gas?

60. A bicycle tire has an internal volume of 1.52 L and contains 0.406 mol of air. The tire will burst if its internal pressure reaches 7.25 atm. To what temperature, in degrees Celsius, does the air in the tire need to be heated to cause a blowout?

61. The temperature of the atmosphere on Mars can be as high as 27 °C at the equator at noon, and the atmospheric pressure is about 8 mm Hg. If a spacecraft could collect 10. m^3 of this atmosphere, compress it to a small volume, and send it back to Earth, how many moles would the sample contain?

62. If you place 2.25 g of solid silicon in a 6.56-L flask that contains CH_3Cl with a pressure of 585 mm Hg at 25 °C, what mass of dimethyldichlorosilane, $(CH_3)_2SiCl_2(g)$, can be formed?

$$Si(s) + 2\ CH_3Cl(g) \longrightarrow (CH_3)_2SiCl_2(g)$$

What pressure of $(CH_3)_2SiCl_2(g)$ would you expect in this same flask at 95 °C on completion of the reaction? (Dimethyldichlorosilane is one starting material used to make silicones, polymeric substances used as lubricants, antistick agents, and water-proofing caulk.)

63. $Ni(CO)_4$ can be made by reacting finely divided nickel with gaseous CO. If you have CO in a 1.50-L flask at a pressure of 418 mm Hg at 25.0 °C, along with 0.450 g of Ni powder, what is the theoretical yield of $Ni(CO)_4$?

64. The gas B_2H_6 burns in air to give H_2O and B_2O_3.

$$B_2H_6(g) + 3\,O_2(g) \longrightarrow B_2O_3(s) + 3\,H_2O(g)$$

(a) Three gases are involved in this reaction. Place them in order of increasing molecular speed. (Assume all are at the same temperature.)

(b) A 3.26-L flask contains B_2H_6 at a pressure of 256 mm Hg and a temperature of 25 °C. Suppose O_2 gas is added to the flask until B_2H_6 and O_2 are in the correct stoichiometric ratio for the combustion reaction. At this point, what is the partial pressure of O_2?

65. ■ You have four gas samples:

1. 1.0 L of H_2 at STP

2. 1.0 L of Ar at STP

3. 1.0 L of H_2 at 27 °C and 760 mm Hg

4. 1.0 L of He at 0 °C and 900 mm Hg

(a) Which sample has the largest number of gas particles (atoms or molecules)?

(b) Which sample contains the smallest number of particles?

(c) Which sample represents the largest mass?

66. An automobile tire has a volume of 17 L. What mass of air is contained in the tire at 25 °C and a pressure of 3.2 atm? (Molar mass of air = 28.96 g/mol.)

67. Diborane, B_2H_6, reacts with oxygen to give boric oxide and water vapor.

$$B_2H_6(g) + 3\,O_2(g) \longrightarrow B_2O_3(s) + 3\,H_2O(g)$$

If you mix B_2H_6 and O_2 in the correct stoichiometric ratio, and if the total pressure of the mixture is 228 mm Hg, what are the partial pressures of B_2H_6 and O_2? If the temperature and volume do not change, what is the pressure of the water vapor?

68. Analysis of a gaseous chlorofluorocarbon, CCl_xF_y, shows that it contains 11.79% C and 69.57% Cl. In another experiment you find that 0.107 g of the compound fills a 458-mL flask at 25 °C with a pressure of 21.3 mm Hg. What is the molecular formula of the compound?

69. There are five compounds in the family of sulfur–fluorine compounds with the general formula S_xF_y. One of these compounds is 25.23% S. If you place 0.0955 g of the compound in a 89-mL flask at 45 °C, the pressure of the gas is 83.8 mm Hg. What is the molecular formula of S_xF_y?

70. A miniature volcano can be made in the laboratory with ammonium dichromate. When ignited, it decomposes in a fiery display.

$$(NH_4)_2Cr_2O_7(s) \longrightarrow N_2(g) + 4\,H_2O(g) + Cr_2O_3(s)$$

If 0.95 g of ammonium dichromate is used, and if the gases from this reaction are trapped in a 15.0-L flask at 23 °C, what is the total pressure of the gas in the flask? What are the partial pressures of N_2 and H_2O?

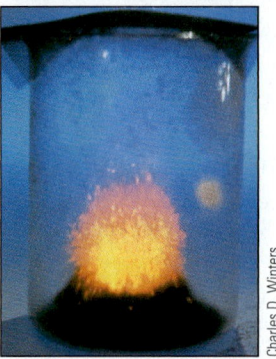

Charles D. Winters

Ammonium dichromate, $(NH_4)_2Cr_2O_7$, decomposes on heating to give nitrogen gas, water vapor, and the green solid, chromium(III) oxide.

71. ■ Iron carbonyl can be made by the direct reaction of iron metal and carbon monoxide.

$$Fe(s) + 5\,CO(g) \longrightarrow Fe(CO)_5(\ell)$$

What is the theoretical yield of $Fe(CO)_5$ if 3.52 g of iron is treated with CO gas having a pressure of 732 mm Hg in a 5.50-L flask at 23 °C?

72. You are given a solid mixture of $NaNO_2$ and NaCl and are asked to analyze it for the amount of $NaNO_2$ present. To do so you allow the mixture to react with sulfamic acid, HSO_3NH_2, in water according to the equation

$$NaNO_2(aq) + HSO_3NH_2(aq) \longrightarrow$$
$$NaHSO_4(aq) + H_2O(\ell) + N_2(g)$$

What is the weight percentage of $NaNO_2$ in 1.232 g of the solid mixture if reaction with sulfamic acid produces 295 mL of N_2 gas with a pressure of 713 mm Hg at 21.0 °C?

73. The density of air 20 km above the earth's surface is 92 g/m^3. The pressure of the atmosphere is 42 mm Hg and the temperature is −63 °C.

(a) What is the average molar mass of the atmosphere at this altitude?

(b) If the atmosphere at this altitude consists of only O_2 and N_2, what is the mole fraction of each gas?

74. ■ A 3.0-L bulb containing He at 145 mm Hg is connected by a valve to a 2.0-L bulb containing Ar at 355 mm Hg. (See the accompanying figure.) Calculate the partial pressure of each gas and the total pressure after the valve between the flasks is opened.

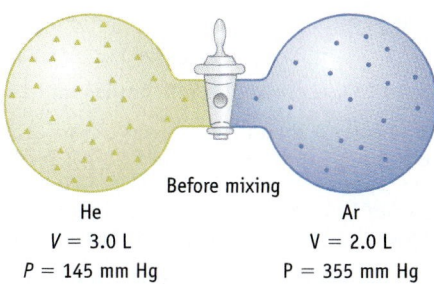

Before mixing

He
V = 3.0 L
P = 145 mm Hg

Ar
V = 2.0 L
P = 355 mm Hg

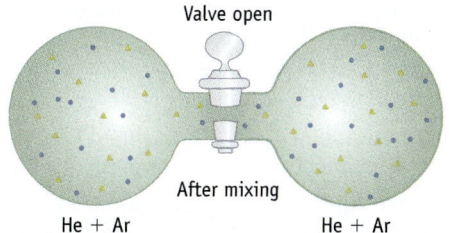

Valve open

After mixing

He + Ar   He + Ar

75. Phosphine gas, PH_3, is toxic when it reaches a concentration of 7×10^{-5} mg/L. To what pressure does this correspond at 25 °C?

76. A xenon fluoride can be prepared by heating a mixture of Xe and F_2 gases to a high temperature in a pressure-proof container. Assume that xenon gas was added to a 0.25-L container until its pressure reached 0.12 atm at 0.0 °C. Fluorine gas was then added until the total pressure reached 0.72 atm at 0.0 °C. After the reaction was complete, the xenon was consumed completely and the pressure of the F_2 remaining in the container was 0.36 atm at 0.0 °C. What is the empirical formula of the xenon fluoride?

77. Chlorine dioxide, ClO_2, reacts with fluorine to give a new gas that contains Cl, O, and F. In an experiment you find that 0.150 g of this new gas has a pressure of 17.2 mm Hg in a 1850-mL flask at 21 °C. What is the identity of the unknown gas?

78. A balloon at the circus is filled with helium gas to a gauge pressure of 22 mm Hg at 25 °C. The volume of the gas is 305 mL, and the barometric pressure is 755 mm Hg. What amount of helium is in the balloon? (Remember that gauge pressure = total pressure − barometric pressure. See page 550.)

79. Acetylene can be made by allowing calcium carbide to react with water:

$$CaC_2(s) + 2\ H_2O(\ell) \longrightarrow C_2H_2(g) + Ca(OH)_2(s)$$

Suppose you react 2.65 g of CaC_2 with excess water. If you collect the acetylene and find that the gas has a volume of 795 mL at 25.2 °C with a pressure of 735.2 mm Hg, what is the percent yield of acetylene?

80. If you have a sample of water in a closed container, some of the water will evaporate until the pressure of the water vapor, at 25 °C, is 23.8 mm Hg. How many molecules of water per cubic centimeter exist in the vapor phase?

81. You are given 1.56 g of a mixture of $KClO_3$ and KCl. When heated, the $KClO_3$ decomposes to KCl and O_2,

$$2\ KClO_3(s) \longrightarrow 2\ KCl(s) + 3\ O_2(g)$$

and 327 mL of O_2 with a pressure of 735 mm Hg is collected at 19 °C. What is the weight percentage of $KClO_3$ in the sample?

82. ▲ A study of climbers who reached the summit of Mount Everest without supplemental oxygen showed that the partial pressures of O_2 and CO_2 in their lungs were 35 mm Hg and 7.5 mm Hg, respectively. The barometric pressure at the summit was 253 mm Hg. Assume the lung gases are saturated with moisture at a body temperature of 37 °C [which means the partial pressure of water vapor in the lungs is $P(H_2O) = 47.1$ mm Hg]. If you assume the lung gases consists of only O_2, N_2, CO_2, and H_2O, what is the partial pressure of N_2?

83. Nitrogen monoxide reacts with oxygen to give nitrogen dioxide:

$$2\ NO(g) + O_2(g) \longrightarrow 2\ NO_2(g)$$

(a) Place the three gases in order of increasing rms speed at 298 K.

(b) If you mix NO and O_2 in the correct stoichiometric ratio, and NO has a partial pressure of 150 mm Hg, what is the partial pressure of O_2?

(c) After reaction between NO and O_2 is complete, what is the pressure of NO_2 if the NO originally had a pressure of 150 mm Hg and O_2 was added in the correct stoichiometric amount?

84. ▲ Ammonia gas is synthesized by combining hydrogen and nitrogen:

$$3\ H_2(g) + N_2(g) \longrightarrow 2\ NH_3(g)$$

(a) If you want to produce 562 g of NH_3, what volume of H_2 gas, at 56 °C and 745 mm Hg, is required?

(b) To produce 562 g of NH_3, what volume of air (the source of N_2) is required if the air is introduced at 29 °C and 745 mm Hg? (Assume the air sample has 78.1 mole % N_2.)

85. ▲ You have a 550-mL tank of gas with a pressure of 1.56 atm at 24 °C. You thought the gas was pure carbon monoxide gas, CO, but you later found it was contaminated by small quantities of gaseous CO_2 and O_2. Analysis shows that the tank pressure is 1.34 atm (at 24 °C) if the CO_2 is removed. Another experiment shows that 0.0870 g of O_2 can be removed chemically. What are the masses of CO and CO_2 in the tank, and what is the partial pressure of each of the three gases at 25 °C?

86. ▲ Methane is burned in a laboratory Bunsen burner to give CO_2 and water vapor. Methane gas is supplied to the burner at the rate of 5.0 L/min (at a temperature of 28 °C and a pressure of 773 mm Hg). At what rate must oxygen be supplied to the burner (at a pressure of 742 mm Hg and a temperature of 26 °C)?

87. ▲ Iron forms a series of compounds of the type $Fe_x(CO)_y$. In air they are oxidized to Fe_2O_3 and CO_2 gas. After heating a 0.142-g sample of $Fe_x(CO)_y$ in air, you isolate the CO_2 in a 1.50-L flask at 25 °C. The pressure of the gas is 44.9 mm Hg. What is the formula of $Fe_x(CO)_y$?

88. ▲ Group 2A metal carbonates are decomposed to the metal oxide and CO_2 on heating:

$$MCO_3(s) \longrightarrow MO(s) + CO_2(g)$$

You heat 0.158 g of a white, solid carbonate of a Group 2A metal (M) and find that the evolved CO_2 has a pressure of 69.8 mm Hg in a 285-mL flask at 25 °C. Identify M.

89. Silane, SiH_4, reacts with O_2 to give silicon dioxide and water vapor:

$$SiH_4(g) + 2\,O_2(g) \longrightarrow SiO_2(s) + 2\,H_2O(g)$$

If you mix SiH_4 with O_2 in the correct stoichiometric ratio, and if the total pressure of the mixture is 120 mm Hg, what are the partial pressures of SiH_4 and O_2? When the reactants have been completely consumed, what is the total pressure in the flask? (Assume T is constant.)

90. Chlorine trifluoride, ClF_3, is a valuable reagent because it can be used to convert metal oxides to metal fluorides:

$$6\,NiO(s) + 4\,ClF_3(g) \longrightarrow 6\,NiF_2(s) + 2\,Cl_2(g) + 3\,O_2(g)$$

(a) What mass of NiO will react with ClF_3 gas if the gas has a pressure of 250 mm Hg at 20 °C in a 2.5-L flask?

(b) If the ClF_3 described in part (a) is completely consumed, what are the partial pressures of Cl_2 and of O_2 in the 2.5-L flask at 20 °C (in mm Hg)? What is the total pressure in the flask?

91. One way to synthesize diborane, B_2H_6, is the reaction

$$2\,NaBH_4(s) + 2\,H_3PO_4(aq) \longrightarrow$$
$$B_2H_6(g) + 2\,NaH_2PO_4(aq) + 2\,H_2(g)$$

(a) If you have 0.136 g of $NaBH_4$ and excess H_3PO_4, and you collect the B_2H_6 in a 2.75 L flask at 25 °C, what is the pressure of the B_2H_6 in the flask?

(b) A byproduct of the reaction is H_2 gas. If both B_2H_6 and H_2 gas come from this reaction, what is the *total* pressure in the 2.75-L flask (after reaction of 0.136 g of $NaBH_4$ with excess H_3PO_4) at 25 °C?

92. Calcium carbide reacts with water to produce acetylene and calcium hydroxide:

$$CaC_2(s) + 2\,H_2O(\ell) \longrightarrow C_2H_2(g) + Ca(OH)_2(s)$$

Suppose you combine 13.0 g of CaC_2 with 4.65 g of water and collect the acetylene in a 4.66-L flask. What is the pressure of the acetylene at 23 °C?

93. ▲ You have 1.249 g of a mixture of $NaHCO_3$ and Na_2CO_3. You find that 12.0 mL of 1.50 M HCl is required to convert the sample completely to NaCl, H_2O, and CO_2.

$$NaHCO_3(aq) + HCl(aq) \longrightarrow$$
$$NaCl(aq) + H_2O(\ell) + CO_2(g)$$

$$Na_2CO_3(aq) + 2\,HCl(aq) \longrightarrow$$
$$2\,NaCl(aq) + H_2O(\ell) + CO_2(g)$$

What volume of CO_2 is evolved at 745 mm Hg and 25 °C?

94. ▲ A mixture of $NaHCO_3$ and Na_2CO_3 has a mass of 2.50 g. When treated with HCl(aq), 665 mL of CO_2 gas is liberated with a pressure of 735 mm Hg at 25 °C. What is the weight percent of $NaHCO_3$ and Na_2CO_3 in the mixture? (See Study Question 93 for the reactions that occur.)

95. ▲ Relative humidity is the ratio of the partial pressure of water in air at a given temperature to the vapor pressure of water at that temperature. Calculate the mass of water per liter of air under the following conditions.

(a) at 20 °C and 45% relative humidity
(b) at 0 °C and 95% relative humidity

Under which circumstances is the mass of H_2O per liter greater? (See Appendix G for the vapor pressure of water.)

96. How much water vapor is present in a dormitory room when the relative humidity is 55% and the temperature is 23 °C? The dimensions of the room are 4.5 m² floor area and 3.5 m ceiling height. (See Study Question 95 for a definition of relative humidity and Appendix G for the vapor pressure of water.)

Summary and Conceptual Questions

The following questions may use concepts from the preceding chapters.

97. A 1.0-L flask contains 10.0 g each of O_2 and CO_2 at 25 °C.

(a) Which gas has the greater partial pressure, O_2 or CO_2, or are they the same?

(b) Which molecules have the greater average speed, or are they the same?

(c) Which molecules have the greater average kinetic energy, or are they the same?

98. If equal masses of O_2 and N_2 are placed in separate containers of equal volume at the same temperature, which of the following statements is true? If false, tell why it is false.

(a) The pressure in the flask containing N_2 is greater than that in the flask containing O_2.

(b) There are more molecules in the flask containing O_2 than in the flask containing N_2.

99. You have two pressure-proof steel cylinders of equal volume, one containing 1.0 kg of CO and the other containing 1.0 kg of acetylene, C_2H_2.

(a) In which cylinder is the pressure greater at 25 °C?

(b) Which cylinder contains the greater number of molecules?

▲ More challenging ■ In General ChemistryNow Blue-numbered questions answered in Appendix O

100. Two flasks, each with a volume of 1.00 L, contain O_2 gas with a pressure of 380 mm Hg. Flask A is at 25 °C, and flask B is at 0 °C. Which flask contains the greater number of O_2 molecules?

101. ▲ State whether each of the following samples of matter is a gas. If there is not enough information for you to decide, write "insufficient information."

   (a) A material is in a steel tank at 100 atm pressure. When the tank is opened to the atmosphere, the material suddenly expands, increasing its volume by 10%.

   (b) A 1.0-mL sample of material weighs 8.2 g.

   (c) The material is transparent and pale green in color.

   (d) One cubic meter of material contains as many molecules as 1.0 m^3 of air at the same temperature and pressure.

102. Each of the four tires of a car is filled with a different gas. Each tire has the same volume, and each is filled to the same pressure, 3.0 atm, at 25 °C. One tire contains 116 g of air, another tire has 80.7 g of neon, another tire has 16.0 g of helium, and the fourth tire has 160. g of an unknown gas.

   (a) Do all four tires contain the same number of gas molecules? If not, which one has the greatest number of molecules?

   (b) How many times heavier is a molecule of the unknown gas than an atom of helium?

   (c) In which tire do the molecules have the largest kinetic energy? The highest average speed?

103. The sodium azide required for automobile air bags is made by the reaction of sodium metal with dinitrogen oxide in liquid ammonia:

$$3\ N_2O(g) + 4\ Na(s) + NH_3(\ell) \longrightarrow$$
$$NaN_3(s) + 3\ NaOH(s) + 2\ N_2(g)$$

   (a) You have 65.0 g of sodium and a 35.0-L flask containing N_2O gas with a pressure of 2.12 atm at 23 °C. What is the theoretical yield (in grams) of NaN_3?

   (b) Draw a Lewis structure for the azide ion. Include all possible resonance structures. Which resonance structure is most likely?

   (c) What is the shape of the azide ion?

104. ▲ Chlorine gas (Cl_2) is used as a disinfectant in municipal water supplies, although chlorine dioxide (ClO_2) and ozone are becoming more widely used. ClO_2 is a better choice than Cl_2 in this application because it leads to fewer chlorinated byproducts, which are themselves pollutants.

   (a) How many valence electrons are in ClO_2?

   (b) The chlorite ion, ClO_2^-, is obtained by reducing ClO_2. Draw a possible electron dot structure for ClO_2^-. (Cl is the central atom.)

   (c) What is the hybridization of the central Cl atom in ClO_2^-? What is the shape of the ion?

   (d) Which species has the larger bond angle, O_3 or ClO_2^-? Explain briefly.

   (e) Chlorine dioxide, ClO_2, a yellow-green gas, can be made by the reaction of chlorine with sodium chlorite:

$$2\ NaClO_2(s) + Cl_2(g) \longrightarrow 2\ NaCl(s) + 2\ ClO_2(g)$$

   Assume you react 15.6 g of $NaClO_2$ with chlorine gas, which has a pressure of 1050 mm Hg in a 1.45-L flask at 22 °C. What mass of ClO_2 can be produced?

105. If the absolute temperature of a gas doubles, by how much does the average speed of the gaseous molecules increase? (*See General ChemistryNow Screen 12.9.*)

106. Screen 12.10 of the General ChemistryNow CD-ROM or website shows animations describing the following relationships on the molecular scale: *P* versus *n*, *P* versus *T*, and *P* versus *V*. Sketch a molecular-scale animation for the relationship between *n* and *V*.

Do you need a live tutor for homework problems?
Access vMentor at General ChemistryNow at
http://now.brookscole.com/kotz6e
for one-on-one tutoring from a chemistry expert

13—Intermolecular Forces, Liquids, and Solids

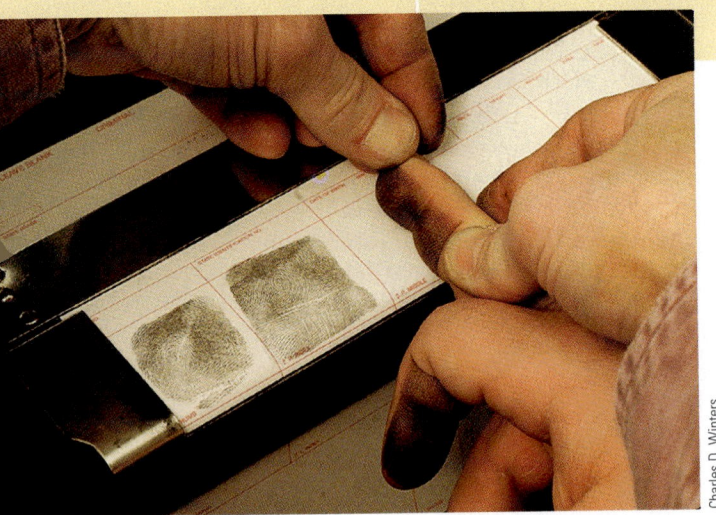

Taking an official fingerprint at the local police station.

Charles D. Winters

The Mystery of the Disappearing Fingerprints

The events of September 11, 2001, are etched in everyone's memory. The specter of domestic terrorism, however, was first raised almost two years earlier. That's when a man was apprehended in December 1999 at the U.S.–Canadian border with bomb materials and a map of Los Angeles International Airport. Although he claimed innocence, his fingerprints were on the bomb materials, and he was convicted of an attempt to bomb the airport.

Each of us has a unique fingerprint pattern, as first described by John Purkinji in 1823. Not long after his discovery, English colonists in India began using fingerprints on contracts because they believed it made the contract appear more binding. Not until late in the 19th century, however, was fingerprinting used as an identifier. Sir Francis Galton, a British anthropologist and a cousin of Charles Darwin, established that a person's fingerprints do not

Dusting for fingerprints on a glass coffee mug.

Charles D. Winters

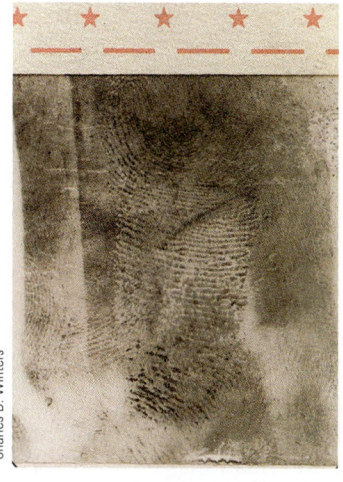

Charles D. Winters

Close-up of a fingerprint.

change over the course of a lifetime and that no two prints are exactly the same.

Fingerprinting has since become an accepted tool in forensic science. In 1993 in Knoxville, Tennessee, Art Bohanan thought he could use it to solve the case of the kidnapping of a young girl. The girl had been taken from her home and driven away in a green car. The girl soon managed to escape from her attacker and was able to describe the car to the police. After four days the police found the car and arrested its owner. But had the girl been in that car? Art Bohanan inspected the car for her fingerprints and even used the latest technique, fuming with superglue. No prints were found.

The abductor of the girl was eventually convicted on the basis of other evidence, but Bohanan wondered why he had never found her prints in the car. He decided to test the permanence of children's fingerprints compared with adults' fingerprints. To his amazement, he found that children's prints disappear in a few hours, whereas an adult's prints can last for days. Bohanan said, "It sounded like the compounds in children's fingerprints might simply be evaporating faster than adult's."

The residue deposited by fingerprints is 99% water. The other 1% contains oils, fatty acids, esters, salts, urea, and amino acids.

Scientists at Oak Ridge National Laboratory studied the fingerprints of 50 child and adult volunteers, identifying the compounds present by such techniques as mass spectrometry [◀ page 127]. What they found clarified the mystery of the disappearing fingerprints.

Children's fingerprints contain more low-molecular-weight fatty acids than do adult fingerprints. (Fatty acids consist of a carbon–hydrogen chain with a carboxylic acid group, $-CO_2H$, at one end. See page 505.) Due to the relatively low molecular weight and low polarity of these acids, their intermolecular forces are weak and the compounds are volatile. As a consequence, children's fingerprints simply evaporate.

In contrast, adult fingerprints contain esters of long-chain fatty acids with long-chain alcohols. These are waxes, semisolid or solid organic compounds with high molecular weights and low volatility. Examples of waxes are lanolin, a component of wool, and the carnuba wax used in furniture polish.

Carnuba wax

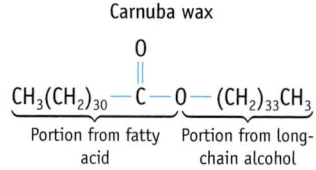

Before puberty, children do not produce waxy compounds in their skin. However, sebaceous glands in adult skin produce sebum, a complex mixture of organic compounds (triglycerides, fatty acids, cholesterol, and waxes). Only a few of these glands are found on the hands; most are located on the mid-back, forehead, and chin. When you touch your face, this mixture of compounds is transferred to your fingers, and you can leave a fingerprint that is unique to you.

589

GENERAL
Chemistry•🔧•Now™

Throughout the chapter this icon introduces a list of resources on the General ChemistryNow CD-ROM or website (http://now .brookscole.com/kotz6e) that will:

- help you evaluate your knowledge of the material
- provide homework problems
- allow you to take an exam-prep quiz
- provide a personalized Learning Plan targeting resources that address areas you should study

The vast majority of the known chemical elements are solids at 25 °C and 1 atm of pressure. Only 11 elements occur as gases under these conditions (H_2, N_2, O_2, F_2, Cl_2, and the six noble gases), and only two elements occur as liquids (Hg and Br_2). Many common compounds are gases (such as CO_2 and CH_4) or liquids (H_2O) at standard temperature and pressure, but, as is the case with the elements, the largest number of compounds are solids.

The primary objective in this chapter is to elucidate the macroscopic properties of the liquid and solid states by looking at the particulate level—that is, the level of atoms, molecules, and ions. You will find this a useful chapter because it explains, among other things, why your body is cooled when you sweat, how bodies of water can influence local climate, why one form of pure carbon (diamond) is hard and another (graphite) is slippery, and why many solid compounds form beautiful crystals.

13.1—States of Matter and the Kinetic-Molecular Theory

The kinetic-molecular theory of gases [◀ Section 12.6] assumes that gas molecules or atoms are widely separated and that these particles can be considered to be independent of one another. Consequently, we can relate the properties of gases under most conditions by a simple mathematical equation, $PV = nRT$, known as the ideal gas law equation (Equation 12.4). Liquids and solids present a more complicated picture, however. In these states, the particles are close enough together that attractive forces between them can have a considerable effect. The ideal gas law is valid because, at the temperatures and pressures at which gases exist, we can usually ignore these forces. In contrast, when these forces are introduced in liquids and solids, it is not possible to create a simple "ideal liquid equation" or "ideal solid equation."

How different are the states of matter at the particulate level? We can get a sense of this by comparing volumes occupied by equal numbers of molecules of a material in different states. Figure 13.1a shows a flask containing about 300 mL of liquid nitrogen. If all of the liquid were allowed to evaporate, the gaseous nitrogen would fill a large balloon (more than 200 L volume) to a pressure of 1 atm at room temperature. A large amount of space exists between molecules in a gas, whereas in liquids the molecules are close together.

The increase in volume when converting liquids to gases is strikingly large. In contrast, no dramatic change in volume occurs when a solid is converted to a liquid. Figure 13.1b shows the same amount of liquid and solid benzene side by side. As you see, they are not appreciably different in volume. This means that the atoms in the liquid are packed together about as tightly as the atoms in the solid phase.

We know that gases can be compressed easily, a process that involves forcing the gas molecules closer together. The air–fuel mixture in your car's engine, for example, is routinely compressed by a factor of about 10 before it is ignited. In contrast, the molecules, ions, or atoms in liquid or solid phases strongly resist forces that

Nitrogen gas

Liquid nitrogen

Photos: Charles. D. Winters

(a)

Liquid benzene   Solid benzene

(b)

Figure 13.1 **Contrasting gases, liquids, and solids.**
(a) When a 300-mL sample of liquid nitrogen evaporates, it will produce more than 200 L of gas at 25 °C and 1.0 atm. In the liquid phase, the molecules of N_2 are close together; in the gas phase, they are far apart.

(b) The same volume of liquid benzene, C_6H_6, is placed in two test tubes, and one tube (*right*) is cooled, freezing the liquid. The solid and liquid states have almost the same volume, showing that the molecules are packed together almost as tightly in the liquid state as they are in the solid state.

would push them closer together. Thus, a characteristic of liquids and solids is a lack of compressibility. For example, the volume of liquid water changes only by 0.005% per atmosphere of pressure applied to it.

In the gaseous state, atoms or molecules are relatively far apart because the forces between the particles are not strong enough to pull them together and overcome their kinetic energy. In liquids and solids, much stronger forces pull the particles together and limit their motion. In solid ionic compounds, the positively and negatively charged ions are held together by electrostatic attraction [◀ page 112]. In molecular solids and liquids, the forces *between* molecules, called **intermolecular forces**, are based on various electrostatic attractions that are weaker than the forces between oppositely charged ions. By comparison, the attractive forces between the ions in ionic compounds are usually in the range of 700 to 1100 kJ/mol, and most covalent bond energies are in the range of 100 to 400 kJ/mol (Table 9.10). As a rough guideline, intermolecular forces are generally 15% (or less) of the values of bond energies.

GENERAL
Chemistry ⚛ Now™

See the General ChemistryNow CD-ROM or website:
- **Screen 13.2 States of Matter,** to view an animation of gases, liquids, and solids at the molecular level

---

13.2—Intermolecular Forces

Intermolecular forces influence chemistry in many ways:

- They are directly related to properties such as melting point, boiling point, and the energy needed to convert a solid to a liquid or a liquid to a vapor.

- They are important in determining the solubility of gases, liquids, and solids in various solvents.

- They are crucial in determining the structures of biologically important molecules such as DNA and proteins.

You will encounter examples of these relationships in this and subsequent chapters.

Bonding in ionic compounds depends on the electrostatic forces of attraction between oppositely charged ions. Similarly, the intermolecular forces attracting one molecule to another are electrostatic. Recall that molecules can have polar bonds owing to the differences in electronegativity of the bonded atoms [◀ Section 9.8]. Depending on the orientation of these polar bonds, an entire molecule can be polar, with one portion of the molecule being negatively charged and another portion being positively charged. Interactions between the polar molecules can have a profound effect on molecular properties and are the subject of this section.

GENERAL
Chemistry ⚛ Now™

See the General ChemistryNow CD-ROM or website:
- **Screen 13.3 Intermolecular Forces,** for an outline of the important intermolecular forces

Interactions Between Ions and Molecules with a Permanent Dipole

The distribution of bonding electrons in a molecule often results in a permanent dipole moment (Section 9.9). Because polar molecules have positive and negative ends, if a polar molecule and an ionic compound are mixed, the negative end of the dipole will be attracted to a positive cation (Figure 13.2). Similarly, the positive end of the dipole will be attracted to a negative anion. Forces involved in the attraction between a positive or negative ion and polar molecules are less than those for ion–ion attractions, but they are greater than other forces between molecules, whether polar or nonpolar. Ion–dipole attraction can be evaluated based on the equation describing the attraction between opposite charges, Coulomb's law (Equation 3.1). It informs us that the force of attraction between two charged objects depends on the product of their charges divided by the square of the distance between them (see Section 3.3). Therefore, when a polar molecule encounters an ion, the attractive forces depend on three factors:

- *The distance between the ion and the dipole.* The closer the ion and dipole, the stronger the attraction.
- *The charge on the ion.* The higher the ion charge, the stronger the attraction.
- *The magnitude of the dipole.* The greater the magnitude of the dipole, the stronger the attraction.

The formation of hydrated ions in aqueous solution is one of the most important examples of the interaction between an ion and a polar molecule. (See Figure 13.2 and "A Closer Look: Hydrated Salts.") The energy associated with the hydration of ions—which is generally called the **solvation energy** or, for ions in water, the **enthalpy of hydration**—can be substantial. The solvation energy or enthalpy for an individual ion cannot be measured directly, but values can be estimated. For example, the solvation or hydration of sodium ions is described by the following reaction:

$$Na^+(g) + x\ H_2O(\ell) \longrightarrow [Na(H_2O)_x]^+(aq)(x \text{ probably} = 6) \qquad \Delta H_{rxn} = -405 \text{ kJ}$$

The energy of attraction depends on $1/d$, where d is the distance between the center of the ion and the oppositely charged "pole" of the dipole.

As the ion radius becomes larger, d increases and the enthalpy of hydration becomes less exothermic. The trend in the enthalpy of hydration of the alkali metal

---

■ **Coulomb's Law**
The **force** of attraction between oppositely charged ions depends directly on the product of the ion charges and inversely on the square of the distance between the ions (Equation 3.1, page 112). The **energy** of the attraction is also proportional to the ion charge product, but it is inversely proportional to the distance between them.

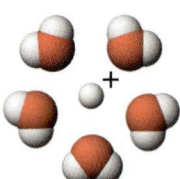

Water surrounding
a cation

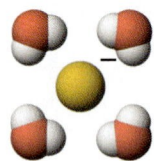

Water surrounding
an anion

Active Figure 13.2   **Ion–dipole interactions.** When an ionic compound such as NaCl is placed in water, the polar water molecules surround the cations and anions.

GENERAL
Chemistry ⚛ Now™ See the General ChemistryNow CD-ROM or website to explore an interactive version of this figure accompanied by an exercise.

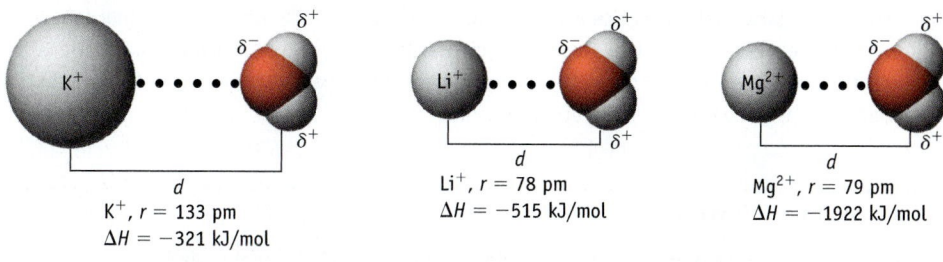

Increasing force of attraction; more exothermic enthalpy of hydration

Figure 13.3   Enthalpy of hydration. The energy evolved when an ion is hydrated depends on the ion charge and the distance d between the ion and the polar water molecule. The distance d increases as ion size increases.

cations illustrates this property, as do the hydration enthalpy values for Mg^{2+}, Li^+, and K^+ in Figure 13.3.

Cation	Ion Radius (pm)	Enthalpy of Hydration (kJ/mol)
Li^+	78	−515
Na^+	98	−405
K^+	133	−321
Rb^+	149	−296
Cs^+	165	−263

It is interesting to compare these values with the enthalpy of hydration of the H^+ ion, estimated to be −1090 kJ/mol. This extraordinarily large value is due to the tiny size of the H^+ ion.

GENERAL
Chemistry ⊹ Now™

See the General ChemistryNow CD-ROM or website:
• **Screen 13.4 Intermolecular Forces (2),** to view an animation of ion–dipole forces

Example 13.1—Hydration Energy

Problem Explain why the enthalpy of hydration of Na^+ (−405 kJ/mol) is somewhat more exothermic than that of Cs^+ (−263 kJ/mol), whereas that of Mg^{2+} is much more exothermic (−1922 kJ/mol) than that of either Na^+ or Cs^+.

Strategy The strength of ion–dipole attractions depends directly on the size of the ion charge and the magnitude of the dipole, and inversely on the distance between them. To judge the ion–dipole distance, we need ion sizes from Figure 8.15.

Solution The relevant ion sizes are $Na^+ = 98$ pm, $Cs^+ = 165$ pm, and $Mg^{2+} = 79$ pm. From these values we can predict that the distances between the center of the positive charge on the metal ion and the negative side of the water dipole will vary in this order: $Mg^{2+} < Na^+ <$ Cs^+. The hydration energy varies in the reverse order (with the hydration energy of Mg^{2+} being

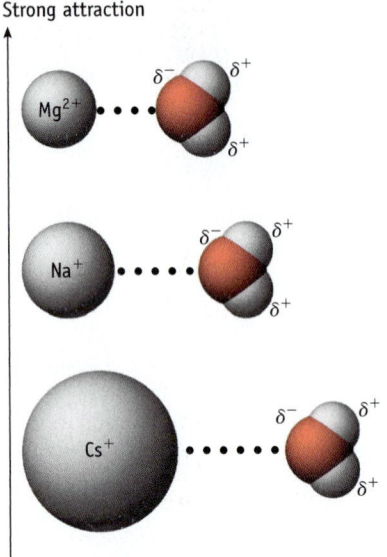

Strong attraction

Weak attraction

the most negative value). Notice also that Mg^{2+} has a 2+ charge, whereas the other ions are 1+. The greater charge on Mg^{2+} leads to a greater force of ion–dipole attraction than for the other two ions, which have only a 1+ charge. As a result, the hydration energy for MG^{2+} is much more negative than for the other two ions.

Exercise 13.1—Hydration Energy

Which should have the more negative hydration energy, F^- or Cl^-? Explain briefly.

Interactions Between Molecules with Permanent Dipoles

When a polar molecule encounters another polar molecule, of the same or a different kind, the positive end of one molecule is attracted to the negative end of the other polar molecule.

Many molecules have dipoles, and their interactions occur by **dipole–dipole attraction**.

For polar molecules, dipole–dipole attractions influence the evaporation of a liquid and the condensation of a gas (Figure 13.4). An energy change occurs in both processes. Evaporation requires the addition of heat, specifically the enthalpy of vaporization (ΔH°_{vap}) [◄ Section 6.3; see also Section 13.5]. The value for the enthalpy of vaporization has a positive sign, indicating that evaporation is an endothermic process. The enthalpy change for the condensation process—the reverse of evaporation—has a negative value, because heat is transferred out of the system.

The greater the forces of attraction between molecules in a liquid, the greater the energy that must be supplied to separate them. Thus, we expect polar compounds to have a higher value for their enthalpy of vaporization than nonpolar compounds with similar molar masses. Comparisons between a few polar and nonpolar molecules that illustrate this trend appear in Table 13.1. For example, notice that ΔH°_{vap} for polar molecules is greater than for nonpolar molecules of approximately the same size and mass.

Figure 13.4 Evaporation at the molecular level. Energy must be supplied to separate molecules in the liquid state against intermolecular forces of attraction.

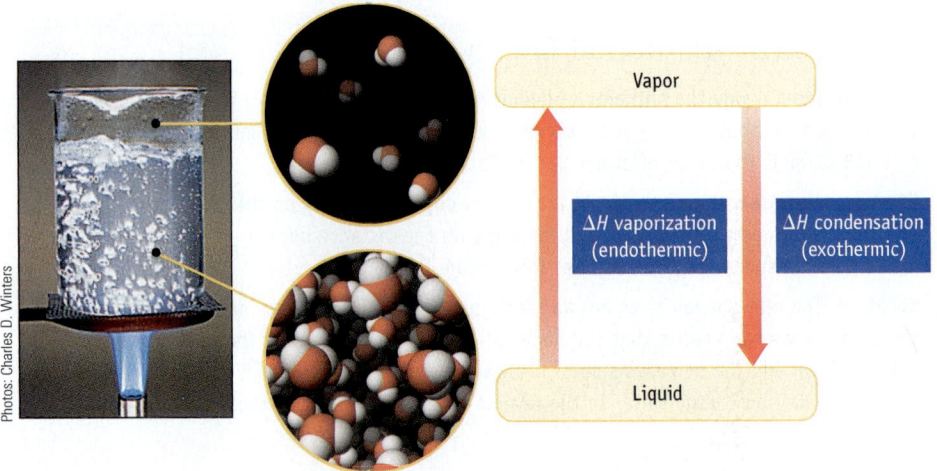

Photos: Charles D. Winters

A Closer Look

Hydrated Salts

Solid salts with waters of hydration are common. The formulas of these compounds are given by appending a specific number of water molecules to the end of the formula, as in $BaCl_2 \cdot 2\,H_2O$. Sometimes the water molecules simply fill in empty spaces in a crystalline lattice, but often the cation in these salts is directly associated with water molecules. For example, the compound $CrCl_3 \cdot 6\,H_2O$ is better written as $[Cr(H_2O)_4Cl_2]Cl \cdot 2\,H_2O$. Four of the six water molecules are associated with the Cr^{3+} ion by ion–dipole attractive forces; the remaining two water molecules are in the lattice. Common examples of hydrated salts are listed in the table.

Compound	Common Name	Users
$Na_2CO_3 \cdot 10\,H_2O$	Washing soda	Water softener
$Na_2S_2O_3 \cdot 5\,H_2O$	Hypo	Photography
$MgSO_4 \cdot 7\,H_2O$	Epsom salt	Cathartic, dyeing and tanning
$CaSO_4 \cdot 2\,H_2O$	Gypsum	Wallboard
$CuCO_3 \cdot 5\,H_2O$	Blue vitriol	Biocide

Hydrated cobalt(II) chloride, $CoCl_2 \cdot 6\,H_2O$. In the solid state the compound is best described by the formula $[Co(H_2O)_4Cl_2] \cdot 2\,H_2O$. The cobalt(II) ion is surrounded by four water molecules and two chloride ions in an octahedral arrangement. In water, the ion is completely hydrated, now being surrounded by six water molecules. Cobalt(II) ions and water molecules interact by ion–dipole forces. This is an example of a coordination compound, a class of compounds discussed in detail in Chapter 22.

Charles D. Winters

The boiling point of a liquid is also related to intermolecular forces of attraction. As the temperature of a substance is raised, its molecules gain kinetic energy. Eventually, when the boiling point is reached, the molecules have sufficient kinetic energy to escape the forces of attraction of their neighbors. The higher the forces of attraction, the higher the boiling point. In Table 13.1 you see that the boiling point for polar ICl is greater than that for nonpolar Br_2, for example.

Intermolecular forces also influence solubility. A qualitative observation on solubility is that "like dissolves like." In other words, *polar molecules are likely to dissolve in a polar solvent, and nonpolar molecules are likely to dissolve in a nonpolar solvent* (Figure 13.5) [▶ Chapter 14]. The converse is also true; that is, it is unlikely that polar molecules will dissolve in nonpolar solvents or that nonpolar molecules will dissolve in polar solvents.

■ **Dissolving Substances**
Other factors besides energy are also important in controlling the mixing of substances. See Section 14.2.

Table 13.1   Molar Masses and Boiling Points of Nonpolar and Polar Substances

	Nonpolar				Polar		
	M (g/mol)	BP (°C)	ΔH°_{vap} (kJ/mol)		M (g/mol)	BP (°C)	ΔH°_{vap} (kJ/mol)
N_2	28	−196	5.57	CO	28	−192	6.04
SiH_4	32	−112	12.10	PH_3	34	−88	14.06
GeH_4	77	−90	14.06	AsH_3	78	−62	16.69
Br_2	160	59	29.96	ICl	162	97	—

Ethylene glycol

Hydrocarbon

(a) Ethylene glycol (HOCH₂CH₂OH), a polar compound used as antifreeze in automobiles, dissolves in water.

(b) Nonpolar motor oil (a hydrocarbon) dissolves in nonpolar solvents such as gasoline or CCl₄. It will not dissolve in a polar solvent such as water, however. Commercial spot removers use nonpolar solvents to dissolve oil and grease from fabrics.

Figure 13.5   "Like dissolves like."

For example, water and ethanol (C₂H₅OH) can be mixed in any ratio to give a homogeneous mixture. In contrast, water does not dissolve in gasoline to an appreciable extent. The difference in these two situations is that ethanol and water are polar molecules, whereas the hydrocarbon molecules in gasoline (e.g., octane, C₈H₁₈) are nonpolar. The water–ethanol interactions are strong enough that the energy expended in pushing water molecules apart to make room for ethanol molecules is compensated for by the energy of attraction between the two kinds of polar molecules. In contrast, water–hydrocarbon attractions are weak. The hydrocarbon molecules cannot disrupt the stronger water–water attractions.

GENERAL Chemistry⚛Now™

See the General ChemistryNow CD-ROM or website:
• **Screen 13.4 Intermolecular Forces (2),** to view an animation of dipole–dipole forces

Interactions Involving Nonpolar Molecules

Many important molecules such as O₂, N₂, and the halogens are not polar. Why, then, does O₂ dissolve in polar water? Perhaps even more difficult to imagine is how the N₂ of the atmosphere can be liquefied (see Figure 13.1). Some intermolecular forces must be acting between O₂ and water and between N₂ molecules, but what is their nature?

Dipole/Induced Dipole Forces

Polar molecules such as water can *induce*, or create, a dipole in molecules that do not have a permanent dipole. To see how this situation can occur, picture a polar water molecule approaching a nonpolar molecule such as O₂ (Figure 13.6). The electron cloud of an isolated (gaseous) O₂ molecule is symmetrically distributed be-

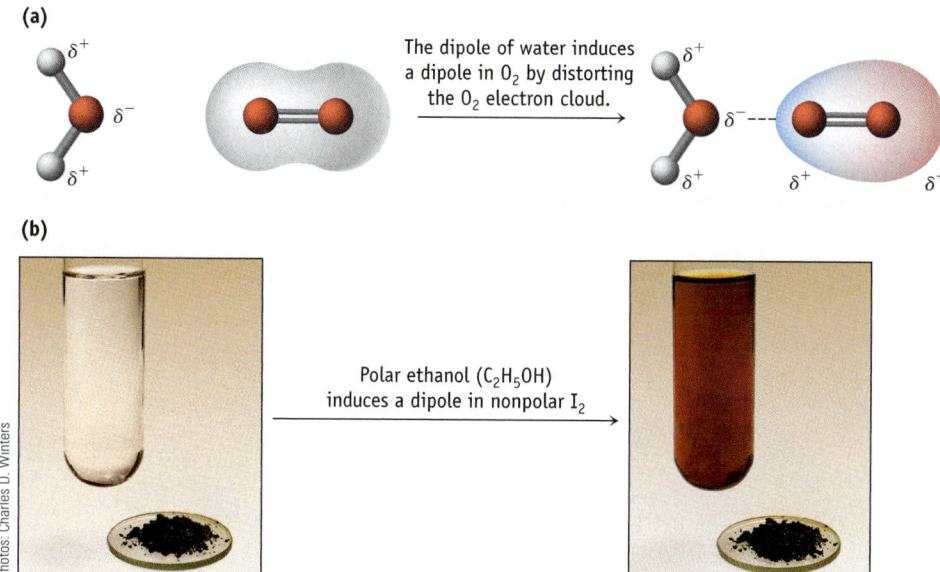

(a)

The dipole of water induces a dipole in O₂ by distorting the O₂ electron cloud.

(b)

Polar ethanol (C₂H₅OH) induces a dipole in nonpolar I₂

Photos: Charles D. Winters

Figure 13.6 Dipole/induced dipole interaction. (a) A polar molecule such as water can induce a dipole in nonpolar O_2 by distorting the molecule's electron cloud. (b) Nonpolar I_2 dissolves in polar ethanol (C_2H_5OH). The intermolecular force involved is a dipole/induced dipole force.

tween the two oxygen atoms. As the negative end of the polar H_2O molecule approaches, however, the O_2 electron cloud becomes distorted. In this process, the O_2 molecule itself becomes polar; that is, a dipole is induced in the otherwise nonpolar O_2 molecule. The result is that H_2O and O_2 molecules are now attracted to one another, albeit only weakly. Oxygen can dissolve in water because a force of attraction exists between water's permanent dipole and the induced dipole in O_2. Chemists refer to such interactions as **dipole/induced dipole interactions**.

The process of inducing a dipole is called **polarization**, and the degree to which the electron cloud of an atom or a molecule can be distorted depends on the **polarizability** of that atom or molecule. This property is difficult to measure experimentally. It makes sense, however, that the electron cloud of an atom or molecule with a large, extended electron cloud, such as I_2, can be polarized more readily than the electron cloud in a much smaller atom or molecule, such as He or H_2, in which the valence electrons are close to the nucleus and more tightly held. In general, for an analogous series of compounds, say the halogens or alkanes (such as CH_4, C_2H_6, C_3H_8, and so on), *the higher the molar mass, the greater the polarizability of the molecule.*

The solubilities of common gases in water illustrate the effect of interactions between a dipole and an induced dipole. Table 13.2 reveals a trend toward higher solubility with increasing mass of the nonpolar gas. As the molar mass of the gas increases, the polarizability of the electron cloud increases, and the strength of the dipole/induced dipole interaction increases.

London Dispersion Forces: Induced Dipole/Induced Dipole Forces

Iodine, I_2, is a solid and not a gas at room temperatures and pressures, proving that nonpolar molecules must also experience intermolecular forces. An estimate of these forces is provided by the enthalpy of vaporization of the substance at its boiling point. The data in the following table suggest that the forces in this case can range from very weak (N_2, O_2, and CH_4 have low enthalpies of vaporization and very low boiling points) to more substantial (I_2 and benzene).

Table 13.2 The Solubility of Some Gases in Water*

Gas	Molar Mass (g/mol)	Solubility at 20 °C (g gas/100 g water)[†]
H_2	2.01	0.000160
N_2	28.0	0.00190
O_2	32.0	0.00434

*Data taken from J. A. Dean: *Lange's Handbook of Chemistry*, 14th ed., pp. 5.3–5.8. New York, McGraw-Hill, 1992.
[†]Measured under conditions where pressure of gas + pressure of water vapor = 760 mm Hg.

■ **Dissolving O₂ In Water**
Oxygen dissolves in water to the extent of about 43 ppm. This property is important because microorganisms use dissolved oxygen to convert the organic substances dissolved in water to simpler compounds. The quantity of oxygen required to oxidize a given quantity of organic material is called the biological oxygen demand (BOD). Highly polluted water often has a high concentration of organic matter and so has a high BOD.

Figure 13.7   Induced dipole interactions. Momentary attractions and repulsions between nuclei and electrons create induced dipoles and lead to a net stabilization due to attractive forces.

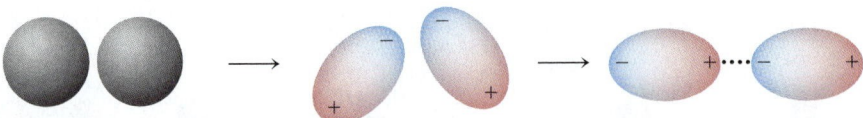

Two nonpolar atoms or molecules (depicted as having an electron cloud that has a time-averaged spherical shape).

Momentary attractions and repulsions between nuclei and electrons in neighboring molecules lead to induced dipoles.

Correlation of the electron motions between the two atoms or molecules (which are now dipolar) leads to a lower energy and stabilizes the system.

	ΔH_{vap} (kJ/mol)	Element/Compound BP (°C)
N_2	5.57	−196
O_2	6.82	−183
CH_4 (methane)	8.2	−161.5
Br_2	29.96	+58.8
C_6H_6 (benzene)	30.7	+80.1
I_2	41.95	+185

■ **Van der Waals Forces**
The name "van der Waals forces" is a general term applied to intermolecular interactions. P. W. Atkins: *Quanta: A Handbook of Concepts,* 2nd ed., p. 187, Oxford, Oxford University Press, 2000.

To understand how two nonpolar molecules can attract each other, recall that the electrons in atoms or molecules are in a state of constant motion (Figure 13.7). When two atoms or nonpolar molecules approach each other, attractions or repulsions between their electrons and nuclei can lead to distortions in these electron clouds. That is, dipoles can be induced momentarily in neighboring atoms or molecules, and these induced dipoles lead to intermolecular attractions. Thus, the intermolecular force of attraction in liquids and solids composed of nonpolar molecules is an **induced dipole/induced dipole force**. Chemists often call them **London dispersion forces**.

GENERAL
Chemistry⚛**Now**™

See the General ChemistryNow CD-ROM or website:
- **Screen 13.5 Intermolecular Forces (3),** to view an animation of induced dipole forces and for an exercise and tutorial on intermolecular forces

---

Example 13.2—Intermolecular Forces

Problem Suppose you have a mixture of solid iodine, I_2, and the liquids water and carbon tetrachloride (CCl_4). What intermolecular forces exist between each possible pair of compounds? Describe what you might see when these compounds are mixed.

Strategy First decide whether each substance is polar or nonpolar. Next, use the "like dissolves like" guideline (Figure 13.5) to decide whether iodine will dissolve in water or CCl_4 and whether CCl_4 will dissolve in water.

Solution Iodine, I_2, is nonpolar. As a molecule composed of large iodine atoms, it has an extensive electron cloud. Thus, the molecule is easily polarized. Iodine could interact with water, a polar molecule, by induced dipole/dipole forces.

Carbon tetrachloride, a tetrahedral molecule, is not polar [◄ Section 9.9]. As a consequence, it can interact with iodine only by dispersion forces. Water and CCl₄ could interact by dipole/induced dipole forces, but the interaction is expected to be weak.

The photo here shows the result of mixing these three compounds. Iodine does dissolve to a small extent in water to give a brown solution. When this brown solution is added to a test tube containing CCl₄, the liquid layers do not mix. (Polar water does not dissolve in nonpolar CCl₄.) When the test tube is shaken, however, nonpolar I₂ dissolves preferentially in nonpolar CCl₄, as evidenced by the disappearance of the color of I₂ in the water layer (top) and the appearance of the purple I₂ color in the CCl₄ layer (bottom).

Br₂   I₂

Induced dipole/induced dipole forces. The molecules Br₂ (*left*) and I₂ (*right*) are both nonpolar. They are a liquid and a solid, respectively, implying that there are forces between the molecules sufficient to cause them to be in a condensed phase. Forces between nonpolar substances are known as London dispersion forces or induced dipole/induced dipole forces.

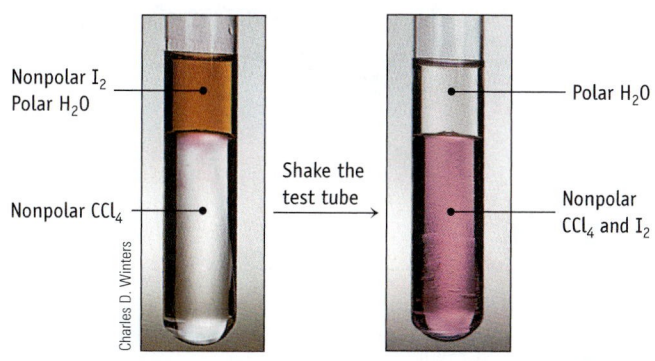

Nonpolar I₂
Polar H₂O

Nonpolar CCl₄

Shake the test tube

Polar H₂O

Nonpolar CCl₄ and I₂

Exercise 13.2—Intermolecular Forces

You mix water, CCl₄, and hexane (CH₃CH₂CH₂CH₂CH₂CH₃). What type of intermolecular forces can exist between each pair of these compounds? If you mix the three liquids, describe what observations you might make.

13.3—Hydrogen Bonding

Hydrogen fluoride and many other compounds with O—H and N—H bonds have exceptional properties. We can see this by examining the boiling points for hydrogen compounds of elements in Groups 4A through 7A (Figure 13.8). Generally, the boiling points of related compounds increase with molar mass because of increasing dispersion forces. This trend is seen in the boiling points of the hydrogen compounds of Group 4A elements, for example (CH₄ < SiH₄ < GeH₄ < SnH₄). The same effect is also operating for the heavier molecules of the hydrogen compounds of elements of Groups 5A, 6A, and 7A. The boiling points of NH₃, H₂O, and HF, however, are greatly out of line with what might be expected based on molar mass alone. If we were to extrapolate the curve for the boiling points of H₂Te, H₂Se, and H₂S to the expected boiling point of water, water should boil around −90 °C. *The boiling point of water is almost 200 °C higher than the expected value!* Similarly, the boiling points of NH₃ and HF are much higher than would be expected based on molar mass. Why do the properties of water, ammonia, and hydrogen fluoride differ so from the extrapolated values?

Because the temperature at which a substance boils depends on the attractive forces between molecules, the boiling points of H₂O, HF, and NH₃ clearly indicate strong intermolecular attractions. The unusually high boiling points in these compounds are due to hydrogen bonding. A **hydrogen bond** is an attraction between the hydrogen atom of an X—H bond and Y, where X and Y are atoms of highly electronegative elements and Y has a lone pair of electrons. Hydrogen bonds are an

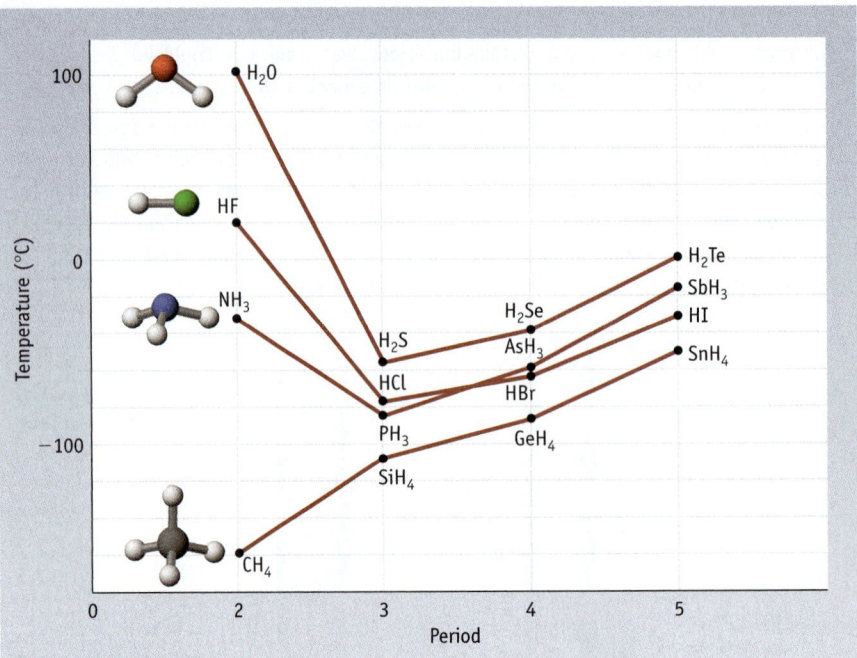

Active Figure 13.8   **The boiling points of some simple hydrogen compounds.** The effect of hydrogen bonding is apparent in the unusually high boiling points of H_2O, HF, and NH_3.

GENERAL
Chemistry-⚛-Now™ See the General ChemistryNow CD-Rom or website to explore an interactive version of this figure accompanied by an exercise.

extreme form of dipole–dipole interaction where one atom involved is always H and the other atom is most often O, N, or F.

A bond dipole arises as a result of a difference in electronegativity between bonded atoms [◀ Section 9.8]. The electronegativities of N (3.0), O (3.5), and F (4.0) are among the highest of all the elements, whereas the electronegativity of hydrogen is much lower (2.2). The large difference in electronegativity means that N—H, O—H, and F—H bonds are very polar. In bonds between H and N, O, or F, the more electronegative element takes on a significant negative charge and the hydrogen atom acquires a significant positive charge.

In hydrogen bonding, there is an unusually strong attraction between an electronegative atom with a lone pair of electrons (an N, O, or F atom in another molecule or even in the same molecule) and the hydrogen atom of the N—H, O—H, or F—H bond. A hydrogen bond can be represented as

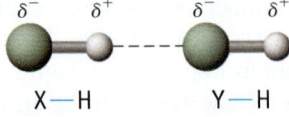

The hydrogen atom becomes a bridge between the two electronegative atoms X and Y, and the dashed line represents the hydrogen bond. The most pronounced effects of hydrogen bonding occur where X and Y are N, O, or F. Energies associated with most hydrogen bonds involving these elements are in the range of 5 to 30 kJ/mol.

Types of Hydrogen Bonds [X—H - - - :Y]

N—H - - - :N—	O—H - - - :N—	F—H - - - :N—
N—H - - - :O—	O—H - - - :O—	F—H - - - :O—
N—H - - - :F—	O—H - - - :F—	F—H - - - :F—

Hydrogen bonding has important implications for any property of a compound that is influenced by intermolecular forces of attraction. For example, it is important in determining structures of molecular solids, one example of which is acetic acid. In the solid state, two molecules of CH_3CO_2H are joined to one another by hydrogen bonding (Figure 13.9).

Hydrogen bonding is also an important factor in the structure of some synthetic polymers. In nylon, for example, the N—H unit of the amide interacts with a carbonyl oxygen on an adjacent polymer chain [◄ Figure 11.19]. Hydrogen bonding in Kevlar, also a polyamide, gives this material the exceptional strength-to-weight ratio needed for its use in making canoes, ski equipment, and bullet-proof vests.

Photo: Charles D. Winters

Figure 13.9 Hydrogen bonding. Two acetic acid molecules can interact through hydrogen bonds. This photo shows partly solid glacial acetic acid. Notice that the solid is denser than the liquid, a property shared by virtually all substances. The notable exception is water.

GENERAL
Chemistry⚛Now™

See the General ChemistryNow CD-ROM or website:
- **Screen 13.6 Hydrogen Bonding,** for a description of hydrogen bonding

Example 13.3—The Effect of Hydrogen Bonding

Problem Ethanol, CH_3CH_2OH, and dimethyl ether, CH_3OCH_3, have the same formula but a different arrangement of atoms (they are isomers). Predict which of these compounds has the higher boiling point.

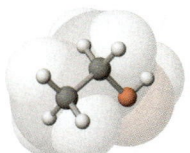   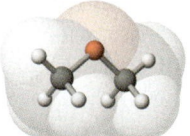

Ethanol, CH_3CH_2OH Dimethyl ether, CH_3OCH_3

Strategy Inspect the structure of each molecule to decide whether each is polar and, if polar, whether hydrogen bonding is possible.

Solution Although these two compounds have identical masses, they have different structures. Ethanol possesses an O—H group, so hydrogen bonding between ethanol molecules makes an important contribution to its intermolecular forces.

$$CH_3CH_2 - \ddot{O}: \cdots H - \ddot{O}:$$
$$\qquad\quad | \qquad\qquad |$$
$$\qquad\quad H \qquad\quad CH_2CH_3$$

hydrogen bonding in ethanol, CH_3CH_2OH

In contrast, dimethyl ether, although a polar molecule, presents no opportunity for hydrogen bonding because there is no O—H bond. We can predict, therefore, that intermolecular forces

Charles D. Winters

Tooth whiteners and hydrogen bonding. Most tooth-whitening products contain urea, $(NH_2)_2CO$, and hydrogen peroxide, H_2O_2 (a mixture sometimes referred to as carbamide peroxide). Hydrogen peroxide, the active ingredient, is stabilized by hydrogen bonding with urea.

will be larger in ethanol than in dimethyl ether and that ethanol will have the higher boiling point . Indeed, ethanol boils at 78.3 °C, whereas dimethyl ether has a boiling point of −24.8 °C, more than 100 °C lower. Dimethyl ether is a gas, whereas ethanol is a liquid under standard conditions.

Exercise 13.3—Hydrogen Bonding

Using structural formulas, describe the hydrogen bonding between methanol (CH_3OH) molecules. What physical properties of methanol are likely to be affected by hydrogen bonding?

Hydrogen Bonding and the Unusual Properties of Water

One of the most striking differences between our planet and others in our solar system is the presence of large amounts of water on earth. Three fourths of the planet is covered by oceans, the polar regions are vast ice fields, and even soil and rocks hold large amounts of water. Although we tend to take water for granted, almost no other substance behaves in a similar manner. Water's unique features reflect the ability of H_2O molecules to cling tenaciously to one another by hydrogen bonding.

■ **Energy of Hydrogen Bonding**
A hydrogen bond between water molecules has an estimated energy of 22 kJ/mol. For comparison, the O—H covalent bond energy is 463 kJ/mol.

One reason for ice's unusual structure, and water's unusual properties, is that each hydrogen atom of a water molecule can form a hydrogen bond to a lone pair of electrons on the oxygen atom of an adjacent water molecule. In addition, because the oxygen atom in water has two lone pairs of electrons, it can form two more hydrogen bonds with hydrogen atoms from adjacent molecules (Figure 13.10a). The result is a tetrahedral arrangement for the hydrogen atoms around each oxygen, involving two covalently bonded hydrogen atoms and two hydrogen-bonded hydrogen atoms.

To achieve the regular arrangement of hydrogen-bonded water molecules linked by hydrogen bonding, ice has an open-cage structure with lots of empty space (Figure 13.10). The result is that ice has a density about 10% less than that of liquid water, which explains why ice floats. (In contrast, virtually all other solids sink in their liquid phase.) We can also see in this structure that the oxygen atoms are arranged at the corners of puckered, six-sided rings, or hexagons. Snowflakes are always based on six-sided figures [◀ page 101], a reflection of the internal molecular structure of ice.

Figure 13.10 **The structure of ice.** (a) The oxygen atom of a water molecule attaches itself to four other water molecules by hydrogen bonds. Notice that the four groups that surround an oxygen atom are arranged tetrahedrally. Each oxygen atom is covalently bonded to two hydrogen atoms and hydrogen-bonded to hydrogen atoms from two other molecules. The hydrogen bonds are longer than the covalent bonds.

(b) In ice, the structural unit shown in part (a) is repeated in the crystalline lattice. This computer-generated structure shows a small portion of the extensive lattice. Notice the six-member, hexagonal rings. The corners of each hexagon are O atoms, and each side is composed of a normal O—H bond and a longer hydrogen bond. (This structure is found on the General ChemistryNow CD-ROM or website.)

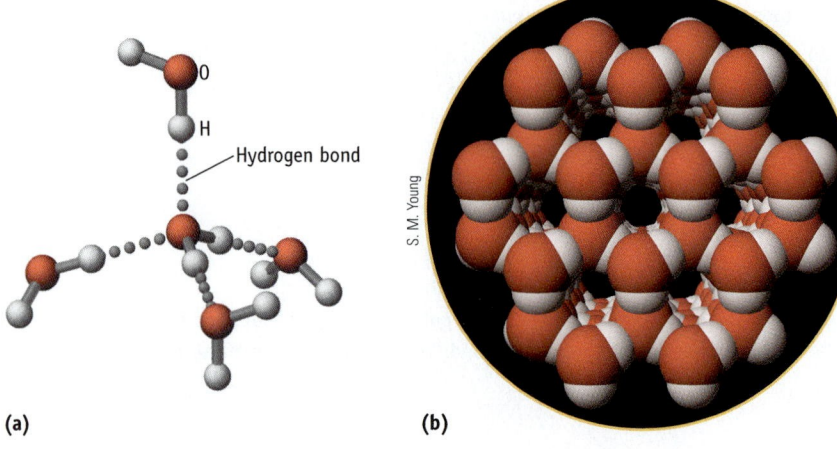

(a)   (b)

S. M. Young

When ice melts at 0 °C, a relatively large increase in density occurs (Figure 13.11), as a result of the breakdown of the regular structure imposed on the solid state by hydrogen bonding. When the temperature of liquid water is raised from 0 °C to 4 °C, another surprising thing occurs: The density of water *increases*. For almost every other substance known, density decreases as its temperature is raised. Once again, hydrogen bonding is the reason for water's seemingly odd behavior. At a temperature just above the melting point, some of the water molecules continue to cluster in ice-like arrangements, which require extra space. As the temperature is raised from 0 °C to 4 °C, the final vestiges of the ice structure disappear and the volume contracts further, giving rise to the increase in density. Water's density reaches a maximum at about 4 °C. From this point, the density declines with increasing temperature in the normal fashion.

Because of the way that water's density changes as the temperature approaches the freezing point, lakes do not freeze solidly from the bottom up in the winter. When lake water cools with the approach of winter, its density increases, the cooler water sinks, and the warmer water rises. This "turnover" process continues until all of the water reaches 4 °C, the maximum density. (This is the way oxygen-rich water moves to the lake bottom to restore the oxygen used during the summer and nutrients are brought to the top layers of the lake.) As the temperature decreases further, the colder water stays on the top of the lake, because water cooler than 4 °C is less dense than water at 4 °C. With further heat loss, ice can then begin to form on the surface, floating there and protecting the underlying water and aquatic life from additional heat loss.

Extensive hydrogen bonding is also the origin of the extraordinarily high heat capacity of water. Although liquid water does not have the regular structure of ice, hydrogen bonding still occurs. With a rise in temperature, the extent of hydrogen bonding diminishes. Disrupting hydrogen bonds requires heat. The high heat capacity of water explains, in large part, why oceans and lakes have such an enormous effect on weather. In autumn, when the temperature of the air is lower than the temperature of the ocean or lake, the ocean or lake gives up heat to the atmosphere, moderating the drop in air temperature. So much heat is available to be given off for each one-degree drop in temperature that the decline in water temperature is gradual. For this reason the temperature of the ocean or of a large lake generally remains higher than the average air temperature until late in the autumn.

Hydrogen bonds involving water are also responsible for the structure and properties of one of the strangest substances on earth: methane hydrate (Figure 13.12). When methane is mixed with water at high pressures and low temperatures, solid methane hydrate forms. Although the substance has been known for years, vast deposits of methane hydrate were recently discovered deep within sediments on the floor of the world's oceans. How they formed remains a mystery, but what is important is their size. It is estimated that the global methane hydrate deposits contain approximately 10^{13} tons of carbon, or about twice the combined amount in all known reserves of coal, oil, and conventional gas.

GENERAL
Chemistry ⚛ Now™

See the General ChemistryNow CD-ROM or website:

Screen 13.7 The Weird Properties of Water, to view an animation of the transformation of ice to water and for a table listing all of the unusual properties of water

Active Figure 13.11 **The temperature dependence of the densities of ice and water.**

GENERAL
Chemistry ⚛ Now™ See the General ChemistryNow CD-ROM or website to explore an interactive version of this figure accompanied by an exercise.

■ **Hydrogen Bonding in DNA**
Hydrogen bonding occurs extensively in biochemical systems. For example, hydrogen bonding takes place between the two bases thymine and adenine in opposing chains of the DNA molecule.

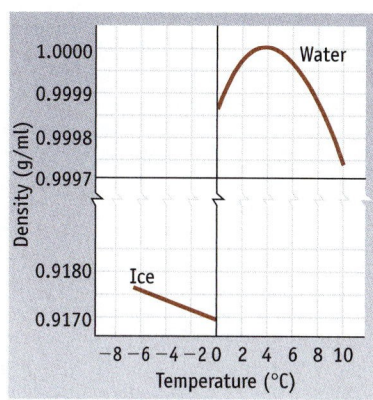

See "The Chemistry of Life: Biochemistry," pages 530–545.

Figure 13.12 Methane hydrate. (a) This interesting substance is found in huge deposits hundreds of feet down on the floor of the ocean. When a sample is brought to the surface, the methane oozes out of the solid, and the gas readily burns as it escapes from the solid hydrate.
(b) The structure of solid methane hydrate consists of methane molecules trapped in a lattice of water molecules. Each point of the lattice shown here is an O atom of an H_2O molecule. The edges are O—H—O hydrogen bonds. Such structures are often called "clathrates." See E. Suess, G. Bohrmann, J. Greinert, and E. Lausch: *Scientific American*, pp. 76–83, November 1999. See also "The Chemistry of Energy and Fuels," page 287.

Photo: John Pinkston and Laura Stern/U.S. Geological Survey/Science News, 11-9-96

(a) Methane hydrate burns as methane gas escapes from the solid hydrate.

(b) Methane hydrate consists of a lattice of water molecules with methane molecules trapped in the cavity.

13.4—Summary of Intermolecular Forces

Intermolecular forces involve molecules that are polar or those in which polarity can be induced (Table 13.3). London dispersion forces are found in all molecules, both nonpolar and polar, but dispersion forces are the only intermolecular forces that allow nonpolar molecules to interact. Furthermore, several types of intermolecular forces can be at work in a single type of molecule (Figure 13.13). A very large molecule, for example, can have polar or nonpolar regions. In general, the strength of intermolecular forces is in the order

$$\text{dipole–dipole (including H-bonding)} > \text{dipole/induced dipole} > \text{induced dipole/induced dipole}$$

Table 13.3 Summary of Intermolecular Forces

Type of Interaction	Factors Responsible for Interaction	Approximate Energy (kJ/mol)	Example
Ion–dipole	Ion charge, magnitude of dipole	40–600	$Na^+ \ldots H_2O$
Dipole–dipole	Dipole moment (depends on atom electronegativities and molecular structure)	20–30	H_2O, HCl
Hydrogen bonding, X—H . . . :Y	Very polar X—H bond (where X = F, N, O) and atom Y with lone pair of electrons An extreme form of dipole–dipole interaction.	5–30	$H_2O \ldots H_2O$
Dipole–induced dipole	Dipole moment of polar molecule and polarizability of nonpolar molecule	2–10	$H_2O \ldots I_2$
Induced dipole–induced dipole (London dispersion forces)	Polarizability	0.05–40	$I_2 \ldots I_2$

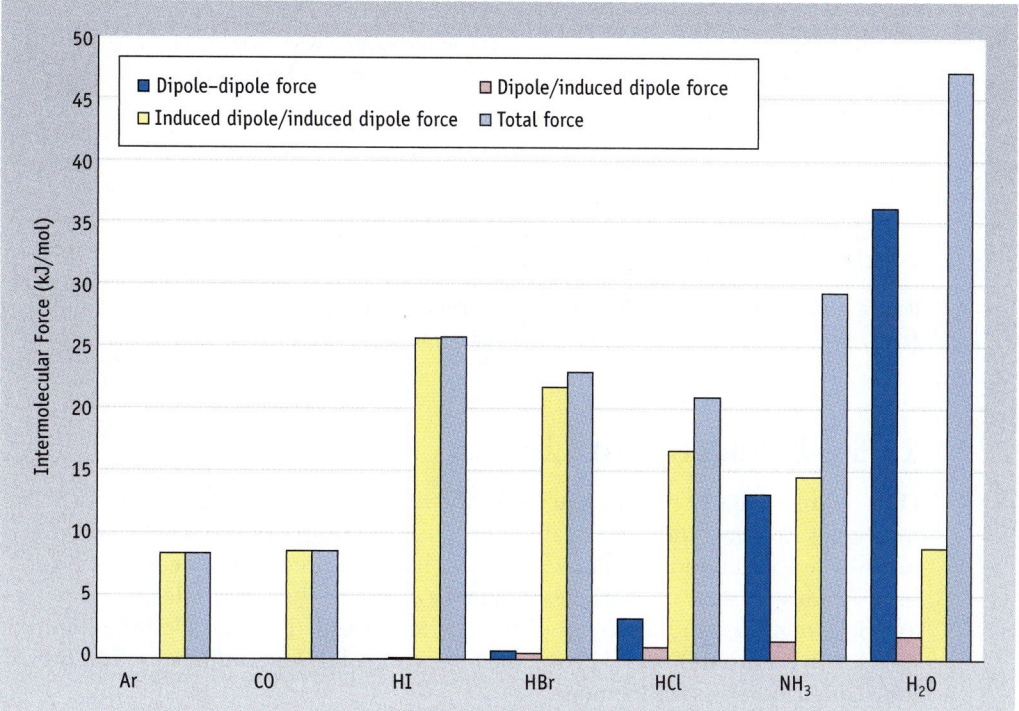

Figure 13.13   Intermolecular forces for various molecules. (Forces are reported in terms of energies in kJ/mol.) The total intermolecular force for atomic argon and weakly polar CO is small (8–9 kJ/mol) and consists entirely of dispersion forces. The polar molecules HI, HBr, and HCl have larger intermolecular forces (21–26 kJ/mol), but dispersion forces dominate in every case. For HCl, dipole–dipole forces contribute 3.3 kJ/mol to the total force of 21.1 kJ/mol. Highly polar water molecules have the largest intermolecular forces (47.2 kJ/mol). Water molecules interact primarily through dipole forces, but induced forces are also present.

Example 13.4—Intermolecular Forces

Problem Decide which are the most important intermolecular forces involved in each of the following and place them in order of increasing strength of interaction: (a) liquid methane, CH_4; (b) a mixture of water and methanol (CH_3OH); and (c) a solution of bromine in water.

Strategy For each molecule we consider its structure and then decide whether it is polar. If polar, consider the possibility of hydrogen bonding.

Solution (a) Methane is a covalently bonded molecule. Based on the Lewis structure we can conclude that it must be a tetrahedral molecule and that it cannot be polar. The only way methane molecules can interact with one another is through induced dipole/induced dipole forces.

(b) Both water and methanol are covalently bonded molecules, both are polar, and both have an O—H bond. They therefore interact through the special dipole–dipole force called hydrogen bonding.

(c) Nonpolar molecules of bromine, Br_2, interact by induced dipole forces, whereas water is a polar molecule. Therefore, dipole/induced dipole forces are involved when Br_2 molecules interact with water. (This is similar to the I_2–ethanol interaction in Figure 13.6.)

In order of increasing strength, the likely order of interactions is

$$\text{liquid } CH_4 < H_2O \text{ and } Br_2 < H_2O \text{ and } CH_3OH$$

Exercise 13.4—Intermolecular Forces

Decide which type of intermolecular force is involved in (a) liquid O_2; (b) liquid CH_3OH; and (c) O_2 dissolved in H_2O. Place the interactions in order of increasing strength.

13.5—Properties of Liquids

Of the three states of matter, liquids are the most difficult to describe precisely. The molecules in a gas under normal conditions are far apart and may be considered more or less independent of one another. The structures of solids can be described more readily because the particles that make up solids—atoms, molecules, or ions—are close together and are in an orderly arrangement. The particles of a liquid interact with their neighbors, like the particles in a solid, but, unlike in solids, there is little order in their arrangement.

In spite of a lack of precision in describing liquids, we can still consider the behavior of liquids at the molecular level. In the following sections we will look further at the process of vaporization, at the vapor pressure of liquids, at their boiling points and critical properties, and at the behavior that results in their surface tension, capillary action, and viscosity.

Vaporization

Vaporization or **evaporation** is the process in which a substance in the liquid state becomes a gas. In this process, molecules escape from the liquid surface and enter the gaseous state.

To understand evaporation, we have to look at molecular energies. Molecules in a liquid have a range of energies (Figure 13.14) that closely resembles the distribution of energies for molecules of a gas (see Figure 12.14). As with gases, the average energy for molecules in a liquid depends only on temperature: The higher the temperature, the higher the average energy and the greater the relative number of molecules with high kinetic energy. In a sample of a liquid, at least a few molecules

Figure 13.14 The distribution of energy among molecules in a liquid sample. T_2 is a higher temperature than T_1, and more molecules have an energy greater than the value marked E in the diagram at T_2 than at T_1.

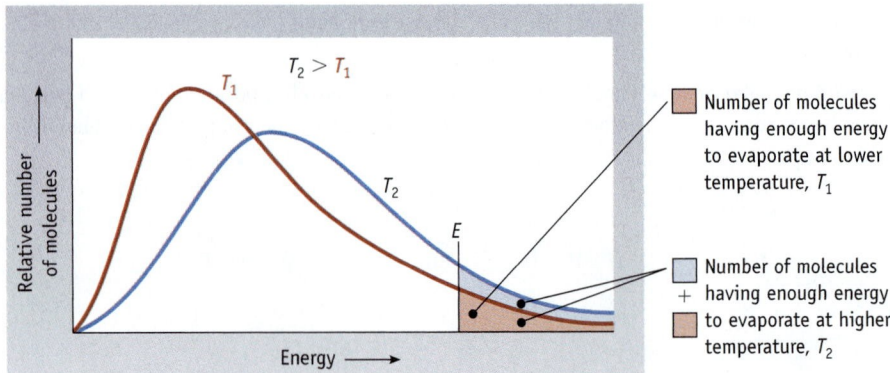

Number of molecules having enough energy to evaporate at lower temperature, T_1

Number of molecules having enough energy to evaporate at higher temperature, T_2

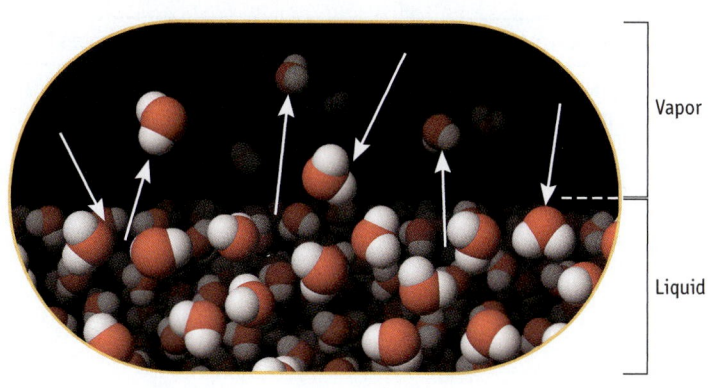

Figure 13.15 **Evaporation.** Some molecules at the surface of a liquid have enough energy to escape the attractions of their neighbors and enter the gaseous state. At the same time, some molecules in the gaseous state can reenter the liquid.

Vapor

Liquid

have very high energy; that is, they may have more kinetic energy than the potential energy of the intermolecular attractive forces holding the liquid molecules to one another. If these high-energy molecules find themselves at the surface of the liquid, and if they are moving in the right direction, they can break free of their neighbors and enter the gas phase (Figure 13.15).

Vaporization is an endothermic process because energy must be added to the system to break the intermolecular forces of attraction holding the molecules together. The heat energy required to vaporize a sample is often given as the **standard molar enthalpy of vaporization, ΔH°_{vap}** (in units of kilojoules per mole; see Tables 13.1 and 13.4 and Figure 13.4).

$$\text{Liquid} \xrightarrow[\substack{\text{heat energy absorbed} \\ \text{by liquid}}]{\text{vaporization}} \text{Vapor} \qquad \Delta H^{\circ}_{vap} = \text{molar heat of vaporization}$$

A molecule in the gas phase will eventually transfer some of its kinetic energy by colliding with slower gaseous molecules and solid objects. If this molecule comes in contact with the surface of the liquid again, it can reenter the liquid phase in the process called **condensation.**

$$\text{Vapor} \xrightarrow[\substack{\text{heat energy released} \\ \text{by vapor}}]{\text{condensation}} \text{Liquid}$$

Condensation is the opposite of vaporization. Condensation is exothermic, so energy is transferred to the surroundings. The enthalpy change for condensation is equal but opposite in sign to the enthalpy of vaporization. For example, the enthalpy change for the vaporization of 1.00 mol of water at 100 °C is +40.7 kJ. On condensing 1.00 mol of water vapor to liquid water at 100 °C, the enthalpy change is −40.7 kJ.

In the discussion of intermolecular forces for polar and nonpolar molecules, we pointed out the relationship between the ΔH°_{vap} values for various substances and the temperatures at which they boil (Table 13.4). Both properties reflect the attractive forces between particles in the liquid. The boiling points of nonpolar liquids (e.g., the hydrocarbons, atmospheric gases, and the halogens) increase with increasing atomic or molecular mass, a reflection of increased intermolecular dispersion forces. The alkanes listed in Table 13.4 show this trend clearly. Similarly, the boiling points and enthalpies of vaporization of the heavier hydrogen halides (HX, where X = Cl, Br, and I) increase with increasing molecular mass. For these molecules, hydrogen bonding is not as important as it is in HF, so dispersion forces and ordinary dipole–dipole forces account for their intermolecular attractions (see

Table 13.4   Molar Enthalpy of Vaporization and Boiling Points for Common Substances*

Compound	Molar Mass (g/mol)	ΔH°_{vap} (kJ/mol)[†]	Boiling Point (°C) (Vapor pressure = 760 mm Hg)
Polar Compounds			
HF	20.0	25.2	19.7
HCl	36.5	16.2	−84.8
HBr	80.9	19.3	−66.4
HI	127.9	19.8	−35.6
NH_3	17.0	23.3	−33.3
H_2O	18.0	40.7	100.0
SO_2	64.1	24.9	−10.0
Nonpolar Compounds			
CH_4 (methane)	16.0	8.2	−161.5
C_2H_6 (ethane)	30.1	14.7	−88.6
C_3H_8 (propane)	44.1	19.0	−42.1
C_4H_{10} (butane)	58.1	22.4	−0.5
Monatomic Elements			
He	4.0	0.08	−268.9
Ne	20.2	1.7	−246.1
Ar	39.9	6.4	−185.9
Xe	131.3	12.6	−108.0
Diatomic Elements			
H_2	2.0	0.90	−252.9
N_2	28.0	5.6	−195.8
O_2	32.0	6.8	−183.0
F_2	38.0	6.6	−188.1
Cl_2	70.9	20.4	−34.0
Br_2	159.8	30.0	58.8

*Data taken from D. R. Lide: *Basic Laboratory and Industrial Chemicals*, Boca Raton, FL, CRC Press, 1993.

[†]ΔH°_{vap} is measured at the normal boiling point of the liquid.

Figure 13.13). Because dispersion forces become increasingly important with increasing mass (Figure 13.13), the boiling points are in the order HCl < HBr < HI. Also notice in Table 13.4 the very high heats of vaporization of water and hydrogen fluoride that result from extensive hydrogen bonding.

GENERAL
Chemistry·Now™

See the General ChemistryNow CD-ROM or website:
- **Screen 13.8 Properties of Liquids (1): Enthalpy of Vaporization,** to view an animation of the vaporization process and for a table of ΔH°_{vap} values

Example 13.5—Enthalpy of Vaporization

Problem You put 1.00 L of water (about 4 cups) in a pan at 100 °C, and the water slowly evaporates. How much heat must have been supplied to vaporize the water?

Strategy Three pieces of information are needed to solve this problem:

1. ΔH°_{vap} for water = +40.7 kJ/mol at 100 °C

2. The density of water at 100 °C = 0.958 g/cm³ (This is needed because ΔH°_{vap} has units of kilojoules per mole, so you first must find the mass of water and then the amount.)

3. Molar mass of water = 18.02 g/mol

Solution A volume of 1.00 L (or 1.00×10^3 cm³) is equivalent to 958 g, and this mass is in turn equivalent to 53.2 mol of water.

$$1.00 \text{ L}\left(\frac{1000 \text{ mL}}{1 \text{ L}}\right)\left(\frac{0.958 \text{ g}}{1 \text{ mL}}\right)\left(\frac{1 \text{ mol H}_2\text{O}}{18.02 \text{ g}}\right) = 53.2 \text{ mol H}_2\text{O}$$

Therefore, the amount of energy required is

$$53.2 \text{ mol H}_2\text{O}\left(\frac{40.7 \text{ kJ}}{\text{mol}}\right) = 2.16 \times 10^3 \text{ kJ}$$

2160 kJ is equivalent to about one quarter of the energy in your daily food intake.

Exercise 13.5—Enthalpy of Vaporization

The molar enthalpy of vaporization of methanol, CH_3OH, is 35.2 kJ/mol at 64.6 °C. How much energy is required to evaporate 1.00 kg of this alcohol at 64.6 °C?

Figure 13.16 Rainstorms release an enormous quantity of energy. When water vapor condenses, energy is evolved to the surroundings. The enthalpy of condensation of water is large, so a large quantity of heat is released in a rainstorm.

Water is exceptional among the liquids listed in Table 13.4 in that an enormous amount of heat is required to convert liquid water to water vapor. This fact is important to your environment and your own physical well-being. When you exercise vigorously, your body responds by sweating to rid itself of the excess heat. Heat from your body is consumed in the process of evaporation, and your body is cooled.

Heats of vaporization and condensation of water also play an important role in weather (Figure 13.16). For example, if enough water condenses from the air to fall as an inch of rain on an acre of ground, the heat released exceeds 2.0×10^8 kJ! This is equivalent to about 50 tons of exploded dynamite, or the energy released by a small bomb.

Vapor Pressure

If you put some water in an open beaker, it will eventually evaporate completely. If you put water in a sealed flask (Figure 13.17), however, the liquid will evaporate only until the rate of vaporization equals the rate of condensation. At this point, no further change will be observed in the system. The situation is an example of what chemists call a **dynamic equilibrium**.

$$\text{Liquid} \rightleftharpoons \text{Vapor}$$

Molecules move continuously from the liquid phase to the vapor phase, and from the vapor phase back to the liquid phase. Even though these changes occur on the molecular level, no change can be detected on the macroscopic level. The rate at which molecules move from liquid to vapor is the same as the rate at which they move from vapor to liquid; thus, there is no net change in the masses of the two phases.

In contrast to a closed flask, water in an open beaker does not reach an equilibrium with gas-phase water molecules. Instead, air movement and gas diffusion remove the water vapor from the vicinity of the liquid surface, so many water molecules are not able to return to the liquid.

■ **Equilibrium**
Equilibrium is a concept used throughout chemistry and one to which we shall return often. This situation is signaled by connecting the two states or the reactants and products by a set of double arrows ($\rightleftharpoons$). See Chapters 16–18 in particular.

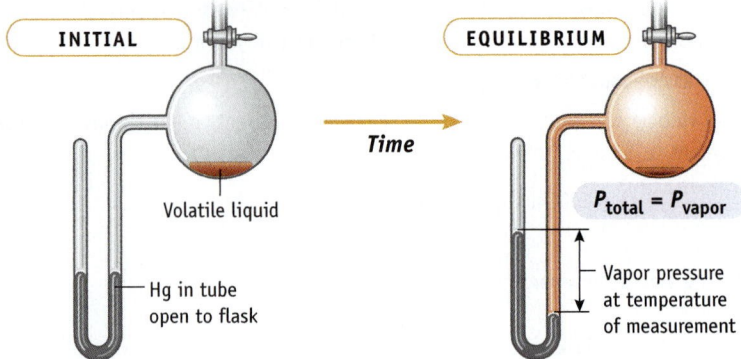

Active Figure 13.17   **Vapor pressure.** A volatile liquid is placed in an evacuated flask (*left*). At the beginning, no molecules of the liquid are in the vapor phase. After a short time, however, some of the liquid evaporates, and the molecules now in the vapor phase exert a pressure. The pressure of the vapor measured when the liquid and the vapor are in equilibrium is called the *equilibrium vapor pressure* (*right*).

GENERAL
Chemistry‑‑Now™ See the General ChemistryNow CD-ROM or website to explore an interactive version of this figure accompanied by an exercise.

■ **Equilibrium Vapor Pressure**
At the conditions of *T* and *P* given by any point on a curve in Figure 13.18, the pure liquid and its vapor are in dynamic equilibrium. If *T* and *P* define a point not on the curve, the system is not at equilibrium. See Appendix G for the vapor pressures of water at various temperatures.

   When a liquid–vapor equilibrium has been established, the **equilibrium vapor pressure** (often just called the *vapor pressure*) can be measured. The equilibrium vapor pressure of any substance is a measure of the tendency of its molecules to escape from the liquid phase and enter the vapor phase at a given temperature. This tendency is referred to qualitatively as the **volatility** of the compound. The higher the equilibrium vapor pressure at a given temperature, the more volatile the compound.
   As described previously (see Figure 13.14), the distribution of molecular energies in the liquid phase is a function of temperature. At a higher temperature, more

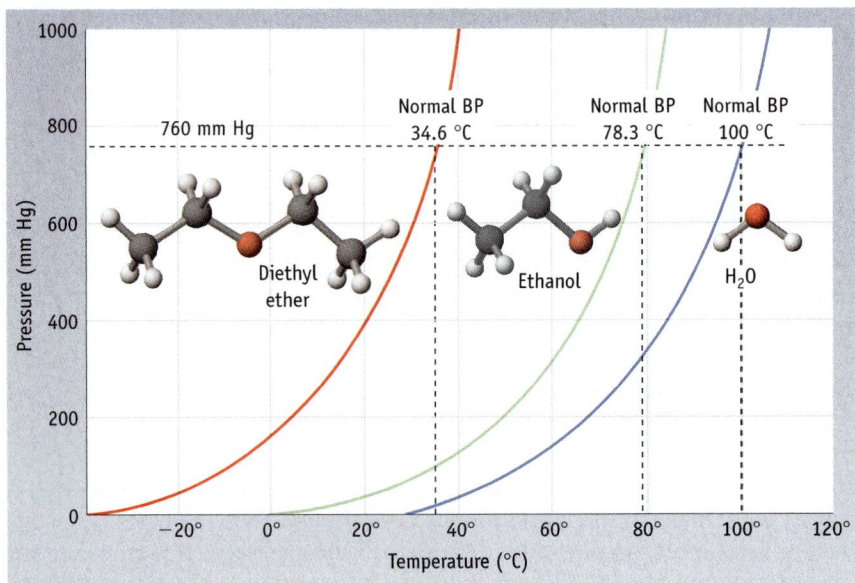

Active Figure 13.18   **Vapor pressure curves for diethyl ether [$(C_2H_5)_2O$], ethanol (C_2H_5OH), and water.** Each curve represents conditions of *T* and *P* at which the two phases, liquid and vapor, are in equilibrium. These compounds exist as liquids for temperatures and pressures to the left of the curve and as gases under conditions to the right of the curve.

GENERAL
Chemistry‑‑Now™ See the General ChemistryNow CD-ROM or website to explore an interactive version of this figure accompanied by an exercise.

molecules have sufficient energy to escape the surface of the liquid. The equilibrium vapor pressure must, therefore, increase with temperature (Figure 13.18). All points along the vapor-pressure-versus-temperature curves in Figure 13.18 represent conditions of pressure and temperature at which liquid and vapor are in equilibrium. For example, at 60 °C the vapor pressure of water is 149 mm Hg (Appendix G). If water is placed in an evacuated flask that is maintained at 60 °C, liquid will evaporate until the pressure exerted by the vapor is 149 mm Hg (assuming enough water is in the flask so that some liquid remains when equilibrium is reached).

GENERAL
Chemistry⋅⚛⋅Now™

See the General ChemistryNow CD-ROM or website:
- **Screen 13.9 Properties of Liquids (2): Vapor Pressure**, to view an animation of equilibrium vapor pressure and for a simulation of vapor pressure curves

---

Example 13.6—Vapor Pressure

Problem You place 2.00 L of water in your dormitory room, which has a volume of 4.25×10^4 L. You seal the room and wait for the water to evaporate. Will all of the water evaporate at 25 °C? (At 25 °C the density of water is 0.997 g/mL, and its vapor pressure is 23.8 mm Hg.)

Strategy One approach to solving this problem is to calculate the quantity of water that must evaporate to exert a pressure of 23.8 mm Hg in a volume of 4.25×10^4 L at 25 °C. Because water vapor is like any other gas, we use the ideal gas law for the calculation.

Solution Calculate the amount and then mass and volume of water that fulfills the following conditions: $P = 23.8$ mm Hg, $V = 4.25 \times 10^4$ L, $T = 25°C$ (or 298 K).

$$P = 23.8 \text{ mm Hg} \times \frac{1 \text{ atm}}{760 \text{ mm Hg}} = 0.0313 \text{ atm}$$

$$n = \frac{PV}{RT} = \frac{(0.0313 \text{ atm})(4.25 \times 10^4 \text{ L})}{\left(0.082057 \dfrac{\text{L} \cdot \text{atm}}{\text{K} \cdot \text{mol}}\right)(298 \text{ K})} = 54.4 \text{ mol}$$

$$54.4 \text{ mol H}_2\text{O} \times \frac{18.02 \text{ g}}{1 \text{ mol H}_2\text{O}} = 980. \text{ g H}_2\text{O}$$

$$980. \text{ g H}_2\text{O} \times \frac{1 \text{ mL}}{0.997 \text{ g}} = 983 \text{ mL}$$

Only about half of the available water needs to evaporate to achieve an equilibrium water vapor pressure of 23.8 mm Hg at 25 °C in the dorm room.

Exercise 13.6—Vapor Pressure Curves

Examine the vapor pressure curve for ethanol in Figure 13.18.

(a) What is the approximate vapor pressure of ethanol at 40 °C?

(b) Are liquid and vapor in equilibrium when the temperature is 60 °C and the pressure is 600 mm Hg? If not, does liquid evaporate to form more vapor, or does vapor condense to form more liquid?

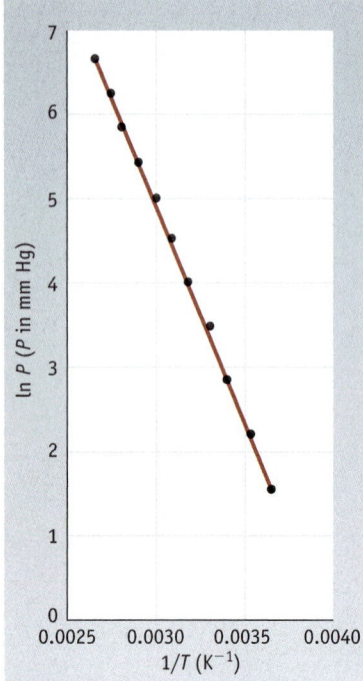

Figure 13.19 **Clausius-Clapeyron equation.** When the natural logarithm of the vapor pressure (ln P) of water at various temperatures (T) is plotted against $1/T$, a straight line is obtained. The slope of the line equals $-\Delta H^{\circ}_{vap}/R$. Values of T and P are from Appendix G.

Exercise 13.7—Vapor Pressure

If 0.50 g of pure water is sealed in an evacuated 5.0-L flask and the whole assembly is heated to 60 °C, will the pressure be equal to or less than the equilibrium vapor pressure of water at this temperature? What if you use 2.0 g of water? Under either set of conditions, is any liquid water left in the flask, or does all of the water evaporate?

Vapor Pressure, Enthalpy of Vaporization, and the Clausius-Clapeyron Equation

Plotting the vapor pressure for a liquid at a series of temperatures results in a curved line (Figure 13.18). However, the German physicist R. Clausius (1822–1888) and the Frenchman B. P. E. Clapeyron (1799–1864) showed that, for a pure liquid, a *linear* relationship exists between the reciprocal of the Kelvin temperature ($1/T$) and the natural logarithm of vapor pressure (ln P).

$$\ln P = -(\Delta H^{\circ}_{vap}/RT) + C \tag{13.1}$$

Here ΔH°_{vap} is the enthalpy of vaporization of the liquid, R is the ideal gas constant (8.314472 J/K · mol), and C is a constant characteristic of the compound in question. This equation, now called the **Clausius-Clapeyron equation**, provides a method of obtaining values for ΔH°_{vap}. The equilibrium vapor pressure of a liquid can be measured at several different temperatures, and the logarithm of these pressures is plotted versus $1/T$. The result is a straight line with a slope of $-\Delta H^{\circ}_{vap}/R$. For example, plotting data for water (Figure 13.19), we find the slope of the line is -4.90×10^3, which gives $\Delta H^{\circ}_{vap} = 40.7$ kJ/mol.

As an alternative to plotting ln P versus $1/T$, we can write the following equation that allows us to calculate ΔH°_{vap} knowing the vapor pressure of a liquid at two different temperatures.

$$\ln P_2 - \ln P_1 = \left[\frac{-\Delta H^{\circ}_{vap}}{RT_2} + C\right] - \left[\frac{-\Delta H^{\circ}_{vap}}{RT_1} + C\right]$$

or

$$\ln\frac{P_2}{P_1} = \frac{\Delta H^{\circ}_{vap}}{R}\left[\frac{1}{T_1} - \frac{1}{T_2}\right] \tag{13.2}$$

For example, ethylene glycol has a vapor pressure of 14.9 mm Hg (P_1) at 373 K (T_1), and a vapor pressure of 49.1 mm Hg (P_2) at 398 K (T_2).

$$\ln\left(\frac{49.1\ \text{mm Hg}}{14.9\ \text{mm Hg}}\right) = \frac{\Delta H^{\circ}_{vap}}{0.0083145\ \text{kJ/K} \cdot \text{mol}}\left[\frac{1}{373\ \text{K}} - \frac{1}{398\ \text{K}}\right]$$

$$1.192 = \frac{\Delta H^{\circ}_{vap}}{0.0083145\ \text{kJ/K} \cdot \text{mol}}(0.000168)\frac{1}{\text{K}}$$

$$\Delta H^{\circ}_{vap} = 59.0\ \text{kJ/mol}$$

GENERAL
Chemistry · Now ™

See the General ChemistryNow CD-ROM or website:
• **Screen 13.9 Properties of Liquids (2): Vapor Pressure,** for three tutorials on using the Clausius-Clapeyron equation

Exercise 13.8—Clausius-Clapeyron Equation

Calculate the enthalpy of vaporization of diethyl ether, $(C_2H_5)_2O$ (see Figure 13.8). This compound has vapor pressures of 57.0 mm Hg and 534 mm Hg at -22.8 °C and 25.0 °C, respectively.

Boiling Point

If you have a beaker of water open to the atmosphere, the mass of the atmosphere presses down on the surface. As heat is added, more and more water evaporates, pushing the molecules of the atmosphere aside. If enough heat is added, a temperature is eventually reached at which the vapor pressure of the liquid equals the atmospheric pressure. Larger and larger bubbles of vapor form in the liquid and rise to the surface; the liquid boils (Figure 13.20).

The boiling point of a liquid is the temperature at which its vapor pressure is equal to the external pressure. If the external pressure is 760 mm Hg, this temperature is designated as the **normal boiling point**. This point is highlighted on the vapor pressure curves for several substances in Figure 13.18. Normal boiling points of other liquids are included in Table 13.4, where you can also see the relationship between normal boiling point and enthalpy of vaporization.

The normal boiling point of water is 100 °C, and in a great many places in the United States, water boils at or near this temperature. If you live at higher altitudes, however, such as in Salt Lake City, Utah, where the barometric pressure is about 650 mm Hg, water will boil at a noticeably lower temperature. The curve for the equilibrium vapor pressure of water in Figure 13.18 shows that a pressure of 650 mm Hg corresponds to a boiling temperature of about 95 °C. Cooks know that food has to be cooked a little longer in Salt Lake City or Denver to achieve the same result as in New York City at sea level.

Figure 13.20 **Vapor pressure and boiling.** When the vapor pressure of the liquid equals the atmospheric pressure, bubbles of vapor begin to form within the body of liquid, and the liquid boils.

Charles D. Winters

GENERAL
Chemistry⚛Now™

See the General ChemistryNow CD-ROM or website:

- **Screen 13.10 Properties of Liquids (3): Boiling Point,** to watch a video of the boiling of a liquid in a partial vacuum and to explore the relationship of molecular composition and structure and boiling point in two simulations

■ **Cooking under Pressure**
To shorten cooking time, a pressure cooker can be used. This sealed pot allows water vapor to build up to pressures somewhat greater than the external or atmospheric pressure. At pressures greater than 760 mm Hg, the boiling point of the water is higher than 100 °C, and foods cook faster.

Critical Temperature and Pressure

The vapor pressure of a liquid will continue to increase if the temperature is raised above the normal boiling point. On first thought it might seem that vapor pressure–temperature curves (such as shown in Figure 13.18) should continue upward without limit, but this is not so. Instead, when a specific temperature and pressure are reached, the interface between the liquid and the vapor disappears. This point is called the **critical point**. The temperature at which this phenomenon occurs is the **critical temperature**, T_c, and the corresponding pressure is the **critical pressure**, P_c (Figure 13.21). The substance that exists under these conditions is called a **supercritical fluid**. It is like a gas under such a high pressure that its density resembles that of a liquid, while its viscosity (ability to flow) remains close to that of a gas.

For most substances the critical point is at a very high temperature and pressure (Table 13.5). Water, for instance, has a critical temperature of 374 °C and a critical pressure of 217.7 atm. Consider what the substance might look like at the molecular

Figure 13.21
Critical temperature and pressure for water. The curve representing equilibrium conditions for liquid and gaseous water ends at the critical point; above that temperature and pressure, water becomes a supercritical fluid.

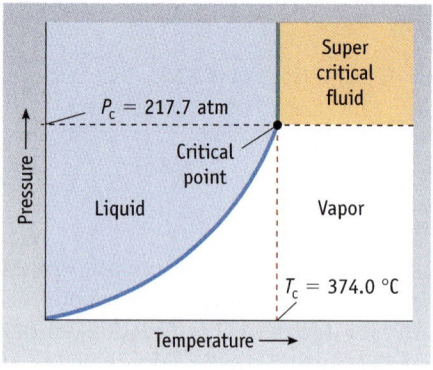

Table 13.5   Critical Temperatures and Pressures for Common Compounds*

Compound	T_c (°C)	P_c (atm)
CH_4 (methane)	−82.6	45.4
C_2H_6 (ethane)	32.3	49.1
C_3H_8 (propane)	96.7	41.9
C_4H_{10} (butane)	152.0	37.3
CCl_2F_2 (CFC-12)	111.8	40.9
NH_3	132.4	112.0
H_2O	374.0	217.7
CO_2	30.99	72.8
SO_2	157.7	77.8

*Data taken from D. R. Lide: *Basic Laboratory and Industrial Chemicals*, Boca Raton FL, CRC Press, 1993.

■ **Supercritical CO_2 and the Environment**
Supercritical CO_2 has properties that make it attractive as a solvent, so not surprisingly other uses are being sought for this substance. For example, more than 10 billion kilograms of organic and halogenated solvents is used worldwide every year in cleaning applications. These cleaning agents can have deleterious effects on the environment, so it is hoped that many can be replaced by supercritical CO_2. (For more about supercritical CO_2 see page 632.)

level under these conditions. At this high pressure, water molecules have been forced almost as close together as they are in the liquid state. The high temperature, however, means that each molecule has enough kinetic energy to exceed the forces holding molecules together. As a result, the supercritical fluid has a tightly packed molecular arrangement like a liquid, but the intermolecular forces of attraction that characterize the liquid state are less than the kinetic energy of the particles.

Supercritical fluids can have unexpected properties, such as the ability to dissolve normally insoluble materials. Supercritical CO_2 is especially useful. Carbon dioxide is widely available, essentially nontoxic, nonflammable, and inexpensive. It is relatively easy to reach its critical temperature of 30.99 °C and critical pressure of 72.8 atm. The material is also easy to handle. CO_2 is highly useful because it does not dissolve water or polar compounds such as sugar, but it does dissolve nonpolar oils, which constitute many of the flavoring or odor-causing compounds in foods. As a result, food companies now use supercritical CO_2 to extract caffeine from coffee, for example.

To decaffeinate coffee, the beans are treated with steam to bring the caffeine to the surface. The beans are then immersed in supercritical CO_2, which selectively dissolves the caffeine but leaves intact the compounds that give flavor to coffee. (Decaffeinated coffee contains less than 3% of the original caffeine.) The solution of caffeine in supercritical CO_2 is poured off, and the CO_2 is evaporated, trapped, and reused.

Surface Tension, Capillary Action, and Viscosity

Molecules in the interior of a liquid interact with molecules all around them (Figure 13.22). In contrast, surface molecules are affected only by those molecules located below the surface layer. This phenomenon leads to a net inward force of attraction on the surface molecules, contracting the surface area and making the liquid behave as though it had a skin. The toughness of the skin of a liquid is measured by its **surface tension**—the energy required to break through the surface or to disrupt a liquid drop and spread the material out as a film. Surface tension causes water drops to be spheres and not little cubes, for example (Figure 13.23a), because the sphere has a smaller surface area than any other shape of the same volume.

Capillary action is closely related to surface tension. When a small-diameter glass tube is placed in water, the water rises in the tube, just as water rises in a piece of paper in water (Figure 13.23b). Because polar Si—O bonds are present on the surface of glass, polar water molecules are attracted by **adhesive forces** between the two different substances. These forces are strong enough that they can compete

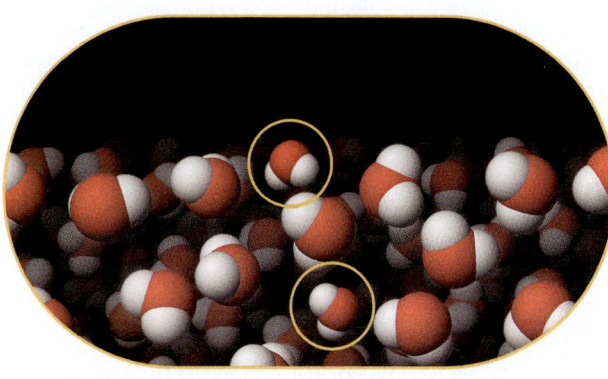

Water molecules on the surface are not completely surrounded by other water molecules.

Water molecules under the surface are completely surrounded by other water molecules.

Figure 13.22   Intermolecular forces in a liquid. Forces acting on a molecule at the surface of a liquid are different than those acting on a molecule in the interior of a liquid.

with the **cohesive forces** between the water molecules themselves. Thus, some water molecules can adhere to the walls; other water molecules are attracted to them and build a "bridge" back into the liquid. The surface tension of the water (from cohesive forces) is great enough to pull the liquid up the tube, so the water level rises in the tube. The rise will continue until the attractive forces—adhesion between water and glass, cohesion between water molecules—are balanced by the force of gravity pulling down on the water column. These forces lead to the characteristic concave, or downward-curving, *meniscus* seen with water in a drinking glass or in a laboratory test tube (Figure 13.23c).

In some liquids, cohesive forces (high surface tension) are much greater than adhesive forces with glass. Mercury is one example. Mercury does not climb the walls of a glass capillary. In fact, when it is in a glass tube, mercury will form a convex, or upward-curving, meniscus (Figure 13.23c).

a, S. R. Nagel, James Frank Institute, University of Chicago; *b* and *c*, Charles D. Winters

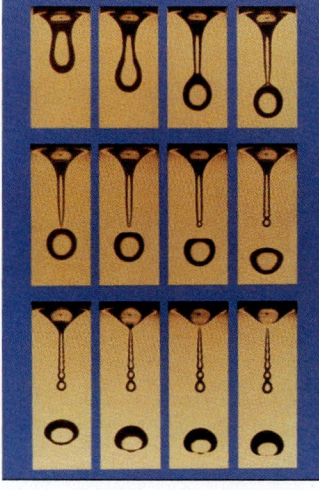

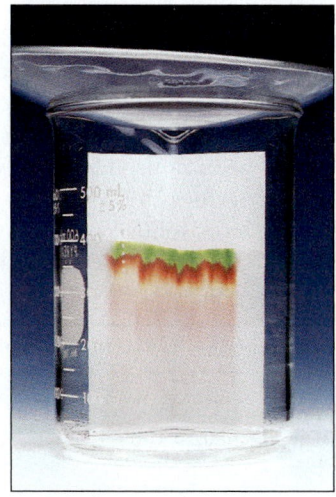

(a) A series of photographs showing the different stages when a water drop falls. The drop was illuminated by a strobe light of 5-ms duration. (The total time for this sequence was 0.05 s.) Water droplets take a spherical shape because of surface tension.

(b) Capillary action. Polar water molecules are attracted to the —OH bonds in paper fibers, and water rises in the paper. If a line of ink is placed in the path of the rising water, the different components of the ink are attracted differently to the water and paper and are separated in a process called chromatography.

(c) Water (top layer) forms a concave meniscus, while mercury (bottom layer) forms a convex meniscus. The different shapes are determined by the adhesive forces of the molecules of the liquid with the walls of the tube and the cohesive forces between molecules of the liquid.

Figure 13.23   Adhesive and cohesive forces in liquids.

One other important property of liquids in which intermolecular forces play a role is their **viscosity**, the resistance of liquids to flow. When you turn over a glassful of water, it empties quickly. In contrast, it takes much more time to empty a glassful of olive oil or honey. Olive oil consists of molecules with long chains of carbon atoms (see Chapter 11), and it is about 70 times more viscous than ethanol, a small molecule with only two carbons and one oxygen. Longer chains have greater intermolecular forces because there are more atoms to attract one another, with each atom contributing to the total force. In contrast, honey is a concentrated solution of smaller sugar molecules. However, these molecules have numerous —OH groups and are thus capable of hydrogen bonding.

■ **Viscosity**
Another factor in determining viscosity may be the presence of long chains of atoms in substances such as oils. These long chains are floppy and could become entangled with one another in the liquid; the longer the chain, the greater the tangling and the greater the viscosity.

GENERAL
Chemistry⚛Now™

See the General ChemistryNow CD-ROM or website:
- **Screen 13.11 Properties of Liquids (4): Surface Tension, Capillary Action, and Viscosity,** to watch videos on these three topics

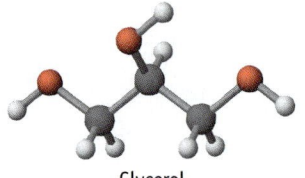

Glycerol

Exercise 13.9—Viscosity

Glycerol (HOCH$_2$CHOHCH$_2$OH) is used in cosmetics. Do you expect its viscosity to be larger or smaller than the viscosity of ethanol, CH$_3$CH$_2$OH? Explain briefly.

13.6—The Solid State: Metals

Many kinds of solids exist in the world around us (Figure 13.24). Solid-state chemistry is one of the booming areas of science, especially because it relates to the development of interesting new materials. As we describe various kinds of solids, we hope to provide a glimpse of the reasons this area is exciting.

Solid-state chemistry can be organized by classifying the common types of solids (Table 13.6). This section describes the solid-state structures of common metals, and the next section takes up ionic solids. Next, we examine molecular and network solids in Section 13.8. Finally, Section 13.9 outlines important properties of solids.

■ **Chemistry of Materials**
For a glimpse into scientists' latest efforts to create new materials and to find new uses for old materials, see "The Chemistry of Modern Materials," page 642.

Table 13.6   Structures and Properties of Various Types of Solid Substances

Type	Examples	Structural Units
Ionic	NaCl, K$_2$SO$_4$, CaCl$_2$, (NH$_4$)$_3$PO$_4$	Positive and negative ions; no discrete molecules
Metallic	Iron, silver, copper, other metals and alloys	Metal atoms (positive metal ions with delocalized electrons)
Molecular	H$_2$, O$_2$, I$_2$, H$_2$O, CO$_2$, CH$_4$, CH$_3$OH, CH$_3$CO$_2$H	Molecules
Network	Graphite, diamond, quartz, feldspars, mica	Atoms held in an infinite two-, or three-dimensional network
Amorphous	Glass, polyethylene, nylon	Covalently bonded networks with no long-range regularity

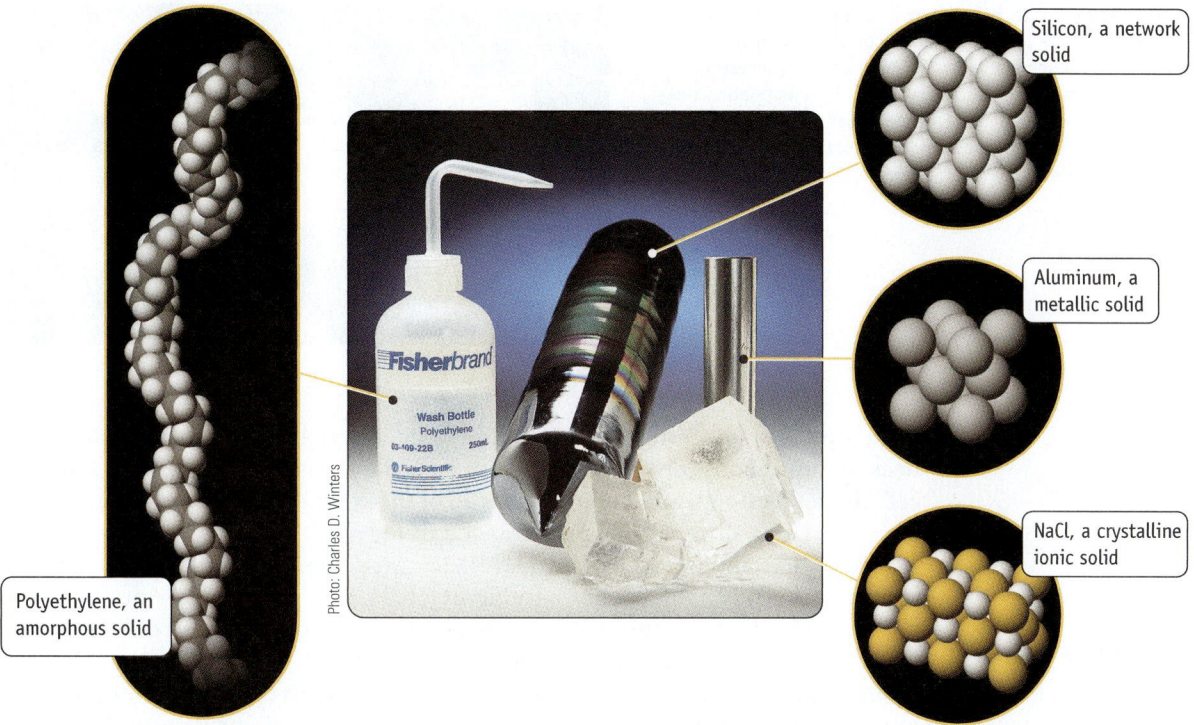

Figure 13.24 **Some common solids.**

Crystal Lattices and Unit Cells

In both gases and liquids, molecules move continually and randomly, and they rotate and vibrate as well. Because of this movement, an orderly arrangement of molecules in the gaseous or liquid state is not possible. In solids, however, the molecules, atoms, or ions cannot change their relative positions (although they vibrate and occasionally rotate). Thus, a regular, repeating pattern of atoms or molecules within the structure—a long-range order—is characteristic of the solid state.

The beautiful, external (macroscopic) regularity of a crystal of salt (see Figure 13.24) suggests that it has an internal symmetry, a symmetry involving the ions that

Forces Holding Units Together	Typical Properties
Ionic; attractions among charges on positive and negative ions	Hard; brittle; high melting point; poor electric conductivity as solid, good as liquid; often water-soluble
Metallic; electrostatic attraction among metal ions and electrons	Malleable; ductile; good electric conductivity in solid and liquid; good heat conductivity; wide range of hardness and melting points
Dispersion forces, dipole–dipole forces, hydrogen bonds	Low to moderate melting points and boiling points; soft; poor electric conductivity in solid and liquid
Covalent; directional electron-pair bonds	Wide range of hardness and melting points (three-dimensional bonding > two-dimensional bonding); poor electric conductivity, with some exceptions
Covalent; directional electron-pair bonds	Noncrystalline; wide temperature range for melting; poor electric conductivity, with some exceptions

Figure 13.25 **Unit cells for a flat, two-dimensional solid made from circular "atoms."** Many unit cells are possible, with two of the most obvious being squares. The lattice can be represented as being made up of repeating unit cells. (That is, this two-dimensional lattice can be built by translating the unit cells throughout the plane of the figure. Each cell must move by the length of one side of the unit cell.) In this figure all unit cells contain a net of one large sphere and one small sphere.

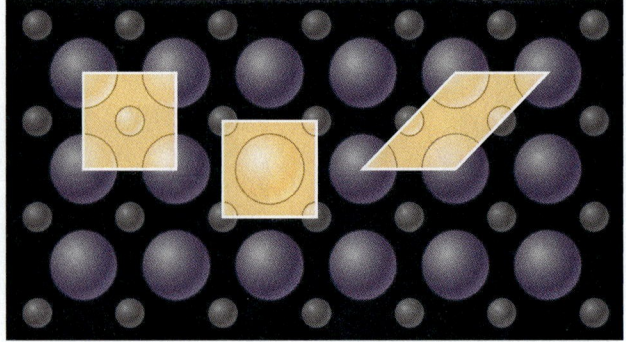

make up the solid. Structures of solids can be described as three-dimensional lattices of atoms, ions, or molecules. For a crystalline solid, we can identify the **unit cell**, the smallest repeating unit that has all of the symmetry characteristic of the way the atoms, ions, or molecules are arranged in the solid.

To understand unit cells, consider first a two-dimensional lattice model, the repeating pattern of spheres shown in Figure 13.25. The yellow square at the left is a unit cell. The overall pattern can be created from a group of these cells by joining them edge to edge. It is also a requirement that unit cells reflect the stoichiometry of the solid. Here the square unit cell at the left contains one-fourth of each of the four larger spheres and one smaller sphere, giving a total of one large and one small sphere per two-dimensional unit cell.

You may recognize that it is possible to draw other unit cells for this two-dimensional lattice. One option is the square in the middle of Figure 13.25 that fully encloses a single large sphere and parts of small spheres that add up to one net small sphere. Yet another possible unit cell is the parallelogram at the right. Other unit cells are possible, but it is conventional to try to draw unit cells in which atoms or ions are placed at the **lattice points**, that is, at the corners of the cube or other geometric object that constitutes the unit cell.

The three-dimensional lattices of solids can be built by assembling three-dimensional unit cells much like building blocks (Figure 13.26). The assemblage of these three-dimensional unit cells defines the **crystal lattice**.

To construct crystal lattices, nature uses seven three-dimensional unit cells. They differ from one another in that their sides have different relative lengths and their edges meet at different angles. The simplest of the seven crystal lattices is the **cubic unit cell**, a cell with edges of equal length that meet at 90° angles. We shall

Figure 13.26 **Cubic unit cells.** (a) The cube, one of the seven basic unit cells that describe crystal systems. (b) Stacking cubes to build a crystal lattice. Each crystal face is part of two cubes, each edge is part of four cubes, and each corner is part of eight cubes.

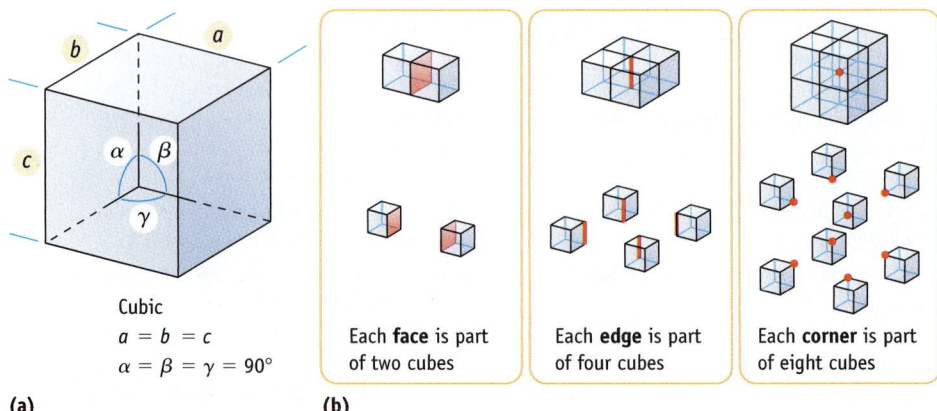

Cubic
$a = b = c$
$\alpha = \beta = \gamma = 90°$

Each **face** is part of two cubes

Each **edge** is part of four cubes

Each **corner** is part of eight cubes

(a)   (b)

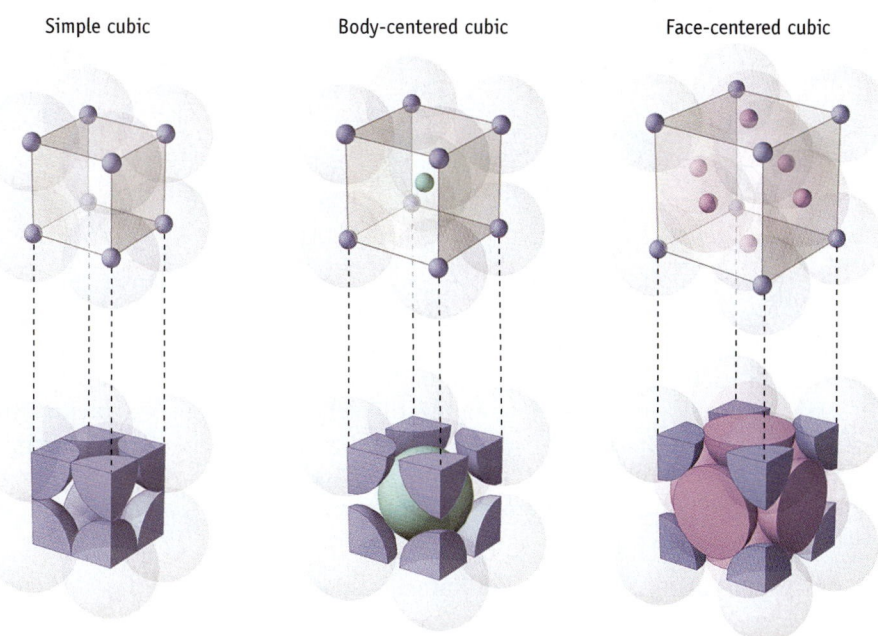

Simple cubic Body-centered cubic Face-centered cubic

Figure 13.27 The three cubic unit cells. The top row shows the lattice points of the three cells, and the bottom row shows the same cells using space-filling spheres. The spheres in each figure represent *identical* atoms (or ions) centered on the lattice points. Because eight unit cells share a corner atom, only 1/8 of each corner atom lies within a given unit cell; the remaining 7/8 lies in seven other unit cells. Because each face of a fcc unit cell is shared with another unit cell, one half of each atom in the face of a face-centered cube lies in a given unit cell, and the other half lies in the adjoining cell.

look in detail at just this structure, not only because cubic unit cells are easily visualized but also because they are commonly encountered.

Within the cubic class, three cell symmetries occur: **primitive** or **simple cubic (sc)**, **body-centered cubic (bcc)**, and **face-centered cubic (fcc)** (Figure 13.27). All three have identical atoms, molecules, or ions at the corners of the cubic unit cell. The bcc and fcc arrangements, however, differ from the primitive cube in that they have additional particles at other locations. The bcc structure is called "body-centered" because it has an additional particle, of the same type as those at the corners, at the center of the cube. The fcc arrangement is called "face-centered" because it has a particle, of the same type as the corner atoms, in the center of each of the six faces of the cube. Metals may assume any of these structures. The alkali metals, for example, are body-centered cubic, whereas nickel, copper, and aluminum are face-centered cubic (Figure 13.28).

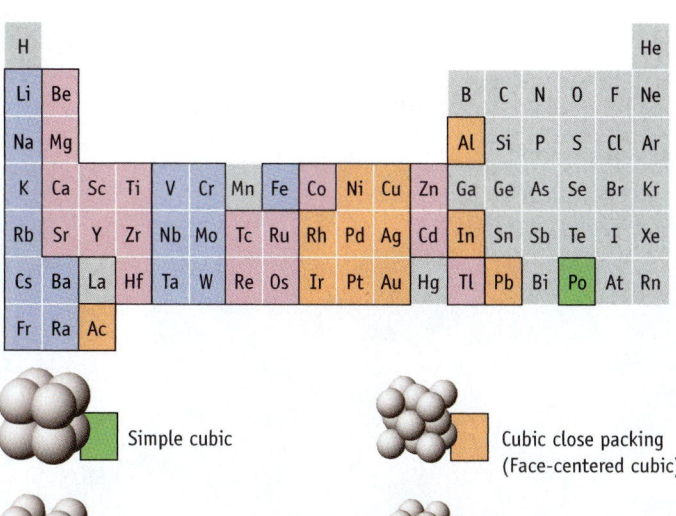

Simple cubic

Body-centered cubic

Cubic close packing (Face-centered cubic)

Hexagonal close packing

Figure 13.28 Metals use four different unit cells. Three are based on the cube, and the fourth is the hexagonal unit cell. See "A Closer Look: Packing Oranges," page 621.

Figure 13.29 Atom sharing at cube corners and faces. (a) In any cubic lattice, each corner particle is shared equally among eight cubes, so one eighth of the particle is within a particular cubic unit cell. (b) In a face-centered lattice, each particle on a cube face is shared equally between two unit cells. One half of each particle of this type is within the given unit cell.

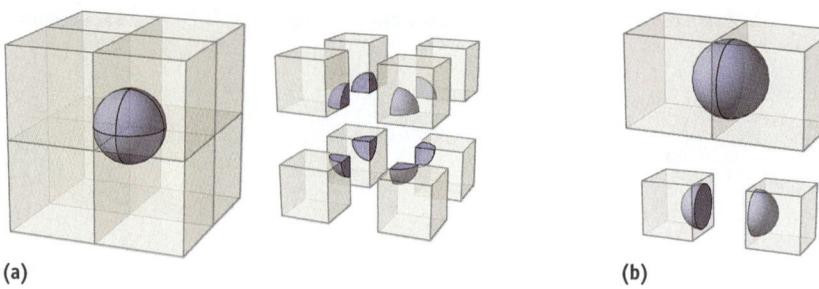

(a)   (b)

When the cubes pack together to make a three-dimensional crystal of a metal, the atom at each corner is shared among eight cubes (Figures 13.26, 13.27, and 13.29a). Because of this, only one eighth of each corner atom is actually within a given unit cell. Furthermore, because a cube has eight corners, and because one eighth of the atom at each corner "belongs to" a particular unit cell, the corner atoms contribute a net of one atom to a given unit cell. Thus, *the primitive or simple cubic arrangement has one net atom within the unit cell.*

(8 corners of a cube)(1/8 of each corner atom within a unit cell) =

1 net atom per unit cell

■ **Atoms per Unit Cell**

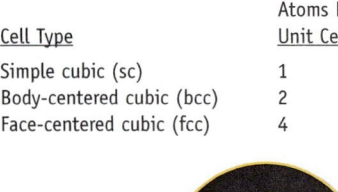

Cell Type	Atoms Per Unit Cell
Simple cubic (sc)	1
Body-centered cubic (bcc)	2
Face-centered cubic (fcc)	4

In contrast to the simple cube, a body-centered cube has an additional atom wholly within the unit cell at the cube's center. The center particle is present in addition to those at the cube corners, so *the body-centered cubic arrangement has a net of two atoms within the unit cell.*

In a face-centered cubic arrangement, there is an atom in each of the six faces of the cube in addition to those at the cube corners. One half of each atom on a face belongs to the given unit cell (Figure 13.29b). Three net particles are therefore contributed by the particles on the faces of the cube:

(6 faces of a cube)(1/2 of an atom within a unit cell) =

3 net face-centered atoms within a unit cell

Aluminum metal. The metal has a face-centered cubic unit cell with a net of four Al atoms in each unit cell.

Thus, *the face-centered cubic arrangement has a net of four atoms within the unit cell,* one contributed by the corner atoms and another three contributed by the atoms centered in the six faces.

An experimental technique, x-ray crystallography, can be used to determine the structure of a crystalline substance (Figure 13.30). Once the structure is known, this information can be combined with other experimental information to calculate such useful parameters as the radius of an atom (Study Questions 13.54–13.55).

Figure 13.30 X-ray crystallography. In the x-ray diffraction experiment, a beam of x-rays is directed at a crystalline solid. The photons of the x-ray beam are scattered by the atoms of the solid. The scattered x-rays are detected by a photographic film or an electronic detector, and the pattern of scattered x-rays is related to the locations of the atoms or ions in the crystal.

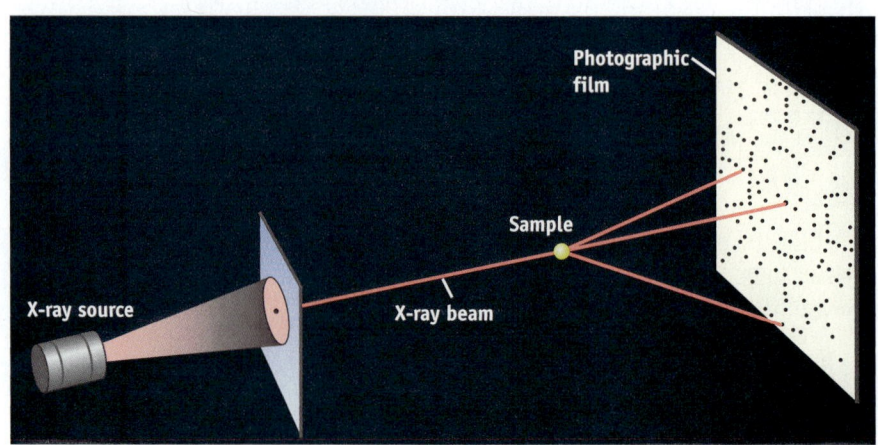

A Closer Look

Packing Oranges

It is a "rule" that nature does things as efficiently as possible. You know this if you have ever tried to stack some oranges into a pile that doesn't fall over and that takes up as little space as possible. How did you do it? Clearly, the pyramid arrangement below on the right works, whereas the cubic one on the left does not.

Open space between balls

If you could look inside the pile, you would find that less open space is left in the pyramid stacking than in the cube stacking. Only 52% of the space is filled in the cubic packing arrangement. (If you could stack oranges as a body-centered cube, that would be slightly better; 68% of the space is used.) However, the best method is the pyramid stack, which is really a face-centered cubic arrangement. Oranges, atoms, or ions packed this way occupy 74% of the available space.

To fill three-dimensional space, the most efficient way to pack oranges or atoms is to begin with a hexagonal arrangement of spheres, as in this arrangement of marbles.

Succeeding layers of atoms or ions are then stacked one on top of the other in two different ways. Depending on the stacking pattern (Figure 1), you will get either a **cubic close-packed (ccp)** or **hexagonal close-packed (hcp)** arrangement.

In the hcp arrangement, additional layers of particles are placed above and below a given layer, fitting into the same depressions on either side of the middle layer. In a three-dimensional

crystal, the layers repeat their pattern in the manner ABABAB. . . . Atoms in each A layer are directly above the ones in another A layer; the same holds true for the B layers.

In the ccp arrangement, the atoms of the "top" layer (A) rest in depressions in the middle layer (B), and those of the "bottom" layer (C) are oriented opposite to those in the top layer. In a crystal, the pattern is repeated ABCABCABC. . . . By turning the whole crystal, you can see that the ccp arrangement is the face-centered cubic structure (Figure 2).

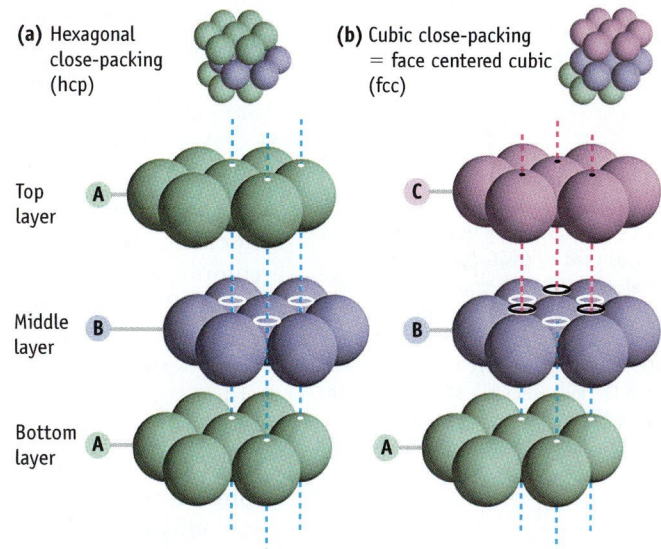

(a) Hexagonal close-packing (hcp)

(b) Cubic close-packing = face centered cubic (fcc)

Top layer — A / C

Middle layer — B / B

Bottom layer — A / A

Figure 1 Efficient packing. The most efficient ways to pack atoms or ions in crystalline materials are hexagonal close packing (hcp) and cubic close packing (ccp).

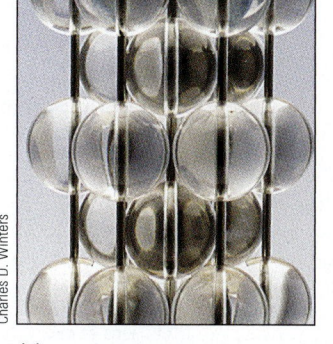

(a)

(b)

Figure 2 Models of close packing. (a) A model of hexagonal close-packing, where the layers repeat in the order ABABAB. . . . (b) A face-centered unit cell (cubic close-packing), where the layers repeat in the order ABCABC. . . . (A kit from which these models can be built is available from the Institute for Chemical Education at the University of Wisconsin at Madison.)

See the General ChemistryNow CD-ROM or website:
- **Screen 13.12 Solid Structures (1): Crystalline and Amorphous Solids,** to view an animation of unit cells

13.7—The Solid State: Structures and Formulas of Ionic Solids

The lattices of many ionic compounds are built by taking a simple cubic or face-centered cubic lattice of spherical ions of one type and placing ions of opposite charge in the holes within the lattice. This produces a three-dimensional lattice of regularly placed ions. The smallest repeating unit in these structures is, by definition, the unit cell for the ionic compound.

The choice of the lattice and the number and location of the holes that are filled are the keys to understanding the relationship between the lattice structure and the formula of a salt. This is illustrated with the ionic compound cesium chloride, $CsCl$ (Figure 13.31). The structure of $CsCl$ has a primitive cubic unit cell of chloride ions. The cesium ion fits into a hole in the center of the cube. (An equivalent unit cell has a primitive cubic unit cell of Cs^+ ions with a Cl^- ion in the center of the cube.)

Next consider the structure for $NaCl$. An extended view of the lattice and one unit cell are illustrated in Figures 13.32a and 13.32b, respectively. The Cl^- ions are arranged in a face-centered cubic unit cell, and the Na^+ ions are arranged in a regular manner between these ions. Notice that each Na^+ ion is surrounded by six Cl^- ions. An octahedral geometry is assumed by the ions surrounding an Na^+ ion, so the Na^+ ions are said to be in **octahedral holes** (Figure 13.32c).

The formula for $NaCl$ can be related to this structure by counting the number of cations and anions contained in one unit cell. A face-centered cubic lattice of Cl^- ions has a net of four Cl^- ions within the unit cell. There is one Na^+ ion in the center of the unit cell, contained totally within the unit cell. In addition, there are 12 Na^+ ions along the edges of the unit cell. Each of these Na^+ ions is shared among four unit cells, so each contributes one fourth of an Na^+ ion to the unit cell, giving three additional Na^+ ions within the unit cell.

(1 Na^+ ion in the center of the unit cell) + (1/4 of Na^+ ion in each edge $\times$ 12 edges)
= net of 4 Na^+ ions in $NaCl$ unit cell

■ Lattice Ions and Holes
When ions are packed into a lattice, the holes in the lattice are usually smaller than the ions used to create the lattice. Therefore, a lattice is usually built out of the larger anions, and the smaller cations are placed into the holes in the lattice. For NaCl, for example, an fcc lattice is built out of the Cl^- ions (radius = 181 pm), and the smaller Na^+ cations (radius = 98 pm) are placed in appropriate holes in the lattice.

Figure 13.31 Cesium chloride (CsCl) unit cell. The unit cell of CsCl may be viewed in two ways. The only requirement is that the unit cell must have a net of one Cs^+ ion and one Cl^- ion. Either way, it is a simple cubic unit cell of ions of one type (Cl^- on the left or Cs^+ on the right). Generally, ionic lattices are assembled by placing the larger ions (here Cl^-) at the lattice points and placing the smaller ions (here Cs^+) in the lattice holes.

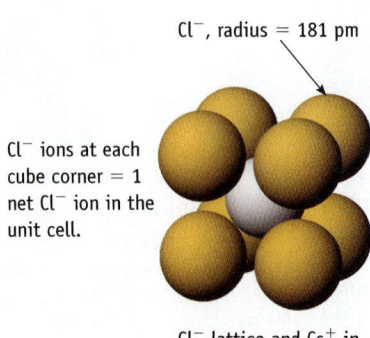

Cl^-, radius = 181 pm

Cl^- ions at each cube corner = 1 net Cl^- ion in the unit cell.

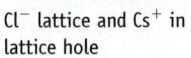

Cl^- lattice and Cs^+ in lattice hole

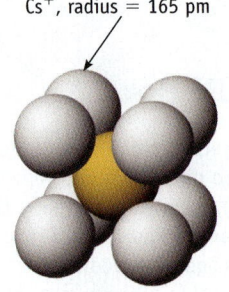

Cs^+, radius = 165 pm

One Cs^+ ion at each cube corner = 1 net Cs^+ ion in the unit cell.

Cs^+ lattice and Cl^- in lattice hole

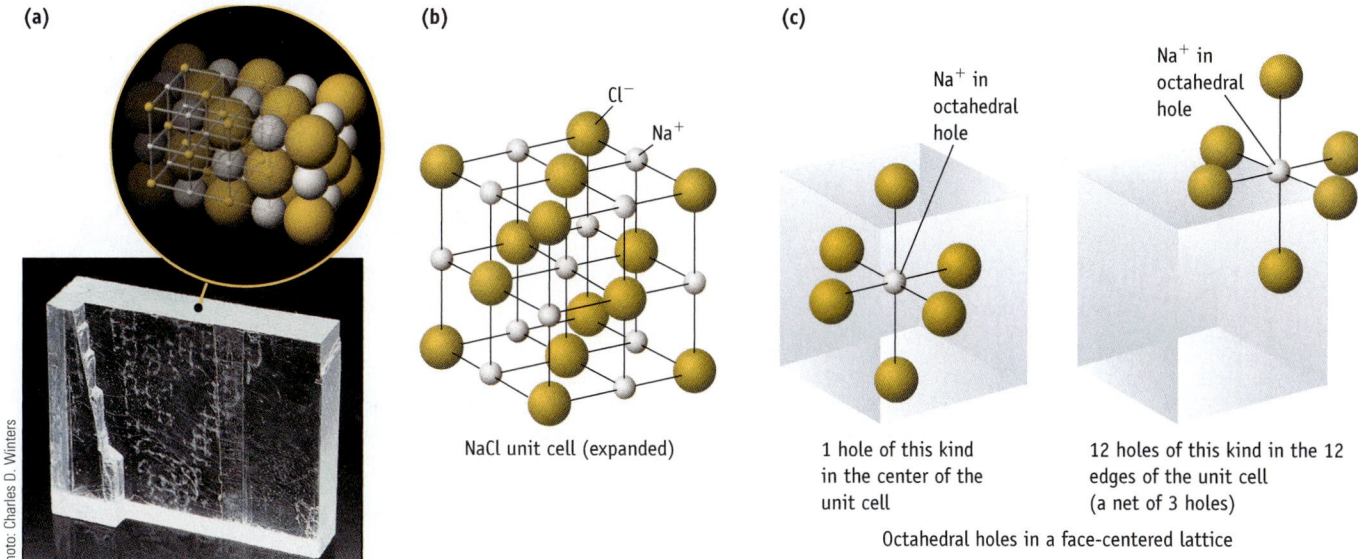

Figure 13.32 Sodium chloride. (a) Cubic NaCl is based on a face-centered cubic unit cell of Na^+ and Cl^- ions. (b) An expanded view of a sodium chloride lattice. (The lines represent the connections between lattice points.) The smaller Na^+ ions (silver) are packed into a face-centered cubic lattice of larger Cl^- ions (yellow). (c) A close-up view of the octahedral holes in the lattice.

This accounts for all of the ions contained in the unit cell: four Cl^- and four Na^+ ions. Thus, a unit cell of NaCl has a 1:1 ratio of Na^+ and Cl^- ions, as the formula requires.

Another common unit cell has ions of one type in a face-centered cubic unit cell. Ions of the other type are located in **tetrahedral holes**, wherein each ion is surrounded by four oppositely charged ions. As illustrated in Figure 13.33, there are eight tetrahedral holes in a face-centered unit cell. In ZnS (zinc blende), the sulfide ions (S^{2-}) form a face-centered cubic unit cell. The zinc ions (Zn^{2+}) then occupy one half of the tetrahedral holes, and each Zn^{2+} ion is surrounded by four S^{2-} ions. The unit cell consists of a net of four S^{2-} ions and four Zn^{2+} ions, which are contained wholly within the unit cell.

In summary, compounds with the formula MX commonly form one of three possible crystal structures:

1. M^{n+} ions occupying all the cubic holes of a simple cubic X^{n-} lattice. Example: CsCl.

2. M^{n+} ions in all the octahedral holes in a face-centered cubic X^{n-} lattice. Example: NaCl.

3. M^{n+} ions occupying half of the tetrahedral holes in a face-centered cube lattice of X^{n-} ions. Example: ZnS.

Chemists and geologists, in particular, have observed that the sodium chloride or "rock salt" structure is adopted by many ionic compounds, most especially by the alkali metal halides (except CsCl, CsBr, and CsI), the oxides and sulfides of the alkaline earth metals, and the oxides of formula MO of the transition metals of the fourth period. Finally, the formulas of compounds must always be reflected in the structures of their unit cells; therefore, the formula can always be derived from the unit cell structure.

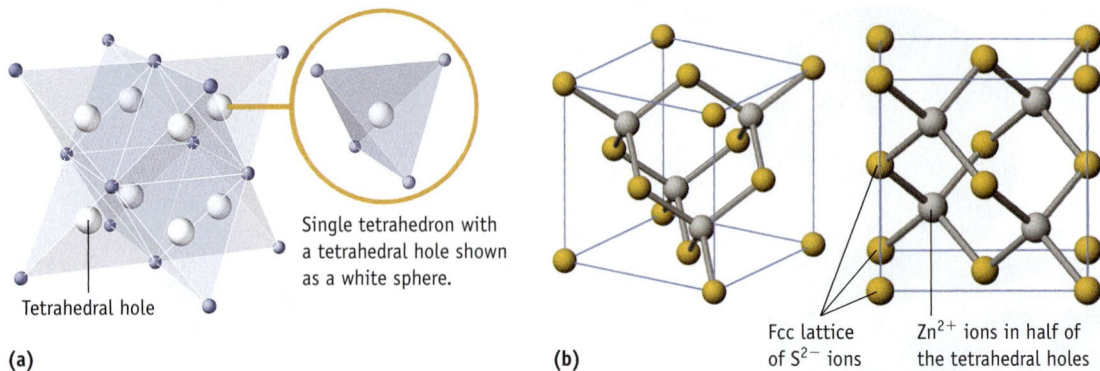

Figure 13.33   Tetrahedral holes and two views of the ZnS (zinc blende) unit cell. (a) The tetrahedral holes in a face-centered cubic lattice. (b) This unit cell is an example of a face-centered cubic lattice of ions of one type with ions of the opposite type in one half of the tetrahedral holes.

See the General ChemistryNow CD-ROM or website:
- **Screen 13.13 Solid Structures (2): Ionic Solids,** to view an animation of ionic unit cells

Example 13.7—Ionic Structure and Formula

Problem One unit cell of the mineral perovskite is illustrated here. This compound is composed of calcium and titanium cations and oxide anions. Based on the unit cell, what is the formula of perovskite?

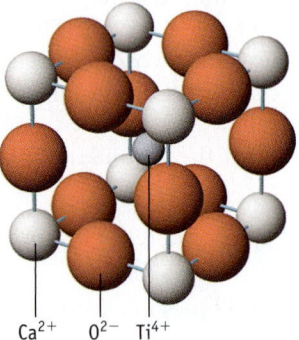

Ca^{2+} O^{2-} Ti^{4+}

Strategy Identify the ions present in the unit cell and their locations within the unit cell. Decide on the net number of ions of each kind in the cell.

Solution The unit cell has Ca^{2+} ions at the corners of the cubic unit cell, a titanium ion in the center of the cell, and oxide ions on the cube edges.

Number of Ca^{2+} ions:

(8 Ca^{2+} ions at cube corners) $\times$ (1/8 of each ion inside unit cell) = 1 net Ca^{2+} ion

Number of Ti^{4+} ions:

One ion is in the cube center = 1 net Ti^{4+} ion

Number of O^{2-} ions:

(12 O^{2-} ions in cube edges) × (1/4 of each ion inside cell) = 3 net O^{2-} ions

Thus, the formula of perovskite is CaTiO₃.

Comment This is a reasonable formula. A Ca^{2+} ion and three O^{2-} ions would require a titanium ion with a 4+ charge. Titanium is in Group 4B of the periodic table, so Ti^{4+} is a predictable ion.

Exercise 13.10—Structure and Formula

If an ionic solid has a fcc lattice of anions (X) and *all* of the tetrahedral holes are occupied by metal cations (M), is the formula of the compound MX, MX_2, or M_2X?

13.8—Other Kinds of Solid Materials

So far we have described the structures of metals and simple ionic solids. Now we will look briefly at the other categories of solids: molecular solids, network solids, and amorphous solids (Table 13.6).

Molecular Solids

Molecules such as H_2O and CO_2 are found in the solid state under appropriate conditions. In these cases, it is molecules, rather than atoms or ions, that pack in a regular fashion in a three-dimensional lattice. We have commented already on one such structure, the structure of ice (Section 13.3, Figure 13.10).

The way molecules are arranged in a crystalline lattice depends on the shape of the molecules and the types of intermolecular forces. Molecules tend to pack in the most efficient manner and to align in ways that maximize intermolecular forces of attraction. Thus, the water structure was established to gain the maximum intermolecular attraction through hydrogen bonding. As illustrated in Figure 13.9, organic acid molecules often assemble in the solid state as dimers, with two molecules being linked by hydrogen bonding.

The greatest interest in the structures of molecular solids focuses on the structures of the molecules themselves and not on the way they pack into the solid. It is from structural studies on molecular solids that most of the information on molecular geometries, bond lengths, and bond angles discussed in Chapters 9 through 11 was assembled.

GENERAL
Chemistry ⚛ Now™

See the General ChemistryNow CD-ROM or website:
- **Screen 13.14 Solid Structures (3): Molecular Solids,** to view an animation of the packing of molecules in a molecular solid

---

Network Solids

Network solids are composed entirely of a three-dimensional array of covalently bonded atoms. Common examples include two allotropes of carbon: graphite and diamond. Elemental silicon is a network solid with a diamond-like structure.

Figure 13.34   A mixture of natural and synthetic industrial diamonds. The colors of diamonds may range from colorless to yellow, brown, or black. Poorer-quality diamonds are used extensively in industry, mainly for cutting or grinding tools. Industrial-quality diamonds are produced synthetically by heating graphite, along with a metal catalyst, to 1200–1500 °C and a pressure of 65–90 kilobars.

Diamonds have a low density ($d = 3.51 \text{ g/cm}^3$), but they are also the hardest material and the best conductor of heat known. They are transparent to visible light, as well as to infrared and ultraviolet radiation. Diamonds are electrically insulating but behave as semiconductors with some advantages over silicon. What more could a scientist want in a material—except a cheap, practical way to make it!

In the 1950s, scientists at General Electric in Schenectady, New York, achieved something alchemists had sought for centuries: the synthesis of diamonds from carbon-containing materials, including wood or peanut butter. Their technique was to heat graphite to a temperature of 1500 °C in the presence of a metal, such as nickel or iron, and under a pressure of 65–90 kilobars. Under these conditions, the carbon dissolves in the metal and slowly forms diamonds (Figure 13.34). More than $500 million worth of diamonds are made this way annually. Most are used for abrasives and diamond-coated cutting tools.

Silicates, compounds composed of silicon and oxygen, represent an enormous class of chemical compounds. You know them in the form of sand, quartz, talc, and mica, or as a major constituent of rocks such as granite. The structure of quartz is illustrated in Figure 13.35, and other silicates are described in Chapter 21.

Most network solids are hard and rigid and are characterized by high melting and boiling points. These characteristics reflect the fact that a great deal of energy must be provided to break the covalent bonds in the lattice. For example, silicon dioxide melts at temperatures higher than 1600 °C. The solid consists of tetrahedral silicon atoms covalently bonded to oxygen atoms in a giant three-dimensional lattice (Figure 13.35). A high temperature is required to break the covalent bonds between silicon and oxygen and thereby disrupt this stable structure.

GENERAL
Chemistry ⚛ Now™

See the General ChemistryNow CD-ROM or website:
- **Screen 13.15 Solid Structures (4): Network Solids,** to view an animation of the structures of network solids.

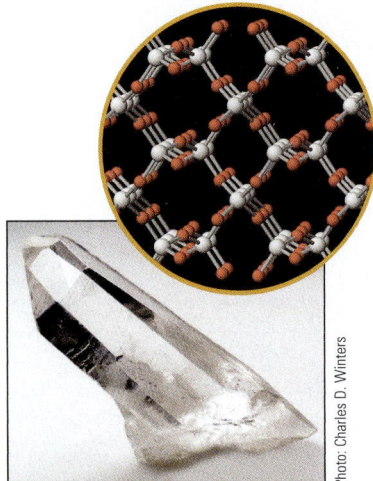

Figure 13.35   Silicon dioxide. Common quartz, SiO$_2$, is a network solid consisting of silicon and oxygen atoms.

Amorphous Solids

A characteristic property of pure crystalline solids—whether metals, ionic solids, or molecular solids—is that they melt at a specific temperature. For example, water melts at 0 °C, aspirin at 135 °C, lead at 327.5 °C, and NaCl at 801 °C. Because they are specific and reproducible values, melting points are often used as a means of identifying chemical compounds.

Another property of crystalline solids is that they form well-defined crystals, with smooth, flat faces. When a sharp force is applied to a crystal, it will most often cleave to give smooth, flat faces. The resulting solid particles are smaller versions of the original crystal (Figure 13.36).

Many common solids, including ones that we encounter everyday, do not have these properties, however. Glass is a good example. When glass is heated it softens over a wide temperature range, a property useful for artisans and craftsmen who can create beautiful and functional products for our enjoyment and use. Glass also possesses a property that we would rather it not have: When glass breaks, it leaves randomly shaped pieces. Other materials that behave similarly include common polymers such as polyethylene, nylon, and other plastics.

(a) A salt crystal can be cleaved cleanly into smaller and smaller crystals that are duplicates of the larger crystal.

(b) Glass is an amorphous solid composed of silicon and oxygen atoms. It has, however, no long-range order as in crystalline quartz.

(c) Glass can be molded and shaped into beautiful forms and, by adding metal oxides, can take on wonderful colors.

Figure 13.36 Crystalline and amorphous solids.

The characteristics of these **amorphous solids** relate to their molecular structure. At the particulate level, amorphous solids do not have a regular structure. In fact, in many ways these substances look a lot like liquids. Unlike liquids, however, the forces of attraction are strong enough that movement of the molecules or ions is restricted.

13.9—The Physical Properties of Solids

The shape of a crystalline solid is a reflection of its internal structure. But what about physical properties such as hardness and the temperatures at which solids melt? These and many other physical properties of solids are of interest to chemists, geologists, engineers, and others.

Melting: Conversion of Solid to Liquid

The melting point of a solid is the temperature at which the lattice collapses and the solid is converted to a liquid. Like the liquid-to-vapor transformation, melting requires energy, called the **enthalpy of fusion** (given in kilojoules per mole) (Chapter 6).

$$\text{Heat energy absorbed on melting} = \text{enthalpy of fusion} = \Delta H^\circ_{\text{fusion}} \text{ (kJ/mol)}$$
$$\text{Heat energy evolved on freezing} = \text{enthalpy of crystallization} = -\Delta H^\circ_{\text{fusion}} \text{ (kJ/mol)}$$

Enthalpies of fusion can range from just a few thousand joules per mole to many thousands of joules per mole (Table 13.7). A low melting temperature will certainly mean a low value for the enthalpy of fusion, whereas high melting points are associated with high enthalpies of fusion. Figure 13.37 shows the enthalpies of fusion for the metals of the fourth through the sixth periods. Based on this figure and Table 13.7 we can make two statements: (1) Metals that have notably low melting points, such as the alkali metals and mercury (mp = −39 °C), also have low enthalpies of fusion; and (2) transition metals have high heats of fusion, with those of the third transition series being extraordinarily high. This trend parallels the trend seen with the melting points for these elements. Tungsten, which has the highest

Table 13.7   Melting Points and Enthalpies of Fusion of Some Elements and Compounds

Compound	Melting Point (°C)	Enthalpy of Fusion (kJ/mol)	Type of Interparticle Forces
Metals			
Hg	−39	2.29	Metal bonding; see page 643
Na	98	2.60	
Al	660	10.7	
Ti	1668	20.9	
W	3422	35.2	
Molecular Solids: Nonpolar Molecules			
O_2	−219	0.440	Dispersion forces only
F_2	−220	0.510	
Cl_2	−102	6.41	
Br_2	−7.2	10.8	
Molecular Solids: Polar Molecules			
HCl	−114	1.99	All three HX molecules have
HBr	−87	2.41	dipole–dipole forces.
HI	−51	2.87	Dispersion forces increase with size and molar mass.
H_2O	0	6.02	Hydrogen bonding
Ionic Solids			
NaF	996	33.4	All ionic solids have extended
NaCl	801	28.2	ion–ion interactions. Note the
NaBr	747	26.1	general trend is the same as for lattice energies (see Section 9.3
NaI	660	23.6	and Figure 9.3).

melting point of all the known elements except for carbon, also has the highest enthalpy of fusion among the transition metals. These properties affect the uses of this metal. For example, tungsten is used for the filaments in lightbulbs; no other material has been found to work better since the invention of the lightbulb in 1908.

The melting temperature of a solid can convey a great deal of information. Table 13.7 provides some data for several basic types of substances: metals, polar and nonpolar molecules, and ionic solids. In general, nonpolar substances that form molecular solids have low melting points. Melting points increase within a series of related molecules, however, as the size and molar mass increase. This happens because dispersion forces are generally larger when the molar mass is larger. Thus, increasing amounts of energy are required to break down the intermolecular forces in the solid, a principle that is reflected in an increasing enthalpy of fusion.

The ionic compounds in Table 13.7 have higher melting points and higher enthalpies of fusion than do molecular solids. This trend is due to the strong ion–ion forces present in ionic solids, forces that are reflected in high lattice energies (see Section 9.3). Because ion–ion forces depend on ion size (as well as ion charge), there is a good correlation between lattice energy and the position of the metal or halogen in the periodic table. For example, the data in Table 13.7 show a decrease in melting point and enthalpy of fusion for sodium salts as the halide ion increases in size. This parallels the decrease in lattice energy seen with increasing ion size illustrated in Figure 9.3.

Sublimation: Conversion of Solid to Vapor

Molecules can escape directly from the solid to the gas phase by **sublimation** (Figure 13.38).

$$\text{Solid} \longrightarrow \text{gas} \qquad \text{heat energy required} = \Delta H^{\circ}_{\text{sublimation}}$$

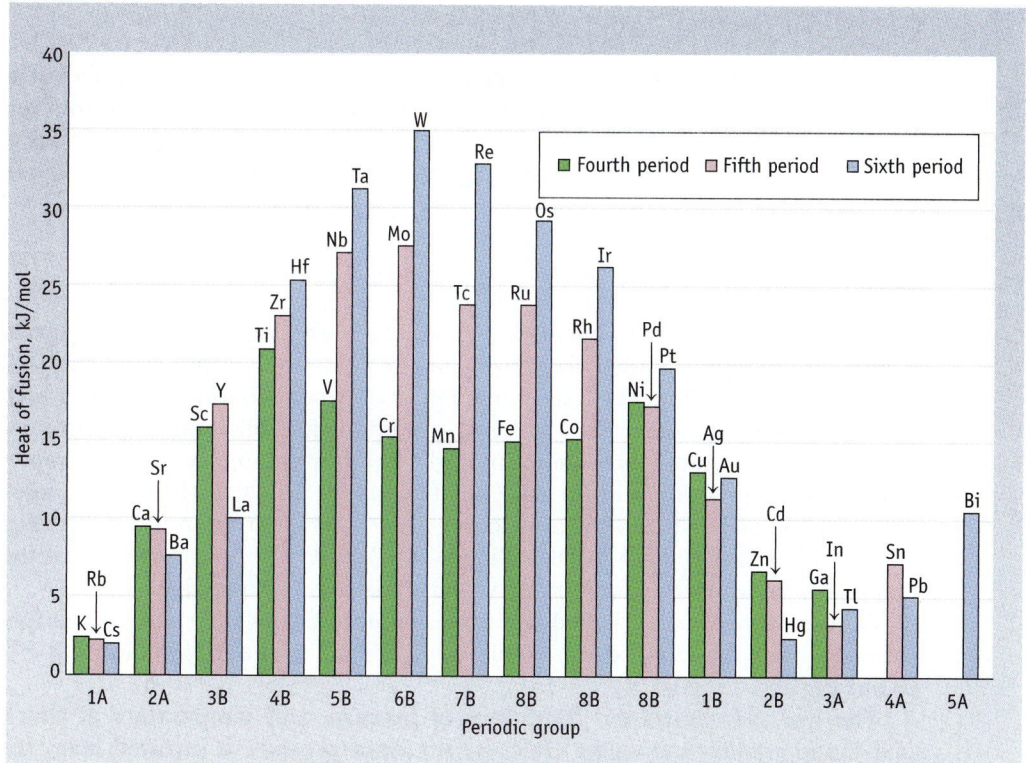

Figure 13.37 **Heat of fusion of fourth-, fifth-, and sixth-period metals.** Heats of fusion range from 2–5 kJ/mol for Group 1A elements to 35.2 kJ/mol for tungsten. Notice that heats of fusion generally increase for B-group metals on descending the periodic table.

Sublimation, like fusion and evaporation, is an endothermic process. The heat energy required is called the *enthalpy of sublimation*. Water, which has a molar enthalpy of sublimation of 51 kJ/mol, can be converted from solid ice to water vapor quite readily. A good example of this phenomenon is the sublimation of frost from grass and trees as night turns to day on a cold morning in the winter.

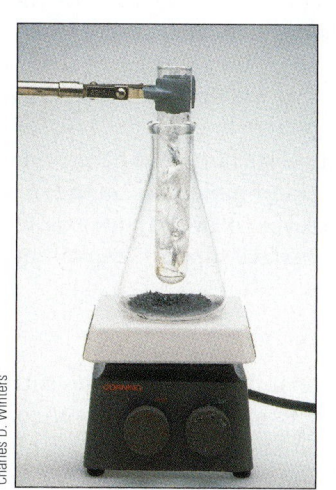 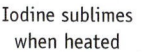 

Iodine sublimes when heated

Figure 13.38 **Sublimation.** Sublimation entails the conversion of a solid directly to its vapor. Here, iodine (I_2) sublimes when heated in warm water. If an ice-filled test tube is inserted into the flask, the vapor condenses on the cold surface.

13.10—Phase Diagrams

Depending on the conditions of temperature and pressure, a substance can exist as a gas, a liquid, or a solid. In addition, under certain specific conditions, two (or even three) states can coexist in equilibrium. It is possible to summarize this information in the form of a graph called a **phase diagram**. Phase diagrams are used to illustrate the relationship between phases of matter and the pressure and temperature.

Water

A phase diagram for water appears in Figure 13.39. The lines in a phase diagram identify the conditions under which two phases exist at equilibrium. Conversely, all points that do not fall on the lines in the figure represent conditions under which only one state exists. Line A-B represents conditions for solid–vapor equilibrium, and line A-C for liquid–solid equilibrium. The line from point A to point D, representing the temperature and pressure combination at which the liquid and vapor phases are in equilibrium, is the same curve plotted for water vapor pressure in Figure 13.18. Recall that the normal boiling point, 100 °C in the case of water, is the temperature at which the equilibrium vapor pressure is 760 mm Hg.

Point A, appropriately called the **triple point**, indicates the conditions under which all three phases coexist in equilibrium. For water, the triple point is at $P = 4.6$ mm Hg and $T = 0.01$ °C.

The line A-C shows the conditions of pressure and temperature at which solid–liquid equilibrium exists. (Because no vapor pressure is involved here, the pressure referred to is the external pressure on the liquid.) *For water, this line has a negative slope;* the change for water is approximately −0.01 °C for each one-atmosphere increase in pressure. That is, the higher the external pressure, the lower the melting point.

The negative slope of the water solid–liquid equilibrium line can be explained from our knowledge of the structure of water and ice. When the pressure on an object increases, common sense tells us that the volume of the object will become smaller, giving the substance a higher density. Because ice is less dense than liquid water (due to the open lattice structure of ice), ice and water in equilibrium respond to increased pressure (at constant T) by melting ice to form more water because the same mass of water requires less volume.

Ice Skating and the Ice-Liquid Equilibrium

Ice is slippery stuff. It was long assumed that you can ski or skate on ice because the surface melts slightly from the pressure of a skate blade or ski, or that surface melting occurs because of frictional heating. This has always seemed an unsatisfying explanation, however, because it does not seem possible that just standing or sliding on a piece of ice could produce a pressure or temperature high enough to cause sufficient melting. Recently, surface chemists have studied ice surfaces and have come up with a better explanation. They have concluded that water molecules on the surface of ice are vibrating rapidly. In fact, the outermost layer or two of water molecules is almost liquid-like. This arrangement makes the surface slippery, explaining why we can ski on snow and skate on ice.

Phase Diagrams and Thermodynamics

Let us explore the water phase diagram further by correlating phase changes with thermodynamic data. Suppose we begin with ice at −10 °C and under a pressure of 500 mm Hg (point *a* on Figure 13.39). As ice is heated, it absorbs about 2.1 J/g · K in warming from point *a* to point *b* at a temperature between 0 °C and 0.01 °C. At

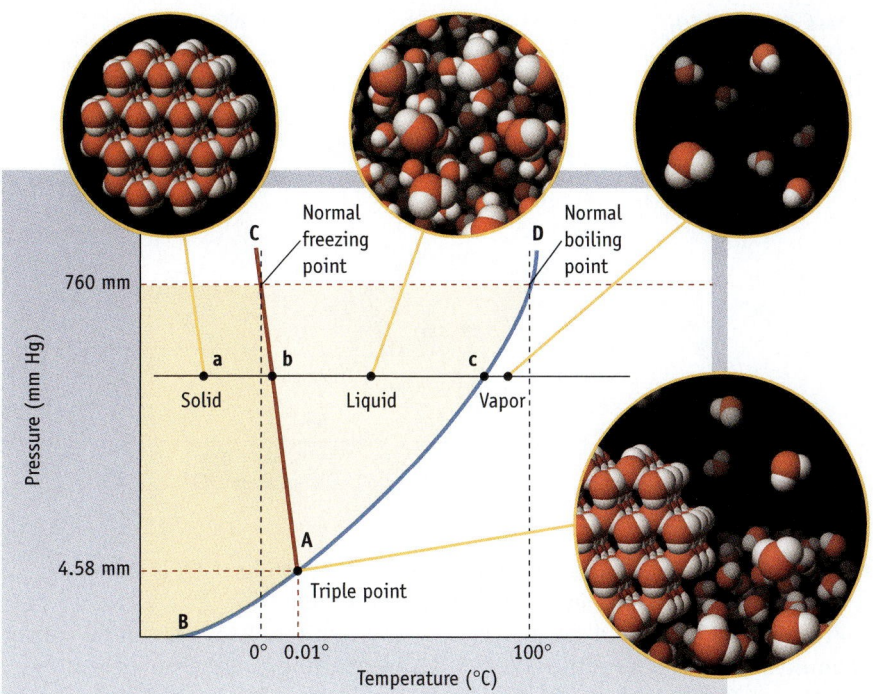

this point the solid is in equilibrium with liquid water. Solid–liquid equilibrium is maintained until 333 J/g has been transferred to the sample and it has become liquid water at a temperature slightly greater than 0 °C. If the liquid, still under a pressure of 500 mm Hg, now absorbs 4.184 J/g, it warms to point *c*. The temperature at point *c* is about 89 °C, and equilibrium is established between liquid water and water vapor. The equilibrium vapor pressure of the liquid water is 500 mm Hg. If 2260 J/g is transferred to the liquid–vapor sample, the equilibrium vapor pressure remains 500 mm Hg until the liquid is completely converted to vapor at 89 °C.

Carbon Dioxide

The basic features of the phase diagram for CO_2 (Figure 13.40) are the same as those for water. There are some important differences, however.

In contrast to water, the CO_2 solid–liquid equilibrium line has a *positive* slope. That is, increasing pressure on solid CO_2 in equilibrium with liquid CO_2 will shift the equilibrium to solid CO_2. Thus, adding pressure to solid CO_2 will cause it to move to the more dense phase, the solid phase. Because solid CO_2 is denser than the liquid, the newly formed solid CO_2 sinks to the bottom in a container of liquid CO_2.

Another feature of the CO_2 phase diagram is the triple point that occurs at a pressure of 5.19 atm (3940 mm Hg) and 216.6 K (−56.6 °C). CO_2 cannot be a liquid at pressures lower than this. Thus, at a pressure of 1 atm, solid CO_2 is in equilibrium with the gas at a temperature of 197.5 K (−78.7 °C). As a result, as solid CO_2

Figure 13.40   The phase diagram of CO₂.
Notice in particular the positive slope of the solid–liquid equilibrium line.

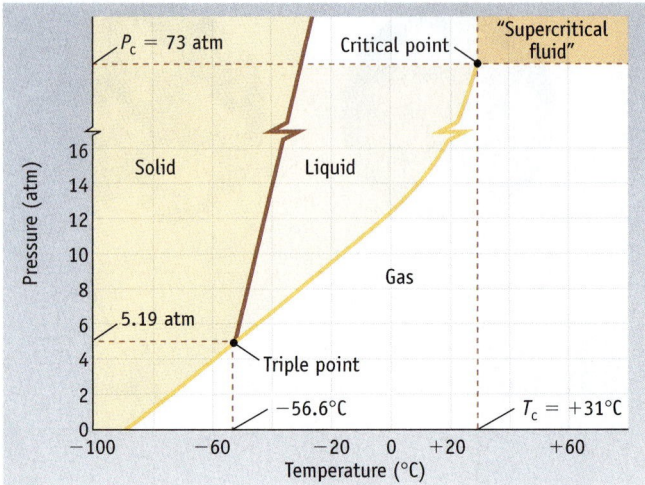

warms to room temperature, it sublimes rather than melts. CO_2 is called *dry ice* for this reason; it looks like water ice, but it does not melt.

From the CO_2 phase diagram we can also learn that CO_2 gas can be converted to a liquid at room temperature (20–25 °C) by exerting a moderate pressure on the gas. In fact, CO_2 is regularly shipped in tanks as a liquid to laboratories and industrial companies.

Finally, the critical pressure and temperature for CO_2 are 73 atm and 31 °C, respectively. Because the critical temperature and pressure are easily attained in the laboratory, it is possible to observe the transformation to supercritical CO_2 (Figure 13.41).

GENERAL
Chemistry∙Now™

See the General ChemistryNow CD-ROM or website:

● **Screen 13.17 Phase Changes,** to view animations of phase changes and to do an exercise on phase diagrams

---

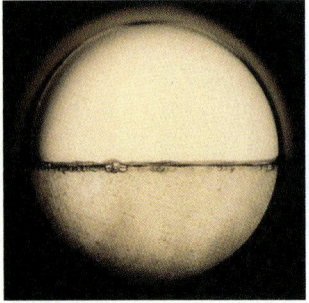

The separate phases of CO_2 are seen through the window in a high-pressure vessel.

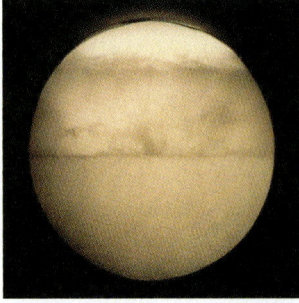

As the sample warms and the pressure increases, the meniscus becomes less distinct.

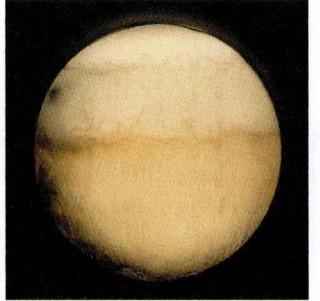

As the temperature continues to increase, it is more difficult to distinguish the liquid and vapor phases.

Once the critical T and P are reached, distinct liquid and vapor phases are no longer in evidence. This homogenous phase is "supercritical CO_2."

Dr. Christopher M. Rayner/University of Leeds.

Figure 13.41   Phase changes for CO₂.

Chapter Goals Revisited

When you have finished studying this chapter, you should ask if you have met the chapter goals. In particular you should be able to

Describe intermolecular forces and their effects

a. Describe the various intermolecular forces found in liquids and solids (Sections 13.2 and 13.4). General ChemistryNow homework: Study Question(s) 4, 6

b. Tell when two molecules can interact through a dipole–dipole attraction and when hydrogen bonding may occur. The latter occurs most strongly when H is attached to O, N, or F (Sections 13.2 and 13.3). General ChemistryNow homework: SQ(s) 8, 87

c. Identify instances in which molecules interact by induced dipoles (dispersion forces) (Section 13.2).

Understand the importance of hydrogen bonding

a. Explain how hydrogen bonding affects the properties of water (Section 13.3).

Understand the properties of liquids

a. Explain the processes of evaporation and condensation, and use the enthalpy of vaporization in calculations (Section 13.5). General ChemistryNow homework: SQ(s) 12, 53

b. Define the equilibrium vapor pressure of a liquid, and explain the relationship between the vapor pressure and boiling point of a liquid (Section 13.5). General ChemistryNow homework: SQ(s) 14, 17, 19, 20, 43, 47, 79

c. Describe the phenomena of the critical temperature, T_c, and critical pressure, P_c, of a substance (Section 13.5).

d. Describe how intermolecular interactions affect the cohesive forces between identical liquid molecules, the energy necessary to break through the surface of a liquid (surface tension), and the resistance to flow, or viscosity, of liquids (Section 13.5).

e. Use the Clausius-Clapeyron equation, which connects temperature, vapor pressure, and enthalpy of vaporization for liquids (Section 13.5). General ChemistryNow homework: SQ(s) 22

Understand cubic unit cells

a. Describe the three types of cubic unit cells: primitive or simple cubic (sc), body-centered cubic (bcc), and face-centered cubic (fcc) (Section 13.5).

b. Relate atom size and unit cell dimensions. General ChemistryNow homework: SQ(s) 54

Relate unit cells for ionic compounds to formulas (Section 13.7) General ChemistryNow homework: SQ(s) 26, 28, 83, 85

Describe the properties of solids

a. Characterize different types of solids: metallic (e.g., copper), ionic (e.g., NaCl and CaF_2), molecular (e.g., water and I_2), network (e.g., diamond), and amorphous (e.g., glass and many synthetic polymers) (Sections 13.6–13.8 and Table 13.6).

b. Define the enthalpy of fusion and use it in a calculation (Section 13.9).

GENERAL
Chemistry ⚛ Now™

See the General ChemistryNow CD-ROM or website to:

- Assess your understanding with homework questions keyed to each goal

- Check your readiness for an exam by taking the exam-prep quiz and exploring the resources in the personalized Learning Plan it provides

For additional preparation for an examination on this chapter see the **Let's Review** section on pages 642–655.

Understand the nature of phase diagrams

a. Identify the different points (triple point, normal boiling point, freezing point) and regions (solid, liquid, vapor) of a phase diagram, and use the diagram to evaluate the vapor pressure of a liquid and the relative densities of a liquid and a solid (Section 13.10). General ChemistryNow homework: SQ(s) 34

Key Equation

Equation 13.2 (page 612)

The Clausius-Clapeyron equation relates the equilibrium vapor pressure, P, of a volatile liquid to the molar enthalpy of vaporization (ΔH°_{vap}) at a given temperature, T. (R is the universal constant, $8.314472 \, J/K \cdot mol$.) Equation 13.2 allows you to calculate ΔH°_{vap} if you know the vapor pressures at two different temperatures. Alternatively, you may plot ln P versus $1/T$; the slope of the line is $-\Delta H^\circ_{vap}/R$.

$$\ln\frac{P_2}{P_1} = \frac{\Delta H^\circ_{vap}}{R}\left[\frac{1}{T_1} - \frac{1}{T_2}\right]$$

Study Questions

▲ denotes more challenging questions.

■ denotes questions available in the Homework and Goals section of the General ChemistryNow CD-ROM or website.

Blue numbered questions have answers in Appendix O and fully worked solutions in the *Student Solutions Manual*.

Structures of many of the compounds used in these questions are found on the General ChemistryNow CD-ROM or website in the Models folder.

Chemistry ⚛ Now™ Assess your understanding of this chapter's topics with additional quizzing and conceptual questions at **http://now.brookscole.com/kotz6e**

Practicing Skills

Intermolecular Forces
(See Examples 13.2–13.4 and General ChemistryNow Screens 13.3–13.7.)

1. What intermolecular force(s) must be overcome to
   (a) Melt ice?
   (b) Sublime solid I_2?
   (c) Convert liquid NH_3 to NH_3 vapor?

2. What type of forces must be overcome within the solid I_2 when I_2 dissolves in methanol, CH_3OH? What type of forces must be disrupted between CH_3OH molecules when I_2 dissolves? What type of forces exist between I_2 and CH_3OH molecules in solution?

3. What type of intermolecular force must be overcome in converting each of the following from a liquid to a gas?
   (a) liquid O_2 (c) CH_3I (methyl iodide)
   (b) mercury   (d) CH_3CH_2OH (ethanol)

4. ■ What type of intermolecular forces must be overcome in converting each of the following from a liquid to a gas?
   (a) CO_2 (c) $CHCl_3$
   (b) NH_3 (d) CCl_4

5. Rank the following atoms or molecules in order of increasing strength of intermolecular forces in the pure substance. Which exist as gases at 25 °C and 1 atm?
   (a) Ne   (c) CO
   (b) CH_4 (d) CCl_4

6. ■ Rank the following in order of increasing strength of intermolecular forces in the pure substances. Which exist as gases at 25 °C and 1 atm?
   (a) $CH_3CH_2CH_2CH_3$ (butane)
   (b) CH_3OH (methanol)
   (c) He

7. Which of the following compounds would be expected to form intermolecular hydrogen bonds in the liquid state?
   (a) CH_3OCH_3 (dimethyl ether)
   (b) CH_4
   (c) HF
   (d) CH_3CO_2H (acetic acid)
   (e) Br_2
   (f) CH_3OH (methanol)

8. ■ Which of the following compounds would be expected to form intermolecular hydrogen bonds in the liquid state?

(a) H_2Se

(b) HCO_2H (formic acid)

(c) HI   (d) acetone

$$H_3C - \overset{\overset{\displaystyle O}{\|}}{C} - CH_3$$

9. In each pair of ionic compounds, which is more likely to have the greater heat of hydration? Briefly explain your reasoning in each case.

(a) LiCl or CsCl   (c) RbCl or $NiCl_2$

(b) $NaNO_3$ or $Mg(NO_3)_2$

10. When salts of Mg^{2+}, Na^+, and Cs^+ are placed in water, the positive ion is hydrated (as is the negative ion). Which of these three cations is most strongly hydrated? Which one is least strongly hydrated?

Liquids

(See Examples 13.5 and 13.6 and General ChemistryNow Screens 13.8–13.11.)

11. Ethanol, CH_3CH_2OH, has a vapor pressure of 59 mm Hg at 25 °C. What quantity of heat energy is required to evaporate 125 mL of the alcohol at 25 °C? The enthalpy of vaporization of the alcohol at 25 °C is 42.32 kJ/mol. The density of the liquid is 0.7849 g/mL.

12. ■ The enthalpy of vaporization of liquid mercury is 59.11 kJ/mol. What quantity of heat is required to vaporize 0.500 mL of mercury at 357 °C, its normal boiling point? The density of mercury is 13.6 g/mL.

13. Answer the following questions using Figure 13.18.

(a) What is the approximate equilibrium vapor pressure of water at 60 °C? Compare your answer with the data in Appendix G.

(b) At what temperature does water have an equilibrium vapor pressure of 600 mm Hg?

(c) Compare the equilibrium vapor pressures of water and ethanol at 70 °C. Which is higher?

14. ■ Answer the following questions using Figure 13.18.

(a) What is the equilibrium vapor pressure of diethyl ether at room temperature (approximately 20 °C)?

(b) Place the three compounds in Figure 13.18 in order of increasing intermolecular forces.

(c) If the pressure in a flask is 400 mm Hg and the temperature is 40 °C, which of the three compounds (diethyl ether, ethanol, and water) are liquids and which are gases?

15. Assume you seal 1.0 g of diethyl ether (see Figure 13.18) in an evacuated 100.-mL flask. If the flask is held at 30 °C, what is the approximate gas pressure in the flask? If the flask is placed in an ice bath, does additional liquid ether evaporate or does some ether condense to a liquid?

16. Refer to Figure 13.18 as an aid in answering these questions:

(a) You put some water at 60 °C in a plastic milk carton and seal the top very tightly so that gas cannot enter or leave the carton. What happens when the water cools?

(b) If you put a few drops of liquid diethyl ether on your hand, does it evaporate completely or remain a liquid?

17. ■ Which member of each of the following pairs of compounds has the higher boiling point?

(a) O_2 or N_2 (c) HF or HI

(b) SO_2 or CO_2 (d) SiH_4 or GeH_4

18. Place the following four compounds in order of increasing boiling point.

(a) SCl_2 (b) NH_3 (c) CH_4 (d) CO

19. ■ Vapor pressure curves for CS_2 (carbon disulfide) and CH_3NO_2 (nitromethane) are drawn here.

(a) What are the approximate vapor pressures of CS_2 and CH_3NO_2 at 40 °C?

(b) What types of intermolecular forces exist in the liquid phase of each compound?

(c) What is the normal boiling point of CS_2? Of CH_3NO_2?

(d) At what temperature does CS_2 have a vapor pressure of 600 mm Hg?

(e) At what temperature does CH_3NO_2 have a vapor pressure of 60 mm Hg?

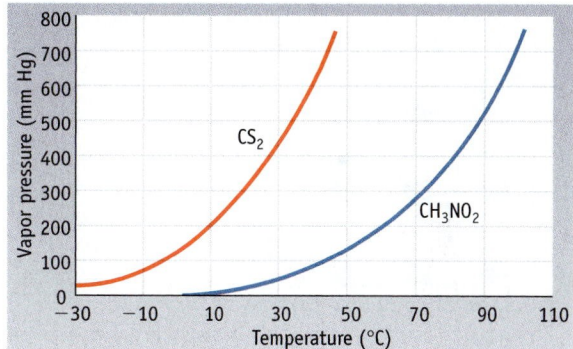

20. ■ Answer each of the following questions with *increases, decreases,* or *does not change.*

(a) If the intermolecular forces in a liquid increase, the normal boiling point of the liquid _____.

(b) If the intermolecular forces in a liquid decrease, the vapor pressure of the liquid _____.

(c) If the surface area of a liquid decreases, the vapor pressure _____.

(d) If the temperature of a liquid increases, the equilibrium vapor pressure _____.

21. The following table gives the equilibrium vapor pressure of benzene, C_6H_6, at various temperatures.

Temperature (°C)	Vapor Pressure (mm Hg)
7.6	40.
26.1	100.
60.6	400.
80.1	760.

(a) What is the normal boiling point of benzene?

(b) Plot these data so that you have plot resembling the one in Figure 13.19. At what temperature does the liquid have an equilibrium vapor pressure of 250 mm Hg? At what temperature is this pressure 650 mm Hg?

(c) Calculate the molar enthalpy of vaporization for benzene using the the Clausius–Clapeyron equation (Equation 13.1, page 612).

22. ■ Vapor pressure data are given here for octane, C_8H_{18}.

Temperature (°C)	Vapor Pressure (mm Hg)
25	13.6
50.	45.3
75	127.2
100.	310.8

Use the Clausius-Clapeyron equation (Equation 13.1, page 612) to calculate the molar enthalpy of vaporization of octane and its normal boiling point.

Metallic and Ionic Solids

(See Example 13.7, Exercise 13.10, and the General ChemistryNow Screens 13.12–13.13.)

23. Outline a two-dimensional unit cell for the pattern shown here. If the black squares are labeled A and the white squares are B, what is the simplest formula for a "compound" based on this pattern?

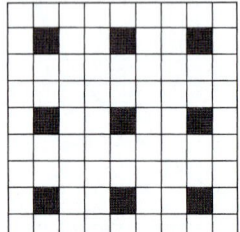

24. Outline a two-dimensional unit cell for the pattern shown here. If the black squares are labeled A and the white squares are B, what is the simplest formula for a "compound" based on this pattern?

25. One way of viewing the unit cell of perovskite was illustrated in Example 13.7. Another way is shown here. Prove that this view also leads to a formula of $CaTiO_3$.

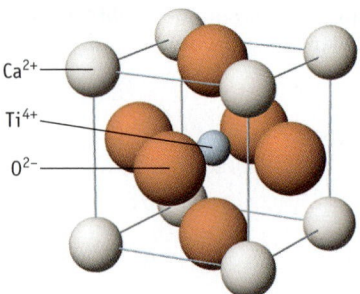

26. ■ Rutile, TiO_2, crystallizes in a structure characteristic of many other ionic compounds. How many formula units of TiO_2 are in the unit cell illustrated here? (The oxide ions marked by an *x* are wholly within the cell; the others are in the cell faces.)

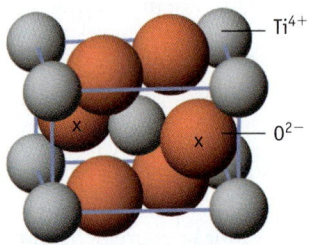

27. Cuprite is a semiconductor. Oxide ions are at the cube corners and in the cube center. Copper ions are wholly within the unit cell.

(a) What is the formula of cuprite?

(b) What is the oxidation number of copper?

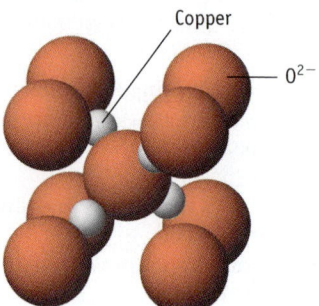

28. ■ The mineral fluorite, which is composed of calcium ions and fluoride ions, has the unit cell shown here.

(a) What type of unit cell is described by the Ca^{2+} ions?

(b) Where are the F^- ions located, in octahedral holes or tetrahedral holes?

(c) Based on this unit cell, what is the formula of fluorite?

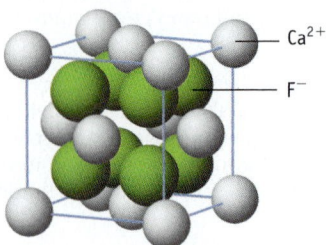

Other Types of Solids

(See the General ChemistryNow Screens 13.14–13.16.)

29. A diamond unit cell is shown here.
    (a) How many carbon atoms are in one unit cell?
    (b) The unit cell can be considered as a cubic unit cell of C atoms with other C atoms in holes in the lattice. What type of unit cell is this (sc, bcc, fcc)? In what holes are other C atoms located, octahedral or tetrahedral holes?

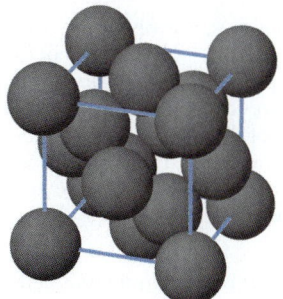

30. The structure of graphite is given in Figure 2.15.
    (a) What type of intermolecular bonding forces exist between the layers of six-member carbon rings?
    (b) Account for the lubricating ability of graphite. That is, why does graphite feel slippery? Why does pencil lead (which is really graphite in clay) leave black marks on paper?

Physical Properties of Solids

31. Benzene, C_6H_6, is an organic liquid that freezes at 5.5 °C (see Figure 13.1) to form beautiful, feather-like crystals. How much heat is evolved when 15.5 g of benzene freezes at 5.5 °C? (The heat of fusion of benzene is 9.95 kJ/mol.) If the 15.5-g sample is remelted, again at 5.5 °C, what quantity of heat is required to convert it to a liquid?

32. The specific heat capacity of silver is 0.235 J/g · K. Its melting point is 962 °C, and its heat of fusion is 11.3 kJ/mol. What quantity of heat, in joules, is required to change 5.00 g of silver from a solid at 25 °C to a liquid at 962 °C?

Phase Diagrams and Phase Changes

(See the General ChemistryNow Screen 13.17.)

33. Consider the phase diagram of CO_2 in Figure 13.40.
    (a) Is the density of liquid CO_2 greater or less than that of solid CO_2?
    (b) In what phase do you find CO_2 at 5 atm and 0 °C?
    (c) Can CO_2 be liquefied at 45 °C?

34. ■ Use the phase diagram given here to answer the following questions:

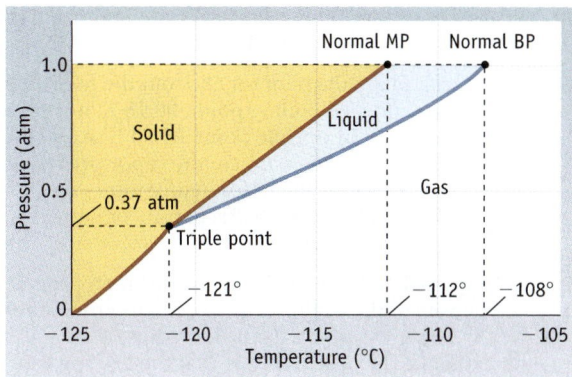

   (a) In what phase is the substance found at room temperature and 1.0 atm pressure?
   (b) If the pressure exerted on a sample is 0.75 atm and the temperature is −114 °C, in what phase does the substance exist?
   (c) If you measure the vapor pressure of a liquid sample and find it to be 380 mm Hg, what is the temperature of the liquid phase?
   (d) What is the vapor pressure of the solid at −122 °C?
   (e) Which is the denser phase—solid or liquid? Explain briefly.

35. Liquid ammonia, $NH_3(\ell)$, was once used in home refrigerators as the heat transfer fluid. The specific heat of the liquid is 4.7 J/g · K and that of the vapor is 2.2 J/g · K. The enthalpy of vaporization is 23.33 kJ/mol at the boiling point. If you heat 12 kg of liquid ammonia from −50.0 °C to its boiling point of −33.3 °C, allow it to evaporate, and then continue warming to 0.0 °C, how much heat energy must you supply?

36. If your air conditioner is more than several years old, it may use the chlorofluorocarbon CCl_2F_2 as the heat transfer fluid. The normal boiling point of CCl_2F_2 is −29.8 °C, and the enthalpy of vaporization is 20.11 kJ/mol. The gas and the liquid have specific heats of 117.2 J/mol · K and 72.3 J/mol · K, respectively. How much heat is evolved when 20.0 g of CCl_2F_2 is cooled from +40 °C to −40 °C?

37. The critical temperature and pressure of chloromethane are 416 K and 66.1 atm, respectively. (Chloromethane's triple point is at 175.4 K and 0.0086 atm.) Can CH_3Cl be liquefied at or above room temperature? Explain briefly.

38. Methane (CH_4) cannot be liquefied at room temperature, no matter how high the pressure. Propane (C_3H_8), another alkane, has a critical pressure of 41.8 atm and a critical temperature of 369.9 K. (The triple point for propane is at 85 K and 1.7×10^{-9} atm.) Can propane be liquefied at room temperature?

General Questions

These questions are not designated as to type or location in the chapter. They may combine several concepts.

39. Rank the following substances in order of increasing strength of intermolecular forces: Ar, CH_3OH, CO_2.

40. What types of intermolecular forces are important in the liquid phase of (a) C_2H_6 and (b) $(CH_3)_2CHOH$?

41. Construct a phase diagram for O_2 from the following information: normal boiling point, 90.18 K; normal melting point, 54.8 K; and triple point, 54.34 K at a pressure of 2 mm Hg. Very roughly estimate the vapor pressure of liquid O_2 at $-196\ °C$, the lowest temperature easily reached in the laboratory. Is the density of liquid O_2 greater or less than that of solid O_2?

42. ▲ A unit cell of cesium chloride is shown on page 622. The density of the solid is 3.99 g/cm³, and the radius of the Cl^- ion is 181 pm. What is the radius of the Cs^+ ion in the center of the cell? (Assume that the Cs^+ ion touches all of the corner Cl^- ions.)

43. ■ If you place 1.0 L of ethanol (C_2H_5OH) in a room that is 3.0 m long, 2.5 m wide, and 2.5 m high, will all of the alcohol evaporate? If some liquid remains, how much will there be? The vapor pressure of ethanol at 25 °C is 59 mm Hg, and the density of the liquid at this temperature is 0.785 g/cm³.

44. Select the substance in each of the following pairs that should have the higher boiling point.
   (a) Br_2 or ICl
   (b) neon or krypton
   (c) CH_3CH_2OH (ethanol) or C_2H_4O (ethylene oxide, structure below)

$$H_2C — CH_2$$
$$\diagdown\ \diagup$$
$$O$$

45. Which salt, Li_2SO_4 or Cs_2SO_4, is expected to have the more exothermic enthalpy of hydration?

46. In which salts does the cation bind most strongly to water molecules? In which is the binding less strong in comparison? Explain your reasoning.
   (a) $Fe(NO_3)_3$ (c) NaCl
   (b) $CoCl_2$ (d) $Al(NO_3)_3$

47. ■ Use the vapor pressure curves illustrated here to answer the questions that follow.

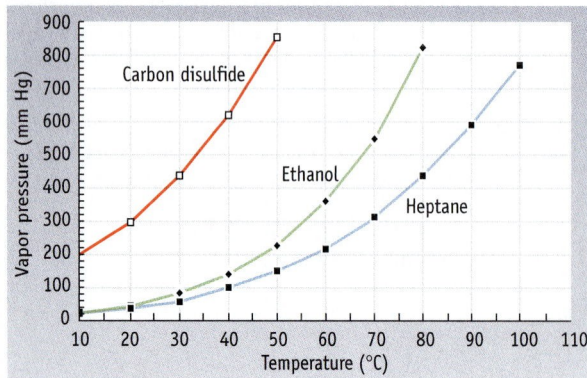

(a) What is the vapor pressure of ethanol (C_2H_5OH) at 60°C?

(b) Considering only carbon disulfide (CS_2) and ethanol, which has the stronger intermolecular forces in the liquid state?

(c) At what temperature does heptane (C_7H_{16}) have a vapor pressure of 500 mm Hg?

(d) What are the approximate normal boiling points of each of the three substances?

(e) At a pressure of 400 mm Hg and a temperature of 70 °C, is each substance a liquid, a gas, or a mixture of liquid and gas?

48. What quantity of energy, in joules, is evolved when 1.00 mol of liquid ammonia cools from $-33.3\ °C$ (its boiling point) to $-43.3\ °C$? (The specific heat capacity of liquid NH_3 is 4.70 J/g · K.) Compare this value with the quantity of heat evolved by 1.00 mol of liquid water cooling by exactly 10 °C. Which evolves more heat on cooling 10 °C, liquid water or liquid ammonia? Using intermolecular forces, explain briefly why one liquid should evolve more energy than the other.

49. ▲ Silver crystallizes in a face-centered cubic unit cell. Each side of the unit cell has a length of 409 pm. What is the radius of a silver atom? (*Hint:* Assume the atoms just touch each other on the diagonal across the face of the unit cell. That is, each face atom is touching the four corner atoms.)

50. ▲ Tungsten crystallizes in the unit cell shown here.

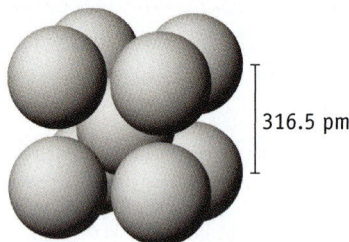

316.5 pm

(a) What type of unit cell is this?
(b) How many tungsten atoms occur per unit cell?
(c) If the edge of the unit cell is 316.5 pm, what is the radius of a tungsten atom? (*Hint:* The W atoms touch each other along the diagonal line from one corner of the unit cell to the opposite corner of the unit cell.)

51. ▲ The unit cell shown here is for calcium carbide. How many calcium atoms and how many carbon atoms are in each unit cell? What is the formula of calcium carbide? (Calcium ions are silver in color and carbon atoms are gray.)

52. Rank the following molecules in order of *increasing* intermolecular forces: CH_3Cl, HCO_2H (formic acid), and CO_2.

53. ■ Rank the following compounds in order of increasing molar enthalpy of vaporization: CH_3OH, C_2H_6, HCl.

54. ■ Calcium metal crystallizes in a face-centered cubic unit cell. The density of the solid is $1.54\ g/cm^3$. What is the radius of a calcium atom?

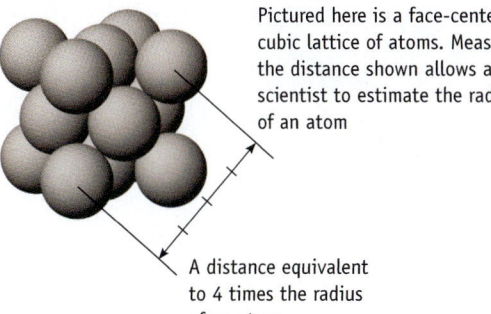

Pictured here is a face-centered cubic lattice of atoms. Measuring the distance shown allows a scientist to estimate the radius of an atom

A distance equivalent to 4 times the radius of an atom

55. ▲ The very dense metal iridium has a face-centered cubic unit cell and a density of $22.56\ g/cm^3$. Use this information to calculate the radius of an atom of the element.

56. ▲ The density of copper metal is $8.95\ g/cm^3$. If the radius of a copper atom is 127.8 pm, is the copper unit cell simple cubic, body-centered cubic, or face-centered cubic?

57. ▲ Vanadium metal has a density of $6.11\ g/cm^3$. Assuming the vanadium atomic radius is 132 pm, is the vanadium unit cell simple cubic, body-centered cubic, or face-centered cubic?

58. ▲ Iron has a body-centered cubic unit cell with a cell dimension of 286.65 pm. The density of iron is $7.874\ g/cm^3$. Use this information to calculate Avogadro's number.

59. ▲ Calcium fluoride is the well-known mineral fluorite. It is known that each unit cell contains four Ca^{2+} ions and eight F^- ions and that the Ca^{2+} ions are arranged in a fcc lattice. The F^- ions fill all the tetrahedral holes in a face-centered cubic lattice of Ca^{2+} ions. The edge of the CaF_2 unit cell is 5.46295×10^{-8} cm in length. The density of the solid is $3.1805\ g/cm^3$. Use this information to calculate Avogadro's number.

60. Mercury and many of its compounds are dangerous poisons if breathed, swallowed, or even absorbed through the skin. The liquid metal has a vapor pressure of 0.00169 mm Hg at 24 °C. If the air in a small room is saturated with mercury vapor, how many atoms of mercury vapor occur per cubic meter?

61. ▲ You can get some idea of how efficiently spherical atoms or ions are packed in a three-dimensional solid by seeing how well circular atoms pack in two dimensions. Using the drawings shown here, prove that B is a more efficient way to pack circular atoms than A. A unit cell of A contains portions of four circles and one hole. In B, packing coverage can be calculated by looking at a triangle that contains portions of three circles and one hole. Show that A fills about 80% of the available space, whereas B fills closer to 90% of the available space.

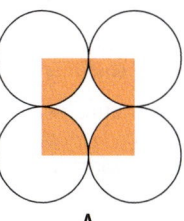 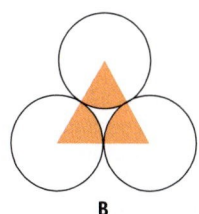

A   B

62. ▲ Assuming that in a simple cubic unit cell the spherical atoms or ions just touch along the cube's edges, calculate the percentage of empty space within the unit cell. (Recall that the volume of a sphere is $(4/3)\pi r^3$, where r is the radius of the sphere.)

63. Equilibrium vapor pressures of dichlorodimethylsilane, $SiCl_2(CH_3)_2$, are given below. (The compound is a starting material to making silicone polymers.)

Temperature (°C)	Vapor Pressure (mm Hg)
−0.4	40.
+17.5	100.
+51.9	400.
+70.3	760.

(a) What is the normal boiling point of dichlorodimethylsilane?

(b) Plot these data as ln P versus $1/T$ so that you have a plot resembling the one in Figure 13.19. At what temperature does the liquid have an equilibrium vapor pressure of 250 mm Hg? At what temperature is this pressure 650 mm Hg?

(c) Calculate the molar enthalpy of vaporization for dichlorodimethylsilane using the the Clausius-Clapeyron equation (Equation 13.1).

64. ▲ The following data are the equilibrium vapor pressures of limonene, $C_{10}H_{16}$, at various temperatures. (Limonene is used as a scent in commercial products.)

Temperature (°C)	Vapor Pressure (mm Hg)
14.0	1.0
53.8	10.
84.3	40.
108.3	100.
151.4	400.

(a) Plot these data as ln P versus $1/T$ so that you have a plot resembling the one in Figure 13.19.

(b) At what temperature does the liquid have an equilibrium vapor pressure of 250 mm Hg? At what temperature is this pressure 650 mm Hg?

(c) What is the normal boiling point of limonene?

(d) Calculate the molar enthalpy of vaporization for limonene using the the Clausius-Clapeyron equation (Equation 13.1).

Summary and Conceptual Questions

The following questions may use concepts from the preceding chapters.

65. Acetone, CH_3COCH_3, is a common laboratory solvent. It is usually contaminated with water, however. Why does acetone absorb water so readily? Draw molecular structures showing how water and acetone can interact. What intermolecular force(s) is (are) involved in the interaction?

66. Cooking oil is not miscible with water. From this observation, what conclusions can you draw regarding the polarity or hydrogen-bonding ability of molecules found in cooking oil?

67. Liquid ethylene glycol, $HOCH_2CH_2OH$, is one of the main ingredients in commercial antifreeze. Do you predict its viscosity to be greater or less than that of ethanol, CH_3CH_2OH?

68. Liquid methanol, CH_3OH, is placed in a glass tube. Predict whether the meniscus of the liquid is concave or convex.

69. Account for these facts:
  (a) Although ethanol (C_2H_5OH) (bp, 80 °C) has a higher molar mass than water (bp, 100 °C), the alcohol has a lower boiling point.
  (b) Mixing 50 mL of ethanol with 50 mL of water produces a solution with a volume slightly less than 100 mL.

70. ▲ Why is it not possible for a salt with the formula M_3X (Na_3PO_4, for example) to have a face-centered cubic lattice of X anions with M cations in octahedral holes?

71. Can $CaCl_2$ have a unit cell like that of sodium chloride? Explain.

72. Rationalize the observation that $CH_3CH_2CH_2OH$, 1-propanol, has a boiling point of 97.2 °C, whereas a compound with the same empirical formula, methyl ethyl ether ($CH_3CH_2OCH_3$) boils at 7.4 °C.

73. Cite two pieces of evidence to support the statement that water molecules in the liquid state exert considerable attractive force on one another.

74. During thunderstorms in the Midwest, very large hailstones can fall from the sky. (Some are the size of golf balls!) To preserve some of these stones, we put them in the freezer compartment of a frost-free refrigerator. Our friend, who is a chemistry student, tells us to use an older model that is not frost-free. Why?

75. Refer to Figure 13.13 to answer the following questions.
  (a) Of the three hydrogen halides (HX), which has the largest total intermolecular force?
  (b) Why are the dispersion forces greater for HI than for HCl?
  (c) Why are the dipole–dipole forces greater for HCl than for HI?
  (d) Of the seven molecules in Figure 13.13, which involves the largest dispersion forces? Explain why this is reasonable.

76. A "hand boiler" can be purchased in toy stores or at science supply companies. If you cup your hand around the bottom bulb, the volatile liquid in the boiler boils and the liquid moves to the upper chamber. Using your knowledge of kinetic-molecular theory and intermolecular forces, explain how the hand boiler works.

Charles D. Winters

77. ▲ The photos illustrate an experiment you can do yourself. Place 10 mL of water in an empty soda can and heat the water to boiling. Using tongs or pliers, quickly turn the can over in a pan of cold water, making sure the opening in the can is below the water level in the pan.
  (a) Describe what happens and explain it in terms of the subject of this chapter.

(a)   **(b)**

Charles D. Winters

  (b) Prepare a molecular-level sketch of the situation inside the can before heating and after heating (but prior to inverting the can).

78. A fluorocarbon, CF_4, has a critical temperature of −45.7 °C, a critical pressure of 37 atm, and a normal boiling point of −128 °C. Are there any conditions under which this compound can be a liquid at room temperature? Explain briefly.

79. ■ ▲ The following figure is a plot of vapor pressure versus temperature for dichlorodifluoromethane, CCl_2F_2. The heat of vaporization of the liquid is 165 kJ/g, and the specific heat capacity of the liquid is about 1.0 J/g · K.

▲ More challenging   ■ In General ChemistryNow   **Blue-numbered questions** answered in Appendix O

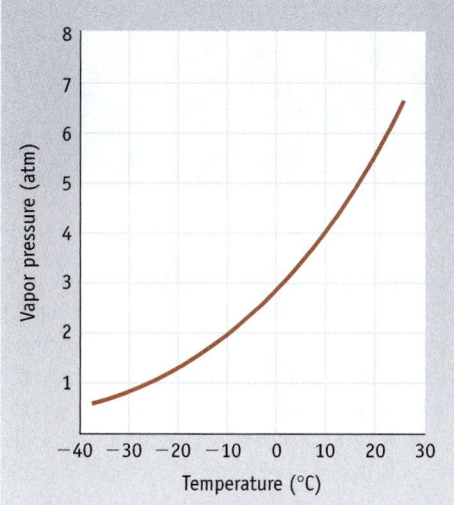

(a) What is the normal boiling point of CCl_2F_2?

(b) A steel cylinder containing 25 kg of CCl_2F_2 in the form of liquid and vapor is set outdoors on a warm day (25 °C). What is the approximate pressure of the vapor in the cylinder?

(c) The cylinder valve is opened, and CCl_2F_2 vapor gushes out of the cylinder in a rapid flow. Soon, however, the flow becomes much slower, and the outside of the cylinder is coated with ice frost. When the valve is closed and the cylinder is reweighed, it is found that 20 kg of CCl_2F_2 is still in the cylinder. Why is the flow fast at first? Why does it slow down long before the cylinder is empty? Why does the outside become icy?

(d) Which of the following procedures would be effective in emptying the cylinder rapidly (and safely)?

  (1) Turn the cylinder upside down and open the valve.

  (2) Cool the cylinder to −78 °C in dry ice and open the valve.

  (3) Knock off the top of the cylinder, valve and all, with a sledge hammer.

80. ▲ Two identical swimming pools are filled with uniform spheres of ice packed as closely as possible. The spheres in the first pool are the size of grains of sand; those in the second pool are the size of oranges. The ice in both pools melts. In which pool, if either, will the water level be higher? (Ignore any differences in filling space at the planes next to the walls and bottom.)

81. Figure 13.41 is a series of photos of CO_2 as it changes from a mixture of liquid and vapor at equilibrium to the supercritical fluid. Draw a representation of the situation at the molecular level of the liquid–vapor equilibrium in the photo at the left and of the supercritical fluid at the right.

Using Molecular Models to Explore Intermolecular Forces and the Solid State

On any screen of the General ChemistryNow CD-ROM or website, click the Molecular Models menu item. For each of the following questions, locate the required model in the indicated folder.

82. ▲ Examine the model of the structure of CaO (Inorganic folder, Ionic Solids).

  (a) Describe the structure. What type of lattice is described by the Ca^{2+} ions? What type of hole do the O^{2-} ions occupy in the lattice of Ca^{2+} ions?

  (b) How is the formula of CaO related to its unit cell structure? (How many Ca^{2+} ions and how many O^{2-} ions are in one unit cell?)

  (c) How is the CaO structure related to the NaCl structure?

83. ■ Examine the structure of ZnS using the model in Inorganic: Ionic Solids.

  (a) Describe the structure. What type of lattice is described by the Zn^{2+} ions (silver color)? What type of holes do the S^{2-} ions (yellow color) occupy in the lattice of Zn^{2+} ions?

  (b) How is the formula of ZnS related to its unit cell structure?

84. Lead sulfide, PbS (commonly called galena), has the same formula as ZnS. Does it have the same solid structure? If different, how it is different? How is its unit cell related to its formula? (The model is found in Inorganic: Ionic Solids.)

85. ■ Examine the structure of CaF_2 using the model in Inorganic: Ionic Solids.

  (a) Describe the structure. What type of lattice is described by the Ca^{2+} ions? What type of holes do the F^- ions occupy in the lattice of Ca^{2+} ions?

  (b) How is the formula of CaF_2 related to its unit cell structure? (How many Ca^{2+} ions and how many F^- ions are in one unit cell?)

  (c) How is the CaF_2 structure related to the ZnS structure?

86. Acetaminophen is used in analgesics. (A model is in Organic: Alcohols.)

  (a) Draw the structure of acetaminophen.

  (b) Is the molecule capable of hydrogen bonding? If so, what are the sites of hydrogen bonding?

87. ■ Aspartame is an artificial sweetener. (A model is in Organic: Amines.)

  (a) Draw the structure of aspartame.

  (b) Is the molecule capable of hydrogen bonding? If so, what are the sites of hydrogen bonding?

Do you need a live tutor for homework problems?
Access vMentor at General ChemistryNow at
http://now.brookscole.com/kotz6e
for one-on-one tutoring from a chemistry expert

Let's Review Chapters 12–13

Photo: John C. Kotz

Many of us love the beach, and of course there are things of interest in chemistry there. The sand on the beach consists of many different substances, among them pulverized corals and sea shells. Another ingredient is quartz sand, and a tiny portion of the structure of quartz is shown in the inset in the photograph. The water washing onto the sand is also a most peculiar substance, and one of its properties is that it can dissolve substances of many kinds—ionic salts and gases, among others. One substance it does not dissolve is the sand itself, although the sand does become wet, and wet sand makes wonderful sand castles.

The Purpose of *Let's Review*

- *Let's Review* provides additional homework questions for Chapters 12 and 13. Routine questions covering each of the concepts in a chapter are found in that chapter and are those usually identified by topic. In contrast, many of the *Let's Review* questions combine several concepts from one or more chapters. Many come from the examinations given by the authors and others are based often on actual experiments or processes in chemical research or in the chemical industry.

- *Let's Review* provides some guidance for Chapters 12 and 13 as you prepare for an exam covering material in these chapters. Although this is designated for Chapters 12 and 13 you may choose only material appropriate to the exam in your specific course.

- Each Comprehensive Question is correlated with tutorials and exercises on the General ChemistryNow website and with the OWL online homework system to which you may have purchased access. Some questions may include a screen shot of a tutorial so you see what resources are available to help you review.

Preparing for an Examination on Chapters 12–13

1. **GENERAL Chemistry☼Now™** Use General ChemistryNow online to take a chapter **Pre-Test.** A **Personalized Learning Plan** will indicate sections of the textbook that should be reviewed.

2. Use General ChemistryNow online to work though media-based **Exercises, Guided Simulations,** and **Intelligent Tutors.**

3. If you subscribe to OWL, use the **Tutorials** in that system.

4. Work on the questions below that are relevant to a particular chapter or chapters. See the solutions to those questions at the end of this section.

5. For background and help with a question, use the General ChemistryNow information or OWL questions that are correlated with the question.

Key Points to Know for Chapters 12–13

Here are some of the key points you must know to be successful in Chapters 12 and 13.

- The gas laws: Boyle's and Charles's laws and Avogadro's hypothesis.
- The ideal gas law, $PV = nRT$
- Reaction stoichiometry and the gas laws.
- Kinetic molecular theory and its connection to the gas laws and gas behavior.
- The non-ideal behavior of gases.
- Intermolecular forces and their effect.
- The importance of hydrogen bonding and the circumstances under which it occurs.
- The properties of liquids (melting and boiling point, vapor pressure)
- Unit cells of solids, especially cubic unit cells.
- Interpretation of phase diagrams.
- The critical temperature and pressure of a substance.

Examination Preparation Questions

▲ denotes more challenging questions.

Important information about the questions that follow:

- See the Study Questions in each chapter for questions on basic concepts.

- Some questions arise from recent research in chemistry and the other sciences. They often involve concepts from more than one chapter and may be more challenging than those in earlier chapters. Not all chapter goals or concepts are necessarily addressed in these questions.

- Assessing Key Points questions are short-answer questions covering the key points on this page.

- Each Comprehensive Question is correlated with the text section covering that topic, with General ChemistryNow material, and with questions in OWL that may provide additional background.

- The computer screens are largely taken from General ChemistryNow but all are also contained in the OWL system if you subscribe to it. They illustrate **Tutorials** or other resources that may help in answering the questions.

Assessing the Key Points

1. You have a 5.0 L flask filled with N_2 at a pressure of 735 mm Hg at 21 °C.
   (a) How many N_2 molecules are in the flask?
   (b) If the N_2 originally in the 5.0-L flask is placed in a 2.5-L flask and cooled to 0.0 °C, what is the pressure under the new conditions?

2. Consider the following gases: He, SO_2, CO_2, and Cl_2.
   (a) Which has the largest density (assuming that all gases are at the same T and P)?
   (b) Which gas will effuse fastest through a porous plate?

3. What volume (in liters) of O_2, measured at standard temperature and pressure, is required to oxidize 0.400 mol of phosphorus (P_4)?

$$P_4(s) + 5\ O_2(g) \longrightarrow P_4O_{10}(s)$$

4. Several small molecules (besides water) are important in biochemical systems: O_2, CO, CO_2, and NO. You have isolated one of these and to identify it you determine its molar mass. You release 0.37 g of the gas into a flask with a volume of 732 mL volume at 21 °C. The gas pressure in the flask is 209 mm Hg. What is the unknown gas?

5. Under which set of conditions will CO_2 deviate most from ideal gas behavior?
   (a) 1 atm, 0 °C   (c) 10 atm, 0 °C
   (b) 0.1 atm, 100 °C   (d) 1 atm, 100 °C

6. A 10.0-L flask contains the following mixture of gases at 25 °C: 16 g of O_2, 28 g of N_2, 2.0 g of He, and 4.0 g of H_2.

   (a) What is the total pressure in the flask?

   (b) Which gas exerts the largest partial pressure?

7. Air contains 21 mol % O_2. If the pressure of air in a room is 745 mm Hg, what is the partial pressure of O_2?

8. Two flasks, each with a volume of 1.00 L, contain O_2 with a pressure of 380 mm Hg. Flask A is inside the Chemistry Building where the temperature is 25 °C. Flask B is outside in the cold winter air at 0 °C.

   (a) Which flask contains the greater number of O_2 molecules?

   (b) Which flask contains O_2 molecules with greater average speed? With greater average kinetic energy?

9. Rank the following gases in order of increasing molecular speed at a given temperature: CO, CO_2, O_2, and NH_3. (The relation between molar mass and molecular speed can be explored using the Simulation on Screen 12.11.)

12.11 Boltzman Distributions

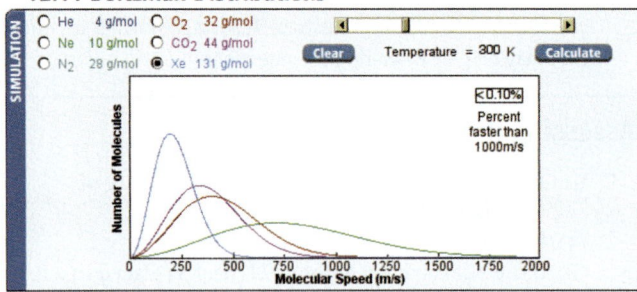

General ChemistryNow Screen 12.11 A Simulation of the Boltzmann distribution of molecular speeds for a variety of molecules.

10. Propane reacts with O_2 according to the equation

$$C_3H_8(g) + 5\,O_2(g) \longrightarrow 3\,CO_2(g) + 4\,H_2O(g)$$

   You mix C_3H_8 with O_2 in the correct stoichiometric ratio, and the total pressure of the mixture is 120 mm Hg at some temperature.

   (a) What are the partial pressures of C_3H_8 and O_2?

   (b) What will be the pressure of water vapor, at the same temperature, after complete reaction?

11. For which of the following will there be hydrogen bonding between molecules: H_2O, CH_3OCH_3, CH_3OH, CH_4, H_2, HNO_3?

12. Which of the following compounds would be expected to form intermolecular hydrogen bonds with water?

   (a) CH_3OCH_3 (dimethyl ether)

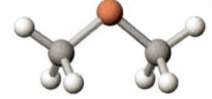

(b) C_3H_8 (propane)

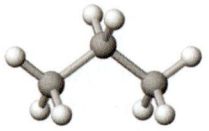

(c) CH_3CO_2H (acetic acid)

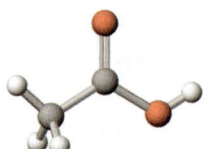

(d) Aspirin $C_6H_4(CO_2H)(CO_2CH_3)$

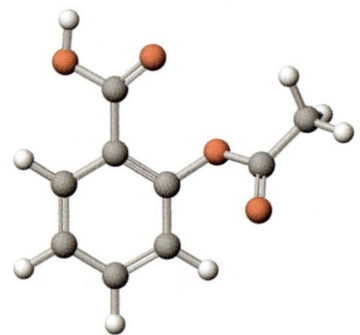

13. Rank the following compounds in terms of increasing strength of intermolecular forces in the pure substances: Ar, CH_3OH, CO_2, CaO.

14. The following types of intermolecular forces are generally found to be important:

   (i)   ion-dipole

   (ii) dipole-induced dipole

   (iii) dipole-dipole

   (iv) induced dipole-induced dipole

   (v)   hydrogen bonds

   Decide which type of intermolecular force is most important for each of the following systems.

   (a) Between methane (CH_4) molecules in liquid methane.

   (b) Between H_2O and CH_3OH molecules in a mixture of the liquids.

   (c) Between Li^+ and Cl^- ions and water in aqueous lithium chloride.

   (d) Between O_2 and H_2O molecules when O_2 is dissolved in water.

13.5 Determining Intermolecular Forces

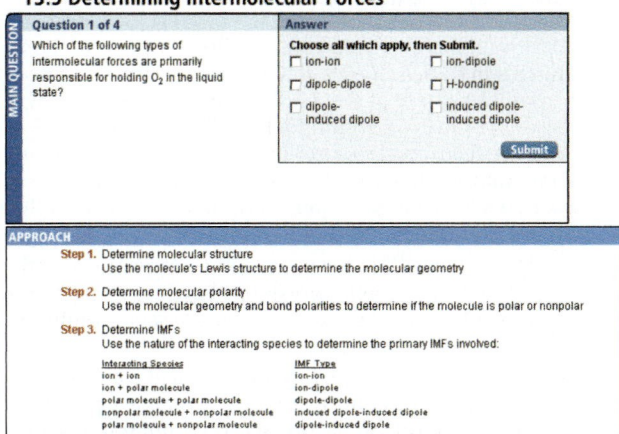

General ChemistryNow Screen 13.5 Exploring intermolecular forces

15. Rank the following substances in order of increasing boiling point: (a) Ne, (b) CH_3OH, and (c) SO_2.

16. Use the vapor pressure curves for carbon disulfide (CS_2), ethanol (CH_3CH_2OH), and heptane (C_7H_{16}) below to answer the following questions:

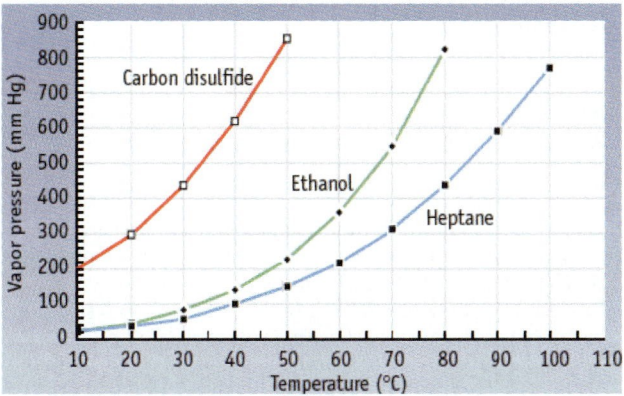

Vapor pressure curves for carbon disulfide (CS_2), ethanol (CH_3CH_2OH), and heptane (C_7H_{16}).

(a) What is the vapor pressure of heptane at 70 °C?

(b) What is the normal boiling point of ethanol?

(c) What type of intermolecular forces exist

    (i)   between two CS_2 molecules

    (ii) between two heptane molecules

    (iii) between two ethanol molecules

(d) Decide if each substance is a liquid or vapor under those conditions if the pressure is 600 mm Hg and the temperature is 50 °C.

(e) Which one or ones of these molecules is expected to dissolve in water?

(f) Which has the higher normal boiling point, ethanol or CS_2? Explain.

(g) Which has the higher normal boiling point, heptane or CS_2? Explain.

13.10 Vapor Pressure and Boiling Point

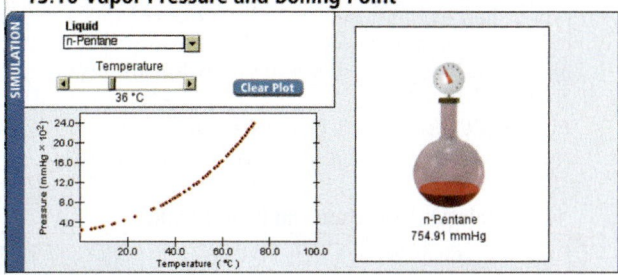

General ChemistryNow Screen 13.10 Vapor pressures and boiling points of various volatile liquids

17. Types of solids include molecular solids, metallic solids, network solids, amorphous solids, and ionic solids. Classify each of the following solids:

(a) $CaCO_3$ (c) tungsten

(b) polyethylene   (d) solid CO_2 (dry ice)

18. Magnesium oxide has a NaCl-like crystal structure.

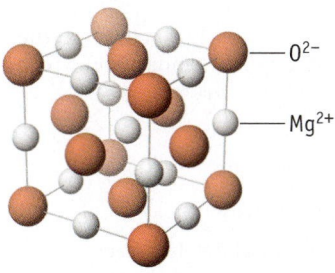

Unit cell of MgO

(a) In which type of unit cell are the O^{2-} ions arranged?

(b) How many Mg^{2+} ions are there per unit cell?

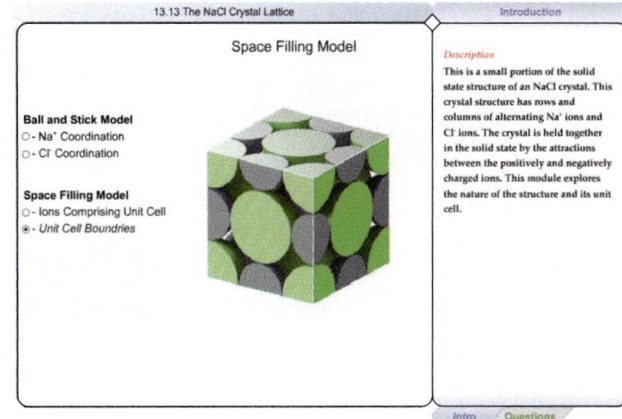

General ChemistryNow Screen 13.13 Tutorial on crystal lattices.

19. List the following compounds in order of increased boiling point, then explain your choice in terms of intermolecular forces. For part (b) you can predict the answer by examining the boiling points of a series of hydrocarbons on Screen 13.10.

    (a) CH_3CO_2H, CH_3OCH_3, CH_3CH_2OH

    (b) hexane (C_6H_{14}), octane (C_8H_{18}), propane (C_3H_8)

13.10 Molecular Structure and Boiling Point

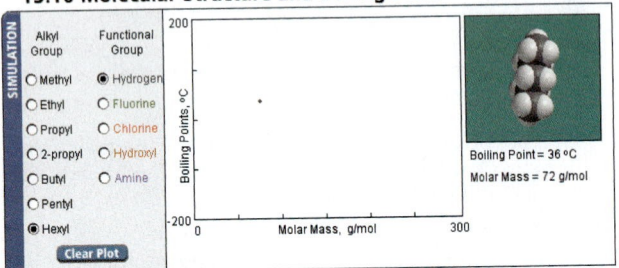

General ChemistryNow Screen 13.10 The relation of molecular structure and boiling point.

20. The following data are available for methane at the NIST website (National Institutes for Standards and Technology, http://webbook.nist.gov/chemistry/):

    Triple point = 90.67 K at 0.117 bar

    Normal melting point = 90.69 K

    Normal boiling point = 111.2 K

    Critical T = 190.6 K at critical P = 46.1 bar

    (a) Is solid methane more or less dense than liquid methane?

    (b) Can methane be liquefied at 0 °C?

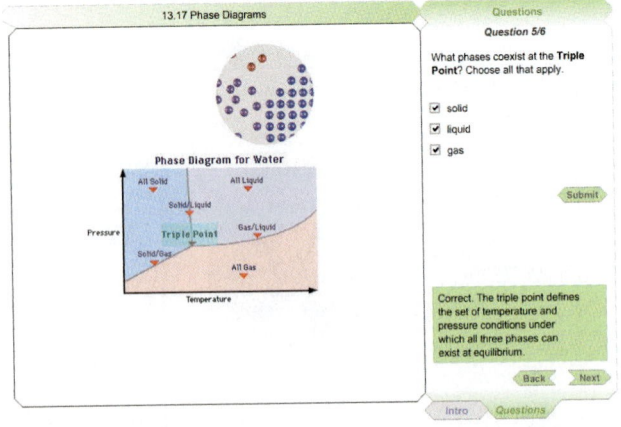

General ChemistryNow Screen 13.17 Exploring phase diagrams

Comprehensive Questions

21. (Chapters 12 and 13) It was recently discovered at the Fred Hutchison Cancer Research Center in Seattle that mice can be put into a state of suspended animation by applying a low dose of hydrogen sulfide, H_2S. The breathing rate of the mice fell from 120 to 10 breaths a minute and their temperature fell to just 2 °C above ambient temperature. Six hours later the mice were revived and seemed to show no negative effects.

    (a) Hydrogen sulfide is a gas at room temperature and normal atmospheric pressure whereas water is a liquid with a low vapor pressure under the same conditions. Explain this observation.

    (b) The H_2S gas delivered to the mice had a concentration of 80 ppm. (A concentration of 1 ppm is 1 part per million, or one molecule in every 1 million molecules.) If you deliver 1.0 L of gas (a mixture of O_2, N_2, and H_2S) at a total pressure of 725 mm Hg at a temperature of 22 °C, what is the partial pressure of the H_2S gas?

    (c) Hydrogen sulfide can be converted to sulfuric acid. If 5.2 L of H_2S gas at 130 mm Hg pressure and 25 °C is allowed to react with O_2 gas, how many liters of O_2 gas, also at 130 mm Hg pressure and 25 °C, are required for complete reaction? Assume the following reaction occurs.

    $$H_2S(g) + 2\,O_2(g) \longrightarrow H_2SO_4(liq)$$

    Text Sections: 12.3–12.5, 13.2–13.4

    General ChemistryNow Screens: 12.4, 12.7, 12.8, 13.3–13.6

    OWL Questions: 12.1, 12.3, 12.4, 13.1

22. (Chapter 12) A common industrial chemical, butyl acrylate, is made by the Reppe process. This combines acetylene (C_2H_2), butanol (C_4H_9OH), and CO in the presence of a catalyst. If you release 0.370 g of the volatile liquid product of the reaction into a flask with a volume of 732 mL at 21 °C, the pressure is 72.4 mm Hg. What is the molar mass of the gas? Based on the reactants in the Reppe reaction, and the molar mass of the product, what is its formula?

    Text Section: 12.3

    General ChemistryNow Screens: 12.5, 12.6

    OWL Question 12.2

23. (Chapters 4 and 12) Jacques Charles used 225 kg of H_2SO_4 and 450. kg of Fe to prepare H_2 gas for one of his earliest lighter-than-air balloon flights (in August, 1783; see text page 546). Assume that the atmospheric pressure that day was 735 mm Hg, and the temperature was 18 °C. What was the volume of the balloon (that is, the volume needed to contain this gas)? The hydrogen-forming reaction is

    $$Fe(s) + H_2SO_4(aq) \longrightarrow FeSO_4(aq) + H_2(g)$$

24. (Chapter 12) Mercury has been poured into the U-tube, as shown below, trapping air in the right side. What is the pressure of the trapped air? The atmospheric pressure is 743 mm.

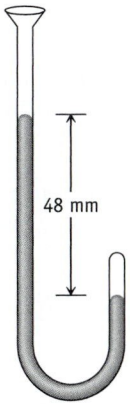

48 mm

25. ▲ (Chapter 12) Unlike balloons made of rubber, balloons made of Mylar do not stretch. If you add gas to a Mylar balloon, the volume increases and, while filling, the pressure inside equals the pressure outside. This persists until the balloon is filled. After that, addition of gas will cause the pressure inside the balloon to increase above that of the outside atmosphere, and the volume won't change. At some point, you will exceed the tensile strength of the plastic film and the balloon will break.

(a) You place 4.8 g of dry ice [$CO_2(s)$] in a Mylar balloon whose volume when filled is 2.33 L. The dry ice sublimes and the gaseous CO_2 warms to room temperature, 18 °C. Atmospheric pressure that day is 0.98 atm. Assume there is no air inside the balloon at the start of this experiment. First, determine if the balloon is underfilled and the pressure inside is 0.98 atm or if the balloon is overfilled and the pressure inside is greater than 0.98 atm. Then, determine the actual pressure of gas inside the balloon.

(b) Compare the number of molecules of CO_2 used to fill the balloon in part (a) with the number of atoms of helium needed to fill the same balloon to the pressure calculated in (a).

(c) In which balloon (the one with He or the one with CO_2) are the gas molecules or atoms traveling at a greater average speed? What are their relative average kinetic energies at 18 °C?

26. (Chapter 12) Carbon dioxide, CO_2, was determined to effuse through a porous plate at the rate of 0.0330 mol/min. The same amount of an unknown gas, 0.0330 moles, is found to effuse through the same porous barrier in 104 seconds. Calculate the molar mass of the unknown gas.

27. (Chapters 9, 10, and 13) Bonding, structure, and properties of formamide, $HCONH_2$.

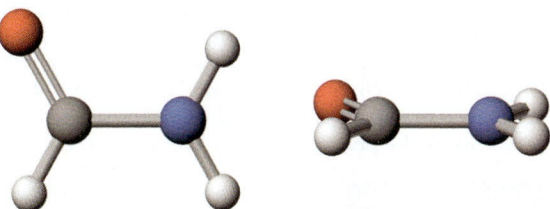

Front and side views of formamide, $HCONH_2$

(a) Draw an electron dot structure for formamide. Include all resonance structures.

(b) What is the C atom hybridization?

(c) Is the molecule polar?

(d) Can two formamide molecules interact by hydrogen bonding? Can formamide interact with water by hydrogen bonding?

(e) As indicated by the molecular model, the molecule is flat. Suggest a reason for the planar structure of the molecule.

(f) The enthalpy of vaporization of formamide at 25 °C is 60.15 kJ/mol. What amount of heat is required to vaporize 10.0 g of the compound at 25 °C?

28. (Chapter 13) The triple point of ammonia, NH_3, is 195.4 K at 45.9 mm Hg. The normal boiling point is 239.8 K. Use the Clausius-Clapeyron equation to calculate the enthalpy of vaporization for ammonia.

13.9 Heat of Vaporization from Graph

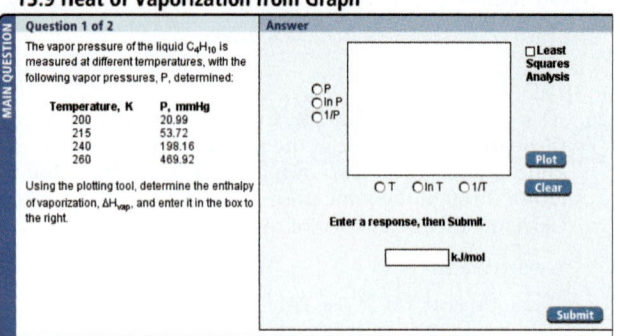

General ChemistryNow Screen 13.9 Tutorial: Using the Clausius-Clapeyron equation.

29. (Chapter 13) A "hand boiler" can be purchased in toy and novelty stores. Holding your hand under the bottom chamber raises the liquid vapor pressure until the liquid boils. The increase in vapor pressure forces the liquid into the upper chamber. (In Study Question 76 on page 640 you were asked to explain how it works.) Which of the following liquids would be best to use in the hand boiler? Explain.

Charles D. Winters

Compound	Temperature at which the vapor pressure is 100 mm Hg	Normal Boiling Point
Ethanol	34.9	78.4 °C
Chloroform	10.4	61.3 °C
CCl$_3$F	−23.0	23.7 °C

Text Section: 13.5

General ChemistryNow Screen: 13.9

OWL Question: 13.2

30. (Chapters 12 and 13) Acetone is a common solvent.

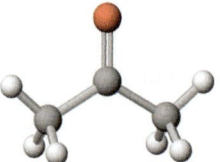

Acetone, $(CH_3)_2CO$

(a) Allyl alcohol, CH_2=CH—CH_2OH, is an isomer of acetone. Acetone has a vapor pressure of 100 mm Hg at +7.7 °C. Predict whether the vapor pressure of allyl alcohol is higher or lower than 100 mm Hg at this temperature? Explain.

(b) Use the Clausius-Clapeyron equation to calculate the enthalpy of vaporization of acetone from the following data.

Temperature (°C)	Vapor Pressure (mm Hg)
−9.4	40.
+7.7	100.
39.5	400.
56.5	760.

(c) Fluorination of acetone, C_3H_6O (substitution of fluorine for H) produces a gaseous compound with the formula $C_3H_{6-x}F_xO$. To identify this compound its molar mass was determined by measuring the gas density. The following data were obtained: Mass of gas, 1.53 g; volume of flask = 264 mL; pressure exerted by gas, 722 mm Hg; temperature, 22 °C. Calculate the molar mass from this information, then identify the molecular formula.

Text Sections: 12.3, 13.5

General ChemistryNow Screens: 12.6, 13.9

OWL Questions: 12.2, 13.3

31. (Chapter 13) Butane, the fuel in lighters and camping stoves, has the following physical properties:

Normal melting point = 136 K

Normal boiling point = 273 K

Triple point = 135 K at 7×10^{-6} bar

Construct a phase diagram for butane based on this information. Be sure to label the axes. Use S, L, and G to represent conditions for the three states of matter. Finally, use this drawing to answer the following questions:

(a) Is butane a gas, liquid or solid at 0 °C and 400 mm Hg?

(b) Will an increase in pressure result in an increase, a decrease, or no change in the melting point of butane?

(c) Between what temperatures is butane a liquid at 1.0 atm pressure?

Text Section: 13.10

General ChemistryNow Screens: 13.8–13.10, 13.17

OWL Question: 13.7

32. The unit cell of one of several known molybdenum fluorides is pictured here.

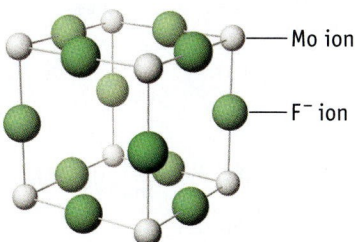

A unit cell of molybdenum fluoride

How many molybdenum ions and how many fluoride ions are in each unit cell? What is the formula and name of this molybdenum fluoride?

Text Section: 13.7

General ChemistryNow Screen: 13.13

OWL Question: 13.5

33. (Chapter 13) The unit cell of sodium chloride is pictured below. The length of each side of the unit cell is 562.8 pm.

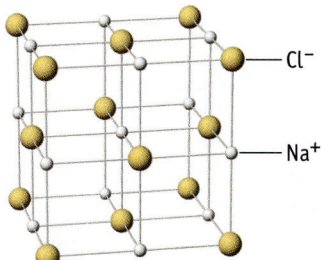

NaCl unit cell (expanded)

(a) What is the mass of one unit cell?

(b) What is the volume of one unit cell?

(c) What is the density of NaCl based on the mass and volume of the unit cell? (Compare with a literature density of 2.17 g/cm³.)

Text Section: 13.7

General ChemistryNow Screen: 13.13

OWL Question: 13.5

34. ▲ (Chapter 13) The unit cell of silver iodide is illustrated below.

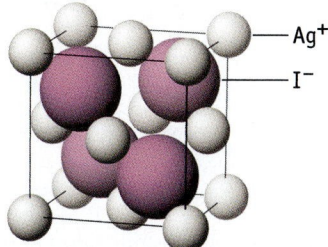

Silver iodide unit cell

(a) How many silver ions and how many iodide ions are there per unit cell?

(b) Sides of the unit cell are 649.6 pm. The density is 5.68 g/cm³. What is the mass of one unit cell?

(c) Based on the unit cell mass, what is the mass of one silver iodide "molecule" and what is its molar mass? Compare this with the molar mass of silver iodide calculated from atomic weights.

Text Section: 13.7

General ChemistryNow Screen: 13.13

OWL Question: 13.5

13.13 Calculating Density from Ion Radii and Structure

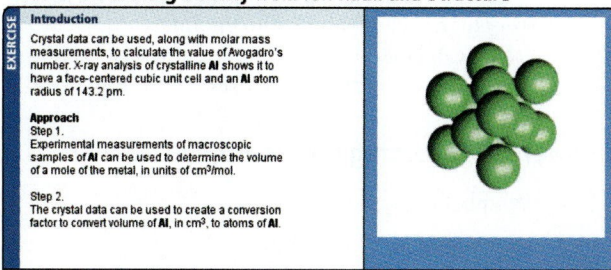

Introduction

Crystal data can be used, along with molar mass measurements, to calculate the value of Avogadro's number. X-ray analysis of crystalline **Al** shows it to have a face-centered cubic unit cell and an **Al** atom radius of 143.2 pm.

Approach
Step 1.
Experimental measurements of macroscopic samples of **Al** can be used to determine the volume of a mole of the metal, in units of cm³/mol.

Step 2.
The crystal data can be used to create a conversion factor to convert volume of **Al**, in cm³, to atoms of **Al**.

General ChemistryNow Screen 13.13 Exercise: Calculating the density of a solid from unit cell information.

35. ▲ (Chapter 13) Units cells and close packing. Prove that the atoms packed in an idealized face-centered cubic unit cell occupy about 74% of the available space.

Text Section: 13.7

General ChemistryNow Screen: 13.13

OWL Question: 13.5

36. (Chapters 13) Below you see the Simulation from Screen 13.9 on vapor pressures. Observe the behavior of the vapor pressure of pentane (C_5H_{12}) and hexane (C_6H_{14}) at various temperatures.

(a) Which has the higher vapor pressure at all temperatures and why?

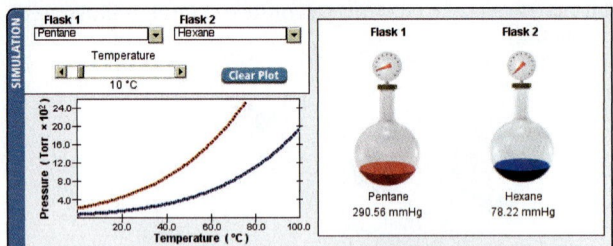

General ChemistryNow Screen 13.9 Simulation on vapor pressures.

(b) What is the approximate normal boiling point of these two hydrocarbons?

(c) Use vapor pressure-temperature data from this simulation to calculate the molar enthalpy of vaporization of hexane.

Text Section: 13.5

General ChemistryNow Screens: 13.8–13.10

OWL Questions: 13.3, 13.4

37. See the photograph and structure of quartz on page 642.

(a) What are some interesting structural features of quartz? What kind of solid does it represent?

(b) How does water interact with ordinary salt and oxygen gas so that they dissolve in water?

(c) Why does water have such peculiar properties? For example, for such a small molecule why does it have a very high melting and boiling point?

(d) ▲ Why does water interact with sand—the sand gets wet—but it does not dissolve the sand?

Answers to Assessing Key Points Questions

1. (a) Number of N_2 molecules

Amount of $N_2 = n = PV/RT$

$$= \frac{\left[735 \text{ mm Hg}\left(\dfrac{1 \text{ atm}}{760 \text{ mm Hg}}\right)\right](5.0 \text{ L})}{\left(0.08206 \dfrac{\text{L} \cdot \text{atm}}{\text{K} \cdot \text{mol}}\right)(294 \text{ K})}$$

$$= 0.20 \text{ mol}$$

Number of molecules =

$$= (0.20 \text{ mol})\left(\frac{6.022 \times 10^{23} \text{ molecules}}{\text{mol}}\right)$$

$$= 1.2 \times 10^{23} \text{ molecules}$$

(b) Pressure under new conditions. Use the general gas law, Equation 12.3, on page 554

$$P_{new} = 735 \text{ mm Hg}\left(\frac{5.0 \text{ L}}{2.5 \text{ L}}\right)\left(\frac{273 \text{ K}}{294 \text{ K}}\right) = 1400 \text{ mm Hg}$$

2. (a) Equation 12.5 (page 559) states that the density of a gas is proportional to the molar mass at a given P and T. The masses are in the order $He < CO_2 < SO_2 < Cl_2$. Therefore, Cl_2 has the greatest density.

(b) The rate of effusion is inversely proportional to the molar mass (page 572). Therefore, He will diffuse most rapidly.

3. (a) Amount of O_2 required

$$0.400 \text{ mol P}_4\left(\frac{5 \text{ mol O}_2}{1 \text{ mol P}_4}\right) = 2.00 \text{ mol O}_2$$

Volume of O_2 required.

$$V = nRT/P$$

$$= \frac{(2.00 \text{ mol O}_2)\left(0.08206 \dfrac{\text{L} \cdot \text{atm}}{\text{K} \cdot \text{mol}}\right)(273 \text{ K})}{1.00 \text{ atm}}$$

$$= 44.8 \text{ L}$$

Remember that the molar volume of an ideal gas is 22.4 L/mol at STP.

4. Calculate the molar mass.

Amount of gas $= n = \dfrac{PV}{RT}$

$$= \frac{(209 \text{ mm Hg})\left(\dfrac{1 \text{ atm}}{760 \text{ mm Hg}}\right)(0.732 \text{ L})}{\left(0.08206 \dfrac{\text{L} \cdot \text{atm}}{\text{K} \cdot \text{mol}}\right)(294 \text{ K})}$$

$$= 0.00834 \text{ mol}$$

$$\text{Molar mass} = \frac{0.37 \text{ g}}{0.00834 \text{ mol}} = 44 \text{ g/mol}$$

The gas is CO_2.

5. (c) Gases at higher pressures and lower temperatures usually deviate most from ideality.

6. (a) The total pressure depends on the total number of molecules (moles) of gas in the flask (and T and V).

$n(O_2) = 0.50 \text{ mol}$

$n(N_2) = 1.0 \text{ mol}$

$n(He) = 0.50 \text{ mol}$

$n(H_2) = 2.0 \text{ mol}$

Total amount $= n_{total} = 4.0 \text{ mol}$

$$P = \frac{n_{total}RT}{V} = \frac{(4.0 \text{ mol})\left(0.08206 \dfrac{\text{L} \cdot \text{atm}}{\text{K} \cdot \text{mol}}\right)(298 \text{ K})}{10.0 \text{ L}}$$

$$= 9.8 \text{ atm}$$

See Equation 12.7 on page 565.

(b) The gas with the highest partial pressure is that present in the largest amount, here H_2.

7. The partial pressure of a gas depends on its mole fraction, here 0.21 for O_2. (See Equation 12.8 on page 565.) Thus, the O_2 partial pressure is 160 mm Hg (= 0.21 × 745 mm Hg).

8. (a) Both flasks have the same P and same V, but B has a lower T. Therefore, B must have more molecules because more molecules are needed to produce the same P at a lower T. That is, given that $n = PV/RT$, when P, V, and R are constant, n must be greater for a lower T.

   (b) Flask A contains molecules with a greater average speed and greater average kinetic energy.

9. Molecules in order of increasing molar mass and decreasing speed: NH_3 ($M = 17$ g/mol), CO ($M = 28$ g/mol), O_2 ($M = 32$ g/mol), CO_2 ($M = 44$ g/mol)

10. (a) C_3H_8 and O_2 are mixed in a ratio of 1 mol to 5 mol, which reflects their mole fraction.

    $X_{propane} = 1\ mol/(1\ mol + 5\ mol) = 1/6$

    $P_{propane} = X_{propane}P_{total} = (1/6)(120\ mm\ Hg) = 20.\ mm\ Hg$

    $P_{oxygen} = (120 - 20.)\ mm\ Hg = 100\ mm\ Hg$

    The pressure of O_2 is 5 times that of C_3H_8 as required by stoichiometry.

    (b) Pressure of water vapor

    $$P_{H_2O} = P_{propane}\left(\frac{4\ mm\ Hg\ H_2O}{1\ mm\ Hg\ propane}\right) = 80.\ mm\ Hg$$

11. H_2O, CH_3OH, and HNO_3. Note that HNO_3 has an OH bond and can H-bond to neighboring nitric acid molecules.

12. A H-bond could form between H_2O and the O atom of the polar molecule CH_3OCH_3. Both acetic acid and aspirin can readily H-bond to water.

13. Intermolecular forces:

    Atoms of Ar < nonpolar CO_2 < polar CH_3OH with H-bonding < ionic solid CaO

14. (a) Induced dipole-induced dipole forces between nonpolar CH_4 molecules

    (b) Dipole-dipole and hydrogen bonding between H_2O and CH_3OH

    (c) Ion-dipole forces between Li^+ and Cl^- ions and water

    (d) Dipole-induced dipole forces between polar water and nonpolar O_2

15. Ne (−246 °C) < SO_2 (−10.05 °C) < CH_3OH (+64.6 °C)

    Neon atoms interact only by induced dipole forces, whereas SO_2 is weakly polar. Methanol molecules interact by dipole-dipole forces and hydrogen bonding.

16. (a) 300 mm Hg

    (b) About 75 °C (literature value, 78.3 °C)

(c) Nonpolar CS_2 molecules interact by induced dipole forces as do heptane molecules. Ethanol molecules interact through dipole-dipole forces and hydrogen bonds.

(d) CS_2 is a vapor whereas both ethanol and heptane are liquids.

(e) Ethanol can interact with water through dipole forces and hydrogen bonding, so it dissolves readily in water. Neither of the nonpolar molecules, CS_2 or heptane, are miscible with water.

(f) The vapor pressure curve indicates CS_2 has a normal boiling point of about 45 °C, whereas that of ethanol is 78 °C. This makes sense because we predict that ethanol molecules interact through dipole-dipole forces and extensive hydrogen bonding, whereas CS_2 molecules interact through much weaker induced dipole forces.

(g) Boiling points depend on only intermolecular forces for molecules of similar molar mass. Both CS_2 and heptane are nonpolar, but heptane molecules are significantly heavier than CS_2 molecules (100.2 u versus 76.1 u) so its boiling point is predicted to be much higher. This is confirmed by the vapor pressure curves, which indicate that heptane has a normal boiling point of about 100 °C and that the CS_2 normal boiling point is only about 45 °C.

17. (a) $CaCO_3$ is an ionic solid.

    (b) Polyethylene is an amorphous solid (although crystalline polyethylene does exist).

    (c) Tungsten is a metallic solid.

    (d) Solid CO_2 is a molecular solid.

18. (a) The oxide ions are arranged in a face-centered cubic unit cell.

    (b) There are four Mg^{2+} (and four O^{2-} ions) per unit cell.

19. (a) $CH_3OCH_3 < CH_3CH_2OH < CH_3CO_2H$

    The molecule CH_3OCH_3 (dimethyl ether) is polar. However, the other two molecules are not only polar but can interact through hydrogen bonds. This is especially true for CH_3CO_2H (acetic acid), which has not only an OH group but a very polar C=O group. (See Figure 13.9 on page 601.)

    (b) propane (C_3H_8) < hexane (C_6H_{14}) < octane (C_8H_{18})

    The boiling points of the nonpolar alkanes (a class of hydrocarbons with the general formula C_nH_{2n+2}) increase with increasing molar mass.

20. Phase diagram for methane, CH_4.

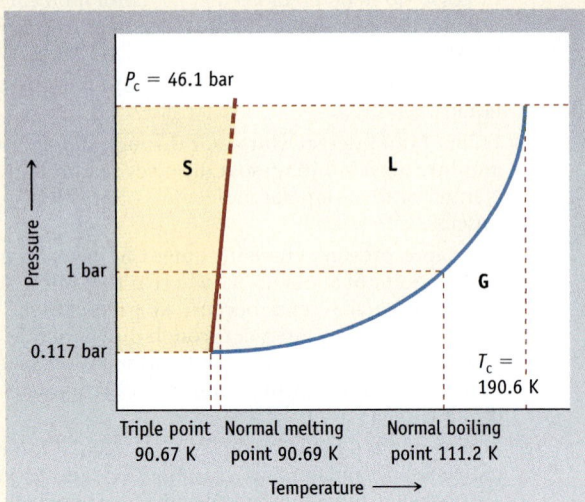

(a) Solid methane is slightly more dense than liquid methane.

(b) No, this is above the critical temperature.

Solutions To Comprehensive Questions

21. (a) H_2S, unlike water, does not exhibit significant hydrogen bonding.

(b) Partial pressure of H_2S depends on the mole fraction of H_2S in the mixture (Equation 12.8). Here the mole fraction is the number of moles of H_2S divided by total moles of all gases, which is the same as the ratio of H_2S molecules relative to the total number of molecules.

P_{H_2S} = (mole fraction H_2S)(P_{total})

$$= \left(\frac{80 \text{ molecules}}{1 \times 10^6 \text{ molecules}}\right)(725 \text{ mm Hg})$$

$$= 0.06 \text{ mm Hg}$$

(c) For gases at the same T and P, their volumes are directly proportional to the amount of gas. (This is Avogadro's Hypothesis; page 556. See also Example 12.5.)

$$5.2 \text{ L } H_2S \left(\frac{2 \text{ L } O_2}{1 \text{ L } H_2S}\right) = 10. \text{ L } O_2 \text{ required}$$

22. This is similar to Example 12.8.

$$n = \frac{PV}{RT} = \frac{72.4 \text{ mm Hg}\left(\dfrac{1 \text{ atm}}{760 \text{ mm Hg}}\right)(0.732 \text{ L})}{\left(0.08206 \dfrac{\text{L} \cdot \text{atm}}{\text{K} \cdot \text{mol}}\right)294 \text{ K}}$$

$$n = 0.00289 \text{ mol}$$

$$\text{Molar mass} = \frac{0.370 \text{ g}}{0.00289 \text{ mol}} = 128 \text{ g/mol}$$

The product of the reaction, butyl acrylate, has a formula of $C_7H_{12}O_2$, formed by combining the reactants in a $1:1:1$ ratio.

23. The volume of H_2 depends on its amount, and that in turn depends on which is the limiting reactant, Fe or H_2SO_4.

$$450. \times 10^3 \text{ g Fe}\left(\frac{1 \text{ mol Fe}}{55.85 \text{ g Fe}}\right) = 8.06 \times 10^3 \text{ mol Fe}$$

$$225 \times 10^3 \text{ } H_2SO_4\left(\frac{1 \text{ mol } H_2SO_4}{98.08 \text{ g } H_2SO_4}\right) = 2.29 \times 10^3 \text{ mol } H_2SO_4$$

The limiting reactant is H_2SO_4, which produces 2.29×10^3 mol H_2.

$$V = \frac{nRT}{P} = \frac{(2.29 \times 10^3 \text{ mol } H_2)\left(0.08206 \dfrac{\text{L} \cdot \text{atm}}{\text{K} \cdot \text{mol}}\right)(291 \text{ K})}{735 \text{ mm Hg}\left(\dfrac{1 \text{ atm}}{760 \text{ mm Hg}}\right)}$$

$$V = 5.65 \times 10^4 \text{ L}$$

24. Pressure = 743 mm Hg + 48 mm Hg = 791 mm Hg.

25. (a) Amount of CO_2 in the balloon:

$$\text{Amount of } CO_2 = 4.8 \text{ g } CO_2\left(\frac{1 \text{ mol } CO_2}{44.0 \text{ g } CO_2}\right) = 0.11 \text{ mol}$$

Pressure in the balloon with $n = 0.11$ mol, $T = 291$ K, and $V = 2.33$ L.

$$P = \frac{(0.11 \text{ mol})\left(0.08206 \dfrac{\text{L} \cdot \text{atm}}{\text{K} \cdot \text{mol}}\right)(291 \text{ K})}{2.33 \text{ L}} = 1.1 \text{ atm}$$

The balloon is overfilled because the pressure generated by 0.11 mol of gas under these conditions is greater than the atmospheric pressure. The actual pressure in the balloon is 1.1 atm.

(b) The number of molecules is the same. The pressure of a gas (at a given T and V) depends only on the number of particles, not their identity.

(c) He atoms travel faster at a given T than CO_2 molecules (see Figure 12.15 and Screen 12.11) but their average kinetic energies are identical.

26. This problem on gas effusion uses Graham's law, Equation 12.10, page 572.

$$\frac{\text{Rate of effusion of } CO_2}{\text{Rate of effusion of unknown}} = \sqrt{\frac{\text{Molar mass unknown}}{\text{Molar mass } CO_2}}$$

$$\frac{0.0330 \text{ mol/min}}{0.0330 \text{ mol/1.73 min}} = \sqrt{\frac{\text{Unknown}}{44.0 \text{ g/mol}}}$$

$$1.73 = \sqrt{\frac{\text{Unknown}}{44.0 \text{ g/mol}}}$$

$$2.99 = \frac{\text{Unknown}}{44.0 \text{ g/mol}}$$

$$\text{Unknown} = 132 \text{ g/mol}$$

12.12 Effusion

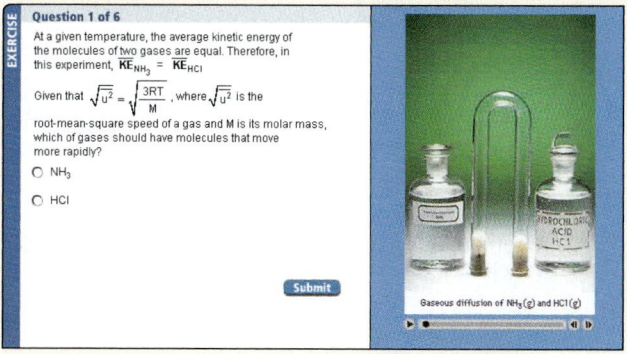

General ChemistryNow Screen 12.12a Tutorial on using Graham's law to estimate a molar mass.

12.12 Diffusion

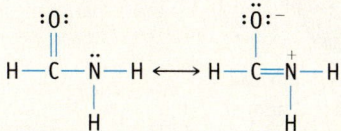

General ChemistryNow Screen 12.12b This exercise gives further insight into diffusion.

27. (a) Electron dot structure

$$:\!O\!:\qquad\qquad :\!\ddot{O}\!:^-$$
$$\|\qquad\qquad\qquad\qquad$$
$$H\!-\!C\!-\!\ddot{N}\!-\!H \longleftrightarrow H\!-\!C\!=\!\overset{+}{N}\!-\!H$$
$$\quad|\qquad\qquad\qquad\qquad|$$
$$\quad H\qquad\qquad\qquad H$$

(b) The C atom is sp^2 hybridized.

(c) The molecule is polar.

(d) Yes, the N—H of one molecule could interact with an N atom or O atom on a neighboring molecule. The formamide molecule can also hydrogen bond to water.

(e) The molecule is flat owing to the importance of the resonance structure with a N=C double bond.

(f) Vaporizing 10.0 g of formamide:

Heat required =

$$10.0 \text{ g HCONH}_2\left(\frac{1 \text{ mol HCONH}_2}{45.04 \text{ g HCONH}_2}\right)\left(\frac{60.15 \text{ kJ}}{1 \text{ mol HCONH}_2}\right)$$
$$= 13.4 \text{ kJ}$$

28. Using the Clausius-Clapeyron equation:

$$\ln\frac{P_2}{P_1} = \frac{\Delta H^\circ_{vap}}{R}\left[\frac{1}{T_1} - \frac{1}{T_2}\right]$$

$$\ln\frac{760.}{45.9} = \frac{\Delta H^\circ_{vap}}{8.314 \text{ J/K}\cdot\text{mol}}\left[\frac{1}{195.4} - \frac{1}{239.8}\right]\frac{1}{\text{K}}$$

$$\ln 16.558 = \frac{\Delta H^\circ_{vap}}{8.314 \text{ J/K}\cdot\text{mol}}[9.476\times10^{-4}]1/\text{K}$$

$$\Delta H^\circ_{vap} = 2.46\times10^4 \text{ J/mol} \quad\text{or}\quad 24.6 \text{ kJ/mol}$$

29. The best liquid to use would be one that boils close to room temperature or to the temperature of a hand (37 °C). The chlorofluorocarbon CCl_3F is the best choice here.

30. (a) Allyl alcohol is an alcohol (like ethanol) and has an OH group capable of hydrogen bonding. This should lower its vapor pressure at a given temperature relative to that of acetone, which, while polar, is not capable of hydrogen bonding.

(b) One can either choose a set of temperatures and pressures and use Equation 13.2 or plot the data (as in Figure 13.19) and derive the heat of vaporization from the slope of the line. Here a plot of ln P versus $1/T$ gives a slope of -3875.3.

Slope $= -3875.3 = -\Delta H^\circ_{vap}/R$

Using $R = 8.314$ J/K·mol, and converting to kilojoules, we obtain $\Delta H^\circ_{vap} = 32.2$ kJ/mol. (An identical result is obtained using the Clausius-Clapeyron equation.)

(c) Calculate the molar mass of the fluorination product.

$$n = \frac{PV}{RT} = \frac{722 \text{ mm Hg}\left(\dfrac{1 \text{ atm}}{760 \text{ mm Hg}}\right)(0.264 \text{ L})}{\left(0.08206 \dfrac{\text{L}\cdot\text{atm}}{\text{K}\cdot\text{mol}}\right)295 \text{ K}}$$

$$n = 0.0104 \text{ mol}$$

$$\text{Molar mass} = \frac{1.53 \text{ g}}{0.0104 \text{ mol}} = 148 \text{ g/mol}$$

The formula of the product must be C_3HF_5O. (The 3 C atoms and 1 O atom account for a mass of 52 g/mol of the 148 g/mol. The difference, 96, must be due to F and H atoms. This is very close to 5 F atoms.)

31. The approximate phase diagram for butane is illustrated here. Note that it is not to scale. Also note that the triple point is at a very low pressure.

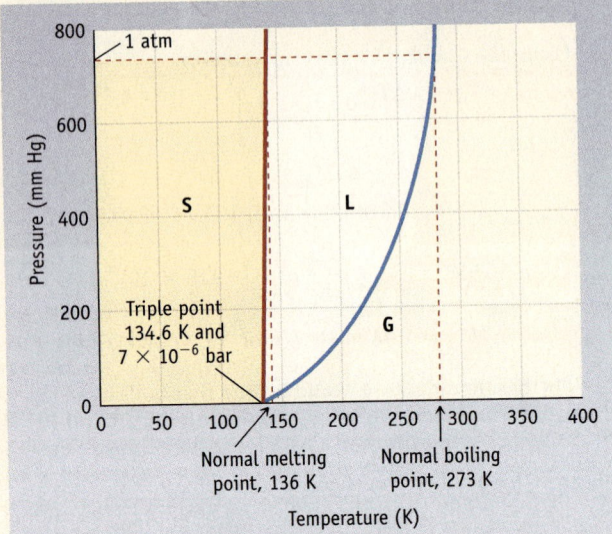

- Y-axis: Pressure (mm Hg), values 0, 200, 400, 600, 800
- 1 atm (dashed line near 800)
- S (solid region), L (liquid region), G (gas region)
- Triple point 134.6 K and 7×10^{-6} bar
- Normal melting point, 136 K
- Normal boiling point, 273 K
- X-axis: Temperature (K), values 0, 50, 100, 150, 200, 250, 300, 350, 400

(a) At 0 °C and 400 mm Hg the butane is a gas.

(b) An increase in pressure will result in a higher melting point for butane.

(c) Between the melting point (136 K) and the boiling point (273 K).

32. Fluoride ions occupy each of 12 edges and contribute $\frac{1}{4}$ of each ion to the unit cell for a total of three F^- ions. Mo ions are located at each corner and contribute a net of one Mo ion to the unit cell. Therefore, the formula is MoF_3 and its name is molybdenum(III) fluoride.

33. Sodium chloride unit cell.

(a) Mass of one unit cell (which contains 4 NaCl units).

$$\left(\frac{58.44 \text{ g NaCl}}{1 \text{ mol NaCl}}\right)\left(\frac{1 \text{ mol NaCl}}{6.022 \times 10^{23} \text{ NaCl units}}\right)\left(\frac{4 \text{ NaCl units}}{1 \text{ unit cell}}\right)$$

$$= 3.882 \times 10^{-22} \text{ g NaCl/unit cell}$$

(b) Volume of one unit cell

Length of unit cell edge =

$$562.8 \text{ pm}\left(\frac{1 \text{ m}}{1 \times 10^{12} \text{ pm}}\right)\left(\frac{100 \text{ cm}}{1 \text{ m}}\right)$$

$$= 5.628 \times 10^{-8} \text{ cm}$$

Volume = $(5.628 \times 10^{-8} \text{ cm})^3 = 1.783 \times 10^{-22} \text{ cm}^3$

(c) Density of NaCl

$$\text{Density} = \frac{3.882 \times 10^{-22} \text{ g NaCl}}{1.783 \times 10^{-22} \text{ cm}^3} = 2.178 \text{ g/cm}^3$$

This calculated value agrees with the literature value.

34. (a) There are 4 Ag^+ ions (as a face-centered cube) in the unit cell. There are 4 I^- ions in tetrahedral holes.

(b) First find the volume of the unit cell.

Length of unit cell edge =

$$(649.6 \text{ pm})\left(\frac{1 \text{ m}}{1 \times 10^{12} \text{ pm}}\right)\left(\frac{100 \text{ cm}}{1 \text{ m}}\right)$$

$$= 6.496 \times 10^{-8} \text{ cm}$$

Volume = $(6.496 \times 10^{-8} \text{ cm})^3 = 2.741 \times 10^{-22} \text{ cm}^3$

Use this volume with the density to derive the mass of the unit cell.

$$\left(\frac{2.741 \times 10^{-22} \text{ cm}^3}{1 \text{ unit cell}}\right)\left(\frac{5.68 \text{ g}}{1 \text{ cm}^3}\right) = \frac{1.557 \times 10^{-21} \text{ g}}{\text{unit cell}}$$

(c) There are 4 AgI units per unit cell, so the mass of 1 AgI is $\frac{1}{4}$ of 1.557×10^{-21} g or 3.893×10^{-22} g. Multiplying by Avogadro's number gives 234.4 g/mol, very close to the value calculated from atomic weights (234.8 g/mol).

35. (a) The key assumption here is that the atoms in a face-centered cube just touch along the diagonal dimension in each face.

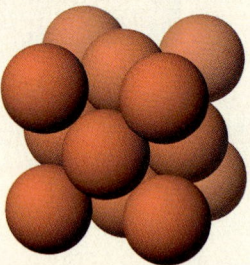

Atoms of a face-centered unit cell just touch along the face diagonal.

Let us assume the cell has an edge dimension of e, so the volume of the unit cell is e^3.

First solve for the volume of an individual atom.

Length of face diagonal = $\sqrt{2}\,e$

Diagonal = 4 × radius of an atom

$$\text{Radius of an atom} = \frac{\sqrt{2}}{4}e$$

$$\text{Volume of an atom} = \frac{4}{3}\pi\left(\frac{\sqrt{2}}{4}e\right)^3$$

$$= 0.1852\,e^3$$

There are four atoms in each unit cell, so the total volume of the atoms is $4 \times 0.1852 \, e^3 = 0.7408 \, e^3$. This is equivalent to 74% of the volume of the unit cell.

36. (a) Pentane has the higher vapor pressure because it has the smaller molar mass.

(b) Approximate normal boiling points: pentane, 36 °C, and hexane, 69 °C.

(c) At $T_1 = 0$ °C or 273 K, $P_1 = 48.3$ mm Hg
At $T_2 = 50$ °C or 323 K, $P_2 = 399.20$ mm Hg

$$\ln \frac{P_2}{P_1} = \frac{\Delta H^\circ_{vap}}{R}\left[\frac{1}{T_1} - \frac{1}{T_2}\right]$$

$$\ln \frac{399.2}{48.3} = \frac{\Delta H^\circ_{vap}}{8.314 \, \text{J/K} \cdot \text{mol}}\left[\frac{1}{273} - \frac{1}{323}\right]\frac{1}{\text{K}}$$

$$\ln 8.265 = \frac{\Delta H^\circ_{vap}}{8.314 \, \text{J/K} \cdot \text{mol}}[5.67 \times 10^{-4}]1/\text{K}$$

$$\Delta H^\circ_{vap} = 3.10 \times 10^4 \, \text{J/mol} \quad \text{or} \quad 31.0 \, \text{kJ/mol}$$

37. (a) The quartz structure consists of tetrahedral sp^3 hybridized Si atoms bridged by O atoms. Also notice that it forms six-member rings. Quartz is a covalent, network solid.

(b) The significant polarity of water molecules allows them to interact with the cations and anions of salts and bring the ions into solution. (See pages 592–593.) Water molecules can also induce a dipole in nonpolar molecules such as O_2 and N_2, allowing them to dissolve to a small extent in water. (See pages 596–597.)

(c) For such a small molecule, water has unusual properties (such as its high boiling point, low vapor pressure, and high specific heat capacity). Also, it is one of only two or three substances known to have a solid phase of lower density than the liquid phase. Much of this can be explained by the hydrogen bonding capability of the molecule. (See Section 13.3.)

(d) The quartz structure on page 642 shows the internal structure of the crystal. However, on the surface of the solid there are O atoms attached to only one Si atom, and these atoms bear a negative charge. (This is similar to the O atoms in anions such as SO_4^{2-} and PO_4^{3-}.) Polar water molecules interact with the negative O atoms on the surface of sand particles and wet the sand. However, the interaction is not strong enough to break down Si—O covalent bonds, and so the quartz sand does not dissolve in water.

14— Solutions and Their Behavior

Lake Nyos in Cameroon (western Africa), the site of a natural disaster. In 1986, a huge bubble of CO_2 escaped from the lake and asphyxiated more than 1700 people.

The Killer Lakes of Cameroon

It was evening on Thursday, August 21, 1986. Suddenly people and animals around Lake Nyos in Cameroon, a small nation on the west coast of Africa, collapsed and died. By the next morning, 1700 people and hundreds of animals were dead. The calamity had no apparent cause—no fire, no earthquake, no storm. What had brought on this disaster?

Some weeks later, the mystery was solved. Lake Nyos and nearby Lake Monoun are crater lakes, which formed when cooled volcanic craters filled with water. Lake Nyos was lethal because it contains an enormous amount of dissolved carbon dioxide. The CO_2 in the lake was generated as a result of volcanic activity deep in the earth. Under the high pressures found at the bottom of the lake, a very large amount of CO_2 dissolved in the water.

On that fateful evening in 1986, something happened to disturb the lake. The CO_2-saturated water at the bottom of the lake was carried to the surface where, under lower pressure, the gas was much less soluble. Approximately one cubic kilometer of carbon dioxide was released into the atmosphere, much like the explosive release of CO_2 from a can of carbonated beverage that has been shaken. The CO_2 shot up about 260 feet; then, because this gas is more dense than air, it hugged the ground and began to move with the prevailing breeze along the ground at about 45 miles per hour. When it reached the

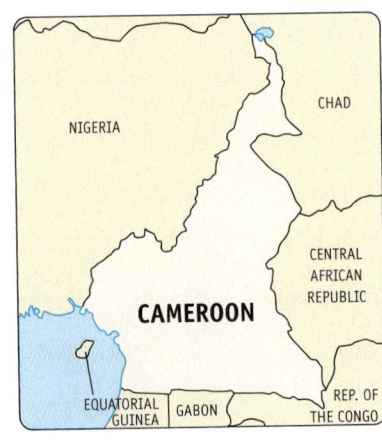

See Chapter Goals Revisited (page 690). Test your knowledge of these goals by taking the exam-prep quiz on the General ChemistryNow CD-ROM or website.

Chapter Goals

- Learn additional methods of expressing solution concentration.
- Understand the solution process.
- Understand and use the colligative properties of solutions.
- Describe colloids and their applications.

Charles D. Winters

A soft drink is saturated with CO₂ gas. If the equilibrium is disturbed, the gas erupts from the solution.

villages 12 miles away, vital oxygen was displaced. The result—both people and animals were asphyxiated.

In most lakes this situation would not occur because lake water "turns over" as the seasons change. In the autumn the top layer of water in a lake cools, its density increases, and the water sinks. This process continues, with warmer water coming to the surface and cooler water sinking. Dissolved CO₂ at the bottom of a lake would normally be expelled in this turnover process, but geologists found that the lakes in Cameroon are different. The chemocline, the boundary between deep water, rich in gas and minerals, and the upper layer, full of fresh water, stays intact. As carbon dioxide continues to enter the lake through vents in the bottom of the lake, the water becomes saturated with this gas. It is presumed that a minor disturbance—perhaps a small earthquake, a strong wind, or an underwater landslide—caused the lake water to turn over and led to the explosive and deadly release of CO₂.

Lake Nyos remains potentially deadly. Geologists estimate that the lake contains 10.6 to 14.1 billion cubic feet (300–400 million cubic meters) of carbon dioxide. This is about 16,000 times the amount found in an average lake that size.

A team of geologists from France and the United States has been working to resolve this potential threat. In early 2001 scientists lowered a pipe, about 200 meters long, into the lake. Now the pressure of escaping carbon dioxide causes a jet of water to rise as high as 165 feet in the air. Over the course of a year, about 20 million cubic meters of gas will be released. While this has been a successful first step, more gas must be removed to make the lake entirely safe, so additional vents are planned.

Courtesy of George Kling

Venting CO₂ gas. A pipe, extending into the depths of Lake Nyos, lets gas-rich water spout into the air.

To Review Before You Begin

- Review solution concentrations (Section 5.8)
- Review thermochemistry and enthalpy of reaction (Chapter 6)
- Review intermolecular forces (Section 13.2)
- Review properties of liquids (Section 13.5)

GENERAL
Chemistry◆Now™

Throughout the chapter this icon introduces a list of resources on the General ChemistryNow CD-ROM or website (http://now .brookscole.com/kotz6e) that will:

- help you evaluate your knowledge of the material
- provide homework problems
- allow you to take an exam-prep quiz
- provide a personalized Learning Plan targeting resources that address areas you should study

We come into contact with solutions every day: aqueous solutions of ionic salts, gasoline with additives to improve its properties, and household cleaners such as ammonia in water. We purposely make solutions. Adding sugar, flavoring, and sometimes CO_2 to water produces a palatable soft drink. Athletes drink commercial beverages with dissolved salts to match salt concentrations in body fluids precisely, thus allowing the fluid to be taken into the body more rapidly. In medicine, saline solutions (aqueous solutions containing NaCl and other soluble salts) are infused into the body to replace lost fluids.

A **solution** is a homogeneous mixture of two or more substances in a single phase. By convention, the component present in largest amount is considered as the **solvent** and the other component as the **solute** (Figure 14.1). When you think of solutions, those that occur to you first probably involve a liquid as solvent. Some solutions, however, do not involve a liquid solvent at all; examples include the air you breathe (a solution of nitrogen, oxygen, carbon dioxide, water vapor, and other gases) and solid solutions such as 18-carat gold, brass, bronze, and pewter. Although many types of solutions exist, the objective in this chapter is to develop an understanding of gases, liquids, and solids dissolved in liquid solvents.

Experience tells you that adding a solute to a pure liquid will change the properties of the liquid. Indeed, that is the reason some solutions are made. For instance, adding antifreeze to your car's radiator prevents the coolant from boiling in the summer and freezing in the winter. The changes that occur in the freezing and boiling points when a substance is dissolved in a pure liquid are two properties that we will examine in detail. These properties, as well as the osmotic pressure of a so-

Figure 14.1 Making a solution of copper(II) chloride (the solute) in water (the solvent). When ionic compounds dissolve in water, each ion is surrounded by water molecules. The number of water molecules is usually six, but fewer are possible.

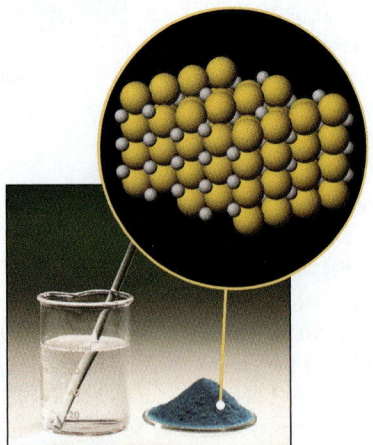

(a) Copper(II) chloride, the solute, is added to water, the solvent.

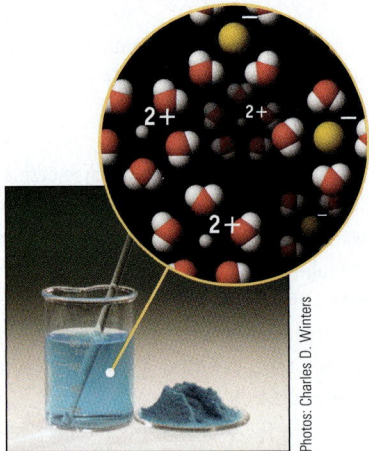

(b) Interactions between water molecules and Cu^{2+} and Cl^- ions allow the solid to dissolve. The ions are now sheathed with water molecules.

Photos: Charles D. Winters

lution and changes in vapor pressure, are examples of colligative properties. **Colligative properties** depend only on the number of solute particles per solvent molecule and not on the identity of the solute.

This chapter covers four major topics. First, because colligative properties depend on the relative number of solvent and solute particles in solution, we discuss convenient ways of describing solution concentrations in these terms. Second, we consider how and why solutions form on the molecular and ionic levels. This examination gives us some insight into the third topic, the colligative properties themselves. The chapter concludes with a brief discussion of colloids, mixtures that have properties intermediate between solutions and suspensions and that are important in many biological systems.

14.1—Units of Concentration

To analyze the colligative properties of a solution, we need ways of measuring solute concentrations that reflect the number of molecules or ions of solute per molecule of solvent. Molarity, a concentration unit useful in stoichiometry calculations, does not work when dealing with colligative properties. Recall that molarity (M) is defined as the number of moles of solute per liter of solution:

$$\text{Molar concentration of solute A} = \frac{\text{amount of A (mol)}}{\text{volume of solution (L)}}$$

Molarity does not allow us to identify the exact amount of solvent used to make the solution. This problem is illustrated in Figure 14.2. The flask on the right contains a 0.100 M aqueous solution of potassium chromate. It was made by adding enough water to 0.100 mol of K_2CrO_4 to make 1.00 L of solution. There is no way to identify the amount of solvent (water) that was actually added. If 1.00 L of water had been added to 0.100 mol of K_2CrO_4, as illustrated with the flask on the left in Figure 14.2, the volume of solution would not be 1.00 L. It is slightly greater than a liter, a result that might have been anticipated because the solute molecules or ions are expected to occupy some volume.

Four concentration units are described here that reflect the number of molecules or ions of solute per solvent molecule: molality, mole fraction, weight percent, and parts per million.

The **molality**, *m*, of a solution is defined as the amount of solute (mol) per kilogram of solvent.

$$\text{Molality of solute } (m) = \frac{\text{amount of solute (mol)}}{\text{mass of solvent (kg)}} \tag{14.1}$$

The concentration of the K_2CrO_4 solution in the flask on the left side of Figure 14.2 is 0.100 molal (0.100 *m*). It was prepared from 0.100 mol (19.4 g) of K_2CrO_4 and 1.00 kg (1.00 L × 1.00 kg/L) of water:

$$\text{Molality of } K_2CrO_4 = \frac{0.100 \text{ mol}}{1.00 \text{ kg water}} = 0.100 \ m$$

Notice that different quantities of water were used to make the 0.100 M (0.100 molar) and 0.100 *m* (0.100 molal) solutions of K_2CrO_4. This means the *molarity and the molality of a given solution cannot be the same* (although the difference may be negligibly small when the solution is quite dilute).

$V_{\text{soln}} > 1.00 \text{ L}$ $V_{\text{soln}} = 1.00 \text{ L}$
$V_{\text{H}_2\text{O}} \text{ added} = 1.00 \text{ L}$ $V_{\text{H}_2\text{O}} \text{ added} < 1.00 \text{ L}$
0.100 molal solution   0.100 molar solution

Figure 14.2 **Preparing 0.100 molal and 0.100 molar solutions.** In the flask on the right, 0.100 mol (19.4 g) of K_2CrO_4 was mixed with enough water to make 1.00 L of solution. (The volumetric flask was filled to the mark on its neck, indicating that the volume is exactly 1.00 L. Slightly less than 1.00 L of water was added.) Exactly 1.00 kg of water was added to 0.100 mol of K_2CrO_4 in the flask on the left. Notice that the volume of solution is greater than 1.00 L. (The small pile of yellow solid in front of the flask is 0.100 mol of K_2CrO_4.)

■ **Symbols, *m*, M, and *M***
The symbol for molality is a small italicized *m*. The symbol for molarity is a regular capital M. You will also see an italicized *M*, which stands for molar mass.

■ **Mole Fraction**
Mole fraction was introduced for mixtures of gases, page 565.

The **mole fraction**, X, of a solution component is defined as the amount of that component divided by the total amount of all of the components of the mixture. Mathematically it is represented as

$$\text{Mole fraction of A } (X_A) = \frac{n_A}{n_A + n_B + n_C + \cdots} \tag{14.2}$$

Consider a solution that contains 1.00 mol (46.1 g) of ethanol, C_2H_5OH, in 9.00 mol (162 g) of water. Here the mole fraction of alcohol is 0.100 and that of water is 0.900:

$$X_{ethanol} = \frac{1.00 \text{ mol ethanol}}{1.00 \text{ mol ethanol} + 9.00 \text{ mol water}} = 0.100$$

$$X_{water} = \frac{9.00 \text{ mol water}}{1.00 \text{ mol ethanol} + 9.00 \text{ mol water}} = 0.900$$

Notice that the sum of the mole fractions of the components in the solution equals 1.000, a relationship that holds true for the solute and solvent in all solutions:

$$X_{water} + X_{ethanol} = 1.000$$

Weight percent is the mass of one component divided by the total mass of the mixture, multiplied by 100%:

$$\text{Weight \% A} = \frac{\text{mass of A}}{\text{mass of A} + \text{mass of B} + \text{mass of C} + \cdots} \times 100\% \tag{14.3}$$

The alcohol–water mixture has 46.1 g of ethanol and 162 g of water, so the total mass of solution is 208 g, and the weight % of alcohol is

$$\text{Weight \% ethanol} = \frac{46.1 \text{ g ethanol}}{46.1 \text{ g ethanol} + 162 \text{ g water}} \times 100\% = 22.2\%$$

Notice that if you know the weight percent of a solute, you can determine its mole fraction or molality (or vice versa) because the masses of solute and solvent are known.

Weight percent is a common unit in consumer products (Figure 14.3). Vinegar, for example, is an aqueous solution containing approximately 5% acetic acid and 95% water. The label on a common household bleach lists its active ingredient as 6.00% sodium hypochlorite (NaOCl) and 94.00% inert ingredients.

Naturally occurring solutions are often very dilute. Environmental chemists, biologists, geologists, oceanographers, and others frequently use **parts per million (ppm)** to express their concentrations. The unit ppm refers to relative amounts by mass; 1.0 ppm represents 1.0 g of a substance in a sample with a total mass of 1.0 million g. Because water at 25 °C has a density of 1.0 g/mL, a concentration of 1.0 mg/L is equivalent to 1.0 mg of solute in 1000 g of water or to 1.0 g of solute in 1,000,000 g of water; that is, units of ppm and mg/L are approximately equivalent.

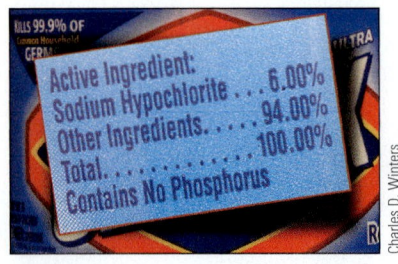

Figure 14.3   Weight percent. The composition of many common products is often given in terms of weight percent. Here the label on household bleach indicates that it contains 6.00% sodium hypochlorite.

GENERAL
Chemistry·⚛·Now™

See the General ChemistryNow CD-ROM or website:

- **Screen 14.2 Solubility,** for an exercise on calculating solution concentrations in various units

Commercial antifreeze. This solution contains ethylene glycol, $HOCH_2CH_2OH$, an organic alcohol that is readily soluble in water. Regulations specify that ethylene glycol–based antifreeze must contain a minimum of 75 weight percent of the glycol. (The remainder of the solution can be other glycols and water.)

Example 14.1—Calculating Mole Fractions, Molality, and Weight Percent

Problem Assume you add 1.2 kg of ethylene glycol, $HOCH_2CH_2OH$, as an antifreeze to 4.0 kg of water in the radiator of your car. What are the mole fraction, molality, and weight percent of the ethylene glycol?

Strategy Calculate the amount of ethylene glycol and water and then use Equations 14.1–14.3.

Solution The 1.2 kg of ethylene glycol (molar mass = 62.1 g/mol) is equivalent to 19 mol, and 4.0 kg of water represents 220 mol.

Mole fraction:

$$X_{glycol} = \frac{19 \text{ mol glycol}}{19 \text{ mol glycol} + 220 \text{ mol water}} = \boxed{0.080}$$

Molality:

$$\text{Molality} = \frac{19 \text{ mol glycol}}{4.0 \text{ kg}} = \boxed{4.8 \text{ m}}$$

Weight percent:

$$\text{Weight \%} = \frac{1.2 \times 10^3 \text{ g glycol}}{1.2 \times 10^3 \text{ g glycol} + 4.0 \times 10^3 \text{ g water}} \times 100\% = \boxed{23\%}$$

Example 14.2—Parts per Million

Problem You dissolve 560 g of $NaHSO_4$ in a swimming pool that contains 4.5×10^5 L of water at 25 °C. What is the sodium ion concentration in parts per million? [Sodium hydrogen sulfate is used to adjust the pH of the pool water because the anion, HSO_4^-, can furnish H^+(aq) to the solution.]

Strategy First calculate the quantity of sodium ions (in grams) in 560 g of $NaHSO_4$. Then use this mass of sodium and the volume to calculate milligrams per liter, which is equivalent to parts per million.

Solution

Mass of Na^+:

$$560 \text{ g NaHSO}_4 \left(\frac{1 \text{ mol NaHSO}_4}{120 \text{ g NaHSO}_4}\right)\left(\frac{1 \text{ mol Na}^+}{1 \text{ mol NaHSO}_4}\right)\left(\frac{23.0 \text{ g}}{1 \text{ mol Na}^+}\right) = 110 \text{ g Na}^+$$

Concentration of Na^+:

$$\frac{110 \text{ g } (1000 \text{ mg/g})}{4.5 \times 10^5 \text{ L}} = \boxed{0.24 \text{ mg/L or 0.24 ppm}}$$

Exercise 14.1—Mole Fraction, Molality, and Weight Percent

If you dissolve 10.0 g (about one heaping teaspoonful) of sugar (sucrose, $C_{12}H_{22}O_{11}$), in a cup of water (250. g), what are the mole fraction, molality, and weight percent of sugar?

Exercise 14.2—Parts per Million

Sea water has a sodium ion concentration of 1.08×10^4 ppm. If the sodium is present in the form of dissolved sodium chloride, what mass of NaCl is in each liter of sea water? Sea water is denser than pure water because of dissolved salts. Its density is 1.05 g/mL.

14.2—The Solution Process

If solid $CuCl_2$ is added to a beaker of water, the salt will begin to dissolve (see Figure 14.1). The amount of solid diminishes, and the concentration of Cu^{2+}(aq) and Cl^-(aq) in the solution increases. If we continue to add $CuCl_2$, however, we will eventually reach a point when no additional $CuCl_2$ seems to dissolve. The concentrations of Cu^{2+}(aq) and Cl^-(aq) will not increase further, and any additional solid $CuCl_2$ added after this point will simply remain as a solid at the bottom of the beaker. We say that such a solution is **saturated**.

Although no change is observed on the macroscopic level, it is a different matter on the particulate level. The process of dissolving continues, with Cu^{2+}(aq) and Cl^-(aq) ions leaving the solid state and entering solution. Concurrently, a second process is occurring: the formation of solid $CuCl_2$(s) from Cu^{2+}(aq) and Cl^-(aq). The rates at which $CuCl_2$ is dissolving and reprecipitating are equal in a saturated solution, so that no net change is observed on the macroscopic level.

This reaction is an another example of equilibrium in chemistry, and we can describe the situation in terms of an equation with substances linked by a set of double arrows ($\rightleftharpoons$):

$$CuCl_2(s) \overset{H_2O}{\rightleftharpoons} Cu^{2+}(aq) + 2\ Cl^-(aq)$$

Recall that we encountered equilibrium systems earlier when changes of state were introduced [◄ Section 13.5]. In equilibria involving two states of matter, the description was very similar to what we are seeing for a saturated solution: No changes are observable at the macroscopic level, but two opposing processes go on at the particulate level at the same rate.

A saturated solution gives us a way to define precisely the solubility of a solid in a liquid. **Solubility** is the concentration of solute in equilibrium with undissolved solute in a saturated solution. The solubility of $CuCl_2$, for example, is 70.6 g in 100 mL of water at 0 °C. If we added 100.0 g of $CuCl_2$ to 100 mL of water at 0 °C, we can expect 70.6 g to dissolve, and 29.4 g of solid to remain.

Liquids Dissolving in Liquids

If two liquids mix to an appreciable extent to form a solution, they are said to be **miscible**. In contrast, **immiscible** liquids do not mix to form a solution; they exist in contact with each other as separate layers [◄ Figure 13.5].

The polar compound ethanol (C_2H_5OH) dissolves in water, which is also a polar compound. In fact, ethanol and water are miscible in all proportions. The nonpolar liquids octane (C_8H_{18}) and carbon tetrachloride (CCl_4) are also miscible in all proportions. On the other hand, neither C_8H_{18} nor CCl_4 is miscible with water. Ob-

■ **Unsaturated**
The term *unsaturated* is used when referring to solutions with concentrations of solute that are less than that of a saturated solution.

■ **Solubility Data**
Quantitative data on solubility for many compounds are listed in chemical handbooks. It is from these data that solubility rules such as those given in Figure 5.3 were created.

A Closer Look

Supersaturated Solutions

Although at first glance it may seem a contradiction, it is possible for a solution to hold *more* dissolved solute than the amount in a saturated solution. Such solutions are referred to as *supersaturated* solutions. Supersaturated solutions are unstable, and the excess solid eventually crystallizes from the solution until the equilibrium concentration of the solute is reached.

The solubility of substances often decreases if the temperature is lowered. Supersaturated solutions are usually made by preparing a saturated solution at a given temperature and then carefully cooling it. If the rate of crystallization is slow, the solid may not precipitate when the solubility is exceeded. Going to still lower temperatures results in a solution that has more solute than the amount defined by equilibrium conditions; it is supersaturated.

When disturbed in some manner, a supersaturated solution moves toward equilibrium by precipitating solute. This change can occur rapidly and often with

Supersaturated solutions. When a supersaturated solution is disturbed, the dissolved salt (here sodium acetate, $NaCH_3CO_2$) rapidly crystallizes. (*See General ChemistryNow Screen 14.2 Solubility, to watch a video of this process.*)

the evolution of heat. In fact, supersaturated solutions are used in "heat packs" to apply heat to injured muscles. When crystallization of sodium acetate ($NaCH_3CO_2$)

from a supersaturated solution in a heat pack is initiated, the temperature of the heat pack rises to close to 50 °C, and crystals of solid sodium acetate are detectable inside the bag.

Heat of crystallization. A heat pack relies on the heat evolved by the crystallization of sodium acetate. (*See General ChemistryNow Screen 14.6 Factors Affecting Solubility (2), to watch a video of a heat pack.*)

---

servations like these have led to a familiar rule of thumb described in Section 13.2: *Like dissolves like.* That is, two or more nonpolar liquids frequently are miscible, just as are two or more polar liquids.

What is the molecular basis for the "like dissolves like" guideline? In pure water and pure ethanol, the major force between molecules is hydrogen bonding involving O—H groups. When the two liquids are mixed, hydrogen bonding between ethanol and water molecules also occurs and assists in the solution process. Molecules of pure octane or pure CCl_4, both of which are nonpolar, are held together in the liquid phase by dispersion forces [◄ Section 13.2]. When these nonpolar liquids are mixed, the energy associated with these forces of attraction is similar in value to the energy due to the forces of attraction between octane and CCl_4 molecules. Thus, little or no energy change occurs when octane–octane and CCl_4–CCl_4 attractive forces are replaced with octane–CCl_4 forces. The solution process is expected to be nearly energy-neutral. So, why do the liquids mix? The answer lies deeper in thermodynamics. As you shall see in Chapter 19, processes that move to less orderly arrangements tend to occur. The tendency is measured by a thermodynamic function called *entropy* (Figure 14.4).

In contrast, polar and nonpolar liquids usually do not mix to an appreciable degree; when placed together in a container, they separate into two distinct layers (Figure 14.5). The rationale for this behavior is complicated. Experimental data show that the enthalpy of mixing of dissimilar liquids is also near zero, so the energetics of the process is not the primary factor. Apparently interposing nonpolar molecules

■ **Entropy and the Solution Process**
Although the energetics of solution formation are important, it is generally accepted that the more important contributor is the entropy of mixing. As you shall see in Chapter 19, entropy favors the formation of less ordered systems such as solutions. See also T. P. Silverstein: "The real reason why oil and water don't mix." *Journal of Chemical Education*, Vol. 75, pp. 116–118, 1998.

■ **Alcohol–Water Solutions**
Beer, wine, and other alcoholic beverages contain amounts of alcohol ranging from just a few percent to more than 50%. Ethanol commonly used in laboratories has a concentration of 95% ethanol and 5% water (and other ingredients).

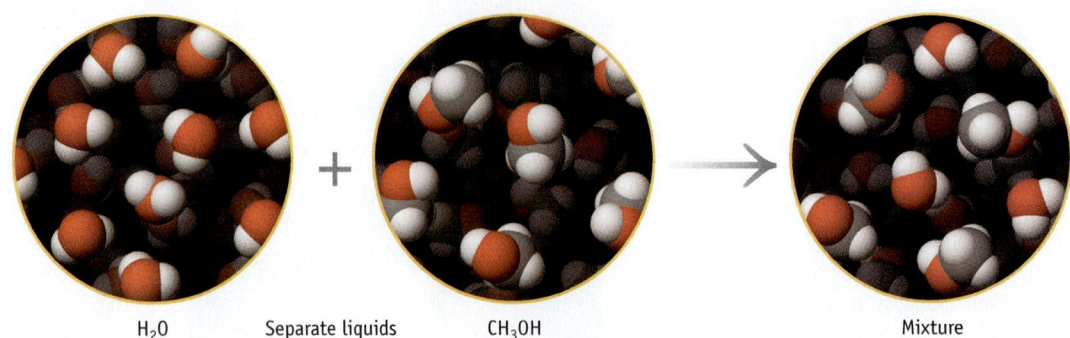

H₂O   Separate liquids   CH₃OH   Mixture

Figure 14.4   Driving the solution process—entropy. When two similar liquids—here water and methanol—are mixed, the molecules are intermingled. The mixture has a less ordered arrangement of molecules than the separate liquids. This disordering process largely drives solution formation. For more on this thermodynamic quantity, see Chapter 19.

into a polar solvent such as water causes the internal structure of water at the molecular level to become more ordered. Forming a more ordered system disfavors the process of mixing.

Solids Dissolving in Water

The "like dissolves like" guideline also holds for molecular solids dissolving in liquids. Nonpolar organic solids such as naphthalene, $C_{10}H_8$, dissolve readily in solvents such as benzene, C_6H_6, and hexane, C_6H_{14}. Iodine, I_2, a nonpolar inorganic solid, dissolves in water to some extent but given a choice, prefers to be in solution in nonpolar liquid such as CCl_4 (Figure 14.6). Sucrose (sugar), a polar molecular solid, is not very soluble in nonpolar solvents but it is readily soluble in water, a fact that we know well because of its use to sweeten beverages. The presence of O—H groups in the structure of sugar and other substances such as glucose allows these molecules to interact with polar water molecules through strong hydrogen bonding.

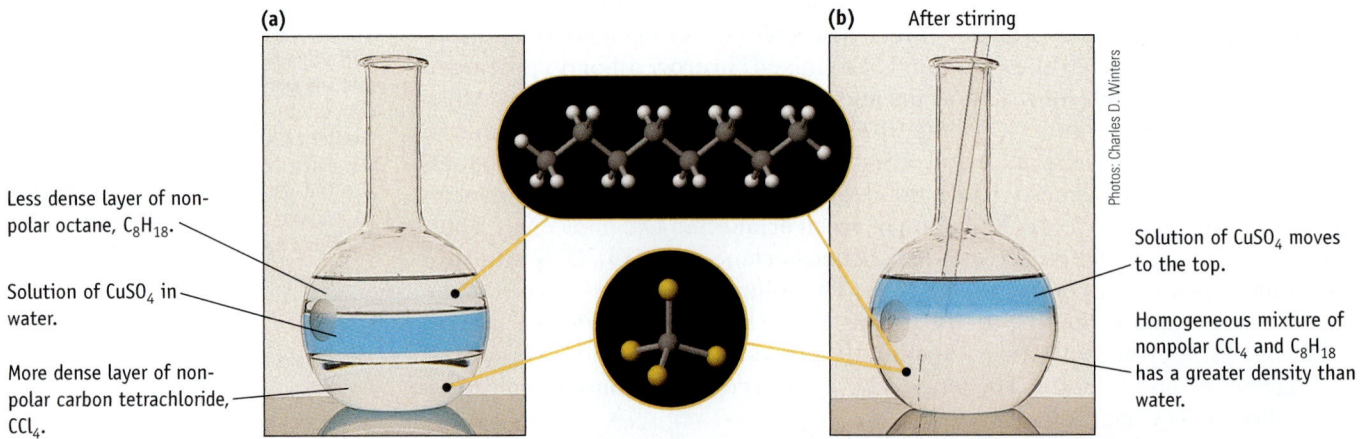

(a)

Less dense layer of non-polar octane, C_8H_{18}.

Solution of $CuSO_4$ in water.

More dense layer of non-polar carbon tetrachloride, CCl_4.

(b)   After stirring

Photos: Charles D. Winters

Solution of $CuSO_4$ moves to the top.

Homogeneous mixture of nonpolar CCl_4 and C_8H_{18} has a greater density than water.

Figure 14.5   Miscibility. (a) The colorless, denser bottom layer is nonpolar carbon tetrachloride, CCl_4. The blue middle layer is a solution of $CuSO_4$ in water, and the colorless, less dense top layer is nonpolar octane, C_8H_{18}. This mixture was prepared by carefully layering one liquid on top of another, without mixing. (b) After stirring the mixture, the two nonpolar liquids form a homogeneous mixture. This layer of mixed liquids is under the water layer because the mixture of CCl_4 and C_8H_{18} has a greater density than water.

Active Figure 14.6 **Solubility of nonpolar iodine in polar water and nonpolar carbon tetrachloride.** When a solution of nonpolar I_2 in water (the brown layer on top in the left test tube) is shaken with non-polar CCl_4 (the clear bottom layer in the left test tube), the I_2 transfers preferentially to the nonpolar solvent. Evidence for this is the purple color of the bottom CCl_4 layer in the test tube on the right.

GENERAL
Chemistry ☆ Now™ See the General ChemistryNow CD-ROM or website to explore an interactive version of this figure accompanied by an exercise.

"Like dissolves like" is a somewhat less effective but still useful guideline when considering ionic solids. Thus, we can reasonably predict that ionic compounds, which can be considered extreme examples of polar compounds, will not dissolve in nonpolar solvents. This fact is amply borne out by observation. Sodium chloride, for example, will not dissolve in liquids such as hexane or CCl_4 but this common ionic compound does have a significant solubility in water. Many other ionic compounds such as $CuCl_2$ (see Figure 14.1) are soluble in water, but also many other ionic solids are not. Recall the solubility rules (page 179) from which many insoluble ionic compounds can be identified.

Predicting the solubility of ionic compounds in water is a complicated business. As mentioned earlier, two factors—enthalpy and entropy—together determine the extent to which one substance dissolves in another. For ionic compounds dissolving in water, entropy usually (but not always) favors solution. A favorable enthalpy factor generally leads to a compound being soluble, but the reverse is not true. We can see this clearly with an example. Both ammonium nitrate and sodium hydroxide dissolve readily in water, but the solution becomes colder when NH_4NO_3 dissolves, and it warms up when NaOH dissolves (Figure 14.7). Heat is evolved if NaOH dissolves,

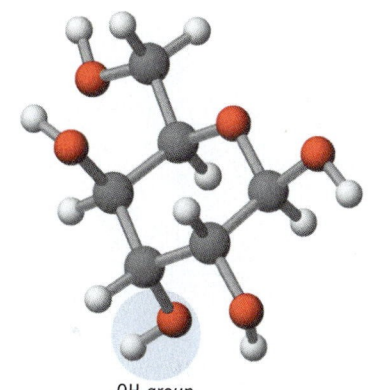

OH group

Like dissolves like. Glucose has five —OH groups on each molecule, groups that allow it to form hydrogen bonds with water molecules. As a result, glucose dissolves readily in water.

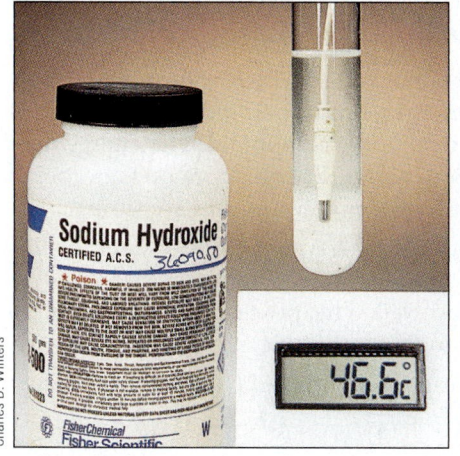

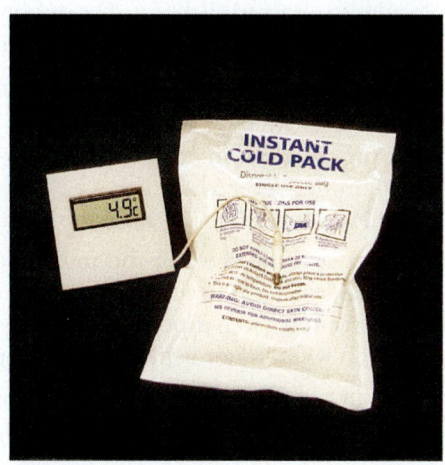

Figure 14.7 **Dissolving ionic solids and heat of solution.** (a) Dissolving NaOH in water is a strongly exothermic process. (b) A "cold pack" contains solid ammonium nitrate, NH_4NO_3, and a package of water. When the water and NH_4NO_3 are mixed, and the salt dissolves, the temperature of the system drops owing to the endothermic heat of the solution of ammonium nitrate ($\Delta H°_{soln} = +25.7$ kJ/mol).

(a)   (b)

Figure 14.8 **Model for energy changes on dissolving KF.** An estimate of the magnitude of the energy change on dissolving an ionic compound in water is achieved by imagining it as occurring in two steps at the particulate level. Here KF is first separated into cations and anions in the gas phase with an expenditure of 821 kJ per mol of KF. These ions are then hydrated, with $\Delta H_{hydration}$ estimated to be -837 kJ. Thus, the net energy change is -16 kJ, a slightly exothermic heat of solution. (*See the General ChemistryNow Screen 14.4 Energetics of Solution Formation, to view an animation of this diagram.*)

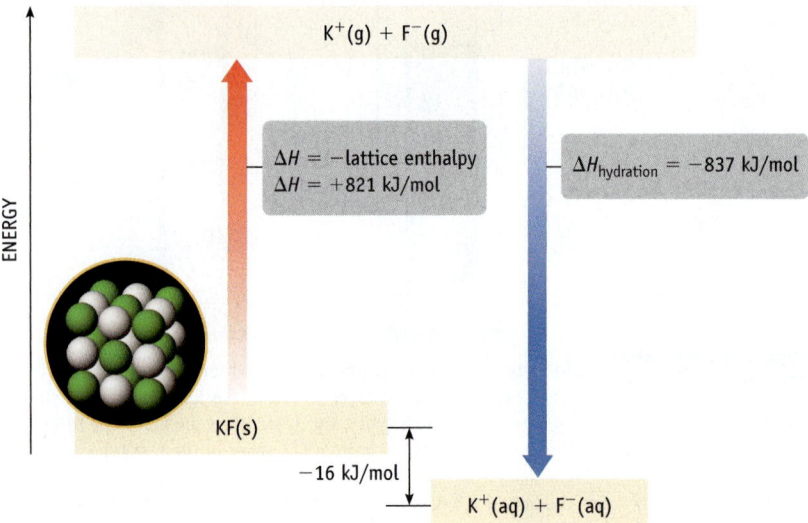

but addition of heat is required for the solution of NH_4NO_3 to form. Nonetheless, both compounds are soluble.

Network solids, including graphite, diamond, and quartz sand (SiO_2), do not dissolve in water. Indeed, where would all the beaches be if sand dissolved in water? What if the diamond in a ring dissolved when you washed your hands? The normal covalent chemical bonding in network solids is simply too strong to be broken; the lattice remains intact when in contact with water.

Heat of Solution

To understand the energetics of the solution process, let us view this process at the molecular level. We will use the process of dissolving potassium fluoride, KF, in water to illustrate what occurs, and the energy-level diagram in Figure 14.8 will assist us in following the changes.

Solid potassium fluoride has an ionic lattice structure like that of NaCl (Figure 13.24). That is, solid KF has alternating K^+ and F^- ions held in place by the attractive forces due to their opposite charges. As described in Chapter 13 (page 592) and in Figure 14.1, in water these ions are separated from each other and hydrated; that is, they are surrounded by water molecules. Ion–dipole forces of attraction bind water molecules strongly to each ion. The energy change to go from the reactant, KF(s), to the products, K^+(aq) and F^-(aq), thus can be considered to take place in two stages:

1. Energy must be supplied to separate the ions in the lattice against their attractive forces. We have encountered this energy quantity before; it is the reverse of the process defining the lattice energy of an ionic compound. The energy here is identified as $-\Delta H_{lattice}$ (page 379). Separating the ions from one another is highly endothermic because the attractive forces between ions are strong.

2. Energy is evolved when the individual ions are transferred into water. In this process, each ion becomes surrounded by water molecules. Again, strong forces of attraction (ion–dipole forces) are involved. This process, referred to as hydration when water is the solvent, is strongly exothermic.

We can represent the process of dissolving KF in terms of chemical equations:

Step 1	$KF(s) \longrightarrow K^+(g) + F^-(g)$	$-\Delta H_{lattice}$
Step 2	$K^+(g) + F^-(g) \longrightarrow K^+(aq) + F^-(aq)$	$\Delta H_{hydration}$

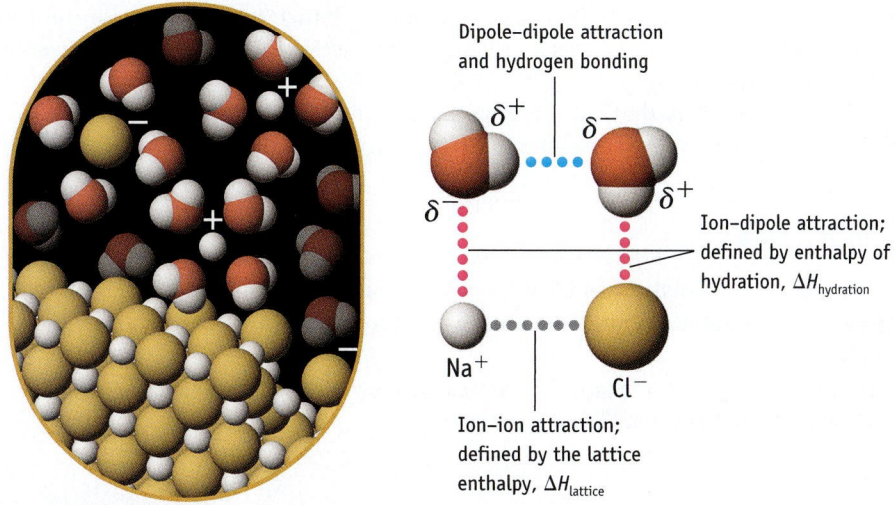

Dipole–dipole attraction and hydrogen bonding

δ^+ δ^-

δ^- δ^+

Ion–dipole attraction; defined by enthalpy of hydration, $\Delta H_{hydration}$

Na$^+$ Cl$^-$

Ion–ion attraction; defined by the lattice enthalpy, $\Delta H_{lattice}$

Active Figure 14.9   **Dissolving an ionic solid in water.** This process is a balance of forces. There are intermolecular forces between water molecules, and ion–ion forces are at work in the ionic crystal lattice. To dissolve, the ion–dipole forces between water and the ions (as measured by $\Delta H_{hydration}$) must overcome the ion-ion forces (as measured by $\Delta H_{lattice}$) and the intermolecular forces in water.

GENERAL
Chemistry ❖ Now™ See the General ChemistryNow CD-ROM or website to explore an interactive version of this figure accompanied by an exercise.

The overall reaction is the sum of these two steps. The energy of the reaction, called the **heat of solution**, is the sum of the two energy quantities.

$$\text{Overall} \qquad KF(s) \longrightarrow K^+(aq) + F^-(aq) \qquad \Delta H_{soln} = -\Delta H_{lattice} + \Delta H_{hydration}$$

The lattice energy for an ionic compound is estimated with reasonable accuracy from thermochemical data using a Born-Haber cycle calculation (page 379). The lattice energy for KF is -821 kJ/mol, so to separate one mole of $K^+(g)$ from one mole of $F^-(g)$ requires 821 kJ. The heat of solution, ΔH_{soln}, can be measured using a calorimeter. For KF, the heat of solution is -16.4 kJ. From these two values, we can determine the hydration energy, $\Delta H_{hydration}$, which cannot be measured directly, to be -837 kJ/mol.

Several aspects of this analysis are of interest. We see that the energy required to break down the KF crystal lattice is returned by the energy of attraction between the ions and the polar water molecules. Indeed, this will necessarily be the case if an ionic compound is to be soluble in water.

As a general rule, to be soluble, salts dissolving are exothermic or only slightly endothermic (Figure 14.9). In the latter instance, it is assumed that the disfavoring of the solution process due to its endothermic character will be balanced by a favorable entropy of solution. If the solution process is very endothermic—because of a low solvation energy, for example—then the compound is unlikely to be soluble. We can reasonably speculate that nonpolar solvents would not solvate ions strongly, and that the process of solution would thus be energetically unfavorable.

It is also useful to recognize that the heat of solution is the difference between two very large numbers. Small variations in either lattice energy or hydration energy can determine whether a salt dissolves endothermically or exothermically.

Finally, notice that the two energy quantities, $\Delta H_{lattice}$ and $\Delta H_{hydration}$, are both affected by ion sizes and ion charges (pages 379 and 592). A salt composed of smaller ions is expected to have a higher (more negative) lattice energy because the ions can be closer together and experience higher attractive forces. However, the small size will also allow a closer approach of solvent molecules and a higher solvation energy. The net result is that simple correlations of solubility with structure (ionic radii) or thermodynamic parameters ($\Delta H_{\text{lattice}}$) are generally not successfully made.

GENERAL
Chemistry·⚛·Now™

See the General ChemistryNow CD-ROM or website:
- **Screen 14.3 The Solution Process: Intermolecular Forces,** for a visualization of the process and for a problem
- **Screen 14.4 Energetics of Solution Formation—Dissolving Ionic Compounds,** for an analysis of the dissolution of KF and an exercise

---

■ **Heat of Solution and Preparing Solutions of H₂SO₄**
The heat generated when H_2SO_4 is dissolved in water is so large that the water will boil, if care is not taken. Students are advised to add H_2SO_4 in small increments to water when preparing a dilute solution of this acid; the opposite addition (water to acid) may produce enough heat to result in acid spattering onto skin, clothing, and lab equipment.

Heat of Solution: Thermodynamic Data

As mentioned earlier, the enthalpy of solution for a salt can be measured using a calorimeter. This is usually done in an open system such as the coffee-cup calorimeter described in Section 6.6. For an experiment run under standard conditions, the resulting measurement produces a value for the standard enthalpy of solution, $\Delta H^\circ_{\text{soln}}$, where standard conditions refer to a concentration of 1 molal.

Tables of thermodynamic values often include values for the heats of formation of aqueous solutions of salts. For example, a value of ΔH°_f for NaCl(aq) of -407.3 kJ/mol is listed in Appendix L. This value refers to the formation of a 1 m solution of NaCl from the elements. It may be considered to involve the enthalpies of two steps: (1) the formation of NaCl(s) from the elements Na(s) and Cl_2(g) in their standard states, and (2) the formation of a 1 m solution by dissolving solid NaCl in water:

Formation of NaCl(s):	$Na(s) + \frac{1}{2}Cl_2(g) \longrightarrow NaCl(s)$	$\Delta H^\circ_f = -411.1$ kJ/mol
Dissolving NaCl:	$NaCl(s) \longrightarrow NaCl(aq, 1\ m)$	$\Delta H^\circ_{\text{soln}} = +3.9$ kJ/mol
Net process:	$Na(s) + \frac{1}{2}Cl_2(g) \longrightarrow NaCl(aq, 1\ m)$	$\Delta H^\circ_f = -407.3$ kJ/mol

Heats of solution, $\Delta H^\circ_{\text{soln}}$ (Table 14.1), can be calculated using ΔH°_f data. The solution process for NaCl(s), for example, is represented by the equation

$$NaCl(s) \longrightarrow NaCl(aq, 1\ m)$$

The energy of this process is calculated using Equation 6.6:

$$\begin{aligned}
\Delta H^\circ_{\text{soln}} &= \sum \left[\Delta H^\circ_f(\text{products})\right] - \sum \left[\Delta H^\circ_f(\text{reactants})\right] \\
&= \Delta H^\circ_f[\text{NaCl(aq)}] - \Delta H^\circ_f[\text{NaCl(s)}] \\
&= -407.3 \text{ kJ/mol} - (-411.1 \text{ kJ/mol}) = +3.8 \text{ kJ/mol}
\end{aligned}$$

Table 14.1   Data for Calculating Enthalpy of Solution

Compound	$\Delta H^\circ_f(s)$ (kJ/mol)	ΔH°_f(aq, 1 m) (kJ/mol)
LiF	−616.9	−611.1
NaF	−573.6	−572.8
KF	−568.6	−585.0
RbF	−557.7	−583.8
LiCl	−408.7	−445.6
NaCl	−411.1	−407.3
KCl	−436.7	−419.5
RbCl	−435.4	−418.3
NaOH	−425.9	−469.2
NH₄NO₃	−365.6	−339.9

Example 14.3—Calculating an Enthalpy of Solution

Problem Use the data given in Table 14.1 to determine the heat of solution for NH_4NO_3, the compound used in cold packs (see page 665).

Strategy Use Equation 6.6 and data from Table 14.1 for reactants and products.

Solution The solution process for NH_4NO_3 is represented by the equation

$$NH_4NO_3(s) \longrightarrow NH_4NO_3(aq)$$

The energy of this process is calculated using heats of formation given in Table 14.1:

$$\Delta H°_{soln} = \sum [\Delta H°_f(\text{product})] - \sum [\Delta H°_f(\text{reactant})]$$
$$= \Delta H°_f[NH_4NO_3(aq)] - \Delta H°_f[NH_4NO_3(s)]$$
$$= -339.9 \text{ kJ/mol} - (-365.6 \text{ kJ/mol}) = +25.7 \text{ kJ/mol}$$

The process is endothermic, as indicated by the fact that $\Delta H°_{soln}$ has a positive value.

Table 14.2 Henry's Law Constants (25 °C)*

Gas	k_H (M/mm Hg)
N_2	8.42×10^{-7}
O_2	1.66×10^{-6}
CO_2	4.48×10^{-5}

*From W. Stumm and J. J. Morgan: *Aquatic Chemistry*, p. 109. New York, Wiley, 1981.

Exercise 14.3—Calculating an Enthalpy of Solution

Use the data in Table 14.1 to calculate the enthalpy of solution for NaOH.

14.3—Factors Affecting Solubility: Pressure and Temperature

Biochemists and physicians, among others, are interested in the solubility of gases such as CO_2 and O_2 in water or body fluids, and scientists and engineers need to know about the solubility of solids in various solvents. Pressure and temperature are two external factors that influence solubility. Both affect the solubility of gases in liquids, whereas only temperature is an important factor in the solubility of solids in liquids.

Dissolving Gases in Liquids: Henry's Law

The solubility of a gas in a liquid is directly proportional to the gas pressure. This is a statement of **Henry's law**,

$$S_g = k_H P_g \tag{14.4}$$

where S_g is the gas solubility, P_g is the partial pressure of the gaseous solute, and k_H is Henry's law constant (Table 14.2), a constant characteristic of the solute and solvent.

Carbonated soft drinks illustrate how Henry's law works. These beverages are packed under pressure in a chamber filled with carbon dioxide gas, some of which dissolves in the drink. When the can or bottle is opened, the partial pressure of CO_2 above the solution drops, which causes the solubility of CO_2 to drop, allowing gas to bubble out of the solution (Figure 14.10).

Henry's law has important consequences in SCUBA diving (Figure 14.11). When you dive, the pressure of the air you breathe must be balanced against the external pressure of the water. In deeper dives, the pressure of the gases in the SCUBA gear must be several atmospheres and, as a result, more gas dissolves in the blood. This can lead to a problem. If you ascend too rapidly, you can experience a painful and potentially lethal condition referred to as "the bends," in which nitrogen gas bubbles form in the blood as the solubility of nitrogen decreases with decreasing pressure. In an effort to prevent the bends, divers may use a helium–oxygen mixture (rather than nitrogen–oxygen) because helium is not as soluble in aqueous media as nitrogen is.

We can better understand the effect of pressure on solubility by examining the system at the particulate level. The solubility of a gas is defined as the concentration of the dissolved gas in equilibrium with the substance in the gaseous state. At equilibrium, the rate at which solute gas molecules escape the solution and enter the

Charles D. Winters

Figure 14.10 Gas solubility and pressure. Carbonated beverages are bottled under CO_2 pressure. When the bottle is opened, the pressure is released and bubbles of CO_2 form within the liquid and rise to the surface. After some time, an equilibrium between dissolved CO_2 and atmospheric CO_2 is reached. Because CO_2 provides some of the taste in the beverage, the beverage tastes flat when most of its dissolved CO_2 is lost.

Figure 14.11 Illustrations of Henry's law.

(a) SCUBA divers must pay attention to the solubility of gases in the blood and the fact that solubility increases with pressure.

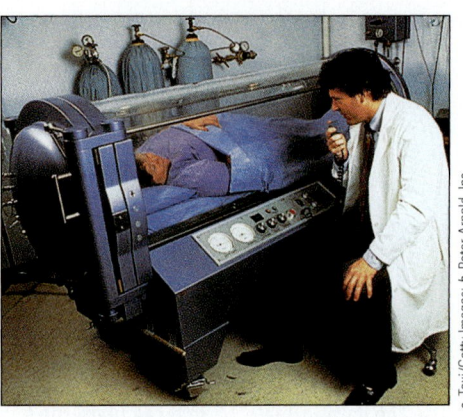

(b) A hyperbaric chamber. People who have problems breathing can be placed in a hyperbaric chamber where they are exposed to a higher partial pressure of oxygen.

a, Taxi/Getty Images; b, Peter Arnold, Inc.

gaseous state equals the rate at which gas molecules reenter the solution. An increase in pressure results in more molecules of gas striking the surface of the liquid and entering solution in a given time. The solution eventually reaches equilibrium when the concentration of gas dissolved in the solvent is high enough that the rates of gas molecules escaping and entering the solution are the same.

GENERAL
Chemistry✦Now™

See the General ChemistryNow CD-ROM or website:
- **Screen 14.5 Factors Affecting Solubility (1)—Henry's Law and Gas Pressure,** for an exercise and tutorial on Henry's law

Example 14.4—Using Henry's Law

Problem What is the concentration of O_2 in a fresh water stream in equilibrium with air at 25 °C and 1.0 atm? Express the answer in grams of O_2 per liter of solution.

Strategy To use Henry's law to calculate the molar solubility of oxygen, the partial pressure of O_2 in air must first be calculated. Recall that the mole fraction of O_2 in air is 0.21 (Table 12.1).

Solution The mole fraction of O_2 in air is 0.21, and, assuming the total pressure is 1.0 atm, the partial pressure of O_2 is 160 mm Hg ($= 0.21 \times 760$ mm Hg). Using this pressure for P_g in Henry's law we have

$$\text{Solubility of } O_2 = \left(\frac{1.66 \times 10^{-6} \text{ M}}{\text{mm Hg}} \right) (160 \text{ mm Hg}) = 2.7 \times 10^{-4} \text{ M}$$

This concentration, in grams per liter, can then be calculated using the molar mass of O_2:

$$\text{Solubility of } O_2 = \left(\frac{2.7 \times 10^{-4} \text{ mol}}{\text{L}} \right) \left(\frac{32.0 \text{ g}}{\text{mol}} \right) = \boxed{0.0085 \text{ g/L}}$$

This concentration of O_2 (8.5 mg/L) is quite low, but it is sufficient to provide the oxygen required by aquatic life.

■ **Limitations of Henry's Law**
Henry's law holds quantitatively only for gases that do not interact chemically with the solvent. It does not work perfectly for NH_3, for example, because this compound gives small concentrations of NH_4^+ and OH^- in water.

Exercise 14.4—Using Henry's Law

The Henry's law constant for CO_2 in water at 25 °C is 4.48×10^{-5} M/mm Hg. What is the concentration of CO_2 in water when the partial pressure is 0.33 atm? (Although CO_2 reacts with water to give traces of H^+ and HCO_3^-, the reaction occurs to such a small extent that Henry's law is obeyed at low CO_2 partial pressures.)

Temperature Effects on Solubility: Le Chatelier's Principle

The solubility of all gases in water decreases with increasing temperature. You may realize this from everyday observations such as the appearance of bubbles as water is heated below the boiling point.

Decreased solubility of gases with increasing temperature also has environmental consequences. Fish often seek lower depths of water in summer because the warmer surface layers of lakes and rivers have lower oxygen concentrations. Thermal pollution, resulting from the use of surface water as a coolant for various industries, can pose a special problem for marine life that requires oxygen to survive. Effluent water returned to a natural water source at a warmer temperature will be depleted of oxygen.

To understand the effect of temperature on the solubility of gases, let us reexamine the heat of solution. Gases that dissolve to an appreciable extent in water usually do so in an exothermic process.

$$\text{Gas} + \text{liquid solvent} \xrightleftharpoons{\Delta H_{soln} < 0} \text{saturated solution} + \text{heat}$$

The reverse process, loss of dissolved gas molecules from a solution, requires heat to occur. These two processes can reach equilibrium, as depicted in an equation where products and reactants are connected by the symbol for an equilibrium process, a double arrow ($\rightleftharpoons$).

To understand how temperature affects solubility, we turn to **Le Chatelier's principle**, which states that a change in any of the factors determining an equilibrium causes the system to adjust so as to reduce or counteract the effect of the change. If a solution of a gas in a liquid is heated, for example, the equilibrium will shift to absorb some of the added heat energy. That is, the reaction

$$\text{Gas} + \text{liquid solvent} \underset{\substack{\text{Exothermic process}\\ \Delta H_{soln} \text{ is negative.}}}{\rightleftharpoons} \text{saturated solution} + \text{heat}$$

Add heat energy. Equilibrium shifts left.

shifts back to the left if the temperature is raised because heat energy can be consumed in the process that gives free gas molecules and pure solvent. This shift corresponds to less gas dissolved, or a lower solubility, at higher temperature—the observed result.

The solubility of solids in water is also affected by temperature, but unlike the situation involving solutions of gases no general pattern of behavior is observed. In Figure 14.12, the solubilities of several salts are plotted versus temperature. The solubility of many salts increases with increasing temperature, but there are notable exceptions. Predictions based on whether the heat of solution is positive or negative work most of the time, but exceptions do occur.

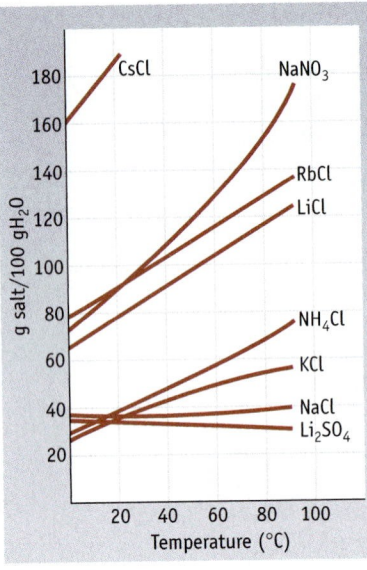

(a) Temperature dependence of the solubility of some ionic compounds.

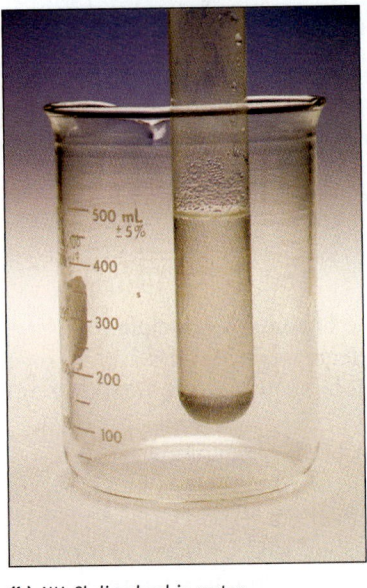

(b) NH₄Cl dissolved in water.

(c) NH₄Cl precipitates when the solution is cooled in ice.

Photos: Charles D. Winters

Figure 14.12 **The temperature dependence of the solubility of some ionic compounds in water.** Most compounds increase in solubility with increasing temperature.

Chemists take advantage of the variation of solubility with temperature to purify compounds. An impure sample of a given compound is dissolved in a solvent at high temperature, a condition under which it is more soluble. The solution is cooled to decrease the solubility. When the limit of solubility is reached at the lower temperature, crystals of the pure compound form. If the process is done slowly and carefully, it is sometimes possible to obtain very large crystals (Figure 14.13).

Charles D. Winters

Figure 14.13 **Giant crystals of potassium dihydrogen phosphate.** The crystal being measured by this researcher at Lawrence Livermore Laboratory in California weighs 318 kg and measures 66 × 53 × 58 cm. The crystals were grown by suspending a thumbnail-sized seed crystal in a 6-foot tank of saturated KH₂PO₄. The temperature of the solution was gradually reduced from 65 °C over a period of about 50 days. The crystals are sliced into thin plates, which are used to convert light from a giant laser from infrared to ultraviolet.

GENERAL
Chemistry ⚛ Now™

See the General ChemistryNow CD-ROM or website:
• **Screen 14.6 Factors Affecting Solubility (2)—Temperature and Le Chatelier's Principle,** for a tutorial on enthalpy of solution

14.4—Colligative Properties

When water contains dissolved sodium chloride, the vapor pressure of water over the solution is different from that over pure water, as is the freezing point of the solution, its boiling point, and its osmotic pressure. These *colligative properties* depend on the relative numbers of solute and solvent particles.

Changes in Vapor Pressure: Raoult's Law

The equilibrium vapor pressure at a particular temperature is the pressure of the vapor when the liquid and the vapor are in equilibrium. When the vapor pressure of the solvent over a solution is measured at a given temperature, it is experimentally observed that

- The vapor pressure of the solvent over the solution is lower than the vapor pressure of the pure solvent, and

- The vapor pressure of the solvent, $P_{solvent}$, is proportional to the relative number of solvent molecules in the solution; that is, the solvent vapor pressure is proportional to the solvent mole fraction, $P_{solvent} \propto X_{solvent}$.

Because solvent vapor pressure and the relative number of solvent molecules are proportional, we can write the following equation for the equilibrium vapor pressure of the solvent over a solution:

$$P_{solvent} = X_{solvent}\, P°_{solvent}$$

(14.5)

This equation, called **Raoult's law**, tells us that the vapor pressure of solvent over a solution ($P_{solvent}$) is some fraction of the pure solvent equilibrium vapor pressure ($P°_{solvent}$). For example, if 95% of the molecules in a solution are solvent molecules ($X_{solvent} = 0.95$), then the vapor pressure of the solvent ($P_{solvent}$) is 95% of $P°_{solvent}$.

Like the ideal gas law, Raoult's law describes a simplified model of a solution. We say that an **ideal solution** is one that obeys Raoult's law. No solution is ideal, of course, just as no gas is truly ideal. Nevertheless, Raoult's law is a good approximation of solution behavior in many instances, especially at low solute concentration.

When will a solution not be ideal? This question brings us to another effect of dissolved solutes—the forces of attraction between solute and solvent molecules. For Raoult's law to hold, the forces between solute and solvent molecules must be the same as those between solvent molecules in the pure solvent. This is frequently the case when molecules with similar structures are involved. Solutions of one hydrocarbon in another (hexane, C_6H_{14}, dissolved in octane, C_8H_{18}, for example) usually follow Raoult's law quite closely. If solvent–solute interactions are stronger than solvent–solvent interactions, the actual vapor pressure will be lower than calculated by Raoult's law. If the solvent–solute interactions are weaker than solvent–solvent interactions, the vapor pressure will be higher.

■ **Raoult's Law**
Raoult's law is named for Francois M. Raoult (1830–1901), a professor of chemistry at the University of Grenoble in France, who did the pioneering studies in this area.

GENERAL
Chemistry ⚛ Now™

See the General ChemistryNow CD-ROM or website
- **Screen 14.7 Colligative Properties (1)—Vapor Pressure and Raoult's Law,** for an exercise and a tutorial on Raoult's law

---

Example 14.5—Using Raoult's Law

Problem Suppose 651 g of ethylene glycol, $HOCH_2CH_2OH$, is dissolved in 1.50 kg of water. What is the vapor pressure of the water over the solution at 90 °C? Assume ideal behavior for the solution.

Strategy To use Raoult's law (Equation 14.5), we first must calculate the mole fraction of the solvent (water). We also need the vapor pressure of pure water at 90 °C (= 525.8 mm Hg, Appendix G).

Solution We first calculate the amount of water and ethylene glycol and, from these, the mole fraction of water.

$$\text{Amount of water} = 1.50 \times 10^3 \text{ g} \left(\frac{1 \text{ mol}}{18.02 \text{ g}}\right) = 83.2 \text{ mol water}$$

$$\text{Amount of ethylene glycol} = 651 \text{ g} \left(\frac{1 \text{ mol}}{62.07 \text{ g}}\right) = 10.5 \text{ mol glycol}$$

$$X_{water} = \frac{83.2 \text{ mol water}}{83.2 \text{ mol water} + 10.5 \text{ mol glycol}} = 0.888$$

Next we apply Raoult's law, calculating the vapor pressure from the mole fraction of water and the vapor pressure of pure water:

$$P_{water} = X_{water}P^{\circ}_{water} = (0.888)(525.8 \text{ mm Hg}) = 467 \text{ mm Hg}$$

The dissolved solute decreases the vapor pressure by 59 mm Hg, or about 11%:

$$\Delta P_{water} = P_{water} - P^{\circ}_{water} = 467 \text{ mm Hg} - 525.8 \text{ mm Hg} = -59 \text{ mm Hg}$$

Comment Ethylene glycol dissolves easily in water, is noncorrosive, and is relatively inexpensive. Because of its high boiling point, it will not boil off. These features make it ideal for use as antifreeze. It is, however, toxic to animals, so it is being replaced by less toxic propylene glycol for this application.

Exercise 14.5—Using Raoult's Law

Assume you dissolve 10.0 g of sucrose ($C_{12}H_{22}O_{11}$) in 225 mL (225 g) of water and warm the water to 60 °C. What is the vapor pressure of the water over this solution? [Appendix G lists $P^{\circ}(H_2O)$ at various temperatures.]

Adding a nonvolatile solute to a solvent lowers the vapor pressure of the solvent (Example 14.5). Raoult's law can be modified to calculate directly the lowering of the vapor pressure, $\Delta P_{solvent}$, as a function of the mole fraction of the solute.

$$\Delta P_{solvent} = P_{solvent} - P^{\circ}_{solvent}$$

Substituting Raoult's law for $P_{solvent}$, we have

$$\Delta P_{solvent} = (X_{solvent} P^{\circ}_{solvent}) - P^{\circ}_{solvent} = -(1 - X_{solvent})P^{\circ}_{solvent}$$

In a solution that has only the volatile solvent and one nonvolatile solute, the sum of the mole fraction of solvent and solute must be 1:

$$X_{solvent} + X_{solute} = 1$$

Therefore, $1 - X_{solvent} = X_{solute}$, and the equation for $\Delta P_{solvent}$ can be rewritten as

$$\Delta P_{solvent} = -X_{solute}P^{\circ}_{solvent} \tag{14.6}$$

Thus, the change in the vapor pressure of the solvent is *proportional to the mole fraction (the relative number of particles) of solute.*

Boiling Point Elevation

Suppose you have a solution of a nonvolatile solute in the volatile solvent benzene. If the solute concentration is 0.200 mol in 100. g of benzene (C_6H_6) (= 2.00 *m*), this

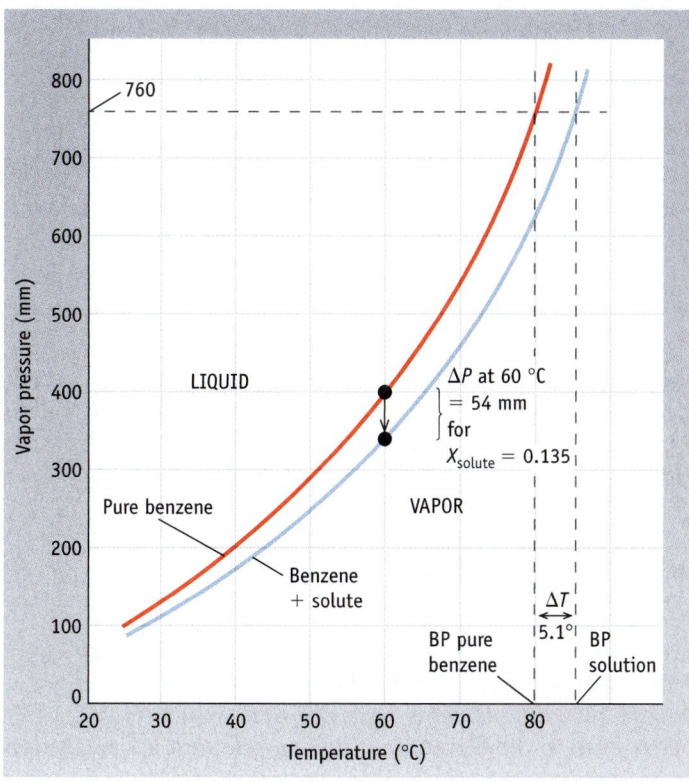

The curve drawn in red represents the vapor pressure of pure benzene, and the curve in blue represents the vapor pressure of a solution containing 0.200 mol of a solute dissolved in 0.100 kg of solvent (2.00 *m*). This graph was created using a series of calculations such as those shown in the text. As an alternative, the graph could be created by measuring various vapor pressures for the solution in a laboratory experiment. (*See General ChemistryNow Screen 14.8 Colligative Properties (2), to view an animation of this vapor pressure lowering.*)

means that $X_{benzene} = 0.865$. Using $X_{benzene}$ and applying Raoult's law, we can calculate that the vapor pressure of the solvent at 60 °C will drop from 400. mm Hg for the pure solvent to 346 mm Hg for the solution:

$$X_{benzene} = 0.865$$
$$P_{benzene} = X_{benzene}\, P^{\circ}_{benzene} = (0.865)(400.\text{ mm Hg}) = 346\text{ mm Hg}$$

This point is marked on the vapor pressure graph in Figure 14.14. Now, what is the vapor pressure when the temperature of the solution is raised another 10 °C? $P^{\circ}_{benzene}$ becomes larger with increasing temperature, so $P_{benzene}$ for the solution must also become larger. This new point, and additional ones calculated in the same way for other temperatures, define the vapor pressure curve for the solution (the lower curve in Figure 14.14).

An important observation we can make in Figure 14.14 is that *the vapor pressure lowering caused by the nonvolatile solute leads to an increase in the boiling point.* The normal boiling point of a liquid is the temperature at which its vapor pressure is equal to 1 atm or 760 mm Hg [◄ Section 13.5]. In Figure 14.14 we see that the normal boiling point of pure benzene (at 760 mm Hg) is about 80 °C. Tracing the vapor pressure curve for the solution, we also see that the vapor pressure reaches 760 mm Hg at a temperature about 5 °C higher than this value.

The vapor pressure curve and increase in the boiling point shown in Figure 14.14 refer specifically to a 2.00 *m* solution. We might wonder how the boiling point of the solution would vary with solute concentration, and it is possible to reason out the answer. Recall that the change in vapor pressure is directly proportional to the concentration of solute ($\Delta P_{benzene} = -X_{solute}\, P^{\circ}_{benzene}$, Equation 14.6). Concentrations of solute greater than 2.00 *m* lead to a larger decrease in vapor pressure and

Table 14.3   Some Boiling Point Elevation and Freezing Point Depression Constants

Solvent	Normal Boiling Point (°C) Pure Solvent	K_{bp} (°C/m)	Normal Freezing Point (°C) Pure Solvent	K_{fp} (°C/m)
Water	100.00	+0.5121	0.0	−1.86
Benzene	80.10	+2.53	5.50	−5.12
Camphor	207.4	+5.611	179.75	−39.7
Chloroform ($CHCl_3$)	61.70	+3.63	—	—

consequently to a higher boiling point. Conversely, solute concentrations less than 2.00 m show a smaller decrease in vapor pressure and a smaller increase in boiling point. In fact, a simple relationship exists between boiling point elevation and molal concentration: *The boiling point elevation, ΔT_{bp}, is directly proportional to the molality of the solute.*

$$\text{Elevation in boiling point} = \Delta T_{bp} = K_{bp}m_{solute} \qquad (14.7)$$

In this equation, K_{bp} is a proportionality constant called the **molal boiling point elevation constant**. It has the units of degrees/molal (°C/m). Values for K_{bp} are determined experimentally, and different solvents have different values (Table 14.3). Formally, the value corresponds to the elevation in boiling point for a 1 m solution.

GENERAL
Chemistry⋅⚛⋅Now™

See the General ChemistryNow CD-ROM or website:

- **Screen 14.8, Colligative Properties (2)—Boiling Point and Freezing Point,** for an exercise and a tutorial on the effect of a solute on the solution boiling point

Example 14.6—Boiling Point Elevation

Problem Eugenol, the active ingredient in cloves, has a formula of $C_{10}H_{12}O_2$ (page 123). What is the boiling point of a solution containing 0.144 g of this compound dissolved in 10.0 g of benzene?

Strategy We can use Equation 14.7 to calculate the change in boiling point. This value is then added to the boiling point of pure benzene to provide the answer. To use Equation 14.7 you need a value of K_{bp} and the molality of the solution. The K_{bp} value for benzene is given in Table 14.3, but you need to calculate the molality, m, first.

Solution

$$0.144 \text{ g eugenol} \left(\frac{1 \text{ mol eugenol}}{164.2 \text{ g}} \right) = 8.77 \times 10^{-4} \text{ mol eugenol}$$

$$\frac{8.77 \times 10^{-4} \text{ mol eugenol}}{0.0100 \text{ kg benzene}} = 8.77 \times 10^{-2} \ m$$

Use the value for the molality to calculate the boiling point elevation and then the boiling point:

$$\Delta T_{bp} = (2.53 \text{ °C/}m)(0.0877 \ m) = 0.222 \text{ °C}$$

Because the boiling point *rises* relative to that of the pure solvent, the boiling point of the solution is

$$80.10\ °C + 0.222\ °C = \boxed{80.32\ °C}$$

Exercise 14.6—Boiling Point Elevation

What quantity of ethylene glycol, $HOCH_2CH_2OH$, must be added to 125 g of water to raise the boiling point by 1.0 °C? Express the answer in grams.

The elevation of the boiling point of a solvent on adding a solute has many practical consequences. One of them is the summer protection your car's engine receives from "all-season" antifreeze. The main ingredient of commercial antifreeze is ethylene glycol, $HOCH_2CH_2OH$. The car's radiator and cooling system are sealed to keep the coolant under pressure, ensuring that it will not vaporize at normal engine temperatures. When the air temperature is high in the summer, however, the radiator could "boil over" if it were not protected with "antifreeze." By adding this nonvolatile liquid, the solution in the radiator has a higher boiling point than that of pure water.

Freezing Point Depression

Another consequence of dissolving a solute in a solvent is that the freezing point of the solution is lower than that of the pure solvent (Figure 14.15). For an ideal solution, the depression of the freezing point is given by an equation similar to that for the elevation of the boiling point:

$$\text{Freezing point depression} = \Delta T_{fp} = K_{fp}m_{solute} \tag{14.8}$$

where K_{fp} is the **freezing point depression constant** in degrees per molal ($°C/m$). Values of K_{fp} for a few common solvents are given in Table 14.3. The values are negative quantities, so the result of the calculation is a negative value for ΔT_{fp}, signifying a decrease in temperature.

The practical aspects of freezing point changes from pure solvent to solution are similar to those for boiling point elevation. The very name of the liquid you add to the radiator in your car, *antifreeze*, indicates its purpose (see Figure 14.15a). The

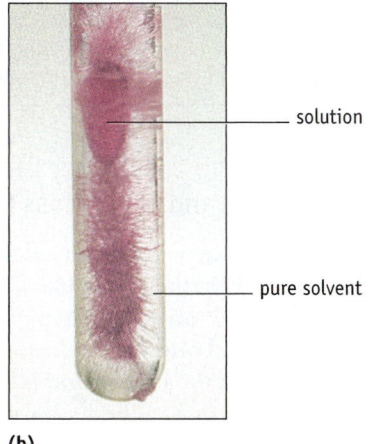

Figure 14.15 Freezing a solution. (a) Adding antifreeze to water prevents the water from freezing. Here a jar of pure water (*left*) and a jar of water to which automobile antifreeze had been added (*right*) were kept overnight in the freezing compartment of a home refrigerator.

(b) When a solution freezes, it is pure solvent that solidifies. To take this photo, a purple dye was dissolved in water, and the solution was frozen slowly. Pure ice formed along the walls of the tube, and the dye stayed in solution. The concentration of the solute increased as more and more solvent was frozen out, and the resulting solution had a lower and lower freezing point. At equilibrium, the system contains pure, colorless ice that formed along the walls of the tube and a concentrated solution of dye in the center of the tube.

solution

pure solvent

Charles D. Winters

(a)　　　　(b)

label on the container of antifreeze tells you, for example, to add 6 qt (5.7 L) of antifreeze to a 12-qt (11.4-L) cooling system to lower the freezing point to $-34\,°C$ and to raise the boiling point to $+109\,°C$.

GENERAL
Chemistry• •Now™

See the General ChemistryNow CD-ROM or website:

- **Screen 14.8 Colligative Properties (2)—Boiling Point and Freezing Point,** for an exercise and a tutorial on the effect of a solute on the solution freezing point

---

Example 14.7—Freezing Point Depression

Problem What mass of ethylene glycol, $HOCH_2CH_2OH$, must be added to 5.50 kg of water to lower the freezing point of the water from $0.0\,°C$ to $-10.0\,°C$?

Strategy To use Equation 14.8, you need K_{fp} (Table 14.3). You can then calculate the molality of the solution and, from this value, the amount and quantity of ethylene glycol required.

Solution The solute concentration (molality) in a solution with a freezing point depression of $-10.0\,°C$ is

$$\text{Solute concentration }(m) = \frac{\Delta T_{fp}}{K_{fp}} = \frac{-10.0\,°C}{-1.86\,°C/m} = 5.38\ m$$

Because the radiator contains 5.50 kg of water, we need 29.6 mol of ethylene glycol:

$$\left(\frac{5.38\text{ mol glycol}}{1.00\text{ kg water}}\right)(5.50\text{ kg water}) = 29.6\text{ mol glycol}$$

The molar mass of ethylene glycol is 62.07 g/mol, so the mass required is

$$29.6\text{ mol glycol}\left(\frac{62.07\text{ g}}{1\text{ mol}}\right) = \boxed{1840\text{ g glycol}}$$

Comment The density of ethylene glycol is 1.11 kg/L, so the volume of antifreeze to be added is 1.84 kg (1 L/1.11 kg) = 1.66 L.

Exercise 14.7—Freezing Point Depression

In the northern United States, summer cottages are usually closed up for the winter. When doing so, the owners "winterize" the plumbing by putting antifreeze in the toilet tanks, for example. Will adding 525 g of $HOCH_2CH_2OH$ to 3.00 kg of water ensure that the water will not freeze at $-25\,°C$?

Colligative Properties and Molar Mass Determination

Early in this book you learned how to calculate a molecular formula from an empirical formula when given the molar mass. But how do you know the molar mass of an unknown compound? An experiment must be carried out to find this crucial piece of information, and one way to do so is to use a colligative property of a solution of the compound. If the compound is soluble in a solvent of appreciable vapor pressure or a known K_{bp} or K_{fp}, the molar mass can be determined. All approaches use the same basic logic:

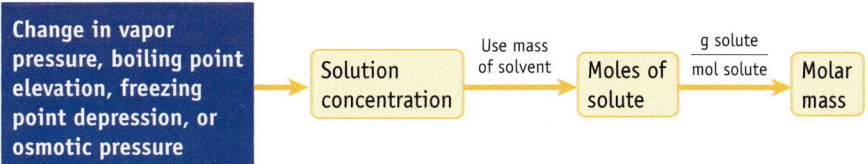

Example 14.8—Determining Molar Mass from Boiling Point Elevation

Problem A solution prepared from 1.25 g of oil of wintergreen (methyl salicylate) in 99.0 g of benzene has a boiling point of 80.31 °C. Determine the molar mass of this compound.

Strategy Calculations using colligative properties to determine a molar mass always follow the pattern outlined in the text.

Solution We first use the boiling point elevation to calculate the solution concentration:

$$\text{Boiling point elevation } (\Delta T_{bp}) = 80.31 \text{ °C} - 80.10 \text{ °C} = 0.21 \text{ °C}$$

and then calculate the molality:

$$\text{Molality of solution} = \frac{\Delta T_{bp}}{K_{bp}} = \frac{0.21 \text{ °C}}{2.53 \text{ °C}/m} = 0.083 \text{ } m$$

The amount of solute in the solution is calculated from the solution concentration:

$$\text{Amount of solute} = \left(\frac{0.083 \text{ mol}}{1.00 \text{ kg}}\right)(0.099 \text{ kg solvent}) = 0.0082 \text{ mol solute}$$

Now we can combine the amount of solute with its mass:

$$\frac{1.25 \text{ g}}{0.0083 \text{ mol}} = \boxed{150 \text{ g/mol}}$$

Comment Methyl salicylate has the formula $C_8H_8O_3$ and a molar mass of 152.14 g/mol.

Exercise 14.8—Determining Molar Mass from Boiling Point Elevation

Crystals of the beautiful blue hydrocarbon, azulene (mass, 0.640 g), which has an empirical formula of C_5H_4, are dissolved in 99.0 g of benzene. The boiling point of the solution is 80.23 °C. What is the molecular formula of azulene?

Colligative Properties of Solutions Containing Ions

In the northern United States it is common practice to scatter salt on snowy or icy roads or sidewalks. When the sun shines on the snow or patch of ice, a small amount melts, and the water dissolves some of the salt. As a result of the dissolved solute, the freezing point of the solution is lower than 0 °C. The solution "eats" its way through the ice, breaking it up, and the icy patch is no longer dangerous for drivers or for people walking.

Salt (NaCl) is the most common substance used on roads because it is inexpensive and dissolves readily in water. Its relatively low molar mass means that the effect per gram is large. In addition, salt is especially effective because it is an electrolyte. That is, it dissolves to give ions in solution:

$$NaCl(s) \longrightarrow Na^+(aq) + Cl^-(aq)$$

Putting salt on ice assists in melting the ice.

Charles D. Winters

Table 14.4   Freezing Point Depressions of Some Ionic Solutions

Mass %	m (mol/kg)	ΔT_{fp} (measured, °C)	ΔT_{fp} (calculated, °C)	$\dfrac{\Delta T_{fp}, \text{ measured}}{\Delta T_{fp}, \text{ calculated}}$
NaCl				
0.00700	0.0120	−0.0433	−0.0223	1.94
0.500	0.0860	−0.299	−0.160	1.87
1.00	0.173	−0.593	−0.322	1.84
2.00	0.349	−1.186	−0.649	1.83
Na₂SO₄				
0.00700	0.00493	−0.0257	−0.00917	2.80
0.500	0.0354	−0.165	−0.0658	2.51
1.00	0.0711	−0.320	−0.132	2.42
2.00	0.144	−0.606	−0.268	2.26

Recall that colligative properties depend not on what is dissolved but *only on the number of particles of solute per solvent particle.* When 1 mol of NaCl dissolves, 2 mol of ions forms, which means that the effect on the freezing point of water should be twice as large as that expected for a mole of sugar. This peculiarity was discovered by Raoult in 1884 and studied in detail by Jacobus Henrikus van't Hoff (1852–1911) in 1887. Later in that same year Svante Arrhenius (1859–1927) provided the explanation for the behavior of electrolytes based on ions in solution. A 0.100 m solution of NaCl really contains two solutes, 0.100 m Na$^+$ and 0.100 m Cl$^-$. What we should use to estimate the freezing point depression is the *total* molality of solute particles:

$$m_{total} = m(Na^+) + m(Cl^-) = (0.100 + 0.100) \text{ mol/kg} = 0.200 \text{ mol/kg}$$
$$\Delta T_{fp} = (-1.86 \text{ °C}/m)(0.200 \ m) = -0.372 \text{ °C}$$

To estimate the freezing point depression for an ionic compound, first find the molality of solute from the mass and molar mass of the compound. Then, multiply the number you get by the number of ions in the formula: two for NaCl, three for Na₂SO₄, four for LaCl₃, five for Al₂(SO₄)₃, and so on.

As it turns out, this model gives a reasonable estimate of the effect of the ionization of an electrolyte on colligative properties, but it is not exact. Let us look at some experimental data (Table 14.4) for the effect of the dissociation of two ionic compounds, NaCl and Na₂SO₄, on the solution freezing point. The measured freezing point depression is larger than that calculated from Equation 14.9, assuming no ionization. As seen in the last column of the table, however, ΔT_{fp} is not twice the value expected for NaCl, but only about 1.8 times larger. Likewise, ΔT_{fp} for Na₂SO₄ approaches but does not reach a value that is 3 times larger than the value assuming no ionization. The ratio of the experimentally observed value of ΔT_{fp} to the value calculated, assuming no ionization, is called the **van't Hoff factor** and is represented by i.

$$i = \frac{\Delta T_{fp}, \text{ measured}}{\Delta T_{fp}, \text{ calculated}} = \frac{\Delta T_{fp}, \text{ measured}}{K_{fp} \ m}$$

or

$$\Delta T_{fp}, \text{ measured} = K_{fp} \times m \times i \qquad (14.9)$$

The numbers in the last column of Table 14.4 are van't Hoff factors. These values can be used in calculations of any colligative property. Vapor pressure lowering, boiling point elevation, freezing point depression, and osmotic pressure are all larger for electrolytes than for nonelectrolytes of the same molality.

The van't Hoff factor approaches a whole number (2, 3, and so on) only with very dilute solutions. In more concentrated solutions, the experimental freezing point depressions tell us that there are fewer ions in solution than expected. This behavior, which is typical of all ionic compounds, is a consequence of the strong attractions between ions. The result is as if some of the positive and negative ions are paired, decreasing the total molality of particles. Indeed, in more concentrated solutions, and especially in solvents less polar than water, ions are extensively associated in ion pairs and in even larger clusters.

Example 14.9—Freezing Point and Ionic Solutions

Problem A 0.00200 m aqueous solution of an ionic compound, $Co(NH_3)_5(NO_2)Cl$, freezes at -0.00732 °C. How many moles of ions does 1.0 mol of the salt produce on being dissolved in water?

Strategy The solution is to calculate ΔT_{fp} of the solution assuming no ions are produced. Compare this value with the actual value of ΔT_{fp}. The ratio will reflect the number of ions produced.

Solution The freezing-point depression expected for a 0.00200 m solution assuming that the salt does not dissociate into ions is

$$\Delta T_{fp}, \text{ calculated} = K_{fp}m = (-1.86 \text{ °C}/m)(0.00200 \text{ } m) = -3.72 \times 10^{-3} \text{ °C}$$

Now compare the calculated freezing point depression with the measured depression. This gives us the van't Hoff factor:

$$i = \frac{\Delta T_{fp}, \text{ measured}}{\Delta T_{fp}, \text{ calculated}} = \frac{-7.32 \times 10^{-3} \text{ °C}}{-3.72 \times 10^{-3} \text{ °C}} = 1.97 \approx 2$$

It appears that 1 mol of this compound gives 2 mol of ions. In this case, the ions are $[Co(NH_3)_5(NO_2)]^+$ and Cl^-.

Exercise 14.9—Freezing Point and Ionic Compounds

Calculate the freezing point of 525 g of water that contains 25.0 g of NaCl. Assume i, the van't Hoff factor, is 1.85 for NaCl.

Osmosis

Osmosis is the movement of solvent molecules through a semipermeable membrane from a region of lower solute concentration to a region of higher solute concentration. This movement can be demonstrated with a simple experiment. The beaker in Figure 14.16 contains pure water, and the bag and tube hold a concentrated sugar solution. The liquids are separated by a semipermeable membrane, a thin sheet of material (such as a vegetable tissue or cellophane) through which only certain types of molecules can pass. Here, water molecules can pass through the membrane but larger sugar molecules (or hydrated ions) cannot (Figure 14.17). When the experiment is begun, the liquid levels in the beaker and the tube are the same. Over time, however, the level of the sugar solution inside the tube rises, the level of pure water in the beaker falls, and the sugar solution becomes steadily more dilute. After a while, no further net change occurs; equilibrium is reached.

Figure 14.16 The process of osmosis. (a) The bag attached to the tube contains a solution that is 5% sugar and 95% water. The beaker contains pure water. The bag is made of a material that is semipermeable, meaning that it allows water, but not sugar molecules, to pass through.
(b) Over time, water flows from the region of low solute concentration (pure water) to the region of higher solute concentration (the sugar solution). Flow continues until the pressure exerted by the column of solution in the tube above the water level in the beaker is great enough to result in equal rates of passage of water molecules in both directions. The height of the column of solution (b) is a measure of the osmotic pressure, Π. (*See the General ChemistryNow Screen 14.9 Colligative Properties (3), for an animation of osmosis.*)

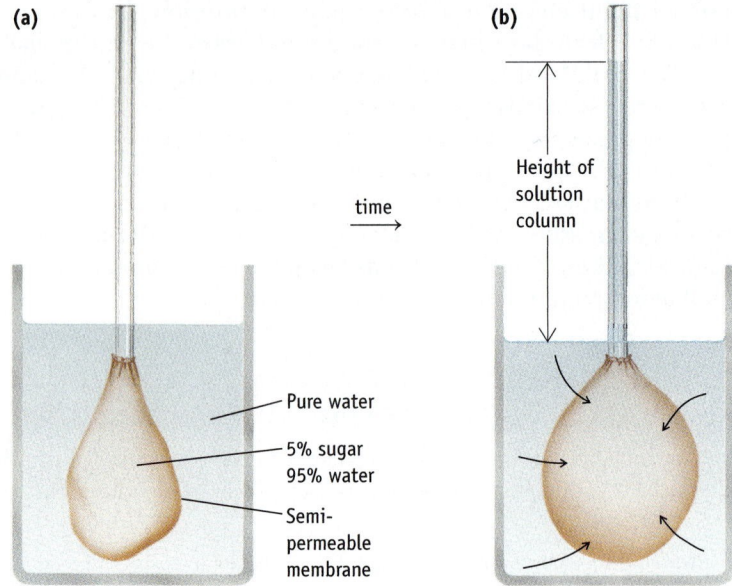

From a molecular point of view, the semipermeable membrane does not present a barrier to the movement of water molecules, so they move through the membrane in both directions. Over time, more water molecules pass through the membrane from the pure water side to the solution side than in the opposite direction. In effect, water molecules tend to move from regions of low solute concentration to regions of high solute concentration. The same is true for any solvent, as long as the membrane allows solvent molecules but not solute molecules or ions to pass through.

Why does the system eventually reach equilibrium? Clearly, the solution in the tube in Figure 14.16 can never reach zero sugar or salt concentration, which would be required to equalize the number of water molecules moving through the membrane in each direction in a given time. The answer lies in the fact that the solution moves higher and higher in the tube as osmosis continues and water moves into the

Figure 14.17 Osmosis at the particulate level. Osmotic flow through a membrane that is selectively permeable (semipermeable) to water. Dissolved substances such as hydrated ions or large sugar molecules cannot diffuse through the membrane. The membrane acts as a "molecular sieve."

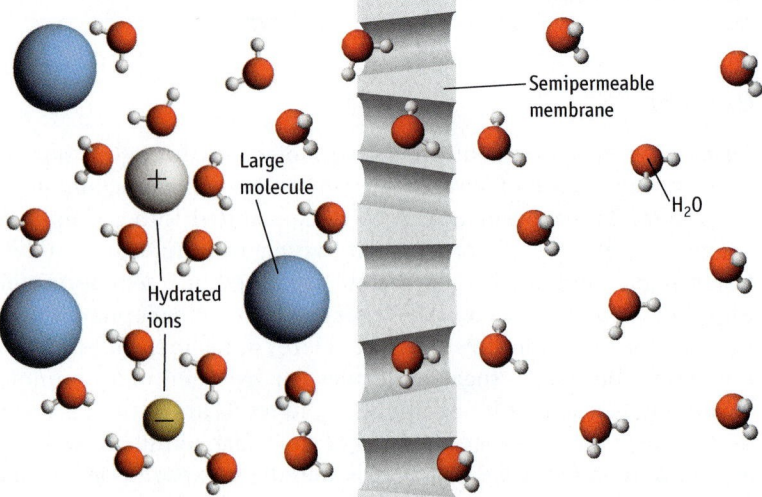

sugar solution. Eventually the pressure exerted by this column of solution counter-balances the pressure exerted by the water moving through the membrane from the pure water side, and no further net movement of water occurs. An equilibrium of forces is achieved. The pressure created by the column of solution for the system at equilibrium is called the **osmotic pressure**. A measure of this pressure is the difference in the height of the solution in the tube and the level of pure water in the beaker.

From experimental measurements on dilute solutions, it is known that osmotic pressure (Π) and concentration (c) are related by the equation

$$\Pi = cRT \tag{14.10}$$

In this equation, c is the molar concentration (in moles per liter), R is the gas constant, and T is the absolute temperature (in kelvins). Using a value for the gas law constant of 0.082057 L · atm/K · mol allows calculation of the osmotic pressure Π in atmospheres. This equation is analogous to the ideal gas law ($PV = nRT$), with Π taking the place of P and c being equivalent to n/V.

According to the osmotic pressure equation, the pressure exerted by a 0.10 M solution of particles at 25 °C is

$$\Pi = (0.10 \text{ mol/L})(0.0821 \text{ L} \cdot \text{atm/K} \cdot \text{mol})(298 \text{ K}) = 2.4 \text{ atm}$$

Because pressures on the order of 10^{-3} atm are easily measured, concentrations of about 10^{-4} M can be determined through measurements of osmotic pressure. Polymers and biologically important molecules often have a large number of atoms, so osmosis is an ideal method for measuring the molar masses of such molecules.

GENERAL
Chemistry⚛Now™

See the General ChemistryNow CD-ROM or website:

- **Screen 14.9 Colligative Properties (3),** for a tutorial on osmosis and to watch a video of the effect of different solution concentrations on an egg

---

Example 14.10—Osmotic Pressure and Molar Mass

Problem Beta-carotene is the most important of the A vitamins. Its molar mass can be determined by measuring the osmotic pressure generated by a given mass of β-carotene dissolved in the solvent chloroform. Calculate the molar mass of β-carotene if 10.0 mL of a solution containing 7.68 mg of β-carotene has an osmotic pressure of 26.57 mm Hg at 25.0 °C.

Strategy First, use Equation 14.10 to calculate the solution concentration from the osmotic pressure. Then, use the volume and concentration of the solution to calculate the amount of solute. Finally, find the molar mass of the solute from its mass and amount.

Solution The osmotic pressure can be used to calculate the concentration of β-carotene:

$$\text{Concentration (M)} = \frac{\Pi}{RT} = \frac{(26.57 \text{ mm Hg})\left(\dfrac{1 \text{ atm}}{760 \text{ mm Hg}}\right)}{(0.082057 \text{ L} \cdot \text{atm/K} \cdot \text{mol})(298.2 \text{ K})} = 1.429 \times 10^{-3} \text{ mol/L}$$

Now the amount of β-carotene dissolved in 10.0 mL of solvent can be calculated:

$$(1.429 \times 10^{-3} \text{ mol/L})(0.0100 \text{ L}) = 1.43 \times 10^{-5} \text{ mol}$$

(a) A fresh egg is placed in dilute acetic acid. The acid reacts with the $CaCO_3$ of the shell but leaves the egg membrane intact.

(b) If the egg, with its shell removed, is placed in pure water, the egg swells.

(c) If the egg, with its shell removed, is placed in a concentrated sugar solution, the egg shrivels.

An experiment to observe osmosis. You can try this experiment in your kitchen. In the first step use vinegar as a source of acetic acid. (*See the General ChemistryNow Screen 14.1 Puzzler, and Screen 14.9 Colligative Properties (3), to watch a video of this experiment.*)

This amount of β-carotene (1.43×10^{-5} mol) is equivalent to 7.68 mg (7.68×10^{-3} g). This gives us a way to calculate the molar mass:

$$\frac{7.68 \times 10^{-3} \text{ g}}{1.43 \times 10^{-5} \text{ mol}} = 538 \text{ g/mol}$$

Comment Beta-carotene is a hydrocarbon with the formula $C_{40}H_{56}$.

Exercise 14.10—Osmotic Pressure and Molar Mass

A 1.40-g sample of polyethylene, a common plastic, is dissolved in enough benzene to give exactly 100 mL of solution. The measured osmotic pressure of the solution is 1.86 mm Hg at 25 °C. Calculate the average molar mass of the polymer.

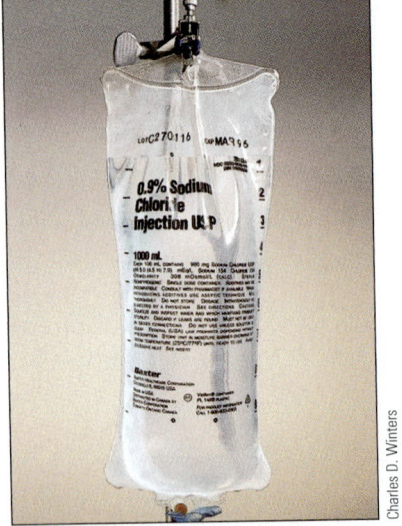

An isotonic saline solution.

Osmosis is of practical significance for people in the health professions. Patients who become dehydrated through illness often need to be given water and nutrients intravenously. Water cannot simply be dripped into a patient's vein, however. Rather, the intravenous solution must have the same overall solute concentration as the patient's blood: The solution must be iso-osmotic or **isotonic**. If pure water were used, the inside of a blood cell would have a higher solute concentration (lower water concentration), and water would flow into the cell. This *hypotonic* situation would cause red blood cells to burst (lyse) (Figure 14.18). The opposite situation, *hypertonicity*, occurs if the intravenous solution is more concentrated than the contents of the blood cell. In this case the cell would lose water and shrivel up (crenate). To combat this problem, a dehydrated patient is rehydrated in the hospital with a sterile saline solution that is 0.154 M NaCl. This solution is isotonic with the cells of the body.

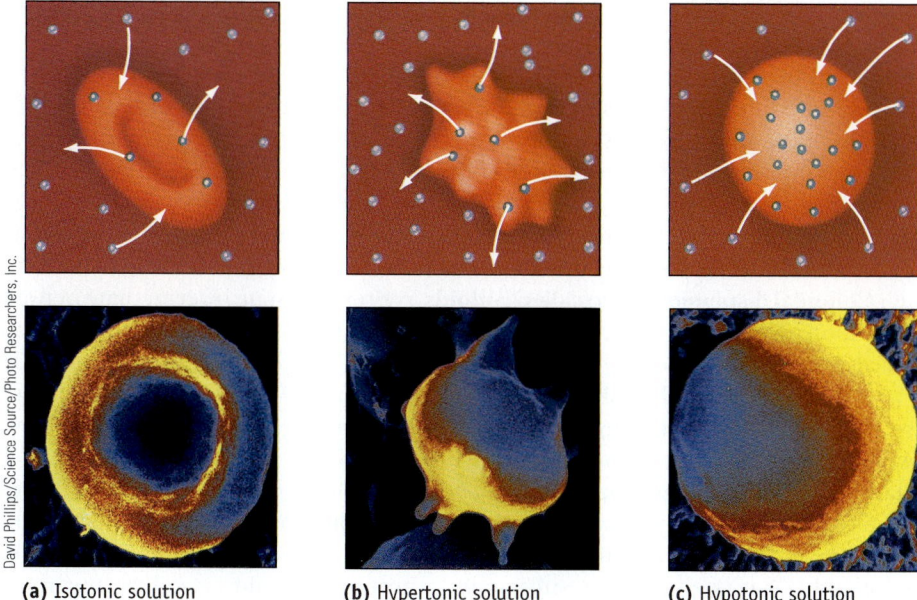

David Phillips/Science Source/Photo Researchers, Inc.

(a) Isotonic solution   **(b)** Hypertonic solution   **(c)** Hypotonic solution

Figure 14.18 **Osmosis and living cells.** (a) A cell placed in an isotonic solution. The net movement of water into and out of the cell is zero because the concentration of solutes inside and outside the cell is the same. (b) In a hypertonic solution, the concentration of solutes outside the cell is greater than that inside. There is a net flow of water out of the cell, causing the cell to dehydrate, shrink, and perhaps die. (c) In a hypotonic solution, the concentration of solutes outside the cell is less than that inside. There is a net flow of water into the cell, causing the cell to swell and perhaps to burst (or lyse).

A Closer Look

Reverse Osmosis in Tampa Bay

Finding sources of fresh water for human and agricultural use has been a constant battle for centuries. Water has been the cause of numerous conflicts and a target in war. Although the earth has abundant water, 97% of it is too salty to drink or to irrigate crops. A large portion of the remaining 3% is in the form of ice in the polar regions.

One of the oldest ways to obtain fresh water from the oceans is evaporation. Heating water, however, requires large quantities of heat, and the salt left behind may not be suitable for consumption.

A more modern way to extract fresh water from the oceans is desalination by reverse osmosis. The best estimate is that about 1% of the world's drinking water is supplied by approximately 12,500 desalina-

Courtesy of Tampa Bay Water

Tampa Bay reverse osmosis plant. This plant will eventually produce about 95 million liters of potable water from sea water each day.

tion plants. One of the latest, and one of the largest, is operating in Tampa, Florida. This plant uses sea water and will eventually produce 25 million gallons of water per day (about 95 million liters), or roughly 10% of the region's water needs.

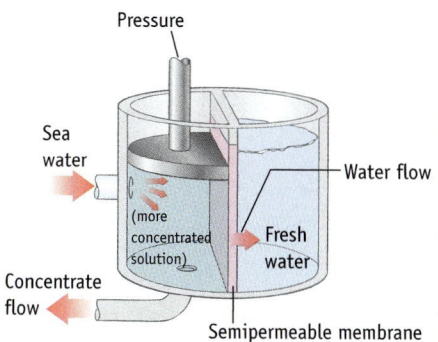

Reverse osmosis. Drinking water can be produced from sea water by reverse osmosis. The osmotic pressure of sea water is approximately 27 atm. To obtain fresh water at a reasonable rate, reverse osmosis requires a pressure of about 50 atm. For comparison, bicycle tires are usually pumped up to about 2–3 atm of pressure.

Gold colloid. Aqueous $AuCl_4^-$ is reduced to give colloidal gold metal. Since the days of alchemy some have claimed that drinking a colloidal gold solution "cleared the mind, increased intelligence and will power, and balanced the emotions."

Charles D. Winters

14.5—Colloids

Earlier in this chapter, we defined a solution broadly as a homogeneous mixture of two or more substances in a single phase (page 658). To this definition we should add that, in a true solution, no settling of the solute should be observed and the solute particles should be in the form of ions or relatively small molecules. Thus, NaCl and sugar form true solutions in water. You are also familiar with suspensions, which result, for example, if a handful of fine sand is added to water and shaken vigorously. Sand particles are still visible and gradually settle to the bottom of the beaker or bottle. **Colloidal dispersions**, also called **colloids**, represent a state intermediate between a solution and a suspension. Colloids include many of the foods you eat and the materials around you; among them are Jello, milk, fog, and porcelain (see Table 14.5).

Around 1860, the British chemist Thomas Graham (1805–1869) found that substances such as starch, gelatin, glue, and albumin from eggs diffused only very slowly when placed in water, compared with sugar or salt. In addition, the former substances differ significantly in their ability to diffuse through a thin membrane: Sugar molecules can diffuse through many membranes, but the very large molecules that make up starch, gelatin, glue, and albumin do not. Moreover, Graham found that he could not crystallize these substances, whereas he could crystallize sugar, salt, and other materials that form true solutions. Graham coined the word "colloid" (from the Greek, meaning "glue") to describe this class of substances that are distinctly different from true solutions and suspensions.

We now know that it is possible to crystallize some colloidal substances, albeit with difficulty, so there really is no sharp dividing line between these classes based on this property. Colloids do, however, have two distinguishing characteristics. First, colloids generally have high molar masses; this is true of proteins such as hemoglobin that have molar masses in the thousands. Second, the particles of a colloid are relatively large (say, 1000 nm in diameter). As a consequence, they exhibit the **Tyndall effect**; they scatter visible light when dispersed in a solvent, making the mixture appear cloudy (Figure 14.19). Third, even though colloidal particles are large, they are not so large that they settle out.

Graham also gave us the words **sol** for a dispersion of a solid substance in a fluid medium, and **gel** for a dispersion that has a structure that prevents it from being mobile. Jello is a sol when the solid is first mixed with boiling water, but it becomes a gel when cooled. Other examples of gels are the gelatinous precipitates of $Al(OH)_3$, $Fe(OH)_3$, and $Cu(OH)_2$ (Figure 14.20).

Colloidal dispersions consist of finely divided particles that, as a result, have a very high surface area. For example, if you have one millionth of a mole of colloidal particles, each assumed to be a sphere with a diameter of 200 nm, the total surface

Figure 14.19 The Tyndall effect. Colloidal dispersions scatter light, a phenomenon known as the Tyndall effect. (a) Dust in the air scatters the light coming through the trees in a forest along the Oregon coast. (b) A narrow beam of light from a laser is passed through an NaCl solution (*left*) and then a colloidal mixture of gelatin and water (*right*).

(a)

(b)

Charles D. Winters

Table 14.5 Types of Colloids

Type	Dispersing Medium	Dispersed Phase	Examples
Aerosol	Gas	Liquid	Fog, clouds, aerosol sprays
Aerosol	Gas	Solid	Smoke, airborne viruses, automobile exhaust
Foam	Liquid	Gas	Shaving cream, whipped cream
Foam	Solid	Gas	Styrofoam, marshmallow
Emulsion	Liquid	Liquid	Mayonnaise, milk, face cream
Gel	Solid	Liquid	Jelly, Jello, cheese, butter
Sol	Liquid	Solid	Gold in water, milk of magnesia, mud
Solid sol	Solid	Solid	Milkglass

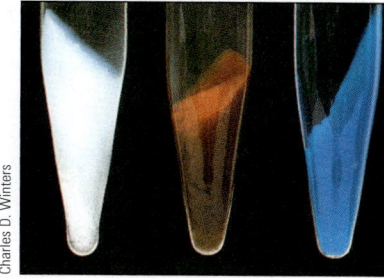

Charles D. Winters

Figure 14.20 **Gelatinous precipitates.** (*left*) $Al(OH)_3$, (*center*) $Fe(OH)_3$, and (*right*) $Cu(OH)_2$.

area of the particles would be on the order of 200 million cm^2, or the size of several football fields. It is not surprising, therefore, that many of the properties of colloids depend on the properties of surfaces.

Types of Colloids

Colloids are classified according to the state of the dispersed phase and the dispersing medium. Table 14.5 lists several types of colloids and gives examples of each.

Colloids with water as the dispersing medium can be classified as **hydrophobic** (from the Greek, meaning "water-fearing") or **hydrophilic** ("water-loving"). A hydrophobic colloid is one in which only weak attractive forces exist between the water and the surfaces of the colloidal particles. Examples include dispersions of metals and of nearly insoluble salts in water. When compounds like AgCl precipitate, the result is often a colloidal dispersion. The precipitation reaction occurs too rapidly for ions to gather from long distances and make large crystals, so the ions aggregate to form small particles that remain suspended in the liquid.

Why don't the particles come together (coagulate) and form larger particles? The answer is that the colloidal particles carry electric charges. An AgCl particle, for example, will absorb Ag^+ ions if the ions are present in substantial concentration; an attraction occurs between Ag^+ ions in solution and Cl^- ions on the surface of the particle. In this way, the colloidal particles become positively charged, allowing them to attract a secondary layer of anions. The particles, now surrounded by layers of ions, repel one another and are prevented from coming together to form a precipitate (Figure 14.21).

A stable *hydrophobic colloid* can be made to coagulate by introducing ions into the dispersing medium. Consider milk, which contains a colloidal suspension of protein-rich casein micelles with a hydrophobic core. When milk ferments, lactose (milk sugar) is converted to lactic acid, which forms lactate ions and hydrogen ions. The protective charges on the surfaces of the colloidal particles are overcome, and the milk coagulates; the milk solids come together in clumps called "curds."

Soil particles are often carried by water in rivers and streams as hydrophobic colloids. When river water carrying large amounts of colloidal particles meets sea water with its high concentration of salts, the particles coagulate to form the silt seen at the mouth of the river (Figure 14.22). Municipal water treatment plants often add salts such as $Al_2(SO_4)_3$ to clarify water. In aqueous solution, aluminum ions exist as $[Al(H_2O)_6]^{3+}$ cations, which neutralize the charge on the hydrophobic colloidal soil particles, causing these particles to aggregate and settle out.

Figure 14.21   Hydrophobic colloids. A hydrophobic colloid is stabilized by positive ions absorbed onto each particle and a secondary layer of negative ions. Because the particles bear similar charges, they repel one another and precipitation is prevented.

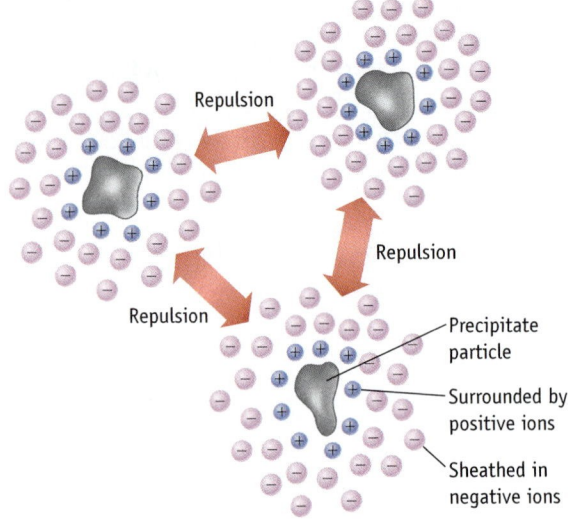

Repulsion

Repulsion

Repulsion

Precipitate particle

Surrounded by positive ions

Sheathed in negative ions

NASA/Peter Arnold, Inc.

Figure 14.22   Formation of silt. Silt forms at a river delta as colloidal soil particles come in contact with salt water in the ocean. Here the Ashley and Cooper Rivers empty into the Atlantic Ocean at Charleston, South Carolina. The high concentration of ions in sea water causes the colloidal soil particles to coagulate.

Hydrophilic colloids are strongly attracted to water molecules. They often have groups such as —OH and —NH_2 on their surfaces. These groups form strong hydrogen bonds to water, thereby stabilizing the colloid. Proteins and starch are important examples of hydrophilic colloids, and homogenized milk is the most familiar example.

Emulsions are colloidal dispersions of one liquid in another, such as oil or fat in water. Familiar examples include salad dressing, mayonnaise, and milk. If vegetable oil and vinegar are mixed to make a salad dressing, the mixture quickly separates into two layers because the nonpolar oil molecules do not interact with the polar water and acetic acid (CH_3CO_2H) molecules. So why are milk and mayonnaise apparently homogeneous mixtures that do not separate into layers? The answer is that they contain an **emulsifying agent** such as soap or a protein. Lecithin is a phospholipid found in egg yolks, so mixing egg yolks with oil and vinegar stabilizes the colloidal dispersion known as mayonnaise. To understand this process further, let us look into the functioning of soaps and detergents, substances known as surfactants.

Surfactants

Soaps and detergents are emulsifying agents. Soap is made by heating a fat with sodium or potassium hydroxide (see page 510), which produces the anion of a fatty acid.

$$H_3C(CH_2)_{16} - \overset{\overset{\textstyle O}{\|}}{C} - O^- \ Na^+$$

Hydrocarbon tail   Polar head
Soluble in oil   Soluble in water

sodium stearate, a soap

■ **Soaps and Surfactants**
A sodium soap is a solid at room temperature, whereas potassium soaps are usually liquids. About 30 million tons of household and toilet soap, and synthetic and soap-based laundry detergents, are produced annually worldwide.

The fatty acid anion has a split personality: It has a nonpolar, hydrophobic hydrocarbon tail that is soluble in other similar hydrocarbons and a polar, hydrophilic head that is soluble in water.

Oil cannot be readily washed away from dishes or clothing with water because oil is nonpolar and thus insoluble in water. Instead, we add soap to the water to clean away the oil. The nonpolar molecules of the oil interact with the nonpolar hy-

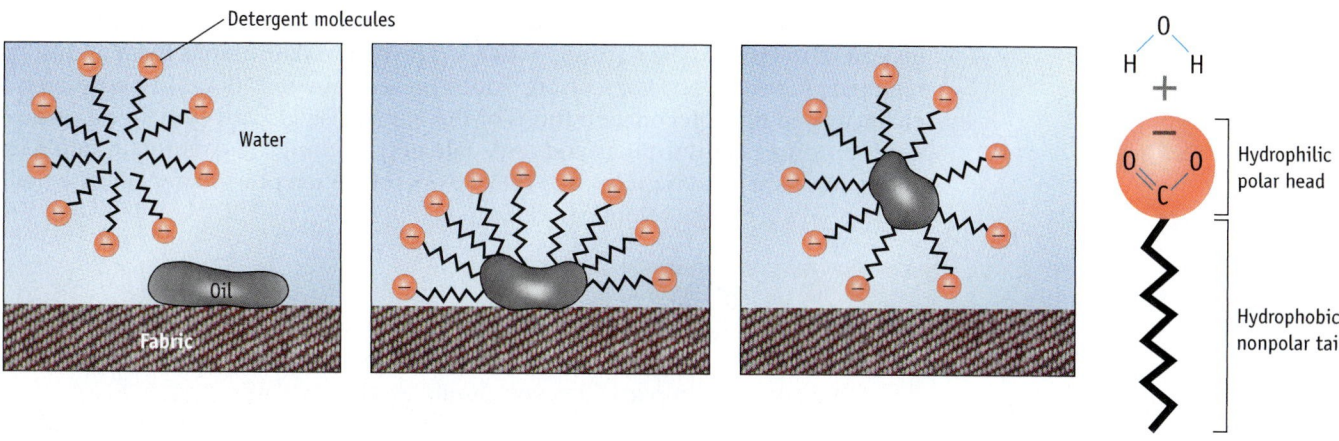

Detergent molecules

Water

Oil

Fabric

Hydrophilic polar head

Hydrophobic nonpolar tail

Figure 14.23 **The cleaning action of soap.** Soap molecules interact with water through the charged, hydrophilic end of the molecule. The long, hydrocarbon end of the molecule is hydrophobic, but it can bind through dispersion forces with hydrocarbons and other nonpolar substances.

drocarbon tails of the soap molecules, leaving the polar heads of the soap to interact with surrounding water molecules. The oil and water then mix (Figure 14.23). If the oily material on a piece of clothing or a dish also contains some dirt particles, that dirt can now be washed away.

Substances such as soaps that affect the properties of surfaces, and therefore affect the interaction between two phases, are called surface-active agents, or **surfactants**, for short. A surfactant used for cleaning is called a **detergent**. One function of a surfactant is to lower the surface tension of water, which enhances the cleansing action of the detergent (Figure 14.24).

Many detergents used in the home and industry are synthetic. One example is sodium laurylbenzenesulfonate, a biodegradable compound.

$$CH_3CH_2CH_2CH_2CH_2CH_2CH_2CH_2CH_2CH_2CH_2CH_2 - \bigcirc - SO_3^- \ Na^+$$

sodium laurylbenzenesulfonate

add surfactant →

Figure 14.24 **Effect of a detergent on the surface tension of water.** Sulfur (density = 2.1 g/cm³) is carefully placed on the surface of water (density, 1.0 g/cm³) (*left*). The surface tension of the water keeps the denser sulfur afloat. Several drops of detergent are then placed on the surface of the water (*right*). The surface tension of the water is reduced, and the sulfur sinks to the bottom of the beaker.

In general, synthetic detergents use the sulfonate group, $-SO_3^-$, as the polar head instead of the carboxylate group, $-CO_2^-$. The carboxylate anions form an insoluble precipitate with any Ca^{2+} or Mg^{2+} ions present in water. Because hard water is characterized by high concentrations of these ions, using soaps containing carboxylates produces bathtub rings and tattle-tale gray clothing. The synthetic sulfonate detergents have the advantage that they do not form such precipitates because their calcium salts are more soluble in water.

GENERAL
Chemistry ᴓ Now™

See the General ChemistryNow CD-ROM or website to:
- Assess your understanding with homework questions keyed to each goal
- Check your readiness for an exam by taking the exam-prep quiz and exploring the resources in the personalized Learning Plan it provides

For additional preparation for an examination on this chapter see the **Let's Review** section on pages 998–1011.

Chapter Goals Revisited

Now that you have studied this chapter, you should ask whether you have met the chapter goals. In particular, you should be able to

Learn additional methods of expressing solution concentration

a. Define the terms *solution, solvent, solute,* and *colligative properties* (Section 14.1).

b. Use the following concentration units: molality, mole fraction, weight percent, and parts per million (Section 14.1). General ChemistryNow homework: Study Question(s) 2, 6, 10, 12, 62

c. Understand the distinctions between saturated, unsaturated, and supersaturated solutions (Section 14.2).

d. Define and illustrate the terms *miscible* and *immiscible* (Section 14.2).

Understand the solution process

a. Describe the process of dissolving a solute in a solvent, including the energy changes that may occur (Section 14.2). General ChemistryNow homework: SQ(s) 16

b. Understand the relationship of lattice energy and enthalpy of hydration to the enthalpy of solution for an ionic solute (Section 14.2). General ChemistryNow homework: SQ(s) 20

c. Describe the effect of pressure and temperature on the solubility of a solute (Section 14.2).

d. Use Henry's law to calculate the solubility of a gas in a solvent (Section 14.2). General ChemistryNow homework: SQ(s) 24

e. Apply Le Chatelier's principle to the change in solubility of gases with pressure and temperature changes (Section 14.2).

Understand and use the colligative properties of solutions

a. Calculate the mole fraction of a solute or solvent ($X_{solvent}$) and the effect of a solute on solvent vapor pressure ($P_{solvent}$) using Raoult's law (Section 14.4). General ChemistryNow homework: SQ(s) 26

b. Calculate the boiling point elevation or freezing point depression caused by a solute in a solvent (Section 14.4). General ChemistryNow homework: SQ(s) 30, 34, 36, 51

c. Use colligative properties to determine the molar mass of a solute (Section 14.4). General ChemistryNow homework: SQ(s) 39, 43, 67

d. Characterize the effect of ionic solutes on colligative properties (Section 14.4). General ChemistryNow homework: SQ(s) 49

e. Use the van't Hoff factor, *i*, in calculations involving colligative properties (Section 14.4). General ChemistryNow homework: SQ(s) 47

f. Calculate the osmotic pressure (Π) for solutions, and use the equation defining osmotic pressure to determine the molar mass of a solute (Section 14.4). General ChemistryNow homework: SQ(s) 53, 88

Describe colloids and their applications

a. Recognize the difference among a homogeneous solution, a suspension, and a colloid (or colloidal dispersion) (Section 14.5).

b. Recognize hydrophobic and hydrophilic colloids (Section 14.5).

c. Describe the action of a surfactant (Section 14.5).

Key Equations

Equation 14.1 (page 659)—Molality is defined as the amount of solute per kilogram of solvent.

$$\text{Molality of solute } (m) = \frac{\text{amount of solute (mol)}}{\text{mass of solvent (kg)}}$$

Equation 14.2 (page 660)—The mole fraction, X, of a solution component is defined as the number of moles of a given component of a mixture (n, mol) divided by the total number of moles of all of the components of the mixture.

$$\text{Mole fraction of A } (X_A) = \frac{n_A}{n_A + n_B + n_C + \cdots}$$

Equation 14.3 (page 660)—Weight percent is the mass of one component divided by the total mass of the mixture (multiplied by 100%).

$$\text{Weight \% A} = \frac{\text{mass of A}}{\text{mass of A} + \text{mass of B} + \text{mass of C} + \cdots} \times 100\%$$

Equation 14.4 (page 669)—Henry's law states that the solubility of a gas, S_g, is equal to the product of the partial pressure of the gaseous solute (P_g) and a constant (k_H) characteristic of the solute and solvent.

$$S_g = k_H P_g$$

Equation 14.5 (page 673)—Raoult's law states that the equilibrium vapor pressure of a solvent over a solution at a given temperature, P_{solvent}, is the product of the mole fraction of the solvent (X_{solvent}) and the vapor pressure of the pure solvent (P°_{solvent}).

$$P_{\text{solvent}} = X_{\text{solvent}} P^\circ_{\text{solvent}}$$

Equation 14.6 (page 674)—The change in the equilibrium vapor pressure of the solvent at a given temperature, $\Delta P_{\text{solvent}}$, is the product of the mole fraction of the solute, X_{solute}, and the vapor pressure of the pure solvent.

$$\Delta P_{\text{solvent}} = -X_{\text{solute}} P^\circ_{\text{solvent}}$$

Equation 14.7 (page 676)—The elevation in boiling point of the solvent in a solution, ΔT_{bp}, is the product of the molality of the solute, m_{solute}, and a constant characteristic of the solvent, K_{bp}.

$$\text{Elevation in boiling point: } \Delta T_{\text{bp}} = K_{\text{bp}} m_{\text{solute}}$$

Equation 14.8 (page 677)—The depression of the freezing point of the solvent in a solution, ΔT_{fp}, is the product of the molality of the solute, m_{solute}, and a constant characteristic of the solvent, K_{fp}.

$$\text{Depression of freezing point, } \Delta T_{\text{fp}} = K_{\text{fp}} m_{\text{solute}}$$

Equation 14.9 (page 680)—This equation takes into account the possible dissociation of the solute. Here i, the van't Hoff factor, is the ratio of the measured freezing point depression and the freezing point depression calculated assuming no solute dissociation. As such, it is related to the relative number of particles produced by a solute.

$$\Delta T_{fp}, \text{ measured} = K_{fp} \times m \times i$$

Equation 14.10 (page 683)—The osmotic pressure, Π, is the product of the solute concentration c (in mol/L), the universal gas constant R (0.0821 L · atm/K · mol), and the temperature T (in kelvins).

$$\Pi = cRT$$

Study Questions

▲ denotes more challenging questions.

■ denotes questions available in the Homework and Goals section of the General ChemistryNow CD-ROM or website.

Blue numbered questions have answers in Appendix O and fully worked solutions in the *Student Solutions Manual*.

Structures of many of the compounds used in these questions are found on the General ChemistryNow CD-ROM or website in the Models folder.

GENERAL
Chemistry ⚛ Now™ Assess your understanding of this chapter's topics with additional quizzing and conceptual questions at **http://now.brookscole.com/kotz6e**

Practicing Skills

Concentration
(See Examples 14.1 and 14.2 and General ChemistryNow Screen 14.2.)

1. Suppose you dissolve 2.56 g of succinic acid, $C_2H_4(CO_2H)_2$, in 500. mL of water. Assuming that the density of water is 1.00 g/cm^3, calculate the molality, mole fraction, and weight percentage of acid in the solution.

2. ■ Assume you dissolve 45.0 g of camphor, $C_{10}H_{16}O$, in 425 mL of ethanol, C_2H_5OH. Calculate the molality, mole fraction, and weight percent of camphor in this solution. (The density of ethanol is 0.785 g/mL.)

3. Fill in the blanks in the table. Aqueous solutions are assumed.

Compound	Molality	Weight Percent	Mole Fraction
NaI	0.15	_____	_____
C_2H_5OH	_____	5.0	_____
$C_{12}H_{22}O_{11}$	0.15	_____	_____

4. Fill in the blanks in the table. Aqueous solutions are assumed.

Compound	Molality	Weight Percent	Mole Fraction
KNO_3	_____	10.0	_____
CH_3CO_2H	0.0183	_____	_____
$HOCH_2CH_2OH$	_____	18.0	_____

5. What mass of Na_2CO_3 must you add to 125 g of water to prepare 0.200 m Na_2CO_3? What is the mole fraction of Na_2CO_3 in the resulting solution?

6. ■ You want to prepare a solution that is 0.0512 m in $NaNO_3$. What mass of $NaNO_3$ must be added to 500. g of water? What is the mole fraction of $NaNO_3$ in the solution?

7. You wish to prepare an aqueous solution of glycerol, $C_3H_5(OH)_3$, in which the mole fraction of the solute is 0.093. What mass of glycerol must you add to 425 g of water to make this solution? What is the molality of the solution?

8. You want to prepare an aqueous solution of ethylene glycol, $HOCH_2CH_2OH$, in which the mole fraction of solute is 0.125. What mass of ethylene glycol, in grams, should you combine with 955 g of water? What is the molality of the solution?

9. Hydrochloric acid is sold as a concentrated aqueous solution. If the molarity of commercial HCl is 12.0 and its density is 1.18 g/cm^3, calculate the following:
   (a) the molality of the solution
   (b) the weight percent of HCl in the solution

10. ■ Concentrated sulfuric acid has a density of 1.84 g/cm^3 and is 95.0% by weight H_2SO_4. What is the molality of this acid? What is its molarity?

11. The average lithium ion concentration in sea water is 0.18 ppm. What is the molality of Li^+ in sea water?

12. ■ Silver ion has an average concentration of 28 ppb (parts per billion) in U.S. water supplies.
   (a) What is the molality of the silver ion?
   (b) If you wanted 1.0×10^2 g of silver and could recover it chemically from water supplies, what volume of water in liters, would you have to treat? (Assume the density of water is 1.0 g/cm^3.)

The Solution Process

(See Example 14.3 and General ChemistryNow Screens 14.3 and 14.4.)

13. Which pairs of liquids will be miscible?
   (a) H_2O and $CH_3CH_2CH_2CH_3$
   (b) C_6H_6 (benzene) and CCl_4
   (c) H_2O and CH_3CO_2H

14. Acetone, CH_3COCH_3, is quite soluble in water. Explain why this should be so.

15. Use the data of Table 14.1 to calculate the enthalpy of solution of LiCl.

16. ■ Use the following data to calculate the enthalpy of solution of sodium perchlorate, $NaClO_4$:

$$\Delta H_f^\circ(s) = -382.9 \text{ kJ/mol}$$

$$\Delta H_f^\circ(aq, 1 \; m) = -369.5 \text{ kJ/mol}$$

17. You make a saturated solution of NaCl at 25 °C. No solid is present in the beaker holding the solution. What can be done to increase the amount of dissolved NaCl in this solution? (See Figure 14.12.)
   (a) Add more solid NaCl.
   (b) Raise the temperature of the solution.
   (c) Raise the temperature of the solution and add some NaCl.
   (d) Lower the temperature of the solution and add some NaCl.

18. Some lithium chloride, LiCl, is dissolved in 100 mL of water in one beaker and some Li_2SO_4 is dissolved in 100 mL of water in another beaker. Both are at 10 °C and both are saturated solutions; some solid remains undissolved in each beaker. Describe what you would observe as the temperature is raised. The following data are available to you from a handbook of chemistry:

	Solubility (g/100 mL)	
Compound	10 °C	40 °C
Li_2SO_4	35.5	33.7
LiCl	74.5	89.8

19. In each pair of ionic compounds, which is more likely to have the more negative heat of hydration? Briefly explain your reasoning in each case.
   (a) LiF or RbF
   (b) KNO_3 or $Ca(NO_3)_2$
   (c) CsBr or $CuBr_2$

20. ■ When salts of Mg^{2+}, Ca^{2+}, and Be^{2+} are placed in water, the positive ion is hydrated (as is the negative ion). Which of these three cations is most strongly hydrated? Which one is least strongly hydrated?

Henry's Law

(See Example 14.4 and General ChemistryNow Screen 14.5.)

21. The partial pressure of O_2 in your lungs varies from 25 mm Hg to 40 mm Hg. What mass of O_2 can dissolve in 1.0 L of water at 25 °C if the partial pressure of O_2 is 40 mm Hg?

22. The Henry's law constant for O_2 in water at 25 °C is 1.66×10^{-6} M/mm Hg. Which of the following is a reasonable constant when the temperature is 50 °C? Explain the reason for your choice.
   (a) 8.80×10^{-7} M/mm Hg   (c) 1.66×10^{-6} M/mm Hg
   (b) 3.40×10^{-6} M/mm Hg   (d) 8.40×10^{-5} M/mm Hg

23. An unopened soda can has an aqueous CO_2 concentration of 0.0506 M at 25 °C. What is the pressure of CO_2 gas in the can?

24. ■ Hydrogen gas has a Henry's law constant of 1.07×10^{-6} M/mm Hg at 25 °C when dissolving in water. If the total pressure of gas (H_2 gas plus water vapor) over water is 1.0 atm, what is the concentration of H_2 in the water in grams per milliliter? (See Appendix G for the vapor pressure of water.)

Raoult's Law

(See Example 14.5 and General ChemistryNow Screen 14.7.)

25. A 35.0-g sample of ethylene glycol, $HOCH_2CH_2OH$, is dissolved in 500.0 g of water. The vapor pressure of water at 32 °C is 35.7 mm Hg. What is the vapor pressure of the water–ethylene glycol solution at 32 °C? (Ethylene glycol is nonvolatile.)

26. ■ Urea, $(NH_2)_2CO$, which is widely used in fertilizers and plastics, is quite soluble in water. If you dissolve 9.00 g of urea in 10.0 mL of water, what is the vapor pressure of the solution at 24 °C? Assume the density of water is 1.00 g/mL.

27. Pure ethylene glycol, $HOCH_2CH_2OH$, is added to 2.00 kg of water in the cooling system of a car. The vapor pressure of the water in the system when the temperature is 90 °C is 457 mm Hg. What mass of glycol was added? (Assume the solution is ideal. See Appendix G for the vapor pressure of water.)

28. Pure iodine (105 g) is dissolved in 325 g of CCl_4 at 65 °C. Given that the vapor pressure of CCl_4 at this temperature is 531 mm Hg, what is the vapor pressure of the CCl_4–I_2 solution at 65 °C? (Assume that I_2 does not contribute to the vapor pressure.)

Boiling Point Elevation

(See Example 14.6 and General ChemistryNow Screen 14.8.)

29. Verify that 0.200 mol of a nonvolatile solute in 125 g of benzene (C_6H_6) produces a solution whose boiling point is 84.2 °C.

30. ■ What is the boiling point of a solution composed of 15.0 g of urea, $(NH_2)_2CO$, in 0.500 kg of water?

---

31. What is the boiling point of a solution composed of 15.0 g of $CHCl_3$ and 0.515 g of the nonvolatile solute acenaphthene, $C_{12}H_{10}$, a component of coal tar?

32. What is the boiling point of a solution composed of 0.755 g of caffeine, $C_8H_{10}O_2N_4$, in 95.6 g of benzene, C_6H_6?

33. Phenanthrene, $C_{14}H_{10}$, is an aromatic hydrocarbon. If you dissolve some phenanthrene in 50.0 g of benzene, the boiling point of the solution is 80.51 °C. What mass of the hydrocarbon must have been dissolved?

34. ■ A solution of glycerol, $C_3H_5(OH)_3$, in 735 g of water has a boiling point of 104.4 °C at a pressure of 760 mm Hg. What is the mass of glycerol in the solution? What is the mole fraction of the solute?

Freezing Point Depression

(See Example 14.7 and General ChemistryNow Screen 14.8.)

35. A mixture of ethanol, C_2H_5OH, and water has a freezing point of −16.0 °C.
    (a) What is the molality of the alcohol?
    (b) What is the weight percent of alcohol in the solution?

36. ■ Some ethylene glycol, $HOCH_2CH_2OH$, is added to your car's cooling system along with 5.0 kg of water. If the freezing point of the water–glycol solution is −15.0 °C, what mass of $HOCH_2CH_2OH$ must have been added?

37. You dissolve 15.0 g of sucrose, $C_{12}H_{22}O_{11}$, in a cup of water (225 g). What is the freezing point of the solution?

38. Assume a bottle of wine consists of an 11 weight percent solution of ethanol (C_2H_5OH) in water. If the bottle of wine is chilled to −20 °C, will the solution begin to freeze?

Colligative Properties and Molar Mass Determination

(See Example 14.8.)

39. ■ You add 0.255 g of an orange, crystalline compound whose empirical formula is $C_{10}H_8Fe$ to 11.12 g of benzene. The boiling point of the benzene rises from 80.10 °C to 80.26 °C. What is the molar mass and molecular formula of the compound?

40. Butylated hydroxyanisole (BHA) is used as an antioxidant in margarine and other fats and oils; it prevents oxidation and prolongs the shelf life of the food. What is the molar mass of BHA if 0.640 g of the compound, dissolved in 25.0 g of chloroform, produces a solution whose boiling point is 62.22 °C?

41. Benzyl acetate is one of the active components of oil of jasmine. If 0.125 g of the compound is added to 25.0 g of chloroform ($CHCl_3$), the boiling point of the solution is 61.82 °C. What is the molar mass of benzyl acetate?

42. Anthracene, a hydrocarbon obtained from coal, has an empirical formula of C_7H_5. To find its molecular formula you dissolve 0.500 g in 30.0 g of benzene. The boiling point of the pure benzene is 80.10 °C, whereas the solution has a boiling point of 80.34 °C. What is the molecular formula of anthracene?

43. ■ An aqueous solution contains 0.180 g of an unknown, nonionic solute in 50.0 g of water. The solution freezes at −0.040 °C. What is the molar mass of the solute?

44. The organic compound called aluminon is used as a reagent to test for the presence of the aluminum ion in aqueous solution. A solution of 2.50 g of aluminon in 50.0 g of water freezes at −0.197 °C. What is the molar mass of aluminon?

45. ■ The melting point of pure biphenyl ($C_{12}H_{10}$) is found to be 70.03 °C. If 0.100 g of naphthalene is added to 10.0 g of biphenyl, the freezing point of the mixture is 69.40 °C. If K_{fp} for biphenyl is −8.00 °C/m, what is the molar mass of naphthalene?

46. Phenylcarbinol is used in nasal sprays as a preservative. A solution of 0.52 g of the compound in 25.0 g of water has a melting point of −0.36 °C. What is the molar mass of phenylcarbinol?

Colligative Properties of Ionic Compounds

(See Example 14.9 and General ChemistryNow Screen 14.8.)

47. ■ If 52.5 g of LiF is dissolved in 306 g of water, what is the expected freezing point of the solution? (Assume the van't Hoff factor, i, for LiF is 2.)

48. To make homemade ice cream, you cool the milk and cream by immersing the container in ice and a concentrated solution of rock salt (NaCl) in water. If you want to have a water–salt solution that freezes at −10. °C, what mass of NaCl must you add to 3.0 kg of water? (Assume the van't Hoff factor, i, for NaCl is 1.85.)

49. ■ List the following aqueous solutions in order of increasing melting point. (The last three are all assumed to dissociate completely into ions in water.)
    (a) 0.1 m sugar   (c) 0.08 m $CaCl_2$
    (b) 0.1 m NaCl   (d) 0.04 m Na_2SO_4

50. Arrange the following aqueous solutions in order of decreasing freezing point. (The last three are all assumed to dissociate completely into ions in water.)
    (a) 0.20 m ethylene glycol (nonvolatile, nonelectrolyte)
    (b) 0.12 m K_2SO_4
    (c) 0.10 m $MgCl_2$ (d) 0.12 m KBr

Osmosis

(See Example 14.10 and General ChemistryNow Screen 14.9.)

51. ■ An aqueous solution contains 3.00% phenylalanine ($C_9H_{11}NO_2$) by mass. Assume the phenylalanine is nonionic and nonvolatile. Find the following:
    (a) the freezing point of the solution
    (b) the boiling point of the solution
    (c) the osmotic pressure of the solution at 25 °C

    In your view, which of these values is most easily measurable in the laboratory?

52. Estimate the osmotic pressure of human blood at 37 °C. Assume blood is isotonic with a 0.154 M NaCl solution, and assume the van't Hoff factor, i, is 1.9 for NaCl.

▲ More challenging   ■ In General ChemistryNow   Blue-numbered questions answered in Appendix O

53. ■ An aqueous solution containing 1.00 g of bovine insulin (a protein, not ionized) per liter has an osmotic pressure of 3.1 mm Hg at 25 °C. Calculate the molar mass of bovine insulin.

54. Calculate the osmotic pressure of a 0.0120 M solution of NaCl in water at 0 °C. Assume the van't Hoff factor, i, is 1.94 for this solution.

Colloids

(See Section 14.5 and General ChemistryNow Screen 14.10.)

55. When solutions of $BaCl_2$ and Na_2SO_4 are mixed, the mixture becomes cloudy. After a few days, a white solid is observed on the bottom of the beaker with a clear liquid above it.
    (a) Write a balanced equation for the reaction that occurs.
    (b) Why is the solution cloudy at first?
    (c) What happens during the few days of waiting?

56. The dispersed phase of a certain colloidal dispersion consists of spheres of diameter 1.0×10^2 nm.
    (a) What is the volume ($V = \frac{4}{3}\pi r^3$) and surface area ($A = 4\pi r^2$) of each sphere?
    (b) How many spheres are required to give a total volume of 1.0 cm^3? What is the total surface area of these spheres in square meters?

General Questions

These questions are not designated as to type or location in the chapter. They may combine several concepts.

57. Which salt, Li_2SO_4 or Cs_2SO_4, is expected to have the more exothermic heat of hydration? Explain briefly.

58. (a) Which aqueous solution is expected to have the higher boiling point: 0.10 m Na_2SO_4 or 0.15 m sugar?
    (b) For which aqueous solution is the vapor pressure of water higher: 0.30 m NH_4NO_3 or 0.15 m Na_2SO_4?

59. Arrange the following aqueous solutions in order of (i) increasing vapor pressure of water and (ii) increasing boiling point.
    (a) 0.35 m $HOCH_2CH_2OH$ (a nonvolatile solute)
    (b) 0.50 m sugar
    (c) 0.20 m KBr (a strong electrolyte)
    (d) 0.20 m Na_2SO_4 (a strong electrolyte)

60. Making homemade ice cream is one of life's great pleasures. Fresh milk and cream, sugar, and flavorings are churned in a bucket suspended in an ice–water mixture, the freezing point of which has been lowered by adding rock salt. One manufacturer of home ice cream freezers recommends adding 2.50 lb (1130 g) of rock salt (NaCl) to 16.0 lb of ice (7250 g) in a 4-qt freezer. For the solution when this mixture melts, calculate the following:
    (a) the weight percent of NaCl
    (b) the mole fraction of NaCl
    (c) the molality of the solution

61. Dimethylglyoxime [DMG, $(CH_3CNOH)_2$] is used as a reagent to precipitate nickel ion. Assume that 53.0 g of DMG has been dissolved in 525 g of ethanol (C_2H_5OH).

Charles D. Winters

The red, insoluble compound formed between nickel(II) ion and dimethylglyoxime (DMG) is precipitated when DMG is added to a basic solution of Ni^{2+}(aq).

    (a) What is the mole fraction of DMG?
    (b) What is the molality of the solution?
    (c) What is the vapor pressure of the ethanol over the solution at ethanol's normal boiling point of 78.4 °C?
    (d) What is the boiling point of the solution? (DMG does not produce ions in solution.) (K_{bp} for ethanol = + 1.22 °C/m)

62. ■ A 10.7 m solution of NaOH has a density of 1.33 g/cm^3 at 20 °C. Calculate the following:
    (a) the mole fraction of NaOH
    (b) the weight percent of NaOH
    (c) the molarity of the solution

63. Concentrated aqueous ammonia has a molarity of 14.8 and a density of 0.90 g/cm^3. What is the molality of the solution? Calculate the mole fraction and weight percent of NH_3.

64. If you dissolve 2.00 g of $Ca(NO_3)_2$ in 750 g of water, what is the molality of $Ca(NO_3)_2$? What is the total molality of ions in solution? (Assume total dissociation of the ionic solid.)

65. If you want a solution that is 0.100 m in ions, what mass of Na_2SO_4 must you dissolve in 125 g of water? (Assume total dissociation of the ionic solid.)

66. Consider the following aqueous solutions: (i) 0.20 m $HOCH_2CH_2OH$ (nonvolatile, nonelectrolyte); (ii) 0.10 m $CaCl_2$; (iii) 0.12 m KBr; and (iv) 0.12 m Na_2SO_4.
    (a) Which solution has the highest boiling point?
    (b) Which solution has the lowest freezing point?
    (c) Which solution has the highest water vapor pressure?

67. ■ (a) Which solution is expected to have the higher boiling point: 0.20 m KBr or 0.30 m sugar?
    (b) Which aqueous solution has the lower freezing point: 0.12 m NH_4NO_3 or 0.10 m Na_2CO_3?

68. The solubility of NaCl in water at 100 °C is 39.1 g/100. g of water. Calculate the boiling point of this solution. (Assume $i = 1.85$ for NaCl.)

▲ More challenging   ■ In General ChemistryNow   **Blue-numbered questions** answered in Appendix O

69. Instead of using NaCl to melt the ice on your sidewalk, you decide to use $CaCl_2$. If you add 35.0 g of $CaCl_2$ to 150. g of water, what is the freezing point of the solution? (Assume $i = 2.7$ for $CaCl_2$.)

70. The smell of ripe raspberries is due to *p*-hydroxyphenyl-2-butanone, which has the empirical formula C_5H_6O. To find its molecular formula, you dissolve 0.135 g in 25.0 g of chloroform, $CHCl_3$. The boiling point of the solution is 61.82 °C. What is the molecular formula of the solute?

71. Hexachlorophene has been used in germicidal soap. What is its molar mass if 0.640 g of the compound, dissolved in 25.0 g of chloroform, produces a solution whose boiling point is 61.93 °C?

72. The solubility of ammonium formate, NH_4CHO_2, in 100 g of water is 102 g at 0 °C and 546 g at 80 °C. A solution is prepared by dissolving NH_4CHO_2 in 200 g of water until no more will dissolve at 80 °C. The solution is then cooled to 0 °C. What mass of NH_4CHO_2 precipitates? (Assume that no water evaporates and that the solution is not super-saturated.)

73. How much N_2 can dissolve in water at 25 °C if the N_2 partial pressure is 585 mm Hg?

74. Cigars are best stored in a "humidor" at 18 °C and 55% relative humidity. This means the pressure of water vapor should be 55% of the vapor pressure of pure water at the same temperature. The proper humidity can be maintained by placing a solution of glycerol $[C_3H_5(OH)_3]$ and water in the humidor. Calculate the percent by mass of glycerol that will lower the vapor pressure of water to the desired value. (The vapor pressure of glycerol is zero.)

75. An aqueous solution containing 10.0 g of starch per liter has an osmotic pressure of 3.8 mm Hg at 25 °C.
   (a) What is the molar mass of starch? (Because not all starch molecules are identical, the result will be an average.)
   (b) What is the freezing point of the solution? Would it be easy to determine the molecular weight of starch by measuring the freezing point depression? (Assume that the molarity and molality are the same for this solution.)

76. Vinegar is a 5% solution (by weight) of acetic acid in water. Determine the mole fraction and molality of acetic acid. What is the concentration of acetic acid in parts per million (ppm)? Explain why it is not possible to calculate the molarity of this solution from the information provided.

77. ▲ A solution of 5.00 g of acetic acid in 100. g of benzene freezes at 3.37 °C. A solution of 5.00 g of acetic acid in 100. g of water freezes at −1.49 °C. Find the molar mass of acetic acid from each of these experiments. What can you conclude about the state of the acetic acid molecules dissolved in each of these solvents? Recall the discussion of hydrogen bonding in Section 13.3 (and see Figure 13.9), and propose a structure for the species in benzene solution.

78. Some ethylene glycol, $HOCH_2CH_2OH$, is added to your car's cooling system along with 5.0 kg of water. If the freezing point of the water–glycol solution is −15.0 °C, what is the boiling point of the solution?

79. Calculate the enthalpies of solution for Li_2SO_4 and K_2SO_4. Are the solution processes exothermic or endothermic? Compare them with LiCl and KCl. What similarities or differences do you find?

Compound	ΔH_f° (s) (kJ/mol)	ΔH_f° (aq, 1 *m*) (kJ/mol)
Li_2SO_4	−1436.4	−1464.4
K_2SO_4	−1437.7	−1414.0

80. ▲ Water at 25 °C has a density of 0.997 g/cm^3. Calculate the molality and molarity of pure water at this temperature.

81. ▲ If a volatile solute is added to a volatile solvent, both substances contribute to the vapor pressure over the solution. Assuming an ideal solution, the vapor pressure of each is given by Raoult's law, and the total vapor pressure is the sum of the vapor pressures for each component. A solution, assumed to be ideal, is made from 1.0 mol of toluene ($C_6H_5CH_3$) and 2.0 mol of benzene (C_6H_6). The vapor pressures of the pure solvents are 22 mm Hg and 75 mm Hg, respectively, at 20 °C. What is the total vapor pressure of the mixture? What is the mole fraction of each component in the liquid and in the vapor?

82. A solution is made by adding 50.0 mL of ethanol (C_2H_5OH, $d = 0.789$ g/ml) to 50.0 mL of water ($d = 0.998$ g/mL). What is the total vapor pressure over the solution at 20 °C? (See Study Question 81.) The vapor pressure of ethanol at 20 °C is 43.6 mm Hg.

83. ▲ A solution of benzoic acid in benzene has a freezing point of 3.1 °C and a boiling point of 82.6 °C. (The freezing point of pure benzene is 5.12 °C and its boiling point is 80.1 °C.) The structure of benzoic acid is

What can you conclude about the state of the benzoic acid molecules at the two different temperatures? Recall the discussion of hydrogen bonding in Section 13.3 and see Figure 13.9.

84. ▲ You dissolve 5.0 mg of iodine, I_2, in 25 mL of water. You then add 10.0 mL of CCl_4 and shake the mixture. If I_2 is 85 times more soluble in CCl_4 than in H_2O (on a volume basis), what are the masses of I_2 in the water and CCl_4 layers after shaking? (See Figure 14.6.)

85. A 2.0% (by mass) aqueous solution of novocainium chloride ($C_{13}H_{21}ClN_2O_2$) freezes at −0.237 °C. Calculate the van't Hoff factor, *i*. How many moles of ions are in the solution per mole of compound?

86. A solution is 4.00% (by mass) maltose and 96.00% water. It freezes at −0.229 °C.
   (a) Calculate the molar mass of maltose (which is not an ionic compound).
   (b) The density of the solution is 1.014 g/mL. Calculate the osmotic pressure of the solution.

▲ More challenging   ■ In General ChemistryNow   Blue-numbered questions answered in Appendix O

87. ▲ The following table lists the concentrations of the principal ions in sea water:

Ion	Concentration (ppm)
Cl^-	1.95×10^4
Na^+	1.08×10^4
Mg^{2+}	1.29×10^3
SO_4^{2-}	9.05×10^2
Ca^{2+}	4.12×10^2
K^+	3.80×10^2
Br^-	67

(a) Calculate the freezing point of water.

(b) Calculate the osmotic pressure of sea water at 25 °C. What is the minimum pressure needed to purify sea water by reverse osmosis?

88. ■ ▲ A tree is exactly 10 m tall.

(a) What must be the total molarity of the solutes if sap rises to the top of the tree by osmotic pressure at 20 °C? Assume the groundwater outside the tree is pure water and that the density of the sap is 1.0 g/mL. (1 mm Hg = 13.6 mm H_2O.)

(b) If the only solute in the sap is sucrose, $C_{12}H_{22}O_{11}$, what is its percent by mass?

89. A 2.00% solution of H_2SO_4 in water freezes at −0.796 °C.

(a) Calculate the van't Hoff factor, i.

(b) Which of the following best represents sulfuric acid in a dilute aqueous solution: H_2SO_4, $H^+ + HSO_4^-$, or $2\,H^+ + SO_4^{2-}$?

90. A compound is known to be a potassium halide, KX. If 4.00 g of the salt is dissolved in exactly 100 g of water, the solution freezes at −1.28 °C. Identify the halide ion in this formula.

Summary and Conceptual Questions

The following questions may use concepts from the preceding chapters.

91. A newly synthesized compound containing boron and fluorine is 22.1% boron. Dissolving 0.146 g of the compound in 10.0 g of benzene gives a solution with a vapor pressure of 94.16 mm Hg at 25 °C. (The vapor pressure of pure benzene at this temperature is 95.26 mm Hg.) In a separate experiment, it is found that the compound does not have a dipole moment.

(a) What is the molecular formula for the compound?

(b) Draw a Lewis structure for the molecule, and suggest a possible molecular structure. Give the bond angles in the molecule and the hybridization of the boron atom.

92. In chemical research we often send newly synthesized compounds to commercial laboratories for analysis. These laboratories determine the weight percent of C and H by burning the compound and collecting the evolved CO_2 and H_2O. They determine the molar mass by measuring the osmotic pressure of a solution of the compound. Calculate the empirical and molecular formulas of a compound, C_xH_yCr, given the following information:

(a) The compound contains 73.94% C and 8.27% H; the remainder is chromium.

(b) At 25 °C, the osmotic pressure of a solution containing 5.00 mg of the unknown dissolved in exactly 100 mL of chloroform solution is 3.17 mm Hg.

93. If you dissolve equal molar amounts of NaCl and $CaCl_2$ in water, the $CaCl_2$ lowers the freezing point of the water almost 1.5 times as much as the NaCl. Why?

94. Explain why a cucumber shrivels up when it is placed in a concentrated solution of salt.

95. A 100.-gram sample of sodium chloride (NaCl) is added to 100. mL of water at 0 °C. After equilibrium is reached, about 64 g of solid remains undissolved. Describe the equilibrium that exists in this system at the particulate level.

96. In words, describe an experimental procedure using freezing point depression to determine the molar mass of an unknown compound.

97. Which of the following substances is likely to dissolve in water, and which is likely to dissolve in benzene (C_6H_6)?

(a) $NaNO_3$

(b) diethyl ether, $CH_3CH_2OCH_2CH_3$

(c) naphthalene, $C_{10}H_8$

(d) NH_4Cl

98. Account for the fact that alcohols such as methanol (CH_3OH) and ethanol (C_2H_5OH) are quite miscible with water, whereas an alcohol with a long-carbon chain, such as octanol ($C_8H_{17}OH$), is poorly soluble in water.

99. Starch contains C—C, C—H, C—O, and O—H bonds. Hydrocarbons have only C—C and C—H bonds. Both starch and hydrocarbons can form colloidal dispersions in water. Which dispersion is classified as hydrophobic? Which is hydrophilic? Explain briefly.

100. Which substance would have the greater influence on the vapor pressure of water when added to 1000. g of the liquid: 10.0 g of sucrose ($C_{12}H_{22}O_{11}$) or 10.0 g of ethylene glycol [$HOCH_2CH_2OH$]?

101. Suppose you have two aqueous solutions separated by a semipermeable membrane. One contains 5.85 g of NaCl dissolved in 100. mL of solution, and the other contains 8.88 g of KNO_3 dissolved in 100. mL of solution. In which direction will solvent flow: from the NaCl solution to the KNO_3 solution, or from KNO_3 to NaCl? Explain briefly.

102. A protozoan (single-celled animal) that normally lives in the ocean is placed in fresh water. Will it shrivel or burst? Explain briefly.

Do you need a live tutor for homework problems?
Access vMentor at General ChemistryNow at
http://now.brookscole.com/kotz6e
for one-on-one tutoring from a chemistry expert

▲ More challenging   ■ In General ChemistryNow   **Blue-numbered questions** answered in Appendix O

15—Principles of Reactivity: Chemical Kinetics

Hard-to-digest foods. Foods such as beans, cabbage, and broccoli are known to produce flatulence in some people due to incomplete digestion of complex sugars. However, an enzyme, when taken with food, can help break down these complex sugars and avoid "problem gas."

Faster and Faster

Certain foods such as beans, cabbage, and broccoli contain complex sugars known as oligosaccharides. Although these compounds are broken down to simple sugars during the digestive process, some people have a problem breaking them down completely. This failure can lead to a condition known politely as flatulence, because the undigested material is eventually fermented by anaerobic organisms in the colon to produce gases such as CO_2, H_2, CH_4, and small amounts of smelly compounds.

To help people who have this problem, a commercial product called Beano was developed. Its maker's advertising material states that Beano "is a food enzyme from a natural source that breaks down the complex sugars in gassy foods, making them more digestible."

As you will learn in this chapter, which describes the factors affecting the speed of chemical reactions, enzymes are biological catalysts. Their role is to speed up chemical reactions. One of the enzymes in Beano, galactosidase, accelerates the breakdown of the

(a) $t = 0$

(b) $t = 9$ sec

(c) $t = 28$ sec

(d) $t = 34$ sec

(e) $t = 37$ sec

Figure A CO_2 in water. (a) A cold solution of CO_2 in water. (b) A few drops of a dye (bromthymol blue) are added to the cold solution. The yellow color of the dye indicates an acidic solution. (c) A less than stoichiometric amount of sodium hydroxide is added, converting H_2CO_3 to HCO_3^- (and CO_3^{2-}). (d) The blue color of the dye indicates a basic solution. (e) The blue color begins to fade after some seconds as CO_2 slowly forms more H_2CO_3. The amount of H_2CO_3 formed is finally sufficient to consume the added NaOH and the solution is again acidic.

Chapter Goals

See Chapter Goals Revisited (page 741). Test your knowledge of these goals by taking the exam-prep quiz on the General ChemistryNow CD-ROM or website.

- Understand rates of reaction and the conditions affecting rates.
- Derive the rate equation, rate constant, and reaction order from experimental data.
- Use integrated rate laws.
- Understand the collision theory of reaction rates and the role of activation energy.
- Relate reaction mechanisms and rate laws.

Chapter Outline

oligosaccharides in certain foods to the simple sugars galactose and glucose.

$$\text{Oligosaccharide} + H_2O \xrightarrow{\text{galactosidase}} \text{galactose} + \text{glucose}$$

Carbonic anhydrase is one of the many enzymes that play important roles in biological processes (see page 732). Carbon dioxide dissolves in water to a small extent to produce carbonic acid, which ionizes to give H^+ and HCO_3^- ions.

$$CO_2(g) \longrightarrow CO_2(aq) \tag{1}$$

$$CO_2(aq) + H_2O(\ell) \longrightarrow H_2CO_3(aq) \tag{2}$$

$$H_2CO_3(aq) \longrightarrow H^+(aq) + HCO_3^-(aq) \tag{3}$$

Carbonic anhydrase speeds up reactions 1 and 2. Many of the H^+ ions produced by ionization of H_2CO_3 (reaction 3) are picked up by hemoglobin in the blood as hemoglobin loses O_2. The resulting HCO_3^- ions are transported back to the lungs. When hemoglobin again takes on O_2, it releases H^+ ions. These ions and HCO_3^- re-form H_2CO_3, from which CO_2 is liberated and exhaled.

A simple experiment illustrates the effect of carbonic anhydrase. First, add a small amount of NaOH to a cold, aqueous solution of CO_2 (Figure A). The solution becomes basic immediately, because there is not enough H_2CO_3 in the solution to use up the NaOH. After some seconds, however, dissolved CO_2 slowly produces more H_2CO_3, which consumes NaOH, and the solution again becomes acidic.

Now try the experiment again, this time adding a few drops of blood to the solution (Figure B). Carbonic anhydrase in blood speeds up reactions 1 and 2 by a factor of about 10^7, as evidenced by the more rapid reaction that occurs under these conditions.

To learn more about Beano and enzymes, see J. R. Hardee, T. M. Montgomery, and W. H. Jones: *Journal of Chemical Education*, Vol. 77, p. 498, 2000. For a more detailed discussion of the rate of conversion of aqueous CO_2 to H_2CO_3 and HCO_3^-, see J. Bell: *Journal of Chemical Education*, Vol. 77, p. 1098, 2000.

Photos: Charles D. Winters

(a) $t = 0$ (b) $t = 3$ sec (c) $t = 15$ sec (d) $t = 17$ sec (e) $t = 21$ sec

Figure B   Action of carbonic anhydrase. (a) A few drops of blood are added to a cold solution of CO_2 in water. (b) The dye indicates an acidic solution. (c, d) A less than stoichiometric amount of sodium hydroxide is added, converting H_2CO_3 to HCO_3^- (and CO_3^{2-}). The dye's blue color indicates a basic solution. (e) The blue color begins to fade after a few seconds as more H_2CO_3 forms, and the solution again becomes acidic. The formation of H_2CO_3 is more rapid in the presence of an enzyme.

When carrying out a chemical reaction, chemists are concerned with two issues: the *rate* at which the reaction proceeds and the *extent* to which the reaction is product-favored. Chapter 6 began to address the second question, and Chapters 16 and 19 will develop that topic further. In this chapter we turn to the other part of our question, **chemical kinetics**, the study of the rates of chemical reactions.

The study of kinetics is divided into two parts. The first part concerns the *macroscopic level*, which addresses rates of reactions: what the reaction rate means, how to determine a reaction rate experimentally, and how factors such as temperature and the concentrations of reactants influence rates. The second part considers chemical reactions at the *particulate level*. Here, the concern is with the **reaction mechanism,** the detailed pathway taken by atoms and molecules as a reaction proceeds. The goal is to reconcile data in the macroscopic world of chemistry with an understanding of how and why chemical reactions occur at the particulate level—and then to apply this information to control important reactions.

15.1—Rates of Chemical Reactions

The concept of rate is encountered in many nonchemical circumstances. Common examples are the speed of an automobile given in terms of the distance traveled per unit time (e.g., kilometers per hour) and the rate of flow of water from a faucet given as volume per unit time (liters per minute). In each case a change is measured over an interval of time. The **rate of a chemical reaction** refers to the change in concentration of a substance per unit of time.

$$\text{Rate of reaction} = \frac{\text{change in concentration}}{\text{change in time}}$$

An easy way to gauge the speed of an automobile is to measure how far it travels during a specified time interval. Two measurements are made: distance traveled and time elapsed. The speed is the distance traveled divided by the time elapsed, or $\Delta(\text{distance})/\Delta(\text{time})$. If an automobile travels 2.4 mi in 4.5 min (0.075 h), its average speed is (2.4 mi/0.075 h), or 32 mph.

Chemical reaction rates are determined in a similar manner. Two quantities, concentration and time, must be measured. The rate of the reaction can then be described as the change in the concentration of a reactant or a product per unit time—that is, $\Delta(\text{concentration})/\Delta(\text{time})$.

For a rate study, the concentration of a substance undergoing reaction can be determined by a variety of methods. Concentrations can sometimes be measured directly, by using a pH meter, for example. Often, concentrations are obtained by measuring a property such as the absorbance of light that is related to concentration (Figure 15.1).

During a chemical reaction, amounts of reactants decrease with time, and amounts of products increase. It is possible to describe the rate of reaction based on either the increase in concentration of a product or the decrease in concentration of a reactant per unit of time. Consider the decomposition of N_2O_5 in a solvent. This reaction occurs according to the following equation:

$$2\ N_2O_5 \longrightarrow 4\ NO_2 + O_2$$

(a)   (b)   (c)

Figure 15.1 An experiment to measure rate of reaction. (a) A few drops of blue food dye were added to water, followed by a solution of bleach. Initially, the concentration of dye was about 3.4×10^{-5} M, and the bleach (NaOCl) concentration was about 0.034 M. (b, c) The dye faded as it reacted with the bleach. The absorbance of the solution can be measured at various times using a spectrophotometer, and these values can be used to determine the concentration of the dye.

The progress of this reaction can be followed in a number of ways, including monitoring the increase in O_2 pressure. The amount of O_2 that is formed (calculated from measured values of P, V, and T) is related to the amount of N_2O_5 that has decomposed: For every 1 mol of O_2 formed, 2 mol of N_2O_5 decomposed. The amount of N_2O_5 in solution at a given time equals the initial amount of N_2O_5 minus the amount decomposed. (If the volume of the solution is known, the concentration can be determined from that amount.) Data for a typical experiment done at 30.0 °C are presented as a graph of concentration of N_2O_5 versus time in Figure 15.2.

The rate of this reaction for any interval of time can be expressed as the change in concentration of N_2O_5 divided by the change in time:

$$\text{Rate of reaction} = \frac{\text{change in } [N_2O_5]}{\text{change in time}} = -\frac{\Delta [N_2O_5]}{\Delta t}$$

The minus sign is required because the concentration of N_2O_5 decreases with time, and the rate is always expressed as a positive quantity.

The rate could also be expressed in terms of the rate of formation of NO_2 or the rate of formation of O_2. Rates expressed in these ways will have a positive sign because the concentration is increasing. Furthermore, the rate of formation of NO_2 is twice the rate of decomposition of N_2O_5 because the balanced chemical equation tells us that 2 mol of NO_2 form when 1 mol of N_2O_5 decomposes. The rate of formation of O_2 is one half of the rate of decomposition of N_2O_5 because one mole of O_2 is formed per two moles of N_2O_5 decomposed. For example, the rate of disappearance of N_2O_5 between 40 min and 55 min (see Figure 15.2) is given by

$$-\frac{\Delta [N_2O_5]}{\Delta t} = -\frac{(1.10 \text{ mol/L}) - (1.22 \text{ mol/L})}{55 \text{ min} - 40 \text{ min}} = +\frac{0.12 \text{ mol/L}}{15 \text{ min}}$$

$$= 0.0080 \frac{\text{mol } N_2O_5 \text{ consumed}}{\text{L} \cdot \text{min}}$$

■ **Representing Concentration**
Recall that square brackets around a formula indicate its concentration in mol/L (Section 5.8).

■ **Calculating Changes**
Recall that when we calculate a change in a quantity, we always do so by subtracting the initial quantity from the final quantity: $\Delta X = X_{\text{final}} - X_{\text{initial}}$. Therefore, ΔX will be negative for the disappearance of a reactant.

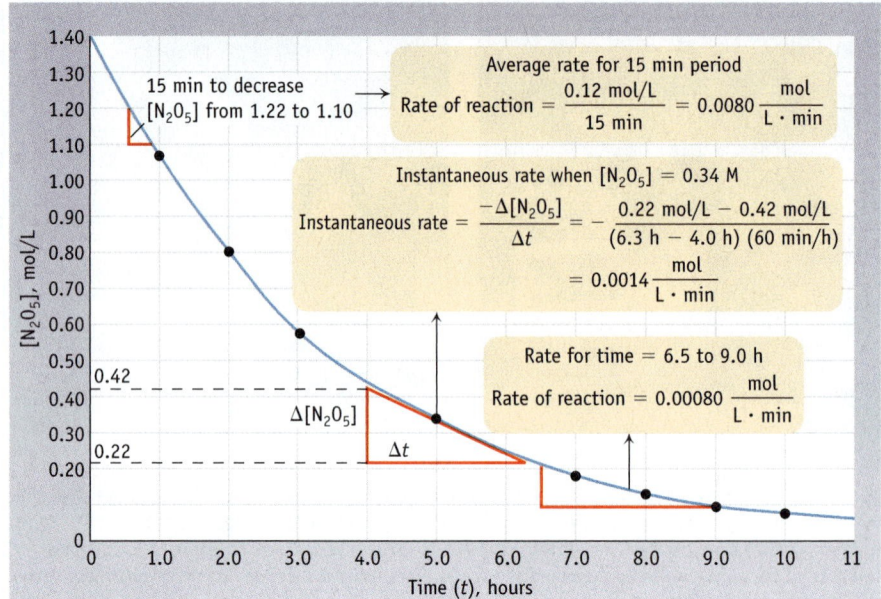

Active Figure 15.2   **A plot of reactant concentration versus time for the decomposition of N_2O_5.** The average rate for a 15-min interval from 40 min to 55 min is 0.0080 mol/L · min. The instantaneous rate calculated when $[N_2O_5] = 0.34$ M is 0.0014 mol/L · min.

GENERAL
Chemistry⚛Now™ See the General ChemistryNow CD-ROM or website to explore an interactive version of this figure accompanied by an exercise.

Expressing the rate in terms of the rate of appearance of NO_2 produces a rate that is twice the rate of disappearance of N_2O_5.

■ **Units of Reaction Rates**
Notice that the units used to describe reaction rates are mol/L · time.

$$\text{Rate} = \frac{\Delta[NO_2]}{\Delta t} = \frac{0.0080 \text{ mol } N_2O_5 \text{ consumed}}{L \cdot min} \times \frac{4 \text{ mol } NO_2 \text{ formed}}{2 \text{ mol } N_2O_5 \text{ consumed}}$$

$$= 0.016 \frac{\text{mol } NO_2 \text{ formed}}{L \cdot min}$$

The rate of the reaction in terms of the rate at which O_2 is formed is

$$\text{Rate} = \frac{\Delta[O_2]}{\Delta t} = \frac{0.0080 \text{ mol } N_2O_5 \text{ consumed}}{L \cdot min} \times \frac{1 \text{ mol } O_2 \text{ formed}}{2 \text{ mol } N_2O_5 \text{ consumed}}$$

$$= 0.0040 \frac{\text{mol } O_2 \text{ formed}}{L \cdot min}$$

■ **Summarizing Rate Expressions**

$$-\frac{1}{2}\frac{\Delta[N_2O_5]}{\Delta t} = \frac{1}{4}\frac{\Delta[NO_2]}{\Delta t} = \frac{\Delta[O_2]}{\Delta t}$$

for the reaction $2\,N_2O_5 \longrightarrow 4\,NO_2 + O_2$. To equate rates of disappearance or appearance, you should divide $\Delta[\text{reagent}]/\Delta t$ by the stoichiometric coefficient in the balanced equation.

Graphing concentration versus time in Figure 15.2 does not give a straight line because the rate of the reaction changes during the course of the reaction. The concentration of N_2O_5 decreases rapidly at the beginning of the reaction but more slowly near the end. We can verify this by comparing the rate of disappearance of N_2O_5 calculated previously (the concentration decreased by 0.12 mol/L in 15 min) with the rate of reaction calculated for the time interval from 6.5 h to 9.0 h (when the concentration drops by 0.12 mol/L in 150 min).

$$-\frac{\Delta[N_2O_5]}{\Delta t} = -\frac{(0.1 \text{ mol/L}) - (0.22 \text{ mol/L})}{540 \text{ min} - 390 \text{ min}} = +\frac{0.12 \text{ mol/L}}{150 \text{ min}}$$

$$= 0.00080 \frac{\text{mol}}{L \cdot min}$$

The rate in this later stage of this reaction is only one tenth of the previous value.

The procedure we have used to calculate the reaction rate gives the **average rate** over the chosen time interval. We might also ask what the **instantaneous rate** is at a single point in time. The instantaneous rate is determined by drawing a line tangent to the concentration–time curve at a particular time (see Figure 15.2) and obtaining the rate from the slope of this line. For example, when $[N_2O_5] = 0.34$ mol/L and $t = 5.0$ h, the rate is

$$\text{Rate when } [N_2O_5] \text{ is } 0.34 \text{ M} = -\frac{\Delta[N_2O_5]}{\Delta t} = +\frac{0.20 \text{ mol/L}}{140 \text{ min}} = 1.4 \times 10^{-3} \frac{\text{mol}}{\text{L} \cdot \text{min}}$$

At that particular moment in time ($t = 5.0$ h), N_2O_5 is being consumed at a rate of 0.0014 mol/L · min.

The difference between an average rate and an instantaneous rate has an analogy in the speed of an automobile. In the previous example, the car traveled 2.4 mi in 4.5 min for an average speed of 32 mph. At any instant in time, however, the car may be moving much slower or much faster. The instantaneous speed at any instant is indicated by the car's speedometer.

■ **The Slope of a Line**
The instantaneous rate in Figure 15.2 can be determined from an analysis of the slope of the line. See pages 43–44 for more on finding the slope of a line.

GENERAL
Chemistry ⋅ Now™

See the General ChemistryNow CD-ROM or website:
• **Screen 15.2 Rates of Chemical Reactions,** for a visualization of ways to express reaction rates

Example 15.1—Relative Rates and Stoichiometry

Problem Give the relative rates for the disappearance of reactants and formation of products for the following reaction:

$$4 \text{ PH}_3(g) \longrightarrow P_4(g) + 6 \text{ H}_2(g)$$

Strategy In this reaction PH_3 disappears and P_4 and H_2 are formed. Consequently, the value of $\Delta[PH_3]/\Delta t$ will be negative, whereas $\Delta[P_4]/\Delta t$ and $\Delta[H_2]/\Delta t$ will be positive. To equate rates, we divide $\Delta[\text{reagent}]/\Delta t$ by its stoichiometric coefficient in the balanced equation.

Solution Because four moles of PH_3 disappear for every one mole of P_4 formed, the numerical value of the rate of formation of P_4 can only be one fourth of the rate of disappearance of PH_3. Similarly, P_4 is formed at only one sixth of the rate that H_2 is formed.

$$-\frac{1}{4}\left(\frac{\Delta[PH_3]}{\Delta t}\right) = +\frac{\Delta[P_4]}{\Delta t} = +\frac{1}{6}\left(\frac{\Delta[H_2]}{\Delta t}\right)$$

Example 15.2—Rate of Reaction

Problem Data collected on the concentration of dye as a function of time (see Figure 15.1) are given in the graph below. What is the average rate of change of the dye concentration over the first 2 min? What is the average rate of change during the fifth minute (from $t = 4$ to $t = 5$)? Estimate the instantaneous rate at 4 min.

Strategy To find the average rate, calculate the difference in concentration at the beginning and end of a time period ($\Delta c = c_{\text{final}} - c_{\text{initial}}$) and divide by the elapsed time. To find the

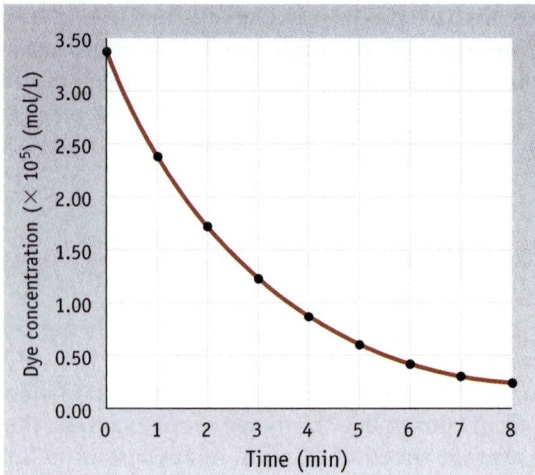

instantaneous rate, draw a line tangent to the graph at the given time. The slope of the line (page 43) is the instantaneous rate. (*See also General ChemistryNow Screen 15.2.*)

Solution The concentration of dye decreases from 3.4×10^{-5} M at $t = 0$ min to 1.7×10^{-5} M at $t = 2.0$ min. The average rate of the reaction in this interval of time is

$$-\frac{\Delta[\text{Dye}]}{\Delta t} = -\frac{(1.7 \times 10^{-5}\ \text{mol/L}) - (3.4 \times 10^{-5}\ \text{mol/L})}{2.0\ \text{min}} = +\frac{8.5 \times 10^{-6}\ \text{mol}}{\text{L} \cdot \text{min}}$$

The concentration of dye decreases from 0.90×10^{-5} M at $t = 4.0$ min to 0.60×10^{-5} M at $t = 5.0$ min. The average rate of the reaction in this interval of time is

$$-\frac{\Delta[\text{Dye}]}{\Delta t} = -\frac{(0.60 \times 10^{-5}\ \text{mol/L}) - (0.90 \times 10^{-5}\ \text{mol/L})}{1.0\ \text{min}} = +\frac{3.0 \times 10^{-6}\ \text{mol}}{\text{L} \cdot \text{min}}$$

Finally, from the slope of the line tangent to the curve, the instantaneous rate at 4 min is $+3.5 \times 10^{-6}$ mol/L · min.

Comment Notice that the average rate of reaction in the 4- to 5-min interval is less than half the value in the first minute.

Exercise 15.1—Reaction Rates and Stoichiometry

What are the relative rates of appearance or disappearance of each product and reactant, respectively, in the decomposition of nitrosyl chloride, NOCl?

$$2\ \text{NOCl}(g) \longrightarrow 2\ \text{NO}(g) + \text{Cl}_2(g)$$

Exercise 15.2—Rate of Reaction

Sucrose decomposes to fructose and glucose in acid solution. A plot of the concentration of sucrose as a function of time is given here. What is the rate of change of the sucrose concentration over the first 2 h? What is the rate of change over the last 2 h? Estimate the instantaneous rate at 4 h.

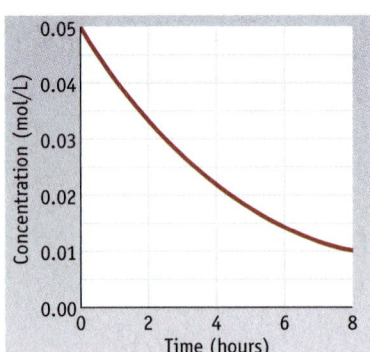

Concentration versus time for the decomposition of sucrose. See Exercise 15.2.

15.2—Reaction Conditions and Rate

For a chemical reaction to occur, molecules of the reactants must come together so that atoms can be exchanged or rearranged. Atoms and molecules are mobile in the gas phase or in solution, so reactions are often carried out using a mixture of gases or using solutions of reactants. Under these circumstances, several factors—reactant concentrations, temperature, and presence of catalysts—affect the rate of a reaction. If the reactant is a solid, the surface area available for reaction will also affect the rate of reaction.

The "iodine clock reaction" in Figure 15.3 illustrates the effect of concentration and temperature. The reaction mixture contains hydrogen peroxide (H_2O_2), iodide ion (I^-), vitamin C (ascorbic acid), and starch (which is an indicator of the presence of iodine, I_2). A sequence of reactions begins with the slow oxidation of iodide ion to I_2 by H_2O_2.

$$H_2O_2(aq) + 2\ I^-(aq) + 2\ H^+(aq) \longrightarrow 2\ H_2O(\ell) + I_2(aq)$$

As soon as I_2 is formed in the solution, vitamin C rapidly reduces it to I^-.

$$I_2(aq) + C_6H_8O_6(aq) \longrightarrow C_6H_6O_6(aq) + 2\ H^+(aq) + 2\ I^-(aq)$$

When all of the vitamin C has been consumed, I_2 remains in solution and forms a blue-black complex with starch. The time measured represents how long it has taken for the given amount of vitamin C to react. For the first experiment (A), the time required is 51 seconds. When the concentration of iodide ion is smaller (B), the time required for the vitamin C to be consumed is longer, 1 minute and 33 seconds. Finally, when the concentrations are again the same as in experiment B but the reaction mixture is heated, the reaction occurs more rapidly (56 seconds).

Catalysts are substances that accelerate chemical reactions but are not themselves transformed. For example, hydrogen peroxide, H_2O_2, decomposes to water and oxygen,

$$2\ H_2O_2(aq) \longrightarrow O_2(g) + 2\ H_2O(\ell)$$

■ **Effect of Temperature on Reaction Rate**
Cooking involves chemical reactions, and a higher temperature results in foods cooking faster. In the laboratory, reaction mixtures are often heated to make reactions occur faster.

(a) Initial Experiment.
The blue color of the starch-iodine complex develops in 51 seconds.

(b) Change Concentration.
The blue color of starch-iodine complex develops in 1 minute, 33 seconds when the solution is less concentrated than A.

(c) Change Temperature.
The blue color of the starch-iodine complex develops in 56 seconds when the solution is less concentrated than A but at a higher temperature.

Hot bath

Solutions containing vitamin C, H_2O_2, I^-, and starch are mixed.

Smaller concentration of I^- than in Experiment A.

Same concentrations as in Experiment B, but at a higher temperature.

Figure 15.3 The iodine clock reaction. This reaction illustrates the effects of concentration and temperature on reaction rate. (You can do these experiments yourself with reagents available in the supermarket. For details see S. W. Wright: "The vitamin C clock reaction," *Journal of Chemical Education*, Vol. 79, p. 41, 2002.) (*See also General ChemistryNow Screen 15.11.*)

Photos: Charles D. Winters

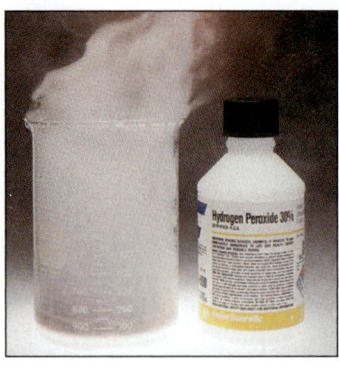  

(a)   (b)   (c)

a and c, Charles D. Winters; b, Thomas Eisner with Daniel Aneshansley, Cornell University

Figure 15.4 **Catalyzed decomposition of H_2O_2.** (a) The rate of decomposition of hydrogen peroxide is increased by the catalyst MnO_2. Here a 30% solution of H_2O_2, poured onto the black solid MnO_2, rapidly decomposes to O_2 and H_2O. Steam forms because of the high heat of reaction. (b) A bombardier beetle uses the catalyzed decomposition of H_2O_2 as a defense mechanism. The heat of the reaction lets the insect eject hot water and other irritating chemicals with explosive force. (c) A naturally occurring catalyst, called an enzyme, decomposes hydrogen peroxide. Here the enzyme found in a potato is used to catalyze H_2O_2 decomposition, and bubbles of O_2 gas are seen rising in the solution.

(a)

(b)

Figure 15.5 **The combustion of lycopodium powder.** (a) The spores of this common fern burn only with difficulty when piled in a dish. (b) If the spores are ground to a fine powder and sprayed into a flame, combustion is rapid.

Charles D. Winters

but a solution of H_2O_2 can be stored for many months because the rate of the decomposition reaction is extremely slow. Adding a manganese salt, an iodide-containing salt, or a biological substance called an *enzyme* causes this reaction to occur rapidly, as shown by vigorous bubbling as gaseous oxygen escapes from the solution (Figure 15.4).

The surface area of a solid reactant can also affect the reaction rate. Only molecules at the surface of a solid can come in contact with other reactants. The smaller the particles of a solid, the more molecules found on the solid's surface. With very small particles, the effect of surface area on rate can be quite dramatic (Figure 15.5). Farmers know that explosions of fine dust particles (suspended in the air in an enclosed silo or at a feed mill) represent a major hazard.

GENERAL
Chemistry⚛Now™

See the General ChemistryNow CD-ROM or website:
- **Screens 15.3 and 15.4 Control of Reaction Rates,** for a visualization of the factors controlling rates and for a simulation of the effect of concentration on rate

15.3—Effect of Concentration on Reaction Rate

One important goal in studying kinetics is to determine how concentrations of reactants affect the reaction rate. The effects can be determined by evaluating the rate of a reaction using different concentrations of each reactant (with the temperature held constant). Consider, for example, the decomposition of N_2O_5 to NO_2 and O_2. Figure 15.2 presents data on the concentration of N_2O_5 as a function of time. We previously calculated that, when $[N_2O_5] = 0.34$ mol/L, the instantaneous rate of disappearance of N_2O_5 is 0.0014 mol/L · min. An evaluation of the instantaneous rate of the reaction when $[N_2O_5] = 0.68$ mol/L reveals a rate of 0.0028 mol/L · min. That is, doubling the concentration of N_2O_5 doubles the reaction rate. A similar exercise shows that if $[N_2O_5]$ is halved to 0.17 mol/L, the reaction rate is also halved.

These results tell us that the reaction rate is directly proportional to the reactant concentration for this reaction:

$$\text{Rate of reaction} \propto [N_2O_5]$$

where the symbol $\propto$ means "proportional to."

Different relationships between reaction rate and reactant concentration are encountered in other reactions. For example, the reaction rate could be independent of concentration, or it may depend on the reactant concentration raised to some power (that is, $[\text{reactant}]^n$). If the reaction involves several reactants, the reaction rate may depend on the concentrations of each of them or on only one of them. Finally, if a catalyst is involved, its concentration may also affect the rate.

Rate Equations

The relationship between reactant concentration and reaction rate is expressed by an equation called a **rate equation**, or **rate law**. For the decomposition of N_2O_5 the rate equation is

$$\text{Rate of reaction} = k[N_2O_5]$$

where the proportionality constant, k, is called the **rate constant**. This rate equation tells us that this reaction rate is proportional to the concentration of the reactant. That is, when $[N_2O_5]$ is doubled, the reaction rate doubles, for example.

In general, for a reaction such as

$$a\,A + b\,B \longrightarrow x\,X$$

the rate equation has the form

$$\text{Rate} = k[A]^m[B]^n$$

The rate equation expresses the fact that the rate of reaction is proportional to the reactant concentrations, each concentration being raised to some power. It is important to recognize that the exponents m and n are *not necessarily the stoichiometric coefficients* (a and b) for the balanced chemical equation. *The exponents must be determined by experiment.* They are often positive whole numbers, but they can also be negative numbers, fractions, or zero.

If a homogeneous catalyst is present, its concentration might also be included in the rate equation, even though the catalytic species does not appear in the balanced, overall equation for the reaction. Consider, for example, the decomposition of hydrogen peroxide in the presence of a catalyst such as iodide ion.

$$2\,H_2O_2(aq) \xrightarrow{\;I^-(aq)\;} 2\,H_2O(\ell) + O_2(g)$$

Experiments show that this reaction has the following rate equation:

$$\text{Reaction rate} = k[H_2O_2][I^-]$$

Here the exponent on each concentration term is 1, even though the stoichiometric coefficient of H_2O_2 is 2 and I^- does not appear in the balanced equation.

The Order of a Reaction

The **order** of a reaction with respect to a particular reactant is the exponent of its concentration term in the rate expression, and the total reaction order is the sum

■ **The Nature of Catalysts**
A catalyst does not appear as a reactant in the balanced, overall equation for the peroxide decomposition reaction, but it may appear in the rate expression. It is common practice to indicate catalysts by placing them above the reaction arrow, as shown in the example. A homogeneous catalyst is one in the same phase as the reactants. For example, both H_2O_2 and I^- are dissolved in water.

of the exponents on all concentration terms. For example, the rate equation for the decomposition of H_2O_2 in the presence of iodide ion,

$$\text{Reaction rate} = k[H_2O_2][I^-]$$

shows that the reaction is first order with respect to H_2O_2 and also with respect to I^-; it is second order overall. This tells us that the rate doubles if either $[H_2O_2]$ or $[I^-]$ is doubled and that the rate increases by a factor of 4 if both concentrations are doubled.

Consider another example, the reaction of NO and Cl_2:

$$2\,NO(g) + Cl_2(g) \longrightarrow 2\,NOCl(g)$$

The experimentally determined rate equation for this reaction is

$$\text{Rate} = k[NO]^2[Cl_2]$$

This reaction is second order in NO, first order in Cl_2, and third order overall. We can see how this rate equation is related to experimental data by examining some data for the rate of disappearance of NO.

Experiment	[NO] mol/L	[Cl₂] mol/L	Rate mol/L · s
1	0.250	0.250	1.43×10^{-6}
	↓ × 2	↓ no change	↓ × 4
2	0.500	0.250	5.72×10^{-6}
3	0.250	0.500	2.86×10^{-6}
4	0.500	0.500	11.4×10^{-6}

- *Experiments 1 and 2:* If $[Cl_2]$ is held constant and $[NO]$ is doubled from 0.250 mol/L to 0.500 mol/L, the reaction rate increases by a factor of 4 (from 1.43×10^{-6} mol/L · s to 5.72×10^{-6} mol/L · s); that is, Rate $\propto [NO]^2$.

$$\frac{\text{Rate for experiment 2}}{\text{Rate for experiment 1}} = \frac{5.72 \times 10^{-6} \text{ mol/L} \cdot \text{s}}{1.43 \times 10^{-6} \text{ mol/L} \cdot \text{s}} = \frac{4}{1}$$

- *Experiments 1, 3, and 4:* Comparing experiments 1 and 3, we see that, when $[NO]$ is held constant and $[Cl_2]$ is doubled from 0.250 mol/L to 0.500 mol/L, the rate is doubled. Comparing experiments 1 and 4, we see that if both $[NO]$ and $[Cl_2]$ are doubled from 0.250 M to 0.500 M, then the rate (11.4×10^{-5} mol/L · s) is 8 times the original value.

The decomposition of ammonia on a platinum surface at 856 °C is interesting because it is zero order.

$$2\,NH_3(g) \longrightarrow N_2(g) + 3\,H_2(g)$$

This means that the reaction rate is independent of NH_3 concentration.

$$\text{Rate} = k[NH_3]^0 = k$$

The reaction order is important because it gives some insight into the most interesting question of all—how the reaction occurs. This is described further in Section 15.6.

The Rate Constant, k

The rate constant, k, is a proportionality constant that relates rate and concentration *at a given temperature*. It is an important quantity because it enables you to find the reaction rate for a new set of concentrations. To see how to use k, consider the substitution of Cl^- ion by water in the cancer chemotherapy agent cisplatin, $Pt(NH_3)_2Cl_2$.

$$Pt(NH_3)_2Cl_2(aq) \quad + \quad H_2O(\ell) \quad \longrightarrow \quad [Pt(NH_3)_2(H_2O)Cl]^+(aq) \quad + \quad Cl^-(aq)$$

■ **Time and Rate Constants**
The time in a rate constant can be seconds, minutes, hours, days, years, or whatever time unit is appropriate.

The rate expression for this reaction is

$$Rate = k[Pt(NH_3)_2Cl_2]$$

and the rate constant, k, is 0.090/h at 25 °C. Knowing k allows you to calculate the rate at a particular reactant concentration—for example, when $[Pt(NH_3)_2Cl_2] = 0.018$ mol/L:

$$Rate = (0.090/h)(0.018\ mol/L) = 0.0016\ mol/L \cdot h$$

Reaction rates $(\Delta[R]/\Delta t)$ have units of mol/L · time when concentrations are given as moles per liter. Rate constants must have units consistent with the units for the other terms in the rate equation.

■ **Expressing Time and Rate**
The fraction 1/time can also be written as time^{-1}. For example, 1/y is equivalent to y^{-1}, and 1/s is equivalent to s^{-1}.

- First-order reactions: the units of k are time^{-1}.
- Second-order reactions: the units of k are L/mol · time.
- Zero-order reaction: the units of k are mol/L · time.

Determining a Rate Equation

A rate equation must be determined experimentally. One way to do so is by using the "method of initial rates." The **initial rate** is the instantaneous reaction rate at the start of the reaction (the rate at $t = 0$). An approximate value of the initial rate can be obtained by mixing the reactants and determining $\Delta[product]/\Delta t$ or $-\Delta[reactant]/\Delta t$ after 1% to 2% of the limiting reactant has been consumed. Measuring the rate during the initial stage of a reaction is convenient because initial concentrations are known, and this method avoids possible complications arising from interference by reaction products or the occurrence of side reactions.

As an example of the determination of a reaction rate by this method, let us look at the reaction of sodium hydroxide with methyl acetate to produce acetate ion and methanol.

$$CH_3CO_2CH_3(aq) \quad + \quad OH^-(aq) \quad \longrightarrow \quad CH_3CO_2^-(aq) \quad + \quad CH_3OH(aq)$$

Data in the table were collected for several experiments at 25 °C:

Experiment	Initial Concentrations		Initial Reaction Rate (mol/L · s) at 25 °C
	[CH₃CO₂CH₃]	[OH⁻]	
1	0.050 M	0.050 M	0.00034
	↓ no change	↓ × 2	↓ × 2
2	0.050 M	0.10 M	0.00069
	↓ × 2	↓ no change	↓ × 2
3	0.10 M	0.10 M	0.00137

This table shows that when the initial concentration of one reactant (either $CH_3CO_2CH_3$ or OH^-) is doubled while the concentration of the other reactant is held constant, the initial reaction rate doubles. This rate doubling shows that the rate for the reaction is directly proportional to the concentrations of *both* $CH_3CO_2CH_3$ and OH^-; thus, the reaction is first order in each of these reactants. The rate equation that reflects these experimental observations is

$$\text{Rate} = k[CH_3CO_2CH_3][OH^-]$$

Using this equation we can predict that doubling *both* concentrations at the same time should cause the rate to go up by a factor of 4. What happens, however, if one concentration is doubled and the other is halved? The rate equation tells us the rate should not change!

If the rate equation is known, the value of k, the rate constant, can be found by substituting values for the rate and concentration into the rate equation. To find k for the methyl acetate/hydroxide ion reaction, for example, data from one of the experiments are substituted into the rate equation. Using the data from the first experiment, we have

$$\text{Rate} = 0.00034 \text{ mol/L} \cdot \text{s} = k(0.050 \text{ mol/L})(0.050 \text{ mol/L})$$

$$k = \frac{0.00034 \text{ mol/L} \cdot \text{s}}{(0.050 \text{ mol/L})(0.050 \text{ mol/L})} = 0.14 \text{ L/mol} \cdot \text{s}$$

GENERAL
Chemistry ⟨Now™

See the General ChemistryNow CD-ROM or website:
- **Screen 15.4 Control of Reaction Rates (2) and Screen 15.5 Determination of the Rate Equation (1),** for a simulation, tutorial, and exercise on determining rate equations from a study of the effect of concentration on reaction rate

Example 15.3—Determining a Rate Equation

Problem The rate of the reaction between CO and NO_2

$$CO(g) + NO_2(g) \longrightarrow CO_2(g) + NO(g)$$

was studied at 540 K starting with various concentrations of CO and NO_2, and the data in the table were collected. Determine the rate equation and the value of the rate constant.

| Experiment | Initial Concentrations | | Initial Rate (mol/L · h) |
	[CO], mol/L	[NO$_2$], mol/L	
1	5.10×10^{-4}	0.350×10^{-4}	3.4×10^{-8}
2	5.10×10^{-4}	0.700×10^{-4}	6.8×10^{-8}
3	5.10×10^{-4}	0.175×10^{-4}	1.7×10^{-8}
4	1.02×10^{-3}	0.350×10^{-4}	6.8×10^{-8}
5	1.53×10^{-3}	0.350×10^{-4}	10.2×10^{-8}

Strategy For a reaction involving several reactants, the general approach is to keep the concentration of one reactant constant and then decide how the rate of reaction changes as the concentration of the other reagent is varied. Because the rate is proportional to the concentration of a reactant, say A, raised to some power n (the reaction order)

$$\text{Rate} \propto [A]^n$$

we can write the general equation

$$\frac{\text{Rate in experiment 2}}{\text{Rate in experiment 1}} = \frac{[A_2]^n}{[A_1]^n} = \left(\frac{[A_2]}{[A_1]}\right)^n$$

If $[A]$ doubles ($[A_2] = 2[A_1]$), and the rate doubles from experiment 1 to experiment 2, then $n = 1$. If $[A]$ doubles, and the rate goes up by 4, then $n = 2$.

Solution In the first three experiments, the concentration of CO remains constant. In the second experiment, the NO$_2$ concentration has been doubled relative to Experiment 1, leading to a twofold increase in the rate. Thus, $n = 1$ and the reaction is first order in NO$_2$.

$$\frac{6.8 \times 10^{-8} \text{ mol/L} \cdot \text{h}}{3.4 \times 10^{-8} \text{ mol/L} \cdot \text{h}} = \left(\frac{0.700 \times 10^{-4}}{0.350 \times 10^{-4}}\right)^n$$

$$2 = (2)^n$$

and so $n = 1$.

This finding is confirmed by experiment 3. Decreasing $[NO_2]$ to half its original value in experiment 3 causes the rate to decrease by half.

The data in experiments 1 and 4 (with constant $[NO_2]$) show that doubling $[CO]$ doubles the rate, and the data from experiments 1 and 5 show that tripling the concentration of CO triples the rate. These results mean that the reaction is also first order in $[CO]$. We now know the rate equation is

$$\text{Rate} = k[CO][NO_2]$$

The rate constant, k, can be found by inserting data for one of the experiments into the rate equation. Using data from experiment 1, for example,

$$\text{Rate} = 3.4 \times 10^{-8} \text{ mol/L} \cdot \text{h} = k(5.10 \times 10^{-4} \text{ mol/L})(0.350 \times 10^{-4} \text{ mol/L})$$

$$k = 1.9 \text{ L/mol} \cdot \text{h}$$

Example 15.4—Using a Rate Equation to Determine Rates

Problem Using the rate equation and rate constant determined for the reaction of CO and NO$_2$ at 540 K in Example 15.3, determine the initial rate of the reaction when $[CO] = 3.8 \times 10^{-4}$ and $[NO_2] = 0.650 \times 10^{-4}$.

Strategy A rate equation consists of three parts: a rate, a rate constant (k), and the concentration terms. If two of these parts are known (here k and the concentrations), the third can be calculated.

Solution Substitute k ($= 1.9$ L/mol · h) and the concentration of each reactant into the rate law determined in Example 15.3.

$$\text{Rate} = k[CO][NO_2] = (1.9 \text{ L/mol} \cdot \text{h})(3.8 \times 10^{-4} \text{ mol/L})(0.650 \times 10^{-4} \text{ mol/L})$$

$$\text{Rate} = 4.7 \times 10^{-8} \text{ mol/L} \cdot \text{h}$$

Comment As a check on the calculated result, it is sometimes useful to make an educated guess at the answer before carrying out the mathematical solution. We know that the reaction here is first order in both reactants. Comparing the concentration values given in this problem with the concentration values in found experiment 1 in Example 15.3, we notice that $[CO]$ is about three fourths of the concentration value, whereas $[NO_2]$ is almost twice the value. The effects do not precisely offset each other, but we might predict that the difference in rates between this experiment and experiment 1 will be fairly small, with the rate of this experiment being just a little greater. The calculated value bears this out.

Exercise 15.3—Determining a Rate Equation

The initial rate of the reaction of nitrogen monoxide and oxygen

$$2 \text{ NO}(g) + O_2(g) \longrightarrow 2 \text{ NO}_2(g)$$

was measured at 25 °C for various initial concentrations of NO and O_2. Data are collected in the table. Determine the rate equation from these data. What is the value of the rate constant, k, and what are its units?

	Initial Concentrations (mol/L)		Initial Rate
Experiment	[NO]	[O₂]	(mol/L · s)
1	0.020	0.010	0.028
2	0.020	0.020	0.057
3	0.020	0.040	0.114
4	0.040	0.020	0.227
5	0.010	0.020	0.014

Exercise 15.4—Using Rate Laws

The rate constant, k, is 0.090 h^{-1} for the reaction

$$\text{Pt}(NH_3)_2Cl_2(aq) + H_2O(\ell) \longrightarrow [\text{Pt}(NH_3)_2(H_2O)Cl]^+(aq) + Cl^-(aq)$$

and the rate equation is

$$\text{Rate} = k[\text{Pt}(NH_3)_2Cl_2]$$

Calculate the rate of reaction when the concentration of $\text{Pt}(NH_3)_2Cl_2$ is 0.020 M. What is the rate of change in the concentration of Cl^- under these conditions?

15.4—Concentration–Time Relationships: Integrated Rate Laws

It is often useful or important to know how long a reaction must proceed to reach a predetermined concentration of some reactant or product, or what the reactant and product concentrations will be after some time has elapsed. One way to make this determination is to use a mathematical equation that relates time and concentration. That is, we would like to have an equation that will describe concentration–time curves like the one shown in Figure 15.2. With such an equation we could calculate a concentration at any given time or the length of time needed for a given amount of reactant to react.

First-Order Reactions

Suppose the reaction "R $\longrightarrow$ products" is first order. This means the reaction rate is directly proportional to the concentration of R raised to the first power, or, mathematically,

$$-\frac{\Delta[R]}{\Delta t} = k[R]$$

Using calculus, this relationship can be transformed into a very useful equation called an **integrated rate equation** (because integral calculus is used in its derivation).

$$\ln \frac{[R]_t}{[R]_0} = -kt \qquad (15.1)$$

Here $[R]_0$ and $[R]_t$ are concentrations of the reactant at time $t = 0$ and at a later time, t, respectively. The *ratio* of concentrations, $[R]_t/[R]_0$, is the fraction of reactant that *remains* after a given time has elapsed. In words, the equation says

$$\text{Natural logarithm} \left(\frac{\text{concentration of R after some time}}{\text{concentration of R at start of experiment}} \right)$$

$$= \ln (\text{fraction remaining at time, } t)$$

$$= -(\text{rate constant})(\text{elapsed time})$$

Notice the negative sign in the equation. The ratio $[R]_t/[R]_0$ is less than 1 because $[R]_t$ is always less than $[R]_0$; the reactant R is consumed during the reaction. This means the logarithm of $[R]_t/[R]_0$ is negative, so the other side of the equation must also bear a negative sign.

Equation 15.1 is useful in three ways:

- If $[R]_t/[R]_0$ is measured in the laboratory after some amount of time has elapsed, then k can be calculated.

- If $[R]_0$ and k are known, then the concentration of material expected to remain after a given amount of time ($[R]_t$) can be calculated.

- If k is known, then the time elapsed until a specific fraction ($[R]_t/[R]_0$) remains can be calculated.

Finally, notice that k for a first-order reaction is independent of concentration; k has units of time^{-1} (y^{-1} or s^{-1}, for example). This means we can choose any convenient unit for $[R]_t$ and $[R]_0$: moles per liter, moles, grams, number of atoms, number of molecules, or pressure.

■ **Initial and Final Time, *t***
The time $t = 0$ does not need to correspond to the actual beginning of the experiment. It can be the time when instrument readings were started, for example.

GENERAL
Chemistry ·Now™

See the General ChemistryNow CD-ROM or website:

- **Screen 15.6 Concentration–Time Relationships,** for a tutorial on the use of the integrated first-order rate equation

Example 15.5—The First-Order Rate Equation

Problem In the past cyclopropane, C_3H_6, was used in a mixture with oxygen as an anesthetic. (This practice has almost ceased today, because the compound is very flammable.) When heated, cyclopropane rearranges to propene in a first-order process.

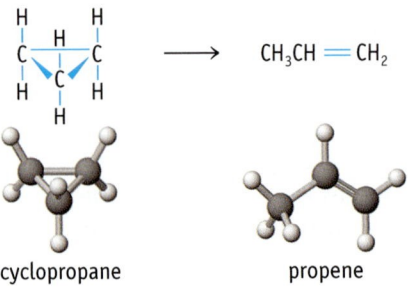

cyclopropane   propene

$$\text{Rate} = k[\text{cyclopropane}] \qquad k = 5.4 \times 10^{-2} \text{ h}^{-1}$$

If the initial concentration of cyclopropane is 0.050 mol/L, how much time (in hours) must elapse for its concentration to drop to 0.010 mol/L?

Strategy The reaction is first order in cyclopropane. You know the rate constant, k, and the concentrations at $t = 0$ and after some time has elapsed. Use Equation 15.1 to calculate the time (t) elapsed to reach a concentration of 0.010 mol/L.

Solution The first-order rate equation applied to this reaction is

$$\ln \frac{[\text{cyclopropane}]_t}{[\text{cyclopropane}]_0} = -kt$$

Values for $[\text{cyclopropane}]_t$, $[\text{cyclopropane}]_0$, and k are given:

$$\ln \frac{[0.010]}{[0.050]} = -(5.4 \times 10^{-2} \text{ h}^{-1})t$$

$$t = \frac{-\ln(0.20)}{5.4 \times 10^{-2} \text{ h}^{-1}} = \frac{-(-1.61)}{5.4 \times 10^{-2} \text{ h}^{-1}} = \boxed{30. \text{ h}}$$

Comment Cycloalkanes with fewer than five carbon atoms are strained because the C—C—C bond angles cannot match the preferred 109.5°. As a consequence of this ring strain, the ring opens readily to form propene.

Example 15.6—Using the First-Order Rate Equation

Problem Hydrogen peroxide decomposes in a dilute sodium hydroxide solution at 20 °C in a first-order reaction:

$$2\ H_2O_2(aq) \longrightarrow 2\ H_2O(\ell) + O_2(g)$$

$$\text{Rate} = k[H_2O_2] \qquad k = 1.06 \times 10^{-3} \text{ min}^{-1}$$

What is the fraction remaining after exactly 100 min if the initial concentration of H_2O_2 is 0.020 mol/L? What is the concentration of the peroxide after exactly 100 min?

Strategy Because the reaction is first order in H_2O_2, we use Equation 15.1. Here $[H_2O_2]_0$, k, and t are known, and we are asked to find $[H_2O_2]_t$. Recall that

$$\frac{[R]_t}{[R]_0} = \text{fraction remaining}$$

Therefore, once this value is known, and knowing $[H_2O_2]_0$, we can calculate $[H_2O_2]_t$. *(See General ChemistryNow Screen 15.6.)*

Solution Substitute the known values into Equation 15.1.

$$\ln \frac{[H_2O_2]_t}{[H_2O_2]_0} = -kt = -(1.06 \times 10^{-3} \text{ min}^{-1})(100 \text{ min})$$

$$\ln \frac{[H_2O_2]_t}{[H_2O_2]_0} = -0.106$$

Taking the antilogarithm of -0.106 [i.e., the inverse of ln (-0.106) or $e^{-0.106}$], we find the fraction remaining to be 0.90.

$$\text{Fraction remaining} = \frac{[H_2O_2]_t}{[H_2O_2]_0} = 0.90$$

Because $[H_2O_2]_0 = 0.020$ mol/L, this gives

$$[H_2O_2]_t = 0.018 \text{ mol/L}$$

Exercise 15.5—Using the First-Order Rate Equation

Sucrose, a sugar, decomposes in acid solution to give glucose and fructose. The reaction is first order in sucrose, and the rate constant at 25 °C is $k = 0.21$ h^{-1}. If the initial concentration of sucrose is 0.010 mol/L, what is its concentration after 5.0 h?

Exercise 15.6—Using the First-Order Rate Equation

Gaseous NO_2 decomposes when heated:

$$2 \text{ NO}_2(g) \longrightarrow 2 \text{ NO}(g) + O_2(g)$$

The disappearance of NO_2 is a first-order reaction with $k = 3.6 \times 10^{-3}$ s^{-1} at 300 °C.

(a) A sample of gaseous NO_2 is placed in a flask and heated at 300 °C for 150 s. What fraction of the initial sample remains after this time?

(b) How long must a sample be heated so that 99% of the sample has decomposed?

Second-Order Reactions

Suppose the reaction "R $\longrightarrow$ products" is second order. The rate equation is

$$-\frac{\Delta[R]}{\Delta t} = k[R]^2$$

Using the methods of calculus, this relationship can be transformed into the following equation that relates reactant concentration and time:

$$\frac{1}{[R]_t} - \frac{1}{[R]_0} = kt \qquad (15.2)$$

The same symbolism used with first-order reactions applies: $[R]_0$ is the concentration of reactant at the time $t = 0$, $[R]_t$ is the concentration at a later time, and k is the second-order rate constant (with units of L/mol · time).

Example 15.7—Using the Second-Order Integrated Rate Equation

Problem The gas-phase decomposition of HI

$$HI(g) \longrightarrow \tfrac{1}{2} H_2(g) + \tfrac{1}{2} I_2(g)$$

has the rate equation

$$-\frac{\Delta[HI]}{\Delta t} = k[HI]^2$$

where $k = 30.$ L/mol · min at 443 °C. How much time does it take for the concentration of HI to drop from 0.010 mol/L to 0.0050 mol/L at 443 °C?

Strategy Substitute the values of $[HI]_0$, $[HI]_t$, and k into Equation 15.2.

Solution Here $[HI]_0 = 0.010$ mol/L and $[HI]_t = 0.0050$ mol/L. Using Equation 15.2, we have

$$\frac{1}{0.0050 \text{ mol/L}} - \frac{1}{0.010 \text{ mol/L}} = (30. \text{ L/mol} \cdot \text{min})t$$

$$(2.0 \times 10^2 \text{ L/mol}) - (1.0 \times 10^2 \text{ L/mol}) = (30. \text{ L/mol} \cdot \text{min})t$$

$$t = 3.3 \text{ min}$$

Exercise 15.7—Using the Second-Order Concentration–Time Equation

Using the rate constant for HI decomposition given in Example 15.7, calculate the concentration of HI after 12 min if $[HI]_0 = 0.010$ mol/L.

Zero-Order Reactions

If a reaction (R $\longrightarrow$ products) is zero order, the rate equation is

$$-\frac{\Delta[R]}{\Delta t} = k[R]^0$$

This equation leads to the integrated rate equation

$$[R]_0 - [R]_t = kt \tag{15.3}$$

where the units of k are mol/L · s.

Graphical Methods for Determining Reaction Order and the Rate Constant

Equations 15.1, 15.2, and 15.3 relating concentration and time for first-, second-, and zero-order reactions, respectively, suggest a convenient way to determine the order of a reaction and its rate constant. Rearranged slightly, each of these equations has the form $y = mx + b$. This is the equation for a straight line, where m is the slope of the line and b is the y-intercept (the value of y when x is zero) (page 43). As illustrated here, $x = t$ in each case.

Zero order	First order	Second order
$[R]_t = -kt + [R]_0$	$\ln[R]_t = -kt + \ln[R]_0$	$\dfrac{1}{[R]_t} = +kt + \dfrac{1}{[R]_0}$
$\downarrow \quad\quad \downarrow \quad\quad \downarrow$	$\downarrow \quad\quad \downarrow \quad\quad \downarrow$	$\downarrow \quad\quad \downarrow \quad\quad \downarrow$
$y \quad\quad mx \quad\quad b$	$y \quad\quad mx \quad\quad b$	$y \quad\quad mx \quad\quad b$

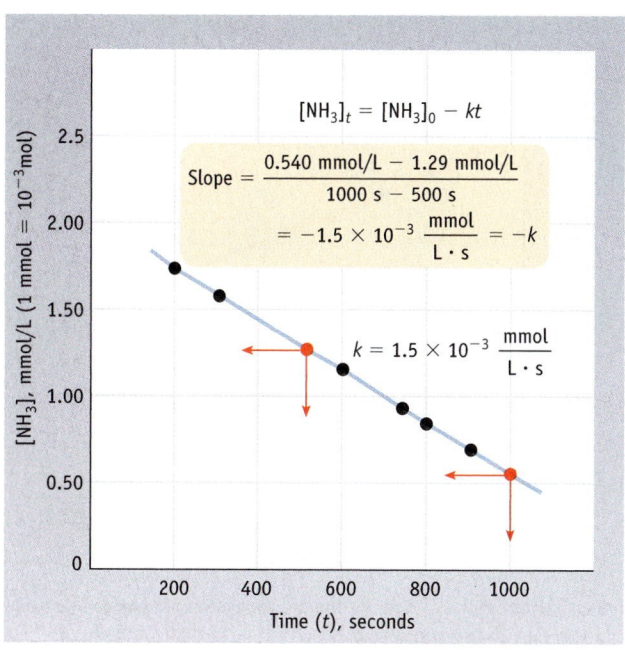

Figure 15.6 Plot of a zero-order reaction. A graph of the concentration of ammonia, $[NH_3]_t$, against time for the decomposition of NH_3

$$2 NH_3(g) \longrightarrow N_2(g) + 3 H_2(g)$$

on a metal surface at 856 °C is a straight line, indicating that this is a zero-order reaction. The rate constant, k, for this reaction is found from the slope of the line; $k = -$slope. (The points chosen to calculate the slope are given in red.)

As an example of the use of a concentration/time equation, consider the zero-order decomposition of ammonia on a platinum surface.

$$2 NH_3(g) \longrightarrow N_2(g) + 3 H_2(g) \qquad \text{Rate} = k[NH_3]^0 = k$$

The rate here is proportional to the ammonia concentration to the zero power, which is 1. That is, the reaction rate is independent of NH_3 concentration. The straight line, obtained when the concentration at time t, $[R]_t$, is plotted against time (Figure 15.6), is proof that this reaction is zero order in NH_3 concentration. The rate constant, k, can be determined from the slope of the line. Here the slope $= -k$, so in this case

$$\text{slope} = -k = -1.5 \times 10^{-3} \text{ mmol/L} \cdot \text{s}$$
$$k = 1.5 \times 10^{-3} \text{ mmol/L} \cdot \text{s}$$

The y-intercept of the line at $t = 0$ is equal to $[R]_0$.

A plot of concentration versus time for a first-order reaction is always a curved line (see Figure 15.2). Plotting ln [reactant] versus time, however, produces a straight line with a negative slope when the reaction is first order in that reactant. Consider the decomposition of hydrogen peroxide, a first-order reaction referred to earlier in Example 15.6.

$$2 H_2O_2(aq) \longrightarrow 2 H_2O(\ell) + O_2(g)$$
$$\text{Rate} = k [H_2O_2]$$

Values of the concentration of H_2O_2 as a function of time for a typical experiment are given as the first two columns of numbers in Figure 15.7. The third column lists values of ln $[H_2O_2]$. A graph of ln $[H_2O_2]$ versus time produces a straight line, showing that the reaction is first order in H_2O_2. The negative of the slope of the line equals the rate constant for the reaction, $1.05 \times 10^{-3} \text{ min}^{-1}$.

■ **Finding the Slope of a Line**
See Section 1.8 for a description of methods for finding the slope of a line. The graphing program on the General ChemistryNOW CD-ROM or website will also give the slope of a line from experimental data.

Time (min)	[H₂O₂] mol/L	ln [H₂O₂]
0	0.0200	−3.912
200	0.0160	−4.135
400	0.0131	−4.335
600	0.0106	−4.547
800	0.0086	−4.76
1000	0.0069	−4.98
1200	0.0056	−5.18
1600	0.0037	−5.60
2000	0.0024	−6.03

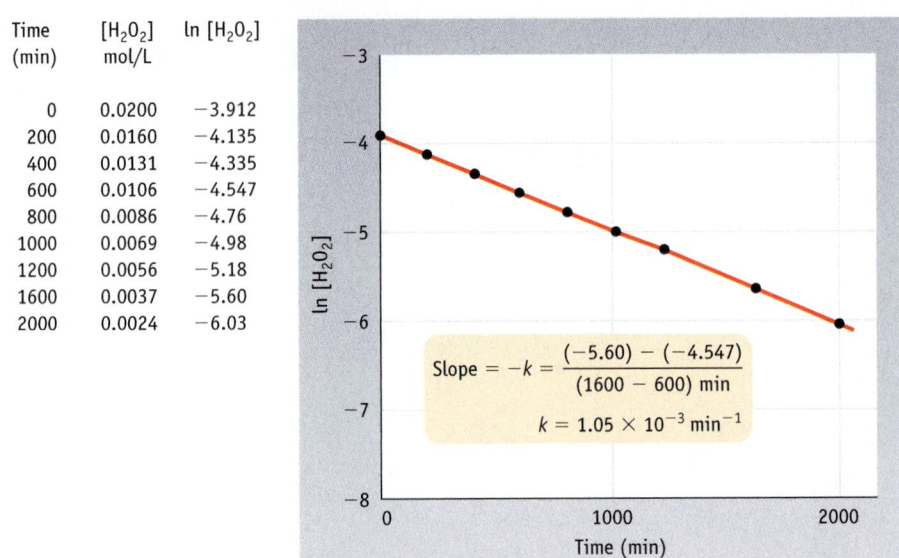

$$\text{Slope} = -k = \frac{(-5.60) - (-4.547)}{(1600 - 600)\ \text{min}}$$

$$k = 1.05 \times 10^{-3}\ \text{min}^{-1}$$

Active Figure 15.7   **The decomposition of H₂O₂.** If data for the decomposition of hydrogen peroxide,

$$2\ H_2O_2(aq) \longrightarrow 2\ H_2O(\ell) + O_2(g)$$

are plotted as the natural logarithm of the H₂O₂ concentration versus time, the result is a straight line with a negative slope. This indicates a first-order reaction. The rate constant $k = -$slope.

GENERAL
Chemistry✦Now™ **See the General ChemistryNow CD-ROM or website to explore an interactive version of this figure accompanied by an exercise.**

The decomposition of NO_2 is a second-order process.

$$NO_2(g) \longrightarrow NO(g) + \tfrac{1}{2}\ O_2(g)$$

$$\text{Rate} = k[NO_2]^2$$

This fact can be verified by showing that a plot of $1/[NO_2]$ versus time is a straight line (Figure 15.8). Here the slope of the line is equal to k.

To determine the reaction order, therefore, a chemist will plot the experimental concentration–time data in different ways until a straight-line plot is achieved. The mathematical relationships for zero-, first-, and second-order reactions are summarized in Table 15.1.

Figure 15.8   A second-order reaction. A plot of $1/[NO_2]$ versus time for the decomposition of NO_2,

$$NO_2(g) \longrightarrow NO(g) + \tfrac{1}{2}\ O_2(g)$$

results in a straight line. This confirms that this is a second-order reaction. The slope of the line equals the rate constant for this reaction.

Time (min)	[NO₂] (M)	1/[NO₂] (M⁻¹)
0	0.020	50
0.50	0.015	67
1.0	0.012	83
1.5	0.010	100
2.0	0.0087	115

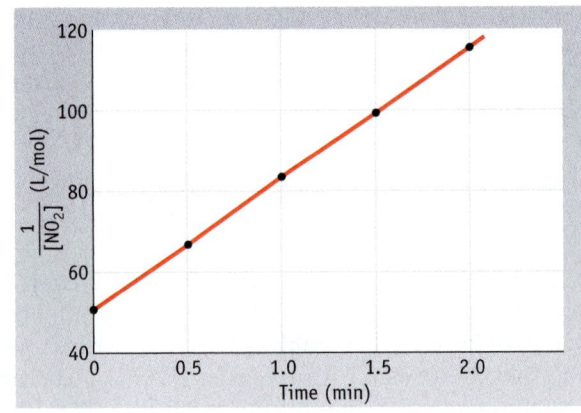

Table 15.1 Characteristic Properties of Reactions of the Type "R $\longrightarrow$ Products"

Order	Rate Equation	Integrated Rate Equation	Straight-Line Plot	Slope	k Units
0	$-\Delta[R]/\Delta T = k[R]^0$	$[R]_0 - [R]_t = kt$	$[R]_t$ vs. t	$-k$	mol/L · time
1	$-\Delta[R]/\Delta T = k[R]^1$	$\ln([R]_t/[R]_0) = -kt$	$\ln[R]_t$ vs. t	$-k$	time^{-1}
2	$-\Delta[R]/\Delta T = k[R]^2$	$(1/[R]_t) - (1/[R]_0) = kt$	$1/[R]_t$ vs. t	k	L/mol · time

GENERAL
Chemistry Now™

See the General ChemistryNow CD-ROM or website
• **Screen 15.7 Determination of Rate Equation (2),** for a tutorial on graphical methods

Exercise 15.8—Using Graphical Methods

Data for the decomposition of N_2O_5 in a particular solvent at 45 °C are as follows:

$[N_2O_5]$, mol/L	t, min
2.08	3.07
1.67	8.77
1.36	14.45
0.72	31.28

Plot $[N_2O_5]$, $\ln[N_2O_5]$, and $1/[N_2O_5]$ versus time, t. What is the order of the reaction? What is the rate constant for the reaction?

Half-Life and First-Order Reactions

The **half-life, $t_{1/2}$,** of a reaction is the time required for the concentration of a reactant to decrease to one-half its initial value. It indicates the rate at which a reactant is consumed in a chemical reaction: The longer the half-life, the slower the reaction. Half-life is used primarily when dealing with first-order processes.

The half-life, $t_{1/2}$, is the time when the fraction of the reactant R remaining is $\frac{1}{2}$.

$$[R]_t = \frac{1}{2}[R]_0 \quad \text{or} \quad \frac{[R]_t}{[R]_0} = \frac{1}{2}$$

Here $[R]_0$ is the initial concentration, and $[R]_t$ is the concentration after the reaction is half completed. To evaluate $t_{1/2}$ for a first-order reaction, we substitute $[R]_t/[R]_0 = \frac{1}{2}$ and $t = t_{1/2}$ into the integrated first-order rate equation (Equation 15.1),

$$\ln\frac{[R]_t}{[R]_0} = -kt$$

$$\ln\left(\tfrac{1}{2}\right) = -kt_{1/2}$$

Rearranging this equation (and knowing that $\ln 2 = 0.693$), provides a useful equation that relates half-life and the first-order rate constant:

$$t_{1/2} = \frac{0.693}{k} \tag{15.4}$$

■ **Half-Life and Radioactivity**
Half-life is a term often encountered when dealing with radioactive elements. Radioactive decay is a first-order process, and half-life is commonly used to describe how rapidly a radioactive element decays. See Chapter 23 and Example 15.9.

■ **Half-Life Equations for Other Reaction Orders**

For a zero-order reaction,

$$t_{1/2} = \frac{[R]_0}{2k}$$

For a second-order reaction,

$$t_{1/2} = \frac{1}{k[R]_0}$$

This equation identifies an important feature of first-order reactions: $t_{1/2}$ is *independent* of concentration.

To illustrate the concept of half-life, consider the first-order decomposition of H_2O_2:

$$2 \; H_2O_2(aq) \longrightarrow 2 \; H_2O(\ell) + O_2(g)$$

The data provided in Figure 15.7 allowed us to determine that the rate constant, k, for this reaction is $1.05 \times 10^{-3} \; min^{-1}$. Using Equation 15.4, the half-life of H_2O_2 in this reaction can be calculated from the rate constant.

$$t_{1/2} = \frac{0.693}{k} = \frac{0.693}{1.05 \times 10^{-3} \; min^{-1}} = 660. \; min$$

In Figure 15.9, the concentration of H_2O_2 has been plotted as a function of time. This graph shows that $[H_2O_2]$ decreases by half within each 660-min period. The initial concentration of H_2O_2 is 0.020 M, but it drops to 0.010 M after 660 min. The concentration drops again by half (to 0.0050 M) after another 660 min. That is, after two half-lives (1320 min), the concentration is $(\frac{1}{2}) \times (\frac{1}{2}) = (\frac{1}{2})^2 = \frac{1}{4}$, or 25% of the initial concentration. After three half-lives (1980 min), the concentration has dropped to $(\frac{1}{2}) \times (\frac{1}{2}) \times (\frac{1}{2}) = (\frac{1}{2})^3 = \frac{1}{8}$, or 12.5% of the initial value; here $[H_2O_2] = 0.0025$ M.

It is hard to visualize whether a reaction is fast or slow from the value of the rate constant. Can you tell from the value of the rate constant, $k = 1.05 \times 10^{-3}$ min,

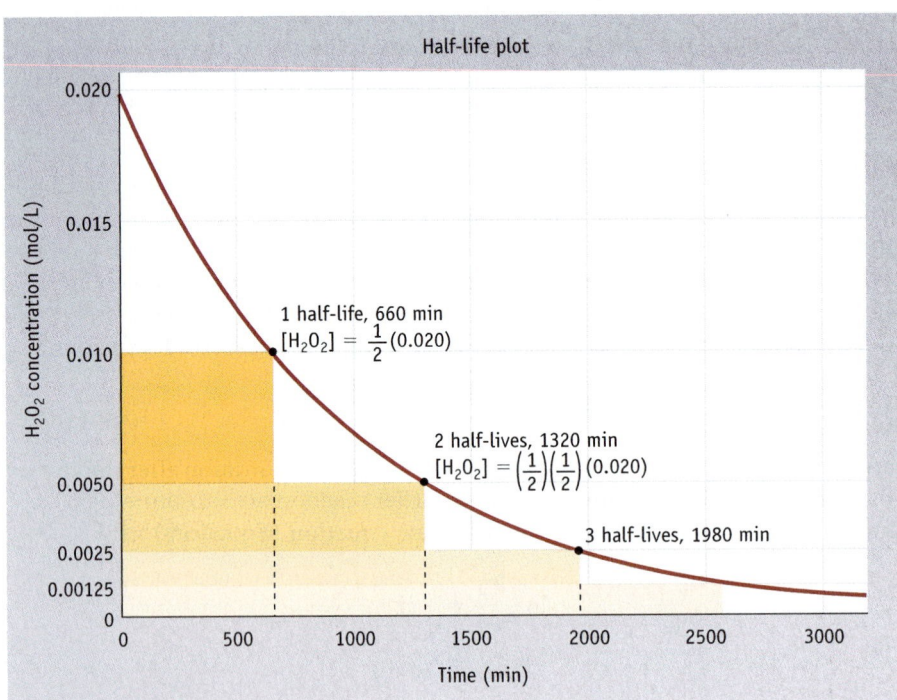

Active Figure 15.9   **Half-life of a first-order reaction.** This concentration-versus-time curve shows the disappearance of H_2O_2 (where $k = 1.05 \times 10^{-3} \; min^{-1}$). The concentration of H_2O_2 is halved every 660 min. (This plot of concentration versus time is similar in shape to those for all other first-order reactions.)

GENERAL

Chemistry ⚛ Now™ See the General ChemistryNow CD-ROM or website to explore an interactive version of this figure accompanied by an exercise.

whether the decomposition of H_2O_2 will require seconds, minutes, hours, or days to reach completion? Probably not, but this is easily assessed from the value of half-life for this reaction, 660 min. The half-life is 11 h, so you will have to wait several days for most of the H_2O_2 in a sample to decompose.

GENERAL
Chemistry⚛Now™

See the General ChemistryNow CD-ROM or website:
• **Screen 15.8 Half-Life,** for tutorials on using half-life

---

Example 15.8—Half-Life and a First-Order Process

Problem Sucrose, $C_{12}H_{22}O_{11}$, decomposes to fructose and glucose in acid solution with the rate law

$$\text{Rate} = k[\text{sucrose}] \qquad k = 0.208 \text{ h}^{-1} \text{ at 25 °C}$$

What amount of time is required for 87.5% of the initial concentration of sucrose to decompose?

Strategy After 87.5% of the sucrose has decomposed, 12.5% remains. That is, the fraction remaining is 0.125. To reach this point, three half-lives are required.

Half-Life	Fraction Remaining
1	0.5
2	0.25
3	0.125

Therefore, we calculate the half-life from Equation 15.4 and then multiply by 3.

Solution The half-life for the reaction is

$$t_{1/2} = \frac{0.693}{k} = \frac{0.693}{0.208 \text{ h}^{-1}} = 3.33 \text{ h}$$

Three half-lives must elapse before the fraction remaining is 0.125, so

$$\text{Time elapsed} = 3 \times 3.33 \text{ h} = \boxed{9.99 \text{ h}}$$

Example 15.9—Half-Life and First-Order Processes

Problem Radioactive radon-222 gas (^{222}Rn) from natural sources can seep into the basement of a home. The half-life of ^{222}Rn is 3.8 days. If a basement has 4.0×10^{13} atoms of ^{222}Rn per liter of air, and the radon gas is trapped in the basement, how many atoms of ^{222}Rn will remain after one month (30 days)?

Strategy Using Equation 15.1, and knowing the number of atoms at the beginning ($= [R]_0$), the elapsed time (30 days), and the rate constant, we can calculate the number of atoms remaining ($= [R]_t$). First, the rate constant, k, must be found from the half-life using Equation 15.4. (*See General ChemistryNow Screen 15.8.*)

Solution The rate constant, k, is

$$k = \frac{0.693}{t_{1/2}} = \frac{0.693}{3.8 \text{ d}} = 0.18 \text{ d}^{-1}$$

Now use Equation 15.1 to calculate the number of atoms remaining after 30. days.

$$\ln \frac{[\text{Rn}]_t}{4.0 \times 10^{13} \text{ atom/L}} = -(0.18 \text{ d}^{-1})(30. \text{ d}) = -5.5$$

$$\frac{[\text{Rn}]_t}{4.0 \times 10^{13} \text{ atom/L}} = e^{-5.5} = 0.0042$$

$$[\text{Rn}]_t = \boxed{1.7 \times 10^{11} \text{ atom/L}}$$

Exercise 15.9—Half-Life and a First-Order Process

Americium is used in smoke detectors and in medicine for the treatment of certain malignancies. One isotope of americium, ^{241}Am, has a rate constant, k, for radioactive decay of 0.0016 y^{-1}. In contrast, radioactive iodine-125, which is used for studies of thyroid functioning, has a rate constant for decay of 0.011 d^{-1}.

(a) What are the half-lives of these isotopes?
(b) Which element decays faster?
(c) If you are given a dose of iodine-125, and have 1.6×10^{15} atoms, how many atoms remain after 2.0 days?

15.5—A Microscopic View of Reaction Rates

Throughout this book we have turned to the particulate level of chemistry to understand chemical phenomena. The rate of a reaction is no exception. Looking at the way reactions occur at the atomic and molecular levels provides some insight into the various influences on rates of reactions.

Let us review the macroscopic observations we have made so far concerning reaction rates. We know the wide difference in rates of reaction relates to the specific compounds involved—from very fast reactions like the explosion that occurs when hydrogen and oxygen are exposed to a spark or flame (Figure 1.13), to slow reactions like the formation of rust that occur over days, weeks, or years. For a specific reaction, factors that influence reaction rate include the concentration of the reactants, the temperature of the reaction system, and the presence of catalysts. Let us next look at each of these influences in more depth.

Concentration, Reaction Rate, and Collision Theory

Consider the gas-phase reaction of nitric oxide and ozone:

$$\text{NO(g)} + \text{O}_3\text{(g)} \longrightarrow \text{NO}_2\text{(g)} + \text{O}_2\text{(g)}$$

The rate law for this product-favored reaction is first order in each reactant: Rate = $k[\text{NO}][\text{O}_3]$. How can this reaction have this rate law?

Let us consider the reaction at the particulate level and imagine a flask containing a mixture of NO and O_3 molecules in the gas phase. Both kinds of molecules are in rapid and random motion within the flask. They strike the walls of the vessel and collide with other molecules. For this or any other reaction to occur, the **collision theory** of reaction rates states that three conditions must be met:

1. The reacting molecules must collide with one another.
2. The reacting molecules must collide with sufficient energy to break bonds.

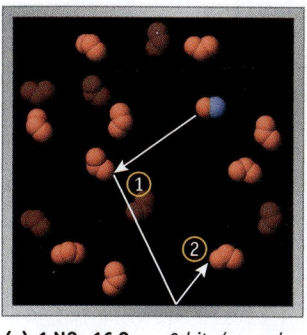 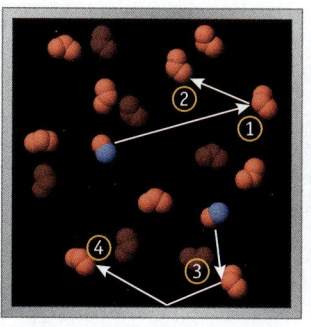 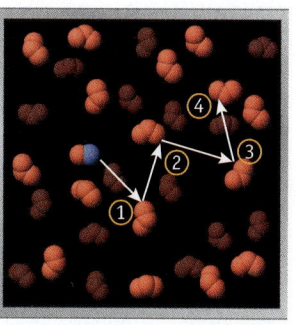

(a) 1 NO : 16 O$_3$ — *2 hits/second*   **(b) 2 NO : 16 O$_3$** — *4 hits/second*   **(c) 1 NO : 32 O$_3$** — *4 hits/second*

Figure 15.10 **The effect of concentration on the frequency of molecular collisions.** (a) A single NO molecule, moving among sixteen O$_3$ molecules, is shown colliding with two of them per second. (b) If two NO molecules move among 16 O$_3$ molecules, we would predict that four NO–O$_3$ collisions would occur per second. (c) If the number of O$_3$ molecules is doubled (to 32), the frequency of NO–O$_3$ collisions is also doubled, to four per second.

3. The molecules must collide in an orientation that can lead to rearrangement of the atoms.

We shall discuss each of these conditions within the context of the effects of concentration and temperature on reaction rate.

To react, molecules must collide with one another. The rate of their reaction is primarily related to the number of collisions, which is in turn related to their concentrations (Figure 15.10). Doubling the concentration of one reagent in the NO + O$_3$ reaction, say NO, will lead to twice the number of molecular collisions. Figure 15.10a shows a single molecule of one of the reactants (NO) moving randomly among sixteen O$_3$ molecules. In a given time period, it might collide with two O$_3$ molecules. The number of NO—O$_3$ collisions will double, however, if the concentration of NO molecules is doubled (to 2, as shown in Figure 15.10b) or if the number of O$_3$ molecules is doubled (to 32, as in Figure 15.10c). Thus we can explain the dependence of reaction rate on concentration: The number of collisions between the two reactant molecules is directly proportional to the concentration of each reactant, and the rate of the reaction shows a first-order dependence on each reactant.

GENERAL
Chemistry ⚛ Now™

See the General ChemistryNow CD-ROM or website:
• **Screen 15.9 Microscopic View of Reactions (1),** for a visualization of collision theory

---

Temperature, Reaction Rate, and Activation Energy

In a laboratory or in the chemical industry, a chemical reaction is often carried out at elevated temperature because this allows the reaction to occur more rapidly. Conversely, it is sometimes desirable to lower the temperature to slow down a chemical reaction (to avoid an uncontrollable reaction or a potentially dangerous explosion). Chemists are very aware of the effect of temperature on the rate of a reaction. But how and why does temperature influence reaction rate?

A discussion of the effect of temperature on reaction rate goes back to the distribution of energies for molecules in a sample of a gas or liquid. Recall from studying gases and liquids that the molecules in a sample have a wide range of energies, described earlier as a Boltzmann distribution of energies [◄ Figure 12.14 and Figure 13.14]. That is, in any sample of a gas or liquid, some molecules have very low energies, others have very high energies, but most have some intermediate energy. As

Figure 15.11 Kinetic-energy distribution curve. The vertical axis gives the relative number of molecules possessing the energy indicated on the horizontal axis. The graph indicates the minimum energy required for an arbitrary reaction. At a higher temperature, a larger fraction of the molecules have sufficient energy to react. (Recall Figure 12.14, the Boltzmann distribution function, for a collection of gas molecules.)

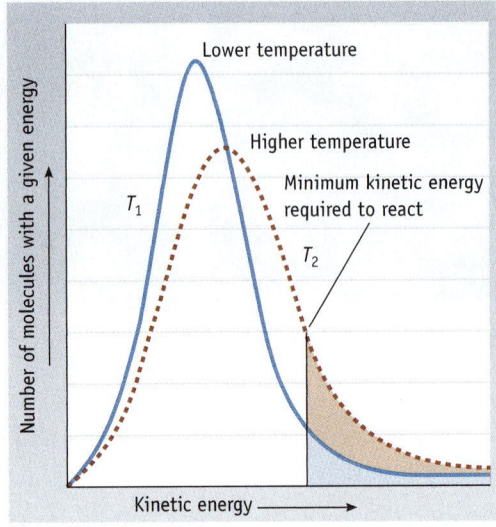

the temperature increases, the average energy of the molecules in the sample increases, as does the fraction having higher energies (Figure 15.11).

Activation Energy

Molecules require some minimum energy to react. Chemists visualize this as an energy barrier that must be surmounted by the reactants for a reaction to occur (Figure 15.12). The energy required to surmount the barrier is called the **activation energy, E_a**. If the barrier is low, the energy required is low, and a high proportion of the molecules in a sample may have sufficient energy to react. In such a case, the reaction will be fast. If the barrier is high, the activation energy is high, and only a few reactant molecules in a sample may have sufficient energy. In this case, the reaction will be slow.

As an illustration of an activation energy barrier, consider the conversion of NO_2 and CO to NO and CO_2 or the reverse reaction (Figure 15.13). At the molecular level we imagine that the reaction involves the transfer of an O atom from an NO_2 molecule to a CO molecule (or, in the reverse reaction, the transfer of an O atom from CO_2 to NO).

$$NO_2(g) + CO(g) \rightleftharpoons NO(g) + CO_2(g)$$

These reactions cannot occur, however, without the initial input of energy, the activation energy. For example, for NO_2 to transfer an O atom to CO, the N—O bond must be broken. We show this process by using an energy diagram or *reaction coordinate diagram*. The horizontal axis represents the nature of the reactants and products as the reaction proceeds, and the vertical axis represents the potential energy of the system during the reaction. When NO_2 and CO approach and O atom transfer begins, an N—O bond is being broken and a C=O bond is forming. The energy of the system reaches a maximum at the **transition state**. At the transition state, sufficient energy has been concentrated in the appropriate bonds; bonds in the reactants can now break and new bonds can form to give products. The system is poised to go on to products, or it can return to the reactants. Because the transition state is at a maximum in potential energy, it cannot be isolated. However, chemists can

Figure 15.12 An analogy to chemical activation energy. For the volleyball to go over the net, the player must give it sufficient energy.

Charles D. Winters

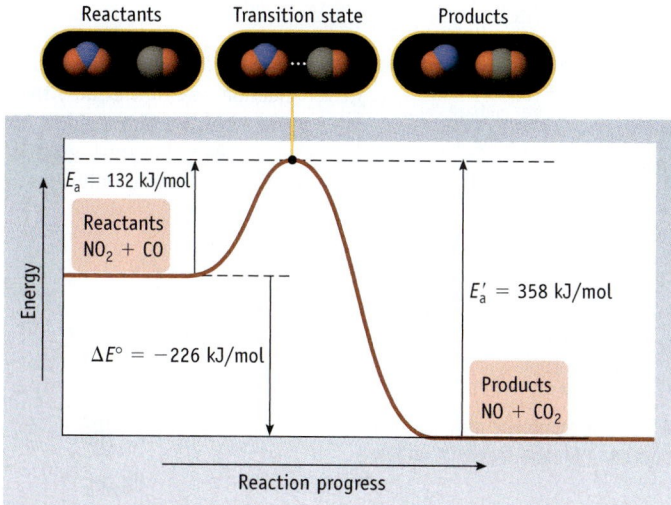

Active Figure 15.13 **Activation energy.** The reaction of NO_2 and CO (to give NO and CO_2) has an activation energy barrier of 132 kJ/mol. The reverse reaction (NO + CO_2 ⟶ NO_2 + CO) requires 358 kJ/mol. The net energy change for the reaction of NO_2 and CO is −226 kJ/mol.

GENERAL
Chemistry⋅Now™ See the General ChemistryNow CD-ROM or website to explore an interactive version of this figure accompanied by an exercise.

sometimes get a notion of the transition state using computer molecular modeling techniques.

In the NO_2 + CO reaction, 132 kJ/mol is required to reach the transition state at the top of the energy barrier. As the reaction "slides down" the other side of the barrier—as the N—O bond is finally broken and a C=O bond forms—the reaction evolves energy, 358 kJ/mol. The net energy involved in the reaction is

$$\text{Net energy} = +132 \text{ kJ/mol} + (-358 \text{ kJ/mol}) = -226 \text{ kJ/mol}$$

Overall, the reaction is exothermic by 226 kJ/mol.

What happens if NO and CO_2 are mixed to form NO_2 and CO? Now the reaction requires 358 kJ/mol to reach the transition state, and 132 kJ/mol is evolved on proceeding to the product, NO_2 and CO. The reaction in this direction is endothermic.

GENERAL
Chemistry⋅Now™

See the General ChemistryNow CD-ROM or website:
- **Screen 15.10 Microscopic View of Reactions (2),** for a simulation of reaction coordinate diagrams

Effect of a Temperature Increase

The conversion of NO_2 and CO to products at room temperature is slow because only a few of the molecules have enough energy to undergo this reaction. The rate can be increased, by heating the sample, which has the effect of increasing the

A Closer Look

Reaction Coordinate Diagrams

Reaction coordinate diagrams (Figure 15.13) can convey a great deal of information. Another reaction that would have an energy diagram like that in Figure 15.13 is the substitution of a halogen atom of CH_3Cl by an ion such as F^-. Here the F^- ion attacks the molecule from the side opposite the Cl substituent. As F^- begins to form a bond to carbon, the C—Cl bond weakens and the CH_3 portion of the molecule changes shape. As time progresses, the products CH_3F and Cl^- are formed.

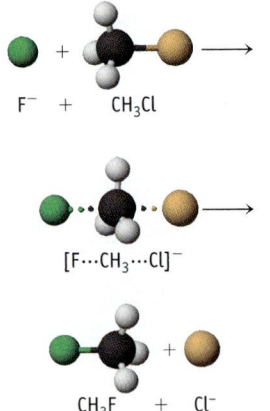

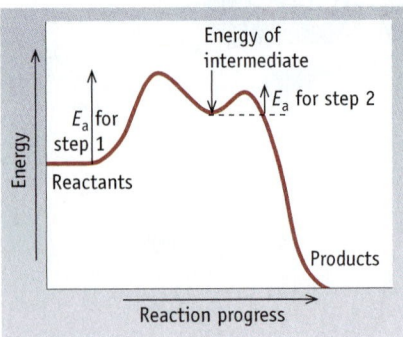

Figure A A reaction coordinate diagram for a two-step reaction, a process involving an intermediate.

The diagram in Figure A shows a different type of reaction, a two-step reaction that involves a reaction intermediate. An example would be the substitution of the —OH group on methanol by a halide ion in the presence of acid. In the first step, an H^+ ion attaches to the O of the C—O—H group in a rapid, reversible reaction. Activation energy is required to reach this state. The energy of this protonated species, $CH_3OH_2^+$, a reaction intermediate, is higher than the energies of the reactants and is represented by the dip in the curve

shown in Figure A. In the second step, a halide ion, say Br^-, attacks the intermediate in a process that requires further activation energy. The final result is methyl bromide, CH_3Br, and water.

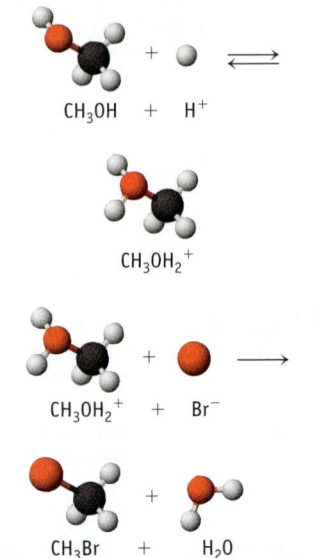

Notice in Figure A, as in Figure 15.13, that the energy of the products is lower than the energy of the reactants. The reaction is exothermic.

---

fraction of molecules having higher energies (Figure 15.11). Raising the temperature always increases the reaction rate by increasing the fraction of molecules with enough energy to surmount the activation energy barrier.

Effect of Molecular Orientation on Reaction Rate

Not only must the NO_2 and CO molecules collide with sufficient energy, but they must also come together in the correct orientation. Having a sufficiently high energy is necessary, but it is not sufficient to ensure that reactants will form products. For the reaction of NO_2 and CO, we can imagine that the transition state structure has one of the O atoms of NO_2 beginning to bind to the C atom of CO in preparation for O atom transfer (Figure 15.13). This "steric factor" is important in determining the rate of the reaction and affects the value of the rate constant, k. The lower the probability of achieving the proper alignment, the smaller the value of k, and the slower the reaction.

Imagine what happens when two or more complicated molecules collide. In only a small fraction of the collisions will the molecules come together in exactly the right orientation. Thus, only a tiny fraction of the collisions can be effective. No wonder some reactions are so slow. Conversely, it is amazing that so many are so fast!

The Arrhenius Equation

The observation that reaction rates depend on the energy and frequency of collisions between reacting molecules, on the temperature, and on whether the collisions have the correct geometry is summarized by the **Arrhenius equation**:

$$k = \text{rate constant} = Ae^{-E_a/RT}$$

Frequency factor ⟶   ⟵ Fraction of molecules with minimum energy for reaction   (15.5)

where R is the gas constant with a value of 8.314510×10^{-3} kJ/K · mol. The parameter A is called the *frequency factor*, and it has units of L/mol · s. It is related to the number of collisions and to the fraction of collisions that have the correct geometry; A is specific to each reaction and is temperature-dependent. The factor $e^{-E_a/RT}$ is interpreted as the *fraction of molecules having the minimum energy required for reaction*; its value is always less than 1. As the table in the margin shows, this fraction changes significantly with temperature.

The Arrhenius equation is valuable because it can be used to (1) calculate the value of the activation energy from the temperature dependence of the rate constant and (2) calculate the rate constant for a given temperature if the activation energy and A are known. Taking the natural logarithm of each side of Equation 15.5, we have

$$\ln k = \ln A + \left(-\frac{E_a}{RT} \right)$$

If we rearrange this expression slightly, it becomes an equation for a straight line relating $\ln k$ to $(1/T)$:

$$\ln k = -\frac{E_a}{R}\left(\frac{1}{T}\right) + \ln A \quad \longleftarrow \text{Arrhenius equation}$$

$$\downarrow \qquad\qquad \downarrow \qquad\quad \downarrow$$

$$y \quad = \quad mx \quad + b \quad \longleftarrow \text{Equation for straight line}$$

(15.6)

This means that, if the natural logarithm of k ($\ln k$) is plotted versus $1/T$, the result is a downward-sloping line with a slope of $(-E_a/R)$. Now we have a way to calculate E_a from experimental values of k at several temperatures, a calculation illustrated in Example 15.10 and in Figure 15.14.

■ **Interpreting the Arrhenius Equation**
(a) The exponential term gives the fraction of molecules having sufficient energy for reaction as a function of T.

Temperature (K)	Value of $e^{-E_a/RT}$ for $E_a = 40$ kJ
298	9.7×10^{-8}
400	5.9×10^{-6}
600	3.3×10^{-4}

(b) Significance of A. Although a complete understanding of A goes beyond the level of this text, it can be noted that A becomes smaller as the reactants become larger, a reflection of the "steric effect."

GENERAL
Chemistry ❖ Now™

See the General ChemistryNow CD-ROM or website:

- **Screen 15.11 Control of Reaction Rates (3),** for a simulation and three tutorials on the temperature dependence of reaction rates and the Arrhenius equation

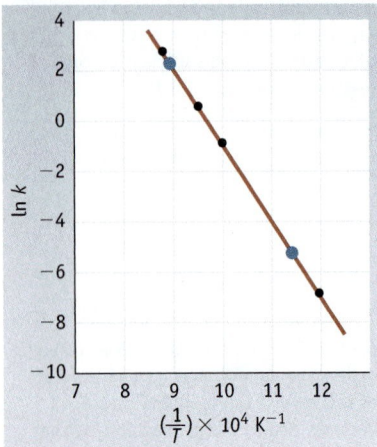

Active Figure 15.14 **Arrhenius plot.**
A plot of ln k versus $1/T$ for the reaction
$2 N_2O(g) \longrightarrow 2 N_2(g) + O_2(g)$. The slope of
the line gives E_a. See Example 15.10.

GENERAL
Chemistry Now™ See the General
ChemistryNow CD-ROM or website to explore
an interactive version of this figure accompanied by an exercise.

Example 15.10—Determination of E_a from the Arrhenius Equation

Problem Using the experimental data shown in the table, calculate the activation energy E_a for the reaction

$$2 N_2O(g) \longrightarrow 2 N_2(g) + O_2(g)$$

Experiment	Temperature (K)	k (L/mol · s)
1	1125	11.59
2	1053	1.67
3	1001	0.380
4	838	0.0011

Strategy To use the Arrhenius equation (Equation 15.6), we first need to calculate ln k and $1/T$ for each data point. These data are then plotted, and E_a is calculated from the resulting straight line (slope $= -E_a/R$).

Solution The data are expressed as $1/T$ and ln k.

Experiment	$1/T$ (K^{-1})	ln k
1	8.889×10^{-4}	2.4501
2	9.497×10^{-4}	0.513
3	9.990×10^{-4}	-0.968
4	11.9×10^{-4}	-6.81

Plotting these data gives the graph shown here. Choosing the large blue points on the graph in Figure 15.14, the slope is found to be

$$\text{Slope} = \frac{\Delta \ln k}{\Delta(1/T)} = \frac{2.0 - (-5.6)}{(9.0 - 11.5)(10^{-4})/K} = -3.0 \times 10^4 \text{ K}$$

The activation energy is evaluated from

$$\text{Slope} = -\frac{E_a}{R}$$

$$-3.0 \times 10^4 \text{ K} = -\frac{E_a}{8.31 \times 10^{-3} \text{ kJ/K} \cdot \text{mol}}$$

$$E_a = 250 \text{ kJ/mol}$$

In addition to the graphical method for evaluating E_a used in Example 15.10, E_a can be obtained algebraically. Knowing k at two different temperatures, we can write an equation for each of these conditions:

$$\ln k_1 = -\left(\frac{E_a}{RT_1}\right) + \ln A \quad \text{or} \quad \ln k_2 = -\left(\frac{E_a}{RT_2}\right) + \ln A$$

If one of these equations is subtracted from the other, we have

■ **E_a, Reaction Rates, and Temperature**
A good rule of thumb is that reaction rates double for every 10 °C rise in temperature in the vicinity of room temperature.

$$\ln k_2 - \ln k_1 = \ln \frac{k_2}{k_1} = -\frac{E_a}{R}\left[\frac{1}{T_2} - \frac{1}{T_1}\right] \qquad (15.7)$$

Example 15.11 demonstrates the use of this equation.

Example 15.11—Calculating E_a from the Temperature Dependence of k

Problem Using values of k determined at two different temperatures, calculate the value of E_a for the decomposition of HI:

$$2\ HI(g) \longrightarrow H_2(g) + I_2(g)$$
$$k = 2.15 \times 10^{-8}\ L/(mol \cdot s)\ \text{at}\ 6.50 \times 10^2\ K$$
$$k = 2.39 \times 10^{-7}\ L/(mol \cdot s)\ \text{at}\ 7.00 \times 10^2\ K$$

Strategy Here we are given values of k_1, T_1, k_2, and T_2, so we use Equation 15.7.

Solution

$$\ln \frac{2.39 \times 10^{-7}\ L/(mol \cdot s)}{2.15 \times 10^{-8}\ L/(mol \cdot s)} = -\frac{E_a}{8.315 \times 10^{-3}\ kJ/K \cdot mol} \left[\frac{1}{7.00 \times 10^2 K} - \frac{1}{6.50 \times 10^2\ K} \right]$$

Solving this equation for E_a, we find $E_a = 180\ kJ/mol$.

Comment When using Equation 15.7, be aware that another way to write the difference in fractions in brackets is

$$\left[\frac{1}{T_2} - \frac{1}{T_1} \right] = \frac{T_1 - T_2}{T_1 T_2}$$

Also be very careful of significant figures.

Exercise 15.10—Calculating E_a from the Temperature Dependence of k

The colorless gas N_2O_4 decomposes to the brown gas NO_2 in a first-order reaction:

$$N_2O_4(g) \longrightarrow 2\ NO_2(g)$$

The rate constant $k = 4.5 \times 10^3\ s^{-1}$ at 274 K and $k = 1.00 \times 10^4\ s^{-1}$ at 283 K. What is the activation energy, E_a?

Effect of Catalysts on Reaction Rate

Catalysts are substances that speed up the rate of a chemical reaction, and we have seen several examples of catalysts in earlier discussions in this chapter: MnO_2 (Figure 15.4), iodide ion (page 705), an enzyme in a potato (page 706), and hydroxide ion (page 714) all catalyze the decomposition of hydrogen peroxide. In biological systems, catalysts called *enzymes* influence the rates of most reactions (page 699).

Catalysts are not consumed in a chemical reaction. They are, however, intimately involved in the details of the reaction at the particulate level. Their function is to provide a different pathway with a lower activation energy for the reaction.

To illustrate how a catalyst participates in a reaction, let us consider the first-order interconversion of the butene isomer, *cis*-2-butene, to the slightly more stable isomer, *trans*-2-butene.

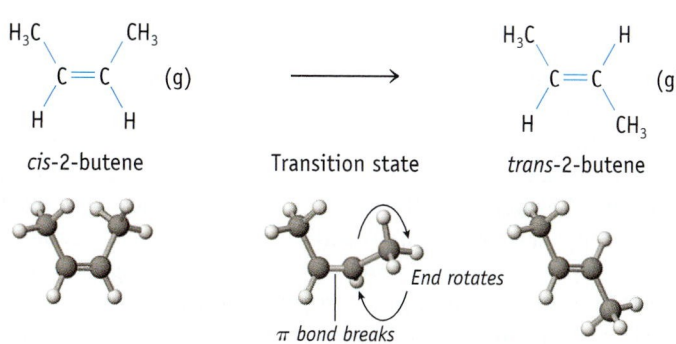

cis-2-butene   Transition state   trans-2-butene

■ **Enzymes: Biological Catalysts**
Catalase is an enzyme whose function is to speed up the decomposition of hydrogen peroxide. This enzyme ensures that hydrogen peroxide, which is highly toxic, does not build up in the body.

■ **Butene Isomerization**
Isomerization of *cis*-2-butene has the rate law "Rate = k[*cis*-2-butene]." In a large collection of *cis*-2-butene molecules, the probability that a molecule will isomerize is related to the fraction of molecules that have a high enough energy. The rate of such a reaction would have a first-order dependence on concentration. (*See Screen 10.8 of the General ChemistryNow* to view an animation of the interconversion of butene isomers and the energy barrier to the process.)

Figure 15.15   The mechanism of the iodine-catalyzed isomerization of *cis*-2-butene. *Cis*-2-butene is converted to *trans*-2-butene in the presence of a catalytic amount of iodine. Catalyzed reactions are often pictured in such diagrams to emphasize what chemists refer to as a "catalytic cycle."

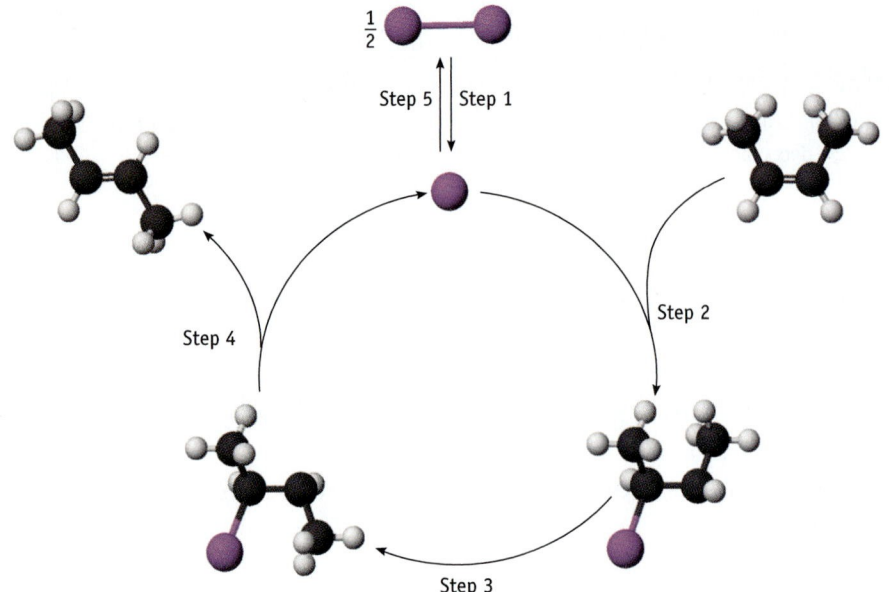

The activation energy for the uncatalyzed conversion is relatively large—264 kJ/mol—because the π bond must be broken to allow one end of the molecule to rotate into a new position. Because of the high activation energy this is a slow reaction, and rather high temperatures are required for it to occur at a reasonable rate.

The *cis*- to *trans*-2-butene reaction is greatly accelerated by a catalyst, iodine. The presence of iodine allows the isomerization reaction to be carried out at a temperature several hundred degrees lower than the uncatalyzed reaction. Iodine is not consumed (nor is it a product), and it does not appear in the overall balanced equation. It does appear in the reaction rate law, however; the rate of the reaction depends on the square root of the iodine concentration:

$$\text{Rate} = k\,[\textit{cis}\text{-2-butene}]\,[\text{I}_2]^{1/2}$$

The rate of the *cis–trans* conversion changes because the presence of I_2 changes the way the reaction occurs. That is, it changes the *mechanism* of the reaction (see Section 15.6 and Figure 15.15). The best hypothesis is that iodine molecules first dissociate to form iodine atoms (Step 1). An I atom then adds to one of the C atoms of the C=C double bond (Step 2). This converts the double bond between the carbon atoms to a single bond (the π bond is broken) and allows the ends of the molecule to twist freely relative to each other (Step 3). If the I atom then dissociates from the intermediate, the double bond can re-form in the *trans* configuration (Step 4).

The iodine atom catalyzing the rotation is now free to add to another molecule of *cis*-2-butene. The result is a kind of chain reaction, as one molecule of *cis*-2-butene after another is converted to the *trans* isomer. The chain is broken if the iodine atom recombines with another iodine atom to re-form molecular iodine.

An energy profile for the catalyzed reaction (Figure 15.16) shows that the overall energy barrier has been greatly lowered from the situation in the uncatalyzed reaction. In addition, the energy profile for the reaction includes several steps (a total of five), representing each step in the reaction. This proposed mechanism includes a series of chemical species called **reaction intermediates**, species formed in one step of the reaction and consumed in a later step. Iodine atoms are intermediates, as are the free radical species formed when an iodine atom adds to *cis*-2-butene.

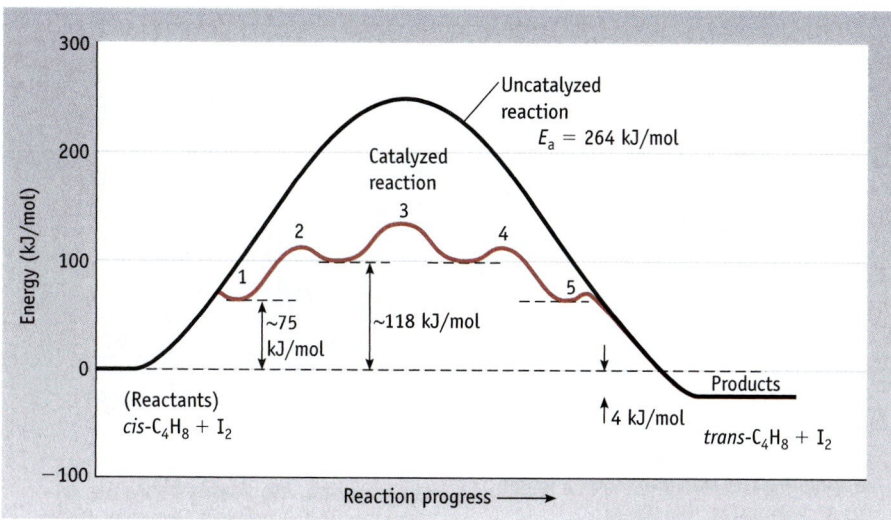

Figure 15.16 **Energy profile for the iodine-catalyzed reaction of *cis*-2-butene.** A catalyst accelerates a reaction by altering the mechanism so that the activation energy is lowered. With a smaller barrier to overcome, more reacting molecules have sufficient energy to surmount the barrier, and the reaction occurs more rapidly. The energy profile for the uncatalyzed conversion of *cis*-2-butene to *trans*-2-butene is shown by the black curve, and that for the iodine-catalyzed reaction is represented by the red curve. Notice that the shape of the barrier has changed because the mechanism has changed.

Five important points are associated with this mechanism:

- Iodine molecules, I_2, dissociate to atoms and then re-form. On the macroscopic level, the concentration of I_2 is unchanged. Iodine does not appear in the balanced, stoichiometric equation even though it appears in the rate equation. This is generally true of catalysts.

- Both the catalyst I_2 and the reactant *cis*-2-butene are in the gas phase. If a catalyst is present in the same phase as the reacting substance, it is called a **homogeneous catalyst**.

- Iodine atoms and the radical species formed by addition of an I atom to a 2-butene molecule are intermediates.

- The activation energy barrier to reaction is significantly lower because the mechanism changed. Dropping the activation energy from 264 kJ/mol for the uncatalyzed reaction to about 150 kJ/mol for the catalyzed process makes the catalyzed reaction 10^{15} times faster!

- The diagram of energy-versus-reaction progress has five energy barriers (five humps appear in the curve). This feature in the diagram means that the reaction occurs in a series of five steps.

What we have described here is a reaction mechanism. The uncatalyzed isomerization reaction of *cis*-2-butene is a one-step reaction mechanism, whereas the catalyzed mechanism involves a series of steps. We shall discuss reaction mechanisms in more detail in the next section.

See the General ChemistryNow CD-ROM or website:

- **Screen 15.14 Catalysis and Reaction Rate,** for a description of various catalysts, a visualization of the effect of a catalyst on activation energy, an interview of a scientist describing catalyst use in industry, two exercises on reaction mechanisms and the effect of catalysts, and a video exercise on catalysis

A Closer Look

Enzymes: Nature's Catalysts

Enzymes are powerful catalysts, typically producing a reaction rate that is 10^7 to 10^{14} times faster than the uncatalyzed rate. Metal ions are often part of an enzyme. Carboxypeptidase, for example, contains Zn^{2+} ions at the active site.

In 1913 Leonor Michaelis and Maud L. Menten proposed a general theory of enzyme action based on kinetic observations. They assumed that the substrate, S (the reactant), and the enzyme, E, form a complex, ES. This complex then breaks down, releasing the enzyme and the product, P.

$$E + S \rightleftarrows ES \rightleftarrows E + P$$

The table below lists a few important enzymes. One, carbonic anhydrase, was mentioned at the beginning of the chapter (page 699), where a simple experiment showed the rate-enhancing ability of this enzyme.

Here is another experiment you can do with carbonic anhydrase. Take a sip of very cold carbonated beverage. The tingling sensation you feel on your tongue and in your mouth is not from the CO_2 bubbles. Rather, it comes from the protons released when carbonic anhydrase accelerates the formation of H^+ ions from dissolved H_2CO_3 (see page 699). Acidification of nerve endings creates the tingling feeling.

The enzymes trypsin, chymotrypsin, and elastase are digestive enzymes, catalyzing

the hydrolysis of peptide bonds [◀ page 532]. They are synthesized in the pancreas and secreted into the digestive tract.

Acetylcholinesterase is involved in transmission of nerve impulses. Many pesticides interfere with this enzyme, so farm workers are often tested to be sure they have not been overexposed to agricultural toxins.

The liver has the primary role in maintaining blood glucose levels. This organ produces glucose with phosphate groups attached (PO_4^{3-}). The enzyme glucose phosphatase in the liver has the function of removing the phosphate group before the glucose enters the blood.

See also "The Chemistry of Life: Biochemistry," page 530.

Charles D. Winters

Enzyme action. The tingling feeling you get when you drink a carbonated beverage comes from the H^+ ions released by H_2CO_3. The acid is formed rapidly from dissolved CO_2 in presence of the enzyme carbonic anhydrase.

Biologically Important Reactions Catalyzed by Enzymes

Enzyme	Enzyme Function or Reaction Catalyzed
Carbonic anhydrase	$CO_2 + H_2O \longrightarrow H_2CO_3$
Chymotrypsin	Cleavage of peptide linkages in proteins
Urease	$(H_2N)_2CO + 2\ H_2O + H^+ \longrightarrow 2\ NH_4^+ + HCO_3^-$
Catalase	$2\ H_2O_2 \longrightarrow 2\ H_2O + O_2$
Acetylcholinesterase	Regenerates acetylcholine, an important substance in the transmission of nerve impulses, from acetate and choline
Hexokinase and glucokinase	Both enzymes catalyze the formation of a phosphate ester linkage to a —OH group of a sugar. Glucokinase is a liver-specific enzyme, and the liver is the major organ for the storage of excess dietary sugar as glycogen.

15.6—Reaction Mechanisms

■ **Rate Laws and Mechanisms**
Rate laws are macroscopic observations. Mechanisms analyze how reactions occur at the particulate level.

One of the most important reasons to study reaction rates is the fact that rate laws help us to understand **reaction mechanisms**, the sequence of bond-making and bond-breaking steps that occurs during the conversion of reactants to products. The study of reaction mechanisms places us squarely within the realm of the particulate level of chemistry. We want to analyze the changes that atoms and molecules undergo when they react. We then want to relate this description back to the macroscopic world, to the experimental observations of reaction rates.

Based on the rate equation for a reaction, and by applying chemical intuition, chemists can often make an educated guess about the mechanism for the reaction. In some reactions, the conversion of reactants to products in a single step is envisioned. For example, nitrogen dioxide and carbon monoxide react in a single-step reaction, with the reaction occurring as a consequence of a collision between reac-

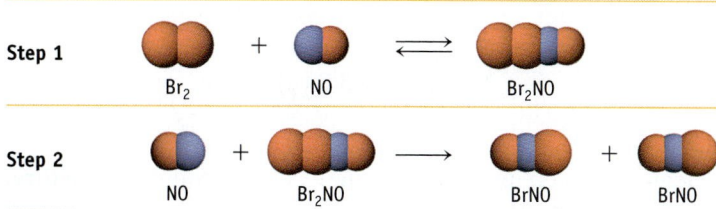

Figure 15.17   A reaction mechanism. A representation of the proposed two-step mechanism by which NO and Br_2 are converted to NOBr.

tant molecules (Figure 15.13). The uncatalyzed isomerization of *cis*-2-butene to *trans*-2-butene is also best described as a single-step reaction (Figure 15.16).

Most chemical reactions occur in a sequence of steps, however. We saw an example with the iodine-catalyzed 2-butene isomerization reaction. Another example of a reaction that occurs in several steps is the reaction of bromine and NO:

$$Br_2(g) + 2\,NO(g) \longrightarrow 2\,BrNO(g)$$

A single-step reaction would require that three reactant molecules collide simultaneously in just the right orientation to be productive. Clearly, such an event has a low probability of occurring. Thus, for this reaction it would be reasonable to look for a mechanism that occurs in a series of steps, with each step involving only one or two molecules. For example, in one possible mechanism Br_2 and NO might combine in an initial step to produce an intermediate species, Br_2NO (Figure 15.17). This intermediate would then react with another NO molecule to give the reaction products. The equation for the overall reaction is obtained by adding the equations for these two steps:

Step 1.	$Br_2(g) + NO(g) \rightleftharpoons Br_2NO(g)$
Step 2.	$\underline{Br_2NO(g) + NO(g) \longrightarrow 2\,BrNO(g)}$
Overall Reaction:	$Br_2(g) + 2\,NO(g) \longrightarrow 2\,BrNO(g)$

Each step in a multistep reaction sequence is an **elementary step**, which is defined as a chemical equation that describes a single molecular event such as the formation or rupture of a chemical bond or the displacement of atoms as a result of a molecular collision. Each step has its own activation energy barrier, E_a, and rate constant, k. The steps must add up to give the balanced equation for the overall reaction, and the time required to complete all of the steps defines the overall reaction rate. A series of steps that satisfactorily explains the kinetic properties of a chemical reaction constitutes a possible reaction mechanism.

Mechanisms of reactions are postulated starting with experimental data. To see how this is done, we first describe three types of elementary steps in terms of the concept of *molecularity*.

Molecularity of Elementary Steps

Elementary steps are classified by the number of reactant molecules (or ions, atoms, or free radicals) that come together. This whole, positive number is called the **molecularity** of the elementary step. When one molecule is the only reactant in an elementary step, the reaction is a **unimolecular** process. A **bimolecular** elementary process involves two molecules, which may be identical ($A + A \longrightarrow$ products) or different ($A + B \longrightarrow$ products). The mechanism proposed for the decomposition of ozone in the stratosphere is an example of the use of these terms.

Step 1.	Unimolecular	$O_3(g) \longrightarrow O_2(g) + O(g)$
Step 2.	Bimolecular	$O_3(g) + O(g) \longrightarrow 2\,O_2(g)$
Overall reaction:		$2\,O_3(g) \longrightarrow 3\,O_2(g)$

Here an initial unimolecular step is followed by a bimolecular step.

A **termolecular** elementary step involves three molecules. It could involve three molecules of the same or a different type ($3\,A \longrightarrow$ products; $2\,A + B \longrightarrow$ products; or $A + B + C \longrightarrow$ products). As you might suspect, the simultaneous collision of three molecules is not likely, unless one of the molecules involved is in high concentration, such as a solvent molecule. In fact, most termolecular processes involve the collision of two reactant molecules and a third, inert molecule. The function of the inert molecule is to absorb the excess energy produced when a new chemical bond is formed by the first two molecules. For example, N_2 is unchanged in a termolecular reaction between oxygen molecules and oxygen atoms that produces ozone in the upper atmosphere:

$$O(g) + O_2(g) + N_2(g) \longrightarrow O_3(g) + \text{energetic } N_2(g)$$

The probability that four or more molecules will simultaneously collide with sufficient kinetic energy and proper orientation to react is so small that reaction molecularities greater than three are never proposed.

Rate Equations for Elementary Steps

As you have already seen, the experimentally determined rate equation for a reaction cannot be predicted from its overall stoichiometry. In contrast, *the rate equation for any elementary step is defined by the reaction stoichiometry. The rate equation of an elementary step is given by the product of the rate constant and the concentrations of the reactants in that step.* We can therefore write the rate equation for any elementary step, as shown by examples in the following table:

Elementary Step	Molecularity	Rate Equation
$A \longrightarrow$ product	unimolecular	Rate $= k[A]$
$A + B \longrightarrow$ product	bimolecular	Rate $= k[A][B]$
$A + A \longrightarrow$ product	bimolecular	Rate $= k[A]^2$
$2\,A + B \longrightarrow$ product	termolecular	Rate $= k[A]^2[B]$

For example, the rate laws for each of the two steps in the decomposition of ozone are

$$\text{Rate for (unimolecular) Step 1} = k[O_3]$$
$$\text{Rate for (bimolecular) Step 2} = k'[O_3][O]$$

When a reaction mechanism consists of two elementary steps, the two steps will likely occur at different rates. The two rate constants (*k* and *k'* in this example) are not expected to have the same value (nor the same units, if the two steps have different molecularities).

Molecularity and Reaction Order

The molecularity of an elementary step and its order are the same. A unimolecular elementary step must be first order, a bimolecular elementary step must be second order, and a termolecular elementary step must be third order. Such a direct relation between molecularity and order is emphatically *not* true for the *overall* reaction. If you discover experimentally that a reaction is first order, you cannot conclude that it occurs in a single, unimolecular elementary step. Similarly, a second-order rate equation does not imply that the reaction occurs in a single, bimolecular elementary step. An example illustrating this is the decomposition of N_2O_5:

$$2\ N_2O_5(g) \longrightarrow 4\ NO_2(g) + O_2(g)$$

Here the rate equation is "Rate = $k[N_2O_5]$," but chemists are fairly certain that the mechanism involves a series of unimolecular and bimolecular steps.

To see how the experimentally observed rate equation for the *overall reaction* is connected with a possible mechanism or sequence of elementary steps requires some chemical intuition. We will provide only a glimpse of the subject in the next section.

GENERAL
Chemistry Now™

See the General ChemistryNow CD-ROM or website:
- **Screen 15.12 Reaction Mechanisms and Screen 15.13 Reaction Mechanisms and Rate Equations,** for exercises on reaction mechanisms

---

Example 15.12—Elementary Steps

Problem The hypochlorite ion undergoes self-oxidation–reduction to give chlorate, ClO_3^-, and chloride ions.

$$3\ ClO^-(aq) \longrightarrow ClO_3^-(aq) + 2\ Cl^-(aq)$$

This reaction is thought to occur in two steps:

Step 1. $ClO^-(aq) + ClO^-(aq) \longrightarrow ClO_2^-(aq) + Cl^-(aq)$

Step 2. $ClO_2^-(aq) + ClO^-(aq) \longrightarrow ClO_3^-(aq) + Cl^-(aq)$

What is the molecularity of each step? Write the rate equation for each reaction step. Show that the sum of these reactions gives the equation for the net reaction.

Strategy The molecularity is the number of ions or molecules involved in a reaction step. The rate equation involves the concentration of each ion or molecule in an elementary step, raised to the power of its stoichiometric coefficient.

Solution Because two ions are involved in each elementary step, each step is bimolecular. The rate equation for any elementary step involves the product of the concentrations of the reactants. Thus, in this case, the rate equations are

Step 1. Rate = $k[ClO^-]^2$

Step 2. Rate = $k[ClO^-][ClO_2^-]$

On adding the equations for the two elementary steps, we see that the ClO_2^- ion is an intermediate, a product of the first step and a reactant in the second step. It therefore cancels out, and we are left with the stoichiometric equation for the overall reaction:

Step 1. $ClO^-(aq) + ClO^-(aq) \longrightarrow ClO_2^-(aq) + Cl^-(aq)$

Step 2. $\underline{ClO_2^-(aq) + ClO^-(aq) \longrightarrow ClO_3^-(aq) + Cl^-(aq)}$

Sum of steps: $3\ ClO^-(aq) \longrightarrow ClO_3^-(aq) + 2\ Cl^-(aq)$

Exercise 15.11—Elementary Steps

Nitrogen monoxide is reduced by hydrogen to give nitrogen and water:

$$2\ NO(g) + 2\ H_2(g) \longrightarrow N_2(g) + 2\ H_2O(g)$$

One possible mechanism for this reaction is

$$2\ NO(g) \longrightarrow N_2O_2(g)$$

$$N_2O_2(g) + H_2(g) \longrightarrow N_2O(g) + H_2O(g)$$

$$N_2O(g) + H_2(g) \longrightarrow N_2(g) + H_2O(g)$$

What is the molecularity of each of the three steps? What is the rate equation for the third step? Show that the sum of these elementary steps gives the net reaction.

Reaction Mechanisms and Rate Equations

The dependence of rate on concentration is an experimental fact. Mechanisms, by contrast, are constructs of our imagination, intuition, and good "chemical sense." To describe a mechanism, we need to make a guess (a *good* guess, we hope) about how the reaction occurs at the particulate level. Several mechanisms can often be proposed that correspond to the observed rate equation, and a postulated mechanism can be wrong. A good mechanism is a worthy goal because it allows us to understand the chemistry better. A practical consequence of a good mechanism is that it allows us to predict important things, such as how to control a reaction better and how to design new experiments.

One of the important guidelines of kinetics is that *products of a reaction can never be produced at a rate faster than the rate of the slowest step*. If one step in a multistep reaction is slower than the others, then the rate of the overall reaction is limited by the combined rates of all elementary steps up through the slowest step in the mechanism. Often the overall reaction rate and the rate of the slow step are nearly the same. If the slow step determines the rate of the reaction, it is called the **rate-determining step**, or rate-limiting step. You are already familiar with rate-determining steps. No matter how fast you shop in the supermarket, it always seems that the time it takes to finish is determined by the wait in the checkout line.

Imagine that a reaction takes place with a mechanism involving two sequential steps, and assume that we know the rates of both steps. The first step is slow and the second is fast:

Elementary Step 1 $A + B \xrightarrow[\text{Slow, } E_a \text{ large}]{k_1} X + M$

Elementary Step 2 $M + A \xrightarrow[\text{Fast, } E_a \text{ small}]{k_2} Y$

Overall Reaction $2\ A + B \longrightarrow X + Y$

In the first step, A and B come together and slowly react to form one of the products (X) plus another reactive species, M. Almost as soon as M is formed, however, it is rapidly consumed by reaction with an additional molecule of A to form the second product Y. The products X and Y are the result of two elementary steps. The rate-determining elementary step is the first step. That is, the rate of the first step

■ **Can You Derive a Mechanism?**
At this introductory level you cannot be expected to derive reaction mechanisms. Given a mechanism, however, you can decide whether it agrees with experimental rate laws.

is equal to the rate of the overall reaction. This step is bimolecular and so has the rate equation

$$Rate = k_1[A][B]$$

where k_1 is the rate constant for that step. The overall reaction is expected to follow this same second-order rate equation.

Let us apply these ideas to the mechanism of a real reaction. Experiment shows that the reaction of nitrogen dioxide with fluorine has a second-order rate equation:

Overall Reaction $2 NO_2(g) + F_2(g) \longrightarrow 2 FNO_2(g)$

$$Rate = k[NO_2][F_2]$$

The experimental rate equation immediately rules out the possibility that the reaction occurs in a single step. If the equation for the reaction represented an elementary step, the rate law would have a second-order dependence on $[NO_2]$. Because a single-step reaction is ruled out, it follows that the mechanism must include at least two steps. We can also conclude from the rate law that the rate-determining elementary step must involve NO_2 and F_2 in a $1:1$ ratio. The simplest possible mechanism is as follows:

Elementary Step 1 Slow $NO_2(g) + F_2(g) \xrightarrow{k_1} FNO_2(g) + F(g)$

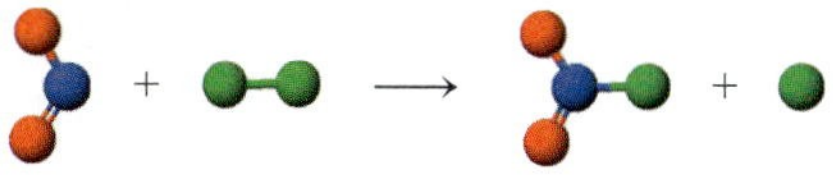

Elementary Step 2 Fast $NO_2(g) + F(g) \xrightarrow{k_2} FNO_2(g)$

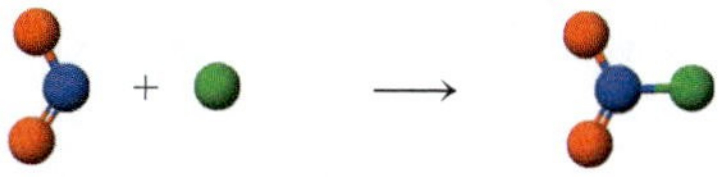

Overall Reaction $2 NO_2(g) + F_2(g) \longrightarrow 2 FNO_2(g)$

This proposed mechanism suggests that molecules of NO_2 and F_2 first react to produce one molecule of the product (FNO_2) plus one F atom. In a second step, the F atom produced in the first step reacts with additional NO_2 to give a second molecule of product. If we assume that the first, bimolecular step is rate-determining, its rate equation would be "Rate = $k_1[NO_2][F_2]$," the same as the experimentally observed rate equation. The experimental rate constant is, therefore, the same as k_1.

The F atom formed in the first step of the NO_2/F_2 reaction is a reaction intermediate. It does not appear in the equation describing the overall reaction. Reaction intermediates usually have only a fleeting existence, but occasionally they have long enough lifetimes to be observed. One of the tests of a proposed mechanism is the detection of an intermediate.

Example 15.13—Elementary Steps and Reaction Mechanisms

Problem Oxygen atom transfer from nitrogen dioxide to carbon monoxide produces nitrogen monoxide and carbon dioxide (Figure 15.13):

$$NO_2(g) + CO(g) \longrightarrow NO(g) + CO_2(g)$$

This reaction has the following rate equation at temperatures less than 500 K:

$$\text{Rate} = k[NO_2]^2$$

Can this reaction occur in one bimolecular step whose stoichiometry is the same as the overall reaction?

Strategy Write the rate law based on the equation for the $NO_2 + CO$ reaction occurring as an elementary step. If this rate law corresponds to the observed rate law, then the overall equation may reflect the way the reaction occurs.

Solution If the reaction occurs by the collision of one NO_2 molecule with one CO molecule, the rate equation would be

$$\text{Rate} = k[NO_2][CO]$$

This does not agree with experiment, so the mechanism must involve more than a single step. In one possible mechanism, the reaction occurs in two, bimolecular steps, the first one slow and the second one fast:

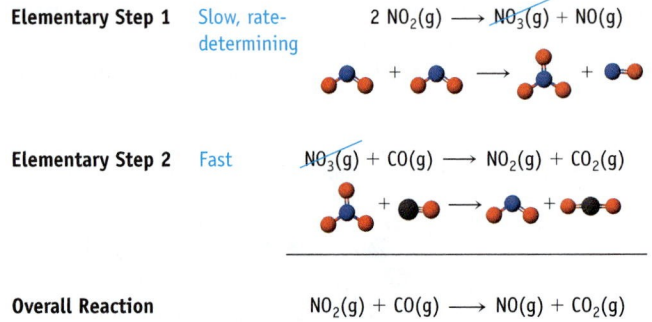

Elementary Step 1	Slow, rate-determining	$2\,NO_2(g) \longrightarrow NO_3(g) + NO(g)$
Elementary Step 2	Fast	$NO_3(g) + CO(g) \longrightarrow NO_2(g) + CO_2(g)$
Overall Reaction		$NO_2(g) + CO(g) \longrightarrow NO(g) + CO_2(g)$

The first (rate-determining) step has a rate equation that agrees with experiment, so this mechanism may be the way the reaction actually occurs.

Exercise 15.12—Elementary Steps and Reaction Mechanisms

The Raschig reaction produces hydrazine, N_2H_4, an industrially important reducing agent, from NH_3 and OCl^- in basic, aqueous solution. A proposed mechanism is

Step 1	Fast	$NH_3(aq) + OCl^-(aq) \longrightarrow NH_2Cl(aq) + OH^-(aq)$
Step 2	Slow	$NH_2Cl(aq) + NH_3(aq) \longrightarrow N_2H_5^+(aq) + Cl^-(aq)$
Step 3	Fast	$N_2H_5^+(aq) + OH^-(aq) \longrightarrow N_2H_4(aq) + H_2O(\ell)$

(a) What is the overall stoichiometric equation?
(b) Which step of the three is rate-determining?
(c) Write the rate equation for the rate-determining elementary step.
(d) What reaction intermediates are involved?

Another common two-step reaction mechanism involves an initial fast reaction that produces an intermediate, followed by a slower second step in which the intermediate is converted to the final product. The rate of the reaction is determined by the second step, for which a rate law can be written. The rate of that step, however, depends on the concentration of the intermediate. An important thing to remember, though, is that *the rate law must be written with respect to the reactants only.* An intermediate, whose concentration will probably not be measurable, cannot appear as a term in the overall rate equation.

The reaction of nitrogen monoxide and oxygen is an example of a two-step reaction where the first step is fast and the second step is rate-determining.

$$2\ NO(g) + O_2(g) \longrightarrow 2\ NO_2(g)$$
$$Rate = k[NO]^2[O_2]$$

The experimentally determined rate law shows second-order dependence on NO and first-order dependence on O_2. Although this rate law would be correct for a termolecular reaction, experimental evidence indicates that an intermediate is formed in this reaction. A possible two-step mechanism that proceeds through an intermediate is

Elementary Step 1. Fast, equilibrium $\qquad NO(g) + O_2(g) \underset{k_{-1}}{\overset{k_1}{\rightleftharpoons}} \underset{\text{intermediate}}{OONO(g)}$

Elementary Step 2. Slow, rate-determining $\quad \underline{NO(g) + OONO(g) \overset{k_2}{\longrightarrow} 2\ NO_2(g)}$

Overall Reaction $\qquad\qquad\qquad\qquad NO(g) + O_2(g) \longrightarrow 2\ NO_2(g)$

The second step of this reaction is the slow step, and the overall rate depends on it. We can write a rate law for the second step:

$$Rate = k_2[NO][OONO]$$

This rate law cannot be compared directly with the experimental rate law because it contains the concentration of an intermediate, OONO. Recall that the experimental rate law must be written only in terms of compounds appearing in the overall equation. We therefore need to express the postulated rate law in a way that eliminates the intermediate. To do so, we look at the rapid first step in this reaction sequence.

At the beginning of the reaction, NO and O_2 react rapidly and produce the intermediate OONO. The rate of formation can be defined by a rate law with a rate constant k_1:

$$\text{Rate of production of OONO} = k_1[NO][O_2]$$

Because the intermediate is consumed only very slowly in the second step, it is possible for the OONO to revert to NO and O_2 before it reacts further:

$$\text{Rate of reverse reaction (OONO to NO and } O_2) = k_{-1}[OONO]$$

As NO and O_2 form OONO, their concentrations drop, so the rate of the forward reaction decreases. At the same time, the concentration of OONO builds up, so the rate of the reverse reaction increases. Eventually, the rates of the forward and reverse reactions become the same, and the first elementary step reaches a *state of equilibrium*. The forward and reverse reactions in the first elementary step are so much faster than the second elementary step that equilibrium is established before any significant amount of OONO is consumed by NO to give NO_2. The state of equilibrium for the first step remains throughout the lifetime of the overall reaction.

Because equilibrium is established when the rates of the forward and reverse reactions are the same, this means

$$\text{Rate of forward reaction} = \text{rate of reverse reaction}$$
$$k_1[NO][O_2] = k_{-1}[OONO]$$

Rearranging this equation, we find

$$\frac{k_1}{k_{-1}} = \frac{[OONO]}{[NO][O_2]} = K$$

Both k_1 and k_{-1} are constants (they will change only if the temperature changes). We can define a new constant K equal to the ratio of these two constants and called the *equilibrium constant*, which is equal to the quotient $[OONO]/[NO][O_2]$. From this we can come up with an expression for the concentration of OONO:

$$[OONO] = K[NO][O_2]$$

If $K[NO][O_2]$ is substituted for $[OONO]$ in the rate law for the rate-determining elementary step, we have

$$\text{Rate} = k_2[NO][OONO] = k_2[NO]\{K[NO][O_2]\}$$
$$= k_2 K[NO]^2[O_2]$$

Because both k_2 and K are constants, their product is another constant k', and we have

$$\text{Rate} = k'[NO]^2[O_2]$$

This is exactly the rate law derived from experiment. Thus, the sequence of reactions on which the rate law is based may be a reasonable mechanism for this reaction. It is *not* the only possible mechanism, however. This rate equation is also consistent with the reaction occurring in a single termolecular step. Another possible mechanism is illustrated in Example 15.14.

Example 15.14—Reaction Mechanism Involving an Equilibrium Step

Problem The $NO + O_2$ reaction described in the text could also occur by the following mechanism:

Elementary Step 1:   Fast, equilibrium

$$NO(g) + NO(g) \underset{k_{-1}}{\overset{k_1}{\rightleftharpoons}} N_2O_2(g) \text{ intermediate}$$

Elementary Step 2:   Slow, rate-determining

$$N_2O_2(g) + O_2(g) \xrightarrow{k_2} NO_2(g)$$

Overall Reaction: $2\, NO(g) + O_2(g) \longrightarrow 2\, NO_2(g)$

Show that this mechanism leads to the following experimental rate law: Rate = $k[NO]^2[O_2]$.

Strategy The rate law for the rate-determining elementary step is

$$\text{Rate} = k_2[N_2O_2][O_2]$$

The compound N_2O_2 is an intermediate and cannot appear in the final derived rate law. (A postulated rate law must not include an intermediate.) To obtain the rate law we use the equilibrium constant expression for the first step.

Solution $[N_2O_2]$ and $[NO]$ are related by the equilibrium constant.

$$\frac{k_1}{k_{-1}} = \frac{[N_2O_2]}{[NO]^2} = K$$

Problem-Solving Tip

Relating Rate Equations and Reaction Mechanisms

The connection between an experimental rate equation and the proposed reaction mechanism is important in chemistry.

1. Experiments must first be performed that define the effect of reactant concentrations on the rate of the reaction. This gives the experimental rate equation.

2. A mechanism for the reaction is proposed on the basis of the experimental rate equation, the principles of stoichiometry and molecular structure and bonding, general chemical experience, and intuition.

3. The *proposed* reaction mechanism is used to *derive* a rate equation. This rate equation must contain only those species present in the overall chemical reaction but not any reaction intermediates. If the derived and experimental rate equations are the same, the postulated mechanism *may* be a reasonable hypothesis of the reaction sequence.

4. If more than one mechanism can be proposed, and they all predict derived rate equations in agreement with experiment, then more experiments must be done.

If we solve this equation for $[N_2O_2]$, we have $[N_2O_2] = K[NO]^2$. When this is substituted into the derived rate law

$$Rate = k_2\{K[NO]^2\}[O_2]$$

the resulting equation is identical with the experimental rate law where $k_2K = k$.

Comment The $NO + O_2$ reaction has an experimental rate law for which at least three mechanisms can be proposed. The challenge is to decide which is correct. In this case further experimentation detected the species OONO as a short-lived intermediate, confirming the mechanism involving this intermediate.

Exercise 15.13—Reaction Mechanism Involving a Fast Initial Step

One possible mechanism for the decomposition of nitryl chloride, NO_2Cl, is

Elementary Step 1:   Fast, Equilibrium $NO_2Cl(g) \underset{k_{-1}}{\overset{k_1}{\rightleftharpoons}} NO_2(g) + Cl(g)$

Elementary Step 2:   Slow $NO_2Cl(g) + Cl(g) \xrightarrow{k_2} NO_2(g) + Cl_2(g)$

What is the overall reaction? What rate law would be derived from this mechanism? What effect does increasing the concentration of the product NO_2 have on the reaction rate?

Chapter Goals Revisited

Now that you have studied this chapter, you should ask whether you have met the chapter goals. In particular you should be able to:

Understand rates of reaction and the conditions affecting rates
  a. Explain the concept of reaction rate (Section 15.1).
  b. Derive the average and instantaneous rates of a reaction from experimental information (Section 15.1). General ChemistryNow homework: Study Question(s) 5

GENERAL
Chemistry Now™

See the General ChemistryNow CD-ROM or website to:
• Assess your understanding with homework questions keyed to each goal
• Check your readiness for an exam by taking the exam-prep quiz and exploring the resources in the personalized Learning Plan it provides

For additional preparation for an examination on this chapter see the **Let's Review** section on pages 998–1011.

c. Describe factors that affect reaction rate (i.e., reactant concentrations, temperature, presence of a catalyst, and the state of the reactants) (Section 15.2). General ChemistryNow homework: SQ(s) 8, 10

Derive the rate equation, rate constant, and reaction order from experimental data

a. Define the various parts of a rate equation (the rate constant and order of reaction) and understand their significance (Section 15.3). General ChemistryNow homework: SQ(s) 12, 14

b. Derive a rate equation from experimental information (Section 15.3).

Use integrated rate laws

a. Describe and use the relationships between reactant concentration and time for zero-order, first-order, and second-order reactions (Section 15.4 and Table 15.1). General ChemistryNow homework: SQ(s) 18, 20, 22, 23, 82

b. Apply graphical methods for determining reaction order and the rate constant from experimental data (Section 15.4 and Table 15.1). General ChemistryNow homework: SQ(s) 36, 38

c. Use the concept of half-life ($t_{1/2}$), especially for first-order reactions (Section 15.4). General ChemistryNow homework: SQ(s) 26, 30, 81

Understand the collision theory of reaction rates and the role of activation energy

a. Describe the collision theory of reaction rates (Section 15.5).

b. Relate activation energy (E_a) to the rate and thermodynamics of a reaction (Section 15.5). General ChemistryNow homework: SQ(s) 44, 79

c. Use collision theory to describe the effect of reactant concentration on reaction rate (Section 15.5).

d. Understand the effect of molecular orientation on reaction rate (Section 15.5).

e. Describe the effect of temperature on reaction rate using the collision theory of reaction rates and the Arrhenius equation (Equation 15.7 and Section 15.5).

f. Use Equations 15.5, 15.6, and 15.7 to calculate the activation energy from experimental data (Section 15.5).

Relate reaction mechanisms and rate laws

a. Describe the functioning of a catalyst and its effect on the activation energy and mechanism of a reaction (Section 15.5).

b. Understand reaction coordinate diagrams (Section 15.5).

c. Understand the concept of a reaction mechanism (the sequence of bond-making and bond-breaking steps that occurs during the conversion of reactants to products) and the relation of the mechanism to the overall, stoichiometric equation for a reaction (Section 15.6).

d. Describe the elementary steps of a mechanism and give their molecularity (Section 15.6). General ChemistryNow homework: SQ(s) 48, 50

e. Define the rate-determining step in a mechanism and identify any reaction intermediates (Section 15.6). General ChemistryNow homework: SQ(s) 52

Key Equations

Equation 15.1 (page 713)
Integrated rate equation for a first-order reaction (in which $-\Delta[R]/\Delta t = k[R]$).

$$\ln \frac{[R]_t}{[R]_0} = -kt$$

Here $[R]_0$ and $[R]_t$ are concentrations of the reactant at time $t = 0$ and at a later time, t. The *ratio* of concentrations, $[R]_t/[R]_0$, is the fraction of reactant that *remains* after a given time has elapsed.

Equation 15.2 (page 715)
Integrated rate equation for a second-order reaction (in which $-\Delta[R]/\Delta t = k[R]^2$).

$$\frac{1}{[R]_t} - \frac{1}{[R]_0} = kt$$

Equation 15.3 (page 716)
Integrated rate equation for a zero-order reaction (in which $-\Delta[R]/\Delta t = k[R]^0$).

$$[R]_0 - [R]_t = kt$$

Equation 15.4 (page 719)
The relation between the half-life $(t_{1/2})$ and the rate constant (k) for a first-order reaction.

$$t_{1/2} = \frac{0.693}{k}$$

Equation 15.5 (page 727)
Arrhenius equation in exponential form

$$k = \text{rate constant} = Ae^{-E_a/RT}$$

Frequency factor   Fraction of molecules with minimum energy for reaction

A is the frequency factor, E_a is the activation energy, T is the temperature (in kelvins), and R is the gas constant ($= 8.314510 \times 10^{-3}$ kJ/K · mol).

Equation 15.6 (page 727)
Expanded Arrhenius equation in logarithmic form.

$$\ln k = -\frac{E_a}{R}\left(\frac{1}{T}\right) + \ln A \quad \longleftarrow \text{Arrhenius equation}$$

$$\begin{array}{ccc} \downarrow & \downarrow & \downarrow \\ y & = mx & + b \end{array} \quad \longleftarrow \text{Equation for straight line}$$

Equation 15.7 (page 728)
A version of the Arrhenius equation used to calculate the activation energy for a reaction when you know the values of the rate constant at two temperatures (in kelvins).

$$\ln k_2 - \ln k_1 = \ln \frac{k_2}{k_1} = -\frac{E_a}{R}\left[\frac{1}{T_2} - \frac{1}{T_1}\right]$$

Study Questions

▲ denotes more challenging questions.

■ denotes questions available in the Homework and Goals section of the General ChemistryNow CD-ROM or website.

Blue numbered questions have answers in Appendix O and fully worked solutions in the *Student Solutions Manual*.

Structures of many of the compounds used in these questions are found on the General ChemistryNow CD-ROM or website in the Models folder.

GENERAL
Chemistry Now™ Assess your understanding of this chapter's topics with additional quizzing and conceptual questions at http://now.brookscole.com/kotz6e

Practicing Skills

Reaction Rates

(See Examples 15.1–15.2, Exercises 15.1–15.2, and General ChemistryNow Screen 15.2.)

1. Give the relative rates of disappearance of reactants and formation of products for each of the following reactions.
   (a) $2 O_3(g) \longrightarrow 3 O_2(g)$
   (b) $2 HOF(g) \longrightarrow 2 HF(g) + O_2(g)$

2. Give the relative rates of disappearance of reactants and formation of products for each of the following reactions.
   (a) $2 NO(g) + Br_2(g) \longrightarrow 2 NOBr(g)$
   (b) $N_2(g) + 3 H_2(g) \longrightarrow 2 NH_3(g)$

3. In the reaction $2 O_3(g) \longrightarrow 3 O_2(g)$, the rate of formation of O_2 is 1.5×10^{-3} mol/L · s. What is the rate of decomposition of O_3?

4. In the synthesis of ammonia, if $-\Delta[H_2]/\Delta t = 4.5 \times 10^{-4}$ mol/ L · min, what is $\Delta[NH_3]/\Delta t$?
   $N_2(g) + 3 H_2(g) \longrightarrow 2 NH_3(g)$

5. ■ Experimental data are listed here for the reaction $A \longrightarrow 2 B$.

Time (s)	[B] (mol/L)
0.00	0.000
10.0	0.326
20.0	0.572
30.0	0.750
40.0	0.890

(a) Prepare a graph from these data, connect the points with a smooth line, and calculate the rate of change of [B] for each 10-s interval from 0.0 to 40.0 s. Does the rate of change decrease from one time interval to the next? Suggest a reason for this result.

(b) How is the rate of change of [A] related to the rate of change of [B] in each time interval? Calculate the rate of change of [A] for the time interval from 10.0 to 20.0 s.

(c) What is the instantaneous rate when [B] = 0.750 mol/L?

6. Phenyl acetate, an ester, reacts with water according to the equation

$$
\underset{\text{phenyl acetate}}{CH_3\overset{\displaystyle O}{\overset{\|}{C}}OC_6H_5} + H_2O \longrightarrow \underset{\text{acetic acid}}{CH_3\overset{\displaystyle O}{\overset{\|}{C}}OH} + \underset{\text{phenol}}{C_6H_5OH}
$$

The data in the table were collected for this reaction at 5 °C.

Time (s)	[Phenyl acetate] (mol/L)
0	0.55
15.0	0.42
30.0	0.31
45.0	0.23
60.0	0.17
75.0	0.12
90.0	0.085

(a) Plot the phenyl acetate concentration versus time, and describe the shape of the curve observed.

(b) Calculate the rate of change of the phenyl acetate concentration during the period 15.0 s to 30.0 s and also during the period 75.0 s to 90.0 s. Compare the values, and suggest a reason why one value is smaller than the other.

(c) What is the rate of change of the phenol concentration during the time period 60.0 s to 75.0 s?

(d) What is the instantaneous rate at 15.0 s?

Concentration and Rate Equations

(See Examples 15.3–15.4, Exercises 15.3–15.4, and General ChemistryNow Screens 15.4 and 15.5.)

7. Using the rate equation "Rate = $k[A]^2[B]$," define the order of the reaction with respect to A and B. What is the total order of the reaction?

8. ■ A reaction has the experimental rate equation "Rate = $k[A]^2$." How will the rate change if the concentration of A is tripled? If the concentration of A is halved?

9. The reaction between ozone and nitrogen dioxide at 231 K is first order in both $[NO_2]$ and $[O_3]$.

$$2 NO_2(g) + O_3(g) \longrightarrow N_2O_5(s) + O_2(g)$$

(a) Write the rate equation for the reaction.

(b) If the concentration of NO_2 is tripled, what is the change in the reaction rate?

(c) What is the effect on reaction rate if the concentration of O_3 is halved?

10. ■ Nitrosyl bromide, NOBr, is formed from NO and Br_2:

$$2\,NO(g) + Br_2(g) \longrightarrow 2\,NOBr(g)$$

Experiments show that this reaction is second order in NO and first order in Br_2.

(a) Write the rate equation for the reaction.

(b) How does the initial reaction rate change if the concentration of Br_2 is changed from 0.0022 mol/L to 0.0066 mol/L?

(c) What is the change in the initial rate if the concentration of NO is changed from 0.0024 mol/L to 0.0012 mol/L?

11. The data in the table are for the reaction of NO and O_2 at 660 K.

$$2\,NO(g) + O_2(g) \longrightarrow 2\,NO_2(g)$$

Reactant Concentration (mol/L)		Rate of Disappearance of NO (mol/L · s)
[NO]	[O_2]	
0.010	0.010	2.5×10^{-5}
0.020	0.010	1.0×10^{-4}
0.010	0.020	5.0×10^{-5}

(a) Determine the order of the reaction for each reactant.

(b) Write the rate equation for the reaction.

(c) Calculate the rate constant.

(d) Calculate the rate (in mol/L · s) at the instant when $[NO] = 0.015$ mol/L and $[O_2] = 0.0050$ mol/L.

(e) At the instant when NO is reacting at the rate 1.0×10^{-4} mol/L · s, what is the rate at which O_2 is reacting and NO_2 is forming?

12. ■ The reaction

$$2\,NO(g) + 2\,H_2(g) \longrightarrow N_2(g) + 2\,H_2O(g)$$

was studied at 904 °C, and the data in the table were collected.

Reactant Concentration (mol/L)		Rate of Appearance of N_2 (mol/L · s)
[NO]	[H_2]	
0.420	0.122	0.136
0.210	0.122	0.0339
0.210	0.244	0.0678
0.105	0.488	0.0339

(a) Determine the order of the reaction for each reactant.

(b) Write the rate equation for the reaction.

(c) Calculate the rate constant for the reaction.

(d) Find the rate of appearance of N_2 at the instant when $[NO] = 0.350$ mol/L and $[H_2] = 0.205$ mol/L.

13. Data for the reaction $2\,NO(g) + O_2(g) \longrightarrow 2\,NO_2(g)$ are given in the table.

Experiment	Concentration (mol/L)		Initial Rate (mol/L · h)
	[NO]	[O_2]	
1	3.6×10^{-4}	5.2×10^{-3}	3.4×10^{-8}
2	3.6×10^{-4}	1.04×10^{-2}	6.8×10^{-8}
3	1.8×10^{-4}	1.04×10^{-2}	1.7×10^{-8}
4	1.8×10^{-4}	5.2×10^{-3}	?

(a) What is the rate law for this reaction?

(b) What is the rate constant for the reaction?

(c) What is the initial rate of the reaction in experiment 4?

14. ■ Data for the following reaction are given in the table below.

$$CO(g) + NO_2(g) \longrightarrow CO_2(g) + NO(g)$$

Experiment	Concentration (mol/L)		Initial Rate (mol/L · h)
	[CO]	[NO_2]	
1	5.0×10^{-4}	0.36×10^{-4}	3.4×10^{-8}
2	5.0×10^{-4}	0.18×10^{-4}	1.7×10^{-8}
3	1.0×10^{-3}	0.36×10^{-4}	6.8×10^{-8}
4	1.5×10^{-3}	0.72×10^{-4}	?

(a) What is the rate law for this reaction?

(b) What is the rate constant for the reaction?

(c) What is the initial rate of the reaction in experiment 4?

15. Carbon monoxide reacts with O_2 to form CO_2:

$$2\,CO(g) + O_2(g) \longrightarrow 2\,CO_2(g)$$

Information on this reaction is given in the table below.

[CO] (mol/L)	[O_2] (mol/L)	Rate (mol/L · min)
0.02	0.02	3.68×10^{-5}
0.04	0.02	1.47×10^{-4}
0.02	0.04	7.36×10^{-5}

(a) What is the rate law for this reaction?

(b) What is the order of the reaction with respect to CO? What is the order with respect O_2? What is the overall order of the reaction?

(c) What is the value for the rate constant, k?

16. Data for the reaction

$$H_2PO_4^-(aq) + OH^-(aq) \longrightarrow HPO_4^{2-}(aq) + H_2O(\ell)$$

are provided in the table.

Experiment	[$H_2PO_4^-$] (M)	[OH^-] (M)	Initial Rate (mol/L · min)
1	0.0030	0.00040	0.0020
2	0.0030	0.00080	0.0080
3	0.0090	0.00040	0.0060
4	?	0.00033	0.0020

(a) What is the rate law for this reaction?

(b) What is the value of k?

(c) What is the concentration of $H_2PO_4^-$ in experiment 4?

Concentration–Time Relationships

(See Examples 15.5–15.7, Exercises 15.5–15.7, and General ChemistryNow Screen 15.6.)

17. The rate equation for the hydrolysis of sucrose to fructose and glucose

$$C_{12}H_{22}O_{11}(aq) + H_2O(\ell) \longrightarrow 2\ C_6H_{12}O_6(aq)$$

is "$-\Delta[\text{sucrose}]/\Delta t = k[C_{12}H_{22}O_{11}]$." After 2.57 h at 27 °C, the sucrose concentration decreased from 0.0146 M to 0.0132 M. Find the rate constant, k.

18. ■ The decomposition of N_2O_5 in CCl_4 is a first-order reaction. If 2.56 mg of N_2O_5 is present initially, and 2.50 mg is present after 4.26 min at 55 °C, what is the value of the rate constant, k?

19. The decomposition of SO_2Cl_2 is a first-order reaction:

$$SO_2Cl_2(g) \longrightarrow SO_2(g) + Cl_2(g)$$

The rate constant for the reaction is 2.8×10^{-3} min^{-1} at 600 K. If the initial concentration of SO_2Cl_2 is 1.24×10^{-3} mol/L, how long will it take for the concentration to drop to 0.31×10^{-3} mol/L?

20. ■ The conversion of cyclopropane to propene, described in Example 15.5, occurs with a first-order rate constant of 5.4×10^{-2} h^{-1}. How long will it take for the concentration of cyclopropane to decrease from an initial concentration 0.080 mol/L to 0.020 mol/L?

21. Ammonium cyanate, NH_4NCO, rearranges in water to give urea, $(NH_2)_2CO$:

$$NH_4NCO(aq) \longrightarrow (NH_2)_2CO(aq)$$

The rate equation for this process is "Rate = k $[NH_4NCO]^2$," where $k = 0.0113$ L/mol · min. If the original concentration of NH_4NCO in solution is 0.229 mol/L, how long will it take for the concentration to decrease to 0.180 mol/L?

22. ■ The decomposition of nitrogen dioxide at a high temperature

$$NO_2(g) \longrightarrow NO(g) + \tfrac{1}{2} O_2(g)$$

is second order in this reactant. The rate constant for this reaction is 3.40 L/mol · min. Determine the time needed for the concentration of NO_2 to decrease from 2.00 mol/L to 1.50 mol/L.

23. ■ Hydrogen peroxide, $H_2O_2(aq)$, decomposes to $H_2O(\ell)$ and $O_2(g)$ in a reaction that is first order in H_2O_2 and has a rate constant $k = 1.06 \times 10^{-3}$ min^{-1}.

(a) How long will it take for 15% of a sample of H_2O_2 to decompose?

(b) How long will it take for 85% of the sample to decompose?

24. The thermal decomposition of HCO_2H is a first-order reaction with a rate constant of 2.4×10^{-3} s^{-1} at a given temperature. How long will it take for three fourths of a sample of HCO_2H to decompose?

Half-Life

(See Examples 15.8 and 15.9, Exercise 15.9, and General ChemistryNow Screen 15.8.)

25. The rate equation for the decomposition of N_2O_5 (giving NO_2 and O_2) is "$-\Delta[N_2O_5]/\Delta t = k[N_2O_5]$." The value of k is 5.0×10^{-4} s^{-1} for the reaction at a particular temperature.

(a) Calculate the half-life of N_2O_5.

(b) How long does it take for the N_2O_5 concentration to drop to one tenth of its original value?

26. ■ The decomposition of SO_2Cl_2

$$SO_2Cl_2(g) \longrightarrow SO_2(g) + Cl_2(g)$$

is first order in SO_2Cl_2, and the reaction has a half-life of 245 min at 600 K. If you begin with 3.6×10^{-3} mol of SO_2Cl_2 in a 1.0-L flask, how long will it take for the amount of SO_2Cl_2 to decrease to 2.00×10^{-4} mol?

27. Gaseous azomethane, $CH_3N{=}NCH_3$, decomposes in a first-order reaction when heated:

$$CH_3N{=}NCH_3(g) \longrightarrow N_2(g) + C_2H_6(g)$$

The rate constant for this reaction at 425 °C is 40.8 min^{-1}. If the initial quantity of azomethane in the flask is 2.00 g, how much remains after 0.0500 min? What quantity of N_2 is formed in this time?

28. The compound $Xe(CF_3)_2$ decomposes in a first-order reaction to elemental Xe with a half-life of 30. min. If you place 7.50 mg of $Xe(CF_3)_2$ in a flask, how long must you wait until only 0.25 mg of $Xe(CF_3)_2$ remains?

29. The radioactive isotope ^{64}Cu is used in the form of copper(II) acetate to study Wilson's disease. The isotope has a half-life of 12.70 h. What fraction of radioactive copper(II) acetate remains after 64 h?

30. ■ Radioactive gold-198 is used in the diagnosis of liver problems. The half-life of this isotope is 2.7 days. If you begin with a 5.6-mg sample of the isotope, how much of this sample remains after 1.0 day?

31. Formic acid decomposes at 550 °C according to the equation

$$HCO_2H(g) \longrightarrow CO_2(g) + H_2(g)$$

The reaction follows first-order kinetics. In an experiment, it is determined that 75% of a sample of HCO_2H has decomposed in 72 seconds. Determine $t_{1/2}$ for this reaction.

32. The decomposition of SO_2Cl_2 to SO_2 and Cl_2 at high temperature is a first-order reaction with a half-life of 2.5×10^3 min. What fraction of SO_2Cl_2 will remain after 750 min?

▲ More challenging   ■ In General ChemistryNow   **Blue-numbered questions** answered in Appendix O

Graphical Analysis: Rate Equations and *k*

(See Exercise 15.8 and General ChemistryNow Screen 15.7.)

33. Common sugar, sucrose, breaks down in dilute acid solution to form glucose and fructose. Both products have the same formula, $C_6H_{12}O_6$.

$$C_{12}H_{22}O_{11}(aq) + H_2O(\ell) \longrightarrow 2\ C_6H_{12}O_6(aq)$$

The rate of this reaction has been studied in acid solution, and the data in the table were obtained.

Time (min)	$[C_{12}H_{22}O_{11}]$ (mol/L)
0	0.316
39	0.274
80	0.238
140	0.190
210	0.146

(a) Plot ln [sucrose] versus time and 1/[sucrose] versus time. What is the order of the reaction?

(b) Write the rate equation for the reaction, and calculate the rate constant, *k*.

(c) Estimate the concentration of sucrose after 175 min.

34. Data for the reaction of phenyl acetate with water are given in Study Question 6. Plot these data as ln [phenyl acetate] and 1/[phenyl acetate] versus time. Based on the appearance of the two graphs, what can you conclude about the order of the reaction with respect to phenyl acetate? Working from the data and the rate law, determine the rate constant for the reaction.

35. Data for the decomposition of dinitrogen oxide

$$2\ N_2O(g) \longrightarrow 2\ N_2(g) + O_2(g)$$

on a gold surface at 900 °C are given below. Verify that the reaction is first order by preparing a graph of ln [N_2O] versus time. Derive the rate constant from the slope of the line in this graph. Using the rate law and value of *k*, determine the decomposition rate at 900 °C when [N_2O] = 0.035 mol/L.

Time (min)	$[N_2O]$ (mol/L)
15.0	0.0835
30.0	0.0680
80.0	0.0350
120.0	0.0220

36. ■ Ammonia decomposes when heated according to the equation

$$NH_3(g) \longrightarrow NH_2(g) + H(g)$$

The data in the table for this reaction were collected at a high temperature.

Time (h)	$[NH_3]$ (mol/L)
0	8.00×10^{-7}
25	6.75×10^{-7}
50	5.84×10^{-7}
75	5.15×10^{-7}

Plot ln [NH_3] versus time and 1/[NH_3] versus time. What is the order of this reaction with respect to NH_3? Find the rate constant for the reaction from the slope.

37. Gaseous [NO_2] decomposes at 573 K.

$$2\ NO_2(g) \longrightarrow 2\ NO(g) + O_2(g)$$

The concentration of NO_2 was measured as a function of time. A graph of 1/[NO_2] versus time gives a straight line with a slope of 1.1 L/mol · s. What is the rate law for this reaction? What is the rate constant?

38. ■ The decomposition of HOF occurs at 25 °C.

$$2\ HOF(g) \longrightarrow 2\ HF(g) + O_2(g)$$

Using the data in the table below, determine the rate law and then calculate the rate constant.

[HOF] (mol/L)	Time (min)
0.850	0
0.810	2.00
0.754	5.00
0.526	20.0
0.243	50.0

39. For the reaction $2\ C_2F_4 \longrightarrow C_4F_8$, a graph of 1/[$C_2F_4$] versus time gives a straight line with a slope of +0.04 L/mol · s. What is the rate law for this reaction?

40. Butadiene, $C_4H_6(g)$, dimerizes when heated, forming 1,5-cyclooctadiene, C_8H_{12}. The data in the table were collected.

$[C_4H_6]$ (mol/L)	Time (s)
1.0×10^{-2}	0
8.7×10^{-3}	200.
7.7×10^{-3}	500.
6.9×10^{-3}	800.
5.8×10^{-3}	1200.

▲ More challenging   ■ In General ChemistryNow   **Blue-numbered questions** answered in Appendix O

(a) Use a graphical method to verify that this is a second-order reaction.

(b) Calculate the rate constant for this reaction.

Kinetics and Energy

(See Examples 15.10 and 15.11, Exercise 15.10, and General ChemistryNow Screens 15.9 and 15.10.)

41. Calculate the activation energy, E_a, for the reaction

$$N_2O_5(g) \longrightarrow 2\,NO_2(g) + \tfrac{1}{2}O_2(g)$$

from the observed rate constants: k at 25 °C = $3.46 \times 10^{-5}\,s^{-1}$ and k at 55 °C = $1.5 \times 10^{-3}\,s^{-1}$.

42. If the rate constant for a reaction triples when the temperature rises from 3.00×10^2 K to 3.10×10^2 K, what is the activation energy of the reaction?

43. When heated to a high temperature, cyclobutane, C_4H_8, decomposes to ethylene:

$$C_4H_8(g) \longrightarrow 2\,C_2H_4(g)$$

The activation energy, E_a, for this reaction is 260 kJ/mol. At 800 K, the rate constant $k = 0.0315\,s^{-1}$. Determine the value of k at 850 K.

44. ■ When heated, cyclopropane is converted to propene (see Example 15.5). Rate constants for this reaction at 470 °C and 510 °C are $k = 1.10 \times 10^{-4}\,s^{-1}$ and $k = 1.02 \times 10^{-3}\,s^{-1}$, respectively. Determine the activation energy, E_a, from these data.

45. The reaction of H_2 molecules with F atoms

$$H_2(g) + F(g) \longrightarrow HF(g) + H(g)$$

has an activation energy of 8 kJ/mol and an energy change of −133 kJ/mol. Draw a diagram similar to Figure 15.13 for this process. Indicate the activation energy and enthalpy of reaction on this diagram.

46. Answer questions (a) and (b) based on the accompanying reaction coordinate diagram.

(a) Is the reaction exothermic or endothermic?

(b) Does the reaction occur in more than one step? If so, how many?

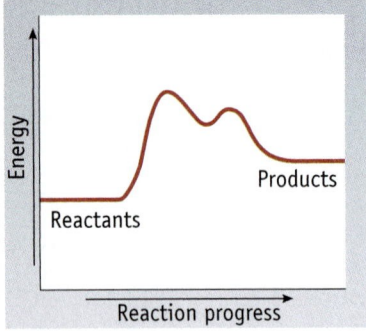

Reaction Mechanisms

(See Examples 15.12–15.14, Exercises 15.11–15.13, and General ChemistryNow Screens 15.12 and 15.13.)

47. What is the rate law for each of the following *elementary* reactions?

(a) $NO(g) + NO_3(g) \longrightarrow 2\,NO_2(g)$

(b) $Cl(g) + H_2(g) \longrightarrow HCl(g) + H(g)$

(c) $(CH_3)_3CBr(aq) \longrightarrow (CH_3)_3C^+(aq) + Br^-(aq)$

48. ■ What is the rate law for each of the following *elementary* reactions?

(a) $Cl(g) + ICl(g) \longrightarrow I(g) + Cl_2(g)$

(b) $O(g) + O_3(g) \longrightarrow 2\,O_2(g)$

(c) $2\,NO_2(g) \longrightarrow N_2O_4(g)$

49. Ozone, O_3, in the earth's upper atmosphere decomposes according to the equation

$$2\,O_3(g) \longrightarrow 3\,O_2(g)$$

The mechanism of the reaction is thought to proceed through an initial fast, reversible step followed by a slow, second step.

Step 1   Fast, reversible $O_3(g) \rightleftharpoons O_2(g) + O(g)$

Step 2   Slow $O_3(g) + O(g) \longrightarrow 2\,O_2(g)$

(a) Which of the steps is rate-determining?

(b) Write the rate equation for the rate-determining step.

50. ■ The reaction of $NO_2(g)$ and $CO(g)$ is thought to occur in two steps:

Step 1   Slow $NO_2(g) + NO_2(g) \longrightarrow NO(g) + NO_3(g)$

Step 2   Fast $NO_3(g) + CO(g) \longrightarrow NO_2(g) + CO_2(g)$

(a) Show that the elementary steps add up to give the overall, stoichiometric equation.

(b) What is the molecularity of each step?

(c) For this mechanism to be consistent with kinetic data, what must be the experimental rate equation?

(d) Identify any intermediates in this reaction.

51. Iodide ion is oxidized in acid solution by hydrogen peroxide (Figure 15.3).

$$H_2O_2(aq) + 2\,H^+(aq) + 2\,I^-(aq) \longrightarrow I_2(aq) + 2\,H_2O(\ell)$$

A proposed mechanism is

Step 1   Slow $H_2O_2(aq) + I^-(aq) \longrightarrow$
$$H_2O(\ell) + OI^-(aq)$$

Step 2   Fast $H^+(aq) + OI^-(aq) \longrightarrow HOI(aq)$

Step 3   Fast $HOI(aq) + H^+(aq) + I^-(aq) \longrightarrow$
$$I_2(aq) + H_2O(\ell)$$

(a) Show that the three elementary steps add up to give the overall, stoichiometric equation.

(b) What is the molecularity of each step?

(c) For this mechanism to be consistent with kinetic data, what must be the experimental rate equation?

(d) Identify any intermediates in the elementary steps in this reaction.

---

52. ■ The mechanism for the reaction of CH_3OH and HBr is believed to involve two steps. The overall reaction is exothermic.

    Step 1   Fast, endothermic
    $$CH_3OH + H^+ \rightleftharpoons CH_3OH_2^+$$
    Step 2   Slow
    $$CH_3OH_2^+ + Br^- \longrightarrow CH_3Br + H_2O$$

    (a) Write an equation for the overall reaction.
    (b) Draw a reaction coordinate diagram for this reaction.
    (c) Show that the rate law for this reaction is
    $-\Delta[CH_3OH]/\Delta t = k[CH_3OH][H^+][Br^-]$.

53. A proposed mechanism for the reaction of NO_2 and CO is

    Step 1   Slow, endothermic
    $$2\,NO_2(g) \longrightarrow NO(g) + NO_3(g)$$
    Step 2   Fast, exothermic
    $$NO_3(g) + CO(g) \longrightarrow NO_2(g) + CO_2(g)$$
    Overall Reaction   Exothermic
    $$NO_2(g) + CO(g) \longrightarrow NO(g) + CO_2(g)$$

    (a) Identify each of the following as a reactant, product, or intermediate: $NO_2(g)$, $CO(g)$, $NO_3(g)$, $CO_2(g)$, $NO(g)$.
    (b) Draw a reaction coordinate diagram for this reaction. Indicate on this drawing the activation energy for each step and the overall reaction enthalpy.

54. A three-step mechanism for the reaction of $(CH_3)_3CBr$ and H_2O is proposed:

    Step 1   Slow
    $$(CH_3)_3CBr \longrightarrow (CH_3)_3C^+ + Br^-$$
    Step 2   Fast
    $$(CH_3)_3C^+ + H_2O \longrightarrow (CH_3)_3COH_2^+$$
    Step 3   Fast
    $$(CH_3)_3COH_2^+ + Br^- \longrightarrow (CH_3)_3COH + HBr$$

    (a) Write an equation for the overall reaction.
    (b) Which step is rate-determining?
    (c) What rate law is expected for this reaction?

General Questions

These questions are not designated as to type or location in the chapter. They may contain several concepts.

55. A reaction has the following experimental rate equation: Rate = $k[A]^2[B]$. If the concentration of A is doubled and the concentration of B is halved, what happens to the reaction rate?

56. After five half-life periods for a first-order reaction, what fraction of reactant remains?

57. To determine the concentration dependence of the rate of the reaction
    $$H_2PO_3^-(aq) + OH^-(aq) \longrightarrow HPO_3^{2-}(aq) + H_2O(\ell)$$
    you might measure $[OH^-]$ as a function of time using a pH meter. (To do so, you would set up conditions under which $[H_2PO_3^-]$ remains constant, by using a large excess of this reactant.) How would you prove a second-order rate dependence for $[OH^-]$?

58. Gaseous ammonia is made by the reaction
    $$N_2(g) + 3\,H_2(g) \longrightarrow 2\,NH_3(g)$$
    Use the information on the formation of NH_3 given in the table to answer the questions that follow.

$[N_2]$ (M)	$[H_2]$ (M)	Rate (mol/L · min)
0.030	0.010	4.21×10^{-5}
0.060	0.010	1.68×10^{-4}
0.030	0.020	3.37×10^{-4}

    (a) Determine n and m in the rate equation: Rate = $k[N_2]^n[H_2]^m$.
    (b) Calculate the value of the rate constant.
    (c) What is the order of the reaction with respect to $[H_2]$?
    (d) What is the overall order of the reaction?

59. The decomposition of ammonia is first order with respect to NH_3. (Compare with Study Question 58.)
    $$2\,NH_3(g) \longrightarrow N_2(g) + 3\,H_2(g)$$

    (a) What is the rate equation for this reaction?
    (b) Calculate the rate constant, k, given the following data:

$[NH_3]$ (mol/L)	Time (s)
0.67	0
0.26	19

    (c) Determine the half-life of NH_3.

60. Data for the following reaction are given in the table.
    $$2\,NO(g) + Br_2(g) \longrightarrow 2\,NOBr(g)$$

Experiment	[NO] (M)	$[Br_2]$ (M)	Initial Rate (mol/L · s)
1	1.0×10^{-2}	2.0×10^{-2}	2.4×10^{-2}
2	4.0×10^{-2}	2.0×10^{-2}	0.384
3	1.0×10^{-2}	5.0×10^{-2}	6.0×10^{-2}

    (a) What is the order of the reaction with respect to [NO]?
    (b) What is the order with respect to $[Br_2]$?
    (c) What is the overall order of the reaction?

▲ More challenging   ■ In General ChemistryNow   Blue-numbered questions answered in Appendix O

61. The decomposition of CO_2 is first order with respect to the concentration of CO_2.

$$2\,CO_2(g) \longrightarrow 2\,CO(g) + O_2(g)$$

Data on this reaction are provided in the table below.

$[CO_2]$ (mol/L)	Time (s)
0.38	0
0.27	12

(a) Write the rate equation for this reaction.

(b) Use the data to determine the value of k.

(c) What is the half-life of CO_2 under these conditions?

62. The isomerization of CH_3NC occurs slowly when CH_3NC is heated.

$$CH_3NC(g) \longrightarrow CH_3CN(g)$$

To study the rate of this reaction at 488 K, data on $[CH_3NC]$ were collected at various times. Analysis led to the graph below.

(a) What is the rate law for this reaction?

(b) What is the equation for the straight line in this graph?

(c) Calculate the rate constant for this reaction, giving the correct units.

(d) How long does it take for half of the sample to isomerize?

(e) What is the concentration of CH_3NC after 10,000 s?

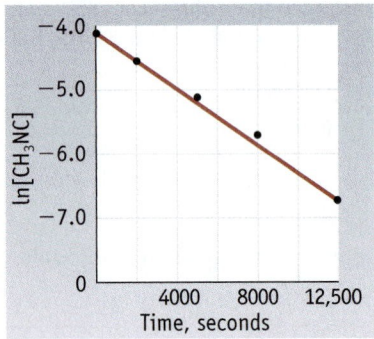

63. When heated, tetrafluoroethylene dimerizes to form octafluorocyclobutane.

$$2\,C_2F_4(g) \longrightarrow C_4F_8(g)$$

To determine the rate of this reaction at 488 K, the data in the table were collected. Analysis was done graphically, as shown below:

$[C_2F_4]$ (M)	Time (s)
0.100	0
0.080	56
0.060	150.
0.040	335
0.030	520.

(a) What is the rate law for this reaction?

(b) What is the value of the rate constant?

(c) What is the concentration of C_2F_4 after 600 s?

(d) How long will it take until the reaction is 90% complete?

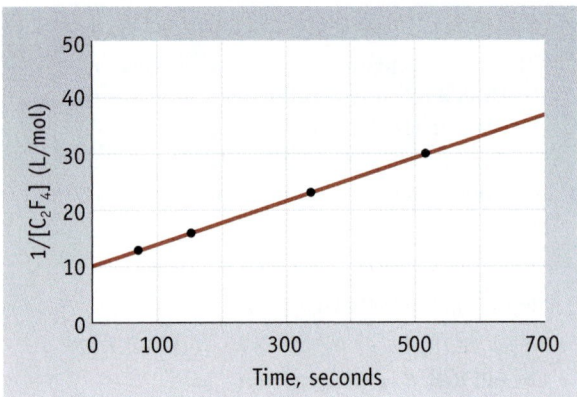

64. Data in the table were collected at 540 K for the following reaction:

$$CO(g) + NO_2(g) \longrightarrow CO_2(g) + NO(g)$$

(a) Derive the rate equation.

(b) Determine the reaction order with respect to each reactant.

(c) Calculate the rate constant, giving the correct units for k.

Initial Concentration (mol/L)		Initial Rate
[CO]	[NO₂]	(mol/L · h)
5.1×10^{-4}	0.35×10^{-4}	3.4×10^{-8}
5.1×10^{-4}	0.70×10^{-4}	6.8×10^{-8}
5.1×10^{-4}	0.18×10^{-4}	1.7×10^{-8}
1.0×10^{-3}	0.35×10^{-4}	6.8×10^{-8}
1.5×10^{-3}	0.35×10^{-4}	10.2×10^{-8}

65. Ammonium cyanate, NH_4NCO, rearranges in water to give urea, $(NH_2)_2CO$:

$$NH_4NCO(aq) \longrightarrow (NH_2)_2CO(aq)$$

Time (min)	$[NH_4NCO]$ (mol/L)
0	0.458
4.50×10^1	0.370
1.07×10^2	0.292
2.30×10^2	0.212
6.00×10^2	0.114

Using the data in the table:

(a) Decide whether the reaction is first order or second order.

(b) Calculate k for this reaction.

(c) Calculate the half-life of ammonium cyanate under these conditions.

(d) Calculate the concentration of NH_4NCO after 12.0 h.

66. Nitrogen oxides, NO_x (a mixture of NO and NO_2 collectively designated as NO_x), play an essential role in the production of pollutants found in photochemical smog. The NO_x in the atmosphere is slowly broken down to N_2 and O_2 in a first-order reaction. The average half-life of NO_x in the smokestack emissions in a large city during daylight is 3.9 h.

(a) Starting with 1.50 mg in an experiment, what quantity of NO_x remains after 5.25 h?

(b) How many hours of daylight must have elapsed to decrease 1.50 mg of NO_x to 2.50×10^{-6} mg?

67. At temperatures below 500 K, the reaction between carbon monoxide and nitrogen dioxide

$$NO_2(g) + CO(g) \longrightarrow CO_2(g) + NO(g)$$

has the following rate equation: Rate = $k[NO_2]^2$. Which of the three mechanisms suggested here best agrees with the experimentally observed rate equation?

Mechanism 1   *Single, elementary step*

$$NO_2 + CO \longrightarrow CO_2 + NO$$

Mechanism 2   *Two steps*

Slow $NO_2 + NO_2 \longrightarrow NO_3 + NO$

Fast $NO_3 + CO \longrightarrow NO_2 + CO_2$

Mechanism 3   *Two steps*

Slow $NO_2 \longrightarrow NO + O$

Fast $CO + O \longrightarrow CO_2$

68. Nitryl fluoride can be made by treating nitrogen dioxide with fluorine:

$$2 NO_2(g) + F_2(g) \longrightarrow 2 NO_2F(g)$$

Use the rate data in the table to do the following:

(a) Write the rate equation for the reaction.

(b) Indicate the order of reaction with respect to each component of the reaction.

(c) Find the numerical value of the rate constant, k.

Experiment	Initial Concentrations(mol/L)			Initial Rate (mol/L · s)
	[NO₂]	[F₂]	[NO₂F]	
1	0.001	0.005	0.001	2×10^{-4}
2	0.002	0.005	0.001	4×10^{-4}
3	0.006	0.002	0.001	4.8×10^{-4}
4	0.006	0.004	0.001	9.6×10^{-4}
5	0.001	0.001	0.001	4×10^{-5}
6	0.001	0.001	0.002	4×10^{-5}

69. ▲ The decomposition of dinitrogen pentaoxide

$$2 N_2O_5(g) \longrightarrow 4 NO_2(g) + O_2(g)$$

has the following rate equation: $-\Delta[N_2O_5]/\Delta t = k[N_2O_5]$. It has been found experimentally that the decomposition is 20% complete in 6.0 h at 300 K. Calculate the rate constant and the half-life at 300 K.

70. The data in the table give the temperature dependence of the rate constant for the reaction $N_2O_5(g) \longrightarrow 2 NO_2(g) + \frac{1}{2} O_2(g)$. Plot these data in the appropriate way to derive the activation energy for the reaction.

T(K)	$k(s^{-1})$
338	4.87×10^{-3}
328	1.50×10^{-3}
318	4.98×10^{-4}
308	1.35×10^{-4}
298	3.46×10^{-5}
273	7.87×10^{-7}

71. The decomposition of gaseous dimethyl ether at ordinary pressures is first order. Its half-life is 25.0 min at 500 °C:

$$CH_3OCH_3(g) \longrightarrow CH_4(g) + CO(g) + H_2(g)$$

(a) Starting with 8.00 g of dimethyl ether, what mass remains (in grams) after 125 min and after 145 min?

(b) Calculate the time in minutes required to decrease 7.60 ng (nanograms) to 2.25 ng.

(c) What fraction of the original dimethyl ether remains after 150 min?

72. The decomposition of phosphine, PH_3, proceeds according to the equation

$$4 PH_3(g) \longrightarrow P_4(g) + 6 H_2(g)$$

It is found that the reaction has the following rate equation: Rate = $k[PH_3]$. The half-life of PH_3 is 37.9 s at 120 °C.

(a) How much time is required for three fourths of the PH_3 to decompose?

(b) What fraction of the original sample of PH_3 remains after 1 min?

73. Three mechanisms are proposed for the gas-phase reaction of NO with Br_2 to give BrNO:

Mechanism 1

$$NO(g) + NO(g) + Br_2(g) \longrightarrow 2 BrNO(g)$$

Mechanism 2

Step 1 Slow $NO(g) + Br_2(g) \longrightarrow Br_2NO(g)$

Step 2 Fast $Br_2NO(g) + NO(g) \longrightarrow 2 BrNO(g)$

Mechanism 3

Step 1 Slow $NO(g) + NO(g) \longrightarrow N_2O_2(g)$

Step 2 Fast $N_2O_2(g) + Br_2(g) \longrightarrow 2 BrNO(g)$

(a) Write the balanced equation for the net reaction.

▲ More challenging   ■ In General ChemistryNow   **Blue-numbered questions** answered in Appendix O

(b) What is the molecularity for each step in each mechanism?

(c) What are the intermediates formed in mechanisms 2 and 3?

(d) Compare the rate laws that are derived from these three mechanisms. How could you differentiate them experimentally?

74. Radioactive iodine-131, which has a half-life of 8.04 days, is used in the form of sodium iodide to treat cancer of the thyroid. If you begin with 25.0 mg of $Na^{131}I$, what quantity of the material remains after 31 days?

75. The ozone in the earth's ozone layer decomposes according to the equation

$$2\,O_3(g) \longrightarrow 3\,O_2(g)$$

The mechanism of the reaction is thought to proceed through an initial fast equilibrium and a slow step:

Step 1 Fast, Reversible $O_3(g) \rightleftharpoons O_2(g) + O(g)$
Step 2 Slow $O_3(g) + O(g) \longrightarrow 2\,O_2(g)$

Show that the mechanism agrees with this experimental rate law: $-\Delta[O_3]/\Delta t = k[O_3]^2/[O_2]$.

76. Hundreds of different reactions can occur in the stratosphere, among them reactions that destroy the earth's ozone layer. The table below lists several (second-order) reactions of Cl atoms with ozone and organic compounds; each is given with its rate constant.

Reaction	Rate Constant (298 K, cm³/molecule · s)
(a) $Cl + O_3 \longrightarrow ClO + O_2$	1.2×10^{-11}
(b) $Cl + CH_4 \longrightarrow HCl + CH_3$	1.0×10^{-13}
(c) $Cl + C_3H_8 \longrightarrow HCl + C_3H_7$	1.4×10^{-10}
(d) $Cl + CH_2FCl \longrightarrow HCl + CHFCl$	3.0×10^{-18}

For equal concentrations of Cl and the other reactant, which is the slowest reaction? Which is the fastest reaction?

77. Data for the reaction

$$[Mn(CO)_5(CH_3CN)]^+ + NC_5H_5 \longrightarrow$$
$$[Mn(CO)_5(NC_5H_5)]^+ + CH_3CN$$

are given in the table. Calculate E_a from a plot of ln k versus $1/T$.

T(K)	k(min^{-1})
298	0.0409
308	0.0818
318	0.157

78. The gas-phase reaction

$$2\,N_2O_5(g) \longrightarrow 4\,NO_2(g) + O_2(g)$$

has an activation energy of 103 kJ, and the rate constant is 0.0900 min^{-1} at 328.0 K. Find the rate constant at 318.0 K.

79. ■ ▲ Egg protein albumin is precipitated when an egg is cooked in boiling (100 °C) water. E_a for this first-order reaction is 52.0 kJ/mol. Estimate the time to prepare a 3-min egg at an altitude at which water boils at 90 °C.

80. ▲ Two molecules of the unsaturated hydrocarbon 1,3-butadiene (C_4H_6) form the "dimer" C_8H_{12} at higher temperatures.

$$2\,C_4H_6(g) \longrightarrow C_8H_{12}(g)$$

Use the following data to determine the order of the reaction and the rate constant, k. (Note that the total pressure is the pressure of the unreacted C_4H_6 at any time and the pressure of the C_8H_{12}.)

Time (min)	Total Pressure (mm Hg)
0	436
3.5	428
11.5	413
18.3	401
25.0	391
32.0	382
41.2	371

81. ■ ▲ Hypofluorous acid, HOF, is very unstable, decomposing in a first-order reaction to give HF and O_2, with a half-life of only 30 min at room temperature:

$$HOF(g) \longrightarrow HF(g) + \tfrac{1}{2}\,O_2(g)$$

If the partial pressure of HOF in a 1.00-L flask is initially 1.00×10^2 mm Hg at 25 °C, what is the total pressure in the flask and the partial pressure of HOF after exactly 30 min? After 45 min?

82. ■ ▲ We know that the decomposition of SO_2Cl_2 is first order in SO_2Cl_2,

$$SO_2Cl_2(g) \longrightarrow SO_2(g) + Cl_2(g)$$

with a half-life of 245 min at 600 K. If you begin with a partial pressure of SO_2Cl_2 of 25 mm Hg in a 1.0-L flask, what is the partial pressure of each reactant and product after 245 min? What is the partial pressure of each reactant after 12 h?

83. The substitution of CO in $Ni(CO)_4$ by another group L [where L is an electron-pair donor such as $P(CH_3)_3$] was studied some years ago and led to an understanding of some of the general principles that govern the chemistry of compounds having metal–CO bonds. (See J. P. Day, F. Basolo, and R. G. Pearson: *Journal of the American Chemical Society,* Vol. 90, p. 6927, 1968.) A detailed study of the kinetics of the reaction led to the following mechanism:

Slow　　$Ni(CO)_4 \longrightarrow Ni(CO)_3 + CO$

Fast　　$Ni(CO)_3 + L \longrightarrow Ni(CO)_3L$

(a) What is the molecularity of each of the elementary reactions?

(b) Doubling the concentration of $Ni(CO)_4$ increased the reaction rate by a factor of 2. Doubling the concentration of L had no effect on the reaction rate. Based on this information, write the rate equation for the reaction. Does this agree with the mechanism described?

(c) The experimental rate constant for the reaction, when $L = P(C_6H_5)_3$, is $9.3 \times 10^{-3}\ s^{-1}$ at 20 °C. If the initial concentration of $Ni(CO)_4$ is 0.025 M, what is the concentration of the product after 5.0 min?

84. Screen 15.5 of the General ChemistryNow CD-ROM or website describes how to determine experimentally a rate law using the method of initial rates.

(a) Why is it best to measure the rate of reaction at the beginning of the process for this method to be valid?

(b) The first experiment shows that the initial rate of NH_4NCO degradation is $2.2 \times 10^{-4}\ mol/L \cdot s$ when $[NH_4NCO] = 0.14$ M. Using the rate law determined on this screen, predict what the rate would be if $[NH_4NCO] = 0.18$ M.

Summary and Conceptual Questions

The following questions may use concepts from the preceding chapters.

85. Hydrogenation reactions, processes wherein H_2 is added to a molecule, are usually catalyzed. An excellent catalyst is a very finely divided metal suspended in the reaction solvent. Tell why finely divided rhodium, for example, is a much more efficient catalyst than a small block of the metal.

86. ▲ It is instructive to use a mathematical model in connection with Study Question 85. Suppose you have 1000 blocks, each of which is 1.0 cm on a side. If all 1000 of these blocks are stacked to give a cube that is 10. cm on a side, what fraction of the 1000 blocks have at least one surface on the outside surface of the cube? Next divide the 1000 blocks into eight equal piles of blocks and form them into eight cubes, 5.0 cm on a side. What fraction of the blocks now have at least one surface on the outside of the cubes? How does this mathematical model pertain to Study Question 85?

87. The following statements relate to the reaction with the following rate law: Rate = $k[H_2][I_2]$.

$$H_2(g) + I_2(g) \longrightarrow 2\ HI(g)$$

Determine which of the following statements are true. If a statement is false, indicate why it is incorrect.

(a) The reaction must occur in a single step.

(b) This is a second-order reaction overall.

(c) Raising the temperature will cause the value of k to decrease.

(d) Raising the temperature lowers the activation energy for this reaction.

(e) If the concentrations of both reactants are doubled, the rate will double.

(f) Adding a catalyst in the reaction will cause the initial rate to increase.

88. Chlorine atoms contribute to the destruction of the earth's ozone layer by the following sequence of reactions:

$$Cl + O_3 \longrightarrow ClO + O_2$$

$$ClO + O \longrightarrow Cl + O_2$$

where the O atoms in the second step come from the decomposition of ozone by sunlight:

$$O_3(g) \longrightarrow O(g) + O_2(g)$$

What is the net equation on summing these three equations? Why does this lead to ozone loss in the stratosphere? What is the role played by Cl in this sequence of reactions? What name is given to species such as ClO?

89. Describe each of the following statements as true or false. If false, rewrite the sentence to make it correct.

(a) The rate-determining elementary step in a reaction is the slowest step in a mechanism.

(b) It is possible to change the rate constant by changing the temperature.

(c) As a reaction proceeds at constant temperature, the rate remains constant.

(d) A reaction that is third order overall must involve more than one step.

90. Identify which of the following statements are incorrect. If the statement is incorrect, rewrite it to be correct.

(a) Reactions are faster at a higher temperature because activation energies are lower.

(b) Rates increase with increasing concentration of reactants because there are more collisions between reactant molecules.

(c) At higher temperatures a larger fraction of molecules have enough energy to get over the activation energy barrier.

(d) Catalyzed and uncatalyzed reactions have identical mechanisms.

---

91. The reaction cyclopropane $\longrightarrow$ propene occurs on a platinum metal surface at 200 °C. (The platinum is a catalyst.) The reaction is first order in cyclopropane. Indicate how the following quantities change (increase, decrease, or no change) as this reaction progresses, assuming constant temperature.

(a) [cyclopropane]

(b) [propene]

(c) [catalyst]

(d) the rate constant, k

(e) the order of the reaction

(f) the half-life of cyclopropane

92. Isotopes are often used as "tracers" to follow an atom through a chemical reaction, and the following is an example. Acetic acid reacts with methanol by eliminating a molecule of water and forming methyl acetate (see Chapter 11).

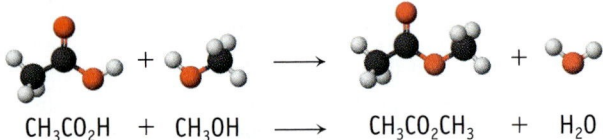

$$CH_3CO_2H \; + \; CH_3OH \; \longrightarrow \; CH_3CO_2CH_3 \; + \; H_2O$$

Explain how you could use the isotope ^{18}O to show whether the oxygen atom in the water comes from the —OH of the acid or the —OH of the alcohol.

93. Examine the reaction coordinate diagram given here.

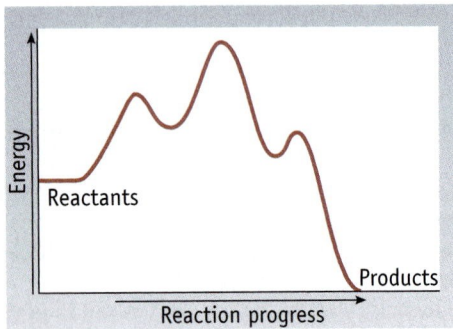

(a) How many steps are in the mechanism for the reaction described by this diagram?

(b) Is the reaction overall exothermic or endothermic?

94. Draw a reaction coordinate diagram for an exothermic reaction that occurs in a single step. Mark the activation energy, and identify the net energy change for the reaction on this diagram. Draw a second diagram that represents the same reaction in the presence of a catalyst. Identify the activation energy of this reaction and the energy change. Is the activation energy in the two drawings different? Does the energy evolved in the two reactions differ?

95. Screen 15.2 of the General ChemistryNow CD-ROM or website (Rates of Chemical Reactions) illustrates the rate at which a blue dye is bleached.

(a) What is the difference between an instantaneous rate and an average rate?

(b) Observe the graph of food dye concentration versus time on this screen. (Click the "tool" icon on this screen.) The plot shows the concentration of dye as the reaction progresses. What does the steepness of the plot at any particular time tell you about the rate of the reaction at that time?

(c) As the reaction progresses, the concentration of dye decreases as it is consumed. What happens to the reaction rate as this occurs? What is the relationship between reaction rate and dye concentration?

96. Watch the video on Screen 15.4 of the General ChemistryNow CD-ROM or website (Control of Reaction Rates—Concentration Dependence).

(a) How does an increase in HCl concentration affect the rate of the reaction of the acid with magnesium metal?

(b) On the second portion of this screen are data for the rate of decomposition of N_2O_5 (click "More"). The initial reaction rate is given for three separate experiments, each beginning with a different concentration of N_2O_5. How is the initial reaction rate related to $[N_2O_5]$?

97. The "Microscopic View of Reactions" is described on Screen 15.9 of the General ChemistryNow CD-ROM or website.

(a) According to collision theory, what three conditions must be met for two molecules to react?

(b) Examine the animations that play when numbers 1 and 2 are selected. One of these occurs at a higher temperature than the other. Which one? Explain briefly.

(c) Examine the animations that play when numbers 2 and 3 are selected. Would you expect the reaction of O_3 with N_2,

$$O_3(g) + N_2(g) \longrightarrow O_2(g) + ONN(g)$$

to be more or less sensitive to requiring a proper orientation for reaction than the reaction displayed on this screen? Explain briefly.

98. "Reaction Mechanisms and Rate Equations" are described on Screen 15.13 of the General ChemistryNow CD-ROM or website.

(a) What is the relationship between the stoichiometric coefficients of the reactants in an elementary step and the rate law for that step?

(b) What is the rate law for Step 2 of mechanism 2?

(d) Examine the "Isotopic Labeling" sidebar to this screen. If the transfer of an oxygen atom from NO_2 to CO occurred in a single-step, would any $N^{16}O^{18}O$ be found if the reaction is started using a mixture of $N^{16}O_2$ and $N^{18}O_2$? Why or why not?

99. The mechanism for the iodide ion–catalyzed decomposition of H_2O_2 is described on Screen 15.14 (Catalysis and Reaction Rate) of the General ChemistryNow CD-ROM or website.

(a) Examine the mechanism for the iodide ion–catalyzed decomposition of H_2O_2. Explain how the mechanism shows that I^- is a catalyst.

(b) How does the reaction coordinate diagram show that the catalyzed reaction is expected to be faster than the uncatalyzed reaction?

100. Many biochemical reactions are catalyzed by acids. A typical mechanism consistent with the experimental results (in which HA is the acid and X is the reactant) is

Step 1 Fast, reversible $HA \rightleftharpoons H^+ + A^-$

Step 2 Fast, reversible $X + H^+ \rightleftharpoons XH^+$

Step 3 Slow $XH^+ \longrightarrow products$

What rate law is derived from this mechanism? What is the order of the reaction with respect to HA? How would doubling the concentration of HA affect the reaction?

Do you need a live tutor for homework problems?
Access vMentor at General ChemistryNow at
http://now.brookscole.com/kotz6e
for one-on-one tutoring from a chemistry expert

16—Principles of Reactivity: Chemical Equilibria

Ammonia gas is "drilled" into the soil of a farm field. Most of the ammonia manufactured in the world is used as a fertilizer because ammonia supplies the nitrogen needed by green plants. Some ammonia is converted to nitric acid, and ammonia and the acid are combined to give ammonium nitrate, another important industrial chemical.

Arthur C. Smith III. from Grant Heilman

Fertilizer and Poison Gas

Nitrogen-containing substances are used around the world to stimulate the growth of field crops. Farmers from Portugal to Tibet have used animal waste for centuries as a "natural" fertilizer. In the 19th century industrialized countries imported nitrogen-rich marine bird manure from Peru, Bolivia, and Chile, but the supply of this material was clearly limited. In 1898 William Ramsay (the discoverer of the noble gases) pointed out that the amount of "fixed nitrogen" in the world was being depleted and predicted that world food shortages would occur by the mid-20th century as a result. That Ramsay's prediction failed to materialize was due in part to the work of Fritz Haber. His method of making ammonia from nitrogen and hydrogen is a practical example of the importance of understanding chemical equilibria, the subject of this chapter.

Fritz Haber was born in 1868 in Germany. When he was a young man, his father insisted that Fritz join his business of selling dyes and chemicals. Fritz was restless, however, and spent some years at various universities before earning his Ph.D. in 1891 from the University of Berlin. What he truly wanted was to become a professor. One problem facing him in this quest was the fact that his scientific training was considered unsound. A more important problem was that he was Jewish, and few Jews were given professorships during that era. Haber "solved" this issue by converting to Christianity, but his Jewish roots followed him for the rest of his life.

Haber did finally join the family business, but, after losing a large amount of the firm's money, he found a home at the famous Karlsruhe Institute of Technology. There his work ranged widely across industrial problems and studies in electrochemistry. In 1906 he achieved his goal of a professorship, and his interest turned to a fundamental chemical problem: how to turn atmospheric N_2 into a form usable by plants.

Haber was an ideal person to enter the race to produce ammonia. He had practical industrial experience and understood the problem theoretically. And he was tireless. Haber and his assistant, Robert Le Rossignol, discovered that the best way to accomplish the union of N_2 and H_2 was at high temperatures (over 200 °C) and high pressures (200 atm). At that time no one had yet developed methods of achieving those conditions in the laboratory. However, Haber and his coworkers soon did, and they produced ammonia—albeit very slowly. To speed up the process, they knew they needed a catalyst. After testing many substances, Haber's team found that osmium and uranium metals accelerated the reaction. Haber excitedly told his colleagues, "You have to see how liquid ammonia is pouring out."

German industry was skeptical that an industrial process could run under the severe conditions that Haber prescribed and a uranium catalyst was out of the question. Nonetheless, the process appeared so promising that Carl Bosch (1874–1940) and chemists at a large German chemical company conducted more than 10,000 experiments and tested 2000 catalysts. A suitable catalyst—based on iron oxide—was finally found, and the process was patented in 1913. The *Haber-Bosch process* is still used today and remains the cheapest way to "fix" atmospheric nitrogen. Because Haber got 1 penny per kilogram of

Fritz Haber (1868–1934). Haber developed a method for combining nitrogen from the air with hydrogen to make ammonia, a valuable agricultural chemical. A biographer described him as "verbally and action-oriented rather than contemplative," and a contemporary said that his talent for gaiety and laughter was enormously appealing.

Oesper Collection in the History of Chemistry, University of Cincinnati.

ammonia, he soon became not only famous but also rich from his discovery.

Haber demonstrated the intense nationalistic spirit prevalent in Germany at the time. When Germany became embroiled in World War I, he turned his scientific expertise to chemical problems of warfare. He became the director of the German Chemical Warfare Service, whose primary mission was to develop gas warfare. In 1915 Haber supervised the first use of chlorine gas against an opposing army at the infamous battle of Ypres in Belgium. Not only was this development a tragedy of modern warfare but it also proved to be a personal tragedy for Haber. His wife pleaded with him to stop his work on gas warfare, and, when he refused, she committed suicide.

Haber received the Nobel Prize in 1918 for his work on ammonia synthesis, although the award was widely criticized because of his wartime activities. After the war he did some of his best work, continuing his studies of thermodynamics (page 379). Because he had a Jewish background, however, Haber was forced to leave Germany in 1933. He worked for a time in England and died in Switzerland in 1934.

Haber's process to synthesize ammonia represents a triumph of chemistry. To understand it requires some knowledge of the principles of chemical equilibria, the subject of this chapter.

To Review Before You Begin
- Review reaction stoichiometry (Chapters 4 and 5)
- Review the principles of chemical kinetics (Chapter 15)
- Review the behavior of gases (Chapter 12)

GENERAL
Chemistry Now™

Throughout the chapter this icon introduces a list of resources on the General ChemistryNow CD-ROM or website (http://now .brookscole.com/kotz6e) that will:
- help you evaluate your knowledge of the material
- provide homework problems
- allow you to take an exam-prep quiz
- provide a personalized Learning Plan targeting resources that address areas you should study

■ **Reversibility of Reactions**
All chemical reactions are reversible, in theory. Practically speaking, some reactions cannot be reversed. Frying an egg, for example, is not a reversible process in practical terms.

The concept of equilibrium is fundamental in chemistry. Our goal in this and the next two chapters is to explore the consequences of the facts that chemical reactions are reversible, that in a closed system a state of equilibrium is achieved eventually between reactants and products, and that outside forces can affect the equilibrium. A major result of this exploration will be an ability to describe chemical reactions in quantitative terms and further describe the concept of "chemical reactivity."

16.1—The Nature of the Equilibrium State

If you have ever visited a limestone cave, you were surely impressed with the beautiful limestone stalactites and stalagmites, which are made chiefly of calcium carbonate (Figure 16.1). How did these structures evolve?

The formation of stalactites and stalagmites depends on the reversibility of a chemical reaction. Calcium carbonate is found in underground deposits in the form of limestone, a leftover from ancient oceans. If water seeping through the limestone contains dissolved CO_2, a reaction occurs in which the mineral dissolves, giving an aqueous solution of Ca^{2+} and HCO_3^- ions.

$$CaCO_3(s) + CO_2(aq) + H_2O(\ell) \longrightarrow Ca^{2+}(aq) + 2\ HCO_3^-(aq)$$

When the mineral-laden water reaches a cave, the reverse reaction occurs, with CO_2 being evolved into the cave and solid $CaCO_3$ being deposited.

$$Ca^{2+}(aq) + 2\ HCO_3^-(aq) \longrightarrow CaCO_3(s) + CO_2(g) + H_2O(\ell)$$

Dissolving and reprecipitating $CaCO_3$ can be illustrated by a laboratory experiment with soluble salts containing the Ca^{2+} and HCO_3^- ions (say, $CaCl_2$ and $NaHCO_3$, respectively). If you dissolve these salts in a beaker of water, you will soon see bubbles of CO_2 gas and a precipitate of solid $CaCO_3$ (Figure 16.2). If you then

Figure 16.1 Cave chemistry. Calcium carbonate stalactites cling to the roof of a cave, and stalagmites grow up from the cave floor. The chemistry producing these formations is a good example of the reversibility of chemical reactions.

Dr. Arthur N. Palmer

bubble CO_2 into the solution, the solid $CaCO_3$ dissolves. This experiment illustrates an important feature of chemical reactions: All chemical reactions are reversible.

It is informative to ask what happens if the initial solution of Ca^{2+} and HCO_3^- ions is in a closed container (unlike the reaction in Figure 16.2, which is done in an open test tube). As the reaction begins, Ca^{2+} and HCO_3^- react to give the products at some rate [◀ Section 15.1]. As the reactants are used up, the rate of this reaction

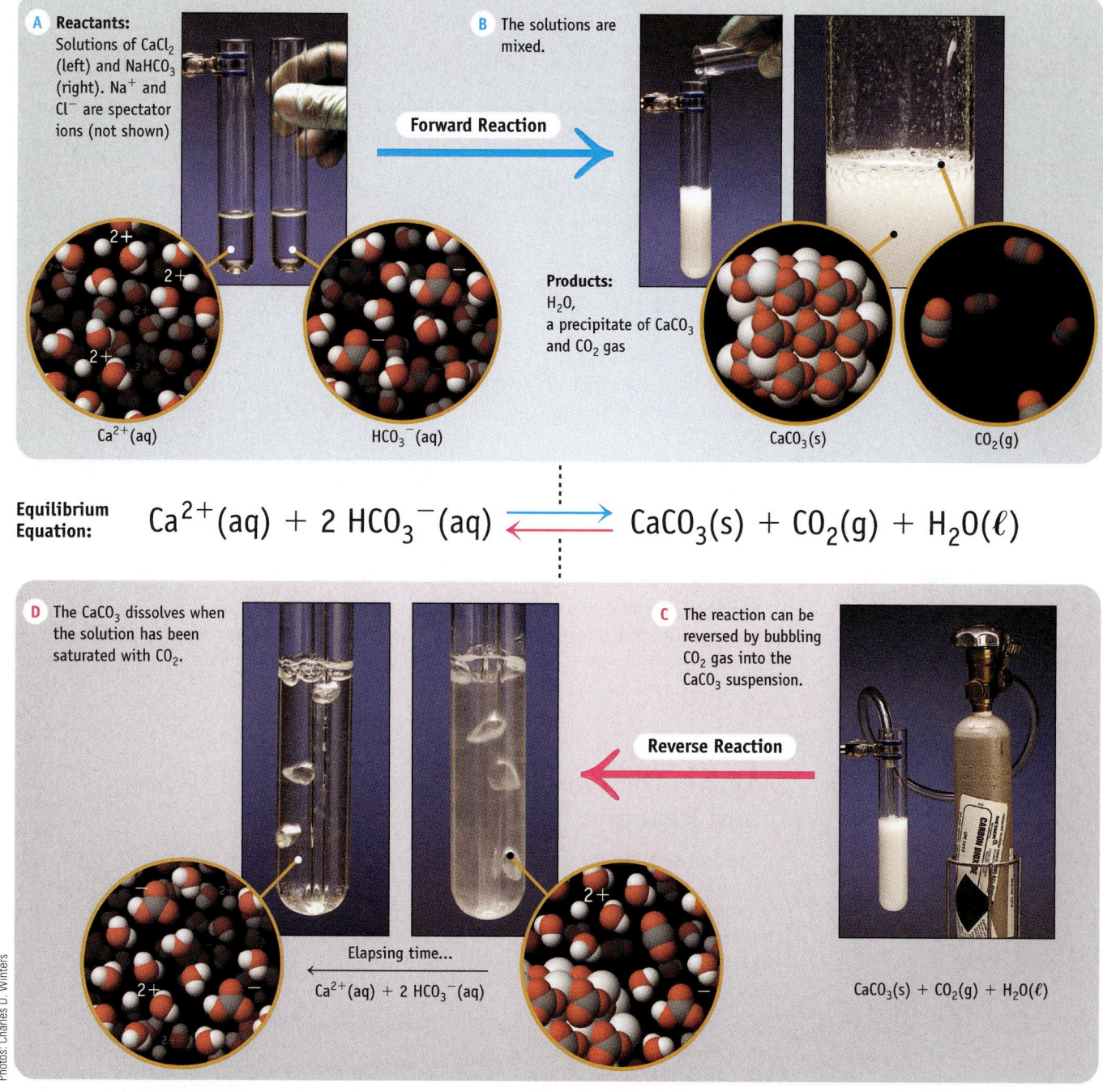

A **Reactants:** Solutions of $CaCl_2$ (left) and $NaHCO_3$ (right). Na^+ and Cl^- are spectator ions (not shown)

$Ca^{2+}(aq)$ $HCO_3^-(aq)$

B The solutions are mixed.

Forward Reaction

Products: H_2O, a precipitate of $CaCO_3$ and CO_2 gas

$CaCO_3(s)$ $CO_2(g)$

Equilibrium Equation:

$$Ca^{2+}(aq) + 2\,HCO_3^-(aq) \rightleftharpoons CaCO_3(s) + CO_2(g) + H_2O(\ell)$$

D The $CaCO_3$ dissolves when the solution has been saturated with CO_2.

Elapsing time...

$Ca^{2+}(aq) + 2\,HCO_3^-(aq)$

Reverse Reaction

C The reaction can be reversed by bubbling CO_2 gas into the $CaCO_3$ suspension.

$CaCO_3(s) + CO_2(g) + H_2O(\ell)$

Photos: Charles D. Winters

Figure 16.2   The nature of chemical equilibrium. The experiments pictured here demonstrate the reversibility of chemical reactions. All chemical reactions are, in principle, reversible and given enough time and the proper conditions, will achieve a state of dynamic equilibrium.

■ **Double Arrows,** ⇌
The set of double arrows, ⇌, in an equation indicates that the reaction is reversible. It signals that the reaction will be studied using the concepts of chemical equilibria.

slows. At the same time, however, the reaction products ($CaCO_3$, CO_2, and H_2O) begin to combine to reform Ca^{2+} and HCO_3^-, at a rate that increases as the amounts of $CaCO_3$ and CO_2 increase. Eventually the rate of the forward reaction, the formation of $CaCO_3$, and the rate of the reverse reaction, the redissolving of $CaCO_3$, become equal. With $CaCO_3$ being formed and redissolving at the same rate, no further *macroscopic* change is observed. The system is at **equilibrium**, a state in which both the forward and the reverse reactions continue to occur at equal rates so that no *net* change is observed. We depict this situation by writing a balanced equation with reactants and products connected with double arrows.

$$Ca^{2+}(aq) + 2\ HCO_3^-(aq) \rightleftharpoons CaCO_3(s) + CO_2(g) + H_2O(\ell)$$

Another example of a chemical equilibrium is the ionization of acetic acid, the reaction responsible for the acidity of vinegar.

$$CH_3CO_2H(aq) + H_2O(\ell) \rightleftharpoons CH_3CO_2^-(aq) + H_3O^+(aq)$$

Acetic acid   Acetate ion   Hydronium ion

Dissolving 1.0 mol of CH_3CO_2H in enough water to make a 1.0 M solution will produce a solution having 0.0042 M $CH_3CO_2^-$ and 0.0042 M H_3O^+ at equilibrium at 25 °C. An identical set of concentrations can be achieved by dissolving 1.0 mol of a source of $CH_3CO_2^-$ ions (say, $NaCH_3CO_2$) and 1.0 mol of a source of H_3O^+ ions (say, HCl) in the same volume of water at 25 °C.

GENERAL
Chemistry·̇·Now™

See the General ChemistryNow CD-ROM or website:
• **Screen 16.2 The Principle of Microscopic Reversibility and Screen 16.3 The Equilibrium State**, to watch a video of a reversible reaction and for a simulation of a chemical equilibrium

16.2—The Equilibrium Constant and Reaction Quotient

The concentrations of reactants and products when a reaction has reached equilibrium are related. For the reaction of hydrogen and iodine to produce hydrogen iodide, for example, a very large number of experiments have shown that *at equilibrium* the ratio of the square of the HI concentration to the product of the H_2 and I_2 concentrations is a constant.

$$H_2(g) + I_2(g) \rightleftharpoons 2\ HI(g)$$

$$\frac{[HI]^2}{[H_2][I_2]} = \text{constant } (K) \text{ at equilibrium}$$

This constant is always the same within experimental error for all experiments done at a given temperature. Suppose, for example, the concentrations of H_2 and I_2 in a

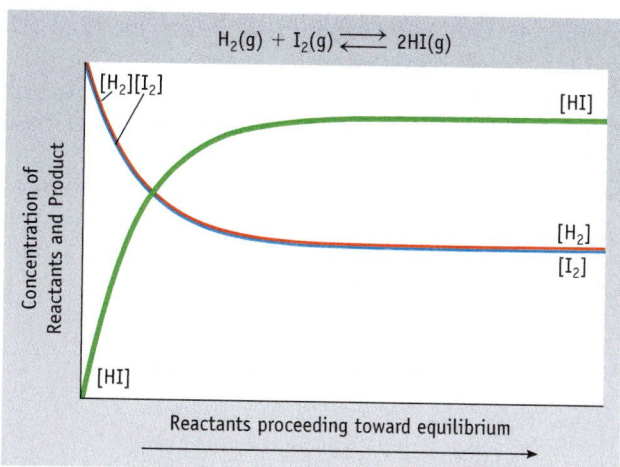

Active Figure 16.3   **The reaction of H_2 and I_2 reaches equilibrium.** The final concentrations of H_2, I_2, and HI depend on the initial concentrations of H_2 and I_2. If one begins with a different set of initial concentrations, the equilibrium concentrations will be different but the quotient $[HI]^2/[H_2][I_2]$ will always be the same at a given temperature.

GENERAL
Chemistry ⚛ Now™ **See the General ChemistryNow CD-ROM or website to explore an interactive version of this figure accompanied by an exercise.**

flask are each initially 0.0175 mol/L at 425 °C; no HI is present. Over time, the concentrations of H_2 and I_2 will decrease and the concentration of HI will increase until a state of equilibrium is reached (Figure 16.3). If the gases in the flask are then analyzed, the observed concentrations would be $[H_2] = [I_2] = 0.0037$ mol/L and $[HI] = 0.0276$ mol/L. The following table—which we call an ICE table for *initial, change,* and *equilibrium* concentrations—summarizes these results:

Equation	$H_2(g)$	+	$I_2(g)$	⇌	2 HI(g)
I = *Initial* concentration (M)	0.0175		0.0175		0
C = *Change* in concentration as reaction proceeds to equilibrium	−0.0138		−0.0138		+0.0276
E = *Equilibrium* concentration (M)	0.0037		0.0037		0.0276

■ **ICE Table: Initial, Change, and Equilibrium**
Throughout our discussions of chemical equilibria we shall express the quantitative information for reactions in an amounts table or ICE table (see Chapter 4, page 148). These tables show what the *initial* (I) concentrations are, how those concentrations *change* (C) on proceeding to equilibrium, and what the concentrations are at *equilibrium* (E).

The second line in the table gives the *change* in concentration of a reactant or product on proceeding to equilibrium. Changes are always equal to the difference between the experimentally observed equilibrium and initial concentrations.

Change in concentration = Equilibrium concentration − Initial concentration

Putting the equilibrium concentration values from the ICE table into the expression for the constant (K) gives a value of about 56 (or 55.64 if the experimental information contains more significant figures).

$$\frac{[HI]^2}{[H_2][I_2]} = \frac{(0.0276)^2}{(0.0037)(0.0037)} = 56$$

Other experiments can be done on the H_2/I_2 reaction with different concentrations of reactants, or done using mixtures of reactants and products. Regardless of the initial conditions, when equilibrium is achieved, the ratio $[HI]^2/[H_2][I_2]$ is always the same (at the same temperature).

The observation that the product and reactant concentrations for the H_2 and I_2 reaction are always in the same ratio can be generalized to other reactions. For the general chemical reaction

$$a\,A + b\,B \rightleftharpoons c\,C + d\,D$$

we can define the **equilibrium constant, K**.

When reaction is at equilibrium

$$\text{Equilibrium constant} = K = \frac{[C]^c[D]^d}{[A]^a[B]^b}$$

(16.1)

Equation 16.1 is called the **equilibrium constant expression**. In an equilibrium constant expression,

- All concentrations are equilibrium values.
- Product concentrations appear in the numerator and reactant concentrations appear in the denominator.
- Each concentration is raised to the power of its stoichiometric coefficient in the balanced chemical equation.
- The value of the constant K depends on the particular reaction and on the temperature.
- Units are never given with K.

The equilibrium constant for a chemical reaction is very useful.

- If the ratio of products to reactants as defined by Equation 16.1 matches the equilibrium constant value, the system is known to be at equilibrium. Conversely, if the ratio has a different value, the system is not at equilibrium, and we can predict in which direction the reaction will proceed to reach equilibrium.
- The equilibrium constant value indicates whether a reaction is product- or reactant-favored (page 197).
- If some concentrations of reactants or products are known at equilibrium, others can be calculated from Equation 16.1 (Section 16.4).
- If the initial concentrations and the value of K are known, the concentrations of reactants and products at equilibrium can be calculated (Section 16.4).

See the General ChemistryNow CD-ROM or website:

- **Screen 16.4 The Equilibrium Constant,** for a simulation of equilibrium and determination of the constant

Writing Equilibrium Constant Expressions

Reactions Involving Solids

The oxidation of solid, yellow sulfur produces colorless sulfur dioxide gas in a product-favored reaction (Figure 16.4).

$$S(s) + O_2(g) \rightleftharpoons SO_2(g)$$

The general principle when writing an equilibrium constant expression is to place product concentrations in the numerator and reactant concentrations in the denominator. In reactions involving solids, however, experiments show that the equilibrium concentrations of other reactants or products—here O_2 and SO_2—do not depend on the amount of solid present (as long as some solid is present at equilibrium). The concentration of a solid such as sulfur is determined by its density, and the density is a fixed value. Therefore, the concentration of sulfur is essentially constant; it is unchanged by addition or removal of the solid. Because it is constant, we do not need to include sulfur in the equilibrium constant expression.

$$K = \frac{[SO_2]}{[O_2]}$$

In general, *the concentrations of any solid reactants and products are not included in the equilibrium constant expression.*

Figure 16.4 Burning sulfur. Elemental sulfur burns in oxygen with a beautiful blue flame to give SO_2 gas.

Charles D. Winters

Reactions in Aqueous Solution

There are special considerations for reactions occurring in aqueous solution. Consider ammonia, which is a weak base owing to its reaction with water (Figure 5.7).

$$NH_3(aq) + H_2O(\ell) \rightleftharpoons NH_4^+(aq) + OH^-(aq)$$

Because the water concentration is very high in a dilute ammonia solution, the concentration of water is essentially unchanged by the reaction. The general rule for reactions in aqueous solution is that *the molar concentration of water is not included in the equilibrium constant expression.* Thus, for aqueous ammonia we write

$$K = \frac{[NH_4^+][OH^-]}{[NH_3]}$$

Reactions Involving Gases: K_c and K_p

Concentration data can be used to calculate equilibrium constants for both aqueous and gaseous systems. In these cases, the symbol K is sometimes given the subscript "c" for "concentration," as in K_c. For gases, however, equilibrium constant expressions can be written in another way—in terms of partial pressures of reactants and products. If you rearrange the ideal gas law [$PV = nRT$; Chapter 12], and recognize that the "gas concentration," (n/V), is equivalent to P/RT, you see that the partial pressure of a gas is proportional to its concentration [$P = (n/V)RT$]. If reactant and product quantities are given in partial pressures, then K is given the subscript "p," as in K_p.

$$H_2(g) + I_2(g) \rightleftharpoons 2\ HI(g)$$

$$K_p = \frac{P_{HI}^2}{P_{H_2}P_{I_2}}$$

■ **K_c and K_p**
The subscript "c" (K_c) indicates that the numerical values of concentrations in the equilibrium constant expression have units of mol/L. A subscript "p" (K_p) indicates values in units of pressure. In Chapter 16 we use K for K_c and use K_p only when equilibrium values are in units of pressure.

A Closer Look

Equilibrium Constant Expressions for Gases—K_c and K_p

Many metal carbonates, such as limestone, decompose on heating to give the metal oxide and CO_2 gas.

$$CaCO_3(s) \rightleftharpoons CaO(s) + CO_2(g)$$

The equilibrium condition for this reaction can be expressed either in terms of the number of moles per liter of CO_2, $K_c = [CO_2]$, or in terms of the pressure of CO_2, $K_p = P_{CO_2}$. From the ideal gas law, you know that

$$P = (n/V)RT = (\text{concentration in mol/L}) \times RT$$

For this reaction, we can therefore say that $P_{CO_2} = [CO_2]RT = K_p$. Because $K_c = [CO_2]$, we find that $K_p = K_c(RT)$. That is, the *values* of K_p and K_c are *not* the same; for the decomposition of calcium carbonate, K_p is the product of K_c and the factor RT.

Consider the equilibrium constant for the reaction of N_2 and H_2 to produce ammonia in terms of partial pressures, K_p.

$$N_2(g) + 3 H_2(g) \rightleftharpoons 2 NH_3(g)$$

$$K_p = \frac{(P_{NH_3})^2}{(P_{N_2})(P_{H_2})^3} = 5.8 \times 10^5 \text{ at } 25 \text{ °C}$$

If pressures are in atmospheres, K_p has the value 5.8×10^5 at 25 °C. Does K_c, the equilibrium constant in terms of concentrations, have the same value as or a different value than K_p? We can

answer this question by substituting for each pressure in K_p the equivalent expression $[C](RT)$. That is,

$$K_p = \frac{\{[NH_3](RT)\}^2}{\{[N_2](RT)\}\{[H_2](RT)\}^3} = \frac{[NH_3]^2}{[N_2][H_2]^3} \times \frac{1}{(RT)^2} = \frac{K_c}{(RT)^2}$$

Solving for K_c, we find

$$K_p = 5.8 \times 10^5 = \frac{K_c}{[(0.08206)(298)]^2}$$

$$K_c = 3.5 \times 10^8$$

Once again you see that K_p and K_c are not the same but are related by some function of RT.

Looking carefully at these examples, we find that, in general,

$$K_p = K_c(RT)^{\Delta n}$$

where Δn is the change in the number of moles of gas on going from reactants to products.

$$\Delta n = \text{total moles of gaseous products}$$
$$- \text{ total moles of gaseous reactants}$$

For the decomposition of $CaCO_3$,

$$\Delta n = 1 - 0 = 1$$

whereas the value of Δn for the ammonia synthesis is

$$\Delta n = 2 - 4 = -2$$

Notice that the basic form of the equilibrium constant expression is the same as for K_c.

In some cases the numerical values of K_c and K_p may be the same, but they are different when the numbers of moles of gaseous reactants and products are different. "A Closer Look: Equilibrium Constant Expressions for Gases—K_c and K_p," shows how K_c and K_p are related and how to convert from one to the other if necessary.

GENERAL
Chemistry·⚛·Now™

See the General ChemistryNow CD-ROM or website:
- **Screen 16.5 Writing Equilibrium Expressions,** for a simulation and a tutorial

Example 16.1—Writing Equilibrium Constant Expressions

Problem Write the equilibrium constant expressions for the following reactions.

(a) $N_2(g) + 3 H_2(g) \rightleftharpoons 2 NH_3(g)$

(b) $H_2CO_3(aq) + H_2O(\ell) \rightleftharpoons HCO_3^-(aq) + H_3O^+(aq)$

Strategy Remember that product concentrations always appear in the numerator and reactant concentrations appear in the denominator. Each concentration should be raised to a power equal to the stoichiometric coefficient in the balanced equation. In reaction (b) the water concentration does not appear in the equilibrium constant expression.

Solution

(a) $K = \dfrac{[NH_3]^2}{[N_2][H_2]^3}$ (b) $K = \dfrac{[HCO_3^-][H_3O^+]}{[H_2CO_3]}$

Exercise 16.1—Writing Equilibrium Constant Expressions

Write the equilibrium constant expression for each of the following reactions in terms of concentrations.

(a) $PCl_5(g) \rightleftharpoons PCl_3(g) + Cl_2(g)$

(b) $CO_2(g) + C(s) \rightleftharpoons 2\ CO(g)$

(c) $Cu(NH_3)_4^{2+}(aq) \rightleftharpoons Cu^{2+}(aq) + 4\ NH_3(aq)$

(d) $CH_3CO_2H(aq) + H_2O(\ell) \rightleftharpoons CH_3CO_2^-(aq) + H_3O^+(aq)$

The Meaning of the Equilibrium Constant, K

Table 16.1 lists a few equilibrium constants for different kinds of reactions. A large value of K means that the concentration of the products is higher than the concentration of the reactants at equilibrium. That is, the products are favored over the reactants at equilibrium.

$K \gg 1$: **Reaction is product-favored.** Concentrations of products are greater than concentrations of reactants at equilibrium.

Table 16.1 Selected Equilibrium Constant Values

Reaction	Equilibrium Constant, K (at 25 °C)	Product- or Reactant-Favored
Combination Reaction of Nonmetals		
$S(s) + O_2(g) \rightleftharpoons SO_2(g)$	4.2×10^{52}	$K > 1$; product-favored
$2\ H_2(g) + O_2(g) \rightleftharpoons 2\ H_2O(g)$	3.2×10^{81}	$K > 1$; product-favored
$N_2(g) + 3\ H_2(g) \rightleftharpoons 2\ NH_3(g)$	3.5×10^8	$K > 1$; product-favored
$N_2(g) + O_2(g) \rightleftharpoons 2\ NO(g)$	1.7×10^{-3} (at 2300 K)	$K < 1$; reactant-favored
Ionization of Weak Acids and Bases		
$HCO_2H(aq) + H_2O(\ell) \rightleftharpoons HCO_2^-(aq) + H_3O^+(aq)$ formic acid	1.8×10^{-4}	$K < 1$; reactant-favored
$CH_3CO_2H(aq) + H_2O(\ell) \rightleftharpoons CH_3CO_2^-(aq) + H_3O^+(aq)$ acetic acid	1.8×10^{-5}	$K < 1$; reactant-favored
$H_2CO_3(aq) + H_2O(\ell) \rightleftharpoons HCO_3^-(aq) + H_3O^+(aq)$ carbonic acid	4.2×10^{-7}	$K < 1$; reactant-favored
$NH_3(aq) + H_2O(\ell) \rightleftharpoons NH_4^+(aq) + OH^-(aq)$ ammonia	1.8×10^{-5}	$K < 1$; reactant-favored
Dissolution of "Insoluble" Solids		
$CaCO_3(s) \rightleftharpoons Ca^{2+}(aq) + CO_3^{2-}(aq)$	3.8×10^{-9}	$K < 1$; reactant-favored
$AgCl(s) \rightleftharpoons Ag^+(aq) + Cl^-(aq)$	1.8×10^{-10}	$K < 1$; reactant-favored

An example is the reaction of nitrogen monoxide and ozone.

$$NO(g) + O_3(g) \rightleftharpoons NO_2(g) + O_2(g)$$

$$K = \frac{[NO_2][O_2]}{[NO][O_3]} = 6 \times 10^{34} \text{ at 25 °C}$$

The very large value of K indicates that, at equilibrium, $[NO_2][O_2] \gg [NO][O_3]$. If stoichiometric amounts of NO and O_3 are mixed and allowed to come to equilibrium, virtually none of the reactants will be found (Figure 16.5a). Essentially all will have been converted to NO_2 and O_2. A chemist would say that "the reaction has gone to completion."

Conversely, a small value of K means that very little of the products exist when equilibrium has been achieved (Figure 16.5b). In other words, the reactants are favored over the products at equilibrium.

$K \ll 1$: Reaction is reactant-favored. Concentrations of reactants are greater than concentrations of products at equilibrium.

This is true for the formation of ozone from oxygen.

$$\tfrac{3}{2} O_2(g) \rightleftharpoons O_3(g)$$

$$K = \frac{[O_3]}{[O_2]^{3/2}} = 2.5 \times 10^{-29} \text{ at 25 °C}$$

The very small value of K indicates that, at equilibrium, $[O_3] \ll [O_2]^{3/2}$. If O_2 is placed in a flask, very little O_2 will have been converted to O_3 when equilibrium has been achieved.

When K is close to 1, it may not be immediately clear whether the reactant concentrations are larger than the product concentrations, or *vice versa*. It will depend on the form of K and thus on the reaction stoichiometry. Calculations of the concentrations will have to be done.

GENERAL
Chemistry Now™

See the General ChemistryNow CD-ROM or website:
• **Screen 16.6 The Meaning of the Equilibrium Constant,** for a simulation of an equilibrium

---

Figure 16.5   The difference between product- and reactant-favored reactions. (a) When $K > 1$, there is much more product than reactant at equilibrium. (b) When $K < 1$, there is much more reactant than product at equilibrium.

(a)

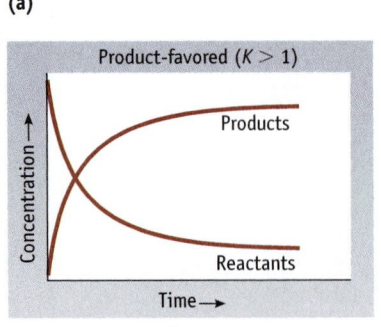

(b)

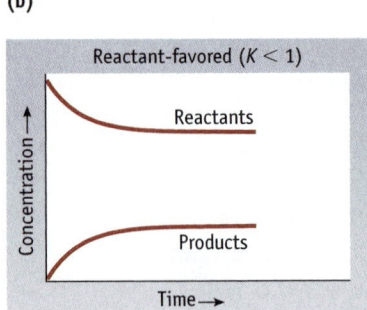

Exercise 16.2—The Equilibrium Constant and Extent of Reaction

Are the following reactions product- or reactant-favored?

(a) $Cu(NH_3)_4^{2+} \rightleftharpoons Cu^{2+}(aq) + 4\ NH_3(aq)$ $K = 1.5 \times 10^{-13}$
(b) $Cd(NH_3)_4^{2+} \rightleftharpoons Cd^{2+}(aq) + 4\ NH_3(aq)$ $K = 1.0 \times 10^{-7}$

If each reaction has a reactant concentration of 0.10 M, in which solution is the NH_3 concentration greater?

The Reaction Quotient, Q

The equilibrium constant, K, for a reaction has a particular numerical value when the reactants and products are at equilibrium. When the reactants and products in a reaction are not at equilibrium, however, it is convenient to calculate the **reaction quotient, Q**. For the general reaction of A and B to give C and D,

$$a\ A + b\ B \rightleftharpoons c\ C + d\ D$$

the reaction quotient is defined as

$$\text{Reaction quotient} = Q = \frac{[C]^c[D]^d}{[A]^a[B]^b} \tag{16.2}$$

This expression *appears* to be just like Equation 16.1, but it is not. The concentrations of reactants and products in the expression for Q are those that occur *at any point* as the reaction proceeds from reactants to an equilibrium mixture. *Only when the system is at equilibrium does $Q = K$.* For the reaction of H_2 and I_2 to give HI (Figure 16.3), any combination of reactant and product concentrations before equilibrium is achieved will give a value of Q different than K.

Determining a reaction quotient is useful for two reasons. First, it will tell you whether a system is at equilibrium (when $Q = K$) or is not at equilibrium (when $Q \neq K$). Second, by comparing Q and K, we can predict what changes will occur in reactant and product concentrations as the reaction proceeds to equilibrium.

- **$Q < K$** If Q is less than K, some reactants must be converted to products for the reaction to reach equilibrium. This will decrease the reactant concentrations and increase the product concentrations. (This is the case for the system at the left side of Figure 16.3.)

- **$Q > K$** If Q is greater than K, some products must be converted to reactants for the reaction to reach equilibrium. This will increase the reactant concentrations and decrease the product concentrations.

To illustrate these points, let us consider a reaction such as the transformation of butane to isobutane (2-methylpropane).

■ **Comparing Q and K**

Relative Magnitude	Direction of Reaction
$Q < K$	Reactants $\longrightarrow$ Products
$Q = K$	Reaction at equilibrium
$Q > K$	Reactants $\longleftarrow$ Products

Butane $\rightleftharpoons$ Isobutane

$$CH_3CH_2CH_2CH_3 \quad \rightleftharpoons \quad CH_3\overset{\displaystyle CH_3}{\underset{|}{C}}HCH_3$$

$$K_c = \frac{[\text{isobutane}]}{[\text{butane}]} = 2.50 \text{ at } 298 \text{ K}$$

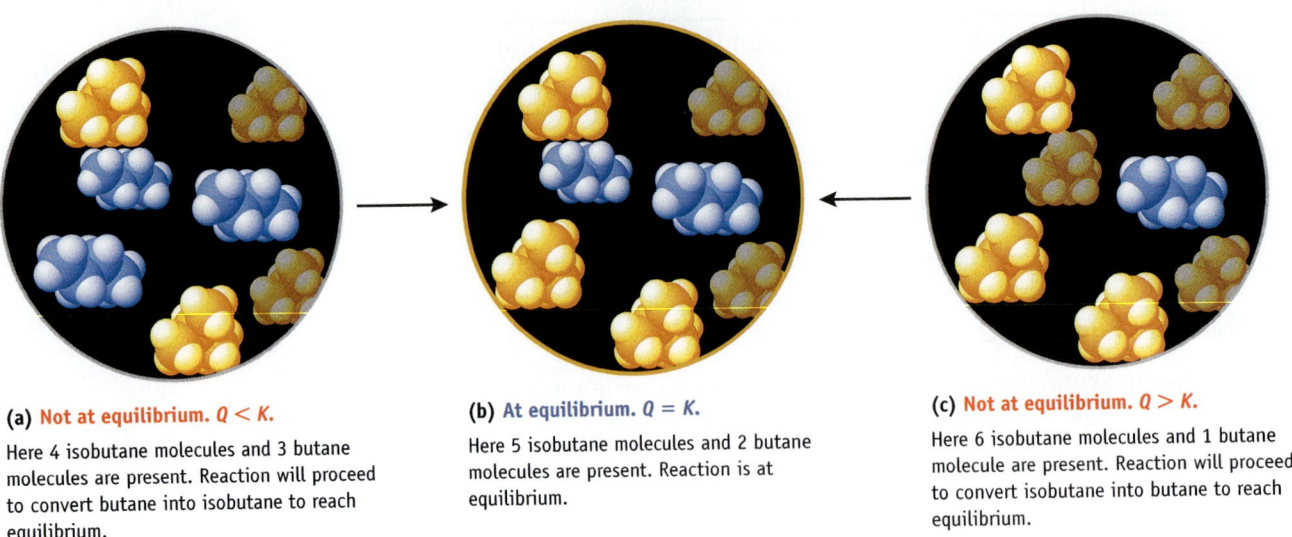

(a) Not at equilibrium. $Q < K$.

Here 4 isobutane molecules and 3 butane molecules are present. Reaction will proceed to convert butane into isobutane to reach equilibrium.

(b) At equilibrium. $Q = K$.

Here 5 isobutane molecules and 2 butane molecules are present. Reaction is at equilibrium.

(c) Not at equilibrium. $Q > K$.

Here 6 isobutane molecules and 1 butane molecule are present. Reaction will proceed to convert isobutane into butane to reach equilibrium.

Figure 16.6   The interconversion of isobutane and butane. Only when the concentrations of isobutane and butane are in the ratio [isobutane/butane] = 2.5 is the system at equilibrium (b) at 25 °C. With any other ratio of concentration, one molecule will be converted into another until equilibrium is achieved.

If the concentration of one of the compounds is known, then only one value of the other concentration will satisfy the equation for the equilibrium constant. For example, if [butane] is 1.0 mol/L, then the equilibrium concentration of isobutane, [isobutane], must be 2.5 mol/L. If [butane] is 0.80 M, then [isobutane] at equilibrium is 2.0 M.

$$[isobutane] = K[butane] = (2.50)(0.80 \text{ M}) = 2.0 \text{ M}$$

Any mixture of butane and isobutane, whether at equilibrium or not, can be represented by the reaction quotient Q (= [isobutane]/[butane]). Suppose you have a mixture composed of 3 mol/L of butane and 4 mol/L of isobutane (at 298 K) (Figure 16.6a). This means that the reaction quotient, Q, is

$$Q = \frac{[isobutane]}{[butane]} = \frac{4.0}{3.0} = 1.3$$

This set of concentrations clearly does *not* represent an equilibrium system because $Q < K$. To reach equilibrium, some butane molecules must be transformed into molecules of isobutane, thereby lowering [butane] and raising [isobutane]. This transformation will continue until the ratio [isobutane]/[butane] = 2.5, that is, until $Q = K$ (Figure 16.6b).

What happens when there is too much isobutane in the system relative to the amount of butane? Suppose [isobutane] = 6.0 M but [butane] is only 1.0 M (Figure 16.6c). Now the reaction quotient Q is greater than K ($Q > K$), and the system is again not at equilibrium. It can proceed to equilibrium by converting isobutane molecules to butane molecules.

GENERAL
Chemistry Now™

See the General ChemistryNow CD-ROM or website:

- **Screen 16.9 Systems at Equilibrium,** for a simulation and tutorial on Q

Example 16.2—The Reaction Quotient

Problem The brown gas nitrogen dioxide, NO_2, will exist in equilibrium with the colorless gas N_2O_4. $K = 170$ at 298 K.

$$2\ NO_2(g) \rightleftharpoons N_2O_4(g) \qquad K = 170$$

Suppose that, at a specific time, the concentration of NO_2 is 0.015 M, and the concentration of N_2O_4 is 0.025 M. Is Q larger than, smaller than, or equal to K? If the system is not at equilibrium, in which direction will the reaction proceed to achieve equilibrium?

Strategy Write the expression for Q and substitute the numerical values into the equation. Decide whether Q is less than, equal to, or greater than K.

Solution When the reactant and product concentrations are substituted into the reaction quotient expression, we have

$$Q = \frac{[N_2O_4]}{[NO_2]^2} = \frac{(0.025)}{(0.015)^2} = 110$$

The value of Q is less than the value of K ($Q < K$), so the reaction is not at equilibrium. The system proceeds to equilibrium by converting more NO_2 to N_2O_4, increasing $[N_2O_4]$ and decreasing $[NO_2]$ until $Q = K$.

Comment When writing Q, make sure that you raise each concentration to the power of the stoichiometric coefficient.

Exercise 16.3—The Reaction Quotient

Answer the following questions regarding the butane $\rightleftharpoons$ isobutane equilibrium ($K = 2.50$ at 298 K).

(a) Is the system at equilibrium when [butane] = 0.97 M and [isobutane] = 2.18 M? If it is not at equilibrium, in which direction will the reaction proceed to achieve equilibrium?

(b) Is the system at equilibrium when [butane] = 0.75 M and [isobutane] = 2.60 M? If it is not at equilibrium, in which direction will the reaction proceed to achieve equilibrium?

Exercise 16.4—The Reaction Quotient

At 2000 K the equilibrium constant, K, for the formation of $NO(g)$,

$$N_2(g) + O_2(g) \rightleftharpoons 2\ NO(g)$$

is 4.0×10^{-4}. You have a flask in which, at 2000 K, the concentration of N_2 is 0.50 mol/L, that of O_2 is 0.25 mol/L, and that of NO is 4.2×10^{-3} mol/L. Is the system at equilibrium? If not, predict which way the reaction will proceed to achieve equilibrium.

16.3—Determining an Equilibrium Constant

When the experimental values of the concentrations of all of the reactants and products are known *at equilibrium*, an equilibrium constant can be calculated by substituting the data into the equilibrium constant expression. Consider this concept as it applies to the oxidation of sulfur dioxide.

$$2 \, SO_2(g) + O_2(g) \rightleftharpoons 2 \, SO_3(g)$$

In an experiment done at 852 K, the equilibrium concentrations are found to be $[SO_2] = 3.61 \times 10^{-3}$ mol/L, $[O_2] = 6.11 \times 10^{-4}$ mol/L, and $[SO_3] = 1.01 \times 10^{-2}$ mol/L. Substituting these data into the equilibrium constant expression we can determine the value of *K*.

$$K = \frac{[SO_3]^2}{[SO_2]^2[O_2]} = \frac{(1.01 \times 10^{-2})^2}{(3.61 \times 10^{-3})^2(6.11 \times 10^{-4})} = 1.28 \times 10^4 \text{ at 852 K}$$

(Notice that *K* has a large value; the oxidation of sulfur dioxide is clearly product-favored at 852 K.)

More commonly, an experiment will provide information on the initial quantities of reactants and the concentration at equilibrium of only one of the reactants or of one of the products. The equilibrium concentrations of the rest of the reactants and products must then be inferred from the balanced chemical equation. As an example, consider again the oxidation of sulfur dioxide to sulfur trioxide. Suppose that 1.00 mol of SO_2 and 1.00 mol of O_2 are placed in a 1.00-L flask, this time at 1000 K. When equilibrium has been achieved, 0.925 mol of SO_3 has been formed. Let us use these data to calculate the equilibrium constant for the reaction. After writing the equilibrium constant expression in terms of concentrations, we set up an ICE table (page 761) showing the initial concentrations, the changes in those concentrations on proceeding to equilibrium, and the concentrations at equilibrium.

Equation	2 SO_2(g) +	O_2(g) ⇌	2 SO_3(g)
Initial (M)	1.00	1.00	0
Change (M)	−0.925	−0.925/2	+0.925
Equilibrium (M)	1.00 − 0.925	1.00 − 0.925/2	0.925
	= 0.075	= 0.54	

The quantities in the ICE table result from the following analysis:

- The amount of SO_2 consumed on proceeding to equilibrium is equal to the amount of SO_3 produced (= 0.925 mol because the stoichiometric factor is [2 mol SO_2 consumed/2 mol SO_3 produced]). Because SO_2 is consumed, the *change* in SO_2 concentration is −0.925 M.

- The amount of O_2 consumed is half of the amount of SO_3 produced (= 0.463 mol because the stoichiometric factor is [1 mol O_2 consumed/2 mol SO_3 produced]). The amount of O_2 remaining is 0.54 M.

- The equilibrium concentration of a reactant is always the initial concentration minus the quantity consumed or produced on proceeding to equilibrium.

With the equilibrium concentrations now known, it is possible to calculate K.

$$K = \frac{[SO_3]^2}{[SO_2]^2[O_2]} = \frac{(0.925)^2}{(0.075)^2(0.54)} = 2.8 \times 10^2 \text{ at } 1000 \text{ K}$$

GENERAL

Chemistry✦Now™

See the General ChemistryNow CD-ROM or website:

• **Screen 16.4 The Equilibrium Constant and Screen 16.8 Determining an Equilibrium Constant,** for a simulation and a tutorial on calculating an equilibrium constant

Example 16.3—Calculating an Equilibrium Constant

Problem An aqueous solution of ethanol and acetic acid, each at an initial concentration of 0.810 M, is heated to 100 °C. At equilibrium, the acetic acid concentration is 0.748 M. Calculate K for the reaction

$$C_2H_5OH(aq) + CH_3CO_2H(aq) \rightleftharpoons CH_3CO_2C_2H_5(aq) + H_2O(\ell)$$
$$\quad\text{ethanol} \qquad\quad \text{acetic acid} \qquad\qquad \text{ethyl acetate}$$

Strategy Always focus on defining equilibrium concentrations. The amount of acetic acid remaining is known, so the amount consumed is given by [initial concentration of reactant − concentration of reactant remaining]. Because the balanced chemical equation tells us that 1 mol of ethanol reacts per 1 mol of acetic acid, the concentration of ethanol is also known at equilibrium. Finally, the concentration of the product formed upon reaching equilibrium is equivalent to the amount of reactant consumed.

Solution The amount of acetic acid consumed is 0.810 M − 0.748 M = 0.062 M. This is the same as the amount of ethanol consumed and the same as the amount of ethyl acetate produced. The ICE table for this reaction is therefore

Equation	C_2H_5OH	+ CH_3CO_2H	$\rightleftharpoons$ $CH_3CO_2C_2H_5$	+ H_2O
Initial (M)	0.810	0.810	0	
Change (M)	−0.062	−0.062	+0.062	
Equilibrium (M)	0.748	0.748	0.062	

The concentration of each substance at equilibrium is now known, and K can be calculated.

$$K = \frac{[CH_3CO_2C_2H_5]}{[C_2H_5OH][CH_3CO_2H]} = \frac{0.062}{(0.748)(0.748)} = 0.11$$

Comment Notice that water does not appear in the equilibrium expression.

Example 16.4—Calculating an Equilibrium Constant (K_p) Using Partial Pressures

Problem Suppose a tank initially contains H_2S at a pressure of 10.00 atm and a temperature of 800 K. When the reaction

$$2 \text{ H}_2S(g) \rightleftharpoons 2 \text{ H}_2(g) + S_2(g)$$

has come to equilibrium, the partial pressure of S_2 vapor is 0.020 atm. Calculate K_p.

Strategy Recall from page 763 that the equilibrium constant expression can be written in terms of gas partial pressures or concentrations. The equilibrium pressure of a gaseous reactant is $P_{initial} - P_{gas\ consumed}$.

Solution The equilibrium constant expression that we want to evaluate is

$$K_p = \frac{(P_{H_2})^2 P_{S_2}}{(P_{H_2S})^2}$$

We know that $P(H_2S)_{initial} = 10.00$ atm and that $P(S_2)_{equilibrium} = 0.020$ atm, so we can set up an ICE table that expresses the equilibrium partial pressures of each gas.

Equation	2 H₂S(g)	⇌	2 H₂(g)	+	S₂(g)
Initial (atm)	10.00		0		0
Change (atm)	−2(0.020)		+2(0.020)		+0.020
Equilibrium (atm)	9.96		0.040		0.020

The balanced equation informs us that for every mole (or atmosphere of pressure) of S_2 formed, two moles (or 2 atm) of H_2 are also formed, and two moles (or 2 atm) of H_2S are consumed. Because the quantity of S_2 present at equilibrium is known from experiment to be 0.020 atm, the partial pressure of each gas is known at equilibrium, and K_p can be calculated.

$$K_p = \frac{(P_{H_2})^2 P_{S_2}}{(P_{H_2S})^2} = \frac{(0.040)^2(0.020)}{(9.96)^2} = 3.2 \times 10^{-7}$$

Comment The value of K_p will be the same as the value of K_c only when the number of moles of gaseous reactants is the same as the number of moles of gaseous products. This is not true here, so $K_p \neq K_c$. See "A Closer Look" (page 764).

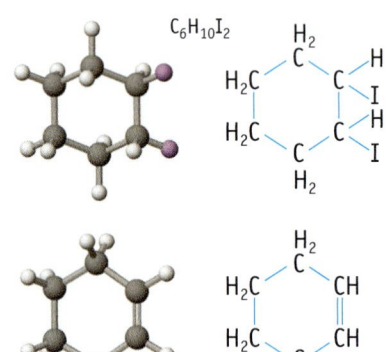

$C_6H_{10}I_2$

C_6H_{10}

Exercise 16.5—Calculating an Equilibrium Constant, K

A solution is prepared by dissolving 0.050 mol of diiodocyclohexane, $C_6H_{10}I_2$, in the solvent CCl_4. The total solution volume is 1.00 L. When the reaction

$$C_6H_{10}I_2 \rightleftharpoons C_6H_{10} + I_2$$

has come to equilibrium at 35 °C, the concentration of I_2 is 0.035 mol/L.

(a) What are the concentrations of $C_6H_{10}I_2$ and C_6H_{10} at equilibrium?
(b) Calculate K, the equilibrium constant.

16.4—Using Equilibrium Constants in Calculations

In many cases the value of K and the initial amounts of reactants are known, and you want to know the amounts present at equilibrium. As we look at several examples of this situation, we will again use ICE tables that summarize initial conditions, final conditions, and changes on proceeding to equilibrium.

GENERAL
Chemistry ☆ Now™

See the General ChemistryNow CD-ROM or website:
- **Screen 16.10 Estimating Equilibrium Concentrations,** for a tutorial

Example 16.5—Calculating Equilibrium Concentrations

Problem The equilibrium constant K ($= 55.64$) for

$$H_2(g) + I_2(g) \rightleftharpoons 2\,HI(g)$$

has been determined at 425 °C. If 1.00 mol each of H_2 and I_2 are placed in a 0.500-L flask at 425 °C, what are the concentrations of H_2, I_2, and HI when equilibrium has been achieved?

Strategy Because we know the value of K and the initial concentrations, we can set up the equilibrium constant expression and an ICE table. We will use the equilibrium constant expression to solve for the unknown values in the table.

Solution Having written the balanced chemical equation, the next step is to write the equilibrium constant expression.

$$K = \frac{[HI]^2}{[H_2][I_2]} = 55.64$$

Next set up an ICE table to express the concentrations of H_2, I_2, and HI before reaction and upon reaching equilibrium. Here, however, we do not know the numerical values of the changes in the H_2 and I_2 concentrations on proceeding to equilibrium. Because the change in $[H_2]$ is the same as the change in $[I_2]$, we express these changes as the unknown quantity x. It follows that $2x$ is the quantity of HI produced (because the stoichiometric factor is [2 mol HI produced/1 mol H_2 consumed]).

Equation	$H_2(g)$	$+$	$I_2(g)$	$\rightleftharpoons$	$2\,HI(g)$
Initial (M)	$\dfrac{1.00\ mol}{0.500\ L}$		$\dfrac{1.00\ mol}{0.500\ L}$		0
	$= 2.00$ M		$= 2.00$ M		0
Change (M)	$-x$		$-x$		$+2x$
Equilibrium (M)	$2.00 - x$		$2.00 - x$		$2x$

Now the expressions for the equilibrium concentrations can be substituted into the equilibrium constant expression.

$$55.64 = \frac{(2x)^2}{(2.00 - x)(2.00 - x)} = \frac{(2x)^2}{(2.00 - x)^2}$$

In this case, the unknown quantity x can be found by taking the square root of both sides of the equation,

$$\sqrt{K} = 7.459 = \frac{2x}{2.00 - x}$$

$$7.459(2.00 - x) = 14.9 - 7.459x = 2x$$

$$14.9 = 9.459x$$

$$x = 1.58$$

With x known, we can solve for the equilibrium concentrations of the reactants and products.

$$[H_2] = [I_2] = 2.00 - x = 0.42\ M$$

$$[HI] = 2x = 3.16\ M$$

Comment It is always wise to verify the answer by substituting the values back into the equilibrium expression to see if your calculated K agrees with the one given in the problem. In this case $(3.16)^2/(0.42)^2 = 57$. The slight discrepancy with the given value, $K = 55.64$, is because we know $[H_2]$ and $[I_2]$ to only two significant figures.

Exercise 16.6—Calculating Equilibrium Concentrations

At some temperature, $K = 33$ for the reaction

$$H_2(g) + I_2(g) \rightleftharpoons 2\,HI(g)$$

Assume the initial concentrations of both H_2 and I_2 are 6.00×10^{-3} mol/L. Find the concentration of each reactant and product at equilibrium.

Calculations Where the Solution Involves a Quadratic Expression

Suppose you are studying the decomposition of PCl_5 to form PCl_3 and Cl_2. You know that $K = 1.20$ at a given temperature.

$$PCl_5(g) \rightleftharpoons PCl_3(g) + Cl_2(g)$$

If the initial concentration of PCl_5 is 1.60 M, what will be the concentrations of reactant and product when the system reaches equilibrium? Following the procedures outlined in Example 16.5, you would set up an ICE table to define the equilibrium concentrations of reactants and products.

Reaction	$PCl_5(g)$	$\rightleftharpoons$	$PCl_3(g)$	+ $Cl_2(g)$
Initial (M)	1.60		0	0
Change (M)	$-x$		$+x$	$+x$
Equilibrium (M)	$1.60 - x$		x	x

Substituting into the equilibrium constant expression, we have

$$K = 1.20 = \frac{[PCl_3][Cl_2]}{[PCl_5]} = \frac{(x)(x)}{1.60 - x}$$

Expanding the algebraic expression results in a quadratic equation,

$$x^2 + 1.20x - 1.92 = 0$$

Using the quadratic formula (Appendix A; $a = 1$, $b = 1.20$, and $c = -1.92$), we find two roots to the equation: $x = 0.91$ and -2.11. Because a negative value of x (which represents a negative concentration) is not chemically meaningful, the answer is $x = 0.91$. Therefore, we have, at equilibrium,

$$[PCl_5] = 1.60 - 0.91 = 0.69\,M$$
$$[PCl_3] = [Cl_2] = 0.91\,M$$

Although a solution to a quadratic equation can always be obtained using the quadratic formula, in many instances an acceptable answer can be obtained by using a realistic approximation to simplify the equation. To illustrate this situation, let us consider another equilibrium, the dissociation of I_2 molecules to form I atoms, for which $K = 5.6 \times 10^{-12}$ at 500 K.

$$I_2(g) \rightleftharpoons 2\,I(g)$$

$$K = \frac{[I]^2}{[I_2]} = 5.6 \times 10^{-12}$$

Problem-Solving Tip 16.1

When Do You Need to Use the Quadratic Formula?

In most equilibrium calculations, the quantity x may be neglected in the denominator of the equation $K = x^2/([A]_0 - x)$ if x is less than 10% of the quantity of reactant initially present. The guideline presented in the text for making the approximation that $[A]_0 - x \approx [A]_0$ when $100 \times K < [A]_0$ reflects this fact.

In general, when K is about 1 or greater, the approximation *cannot* be made. If K is much less than 1 and $100 \times K < [A]_0$ (you will see many such cases in Chapter 17), the approximate expression $(K = x^2/[A]_0)$ gives an acceptable answer.

If you are not certain, then first make the assumption that the unknown (x) is small and solve the approximate expression (Equation 16.3). Next compare the "approximate" value of x with $[A]_0$. If x has a value equal to or less than 10% of $[A]_0$, then there is no need to solve the full equation using the quadratic formula.

Assuming the initial I_2 concentration is 0.45 M, and setting up the ICE table in the usual manner, we have

Reaction	$I_2(g)$	$\rightleftharpoons$	$2 I(g)$
Initial (M)	0.45		0
Change (M)	$-x$		$+2x$
Equilibrium (M)	$0.45 - x$		$2x$

For the equilibrium constant expression, we again arrive at a quadratic equation.

$$K = 5.6 \times 10^{-12} = \frac{(2x)^2}{(0.45 - x)}$$

Although we could solve this equation using the quadratic formula, there is a simpler way to reach an answer. Notice that the value of K is very small, indicating that the amount of I_2 that will be dissociated $(= x)$ is very small. In fact, K is so small that subtracting x from the original reactant concentration (0.45 mol/L) in the denominator of the equilibrium constant expression will leave the denominator essentially unchanged. That is, $(0.45 - x)$ is essentially equal to 0.45. Thus, we drop x in the denominator and have a simpler equation to solve.

$$K = 5.6 \times 10^{-12} = \frac{(2x)^2}{0.45}$$

The solution to this equation gives $x = 7.9 \times 10^{-7}$. From this value we can determine that $[I_2] = 0.45 - x \approx 0.45$ M and $[I] = 2x \approx 1.6 \times 10^{-6}$ M. Notice that the answer to the I_2 dissociation problem confirms the assumption that the dissociation of I_2 is so small that $[I_2]$ at equilibrium is essentially equal to the initial concentration.

When is it possible to simplify a quadratic equation? The decision depends on both the value of the initial concentration and the value of x, which is in turn related to the value of K. For the general reaction

$$A \rightleftharpoons B + C$$

the equilibrium constant expression is

$$K = \frac{[B][C]}{[A]} = \frac{(x)(x)}{[A]_0 - x}$$

where K and the initial concentration of A ($= [A]_0$) are known. We wish to find $[B]$ and $[C]$, and we say they are both equal to the unknown quantity x. When K is very small, the value of x will be *much* less than $[A]_0$, so $[A]_0 - x \approx [A]_0$. That is, the equilibrium concentration of A is essentially equal to $[A]_0$, so

$$K = \frac{[B][C]}{[A]} \approx \frac{(x)(x)}{[A]_0} \tag{16.3}$$

Our guideline for the use of Equation 16.3 is

If $100 \times K < [A]_0$ the approximate expression, Equation 16.3, will give acceptable values of equilibrium concentrations.

This guideline will lead to acceptable values of equilibrium concentrations (to two significant figures). For more about this useful guideline, see Problem-Solving Tip, 16.1.

Example 16.6—Calculating an Equilibrium Concentration Using an Equilibrium Constant

Problem The reaction

$$N_2(g) + O_2(g) \rightleftharpoons 2\,NO(g)$$

contributes to air pollution whenever a fuel is burned in air at a high temperature, as in a gasoline engine. At 1500 K, $K = 1.0 \times 10^{-5}$. Suppose a sample of air has $[N_2] = 0.80$ mol/L and $[O_2] = 0.20$ mol/L before any reaction occurs. Calculate the equilibrium concentrations of reactants and products after the mixture has been heated to 1500 K.

Strategy Set up an ICE table of equilibrium concentrations and then substitute these concentrations into the equilibrium constant expression. The result will be a quadratic equation. This expression can be solved by using the methods outlined in Appendix A or by using the guideline in the text to derive an acceptable, approximate answer.

Solution We first set up an ICE table of equilibrium concentrations.

Equation	$N_2(g)$ +	$O_2(g)$ $\rightleftharpoons$	$2\,NO(g)$
Initial (M)	0.80	0.20	0
Change (M)	$-x$	$-x$	$+2x$
Equilibrium (M)	$0.80 - x$	$0.20 - x$	$2x$

Next, the equilibrium concentrations are substituted into the equilibrium constant expression.

$$K = 1.0 \times 10^{-5} = \frac{[NO]^2}{[N_2][O_2]} = \frac{(2x)^2}{(0.80 - x)(0.20 - x)}$$

We refer to our guideline (Equation 16.3) to decide whether an approximate solution is possible. Here $100 \times K$ ($= 1.0 \times 10^{-3}$) is smaller than either of the initial reactant concentrations (0.80 and 0.20). This means we can use the approximate expression

$$K = 1.0 \times 10^{-5} = \frac{[NO]^2}{[N_2][O_2]} = \frac{(2x)^2}{(0.80)(0.20)}$$

Solving this expression, we find

$$1.6 \times 10^{-6} = 4x^2$$

$$x = 6.3 \times 10^{-4}$$

Therefore, the reactant and product concentrations at equilibrium are

$$[N_2] = 0.80 - 6.3 \times 10^{-4} \approx 0.80 \text{ M}$$

$$[O_2] = 0.20 - 6.3 \times 10^{-4} \approx 0.20 \text{ M}$$

$$[NO] = 2x \approx 1.3 \times 10^{-3} \text{ M}$$

Comment The value of x obtained using the approximation is the same as that obtained from the quadratic formula. If the full equilibrium constant expression is expanded, we have

$$(1.0 \times 10^{-5})(0.80-x)(0.20-x) = 4x^2$$

$$(1.0 \times 10^{-5})(0.16-1.00x + x^2) = 4x^2$$

$$\underset{ax^2}{(4-1.0 \times 10^{-5})x^2} + \underset{bx}{(1.0 \times 10^{-5})x} - \underset{c}{0.16 \times 10^{-5}} = 0$$

The two roots to this equation are

$$x = 6.3 \times 10^{-4} \quad \text{or} \quad x = -6.3 \times 10^{-4}$$

One root is identical to the approximate answer obtained above. The approximation is indeed valid in this case.

Exercise 16.7—Calculating an Equilibrium Partial Pressure Using an Equilibrium Constant

Graphite and carbon dioxide are kept at constant volume at 1000 K until the reaction

$$C(\text{graphite}) + CO_2(g) \rightleftharpoons 2\ CO(g)$$

has come to equilibrium. At this temperature, $K = 0.021$. The initial concentration of CO_2 is 0.012 mol/L. Calculate the equilibrium concentration of CO.

16.5—More About Balanced Equations and Equilibrium Constants

Chemical equations can be balanced using different sets of stoichiometric coefficients. For example, the equation for the oxidation of carbon to give carbon monoxide can be written

$$C(s) + \tfrac{1}{2} O_2(g) \rightleftharpoons CO(g)$$

In this case the equilibrium constant expression would be

$$K_1 = \frac{[CO]}{[O_2]^{1/2}} = 4.6 \times 10^{23} \text{ at 25 °C}$$

You can write the chemical equation equally well as

$$2\ C(s) + O_2(g) \rightleftharpoons 2\ CO(g)$$

The equilibrium constant expression would then be

$$K_2 = \frac{[CO]^2}{[O_2]} = 2.1 \times 10^{47} \text{ at 25 °C}$$

When you compare the two equilibrium constant expressions you find that $K_2 = (K_1)^2$; that is,

$$K_2 = \frac{[CO]^2}{[O_2]} = \left\{ \frac{[CO]}{[O_2]^{1/2}} \right\}^2 = K_1^2$$

When the stoichiometric coefficients of a balanced equation are multiplied by some factor, the equilibrium constant for the new equation (K_{new}) is the old equilibrium constant (K_{old}) raised to the power of the multiplication factor.

In the case of the oxidation of carbon, the second equation was obtained by multiplying the first equation by 2. Therefore, K_2 is the *square* of K_1 ($K_2 = K_1^2$).

What happens if a chemical equation is reversed? Let us compare the value of K for formic acid transferring a H^+ ion to water

$$HCO_2H(aq) + H_2O(\ell) \rightleftharpoons HCO_2^-(aq) + H_3O^+(aq)$$

$$K_1 = \frac{[HCO_2^-][H_3O^+]}{[HCO_2H]} = 1.8 \times 10^{-4} \text{ at } 25 \text{ °C}$$

with the opposite reaction, the gain of a H^+ ion by the formate ion, HCO_2^-.

$$HCO_2^-(aq) + H_3O^+(aq) \rightleftharpoons HCO_2H(aq) + H_2O(\ell)$$

$$K_2 = \frac{[HCO_2H]}{[HCO_2^-][H_3O^+]} = 5.6 \times 10^3 \text{ at } 25 \text{ °C}$$

Here $K_2 = 1/K_1$.

The equilibrium constants for a reaction and its reverse are the reciprocals of each other.

It is often useful to add two equations to obtain the equation for a net process. As an example, consider the reactions that take place when silver chloride dissolves in water (to a *very* small extent) and ammonia is added to the solution. The ammonia reacts with the silver ion to form a water-soluble compound, $Ag(NH_3)_2Cl$ (Figure 16.7). Adding the equation for dissolving solid AgCl to the equation for the

AgCl(s) in water   After adding NH₃(aq)

Figure 16.7   Dissolving silver chloride in aqueous ammonia. (*left*) A precipitate of AgCl(s) is suspended in water. (*right*) When aqueous ammonia is added, the ammonia reacts with the trace of silver ion in solution, the equilibrium shifts, and the silver chloride dissolves. (*See Figure 5.3 and the General ChemistryNow Screen 18.12 for the process of forming a precipitate of AgCl.*)

reaction of Ag^+ ion with ammonia gives the equation for the net reaction, dissolving solid AgCl in aqueous ammonia. (All equilibrium constants are given at 25 °C.)

$$AgCl(s) \rightleftharpoons Ag^+(aq) + Cl^-(aq) \qquad K_1 = [Ag^+][Cl^-] = 1.8 \times 10^{-10}$$

$$Ag^+(aq) + 2\ NH_3(aq) \rightleftharpoons Ag(NH_3)_2^+(aq) \qquad K_2 = \frac{[Ag(NH_3)_2^+]}{[Ag^+][NH_3]^2} = 1.6 \times 10^7$$

Net reaction: $AgCl(s) + 2\ NH_3(aq) \rightleftharpoons Ag(NH_3)_2^+(aq) + Cl^-(aq)$

To obtain the equilibrium constant for the net reaction, K_{net}, we *multiply* the equilibrium constants for the two reactions, $K_1 \times K_2$.

$$K_{net} = K_1 \times K_2 = [Ag^+][Cl^-] \times \frac{[Ag(NH_3)_2^+]}{[Ag^+][NH_3]^2} = \frac{[Ag(NH_3)_2^+][Cl^-]}{[NH_3]^2}$$

$$K_{net} = K_1 \times K_2 = 2.9 \times 10^{-3}$$

When two or more chemical equations are added to produce a net equation, the equilibrium constant for the net equation is the product of the equilibrium constants for the added equations.

Example 16.7—Balanced Equations and Equilibrium Constants

Problem A mixture of nitrogen, hydrogen, and ammonia is brought to equilibrium. When the equation is written using whole-number coefficients, as follows, the value of K is 3.5×10^8 at 25 °C.

Equation 1: $N_2(g) + 3\ H_2(g) \rightleftharpoons 2\ NH_3(g) \qquad K_1 = 3.5 \times 10^8$

However, the equation can also be written as in Equation 2. What is the value of K_2?

Equation 2: $\frac{1}{2} N_2(g) + \frac{3}{2} H_2(g) \rightleftharpoons NH_3(g)$ $K_2 = ?$

The decomposition of ammonia to the elements is the reverse of its formation (Equation 3). What is the value of K_3?

Equation 3: $2 NH_3(g) \rightleftharpoons N_2(g) + 3 H_2(g)$ $K_3 = ?$

Strategy Review what happens to the value of K when the stoichiometric coefficients are changed or the reaction is reversed. See Problem-Solving Tip, 16.2.

Solution To see the relation between K_1 and K_2, first write the equilibrium constant expressions for these two balanced equations.

$$K_1 = \frac{[NH_3]^2}{[N_2][H_2]^3} \qquad K_2 = \frac{[NH_3]}{[N_2]^{1/2}[H_2]^{3/2}}$$

Writing these expressions makes it clear that K_2 is the square root of K_1.

$$K_2 = \sqrt{K_1} = \sqrt{3.5 \times 10^8} = 1.9 \times 10^4$$

Equation 3 is the reverse of Equation 1, and its equilibrium constant expression is

$$K_3 = \frac{[N_2][H_2]^3}{[NH_3]^2}$$

In this case, K_3 is the reciprocal of K_1. That is, $K_3 = 1/K_1$.

$$K_3 = \frac{1}{K_1} = \frac{1}{3.5 \times 10^8} = 2.9 \times 10^{-9}$$

Comment Notice that the production of ammonia from the elements has a large equilibrium constant and is product-favored (see Section 16.2). As expected, the reverse reaction, the decomposition of ammonia to its elements, has a small equilibrium constant and is reactant-favored.

Exercise 16.8—Manipulating Equilibrium Constant Expressions

The conversion of oxygen to ozone has a very small equilibrium constant.

$$\frac{3}{2} O_2(g) \rightleftharpoons O_3(g) \qquad K = 2.5 \times 10^{-29}$$

(a) What is the value of K when the equation is written using whole-number coefficients?

$$3 O_2(g) \rightleftharpoons 2 O_3(g)$$

(b) What is the value of K for the conversion of ozone to oxygen?

$$2 O_3(g) \rightleftharpoons 3 O_2(g)$$

Exercise 16.9—Manipulating Equilibrium Constant Expressions

The following equilibrium constants are given at 500 K:

$$H_2(g) + Br_2(g) \rightleftharpoons 2 HBr(g) \qquad K_p = 7.9 \times 10^{11}$$
$$H_2(g) \rightleftharpoons 2 H(g) \qquad K_p = 4.8 \times 10^{-41}$$
$$Br_2(g) \rightleftharpoons 2 Br(g) \qquad K_p = 2.2 \times 10^{-15}$$

Calculate K_p for the reaction of H and Br atoms to give HBr.

$$H(g) + Br(g) \rightleftharpoons HBr(g) \qquad K_p = ?$$

16.6—Disturbing a Chemical Equilibrium

The equilibrium between reactants and products may be disturbed in three ways: (1) by changing the temperature, (2) by changing the concentration of a reactant or product, or (3) by changing the volume (for systems involving gases) (Table 16.2). *A change in any of the factors that determine the equilibrium conditions of a system will cause the system to change in such a manner as to reduce or counteract the effect of the change.* This statement is often referred to as *Le Chatelier's principle* [◀ see page 671]. It is a shorthand way of saying how a reversible reaction will attempt to adjust the quantities of reactants and products so that equilibrium is restored—that is, so that the reaction quotient is once again equal to the equilibrium constant.

■ **Le Chatelier's Principle**
Phase changes (Section 13.5) and the solubility of gases in water (Section 14.3) are affected by temperature changes. You can use Le Chatelier's principle to predict the consequences of these changes.

GENERAL
Chemistry·꙰·Now™

See the General ChemistryNow CD-ROM or website:
• **Screen 16.11 Disturbing a Chemical Equilibrium (1),** to view an animation of Le Chatelier's principle

Effect of Temperature Changes on Equilibrium Composition

You can make a qualitative prediction about the effect of a temperature change on the equilibrium composition of a chemical reaction if you know whether the reaction is exothermic or endothermic. As an example, consider the endothermic reaction of N_2 with O_2 to give NO.

$$N_2(g) + O_2(g) \rightleftharpoons 2\ NO(g) \qquad \Delta H^\circ_{rxn} = +180.6\ kJ$$

$$K = \frac{[NO]^2}{[N_2][O_2]}$$

Table 16.2   Effects of Disturbances on Equilibrium Composition

Disturbance	Change as Mixture Returns to Equilibrium	Effect on Equilibrium	Effect on K
Reactions Involving Solids, Liquids, or Gases			
Rise in temperature	Heat energy is consumed by system	Shift in endothermic direction	Change
Drop in temperature	Heat energy is generated by system	Shift in exothermic direction	Change
Addition of reactant*	Some of added reactant is consumed	Product concentration increases	No change
Addition of product*	Some of added product is consumed	Reactant concentration increases	No change
Reactions Involving Gases			
Decrease in volume, increase in pressure	Pressure decreases	Composition changes to reduce total number of gas molecules	No change
Increase in volume, decrease in pressure	Pressure increases	Composition changes to increase total number of gas molecules	No change

* Does not apply when an insoluble solid or pure liquid reactant or product is added. Recall that their "concentrations" do not appear in the reaction quotient.

Equilibrium Constant, K	Temperature
4.5×10^{-31}	298 K
6.7×10^{-10}	900 K
1.7×10^{-3}	2300 K

■ K for the N_2/O_2 Reaction
We are surrounded by N_2 and O_2, but you know that they do not react appreciably at room temperature. However, if a mixture of N_2 and O_2 is heated above 700 °C, as in an automobile engine, an equilibrium mixture will contain NO.

Notice that the equilibrium constant varies with temperature. The values of the experimental equilibrium constants indicate that the proportion of NO in the equilibrium mixture increases with temperature.

Le Chatelier's principle allows us to predict how the value of K will vary with temperature. The formation of NO from N_2 and O_2 is endothermic; that is, heat must be provided for the reaction to occur. We might imagine that heat is a "reactant." If the system is at equilibrium, and the temperature then increases, the system will adjust to alleviate this "stress." The way to counteract the energy input is to use up some of the added heat by consuming N_2 and O_2 and producing more NO as the system returns to equilibrium. This raises the value of the numerator ($[NO]^2$) and lowers the value of the denominator ($[N_2][O_2]$) in the reaction quotient, Q, resulting in a higher value of K.

As another example, consider the combination of molecules of the brown gas NO_2 to form colorless N_2O_4. An equilibrium between these compounds is readily achieved in a closed system (Figure 16.8).

$$2\ NO_2(g) \rightleftharpoons N_2O_4(g) \qquad \Delta H° = -57.1\ kJ$$

$$K = \frac{[N_2O_4]}{[NO_2]^2}$$

Equilibrium Constant, K	Temperature
1300	273 K
170	298 K

Here the reaction is exothermic, so we might imagine heat as being a reaction "product." By lowering the temperature of the system, as in Figure 16.8, some heat

Figure 16.8   Effect of temperature on an equilibrium. The tubes in the photograph both contain gaseous NO_2 (brown) and N_2O_4 (colorless) at equilibrium. K is larger at the lower temperature because the equilibrium favors colorless N_2O_4. This is clearly seen in the tube at the right, where the gas in the ice bath at 0 °C is only slightly brown because the brown gas NO_2 exerts only a small partial pressure. At 50 °C (the tube at the left), the equilibrium is shifted toward NO_2, as indicated by the darker brown color.

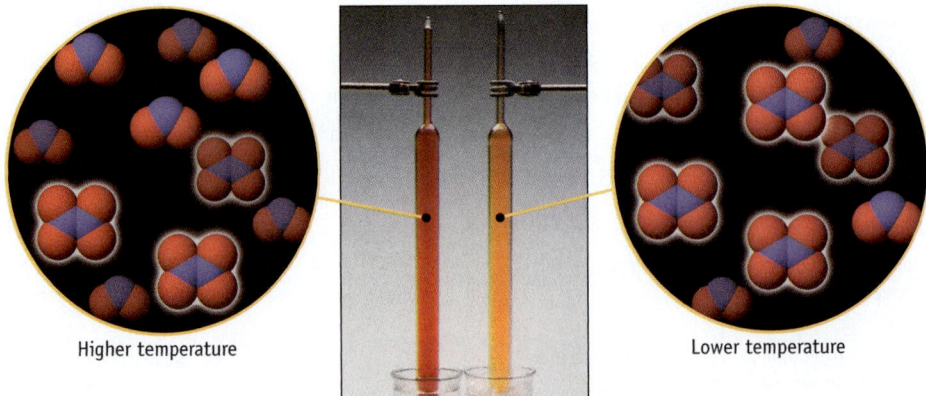

Higher temperature

Lower temperature

is removed. The removal of heat can be counteracted if the reaction produces heat by the combination of NO_2 molecules to give more N_2O_4. Thus, the equilibrium concentration of NO_2 decreases, the concentration of N_2O_4 increases, and the value of K is larger at lower temperatures.

In summary,

- When the temperature of a system at equilibrium increases, the equilibrium will shift in the direction that absorbs heat energy (Table 16.2)—that is, in the endothermic direction.

- If the temperature decreases, the equilibrium will shift in the direction that releases heat energy—that is, in the exothermic direction.

- Changing the temperature changes the equilibrium composition, and the value of K will change.

GENERAL
Chemistry**Now**™

See the General ChemistryNow CD-ROM or website:
- **Screen 16.12 Disturbing a Chemical Equilibrium (2),** for a tutorial on the effect of temperature changes

---

Exercise 16.10—Disturbing a Chemical Equilibrium

Consider the effect of temperature changes on the following equilibria.

(a) Does the equilibrium concentration of NOCl increase or decrease as the temperature of the system is increased?

$$2 \text{ NOCl(g)} \rightleftharpoons 2 \text{ NO(g)} + \text{Cl}_2\text{(g)} \qquad \Delta H^\circ_{rxn} = +77.1 \text{ kJ}$$

(b) Does the equilibrium concentration of SO_3 increase or decrease when the temperature increases?

$$2 \text{ SO}_2\text{(g)} + \text{O}_2\text{(g)} \rightleftharpoons 2 \text{ SO}_3\text{(g)} \qquad \Delta H^\circ_{rxn} = -198 \text{ kJ}$$

Effect of the Addition or Removal of a Reactant or Product

If the concentration of a reactant or product is changed from its equilibrium value *at a given temperature*, equilibrium will be reestablished eventually. The new equilibrium concentrations of reactants and products will be different, but the value of the equilibrium constant expression will still equal K (Table 16.2). To illustrate this, let us return to the butane/isobutane equilibrium (with $K = 2.5$).

$$\underset{\text{butane}}{\text{CH}_3\text{CH}_2\text{CH}_2\text{CH}_3} \rightleftharpoons \underset{\text{isobutane}}{\overset{\overset{\displaystyle \text{CH}_3}{\displaystyle |}}{\text{CH}_3\text{CHCH}_3}} \qquad K = 2.5$$

Suppose the equilibrium mixture consists of two molecules of butane and five molecules of isobutane (Figure 16.9). The ratio of molecules is 5/2 (or 2.5/1), the value of the equilibrium constant for the reaction. Now add seven more molecules of isobutane to the mixture to give a ratio of twelve isobutane molecules to two butane molecules. The ratio or reaction quotient, Q, is now 6/1. Q is greater than K, so the

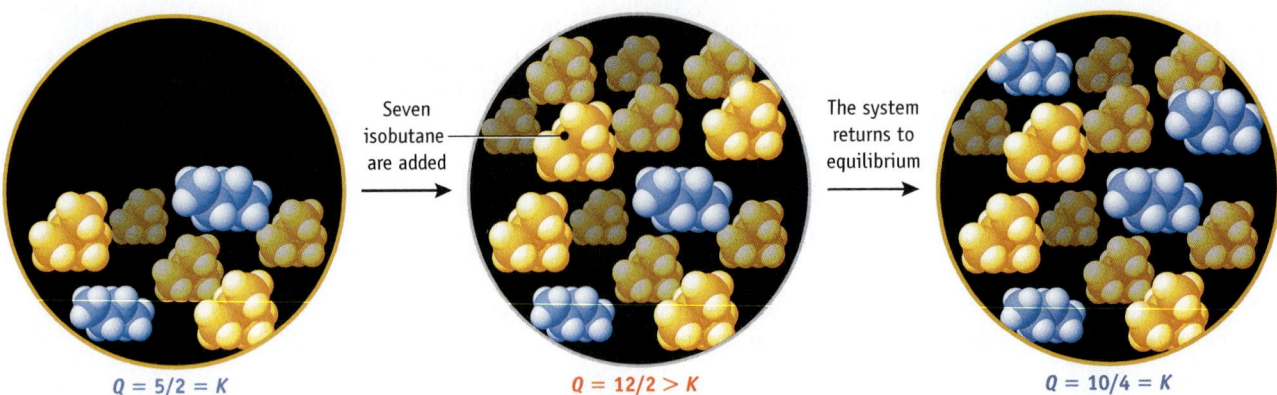

$Q = 5/2 = K$

An equilibrium mixture of 5 isobutane molecules and 2 butane molecules.

$Q = 12/2 > K$

Seven isobutane molecules are added, so the system is no longer at equilibrium.

$Q = 10/4 = K$

A net of 2 isobutane molecules has changed to butane molecules, to once again give an equilibrium mixture where the ratio of isobutane to butane is 5 to 2 (or 2.5/1).

Seven isobutane are added

The system returns to equilibrium

Active Figure 16.9   Addition of more reactant or product to an equilibrium system.

GENERAL
Chemistry⋅⚛⋅**Now**™ See the General ChemistryNow CD-ROM or website to explore an interactive version of this figure accompanied by an exercise.

system will change to reestablish equilibrium. To do so, some molecules of isobutane must be changed into butane molecules, a process that continues until the ratio [isobutane]/[butane] is once again 5/2 or 2.5/1. In this particular case, if two of the twelve isobutane molecules change to butane, the *ratio* of isobutane to butane is again equal to $K (= 10/4 = 2.5/1)$ and equilibrium is reestablished.

GENERAL
Chemistry⋅⚛⋅**Now**™

See the General ChemistryNow CD-ROM or website:

• **Screen 16.13 Disturbing a Chemical Equilibrium (3),** for a simulation and a tutorial on the effect of concentration changes on an equilibrium

Example 16.8—Effect of Concentration Changes on Equilibrium

Problem Assume equilibrium has been established in a 1.00-L flask with [butane] = 0.500 mol/L and [isobutane] = 1.25 mol/L.

$$\text{Butane} \rightleftharpoons \text{Isobutane} \qquad K = 2.50$$

Then 1.50 mol of butane is added. What are the concentrations of butane and isobutane when equilibrium is reestablished?

Strategy After adding excess butane, $Q < K$. To reestablish equilibrium, the concentration of butane must decrease, and that of isobutane must increase. Use an ICE table to track the changes. The decrease in butane concentration and the increase in isobutane concentration are both designated as x.

Solution First organize the information in a modified ICE table.

Equation	Butane	$\rightleftharpoons$	Isobutane
Initial (M)	0.500		1.25
Concentration immediately on adding butane (M)	0.500 + 1.50		1.25
Change in concentration to reestablish equilibrium	$-x$		$+x$
Equilibrium (M)	0.500 + 1.50 − x		1.25 + x

The entries in this table were arrived at as follows:

(a) The concentration of butane when equilibrium is reestablished will be the original equilibrium concentration plus what was added (1.50 mol/L) minus the concentration of butane that is converted to isobutane to reestablish equilibrium. The quantity of butane converted to isobutane is unknown and so is designated as *x*.

(b) The concentration of isobutane when equilibrium is reestablished is the concentration that was already present (1.25 mol/L) plus the concentration formed (*x* mol/L) on reestablishing equilibrium.

Having defined [butane] and [isobutane] when equilibrium is reestablished, and remembering that *K* is a constant (= 2.50), we can write

$$K = 2.50 = \frac{[\text{isobutane}]}{[\text{butane}]}$$

$$2.50 = \frac{1.25 + x}{0.500 + 1.50 - x} = \frac{1.25 + x}{2.00 - x}$$

$$2.50\,(2.00 - x) = 1.25 + x$$

$$x = 1.07 \text{ mol/L}$$

We now know the new equilibrium composition:

$$[\text{butane}] = 0.500 + 1.50 - x = 0.93 \text{ mol/L}$$

$$[\text{isobutane}] = 1.25 + x = 2.32 \text{ mol/L}$$

Comment Check your answer to verify that [isobutane]/[butane] = 2.32/0.93 = 2.5.

Exercise 16.11—Effect of Concentration Changes on Equilibrium

Equilibrium exists between butane and isobutane when [butane] = 0.20 M and [isobutane] = 0.50 M. An additional 2.00 mol/L of isobutane is added to the mixture. What are the concentrations of butane and isobutane after equilibrium has again been attained?

Effect of Volume Changes on Gas-Phase Equilibria

For a reaction that involves gases, what happens to equilibrium concentrations or pressures if the size of the container is changed? (Such a change occurs, for example, when fuel and air are compressed in an automobile engine.) To answer this question, recall that concentrations are in moles per liter. If the volume of a gas changes, its concentration therefore must also change, and the equilibrium composition can change. As an example, again consider the following equilibrium (Figure 16.8):

$$2\ NO_2(g) \rightleftharpoons N_2O_4(g)$$

brown gas   colorless gas

$$K = \frac{[N_2O_4]}{[NO_2]^2} = 170 \text{ at } 298\ K$$

What happens to this equilibrium if the volume of the flask holding the gases is suddenly halved? The immediate result is that the concentrations of both gases will double. For example, assume equilibrium is established when $[N_2O_4]$ is 0.0280 mol/L and $[NO_2]$ is 0.0128 mol/L. When the volume is halved, $[N_2O_4]$ becomes 0.0560 mol/L and $[NO_2]$ is 0.0256 mol/L. The reaction quotient, Q, under these circumstances is $(0.0560)/(0.0256)^2 = 85.5$, a value clearly less than K. Because Q is less than K, the quantity of product must increase at the expense of the reactants to return to equilibrium, and the new equilibrium composition will have a higher concentration of N_2O_4 than before the volume change.

$$2\ NO_2(g) \rightleftharpoons N_2O_4(g)$$

decrease volume of container

new equilibrium favors product

This means that one molecule of N_2O_4 is formed by consuming two molecules of NO_2. The concentration of NO_2 decreases twice as fast as the concentration of N_2O_4 increases until the reaction quotient, $Q = [N_2O_4]/[NO_2]^2$, is once again equal to K. The conclusions for the NO_2/N_2O_4 equilibrium can be generalized:

- For reactions involving gases, the stress of a volume decrease (a pressure increase) will be counterbalanced by a change in the equilibrium composition to one having a smaller number of molecules.

- For a volume increase (a pressure decrease), the equilibrium composition will favor the side of the reaction with the larger number of molecules.

- For a reaction in which there is no change in the number of molecules, such as in the reaction of H_2 and I_2 to produce HI $[H_2(g) + I_2(g) \rightleftharpoons 2\ HI(g)]$, a volume change will have no effect.

GENERAL
Chemistry  **Now**™

See the General ChemistryNow CD-ROM or website:
- **Screen 16.14 Disturbing a Chemical Equilibrium (4)**, for a simulation of the effect of a volume change for a reaction involving gases

---

Exercise 16.12—Effect of Concentration and Volume Changes on Equilibria

The formation of ammonia from its elements is an important industrial process.

$$3\ H_2(g) + N_2(g) \rightleftharpoons 2\ NH_3(g)$$

(a) How does the equilibrium composition change when extra H_2 is added? When extra NH_3 is added?
(b) What is the effect on the equilibrium when the volume of the system is increased? Does the equilibrium composition change or is the system unchanged?

16.7—Applying the Principles of Chemical Equilibrium

The Haber-Bosch Process

As described on pages 756–757, one of the greatest advances in agriculture has been the use of manufactured nitrogen-containing fertilizers, largely ammonia and ammonium salts. This industry is based on the Haber-Bosch process, the synthesis of ammonia from its elements, nitrogen and hydrogen.

$$N_2(g) + 3\ H_2(g) \rightleftharpoons 2\ NH_3(g)$$

At 25 °C, K (calculated value) $= 3.5 \times 10^8$ and $\Delta H^\circ_{rxn} = -91.8$ kJ/mol

At 450 °C, K (experimental value) $= 0.16$ and $\Delta H^\circ_{rxn} = -111.3$ kJ/mol

Ammonia is now made for pennies per kilogram and is consistently ranked in the top five chemicals produced in the United States, with 15–20 billion kilograms being produced annually. Not only is it used as fertilizer, but it also serves as a starting material for making nitric acid and ammonium nitrate, among other things.

The manufacture of ammonia (Figure 16.10) is a good example of the role that kinetics and chemical equilibria play in practical chemistry.

- The $N_2 + H_2$ reaction is exothermic and product-favored ($K > 1$ at 25 °C). The reaction at 25 °C is slow, however, so it is carried out at a higher temperature to increase the reaction rate.

- Although the reaction rate increases with temperature, the equilibrium constant decreases, as predicted by Le Chatelier's principle. Thus, for a given concentration of starting material, the equilibrium concentration of NH_3 is smaller at higher temperatures. In an industrial ammonia plant, it is necessary to balance reaction rate (improved at higher temperature) with product yield (K is smaller at higher temperatures).

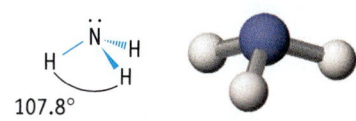

107.8°

Properties of Ammonia
Melting point (K) = 195.42
Boiling point (K) = 239.74
Liquid density = 0.6826 g/cm³
Dipole moment = 1.46 Debye

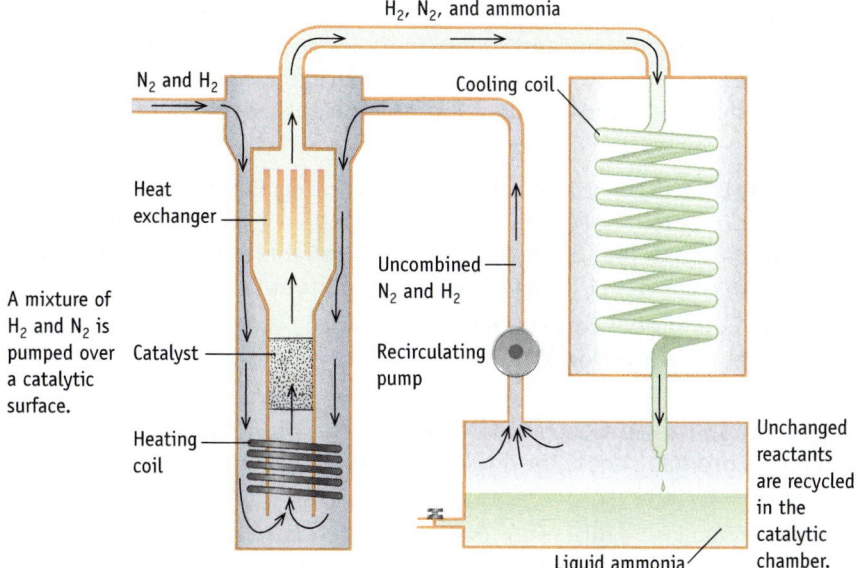

H₂, N₂, and ammonia

N₂ and H₂

Cooling coil

Heat exchanger

Uncombined N₂ and H₂

A mixture of H₂ and N₂ is pumped over a catalytic surface.

Catalyst

Recirculating pump

Heating coil

Unchanged reactants are recycled in the catalytic chamber.

Liquid ammonia

Figure 16.10 The Haber-Bosch process for ammonia synthesis. A mixture of H_2 and N_2 is pumped over a catalytic surface. The NH_3 is collected as a liquid (at -33 °C), and unchanged reactants are recycled in the catalytic chamber.

- To increase the equilibrium concentration of NH_3, the reaction is carried out at a higher pressure. This does not change the value of K, but an increase in pressure can be compensated for by converting 4 mol of reactants to 2 mol of product.

- A catalyst is used to increase the reaction rate even more. An effective catalyst for the Haber-Bosch process is Fe_3O_4 mixed with KOH, SiO_2, and Al_2O_3 (all inexpensive chemicals). The catalyst is not effective at temperatures less than 400 °C, however, and the optimal temperature is about 450 °C.

Chapter Goals Revisited

Now that you have studied this chapter, you should ask whether you have met the chapter goals. In particular you should be able to:

Understand the nature and characteristics of chemical equilibria

a. Chemical reactions are reversible and equilibria are dynamic (Section 16.1).

Understand the significance of the equilibrium constant, K, and reaction quotient, Q.

a. Write the reaction quotient, Q, for a chemical reaction (Section 16.2). When the system is at equilibrium, the reaction quotient is called the equilibrium constant expression and has a constant value called the equilibrium constant, which is symbolized by K (Equation 16.2). General ChemistryNow homework: Study Question(s) 2, 4

b. Recognize that the concentrations of solids, pure liquids, and solvents (e.g., water) are not included in the equilibrium constant expression (Equation 16.1, Section 16.2).

c. Recognize that a large value of K ($K \gg 1$) means the reaction is product-favored, and the product concentrations are greater than the reactant concentrations at equilibrium. A small value of K ($K \ll 1$) indicates a reactant-favored reaction in which the product concentrations are smaller than the reactant concentrations at equilibrium (Section 16.2).

d. Appreciate the fact that equilibrium concentrations may be expressed in terms of reactant and product concentrations (in moles per liter), and K is then sometimes designated as K_c. Alternatively, concentrations of gases may be represented by partial pressures, and K for such cases is designated K_p (Section 16.2).

Understand how to use K in quantitative studies of chemical equilibria

a. Use the reaction quotient (Q) to decide whether a reaction is at equilibrium ($Q = K$), or if there will be a net conversion of reactants to products ($Q < K$) or products to reactants ($Q > K$) to attain equilibrium (Section 16.2).

b. Calculate an equilibrium constant given the reactant and product concentrations at equilibrium (Section 16.3). General ChemistryNow homework: SQ(s) 8, 11, 33, 34, 44, 61a

c. Use equilibrium constants to calculate the concentration (or pressure) of a reactant or a product at equilibrium (Section 16.4). General ChemistryNow homework: SQ(s) 16, 32, 42, 47, 56, 59, 62

d. Know how K changes as different stoichiometric coefficients are used in a balanced equation, if the equation is reversed, or if several equations are added to give a new net equation (Section 16.5). General ChemistryNow homework: SQ(s) 21

e. Know how to predict, using Le Chatelier's principle, the effect of a disturbance on a chemical equilibrium—a change in temperature, a change in concentrations, or a change in volume or pressure for a reaction involving gases (Section 16.6 and Table 16.2). General ChemistryNow homework: SQ(s) 26, 28, 39

Key Equations

Equation 16.1 (page 762)

The equilibrium constant expression. At equilibrium the ratio of products to reactants has a constant value, K (at a particular temperature). For the general reaction $a\,A + b\,B \rightleftharpoons c\,C + d\,D$,

$$\text{Equilibrium constant} = K = \frac{[C]^c[D]^d}{[A]^a[B]^b}$$

Equation 16.2 (page 767)

For the general reaction $a\,A + b\,B \rightleftharpoons c\,C + d\,D$, the ratio of product to reactant concentrations at any point in the reaction is the reaction quotient.

$$\text{Reaction quotient} = Q = \frac{[C]^c[D]^d}{[A]^a[B]^b}$$

Study Questions

▲ denotes more challenging questions.

■ denotes questions available in the Homework and Goals section of the General ChemistryNow CD-ROM or website.

Blue numbered questions have answers in Appendix O and fully worked solutions in the *Student Solutions Manual*.

Structures of many of the compounds used in these questions are found on the General ChemistryNow CD-ROM or website in the Models folder.

GENERAL
Chemistry ⚛ Now™ Assess your understanding of this chapter's topics with additional quizzing and conceptual questions at **http://now.brookscole.com/kotz6e**

Practicing Skills

Writing Equilibrium Constant Expressions

(See Example 16.1 and General ChemistryNow Screens 16.3, 16.4, and 16.5.)

1. Write equilibrium constant expressions for the following reactions. For gases use either pressures or concentrations.
   - (a) $2\,H_2O_2(g) \rightleftharpoons 2\,H_2O(g) + O_2(g)$
   - (b) $CO(g) + \frac{1}{2}\,O_2(g) \rightleftharpoons CO_2(g)$
   - (c) $C(s) + CO_2(g) \rightleftharpoons 2\,CO(g)$
   - (d) $NiO(s) + CO(g) \rightleftharpoons Ni(s) + CO_2(g)$

2. ■ Write equilibrium constant expressions for the following reactions. For gases use either pressures or concentrations.
   - (a) $3\,O_2(g) \rightleftharpoons 2\,O_3(g)$
   - (b) $Fe(s) + 5\,CO(g) \rightleftharpoons Fe(CO)_5(g)$
   - (c) $(NH_4)_2CO_3(s) \rightleftharpoons 2\,NH_3(g) + CO_2(g) + H_2O(g)$
   - (d) $Ag_2SO_4(s) \rightleftharpoons 2\,Ag^+(aq) + SO_4^{2-}(aq)$

The Equilibrium Constant and Reaction Quotient

(See Example 16.2 and General ChemistryNow Screen 16.9.)

3. $K = 5.6 \times 10^{-12}$ at 500 K for the dissociation of iodine molecules to iodine atoms.

$$I_2(g) \rightleftharpoons 2\,I(g)$$

A mixture has $[I_2] = 0.020$ mol/L and $[I] = 2.0 \times 10^{-8}$ mol/L. Is the reaction at equilibrium (at 500 K)? If not, which way must the reaction proceed to reach equilibrium?

4. ■ The reaction

$$2\,NO_2(g) \rightleftharpoons N_2O_4(g)$$

has an equilibrium constant, K, of 170 at 25 °C. If 2.0×10^{-3} mol of NO_2 is present in a 10.-L flask along with 1.5×10^{-3} mol of N_2O_4, is the system at equilibrium? If it is not at equilibrium, does the concentration of NO_2 increase or decrease as the system proceeds to equilibrium?

5. A mixture of SO_2, O_2, and SO_3 at 1000 K contains the gases at the following concentrations: $[SO_2] = 5.0 \times 10^{-3}$ mol/L, $[O_2] = 1.9 \times 10^{-3}$ mol/L, and $[SO_3] = 6.9 \times 10^{-3}$ mol/L. Is the reaction at equilibrium? If not, which way will the reaction proceed to reach equilibrium?

$$2\,SO_2(g) + O_2(g) \rightleftharpoons 2\,SO_3(g) \quad K = 279$$

6. The equilibrium constant, K, for the reaction

$$2\, NOCl(g) \rightleftharpoons 2\, NO(g) + Cl_2(g)$$

is 3.9×10^{-3} at 300 °C. A mixture contains the gases at the following concentrations: $[NOCl] = 5.0 \times 10^{-3}$ mol/L, $[NO] = 2.5 \times 10^{-3}$ mol/L, and $[Cl_2] = 2.0 \times 10^{-3}$ mol/L. Is the reaction at equilibrium at 300 °C? If not, in which direction does the reaction proceed to come to equilibrium?

Calculating an Equilibrium Constant
(See Examples 16.3 and 16.4 and General ChemistryNow Screens 16.4 and 16.8.)

7. The reaction

$$PCl_5(g) \rightleftharpoons PCl_3(g) + Cl_2(g)$$

was examined at 250 °C. At equilibrium, $[PCl_5] = 4.2 \times 10^{-5}$ mol/L, $[PCl_3] = 1.3 \times 10^{-2}$ mol/L, and $[Cl_2] = 3.9 \times 10^{-3}$ mol/L. Calculate K for the reaction.

8. ■ An equilibrium mixture of SO_2, O_2, and SO_3 at 1000 K contains the gases at the following concentrations: $[SO_2] = 3.77 \times 10^{-3}$ mol/L, $[O_2] = 4.30 \times 10^{-3}$ mol/L, and $[SO_3] = 4.13 \times 10^{-3}$ mol/L. Calculate the equilibrium constant, K, for the reaction.

$$2\, SO_2(g) + O_2(g) \rightleftharpoons 2\, SO_3(g)$$

9. The reaction

$$C(s) + CO_2(g) \rightleftharpoons 2\, CO(g)$$

occurs at high temperatures. At 700 °C, a 2.0-L flask contains 0.10 mol of CO, 0.20 mol of CO_2, and 0.40 mol of C at equilibrium.

(a) Calculate K for the reaction at 700 °C.

(b) Calculate K for the reaction, also at 700 °C, if the amounts at equilibrium in the 2.0-L flask are 0.10 mol of CO, 0.20 mol of CO_2, and 0.80 mol of C.

(c) Compare the results of (a) and (b). Does the quantity of carbon affect the value of K? Explain.

10. Hydrogen and carbon dioxide react at a high temperature to give water and carbon monoxide.

$$H_2(g) + CO_2(g) \rightleftharpoons H_2O(g) + CO(g)$$

(a) Laboratory measurements at 986 °C show that there are 0.11 mol each of CO and H_2O vapor and 0.087 mol each of H_2 and CO_2 at equilibrium in a 1.0-L container. Calculate the equilibrium constant for the reaction at 986 °C.

(b) Suppose 0.050 mol each of H_2 and CO_2 are placed in a 2.0-L container. When equilibrium is achieved at 986 °C, what amounts of CO(g) and H_2O(g), in moles, would be present? [Use the value of K from part (a).]

11. ■ A mixture of CO and Cl_2 is placed in a reaction flask: $[CO] = 0.0102$ mol/L and $[Cl_2] = 0.00609$ mol/L. When the reaction

$$CO(g) + Cl_2(g) \rightleftharpoons COCl_2(g)$$

has come to equilibrium at 600 K, $[Cl_2] = 0.00301$ mol/L.

(a) Calculate the concentrations of CO and $COCl_2$ at equilibrium.

(b) Calculate K.

12. You place 3.00 mol of pure SO_3 in an 8.00-L flask at 1150 K. At equilibrium, 0.58 mol of O_2 has been formed. Calculate K for the reaction at 1150 K.

$$2\, SO_3(g) \rightleftharpoons 2\, SO_2(g) + O_2(g)$$

Using Equilibrium Constants
(See Examples 16.5 and 16.6 and General ChemistryNow Screen 16.10.)

13. The value of K for the interconversion of butane and isobutane is 2.5 at 25 °C.

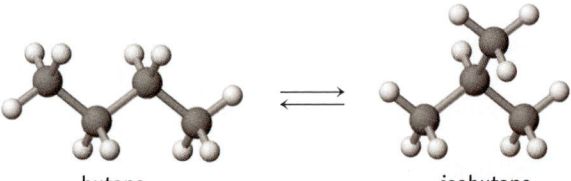

butane   isobutane

If you place 0.017 mol of butane in a 0.50-L flask at 25 °C and allow equilibrium to be established, what will be the equilibrium concentrations of the two forms of butane?

14. Cyclohexane, C_6H_{12}, a hydrocarbon, can isomerize or change into methylcyclopentane, a compound of the same formula ($C_5H_9CH_3$) but with a different molecular structure.

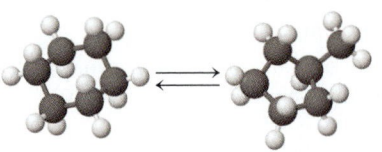

cyclohexane   methylcyclopentane

The equilibrium constant has been estimated to be 0.12 at 25 °C. If you had originally placed 0.045 mol of cyclohexane in a 2.8-L flask, what would be the concentrations of cyclohexane and methylcyclopentane when equilibrium is established?

15. The equilibrium constant for the dissociation of iodine molecules to iodine atoms

$$I_2(g) \rightleftharpoons 2\, I(g)$$

is 3.76×10^{-3} at 1000 K. Suppose 0.105 mol of I_2 is placed in a 12.3-L flask at 1000 K. What are the concentrations of I_2 and I when the system comes to equilibrium?

16. ■ The equilibrium constant for the reaction

$$N_2O_4(g) \rightleftharpoons 2\, NO_2(g)$$

at 25 °C is 5.88×10^{-3}. Suppose 15.6 g of N_2O_4 is placed in a 5.00-L flask at 25 °C. Calculate the following:

(a) the amount of NO_2 (mol) present at equilibrium

(b) the percentage of the original N_2O_4 that is dissociated

17. Carbonyl bromide decomposes to carbon monoxide and bromine.

$$COBr_2(g) \rightleftharpoons CO(g) + Br_2(g)$$

K is 0.190 at 73 °C. If you place 0.500 mol of $COBr_2$ in a 2.00-L flask and heat it to 73 °C, what are the equilibrium concentrations of $COBr_2$, CO, and Br_2? What percentage of the original $COBr_2$ decomposed at this temperature?

18. Iodine dissolves in water, but its solubility in a nonpolar solvent such as CCl_4 is greater.

Nonpolar I_2
Polar H_2O

Nonpolar CCl_4

Shake the test tube →

Polar H_2O

Nonpolar CCl_4 and I_2

Extracting iodine (I_2) from water with the nonpolar solvent CCl_4. I_2 is more soluble in CCl_4 and, after shaking a mixture of water and CCl_4, the I_2 has accumulated in the more dense CCl_4 layer.

The equilibrium constant is 85.0 for the reaction

$$I_2(aq) \rightleftharpoons I_2(CCl_4)$$

You place 0.0340 g of I_2 in 100.0 mL of water. After shaking it with 10.0 mL of CCl_4, how much I_2 remains in the water layer?

Manipulating Equilibrium Constant Expressions
(See Example 16.7 and General ChemistryNow Screen 16.7.)

19. Which of the following correctly relates the equilibrium constants for the two reactions shown?

$$A + B \rightleftharpoons 2 C \qquad K_1$$
$$2 A + 2 B \rightleftharpoons 4 C \qquad K_2$$

(a) $K_2 = 2K_1$ (c) $K_2 = 1/K_1$
(b) $K_2 = K_1^2$ (d) $K_2 = 1/K_1^2$

20. Which of the following correctly relates the equilibrium constants for the two reactions shown?

$$A + B \rightleftharpoons 2 C \qquad K_1$$
$$C \rightleftharpoons \tfrac{1}{2} A + \tfrac{1}{2} B \qquad K_2$$

(a) $K_2 = 1/(K_1)^{1/2}$ (c) $K_2 = K_1^2$
(b) $K_2 = 1/K_1$ (d) $K_2 = -K_1^{1/2}$

21. ■ Consider the following equilibria involving $SO_2(g)$ and their corresponding equilibrium constants.

$$SO_2(g) + \tfrac{1}{2} O_2(g) \rightleftharpoons SO_3(g) \qquad K_1$$
$$2 SO_3(g) \rightleftharpoons 2 SO_2(g) + O_2(g) \qquad K_2$$

Which of the following expressions relates K_1 to K_2?

(a) $K_2 = K_1^2$ (d) $K_2 = 1/K_1$
(b) $K_2^2 = K_1$ (e) $K_2 = 1/K_1^2$
(c) $K_2 = K_1$

22. The equilibrium constant K for the reaction

$$CO_2(g) \rightleftharpoons CO(g) + \tfrac{1}{2} O_2(g)$$

is 6.66×10^{-12} at 1000 K. Calculate K for the reaction

$$2 CO(g) + O_2(g) \rightleftharpoons 2 CO_2(g)$$

23. Calculate K for the reaction

$$SnO_2(s) + 2 CO(g) \rightleftharpoons Sn(s) + 2 CO_2(g)$$

given the following information:

$$SnO_2(s) + 2 H_2(g) \rightleftharpoons Sn(s) + 2 H_2O(g) \qquad K = 8.12$$
$$H_2(g) + CO_2(g) \rightleftharpoons H_2O(g) + CO(g) \qquad K = 0.771$$

24. Calculate K for the reaction

$$Fe(s) + H_2O(g) \rightleftharpoons FeO(s) + H_2(g)$$

given the following information:

$$H_2O(g) + CO(g) \rightleftharpoons H_2(g) + CO_2(g) \qquad K = 1.6$$
$$FeO(s) + CO(g) \rightleftharpoons Fe(s) + CO_2(g) \qquad K = 0.67$$

Disturbing a Chemical Equilibrium
(See Example 16.8 and General ChemistryNow Screens 16.11–16.14.)

25. Dinitrogen trioxide decomposes to NO and NO_2 in an endothermic process ($\Delta H = 40.5$ kJ/mol).

$$N_2O_3(g) \rightleftharpoons NO(g) + NO_2(g)$$

Predict the effect of the following changes on the position of the equilibrium; that is, state which way the equilibrium will shift (left, right, or no change) when each of the following changes is made.
(a) adding more $N_2O_3(g)$
(b) adding more $NO_2(g)$
(c) increasing the volume of the reaction flask
(d) lowering the temperature

26. ■ K_p for the following reaction is 0.16 at 25 °C:

$$2 NOBr(g) \rightleftharpoons 2 NO(g) + Br_2(g)$$

The enthalpy change for the reaction at standard conditions is +16.3 kJ. Predict the effect of the following changes on the position of the equilibrium; that is, state which way the equilibrium will shift (left, right, or no change) when each of the following changes is made.
(a) adding more $Br_2(g)$
(b) removing some NOBr(g)
(c) decreasing the temperature
(d) increasing the container volume

27. Consider the isomerization of butane with an equilibrium constant of $K = 2.5$. (See Study Question 13.) The system is originally at equilibrium with [butane] = 1.0 M and [isobutane] = 2.5 M.
(a) If 0.50 mol/L of isobutane is suddenly added and the system shifts to a new equilibrium position, what is the equilibrium concentration of each gas?
(b) If 0.50 mol/L of butane is added and the system shifts to a new equilibrium position, what is the equilibrium concentration of each gas?

▲ More challenging   ■ In General ChemistryNow   Blue-numbered questions answered in Appendix O

28. ■ The decomposition of NH_4HS

$$NH_4HS(s) \rightleftharpoons NH_3(g) + H_2S(g)$$

is an endothermic process. Using Le Chatelier's principle, explain how increasing the temperature would affect the equilibrium. If more NH_4HS is added to a flask in which this equilibrium exists, how is the equilibrium affected? What if some additional NH_3 is placed in the flask? What will happen to the pressure of NH_3 if some H_2S is removed from the flask?

General Questions

These questions are not designated as to type or location in the chapter. They may combine several concepts.

29. Suppose 0.086 mol of Br_2 is placed in a 1.26-L flask and heated to 1756 K, a temperature at which the halogen dissociates to atoms

$$Br_2(g) \rightleftharpoons 2\ Br(g)$$

If Br_2 is 3.7% dissociated at this temperature, calculate K.

30. The equilibrium constant for the reaction

$$N_2(g) + O_2(g) \rightleftharpoons 2\ NO(g)$$

is 1.7×10^{-3} at 2300 K.

(a) What is K for the reaction when written as follows?

$$\tfrac{1}{2}\ N_2(g) + \tfrac{1}{2}\ O_2(g) \rightleftharpoons NO(g)$$

(b) What is K for the following reaction?

$$2\ NO(g) \rightleftharpoons N_2(g) + O_2(g)$$

31. K_p for the formation of phosgene, $COCl_2$, is 6.5×10^{11} at 25 °C.

$$CO(g) + Cl_2(g) \rightleftharpoons COCl_2(g)$$

What is the value of K_p for the dissociation of phosgene?

$$COCl_2(g) \rightleftharpoons CO(g) + Cl_2(g)$$

32. ■ The equilibrium constant, K_c, for the following reaction is 1.05 at 350 K.

$$2\ CH_2Cl_2(g) \rightleftharpoons CH_4(g) + CCl_4(g)$$

If an equilibrium mixture of the three gases at 350 K contains 0.0206 M $CH_2Cl_2(g)$ and 0.0163 M CH_4, what is the equilibrium concentration of CCl_4?

33. ■ Carbon tetrachloride can be produced by the following reaction:

$$CS_2(g) + 3\ Cl_2(g) \rightleftharpoons S_2Cl_2(g) + CCl_4(g)$$

Suppose 1.2 mol of CS_2 and 3.6 mol of Cl_2 were placed in a 1.00-L flask. After equilibrium has been achieved, the mixture contains 0.90 mol CCl_4. Calculate K.

34. ■ Equal numbers of moles of H_2 gas and I_2 vapor are mixed in a flask and heated to 700 °C. The initial concentration of each gas is 0.0088 mol/L, and 78.6% of the I_2 is consumed when equilibrium is achieved according to the equation

$$H_2(g) + I_2(g) \rightleftharpoons 2\ HI(g)$$

Calculate K for this reaction.

35. The equilibrium constant for the butane $\rightleftharpoons$ isobutane isomerization reaction is 2.5 at 25 °C. If 1.75 mol of butane and 1.25 mol of isobutane are mixed, is the system at equilibrium? If not, when it proceeds to equilibrium, which reagent increases in concentration? Calculate the concentrations of the two compounds when the system reaches equilibrium.

36. At 2300 K the equilibrium constant for the formation of $NO(g)$ is 1.7×10^{-3}.

$$N_2(g) + O_2(g) \rightleftharpoons 2\ NO(g)$$

(a) Analysis shows that the concentrations of N_2 and O_2 are both 0.25 M, and that of NO is 0.0042 M under certain conditions. Is the system at equilibrium?

(b) If the system is not at equilibrium, in which direction does the reaction proceed?

(c) When the system is at equilibrium, what are the equilibrium concentrations?

37. Which of the following correctly relates the two equilibrium constants for the two reactions shown?

$$NOCl(g) \rightleftharpoons NO(g) + \tfrac{1}{2}\ Cl_2(g) \qquad K_1$$
$$2\ NO(g) + Cl_2(g) \rightleftharpoons 2\ NOCl(g) \qquad K_2$$

(a) $K_2 = -K_1^2$ (c) $K_2 = 1/K_1^2$

(b) $K_2 = 1/(K_1)^{1/2}$ (d) $K_2 = 2K_1$

38. Sulfur dioxide is readily oxidized to sulfur trioxide.

$$2\ SO_2(g) + O_2(g) \rightleftharpoons 2\ SO_3(g) \qquad K = 279$$

If we add 3.00 g of SO_2 and 5.00 g of O_2 to a 1.0-L flask, approximately what quantity of SO_3 will be in the flask once the reactants and the product reach equilibrium?

(a) 2.21 g   (c) 3.61 g

(b) 4.56 g   (d) 8.00 g

(Note: The full solution to this problem results in a cubic equation. Do not try to solve it exactly. Decide only which of the answers is most reasonable.)

39. ■ Heating a metal carbonate leads to decomposition.

$$BaCO_3(s) \rightleftharpoons BaO(s) + CO_2(g)$$

Predict the effect on the equilibrium of each change listed below. Answer by choosing (i) no change, (ii) shifts left, or (iii) shifts right.

(a) add $BaCO_3$ (c) add BaO

(b) add CO_2 (d) raise the temperature

(e) increase the volume of the flask containing the reaction

40. Carbonyl bromide decomposes to carbon monoxide and bromine.

$$COBr_2(g) \rightleftharpoons CO(g) + Br_2(g)$$

K is 0.190 at 73 °C. Suppose you placed 0.500 mol of $COBr_2$ in a 2.00-L flask and heated it to 73 °C (Study Question 17). After equilibrium had been achieved, you added an additional 2.00 mol of CO.

(a) How is the equilibrium mixture affected by adding more CO?

(b) When equilibrium is reestablished, what are the new equilibrium concentrations of $COBr_2$, CO, and Br_2?

(c) How has the addition of CO affected the percentage of $COBr_2$ that decomposed?

41. Phosphorus pentachloride decomposes at higher temperatures.

$$PCl_5(g) \rightleftharpoons PCl_3(g) + Cl_2(g)$$

An equilibrium mixture at some temperature consists of 3.120 g of PCl_5, 3.845 g of PCl_3, and 1.787 g of Cl_2 in a 1.00-L flask. If you add 1.418 g of Cl_2, how will the equilibrium be affected? What will the concentrations of PCl_5, PCl_3, and Cl_2 be when equilibrium is reestablished?

42. ■ Ammonium hydrogen sulfide decomposes on heating.

$$NH_4HS(s) \rightleftharpoons NH_3(g) + H_2S(g)$$

If K_p for this reaction is 0.11 at 25 °C (when the partial pressures are measured in atmospheres), what is the total pressure in the flask at equilibrium?

43. Ammonium iodide dissociates reversibly to ammonia and hydrogen iodide if the salt is heated to a sufficiently high temperature.

$$NH_4I(s) \rightleftharpoons NH_3(g) + HI(g)$$

Some ammonium iodide is placed in a flask, which is then heated to 400 °C. If the total pressure in the flask when equilibrium has been achieved is 705 mm Hg, what is the value of K_p (when partial pressures are in atmospheres)?

44. ■ When solid ammonium carbamate sublimes, it dissociates completely into ammonia and carbon dioxide according to the following equation:

$$(NH_4)(H_2NCO_2)(s) \rightleftharpoons 2 NH_3(g) + CO_2(g)$$

At 25 °C, experiment shows that the total pressure of the gases in equilibrium with the solid is 0.116 atm. What is the equilibrium constant, K_p?

45. The equilibrium constant, K_p, for $N_2O_4(g) \rightleftharpoons 2 NO_2(g)$ is 0.15 at 25 °C. If the pressure of N_2O_4 at equilibrium is 0.85 atm, what is the total pressure of the gas mixture ($N_2O_4 + NO_2$) at equilibrium?

46. In the gas phase, acetic acid exists as an equilibrium of monomer and dimer molecules. (The dimer consists of two molecules linked through hydrogen bonds.)

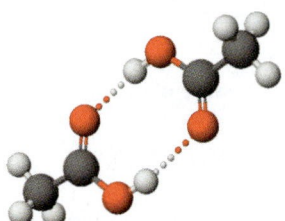

The equilibrium constant, K, at 25 °C for the monomer–dimer equilibrium

$$2 CH_3CO_2H \rightleftharpoons (CH_3CO_2H)_2$$

has been determined to be 3.2×10^4. Assume that acetic acid is present initially at a concentration of 5.4×10^{-4} mol/L at 25 °C and that no dimer is present initially.

(a) What percentage of the acetic acid is converted to dimer?

(b) As the temperature increases, in which direction does the equilibrium shift? (Recall that hydrogen-bond formation is an exothermic process.)

47. ■ At 450 °C, 3.60 mol of ammonia is placed in a 2.00-L vessel and allowed to decompose to the elements.

$$2 NH_3(g) \rightleftharpoons N_2(g) + 3 H_2(g)$$

If the experimental value of K is 6.3 for this reaction at this temperature, calculate the equilibrium concentration of each reagent. What is the total pressure in the flask?

48. The total pressure for a mixture of N_2O_4 and NO_2 is 1.5 atm. If $K_p = 6.75$ (at 25 °C), calculate the partial pressure of each gas in the mixture.

$$2 NO_2(g) \rightleftharpoons N_2O_4(g)$$

49. K_c for the decomposition of ammonium hydrogen sulfide is 1.8×10^{-4} at 25 °C.

$$NH_4HS(s) \rightleftharpoons NH_3(g) + H_2S(g)$$

(a) When the pure salt decomposes in a flask, what are the equilibrium concentrations of NH_3 and H_2S?

(b) If NH_4HS is placed in a flask already containing 0.020 mol/L of NH_3 and then the system is allowed to come to equilibrium, what are the equilibrium concentrations of NH_3 and H_2S?

50. ▲ A reaction important in smog formation is

$$O_3(g) + NO(g) \rightleftharpoons O_2(g) + NO_2(g) \qquad K = 6.0 \times 10^{34}$$

(a) If the initial concentrations are $[O_3] = 1.0 \times 10^{-6}$ M, $[NO] = 1.0 \times 10^{-5}$ M, $[NO_2] = 2.5 \times 10^{-4}$ M, and $[O_2] = 8.2 \times 10^{-3}$ M, is the system at equilibrium? If not, in which direction does the reaction proceed?

(b) If the temperature is increased, as on a very warm day, will the concentrations of the products increase or decrease? (*Hint:* You may have to calculate the enthalpy change for the reaction to find out if it is exothermic or endothermic.)

51. The equilibrium reaction $N_2O_4(g) \rightleftharpoons 2 NO_2(g)$ has been thoroughly studied (see Figures 16.6 and 16.8).

(a) If the total pressure in a flask containing NO_2 and N_2O_4 gas at 25 °C is 1.50 atm, and the value of K_p at this temperature is 0.148, what fraction of the N_2O_4 has dissociated to NO_2?

(b) What happens to the fraction dissociated if the volume of the container is increased so that the total equilibrium pressure falls to 1.00 atm?

52. ▲ Lanthanum oxalate decomposes when heated to lanthanum oxide, CO, and CO_2.

$$La_2(C_2O_4)_3(s) \rightleftharpoons La_2O_3(s) + 3 CO(g) + 3 CO_2(g)$$

(a) If, at equilibrium, the total pressure in a 10.0-L flask is 0.200 atm, what is the value of K_p?

(b) Suppose 0.100 mol of $La_2(C_2O_4)_3$ was originally placed in the 10.0-L flask. What quantity of $La_2(C_2O_4)_3$ remains unreacted at equilibrium?

53. ▲ The ammonia complex of trimethylborane, $(NH_3)B(CH_3)_3$, dissociates at 100 °C to its components with $K_p = 4.62$ (when the pressures are in atmospheres).

$$(NH_3)B(CH_3)_3(g) \rightleftharpoons B(CH_3)_3(g) + NH_3(g)$$

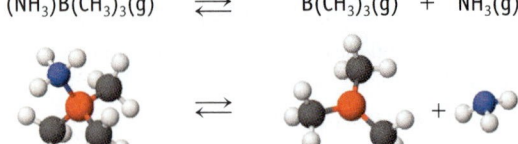

If NH_3 is changed to some other molecule, the equilibrium constant is different.

For $[(CH_3)_3P]B(CH_3)_3$ $K_p = 0.128$

For $[(CH_3)_3N]B(CH_3)_3$ $K_p = 0.472$

For $(NH_3)B(CH_3)_3$ $K_p = 4.62$

(a) If you begin an experiment by placing 0.010 mol of each complex in a flask, which would have the largest partial pressure of $B(CH_3)_3$ at 100 °C?

(b) If 0.73 g (0.010 mol) of $(NH_3)B(CH_3)_3$ is placed in a 100.-mL flask and heated to 100 °C, what is the partial pressure of each gas in the equilibrium mixture and what is the total pressure? What is the percent dissociation of $(NH_3)B(CH_3)_3$?

54. Sulfuryl chloride, SO_2Cl_2, is a compound with very irritating vapors; it is used as a reagent in the synthesis of organic compounds. When heated to a sufficiently high temperature it decomposes to SO_2 and Cl_2.

$$SO_2Cl_2(g) \rightleftharpoons SO_2(g) + Cl_2(g) \qquad K = 0.045 \text{ at } 375 °C$$

(a) Suppose 6.70 g of SO_2Cl_2 is placed in a 1.00-L flask and then heated to 375 °C. What is the concentration of each of the compounds in the system when equilibrium is achieved? What fraction of SO_2Cl_2 has dissociated?

(b) What are the concentrations of SO_2Cl_2, SO_2, and Cl_2 at equilibrium in the 1.00-L flask at 375 °C if you begin with a mixture of SO_2Cl_2 (6.70 g) and Cl_2 (1.00 atm)? What fraction of SO_2Cl_2 has dissociated?

(c) Compare the fractions of SO_2Cl_2 in parts (a) and (b). Do they agree with your expectations based on Le Chatelier's principle?

55. Hemoglobin (Hb) can form a complex with both O_2 and CO. For the reaction

$$HbO_2(aq) + CO(g) \rightleftharpoons HbCO(aq) + O_2(g)$$

at body temperature, K is about 200. If the ratio $[HbCO]/[HbO_2]$ comes close to 1, death is probable. What partial pressure of CO in the air is likely to be fatal? Assume the partial pressure of O_2 is 0.20 atm.

56. ■ ▲ Limestone decomposes at high temperatures.

$$CaCO_3(s) \rightleftharpoons CaO(s) + CO_2(g)$$

At 1000 °C, $K_p = 3.87$. If pure $CaCO_3$ is placed in a 5.00-L flask and heated to 1000 °C, what quantity of $CaCO_3$ must decompose to achieve the equilibrium pressure of CO_2?

57. At 1800 K, oxygen dissociates very slightly into its atoms.

$$O_2(g) \rightleftharpoons 2\,O(g) \qquad K_p = 1.2 \times 10^{-10}$$

If you place 1.0 mol of O_2 in a 10.-L vessel and heat it to 1800 K, how many O atoms are present in the flask?

58. ▲ Nitrosyl bromide, NOBr, is prepared by the direct reaction of NO and Br_2.

$$2\,NO(g) + Br_2(g) \longrightarrow 2\,NOBr(g)$$

The compound dissociates readily at room temperature, however.

$$NOBr(g) \rightleftharpoons NO(g) + \tfrac{1}{2}\,Br_2(g)$$

Some NOBr is placed in a flask at 25 °C and allowed to dissociate. The total pressure at equilibrium is 190 mm Hg and the compound is found to be 34% dissociated. What is the value of K_p?

59. ■ ▲ Boric acid and glycerin form a complex

$$B(OH)_3(aq) + glycerin(aq) \rightleftharpoons B(OH)_3 \cdot glycerin(aq)$$

with an equilibrium constant of 0.90. If the concentration of boric acid is 0.10 M, how much glycerin should be added, per liter, so that 60.% of the boric acid is in the form of the complex?

60. ▲ The dissociation of calcium carbonate has an equilibrium constant of $K_p = 1.16$ at 800 °C.

$$CaCO_3(s) \rightleftharpoons CaO(s) + CO_2(g)$$

(a) What is K_c for the reaction?

(b) If you place 22.5 g of $CaCO_3$ in a 9.56-L container at 800 °C, what is the pressure of CO_2 in the container?

(c) What percentage of the original 22.5-g sample of $CaCO_3$ remains undecomposed at equilibrium?

61. ▲ A sample of N_2O_4 gas with a pressure of 1.00 atm is placed in a flask. When equilibrium is achieved, 20.0% of the N_2O_4 has been converted to NO_2 gas.

(a) ■ Calculate K_p.

(b) If the original pressure of N_2O_4 is 0.10 atm, what is the percent dissociation of the gas? Is the result in agreement with Le Chatelier's principle?

62. ■ ▲ The equilibrium constant, K_p, is 0.15 at 25 °C for the following reaction:

$$N_2O_4(g) \rightleftharpoons 2\,NO_2(g)$$

If the total pressure of the gas mixture is 2.5 atm at equilibrium, what is the partial pressure of each gas?

Summary and Conceptual Questions

The following questions may use concepts from preceding chapters.

63. Decide whether each of the following statements is true or false. If false, change the wording to make it true.

(a) The magnitude of the equilibrium constant is always independent of temperature.

(b) When two chemical equations are added to give a net equation, the equilibrium constant for the net equation is the product of the equilibrium constants of the summed equations.

(c) The equilibrium constant for a reaction has the same value as K for the reverse reaction.

(d) Only the concentration of CO_2 appears in the equilibrium constant expression for the reaction
$CaCO_3(s) \rightleftharpoons CaO(s) + CO_2(g)$.

(e) For the reaction $CaCO_3(s) \rightleftharpoons CaO(s) + CO_2(g)$, the value of K is numerically the same no matter whether the amount of CO_2 is expressed as moles/liter or as gas pressure.

64. Neither $PbCl_2$ nor PbF_2 is appreciably soluble in water. If solid $PbCl_2$ and solid PbF_2 are placed in equal amounts of water in separate beakers, in which beaker is the concentration of Pb^{2+} greater? Equilibrium constants for these solids dissolving in water are as follows:

$PbCl_2(s) \rightleftharpoons Pb^{2+}(aq) + 2\ Cl^-(aq)$ $K = 1.7 \times 10^{-5}$
$PbF_2(s) \rightleftharpoons Pb^{2+}(aq) + 2\ F^-(aq)$ $K = 3.7 \times 10^{-8}$

65. Characterize each of the following as product- or reactant-favored.

(a) $CO(g) + \frac{1}{2} O_2(g) \rightleftharpoons CO_2(g)$ $K_p = 1.2 \times 10^{45}$
(b) $H_2O(g) \rightleftharpoons H_2(g) + \frac{1}{2} O_2(g)$ $K_p = 9.1 \times 10^{-41}$
(c) $CO(g) + Cl_2(g) \rightleftharpoons COCl_2(g)$ $K_p = 6.5 \times 10^{11}$

66. A sample of liquid water is sealed in a container. Over time some of the liquid evaporates, and equilibrium is reached eventually. At this point you can measure the equilibrium vapor pressure of the water. Is the process $H_2O(g) \rightleftharpoons H_2O(\ell)$ a dynamic equilibrium? Explain the changes that take place in reaching equilibrium in terms of the rates of the competing processes of evaporation and condensation.

67. ▲ The reaction of hydrogen and iodine to give hydrogen iodide has an equilibrium constant, K_c, of 56 at 435 °C.

(a) What is the value of K_p?
(b) Suppose you mix 0.45 mol of H_2 and 0.45 mol of I_2 in a 10.0-L flask at 425 °C. What is the total pressure of the mixture before and after equilibrium is achieved?
(c) What is the partial pressure of each gas at equilibrium?

68. An ice cube is placed in a beaker of water at 20 °C. The ice cube partially melts, and the temperature of the water is lowered to 0 °C. At this point, both ice and water are at 0 °C, and no further change is apparent. Is the system at equilibrium? Is this a dynamic equilibrium? That is, are events still occurring at the molecular level? Suggest an experiment to test whether this is so. (*Hint:* Consider using D_2O.)

69. The photo shows the result of heating and cooling a solution of Co^{2+} ions in water containing hydrochloric acid. The equation for the equilibrium existing in these solutions is

$Co(H_2O)_6^{2+}(aq) + 4\ Cl^-(aq) \rightleftharpoons CoCl_4^{2-}(aq) + 6\ H_2O(\ell)$

The $Co(H_2O)_6^{2+}$ ion is pink, whereas the $CoCl_4^{2-}$ ion is blue. Is the transformation of $Co(H_2O)_6^{2+}$ to $CoCl_4^{2-}$ exothermic or endothermic?

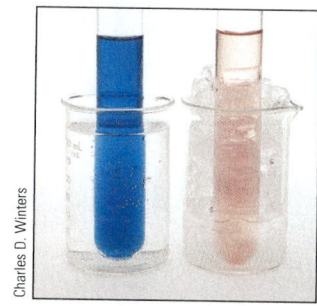

Charles D. Winters

Solution of Co^{2+} ion in water containing HCl(aq). The solution at the left is in a beaker of hot water, whereas the solution at the right is in a beaker of ice water.

70. Consider a gas-phase reaction where a colorless compound C produces a blue compound B.

$$2\ C \rightleftharpoons B$$

After reaching equilibrium, the size of the flask is halved.

(a) What color change (if any) is observed immediately upon halving the flask size?
(b) What color change (if any) is observed after equilibrium has been reestablished in the flask?

71. The "principle of microscopic reversibility" is a useful concept when describing chemical equilibria. To learn more about it, See the General ChemistryNow CD-ROM or website Screen 16.2. How is this principle illustrated by the following equilibrium?

$$PbCl_2(s) \rightleftharpoons Pb^{2+}(aq) + 2\ Cl^-(aq)$$

72. See the simulation on General ChemistryNow CD-ROM or website Screen 16.4.

(a) Set the concentration of Fe^{3+} at 0.0050 M and that of SCN^- at 0.0070 M. Click the "React" button. Does the concentration of Fe^{3+} go to zero? When equilibrium is reached, what are the concentrations of the reactants and the products? What is the equilibrium constant?
(b) Begin with $[Fe^{3+}] = [SCN^-] = 0.0$ M and $[FeSCN^{2+}] = 0.0080$ M. When equilibrium is reached, which ion has the largest concentration in solution?
(c) Begin with $[Fe^{3+}] = 0.0010$ M, $[SCN^-] = 0.0020$ M, and $[FeSCN^{2+}] = 0.0030$ M. Describe the result of allowing this system to come to equilibrium.

Do you need a live tutor for homework problems?
Access vMentor at General ChemistryNow at
http://now.brookscole.com/kotz6e
for one-on-one tutoring from a chemistry expert

17—Principles of Reactivity: The Chemistry of Acids and Bases

Rhubarb, a source of many organic acids.
The leaves and stalks of rhubarb are the source of acids such as oxalic, citric, acetic, and succinic, among others.

David Turnley/2002 Corbis Images

Nature's Acids

Many people grow rhubarb in their gardens because the stalks of the plant, when stewed with sugar, make a wonderful dessert or filling for a pie or tart. But the leaves can make us sick! Why?

Rhubarb leaves are a source of oxalic acid, $H_2C_2O_4$, an organic acid.

$$H_2O(\ell) + H_2C_2O_4(aq) \rightleftharpoons H_3O^+(aq) + HC_2O_4^-(aq)$$

$$H_2O(\ell) + HC_2O_4^-(aq) \rightleftharpoons H_3O^+(aq) + C_2O_4^{2-}(aq)$$

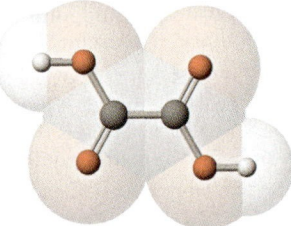

Oxalic acid, $H_2C_2O_4$

Rhubarb leaves contain a very large amount of this acid, 3000–11,000 parts per million. The problem with ingesting oxalic acid is that it interferes with essential elements in the body such as iron, magnesium, and especially calcium. The Ca^{2+} ion and oxalic acid react to form insoluble calcium oxalate, CaC_2O_4.

$$Ca^{2+}(aq) + H_2C_2O_4(aq) \longrightarrow CaC_2O_4(s) + 2\ H^+(aq)$$

Not only does this reaction effectively remove calcium ions from the body, but calcium oxalate crystals can also grow into painful kidney and bladder stones. For this reason, people who are susceptible to kidney stones are put on a diet that is low in oxalic acid. Such people also have to be careful not to take too much vitamin C, a compound that can be turned into oxalic acid in the body. A few people have died from accidentally drinking antifreeze because the ethylene glycol in the antifreeze is converted to oxalic acid in the body. Symptoms of oxalic acid poisoning include nausea, vomiting, abdominal pain, and hemorrhaging.

Oxalic acid is found in the stems and leaves of many plants other than rhubarb, such as cabbage, spinach, and beets. Because it occurs in so many other edible substances, including cocoa, peanuts, and tea, the average person consumes about 150 mg of oxalic acid per day. But will it kill you? For a person weighing about 145 pounds (65.8 kg), the lethal dose is approximately 24 g of pure oxalic acid. You would have to eat a field of rhubarb leaves or drink an ocean of tea to come close to ingesting a fatal dose of oxalic acid. What would happen first, however, is that you might develop severe diarrhea. Your gut knows oxalic acid is a natural toxin and is stimulated to get rid of it.

Despite the minor health risk from eating too much rhubarb, this plant has been cultivated for thousands of years for its healthful properties. In particular, Chinese herbalists have used rhubarb in traditional medicine for centuries. Indeed, it was considered so important that emperors of China in the 18th and 19th centuries forbade its export. Rhubarb was also cultivated in Russia and later in England. It made its first appearance in the United States in about 1800.

Charles D. Winters

Calcium oxalate. Calcium oxalate precipitates when solutions of calcium chloride and the organic acid oxalic acid are mixed.

To Review Before You Begin

- Review the pH scale in Section 5.9
- Review the discussion of acids and bases and their chemistry in Chapter 5 (pages 185–194)
- Review the principles of chemical equilibria in Chapter 16

GENERAL
Chemistry ⚛ Now™

Throughout the chapter this icon introduces a list of resources on the General ChemistryNow CD-ROM or website (http://now .brookscole.com/kotz6e) that will:

- help you evaluate your knowledge of the material
- provide homework problems
- allow you to take an exam-prep quiz
- provide a personalized Learning Plan targeting resources that address areas you should study

Acids and bases are among the most common substances in nature. Amino acids, for example, are the building blocks of proteins. The pH of lakes, rivers, and oceans is affected by dissolved acids and bases, and your bodily functions depend on acids and bases. This chapter and the next take up the detailed chemistry of these substances.

17.1—Acids, Bases, and the Equilibrium Concept

An acid was described in Chapter 5 as any substance that, when dissolved in water, increases the concentration of hydrogen ions, H^+ [◀ page 186]. A base was defined as any substance that increases the concentration of hydroxide ions, OH^-, when dissolved in water. Two other features of acids and bases were also introduced.

- Acids and bases can be divided roughly into those that are strong electrolytes (such as HCl, HNO_3, and NaOH) and those that are weak electrolytes (such as CH_3CO_2H and NH_3) [◀ Table 5.2, Common Acids and Bases, page 187].

- A H^+ ion—the nucleus of the hydrogen atom—cannot exist in water. When an acid is dissolved in water, the proton donated by the acid combines with water to produce the hydronium ion, H_3O^+, and similar ions [◀ "A Closer Look: H^+ Ions in Water," page 188].

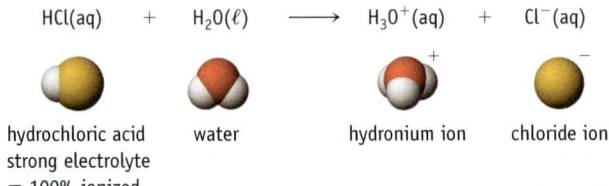

$$HCl(aq) \quad + \quad H_2O(\ell) \quad \longrightarrow \quad H_3O^+(aq) \quad + \quad Cl^-(aq)$$

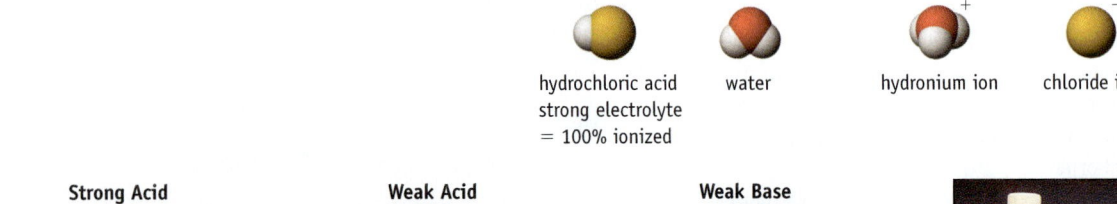

| hydrochloric acid strong electrolyte = 100% ionized | water | hydronium ion | chloride ion |

Strong Acid   **Weak Acid**   **Weak Base**

(a) HCl completely ionizes in aqueous solution.

(b) Acetic acid, CH_3CO_2H, ionizes only slightly in water.

(c) The weak base ammonia reacts to a small extent with water to give a weakly basic solution.

HCl CH_3CO_2H NH_3

Photo: Charles D. Winters

Figure 17.1 Acids and bases. (a) Hydrochloric acid, a strong acid, is sold for household use as "muriatic acid." The acid completely ionizes in water. (b) Vinegar is a solution of acetic acid, a weak acid that ionizes to only a small extent in water. (c) Ammonia is a weak base, ionizing to a small extent in water.

Now let us take a closer look at what is meant by a *strong* or *weak* electrolyte (Figure 17.1). Hydrochloric acid is a strong acid, so 100% of the acid ionizes to produce hydronium and chloride ions. In contrast, acetic acid and ammonia are weak electrolytes. They ionize to only a very small extent in water. For example, for acetic acid, the acid, its anion, and the hydronium ion are all present at equilibrium in solution, but the product ions are present in very low concentration relative to the acid concentration. This chapter describes the extent to which acids or bases ionize in terms of the equilibrium constant for the ionization process.

| acetic acid | water | acetate ion | hydronium ion |

$$K = \frac{[CH_3CO_2^-][H_3O^+]}{[CH_3CO_2H]} = 1.8 \times 10^{-5}$$

The equilibrium constants for the ionization of many weak acids and bases, often called **ionization constants,** are a measure of the extent to which these substances ionize in water. Thus, the ionization constants are a reflection of acid and base strength.

- Acids or bases that ionize extensively, with $K > 1$, are referred to as *strong acids or bases.*

- Acids or bases that do not ionize extensively, with $K < 1$, are referred to as *weak acids or bases.*

17.2—The Brønsted-Lowry Concept of Acids and Bases

In 1923, Johannes N. Brønsted (1879–1947) in Copenhagen, Denmark, and Thomas M. Lowry (1874–1936) in Cambridge, England, independently suggested a new concept of acid and base behavior. They proposed that an *acid* is any substance that can *donate a proton* to any other substance. **Brønsted acids** can be molecular compounds such as nitric acid,

$$HNO_3(aq) + H_2O(\ell) \longrightarrow NO_3^-(aq) + H_3O^+(aq)$$

Acid

■ **Brønsted-Lowry Theory**
This theory broadens the definition of acids and bases first given in Chapter 5 (pages 186–188). Note that the theory is not restricted to compounds in water.

cations such as NH_4^+,

$$NH_4^+(aq) + H_2O(\ell) \rightleftharpoons NH_3(aq) + H_3O^+(aq)$$
Acid

hydrated metal cations,

$$[Fe(H_2O)_6]^{3+}(aq) + H_2O(\ell) \rightleftharpoons [Fe(H_2O)_5(OH)]^{2+}(aq) + H_3O^+(aq)$$
Acid

or anions

$$H_2PO_4^-(aq) + H_2O(\ell) \rightleftharpoons HPO_4^{2-}(aq) + H_3O^+(aq)$$
Acid

According to Brønsted and Lowry, a **Brønsted base** is a substance that can *accept a proton* from any other substance. These can also be molecular compounds,

$$NH_3(aq) + H_2O(\ell) \rightleftharpoons NH_4^+(aq) + OH^-(aq)$$
Base

anions,

$$CO_3^{2-}(aq) + H_2O(\ell) \rightleftharpoons HCO_3^-(aq) + OH^-(aq)$$
Base

or cations

$$[Al(H_2O)_5(OH)]^{2+}(aq) + H_2O(\ell) \rightleftharpoons [Al(H_2O)_6]^{3+}(aq) + OH^-(aq)$$

A wide variety of Brønsted acids are known, and you are familiar with many of them [◀ Table 5.2, page 187]. Acids such as HF, HCl, HNO_3, and CH_3CO_2H (acetic acid) are all capable of donating one proton and so are called **monoprotic acids**. Other acids, called **polyprotic acids** (Table 17.1), are capable of donating two or more protons. An example is sulfuric acid.

$$H_2SO_4(aq) + H_2O(\ell) \longrightarrow HSO_4^-(aq) + H_3O^+(aq)$$

$$HSO_4^-(aq) + H_2O(\ell) \rightleftharpoons SO_4^{2-}(aq) + H_3O^+(aq)$$

Table 17.1   Polyprotic Acids and Bases

Acid Form	Amphiprotic Form	Base Form
H_2S (hydrosulfuric acid or hydrogen sulfide)	HS^- (hydrogen sulfide ion)	S^{2-} (sulfide ion)
H_3PO_4 (phosphoric acid)	$H_2PO_4^-$ (dihydrogen phosphate ion) HPO_4^{2-} (hydrogen phosphate ion)	PO_4^{3-} (phosphate ion)
H_2CO_3 (carbonic acid)	HCO_3^- (hydrogen carbonate ion or bicarbonate ion)	CO_3^{2-} (carbonate ion)
$H_2C_2O_4$ (oxalic acid)	$HC_2O_4^-$ (hydrogen oxalate ion)	$C_2O_4^{2-}$ (oxalate ion)

Just as there are acids that can donate more than one proton, so there are **polyprotic bases** that can accept more than one proton. The anions of polyprotic acids are polyprotic bases; examples include SO_4^{2-}, PO_4^{3-}, CO_3^{2-}, or $C_2O_4^{2-}$. This behavior is illustrated by the carbonate and bicarbonate ions.

$$\underset{\text{Base}}{CO_3^{2-}(aq)} + H_2O(\ell) \rightleftharpoons HCO_3^-(aq) + OH^-(aq)$$

$$\underset{\text{Base}}{HCO_3^-(aq)} + H_2O(\ell) \rightleftharpoons H_2CO_3(aq) + OH^-(aq)$$

Some molecules or ions can behave either as Brønsted acids or bases. These species are called **amphiprotic**, and one example is the hydrogen phosphate anion (see Table 17.1).

$$\underset{\text{Acid}}{HPO_4^{2-}(aq)} + H_2O(\ell) \rightleftharpoons H_3O^+(aq) + PO_4^{3-}(aq)$$

$$\underset{\text{Base}}{HPO_4^{2-}(aq)} + H_2O(\ell) \rightleftharpoons H_2PO_4^-(aq) + OH^-(aq)$$

There is a final, important point illustrated by the chemical equations written above: *Water is amphiprotic.* It can accept a proton to form H_3O^+,

$$\underset{\text{Base}}{H_2O(\ell)} + \underset{\text{Acid}}{HCl(aq)} \rightleftharpoons H_3O^+(aq) + Cl^-(aq)$$

or it can donate a proton to form the OH^- ion

$$\underset{\text{Acid}}{H_2O(\ell)} + \underset{\text{Base}}{NH_3(aq)} \rightleftharpoons NH_4^+(aq) + OH^-(aq)$$

Charles D. Winters

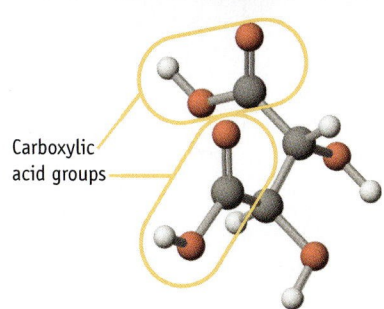

Carboxylic acid groups

Tartaric acid, $H_2C_4H_4O_6$, is a naturally occurring diprotic acid. Tartaric acid and its potassium salt are found in many fruits.

Exercise 17.1—Brønsted Acids and Bases

(a) Write a balanced equation for the reaction that occurs when H_3PO_4, phosphoric acid, donates a proton to water to form the dihydrogen phosphate ion. Is the dihydrogen phosphate ion an acid, a base, or amphiprotic?

(b) Write a balanced equation for the reaction that occurs when the cyanide ion, CN^-, accepts a proton from water to form HCN. Is CN^- a Brønsted acid or base?

Conjugate Acid–Base Pairs

In each of the chemical equations written so far, *a proton has been transferred* to or from water. For example, a reaction important in the control of acidity in biological systems involves the hydrogen carbonate ion, which can act as a Brønsted base (page 801) or acid in water.

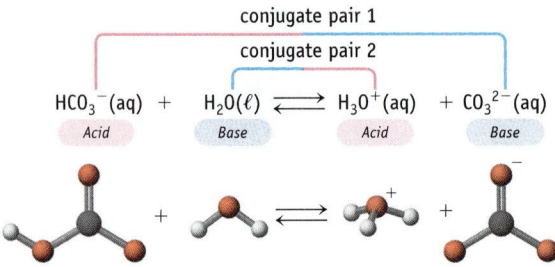

This equation for HCO_3^- as an acid exemplifies a feature of all reactions involving Brønsted acids and bases. The HCO_3^- and CO_3^{2-} ions are related by the loss or gain of H^+, as are H_2O and H_3O^+. A pair of compounds or ions that differ by the presence of one H^+ ion is called a **conjugate acid–base pair**. We say that HCO_3^- is the conjugate acid of the base CO_3^{2-} or that CO_3^{2-} is the conjugate base of the acid HCO_3^-. *Every reaction between a Brønsted acid and Brønsted base involves H^+ transfer and has two conjugate acid–base pairs.* To convince yourself of this fact, look at the reactions above and those in Table 17.2.

GENERAL
Chemistry·⚛·Now™

See the General ChemistryNow CD-ROM or website:
- **Screen 17.2 Brønsted Acids and Bases,** for an exercise and tutorial on acids, bases, and their conjugates

---

Exercise 17.2—Conjugate Acids and Bases

In the following reaction, identify the acid on the left and its conjugate base on the right. Similarly, identify the base on the left and its conjugate acid on the right.

$$HNO_3(aq) + NH_3(aq) \rightleftharpoons NH_4^+(aq) + NO_3^-(aq)$$

17.3—Water and the pH Scale

The properties of water are a recurring topic in this book [◀ Sections 5.1 and 13.3]. Because we generally use aqueous solutions of acids and bases, and because the acid–base reactions in your body occur in your aqueous interior, we come again to the behavior of water.

Table 17.2 Conjugate Acid–Base Pairs*

Name	Acid 1		Base 2			Base 1		Acid 2
Hydrochloric acid	HCl	+	H_2O	$\longrightarrow$		Cl^-	+	H_3O^+
Nitric acid	HNO_3	+	H_2O	$\longrightarrow$		NO_3^-	+	H_3O^+
Hydrogen carbonate	HCO_3^-	+	H_2O	$\rightleftharpoons$		CO_3^{2-}	+	H_3O^+
Acetic acid	CH_3CO_2H	+	H_2O	$\rightleftharpoons$		$CH_3CO_2^-$	+	H_3O^+
Hydrocyanic acid	HCN	+	H_2O	$\rightleftharpoons$		CN^-	+	H_3O^+
Hydrogen sulfide	H_2S	+	H_2O	$\rightleftharpoons$		HS^-	+	H_3O^+
Ammonia	H_2O	+	NH_3	$\rightleftharpoons$		OH^-	+	NH_4^+
Carbonate ion	H_2O	+	CO_3^{2-}	$\rightleftharpoons$		OH^-	+	HCO_3^-
Water	H_2O	+	H_2O	$\rightleftharpoons$		OH^-	+	H_3O^+

*Acid 1 and base 1 are a conjugate pair, as are base 2 and acid 2.

Water Autoionization and the Water Ionization Constant, K_w

An acid such as HCl does not need to be present for the hydronium ion to exist in water. In fact, two water molecules interact with each other to produce a hydronium ion and a hydroxide ion by proton transfer from one water molecule to the other.

$$2\ H_2O(\ell) \rightleftharpoons H_3O^+(aq)\ +\ OH^-(aq)$$

This **autoionization** reaction of water was demonstrated many years ago by Friedrich Kohlrausch (1840–1910). He found that, even after water is painstakingly purified, it still conducts electricity to a very small extent because autoionization produces very low concentrations of H_3O^+ and OH^- ions. Water autoionization is the cornerstone of our concepts of aqueous acid–base behavior.

When water autoionizes, the equilibrium lies far to the left side. In fact, in pure water at 25 °C only about two out of a billion (10^9) water molecules are ionized at any instant. To express this idea more quantitatively, we can write the equilibrium constant expression for autoionization.

$$K = \frac{[H_3O^+][OH^-]}{[H_2O]^2}$$

Recall that in pure water or in dilute aqueous solutions [◀ Section 16.2], the concentration of water can be considered to be a constant (55.5 M). For this reason

$[H_2O]^2$ is included in the constant K and the equilibrium constant expression becomes

$$K[H_2O]^2 = [H_3O^+][OH^-] = K_w$$

This equilibrium constant is given a special symbol, K_w, and is known as the **ionization constant for water**. In pure water, the transfer of a proton between two water molecules leads to one H_3O^+ ion and one OH^- ion. Because this is the only source of these ions in pure water, we know that $[H_3O^+]$ must equal $[OH^-]$. Electrical conductivity measurements of pure water show that $[H_3O^+] = [OH^-] = 1.0 \times 10^{-7}$ M at 25 °C, so K_w has a value of 1.0×10^{-14} at 25 °C.

$$K_w = [H_3O^+][OH^-] = 1.0 \times 10^{-14} \text{ at 25 °C} \tag{17.1}$$

In pure water the hydronium ion and hydroxide ion concentrations are equal and the water is said to be *neutral*. If some acid or base is added to pure water, however, the equilibrium

$$2 H_2O(\ell) \rightleftharpoons H_3O^+(aq) + OH^-(aq)$$

is disturbed. Adding acid raises the concentration of the H_3O^+ ions, so the solution is *acidic*. To oppose this increase, Le Chatelier's principle [◄ Section 16.6] predicts that a small fraction of the H_3O^+ ions will react with OH^- ions from water autoionization to form water. This lowers $[OH^-]$ until the product of $[H_3O^+]$ and $[OH^-]$ is again equal to 1.0×10^{-14} at 25 °C. Similarly, adding a base to pure water gives a *basic* solution because the OH^- ion concentration has increased. Le Chatelier's principle predicts that some of the added OH^- ions will react with H_3O^+ ions present in the solution from water autoionization, thereby lowering $[H_3O^+]$ until the value of the product $[H_3O^+] \times [OH^-]$ equals 1.0×10^{-14} at 25 °C.

Thus, for aqueous solutions at 25 °C, we can say that

- In a **neutral** solution, $[H_3O^+] = [OH^-]$.
  Both are equal to 1.0×10^{-7} M.

- In an **acidic** solution, $[H_3O^+] > [OH^-]$.
  $[H_3O^+] > 1.0 \times 10^{-7}$ M and $[OH^-] < 1.0 \times 10^{-7}$ M.

- In a **basic** solution, $[H_3O^+] < [OH^-]$.
  $[H_3O^+] < 1.0 \times 10^{-7}$ M and $[OH^-] > 1.0 \times 10^{-7}$ M

■ **K_w and Temperature**
The equation $K_w = [H_3O^+][OH^-]$ is valid for pure water and for any aqueous solution. K_w is temperature dependent because the autoionization reaction is endothermic. K_w increases with temperature.

°C	K_w
10	0.29×10^{-14}
15	0.45×10^{-14}
20	0.68×10^{-14}
25	1.01×10^{-14}
30	1.47×10^{-14}
50	5.48×10^{-14}

GENERAL
Chemistry⋅❀⋅Now™

See the General ChemistryNow CD-ROM or website:
- **Screen 17.3 The Acid–Base Properties of Water,** for a simulation of the effect of temperature on K_w

Example 17.1—Ion Concentrations in a Solution of a Strong Base

Problem What are the hydroxide and hydronium ion concentrations in a 0.0012 M solution of NaOH at 25 °C?

Strategy NaOH, a strong base, is 100% dissociated into ions in water, so we assume that the OH^- ion concentration is the same as the NaOH concentration. The H_3O^+ ion concentration can then be calculated using Equation 17.1.

Solution The initial concentration of OH^- is 0.0012 M.

$$0.0012 \text{ mol NaOH per liter} \longrightarrow 0.0012 \text{ M Na}^+(aq) + 0.0012 \text{ M OH}^-(aq)$$

Substituting the OH^- concentration into Equation 17.1, we have

$$K_w = 1.0 \times 10^{-14} = [H_3O^+][OH^-] = [H_3O^+](0.0012)$$

and so

$$[H_3O^+] = \frac{1.0 \times 10^{-14}}{0.0012} = \boxed{8.3 \times 10^{-12} \text{ M}}$$

Comment Why didn't we take into account the ions produced by water autoionization? It is expected to add OH^- and H_3O^+ ions to the solution (with concentrations of x mol/L). When equilibrium is achieved in this case, it means that the H_3O^+ concentration is x, and

$$[OH^-] = (0.0012 \text{ M} + OH^- \text{ from water autoionization})$$

$$[OH^-] = (0.0012 \text{ M} + x)$$

In pure water, the amount of OH^- ion generated is 1.0×10^{-7} M. Le Chatelier's principle [◀ Section 16.6] suggests that the amount should be even smaller when OH^- ions are already present in solution from NaOH; that is, x should be $\ll 1.0 \times 10^{-7}$ M. This means x in the term $(0.0012 + x)$ is insignificant compared with 0.0012. (Following the rules for significant figures, the sum of 0.0012 and a number even smaller than 1.0×10^{-7} is 0.0012.) Thus, the equilibrium concentration of OH^- is just equivalent to the quantity of NaOH added.

What about the Na^+ ion? As described later (see page 810), alkali metal ions have no effect on the acidity or basicity of a solution.

Exercise 17.3—Hydronium Ion Concentration in a Solution of a Strong Acid

A solution of the strong acid HCl has $[HCl] = 4.0 \times 10^{-3}$ M. What are the concentrations of H_3O^+ and OH^- in this solution at 25 °C? (Recall that because HCl is a strong acid, it is 100% ionized in water.)

The pH Scale

The **pH** of a solution is defined as the negative of the base-10 logarithm (log) of the hydronium ion concentration [◀ Section 5.9, page 212].

$$pH = -\log [H_3O^+] \qquad \text{(5.2 and 17.2)}$$

In a similar way, we can now define the pOH of a solution as the negative of the base-10 logarithm of the hydroxide ion concentration.

$$pOH = -\log[OH^-] \qquad \text{(17.3)}$$

■ **The pK Scale**
In general, $-\log X = pX$, so $-\log K = pK$
$-\log[H_3O^+] = pH$
$-\log[OH^-] = pOH$

In pure water, the hydronium and hydroxide ion concentrations are both 1.0×10^{-7} M. Therefore, for pure water at 25 °C

$$pH = -\log (1.0 \times 10^{-7}) = 7.00$$

In the same way, you can show that the pOH of pure water is also 7.00 at 25 °C.

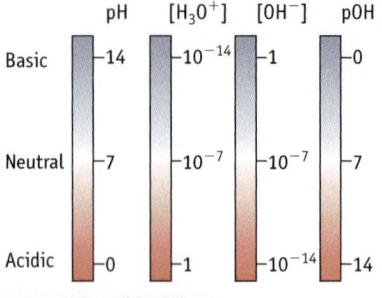

Active Figure 17.2   **pH and pOH.**
This figure shows the relationship of hydronium ion and hydroxide ion concentrations and of pH and pOH.

GENERAL
Chemistry ⚛ **Now**™ See the General ChemistryNow CD-ROM or website to explore an interactive version of this figure accompanied by an exercise.

■ **pH Calculations**
Because we make pH measurements to determine solution H_3O^+ and OH^- concentrations, it is useful to be able to convert experimental pH readings to concentrations. Review Example 5.11, pH of Solutions, and check yourself with Exercise 17.4.

If we take the negative logarithms of both sides of the expression $K_w = [H_3O^+][OH^-]$, we obtain another useful equation.

$$K_w = 1.0 \times 10^{-14} = [H_3O^+][OH^-]$$
$$-\log K_w = -\log(1.0 \times 10^{-14}) = -\log([H_3O^+][OH^-])$$
$$pK_w = 14.00 = -\log([H_3O^+]) + (-\log[OH^-])$$

$$pK_w = 14.00 = pH + pOH \qquad (17.4)$$

The sum of the pH and pOH of a solution must be equal to 14.00 at 25 °C.

As illustrated in Figures 5.20 and 17.2, solutions with pH less than 7.00 (at 25 °C) are acidic, whereas solutions with pH greater than 7.00 are basic. Solutions with pH = 7.00 at 25 °C are neutral.

Determining and Calculating pH

Common litmus paper will show us whether a solution is acidic or basic, and a wide variety of dyes called acid–base indicators that change color in some known pH range are available (see Section 18.3). The indicators we use in the laboratory, such as phenolphthalein, are Brønsted acids or bases that have the property that the acid and its conjugate base have different colors.

The calculation of pH from the hydronium ion concentration, or the concentration of hydronium ion concentration from pH, was introduced in Chapter 5 (page 212). Exercise 17.4 reviews those calculations.

GENERAL
Chemistry ⚛ **Now**™

See the General ChemistryNow CD-ROM or website:
• **Screen 17.4 The pH Scale,** for a simulation and tutorial on using pH and pOH

Exercise 17.4—Reviewing pH Calculations

(a) What is the pH of a 0.0012 M NaOH solution at 25 °C?
(b) The pH of a diet soda is 4.32 at 25 °C. What are the hydronium and hydroxide ion concentrations in the soda?
(c) If the pH of a solution containing the strong base $Sr(OH)_2$ is 10.46 at 25 °C, what is the concentration of $Sr(OH)_2$?

■ **Weak Acid or Weak Base**
If an acid or base is weak, a dilute aqueous solution of the acid or base (say 0.1 M) will have pH values in the following ranges.

| Weak acid | Small [H_2O] (~10^{-2} to < 10^{-7} M) pH ≈ 2 to < 7 |
| Weak base | Small [OH^-] (~10^{-2} to < 10^{-7} M) pH ≈ 12 to > 7 |

17.4—Equilibrium Constants for Acids and Bases

How can we define quantitatively the extent to which an acid or a base reacts with water? That is, how can we define the relative strengths of acids and bases?

One way to define the relative strengths of a series of acids would be to measure the pH of solutions of acids of equal concentration: The lower the pH, the stronger the acid.

• For a strong acid, $[H_3O^+]$ in solution will be equal to the original acid concentration. Similarly, for a strong base, $[OH^-]$ will be equal to the original base concentration.

- For a weak acid, $[H_3O^+]$ will be much less than the original acid concentration. That is, $[H_3O^+]$ will be smaller than if the acid were a strong acid of the same concentration. Similarly, a weak base will give a smaller $[OH^-]$ than if the base were a strong base of the same concentration.

- For a series of weak monoprotic acids (of the type HA) of the same concentration, $[H_3O^+]$ will increase (and the pH will decrease) as the acids become stronger. Similarly, for a series of weak bases, $[OH^-]$ will increase (and the pH will increase) as the bases become stronger.

The relative strength of an acid or base can be expressed quantitatively with an equilibrium constant. For the general acid HA, we can write

$$HA(aq) + H_2O(\ell) \rightleftharpoons H_3O^+(aq) + A^-(aq)$$

$$K_a = \frac{[H_3O^+][A^-]}{[HA]} \qquad (17.5)$$

where the equilibrium constant, K, has a subscript "a" to indicate that it is an equilibrium constant for an *acid* in water. For weak acids, the value of K_a is less than 1 because the product $[H_3O^+][A^-]$ is less than the equilibrium concentration of the weak acid, $[HA]$. For a series of acids, the acid strength increases as the value of K_a increases.

Similarly, we can write the equilibrium expression for a *weak base* B in water. Here we label K with a subscript "b." Its value is also less than 1 for weak bases.

$$B(aq) + H_2O(\ell) \rightleftharpoons BH^+(aq) + OH^-(aq)$$

$$K_b = \frac{[BH^+][OH^-]}{[B]} \qquad (17.6)$$

Some acids and bases are listed in Table 17.3, each with its value of K_a or K_b. The following are important ideas concerning this table.

- Acids are listed in Table 17.3 at the left and their conjugate bases are on the right.

- A large value of K indicates that ionization products are strongly favored, whereas a small value of K indicates that reactants are favored.

- The strongest acids are at the upper left. They have the largest K_a values. K_a values become smaller on descending the chart as the acid strength declines.

- The strongest bases are at the lower right. They have the largest K_b values. K_b values become larger on descending the chart as base strength increases.

- The weaker the acid, the stronger its conjugate base. That is, the smaller the value of K_a, the larger the value of K_b.

- Some acids or bases are listed as having K_a or K_b values that are large or very small. Acids that are stronger than H_3O^+ are completely ionized, so their K_a values are "large." Their conjugate bases do not produce meaningful concentrations of OH^- ions, so their K_b values are "very small." Similar arguments follow for strong bases and their conjugate acids.

To illustrate these ideas, let us compare some common acids and bases. For example, nitric acid, a strong acid, is *much* stronger than the related weak acid nitrous acid.

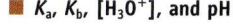

■ K_a, K_b, $[H_3O^+]$, and pH

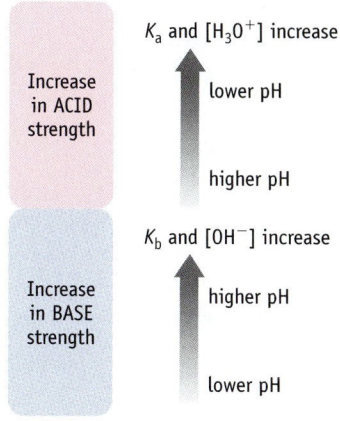

Table 17.3   Ionization Constants for Some Acids and Their Conjugate Bases at 25 °C

Acid Name	Acid	K_a	Base	K_b	Base Name
Perchloric acid	$HClO_4$	large	ClO_4^-	very small	perchlorate ion
Sulfuric acid	H_2SO_4	large	HSO_4^-	very small	hydrogen sulfate ion
Hydrochloric acid	HCl	large	Cl^-	very small	chloride ion
Nitric acid	HNO_3	large	NO_3^-	very small	nitrate ion
Hydronium ion	H_3O^+	1.0	H_2O	1.0×10^{-14}	water
Sulfurous acid	H_2SO_3	1.2×10^{-2}	HSO_3^-	8.3×10^{-13}	hydrogen sulfite ion
Hydrogen sulfate ion	HSO_4^-	1.2×10^{-2}	SO_4^{2-}	8.3×10^{-13}	sulfate ion
Phosphoric acid	H_3PO_4	7.5×10^{-3}	$H_2PO_4^-$	1.3×10^{-12}	dihydrogen phosphate ion
Hexaaquairon(III) ion	$[Fe(H_2O)_6]^{3+}$	6.3×10^{-3}	$[Fe(H_2O)_5OH]^{2+}$	1.6×10^{-12}	pentaaquahydroxoiron(III) ion
Hydrofluoric acid	HF	7.2×10^{-4}	F^-	1.4×10^{-11}	fluoride ion
Nitrous acid	HNO_2	4.5×10^{-4}	NO_2^-	2.2×10^{-11}	nitrite ion
Formic acid	HCO_2H	1.8×10^{-4}	HCO_2^-	5.6×10^{-11}	formate ion
Benzoic acid	$C_6H_5CO_2H$	6.3×10^{-5}	$C_6H_5CO_2^-$	1.6×10^{-10}	benzoate ion
Acetic acid	CH_3CO_2H	1.8×10^{-5}	$CH_3CO_2^-$	5.6×10^{-10}	acetate ion
Propanoic acid	$CH_3CH_2CO_2H$	1.3×10^{-5}	$CH_3CH_2CO_2^-$	7.7×10^{-10}	propanoate ion
Hexaaquaaluminum ion	$[Al(H_2O)_6]^{3+}$	7.9×10^{-6}	$[Al(H_2O)_5OH]^{2+}$	1.3×10^{-9}	pentaaquahydroxoaluminum ion
Carbonic acid	H_2CO_3	4.2×10^{-7}	HCO_3^-	2.4×10^{-8}	hydrogen carbonate ion
Hexaaquacopper(II) ion	$[Cu(H_2O)_6]^{2+}$	1.6×10^{-7}	$[Cu(H_2O)_5OH]^+$	6.3×10^{-8}	pentaaquahydroxocopper(II) ion
Hydrogen sulfide	H_2S	1×10^{-7}	HS^-	1×10^{-7}	hydrogen sulfide ion
Dihydrogen phosphate ion	$H_2PO_4^-$	6.2×10^{-8}	HPO_4^{2-}	1.6×10^{-7}	hydrogen phosphate ion
Hydrogen sulfite ion	HSO_3^-	6.2×10^{-8}	SO_3^{2-}	1.6×10^{-7}	sulfite ion
Hypochlorous acid	$HClO$	3.5×10^{-8}	ClO^-	2.9×10^{-7}	hypochlorite ion
Hexaaqualead(II) ion	$[Pb(H_2O)_6]^{2+}$	1.5×10^{-8}	$[Pb(H_2O)_5OH]^+$	6.7×10^{-7}	pentaaquahydroxolead(II) ion
Hexaaquacobalt(II) ion	$[Co(H_2O)_6]^{2+}$	1.3×10^{-9}	$[Co(H_2O)_5OH]^+$	7.7×10^{-6}	pentaaquahydroxocobalt(II) ion
Boric acid	$B(OH)_3(H_2O)$	7.3×10^{-10}	$B(OH)_4^-$	1.4×10^{-5}	tetrahydroxoborate ion
Ammonium ion	NH_4^+	5.6×10^{-10}	NH_3	1.8×10^{-5}	ammonia
Hydrocyanic acid	HCN	4.0×10^{-10}	CN^-	2.5×10^{-5}	cyanide ion
Hexaaquairon(II) ion	$[Fe(H_2O)_6]^{2+}$	3.2×10^{-10}	$[Fe(H_2O)_5OH]^+$	3.1×10^{-5}	pentaaquahydroxoiron(II) ion
Hydrogen carbonate ion	HCO_3^-	4.8×10^{-11}	CO_3^{2-}	2.1×10^{-4}	carbonate ion
Hexaaquanickel(II) ion	$[Ni(H_2O)_6]^{2+}$	2.5×10^{-11}	$[Ni(H_2O)_5OH]^+$	4.0×10^{-4}	pentaaquahydroxonickel(II) ion
Hydrogen phosphate ion	HPO_4^{2-}	3.6×10^{-13}	PO_4^{3-}	2.8×10^{-2}	phosphate ion
Water	H_2O	1.0×10^{-14}	OH^-	1.0	hydroxide ion
Hydrogen sulfide ion*	HS^-	1×10^{-19}	S^{2-}	1×10^5	sulfide ion
Ethanol	C_2H_5OH	very small	$C_2H_5O^-$	large	ethoxide ion
Ammonia	NH_3	very small	NH_2^-	large	amide ion
Hydrogen	H_2	very small	H^-	large	hydride ion

Increasing Acid Strength (left margin, upward arrow)

Increasing Base Strength (right margin, downward arrow)

*The values of K_a for HS^- and K_b for S^{2-} are estimates.

$$HNO_3, K_a \gg 1 \quad \gg \quad HNO_2, K_a = 4.5 \times 10^{-4}$$

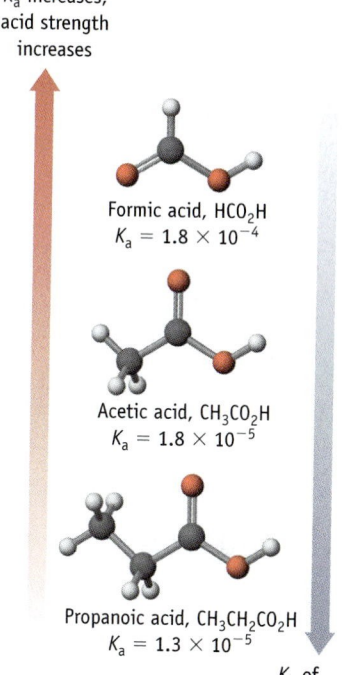

■ **Relative Strengths of Some Organic Acids and Bases**

K_a increases; acid strength increases

Formic acid, HCO_2H
$K_a = 1.8 \times 10^{-4}$

Acetic acid, CH_3CO_2H
$K_a = 1.8 \times 10^{-5}$

Propanoic acid, $CH_3CH_2CO_2H$
$K_a = 1.3 \times 10^{-5}$

K_b of conjugate base increases

Their conjugate bases, however, are reversed in their relative strength. Indeed, the NO_3^- ion is such a weak base that it has no effect on solution pH.

The three organic acids pictured in the margin decline in strength as more carbon atoms are added to the carboxylic acid group. The opposite ordering occurs for their conjugate bases, however. That is, the propanoate ion, $CH_3CH_2CO_2^-$ ($K_b = 7.7 \times 10^{-10}$) is a stronger base than the formate ion (HCO_2^-, $K_b = 5.6 \times 10^{-11}$).

Nature abounds in weak bases as well as weak acids (Figure 17.3). Ammonia and its conjugate acid, the ammonium ion, are part of the nitrogen cycle in the environment. Biological systems reduce nitrate ion to NH_3 and NH_4^+ and incorporate nitrogen into amino acids and proteins. Many organic bases are derived from NH_3 by replacement of the H atoms with organic groups.

Ammonia
$K_b = 1.8 \times 10^{-5}$

Methylamine
$K_b = 5.0 \times 10^{-4}$

Aniline
$K_b = 4.0 \times 10^{-10}$

Ammonia is a weaker base than methylamine (K_b for $NH_3 < K_b$ for CH_3NH_2). However, the conjugate acid of ammonia, NH_4^+ ($K_a = 5.6 \times 10^{-10}$) is stronger than the conjugate acid of methylamine ($CH_3NH_3^+$, $K_a = 2.0 \times 10^{-11}$).

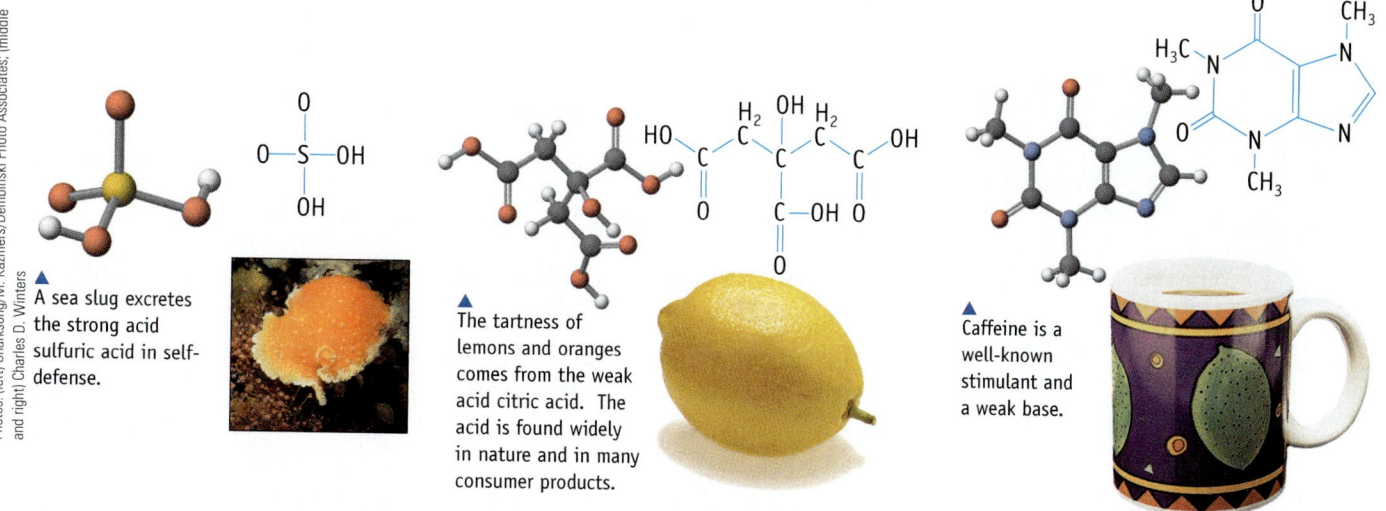

A sea slug excretes the strong acid sulfuric acid in self-defense.

The tartness of lemons and oranges comes from the weak acid citric acid. The acid is found widely in nature and in many consumer products.

Caffeine is a well-known stimulant and a weak base.

Figure 17.3 Natural acids and bases. Hundreds of acids and bases are found in nature. Our foods contain a wide variety, and biochemically important molecules are acids and bases.

Problem-Solving Tip 17.1

Strong or Weak?

How can you tell whether an acid or a base is weak? The easiest way is to remember those few that are strong (see Table 5.2 and the information here). All others are probably weak.

Strong acids are:

Hydrohalic acids: HCl, HBr, and HI (but *not* HF)

Nitric acid: HNO_3

Sulfuric acid: H_2SO_4 (for loss of first H^+ only)

Perchloric acid: $HClO_4$

Some common strong bases include the following:

All Group 1A hydroxides: LiOH, NaOH, KOH, RbOH, CsOH

Group 2A hydroxides: $Sr(OH)_2$ and $Ba(OH)_2$ [$Mg(OH)_2$ and $Ca(OH)_2$ are not considered strong bases because they do not dissolve appreciably in water.]

GENERAL
Chemistry ⚛ Now™

See the General ChemistryNow CD-ROM or website:
- **Screen 17.5 Strong Acids and Bases**, for tutorials on the pH of solutions of acids and bases
- **Screen 17.6 Weak Acids and Bases**, for a table of K_a and K_b values

---

Exercise 17.5—Strengths of Acids and Bases

Use Table 17.3 to answer the following questions.

(a) Which is the stronger acid, H_2SO_4 or H_2SO_3?
(b) Is benzoic acid, $C_6H_5CO_2H$, stronger or weaker than acetic acid?
(c) Which has the stronger conjugate base, acetic acid or boric acid?
(d) Which is the stronger base, ammonia or the acetate ion?
(e) Which has the stronger conjugate acid, ammonia or the acetate ion?

Aqueous Solutions of Salts

A number of the acids and bases listed in Table 17.3 are cations or anions. As described earlier, anions in particular can act as Brønsted bases because they can accept a proton from an acid to form the conjugate acid of the base.

$$CO_3^{2-}(aq) + H_2O(\ell) \rightleftharpoons HCO_3^-(aq) + OH^-(aq) \qquad K_b = 2.1 \times 10^{-4}$$

You should also notice that many metal cations in water are effective Brønsted acids.

$$[Al(H_2O)_6]^{3+}(aq) + H_2O(\ell) \rightleftharpoons [Al(H_2O)_5(OH)]^{2+}(aq) + H_3O^+(aq) \qquad K_a = 7.9 \times 10^{-6}$$

Table 17.4 summarizes the acid–base properties of some of the common cations and anions. As you look over this table, notice the following points:

- Anions that are conjugate bases of strong acids (for example, Cl^- and NO_3^-) are such weak bases that they *have no effect on solution pH*.
- There are numerous basic anions (such as CO_3^{2-}). All are the conjugate bases of weak acids.
- Anions from polyprotic acids can be either acidic or basic.
- Alkali metal and alkaline earth cations have no measurable effect on solution pH.
- Basic cations are conjugate bases of acidic cations such as $[Al(H_2O)_6]^{3+}$.

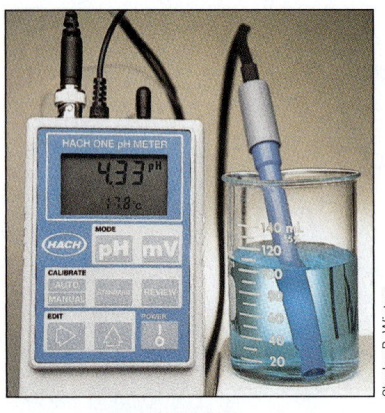

Charles D. Winters

Many aqueous metal cations are Brønsted acids. A pH measurement of a dilute solution of copper (II) sulfate shows that the solution is clearly acidic. Among the common cations, Al^{3+} and transition metal ions form acidic solutions in water.

Table 17.4   Acid and Base Properties of Some Ions in Aqueous Solution

Neutral			Basic			Acidic
Anions	Cl^- Br^- I^-	NO_3^- ClO_4^-	$CH_3CO_2^-$ HCO_2^- CO_3^{2-} S^{2-} F^-	CN^- PO_4^{3-} HCO_3^- HS^- NO_2^-	SO_4^{2-} HPO_4^{2-} SO_3^{2-} OCl^-	HSO_4^- $H_2PO_4^-$ HSO_3^-
Cations	Li^+ Na^+ K^+	Ca^{2+} Ba^{2+}	$[Al(H_2O)_5(OH)]^{2+}$ (for example)			$[Al(H_2O)_6]^{3+}$ and hydrated transition metal cations (such as $[Fe(H_2O)_6]^{3+}$) NH_4^+

- Acidic cations are limited to metal cations with 2+ and 3+ charges and to ammonium ions (and their organic derivatives).

- All metal cations are hydrated in water. That is, they form ions such as $[M(H_2O)_6]^{n+}$. However, only when M is a 2+ or 3+ ion, and particularly a transition metal ion, does the ion act as an acid.

GENERAL
Chemistry ∙ Now™

See the General ChemistryNow CD-ROM or website:
- **Screen 17.11 Acid–Base Properties of Salts,** for a simulation showing the pH of a number of cation/anion combinations

■ **Hydrolysis Reactions**
Chemists often say that, when ions interact with water to produce acidic or basic solutions, the ions "hydrolyze" in water or they undergo "hydrolysis." Thus, some books refer to the K_a and K_b values of ions as "hydrolysis constants," K_h.

Example 17.2—Acid–Base Properties of Salts

Problem Decide whether each of the following will give rise to an acidic, basic, or neutral solution in water.

(a) $NaNO_3$ **(d)** $NaHCO_3$

(b) K_3PO_4 **(e)** NH_4F

(c) $FeCl_2$

Problem-Solving Tip 17.2

Aqueous Solutions of Salts

Because aqueous solutions of salts are found in our bodies and throughout our economy and environment, it is important to know how to predict their acid and base properties. Information on the pH of an aqueous solution of a salt is summarized in Table 17.4. Consider also the following examples:

Cation	Anion	pH of the Solution
From strong base (Na^+)	From strong acid (Cl^-)	$= 7$ (neutral)
From strong base (K^+)	From weak acid ($CH_3CO_2^-$)	> 7 (basic)
From weak base (NH_4^+)	From strong acid (Cl^-)	< 7 (acidic)
From any weak base (BH^+)	From any weak acid (A^-)	Depends on relative strengths of BH^+ and A^-

Strategy First, decide on the cation and anion in each salt. Next, use Tables 17.3 and 17.4 to describe the properties of each ion.

Solution

(a) **NaNO₃:** This salt gives a neutral, aqueous solution (pH = 7). The Na^+ ion does not react with water to an appreciable extent. The nitrate ion, NO_3^-, is the *very* weak conjugate base of a strong acid, so it does not affect the solution pH.

(b) **K₃PO₄:** An aqueous solution of K_3PO_4 should be basic (pH > 7) because PO_4^{3-} is the conjugate base of the weak acid HPO_4^{2-}. In contrast, the K^+ ion, like the Na^+ ion, does not react with water appreciably.

(c) **FeCl₂:** An aqueous solution of $FeCl_2$ should be weakly acidic (pH < 7). The Fe^{2+} ion in water, $[Fe(H_2O)_6]^{2+}$, is a Brønsted acid. In contrast, Cl^- is the *very* weak conjugate base of the strong acid HCl, so it does not contribute excess OH^- ions to the solution.

(d) **NaHCO₃:** Some additional information is needed concerning salts of amphiprotic anions such as HCO_3^- and $H_2PO_4^-$. Because they have an ionizable hydrogen, they can act as acids.

$$HCO_3^-(aq) + H_2O(\ell) \rightleftharpoons CO_3^{2-}(aq) + H_3O^+(aq) \qquad K_a = 4.8 \times 10^{-11}$$

They are also the conjugate bases of weak acids.

$$HCO_3^-(aq) + H_2O(\ell) \rightleftharpoons H_2CO_3(aq) + OH^-(aq) \qquad K_b = 2.4 \times 10^{-8}$$

Whether the solution is acidic or basic will depend on the *relative* magnitude of K_a and K_b. In the case of the hydrogen carbonate anion, K_b is larger than K_a, so $[OH^-]$ is larger than $[H_3O^+]$, and an aqueous solution of NaHCO₃ will be slightly basic.

(e) **NH₄F:** What happens if you have a salt based on an acidic cation and a basic anion? One example is ammonium fluoride. Here the ammonium ion would decrease the pH, and the fluoride ion would increase the pH.

$$NH_4^+(aq) + H_2O(\ell) \rightleftharpoons H_3O^+(aq) + NH_3(aq) \qquad K_a(NH_4^+) = 5.6 \times 10^{-10}$$

$$F^-(aq) + H_2O(\ell) \rightleftharpoons HF(aq) + OH^-(aq) \qquad K_b(F^-) = 1.4 \times 10^{-11}$$

Because $K_a(NH_4^+) > K_b(F^-)$, the ammonium ion is a stronger acid than the fluoride ion is a base. The resulting solution should be slightly acidic.

Comment There are two important points to notice here:

- Anions that are conjugate bases of strong acids—such as Cl^- and NO_3^-—have no effect on solution pH.

- In general, *for a salt that has an acidic cation and a basic anion, the pH of the solution will be determined by the ion that is the stronger acid or base of the two.*

Exercise 17.6—Acid-Base Properties of Salts in Aqueous Solution

For each of the following salts in water, predict whether the pH will be greater than, less than, or equal to 7.

(a) KBr   (b) NH₄NO₃   (c) AlCl₃   (d) Na₂HPO₄

A Logarithmic Scale of Relative Acid Strength, pK_a

Many chemists and biochemists use a logarithmic scale to report and compare relative acid strengths.

$$pK_a = -\log K_a \tag{17.7}$$

The pK_a of an acid is the negative log of the K_a value (just as pH is the negative log of the hydronium ion concentration). For example, acetic acid has a pK_a value of 4.74.

$$pK_a = -\log (1.8 \times 10^{-5}) = 4.74$$

The pK_a value becomes smaller as the acid strength increases.

acid strength increases →

Propanoic acid	Acetic acid	Formic acid
$CH_3CH_2CO_2H$	CH_3CO_2H	HCO_2H
$K_a = 1.3 \times 10^{-5}$	$K_a = 1.8 \times 10^{-5}$	$K_a = 1.8 \times 10^{-4}$
$pK_a = 4.89$	$pK_a = 4.74$	$pK_a = 3.74$

← *pK_a increases*

Exercise 17.7—A Logarthimic Scale for Acid Strength, pK_a

(a) What is the pK_a value for benzoic acid, $C_6H_5CO_2H$?

(b) Is chloroacetic acid ($ClCH_2CO_2H$), $pK_a = 2.87$, a stronger or weaker acid than benzoic acid?

(c) What is the pK_a for the conjugate acid of ammonia? Is this acid stronger or weaker than acetic acid?

Relating the Ionization Constants for an Acid and Its Conjugate Base

Let us look again at Table 17.3. From the top of the table to the bottom, the strengths of the acids decline (K_a becomes smaller) and the strengths of their conjugate bases increase (the values of K_b increase). Examining a few cases shows that the product of K_a for an acid and K_b for its conjugate base is equal to a constant, specifically K_w.

$$K_a \times K_b = K_w \tag{17.8}$$

■ **A Relation Among pK Values**
A useful relationship for an acid–conjugate base pair can be derived from Equation 17.8.

$$pK_w = pK_a + pK_b$$

Consider the specific case of the ionization of a weak acid, say HCN, and the interaction of its conjugate base, CN^-, with H_2O.

Weak acid:	$HCN(aq) + H_2O(\ell) \rightleftharpoons H_3O^+(aq) + CN^-(aq)$	$K_a = 4.0 \times 10^{-10}$
Conjugate base:	$\underline{CN^-(aq) + H_2O(\ell) \rightleftharpoons HCN(aq) + OH^-(aq)}$	$K_b = 2.5 \times 10^{-5}$
	$2\,H_2O(\ell) \rightleftharpoons H_3O^+(aq) + OH^-(aq)$	$K_w = 1.0 \times 10^{-14}$

Adding the equations gives the chemical equation for the autoionization of water, and the numerical value is indeed 1.0×10^{-14}. That is,

$$K_a \times K_b = \left(\frac{[H_3O^+][CN^-]}{[HCN]} \right)\left(\frac{[HCN][OH^-]}{[CN^-]} \right) = [H_3O^+][OH^-] = K_w$$

Equation 17.8 is useful because K_b can be calculated from a knowledge of K_a. The value of K_b for the cyanide ion, for example, is

$$K_b \text{ for } CN^- = \frac{K_w}{K_a \text{ for } HCN} = \frac{1.0 \times 10^{-14}}{4.0 \times 10^{-10}} = 2.5 \times 10^{-5}$$

Exercise 17.8—Using the Equation $K_a \times K_b = K_w$

K_a for lactic acid, $CH_3CHOHCO_2H$, is 1.4×10^{-4}. What is K_b for the conjugate base of this acid, $CH_3CHOHCO_2^-$? Where does this base fit in Table 17.3?

17.5—Equilibrium Constants and Acid–Base Reactions

According to the Brønsted-Lowry theory, all acid–base reactions can be written as equilibria involving the acid and base and their conjugates.

$$\text{Acid} + \text{Base} \rightleftharpoons \text{Conjugate base of the acid} + \text{Conjugate acid of the base}$$

■ **K and product- and reactant-favored reactions**
Reactions with an equilibrium constant greater than 1 are said to be product-favored. Those with K < 1 are reactant-favored.

In Section 17.4, we used equilibrium constants to provide quantitative information about the relative strengths of acids and bases. Now we want to show how the constants can be used to decide whether a particular acid–base reaction is product- or reactant-favored. If the reaction is product-favored, what is the nature of the solution when the reaction is complete?

Predicting the Direction of Acid–Base Reactions

Hydrochloric acid is a strong Brønsted acid. Its equilibrium constant for reaction with water is very large, with the equilibrium effectively lying completely to the right.

$$HCl(aq) + H_2O(\ell) \longrightarrow H_3O^+(aq) + Cl^-(aq)$$

Strong acid ($\approx$ 100%   [H_3O^+] $\approx$ initial
ionized), $K \gg 1$ concentration of the acid

For all strong acids, the acid on the reactant side of the balanced equation is stronger than the acid on the product side (and the base on the reactant side is stronger than the base on the product side).

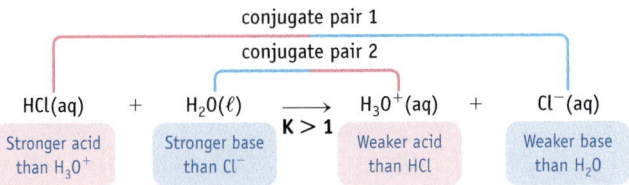

Of the two acids here, HCl is stronger than H_3O^+. Of the two bases, H_2O and Cl^-, water is the stronger base and wins out in the competition for the proton. The equilibrium lies to the side of the chemical equation having the weaker acid and base.

In contrast to HCl and other strong acids, acetic acid, a *weak* Brønsted acid, ionizes to only a very small extent (Table 17.3).

$$CH_3CO_2H(aq) + H_2O(\ell) \rightleftharpoons H_3O^+(aq) + CH_3CO_2^-(aq)$$

Weak acid (< 100%   [H_3O^+] $\ll$ initial
ionized), $K = 1.8 \times 10^{-5}$ concentration of the acid

When equilibrium is achieved in a 0.1 M aqueous solution of CH_3CO_2H, the concentrations of $H_3O^+(aq)$ and $CH_3CO_2^-(aq)$ are each only about 0.001 M. Approximately 99% of the acetic acid is *not* ionized.

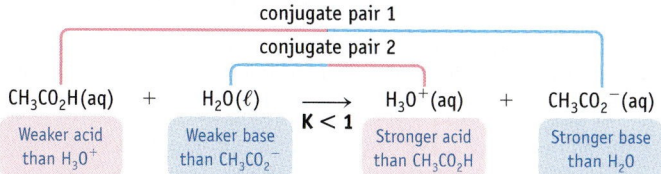

Again, the equilibrium lies toward the side of the reaction having the weaker acid and base.

These two examples of the relative extent of acid–base reactions illustrate another general principle: *All proton transfer reactions proceed from the stronger acid and base to the weaker acid and base.* Using this principle and Table 17.3, you can predict which reactions are product-favored and which are reactant-favored. Consider the possible reaction of phosphoric acid and acetate ion to give acetic acid and the dihydrogen phosphate ion. Table 17.3 informs us that H_3PO_4 is a stronger acid ($K_a = 7.5 \times 10^{-3}$) than acetic acid ($K_a = 1.8 \times 10^{-5}$), and the acetate ion ($K_b = 5.6 \times 10^{-10}$) is a stronger base than the dihydrogen phosphate ion ($K_b = 1.3 \times 10^{-12}$).

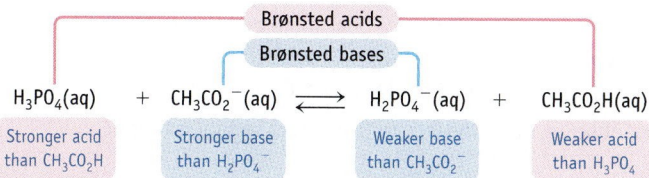

Thus, mixing phosphoric acid with sodium acetate would produce a significant amount of dihydrogen phosphate ion and acetic acid. That is, the equilibrium is predicted to lie to the right because the reaction has proceeded from the stronger acid–base combination to the weaker acid–base combination.

GENERAL
Chemistry • Now™

See the General ChemistryNow CD-ROM or website:
• **Screen 17.7 Acid–Base Reactions,** for a simulation on predicting the direction of acid–base reactions

---

Example 17.3—Reactions of Acids and Bases

Problem Write a balanced, net ionic equation for the reaction that occurs between acetic acid and sodium bicarbonate. Decide whether the equilibrium lies predominantly to the left or to the right.

Strategy First, identify the products of the acid–base reaction (which arise by H^+ transfer from the acid to the base). Next, identify the two acids (or the two bases) in the reaction. Finally, use Table 17.3 to decide which is the weaker of the two acids (or the weaker of the two bases). The reaction will proceed from the stronger acid (or base) to the weaker acid (or base).

Solution Acetic acid is clearly one acid involved (and its conjugate base is the acetate ion, $CH_3CO_2^-$). The other reactant, $NaHCO_3$, is a water-soluble salt that forms Na^+ and HCO_3^- ions in water. Because acetic acid can function only as an acid, the HCO_3^- ion in this case must be the Brønsted base. Thus, hydrogen ion transfer from the acid to the base (HCO_3^- ion) could lead to the following net ionic equation:

$$CH_3CO_2H(aq) + HCO_3^-(aq) \rightleftharpoons CH_3CO_2^-(aq) + H_2CO_3(aq)$$

Reaction of vinegar and baking soda.
This reaction involves the weak acid acetic acid and the weak base HCO_3^- from sodium hydrogen carbonate. Based on the values of the equilibrium constants, the reaction is predicted to proceed to the right.

According to Table 17.3, H_2CO_3 is a weaker acid ($K_a = 4.2 \times 10^{-7}$) than CH_3CO_2H ($K_a = 1.8 \times 10^{-5}$), and $CH_3CO_2^-$ is a weaker base ($K_b = 5.6 \times 10^{-10}$) than HCO_3^- ($K_b = 2.4 \times 10^{-8}$). The reaction favors the side having the weaker acid and base—that is, the right side.

Comment The reaction of acetic acid and $NaHCO_3$ favors the weaker acid (H_2CO_3) and base ($CH_3CO_2^-$). In the photograph you see that the product, H_2CO_3, must have dissociated into CO_2 and H_2O because the CO_2 bubbles out of the solution: the equilibrium lies far to the right.

$$H_2CO_3(aq) \rightleftharpoons CO_2(g) + H_2O(\ell)$$

See the discussion of gas-forming reactions in Chapter 5 and of Le Chatelier's principle in Section 16.6.

Exercise 17.9—Relative Strengths of Acids and Bases—Predicting the Direction of an Acid–Base Reaction

(a) Which is the stronger Brønsted acid, HCO_3^- or NH_4^+? Which has the stronger conjugate base?

(b) Is a reaction between HCO_3^- ions and NH_3 product- or reactant-favored?

$$HCO_3^-(aq) + NH_3(aq) \rightleftharpoons CO_3^{2-}(aq) + NH_4^+(aq)$$

(c) You mix solutions of sodium hydrogen phosphate and ammonia. The net ionic equation for a possible reaction is

$$HPO_4^{2-}(aq) + NH_3(aq) \rightleftharpoons PO_4^{3-}(aq) + NH_4^+(aq)$$

Does the equilibrium lie to the left or to the right in this reaction?

Exercise 17.10—Reaction of an Acid and a Base

Write the net ionic equation for the possible reaction between acetic acid and sodium hydrogen sulfate, $NaHSO_4$. Does the equilibrium lie to the left or right?

17.6—Types of Acid–Base Reactions

The reaction of hydrochloric acid and sodium hydroxide is the classic example of a strong acid–strong base reaction, whereas citric acid and bicarbonate ion represent the reaction of a weak acid and weak base (Figure 17.4). There are two other types of acid–base reactions.

Type of Acid–Base Reaction	Example
Strong acid + strong base	HCl and NaOH
Strong acid + weak base	HCl and NH_3
Weak acid + strong base	CH_3CO_2H and NaOH
Weak acid + weak base	Citric acid and HCO_3^-

Because acid–base reactions are among the most important classes of chemical reactions, it is useful for you to know the outcome of the various types of these reactions (Table 17.5).

Figure 17.4   Reaction of a weak acid with a weak base. The bubbles coming from the tablet are carbon dioxide. This gas arises from the reaction of a weak Brønsted acid (citric acid) with a weak Brønsted base (HCO_3^-). The reaction is driven to completion by gas evolution.

The Reaction of a Strong Acid with a Strong Base

Strong acids and bases are effectively 100% ionized in solution. Therefore, the total ionic equation for the reaction of HCl (strong acid) and NaOH (strong base) is

$$H_3O^+(aq) + Cl^-(aq) + Na^+(aq) + OH^-(aq) \longrightarrow 2\ H_2O(\ell) + Na^+(aq) + Cl^-(aq)$$

Table 17.5 Characteristics of Acid–Base Reactions

Type	Example	Net Ionic Equation	Species Present After Equal Molar Amounts are Mixed; pH
Strong acid + strong base	HCl + NaOH	$H_3O^+(aq) + OH^-(aq) \longrightarrow 2\ H_2O(\ell)$	Cl^-, Na^+, pH = 7
Strong acid + weak base	HCl + NH$_3$	$H_3O^+(aq) + NH_3(aq) \rightleftharpoons NH_4^+(aq) + H_2O(\ell)$	Cl^-, NH_4^+, pH < 7
Weak acid + strong base	HCO$_2$H + NaOH	$HCO_2H(aq) + OH^-(aq) \rightleftharpoons HCO_2^-(aq) + H_2O(\ell)$	HCO_2^-, Na^+, pH > 7
Weak acid + weak base	HCO$_2$H + NH$_3$	$HCO_2H(aq) + NH_3(aq) \rightleftharpoons HCO_2^-(aq) + NH_4^+(aq)$	HCO_2^-, NH_4^+, pH dependent on K_a and K_b of conjugate acid and base

which leads to the following net ionic equation:

$$H_3O^+(aq) + OH^-(aq) \rightleftharpoons 2\ H_2O(\ell) \qquad K = 1/K_w = 1.0 \times 10^{14}$$

The net ionic equation for the reaction of any strong acid with any strong base is always simply the union of hydronium ion and hydroxide ion to give water [◀ Section 5.4]. Because this reaction is the reverse of the autoionization of water, it has an equilibrium constant of $1/K_w$. This very large value of K shows that, for all practical purposes, the reactants are completely consumed to form products. Thus, if equal numbers of moles of NaOH and HCl are mixed, the result is just a solution of NaCl in water. The constituents of NaCl, the Na^+ and Cl^- ions, which arise from a strong base and a strong acid, respectively, produce a neutral aqueous solution. For this reason reactions of strong acids and bases are often called "neutralizations."

> Mixing equal molar quantities of a strong base with a strong acid produces a neutral solution (pH = 7 at 25 °C).

The Reaction of a Weak Acid with a Strong Base

Consider the reaction of the naturally occurring weak acid formic acid, HCO$_2$H, with sodium hydroxide. The net ionic equation is

$$HCO_2H(aq) + OH^-(aq) \rightleftharpoons H_2O(\ell) + HCO_2^-(aq)$$

In the reaction of formic acid with NaOH, OH^- is a much stronger base than HCO_2^- ($K_b = 5.6 \times 10^{-11}$), and the reaction is predicted to proceed to the right. If equal molar quantities of weak acid and base are mixed, the final solution will contain sodium formate (NaHCO$_2$), a salt that is 100% dissociated in water. The Na^+ ion is the cation of a strong base, so it gives a neutral solution. The formate ion, however, is the conjugate base of a weak acid (Table 17.3), so the solution is basic. This example leads to a useful general conclusion:

> Mixing equal molar quantities of a strong base with a weak acid produces a salt whose anion is the conjugate base of the weak acid. The solution is basic, with the pH depending on the value of K_b for the anion.

The Reaction of a Strong Acid with a Weak Base

The net ionic equation for the reaction of the strong acid HCl and the weak base NH$_3$ is

$$H_3O^+(aq) + NH_3(aq) \rightleftharpoons H_2O(\ell) + NH_4^+(aq)$$

■ **Formic Acid + NaOH**
The equilibrium constant for the reaction of formic acid and sodium hydroxide is 1.8×10^{10}. Can you confirm this? (See Study Question 17.97.)

■ **Ammonia + HCl**
The equilibrium constant for the reaction of a strong acid with aqueous ammonia is 1.8×10^9. Can you confirm this? (See Study Question 17.102.)

A weak acid reacting with a weak base. Baking powder contains the weak acid calcium dihydrogen phosphate, $Ca(H_2PO_4)_2$. It can react with the basic HCO_3^- ion in baking soda to give HPO_4^{2-}, CO_2 gas, and water.

■ **K for Reaction of a Weak Acid and a Weak Base** The equilibrium constant for the reaction between a weak acid and a weak base is $K_{net} = K_w/K_aK_b$. Can you confirm this? (See Study Question 17.119.)

The hydronium ion, H_3O^+, is a much stronger acid than NH_4^+ ($K_a = 5.6 \times 10^{-10}$), and NH_3 ($K_b = 1.8 \times 10^{-5}$) is a stronger base than H_2O. Therefore, the reaction is predicted to proceed to the right and essentially to completion. Thus, after mixing equal molar quantities of HCl and NH_3, the solution contains the salt ammonium chloride, NH_4Cl. The Cl^- ion has no effect on the solution pH (Tables 17.3 and 17.4). However, the NH_4^+ ion is the conjugate acid of the weak base NH_3, so the solution is acidic at the conclusion of the reaction. In general, we can draw the following conclusion:

> Mixing equal molar quantities of a strong acid and a weak base produces a salt whose cation is the conjugate acid of the weak base. The solution is acidic, with the pH depending on the value of K_a for the cation.

The Reaction of a Weak Acid with a Weak Base

If acetic acid, a weak acid, is mixed with ammonia, a weak base, the following reaction occurs.

$$CH_3CO_2H(aq) + NH_3(aq) \rightleftharpoons NH_4^+(aq) + CH_3CO_2^-(aq)$$

You know that this reaction is product-favored because CH_3CO_2H is a stronger acid than NH_4^+ and NH_3 is a stronger base than $CH_3CO_2^-$ (Table 17.3). Thus, if equal molar quantities of the acid and the base are mixed, the resulting solution contains ammonium acetate, $NH_4CH_3CO_2$. Is this solution acidic or basic? The answer depends on the relative values of K_a for the conjugate acid (here NH_4^+; $K_a = 5.6 \times 10^{-10}$) and K_b for the conjugate base (here $CH_3CO_2^-$; $K_b = 5.6 \times 10^{-10}$). In this case the values of K_a and K_b are the same, so the solution is predicted to be neutral.

> Mixing equal molar quantities of a weak acid and a weak base produces a salt whose cation is the conjugate acid of the weak base and whose anion is the conjugate base of the weak acid. The solution pH depends on the relative K_a and K_b values.

Exercise 17.11—Acid–Base Reactions

(a) Equal molar quantities of HCl(aq) and NaCN(aq) are mixed. Is the resulting solution acidic, basic, or neutral?

(b) Equal molar quantities of acetic acid and sodium sulfite, Na_2SO_3, are mixed. Is the resulting solution acidic, basic, or neutral?

17.7—Calculations with Equilibrium Constants

Determining K from Initial Concentrations and Measured pH

The K_a and K_b values found in Table 17.3 and in the more extensive tables in Appendices H and I were all determined by experiment. Several experimental methods are available, but one approach, illustrated by the following example, is to determine the pH of the solution.

See the General ChemistryNow CD-ROM or website:

• **Screen 17.8 Determining K_a and K_b Values,** for a tutorial on calculating K_a or K_b from experimental data

Example 17.4—Calculating a K_a Value from a Measured pH

Problem A 0.10 M aqueous solution of lactic acid, $CH_3CHOHCO_2H$, has a pH of 2.43. What is the value of K_a for lactic acid?

Strategy To calculate K_a, we must know the equilibrium concentration of each species. The pH of the solution directly tells us the equilibrium concentration of H_3O^+, and we can derive the other equilibrium concentrations from this value. These concentrations are used to calculate K_a.

Solution The equation for the equilibrium interaction of lactic acid with water is

$$CH_3CHOHCO_2H(aq) + H_2O(\ell) \rightleftharpoons CH_3CHOHCO_2^-(aq) + H_3O^+(aq)$$

      Lactic acid   Lactate ion

The equilibrium constant expression is

$$K_a \text{ (lactic acid)} = \frac{[H_3O^+][CH_3CHOHCO_2^-]}{[CH_3CHOHCO_2H]}$$

We begin by converting the pH to $[H_3O^+]$.

$$[H_3O^+] = 10^{-pH} = 10^{-2.43} = 3.7 \times 10^{-3} \text{ M}$$

Next, prepare an ICE table of the concentrations in the solution before equilibrium is established, the change occurring as the reaction proceeds to equilibrium, and the concentrations when equilibrium has been achieved [◄ Examples 16.4–16.6].

Equilibrium	$CH_3CHOHCO_2H + H_2O \rightleftharpoons CH_3CHOHCO_2^- + H_3O^+$		
Initial (M)	0.10	0	0
Change (M)	$-x$	$+x$	$+x$
Equilibrium (M)	$(0.10 - x)$	x	x

The following points can be made concerning the ICE table.

• The quantity x represents the equilibrium concentrations of hydronium ion and lactate ion. That is, at equilibrium $x = [H_3O^+] = [CH_3CHOHCO_2^-] = 3.7 \times 10^{-3}$ M.

• By stoichiometry, x is also the quantity of acid that ionized on proceeding to equilibrium.

With these points in mind, we can calculate K_a for lactic acid.

$$K_a \text{ (lactic acid)} = \frac{[H_3O^+][CH_3CHOHCO_2^-]}{[CH_3CHOHCO_2H]}$$

$$= \frac{(3.7 \times 10^{-3})(3.7 \times 10^{-3})}{0.10 - 0.0037} = 1.4 \times 10^{-4}$$

Comparing this value of K_a with others in Table 17.3, we see it is similar to formic acid in its strength.

Comment Hydronium ion, H_3O^+, is present in solution from lactic acid ionization *and* from water autoionization. Le Chatelier's principle informs us that the H_3O^+ added to the water by lactic acid will suppress the H_3O^+ coming from the water autoionization. However, because $[H_3O^+]$ from water must be less than 10^{-7} M, the pH is almost completely a reflection of H_3O^+ from lactic acid [◄ Example 17.1, page 804].

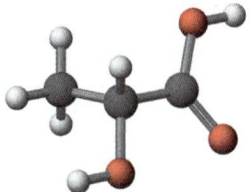

Lactic acid, $CH_3CHOHCO_2H$

Lactic acid, $CH_3CHOHCO_2H$. Lactic acid is a weak monoprotic acid that occurs naturally in sour milk and arises from metabolism in the human body.

Exercise 17.12—Calculating a K_a Value from a Measured pH

A solution prepared from 0.055 mol of butanoic acid dissolved in sufficient water to give 1.0 L of solution has a pH of 2.72 at 25 °C. Determine K_a for butanoic acid. The acid ionizes according to the balanced equation

$$CH_3CH_2CH_2CO_2H(aq) + H_2O(\ell) \rightleftharpoons H_3O^+(aq) + CH_3CH_2CH_2CO_2^-(aq)$$

There is an important point to notice in Example 17.4. The lactic acid concentration at equilibrium was given by $(0.10 - x)$, where x was found to be 3.7×10^{-3} M. By the usual rules governing significant figures, $(0.10 - 0.0037)$ is equal to 0.10. The acid is weak, so very little of it ionizes (approximately 4%). Thus, the equilibrium concentration of lactic acid is essentially equal to the initial acid concentration. Neglecting to subtract 0.0037 from 0.10 has little effect on the answer.

Like lactic acid, most weak acids (HA) are so weak that the equilibrium concentration of the acid, $[HA]$, is effectively its initial concentration ($\approx [HA]_0$). This fact leads to the useful conclusion that the denominator in the equilibrium constant expression for dilute solutions of most weak acids is simply $[HA]_0$, the original or initial concentration of the weak acid.

$$HA(aq) + H_2O(\ell) \rightleftharpoons H_3O^+(aq) + A^-(aq)$$

$$K_a = \frac{[H_3O^+][A^-]}{[HA]_0 - [H_3O^+]} \approx \frac{[H_3O^+][A^-]}{[HA]_0}$$

Analysis shows that

> The approximation that $[HA]_{equilibrium}$ is effectively equal to $[HA]_0$
> $$[HA]_{equilibrium} = [HA]_0 - [H_3O^+] \approx [HA]_0$$
> is valid whenever $[HA]_0$ is greater than or equal to $100 \times K_a$.

This is the same approximation we derived in Chapter 16 when deciding whether we needed to solve quadratic equations exactly [◄ Problem-Solving Tip 16.1, page 775].

What Is the pH of an Aqueous Solution of a Weak Acid or Base?

Knowing the values of the equilibrium constants for weak acids and bases enables us to calculate the pH of a solution of a weak acid or base.

GENERAL
Chemistry ⚛ Now™

See the General ChemistryNow CD-ROM or website:
- **Screen 17.9 Estimating the pH of Weak Acid Solutions,** for a tutorial on calculating the pH of solutions

Example 17.5—Calculating Equilibrium Concentrations and pH from K_a

Problem Calculate the pH of a 0.020 M solution of benzoic acid ($C_6H_5CO_2H$) if $K_a = 6.3 \times 10^{-5}$ for the acid.

$$C_6H_5CO_2H(aq) + H_2O(\ell) \rightleftharpoons H_3O^+(aq) + C_6H_5CO_2^-(aq)$$

Strategy This is similar to Examples 16.5 and 16.6 where we wanted to find the concentration of a reaction product. The strategy is the same: Designate the quantity of product (here $[H_3O^+]$) by x and derive the other concentrations from that starting point.

Solution Organize the information in an ICE table.

Equilibrium	$C_6H_5CO_2H + H_2O$	$\rightleftharpoons$	$C_6H_5CO_2^-$	$+$	H_3O^+
Initial (M)	0.020		0		0
Change (M)	$-x$		$+x$		$+x$
Equilibrium (M)	$(0.020 - x)$		x		x

According to the reaction stoichiometry,

$$[H_3O^+] = [C_6H_5CO_2^-] = x \text{ at equilibrium}$$

Stoichiometry also tells us that the quantity of acid ionized is x. Thus, the benzoic acid concentration at equilibrium is

$$[C_6H_5CO_2H] = \text{initial acid concentration} - \text{quantity of acid that ionized}$$
$$[C_6H_5CO_2H] = [C_6H_5CO_2H]_0 - x$$
$$[C_6H_5CO_2H] = 0.020 - x$$

Substituting these equilibrium concentrations into the K_a expression, we have

$$K_a = \frac{[H_3O^+][C_6H_5CO_2^-]}{[C_6H_5CO_2H]}$$

$$= 6.3 \times 10^{-5} = \frac{(x)(x)}{0.020 - x}$$

The value of x is small compared with 0.020 (because $[HA]_0 > 100 \times K_a$; 0.020 M $> 6.3 \times 10^{-3}$). Therefore,

$$K_a = 6.3 \times 10^{-5} = \frac{x^2}{0.020}$$

Solving for x, we have

$$x = \sqrt{K_a \times (0.020)} = 0.0011 \text{ M}$$

and we find that

$$[H_3O^+] = [C_6H_5CO_2^-] = 0.0011 \text{ M}$$

and

$$[C_6H_5CO_2H] = (0.020 - x) = 0.019 \text{ M}$$

Finally, the pH of the solution is found to be

$$pH = -\log(1.1 \times 10^{-3}) = \boxed{2.96}$$

Comment Let us think again about the result. Because benzoic acid is weak, we made the approximation that $(0.020 - x) \approx 0.020$. If we do *not* make the approximation and instead solve the exact expression, $x = [H_3O^+] = 1.1 \times 10^{-3}$ M. This is the same answer to two significant figures that we obtained from the "approximate" expression. Finally, notice that we again ignored any H_3O^+ that arises from water ionization.

Example 17.6—Calculating Equilibrium Concentrations and pH from K_a and Using the Method of Successive Approximations

Problem What is the pH of a 0.0010 M solution of formic acid? What is the concentration of formic acid at equilibrium? The acid is moderately weak, with $K_a = 1.8 \times 10^{-4}$.

$$HCO_2H(aq) + H_2O(\ell) \rightleftharpoons HCO_2^-(aq) + H_3O^+(aq)$$

Strategy This problem is similar to Example 17.5 except that an approximate solution will not be possible.

Solution The ICE table is shown here.

Equilibrium	HCO_2H	$+ H_2O$	$\rightleftharpoons$	HCO_2^-	$+$	H_3O^+
Initial (M)	0.0010			0		0
Change (M)	$-x$			$+x$		$+x$
Equilibrium (M)	$(0.0010 - x)$			x		x

Substituting the values from the table into the K_a expression, we have

$$K_a = \frac{[H_3O^+][HCO_2^-]}{[HCO_2H]} = 1.8 \times 10^{-4} = \frac{(x)(x)}{0.0010 - x}$$

Formic acid is a weak acid because it has a value of K_a much less than 1. In this example, however, $[HA]_0$ ($= 0.0010$ M) is *not* greater than $100 \times K_a$ ($= 1.8 \times 10^{-2}$), so the usual approximation is not reasonable. Thus, we have to find the equilibrium concentrations by solving the "exact" expression. This can be solved with the quadratic formula (page 775) or by successive approximations (Appendix A). Let us use the successive approximation method here.

Begin by solving the approximate expression for x.

$$1.8 \times 10^{-4} = \frac{(x)(x)}{0.0010}$$

Solving this, $x = 4.2 \times 10^{-4}$. Put this value into the expression for x in the denominator of the exact expression.

$$1.8 \times 10^{-4} = \frac{(x)(x)}{0.0010 - x} = \frac{(x)(x)}{0.0010 - 4.2 \times 10^{-4}}$$

Solving this equation for x, we find $x = 3.2 \times 10^{-4}$. Again put this value into the denominator and solve for x.

$$1.8 \times 10^{-4} = \frac{(x)(x)}{0.0010 - x} = \frac{(x)(x)}{0.0010 - 3.2 \times 10^{-4}}$$

Continue this procedure until the value of x does not change from one cycle to the next. In this case, two more steps give us the result that

$$x = [H_3O^+] = [HCO_2^-] = 3.4 \times 10^{-4} \text{ M}$$

Thus,

$$[HCO_2H] = 0.0010 - x \approx 0.0007 \text{ M}$$

and the pH of the formic acid solution is

$$pH = -\log(3.4 \times 10^{-4}) = \boxed{3.47}$$

Comment If we had used the approximate expression to find the H_3O^+ concentration, we would have obtained a value of $[H_3O^+] = 4.2 \times 10^{-4}$ M. A simplifying assumption led to a large error, about 24%. The approximate solution fails in this case because (1) the acid concentration is small and (2) the acid is not all that weak. These facts made invalid the approximation that $[HA]_{equilibrium} \approx [HA]_0$.

Exercise 17.13—Calculating Equilibrium Concentrations and pH from K_a

What are the equilibrium concentrations of acetic acid, the acetate ion, and H_3O^+ for a 0.10 M solution of acetic acid ($K_a = 1.8 \times 10^{-5}$)? What is the pH of the solution?

Ammonia, NH_3
$K_b = 1.8 \times 10^{-5}$

Caffeine, $C_8H_{10}N_4O_2$
$K_b = 2.5 \times 10^{-4}$

Benzoate ion, $C_6H_5CO_2^-$
$K_b = 1.6 \times 10^{-10}$

Phosphate ion, PO_4^{3-}
$K_b = 2.8 \times 10^{-2}$

Photo: Charles D. Winters

Figure 17.5   Examples of weak bases. Weak bases in water include molecules having one or more N atoms capable of accepting a H^+ ion. Anionic bases are conjugate bases of weak acids.

Exercise 17.14—Calculating Equilibrium Concentrations and pH from K_a

What are the equilibrium concentrations of HF, F^- ion, and H_3O^+ ion in a 0.015 M solution of HF? What is the pH of the solution?

Just as acids can be molecular species or ions, so too can bases be molecular or ionic (Figures 17.3–17.5). Many molecular bases are based on nitrogen, with ammonia being the simplest. Many other nitrogen-containing bases occur naturally; caffeine and nicotine are two well-known examples. The anionic conjugate bases of weak acids make up another group of bases. The following example describes the calculation of the pH for a solution of sodium acetate.

GENERAL
Chemistry.⚛.Now™

See the General ChemistryNow CD-ROM or website:

• **Screen 17.10 Estimating the pH of Weak Base Solutions,** for a simulation on predicting the pH of weak bases in water

• **Screen 17.11 Acid–Base Properties of Salts,** for a simulation on predicting the pH of salt solutions

Example 17.7—The pH of a Solution of a Weakly Basic Salt, Sodium Acetate

Problem What is the pH of a 0.015 M solution of sodium acetate, $NaCH_3CO_2$ at 25 °C?

Strategy Sodium acetate will be basic in water because the acetate ion, the conjugate base of a weak acid, acetic acid, reacts with water to form OH^- (Tables 17.3 and 17.4). (Note that the sodium ion of sodium acetate does not affect the solution pH.) We shall calculate the hydroxide ion concentration in a manner parallel to that in Example 17.5.

Solution The value of K_b for the acetate ion is 5.6×10^{-10} (Table 17.3).

$$CH_3CO_2^-(aq) + H_2O(\ell) \rightleftharpoons CH_3CO_2H(aq) + OH^-(aq)$$

Set up an ICE table to summarize the initial and equilibrium concentrations of the species in solution.

Equilibrium	$CH_3CO_2^- + H_2O$ $\rightleftharpoons$	CH_3CO_2H +	OH^-
Initial (M)	0.015	0	0
Change (M)	$-x$	$+x$	$+x$
Equilibrium (M)	$(0.015 - x)$	x	x

Next substitute the values from the table into the K_b expression.

$$K_b = 5.6 \times 10^{-10} = \frac{[CH_3CO_2H][OH^-]}{[CH_3CO_2^-]} = \frac{x^2}{0.015 - x}$$

The acetate ion is a very weak base, as reflected by the very small value of K_b. Therefore, we assume that x, the concentration of hydroxide ion generated by the reaction of acetate ion with water, is very small, and we use the approximate expression to solve for x.

$$K_b = 5.6 \times 10^{-10} = \frac{x^2}{0.015}$$

$$x = [OH^-] = [CH_3CO_2H] = \sqrt{(5.6 \times 10^{-10})(0.015)}$$

$$= [OH^-] = [CH_3CO_2H] = 2.9 \times 10^{-6} \text{ M}$$

To calculate the pH of the solution, we need the hydronium ion concentration. In aqueous solutions at 25 °C, it is always true that

$$K_w = 1.0 \times 10^{-14} = [H_3O^+][OH^-]$$

Therefore,

$$[H_3O^+] = \frac{K_w}{[OH^-]} = \frac{1.0 \times 10^{-14}}{2.9 \times 10^{-6}} = 3.5 \times 10^{-9} \text{ M}$$

$$pH = -\log(3.5 \times 10^{-9}) = 8.46$$

The acetate ion does indeed give rise to a weakly basic solution.

Comment The hydroxide ion concentration (x) is quite small relative to the initial acetate ion concentration. (We would have predicted this from our "rule of thumb": that $100 \times K_b$ should be less than the initial base concentration if we wish to use the approximate expression.)

Exercise 17.15—The pH of a Solution of a Conjugate Base of a Weak Acid

Sodium hypochlorite, NaClO, is used as a disinfectant in swimming pools and water treatment plants. What are the concentrations of HClO and OH^- and the pH of a 0.015 M solution of NaClO at 25 °C?

What Is the pH of a Solution After an Acid–Base Reaction?

In Section 17.6 you learned how to predict the relative pH of the solution resulting from an acid–base reaction. Whether a solution will be acidic, basic, or neutral depends on the reactants, and the results are summarized in Table 17.5. Let us turn now to the way in which you can calculate a value for the pH after such a reaction.

See the General ChemistryNow CD-ROM or website:

• **Screen 17.10 Estimating the pH of Weak Base Solutions,** for a tutorial on estimating the pH following an acid-base reaction

Example 17.8—Calculating the pH After the Reaction of a Base with an Acid

Problem What is the pH of the solution that results from mixing 25 mL of 0.016 M NH_3 and 25 mL of 0.016 M HCl?

Strategy This question involves three problems in one:

(a) *Writing a Balanced Equation.* We first have to write a balanced equation for the reaction that occurs and then decide whether the reaction products are acidic or basic. Here NH_4^+ is the product of interest, and it is a weak acid.

(b) *Stoichiometry Problem.* To find the "initial" NH_4^+ concentration is a stoichiometry problem: What amount of NH_4^+ (in moles) is produced in the HCl + NH_3 reaction, and in what volume of solution is the NH_4^+ ion found?

(c) *Equilibrium Problem.* Calculating the pH involves solving an equilibrium problem. The crucial piece of information needed here is the "initial" concentration of NH_4^+ from part (b).

Solution If equal molar quantities of base (NH_3) and acid (HCl) are mixed, the result should be an acidic solution because the significant species remaining in solution upon completion of the reaction is NH_4^+, the conjugate acid of the weak base NH_3 (see Tables 17.3 and 17.5). The chemistry can be summarized by writing the following net ionic equations.

(a) *Writing a Balanced Equation*

Reaction of HCl (the supplier of hydronium ion) with NH_3 to give NH_4^+:

$$NH_3(aq) + H_3O^+(aq) \longrightarrow NH_4^+(aq) + H_2O(\ell)$$

Reaction of NH_4^+ with water:

$$NH_4^+(aq) + H_2O(\ell) \rightleftharpoons H_3O^+(aq) + NH_3(aq)$$

(b) *Stoichiometry Problem*

Amount of HCl and NH_3 consumed:

$$(0.025 \text{ L HCl})(0.016 \text{ mol/L}) = 4.0 \times 10^{-4} \text{ mol HCl}$$

$$(0.025 \text{ L } NH_3)(0.016 \text{ mol/L}) = 4.0 \times 10^{-4} \text{ mol } NH_3$$

Amount of NH_4^+ produced upon completion of the reaction:

$$4.0 \times 10^{-4} \text{ mol } NH_3 \left(\frac{1 \text{ mol } NH_4^+}{1 \text{ mol } NH_3} \right) = 4.0 \times 10^{-4} \text{ mol } NH_4^+$$

Concentration of NH_4^+: Combining 25 mL each of HCl and NH_3 gives a total solution volume of 50. mL. Therefore, the concentration of NH_4^+ is

$$[NH_4^+] = \frac{4.0 \times 10^{-4} \text{ mol}}{0.050 \text{ L}} = 8.0 \times 10^{-3} \text{ M}$$

(c) *Acid–Base Equilibrium Problem*

With the initial concentration of ammonium ion known, set up an ICE table to find the equilibrium concentration of hydronium ion.

Problem-Solving Tip 17.3

What Is the pH After Mixing Equal Molar Amounts of an Acid and a Base?

Table 17.5 summarizes the outcome of mixing various types of acids and bases. But how do you calculate a numerical value for the pH, particularly in the case of mixing a weak acid with a strong base or a weak base with a strong acid? The strategy (Example 17.8) is to recognize that this problem's solution involves two related

calculations: a stoichiometry calculation and an equilibrium calculation. The key is that you need to know the concentration of the weak acid or weak base produced when the acid and base are mixed. You should ask yourself the following questions:

1. What amounts of acid and base are used (in moles)? (This is a stoichiometry problem.)

2. What is the total volume of the solution after mixing the acid and base solutions?

3. What is the concentration of the weak acid or base produced on mixing the acid and base solutions?

4. Using the concentration found in Step 3, what is the hydronium ion concentration in the solution? (This is an equilibrium problem.)

5. Calculate the pH of the solution from $[H_3O^+]$.

Equilibrium	$NH_4^+ + H_2O$	$\rightleftharpoons$	NH_3	$+$	H_3O^+
Initial (M)	0.0080		0		0
Change (M)	$-x$		$+x$		$+x$
Equilibrium (M)	$(0.0080 - x)$		x		x

Next, substitute the values in the table into the K_a expression for the ammonium ion. Thus, we have

$$K_a = 5.6 \times 10^{-10} = \frac{[H_3O^+][NH_3]}{[NH_4^+]} = \frac{(x)(x)}{0.0080 - x}$$

The ammonium ion is a very weak acid, as reflected by the very small value of K_a. Therefore, x, the concentration of hydronium ion generated by reaction of ammonium ion with water, is assumed to be very small, and the approximate expression is used to solve for x. (Here $100 \times K_a$ is much less than the original acid concentration.)

$$K_a = 5.6 \times 10^{-10} \approx \frac{x^2}{0.0080}$$

$$x = \sqrt{(5.6 \times 10^{-10})(0.0080)}$$

$$= [H_3O^+] = [NH_3] = 2.1 \times 10^{-6} \text{ M}$$

$$pH = -\log(2.1 \times 10^{-6}) = 5.67$$

Comment As predicted (Table 17.5), the solution after mixing equal molar quantities of a strong acid and weak base is weakly acidic.

Exercise 17.16—What Is the pH After the Reaction of a Weak Acid and a Strong Base?

Calculate the pH after mixing 15 mL of 0.12 M acetic acid with 15 mL of 0.12 M NaOH. What are the major species in solution at equilibrium (besides water) and what are their concentrations?

17.8—Polyprotic Acids and Bases

Polyprotic acids are capable of donating more than one proton (Table 17.1). Many of these acids occur in nature, such as oxalic acid in rhubarb (page 796), citric acid in citrus fruit, malic acid in apples, and tartaric acid in grapes (page 801).

Phosphoric acid ionizes in three steps:

First ionization step:

$$H_3PO_4(aq) + H_2O(\ell) \rightleftharpoons H_3O^+(aq) + H_2PO_4^-(aq) \qquad K_{a_1} = 7.5 \times 10^{-3}$$

Second ionization step:

$$H_2PO_4^-(aq) + H_2O(\ell) \rightleftharpoons H_3O^+(aq) + HPO_4^{2-}(aq) \qquad K_{a_2} = 6.2 \times 10^{-8}$$

Third ionization step:

$$HPO_4^{2-}(aq) + H_2O(\ell) \rightleftharpoons H_3O^+(aq) + PO_4^{3-}(aq) \qquad K_{a_3} = 3.6 \times 10^{-13}$$

Notice that the K_a value for each successive step becomes smaller and smaller because it is more difficult to remove H^+ from a negatively charged ion, such as $H_2PO_4^-$, than from a neutral molecule, such as H_3PO_4. Furthermore, the larger the negative charge of the anionic acid, the more difficult it is to remove H^+.

For many inorganic polyprotic acids, such as phosphoric acid, carbonic acid, and hydrogen sulfide, each successive loss of a proton is about 10^4 to 10^6 times more difficult than the previous ionization step. As a consequence, the first ionization step of a polyprotic acid produces up to about a million times more H_3O^+ ions than the second step. For this reason, *the pH of many inorganic polyprotic acids depends primarily on the hydronium ion generated in the first ionization step; the hydronium ion produced in the second step can be neglected.* The same principle applies to the conjugate bases of polyprotic acids. It is illustrated by the calculation of the pH of a solution of carbonate ion, an important basic anion in our environment (Example 17.9).

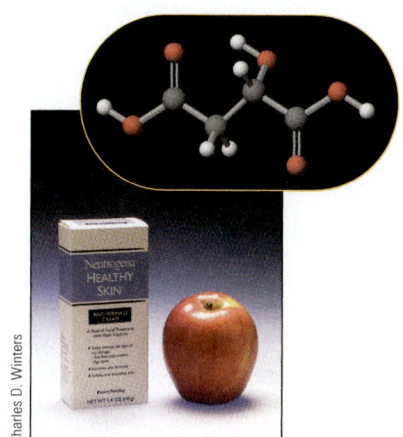

A polyprotic acid. Malic acid is a diprotic acid occurring in apples. It is also classified as an alpha-hydroxy acid because it has an —OH group on the C atom next to the —CO_2H (in the alpha position). It is one of a larger group of natural acids such as lactic acid, citric acid, and ascorbic acid. Alpha-hydroxy acids have been touted as an ingredient in "anti-aging" skin creams. They work by accelerating the natural process by which skin replaces the outer layer of cells with new cells.

Charles D. Winters

Example 17.9—Calculating the pH of the Solution of a Polyprotic Base

Problem The carbonate ion, CO_3^{2-}, is a base in water, forming the hydrogen carbonate ion, which in turn can form carbonic acid.

$$CO_3^{2-}(aq) + H_2O(\ell) \rightleftharpoons HCO_3^-(aq) + OH^-(aq) \qquad K_{b1} = 2.1 \times 10^{-4}$$
$$HCO_3^-(aq) + H_2O(\ell) \rightleftharpoons H_2CO_3(aq) + OH^-(aq) \qquad K_{b2} = 2.4 \times 10^{-8}$$

What is the pH of a 0.10 M solution of Na_2CO_3 at 25 °C?

Strategy The second ionization constant, K_{b2}, is much smaller than the first ionization constant, K_{b1}, so the hydroxide ion concentration in the solution results almost entirely from the first step. Therefore, let us calculate the OH^- concentration produced in the first ionization step but test the conclusion that OH^- produced in the second step is negligible.

Solution Set up an ICE table for the reaction of the carbonate ion (Equilibrium Table 1).

Equilibrium Table 1: Reaction of CO_3^{2-} Ion

Equilibrium	CO_3^{2-} + H_2O	$\rightleftharpoons$	HCO_3^-	+	OH^-
Initial (M)	0.10		0		0
Change (M)	−x		+x		+x
Equilibrium (M)	(0.10 − x)		x		x

Based on this table, the equilibrium concentration of OH^- ($= x$) can be calculated.

$$K_{b1} = 2.1 \times 10^{-4} = \frac{[HCO_3^-][OH^-]}{[CO_3^{2-}]} = \frac{x^2}{0.10 - x}$$

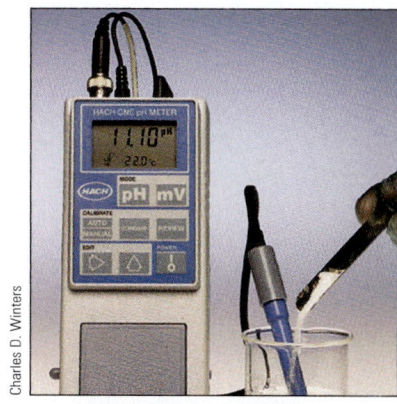

Sodium carbonate, a polyprotic base. This common substance is a base in aqueous solution. Its primary use is in the glass industry. Although it used to be manufactured, it is now mined as the mineral trona, $Na_2CO_3 \cdot NaHCO_3 \cdot 2\ H_2O$.

Charles D. Winters

Because K_{b1} is relatively small, it is reasonable to make the approximation that $(0.10 - x) \approx 0.10$. Therefore,

$$x = [HCO_3^-] = [OH^-] = \sqrt{(2.1 \times 10^{-4})(0.10)} = 4.6 \times 10^{-3} \text{ M}$$

Using this value of $[OH^-]$, we first calculate the pOH of the solution.

$$pOH = -\log (4.6 \times 10^{-3}) = 2.34$$

We then use the relationship $pH + pOH = 14.00$ (at 25 °C) to calculate the pH.

$$pH = 14.00 - pOH = \boxed{11.66}$$

Comment It is instructive to ask what the concentration of H_2CO_3 in the solution might be. If HCO_3^- were to react significantly with water to produce H_2CO_3, the pH of the solution would be affected. Let us set up a second ICE Table.

Equilibrium Table 2: Reaction of HCO_3^- Ion

Equilibrium	$HCO_3^- + H_2O \rightleftharpoons$	H_2CO_3	$+$	OH^-
Initial (M)	4.6×10^{-3}	0		4.6×10^{-3}
Change (M)	$-y$	$+y$		$+y$
Equilibrium (M)	$(4.6 \times 10^{-3} - y)$	y		$(4.6 \times 10^{-3} + y)$

Because K_{b2} is so small, the second step occurs to a *much* smaller extent than the first step. Thus, the amount of H_2CO_3 and OH^- produced in the second step $(= y)$ is *much* smaller than 10^{-3} M. It is therefore reasonable to assume that both $[HCO_3^-]$ and $[OH^-]$ are very close to 4.6×10^{-3} M.

$$K_{b2} = 2.4 \times 10^{-8} = \frac{[H_2CO_3][OH^-]}{[HCO_3^-]} = \frac{(y)(4.6 \times 10^{-3})}{4.6 \times 10^{-3}}$$

Because $[HCO_3^-]$ and $[OH^-]$ (from step 2) have nearly identical values, they cancel from the expression, and we find that $[H_2CO_3]$ is simply equal to K_{b2}.

$$y = [H_2CO_3] = K_{b2} \approx 2.4 \times 10^{-8} \text{ M}$$

For the carbonate ion, where K_1 and K_2 differ by about 10^4, essentially all of the hydroxide ion is produced in the first equilibrium process.

Exercise 17.17—Calculating the pH of the Solution of a Polyprotic Acid

What is the pH of a 0.10 M solution of oxalic acid, $H_2C_2O_4$? What are the concentrations of H_3O^+, $HC_2O_4^-$, and the oxalate ion, $C_2O_4^{2-}$?

17.9—The Lewis Concept of Acids and Bases

The concept of acid–base behavior advanced by Brønsted and Lowry in the 1920s works well for reactions involving proton transfer. A more general acid–base concept, however, was developed by Gilbert N. Lewis in the 1930s. This concept is based on the sharing of electron pairs between an acid and a base. A **Lewis acid** is a substance that can accept a pair of electrons from another atom to form a new bond, and a **Lewis base** is a substance that can donate a pair of electrons to another atom to form a new bond. Thus, an acid–base reaction in the Lewis sense occurs when a molecule (or ion) donates a pair of electrons to another molecule (or ion).

$$A + \quad B: \quad \longrightarrow \quad B{\rightarrow}A$$

$$\text{Acid} \qquad \text{Base} \qquad\qquad \text{Adduct}$$

The product is often called an **acid–base adduct**. In Section 9.6 this type of chemical bond was called a *coordinate covalent bond.*

Formation of a hydronium ion from H^+ and water is a good example of a Lewis acid–base reaction. The H^+ ion has no electrons in its valence ($1s$) orbital, and the water molecule has two unshared pairs of electrons (located in sp^3 hybrid orbitals). One of the O atom lone pairs of a water molecule can be shared with an H^+ ion, thus forming an H—O bond in an H_3O^+ ion. A similar interaction occurs between H^+ and the base ammonia to form the ammonium ion.

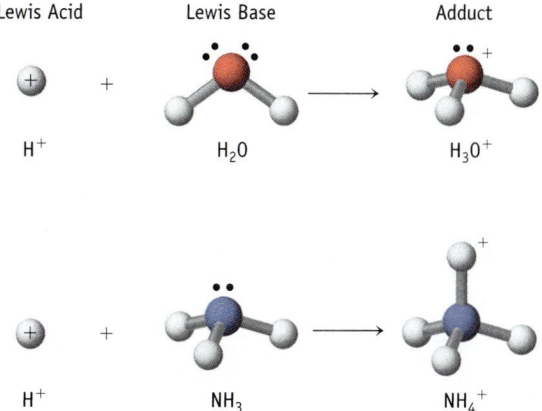

Such reactions are very common. In general, they involve Lewis acids that are cations or neutral molecules with an available, empty valence orbital and bases that are anions or neutral molecules with a lone electron pair.

Cationic Lewis Acids

Metal cations interact with water molecules to form hydrated cations, ions in which the metal ion is surrounded by water molecules (Figure 17.6 and page 658). In these species, coordinate covalent bonds form between the metal cation and a lone pair of electrons on the O atom of each water. For example, an iron(II) ion, Fe^{2+}, forms six coordinate covalent bonds to water.

$$Fe^{2+}(aq) + 6\ H_2O(\ell) \longrightarrow [Fe(H_2O)_6]^{2+}(aq)$$

Similar structures formed by transition metal cations are generally very colorful (Figure 17.6 and Section 22.3). Chemists call them **complex ions** or, because of the coordinate covalent bond, **coordination complexes**. Several are listed in Table 17.3 as acids, and their behavior is described further in Section 17.10 and Chapter 22.

Like water, ammonia is an excellent Lewis base and combines with metal cations to give adducts (complex ions), which are often very colorful. For example, copper(II) ions, which are light blue in aqueous solution (Figure 17.6), react with ammonia to give a deep blue adduct with four ammonia molecules surrounding each Cu^{2+} ion (Figure 17.7).

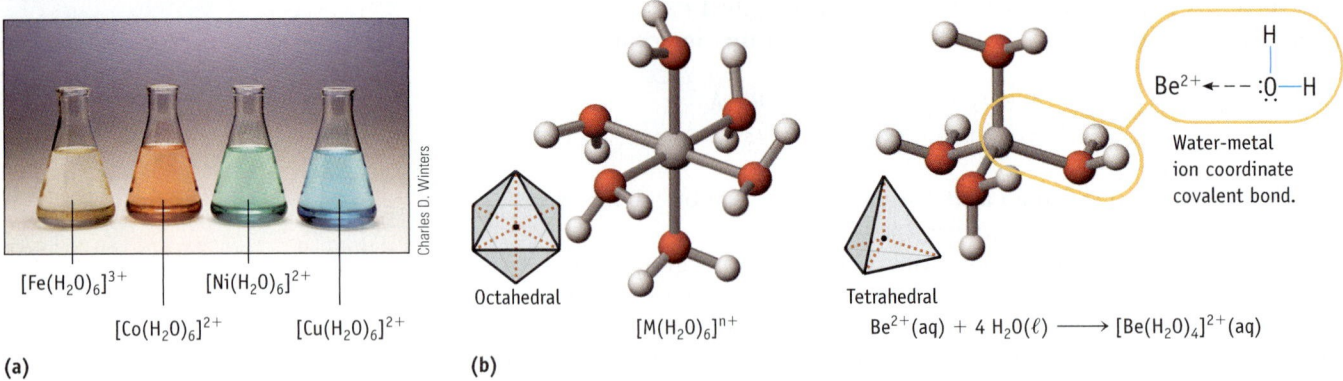

(a)

(b)

Figure 17.6   Metal cations in water. (a) Solutions of the nitrate salts of iron(III), cobalt(II), nickel(II), and copper(II). All have characteristic colors. (b) Models of complex ions (Lewis acid–base adducts) formed between a metal cation and water molecules. Such complexes often have six water molecules arranged octahedrally around the metal cation.

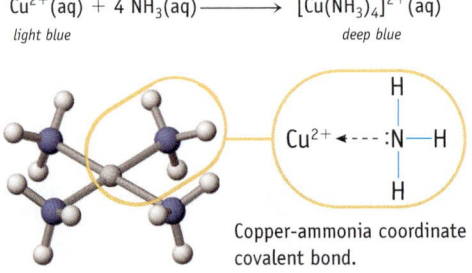

Hydroxide ion, OH^-, is an excellent Lewis base and binds readily to metal cations to give metal hydroxides. An important feature of the chemistry of some metal hydroxides is that they are **amphoteric**. An amphoteric metal hydroxide can behave as an acid or a base (Table 17.6). One of the best examples of this behavior is provided by aluminum hydroxide, $Al(OH)_3$ (Figure 17.8). Adding OH^- to a precipitate of $Al(OH)_3$ produces the water-soluble $[Al(OH)_4]^-$ ion.

$$Al(OH)_3(s) + OH^-(aq) \longrightarrow [Al(OH)_4]^-(aq)$$
Acid   Base

If acid is added to the $Al(OH)_3$ precipitate, it again dissolves. In this reaction, however, aluminum hydroxide acts as a base.

$$Al(OH)_3(s) + 3\ H_3O^+(aq) \longrightarrow Al^{3+}(aq) + 6\ H_2O(\ell)$$
Base   Acid

Molecular Lewis Acids

Lewis's acid–base concept accounts nicely for the fact that oxides of nonmetals behave as acids [◀ Section 5.3]. Two important examples of acidic oxides are carbon dioxide and sulfur dioxide.

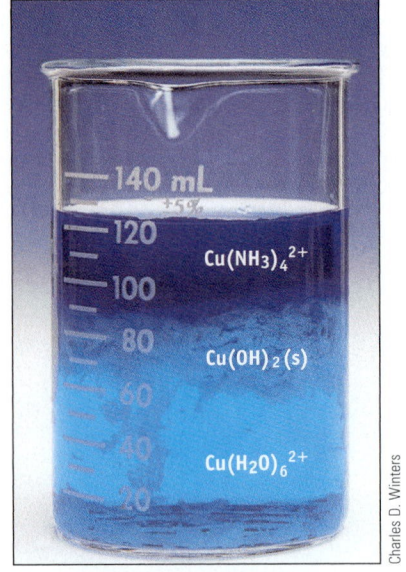

Figure 17.7   The Lewis acid–base complex ion $[Cu(NH_3)_4]^{2+}$. Here aqueous ammonia was added to aqueous $CuSO_4$ (the light blue solution at the bottom of the beaker). The small concentration of OH^- in $NH_3(aq)$ first formed insoluble blue-white $Cu(OH)_2$ (the solid in the middle of the beaker). With additional NH_3, however, the deep blue, soluble complex ion is formed (the solution at the top of the beaker). The model in the text shows the copper(II)–ammonia complex ion.

Table 17.6   Some Common Amphoteric Metal Hydroxides*

Hydroxide	Reaction as a Base	Reaction as an Acid
$Al(OH)_3$	$Al(OH)_3(s) + 3\ H_3O^+(aq) \rightleftharpoons Al^{3+}(aq) + 6\ H_2O(\ell)$	$Al(OH)_3(s) + OH^-(aq) \rightleftharpoons [Al(OH)_4]^-(aq)$
$Zn(OH)_2$	$Zn(OH)_2(s) + 2\ H_3O^+(aq) \rightleftharpoons Zn^{2+}(aq) + 4\ H_2O(\ell)$	$Zn(OH)_2(s) + 2\ OH^-(aq) \rightleftharpoons [Zn(OH)_4]^{2-}(aq)$
$Sn(OH)_4$	$Sn(OH)_4(s) + 4\ H_3O^+(aq) \rightleftharpoons Sn^{4+}(aq) + 8\ H_2O(\ell)$	$Sn(OH)_4(s) + 2\ OH^-(aq) \rightleftharpoons [Sn(OH)_6]^{2-}(aq)$
$Cr(OH)_3$	$Cr(OH)_3(s) + 3\ H_3O^+(aq) \rightleftharpoons Cr^{3+}(aq) + 6\ H_2O(\ell)$	$Cr(OH)_3(s) + OH^-(aq) \rightleftharpoons [Cr(OH)_4]^-(aq)$

* The aqueous metal cations are best described as $[M(H_2O)_6]^{n+}$.

Because oxygen is electronegative, the C—O bonding electrons in CO_2 are polarized away from carbon and toward oxygen. This causes the carbon atom to be slightly positive. The negatively charged Lewis base OH^- can then attack this atom to give, ultimately, the bicarbonate ion.

■ **CO_2 in Basic Solution**
This reaction of CO_2 with OH^- is the first step in the precipitation of $CaCO_3$ when CO_2 is bubbled into a solution of $Ca(OH)_2$ (Figure 16.2).

Similarly, SO_2 reacts with aqueous OH^- to form the HSO_3^- ion.

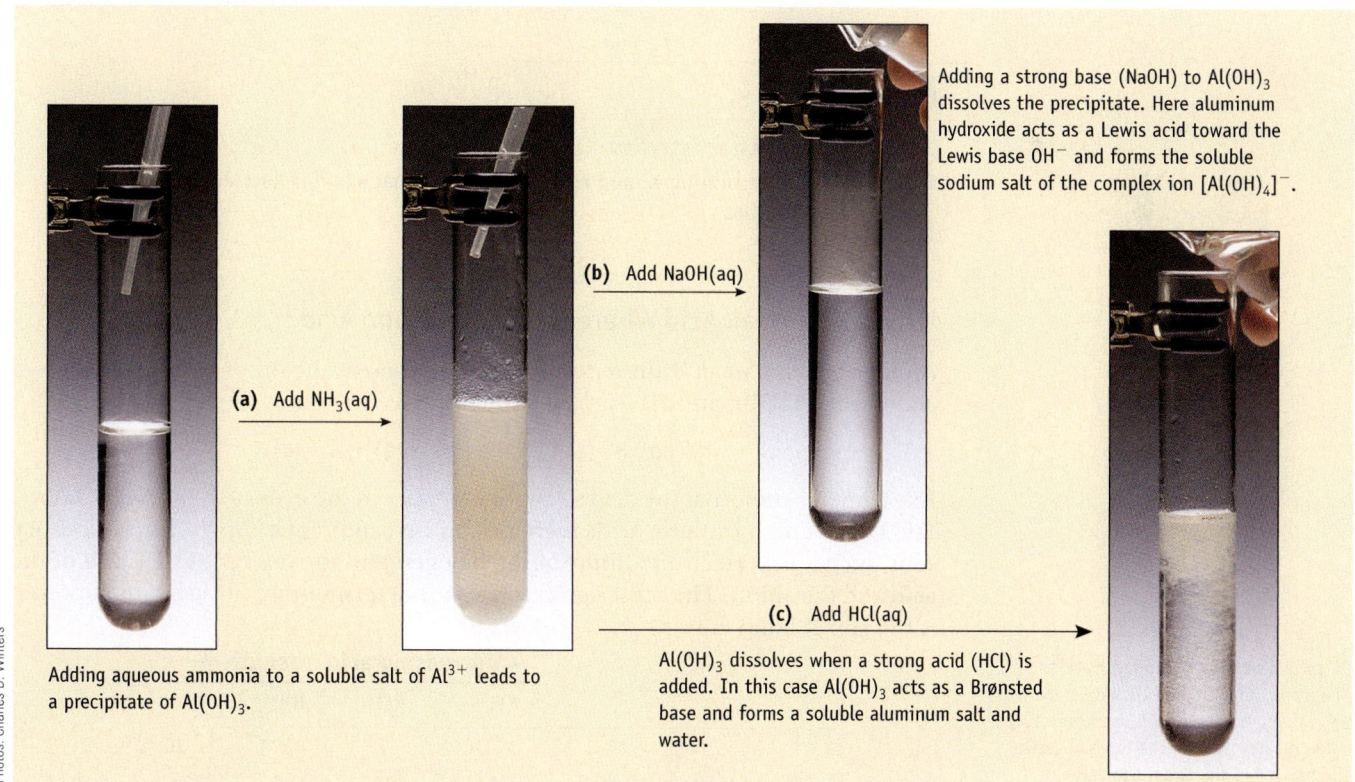

Adding a strong base (NaOH) to $Al(OH)_3$ dissolves the precipitate. Here aluminum hydroxide acts as a Lewis acid toward the Lewis base OH^- and forms the soluble sodium salt of the complex ion $[Al(OH)_4]^-$.

(a) Add NH_3(aq)

(b) Add NaOH(aq)

(c) Add HCl(aq)

Adding aqueous ammonia to a soluble salt of Al^{3+} leads to a precipitate of $Al(OH)_3$.

$Al(OH)_3$ dissolves when a strong acid (HCl) is added. In this case $Al(OH)_3$ acts as a Brønsted base and forms a soluble aluminum salt and water.

Photos: Charles D. Winters

Figure 17.8   The amphoteric nature of $Al(OH)_3$. Aluminum hydroxide is formed by the reaction of aqueous Al^{3+} and ammonia.

$$Al^{3+}(aq) + 3\ NH_3(aq) + 3\ H_2O(\ell) \rightleftharpoons Al(OH)_3(s) + 3\ NH_4^+(aq)$$

Reactions of solid $Al(OH)_3$ with aqueous NaOH and HCl demonstrate that aluminum hydroxide is amphoteric.

GENERAL
Chemistry-᠅-Now™

See the General ChemistryNow CD-ROM or website:
- **Screen 17.12 Lewis Acids and Bases,** for a tutorial on identifying Lewis base sites
- **Screen 17.14 Neutral Lewis Acids,** for a tutorial on identifying Lewis acid sites

---

Exercise 17.18—Lewis Acids and Bases

Describe each of the following as a Lewis acid or a Lewis base.

(a) PH_3 **(b)** BCl_3 **(c)** H_2S **(d)** HS^-

Hint: In each case draw the Lewis electron dot structure of the molecule or ion. Does the central atom have lone pairs of electrons? If so, it can be a Lewis base. Does the central atom lack an electron pair? If so, it can behave as a Lewis acid.

17.10—Molecular Structure, Bonding, and Acid–Base Behavior

One of the most interesting aspects of chemistry is the correlation between molecular structure, bonding, and observed properties. Here it is useful to analyze the connection between the structure and bonding in some acids and their relative strengths.

GENERAL
Chemistry-᠅-Now™

See the General ChemistryNow CD-ROM or website:
- **Screen 17.8 Determining K_a and K_b Values,** for information on molecular interpretation of acid–base properties

Why Is HF a Weak Acid Whereas HCl Is a Strong Acid?

Aqueous HF is a weak Brønsted acid in water, whereas the other hydrohalic acids—aqueous HCl, HBr, and HI—are all strong acids.

$$HX(aq) + H_2O(\ell) \longrightarrow H_3O^+(aq) + X^-(aq)$$

Experiments show that the acid strength increases in the order HF $\ll$ HCl $<$ HBr $<$ HI. The strength of these acids increases on descending Group 7A for several reasons, such as the electron affinity of the halogen and the energy of solvation of the acid and the anion. The most significant factor determining acid strength, however, is the H—X bond energy.

■ **pK_a Values for Hydrogen Halides**
The acids HCl, HBr, and HI have negative pK_a values. A negative pK_a indicates a K_a value greater than 1. The more negative the pK_a value, the larger the value of K_a and the stronger the acid.

	—Increasing acid strength ⟶			
	HF	**HCl**	**HBr**	**HI**
pK_a	+3.14	−7	−9	−10
H—X bond strength (kJ/mol)	565	432	366	299

The weakest acid, HF, has the strongest H—X bond, whereas the strongest acid, HI, has the weakest H—X bond.

Chemical Perspectives

Lewis and Brønsted Bases: Adrenaline and Serotonin

You are going to take a chemistry exam—your heart races and you begin to sweat. These actions of your nervous system are affected by a chemical compound called *epinephrine*, which is also known as *adrenaline*. This compound has a basic NH_2 group that is protonated in an acid solution. Epinephrine, which is produced in the body through a chain of reactions starting with the amino acid phenylalanine, is known as the "flight or fight" hormone. It causes the release of glucose and other nutrients into the blood and stimulates brain function. Currently, epinephrine is used as a bronchodilator by people with asthma. It is also used to treat glaucoma.

Epinephrine is a member of a class of compounds called neurotransmitters. This class includes serotonin, another Lewis and Brønsted base. Very low levels of serotonin are associated with depression, whereas very high levels can produce a manic state.

Serotonin is derived from the amino acid tryptophan. Some people take tryptophan because they believe that it makes them feel good and helps them sleep at night. Milk proteins have a high level of tryptophan, which may explain why you enjoy a glass of milk or dish of ice cream before you go to bed at night.

See J. Mann: *Murder, Magic, and Medicine*, New York, Oxford University Press, 1994.

Phenylalanine

Epinephrine · HCl

Serotonin

Why Is HNO₂ a Weak Acid Whereas HNO₃ Is a Strong Acid?

Nitrous acid (HNO_2) and nitric acid (HNO_3) are examples of several series of **oxoacids**. Oxoacids contain an atom (usually a nonmetal atom) bonded to one or more oxygen atoms, some with hydrogen atoms attached. Besides those oxoacids based on N, you are familiar with the sulfur- and chlorine-based oxoacids (Table 17.7). In all of these series of related compounds, the acid strength increases as the number of oxygen atoms bonded to the central element increases.

HNO_3, strong acid, $pK_a = -1.4$

$>>$

HNO_2, weak acid, $pK_a = +3.35$

Thus, nitric acid (HNO_3) is a stronger acid than nitrous acid (HNO_2), and the order of acid strength for the chlorine-based oxoacids is $HOCl < HOClO < HOClO_2 < HOClO_3$ (Table 17.7).

Table 17.7 Oxoacids

Acid	pK_a
Cl-Based Oxoacids	
HOCl	7.46
HOClO ($HClO_2$)	~ 2
HOClO₂ ($HClO_3$)	~ −1
HOClO₃ ($HClO_4$)	~ −10
S-Based Oxoacids	
$(HO)_2SO$ [H_2SO_3]	1.92, 7.21
$(HO)_2SO_2$ [H_2SO_4]	~ −3, 1.92

According to Linus Pauling, for oxoacids with the general formula $(HO)_nE(0)_m$, the value of pK_a is about 8–5m. When $n > 1$, the pK_a increases by about 5 for each successive loss of a proton.

Acid strength is determined by the extent to which the ionization reaction is favored—that is, the extent to which the reaction of an acid HX to form H^+ and the conjugate base (X^-) is product-favored. It is necessary to assume that acid strength will be related to characteristics of *both* the reactant (HX) *and* the products (H^+ and X^-).

In an oxoacid molecule, attention focuses on the H—O bond and the influence of other atoms in the molecule on this bond. Because of the difference in the electronegativities of O and H, the H—O bond is polar ($O^{\delta-}$—$H^{\delta+}$). But the question is how the other atoms in the molecule affect the H—O bond polarity. One view is that electrons in the H—O bond are attracted by other electronegative atoms or groups in the molecule and are drawn away from the H—O hydrogen. This increases the H—O bond polarity and makes hydrogen easier to separate from the molecule as H^+.

The extent to which adjacent atoms or groups of atoms attract electrons from another part of a molecule is called the **inductive effect**; it comprises the attraction of electrons from adjacent bonds by more electronegative atoms. Inductive effects are used to explain many properties of molecules. The analysis presented here, related to acid strength, is merely one example. With nitric and nitrous acid, we are comparing the ability of an NO group (in HONO) and an NO_2 group (in $HONO_2$) to attract electrons and increase the polarity of the H—O bond. There are two oxygen atoms bonded to the central nitrogen atom in the NO_2 group but only one oxygen bonded to the nitrogen atom in the NO group. By attaching more oxygen atoms to nitrogen, the inductive effect is greater, and the H—O bond becomes more polarized. Thus, the H atom in the H—O group is more positive in HNO_3 than in HNO_2, and HNO_3 is a stronger acid.

Hydrogen bond to H atom assists in O—H bond breaking in HNO_3.

Electrons in H—O bond flow toward electronegative O atoms owing to inductive effect.

Additional oxygen atoms in an oxoacid also have the effect of stabilizing the anion formed by removal of H^+ from the oxoacid. This greater stability arises because the negative charge on the anion can be dispersed over more atoms. In the nitrate ion, for example, the negative charge is shared equally over the three oxygen atoms. This situation is represented symbolically in the three resonance structures for this ion.

In the nitrite ion, only two atoms share the negative charge. The greater stability of the products formed by ionizing the acid contributes to increased acidity.

In summary, a molecule can behave as a Brønsted acid if electronegative atoms increase the polarization of the H—O bond. In addition, the anion created by loss of H^+ should be stable and able to accommodate the negative charge. These conditions are promoted by two factors:

- The presence of electronegative atoms attached to the central atom
- The possibility of resonance structures for the anion, which lead to delocalization of the negative charge over the anion and thus to a stable ion

■ **Anion Solvation**
The solvation of the anion is another contributing factor in determining the relative strength of an acid.

Why Are Carboxylic Acids Brønsted Acids?

Other important questions are why carboxylic acids (such as acetic acid, CH_3CO_2H) are Brønsted acids and which H atom is lost as an H^+ ion. (A related question is why so few substances behave as Brønsted acids, even though hundreds of molecules have some type of E—H bond.)

The arguments used to explain the acidity of oxoacids can also be applied to carboxylic acids. The H—O bond in these compounds is polar, a prerequisite for ionization.

Polar O—H bond broken by interaction of positively charged H atom with hydrogen-bonded H_2O

In addition, carboxylate anions are stabilized by delocalizing the negative charge over the two oxygen atoms.

The simple carboxylic acids, RCO_2H, in which R is a hydrocarbon group [◀ Section 11.4], do not differ markedly in acid strength (compare acetic acid, $pK_a = 4.74$, and propanoic acid, $pK_a = 4.89$; Table 17.3). The acidity of carboxylic acids is enhanced, however, if electronegative substituents replace the hydrogens in the alkyl group. Compare for example, the pK_a values of a series of acetic acids in which hydrogen is replaced sequentially by the more electronegative element chlorine.

Acid		pK_a Value	
CH_3CO_2H	Acetic acid	4.74	
$ClCH_2CO_2H$	Chloroacetic acid	2.85	increasing
Cl_2CHCO_2H	Dichloracetic acid	1.49	acid
Cl_3CCO_2H	Trichloroacetic acid	0.7	strength

We can again rationalize the trend in acidity based on the increasing inductive effect when electronegative Cl atoms are substituted for H atoms.

Finally, why are the C—H hydrogens of carboxylic acids not dissociated as H^+ instead of (or in addition to) the O—H hydrogen atom? Recall that the stability of the product ion is an important part of promoting ionization. In carboxylic acids, the adjacent carbon atom is not sufficiently electronegative to stabilize the negative charge left if the terminal bond breaks as C—H $\longrightarrow$ $C:^- + H^+$ (Figure 17.9).

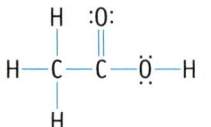

(a) Lewis electron dot structure of acetic acid, CH_3CO_2H

(b) Ball-and-stick model of acetic acid

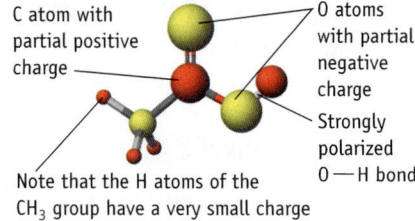

C atom with partial positive charge

O atoms with partial negative charge

Strongly polarized O—H bond

Note that the H atoms of the CH_3 group have a very small charge

(c) Computer-calculated partial charges on atoms of CH_3CO_2H

Figure 17.9   Acetic acid, a carboxylic acid. The model in part (c) was generated by computer to show the partial + and − charges on the H, C, and O atoms. Atoms with a positive charge are red, and atoms with a negative charge are yellow. The relative size of the partial charge is reflected by the relative size of the red or yellow sphere. Bonding electrons in the molecule flow toward O atoms and away from the H atom, leaving the H atom of the −CO_2H group positively charged and readily removed by interaction with a polar water molecule.

Why Are Hydrated Metal Cations Brønsted Acids?

When a coordinate covalent bond is formed between a metal cation (a Lewis acid) and a water molecule (a Lewis base), the positive charge of the metal ion and its small size mean that the electrons of the $H_2O—M^{n+}$ bond are very strongly attracted to the metal. As a result of this inductive effect, the H—O bonds of the bound water molecules are polarized, just as in oxoacids and carboxylic acids. The net effect is that an H atom of a coordinated water molecule is removed as H^+ more readily than in an uncoordinated water molecule. Thus, a hydrated metal cation functions as a Brønsted acid or proton donor (Figure 17.6).

$$[Cu(H_2O)_6]^{2+}(aq) + H_2O(\ell) \rightleftharpoons [Cu(H_2O)_5(OH)]^+(aq) + H_3O^+(aq)$$

The inductive effect of a metal ion increases with increasing charge. Consulting Table 17.3, you see that the Brønsted acidity of 3+ cations (for example, $[Al(H_2O)_6]^{3+}$ and $[Fe(H_2O)_6]^{3+}$ is greater than that of 2+ cations ($[Cu(H_2O)_6]^{2+}$ and related, aquated ions such as Pb^{2+}, Co^{2+}, Fe^{2+}, Ni^{2+}). Ions with a single positive charge such as Na^+ and K^+ are not acidic.

Why Are Anions Brønsted Bases?

Anions, particularly oxoanions such as PO_4^{3-}, are Brønsted bases. The negatively charged anion interacts with the positively charged H atom of a polar water molecule, and an H^+ ion is transferred to the anion.

■ **Polarization of O—H Bonds**
Water molecules attached to a metal cation have strongly polarized H—O bonds.

$$(H_2O)_5M^{n+} \longleftarrow :\overset{..}{\underset{\delta-}{O}}—H^{\delta+}$$
$$H^{\delta+}$$

The data in Table 17.8 show that, in a series of related anions, the basicity of an anionic base increases substantially as the negative charge of the anion increases.

Why Are Organic Amines Brønsted and Lewis Bases?

Ammonia is the parent compound of an enormous number of compounds that behave as Brønsted and Lewis bases. These molecules have an N atom surrounded by three other atoms as well as a lone pair of electrons.

Table 17.8   Basic Oxoanions

Anion	pK_b
PO_4^{3-}	1.55
HPO_4^{2-}	6.80
$H_2PO_4^-$	11.89
CO_3^{2-}	3.68
HCO_3^-	7.62
SO_3^{2-}	6.80
HSO_3^-	12.08

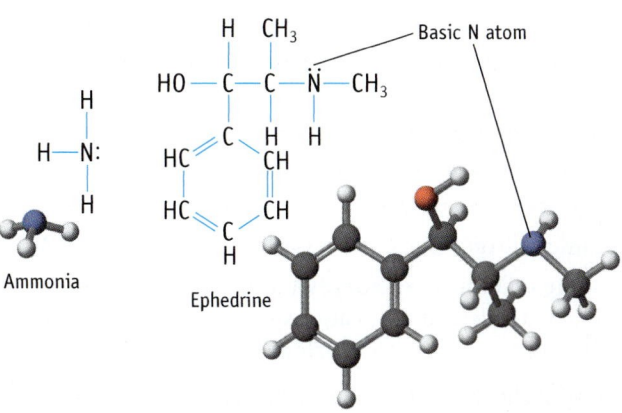

Ammonia

Ephedrine

Basic N atom

■ **Ephedra**
Ma Huang, an extract from the *ephedra* species of plants, contains ephedrine. Chinese herbalists have used it for more than 5000 years to treat asthma. Recently, however, the substance has been used in diet pills that can be purchased over the counter in herbal medicine shops. Very serious concerns about these pills have arisen following reports of serious heart problems in users, and the substance is now banned in the U.S.

In each case, the positively charged H atom of a polar water molecule can interact with the lone pair of electrons on the electronegative N atom. An H^+ ion transfers from water to the nitrogen atom, and an OH^- ion enters the solution.

Hydrogen bond with water.
H^+ ion moves to N atom.

Exercise 17.19—Molecular Structure, Acids, and Bases

(a) Which should be the stronger acid, H_2SeO_4 or H_2SeO_3?

(b) Which should be the stronger acid, $[Fe(H_2O)_6]^{2+}$ or $[Fe(H_2O)_6]^{3+}$?

(c) Which should be the stronger acid, HOCl or HOBr?

(d) The molecule whose structure is illustrated here is amphetamine, a stimulant. Is the compound a Brønsted acid, a Lewis acid, a Brønsted base, a Lewis base, or some combination of these?

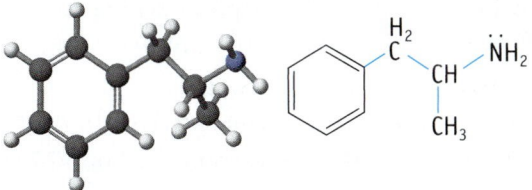

Chapter Goals Revisited

When you have finished studying this chapter, you should ask whether you have met the chapter goals. In particular you should be able to

Use the Brønsted-Lowry and Lewis theories of acids and bases

a. Define and use the Brønsted concept of acids and bases (Section 17.2).

b. Recognize common monoprotic and polyprotic acids and bases and write balanced equations for their ionization in water (Section 17.2).

For additional preparation for an examination on this chapter see the **Let's Review** section on pages 998–1011.

c. Appreciate when a substance can be amphiprotic (Section 17.2).

d. Recognize the Brønsted acid and base in a reaction and identify the conjugate partner of each (Section 17.2). General ChemistryNow homework: Study Question(s) 2, 4, 8

e. Understand the concept of water autoionization and its role in Brønsted acid–base chemistry. Use the water ionization constant, K_w (Section 17.3).

f. Use the pH concept (Section 17.3). General ChemistryNow homework: SQ(s) 10, 12

g. Identify common strong acids and bases (Tables 5.2 and 17.3).

h. Recognize some common weak acids and understand that they can be neutral molecules (such as acetic acid), cations (such as NH_4^+ or hydrated metal ions such as $[Fe(H_2O)_6]^{2+}$, or anions (such as HCO_3^-) (Table 17.3).

Apply the principles of chemical equilibrium to acids and bases in aqueous solution

a. Write equilibrium constant expressions for weak acids and bases (Section 17.4).

b. Calculate pK_a from K_a (or K_a from pK_a) and understand how pK_a is correlated with acid strength (Section 17.4). General ChemistryNow homework: SQ(s) 26, 28, 30

c. Understand the relationship between K_a for a weak acid and K_b for its conjugate base (Section 17.4). General ChemistryNow homework: SQ(s) 18

d. Write equations for acid–base reactions and decide whether they are product- or reactant-favored (Sections 17.5 and 17.6 and Table 17.5). General ChemistryNow homework: SQ(s) 36, 38

e. Calculate the equilibrium constant for a weak acid (K_a) or a weak base (K_b) from experimental information (such as pH, $[H_3O^+]$, or $[OH^-]$) (Section 17.7 and Example 17.4). General ChemistryNow homework: SQ(s) 42

f. Use the equilibrium constant and other information to calculate the pH of a solution of a weak acid or weak base (Section 17.7 and Examples 17.5 and 17.6). General ChemistryNow homework: SQ(s) 48, 50, 56

g. Describe the acid–base properties of salts and calculate the pH of a solution of a salt of a weak acid or of a weak base (Section 17.7 and Example 17.7). General ChemistryNow homework: SQ(s) 94, 103

h. Calculate the pH of a solution of a polyprotic acid or base (Section 17.8 and Example 17.9). General ChemistryNow homework: SQ(s) 66

Predict the outcome of reactions of acids and bases

a. Recognize the type of acid–base reaction and describe its result (Section 17.6).

b. Calculate the pH after an acid–base reaction (Section 17.7 and Example 17.8). General ChemistryNow homework: SQ(s) 62

Understand the influence of structure and bonding on acid–base properties

a. Characterize a compound as a Lewis base (an electron-pair donor) or a Lewis acid (an electron-pair acceptor) (Section 17.9). General ChemistryNow homework: SQ(s) 70

b. Appreciate the connection between the structure of a compound and its acidity or basicity (Section 17.10).

Key Equations

Equation 17.1 (page 804):
Water ionization constant.

$$K_w = [H_3O^+][OH^-] = 1.0 \times 10^{-14} \text{ at 25 °C}$$

Equation 17.2 (page 805):
Definition of pH (see also Equation 5.2).

$$pH = -\log [H_3O^+]$$

Equation 17.3 (page 805):
Definition of pOH

$$pOH = -\log [OH^-]$$

Equation 17.4 (page 806):
Definition of pK_w

$$pK_w = 14.00 = pH + pOH$$

Equation 17.5 (page 807):
Equilibrium expression for a general acid, HA, in water.

$$HA(aq) + H_2O(\ell) \rightleftharpoons H_3O^+(aq) + A^-(aq)$$

$$K_a = \frac{[H_3O^+][A^-]}{[HA]}$$

Equation 17.6 (page 807):
Equilibrium expression for a general base, B, in water.

$$B(aq) + H_2O(\ell) \rightleftharpoons BH^+(aq) + OH^-(aq)$$

$$K_b = \frac{[BH^+][OH^-]}{[B]}$$

Equation 17.7 (page 812):
Definition of pK_a.

$$pK_a = -\log K_a$$

Equation 17.8 (page 813):
Relationship of K_a, K_b, and K_w where K_a and K_b are for a conjugate acid–base pair.

$$K_a \times K_b = K_w$$

Study Questions

▲ denotes more challenging questions.

■ denotes questions available in the Homework and Goals section of the General ChemistryNow CD-ROM or website.

Blue numbered questions have answers in Appendix O and fully worked solutions in the *Student Solutions Manual*.

Structures of many of the compounds used in these questions are found on the General ChemistryNow CD-ROM or website in the Models folder.

GENERAL
Chemistry **Now**™ Assess your understanding of this chapter's topics with additional quizzing and conceptual questions at **http://now.brookscole.com/kotz6e**

Practicing Skills

The Brønsted Concept
(See Exercises 17.1 and 17.2 and General ChemistryNow Screen 17.2.)

1. Write the formula and give the name of the conjugate base of each of the following acids.
   (a) HCN
   (b) HSO_4^-
   (c) HF

2. ■ Write the formula and give the name of the conjugate acid of each of the following bases.
   (a) NH_3
   (b) HCO_3^-
   (c) Br^-

3. What are the products of each of the following acid–base reactions? Indicate the acid and its conjugate base, and the base and its conjugate acid.

(a) $HNO_3 + H_2O \longrightarrow$

(b) $HSO_4^- + H_2O \longrightarrow$

(c) $H_3O^+ + F^- \longrightarrow$

4. ■ What are the products of each of the following acid–base reactions? Indicate the acid and its conjugate base, and the base and its conjugate acid.

(a) $HClO_4 + H_2O \longrightarrow$

(b) $NH_4^+ + H_2O \longrightarrow$

(c) $HCO_3^- + OH^- \longrightarrow$

5. Write balanced equations showing how the hydrogen oxalate ion, $HC_2O_4^-$, can be both a Brønsted acid and a Brønsted base.

6. Write balanced equations showing how the HPO_4^{2-} ion of sodium hydrogen phosphate, Na_2HPO_4, can be a Brønsted acid or a Brønsted base.

7. In each of the following acid–base reactions, identify the Brønsted acid and base on the left and their conjugate partners on the right.

(a) $HCO_2H(aq) + H_2O(\ell) \rightleftharpoons HCO_2^-(aq) + H_3O^+(aq)$

(b) $NH_3(aq) + H_2S(aq) \rightleftharpoons NH_4^+(aq) + HS^-(aq)$

(c) $HSO_4^-(aq) + OH^-(aq) \rightleftharpoons SO_4^{2-}(aq) + H_2O(\ell)$

8. ■ In each of the following acid–base reactions, identify the Brønsted acid and base on the left and their conjugate partners on the right.

(a) $C_5H_5N(aq) + CH_3CO_2H(aq) \rightleftharpoons$
$C_5H_5NH^+(aq) + CH_3CO_2^-(aq)$

(b) $N_2H_4(aq) + HSO_4^-(aq) \rightleftharpoons N_2H_5^+(aq) + SO_4^{2-}(aq)$

(c) $[Al(H_2O)_6]^{3+}(aq) + OH^-(aq) \rightleftharpoons$
$[Al(H_2O)_5OH]^{2+}(aq) + H_2O(\ell)$

pH Calculations

(See Examples 5.11 and 17.1, Exercise 17.4, and General ChemistryNow Screens 5.17 and 17.3–17.4.)

9. An aqueous solution has a pH of 3.75. What is the hydronium ion concentration of the solution? Is it acidic or basic?

10. ■ A saturated solution of milk of magnesia, $Mg(OH)_2$, has a pH of 10.52. What is the hydronium ion concentration of the solution? What is the hydroxide ion concentration? Is the solution acidic or basic?

11. What is the pH of a 0.0075 M solution of HCl? What is the hydroxide ion concentration of the solution?

12. ■ What is the pH of a 1.2×10^{-4} M solution of KOH? What is the hydronium ion concentration of the solution?

13. What is the pH of a 0.0015 M solution of $Ba(OH)_2$?

14. The pH of a solution of $Ba(OH)_2$ is 10.66 at 25 °C. What is the hydroxide ion concentration in the solution? If the solution volume is 125 mL, how many grams of $Ba(OH)_2$ must have been dissolved?

Equilibrium Constants for Acids and Bases

(See Example 17.2, Exercise 17.5, and General ChemistryNow Screen 17.6.)

15. Several acids are listed here with their respective equilibrium constants:

$$C_6H_5OH(aq) + H_2O(\ell) \rightleftharpoons H_3O^+(aq) + C_6H_5O^-(aq)$$

$$K_a = 1.3 \times 10^{-10}$$

$$HCO_2H(aq) + H_2O(\ell) \rightleftharpoons H_3O^+(aq) + HCO_2^-(aq)$$

$$K_a = 1.8 \times 10^{-4}$$

$$HC_2O_4^-(aq) + H_2O(\ell) \rightleftharpoons H_3O^+(aq) + C_2O_4^{2-}(aq)$$

$$K_a = 6.4 \times 10^{-5}$$

(a) Which is the strongest acid? Which is the weakest acid?

(b) Which acid has the weakest conjugate base?

(c) Which acid has the strongest conjugate base?

16. Several acids are listed here with their respective equilibrium constants.

$$HF(aq) + H_2O(\ell) \rightleftharpoons H_3O^+(aq) + F^-(aq)$$
$$K_a = 7.2 \times 10^{-4}$$
$$HPO_4^-(aq) + H_2O(\ell) \rightleftharpoons H_3O^+(aq) + PO_4^{3-}(aq)$$
$$K_a = 3.6 \times 10^{-13}$$
$$CH_3CO_2H(aq) + H_2O(\ell) \rightleftharpoons H_3O^+(aq) + CH_3CO_2^-(aq)$$
$$K_a = 1.8 \times 10^{-5}$$

(a) Which is the strongest acid? Which is the weakest acid?

(b) What is the conjugate base of the acid HF?

(c) Which acid has the weakest conjugate base?

(d) Which acid has the strongest conjugate base?

17. State which of the following ions or compounds has the strongest conjugate base and briefly explain your choice.

(a) HSO_4^-

(b) CH_3CO_2H

(c) HClO

18. ■ Which of the following compounds or ions has the strongest conjugate acid? Briefly explain your choice.

(a) CN^-

(b) NH_3

(c) SO_4^{2-}

19. Dissolving K_2CO_3 in water gives a basic solution. Write a balanced equation showing how the carbonate ion is responsible for this effect.

20. Dissolving ammonium bromide in water gives an acidic solution. Write a balanced equation showing how this reaction can occur.

21. If each of the salts listed here were dissolved in water to give a 0.10 M solution, which solution would have the highest pH? Which would have the lowest pH?

(a) Na_2S (d) NaF

(b) Na_3PO_4 (e) $NaCH_3CO_2$

(c) NaH_2PO_4 (f) $AlCl_3$

▲ More challenging   ■ In General ChemistryNow   **Blue-numbered questions** answered in Appendix O

22. Which of the following common food additives would give a basic solution when dissolved in water?
    (a) $NaNO_3$ (used as a meat preservative)
    (b) $NaC_6H_5CO_2$ (sodium benzoate; used as a soft-drink preservative)
    (c) Na_2HPO_4 (used as an emulsifier in the manufacture of pasteurized cheese)

pK_a: A Logarithmic Scale of Acid Strength

(See Exercise 17.7 and General ChemistryNow Screen 17.6.)

23. A weak acid has a K_a of 6.5×10^{-5}. What is the value of pK_a for the acid?

24. If K_a for a weak acid is 2.4×10^{-11}, what is the value of pK_a?

25. Epinephrine hydrochloride (page 833) has a pK_a value of 9.53. What is the value of K_a? Where does the acid fit in Table 17.3?

26. ■ An organic acid has $pK_a = 8.95$. What is its K_a value? Where does the acid fit in Table 17.3?

27. Which is the stronger of the following two acids?
    (a) benzoic acid, $C_6H_5CO_2H$, $pK_a = 4.20$
    (b) 2-chlorobenzoic acid, $ClC_6H_4CO_2H$, $pK_a = 2.88$

28. ■ Which is the stronger of the following two acids?
    (a) acetic acid, CH_3CO_2H, $K_a = 1.8 \times 10^{-5}$
    (b) chloroacetic acid, $ClCH_2CO_2H$, $pK_a = 2.87$

Ionization Constants for Weak Acids and Their Conjugate Bases

(See Exercise 17.8 and General ChemistryNow Screen 17.6.)

29. Chloroacetic acid ($ClCH_2CO_2H$) has $K_a = 1.36 \times 10^{-3}$. What is the value of K_b for the chloroacetate ion ($ClCH_2CO_2^-$)?

30. ■ A weak base has $K_b = 1.5 \times 10^{-9}$. What is the value of K_a for the conjugate acid?

31. The trimethylammonium ion, $(CH_3)_3NH^+$, is the conjugate acid of the weak base trimethylamine, $(CH_3)_3N$. A chemical handbook gives 9.80 as the pK_a value for $(CH_3)_3NH^+$. What is the value of K_b for $(CH_3)_3N$?

32. The chromium(III) ion in water, $[Cr(H_2O)_6]^{3+}$, is a weak acid with $pK_a = 3.95$. What is the value of K_b for its conjugate base, $[Cr(H_2O)_5OH]^{2+}$?

Predicting the Direction of Acid–Base Reactions

(See Example 17.3 and General ChemistryNow Screen 17.7.)

33. Acetic acid and sodium hydrogen carbonate, $NaHCO_3$, are mixed in water. Write a balanced equation for the acid–base reaction that could, in principle, occur. Using Table 17.3, decide whether the equilibrium lies predominantly to the right or to the left.

34. Ammonium chloride and sodium dihydrogen phosphate, NaH_2PO_4, are mixed in water. Write a balanced equation for the acid–base reaction that could, in principle, occur. Using Table 17.3, decide whether the equilibrium lies predominantly to the right or to the left.

35. For each of the following reactions, predict whether the equilibrium lies predominantly to the left or to the right. Explain your prediction briefly.
    (a) $NH_4^+(aq) + Br^-(aq) \rightleftharpoons NH_3(aq) + HBr(aq)$
    (b) $HPO_4^{2-}(aq) + CH_3CO_2^-(aq) \rightleftharpoons$
    $PO_4^{3-}(aq) + CH_3CO_2H(aq)$
    (c) $[Fe(H_2O)_6]^{3+}(aq) + HCO_3^-(aq) \rightleftharpoons$
    $[Fe(H_2O)_5(OH)]^{2+}(aq) + H_2CO_3(aq)$

36. ■ For each of the following reactions, predict whether the equilibrium lies predominantly to the left or to the right. Explain your prediction briefly.
    (a) $H_2S(aq) + CO_3^{2-}(aq) \rightleftharpoons HS^-(aq) + HCO_3^-(aq)$
    (b) $HCN(aq) + SO_4^{2-}(aq) \rightleftharpoons CN^-(aq) + HSO_4^-(aq)$
    (c) $SO_4^{2-}(aq) + CH_3CO_2H(aq) \rightleftharpoons$
    $HSO_4^-(aq) + CH_3CO_2^-(aq)$

Types of Acid–Base Reactions

(See Exercise 17.11 and General ChemistryNow Screen 17.7.)

37. Equal molar quantities of sodium hydroxide and sodium hydrogen phosphate (Na_2HPO_4) are mixed.
    (a) Write the balanced, net ionic equation for the acid–base reaction that can, in principle, occur.
    (b) Does the equilibrium lie to the right or left?

38. ■ Equal molar quantities of hydrochloric acid and sodium hypochlorite (NaClO) are mixed.
    (a) Write the balanced, net ionic equation for the acid–base reaction that can, in principle, occur.
    (b) Does the equilibrium lie to the right or left?

39. Equal molar quantities of acetic acid and sodium hydrogen phosphate (Na_2HPO_4) are mixed.
    (a) Write a balanced, net ionic equation for the acid–base reaction that can, in principle, occur.
    (b) Does the equilibrium lie to the right or left?

40. Equal molar quantities of ammonia and sodium dihydrogen phosphate (NaH_2PO_4) are mixed.
    (a) Write a balanced, net ionic equation for the acid–base reaction that can, in principle, occur.
    (b) Does the equilibrium lie to the right or left?

Using pH to Calculate Ionization Constants

(See Example 17.4 and General ChemistryNow Screen 17.8.)

41. A 0.015 M solution of hydrogen cyanate, HOCN, has a pH of 2.67.
    (a) What is the hydronium ion concentration in the solution?
    (b) What is the ionization constant, K_a, for the acid?

42. ■ A 0.10 M solution of chloroacetic acid, $ClCH_2CO_2H$, has a pH of 1.95. Calculate K_a for the acid.

43. A 0.025 M solution of hydroxylamine has a pH of 9.11. What is the value of K_b for this weak base?

$$H_2NOH(aq) + H_2O(\ell) \rightleftharpoons H_3NOH^+(aq) + OH^-(aq)$$

44. Methylamine, CH_3NH_2, is a weak base.

$$CH_3NH_2(aq) + H_2O(\ell) \rightleftharpoons CH_3NH_3^+(aq) + OH^-(aq)$$

If the pH of a 0.065 M solution of the amine is 11.70, what is the value of K_b?

45. A 2.5×10^{-3} M solution of an unknown acid has a pH of 3.80 at 25 °C.
    (a) What is the hydronium ion concentration of the solution?
    (b) Is the acid a strong acid, a moderately weak acid (K_a of about 10^{-5}), or a very weak acid (K_a of about 10^{-10})?

46. A 0.015 M solution of a base has a pH of 10.09.
    (a) What are the hydronium and hydroxide ion concentrations of this solution?
    (b) Is the base a strong base, a moderately weak base (K_b of about 10^{-5}), or a very weak base (K_b of about 10^{-10})?

Using Ionization Constants

(See Examples 17.5–17.7 and General ChemistryNow Screens 17.9–17.11.)

47. What are the equilibrium concentrations of hydronium ion, acetate ion, and acetic acid in a 0.20 M aqueous solution of acetic acid?

48. ■ The ionization constant of a very weak acid, HA, is 4.0×10^{-9}. Calculate the equilibrium concentrations of H_3O^+, A^-, and HA in a 0.040 M solution of the acid.

49. What are the equilibrium concentrations of H_3O^+, CN^-, and HCN in a 0.025 M solution of HCN? What is the pH of the solution?

50. Phenol (C_6H_5OH), commonly called carbolic acid, is a weak organic acid.

$$C_6H_5OH(aq) + H_2O(\ell) \rightleftharpoons C_6H_5O^-(aq) + H_3O^+(aq)$$
$$K_a = 1.3 \times 10^{-10}$$

If you dissolve 0.195 g of the acid in enough water to make 125 mL of solution, what is the equilibrium hydronium ion concentration? What is the pH of the solution?

51. What are the equilibrium concentrations of NH_3, NH_4^+, and OH^- in a 0.15 M solution of ammonia? What is the pH of the solution?

52. ■ A hypothetical weak base has $K_b = 5.0 \times 10^{-4}$. Calculate the equilibrium concentrations of the base, its conjugate acid, and OH^- in a 0.15 M solution of the base.

53. The weak base methylamine, CH_3NH_2, has $K_b = 4.2 \times 10^{-4}$. It reacts with water according to the equation

$$CH_3NH_2(aq) + H_2O(\ell) \rightleftharpoons CH_3NH_3^+(aq) + OH^-(aq)$$

Calculate the equilibrium hydroxide ion concentration in a 0.25 M solution of the base. What are the pH and pOH of the solution?

54. Calculate the pH of a 0.12 M aqueous solution of the base aniline, $C_6H_5NH_2$ ($K_b = 4.0 \times 10^{-10}$).

$$C_6H_5NH_2(aq) + H_2O(\ell) \rightleftharpoons C_6H_5NH_3^+(aq) + OH^-(aq)$$

55. Calculate the pH of a 0.0010 M aqueous solution of HF.

56. ■ A solution of hydrofluoric acid, HF, has a pH of 2.30. Calculate the equilibrium concentrations of HF, F^-, and H_3O^+, and calculate the amount of HF originally dissolved per liter.

Acid–Base Properties of Salts

(See Example 17.7 and General ChemistryNow Screen 17.11.)

57. Calculate the hydronium ion concentration and pH in a 0.20 M solution of ammonium chloride, NH_4Cl.

58. ▲ Calculate the hydronium ion concentration and pH for a 0.015 M solution of sodium formate, $NaHCO_2$.

59. Sodium cyanide is the salt of the weak acid HCN. Calculate the concentration of H_3O^+, OH^-, HCN, and Na^+ in a solution prepared by dissolving 10.8 g of NaCN in enough water to make 5.00×10^2 mL of solution at 25 °C.

60. The sodium salt of propanoic acid, $NaCH_3CH_2CO_2$, is used as an antifungal agent by veterinarians. Calculate the equilibrium concentrations of H_3O^+ and OH^-, and the pH, for a solution of 0.10 M $NaCH_3CH_2CO_2$.

pH After an Acid–Base Reaction

(See Example 17.8 and General ChemistryNow Screens 17.7 and 17.10.)

61. Calculate the hydronium ion concentration and pH of the solution that results when 22.0 mL of 0.15 M acetic acid, CH_3CO_2H, is mixed with 22.0 mL of 0.15 M NaOH.

62. ■ Calculate the hydronium ion concentration and the pH when 50.0 mL of 0.40 M NH_3 is mixed with 50.0 mL of 0.40 M HCl.

63. For each of the following cases, decide whether the pH is less than 7, equal to 7, or greater than 7.
    (a) equal volumes of 0.10 M acetic acid, CH_3CO_2H, and 0.10 M KOH are mixed
    (b) 25 mL of 0.015 M NH_3 is mixed with 25 mL of 0.015 M HCl
    (c) 150 mL of 0.20 M HNO_3 is mixed with 75 mL of 0.40 M NaOH

64. For each of the following cases, decide whether the pH is less than 7, equal to 7, or greater than 7.
    (a) 25 mL of 0.45 M H_2SO_4 is mixed with 25 mL of 0.90 M NaOH
    (b) 15 mL of 0.050 M formic acid, HCO_2H, is mixed with 15 mL of 0.050 M NaOH
    (c) 25 mL of 0.15 M $H_2C_2O_4$ (oxalic acid) is mixed with 25 mL of 0.30 M NaOH (Both H^+ ions of oxalic acid are removed with NaOH.)

Polyprotic Acids and Bases

(See Example 17.9.)

65. Sulfurous acid, H_2SO_3, is a weak acid capable of providing two H^+ ions.
    (a) What is the pH of a 0.45 M solution of H_2SO_3?

▲ More challenging   ■ In General ChemistryNow   Blue-numbered questions answered in Appendix O

(b) What is the equilibrium concentration of the sulfite ion, SO_3^{2-}, in the 0.45 M solution of H_2SO_3?

66. ■ Ascorbic acid (vitamin C, $C_6H_8O_6$) is a diprotic acid ($K_{a1} = 6.8 \times 10^{-5}$ and $K_{a2} = 2.7 \times 10^{-12}$). What is the pH of a solution that contains 5.0 mg of acid per milliliter of solution?

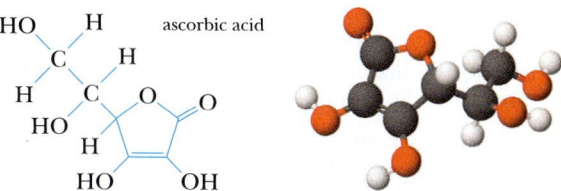

67. Hydrazine, N_2H_4, can interact with water in two steps.

$$N_2H_4(aq) + H_2O(\ell) \rightleftharpoons N_2H_5^+(aq) + OH^-(aq)$$
$$K_{b1} = 8.5 \times 10^{-7}$$

$$N_2H_5^+(aq) + H_2O(\ell) \rightleftharpoons N_2H_6^{2+}(aq) + OH^-(aq)$$
$$K_{b2} = 8.9 \times 10^{-16}$$

(a) What is the concentration of OH^-, $N_2H_5^+$, and $N_2H_6^{2+}$ in a 0.010 M aqueous solution of hydrazine?

(b) What is the pH of the 0.010 M solution of hydrazine?

68. Ethylenediamine, $H_2NCH_2CH_2NH_2$, can interact with water in two steps, forming OH^- in each step (see Appendix I). If you have a 0.15 M aqueous solution of the amine, calculate the concentration of $[H_3NCH_2CH_2NH_3]^{2+}$ and OH^-.

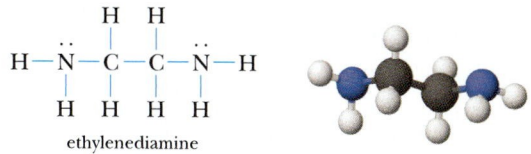

ethylenediamine

Lewis Acids and Bases

(See Exercise 17.18 and General ChemistryNow Screens 17.12–17.14.)

69. Decide whether each of the following substances should be classified as a Lewis acid or a Lewis base.

(a) H_2NOH in the reaction

$$H_2NOH(aq) + HCl(aq) \longrightarrow [H_3NOH]Cl(aq)$$

(b) Fe^{2+}

(c) CH_3NH_2 (Hint: Draw the electron dot structure.)

70. ■ Decide whether each of the following substances should be classified as a Lewis acid or a Lewis base.

(a) BCl_3 (Hint: Draw the electron dot structure.)

(b) H_2NNH_2, hydrazine (Hint: Draw the electron dot structure.)

(c) the reactants in the reaction

$$Ag^+(aq) + 2\,NH_3(aq) \rightleftharpoons [Ag(NH_3)_2]^+(aq)$$

71. Carbon monoxide forms complexes with low-valent metals. For example, $Ni(CO)_4$ and $Fe(CO)_5$ are well known. CO also forms complexes with the iron(II) ion in hemoglobin, which prevents the hemoglobin from acting in its normal way. Is CO a Lewis acid or a Lewis base?

72. Trimethylamine, $(CH_3)_3N$, is a common reagent. It interacts readily with diborane gas, B_2H_6. The latter dissociates to BH_3, and this forms a complex with the amine, $(CH_3)_3N{\rightarrow}BH_3$. Is the BH_3 fragment a Lewis acid or a Lewis base?

Molecular Structure, Bonding, and Acid–Base Behavior
(See Section 17.9 and Exercise 17.19.)

73. Which should be the stronger acid, HOCN or HCN? Explain briefly. (In HOCN, the H^+ ion is attached to the O atom of the OCN^- ion.)

74. Which should be the stronger Brønsted acid, $[V(H_2O)_6]^{2+}$ or $[V(H_2O)_6]^{3+}$?

75. Explain why benzenesulfonic acid is a Brønsted acid.

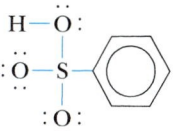

benzenesulfonic acid

76. The structure of ethylenediamine is illustrated in Study Question 68. Is this compound a Brønsted acid, a Brønsted base, a Lewis acid, or a Lewis base, or some combination of these?

General Questions on Acids and Bases

These questions are not designated as to type or location in the chapter. They may combine several concepts.

77. About this time, you may be wishing you had an aspirin. Aspirin is an organic acid (page 507) with a K_a of 3.27×10^{-4} for the reaction

$$HC_9H_7O_4(aq) + H_2O(\ell) \rightleftharpoons C_9H_7O_4^-(aq) + H_3O^+(aq)$$

If you have two tablets, each containing 0.325 g of aspirin (mixed with a neutral "binder" to hold the tablet together), and you dissolve them in a glass of water to give 225 mL of solution, what is the pH of the solution?

78. Consider the following ions: NH_4^+, CO_3^{2-}, Br^-, S^{2-}, and ClO_4^-.

(a) Which of these ions might lead to an acidic solution and which might lead to a basic solution?

(b) Which of these anions will have no effect on the pH of an aqueous solution?

(c) Which ion is the strongest base?

(d) Write a chemical equation for the reaction of each basic anion with water.

79. You have 0.010 M solutions of benzoic acid, $C_6H_5CO_2H$ ($K_a = 6.3 \times 10^{-5}$) and 4-chlorobenzoic acid, $ClC_6H_4CO_2H$ ($K_a = 1.0 \times 10^{-4}$). Which solution will have the higher pH?

80. Place the following acids in order of (i) increasing strength and (ii) increasing pH. Assume you have a 0.10 M solution of each acid.
    (a) 4-chlorobenzoic acid, $ClC_6H_4CO_2H$, $K_a = 1.0 \times 10^{-4}$
    (b) bromoacetic acid, $BrCH_2CO_2H$, $K_a = 1.3 \times 10^{-3}$
    (c) trimethylammonium ion, $(CH_3)_3NH^+$, $K_a = 1.6 \times 10^{-10}$

81. Hydrogen sulfide, H_2S, and sodium acetate, $Na(CH_3CO_2)$, are mixed in water. Using Table 17.3, write a balanced equation for the acid–base reaction that could, in principle, occur. Does the equilibrium lie toward the products or the reactants?

82. For each of the following reactions, predict whether the equilibrium lies predominantly to the left or to the right. Explain your prediction briefly.
    (a) $HCO_3^-(aq) + SO_4^{2-}(aq) \rightleftharpoons CO_3^{2-}(aq) + HSO_4^-(aq)$
    (b) $HSO_4^-(aq) + CH_3CO_2^-(aq) \rightleftharpoons$
    $$SO_4^{2-}(aq) + CH_3CO_2H(aq)$$
    (c) $[Co(H_2O)_6]^{2+}(aq) + CH_3CO_2^-(aq) \rightleftharpoons$
    $$[Co(H_2O)_5(OH)]^+(aq) + CH_3CO_2H(aq)$$

83. A monoprotic acid HX has $K_a = 1.3 \times 10^{-3}$. Calculate the equilibrium concentration of HX and H_3O^+ and the pH for a 0.010 M solution of the acid.

84. Calcium hydroxide, $Ca(OH)_2$, is almost insoluble in water; only 0.50 g can be dissolved in 1.0 L of water at 25 °C. If the dissolved substance is completely dissociated into its constituent ions, what is the pH of a saturated solution?

85. *m*-Nitrophenol, a weak acid, can be used as a pH indicator because it is yellow at a pH above 8.6 and colorless at a pH below 6.8. If the pH of a 0.010 M solution of the compound is 3.44, calculate its pK_a.

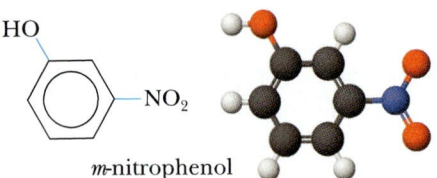

m-nitrophenol

86. The butylammonium ion, $C_4H_9NH_3^+$, has a K_a of 2.3×10^{-11}.

    $$C_4H_9NH_3^+(aq) + H_2O(\ell) \rightleftharpoons H_3O^+(aq) + C_4H_9NH_2(aq)$$

    (a) Calculate K_b for the conjugate base, $C_4H_9NH_2$ (butylamine).
    (b) Place the butylammonium ion and its conjugate base in Table 17.3. Name an acid weaker than $C_4H_9NH_3^+$ and a base stronger than $C_4H_9NH_2$.
    (c) What is the pH of a 0.015 M solution of the butylammonium ion?

87. The local anesthetic novocaine is the hydrogen chloride salt of an organic base, procaine.

    $$\underset{\text{procaine}}{C_{13}H_{20}N_2O_2(aq)} + HCl(aq) \longrightarrow \underset{\text{novocaine}}{[HC_{13}H_{20}N_2O_2]^+Cl^-(aq)}$$

    The pK_a for novocaine is 8.85. What is the pH of a 0.0015 M solution of novocaine?

88. The anilinium ion, $C_6H_5NH_3^+$, is the conjugate acid of the weak organic base aniline. If the anilinium ion has a pK_a of 4.60, what is the pH of a 0.080 M solution of anilinium hydrochloride, $C_6H_5NH_3Cl$?

89. The base ethylamine ($CH_3CH_2NH_2$) has a K_b of 4.3×10^{-4}. A closely related base, ethanolamine ($HOCH_2CH_2NH_2$), has a K_b of 3.2×10^{-5}.
    (a) Which of the two bases is stronger?
    (b) Calculate the pH of a 0.10 M solution of the stronger base.

90. Chloroacetic acid, $ClCH_2CO_2H$, is a moderately weak acid ($K_a = 1.40 \times 10^{-3}$). If you dissolve 94.5 mg of the acid in water to give 125 mL of solution, what is the pH of the solution?

91. Pyridine is a weak organic base and readily forms a salt with hydrochloric acid.

    $$\underset{\text{pyridine}}{C_5H_5N(aq)} + HCl(aq) \longrightarrow \underset{\text{pyridinium ion}}{C_5H_5NH^+(aq)} + Cl^-(aq)$$

    What is the pH of a 0.025 M solution of pyridinium hydrochloride, $[C_5H_5NH^+]Cl^-$?

92. Saccharin ($HC_7H_4NO_3S$) is a weak acid with $pK_a = 2.32$ at 25 °C. It is used in the form of sodium saccharide, $NaC_7H_4NO_3S$. What is the pH of a 0.10 M solution of sodium saccharide at 25 °C?

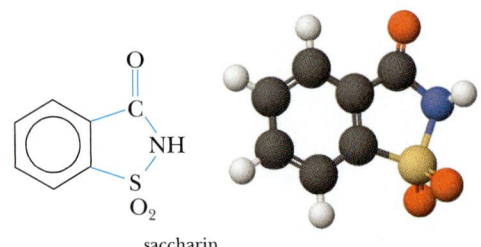

saccharin

93. For each of the following salts, predict whether a 0.1 M solution has a pH less than, equal to, or greater than 7.
    (a) $NaHSO_4$
    (b) NH_4Br
    (c) $KClO_4$
    (d) Na_2CO_3
    (e) $(NH_4)_2S$
    (f) $NaNO_3$
    (g) Na_2HPO_4
    (h) $LiBr$
    (i) $FeCl_3$

94. ■ Given the following solutions:
    (a) 0.1 M NH_3 (e) 0.1 M NH_4Cl
    (b) 0.1 M Na_2CO_3 (f) 0.1 M $NaCH_3CO_2$
    (c) 0.1 M $NaCl$ (g) 0.1 M $NH_4CH_3CO_2$
    (d) 0.1 M CH_3CO_2H
    (i) Which of the solutions are acidic?
    (ii) Which of the solutions are basic?
    (iii) Which of the solutions is most acidic?

95. Oxalic acid is a relatively weak diprotic acid. Calculate the equilibrium constant for the reaction shown below from K_{a1} and K_{a2}. (See Appendix H for the required K_a values.)

    $$H_2C_2O_4(aq) + 2\ H_2O(\ell) \rightleftharpoons C_2O_4^{2-}(aq) + 2\ H_3O^+(aq)$$

96. Nicotinic acid, $C_6H_5NO_2$, is found in minute amounts in all living cells, but appreciable amounts occur in liver, yeast, milk, adrenal glands, white meat, and corn. Whole-wheat flour contains about 60. μg per 1g of flour. One gram (1.00 g) of the acid dissolves in water to give 60. mL of solution having a pH of 2.70. What is the approximate value of K_a for the acid?

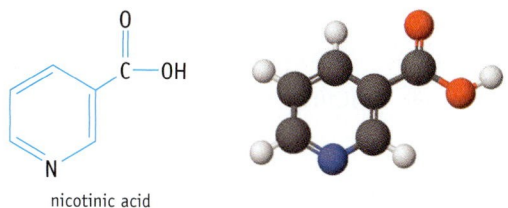

nicotinic acid

97. ▲ The equilibrium constant for the reaction of formic acid and sodium hydroxide is 1.8×10^{10} (page 817). Confirm this value.

98. Nicotine, $C_{10}H_{14}N_2$, has two basic nitrogen atoms (page 501), and both can react with water in two steps.

    $$Nic(aq) + H_2O(\ell) \rightleftharpoons NicH^+(aq) + OH^-(aq)$$

    $$NicH^+(aq) + H_2O(\ell) \rightleftharpoons NicH_2^{2+}(aq) + OH^-$$

    K_{b1} is 7.0×10^{-7} and K_{b2} is 1.1×10^{-10}. Calculate the approximate pH of the 0.020 M solution.

99. ▲ To what volume should 1.00×10^2 mL of any weak acid, HA, with a concentration 0.20 M be diluted to double the percentage ionization?

100. ▲ Equilibrium constants can be measured for the dissociation of Lewis acid–base complexes such as the dimethyl ether complex of BF_3, $(CH_3)_2O{\rightarrow}BF_3$. The value of K (here K_p) for the reaction is 0.17.

    $$(CH_3)_2O{-}BF_3(g) \rightleftharpoons BF_3(g) + (CH_3)_2O(g)$$

    (a) Describe each product as a Lewis acid or a Lewis base.
    (b) If you place 1.00 g of the complex in a 565-mL flask at 25 °C, what is the total pressure in the flask? What are the partial pressures of the Lewis acid, the Lewis base, and the complex?

101. ▲ Sulfanilic acid, which is used in making dyes, is made by reacting aniline with sulfuric acid.

    $$H_2SO_4(aq) + C_6H_5NH_2(aq) \longrightarrow \qquad + H_2O(\ell)$$
    aniline

    SO$_3$H
    NH$_2$
    sulfanilic acid

    The acid has a pK_a value of 3.23. The sodium salt of the acid, $Na(H_2NC_6H_4SO_3)$, is quite soluble in water. If you dissolve 1.25 g of the salt in water to give 125 mL of solution, what is the pH of the solution?

102. ▲ The equilibrium constant for the reaction of hydrochloric acid and ammonia is 1.8×10^9 (page 817). Confirm this value.

103. ■ Arrange the following 0.10 M solutions in order of increasing pH.
    (a) $NaCl$ (d) $NaCH_3CO_2$
    (b) NH_4Cl (e) KOH
    (c) HCl

104. ▲ Calculate the pH of the solution that results from mixing 25.0 mL of 0.14 M formic acid and 50.0 mL of 0.070 M sodium hydroxide.

105. ▲ The hydrogen phthalate ion, $C_8H_5O_4^-$, is a weak acid with $K_a = 3.91 \times 10^{-6}$.

    $$C_8H_5O_4^-(aq) + H_2O(\ell) \rightleftharpoons C_8H_4O_4^{2-}(aq) + H_3O^+(aq)$$

    What is the pH of a 0.050 M solution of potassium hydrogen phthalate, $KC_8H_5O_4$? *Note:* To find the pH for a solution of the anion, we must take into account that the ion is amphiprotic. It can be shown that, for most cases of amphiprotic ions, the H_3O^+ concentration is

    $$[H_3O^+] = \sqrt{K_1 \times K_2}$$

    where K_1 here is for phthalic acid, $C_8H_6O_4$ ($= 1.12 \times 10^{-3}$), and K_2 is for the hydrogen phthalate anion ($= 3.91 \times 10^{-6}$).

106. Iodine, I_2, is much more soluble in a water solution of potassium iodide, KI, than it is in pure water. The anion found in solution is I_3^-.
    (a) Draw an electron dot structure for I_3^-.
    (b) Write an equation for this reaction, indicating the Lewis acid and the Lewis base.

Summary and Conceptual Questions

The following questions use concepts from the preceding chapters.

107. Why can water be both a Brønsted base and a Lewis base? Can water be a Brønsted acid? A Lewis acid?

108. The nickel(II) ion exists as $[Ni(H_2O)_6]^{2+}$ in aqueous solution. Why is such a solution acidic? As part of your answer include a balanced equation depicting what happens when $[Ni(H_2O)_6]^{2+}$ interacts with water.

109. Describe an experiment that will allow you to place the following three bases in order of increasing base strength: $NaCN$, CH_3NH_2, Na_2CO_3.

110. Which should be the stronger acid: H_2SeO_4 or H_2SeO_3? Why? Describe an experiment by which you could confirm your prediction.

111. ▲ You prepare a 0.10 M solution of HCN. What molecules and ions exist in this solution? List them in order of decreasing concentration.

112. ▲ You prepare a 0.10 M solution of oxalic acid, $H_2C_2O_4$. What molecules and ions exist in this solution? List them in order of decreasing concentration.

113. ▲ You mix 30.0 mL of 0.15 M NaOH with 30.0 mL of 0.15 M acetic acid. What molecules and ions exist in this solution? List them in order of decreasing concentration.

114. The data below compare the strength of acetic acid with a related series of acids where the H atoms of the CH_3 group in acetic acid are successively replaced by Br.

Acid	pK_a
CH_3CO_2H	4.74
$BrCH_2CO_2H$	2.90
Br_2CHCO_2H	1.39
Br_3CCO_2H	−0.147

(a) What trend in acid strength do you observe as H is successively replaced by Br? Can you suggest a reason for this trend?

(b) Suppose each of the acids above was present as a 0.10 M aqueous solution. Which would have the highest pH? The lowest pH?

115. Perchloric acid behaves as an acid, even when it is dissolved in sulfuric acid.

(a) Write a balanced equation showing how perchloric acid can transfer a proton to sulfuric acid.

(b) Draw a Lewis electron dot structure for sulfuric acid. How can sulfuric acid function as a base?

116. You purchase a bottle of water. On checking its pH, you find that it is not neutral as you might have expected. Instead, it is slightly acidic. Why?

117. ▲ You have three solutions labeled A, B, and C. You know only that each contains a different cation—Na^+, NH_4^+, or H^+. Each has an anion that does not contribute to the solution pH (e.g., Cl^-). You also have two other solutions, Y and Z, each containing a different anion, Cl^- or OH^-, with a cation that does not influence solution pH (e.g., K^+). If equal amounts of B and Y are mixed, the result is an acidic solution. Mixing A and Z gives a neutral solution, whereas B and Z give a basic solution. Identify the five unknown solutions. (Adapted from D. H. Barouch: *Voyages in Conceptual Chemistry*, Boston, Jones and Bartlett, 1997.)

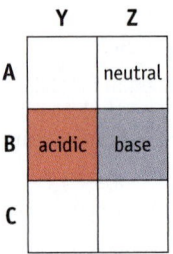

118. A hydrogen atom in the organic base pyridine, C_5H_5N, can be substituted by various atoms or groups to give XC_5H_4N, where X is an atom such as Cl or a group such as CH_3. The following table gives K_a values for the conjugate acids of a variety of substituted pyridines.

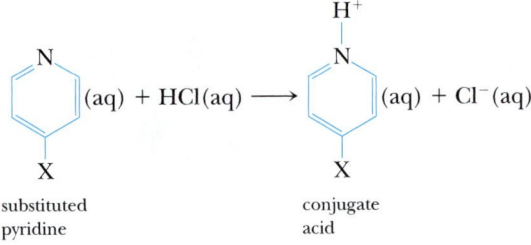

substituted pyridine   conjugate acid

Atom or Group X	K_a of Conjugate Acid
NO_2	5.9×10^{-2}
Cl	1.5×10^{-4}
H	6.8×10^{-6}
CH_3	1.0×10^{-6}

(a) Suppose each conjugate acid is dissolved in sufficient water to give a 0.050 M solution. Which solution would have the highest pH? The lowest pH?

(b) Which of the substituted pyridines is the strongest Brønsted base? Which is the weakest Brønsted base?

119. ▲ Consider a salt of a weak base and a weak acid such as ammonium cyanide. Both the NH_4^+ and CN^- ions interact with water in aqueous solution, but the net reaction can be considered as a proton transfer from NH_4^+ to CN^-.

$$NH_4^+(aq) + CN^-(aq) \rightleftharpoons NH_3(aq) + HCN(aq)$$

(a) Show that the equilibrium constant for this reaction, K_{net}, is

$$K_{net} = \frac{K_w}{K_a K_b}$$

where K_a is the ionization constant for the weak acid HCN and K_b is the constant for the weak base NH_3.

(b) Prove that the hydronium ion concentration in this solution must be given by

$$[H_3O^+] = \sqrt{\frac{K_w K_a}{K_b}}$$

(c) What is the pH of a 0.15 M solution of ammonium cyanide?

120. Listed below are values of pK_a for some compounds.

Acid	pK_a
Benzoic acid, $C_6H_5CO_2H$	4.20
Benzylammonium ion, $C_6H_5CH_2NH_3^+$	9.35
Chloroacetic acid, $ClCH_2CO_2H$	2.87
Conjugate acid of Cocaine	8.41
Thioacetic acid, $HSCH_2CO_2H$	3.33

(a) Which is the strongest acid?
(b) Which acid has the strongest conjugate base?
(c) List the acids in order of increasing strength.

Mentor

Do you need a live tutor for homework problems?
Access vMentor at General ChemistryNow at
http://now.brookscole.com/kotz6e
for one-on-one tutoring from a chemistry expert

18—Principles of Reactivity: Other Aspects of Aqueous Equilibria

Hydrangea blossoms in a Buddhist temple in Kunming, China.

Roses Are Red, Violets Are Blue, and Hydrangeas Are Red or Blue

Like the colors of some dyes, the colors of many flowers change to reflect the pH of their environment. This is certainly true for the beautiful, showy flowers of the hydrangea bush.

According to a gardening book, "Hydrangea flowers need aluminum to maintain a good blue. Aluminum is easily absorbed from acidic, lower pH soils. . . . The quickest way is to add aluminum sulfate to the soil." What does pH have to do with the color? What role does aluminum play? And what does this have to do with chemical equilibria?

The pigment in hydrangea blossoms belongs to the class of molecules called anthocyanins; more precisely, it is a cyanidin. Cyanidins are responsible for the red color of roses, strawberries, raspberries, apple skins, rhubarb, and cherries and for the purple color of blueberries. What is more, the color of cyanidins depends on the pH. Red cabbage juice is only red in acid; it is purple in a solution closer to a pH of 7 [◄ Figure 5.6]. An extract of red rose petals is red in acid but blue in a basic solution. The reason for this shift in color with pH is that the pigment is an acid and can donate an H^+ ion. As the pH increases, the conjugate base is formed, and its formation is accompanied by a significant color change. If the pH could be controlled, then the color could be controlled.

But why does the color of hydrangea blossoms depend on the

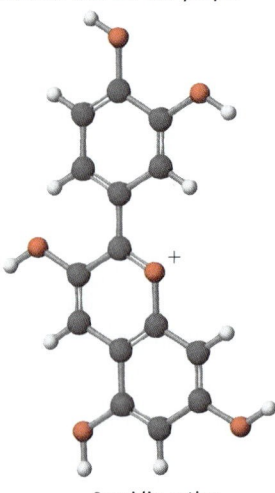

Cyanidin cation
$C_{15}H_{11}O_6{}^+$

Chapter Goals

See Chapter Goals Revisited (page 892). Test your knowledge of these goals by taking the exam-prep quiz on the General ChemistryNow CD-ROM or website.

- Understand the common ion effect.
- Understand the control of pH in aqueous solutions with buffers.
- Evaluate the pH in the course of acid–base titrations.
- Apply chemical equilibrium concepts to the solubility of ionic compounds.

Cyanidin chloride in acidic solution

Cyanidin in basic solution

$-HCl$ $\rightleftarrows$ $+HCl$

Low pH

High pH

When cyanidin functions as an acid and loses an H^+ ion, a color shift can occur.

presence of aluminum ions? One idea is that, at lower soil pH values, cyanidin reacts with Al^{3+} ions (a Lewis acid-base interaction), and the color of this complex ion shifts to blue. Aluminum is an avid seeker of oxygen atoms, so it can not only interact with the O atoms of cyanidin but also form $Al(H_2O)_6^{3+}$ in acidic solutions and insoluble $Al(OH)_3$ in basic solutions.

$$Al(H_2O)_6^{3+}(aq) + 3\ OH^-(aq) \rightleftarrows Al(OH)_3(s) + 6\ H_2O(\ell)$$

Thus, soil should be on the acid side if Al^{3+} ions are to be available to bind to the cyanidin pigment and give blue flowers. Indeed, the ideal pH range to promote blue hydrangeas is 5 to 5.5, whereas a pH of 6 to 6.5 is better suited to pink blossoms.

Al^{3+} ions stabilize the low pH form of cyanidin, and shift the color to blue.

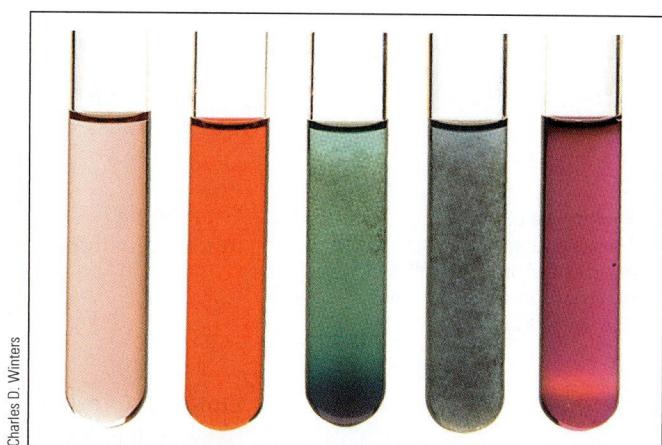

Charles D. Winters

Cyanidin with added acid, base, and aluminum ions. *(from left to right)* The pigment in red rose petals was extracted with ethanol; the extract was a faint red. After adding one drop of 6 M HCl, the color changed to a vivid red. Adding two drops of 6 M NH_3 produced a green color, and adding 1 drop each of HCl and NH_3 gave a blue solution. Finally, adding a few milligrams of $Al(NO_3)_3$ turned the solution deep purple. (The deep purple color with aluminum ions was so intense that the solution had to be diluted significantly to take the photo.)

To Review Before You Begin

- Review the principles of acid–base equilibria and ICE tables (Chapter 17)
- Review the solubility rules (Chapter 5)
- Review the stoichiometry of acid–base reactions (Chapter 5)

GENERAL
Chemistry ⚛ Now™

Throughout the chapter this icon introduces a list of resources on the General ChemistryNow CD-ROM or website (http://now .brookscole.com/kotz6e) that will:

- help you evaluate your knowledge of the material
- provide homework problems
- allow you to take an exam-prep quiz
- provide a personalized Learning Plan targeting resources that address areas you should study

In Chapter 5 we described four fundamental types of chemical reactions: acid–base reactions, precipitation reactions, gas-forming reactions, and oxidation–reduction reactions [◀ Sections 5.2–5.7]. In this chapter we want to apply the principles of chemical equilibria to develop a further understanding of the first two kinds of reactions.

With regard to acid–base reactions, we are looking for answers to the following questions:

- How can we control the pH in a solution?
- What happens when an acid and a base are mixed in any amount?

Precipitation reactions can also be understood in terms of chemical equilibria. The following questions are discussed in this chapter:

- If aqueous solutions of two ionic compounds are mixed, will precipitation occur?
- To what extent does an insoluble substance actually dissolve?
- What chemical reactions can be used to redissolve a precipitate?

18.1—The Common Ion Effect

Lactic acid is a weak acid found in sour milk, apples, beer, wine, and sore muscles.

lactic acid ($HC_3H_5O_3$)
$K_a = 1.4 \times 10^{-4}$

lactate ion ($C_3H_5O_3^-$)

■ **Equilibrium constants and temperature.** Unless specified otherwise, all equilibrium constants and all calculations in this chapter are at 25 °C.

Suppose you add 0.025 mol of sodium hydroxide, a strong base, to an aqueous solution containing 0.100 mol of the acid.

$$HC_3H_5O_3(aq) + OH^-(aq) \rightleftharpoons C_3H_5O_3^-(aq) + H_2O(\ell)$$

lactic acid   lactate ion

Only one-fourth as much base as is required to completely consume the acid has been added. Therefore, when reaction is complete, the solution contains both a weak acid (unreacted lactic acid, 0.075 mol) and a weak base (the product, lactate ion, 0.025 mol). Le Chatelier's principle informs us that the extent to which lactic acid can ionize under these conditions—and thus affect the pH of the solution—is influenced by the presence of the lactate ion. A 0.100 M solution of lactic acid will have a pH of about 2.43, whereas the pH is higher, 3.38, after converting some of the lactic acid to the conjugate base, lactate ion.

Figure 18.1 **The common ion effect.**
Approximately equal amounts of acetic acid (*left flask*, pH about 2.7) and sodium acetate, a base (*right flask*, pH about 9), were mixed in the beaker. The pH meter shows that the resulting solution in the beaker has a lower hydronium ion concentration (pH about 5) than the acetic acid solution owing to the presence of acetate ion, the conjugate base of the acid and an ion common to the ionization reaction of the acid. (The acid and base solutions each had about the same concentration. Each solution contains universal indicator. This dye is red in low pH, yellow in slightly acidic media, and green in neutral to weakly basic media.)

Aqueous acetic acid pH 2.7

Aqueous sodium acetate pH 9

Mixture of acetic acid and sodium acetate

The interaction between lactic acid and the lactate ion is an example of the **common ion effect**: The ionization of an acid or a base is limited by the presence of its conjugate base or acid (Figure 18.1).

$$HC_3H_5O_3(aq) + H_2O(\ell) \rightleftharpoons H_3O^+(aq) + C_3H_5O_3^-(aq)$$

lactic acid   lactate ion

Ionization of the acid is affected by the presence of its conjugate base

Let us see exactly how the common ion effect works. If 1.0 L of a 0.25-M acetic acid solution has a pH of 2.67, what is the pH after adding 0.10 mol of sodium acetate? Sodium acetate, $NaCH_3CO_2$, is 100% dissociated into its ions, Na^+ and $CH_3CO_2^-$, in water. Sodium ion has no effect on the pH of a solution (see Table 17.4 and Example 17.2). Thus, the important components of the solution are a weak acid (CH_3CO_2H) and its conjugate base ($CH_3CO_2^-$); the latter is an ion "common" to the ionization equilibrium reaction of acetic acid.

$$CH_3CO_2H(aq) + H_2O(\ell) \rightleftharpoons H_3O^+(aq) + CH_3CO_2^-(aq)$$

Assume the acid ionizes to give H_3O^+ and $CH_3CO_2^-$, both by x mol/L. This means that, relative to their initial concentrations, CH_3CO_2H decreases in concentration slightly (by x) and $CH_3CO_2^-$ increases slightly (by x).

Equation	$CH_3CO_2H + H_2O \rightleftharpoons H_3O^+ + CH_3CO_2^-$		
Initial (M)	0.25	0	0.10
Change (M)	$-x$	$+x$	$+x$
Equilibrium (M)	$(0.25 - x)$	x	$0.10 + x$

■ **The Common Ion Effect**
In this ICE table the first row (Initial) reflects the assumption that no ionization of the acid (or hydrolysis of the conjugate base) has yet occurred. Ionization of the acid in the presence of the conjugate base then produces x mol/L of hydronium ion and x mol/L more of the conjugate base.

Because we have been able to define the equilibrium concentrations of acid and conjugate base, and we know K_a, the hydronium ion concentration $(= x)$ can be calculated from the usual equilibrium constant expression.

$$K_a = 1.8 \times 10^{-5} = \frac{[H_3O^+][CH_3CO_2^-]}{[CH_3CO_2H]} = \frac{(x)(0.10 + x)}{0.25 - x}$$

Now, because acetic acid is a weak acid, and because it is ionizing in the presence of a significant concentration of its conjugate base, let us assume x is quite small. That is, it is reasonable to assume that $(0.10 + x)$ M ≈ 0.10 M and that $(0.25 - x)$ M ≈ 0.25 M. This leads to the "approximate" expression.

$$K_a = 1.8 \times 10^{-5} = \frac{[H_3O^+][CH_3CO_2^-]}{[CH_3CO_2H]} = \frac{(x)(0.10)}{0.25}$$

Solving this expression, we find that $x = [H_3O^+] = 4.5 \times 10^{-5}$ M and the pH is 4.35.

Without added $NaCH_3CO_2$, which provides the "common ion" $CH_3CO_2^-$, ionization of 0.25 M acetic acid will produce H_3O^+ and $CH_3CO_2^-$ ions in a concentration of 0.0021 M (to give a pH of 2.67). Le Chatelier's principle, however, predicts that the added common ion will cause the reaction to proceed less far to the right. Hence, as we have found, $x = [H_3O^+]$ is less than 0.0021 M in the presence of added acetate ion.

This section began with a discussion of the reaction of lactic acid with sodium hydroxide. Example 18.1 describes this in more detail.

GENERAL
Chemistry·•·Now™

See the General ChemistryNow CD-ROM or website:

• **Screen 18.2 Common Ion Effect**, for a simulation on the effect of a common ion on pH

Example 18.1—Reaction of Lactic Acid with a Deficiency of Sodium Hydroxide

Problem What is the pH of the solution that results from adding 25.0 mL of 0.0500 M NaOH to 25.0 mL of 0.100 M lactic acid? (K_a for lactic acid = 1.4×10^{-4}.)

$$HC_3H_5O_3(aq) + OH^-(aq) \rightleftharpoons C_3H_5O_3^-(aq) + H_2O(\ell)$$
lactic acid   lactate ion

Strategy There are two parts to this problem: a stoichiometry problem and an equilibrium problem. We first calculate the concentrations of lactic acid and lactate ion that are present following the reaction of lactic acid with NaOH. Then, with the acid and conjugate base concentrations known, we follow the strategy described in the text to determine the pH.

Solution

Part 1: Stoichiometry Problem

First, consider what species remain in solution after the acid–base reaction, and what the concentrations of those species are.

(a) Amounts of NaOH and lactic acid:

$$(0.0250 \text{ L NaOH})(0.0500 \text{ mol/L}) = 1.25 \times 10^{-3} \text{ mol NaOH}$$

$$(0.0250 \text{ L lactic acid})(0.100 \text{ mol/L}) = 2.50 \times 10^{-3} \text{ mol lactic acid}$$

(b) Amount of lactate ion produced:

Recognizing that NaOH is the limiting reactant, we have

$$(1.25 \times 10^{-3}\,\text{mol NaOH})\left(\frac{1\,\text{mol lactate ion}}{1\,\text{mol NaOH}}\right) = 1.25 \times 10^{-3}\,\text{mol lactate ion produced}$$

(c) Amount of lactic acid consumed:

$$(1.25 \times 10^{-3}\,\text{mol NaOH})\left(\frac{1\,\text{mol lactic acid}}{1\,\text{mol NaOH}}\right) = 1.25 \times 10^{-3}\,\text{mol lactic acid consumed}$$

(d) Amount of lactic acid remaining when reaction is complete:

$$2.50 \times 10^{-3}\,\text{mol lactic acid available} - 1.25 \times 10^{-3}\,\text{mol lactic acid consumed}$$

$$= 1.25 \times 10^{-3}\,\text{mol lactic acid remaining}$$

(e) Concentrations of lactic acid and lactate ion after reaction:

Note that the total solution volume after reaction is 50.0 mL or 0.0500 L.

$$[\text{lactic acid}] = \frac{1.25 \times 10^{-3}\,\text{mol lactic acid}}{0.0500\,\text{L}} = 2.50 \times 10^{-2}\,\text{M}$$

Because the amount of lactic acid remaining is the same as the amount of lactate ion produced, we have

$$[\text{lactic acid}] = [\text{lactate ion}] = 2.50 \times 10^{-2}\,\text{M}$$

Part 2: Equilibrium Calculation

With the "initial" concentrations known, construct a table summarizing the equilibrium concentrations.

Equilibrium	$HC_3H_5O_3 + H_2O \rightleftharpoons H_3O^+ + C_3H_5O_3^-$		
Initial (M)	0.0250	0	0.0250
Change (M)	$-x$	$+x$	$+x$
Equilibrium (M)	$(0.0250-x)$	x	$(0.0250 + x)$

Substituting the concentrations into the equilibrium expression, we have

$$K_a(\text{lactic acid}) = 1.4 \times 10^{-4} = \frac{[H_3O^+][C_3H_5O_3^-]}{[HC_3H_5O_3]} = \frac{(x)(0.0250 + x)}{0.0250 - x}$$

Making the assumption that x is small with respect to 0.0250 M, we see that

$$x = [H_3O^+] = K_a = 1.4 \times 10^{-4}\,\text{M}$$

which gives a pH of 3.85 .

Comment There are two final points to be made:

• Our assumption that $x \ll 0.0250$ is valid.

• The pH of a solution containing only 0.100 M lactic acid solution is 2.43. Adding a base (lactate ion) increases the pH.

Exercise 18.1—Common Ion Effect

Assume you have a 0.30 M solution of formic acid (HCO_2H) and have added enough sodium formate ($NaHCO_2$) to make the solution 0.10 M in the salt. Calculate the pH of the formic acid solution before and after adding sodium formate.

Exercise 18.2—Mixing an Acid and a Base

What is the pH of the solution that results from adding 30.0 mL of 0.100 M NaOH to 45.0 mL of 0.100 M acetic acid?

18.2—Controlling pH: Buffer Solutions

The normal pH of human blood is 7.4. However, the addition of a small quantity of strong acid or base, say 0.01 mol, to a liter of blood, leads to a change in pH of only about 0.1 pH unit. In comparison, if you add 0.01 mol of HCl to 1.0 L of pure water, the pH drops from 7.00 to 2.0. Addition of 0.01 mol of NaOH to pure water increases the pH from 7.00 to 12.0. Blood, and many other body fluids, are said to be buffered. A **buffer** causes solutions to resist a change in pH when a strong acid or base is added (Figure 18.2).

There are two requirements for a buffer:

- Two substances are needed: an acid that is capable of reacting with added OH^- ions and a base that can consume added H_3O^+ ions.

- The acid and the base must not react with each other.

These requirements mean a buffer is usually prepared from a conjugate acid–base pair: (1) a weak acid and its conjugate base (acetic acid and acetate ion, for example) or (2) a weak base and its conjugate acid (ammonia and ammonium ion, for example). Some buffers commonly used in the laboratory are given in Table 18.1.

To see how a buffer works, let us consider an acetic acid/acetate ion buffer. Acetic acid, the weak acid, is needed to consume any added hydroxide ion.

$$CH_3CO_2H(aq) + OH^-(aq) \rightleftharpoons CH_3CO_2^-(aq) + H_2O(\ell) \qquad K = 1.8 \times 10^9$$

The equilibrium constant for the reaction is very large because OH^- is a much stronger base than acetate ion, $CH_3CO_2^-$ (see Section 17.5 and Table 17.3). This means that any OH^- entering the solution from an outside source is consumed completely. In a similar way, any hydronium ion added to the solution reacts with the acetate ion present in the buffer.

$$H_3O^+(aq) + CH_3CO_2^-(aq) \rightleftharpoons H_2O(\ell) + CH_3CO_2H(aq) \qquad K = 5.6 \times 10^4$$

■ **Buffers and the Common Ion Effect**
The common ion effect is observed for an acid (or a base) ionizing in the presence of its conjugate base (or acid). A buffer is a solution of an acid, for example, and its conjugate base. Thus, buffers are just a particular type of the common ion effect.

Before

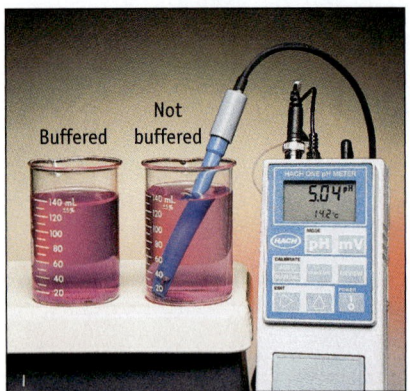

After adding 0.10 M HCl

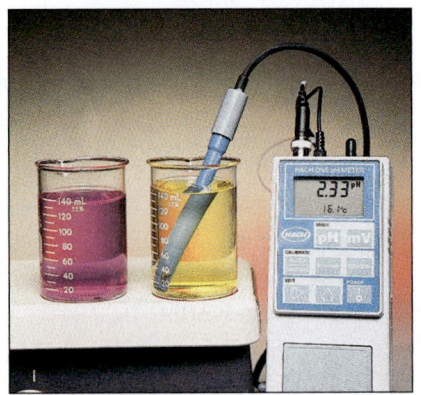

(a) The pH electrode is indicating the pH of water that contains a trace of acid (and bromphenol blue acid-base indicator. The solution at the left is a buffer solution with a pH of about 7. (It also contains bromphenol blue dye.)

(b) When 5 mL of 0.10 M HCl is added to each solution, the pH of the water drops several units, whereas the pH of the buffer stays constant as implied by the fact that the indicator color did not change.

Active Figure 18.2   **Buffer solutions.**

GENERAL
Chemistry ⚛ Now™ See the General ChemistryNow CD-ROM or website to explore an interactive version of this figure accompanied by an exercise.

Table 18.1 Some Commonly Used Buffer Systems

Weak Acid	Conjugate Base	Acid K_a (pK_a)	Useful pH Range
Phthalic acid, $C_6H_4(CO_2H)_2$	Hydrogen phthalate ion $C_6H_4(CO_2H)(CO_2)^-$	1.3×10^{-3} (2.89)	1.9–3.9
Acetic acid, CH_3CO_2H	Acetate ion, $CH_3CO_2^-$	1.8×10^{-5} (4.74)	3.7–5.8
Dihydrogen phosphate ion, $H_2PO_4^-$	Hydrogen phosphate ion, HPO_4^{2-}	6.2×10^{-8} (7.21)	6.2–8.2
Hydrogen phosphate ion, HPO_4^{2-}	Phosphate ion, PO_4^{3-}	3.6×10^{-13} (12.44)	11.3–13.3

The equilibrium constant for this reaction is also quite large because H_3O^+ is a much stronger acid than CH_3CO_2H.

The next several examples illustrate how to calculate the pH of a buffer solution, how to prepare a buffer, and how a buffer can control the pH of a solution.

GENERAL
Chemistry⋅⚛⋅Now™

See the General ChemistryNow CD-ROM or website:
- **Screen 18.3 Buffer Solutions,** for a simulation and tutorial on buffer solutions
- **Screen 18.4 pH of Buffer Solutions,** for a simulation and tutorial to study buffers

Example 18.2—pH of a Buffer Solution

Problem What is the pH of an acetic acid/sodium acetate buffer with $[CH_3CO_2H] = 0.700$ M and $[CH_3CO_2^-] = 0.600$ M?

Strategy Knowing the concentrations of the weak acid and its conjugate base, as well as K_a, we can calculate the hydronium ion concentration.

Solution Write a balanced equation for the ionization of acetic acid and set up an ICE table.

Equilibrium	$CH_3CO_2H + H_2O \rightleftharpoons H_3O^+ + CH_3CO_2^-$		
Initial (M)	0.700	0	0.600
Change (M)	$-x$	$+x$	$+x$
Equilibrium (M)	$0.700 - x$	x	$0.600 + x$

The appropriate equilibrium constant expression is

$$K_a = 1.8 \times 10^{-5} = \frac{[H_3O^+][CH_3CO_2^-]}{[CH_3CO_2H]} = \frac{(x)(0.600 + x)}{0.700 - x}$$

As explained in Example 18.1, the value of x will be very small compared with 0.700 or 0.600, so we can use the "approximate expression" to find x, the hydronium ion concentration.

$$K_a = 1.8 \times 10^{-5} = \frac{[H_3O^+][CH_3CO_2^-]}{[CH_3CO_2H]} = \frac{(x)(0.600)}{0.700}$$

$$x = 2.1 \times 10^{-5} \text{ M}$$

$$pH = -\log(2.1 \times 10^{-5}) = \boxed{4.68}$$

Comment The pH of the buffer has a value between the pH of 0.700 M acetic acid (2.45) and 0.600 M sodium acetate (9.26).

Exercise 18.3—pH of a Buffer Solution

What is the pH of a buffer solution composed of 0.50 M formic acid (HCO_2H) and 0.70 M sodium formate ($NaHCO_2$)?

General Expressions for Buffer Solutions

In Example 18.2 we found the hydronium ion concentration of the acetic acid/acetate ion buffer solution by solving for x in the equation

$$K_a = 1.8 \times 10^{-5} = \frac{[H_3O^+][CH_3CO_2^-]}{[CH_3CO_2H]} = \frac{(x)(0.600)}{0.700}$$

If we rearrange this equation, we obtain a very useful equation that can help you understand better how a buffer works.

$$[H_3O^+] = \frac{[CH_3CO_2H]}{[CH_3CO_2^-]} \times K_a$$

That is, the hydrogen ion concentration in the acetic acid/acetate ion buffer is given by the ratio of the acid and conjugate base concentrations multiplied by the acid ionization constant. Indeed, this expression holds true for all buffer solutions based on *a weak acid and its conjugate base*.

$$[H_3O^+] = \frac{[acid]}{[conjugate\ base]} \times K_a \qquad (18.1)$$

■ **Buffer Solutions**
You will find it generally useful to consider all buffer solutions as being composed of a weak acid and its conjugate base. Suppose, for example, a buffer is composed of the weak base ammonia and its conjugate acid ammonium ion. The hydronium ion concentration can be found from Equation 18.1 by assuming the buffer is composed of the weak acid NH_4^+ and its conjugate base, NH_3.

It is often convenient to use Equation 18.1 in a different form. If we take the negative logarithm of each side of the equation, we have

$$-\log[H_3O^+] = \left\{ -\log \frac{[acid]}{[conjugate\ base]} \right\} + (-\log K_a)$$

You know that $-\log[H_3O^+]$ is defined as pH, and $-\log K_a$ is defined as pK_a [◀ Sections 17.3 and 17.4]. Furthermore, because

$$-\log \frac{[acid]}{[conjugate\ base]} = +\log \frac{[conjugate\ base]}{[acid]}$$

the preceding equation can be rewritten as

$$pH = pK_a + \log \frac{[conjugate\ base]}{[acid]} \qquad (18.2)$$

■ **The Henderson-Hasselbalch Equation**
Many chemistry handbooks list acid ionization constants in terms of pK_a values, so the approximate pH values of possible buffer solutions are readily apparent.

This equation is known as the **Henderson-Hasselbalch equation**.

Both Equations 18.1 and 18.2 show that the pH of a buffer solution is controlled by two factors: the strength of the acid (as expressed by K_a or pK_a) and the relative amounts of acid and conjugate base. The buffer's pH is established primarily by the value of K_a or pK_a, and the pH is fine-tuned by adjusting the acid-to-conjugate base ratio.

When the concentrations of conjugate base and acid are the same in a solution, the ratio [conjugate base]/[acid] is 1. The log of 1 is zero, so $pH = pK_a$ under these circumstances. If there is more of the conjugate base in the solution than acid, for example, then $pH > pK_a$. Conversely, if there is more acid than conjugate base in solution, then $pH < pK_a$.

GENERAL
Chemistry·⚛·Now™

See the General ChemistryNow CD-ROM or website:

- **Screen 18.4 pH of Buffer Solutions,** for a simulation and tutorial that uses the Henderson-Hasselbalch equation

Example 18.3—Using the Henderson-Hasselbalch Equation

Problem Benzoic acid ($C_6H_5CO_2H$, 2.00 g) and sodium benzoate ($NaC_6H_5CO_2$, 2.00 g) are dissolved in enough water to make 1.00 L of solution. Calculate the pH of the solution using the Henderson-Hasselbalch equation.

Strategy The Henderson-Hasselbalch equation requires the pK_a of the acid, which is obtained from the K_a for the acid (see Table 17.3 or Appendix H). You will also need the acid and conjugate base concentrations.

Solution K_a for benzoic acid is 6.3×10^{-5}. Therefore,

$$pK_a = -\log (6.3 \times 10^{-5}) = 4.20$$

Next, we need the concentrations of the acid (benzoic acid) and its conjugate base (benzoate ion).

$$2.00 \text{ g benzoic acid} \left(\frac{1 \text{ mol}}{122.1 \text{ g}} \right) = 0.0164 \text{ mol benzoic acid}$$

$$2.00 \text{ g sodium benzoate} \left(\frac{1 \text{ mol}}{144.1 \text{ g}} \right) = 0.0139 \text{ mol sodium benzoate}$$

Because the solution volume is 1.00 L, the concentrations are [benzoic acid] = 0.0164 M and [sodium benzoate] = 0.0139 M. Using Equation 18.2, we have

$$pH = 4.20 + \log \frac{0.0139}{0.0164} = 4.20 + \log(0.848) = \boxed{4.13}$$

Comment The pH is less than the pK_a because the concentration of the acid is greater than the concentration of the conjugate base (the ratio of conjugate base to acid concentration is less than 1).

Exercise 18.4—Using the Henderson-Hasselbalch Equation

Use the Henderson-Hasselbalch equation to calculate the pH of 1.00 L of a buffer solution containing 15.0 g of $NaHCO_3$ and 18.0 g of Na_2CO_3. (Consider this buffer to be a solution of the weak acid HCO_3^- with CO_3^{2-} as its conjugate base.)

Preparing Buffer Solutions

To be useful, a buffer solution must have two characteristics:

- *pH control:* It should control the pH at the desired value. The Henderson-Hasselbalch equation shows us how this can be done.

$$pH = pK_a + \log \frac{[\text{conjugate base}]}{[\text{acid}]}$$

First, an acid is chosen whose pK_a (or K_a) is near the intended value of pH (or $[H_3O^+]$). Second, the exact value of pH (or $[H_3O^+]$) is obtained by adjusting the acid-to-conjugate base ratio. (Example 18.4 illustrates this approach.)

- *Buffer capacity:* The buffer should have the capacity to control the pH after the addition of reasonable amounts of acid and base. For example, the concentration of acetic acid in an acetic acid/acetate ion buffer must be sufficient to consume all of the hydroxide ion that may be added and still control the pH (see Example 18.4). Buffers are usually prepared as 0.10 M to 1.0 M solutions of reagents. However, any buffer will lose its capacity if too much strong acid or base is added.

GENERAL
Chemistry⋅Now™

See the General ChemistryNow CD-ROM or website:
- **Screen 18.5 Preparing Buffer Solutions,** for a simulation and tutorial on the preparation of a buffer

Example 18.4—Preparing a Buffer Solution

Problem You wish to prepare 1.0 L of a buffer solution with a pH of 4.30. A list of possible acids (and their conjugate bases) follows:

Acid	Conjugate Base	K_a	pK_a
HSO_4^-	SO_4^{2-}	1.2×10^{-2}	1.92
CH_3CO_2H	$CH_3CO_2^-$	1.8×10^{-5}	4.74
HCO_3^-	CO_3^{2-}	4.8×10^{-11}	10.32

Which combination should be selected, and what should the ratio of acid to conjugate base be?

Strategy Use either the general equation for a buffer (Equation 18.1) or the Henderson-Hasselbalch equation (Equation 18.2). Equation 18.1 informs you that $[H_3O^+]$ should be close to the acid K_a value, and Equation 18.2 tells you that the pH should be close to the acid pK_a value. This will establish which acid you will use.

Having decided which acid to use, convert pH to $[H_3O^+]$ to use Equation 18.1. If you use Equation 18.2, use the pK_a value in the table. Finally, calculate the ratio of acid to conjugate base.

Solution The hydronium ion concentration for the buffer is found from the targeted pH.

$$pH = 4.30, \text{ so } [H_3O^+] = 10^{-pH} = 10^{-4.30} = 5.0 \times 10^{-5} \text{ M}$$

Of the acids given, only acetic acid (CH_3CO_2H) has a K_a value close to that of the desired $[H_3O^+]$ (or a pK_a close to a pH of 4.30). Now you need merely to adjust the ratio $[CH_3CO_2H]/[CH_3CO_2^-]$ to achieve the desired hydronium ion concentration.

$$[H_3O^+] = 5.0 \times 10^{-5} = \frac{[CH_3CO_2H]}{[CH_3CO_2^-]}(1.8 \times 10^{-5})$$

Rearrange this equation to find the ratio $[CH_3CO_2H]/[CH_3CO_2^-]$.

$$\frac{[CH_3CO_2H]}{[CH_3CO_2^-]} = \frac{[H_3O^+]}{K_a} = \frac{5.0 \times 10^{-5}}{1.8 \times 10^{-5}} = \frac{2.8 \text{ mol/L}}{1.0 \text{ mol/L}}$$

If you add 0.28 mol of acetic acid and 0.10 mol of sodium acetate (or any other pair of molar quantities in the ratio 2.8/1) to enough water to make 1.0 L of solution, the buffer solution will have a pH of 4.30.

Comment If you prefer to use the Henderson-Hasselbalch equation, you would have

$$pH = 4.30 = 4.74 + \log\frac{[CH_3CO_2^-]}{[CH_3CO_2H]}$$

$$\log\frac{[CH_3CO_2^-]}{[CH_3CO_2H]} = 4.30 - 4.74 = -0.44$$

$$\frac{[CH_3CO_2^-]}{[CH_3CO_2H]} = 10^{-0.44} = 0.36$$

The ratio of conjugate base to acid, $[CH_3CO_2^-]/[CH_3CO_2H]$, is 0.36. The reciprocal of this ratio $\{= [CH_3CO_2H]/[CH_3CO_2^-] = 1/0.36)\}$ is 2.8. This is the same result obtained above using Equation 18.1.

Exercise 18.5—Preparing a Buffer Solution

Using an acetic acid/sodium acetate buffer solution, what ratio of acid to conjugate base will you need to maintain the pH at 5.00? Describe how you would make up such a solution.

Example 18.4 illustrates several important points concerning buffer solutions. The hydronium ion concentration depends not only on the K_a value of the acid but also on the ratio of acid and conjugate base concentrations. However, even though we write these ratios in terms of reagent concentrations, it is *the relative number of moles of acid and conjugate base that is important in determining the pH of a buffer solution.* Because both reagents are dissolved in the same solution, their concentrations depend on the same solution volume. In Example 18.4, the ratio 2.8/1 for acetic acid and sodium acetate implies that 2.8 times as many moles of acid were dissolved per liter as moles of sodium acetate.

$$\frac{[CH_3CO_2H]}{[CH_3CO_2^-]} = \frac{2.8 \text{ mol } CH_3CO_2H/L}{1.0 \text{ mol } CH_3CO_2^-/L} = \frac{2.8 \text{ mol } CH_3CO_2H}{1.0 \text{ mol } CH_3CO_2^-}$$

Problem-Solving Tip 18.1

Buffer Solutions

The following summary highlights important aspects of buffer solutions.

- A buffer resists changes in pH on adding small quantities of strong acid or base.

- A buffer contains a weak acid and its conjugate base.

- The hydronium ion concentration of a buffer solution can be calculated from Equation 18.1,

$$[H_3O^+] = \frac{[\text{acid}]}{[\text{conjugate base}]} \times K_a$$

or the pH can be calculated from the Henderson-Hasselbalch equation (Equation 18.2).

$$pH = pK_a + \log\frac{[\text{conjugate base}]}{[\text{acid}]}$$

- The pH depends primarily on the K_a value of the weak acid and secondarily on the relative amounts of acid and conjugate base.

- The function of the weak acid in a buffer is to consume added base; the conjugate base consumes added acid. Such reactions affect the relative quantities of the weak acid and its conjugate base. Because this ratio of acid to its conjugate base has only a secondary effect on the pH, the pH can be maintained at a relatively constant level.

- The buffer must have sufficient capacity to react with reasonable quantities of added acid or base.

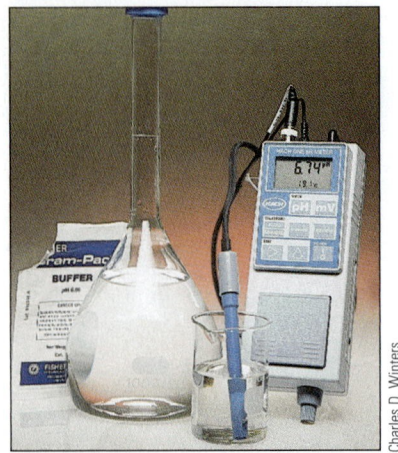

Figure 18.3   A commercial buffer solution. The solid acid and conjugate base in the packet are mixed with water to give a solution with the indicated pH. The quantity of water used does not affect the pH because the ratio [acid]/[conjugate base] does not depend on the solution volume. (However, if too much water is added, the acid and conjugate base concentrations will be low and the buffer capacity could be exceeded. Again, buffer solutions usually have solute concentrations around 0.1 M to 1.0 M.)

Notice that on dividing one concentration by the other, the volumes "cancel." This means that we only need to ensure that the ratio of moles of acid to moles of conjugate base is 2.8 to 1.0 in this example. The acid and its conjugate base could have been dissolved in any reasonable amount of water. It also means that *diluting a buffer solution will not change its pH*. Commercially available buffers are often sold as premixed, dry ingredients. To use them, you simply mix the ingredients in some volume of pure water (Figure 18.3).

How Does a Buffer Maintain pH?

Now let us explore quantitatively how a given buffer solution can maintain the pH of a solution on adding a small amount of strong acid.

GENERAL
Chemistry ⚛ Now™

See the General ChemistryNow CD-ROM or website:
- **Screen 18.6 Adding Reagents to a Buffer Solution,** for a tutorial on adding acids and bases to buffers

---

Example 18.5—How Does a Buffer Maintain a Constant pH?

Problem What is the change in pH when 1.00 mL of 1.00 M HCl is added to (1) 1.000 L of pure water and to (2) 1.000 L of acetic acid/sodium acetate buffer with $[CH_3CO_2H] = 0.700$ M and $[CH_3CO_2^-] = 0.600$ M. (See Example 18.2, where the pH of this acetic acid/acetate ion buffer was found to be 4.68.)

Strategy HCl is a strong acid, so it ionizes completely to supply H_3O^+ ions. Part 1 involves two steps: (a) Find the H_3O^+ concentration when adding 1.00 mL of acid to 1.000 L. (b) Convert the value of $[H_3O^+]$ for the dilute solution to pH.

Part 2 involves three steps: (a) A stoichiometry calculation finds how the concentrations of acid and conjugate base change on adding H_3O^+. (b) An equilibrium calculation to find $[H_3O^+]$ for a buffer solution in which the concentrations of CH_3CO_2H and $CH_3CO_2^-$ are slightly altered owing to the reaction of $CH_3CO_2^-$ with added H_3O^+. (c) Convert $[H_3O^+]$ to pH.

Solution

Part 1. *Adding Acid to Pure Water*

Here 1.00 mL of 1.00 M HCl represents 0.00100 mol of acid. If this amount is added to 1.000 L of pure water, the H_3O^+ concentration of the water changes from 10^{-7} M to 10^{-3} M,

$$c_1 \times V_1 = c_2 \times V_2$$

$$(1.00 \text{ M})(0.00100 \text{ L}) = c_2 \times (1.001 \text{ L})$$

$$c_2 = [H_3O^+] \text{ in diluted solution} = 1.00 \times 10^{-3} \text{ M}$$

The pH falls from 7.00 to 3.00.

Part 2. *Adding Acid to an Acetic Acid/Acetate Buffer Solution*

HCl is a strong acid that is 100% ionized in water and supplies H_3O^+, which reacts completely with the base (acetate ion) in the buffer solution according to the following equation:

$$H_3O^+(aq) + CH_3CO_2^-(aq) \longrightarrow H_2O(\ell) + CH_3CO_2H(aq)$$

	$[H_3O^+]$ from added HCl	$[CH_3CO_2^-]$ from buffer	$[CH_3CO_2H]$ from buffer
Initial amount of acid or base (mol $= c \times V$)	0.00100	0.600	0.700
Change (mol)	−0.00100	−0.00100	+0.00100
Equilibrium (mol)	0	0.599	0.701
Concentrations after reaction ($c =$ mol/V)	0	0.598	0.700

Because the added HCl reacts *completely* with the acetate ion to produce acetic acid, the solution after this reaction (with $V = 1.001$ L) is once again a buffer containing only the weak acid and its salt. Now we can use Equation 18.1 (or the Henderson-Hasselbalch equation) to find $[H_3O^+]$ and the pH in the buffer solution as in Examples 18.2 and 18.3.

Equilibrium	CH_3CO_2H + H_2O $\rightleftharpoons$ H_3O^+ + $CH_3CO_2^-$		
Initial (M)	0.701	0	0.598
Change (M)	−x	+x	+x
Equilibrium (M)	0.701−x	x	0.598 + x

As usual, we make the approximation that x, the amount of H_3O^+ formed by ionizing acetic acid in the presence of acetate ion, is very small compared with 0.701 M or 0.598 M. Using Equation 18.1, we calculate a pH of 4.68.

$$[H_3O^+] = x = \frac{[CH_3CO_2H]}{[CH_3CO_2^-]} \times K_a = \left(\frac{0.701 \text{ mol/L}}{0.598 \text{ mol/L}}\right)(1.8 \times 10^{-5}) = 2.1 \times 10^{-5} \text{ M}$$

$$pH = 4.68$$

Comment Within the number of significant figures allowed, the *pH of the buffer solution does not change* after adding HCl. In contrast, it changed by 4 units when 1.0 mL of 1.0 M HCl was added to 1.0 L of pure water.

Exercise 18.6—Buffer Solutions

Calculate the pH of 0.500 L of a buffer solution composed of 0.50 M formic acid (HCO_2H) and 0.70 M sodium formate ($NaHCO_2$) before and after adding 10.0 mL of 1.0 M HCl.

18.3—Acid–Base Titrations

A *titration* is one of the most important ways of determining accurately the quantity of an acid, a base, or some other substance in a mixture or of ascertaining the purity of a substance. You learned how to perform the stoichiometry calculations involved in titrations in Chapter 5 [◀ Section 5.10]. In Chapter 17, we described the following important points regarding acid–base reactions [◀ Section 17.6]:

- The pH at the equivalence point of a strong acid–strong base titration is 7. The solution at the equivalence point is truly "neutral" *only* when a strong acid is titrated with a strong base, and vice versa.

- If the substance being titrated is a weak acid or base, then the pH at the equivalence point is not 7 [◀ Table 17.5].

  (a) A weak acid titrated with a strong base leads to a pH $>$ 7 at the equivalence point due to the conjugate base of the weak acid.

■ **Equivalence Point**
The equivalence point for a reaction is the point at which one reactant has been completely consumed by addition of another reactant. See page 217.

Chemical Perspectives

Buffers in Biochemistry

Maintenance of pH is vital to the cells of all living organisms, because enzyme activity is influenced by pH. The primary protection against harmful pH changes in cells is provided by buffer systems. The intracellular pH of most cells is maintained within the range 6.9 to 7.4. Two important biological buffer systems control pH in this range: the phosphate system ($HPO_4^{2-}/H_2PO_4^-$) and the bicarbonate/carbonic acid system (HCO_3^-/H_2CO_3).

Phosphate ions are abundant in cells, both as the ions themselves and as important substituents on organic molecules. Most importantly, the pK_a for the $H_2PO_4^-$ ion is 7.21, which is very close to the high end of the normal pH range.

$$H_2PO_4^-(aq) + H_2O(\ell) \rightleftharpoons$$
$$H_3O^+(aq) + HPO_4^{2-}(aq)$$

If the buffer is to control the pH at about 7.4, the ratio of HPO_4^{2-} to $H_2PO_4^-$ must be 1.58.

$$pH = pK_a + \log\frac{[HPO_4^{2-}]}{[H_2PO_4^-]}$$

$$7.4 = 7.21 + \log\frac{[HPO_4^{2-}]}{[H_2PO_4^-]}$$

$$\frac{[HPO_4^{2-}]}{[H_2PO_4^-]} \approx 1.5$$

Charles D. Winters

Alkalosis. If blood pH is too high, alkalosis results. It can be reversed by breathing into a bag, an action that recycles exhaled CO_2. The returned CO_2 affects the carbonic acid buffer system in the body, raising the blood hydronium ion concentration. The blood pH drops back to a more normal level of 7.4.

A typical total phosphate concentration in a cell ($[HPO_4^{2-}] + [H_2PO_4^-]$) is 2.0×10^{-2} M. You can calculate that $[HPO_4^{2-}]$ should be about 1.2×10^{-2} M and $[H_2PO_4^-]$ should be about 7.7×10^{-3} M.

The bicarbonate/carbonic acid buffer is important in blood plasma. Three equilibria are important here.

$$CO_2(g) \rightleftharpoons CO_2(aq)$$
$$CO_2(aq) + H_2O(\ell) \rightleftharpoons H_2CO_3(aq)$$
$$H_2CO_3(aq) + H_2O(\ell) \rightleftharpoons$$
$$H_3O^+(aq) + HCO_3^-(aq)$$

The overall equilibrium constant for the second and third steps is $pK_{overall} = 6.3$ at 37 °C, the temperature of the human body. Thus,

$$7.4 = 6.3 + \log\frac{[HCO_3^-]}{[CO_2(aq)]}$$

Although the value of $pK_{overall}$ is about 1 pH unit away from the blood pH, the natural partial pressure of CO_2 in the alveoli of the lungs (about 40 mm Hg) is sufficient to keep $[CO_2(aq)]$ at about 1.2×10^{-3} M and $[HCO_3^-]$ at about 1.5×10^{-2} M.

If blood pH rises above 7.45, you can develop a condition called *alkalosis*. This problem can arise from hyperventilation, from severe anxiety, or from an oxygen deficiency at high altitude. It can ultimately lead to overexcitation of the central nervous system, muscle spasms, convulsions, and death. One way to treat acute alkalosis is to breathe into a paper bag. The CO_2 you exhale is recycled, so it raises the blood CO_2 level and causes the equilibria above to shift to the right, thus raising the hydronium ion concentration.

Acidosis is the opposite of alkalosis. It can arise from inadequate exhalation of CO_2. Acidosis can be reversed by breathing rapidly and deeply. Doubling the breathing rate increases the blood pH by about 0.23 units.

(b) A weak base titrated with a strong acid leads to a pH < 7 at the equivalence point due to the conjugate acid of the weak base.

A knowledge of buffer solutions and how they work will now allow us to more fully understand how the pH changes in the course of an acid–base reaction.

■ **Weak Acid–Weak Base Titrations** Titrations combining a weak acid and a weak base are generally not done because the equivalence point often cannot be judged accurately.

Titration of a Strong Acid with a Strong Base

Figure 18.4 illustrates what happens to the pH as 0.100 M NaOH is slowly added to 50.0 mL of 0.100 M HCl.

$$HCl(aq) + NaOH(aq) \longrightarrow NaCl(aq) + H_2O(\ell)$$
$$\text{Net ionic equation: } H_3O^+(aq) + OH^-(aq) \longrightarrow 2\,H_2O(\ell)$$

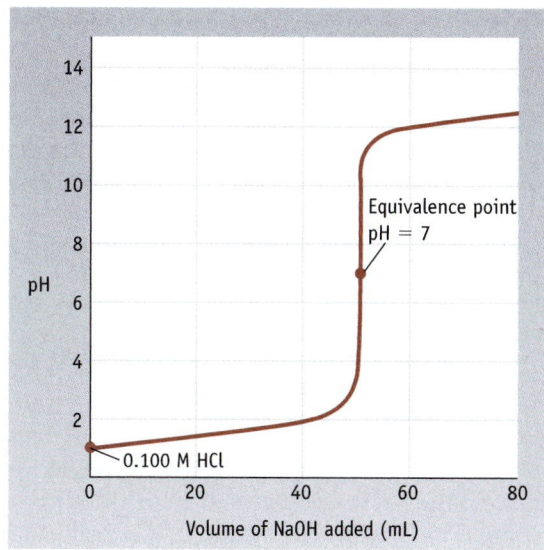

50.0 mL of 0.100 M HCl titrated with 0.100 M NaOH

Volume of base added	pH
100.0	12.52
80.0	12.36
60.0	11.96
55.0	11.68
51.0	11.00
50.0	7.00
49.0	3.00
48.0	2.69
45.0	2.28
40.0	1.95
20.0	1.37
10.0	1.18
0.0	1.00
very large amount	13.00 (maximum)

Figure 18.4 **The change in pH as a strong acid is titrated with a strong base.** Here 50.0 mL of 0.100 M HCl is titrated with 0.100 M NaOH. The pH at the equivalence point is 7.0 for the reaction of a strong acid with a strong base.

Let us focus on four regions on this plot.

- pH of the initial solution
- pH as NaOH is added to the HCl solution before the equivalence point
- pH at the equivalence point
- pH after the equivalence point

Before beginning the titration, the 0.100 M solution of HCl has a pH of 1.000. As NaOH is added to the acid solution, the amount of HCl declines, and the acid remaining is dissolved in an ever-increasing volume of solution. Thus, $[H_3O^+]$ decreases, and the pH slowly increases. As an example, let us find the pH of the solution after 10.0 mL of 0.100 M NaOH has been added to 50.0 mL of 0.100 M HCl. We will set up a table to list the amounts of acid and base before reaction, the changes in those amounts, and the amounts remaining after reaction. Notice that the volume of the solution after reaction is the sum of the combined volumes of NaOH and HCl (60.0 mL or 0.0600 L in this case).

	$H_3O^+(aq)$	+	$OH^-(aq) \longrightarrow 2 H_2O(\ell)$	
Initial amount (mol = $c \times V$)	0.00500		0.00100	
Change (mol)	−0.00100		−0.00100	
After reaction (mol)	0.00400		0	
After reaction (c = mol/V)	0.00400 mol/0.0600 L		0	
	=0.0667 M			

After addition of 10.0 mL of NaOH, the final solution has a hydronium ion concentration of 0.0667 M. The pH is

$$pH = -\log [H_3O^+] = -\log (0.0667) = 1.176$$

After 49.5 mL of base has been added—that is, just before the equivalence point—we can use the same approach to show that

$$pH = -\log [H_3O^+] = -\log (5.0 \times 10^{-4}) = 3.3$$

The solution being titrated is still quite acidic even very close to the equivalence point.

The pH of the equivalence point in an acid–base titration is taken as the mid-point in the vertical portion of the pH versus volume of titrant curve. (The **titrant** is the substance being added during the titration.) In the HCl/NaOH titration illustrated in Figure 18.4, the pH increases very rapidly near the equivalence point. In fact, in this case the pH rises 7 units (the H_3O^+ concentration decreases by a factor of 10 million!) when only a drop or two of the NaOH solution is added, and the mid-point of the vertical portion of the curve is at a pH of 7.00.

> The pH of the solution at the equivalence point in a strong acid–strong base reaction is always 7.00 (at 25 °C) because the solution contains a neutral salt.

After all of the HCl has been consumed, and the slightest excess of NaOH has been added, the solution will be basic, and the pH will continue to increase as more NaOH is added (and the solution volume increases). For example, if we calculate the pH of the solution after 55.0 mL of 0.100 M NaOH has been added to 50.0 mL of 0.100 M HCl, we find

	$H_3O^+(aq)$ +	$OH^-(aq)$ $\longrightarrow$	$2\ H_2O(\ell)$
Initial amount (mol = $c \times V$)	0.00500	0.00550	
Change (mol)	−0.00500	−0.00500	
After reaction (mol)	0	0.00050	
After reaction (c = mol/V)	0	0.00050 mol/0.1050 L	
		= 0.0048 M	

At this point, the solution has a hydroxide ion concentration of 0.0048 M. The pH is

$$pH = 14.00 - pOH = 14.00 - \log (0.0048)$$
$$pH = 14.00 - 2.32 = 11.68$$

Exercise 18.7—Titration of a Strong Acid with a Strong Base

What is the pH after 25.0 mL of 0.100 M NaOH has been added to 50.0 mL of 0.100 M HCl? What is the pH after 50.50 mL of NaOH has been added?

Titration of a Weak Acid with a Strong Base

The titration of a weak acid with a strong base is somewhat different from the strong acid–strong base titration. Look carefully at the curve for the titration of 100.0 mL of 0.100 M acetic acid with 0.100 M NaOH (Figure 18.5).

$$CH_3CO_2H(aq) + NaOH(aq) \longrightarrow NaCH_3CO_2(aq) + H_2O(\ell)$$

Let us focus on three important points on this curve:

- *The pH before titration begins.* The pH before any base is added can be calculated from the weak acid K_a value and the acid concentration [◀ Example 17.5].
- *The pH at the equivalence point.* At the equivalence point the solution contains only sodium acetate, with the CH_3CO_2H and NaOH having been completely

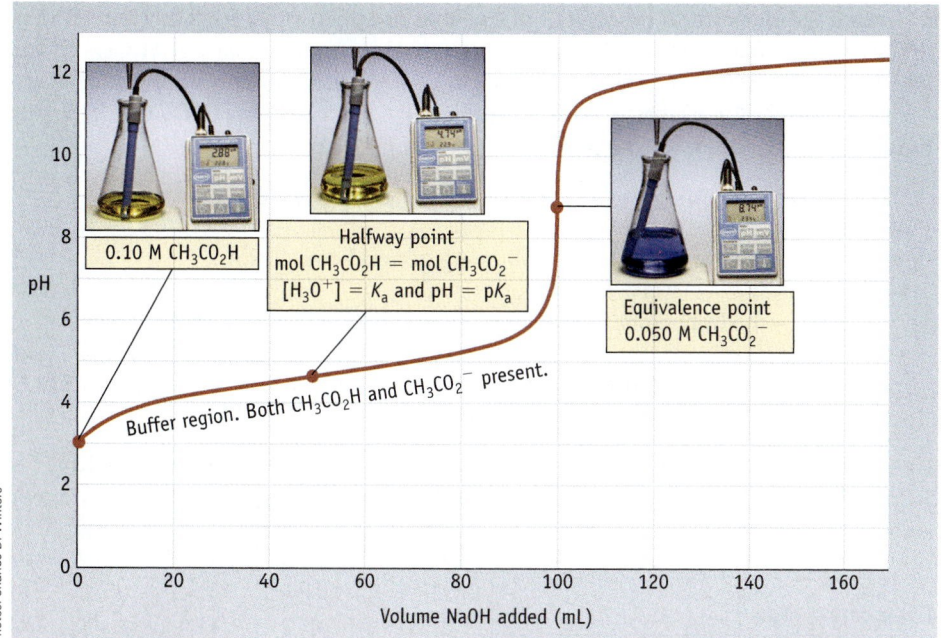

Active Figure 18.5   **The change in pH during the titration of a weak acid with a strong base.** Here 100.0 mL of 0.100 M acetic acid is titrated with 0.100 M NaOH. Note especially the following: (1) Acetic acid is a weak acid. The pH of the original solution is 2.87. (2) The pH at the point at which half the acid has reacted with the base is equal to the pK_a value for the acid (pH = pK_a = 4.74). (3) At the equivalence point, the solution consists of acetate ion, a weak base, and the pH is 8.72.

GENERAL

Chemistry ⚛ **Now**™ See the General ChemistryNow CD-ROM or website to explore an interactive version of this figure accompanied by an exercise.

consumed. The pH is controlled by the acetate ion, the conjugate base of acetic acid [◄ Table 17.5, page 817].

- *The pH at the halfway point (half-equivalence point) of the titration.* Here the pH is equal to the pK_a of the weak acid, a conclusion that is discussed in more detail below.

As NaOH is added to acetic acid, for example, the base is consumed and sodium acetate is produced. Thus, at every point between the beginning of the titration (when only acetic acid is present) and the equivalence point (when only sodium acetate is present), the solution contains both acetic acid and its salt, sodium acetate. These are the components of a *buffer solution,* and the hydronium ion concentration can be found from Equation 18.1 or 18.2. Therefore, between the beginning point and the equivalence point,

$$[H_3O^+] = \frac{[\text{weak acid remaining}]}{[\text{conjugate base produced}]} \times K_a \qquad (18.3)$$

or

$$pH = pK_a + \log \frac{[\text{conjugate base produced}]}{[\text{weak acid remaining}]} \qquad (18.4)$$

The fact that a buffer is present at any point between the beginning of the titration and the equivalence point is the reason that the pH of the solution rises more slowly after a few milliliters of titrant have been added.

What happens when *exactly* half of the acid has been consumed by base? Half of the acid (say, CH_3CO_2H) has been converted to the conjugate base ($CH_3CO_2^-$), and half remains. Therefore, the concentration of weak acid remaining is equal to the concentration of conjugate base produced $\{[CH_3CO_2H] = [CH_3CO_2^-]\}$. Using Equation 18.3 or 18.4, we see that

$$[H_3O^+] = (1) \times K_a \quad \text{or} \quad pH = pK_a + \log(1)$$

Because $\log 1 = 0$, we come to the following general conclusion:

> At the halfway point in the titration of a weak acid with a strong base
>
> $$[H_3O^+] = K_a \quad \text{and} \quad pH = pK_a \qquad (18.5)$$

In the particular case of the titration of acetic acid with a strong base, $[H_3O^+] = 1.8 \times 10^{-5}$ M at the halfway point, so the pH is 4.74. This is equal to the pK_a of acetic acid.

GENERAL
Chemistry⚛Now™

See the General ChemistryNow CD-ROM or website:
- **Screen 18.7 Titration Curves,** for a simulation and tutorial on titration curves for strong and weak acids

Example 18.6—Titration of Acetic Acid with Sodium Hydroxide

Problem What is the pH of the solution when 90.0 mL of 0.100 M NaOH has been added to 100.0 mL of 0.100 M acetic acid (see Figure 18.5)?

Strategy This problem is like Example 18.1. It involves two major steps: (a) a stoichiometry calculation to find the amount of acid remaining and amount of conjugate base formed after adding NaOH, and (b) an equilibrium calculation to find $[H_3O^+]$ for a buffer solution where the amounts of CH_3CO_2H and $CH_3CO_2^-$ are known from the first part of the calculation.

Solution Let us first calculate the amounts of reactants before reaction (= concentration × volume) and then use the principles of stoichiometry to calculate the amounts of reactants and products after reaction.

Equation	CH_3CO_2H +	OH^-	$\rightleftharpoons$	$CH_3CO_2^-$ + H_2O
Initial (mol)	0.0100	0.00900		0
Change (mol)	−0.00900	−0.00900		+0.00900
After reaction (mol)	0.0010	0		0.00900

The ratio of amounts (moles) of acid and conjugate base is the same as the ratio of their concentrations. Therefore, we can use the amounts of weak acid remaining and conjugate base formed to find the pH from Equation 18.3 (where we use amounts and not concentrations).

$$[H_3O^+] = \frac{\text{mol } CH_3CO_2H}{\text{mol } CH_3CO_2^-} \times K_a = \left(\frac{0.0010 \text{ mol}}{0.0090 \text{ mol}}\right)(1.8 \times 10^{-5}) = 2.0 \times 10^{-6} \text{ M}$$

$$pH = -\log(2.0 \times 10^{-6}) = 5.70$$

The pH is 5.70, in agreement with Figure 18.5.

Comment If you use the Henderson-Hasselbalch equation (18.2), the solution is

$$pH = pK_a + \log \frac{\text{mol } CH_3CO_2^-}{\text{mol } CH_3CO_2H} = 4.75 + \log \frac{0.0090}{0.0010}$$

$$pH = 4.74 + 0.95 = \boxed{5.69}$$

Finally, notice that the pH obtained (5.69) is appropriate for a point after the halfway point (4.74) but before the equivalence point (8.72).

Exercise 18.8—Titration of a Weak Acid with a Strong Base

The titration of 0.100 M acetic acid with 0.100 M NaOH is described in the text. What is the pH of the solution when 35.0 mL of the base has been added to 100.0 mL of 0.100 M acetic acid?

Titration of Weak Polyprotic Acids

The titrations illustrated thus far have been for the reaction of a monoprotic acid (HA) with a strong base such as NaOH. It is possible to extend the discussion of titrations to polyprotic acids such as oxalic acid, $H_2C_2O_4$.

$$H_2C_2O_4(aq) + H_2O(\ell) \rightleftharpoons HC_2O_4^-(aq) + H_3O^+(aq) \qquad K_{a_1} = 5.9 \times 10^{-2}$$

$$HC_2O_4^-(aq) + H_2O(\ell) \rightleftharpoons C_2O_4^{2-}(aq) + H_3O^+(aq) \qquad K_{a_2} = 6.4 \times 10^{-5}$$

Figure 18.6 illustrates the curve for the titration of 100. mL of 0.100 M oxalic acid with 0.100 M NaOH. The first significant rise in pH is experienced after 100 mL of base has been added, indicating that the first proton of the acid has been titrated.

$$H_2C_2O_4(aq) + OH^-(aq) \rightleftharpoons HC_2O_4^-(aq) + H_2O(\ell)$$

When the second proton of oxalic acid is titrated, the pH again rises significantly.

$$HC_2O_4^-(aq) + OH^-(aq) \rightleftharpoons C_2O_4^{2-}(aq) + H_2O(\ell)$$

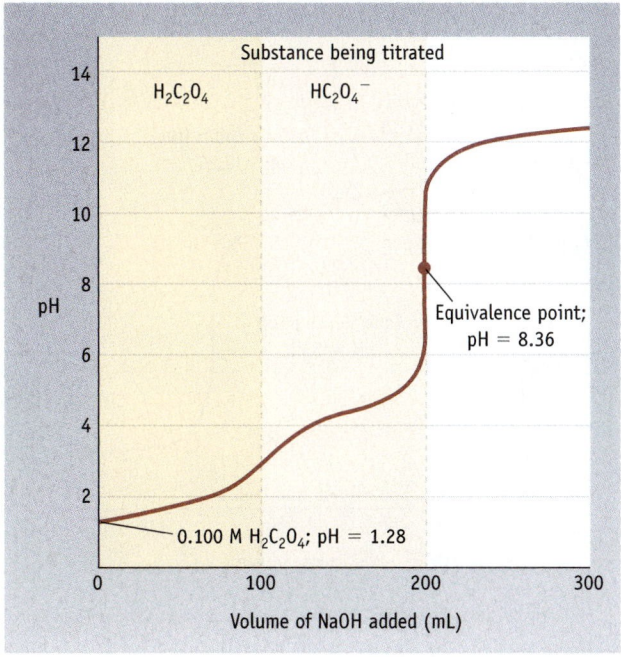

Figure 18.6 Titration curve for a diprotic acid. The curve for the titration of 100. mL of 0.100 M oxalic acid ($H_2C_2O_4$, a weak diprotic acid) with 0.100 M NaOH. The first equivalence point (at 100 mL) occurs when the first hydrogen ion of $H_2C_2O_4$ is titrated, and the second (at 200 mL) occurs at the completion of the reaction. The curve for pH versus volume of NaOH added shows an initial rise at the first equivalence point and then another rise at the second equivalence point.

The pH at this second equivalence point is controlled by the oxalate ion, $C_2O_4^{2-}$.

$$C_2O_4^{2-}(aq) + H_2O(\ell) \rightleftharpoons HC_2O_4^-(aq) + OH^-(aq) \qquad K_b = K_w/K_{a_2} = 1.6 \times 10^{-10}$$

Calculation of the pH at the equivalence point indicates that it should be about 8.4, as observed.

GENERAL
Chemistry Now™

See the General ChemistryNow CD-ROM or website:
- **Screen 18.7 Titration curves,** for a simulation of the titration of a polyprotic acid

Titration of a Weak Base with a Strong Acid

Finally, it is useful to consider the titration of a weak base with a strong acid. Figure 18.7 illustrates the pH curve for the titration of 100.0 mL of 0.100 M NH_3 with 0.100 M HCl.

$$NH_3(aq) + H_3O^+(aq) \rightleftharpoons NH_4^+(aq) + H_2O(\ell)$$

The initial pH for a 0.100 M NH_3 solution is 11.12. As the titration progresses, the important species in solution are the weak acid NH_4^+ and its conjugate base NH_3.

$$NH_4^+(aq) + H_2O(\ell) \rightleftharpoons NH_3(aq) + H_3O^+(aq) \qquad K_a = 5.6 \times 10^{-10}$$

At the halfway point, the concentrations of NH_4^+ and NH_3 are the same, so

Figure 18.7 Titration of a weak base with a strong acid. The graph shows the change in pH during the titration of 100.0 mL of 0.100 M NH_3 (a weak base) with 0.100 M HCl (a strong acid). The pH at the halfway point is equal to the pK_a for the conjugate acid (NH_4^+) of the weak base (NH_3); pH = pK_a = 9.26. At the equivalence point the solution contains the NH_4^+ ion, a weak acid, and the pH is about 5.

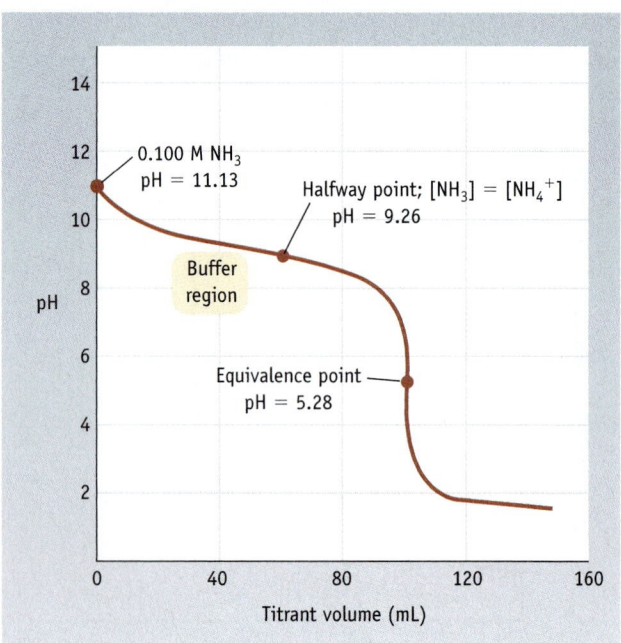

$$[H_3O^+] = \frac{[NH_4^+]}{[NH_3]} \times K_a = 5.6 \times 10^{-10}$$

$$[H_3O^+] = K_a$$

$$pH = pK_a = -\log(5.6 \times 10^{-10}) = 9.25$$

As the addition of HCl to NH_3 continues, the pH declines slowly because of the buffering action of the NH_3/NH_4^+ combination. Near the equivalence point, however, the pH drops rapidly. At the equivalence point, the solution contains only ammonium chloride, a weak Brønsted acid, and the solution is weakly acidic.

GENERAL
Chemistry⋅Now™

See the General ChemistryNow CD-ROM or website:
- **Screen 18.7 Titration of a Weak Base with a Strong Acid,** for a simulation of this titration

Example 18.7—Titration of Ammonia with HCl

Problem What is the pH of the solution at the equivalence point in the titration of 100.0 mL of 0.100 M ammonia with 0.100 M HCl (see Figure 18.7)?

Strategy This problem has two steps: (a) a stoichiometry calculation to find the concentration of NH_4^+ at the equivalence point, and (b) an equilibrium calculation to find $[H_3O^+]$ for a solution of the weak acid NH_4^+.

Solution

Part 1: Stoichiometry Problem

Here we are titrating 0.0100 mol of NH_3 ($= c \times V$), so 0.0100 mol of HCl is required. Thus, 100.0 mL of 0.100 M HCl ($= 0.0100$ mol HCl) must be used in the titration.

Equation	NH_3	+	H_3O^+	⇌	NH_4^+	+ H_2O
Initial (mol = $c \cdot V$)	0.0100		0		0	
Amount HCl added (mol)	—		0.0100		—	
Change on reaction (mol)	−0.0100		−0.0100		+0.0100	
After reaction (mol)	0		0		0.0100	
Concentration (M)	0		0		$\dfrac{0.0100 \text{ mol}}{0.200 \text{ L}}$ $=0.0500$ M	

Part 2: Equilibrium Problem

When the equivalence point is reached, the solution consists of 0.0500 M NH_4^+. The pH is determined by the ionization of this weak acid.

Equation	$NH_4^+ + H_2O$ ⇌	NH_3	+ H_3O^+
Initial (M)	0.0500	0	0
Change (M)	−x	+x	+x
Equilibrium (M)	0.0500−x	x	x

Using K_a for the weak acid NH_4^+, we have

$$K_a = 5.6 \times 10^{-10} = \frac{[NH_3][H_3O^+]}{[NH_4^+]}$$

$$[H_3O^+] = \sqrt{(5.6 \times 10^{-10})(0.0500)} = 5.3 \times 10^{-6}\ M$$

$$pH = 5.28$$

The pH at the equivalence point in this weak base–strong acid titration is, indeed, slightly acidic as expected from Figure 18.7.

Exercise 18.9—Titration of a Weak Base with a Strong Acid

Calculate the pH after 75.0 mL of 0.100 M HCl has been added to 100.0 mL of 0.100 M NH_3. See Figure 18.7.

---

pH Indicators

The colors of flowers often depend on the pH of the medium (page 848). Indeed, many organic compounds (Figure 18.8), both natural and synthetic, have a color that changes with pH. Not only does this add beauty and variety to our world, but it is a useful property in chemistry.

You have likely carried out an acid–base titration in the laboratory, and, before starting the titration, you would have added an **indicator**. The acid–base indicator is usually an organic compound that is itself a weak acid or weak base [similar to the anthocyanin dyes of flowers (page 848)]. In aqueous solution, the acid form is in equilibrium with its conjugate base. Abbreviating the indicator's acid formula as HInd and the formula of its conjugate base as Ind^-, we can write the equilibrium equation

$$HInd(aq) + H_2O(\ell) \rightleftharpoons H_3O^+(aq) + Ind^-(aq)$$

The important characteristic of acid–base indicators is that the acid form of the compound (HInd) has one color, and the conjugate base (Ind^-) has another. To see how such compounds can be used as equivalence point indicators, let us write the

---

Problem-Solving Tip 18.2

Calculating the pH at Various Stages of an Acid–Base Reaction

Finding the pH at or before the equivalence point for an acid–base reaction always involves several calculation steps. There are no shortcuts. Consider the *titration of a weak base, B, with a strong acid* as in Example 18.7. (The same principles apply to other acid–base titrations.)

$$H_3O^+(aq) + B(aq) \rightleftharpoons BH^+(aq) + H_2O(\ell)$$

Step 1. *Solve the stoichiometry problem.* Up to the equivalence point, acid is consumed completely to leave a solution containing some base (B) and its conjugate acid (BH^+). Use the principles of stoichiometry to calculate (a) the amount of acid added, (b) the amount of base consumed, and (c) the amount of conjugate base (BH^+) formed.

Step 2. *Calculate the concentrations of base, [B], and conjugate acid, [BH^+].* Recognize that the volume of the solution at any point is the sum of the original volume of base solution plus the volume of acid solution added.

Step 3. *Calculate the pH before the equivalence point.* At any point before the equivalence point, the solution is a buffer solution because both the base and its conjugate acid are present. Calculate [H_3O^+] using the concentrations of Step 2 and the value of K_a for the conjugate acid of the weak base.

Step 4. *Calculate the pH at the equivalence point.* Calculate the concentration of the conjugate acid using the procedure of Steps 1 and 2. Use the value of K_a for the conjugate acid of the weak base and the procedure outlined in Example 18.7.

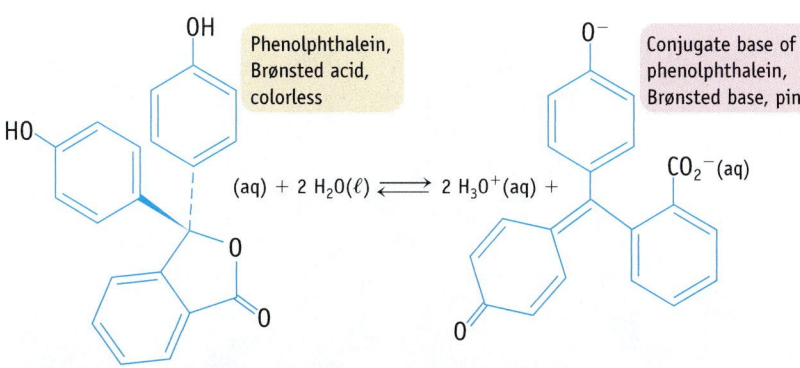

Figure 18.8   Phenolphthalein, a common acid–base indicator. Phenolphthalein, a weak acid, is colorless. As the pH increases, the pink conjugate base form predominates, and the color of the solution changes. The change in color is most noticeable around pH 9. The dye is commonly used for strong acid + strong base titrations because the change in color (appearance of a tinge of red) is noticeably slightly above pH = 7 (Figure 18.4). For other suitable indicator dyes, see Figure 18.10.

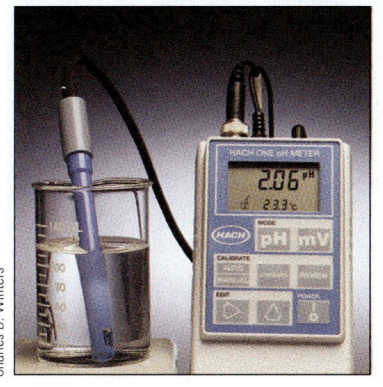
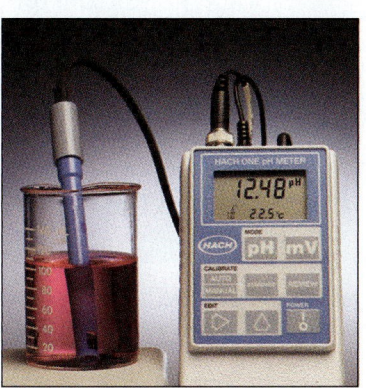

Charles D. Winters

usual equilibrium constant expressions for the dependence of hydronium ion concentration or pH on the indicator's ionization constant (K_a) and on the relative quantities of the acid and conjugate base.

$$[H_3O^+] = \frac{[HInd]}{[Ind^-]} \times K_a \quad \text{or} \quad pH = pK_a + \log\frac{[Ind^-]}{[HInd]} \qquad (18.6)$$

These equations inform us that

- When the hydronium ion concentration is equivalent to the value of K_a (or when pH = pK_a), then $[HInd] = [Ind^-]$.
- When $[H_3O^+] > K_a$ (or pH < pK_a), then $[HInd] > [Ind^-]$.
- When $[H_3O^+] < K_a$ (or pH > pK_a), then $[HInd] < [Ind^-]$.

Now let us apply these conclusions to, for example, the titration of an acid with a base using an indicator whose pK_a value is in the same range as the pH at the equivalence point in the titration (Figure 18.9). At the beginning of the titration, the pH is low and $[H_3O^+]$ is high; the acid form of the indicator (HInd) predominates. Its color is the one observed. As the titration progresses and the pH increases ($[H_3O^+]$ decreases), less of the acid HInd and more of its conjugate base exist in solution. Finally, just after we reach the equivalence point, $[Ind^-]$ is much larger than $[HInd]$, and the color of $[Ind^-]$ is observed.

Several obvious questions remain to be answered. If you are trying to analyze for an acid, and add an indicator that is a weak acid, won't this choice affect the analysis? Recall that you used only a tiny amount of an indicator in a titration. Although the acidic indicator molecules also react with the base as the titration progresses, so little indicator is present that the analysis will not be significantly in error.

■ **Mauve, a Synthetic Dye**
Mauve, the synthetic dye discovered by Perkin, is a weak acid. See Chapter 11, page 474.

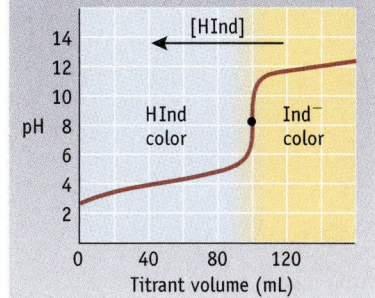

Figure 18.9 Indicator color change in the course of a titration when the pK_a of HInd is about 8.

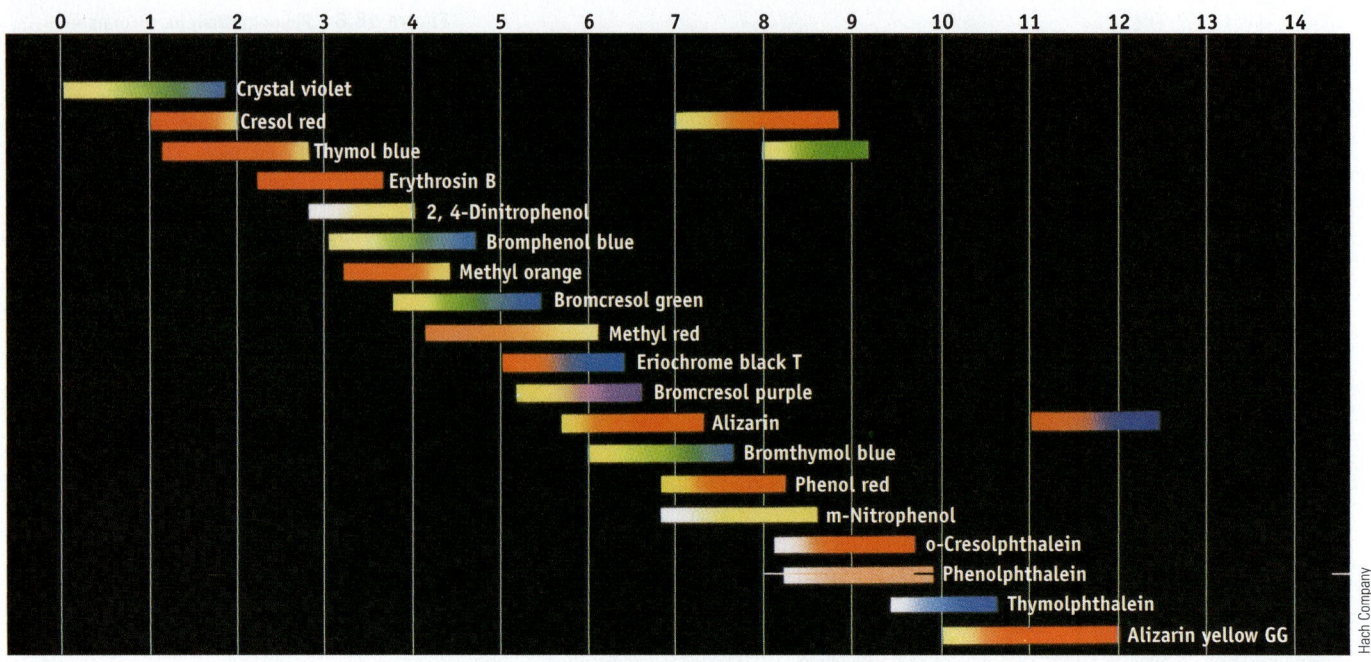

Figure 18.10 Common acid–base indicators. The color changes occur over a range of pH values. Notice that a few indicators have color changes over two different pH ranges.

Another question is whether you could accurately determine the pH by observing the color change of an indicator. In practice, your eyes are not quite that good. Usually, you see the color of HInd when $[HInd]/[Ind^-]$ is about 10/1, and the color of Ind^- when $[HInd]/[Ind^-]$ is about 1/10. Thus, the color change is observed over a hydronium ion concentration interval of about 2 pH units. However, as you can see in Figures 18.4–18.7, on passing through the equivalence point of these titrations, the pH changes by as many as 7 pH units.

As Figure 18.10 shows, a variety of indicators are available, each changing color in a different pH range. If you are analyzing a weak acid or base by titration, you must choose an indicator that changes color in a range that includes the pH to be observed at the equivalence point. For example, an indicator that changes color in the pH range 7 ± 2 should be used for a strong acid–strong base titration. On the other hand, the pH at the equivalence point in the titration of a weak acid with a strong base is greater than 7, so you should choose an indicator that changes color at a pH near the anticipated equivalence point.

GENERAL
Chemistry ⚛ Now™

See the General ChemistryNow CD-ROM or website:
• **Screen 18.7 Acid-Base Indicators,** for a simulation of indicator chemistry

Exercise 18.10—pH Indicators

Use Figure 18.10 to decide which indicator is best to use in the titration of NH_3 with HCl shown in Figure 18.7.

18.4—Solubility of Salts

Precipitation reactions [◀ Section 5.2] are exchange reactions in which one of the products is a water-insoluble compound such as $CaCO_3$,

$$CaCl_2(aq) + Na_2CO_3(aq) \longrightarrow CaCO_3(s) + 2\ NaCl(aq)$$

that is, a compound having a water solubility of less than about 0.01 mole of dissolved material per liter of solution (Figure 18.11).

How do you know when to predict an insoluble compound as the product of a reaction? In Chapter 5 we listed some guidelines for predicting solubility (Figure 5.3) and mentioned a few important minerals that are insoluble in water. Now we want to make our estimates of solubility more quantitative and to explore conditions under which some compounds precipitate and others do not.

The Solubility Product Constant, K_{sp}

Silver bromide, AgBr, is used in photographic film (Figure 18.11c). If some AgBr is placed in pure water, a tiny amount of the compound dissolves, and an equilibrium is established.

$$AgBr(s) \rightleftharpoons Ag^+(aq, 7.35 \times 10^{-7}\ M) + Br^-(aq, 7.35 \times 10^{-7}\ M)$$

When the AgBr has dissolved to the greatest extent possible, the solution is said to be saturated [◀ Section 14.2], and experiments show that the concentrations of the silver and bromide ions in the solution are each about 7.35×10^{-7} M at 25 °C. The extent to which an insoluble salt dissolves can be expressed in terms of the equilibrium constant for the dissolving process. In this case the appropriate expression is

$$K_{sp} = [Ag^+][Br^-]$$

a, National Oceanic and Atmospheric Administration/NOAA; b, Arthur Palmer; c, Charles D. Winters.

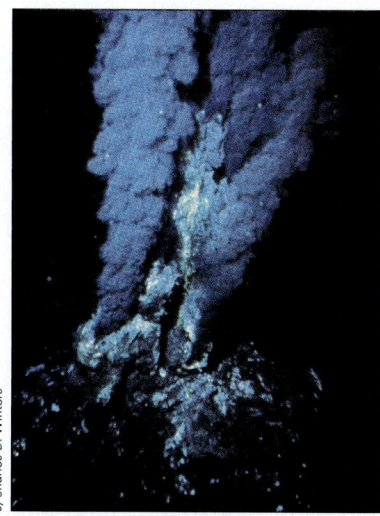

(a) Metal sulfides (and hydroxides) in a black smoker (page 140).

(b) $CaCO_3$ stalactites.

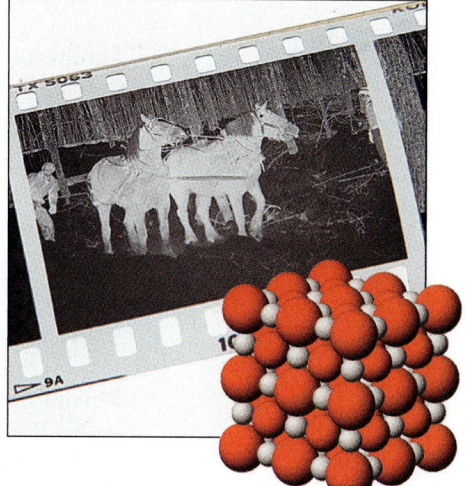

(c) Black-and-white film is coated with water-insoluble silver bromide. The image is formed by metallic silver particles.

Figure 18.11   Some insoluble substances.

Table 18.2   Some Common Insoluble Compounds and Their K_{sp} Values*

Formula	Name	K_{sp} (25 °C)	Common Names/Uses
$CaCO_3$	Calcium carbonate	3.4×10^{-9}	Calcite, iceland spar
$MnCO_3$	Manganese(II) carbonate	2.3×10^{-11}	Rhodochrosite (forms rose-colored crystals)
$FeCO_3$	Iron(II) carbonate	3.1×10^{-11}	Siderite
CaF_2	Calcium fluoride	5.3×10^{-11}	Fluorite (source of HF and other inorganic fluorides)
$AgCl$	Silver chloride	1.8×10^{-10}	Chlorargyrite
$AgBr$	Silver bromide	5.4×10^{-13}	Used in photographic film
$CaSO_4$	Calcium sulfate	4.9×10^{-5}	The hydrated form is commonly called gypsum
$BaSO_4$	Barium sulfate	1.1×10^{-10}	Barite (used in "drilling mud" and as a component of paints)
$SrSO_4$	Strontium sulfate	3.4×10^{-7}	Celestite
$Ca(OH)_2$	Calcium hydroxide	5.5×10^{-5}	Slaked lime

* The values in this table were taken from *Lange's Handbook of Chemistry,* 15th ed., New York, NY, McGraw-Hill Publishers, 1999. Additional K_{sp} values are given in Appendix J.

The value of the equilibrium constant that reflects the solubility of a compound is often referred to as its **solubility product constant**. Chemists often use the notation $\boldsymbol{K_{sp}}$ for such constants, with the subscript "sp" denoting a "solubility product."

The water solubility of a compound, and thus its K_{sp} value, can be estimated by determining the concentration of the cation or anion when the compound dissolves. For example, if we find that AgBr dissolves to give a silver ion concentration of 7.35×10^{-7} mol/L, we know that 7.35×10^{-7} mol of AgBr must have dissolved per liter of solution (and that the bromide ion concentration also equals 7.35×10^{-7} M). Therefore, the calculated value of the equilibrium constant for the dissolving process is

$$K_{sp} = [Ag^+][Br^-] = (7.35 \times 10^{-7})(7.35 \times 10^{-7}) = 5.40 \times 10^{-13} \text{ (at 25 °C)}$$

Equilibrium constants for the dissolving of other insoluble salts can be calculated in the same manner.

The solubility product constant, K_{sp}, for any salt always has the form

$$A_xB_y(s) \rightleftharpoons x\, A^{y+}(aq) + y\, B^{x-}(aq) \qquad K_{sp} = [A^{y+}]^x[B^{x-}]^y$$

For example,

$$CaF_2(s) \rightleftharpoons Ca^{2+}(aq) + 2\, F^-(aq) \qquad K_{sp} = [Ca^{2+}][F^-]^2 = 5.3 \times 10^{-11}$$
$$Ag_2SO_4(s) \rightleftharpoons 2\, Ag^+(aq) + SO_4{}^{2-}(aq) \qquad K_{sp} = [Ag^+]^2[SO_4{}^{2-}] = 1.2 \times 10^{-5}$$

The numerical values of K_{sp} for a few salts are given in Table 18.2, and more values are collected in Appendix J.

Do not confuse the *solubility* of a compound with its *solubility product constant.* The *solubility* of a salt is the quantity present in some volume of a saturated solution, expressed in moles per liter, grams per 100 mL, or other units. The *solubility product constant* is an equilibrium constant. Nonetheless, there is a connection between them: if one is known, the other can be calculated.

GENERAL
Chemistry Now™

See the General ChemistryNow CD-ROM or website:
- **Screen 18.8 Precipitation Reactions,** for a review of these reactions
- **Screen 18.9 Solubility Product Constant,** for a simulation and tutorial on using K_{sp}

Exercise 18.11—Writing K_{sp} Expressions

Write K_{sp} expressions for the following insoluble salts and look up numerical values for the constant in Appendix J.

(a) AgI   (b) BaF_2 (c) Ag_2CO_3

Relating Solubility and K_{sp}

Solubility product constants are determined by careful laboratory measurements of the concentrations of ions in solution.

GENERAL
Chemistry Now™

See the General ChemistryNow CD-ROM or website:
- **Screen 18.10 Determining K_{sp} Experimentally,** for a tutorial on determining K_{sp}

Example 18.8—K_{sp} from Solubility Measurements

Problem Calcium fluoride, the main component of the mineral fluorite, dissolves to a slight extent in water.

$$CaF_2(s) \rightleftharpoons Ca^{2+}(aq) + 2\ F^-(aq) \qquad K_{sp} = [Ca^{2+}][F^-]^2$$

Calculate the K_{sp} value for CaF_2 if the calcium ion concentration has been found to be 2.4×10^{-4} mol/L.

Strategy We first write the K_{sp} expression for CaF_2 and then substitute the numerical values for the equilibrium concentrations of the ions.

Solution When CaF_2 dissolves to a small extent in water, the balanced equation shows that the concentration of F^- ion must be twice the Ca^{2+} ion concentration.

$$\text{If } [Ca^{2+}] = 2.4 \times 10^{-4}\ M, \text{ then } [F^-] = 2 \times [Ca^{2+}] = 4.8 \times 10^{-4}\ M$$

This means the solubility product constant is

$$K_{sp} = [Ca^{2+}][F^-]^2 = (2.4 \times 10^{-4})(4.8 \times 10^{-4})^2 = 5.5 \times 10^{-11}$$

Exercise 18.12—K_{sp} from Solubility Measurements

The barium ion concentration, $[Ba^{2+}]$, in a saturated solution of barium fluoride is 3.6×10^{-3} M. Calculate the value of K_{sp} for BaF_2.

$$BaF_2(s) \rightleftharpoons Ba^{2+}(aq) + 2\ F^-(aq)$$

Charles D. Winters

Fluorite. The mineral fluorite is water-insoluble calcium fluoride. The mineral can vary widely in color from purple to green to colorless. The colors are likely due to impurities. The K_{sp} of CaF_2 is 5.3×10^{-11}.

Figure 18.12   Barium sulfate. Barium sulfate, a white solid, is quite insoluble in water ($K_{sp} = 1.1 \times 10^{-10}$) (see Example 18.9). (a) A sample of the mineral barite, which is mostly barium sulfate. (b) Barium sulfate is opaque to x-rays, so physicians use it to examine the digestive tract. A patient drinks a "cocktail" containing $BaSO_4$, and the progress of the $BaSO_4$ through the digestive organs can be followed by x-ray analysis. This photo is an x-ray of a gastrointestinal tract after a person ingested barium sulfate. It is fortunate that $BaSO_4$ is so insoluble, because water- and acid-soluble barium salts are toxic.

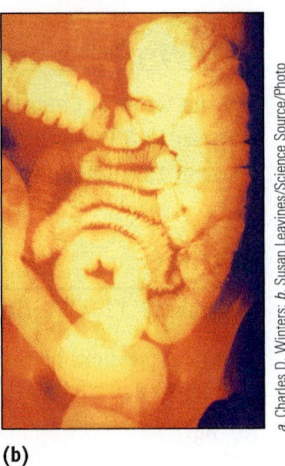

(a)   (b)

K_{sp} values for insoluble salts can be used to estimate the solubility of a solid salt or to determine whether a solid will precipitate when solutions of its anion and cation are mixed. Let us first look at an example of the estimation of the solubility of a salt from its K_{sp} value. Later we will see how to use these predictions to plan the separation of ions that are mixed in solution (page 890).

(page 890)

GENERAL
Chemistry•⚛•Now™

See the General ChemistryNow CD-ROM or website:
- **Screen 18.11 Estimating Salt Solubility: Using K_{sp},** for a tutorial on the relationship of K_{sp} and solubility

---

Example 18.9—Solubility from K_{sp}

Problem The K_{sp} value for $BaSO_4$ (as the mineral barite; see Figure 18.12) is 1.1×10^{-10} at 25 °C. Calculate the solubility of barium sulfate in pure water in (a) moles per liter and (b) grams per liter.

Strategy When 1 mol of $BaSO_4$ dissolves, 1 mol of Ba^{2+} ions and 1 mol of SO_4^{2-} ions are produced. Thus, the solubility of $BaSO_4$ can be estimated by calculating the concentration of either Ba^{2+} or SO_4^{2-} from the solubility product constant.

Solution The equation for the solubility of $BaSO_4$ is

$$BaSO_4(s) \rightleftharpoons Ba^{2+}(aq) + SO_4^{2-}(aq) \qquad K_{sp} = [Ba^{2+}][SO_4^{2-}] = 1.1 \times 10^{-10}$$

Let us denote the solubility of $BaSO_4$ (in moles per liter) by x; that is, x moles of $BaSO_4$ dissolve per liter. Therefore, both $[Ba^{2+}]$ and $[SO_4^{2-}]$ must also equal x at equilibrium.

Equation	$BaSO_4(s) \rightleftharpoons Ba^{2+}(aq) + SO_4^{2-}(aq)$	
Initial (M)	0	0
Change (M)	$+x$	$+x$
Equilibrium (M)	x	x

Because K_{sp} is the product of the barium and sulfate ion concentrations, K_{sp} is the square of the solubility, x:

$$K_{sp} = [Ba^{2+}][SO_4^{2-}] = 1.1 \times 10^{-10} = (x)(x) = x^2$$

A Closer Look

Solubility Calculations

The K_{sp} value reported for lead(II) chloride, $PbCl_2$, is 1.7×10^{-5}. If we assume the appropriate equilibrium in solution is

$$PbCl_2(s) \rightleftharpoons Pb^{2+}(aq) + 2\ Cl^-(aq)$$

the calculated solubility of $PbCl_2$ is 0.016 M. The experimental value for the solubility of the salt, however, is 0.036 M, more than twice the calculated value! What is the problem? There are several, as summarized by the diagram below.

The main problem in the lead(II) chloride case, and in many others, is that the compound dissolves but is not 100% dissociated into its constituent ions. Instead, it dissolves as the undissociated salt or forms ion pairs.

Other problems that lead to discrepancies between calculated and experimental solubilities are the reactions of ions with water (particularly anions) and complex ion formation. An example of the former effect is the reaction of sulfide ion with water, a reaction that is strongly product-favored.

$$S^{2-}(aq) + H_2O(\ell) \rightleftharpoons HS^-(aq) + OH^-(aq)$$

$$K_b = 1 \times 10^5$$

This means that the solubility of a metal sulfide is better described by a chemical equation such as

$$NiS(s) + H_2O(\ell) \rightleftharpoons Ni^{2+}(aq) + HS^-(aq) + OH^-(aq)$$

Complex ion formation is illustrated by the fact that lead chloride is more soluble in the presence of *excess* chloride ion, owing to the formation of the complex ion $PbCl_4{}^{2-}$.

$$PbCl_2(s) + 2\ Cl^-(aq) \rightleftharpoons PbCl_4{}^{2-}(aq)$$

Both hydrolysis and complex ion formation are discussed further on pages 882–883 and 887–890.

For further information on these issues, see:

a. L. Meites, J. S. F. Pode, and H. C. Thomas: *Journal of Chemical Education,* Vol. 43, pp. 667–672, 1966.

b. S. J. Hawkes: *Journal of Chemical Education,* Vol. 75, pp. 1179–1181, 1998.

c. R. W. Clark and J. M. Bonicamp: *Journal of Chemical Education,* Vol. 75, pp. 1182–1185, 1998.

d. R. J. Myers: *Journal of Chemical Education,* Vol. 63, pp. 687–690, 1986.

The value of x is

$$x = [Ba^{2+}] = [SO_4{}^{2-}] = \sqrt{1.1 \times 10^{-10}} = 1.0 \times 10^{-5}\ M$$

The solubility of $BaSO_4$ in pure water is 1.0×10^{-5} mol/L. To find its solubility in grams per liter, we need just to multiply by the molar mass of $BaSO_4$.

$$\text{Solubility in g/L} = (1.0 \times 10^{-5}\ \text{mol/L})(233\ \text{g/mol}) = 0.0024\ \text{g/L}$$

Example 18.10—Solubility from K_{sp}

Problem Knowing that the K_{sp} value for MgF_2 is 5.2×10^{-11}, calculate the solubility of the salt in (a) moles per liter and (b) grams per liter.

Strategy The problem is to define the salt solubility in terms that will allow us to solve the K_{sp} expression for this value. We know that, if 1 mol of MgF_2 dissolves, 1 mol of Mg^{2+} and 2 mol of F^- appear in the solution. This means the MgF_2 solubility (in moles dissolved per liter) is equivalent to the concentration of Mg^{2+} ion in the solution. Thus, if the solubility of MgF_2 is x mol/L, then $[Mg^{2+}] = x$ and $[F^-] = 2x$.

Solution We begin by writing the equilibrium equation and the K_{sp} expression.

$$MgF_2(s) \rightleftharpoons Mg^{2+}(aq) + 2\ F^-(aq) \qquad K_{sp} = [Mg^{2+}][F^-]^2 = 5.2 \times 10^{-11}$$

We then set up an ICE table.

Equation	$MgF_2(s) \rightleftharpoons Mg^{2+}(aq) + 2\ F^-(aq)$	
Initial (M)	0	0
Change (M)	$+x$	$+2x$
Equilibrium (M)	x	$2x$

Substituting these values into the K_{sp} expression, we find

$$K_{sp} = [Mg^{2+}][F^-]^2 = (x)(2x)^2 = 4\ x^3$$

Solving the equation for x,

$$x = \sqrt[3]{\frac{K_{sp}}{4}} = \sqrt[3]{\frac{5.2 \times 10^{-11}}{4}} = 2.4 \times 10^{-4}\ M$$

we find that 2.4×10^{-4} mol of MgF_2 dissolves per liter.

The solubility of MgF_2 in grams per liter is

$$(2.4 \times 10^{-4}\ mol/L)(62.3\ g/mol) = 0.015\ g\ MgF_2/L$$

Comment Problems like this one often provoke students to ask such questions as "Aren't you counting things twice when you multiply x by 2 and then square it as well?" in the expression $K_{sp} = (x)(2x)^2$. The answer is no. The 2 in the $2x$ term is based on the stoichiometry of the compound. The exponent of 2 on the F^- ion concentration arises from the rules for writing equilibrium expressions.

Exercise 18.13—Salt Solubility from K_{sp}

Calculate the solubility of $Ca(OH)_2$ in moles per liter and grams per liter using the value of K_{sp} in Appendix J.

The *relative* solubilities of salts can often be deduced by comparing values of solubility product constants, but you must be careful! For example, the K_{sp} value for silver chloride is

$$AgCl(s) \rightleftharpoons Ag^+(aq) + Cl^-(aq) \qquad K_{sp} = 1.8 \times 10^{-10}$$

whereas that for silver chromate is

$$Ag_2CrO_4(s) \rightleftharpoons 2\ Ag^+(aq) + CrO_4^{2-}(aq) \qquad K_{sp} = 1.1 \times 10^{-12}$$

In spite of the fact that Ag_2CrO_4 has a numerically smaller K_{sp} value than AgCl, the chromate salt is about 5 times *more* soluble than the chloride salt. If you determine solubilities from K_{sp} values as in the preceding examples, you would find that the solubility of AgCl is 1.3×10^{-5} mol/L, whereas that of Ag_2CrO_4 is 6.5×10^{-5} mol/L. From this example and countless others, we conclude that

Direct comparisons of the solubility of two salts on the basis of their K_{sp} values can only be made for salts having the same cation-to-anion ratio.

This means, for example, that you can directly compare solubilities of 1:1 salts such as the silver halides by comparing their K_{sp} values.

$$AgI(K_{sp} = 8.5 \times 10^{-17}) < AgBr(K_{sp} = 5.4 \times 10^{-13}) < AgCl(K_{sp} = 1.8 \times 10^{-10})$$
$$\xrightarrow{\text{increasing } K_{sp} \text{ and increasing solubility}}$$

Similarly, you could compare 1:2 salts such as the lead halides.

$$PbI_2(K_{sp} = 9.8 \times 10^{-9}) < PbBr_2(K_{sp} = 6.6 \times 10^{-6}) < PbCl_2(K_{sp} = 1.7 \times 10^{-5})$$
$$\xrightarrow{\text{increasing } K_{sp} \text{ and increasing solubility}}$$

but you cannot directly compare a 1:1 salt (AgCl) with a 2:1 salt (Ag$_2$CrO$_4$).

Exercise 18.14—Comparing Solubilities

Using K_{sp} values, predict which salt in each pair is more soluble in water.

(a) AgCl or AgCN
(b) Mg(OH)$_2$ or Ca(OH)$_2$
(c) Ca(OH)$_2$ or CaSO$_4$

Solubility and the Common Ion Effect

The test tube on the left in Figure 18.13 contains a precipitate of silver acetate, AgCH$_3$CO$_2$, in water. The solution is saturated, and the silver ions and acetate ions in the solution are in equilibrium with solid silver acetate.

$$AgCH_3CO_2(s) \rightleftharpoons Ag^+(aq) + CH_3CO_2^-(aq)$$

But what would happen if the silver ion concentration is increased, such as by adding silver nitrate? Le Chatelier's principle [◀ Section 16.6] suggests—and we observe—that more silver acetate precipitate should form because a product ion has been added, causing the equilibrium to shift to form more silver acetate.

The ionization of weak acids and bases is affected by the presence of an ion common to the equilibrium process (Section 18.1), and the effect of adding silver ions to a saturated silver acetate solution is another example of the common ion effect. Adding a common ion to a saturated solution of a salt will always lower the salt solubility.

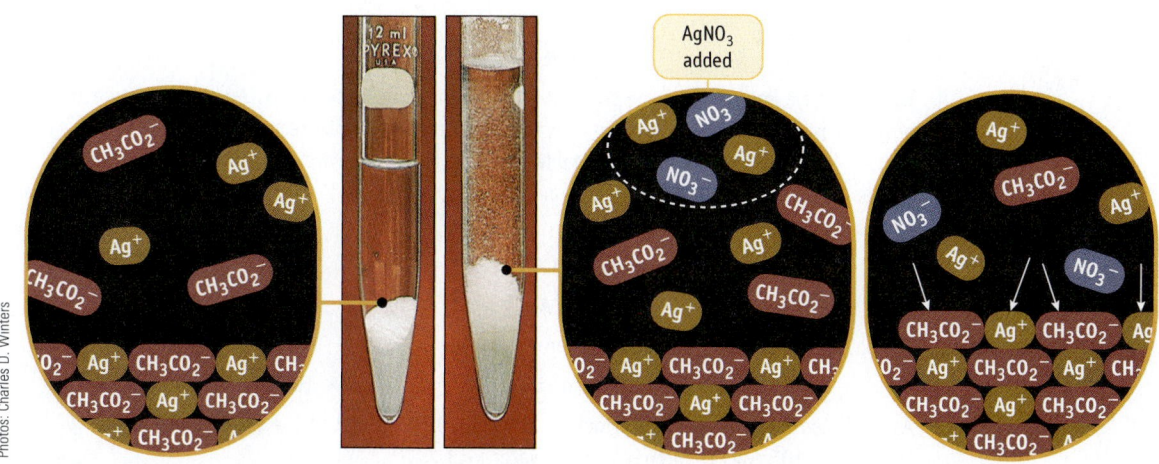

Photos: Charles D. Winters

Figure 18.13 The common ion effect. The tube at the left contains a saturated solution of silver acetate, AgCH$_3$CO$_2$. When 1.0 M AgNO$_3$ is added to the tube (right), more solid silver acetate forms.

GENERAL
Chemistry••••Now™

See the General ChemistryNow CD-ROM or website:

• **Screen 18.12 Common Ion Effect,** for a simulation and tutorial on the effect of a common ion on solubility

---

Example 18.11—The Common Ion Effect and Salt Solubility

Problem If solid AgCl is placed in 1.00 L of 0.55 M NaCl, what mass of AgCl will dissolve?

Strategy The presence of an ion common to the equilibrium suppresses the solubility of a salt. To determine the solubility of the salt under these circumstances, calculate the concentration of the ion other than the common ion (Ag^+ ion in this case).

Solution In pure water, the solubility of AgCl is equal to *either* $[Ag^+]$ *or* $[Cl^-]$.

$$AgCl(s) \rightleftharpoons Ag^+(aq) + Cl^-(aq)$$

Solubility of AgCl in pure water = $[Ag^+]$ or $[Cl^-]$ = $\sqrt{K_{sp}}$ = 1.3×10^{-5} mol/L or 0.0019 g/L

However, in water already containing a common ion (here the Cl^- ion), Le Chatelier's principle predicts that the solubility will be less than 1.3×10^{-5} mol/L. In this case the solubility of AgCl is equivalent to the concentration of Ag^+ ion in solution, so we set up an ICE table to show the concentrations of Ag^+ and Cl^- when equilibrium is attained.

Equation	$AgCl(s) \rightleftharpoons Ag^+(aq) + Cl^-(aq)$	
Initial (M)	0	0.55
Change (M)	$+x$	$+x$
Equilibrium (M)	x	$0.55 + x$

Some AgCl dissolves in the presence of chloride ion and produces Ag^+ and Cl^- ion concentrations of x mol/L. Because some chloride ion was already present, the total chloride ion concentration is what was already there (0.55 M) *plus* the amount supplied by AgCl dissociation ($= x$).

The equilibrium concentrations from the table are substituted into the K_{sp} expression,

$$K_{sp} = 1.8 \times 10^{-10} = [Ag^+][Cl^-] = (x)(0.55 + x)$$

and rearranged to

$$x^2 + 0.55\, x - K_{sp} = 0$$

This quadratic equation can be solved by the methods in Appendix A. An easier approach, however, is to make the approximation that x is *very* small with respect to 0.55 [so $(0.55 + x) \approx 0.55$]. This is a reasonable assumption because we know that the solubility equals 1.3×10^{-5} M without the common ion Cl^-, and that it will be even smaller in the presence of added Cl^-. Therefore,

$$K_{sp} = 1.8 \times 10^{-10} \approx (x)(0.55)$$
$$x = [Ag^+] \approx 3.3 \times 10^{-10} \text{ M}$$

The solubility in grams per liter is then

$$(3.3 \times 10^{-10} \text{ mol/L})(143 \text{ g/mol}) = 4.7 \times 10^{-8} \text{ g/L}$$

As predicted by Le Chatelier's principle, the solubility of AgCl in the presence of added Cl^- is less (3.3×10^{-10} M) than that in pure water (1.3×10^{-5} M).

Comment As a final step, check the approximation by substituting the calculated value of x into the exact expression $K_{sp} = (x)(0.55 + x)$. If the product $(x)(0.55 + x)$ is the same as the given value of K_{sp}, the approximation is valid.

$$K_{sp} = (x)(0.55 + x) = (3.3 \times 10^{-10})(0.55 + 3.3 \times 10^{-10}) \approx 1.8 \times 10^{-10}$$

The approximation we made here is similar to the approximations we make in acid–base equilibrium problems [see Example 17.5].

Example 18.12—The Common Ion Effect and Salt Solubility

Problem Calculate the solubility of silver chromate, Ag_2CrO_4, at 25 °C in the presence of 0.0050 M K_2CrO_4 solution.

$$Ag_2CrO_4(s) \rightleftharpoons 2\ Ag^+(aq) + CrO_4^{2-}(aq) \qquad K_{sp} = [Ag^+]^2[CrO_4^{2-}] = 1.1 \times 10^{-12}$$

For comparison, the solubility of Ag_2CrO_4 in pure water is 1.3×10^{-4} mol/L.

Strategy In the presence of chromate ion from the water-soluble salt K_2CrO_4, the concentration of Ag^+ ions produced by Ag_2CrO_4 will be less than in pure water. Assume the solubility of Ag_2CrO_4 is x mol/L. This means the concentration of Ag^+ ions will be $2x$ mol/L, whereas the concentration of CrO_4^{2-} ions will be x mol/L *plus* the amount of CrO_4^{2-} already in the solution.

Solution

Equation	$Ag_2CrO_4(s) \rightleftharpoons 2\ Ag^+(aq)$	$+ CrO_4^{2-}(aq)$
Initial (M)	0	0.0050
Change (M)	$+2x$	$+x$
Equilibrium (M)	$2x$	$0.0050 + x$

Substituting the equilibrium amounts into the K_{sp} expression, we have

$$K_{sp} = 1.1 \times 10^{-12} = [Ag^+]^2[CrO_4^{2-}]$$
$$= (2x)^2(0.0050 + x)$$

As in Example 18.11 you can make the approximation that x is very small with respect to 0.0050, so $(0.0050 + x) \approx 0.0050$. (This assumption is reasonable because $[CrO_4^{2-}]$ is 0.00013 M without added chromate ion, and it is certain that x is even smaller in the presence of extra chromate ion.) Therefore, the approximate expression is

$$K_{sp} = 1.1 \times 10^{-12} = [Ag^+]^2[CrO_4^{2-}] = (2x)^2(0.0050)$$

Solving, we find x, the solubility of silver chromate in the presence of excess chromate ion, is

$$x = \text{solubility of } Ag_2CrO_4 = 7.4 \times 10^{-6} \text{ M}$$

Comment The silver ion concentration in the presence of the common ion is

$$[Ag^+] = 2x = 1.5 \times 10^{-5} \text{ M}$$

This silver ion concentration is, indeed, less than its value in pure water (1.3×10^{-4} M) owing to the presence of an ion "common" to the equilibrium.

Exercise 18.15—The Common Ion Effect and Salt Solubility

Calculate the solubility of $BaSO_4$ (a) in pure water and (b) in the presence of 0.010 M $Ba(NO_3)_2$. K_{sp} for $BaSO_4$ is 1.1×10^{-10}.

Figure 18.14   Lead sulfide. This and other metal sulfides dissolve in water to a greater extent than expected because the sulfide ion reacts with water to form the very stable species HS^- and OH^-.

$$PbS(s) + H_2O(\ell) \rightleftharpoons$$
$$Pb^{2+}(aq) + HS^-(aq) + OH^-(aq)$$

(The model of PbS shows that the unit cell is cubic, a feature reflected by the cubic crystals of the mineral galena.)

Charles D. Winters

Exercise 18.16—The Common Ion Effect and Salt Solubility

Calculate the solubility of $Zn(CN)_2$ at 25 °C (a) in pure water and (b) in the presence of 0.10 M $Zn(NO_3)_2$. K_{sp} for $Zn(CN)_2$ is 8.0×10^{-12}.

Examples 18.11 and 18.12 allow us to propose two important general ideas:

- The solubility of a salt will always be reduced by the presence of a common ion, in accordance with Le Chatelier's principle.

- We made the approximation that the amount of common ion added to the solution was very large in comparison with the amount of that ion coming from the insoluble salt, which allowed us to simplify our calculations. This is almost always the case, but you should check to be sure.

The Effect of Basic Anions on Salt Solubility

The next time you are tempted to wash a supposedly insoluble salt down the kitchen or laboratory drain, stop and consider the consequences. Many metal ions such as lead, chromium, and mercury are toxic in the environment. Even if a so-called insoluble salt of one of these cations does not appear to dissolve, its solubility in water may be greater than you think, in part owing to the possibility that the anion of the salt is a weak base or the cation is a weak acid.

Lead sulfide, PbS, which is found in nature as the mineral galena (Figure 18.14), provides an example of the effect of the acid–base properties of an ion on salt solubility. When placed in water, a trace amount dissolves,

$$PbS(s) \rightleftharpoons Pb^{2+}(aq) + S^{2-}(aq)$$

and one product of the reaction is the sulfide ion. This anion is a strong base,

$$S^{2-}(aq) + H_2O(\ell) \rightleftharpoons HS^-(aq) + OH^-(aq) \qquad K_b = 1 \times 10^5$$

and it undergoes extensive hydrolysis (reaction with water) [◀ Table 17.3]. The equilibrium process for dissolving PbS thus shifts to the right, and the lead ion concentration in solution is greater than expected from the simple ionization of the salt.

The lead sulfide example leads to the following general observation:

> Any salt containing an anion that is the conjugate base of a weak acid will dissolve in water to a greater extent than given by K_{sp}.

This means that phosphate, acetate, carbonate, and cyanide salts, as well as sulfide salts, can be affected, because all of these anions undergo the general hydrolysis reaction

$$X^{n-}(aq) + H_2O(\ell) \rightleftharpoons HX^{(n-1)}(aq) + OH^-(aq)$$

The observation that ions from insoluble salts can undergo hydrolysis is related to another useful, general conclusion:

> Insoluble salts in which the anion is the conjugate base of a weak acid dissolve in strong acids.

■ **Metal Sulfide Solubility**
The solubility of a metal sulfide is better represented by a modified solubility product constant, K_{spa}, which is defined as follows:

$$MS(s) \rightleftharpoons M^{2+}(aq) + S^{2-}(aq)$$
$$K_{sp} = [M^{2+}][S^{2-}]$$

$$S^{2-}(aq) + H_2O(\ell) \rightleftharpoons HS^-(aq) + OH^-(aq)$$
$$K_b = [HS^-][OH^-]/[S^{2-}]$$

Net reaction:

$$MS(s) + H_2O(\ell) \rightleftharpoons$$
$$HS^-(aq) + M^{2+}(aq) + OH^-(aq)$$
$$K_{spa} = [M^{2+}][HS^-][OH^-] = K_{sp} \times K_b$$

Values for K_{spa} for several metal sulfides are included in Appendix J.

Insoluble salts containing such anions as acetate, carbonate, hydroxide, phosphate, and sulfide dissolve in strong acids. For example, you know that if a strong acid is added to a water-insoluble metal carbonate such as $CaCO_3$, the salt dissolves [◀ Section 5.5].

$$CaCO_3(s) + 2\ H_3O^+(aq) \longrightarrow Ca^{2+}(aq) + 3\ H_2O(\ell) + CO_2(g)$$

You can think of this as the result of a series of reactions.

$$CaCO_3(s) \rightleftharpoons Ca^{2+}(aq) + CO_3^{2-}(aq) \qquad K_{sp} = 3.4 \times 10^{-9}$$
$$CO_3^{2-}(aq) + H_2O(\ell) \rightleftharpoons HCO_3^-(aq) + OH^-(aq) \qquad K_{b_1} = 2.1 \times 10^{-4}$$
$$HCO_3^-(aq) + H_2O(\ell) \rightleftharpoons H_2CO_3(aq) + OH^-(aq) \qquad K_{b_2} = 2.4 \times 10^{-8}$$
$$2[OH^-(aq) + H_3O^+(aq) \rightleftharpoons 2\ H_2O(\ell)] \qquad K = (1/K_w)^2 = [1/(1 \times 10^{-14})]^2$$
Overall: $CaCO_3(s) + 2\ H_3O^+(aq) \rightleftharpoons Ca^{2+}(aq) + 2\ H_2O(\ell) + H_2CO_3(aq)$
$$K_{net} = (K_{sp})(K_{b_1})(K_{b_2})(1/K_w)^2 = 1.7 \times 10^8$$

Carbonic acid, a product of this reaction, is not stable,

$$H_2CO_3(aq) \rightleftharpoons CO_2(g) + H_2O(\ell) \qquad K \approx 10^5$$

so you see CO_2 bubbling out of the solution, a process that moves the $CaCO_3$ + H_3O^+ equilibrium even farther to the right. Calcium carbonate dissolves completely in strong acid when the system is open and the CO_2 can escape!

Many metal sulfides are also soluble in strong acids,

$$FeS(s) + 2\ H_3O^+(aq) \rightleftharpoons Fe^{2+}(aq) + H_2S(aq) + 2\ H_2O(\ell)$$

as are metal phosphates (Figure 18.15),

$$Ag_3PO_4(s) + 3\ H_3O^+(aq) \rightleftharpoons 3\ Ag^+(aq) + H_3PO_4(aq) + 3\ H_2O(\ell)$$

and metal hydroxides.

$$Mg(OH)_2(s) + 2\ H_3O^+(aq) \rightleftharpoons Mg^{2+}(aq) + 4\ H_2O(\ell)$$

In general, the solubility of a salt containing the conjugate base of a weak acid is increased by addition of a stronger acid to the solution. In contrast, salts are not soluble in strong acid if the anion is the conjugate base of a strong acid. For example, AgCl is not soluble in strong acid

$$AgCl(s) \rightleftharpoons Ag^+(aq) + Cl^-(aq) \qquad K_{sp} = 1.8 \times 10^{-10}$$
$$H_3O^+(aq) + Cl^-(aq) \rightleftharpoons HCl(aq) + H_2O(\ell) \qquad K \ll 1$$

because Cl^- is a very weak base (Table 17.3), so its concentration is not lowered by a reaction with the strong acid H_3O^+ (Figure 18.15). This same conclusion would also apply to insoluble salts of Br^- and I^-.

GENERAL
Chemistry ⚛ Now™

See the General ChemistryNow CD-ROM or website:
- **Screen 18.13 Solubility and pH,** for a simulation and tutorial on the effect of pH on solubility

Figure 18.15   The effect of the anion on salt solubility in acid. (*left*) A precipitate of AgCl (white) and Ag_3PO_4 (yellow). (*right*) Adding a strong acid (HNO_3) dissolves Ag_3PO_4 (and leaves insoluble AgCl). The basic anion PO_4^{3-} reacts with acid to give H_3PO_4, whereas Cl^- is too weakly basic to form HCl.

Add strong acid →

Photos: Charles D. Winters

Precipitate of
AgCl and Ag_3PO_4

Precipitate of
AgCl

18.5—Precipitation Reactions

Metal-bearing ores contain the metal in the form of an insoluble salt (Figure 18.16). To complicate matters further, ores often contain several such metal salts. Many industrial methods for separating metals from their ores involve dissolving metal salts to obtain the metal ion or ions in solution. The solution is then concentrated in some manner, and a precipitating agent is added to precipitate selectively only one type of metal ion as an insoluble salt. In the case of nickel, for example, the Ni^{2+} ion can be precipitated as insoluble nickel(II) sulfide or nickel(II) carbonate.

$$Ni^{2+}(aq) + HS^-(aq) + H_2O(\ell) \rightleftharpoons NiS(s) + H_3O^+(aq) \quad K = 0.3$$
$$Ni^{2+}(aq) + CO_3^{2-}(aq) \rightleftharpoons NiCO_3(s) \quad K = 7.1 \times 10^6$$

The final step in obtaining the metal itself is to reduce the metal cation to the metal either chemically or electrochemically (Chapter 20).

    Our immediate goal is to work out methods to determine whether a precipitate will form under a given set of conditions. For example, if Ag^+ and Cl^- are present at some given concentrations, will AgCl precipitate from the solution?

K_{sp} and the Reaction Quotient, Q

Silver chloride, like silver bromide, is used in photographic films. It dissolves to a very small extent in water and has a correspondingly small value of K_{sp}.

$$AgCl(s) \rightleftharpoons Ag^+(aq) + Cl^-(aq) \quad K_{sp} = [Ag^+][Cl^-] = 1.8 \times 10^{-10}$$

But let us look at this problem from the other direction: If a solution contains Ag^+ and Cl^- ions at some concentration, will AgCl precipitate from solution? This is the same question we asked in Section 16.3 when we wanted to know if a given mixture of reactants and products was an equilibrium mixture, if the reactants continued to form products, or if products would revert to reactants. The procedure there was to calculate the reaction quotient, Q.

    For silver chloride, the expression for the reaction quotient, Q, is

$$Q = [Ag^+][Cl^-]$$

Recall that *the difference between Q and K is that the concentrations required in the reaction quotient expression may or may not be those at equilibrium.* For the case of a slightly soluble salt such as AgCl, we can reach the following conclusions [◀ Section 16.3].

1. If $Q = K_{sp}$, the solution is saturated and at equilibrium.

Charles D. Winters

Figure 18.16   Minerals. Minerals are often insoluble salts. The minerals shown here are light purple fluorite (calcium fluoride), black hematite [iron(III) oxide], and rust brown goethite, a mixture of iron(III) oxide and iron(III) hydroxide.

When the product of the ion concentrations is equal to K_{sp}, the ion concentrations must be their equilibrium values.

2. If $Q < K_{sp}$, the solution is not saturated.

This can mean two things: (i) If solid AgCl is present, more will dissolve until equilibrium is achieved (when $Q = K_{sp}$). (ii) If solid AgCl is not already present, more $Ag^+(aq)$ or more $Cl^-(aq)$ (or both) could be added to the solution until precipitation of solid AgCl begins (when $Q > K_{sp}$).

3. If $Q > K_{sp}$, the system is not at equilibrium; the solution is supersaturated.

The concentrations of Ag^+ and Cl^- in solution are too high, and AgCl will precipitate until $Q = K_{sp}$.

GENERAL
Chemistry⚛Now™

See the General ChemistryNow CD-ROM or website:
• **Screen 18.14 Can a Precipitation Reaction Occur?**, for a simulation and tutorial

---

Example 18.13—Solubility and the Reaction Quotient

Problem Solid AgCl has been placed in a beaker of water. After some time, the concentrations of Ag^+ and Cl^- are each 1.2×10^{-5} mol/L. Has the system reached equilibrium? If not, will more AgCl dissolve?

Strategy Use the experimental ion concentrations to calculate the reaction quotient, Q. Compare Q and K_{sp} to decide whether the system is at equilibrium (if $Q = K_{sp}$).

Solution For this AgCl case,

$$Q = [Ag^+][Cl^-] = (1.2 \times 10^{-5})(1.2 \times 10^{-5}) = 1.4 \times 10^{-10}$$

Here Q is less than K_{sp} (1.8×10^{-10}). The solution is not yet saturated, and AgCl will continue to dissolve until $Q = K_{sp}$, at which point $[Ag^+] = [Cl^-] = 1.3 \times 10^{-5}$ M. That is, an additional 0.1×10^{-5} mol of AgCl (about 0.14 mg) will dissolve per liter.

Exercise 18.17—Solubility and the Reaction Quotient

Solid PbI_2 ($K_{sp} = 9.8 \times 10^{-9}$) is placed in a beaker of water. After a period of time, the lead(II) concentration is measured and found to be 1.1×10^{-3} M. Has the system reached equilibrium? That is, is the solution saturated? If not, will more PbI_2 dissolve?

K_{sp}, the Reaction Quotient, and Precipitation Reactions

With some knowledge of the reaction quotient, we can decide (i) whether a precipitate will form when the ion concentrations are known or (ii) what concentrations of ions are required to begin the precipitation of an insoluble salt.

Suppose the concentration of aqueous magnesium ion in a solution is 1.5×10^{-6} M. If enough NaOH is added to make the solution 1.0×10^{-4} M in hydroxide ion, OH^-, will precipitation of $Mg(OH)_2$ occur ($K_{sp} = 5.6 \times 10^{-12}$)? If not, will it occur if the concentration of OH^- is increased to 1.0×10^{-2} M?

Our strategy will be as in Example 18.13. That is, use the ion concentrations to calculate the value of Q and then compare Q with K_{sp} to decide whether the system is at equilibrium. Let us begin with the equation for the dissolution of insoluble $Mg(OH)_2$.

$$Mg(OH)_2(s) \rightleftharpoons Mg^{2+}(aq) + 2\ OH^-(aq)$$

When the concentrations of magnesium and hydroxide ions are those stated above, we find that Q is less than K_{sp}.

$$Q = [Mg^{2+}][OH^-]^2 = (1.5 \times 10^{-6})(1.0 \times 10^{-4})^2 = 1.5 \times 10^{-14}$$
$$Q\,(1.5 \times 10^{-14}) < K_{sp}\,(5.6 \times 10^{-12})$$

This means the solution is not yet saturated, and precipitation does not occur.

When $[OH^-]$ is increased to 1.0×10^{-2} M, the reaction quotient is 1.5×10^{-10},

$$Q = (1.5 \times 10^{-6})(1.0 \times 10^{-2})^2 = 1.5 \times 10^{-10}$$
$$Q\,(1.5 \times 10^{-10}) > K_{sp}\,(5.6 \times 10^{-12})$$

The reaction quotient is now *larger* than K_{sp}. Precipitation of $Mg(OH)_2$ occurs, and will continue until the Mg^{2+} and OH^- ion concentrations have declined to the point where their product is equal to K_{sp}.

Exercise 18.18—Deciding Whether a Precipitate Will Form

Will $SrSO_4$ precipitate from a solution containing 2.5×10^{-4} M strontium ion, Sr^{2+}, if enough of the soluble salt Na_2SO_4 is added to make the solution 2.5×10^{-4} M in SO_4^{2-}? (K_{sp} for $SrSO_4$ is 3.4×10^{-7}.)

Now that we know how to decide whether a precipitate will form when the concentration of each ion is known, let us turn to the problem of deciding how much of the precipitating agent is required to begin the precipitation of an ion at a given concentration level.

Example 18.14—Ion Concentrations Required to Begin Precipitation

Problem The concentration of barium ion, Ba^{2+}, in a solution is 0.010 M.

(a) What concentration of sulfate ion, SO_4^{2-}, is required to just begin precipitating $BaSO_4$?

(b) When the concentration of sulfate ion in the solution reaches 0.015 M, what concentration of barium ion will remain in solution?

Strategy There are three variables in the K_{sp} expression: K_{sp} and the anion and cation concentrations. Here we know K_{sp} (1.1×10^{-10}) and one of the ion concentrations. We can calculate the other ion concentration.

Solution Let us begin by writing the balanced equation for the equilibrium that will exist when $BaSO_4$ has been precipitated.

$$BaSO_4(s) \rightleftharpoons Ba^{2+}(aq) + SO_4^{2-}(aq) \qquad K_{sp} = [Ba^{2+}][SO_4^{2-}] = 1.1 \times 10^{-10}$$

(a) When the product of the ion concentrations exceeds K_{sp} ($= 1.1 \times 10^{-10}$)—that is, when $Q > K_{sp}$—precipitation will occur. The Ba^{2+} ion concentration is known (0.010 M), so the SO_4^{2-} ion concentration necessary for precipitation can be calculated.

$$[SO_4^{2-}] = \frac{K_{sp}}{[Ba^{2+}]} = \frac{1.1 \times 10^{-10}}{0.010} = 1.1 \times 10^{-8} \, M$$

The result tells us that if the sulfate ion is just slightly greater than 1.1×10^{-8} M, $BaSO_4$ will begin to precipitate; $Q = [Ba^{2+}][SO_4^{2-}]$ would then be greater than K_{sp}.

(b) If the sulfate ion concentration is increased to 0.015 M, the maximum concentration of Ba^{2+} ion that can exist in solution (in equilibrium with solid $BaSO_4$) is

$$[Ba^{2+}] = \frac{K_{sp}}{[SO_4^{2-}]} = \frac{1.1 \times 10^{-10}}{0.015} = 7.3 \times 10^{-9} \, M$$

Comment The fact that the barium ion concentration is so small under these circumstances means that the Ba^{2+} ion has been essentially completely removed from solution. (Its concentration began at 0.010 M and has dropped by a factor of about 1 million.) You would say that Ba^{2+} ion precipitation is, for all practical purposes, complete.

Example 18.15—K_{sp} and Precipitation

Problem Suppose you mix 100. mL of 0.0200 M $BaCl_2$ with 50.0 mL of 0.0300 M Na_2SO_4. Will $BaSO_4$ ($K_{sp} = 1.1 \times 10^{-10}$) precipitate?

Strategy Here we mix two solutions, one containing Ba^{2+} ions and the other containing SO_4^{2-} ions. First, find the concentration of each of these ions *after* mixing. Then, knowing the ion concentrations in the diluted solution, calculate Q and compare it with the K_{sp} value for $BaSO_4$.

Solution First use the equation $c_1V_1 = c_2V_2$ (see Section 5.8) to calculate c_2, the concentration of the Ba^{2+} or SO_4^{2-} ions after mixing to give a new solution with a volume of 150. mL ($= V_2$).

$$[Ba^{2+}] \text{ after mixing} = \frac{(0.0200 \text{ mol/L})(0.100 \text{ L})}{0.1500 \text{ L}} = 0.0133 \, M$$

$$[SO_4^{2-}] \text{ after mixing} = \frac{(0.0300 \text{ mol/L})(0.0500 \text{ L})}{0.150 \text{ L}} = 0.0100 \, M$$

Now the reaction quotient can be calculated.

$$Q = [Ba^{2+}][SO_4^{2-}] = (0.0133)(0.0100) = 1.33 \times 10^{-4}$$

Q is much larger than K_{sp}, so $BaSO_4$ precipitates.

Exercise 18.19—Ion Concentrations Required to Begin Precipitation

What is the minimum concentration of I^- that can cause precipitation of PbI_2 from a 0.050 M solution of $Pb(NO_3)_2$? K_{sp} for PbI_2 is 9.8×10^{-9}. What concentration of Pb^{2+} ions remains in solution when the concentration of I^- is 0.0015 M?

Exercise 18.20—K_{sp} and Precipitation

You have 100.0 mL of 0.0010 M silver nitrate. Will AgCl precipitate if you add 5.0 mL of 0.025 M HCl?

18.6—Solubility and Complex Ions

Calcium carbonate does not dissolve in water, but it does dissolve in strong acids. In contrast, silver chloride dissolves in neither water nor strong acid—but it does dissolve in ammonia. How do we explain these observations, and how can we use them in a practical way?

Dimethylglyoxime complex of Ni^{2+} ion

Photo: Charles D. Winters

[Ni(NH$_3$)$_6$]$^{2+}$

[Ni(H$_2$O)$_6$]$^{2+}$

Figure 18.17   Complex ions. The green solution contains soluble [Ni(H$_2$O)$_6$]$^{2+}$ ions in which water molecules are bound to Ni^{2+} ions by ion–dipole forces. This complex ion gives the solution its green color. The red, insoluble solid is the dimethylglyoximate complex of the Ni^{2+} ion [Ni(C$_4$H$_7$O$_2$N$_2$)$_2$] (*model at top right*). Formation of this beautiful red insoluble compound is the classical test for the presence of the aqueous Ni^{2+} ion.

■ **Successive Equilibria**

Dissolving AgCl in NH$_3$ is an example of what chemists often refer to as "successive equilibria" where one product-favored reaction affects an overall equilibrium. Another example is the increase in solubility of an insoluble salt owing to the reaction of its anion with water (page 882), and yet another is dissolving CaCO$_3$ in strong acid (page 883).

Metal ions exist in aqueous solution as complex ions [◄ Section 17.9] (Figure 18.17). Complex ions consist of the metal ion and other molecules or ions, bound into a single entity. In water, metal ions are always surrounded by water molecules; the negative end of the polar water molecule, the oxygen atom, is attracted to the positive metal ion. Indeed, any negative ion or polar molecule, such as NH$_3$, can be attracted to a metal ion.

As you shall see in Chapter 22, complex ions are prevalent in chemistry, and are the basis of such biologically important substances as hemoglobin and vitamin B$_{12}$. For our present purposes, they are important because a water-soluble complex ion can often be formed in preference to an insoluble salt. For example, adding sufficient aqueous ammonia to solid AgCl will cause the insoluble salt to dissolve owing to the formation of the water-soluble complex ion Ag(NH$_3$)$_2$$^+$ (Figure 18.18).

$$\text{AgCl(s)} + 2\text{ NH}_3\text{(aq)} \rightleftharpoons [\text{Ag(NH}_3)_2]^+\text{(aq)} + \text{Cl}^-\text{(aq)}$$

We can view dissolving AgCl(s) in this way as a two-step process. First, AgCl dissolves minimally in water, giving Ag$^+$(aq) and Cl$^-$(aq) ion. Second, the Ag$^+$(aq) ion combines with NH$_3$ to give the ammonia complex. Lowering the Ag$^+$(aq) concentration through complexation with NH$_3$ shifts the solubility equilibrium to the right, so more solid AgCl dissolves.

$$\text{AgCl(s)} \rightleftharpoons \text{Ag}^+\text{(aq)} + \text{Cl}^-\text{(aq)} \qquad K_{sp} = 1.8 \times 10^{-10}$$
$$\text{Ag}^+\text{(aq)} + 2\text{ NH}_3\text{(aq)} \rightleftharpoons [\text{Ag(NH}_3)_2]^+\text{(aq)} \qquad K_{formation} = 1.6 \times 10^7$$

This is an example of combining two (or more) equilibria where one is a reactant-favored reaction and the other is a product-favored reaction.

The equilibrium constant for the formation of a complex ion such as [Ag(NH$_3$)$_2$]$^+$ is called the **formation constant**, $K_{formation}$. The large value of this equilibrium constant means that the equilibrium lies well to the right and provides the driving force for dissolving AgCl in the presence of NH$_3$. If we combine $K_{formation}$ with K_{sp}, we obtain the net equilibrium constant for dissolving AgCl in aqueous ammonia.

$$K_{net} = K_{sp} \times K_{formation} = (1.8 \times 10^{-10})(1.6 \times 10^7) = 2.9 \times 10^{-3}$$
$$= 2.9 \times 10^{-3} = \frac{[\text{Ag(NH}_3)_2^+][\text{Cl}^-]}{[\text{NH}_3]^2}$$

Even though the value of K_{net} seems small, we use a large concentration of NH$_3$, so the concentration of [Ag(NH$_3$)$_2$]$^+$ in solution can be quite high. Silver chloride is thus much more soluble in the presence of ammonia than in pure water.

Formation constants have been measured for many complex ions (Appendix K). As a consequence, it is possible to compare the stabilities of various complex ions by comparing the values of their formation constants. For the silver(I) ion, a few other values are given here:

Formation Equilibrium	$K_{formation}$
Ag$^+$(aq) + 2 Cl$^-$(aq) $\rightleftharpoons$ [AgCl$_2$]$^-$(aq)	2.5×10^5
Ag$^+$(aq) + 2 S$_2$O$_3$$^{2-}$(aq) $\rightleftharpoons$ [Ag(S$_2$O$_3$)$_2$]$^{3-}$(aq)	2.0×10^{13}
Ag$^+$(aq) + 2 CN$^-$(aq) $\rightleftharpoons$ [Ag(CN)$_2$]$^-$(aq)	5.6×10^{18}

The formation of all three silver complexes is strongly product-favored. The cyanide complex ion [Ag(CN)$_2$]$^-$ is the most stable of the three products.

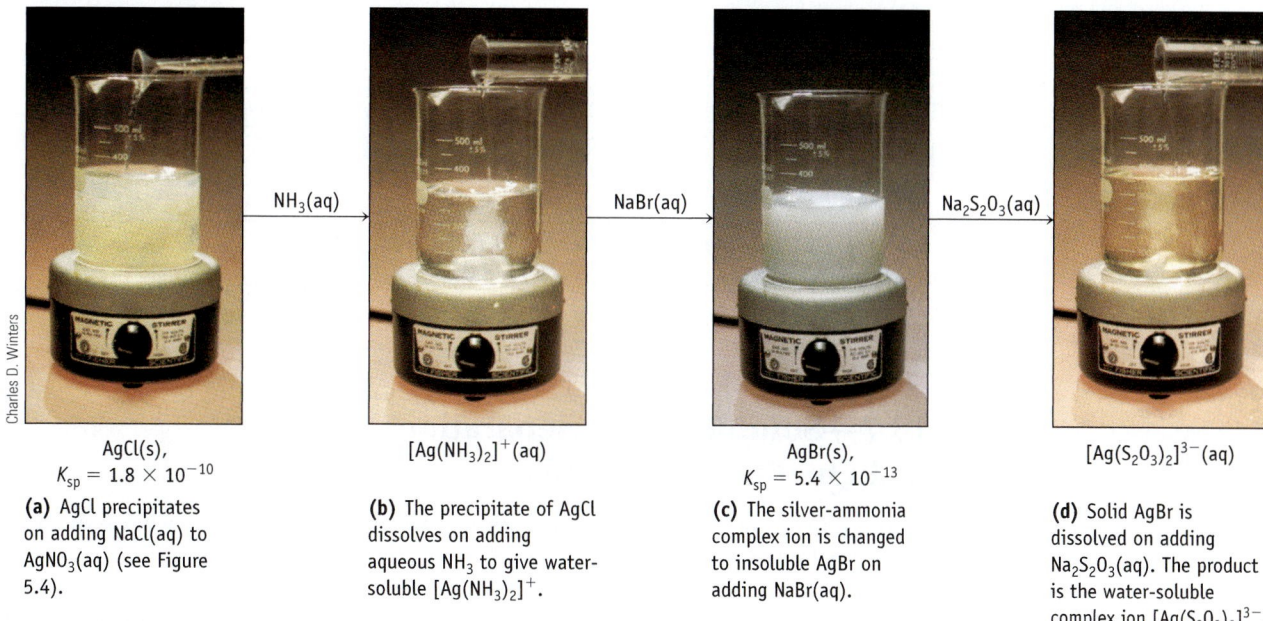

(a) AgCl(s),
$K_{sp} = 1.8 \times 10^{-10}$

(a) AgCl precipitates on adding NaCl(aq) to AgNO₃(aq) (see Figure 5.4).

NH₃(aq) →

[Ag(NH₃)₂]⁺(aq)

(b) The precipitate of AgCl dissolves on adding aqueous NH₃ to give water-soluble [Ag(NH₃)₂]⁺.

NaBr(aq) →

AgBr(s),
$K_{sp} = 5.4 \times 10^{-13}$

(c) The silver-ammonia complex ion is changed to insoluble AgBr on adding NaBr(aq).

Na₂S₂O₃(aq) →

[Ag(S₂O₃)₂]³⁻(aq)

(d) Solid AgBr is dissolved on adding Na₂S₂O₃(aq). The product is the water-soluble complex ion [Ag(S₂O₃)₂]³⁻.

Figure 18.18 **Forming and dissolving precipitates.** Insoluble compounds often dissolve upon addition of a complexing agent.

Figure 18.18 shows what happens as complex ions form. Beginning with a precipitate of AgCl, adding aqueous ammonia dissolves the precipitate to give the soluble complex ion $[Ag(NH_3)_2]^+$. Silver bromide is even more stable than $[Ag(NH_3)_2]^+$, so AgBr forms in preference to the complex ion on adding bromide ion. If thiosulfate ion, $S_2O_3^{2-}$, is then added, AgBr dissolves due to the formation of $[Ag(S_2O_3)_2]^{3-}$, a complex ion with a large formation constant.

See the General ChemistryNow CD-ROM or website:

- **Screen 18.15 Simultaneous Equilibria,** for more information on combining equilibria
- **Screen 18.16 Complex Ion Formation and Solubility,** for a tutorial on the effect of complexation on solubility

Example 18.16—Complex Ions and Solubility

Problem What is the value of the equilibrium constant, K_{net}, for dissolving AgBr in a solution containing the thiosulfate ion, $S_2O_3^{2-}$ (see Figure 18.18)? Explain why AgBr dissolves readily on adding aqueous sodium thiosulfate to the solid.

Strategy Summing several equilibrium processes gives the net chemical equation. K_{net} is the product of the values of K of the summed chemical equations. (See the preceding text and Section 16.10.)

Solution The overall reaction for dissolving AgBr in the presence of the thiosulfate anion is the sum of two equilibrium processes.

$$AgBr(s) \rightleftharpoons Ag^+(aq) + Br^-(aq) \qquad K_{sp} = 5.4 \times 10^{-13}$$

$$Ag^+(aq) + 2\,S_2O_3^{2-}(aq) \rightleftharpoons Ag(S_2O_3)_2^{3-}(aq) \qquad K_{formation} = 2.0 \times 10^{13}$$

Net chemical equation:

$$AgBr(s) + 2\,S_2O_3{}^{2-}(aq) \rightleftharpoons Ag(S_2O_3)_2{}^{3-}(aq) + Br^-(aq) \qquad K_{net} = K_{sp} \cdot K_{formation} = 1.0 \times 10^1$$

The value of K_{net} is greater than 1, indicating a decidedly product-favored reaction. AgBr is predicted to dissolve readily in aqueous $Na_2S_2O_3$, as observed (Figure 18.18).

Exercise 18.21—Complex Ions and Solubility

Calculate the value of the equilibrium constant, K_{net}, for dissolving $Cu(OH)_2$ in aqueous ammonia to form the complex ion $[Cu(NH_3)_4]^{2+}$ (see Figure 17.7).

18.7—Solubility, Ion Separations, and Qualitative Analysis

In many introductory chemistry courses, a portion of the laboratory work is devoted to the qualitative analysis of aqueous solutions, focusing on the identification of anions and metal cations. The purpose of such laboratory work is (1) to introduce some basic chemistry of various ions and (2) to illustrate how the principles of chemical equilibria can be applied.

Assume you have an aqueous solution that contains the metal ions Ag^+, Pb^{2+}, and Cu^{2+}. Your objective is to separate the ions so that each type of ion ends up in a separate test tube; the identity of each ion can then be confirmed. As a first step in this process, you want to find one reagent that will form a precipitate with one or more of the cations and leave the others in solution. This is done by comparing K_{sp} values for salts of cations with various anions (say, S^{2-}, OH^-, Cl^-, or $SO_4{}^{2-}$), looking for an anion that gives insoluble salts for some cations but not others.

Looking over the list of solubility products in Appendix J, you notice that the ions in our example solution all form very insoluble sulfides (Ag_2S, PbS, and CuS). However, only two of them form insoluble chlorides, $AgCl$ and $PbCl_2$. Thus, your "magic reagent" for partial cation separation could be aqueous HCl, which will form precipitates with Ag^+ and Pb^{2+} ($AgCl$ and $PbCl_2$) while Cu^{2+} ions remain in solution (Figure 18.19).

Some Insoluble Sulfides and Chlorides

Compound	K_{sp} or K_{spa} at 25 °C*
Ag_2S	6×10^{-51}
PbS	3×10^{-28}
CuS	6×10^{-37}
$AgCl$	1.8×10^{-10}
$PbCl_2$	1.7×10^{-5}

* See Appendix J

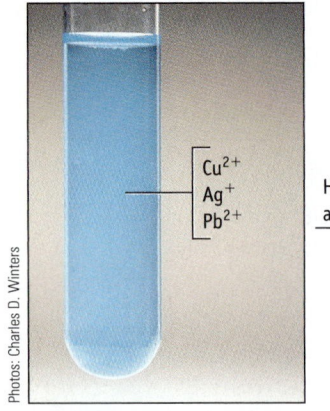

(a) The solution contains nitrate salts of Ag^+, Pb^{2+}, and Cu^{2+}. (The Cu^{2+} ion in water is light blue; the others are colorless.)

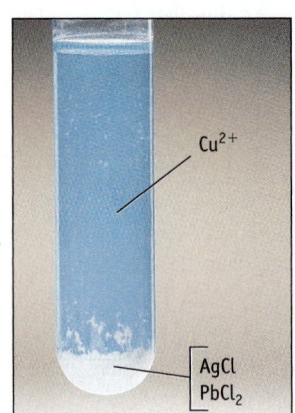

(b) Aqueous HCl is added in an amount sufficient to precipitate completely the white solids $AgCl$ and $PbCl_2$.

(c) The blue solution containing the Cu^{2+} ion is poured carefully into another test tube, leaving the white precipitates in the first test tube.

Figure 18.19   Ion separations by solubility difference.

Now we have four problems to solve. The first is how to separate AgCl and $PbCl_2$, which are now present in the same test tube (panel *c* of Figure 18.19) into separate test tubes. This turns out to be relatively easy: $PbCl_2$ (with a value of K_{sp} of 1.7×10^{-5}) is much more soluble than AgCl, and the lead salt dissolves readily in hot water whereas AgCl remains insoluble.

We now have three test tubes: one containing $PbCl_2$, another with AgCl, and a third with Cu^{2+} ions. How can we verify the presence of $PbCl_2$ in one of the test tubes? The answer is to convert it to an even more insoluble salt with a distinctive appearance. Consider two common lead compounds, white lead(II) chloride and bright yellow lead(II) chromate.

$$PbCl_2, K_{sp} = 1.7 \times 10^{-5} \quad \text{and} \quad PbCrO_4, K_{sp} = 2.8 \times 10^{-13}$$

If you add a few drops of K_2CrO_4 to a small amount of the white precipitate, $PbCl_2$, and shake the mixture, the solid will change to yellow $PbCrO_4$ (Figure 18.20). That this reaction is possible is evident from the equilibrium constant, K_{net}, for the process

$$PbCl_2(s) \rightleftharpoons Pb^{2+}(s) + 2\ Cl^-(aq) \qquad K_1 = K_{sp} = 1.7 \times 10^{-5}$$
$$Pb^{2+}(aq) + CrO_4{}^{2-}(aq) \rightleftharpoons PbCrO_4(s) \qquad K_2 = 1/K_{sp} = 1/(2.8 \times 10^{-13})$$

Net chemical equation:

$$PbCl_2(s) + CrO_4{}^{2-}(aq) \rightleftharpoons PbCrO_4(s) + 2\ Cl^-(aq) \qquad K_{net} = 6.1 \times 10^7$$

The value of K_{net} ($= K_1 \times K_2$) is very large, indicating that the reaction should proceed from left to right, as observed (Figure 18.20).

How can we verify the presence of AgCl in one of the test tubes? One way is to attempt to dissolve the solid AgCl in aqueous ammonia (Figure 18.18), a characteristic reaction of AgCl. In contrast, $PbCl_2$ does not dissolve in aqueous ammonia.

Finally, how might we verify that the blue solution in the third test tube really contains aqueous Cu^{2+} ions? The classical test is to add aqueous ammonia. This first precipitates blue-white copper(II) hydroxide (owing to the presence of OH^- ion in solution of aqueous ammonia).

$$Cu^{2+}(aq) + 2\ NH_3(aq) + 2\ H_2O(\ell) \rightleftharpoons Cu(OH)_2(s) + 2\ NH_4{}^+(aq)$$

As more ammonia is added, however, copper(II) hydroxide is converted to the very stable complex ion $[Cu(NH_3)_4]^{2+}$, which has a very distinctive deep blue color (Figure 17.7).

$$Cu(OH)_2(s) + 4\ NH_3(aq) \rightleftharpoons [Cu(NH_3)_4]^{2+}(aq) + 2\ OH^-(aq)$$

K₂CrO₄ added

Stirred

PbCl₂ precipitate

Photos: Charles D. Winters

Figure 18.20 Lead(II) chloride and lead(II) chromate. The test tube on the left contains a precipitate of white $PbCl_2$. The test tube on the right shows what happens when the $PbCl_2$ precipitate is stirred in the presence of K_2CrO_4. The white solid $PbCl_2$ has been transformed into the less soluble yellow $PbCrO_4$.

Exercise 18.22—Ion Separation

The cations of each pair given below appear together in one solution.

(a) Ag^+ and Ca^{2+} **(b)** Fe^{2+} and K^+

You may add only one reagent to precipitate one cation and not the other. Consult the solubility product table in Appendix J and tell whether you would use Cl^-, S^{2-}, or OH^- as the precipitating ion in each case. [The precipitating ions are introduced in the form of HCl, $(NH_4)_2S$, or NaOH, for example.]

Chapter Goals Revisited

When you have finished reading this chapter, ask whether you have met the chapter goals. In particular, you should be able to

Understand the common ion effect
a. Predict the effect of the addition of a "common ion" on the pH of the solution of a weak acid or base (Section 18.1). General ChemistryNow homework: Study Question(s) 2, 4

Understand the control of pH in aqueous solutions with buffers (Section 18.2)
a. Describe the functioning of buffer solutions.

b. Use the Henderson-Hasselbalch equation (Equation 18.2) to calculate the pH of a buffer solution of given composition. General ChemistryNow homework: SQ(s) 6, 7, 14

c. Describe how a buffer solution of a given pH can be prepared. General ChemistryNow homework: SQ(s) 9, 16

d. Calculate the pH of a buffer solution before and after adding acid or base. General ChemistryNow homework: SQ(s) 20, 22

e. Predict the pH of an acid–base reaction at its equivalence point (see also Sections 17.5 and 17.6).

Acid	Base	pH at Equivalence Point
Strong	Strong	= 7 (neutral)
Strong	Weak	< 7 (acidic)
Weak	Strong	> 7 (basic)

Evaluate the pH in the course of acid–base titrations (Section 18.3)
a. Calculate the pH at the equivalence point in the reaction of a strong acid with a weak base, or in the reaction of a strong base with a weak acid. General ChemistryNow homework: SQ(s) 24

b. Understand the differences between the titration curves for a strong acid–strong base titration versus cases in which one of the substances is weak.

c. Describe how an indicator functions in an acid–base titration. General ChemistryNow homework: SQ(s) 32

Apply chemical equilibrium concepts to the solubility of ionic compounds.

a. Write the equilibrium constant expression—the solubility product constant, K_{sp}—for any insoluble salt (Section 18.4)

b. Calculate K_{sp} values from experimental data (Section 18.4). General ChemistryNow homework: SQ(s) 40, 42

c. Estimate the solubility of a salt from the value of K_{sp} (Section 18.4). General ChemistryNow homework: SQ(s) 46, 48

d. Calculate the solubility of a salt in the presence of a common ion (Section 18.4). General ChemistryNow homework: SQ(s) 54

e. Understand the effect of basic anions on the solubility of a salt (Section 18.4). General ChemistryNow homework: SQ(s) 58

f. Decide whether a precipitate will form when the ion concentrations are known (Section 18.5). General ChemistryNow homework: SQ(s) 36, 64

g. Calculate the ion concentrations that are required to begin the precipitation of an insoluble salt (Section 18.5).

h. Understand that the formation of a complex ion can increase the solubility of an insoluble salt (Section 18.6). General ChemistryNow homework: SQ(s) 66

i. Use K_{sp} values to devise a method of separating ions in solution from one another (Section 18.7).

Key Equations

Equation 18.1 (page 856):
Hydronium ion concentration in a buffer solution composed of a weak acid and its conjugate base.

$$[H_3O^+] = \frac{[\text{acid}]}{[\text{conjugate base}]} \times K_a$$

Equation 18.2 (page 856):
Henderson-Hasselbalch equation. To calculate the pH of a buffer solution composed of a weak acid and its conjugate base.

$$pH = pK_a + \log \frac{[\text{conjugate base}]}{[\text{acid}]}$$

Equation 18.3 (page 865):
Equation to calculate the hydronium ion concentration before the equivalence point in the titration of a weak acid with a strong base. See also Equation 18.4 for the version of this equation based on the Henderson-Hasselbalch equation.

$$[H_3O^+] = \frac{[\text{weak acid remaining}]}{[\text{conjugate base produced}]} \times K_a$$

Equation 18.5 (page 866):
The relationship between the pH of the solution and the pK_a of the weak acid at the halfway (or half-neutralization) point in the titration of a weak acid with a strong base (or of a weak base with a strong acid).

$$pH = pK_a$$

Study Questions

▲ denotes more challenging questions.

■ denotes questions available in the Homework and Goals section of the General ChemistryNow CD-ROM or website.

Blue numbered questions have answers in Appendix O and fully worked solutions in the *Student Solutions Manual*.

Structures of many of the compounds used in these questions are found on the General ChemistryNow CD-ROM or website in the Models folder.

GENERAL
Chemistry⚛Now™ Assess your understanding of this chapter's topics with additional quizzing and conceptual questions at http://now.brookscole.com/kotz6e

Practicing Skills

The Common Ion Effect
(See Example 18.1 and General ChemistryNow Screen 18.2.)

1. Does the pH of the solution increase, decrease, or stay the same when you
   (a) Add solid ammonium chloride to a dilute aqueous solution of NH_3?
   (b) Add solid sodium acetate to a dilute aqueous solution of acetic acid?
   (c) Add solid NaCl to a dilute aqueous solution of NaOH?

2. ■ Does the pH of the solution increase, decrease, or stay the same when you
   (a) Add solid sodium oxalate, $Na_2C_2O_4$, to 50.0 mL of 0.015 M oxalic acid, $H_2C_2O_4$?
   (b) Add solid ammonium chloride to 75 mL of 0.016 M HCl?
   (c) Add 20.0 g of NaCl to 1.0 L of 0.10 M sodium acetate, $NaCH_3CO_2$?

3. What is the pH of a solution that consists of 0.20 M ammonia, NH_3, and 0.20 M ammonium chloride, NH_4Cl?

4. ■ What is the pH of 0.15 M acetic acid to which 1.56 g of sodium acetate, $NaCH_3CO_2$ has been added?

5. What is the pH of the solution that results from adding 30.0 mL of 0.015 M KOH to 50 mL of 0.015 M benzoic acid?

6. ■ What is the pH of the solution that results from adding 25.0 mL of 0.12 M HCl to 25.0 mL of 0.43 M NH_3?

Buffer Solutions
(See Example 18.2 and General ChemistryNow Screens 18.3 and 18.4.)

7. ■ What is the pH of the buffer solution that contains 2.2 g of NH_4Cl in 250 mL of 0.12 M NH_3? Is the final pH lower or higher than the pH of the original ammonia solution?

8. Lactic acid ($CH_3CHOHCO_2H$) is found in sour milk, in sauerkraut, and in muscles after activity (see page 479). (K_a for lactic acid = 1.4×10^{-4}.)
   (a) If 2.75 g of $NaCH_3CHOHCO_2$, sodium lactate, is added to 5.00×10^2 mL of 0.100 M lactic acid, what is the pH of the resulting buffer solution?
   (b) Is the final pH lower or higher than the pH of the lactic acid solution?

9. ■ What mass of sodium acetate, $NaCH_3CO_2$, must be added to 1.00 L of 0.10 M acetic acid to give a solution with a pH of 4.50?

10. What mass of ammonium chloride, NH_4Cl, must be added to exactly 5.00×10^2 mL of 0.10 M NH_3 solution to give a solution with a pH of 9.00?

Using the Henderson-Hasselbalch Equation
(See Example 18.3 and General ChemistryNow Screen 18.4.)

11. Calculate the pH of a solution that has an acetic acid concentration of 0.050 M and a sodium acetate concentration of 0.075 M.

12. Calculate the pH of a solution that has an ammonium chloride concentration of 0.050 M and an ammonia concentration of 0.045 M.

13. A buffer is composed of formic acid and its conjugate base, the formate ion.
    (a) What is the pH of a solution that has a formic acid concentration of 0.050 M and a sodium formate concentration of 0.035 M?
    (b) What must the ratio of acid to conjugate base be to increase the pH by 0.5 unit?

14. ■ A buffer solution is composed of 1.360 g of KH_2PO_4 and 5.677 g of Na_2HPO_4.
    (a) What is the pH of the buffer solution?
    (b) What mass of KH_2PO_4 must be added to decrease the buffer solution pH by 0.5 unit?

Preparing a Buffer Solution
(See Example 18.4 and General ChemistryNow Screen 18.5.)

15. Which of the following combinations would be the best to buffer the pH of a solution at approximately 9?
    (a) HCl and NaCl
    (b) NH_3 and NH_4Cl
    (c) CH_3CO_2H and $NaCH_3CO_2$

16. ■ Which of the following combinations would be the best choice to buffer the pH of a solution at approximately 7?
    (a) H_3PO_4 and NaH_2PO_4
    (b) NaH_2PO_4 and Na_2HPO_4
    (c) Na_2HPO_4 and Na_3PO_4

17. Describe how to prepare a buffer solution from NaH_2PO_4 and Na_2HPO_4 to have a pH of 7.5.

18. Describe how to prepare a buffer solution from NH_3 and NH_4Cl to have a pH of 9.5.

Adding an Acid or a Base to a Buffer Solution

(See Example 18.5 and General ChemistryNow Screen 18.6.)

19. A buffer solution was prepared by adding 4.95 g of sodium acetate, $NaCH_3CO_2$, to 2.50×10^2 mL of 0.150 M acetic acid, CH_3CO_2H.
    (a) What is the pH of the buffer?
    (b) What is the pH of 1.00×10^2 mL of the buffer solution if you add 82 mg of NaOH to the solution?

20. ■ You dissolve 0.425 g of NaOH in 2.00 L of a buffer solution that has $[H_2PO_4^-] = [HPO_4^{2-}] = 0.132$ M. What is the pH of the solution before adding NaOH? After adding NaOH?

21. A buffer solution is prepared by adding 0.125 mol of ammonium chloride to 5.00×10^2 mL of 0.500 M solution of ammonia.
    (a) What is the pH of the buffer?
    (b) If 0.0100 mol of HCl gas is bubbled into 5.00×10^2 mL of the buffer, what is the new pH of the solution?

22. ■ What will be the pH change when 20.0 mL of 0.100 M NaOH is added to 80.0 mL of a buffer solution consisting of 0.169 M NH_3 and 0.183 M NH_4Cl?

More about Acid–Base Reactions: Titrations

(See Examples 18.6 and 18.7, and General ChemistryNow Screen 18.7.)

23. Phenol, C_6H_5OH, is a weak organic acid. Suppose 0.515 g of the compound is dissolved in exactly 125 mL of water. The resulting solution is titrated with 0.123 M NaOH.

    $$C_6H_5OH(aq) + OH^-(aq) \rightleftharpoons C_6H_5O^-(aq) + H_2O(\ell)$$

    (a) What is the pH of the original solution of phenol?
    (b) What are the concentrations of all of the following ions at the equivalence point: Na^+, H_3O^+, OH^-, and $C_6H_5O^-$?
    (c) What is the pH of the solution at the equivalence point?

24. ■ Assume you dissolve 0.235 g of the weak acid benzoic acid, $C_6H_5CO_2H$, in enough water to make 1.00×10^2 mL of solution and then titrate the solution with 0.108 M NaOH.

    $$C_6H_5CO_2H(aq) + OH^-(aq) \rightleftharpoons C_6H_5CO_2^-(aq) + H_2O(\ell)$$

    (a) What was the pH of the original benzoic acid solution?
    (b) What are the concentrations of all of the following ions at the equivalence point: Na^+, H_3O^+, OH^-, and $C_6H_5CO_2^-$?
    (c) What is the pH of the solution at the equivalence point?

25. You require 36.78 mL of 0.0105 M HCl to reach the equivalence point in the titration of 25.0 mL of aqueous ammonia.
    (a) What was the concentration of NH_3 in the original ammonia solution?
    (b) What are the concentrations of H_3O^+, OH^-, and NH_4^+ at the equivalence point?
    (c) What is the pH of the solution at the equivalence point?

26. A solution of the weak base aniline, $C_6H_5NH_2$, in 25.0 mL of water requires 25.67 mL of 0.175 M HCl to reach the equivalence point.

    $$C_6H_5NH_2(aq) + H_3O^+(aq) \longrightarrow C_6H_5NH_3^+(aq) + H_2O(\ell)$$

    (a) What was the concentration of aniline in the original solution?
    (b) What are the concentrations of H_3O^+, OH^-, and $C_6H_5NH_3^+$ at the equivalence point?
    (c) What is the pH of the solution at the equivalence point?

Titration Curves and Indicators

(See Figures 18.4–18.10 and General ChemistryNow Screen 18.7.)

27. Without doing detailed calculations, sketch the curve for the titration of 30.0 mL of 0.10 M NaOH with 0.10 M HCl. Indicate the approximate pH at the beginning of the titration and at the equivalence point. What is the total solution volume at the equivalence point?

28. Without doing detailed calculations, sketch the curve for the titration of 50 mL of 0.050 M pyridine, C_5H_5N (a weak base), with 0.10 M HCl. Indicate the approximate pH at the beginning of the titration and at the equivalence point. What is the total solution volume at the equivalence point?

29. You titrate 25.0 mL of 0.10 M NH_3 with 0.10 M HCl.
    (a) What is the pH of the NH_3 solution before the titration begins?
    (b) What is the pH at the equivalence point?
    (c) What is the pH at the halfway point of the titration?
    (d) What indicator in Figure 18.10 could be used to detect the equivalence point?
    (e) Calculate the pH of the solution after adding 5.00, 15.0, 20.0, 22.0, and 30.0 mL of the acid. Combine this information with that in parts (a)–(c) and plot the titration curve.

30. Construct a rough plot of pH versus volume of base for the titration of 25.0 mL of 0.050 M HCN with 0.075 M NaOH.
    (a) What is the pH before any NaOH is added?
    (b) What is the pH at the halfway point of the titration?
    (c) What is the pH when 95% of the required NaOH has been added?
    (d) What volume of base, in milliliters, is required to reach the equivalence point?
    (e) What is the pH at the equivalence point?
    (f) What indicator would be most suitable for this titration? (See Figure 18.10.)
    (g) What is the pH when 105% of the required base has been added?

31. Using Figure 18.10, suggest an indicator to use in each of the following titrations:
    (a) the weak base pyridine is titrated with HCl
    (b) formic acid is titrated with NaOH
    (c) ethylenediamine, a weak diprotic base, is titrated with HCl

---

▲ More challenging ■ In General ChemistryNow **Blue-numbered questions** answered in Appendix 0

32. ■ Using Figure 18.10, suggest an indicator to use in each of the following titrations.
    (a) $NaHCO_3$ is titrated to CO_3^- with NaOH
    (b) hypochlorous acid is titrated with NaOH
    (c) trimethylamine is titrated with HCl

Solubility Guidelines

(Review Section 5.1, Figure 5.3, and Example 5.1; also see General ChemistryNow.)

33. Name two insoluble salts of each of the following ions.
    (a) Cl^-
    (b) Zn^{2+}
    (c) Fe^{2+}

34. Name two insoluble salts of each of the following ions.
    (a) SO_4^{2-}
    (b) Ni^{2+}
    (c) Br^-

35. Using the solubility guidelines (Figure 5.3), predict whether each of the following is insoluble or soluble in water.
    (a) $(NH_4)_2CO_3$
    (b) $ZnSO_4$
    (c) NiS
    (d) $BaSO_4$

36. ■ Predict whether each of the following is insoluble or soluble in water.
    (a) $Pb(NO_3)_2$
    (b) $Fe(OH)_3$
    (c) $ZnCl_2$
    (d) CuS

Writing Solubility Product Constant Expressions

(See Exercise 18.11 and General ChemistryNow Screen 18.9.)

37. For each of the following insoluble salts, (i) write a balanced equation showing the equilibrium occurring when the salt is added to water and (ii) write the K_{sp} expression.
    (a) AgCN
    (b) $NiCO_3$
    (c) $AuBr_3$

38. For each of the following insoluble salts, (i) write a balanced equation showing the equilibrium occurring when the salt is added to water and (ii) write the K_{sp} expression.
    (a) $PbSO_4$
    (b) BaF_2
    (c) Ag_3PO_4

Calculating K_{sp}

(See Example 18.8 and General ChemistryNow Screen 18.10.)

39. When 1.55 g of solid thallium(I) bromide is added to 1.00 L of water, the salt dissolves to a small extent.

$$TlBr(s) \rightleftharpoons Tl^+(aq) + Br^-(aq)$$

The thallium(I) and bromide ions in equilibrium with TlBr each have a concentration of 1.9×10^{-3} M. What is the value of K_{sp} for TlBr?

40. ■ At 20 °C, a saturated aqueous solution of silver acetate, $AgCH_3CO_2$, contains 1.0 g of the silver compound dissolved in 100.0 mL of solution. Calculate K_{sp} for silver acetate.

$$AgCH_3CO_2(s) \rightleftharpoons Ag^+(aq) + CH_3CO_2^-(aq)$$

41. When 250 mg of SrF_2, strontium fluoride, is added to 1.00 L of water, the salt dissolves to a very small extent.

$$SrF_2(s) \rightleftharpoons Sr^{2+}(aq) + 2 F^-(aq)$$

At equilibrium, the concentration of Sr^{2+} is found to be 1.0×10^{-3} M. What is the value of K_{sp} for SrF_2?

42. ■ Calcium hydroxide, $Ca(OH)_2$, dissolves in water to the extent of 1.3 g per liter. What is the value of K_{sp} for $Ca(OH)_2$?

$$Ca(OH)_2(s) \rightleftharpoons Ca^{2+}(aq) + 2 OH^-(aq)$$

43. You add 0.979 g of $Pb(OH)_2$ to 1.00 L of pure water at 25 °C. The pH is 9.15. Estimate the value of K_{sp} for $Pb(OH)_2$.

44. You place 1.234 g of solid $Ca(OH)_2$ in 1.00 L of pure water at 25 °C. The pH of the solution is found to be 12.68. Estimate the value of K_{sp} for $Ca(OH)_2$.

Estimating Salt Solubility from K_{sp}

(See Examples 18.9 and 18.10, Exercise 8.14, and General ChemistryNow Screen 18.11.)

45. Estimate the solubility of silver iodide in pure water at 25 °C (a) in moles per liter and (b) in grams per liter.

$$AgI(s) \rightleftharpoons Ag^+(aq) + I^-(aq)$$

46. ■ What is the molar concentration of $Au^+(aq)$ in a saturated solution of AuCl in pure water at 25 °C?

$$AuCl(s) \rightleftharpoons Au^+(aq) + Cl^-(aq)$$

47. Estimate the solubility of calcium fluoride, CaF_2, (a) in moles per liter and (b) in grams per liter of pure water.

$$CaF_2(s) \rightleftharpoons Ca^{2+}(aq) + 2 F^-(aq)$$

48. ■ Estimate the solubility of lead(II) bromide (a) in moles per liter and (b) in grams per liter of pure water.

49. The K_{sp} value for radium sulfate, $RaSO_4$, is 4.2×10^{-11}. If 25 mg of radium sulfate is placed in 1.00×10^2 mL of water, does all of it dissolve? If not, how much dissolves?

50. If 55 mg of lead(II) sulfate is placed in 250 mL of pure water, does all of it dissolve? If not, how much dissolves?

51. Use K_{sp} values to decide which compound in each of the following pairs is the more soluble.
    (a) $PbCl_2$ or $PbBr_2$
    (b) HgS or FeS
    (c) $Fe(OH)_2$ or $Zn(OH)_2$

▲ More challenging   ■ In General ChemistryNow   **Blue-numbered questions** answered in Appendix O

52. Use K_{sp} values to decide which compound in each of the following pairs is the more soluble.
    (a) AgBr or AgSCN
    (b) $SrCO_3$ or $SrSO_4$
    (c) AgI or PbI_2
    (d) MgF_2 or CaF_2

The Common Ion Effect and Salt Solubility

(See Examples 18.11 and 18.12, and General ChemistryNow Screen 18.12.)

53. Calculate the molar solubility of silver thiocyanate, AgSCN, in pure water and in water containing 0.010 M NaSCN.

54. ■ Calculate the solubility of silver bromide, AgBr, in moles per liter, in pure water. Compare this value with the molar solubility of AgBr in 225 mL of water to which 0.15 g of NaBr has been added.

55. Compare the solubility, in milligrams per milliliter, of silver iodide, AgI, (a) in pure water and (b) in water that is 0.020 M in $AgNO_3$.

56. What is the solubility, in milligrams per milliliter, of BaF_2, (a) in pure water and (b) in water containing 5.0 mg/mL KF?

The Effect of Basic Anions on Salt Solubility

(See pages 882–883, and General ChemistryNow Screen 18.13.)

57. Which insoluble compound in each pair should be more soluble in nitric acid than in pure water?
    (a) $PbCl_2$ or PbS
    (b) Ag_2CO_3 or AgI
    (c) $Al(OH)_3$ or AgCl

58. ■ Which compound in each pair is more soluble in water than is predicted by a calculation from K_{sp}?
    (a) AgI or Ag_2CO_3
    (b) $PbCO_3$ or $PbCl_2$
    (c) AgCl or AgCN

Precipitation Reactions

(See Examples 18.13–18.15 and General ChemistryNow Screen 18.14.)

59. You have a solution that has a lead(II) concentration of 0.0012 M.

$$PbCl_2(s) \rightleftharpoons Pb^{2+}(aq) + 2\,Cl^-(aq)$$

If enough soluble chloride-containing salt is added so that the Cl^- concentration is 0.010 M, will $PbCl_2$ precipitate?

60. Sodium carbonate is added to a solution in which the concentration of Ni^{2+} ion is 0.0024 M.

$$NiCO_3(s) \rightleftharpoons Ni^{2+}(aq) + CO_3{}^{2-}(aq)$$

Will precipitation of $NiCO_3$ occur (a) when the concentration of the carbonate ion is 1.0×10^{-6} M or (b) when it is 100 times greater (or 1.0×10^{-4} M)?

61. If the concentration of Zn^{2+} in 10.0 mL of water is 1.6×10^{-4} M, will zinc hydroxide, $Zn(OH)_2$, precipitate when 4.0 mg of NaOH is added?

62. You have 95 mL of a solution that has a lead(II) concentration of 0.0012 M. Will $PbCl_2$ precipitate when 1.20 g of solid NaCl is added?

63. If the concentration of Mg^{2+} ion in seawater is 1350 mg per liter, what OH^- concentration is required to precipitate $Mg(OH)_2$?

64. ■ Will a precipitate of $Mg(OH)_2$ form when 25.0 mL of 0.010 M NaOH is combined with 75.0 mL of a 0.10 M solution of magnesium chloride?

Solubility and Complex Ions

(See Example 18.16 and General ChemistryNow Screen 18.16.)

65. Solid gold(I) chloride, AuCl, dissolves when excess cyanide ion, CN^-, is added to give a water-soluble complex ion.

$$AuCl(s) + 2\,CN^-(aq) \rightleftharpoons [Au(CN)_2]^-(aq) + Cl^-(aq)$$

Show that this equation is the sum of two other equations, one for dissolving AuCl to give its ions and the other for the formation of the $Au(CN)_2{}^-$ ion from Au^+ and CN^-. Calculate K_{net} for the overall reaction.

66. ■ Solid silver iodide, AgI, can be dissolved by adding aqueous sodium cyanide to it.

$$AgI(s) + 2\,CN^-(aq) \rightleftharpoons [Ag(CN)_2]^-(aq) + I^-(aq)$$

Show that this equation is the sum of two other equations, one for dissolving AgI to give its ions and the other for the formation of the $[Ag(CN)_2]^-$ ion from Ag^+ and CN^-. Calculate K_{net} for the overall reaction.

Separations

(See Exercise 18.22.)

67. Each pair of ions below is found together in aqueous solution. Using the table of solubility product constants in Appendix J, devise a way to separate these ions by precipitating one of them as an insoluble salt and leaving the other in solution.
    (a) Ba^{2+} and Na^+
    (b) Ni^{2+} and Pb^{2+}

68. ■ Each pair of ions below is found together in aqueous solution. Using the table of solubility product constants in Appendix J, devise a way to separate these ions by adding one reagent to precipitate one of them as an insoluble salt and leave the other in solution.
    (a) Cu^{2+} and Ag^+
    (b) Al^{3+} and Fe^{3+}

General Questions

These questions are not designated as to type or location in the chapter. They may combine several concepts.

69. In each of the following cases, decide whether a precipitate will form when mixing the indicated reagents, and write a balanced equation for the reaction.
    (a) $NaBr(aq) + AgNO_3(aq)$
    (b) $KCl(aq) + Pb(NO_3)_2(aq)$

---

70. In each of the following cases, decide whether a precipitate will form when mixing the indicated reagents, and write a balanced equation for the reaction.

    (a) $Na_2SO_4(aq) + Mg(NO_3)_2(aq)$

    (b) $K_3PO_4(aq) + FeCl_3(aq)$

71. If you mix 48 mL of 0.0012 M $BaCl_2$ with 24 mL of 1.0×10^{-6} M Na_2SO_4 will a precipitate of $BaSO_4$ form?

72. Calculate the hydronium ion concentration and the pH of the solution that results when 20.0 mL of 0.15 M acetic acid, CH_3CO_2H, is mixed with 5.0 mL of 0.17 M NaOH.

73. Calculate the hydronium ion concentration and the pH of the solution that results when 50.0 mL of 0.40 M NH_3 is mixed with 25.0 mL of 0.20 M HCl.

74. For each of the following cases, decide whether the pH is less than 7, equal to 7, or greater than 7.

    (a) equal volumes of 0.10 M acetic acid, CH_3CO_2H, and 0.10 M KOH are mixed

    (b) 25 mL of 0.015 M NH_3 is mixed with 12 mL of 0.015 M HCl

    (c) 150 mL of 0.20 M HNO_3 is mixed with 75 mL of 0.40 M NaOH

    (d) 25 mL of 0.45 M H_2SO_4 is mixed with 25 mL 0.90 M NaOH

75. Rank the following compounds in order of increasing solubility in water: Na_2CO_3, $BaCO_3$, Ag_2CO_3.

76. A sample of hard water contains about 2.0×10^{-3} M Ca^{2+}. A soluble fluoride-containing salt such as NaF is added to "fluoridate" the water (to aid in the prevention of dental caries). What is the maximum concentration of F^- that can be present without precipitating CaF_2?

Dietary sources of fluoride ion. Adding fluoride ion to drinking water (or toothpaste) prevents the formation of dental caries.

77. What is the pH of a buffer solution prepared from 5.15 g of NH_4NO_3 and 0.10 L of 0.15 M NH_3? What is the new pH if the solution is diluted with pure water to a volume of 5.00×10^2 mL?

78. The weak base ethanolamine, $HOCH_2CH_2NH_2$, can be titrated with HCl.

$$HOCH_2CH_2NH_2(aq) + H_3O^+(aq) \longrightarrow$$
$$HOCH_2CH_2NH_3^+(aq) + H_2O(\ell)$$

Assume you have 25.0 mL of a 0.010 M solution of ethanolamine and titrate it with 0.0095 M HCl. (K_b for ethanolamine is 3.2×10^{-5}.)

    (a) What is the pH of the ethanolamine solution before the titration begins?

    (b) What is the pH at the equivalence point?

    (c) What is the pH at the halfway point of the titration?

    (d) Which indicator in Figure 18.10 would be the best choice to detect the equivalence point?

    (e) Calculate the pH of the solution after adding 5.00, 10.0, 20.0, and 30.0 mL of the acid.

    (f) Combine the information in parts (a), (b), and (e) and plot an approximate titration curve.

79. Aniline hydrochloride, $(C_6H_5NH_3)Cl$, is a weak acid. (Its conjugate base is the weak base aniline, $C_6H_5NH_2$.) The acid can be titrated with a strong base such as NaOH.

$$C_6H_5NH_3^+(aq) + OH^-(aq) \longrightarrow C_6H_5NH_2(aq) + H_2O(\ell)$$

Assume 50.0 mL of 0.100 M aniline hydrochloride is titrated with 0.185 M NaOH. (K_a for aniline hydrochloride is 2.4×10^{-5}.)

    (a) What is the pH of the $(C_6H_5NH_3)Cl$ solution before the titration begins?

    (b) What is the pH at the equivalence point?

    (c) What is the pH at the halfway point of the titration?

    (d) Which indicator in Figure 18.10 could be used to detect the equivalence point?

    (e) Calculate the pH of the solution after adding 10.0, 20.0, and 30.0 mL of base.

    (f) Combine the information in parts (a), (b), and (e) and plot an approximate titration curve.

80. If you place 5.0 mg of $SrCO_3$ in 1.0 L of pure water, will all of the salt dissolve before equilibrium is established, or will some salt remain undissolved?

81. To have a buffer with a pH of 2.50, what volume of 0.150 M NaOH must be added to 100. mL of 0.230 M H_3PO_4?

82. ▲ What mass of Na_3PO_4 must be added to 80.0 mL of 0.200 M HCl to obtain a buffer with a pH of 7.75?

83. For the titration of 50.0 mL of 0.150 M ethylamine, $C_2H_5NH_2$, with 0.100 M HCl, find the pH at each of the following points and then use that information to sketch the titration curve and decide on an appropriate indicator.

    (a) at the beginning, before HCl is added

    (b) at the halfway point in the titration

(c) when 75% of the required acid has been added

(d) at the equivalence point

(e) when 10.0 mL more HCl has been added than is required

(f) Sketch the titration curve.

(g) Suggest an appropriate indicator for this titration.

84. What volume of 0.120 M NaOH must be added to 100. mL of 0.100 M $NaHC_2O_4$ to reach a pH of 4.70?

85. Describe the effect on the pH of the following actions:

(a) adding sodium acetate, $NaCH_3CO_2$, to 0.100 M CH_3CO_2H

(b) adding $NaNO_3$ to 0.100 M HNO_3

(c) Explain why there is or is not an effect in each case.

86. ▲ A buffer solution is prepared by dissolving 1.50 g each of benzoic acid, $C_6H_5CO_2H$, and sodium benzoate, $NaC_6H_5CO_2$, in 150.0 mL of solution.

(a) What is the pH of this buffer solution?

(b) Which buffer component must be added and what quantity is needed to change the pH to 4.00?

(c) What quantity of 2.0 M NaOH or 2.0 M HCl must be added to the buffer to change the pH to 4.00?

87. A buffer solution with a pH of 12.00 consists of Na_3PO_4 and Na_2HPO_4. The volume of solution is 200.0 mL.

(a) Which component of the buffer is present in a larger amount?

(b) If the concentration of Na_3PO_4 is 0.400 M, what mass of Na_2HPO_4 is present?

(c) Which component of the buffer must be added to change the pH to 12.25? What mass of that component is required?

88. What volume of 0.200 M HCl must be added to 500.0 mL of 0.250 M NH_3 to have a buffer with a pH of 9.00?

89. ▲ The cations Ba^{2+} and Sr^{2+} can be precipitated as very insoluble sulfates.

(a) If you add sodium sulfate to a solution containing these metal cations, each with a concentration of 0.10 M, which is precipitated first, $BaSO_4$ or $SrSO_4$?

(b) What will be the concentration of the first ion that precipitates (Ba^{2+} or Sr^{2+}) when the second, more soluble salt begins to precipitate?

90. ▲ You will often work with salts of Fe^{3+}, Pb^{2+}, and Al^{3+} in the laboratory. (All are found in nature, and all are important economically.) If you have a solution containing these three ions, each at a concentration of 0.10 M, what is the order in which their hydroxides precipitate as aqueous NaOH is slowly added to the solution?

91. What is the equilibrium constant for the following reaction?

$$AgCl(s) + I^-(aq) \rightleftharpoons AgI(s) + Cl^-(aq)$$

Does the equilibrium lie predominantly to the left or to the right? Will AgI form if iodide ion, I^-, is added to a saturated solution of AgCl?

92. Calculate the equilibrium constant for the following reaction.

$$Zn(OH)_2(s) + 2 CN^-(aq) \rightleftharpoons Zn(CN)_2(s) + 2 OH^-(aq)$$

Does the equilibrium lie predominantly to the left or to the right? Can zinc hydroxide be transformed into zinc cyanide by adding a soluble salt of the cyanide ion?

93. ▲ In principle, the ions Ba^{2+} and Ca^{2+} can be separated by the difference in solubility of their fluorides, BaF_2 and CaF_2. If you have a solution that is 0.10 M in both Ba^{2+} and Ca^{2+}, CaF_2 will begin to precipitate first as fluoride ion is added slowly to the solution.

(a) What concentration of fluoride ion will precipitate the maximum amount of Ca^{2+} ion without precipitating BaF_2?

(b) What concentration of Ca^{2+} remains in solution when BaF_2 just begins to precipitate?

94. ▲ A solution contains 0.10 M iodide ion, I^-, and 0.10 M carbonate ion, CO_3^{2-}.

(a) If solid $Pb(NO_3)_2$ is slowly added to the solution, which salt will precipitate first, PbI_2 or $PbCO_3$?

(b) What will be the concentration of the first ion that precipitates (CO_3^{2-} or I^-) when the second, more soluble salt begins to precipitate?

95. ▲ A solution contains Ca^{2+} and Pb^{2+} ions, both at a concentration of 0.010 M. You wish to separate the two ions from each other as completely as possible by precipitating one but not the other using aqueous Na_2SO_4 as the precipitating agent.

(a) Which will precipitate first as sodium sulfate is added, $CaSO_4$ or $PbSO_4$?

(b) What will be the concentration of the first ion that precipitates (Ca^{2+} or Pb^{2+}) when the second, more soluble salt begins to precipitate?

96. Buffer capacity is defined as the number of moles of a strong acid or strong base that are required to change the pH of one liter of the buffer solution by one unit. What is the buffer capacity of a solution that is 0.10 M in acetic acid and 0.10 M in sodium acetate?

Summary and Conceptual Questions

The following questions may use concepts from the preceding chapters.

97. Suggest a method for separating a precipitate consisting of a mixture of solid CuS and solid $Cu(OH)_2$.

98. Which of the following barium salts should dissolve in a strong acid such as HCl: $Ba(OH)_2$, $BaSO_4$, or $BaCO_3$?

99. Describe how a buffer solution can control the pH of a solution when a strong base is added. Use a solution of acetic acid and sodium acetate as an example, and include balanced chemical equations in your answer.

100. Use the Henderson-Hasselbalch equation to explain how the pH of a buffer solution based on a weak acid and its conjugate base changes (a) when the ionization constant of the weak acid increases and (b) when the acid concentration is decreased relative to the concentration of its conjugate base.

101. Explain why the solubility of Ag_3PO_4 can be greater in water than is calculated from the K_{sp} value of the salt.

102. Two acids, each approximately 0.01 M in concentration, are titrated separately with a strong base. The acids show the following pH values at the equivalence point: HA, pH = 9.5, and HB, pH = 8.5.
   (a) Which is the stronger acid, HA or HB?
   (b) Which of the conjugate bases, A^- or B^-, is the stronger base?

103. Composition diagrams, commonly known as "alpha plots," are often used to visualize the species in a solution of an acid or base as the pH is varied. The diagram for 0.100 M acetic acid is shown here.

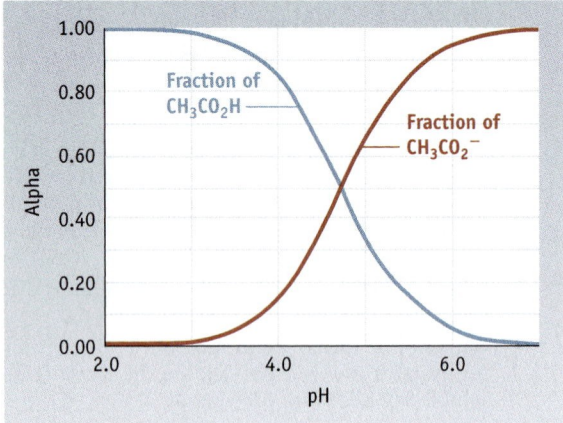

The plot shows how the fraction [= alpha (α)] of acetic acid in solution, $[CH_3CO_2H]/([CH_3CO_2H] + [CH_3CO_2^-])$, changes as the pH increases (blue curve). (The red curve shows how the fraction of acetate ion, $CH_3CO_2^-$, changes as the pH increases.) Alpha plots are another way of viewing the relative concentrations of acetic acid and acetate ion as a strong base is added to a solution of acetic acid in the course of a titration.
   (a) Explain why the fraction of acetic acid declines and that of acetate ion increases as the pH increases.
   (b) Which species predominates at a pH of 4, acetic acid or acetate ion? What is the situation at a pH of 6?
   (c) Consider the point where the two lines cross. The fraction of acetic acid in the solution is 0.5, and so is that of acetate ion. That is, the solution is half acid and half conjugate base; their concentrations are equal. At this point the graph shows the pH is 4.74. Explain why the pH at this point is 4.74.

104. The composition diagram, or alpha plot, for the important acid–base system of carbonic acid, H_2CO_3, is illustrated below. (See Study Question 103 for more information on such diagrams.)

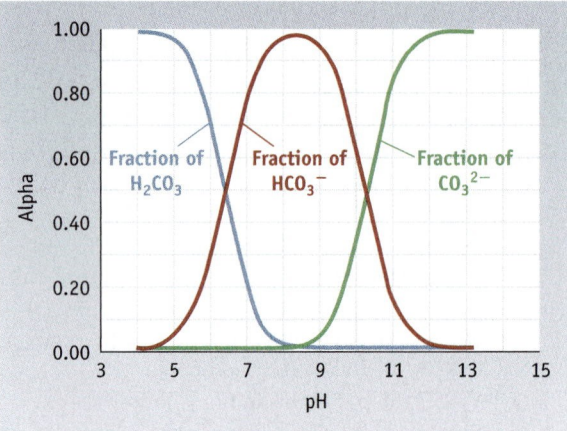

   (a) Explain why the fraction of bicarbonate ion, HCO_3^-, rises and then falls as the pH increases.
   (b) What is the composition of the solution when the pH is 6.0? When the pH is 10.0?
   (c) If you wanted to buffer a solution at a pH of 11.0, what should be the ratio of HCO_3^- to CO_3^{2-}?

105. The chemical name for aspirin is acetylsalicylic acid. It is believed that the analgesic and other desirable properties of aspirin are due not to the aspirin itself but rather to the simpler compound salicylic acid, $C_6H_4(OH)CO_2H$, that results from the breakdown of aspirin in the stomach.

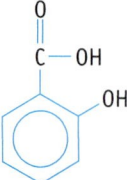

salicylic acid

   (a) Give approximate values for the following bond angles in the acid: (i) C—C—C in the ring; (ii) O—C=O; (iii) either of the C—O—H angles; and (iv) C—C—H.
   (b) What is the hybridization of the C atoms of the ring? Of the C atom in the —CO_2H group?
   (c) Experiment shows that 1.00 g of the acid will dissolve in 460 mL of water. If the pH of this solution is 2.4, what is K_a for the acid?
   (d) If you have salicylic acid in your stomach, and if the pH of gastric juice is 2.0, calculate the percentage of salicylic acid that will be present in the stomach in the form of the salicylate ion, $C_6H_4(OH)CO_2^-$.
   (e) Assume you have 25.0 mL of a 0.014 M solution of salicylic acid and titrate it with 0.010 M NaOH. What is the pH at the halfway point of the titration? What is the pH at the equivalence point?

106. Observe the titration curves on the General ChemistryNow CD-ROM or website Screen 18.7 simulation. Titrate 25.0 mL of 0.30 M acetic acid with 0.50 M NaOH.

    (a) What volume of NaOH is required?

    (b) What is the pH at the equivalence point? Explain why the pH has this value.

    (c) Which of the three indicators available are best used in this titration?

107. Explore the common ion effect on Screen 18.13 of the General ChemistryNow CD-ROM or website. The animation on the Description screen illustrates the common ion effect for the case of adding extra chloride ion to an equilibrium system containing $PbCl_2(s)$, $Pb^{2+}(aq)$, and $Cl^-(aq)$. Explain the changes you see in terms of the solubility product constant expression for this system.

108. Simultaneous equilibria are explored on Screen 18.15 of the General ChemistryNow CD-ROM or website. Why is the experiment on this screen good evidence that chemical equilibria are dynamic as opposed to static?

109. Examine the pH and Solubility Table on Screen 18.16 of the General ChemistryNow CD-ROM or website. Explain why the solubility of $Co(OH)_2$ increases by 100 for each 1.0 unit decrease in pH.

110. Examine the sidebar on Screen 18.17 of the General ChemistryNow CD-ROM or website. How does the chemistry of floor wax support the idea that reactions can be reversible?

Do you need a live tutor for homework problems?
Access vMentor at General ChemistryNow at
http://now.brookscole.com/kotz6e
for one-on-one tutoring from a chemistry expert

19—Principles of Reactivity: Entropy and Free Energy

Waterfall by M. C. Escher, 1961. This drawing is reminiscent of Fludd's perpetual motion machine.

Perpetual Motion Machines

"You can't get something for nothing."

Economists have often said, "There is no free lunch," meaning that every desirable thing requires the expenditure of some effort, money, or energy. But people have certainly tried.

Wouldn't it be wonderful if someone could design a perpetual motion machine, one that produces energy but requires no energy itself? Over the centuries many people have attempted this feat. But, as you will see in this chapter, such a machine violates the fundamental laws of thermodynamics.

The first perpetual motion machine was apparently proposed by Villand de Honnecourt in the 13th century. Later, Leonardo da Vinci made a number of drawings of machines that would produce energy at no cost.

Among the best-known early machines was one proposed by Robert Fludd in the 1600s. Fludd, a well-known scientist, also suggested that the sun and not the earth was the center of our universe, and that blood carries the gases important to life throughout the body. But, he also believed that lightning was simply an act of God.

In 1812 Charles Redheffer of Philadelphia set up a machine that, he claimed, required no source of energy to run. He applied to the city government for funds to build an even larger version, but did not allow the city commissioners to get too close to the machine. The commissioners were suspicious, however, and asked a local engineer, Isaiah Lukens, to build a machine that worked on the same principle as Redheffer's machine. When Redheffer saw Lukens's replica, Redheffer apparently realized his fraud was exposed. He soon left town for New York City, where he tried anew to interest investors in his machine.

In New York Redheffer was again exposed as a charlatan, this time by the inventor Robert Fulton. When Fulton was invited to view the machine, he noticed that it was operating in a wobbly manner. Fulton challenged Redheffer, stating that he could expose the secret source of energy. Fulton also offered to compensate Redheffer for damages if the accusations proved to be false. Redheffer should never have agreed. When Fulton removed some boards in a wall near the machine, a thin cord of cat gut was discovered to lead to yet another room. There Fulton found an elderly gentleman, eating a crust of bread with one hand and turning a crank with the other hand—not so smoothly—to run Redheffer's machine! Upon the discovery Redheffer disappeared.

The vast majority of the so-called perpetual motion machines violate the first law of thermodynamics. Falling water in Fludd's machine can, indeed, turn a crank to perform useful work. However, then there is insufficient energy available to lift the water to the reservoir to begin the cycle again. Besides, energy losses due to friction occur in any machine.

In the 1880s John Gamgee invented an "ammonia engine" and tried to persuade the U. S. Navy to use it for ship propulsion. Even President Garfield took the time to inspect Gamgee's engine. The engine worked by using the heat of ocean water to evaporate liquid ammonia. The expanding ammonia vapor was supposed to drive a piston, in the same way that the combustion of gasoline in an automobile's engine drives its pistons. In the ammonia engine, the ammonia vapor cools when it expands, which would lead to the condensation of the ammonia. The reservoir of liquid ammonia is thereby replenished, and the cycle begins anew.

Gamgee's engine sounded good—but it violates the second law of thermodynamics, the subject of this chapter.

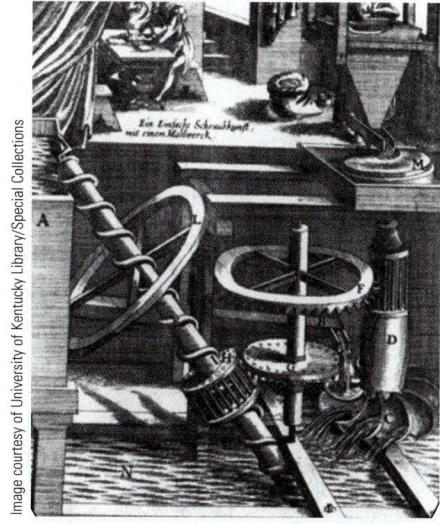

Image courtesy of University of Kentucky Library/Special Collections

A late-17th-century version of Fludd's proposed "perpetual motion machine." Water flows from a reservoir at the right, turning a water wheel. Work could be extracted from this motion, and it was also used to turn an Archimedes screw, the long rod to the left, that transported the water back to the top where it could again flow downward onto the water wheel. Can you figure out what is wrong with this device?

Chapter 6 (Energy and Chemical Reactions) and Chapter 19 (Entropy and Free Energy) together provide an introduction to the subject of thermodynamics. Chapter 6 focused on heat and heat transfer, with the discussion being guided by the first law of thermodynamics, the law of conservation of energy. In Chapter 19, we encounter two more laws of thermodynamics. These laws determine the spontaneity of chemical and physical processes, explain the driving forces that compel the changes that a system undergoes to achieve equilibrium, and guide us to understand when reactions are product-favored or reactant-favored.

19.1—Spontaneous Change and Equilibrium

Change is central to chemistry, so it is important to understand the factors that determine whether a change will occur. In chemistry, we encounter many examples of both chemical changes (chemical reactions) and physical changes (the formation of mixtures, expansion of gases, and changes of state, to name a few). When describing changes, chemists use the term **spontaneous**. A spontaneous change is one that occurs without outside intervention. This statement does not say anything about the rate of the change, merely that a spontaneous change is naturally occurring and unaided. Furthermore, *a spontaneous change leads inexorably to equilibrium.*

If a piece of hot metal is placed in a beaker of cold water, heat transfer from the hot metal to the cooler water occurs spontaneously. The process of heat transfer continues until the two objects are at the same temperature and thermal equilibrium is attained. Similarly, many chemical reactions proceed spontaneously until equilibrium is reached. Some chemical reactions greatly favor products at equilibrium, as in the reaction of sodium and chlorine (page 19). In Chapters 16–18 we described them as *product-favored* reactions (page 765). In other instances, the position of the equilibrium favors the reactants. One example of such a *reactant-favored* process would be the dissolution of an insoluble substance like limestone. If you place a handful of $CaCO_3$ in a small amount of water, only a tiny fraction of the sample will dissolve spontaneously until equilibrium is reached; most of the salt will remain undissolved.

When considering both physical and chemical processes, the focus is always on the changes that must occur to achieve equilibrium. The factors that determine the directionality of the change are the topic of this chapter. Given two objects at the same temperature but isolated thermally from their surroundings, it will never happen that one will heat up while the other becomes colder. Gas molecules will never spontaneously congregate at one end of a flask. A chemical system at equilibrium will not spontaneously change in a way that results in the system no longer being in equilibrium. Neither will a chemical system not at equilibrium change in the direction that takes it farther from the equilibrium condition.

19.2—Heat and Spontaneity

In previous chapters, we have encountered many spontaneous chemical reactions: hydrogen and oxygen combine to form water; methane burns to give CO_2 and H_2O; Na and Cl_2 react to form NaCl; HCl(aq) and NaOH(aq) react to form H_2O

A spontaneous process. Heat transfers spontaneously from a hotter object to a cooler object.

Charles D. Winters

A Review of Concepts of Thermodynamics

To understand the thermodynamic concepts introduced in this chapter, be sure to review the ideas of Chapter 6.

System: The part of the universe under study.

Surroundings: The rest of the universe exclusive of the system.

Exothermic: Thermal energy transfer occurs from the system to the surroundings.

Endothermic: Thermal energy transfer occurs from the surroundings to the system.

First law of thermodynamics: The law of the conservation of energy, $\Delta E = q + w$. The change in internal energy of a system is the sum of heat transferred to or from the system and the work done on or by the system. Energy can be neither created nor destroyed.

Enthalpy change: The thermal energy transferred at constant pressure.

State function: A quantity whose value is determined only by the initial and final states of a system.

Standard conditions: Pressure of 1 bar (1 bar = 0.98692 atm) and solution concentration of 1 m.

Standard enthalpy of formation, ΔH_f°: The enthalpy change occurring when a compound is formed from its elements in their standard states.

---

and NaCl(aq). These reactions proceed spontaneously from reactants to products and have gone substantially to completion when equilibrium is reached. These chemical reactions and many others share a common feature: They are exothermic. Therefore, it might be tempting to conclude that evolution of heat is the criterion that determines whether a reaction or process is spontaneous. Further inspection, however, will reveal flaws in this reasoning. This is especially evident with the inclusion of some common physical changes, changes that are spontaneous but that are endothermic or energy-neutral:

- *Dissolving NH_4NO_3.* The ionic compound NH_4NO_3 dissolves spontaneously in water in an endothermic process with $\Delta H° = +25.7$ kJ/mol (page 665).

- *Expansion of a gas into a vacuum.* A system is set up with two flasks connected by a valve (Figure 19.1). One flask is filled with a gas and the other is evacuated. When the valve is opened, the gas will flow spontaneously from one flask to the other until the pressure is the same throughout. The expansion of an ideal gas is energy-neutral, with heat being neither evolved nor required.

- *Phase changes.* Melting of ice is an endothermic process; to melt one mole of H_2O requires about 6 kJ. At temperatures above 0 °C, the melting of ice is spontaneous. Below 0 °C, however, ice does not melt; melting is not a spontaneous process under these conditions. At 0 °C, liquid water and ice coexist at equilibrium, and no net change occurs. This example illustrates that temperature can have a role in determining whether a process is spontaneous. We will return to this important issue later in this chapter.

- *Heat transfer.* The temperature of a cold soft drink sitting in a warm environment will rise until the beverage reaches the ambient temperature. The heat required for this process comes from the surroundings. The endothermic process illustrates that heat transfer from a hotter object (the surroundings) to a cooler object (the soft drink) is spontaneous.

We can gain further insight into spontaneity if we think about a specific chemical system—for example, the reaction $H_2(g) + I_2(g) \rightleftharpoons 2\,HI(g)$. Equilibrium in this system can be approached from either direction. The reaction of H_2 and I_2 is endothermic, but the reaction occurs spontaneously until an equilibrium mixture containing H_2, I_2, and HI forms. The reverse reaction, the decomposition of HI to form H_2 and I_2

■ **Spontaneous Reactions**
A spontaneous reaction proceeds to equilibrium without outside intervention. Such a reaction may or may not be product-favored.

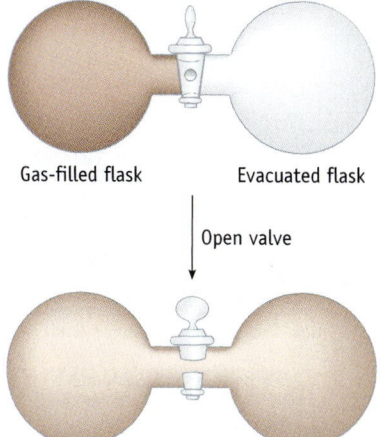

Gas-filled flask   Evacuated flask

Open valve

When the valve is opened the gas expands irreversibly to fill both flasks.

Figure 19.1 **Spontaneous expansion of a gas.** *(See General ChemistryNow Screen 19.3 Directionality of Reactions, to view an animation of this figure.)*

until equilibrium is achieved, is also spontaneous, but in this case the process evolves heat. Equilibrium can be approached spontaneously from either direction.

From these examples, and many others, we must conclude that evolution of heat is not a sufficient criterion to determine whether a process is spontaneous. In retrospect, this conclusion makes sense. The first law of thermodynamics tells us that in any process energy must be conserved. If energy is evolved by a system, then the same amount of energy must be absorbed by the surroundings. The exothermicity of the system must be balanced by the endothermicity of the surroundings so that the energy content of the universe remains unchanged. If energy evolution were the only factor determining whether a change is spontaneous, then for every spontaneous process there would be a corresponding nonspontaneous change in the surroundings. We must search further than heat evolution and the first law to determine whether a change is spontaneous.

GENERAL
Chemistry ⚛ Now™

See the General ChemistryNow CD-ROM or website:
- **Screen 19.2 Reaction Spontaneity,** for a description of kinetic and thermodynamic control of reactions

19.3—Dispersal of Energy and Matter

A better way to predict whether a process will be spontaneous is with a thermodynamic function called **entropy, S**. Entropy is tied to the **second law of thermodynamics**, which states that *in a spontaneous process, the entropy of the universe increases.* Ultimately, this law allows us to predict the conditions at equilibrium as well as the direction of spontaneous change toward equilibrium.

The concept of entropy is built around the idea that *spontaneous change results in dispersal of energy*. Many times, a dispersal of matter is also involved, and it can contribute to energy dispersal in some systems. Because the logic underlying these ideas is statistical, let us examine the statistical nature of entropy more closely.

Dispersal of Energy

The dispersal of energy over as many different energy states as possible is the key contribution to entropy. We can explore a simple example of this phenomenon in which heat flows from a hot object to a cold object until both have the same temperature. A model involving gaseous atoms provides a basis for this analysis (Figure 19.2). Start by placing a sample of hot atoms in contact with a sample of cold atoms. The atoms move randomly in each container and collide with the walls. When the containers come in contact with each other, thermal energy is transferred from the warmer container to the cooler one, and the energy is thus transferred through wall collisions from the warmer atoms to the cooler atoms. Eventually the system stabilizes at an average temperature so that each sample of gas has the same molecular distribution of energies [◀ Section 12.6].

We can also use a statistical explanation to show how energy is dispersed in a system. With statistical arguments, systems must include large numbers of particles for the arguments to be accurate. We will look first at a simple example to understand the underlying concept, and then build up to a larger system.

Consider a system in which there are two atoms (1 and 2) with one discrete packet, or quantum, of energy each, and two other atoms (3 and 4) with no energy

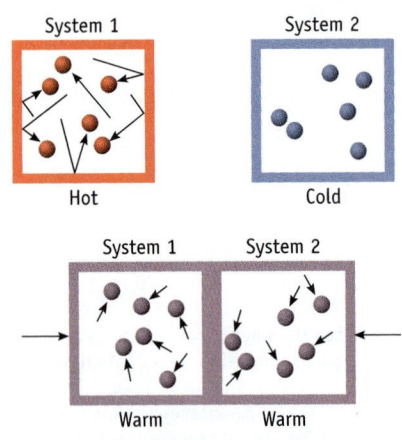

Figure 19.2 Energy transfer between molecules in the gas phase.

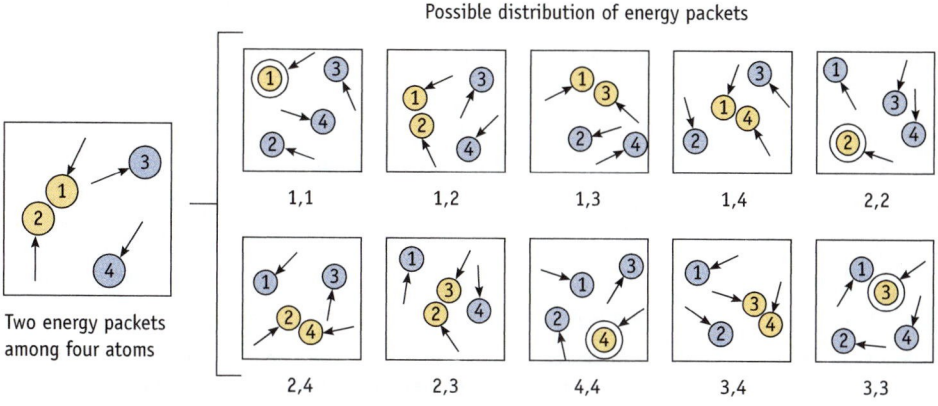

Possible distribution of energy packets

Two energy packets among four atoms

1,1   1,2   1,3   1,4   2,2

2,4   2,3   4,4   3,4   3,3

Figure 19.3 **Energy dispersal.** Possible ways of distributing two packets of energy among four atoms. Here there are two atoms (1 and 2) with one quantum of energy apiece, and two other atoms (3 and 4) that have no energy initially. The figure shows there are 10 different ways to distribute the two quanta of energy over the four atoms.

initially (Figure 19.3). When these four atoms are brought together, the total energy in the system is simply the total of the two quanta. Collisions among the atoms allow energy to be transferred so that all distributions of the two packets of energy over the four atoms are eventually achieved. There are 10 different ways to distribute these two quanta of energy over the four atoms. In only 4 of the 10 cases [1, 1; 2, 2; 3, 3; and 4,4], is the energy concentrated on a single particle. In the majority of the cases, 6 out of 10, the energy is distributed to two different particles. Thus, even in a simple example with only two packets of energy to consider, it is more likely that the energy will be found distributed over multiple particles than concentrated in one place.

Now let's consider a slightly larger sample. Assume we have four particles again, but this time two of them each begin with three quanta of energy (Figure 19.4). The other two particles initially have zero energy. When these four particles are brought together, the total energy of the system is six quanta. There are nine different arrangements for distributing six quanta of energy among four particles (Figure 19.4). For example, one particle can have all six quanta of energy, and the others can have none. In a different arrangement, two particles can have two quanta of energy each and the other two can have one each.

As in our previous example (Figure 19.3), if we label the particles we will find that some arrangements are more likely than others. Figure 19.4 shows the number of different ways to distribute four particles with a total of six quanta of energy in each of the arrangements indicated. The arrangement that occurs most often is one in which the energy is distributed over all four particles and to a large number of states (four distinct levels are occupied). *As the number of particles and the number of energy levels grows, one arrangement turns out to be vastly more probable than all the others.*

Dispersal of Matter

It is rarely obvious how to calculate the different energy levels of a system and how to discern the distribution of the total energy among them. Therefore, it is useful to look at the dispersal of matter, because matter dispersal often contributes to energy dispersal. Let us examine a system qualitatively to gain insight into the likely distribution of energy from the distribution of matter.

Matter dispersal is illustrated in Figure 19.1 by the expansion of a gas into a vacuum. Let us consider this same arrangement—a flask containing two molecules of a gas connected by a valve to an evacuated flask having the same volume (Figure 19.5). Assume that, when the valve connecting the flasks is opened, the two molecules originally in flask A can move randomly throughout flasks A and B. At any given instant, the molecules will be in one of four possible configurations: two molecules in flask A,

Number of different ways to achieve this arrangement

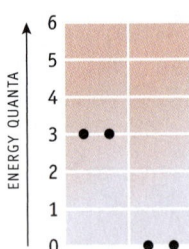

(a) Initially four particles are separated from each other. Two particles each have three quanta of energy, and the other two have none. A total of six quanta of energy will be distributed once the particles interact.

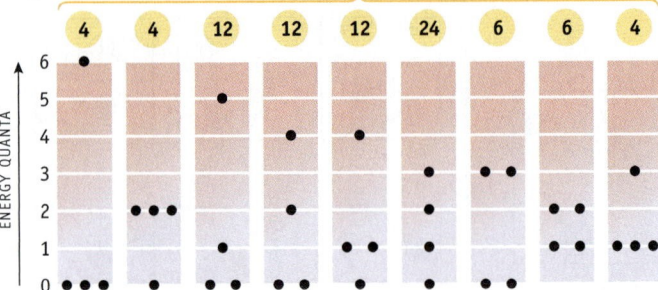

(b) Once the particles begin to interact, there are nine ways to distribute the six available quanta. Each of these arrangements will have multiple ways of distributing the energy among the four atoms. Part (c) shows why the arrangement on the right can be achieved four ways.

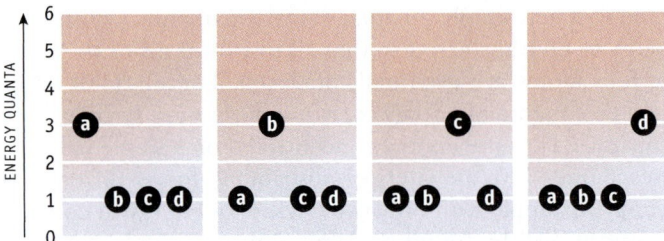

(c) There are four different ways to arrange four particles (a, b, c, and d) such that one particle has three quanta of energy and the other three each have one quantum of energy.

Figure 19.4   Energy dispersal. Possible ways of distributing six quanta of energy among four atoms.

two molecules in flask B, or one molecule in each flask. The probability of having one molecule in each flask is 50%. There is a 25% probability that the two molecules will be simultaneously in flask A and a 25% probability that they will be in flask B.

If we next consider a system having three molecules originally in flask A, we would find there is only a one in eight chance that the three molecules will remain in the original flask. With 10 molecules, there is only one chance in 1024 of finding all the molecules in flask A. The probability of n molecules remaining in the initial flask in this two-flask system is $(\frac{1}{2})^n$, where n is the number of molecules. If flask A contained one mole of a gas, the probability of all the molecules being found in that flask when the connecting valve is opened is $(\frac{1}{2})^N$, where N is Avogadro's number—a probability almost too small to comprehend! If we calculate the probabilities for all the other possible arrangements of the mole of molecules in this scenario, we would find that the most probable arrangement, by a huge margin, is the one in which the molecules are, on average, evenly distributed over the entire two-flask volume.

This analysis shows that, in a system such as is depicted in Figure 19.1 or 19.5, it is highly probable that gas molecules will flow from one flask into an evacuated flask until the pressures in the two flasks are equal. Conversely, the opposite process, in which all the gas molecules in the apparatus congregate in one of the two flasks, is highly improbable.

As stated earlier, matter dispersal contributes to energy dispersal, so let us next describe the experiment in Figure 19.5 in terms of energy dispersal. To do so, we need to remember an idea introduced in Chapter 7, that all energy is quantized. Schrödinger's model [◄ Section 7.5], and the equation that he developed, was applied to the atom to derive the three quantum numbers and the orbitals for elec-

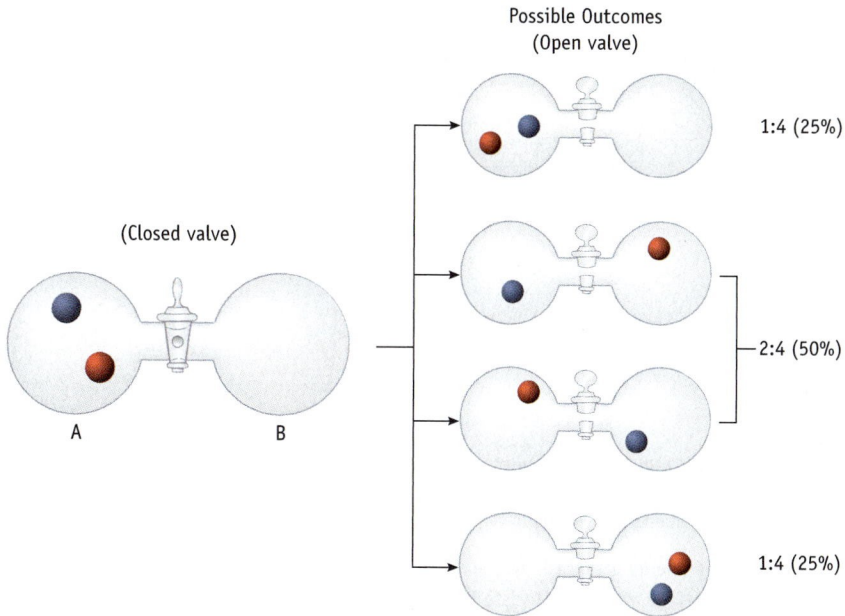

Possible Outcomes
(Open valve)

(Closed valve)

A   B

1:4 (25%)

2:4 (50%)

1:4 (25%)

Figure 19.5   Dispersal of matter. The expansion of a gas into a vacuum. There are four possible arrangements for two molecules in this apparatus. There is a 50% probability that there will be one molecule in each flask at any instant of time. *(See General ChemistryNow Screen 19.3 Reaction Directionality, to view an animation of the expansion of a gas.)*

trons around a nucleus. It turns out that Schrödinger's equation is universal and can be applied to any system, including gas molecules in a room or in a reaction flask. When we do this, we find that *all* systems have quantized energies, just as we found for electrons in an atom.

Recall from Chapter 7 that the energy levels in an atom get closer together as the principal quantum number, *n*, gets larger (Figure 7.12). It is also true that the average radius of the atom increases with the primary quantum number (page 321). If we apply the Schrödinger equation to other systems, we find the same general trends: the size of the system affects the size of *n*, and energy levels get closer together as *n* gets larger.

In a macroscopic system such as the example of a gas inside a flask, the principal quantum number is very large (because the system is much larger than the size of an atom), and the energy levels are infinitesimally close together as a result of *n* being so large. Nonetheless, they are still separate, quantized, energy levels. However, because they are so close together, we often simply treat gas samples as though the energies are continuous, such as when we discuss the Maxwell-Boltzmann distribution (see Section 12.6). If we increase the size of the vessel holding the gas, the corresponding quantum number will increase, and the energies will get even closer together. In our statistical discussion of matter and energy dispersal, a decreased spacing between energy levels means a larger number of energy levels is available to our system. Because the total energy available to our system has not changed (only the volume has changed), the number of ways of distributing the total energy within those energy levels increases (Figure 19.6). That is, when matter is dispersed into a larger volume, energy is dispersed over more energy levels.

Applications of the Dispersal of Matter

The logic applied to the expansion of a gas into a vacuum can be used to rationalize the mixing of two gases. If flasks containing O_2 and N_2 are connected (in an experimental set-up like that in Figure 19.5), the two gases will diffuse together, eventually leading to a mixture in which molecules of O_2 and N_2 are evenly distributed throughout the total volume. A mixture of N_2 and O_2 will never separate into samples

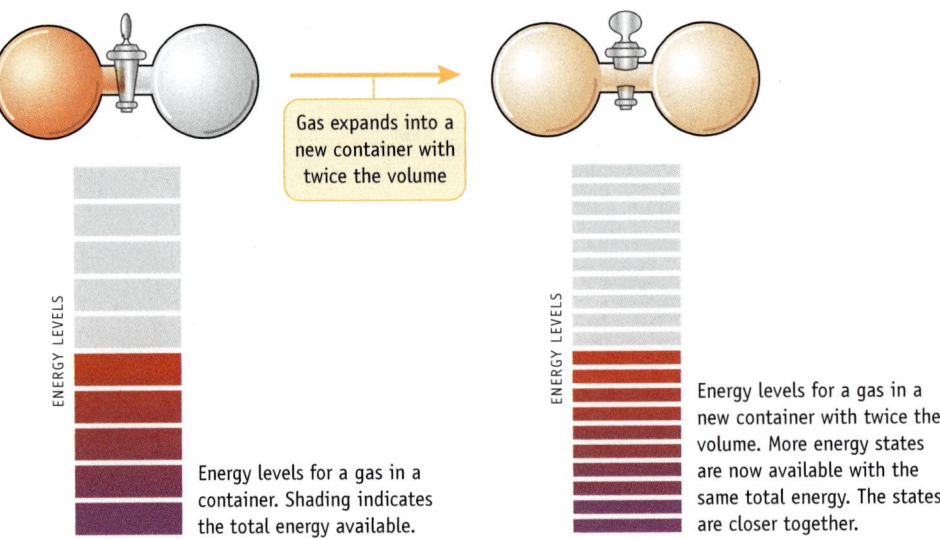

Gas expands into a new container with twice the volume

ENERGY LEVELS

ENERGY LEVELS

Energy levels for a gas in a container. Shading indicates the total energy available.

Energy levels for a gas in a new container with twice the volume. More energy states are now available with the same total energy. The states are closer together.

Figure 19.6 Matter and energy dispersal. As the size of the container for the chemical or physical change increases, the number of energy states accessible to the molecules of the system increases and the states come closer together.

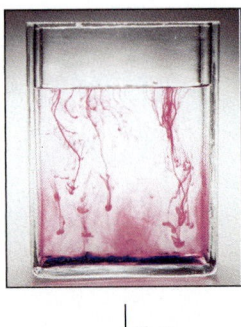

↓ Time

Photos: Charles D. Winters

Figure 19.7 Matter dispersal. A small quantity of purple $KMnO_4$ is added to water (*top*). With time, the solid dissolves and the highly colored MnO_4^- ion (and the K^+ ions) become dispersed throughout the solution (*bottom*).

of each component of its own accord. The important point is that what began as a relatively orderly system with N_2 and O_2 molecules in separate flasks spontaneously moves toward a system in which each gas is maximally dispersed. *For gases at room temperature, the entropy-driven dispersal of matter is equivalent to an increase in disorder of the system.*

The equivalence of matter and energy dispersal with disorder is true for some solutions as well. For example, when a water-soluble compound is placed in water, it is highly probable that the molecules or ions of the compound will ultimately become distributed evenly throughout the solution. The tendency to move from order to disorder explains why the $KMnO_4$ sample in Figure 19.7 eventually disperses over the entire container. The process leads to a mixture, a system in which the solute and solvent molecules are more widely dispersed.

It is also evident from these examples of matter dispersal that if we wanted to put all of the gas molecules in Figure 19.5 back into one flask, or recover the $KMnO_4$ crystals, we would have to intervene in some way. For example, we could use a pump to force all of the gas molecules from one side to the other or we could lower the temperature drastically (Figure 19.8). In the latter case, the molecules would move "downhill" energetically to a place of very low kinetic energy.

Does the formation of a mixture always lead to greater disorder? It does when gases are mixed. With liquids and solids this is usually the case as well, but not always. Exceptions can be found, especially when considering aqueous solutions. For example, Li^+ and OH^- ions are highly ordered in the solid lattice [◄ Section 13.6], but they become more disordered when they enter into solution. However, the solution process is accompanied by solvation, a process in which water molecules become tightly bound to the ions. As a result, the water molecules are constrained to a more ordered arrangement than in pure water. Thus, two opposing effects are at work here: a decrease in order for the Li^+ and OH^- ions when dispersed in the water, and an increased order for water. The higher degree of ordering due to solvation dominates, and the result is a higher degree of order overall in the system.

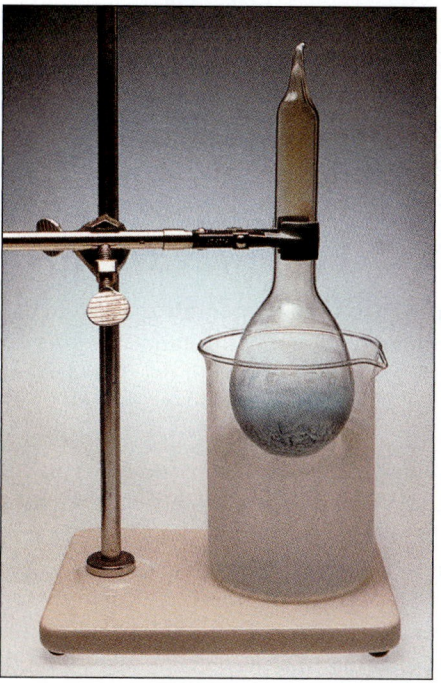

(a)   (b)

Charles D. Winters

Figure 19.8 **Reversing the process of matter dispersal.** (a) Brown NO_2 gas is dispersed evenly throughout the flask. (b) If the flask is immersed in liquid nitrogen at $-196\ °C$, the kinetic energy of the NO_2 molecules is reduced to the point that they become a solid and collect on the cold walls of the flask.

GENERAL
Chemistry⚛Now™

See the General ChemistryNow CD-ROM or website:
- **Screen 19.3 Directionality of Reactions,** for an illustration of matter and energy dispersal

The Boltzmann Equation for Entropy

Ludwig Boltzmann (1844–1906) developed the idea of looking at the distribution of energy over different energy states as a way to calculate entropy. At the time of his death, the scientific world had not yet accepted his ideas. In spite of this, he was firmly committed to his theories and had his equation for entropy engraved on his tombstone in Vienna, Austria. The equation is

$$S = k \log W$$

where k is the Boltzmann constant, and W represents the number of different ways that the energy can be distributed over the available energy levels. Boltzmann concluded that the maximum entropy will be achieved at equilibrium, a state in which W has the maximum value.

A Summary: Matter and Energy Dispersal

In summary, the final state of a system can be more probable than the initial state in either or both of two ways: (1) the atoms and molecules can be more disordered and (2) energy can be dispersed over a greater number of atoms and molecules.

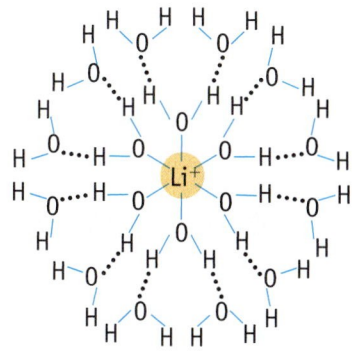

Order and disorder in solid and solution. Lithium hydroxide is a crystalline solid with an orderly arrangement of Li^+ and OH^- ions. When placed in water, it dissolves, and water molecules interact with the Li^+ cations and OH^- anions. Water molecules in pure water have a highly disordered arrangement. However, the attraction between Li^+ ions and water molecules leads to a net ordering of the system when LiOH dissolves.

- If energy and matter are both dispersed in a process, it is spontaneous.
- If only matter is dispersed, then quantitative information is needed to decide whether the process is spontaneous.
- If energy is not dispersed after a process occurs, then that process will never be spontaneous.

19.4—Entropy and the Second Law of Thermodynamics

Entropy is used to quantify the extent of disorder resulting from dispersal of energy and matter. For any substance under a given set of conditions, a numerical value for entropy can be determined. The greater the disorder in a system, the greater the entropy and the larger the value of S.

Like internal energy (E) and enthalpy (H), entropy is a state function. This means that the *change in entropy* for any process depends only on the initial and final states of the system, and not on the pathway by which the process occurs [◀ Section 6.4]. This point is important because it has a bearing on how the value of S is determined, as we will see later.

The point of reference for entropy values is established by the **third law of thermodynamics**. Defined by Ludwig Boltzmann, the third law states that there is no disorder in a perfect crystal at 0 K; that is, $S = 0$. The entropy of an element or compound under any set of conditions is the entropy gained by converting the substance from 0 K to the defined conditions. The entropy of a substance at any temperature can be obtained by measuring the heat required to raise the temperature from 0 K, but with a specific provision that the conversion must be carried out by a reversible process. Adding heat slowly and in very small increments approximates a reversible process. The entropy added by each incremental change is calculated using Equation 19.1:

$$\Delta S = \frac{q_{rev}}{T} \tag{19.1}$$

In this equation, q_{rev} is the heat absorbed and T is the Kelvin temperature at which the change occurs. Adding the entropy changes for the incremental steps gives the total entropy of a substance. (Because it is necessary to add heat to raise the temperature, *all substances have positive entropy values at temperatures above 0 K*. Based on the third law of thermodynamics, negative values of entropy cannot occur.)

The standard entropy, $S°$, of a substance, is the entropy gained by converting it from a perfect crystal at 0 K to standard state conditions (1 bar, 1 molal solution). The units for entropy are J/K · mol. A few values of $S°$ are given in Table 19.1. Generally, values of $S°$ found in tables of data refer to a temperature of 298 K.

Some interesting and useful generalizations can be drawn from the data given in Table 19.1 (and Appendix L).

- *When comparing the same or similar substances, entropies of gases are much larger than those for liquids, and entropies of liquids are larger than those for solids.* In a solid the particles have fixed positions in the solid lattice. When a solid melts, its particles have more freedom to assume different positions, resulting in an increase in disorder (Figure 19.9). When a liquid evaporates, restrictions due to forces between the particles nearly disappear, and another large entropy increase occurs. For example, the standard entropies of $I_2(s)$, $Br_2(\ell)$, and $Cl_2(g)$ are 116.1, 152.2, and 223.1 J/K · mol, respectively, and the standard entropies of C(s, graphite) and C(g) are 5.6 and 158.1 J/K · mol, respectively.

$S°$ (J/K · mol)

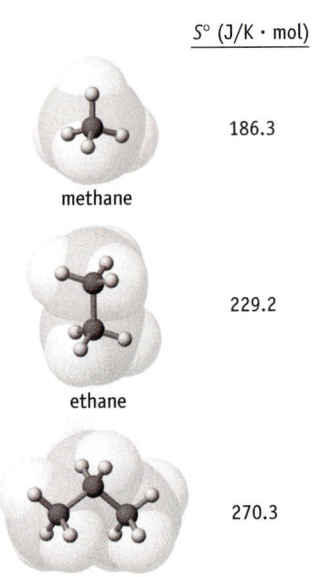

methane   186.3

ethane   229.2

propane   270.3

Table 19.1   Some Standard Molar Entropy Values at 298 K

Element	Entropy, $S°$ (J/K mol)	Compound	Entropy, $S°$ (J/K mol)
C(graphite)	5.6	$CH_4(g)$	186.3
C(diamond)	2.377	$C_2H_6(g)$	229.2
C(vapor)	158.1	$C_3H_8(g)$	270.3
Ca(s)	41.59	$CH_3OH(\ell)$	127.2
Ar(g)	154.9	$CO(g)$	197.7
$H_2(g)$	130.7	$CO_2(g)$	213.7
$O_2(g)$	205.1	$H_2O(g)$	188.84
$N_2(g)$	191.6	$H_2O(\ell)$	69.95
$F_2(g)$	202.8	$HCl(g)$	186.2
$Cl_2(g)$	223.1	NaCl(s)	72.11
$Br_2(\ell)$	152.2	MgO(s)	26.85
$I_2(s)$	116.1	$CaCO_3(s)$	91.7

■ **Entropy Values**
A longer list of entropy values appears in Appendix L. Extensive lists of $S°$ values can be found in standard chemical reference sources such as the NIST tables (**http://webbook.nist.gov**).

- As a general rule, *larger molecules have a larger entropy than smaller molecules, and molecules with more complex structures have larger entropies than simpler molecules.* These generalizations work best in series of related compounds. With a more complicated molecule, there are more ways for the molecule to rotate, twist, and vibrate in space (which provides a larger number of internal energy states over which energy can be distributed). Entropies for methane (CH_4), ethane (C_2H_6), and propane (C_3H_8) are 186.3, 229.2, and 270.3 J/K · mol, respectively. The effect of molecular structure can be seen with atoms or molecules of similar molar mass: Ar, CO_2, and C_3H_8 have entropies of 154.9, 213.7, and 270.3 J/K · mol, respectively.

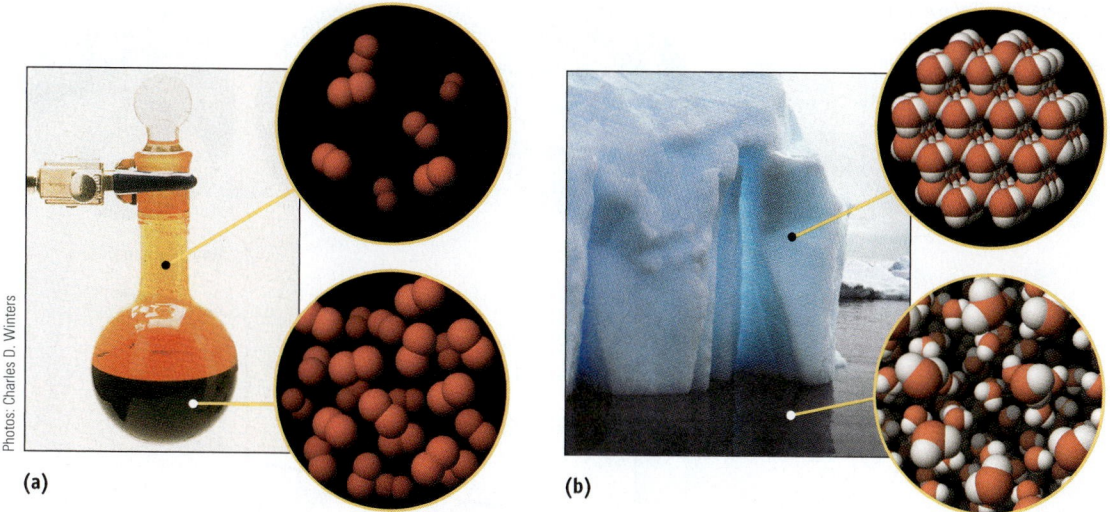

Photos: Charles D. Winters

(a)   (b)

Figure 19.9   Entropy and states of matter. (a) The entropy of liquid bromine, $Br_2(\ell)$, is 152.2 J/K · mol, and that for more disordered bromine vapor is 245.47 J/K · mol. (b) The entropy of ice, which has a highly ordered molecular arrangement, is smaller than that for disordered liquid water. (*See General ChemistryNow Screen 19.4 Entropy, to view animations of the relationship of entropy and molecular state or molecular properties.*)

A Closer Look

Reversible and Irreversible Processes

To determine the entropy change experimentally, the heat transfer must be measured for a reversible process. But what is a reversible process, and why is this constraint important in this discussion?

The melting of ice/freezing of water at 0 °C is an example of a reversible process. Given a mixture of ice and water at equilibrium, adding heat in small increments will convert ice to water; removing heat in small increments will convert water back to ice. The test for reversibility is that after carrying out a change along a given path (in this instance, heat added), it must be possible to return to the starting point by

the same path (heat taken away) without altering the surroundings.

Reversibility is closely associated with equilibrium. Assume that we have a system at equilibrium. Changes can then be made by slightly perturbing the equilibrium and letting the system readjust. Melting (or freezing) water is carried out by adding (or removing) heat in small increments.

Spontaneous processes are not reversible: The process occurs in one direction. Suppose a gas is allowed to expand into a vacuum, which is clearly a spontaneous process (Figure 19.1). No work is done in this process because no force resists this expansion. To return the system to its original state, it will be necessary to compress the gas, but doing so means doing work on the system. In this process, the

energy content of the surroundings will decrease by the amount of work expended by the surroundings. The system can be restored to its original state, but the surroundings will be altered in the process.

In summary, there are two important points concerning reversibility:

- At every step along a reversible pathway between two states, the system remains at equilibrium.
- Spontaneous processes follow irreversible pathways and involve non-equilibrium conditions.

To determine the entropy change for a process, it is necessary to identify a reversible pathway. Only then can an entropy change for the process be calculated from the measured heat change for the process, q_{rev}, and the temperature at which it occurs.

- For a given substance, entropy increases as the temperature is raised. Large increases in entropy accompany changes of state (Figure 19.10).

GENERAL
Chemistry ⚛ Now™

See the General ChemistryNow CD-ROM or website:
- **Screen 19.4 Entropy: Matter Dispersal and Disorder,** to view animations of the effects of molecular properties on entropy

Figure 19.10 Entropy and temperature. For each of the three states of matter, entropy increases with increasing temperature. An especially large increase in entropy accompanies a phase change from a solid to a liquid to a gas.

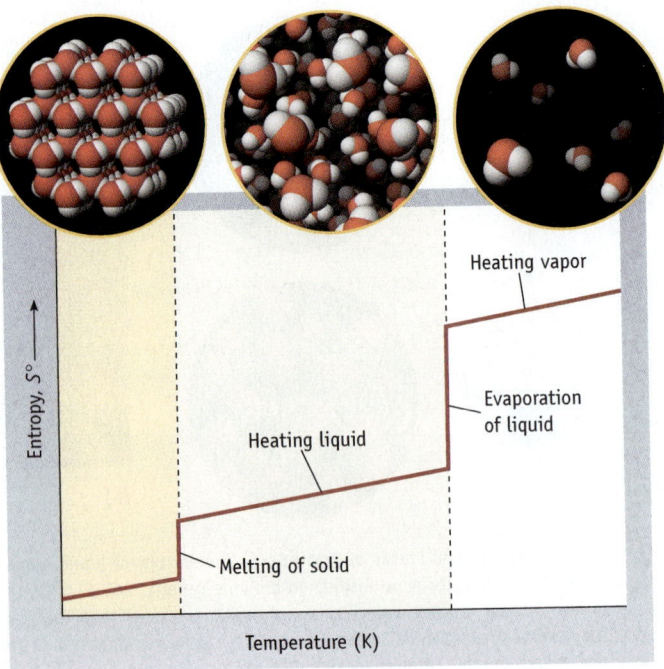

Problem-Solving Tip 19.2

Summary of Common Entropy-Favored Processes

The discussion to this point, along with the examples and exercises, allows the listing of several general principles involving entropy changes:

- A substance becomes increasingly disordered going from a solid to a liquid to a gas. Significant increases in entropy correspond to these phase changes.

- Entropy of any substance increases with temperature (Figure 19.10). Heat must be added to a system to increase its temperature (that is, $q > 0$), so q_{rev}/T is necessarily positive.

- Entropy of a gas increases with an increase in volume. A larger volume provides a larger number of energy states over which to disperse energy.

- Reactions that increase the number of moles of gases in a system are accompanied by an increase in entropy.

Example 19.1—Entropy Comparisons

Problem Which substance has the higher entropy under standard conditions? Explain your reasoning. Check your answer against data in Appendix L.

(a) $NO_2(g)$ or $N_2O_4(g)$

(b) $I_2(g)$ or $I_2(s)$

Strategy Use the general guidelines on entropy listed in the text: Entropy decreases in the order gas > liquid > solid; larger molecules have greater entropy than smaller molecules.

Solution

(a) Both NO_2 and N_2O_4 are gases. Dinitrogen tetraoxide, N_2O_4, the larger molecule, is expected to have the higher standard entropy. $S°$ values (Appendix L) confirm this prediction: $S°$ for $NO_2(g)$ is 240.04 J/K · mol; $S°$ for $N_2O_4(g)$ is 304.38 J/K · mol.

(b) Gases have higher entropies than solids. $S°$ for $I_2(g)$ is 260.69 J/K · mol; $S°$ for $I_2(s)$ is 116.135 J/K · mol

Comment A prediction is a useful check on your numerical result. If an error is inadvertently made, then the prediction will alert you to reconsider your work.

Exercise 19.1—Entropy Comparisons

Predict which substance has the higher entropy and explain your reasoning.

(a) $O_2(g)$ or $O_3(g)$
(b) $SnCl_4(\ell)$ or $SnCl_4(g)$

Entropy Changes in Physical and Chemical Processes

The entropy changes ($\Delta S°$) for chemical and physical changes under standard conditions can be calculated from values of $S°$. The procedure used to calculate $\Delta S°$ is similar to that used to obtain values of $\Delta H°$ [◀ Equation 6.6]. The entropy change is the sum of the entropies of the products minus the sum of the entropies of reactants:

$$\Delta S°_{system} = \sum S°(products) - \sum S°(reactants) \qquad (19.2)$$

To illustrate, let us calculate $\Delta S°_{rxn}$ ($=\Delta S°_{system}$) for the oxidation of NO with O_2.

$$2\ NO(g) + O_2(g) \longrightarrow 2\ NO_2(g)$$

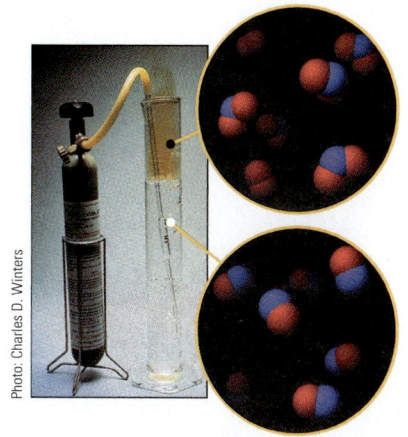

Photo: Charles D. Winters

The reaction of NO with O_2. The entropy of the system decreases when two molecules of gas are produced from three molecules of gaseous reactants.

Here we subtract the entropies of the reactants (2 mol NO and 1 mol O_2) from the entropy of the products (2 mol NO_2).

$$\Delta S^\circ_{rxn} = (2 \text{ mol } NO_2)(240.0 \text{ J/K} \cdot \text{mol})$$
$$- [(2 \text{ mol } NO)(210.8 \text{ J/K mol}) + (1 \text{ mol } O_2)(205.1 \text{ J/K} \cdot \text{mol})]$$
$$= -146.7 \text{ J/K}$$

or -73.35 J/K for 1 mol of NO_2 formed. Notice that the entropy of the system decreases, as predicted for a reaction that converts three molecules of gas into two molecules of another gas.

GENERAL
Chemistry ⚛ Now™

See the General ChemistryNow CD-ROM or website:

- **Screen 19.5 Calculating ΔS° for a Chemical Reaction,** for a tutorial on entropy calculations

---

Example 19.2—Predicting and Calculating ΔS°

Problem Calculate the standard entropy changes for the following processes. Do the calculations match prediction?

(a) Evaporation of 1.00 mol of liquid ethanol to ethanol vapor

$$C_2H_5OH(\ell) \longrightarrow C_2H_5OH(g)$$

(b) Oxidation of 1 mol of ethanol vapor

$$C_2H_5OH(g) + 3 O_2(g) \longrightarrow 2 CO_2(g) + 3 H_2O(g)$$

Strategy Entropy changes are calculated from values of standard entropies (Appendix L) using Equation 19.2. Predictions are made using the guidelines given in the text: Entropy increases going from solid to liquid to gas (see Figure 19.10), and entropy increases in a chemical reaction if there is an increase in the number of moles of gases in the system.

Solution

(a) Evaporating ethanol

$$\Delta S^\circ = \sum S^\circ(\text{products}) - \sum S^\circ(\text{reactants})$$
$$= S^\circ[C_2H_5OH(g)] - S^\circ[C_2H_5OH(\ell)]$$
$$= 1 \text{ mol } (282.70 \text{ J/K} \cdot \text{mol}) - 1 \text{ mol } (160.7 \text{ J/K} \cdot \text{mol})$$
$$= +122.0 \text{ J/K}$$

The large positive value for the entropy change is expected because the process converts ethanol from a more ordered state (liquid) to a less ordered state (vapor).

(b) Oxidation of ethanol vapor

$$\Delta S^\circ = 2 S^\circ[CO_2(g)] + 3 S^\circ[H_2O(g)] - \{S^\circ[C_2H_5OH(g)] + 3 S^\circ[O_2(g)]\}$$
$$= 2 \text{ mol } (213.74 \text{ J/K} \cdot \text{mol}) + 3 \text{ mol } (188.84 \text{ J/K} \cdot \text{mol})$$
$$- \{1 \text{ mol } (282.70 \text{ J/K} \cdot \text{mol}) + 3 \text{ mol } (205.07 \text{ J/K} \cdot \text{mol})\}$$
$$= +96.09 \text{ J/K}$$

An increase in entropy is predicted for this reaction because the number of moles of gases increases from four to five.

Comment Values of entropies in tables are based on 1 mol of the compound. In part (b), the number of moles of reactants and products is defined by the stoichiometric coefficients in the balanced chemical equation.

Exercise 19.2—Calculating $\Delta S°$

Calculate the standard entropy changes for the following processes using the entropy values in Appendix L. Do the calculated values of $\Delta S°$ match predictions?

(a) Dissolving 1 mol of $NH_4Cl(s)$ in water: $NH_4Cl(s) \longrightarrow NH_4Cl(aq)$

(b) Forming 2.0 mol of $NH_3(g)$ from $N_2(g)$ and $H_2(g)$: $N_2(g) + 3\ H_2(g) \longrightarrow 2\ NH_3(g)$

19.5—Entropy Changes and Spontaneity

Which processes are spontaneous, and how is spontaneity predicted? Entropy is the basis for this determination. But how can it be useful when, as you have seen earlier, the entropy of a system may either decrease (oxidation of NO) or increase (evaporation and oxidation of ethanol). The second law of thermodynamics provides an answer. The **second law of thermodynamics** states that *a spontaneous process is one that results in an increase of entropy in the universe.* This criterion requires assessing entropy changes in both the system under study and the surroundings.

At first glance, the statement of the second law may seem curious. Ordinarily our thinking is not so expansive as to consider the whole universe; we think mainly about a given system. Recall that the term "system" is defined as "that part of the universe being studied." As we continue, however, it will become apparent that this view is not so complicated as it might first sound.

The "universe" (= univ) has two parts: the system (= sys) and its surroundings (= surr) [◀ Section 6.1]. The entropy change for the universe is the sum of the entropy changes for the system and the surroundings:

$$\Delta S_{univ} = \Delta S_{sys} + \Delta S_{surr} \tag{19.3}$$

The second law states that ΔS_{univ} is positive for a spontaneous process. Conversely, a negative value of ΔS_{univ} means the process cannot be spontaneous as written. If $\Delta S_{univ} = 0$, the system is at equilibrium.

A similar equation can be written for the entropy change for a process under standard conditions:

$$\Delta S°_{univ} = \Delta S°_{sys} + \Delta S°_{surr} \tag{19.4}$$

The value of $\Delta S°_{univ}$ represents the entropy change for a process in which all of the reactants and products are in their standard states.

The standard entropy change for a chemical reaction can be calculated from standard entropy values found in tables such as Appendix L. The value of $\Delta S°_{univ}$ calculated in this way is the entropy change when reactants are converted *completely* to products, with all species at standard conditions.

As an example of the calculation of $\Delta S°_{univ}$, consider the reaction currently used to manufacture methanol, CH_3OH.

$$CO(g) + 2\ H_2(g) \longrightarrow CH_3OH(\ell)$$

■ **Spontaneity and the Second Law**
Spontaneous change is always accompanied by an increase in entropy in the universe. This is in contrast to enthalpy and internal energy. According to the first law, the energy contained in the universe is constant.

■ **Using $\Delta S°$ universe**
For a spontaneous process:

$\Delta S°_{univ} > 0$

For a system at equilibrium:

$\Delta S°_{univ} = 0$

For a nonspontaneous process:

$\Delta S°_{univ} < 0$

If the change in entropy for the universe is positive, the conversion of 1 mol of CO and 2 mol of H_2 to 1 mol of CH_3OH will be spontaneous under standard conditions.

Calculating ΔS°_{sys}, the Entropy Change for the System

To calculate ΔS°_{sys}, we start by defining the system to include the reactants and products. This means that dispersal of matter in this process occurs entirely within the system. That is, to evaluate the entropy change for matter dispersal, we need to look only at ΔS°_{sys}. Calculation of the entropy change follows the procedure given in Example 19.2.

$$\Delta S^\circ_{sys} = \sum S^\circ(\text{products}) - \sum S^\circ(\text{reactants})$$
$$= S^\circ[CH_3OH(\ell)] - \{S^\circ[CO(g)] + 2\ S^\circ[H_2(g)]\}$$
$$= (1\ \text{mol})(127.2\ \text{J/K} \cdot \text{mol}) - \{(1\ \text{mol})(197.7\ \text{J/K mol}) + (2\ \text{mol})(130.7\ \text{J/K} \cdot \text{mol})\}$$
$$= -331.9\ \text{J/K}$$

A decrease in entropy for the system is expected because three moles of gaseous reactants are converted to one mole of a liquid product.

Calculating ΔS°_{surr}, the Entropy Change for the Surroundings

The entropy change resulting from the dispersal of energy produced in this exothermic chemical reaction is evaluated from the enthalpy change for the reaction. The heat evolved in this reaction is transferred to the surroundings, so $q_{surr} = -\Delta H^\circ_{sys}$, and

$$\Delta S^\circ_{surr} = \frac{q_{surr}}{T} = -\frac{\Delta H^\circ_{sys}}{T}$$

According to this equation, an exothermic reaction ($\Delta H^\circ_{sys} < 0$) is accompanied by an increase in entropy in the surroundings. For the synthesis of methanol, the enthalpy change is -127.9 kJ.

$$\Delta H^\circ_{sys} = \sum \Delta H^\circ_f(\text{products}) - \sum \Delta H^\circ_f(\text{reactants})$$
$$= \Delta H^\circ_f[CH_3OH(\ell)] - \{\Delta H^\circ_f[CO(g)] + 2\ \Delta H^\circ_f[H_2(g)]\}$$
$$= (1\ \text{mol})(-238.4\ \text{kJ/mol}) - \{(1\ \text{mol})(-110.5\ \text{kJ/mol}) + (2\ \text{mol})(0)\}$$
$$= -127.9\ \text{kJ}$$

Therefore, if we make the simplifying assumption that the process is reversible and occurs at a constant temperature, the entropy change for the surroundings in the methanol synthesis is $+429.2$ J/K.

$$\Delta S^\circ_{surr} = -\frac{\Delta H^\circ_{sys}}{T} = -\frac{-127.9\ \text{kJ}}{298\ \text{K}}(1000\ \text{J/kJ}) = +429.2\ \text{J/K}$$

Calculating ΔS°_{univ}, the Total Entropy Change for the System and Surroundings

The pieces are now in place to calculate the entropy change in the universe. For the formation of $CH_3OH(\ell)$ from $CO(g)$ and $H_2(g)$, ΔS°_{univ} is

$$\Delta S^\circ_{univ} = \Delta S^\circ_{sys} + \Delta S^\circ_{surr}$$
$$= -331.9\ \text{J/K} + 429.2\ \text{J/K}$$
$$= +97.3\ \text{J/K}$$

The positive value indicates an increase in the entropy of the universe. It follows from the second law of thermodynamics that this reaction is spontaneous.

GENERAL
Chemistry Now™

See the General ChemistryNow CD-ROM or website:
- **Screen 19.6 The Second Law of Thermodynamics,** for a simulation and tutorial to explore entropy changes in reactions

---

Example 19.3—Determining whether a Process Is Spontaneous

Problem Show that ΔS°_{univ} is positive (> 0) for dissolving NaCl in water.

Strategy The problem is divided into two parts: determining ΔS°_{sys} and determining ΔS°_{surr} (calculated from ΔH° for the process). The sum of these two entropy changes is ΔS°_{univ}. Values of ΔH°_f and S° for NaCl(s) and NaCl(aq) are obtained from Appendix L.

Solution The process occurring is NaCl(s) $\longrightarrow$ NaCl(aq). Its entropy change, ΔS°_{sys}, can be calculated from values of S° for the two species using Equation 19.2. The calculation is based on 1 mol of NaCl.

$$\Delta S^\circ_{sys} = S^\circ[NaCl(aq)] - S^\circ[NaCl(s)]$$
$$= (1\ mol)(115.5\ J/K \cdot mol) - (1\ mol)(72.11\ J/K \cdot mol)$$
$$= +43.4\ J/K$$

The heat of solution is determined from values of ΔH°_f for solid and aqueous sodium chloride. The solution process is slightly endothermic, indicating heat transfer occurs from the surroundings to the system:

$$\Delta H^\circ_{sys} = \Delta H^\circ_f[NaCl(aq)] - \Delta H^\circ_f[NaCl(s)]$$
$$= (1\ mol)(-407.27\ kJ/mol) - (1\ mol)(-411.12\ kJ/mol)$$
$$= +3.85\ kJ$$

The enthalpy change of the surroundings has the same numerical value but is opposite in sign:
$$q_{surr} = -\Delta H^\circ_{sys} = -3.85\ kJ.$$

The entropy change of the surroundings is determined by dividing q_{surr} by the Kelvin temperature.

$$\Delta S^\circ_{surr} = \frac{q_{surr}}{T} = \frac{-3.85\ kJ}{298\ K}(1000\ J/kJ) = -12.9\ J/K$$

We see that ΔS°_{surr} has a negative sign. Heat was transferred from the surroundings to the system in this endothermic process.

The overall entropy change—the change of entropy in the universe—is the sum of the values for the system and the surroundings.

$$\Delta S^\circ_{univ} = \Delta S^\circ_{sys} + \Delta S^\circ_{surr} = (+43.4\ J/K) + (-12.9\ J/K) = 30.5\ J/K$$

Comment The sum of the two entropy quantities is positive, indicating that, overall, entropy in the universe increases. Dissolving NaCl is spontaneous. Notice that the process is favored based on dispersal of matter ($\Delta S^\circ_{sys} > 0$) and disfavored based on dispersal of energy ($\Delta S^\circ_{surr} < 0$).

In Summary: Spontaneous or Not?

In the preceding examples, predictions were made using values of ΔS°_{sys} and ΔH°_{sys} calculated from tables of thermodynamic data. It will be useful to look at the possibilities that result from the interplay of these two quantities. There are four possible

Table 19.2   Predicting Whether a Process Will Be Spontaneous

Type	ΔH°_{sys}	ΔS°_{sys}	Spontaneous Process?
1	Exothermic process	Less order	Spontaneous under all conditions
	$\Delta H^\circ_{sys} < 0$	$\Delta S^\circ_{sys} > 0$	$\Delta S^\circ_{univ} > 0$.
2	Exothermic process	more order	Depends on relative magnitudes of ΔH and ΔS.
	$\Delta H^\circ_{sys} < 0$	$\Delta S^\circ_{sys} < 0$	More favorable at *lower* temperatures.
3	Endothermic process	Less order	Depends on relative magnitudes of ΔH and ΔS.
	$\Delta H^\circ_{sys} > 0$	$\Delta S^\circ_{sys} > 0$	More favorable at *higher* temperatures.
4	Endothermic process	More order	Not spontaneous under any conditions
	$\Delta H^\circ_{sys} > 0$	$\Delta S^\circ_{sys} < 0$	$\Delta S^\circ_{univ} < 0$

outcomes when these two quantities are matched (Table 19.2). In two, ΔH°_{sys} and ΔS°_{sys} work in concert (Types 1 and 4 in Table 19.2). In the other two, the two quantities are opposed (Types 2 and 3).

Processes in which both enthalpy and entropy favor energy dispersal (Type 1) are always spontaneous. Processes disfavored by both enthalpy and entropy (Type 4) can never be spontaneous. Let us consider examples that illustrate each situation.

Combustion reactions are always exothermic and often produce a larger number of product molecules from a few reactant molecules. They are Type 1 reactions. The equation for the combustion of butane is an example.

$$2\ C_4H_{10}(g) + 13\ O_2(g) \longrightarrow 8\ CO_2(g) + 10\ H_2O(g)$$

For this reaction $\Delta H^\circ = -5315.1$ kJ and $\Delta S^\circ = 312.4$ J/K. Both values indicate that this reaction, like all combustion reactions, is spontaneous.

Hydrazine, N_2H_4, is used as a high-energy rocket fuel. Synthesis of N_2H_4 from gaseous N_2 and H_2 would be attractive because these reactants are inexpensive.

$$N_2(g) + 2\ H_2(g) \longrightarrow N_2H_4(\ell)$$

However, this reaction fits into the Type 4 category. The reaction is endothermic ($\Delta H^\circ = +50.63$ kJ/mol), and the entropy change ΔS° is negative ($\Delta S^\circ = -331.4$ J/K (1 mol of liquid is produced from 3 mol of gases).

In the two other possible outcomes, entropy and enthalpy changes oppose each other. A process could be favored by the enthalpy change but disfavored by the entropy change (Type 2), or vice versa (Type 3). In either instance, whether a process is spontaneous depends on which factor is more important.

Temperature also influences the value of ΔS°_{univ} because ΔS°_{surr} (the dispersal of energy) varies with temperature. Because the enthalpy change for the surroundings is divided by the temperature to obtain ΔS°_{surr}, the numerical value of ΔS°_{surr} is smaller (either less positive or less negative) at higher temperatures. In contrast, ΔS°_{sys} does not depend on temperature. Thus, the effect of ΔS°_{surr} relative to ΔS°_{sys} diminishes at higher temperature. Stated another way, at higher temperatures the enthalpy change becomes less of a factor relative to the entropy change. Consider the two cases where ΔH°_{sys} and ΔS°_{sys} are in opposition (Table 19.2):

- Type 2: *Exothermic processes that are entropy-disfavored.* Such processes become less favorable with an increase in temperature.

- Type 3: *Endothermic processes that are entropy-favored.* These processes become more favorable as the temperature increases.

The effect of temperature is illustrated by two examples. The first is the reaction of N_2 and H_2 to form NH_3, one of the most important industrial chemical processes. The reaction is exothermic; that is, it is favored by energy dispersal. The entropy change for the system is unfavorable, however, because the reaction, $N_2(g) + 3\ H_2(g) \longrightarrow 2\ NH_3(g)$, converts four moles of gaseous reactants to two moles of gaseous products. To maximize the yield of ammonia, the lowest possible temperature should be used [◄ Section 16.7].

The second example considers the thermal decomposition of NH_4Cl (Figure 19.11). At room temperature, NH_4Cl is a stable, white, crystalline salt. When heated strongly, it decomposes to $NH_3(g)$ and $HCl(g)$. The reaction is endothermic (enthalpy-disfavored) but entropy-favored because of the formation of two moles of gas from one mole of a solid reactant.

Exercise 19.3—Is a Reaction Spontaneous?

Classify the following reactions as one of the four types of reactions summarized in Table 19.2.

Reaction	ΔH°_{rxn} (at 298 K) (kJ)	ΔS°_{sys} (at 298 K) (J/K)
(a) $CH_4(g) + 2\ O_2(g) \longrightarrow 2\ H_2O(\ell) + CO_2(g)$	−890.6	−242.8
(b) $2\ Fe_2O_3(s) + 3\ C(graphite) \longrightarrow 4\ Fe(s) + 3\ CO_2(g)$	+467.9	+560.7
(c) $C(graphite) + O_2(g) \longrightarrow CO_2(g)$	−393.5	+3.1
(d) $N_2(g) + 3\ F_2(g) \longrightarrow 2\ NF_3(g)$	−264.2	−277.8

Exercise 19.4—Is a Reaction Spontaneous?

Is the direct reaction of hydrogen and chlorine to give hydrogen chloride gas predicted to be spontaneous?

$$H_2(g) + Cl_2(g) \longrightarrow 2\ HCl(g)$$

Answer the question by calculating the values for ΔS°_{sys} and ΔS°_{surr} (at 298 K) and then summing them to determine ΔS°_{univ}.

Exercise 19.5—Effect of Temperature on Spontaneity

Iron is produced in a blast furnace by reducing iron oxide using carbon. For this reaction, $2\ Fe_2O_3(s) + 3\ C(graphite) \longrightarrow 4\ Fe(s) + 3\ CO_2(g)$, the following parameters are determined: $\Delta H^\circ_{rxn} = +467.9$ kJ and $\Delta S^\circ_{rxn} = +560.7$ J/K. Show that it is necessary that this reaction be carried out at a high temperature.

19.6—Gibbs Free Energy

The method used so far to determine whether a process is spontaneous required evaluation of two quantities, ΔS°_{sys} and ΔS°_{surr}. Wouldn't it be convenient to have a single thermodynamic function that serves the same purpose? A function associated with a system only—one that does not require assessment of the surroundings—would be even better. In fact, such a function exists. It is called Gibbs free energy, with the name honoring J. Willard Gibbs (1839–1903). **Gibbs free energy**, *G*, often referred to simply as "free energy," is defined mathematically as

$$G = H - TS$$

Figure 19.11 Thermal decomposition of $NH_4Cl(s)$. White, solid ammonium chloride is heated in a spoon. At high temperatures, $NH_4Cl(s)$ decomposes to form $NH_3(g)$ and $HCl(g)$ in a spontaneous reaction. At lower temperatures, the reverse reaction, forming $NH_4Cl(s)$, is spontaneous. As $HCl(g)$ and $NH_3(g)$ above the heated solid cool, they recombine to form solid NH_4Cl, the white "smoke" seen in this photo.

A reaction requiring a higher temperature so that ΔS°_{univ} is positive. Iron is obtained in a blast furnace by heating iron ore and coke (carbon). See Exercise 19.5.

J. Willard Gibbs (1839–1903) Gibbs received a Ph.D. from Yale University in 1863. His was the first Ph.D. in science awarded from an American university.

Burndy Library/courtesy AIP Emilio Segre Visual Archives

where H is enthalpy, T is the Kelvin temperature, and S is entropy. In this equation, G, H, and S all refer to a system. Because enthalpy and entropy are state functions [◀ Section 6.4], free energy is also a state function.

Every substance possesses a specific quantity of free energy. The actual quantity of free energy is seldom known, however, or even of interest. Instead, we are concerned with *changes* in free energy, ΔG, in the course of a chemical or physical process. We do not need to know the free energy of a substance to determine ΔG. In this sense, free energy and enthalpy are similar. A substance possesses some amount of enthalpy, but we do not have to know what the actual value of H is to obtain or use ΔH.

Let us see first how to use free energy as a way to determine whether a reaction is spontaneous. We can then ask further questions about the meaning of the term "free energy" and its use in deciding whether a reaction is product- or reactant-favored.

$\Delta G°$ and Spontaneity

Recall the equation defining the entropy change for the universe (Equation 19.3):

$$\Delta S_{univ} = \Delta S_{surr} + \Delta S_{sys}$$

On page 918 we noted that the entropy change of the surroundings equals the negative of the change in enthalpy of the system divided by T. Thus

$$\Delta S_{univ} = (-\Delta H_{sys}/T) + \Delta S_{sys}$$

Multiplying through this equation by $-T$, we have

$$-T\Delta S_{univ} = \Delta H_{sys} - T\Delta S_{sys}$$

Gibbs defined the free energy function so that

$$\Delta G_{sys} = -T\Delta S_{univ}$$

Therefore, we have

$$\Delta G_{sys} = \Delta H_{sys} - T\Delta S_{sys}$$

The connection between ΔG_{rxn} ($= \Delta G_{sys}$) and spontaneity is the following:

- If $\Delta G_{rxn} < 0$, a reaction is spontaneous.
- If $\Delta G_{rxn} = 0$, the reaction is at equilibrium.
- If $\Delta G_{rxn} > 0$, the reaction is not spontaneous.

The free energy change can also be defined under standard conditions.

$$\Delta G°_{sys} = \Delta H°_{sys} - T\Delta S°_{sys} \qquad (19.5)$$

$\Delta G°_{sys}$ is generally used as a criterion of reaction spontaneity, and, as you shall see, *it is directly related to the value of the equilibrium constant and hence to product favorability.*

What Is "Free" Energy?

The term "free energy" was not arbitrarily chosen. In any given process, *the free energy represents the maximum energy available to do useful work* (or, mathematically, $\Delta G = w_{max}$). In this context, the word "free" means "available."

To illustrate the reasoning behind this relationship, consider a reaction in which heat is evolved ($\Delta H^\circ_{rxn} < 0$) and entropy decreases ($\Delta S^\circ_{rxn} < 0$).

$$C(graphite) + 2\ H_2(g) \longrightarrow CH_4(g)$$

$$\Delta H^\circ_{rxn} = -74.9\ kJ \quad \text{and} \quad \Delta S^\circ_{rxn} = -80.7\ J/K$$

$$\Delta G^\circ_{rxn} = -50.8\ kJ$$

At first glance, it might seem that all the energy available from the reaction could be transferred to the surroundings and would thus be available to do work, but this is not the case. The negative entropy change means that the system is becoming more ordered. A portion of the energy from the reaction was used to create this more ordered system, so this energy is not available to do work. The energy left over is "free," or available, to do work. Here that free energy amounts to 50.8 kJ, as shown in Example 19.4.

The analysis of the enthalpy and entropy changes for the carbon and hydrogen reaction applies to any combination of ΔH° and ΔS°. The "free" energy is the sum of the energies available from dispersal of energy and matter. In a process that is favored by both the enthalpy and entropy changes, the free energy available exceeds what is available based on enthalpy alone.

Calculating ΔG°_{rxn}, the Free Energy Change for a Reaction

Enthalpy and entropy changes at standard conditions can be calculated for chemical reactions using values of ΔH°_f and S° for substances in the reaction. Then, ΔG°_{rxn} ($= \Delta G^\circ_{sys}$) can be found from the resulting values of ΔH°_{rxn} and ΔS°_{rxn} using Equation 19.5, as illustrated in the following example and exercise.

GENERAL
Chemistry ⟵⟶ Now™

See the General ChemistryNow CD-ROM or website:
- **Screen 19.7 Gibbs Free Energy,** for tutorials on calculating ΔG°_{rxn}

Example 19.4—Calculating ΔG°_{rxn} from ΔH°_{rxn} and ΔS°_{rxn}

Problem Calculate the standard free energy change, ΔG°, for the formation of methane at 298 K:

$$C(graphite) + 2\ H_2(g) \longrightarrow CH_4(g)$$

Strategy Values for ΔH°_f and S° are provided in Appendix L. These are first combined to find ΔH°_{rxn} and ΔS°_{rxn}. With these values known, ΔG°_{rxn} can be calculated using Equation 19.5. When doing so, recall that S° values in tables are given in units of J/K · mol, whereas ΔH° values are given in units of kJ/mol.

Solution

	C(graphite)	$H_2(g)$	$CH_4(g)$
ΔH°_f (kJ/mol)	0	0	−74.9
S° (J/K · mol)	+5.6	+130.7	+186.3

From these values, we can find both $\Delta H°$ and $\Delta S°$ for the reaction:

$$\Delta H°_{rxn} = \Delta H°_f[CH_4(g)] - \{\Delta H°_f[C(graphite)] + 2\Delta H°_f[H_2(g)]\}$$

$$= (1 \text{ mol})(-74.9 \text{ kJ/mol}) - (0 + 0)$$

$$= -74.9 \text{ kJ}$$

$$\Delta S°_{rxn} = S°[CH_4(g)] - \{S°[C(graphite)] + 2S°[H_2(g)]\}$$

$$= (1 \text{ mol})186.3 \text{ J/K} \cdot \text{mol} - [(1 \text{ mol})(5.6 \text{ J/K} \cdot \text{mol}) + (2 \text{ mol})(130.7 \text{ J/K} \cdot \text{mol})]$$

$$= -80.7 \text{ J/K}$$

Both the enthalpy change and the entropy change for this reaction are negative. In this case, the reaction is predicted to be spontaneous at "low temperature" (see Table 19.2). These values alone do not tell us whether the temperature is low enough, however. By combining them in the Gibbs free energy equation, and calculating $\Delta G°_{rxn}$ for a temperature of 298 K, we can predict with certainty the outcome of the reaction.

$$\Delta G°_{rxn} = \Delta H°_{rxn} - T\Delta S°_{rxn}$$

$$= -74.9 \text{ kJ} - (298 \text{ K})(-80.7 \text{ J/K})(1 \text{ kJ/1000 J})$$

$$= -74.9 \text{ kJ} - (-24.1 \text{ kJ})$$

$$= -50.8 \text{ kJ}$$

$\Delta G°_{rxn}$ is negative at 298 K, so the reaction is predicted to be spontaneous.

Comment In this case the product $T\Delta S°$ is negative and, because the entropy change is relatively small, $T\Delta S°$ is less negative than $\Delta H°_{rxn}$. Chemists call this situation an "enthalpy-driven reaction" because the exothermic nature of the reaction overcomes the decrease in entropy of the system.

Exercise 19.6—Calculating $\Delta G°_{rxn}$ from $\Delta H°_{rxn}$ and $\Delta S°_{rxn}$

Using values of $\Delta H°_f$ and $S°$ to find $\Delta H°_{rxn}$ and $\Delta S°_{rxn}$, respectively, calculate the free energy change, $\Delta G°$, for the formation of 2 mol of $NH_3(g)$ from the elements at standard conditions (and 25 °C): $N_2(g) + 3 H_2(g) \longrightarrow 2 NH_3(g)$.

Standard Free Energy of Formation

The **standard free energy of formation** of a compound, $\Delta G°_f$, is the free energy change when forming one mole of the compound from the component elements, with products and reactants in their standard states. By defining $\Delta G°_f$ in this way, *the free energy of formation of an element in its standard states is zero* (Table 19.3).

Table 19.3   Standard Molar Free Energies of Formation of Some Substances at 298 K

Element/Compound	$\Delta G°_f$ (kJ/mol)	Element/Compound	$\Delta G°_f$ (kJ/mol)
$H_2(g)$	0	$CO_2(g)$	−394.4
$O_2(g)$	0	$CH_4(g)$	−50.8
$N_2(g)$	0	$H_2O(g)$	−228.6
C(graphite)	0	$H_2O(\ell)$	−237.2
C(diamond)	2.900	$NH_3(g)$	−16.4
CO(g)	−137.2	$Fe_2O_3(s)$	−742.2

Just as the standard enthalpy change for a reaction can be calculated using values of ΔH_f°, the standard free energy change for a reaction can be calculated from values of ΔG_f°:

$$\Delta G^\circ = \sum \Delta G_f^\circ(\text{products}) - \sum \Delta G_f^\circ(\text{reactants}) \qquad (19.6)$$

Example 19.5—Calculating $\Delta G_{\text{rxn}}^\circ$ from ΔG_f°

Problem Calculate the standard free energy change for the combustion of one mole of methane from the standard free energies of formation of the products and reactants.

Strategy Use Equation 19.6 with values obtained from Table 19.3 or Appendix L.

Solution First, write a balanced equation for the reaction. Then, find values of ΔG_f° for each reactant and product (in Appendix L).

$$CH_4(g) + 2\ O_2(g) \longrightarrow 2\ H_2O(g) + CO_2(g)$$

ΔG_f° (kJ/mol) $\quad$ −50.8	0	−228.6	−394.4

Because ΔG_f° values are given for 1 mol of each substance (the units are kJ/mol), each value of ΔG_f° must be multiplied by the number of moles defined by the stoichiometric coefficient in the balanced chemical equation.

$$\Delta G_{\text{rxn}}^\circ = 2\Delta G_f^\circ[H_2O(g)] + \Delta G_f^\circ[CO_2(g)] - \{\Delta G_f^\circ[CH_4(g)] + 2\Delta G_f^\circ[O_2(g)]\}$$

$$= (2\ \text{mol})(-228.6\ \text{kJ/mol}) + (1\ \text{mol})(-394.4\ \text{kJ/mol})$$

$$- [(1\ \text{mol})(-50.8\ \text{kJ/mol}) + (2\ \text{mol})(0\ \text{kJ/mol})]$$

$$= -801.0\ \text{kJ}$$

The large negative value of $\Delta G_{\text{rxn}}^\circ$ indicates that the reaction is spontaneous under standard conditions.

Comment The most common errors made by students in this calculation are (1) ignoring the stoichiometric coefficients in the equation and (2) confusing the signs for the terms when using Equation 19.6.

Exercise 19.7—Calculating $\Delta G_{\text{rxn}}^\circ$ From ΔG_f°

Calculate the standard free energy change for the oxidation of 1.00 mol of $SO_2(g)$ to form $SO_3(g)$.

Free Energy and Temperature

The definition for free energy, $G = H - TS$, states that free energy is a function of temperature, so ΔG will change as the temperature changes (Figure 19.12).

A consequence of this dependence on temperature is that, in certain instances, reactions can be spontaneous at one temperature and not spontaneous at another. Those instances arise when the ΔH° and $T\Delta S^\circ$ terms work in opposite directions:

- Processes that are entropy-favored $(\Delta S^\circ > 0)$ and enthalpy-disfavored $(\Delta H^\circ > 0)$

- Processes that are enthalpy-favored $(\Delta H^\circ < 0)$ and entropy-disfavored $(\Delta S^\circ < 0)$

Let us explore the relationship of ΔG° and T further and illustrate how it can be used to advantage.

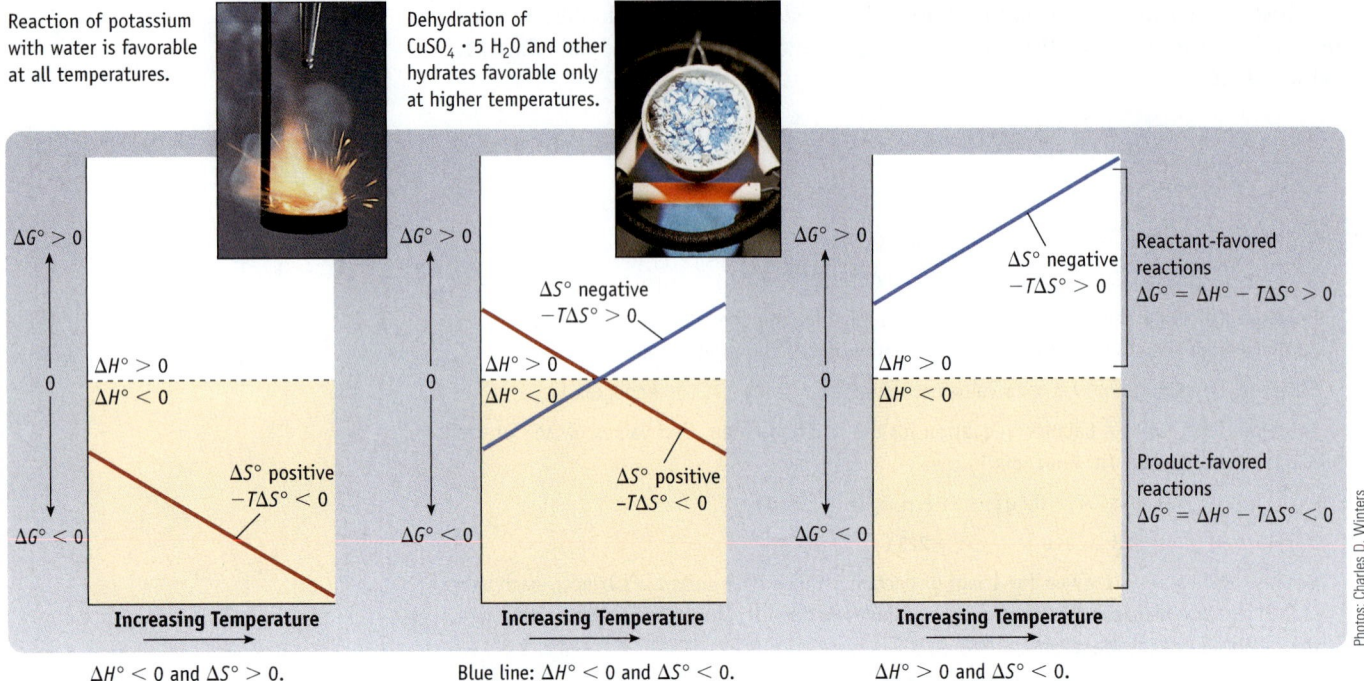

Reaction of potassium with water is favorable at all temperatures.

Dehydration of $CuSO_4 \cdot 5 H_2O$ and other hydrates favorable only at higher temperatures.

$\Delta H^\circ < 0$ and $\Delta S^\circ > 0$. Product-favored at all temperatures.

Blue line: $\Delta H^\circ < 0$ and $\Delta S^\circ < 0$. Favored at low T. Red line: $\Delta H^\circ > 0$ and $\Delta S^\circ > 0$. Favored at high T.

$\Delta H^\circ > 0$ and $\Delta S^\circ < 0$. Reactant-favored at all temperatures.

Active Figure 19.12   Changes in ΔG° with temperature.

GENERAL
Chemistry ·⚛· Now™ See the General ChemistryNow CD-ROM or website to explore an interactive version of this figure accompanied by an exercise.

Calcium carbonate is a common substance. Among other things, it is the primary component of limestone, marble, and seashells. Heating $CaCO_3$ produces lime, CaO, along with gaseous CO_2. The data below are from Appendix L.

	$CaCO_3(s)$	$\longrightarrow$	$CaO(s)$	+	$CO_2(g)$
ΔG_f° (kJ/mol)	−1129.16		−603.42		−394.36
ΔH_f° (kJ/mol)	−1207.6		−635.09		−393.51
S° (J/K · mol)	91.7		38.2		213.74

For the conversion of limestone, $CaCO_3(s)$, to lime, $CaO(s)$, $\Delta G_{rxn}^\circ = +131.38$ kJ, $\Delta H_{rxn}^\circ = +179.0$ kJ, and $\Delta S_{rxn}^\circ = +160.2$ kJ/K. Although the reaction is entropy-favored, the large positive and unfavorable enthalpy change dominates at this temperature. Thus, the free energy change is positive at 298 K and 1 bar, indicating that overall the reaction is not spontaneous under these conditions.

Although the formation of CaO from $CaCO_3$ is unfavorable at 298 K, the temperature dependence of ΔG° provides a means to turn it into a product-favored reaction. Notice that the entropy change in the reaction is positive as a result of the formation of CO_2 gas in the reaction. Thus, raising the temperature results in the value of $T\Delta S^\circ$ becoming increasingly large. At a high enough temperature, the effect of ΔS° will outweigh the enthalpy effect and the process will become favorable.

How high must the temperature be for this reaction to become spontaneous? An estimate of the temperature can be obtained using Equation 19.5. To do this, let us calculate the temperature at which $\Delta G^\circ = 0$. Above that temperature, ΔG° will have a negative value. (Note that in this calculation the enthalpy must be in joules, not kilojoules, to match the units used in the entropy term. Keeping track of the units will show that the temperature will have the units of kelvins.)

$$\Delta G° = \Delta H° - T\Delta S°$$
$$0 = 179.0 \text{ kJ } (1000 \text{ J/K}) - T(160.2 \text{ J/K})$$
$$T = 1117 \text{ K (or 844 °C)}$$

How accurate is this result? As noted earlier, we can obtain only an approximate answer from this calculation. The biggest source of error is the assumption that $\Delta H°$ and $\Delta S°$ do not vary with temperature, which is not strictly true. There is usually a small variation in these values when the temperature changes—not too important if the temperature range is narrow, but potentially creating a problem over wider temperature ranges. As an estimate, however, a temperature in the 850 °C range for this reaction is reasonable. In fact, the pressure of CO_2 in an equilibrium system $[CaCO_3(s) \longrightarrow CaO(s) + CO_2(g); \Delta G° = 0]$ is 1 bar at about 900 °C, very close to our estimated temperature.

GENERAL
Chemistry ·.·.Now™

See the General ChemistryNow CD-ROM or website:

• **Screen 19.8 Free Energy and Temperature,** for a simulation and tutorial on the relationship of $\Delta H°$, $\Delta S°$, and T.

Example 19.6—Effect of Temperature on $\Delta G°$

Problem Use thermodynamic parameters to estimate the boiling point of methanol.

Strategy At the boiling point, liquid and gas exist at equilibrium, and the condition for equilibrium is $\Delta G° = 0$. Values of $\Delta H_f°$ and $S°$ from Appendix L for the process $CH_3OH(\ell) \longrightarrow CH_3OH(g)$, are used to calculate $\Delta H°$ and $\Delta S°$. T is the unknown in Equation 19.5.

Solution Values for $\Delta G_f°$, $\Delta H_f°$, and $S°$ are obtained from Appendix L for CH_3OH liquid and vapor.

	$CH_3OH(\ell)$	$CH_3OH(g)$
$\Delta G_f°$ (kJ/mol)	-166.14	-162.5
$\Delta H_f°$ (kJ/mol)	-238.4	-201.0
$S°$ (J/K · mol)	$+127.19$	$+239.7$

For a process in which 1 mol of liquid is converted to 1 mol of gas, $\Delta G°_{rxn} = +3.6$ kJ, $\Delta H°_{rxn} = +37.4$ kJ, and $\Delta S°_{rxn} = +112.5$ J/K. The process is endothermic (as expected, heat being required to convert a liquid to a gas), and entropy favors the process (the vapor has a higher degree of disorder than the liquid). These two quantities oppose each other. The free energy change per mole for this process under standard conditions is +3.6 kJ. The positive value indicates that this process is not spontaneous under standard conditions (298 K and 1 bar).

If we use the values of $\Delta H°_{rxn}$ and $\Delta S°_{rxn}$, along with the criterion that $\Delta G° = 0$ at equilibrium (at the boiling point), we have

$$\Delta G° = \Delta H° - T\Delta S°$$
$$0 = +37,400 \text{ J} - [T(+112.5 \text{ J/K})]$$
$$T = 332 \text{ K (or 59 °C)}$$

Comment The calculated boiling temperature is close to the observed value of 65.0 °C.

Exercise 19.8—Effect of Temperature on $\Delta G°$

Oxygen was first prepared by Joseph Priestley (1733–1804), by heating HgO. Use the thermodynamic data in Appendix L to estimate the temperature required to decompose HgO(s) into Hg(ℓ) and O_2(g).

19.7—$\Delta G°$, K, and Product Favorability

The terms *product-favored* and *reactant-favored* were introduced in Chapter 5, and in Chapter 16 we described how these terms are related to the values of equilibrium constants. *Reactions for which K is large are product-favored and those for which K is small are reactant-favored.* We now return to this important topic, to relate the value of the equilibrium constant K, and thus the product or reactant favorability, to the standard free energy change, $\Delta G°$, for a chemical reaction.

The standard free energy change for a reaction, $\Delta G°$, is the increase or decrease in free energy as the reactants in their standard states are converted *completely* to the products in their standard states. But complete conversion is not always observed in practice. A product-favored reaction proceeds largely to products, but some reactants may remain when equilibrium is achieved. A reactant-favored reaction proceeds only partially to products before achieving equilibrium.

To discover the relationship of $\Delta G°$ and the equilibrium constant K, let us use Figure 19.13. The free energy of the pure reactants in their standard states is indicated at the left, and the free energy of the pure products in their standard states is at the right. The difference in these values is $\Delta G°$. In this example $\Delta G° = G°_{products} - G°_{reactants}$ has a negative value, and the reaction is product-favored.

When the reactants are mixed in a chemical system, the system will proceed spontaneously to a position of lower free energy, and the system will eventually achieve equilibrium. At any point along the way from the pure reactants to equilibrium, the reactants are not at standard conditions. The change in free energy under these nonstandard conditions, ΔG, is related to $\Delta G°$ by the equation

$$\Delta G = \Delta G° + RT \ln Q \tag{19.7}$$

where R is the universal gas constant, T is the temperature in kelvins, and Q is the reaction quotient [◀ Section 16.2]. Recall that, for the general reaction of A and B giving products C and D,

$$a\,A + b\,B \longrightarrow c\,C + d\,D$$

the reaction quotient, Q, is

$$Q = \frac{[C]^c[D]^d}{[A]^a[B]^b}$$

Equation 19.7 informs us that, at a given temperature, the difference in free energy between that of the pure reactants and that of a mixture of reactants and products is determined by the values of $\Delta G°$ and Q. Further, as long as ΔG is negative—that is, the reaction is "descending" from the free energy of the pure reactants to the equilibrium position—the reaction is spontaneous.

Eventually the system reaches equilibrium. Because no further change in concentration of reactants and products is seen at this point, ΔG must be zero; that is,

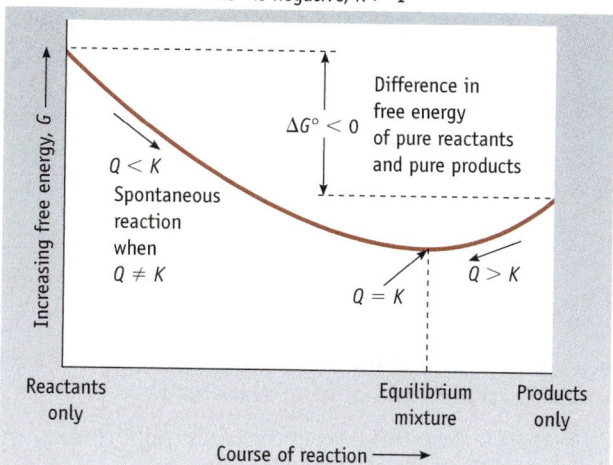

Reaction is product-favored
ΔG° is negative, K > 1

Active Figure 19.13 **Free energy changes as a reaction approaches equilibrium.** Equilibrium represents a minimum in the free energy for a system. The reaction portrayed here has ΔG° < 0, K > 1, and is product-favored.

GENERAL
Chemistry·₊·Now™ See the General ChemistryNow CD-ROM or website to explore an interactive version of this figure accompanied by an exercise.

there is no further change in free energy in the system. Substituting $\Delta G = 0$ and $Q = K$ into Equation 19.7 gives

$$0 = \Delta G° + RT \ln K \text{ (at equilibrium)}$$

Rearranging this equation leads to a useful relationship between the standard free energy change for a reaction and the equilibrium constant, K:

$$\Delta G°_{rxn} = -RT \ln K \qquad (19.8)$$

From Equation 19.8 we learn that, *when $\Delta G°_{rxn}$ is negative, K must be greater than 1, and the reaction is product-favored.* The more negative the value of $\Delta G°$, the larger the equilibrium constant. This makes sense because, as described in Chapter 16, large equilibrium constants are associated with product-favored reactions. *For reactant-favored reactions, $\Delta G°$ is positive and K is less than 1.* If $K = 1$, then $\Delta G° = 0$. (This is a rare situation, requiring that $[C]^c[D]^d/[A]^a[B]^b = 1$ for the reaction $a\,A + b\,B \longrightarrow c\,C + d\,D$.)

Free Energy, the Reaction Quotient, and the Equilibrium Constant

Let us summarize the relationships among $\Delta G°$, ΔG, Q, and K.

- The solid line in Figure 19.13 shows how free energy decreases to a minimum as a system approaches equilibrium. *The free energy at equilibrium is lower than the free energy of the pure reactants and of the pure products.*

Charles D. Winters

Spontaneous but not product-favored. If a sample of yellow lead iodide is placed in pure water, the compound will begin to dissolve. The dissolving process will be spontaneous ($\Delta G < 0$) until equilibrium is reached. However, because PbI_2 is insoluble, with $K_{sp} = 9.8 \times 10^{-9}$, the process of dissolving the compound is strongly reactant-favored, and the value of $\Delta G°$ is positive.

- ΔG°_{rxn} gives the position of equilibrium and may be calculated from

$$\Delta G^{\circ}_{rxn} = \sum \Delta G^{\circ}_f(\text{products}) - \sum \Delta G^{\circ}_f(\text{reactants})$$
$$= \Delta H^{\circ}_{rxn} - T\Delta S^{\circ}_{rxn}$$
$$= -RT \ln K$$

- ΔG_{rxn} describes the direction in which a reaction proceeds to reach equilibrium and may be calculated from

$$\Delta G_{rxn} = \Delta G^{\circ}_{rxn} + RT \ln Q$$

When $\Delta G_{rxn} < 0$, $Q < K$ and the reaction proceeds spontaneously to convert reactants to products until equilibrium is attained.

When $\Delta G_{rxn} > 0$, $Q > K$ and the reaction proceeds spontaneously to convert products to reactants until equilibrium is attained.

When $\Delta G_{rxn} = 0$, $Q = K$, and the reaction is at equilibrium.

Using the Relationship Between ΔG°_{rxn} and K

Equation 19.8 provides a direct route to determine the standard free energy change from experimentally determined equilibrium constants. Alternatively, it allows calculation of an equilibrium constant from thermochemical data contained in tables or obtained from an experiment.

GENERAL
Chemistry ⚛ Now™

See the General ChemistryNow CD-ROM or website:
- **Screen 19.9 Thermodynamics and the Equilibrium Constant,** for a simulation and tutorial on $\Delta G°$ and K

Example 19.7—Calculating K_p from ΔG°_{rxn}

Problem Determine the standard free energy change, ΔG°_{rxn}, for the formation of 1.00 mol of ammonia from nitrogen and hydrogen, and use this value to calculate the equilibrium constant for this reaction at 25 °C.

Strategy The free energy of formation of ammonia represents the free energy change to form 1.00 mol of $NH_3(g)$ from the elements. The equilibrium constant for this reaction is calculated from $\Delta G°$ using Equation 19.8. Because the reactants and products are gases, the calculated value will be K_p.

Solution Begin by specifying a balanced equation for the chemical reaction under investigation.

$$\tfrac{1}{2} N_2(g) + \tfrac{3}{2} H_2(g) \longrightarrow NH_3(g)$$

The free energy change for this reaction is -16.37 kJ/mol ($\Delta G^{\circ}_{rxn} = \Delta G^{\circ}_f$ for $NH_3(g)$; Appendix L). In a calculation of K_p using Equation 19.8, we will need consistent units. The gas constant, R, is 8.3145 J/K · mol, so the value of $\Delta G°$ must be in joules/mol (not kilojoules/mol). The temperature is 298 K.

$$\Delta G° = -RT \ln K$$
$$-16{,}370 \text{ J/mol} = -(8.3145 \text{ J/K} \cdot \text{mol})(298.15 \text{ K}) \ln K_p$$
$$\ln K_p = 6.604$$
$$\boxed{K_p = 7.38 \times 10^2}$$

Comment As illustrated by this example, it is possible to calculate equilibrium constants from thermodynamic data. This gives us another use for these valuable tables. For the reaction of N_2 and H_2 to form NH_3, at 298 K (25 °C), the value of the equilibrium constant is quite large, indicating that the reaction is product-favored.

Example 19.8—Calculating ΔG°_{rxn} from K_c

Problem The value of K_{sp} for AgCl(s) at 25 °C is 1.8×10^{-10}. Use this value in Equation 19.8 to determine ΔG° for the process $Ag^+(aq) + Cl^-(aq) \longrightarrow AgCl(s)$ at 25 °C.

Strategy The chemical equation given is the opposite of the equation used to define K_{sp}; therefore, its equilibrium constant is $1/K_{sp}$. This value is used to calculate ΔG°.

Solution For $Ag^+(aq) + Cl^-(aq) \longrightarrow AgCl(s)$,

$$K = \frac{1}{K_{sp}} = \frac{1}{1.8 \times 10^{-10}} = 5.6 \times 10^9$$

$$\Delta G^\circ = -RT \ln K = -(8.3145 \text{ J/K} \cdot \text{mol})(298.15 \text{ K}) \ln (5.6 \times 10^9)$$

$$= -56 \text{ kJ/mol (to two significant figures)}$$

Comment The negative value of ΔG° indicates that the precipitation of AgCl from $Ag^+(aq)$ and $Cl^-(aq)$ is a product-favored process. Earlier we described the experimental determination of ΔG° values from thermochemical measurements—that is, determining ΔH and ΔS values from the measurement of heat gain or loss. Here is a second method to determine ΔG°.

Exercise 19.9—Calculating K_p from ΔG°_{rxn}

Determine the value of ΔG°_{rxn} for the reaction $C(s) + CO_2(g) \longrightarrow 2\ CO(g)$ from the thermodynamic data in Appendix L. Use this result to calculate the equilibrium constant.

Exercise 19.10—Calculating ΔG°_{rxn} from K_c

The formation constant for $[Ag(NH_3)_2]^+$ is 1.6×10^7. Use this value to calculate ΔG° for the reaction $Ag^+(aq) + 2\ NH_3(aq) \longrightarrow [Ag(NH_3)_2]^+(aq)$.

19.8—Thermodynamics, Time, and Life

Chapters 6 and 19 have brought together the three laws of thermodynamics.

First law:	The total energy of the universe is a constant.
Second law:	The total entropy of the universe is always increasing.
Third law:	The entropy of a pure, perfectly formed crystalline substance at 0 K is zero.

■ **Time's Arrow**
If you are interested in the theories of the origin of the universe, and in "time's arrow," read *A Brief History of Time, From the Big Bang to Black Holes* by Stephen W. Hawking, New York, Bantam Books, 1988.

Some cynic long ago paraphrased the first two laws. The first law was transmuted into "You can't win!," referring to the fact that energy will always be conserved, so a process in which you get back more energy than you put in is impossible. The paraphrase of the second law? "You can't break even!" The Gibbs free energy provides a rationale for this interpretation. Only part of the energy from a chemical reaction can be converted to useful work; the rest will be committed to the redistribution of matter or energy.

Chemical Perspectives

Thermodynamics and Speculation on the Origin of Life

Early Earth was very different than it is today. The atmosphere was made up of simple molecular substances such as N_2, H_2, CO_2, CO, NH_3, CH_4, H_2S, and H_2O. Elemental oxygen was not originally present; this atmospheric gas, now necessary for most life, would eventually be produced by plants via photosynthesis more than a billion years after Earth was formed.

The crust of the planet was rocky, consisting mostly of silicon and aluminum oxides. What was available was an abundance of energy, from solar radiation (without O_2 and O_3, a high level of ultraviolet radiation reached the surface), from lightning, and from the heat in Earth's core. How did the molecules found in living organisms form under these conditions?

A classic experiment to probe the question of how more complex molecules arose was performed in the late 1950s. In the Miller–Urey experiment, a mixture of these gases was subjected to an electric discharge, where the spark simulated lightning. After a few days HCN and

formaldehyde (CH_2O), along with amino acids and other organic molecules, had accumulated in this system. Although scientists no longer believe life-giving molecules may have been formed this way, it is nonetheless a starting point for new experiments in this direction.

Other theories on the formation of organic molecules have their own supporters. One theory suggests that organic molecules formed in aqueous solution on the surface of clay particles. Another argues that important molecules formed near cracks in Earth's crust on the floor of the oceans. Such processes have also been shown to occur in the laboratory.

It is, of course, a long way from small organic molecules to larger molecules, and the conversion of inanimate molecules to living beings is an even greater leap. Creating larger and larger molecules from small molecules requires energy and means that entropy must decrease in the "system." For the process to occur, however, the entropy of the universe must increase. Thus the creation of living systems means there must be a corresponding increase in the entropy of the "surroundings" elsewhere in the universe. One thing remains clear: the laws of thermodynamics must be obeyed.

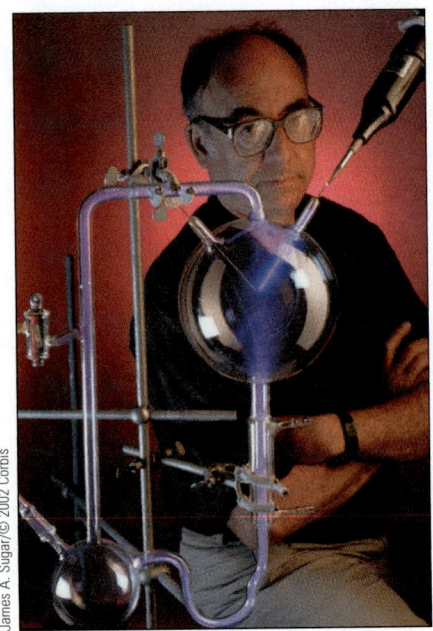

James A. Sugar/© 2002 Corbis

The Miller–Urey experiment. An electric discharge through a mixture of gases leads to formation of organic molecules, including formaldehyde and amino acids.

■ **Entropy and Time**
The second law of thermodynamics requires that disorder increase with time. Because all natural processes that occur do so as time progresses and result in increased disorder, it is evident that increasing entropy and time "point" in the same direction.

The second law tells us that the entropy of the universe is continually increasing. A snowflake will melt in a warm room, but you won't see a glassful of water molecules reassemble themselves into snowflakes at any temperature above 0 °C. Molecules of perfume will diffuse throughout a room, but they won't collect again on your body. All spontaneous processes result in energy becoming more dispersed throughout the universe. This is what scientists mean when they say that the second law is an expression of time in a physical—as opposed to psychological—form. In fact, entropy has been called "time's arrow."

Neither the first nor the second law of thermodynamics has ever been proven. It is just that there never has been a single example demonstrating that either is false. Albert Einstein once remarked that thermodynamic theory "is the only physical theory of the universe content [which], within the framework of applicability of its basic concepts, will never be overthrown." Einstein's statement does not mean that people have not tried (and are continuing to try) to disprove the laws of thermodynamics. Claims to have invented machines that perform useful work without expending energy—perpetual motion machines—are frequently made. However, no perpetual motion machine has ever been shown to work (page 902). We can feel safe with the assumption that such a machine never will be built.

Roald Hoffmann, a chemist who shared the 1981 Nobel Prize in chemistry, has said, "One amusing way to describe synthetic chemistry, the making of molecules that is at the intellectual and economic center of chemistry, is that it is the local defeat of entropy" [*American Scientist*, pp. 619–621, Nov–Dec 1987]. "Local defeat" in this context refers to an unfavorable entropy change in a system because, of course, the entropy of the universe must increase if a process is to occur. An important point to make is that chemical syntheses are often entropy-disfavored. Chemists find ways to accomplish them by balancing the unfavorable changes in the system with favorable changes in the surroundings to make these reactions occur.

Finally, thermodynamics speaks to one of the great mysteries: the origin of life. Life as we know it requires the creation of extremely complex molecules such as proteins and nucleic acids. Their formation must have occurred from atoms and small molecules. Assembling thousands of atoms into a highly ordered state in biochemical compounds clearly requires a local decrease in entropy. Some have said that life is a violation of the second law of thermodynamics. A more logical view, however, is that the local decrease is offset by an increase in entropy in the rest of the universe. Here again, thermodynamics is unchallenged.

Chapter Goals Revisited

When you have finished studying this chapter, you should ask whether you have met the chapter goals. In particular, you should be able to:

Understand the concept of entropy and its relationship to spontaneity

a. Understand that entropy is a measure of matter and energy dispersal or disorder (Section 19.3). General ChemistryNow homework: Study Question(s) 2

b. Recognize that entropy can be determined experimentally as the heat change of a reversible process (A Closer Look, Section 19.4). General ChemistryNow homework: SQ(s) 44

c. Know how to calculate entropy changes from tables of entropy values for compounds (Section 19.4). General ChemistryNow homework: SQ(s) 6, 10, 11

d. Identify common processes that are entropy-favored (Section 19.4).

Predict whether a process will be spontaneous

a. Use entropy and enthalpy changes to predict whether a reaction will be spontaneous (Section 19.5 and Table 19.2). General ChemistryNow homework: SQ(s) 14, 16, 55, 63

b. Recognize how temperature influences whether a reaction is spontaneous (Section 19.5). General ChemistryNow homework: SQ(s) 18, 19, 20

Understand and use a new thermodynamic function, Gibbs free energy

a. Understand the connection between enthalpy and entropy changes and the Gibbs free energy change for a process (Section 19.6).

b. Calculate the change in free energy at standard conditions for a reaction from the enthalpy and entropy changes or from the standard free energy of formation of reactants and products (ΔG_f°) (Section 19.6). General ChemistryNow homework: SQ(s) 22, 24, 26

c. Know how free energy changes with temperature (Section 19.6). General ChemistryNow homework: SQ(s) 30, 72

Understand the relationship of a free energy change for a reaction, its equilibrium constant, and whether the reaction is product- or reactant-favored

a. Describe the relationship between the free energy change and equilibrium constants and calculate K from ΔG_{rxn}° (Section 19.7) General ChemistryNow homework: SQ(s) 48, 50

GENERAL
Chemistry ⚛ Now™

See the General ChemistryNow CD-ROM or website to:
- Assess your understanding with homework questions keyed to each goal
- Check your readiness for an exam by taking the exam-prep quiz and exploring the resources in the personalized Learning Plan it provides

For additional preparation for an examination on this chapter see the **Let's Review** section on pages 998–1011.

Key Equations

Equation 19.1 (page 912):
Calculate entropy change from the heat of the process and the temperature at which it occurs.

$$\Delta S = q_{rev}/T$$

Equation 19.2 (page 915):
Calculate the standard entropy change for a process from the tabulated entropies of the products and reactants.

$$\Delta S^\circ_{system} = \sum S^\circ(\text{products}) - \sum S^\circ(\text{reactants})$$

Equation 19.4 (page 917):
Calculate the total entropy change for a system and its surroundings, to determine whether a process is product-favored.

$$\Delta S^\circ_{univ} = \Delta S^\circ_{sys} + \Delta S^\circ_{surr}$$

Equation 19.5 (page 922):
Calculate the free energy change for a process from the enthalpy and entropy change for the process.

$$\Delta G^\circ = \Delta H^\circ - T\Delta S^\circ$$

Equation 19.6 (page 925):
Calculate the standard free energy change for a reaction using tabulated values of ΔG°_f, the standard free energy of formation.

$$\Delta G^\circ_{rxn} = \sum \Delta G^\circ_f(\text{products}) - \sum \Delta G^\circ_f(\text{reactants})$$

Equation 19.7 (page 928):
The relationship between the free energy change under nonstandard conditions and ΔG° and the reaction quotient Q.

$$\Delta G = \Delta G^\circ + RT \ln Q$$

Equation 19.8 (page 929):
The relationship between the standard free energy change for a reaction and its equilibrium constant.

$$\Delta G^\circ = -RT \ln K$$

Study Questions

▲ denotes more challenging questions.

■ denotes questions available in the Homework and Goals section of the General ChemistryNow CD-ROM or website.

Blue numbered questions have answers in Appendix O and fully worked solutions in the *Student Solutions Manual*.

Structures of many of the compounds used in these questions are found on the General ChemistryNow CD-ROM or website in the Models folder.

GENERAL
Chemistry⚛Now™ Assess your understanding of this chapter's topics with additional quizzing and conceptual questions at http://now.brookscole.com/kotz6e

Practicing Skills

Entropy
(See Example 19.1 and General ChemistryNow Screen 19.4.)

1. Which substance has the higher entropy in each of the following pairs?
   (a) dry ice (solid CO_2) at -78 °C or CO_2 (g) at 0 °C
   (b) liquid water at 25 °C or liquid water at 50 °C
   (c) pure alumina, Al_2O_3(s), or ruby (Ruby is Al_2O_3 in which some of the Al^{3+} ions in the crystalline lattice are replaced with Cr^{3+} ions.)
   (d) one mole of N_2(g) at 1 bar pressure or one mole of N_2(g) at 10 bar pressure (both at 298 K)

2. ■ Which substance has the higher entropy in each of the following pairs?

  (a) a sample of pure silicon (to be used in a computer chip) or a piece of silicon containing a trace of some other elements such as boron or phosphorus

  (b) $O_2(g)$ at 0 °C or $O_2(g)$ at −50 °C

  (c) $I_2(s)$ or $I_2(g)$, both at room temperature

  (d) one mole of $O_2(g)$ at 1 bar pressure or one mole of $O_2(g)$ at 0.01 bar pressure (both at 298 K)

3. By comparing the formulas for each pair of compounds, decide which is expected to have the higher entropy. Assume both are at the same temperature. Check your answers using data in Appendix L.

  (a) $O_2(g)$ or $CH_3OH(g)$ (two substances with the same molar mass)

  (b) $HF(g)$, $HCl(g)$, or $HBr(g)$

  (c) $NH_4Cl(s)$ or $NH_4Cl(aq)$

  (d) $HNO_3(g)$, $HNO_3(\ell)$, or $HNO_3(aq)$

4. By comparing the formulas for each pair of compounds, decide which is expected to have the higher entropy. Assume both are at the same temperature. Check your answers using data in Appendix L.

  (a) $NaCl(s)$, $NaCl(g)$, or $NaCl(aq)$

  (b) $H_2O(g)$ or $H_2S(g)$

  (c) $C_2H_4(g)$ or $N_2(g)$ (two substances with the same molar mass)

  (d) $H_2SO_4(\ell)$ or $H_2SO_4(aq)$

Predicting and Calculating Entropy Changes

(See Example 19.2 and General ChemistryNow Screen 19.5.)

5. Use $S°$ values to calculate the entropy change, $\Delta S°$, for each of the following processes and comment on the sign of the change.

  (a) $KOH(s) \longrightarrow KOH(aq)$

  (b) $Na(g) \longrightarrow Na(s)$

  (c) $Br_2(\ell) \longrightarrow Br_2(g)$

  (d) $HCl(g) \longrightarrow HCl(aq)$

6. ■ Use $S°$ values to calculate the entropy change, $\Delta S°$, for each of the following changes and comment on the sign of the change.

  (a) $NH_4Cl(s) \longrightarrow NH_4Cl(aq)$

  (b) $C_2H_5OH(\ell) \longrightarrow C_2H_5OH(g)$

  (c) $CCl_4(g) \longrightarrow CCl_4(\ell)$

  (d) $NaCl(s) \longrightarrow NaCl(g)$

7. Calculate the standard entropy change for the formation of one mole of gaseous ethane (C_2H_6) at 25 °C.

$$2\,C(\text{graphite}) + 3\,H_2(g) \longrightarrow C_2H_6(g)$$

8. Using standard entropy values, calculate the standard entropy change for a reaction forming one mole of $NH_3(g)$ from $N_2(g)$ and $H_2(g)$.

$$\tfrac{1}{2}\,N_2(g) + \tfrac{3}{2}\,H_2(g) \longrightarrow NH_3(g)$$

9. Calculate the standard molar entropy change for the formation of each of the following compounds from the elements at 25 °C.

  (a) $HCl(g)$ (b) $Ca(OH)_2(s)$

10. ■ Calculate the standard molar entropy change for the formation of each of the following compounds from the elements at 25 °C.

  (a) $H_2S(g)$ (b) $MgCO_3(s)$

11. ■ Calculate the standard molar entropy change for each of the following reactions at 25 °C. Comment on the sign of $\Delta S°$.

  (a) $2\,Al(s) + 3\,Cl_2(g) \longrightarrow 2\,AlCl_3(s)$

  (b) $2\,CH_3OH(\ell) + 3\,O_2(g) \longrightarrow 2\,CO_2(g) + 4\,H_2O(g)$

12. Calculate the standard molar entropy change for each of the following reactions at 25 °C. Comment on the sign of $\Delta S°$.

  (a) $2\,Na(s) + 2\,H_2O(\ell) \longrightarrow 2\,NaOH(aq) + H_2(g)$

  (b) $Na_2CO_3(s) + 2\,HCl(aq) \longrightarrow$
$$2\,NaCl(aq) + H_2O(\ell) + CO_2(g)$$

$\Delta S°_{univ}$ and Spontaneity

(See Example 19.3 and General ChemistryNow Screen 19.6.)

13. Is the reaction

$$Si(s) + 2\,Cl_2(g) \longrightarrow SiCl_4(g)$$

spontaneous? Answer this question by calculating $\Delta S°_{sys}$, $\Delta S°_{surr}$, and $\Delta S°_{univ}$. (The reactants and products are defined as the system and T = 298 K.)

14. ■ Is the reaction

$$Si(s) + 2\,H_2(g) \longrightarrow SiH_4(g)$$

spontaneous? Answer this question by calculating $\Delta S°_{sys}$, $\Delta S°_{surr}$, and $\Delta S°_{univ}$. (The reactants and products are defined as the system and T = 298 K.)

15. Calculate the standard enthalpy and entropy changes for the decomposition of liquid water to form gaseous hydrogen and oxygen. Is this reaction spontaneous? Explain your answer briefly.

16. ■ Calculate the standard enthalpy and entropy changes for the formation of $HCl(g)$ from gaseous hydrogen and chlorine. Is this reaction spontaneous? Explain your answer briefly.

17. Classify each of the reactions according to one of the four reaction types summarized in Table 19.2.

  (a) $Fe_2O_3(s) + 2\,Al(s) \longrightarrow 2\,Fe(s) + Al_2O_3(s)$
  $$\Delta H° = -851.5 \text{ kJ}; \qquad \Delta S° = -375.2 \text{ J/K}$$

  (b) $N_2(g) + 2\,O_2(g) \longrightarrow 2\,NO_2(g)$
  $$\Delta H° = 66.2 \text{ kJ}; \qquad \Delta S° = -121.6 \text{ J/K}$$

18. ■ Classify each of the reactions according to one of the four reaction types summarized in Table 19.2.
    (a) $C_6H_{12}O_6(s) + 6\ O_2(g) \longrightarrow 6\ CO_2(g) + 6\ H_2O(\ell)$

    $\Delta H° = -673\ \text{kJ};\qquad \Delta S° = 60.4\ \text{J/K}$

    (b) $MgO(s) + C(graphite) \longrightarrow Mg(s) + CO(g)$

    $\Delta H° = 490.7\ \text{kJ};\qquad \Delta S° = 197.9\ \text{J/K}$

Effect of Temperature on Reactions

(See Example 19.4.)

19. ■ Heating some metal carbonates, among them magnesium carbonate, leads to their decomposition.

    $$MgCO_3(s) \longrightarrow MgO(s) + CO_2(g)$$

    (a) Calculate $\Delta H°$ and $\Delta S°$ for the reaction.
    (b) Is the reaction spontaneous at 298 K?
    (c) Is the reaction predicted to be spontaneous at higher temperatures?

20. ■ Calculate $\Delta H°$ and $\Delta S°$ for the reaction of tin(IV) oxide with carbon.

    $$SnO_2(s) + C(s) \longrightarrow Sn(s) + CO_2(g)$$

    (a) Is the reaction spontaneous at 298 K?
    (b) Is the reaction predicted to be spontaneous at higher temperatures?

Changes in Free Energy

(See Example 19.4; use $\Delta G° = \Delta H° - T\Delta S°$; see General ChemistryNow Screen 19.7.)

21. Using values of $\Delta H_f°$ and $S°$, calculate $\Delta G_{rxn}°$ for each of the following reactions.
    (a) $2\ Pb(s) + O_2(g) \longrightarrow 2\ PbO(s)$
    (b) $NH_3(g) + HNO_3(aq) \longrightarrow NH_4NO_3(aq)$

    Which of these reactions is (are) predicted to be product-favored? Are the reactions enthalpy- or entropy-driven?

22. ■ Using values of $\Delta H_f°$ and $S°$, calculate $\Delta G_{rxn}°$ for each of the following reactions.
    (a) $Ca(s) + 2\ H_2O(\ell) \longrightarrow Ca(OH)_2(aq) + 2\ H_2(g)$
    (b) $6\ C(graphite) + 3\ H_2(g) \longrightarrow C_6H_6(\ell)$

    Which of these reactions is (are) predicted to be product-favored? Are the reactions enthalpy- or entropy-driven?

23. Using values of $\Delta H_f°$ and $S°$, calculate the standard molar free energy of formation, $\Delta G_f°$, for each of the following compounds:
    (a) $CS_2(g)$
    (b) $NaOH(s)$
    (c) $ICl(g)$

    Compare your calculated values of $\Delta G_f°$ with those listed in Appendix L. Which compounds are predicted to be formed spontaneously?

24. ■ Using values of $\Delta H_f°$ and $S°$, calculate the standard molar free energy of formation, $\Delta G_f°$, for each of the following:
    (a) $Ca(OH)_2(s)$
    (b) $Cl(g)$
    (c) $Na_2CO_3(s)$

    Compare your calculated values of $\Delta G_f°$ with those listed in Appendix L. Which compounds are predicted to be formed spontaneously?

Free Energy of Formation

(See Example 19.5; use $\Delta G_{rxn}° = \Sigma\Delta G_f°\ (products) - \Sigma\Delta G_f°\ (reactants)$; see General ChemistryNow Screen 19.7.)

25. Using values of $\Delta G_f°$, calculate $\Delta G_{rxn}°$ for each of the following reactions. Which are product-favored?
    (a) $2\ K(s) + Cl_2(g) \longrightarrow 2\ KCl(s)$
    (b) $2\ CuO(s) \longrightarrow 2\ Cu(s) + O_2(g)$
    (c) $4\ NH_3(g) + 7\ O_2(g) \longrightarrow 4\ NO_2(g) + 6\ H_2O(g)$

26. ■ Using values of $\Delta G_f°$, calculate $\Delta G_{rxn}°$ for each of the following reactions. Which are product-favored?
    (a) $HgS(s) + O_2(g) \longrightarrow Hg(\ell) + SO_2(g)$
    (b) $2\ H_2S(g) + 3\ O_2(g) \longrightarrow 2\ H_2O(g) + 2\ SO_2(g)$
    (c) $SiCl_4(g) + 2\ Mg(s) \longrightarrow 2\ MgCl_2(s) + Si(s)$

27. For the reaction $BaCO_3(s) \longrightarrow BaO(s) + CO_2(g)$, $\Delta G_{rxn}° = +219.7\ \text{kJ}$. Using this value and other data available in Appendix L, calculate the value of $\Delta G_f°$ for $BaCO_3(s)$.

28. For the reaction $TiCl_2(s) + Cl_2(g) \longrightarrow TiCl_4(\ell)$, $\Delta G_{rxn}° = -272.8\ \text{kJ}$. Using this value and other data available in Appendix L, calculate the value of $\Delta G_f°$ for $TiCl_2(s)$.

Effect of Temperature on ΔG

(See Example 19.6 and General ChemistryNow Screen 19.8.)

29. Determine whether each of the reactions listed below is entropy-favored or -disfavored under standard conditions. Predict how an increase in temperature will affect the value of $\Delta G_{rxn}°$.
    (a) $N_2(g) + 2\ O_2(g) \longrightarrow 2\ NO_2(g)$
    (b) $2\ C(s) + O_2(g) \longrightarrow 2\ CO(g)$
    (c) $CaO(s) + CO_2(g) \longrightarrow CaCO_3(s)$
    (d) $2\ NaCl(s) \longrightarrow 2\ Na(s) + Cl_2(g)$

30. ■ Determine whether each of the reactions listed below is entropy-favored or -disfavored under standard conditions. Predict how an increase in temperature will affect the value of $\Delta G_{rxn}°$.
    (a) $I_2(g) \longrightarrow 2\ I(g)$
    (b) $2\ SO_2(g) + O_2(g) \longrightarrow 2SO_3(g)$
    (c) $SiCl_4(g) + 2\ H_2O(\ell) \longrightarrow SiO_2(s) + 4\ HCl(g)$
    (d) $P_4(s, white) + 6\ H_2(g) \longrightarrow 4\ PH_3(g)$

31. Estimate the temperature required to decompose $HgS(s)$ into $Hg(\ell)$ and $S(g)$.

32. Estimate the temperature required to decompose $CaSO_4(s)$ into $CaO(s)$ and $SO_3(g)$.

▲ More challenging   ■ In General ChemistryNow   **Blue-numbered questions** answered in Appendix O

Free Energy and Equilibrium Constants

(See Example 19.8; use $\Delta G° = -RT \ln K$; see General ChemistryNow Screen 19.9.)

33. The formation of NO(g) from its elements

$$\tfrac{1}{2} N_2(g) + \tfrac{1}{2} O_2(g) \longrightarrow NO(g)$$

has a standard free energy change, $\Delta G°$, of +86.58 kJ/mol at 25 °C. Calculate K_p at this temperature. Comment on the connection between the sign of $\Delta G°$ and the magnitude of K_p.

34. The formation of $O_3(g)$ from $O_2(g)$ has a standard free energy change, $\Delta G°$, of +163.2 kJ/mol at 25 °C. Calculate K_p at this temperature. Comment on the connection between the sign of $\Delta G°$ and the magnitude of K_p.

35. Calculate $\Delta G°$ and K_p at 25 °C for the reaction

$$C_2H_4(g) + H_2(g) \longrightarrow C_2H_6(g)$$

Comment on the sign of $\Delta G°$ and the magnitude of K_p.

36. Calculate $\Delta G°$ and K_p at 25 °C for the reaction

$$2 HBr(g) + Cl_2(g) \longrightarrow 2 HCl(g) + Br_2(\ell)$$

Is the reaction predicted to be product-favored under standard conditions? Comment on the sign of $\Delta G°$ and the magnitude of K_p.

General Questions

These questions are not designated as to type or location in the chapter. They may combine several concepts.

37. Calculate the standard molar entropy change, $\Delta S°$, for each of the following reactions:
    1. $C(s) + 2 H_2(g) \longrightarrow CH_4(g)$
    2. $CH_4(g) + \tfrac{1}{2} O_2(g) \longrightarrow CH_3OH(\ell)$
    3. $C(s) + 2 H_2(g) + \tfrac{1}{2} O_2(g) \longrightarrow CH_3OH(\ell)$

    Verify that these values are related by the equation $\Delta S°_1 + \Delta S°_2 = \Delta S°_3$. What general principle is illustrated here?

38. Hydrogenation, the addition of hydrogen to an organic compound, is a reaction of considerable industrial importance. Calculate $\Delta H°$, $\Delta S°$, and $\Delta G°$ at 25 °C for the hydrogenation of octene, C_8H_{16}, to give octane, C_8H_{18}. Is the reaction product- or reactant-favored under standard conditions?

    $$C_8H_{16}(g) + H_2(g) \longrightarrow C_8H_{18}(g)$$

    The following information is required, in addition to data in Appendix L.

Compound	$\Delta H°_f$ (kJ/mol)	S° (J/K · mol)
Octene	−82.93	462.8
Octane	−208.45	463.639

39. Is the combustion of ethane, C_2H_6, a spontaneous reaction?

    $$C_2H_6(g) + \tfrac{7}{2} O_2(g) \longrightarrow 2 CO_2(g) + 3 H_2O(g)$$

    Answer the question by calculating the value of $\Delta S°_{univ}$ at 298 K, using values of $\Delta H°_f$ and $S°$ in Appendix L. Does your calculated answer agree with your preconceived idea of this reaction?

40. Write a balanced equation that depicts the formation of 1 mol of $Fe_2O_3(s)$ from its elements. What is the standard free energy of formation of 1.00 mol of $Fe_2O_3(s)$? What is the value of $\Delta G°_{rxn}$ when 454 g (1 lb) of $Fe_2O_3(s)$ is formed from the elements?

41. When vapors from hydrochloric acid and aqueous ammonia come in contact, the following reaction takes place, producing a white "cloud" of solid NH_4Cl (Figure 19.11).

    $$HCl(g) + NH_3(g) \longrightarrow NH_4Cl(s)$$

    Defining this as the system under study:
    (a) Predict whether the signs of $\Delta S°_{sys}$, $\Delta S°_{surr}$, $\Delta S°_{univ}$, $\Delta H°$, and $\Delta G°$ are greater than zero, equal to zero, or less than zero, and explain your prediction. Verify your predictions by calculating values for each of these quantities.
    (b) Calculate a value of K_p for this reaction at 298 K.

42. Calculate $\Delta S°_{sys}$, $\Delta S°_{surr}$, $\Delta S°_{univ}$ for each of the following processes at 298 K.
    (a) $NaCl(s) \longrightarrow NaCl(aq)$
    (b) $NaOH(s) \longrightarrow NaOH(aq)$

    Comment on how these systems differ.

43. Methanol is now widely used as a fuel in race cars such as those that compete in the Indianapolis 500 race (see page 496). Consider the following reaction as a possible synthetic route to methanol.

    $$C(graphite) + \tfrac{1}{2} O_2(g) + 2 H_2(g) \longrightarrow CH_3OH(\ell)$$

    Calculate K_p for the formation of methanol at 298 K using this reaction. Would a different temperature be better suited to this reaction?

44. ■ The enthalpy of vaporization of liquid diethyl ether, $(C_2H_5)_2O$, is 26.0 kJ/mol at the boiling point of 35.0 °C. Calculate $\Delta S°$ for (a) a liquid to vapor transformation and (b) a vapor to liquid transformation at 35.0 °C.

45. Calculate the entropy change, $\Delta S°$, for the vaporization of ethanol, C_2H_5OH, at its normal boiling point, 78.0 °C. The enthalpy of vaporization of ethanol is 39.3 kJ/mol.

46. If gaseous hydrogen can be produced cheaply, it could be burned directly as a fuel or converted to another fuel, methane (CH_4), for example.

    $$3 H_2(g) + CO(g) \longrightarrow CH_4(g) + H_2O(g)$$

    Calculate $\Delta G°$, $\Delta S°$, and $\Delta H°$ at 298 K for the reaction. Is it predicted to be product- or reactant-favored under standard conditions?

---

47. Using thermodynamic data, estimate the normal boiling point of ethanol. (Recall that liquid and vapor are in equilibrium at 1.0 atm pressure at the normal boiling point.) The actual normal boiling point is 78 °C. How well does your calculated result agree with the actual value?

48. ■ ▲ Estimate the vapor pressure of ethanol at 37 °C using thermodynamic data. Express the result in millibars of mercury.

49. The following reaction is not spontaneous at room temperature.

$$COCl_2(g) \longrightarrow CO(g) + Cl_2(g)$$

Would you have to raise or lower the temperature to make it spontaneous?

50. ■ When calcium carbonate is heated strongly, CO_2 gas is evolved. The equilibrium pressure of the gas is 1.00 bar at 897 °C, and $\Delta H°_{rxn}$ at 298 K = 179.0 kJ.

$$CaCO_3(s) \longrightarrow CaO(s) + CO_2(g)$$

Estimate the value of $\Delta S°$ at 897 °C for the reaction.

51. The reaction used by Joseph Priestley to prepare oxygen is

$$2 HgO(s) \longrightarrow 2 Hg(\ell) + O_2(g)$$

Defining this reaction as the system under study:

(a) Predict whether the signs of $\Delta S°_{sys}$, $\Delta S°_{surr}$, $\Delta S°_{univ}$, $\Delta H°$, and $\Delta G°$ are greater than zero, equal to zero, or less than zero, and explain your prediction. Using data from Appendix L, calculate the value of each of these quantities to verify your prediction.

(b) Calculate K_p at 298 K. Is the reaction product-favored?

52. ▲ Estimate the boiling point of water in Denver, Colorado (where the altitude is 1.60 km and the atmospheric pressure is 630 mm Hg).

53. Sodium reacts violently with water according to the equation

$$Na(s) + H_2O(\ell) \longrightarrow NaOH(aq) + \tfrac{1}{2} H_2(g)$$

Without doing calculations, predict the signs of $\Delta H°$ and $\Delta S°$ for the reaction. Verify your prediction with a calculation.

54. Yeast can produce ethanol by the fermentation of glucose ($C_6H_{12}O_6$), which is the basis for the production of most alcoholic beverages.

$$C_6H_{12}O_6(aq) \longrightarrow 2 C_2H_5OH(\ell) + 2 CO_2(g)$$

Calculate $\Delta H°$, $\Delta S°$, and $\Delta G°$ for the reaction. Is the reaction product- or reactant-favored? (In addition to the thermodynamic values in Appendix L, you will need the following data for $C_6H_{12}O_6(aq)$: $\Delta H°_f = -1260.0$ kJ/mol; $S° = 289$ J/K · mol; and $\Delta G°_f = -918.8$ kJ/mol.)

55. ■ Elemental boron, in the form of thin fibers, can be made by reducing a boron halide with H_2.

$$BCl_3(g) + \tfrac{3}{2} H_2(g) \longrightarrow B(s) + 3 HCl(g)$$

Calculate $\Delta H°$, $\Delta S°$, and $\Delta G°$ at 25 °C for this reaction. Is the reaction predicted to be spontaneous under standard conditions? If spontaneous, is it enthalpy-driven or entropy-driven? [$S°$ for B(s) is 5.86 J/K · mol.]

56. The equilibrium constant, K_p, for $N_2O_4(g) \rightleftharpoons 2 NO_2(g)$ is 0.14 at 25 °C. Calculate $\Delta G°$ from this constant, and compare your calculated value with that determined from the $\Delta G°_f$ values in Appendix L.

57. Most metal oxides can be reduced with hydrogen to the pure metal. (Although such reactions work well, this expensive method is not used often for large-scale preparations.) The equilibrium constant for the reduction of iron(II) oxide is 0.422 at 700 °C. Estimate $\Delta G°_{rxn}$.

$$FeO(s) + H_2(g) \longrightarrow Fe(s) + H_2O(g)$$

58. The equilibrium constant for the butane $\rightleftharpoons$ isobutane equilibrium at 25 °C is 2.50.

$$K_c = \frac{[isobutane]}{[butane]} = 2.50 \text{ at } 298 \text{ K}$$

Calculate $\Delta G°_{rxn}$ at this temperature in units of kJ/mol.

59. Almost 5 billion kilograms of benzene, C_6H_6, are made each year. Benzene is used as a starting material for many other compounds and as a solvent (although it is also a carcinogen, and its use is restricted). One compound that can be made from benzene is cyclohexane, C_6H_{12}.

$$C_6H_6(\ell) + 3 H_2(g) \longrightarrow C_6H_{12}(\ell)$$
$$\Delta H°_{rxn} = -206.7 \text{ kJ}; \quad \Delta S°_{rxn} = -361.5 \text{ J/K}$$

Is this reaction predicted to be spontaneous under standard conditions at 25 °C? Is the reaction enthalpy- or entropy-driven?

60. A crucial reaction for the production of synthetic fuels is the conversion of coal to H_2 with steam. The chemical reaction is

$$C(s) + H_2O(g) \longrightarrow CO(g) + H_2(g)$$

(a) Calculate $\Delta G°_{rxn}$ for this reaction at 25 °C assuming C(s) is graphite.

(b) Calculate K_p for the reaction at 25 °C.

(c) Is the reaction predicted to be spontaneous under standard conditions? If not, at what temperature will it become so?

61. Calculate $\Delta G°_{rxn}$ for the decomposition of sulfur trioxide to sulfur dioxide and oxygen.

$$2 SO_3(g) \longrightarrow 2 SO_2(g) + O_2(g)$$

(a) Is the reaction spontaneous under standard conditions at 25 °C?

(b) If the reaction is not spontaneous at 25 °C, is there a temperature at which it will become so? Estimate this temperature.

(c) What is the equilibrium constant for the reaction at 1500 °C?

▲ More challenging   ■ In General ChemistryNow   Blue-numbered questions answered in Appendix O

62. Methanol is relatively inexpensive to produce. Much consideration has been given to using it as a precursor to other fuels such as methane, which could be obtained by the decomposition of the alcohol.

$$CH_3OH(\ell) \longrightarrow CH_4(g) + \tfrac{1}{2} O_2(g)$$

(a) What are the sign and magnitude of the entropy change for the reaction? Does the sign of $\Delta S°$ agree with your expectation? Explain briefly.

(b) Is the reaction spontaneous under standard conditions at 25 °C? Use thermodynamic values to prove your answer.

(c) If it is not spontaneous at 25 °C, at what temperature does the reaction become spontaneous?

63. ■ A cave in Mexico was recently discovered to have some interesting chemistry (see Chapter 21). Hydrogen sulfide, H_2S, reacts with oxygen in the cave to give sulfuric acid, which drips from the ceiling in droplets with a pH less than 1. If the reaction occurring is

$$H_2S(g) + 2 O_2(g) \longrightarrow H_2SO_4(\ell)$$

calculate $\Delta H°$, $\Delta S°$, and $\Delta G°$. Is the reaction spontaneous? Is it enthalpy- or entropy-driven?

64. Wet limestone is used to scrub SO_2 gas from the exhaust gases of power plants. One possible reaction gives hydrated calcium sulfite:

$$CaCO_3(s) + SO_2(g) + \tfrac{1}{2} H_2O(\ell) \rightleftharpoons$$
$$CaSO_3 \cdot \tfrac{1}{2} H_2O(s) + CO_2(g)$$

Another reaction gives hydrated calcium sulfate:

$$CaCO_3(s) + SO_2(g) + \tfrac{1}{2} H_2O(\ell) + \tfrac{1}{2} O_2(g) \rightleftharpoons$$
$$CaSO_4 \cdot \tfrac{1}{2} H_2O(s) + CO_2(g)$$

Which is the more product-favored reaction? Use the data in the table below and any other needed in Appendix L.

	$CaSO_3 \cdot \tfrac{1}{2} H_2O(s)$	$CaSO_4 \cdot \tfrac{1}{2} H_2O(s)$
$\Delta H_f°$ (kJ/mol)	−1311.7	−1574.65
$S°$ (J/K mol)	121.3	134.8

65. Sulfur undergoes a phase transition between between 80 and 100 °C.

$$S_8(\text{rhombic}) \longrightarrow S_8(\text{monoclinic})$$
$$\Delta H°_{\text{rxn}} = 3.213 \text{ kJ/mol} \qquad \Delta S°_{\text{rxn}} = 8.7 \text{ J/K}$$

(a) Estimate $\Delta G°$ for the transition at 80.0 °C and 110.0 °C. What do these results tell you about the stability of the two forms of sulfur at each of these temperatures?

(b) Calculate the temperature at which $\Delta G° = 0$. What is the significance of this temperature?

66. Copper(II) oxide, CuO, can be reduced to copper metal with hydrogen at higher temperatures.

$$CuO(s) + H_2(g) \longrightarrow Cu(s) + H_2O(g)$$

Is this reaction product- or reactant-favored under standard conditions at 298 K?

If copper metal is heated in air, a black film of CuO forms on the surface. In this photo the heated bar, covered with a black CuO film, has been bathed in hydrogen gas. Black, solid CuO is reduced rapidly to copper at higher temperatures.

67. ▲ Consider the formation of NO(g) from its elements.

$$N_2(g) + O_2(g) \longrightarrow 2 NO(g)$$

(a) Calculate K_p at 25 °C. Is the reaction product-favored at this temperature?

(b) Assuming $\Delta H°_{\text{rxn}}$ and $\Delta S°_{\text{rxn}}$ are nearly constant with temperature, calculate $\Delta G°_{\text{rxn}}$ at 700 °C. Estimate K_p from the new value of $\Delta G°_{\text{rxn}}$ at 700 °C. Is the reaction product-favored at 700 °C?

(c) Using K_p at 700 °C, calculate the equilibrium partial pressures of the three gases if you mix 1.00 bar each of N_2 and O_2.

68. ▲ Silver(I) oxide can be formed by the reaction of silver metal and oxygen.

$$4 Ag(s) + O_2(g) \longrightarrow 2 Ag_2O(s)$$

(a) Calculate $\Delta H°_{\text{rxn}}$, $\Delta S°_{\text{rxn}}$, and $\Delta G°_{\text{rxn}}$ for the reaction.

(b) What is the pressure of O_2 in equilibrium with Ag and Ag_2O at 25 °C?

(c) At what temperature would the pressure of O_2 in equilibrium with Ag and Ag_2O become equal to 1.00 bar?

69. Calculate $\Delta G_f°$ for HI(g) at 350 °C, given the following equilibrium partial pressures: $P(H_2) = 0.132$ bar, $P(I_2) = 0.295$ bar, and $P(HI) = 1.61$ bar. At 350 °C and 1 bar, I_2 is a gas.

$$\tfrac{1}{2} H_2(g) + \tfrac{1}{2} I_2(g) \rightleftharpoons HI(g)$$

70. Calculate the entropy change for dissolving HCl gas in water. Is the sign of $\Delta S°$ what you expected? Why or why not?

▲ More challenging   ■ In General ChemistryNow   Blue-numbered questions answered in Appendix O

71. ▲ Mercury vapor is dangerous because it can be breathed into the lungs. We wish to estimate the vapor pressure of mercury at two different temperatures from the following data:

	ΔH_f° (kJ/mol)	S° (J/K · mol)	ΔG_f° (kJ/mol)
$Hg(\ell)$	0	76.02	0
$Hg(g)$	61.38	174.97	31.88

Estimate the temperature at which K_p for the process $Hg(\ell) \longrightarrow Hg(g)$ is equal to (a) 1.00 and (b) 1/760. What is the vapor pressure at each of these temperatures? (Experimental vapor pressures are 1.00 mm Hg at 126.2 °C and 1.00 bar at 356.6 °C.) (Note: The temperature at which $P = 1.00$ bar can be calculated from thermodynamic data. To find the other temperature, you will need to use the temperature for $P = 1.00$ bar and the Clausius-Clapeyron equation on page 612.)

72. ■ Some metal oxides can be decomposed to the metal and oxygen under reasonable conditions. Is the decomposition of silver(I) oxide product-favored at 25 °C?

$$2 Ag_2O(s) \longrightarrow 4 Ag(s) + O_2(g)$$

If not, can it become so if the temperature is raised? At what temperature is the reaction product-favored?

Summary and Conceptual Questions

The following questions may use concepts from the preceding chapters.

73. Based on your experience and common sense, which of the following processes would you describe as product-favored and which as reactant-favored under standard conditions?
    (a) $Hg(\ell) \longrightarrow Hg(s)$
    (b) $H_2O(g) \longrightarrow H_2O(\ell)$
    (c) $2 HgO(s) \longrightarrow Hg(\ell) + O_2(g)$
    (d) $C(s) + O_2(g) \longrightarrow CO_2(g)$
    (e) $NaCl(s) \longrightarrow NaCl(aq)$
    (f) $CaCO_3(s) \longrightarrow Ca^{2+}(aq) + CO_3^{2-}(aq)$

74. Explain why each of the following statements is incorrect.
    (a) Entropy increases in all spontaneous reactions.
    (b) Reactions with a negative free energy change ($\Delta G_{rxn}^\circ < 0$) are product-favored and occur with rapid transformation of reactants to products.
    (c) All spontaneous processes are exothermic.
    (d) Endothermic processes are never spontaneous.

75. Decide whether each of the following statements is true or false. If false, rewrite it to make it true.
    (a) The entropy of a substance increases on going from the liquid to the vapor state at any temperature.
    (b) An exothermic reaction will always be spontaneous.
    (c) Reactions with a positive ΔH_{rxn}° and a positive ΔS_{rxn}° can never be product-favored.
    (d) If ΔG_{rxn}° for a reaction is negative, the reaction will have an equilibrium constant greater than 1.

76. Under what conditions is the entropy of a pure substance 0 J/K · mol? Could a substance at standard conditions have a value of 0 J/K · mol? A negative entropy value? Are there any conditions under which a substance will have negative entropy? Explain your answer.

77. In Chapter 14 you learned that entropy, as well as enthalpy, plays a role in the solution process. If ΔH° for the solution process is zero, explain how the process can be driven by entropy.

78. Draw a diagram like that in Figure 19.13 for a reactant-favored process.

79. Write a chemical equation for the oxidation of $C_2H_6(g)$ by $O_2(g)$ to form $CO_2(g)$ and $H_2O(g)$. Defining this as the system:
    (a) Predict whether the signs of ΔS_{sys}°, ΔS_{surr}°, and ΔS_{univ}° will be greater than zero, equal to zero, or less than zero. Explain your prediction.
    (b) Predict the signs of ΔH° and ΔG°. Explain how you made this prediction.
    (c) Will the value of K_p be very large, very small, or near 1? Will the equilibrium constant, K_p, for this system be larger or smaller at temperatures greater than 298 K? Explain how you made this prediction.

80. The normal melting point of benzene, C_6H_6, is 5.5 °C. For the process of melting, what is the sign of each of the following?
    (a) ΔH° (c) ΔG° at 5.5 °C   (e) ΔG° at 25.0°C
    (b) ΔS° (d) ΔG° at 0.0 °C

81. Explain why the entropy of the system increases on dissolving solid NaCl in water {$S^\circ[NaCl(s)] = 72.1$ J/K · mol and $S^\circ[NaCl(aq)] = 115.5$ J/K · mol}.

82. For each of the following processes, give the algebraic sign of ΔH°, ΔS°, and ΔG°. No calculations are necessary; use your common sense.
    (a) The decomposition of liquid water to give gaseous oxygen and hydrogen, a process that requires a considerable amount of energy.
    (b) Dynamite is a mixture of nitroglycerin, $C_3H_5N_3O_9$, and diatomaceous earth. The explosive decomposition of nitroglycerin gives gaseous products such as water, CO_2, and others; much heat is evolved.
    (c) The combustion of gasoline in the engine of your car, as exemplified by the combustion of octane.

$$2 C_8H_{18}(g) + 25 O_2(g) \longrightarrow 16 CO_2(g) + 18 H_2O(g)$$

83. Iodine, I_2, dissolves readily in carbon tetrachloride. For this process, $\Delta H^\circ = 0$ kJ/mol.

$$I_2(s) \longrightarrow I_2 \text{ (in } CCl_4 \text{ solution)}$$

What is the sign of ΔG_{rxn}°? Is the dissolving process entropy-driven or enthalpy-driven? Explain briefly.

84. *Abba's Refrigerator* was described on page 232 as an example of the validity of the second law of thermodynamics. Explain how the second law applies to this simple but useful device.

---

85. "Heater Meals" are food packages that contain their own heat source. Just pour water into the heater unit, wait a few minutes, and voila! You have a hot meal.

The heat for the heater unit is produced by the reaction of magnesium with water.

$$Mg(s) + 2 H_2O(\ell) \longrightarrow Mg(OH)_2(s) + H_2(g)$$

(a) Confirm that this is a spontaneous reaction.

(b) What mass of magnesium is required to produce sufficient energy to heat 225 mL of water (density = 0.995 g/mL) from 25 °C to the boiling point?

86. The formation of diamond from graphite is a process of considerable importance.

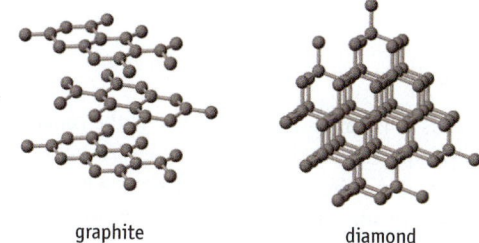

graphite   diamond

(a) Using data in Appendix L, calculate ΔS°_{univ}, ΔH°, and ΔG° for this process.

(b) The calculations will suggest that this process is not possible under any conditions. However, the synthesis of diamonds by this reaction is a commercial process. How can this contradiction be rationalized? (Note: In the synthesis, high pressures and temperatures are used.)

87. Oxygen dissolved in water can cause corrosion in hot-water heating systems. To remove oxygen, hydrazine (N_2H_4) is often added. Hydrazine reacts with dissolved O_2 to form water and N_2.

(a) Write a balanced chemical equation for the reaction of hydrazine and oxygen. Identify the oxidizing and reducing agents in this redox reaction.

(b) Calculate ΔH°, ΔS°, and ΔG° for this reaction.

(c) Because this is an exothermic reaction, heat is evolved. What temperature change is expected in a heating system containing 5.5×10^4 L of water? (Assume no heat is lost to the surroundings.)

(d) The mass of a hot-water heating system is 5.5×10^4 Kg. What amount of O_2 (in moles) would be present in this system if it is filled with water saturated with O_2? (The solubility of O_2 in water at 25 °C is 0.000434 g per 100 g of water.)

(e) Assume hydrazine is available as a 5.0% solution in water. What mass of this solution should be added to totally consume the dissolved O_2 (described in part d)?

(f) Assuming the N_2 escapes as a gas, calculate the volume of $N_2(g)$ (measured at 273 K and 1.00 bar) that will be produced.

88. If gaseous H_2 and O_2 are carefully mixed and left alone, they can remain intact for millions of years. Is this "stability" a function of thermodynamics or of kinetics? (You may wish to review the General ChemistryNow CD-ROM or website Screen 6.3.)

89. Use the simulation on the General ChemistryNow CD-ROM or website Screen 19.6 to answer the following questions regarding the reaction of NO and Cl_2 to produce NOCl.

(a) What is ΔS°_{sys} at 400 K for this reaction?

(b) Does ΔS°_{sys} change with temperature?

(c) Does ΔS°_{surr} change with temperature?

(d) Does ΔS°_{surr} always change with an increase in temperature?

(e) Do exothermic reactions always lead to positive values of ΔS°_{surr}?

(f) Is the NO + Cl_2 reaction spontaneous at 400 K? At 700 K?

90. Use the simulation on the General ChemistryNow CD-ROM or website Screen 19.6 to answer the following questions.

(a) Does the spontaneity of the decomposition of CH_3OH change as the temperature increases?

(b) Is there a temperature between 400 K and 1000 K at which the decomposition is spontaneous?

91. Use the simulation on the General ChemistryNow CD-ROM or website Screen 19.8 to answer the following questions.

(a) Consider the reaction of Fe_2O_3 and C. How does ΔG°_{rxn} vary with temperature? Is there a temperature at which the reaction is spontaneous?

(b) Consider the reaction of HCl and Na_2CO_3. Is there a temperature at which the reaction is no longer spontaneous?

(c) Is the spontaneity of a reaction dependent or independent of temperature?

92. The General ChemistryNow CD-ROM or website Screen 19.9 considers the relationships among thermodynamics, the equilibrium constant, and kinetics. Use the simulation to consider the outcome of the reaction sequence

$$A \rightleftharpoons B \quad \text{and} \quad B \rightleftharpoons C$$

(a) Set up the activation energy diagram with $\Delta E_a(1) = 18$ kJ, $\Delta G^\circ(1) = 7$ kJ, $\Delta E_a(2) = 13$ kJ, and $\Delta G^\circ(2) = -22$ kJ. (Energies need only be approximate.) When the system reaches equilibrium, will there be an appreciable concentration of B? Which species predominates? Explain the results.

(b) Leave the activation energy values the same as in (a) but make $\Delta G^\circ(1) = 10$ kJ and $\Delta G^\circ(2) = -10$ kJ. When equilibrium is attained, what are the relative amounts of A, B, and C? Explain the results.

Do you need a live tutor for homework problems?

Access vMentor at General ChemistryNow at

http://now.brookscole.com/kotz6e

for one-on-one tutoring from a chemistry expert

▲ More challenging   ■ In General ChemistryNow   Blue-numbered questions answered in Appendix O

20 — Principles of Reactivity: Electron Transfer Reactions

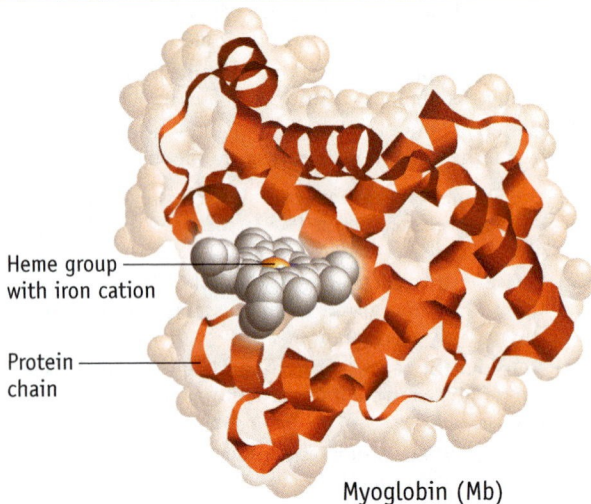

Heme group
with iron cation

Protein
chain

Myoglobin (Mb)

Myoglobin, the protein that binds O_2 in muscles. At the center of the molecule is a heme group. The oxygen binds to the iron in the center of this group. The red color of muscle comes from the oxygenated protein.

Blood Gases

Today, in hundreds of hospitals and clinics around the country, doctors will order up a BGA, a blood gas analysis. A complete analysis will give the physician information on pH and the partial pressures of O_2 and CO_2 in arterial blood. The pH is a measure of the body's acid–base equilibrium, arterial O_2 pressure measures oxygenation in the blood, and arterial CO_2 pressure measures the body's ability to excrete CO_2. An elevated CO_2 pressure may indicate a lung dysfunction, whereas a low pressure may suggest a metabolic problem.

Because the solubility of O_2 in water is low, the blood must contain a substance that transports the gas to tissues. That role is filled by hemoglobin in blood and by myoglobin in muscles. Both are proteins that have at their center a heme group, and it is the iron ion at the center of the heme group that binds O_2.

Myoglobin and hemoglobin have different affinities for O_2, reflecting their different physiological functions. Myoglobin, the oxygen delivery system in muscles, has a higher affinity for O_2 than hemoglobin at all O_2 pressures. Hemoglobin, the oxygen carrier, becomes saturated with O_2 in the lungs, where the O_2 partial pressure is about 100 mm Hg. In tissue capillaries, however, the O_2 partial pressure drops, and hemoglobin releases O_2. Some is stored in myoglobin for use in times of severe oxygen deprivation, such as during strenuous exercise.

Monitoring blood O_2 levels is important, especially in a newborn child. Oxygen is a toxic substance; high concentrations can damage lungs and eyes. For this reason, blood O_2 levels must be kept within certain limits.

Because the level of O_2 in blood is so important, a number of devices have been developed for its measurement. Among them is a tiny electrochemical cell. One half of this cell is a reference electrode; the second half is an electrode that detects and measures

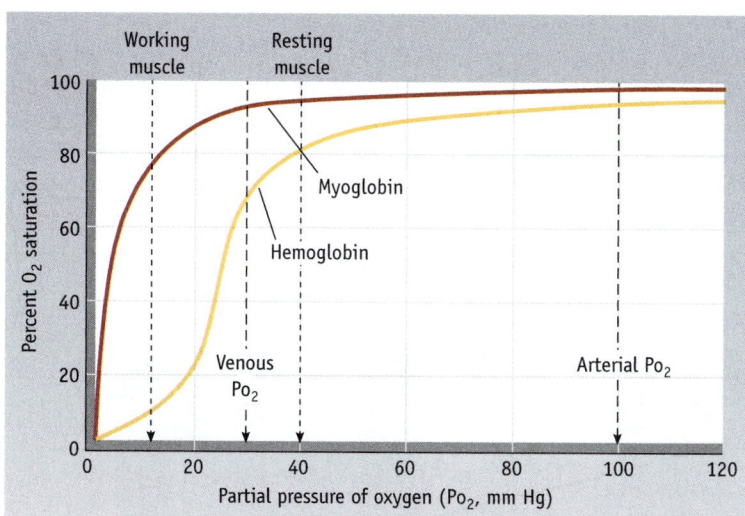

O₂ binding by myoglobin and hemoglobin at a pH of 7.4 in a normal adult. Interestingly, fetal blood has a low Po₂, about 30 mm Hg, equivalent to living at an altitude of 26,000 feet.

Measurement of oxygen concentration using electrochemistry has been a well-established laboratory procedure for some time, but applying this technology to a mixture as complicated as blood did not work well because many other substances in blood interfered with the measurement. The key development that led to the new device was incredibly simple. Biochemist L. C. Clark, Jr., modified the laboratory device, covering the electrode with a polyethylene membrane. The membrane screens most of the components of blood from the electrode, but it is permeable to oxygen. Thus, oxygen passes through the membrane to reach the electrode where it is detected. Clark obtained a patent for his invention in 1956, and the Clark electrode remains the basis for this important medical diagnostic procedure to this day.

oxygen concentration. The sensing electrode is placed in a sample of arterial blood, and a small voltage applied to the cell causes the reduction of dissolved oxygen in the sample.

$$O_2(aq) + 2 H_2O(\ell) + 4 e^- \longrightarrow 4 OH^-(aq)$$

The device "counts" the number of electrons moving through the circuit and from that value determines the amount of dissolved oxygen reduced. This device has been miniaturized to fit into the tip of a narrow catheter so that it can be inserted into an artery, allowing for direct measurement of the oxygen concentration within the body.

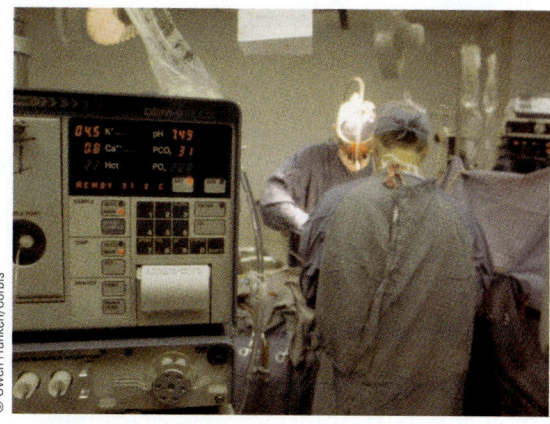

Blood gas analyzer.

© Owen Franken/Corbis

To Review Before You Begin

- Review the definitions of oxidation, reduction, reducing agent, oxidizing agent, and oxidation number (Section 5.7)
- Review common redox reactions and common oxidizing and reducing agents (Section 5.7)
- Review the concepts of thermodynamics (Chapters 6 and 19)

GENERAL
Chemistry Now™

Throughout the chapter this icon introduces a list of resources on the General ChemistryNow CD-ROM or website (**http://now .brookscole.com/kotz6e**) that will:

- help you evaluate your knowledge of the material
- provide homework problems
- allow you to take an exam-prep quiz
- provide a personalized Learning Plan targeting resources that address areas you should study

Let us introduce you to electrochemistry and electron transfer reactions with a simple chemistry experiment. Place a piece of copper in an aqueous solution of silver nitrate. After a short time, metallic silver deposits on the copper and the solution takes on the blue color typical of aqueous Cu^{2+} ions (Figure 20.1). The following oxidation–reduction (redox) reaction has occurred:

$$Cu(s) + 2\ Ag^{+}(aq) \longrightarrow Cu^{2+}(aq) + 2\ Ag(s)$$

At the submicroscopic level, Ag^{+} ions in solution come into direct contact with the copper surface where the transfer of electrons occurs. Two electrons are transferred from a Cu atom to two Ag^{+} ions. Copper ions, Cu^{2+}, enter the solution and silver atoms are deposited on the copper surface. This product-favored reaction continues until one or both of the reactants is consumed.

The reaction between copper metal and silver ions can be used to generate an electric current. The experiment must be carried out in a different way, however. If the reactants, Cu(s) and Ag^{+}(aq), are in direct contact, electrons will be transferred directly from copper atoms to silver ions, and the energy produced will take the form of heat rather than electricity. Instead, the reaction is done in an appara-

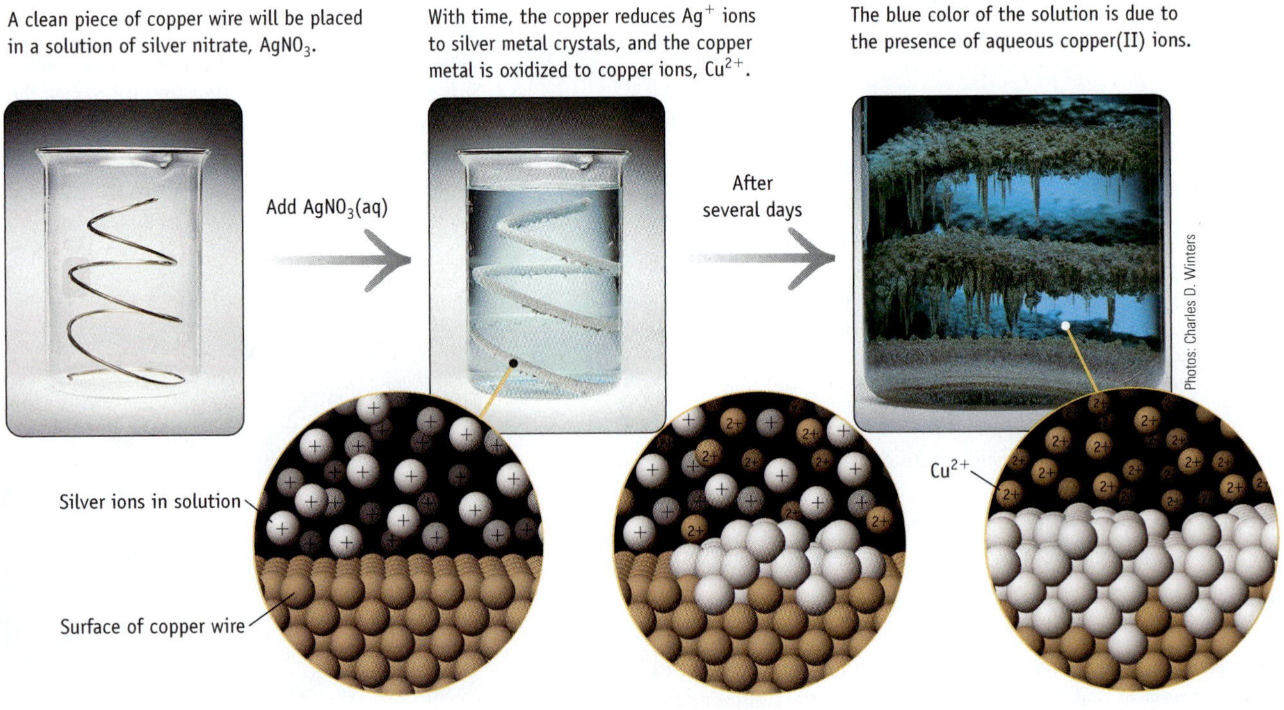

A clean piece of copper wire will be placed in a solution of silver nitrate, $AgNO_3$.

Add $AgNO_3$(aq)

With time, the copper reduces Ag^+ ions to silver metal crystals, and the copper metal is oxidized to copper ions, Cu^{2+}.

After several days

The blue color of the solution is due to the presence of aqueous copper(II) ions.

Cu^{2+}

Silver ions in solution

Surface of copper wire

Photos: Charles D. Winters

Figure 20.1 **The oxidation of copper by silver ions.** *(See the General ChemistryNow Screen 5.12 for photographs of this reaction sequence.)*

tus that allows electrons to be transferred from one reactant to the other through an electrical circuit. The movement of electrons through the circuit constitutes an electric current that can be used to light a light bulb or run a motor.

Devices that use chemical reactions to produce an electric current are called **voltaic cells** or **galvanic cells**, names honoring Count Alessandro Volta (1745–1827) and Luigi Galvani (1737–1798). All voltaic cells work in the same general way. They use product-favored redox reactions composed of an oxidation and a reduction. The cell is constructed so that electrons produced by the reducing agent are transferred through an electric circuit to the oxidizing agent.

Chemical energy is converted to electrical energy in a voltaic cell. The opposite process, the use of electric energy to effect a chemical change, occurs in **electrolysis**. An example is the electrolysis of water (Figure 1.6), in which electrical energy is used to split water into its component elements, hydrogen and oxygen. Electrolysis is also used to electroplate one metal onto another, to obtain aluminum from its common ore (bauxite, mostly Al_2O_3), and to prepare important chemicals such as chlorine (see Chapter 21).

Electrochemistry is the field of chemistry that considers chemical reactions that produce or are caused by electrical energy. Because all electrochemical reactions are oxidation–reduction (*redox*) reactions, we begin our exploration of this subject by first describing electron transfer reactions in more detail.

■ **Two Types of Electrochemical Processes**
- Chemical change can produce an electric current in a voltaic cell.
- Electric energy can cause chemical change in the process of electrolysis.

20.1—Oxidation–Reduction Reactions

In an oxidation–reduction reaction electron transfer occurs between a reducing agent and an oxidizing agent [◀ Section 5.7]. The essential features of all electron transfer reactions are as follows:

- One reactant is oxidized and one is reduced.
- The extent of oxidation and reduction must balance.
- The oxidizing agent (the chemical species causing oxidation) is reduced.
- The reducing agent (the chemical species causing reduction) is oxidized.
- Oxidation numbers (page 200) can be used to determine whether a substance is oxidized or reduced. An element is oxidized if its oxidation number increases. The oxidation number decreases in a reduction.

These aspects of oxidation–reduction or redox reactions are illustrated for the reaction of copper metal and silver ion (Figure 20.1).

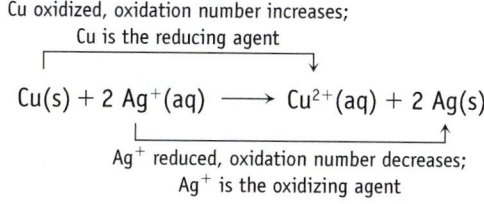

Cu oxidized, oxidation number increases;
Cu is the reducing agent

$$Cu(s) + 2\ Ag^+(aq) \longrightarrow Cu^{2+}(aq) + 2\ Ag(s)$$

Ag^+ reduced, oxidation number decreases;
Ag^+ is the oxidizing agent

GENERAL
Chemistry Now™

See the General ChemistryNow CD-ROM or website:
- **Screen 20.2 Redox Reactions: Electron Transfer,** to view photographs and an animation of different types of oxidation–reduction reactions

Balancing Oxidation–Reduction Equations

All equations for oxidation–reduction reactions must be balanced for both mass and charge. The same number of atoms appear in the products and reactants of an equation, and the sum of electric charges of all the species on each side of the equation arrow must be the same. Charge balance guarantees that the number of electrons produced in oxidation equals the number of electrons consumed in reduction.

Balancing redox equations can be complicated, but fortunately some systematic procedures can be used in these cases. Here, we describe the half-reaction method, a procedure that involves writing separate, balanced equations for the oxidation and reduction processes called **half-reactions**. One half-reaction describes the oxidation part of the reaction, and a second half-reaction describes the reduction part. The equation for the overall reaction is the sum of the two half-reactions, after adjustments have been made (if necessary) in one or both half-reaction equations so that the numbers of electrons transferred from reducing agent to oxidizing agent balance. For example, the half-reactions for the reaction of copper metal with silver ions are

Reduction half-reaction: $\quad Ag^+(aq) + e^- \longrightarrow Ag(s)$

Oxidation half-reaction: $\quad Cu(s) \longrightarrow Cu^{2+}(aq) + 2\,e^-$

Notice that the equations for the half-reactions are themselves balanced for mass and charge. In the copper half-reaction there is one Cu atom on each side of the equation (*mass balance*). The electric charge on the right side of the equation is 0 (the sum of +2 for the ion and −2 for two electrons), as it is on the left side (*charge balance*).

To produce the net chemical equation, we add the two half-reactions. First, however, we must multiply the silver half-reaction by 2.

$$2\,Ag^+(aq) + 2\,e^- \longrightarrow 2\,Ag(s)$$

Each mole of copper atoms produces two moles of electrons, and two moles of Ag^+ ions are required to consume those electrons.

Finally, adding the two half-reactions, and canceling electrons from both sides, leads to the net ionic equation for the reaction.

Reduction half-reaction: $\quad 2\,Ag^+(aq) + 2\,e^- \longrightarrow 2\,Ag(s)$

Oxidation half-reaction: $\quad Cu(s) \longrightarrow Cu^{2+}(aq) + 2\,e^-$

Balanced net ionic equation: $\quad Cu(s) + 2\,Ag^+(aq) \longrightarrow Cu^{2+}(aq) + 2\,Ag(s)$

The net ionic equation is likewise balanced for mass and charge.

GENERAL
Chemistry ⚛ Now™

See the General ChemistryNow CD-ROM or website:

- **Screen 20.3 Balancing Equations for Redox Reactions,** for methods of balancing redox reactions in acidic solution

Figure 20.2 Reduction of Cu²⁺ by Al. (a) A ball of aluminum foil is added to a solution of $Cu(NO_3)_2$ and NaCl. (b) A coating of copper is soon seen on the surface of the aluminum, and the reaction generates a significant amount of heat. (A coating of Al_2O_3 on the surface of aluminum protects the metal from reaction. However, in the presence of Cl^- ion, the coating is breached, and reaction occurs.) See Example 20.1.

(a)   (b)

Example 20.1—Balancing Oxidation–Reduction Equations

Problem Balance the equation

$$Al(s) + Cu^{2+}(aq) \longrightarrow Al^{3+}(aq) + Cu(s)$$

Identify the oxidizing agent, the reducing agent, the substance oxidized, and the substance reduced. Write balanced half-reactions and the balanced net ionic equation. See Figure 20.2 for photographs of this reaction.

Strategy First, make sure the reaction is an oxidation–reduction reaction by checking each element to see whether the oxidation numbers change. Next, separate the equation into half-reactions, identifying what has been reduced (oxidizing agent) and oxidized (reducing agent). Then balance the half-reactions, first for mass and then for charge. Finally, add the two half-reactions, after ensuring that the reducing agent half-reaction involves the same number of moles of electrons as the oxidizing agent half-reaction.

Solution

Step 1. Recognize the reaction as an oxidation–reduction reaction.

Here the oxidation number for aluminum changes from 0 to +3 and the oxidation number of copper changes from +2 to 0. Aluminum is oxidized and serves as the reducing agent. Copper(II) ions are reduced, and Cu^{2+} is the oxidizing agent.

Step 2. Separate the process into half-reactions.

Reduction: $Cu^{2+}(aq) \longrightarrow Cu(s)$ (*oxidation number of Cu decreases*)

Oxidation: $Al(s) \longrightarrow Al^{3+}(aq)$ (*oxidation number of Al increases*)

Step 3. Balance each half-reaction for mass.

Both half-reactions are already balanced for mass.

Step 4. Balance each half-reaction for charge.

To balance the equations for charge add electrons to the more positive side of each half-reaction.

Reduction: $2\ e^- + Cu^{2+}(aq) \longrightarrow Cu(s)$

   Each Cu^{2+} ion requires two electrons

Oxidation: $Al(s) \longrightarrow Al^{3+}(aq) + 3\ e^-$

   Each Al atom releases three electrons

Step 5. Multiply each half-reaction by an appropriate factor.

The reducing agent must donate as many electrons as the oxidizing agent acquires. Three Cu^{2+} ions are required to take on the six electrons produced by two Al atoms. Thus, we multiply the Cu^{2+}/Cu half-reaction by 3 and the Al/Al^{3+} half-reaction by 2.

Reduction: $3[2\ e^- + Cu^{2+}(aq) \longrightarrow Cu(s)]$

Oxidation: $2[Al(s) \longrightarrow Al^{3+}(aq) + 3\ e^-]$

Step 6. Add the half-reactions to produce the overall balanced equation.

Reduction: $6\ e^- + 3\ Cu^{2+}(aq) \longrightarrow 3\ Cu(s)$

Oxidation: $2\ Al(s) \longrightarrow 2\ Al^{3+}(aq) + 6\ e^-$

Net ionic equation: $3\ Cu^{2+}(aq) + 2\ Al(s) \longrightarrow 3\ Cu(s) + 2\ Al^{3+}(aq)$

Step 7. Simplify by eliminating reactants and products that appear on both sides.

This step is not required here.

Comment You should always check the overall equation to ensure there is a mass and charge balance. In this case three Cu atoms and two Al atoms appear on each side. The net electric charge on each side is $+6$. The equation is balanced.

Exercise 20.1—Balancing Oxidation–Reduction Equations

Aluminum reacts with nonoxidizing acids to give $Al^{3+}(aq)$ and $H_2(g)$. The (unbalanced) equation is

$$Al(s) + H^+(aq) \longrightarrow Al^{3+}(aq) + H_2(g)$$

Write balanced equations for the half-reactions and the balanced net ionic equation. Identify the oxidizing agent, the reducing agent, the substance oxidized, and the substance reduced.

When balancing equations for redox reactions in aqueous solution, it is sometimes necessary to add water molecules (H_2O) and either $H^+(aq)$ in acidic solution or $OH^-(aq)$ in basic solution to the equation. Equations that include oxoanions such as $SO_4{}^{2-}$, $NO_3{}^-$, ClO^-, $CrO_4{}^{2-}$, and $MnO_4{}^-$ fall into this category. The process is outlined in Example 20.2 for the reduction of an oxocation in acid solution and in Example 20.3 for a reaction in a basic solution.

Example 20.2—Balancing Equations for Oxidation–Reduction Reactions in Acid Solution

Problem Balance the net ionic equation for the reaction of the dioxovanadium(V) ion, $VO_2{}^+$, with zinc in acid solution to form VO^{2+} (see Figure 20.3).

$$VO_2{}^+(aq) + Zn(s) \longrightarrow VO^{2+}(aq) + Zn^{2+}(aq)$$

Strategy Follow the strategy outlined in Example 20.1. One exception is that water and H^+ ions will appear as product and reactant, respectively, in the half-reaction for the reduction of $VO_2{}^+$ ion.

Solution

Step 1. Recognize the reaction as an oxidation–reduction reaction.

The oxidation number of V changes from $+5$ in $VO_2{}^+$ to $+4$ in VO^{2+}. The oxidation number of Zn changes from 0 in the metal to $+2$ in Zn^{2+}.

Step 2. Separate the process into half-reactions.

Oxidation: $Zn(s) \longrightarrow Zn^{2+}(aq)$

           Zn(s) is oxidized and is the reducing agent

Reduction: $VO_2{}^+(aq) \longrightarrow VO^{2+}(aq)$

           $VO_2{}^+(aq)$ is reduced and is the oxidizing agent

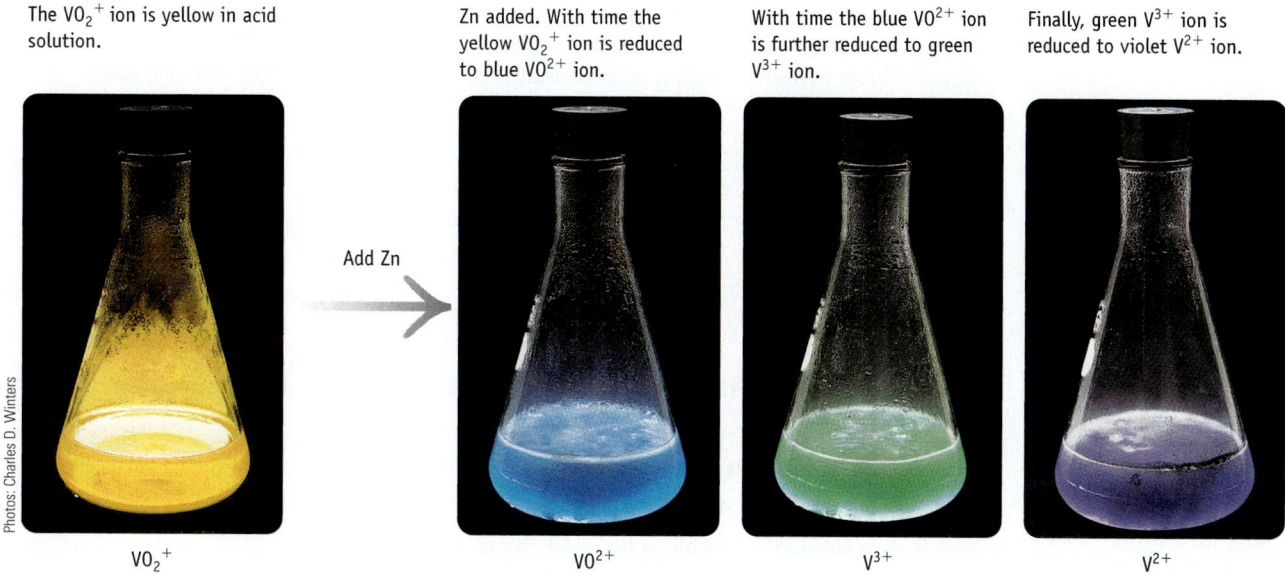

The VO₂⁺ ion is yellow in acid solution.

Zn added. With time the yellow VO₂⁺ ion is reduced to blue VO²⁺ ion.

With time the blue VO²⁺ ion is further reduced to green V³⁺ ion.

Finally, green V³⁺ ion is reduced to violet V²⁺ ion.

Add Zn

VO_2^+ VO^{2+} V^{3+} V^{2+}

Photos: Charles D. Winters

Figure 20.3 Reduction of vanadium(V) with zinc. See Example 20.2 for the balanced equation for the first stage of the reduction.

Step 3. Balance the half-reactions for mass.

Begin by balancing all atoms except H and O. (These atoms are always the last to be balanced because they often appear in more than one reactant or product).

Zinc half-reaction: $Zn(s) \longrightarrow Zn^{2+}(aq)$

This half-reaction is already balanced for mass.

Vanadium half-reaction: $VO_2^+(aq) \longrightarrow VO^{2+}(aq)$

The V atoms in this half-reaction are already balanced. An oxygen-containing species must be added to the right side of the equation to achieve an O atom balance, however.

$$VO_2^+(aq) \longrightarrow VO^{2+}(aq) + (\text{need 1 O atom})$$

In acid solution, add H_2O to the side requiring O atoms, one H_2O molecule for each O atom required.

$$VO_2^+(aq) \longrightarrow VO^{2+}(aq) + H_2O(\ell)$$

There are now two unbalanced H atoms on the right. Because the reaction occurs in an acidic solution, H^+ ions are present. Therefore, a mass balance for H can be achieved by adding H^+ to the side of the equation deficient in H atoms. Here two H^+ ions are added to the left side of the equation.

$$2\ H^+(aq) + VO_2^+(aq) \longrightarrow VO^{2+}(aq) + H_2O(\ell)$$

Step 4. Balance the half-reactions for charge.

Zinc half-reaction: $Zn(s) \longrightarrow Zn^{2+}(aq) + 2\ e^-$

The mass-balanced VO_2^+ equation has a net charge of 3+ on the left side and 2+ on the right. Therefore, 1 e^- is added to the more positive left side.

Vanadium half-reaction: $e^- + 2\ H^+(aq) + VO_2^+(aq) \longrightarrow VO^{2+}(aq) + H_2O(\ell)$

Step 5. Multiply the half-reactions by appropriate factors so that the reducing agent donates as many electrons as the oxidizing agent consumes.

Here the reducing agent half-reaction supplies 2 mol of electrons per 1 mol of Zn and the oxidizing agent half-reaction consumes 1 mol of electrons per 1 mol of VO_2^+. Therefore, the oxidizing agent half-reaction must be multiplied by 2. Now 2 mol of the oxidizing agent (VO^{2+}) consumes the 2 mol of electrons provided per mole of the reducing agent (Zn).

$$Zn(s) \longrightarrow Zn^{2+}(aq) + 2\ e^-$$
$$2[e^- + 2\ H^+(aq) + VO_2^+(aq) \longrightarrow VO^{2+}(aq) + H_2O(\ell)]$$

Step 6. Add the half-reactions to give the balanced, overall equation.

Oxidation:	$Zn(s) \longrightarrow Zn^{2+}(aq) + 2\ e^-$
Reduction:	$2\ e^- + 4\ H^+(aq) + 2\ VO_2^+(aq) \longrightarrow 2\ VO^{2+}(aq) + 2\ H_2O(\ell)$
Net ionic equation:	$Zn(s) + 4\ H^+(aq) + 2\ VO_2^+(aq) \longrightarrow Zn^{2+}(aq) + 2\ VO^{2+}(aq) + 2\ H_2O(\ell)$

Step 7. Simplify by eliminating reactants and products that appear on both sides.

This step is not required here.

Comment Check the overall equation to ensure that there is a mass and charge balance.

Mass balance: 1 Zn, 2 V, 4 H, and 4 O

Charge balance: Each side has a net charge of 6+.

Exercise 20.2—Balancing Equations for Oxidation–Reduction Reactions in Acid Solution

The yellow dioxovanadium (V) ion, $VO_2^+(aq)$, is reduced by zinc metal in three steps. The first step reduces it to blue $VO^{2+}(aq)$ (Example 20.2). This ion is further reduced to green $V^{3+}(aq)$ in the second step, and V^{3+} can be reduced to violet $V^{2+}(aq)$ in a third step. In each step zinc is oxidized to $Zn^{2+}(aq)$. Write balanced net ionic equations for Steps 2 and 3. (This reduction sequence is shown in Figure 20.3.)

Exercise 20.3—Balancing Equations for Oxidation–Reduction Reactions in Acid Solution

The permanganate ion, MnO_4^-, is an oxidizing agent (see Section 5.7). A common laboratory analysis for iron is to titrate aqueous iron(II) ion with a solution of potassium permanganate of precisely known concentration (Figure 20.4; see also Example 5.16, page 220). Use the half-reaction method to write the balanced net ionic equation for the reaction in acid solution.

$$MnO_4^-(aq) + Fe^{2+}(aq) \longrightarrow Mn^{2+}(aq) + Fe^{3+}(aq)$$

Figure 20.4 The reaction of purple permanganate ion (MnO_4^-) with iron(II) ion in aqueous solution. The reaction gives the nearly colorless ions Mn^{2+} and Fe^{3+}. See Exercise 20.3.

Charles D. Winters

Problem-Solving Tip 20.1

Balancing Oxidation–Reduction Equations: A Summary

- Hydrogen balance can be achieved only with H^+/H_2O (in acid) or OH^-/H_2O (in base). Never add H or H_2 to balance hydrogen.

- Use H_2O or OH^- as appropriate to balance oxygen. Never add O atoms, O^{2-} ions, or O_2 for O balance.

- Never include $H^+(aq)$ and $OH^-(aq)$ in the same equation. A solution can be either acidic or basic, never both.

- The number of electrons in a half-reaction reflects to the change in oxidation state of the element being oxidized or reduced.

- Electrons are always a component of half-reactions but should never appear in the overall equation.

- Include charges in the formulas for ions. Omitting the charge, or writing the charge incorrectly, is one of the most common errors seen on student papers.

- The best way to become competent in balancing redox equations is to practice, practice, practice.

Example 20.2 and Exercises 20.2 and 20.3 illustrate the technique of balancing equations for redox reactions involving oxocations and oxoanions that occur in acid solution. Under these conditions, H^+ ion or the H^+/H_2O pair can be used to achieve a balanced equation if required. Conversely, in basic solution, only OH^- ion or the OH^-/H_2O pair can be used.

Example 20.3—Balancing Equations for Oxidation–Reduction Reactions in Basic Solution

Problem Aluminum metal is oxidized in aqueous base, with water serving as the oxidizing agent. The products of the reaction are $Al(OH)_4{}^-(aq)$ and $H_2(g)$. Write a balanced net ionic equation for this reaction.

Strategy First identify the oxidizing and reducing half-reactions, and then balance them for mass and charge. Finally, add the balanced half-reactions to obtain the balanced net ionic equation for the reaction.

Solution

Step 1. Recognize the reaction as an oxidation and a reduction.

The unbalanced equation is

$$Al(s) + H_2O(\ell) \longrightarrow Al(OH)_4{}^-(aq) + H_2(g)$$

Here aluminum is oxidized, with its oxidation number changing from 0 to +3. Hydrogen is reduced, with its oxidation number decreasing from +1 to zero.

Step 2. Separate the process into half-reactions.

Oxidation half-reaction: $\quad Al(s) \longrightarrow Al(OH)_4{}^-(aq)$

$\qquad\qquad\qquad\qquad\qquad$ (*Al oxidation number increases from 0 to +3*)

Reduction half-reaction: $\quad H_2O(\ell) \longrightarrow H_2(g)$ $\quad$ (*H oxidation number decreases from +1 to 0*)

Step 3. Balance the half-reactions for mass.

Addition of OH^- and/or H_2O is required for mass balance in both half-reactions. In the case of the aluminum half-reaction, we simply add OH^- ions to the left side.

Oxidation half-reaction: $\quad Al(s) + 4\ OH^-(aq) \longrightarrow Al(OH)_4{}^-(aq)$

To balance the half-reaction for water reduction, notice that an oxygen-containing species must appear on the right side of the equation. Because H_2O is a reactant, we use OH^-, which is present in this basic solution, as the other product.

Reduction half-reaction: $\quad 2\ H_2O(\ell) \longrightarrow H_2(g) + 2\ OH^-(aq)$

Step 4. Balance the half-reactions for charge.

Electrons are added to balance charge.

Oxidation half-reaction: $\quad Al(s) + 4\ OH^-(aq) \longrightarrow Al(OH)_4{}^-(aq) + 3\ e^-$

Reduction half-reaction: $\quad 2\ H_2O(\ell) + 2\ e^- \longrightarrow H_2(g) + 2\ OH^-(aq)$

Step 5. Balance the oxidation and reduction reactions.

Here electron balance is achieved by using 2 mol of Al to provide 6 mol of e^-, which are then acquired by 6 mol of H_2O.

Oxidation half-reaction: $\quad 2[Al(s) + 4\ OH^-(aq) \longrightarrow Al(OH)_4{}^-(aq) + 3\ e^-]$

Reduction half-reaction: $\quad 3[2\ H_2O(\ell) + 2\ e^- \longrightarrow H_2(g) + 2\ OH^-(aq)]$

Step 6. Add the half-reactions.

$$2\ Al(s) + 8\ OH^-(aq) \longrightarrow 2\ Al(OH)_4{}^-(aq) + \cancel{6\ e^-}$$

$$\underline{6\ H_2O(\ell) + \cancel{6\ e^-} \longrightarrow 3\ H_2(g) + 6\ OH^-(aq)}$$

Net equation: $\quad 2\ Al(s) + 8\ OH^-(aq) + 6\ H_2O(\ell) \longrightarrow 2\ Al(OH)_4{}^-(aq) + 3\ H_2(g) + 6\ OH^-(aq)$

Step 7. Simplify by eliminating reactants and products that appear on both sides.

Six OH^- ions can be canceled from the two sides of the equation:

$$2\ Al(s) + 2\ OH^-(aq) + 6\ H_2O(\ell) \longrightarrow 2\ Al(OH)_4^-(aq) + 3\ H_2(g)$$

Comment The final equation is balanced for mass and charge.

Mass balance: 2 Al, 14 H, and 8 O

Charge balance: There is a net −2 charge on each side.

Exercise 20.4—Balancing Equations for Oxidation–Reduction Reactions in Basic Solution

Voltaic cells based on the reduction of sulfur are under development. One such cell involves the reaction of sulfur with aluminum under basic conditions.

$$Al(s) + S(s) \longrightarrow Al(OH)_3(s) + HS^-(aq)$$

(a) Balance this equation showing each balanced half-reaction.
(b) Identify the oxidizing and reducing agents, the substance oxidized, and the substance reduced.

20.2—Simple Voltaic Cells

Let us use the reaction of copper metal and silver ions (Figure 20.1) as the basis of a voltaic cell. To do so, we place the components of the two half-reactions in separate compartments (Figure 20.5). This prevents the copper metal from transferring electrons directly to silver ions. Instead, electrons are transferred through an external circuit, and useful work can potentially be done.

The copper half-cell (on the left in Figure 20.5) holds copper metal that serves as one electrode and a solution containing copper(II) ions. The half-cell on the right uses a silver electrode and a solution containing silver(I) ions. Important features of this simple cell are as follows:

- *The two half-cells are connected with a **salt bridge** that allows cations and anions to move between the two half-cells.* The electrolyte chosen for the salt bridge should contain ions that will not react with chemical reagents in both half-cells. In the example in Figure 20.5, $NaNO_3$ is used.

- *In all electrochemical cells the **anode** is the electrode at which oxidation occurs. The electrode at which reduction occurs is always the **cathode**.* (In Figure 20.5, the copper electrode is the anode and the silver electrode is the cathode.)

■ **Salt Bridges**
A simple salt bridge can be made by adding gelatin to a solution of an electrolyte. Gelatin makes the contents semi-rigid so that the salt bridge is easier to handle. Porous glass disks and permeable membranes are alternatives to a salt bridge. These devices allow ions to traverse from one half-cell to the other while keeping the two solutions from mixing.

Problem-Solving Tip 20.2

An Alternative Method for Balancing Equations in Basic Solution

Another way to balance equations for reactions in basic solution is to first do so in acidic solution and then add enough OH^- ions to both sides of the equation so that the H^+ ions are converted to water.

Taking the half-reaction for the reduction of ClO^- ion to Cl_2, we have these steps:

1. Balance in acid.
$$4\ H^+(aq) + 2\ ClO^-(aq) + 2\ e^- \longrightarrow Cl_2(g) + 2\ H_2O(\ell)$$

2. Add 4 OH^- ions to both sides.
$$4\ OH^-(aq) + 4\ H^+(aq) + 2\ ClO^-(aq) + 2\ e^-$$
$$\longrightarrow Cl_2(g) + 2\ H_2O(\ell) + 4\ OH^-(aq)$$

3. Combine OH^- and H^+ to form water where appropriate.
$$4\ H_2O(\ell) + 2\ ClO^-(aq) + 2\ e^- \longrightarrow Cl_2(g) + 2\ H_2O(\ell) + 4\ OH^-(aq)$$

4. Simplify.
$$2\ H_2O(\ell) + 2\ ClO^-(aq) + 2\ e^- \longrightarrow Cl_2(g) + 4\ OH^-(aq)$$

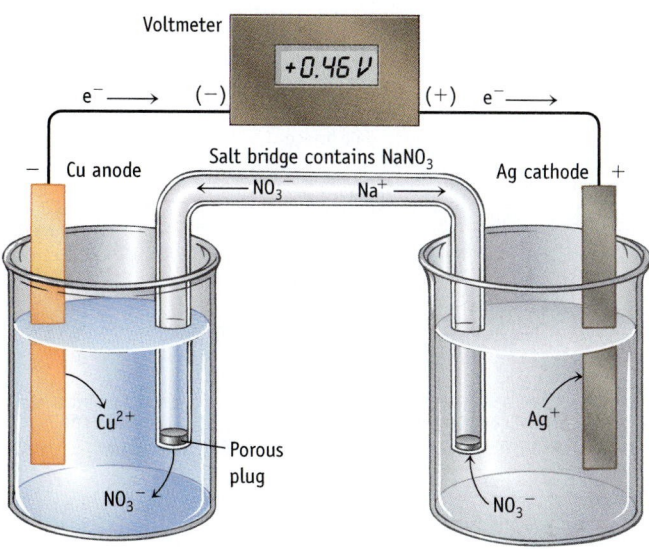

Figure 20.5 **A voltaic cell using $Cu(s)\,|\,Cu^{2+}(aq)$ and $Ag(s)\,|\,Ag^{+}(aq)$ half-cells.** Electrons flow through the external circuit from the anode (the copper electrode) to the cathode (silver electrode). In the salt bridge, which contains aqueous $NaNO_3$, negative $NO_3^{-}(aq)$ ions migrate toward the copper half-cell, and positive $Na^{+}(aq)$ ions migrate toward the silver half-cell. Using 1.0 M $Cu^{2+}(aq)$ and 1.0 M $Ag^{+}(aq)$ solutions, this cell will generate 0.46 volts.

Net reaction: $Cu(s) + 2\,Ag^{+}(aq) \longrightarrow Cu^{2+}(aq) + 2\,Ag(s)$

- *A minus sign can be assigned to the anode in a voltaic cell, and the cathode is marked with a positive sign.* The chemical oxidation occurring at the anode, which produces electrons, gives it a negative charge. Electric current in the external circuit of a voltaic cell consists of electrons moving from the negative to the positive electrode.

- *In all electrochemical cells, electrons flow in the external circuit from the anode to the cathode.*

The chemistry occurring in the cell pictured in Figure 20.5 is summarized by the following half-reactions and net ionic equation:

Cathode (reduction):	$2\,Ag^{+}(aq) + 2\,e^{-} \longrightarrow 2\,Ag(s)$
Anode (oxidation):	$Cu(s) \longrightarrow Cu^{2+}(aq) + 2\,e^{-}$
Net ionic equation:	$Cu(s) + 2\,Ag^{+}(aq) \longrightarrow Cu^{2+}(aq) + 2\,Ag(s)$

The salt bridge is required in a voltaic cell for the reaction to proceed. In the Cu/Ag^{+} voltaic cell, anions move in the salt bridge toward the copper half-cell and cations move toward the silver half-cell (Figure 20.5). As $Cu^{2+}(aq)$ ions are formed in the copper half-cell by oxidation, negative ions enter that cell from the salt bridge (and positive ions leave the cell), so that the numbers of positive and negative charges in the half-cell compartment remain in balance. Likewise, in the silver half-cell, negative ions move out of the half-cell into the salt bridge and positive ions move into the cell as $Ag^{+}(aq)$ ions are reduced to silver metal. A complete circuit is required for current to flow. If the salt bridge is removed, reactions at the electrodes will cease.

In Figure 20.5, the electrodes are connected by wires to a voltmeter. In an alternative set-up, the connections might be to a light bulb or other device that uses electricity. Electrons are produced by oxidation of copper, and $Cu^{2+}(aq)$ ions enter the solution. The electrons traverse the external circuit to the silver electrode, where they reduce $Ag^{+}(aq)$ ions to silver metal. To balance the extent of oxidation and reduction, two $Ag^{+}(aq)$ ions are reduced for every $Cu^{2+}(aq)$ ion formed. The main features of this and of all other voltaic cells are summarized in Figure 20.6.

■ **Electron and Ion Flow**
It is helpful to notice that in an electrochemical cell the negative electrons and negatively charged anions make a "circle." That is, electrons move from anode to cathode in the external circuit, and negative anions move from the cathode compartment, through the salt bridge, to the anode compartment.

Figure 20.6 **Summary of terms used in a voltaic cell.**
Electrons move from the anode, the site of oxidation, through the external circuit to the cathode, the site of reduction. Charge balance in each half-cell is achieved by migration of ions through the salt bridge. Negative ions move from the cathode compartment to the anode compartment, and positive ions move in the opposite direction.

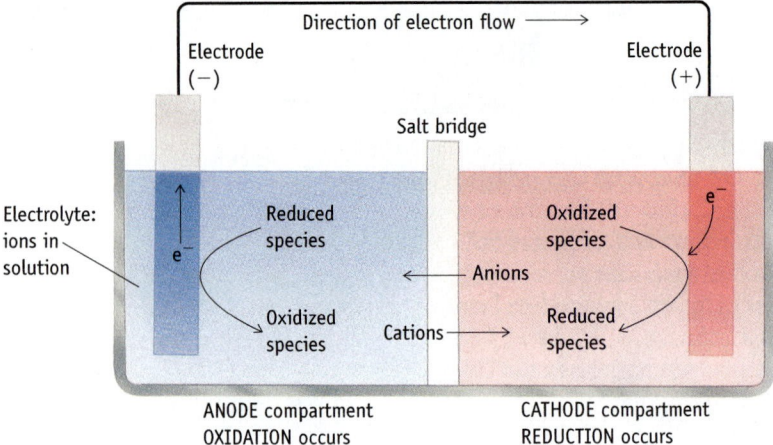

See the General ChemistryNow CD-ROM or website:

• **Screen 20.4 Electrochemical Cells,** to view an animation of a cell based on zinc and copper

Example 20.4—Electrochemical Cells

Problem Describe how to set up a voltaic cell to generate an electric current using the reaction

$$Fe(s) + Cu^{2+}(aq) \longrightarrow Cu(s) + Fe^{2+}(aq)$$

Which electrode is the anode and which is the cathode? In which direction do electrons flow in the external circuit? In which direction do the positive and negative ions flow in the salt bridge? Write equations for the half-reactions that occur at each electrode.

Strategy First, identify the two different half-cells that make up the cell. Next, decide in which half-cell oxidation occurs and in which reduction occurs.

Solution This voltaic cell is similar to the one diagrammed in Figure 20.5. One half-cell contains an iron electrode and a solution of an iron(II) salt such as $Fe(NO_3)_2$. The other half-cell contains a copper electrode and a soluble copper(II) salt such as $Cu(NO_3)_2$. The two half-cells are linked with a salt bridge containing an electrolyte such as KNO_3. Iron is oxidized, so the iron electrode is the anode:

Oxidation, anode: $Fe(s) \longrightarrow Fe^{2+}(aq) + 2 e^-$

Because copper(II) ions are reduced, the copper electrode is the cathode. The cathodic half-reaction is

Reduction, cathode: $Cu^{2+}(aq) + 2 e^- \longrightarrow Cu(s)$

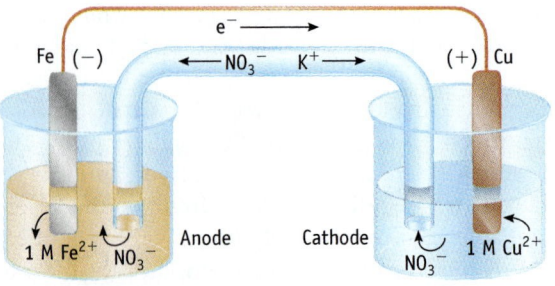

In the external circuit, electrons flow from the iron electrode (anode) to the copper electrode (cathode). In the salt bridge, negative ions flow toward the $Fe|Fe^{2+}(aq)$ half-cell and positive ions flow in the opposite direction.

Comment Recall that oxidation always occurs at the anode, reduction always occurs at the cathode, and electrons always flow from the anode to the cathode. See Figure 20.6.

Exercise 20.5—Electrochemical Cells

Describe how to set up a voltaic cell using the following half-reactions:

Reduction half-reaction: $Ag^+(aq) + e^- \longrightarrow Ag(s)$

Oxidation half-reaction: $Ni(s) \longrightarrow Ni^{2+}(aq) + 2\ e^-$

Which is the anode and which is the cathode? What is the overall cell reaction? What is the direction of electron flow in an external wire connecting the two electrodes? Describe the ion flow in a salt bridge (with $NaNO_3$) connecting the cell compartments.

Voltaic Cells with Inert Electrodes

In the half-cells described so far, the metal used as an electrode is also a reactant or a product in the redox reaction. Not all half-reactions involve a metal as a reactant or product, however. With the exception of carbon in the form of graphite, most nonmetals are unsuitable as electrode materials because they do not conduct electricity. It is not possible to make an electrode from a gas, a liquid, or a solution. Ionic solids do not make satisfactory electrodes because the ions are locked tightly in a crystal lattice, and these materials do not conduct electricity.

In situations where reactants and products cannot serve as the electrode material, an **inert electrode** must be used. Such electrodes are made of materials that conduct an electric current but that are neither oxidized nor reduced in the cell.

Consider constructing a voltaic cell to accommodate the following product-favored reaction:

$$2\ Fe^{3+}(aq) + H_2(g) \longrightarrow 2\ Fe^{2+}(aq) + 2\ H^+(aq)$$

Reduction half-reaction: $Fe^{3+}(aq) + e^- \longrightarrow Fe^{2+}(aq)$

Oxidation half-reaction: $H_2(g) \longrightarrow 2\ H^+(aq) + 2\ e^-$

Neither the reactants nor the products can be used as an electrode material. Therefore, the two half-cells are set up so that the reactants and products come in contact with an electrode such as graphite where they can accept or give up electrons. Graphite is a commonly used electrode material: It is a conductor of electricity, inexpensive (essential in commercial cells), and not readily oxidized under the conditions encountered in most cells. Platinum and gold are also commonly used in laboratory experiments because both are chemically inert under most circumstances. They are generally too costly for commercial cells, however.

The *hydrogen electrode* is particularly important in the field of electrochemistry because it is used as a reference in assigning cell voltages (see Section 20.4) (Figure 20.7). The electrode itself is platinum, chosen because hydrogen adsorbs on the metal's surface. In this cell's operation, hydrogen is bubbled over the electrode and a large surface area maximizes the contact of the gas and the electrode. The aqueous solution contains $H^+(aq)$. The half-reactions involving $H^+(aq)$ and $H_2(g)$,

$$2\ H^+(aq) + 2\ e^- \longrightarrow H_2(g) \quad \text{or} \quad H_2(g) \longrightarrow 2\ H^+(aq) + 2\ e^-$$

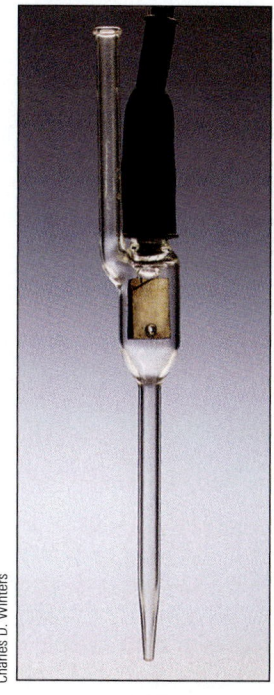

Charles D. Winters

Figure 20.7 Hydrogen electrode. Hydrogen gas is bubbled over a platinum electrode in a solution containing H^+ ions. Such electrodes function best if they have a large surface area. Often platinum wires are woven into a gauze or the metal surface is roughened either by abrasion or by chemical treatment to increase the surface area.

Figure 20.8 A voltaic cell with a hydrogen electrode. This cell has Fe^{2+}(aq, 1.0 M) and Fe^{3+}(aq, 1.0 M) in the cathode compartment and H_2(g) and H^+(aq, 1.0 M) in the anode compartment. At 25 °C, the cell generates 0.77 V.

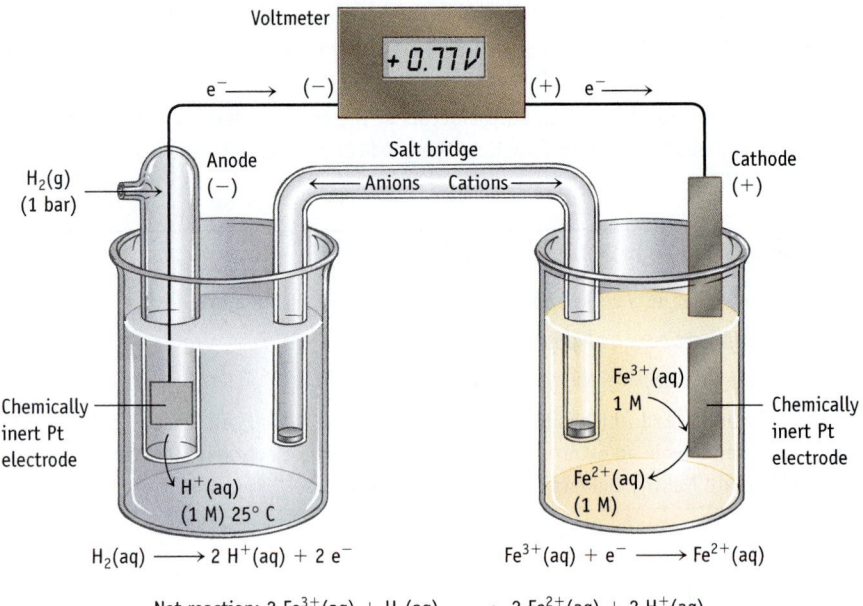

Voltmeter

$+0.77V$

$e^- \longrightarrow$ (−)   (+) $e^- \longrightarrow$

Anode   Salt bridge   Cathode
(−) $\longleftarrow$ Anions   Cations $\longrightarrow$ (+)

H_2(g)
(1 bar)

Chemically
inert Pt
electrode

H^+(aq)
(1 M) 25° C

Fe^{3+}(aq)
1 M

Fe^{2+}(aq)
(1 M)

Chemically
inert Pt
electrode

H_2(aq) $\longrightarrow$ 2 H^+(aq) + 2 e^- Fe^{3+}(aq) + e^- $\longrightarrow$ Fe^{2+}(aq)

Net reaction: 2 Fe^{3+}(aq) + H_2(aq) $\longrightarrow$ 2 Fe^{2+}(aq) + 2 H^+(aq)

take place at the electrode surface, and the electrons involved in the reaction are conducted to or from the reaction site by the metal electrode.

A half-cell using the reduction of Fe^{3+}(aq) to Fe^{2+}(aq) can also be set up with a platinum electrode. In this case the solution surrounding the electrode contains iron ions in two different oxidation states. Transfer of electrons to or from the reactant occurs at the electrode surface.

A voltaic cell involving the reduction of Fe^{3+}(aq, 1.0 M) to Fe^{2+}(aq, 1.0 M) with H_2 gas is illustrated in Figure 20.8. In this cell, the hydrogen electrode is the anode (H_2 is oxidized to H^+), and the iron-containing compartment is the cathode (Fe^{3+} is reduced to Fe^{2+}). The cell produces 0.77 V.

Electrochemical Cell Conventions

Chemists often use a shorthand notation to simplify cell descriptions. For example, the cell involving the reduction of silver ion with copper metal is written as

$$Cu(s)|Cu^{2+}(aq, 1.0\ M)||Ag^+(aq, 1.0\ M)|Ag(s)$$

The cell using H_2 gas to reduce Fe^{3+} ions is written as

$$2\ Fe^{3+}(aq) + H_2(g) \longrightarrow 2\ Fe^{2+}(aq) + 2\ H^+(aq)$$
$$Pt|H_2(P = 1\ bar)|H^+(aq, 1.0\ M)||Fe^{3+}(aq, 1.0\ M),\ Fe^{2+}(aq, 1.0\ M)|Pt$$
Anode information   Cathode information

By convention, the anode and information with respect to the solution with which it is in contact are always written on the left. A single vertical line (|) indicates a phase boundary, and double vertical lines (||) indicate a salt bridge.

Chemical Perspectives

Frogs and Voltaic Piles

Voltaic cells are also called galvanic cells after the Italian physician Luigi Galvani (1737–1798), who carried out early studies of what he called "animal electricity." These studies brought several new words into our language—among them "galvanic" and "galvanize."

Around 1780 Galvani observed that the electric current from a static electricity generator caused the contraction of the muscles in a frog's leg. Investigating this phenomenon further, he found that he could induce contraction when the muscle was in contact with two different metals. Because no external source of electricity was applied to the muscles, Galvani con-

Alessandro Volta, 1745–1827.

© Bettman/Corbis

cluded that the frog's muscles were themselves generating electricity. This was evidence, he believed, of a kind of "vital energy" or "animal electricity," which was related to but different from "natural electricity" generated by machines or lightning.

Alessandro Volta repeated Galvani's experiments, with the same results, but he came to different conclusions. Volta proposed that an electric current was generated by the contact between two different metals—an explanation we now know to be correct—and that the muscle was a detector of the small current generated.

To prove his hypothesis, Volta built the first "electric pile" in 1800. This device comprised a series of metal disks of two kinds (silver and zinc), separated by paper disks soaked in acid or salt solutions. Soon after Volta announced his discovery, Carlisle and Nicholson in England used the

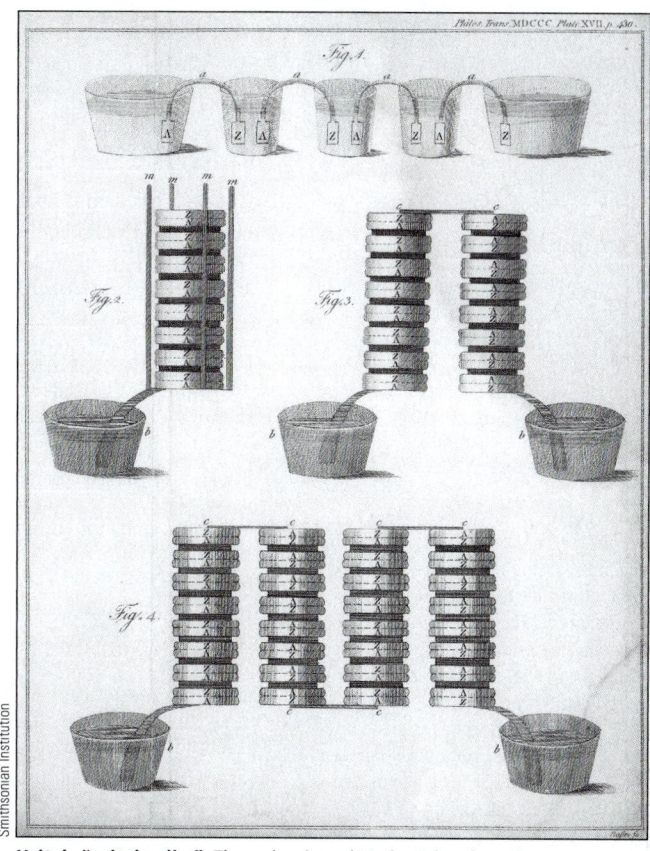

Smithsonian Institution

Volta's "voltaic pile." These drawings done by Volta show the arrangement of silver and zinc disks used to generate an electric current.

electricity from a "pile" to decompose water into hydrogen and oxygen. Within a few years, the great English chemist Humphry Davy used a more powerful voltaic pile to isolate potassium and sodium metals by electrolysis.

20.3—Commercial Voltaic Cells

The cells described so far are unlikely to have practical use. They are neither compact nor robust, high priorities for most applications. In most situations, it is also important that the cell produce a constant voltage, but a problem with the cells described so far is that the voltage produced varies as the concentrations of ions in solution change (see Section 20.5). Also, the rate of current production is low. Attempting to draw a large current results in a drop in voltage because *the current depends on how fast ions in solution migrate to the electrode.* Ion concentrations near the electrode become depleted if current is drawn rapidly, resulting in a decline in voltage.

The amount of current that can be drawn from a voltaic cell depends on the quantity of reagents consumed. A voltaic cell must have a large mass of reactants to

A voltaic cell. A voltaic cell can be made by inserting copper and zinc electrodes into almost any conductive material. Unfortunately, this cell is a "lemon." Although we measure about 1 V, the current is too low to be practical.

■ **Batteries**
The word *battery* has become part of our common language, designating any self-contained device that generates an electric current. The term *battery* has a more precise scientific meaning, however: It refers to a collection of two or more voltaic cells. For example, the 12-volt battery used in automobiles is made up of six voltaic cells. Each voltaic cell develops a voltage of 2 volts. Six cells connected in series produce 12 volts.

produce current over a prolonged period. In addition, a voltaic cell that can be recharged is attractive. Recharging a cell means returning the reagents to their original sites in the cell. In the cells described so far, the movement of ions in the cell mixes the reagents, and they cannot be "unmixed" after the cell has been running.

Batteries can be classified as primary and secondary. **Primary batteries** use redox reactions that cannot be returned to their original state by recharging, so when the reactants are consumed, the battery is "dead" and must be discarded. **Secondary batteries** are often called **storage batteries** or **rechargeable batteries**. The reactions in these batteries can be reversed; thus, the batteries can be recharged.

Years of development have led to many different commercial voltaic cells to meet specific needs (Figure 20.9). Several common ones are described below. All adhere to the principles that have been developed in earlier discussions.

Primary Batteries: Dry Cells and Alkaline Batteries

If you buy an inexpensive flashlight battery or **dry cell battery**, it will probably be a modern version of a voltaic cell invented by George LeClanché in 1866 (Figure 20.10). Zinc serves as the anode, and the cathode is a graphite rod placed down the center of the device. These cells are often called "dry cells" because there is no visible liquid phase. However, the cell contains a moist paste of NH_4Cl, $ZnCl_2$, and MnO_2. The moisture is necessary because the ions present must be in a medium in which they can migrate from one electrode to the other. The cell generates a potential of 1.5 V using the following half-reactions:

Cathode, reduction: $2\ NH_4^+(aq) + 2\ e^- \longrightarrow 2\ NH_3(g) + H_2(g)$
Anode, oxidation: $Zn(s) \longrightarrow Zn^{2+}(aq) + 2\ e^-$

The two gases formed at the cathode will build up pressure and could cause the cell to rupture. This problem is avoided, however, by two other reactions that take place in the cell. Ammonia molecules bind to Zn^{2+} ions, and hydrogen gas is oxidized by MnO_2.

$$Zn^{2+}(aq) + 2\ NH_3(g) + 2\ Cl^-(aq) \longrightarrow Zn(NH_3)_2Cl_2(s)$$
$$2\ MnO_2(s) + H_2(g) \longrightarrow Mn_2O_3(s) + H_2O(\ell)$$

Figure 20.9   Some commercial voltaic cells. Commercial voltaic cells provide energy for a wide range of devices, come in a myriad of sizes and shapes, and produce different voltages. Some are rechargeable; others are thrown away after use. One might think that there is nothing further to learn about batteries, yet research on these devices is being actively pursued in the chemical community.

LeClanché cells are widely used because of their low cost, but they have several disadvantages. If current is drawn from the battery rapidly, the gaseous products cannot be consumed rapidly enough, so the cell resistance rises and the voltage drops. In addition, the zinc electrode and ammonium ions are in contact in the cell, and these chemicals react slowly. Recall that zinc reacts with acid to form hydrogen. The ammonium ion, $NH_4^+(aq)$, is a weak Brønsted acid [◄ Table 17.3] and reacts slowly with zinc. Because of this reaction these voltaic cells cannot be stored indefinitely, a fact you may have learned from experience. When the zinc outer shell deteriorates, the battery can leak acid and perhaps damage the flashlight or other appliance in which it is contained.

Not satisfied with a simple battery? You can spend a little more money to buy **alkaline cells** for your CD player. They generate current up to 50% longer than a dry cell of the same size. The chemistry of alkaline cells is quite similar to that in a LeClanché cell except that the material inside the cell is basic (alkaline). Alkaline cells use the oxidation of zinc and the reduction of MnO_2 to generate a current, but NaOH or KOH is used in the cell instead of the acidic salt NH_4Cl.

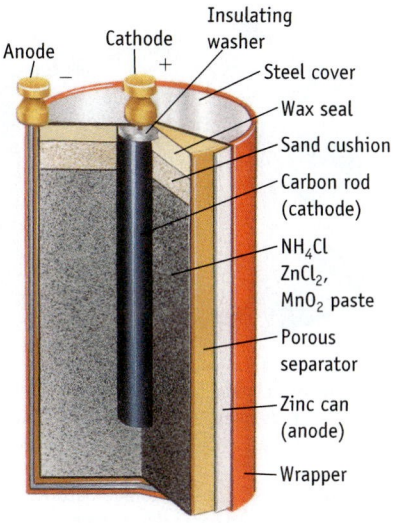

Figure 20.10 **The common LeClanché dry cell battery.**

Cathode, reduction: $2\ MnO_2(s) + H_2O(\ell) + 2\ e^- \longrightarrow Mn_2O_3(s) + 2\ OH^-(aq)$

Anode, oxidation: $Zn(s) + 2\ OH^-(aq) \longrightarrow ZnO(s) + H_2O(\ell) + 2\ e^-$

Alkaline cells, which produce 1.54 V (approximately the same voltage as the LeClanché cell), have the further advantage that the cell potential does not decline under high current loads because no gases are formed.

Prior to 2000, mercury-containing batteries were widely used in calculators, cameras, watches, heart pacemakers, and other devices. However, these small batteries were banned in the United States in the 1990s because of environmental problems. Taking their place have been several other types of batteries, such as silver oxide batteries and zinc-oxygen batteries. Both operate under alkaline conditions and both have zinc anodes. In the silver oxide battery, which produces a voltage of about 1.5 V, the cell reactions are

Cathode, reduction: $Ag_2O(s) + H_2O(\ell) + 2\ e^- \longrightarrow 2\ Ag(s) + 2\ OH^-(aq)$

Anode, oxidation: $Zn(s) + 2\ OH^-(aq) \longrightarrow ZnO(s) + H_2O(\ell) + 2\ e^-$

The zinc-oxygen battery, which produces about 1.15–1.35 V, is unique in that atmospheric oxygen and not a metal oxide is the oxidizing agent.

Cathode, reduction: $O_2(g) + 2\ H_2O(\ell) + 4\ e^- \longrightarrow 4\ OH^-(aq)$

Anode, oxidation: $2\ Zn(s) + 4\ OH^-(aq) \longrightarrow 2\ ZnO(s) + 2\ H_2O(\ell) + 4\ e^-$

These batteries have found use in hearing aids, pagers, and medical devices.

Secondary or Rechargeable Batteries

When a LeClanché cell or an alkaline cell ceases to produce a usable electric current, it is discarded. In contrast, some types of cells can be recharged, often hundreds of times. Recharging requires applying an electric current from an external source to restore the cell to its original state.

An automobile battery—the **lead storage battery**—is probably the best-known rechargeable battery (Figure 20.11). The 12-V version of this battery contains six voltaic cells, each generating 2.04 V. The lead storage battery can produce a large initial current, an essential feature to start an automobile engine.

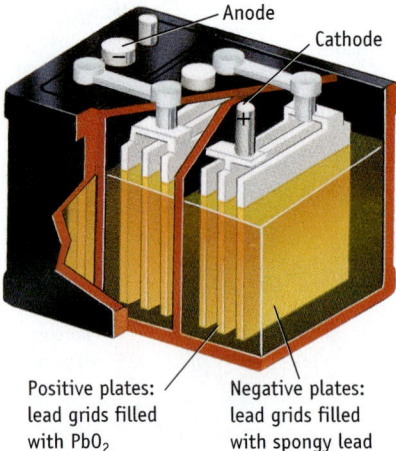

Positive plates: lead grids filled with PbO₂

Negative plates: lead grids filled with spongy lead

Figure 20.11 Lead storage battery, a secondary or rechargeable battery. The negative plates (anode) are lead grids filled with spongy lead. The positive plates (cathode) are lead grids filled with lead(IV) oxide, PbO₂. Each cell of the battery generates 2 V.

The anode of a lead storage battery is metallic lead. The cathode is also made of lead, but it is covered with a layer of compressed, insoluble lead(IV) oxide, PbO_2. The electrodes, arranged alternately in a stack and separated by thin fiberglass sheets, are immersed in aqueous sulfuric acid. When the cell supplies electrical energy, the lead anode is oxidized to lead(II) sulfate, an insoluble substance that adheres to the electrode surface. The two electrons produced per lead atom move through the external circuit to the cathode, where PbO_2 is reduced to Pb^{2+} ions that, in the presence of H_2SO_4, also form lead(II) sulfate.

Cathode, reduction: $PbO_2(s) + 4\ H^+(aq) + SO_4^{2-}(aq) + 2\ e^- \longrightarrow PbSO_4(s) + 2\ H_2O(\ell)$

Anode, oxidation: $Pb(s) + SO_4^{2-}(aq) \longrightarrow PbSO_4(s) + 2\ e^-$

Net ionic equation: $Pb(s) + PbO_2(s) + 2\ H_2SO_4(aq) \longrightarrow 2\ PbSO_4(s) + 2\ H_2O(\ell)$

When current is generated, sulfuric acid is consumed and water is formed. Because water is less dense than sulfuric acid, the density of the solution decreases during this process. Therefore, one way to determine whether a lead storage battery needs to be recharged is to measure the density of the solution.

A lead storage battery is recharged by supplying electrical energy. The $PbSO_4$ coating the surfaces of the electrodes is converted back to metallic lead and PbO_2, and sulfuric acid is regenerated. Recharging this battery is possible because the reactants and products remain attached to the electrode surface. The lifetime of a lead storage battery is limited, however, because the coatings of PbO_2 and $PbSO_4$ flake off of the surface and fall to the bottom of the battery case.

Scientists and engineers would like to find an alternative to lead storage batteries, especially for use in electric cars. Lead storage batteries have the disadvantage of being large and heavy. In addition, lead and its compounds are toxic and their disposal adds a further complication. Nevertheless, at this time, the advantages of lead storage batteries outweigh their disadvantages.

Nickel-cadmium ("Ni-cad") batteries, used in a variety of cordless appliances such as telephones, video camcorders, and cordless power tools, are lightweight and rechargeable. The chemistry of the cell utilizes the oxidation of cadmium and the reduction of nickel(III) oxide under basic conditions. As with the lead storage battery, the reactants and products formed when producing a current are solids that adhere to the electrodes.

Cathode, reduction: $2\ NiO(OH)(s) + 2\ H_2O(\ell) + 2\ e^- \longrightarrow 2\ Ni(OH)_2(s) + 2\ OH^-(aq)$

Anode, oxidation: $Cd(s) + 2\ OH^-(aq) \longrightarrow Cd(OH)_2(s) + 2\ e^-$

Ni-cad batteries produce a nearly constant voltage. However, their cost is relatively high and there are restrictions on their disposal because cadmium compounds are toxic and present an environmental hazard.

Fuel Cells

An advantage of voltaic cells is that they are small and portable, but their size is also a limitation. The amount of electric current produced is limited by the quantity of reagents contained in the cell. When one of the reactants is completely consumed, the cell will no longer generate a current. **Fuel cells** avoid this limitation. The reac-

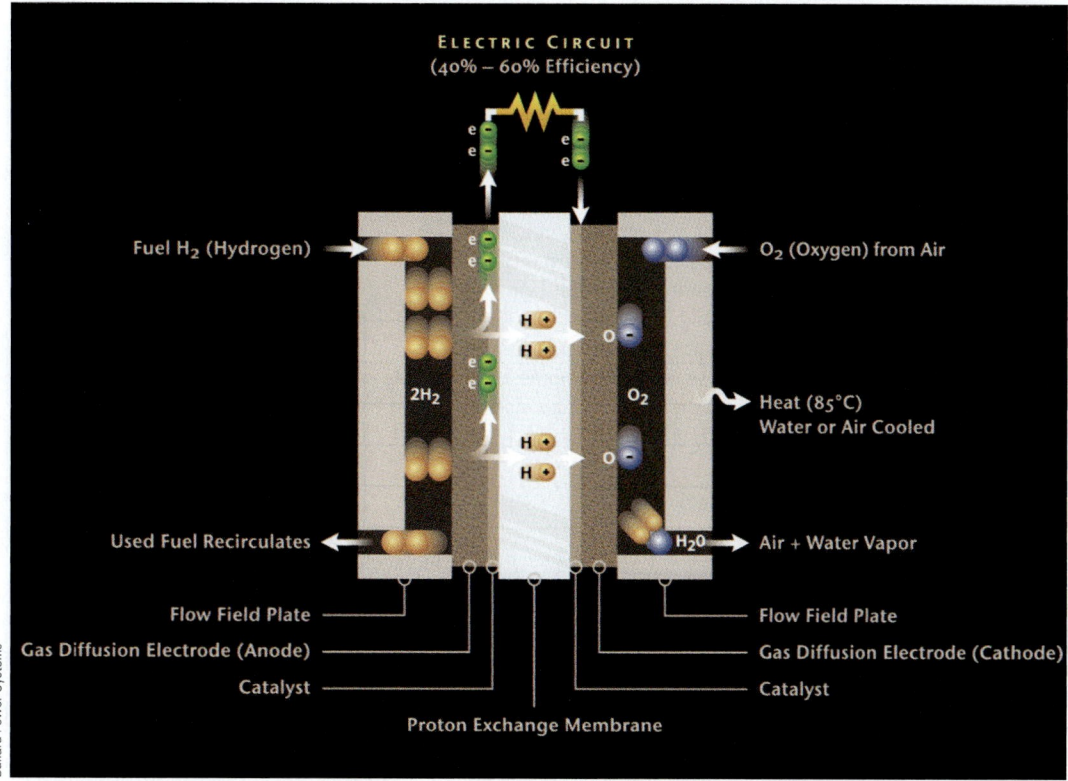

Figure 20.12 **Fuel cell design.** Hydrogen gas is oxidized to $H^+(aq)$ at the anode surface. On the other side of the proton exchange membrane (PEM), oxygen gas is reduced to $OH^-(aq)$. The $H^+(aq)$ ions travel through the PEM and combine with $OH^-(aq)$, forming water.

tants (fuel and oxidant) can be supplied continuously to the cell from an external reservoir.

Although the first fuel cells were constructed more than 150 years ago, little was done to develop this technology until the space program rekindled interest in these devices. Hydrogen-oxygen fuel cells have been used in NASA's Gemini, Apollo, and Space Shuttle programs. Not only are they lightweight and efficient, but they also have the added benefit that they generate drinking water for the ship's crew. The fuel cells on board the Space Shuttle deliver the same power as batteries weighing ten times as much.

In a hydrogen-oxygen fuel cell (Figure 20.12), hydrogen is pumped onto the anode of the cell, and O_2 (or air) is directed to the cathode where the following reactions occur:

Cathode, reduction: $O_2(g) + 2 H_2O(\ell) + 4 e^- \longrightarrow 4 OH^-(aq)$

Anode, oxidation: $2 H_2(g) \longrightarrow 4 H^+(aq) + 4 e^-$

The two halves of the cell are separated by a special material called a proton exchange membrane (PEM). Protons, $H^+(aq)$, formed at the anode traverse the PEM and react with the hydroxide ions produced at the cathode, forming water. The net reaction in the cell is thus the formation of water from H_2 and O_2. Cells currently in use run at temperatures of 70–140 °C and produce about 0.9 V.

Fuel cells for automotive use. Fuel cells are being developed for transportation applications as well as for stationary power sources. Considered promising over the long term, one drawback to hydrogen-based fuel cells is the need to produce hydrogen gas. See "The Chemistry of Fuels and Energy Sources," page 282, for more information.

Chemical Perspectives

Your Next Car?

As a response to federally mandated clean air standards, several major car manufacturers have designed prototype electric cars that use various types of batteries to provide the power to drive the car. The most commonly employed type for automotive use is the lead storage battery, but these devices are problematic owing to their mass. To produce one mole of electrons requires 321 g of reactants in lead storage batteries. As a result, these batteries rank very low among various options in power per kilogram of battery weight. In fact, the power available from any type of battery is much less than that available from an equivalent mass of gasoline. The best values achieved by several high-tech batteries currently being developed are still almost 100 times less than what is available from gasoline.

Chemical System	W · h/kg* (1 W · h = 3600 J)
Lead-acid battery	18—56
Nickel-cadmium battery	33—70
Sodium-sulfur battery	80—140
Lithium polymer battery	150
Gasoline-air combustion engine	12,200

* watt-hour/kilogram

Fuel cells improve the situation. Hydrogen-oxygen fuel cells operate at 40–60% efficiency, and they meet most of the requirements for use in automobiles. They operate at room temperature or slightly above, start rapidly, and develop a high current density. Cost is a serious problem, however, and it appears that a substantial shift away from the internal combustion engine remains a long way off.

The hybrid car appears to offer an interim solution. These vehicles combine a small gasoline-fueled engine with an electric motor and batteries for storage of electric energy.

A 15-kg lead-acid battery has the same amount of stored energy as 59 mL (47 g) of gasoline.

Hybrid cars. These cars combine gasoline-fueled engines with electric motors. Their fuel efficiency is about double that of the current generation of cars using only gasoline engines.

Currently hybrid cars use rechargeable nickel–metal hydride batteries. Electrons are generated when H atoms interact with OH^- ions at the metal alloy anode.

$$Alloy(H) + OH^- \longrightarrow alloy + H_2O + e^-$$

The reaction at the cathode is the same as in ni-cad batteries.

$$NiO(OH) + H_2O + e^- \longrightarrow Ni(OH)_2 + OH^-$$

The alloy used in these batteries is interesting in itself. When researchers first considered the possibility of using hydrogen as a fuel, a problem arose—namely, how to store the element. Certain metallic alloys were found to absorb (and later release) hydrogen in volumes up to 1000 times the alloy volume. Currently alloys based on rare earth elements such as $LaNi_5$ are used in these batteries.

■ **History of Fuel Cells**
William Grove (1811–1896) demonstrated a fuel cell in 1839 at the Royal Institute in London. At the time, Michael Faraday (page 981) directed the institute.

GENERAL
Chemistry ·❀· Now™

See the General ChemistryNow CD-ROM or website:
• **Screen 20.5 Batteries,** to view animations of various types of batteries

20.4—Standard Electrochemical Potentials

Different electrochemical cells produce different voltages: 1.5 V for the LeClanché and alkaline cells, about 1.25 V for a Ni-Cd battery, and about 2.0 V for the individual cells in a lead storage battery. In this section, we want to identify the various factors affecting cell voltages and develop procedures to calculate the voltage of a cell based on the chemistry in the cell and the conditions used.

Electromotive Force

Electrons generated at the anode of an electrochemical cell move through the external circuit toward the cathode, and the force needed to move the electrons arises from a difference in potential energy of electrons at the two electrodes. This difference in potential energy per electrical charge is called the **electromotive force** or **emf**, for which the literal meaning is "force causing electrons to move." Emf has units of volts (V); one volt is the potential difference needed to impart one joule of energy to an electric charge of one coulomb ($1 \, J = 1 \, V \times 1 \, C$). *One coulomb is the quantity of charge that passes a point in an electric circuit when a current of one ampere flows for one second ($1 \, C = 1 \, A \times 1 \, s$).*

It would be possible to calculate cell voltages if a value for the potential energy of the electron for each half-cell were known. The problem is that the potential energy of the electron for a single half-cell (not connected to anything else) is not easily determined. Instead, the emf of a voltaic cell is equated with the cell voltage, E_{cell}, which can be measured.

You may recognize a similarity between emf, enthalpy (H), and free energy (G). Changes in enthalpy and free energy (ΔH and ΔG) can be measured but the value of H or G for a specific substance is generally not known. Values for ΔH_f° and ΔG_f° (Appendix L) were established by choosing a reference point, the elements in their standard states. As you will see next, data on cell potentials are all referenced to the standard hydrogen electrode. All other voltages are measured against this reference cell.

Measuring Standard Potentials

Imagine you planned to study cell voltages in a laboratory with two objectives: (1) to understand the factors that affect these values and (2) to be able to predict the potential of a voltaic cell. You might construct a number of different half-cells, link them together in various combinations to form voltaic cells (as in Figure 20.13), and measure the cell potentials. After a few experiments, it would become apparent that cell voltages depend on a number of factors: the half-cells used (that is, the reaction in each half-cell and the overall or net reaction in the cell), the concentrations of reactants and products in solution, the pressure of gaseous reactants, and the temperature.

So that we can later compare the potential of one half-cell with another, let us measure all cell voltages under **standard conditions**:

- Reactants and products are present in their standard states.
- Solutes in aqueous solution have a concentration of 1.0 M.
- Gaseous reactants or products have a pressure of 1.0 bar.

A cell potential measured under these conditions is called the **standard potential** and is denoted by E_{cell}°. Unless otherwise specified, all values of E_{cell}° refer to measurements at 298 K (25 °C).

Suppose you set up a number of standard half-cells and connect each in turn to a **standard hydrogen electrode (SHE)**. Your apparatus would look like the voltaic cell in Figure 20.13. There are three important things to learn here:

1. *The reaction that occurs.* The reaction occurring in the cell pictured in Figure 20.13 could be *either* the reduction of Zn^{2+} ions with H_2 gas

$$Zn^{2+}(aq) + H_2(g) \longrightarrow Zn(s) + 2 \, H^+(aq)$$

$Zn^{2+}(aq)$ is the oxidizing agent and H_2 is the reducing agent
Standard hydrogen electrode would be the anode or negative electrode

■ **Electrochemical Units**
- The coulomb (abbreviated C) is the standard (SI) unit of electrical charge (Appendix C.3).
- 1 joule = 1 volt × 1 coulomb
- 1 coulomb = 1 ampere × 1 second

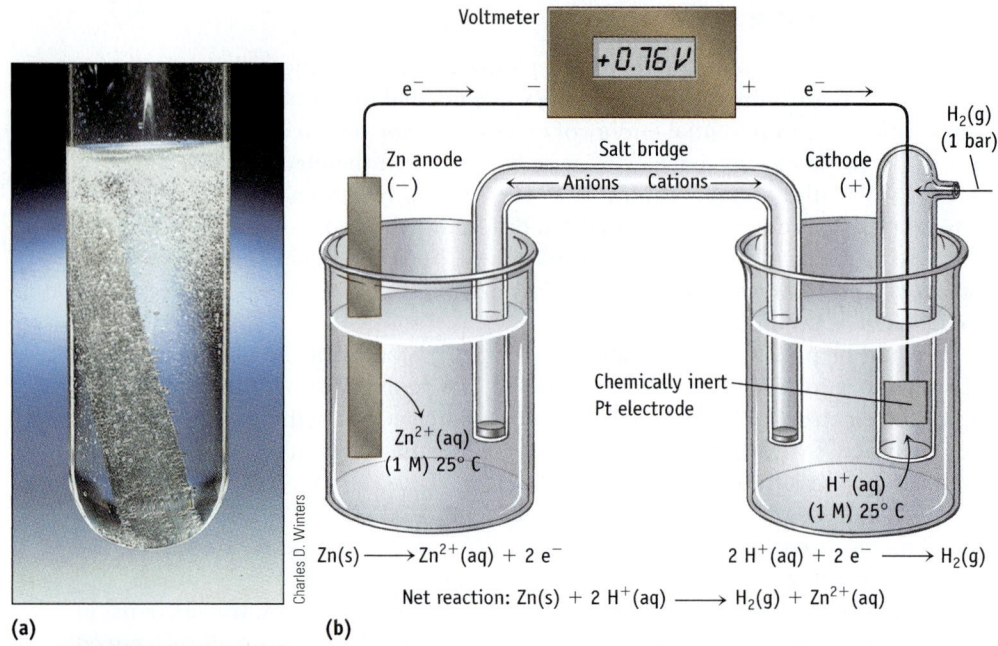

(a)   (b)

Active Figure 20.13 A voltaic cell using Zn|Zn²⁺(aq, 1.0 M) and H₂(1 bar)|H⁺(aq, 1.0 M) half-cells. (a) The reaction of zinc and H⁺ ions is product-favored. This is reflected by a potential of 0.76 V generated by this cell. (b) The electrode in the H₂(1 bar)|H⁺(aq, 1.0 M) half-cell is the cathode, and the Zn electrode is the anode. Electrons flow in the external circuit to the hydrogen half-cell from the zinc half-cell. The positive sign of the measured voltage indicates that the hydrogen electrode is the cathode or positive electrode.

GENERAL
Chemistry Now™ See the General ChemistryNow CD-ROM or website to explore an interactive version of this figure accompanied by an exercise.

or the reduction of H⁺(aq) ions by Zn(s)

$$Zn(s) + 2\,H^+(aq) \longrightarrow Zn^{2+}(aq) + H_2(g)$$

Zn is the reducing agent and H⁺(aq) is the oxidizing agent

Standard hydrogen electrode would be the cathode or positive electrode

All the substances named in these equations are present in the cell. The reaction that actually occurs is the product-favored reaction. That is, *the reaction occurring is the one in which the reactants are the stronger reducing and oxidizing agents.*

2. *Direction of electron flow in the external circuit.* In a voltaic cell, electrons always flow from the anode (negative electrode) to the cathode (positive electrode). That is, *electrons move from the electrode of higher potential energy to the one of lower potential energy.* We can tell the direction of electron movement by placing a voltmeter in the circuit. A positive potential is observed if the voltmeter terminal with a plus sign (+) is connected to the positive electrode [and the terminal with the minus sign (−) is connected to the negative electrode]. Connected in the opposite way (plus to minus and minus to plus) will give a negative value on the digital readout.

3. *Cell potential.* In Figure 20.13 the voltmeter is hooked up with its positive terminal connected to the hydrogen half-cell, and a reading of +0.76 V is observed. The hydrogen electrode is thus the positive electrode or cathode, and the reactions occurring in this cell must be

A Closer Look

EMF, Cell Potential, and Voltage

Emf and *cell potential* (E_{cell}) are often used synonymously but the two are subtly different. E_{cell} is a measured quantity, so its value is affected by how the measurement is made. To understand this point, consider as an analogy water in a pipe under pressure. Water pressure can be viewed as analogous to emf; it represents a force that will cause water in the pipe to move. If we open a faucet, water will flow. Opening the faucet will, however, decrease the pressure in the system.

Emf is the potential difference when no current flows. To measure E_{cell}, a voltmeter is placed in the external circuit. Although voltmeters have high internal resistance to minimize current flow, a small current flows nonetheless. As a result, the value of E_{cell} will be slightly different than the emf.

Finally, there is a difference between a *potential* and a *voltage*. The voltage of a cell has a magnitude but no sign. In contrast, the potential of a half-reaction or a cell has a sign (+ or −) and a magnitude.

Reduction, cathode:	$2\ H^+(aq) + 2\ e^- \longrightarrow H_2(g)$
Oxidation, anode:	$Zn(s) \longrightarrow Zn^{2+}(aq) + 2\ e^-$
Net cell reaction:	$Zn(s) + 2\ H^+(aq) \longrightarrow Zn^{2+}(aq) + H_2(g)$

This result confirms that, of the two oxidizing agents present in the cell, $H^+(aq)$ is better than $Zn^{2+}(aq)$ and that Zn metal is a better reducing agent than H_2 gas.

A potential of +0.76 V was measured for the oxidation of zinc with hydrogen ion. *This value reflects the difference in potential energy of an electron at each electrode.* From the direction of flow of electrons in the external circuit (Zn electrode ⟶ H_2 electrode), we conclude that the potential energy of an electron at the zinc electrode is higher than the potential energy of the electron at the hydrogen electrode.

Hundreds of electrochemical cells like that shown in Figure 20.13 can be set up, allowing us to determine the relative oxidizing or reducing ability of various chemical species and to determine the electrical potential generated by the reaction under standard conditions. A few results are given in Figure 20.14, where half-reactions are listed as reductions. That is, the chemical species on the left is acting as an oxidizing agent. In Figure 20.14, we list them in *descending* ability to act as oxidizing agents.

Standard Reduction Potentials

By doing experiments such as that as illustrated by Figure 20.13, we not only have a notion of the *relative* oxidizing and reducing abilities of various chemical species, but we can also rank them quantitatively.

If E°_{cell} is a measure of the standard potential for the cell, then $E^\circ_{cathode}$ and E°_{anode} can be taken as a measure of electrode potential. Because E°_{cell} reflects the *difference* in electrode potentials, E°_{cell} must be the difference between $E^\circ_{cathode}$ and E°_{anode}.

$$E^\circ_{cell} = E^\circ_{cathode} - E^\circ_{anode} \quad (20.1)$$

Equation 20.1
Equation 20.1 is another example of calculating a change from $X_{final} - X_{initial}$. Electrons move to the cathode (the "final" state) from the anode (the "initial" state). Thus, Equation 20.1 resembles equations you have seen previously in this book (such as Equations 6.6 and 7.5).

Here, $E^\circ_{cathode}$ and E°_{anode} are the *standard reduction potentials* for the half-cell reactions that occur at the cathode and anode, respectively. Equation 20.1 is important for three reasons:

- If we have values for $E^\circ_{cathode}$ and E°_{anode}, we can calculate the standard potential, E°_{cell}, for a voltaic cell.
- *When the calculated value of E°_{cell} is positive, the reaction is predicted to be product-favored as written.* Conversely, if the calculated value of E°_{cell} is negative, the

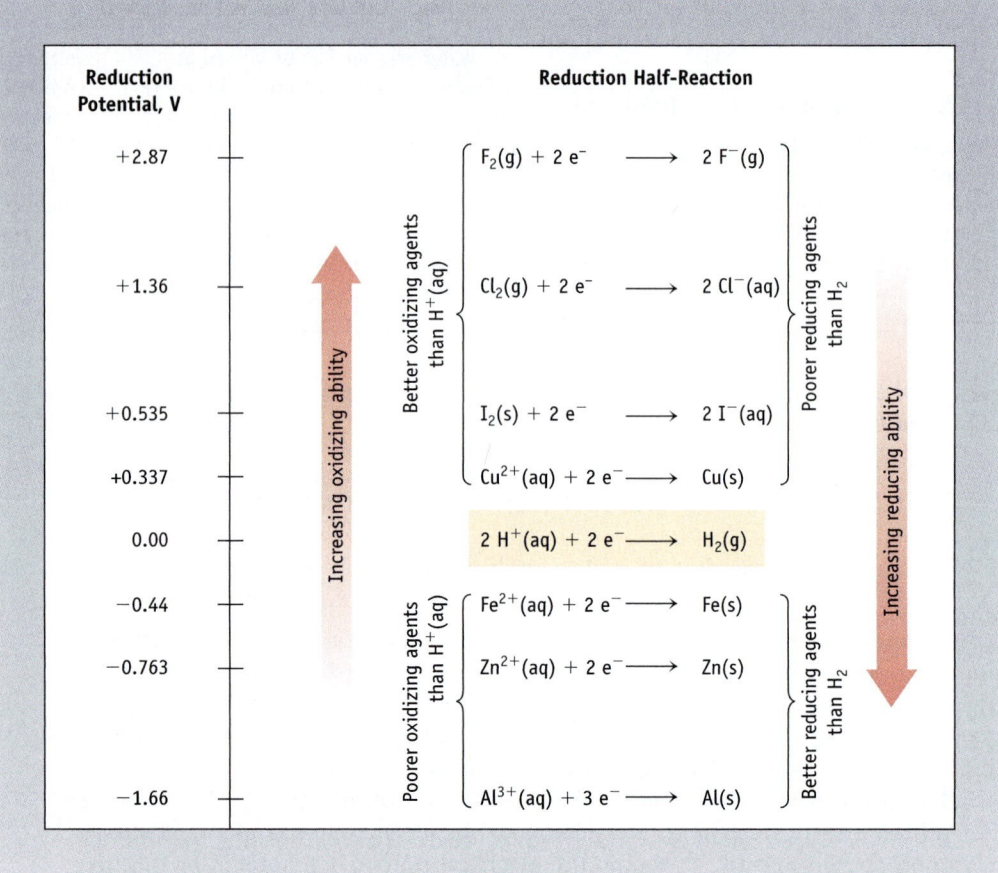

Figure 20.14   A potential ladder for reduction half-reactions. The relative position of a half-reaction on this potential ladder reflects the relative ability of the species at the left to act as an oxidizing agent. The higher the compound or ion is in the list, the better it is as an oxidizing agent. Conversely, the atoms or ions on the right are reducing agents. The lower they are in the list, the better they are as reducing agents. The potential for each half-reaction is given with its reduction potential, $E°_{cathode}$. (For more information see J. R. Runo and D. G. Peters: *Journal of Chemical Education*, Vol. 70, p. 708, 1993.)

reaction is predicted to be reactant-favored. The reaction will be product-favored in a direction opposite to the way it is written.

- If we measure $E°_{cell}$ and know either $E°_{cathode}$ or $E°_{anode}$, we can calculate the other value. This value would tell us how one half-cell reaction compares with others in terms of relative oxidizing or reducing ability.

But here is a dilemma. One cannot measure individual half-cell potentials. Instead, scientists have assigned a potential of 0.00 V to the half-reaction that occurs at a standard hydrogen electrode (SHE).

$$2 \text{ H}^+(\text{aq, 1 M}) + 2 \text{ e}^- \longrightarrow \text{H}_2(\text{g, 1 bar}) \qquad E° = 0.00 \text{ V}$$

With this standard, we can now set up experiments such as those in Figures 20.8 and 20.13 to determine $E°$ for half-cells by measuring $E°_{cell}$ where one of the electrodes is the standard hydrogen electrode. We can then quantify the information in Figure 20.14 and use these values to make predictions about $E°_{cell}$ for new voltaic cells.

Table 20.1 Standard Reduction Potentials in Aqueous Solution at 25 °C*

Reduction Half-Reaction		$E°$ (V)
$F_2(g) + 2 e^-$	$\longrightarrow 2 F^-(aq)$	+2.87
$H_2O_2(aq) + 2 H^+(aq) + 2 e^-$	$\longrightarrow 2 H_2O(\ell)$	+1.77
$PbO_2(s) + SO_4^{2-}(aq) + 4 H^+(aq) + 2 e^-$	$\longrightarrow PbSO_4(s) + 2 H_2O(\ell)$	+1.685
$MnO_4^-(aq) + 8 H^+(aq) + 5 e^-$	$\longrightarrow Mn^{2+}(aq) + 4 H_2O(\ell)$	+1.51
$Au^{3+}(aq) + 3 e^-$	$\longrightarrow Au(s)$	+1.50
$Cl_2(g) + 2 e^-$	$\longrightarrow 2 Cl^-(aq)$	+1.36
$Cr_2O_7^{2-}(aq) + 14 H^+(aq) + 6 e^-$	$\longrightarrow 2 Cr^{3+}(aq) + 7 H_2O (\ell)$	+1.33
$O_2(g) + 4 H^+(aq) + 4 e^-$	$\longrightarrow 2 H_2O(\ell)$	+1.229
$Br_2(\ell) + 2 e^-$	$\longrightarrow 2 Br^-(aq)$	+1.08
$NO_3^-(aq) + 4 H^+(aq) + 3 e^-$	$\longrightarrow NO(g) + 2 H_2O(\ell)$	+0.96
$OCl^-(aq) + H_2O(\ell) + 2 e^-$	$\longrightarrow Cl^-(aq) + 2 OH^-(aq)$	+0.89
$Hg^{2+}(aq) + 2 e^-$	$\longrightarrow Hg(\ell)$	+0.855
$Ag^+(aq) + e^-$	$\longrightarrow Ag(s)$	+0.799
$Hg_2^{2+}(aq) + 2 e^-$	$\longrightarrow 2 Hg(\ell)$	+0.789
$Fe^{3+}(aq) + e^-$	$\longrightarrow Fe^{2+}(aq)$	+0.771
$I_2(s) + 2 e^-$	$\longrightarrow 2 I^-(aq)$	+0.535
$O_2(g) + 2 H_2O(\ell) + 4 e^-$	$\longrightarrow 4 OH^-(aq)$	+0.40
$Cu^{2+}(aq) + 2 e^-$	$\longrightarrow Cu(s)$	+0.337
$Sn^{4+}(aq) + 2 e^-$	$\longrightarrow Sn^{2+}(aq)$	+0.15
$2 H^+(aq) + 2 e^-$	$\longrightarrow H_2(g)$	0.00
$Sn^{2+}(aq) + 2 e^-$	$\longrightarrow Sn(s)$	−0.14
$Ni^{2+}(aq) + 2 e^-$	$\longrightarrow Ni(s)$	−0.25
$V^{3+}(aq) + e^-$	$\longrightarrow V^{2+}(aq)$	−0.255
$PbSO_4(s) + 2 e^-$	$\longrightarrow Pb(s) + SO_4^{2-}(aq)$	−0.356
$Cd^{2+}(aq) + 2 e^-$	$\longrightarrow Cd(s)$	−0.40
$Fe^{2+}(aq) + 2 e^-$	$\longrightarrow Fe(s)$	−0.44
$Zn^{2+}(aq) + 2 e^-$	$\longrightarrow Zn(s)$	−0.763
$2 H_2O(\ell) + 2 e^-$	$\longrightarrow H_2(g) + 2 OH^-(aq)$	−0.8277
$Al^{3+}(aq) + 3 e^-$	$\longrightarrow Al(s)$	−1.66
$Mg^{2+}(aq) + 2 e^-$	$\longrightarrow Mg(s)$	−2.37
$Na^+(aq) + e^-$	$\longrightarrow Na(s)$	−2.714
$K^+(aq) + e^-$	$\longrightarrow K(s)$	−2.925
$Li^+(aq) + e^-$	$\longrightarrow Li(s)$	−3.045

Increasing strength of oxidizing agents (left arrow)

Increasing strength of reducing agents (right arrow)

* In volts (V) versus the standard hydrogen electrode.

Tables of Standard Reduction Potentials

The experimental approach just described leads to lists of $E°$ values such as seen in Figure 20.14 and Table 20.1. Let us list some important points concerning these tables and then illustrate them in the discussion and examples that follow.

1. As in Figure 20.14 reactions are written as "oxidized form + electrons $\longrightarrow$ reduced form." The species on the left side of the reaction arrow is an oxidizing agent, and the species on the right side of the reaction arrow is a reducing agent. Therefore, *all potentials are for reduction reactions.*

■ **$E°$ Values**
An extensive listing of $E°$ values is found in Appendix M, and still larger tables of data can be found in chemistry reference books. A common convention, used in Appendix M, lists standard reduction potentials in two groups, one for acid and neutral solutions and the other for basic solutions.

2. The more positive the value of $E°$ for the reactions in Table 20.1, the better the oxidizing ability of the ion or compound on the left side of the reaction. This means $F_2(g)$ *is the best oxidizing agent in the table.*

$$F_2(g, 1 \text{ atm}) + 2\,e^- \longrightarrow 2\,F^-(aq, 1\,M) \quad E° = +2.87 \text{ V}$$

Lithium ion at the lower-left corner of the table is the poorest oxidizing agent because its $E°$ value is the most negative.

3. The oxidizing agents in the table (the ions, elements, and compounds at the left) increase in strength from the bottom to the top of the table.

4. The more negative the value of the reduction potential, $E°$, the less likely the half-reaction will occur as a reduction, and the more likely the reverse half-reaction will occur (as an oxidation). Thus, Li(s) is the strongest reducing agent in the table, and F^- is the weakest reducing agent. The reducing agents in the table (the ions, elements, and compounds at the right) increase in strength from the top to the bottom.

5. When a reaction is reversed (to give "reduced form $\longrightarrow$ oxidized form + electrons"), the sign of $E°$ is reversed but the value of $E°$ is unaffected.

$$Fe^{3+}(aq, 1\,M) + e^- \longrightarrow Fe^{2+}(aq, 1\,M) \quad E° = +0.771 \text{ V}$$
$$Fe^{2+}(aq, 1\,M) \longrightarrow Fe^{3+}(aq, 1\,M) + e^- \quad E° = -0.771 \text{ V}$$

If a reaction is product-favored in one direction, it is reactant-favored in the opposite direction.

6. The reaction between any substance on the left in this table (an oxidizing agent) with any substance lower than it on the right (a reducing agent) is product-favored under standard conditions. This has been called the *northwest-southeast rule*: Product-favored reactions will always involve a reducing agent that is "southeast" of the proposed oxidizing agent.

■ **Northwest-Southeast Rule**
This guideline reflects the idea of moving down a potential "ladder" in a product-favored reaction.

Reduction Half-Reaction

$$I_2(s) + 2\,e^- \longrightarrow 2\,I^-(aq)$$
$$Cu^{2+}(aq) + 2\,e^- \longrightarrow Cu(s)$$
$$2\,H^+(aq) + 2\,e^- \longrightarrow H_2(g)$$
$$Fe^{2+}(aq) + 2\,e^- \longrightarrow Fe(s)$$
$$Zn^{2+}(aq) + 2\,e^- \longrightarrow Zn(s)$$

The northwest-southeast rule: The reducing agent always lies to the southeast of the oxidizing agent in a product-favored reaction.

For example, Zn can reduce Fe^{2+}, H^+, Cu^{2+}, and I_2, but Cu can reduce only I_2.

7. The algebraic sign of the half-reaction potential is the sign of the electrode when it is attached to the H_2/H^+ standard cell (see Figures 20.8 and 20.13).

8. Electrochemical potentials depend on the nature of the reactants and products and their concentrations, not on the quantities of material used. Therefore, changing the stoichiometric coefficients for a half-reaction does not change the value of $E°$. For example, the reduction of Fe^{3+} has an $E°$ of $+0.771$ V whether the reaction is written as

■ **Changing Stoichiometric Coefficients**
The volt is defined as "energy/charge" (V = J/C). Multiplying a reaction by some number causes both the energy and the charge to be multiplied by that number. Thus, the ratio "energy/charge = volt" does not change.

$$Fe^{3+}(aq, 1\,M) + e^- \longrightarrow Fe^{2+}(aq, 1\,M) \quad E° = +0.771 \text{ V}$$

or as

$$2\,Fe^{3+}(aq, 1\,M) + 2\,e^- \longrightarrow 2\,Fe^{2+}(aq, 1\,M) \quad E° = +0.771 \text{ V}$$

Using Tables of Standard Reduction Potentials

Tables or "ladders" of standard reduction potentials are immensely useful. They allow you to predict the potential of a new voltaic cell, provide information that can be used to balance redox equations, and help predict which redox reactions are product-favored. Let us expand on each of these ideas.

Calculating Cell Potentials, $E°_{cell}$

The standard reduction potentials for half-reactions were obtained by measuring cell potentials. It makes sense, therefore, that these values can be combined to give the potential of some new cell.

The net reaction occurring in a voltaic cell using silver and copper half-cells is

$$2\ Ag^+(aq) + Cu(s) \longrightarrow 2\ Ag(s) + Cu^{2+}(aq)$$

The silver electrode is the cathode and the copper electrode is the anode. We know this because silver ion is reduced (to silver metal) and copper metal is oxidized (to Cu^{2+} ions). (Recall that oxidations always occur at the anode and reductions at the cathode.) Also notice that the $Cu^{2+}|Cu$ half-reaction is "southeast" of the $Ag^+|Ag$ half-reaction in the potential ladder (Table 20.1).

$E°_{cathode} = +0.799\ V \qquad Ag^+(aq) + e^- \longrightarrow Ag(s)$

"Distance" from $E°_{cathode}$ to $E°_{anode}$ is $0.799\ V - 0.337\ V = 0.462\ V$.

Cu is "southeast" of Ag^+

$E°_{anode} = +0.337\ V \qquad Cu^{2+}(aq) + 2e^- \longrightarrow Cu(s)$

The equations for the half-reactions at each electrode and the standard reduction potentials (Table 20.1) are

Reduction, cathode:	$2\ Ag^+(aq) + 2\ e^- \longrightarrow 2\ Ag(s)$
Standard reduction potential for $Ag^+|Ag$:	$E° = +0.799\ V$
Oxidation, anode:	$Cu\ (s) \longrightarrow Cu^{2+}(aq) + 2\ e^-$
Standard reduction potential for $Cu^{2+}|Cu$:	$E° = +0.337\ V$

The potential for the voltaic cell is the difference between the standard reduction potentials.

$$E°_{cell} = E°_{cathode} - E°_{anode}$$
$$E°_{cell} = (+0.799\ V) - (+0.337\ V)$$
$$E°_{cell} = +0.462\ V$$

Notice that the value of $E°_{cell}$ is related to the "distance" between the cathode and anode reactions on the potential ladder. The products have a lower potential energy than the reactants (we have moved down the potential ladder) and the cell potential, $E°_{cell}$, has a positive value.

A positive potential calculated for the $Ag^+|Ag$ and $Cu^{2+}|Cu$ cell ($E°_{cell} = +0.462\ V$) confirms that the reduction of silver ions in water with copper metal is product-favored (Figure 20.1). We might ask, however, about the value of $E°_{cell}$ if a reactant-favored equation had been selected. For example, what is $E°_{cell}$ for the reduction of copper ions with silver metal?

Cathode, reduction: $Cu^{2+}(aq) + 2\,e^- \longrightarrow 2\,Cu(s)$

Anode, oxidation: $2\,Ag(s) \longrightarrow 2\,Ag^+(aq) + 2\,e^-$

Net ionic equation: $2\,Ag(s) + Cu^{2+}(aq) \longrightarrow 2\,Ag^+(aq) + Cu(s)$

Cell Voltage Calculation

$$E^\circ_{cathode} = +0.337\ V \quad \text{and} \quad E^\circ_{anode} = +0.799\ V$$

$$E^\circ_{cell} = E^\circ_{cathode} - E^\circ_{anode} = (+0.337\ V) - (0.799\ V)$$

$$E^\circ_{cell} = -0.462\ V$$

The negative sign for E°_{cell} indicates that the reaction as written is reactant-favored. The products of the reaction (Ag^+ and Cu) have a higher potential energy than the reactants (Ag and Cu^{2+}). We have moved *up* the potential ladder. For the indicated reaction to occur, a potential of at least 0.46 V would have to be imposed on the system by an external source of electricity (see Section 20.7).

GENERAL
Chemistry⚛Now™

See the General ChemistryNow CD-ROM or website:

- **Screen 20.6 Electrochemical Cells and Potentials,** for a demonstration of the potentials of various cells

- **Screen 20.7 Standard Potentials,** for a simulation and tutorial on calculating E°_{cell}

Exercise 20.6—Calculating Standard Cell Potentials

The net reaction that occurs in a voltaic cell is

$$Zn(s) + 2\,Ag^+(aq) \longrightarrow Zn^{2+}(aq) + 2\,Ag$$

Identify the half-reactions that occur at the anode and the cathode, and calculate a potential for the cell assuming standard conditions.

Relative Strengths of Oxidizing and Reducing Agents

Five half-reactions, selected from Table 20.1, are arranged from the half-reaction with the highest (most positive) E° value to the one with the lowest (most negative) value.

E°, V			Reduction Half-Reaction
+1.36			$Cl_2(g) + 2\,e^- \longrightarrow 2\,Cl^-(aq)$
+0.80			$Ag^+(aq) + e^- \longrightarrow Ag(s)$
0.00	Increasing strength	as oxidizing agents	$2\,H^+(aq) + 2\,e^- \longrightarrow H_2(g)$
−0.25			$Ni^{2+}(aq) + 2\,e^- \longrightarrow Ni(s)$
−0.76			$Zn^{2+}(aq) + 2\,e^- \longrightarrow Zn(s)$

Listing half-reactions in this order matches important trends in chemical behavior (see point 3 on page 968).

- The list on the left is headed by Cl_2, an element that is a strong oxidizing agent and thus is easily reduced. At the bottom of the list is $Zn^{2+}(aq)$, an ion not easily reduced and thus a poor oxidizing agent.

- On the right, the list is headed by $Cl^-(aq)$, an ion that can be oxidized to Cl_2 only with difficulty. It is a very poor reducing agent. At the bottom of the list is zinc metal, which is quite easy to oxidize and a good reducing agent.

By arranging these half-reactions based on $E°$ values, we have also arranged the chemical species on the two sides of the equation in order of their strengths as oxidizing or reducing agents. In this list, from strongest to weakest, the order is

Oxidizing agents: $Cl_2 > Ag^+ > H^+ > Ni^{2+} > Zn^{2+}$

strong $\longrightarrow$ weak

Reducing agents: $Cl^- < Ag < H_2 < Ni < Zn$

weak $\longrightarrow$ strong

This example illustrates the use of the table of standard reduction potentials to provide information on relative strengths of oxidizing and reducing agents.

Finally, notice that the value of $E°_{cell}$ is greater the farther apart the oxidizing and reducing agents are on the potential ladder. For example,

$$Zn(s) + Cl_2(g) \longrightarrow Zn^{2+}(aq) + 2\ Cl^-(aq) \qquad E°_{cell} = +2.12\ V$$

is more strongly product-favored than the reduction of hydrogen ions with nickel metal

$$Ni(s) + 2\ H^+(aq) \longrightarrow Ni^{2+}(aq) + H_2(g) \qquad E°_{cell} = +0.25\ V$$

Example 20.5—Ranking Oxidizing and Reducing Agents

Problem Use the table of standard reduction potentials (Table 20.1) to do the following:

(a) Rank the halogens in order of their strength as oxidizing agents.

(b) Decide whether hydrogen peroxide (H_2O_2) in acid solution is a stronger oxidizing agent than Cl_2.

(c) Decide which of the halogens is capable of oxidizing gold metal to $Au^{3+}(aq)$.

Strategy The ability of a species on the left side of Table 20.1 to function as an oxidizing agent declines on descending the list (see points 2–4, page 968).

Solution

(a) *Ranking halogens according to oxidizing ability.* The halogens (F_2, Cl_2, Br_2, and I_2) appear in the upper-left portion of the table, with F_2 being highest, followed in order by the other three species. Their strengths as oxidizing agents are $F_2 > Cl_2 > Br_2 > I_2$. (The ability of bromine to oxidize iodide ions to molecular iodine is illustrated in Figure 20.15.)

(b) *Comparing hydrogen peroxide and chlorine.* H_2O_2 lies just below F_2 but well above Cl_2 in the potential ladder (Table 20.1). Thus, H_2O_2 is a weaker oxidizing agent than F_2 but a stronger one than Cl_2. (Note that the $E°$ value for H_2O_2 refers to an acidic solution and standard conditions.)

(c) *Which halogen will oxidize gold metal to gold(III) ions?* The $Au^{3+}|Au$ half-reaction is listed below the $F_2|F^-$ half-reaction and just above the $Cl_2|Cl^-$ half-reaction. This tells us that, among the halogens, only F_2 is capable of oxidizing Au to Au^{3+} under standard conditions. That is, in the reaction of Au and F_2,

Oxidation, anode: $Au(s) \longrightarrow Au^{3+}(aq) + 3\ e^-$

Reduction, cathode: $F_2(aq) + 2\ e^- \longrightarrow 2\ F^-(aq)$

Net ionic equation: $3\ F_2(aq) + 2\ Au(s) \longrightarrow 6\ F^-(aq) + 2\ Au^{3+}(aq)$

$$E°_{cell} = E°_{cathode} - E°_{anode} = +1.37\ V$$

Figure 20.15 The reaction of bromine and iodide ion. This experiment proves that Br_2 is a better oxidizing agent than I_2.

The test tube contains an aqueous solution of KI (top layer) and immiscible CCl_4 (bottom layer).

Add Br_2 to solution of KI and shake.

After adding a few drops of Br_2 in water the I_2 produced collects in the bottom CCl_4 layer and gives it a purple color. (The top layer contains excess Br_2 in water.)

Charles D. Winters

F_2 is a stronger oxidizing agent than Au^{3+} so the reaction proceeds from left to right as written. (This is confirmed by a positive value of $E°_{cell}$.) For the reaction of Cl_2 and Au, Table 20.1 shows us that Cl_2 is a *weaker* oxidizing agent than Au^{3+}, so the reaction would be expected to proceed in the opposite direction.

Oxidation, anode: $Au(s) \longrightarrow Au^{3+}(aq) + 3 e^-$

Reduction, cathode: $Cl_2(aq) + 2 e^- \longrightarrow 2 Cl^-(aq)$

Net ionic equation: $3 Cl_2(aq) + 2 Au(s) \longrightarrow 6 Cl^-(aq) + 2 Au^{3+}(aq)$

$$E°_{cell} = E°_{cathode} - E°_{anode} = -0.14 \text{ V}$$

This is confirmed by the negative value for $E°_{cell}$.

Comment In part (c) we calculated $E°_{cell}$ for two reactions. To achieve a balanced net ionic equation we added the half-reactions, but only after multiplying the gold half-reaction by 2 and the halogen half-reaction by 3. (This means 6 mol of electrons was transferred from 2 mol Au to 3 mol Cl_2.) Notice that this multiplication does not change the value of $E°$ for the half-reactions because cell potentials do not depend on the quantity of material.

Exercise 20.7—Relative Oxidizing and Reducing Ability

Which metal in the following list is easiest to oxidize: Fe, Ag, Zn, Mg, Au? Which metal is the most difficult to oxidize?

Product- and Reactant-Favored Oxidation–Reduction Reactions

A table of standard reduction potentials (such as Table 20.1 or Appendix M) is a valuable resource for writing redox equations.

The mechanics of writing a redox equation from half-reactions is illustrated by the oxidation of tin(II) ions by permanganate ions in an acidic solution.

$$Sn^{2+}(aq) + MnO_4^-(aq) \longrightarrow Sn^{4+}(aq) + Mn^{2+}(aq) \quad (not \ balanced)$$

1. Select two half-reactions from the table that include the reactants and products shown in the unbalanced equation. (Here the $MnO_4^-|Mn^{2+}$ half-reaction should be taken from the list of reactions occurring in acid.) Use one as it appears in the table (a reduction reaction) and write the second in the reverse direction (an oxidation process).

Chemical Perspectives

An Electrochemical Toothache!

It was recently reported that a 66-year-old woman had intense pain that was traced to her dental work (*New England Journal of* *Medicine*, Vol. 342, p. 2000, 2003). A root canal job had moved a mercury amalgam filling on one tooth slightly closer to a gold alloy crown on an adjacent tooth. Eating acidic foods caused her intense pain. When dental amalgams of dissimilar metals come in contact with saliva, a voltaic cell is formed that generates potentials up to several hundred millivolts—and you feel it! You can do it, too, if you chew a foil gum wrapper with teeth that have been filled with a dental amalgam. Ouch!

Reduction, cathode: $\quad MnO_4^-(aq) + 8\ H^+(aq) + 5\ e^- \longrightarrow Mn^{2+}(aq) + 4\ H_2O(\ell)$

Oxidation, anode: $\quad Sn^{2+}(aq) \longrightarrow Sn^{4+}(aq) + 2\ e^-$

2. Multiply one or both equations by an integer so that when the two equations are added together the electrons will cancel out. Here the reduction half-reaction is multiplied by 2 and the oxidation half-reaction by 5. This means that 10 mol of electrons is transferred to 2 mol of MnO_4^- ions from 5 mol of Sn^{2+} ions.

$$2[MnO_4^-(aq) + 8\ H^+(aq) + 5\ e^- \longrightarrow Mn^{2+}(aq) + 4\ H_2O(\ell)]$$
$$5[Sn^{2+}(aq) \longrightarrow Sn^{4+}(aq) + 2\ e^-]$$

3. Add the two half-reactions together. Simplify, if necessary, and check the result for mass and charge balance.

$$2\ MnO_4^-(aq) + 16\ H^+(aq) + 5\ Sn^{2+}(aq) \longrightarrow 2\ Mn^{2+}(aq) + 8\ H_2O(\ell) + 5\ Sn^{4+}(aq)$$

A redox equation constructed from two randomly chosen half-reactions could be either product- or reactant-favored. We can determine which in several ways. One approach utilizes Equation 20.1 to calculate a cell voltage from the standard reduction potentials of the two half-reactions. For the Sn^{2+}/MnO_4^- reaction,

$$E^\circ_{cell} = E^\circ_{cathode} - E^\circ_{anode} = (+1.51\ V) - (+0.15\ V) = +1.36\ V$$

The positive value indicates that the reaction is product-favored.

Selecting the half-reaction higher on the potential ladder as the oxidizing agent in the overall reaction (the cathode reaction) will automatically assure a product-favored process. To illustrate, $MnO_4^-(aq)$, a strong oxidizing agent, is high on the list of oxidizing agents. It is capable of oxidizing any species below the $MnO_4^-|Mn^{2+}$ half-reaction and on the right side of the table. Thus, permanganate is capable of oxidizing such species as Cl^-, Br^-, Sn^{2+}, Sn, Fe, Zn, Al, and Li. In contrast, a reaction between $MnO_4^-(aq)$ and F^- (a species on the right, *higher* up on the potential ladder) is reactant-favored. The following example further illustrates this point.

Example 20.6—Using a Table of Standard Reduction Potentials to Predict Chemical Reactions

Problem

(a) Which of the following metals will react with $H^+(aq)$ to produce H_2 in a product-favored reaction: Cu, Al, Ag, Fe, Zn?

(b) Select from Table 20.1 three oxidizing agents that are capable of oxidizing $Cl^-(aq)$ to Cl_2.

The reaction of zinc metal with acids is product-favored. See Example 20.6, part (a).

Charles D. Winters

Strategy Recall that the reaction between any substance on the left in this table (an oxidizing agent) with any substance lower than it on the right (a reducing agent) is product-favored under standard conditions.

Solution

(a) Hydrogen ion is an oxidizing agent, so product-favored reactions will occur between H^+ and reducing agents found on the right side of the table and below the $H^+(aq)/H_2(g)$ half-reaction. Of the metals on our list, three meet this criterion: Al, Fe, and Zn. In contrast, Cu and Ag are located above the $H^+(aq)/H_2$ half-reaction in table 20.1. Reactions of these metals and $H^+(aq)$ are not product-favored.

Reaction	E°_{cell} (V)
Product-favored	
$2\ Al(s) + 6\ H^+(aq) \longrightarrow 2\ Al^{3+}(aq) + 3\ H_2(g)$	+1.66
$Zn(s) + 2\ H^+(aq) \longrightarrow Zn^{2+}(aq) + H_2(g)$	+0.763
$Fe(s) + 2\ H^+(aq) \longrightarrow Fe^{2+}(aq) + H_2(g)$	+0.44
Reactant-favored	
$Cu(s) + 2\ H^+(aq) \longrightarrow Cu^{2+}(aq) + H_2(g)$	−0.337
$2\ Ag(s) + 2\ H^+(aq) \longrightarrow 2\ Ag^+(aq) + H_2(g)$	−0.80

(b) Locate the $Cl_2(g)|Cl^-(aq)$ half-reaction in Table 20.1. Chloride ion, $Cl^-(aq)$, is quite high on the list of species that can be oxidized, so only strong oxidizing agents are capable of oxidizing it to Cl_2. (Conversely, Cl_2 is a strong oxidizing agent, so an oxidant even more powerful than Cl_2 is required to convert Cl^- to Cl_2.) Five substances in Table 20.1 can oxidize $Cl^-(aq)$ to Cl_2: F_2, H_2O_2, $PbO_2(s)$, $MnO_4^-(aq)$, and Au^{3+}. [Notice that reactions with H_2O_2, $PbO_2(s)$, or $MnO_4^-(aq)$ require acid conditions.]

$$H_2O_2(aq) + 2\ H^+(aq) + 2\ Cl^-(aq) \longrightarrow 2\ H_2O(\ell) + Cl_2(g)$$

Comment We checked our predictions by calculating E°_{cell}. The reaction of Cl^- with H_2O_2, for example, has a positive potential ($E^\circ_{cell} = +0.41$ V), as expected for a product-favored reaction.

Exercise 20.8—Using a Table of Standard Reduction Potentials to Predict Chemical Reactions

Determine whether the following redox equations are product-favored. Assume standard conditions.

(a) $Ni^{2+}(aq) + H_2(g) \longrightarrow Ni(s) + 2\ H^+(aq)$

(b) $Fe^{3+}(aq) + 2\ I^-(aq) \longrightarrow Fe^{2+}(aq) + I_2(s)$

(c) $Br_2(\ell) + 2\ Cl^-(aq) \longrightarrow 2\ Br^-(aq) + Cl_2(g)$

(d) $Cr_2O_7^{2-}(aq) + 6\ Fe^{2+}(aq) + 14\ H^+(aq) \longrightarrow 2\ Cr^{3+}(aq) + 6\ Fe^{3+}(aq) + 7\ H_2O(\ell)$

20.5—Electrochemical Cells Under Nonstandard Conditions

Electrochemical cells seldom operate under standard conditions in the real world. Even if the cell is constructed with all dissolved species at 1 M, reactant concentrations decrease and product concentrations increase in the course of the reaction. Changing concentrations of reactants and products will affect the cell voltage. Thus, we need to ask what happens to cell potentials under nonstandard conditions.

The Nernst Equation

Based on both theory and experimental results, it has been determined that cell potentials are related to concentrations of reactants and products, and to temperature, as follows:

$$E = E° - (RT/nF) \ln Q \qquad (20.2)$$

In this equation, which is known as the **Nernst equation**, R is the gas constant $(8.314472 \text{ J/K} \cdot \text{mol})$, T is the temperature (K), and n is the number of moles of electrons transferred between oxidizing and reducing agents (as determined by the balanced equation for the reaction). The symbol F represents the **Faraday constant** $(9.6485338 \times 10^4 \text{ C/mol})$. *One Faraday is the quantity of electric charge carried by one mole of electrons.* The term Q is the reaction quotient, an expression relating the concentrations of the products and reactants raised to an appropriate power as defined by the stoichiometric coefficients in the balanced, net equation [◄ Equation 16.2, Section 16.2]. Substituting values for the constants in Equation 20.2, and using 298 K as the temperature, gives

$$E = E° - \frac{0.0257}{n} \ln Q \qquad \text{at 25 °C} \qquad (20.3)$$

In essence this equation "corrects" the standard potential $E°$ for nonstandard conditions or concentrations.

■ **Walter Nernst (1864–1941)** Nernst was a German physicist and chemist known for his work relating to the third law of thermodynamics.

■ **Units of R and F** The gas constant R has units of J/K · mol, and F has units of coulombs per mol (C/mol). Because 1 J = 1 C · V, the factor RT/F has units of volts.

GENERAL
Chemistry ⚛ Now™

See the General ChemistryNow CD-ROM or website:
• **Screen 20.8 Cells at Nonstandard Conditions,** for a tutorial on the Nernst equation

---

Example 20.7—Using the Nernst Equation

Problem A voltaic cell is set up at 25 °C with the following half-cells: Al^{3+}(0.0010 M) | Al and Ni^{2+}(0.50 M) | Ni. Write an equation for the reaction that occurs when the cell generates an electric current and determine the cell potential.

Strategy First, determine which substance is oxidized (Al or Ni) by looking at the appropriate half-reactions in Table 20.1 and deciding which is the better reducing agent (Examples 20.5 and 20.6). Next, add the half-reactions to determine the net ionic equation and calculate $E°$. Finally, use the Nernst equation to calculate E, the nonstandard potential.

Solution Aluminum metal is a stronger reducing agent than Ni metal. (Conversely, Ni^{2+} is a better oxidizing agent than Al^{3+}.) Therefore, Al is oxidized and the Al^{3+} | Al compartment is the anode.

Cathode, reduction:	$Ni^{2+}(aq) + 2 e^- \longrightarrow Ni(s)$
Anode, oxidation:	$Al(s) \longrightarrow Al^{3+}(aq) + 3 e^-$
Net ionic equation:	$2 Al(s) + 3 Ni^{2+}(aq) \longrightarrow 2 Al^{3+}(aq) + 3 Ni(s)$

$$E°_{cell} = E°_{cathode} - E°_{anode}$$

$$E°_{cell} = (-0.25 \text{ V}) - (-1.66 \text{ V}) = 1.41 \text{ V}$$

The expression for Q is written based on the cell reaction. In the net reaction, $Al^{3+}(aq)$ has a coefficient of 2 so this concentration is squared. Similarly, $[Ni^{2+}(aq)]$ is cubed. Solids are not included in the expression for Q [◀ Section 16.2].

$$Q = \frac{[Al^{3+}]^2}{[Ni^{2+}]^3}$$

The net equation requires transfer of 6 electrons from two Al atoms to three Ni^{2+} ions, so $n = 6$. Substituting for $E°$, n, and Q in the Nernst equation gives

$$E_{cell} = E°_{cell} - \frac{0.0257}{n} \ln \frac{[Al^{3+}]^2}{[Ni^{2+}]^3}$$

$$= +1.41 \text{ V} - \frac{0.0257}{6} \ln \frac{[0.0010]^2}{[0.50]^3}$$

$$= +1.41 \text{ V} - 0.00428 \ln (8.0 \times 10^{-6})$$

$$= +1.41 \text{ V} - 0.00428 (-11.7)$$

$$= \boxed{1.46 \text{ V}}$$

Comment Notice that E_{cell} is larger than $E°_{cell}$ because the product concentration, $[Al^{3+}]$, is much smaller than 1.0 M. Generally, when product concentrations are smaller initially than the reactant concentrations in a product-favored reaction, the cell potential is *more* positive than $E°$.

Exercise 20.9—Using the Nernst Equation

The half-cells $Fe^{2+}(aq, 0.024 \text{ M})|Fe(s)$ and $H^+(aq, 0.056 \text{ M})|H_2(1.0 \text{ bar})$ are linked by a salt bridge to create a voltaic cell. Determine the cell potential, E_{cell}, at 298 K.

Example 20.7 demonstrates the calculation of a cell potential if concentrations are known. It is also useful to apply the Nernst equation in the opposite sense, using a measured cell potential to determine an unknown concentration. A device that does just this is the pH meter (Figure 20.16). In an electrochemical cell in which $H^+(aq)$ is a reactant or product, the cell voltage will vary predictably with the hydrogen ion concentration. The cell voltage is measured and the value is used to calculate pH. Example 20.8 illustrates how E_{cell} varies with the hydrogen ion concentration in a simple cell.

Example 20.8—Variation of E_{cell} with Concentration

Problem A voltaic cell is set up with copper and hydrogen half-cells. Standard conditions are employed in the copper half-cell, $Cu^{2+}(aq, 1.00 \text{ M})|Cu(s)$. The hydrogen gas pressure is 1.00 bar, and $[H^+(aq)]$ in the hydrogen half-cell is the unknown. A value of 0.49 V is recorded for E_{cell} at 298 K. Determine the pH of the solution.

Strategy We first decide which is the better oxidizing and reducing agent so as to decide what net reaction is occurring in the cell. With this known, $E°_{cell}$ can be calculated. The only unknown quantity in the Nernst equation is the concentration of hydrogen ion, from which we can calculate the solution pH.

Solution Hydrogen is a better reducing agent than copper metal, so $Cu(s)|Cu^{2+}(aq, 1.0 \text{ M})$ is the cathode, and $H_2(g, 1.00 \text{ bar})|H^+(aq, ? \text{ M})$ is the anode.

Cathode, reduction: $Cu^{2+}(aq) + 2e^- \longrightarrow Cu(s)$

Anode, oxidation: $H_2(g) \longrightarrow 2 H^+(aq) + 2e^-$

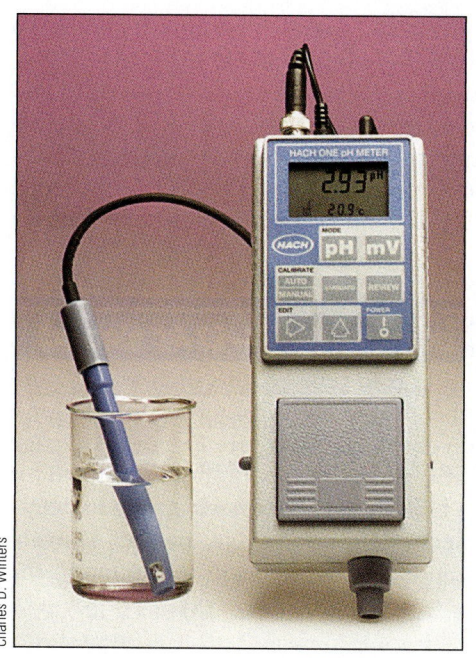

Charles D. Winters

(a)   **(b)**

Figure 20.16 Measuring pH. (a) A portable pH meter that can be used in the field. (b) The tip of a glass electrode for measuring pH. *(See General ChemistryNow Screen 17.4 The pH Scale, to view an animation of the operation of a glass electrode for pH measurement.)*

Net ionic equation: $H_2(g) + Cu^{2+}(aq) \longrightarrow Cu(s) + 2\,H^+(aq)$

$$E^{\circ}_{cell} = E^{\circ}_{cathode} - E^{\circ}_{anode}$$

$$E^{\circ}_{cell} = (+0.37\text{ V}) - (0.00\text{ V}) = +0.37\text{ V}$$

The reaction quotient, Q, is derived from the balanced net ionic equation.

$$Q = \frac{[H^+]^2}{[Cu^{2+}]P_{H_2}}$$

The net equation requires the transfer of two electrons, so $n = 2$. The value of $[Cu^{2+}]$ is 1.00 M, but $[H^+]$ is unknown. Substitute this information into the Nernst equation (and don't overlook the fact that $[H^+]$ is squared in the expression for Q).

$$E_{cell} = E^{\circ}_{cell} - \frac{0.0257}{n} \ln \frac{[H^+]^2}{[Cu^{2+}]P_{H_2}}$$

$$0.49\text{ V} = 0.37\text{ V} - \frac{0.0257}{2} \ln \frac{[H^+]^2}{(1.00)(1.00)}$$

$$-12 = \ln [H^+]^2$$

$$[H^+] = 2.6 \times 10^{-3}\text{ M}$$

$$\boxed{pH = 2.59}$$

Comment The determination of solution pH is clearly an important application of electrochemistry. See Figure 20.16.

Exercise 20.10—Variation of E_{cell} with Concentration

A voltaic cell is set up with an aluminum electrode in a 0.025 M $Al(NO_3)_3(aq)$ solution and an iron electrode in a 0.50 M $Fe(NO_3)_2(aq)$ solution. Calculate the voltage produced by this cell at 25 °C.

In the real world, using a hydrogen electrode in a pH meter is not practical. The apparatus is clumsy, it is anything but robust, and platinum (for the electrode) is costly. Common pH meters use a glass electrode, so-called because it contains a thin glass membrane separating the cell from the solution whose pH is to be measured (Figure 20.16). Inside the glass electrode is a silver wire coated with AgCl and a solution of HCl; outside is the solution of unknown pH to be evaluated. A calomel electrode—a common reference electrode using a mercury(I)–mercury redox couple ($Hg_2Cl_2|Hg$)—serves as the second electrode of the cell. The potential across the glass membrane depends on $[H^+]$. Common pH meters give a direct readout of pH.

Concentrations of ions other than $H^+(aq)$ can also be measured using electrochemistry. Collectively, the electrodes used to measure ion concentrations are known as **ion-selective electrodes**. In many areas of the United States, houses are equipped with water softeners. Their function is to remove ions such as Ca^{2+} and Mg^{2+} from household water and replace the alkaline earth cations with Na^+ ions. They work by using a material called an ion-exchange resin (see Figure 6, on page 1002). To test whether the resin is functioning, the water is periodically sampled for Ca^{2+} ions. One type of water softener has a built-in ion-selective electrode to detect the concentration of this ion. When the electrode indicates that the Ca^{2+} concentration has reached a designated level, it sends a signal to begin regenerating the ion-exchange resin.

20.6—Electrochemistry and Thermodynamics

Work and Free Energy

The first law of thermodynamics [◄ Section 6.4] states that the internal energy change in a system (ΔE) is related to two quantities, heat (q) and work (w): $\Delta E = q + w$. This equation also applies to chemical changes that occur in a voltaic cell. As current flows, energy is transferred from the system (the voltaic cell) to the surroundings.

In a voltaic cell, the decrease in internal energy in the system will manifest itself ideally as electrical work done on the surroundings by the system. In practice, however, some heat evolution by the voltaic cell is usually observed. The maximum work done by an electrochemical system (ideally, assuming no heat is generated) is proportional to the potential difference (volts) and the quantity of charge (coulombs):

$$w_{max} = nFE \tag{20.4}$$

In this equation, E is the cell voltage and nF is the quantity of electric charge transferred from anode to cathode.

The free energy change for a process is, by definition, the maximum amount of work that can be obtained [◄ Section 19.6]. Because the maximum work and the cell potential are related, $E°$ and $\Delta G°$ can be related mathematically (taking care to assign signs correctly). The maximum work done on the surroundings when electricity is produced by a voltaic cell is $+nFE$, with the positive sign denoting an increase in energy in the surroundings. The energy content of the cell decreases by this amount. Thus, ΔG for the voltaic cell has the opposite sign.

$$\Delta G = -nFE \tag{20.5}$$

Under standard conditions, the appropriate equation is

$$\Delta G° = -nFE° \tag{20.6}$$

This expression shows that, the more positive the value of $E°_{cell}$, the larger and more negative the value of $\Delta G°$ for the reaction. That is, the farther apart the half-reactions on the potential ladder, the more strongly product-favored the reaction.

Example 20.9—The Relation Between $E°$ and $\Delta G°$

Problem The standard cell potential, $E°_{cell}$, for the reduction of silver ions with copper metal (Figure 20.5) is $+0.46$ V at 25 °C. Calculate $\Delta G°$ for this reaction.

Strategy We use Equation 20.6, where F is a constant and $E°_{cell}$ is given. The only problem here is to determine the value of n, the number of moles of electrons transferred between copper metal and silver ions in the balanced equation.

Solution In this cell, copper is the anode and silver is the cathode. The overall cell reaction is

$$Cu(s) + 2\ Ag^+(aq) \longrightarrow Cu^{2+}(aq) + 2\ Ag(s)$$

which means that each mole of copper transfers two moles of electrons to two moles of Ag^+ ions. That is, $n = 2$. Now use Equation 20.6.

$$\Delta G° = -nFE° = -(2\ \text{mol e}^-)(96{,}500\ \text{C/mol e}^-)(0.462\ \text{V}) = -89{,}200\ \text{C} \cdot \text{V}$$

Because $1\ \text{C} \cdot \text{V} = 1$ J, we have

$$\Delta G° = -89{,}200\ \text{J} \quad \text{or} \quad -89.2\ \text{kJ}$$

Comment This example demonstrates a very effective method of obtaining thermodynamic values from relatively simple electrochemical experiments.

■ **Units in Equation 20.6**
n has units of mol e$^-$ and F has units of (C/mol e$^-$). Therefore, nF has units of coulombs (C). Because $1\ \text{J} = 1\ \text{C} \cdot \text{V}$, the product nFE will have units of energy (J).

Exercise 20.11—The Relationship Between $E°$ and $\Delta G°_{rxn}$

The following reaction has an $E°_{cell}$ value of -0.76 V:

$$H_2(g) + Zn^{2+}(aq) \longrightarrow Zn(s) + 2\ H^+(aq)$$

Calculate $\Delta G°$ for this reaction. Is the reaction product- or reactant-favored?

$E°$ and the Equilibrium Constant

When a voltaic cell produces an electric current, the reactant concentrations decrease and the product concentrations increase. The cell voltage also changes; as reactants are converted to products, the value of E_{cell} decreases. Eventually the cell potential reaches zero, no further net reaction occurs, and equilibrium is achieved.

This situation can be analyzed using the Nernst equation. When $E_{cell} = 0$, the reactants and products are at equilibrium and the reaction quotient, Q, is equal to the equilibrium constant, K. Substituting the appropriate symbols and values into the Nernst equation,

$$E = 0 = E° - \frac{0.0257}{n} \ln K$$

and collecting terms gives an equation that relates the cell potential and equilibrium constant:

$$\ln K = \frac{nE°}{0.0257} \quad \text{at 25 °C} \tag{20.7}$$

Equation 20.7 can be used to determine values for equilibrium constants, as illustrated in Example 20.10 and Exercise 20.12.

K and E°
The farther apart half-reactions for a product-favored reaction are on the potential ladder, the larger the value of K.

Example 20.10—$E°$ and Equilibrium Constants

Problem Calculate the equilibrium constant for the reaction

$$Fe(s) + Cd^{2+}(aq) \rightleftharpoons Fe^{2+}(aq) + Cd(s)$$

Strategy First determine $E°_{cell}$ from $E°$ values for the two half-reactions (see Examples 20.5 and 20.6). This also gives us the value of n, the other parameter required in Equation 20.7.

Solution The half-reactions and $E°$ values are

Cathode, reduction: $Cd^{2+}(aq) + 2\,e^- \longrightarrow Cd(s)$

Anode, oxidation: $Fe(s) \longrightarrow Fe^{2+}(aq) + 2\,e^-$

Net ionic equation: $Fe(s) + Cd^{2+}(aq) \rightleftharpoons Fe^{2+}(aq) + Cd(s)$

$$E°_{cell} = E°_{cathode} - E°_{anode}$$

$$E°_{cell} = (-0.40 \text{ V}) - (-0.44 \text{ V}) = +0.04 \text{ V}$$

Now substitute $n = 2$ and $E°_{cell}$ into Equation 20.7.

$$\ln K = \frac{nE°}{0.0257} = \frac{(2)(0.04 \text{ V})}{0.0257} = 3.1$$

$$K = 20$$

Comment The relatively small positive voltage (0.040 V) for the standard cell indicates that the cell reaction is only mildly product-favored. A value of 20 for the equilibrium constant is in accord with this observation.

Exercise 20.12—$E°$ and Equilibrium Constants

Calculate the equilibrium constant at 25 °C for the reaction

$$2\,Ag^+(aq) + Hg(\ell) \rightleftharpoons 2\,Ag(s) + Hg^{2+}(aq)$$

The relationship between $E°$ and K can be used to obtain equilibrium constants for many different chemical systems. For example, let us construct an electrode in which an insoluble ionic compound is a component of the half-cell. For this purpose, a silver electrode with a surface layer of AgCl can be prepared. The reaction occurring at this electrode is then

$$AgCl(s) + e^- \longrightarrow Ag(s) + Cl^-(aq)$$

The standard reduction potential for this half-cell (Appendix M) is +0.222 V. When this half-reaction is paired with a standard silver electrode in an electrochemical cell, the cell reactions are

Cathode, reduction: $AgCl(s) + e^- \longrightarrow Ag(s) + Cl^-(aq)$

Anode, oxidation: $Ag(s) \longrightarrow Ag^+(aq) + e^-$

Net ionic equation: $AgCl(s) \rightleftharpoons Ag^+(aq) + Cl^-(aq)$

$$E°_{cell} = E°_{cathode} - E°_{anode} = (+0.222 \text{ V}) - (+0.799 \text{ V}) = -0.577 \text{ V}$$

The equation for the net reaction represents the equilibrium of solid AgCl and its ions. The cell potential is negative, indicating a reactant-favored process, as would

be expected based on the low solubility of AgCl. Using Equation 20.7, the value of the equilibrium constant [◀ K_{sp}, Section 18.4] can be obtained from $E°_{cell}$.

$$\ln K = \frac{nE°}{0.0257\ \text{V}} = \frac{(1)(-0.577\ \text{V})}{0.0257\ \text{V}} = -22.5$$

$$K_{sp} = e^{-22.5} = 1.8 \times 10^{-10}$$

Exercise 20.13—Determining an Equilibrium Constant

In Appendix M the following standard reduction potential is reported:

$$[\text{Zn}(\text{CN})_4]^{2-}(\text{aq}) + 2\ \text{e}^- \longrightarrow \text{Zn}(\text{s}) + 4\ \text{CN}^-(\text{aq}) \qquad E° = -1.26\ \text{V}$$

Use this information, along with the data on the $\text{Zn}^{2+}(\text{aq})|\text{Zn}$ half-cell, to calculate the equilibrium constant for the reaction

$$\text{Zn}^{2+}(\text{aq}) + 4\ \text{CN}^-(\text{aq}) \longrightarrow [\text{Zn}(\text{CN})_4]^{2-}(\text{aq})$$

The value calculated is the formation constant for this complex ion at 25 °C.

20.7—Electrolysis: Chemical Change Using Electrical Energy

Electrolysis of water (Figure 20.17a) is a classic chemistry experiment. An electric current is passed through water containing a small amount of an electrolyte, and gaseous hydrogen and oxygen form at the electrodes. This experiment is used to illustrate stoichiometry and gas laws and to show how an energetically disfavored reaction can be carried out.

Figure 20.17 Electrolysis. (a) Electrolysis of water produces hydrogen and oxygen gas. (b) Electroplating adds a layer of metal to the surface of an object, either to protect the object from corrosion or to improve its physical appearance. The procedure uses an electrolysis cell, set up with the object to be plated as the cathode and a solution containing a salt of the metal to be plated.

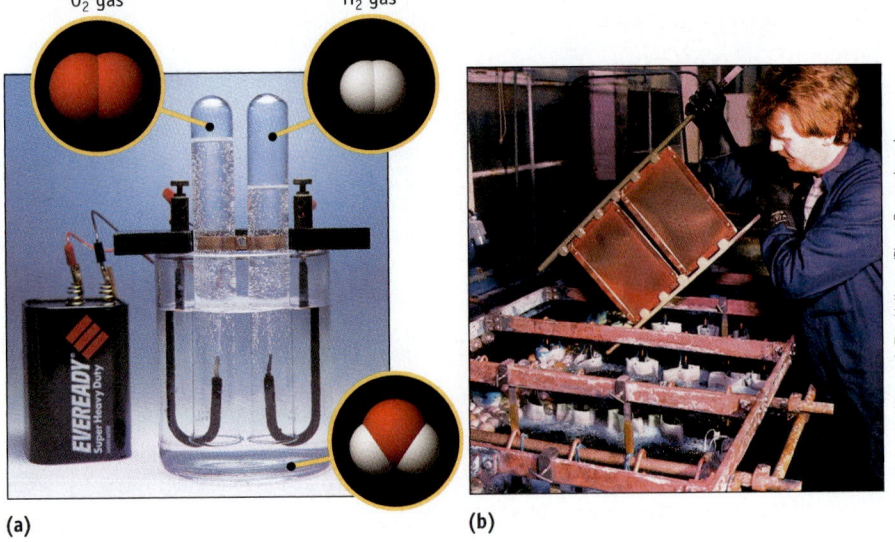

O₂ gas   H₂ gas

(a)   (b)

a, Charles D. Winters; *b*, Tom Hollyman/Photo Researchers, Inc.

Electroplating (Figure 20.17b) is another example of electrolysis. Here, an electric current is passed through a solution containing a salt of the metal to be plated. The object to be plated is the cathode. When metal ions in solution are reduced, the metal deposits on its surface.

Electrolysis is an important procedure because it is widely used in the refining of metals such as aluminum and in the production of chemicals such as chlorine. These important topics are visited in Chapter 21.

Electrolysis of Molten Salts

All electrolysis experiments are the same. The material to be electrolyzed, either a molten salt or a solution, is contained in an electrolysis cell. As was the case with voltaic cells, ions must be present in the liquid or solution for a current to flow. The movement of ions constitutes the electric current within the cell. The cell has two electrodes that are connected to a source of DC (direct-current) voltage. If a high enough voltage is applied, chemical reactions occur at the two electrodes. Reduction occurs at the negatively charged cathode, with electrons being transferred from that electrode to a chemical species in the cell. Oxidation occurs at the positive anode, with electrons from a chemical species being transferred to that electrode.

Let us first focus our attention on the chemical reactions that occur at each electrode in the electrolysis of a molten salt. In molten NaCl (Figure 20.18), sodium ions (Na^+) and chloride ions (Cl^-) are freed from their rigid arrangement in the crystalline lattice at temperatures higher than 800 °C. If a potential is applied to the cell, sodium ions are attracted to the negative electrode and chloride ions are attracted to the positive electrode. If the potential is high enough, chemical reactions occur at each electrode. At the negative cathode, Na^+ ions accept electrons and are reduced to sodium metal (a liquid at this temperature). Simultaneously, at the positive anode, chloride ions give up electrons and form elemental chlorine.

Cathode $(-)$, *reduction:* $2\ Na^+ + 2\ e^- \longrightarrow 2\ Na(\ell)$

Anode $(+)$, *oxidation:* $2\ Cl^- \longrightarrow Cl_2(g) + 2\ e^-$

Net reaction: $2\ Na^+ + 2\ Cl^- \longrightarrow 2\ Na(\ell) + Cl_2(g)$

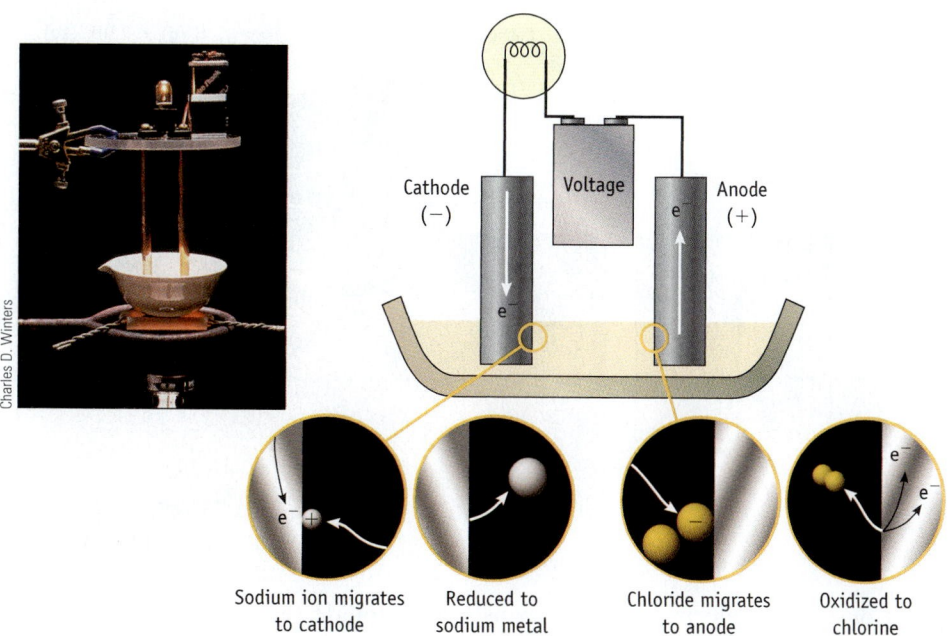

Figure 20.18 **The preparation of sodium and chlorine by the electrolysis of molten NaCl.** In the molten state, sodium ions migrate to the negative cathode, where they are reduced to sodium metal. Chloride ions migrate to the positive anode, where they are oxidized to elemental chlorine.

Electrons move through the external circuit under the force exerted by the applied potential, and the movement of positive and negative ions in the molten salt constitutes the current within the cell. Finally, it is important to recognize that the reaction is *not* product-favored. The energy required for this reaction to occur has been provided by the electric current.

Electrolysis of Aqueous Solutions

Sodium ions (Na^+) and chloride ions (Cl^-) are the primary species present in molten NaCl. Only one of these (Cl^-) can be oxidized, and only one (Na^+) can be reduced. Electrolyses of aqueous solutions are more complicated than electrolyses of molten salts, however, because water is now present. Water is an *electroactive* substance; that is, it can be oxidized or reduced in an electrochemical process.

Problem-Solving Tip 20.3

Electrochemical Conventions: Voltaic Cells and Electrolysis Cells

Whether you are describing a voltaic cell or an electrolysis cell, the terms "anode" and "cathode" always refer to the electrodes at which oxidation and reduction occur, respectively. The polarity of the electrodes is reversed, however.

Type of Cell	Electrode	Function	Polarity
Voltaic	Anode	Oxidation	−
	Cathode	Reduction	+
Electrolysis	Anode	Oxidation	+
	Cathode	Reduction	−

Figure 20.19   Electrolysis of aqueous NaI. A solution of NaI(aq) is electrolyzed, with a potential being applied using an external source of electricity. A drop of phenolphthalein has been added to the solution in this experiment so that the formation of OH⁻(aq) can be detected (by the red color of the indicator in basic solution). Iodine forms at the anode, and H₂ and OH⁻ form at the cathode.

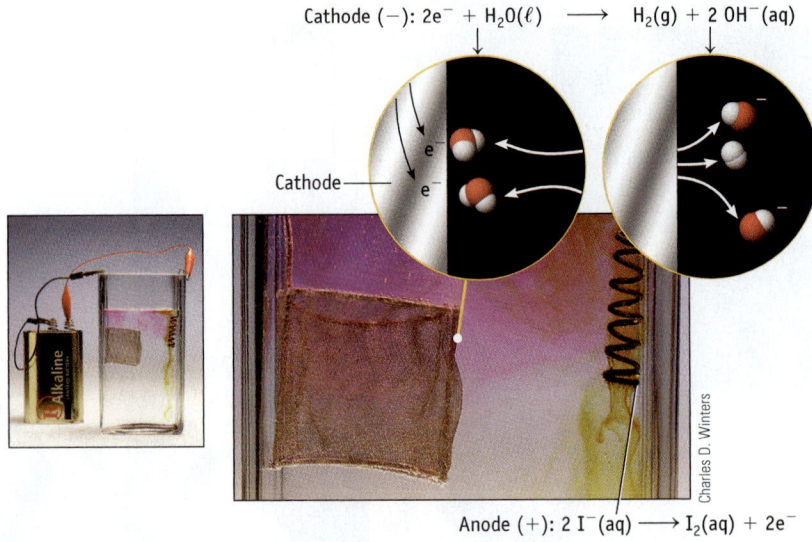

$$\text{Cathode }(-): 2\,e^- + H_2O(\ell) \longrightarrow H_2(g) + 2\,OH^-(aq)$$

Cathode

Charles D. Winters

$$\text{Anode }(+): 2\,I^-(aq) \longrightarrow I_2(aq) + 2\,e^-$$

Consider the electrolysis of aqueous sodium iodide (Figure 20.19). In this experiment the electrolysis cell contains $Na^+(aq)$, $I^-(aq)$, and H_2O molecules. Possible *reduction reactions* at the *cathode* include

$$Na^+(aq) + e^- \longrightarrow Na(s)$$
$$2\,H_2O(\ell) + 2\,e^- \longrightarrow H_2(g) + 2\,OH^-(aq)$$

Possible *oxidation reactions* at the *anode* are

$$2\,I^-(aq) \longrightarrow I_2(aq) + 2\,e^-$$
$$2\,H_2O(\ell) \longrightarrow O_2(g) + 4\,H^+(aq) + 4\,e^-$$

In the electrolysis of aqueous NaI, experiment shows that $H_2(g)$ and $OH^-(aq)$ are formed by water reduction at the cathode, and iodine is formed at the anode (Figure 20.19). Thus, the overall cell process can be summarized by the following equations:

Cathode, reduction: $2\,H_2O(\ell) + 2\,e^- \longrightarrow H_2(g) + 2\,OH^-(aq)$

Anode, oxidation: $2\,I^-(aq) \longrightarrow I_2(aq) + 2\,e^-$

Net ionic equation: $2\,H_2O(\ell) + 2\,I^-(aq) \longrightarrow H_2(g) + 2\,OH^-(aq) + I_2(aq)$

where $E°_{cell}$ has a negative value.

$$E°_{cell} = E°_{cathode} - E°_{anode} = (-0.828\ V) - (+0.621\ V) = -1.449\ V$$

This is a reactant-favored process, which means a potential of at least 1.45 V must be *applied* to the cell for these reactions to occur. If the process had involved the oxidation of water instead of iodide ion at the anode, the required potential would be −2.06 V [$E°_{cathode} - E°_{anode} = (-0.828\ V) - (+1.23\ V)$]. The reaction occurring is the one requiring the smaller applied potential, so the net cell reaction in the electrolysis of NaI(aq) is the oxidation of iodide and reduction of water.

What happens if an aqueous solution of some other metal halide such as $SnCl_2$ is electrolyzed? As before, consult Table 20.1 and consider all possible half-reactions. In this case aqueous Sn^{2+} ion is much more easily reduced ($E° = -0.14\ V$) than water ($E° = -0.83\ V$) at the cathode, so tin metal is produced. At the anode,

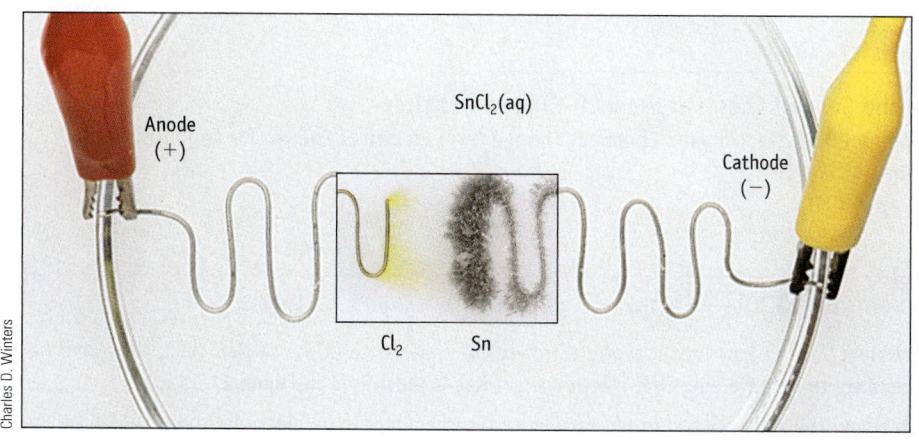

Figure 20.20 **Electrolysis of aqueous tin(II) chloride.** Tin metal collects at the negative cathode. Chlorine gas is formed at the positive anode. Elemental chlorine forms in the cell despite the fact that the potential for the oxidation of Cl^- is more negative than that for the oxidation of water (that is, chlorine should be less easily oxidized than water). The result of chemical kinetics, this process illustrates the complexity of some aqueous electrochemistry.

two oxidations are possible: $Cl^-(aq)$ to $Cl_2(g)$ or H_2O to $O_2(g)$. Experiments show that chloride ion is generally oxidized in preference to water, so the reactions occurring on electrolysis of aqueous tin(II) chloride are (Figure 20.20)

Cathode, reduction:	$Sn^{2+}(aq) + 2\ e^- \longrightarrow Sn(s)$
Anode, oxidation:	$2\ Cl^-(aq) \longrightarrow Cl_2(g) + 2\ e^-$
Net reaction:	$Sn^{2+}(aq) + 2\ Cl^-(aq) \longrightarrow Sn(s) + Cl_2(g)$

$$E^\circ_{cell} = E^\circ_{cathode} - E^\circ_{anode} = (-0.14\ V) - (+1.36\ V) = -1.50\ V$$

Formation of Cl_2 at the anode in the electrolysis of $SnCl_2(aq)$ is contrary to a prediction based on E° values. If the electrode reactions were

Cathode, reduction:	$Sn^{2+}(aq) + 2\ e^- \longrightarrow Sn(s)$
Anode, oxidation:	$2\ H_2O(\ell) \longrightarrow O_2(g) + 4\ H^+(aq) + 4\ e^-$

$$E^\circ_{cell} = (-0.14\ V) - (+1.23\ V) = -1.37\ V$$

a smaller applied potential would seemingly be required. To explain the formation of chlorine instead of oxygen, we must take into account rates of reaction. In the commercially important electrolysis of aqueous NaCl, a voltage high enough to oxidize both Cl^- and H_2O will be used. However, chloride ion is oxidized much faster than H_2O, with the result being that Cl_2 is the major product in this electrolysis.

Another instance in which rates are important concerns electrode materials. Graphite, commonly used to make inert electrodes, can be oxidized. For the half-reaction $CO_2(g) + 4\ H^+(aq) + 4\ e^- \longrightarrow C(s) + 2\ H_2O(\ell)$, E° is $+0.20$ V, indicating that carbon is slightly easier to oxidize than copper ($E^\circ = +0.34$ V). Based on this value, oxidation of a graphite electrode might reasonably be expected to occur during an electrolysis. And indeed it does, albeit slowly; graphite electrodes used in electrolysis cells slowly deteriorate and eventually have to be replaced.

One other factor—the concentration of electroactive species in solution—must be taken into account when discussing electrolyses. As shown in Section 20.6, the potential at which a species in solution is oxidized or reduced depends on concentration. Unless standard conditions are used, predictions based on E° values are merely qualitative. In addition, the rate of a half-reaction depends on the concentration of the electroactive substance at the electrode surface. At a very low concentration, the rate of the redox reaction may depend on the rate at which an ion diffuses from the solution to the electrode surface.

■ **Overvoltage**
Voltages higher than the minimum are typically used to speed up reactions that would otherwise take place only slowly. The term *overvoltage* is often used when describing experiments; it refers to the additional voltage needed to make a reaction occur at a reasonable rate.

GENERAL
Chemistry ·⚛· Now™

See the General ChemistryNow CD-ROM or website:
- **Screen 20.11 Electrolysis: Chemical Change from Electrical Energy,** for an illustration of water electrolysis

---

Example 20.11—Electrolysis of Aqueous Solutions

Problem Predict how products of electrolyses of aqueous solutions of NaF, NaCl, NaBr, and NaI are likely to be different. (The electrolysis of NaI is illustrated in Figure 21.19.)

Strategy The main criterion used to predict the chemistry in an electrolytic cell should be the ease of oxidation and reduction, an assessment based on $E°$ values.

Solution The cathode reaction in all four examples presents no problem—water is reduced to hydroxide ion and H_2 gas in preference to reduction of $Na^+(aq)$ (as in the electrolysis of aqueous NaI). Thus, the primary cathode reaction in all cases is

$$2 H_2O(\ell) + 2 e^- \longrightarrow H_2(g) + 2 OH^-(aq)$$

$$E°_{cathode} = -0.83 \text{ V}$$

At the anode, we need to assess the ease of oxidation of the halide ions relative to water. Based on $E°$ values, the ease of oxidation of halide ions is $I^-(aq) > Br^-(aq) > Cl^-(aq) \gg F^-(aq)$. Fluoride ion is much more difficult to oxidize than water, and electrolysis of an aqueous solution containing this ion results exclusively in O_2 formation. The primary anode reaction for NaF(aq) is

$$2 H_2O(\ell) \longrightarrow O_2(g) + 4 H^+(aq) + 4 e^-$$

Therefore, in this case,

$$E°_{cell} = (-0.83 \text{ V}) - (+1.23 \text{ V}) = -2.06 \text{ V}$$

Bromide and iodide ions are considerably easier to oxidize than chloride ions, however. Recalling that chlorine is the primary product in the electrolysis of chloride salts (Figure 20.20), Br_2 and I_2 may be expected as primary products in the electrolysis of aqueous NaBr and NaI, respectively. For NaBr(aq), the primary anode reaction is

$$2 Br^-(aq) \longrightarrow Br_2(\ell) + 2 e^-$$

so $E°_{cell}$ is

$$E°_{cell} = (-0.83 \text{ V}) - (+1.08 \text{ V}) = -1.91 \text{ V}$$

Exercise 20.14—Electrolysis of Aqueous Solutions

Predict the chemical reactions that will occur at the two electrodes in the electrolysis of an aqueous sodium hydroxide solution. What is the minimum voltage needed to cause this reaction to occur?

20.8—Counting Electrons

Metallic silver is produced at the cathode in the electrolysis of aqueous $AgNO_3$: $Ag^+(aq) + e^- \longrightarrow Ag(s)$. One mole of electrons is required to produce one mole of silver. In contrast, two moles of electrons are required to produce one mole of tin (Figure 20.20):

$$Sn^{2+}(aq) + 2 e^- \longrightarrow Sn(s)$$

It follows that if the number of moles of electrons flowing through the electrolysis cell could be measured, the number of moles of silver or tin produced could be calculated. Conversely, if the amount of silver or tin produced was known, then the number of moles of electrons moving through the circuit could be calculated.

The number of moles of electrons consumed or produced in an electron transfer reaction is obtained by measuring the current flowing in the external electric circuit in a given time. The **current** flowing in an electrical circuit is the amount of charge (in units of coulombs, C) per unit time, and the usual unit for current is the ampere (A). (One ampere equals the passage of one coulomb of charge per second.)

$$\text{Current, } I \text{ (amperes, A)} = \frac{\text{electric charge (coulombs, C)}}{\text{time (seconds, s)}} \qquad (20.8)$$

The current passing through an electrochemical cell and the time for which the current flows are easily measured quantities. Therefore, the charge (in coulombs) that passes through a cell can be obtained by multiplying the current (in amperes) by the time (in seconds). Knowing the charge, and using the Faraday constant as a conversion factor, we can calculate the number of moles of electrons that passed through an electrochemical cell. In turn, we can use this quantity to calculate the quantities of reactants and products. The following examples illustrate this type of calculation.

■ **Faraday Constant**
The Faraday constant is the charge carried by 1 mol of electrons:

$1 F = 9.6485338 \times 10^4$ C/mol e$^-$

GENERAL
Chemistry ⚛ Now™

See the General ChemistryNow CD-ROM or website:

• **Screen 20.12 Coulometry: Counting Electrons,** for a tutorial on quantitative aspects of electrochemistry

Example 20.12—Using the Faraday Constant

Problem A current of 2.40 A is passed through a solution containing Cu^{2+}(aq) for 30.0 minutes, with copper metal being deposited at the cathode. What mass of copper, in grams, is deposited? The reaction at the cathode is Cu^{2+}(aq) $+ 2 \text{ e}^- \longrightarrow Cu(s)$.

Strategy A roadmap for this calculation is as follows:

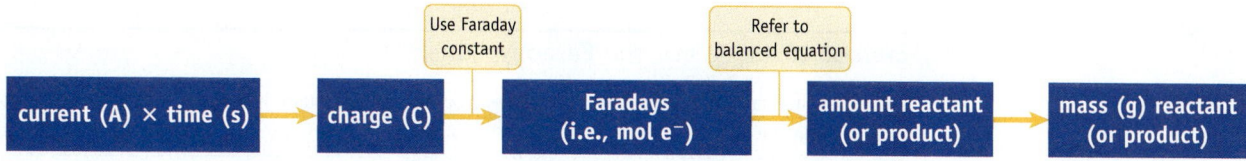

Solution

1. Calculate the charge (number of coulombs) passing through the cell in 30.0 min.

$$\text{Charge (C)} = \text{current (A)} \times \text{time (s)}$$
$$= (2.40 \text{ A})(30.0 \text{ min})(60.0 \text{ s/min})$$
$$= 4.32 \times 10^3 \text{ C}$$

2. Calculate the number of moles of electrons (i.e., the number of Faradays of electricity).

$$\text{mol e}^- = (4.32 \times 10^3 \text{ C})\left(\frac{1 \text{ mol e}^-}{96{,}500 \text{ C}}\right) = 4.48 \times 10^{-2} \text{ mol e}^-$$

3. Calculate the amount of copper and, from this, the mass of copper.

$$\text{Mass of copper} = (4.48 \times 10^{-2} \text{ mol e}^-)\left(\frac{1 \text{ mol Cu}}{2 \text{ mol e}^-}\right)\left(\frac{63.55 \text{ g}}{1 \text{ mol Cu}}\right) = 1.42 \text{ g}$$

Comment The key relation in this calculation is current = charge/time. Most situations will involve knowing two of these three quantities from experiment and calculating the third.

Example 20.13—Using the Faraday Constant

Problem How long must a current of 0.800 A flow to form 2.50 g of silver metal in an electroplating experiment? The cathode reaction is $Ag^+(aq) + e^- \longrightarrow Ag(s)$.

Strategy This problem is the reverse of the problem in Example 20.12. We shall use the roadmap below. Here, we calculate the amount of Ag (mol Ag), then the number of moles of electrons (mol e$^-$), then the charge (C), and finally the time.

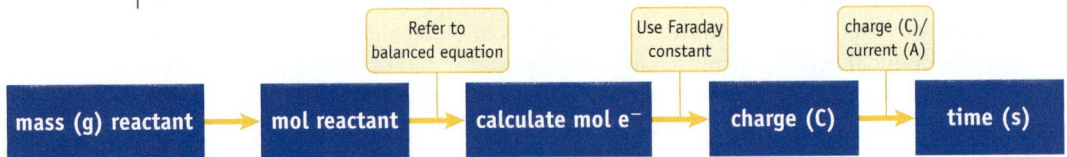

Solution

1. Calculate the moles of electrons required.

$$\text{mol e}^- = (2.50 \text{ g Ag})\left(\frac{1 \text{ mol Ag}}{107.9 \text{ g Ag}}\right)\left(\frac{1 \text{ mol e}^-}{1 \text{ mol Ag}}\right) = 2.32 \times 10^{-2} \text{ mol e}^-$$

2. Calculate the quantity of charge (C).

$$\text{Charge (C)} = (2.32 \times 10^{-2} \text{ mol e}^-)\left(\frac{96,500 \text{ C}}{1 \text{ mol e}^-}\right) = 2240 \text{ C}$$

3. Use the charge and the current (A) to calculate the time required

$$\text{Charge (C)} = \text{current (A)} \times \text{time (s)}$$
$$2240 \text{ C} = (0.800 \text{ A})(\text{time, s})$$
$$\text{Time} = 2.80 \times 10^3 \text{ s (or 46.5 min)}$$

Exercise 20.15—Using the Faraday Constant

Calculate the mass of O_2 produced in the electrolysis of water, using a current of 0.445 A for a period of 45 minutes.

Exercise 20.16—Using the Faraday Constant

In the commercial production of sodium by electrolysis, the cell operates at 7.0 V and a current of 25×10^3 A. What mass of sodium can be produced in one hour?

Chapter Goals Revisited

When you have finished studying this chapter, you should ask whether you have met the chapter goals. In particular, you should be able to

Balance equations for oxidation–reduction reactions in acidic or basic solutions using the half-reaction approach (Section 20.1) General ChemistryNow homework: Study Question(s) 2, 6

Understand the principles underlying voltaic cells

a. In a voltaic cell identify the half-reactions occurring at the anode and the cathode, the polarity of the electrodes, the direction of electron flow in the external circuit, and the direction of ion flow in the salt bridge (Section 20.2). General ChemistryNow homework: SQ(s) 8

b. Appreciate the chemistry and advantages and disadvantages of dry cells, alkaline batteries, lead storage batteries, and Ni-cad batteries (Section 20.3).

c. Understand how fuel cells work, and recognize the difference between batteries and fuel cells (Section 20.3).

Understand how to use electrochemical potentials

a. Understand the process by which standard reduction potentials are determined and identify standard conditions as applied to electrochemistry (Section 20.4).

b. Describe the standard hydrogen electrode ($E° = 0.00$ V) and explain how it is used as the standard to determine the standard potentials of half-reactions (Section 20.4).

c. Know how to use standard reduction potentials to determine cell voltages for cells under standard conditions (Equation 20.1). General ChemistryNow homework: SQ(s) 14, 18, 59

d. Know how to use a table of standard reduction potentials (Table 20.1 and Appendix M) to rank the strengths of oxidizing and reducing agents, to predict which substances can reduce or oxidize another species, and to predict whether redox reactions will be product-favored or reactant-favored (Sections 20.4 and 20.5). General ChemistryNow homework: SQ(s) 20, 22, 57

e. Use the Nernst equation (Equations 20.2 and 20.3) to calculate the cell potential under nonstandard conditions (Section 20.6). General ChemistryNow homework: SQ(s) 26, 30

f. Explain how cell voltage relates to ion concentration, and explain how this allows the determination of pH (Section 20.5) and other ion concentrations.

g. Use the relationships between cell voltage (E_{cell}) and free energy (ΔG) (Equations 20.5 and 20.6) and between $E°_{cell}$ and an equilibrium constant for the cell reaction (Equation 20.7) (Section 20.5). General ChemistryNow homework: SQ(s) 32, 34, 62, 64

Explore electrolysis, the use of electrical energy to produce chemical change

a. Describe the chemical processes occurring in an electrolysis. Recognize the factors that determine which substances are oxidized and reduced at the electrodes (Section 20.7). General ChemistryNow homework: SQ(s) 43

b. Relate the amount of a substance oxidized or reduced to the amount of current and the time the current flows (Section 20.8). General ChemistryNow homework: SQ(s) 46, 48, 50

Key Equations

Equation 20.1 (page 965):
Calculating a standard cell potential, E°_{cell}, from standard half-cell potentials.

$$E^\circ_{cell} = E^\circ_{cathode} - E^\circ_{anode}$$

Equation 20.3 (page 975):
Nernst equation.

$$E = E^\circ - \frac{0.0257}{n} \ln Q \quad \text{at 25 °C}$$

E is the cell potential under nonstandard conditions, n is the number of electrons transferred from the reducing agent to the oxidizing agent (according to the balanced equation), and Q is the reaction quotient.

Equation 20.6 (page 978):
Relationship between standard free energy change and the standard cell potential.

$$\Delta G^\circ = -nFE^\circ$$

F is the Faraday constant, 96,485 C/mol e⁻.

Equation 20.7 (page 979):
Relationship between the equilibrium constant and the standard cell potential for a reaction.

$$\ln K = \frac{nE^\circ}{0.0257} \quad \text{at 25 °C}$$

Equation 20.8 (page 987):
Relationship between current, electric charge, and time.

$$\text{Current, } I \text{ (amperes, A)} = \frac{\text{electric charge (coulombs, C)}}{\text{time (seconds, s)}}$$

Study Questions

▲ denotes more challenging questions.

■ denotes questions available in the Homework and Goals section of the General ChemistryNow CD-ROM or website.

Blue numbered questions have answers in Appendix O and fully worked solutions in the *Student Solutions Manual*.

Structures of many of the compounds used in these questions are found on the General ChemistryNow CD-ROM or website in the Models folder.

GENERAL
Chemistry·⚛·Now™ Assess your understanding of this chapter's topics with additional quizzing and conceptual questions at http://now.brookscole.com/kotz6e

Practicing Skills

Balancing Equations for Oxidation–Reduction Reactions
(See Examples 20.1–20.3 and General ChemistryNow Screen 20.3.)
When balancing the following redox equations, it may be necessary to add $H^+(aq)$ or $H^+(aq)$ plus H_2O for reactions in acid, and $OH^-(aq)$ or $OH^-(aq)$ plus H_2O for reactions in base.

1. Write balanced equations for the following half-reactions. Specify whether each is an oxidation or reduction.
   (a) $Cr(s) \longrightarrow Cr^{3+}(aq)$ (in acid)
   (b) $AsH_3(g) \longrightarrow As(s)$ (in acid)
   (c) $VO_3^-(aq) \longrightarrow V^{2+}(aq)$ (in acid)
   (d) $Ag(s) \longrightarrow Ag_2O(s)$ (in base)

2. ■ Write balanced equations for the following half-reactions. Specify whether each is an oxidation or reduction.
(a) $H_2O_2(aq) \longrightarrow O_2(g)$ (in acid)
(b) $H_2C_2O_4(aq) \longrightarrow CO_2(g)$ (in acid)
(c) $NO_3^-(aq) \longrightarrow NO(g)$ (in acid)
(d) $MnO_4^-(aq) \longrightarrow MnO_2(s)$ (in base)

3. Balance the following redox equations. All occur in acid solution.
(a) $Ag(s) + NO_3^-(aq) \longrightarrow NO_2(g) + Ag^+(aq)$
(b) $MnO_4^-(aq) + HSO_3^-(aq) \longrightarrow$
$$Mn^{2+}(aq) + SO_4^{2-}(aq)$$
(c) $Zn(s) + NO_3^-(aq) \longrightarrow Zn^{2+}(aq) + N_2O(g)$
(d) $Cr(s) + NO_3^-(aq) \longrightarrow Cr^{3+}(aq) + NO(g)$

4. Balance the following redox equations. All occur in acid solution.
(a) $Sn(s) + H^+(aq) \longrightarrow Sn^{2+}(aq) + H_2(g)$
(b) $Cr_2O_7^{2-}(aq) + Fe^{2+}(aq) \longrightarrow Cr^{3+}(aq) + Fe^{3+}(aq)$
(c) $MnO_2(s) + Cl^-(aq) \longrightarrow Mn^{2+}(aq) + Cl_2(g)$
(d) $CH_2O(aq) + Ag^+(aq) \longrightarrow HCO_2H(aq) + Ag(s)$

5. Balance the following redox equations. All occur in basic solution.
(a) $Al(s) + OH^-(aq) \longrightarrow Al(OH)_4^-(aq) + H_2(g)$
(b) $CrO_4^{2-}(aq) + SO_3^{2-}(aq) \longrightarrow$
$$Cr(OH)_3(s) + SO_4^{2-}(aq)$$
(c) $Zn(s) + Cu(OH)_2(s) \longrightarrow [Zn(OH)_4]^{2-}(aq) + Cu(s)$
(d) $HS^-(aq) + ClO_3^-(aq) \longrightarrow S(s) + Cl^-(aq)$

6. ■ Balance the following redox equations. All occur in basic solution.
(a) $Fe(OH)_3(s) + Cr(s) \longrightarrow Cr(OH)_3(s) + Fe(OH)_2(s)$
(b) $NiO_2(s) + Zn(s) \longrightarrow Ni(OH)_2(s) + Zn(OH)_2(s)$
(c) $Fe(OH)_2(s) + CrO_4^{2-}(aq) \longrightarrow$
$$Fe(OH)_3(s) + [Cr(OH)_4]^-(aq)$$
(d) $N_2H_4(aq) + Ag_2O(s) \longrightarrow N_2(g) + Ag(s)$

Constructing Voltaic Cells

(See Example 20.4 and General ChemistryNow Screen 20.4.)

7. A voltaic cell is constructed using the reaction of chromium metal and iron(II) ion.

$$2 Cr(s) + 3 Fe^{2+}(aq) \longrightarrow 2 Cr^{3+}(aq) + 3 Fe(s)$$

Complete the following sentences: Electrons in the external circuit flow from the ___ electrode to the ___ electrode. Negative ions move in the salt bridge from the ___ half-cell to the ___ half-cell. The half-reaction at the anode is ___ and that at the cathode is ___.

8. ■ A voltaic cell is constructed using the reaction

$$Mg(s) + 2 H^+(aq) \longrightarrow Mg^{2+}(aq) + H_2(g)$$

(a) Write equations for the oxidation and reduction half-reactions.
(b) Which half-reaction occurs in the anode compartment and which occurs in the cathode compartment?

(c) Complete the following sentences: Electrons in the external circuit flow from the ___ electrode to the ___ electrode. Negative ions move in the salt bridge from the ___ half-cell to the ___ half-cell. The half-reaction at the anode is ___ and that at the cathode is ___.

9. The half-cells $Fe^{2+}(aq)|Fe(s)$ and $O_2(g)|H_2O$ (in acid solution) are linked to create a voltaic cell.
(a) Write equations for the oxidation and reduction half-reactions and for the overall (cell) reaction.
(b) Which half-reaction occurs in the anode compartment and which occurs in the cathode compartment?
(c) Complete the following sentences: Electrons in the external circuit flow from the ___ electrode to the ___ electrode. Negative ions move in the salt bridge from the ___ half-cell to the ___ half-cell.

10. The half-cells $Ag^+(aq)|Ag(s)$ and $Cl_2(g)|Cl^-(aq)$ are linked to create a voltaic cell.
(a) Write equations for the oxidation and reduction half-reactions and for the overall (cell) reaction.
(b) Which half-reaction occurs in the anode compartment and which occurs in the cathode compartment?
(c) Complete the following sentences: Electrons in the external circuit flow from the ___ electrode to the ___ electrode. Negative ions move in the salt bridge from the ___ half-cell to the ___ half-cell.

Commercial Cells

11. What are the similarities and differences between dry cells, alkaline batteries, and ni-cad batteries?

12. What reactions occur when a lead storage battery is recharged?

Standard Electrochemical Potentials

(See Examples 20.5–20.6 and General ChemistryNow Screens 20.6 and 20.7.)

13. Calculate the value of $E°$ for each of the following reactions. Decide whether each is product-favored in the direction written.
(a) $2 I^-(aq) + Zn^{2+}(aq) \longrightarrow I_2(s) + Zn(s)$
(b) $Zn^{2+}(aq) + Ni(s) \longrightarrow Zn(s) + Ni^{2+}(aq)$
(c) $2 Cl^-(aq) + Cu^{2+}(aq) \longrightarrow Cu(s) + Cl_2(g)$
(d) $Fe^{2+}(aq) + Ag^+(aq) \longrightarrow Fe^{3+}(aq) + Ag(s)$

14. ■ Calculate the value of $E°$ for each of the following reactions. Decide whether each is product-favored in the direction written. [Reaction (d) occurs in basic solution.]
(a) $Br_2(\ell) + Mg(s) \longrightarrow Mg^{2+}(aq) + 2 Br^-(aq)$
(b) $Zn^{2+}(aq) + Mg(s) \longrightarrow Zn(s) + Mg^{2+}(aq)$
(c) $Sn^{2+}(aq) + 2 Ag^+(aq) \longrightarrow Sn^{4+}(aq) + 2 Ag(s)$
(d) $2 Zn(s) + O_2(g) + 2 H_2O(\ell) + 4 OH^-(aq) \longrightarrow$
$$2 [Zn(OH)_4]^{2-}(aq)$$

▲ More challenging   ■ In General ChemistryNow   **Blue-numbered questions** answered in Appendix O

15. Balance each of the following unbalanced equations, then calculate the standard potential, $E°$, and decide whether each is product-favored as written. (All reactions occur in acid solution.)

    (a) $Sn^{2+}(aq) + Ag(s) \longrightarrow Sn(s) + Ag^+(aq)$
    (b) $Al(s) + Sn^{4+}(aq) \longrightarrow Sn^{2+}(aq) + Al^{3+}(aq)$
    (c) $ClO_3^-(aq) + Ce^{3+}(aq) \longrightarrow Cl^-(aq) + Ce^{4+}(aq)$
    (d) $Cu(s) + NO_3^-(aq) \longrightarrow Cu^{2+}(aq) + NO(g)$

16. Balance each of the following unbalanced equations, then calculate the standard potential, $E°$, and decide whether each is product-favored as written. (All reactions occur in acid solution.)

    (a) $I_2(s) + Br^-(aq) \longrightarrow I^-(aq) + Br_2(\ell)$
    (b) $Fe^{2+}(aq) + Cu^{2+}(aq) \longrightarrow Cu(s) + Fe^{3+}(aq)$
    (c) $Fe^{2+}(aq) + Cr_2O_7^{2-}(aq) \longrightarrow Fe^{3+}(aq) + Cr^{3+}(aq)$
    (d) $MnO_4^-(aq) + HNO_2(aq) \longrightarrow Mn^{2+}(aq) + NO_3^-(aq)$

17. Consider the following half-reactions:

Half-Reaction	$E°(V)$
$Cu^{2+}(aq) + 2\,e^- \longrightarrow Cu(s)$	$+0.34$
$Sn^{2+}(aq) + 2\,e^- \longrightarrow Sn(s)$	-0.14
$Fe^{2+}(aq) + 2\,e^- \longrightarrow Fe(s)$	-0.44
$Zn^{2+}(aq) + 2\,e^- \longrightarrow Zn(s)$	-0.76
$Al^{3+}(aq) + 3\,e^- \longrightarrow Al(s)$	-1.66

    (a) Based on $E°$ values, which metal is the most easily oxidized?
    (b) Which metals on this list are capable of reducing $Fe^{2+}(aq)$ to Fe?
    (c) Write a balanced chemical equation for the reaction of $Fe^{2+}(aq)$ with $Sn(s)$. Is this reaction product-favored or reactant-favored?
    (d) Write a balanced chemical equation for the reaction of $Zn^{2+}(aq)$ with $Sn(s)$. Is this reaction product-favored or reactant-favored?

18. ■ Consider the following half-reactions:

Half-Reaction	$E°(V)$
$MnO_4^-(aq) + 8\,H^+(aq) + 5\,e^- \longrightarrow$ $Mn^{2+}(aq) + 4\,H_2O(\ell)$	$+1.51$
$BrO_3^-(aq) + 6\,H^+(aq) + 6\,e^- \longrightarrow$ $Br^-(aq) + 3\,H_2O(\ell)$	$+1.47$
$Cr_2O_7^{2-}(aq) + 14\,H^+(aq) + 6\,e^- \longrightarrow$ $2\,Cr^{3+}(aq) + 7\,H_2O(\ell)$	$+1.33$
$NO_3^-(aq) + 4\,H^+(aq) + 3\,e^- \longrightarrow$ $NO(g) + 2\,H_2O(\ell)$	$+0.96$
$SO_4^{2-}(aq) + 4\,H^+(aq) + 2\,e^- \longrightarrow$ $SO_2(g) + 2\,H_2O(\ell)$	$+0.20$

    (a) Choosing from among the reactants in these half-reactions, identify the strongest and weakest oxidizing agents.

    (b) Which of the oxidizing agents listed is (are) capable of oxidizing $Br^-(aq)$ to $BrO_3^-(aq)$ (in acid solution)?
    (c) Write a balanced chemical equation for the reaction of $Cr_2O_7^{2-}(aq)$ with $SO_2(g)$ in acid solution. Is this reaction product-favored or reactant-favored?
    (d) Write a balanced chemical equation for the reaction of $Cr_2O_7^{2-}(aq)$ with $Mn^{2+}(aq)$. Is this reaction product-favored or reactant-favored?

Ranking Oxidizing and Reducing Agents

(See Examples 20.5 and 20.6 and General ChemistryNow Screen 20.7.) Use a table of standard reduction potentials (Table 20.1 or Appendix M) to answer Study Questions 19–24.

19. Which of the following elements is the best reducing agent?

    (a) Cu   (d) Ag
    (b) Zn   (e) Cr
    (c) Fe

20. ■ From the following list, identify those elements that are easier to oxidize than $H_2(g)$.

    (a) Cu   (d) Ag
    (b) Zn   (e) Cr
    (c) Fe

21. Which of the following ions is most easily reduced?

    (a) $Cu^{2+}(aq)$ (d) $Ag^+(aq)$
    (b) $Zn^{2+}(aq)$ (e) $Al^{3+}(aq)$
    (c) $Fe^{2+}(aq)$

22. ■ From the following list, identify the ions that are more easily reduced than $H^+(aq)$.

    (a) $Cu^{2+}(aq)$ (d) $Ag^+(aq)$
    (b) $Zn^{2+}(aq)$ (e) $Al^{3+}(aq)$
    (c) $Fe^{2+}(aq)$

23. (a) Which halogen is most easily reduced: F_2, Cl_2, Br_2, or I_2 in acidic solution.
    (b) Identify the halogens that are better oxidizing agents than $MnO_2(s)$ in acidic solution.

24. (a) Which ion is most easily oxidized to the elemental halogen: F^-, Cl^-, Br^-, or I^- in acidic solution.
    (b) Identify the halide ions that are more easily oxidized than $H_2O(\ell)$ in acidic solution.

Electrochemical Cells Under Nonstandard Conditions

(See Examples 20.7 and 20.8 and General ChemistryNow Screen 20.8.)

25. Calculate the voltage delivered by a voltaic cell using the following reaction if all dissolved species are 2.5×10^{-2} M.

    $$Zn(s) + 2\,H_2O(\ell) + 2\,OH^-(aq) \longrightarrow [Zn(OH)_4]^{2-}(aq) + H_2(g)$$

26. ■ Calculate the potential developed by a voltaic cell using the following reaction if all dissolved species are 0.015 M.

    $$2\,Fe^{2+}(aq) + H_2O_2(aq) + 2\,H^+(aq) \longrightarrow 2\,Fe^{3+}(aq) + 2\,H_2O(\ell)$$

27. One half-cell in a voltaic cell is constructed from a silver wire dipped into a 0.25 M solution of $AgNO_3$. The other half-cell consists of a zinc electrode in a 0.010 M solution of $Zn(NO_3)_2$. Calculate the cell potential.

28. One half-cell in a voltaic cell is constructed from a copper wire dipped into a 4.8×10^{-3} M solution of $Cu(NO_3)_2$. The other half-cell consists of a zinc electrode in a 0.40 M solution of $Zn(NO_3)_2$. Calculate the cell potential.

29. One half-cell in a voltaic cell is constructed from a silver wire dipped into a $AgNO_3$ solution of unknown concentration. The other half-cell consists of a zinc electrode in a 1.0 M solution of $Zn(NO_3)_2$. A voltage of 1.48 V is measured for this cell. Use this information to calculate the concentration of $Ag^+(aq)$.

30. ■ One half-cell in a voltaic cell is constructed from an iron wire dipped into an $Fe(NO_3)_2$ solution of unknown concentration. The other half-cell is a standard hydrogen electrode. A voltage of 0.49 V is measured for this cell. Use this information to calculate the concentration of $Fe^{2+}(aq)$.

Electrochemistry, Thermodynamics, and Equilibrium
(See Examples 20.9 and 20.10 and General ChemistryNow Screen 20.9.)

31. Calculate $\Delta G°$ and the equilibrium constant for the following reactions.
(a) $2 Fe^{3+}(aq) + 2 I^-(aq) \longrightarrow 2 Fe^{2+}(aq) + I_2(aq)$
(b) $I_2(aq) + 2 Br^-(aq) \longrightarrow 2 I^-(aq) + Br_2(aq)$

32. ■ Calculate $\Delta G°$ and the equilibrium constant for the following reactions.
(a) $Zn^{2+}(aq) + Ni(s) \longrightarrow Zn(s) + Ni^{2+}(aq)$
(b) $Cu(s) + 2 Ag^+(aq) \longrightarrow Cu^{2+}(aq) + 2 Ag(s)$

33. Use standard reduction potentials (Appendix M) for the half-reactions $AgBr(s) + e^- \longrightarrow Ag(s) + Br^-(aq)$ and $Ag^+(aq) + e^- \longrightarrow Ag(s)$ to calculate the value of K_{sp} for AgBr.

34. ■ Use the standard reduction potentials (Appendix M) for the half-reactions $Hg_2Cl_2(s) + 2 e^- \longrightarrow 2 Hg(\ell) + 2 Cl^-(aq)$ and $Hg_2^{2+}(aq) + 2 e^- \longrightarrow 2 Hg(\ell)$ to calculate the value of K_{sp} for Hg_2Cl_2.

35. Use the standard reduction potentials (Appendix M) for the half-reactions $[AuCl_4]^-(aq) + 3 e^- \longrightarrow Au(s) + 4 Cl^-(aq)$ and $Au^{3+}(aq) + 3 e^- \longrightarrow Au(s)$ to calculate the value of $K_{formation}$ for the complex ion $[AuCl_4]^-(aq)$.

36. Use the standard reduction potentials (Appendix M) for the half-reactions $[Zn(OH)_4]^{2-}(aq) + 2 e^- \longrightarrow Zn(s) + 4 OH^-(aq)$ and $Zn^{2+}(aq) + 2 e^- \longrightarrow Zn(s)$ to calculate the value of $K_{formation}$ for the complex ion $[Zn(OH)_4]^{2-}$.

37. ▲ Iron(II) ion undergoes a disproportionation reaction to give Fe(s) and the iron(III) ion. That is, iron(II) ion is both oxidized and reduced within the same reaction.
$$3 Fe^{2+}(aq) \longrightarrow Fe(s) + 2 Fe^{3+}(aq)$$
(a) What two half-reactions make up the disproportionation reaction?

(b) Use the values of the standard reduction potentials for the two half-reactions in part (a) to determine whether this disproportionation reaction is product-favored.
(c) What is the equilibrium constant for this reaction?

38. ▲ Copper(I) ion disproportionates to copper metal and copper(II) ion. (See Study Question 37.)
$$2 Cu^+(aq) \longrightarrow Cu(s) + Cu^{2+}(aq)$$
(a) What two half-reactions make up the disproportionation reaction?
(b) Use values of the standard reduction potentials for the two half-reactions in part (a) to determine whether this disproportionation reaction is product-favored.
(c) What is the equilibrium constant for this reaction?

Electrolysis
(See Section 20.7, Example 20.11, and General ChemistryNow Screen 20.10.)

39. Diagram the apparatus used to electrolyze molten NaCl. Identify the anode and the cathode. Trace the movement of electrons through the external circuit and the movement of ions in the electrolysis cell.

40. Diagram the apparatus used to electrolyze aqueous $CuCl_2$. Identify the reaction products, the anode, and the cathode. Trace the movement of electrons through the external circuit and the movement of ions in the electrolysis cell.

41. Which product, O_2 or F_2, is more likely to form at the anode in the electrolysis of an aqueous solution of KF? Explain your reasoning.

42. Which product, Ca or H_2, is more likely to form at the cathode in the electrolysis of $CaCl_2$? Explain your reasoning.

43. ■ An aqueous solution of KBr is placed in a beaker with two inert platinum electrodes. When the cell is attached to an external source of electrical energy, electrolysis occurs.
(a) Hydrogen gas and hydroxide ion form at the cathode. Write an equation for the half-reaction that occurs at this electrode.
(b) Bromine is the primary product at the anode. Write an equation for its formation.

44. An aqueous solution of Na_2S is placed in a beaker with two inert platinum electrodes. When the cell is attached to an external battery, electrolysis occurs.
(a) Hydrogen gas and hydroxide ion form at the cathode. Write an equation for the half-reaction that occurs at this electrode.
(b) Sulfur is the primary product at the anode. Write an equation for its formation.

Counting Electrons
(See Examples 20.12 and 20.13 and General ChemistryNow Screen 20.12.)

45. In the electrolysis of a solution containing $Ni^{2+}(aq)$, metallic Ni(s) deposits on the cathode. Using a current of 0.150 A for 12.2 min, what mass of nickel will form?

46. ■ In the electrolysis of a solution containing $Ag^+(aq)$, metallic $Ag(s)$ deposits on the cathode. Using a current of 1.12 A for 2.40 h, what mass of silver forms?

47. Electrolysis of a solution of $CuSO_4(aq)$ to give copper metal is carried out using a current of 0.66 A. How long should electrolysis continue to produce 0.50 g of copper?

48. ■ Electrolysis of a solution of $Zn(NO_3)_2(aq)$ to give zinc metal is carried out using a current of 2.12 A. How long should electrolysis continue to prepare 2.5 g of zinc?

49. A voltaic cell can be built using the reaction between Al metal and O_2 from the air. If the Al anode of this cell consists of 84 g of aluminum, how many hours can the cell produce 1.0 A of electricity, assuming an unlimited supply of O_2?

50. ■ Assume the specifications of a Ni-Cd voltaic cell include delivery of 0.25 A of current for 1.00 h. What is the minimum mass of the cadmium that must be used to make the anode in this cell?

General Questions

These questions are not designated as to type or location in the chapter. They may combine several concepts.

51. Write balanced equations for the following half-reactions.
(a) $UO_2^+(aq) \longrightarrow U^{4+}(aq)$ (acid solution)
(b) $ClO_3^-(aq) \longrightarrow Cl^-(aq)$ (acid solution)
(c) $N_2H_4(aq) \longrightarrow N_2(g)$ (basic solution)
(d) $ClO^-(aq) \longrightarrow Cl^-(aq)$ (basic solution)

52. Balance the following equations.
(a) $Zn(s) + VO^{2+}(aq) \longrightarrow$
   $Zn^{2+}(aq) + V^{3+}(aq)$ (acid solution)
(b) $Zn(s) + VO_3^-(aq) \longrightarrow$
   $V^{2+}(aq) + Zn^{2+}(aq)$ (acid solution)
(c) $Zn(s) + ClO^-(aq) \longrightarrow$
   $Zn(OH)_2(s) + Cl^-(aq)$ (basic solution)
(d) $ClO^-(aq) + [Cr(OH)_4]^-(aq) \longrightarrow$
   $Cl^-(aq) + CrO_4^{2-}(aq)$ (basic solution)

53. Magnesium metal is oxidized and silver ions are reduced in a voltaic cell using $Mg^{2+}(aq, 1 M)|Mg$ and $Ag^+(aq, 1 M)|Ag$ half-cells.

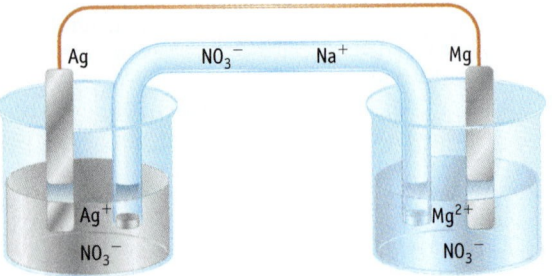

(a) Label each part of the cell.
(b) Write equations for the half-reactions occurring at the anode and the cathode, and write an equation for the net reaction in the cell.
(c) Trace the movement of electrons in the external circuit. Assuming the salt bridge contains $NaNO_3$, trace the movement of the Na^+ and NO_3^- ions in the salt bridge that occurs when a voltaic cell produces current. Why is a salt bridge required in a cell?

54. You want to set up a series of voltaic cells with specific cell voltages. A $Zn^{2+}(aq, 1 M)|Zn(s)$ half-cell is in one compartment. Identify several half-cells that you could use so that the cell voltage will be close to (a) 1.1 V and (b) 0.5 V. Consider cells in which zinc can be either the cathode or the anode.

55. You want to set up a series of voltaic cells with specific cell voltages. The $Ag^+(aq, 1 M)|Ag(s)$ half-cell is one of the compartments. Identify several half-cells that you could use so that the cell voltage will be close to (a) 1.7 V and (b) 0.5 V. Consider cells in which silver can be either the cathode or the anode.

56. Which of the following reactions are product-favored?
(a) $Zn(s) + I_2(s) \longrightarrow Zn^{2+}(aq) + 2 I^-(aq)$
(b) $2 Cl^-(aq) + I_2(s) \longrightarrow Cl_2(g) + 2 I^-(aq)$
(c) $2 Na^+(aq) + 2 Cl^-(aq) \longrightarrow 2 Na(s) + Cl_2(g)$
(d) $2 K(s) + H_2O(\ell) \longrightarrow$
   $2 K^+(aq) + H_2(g) + 2 OH^-(aq)$

57. ■ In the table of standard reduction potentials, locate the half-reactions for the reductions of the following metal ions to the metal: $Sn^{2+}(aq)$, $Au^+(aq)$, $Zn^{2+}(aq)$, $Co^{2+}(aq)$, $Ag^+(aq)$, $Cu^{2+}(aq)$. Among the metal ions and metals that make up these half-reactions:
(a) Which metal ion is the weakest oxidizing agent?
(b) Which metal ion is the strongest oxidizing agent?
(c) Which metal is the strongest reducing agent?
(d) Which metal is the weakest reducing agent?
(e) Will $Sn(s)$ reduce $Cu^{2+}(aq)$ to $Cu(s)$?
(f) Will $Ag(s)$ reduce $Co^{2+}(aq)$ to $Co(s)$?
(g) Which metal ions on the list can be reduced by $Sn(s)$?
(h) What metals can be oxidized by $Ag^+(aq)$?

58. ▲ In the table of standard reduction potentials, locate the half-reactions for the reductions of the following non-metals: F_2, Cl_2, Br_2, I_2 (reduction to halide ions), and O_2, S, Se (reduction to H_2X in aqueous acid). Among the elements, ions, and compounds that make up these half-reactions:
(a) Which element is the weakest oxidizing agent?
(b) Which element is the weakest reducing agent?
(c) Which of the elements listed is (are) capable oxidizing H_2O to O_2?
(d) Which of these elements listed is (are) capable of oxidizing H_2S to S?
(e) Is O_2 capable of oxidizing I^- to I_2, in acid solution?
(f) Is S capable of oxidizing I^- to I_2?

(g) Is the reaction $H_2S(aq) + Se(s) \longrightarrow H_2Se(aq) + S(s)$ product-favored?

(h) Is the reaction $H_2S(aq) + I_2(s) \longrightarrow 2\,H^+(aq) + 2\,I^-(aq) + S(s)$ product-favored?

59. ■ Four voltaic cells are set up. In each, one half-cell contains a standard hydrogen electrode. The second half-cell is one of the following: $Cr^{3+}(aq, 1.0\,M)\,|\,Cr(s)$, $Fe^{2+}(aq, 1.0\,M)\,|\,Fe(s)$, $Cu^{2+}(aq, 1.0\,M)\,|\,Cu(s)$, or $Mg^{2+}(aq, 1.0\,M)\,|\,Mg(s)$.

(a) In which of the voltaic cells does the hydrogen electrode serve as the cathode?

(b) Which voltaic cell produces the highest voltage? Which produces the lowest voltage?

60. The following half-cells are available: $Ag^+(aq, 1.0\,M)\,|\,Ag(s)$, $Zn^{2+}(aq, 1.0\,M)\,|\,Zn(s)$, $Cu^{2+}(aq, 1.0\,M)\,|\,Cu(s)$, and $Co^{2+}(aq, 1.0\,M)\,|\,Co(s)$. Linking any two half-cells makes a voltaic cell. Given four different half-cells, six voltaic cells are possible. These are labeled, for simplicity, Ag-Zn, Ag-Cu, Ag-Co, Zn-Cu, Zn-Co, and Cu-Co.

(a) In which of the voltaic cells does the copper electrode serve as the cathode? In which of the voltaic cells does the cobalt electrode serve as the anode?

(b) Which combination of half-cells generates the highest voltage? Which combination generates the lowest voltage?

61. The reaction occurring in the cell in which Al_2O_3 and aluminum salts are electrolyzed is $Al^{3+}(aq) + 3\,e^- \longrightarrow Al(s)$. If the electrolysis cell operates at 5.0 V and 1.0×10^5 A, what mass of aluminum metal can be produced in a 24-h day?

62. ■ ▲ A potential of $+0.146$ V is recorded (under standard conditions) for a voltaic cell constructed using the following half-reactions:

Anode: $Ag(s) \longrightarrow Ag^+(aq) + e^-$

Cathode: $Ag_2SO_4(s) + 2\,e^- \longrightarrow 2\,Ag(s) + SO_4^{2-}(aq)$

(a) What is the standard reduction potential for the cathode reaction?

(b) Calculate the solubility product, K_{sp}, for Ag_2SO_4.

63. ▲ A potential of 0.142 V is recorded (under standard conditions) for a voltaic cell constructed using the following half reactions:

Cathode: $Pb^{2+}(aq) + 2\,e^- \longrightarrow Pb(s)$

Anode: $PbCl_2(s) + 2\,e^- \longrightarrow Pb(s) + 2\,Cl^-(aq)$

Net: $Pb^{2+}(aq) + 2\,Cl^-(aq) \longrightarrow PbCl_2(s)$

(a) What is the standard reduction potential for the anode reaction?

(b) Estimate the solubility product, K_{sp}, for $PbCl_2$

64. ■ The standard voltage, $E°$, for the reaction of $Zn(s)$ and $Cl_2(g)$ is $+2.12$ V. What is the standard free energy change, $\Delta G°$, for the reaction?

65. The standard potential for the reaction of $Mg(s)$ with $I_2(s)$ is $+2.91$ V. What is the standard free energy change, $\Delta G°$, for the reaction?

66. ▲ An electrolysis cell for aluminum production operates at 5.0 V and a current of 1.0×10^5 A. Calculate the number of kilowatt-hours of energy required to produce 1 metric ton (1.0×10^3 kg) of aluminum. (1 kWh = 3.6×10^6 J and 1 J = 1 C · V.)

67. ▲ Electrolysis of molten NaCl is done in cells operating at 7.0 V and 4.0×10^4 A. What mass of $Na(s)$ and $Cl_2(g)$ can be produced in one day in such a cell? What is the energy consumption in kilowatt-hours? (1 kWh = 3.6×10^6 J and 1 J = 1 C · V.)

68. ▲ A current of 0.0100 A is passed through a solution of rhodium sulfate, causing reduction of the metal ion to the metal. After 3.00 h, 0.038 g of Rh has been deposited. What is the charge on the rhodium ion, Rh^{n+}? What is the formula for rhodium sulfate?

69. ▲ A current of 0.44 A is passed through a solution of ruthenium nitrate causing reduction of the metal ion to the metal. After 25.0 min, 0.345 g of Ru has been deposited. What is the charge on the ruthenium ion, Ru^{n+}? What is the formula for ruthenium nitrate?

70. The total charge that can be delivered by a large dry cell battery before its voltage drops too low is usually about 35 amp-hours. (One amp-hour is the charge that passes through a circuit when 1 A flows for 1 h.) What mass of Zn is consumed when 35 amp-hours is drawn from the cell?

71. Chlorine gas is obtained commercially by electrolysis of brine (a concentrated aqueous solution of NaCl). If the electrolysis cells operate at 4.6 V and 3.0×10^5 A, what mass of chlorine can be produced in a 24-h day?

72. An old method of measuring the current flowing in a circuit was to use a "silver coulometer." The current passed first through a solution of $Ag^+(aq)$ and then into another solution containing an electroactive species. The amount of silver metal deposited at the cathode was weighed. From the mass of silver, the number of atoms of silver was calculated. Since the reduction of a silver ion requires one electron, this value equalled the number of electrons passing through the circuit. If the time was noted, the average current could be calculated. If, in such an experiment, 0.052 g of Ag is deposited during 450 s, what was the current flowing in the circuit?

73. A "silver coulometer" (Study Question 72) was used in the past to measure the current flowing in an electrochemical cell. Suppose you found that the current flowing through an electrolysis cell deposited 0.089 g of Ag metal at the cathode after exactly 10 min. If this same current then passed through a cell containing gold(III) ion in the form of $(AuCl_4)^-$, how much gold was deposited at the cathode in that electrolysis cell?

74. ▲ Write balanced equations for the following reduction half-reactions involving organic compounds.

(a) $HCO_2H \longrightarrow CH_2O$ (acid solution)

(b) $C_6H_5CO_2H \longrightarrow C_6H_5CH_3$ (acid solution)

(c) $CH_3CH_2CHO \longrightarrow CH_3CH_2CH_2OH$ (acid solution)

(d) $CH_3OH \longrightarrow CH_4$ (acid solution)

75. ▲ Balance the following equations involving organic compounds.

(a) $Ag^+(aq) + C_6H_5CHO(aq) \longrightarrow$
$$Ag(s) + C_6H_5CO_2H(aq) \quad \text{(acid solution)}$$

(b) $CH_3CH_2OH + Cr_2O_7^{2-}(aq) \longrightarrow$
$$CH_3CO_2H(aq) + Cr^{3+}(aq) \quad \text{(acid solution)}$$

76. A voltaic cell is constructed in which one half-cell consists of a silver wire in an aqueous solution of $AgNO_3$. The other half-cell consists of an inert platinum wire in an aqueous solution containing $Fe^{2+}(aq)$ and $Fe^{3+}(aq)$.

(a) Calculate the voltage of the cell, assuming standard conditions.

(b) Write the net ionic equation for the reaction occurring in the cell.

(c) In this voltaic cell, which electrode is the anode and which is the cathode?

(d) If $[Ag^+]$ is 0.10 M, and $[Fe^{2+}]$ and $[Fe^{3+}]$ are both 1.0 M, what is the cell voltage? Is the net cell reaction still that used in part (a)? If not, what is the net reaction under the new conditions?

77. ▲ An expensive but lighter alternative to the lead storage battery is the silver-zinc battery.

$$Ag_2O(s) + Zn(s) + H_2O(\ell) \longrightarrow Zn(OH)_2(s) + 2\,Ag(s)$$

The electrolyte is 40% KOH, and silver–silver oxide electrodes are separated from zinc–zinc hydroxide electrodes by a plastic sheet that is permeable to hydroxide ion. Under normal operating conditions, the battery has a potential of 1.59 V.

(a) How much energy can be produced per gram of reactants in the silver-zinc battery? Assume the battery produces a current of 0.10 A.

(b) How much energy can be produced per gram of reactants in the standard lead storage battery? Assume the battery produces a current of 0.10 A at 2.0 V.

(c) Which battery (silver-zinc or lead storage) produces the greater energy per gram of reactants?

78. The specifications for a lead storage battery include delivery of a steady 1.5 A of current for 15 h.

(a) What is the minimum mass of lead that will be used in the anode?

(b) What mass of PbO_2 must be used in the cathode?

(c) Assume that the volume of the battery is 0.50 L. What is the minimum molarity of H_2SO_4 necessary?

Summary and Conceptual Questions

The following questions may use concepts from the preceding chapters.

79. Fluorinated organic compounds are important commercially, as they are used as herbicides, flame retardants, and fire-extinguishing agents, among other things. A reaction such as

$$CH_3SO_2F + 3\,HF \longrightarrow CF_3SO_2F + 3\,H_2$$

is carried out electrochemically in liquid HF as the solvent.

(a) If you electrolyze 150 g of CH_3SO_2F, what mass of HF is required and what mass of each product can be isolated?

(b) Is H_2 produced at the anode or the cathode of the electrolysis cell?

(c) A typical electrolysis cell operates at 8.0 V and 250 A. How many kilowatt-hours of energy does one such cell consume in 24 h?

80. ▲ The free energy change for a reaction, ΔG°_{rxn}, is the maximum energy that can be extracted from the process, whereas ΔH°_{rxn} is the total chemical potential energy change. The efficiency of a fuel cell is the ratio of these two quantities.

$$\text{Efficiency} = \frac{\Delta G^\circ_{rxn}}{\Delta H^\circ_{rxn}} \times 100\%$$

Consider the hydrogen-oxygen fuel cell where the net reaction is

$$H_2(g) + \tfrac{1}{2}\,O_2(g) \longrightarrow H_2O(\ell)$$

(a) Calculate the efficiency of the fuel cell under standard conditions.

(b) Calculate the efficiency of the fuel cell if the product is water vapor instead of liquid water.

(c) Does the efficiency depend on the state of the reaction product? Why or why not?

81. Consider an electrochemical cell based on the half-reactions $Ni^{2+}(aq) + 2\,e^- \longrightarrow Ni(s)$ and $Cd^{2+}(aq) + 2\,e^- \longrightarrow Cd(s)$.

(a) Diagram the cell and label each of the components (including the anode, cathode, and salt bridge).

(b) Use the equations for the half-reactions to write a balanced, net ionic equation for the overall cell reaction.

(c) What is the polarity of each electrode?

(d) What is the value of E°_{cell}?

(e) In which direction do electrons flow in the external circuit?

(f) Assume that a salt bridge containing $NaNO_3$ connects the two half-cells. In which direction do the $Na^+(aq)$ ions move? In which direction do the $NO_3^-(aq)$ ions move?

(g) Calculate the equilibrium constant for the reaction.

(h) If the concentration of Cd^{2+} is reduced to 0.010 M, and $[Ni^{2+}] = 1.0$ M, what is the value of E_{cell}? Is the net reaction still the reaction given in part (b)?

(i) If 0.050 A is drawn from the battery, how long can it last if you begin with 1.0 L of each of the solutions, and each was initially 1.0 M in dissolved species? Each electrode weighs 50.0 g in the beginning.

82. ▲ (a) Is it easier to reduce water in acid or base? To evaluate this, consider the half-reaction

$$2\,H_2O(\ell) + 2\,e^- \longrightarrow 2\,OH^-(aq) + H_2(g)$$
$$E^\circ = -0.83\ \text{V}$$

(b) What is the reduction potential for water for solutions at pH = 7 (neutral) and pH = 1 (acid)? Comment on the value of E° at pH = 1.

▲ More challenging   ■ In General ChemistryNow   **Blue-numbered questions** answered in Appendix O

83. ▲ A solution of KI is added dropwise to a pale blue solution of $Cu(NO_3)_2$. The solution changes to a brown color and a precipitate forms. In contrast, no change is observed if solutions of KCl and KBr are added to aqueous $Cu(NO_3)_2$. Consult the table of standard reduction potentials to explain the dissimilar results seen with the different halides. Write an equation for the reaction that occurs when solutions of KI and $Cu(NO_3)_2$ are mixed.

84. ▲ Four metals, A, B, C, and D, exhibit the following properties:

    (a) Only A and C react with 1.0 M hydrochloric acid to give $H_2(g)$.

    (b) When C is added to solutions of the ions of the other metals, metallic B, D, and A are formed.

    (c) Metal D reduces B^{n+} to give metallic B and D^{n+}.

    Based on this information, arrange the four metals in order of increasing ability to act as reducing agents.

85. A hydrogen-oxygen fuel cell operates on the simple reaction

    $$H_2(g) + \tfrac{1}{2} O_2(g) \longrightarrow H_2O(\ell)$$

    If the cell is designed to produce 1.5 A of current, and if the hydrogen is contained in a 1.0-L tank at 200. atm pressure at 25 °C, how long can the fuel cell operate before the hydrogen runs out? (Assume there is an unlimited supply of O_2.)

86. ▲ Living organisms derive energy from the oxidation of food, typified by glucose.

    $$C_6H_{12}O_6(aq) + 6 O_2(g) \longrightarrow 6 CO_2(g) + 6 H_2O(\ell)$$

    Electrons in this redox process are transferred from glucose to oxygen in a series of at least 25 steps. It is instructive to calculate the total daily current flow in a typical organism and the rate of energy expenditure (power). (See T. P. Chirpich: *Journal of Chemical Education*, Vol. 52, p. 99, 1975.)

    (a) The molar enthalpy of combustion of glucose is −2800 kJ. If you are on a typical daily diet of 2400 Cal (kilocalories), what amount of glucose (in moles) must be consumed in a day if glucose is the only source of energy? What amount of O_2 must be consumed in the oxidation process?

    (b) How many moles of electrons must be supplied to reduce the amount of O_2 calculated in part (a)?

    (c) Based on the answer in part (b), calculate the current flowing, per second, in your body from the combustion of glucose.

    (d) If the average standard potential in the electron transport chain is 1.0 V, what is the rate of energy expenditure in watts?

87. See the General ChemistryNow CD-ROM or website Screen 20.7. Using the reduction potential table in the simulation on this screen, answer these questions:

    (a) What species can be reduced by Cu(s)?

    (b) What metals can be oxidized by Cd^{2+}?

Do you need a live tutor for homework problems?
Access vMentor at General ChemistryNow at
http://now.brookscole.com/kotz6e
for one-on-one tutoring from a chemistry expert

Let's Review Chapters 14–20

Add NH₃

$[Ni(H_2O)_6]^{2+}$

$[Ni(NH_3)_6]^{2+}$

Add
ethylenediamine
$H_2NCH_2CH_2NH_2$

$[Ni(H_2NCH_2CH_2NH_2)_3]^{2+}$

Photos: Charles D. Winters

Chemists and non-chemists alike are drawn to the wonderful colors of many chemical compounds. Here you see an aqueous solution of nickel(II) chloride. When the nickel(II) salt dissolves a green solution is obtained, owing to the presence of the so-called complex ion $[Ni(H_2O)_6]^{2+}$ in solution.

As aqueous ammonia, $NH_3(aq)$, is added to the green solution it turns a characteristic blue color. The color arises from another complex ion, $[Ni(NH_3)_6]^{2+}$, in which ammonia molecules have taken the place of water molecules.

Interestingly, when we add another compound—ethylene-diamine, $H_2NCH_2CH_2NH_2$—the blue color of the ammonia-containing ion is replaced by the purple color of the complex ion, $[Ni(H_2NCH_2CH_2NH_2)_3]^{2+}$.

In this series of chapters you will learn more about the reactivity of molecules and how to predict when one compound can be transformed into another.

The Purpose of *Let's Review*

- *Let's Review* provides additional homework questions for Chapters 14 through 20. Routine questions covering each of the concepts in a chapter are found in that chapter and are those usually identified by topic. In contrast, many of the *Let's Review* questions combine several concepts from one or more chapters. Many come from the examinations given by the authors and others are based often on actual experiments or processes in chemical research or in the chemical industry.

- *Let's Review* provides guidance for Chapters 14 through 20 as you prepare for an exam covering material in these chapters. Although this is designated for Chapters 14 through 20 you may choose only material appropriate to the exam in your specific course.

- Each Comprehensive Question is correlated with tutorials and exercises on the General ChemistryNow website and with the OWL online homework system to which you may have purchased access. Some questions may include a screen shot of a tutorial so you see what resources are available to help you review.

Preparing for an Examination on Chapters 14–20

1. **General Chemistry☆Now™** Use General ChemistryNow online to take a chapter **Pre-Test.** A **Personalized Learning Plan** will indicate sections of the textbook that should be reviewed.

2. Use General ChemistryNow online to work though media-based **Exercises, Guided Simulations,** and **Intelligent Tutors.**

3. If you subscribe to OWL, use the **Tutorials** in that system.

4. Work on the questions below that are relevant to a particular chapter or chapters. See the solutions to those questions at the end of this section.

5. For background and help with a question, use the General ChemistryNow information or OWL questions that are correlated with the question.

Key Points to Know for Chapters 14–20

Here are some of the key points you must know to be successful in Chapters 14 through 20.

- Solution concentrations.
- Colligative properties.
- Rates of reaction and factors affecting reaction rates.
- Rate equations and integrated rate laws.
- Collision theory of reaction rates and activation energy.
- Reaction mechanisms.
- The nature of chemical equlibria.
- The reaction quotient (Q) and equilibrium constant (K).
- The Brønsted-Lowry and Lewis theories of acids and bases.
- Chemical equilibria involving acids, bases, and insoluble salts in aqueous solution.
- The nature of buffer solution.
- The outcome of acid-base reactions.
- Reaction spontaneity, free energy, and entropy.
- The relationship of free energy changes and equilibrium constants and whether a reaction is product-or reactant-favored.
- Balancing equations for redox reactions.
- Voltaic cells and electrochemical potentials.
- Electrolysis and electrical energy.

Examination Preparation Questions

▲ denotes more challenging questions.

Important information about the questions that follow:

- See the Study Questions in each chapter for questions on basic concepts.

- Some questions arise from recent research in chemistry and the other sciences. They often involve concepts from more than one chapter and may be more challenging than those in earlier chapters. Not all chapter goals or concepts are necessarily addressed in these questions.

- Assessing Key Points questions are short-answer questions covering the key points on this page.

- Each Comprehensive Question is correlated with the text section covering that topic, with General ChemistryNow material, and with questions in OWL that may provide additional background.

- The computer screens are largely taken from General ChemistryNow but are also contained in the OWL system if you subscribe to it. They illustrate **Tutorials** or other resources that may help in answering the questions.

Assessing Key Points

1. (Chapter 14) You dissolve 45.0 g of dimethylglyoxime (DMG) ($C_4H_8N_2O_2$) in 500. mL of ethyl alcohol, C_2H_5OH (density of ethanol = 0.790 g/mL). Calculate the molality and weight percent of the glyoxime.

2. (Chapter 14) Which of the following aqueous solutions has the highest boiling point? The lowest freezing point? The highest vapor pressure? The highest osmotic pressure?
   (a) 0.20 m $C_2H_4(OH)_2$ (nonvolatile, nonelectrolyte)
   (b) 0.10 m $MgCl_2$
   (c) 0.12 m KBr
   (d) 0.12 m K_2SO_4

3. (Chapters 13 and 14) Which of the following ions should have the most negative enthalpy of hydration: Rb^+, Mg^{2+}, Na^+, and Ba^{2+}?

4. (Chapter 14) Concentrated salt solutions have boiling points lower than those calculated using the equation $\Delta T_{bp} = K_{bp} \cdot m \cdot i$. Which of the following is a reasonable explanation of this observation?
   (a) Positive ions repel each other more at high concentration.
   (b) In solutions of higher concentration there is ion pairing.
   (c) The water molecules will have a greater attraction for each other.
   (d) Concentrated solutions really have small particles of non-dissolved salt, thus lowering the molality.

5. (Chapter 14) You dissolve 29.3 g of NaCl in 500 g water. What is the freezing point of this solution?

6. (Chapter 14) Erythritol is a compound that occurs naturally in algae and fungi. It is about twice as sweet as sucrose. A solution of 2.50 g of erythritol in 50.0 g of water freezes at $-0.762 \,°C$. What is the molar mass of the compound?

7. (Chapter 15) A few drops of blue food dye were added to water followed by a solution of bleach. (Initially, the concentration of dye was about 3.4×10^{-5} M, and the bleach (NaOCl) concentration was about 0.034 M.) The dye faded as it reacted with the bleach. The color change was followed by a spectrophotometer, and the data are plotted below. (See General ChemistryNow, Screen 15.2)

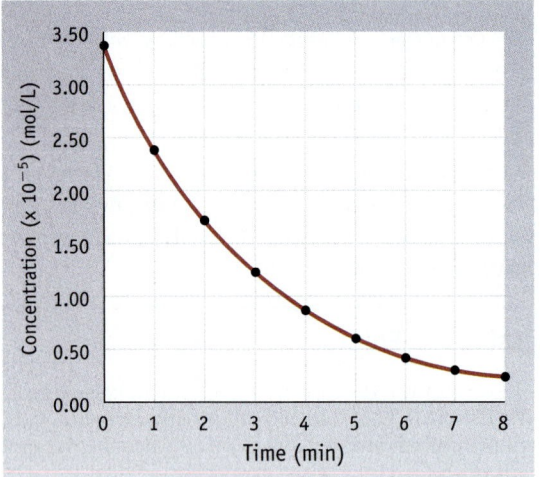

(a) What is the average rate of reaction over the first 2 minutes?

(b) What is the approximate half-life of the reaction?

(c) With time does the rate of reaction decline, stay the same, or increase?

8. (Chapter 15) The reduction of NO with hydrogen produces nitrogen and water.

$$2\,NO(g) + 2\,H_2(g) \longrightarrow N_2(g) + 2\,H_2O(g)$$

(a) The reaction is second order in NO and third order overall. What is the rate law for the reaction?

(b) If [NO] is increased by a factor of 5, how does the rate of the reaction change?

(c) If [H_2] is increased by a factor of 5, how does the rate of the reaction change?

9. (Chapter 15) Given the initial rate data for the reaction A + B $\longrightarrow$ C, determine the rate expression for the reaction. (The reaction is catalyzed by molecule D.)

[A], M	[B], M	[D], M	$\Delta[C]/\Delta t$ (mol/L·s)
0.10	0.20	0.10	2.58
0.10	0.10	0.10	1.29
0.24	0.10	0.20	2.58
0.10	0.20	0.20	5.16

(a) $\Delta[C]/\Delta t = k[A][D]^2$

(b) $\Delta[C]/\Delta t = k[A][B][D]$

(c) $\Delta[C]/\Delta t = k[B][D]$

(d) $\Delta[C]/\Delta t = k[A][B]$

10. (Chapter 15) See the Simulation on General ChemistryNow, Screen 15.4. The three reactions on this screen have the rate law Rate = k[Reactant]^m where m is the order of reaction.

15.4 Concentration Dependence

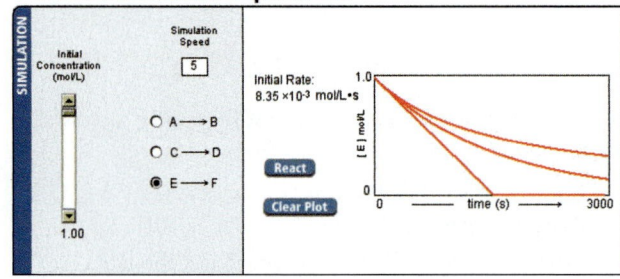

General ChemistryNow Screen 15.4 Concentration Dependence of Rates

(a) For A $\longrightarrow$ B, when [A] = 0.50 M, Rate = 4.18×10^{-3} mol/L·s. When [A] = 1.0 M, Rate = 8.35×10^{-3} mol/L·s. Is m 0, 1, or 2?

(b) For C $\longrightarrow$ D, when [C] = 0.50 M, Rate = 2.09×10^{-3} mol/L·s. When [C] = 1.0 M, Rate = 8.35×10^{-3} mol/L·s. Is m 0, 1, or 2?

(c) For E $\longrightarrow$ F, when [E] = 0.50 M, Rate = 8.35×10^{-3} mol/L·s. When [E] = 1.0 M, Rate = 8.35×10^{-3} mol/L·s. Is m 0, 1, or 2?

(d) Which of the three curves in the figure—top, middle, or bottom—corresponds to the reaction E $\longrightarrow$ F?

11. (Chapter 15) A reaction that occurs in our environment is the oxidation of NO to the brown gas NO_2.

$$2\,NO(g) + O_2(g) \longrightarrow 2\,NO_2(g)$$

The mechanism of the reaction is thought to be

Step 1 $2\,NO(g) \rightleftharpoons N_2O_2(g)$
           rapidly established equilibrium

Step 2 $N_2O_2(g) + O_2(g) \longrightarrow 2\,NO_2(g)$ slow

Which is the rate determining step? Is there an intermediate in the reaction?

12. (Chapter 15) The decomposition of SO_2Cl_2 to SO_2 and Cl_2 is first order in SO_2Cl_2.

$$SO_2Cl_2(g) \longrightarrow SO_2(g) + Cl_2(g)$$
$$\text{Rate} = k\,[SO_2Cl_2] \text{ where } k = 0.17/h$$

(a) What is the rate of decomposition when $[SO_2Cl_2] = 0.010$ M?

(b) What is the half-life of the reaction?

(c) If the initial pressure of SO_2Cl_2 in a flask is 0.050 atm, what is the pressure of all gases (i.e., the total pressure) in the flask after the reaction has proceeded for one half-life period?

13. (Chapter 15) Traces of the radioactive gas radon have been found in homes. (It comes from natural radioactive sources in the soils in and around a house.) The isotope Rn-222 has a half life of 3.82 days. If a sample of Rn-222 is trapped in your basement, how long does it take for the sample to fall to 10.0% of its original activity? (Note that radioactive isotopes always decay in a first-order process.)

14. (Chapter 16) A mixture of SO_2, O_2, and SO_3 at 1000 K contains the gases at the following concentrations: $[SO_2] = 5.0 \times 10^{-3}$ mol/L, $[O_2] = 1.9 \times 10^{-3}$ mol/L, and $[SO_3] = 6.9 \times 10^{-3}$ mol/L. Assuming the reaction is not at equilibrium, which way will the reaction proceed to reach equilibrium?

$$2\,SO_2(g) + O_2(g) \rightleftharpoons 2\,SO_3(g) \quad K_c = 279$$

15. (Chapter 16) You place 0.010 mol of $N_2O_4(g)$ in a 2.0 L flask at 200 °C. After reaching equilibrium, $[N_2O_4] = 0.0042$ M.

$$N_2O_4(g) \rightleftharpoons 2\,NO_2(g)$$

Fill in the table below and then calculate K_c for the reaction at 200 °C.

	$N_2O_4(g)$ $\rightleftharpoons$	2 $NO_2(g)$
Initial concentration (M)	——	0
Change in concentration (M)	——	——
Equilibrium concentration (M)	0.0042	——

16. (Chapter 16) Samples of NO_2 and N_2O_4 are in equilibrium at some temperature. If $[N_2O_4] = 4.6 \times 10^{-5}$ M, what is $[NO_2]$?

$$2\,NO_2(g) \rightleftharpoons N_2O_4(g) \quad K = 240$$

17. (Chapter 16) Calculate K_c for the reaction

$$SnO_2(s) + 2\,CO(g) \rightleftharpoons Sn(s) + 2\,CO_2(g)$$

given the following information:

$$SnO_2(s) + 2\,H_2(g) \rightleftharpoons Sn(s) + 2\,H_2O(g) \quad K_c = 8.12$$
$$H_2(g) + CO_2(g) \rightleftharpoons CO(g) + H_2O(g) \quad K_c = 0.771$$

18. (Chapter 16) Heating ammonium hydrogen sulfide leads to decomposition.

$$\text{heat} + NH_4HS(s) \rightleftharpoons NH_3(g) + H_2S(g)$$

Predict the effect on the equilibrium of each change listed below. Answer by choosing (1) no change, (2) shifts left, or (3) shifts right.

(a) Add solid NH_4HS.

(b) Add NH_3 gas.

(c) Add H_2S gas.

(d) Raise the temperature.

(e) Increase the volume of the flask containing the reaction.

19. (Chapter 17) What is the pH of a 0.020 M NaOH solution?

20. (Chapter 17) Conjugate acids and bases:

(a) What is the conjugate base of formic acid, HCO_2H?

(b) What is the conjugate base of the $H_2PO_4^-$ ion?

(c) What is the conjugate acid of the HCO_3^- ion?

21. (Chapter 17) Which of the following is NOT an acid-base conjugate pair?

(a) HClO and Cl^- (c) HNO_2 and NO_2^-

(b) HF and F^- (d) H_2CO_3 and HCO_3^-

22. (Chapter 17) For each solution below, decide if the pH is less than 7, equal to 7, or greater than 7.

(a) 0.10 M HNO_3 (d) 0.45 M KBr

(b) 0.012 M KOH   (e) 0.25 M Na_3PO_4

(c) 0.15 M formic acid   (f) 0.095 M $AlCl_3$

23. (Chapter 17) What is the pH of a 0.020 M solution of NH_3 (at 25 °C)?

24. (Chapter 17) A 0.040 M solution of a weak acid has a pH of 3.02 at 25 °C. What is the value of K_a for the acid?

25. (Chapter 17) What are the concentrations of H_3O^+ and OH^- and what is the pH of a 0.15 M solution of Na_2CO_3 (at 25 °C)?

26. (Chapter 17) What is the pH of the solution obtained when 12.5 mL of 0.40 M NaOH is added to 40.0 mL of 0.125 M HCl (at 25 °C)?

27. (Chapter 17) Lewis acids and bases. The presence of iron(III) ion in aqueous solution can be detected by adding thiocyanate ion, SCN^- to give a deep red complex ion. Identify the Lewis acids and bases in the following equation.

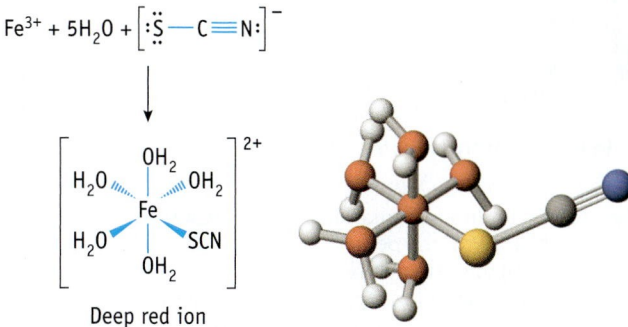

Deep red ion

28. (Chapter 18) Titrations

What is the pH at the half-neutralization point in a titration of formic acid (HCO_2H) with NaOH?

29. (Chapter 18) For each of the following cases, decide whether the pH is less than, equal to, or greater than 7.
    (a) Equal volumes of 0.10 M acetic acid, CH_3CO_2H, and 0.10 M KOH are mixed (at 25 °C).
    (b) 25 mL of 0.015 M NH_3 is mixed with 25 mL of 0.015 M HCl.
    (c) 100. mL of 0.0020 M HNO_3 is mixed with 50. mL of 0.0040 M NaOH.

30. (Chapter 18) Decide which compound in each of the following pairs of salts is the more soluble.
    (a) AgBr or AgSCN   (b) AgCl or Ag_2CO_3

31. (Chapter 18) Rank the following compounds in order of increasing molar solubility in water: Na_2CO_3, CaF_2, $BaCO_3$, Ag_2CO_3.

32. (Chapter 18) Milk of magnesia is an aqueous suspension of $Mg(OH)_2$. What are the concentrations of H_3O^+ and OH^- in solution (at 25 °C)?

33. (Chapter 18) Which of the following insoluble compounds— MnS, $Al(OH)_3$, $Ca_3(PO_4)_2$, $MgCO_3$, AgCl, PbI_2.—will not dissolve in 6 M HCl: ? List all correct answers.

34. (Chapter 18) You have a solution of NH_4Cl. Will the addition of NH_3 increase or decrease the pH or have no effect?

35. (Chapter 18) The Simulation on Screen 18.2 of *General ChemistryNow* explores the common ion effect. Here 0.10 M $HClO_2$ has ionized, about 29% of the molecules having been converted to H_3O^+ and ClO_2^- ions. In this solution

    $[HClO_2] = 0.071$ M, $[H_3O^+] = 0.029$ M, and $[ClO_2^-]$
    $= 0.029$ M.

18.2 Common Ion Effect

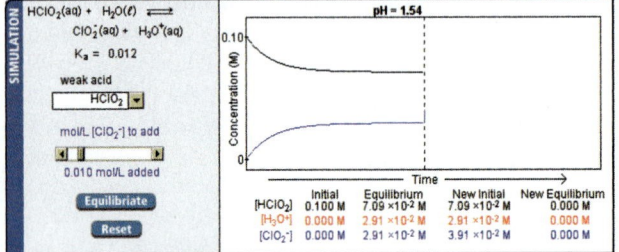

General ChemistryNow Screen 18.2 Common ion effect

    (a) In a solution containing 0.10 N $HClO_2$ and 0.01 M $NaClO_2$, is the percentage ionization of $HClO_2$ larger or smaller than 29%?
    (b) Is the pH of the $NaClO_2$/$HClO_2$ solution in (a) larger or smaller than that of the 0.10 M $HClO_2$ solution?
    (c) ▲ Determine the concentration of each species in the mixture of 0.10 M $HClO_2$ and 0.010 M $NaClO_2$.

36. (Chapter 18) What is the pH of a buffer prepared by mixing 100. mL of 0.20 M NH_4Cl to 200. mL of 0.10 M NH_3? (See General ChemistryNow Screen 18.5).

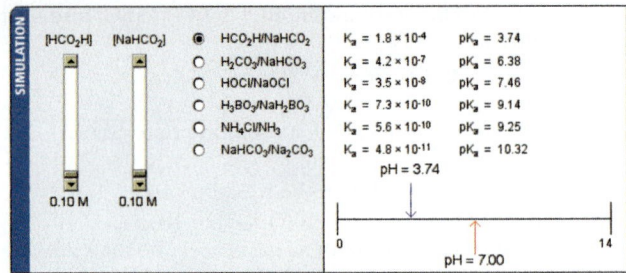

General ChemistryNow Screen 18.5 Simulation on buffers

37. (Chapter 19) Calculate the standard entropy change for the following reaction using data in Appendix L.
    $$F_2(g) + H_2(g) \longrightarrow 2\,HF(g)$$

38. (Chapter 19) Which of the following do you expect to have the largest entropy at 25 °C?
    (a) $H_2O(\ell)$ (c) $O_2(g)$
    (b) $H_2O(s)$ (d) $CCl_4(g)$

39. (Chapter 19) In which of the following physical or chemical changes do you expect to have a significant increase in entropy?
    (a) $Fe(s) \longrightarrow Fe(\ell)$
    (b) $Fe(s) + S(s) \longrightarrow FeS(s)$
    (c) $2\,Fe(s) + \frac{3}{2}\,O_2(g) \longrightarrow Fe_2O_3(s)$
    (d) $HF(\ell) \longrightarrow HF(g)$

40. (Chapter 19) Calculate $\Delta G°$ for the following reaction using data in Appendix L.
    $$N_2(g) + 3\,H_2O(\ell) \longrightarrow 2\,NH_3(g) + \frac{3}{2}\,O_2(g)$$

41. (Chapter 19) For the following general reaction, what can be said about the spontaneity at different temperatures?
    $A + B \longrightarrow C + D$; $\Delta H°$ is positive and $\Delta S°$ is negative
    (a) Product-favored at all temperatures
    (b) Product-favored only at high temperature
    (c) Product-favored only at low temperature
    (d) Not product-favored at any temperature

42. (Chapter 19) Given the following information, calculate $\Delta G°$ for the reaction below at 25 °C. Is the reaction spontaneous under standard conditions?
    $$NiO(s) + 2\,HCl(g) \longrightarrow NiCl_2(s) + H_2O(g)$$
    $$\Delta H° = -122.8 \text{ kJ and } \Delta S° = -123.9 \text{ J/K}$$

43. (Chapter 19) When sodium is added to water, the following reaction takes place readily.
    $$Na(s) + H_2O(\ell) \longrightarrow NaOH(aq) + \tfrac{1}{2}\,H_2(g); \Delta H° < 0$$
    What can be said about this reaction?
    (a) $\Delta G°$ is positive.
    (b) $\Delta S°$ is negative.
    (c) $\Delta G°$ will become less negative with increasing temperature.
    (d) $\Delta G°$ will become more negative with increasing temperature.

44. **(Chapter 19)** Acetic acid has a dissociation constant of 1.8×10^{-5}. What is the predicted sign for $\Delta G°$ for the dissociation of acetic acid at 25 °C? What is the value of $\Delta G°$?

45. **(Chapter 19)** In the reaction of two Cl atoms to give a Cl_2 molecule $[2\ Cl(g) \longrightarrow Cl_2(g)]$, decide if the sign of the enthalpy and entropy changes are positive or negative.

46. **(Chapter 20)** Balance the following oxidation-reduction reaction in acid solution. Designate the oxidizing and reducing agents and the substances oxidized and reduced.

$$MnO_4^-(aq) + H_2C_2O_4(aq) \longrightarrow Mn^{2+}(aq) + CO_2(g)$$

47. **(Chapter 20)** Use the small table of reduction potentials below to answer the questions that follow.

Half Reaction		E° (volts)
$Cl_2(g) + 2e^-$	$\longrightarrow$ $2\ Cl^-(aq)$	+1.36
$Ag^+(aq) + e^-$	$\longrightarrow$ $Ag(s)$	+0.80
$I_2(s) + 2e^-$	$\longrightarrow$ $2\ I^-(aq)$	+0.535
$Cu^{2+}(aq) + 2e^-$	$\longrightarrow$ $Cu(s)$	+0.34
$Pb^{2+}(aq) + 2e^-$	$\longrightarrow$ $Pb(s)$	−0.126
$Ni^{2+}(aq) + 2e^-$	$\longrightarrow$ $Ni(s)$	−0.25
$Zn^{2+}(aq) + 2e^-$	$\longrightarrow$ $Zn(s)$	−0.76
$V^{2+}(aq) + 2e^-$	$\longrightarrow$ $V(s)$	−1.18
$Al^{3+}(aq) + 3e^-$	$\longrightarrow$ $Al(s)$	−1.66

(a) What is the weakest oxidizing agent in the list?
(b) What is the strongest oxidizing agent?
(c) What is the strongest reducing agent?
(d) What is the weakest reducing agent?
(e) Will Pb(s) reduce $V^{2+}(aq)$ to V(s)?
(f) Will $I_2(s)$ oxidize Cl^- to $Cl_2(g)$?
(g) Identify the ions or elements that can be reduced by Pb(s)

48. **(Chapter 20)** Consider the electrochemical cell diagrammed below.

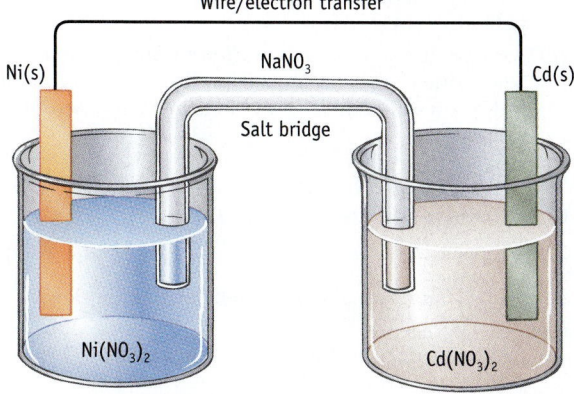

(a) Write the balanced net ionic equation for the reaction occurring in the cell.
(b) Indicate the direction of electron transfer.

(c) Which electrode is the anode and which is the cathode? Indicate the signs of the electrodes.
(d) Indicate the substance oxidized and the substance reduced.
(e) Calculate the standard cell potential, $E°$.
(f) Tell in which direction the Na^+ and NO_3^- ions flow in the salt bridge.

49. **(Chapter 20)** Consider the reaction

$$Cu(s) + 2\ Ag^+(aq) \longrightarrow Cu^{2+}(aq) + 2\ Ag(s)$$

If $[Ag^+] = 1.0$ M and $[Cu^{2+}] = 0.020$ M, is the cell potential E (a) greater than $E°$; (b) less than $E°$; or (c) equal to $E°$?

50. **(Chapter 20)** If you electrolyze a solution of $Ni^{2+}(aq)$ to form Ni(s) and use a current of 0.15 amps for 10 minutes, what mass of Ni(s) is produced?

COMPREHENSIVE QUESTIONS

51. **(Chapters 14 and 17)** Arrange the following aqueous solutions in order of (a) increasing boiling point and (b) increasing vapor pressure at 25 °C:

(i) 0.30 m $C_2H_4(OH)_2$ (ethylene glycol, nonvolatile solute)

(ii) 0.20 m CH_3CO_2H (acetic acid)

(iii) 0.20 m H_3PO_4

(iv) 0.20 m Na_2SO_4

Text Section: 14.4

General General ChemistryNow Screens: 14.7, 14.8

OWL Questions: 14.5, 14.6

52. **(Chapter 14)** You have isolated an unknown compound from treated wastewater, and to help identify the compound you determine its molar mass. A solution of the 0.238 g of the compound in 25.0 g of benzene freezes at 5.11 °C. What is the molar mass of the compound?

Text Section: 14.4

General ChemistryNow Screen: 14.8

OWL Question: 14.6

53. **(Chapter 15)** Data in the table refers to the reaction

$$2\ NO(g) + H_2(g) \longrightarrow N_2O(g) + H_2O(g)$$

Experiment	Initial [NO] (mol/L)	Initial [H₂] (mol/L)	Rate (mol/L · s)
1	2.16×10^{-2}	1.4×10^{-2}	1.2×10^{-2}
2	6.48×10^{-2}	1.4×10^{-2}	1.08×10^{-1}
3	4.32×10^{-2}	2.8×10^{-2}	9.6×10^{-2}

(a) What is the rate law equation for this reaction?
(b) What is the value and units of the rate constant k?

Text Section: 15.3

General ChemistryNow Screen: 15.5

OWL Question: 15.3

54. (Chapter 15) The Simulation from the General ChemistryNow Screen 15.11 for the first-order reaction of A to give B is illustrated here.

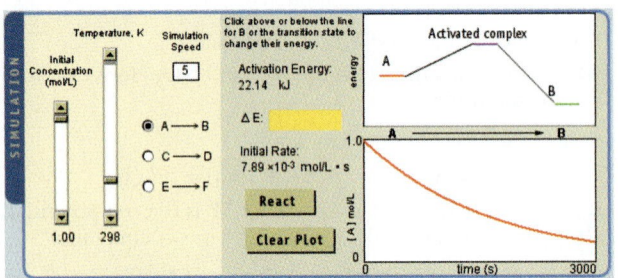

General ChemistryNow Screen 15.11 Activation energy.

(a) Is the reaction of A ⟶ B exothermic or endothermic?

(b) The activation energy for the reaction is 22.14 kJ. If this energy had been higher, say 25 kJ, would the reaction have been faster or slower?

Concentration versus time curves are plotted below for different temperatures or for a change in activation energy.

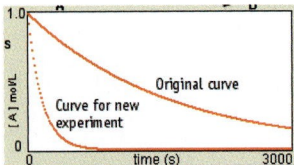 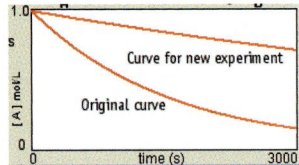

(c) Which diagram—left or right—represents the effect of increasing the temperature?

(d) Which diagram—left or right—represents the effect of having a higher activation energy?

Text Section: 15.5

General ChemistryNow Screens: 15.10, 15.11

OWL Questions: 15.7, 15.8

55. (Chapters 15 and 19) Heating ammonium chloride causes it to decompose to ammonia and hydrogen chloride. (See page 282).

$$NH_4Cl(s) \rightleftharpoons NH_3(g) + HCl(g)$$

(a) When the system has reached equilibrium, what is the effect on the equilibrium of adding more solid NH_4Cl? Of adding NH_3 gas?

(b) Predict the signs of ΔH°_{rxn} and ΔS°_{rxn}. Verify by calculating the enthalpy and entropy changes for the decomposition from data in Appendix L.

(c) Is the reaction predicted to be product- or reactant-favored at 400 °C? Verify the prediction by calculating ΔG°_{rxn} and then calculate the pressure of ammonia in a flask at this temperature.

Text Sections: 16.6, 19.4–19.6

General ChemistryNow Screens: 16.13, 19.5, 19.7

OWL Questions: 16.12, 19.2, 19.6

56. (Chapter 17) A 0.44 M solution of hydroxyacetic acid, $HOCH_2CO_2H$, is 1.8% ionized. Calculate the pH of the solution and the value of the acid dissociation equilibrium constant, K_a.

Text Sections: 17.3–17.5, 17.7

General ChemistryNow Screens: 17.4, 17.6, 17.8

OWL Questions: 17.3, 17.8

57. (Chapter 17) Acid-base chemistry

(a) List the following acids in order of increasing acid strength: HF, NH_4^+, H_3PO_4, CH_3CO_2H, HNO_3.

(b) Use the ranking of relative acid strength to predict whether the equilibrium $HF(aq) + CH_3CO_2^-(aq) \rightleftharpoons F^-(aq) + CH_3CO_2H(aq)$ is product-favored or reactant-favored.

Text Sections: 17.4–17.6

General ChemistryNow Screens: 17.6, 17.7

OWL Questions: 17.5, 17.7

58. (Chapter 18) Refer to the table of K_{sp} values in Appendix J and identify the most and least soluble silver salt, from among the following : AgCl, AgI, $AgCH_3CO_2$, AgCN, AgBr, and AgSCN.

Text Section: 18.4

General ChemistryNow Screens: 18.8–18.11

OWL Questions: 18.7, 18.8

59. (Chapters 17 and 18) You have been given solutions containing the compounds listed below. All concentrations are assumed to be 0.50 M.

    (i) $NH_3 + NH_4Cl$

    (ii) $Na_2HPO_4 + NaH_2PO_4$

    (iii) $HNO_3 + NaNO_3$

    (iv) $H_3PO_4 + NaH_2PO_4$

    (v) $NaF + HF$

(a) Which are buffer solutions?

(b) Which solution, among those listed, has the highest pH? The lowest pH?

(c) Assume that you need a solution buffered at pH = 7.4. Select one of the solutions from the list, then describe how you would adjust its pH to achieve the desired value.

Text Section: 18.2

General ChemistryNow Screens: 18.3–18.5

OWL Questions: 18.1–18.3

60. (Chapter 18) Titration of chloroacetic acid in water with aqueous NaOH:

(a) Chloroacetic acid, $ClCH_2CO_2H$, has a pK_a of 2.867. What is the value of K_a?

(b) Write the net ionic equation for the reaction of chloroacetic acid with NaOH.

(c) What volume of 0.128 M NaOH solution is required to titrate 50.0 mL of 0.154 M acid to the equivalence point?

(d) Which of the following indicators: phenolphthalein, methyl red, or bromthymol blue, would be most suitable for this titration?

(e) What is the pH at the half-neutralization point?

(f) What is the pH at the equivalence point?

Text Section: 18.3

General ChemistryNow Screen: 18.7

OWL Question: 18.5

63. (Chapter 20) Write balanced chemical equations for the following:

(a) $Cr_2O_7^{2-} + I^- \longrightarrow Cr^{3+} + I_2$ (in acid).

(b) $MnO_4^- + HS^- \longrightarrow MnO_2 + SO_4^{2-}$ (in base).

In each case identify the substance reduced, the reducing agent, and the change in oxidation state of chromium and manganese in these reactions.

Text Section: 20.1

General ChemistryNow Screen: 20.3

OWL Questions: 20.2, 20.3

64. (Chapter 20) Electrochemical cells:

(a) Sketch a picture of the voltaic cell $Zn(s)|Zn^{2+}$ $(1.0 M)\|Sn^{4+} (1.0 M)|Sn^{2+}(1.0 M)|Pt(s)$. On the drawing, identify the anode, cathode, and the direction of flow of electrons in the external circuit.

(b) Assume the cell has a salt bridge containing $NaNO_3$. In what direction does each ion in the salt bridge flow?

(c) Write a balanced net ionic equation for the overall cell reaction.

(d) What potential will this cell generate?

Text Sections: 20.2, 20.4

General ChemistryNow Screens: 20.4–20.7

OWL Question: 20.4

65. (Chapter 20) Refer to the table of standard reduction potentials to answer these questions. Questions can have more than one correct answer. List all correct answers.

(a) Which of the following metals react with H^+ (aq) to produce H_2: Ag, Mn, Sn, Cu, Mg?

(b) Which of the following metal ions is most easily reduced: Li^+, Mg^{2+}, Sn^{2+}, Fe^{2+}?

(c) Which of the following metal ions can be reduced to the metal by Sn: Ag^+, Ni^{2+}, Mg^{2+}, Fe^{2+}?

(d) Which of the species listed below is the strongest reducing agent: Cl^-, H_2, Mg^{2+}, Br^-?

Text Section: 20.4

General ChemistryNow Screens: 20.6, 20.7

OWL Question: 20.4

66. (Chapter 20) In the electrolysis of a solution containing Cr^{3+}, how long should one run a current of 2.50 amperes to obtain 5.00 g of Cr?

Text Section: 20.8

General ChemistryNow Screen: 20.12

OWL Question: 20.9

67. The pH curve for the titration of 50.0 mL of 0.10 M diethylamine with 0.10 M aqueous HCl is shown here.

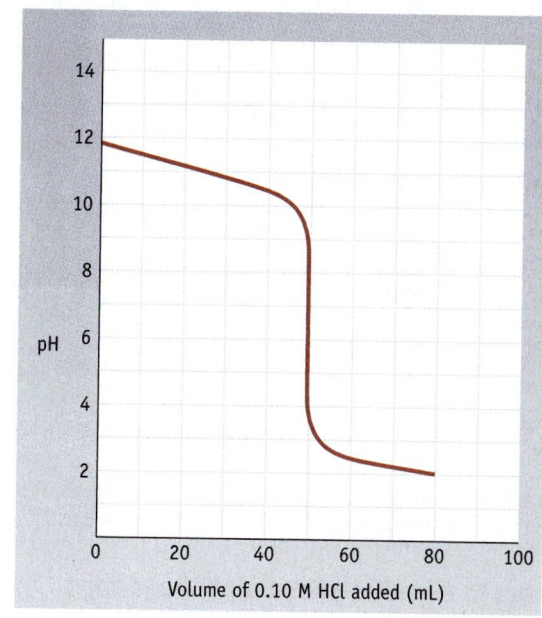

(a) Is diethylamine, $(C_2H_5)_2NH$, a weak acid or weak base?

(b) Write a balanced equation for the reaction of the amine with HCl.

(c) Use the graph of pH versus added HCl to determine the approximate pH at the half-neutralization point? What is the value of K for diethylamine?

(d) What is the approximate pH of the solution at the equivalence point? Verify your observation on the graph with a calculation.

(e) Explain briefly why this solution is acidic or basic at the equivalence point.

Text Section: 18.3

General ChemistryNow Screen: 18.7

OWL Question: 18.5

68. On page 998 you see photos of the reaction of aqueous nickel(II) ion with ammonia to give a complex ion,

$$[Ni(H_2O)_6]^{2+}(aq) + 6\ NH_3(aq)$$
$$\longrightarrow [Ni(NH_3)_6]^{2+}(aq) + 6\ H_2O(aq)$$

and then the reaction of that complex ion with the organic compound ethylenediamine to give another colorful complex ion.

$$[Ni(NH_3)_6]^{2+}(aq) + 3\ H_2NCH_2CH_2NH_2(aq)$$
$$\longrightarrow [Ni(H_2NCH_2CH_2NH_2)_3]^{2+}(aq) + 6\ NH_3(aq)$$

(a) What is the pH of a 0.10 M aqueous solution of nickel(II) chloride? (See Table 17.3, page 808.)

(b) What is the approximate pH of a 0.025 M aqueous solution of ethylenediamine? (See Appendix I.)

(c) In the complex ion $[Ni(NH_3)_6]^{2+}$ what is the Lewis acid and what is the Lewis base?

(d) Is the reaction of $Ni^{2+}(aq)$ with ammonia product- or reactant-favored? (See the discussion of formation constants on page 888 and Appendix K.)

(e) Given the formation constants below, calculate the equilibrium constant and $\Delta G°_{rxn}$ at 298 K for the reaction of $[Ni(NH_3)_6]^{2+}(aq)$ with ethylenediamine.

Complex Ion	Formation Constant
$[Ni(NH_3)_6]^{2+}(aq)$	5.6×10^8
$[Ni(H_2NCH_2CH_2NH_2)_3]^{2+}(aq)$	1.9×10^{18}

Comment on what thermodynamic function controls the outcome of the reaction, $\Delta H°_{rxn}$ or $\Delta S°_{rxn}$.

Answers to Assessing Key Points Questions

1. Molality and weight percent of dimethylglyoxime (DMG)

$$\text{Amount of glyoxime} = 45.0\ \text{g DMG}\left(\frac{1\ \text{mol DMG}}{116.1\ \text{g DMG}}\right)$$
$$= 0.388\ \text{mol DMG}$$

$$\text{Mass of ethanol} = 500.\ \text{mL}\left(\frac{0.790\ \text{g ethanol}}{1\ \text{mL ethanol}}\right) = 395\ \text{g ethanol}$$

$$\text{Molality}\ (m) = \frac{0.388\ \text{mol DMG}}{0.395\ \text{kg ethanol}} = 0.982\ m$$

$$\text{Weight percent} = \left(\frac{45.0\ \text{g DMG}}{45.0\ \text{g} + 395\ \text{g}}\right)100\% = 10.2\%$$

2. The solution with the largest concentration of solute particles, 0.12 m K_2SO_4, has the highest boiling point, the lowest freezing point, and the highest osmotic pressure. The solution with the smallest concentration of solute, 0.20 m $C_2H_4(OH)_2$, has the highest vapor pressure.

3. The answer is Mg^{2+}. The enthalpy of hydration depends directly on the charge on the ion and inversely on its size. Thus, the smallest, most highly charged ion will interact most strongly with water. Ion sizes are in the order $Rb^+ > Ba^{2+} > Na^+ > Mg^{2+}$ (Figure 8.15). (See pages 592–593 and 666–667.)

4. (b). See page 681.

5. Boiling point elevation. See Example 14.6.

$$\text{Amount of NaCl} = 29.3\ \text{g NaCl}\left(\frac{1\ \text{mol NaCl}}{58.45\ \text{g NaCl}}\right)$$
$$= 0.501\ \text{mol NaCl}$$

$$\text{Molality}\ (m) = \frac{0.501\ \text{mol NaCl}}{0.500\ \text{kg}} = 1.00\ m$$

$$\Delta T_{fp} = K_{fp} \cdot m \cdot i = (-1.86\ °C/m)(1.00\ m)(1.84) = -3.42\ °C$$

Here i is 1.84 (see Table 14.4).

6. Calculating molar mass using freezing point depression:

$$\Delta T_{fp} = -0.762\ °C = (-1.86\ °C/m)(C_m)$$
$$\text{Concentration} = C_m = 0.410\ m$$

$$\left(0.410\ \frac{\text{mol}}{\text{kg}}\right)0.0500\ \text{kg} = 0.0205\ \text{mol}$$

$$\text{Molar mass} = \frac{2.50\ \text{g}}{0.0205\ \text{mol}} = 122\ \text{g/mol}$$

Erythritol is $C_4H_{10}O_4$

7. (a) -0.85×10^{-5} mol/L·min. Divide the change in concentration by the time period over which it occurs.

(b) Approximately 2 minutes.

(c) The rate depends on the dye concentration. Therefore, the rate declines with time because the concentration of the dye declines with time.

8. (a) Rate = $k[NO]^2[H_2]$. The sum of the exponents is 3.

(b) The rate will increase by 25-fold (5^2).

(c) The rate increases by 5.

9. Answer (c): $\Delta C/\Delta t = k[B][D]$. Comparing lines 1 and 2 of the data you see that the rate doubles when [B] doubles. Therefore, the reaction is 1st order in B. Comparing lines 1 and 4 you find that the rate doubles when [D] doubles, so the reaction is 1st order in D. Comparing lines 3 and 4 shows that the rate doubles with a doubling of [B], as expected, but the increase in [A] has no effect.

10. (a) $m = 1$; the rate doubles when concentration doubles.

(b) $m = 2$; the rate increases by a factor of 4 when concentration doubles.

(c) $m = 0$; there is no change in rate when the concentration changes.

(d) Bottom. For a zero order reaction Rate = k. The concentration versus time curve is a straight line with a negative slope equal to $-k$.

11. Step 2 is rate determining and N_2O_2 is an intermediate.

12. (a) Rate = $(0.17/h)(0.010\ \text{mol/L}) = 0.0017\ \text{mol/L·h}$

(b) $t_{1/2} = 0.693/k = 4.1\ h$

(c)

	SO_2Cl_2	$\rightleftharpoons$	SO_2	+	Cl_2
Initial P (atm)	0.050		0		0
Change in P (atm)	−0.025		+0.025		0.025
Final P (atm)	0.025		0.025		0.025

Total final pressure = 0.075 atm

13. Rate constant $= 0.693/t_{1/2} = 0.181/d$

$$\ln(0.100/1.00) = -kt = -(0.181/d)(t)$$
$$-2.303 = -(0.181/d)(t)$$
$$t = 12.7\ d$$

14. Calculate Q_c, the reaction quotient.

$$Q_c = \frac{[SO_3]^2}{[SO_2]^2[O_2]} = -\frac{(6.9 \times 10^{-3})^2}{(5.0 \times 10^{-3})^2(1.9 \times 10^{-3})}$$
$$Q_c = 1.0 \times 10^3$$

Because $Q_c > K$, the concentration of product is too large and that of the reactants is too small. The reaction will proceed to the left to attain equilibrium.

15.

	N_2O_4	$\rightleftharpoons$	$2NO_2$
Initial (M)	0.0050		0
Change (M)	**−0.0008**		**+2 × 0.0008**
Equilibrium (M)	0.0042		**2 × 0.0008**

The additions to the table are in bold. The final concentration of NO_2 is derived from the change in N_2O_4 concentration.

$$K_c = \frac{[NO_2]^2}{[N_2O_4]} = \frac{(0.0016)^2}{0.0042} = 6.1 \times 10^{-4}$$

16. $K_c = 240 = \dfrac{[N_2O_4]}{[NO_2]^2} = \dfrac{(4.6 \times 10^{-5})}{[NO_2]^2}$

$$[NO_2] = 4.4 \times 10^{-4}\ M$$

17. See Section 16.5 and Problem-Solving Tip 16.2.

$$SnO_2(s) + 2\ H_2(g) \rightleftharpoons Sn(s) + 2\ H_2O(g)$$
$$K_c = K_1 = 8.12$$

$$2\ [CO(g) + H_2O(g) \rightleftharpoons H_2(g) + CO_2(g)]$$
$$K_c = (1/K_2)^2 = (1/0.771)^2$$

Summed equations:

$$SnO_2(s) + 2\ CO(g) \rightleftharpoons Sn(s) + 2\ CO_2(g)$$
$$K_{net} = K_1 \cdot (1/K_2)^2 = 13.7$$

18. (a) No change   (d) Shifts right
    (b) Shifts left   (e) Shifts right
    (c) Shifts left

19. NaOH is a strong base so $[NaOH] = [OH^-]$

$$pOH = -\log[OH^-] = -\log(0.020) = 1.70$$
$$pH + pOH = 14.00, \text{ so } pH = 12.30$$

Alternatively, we first solve for $[H_3O^+]$

$$K_w = [H_3O^+][OH^-] = 1.00 \times 10^{-14}$$
$$[H_3O^+] = 5.0 \times 10^{-13}\ M$$

and this gives a pH of 12.30.

20. See Table 17.3 on page 808.
    (a) HCO_2^- (b) HPO_4^{2-} (c) H_2CO_3

21. The answer is (a).

22. (a) Strong acid. pH < 7.
    (b) Strong base. pH > 7.
    (c) Weak acid. pH < 7.
    (d) Neutral salt. pH = 7.
    (e) PO_4^{3-} is a relatively strong weak base ($K_b = 2.8 \times 10^{-2}$). pH > 7. See Problem-Solving Tip 17.2, page 811.
    (f) Al^{3+} is a weak acid. pH < 7.

23. Set up an ICE table for the weak base, NH_3.

$$NH_3(aq) + H_2O(\ell) \rightleftharpoons NH_4^+(aq) + OH^-(aq)$$

Initial	0.020	0	0
Change	$-x$	$+x$	$+x$
Equilibrium	$0.020 - x$	x	x

$$K_b = 1.8 \times 10^{-5} = \frac{[NH_4^+][OH^-]}{[NH_3]}$$
$$1.8 \times 10^{-5} = \frac{[x]^2}{0.020 - x}$$

Assuming x is $\ll 0.020$

$$x = [OH^-] = 6.0 \times 10^{-4}\ M$$
$$pOH = -\log[OH^-] = 3.22$$
$$pH + pOH = 14.00$$
$$pH = 10.78$$

24. Set up an ICE table for a weak acid HA where $[H_3O^+] = 10^{-pH} = 9.5 \times 10^{-4}\ M$

$$HA(aq) + H_2O(\ell) \rightleftharpoons H_3O^+(aq) + A^-(aq)$$

Initial	0.040	0	0
Change	$-x$	$+x$	$+x$
Equilibrium	$0.040 - x$	x	x

where $x = 9.5 \times 10^{-4}\ M$

$$K_a = \frac{[H_3O^+][A^-]}{[HA]} = \frac{(9.5 \times 10^{-4})^2}{0.040 - 9.5 \times 10^{-4}} = 2.3 \times 10^{-5}$$

25. Set up an ICE table for the CO_3^{2-} ion behaving as a weak, monoprotic base.

$$CO_3^{2-}(aq) + H_2O(\ell) \rightleftharpoons HCO_3^-(aq) + OH^-(aq)$$

Initial	0.15	0	0
Change	$-x$	$+x$	$+x$
Equilibrium	$0.15 - x$	x	x

$$K_b = \frac{[HCO_3^-][OH^-]}{[CO_3^{2-}]}$$
$$2.1 \times 10^{-4} = \frac{x^2}{0.15 - x}$$

Assuming $x \ll 0.15\ M$

$$x = [OH^-] = 5.6 \times 10^{-3}\ M$$
$$pOH = 2.25 \text{ and so } pH = 11.75$$

The product of CO_3^{2-} hydrolysis, HCO_3^- contributes little to the acidity.

26. Amount of NaOH used = $C \cdot V = 5.0 \times 10^{-3}$ mol.

    Amount of HCl used = $C \cdot V = 5.0 \times 10^{-3}$ mol.

    All of the NaOH and HCl are consumed in the reaction leaving only water and the neutral salt NaCl. The pH of the solution must be 7.00 (at 25 °C).

27. Fe^{3+} is a Lewis acid, an electron pair acceptor (as are all aqueous transition metal cations). Water and SCN^- ion are both electron pair donors, Lewis bases. See Section 17.9.

28. $HCO_2H(aq) + OH^-(aq) \rightleftharpoons HCO_2^-(aq) + H_2O(\ell)$

    At the half-neutralization point of a titration of a weak acid with a strong base the solution contains half of the original acid and an equal amount of weak conjugate base. The solution is a buffer with $[H_3O^+] = K_a$ (and pH = pK_a) Here $[H_3O^+] = 1.8 \times 10^{-4}$ M and pH = 3.74. See page 866.

29. (a) $CH_3CO_2H(aq) + OH^-(aq)$
    $$\rightleftharpoons CH_3CO_2^-(aq) + H_2O(\ell)$$

    This is the reaction of a weak acid with an equal amount of a strong base. The product is the weak conjugate base of the acid, so the pH is greater than 7 after reaction is complete.

    (b) $NH_3(aq) + H_3O^+(aq) \rightleftharpoons NH_4^+(aq) + H_2O(\ell)$

    This is the reaction of a weak base with an equal amount of a strong acid. The product is the weak conjugate acid of the base, so the pH is less than 7 after reaction is complete.

    (c) $H_3O^+(aq)\ OH^-(aq) \rightleftharpoons 2\ H_2O(\ell)$

    This is the reaction of a strong acid with an equal amount of strong base. The solution consists of a neutral salt in water; pH = 7.0.

30. See Section 18.4 and page 878 in particular.

    (a) When two salts have the same formula type (for example, both have one cation and one anion) we can directly compare K_{sp} values to judge their relative solubility.

    $$AgSCN\ (1.0 \times 10^{-12}) > AgBr\ (5.4 \times 10^{-13})$$

    AgSCN is more soluble.

    (b) For AgCl and Ag_2CO_3 we have to calculate their solubilities.

    Solubility of AgCl = $(K_{sp})^{1/2} = (1.8 \times 10^{-10})^{1/2}$
    $$= 1.3 \times 10^{-5}\ M$$

    Solubility of Ag_2CO_3

    $$Ag_2CO_3(s) \rightleftharpoons 2\ Ag^+(aq) + CO_3^{2-}(aq)$$
    $$K_{sp} = [Ag^+]^2[CO_3^{2-}] = (2x)^2(x) = 8.5 \times 10^{-12}$$

    where $x = [CO_3^{2-}]$ and is equivalent to the molar solubility of the salt.

    $$[CO_3^{2-}] = 1.3 \times 10^{-4}\ M = \text{molar solubility of } Ag_2CO_3$$

    Even though Ag_2CO_3 has a smaller K_{sp} value than AgCl, its molar solubility is greater.

31. (a) Na_2CO_3 is a soluble salt.

    (b) $BaCO_3$. Solubility from $K_{sp} = 5.1 \times 10^{-5}$ M.

    (c) CaF_2. Solubility from $K_{sp} = 2.4 \times 10^{-4}$ M.

(d) Ag_2CO_3. Solubility from $K_{sp} = 1.3 \times 10^{-4}$ M.

Order of increasing molar solubility:

$$BaCO_3 < Ag_2CO_3 < CaF_2 < Na_2CO_3$$

32. $Mg(OH)_2(s) \rightleftharpoons Mg^{2+}(aq) + 2\ OH^-(aq)$

    $$K_{sp} = 5.6 \times 10^{-12} = [Mg^{2+}][OH^-]^2 = (x)(2x)^2$$
    $$x = 1.1 \times 10^{-4}\ M$$
    $$[OH^-] = 2x = 2.2 \times 10^{-4}\ M$$
    $$[H_3O^+] = K_w/[OH^-] = 4.5 \times 10^{-11}\ M$$

33. Dissolve in 6 M HCl: MnS, $Ca_3(PO_4)_2$, $Al(OH)_3$, $MgCO_3$. Only these salts contain basic anions (which are conjugate bases of weak acids).

34. $NH_4^+(aq) + H_2O(\ell) \rightleftharpoons NH_3(aq) + H_3O^+(aq)$

    Ammonium ion is in equilibrium with a small amount of NH_3. Adding the weak base NH_3 shifts the equilibrium to the left and raises the pH.

35. (a) Adding ClO_2^- leads to less ionization (about 25%).

    (b) The pH increases (as expected when a weak base is added).

    (c) $[HClO_2] = 0.075$ M, $[H_3O^+] = 0.025$ M, and $[ClO_2^-] = 0.035$ M. Verify using the Simulation on Screen 18.2.

36. The Simulation on Screen 18.5 gives a pH of 9.25.

    $$[H_3O^+] = \frac{[NH_4^+]}{[NH_3]} K_a$$
    $$= \left(\frac{0.020\ mol\ NH_4^+}{0.020\ mol\ NH_3}\right) 5.6 \times 10^{-10} = 5.6 \times 10^{-10}\ M$$

37. $\Delta S° = \Sigma\ S°_{product} - \Sigma\ S°_{reactants}$ (Equation 19.2, page 915)
    $$\Delta S° = 2\ S°(HF) - [S°(H_2) + S°\ (F_2)]$$
    $$\Delta S° = 2\ mol\ (173.78\ J/K \cdot mol)$$
    $$- [1\ mol\ (130.7\ J/K \cdot mol) + 1\ mol\ (202.8\ J/K \cdot mol)]$$
    $$\Delta S° = 14.1\ J/K$$

38. The answer is (d). CCl_4 has more atoms in the molecule than the others and is in the gas phase.

39. (a) and (d)

40. See Equation 19.6, page 925. (Recall that the free energy of formation of elements in their standard states is zero.)
    $$\Delta G°_{rxn} = \Sigma\ \Delta G°_{product} - \Sigma\ \Delta G°_{reactants}$$
    $$\Delta G°_{rxn} = 2\ \Delta G°_f[NH_3(g)] - 3\ \Delta G°_f[H_2O(\ell)]$$
    $$= 2\ mol\ (-16.37\ kJ/mol) - 3\ mol\ (-237.15\ kJ/mol)$$
    $$= +678.71\ kJ$$

41. The answer is (d), not product-favored at any temperature. See Table 19.2, page 920.

42. $\Delta G°_{rxn} = \Delta H°_{rxn} - T\Delta S°_{rxn}$
    $$\Delta G°_{rxn} = -122.8\ kJ - (298\ K)(-0.1239\ kJ/K)$$
    $$\Delta G°_{rxn} = -85.9\ kJ$$
    The reaction is spontaneous.

43. (d) $\Delta G°$ becomes more negative with increasing T because $\Delta S°$ is predicted to be positive (and $\Delta H°$ is negative). See the leftmost panel in Figure 19.12.

44. The sign of $\Delta G°$ is predicted to be positive for a reaction with K less than 1.

$$\Delta G°_{rxn} = -RT\ln K$$

$$= -(8.3145 \text{ J/K} \cdot \text{mol})(298 \text{ K}) \ln (1.8 \times 10^{-5})$$

$$\Delta G°_{rxn} = 2.71 \times 10^4 \text{ J}$$

45. Both ΔH and ΔS are predicted to be negative. The combination of two atoms to form one molecule is always an exothermic process. In addition, two atoms combine to form one molecule, which means $\Delta S° < 0$.

46. MnO_4^- is the oxidizing agent and is reduced.

$$MnO_4^-(aq) + 8 H^+(aq) + 5e^- \longrightarrow Mn^{2+}(aq) + 4 H_2O(\ell)$$

$H_2C_2O_4$ is the reducing agent and is oxidized.

$$H_2C_2O_4(aq) \longrightarrow 2 CO_2(g) + 2H^+(aq) + 2 e^-$$

Complete equation:

$$2 MnO_4^-(aq) + 6 H^+(aq) + 5 H_2C_2O_4(aq)$$
$$\longrightarrow 2 Mn^{2+}(aq) + 8 H_2O(\ell) + 10 CO_2(g)$$

47. (a) Al^{3+}
    (b) Cl_2
    (c) Al
    (d) Cl^-
    (e) No, Pb will not reduce V^{2+}.
    (f) No, I_2 will not oxidize Cl^-.
    (g) Cl_2, Ag^+, I_2, and Cu^{2+}

48. (a) From Table 20.1 we know that Cd will reduce Ni^{2+} to Ni. Therefore,

$$Cd(s) + Ni^{2+}(aq) \longrightarrow Cd^{2+}(aq) + Ni(s)$$

   (b) Electrons transfer in the external wire from Cd to Ni.
   (c) Cd is the site of oxidation and is the anode. The Ni electrode is the cathode.
   (d) Cd is oxidized to Cd^{2+} and Ni^{2+} is reduced to Ni.
   (e) $E°_{cell} = E°_{cathode} - E°_{anode} = -0.25 \text{ V} - (-0.40 \text{ V})$

$$E°_{cell} = +0.15 \text{ V}$$

   (f) Anions flow into the Cd cell (to offset the charge of the Cd^{2+} ions being generated), and the cations flow into the Ni cell.

49. Using the Nernst equation, we have

$$E = E° - \frac{0.0257}{2} \ln \frac{[Cu^{2+}]}{[Ag^+]^2}$$

$$E = +0.462 \text{ V} - 0.0129 \ln \frac{0.020}{(1.0)^2}$$

$$E = 0.512 \text{ V}$$

The electrochemical potential is more positive.

50. The charge passed in the 10.-minute (or 600 second) time period is:

$$(0.15 \text{ amp})(600 \text{ s}) = 90. \text{ coulombs}$$

$$\text{Amount of charge} = 90. \text{ C}\left(\frac{1 \text{ mol electrons}}{96500 \text{ C}}\right)$$
$$= 9.3 \times 10^{-4} \text{ mol } e^-$$

$$9.3 \times 10^{-4} \text{ mol } e^- \left(\frac{1 \text{ mol Ni deposited}}{2 \text{ mol } e^-}\right) = 4.7 \times 10^{-4} \text{ mol Ni}$$

$$4.7 \times 10^{-4} \text{ mol Ni}\left(\frac{58.7 \text{ g Ni}}{1 \text{ mol Ni}}\right) = 0.027 \text{ g Ni}$$

Solutions to Comprehension Questions

51. Boiling point increases in the order:

$$CH_3CO_2H < H_3PO_4 < C_2H_4(OH)_2 < Na_2SO_4.$$

This is in order of the increasing number of particles (ions or molecules) in solution. The increase in vapor pressure occurs in the opposite order.

52. Freezing point depression. Use equation 14.8.

$$\Delta T_{fp} = (5.11 °C - 5.50 °C) = -0.39 °C$$

Solute concentration (m)

$$= \Delta T_{fp}/K_{fp} = -0.39 \text{ C}/(-5.12 °C/m) = 0.076 \text{ } m$$

Amount of solute $= (0.076 \text{ mol}/1.00 \text{ kg})(0.0250 \text{ kg})$
$$= 1.9 \times 10^{-3} \text{ mol}$$

Molar mass $= 0.238 \text{ g}/1.9 \times 10^{-3} \text{ mol} = 1.2 \times 10^2 \text{ g/mol}$

53. (a) Rate $= k[NO]^2[H_2]$. Compare experiments 1 and 2: tripling [NO] increases the rate by a factor of 9 indicating the reaction is second order in [NO]. Compare experiments 1 and 3: doubling [NO] and [H_2] increases the rate by a factor of 8. You know that doubling [NO] increases the rate by 4-fold. To achieve an increase by a factor of 8 after doubling both [NO] and [H_2] must mean the reaction is also first-order in H_2.

   (b) Using the first set of data to calculate k:

$$k = \frac{\text{Rate}}{[NO]^2[H_2]} = \frac{1.2 \times 10^{-2} \text{ mol/L}}{(2.16 \times 10^{-2} \text{ mol/L})^2(1.4 \times 10^{-2} \text{ mol/L})}$$
$$= 1.8 \times 10^3 \text{ L}^2/\text{mol}^2 \cdot \text{s}$$

54. (a) Exothermic. The energy of the products is lower than the energy of the reactants.

   (b) A reaction with a higher activation energy will be slower, assuming the temperature at which the reaction is carried out is the same in each instance.

   (c) Left diagram. The reaction will occur faster at higher temperatures and the concentrations will decrease faster.

   (d) Right diagram. If the activation energy is higher, the reaction will be slower (at the same temperature).

55. (a) Addition of solid NH_4Cl has no effect. Addition of gaseous NH_3 will cause the equilibrium to shift to the left, forming more solid NH_4Cl and consuming some of the HCl and NH_3.

(b) Predictions: ΔH°_{rxn}: positive (heat must be added). ΔS°_{rxn}: positive (conversion of 1 mole of solid to two moles of gases).

$$\Delta H^{\circ}_{rxn} = \Delta H^{\circ}_f[NH_3(g)] + \Delta H^{\circ}_f[HCl(g)] - \Delta H^{\circ}_f[NH_4Cl(s)]$$
$$= 1\ mol(-45.90\ kJ/mol) + 1\ mol(-93.41\ kJ/mol)$$
$$- 1\ mol(-314.55\ kJ/mol)$$
$$= 176.34\ kJ$$

$$\Delta S^{\circ}_{rxn} = S^{\circ}[NH_3(g)] + S^{\circ}[HCl(g)] - S^{\circ}[NH_4Cl(s)]$$
$$= 1\ mol(192.77\ J/mol\ K) + 1\ mol(186.2\ J/mol\ K)$$
$$- 1\ mol(94.85\ J/mol\ K)$$
$$= 284.1\ J/K$$

(c) Prediction: Because of the favorable entropy term, this reaction will become product-favored at higher temperatures. A calculation is needed to determine if 400 °C is a high enough temperature. (Note that 400 °C = 673 K.)

$$\Delta G^{\circ}_{rxn} = \Delta H^{\circ}_{rxn} - T\ \Delta S^{\circ}_{rxn}$$
$$= 176.34\ kJ - (673\ K)(0.28412\ kJ/K)$$
$$= -16\ kJ$$

The negative sign indicates that the reaction is product-favored at this temperature.

To calculate $P(NH_3)$, first determine the equilibrium constant K_p for this reaction, then use this in an equilibrium calculation.

$$\Delta G^{\circ}_{rxn} = -RT\ln K$$

$$\ln K_p = -\frac{(-16000\ J/mol)}{(8.3145\ J/K \cdot mol)(673\ K)} = 2.9$$
$$K_p = 17$$
$$K_p = P_{NH_3}P_{HCl}$$

If $P_{NH_3} = P_{HCl} = x$
then $K_p = 17 = x^2$ and $x = 4.2$
$$P_{NH_3} = P_{HCl} = 4.2\ atm$$

56. Set up a ICE table. Here the acid is ionized to the extent of 1.8% or (0.018)(0.44 M) = 0.0079 M.

	$HOCH_2CO_2H$ $\rightleftharpoons$	H_3O^+	+	$HOCH_2CO_2^-$
Initial (M)	0.44	0		0
Change (M)	−0.0079	+0.0079		+0.0079
Equilib. (M)	0.43	0.0079		0.0079

Because $[H_3O^+] = 0.0079$ M, pH = 2.10

$$K_a = \frac{[H_3O^+][HOCH_2CO_2^-]}{[HOCH_2CO_2H]}$$

$$K_a = \frac{(0.0079)^2}{0.43} = 1.5 \times 10^{-4}$$

57. Refer to K_a values in Table 17.3, page 808.
(a) Acid strength:

$$NH_4^+ < CH_3CO_2H < HF < H_3PO_4 < HNO_3$$

(b) Product-favored. (Brønsted acid–base reactions favor the weaker acid and weaker base.)

58. Most soluble: $AgCH_3CO_2$, least soluble: AgCN.

59. (a) Combinations (i), (ii), (iv), and (v) contain a weak acid and its weak conjugate base and are buffer solutions.
(b) If the concentrations of the two species in the buffer are equal, pH = pK_a. The solution with highest pH is (i) and the solution with the lowest pH is (iii).
(c) Select the buffer system for which the pK_a is closest to 7.40. This is (ii) for which $pK_a = 7.21$. To adjust the pH to 7.40, add base (either additional Na_2HPO_4 or NaOH which would convert some of the NaH_2PO_4 to Na_2HPO_4).

60. (a) $K_a = 10^{-pK_a} = 10^{-2.867} = 1.36 \times 10^{-3}$
(b) $ClCH_2CO_2H(aq) + OH^-(aq)$
$$\rightleftharpoons H_2O(\ell) + ClCH_2CO_2^-(aq)$$
(c) moles acid ($M_{acid} \times V_{acid}$) = moles base ($M_{base} \times V_{base}$)
$$(0.154\ mol/L)(0.0500\ L) = (0.128\ mol/L)(V_{base})$$
$$V_{base} = 0.0602\ L\ (= 60.2\ mL)$$
(d) Phenolphthalein. Choose an indicator that changes color near the pH at the equivalence point of the titration. In this case, the pH at the equivalence point is slightly basic. (Figure 18.2, page 872, shows the ranges of color changes for indicators.)
(e) pH = $pK_a = 2.867$
(f) At the equivalence point, the solution contains 0.00770 moles $ClCH_2CO_2^-$ in a total volume of 0.1102 L (0.0602 L + 0.0500 L). The concentration of this anion is 0.0699 M (0.0077 mol/0.1102 L). The anion is weakly basic:

$$ClCH_2CO_2^-(aq) + H_2O(\ell)$$
$$\rightleftharpoons ClCH_2CO_2H(aq) + OH^-(aq)$$

For this equilibrium
$$K_b = K_w/K_a = (1.0 \times 10^{-14})/(1.36 \times 10^{-3}) = 7.4 \times 10^{-12}$$

$$K_b = \frac{[ClCH_2CO_2H][OH^-]}{[ClCH_2CO_2^-]}$$

$$7.4 \times 10^{-12} = \frac{(x)(x)}{0.0699 - x}$$

$$x = [OH^-] = 7.2 \times 10^{-7}\ M$$
$$pOH = 6.14$$

and so pH = 14.00 − 6.14 = 7.86

63. (a) $Cr_2O_7^{2-} + 14\ H^+ + 6\ e^- \longrightarrow 2\ Cr^{3+} + 7\ H_2O$
$$\underline{3\ [2\ I^- \longrightarrow I_2 + 2\ e^-]}$$
$$Cr_2O_7^{2-} + 14\ H^+ + 6\ I^- \longrightarrow 2\ Cr^{3+} + 7\ H_2O + 3\ I_2$$
$Cr_2O_7^{2-}$ is reduced, I^- is the reducing agent, and the oxidation state of chromium changes from +6 to +3.
(b) $8[MnO_4^- + 2\ H_2O + 3\ e^- \longrightarrow MnO_2 + 4\ OH^-]$
$$\underline{3\ [HS^- + 9\ OH^- \longrightarrow SO_4^{2-} + 5\ H_2O + 8\ e^-]}$$
$$8\ MnO_4^- + 3\ HS^- + H_2O$$
$$\longrightarrow 8\ MnO_2 + 5\ OH^- + 3\ SO_4^{2-}$$
MnO_4^- is reduced, HS^- is the reducing agent, and the oxidation state of the manganese changes from +7 to +4.

64. (a) Cell using Zn as the anode and a chemically inert Pt electrode as the cathode.

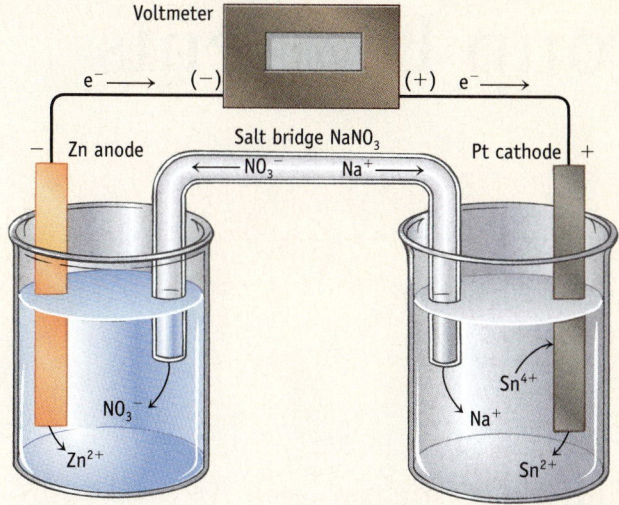

(b) The Na^+ ions move toward the $Sn^{4+}|Sn^{2+}$ (Pt) half-cell. The NO_3^- ions move toward the $Zn|Zn^{2+}$ half-cell.

(c) $Zn + Sn^{4+} \longrightarrow Zn^{2+} + Sn^{2+}$

(d) $E°_{cell} = E°_{cathode} - E°_{anode}$

$= +0.15\ V - (-0.76\ V) = 0.91\ V$

65. For discussion of methodology for part (a) see "Northwest-southeast rule", page 968. For parts (b), (c), and (d) see Example 20.6.

   (a) Sn, Mg, and Mn
   (b) Sn^{2+}
   (c) Ag^+
   (d) H_2

66. First, calculate moles of electrons required, next calculate the quantity of charge, then use charge and current to calculate time required. (See Example 20.13.)

$$\text{Mol } e^- = (5.00\ \text{g Cr})\left(\frac{1\ \text{mol Cr}}{52.0\ \text{g Cr}}\right)\left(\frac{3\ \text{mol } e^-}{1\ \text{mole Cr}}\right) = 0.288\ \text{mol } e^-$$

$$\text{Charge (C)} = (0.288\ \text{mol } e^-)\left(\frac{96,500\ \text{C}}{\text{mol } e^-}\right) = 2.78 \times 10^4\ \text{C}$$

$$\text{Charge (C)} = \text{current (A)} \times \text{time (s)}$$

$$\text{Time} = \frac{2.78 \times 10^4\ \text{C}}{2.50\ \text{A}} = 1.11 \times 10^4\ \text{s}\ (= 186\ \text{min})$$

67. (a) The pH of the original solution is almost 12, so diethylamine is a weak base. Note that $[OH^-]$ for a 0.10 M solution is a little less than 0.010 M, indicating less than 10% of the base reacts with H_2O to form OH^-.

(b) $(C_2H_5)_2NH(aq) + HCl(aq)$
$$\rightleftharpoons (C_2H_5)_2NH_2^+(aq) + Cl^-(aq)$$

(c) At the half-neutralization point (after 25 mL of 0.10 M HCl has been added), pOH = 3.0. (pOH = 14.00 − pH where the pH = 11.0). $pK_b = pOH$, so $K_b = 1.0 \times 10^{-3}$.

(d) The equivalence point occurs when exactly the amount of acid required has been added. Here 50.0 mL of a 0.10 M solution of the weak base requires 50.0 mL of 0.10 M HCl, and you observe a significant drop in the pH at this volume of acid. The pH at the mid-point of the vertical portion of the curve is the pH at the equivalence point, here about 6.2. The solution should be slightly acidic, as observed, because we titrated a weak base and have its weak conjugate acid in solution.

(e) The solution contains $(C_2H_5)_2NH_2^+(aq)$, a weak acid.

68. (a) K_a for $Ni^{2+}(aq) = 2.5 \times 10^{-11}$. pH for a 0.10 M solution is 5.80.

(b) K_{b1} for ethylenediamine = 8.5×10^{-5}. pOH for a 0.025 M solution = 2.84, so pH = 11.16. Ethylenediamine undergoes a second ionization ($K_{b2} = 2.7 \times 10^{-8}$). This will raise the pH slightly.

(c) Ni^{2+} is an electron-pair acceptor, a Lewis acid. NH_3, with its lone pair, is a Lewis base. (See Section 17.9.)

(d) K_f for the formation of $[Ni(NH_3)_6]^{2+}(aq)$ is 5.6×10^8. The reaction is product-favored.

(e) Calculating K_{net}.

$[Ni(NH_3)_6]^{2+}(aq) + 6\ H_2O(aq)$
$$\longrightarrow [Ni(H_2O)_6]^{2+}(aq) + 6\ NH_3(aq)$$
$1/K_{f1} = 1/5.6 \times 10^8$
$[Ni(H_2O)_6]^{2+}(aq) + 3\ H_2NCH_2CH_2NH_2(aq)$
$$\longrightarrow [Ni(H_2NCH_2CH_2NH_2)_3]^{2+}(aq) + 6\ H_2O(aq)$$
$$K_{f2} = 1.9 \times 10^{18}$$

$[Ni(NH_3)_6]^{2+}(aq) + 3\ H_2NCH_2CH_2NH_2(aq)$
$$\longrightarrow [Ni(H_2NCH_2CH_2NH_2)_3]^{2+}(aq) + 6\ NH_3(aq)$$
$$K_{f3} = ?$$

$$K_{f3} = (1/K_{f1})\,K_{f2} = 3.4 \times 10^9$$
$$\Delta G°_{rxn} = -RT\ln K = -54\ \text{kJ/mol at 298 K}$$

This spontaneous (and product-favored) reaction is largely controlled by $\Delta S°_{rxn}$. $\Delta H°_{rxn}$ is likely about 0 because 6 Ni—NH_3 bonds are exchanged for bonds between Ni and the 6 N atoms of the three ethylenediamine molecules. $\Delta S°_{rxn}$ is expected to be large and positive because four ions or molecules are replaced by seven product ions or molecules.

21—The Chemistry of the Main Group Elements

Acid mine drainage. The water flowing from mines is often acidic owing to the production of sulfuric acid from sulfur-bearing minerals.

Simon Fraser/Science Photo Library/Photo Researchers, Inc.

Sulfur Chemistry and Life on the Edge

Whether life exists elsewhere in our solar system is one of the unanswered questions of science. Extremely harsh conditions on planets other than Earth almost surely make human life impossible, but scientists hold out hope that simpler life forms may exist elsewhere. One reason for this belief is that life on Earth has been found to exist under unimaginably severe conditions: extreme heat and cold, high pressure, and highly acidic, highly basic, and highly saline conditions.

How do organisms—often called extremophiles—survive under severe conditions? How and when did they originate, and where do they fit into current ecosystems? Chemistry has a major role in answering these questions, and the nonmetal sulfur is a key player in the existence of some extremophiles.

Exposure of sulfur-containing minerals such as pyrite (FeS_2) to air and water leads to the formation of hydronium and sulfate ions. Sulfide and disulfide ions in minerals are oxidized by air or iron(III) ion, Fe^{3+}. Using pyrite, as an example, one possible reaction is

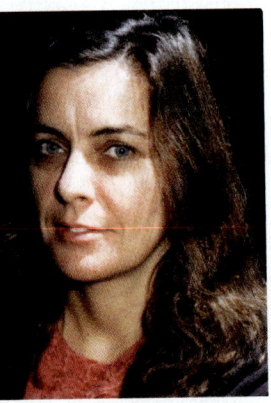

Courtesy of Jillian Banfield

Jillian Banfield, geochemist and MacArthur Fellow. Professor Banfield and her students at the University of California at Berkeley have been at the forefront of studying extremophiles. Findings from these studies have applicability to environmental issues, particularly to the environmental damage caused by drainage from mines. Banfield's work brought her to the attention of the MacArthur Foundation, which awarded her a fellowship, a so-called genius award.

$$FeS_2 + 14\ Fe^{3+} + 8\ H_2O \longrightarrow 15\ Fe^{2+} + 2\ SO_4^{2-} + 16\ H^+$$

Significantly, several species of bacteria (such as *Thiobacillus ferrooxidans*) thrive in highly acidic environments and greatly speed up this mineral degradation. The net result is that water in contact with the waste products of mines is often highly acidic, leading to a serious pollution problem around mines in which sulfide ores are unearthed. In addition to harming plants and animals, the acid can extract arsenic and other toxic elements from minerals that would otherwise remain locked up tight in rocks. The importance of this process is highlighted by the fact that about half of all sulfate ions that enter the oceans are produced in this manner.

Sulfur chemistry can be important in cave formation, as a spectacular example in the jungles of southern Mexico amply demonstrates. Toxic hydrogen sulfide gas spews from the Cueva de Villa Luz along with water that is milky white with suspended sulfur particles. The cave can be followed downward to a large underground stream and a maze of actively enlarging cave passages. Water rises into the cave from underlying sulfur-bearing strata, releasing hydrogen sulfide at concentrations up to 150 ppm. Yellow sulfur crystallizes on the cave walls around the inlets. The sulfur and sulfuric acid are produced by the following reactions:

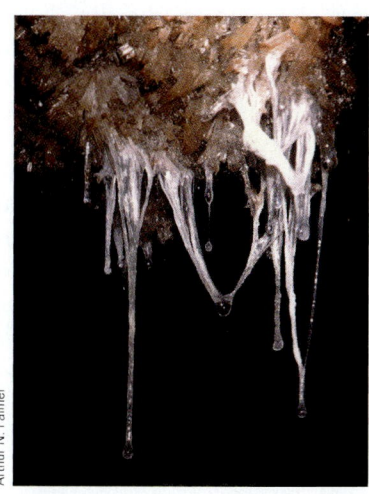

Arthur N. Palmer

Snot-tites. Filaments of sulfur-oxidizing bacteria (dubbed "snot-tites") hang from the ceiling of a Mexican cave containing an atmosphere rich in hydrogen sulfide. The bacteria thrive on the energy released by oxidation of the hydrogen sulfide, forming the base of a complex food chain. Droplets of sulfuric acid on the filaments have the pH of battery acid.

The cave atmosphere is poisonous to humans, so gas masks are essential for would-be explorers. But surprisingly, the cave is teeming with life. Once again, sulfur-oxidizing bacteria speed the chemical reactions and thrive on the large amounts of energy released by them, even in the absence of all other food sources. They use the chemical energy to obtain carbon for their bodies from calcium carbonate and carbon dioxide, both of which are abundant in the cave. Bacterial filaments hang from the walls and ceilings in bundles. (Because the filaments look like something coming from a runny nose, cave explorers refer to them as "snot-tites.") Other microbes feed on the bacteria, and so on up the food chain—which includes spiders, gnats, and pygmy snails—all the way to sardine-like fish that swim in the cave stream. This entire ecosystem is supported by reactions involving sulfur within the cave.

Clearly, this is no ordinary cave environment. Droplets of water seeping in from the surface absorb both hydrogen sulfide and oxygen from the cave air. As illustrated by the reactions shown previously, these dissolved gases react to produce sulfuric acid. This activity depletes the concentrations of both gases in the droplets, allowing more to be absorbed from the air. Meanwhile, the droplets grow more acidic. The longer the droplets remain on the ceiling, the lower their pH becomes. The droplets clinging to the bacterial filaments in the photo had average pH values of 1.4, with some as low as zero! Drops that landed on explorers in the cave burned their skin and disintegrated their clothing.

$$2\ H_2S(g) + O_2(g) \longrightarrow 2\ S(s) + 2\ H_2O(\ell)$$

$$2\ S(s) + 2\ H_2O(\ell) + 3\ O_2(g) \longrightarrow 2\ H_2SO_4(aq)$$

The main group or A-Group elements occupy an important place in the world of chemistry. Eight of the 10 most abundant elements on the earth are in this group. Likewise, the top 10 chemicals produced by the U.S. chemical industry are all main group elements or their compounds.

Because main group elements and their compounds are economically important—and because they have interesting chemistries—we devote this chapter to a brief survey of this group of elements.

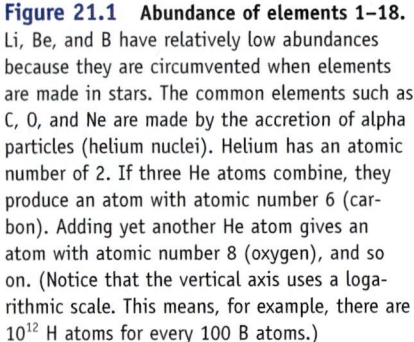

GENERAL
Chemistry Now™

Throughout the chapter this icon introduces a list of resources on the General ChemistryNow CD-ROM or website (http://now.brookscole.com/kotz6e) that will:
- help you evaluate your knowledge of the material
- provide homework problems
- allow you to take an exam-prep quiz
- provide a personalized Learning Plan targeting resources that address areas you should study

21.1—Element Abundances

Figure 21.1 plots the abundance of the first 18 elements in the solar system against their atomic numbers. As you can see, hydrogen and helium are the most abundant by a wide margin because most of the mass of the solar system resides in the sun, and these elements are the sun's primary components. Lithium, beryllium, and boron are low in abundance, but carbon's abundance is very high. From that point on, with the exception of iron and nickel, elemental abundances gradually decline as the atomic number increases.

Data on elemental abundances on the earth are limited to the crust (the outer shell of the planet), the atmosphere, and the hydrosphere. The 10 most abundant elements account for 99% of the aggregate mass of our planet (Table 21.1). Oxygen, silicon, and aluminum represent more than 80% of this mass. Oxygen and nitrogen are the primary components of the atmosphere, and oxygen-containing water is highly abundant on the surface, underground, and as a vapor in the atmosphere. Many common minerals also contain these elements, including limestone ($CaCO_3$) and quartz or sand (SiO_2, Figure 2.13). Aluminum and silicon occur together in many minerals; among the more common ones are feldspar, granite, and clay.

Figure 21.1 Abundance of elements 1–18. Li, Be, and B have relatively low abundances because they are circumvented when elements are made in stars. The common elements such as C, O, and Ne are made by the accretion of alpha particles (helium nuclei). Helium has an atomic number of 2. If three He atoms combine, they produce an atom with atomic number 6 (carbon). Adding yet another He atom gives an atom with atomic number 8 (oxygen), and so on. (Notice that the vertical axis uses a logarithmic scale. This means, for example, there are 10^{12} H atoms for every 100 B atoms.)

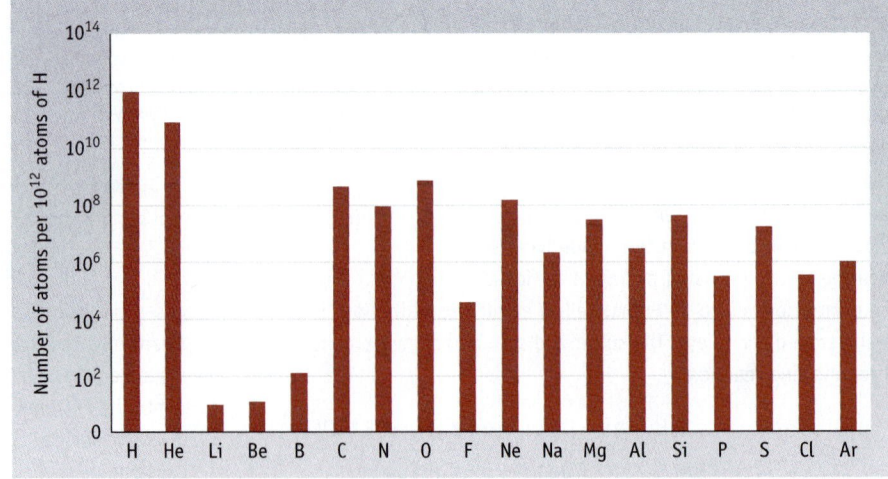

21.2—The Periodic Table: A Guide to the Elements

The similarities in the properties of certain elements guided Mendeleev when he created the first periodic table (page 80). He placed elements in groups based partly on the composition of their common compounds with oxygen and hydrogen, as illustrated in Table 21.2. We now understand that the elements are grouped according to the arrangements of their valence electrons.

Recall that the metallic character of the elements declines on moving from left to right in the periodic table. Elements in Group 1A, the alkali metals, are the most metallic elements in the periodic table. Elements on the far right are nonmetals, and in between are the metalloids. Metallic character increases from the top of a group to the bottom, as illustrated by Group 4A. Carbon, at the top of the group, is a nonmetal; silicon and germanium are metalloids; and tin and lead are metals (Figure 21.2).

Valence Electrons

The ns and np electrons are the valence electrons for main group elements [◀ Section 8.4]; that is, chemistry is determined by these electrons.

A useful reference point is the noble gases (Group 8A), elements having filled electron subshells. Helium has an electron configuration of $1s^2$; the other noble gases have ns^2np^6 valence electron configurations. The dominant characteristic of the noble gases is their lack of reactivity. Indeed, the first two elements in the group do not form any compounds that can be isolated. The other four elements are now known to have limited chemistry, however, and the discovery of xenon compounds in the 1960s ranks as one of the most interesting developments in modern chemistry.

Ionic Compounds of Main Group Elements

Ions with filled s and p subshells are very common—justifying the often-seen statement that elements react in ways that achieve a "noble gas configuration." The elements in Groups 1A and 2A form 1+ and 2+ ions with electron configurations that are the same as those for the previous noble gases. All common compounds of these elements (e.g., NaCl, $CaCO_3$) are ionic (Table 21.3). As expected for ionic compounds, these crystalline solids have high melting points and conduct electricity in the molten state. Many compounds of Group 3A elements (except the metalloid boron) contain 3+ ions. For example, all compounds of aluminum contain the Al^{3+} ion.

Elements of Groups 6A and 7A can achieve a noble gas configuration by adding electrons. Thus, in many reactions, the Group 7A elements (halogens) form anions

Figure 21.2 **Group 4A elements.** A nonmetal, carbon (graphite crucible); a metalloid, silicon (round, lustrous bar); and metals tin (chips of metal) and lead (a bullet, a toy, and a sphere).

Table 21.1 The 10 Most Abundant Elements in Earth's Crust

Rank	Element	Abundance (ppm)*
1	Oxygen	474,000
2	Silicon	277,000
3	Aluminum	82,000
4	Iron	56,300
5	Calcium	41,000
6	Sodium	23,600
7	Magnesium	23,300
8	Potassium	21,000
9	Titanium	5,600
10	Hydrogen	1,520

*ppm = g per 1000 kg. Most abundance data taken from J. Emsley: *The Elements*, New York, Oxford University Press, 3rd edition, 1998.

Table 21.2 Similarities within Periodic Groups*

Group	1A	2A	3A	4A	5A	6A	7A
Common oxide	M_2O	MO	M_2O_3	EO_2	E_4O_{10}	EO_3	E_2O_7
Common hydride	MH	MH_2	MH_3	EH_4	EH_3	EH_2	EH
Highest oxidation state	+1	+2	+3	+4	+5	+6	+7
Common oxoanion			BO_3^{3-}	CO_3^{2-}	NO_3^{-}	SO_4^{2-}	ClO_4^{-}
				SiO_4^{4-}	PO_4^{3-}		

*M denotes a metal and E denotes a nonmetal or metalloid.

Table 21.3   Some Reactions of Group 1A, 2A, and 3A Metals

Metal	+	Nonmetal	⟶	Product
K(s), Group 1A		Br₂(ℓ), Group 7A		KBr(s), ionic
Ba(s), Group 2A		Cl₂(g), Group 7A		BaCl₂(s), ionic
Al(s), Group 3A		F₂(g), Group 7A		AlF₃(s), ionic
Na(s), Group 1A		S₈(s), Group 6A		Na₂S(s), ionic
Mg(s), Group 2A		O₂(g), Group 6A		MgO(s), ionic

with a $1-$ charge (the halide ions, F^-, Cl^-, Br^-, I^-), and the Group 6A elements form anions with a $2-$ charge (O^{2-}, S^{2-}, Se^{2-}, Te^{2-}). In Group 5A chemistry, $3-$ ions with a noble gas configuration (such as the nitride ion, N^{3-}) are encountered. The energy required to form highly charged anions is large, however, which means that other types of chemical behavior will often take precedence.

GENERAL
Chemistry ⚛ Now™

See the General ChemistryNow CD-ROM or website:

- **Screen 21.2 Formation of Ionic Compounds by Main Group Elements,** for a tutorial on ionic compounds

---

Example 21.1—Reactions of Group 1A–3A Elements

Problem Give the formula and name for the product in each of the following reactions. Write a balanced chemical equation for the reaction.

(a) $Ca(s) + S_8(s)$

(b) $Rb(s) + I_2(s)$

(c) lithium and chlorine

(d) aluminum and oxygen

Strategy Predictions are based on the assumption that ions are formed with the electron configuration of the nearest noble gas. Group 1A elements form $1+$ ions, Group 2A elements form $2+$ ions, and metals in Group 3A generally form $3+$ ions. In their reactions with metals, halogen atoms typically add a single electron to give anions with a $1-$ charge; Group 6A elements add two electrons to form anions with a $2-$ charge. For names of products, refer to the nomenclature discussion on page 111.

Solution

Balanced Equation		Product Name
(a)	$8\ Ca(s) + S_8(s) \longrightarrow 8\ CaS(s)$	Calcium sulfide
(b)	$2\ Rb(s) + I_2(s) \longrightarrow 2\ RbI(s)$	Rubidium iodide
(c)	$2\ Li(s) + Cl_2(g) \longrightarrow 2\ LiCl(s)$	Lithium chloride
(d)	$4\ Al(s) + 3\ O_2(g) \longrightarrow 2\ Al_2O_3(s)$	Aluminum oxide

Exercise 21.1—Main Group Element Chemistry

Write a balanced chemical equation for a reaction forming the following compounds from the elements.

(a) NaBr

(b) CaSe

(c) K_2O

(d) $AlCl_3$

Molecular Compounds of Main Group Elements

Many avenues of reactivity are open to main group elements. The metals of Groups 1A–4A are usually encountered in ionic compounds, whereas the metalloids and nonmetals of Groups 3A–7A generally form molecular compounds.

Molecular compounds are encountered with the Group 3A element boron (Figure 21.3), and the chemistry of carbon in Group 4A is dominated by molecular compounds with covalent bonds [◀ Chapter 11]. Similarly, nitrogen chemistry is dominated by molecular compounds. Consider ammonia, NH_3; the various nitrogen oxides; and nitric acid, HNO_3. In each of these species, nitrogen bonds covalently to another nonmetallic element. Also in Group 5A, phosphorus reacts with chlorine to produce the molecular compounds PCl_3 and PCl_5 (page 142).

The valence electron configurations of its elements determine the composition of a molecular compound. Involving all the valence electrons in the formation of a compound is a frequent occurrence in main group element chemistry. We should not be surprised to discover halogen compounds in which the central element has the highest possible oxidation number (such as P in PF_5). The highest oxidation number is readily determined: It has a value equal to the group number. Thus the highest (and only) oxidation number of sodium in its compounds is +1, the highest oxidation number of C is +4, and the highest oxidation number of phosphorus is +5 (Tables 21.2 and 21.4)

Figure 21.3 **Boron halides.** Liquid BBr_3 (*left*) and solid BI_3 (*right*). Formed from a metalloid and a nonmetal, both are molecular compounds. Both are sealed in glass ampules to prevent the boron compound from reacting with H_2O in the air.

Example 21.2—Predicting Formulas for Compounds of Main Group Elements

Problem Predict the formula for each of the following:

(a) the product of the reaction between germanium and excess oxygen

(b) the product of the reaction of arsenic and fluorine

(c) a compound formed from phosphorus and excess chlorine

(d) an anion of selenic acid

Strategy We will predict that in each reaction the element other than the halogen or oxygen in each product achieves its most positive oxidation number, a value equal to the number of its periodic group.

Solution

(a) The Group 4A element germanium should have a maximum oxidation number of +4. Thus, its oxide has the formula GeO_2.

(b) Arsenic, in Group 5A, reacts vigorously with fluorine to form AsF_5, in which arsenic has an oxidation number of +5.

(c) PCl_5 is formed when the Group 5A element phosphorus reacts with excess chlorine.

Table 21.4 **Fluorine Compounds Formed by Main Group Elements**

Group	Compound	Bonding
1A	NaF	Ionic
2A	MgF_2	Ionic
3A	AlF_3	Ionic
4A	SiF_4	Covalent
5A	PF_5	Covalent
6A	SF_6	Covalent
7A	IF_7	Covalent
8A	XeF_4	Covalent

Charles D. Winters

(d) The chemistries of S and Se are similar. Sulfur, in Group 6A, has a maximum oxidation number of +6, so it forms SO_3 and sulfuric acid, H_2SO_4. Selenium, also in Group 6A, has analogous chemistry, forming SeO_3 and selenic acid, H_2SeO_4. The anion of this acid is the selenate ion, SeO_4^{2-}.

Exercise 21.2—Predicting Formulas for Main Group Compounds

Write the formula for each of the following:

(a) hydrogen telluride

(b) sodium arsenate

(c) selenium hexachloride

(d) perbromic acid

You should expect many similarities among elements in the same periodic group. This means you can use compounds of more common elements as examples when you encounter compounds of less common elements. For example, water, H_2O, is the simplest hydrogen compound of oxygen. Therefore, you can reasonably expect the hydrogen compounds of other Group 6A elements to be H_2S, H_2Se, and H_2Te; all are well known.

Example 21.3—Predicting Formulas

Problem Predict the formula for each of the following:

(a) a compound of hydrogen and phosphorus

(b) the hypobromite ion

(c) germane (the simplest hydrogen compound of germanium)

(d) two oxides of tellurium

Strategy Recall as examples some of the compounds of lighter elements in a group and then assume other elements in that group will form analogous compounds.

Solution

(a) Phosphine, PH_3, has a composition analogous to ammonia, NH_3.

(b) Hypobromite ion, BrO^-, is similar to the hypochlorite ion, ClO^-, the anion of hypochlorous acid (HClO).

(c) GeH_4 is analogous to CH_4 and SiH_4, other Group 4A hydrogen compounds.

(d) Te and S are in Group 6A. TeO_2 and TeO_3 are analogs of the oxides of sulfur, SO_2 and SO_3.

Example 21.4—Recognizing Incorrect Formulas

Problem One formula is incorrect in each of the following groups. Pick out the incorrect formula and indicate why it is incorrect.

(a) $CsSO_4$, KCl, $NaNO_3$, Li_2O

(b) MgO, CaI_2, Ba_2SO_4, $CaCO_3$

(c) CO, CO_2, CO_3

(d) PF_5, PF_4^+, PF_2, PF_6^-

Strategy Look for errors such as incorrect charges on ions or an oxidation number exceeding the maximum possible for the periodic group.

Solution

(a) $CsSO_4$. Sulfate ion has a 2− charge, so this formula would require a Cs^{2+} ion. Cesium, in Group 1A, forms only 1+ ions.

(b) Ba_2SO_4. This formula implies a Ba^+ ion (because sulfate is SO_4^{2-}). The ion charge does not equal the group number.

(c) CO_3. Given that O has an oxidation number of −2, carbon would have an oxidation number of +6. Carbon is in Group 4A, however, and can have a maximum oxidation number of +4.

(d) PF_2. This species has an odd number of electrons.

Comment To chemists, this exercise is second nature. Incorrect formulas stand out. You will find that your ability to write and recognize correct formulas will grow as you learn more chemistry.

Exercise 21.3—Predicting Formulas

Identify a compound or ion based on a second-period element that has a formula and Lewis structure analogous to each of the following:

(a) PH_4^+
(b) S_2^{2-}
(c) P_2H_4
(d) PF_3

Exercise 21.4—Recognizing Incorrect Formulas

Explain why compounds with the following formulas would not be expected to exist: ClO, Na_2Cl, $CaCH_3CO_2$, C_3H_7.

21.3—Hydrogen

Chemical and Physical Properties of Hydrogen

Hydrogen has three isotopes, two of them stable (protium and deuterium) and one radioactive (tritium).

Isotopes of Hydrogen		
Isotope Mass (u)	Symbol	Name
1.0078	^1H (H)	Hydrogen (protium)
2.0141	^2H (D)	Deuterium
3.0160	^3H (T)	Tritium

Of the three isotopes, only H and D are found in nature in significant quantities. In contrast, tritium, which is produced by cosmic ray bombardment of nitrogen in the atmosphere, is found to the extent of 1 atom per 10^{18} atoms of ordinary hydrogen. Tritium is radioactive, with half of a sample of the element disappearing in 12.26 years.

A Closer Look

Hydrogen, Helium, and Balloons

In 1783 Jacques Charles first used hydrogen to fill a balloon large enough to float above the French countryside (page 546), a method used in World War I to float observation balloons. The *Graf Zeppelin*, a passenger-carrying dirigible built in Germany in 1928, was also filled with hydrogen. It carried more than 13,000 people between Germany and the United States until 1937, when the dirigible was replaced by the *Hindenburg*. The *Hindenburg* was designed to be filled with helium. At the time World War II was approaching, and the United States, which has much of the world's supply of helium, would not sell the gas to Germany. As a consequence, the

Hindenburg had to use hydrogen. The *Hindenburg* exploded and burned when landing in Lakehurst, New Jersey, in May 1937. Of the 62 people on board, only about half escaped uninjured. As a result of this disaster, hydrogen has acquired a reputation as being a very dangerous substance. Actually, it is as safe to handle as any other fuel, as evidenced by the large quantities used in rockets today.

The Hindenburg. This hydrogen-filled dirigible crashed in Lakehurst, New Jersey, in May 1937. Some have speculated that the aluminum paint coating the skin of the dirigible was involved in sparking the fire.

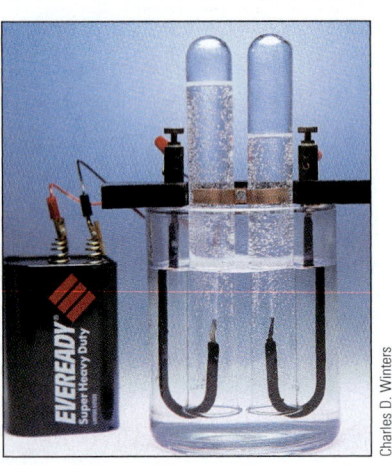

Figure 21.4 Electrolysis of water. Electrolysis of a dilute aqueous solution of H_2SO_4, gives O_2 (*left*) and H_2 (*right*).

Under standard conditions, hydrogen is a colorless gas. Its very low boiling point, 20.7 K, reflects its nonpolar character and low molar mass. As the least dense gas known, it is ideal for filling lighter-than-air craft.

Deuterium compounds have been thoroughly studied. One important observation is that, since D has twice the mass of H, reactions involving D atom transfer are slightly slower than those involving H atoms. This knowledge leads to a way to produce D_2O or "heavy water." Hydrogen can be produced, albeit expensively, by electrolysis of water (Figure 21.4).

$$H_2O(\ell) + \text{electrical energy} \longrightarrow H_2(g) + \tfrac{1}{2} O_2(g)$$

Any sample of natural water always contains a tiny concentration of D_2O. When electrolyzed, H_2O is electrolyzed more rapidly than D_2O. Thus, as the electrolysis proceeds, the liquid remaining is enriched in D_2O. This "heavy water" is valuable as a moderator of some nuclear reactions used in power generation.

Hydrogen combines chemically with virtually every other element except the noble gases. There are many different types of binary hydrogen-containing compounds.

Ionic metal hydrides are formed in the reaction of H_2 with a Group 1A or 2A metal.

$$2\,Na(s) + H_2(g) \longrightarrow 2\,NaH(s)$$
$$Ca(s) + H_2(g) \longrightarrow CaH_2(s)$$

These compounds contain the hydride ion, H^-, in which hydrogen has a -1 oxidation number.

Molecular compounds (such as H_2O, HF, and NH_3) are generally formed by direct combination of the elements (Figure 21.5). In such compounds, covalent bonds to hydrogen are the rule. The oxidation number of the hydrogen atom in these compounds is $+1$.

$$N_2(g) + 3\,H_2(g) \longrightarrow 2\,NH_3(g)$$
$$F_2(g) + H_2(g) \longrightarrow 2\,HF(g)$$

Hydrogen is absorbed by many metals to form *interstitial hydrides*, the third general class of hydrogen compounds. This name refers to the fact that the hydrogen atoms reside in the spaces between the metal atoms (called interstices) in the crystal lattice. For example, when a piece of palladium metal is used as an electrode for the electrolysis of water, the metal can soak up 1000 times its volume of hydrogen (at STP). Most interstitial hydrides are non-stoichiometric; that is, the ratio of metal and hydrogen is not a whole number. When interstitial hydrides are heated, H_2 is driven out. This phenomenon allows these materials to store H_2, just as a sponge can store water. It suggests one way to store hydrogen for use as a fuel in automobiles [◀ page 290].

Preparation of Hydrogen

About 300 billion liters (STP) of hydrogen gas is produced annually worldwide, and virtually all is used immediately in the manufacture of ammonia [▶ Section 21.8], methanol [◀ Section 11.3], or other chemicals.

Some hydrogen is made from coal and steam, a reaction that has been used for more than 100 years.

$$C(s) + H_2O(g) \longrightarrow \underset{\text{water gas or synthesis gas}}{\underline{H_2(g) + CO(g)}}$$

The reaction is carried out by injecting water into a bed of red-hot coke. The mixture of gases produced, called "water gas," was used until about 1950 as a fuel for cooking, heating, and lighting. However, it has serious drawbacks. It produces only about half as much heat as an equal amount of methane does, and the flame is nearly invisible, producing almost no light. Moreover, because it contains carbon monoxide, water gas is toxic.

The largest quantity of hydrogen is now produced by the *catalytic steam re-formation* of hydrocarbons such as methane in natural gas. Methane reacts with steam at high temperature to give H_2 and CO.

$$CH_4(g) + H_2O(g) \longrightarrow 3 H_2(g) + CO(g) \qquad \Delta H^{\circ}_{rxn} = +206 \text{ kJ}$$

The reaction is rapid at 900–1000 °C and goes nearly to completion. More hydrogen can be obtained in a second step in which the CO that is formed in the first step reacts with more water. This so-called *water gas shift reaction* is run at 400–500 °C and is slightly exothermic.

$$H_2O(g) + CO(g) \longrightarrow H_2(g) + CO_2(g) \qquad \Delta H^{\circ} = -41 \text{ kJ}$$

The CO_2 formed in the process is removed by reaction with CaO (to give solid $CaCO_3$), leaving fairly pure hydrogen.

Perhaps the cleanest way to make hydrogen on a relatively large scale is the electrolysis of water (Figure 21.4). This approach provides not only hydrogen gas but also high-purity O_2. Because electricity is quite expensive, however, this method is not generally used commercially.

Table 21.5 and Figure 21.6 give examples of reactions used to produce H_2 gas in the laboratory. The most common is the reaction of a metal with an acid. Alternatively, the reaction of aluminum with NaOH (Figure 21.6b) generates hydrogen as one product. During World War II, this reaction was used to obtain hydrogen to inflate small balloons for weather observation and to raise radio antennas. Metallic aluminum was plentiful at the time because it came from damaged aircraft.

Charles D. Winters

Figure 21.5 **The reaction of H_2 and Br_2.** Hydrogen gas burns in an atmosphere of bromine vapor to give hydrogen bromide.

Roger Ressmeyer/Corbis

Production of water gas. Water gas, also called synthesis gas, is a mixture of CO and H_2. It is produced by treating coal, coke, or some other hydrocarbon with steam at high temperatures in plants such as that pictured here. Methane has the advantage that it gives more total H_2 per gram than other hydrocarbons, and the ratio of the by-product CO_2 to H_2 is lower.

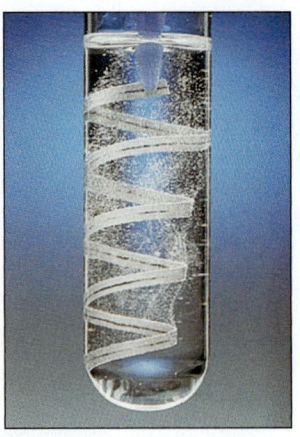

Charles D. Winters

(a) The reaction of magnesium and acid. The products are hydrogen gas and a magnesium salt.

(b) The reaction of aluminum and NaOH. The products of this reaction are hydrogen gas and a solution of Na[Al(OH)$_4$].

(c) The reaction of CaH$_2$ and water. The products are hydrogen gas and Ca(OH)$_2$.

Figure 21.6   Producing hydrogen gas.

Table 21.5   Methods for Preparing H$_2$ in the Laboratory

---

1. Metal + Acid $\longrightarrow$ metal salt + H$_2$

    $Mg(s) + 2\ HCl(aq) \longrightarrow MgCl_2(aq) + H_2(g)$

2. Metal + H$_2$O or base $\longrightarrow$ metal hydroxide or oxide + H$_2$

    $2\ Na(s) + 2\ H_2O(\ell) \longrightarrow 2\ NaOH(aq) + H_2(g)$

    $2\ Fe(s) + 3\ H_2O(\ell) \longrightarrow Fe_2O_3(s) + 3\ H_2(g)$

    $2\ Al(s) + 2\ KOH(aq) + 6\ H_2O(\ell) \longrightarrow 2\ K[Al(OH)_4](aq) + 3\ H_2(g)$

3. Metal hydride + H$_2$O $\longrightarrow$ metal hydroxide + H$_2$

    $CaH_2(s) + 2\ H_2O(\ell) \longrightarrow Ca(OH)_2(s) + 2\ H_2(g)$

The combination of a metal hydride and water (Table 21.5 and Figure 21.6c) is an efficient but expensive way to synthesize H$_2$ in the laboratory. The reaction is more commonly used in laboratories to dry organic solvents because the metal hydride reacts with traces of water present in the solvent.

Exercise 21.5—Hydrogen Chemistry

Use bond energies (page 422) to calculate the enthalpy change for the reaction of methane and water to give hydrogen and carbon monoxide (with all compounds in the gas phase). Considering the bond energies of reactants and products, suggest a reason why the reaction is endothermic.

21.4—The Alkali Metals, Group 1A

Sodium and potassium are the sixth and eighth most abundant elements in the earth's crust by mass. In contrast, lithium is relatively rare (Figure 21.1), as are rubidium and cesium. Only traces of francium occur in nature. All the isotopes of this element are radioactive. Its longest-lived isotope (^{223}Fr) has a half-life of only 22 minutes.

All of the Group 1A elements are metals, and all are highly reactive with oxygen, water, and other oxidizing agents (see Figure 8.9, page 354). In all cases, compounds of the Group 1A metals contain the element as a 1+ ion. None is ever found in nature as the uncombined element.

Most sodium and potassium compounds are water-soluble [◀ solubility rules, Figure 5.3], so it is not surprising that sodium and potassium compounds are found either in the oceans or in underground deposits that are the residue of ancient seas. To a much smaller extent, these elements are also found in minerals, such as Chilean saltpeter ($NaNO_3$).

Despite the fact that sodium is only slightly more abundant than potassium on the earth, sea water contains significantly more sodium than potassium (2.8% NaCl versus 0.8% KCl). Why the great difference? Both are water-soluble, so why didn't the rain dissolve Na- and K-containing minerals over the centuries and carry them down to the sea, so that they appear in the same proportions in the oceans as on land? The answer lies in the fact that potassium is an important factor in plant growth. Most plants contain four to six times as much combined potassium as sodium. Thus most of the potassium ions in groundwater from dissolved minerals are taken up preferentially by plants, whereas sodium salts continue on to the oceans. (Because plants require potassium, fertilizers often contain a significant concentration of potassium salts.)

Some NaCl is essential in the diet of humans and other animals because many biological functions are controlled by the concentrations of Na^+ and Cl^- ions (Figure 21.7). The fact that salt has long been recognized as important is evident in surprising ways. We are paid a "salary" for work done. This word is derived from the Latin *salarium*, which meant "salt money" because Roman soldiers were paid in salt.

Preparation of Sodium and Potassium

Sodium is produced by reducing sodium ions in sodium salts. However, because common chemical reducing agents are not powerful enough to convert sodium ions to sodium metal, an electrolytic method is necessary to accomplish the reduction.

The English chemist Sir Humphry Davy first isolated sodium in 1807 by the electrolysis of molten sodium carbonate. The element remained a laboratory curiosity until 1824, when it was found that sodium could be used to reduce aluminum chloride to aluminum metal. Because aluminum was so valuable, this discovery inspired considerable interest in manufacturing sodium. By 1886, a practical method of sodium production had been devised (the reduction of NaOH with carbon). Unfortunately for sodium producers, in this same year Hall and Heroult invented the electrolytic method for aluminum production, thereby eliminating this market for sodium.

Sodium is currently produced by the electrolysis of molten NaCl [◀ Section 20.7]. The Downs cell for the electrolysis of molten NaCl operates at 7 to 8 V with currents of 25,000 to 40,000 amps (Figure 21.8). The cell is filled with a mixture of dry NaCl, $CaCl_2$, and $BaCl_2$. Adding other salts to NaCl lowers the melting point from that of pure NaCl (800.7 °C) to about 600 °C. [Recall that solutions have lower melting points than pure solvents (Chapter 14).] Sodium is produced at a copper or iron cathode that surrounds a circular graphite anode. Directly over the cathode is an inverted trough in which the low-density, molten sodium (melting point, 97.8 °C) collects. The valuable byproduct, Cl_2 gas, is collected at the anode.

Charles D. Winters

Figure 21.7 **The importance of salt.** All animals, including humans, need a certain amount of salt in their diet. Sodium ions are important in maintaining electrolyte balance and in regulating osmotic pressure. For an interesting account of the importance of salt in society, culture, history, and economy, see *Salt, A World History*, by M. Kurlansky, New York, Penguin Books, 2002.

Group 1A
Alkali metals

Lithium
3
Li
20 ppm
Sodium
11
Na
23,600 ppm
Potassium
19
K
21,000 ppm
Rubidium
37
Rb
90 ppm
Cesium
55
Cs
0.0003 ppm
Francium
87
Fr
trace

Element abundances are in parts per million in the earth's crust.

Figure 21.8. A Downs cell for preparing sodium. A circular iron cathode is separated from the graphite anode by an iron screen. At the temperature of the electrolysis, about 600 °C, sodium is a liquid. It floats to the top and is drawn off periodically. Chlorine gas is produced at the anode and collected inside the inverted cone in the center of the cell.

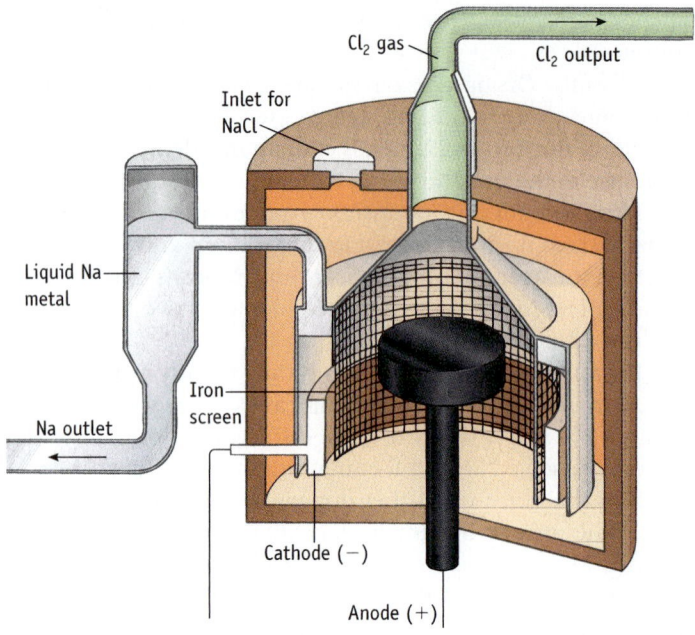

Potassium can also be made by electrolysis. Molten potassium is soluble in molten KCl, however, making separation of the metal difficult. The preferred method for preparation of potassium uses the reaction of sodium vapor with molten KCl, with potassium being continually removed from the equilibrium mixture.

$$Na(g) + KCl(\ell) \longrightarrow K(g) + NaCl(\ell)$$

Properties of Sodium and Potassium

Sodium and potassium are silvery metals that are soft and easily cut with a knife (see Figure 2.12). Their densities are just a bit less than the density of water, and their melting points are quite low (97.8 °C for sodium and 63.7 °C for potassium).

All of the alkali metals are highly reactive. When exposed to moist air, the metal surface quickly becomes coated with a film of oxide or hydroxide. Consequently, the metals must be stored in a way that avoids contact with air, typically by placing them in kerosene or mineral oil.

The high reactivity of Group 1A metals is exemplified by their reaction with water, which generates an aqueous solution of the metal hydroxide and hydrogen (Figure 8.9, page 354),

$$2 Na(s) + 2 H_2O(\ell) \longrightarrow 2 Na^+(aq) + 2 OH^-(aq) + H_2(g)$$

and their reaction with any of the halogens to yield a metal halide (Figure 1.7),

$$2 Na(s) + Cl_2(g) \longrightarrow 2 NaCl(s)$$
$$2 K(s) + Br_2(\ell) \longrightarrow 2 KBr(s)$$

Chemistry sometimes produces surprises. Group 1A metal oxides, M_2O, are known, but they are not the principal products of reactions between the Group 1A elements and oxygen. Instead, the primary product of the reaction of sodium and

Courtesy of the Mine Safety Appliances Company

A closed-circuit breathing apparatus that generates its own oxygen. One source of oxygen is potassium superoxide (KO_2). Both carbon dioxide and moisture exhaled by the wearer into the breathing tube react with the KO_2 to generate oxygen. Because the rate of the chemical reaction is determined by the quantity of moisture and carbon dioxide exhaled, the production of oxygen is regulated automatically. With each exhalation, more oxygen is produced by volume than is required by the user.

A Closer Look

The Reducing Ability of the Alkali Metals

The uses of the Group 1A metals depend on their reducing ability. The values of $E°$ reveal that Li is the best reducing agent in the group, whereas Na is the poorest; the remainder of these metals have roughly comparable reducing ability.

Element	Reduction Potential $E°$ (V)
$Li^+ + e^- \longrightarrow Li$	−3.045
$Na^+ + e^- \longrightarrow Na$	−2.714
$K^+ + e^- \longrightarrow K$	−2.925
$Rb^+ + e^- \longrightarrow Rb$	−2.925
$Cs^+ + e^- \longrightarrow Cs$	−2.92

Analysis of $E°$ is a thermodynamic problem, and to understand it better we can break the process of metal oxidation, $M(s) \longrightarrow M^+(aq) + e^-$, into a series of steps. Here we imagine that the metal

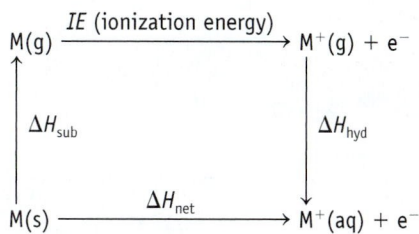

sublimes to vapor, an electron is removed to form the gaseous cation, and the cation is hydrated. The first two steps require energy, but the last is exothermic. From Hess's law we know that the overall energy change should be

$$\Delta H_{net} = \Delta H_{sub} + IE + \Delta H_{hyd}$$

The element that is the best reducing agent should have the most negative (or least positive) value of ΔH_{net}. That is, the best reducing agent should be the metal that has the most exothermic value for its hydration energy because this can offset the energy of the endothermic steps (ΔH_{sub} and IE). For the alkali metals, enthalpies of hydration range from −506 kJ/mol for Li^+ to −180 kJ/mol for Cs^+. The fact that

ΔH_{hyd} is so much greater for Li^+ than for Cs^+ largely accounts for the difference in reducing ability.

While this analysis of the problem gives us a reasonable explanation for the great reducing ability of lithium, recall that $E°$ is directly related to $\Delta G°$ and not to $\Delta H°$. Fortunately, $\Delta G°$ is largely determined by $\Delta H°$ in this case, so it is possible to relate variations in $E°$ to variations in $\Delta H°$.

Potassium is a very good reducing agent and reacts vigorously with water.

---

oxygen is sodium *peroxide*, Na_2O_2, whereas the principal product from the reaction of potassium and oxygen is KO_2, potassium *superoxide*.

$$2 Na(s) + O_2(g) \longrightarrow Na_2O_2(s)$$
$$K(s) + O_2(g) \longrightarrow KO_2(s)$$

Both Na_2O_2 and KO_2 are ionic compounds. The Group 1A cation is paired with either the peroxide ion (O_2^{2-}) or the superoxide ion (O_2^-). These compounds are not merely laboratory curiosities. They are used in oxygen generation devices in places where people are confined, such as submarines, aircraft, and spacecraft, or when an emergency supply is needed. When a person breathes, 0.82 L of CO_2 is exhaled for every 1 L of O_2 inhaled. Thus, a requirement of an O_2 generation system is that it should produce a larger volume of O_2 than the volume of CO_2 taken in. This requirement is met with superoxides. With KO_2 the reaction is

$$4 KO_2(s) + 2 CO_2(g) \longrightarrow 2 K_2CO_3(s) + 3 O_2(g)$$

Important Lithium, Sodium, and Potassium Compounds

Electrolysis of aqueous sodium chloride (*brine*) is the basis of one of the largest chemical industries in the United States.

$$2 NaCl(aq) + 2 H_2O(\ell) \longrightarrow Cl_2(g) + 2 NaOH(aq) + H_2(g)$$

Figure 21.9 Producing soda ash. Trona mined in Wyoming and California is processed into soda ash (Na_2CO_3) and other sodium-based chemicals. Soda ash is the ninth most widely used chemical in the United States. Domestically, about half of all soda ash production is used in making glass. The remainder goes to make chemicals such as sodium silicate, sodium phosphate, and sodium cyanide. Some is also used to make detergents, in the pulp and paper industry, and in water treatment.

(a) (Above) A mine in California. The mineral trona is taken from a mine 1600 feet deep.

(b) (Right) Blocks of trona are cut from the face of the mine.

The products from this process—chlorine, sodium hydroxide, and hydrogen—give the industry its name: the *chlor-alkali industry*. More than 10 billion kilograms of Cl_2 and NaOH is produced annually in the United States.

Sodium carbonate, Na_2CO_3, is another commercially important compound of sodium. It is also known by two common names, soda ash and washing soda. In the past it was largely manufactured by combining NaCl and CO_2 in the Solvay process (which remains the method of choice in many countries). In the United States, however, sodium carbonate is obtained from naturally occurring deposits of the mineral *trona*, $Na_2CO_3 \cdot NaHCO_3 \cdot 2\,H_2O$ (Figure 21.9).

Owing to the environmental problems associated with chlorine and its byproducts, considerable interest has arisen in the possibility of manufacturing sodium hydroxide by methods other than brine electrolysis. This has led to a revival of the old "soda-lime process," which produces NaOH from inexpensive lime (CaO) and soda (Na_2CO_3)

$$Na_2CO_3(aq) + CaO(s) + H_2O(\ell) \longrightarrow 2\,NaOH(aq) + CaCO_3(s)$$

The insoluble calcium carbonate by-product is filtered off and recycled into the process by heating it (calcining) to recover lime

$$CaCO_3(s) \longrightarrow CaO(s) + CO_2(g)$$

Sodium bicarbonate, $NaHCO_3$, also known as *baking soda*, is another common compound of sodium. Not only is $NaHCO_3$ used in cooking, but it is also added in small amounts to table salt. NaCl is often contaminated with small amounts of $MgCl_2$. The magnesium salt is hygroscopic; that is, it picks water up from the air and, in doing so, causes the NaCl to clump. Adding $NaHCO_3$ converts $MgCl_2$ to magnesium carbonate, a non-hygroscopic salt

$$MgCl_2(s) + 2\,NaHCO_3(s) \longrightarrow MgCO_3(s) + 2\,NaCl(s) + H_2O(\ell) + CO_2(g)$$

Large deposits of sodium nitrate, $NaNO_3$, are found in Chile, which explains its common name of "Chile saltpeter." These deposits are thought to have formed by bacterial action on organisms in shallow seas. The initial product was ammonia, which was subsequently oxidized to nitrate ion; combination with sea salt led to sodium nitrate. Because nitrates in general, and alkali metal nitrates in particular, are highly water-soluble, deposits of $NaNO_3$ are found only in areas of very little rainfall.

Sodium nitrate is important because it can be converted to potassium nitrate by an exchange reaction.

$$NaNO_3(aq) + KCl(aq) \rightleftharpoons KNO_3(aq) + NaCl(s)$$

Equilibrium favors the products here because, of the four salts involved in this reaction, NaCl is least soluble in hot water. Sodium chloride precipitates, and the KNO_3 that remains in solution can be recovered by evaporating the water.

Potassium nitrate has been used for centuries as the oxidizing agent in gunpowder. A mixture of KNO_3, charcoal, and sulfur will spontaneously react when ignited.

$$2\ KNO_3(s) + 4\ C(s) \longrightarrow K_2CO_3(s) + 3\ CO(g) + N_2(g)$$
$$2\ KNO_3(s) + 2\ S(s) \longrightarrow K_2SO_4(s) + SO_2(g) + N_2(g)$$

Notice that both reactions (which are doubtless more complex than those written here) produce gases. These gases propel the bullet from a gun or cause a firecracker to explode.

Lithium carbonate, Li_2CO_3, has been used for more than 40 years as a treatment for manic depression, an illness that involves alternating periods of depression and mania or over-excitement that can extend over a few weeks to a year or more. Although the alkali metal salt is efficient in controlling the symptoms of manic depression, its mechanism of action is not understood.

Exercise 21.6—Brine Electrolysis

What current must be used in a Downs cell operating at 7.0 V to produce 1.00 metric ton (exactly 1000 kg) of sodium per day? Assume 100% efficiency.

21.5—The Alkaline Earth Elements, Group 2A

The "earth" part of the name *alkaline earth* dates back to the days of medieval alchemy. To alchemists, any solid that did not melt and was not changed by fire into another substance was called an "earth." Compounds of the Group 2A elements, such as CaO, were alkaline according to experimental tests conducted by the alchemists: They had a bitter taste and neutralized acids. With very high melting points, these compounds were unaffected by fire.

Calcium and magnesium rank fifth and eighth, respectively, in abundance on the earth. Both elements form many commercially important compounds, and we shall focus our attention on this chemistry.

Like the Group 1A elements, the Group 2A elements are very reactive, so they are found in nature only combined with other elements. Unlike Group 1A metals, however, many compounds of the Group 2A elements have low water solubility, which explains their occurrence as various minerals. Common calcium minerals include limestone ($CaCO_3$), gypsum ($CaSO_4 \cdot 2\ H_2O$), and fluorite (CaF_2) (Figure 21.10). Magnesite ($MgCO_3$), talc or soapstone ($3\ MgO \cdot 4\ SiO_2 \cdot H_2O$), and asbestos ($3\ MgO \cdot 4\ SiO_2 \cdot 2\ H_2O$) are common magnesium-containing minerals. The mineral dolomite, $MgCa(CO_3)_2$, contains both magnesium and calcium.

Group 2A
Alkaline earths

Beryllium
4
Be
2.6 ppm

Magnesium
12
Mg
23,000 ppm

Calcium
20
Ca
41,000 ppm

Strontium
38
Sr
370 ppm

Barium
56
Ba
500 ppm

Radium
88
Ra
6×10^{-7} ppm

Element abundances are in parts per million in the earth's crust.

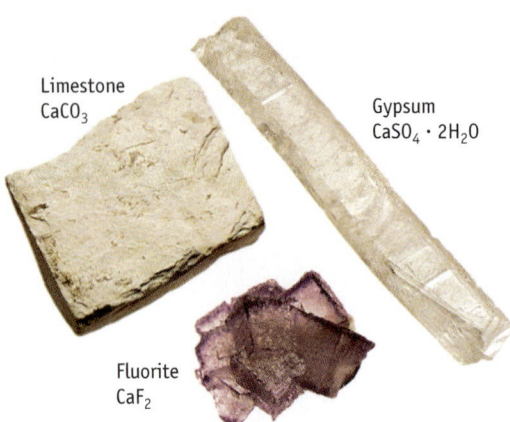

Limestone
CaCO₃

Gypsum
CaSO₄ · 2H₂O

Fluorite
CaF₂

Common minerals of Group 2A elements.

Icelandic spar. This mineral, one of a number of crystalline forms of CaCO₃, displays birefringence, a property in which a double image is formed when light passes through the crystal.

The walls of the Grand Canyon in Arizona are largely limestone or dolomite.

a and *b*, Charles D. Winters; *c*, James Cowlin/Image Enterprises/Phoenix, AZ

Figure 21.10   **Various minerals containing calcium and magnesium.**

Limestone, a sedimentary rock, is found widely on the earth's surface (Figure 21.10). Many of these deposits contain the fossilized remains of marine life. Other forms of calcium carbonate include marble and Icelandic spar, which forms large, clear crystals (Figure 21.10)

Properties of Calcium and Magnesium

Calcium and magnesium are fairly high-melting, silvery metals. The chemical properties of these elements present few surprises. They are oxidized by a wide range of oxidizing agents to form ionic compounds that contain the M^{2+} ion. For example, these elements combine with halogens to form MX_2, with oxygen or sulfur to form MO or MS (Figure 4.3), and with water to form hydrogen and the metal hydroxide, $M(OH)_2$ (Figure 21.11). With acids, hydrogen is evolved (see Figure 21.6 and Table 21.5), and a salt of the metal cation and the anion of the acid results.

Metallurgy of Magnesium

Several hundred thousand tons of magnesium are produced annually, largely for use in lightweight alloys. (Magnesium has a very low density, 1.74 g/cm³.) In fact, most aluminum used today contains about 5% magnesium to improve its mechanical properties and to make it more resistant to corrosion. Other alloys having more magnesium than aluminum are used when a high strength-to-weight ratio is needed and when corrosion resistance is important, such as in aircraft and automotive parts and in lightweight tools.

Interestingly, magnesium-containing minerals are not the source of this element. Most magnesium is obtained from sea water, in which Mg^{2+} ion is present in a concentration of about 0.05 M (Figure 21.12). To obtain magnesium metal, magnesium ions are first precipitated from sea water as the relatively insoluble hydroxide [K_{sp} for $Mg(OH)_2 = 5.6 \times 10^{-12}$]. Calcium hydroxide, the source of OH^- in this reaction, is prepared in a sequence of reactions beginning with CaCO₃, which may be in the form of seashells. Heating CaCO₃ gives CO₂ and CaO, and addition of water to CaO gives calcium hydroxide. When $Ca(OH)_2$ is added to sea water, $Mg(OH)_2$ precipitates:

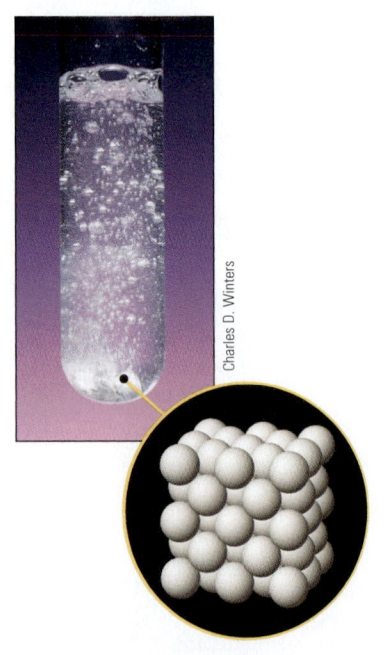

Charles D. Winters

Figure 21.11   **The reaction of calcium and warm water.** Hydrogen bubbles are seen rising from the metal surface. The inset is a model of hexagonal close-packed calcium metal (see Figure 13.28).

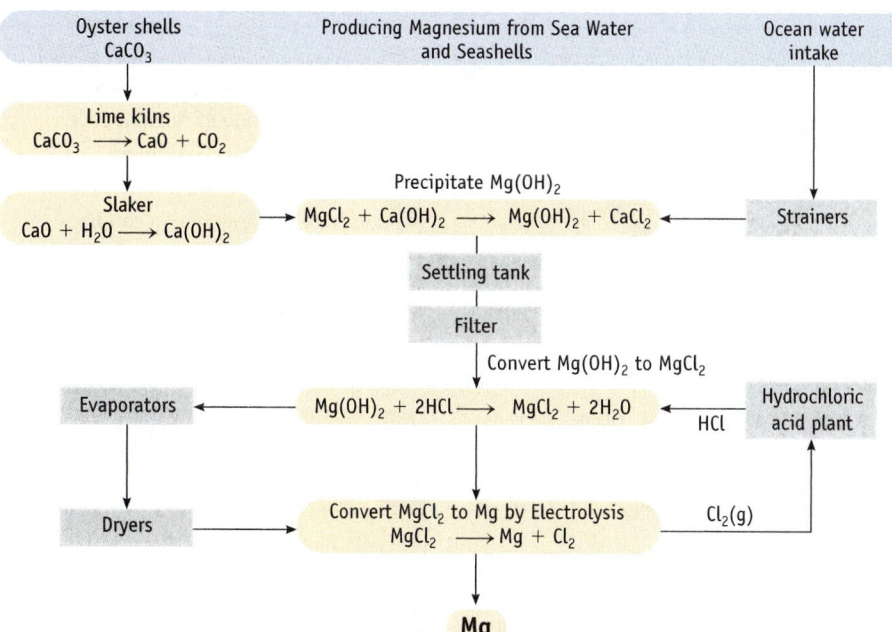

Figure 21.12 The process used to produce magnesium metal from the magnesium in sea water.

$$Mg^{2+}(aq) + Ca(OH)_2(aq) \longrightarrow Mg(OH)_2(s) + Ca^{2+}(aq)$$

Magnesium hydroxide is isolated by filtration and then neutralized with hydrochloric acid.

$$Mg(OH)_2(s) + 2\ HCl(aq) \longrightarrow MgCl_2(aq) + 2\ H_2O(\ell)$$

After evaporating the water, anhydrous magnesium chloride remains. Solid $MgCl_2$ melts at 714 °C, and the molten salt is electrolyzed to give the metal and chlorine.

$$MgCl_2(\ell) \longrightarrow Mg(s) + Cl_2(g)$$

Calcium Minerals and Their Applications

The most common calcium minerals are the fluoride, phosphate, and carbonate salts of the element. Fluorite, CaF_2, and fluorapatite, $Ca_5F(PO_4)_3$, are important as commercial sources of fluorine. Almost half of the CaF_2 mined is used in the steel industry, where it is added to the mixture of materials that is melted to make crude iron. The CaF_2 acts to remove some impurities and improves the separation of molten metal from silicate impurities and other byproducts resulting from the reduction of iron ore to the metal (Chapter 22). A second major application of fluorite is in the manufacture of hydrofluoric acid by a reaction of the mineral with concentrated sulfuric acid.

$$CaF_2(s) + H_2SO_4(\ell) \longrightarrow 2\ HF(g) + CaSO_4(s)$$

Hydrofluoric acid is used to make cryolite, Na_3AlF_6, a material needed in aluminum production [▶ Section 21.6] and in the manufacture of fluorocarbons such as tetrafluoroethylene, the precursor to Teflon (Table 11.12).

Apatites have the general formula $Ca_5X(PO_4)_3$ (X = F, Cl, OH). More than 100 million tons of apatite is mined annually; Florida alone accounts for about one

Chemical Perspectives

Alkaline Earth Metals and Biology

Plants and animals derive energy from the oxidation of a sugar, glucose, with oxygen. Plants are unique, however, in being able to synthesize glucose from CO_2 and H_2O by using sunlight as an energy source. This process is initiated by chlorophyll, a very large, magnesium-based molecule.

In your body the metal ions Na^+, K^+, Mg^{2+}, and Ca^{2+} serve regulatory functions. Although the two alkaline earth metal ions are required by living systems, the other Group 2A elements are toxic. Beryllium compounds are carcinogenic, and soluble barium salts are poisons. You may be concerned if your physician asks you to drink a "barium cocktail" to check the condition of your digestive tract. Don't be afraid, because the "cocktail" contains very insoluble $BaSO_4$ ($K_{sp} = 1.1 \times 10^{-10}$). Barium sulfate is opaque to x-rays, so its path

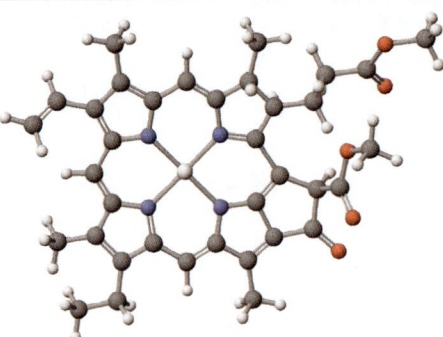

A molecule of chlorophyll. Magnesium is its central element.

through your organs appears on the developed x-ray.

The calcium-containing compound hydroxyapatite is the main component of tooth enamel. Cavities in your teeth form when acids decompose the weakly basic apatite coating.

$$Ca_5(OH)(PO_4)_3(s) + 4\ H^+(aq) \longrightarrow$$
$$5\ Ca^{2+}(aq) + 3\ HPO_4^{2-}(aq) + H_2O(\ell)$$

This reaction can be prevented by converting hydroxyapatite to the much more acid-resistant coating of fluoroapatite.

$$Ca_5(OH)(PO_4)_3(s) + F^-(aq) \longrightarrow$$
$$Ca_5F(PO_4)_3(s) + OH^-(aq)$$

The source of the fluoride ion can be sodium fluoride or sodium monofluorophosphate (Na_2FPO_3, commonly known as MFP) in your toothpaste.

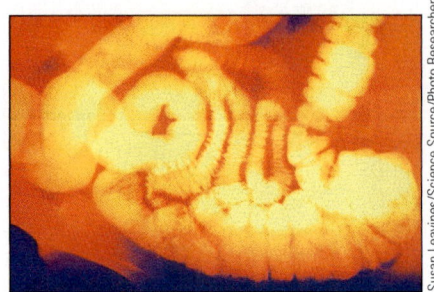

X-ray of a gastrointestinal tract using $BaSO_4$ to make the organs visible.

Apatite. The mineral has the general formula of $Ca_5X(PO_4)_3$ (X = F, Cl, OH). (The apatite is the elongated crystal in the center of a matrix of other rock.)

third of the world's output. Most of this material is converted to phosphoric acid by reaction with sulfuric acid. (Phosphoric acid is needed in the manufacture of a multitude of products, including fertilizers and detergents, baking powder, and various food products; [▶ Section 21.8.]

$$Ca_5F(PO_4)_3(s) + 5\ H_2SO_4(aq) \longrightarrow 5\ CaSO_4(s) + 3\ H_3PO_4(aq) + HF(g)$$
$$\text{fluorapatite}$$

Calcium carbonate and calcium oxide (lime) are of special interest. The thermal decomposition of $CaCO_3$ to lime is one of the oldest chemical reactions known. (Lime is one of the top 10 industrial chemicals produced today, with about 20 billion kilograms produced annually.)

Limestone, which consists mostly of calcium carbonate, has been used in agriculture for centuries. It is spread on fields to neutralize acidic compounds in the soil and to supply the essential nutrient Ca^{2+}. Because magnesium carbonate is often present in limestone, "liming" a field also supplies Mg^{2+}, another important nutrient for plants.

For several thousand years, lime has been used in mortar (a lime, sand, and water paste) to secure stones to one another in building houses, walls, and roads. The Chinese used it to set stones in the Great Wall. The Romans perfected its use, and the fact that many of their constructions still stand today is testament both to their skill and to the usefulness of lime. The famous Appian Way used lime mortar between several layers of its stones.

The utility of mortar depends on some simple chemistry. Mortar consists of one part lime to three parts sand, with water added to make a thick paste. The first reaction that occurs, referred to as *slaking*, produces calcium hydroxide,

Chemical Perspectives

Of Romans, Limestone, and Champagne

The stones of the Appian Way in Italy, a road conceived by the Roman senate in

The Appian Way in Italy.

about 310 B.C., are cemented with mortar made from limestone. The Appian Way was intended to serve as a military road linking Rome to seaports from which soldiers could embark to Greece and other Mediterranean ports. The road stretches 560 kilometers (350 miles) from Rome to Brindisi on the Adriatic Sea (at the heel of the Italian "boot"). It took almost 200 years to construct. The road had a standard width of 14 Roman feet, approximately 20 feet, large enough to allow two chariots to pass, and featured two sidewalks of 4 feet each. Every 10 miles or so there were horse-changing stations with taverns, shops, and latrinae, the famous Roman restrooms.

All over the Roman Empire, buildings, temples, and aqueducts were constructed of limestone and marble. Mortar was made by "burning" chips from the stone cutting. In central France, the Romans dug chalk (also $CaCO_3$) from the ground for cementing sandstone blocks. This activity created huge caves that remain to this day and are used for aging and storing champagne.

Champagne in a limestone cave in France.

which is known as *slaked lime.* When the mortar is placed between bricks or stone blocks, it slowly absorbs CO_2 from the air, and the slaked lime reverts to calcium carbonate.

$$Ca(OH)_2(s) + CO_2(g) \longrightarrow CaCO_3(s) + H_2O(\ell)$$

The sand grains are bound together by the particles of calcium carbonate.

"Hard water" contains dissolved metal ions, chiefly Ca^{2+} and Mg^{2+} (page 1001). These ions are found in water due to the reaction of limestone [or the related mineral dolomite, $CaMg(CO_3)_2$] with water containing dissolved CO_2.

$$CaCO_3(s) + H_2O(\ell) + CO_2(g) \rightleftharpoons Ca^{2+}(aq) + 2 HCO_3^-(aq)$$

This reaction can be reversed. When hard water is heated, the solubility of CO_2 decreases, and the equilibrium shifts to the left. If this happens in a heating system or steam-generating plant, the walls of the hot-water pipes can become coated or even blocked with solid $CaCO_3$. In your house, you may notice a coating of calcium carbonate on the inside of cooking pots.

The previous equations also describe the chemistry occurring inside limestone caves. The acidic oxide CO_2 reacts with $Ca(OH)_2$ to produce white, solid $CaCO_3$. When further CO_2 is available, however, the $CaCO_3$ dissolves due to the formation of aqueous Ca^{2+} and HCO_3^- ions (see page 759).

■ **Dissolving Limestone**
Figure 16.2 illustrates the equilibrium involving $CaCO_3$, CO_2, H_2O, Ca^{2+}, and HCO_3^-.

Exercise 21.7—Beryllium Chemistry

Beryllium, the lightest element in Group 2A, has some important industrial applications, but exposure (by breathing) to some of its compounds can cause berylliosis. Search the World Wide Web for the uses of the element and the causes and symptoms of berylliosis.

21.6—Boron, Aluminum, and the Group 3A Elements

With Group 3A we begin to see the change from metallic behavior of the elements at the left side of the periodic table to nonmetal behavior on the right side of the table. Boron is a metalloid, whereas all the other elements of Group 3A are metals.

The elements of Group 3A vary widely in their relative abundances on earth. Aluminum is the third most abundant element in the earth's crust (82,000 ppm). In contrast, the other elements of the group are all relatively rare and, except for boron compounds, have limited commercial uses.

The General Chemistry of the Group 3A Elements

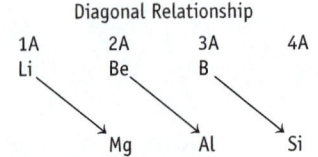

Diagonal Relationship

1A	2A	3A	4A
Li	Be	B	
	Mg	Al	Si

■ **Diagonal relationship.**
The chemistries of elements diagonally situated in the periodic table are often quite similar.

It is generally recognized that a chemical similarity exists between some elements diagonally situated in the periodic table. This *diagonal relationship* means that lithium and magnesium share some chemical properties, as do Be and Al, and B and Si. For example:

- Boric oxide, B_2O_3, and boric acid, $B(OH)_3$, are weakly acidic, as are SiO_2 and its acid, orthosilic acid (H_4SiO_4). In general, boron-oxygen compounds, *borates*, are often chemically similar to silicon-oxygen compounds, *silicates*.
- $Be(OH)_2$ and $Al(OH)_3$ are both amphoteric, dissolving in a strong base such as aqueous NaOH (page 831).
- Chlorides, bromides, and iodides of boron and silicon (such as BCl_3 and $SiCl_4$) react vigorously with water.
- The hydrides of boron and silicon are simple, molecular species; are volatile and flammable; and react readily with water.
- Beryllium hydride and aluminum hydride are colorless, nonvolatile solids that are extensively polymerized through Be—H—Be and Al—H—Al bonds.

Finally, the Group 3A elements of the group are characterized by electron configurations of the type ns^2np^1. This means that each may lose three electrons to have a +3 oxidation number, although the heavier elements, especially thallium, also form compounds with an oxidation number of +1.

Boron Minerals and Production of the Element

Although boron has a low abundance on earth, its minerals are found in concentrated deposits. Large deposits of borax, $Na_2B_4O_7 \cdot 10\ H_2O$, are currently mined in the Mojave Desert near the town of Boron, California (Figures 2.14 and 21.13).

Isolation of pure, elemental boron is extremely difficult and is done in small quantities. Like most metals and metalloids, boron can be obtained by chemically or electrolytically reducing an oxide or halide. Magnesium has often been used for chemical reductions, but the product of this reaction is a noncrystalline boron of low purity.

$$B_2O_3(s) + 3\ Mg(s) \longrightarrow 2\ B(s) + 3\ MgO(s)$$

Boron has several allotropes, all characterized by having the *icosahedron* as one structural element (Figure 21.13c). Partly as a result of extended covalent bonding, elemental boron is very hard, refractory (resistant to heat), and a semiconductor. In this regard, it differs from the other Group 3A elements; Al, Ga, In, and Tl are all relatively low-melting, rather soft metals with high electrical conductivity.

Group 3A

Boron
5
B
10 ppm

Aluminum
13
Al
82,000 ppm

Gallium
31
Ga
18 ppm

Indium
49
In
0.05 ppm

Thallium
81
Tl
0.6 ppm

Element abundances are in parts per million in the earth's crust.

(a) (b) (c)

Figure 21.13 **Boron.** (a) A borax mine near the town of Boron, California. (b) Crystalline borax, $Na_2B_4O_7 \cdot 10\ H_2O$. (c) All allotropes of elemental boron have an icosahedron (a 20-sided polyhedron) of 12 covalently linked boron atoms as a structural element.

Metallic Aluminum and Its Production

The low cost of aluminum and the excellent characteristics of its alloys with other metals (low density, strength, ease of handling in fabrication, and inertness toward corrosion, among others) have led to its widespread use. You know it best in the form of aluminum foil, aluminum cans, and parts of aircraft.

Pure aluminum is soft and weak; moreover, it loses strength rapidly at temperatures higher than 300 °C. What we call "aluminum" is actually aluminum alloyed with small amounts of other elements to strengthen the metal and improve its properties. A typical alloy may contain about 4% copper with smaller amounts of silicon, magnesium, and manganese. Softer, more corrosion-resistant alloys for window frames, furniture, highway signs, and cooking utensils may include only manganese.

The standard reduction potential of aluminum $[Al^{3+}(aq) + 3\ e^- \longrightarrow Al(s);$ $E° = -1.66\ V]$ tells you that aluminum is easily oxidized. From this, we might expect aluminum to be highly susceptible to corrosion but, in fact, it is quite resistant. Aluminum's corrosion resistance is due to the formation of a thin, tough, and transparent skin of Al_2O_3 that adheres to the metal surface. An important feature of the protective oxide layer is that it rapidly self-repairs. If you penetrate the surface coating by scratching it or using some chemical agent, the exposed metal surface immediately reacts with oxygen (or other oxidizing agent) to form a new layer of oxide over the damaged area (Figure 21.14).

Aluminum was first prepared by reducing $AlCl_3$ using sodium or potassium. This was a costly process and, in the 19th century, aluminum was a precious metal. At the 1855 Paris Exposition, in fact, a sample of aluminum was exhibited along with the crown jewels of France. In an interesting coincidence, in 1886 two men, Frenchman Paul Heroult (1863–1914) and American Charles Hall (1863–1914), simultaneously and independently conceived of the electrochemical method used today. The Hall–Heroult method bears the names of the two discoverers.

Aluminum is found in nature as aluminosilicates, minerals such as clay that are based on aluminum, silicon, and oxygen. As these minerals weather, they break down to various forms of hydrated aluminum oxide, $Al_2O_3 \cdot n\ H_2O$, called *bauxite.* Mined in huge quantities, bauxite is the raw material from which aluminum is obtained. The first step is to purify the ore, separating Al_2O_3 from iron and silicon oxides. This is done by the *Bayer process,* which relies on the amphoteric, basic, or acidic

Gallium. Gallium is one of the few metals that can be a liquid at or near room temperature. (Others are Hg and Cs.) Gallium has a melting point of 29.8 °C.

■ **Charles Martin Hall (1863–1914)** Hall was only 22 years old when he worked out the electrolytic process for extracting aluminum from Al_2O_3 in a woodshed behind the family home in Oberlin, Ohio. He went on to found a company that eventually became ALCOA, the Aluminum Corporation of America.

Oesper collection in the History of Chemistry/ University of Cincinatti

Figure 21.14   Corrosion of aluminum.
(a) A ball of aluminum foil is added to a solution of copper(II) nitrate and sodium chloride. Normally, the coating of chemically inert Al_2O_3 on the surface of aluminum protects the metal from further oxidation. (b) In the presence of the Cl^- ion, the coating of Al_2O_3 is breached, and aluminum reduces copper(II) ions to copper metal. The reaction is rapid and so exothermic that the water can boil on the surface of the foil. [The blue color of aqueous copper(II) ions will fade as these ions are consumed in the reaction.]

(a)

(b)

Charles D. Winters

nature of the various oxides. Silica, SiO_2, is an acidic oxide, Al_2O_3 is amphoteric, and Fe_2O_3 is a basic oxide. Silica and Al_2O_3 dissolve in a hot concentrated solution of caustic soda (NaOH), leaving insoluble Fe_2O_3 to be filtered out.

$$Al_2O_3(s) + 2\ NaOH(aq) + 3\ H_2O(\ell) \longrightarrow 2\ Na[Al(OH)_4](aq)$$
$$SiO_2(s) + 2\ NaOH(aq) + 2\ H_2O(\ell) \longrightarrow Na_2[Si(OH)_6](aq)$$

By treating the solution containing aluminate and silicate anions with CO_2, Al_2O_3 precipitates and the silicate ion remains in solution. Recall that CO_2 is an acidic oxide that forms the weak acid H_2CO_3 in water, so the Al_2O_3 precipitation in this step is an acid–base reaction.

$$H_2CO_3(aq) + 2\ Na[Al(OH)_4](aq) \longrightarrow Na_2CO_3(aq) + Al_2O_3(s) + 5\ H_2O(\ell)$$

Metallic aluminum is obtained from purified bauxite by electrolysis (Figure 21.15). Bauxite is first mixed with cryolite, Na_3AlF_6 to give a lower-melting mixture (melting temperature = 980 °C) that is electrolyzed in a cell with graphite electrodes. The cell operates at a relatively low voltage (4.0–5.5 V) but with an extremely high current (50,000–150,000 amps). Aluminum is produced at the cathode and oxygen at the anode. To produce 1 kg of aluminum requires 13 to 16 kilowatt-hours of energy plus the energy required to maintain the high temperature.

Boron Compounds

Borax, $Na_2B_4O_7 \cdot 10\ H_2O$, is the most important boron-oxygen compound and is the form of the element most often found in nature. It has been used for centuries as a low-melting *flux* in metallurgy, because of the ability of molten borax to dissolve other metal oxides. This cleans the surfaces of metals to be joined and permits a good metal-to-metal contact.

The formula given above for borax is misleading. The salt contains an ion better described by the formula $B_4O_5(OH)_4^{2-}$. This ion illustrates two commonly observed structural features in inorganic chemistry. First, many minerals consist of MO_n groups that share O atoms. Second, the sharing of O atoms between two metals or metalloids often leads to MO rings.

After refinement, borax can be treated with sulfuric acid to produce boric acid, $B(OH)_3$.

$$Na_2B_4O_7 \cdot 10\ H_2O(s) + H_2SO_4(aq) \longrightarrow 4\ B(OH)_3(aq) + Na_2SO_4(aq) + 5\ H_2O(\ell)$$

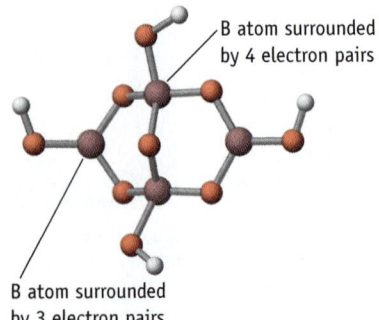

B atom surrounded by 4 electron pairs

B atom surrounded by 3 electron pairs

The borate ion of borax, $B_4O_5(OH)_4^{2-}$.

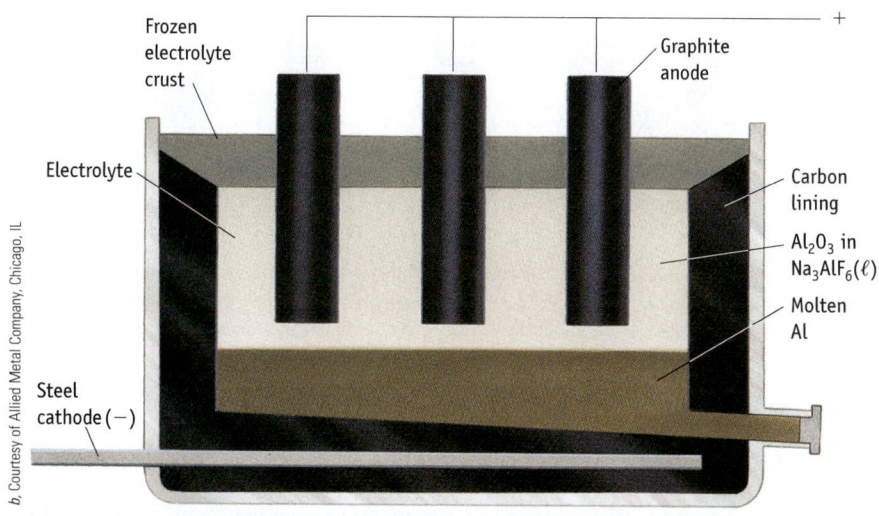

(a) Electrolysis of aluminum oxide to produce aluminum metal.

(b) Molten aluminum from recycled metal.

Active Figure 21.15   **Industrial production of aluminum.** (a) Purified aluminum-containing ore (bauxite), essentially Al_2O_3, is mixed with cryolite (Na_3AlF_6) to give a mixture that melts at a lower temperature
than Al_2O_3 alone. The aluminum-containing substances are reduced at the steel cathode to give molten aluminum. Oxygen is produced at the carbon anode, and the gas reacts slowly with the carbon to give CO_2, leading to eventual destruction of the electrode. (b) Molten aluminum alloy, produced from recycled metal,
at 760 °C, in 1.6×10^4-kg capacity crucibles.

GENERAL
Chemistry·¢·Now™ **See the General ChemistryNow CD-ROM or website to explore an interactive version of this figure accompanied by an exercise.**

The chemistry of boric acid incorporates both Lewis and Brønsted acid behavior.
Hydronium ions are produced by a Lewis acid–base interaction between boric acid
and water.

$$K_a = 7.3 \times 10^{-10}$$

Because of its weak acid properties and slight biological activity, boric acid has been
used for many years as an antiseptic. Furthermore, because the acid is so weak, salts
of borate ions, such as the $B_4O_5(OH)_4^{2-}$ ion in borax, are hydrolyzed in water to
give a basic solution.

Boric acid is dehydrated to boric oxide when strongly heated.

$$2\ B(OH)_3(s) \longrightarrow B_2O_3(s) + 3\ H_2O(\ell)$$

By far the largest use for the oxide is in the manufacture of borosilicate glass. This type
of glass is composed of 76% SiO_2, 13% B_2O_3, and much smaller amounts of Al_2O_3 and
Na_2O. The presence of boric oxide gives the glass a higher softening temperature, im-
parts a better resistance to attack by acids, and makes the glass expand less on heating.

Like its metalloid neighbor silicon, boron forms a series of molecular compounds with hydrogen. Because boron is slightly less electronegative than hydrogen, these compounds are best described as hydrides, in which the H atoms bear a
slight negative charge. More than 20 neutral boron hydrides, or boranes, with the
general formula B_xH_y are known. The simplest of these is *diborane*, B_2H_6, where x is
2 and y is 6. This colorless, gaseous compound has a boiling point of −92.6 °C.

■ **Marco Polo and Boron**
The Venetian adventurer Marco Polo
(1254–1324?) brought borax back from the
Far East, along with gunpowder and
spaghetti.

■ **Borax in Fire Retardants**
The second largest use for boric acid and
borates is as a flame retardant for cellulose
home insulation. Such insulation is often
made of scrap paper, which is inexpensive
but flammable. To control the flammability,
5–10% of the weight of the insulation is
boric acid.

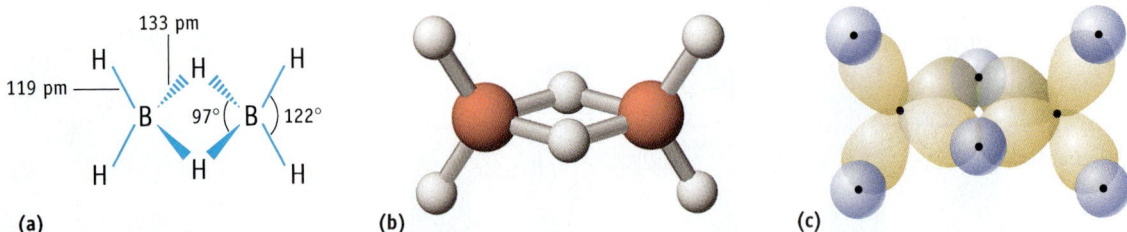

(a)   (b)   (c)

Figure 21.16 **Bonding in diborane.** (a, b) The structure of diborane, B_2H_6. (c) After accounting for bonding to two "terminal" H atoms, two sp^3 hybrid orbitals remain on each B atom. One such orbital from each boron may overlap a hydrogen $1s$ orbital in the bridge to give a three-center bond (three orbitals used), to which two electrons are assigned.

Diborane is an unusual molecule. You might have expected that the simplest boron hydride would have the formula BH_3 and a planar, trigonal geometry like that of the boron trihalides. Diborane seems even more curious if you examine details of its structure (Figure 21.16). Two hydrogen atoms in the structure are bonded not to a single boron atom, but rather to two boron atoms. Furthermore, there appears to be a shortage of electrons for all the bonds. Two boron atoms and six hydrogen atoms bring a total of 12 valence electrons to bind the molecule together. If you take each of the eight lines in the structural diagram as a two-electron bond, 16 electrons would be required. Because there appears not to be enough electrons for all the bonds in the molecule (and the same is true in other boron hydrides), these compounds came to be called *electron-deficient* molecules.

There are several ways to solve the diborane bonding dilemma, but just one is outlined here. The boron atoms are surrounded more or less tetrahedrally by H atoms, so we assume the B atoms are sp^3 hybridized. The "outside" or terminal B—H bonds are assumed to be normal, two-electron bonds formed by the overlap of an H atom's $1s$ orbital with a B atom's sp^3 orbital. The four bonds of this type require eight of the 12 electrons available. Each boron atom has two additional sp^3 hybrid orbitals, which extend into the bridging region (Figure 21.16c). Here a spherical hydrogen $1s$ orbital can overlap with one sp^3 hybrid orbital from each boron, creating a *three-center bond*. This three-center bond can accommodate two electrons, so the two bridges account for the four remaining valence electrons.

Diborane has a very endothermic enthalpy of formation: $\Delta H_f^\circ = +41.0$ kJ/mol. It is not surprising, then, that it and other boron hydrides were once considered as possible rocket fuels. They burn in air to give boric oxide and water vapor in an extremely exothermic reaction.

$$B_2H_6(g) + 3\ O_2(g) \longrightarrow B_2O_3(s) + 3\ H_2O(g) \qquad \Delta H^\circ = -2038\ \text{kJ}$$

Diborane can be synthesized from sodium borohydride, $NaBH_4$, the only B-H compound produced in ton quantities.

$$2\ NaBH_4(s) + I_2(s) \longrightarrow B_2H_6(g) + 2\ NaI(s) + H_2(g)$$

Sodium borohydride, $NaBH_4$, a white, crystalline, water-soluble solid, is made from NaH and a borate.

$$4\ NaH(s) + B(OCH_3)_3(g) \longrightarrow NaBH_4(s) + 3\ NaOCH_3(s)$$

One of the main uses of $NaBH_4$ is as a reducing agent in organic synthesis. We encountered it in Chapter 11, where we described its use to reduce aldehydes, carboxylic acids, and ketones.

Sodium borohydride, NaBH₄, is an excellent reducing agent. Here silver ions are reduced to finely divided silver metal.

Charles D. Winters

See the General ChemistryNow CD-ROM or website:
- **Screen 21.4 Boron Hydrides Structures,** for an exercise on the structures of the family of boron-hydrogen compounds

Aluminum Compounds

Aluminum is an excellent reducing agent, so it reacts readily with hydrochloric acid. In contrast, it does not react with nitric acid. The latter rapidly oxidizes the surface of aluminum, and the resulting film of Al_2O_3 protects the metal from further attack. This protection allows nitric acid to be shipped in aluminum tanks.

Various salts of aluminum dissolve in water, giving the hydrated $Al^{3+}(aq)$ ion. These solutions are acidic [◀ Table 17.3, page 808] because the hydrated ion is a weak Brønsted acid.

$$[Al(H_2O)_6]^{3+}(aq) + H_2O(\ell) \rightleftharpoons [Al(H_2O)_5(OH)]^{2+}(aq) + H_3O^+(aq)$$

Adding acid shifts the equilibrium to the left, whereas adding base causes the equilibrium to shift to the right. Addition of sufficient hydroxide ion results ultimately in precipitation of the hydrated oxide $Al_2O_3 \cdot 3\,H_2O$.

Aluminum oxide, Al_2O_3, formed by dehydrating the hydrated oxide, is quite insoluble in water and generally resistant to chemical attack. In the crystalline form, aluminum oxide is known as corundum. This material is extraordinarily hard, a property that leads to its use as an abrasive in grinding wheels, "sandpaper," and toothpaste.

Some gems are impure aluminum oxide. Rubies, beautiful red crystals prized for jewelry and used in some lasers, are Al_2O_3 contaminated with a small amount of Cr^{3+} (Figure 21.17). The Cr^{3+} ions replace some of the Al^{3+} ions in the crystal lattice. Synthetic rubies were first made in 1902, and the worldwide capacity is now about 200,000 kg/year; much of this production is used for jewel bearings in watches and instruments. Blue sapphires consist of Al_2O_3 with Fe^{2+} and Ti^{4+} impurities in place of Al^{3+} ions.

Boron forms halides such as gaseous BF_3 and BCl_3 that have the expected planar, trigonal molecular geometry of halogen atoms surrounding an sp^2 hybridized boron atom. In contrast, the aluminum halides are all solids and have more interesting structures. Aluminum bromide, which is made by the very exothermic reaction of aluminum metal and bromine (Figure 3.1, page 98),

$$2\,Al(s) + 3\,Br_2(\ell) \longrightarrow Al_2Br_6(s)$$

is composed of two units of $AlBr_3$. That is, Al_2Br_6 is a *dimer* of $AlBr_3$ units. The structure resembles that of diborane in that bridging atoms appear between the two Al atoms. However, Al_2Br_6 is not electron-deficient; the bridge is formed when a Br atom on one Al_3Br uses a lone pair to form a coordinate covalent bond to a neighboring tetrahedral, sp^3-hybridized aluminum atom.

Aluminum does not react with nitric acid. Nitric acid, a strongly oxidizing acid, reacts vigorously with copper (*left*) but aluminum (*right*) is untouched.

Figure 21.17 Sapphires and rubies. Both are minerals based on Al_2O_3 in which a few Al^{3+} ions have been replaced by ions such as Cr^{3+}, Fe^{2+} or Ti^{4+}. (*top*) The Star of Asia sapphire. (*middle*) Various sapphires. (*bottom*) Uncut corundum.

Both aluminum bromide and aluminum iodide have this structure, whereas aluminum chloride exists as a dimer only in the vapor state.

Aluminum chloride can react with a chloride ion to form the simple ion $AlCl_4^-$. Aluminum fluoride, in contrast, can accommodate three additional F^- ions to form an octahedral AlF_6^{3-} ion. This form of aluminum is found in cryolite, Na_3AlF_6, the compound added to aluminum oxide in the electrolytic production of aluminum metal. Apparently, the Al^{3+} ion can bind to six of the smaller F^- ions, whereas only four of the larger Cl^-, Br^-, or I^- ions can surround an Al^{3+} ion.

GENERAL
Chemistry · Now™

See the General ChemistryNow CD-ROM or website:
• **Screen 21.5 Aluminum Compounds,** for an exercise on the chemistry of aluminum compounds

---

Exercise 21.8—Gallium Chemistry

(a) Gallium hydroxide, like aluminum hydroxide, is amphoteric. Write a balanced equation to show how this hydroxide can dissolve in both HCl(aq) and NaOH(aq).

(b) Gallium ion in water, $Ga^{3+}(aq)$, has a K_a value of 1.2×10^{-3}. Is this ion a stronger or a weaker acid than $Al^{3+}(aq)$?

21.7—Silicon and the Group 4A Elements

The elements of Group 4A have the broadest range of chemical behavior of any group in the periodic table. Carbon is distinctly nonmetallic in its chemistry, but silicon and germanium are classed as metalloids, while tin and lead are metals.

All of the Group 4A elements are characterized by half-filled valence shells with two electrons in the ns orbital and two electrons in np orbitals (where n is the period in which the element is found). The bonding in carbon and silicon compounds is largely covalent and involves sharing of four electron pairs with neighboring atoms. In germanium compounds, the +4 oxidation state is common (GeO_2 and $GeCl_4$), but some +2 oxidation state compounds exist (GeI_2). An oxidation number of +2, as well as +4, is even more common for tin and lead (such as $SnCl_2$ and PbO). The increasing importance of the +2 oxidation number for heavier elements in the group illustrates a trend seen in other A-Group elements: Lower oxidation numbers are common for the heavier members of the group (such as Tl^+, Pb^{2+}, and Bi^{3+}).

Group 4A

Carbon
6
C
480 ppm

Silicon
14
Si
277,100 ppm

Germanium
32
Ge
1.8 ppm

Tin
50
Sn
2.2 ppm

Lead
82
Pb
14 ppm

Element abundances are in parts per million in the earth's crust.

Silicon

Silicon is second after oxygen in abundance in the earth's crust, so it is not surprising that we are surrounded by silicon-containing materials: bricks, pottery, porcelain, lubricants, sealants, computer chips, and solar cells. The computer revolution is based on the semiconducting properties of silicon.

Reasonably pure silicon can be made in large quantities by heating pure silica sand with purified coke to approximately 3000 °C in an electric furnace.

$$SiO_2(s) + 2\ C(s) \longrightarrow Si(\ell) + 2\ CO(g)$$

The molten silicon is drawn off the bottom of the furnace and allowed to cool to a shiny blue-gray solid. Because extremely high-purity silicon is needed for the electronics industry, purifying raw silicon requires several steps. First the silicon in the impure sample is allowed to react with chlorine to convert the silicon to liquid silicon tetrachloride.

$$Si(s) + 2\ Cl_2(g) \longrightarrow SiCl_4(\ell)$$

Silicon tetrachloride (boiling point of 57.6 °C) is carefully purified by distillation and then reduced to silicon using magnesium.

$$SiCl_4(g) + 2\ Mg(s) \longrightarrow 2\ MgCl_2(s) + Si(s)$$

The magnesium chloride is washed out with water, and the silicon is remelted and cast into bars. A final purification is carried out by *zone refining*, a process in which a special heating device is used to melt a narrow segment of the silicon rod. The heater is moved slowly down the rod. Impurities contained in the silicon tend to remain in the liquid phase because the melting point of a mixture is lower than that of the pure element (Chapter 14). The silicon that crystallizes above the heated zone is therefore of a higher purity (Figure 21.18).

Silicon Dioxide

The simplest oxide of silicon is SiO_2, commonly called *silica*, a constituent of many rocks such as granite and sandstone. Quartz is a pure crystalline form of silica, but impurities in quartz produce gemstones such as amethyst (Figure 21.19).

Silica and CO_2 are oxides of two elements in the same chemical group, so similarities between them might be expected. In fact, SiO_2 is a high-melting solid (quartz melts at 1610 °C), whereas CO_2 is a gas at room temperature and 1 bar. This great disparity arises from the different structures of the two oxides. Carbon dioxide is a molecular compound, with the carbon atom linked to each oxygen atom by a double bond. In contrast, SiO_2 is a network solid, which is the preferred structure because the energy of two $Si{=}O$ double bonds is much less than the energy of four $Si{-}O$ single bonds. The contrast between SiO_2 and CO_2 exemplifies a more general phenomenon. Multiple bonds, often encountered between second-period elements, are rare among elements in the third and higher periods. There are many compounds with multiple bonds to carbon, but *very* few compounds featuring multiple bonds to silicon.

Figure 21.18 Pure silicon. The manufacture of very pure silicon begins with producing the volatile liquid silanes $SiCl_4$ or $SiHCl_3$. After carefully purifying these by distillation, they are reduced to elemental silicon with extremely pure Mg or Zn. The resulting spongy silicon is heated to produce molten silicon, which is then purified by zone refining. The end result is a cylindrical rod of ultrapure silicon such as those seen in this photograph. Finally, thin wafers of silicon are cut from the bars and are the basis for the semiconducting chips in computers and other devices.

Figure 21.19 Various forms of quartz.

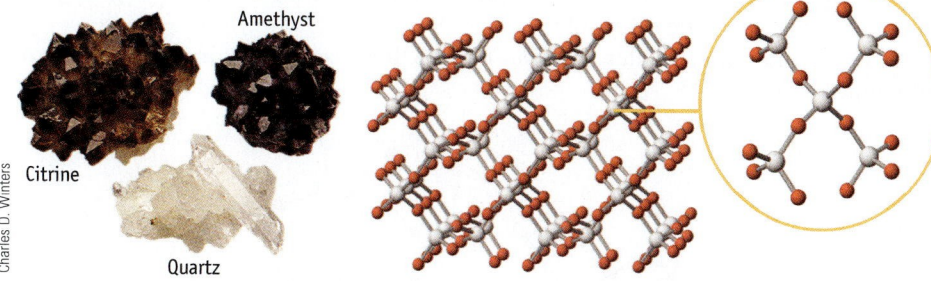

(a) Pure quartz is colorless, but the presence of small amounts of impurities adds color. Purple amethyst and brown citrine crystals are quartz with iron impurities.

(b) Quartz is a network solid in which each Si atom is bound tetrahedrally to four O atoms, each O atom linked to another Si atom. The basic structure consists of a lattice of Si and O atoms. See the Molecular Models folder on the *Interactive General Chemistry CD-ROM*.

Synthetic quartz. These crystals were grown from silica in sodium hydroxide. The colors come from added Co^{2+} ions (blue) or Fe^{2+} ions (brown).

Quartz crystals are used to control the frequency of radio and television transmissions. Because these and related applications use so much quartz, there is not enough natural quartz to fulfill demand, and quartz is therefore synthesized. Noncrystalline, or vitreous, quartz, made by melting pure silica sand, is placed in a steel "bomb" and dilute aqueous NaOH is added. A "seed" crystal is placed in the mixture, just as you might use a seed crystal in a hot sugar solution to grow rock candy. When the mixture is heated above the critical temperature of water (above 400 °C and 1700 atm) over a period of days, pure quartz crystallizes.

Silica is resistant to attack by all acids except HF, with which it reacts to give SiF_4 and H_2O. It also dissolves slowly in hot, molten NaOH or Na_2CO_3 to give Na_4SiO_4, sodium silicate.

$$SiO_2(s) + 4\ HF(\ell) \longrightarrow SiF_4(g) + 2\ H_2O(\ell)$$
$$SiO_2(s) + 2\ Na_2CO_3(\ell) \longrightarrow Na_4SiO_4(s) + 2\ CO_2(g)$$

After the molten mixture has cooled, hot water under pressure is added. This partially dissolves the material to give a solution of sodium silicate. After filtering off insoluble sand or glass, the solvent is evaporated to leave sodium silicate, called *water glass*. The biggest single use of this material is in household and industrial detergents, in which it is included because a sodium silicate solution maintains pH by its buffering ability. Additionally, sodium silicate is used in various adhesives and binders, especially for gluing corrugated cardboard boxes.

If sodium silicate is treated with acid, a gelatinous precipitate of SiO_2 called *silica gel* is obtained. Washed and dried, silica gel is a highly porous material with dozens of uses. It is a drying agent, readily absorbing up to 40% of its own weight of water. Small packets of silica gel are often placed in packing boxes of merchandise during storage. The material is frequently stained with $(NH_4)_2CoCl_4$, a humidity detector that is pink when hydrated and blue when dry.

Silicate Minerals with Chain and Ribbon Structures

Silicate minerals are a world in themselves. All silicates are built from tetrahedral SiO_4 units, but they have different properties owing to the way these tetrahedral SiO_4 units link together.

The simplest silicates, *orthosilicates*, contain SiO_4^{4-} anions. The 4− charge of the anion is balanced by four M^+ ions, two M^{2+} ions, or a combination of ions. Calcium orthosilicate, Ca_2SiO_4, is a component of Portland cement. Olivine, an important mineral in the earth's mantle, contains Mg^{2+} and Fe^{2+}, with the Fe^{2+} ion giving the mineral its characteristic olive color.

A group of minerals called *pyroxenes* have as their basic structural unit a chain of SiO_4 tetrahedra.

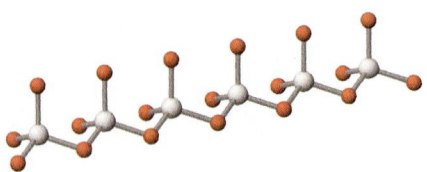

Silica gel. Silica gel is solid, noncrystalline SiO_2. Packages of the material are often used to keep electronic equipment dry when stored. Silica gel is also used to clarify beer; passing beer through a bed of silica gel removes minute particles that would otherwise make the brew cloudy. Yet another use is in kitty litter.

If two such chains are linked together by sharing oxygen atoms, the result is an *amphibole,* of which the asbestos minerals are one example. The molecular chain results in asbestos being a fibrous material.

Silicates with Sheet Structures and Aluminosilicates

Linking many silicate chains together produces a sheet of SiO$_4$ tetrahedra (Figure 21.20). This sheet is the basic structural feature of some of the earth's most important minerals, particularly the clay minerals (such as china clay), mica, and the chrysotile form of asbestos.

The sheet structure leads to the characteristic feature of mica, which is often found as "books" of thin, silicate sheets. Mica is used in furnace windows and as insulation, and flecks of mica give the glitter to "metallic" paints.

Mica, a large family of clays, and asbestos are actually *aluminosilicates*, substances containing both aluminum and silicon. In kaolinite clay, for example, the sheet of SiO$_4$ tetrahedra is bonded to a sheet of AlO$_6$ octahedra. In addition, some Si atoms can be replaced by Al atoms. Because Si^{4+} ions are being replaced by Al^{3+} ions with a smaller charge, nature adds positive ions such as Na$^+$ and Mg^{2+} for every aluminum ion in the lattice. This feature leads to some interesting uses of clays, one being in medicine (Figure 21.21). In certain cultures, clay is eaten for medicinal purposes. Several remedies for the relief of upset stomach contain highly purified clays that absorb excess stomach acid as well as potentially harmful bacteria and their toxins by exchanging the intersheet cations in the clays for the toxins, which are often organic cations.

Other aluminosilicates include the feldspars, common minerals that make up about 60% of the earth's crust, and zeolites (Figure 21.21). Both materials are composed of SiO$_4$ tetrahedra in which some of the Si atoms have been replaced by Al atoms, along with alkali and alkaline earth metal ions for charge balance. The main feature of zeolite structures is their regularly shaped tunnels and cavities. Hole diameters are between 300 and 1000 pm, and small molecules such as water can fit into the cavities of the zeolite. As a result, zeolites can be used as drying agents to selectively absorb water from air or a solvent. Small amounts of zeolites are often sealed into multipane windows to keep the air dry between the panes.

Zeolites are also used as catalysts. ExxonMobil, for example, has patented a process in which methanol, CH$_3$OH, is converted to gasoline in the presence of specially tailored zeolites. In addition, zeolites are added to detergents, where they function as water-softening agents because the sodium ions of the zeolite can be exchanged for Ca^{2+} ions in hard water, effectively removing Ca^{2+} ions from the water.

Kaolinite Clay. The basic structural feature of many clays, and kaolinite in particular, is a sheet of SiO$_4$ tetrahedra (black and red spheres) bonded to a sheet of AlO$_6$ octahedra (gray and green spheres).

■ **A silicate from the center of the earth.** The front cover of this book illustrates the structure of a silicate, MgSiO$_3$. The solid consists of SiO$_6$ octahedra (blue) and magnesium ions (Mg^{2+}; yellow spheres). Each SiO$_6$ octahedron shares the four O atoms in opposite edges with two neighboring octahedra, thus forming a chain of octahedra. These chains are interlinked by sharing the O atoms at the "top" and "bottom" of SiO$_6$ octahedra in neighboring chains. The magnesium ions lie between the layers of interlinked SiO$_6$ chains.

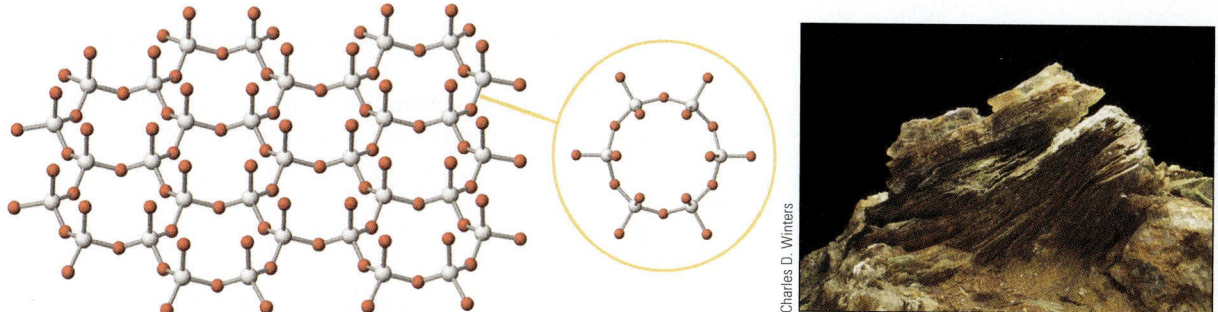

Figure 21.20 Mica, a sheet silicate. The sheet-like structure of mica explains its physical appearance. As in the pyroxenes, each silicon is bonded to four oxygen atoms, but the Si and O atoms form a sheet of six-member rings of Si atoms with O atoms in each edge. The ratio of Si to O is 1 to 2.5. (A formula of SiO$_{2.5}$ requires a positive ion, such as Na$^+$, to counterbalance the charge. Thus, mica and other sheet silicates, and aluminosilicates such as talc and many clays, have positive ions between the sheets.)

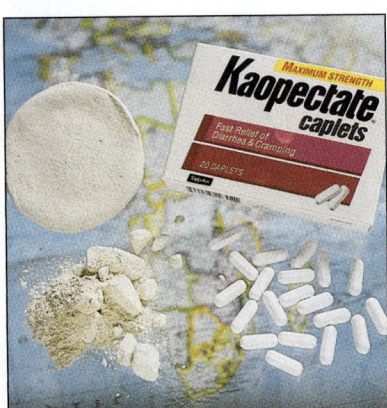

(a) Remedies for stomach upset. One of the ingredients in Kaopectate is kaolin, one form of clay. The off-white objects are pieces of clay purchased in a market in Ghana, West Africa. This clay was made to be eaten as a remedy for stomach ailments. Eating clay is widespread among the world's different cultures.

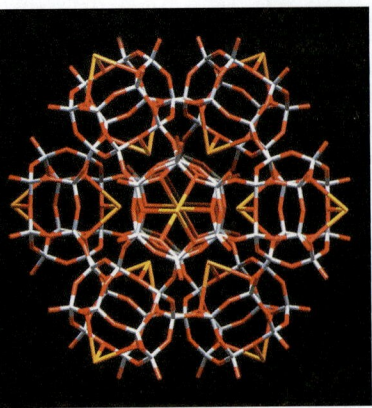

(b) The stucture of a zeolite. Zeolites, which have Si, Al, and O linked in a polyhedral framework, are often portrayed in drawings like this. Each edge consists of a Si-O-Si, Al-O-Si, or Al-O-Al bond. The channels in the framework can selectively capture small molecules or ions or act as catalytic sites.

(c) Apophyllite, a crystalline zeolite.

(d) Consumer products that remove odor-causing molecules from the air often contain zeolites.

a, c, and d, Charles D. Winters; b, Alfred Pasieka/Science Photo Library/Photo Researchers, Inc.

Figure 21.21   Aluminosilicates.

GENERAL
Chemistry ⚛ Now™

See the General ChemistryNow CD-ROM or website:
- **Screen 21.6 Silicon-Oxygen Compounds: Formulas and Structures,** for an exercise on the structural chemistry of silicon-oxygen compounds

---

Silicone Polymers

Silicon and methyl chloride (CH_3Cl) react at 300 °C in the presence of a catalyst, Cu powder. The primary product of this reaction is $(CH_3)_2SiCl_2$.

$$Si(s) + 2\ CH_3Cl(g) \longrightarrow (CH_3)_2SiCl_2(\ell)$$

Halides of Group 4A elements other than carbon hydrolyze readily. The reaction of $(CH_3)_2SiCl_2$ with water, for example, initially produces $(CH_3)_2Si(OH)_2$. On standing, these molecules condense to form a polymer, eliminating water. The polymer is called polydimethylsiloxane, a member of the *silicone* family of polymers.

$$(CH_3)_2SiCl_2 + 2\ H_2O \longrightarrow (CH_3)_2Si(OH)_2 + 2\ HCl$$
$$n\ (CH_3)_2Si(OH)_2 \longrightarrow [-(CH_3)_2SiO-]_n + n\ H_2O$$

Silicone. Some examples of products containing silicones, polymers with repeating —R_2Si—O— units.

Charles D. Winters

Silicone polymers are nontoxic and have good stability to heat, light, and oxygen; they are chemically inert and have valuable antistick and antifoam properties. They can take the form of oils, greases, and resins. Some have rubber-like properties ("Silly Putty," for example, is a silicone polymer). More than 1 million tons of silicone polymers are made worldwide annually. These materials are used in a wide variety of products: lubricants, peel-off labels, lipstick, suntan lotion, car polish, and building caulk.

Chemical Perspectives

Lead Pollution, Old and New

Anchoring the bottom of Group 4A is the element lead. One of a handful of elements known since ancient times, it is also a metal with a variety of modern uses. It is an essential commodity in the modern industrial world, ranking fifth in consumption behind iron, copper, aluminum, and zinc. The major uses of the metal and its compounds are in storage batteries (page 959), pigments, ammunition, solders, plumbing, and bearings. Its chief source is lead sulfide, PbS, known commonly as galena (Figure 18.14).

Lead and its compounds are cumulative poisons, particularly in children. At a blood level as low as 50 ppb (parts per billion), blood pressure is elevated; intelligence is affected at 100 ppb; and blood levels higher than 800 ppb can lead to coma and possible death. Health experts believe that more than 200,000 children become ill from lead poisoning annually. This problem is caused chiefly by children eating paint containing lead pigments. Older homes often contain lead-based paint because white lead [2 $PbCO_3 \cdot$ $Pb(OH)_2$] was the pigment used in white paint until about 40 years ago, when it was replaced by TiO_2. Lead salts have a sweet taste, which may contribute to the tendency of children to chew on painted objects.

Charles D. Winters

Uses of lead. The primary use of lead is in lead storage batteries (where the metal is used as the anode, and PbO_2 is used as the cathode). Other uses include plumbers' lead and as a component of solder. Lead is also used to frame the glass pieces in stained glass windows.

Until just a few years ago, a major use of lead was in tetraethyllead, $Pb(C_2H_5)_4$. This compound was added to gasoline to improve the burning properties of fuel. Its use has since been phased out, in part because of the hazards from tons of lead compounds being spewed into the environment and in part because lead poisons the catalyst in catalytic exhaust systems. All gasoline sold in the United States now bears the label "unleaded" or "no lead."

Lead poisoning is often implicated as one cause of the collapse of the Roman Empire. At the height of the Roman Empire (about 500 A.D.), the production of lead was as much as 80,000 tons per year because lead had so many uses: for roofing, water pipes, kitchenware, and coffins; and for lining the hulls of ships. The Romans used lead to make pipes to carry water, and they cooked in lead vessels. Because many foods are acidic, and lead reacts slowly with acid, lead could be incorporated into food.

Scientists have recently obtained evidence that extensive lead mining and smelting operations during the Roman Empire contributed to atmospheric pollution on a global scale. Drilling in the Greenland ice cap produced an ice core, a cylinder of ice, that is 9938 feet long. The core contains trace remnants of the atmosphere for the last 9000 years, and it is possible to obtain very accurate dates on the various segments of this core.

Among the trace elements in the polar ice core was lead, in concentrations in the parts per trillion (ppt) range. Samples of lead have isotope ratios (such as $^{208}Pb/^{206}Pb$) that depend on the source of the element. (Analysis of isotope ratios is carried out using sensitive mass spectrometric techniques; see Figure 2.8.) In the segment of the ice core between 300 B.C. and 600 A.D., the lead isotope ratios in the samples were identical to those of the lead from Roman mines in southwestern Spain! Lead polluted the earth's atmosphere 2000 years ago.

Exercise 21.9—Silicon Chemistry

Silicon-oxygen rings are a common structural feature in silicate chemistry. Draw the structure for the anion $Si_3O_9^{6-}$, which is found in minerals such as benitoite. The ring has three Si atoms and three O atoms, and there are two other O atoms on each Si atom.

21.8—Nitrogen, Phosphorus, and the Group 5A Elements

Group 5A elements are characterized by the ns^2np^3 configuration with its half-filled np subshell. In compounds of the Group 5A elements, the primary oxidation numbers are +3 and +5, although in a set of common nitrogen compounds a range of oxidation numbers from −3 to +3 and +5 is seen. Once again, as in Groups 3A and 4A, the most positive oxidation number is less stable for the heavier elements. In

Group 5A

Nitrogen
7
N
25 ppm

Phosphorus
15
P
1000 ppm

Arsenic
33
As
1.5 ppm

Antimony
51
Sb
0.2 ppm

Bismuth
83
Bi
0.048 ppm

Element abundances are in parts per million in the earth's crust.

many arsenic and bismuth compounds, the element has an oxidation number of +3 state. In fact, compounds of these elements with oxidation numbers of +5 are powerful oxidizing agents.

This part of our tour of the main group elements will concentrate on the chemistries of nitrogen and phosphorus. Nitrogen is found primarily as N_2 in the atmosphere, where it constitutes 78.1% by volume (75.5% by weight). In contrast, phosphorus occurs in the earth's crust in solids. More than 200 different phosphorus-containing minerals are known; all contain the tetrahedral phosphate ion, PO_4^{3-}, or a derivative of this ion. By far the most abundant minerals are apatites [◄ page 1030].

Nitrogen and its compounds play a key role in our economy, with ammonia making particularly notable contributions. Phosphoric acid is an important commodity chemical; its major use is in fertilizers.

Both phosphorus and nitrogen are part of every living organism. Phosphorus is contained in biochemicals called nucleic acids and phospholipids, and nitrogen occurs in proteins and nucleic acids. Indeed, phosphorus was first derived from human waste (see "A Closer Look: Making Phosphorus").

Properties of Nitrogen and Phosphorus

Nitrogen (N_2) is a colorless gas that liquifies at 77 K (-196 °C) (Figure 13.1, page 591). Its most notable feature is its reluctance to react with other elements or compounds. This is because the $N\equiv N$ triple bond has a large dissociation energy (945 kJ/mol) and because the molecule is nonpolar. Nitrogen does, however, react with hydrogen to give ammonia in the presence of a catalyst (Figure 16.10) and with a few metals to give metal nitrides, compounds containing the N^{3-} ion.

$$3\ Mg(s) + N_2(g) \longrightarrow Mg_3N_2(s)$$
magnesium nitride

Elemental nitrogen is a very useful material. Because of its lack of reactivity, it is used to provide a nonoxidizing atmosphere for packaged foods and wine and to pressurize electric cables and telephone wires. Liquid nitrogen is valuable as a coolant in freezing biological samples such as blood and semen, in freeze-drying food, and for other applications that require extremely low temperatures.

Elemental phosphorus is produced by the reduction of phosphate minerals in an electric furnace.

$$2\ Ca_3(PO_4)_2(s) + 10\ C(s) + 6\ SiO_2(s) \longrightarrow P_4(g) + 6\ CaSiO_3(s) + 10\ CO(g)$$

The most stable allotrope of phosphorus is white phosphorus. Rather than occurring as a diatomic molecule with a triple bond, like its second-period relative nitrogen (N_2), phosphorus is made up of tetrahedral P_4 molecules in which each P atom is joined to three others via single bonds. Red phosphorus is a polymer of P_4 units.

Charles D. Winters

The red and white allotropes of phosphorus.

White phosphorus, P_4

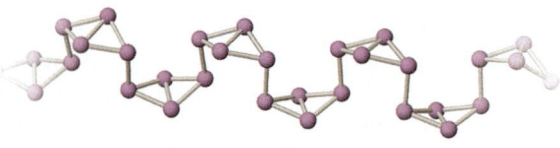

Polymeric red phosphorus

A Closer Look

Making Phosphorus

He stoked his small furnace with more charcoal and pumped the bellows until his retort glowed red hot. Suddenly something strange began to happen. Glowing fumes filled the vessel and from the end of the retort dripped a shining liquid that burst into flames.

J. Emsley: *The 13th Element*, p. 5. New York, John Wiley, 2000.

John Emsley begins his story of phosphorus, its discovery, and its uses, by imagining what the German alchemist Hennig Brandt must have seen in his laboratory that day in 1669. Brandt was in search of the philosopher's stone, the magic elixir that would turn the crudest substance into gold.

Brandt was experimenting with urine, which had served as the source of useful chemicals since Roman times. It is not surprising that phosphorus could be extracted from this source. Humans consume much more phosphorus, in the form of phosphate, than they require, and the excess phosphorus (about 1.4 g per day) is passed in the urine. It is nonetheless extraordinary that Brandt was able to isolate the element. According to an 18th-century chemistry book, about 30 g of phosphorus could be obtained from 60 gallons of urine. And the process was not simple. Another 18th-century recipe states that "50 or 60 pails full" of urine was to be used. "Let it lie steeping . . . till it putrefy and breed worms." The chemist was then to reduce the whole to a paste and finally to heat the paste very strongly in a retort. After some days phosphorus distilled from the mixture and was collected in water. (We know now that carbon from the organic compounds in the urine would have reduced the phosphate to phosphorus.) Phosphorus was made in this manner for more than 100 years.

The Alchymist in Search of the Philosopher's Stone, Discovers Phosphorus. Painted by J. Wright of Derby (1734–1797).

Nitrogen Compounds

A notable feature of the chemistry of nitrogen is the wide diversity of its compounds. Compounds with nitrogen in all oxidation numbers between −3 and +5 are known (Figure 21.22).

Hydrogen Compounds of Nitrogen: Ammonia and Hydrazine

Ammonia is a gas at room temperature and pressure. It has a very penetrating odor and condenses to a liquid at −33 °C under 1 bar of pressure. Solutions in water, often referred to as ammonium hydroxide, are basic due to the reaction of ammonia with water [◀ Section 17.5 and Figure 5.7].

$$NH_3(aq) + H_2O(\ell) \rightleftharpoons NH_4^+(aq) + OH^-(aq) \qquad K_b = 1.8 \times 10^{-5} \text{ at 25 °C}$$

Ammonia is a major industrial chemical and is prepared by the Haber process [◀ page 787], largely for use as a fertilizer.

Hydrazine, N_2H_4, is a colorless, fuming liquid with an ammonia-like odor (mp, 2.0 °C; bp, 113.5 °C). Almost 1 million kilograms of hydrazine is produced annually by the Raschig process—the oxidation of ammonia with alkaline sodium hypochlorite in the presence of gelatin (which is added to suppress metal-catalyzed side reactions that lower the yield of hydrazine).

$$2\ NH_3(aq) + NaClO(aq) \longrightarrow N_2H_4(aq) + NaCl(aq) + H_2O(\ell)$$

Hydrazine, like ammonia, is a Brønsted and Lewis base:

$$N_2H_4(aq) + H_2O(\ell) \rightleftharpoons N_2H_5^+(aq) + OH^-(aq) \qquad K_b = 8.5 \times 10^{-7}$$

It is also a strong reducing agent, as reflected in the reduction potential of N_2 in basic solution:

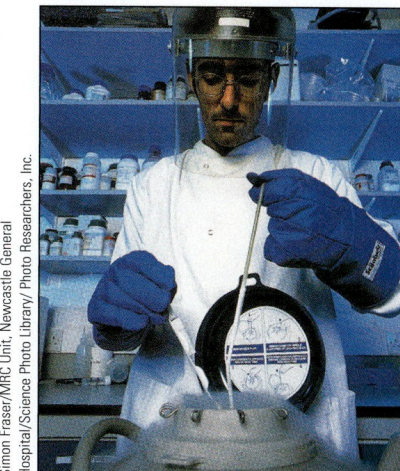

Liquid nitrogen. Biological samples—such as embryos or semen from animals and humans—can be stored in liquid nitrogen for long periods of time.

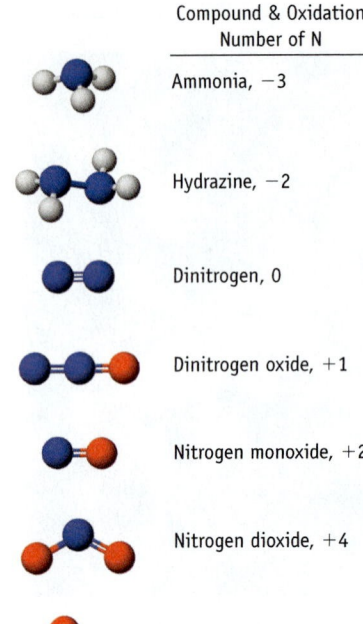

Compound & Oxidation Number of N	
Ammonia, −3	
Hydrazine, −2	
Dinitrogen, 0	
Dinitrogen oxide, +1	
Nitrogen monoxide, +2	
Nitrogen dioxide, +4	
Nitric acid, +5	

Active Figure 21.22 **Compounds and oxidation numbers for nitrogen.** In its compounds, the N atom can have oxidation states ranging from −3 to +5.

GENERAL
Chemistry ❖ Now™ See the General ChemistryNow CD-ROM or website to explore an interactive version of this figure accompanied by an exercise.

$$N_2(g) + 4 H_2O(\ell) + 4 e^- \longrightarrow N_2H_4(aq) + 4 OH^-(aq) \qquad E° = -1.15 \text{ V}$$

Hydrazine's reducing ability is exploited in its use in wastewater treatment for chemical plants. It removes ions such as CrO_4^{2-} by reducing them and thus prevents them from entering the environment. A related use is the treatment of water boilers in large electric-generating plants. Oxygen dissolved in the water presents a serious problem in these plants because the dissolved gas can oxidize the metal of the boiler and pipes and lead to corrosion. Hydrazine reduces the dissolved oxygen to water.

$$N_2H_4(aq) + O_2(g) \longrightarrow N_2(g) + 2 H_2O(\ell)$$

Oxides and Oxoacids of Nitrogen

Nitrogen is unique among elements in the number of binary oxides it forms (Table 21.6). All are thermodynamically unstable with respect to decomposition to N_2 and O_2; that is, all have positive $\Delta G_f°$ values. Most are slow to decompose, however, and so are described as kinetically stable.

Dinitrogen monoxide, N_2O, is a nontoxic, odorless, and tasteless gas in which nitrogen has the lowest oxidation number (+1) among nitrogen oxides. It can be made by the careful decomposition of ammonium nitrate at 250 °C.

$$NH_4NO_3(s) \longrightarrow N_2O(g) + 2 H_2O(g)$$

Dinitrogen monoxide, commonly called nitrous oxide, is used as an anesthetic in minor surgery and has been called "laughing gas" because of its euphoriant effects. Because it is soluble in vegetable fats, the largest commercial use of N_2O is as a propellant and aerating agent in cans of whipped cream.

Table 21.6   Some Oxides of Nitrogen

Formula	Name	Structure	Nitrogen Oxidation Number	Description
N_2O	Dinitrogen monoxide (nitrous oxide)	:N≡N—O: linear	+1	Colorless gas (laughing gas)
NO	Nitrogen monoxide (nitric oxide)	*	+2	Colorless gas; odd-electron molecule (paramagnetic)
N_2O_3	Dinitrogen trioxide	O—N—N—O planar	+3	Blue solid (mp, −100.7 °C); reversibly dissociates to NO and NO_2
NO_2	Nitrogen dioxide	N—O—O	+4	Brown, paramagnetic gas; odd-electron molecule
N_2O_4	Dinitrogen tetraoxide	O—N—N—O planar	+4	Colorless liquid/gas; dissociates to NO_2 (see Figure 16.8)
N_2O_5	Dinitrogen pentaoxide	O—N—O—N—O	+5	Colorless solid

*It is not possible to draw a Lewis structure that accurately represents the electronic structure of NO. See Chapter 10.

Nitrogen monoxide, NO, is an odd-electron molecule. It has 11 valence electrons, giving it one unpaired electron and making it a free radical. The compound has recently been the subject of intense research because it has been found to be important in a number of biochemical processes (page 396).

Nitrogen dioxide, NO₂, is the brown gas you see when a bottle of nitric acid is allowed to stand in the sunlight.

$$2\ HNO_3(aq) \longrightarrow 2\ NO_2(g) + H_2O(\ell) + \tfrac{1}{2}\ O_2(g)$$

Nitrogen dioxide is also a culprit in air pollution. Nitrogen monoxide is often present in urban polluted air and forms when atmospheric nitrogen and oxygen are heated in internal combustion engines. In the presence of excess oxygen, NO rapidly forms NO_2, which, if not removed by a catalytic exhaust system in an automobile, enters the atmosphere.

$$2\ NO(g) + O_2(g) \longrightarrow 2\ NO_2(g)$$

Nitrogen dioxide has 17 valence electrons, so it is also an odd-electron molecule. Because the odd electron largely resides on the N atom, two NO_2 molecules can combine, forming an N—N bond and producing N_2O_4, *dinitrogen tetraoxide*.

$$2\ NO_2(g) \longrightarrow N_2O_4(g)$$

deep brown   colorless
gas   (mp, −11.2 °C)

Solid N_2O_4 (mp, −11.2 °C) is colorless and consists entirely of N_2O_4 molecules. However, as the solid melts and the temperature increases to the boiling point, the color darkens as N_2O_4 dissociates to form brown NO_2. At the normal boiling point (21.5 °C), the distinctly brown gas phase consists of 15.9% NO_2 and 84.1% N_2O_4 [◀ page 911].

When NO_2 is bubbled into water, *nitric acid,* and *nitrous acid* form.

$$2\ NO_2(g) + H_2O(\ell) \longrightarrow HNO_3(aq) + HNO_2(aq)$$

nitric acid   nitrous acid

Nitric acid has been known for centuries and has become an important compound in our modern economy. The oldest way to make the acid is to treat $NaNO_3$ with sulfuric acid (Figure 21.23).

$$2\ NaNO_3(s) + H_2SO_4(\ell) \longrightarrow 2\ HNO_3(\ell) + Na_2SO_4(s)$$

Enormous quantities of nitric acid are now produced industrially by the oxidation of ammonia in the multistep Ostwald process. The acid has many applications, but

Nitrous oxide, N₂O. This oxide readily dissolves in fats, so the gas is added, under pressure, to cans of cream. When the valve is opened, the gas expands, whipping the cream. N_2O is also an anesthetic and is considered safe for medical uses. However, significant dangers arise from using it as a recreational drug. Long-term use can induce nerve damage and cause such problems as weakness and loss of feeling.

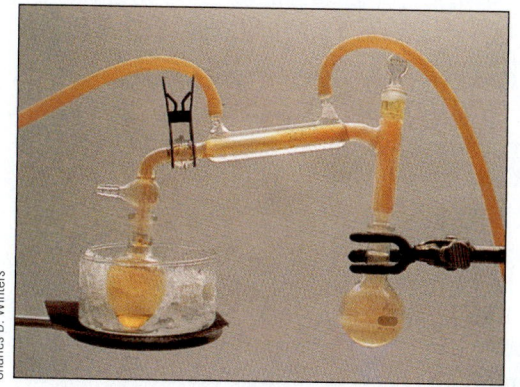

(a) Preparation of nitric acid.

(b) Reaction of HNO₃ with copper.

Figure 21.23 The preparation and properties of nitric acid. (a) Nitric acid is prepared by the reaction of sulfuric acid and sodium nitrate. Pure HNO_3 is colorless, but some acid decomposes to give brown NO_2. This gas fills the apparatus and colors the liquid in the distillation flask. (b) When concentrated nitric acid reacts with copper, the metal is oxidized to copper(II) ions, and NO_2 gas is a reaction product.

by far the greatest amount is turned into ammonium nitrate (for use as a fertilizer) by the reaction of nitric acid and ammonia.

Nitric acid is a powerful oxidizing agent, as the large, positive $E°$ values for the following half-reactions illustrate:

$$NO_3^-(aq) + 4 H^+(aq) + 3 e^- \longrightarrow NO(g) + 2 H_2O(\ell) \qquad E° = + 0.96 V$$
$$NO_3^-(aq) + 2 H^+(aq) + e^- \longrightarrow NO_2(g) + H_2O(\ell) \qquad E° = + 0.80 V$$

Concentrated nitric acid attacks and oxidizes most metals. (Aluminum is an exception; see page 1037.) In this process, the nitrate ion is reduced to one of the nitrogen oxides. Which oxide is formed depends on the metal and on reaction conditions. In the case of copper, for example, either NO or NO_2 is produced, depending on the concentration of the acid (Figure 21.23b).

In dilute acid:

$$3 Cu(s) + 8 H^+(aq) + 2 NO_3^-(aq) \longrightarrow 3 Cu^{2+}(aq) + 4 H_2O(\ell) + 2 NO(g)$$

In concentrated acid:

$$Cu(s) + 4 H^+(aq) + 2 NO_3^-(aq) \longrightarrow Cu^{2+}(aq) + 2 H_2O(\ell) + 2 NO_2(g)$$

Four metals (Au, Pt, Rh, and Ir) that are not attacked by nitric acid are often described as the "noble metals." The alchemists of the 14th century, however, knew that if they mixed HNO_3 with HCl in a ratio of about $1:3$, this *aqua regia*, or "kingly water," would attack even gold, the noblest of metals.

$$10 Au(s) + 6 NO_3^-(aq) + 40 Cl^-(aq) + 36 H^+(aq) \longrightarrow$$
$$10 [AuCl_4]^-(aq) + 3 N_2(g) + 18 H_2O(\ell)$$

Exercise 21.10—Nitrogen Oxide Chemistry

Dinitrogen monoxide can be made by the decomposition of NH_4NO_3.

(a) A Lewis electron dot structure of N_2O is given in Table 21.6. Is it the only possible structure? If other structures are possible, is the one in Table 21.6 the most important?

(b) Is the decomposition of $NH_4NO_3(s)$ to give $N_2O(g)$ and $H_2O(g)$ endothermic or exothermic?

Hydrogen Compounds of Phosphorus and Other Group 5A Elements

The phosphorus analog of ammonia, phosphine (PH_3), is a poisonous, highly reactive gas with a faint garlic-like odor. Industrially, it is made by the alkaline hydrolysis of white phosphorus.

$$P_4(s) + 3 KOH(aq) + 3 H_2O(\ell) \longrightarrow PH_3(g) + 3 KH_2PO_2(aq)$$

The other hydrides of the heavier Group 5A elements are exceedingly toxic and become more unstable as the atomic number of the element increases. Nonetheless, arsine (AsH_3), is used in the semiconductor industry as a starting material in the preparation of gallium arsenide (GaAs), semiconductors.

Phosphorus Oxides and Sulfides

The most important compounds of phosphorus are those with oxygen, and there are at least six simple binary compounds containing just phosphorus and oxygen. All of them can be thought of as being derived structurally from the P_4 tetrahe-

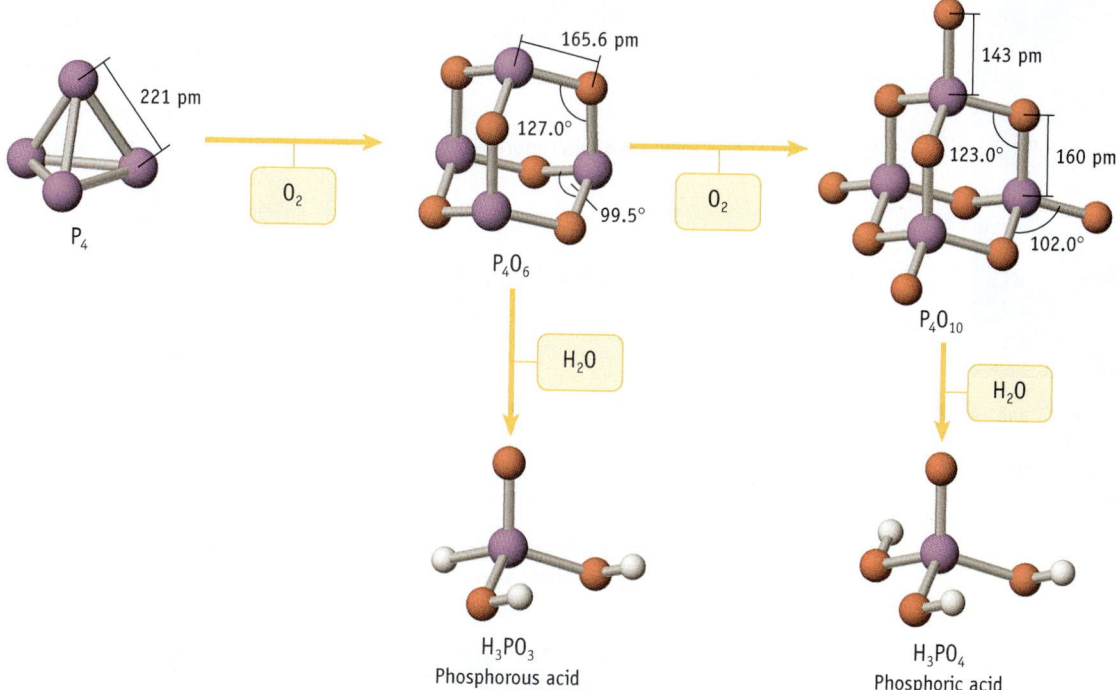

$$P_4 \quad \xrightarrow{O_2} \quad P_4O_6 \quad \xrightarrow{O_2} \quad P_4O_{10}$$

H₃PO₃
Phosphorous acid

H₃PO₄
Phosphoric acid

Figure 21.24 **Phosphorus oxides.** Other binary P—O compounds have formulas between P_4O_6 and P_4O_{10}. They are formed by starting with P_4O_6 and adding O atoms successively to the P atom vertices.

dron of white phosphorus. For example, if P_4 is carefully oxidized, P_4O_6 is formed; an O atom has been inserted into each P—P bond in the tetrahedron (Figure 21.24).

The most common and important phosphorus oxide is P_4O_{10}, a fine white powder commonly called "phosphorus pentaoxide" because its empirical formula is P_2O_5. In P_4O_{10} each phosphorus atom is surrounded tetrahedrally by O atoms.

Unlike nitrogen, phosphorus also forms a series of compounds with sulfur. Of these, the most important is P_4S_3.

$$4\ P(s, \text{red allotrope}) + \tfrac{3}{8}\ S_8(s) \longrightarrow$$

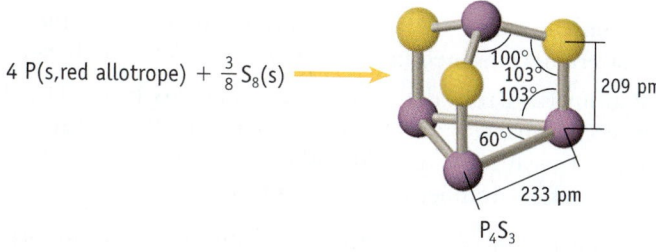

P_4S_3

In this phosphorus sulfide, S atoms are inserted into only three of the P—P bonds. The principal use of P_4S_3 is in "strike anywhere" matches, the kind that light when you rub the head against a rough object. The active ingredients are P_4S_3 and the powerful oxidizing agent potassium chlorate, $KClO_3$. The "safety match" is now more common than the "strike anywhere" match. In safety matches the head is predominantly $KClO_3$, and the material on the match book is red phosphorus (about 50%), Sb_2S_3, Fe_2O_3, and glue.

Charles D. Winters

Matches. The head of a "strike anywhere" match contains P_4S_3 and the oxidizing agent $KClO_3$. (Other components are ground glass, Fe_2O_3, ZnO, and glue.) Safety matches have sulfur (3–5%) and $KClO_3$ (45–55%) in the match head and red phosphorus in the striking strip.

Figure 21.25   Reaction of P₄O₁₀ and water. The white solid oxide reacts vigorously with water to give orthophosphoric acid, H_3PO_4. (The heat generated vaporizes the water, so steam is visible.)

Charles D. Winters

Table 21.7   Phosphorus Oxoacids

Formula	Name	Structure
H_3PO_4	Orthophosphoric acid	
$H_4P_2O_7$	Pyrophosphoric acid (diphosphoric acid)	
$(HPO_3)_3$	Metaphosphoric acid	
H_3PO_3	Phosphorous acid (phosphonic acid)	
H_3PO_2	Hypophosphorous acid (phosphinic acid)	

Phosphorus Oxoacids and Their Salts

A few of the many known phosphorus oxoacids are illustrated in Table 21.7. Indeed, there are so many acids and their salts in this category that structural principles have been developed to organize and understand them.

(a) All P atoms in the oxoacids and their anions (conjugate bases) are four-coordinate and tetrahedral.

(b) All the P atoms in the acids have at least one P—OH group (and this occurs often in the anions as well). In every case the H is ionizable as H^+.

(c) Some oxoacids have one or more P—H bonds. This H atom is not ionizable as H^+.

(d) Polymerization can occur by P—O—P to give both linear and cyclic species. Two P atoms are never joined by more than one P—O—P bridge.

(e) When a P atom is surrounded only by O atoms (as in H_3PO_4), its oxidation number is +5. For each P—OH that is replaced by P—H, the oxidation number drops by 2 (because P is considered more electronegative than H). For example, the oxidation number of P in H_3PO_2 is +1.

Orthophosphoric acid, H_3PO_4, and its salts are far more important commercially than other P-O acids. Millions of tons of phosphoric acid are made annually, some using white phosphorus as the starting material. The phosphorus is burned in oxygen to give P_4O_{10}, and the oxide reacts with water to produce the acid (Figure 21.25).

$$P_4O_{10}(s) + 6\ H_2O(\ell) \longrightarrow 4\ H_3PO_4(aq)$$

This approach gives a pure product, so it is employed to make phosphoric acid for use in food products in particular. When the pure acid is dilute, it is nontoxic and

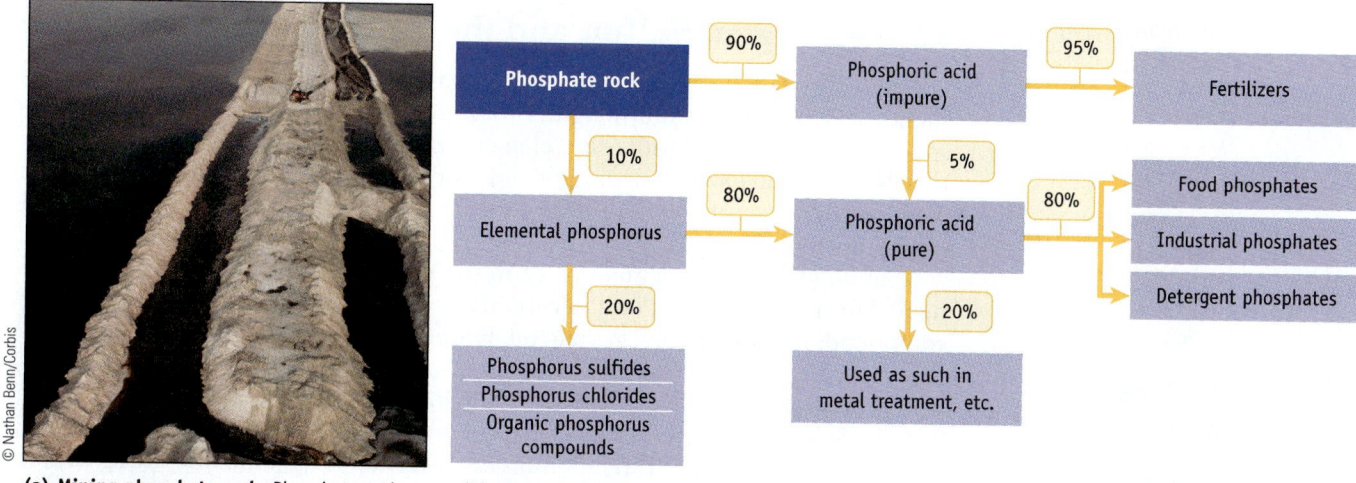

(a) Mining phosphate rock. Phosphate rock is primarily $Ca_3(PO_4)_2$, and most mined in the United States comes from Florida.

(b) Uses of phosphorus and phosphoric acid.

Figure 21.26 Uses of phosphate rock, phosphorus, and phosphoric acid.

gives the tart or sour taste to carbonated "soft drinks," such as various colas (about 0.05% H_3PO_4) or root beer (about 0.01% H_3PO_4).

As illustrated in Figure 21.26b, a major use for phosphoric acid is to impart corrosion resistance to metal objects such as nuts and bolts, tools, and car-engine parts by plunging the object into a hot acid bath. Car bodies are similarly treated with phosphoric acid containing metal ions such as Zn^{2+}, and aluminum trim is "polished" by treating it with the acid.

The protons of H_3PO_4 can be removed to produce salts such as NaH_2PO_4, Na_2HPO_4, and Na_3PO_4. In industry, the monosodium and disodium salts are produced using Na_2CO_3 as the base, but an excess of the stronger (and more expensive) base NaOH is required to remove the third proton to give Na_3PO_4.

Sodium phosphate (Na_3PO_4) is used in scouring powders and paint strippers because the anion PO_4^{3-} is a relatively strong base in water ($K_b = 2.8 \times 10^{-2}$). Sodium monohydrogen phosphate, Na_2HPO_4, which has a less basic anion than PO_4^{3-}, is widely used in food products. Kraft has patented a process using the salt in the manufacture of pasteurized cheese, for example. Thousands of tons of Na_2HPO_4 are still used for this purpose, even though the function of the salt in this process is not completely understood. In addition, a small amount of Na_2HPO_4 in pudding mixes enables the mix to gel in cold water, and the basic anion raises the pH of cereals to provide "quick-cooking" breakfast cereal. (Apparently, the OH^- ion from HPO_4^{2-} hydrolysis accelerates the breakdown of the cellulose material in the cereal.)

Calcium phosphates are used in a broad spectrum of products. For example, the weak acid $Ca(H_2PO_4)_2 \cdot H_2O$ is used as the acid leavening agent in baking powder. A typical baking powder contains (along with inert ingredients) 28% $NaHCO_3$, 10.7% $Ca(H_2PO_4)_2 \cdot H_2O$, and 21.4% $NaAl(SO_4)_2$ (also a weak acid). The weak acids react with sodium bicarbonate to produce CO_2 gas. For example,

$$Ca(H_2PO_4)_2 \cdot H_2O(s) + 2\ NaHCO_3(aq) \longrightarrow$$
$$2\ CO_2(g) + 3\ H_2O(\ell) + Na_2HPO_4(aq) + CaHPO_4(aq)$$

Finally, calcium monophosphate, $CaHPO_4$, is used as an abrasive and polishing agent in toothpaste.

Arsenic in drinking water. In the early 1990s, hundreds of people in the very poor country of Bangladesh became ill with what some thought was leprosy. Soon, however, the problem was recognized as arsenic poisoning. An attempt by world health organizations to alleviate the problem of unsafe drinking water in Bangladesh had gone terribly wrong. Arsenic contamination was found in the shallow "tube wells" dug in the 1980s that served millions of people. The problem and its potential solutions are now the subject of intense study by chemists, geologists, and health workers from around the world.

Group 6A

Oxygen
8
O
474,000 ppm
Sulfur
16
S
260 ppm
Selenium
34
Se
0.5 ppm
Tellurium
52
Te
0.005 ppm
Polonium
84
Po
trace

Element abundances are in parts per million in the earth's crust.

21.9—Oxygen, Sulfur, and the Group 6A Elements

Oxygen is by far the most abundant element in the earth's crust, representing slightly less than 50% of it by weight. It is present as elemental oxygen in the atmosphere and is combined with other elements in water and in many minerals. Scientists believe that elemental oxygen did not appear on this planet until about 2 billion years ago, when it was formed on the planet by plants through the process of photosynthesis.

Sulfur, seventeenth in abundance in the earth's crust, is also found in its elemental form in nature, but only in certain concentrated deposits. Sulfur-containing compounds occur in natural gas and oil. In minerals, sulfur occurs as the sulfide ion (for example, in cinnabar, HgS, and galena, PbS), as the disulfide ion (in iron pyrite, FeS_2, or "fool's gold"), and as sulfate ion (for example, in gypsum, $CaSO_4 \cdot 2 H_2O$). Sulfur oxides (SO_2 and SO_3) also occur in nature, primarily as products of volcanic activity (Figure 21.27). Sulfur chemistry supports some interesting life, as described in the story at the beginning of this chapter.

In the United States, most sulfur—about 10 million tons per year—is obtained from deposits of the element found along the Gulf of Mexico. These deposits occur typically at a depth of 150 to 750 m below the surface in layers about 30 m thick. They are thought to have been formed by anaerobic ("without free oxygen") bacteria acting on sedimentary sulfate deposits such as gypsum.

Preparation and Properties of the Elements

Pure oxygen is obtained by fractionation of air and is among the top five industrial chemicals produced in the United States. Oxygen can be made in the laboratory by electrolysis of water (Figure 21.4) and by the catalyzed decomposition of metal chlorates such as $KClO_3$.

$$2 \ KClO_3(s) \xrightarrow{\text{catalyst}} 2 \ KCl(s) + 3 \ O_2(g)$$

At room temperature and pressure, oxygen is a colorless gas, but it is pale blue when condensed to the liquid at $-183\ °C$ (see Figure 10.16). As described in Section 10.3, diatomic oxygen is paramagnetic because it has two unpaired electrons.

An allotrope of oxygen, ozone (O_3), is a blue, diamagnetic gas with an odor so strong that it can be detected in concentrations as low as 0.05 ppm. Ozone is synthesized by passing O_2 through an electric discharge or by irradiating O_2 with ultraviolet light. Ozone is in the news regularly because of the realization that the earth's protective layer of ozone in the stratosphere is being disrupted by chlorofluorocarbons and other chemicals [◄ page 1008].

Sulfur has numerous allotropes. The most common and most stable allotrope is the yellow, orthorhombic form, which consists of S_8 molecules with the sulfur atoms arranged in a crown-shaped ring (Figure 21.28a). Less stable allotropes are known that have rings of 6 to 20 sulfur atoms. Another form of sulfur, called *plastic sulfur*, has a molecular structure with chains of sulfur atoms (Figure 21.28b).

Sulfur is obtained from underground deposits by a process developed by Herman Frasch (1851–1914) about 1900. Superheated water (at 165 °C) and then air are forced into the deposit. The sulfur melts (mp, 113 °C) and is forced to the surface as a frothy, yellow stream, from which it solidifies.

Selenium and tellurium are comparatively rare on earth, having abundances about the same as those of silver and gold, respectively. Because their chemistry is so similar to that of sulfur, they are often found in minerals associated with the sulfides of copper, silver, iron, and arsenic, and they are recovered as by-products of the industries devoted to those metals.

© Ludovic Maisant/Corbis

Figure 21.27 **Sulfur spewing from a volcano in Indonesia.**

Figure 21.28 Sulfur allotropes. (a) At room temperature, sulfur exists as a bright yellow solid composed of S_8 rings. (b) When heated, the rings break open, and eventually form chains of S atoms in a material described as "plastic sulfur."

(a) (b)

Selenium has a range of uses, including in glass making. A cadmium sulfide/selenide mixture is added to glass to give it a brilliant red color (Figure 21.29a). The most familiar use of selenium is in xerography, a word meaning "dry printing" and a process at the heart of the modern copy machine. Most photocopy machines use an aluminum plate or roller coated with selenium. Light coming from the imaging lens selectively discharges a static electric charge on the selenium film, and the black toner sticks only on the areas that remain charged. A copy is made when the toner is transferred to a sheet of plain paper.

The heaviest element of Group 6A, polonium, is radioactive and found only in trace amounts on earth. It was discovered in Paris, France, in 1898 by Marie Sklodowska Curie (1867–1934) and her husband Pierre Curie (page 65). The Curies painstakingly separated the elements in a large quantity of a uranium-containing ore, pitchblende, and found the new elements radium and polonium.

GENERAL
Chemistry ·ᐡ· Now™

See the General ChemistryNow CD-ROM or website:
• **Screen 21.8 Sulfur Allotropes,** for an exercise on sulfur chemistry

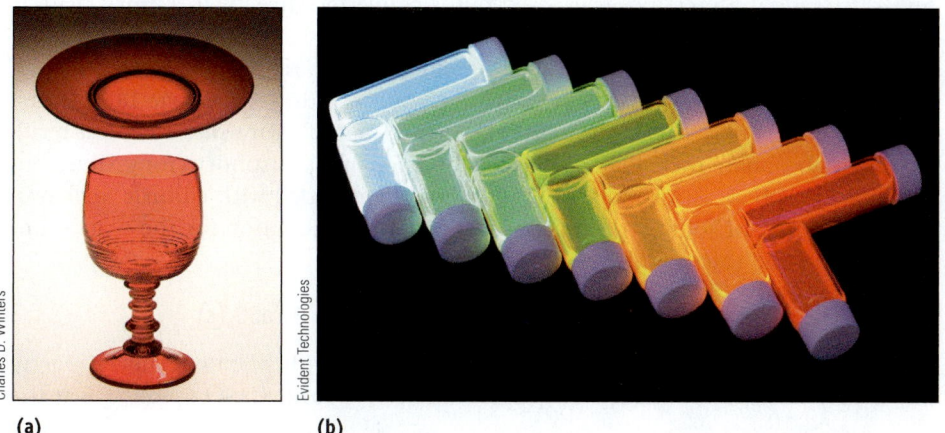

Figure 21.29 Uses of selenium.
(a) Glass takes on a brilliant red color when a mixture of cadmium sulfide/selenide (CdS, CdSe) is added to it. (b) These sample bottles hold suspensions of quantum dots, nonometer-sized crystals of CdSe dispersed in a polymer matrix. The crystals emit light in the visible range when excited by ultraviolet light. Light emission at different wavelengths is achieved by changing the particle size. Crystals of PbS and PbSe can be made that emit light in the infrared range.

(a) (b)

Sulfur Compounds

Hydrogen sulfide, H_2S, has a bent molecular geometry, like water. Unlike water, however, H_2S is a gas under standard conditions (mp, -85.6 °C; bp, -60.3 °C) because its intermolecular forces are weak compared with the strong hydrogen bonding in water (see Figure 13.8). Hydrogen sulfide is poisonous, comparable in toxicity to hydrogen cyanide, but fortunately it has a terrible odor and is detectable in concentrations as low as 0.02 ppm. You must be careful with H_2S, because it has an anesthetic effect; your nose rapidly loses its ability to detect it. Death occurs at H_2S concentrations of 100 ppm.

Sulfur is often found as the sulfide ion in conjunction with metals because all metal sulfides (except those based on Group 1A metals) are water-insoluble. The recovery of metals from their sulfide ores usually begins by heating the ore in air. This converts the metal sulfide to either a metal oxide or the metal itself, with the sulfur appearing as SO_2 as in the reactions of zinc or lead sulfide with oxygen.

$$2\ ZnS(s) + 3\ O_2(g) \longrightarrow 2\ ZnO(s) + 2\ SO_2(g)$$
$$2\ PbO(s) + PbS(s) \longrightarrow 3\ Pb(s) + SO_2(g)$$

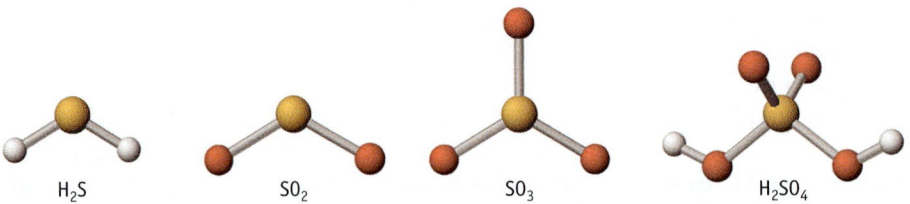

H_2S SO_2 SO_3 H_2SO_4

Models of some common sulfur-containing molecules: H_2S, SO_2, SO_3, and H_2SO_4.

Sulfur dioxide (SO_2) is produced on an enormous scale by the combustion of sulfur and by roasting sulfide ores in air. The combustion of sulfur in sulfur-containing coal and fuel oil creates particularly large environmental problems. It has been estimated that about 2.0×10^8 tons of sulfur oxides are released into the atmosphere each year by human activities, primarily in the form of SO_2; this is more than half of the total emitted by all other natural sources of sulfur in the environment.

Sulfur dioxide is a colorless, toxic gas with a sharp odor. It readily dissolves in water. The most important reaction of this gas is its oxidation to SO_3.

$$SO_2(g) + \tfrac{1}{2} O_2(g) \longrightarrow SO_3(g) \qquad \Delta H° = -98.9 \text{ kJ/mol}$$

Sulfur trioxide is almost never isolated but is converted directly to sulfuric acid by reaction with water in the "contact process."

The largest use of sulfur is the production of sulfuric acid, H_2SO_4, the compound produced in largest quantity by the chemical industry [◀ page 187]. In the United States, roughly 70% of the acid is used to manufacture *superphosphate* fertilizer from phosphate rock. Plants need a soluble form of phosphorus for growth, but calcium phosphate is insoluble. Treating phosphate rock with sulfuric acid produces a mixture of soluble phosphates. The balanced equation for the reaction of sulfuric acid and calcium phosphate is

$$Ca_3(PO_4)_2(s) + 3\ H_2SO_4(\ell) \longrightarrow 2\ H_3PO_4(\ell) + 3\ CaSO_4(s)$$

but it does not tell the whole story. Concentrated superphosphate fertilizer is actually mostly $CaHPO_4$ or $Ca(H_2PO_4)_2$ plus some H_3PO_4 and $CaSO_4$. Notice that the

Common household products containing sulfur or sulfur-based compounds.

Charles D. Winters

chemical principle behind this reaction is that sulfuric acid is a stronger acid than H_3PO_4 (Table 17.3), so the PO_4^{3-} ion is protonated by sulfuric acid.

Smaller amounts of sulfuric acid are used in the conversion of ilmenite, a titanium-bearing ore, to TiO_2, which is then used as a white pigment in paint, plastics, and paper. The acid is also used to make iron and steel, petroleum products, synthetic polymers, and paper.

Exercise 21.11—Sulfur Chemistry

Metal sulfides roasted in air produce metal oxides.

$$2\ ZnS(s) + 3\ O_2(g) \longrightarrow 2\ ZnO(s) + 2\ SO_2(g)$$

Use thermodynamics to decide if the reaction is product- or reactant-favored at 298 K. Will the reaction be more or less product-favored at a high temperature?

21.10—The Halogens, Group 7A

Fluorine and chlorine are the most abundant halogens in the earth's crust, with fluorine somewhat more abundant than chlorine. If their abundance in sea water is measured, however, the situation is quite different. Chlorine has an abundance in sea water of 18,000 ppm, whereas the abundance of fluorine in the same source is only 1.3 ppm. This variation undoubtedly reflects the differences in the solubility of their salts and plays a role in the methods used to recover the elements themselves.

Preparation of the Elements

Fluorine

The water-insoluble mineral *fluorspar*, or calcium fluoride, CaF_2, is one of the many sources of fluorine. Because the mineral was originally used as a flux in metalworking, its name comes from the Latin word meaning "to flow." In the 17th century it was discovered that solid CaF_2 would emit light when heated, and the phenomenon was called *fluorescence*. Thus, in the early 1800s when it was recognized that a new element was contained in fluorspar, A. Ampère suggested that the element be called *fluorine*.

Although fluorine was recognized as an element by 1812, it was not until 1886 that it was isolated by Moisson in elemental form as a colorless gas by the electrolysis of KF dissolved in anhydrous HF. Indeed, because F_2 is such a powerful oxidizing agent, chemical oxidation of F^- to F_2 is not feasible, and electrolysis is the only practical way to obtain gaseous F_2 (Figure 21.30).

The preparation of F_2 is difficult because F_2 oxidizes (corrodes) the equipment and reacts violently with traces of grease or other contaminants. Furthermore, the products of electrolysis, F_2 and H_2, can recombine explosively, so they must be separated carefully. Current U.S. capacity is approximately 5000 metric tons per year.

Group 7A Halogens

| Fluorine 9 **F** 950 ppm |
| Chlorine 17 **Cl** 130 ppm |
| Bromine 35 **Br** 0.37 ppm |
| Iodine 53 **I** 0.14 ppm |
| Astatine 85 **At** trace |

Element abundances are in parts per million in the earth's crust.

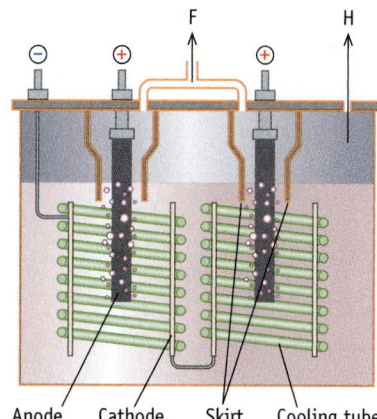

Figure 21.30 Schematic of an electrolysis cell for producing fluorine.

Figure 21.31   Chlorine preparation. Chlorine is prepared by oxidation of chloride ion using a strong oxidizing agent. Here, oxidation of NaCl is accomplished using $K_2Cr_2O_7$ in H_2SO_4. (The Cl_2 gas is bubbled into water in a receiving flask.)

Charles D. Winters

Chlorine

Although chlorine is an excellent oxidizing agent, more powerful oxidizing agents are capable of oxidizing Cl^- ion to Cl_2. In fact, elemental chlorine was first made by the Swedish chemist Karl Wilhelm Scheele (1742–1786) in 1774, who combined sodium chloride with an oxidizing agent in an acidic solution (Figure 21.31).

Industrially, chlorine is made by electrolysis of brine (concentrated aqueous NaCl). The other product of the electrolysis, NaOH, is also a valuable industrial chemical. About 80% of the chlorine produced is made using an electrochemical cell similar to the one depicted in Figure 21.32. Oxidation of chloride ion to Cl_2 gas occurs at the anode and reduction of water occurs at the cathode.

Anode reaction (oxidation) $2\ Cl^-(aq) \longrightarrow Cl_2(g) + 2\ e^-$
Cathode reaction (reduction) $2\ H_2O(\ell) + 2\ e^- \longrightarrow H_2(g) + 2\ OH^-(aq)$

Activated titanium is used for the anode and stainless steel or nickel is preferred for the cathode. The membrane separating the anode and cathode compartments is not permeable to water, but it does allow Na^+ ions to pass so as to maintain the charge balance. Thus the membrane functions as a "salt" bridge between the anode and cathode compartments. The energy consumption of these cells is in the range of 2000–2500 kWh per ton of NaOH produced.

Bromine

All halogens are oxidizing agents, but that ability declines as they become heavier (see Table 20.1).

Half-Reaction	Reduction Potential ($E°$, V)
$F_2(g) + 2\ e^- \longrightarrow 2\ F^-(aq)$	2.87
$Cl_2(g) + 2\ e^- \longrightarrow 2\ Cl^-(aq)$	1.36
$Br_2(\ell) + 2\ e^- \longrightarrow 2\ Br^-(aq)$	1.08
$I_2(s) + 2\ e^- \longrightarrow 2\ I^-(aq)$	0.535

This means that Cl_2 will oxidize Br^- ions to Br_2 in aqueous solution.

$$Cl_2(aq) + 2\ Br^-(aq) \longrightarrow 2\ Cl^-(aq) + Br_2(aq)$$
$$E°_{net} = E°_{cathode} - E°_{anode} = 1.36\ V - (1.08\ V) = +0.28\ V$$

In fact, this is the commercial method of preparing bromine when NaBr is obtained from natural brine wells in Arkansas and Michigan.

Iodine

Iodine is a lustrous, purple-black solid, easily sublimed at room temperature and atmospheric pressure (Figure 13.38). The element was first isolated in 1811 from seaweed and kelp, extracts of which had long been used for treatment of goiter, the enlargement of the thyroid gland. It is now known that the thyroid gland produces a growth-regulating hormone (thyroxine) that contains iodine [▶ page 1108]. Most table salt in the United States has 0.01% NaI added to provide the necessary iodine in the diet.

A laboratory method for preparing I_2 is illustrated in Figure 21.33. The commercial preparation, however, depends on the source of I^- and its concentration. One method is interesting because it involves some chemistry described in this book. Iodide ions are first precipitated with silver ions to give insoluble AgI.

$$I^-(aq) + Ag^+(aq) \longrightarrow AgI(s)$$

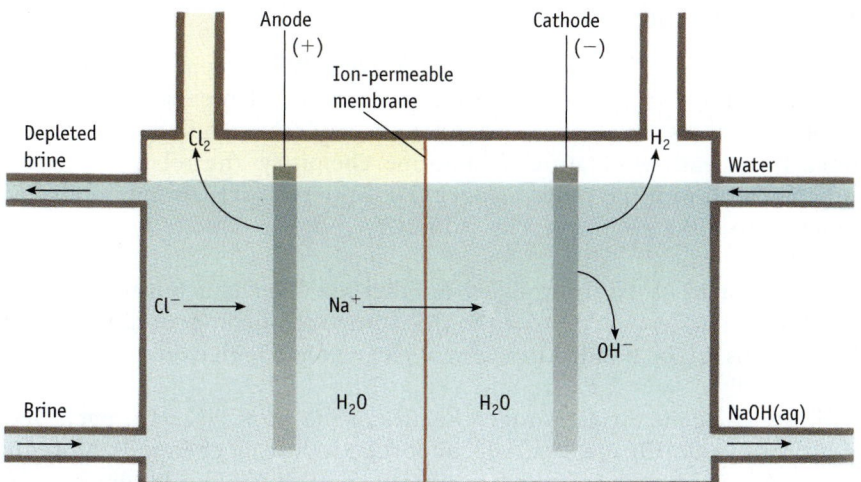

Active Figure 21.32 **A membrane cell for the production of NaOH and Cl₂ gas from a saturated, aqueous solution of NaCl (brine).** Here the anode and cathode compartments are separated by a water-impermeable but ion-conducting membrane. A widely used membrane is made of Nafion, a fluorine-containing polymer that is a relative of polytetrafluoroethylene (Teflon). Brine is fed into the anode compartment and dilute sodium hydroxide or water into the cathode compartment. Overflow pipes carry the evolved gases and NaOH away from the chambers of the electrolysis cell.

GENERAL
Chemistry ·⚛·Now™ See the General ChemistryNow CD-ROM or website to explore an interactive version of this figure accompanied by an exercise.

This is reduced by clean scrap iron to give iron(II) iodide and metallic silver.

$$2\ \text{AgI}(s) + \text{Fe}(s) \longrightarrow \text{FeI}_2(aq) + 2\ \text{Ag}(s)$$

(Because silver is expensive, it is recycled by oxidizing the metal with nitric acid to silver nitrate, which is then reused.) Finally, iodide ion from water-soluble FeI₂ is oxidized to iodine with chlorine [with iron(III) chloride as a by-product].

$$2\ \text{FeI}_2(aq) + 3\ \text{Cl}_2(aq) \longrightarrow 2\ \text{I}_2(s) + 2\ \text{FeCl}_3(aq)$$

Figure 21.33 **The preparation of iodine.** A mixture of sodium iodide and manganese(IV) oxide was placed in the flask (*left*). On adding concentrated sulfuric acid (*right*), brown iodine vapor was evolved.

$$2\ \text{NaI}(s) + 2\ \text{H}_2\text{SO}_4(aq) + \text{MnO}_2(s) \longrightarrow$$
$$\text{Na}_2\text{SO}_4(aq) + \text{MnSO}_4(aq) + 2\ \text{H}_2\text{O}(l) + \text{I}_2(g)$$

Bond Energies of Some Halogen Compounds (kJ/mol)

X	X—X	H—X	C—X (in CX$_4$)
F	155	565	485
Cl	242	432	339
Br	193	366	285
I	151	299	213

Fluorine Compounds

Fluorine is the most reactive of all of the elements, forming compounds with every element except He and Ne. In most cases the elements combine directly, and some reactions can be so vigorous as to be explosive. This reactivity can be explained by at least two features of fluorine chemistry: the relatively weak F—F bond compared with the other halogens, and the relatively strong bonds formed by fluorine to other elements. This is illustrated by the table of bond energies in the margin.

In addition to its oxidizing ability, another notable characteristic of fluorine is its small size. These properties lead to the formation of compounds where a number of F atoms can be bonded to a central element in a high oxidation state. Examples include PtF_6, AgF_2, UF_6, IF_7, and XeF_4.

All the halogens form hydrogen halides of the form HX. Hydrogen fluoride and hydrogen chloride are especially important industrial chemicals. More than 1 million tons of hydrogen fluoride is produced annually worldwide, almost all by the action of concentrated sulfuric acid on fluorspar.

$$CaF_2(s) + H_2SO_4(\ell) \longrightarrow CaSO_4(s) + 2\ HF(g)$$

The U.S. capacity for HF production is approximately 208,000 metric tons, but currently demand is exceeding supply for this chemical. Anhydrous HF is used in a broad range of industries: in the production of refrigerants, herbicides, pharmaceuticals, high-octane gasoline, aluminum, plastics, electrical components, and fluorescent lightbulbs.

The fluorspar used to produce HF must be very pure and free of SiO_2 because HF reacts readily with silicon dioxide.

$$SiO_2(s) + 4\ HF(g) \longrightarrow SiF_4(g) + 2\ H_2O(\ell)$$
$$SiF_4(g) + 2\ HF(aq) \longrightarrow H_2SiF_6(aq)$$

This series of reactions explains why HF can be used to etch or frost glass (such as the inside of fluorescent light bulbs). It also explains why HF is not shipped in glass containers (unlike HCl, for example).

The aluminum industry consumes about 10–40 kg of cryolite, Na_3AlF_6, per metric ton of aluminum produced. The reason is that cryolite must be added to aluminum oxide to produce a lower-melting mixture that can be electrolyzed. Cryolite is found in only small quantities in nature, so it is made in various ways, among them the following reaction:

$$6\ HF(aq) + Al(OH)_3(s) + 3\ NaOH(aq) \longrightarrow Na_3AlF_6(s) + 6\ H_2O(\ell)$$

About 3% of the hydrogen fluoride produced is used in uranium fuel production. Naturally occurring uranium is processed to give UO_2. To separate uranium isotopes in a gas centrifuge (Figure 12.20), the uranium must be in the form of a volatile compound. Therefore, uranium oxide is treated with hydrogen fluoride to give UF_4, which is then reacted with F_2 to produce the volatile solid UF_6.

$$UO_2(s) + 4\ HF(aq) \longrightarrow UF_4(s) + 2\ H_2O(\ell)$$
$$UF_4(s) + F_2(g) \longrightarrow UF_6(s)$$

This last step consumes 70–80% of fluorine production.

Chlorine Compounds

Hydrogen Chloride

Hydrochloric acid, an aqueous solution of hydrogen chloride, is a valuable industrial chemical. Hydrogen chloride gas can be prepared by the reaction of hydrogen and chlorine, but the rapid, exothermic reaction is difficult to control. The classical method of making HCl in the laboratory uses the reaction of NaCl and sulfuric acid, a procedure that takes advantage of the facts that HCl is a gas and that H_2SO_4 will not oxidize the chloride ion.

$$2\ NaCl(s) + H_2SO_4(\ell) \longrightarrow Na_2SO_4(s) + 2\ HCl(g)$$

Hydrogen chloride gas has a sharp, irritating odor. Both gaseous and aqueous HCl react with metals and metal oxides to give metal chlorides and, depending on the reactant, hydrogen or water.

$$Mg(s) + 2\ HCl(aq) \longrightarrow MgCl_2(aq) + H_2(g)$$
$$ZnO(s) + 2\ HCl(aq) \longrightarrow ZnCl_2(aq) + H_2O(g)$$

Oxoacids of Chlorine

Oxoacids of chlorine range from HClO, in which chlorine has an oxidation number of +1, to $HClO_4$, in which the oxidation number is equal to the group number, +7. All are strong oxidizing agents.

Oxoacids of Chlorine			
Acid	Name	Anion	Name
HClO	Hypochlorous	ClO^-	Hypochlorite
$HClO_2$	Chlorous	ClO_2^-	Chlorite
$HClO_3$	Chloric	ClO_3^-	Chlorate
$HClO_4$	Perchloric	ClO_4^-	Perchlorate

Hypochlorous acid, HClO, forms when chlorine dissolves in water. In this reaction, half of the chlorine is oxidized to hypochlorite ion and half is reduced to chloride ion in a **disproportionation reaction.**

$$Cl_2(g) + 2\ H_2O(\ell) \rightleftharpoons H_3O^+(aq) + HClO(aq) + Cl^-(aq)$$

Chlorine, chloride ion, and hypochlorous acid exist in equilibrium. A low pH favors the reactants, whereas a high pH favors the products. Therefore, if Cl_2 is dissolved in cold aqueous NaOH instead of in pure water, hypochlorite ion and chloride ion form.

$$Cl_2(g) + 2\ OH^-(aq) \rightleftharpoons ClO^-(aq) + Cl^-(aq) + H_2O(\ell)$$

Under basic conditions, the equilibrium shifts far to the right. The resulting alkaline solution is the "liquid bleach" used in home laundries. The bleaching action of this solution is a result of the oxidizing ability of ClO^-. Most dyes are colored organic compounds, and hypochlorite ion oxidizes dyes to colorless products.

■ **Disproportionation**
A reaction in which an element or compound is simultaneously oxidized and reduced is called a *disproportionation reaction.* Here Cl_2 is oxidized to ClO^- and reduced to Cl^-.

Use of a perchlorate. The solid-fuel booster rockets of the Space Shuttle utilize a mixture of NH_4ClO_4 (oxidizing agent) and Al powder (reducing agent).

When calcium hydroxide is combined with Cl_2, solid $Ca(ClO)_2$ is the product. This compound is easily handled and is the "chlorine" that is sold for swimming pool disinfection.

When a basic solution of hypochlorite ion is heated, disproportionation again occurs, forming chlorate ion and chloride ion:

$$3\ ClO^-(aq) \longrightarrow ClO_3^-(aq) + 2\ Cl^-(aq)$$

Sodium and potassium chlorates are made in large quantities via this reaction. The sodium salt can be reduced to ClO_2, a compound used for bleaching paper pulp. Some $NaClO_3$ is also converted to potassium chlorate, $KClO_3$, the preferred oxidizing agent in fireworks and a component of safety matches.

Perchlorates, salts containing ClO_4^-, are powerful oxidants. Pure perchloric acid, $HClO_4$, is a colorless liquid that explodes if shocked. It explosively oxidizes organic materials and rapidly oxidizes silver and gold. Dilute aqueous solutions of the acid are safe to handle, however.

Perchlorate salts of most metals are usually relatively stable, albeit unpredictable. Great care should be used when handling any perchlorate salt. Ammonium perchlorate, for example, bursts into flame if heated above 200 °C.

$$2\ NH_4ClO_4(s) \longrightarrow N_2(g) + Cl_2(g) + 2\ O_2(g) + 4\ H_2O(g)$$

The strong oxidizing ability of the ammonium salt accounts for its use as the oxidizer in the solid booster rockets for the Space Shuttle. The solid propellant in these rockets is largely NH_4ClO_4, the remainder being the reducing agent, powdered aluminum. Each launch requires about 750 tons of ammonium perchlorate, and more than half of the sodium perchlorate currently manufactured is converted to the ammonium salt. The process for making this conversion is an exchange reaction that takes advantage of the fact that ammonium perchlorate is less soluble in water than sodium perchlorate:

$$NaClO_4(aq) + NH_4Cl(aq) \longrightarrow NaCl(aq) + NH_4ClO_4(s)$$

Exercise 21.12—Reactions of the Halogen Acids

Metals generally react with hydrogen halides such as HCl to give the metal halide and hydrogen.

$$Ag(s) + HCl(g) \longrightarrow AgCl(s) + \tfrac{1}{2} H_2(g)$$

The reaction is thermodynamically product-favored if ΔG°_{rxn} is negative. Is this true for all of the hydrogen halides reacting with silver? The required free energies of formation are (in kJ/mol)

HX, ΔG°_f (kJ/mol)	AgX, ΔG°_f (kJ/mol)
HF, −273.2	AgF, −193.8
HCl, −95.09	AgCl, −109.76
HBr, −53.45	AgBr, −96.90
HI, +1.56	AgI, −66.19

Chapter Goals Revisited

When you have finished studying this chapter, you should ask whether you have met the chapter goals. In particular, you should be able to

Relate the formulas and properties of compounds to the periodic table

a. Predict several chemical reactions of the Group A elements (Section 21.2).

b. Predict similarities and differences among the elements in a given group, based on the periodic properties (Section 21.2).

c. Know which reactions produce ionic compounds, and predict formulas for common ions and common ionic compounds based on electron configurations (Section 21.2).

d. Recognize when a formula is incorrectly written, based on general principles governing electron configurations (Section 21.2).

Describe the chemistry of the main group or A-Group elements, particularly H; Na and K; Mg and Ca; B and Al; Si; N and P; O and S; and F and Cl

a. Identify the most abundant elements, know how they are obtained, and list some of their common chemical and physical properties.

b. Be able to summarize briefly a series of facts about the most common compounds of main group elements (ionic or covalent bonding, color, solubility, simple reaction chemistry) (Sections 21.3–21.10).

c. Identify uses of common elements and compounds, and understand the chemistry that relates to their usage (Sections 21.3–21.10).

Study Questions

▲ denotes more challenging questions.

■ denotes questions available in the Homework and Goals section of the General ChemistryNow CD-ROM or website.

Blue numbered questions have answers in Appendix O and fully worked solutions in the *Student Solutions Manual*.

Structures of many of the compounds used in these questions are found on the General ChemistryNow CD-ROM or website in the Models folder.

GENERAL
Chemistry⚛Now™ Assess your understanding of this chapter's topics with additional quizzing and conceptual questions at **http://now.brookscole.com/kotz6e**

Practicing Skills

Properties of the Elements

1. Give examples of two basic oxides. Write equations illustrating the formation of each oxide from its component elements. Write another chemical equation that illustrates the basic character of each oxide.

2. Give examples of two acidic oxides. Write equations illustrating the formation of each oxide from its component elements. Write another chemical equation that illustrates the acidic character of each oxide.

3. Give the name and symbol of each element having the valence configuration [noble gas] ns^2np^1.

4. Give symbols and names for four monatomic ions that have the same electron configuration as argon.

5. Select one of the alkali metals and write a balanced chemical equation for its reaction with chlorine. Is the reaction likely to be exothermic or endothermic? Is the product ionic or molecular?

6. Select one of the alkaline earth metals and write a balanced chemical equation for its reaction with oxygen. Is the reaction likely to be exothermic or endothermic? Is the product ionic or molecular?

7. For the product of the reaction you selected in Study Question 5, predict the following physical properties: color, state of matter (s, ℓ, or g), solubility in water.

8. For the product of the reaction you selected in Study Question 6, predict the following physical properties: color, state of matter (s, ℓ, or g), solubility in water.

9. Would you expect to find calcium occurring naturally in the earth's crust as a free element? Why or why not?

10. Which of the first 10 elements in the periodic table are found as free elements in the earth's crust? Which elements in this group occur in the earth's crust only as part of a chemical compound?

11. Place the following oxides in order of increasing basicity: CO_2, SiO_2, SnO_2.

12. Place the following oxides in order of increasing basicity: Na_2O, Al_2O_3, SiO_2, SO_3.

13. Complete and balance the equations for the following reactions. [Assume an excess of oxygen for (d).]
    (a) $Na(s) + Br_2(\ell) \longrightarrow$
    (b) $Mg(s) + O_2(g) \longrightarrow$
    (c) $Al(s) + F_2(g) \longrightarrow$
    (d) $C(s) + O_2(g) \longrightarrow$

14. Complete and balance the equations for the following reactions.
    (a) $K(s) + I_2(g) \longrightarrow$
    (b) $Ba(s) + O_2(g) \longrightarrow$
    (c) $Al(s) + S_8(s) \longrightarrow$
    (d) $Si(s) + Cl_2(g) \longrightarrow$

Hydrogen

15. Write balanced chemical equations for the reaction of hydrogen gas with oxygen, chlorine, and nitrogen.

16. Write an equation for the reaction of potassium and hydrogen. Name the product. Is it ionic or covalent? Predict one physical property and one chemical property of this compound.

17. Write a balanced chemical equation for the preparation of H_2 (and CO) by the reaction of CH_4 and water. Using data in Appendix L, calculate $\Delta H°$, $\Delta G°$, and $\Delta S°$ for this reaction.

18. Using values in Appendix L, calculate $\Delta H°$, $\Delta G°$, and $\Delta S°$ for the reaction of carbon and water to give CO and H_2.

19. A method recently suggested for the preparation of hydrogen (and oxygen) from water proceeds as follows:
    (a) Sulfuric acid and hydrogen iodide are formed from sulfur dioxide, water, and iodine.
    (b) The sulfuric acid from the first step is decomposed by heat to water, sulfur dioxide, and oxygen.
    (c) The hydrogen iodide from the first step is decomposed with heat to hydrogen and iodine.

    Write a balanced equation for each of these steps and show that their sum is the decomposition of water to form hydrogen and oxygen.

20. Compare the mass of H_2 expected from the reaction of steam (H_2O) per mole of methane, petroleum, and coal. (Assume complete reaction in each case. Use CH_2 and CH as representative formulas for petroleum and coal, respectively.)

Alkali Metals

21. Write equations for the reaction of sodium with each of the halogens. Predict at least two physical properties that are common to all of the alkali metal halides.

22. Write balanced equations for the reaction of lithium, sodium, and potassium with O_2. Specify which metal forms an oxide, which forms a peroxide, and which forms a superoxide.

23. The electrolysis of aqueous NaCl gives NaOH, Cl_2, and H_2.
    (a) Write a balanced equation for the process.
    (b) In the United States, 1.19×10^{10} kg of NaOH and 1.14×10^{10} kg of Cl_2 were produced in a recent year. Does the ratio of masses of NaOH and Cl_2 produced agree with the ratio of masses expected from the balanced equation? If not, what does this tell you about the way in which NaOH and Cl_2 are actually produced? Is the electrolysis of aqueous NaCl the only source of these chemicals?

24. (a) Write equations for the half-reactions that occur at the cathode and the anode when an aqueous solution of KCl is electrolyzed. Which chemical species is oxidized, and which chemical species is reduced in this reaction?
    (b) Predict the products formed when an aqueous solution of CsI is electrolyzed.

Alkaline Earth Elements

25. When magnesium burns in air, it forms both an oxide and a nitride. Write balanced equations for the formation of both compounds.

26. Calcium reacts with hydrogen gas at 300–400 °C to form a hydride. This compound reacts readily with water, so it is an excellent drying agent for organic solvents.
    (a) Write a balanced equation showing the formation of calcium hydride from Ca and H_2.
    (b) Write a balanced equation for the reaction of calcium hydride with water (Figure 21.6).

27. Name three uses of limestone. Write a balanced equation for the reaction of limestone with CO_2 in water.

28. Explain what is meant by "hard water." What causes hard water, and what problems are associated with it?

29. Calcium oxide, CaO, is used to remove SO_2 from power plant exhaust. These two compounds react to give solid $CaSO_3$. What mass of SO_2 can be removed using 1.2×10^3 kg of CaO?

30. $Ca(OH)_2$ has a K_{sp} of 5.5×10^{-5}, whereas K_{sp} for $Mg(OH)_2$ is 5.6×10^{-12}. Calculate the equilibrium constant for the reaction

    $$Ca(OH)_2(s) + Mg^{2+}(aq) \rightleftharpoons Ca^{2+}(aq) + Mg(OH)_2(s)$$

    Explain why this reaction can be used in the commercial isolation of magnesium from sea water.

▲ More challenging   ■ In General ChemistryNow   Blue-numbered questions answered in Appendix O

Boron and Aluminum

31. Draw a possible structure for the cyclic anion in the salt $K_3B_3O_6$ and the chain anion in $Ca_2B_2O_5$.

32. The boron trihalides (except BF_3) hydrolyze completely to boric acid and the acid HX.
    (a) Write a balanced equation for the reaction of BCl_3 with water.
    (b) Calculate $\Delta H°$ for the hydrolysis of BCl_3 using data in Appendix L and the following information: $\Delta H_f°$ [$BCl_3(g)$] = -403 kJ/mol; $\Delta H_f°$ [$B(OH)_3(s)$] = -1094 kJ/mol.

33. When boron hydrides burn in air, the reaction is very exothermic.
    (a) Write a balanced equation for the combustion of $B_5H_9(g)$ in air to give $B_2O_3(s)$ and $H_2O(\ell)$.
    (b) Calculate the heat of combustion for $B_5H_9(g)$ ($\Delta H_f° = 73.2$ kJ/mol), and compare it with the heat of combustion of B_2H_6 on page 1036. (The heat of formation of $B_2O_3(s)$ is -1271.9 kJ/mol.)
    (c) Compare the heat of combustion of $C_2H_6(g)$ with that of $B_2H_6(g)$. Which evolves more heat per gram?

34. Diborane can be prepared by the reaction of $NaBH_4$ and I_2. Which substance is oxidized and which is reduced?

35. Write equations for the reactions of aluminum with $HCl(aq)$, Cl_2, and O_2.

36. (a) Write an equation for the reaction of Al and $H_2O(\ell)$ to produce H_2 and Al_2O_3.
    (b) Using thermodynamic data in Appendix L, calculate $\Delta H°$, $\Delta S°$, and $\Delta G°$ for this reaction. Do these data indicate that the reaction should favor the products?
    (c) Why is aluminum metal unaffected by water?

37. Aluminum dissolves readily in hot aqueous NaOH to give the aluminate ion, $Al(OH)_4^-$, and H_2. Write a balanced equation for this reaction. If you begin with 13.2 g of Al, what volume (in milliliters) of H_2 gas is produced when the gas is measured at 735 mm Hg and 22.5 °C?

38. Alumina, Al_2O_3, is amphoteric. Among examples of its amphoteric character are the reactions that occur when Al_2O_3 is heated strongly or "fused" with acidic oxides and basic oxides.
    (a) Write a balanced equation for the reaction of alumina with silica, an acidic oxide, to give aluminum metasilicate, $Al_2(SiO_3)_3$.
    (b) Write a balanced equation for the reaction of alumina with the basic oxide CaO to give calcium aluminate, $Ca(AlO_2)_2$.

39. Aluminum sulfate (1995 worldwide production is about 3×10^9 kg) is the most commercially important aluminum compound, after aluminum oxide and aluminum hydroxide. Write a balanced equation for the reaction of aluminum oxide with sulfuric acid to give aluminum sulfate. To manufacture 1.00 kg of aluminum sulfate, what mass (in kilograms) of aluminum oxide and sulfuric acid must be used?

40. Gallium hydroxide, like aluminum hydroxide, is amphoteric.
    (a) Write balanced equations for the reaction of solid $Ga(OH)_3$ with aqueous HCl and NaOH.
    (b) What volume of 0.0112 M HCl is needed to react completely with 1.25 g of $Ga(OH)_3$?

41. Halides of the Group 3A elements are excellent Lewis acids. When a Lewis base such as Cl^- interacts with $AlCl_3$, the ion $AlCl_4^-$ is formed. Draw a Lewis electron dot structure for this ion. What structure is predicted for $AlCl_4^-$? What hybridization is assigned to the aluminum atom in $AlCl_4^-$?

42. "Aerated" concrete bricks are widely used building materials. They are obtained by mixing gas-forming additives with a moist mixture of lime, sand, and possibly cement. Industrially, the following reaction is important:

    $$2\ Al(s) + 3\ Ca(OH)_2(s) + 6\ H_2O(\ell) \longrightarrow$$
    $$[3\ CaO \cdot Al_2O_3 \cdot 6\ H_2O](s) + 3\ H_2(g)$$

    Assume that the mixture of reactants contains 0.56 g of Al for each brick. What volume of hydrogen gas do you expect at 26 °C and atmospheric pressure (745 mm Hg)?

Silicon

43. Describe the structures of SiO_2 and CO_2. Explain why SiO_2 has a very high melting point, whereas CO_2 is a gas.

44. Describe how ultrapure silicon can be produced from sand.

45. One material needed to make silicones is dichlorodimethylsilane, $(CH_3)_2SiCl_2$. It is made by treating silicon powder at about 300 °C with CH_3Cl in the presence of a copper-containing catalyst.
    (a) Write a balanced equation for the reaction.
    (b) Assume you carry out the reaction on a small scale with 2.65 g of silicon. To measure the CH_3Cl gas, you fill a 5.60-L flask at 24.5 °C. What pressure of CH_3Cl gas must you have in the flask to have the stoichiometrically correct amount of the compound?
    (c) What mass of $(CH_3)_2SiCl_2$ is produced from 2.65 g of Si?

46. Describe the structure of pyroxenes (see page 1040). What is the ratio of silicon to oxygen in this type of silicate?

Nitrogen and Phosphorus

47. Consult the data in Appendix L. Are any of the nitrogen oxides listed there stable with respect to decomposition to N_2 and O_2?

48. Use data in Appendix L to calculate the enthalpy and free energy change for the reaction

    $$2\ NO_2(g) \longrightarrow N_2O_4(g)$$

    Is this reaction exothermic or endothermic? Is the reaction product- or reactant-favored?

49. Use data in Appendix L to calculate the enthalpy and free energy change for the reaction

$$2 NO(g) + O_2(g) \longrightarrow 2 NO_2(g)$$

Is this reaction exothermic or endothermic? Is the reaction product- or reactant-favored?

50. The overall reaction involved in the industrial synthesis of nitric acid is

$$NH_3(g) + 2 O_2(g) \longrightarrow HNO_3(aq) + H_2O(\ell)$$

(a) Calculate $\Delta G°$ for this reaction.
(b) Calculate the equilibrium constant for this reaction at 25 °C.

51. A major use of hydrazine, N_2H_4, is in steam boilers in power plants.

(a) The reaction of hydrazine with O_2 dissolved in water gives N_2 and water. Write a balanced equation for this reaction.
(b) O_2 dissolves in water to the extent of 3.08 cm³ (gas at STP) in 100. mL of water at 20 °C. To consume all of the dissolved O_2 in 3.00×10^4 L of water (enough to fill a small swimming pool), what mass of N_2H_4 is needed?

52. Before hydrazine came into use to remove dissolved oxygen in the water in steam boilers, Na_2SO_3 was commonly used for this purpose:

$$2 Na_2SO_3(aq) + O_2(aq) \longrightarrow 2 Na_2SO_4(aq)$$

What mass of Na_2SO_3 is required to remove O_2 from 3.00×10^4 L of water as outlined in Study Question 51?

53. A common analytical method for hydrazine involves its oxidation with iodate ion, IO_3^-, in acid solution. In the process, hydrazine acts as a four-electron reducing agent.

$$N_2(g) + 5 H_3O^+(aq) + 4 e^- \longrightarrow N_2H_5^+(aq) + 5 H_2O(\ell)$$
$$E° = -0.23 \text{ V}$$

Write the balanced equation for the reaction of hydrazine in acid solution ($N_2H_5^+$) with $IO_3^-(aq)$ to give N_2 and I_2. Calculate $E°$ for this reaction.

54. Unlike carbon, which can form extended chains of atoms, nitrogen can form chains of very limited length. Draw the Lewis electron dot structure of the azide ion, N_3^-.

55. Review the structure of phosphorous acid in Table 21.7.

(a) What is the oxidation number of the phosphorus atom in this acid?
(b) Draw the structure of diphosphorous acid, $H_4P_2O_5$. What is the maximum number of protons this acid can dissociate in water?

56. $CaHPO_4$ is used as an abrasive in toothpaste. Write a balanced equation showing a possible preparation for this compound.

Oxygen and Sulfur

57. In the "contact process" for making sulfuric acid, sulfur is first burned to SO_2. Environmental restrictions allow no more than 0.30% of this SO_2 to be vented to the atmosphere.

(a) If enough sulfur is burned in a plant to produce 1.80×10^6 kg of pure, anhydrous H_2SO_4 per day, what is the maximum amount of SO_2 that is allowed to be exhausted to the atmosphere?
(b) One way to prevent any SO_2 from reaching the atmosphere is to "scrub" the exhaust gases with slaked lime, $Ca(OH)_2$:

$$Ca(OH)_2(s) + SO_2(g) \longrightarrow CaSO_3(s) + H_2O(\ell)$$
$$2 CaSO_3(s) + O_2(g) \longrightarrow 2 CaSO_4(s)$$

What mass of $Ca(OH)_2$ (in kilograms) is needed to remove the SO_2 calculated in part (a)?

58. A sulfuric acid plant produces an enormous amount of heat. To keep costs as low as possible, much of this heat is used to make steam to generate electricity. Some of the electricity is used to run the plant, and the excess is sold to the local electrical utility. Three reactions are important in sulfuric acid production: (1) burning S to SO_2; (2) oxidation of SO_2 to SO_3; and (3) reaction of SO_3 with H_2O:

$$SO_3(g) + H_2O(\text{in } 98\% H_2SO_4) \longrightarrow H_2SO_4(\ell)$$

The enthalpy change of the third reaction is -130 kJ/mol. Estimate the total heat produced when 1.00 mol of S is used to produce 1.00 mol of H_2SO_4 produced. How much heat is produced per metric ton of H_2SO_4?

59. Sulfur forms anionic chains of S atoms called polysulfides. Draw a Lewis electron dot structure for the S_2^{2-} ion. The S_2^{2-} ion is the disulfide ion, an analogue of the peroxide ion. It occurs in iron pyrites, FeS_2.

60. Sulfur forms a range of compounds with fluorine. Draw Lewis electron dot structures for S_2F_2 (connectivity is FSSF), SF_2, SF_4, SF_6, and S_2F_{10}. What is the formal oxidation number of sulfur in each of these compounds?

Fluorine and Chlorine

61. The halogen oxides and oxoanions are good oxidizing agents. For example, the reduction of bromate ion has an $E°$ value of 1.44 V in acid solution:

$$2 BrO_3^-(aq) + 12 H^+(aq) + 10 e^- \longrightarrow Br_2(aq) + 6 H_2O(\ell)$$

Is it possible to oxidize aqueous 1.0 M Mn^{2+} to aqueous MnO_4^- with 1.0 M bromate ion?

62. The hypohalite ions, XO^-, are the anions of weak acids. Calculate the pH of a 0.10 M solution of NaClO. What is the concentration of HClO in this solution?

63. Bromine is obtained from brine wells. The process involves treating water containing bromide ion with Cl_2 and extracting the Br_2 from the solution using an organic solvent. Write a balanced equation for the reaction of Cl_2 and Br^-. What are the oxidizing and reducing agents in this reaction? Using the table of standard reduction potentials (Appendix M), verify that this is a product-favored reaction.

64. To prepare chlorine from chloride ion a strong oxidizing agent is required. The dichromate ion, $Cr_2O_7^{2-}$, is one example (see Figure 21.31). Consult the table of standard reduction potentials (Appendix M) and identify several other oxidizing agents that may be suitable. Write balanced equations for the reactions of these substances with chloride ion.

65. If an electrolytic cell for producing F_2 (Figure 21.30) operates at 5.00×10^3 amps (at 10.0 V), what mass of F_2 can be produced per 24-hour day? Assume the conversion of F^- to F_2 is 100%.

66. Halogens combine with one another to produce *interhalogens* such as BrF_3. Sketch a possible molecular structure for this molecule and decide if the F—Br—F bond angles will be less than or greater than ideal.

General Questions

The questions are not designated as to type or location in the chapter. They may combine several concepts.

67. For each of the third-period elements (Na through Ar), identify the following.
    (a) whether the element is a metal, nonmetal, or metalloid
    (b) the color and appearance of the element
    (c) the state of the element (s, ℓ, or g) under standard conditions

    For help in this question, consult Figure 2.10 or use the periodic table "tool" on the General ChemistryNow CD-ROM or website. The latter provides a picture of each element and a listing of its properties.

68. For each of the second-period elements (Li through Ne), identify the following.
    (a) whether the element is a metal, nonmetal, or metalloid
    (b) the color and appearance of the element
    (c) the state of the element (s, ℓ, or g) under standard conditions

    Consult Figure 2.10 or use the periodic table "tool" on the General ChemistryNow CD-ROM or website.

69. Consider the chemistries of the elements sodium, magnesium, aluminum, silicon, and phosphorus.
    (a) Write a balanced chemical equation depicting the reaction of each element with elemental chlorine.
    (b) Describe the bonding in each of the products of the reactions with chlorine as ionic or covalent.
    (c) Draw Lewis electron dot structures for the products of the reactions of silicon and phosphorus with chlorine. What are their electron-pair and molecular geometries?

70. Consider the chemistries of C, Si, Ge, and Sn.
    (a) Write a balanced chemical equation to depict the reaction of each element with elemental chlorine.

    (b) Describe the bonding in each of the products of the reactions with chlorine as ionic or covalent.

    (You have not seen reactions of some of these elements in the text, but you have been given enough information to be able to predict the reactions that can occur.)

71. Complete and balance the following equations.
    (a) $KClO_3 + heat \longrightarrow$
    (b) $H_2S(g) + O_2(g) \longrightarrow$
    (c) $Na(s) + O_2(g) \longrightarrow$
    (d) $P_4(s) + KOH(aq) + H_2O(\ell) \longrightarrow$
    (e) $NH_4NO_3(s) + heat \longrightarrow$
    (f) $In(s) + Br_2(\ell) \longrightarrow$
    (g) $SnCl_4(\ell) + H_2O(\ell) \longrightarrow$

72. Sodium borohydride, $NaBH_4$, reduces many metal ions to the metal.
    (a) Write a balanced equation for the reaction of $NaBH_4$ with $AgNO_3$ in water to give silver metal, H_2 gas, boric acid, and sodium nitrate (page 1036).
    (b) What mass of silver can be produced from 575 mL of 0.011 M $AgNO_3$ and 13.0 g of $NaBH_4$?

73. When BCl_3 gas is passed through an electric discharge, small amounts of the reactive molecule B_2Cl_4 are produced. (The molecule has a B—B covalent bond.)
    (a) Draw a Lewis electron dot structure for B_2Cl_4.
    (b) Describe the hybridization of the B atoms in the molecule and the geometry around each B atom.

74. Boron carbide, B_4C, is chemically inert. It is used as an abrasive and in body armor. It is synthesized by passing a mixture of $BCl_3(g)$ and H_2 over graphite.
    (a) Write a balanced chemical equation for the reaction of BCl_3, H_2, and C to give B_4C and HCl.
    (b) What mass of B_4C can be produced from 5.45 L of BCl_3 gas at 26.5 °C and 456 mm Hg pressure (and excess H_2 and graphite)?

75. (a) Heating barium oxide in pure oxygen gives barium peroxide. Write a balanced equation for this reaction.
    (b) Barium peroxide is an excellent oxidizing agent. Write a balanced equation for the reaction of iron with barium peroxide to give iron(III) oxide and barium oxide.

76. Worldwide production of silicon carbide, SiC, is several hundred thousand tons annually. If you want to produce 1.0×10^5 metric tons of SiC, what mass (metric tons) of silicon sand (SiO_2) will you use if 70% of the sand is converted to SiC?

77. One of the pieces of evidence relating to the hydride ion in metal hydrides comes from electrochemistry. Predict the reactions that occur at each electrode when molten LiH is electrolyzed.

78. To store 2.88 kg of gasoline with an energy equivalence of 1.43×10^8 J requires a volume of 4.1 L. In comparison, 1.0 kg of H_2 has the same energy equivalence. What volume is required if this quantity of H_2 is to be stored at 25 °C and 1.0 atm of pressure?

79. Using data in Appendix L, calculate $\Delta G°$ values for the decomposition of MCO_3 to MO and CO_2 where M = Mg, Ca, Ba. What is the relative tendency of these carbonates to decompose?

80. Ammonium perchlorate is used as the oxidizer in the solid-fuel booster rockets of the Space Shuttle. Assume that one launch requires 700 tons (6.35×10^5 kg) of the salt, and the salt decomposes according to the equation on page 1060.
    (a) What mass of water is produced? What mass of O_2 is produced?
    (b) If the O_2 produced is assumed to react with the powdered aluminum present in the rocket engine, what mass of aluminum is required to use up all of the O_2?
    (c) What mass of Al_2O_3 is produced?

81. ▲ Metals react with hydrogen halides (such as HCl) to give the metal halide and hydrogen:

    $$M(s) + n\,HX(g) \longrightarrow MX_n(s) + \tfrac{1}{2}n\,H_2(g)$$

    The free energy change for the reaction is $\Delta G°_{rxn} = \Delta G°_f(MX_n) - n\,\Delta G°_f[HX(g)]$
    (a) $\Delta G°_f$ for HCl(g) is -95.1 kJ/mol. What must be the value for $\Delta G°_f$ for MX_n for the reaction to be product-favored?
    (b) Which of the following metals is (are) predicted to have product-favored reactions with HCl(g): Ba, Pb, Hg, Ti?

82. Halogens form polyhalide ions. Sketch Lewis electron dot structures and molecular structures for the following ions:
    (a) I_3^- (b) $BrCl_2^-$ (c) ClF_2^+

83. The standard heat of formation of OF_2 gas is $+24.5$ kJ/mol. Calculate the average O—F bond energy.

84. Calcium fluoride can be used in the fluoridation of municipal water supplies. If you want to achieve a fluoride ion concentration of 2.0×10^{-5} M, what mass of CaF_2 must you use for 1.0×10^6 L of water?

85. The steering rockets in the Space Shuttle use N_2O_4 and a derivative of hydrazine, 1,1-dimethylhydrazine (page 278). This mixture is called a *hypergolic fuel* because it ignites when the reactants come into contact:

    $$H_2NN(CH_3)_2(\ell) + 2\,N_2O_4(\ell) \longrightarrow$$
    $$3\,N_2(g) + 4\,H_2O(g) + 2\,CO_2(g)$$

    (a) Identify the oxidizing agent and the reducing agent in this reaction.
    (b) The same propulsion system was used by the Lunar Lander on moon missions in the 1970s. If the Lander used 4100 kg of $H_2NN(CH_3)_2$, what mass (in kilograms) of N_2O_4 was required to react with it? What mass (in kilograms) of each of the reaction products was generated?

86. ▲ The density of lead is 11.350 g/cm^3, and the metal crystallizes in a face-centered cubic unit cell. Estimate the radius of the lead atom.

87. Dinitrogen trioxide, N_2O_3, has the structure shown here.

    The oxide is unstable, decomposing to NO and NO_2 in the gas phase at 25 °C.

    $$N_2O_3(g) \longrightarrow NO(g) + NO_2(g)$$

    (a) Explain why one N—O bond distance in N_2O_3 is 114.2 pm, whereas the other two bonds are longer (121 pm) and nearly equal to each other.
    (b) For the decomposition reaction, $\Delta H° = +40.5$ kJ/mol and $\Delta G° = -1.59$ kJ/mol. Calculate $\Delta S°$ and K for the reaction at 298 K.
    (c) Calculate $\Delta H°_f$ for $N_2O_3(g)$.

88. ▲ Liquid HCN is dangerously unstable with respect to trimer formation—that is, formation of $(HCN)_3$ with a cyclic structure.
    (a) Propose a structure for this cyclic trimer.
    (b) Estimate the energy of the trimerization reaction using bond energies (Table 9.10).

89. Use $\Delta H°_f$ data in Appendix L to calculate the enthalpy change of the reaction

    $$2\,N_2(g) + 5\,O_2(g) + 2\,H_2O(\ell) \longrightarrow 4\,HNO_3(aq)$$

    Speculate on whether such a reaction could be used to "fix" nitrogen. Would research to find ways to accomplish this reaction be a useful endeavor?

90. (a) Magnesium is obtained from sea water. If the concentration of Mg21 in sea water is 0.050 M, what volume of sea water (in liters) must be treated to obtain 1.00 kg of magnesium metal? What mass of lime (CaO; in kilograms) must be used to precipitate the magnesium in this volume of sea water?
    (b) When 1.2×10^3 kg of molten $MgCl_2$ is electrolyzed to produce magnesium, what mass (in kilograms) of metal is produced at the cathode? What is produced at the anode? What is the mass of this product? What is the total number of faradays of electricity used in the process?
    (c) One industrial process has an energy consumption of 18.5 kWh/kg of Mg. How many joules are required per mole (1 kWh = 1 kilowatt-hour = 3.6×10^6 J)? How does this energy compare with the energy of the following process?

    $$MgCl_2(s) \longrightarrow Mg(s) + Cl_2(g)?$$

91. Assume an electrolysis cell that produces chlorine from aqueous sodium chloride (called "brine") operates at 4.6 V (with a current of 3.0×10^5 amps). Calculate the number of kilowatt-hours of energy required to produce 1.00 kg of chlorine (1 kWh = 1 kilowatt-hour = 3.6×10^6 J).

92. Sodium metal is produced by electrolysis of molten sodium chloride. The cell operates at 7.0 V with a current of 25×10^3 amps.

    (a) What mass of sodium can be produced in 1 h?

    (b) How many kilowatt-hours of electricity are used to produce 1.00 kg of sodium metal (1 kWh = 1 kilowatt-hour = 3.6×10^6 J)?

Summary and Conceptual Questions

The following questions may use concepts from the preceding chapters.

93. You have a 1.0-L flask that contains a mixture of argon and hydrogen. The pressure inside the flask is 745 mm Hg and the temperature is 22 °C. Describe an experiment that you could use to determine the percentage of hydrogen in this mixture.

94. The boron atom in boric acid, $B(OH)_3$, is bonded to three —OH groups. In the solid state, the —OH groups are in turn hydrogen-bonded to —OH groups in neighboring molecules.

    (a) Draw the Lewis structure for boric acid.

    (b) What is the hybridization of the boron atom in the acid?

    (c) Sketch a picture showing how hydrogen bonding can occur between neighboring molecules.

95. How would you extinguish a sodium fire in the laboratory? What is the worst thing you could do?

96. Tin(IV) oxide, cassiterite, is the main ore of tin. It crystallizes in a rutile-like unit cell (page 636).

    (a) How many tin(IV) ions and oxide ions are there per unit cell of this oxide?

    (b) Is it thermodynamically feasible to transform solid SnO_2 into liquid $SnCl_4$ by reaction of the oxide with gaseous HCl? What is the equilibrium constant for this reaction at 25 °C?

97. You are given a stoppered flask that contains either hydrogen, nitrogen, or oxygen. Suggest an experiment to identify the gas.

98. The structure of nitric acid is illustrated on page 1046.

    (a) Why are the N—O bonds the same length, and why are both shorter than the N—OH bond length?

    (b) Rationalize the bond angles in the molecule.

    (c) What is the hybridization of the central N atom? Which orbitals overlap to form the N—O π bond?

99. The reduction potentials for the Group 3A metals, $E°$, are given below. What trend or trends do you observe in these data? What can you learn about the chemistry of the Group 3A elements from these data?

Half-Reaction	Reduction Potential ($E°$, V)
$Al^{3+}(aq) + 3\ e^- \longrightarrow Al(s)$	−1.66
$Ga^{3+}(aq) + 3\ e^- \longrightarrow Ga(s)$	−0.53
$In^{3+}(aq) + 3\ e^- \longrightarrow In(s)$	−0.338
$Tl^{3+}(aq) + 3\ e^- \longrightarrow Tl(s)$	+0.72

100. ▲ You are given air and water for starting materials, along with whatever laboratory equipment you need. Describe how you could synthesize ammonium nitrate from these reagents.

101. ▲ When 1.00 g of a white solid A is strongly heated, you obtain another white solid, B, and a gas. An experiment is carried out on the gas, showing that it exerts a pressure of 209 mm Hg in a 450-mL flask at 25 °C. Bubbling the gas into a solution of $Ca(OH)_2$ gives another white solid, C. If the white solid B is added to water, the resulting solution turns red litmus paper blue. Addition of aqueous HCl to the solution of B and evaporation of the resulting solution to dryness yields 1.055 g of a white solid D. When D is placed in a Bunsen burner flame, it colors the flame green. Finally, if the aqueous solution of B is treated with sulfuric acid, a white precipitate, E, forms. Identify the lettered compounds in the reaction scheme.

The salts $CaCl_2$, $SrCl_2$, and $BaCl_2$ were suspended in methanol. When the methanol is set ablaze, the heat of combustion causes the salts to emit light of characteristic wavelengths: calcium salts are yellow, strontium salts are red, and barium salts are green-yellow.

Charles D. Winters

Do you need a live tutor for homework problems?
Access vMentor at General ChemistryNow at
http://now.brookscole.com/kotz6e
for one-on-one tutoring from a chemistry expert

22—The Chemistry of the Transition Elements

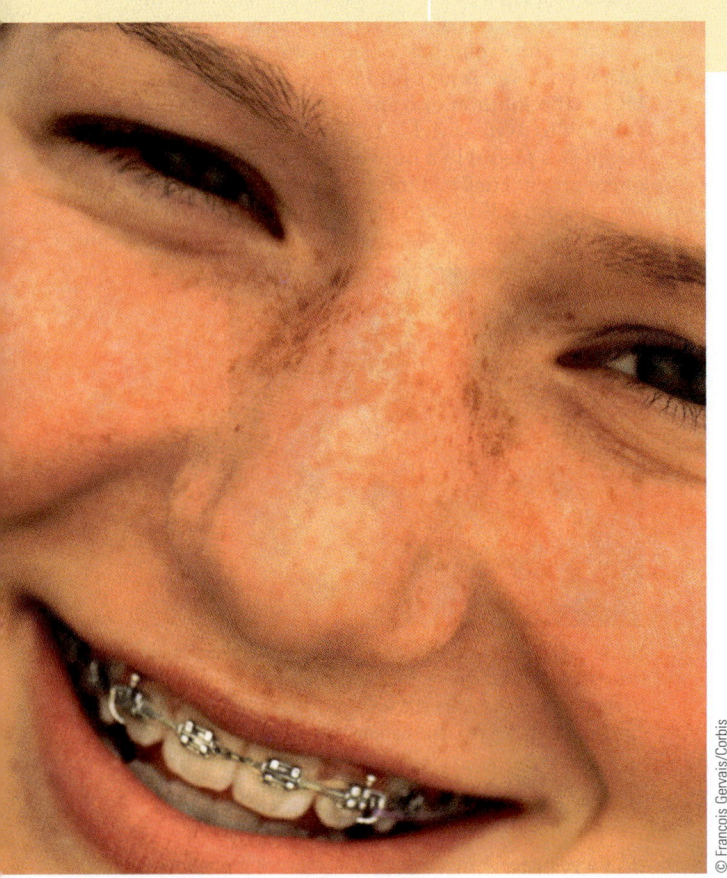

Orthodontic devices from memory metal. Braces for straightening teeth can be made of nitinol.

© François Gervais/Corbis

Memory Metal

In the early 1960s, metallurgical engineer William J. Buehler, a researcher at the Naval Ordnance Laboratory in White Oak, Maryland, was experimenting with alloys made of two metals. He was looking for an impact- and heat-resistant material, intended for use in the nose cone of a Navy missile. It was also important that the material be fatigue resistant; that is, it should not lose its desirable properties when bent and shaped. An alloy of two transition metals, nickel and titanium, appeared to have the desirable properties. Buehler prepared long, thin strips of this alloy to demonstrate that it could be folded and unfolded many times without breaking. At a meeting to discuss this material, one of his associates decided to see what would happen when the strip was heated. He held a cigarette lighter to a folded-up piece of metal and was amazed to observe that the metal strip immediately unfolded and assumed its original shape. Thus, memory metal was discovered. This unusual alloy is now called Nitinol, a name constructed out of "nickel," "titanium," and "Naval Ordnance Laboratory."

Memory metal is an alloy with roughly the same number of Ni and Ti atoms. It remembers its shape because of the arrangement of these atoms in the solid phase. When the atoms are arranged in the highly symmetrical phase (austenite), the alloy is relatively rigid. The metal can be trained to remember its shape

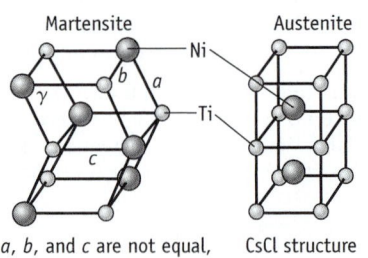

a, *b*, and *c* are not equal, γ about 96°

CsCl structure
$a = b = c$
$\alpha = \beta = \gamma = 90°$

Two phases of nitinol. The austenite form has a structure like CsCl (page 622).

1068

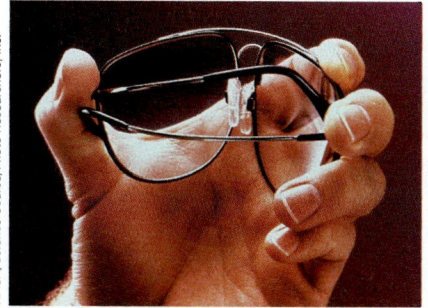

NASA/Science Source/Photo Researchers, Inc.

Nitinol frames for glasses. These sunglass frames are made of nitinol, so they snap back to the proper fit even after being twisted like a pretzel. The metal used in these frames has a critical temperature above room temperature, so it readily returns to its "memorized" shape. Similar alloys are used for wires in dental braces and surgical anchors, which cannot be heated after insertion.

by twisting or bending it to the desired shape when in this phase. When the alloy is cooled below a temperature called its "phase transition temperature," it enters a less symmetrical but flexible phase (martensite). Below its transition temperature, the metal is fairly soft and may be bent and twisted out of shape. When warmed, nitinol returns to its original shape. The temperature at which the change in shape occurs varies with small differences in the nickel-to-titanium ratio. Depending on their composition, materials may change shape at temperatures ranging from −125 °C to about 70 °C, greatly increasing the number of possible uses for this intriguing material.

Memory metal never made it into missile nose cones, but it has found a wide variety of other applications. Some of the most interesting arise in the medical area. One application involves the fabrication of vascular stents to reinforce blood vessels. The stent is crushed and inserted into a blood vessel through a very fine needle. When the nitinol stent warms to body temperature, it returns to its memorized shape and reinforces the walls of the blood vessel.

Nitinol can also be used in orthodontics. Braces made of Nitinol remember their shape and apply a steady, constant pressure to move teeth into position.

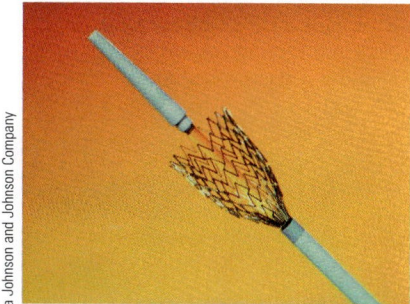

Courtesy of Nitinol Devices & Components, a Johnson and Johnson Company

Nitinol stent. A stent made of nitinol is used to reinforce blood vessels.

To Review Before You Begin

- Recall the position of transition elements in the periodic table (Section 2.7)
- Know the formulas and names for transition metals, their ions, and their compounds (Section 3.3)
- Be able to balance chemical equations (Sections 4.1 and 20.1)
- Review electronic structure, spectra, and magnetism (Chapters 8 and 9)

GENERAL
Chemistry ⚛ Now™

Throughout the chapter this icon introduces a list of resources on the General ChemistryNow CD-ROM or website (http://now.brookscole.com/kotz6e) that will:

- help you evaluate your knowledge of the material
- provide homework problems
- allow you to take an exam-prep quiz
- provide a personalized Learning Plan targeting resources that address areas you should study

The transition elements are the large block of elements in the central portion of the periodic table. All are metals and bridge the *s*-block elements at the left and the *p*-block elements on the right (Figure 22.1). The transition elements are often divided into two groups, depending on the valence electrons involved. Most are **d-block elements,** because their occurrence in the periodic table coincides with the filling of the *d* orbitals. The second group are the **f-block elements,** characterized by filling of the *f* orbitals. Contained within this group of elements are two subgroups: the *lanthanides,* elements that occur between La and Hf, and the *actinides,* elements that occur between Ac and Rf.

This chapter primarily focuses on the *d*-block elements, and within this group we concentrate mainly on the elements in the fourth period, that is, the elements of the first transition series, scandium to zinc.

22.1—Properties of the Transition Elements

The *d*-block metals include elements with a wide range of properties. They encompass the most common metal used in construction and manufacturing (iron), metals that are valued for their beauty (gold, silver, and platinum), and metals used in coins (nickel, copper, and zinc). There are metals used in modern technology (titanium) and metals known and used in early civilizations (copper, silver, gold, and iron). The *d*-block contains the densest elements (osmium, $d = 22.49$ g/cm^3, and iridium, $d = 22.41$ g/cm^3), the metals with the highest and lowest melting points

Fourth-period transition metals:
left to right, Ti, V, Cr, Mn, Fe, Co, Ni, Cu

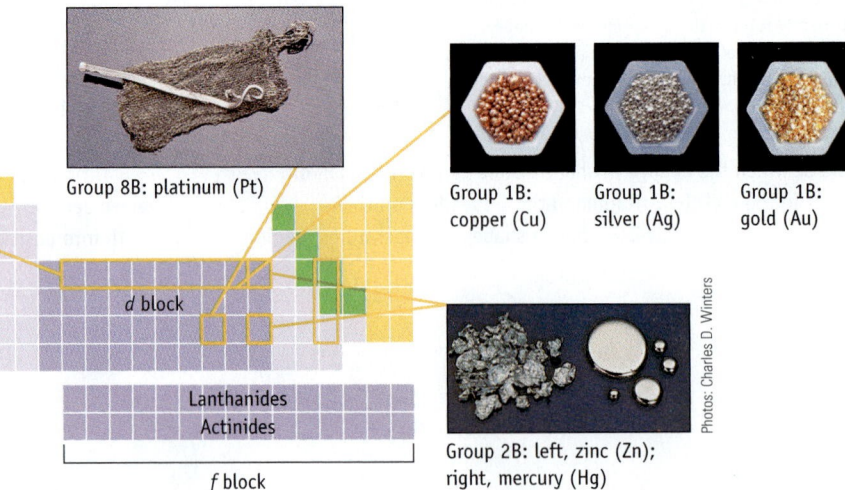

Group 8B: platinum (Pt)

Group 1B: copper (Cu)

Group 1B: silver (Ag)

Group 1B: gold (Au)

d block

Lanthanides
Actinides

f block

Group 2B: left, zinc (Zn); right, mercury (Hg)

Photos: Charles D. Winters

Figure 22.1   **The transition metals.** The *d*-block elements (transition elements) and *f*-block elements are highlighted in a darker shade of purple.

Charles D. Winters

(a) Paint pigments: yellow, CdS; green, Cr_2O_3; white, TiO_2 and ZnO; purple, $Mn_3(PO_4)_2$; blue, Co_2O_3 and Al_2O_3; ochre, Fe_2O_3.

(b) Small amounts of transition metal compounds are used to color glass: blue, Co_2O_3; green, copper or chromium oxides; purple, nickel or cobalt oxides; red, copper oxide; iridescent green, uranium oxide.

(c) Traces of transition metal ions are responsible for the colors of green jade (iron), red corundum (chromium), blue azurite, blue-green turquoise (copper), and purple amethyst (iron).

Figure 22.2 Colorful chemistry. Transition metal compounds are often colored, a property that leads to specific uses.

(tungsten, mp = 3410 °C, and mercury, mp = −38.9 °C), and one of two radioactive elements with atomic numbers less than 83 [technetium (Tc), atomic number 43; the other is promethium (Pm), atomic number 61, in the *f*-block].

With the exception of mercury, the transition elements are solids, often with high melting and boiling points. They have a metallic sheen and conduct electricity and heat. They react with various oxidizing agents to give ionic compounds. There is considerable variation in such reactions among the elements, however. Because silver, gold, and platinum resist oxidation and are resistant to becoming tarnished, for example, they are used for jewelry and decorative items.

Certain *d*-block elements are particularly important in living organisms. Cobalt is the crucial element in vitamin B_{12}, which is part of a catalyst essential for several biochemical reactions. Hemoglobin and myoglobin, oxygen-carrying and -storage proteins, contain iron (see page 942). Molybdenum and iron, together with sulfur, form the reactive portion of nitrogenase, a biological catalyst used by nitrogen-fixing organisms to convert atmospheric nitrogen into ammonia.

Many transition metal compounds are highly colored, which makes them useful as pigments in paints and dyes (Figure 22.2). Prussian blue, $Fe_4[Fe(CN)_6]_3 \cdot 14\,H_2O$ is a "bluing agent" used in engineering blueprints and in the laundry to brighten yellowed white cloth. A common pigment (artist's cadmium yellow) contains cadmium sulfide (CdS), and the white in most white paints is titanium(IV) oxide, TiO_2.

The presence of transition metal ions in crystalline silicates or alumina transforms these common materials into gemstones. Iron(II) ions cause the yellow color in citrine and chromium(III) ions produce the red color of a ruby. Transition metal complexes in small quantities add color to glass. Blue glass contains a small amount of a cobalt(III) oxide, and addition of chromium oxide to glass gives a green color. Old window panes sometimes take on a purple color over time as a consequence of oxidation of traces of manganese(II) ion present in the glass as permanganate ion (MnO_4^-).

In the next few pages we will examine the properties of the transition elements, concentrating on the underlying principles that govern these properties.

Table 22.1 Electron Configurations of the Fourth-Period Transition Elements

	spdf Configu- ration	Box Notation		
		3d		4s
Sc	[Ar]$3d^14s^2$	↑		↑↓
Ti	[Ar]$3d^24s^2$	↑ ↑		↑↓
V	[Ar]$3d^34s^2$	↑ ↑ ↑		↑↓
Cr	[Ar]$3d^54s^1$	↑ ↑ ↑ ↑ ↑		↑
Mn	[Ar]$3d^54s^2$	↑ ↑ ↑ ↑ ↑		↑↓
Fe	[Ar]$3d^64s^2$	↑↓ ↑ ↑ ↑ ↑		↑↓
Co	[Ar]$3d^74s^2$	↑↓ ↑↓ ↑ ↑ ↑		↑↓
Ni	[Ar]$3d^84s^2$	↑↓ ↑↓ ↑↓ ↑ ↑		↑↓
Cu	[Ar]$3d^{10}4s^1$	↑↓ ↑↓ ↑↓ ↑↓ ↑↓		↑
Zn	[Ar]$3d^{10}4s^2$	↑↓ ↑↓ ↑↓ ↑↓ ↑↓		↑↓

Electron Configurations

Because chemical behavior is related to electron structure, it is important to know the electron configurations of the d-block elements (Table 22.1) and their common ions [◀ Section 8.4]. Recall that the configuration of these metals has the general form [noble gas core]$ns^a(n-1)d^b$; that is, valence electrons for the transition elements reside in the ns and $(n-1)d$ subshells.

Oxidation and Reduction

A characteristic chemical property of all metals is that they undergo oxidation by a wide range of oxidizing agents such as oxygen, the halogens, and aqueous acids. Standard reduction potentials for the elements of the first transition series can be used to predict which elements will be oxidized by a given oxidizing agent. For example, all of these metals except vanadium and copper are oxidized by aqueous HCl (Table 22.2). This feature, which dominates the chemistry of these elements, is sometimes highly undesirable (see "Chemical Perspectives: Corrosion of Iron").

When a transition metal is oxidized, the outermost s electrons are removed, followed by one or more d electrons. With a few exceptions, transition metal ions have the electron configuration [noble gas core]$(n-1)d^x$. In contrast to ions formed by main group elements, transition metal cations often possess unpaired electrons, resulting in paramagnetism [◀ page 335]. They are frequently colored as well, due to the absorption of light in the visible region of the electromagnetic spectrum. Color and magnetism figure prominently in a discussion of the properties and bonding of these elements, as you shall see shortly.

In the first transition series, the most commonly encountered metal ions have oxidation numbers of +2 and +3 (Table 22.2). With iron, for example, oxidation converts Fe([Ar]$3d^64s^2$) to either Fe^{2+}([Ar]$3d^6$) or Fe^{3+}([Ar]$3d^5$). Iron reacts with chlorine to give $FeCl_3$, and it reacts with aqueous acids to produce Fe^{2+}(aq) and H_2 (Figure 22.3). Despite the preponderance of 2+ and 3+ ions in compounds of the first transition metal series, the range of possible oxidation states for these compounds is broad (Figure 22.4). Earlier in this text, we encountered chromium with a +6 oxidation number (CrO_4^{2-}, $Cr_2O_7^{2-}$), manganese with an oxidation number of +7 (MnO_4^-), silver and copper as 1+ ions, and vanadium oxidation numbers that can range from +5 to +2 (Figure 20.3).

Charles D. Winters

Prussian blue. When Fe^{3+} ions are added to $[FeCN)_6]^{4-}$ ions in water (or Fe^{2+} ions are added to $[Fe(CN)_6]^{3-}$ ions), a deep blue compound called Prussian blue forms. The formula of the compound is $Fe_4[Fe(CN)_6]_3 \cdot 14 H_2O$. The color arises from the interaction of the Fe(II) and Fe(III) ions in the compound.

Table 22.2 Products from Reactions of the Elements in the First Transition Series with O_2, Cl_2, or Aqueous HCl

Element	Reaction with O_2*	Reaction with Cl_2	Reaction with Aqueous HCl
Scandium	Sc_2O_3	$ScCl_3$	Sc^{3+}(aq)
Titanium	TiO_2	$TiCl_4$	Ti^{3+}(aq)
Vanadium	V_2O_5	VCl_4	NR†
Chromium	Cr_2O_3	$CrCl_3$	Cr^{2+}(aq)
Manganese	MnO_2	$MnCl_2$	Mn^{2+}(aq)
Iron	Fe_2O_3	$FeCl_3$	Fe^{2+}(aq)
Cobalt	Co_2O_3	$CoCl_2$	Co^{2+}(aq)
Nickel	NiO	$NiCl_2$	Ni^{2+}(aq)
Copper	CuO	$CuCl_2$	NR†
Zinc	ZnO	$ZnCl_2$	Zn^{2+}(aq)

* Not all possible compounds are listed.
† NR = no reaction.

(a)

(b)

(c)

Charles D. Winters

Figure 22.3 **Typical reactions of transition metals.** These metals react with oxygen, with halogens, and with acids under appropriate conditions. (a) Steel wool reacts with O_2, (b) steel wool reacts with chlorine gas, Cl_2, and (c) iron chips react with aqueous HCl.

Higher oxidation numbers are more common in compounds of the elements in the second and third transition series. For example, the naturally occurring sources of molybdenum and tungsten are the ores molybdenite (MoS_2) and wolframite (WO_3). This general trend is carried over in the *f*-block. The lanthanides form primarily 3+ ions. In contrast, actinide elements usually have higher oxidation numbers in their compounds; +4 and even +6 are typical. For example, UO_3 is a common oxide of uranium, and UF_6 is a compound important in processing uranium fuel for nuclear reactors [▶ Section 23.6].

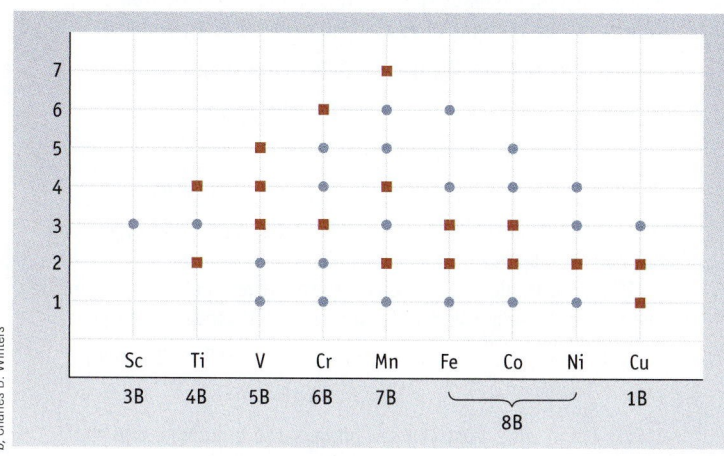

(a)

(b)

b. Charles D. Winters

Figure 22.4 **Oxidation states of the transition elements in the first transition series.** (a) The most common oxidation states are indicated with red squares; less common oxidation states are indicated with blue dots. (b) Aqueous solutions of chromium compounds with two different oxidation numbers: +3 in $Cr(NO_3)_3$ (violet) and $CrCl_3$ (green), and +6 in K_2CrO_4 (yellow) and $K_2Cr_2O_7$ (orange).

Chemical Perspectives

Corrosion of Iron

It is hard not to be aware of corrosion. Those of us who live in the northern part of the United States are well aware of the problems of rust on our automobiles. It is estimated that 20% of iron production each year goes solely to replace iron that has rusted away.

The corrosion or rusting of iron results in major economic loss.

Qualitatively, we describe corrosion as the deterioration of metals by a product-favored oxidation reaction. The corrosion of iron, for example, converts iron metal to red-brown rust, which is hydrated iron(III) oxide, $Fe_2O_3 \cdot H_2O$. This process requires both air and water, and it is enhanced if the water contains dissolved ions and if the metal is stressed (for example, if it has dents, cuts, and scrapes on the surface.)

The corrosion process occurs in what is essentially a small electrochemical cell. There are an anode and a cathode, an electrical connection between the two (the metal itself), and an electrolyte in contact with both anode and cathode. When a metal corrodes, the metal is oxidized on anodic areas of the metal surface.

Anode, oxidation $M(s) \longrightarrow M^{n+} + n\,e^-$

The electrons are consumed by several possible half-reactions in cathodic areas.

Cathode, reduction

$$2\,H^+(aq) + 2\,e^- \longrightarrow H_2(g)$$
$$2\,H_2O(\ell) + 2\,e^- \longrightarrow H_2(g) + 2\,OH^-(aq)$$
$$O_2(g) + 2\,H_2O(\ell) + 4\,e^- \longrightarrow 4\,OH^-(aq)$$

With iron, the rate of the corrosion process is controlled by the rate of the cathodic process. Of the three possible cathodic reactions, the one that is fastest is determined by acidity and the amount of oxygen present. If little or no oxygen is present—as when a piece of iron is buried in soil such as moist clay—hydrogen ion or water is reduced and $H_2(g)$ and hydroxide ions are the products. Iron(II) hydroxide is relatively insoluble and will precipitate on the metal surface, inhibiting the further formation of Fe^{2+}.

Anode	$Fe(s) \longrightarrow Fe^{2+}(aq) + 2\,e^-$
Cathode	$2\,H_2O(\ell) + 2\,e^- \longrightarrow H_2(g) + 2\,OH^-(aq)$
Precipitation	$Fe^{2+}(aq) + 2\,OH^-(aq) \longrightarrow Fe(OH)_2(s)$
Net reaction	$Fe(s) + 2\,H_2O(\ell) \longrightarrow H_2(g) + Fe(OH)_2(s)$

If both water and O_2 are present, the chemistry of iron corrosion is somewhat different, and the corrosion reaction is about 100 times faster than without oxygen.

Anode	$2\,Fe(s) \longrightarrow 2\,Fe^{2+}(aq) + 4\,e^-$
Cathode	$O_2(g) + 2\,H_2O(\ell) + 4\,e^- \longrightarrow 4\,OH^-(aq)$
Precipitation	$2\,Fe^{2+}(aq) + 4\,OH^-(aq) \longrightarrow 2\,Fe(OH)_2(s)$
Net reaction	$2\,Fe(s) + 2\,H_2O(\ell) + O_2(g) \longrightarrow 2\,Fe(OH)_2(s)$

If oxygen is not available in excess, further oxidation of the iron(II) hydroxide leads to the formation of magnetic iron oxide (which can be thought of as a mixed oxide of Fe_2O_3 and FeO).

$$6\,Fe(OH)_2(s) + O_2(g) \longrightarrow 2\,Fe_3O_4 \cdot H_2O(s) + 4\,H_2O(\ell)$$
green hydrated magnetite

$$Fe_3O_4 \cdot H_2O(s) \longrightarrow H_2O(\ell) + Fe_3O_4(s)$$
black magnetite

It is the black magnetite that you find coating an iron object that has corroded by resting in moist soil.

If the iron object has free access to oxygen and water, as in the open or in flowing water, red-brown iron(III) oxide will form.

$$4\,Fe(OH)_2(s) + O_2(g) \longrightarrow 2\,Fe_2O_3 \cdot H_2O(s) + 2\,H_2O(\ell)$$
red-brown

This is the familiar rust you see on cars and buildings, and the substance that colors the water red in some mountain streams or in your home.

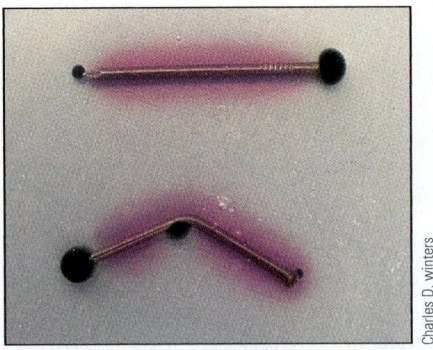

Anode and cathode reactions in iron corrosion. Two iron nails were placed in an agar gel that contains phenolphthalein and $K_3[Fe(CN)_6]$. Iron(II) ion, formed at the tip and where the nail is bent, reacts with $[Fe(CN)_6]^{3-}$ to form blue-green $Fe_4[Fe(CN)_6]_3 \cdot 14\,H_2O$ (Prussian blue). Hydrogen and $OH^-(aq)$ are formed at the other parts of the surface of the nail, the latter being detected by the red color of the acid–base indicator. In this electrochemical cell, regions of stress—the ends and the bent region of the nail—act as anodes, and the remainder of the surface serves as the cathode.

See the General ChemistryNow CD-ROM or website:
- **Screen 22.2 Formulas and Oxidation Numbers in Transition Metal Complexes,** for an exercise on transition metal compounds

Periodic Trends in the *d*-Block: Size, Density, Melting Point

The periodic table is the most useful single reference source for a chemist. Not only does it provide data that have everyday use, but it also organizes the elements with respect to their chemical and physical properties. Let us look at three physical properties of the transition elements that vary periodically: atomic radii, density and melting point.

Metal Atom Radii

The variation in atomic radii for the transition elements in the fourth, fifth, and sixth periods is illustrated in Figure 8.12. The radii of the transition elements vary over a fairly narrow range, with a small decrease to a minimum being observed around the middle of this group of elements. This similarity of radii can be understood based on electron configurations. Atom size is determined by the radius of the outermost orbital, which for these elements is the *ns* orbital ($n = 4$, 5, or 6). Progressing from left to right in the periodic table, the size decline expected from increasing the number of protons in the nucleus is mostly canceled out by an opposing effect, repulsion from additional electrons in the $(n - 1)d$ orbitals.

The radii of the *d*-block elements in the fifth and sixth periods in each group are almost identical. The reason is that the lanthanide elements immediately precede the third series of *d*-block elements. The filling of 4*f* orbitals is accompanied by a steady contraction in size, consistent with the general trend of decreasing size from left to right in the periodic table. At the point where the 5*d* orbitals begin to fill again, the radii have decreased to a size similar to that of elements in the previous period. The decrease in size that results from the filling of the 4*f* orbitals is given a specific name, the **lanthanide contraction**.

The similar sizes of the second- and third-period *d*-block elements have significant consequences for their chemistry. For example, the "platinum group metals" (Ru, Os, Rh, Ir, Pd, and Pt) form similar compounds. Thus, it is not surprising that minerals containing these metals are found in the same geological zones on earth. Nor is it surprising that it is difficult to separate these elements from one another.

Density

The variation in metal radii causes the densities of the transition elements to first increase and then decrease across a period (Figure 22.5a). Although the overall change in radii among these elements is small, the effect is magnified because the volume is actually changing with the cube of the radius $[V = (4/3)\pi r^3]$.

The lanthanide contraction explains why elements in the sixth period have the highest density. The relatively small radii of sixth-period transition metals, combined with the fact that their atomic masses are considerably larger than their counterparts in the fifth period, causes sixth-period metal densities to be very large.

Melting Point

The melting point of any substance reflects the forces of attraction between the atoms, molecules, or ions that compose the solid. With transition elements, the

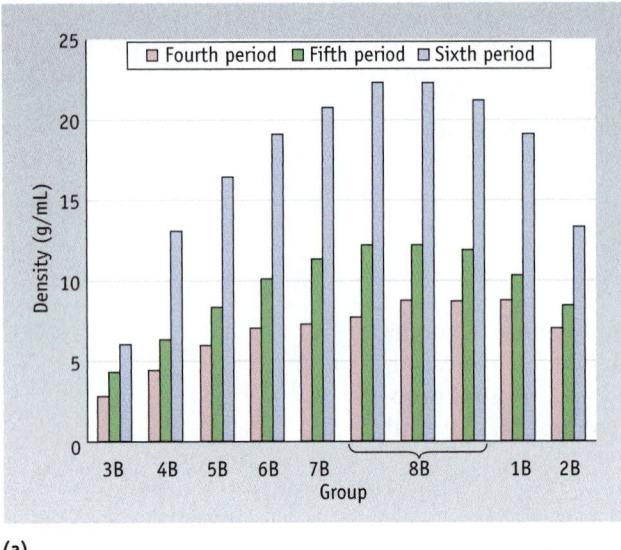

(a)

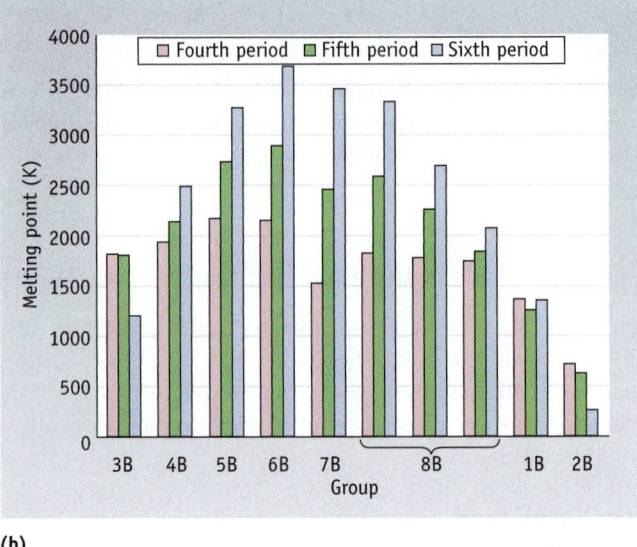

(b)

Figure 22.5 **Periodic properties in the transition series.** Density (a) and melting point (b) of the *d*-block elements.

melting points rise to a maximum around the middle of the series (Figure 22.5b), then descend. Again, these elements' electron configurations provide us with an explanation. The variation in melting point indicates that the strongest metallic bonds occur when the *d* subshell is about half-filled. This is also the point at which the largest number of electrons occupy the bonding molecular orbitals in the metal. (See the discussion of bonding in metals on page 643.)

GENERAL
Chemistry⚛Now™

See the General ChemistryNow CD-ROM or website:
• **Screen 22.3 Periodic Trends for Transition Elements,** for more on transition metal chemistry

---

22.2—Metallurgy

A few metals occur in nature as the free elements. This group includes copper (Figure 22.6), silver, and gold. Most metals, however, are found as oxides, sulfides, halides, carbonates, or other ionic compounds (Figure 22.7). Some metal-containing mineral deposits have little economic value, either because the concentration of the metal is too low or because the metal is difficult to separate from impurities. The relatively few minerals from which elements can be obtained profitably are called ores (Figure 22.7). **Metallurgy** is the general name given to the process of obtaining metals from their ores.

Very few ores are chemically pure substances. Instead, the desired mineral is usually mixed with large quantities of impurities such as sand and clay, called **gangue** (pronounced "gang"). Generally, the first step in a metallurgical process is to separate the mineral from the gangue. Then the ore is converted to the metal, a reduction process. Pyrometallurgy and hydrometallurgy are two methods of recovering metals from their ores. As the names imply, **pyrometallurgy** involves high tem-

Charles D. Winters

Figure 22.6 **Naturally occurring copper.** Copper occurs as the metal (native copper) and as minerals such as blue azurite [2 $CuCO_3 \cdot Cu(OH)_2$] and malachite [$CuCO_3 \cdot Cu(OH)_2$].

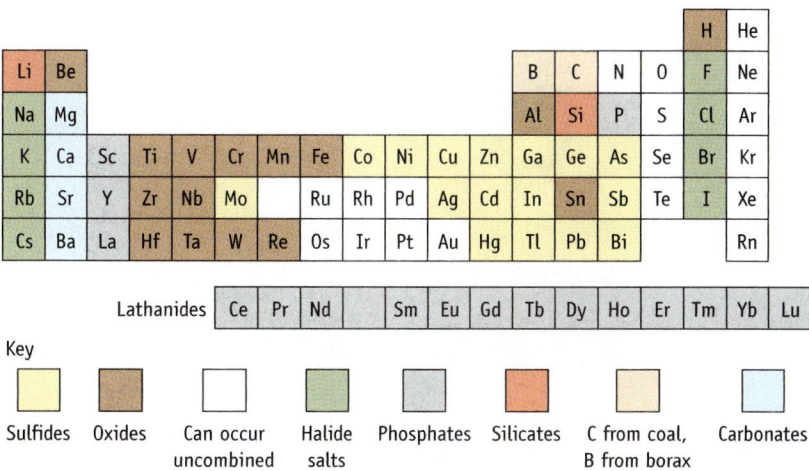

Figure 22.7 Sources of the elements. A few transition metals, such as copper and gold, occur naturally as the metal. Most other elements are found naturally as oxides, sulfides, or other salts.

peratures and **hydrometallurgy** uses aqueous solutions (and thus is limited to the relatively low temperatures at which water is a liquid). Iron and copper metallurgy illustrate these two methods of metal production.

Pyrometallurgy: Iron Production

The production of iron from its ores is carried out in a blast furnace (Figure 22.8). The furnace is charged with a mixture of ore (usually hematite, Fe_2O_3), coke (which is primarily carbon), and limestone ($CaCO_3$). A blast of hot air forced in at the bottom of the furnace causes the coke to burn with such an intense heat that the temperature at the bottom is almost 1500 °C. The quantity of air input is controlled so that carbon monoxide is the primary product. Both carbon and carbon monoxide participate in the reduction of iron(III) oxide to give impure metal.

$$Fe_2O_3(s) + 3\ C(s) \longrightarrow 2\ Fe(\ell) + 3\ CO(g)$$
$$Fe_2O_3(s) + 3\ CO(g) \longrightarrow 2\ Fe(\ell) + 3\ CO_2(g)$$

Much of the carbon dioxide formed in the reduction process (and from heating the limestone) is reduced on contact with unburned coke and produces more reducing agent.

$$CO_2(g) + C(s) \longrightarrow 2\ CO(g)$$

The molten iron flows down through the furnace and collects at the bottom, where it is tapped off through an opening in the side. This impure iron is called *cast iron* or *pig iron*. Usually, the impure metal is either brittle or soft (undesirable properties for most uses) due to the presence of impurities such as elemental carbon, phosphorus, and sulfur.

Iron ores generally contain silicate minerals and silicon dioxide. Lime (CaO), formed when limestone is heated, reacts with these materials to give calcium silicate.

$$SiO_2(s) + CaO(s) \longrightarrow CaSiO_3(\ell)$$

This is an acid–base reaction because CaO is a basic oxide and SiO_2 is an acidic oxide. The calcium silicate, molten at the temperature of the blast furnace and less dense than molten iron, floats on the iron. Other nonmetal oxides dissolve in this layer and the mixture, called *slag*, is easily removed.

■ **Coke: A Reducing Agent**
Coke is made by heating coal in a tall, narrow oven that is sealed to keep out oxygen. Heating drives off volatile chemicals, including benzene and ammonia. What remains is nearly pure carbon.

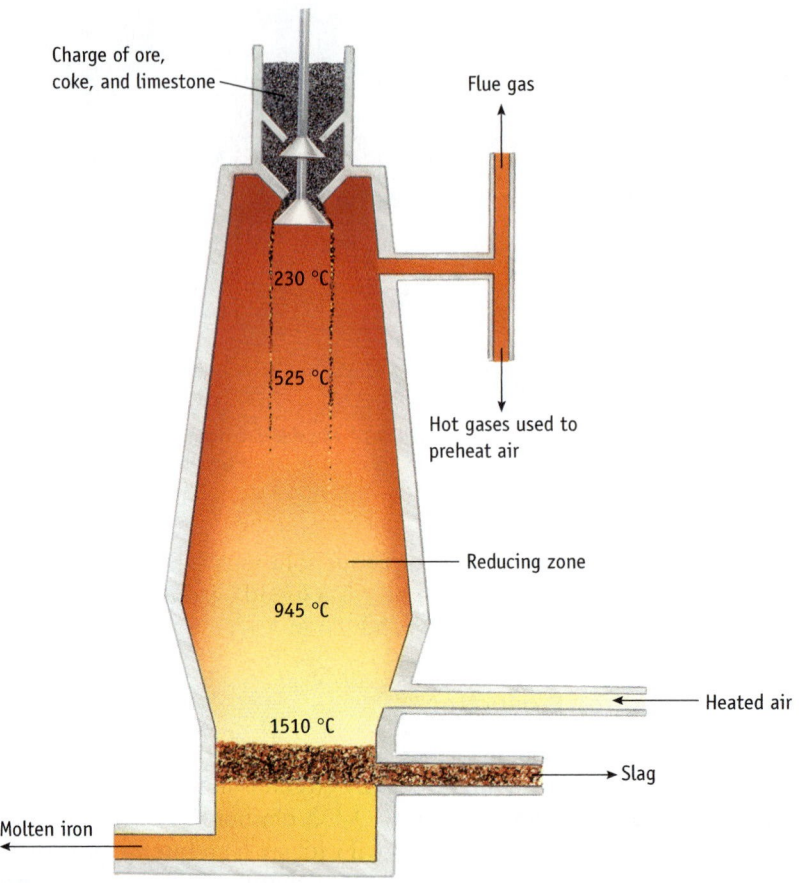

Charge of ore, coke, and limestone

Flue gas

230 °C

525 °C

Hot gases used to preheat air

Reducing zone

945 °C

Heated air

1510 °C

Slag

Molten iron

Active Figure 22.8 **A blast furnace.** The largest modern furnaces have hearths 14 meters in diameter. They can produce as much as 10,000 tons of iron per day.

GENERAL
Chemistry⚛Now™ See the General ChemistryNow CD-ROM or website to explore an interactive version of this figure accompanied by and exercise.

Pig iron from the blast furnace may contain as much as 4.5% carbon, 0.3% phosphorus, 0.04% sulfur, 1.5% silicon, and some other elements as well. The impure iron must be purified to remove these nonmetal components. Several processes are available to accomplish this task, but the most important uses the *basic oxygen furnace* (Figure 22.9). The process in the furnace removes much of the carbon and all of the phosphorus, sulfur, and silicon. Pure oxygen is blown into the molten pig iron and oxidizes phosphorus to P_4O_{10}, sulfur to SO_2, and carbon to CO_2. These nonmetal oxides either escape as gases or react with basic oxides such as CaO that are added or are used to line the furnace. For example,

$$P_4O_{10}(g) + 6\ CaO(s) \longrightarrow 2\ Ca_3(PO_4)_2(\ell)$$

The result is ordinary *carbon steel*. Almost any degree of flexibility, hardness, strength, and malleability can be achieved in carbon steel by reheating and cooling in a process called *tempering*. The resulting material can then be used in a wide variety of applications. The major disadvantages of carbon steel are that it corrodes easily and that it loses its properties when heated strongly.

Other transition metals, such as chromium, manganese, and nickel, can be added during the steel-making process, giving *alloys* (solid solutions of two or more

Figure 22.9   **Molten iron being poured from a basic oxygen furnace.**

metals; see "The Chemistry of Modern Materials," page 642) that have specific physical, chemical, and mechanical properties. One well-known alloy is stainless steel, which contains 18% to 20% Cr and 8% to 12% Ni. Stainless steel is much more resistant to corrosion than carbon steel. Another alloy of iron is Alnico V. Used in loudspeaker magnets because of its permanent magnetism, it contains five elements: Al (8%), Ni (14%), Co (24%), Cu (3%), and Fe (51%).

Hydrometallurgy: Copper Production

In contrast to iron ores, which are mostly oxides, most copper minerals are sulfides. Copper-bearing minerals include chalcopyrite ($CuFeS_2$), chalcocite (Cu_2S), and covellite (CuS) (Figure 22.6). Because ores containing these minerals generally have a very low percentage of copper, enrichment is necessary. This step is carried out by a process known as *flotation*. First, the ore is finely powdered. Next, oil is added and the mixture is agitated with soapy water in a large tank (Figure 22.10). At the same time, compressed air is forced through the mixture, so that the lightweight, oil-covered copper sulfide particles are carried to the top as a frothy mixture. The heavier gangue settles to the bottom of the tank, and the copper-laden froth is skimmed off.

Hydrometallurgy can be used to obtain copper from an enriched ore. In one method, enriched chalcopyrite ore is treated with a solution of copper(II) chloride. A reaction ensues that leaves copper in the form of solid, insoluble CuCl, which is easily separated from the iron that remains in solution as aqueous $FeCl_2$.

$$CuFeS_2(s) + 3\ CuCl_2(aq) \longrightarrow 4\ CuCl(s) + FeCl_2(aq) + 2\ S(s)$$

Aqueous NaCl is then added and CuCl dissolves because of the formation of the complex ion $[CuCl_2]^-$.

$$CuCl(s) + Cl^-(aq) \longrightarrow [CuCl_2]^-(aq)$$

Copper(I) compounds are unstable with respect to Cu(0) and Cu(II). Thus, $[CuCl_2]^-$ disproportionates to the metal and $CuCl_2$, and the latter is used to treat further ore.

$$2\ [CuCl_2]^-(aq) \longrightarrow Cu(s) + CuCl_2(aq) + 2\ Cl^-(aq)$$

Approximately 10% of the copper produced in the United States is obtained with the aid of bacteria. Acidified water is sprayed onto copper-mining wastes that contain low levels of copper. As the water trickles down through the crushed rock, the bacterium *Thiobacillus ferrooxidans* (page 1013) breaks down the iron sulfides in the rock and converts iron(II) to iron(III). Iron(III) ions oxidize the sulfide ion of copper sulfide to sulfate ions, leaving copper(II) ions in solution. Then the copper(II) ion is reduced to metallic copper by reaction with iron.

$$Cu^{2+}(aq) + Fe(s) \longrightarrow Cu(s) + Fe^{2+}(aq)$$

The purity of the copper obtained via these metallurgical processes is about 99%, but this is not acceptable because even traces of impurities greatly diminish the electrical conductivity of the metal. Consequently, a further purification step is needed—one involving electrolysis (Figure 22.11). Thin sheets of pure copper metal and slabs of impure copper are immersed in a solution containing $CuSO_4$ and H_2SO_4. The pure copper sheets serve as the cathode of an electrolysis cell, and the impure slabs are the anode. Copper in the impure sample is oxidized to copper(II) ions at the anode, and copper(II) ions in solution are reduced to pure copper at the cathode.

Figure 22.10 Enriching copper ore by the flotation process. The less dense particles of Cu_2S are trapped in the soap bubbles and float. The denser gangue settles to the bottom.

Figure 22.11   Electrolytic refining of copper. (a) Slabs of impure copper, called "blister copper," form the anode and pure copper is deposited at the cathode. (b) The electrolysis cells at a copper refinery.

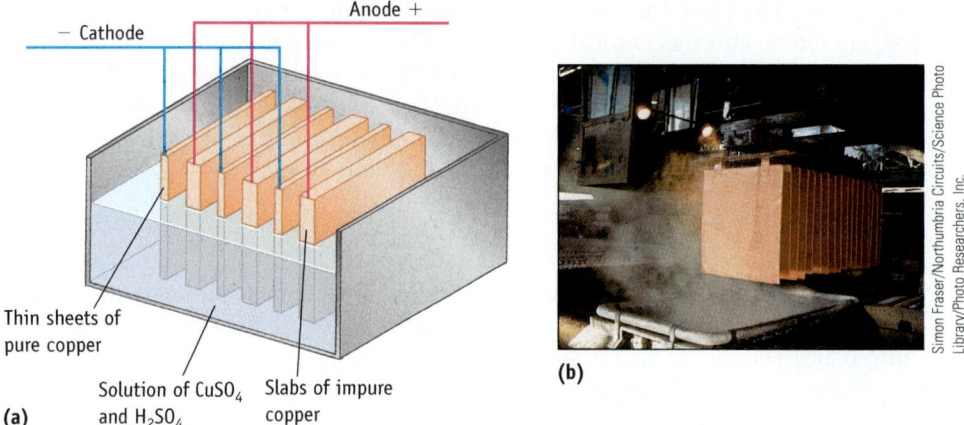

Anode +

− Cathode

Thin sheets of pure copper

Solution of CuSO₄ and H₂SO₄

Slabs of impure copper

(a)

(b)

Simon Fraser/Northumbria Circuits/Science Photo Library/Photo Researchers, Inc.

22.3—Coordination Compounds

When metal salts dissolve, water molecules cluster around the ions (page 177). The negative end of each polar water molecule is attracted to the positively charged metal ion, and the positive end of the water molecule is attracted to the anion. As noted earlier [◀ Section 13.2], the energy of the ion–solvent interaction (solvation energy) is an important aspect of the solution process. But there is much more to this story.

Complexes and Ligands

A green solution formed by dissolving nickel(II) chloride in water contains Ni^{2+} (aq) and Cl^- (aq) ions (Figure 22.12, facing page). If the solvent is removed, a green crystalline solid is obtained. The formula of this solid is often written as $NiCl_2 \cdot 6\,H_2O$, and the compound is called nickel(II) chloride hexahydrate. Addition of ammonia to the aqueous nickel(II) chloride solution gives a lilac-colored solution from which another compound, $NiCl_2 \cdot 6\,NH_3$, can be isolated. This formula looks very similar to the formula for the hydrate, with ammonia substituted for water.

What are these two nickel species? The formulas identify the compounds' compositions but fail to give information about their structures. Because properties of compounds derive from their structures, we need to evaluate the structures in more detail. Typically, metal compounds are ionic, and solid ionic compounds have structures with cations and anions arranged in a regular array. The structure of hydrated nickel chloride contains cations with the formula $[Ni(H_2O)_6]^{2+}$ and chloride anions. The structure of the ammonia-containing compound is similar to the hydrate; it is made up of $[Ni(NH_3)_6]^{2+}$ cations and chloride anions.

Ions such as $[Ni(H_2O)_6]^{2+}$ and $[Ni(NH_3)_6]^{2+}$, in which a metal ion and either water or ammonia molecules compose a single structural unit, are examples of **coordination complexes**, also known as **complex ions** (Figure 22.13). Compounds containing a coordination complex as part of the structure are called **coordination compounds**, and their chemistry is known as coordination chemistry. Although the older "hydrate" formulas are still used, the preferred method of writing the formula for coordination compounds places the metal atom or ion and the molecules or anions directly bonded to it within brackets to show that it is a single structural entity. Thus, the formula for the nickel(II)–ammonia compound is better written as $[Ni(NH_3)_6]Cl_2$.

Sum of metal ion and ligand charges

Coordination complex

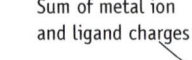

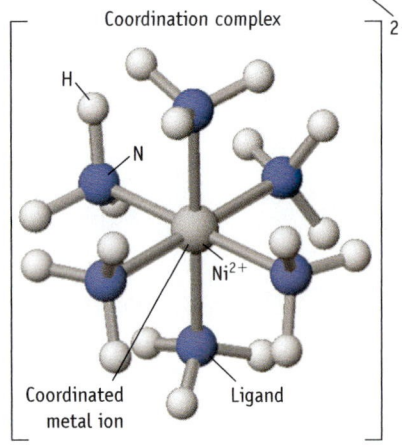

H

N

Ni²⁺

Coordinated metal ion

Ligand

$[Ni(NH_3)_6]^{2+}$

Figure 22.13   A coordination complex. In the $[Ni(NH_3)_6]^{2+}$ ion, the ligands are NH_3 molecules. Because the metal has a 2+ charge, and the ligands have no charge, the charge on the complex ion is 2+.

Figure 22.12 **Coordination compounds of Ni²⁺ ion.** The transition metals and their ions form a wide range of compounds, often with beautiful colors and interesting structures. One purpose of this chapter is to explore some commonly observed structures and explain how these compounds can be so colorful.

All coordination complexes contain a metal atom or ion as the central part of the structure. Bonded to the metal are molecules or ions called **ligands** (from the Latin verb *ligare*, meaning "to bind"). In the preceding examples, water and ammonia are the ligands. The number of ligand atoms attached to the metal defines the **coordination number** of the metal. The geometry described by the attached ligands is called the **coordination geometry**. With structures of coordination compounds, we will also encounter isomerism (see Section 22.4). In the nickel complex ion $[Ni(NH_3)_6]^{2+}$ (Figure 22.12), the six ligands are arranged in a regular octahedral geometry around the central metal ion.

Ligands can be either neutral molecules or anions (or, in rare instances, cations). The characteristic feature of a ligand is that it contains a lone pair of electrons. In the classic description of bonding in a coordination complex, the lone pair of electrons on a ligand is shared with the metal ion. The attachment is a coordinate covalent bond [◀ Section 9.6], because the electron pair being shared was originally on the ligand. The name "coordination complex" derives from the name given to this kind of bonding.

If the ligands are ions, the net charge on a coordination complex is the sum of the charges on the metal and its attached groups. Complexes can be cations (as in the two nickel complexes used as examples here), anions, or uncharged.

Ligands such as H_2O and NH_3, which coordinate to the metal via a single Lewis base atom, are termed **monodentate**. The word "dentate" comes from the Latin *dentis*, meaning "tooth," so NH_3 is a "one-toothed" ligand. Some ligands attach to the metal with more than one donor atom. These ligands are called **polydentate**. Ethylenediamine (1,2-diaminoethane), $H_2NCH_2CH_2NH_2$, often abbreviated as en; oxalate ion, $C_2O_4^{2-}$ (ox^{2-}); and phenanthroline, $C_{12}H_8N_2$ (phen), are examples of the wide variety of bidentate ligands (Figure 22.14). Structures and examples of some complex ions with bidentate ligands are shown in Figure 22.15.

Polydentate ligands are also called **chelating ligands,** or just chelates (pronounced "key-lates"). The name derives from the Greek *chele*, meaning "claw." Because two or more bonds be broken to separate the ligand from the metal, complexes with chelated ligands have greater stability than those with monodentate ligands. Chelated complexes are important in everyday life. One way to clean the rust out of water-cooled automobile engines and steam boilers is to add a solution of oxalic acid. Iron oxide reacts with oxalic acid to give a water-soluble iron oxalate complex ion:

$$Fe_2O_3(s) + 6\ H_2C_2O_4(aq) \longrightarrow 2\ [Fe(C_2O_4)_3]^{3-}(aq) + 6\ H^+(aq) + 3\ H_2O(\ell)$$

■ **Ligands Are Lewis Bases**
Ligands are Lewis bases because they furnish the electron pair; the metal ion is a Lewis acid because it accepts electron pairs (see Section 17.9). Thus, the coordinate covalent bond between ligand and metal results from a Lewis acid–Lewis base interaction.

■ **Bidentate Ligands**
All common bidentate ligands bind to *adjacent* sites on the metal.

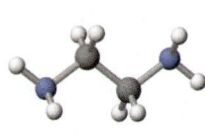

(a) $H_2NCH_2CH_2NH_2$, en

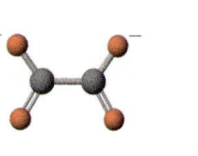

(b) $C_2O_4^{2-}$, ox

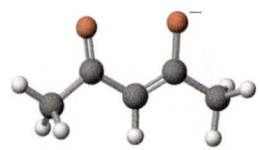

(c) $CH_3COCHCOCH_3^-$, acac$^-$

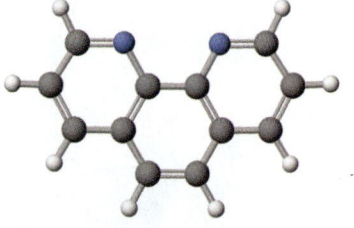

(d) $C_{12}H_8N_2$, phen

Figure 22.14 Common bidentate ligands. (a) Ethylenediamine, $H_2NCH_2CH_2NH_2$; (b) oxalate ion, $C_2O_4^{2-}$; (c) acetylacetonate ion, $CH_3COCHCOCH_3^-$; (d) phenanthroline, $C_{12}H_8N_2$. Coordination of these bidentate ligands to a transition metal ion results in five- or six-member metal-containing rings and no ring strain.

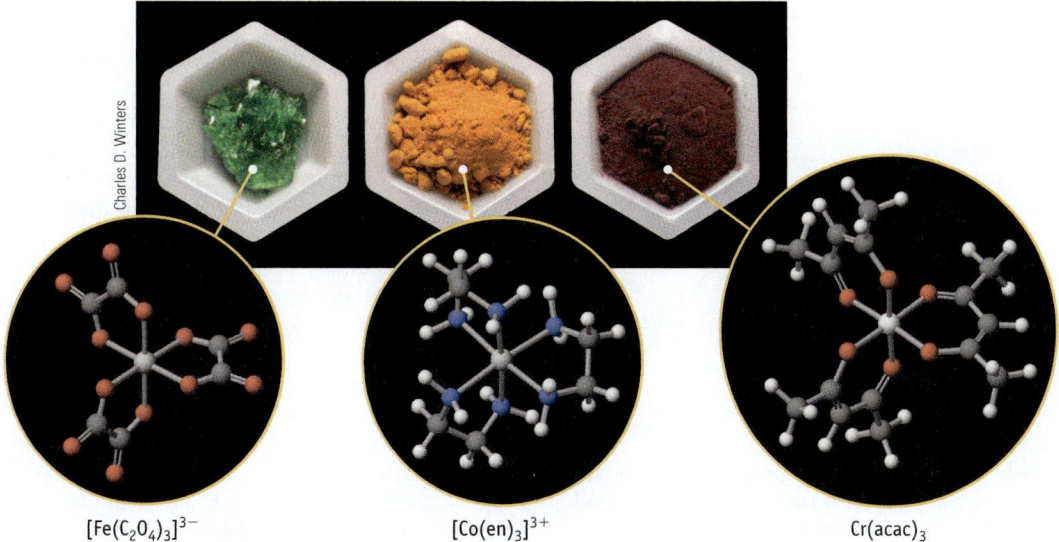

Figure 22.15 Complex ions with bidentate ligands. See Figure 22.14 for abbreviations.

Ethylenediaminetetraacetate ion (EDTA^{4-}), a hexadentate ligand, is an excellent chelating ligand (Figure 22.16). It can wrap around a metal ion, encapsulating it. Salts of this anion are often added to commercial salad dressings to remove traces of free metal ions from solution; otherwise, these metal ions can act as catalysts for the oxidation of the oils in the dressing. Without EDTA^{4-}, the dressing would quickly become rancid. Another use is in bathroom cleansers. The EDTA^{4-} ion removes deposits of $CaCO_3$ and $MgCO_3$ left by hard water by coordinating to Ca^{2+} or Mg^{2+} to create soluble complex ions.

Complexes with polydentate ligands play particularly important roles in biochemistry, as described in "A Closer Look: Hemoglobin."

Formulas of Coordination Compounds

It is useful to be able to predict the formula of a coordination complex, given the metal ion and ligands, and to derive the oxidation number of the coordinated metal ion, given the formula in a coordination compound. The following examples explore these issues.

Figure 22.16 EDTA^{4-}, a hexadentate ligand.
(a) Ethylenediaminetetraacetate, EDTA^{4-}. (b) [Co(EDTA)]$^-$. Notice the five- and six-member rings created when this ligand bonds to the metal.

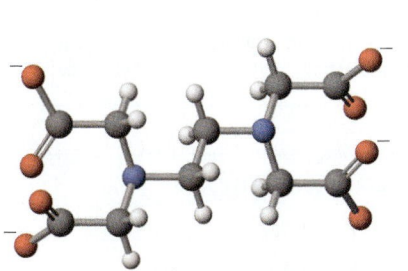

(a) Ethylenediaminetetraacetate, EDTA^{4-}

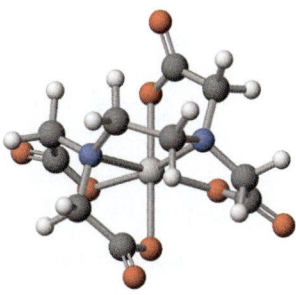

(b) [Co(EDTA)]$^-$

A Closer Look

Hemoglobin

Metal-containing coordination compounds figure prominently in many biochemical reactions. Perhaps the best-known example is hemoglobin, the chemical in the blood responsible for O_2 transport. It is also one of the most thoroughly studied bioinorganic compounds.

As described in "The Chemistry of Life: Biochemistry" (page 534), hemoglobin (Hb) is a large iron-containing protein. It includes four polypeptide segments, each containing an iron(II) ion locked inside a

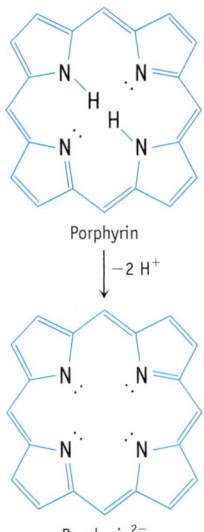

Porphyrin

$$\downarrow -2\ H^+$$

Porphyrin^{2-}

Porphyrin ring of the heme group. The tetradentate ligand surrounding the iron(II) ion in hemoglobin is a dianion of a substituted molecule called a porphyrin. Because of the double bonds in this structure, all of the carbon and nitrogen atoms in the dianion of the porphyrin lie in a plane. In addition, the nitrogen lone pairs are directed toward the center of the molecule and the molecular dimensions are such that a metal ion may fit nicely into the cavity.

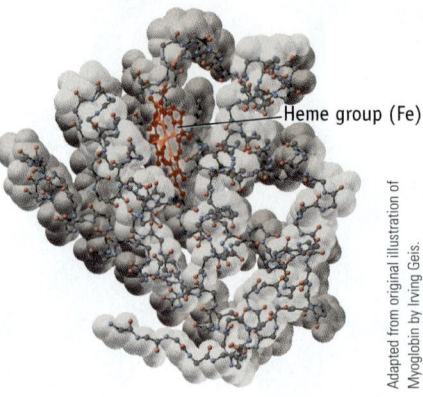

Heme group (Fe)

The heme group in myoglobin. This protein is a close relative of hemoglobin. The heme group with its iron ion is shown.

Adapted from original illustration of Myoglobin by Irving Geis.

porphyrin ring system and coordinated to a nitrogen atom from another part of the protein. A sixth site is available to attach to oxygen.

One segment of the hemoglobin molecule resembles the myoglobin structure shown above. (Myoglobin, the oxygen-storage protein in muscle, has only one polypeptide chain with an enclosed heme group. It is the oxygen storage protein in muscle.) In this case, the iron-containing heme group is enclosed with a polypeptide chain. The iron ion in the porphyrin ring is shown. The first and sixth coordination positions are taken up by nitrogen atoms from amino acids of the polypeptide chain.

Hemoglobin functions by reversibly adding oxygen to the sixth coordination position of each iron, giving a complex called oxyhemoglobin.

Because hemoglobin features four iron centers, a maximum of four molecules of oxygen can bind to the molecule. The binding to oxygen is cooperative; that is, binding one molecule enhances the tendency to bind the second, third, and fourth molecules. Formation of the oxygenated

complex is favored, albeit not too highly, because oxygen must also be released by the molecule to body tissues. Interestingly, an increase in acidity leads to a decrease in the stability of the oxygenated complex. This phenomenon is known as the *Bohr effect*, named for Christian Bohr, Niels Bohr's father. Release of oxygen in tissues is facilitated by an increase in acidity that results from the presence of CO_2 formed by metabolism.

Among the notable properties of hemoglobin is its ability to form a complex with carbon monoxide. This complex is very stable, with the equilibrium constant for the following reaction being about 200 (where Hb is hemoglobin):

$$HbO_2(aq) + CO(g) \rightleftharpoons HbCO(aq) + O_2(g)$$

When CO complexes with iron, the oxygen-carrying capacity of hemoglobin is lost. Consequently, CO is highly toxic to humans. Exposure to even small amounts greatly reduces the capacity of the blood to transport oxygen.

Hemoglobin abnormalities are well known. One of the most common abnormalities causes sickle cell anemia and was described in "The Chemistry of Life: Biochemistry," on page 533.

For more on hemoglobin, see the earlier discussion of "Blood Gas" on page 942.

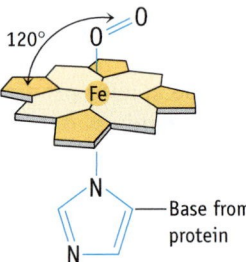

Oxygen binding. Oxygen binds to the iron of the heme group in hemoglobin (and in myoglobin). Interestingly, the Fe—O—O angle is bent.

Example 22.1—Formulas of Coordination Complexes

Problem Give the formulas of the following coordination complexes:

(a) a Ni^{2+} ion is bound to two water molecules and two bidentate oxalate ions

(b) a Co^{3+} ion is bound to one Cl^- ion, one ammonia molecule, and two bidentate ethylenediamine (en) molecules

Strategy The problem requires determining the net charge, which equals the sum of the charges of the various component parts of the complex ion. With that information, the metal and ligands can be assembled in the formula, which is placed in brackets, and the net charge attached.

Solution

(a) This complex ion is constructed from two neutral H_2O molecules, two $C_2O_4^{2-}$ ions, and one Ni^{2+} ion, so the net charge on the complex is -2. The formula for the complex ion is $[Ni(C_2O_4)_2(H_2O)_2]^{2-}$.

(b) This cobalt(III) complex combines two en molecules and one NH_3 molecule, both having no charge, as well as one Cl^- ion and a Co^{3+} ion. The net charge is $+2$. The formula for this complex (writing out the entire formula for ethylenediamine) is $[Co(H_2NCH_2CH_2NH_2)_2(NH_3)Cl]^{2+}$.

Example 22.2—Coordination Compounds

Problem In each of the following complexes, determine the metal's oxidation number and coordination number.

(a) $[Co(en)_2(NO_2)_2]Cl$

(b) $Pt(NH_3)_2(C_2O_4)$

(c) $Pt(NH_3)_2Cl_4$

(d) $[Co(NH_3)_5Cl]SO_4$

Strategy Each formula consists of a complex ion or molecule made up of the metal ion, neutral and/or anionic ligands (the part inside the square brackets), and a counterion (outside the brackets). The oxidation number of the metal is the charge necessary to balance the sum of the negative charges associated with the anionic ligands and counterion. The coordination number is the number of donor atoms in the ligands that are bonded to the metal. Remember that the bidentate ligands in these examples (en, oxalate ion) attach to the metal at two sites and that the counterion is not part of the complex ion—that is, it is not a ligand.

Solution

(a) The chloride ion with a $1-$ charge, outside the brackets, shows that the charge on the complex ion must be $1+$. There are two nitrite ions (NO_2^-) and two neutral bidentate ethylenediamine ligands in the complex. To give a $1+$ charge on the complex ion, the cobalt ion must have a charge of $3+$; that is, the sum of $2-$ (two nitrites), 0 (two en ligands), and $3+$ (the cobalt ion) equals $1+$. Each en ligand fills two coordination positions, and the two nitrite ions fill two more positions. The coordination number of the metal is 6.

(b) There is an oxalate ion ($C_2O_4^{2-}$) and two neutral ammonia ligands. To balance the charge on the oxalate ion, platinum must have a $2+$ charge; that is, it has an oxidation number of $+2$. The coordination number is 4, with an oxalate ligand filling two coordination positions and each ammonia molecule filling one.

(c) There are four chloride ions (Cl^-) and two neutral ammonia ligands. In this complex, the oxidation number of the metal is $+4$ and the coordination number is 6.

(d) There is one chloride ion (Cl^-) and five neutral ammonia ligands. The counter ion is sulfate with a $2-$ charge, so the overall charge on the complex is $2+$. The oxidation number of the metal is $+3$ and the coordination number is 6 (sulfate is not coordinated to the metal).

Exercise 22.1—Formulas of Coordination Compounds

(a) What is the formula of a complex ion composed of one Co^{3+} ion, three ammonia molecules, and three Cl^- ions?

(b) Determine the metal's oxidation number and coordination number in (i) $K_3[Co(NO_2)_6]$ and in (ii) $Mn(NH_3)_4Cl_2$.

Naming Coordination Compounds

Just as rules govern naming of inorganic and organic compounds, coordination compounds are named according to an established system. The three compounds just below are named according to the rules that follow.

Compound	Systematic Name
$[Ni(H_2O)_6]SO_4$	Hexaaquanickel(II) sulfate
$[Cr(en)_2(CN)_2]Cl$	Dicyanobis(ethylenediamine)chromium(III) chloride
$K[Pt(NH_3)Cl_3]$	Potassium amminetrichloroplatinate(II)

1. In naming a coordination compound that is a salt, name the cation first and then the anion. (This is how all salts are commonly named.)

2. When giving the name of the complex ion or molecule, name the ligands first, in alphabetical order, followed by the name of the metal. (When determining alphabetical order, the prefix is not considered part of the name.)

3. Ligands and their names:

   (a) If a ligand is an anion whose name ends in *-ite* or *-ate*, the final *e* is changed to *o* (sulfate $\longrightarrow$ sulfato or nitrite $\longrightarrow$ nitrito).

   (b) If the ligand is an anion whose name ends in *-ide*, the ending is changed to *o* (chloride $\longrightarrow$ chloro, cyanide $\longrightarrow$ cyano).

   (c) If the ligand is a neutral molecule, its common name is usually used with several important exceptions: Water as a ligand is referred to as *aqua*, ammonia is called *ammine*, and CO is called *carbonyl*.

   (d) When there is more than one of a particular monodentate ligand with a simple name, the number of ligands is designated by the appropriate prefix: *di, tri, tetra, penta,* or *hexa*. If the ligand name is complicated (whether monodentate or bidentate), the prefix changes to *bis, tris, tetrakis, pentakis,* or *hexakis*, followed by the ligand name in parentheses.

4. If the coordination complex is an anion, the suffix *-ate* is added to the metal name.

5. Following the name of the metal, the oxidation number of the metal is given in Roman numerals.

Example 22.3—Naming Coordination Compounds

Problem Name the following compounds:

(a) $[Cu(NH_3)_4]SO_4$

(b) $K_2[CoCl_4]$

(c) $Co(phen)_2Cl_2$

(d) $[Co(en)_2(H_2O)Cl]Cl_2$

Strategy Apply the rules for nomenclature given above.

Solution

(a) The complex ion (in square brackets) is composed of four NH_3 molecules (named *ammine* in a complex) and the copper ion. To balance the $2-$ charge on the sulfate counterion, copper must have a $2+$ charge. The compound's name is

tetraamminecopper(II) sulfate

(b) The complex ion $[CoCl_4]^{2-}$ has a $2-$ charge. With four Cl^- ligands, the cobalt ion must have a $+2$ charge, so the sum of charges is $2-$. The name of the compound is

potassium tetrachlorocobaltate(II)

(c) This is a neutral coordination compound. The ligands include two Cl^- ions and two neutral bidentate *phen* (phenanthroline) ligands. The metal ion must have a $2+$ charge (Co^{2+}). The name, listing ligands in alphabetical order, is

dichlorobis(phenanthroline)cobalt(II)

(d) The complex ion has a $2-$ charge because it is paired with two uncoordinated Cl^- ions. The cobalt ion is Co^{3+} because it is bonded to two neutral en ligands, one neutral water, and one Cl^-. The name is

aquachlorobis(ethylenediamine)cobalt(III) chloride

Exercise 22.2—Naming Coordination Compounds

Name the following coordination compounds.

(a) $[Ni(H_2O)_6]SO_4$ **(c)** $K[Pt(NH_3)Cl_3]$

(b) $[Cr(en)_2(CN)_2]Cl$ **(d)** $K[CuCl_2]$

22.4—Structures of Coordination Compounds

Common Coordination Geometries

The geometry of a coordination complex is defined by the arrangement of donor atoms of the ligands around the central metal ion. Metal ions in coordination compounds can have coordination numbers ranging from 2 to 12. Only complexes with coordination numbers of 2, 4, and 6 are common, however, so we will concentrate on species such as $[ML_2]^{n\pm}$, $[ML_4]^{n\pm}$, and $[ML_6]^{n\pm}$, where M is the metal ion and L is a monodentate ligand. Within these stoichiometries, the following geometries are encountered:

- All $[ML_2]^{n\pm}$ complexes are linear. The two ligands are on opposite sides of the metal, and the L—M—L bond angle is 180°. Common examples include $[Ag(NH_3)_2]^+$ and $[CuCl_2]^-$.
- Tetrahedral geometry occurs in many $[ML_4]^{n\pm}$ complexes. Examples include $TiCl_4$, $[CoCl_4]^{2-}$, $[NiCl_4]^{2-}$, and $[Zn(NH_3)_4]^{2+}$.
- The $[ML_4]^{n\pm}$ complexes can also have square planar geometry. This geometry is most often seen with metal ions that have eight *d* electrons. Examples include $Pt(NH_3)_2Cl_2$, $[Ni(CN)_4]^{2-}$, and the nickel complex with the dimethylglyoximate (dmg$^-$) ligand in Figure 22.12.
- Octahedral geometry is found in complexes with the stoichiometry $[ML_6]^{n\pm}$ (Figure 22.12).

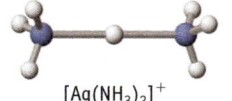

$[Ag(NH_3)_2]^+$ Linear

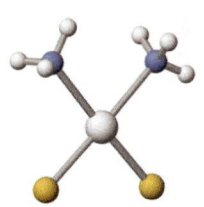

$Pt(NH_3)_2Cl_2$ Square planar

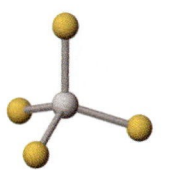

$[NiCl_4]^{2-}$ Tetrahedral

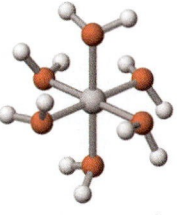

$[M(H_2O)_6]^{n+}$ Octahedral

GENERAL
Chemistry·⚛·Now™

See the General ChemistryNow CD-ROM or website:
• **Screen 22.5 Geometry of Coordination Compounds,** for an exercise on compound geometries

Isomerism

Isomerism is one of the most interesting aspects of molecular structure. Recall that the chemistry of organic compounds is greatly enlivened by the multitude of isomeric compounds that are known.

Isomers are classified as follows:

- *Structural isomers* have the same molecular formula but different bonding arrangements of atoms.

- *Stereoisomers* have the same atom-to-atom bonding sequence, but the atoms differ in their arrangement in space. There are two types of stereoisomers: geometric isomers (such as *cis* and *trans* alkenes, page 477) and optical isomers (non-superimposable mirror images that have the unique property that they rotate planar polarized light, page 478).

All three types of isomerism, structural, geometric, and optical, are encountered in coordination chemistry.

Structural Isomerism

The two most important types of structural isomerism in coordination chemistry are coordination isomerism and linkage isomerism. **Coordination isomerism** occurs when it is possible to exchange a coordinated ligand and the uncoordinated counterion. For example, dark violet $[Co(NH_3)_5Br]SO_4$ and red $[Co(NH_3)_5SO_4]Br$ are coordination isomers. In the first compound, bromide ion is a ligand and sulfate is a counterion; in the second, sulfate is a ligand and bromide is the counterion. A diagnostic test for this kind of isomer is often made based on chemical reactions. For example, these two compounds can be distinguished by precipitation reactions. Addition of $Ag^+(aq)$ to a solution of $[Co(NH_3)_5SO_4]Br$ gives a precipitate of AgBr, indicating the presence of bromide ion in solution. In contrast, no reaction occurs if $Ag^+(aq)$ is added to a solution of $[Co(NH_3)_5Br]SO_4$. In this complex, bromide ion is attached to Co^{3+} and is not a free ion in solution.

$$[Co(NH_3)_5Br]SO_4 + Ag^+(aq) \longrightarrow \text{no reaction}$$
$$[Co(NH_3)_5SO_4]Br + Ag^+(aq) \longrightarrow AgBr(s) + [Co(NH_3)_5SO_4]^{2+}(aq)$$

Linkage isomerism occurs when it is possible to attach a ligand to the metal through different atoms. The two most common ligands with which linkage isomerism arises are thiocyanate, SCN^-, and nitrite, NO_2^-. The Lewis structure of the thiocyanate ion shows that there are lone pairs of electrons on sulfur and nitrogen. The ligand can attach to a metal either through sulfur (called S-bonded thiocyanate) or through nitrogen (called N-bonded thiocyanate). Nitrite ion can attach either at oxygen or at nitrogen. The former are called nitrito complexes; the latter are nitro complexes (Figure 22.17).

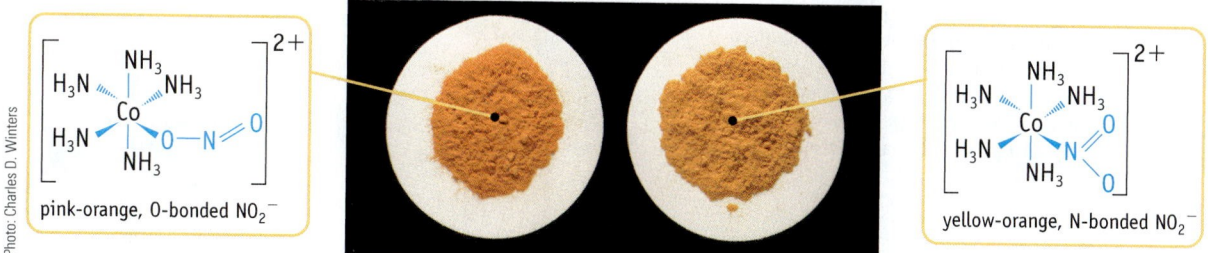

Figure 22.17 **Linkage isomers, [Co(NH₃)₅ONO]²⁺ and [Co(NH₃)₅NO₂]²⁺.** These complexes, whose systematic names are pentaamminenitritocobalt(III) and pentaamminenitrocobalt(III), were the first known examples of this type of isomerism.

Ligands forming linkage isomers

Geometric Isomerism

Geometric isomerism results when the atoms bonded directly to the metal have a different spatial arrangement. The simplest example of geometric isomerism in coordination chemistry is *cis-trans* isomerism, which occurs in both square-planar and octahedral complexes. An example of *cis-trans* isomerism is seen in the square-planar complex $Pt(NH_3)_2Cl_2$ (Figure 22.18a). In this complex, the two Cl^- ions can be either adjacent to each other (*cis*) or on opposite sides of the metal (*trans*). The *cis* isomer is effective in the treatment of testicular, ovarian, bladder, and osteogenic sarcoma cancers, but the *trans* isomer has no effect on these diseases.

Cis-trans isomerism in an octahedral complex is illustrated by $[Co(H_2NCH_2CH_2NH_2)_2Cl_2]^+$, a complex ion with two bidentate ethylenediamine ligands and two chloride ligands. In this complex, the two Cl^- ions occupy positions that are either adjacent (the purple *cis* isomer) or opposite (the green *trans* isomer) (Figure 22.18b).

Another common type of geometric isomerism occurs for octahedral complexes with the general formula MX_3Y_3. A *fac* isomer has three identical ligands

■ ***Cis-Trans* Isomerism**
Cis-trans isomerism is not possible for tetrahedral complexes. All L—M—L angles in tetrahedral geometry are 109.5°, and all positions are equivalent in this three-dimensional structure.

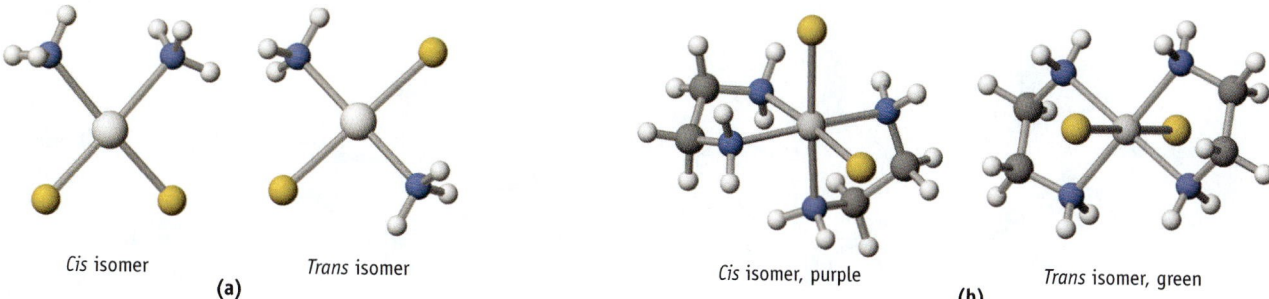

Cis isomer   Trans isomer
                (a)

Cis isomer, purple   Trans isomer, green
                        (b)

Figure 22.18 ***Cis-trans* isomers.** (a) The square planar complex $Pt(NH_3)_2Cl_2$ can exist in two geometries, *cis* and *trans*. (b) Similarly, *cis* and *trans* octahedral isomers are possible for $[Co(en)_2Cl_2]^+$.

Figure 22.19 *Fac* and *mer* isomers of **Cr(NH₃)₃Cl₃.** In the *fac* isomer, the three chloride ligands (and the three ammonia ligands) are arranged at the corners of a triangular face. In the *mer* isomer, the three similar ligands follow a meridian.

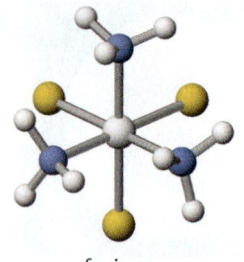
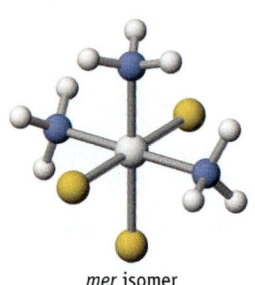

fac isomer

mer isomer

lying at the corners of a triangular face of an octahedron defined by the ligands (*fac* = facial), whereas the ligands follow a meridian in the *mer* isomer (*mer* = meridional). *Fac* and *mer* isomers of Cr(NH₃)₃Cl₃ are shown in Figure 22.19.

Optical Isomerism

Optical isomerism (chirality) occurs for octahedral complexes when the metal ion coordinates to three bidentate ligands or when the metal ion coordinates to two bidentate ligands and two monodentate ligands in a *cis* position. The complexes [Co(en)₃]³⁺ and *cis*-[Co(en)₂Cl₂]⁺, illustrated in Figure 22.20, are examples of chiral complexes. The diagnostic test for chirality is met with both species: Mirror images of these molecules are not superimposable. Solutions of the optical isomers rotate plane-polarized light in opposite directions.

GENERAL
Chemistry⚛Now™

See the General ChemistryNow CD-ROM or website:
- **Screen 22.6 Geometric Isomerism in Coordination Compounds,** for an exercise on isomerism in coordination chemistry

Example 22.4—Isomerism in Coordination Chemistry

Problem For which of the following complexes do isomers exist? If isomers are possible, identify the type of isomerism (structural or geometric). Determine whether the coordination complex is capable of exhibiting optical isomerism.

(a) [Co(NH₃)₄Cl₂]⁺   **(b)** Pt(NH₃)₂(CN)₂ (square planar)

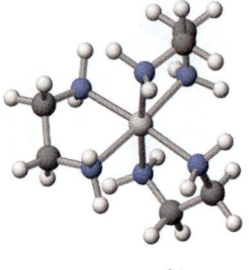
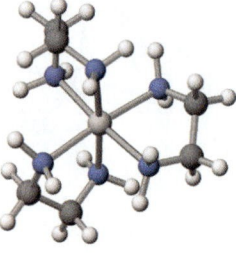
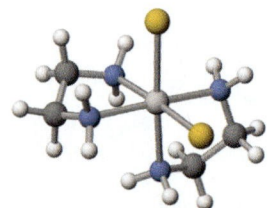
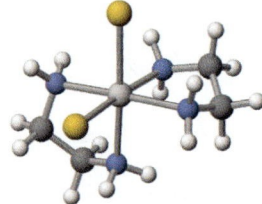

[Co(en)₃]³⁺   [Co(en)₃]³⁺ mirror image   *cis*-[Co(en)₂Cl₂]⁺   *cis*-[Co(en)₂Cl₂]⁺ mirror image

Figure 22.20   Chiral metal complexes. Both [Co(en)₃]³⁺ and *cis*-[Co(en)₂Cl₂]⁺ are chiral. Notice that the mirror images of the two compounds are not superimposable.

(c) $Co(NH_3)_3Cl_3$

(d) $Zn(NH_3)_2Cl_2$ (tetrahedral)

(e) $K_3[Fe(C_2O_4)_3]$

(f) $[Co(NH_3)_5SCN]^{2+}$

Strategy Determine the number of ligands attached to the metal and decide whether the ligands are monodentate or bidentate. Knowing how many donor atoms are coordinated to the metal (the coordination number) will allow you to establish the metal geometry. At that point it is necessary to recall the possible types of isomers for each geometry. The only isomerism possible for square-planar complexes is geometric (*cis* and *trans*). Tetrahedral complexes do not have isomers. Six-coordinate metals of the formula MA_4B_2 can be either *cis* or *trans*. *Mer* and *fac* isomers are possible with a stoichiometry of MA_3B_3. Optical activity arises for metal complexes of the formula *cis*-$M(bidentate)_2X_2$ and $M(bidentate)_3$ (among others). Drawing pictures of the molecules will help you visualize the isomers.

Solution

(a) Two geometric isomers can be drawn for octahedral complexes with a formula of MA_4B_2, such as this one. One isomer has two Cl^- ions in *cis* positions (adjacent positions, at a 90° angle) and the other isomer has the Cl^- ligands in *trans* positions (with a 180° angle between the ligands). Optical isomers are not possible.

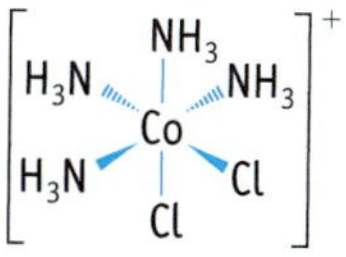

cis isomer 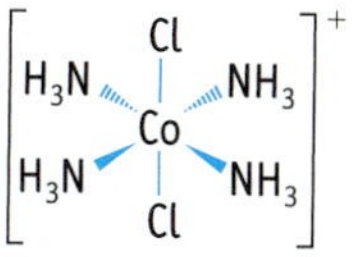   *trans* isomer

(b) In this square-planar complex, the two Cl^- ligands (and the two CN^- ligands) can be either *cis* or *trans*. These are geometric isomers. Optical isomers are not possible.

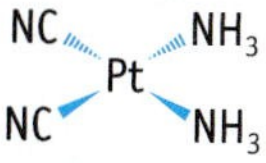

cis isomer 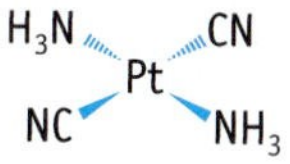   *trans* isomer

(c) Two geometric isomers of this octahedral complex, with chloride ligands either *fac* or *mer*, are possible. In the *fac* isomer, the three Cl^- ligands are all at 90°; in the *mer* isomer, two Cl^- ligands are at 180°, and the third is 90° from the other two. Optical isomers are not possible.

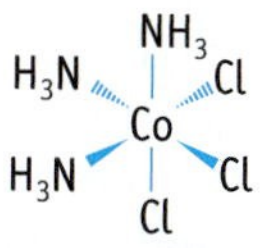

fac isomer 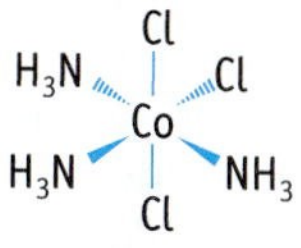   *mer* isomer

(d) Only a single structure is possible for tetrahedral complexes such as $Zn(NH_3)_2Cl_2$.

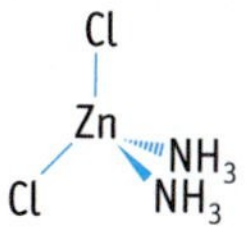

(e) Ignore the counterions, K^+. The anion is an octahedral complex—remember that the bidentate oxalate ion occupies two coordination sites of the metal, and that three oxalate ligands means that the metal has a coordination number of 6. Mirror images of complexes of the stoichiometry M(bidentate)$_3$ are not superimposable; therefore, two optical isomers are possible. (Here the ligands, $C_2O_4^{2-}$, are drawn abbreviated as O—O.)

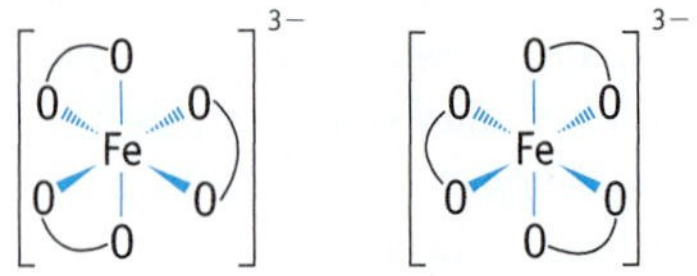

Nonsuperimposable mirror images of $[Fe(ox)_3]^{3-}$

(f) Only linkage isomerism (structural isomerism) is possible for this octahedral cobalt complex. Either the sulfur or the nitrogen of the SCN$^-$ anion can be attached to the cobalt(III) ion in this complex.

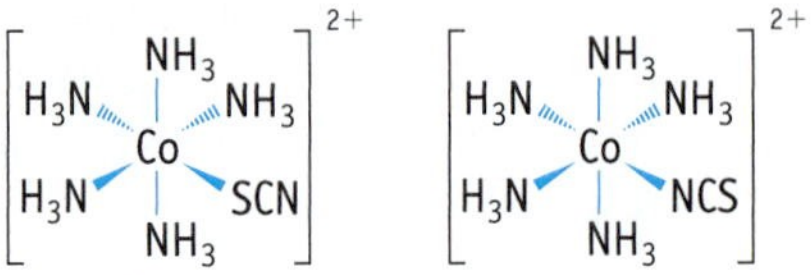

S-bonded SCN$^-$ N-bonded SCN$^-$

Exercise 22.3—Isomers in Coordination Complexes

What types of isomers are possible for the following complexes?

(a) K[Co(NH$_3$)$_2$Cl$_4$]

(b) Pt(en)Cl$_2$ (square planar)

(c) [Co(NH$_3$)$_5$Cl]$^{2+}$

(d) [Ru(phen)$_3$]Cl$_3$

(e) Na$_2$[MnCl$_4$] (tetrahedral)

(f) [Co(NH$_3$)$_5$NO$_2$]$^{2+}$

22.5—Bonding in Coordination Compounds

Metal–ligand bonding in a coordination complex was described earlier in this chapter as being covalent, resulting from the sharing of an electron pair between the metal and the ligand donor atom. Although frequently used, this model is not capable of explaining the color and magnetic behavior of complexes. As a consequence, the covalent bonding picture has now largely been superseded by two other bonding models: molecular orbital theory and ligand field theory.

The bonding model based on molecular orbital theory assumes that the metal and the ligand bond through the molecular orbitals formed by atomic orbital overlap between metal and ligand. The **ligand field model**, in contrast, focuses on repulsion (and destabilization) of electrons in the metal coordination sphere. The ligand field model also assumes that the positive metal ion and the negative ligand lone pair are attracted electrostatically; that is, the bond arises when a positively charged metal ion attracts a negative ion or the negative end of a polar molecule. For the most part, the molecular orbital and ligand field models predict similar, qualitative results regarding color and magnetic behavior. Here we will focus on the ligand field approach and illustrate how it explains color and magnetism of transition metal complexes.

The *d* Orbitals: Ligand Field Theory

To understand ligand field theory, it is necessary to look at the *d* orbitals, particularly with regard to their orientation relative to the positions of ligands in a metal complex.

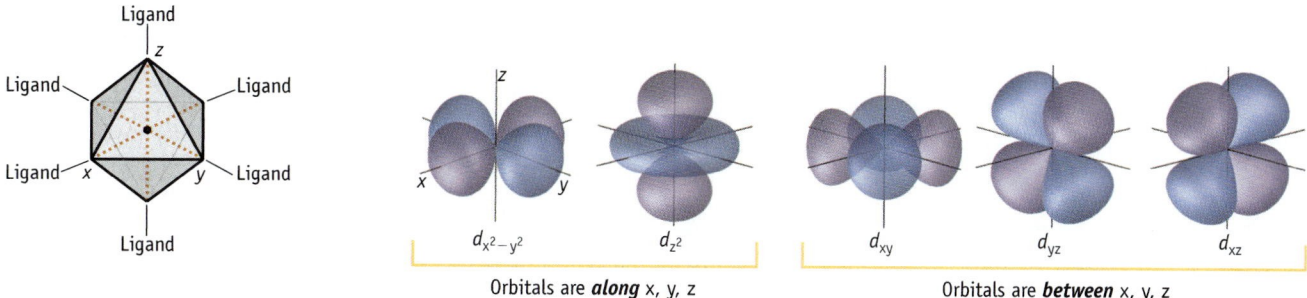

$d_{x^2-y^2}$ d_{z^2}

Orbitals are **along** x, y, z

d_{xy} d_{yz} d_{xz}

Orbitals are **between** x, y, z

Figure 22.21 **The *d* orbitals.** The five *d* orbitals and their spatial relation to the ligands on the *x*-, *y*-, and *z*-axes.

We look first at octahedral complexes. Assume the ligands in an octahedral complex lie along the *x*-, *y*-, and *z*-axes. This results in the five *d* orbitals (Figure 22.21) being subdivided into two sets: the $d_{x^2-y^2}$ and d_{z^2} orbitals in one set and the d_{xy}, d_{xz}, and d_{yz} orbitals in the second. The $d_{x^2-y^2}$ and d_{z^2} orbitals are directed along the *x*-, *y*-, and *z*-axes, whereas the orbitals of the second group are aligned between these axes.

In an isolated atom or ion, the five *d* orbitals have the same energy. For a metal atom or ion in a coordination complex, however, the *d* orbitals have different energies. According to the ligand field model, repulsion between *d* electrons on the metal and electron pairs of the ligands destabilizes electrons that reside in the *d* orbitals; that is, it causes their energy to increase. Electrons in the various *d* orbitals are not affected equally, however, because of their different orientations in space relative to the position of the ligand lone pairs (Figure 22.22). Electrons in the $d_{x^2-y^2}$ and d_{z^2} orbitals experience a larger repulsion because these orbitals point directly at the incoming ligand electron pairs. A smaller repulsive effect is experienced by electrons in the d_{xy}, d_{xz}, and d_{yz} orbitals. The difference in degree of repulsion means that an energy difference exists between the two sets of orbitals. This difference, called the **ligand field splitting** and denoted by the symbol Δ_0, is a function of the metal and the ligands and varies predictably from one complex to another.

A different splitting pattern is encountered with square-planar complexes (Figure 22.23). Assume that the four ligands are along the *x*- and *y*-axes. The $d_{x^2-y^2}$ orbital also points along these axes, so it has the highest energy. The d_{xy} orbital (which also lies in the *xy*-plane, but does not point at the ligands) is next highest in energy,

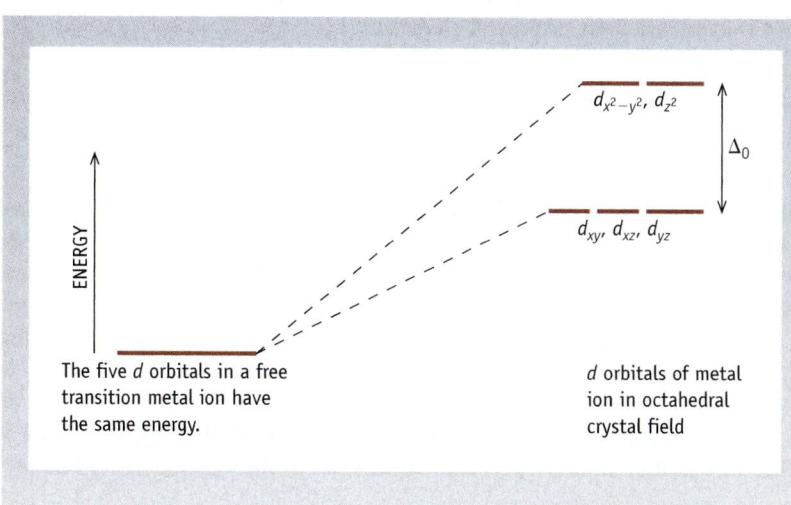

ENERGY

$d_{x^2-y^2}, d_{z^2}$

Δ_0

d_{xy}, d_{xz}, d_{yz}

The five *d* orbitals in a free transition metal ion have the same energy.

d orbitals of metal ion in octahedral crystal field

Figure 22.22 **Ligand field splitting for an octahedral complex.** The *d*-orbital energies increase as the ligands approach the metal along the *x*-, *y*-, and *z*-axes. The d_{xy}, d_{xz}, and d_{yz} orbitals, not pointed toward the ligands, are less destabilized than the $d_{x^2-y^2}$ and d_{z^2} orbitals. Thus, the d_{xy}, d_{xz}, and d_{yz} orbitals are at lower energy.

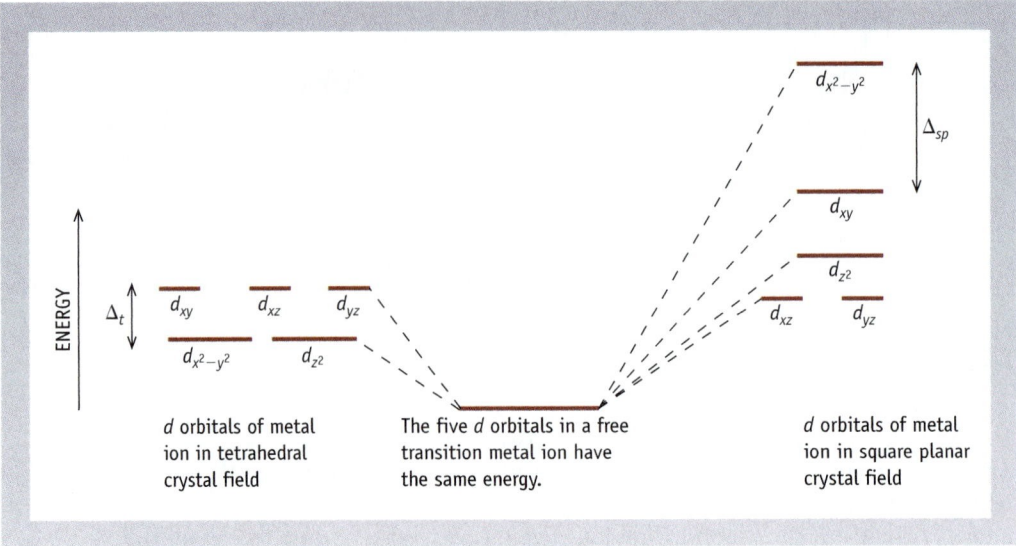

Figure 22.23 **Splitting of the *d* orbitals in (*left*) tetrahedral and (*right*) square planar geometries.**

followed by the d_{z^2} orbital. The d_{xz} and d_{yz} orbitals, both of which partially point in the *z*-direction, have the lowest energy.

The *d*-orbital splitting pattern for a tetrahedral complex is the reverse of the pattern observed for octahedral complexes. Three orbitals (d_{xz}, d_{xy}, d_{yz}) are higher in energy, whereas the $d_{x^2-y^2}$ and d_{z^2} orbitals are below them in energy (Figure 22.23).

Electron Configurations and Magnetic Properties

The *d*-orbital splitting in coordination complexes provides the means to explain both the magnetic behavior and the color of these complexes. To understand this explanation, however, we must first understand how to assign electrons to the various orbitals in each geometry.

A gaseous Cr^{2+} ion has the electron configuration $[Ar]3d^4$. The term "gaseous" in this context is used to denote a single, isolated atom or ion with all other particles located an infinite distance away. In this situation, the five $3d$ orbitals have the same energy. The four electrons reside singly in different *d* orbitals, according to Hund's rule, and the Cr^{2+} ion has four unpaired electrons.

Cr(II) electron configuration

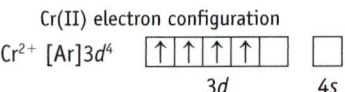

When the Cr^{2+} ion is part of an octahedral complex, the five *d* orbitals do not have identical energies. As illustrated in Figure 22.22, these orbitals divide into two sets, with the d_{xy}, d_{xz}, and d_{yz} orbitals having a lower energy than the $d_{x^2-y^2}$ and d_{z^2} orbitals. Having two sets of orbitals means that two different electron configurations are possible (Figure 22.24). Three of the four *d* electrons in Cr^{2+} are assigned to the lower-energy d_{xy}, d_{xz}, and d_{yz} orbitals. The fourth electron either can be assigned to an orbital in the higher-energy $d_{x^2-y^2}$ and d_{z^2} set or can pair up with an electron already in the lower-energy set. The first arrangement is called **high spin**, because it has the maximum number of unpaired electrons, four in the case of Cr^{2+}. The sec-

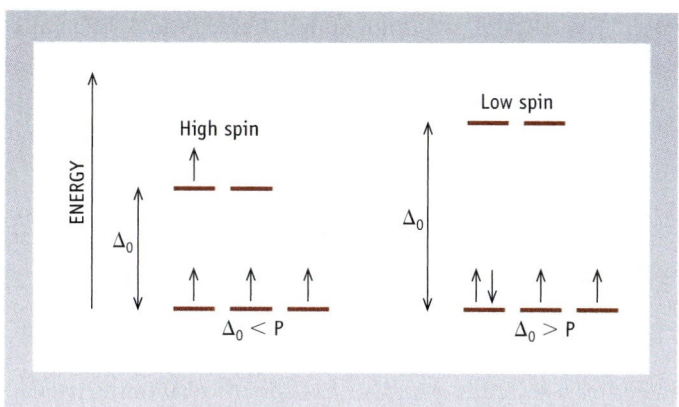

Figure 22.24 **High- and low-spin cases for an octahedral chromium(II) complex.** (*left, high spin*) If the ligand field splitting (Δ_0) is smaller than the pairing energy (P), the electrons are placed in different orbitals and the complex has four unpaired electrons. (*right, low spin*) If the splitting is larger than the pairing energy, all four electrons will be in the lower-energy orbital set. This requires pairing two electrons in one of the orbitals, so the complex will have two unpaired electrons.

ond arrangement is called **low spin**, because it has the minimum number of unpaired electrons possible.

At first glance, a high-spin configuration appears to contradict conventional thinking. It seems logical that the most stable situation would occur when electrons occupy the lowest-energy orbitals. A second factor intervenes, however. Because electrons are negatively charged, repulsion increases when they are assigned to the same orbital. This destabilizing effect bears the name **pairing energy**. The preference for an electron to be in the lowest-energy orbital and the pairing energy have opposing effects (Figure 22.24).

Low-spin complexes arise when the splitting of the d orbitals by the ligand field is large—that is, when Δ_0 has a large value. The energy gained by putting all of the electrons in the lowest-energy level is the dominant effect. In contrast, high-spin complexes occur if the value of Δ_0 is small.

For octahedral complexes, high- and low-spin complexes can occur only for configurations d^4 through d^7 (Figure 22.25). Complexes of the d^6 metal ion, Fe^{2+}, for example, can have either high spin or low spin. The complex formed when the

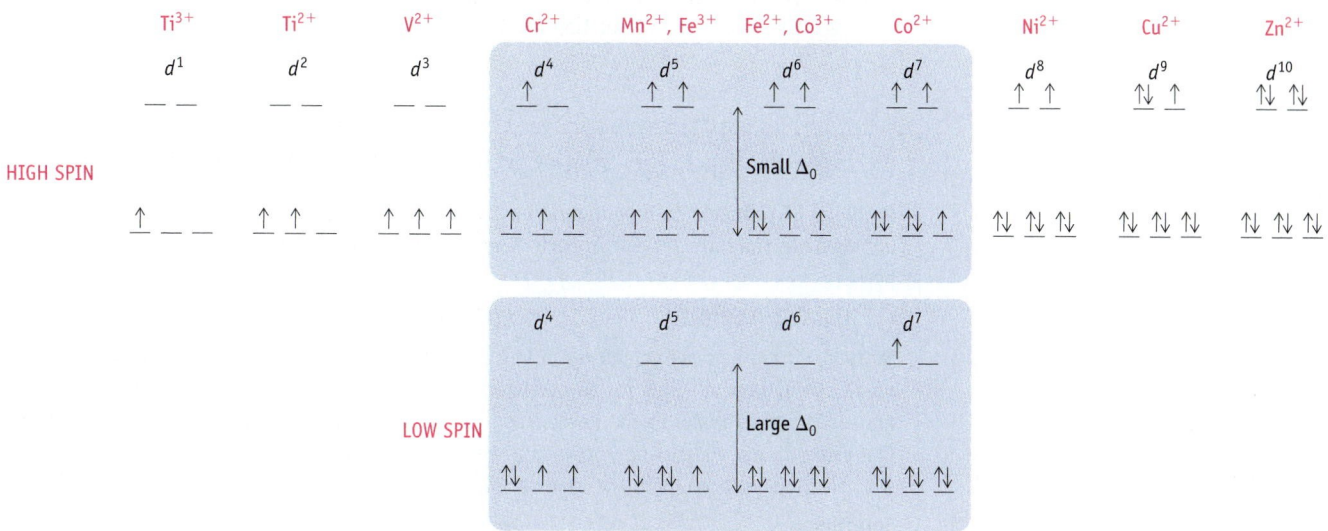

Figure 22.25 **High- and low-spin octahedral complexes.** d-Orbital occupancy for octahedral complexes of metal ions. Only the d^4 through d^7 cases have both high- and low-spin configurations.

Fe^{2+} ion is placed in water, $[Fe(H_2O)_6]^{2+}$, is high spin, whereas the $[Fe(CN)_6]^{4-}$ complex ion is low spin

Electron configuration for Fe^{2+} in an octahedral complex

$d_{x^2-y^2}, d_{z^2}$ ↑ ↑ $\Delta_0(H_2O)$ $d_{x^2-y^2}, d_{z^2}$ — — $\Delta_0(CN^-)$

d_{xy}, d_{xz}, d_{yz} ↑↓ ↑ ↑ d_{xy}, d_{xz}, d_{yz} ↑↓ ↑↓ ↑↓

high spin
$[Fe(H_2O)_6]^{2+}$

low spin
$[Fe(CN)_6]^{4-}$

It is possible to tell whether a complex is high or low spin by examining its magnetic behavior. The high-spin complex $[Fe(H_2O)_6]^{2+}$ has four unpaired electrons and is *paramagnetic* (attracted by a magnet), whereas the low-spin $[Fe(CN)_6]^{4-}$ complex has no unpaired electrons and is *diamagnetic* (repelled by a magnet) [◀ Section 8.1].

Most complexes of Pd^{2+} and Pt^{2+} ions are square planar, the electron configuration of these metals being $[noble\ gas](n-1)d^8$. In a square-planar complex, there are four sets of orbitals (Figure 22.22). All except the highest-energy orbital are filled, and all electrons are paired, resulting in diamagnetic (low-spin) complexes.

Nickel, which is found above palladium in the periodic table, forms both square-planar and tetrahedral complexes. For example, the complex ion $[Ni(CN)_4]^{2-}$ is square planar, whereas the $[NiCl_4]^{2-}$ ion is tetrahedral. Magnetism allows us to differentiate between these two geometries. Based on the ligand field splitting pattern, the cyanide complex is expected to be diamagnetic, whereas the chloride complex is paramagnetic with two unpaired electrons.

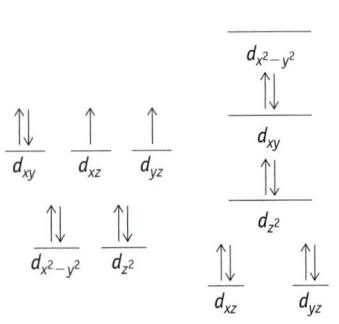

The anion $[NiCl_4]^{2-}$ is a paramagnetic tetrahedral complex. In contrast, $[Ni(CN)_4]^{2-}$ is a diamagnetic square-planar complex.

GENERAL
Chemistry ⚛ Now™

See the General ChemistryNow CD-ROM or website:

- **Screen 21.7 Electronic Structure in Transition Metal Complexes,** for a simulation of bonding and electronic structure in transition metal complexes

---

Example 22.5—High- and Low-Spin Complexes and Magnetism

Problem Give the electron configuration for each of the following complexes. How many unpaired electrons are present in each complex? Are the complexes paramagnetic or diamagnetic?

(a) low-spin $[Co(NH_3)_6]^{3+}$ **(b)** high-spin $[CoF_6]^{3-}$

Strategy These ions are complexes of Co^{3+}, which has a d^6 valence electron configuration. Set up an energy-level diagram for an octahedral complex. In low-spin complexes, the electrons are added preferentially to the lower-energy set of orbitals. In high-spin complexes, the first five electrons are added singly to each of the five orbitals, then additional electrons are paired with electrons in orbitals in the lower-energy set.

Solution

(a) The six electrons of the Co^{3+} ion fill the lower-energy set of orbitals entirely. This d^6 complex ion has no unpaired electrons and is diamagnetic.

(b) To obtain the electron configuration in high-spin $[CoF_6]^{3-}$, place one electron in each of the five d orbitals, and then place the sixth electron in one of the lower-energy orbitals. The complex has four unpaired electrons and is paramagnetic.

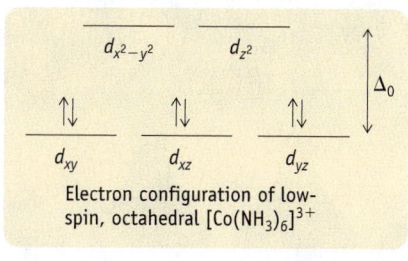

(a)

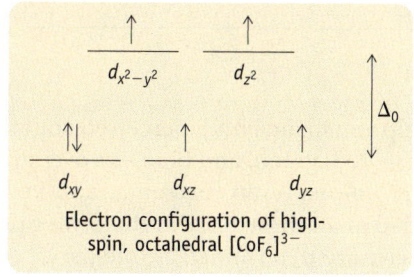

(b)

Exercise 22.4—High- and Low-Spin Configurations and Magnetism

For each of the following complex ions, give the oxidation number of the metal, depict possible low- and high-spin configurations, give the number of unpaired electrons in each configuration, and tell whether each is paramagnetic or diamagnetic.

(a) $[Ru(H_2O)_6]^{2+}$ **(b)** $[Ni(NH_3)_6]^{2+}$

22.6—Colors of Coordination Compounds

One of the most interesting features of compounds of the transition elements is their colors (Figure 22.26). In contrast, compounds of main group metals are usually colorless. The color in these species results from d-orbital splitting. Before discussing how d-orbital splitting is involved, we need to look more closely at what we mean by color.

Color

Visible light, consisting of radiation with wavelengths from 400 to 700 nm [◀ Section 7.1], represents a very small portion of the electromagnetic spectrum. Within this region are all the colors you see when white light passes through a prism: red, orange, yellow, green, blue, indigo, and violet (ROY G BIV). Each color is identified with a portion of the wavelength range.

Isaac Newton did experiments with light and established that the mind's perception of color requires only three colors! When we see white light, we are seeing a mixture of all of the colors—in other words, the superposition of red, green, and blue. If one or more of these colors is absent, the light of the other colors that reaches your eyes is interpreted by your mind as color.

Figure 22.26 **Aqueous solutions of some transition metal ions.** Compounds of transition metal elements are often colored, whereas those of main group metals are usually colorless. Pictured here, from left to right, are solutions of the nitrate salts of Fe^{3+}, Co^{2+}, Ni^{2+}, Cu^{2+}, and Zn^{2+}.

Charles D. Winters

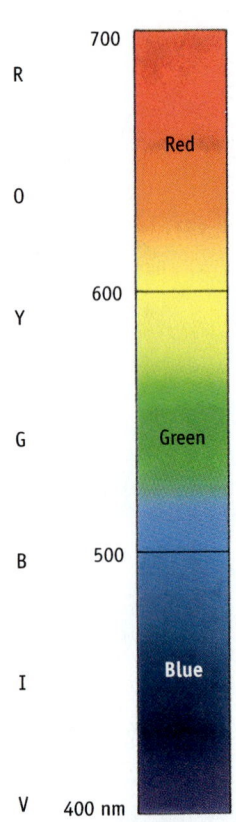

The ROY G BIV spectrum of colors of visible light. The colors used in printing this book are cyan, magenta, yellow, and black. The blue in ROY G BIV is actually cyan, according to color industry standards. Magenta doesn't have its own wavelength region. Rather, it is a mixture of blue and red.

Figure 22.27 will help you in analyzing perceived colors. This color wheel shows the three primary colors—red, green, and blue—as overlapping disks arranged in a triangle. The secondary colors—cyan, magenta, and yellow—appear where two disks overlap. The overlap of all three disks in the center produces white light.

The colors we perceive are determined as follows:

- Light of a single primary color is perceived as that color: Red light is perceived as red, green light as green, blue light as blue.

- Light made up of two primary colors is perceived as the color shown where the disks in Figure 22.27 overlap: Red and green light together appear yellow, green and blue light together are perceived as cyan, and red and blue light are perceived as magenta.

- Light made up of the three primary colors is white (colorless).

In discussing the color of a substance such as a coordination complex *in solution,* these guidelines are turned around.

- Red light is the result of the absence of green and blue light from white light.

- Green light results if red and blue light are absent from white light.

- Blue light results if red and green light are absent.

The secondary colors are rationalized similarly. Absorption of blue light gives yellow (the color across from it in Figure 22.27), absorption of red light results in cyan, and absorption of green light results in magenta.

Now we can apply these ideas to explain colors in transition metal complexes. Focus on what kind of light is *absorbed.* A solution of $[Ni(H_2O)_6]^{2+}$ is green. Green light is the result of removing red and blue light from white light. As white light passes through an aqueous solution of Ni^{2+}, red and blue light are absorbed and green light is allowed to pass (Figure 22.28). Similarly, the $[Co(NH_3)_6]^{3+}$ ion is yellow because blue (B) light has been absorbed and red and green light pass through.

The Spectrochemical Series

Recall that atomic spectra are obtained when electrons are excited from one energy level to another [◀ Section 7.3]. The energy of the light absorbed or emitted is related to the energy levels of the atom or ion under study. The concept that light is absorbed when electrons move from lower to higher energy levels applies to all sub-

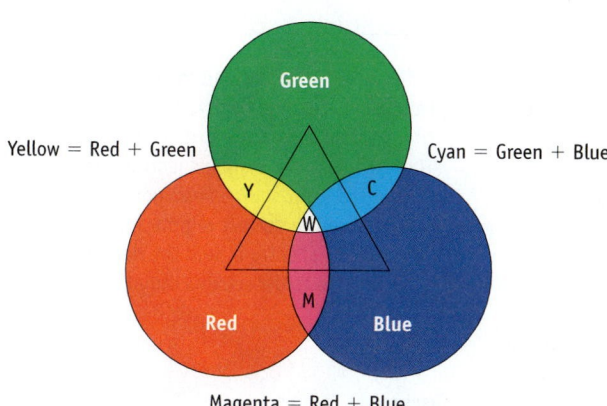

Figure 22.27 **Using color disks to analyze colors.** The three primary colors are red, green, and blue. Adding light of two primary colors gives the secondary colors yellow (= red + green), cyan (= green + blue), and magenta (= red + blue). Adding all three primary colors results in white light.

Yellow = Red + Green

Cyan = Green + Blue

Magenta = Red + Blue

stances, not just atoms. It is the basic premise for the absorption of light for transition metal coordination complexes.

In coordination complexes, the splitting between d orbitals often corresponds to the energy of visible light, so light in the visible region of the spectrum is absorbed when electrons move from a lower-energy d orbital to a higher-energy d orbital. This change, as an electron moves between two orbitals having different energies in a complex, is called a **d-to-d transition.** Qualitatively, such a transition for $[Co(NH_3)_6]^{3+}$ might be represented using an energy-level diagram such as that shown here.

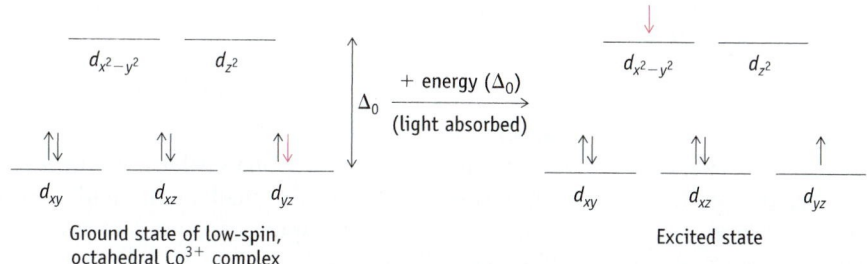

$$d_{x^2-y^2} \qquad d_{z^2}$$

$$\Delta_0 \xrightarrow[\text{(light absorbed)}]{\text{+ energy } (\Delta_0)} \qquad d_{x^2-y^2} \qquad d_{z^2}$$

$$d_{xy} \qquad d_{xz} \qquad d_{yz}$$

Ground state of low-spin, octahedral Co^{3+} complex

$$d_{xy} \qquad d_{xz} \qquad d_{yz}$$

Excited state

Figure 22.28 **Light absorption and color.** The color of a solution is due to the color of the light *not* absorbed by the solution. Here a solution of Ni^{2+} ion in water absorbs red and blue light and so appears green.

A Closer Look

A Spectrophotometer

The qualitative conclusions that we have drawn concerning absorption of light are determined more precisely in the laboratory using an instrument called a *spectrophotometer*. A schematic drawing of a spectrophotometer measuring absorption of visible light is shown in the accompanying figure. White light from a glowing filament first passes through a prism or diffraction grating, both devices that divide the light into its separate frequencies. The instrument selects a specific frequency to pass through a solution of the compound to be studied. After passing through the solution, the intensity of the light is measured by a detector. If light of a given frequency is not absorbed, its intensity remains unchanged when it emerges from the sample. If light is absorbed, the light emerging from the sample has a lower intensity.

An absorption spectrum plots the frequency or wavelength of the light versus the intensity of light absorbed at that frequency or wavelength. When light in a certain frequency range is absorbed, the graph shows an absorption band. For example, $[Co(NH_3)_6]^{3+}$ absorbs in the blue region, allowing its complementary color yellow $(R + G = Y)$ to pass through to the eye.

Scientists are not limited to the visible region of the spectrum when making measurements on absorption of electromagnetic radiation. Ultraviolet and infrared spectrophotometers are common pieces of scientific apparatus in a chemistry laboratory. They differ from a visible spectrophotometer in that the light source must generate radiation in the correct region of the electromagnetic spectrum, and the detector must be able to detect this radiation.

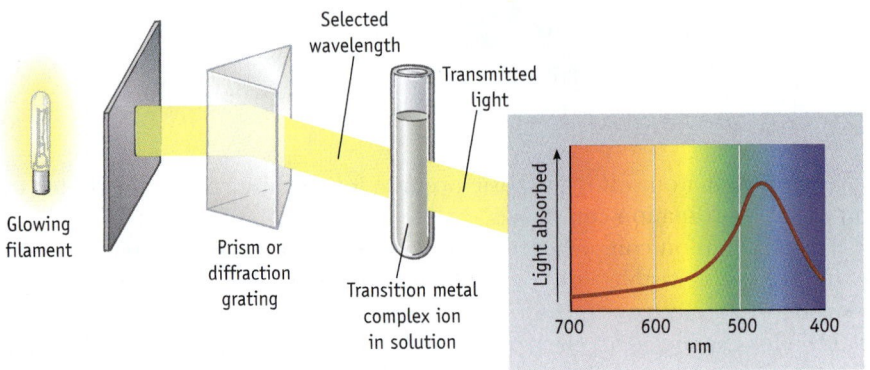

An absorption spectrophotometer. A beam of white light passes through a prism or diffraction grating, which splits the light into its component wavelengths. After passing through a sample, the light reaches a detector. The spectrophotometer "scans" all wavelengths of light and determines whether light is absorbed. Its output is a "spectrum," a graph plotting absorption as a function of wavelength or frequency.

Experiments with coordination complexes reveal that, for a given metal ion, some ligands cause a small energy separation of the *d* orbitals, whereas others cause a large separation. In other words, some ligands create a small ligand field, and others create a large one. An example is seen in the spectroscopic data for several cobalt(III) complexes presented in Table 22.3.

- Both $[Co(NH_3)_6]^{3+}$ and $[Co(en)_3]^{3+}$ are yellow-orange, because they absorb light in the blue portion of the visible spectrum. These compounds have very similar spectra, to be expected because both have six amine-type donor atoms $(H-NH_2$ or $R-NH_2)$.

- Although $[Co(CN)_6]^{3-}$ does not have an absorption band in the visible region, it is pale yellow. Light absorption occurs in the ultraviolet region, but the absorption is broad and extends at least minimally into the visible (blue) region.

- $[Co(C_2O_4)_3]^{3-}$ and $[Co(H_2O)_6]^{3+}$ have similar absorptions, in the yellow and violet regions. Their colors are shades of green with a small difference due to the relative amount of light of each color being absorbed.

The absorption maxima among the listed complexes range from 700 nm for $[CoF_6]^{3-}$ to 310 nm for $[Co(CN)_6]^{3-}$. The ligands change from member to member of this series, and we can conclude that the energy of the light absorbed by the complex is related to the different ligand field splittings, Δ_0, caused by the different ligands. Fluoride ion causes the smallest splitting of the *d* orbitals among the complexes listed in Table 22.3, whereas cyanide causes the largest splitting.

Spectra of complexes of other metals provide similar results. Based on this information, ligands can be listed in order of their ability to split the *d* orbitals. This

Table 22.3 The Colors of Some Co^{3+} Complexes*

Complex Ion	Wavelength of Light Absorbed (nm)	Color of Light Absorbed	Color of Complex
$[CoF_6]^{3-}$	700	Red	Green
$[Co(C_2O_4)_3]^{3-}$	600, 420	Yellow, violet	Dark green
$[Co(H_2O)_6]^{3+}$	600, 400	Yellow, violet	Blue-green
$[Co(NH_3)_6]^{3+}$	475, 340	Blue, ultraviolet	Yellow-orange
$[Co(en)_3]^{3+}$	470, 340	Blue, ultraviolet	Yellow-orange
$[Co(CN)_6]^{3-}$	310	Ultraviolet	Pale yellow

*The complex with fluoride ion, $[CoF_6]^{3-}$, is high spin and has one absorption band. The other complexes are low spin and have two absorption bands. In all but one case, one of these absorptions occurs in the visible region of the spectrum. The wavelengths are measured at the top of that absorption band.

list is called the **spectrochemical series** because it was determined by spectroscopy. A short list, with some of the more common ligands, follows:

Spectrochemical Series

$$Cl^-, Br^-, I^- < C_2O_4^{2-} < H_2O < NH_3 = en < phen < CN^-$$

small orbital splitting
small Δ_0

large orbital splitting
large Δ_0

The spectrochemical series is applicable to a wide range of metal complexes. Indeed, the ability of ligand field theory to explain the differences in the colors of the transition metal complexes is one of the strengths of this theory.

Based on the relative position of a ligand in the series, predictions can be made about a compound's magnetic behavior. Recall that d^4, d^5, d^6, and d^7 complexes can be high or low spin, depending on the ligand field splitting, Δ_0. Complexes formed with ligands near the left end of the spectrochemical series are expected to have small Δ_0 values and, therefore, are likely to be high spin. In contrast, complexes with ligands near the right end are expected to have large Δ_0 values and low-spin configurations. The complex $[CoF_6]^{3-}$ is high spin, whereas $[Co(NH_3)_6]^{3+}$ and the other complexes in Table 22.3 are low spin.

GENERAL
Chemistry-ϕ-Now™

See the General ChemistryNow CD-ROM or website:
- **Screen 21.8 Spectroscopy of Transition Metal Complexes,** for a simulation on the absorption and transmission of light by transition metal complexes

Example 22.6—Spectrochemical Series

Problem An aqueous solution of $[Fe(H_2O)_6]^{2+}$ is light blue-green. Do you expect the d^6 Fe^{2+} ion in this complex to have a high- or low-spin configuration? How would you make this determination by conducting an experiment?

Strategy Use the color wheel in Figure 22.27. The color of the complex, blue-green, tells us what kind of light is transmitted (blue and green), from which we learn what kind of light has been absorbed (red). Red light is at the low-energy end of the visible spectrum. From this fact, we can predict that the d-orbital splitting must be small. Our answer to the question derives from that conclusion.

Problem-Solving Tip 22.1

Ligand Field Theory

This summary of the concepts of ligand field theory may help you to keep the broader picture in mind.

- Ligand–metal bonding results from the electrostatic attraction of a metal cation and an anion or polar molecule.

- The ligands define a coordination geometry. Common geometries are linear (coordination number = 2), tetrahedral and square planar (coordination number = 4), and octahedral (coordination number = 6).

- The placement of the ligands around the metal causes the d orbitals on the metal to have different energies. In an octahedral complex, for example, the d orbitals divide into two groups: a higher-energy group ($d_{x^2-y^2}$ and d_{z^2}) and a lower-energy group (d_{xy}, d_{xz}, and d_{yz}).

- Electrons are placed in the metal d orbitals in a manner that leads to the lowest total energy. Two competing features determine the placement: the relative energy of the sets of orbitals and the electron pairing energy.

- For the electron configurations d^4, d^5, d^6, and d^7, two electron configurations are possible: high spin, which occurs when the orbital splitting is small, and low spin, which occurs with a large orbital splitting. To determine whether a complex is high spin or low spin, measure its magnetism to determine the number of unpaired electrons.

- The d-orbital splitting (the energy difference between the metal d-orbital energies) often corresponds to the energy associated with visible light. As a consequence, many metal complexes absorb visible light and thus are colored.

Solution The low energy of the light absorbed suggests that $[Fe(H_2O)_6]^{2+}$ is likely to be a high-spin complex.

If the complex is high spin, it will have four unpaired electrons and be paramagnetic; if it is low spin it will have no unpaired electrons and be diamagnetic. Identifying the presence of four unpaired electrons by measuring the compound's magnetism can be used to verify the high-spin configuration experimentally.

For additional preparation for an examination on this chapter see the **Let's Review** section on pages 998–1011.

Chapter Goals Revisited

When you have studied this chapter, you should ask whether you have met the chapter goals. In particular, you should be able to

Identify and explain the chemical and physical properties of the transition elements

a. Identify the general classes of elements: transition elements, lanthanides, and actinides (Section 22.1).

b. Identify the more common transition metals from their symbols and positions in the periodic table, and recall some physical and chemical properties (Section 22.1).

c. Understand the electrochemical nature of corrosion (Section 22.1).

d. Know the general features of pyrometallurgy and hydrometallurgy, and describe the metallurgy of iron and copper (Section 22.2).

Understand the composition, structure, and bonding in coordination compounds

a. Given the formula for a coordination complex, be able to identify the metal and its oxidation state, the ligands, the coordination number and coordination geometry, and the overall charge on the complex (Section 22.3).

b. Provide systematic names for complexes from their formulas, and write formulas if the name is given (Section 22.3).

c. Given the formula for a complex, be able to recognize whether isomers will exist, and draw their structures (Section 22.4).

d. Describe the bonding in coordination complexes (Section 22.5).

e. Use ligand field theory to determine the number of unpaired electrons in a complex (Section 22.5).

Show how bonding is used to explain the magnetism and spectra of coordination compounds

a. Understand why substances are colored and be able to predict colors when you know the color of light absorbed (Section 22.6).

b. Understand the relationship between the ligand field splitting, magnetism, and color of metal complexes (Section 22.6).

Study Questions

▲ denotes more challenging questions.

■ denotes questions available in the Homework and Goals section of the General ChemistryNow CD-ROM or website.

Blue numbered questions have answers in Appendix O and fully worked solutions in the *Student Solutions Manual.*

Structures of many of the compounds used in these questions are found on the General ChemistryNow CD-ROM or website in the Models folder.

Chemistry ⚛ Now™ Assess your understanding of this chapter's topics with additional quizzing and conceptual questions at http://now.brookscole.com/kotz6e

Practicing Skills

Properties of Transition Elements
(See Section 22.1 and Example 8.1.)

1. Give the electron configuration for each of the following ions, and tell whether each is paramagnetic or diamagnetic.
(a) Cr^{3+} (b) V^{2+} (c) Ni^{2+} (d) Cu^+

2. Identify two transition metal ions with the following electron configurations.
(a) $[Ar]3d^6$ (b) $[Ar]3d^{10}$ (c) $[Ar]3d^5$ (d) $[Ar]3d^8$

3. Identify an ion of a first series transition metal that is isoelectronic with each of the following.
(a) Fe^{3+} (b) Zn^{2+} (c) Fe^{2+} (d) Cr^{3+}

4. Match up the isoelectronic ions on the following list.
$$Cu^+ \ Mn^{2+} \ Fe^{2+} \ Co^{3+} \ Fe^{3+} \ Zn^{2+} \ Ti^{2+} \ V^{3+}$$

5. The following equations represent various ways of obtaining transition metals from their compounds. Balance each equation.
(a) $Cr_2O_3(s) + Al(s) \longrightarrow Al_2O_3(s) + Cr(s)$
(b) $TiCl_4(\ell) + Mg(s) \longrightarrow Ti(s) + MgCl_2(s)$
(c) $[Ag(CN)_2]^-(aq) + Zn(s) \longrightarrow$
$$Ag(s) + [Zn(CN)_4]^{2-}(aq)$$
(d) $Mn_3O_4(s) + Al(s) \longrightarrow Mn(s) + Al_2O_3(s)$

6. Identify the products of each reaction and balance the equation.
(a) $CuSO_4(aq) + Zn(s) \longrightarrow$
(b) $Zn(s) + HCl(aq) \longrightarrow$
(c) $Fe(s) + Cl_2(g) \longrightarrow$
(d) $V(s) + O_2(g) \longrightarrow$

Formulas of Coordination Compounds
(See Examples 22.1 and 22.2.)

7. Which of the following ligands is expected to be monodentate and which might be polydentate?
(a) CH_3NH_2 (d) en
(b) CH_3CN (e) Br^-
(c) N_3^- (f) phen

8. One of the following nitrogen compounds or ions is not capable of serving as a ligand: NH_4^+, NH_3, NH_2^-. Identify this species and explain your answer.

9. Give the oxidation number of the metal ion in each of the following compounds.
(a) $[Mn(NH_3)_6]SO_4$ (c) $[Co(NH_3)_4Cl_2]Cl$
(b) $K_3[Co(CN)_6]$ (d) $Cr(en)_2Cl_2$

10. Give the oxidation number of the metal ion in each of the following complexes.
    (a) $[Fe(NH_3)_6]^{2+}$
    (b) $[Zn(CN)_4]^{2-}$
    (c) $[Co(NH_3)_5(NO_2)]^+$
    (d) $[Cu(en)_2]^{2+}$

11. Give the formula of a complex constructed from one Ni^{2+} ion, one ethylenediamine ligand, three ammonia molecules, and one water molecule. Is the complex neutral or is it charged? If charged, give the charge.

12. Give the formula of a complex constructed from one Cr^{3+} ion, two ethylenediamine ligands, and two ammonia molecules. Is the complex neutral or is it charged? If charged, give the charge.

Naming Coordination Compounds

(See Example 22.3.)

13. Write formulas for the following ions or compounds.
    (a) dichlorobis(ethylenediamine)nickel(II)
    (b) potassium tetrachloroplatinate(II)
    (c) potassium dicyanocuprate(I)
    (d) tetraamminediaquairon(II)

14. Write formulas for the following ions or compounds.
    (a) diamminetriaquahydroxochromium(II) nitrate
    (b) hexaammineiron(III) nitrate
    (c) pentacarbonyliron(0) (where the ligand is CO)
    (d) ammonium tetrachlorocuprate(II)

15. Name the following ions or compounds.
    (a) $[Ni(C_2O_4)_2(H_2O)_2]^{2-}$
    (b) $[Co(en)_2Br_2]^+$
    (c) $[Co(en)_2(NH_3)Cl]^{2+}$
    (d) $Pt(NH_3)_2(C_2O_4)$

16. Name the following ions or compounds.
    (a) $[Co(H_2O)_4Cl_2]^+$
    (b) $Co(H_2O)_3F_3$
    (c) $[Pt(NH_3)Br_3]^-$
    (d) $[Co(en)(NH_3)_3Cl]^{2+}$

17. Give the name or formula for each ion or compound, as appropriate.
    (a) pentaaquahydroxoiron(III) ion
    (b) $K_2[Ni(CN)_4]$
    (c) $K[Cr(C_2O_4)_2(H_2O)_2]$
    (d) ammonium tetrachloroplatinate(II)

18. Give the name or formula for each ion or compound, as appropriate.
    (a) dichlorotetraaquachromium(III) chloride
    (b) $[Cr(NH_3)_5SO_4]Cl$
    (c) sodium tetrachlorocobaltate(II)
    (d) $[Fe(C_2O_4)_3]^{3-}$

Isomerism

(See Example 22.4.)

19. Draw all possible geometric isomers of the following.
    (a) $Fe(NH_3)_4Cl_2$
    (b) $Pt(NH_3)_2(SCN)(Br)$ (SCN^- is bonded to Pt^{2+} through S)
    (c) $Co(NH_3)_3(NO_2)_3$ (NO_2^- is bonded to Co^{3+} through N)
    (d) $[Co(en)Cl_4]^-$

20. In which of the following complexes are geometric isomers possible? If isomers are possible, draw their structures and label them as *cis* or *trans*, or as *fac* or *mer*.
    (a) $[Co(H_2O)_4Cl_2]^+$
    (b) $Co(NH_3)_3F_3$
    (c) $[Pt(NH_3)Br_3]^-$
    (d) $[Co(en)_2(NH_3)Cl]^{2+}$

21. Determine whether the following complexes have a chiral metal center.
    (a) $[Fe(en)_3]^{2+}$
    (b) *trans*-$[Co(en)_2Br_2]^+$
    (c) *fac*-$[Co(en)(H_2O)Cl_3]$
    (d) square-planar $Pt(NH_3)(H_2O)(Cl)(NO_2)$

22. Four isomers are possible for $[Co(en)(NH_3)_2(H_2O)Cl]^+$. Draw the structures of all four. (Two of the isomers are chiral, meaning that each has a non-superimposable mirror image.)

Magnetic Properties of Complexes

(See Example 22.5.)

23. The following are low-spin complexes. Use the ligand field model to find the electron configuration of each ion. Determine which are diamagnetic. Give the number of unpaired electrons for the paramagnetic complexes.
    (a) $[Mn(CN)_6]^{4-}$
    (b) $[Co(NH_3)_6]Cl_3$
    (c) $[Fe(H_2O)_6]^{3+}$
    (d) $[Cr(en)_3]SO_4$

24. The following are high-spin complexes. Use the ligand field model to find the electron configuration of each ion and determine the number of unpaired electrons in each.
    (a) $K_4[FeF_6]$
    (b) $[MnF_6]^{4-}$
    (c) $[Cr(H_2O)_6]^{2+}$
    (d) $(NH_4)_3[FeF_6]$

25. Determine the number of unpaired electrons in the following tetrahedral complexes. All tetrahedral complexes are high spin.
    (a) $[FeCl_4]^{2-}$
    (b) $Na_2[CoCl_4]$
    (c) $[MnCl_4]^{2-}$
    (d) $(NH_4)_2[ZnCl_4]$

26. Determine the number of unpaired electrons in the following tetrahedral complexes. All tetrahedral complexes are high spin.
    (a) $[Zn(H_2O)_4]^{2+}$
    (b) $VOCl_3$
    (c) $Mn(NH_3)_2Cl_2$
    (d) $[Cu(en)_2]^{2+}$

27. For the high-spin complex $[Fe(H_2O)_6]SO_4$, identify the following.
    (a) the coordination number of iron
    (b) the coordination geometry for iron
    (c) the oxidation number of iron
    (d) the number of unpaired electrons
    (e) whether the complex is diamagnetic or paramagnetic

28. For the low-spin complex $[Co(en)(NH_3)_2Cl_2]ClO_4$, identify the following.
    (a) the coordination number of cobalt
    (b) the coordination geometry for cobalt
    (c) the oxidation number of cobalt
    (d) the number of unpaired electrons
    (e) whether the complex is diamagnetic or paramagnetic
    (f) Draw any geometric isomers.

▲ More challenging ■ In General ChemistryNow Blue-numbered questions answered in Appendix O

29. The coordination compound formed in an aqueous solution of cobalt(III) sulfate is diamagnetic. If NaF is added, the complex in solution becomes paramagnetic. Why does the magnetism change?

30. An aqueous solution of iron(II) sulfate is paramagnetic. If NH_3 is added, the solution becomes diamagnetic. Why does the magnetism change?

Spectroscopy of Complexes

(See Example 22.6.)

31. In water, the titanium(III) ion, $[Ti(H_2O)_6]^{3+}$, has a broad absorption band at about 500 nm. What color light is absorbed by the ion?

32. In water, the chromium(II) ion, $[Cr(H_2O)_6]^{2+}$, absorbs light with a wavelength of about 700 nm. What color is the solution?

General Questions

These questions are not designated as to type or location in the chapter. They may contain several concepts.

33. Describe an experiment that would determine whether nickel in $K_2[NiCl_4]$ is square planar or tetrahedral.

34. Which of the following low-spin complexes has the greatest number of unpaired electrons?
(a) $[Cr(H_2O)_6]^{3+}$
(b) $[Mn(H_2O)_6]^{2+}$
(c) $[Fe(H_2O)_6]^{2+}$
(d) $[Ni(H_2O)_6]^{2+}$

35. How many unpaired electrons are expected for high-spin and low-spin complexes of Fe^{2+}?

36. Excess silver nitrate is added to a solution containing 1.0 mol of $[Co(NH_3)_4Cl_2]Cl$. What amount of AgCl (in moles) will precipitate?

37. Which of the following complexes is (are) square planar?
(a) $[Ti(CN)_4]^{2-}$
(b) $[Ni(CN)_4]^{2-}$
(c) $[Zn(CN)_4]^{2-}$
(d) $[Pt(CN)_4]^{2-}$

38. Which of the following complexes containing the oxalate ion is (are) chiral?
(a) $[Fe(C_2O_4)Cl_4]^{2-}$
(b) *cis*-$[Fe(C_2O_4)_2Cl_2]^{2-}$
(c) *trans*-$[Fe(C_2O_4)_2Cl_2]^{2-}$

39. How many geometric isomers are possible for the square-planar complex $[Pt(NH_3)(CN)Cl_2]^-$?

40. For a tetrahedral complex of a metal in the first transition series, which of the following statements concerning energies of the $3d$ orbitals is correct?
(a) The five d orbitals have the same energy.
(b) The $d_{x^2-y^2}$ and d_{z^2} orbitals are higher in energy than the d_{xz}, d_{yz}, and d_{xy} orbitals.

(c) The d_{xz}, d_{yz}, and d_{xy} orbitals are higher in energy than the $d_{x^2-y^2}$ and d_{z^2} orbitals.
(d) The d orbitals all have different energies.

41. A transition metal complex absorbs 425-nm light. What is its color?
(a) red   (c) yellow
(b) green   (d) blue

42. For the low-spin complex $[Fe(en)_2Cl_2]Cl$, identify the following.
(a) the oxidation number of iron
(b) the coordination number for iron
(c) the coordination geometry for iron
(d) the number of unpaired electrons per metal atom
(e) whether the complex is diamagnetic or paramagnetic
(f) the number of geometric isomers

43. For the high-spin complex $Mn(NH_3)_4Cl_2$, identify the following.
(a) the oxidation number of manganese
(b) the coordination number for manganese
(c) the coordination geometry for manganese
(d) the number of unpaired electrons per metal atom
(e) whether the complex is diamagnetic or paramagnetic
(f) the number of geometric isomers

44. A platinum-containing compound, known as Magnus's green salt, has the formula $[Pt(NH_3)_4][PtCl_4]$ (in which both platinum ions are Pt^{2+}). Name the cation and the anion.

45. Early in the 20th century, complexes sometimes were given names based on their colors. Two compounds with the formula $CoCl_3 \cdot 4 NH_3$ were named praseo-cobalt chloride (*praseo* = green) and violio-cobalt chloride (violet color). We now know that these compounds are octahedral cobalt complexes and that they are *cis* and *trans* isomers. Draw the structures of these two compounds and name them using systematic nomenclature.

46. Give the formula and name of a square-planar complex of Pt^{2+} with one nitrite ion (NO_2^-, which binds to Pt^{2+} through N), one chloride ion, and two ammonia molecules as ligands. Are isomers possible? If so, draw the structure of each isomer, and tell what type of isomerism is observed.

47. Give the formula of the complex formed from one Co^{3+} ion, two ethylenediamine molecules, one water molecule, and one chloride ion. Is the complex neutral or charged? If charged, give the net charge on the ion.

48. ▲ How many geometric isomers of the complex $[Cr(dmen)_3]^{3+}$ can exist? (dmen is the bidentate ligand 1,1-dimethylethylenediamine.)

$$(CH_3)_2\ddot{N}CH_2CH_2\ddot{N}H_2$$

1,1-Dimethylethylenediamine, dmen

49. ▲ Diethylenetriamine (dien) is capable of serving as a tridentate ligand.

$$H_2\overset{..}{N}CH_2CH_2 - \overset{..}{N} - CH_2CH_2\overset{..}{N}H_2$$
$$|$$
$$H$$

Diethylenetriamine, dien

(a) Draw the structures of *fac*-Cr(dien)Cl$_3$ and *mer*-Cr(dien)Cl$_3$.

(b) Two different geometric isomers of *mer*-Cr(dien)Cl$_2$Br are possible. Draw the structure for each.

(c) Three different geometric isomers are possible for [Cr(dien)$_2$]$^{3+}$. Two have the dien ligand in a *fac* configuration, and one has the ligand in a *mer* orientation. Draw the structure of each isomer.

50. From experiment we know that [CoF$_6$]$^{3-}$ is paramagnetic and [Co(NH$_3$)$_6$]$^{3+}$ is diamagnetic. Using the ligand field model, depict the electron configuration for each ion and use this model to explain the magnetic property. What can you conclude about the effect of the ligand on the magnitude of Δ_0?

51. Three geometric isomers are possible for [Co(en)(NH$_3$)$_2$(H$_2$O)$_2$]$^{3+}$. One of the three is chiral; that is, it has a non-superimposable mirror image. Draw the structures of the three isomers. Which one is chiral?

52. The square-planar complex Pt(en)Cl$_2$ has chloride ligands in a *cis* configuration. No *trans* isomer is known. Based on the bond lengths and bond angles of carbon and nitrogen in the ethylenediamine ligand, explain why the *trans* compound is not possible.

53. The complex [Mn(H$_2$O)$_6$]$^{2+}$ has five unpaired electrons, whereas [Mn(CN)$_6$]$^{4-}$ has only one. Using the ligand field model, depict the electron configuration for each ion. What can you conclude about the effects of the different ligands on the magnitude of Δ_0?

54. Experiments show that K$_4$[Cr(CN)$_6$] is paramagnetic and has two unpaired electrons. The related complex K$_4$[Cr(SCN)$_6$] is paramagnetic and has four unpaired electrons. Account for the magnetism of each compound using the ligand field model. Predict where the SCN$^-$ ion occurs in the spectrochemical series relative to CN$^-$.

55. Two different coordination compounds containing one cobalt(III) ion, five ammonia molecules, one bromide ion, and one sulfate ion exist. The dark violet form (A) gives a precipitate upon addition of aqueous BaCl$_2$. No reaction is seen upon addition of aqueous BaCl$_2$ to the violet-red form (B). Suggest structures for these two compounds, and write a chemical equation for the reaction of (A) with aqueous BaCl$_2$.

56. When CrCl$_3$ dissolves in water, three different species can be obtained.

(a) [Cr(H$_2$O)$_6$]Cl$_3$, violet

(b) [Cr(H$_2$O)$_5$Cl]Cl$_2$, pale green

(c) [Cr(H$_2$O)$_4$Cl$_2$]Cl, dark green

If diethylether is added, a fourth complex can be obtained: Cr(H$_2$O)$_3$Cl$_3$ (brown). Describe an experiment that will allow you to differentiate these complexes.

57. ▲ The complex ion [Co(CO$_3$)$_3$]$^{3-}$, an octahedral complex with bidentate carbonate ions as ligands, has one absorption in the visible region of the spectrum at 640 nm. From this information:

(a) Predict the color of this complex, and explain your reasoning.

(b) Is the carbonate ion as weak- or strong-field ligand?

(c) Predict whether [Co(CO$_3$)$_3$]$^{3-}$ will be paramagnetic or diamagnetic.

58. The glycinate ion, H$_2$NCH$_2$CO$_2^-$, formed by deprotonation of the amino acid glycine, can function as a bidentate ligand, coordinating to a metal through the nitrogen of the amino group and one of the oxygen atoms.

Glycinate ion, a bidentate ligand

Site of bonding to transition metal ion

A copper complex of this ligand has the formula Cu(H$_2$NCH$_2$CO$_2$)$_2$(H$_2$O)$_2$. For this complex, determine the following.

(a) the oxidation state of copper

(b) the coordination number of copper

(c) the number of unpaired electrons

(d) whether the complex is diamagnetic or paramagnetic

59. ▲ Draw structures for the five possible geometric isomers of Cu(H$_2$NCH$_2$CO$_2$)$_2$(H$_2$O)$_2$. Are any of these species chiral? (See the structure of the ligand in Study Question 58.)

60. Three different compounds of chromium(III) with water and chloride ion have the same composition: 19.51% Cr, 39.92% Cl, and 40.57% H$_2$O. One of the compounds is violet and dissolves in water to give a complex ion with a 3+ charge and three chloride ions. All three chloride ions precipitate immediately as AgCl on adding AgNO$_3$. Draw the structure of the complex ion and name the compound. Write a net ionic equation for the reaction of this compound with silver nitrate.

61. Titanium is the seventh most abundant metal in the earth's crust. It is strong, lightweight, and resistant to corrosion; these properties lead to its use in aircraft engines. To obtain metallic titanium, ilmenite (FeTiO$_3$), an ore of titanium, is first treated with sulfuric acid to form FeSO$_4$ and Ti(SO$_4$)$_2$. After separating these compounds, the latter substance is converted to TiO$_2$ in basic solution:

$$FeTiO_3(s) + 3\,H_2SO_4(aq) \longrightarrow$$
$$FeSO_4(aq) + Ti(SO_4)_2(aq) + 3\,H_2O(\ell)$$
$$Ti^{4+}(aq) + 4\,OH^-(aq) \longrightarrow TiO_2(s) + 2\,H_2O(\ell)$$

What volume of 18.0 M H_2SO_4 is required to react completely with 1.00 kg of ilmenite? What mass of TiO_2 can theoretically be produced by this sequence of reactions?

62. In the process described in Study Question 61, ilmenite ore is leached with sulfuric acid. This leads to the significant environmental problem of disposal of the iron(II) sulfate (which, in its hydrated form, is commonly called "copperas"). To avoid this dilemma, it has been suggested that HCl be used to leach ilmenite so that the iron-containing product is $FeCl_2$. It can be treated with water and air to give commercially useful iron(III) oxide and regenerate HCl by the reaction

$$2\, FeCl_2(aq) + 2\, H_2O(\ell) + \tfrac{1}{2}\, O_2(g) \longrightarrow Fe_2O_3(s) + 4\, HCl(aq)$$

   (a) Write a balanced equation for the treatment of ilmenite with aqueous HCl to give iron(II) chloride, titanium(IV) oxide, and water.

   (b) If the equation written in part (a) is combined with the preceding equation for oxidation of $FeCl_2$ to Fe_2O_3, is the HCl used in the first step recovered in the second step?

   (c) What mass (in grams) of iron(III) oxide can be obtained from one ton (908 kg) of ilmenite in this process?

63. ▲ A 0.213-g sample of uranyl(VI) nitrate, $UO_2(NO_3)_2$, is dissolved in 20.0 mL of 1.0 M H_2SO_4 and shaken with Zn. The zinc reduces the uranyl ion, UO_2^{2+}, to a uranium ion, U^{n+}. To determine the value of n, this solution is titrated with $KMnO_4$. Permanganate is reduced to Mn^{2+} and U^{n+} is oxidized back to UO_2^{2+}.

   (a) In the titration, 12.47 mL of 0.0173 M $KMnO_4$ was required to reach the equivalence point. Use this information to determine the charge on the ion U^{n+}.

   (b) With the identity of U^{n+} now established, write a balanced net ionic equation for the reduction of UO_2^{2+} by zinc (assume acidic conditions).

   (c) Write a balanced net ionic equation for the oxidation of U^{n+} to UO_2^{2+} by MnO_4^- in acid.

64. ▲ The transition metals form a class of compounds called metal carbonyls, an example of which is the tetrahedral complex $Ni(CO)_4$. Given the following thermodynamic data:

	ΔH_f° (kJ/mol)	S° (J/K · mol)
Ni	0	29.87
CO(g)	−110.525	+197.674
$Ni(CO)_4$(g)	−602.9	+410.6

   (a) Calculate the equilibrium constant for the formation of $Ni(CO)_4$(g) from nickel metal and CO gas.

   (b) Is the reaction of Ni(s) and CO(g) product- or reactant-favored?

   (c) Is the reaction more or less product-favored at higher temperatures? How could this reaction be used in the purification of nickel metal?

Summary and Conceptual Questions

The following questions may use concepts from the preceding chapters.

65. The stability of analogous complexes $[ML_6]^{n+}$ (relative to ligand dissociation) is in the general order Mn^{2+}, Fe^{2+}, Co^{2+}, Ni^{2+}, Cu^{2+}, Zn^{2+}. This order of ions is called the Irving-Williams series. Look up the values of the formation constants for the ammonia complexes of Co^{2+}, Ni^{2+}, Cu^{2+}, and Zn^{2+} in Appendix K and verify this statement.

66. ▲ In this question, we explore the differences between metal coordination by monodentate and bidentate ligands. Formation constants, K_f, for $[Ni(NH_3)_6]^{2+}$(aq) and $[Ni(en)_3]^{2+}$(aq) are as follows:

$$Ni^{2+}(aq) + 6\, NH_3(aq) \longrightarrow [Ni(NH_3)_6]^{2+}(aq) \quad K_f = 10^8$$

$$Ni^{2+}(aq) + 3\, en(aq) \longrightarrow [Ni(en)_3]^{2+}(aq) \quad K_f = 10^{18}$$

The difference in K_f between these complexes indicates a higher thermodynamic stability for the chelated complex, caused by the *chelate effect*. Recall that K is related to the standard free energy of the reaction by $\Delta G^\circ = -RT\ln K$ and $\Delta G^\circ = \Delta H^\circ - T\Delta S^\circ$. We know from experiment that ΔH° for the NH_3 reaction is −109 kJ/mol, and ΔH° for the ethylenediamine reaction is −117 kJ/mol. Is the difference in ΔH° sufficient to account for the 10^{10} difference in K_f? Comment on the role of entropy in the second reaction.

Do you need a live tutor for homework problems?
Access vMentor at General ChemistryNow at
http://now.brookscole.com/kotz6e
for one-on-one tutoring from a chemistry expert

23—Nuclear Chemistry

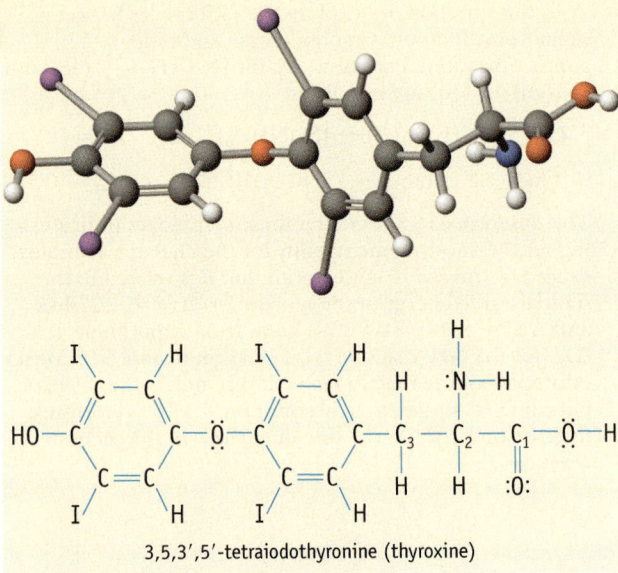

3,5,3′,5′-tetraiodothyronine (thyroxine)

The hormone 3,5,3′,5′-tetraiodothyronine (thyroxine) exerts a stimulating effect on metabolism.

Nuclear Medicine

The primary function of the thyroid gland, located in your neck, is the production of thyroxine (3,5,3′,5′-tetraiodothyronine) and 3,5,3′-triiodothyronine. These chemical compounds are hormones that help to regulate the rate of *metabolism* (page 541), a term that refers to all of the chemical reactions that take place in the body. In particular, the thyroid hormones play an important role in the processes that convert food to energy.

Abnormally low levels of thyroxine result in a condition known as *hypothyroidism*. Its symptoms include lethargy and feeling cold much of the time. The remedy for this condition is medication, consisting of thyroxine pills. The opposite condition, *hyperthyroidism*, also occurs in some people. In this condition, the body produces too much of the hormone. Hyperthyroidism is diagnosed by symptoms such as nervousness, heat intolerance, increased appetite, and muscle weakness and fatigue when blood sugar is too rapidly depleted. The standard remedy for hyperthyroidism is to destroy part of

Most salt contains a small amount of NaI; this supplies the body with sufficient iodine for its needs.

Chapter Goals

See Chapter Goals Revisited (page 1139). Test your knowledge of these goals by taking the exam-prep quiz on the General ChemistryNow CD-ROM or website.

- Identify radioactive elements and describe natural and artificial nuclear reactions.
- Calculate the binding energy and binding energy per nucleon for a particular isotope.
- Understand rates of radioactive decay.
- Understand artificial nuclear reactions.
- Understand issues of health and safety with respect to radioactivity.
- Be aware of some uses of radioactive isotopes in science and medicine.

Chapter Outline

23.1 Natural Radioactivity

23.2 Nuclear Reactions and Radioactive Decay

23.3 Stability of Atomic Nuclei

23.4 Rates of Nuclear Decay

23.5 Artificial Nuclear Reactions

23.6 Nuclear Fission

23.7 Nuclear Fusion

23.8 Radiation Health and Safety

23.9 Applications of Nuclear Chemistry

the thyroid gland, and one way to do so is to use radioactive iodine-131.

To understand this procedure, you need to know something about iodine in the body. Iodine is an essential element. Some diets provide iodine naturally (seaweed, for example, is a good source of

iodine), but in the Western world most iodine taken up by the body comes from iodized salt, NaCl containing a few percent of NaI. An adult man or woman of average size should take in about 150 μg (micrograms) of iodine (1 μg = 10^{-6} g) in the daily diet. In the body, iodide ion is transported to the thyroid, where it serves as one of the raw materials in making thyroxine.

The fact that iodine concentrates in the thyroid tissue is essential to the procedure for using radioiodine therapy as a treatment for hyperthyroidism. Typically, this radioisotope is administered orally as an aqueous NaI solution in which a small fraction of iodide consists of the radioactive isotope iodine-131 or iodine-123, and the rest is nonradioactive iodine-127. The radioactivity destroys thyroid tissue, resulting in a decrease in thyroid activity.

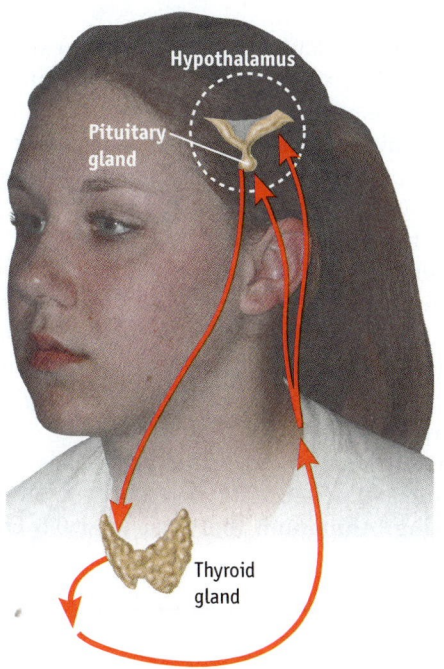

The thyroid gland, located in the neck, is the source of the hormone thyroxine.

To Review Before You Begin

- Review the structure of the atom: protons, neutrons, and electrons (Section 2.1)
- Know the definitions of atomic number and mass number (Section 2.2)
- Understand the mathematics defining first-order kinetics (Section 15.4)
- Be familiar with the units of energy (Appendix B)

GENERAL
Chemistry Now™

Throughout the chapter this icon introduces a list of resources on the General ChemistryNow CD-ROM or website (http://now .brookscole.com/kotz6e) that will:

- help you evaluate your knowledge of the material
- provide homework problems
- allow you to take an exam-prep quiz
- provide a personalized Learning Plan targeting resources that address areas you should study

■ **Discovery of Radioactivity**
The discovery of radioactivity by Henri Becquerel and the isolation of radium and polonium from pitchblende, a uranium ore, by Marie Curie were described in Chapter 2 (page 61).

■ **Common Symbols: α and β**
Symbols used to represent alpha and beta particles do not include a superscript to show that they have a charge.

History of science scholars cite three pillars of modern chemistry: technology, medicine, and alchemy. The third of these pillars, alchemy, was pursued in many cultures on three continents for well over 1000 years. Simply stated, the goal of the ancient alchemists was to turn less valuable materials into gold. We now recognize the futility of these efforts, because this goal is not reachable by chemical processes. We also know that it *is* possible to transmute one element into another. This happens naturally in the decomposition of uranium and other radioactive elements, and scientists can intentionally carry out such reactions in the laboratory. The goal is no longer to make gold, however. Far more important and valuable products of nuclear reactions are possible.

Nuclear chemistry encompasses a wide range of topics that share one thing in common: they involve changes in the nucleus of an atom. While "chemistry" is a major focus in this chapter, the subject cuts across many areas of science. Radioactive isotopes are used in medicine. Nuclear power provides a sizable fraction of energy for modern society. And, then there are nuclear weapons. . . .

23.1—Natural Radioactivity

In the late 19th century, while studying radiation emanating from uranium and thorium, Ernest Rutherford (1871–1937) [◀ page 65] stated, "There are present at least two distinct types of radiation—one that is readily absorbed, which will be termed for convenience *α* **(alpha) radiation,** and the other of a more penetrative character, which will be termed *β* **(beta) radiation.**" Subsequently, charge-to-mass ratio measurements showed that *α* radiation is composed of helium nuclei (He^{2+}) and *β* radiation is composed of electrons (e^-) (Table 23.1).

Rutherford hedged his bet when he said at least two types of radiation existed. A third type was later discovered by the French scientist Paul Villard (1860–1934); he named it *γ* **(gamma) radiation,** using the third letter in the Greek alphabet in keeping with Rutherford's scheme. Unlike *α* and *β* radiation, *γ* radiation is not affected by electric and magnetic fields. Rather, it is a form of electromagnetic radiation like x-rays but even more energetic.

Early studies measured the penetrating power of the three types of radiation (Figure 23.1). Alpha radiation is the least penetrating; it can be stopped by several sheets of ordinary paper or clothing. Aluminum that is at least 0.5 cm thick is

Table 23.1 Characteristics of *α*, *β*, and *γ* Radiation

Name	Symbols	Charge	Mass (g/particle)
Alpha	^4_2He, $^4_2\alpha$	2+	6.65×10^{-24}
Beta	$^0_{-1}e$, $^0_{-1}\beta$	1−	9.11×10^{-28}
Gamma	γ	0	0

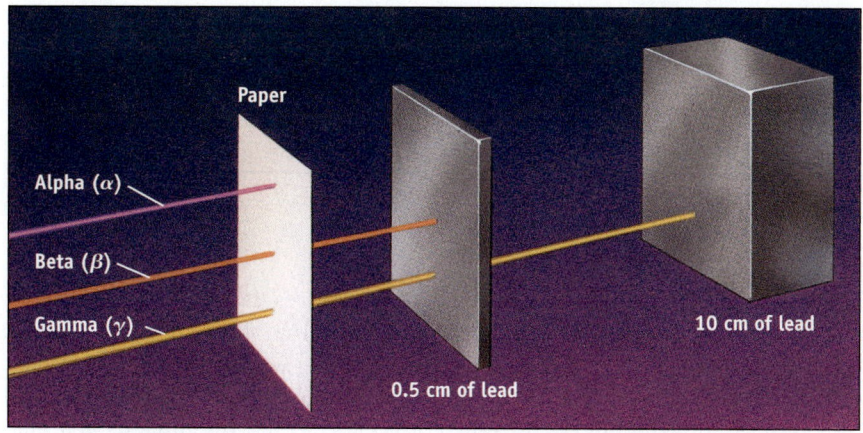

Figure 23.1 **The relative penetrating ability of** α, β, **and** γ **radiation.** Highly charged α particles interact strongly with matter and are stopped by a piece of paper. Beta particles, with less mass and a lower charge, interact to a lesser extent with matter and thus can penetrate farther. Gamma radiation is the most penetrating.

needed to stop β particles; they can penetrate several millimeters of living bone or tissue. Gamma radiation is the most penetrating. Thick layers of lead or concrete are required to shield the body from this radiation, and γ-rays can pass completely through the human body.

Alpha and β particles typically possess high kinetic energies. The energy of γ radiation is similarly very high. The energy associated with this radiation is transferred to any material used to stop the particle or absorb the radiation. This fact is important because the damage caused by radiation is related to the energy absorbed (see Section 23.8).

23.2—Nuclear Reactions and Radioactive Decay

Equations for Nuclear Reactions

In 1903, Rutherford and Frederick Soddy (1877–1956) proposed that radioactivity is the result of a natural change of an isotope of one element into an isotope of a different element. Such processes are called **nuclear reactions.** In a nuclear reaction, there is a change in the atomic number of an atom and often a change in the mass number as well.

Consider a reaction in which radium-226 (the isotope of radium with mass number 226) emits an α particle to form radon-222. The equation for this reaction is

$$^{226}_{88}\text{Ra} \longrightarrow ^{4}_{2}\alpha + ^{222}_{86}\text{Rn}$$

In a nuclear reaction the sum of the mass numbers of reacting particles must equal the sum of the mass numbers of products. Furthermore, to maintain nuclear charge balance, the sum of the atomic numbers of the products must equal the sum of the atomic numbers of the reactants. These principles are illustrated using the preceding nuclear equation:

■ **Symbols Used in Nuclear Equations** The mass number is included as a superscript and the atomic number is included as a subscript preceding the symbols for reactants and products. This is done to facilitate balancing these equations.

	$^{226}_{88}\text{Ra}$	$\longrightarrow$	$^{4}_{2}\alpha$	+	$^{222}_{86}\text{Rn}$
	radium-226	$\longrightarrow$	α particle	+	radon-222
Mass number: (protons + neutrons)	226	=	4	+	222
Atomic number: (protons)	88	=	2	+	86

Alpha particle emission causes a decrease of two units in atomic number and four units in the mass number.

Similarly, nuclear mass and nuclear charge balance accompany β particle emission by uranium-239:

	$^{239}_{92}U$	$\longrightarrow$	$^{0}_{-1}\beta$	+	$^{239}_{93}Np$
	uranium-239	$\longrightarrow$	β particle	+	neptunium-239
Mass number: (protons + neutrons)	239	=	0	+	239
Atomic number: (protons)	92	=	−1	+	93

The β particle has a charge of $1-$. Charge balance requires that the atomic number of the product be one unit greater than the atomic number of the reacting nucleus. The mass number does not change in this process.

How does a nucleus, composed of protons and neutrons, eject an electron? It is a complex process, but the net result is the conversion within the nucleus of a neutron to a proton and an electron.

$$^{1}_{0}n \longrightarrow {}^{0}_{-1}e + {}^{1}_{1}p$$

$$\text{neutron} \qquad \text{electron} \quad \text{proton}$$

Notice that the mass and charge numbers balance in this equation.

What is the origin of the gamma radiation that accompanies most nuclear reactions? Recall that a photon of visible light is emitted when an atom undergoes a transition from an excited electronic state to a lower-energy state [◀ Section 7.3]. Gamma radiation originates from transitions between nuclear energy levels. Nuclear reactions often result in the formation of a product nucleus in an excited nuclear state. One option is to return to the ground state by emitting the excess energy as a photon. The high energy of γ radiation is a measure of the large energy difference between the energy levels in the nucleus.

GENERAL
Chemistry☞Now™

See the General ChemistryNow CD-ROM or website:
- **Screen 23.2 Balancing Nuclear Reaction Equations,** for a tutorial on balancing equations
- **Screen 23.3 Modes of Radioactive Decay,** for a tutorial on modes of radioactive decay

---

Exercise 23.1—Mass and Charge Balance in Nuclear Reactions

Write equations for the following nuclear reactions and confirm that they are balanced with respect to nuclear mass and nuclear charge.

(a) the emission of an α particle by radon-222 to form polonium-218
(b) the emission of a β particle by polonium-218 to form astatine-218

■ **Energy Units**
Gamma ray energies are often reported with the unit *MeV*, which stands for 1 million electron volts. One electron volt (1 eV) is the energy of an electron that has been accelerated by a potential of one volt. The conversion factor between electron volts and joules is 1 eV = 1.60218×10^{-19} J.

Exercise 23.2—Gamma Ray Energies

Calculate the energy per photon, and the energy per mole of photons, for γ radiation with a wavelength of 2.0×10^{-12} m. (*Hint:* Review similar calculations on the energy of photons of visible light, Section 7.2.)

Radioactive Decay Series

Several naturally occurring radioactive isotopes are found to decay to form a product that is also radioactive. When this happens, the initial nuclear reaction is followed by a second nuclear reaction; if the situation is repeated, a third and a fourth nuclear reaction occur; and so on. Eventually, a nonradioactive isotope is formed to end the series. Such a sequence of nuclear reactions is called a **radioactive decay series.** In each step of this nuclear reaction sequence, the reactant nucleus is called the *parent* and the product called the *daughter.*

Uranium-238, the most abundant of three naturally occurring uranium isotopes, heads one of four radioactive decay series. This series begins with the loss of an α particle from $^{238}_{92}U$ to form radioactive $^{234}_{90}Th$. Thorium-234 then decomposes by β emission to $^{234}_{91}Pa$, which emits a β particle to give $^{234}_{92}U$. Uranium-234 is an α emitter, forming $^{230}_{90}Th$. Further α and β emissions follow, until the series ends with formation of the stable, nonradioactive isotope, $^{206}_{82}Pb$. In all, this radioactive decay series converting $^{238}_{92}U$ to $^{206}_{82}Pb$ is made up of 14 reactions, with eight α and six β particles being emitted. The series is portrayed graphically by plotting atomic number versus mass number (Figure 23.2). An equation can be written for each step in the sequence. Equations for the first four steps in the uranium-238 radioactive decay series are

Step 1. $^{238}_{92}U \longrightarrow {}^{234}_{90}Th + {}^{4}_{2}\alpha$

Step 2. $^{234}_{90}Th \longrightarrow {}^{234}_{91}Pa + {}^{0}_{-1}\beta$

Step 3. $^{234}_{91}Pa \longrightarrow {}^{234}_{92}U + {}^{0}_{-1}\beta$

Step 4. $^{234}_{92}U \longrightarrow {}^{230}_{90}Th + {}^{4}_{2}\alpha$

Uranium ore contains trace quantities of the radioactive elements formed in the radioactive decay series. A significant development in nuclear chemistry was

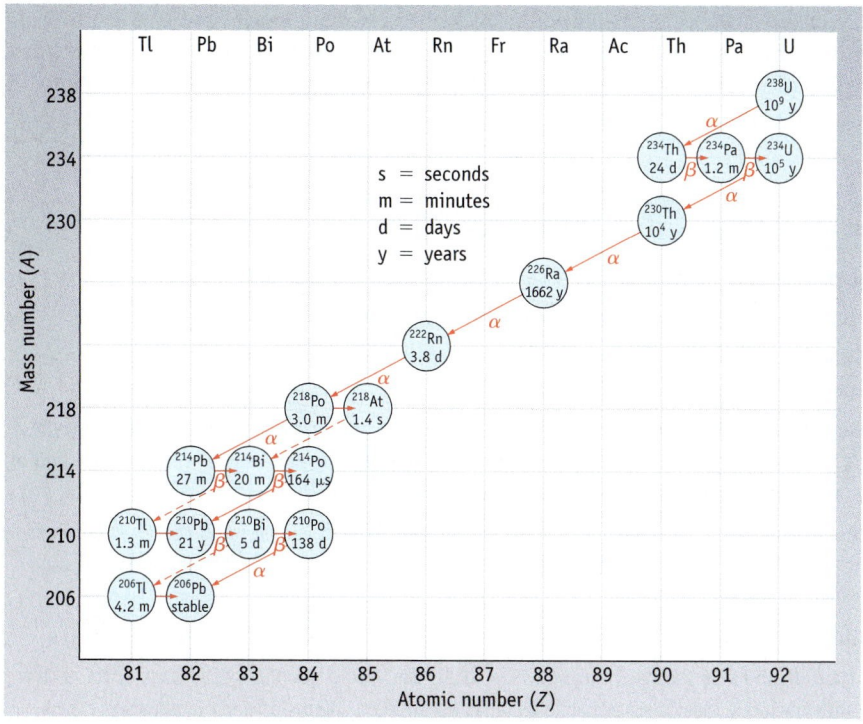

Figure 23.2   The uranium-238 radioactive decay series. The steps in this radioactive decay series are shown graphically in this plot of mass number versus atomic number. Each α decay step lowers the atomic number by two units and the mass number by four units. Beta particle emission does not change the mass but raises the atomic number by one unit. Half-lives of the isotopes are included on the chart.

Charles D. Winters

Radon detector. This kit is intended for use in the home to detect radon gas. The small device is placed in the home's basement for a given time period and is then sent to a laboratory to measure the amount of radon that might be present.

Marie Curie's discovery in 1898 of radium and polonium as trace components of pitchblende, a uranium ore. The amount of each of these elements is small because the isotopes of these elements have short half-lives. It is reported that Curie isolated only a single gram of radium from 7 tons of ore. It is a credit to her skills as a chemist that she extracted sufficient amounts of radium and polonium from uranium ore to identify these elements.

The uranium-238 radioactive decay series is also the source of the environmental hazard radon. Trace quantities of uranium are often present naturally in the soil and rocks in which radon-222 is being continuously formed. Because radon is chemically inert, it is not trapped by chemical processes occurring in soil or water and is free to seep into mines or into homes through pores in cement block walls, through cracks in the basement floor or walls, or around pipes. Because it is more dense than air, radon tends to collect in low spots, so its concentration can build up in a basement if steps are not taken to remove it.

The major health hazard from radon, when it is inhaled by humans, arises not from radon itself but from its decomposition product, polonium.

$$^{222}_{86}Rn \longrightarrow {}^{218}_{84}Po + {}^{4}_{2}\alpha \qquad t_{1/2} = 3.82 \text{ days}$$

$$^{218}_{84}Po \longrightarrow {}^{214}_{82}Pb + {}^{4}_{2}\alpha \qquad t_{1/2} = 3.04 \text{ minutes}$$

Radon does not undergo chemical reactions or form compounds that can be taken up in the body. Polonium, however, is not chemically inert. Polonium-218 can lodge in body tissues, where it undergoes α decay to give lead-214, another radioactive isotope. The range of an α particle in body tissue is quite small, perhaps 0.7 mm. This is approximately the thickness of the epithelial cells of the lungs, however, so α particle radiation can cause serious damage to lung tissues.

Virtually every home in the United States has some level of radon, and kits can be purchased to test for the presence of this gas. If radon gas is detected in your home, you should take corrective actions such as sealing cracks around the foundation and in the basement. It may be reassuring to know that the health risks associated with radon are low. The likelihood of getting lung cancer from exposure to radon is about the same as the likelihood of dying in an accident in your home.

Example 23.1—Radioactive Decay Series

Problem A second radioactive decay series begins with $^{235}_{92}U$ and ends with $^{207}_{82}Pb$.

(a) How many α and β particles are emitted in this series?

(b) The first three steps of this series are (in order) α, β, and α emission. Write an equation for each of these steps.

Strategy First, find the total change in atomic number and mass number. A combination of α and β particles is required that will decrease the total nuclear mass by 28 (235 − 207) and at the same time decrease the atomic number by 10 (92 − 82).

Each equation must give symbols for the parent and daughter nuclei and the emitted particle. In the equations, the sums of the atomic numbers and mass numbers for reactants and products must be equal.

Solution

(a) Mass declines by 28 mass units (235 − 207). Because a decrease of 4 mass units occurs with each α emission, 7 α particles must be emitted. Also, for each α emission, the atomic number decreases by 2. Emission of 7 α particles would cause the atomic number to decrease by 14, but the actual decrease in atomic number is 10 (92 − 82). This means that 4 β particles must also have been emitted because each β emission *increases* the atomic number of the product by one unit. Thus, the radioactive decay sequence involves emission of 7 α and 4 β particles.

(b) Step 1. $^{235}_{92}U \longrightarrow ^{231}_{90}Th + ^{4}_{2}\alpha$

Step 2. $^{231}_{90}Th \longrightarrow ^{231}_{91}Pa + ^{0}_{-1}\beta$

Step 3. $^{231}_{91}Pa \longrightarrow ^{227}_{89}Ac + ^{4}_{2}\alpha$

Comment Notice in Figure 23.2 that all daughter nuclei for the series beginning with $^{238}_{92}U$ have mass numbers differing by four units: 238, 234, 230, . . . , 206. This series is sometimes called the *4n + 2 series* because each mass number (*M*) fits the equation $4n + 2 = M$, where *n* is an integer (*n* is 59 for the first member of this series). For the series headed by $^{235}_{92}U$, the mass numbers are 235, 231, 227, . . . , 207; this is the *4n + 3 series*.

Two other decay series are possible. One, called the *4n series* and beginning with ^{232}Th, is found in nature; the other, the *4n + 1 series*, is not. No member of this series has a very long half-life. During the 5 billion years since this planet was formed, all members of this series have completely decayed.

Exercise 23.3—Radioactive Decay Series

(a) Six α and four β particles are emitted in the thorium-232 radioactive decay series before a stable isotope is reached. What is the final product in this series?

(b) The first three steps in the thorium-232 decay series (in order) are α, β, and β emission. Write an equation for each step.

Other Types of Radioactive Decay

Most naturally occurring radioactive elements decay by emission of α, β, and γ radiation. Other nuclear decay processes became known, however, when new radioactive elements were synthesized by artificial means. These include **positron ($^{0}_{+1}\beta$) emission** and **electron capture.**

Positrons ($^{0}_{+1}\beta$) and electrons have the same mass but opposite charge. The positron is the antimatter analogue to an electron. Positron emission by polonium-207, for example, results in the formation of bismuth-207.

	$^{207}_{84}Po$	$\longrightarrow$	$^{0}_{+1}\beta$	+	$^{207}_{83}Bi$
	polonium-207	$\longrightarrow$	positron	+	bismuth-207
Mass number: (protons + neutrons)	207	=	0	+	207
Atomic number: (protons)	84	=	+1	+	83

To retain charge balance, positron decay results in a decrease in the atomic number.

In *electron capture*, an extranuclear electron is captured by the nucleus. The mass number is unchanged and the atomic number is reduced by 1. (In an old nomenclature, the innermost electron shell was called the K shell, and electron capture was called *K capture.*)

	$^{7}_{4}Be$	+	$^{0}_{-1}e$	$\longrightarrow$	$^{7}_{3}Li$
	beryllium-7	+	electron	$\longrightarrow$	lithium-7
Mass number: (protons + neutrons)	7	+	0	=	7
Atomic number: (protons)	4	+	−1	=	3

■ **Positrons**
Positrons were discovered by Carl Anderson (1905–1991) in 1932. The positron is one of a group of particles that are known as *antimatter*. If matter and antimatter particles collide, mutual annihilation occurs, with energy being emitted.

■ **Neutrinos and Antineutrinos**
Beta particles having a wide range of energies are emitted. To balance the energy associated with β decay, it is necessary to postulate the concurrent emission of another particle, the *antineutrino*. Similarly, neutrino emission accompanies positron emission. Much study has gone into detecting neutrinos and antineutrinos. These massless, chargeless particles are not included when writing nuclear equations.

In summary, most unstable nuclei decay by one of four paths: α or β decay, positron emission, or electron capture. Gamma radiation often accompanies these processes. Section 23.6 introduces a fifth way that nuclei decompose, *fission*.

Example 23.2—Nuclear Reactions

Problem Complete the following equations. Give the symbol, mass number, and atomic number of the product species.

(a) $^{37}_{18}\text{Ar} + ^{0}_{-1}\text{e} \longrightarrow ?$ (c) $^{35}_{16}\text{S} \longrightarrow ^{35}_{17}\text{Cl} + ?$

(b) $^{11}_{6}\text{C} \longrightarrow ^{11}_{5}\text{B} + ?$ (d) $^{30}_{15}\text{P} \longrightarrow ^{0}_{+1}\beta + ?$

Strategy The missing product in each reaction can be determined by recognizing that the sums of mass numbers and atomic numbers for products and reactants must be equal. When you know the nuclear mass and nuclear charge of the product, you can identify it with the appropriate symbol.

Solution

(a) This is an electron capture reaction. The product has a mass number of $37 + 0 = 37$ and an atomic number of $18 - 1 = 17$. Therefore, the symbol for the product is $^{37}_{17}\text{Cl}$.

(b) This missing particle has a mass of zero and a charge of $1+$; these are the characteristics of a positron, $^{0}_{+1}\beta$. If this particle is included in the equation, the sums of the atomic numbers ($6 = 5 + 1$) and the mass numbers (11) on either side of the equation are equal.

(c) A beta particle, $^{0}_{-1}\beta$, is required to balance the mass numbers (35) and atomic numbers ($16 = 17 - 1$) in the equation.

(d) The product nucleus has mass number 30 and atomic number 14. This identifies the unknown as $^{30}_{14}\text{Si}$.

Exercise 23.4—Nuclear Reactions

Indicate the symbol, the mass number, and the atomic number of the missing product in each of the following nuclear reactions.

(a) $^{13}_{7}\text{N} \longrightarrow ^{13}_{6}\text{C} + ?$ (c) $^{90}_{38}\text{Sr} \longrightarrow ^{90}_{39}\text{Y} + ?$

(b) $^{41}_{20}\text{Ca} + ^{0}_{-1}\text{e} \longrightarrow ?$ (d) $^{22}_{11}\text{Na} \longrightarrow ? + ^{0}_{-1}\beta$

23.3—Stability of Atomic Nuclei

We can learn something about nuclear stability from Figure 23.3. In this plot, the horizontal axis represents the number of protons and the vertical axis gives the number of neutrons for known isotopes. Each circle represents an isotope identified by the number of neutrons and protons contained in its nucleus. The black circles represent stable (nonradioactive) isotopes, some 300 in number, and the red circles represent some of the known radioactive isotopes. For example, the three isotopes of hydrogen are $^{1}_{1}\text{H}$ and $^{2}_{1}\text{H}$ (neither is radioactive) and $^{3}_{1}\text{H}$ (tritium, radioactive). For lithium, the third element, isotopes with mass numbers 4, 5, 6, and 7 are known. The isotopes with masses of 6 and 7 (shown in black) are stable, whereas the other two isotopes (in red) are radioactive.

Figure 23.3 contains the following information about nuclear stability:

- Stable isotopes fall in a very narrow range called the **band of stability.** It is remarkable how few isotopes are stable.

- Only two stable isotopes ($^1_1$H and $^3_2$He) have more protons than neutrons.

- Up to calcium ($Z = 20$), stable isotopes often have equal numbers of protons and neutrons or only one or two more neutrons than protons.

- Beyond calcium, the neutron–proton ratio is always greater than 1. As the mass increases, the band of stable isotopes deviates more and more from a line in which $N = Z$.

- Beyond bismuth (83 protons and 126 neutrons), all isotopes are unstable and radioactive. There is apparently no nuclear "superglue" strong enough to hold heavy nuclei together.

- The lifetimes of unstable nuclei are shorter for the heaviest nuclei. For example, half of a sample of $^{238}_{92}$U disintegrates in 4.5 billion years, whereas half of a sample of $^{257}_{103}$Lr is gone in only 0.65 second. Isotopes that fall farther from the band of stability tend to have shorter half-lives than do unstable isotopes nearer to the band of stability.

- Elements of even atomic number have more stable isotopes than do those of odd atomic number. More stable isotopes have an even number of neutrons than have an odd number. Roughly 200 isotopes have an even number of neutrons and an even number of protons. Only about 120 isotopes have an odd number of either protons or neutrons. Only five stable isotopes ($^1_1$H, $^6_3$Li, $^{10}_5$B, $^{14}_7$N, and $^{180}_{73}$Ta) have odd numbers of both protons and neutrons.

The Band of Stability and Radioactive Decay

Besides being a criterion for stability, the neutron–proton ratio can assist in predicting what type of radioactive decay will be observed. Unstable nuclei decay in a

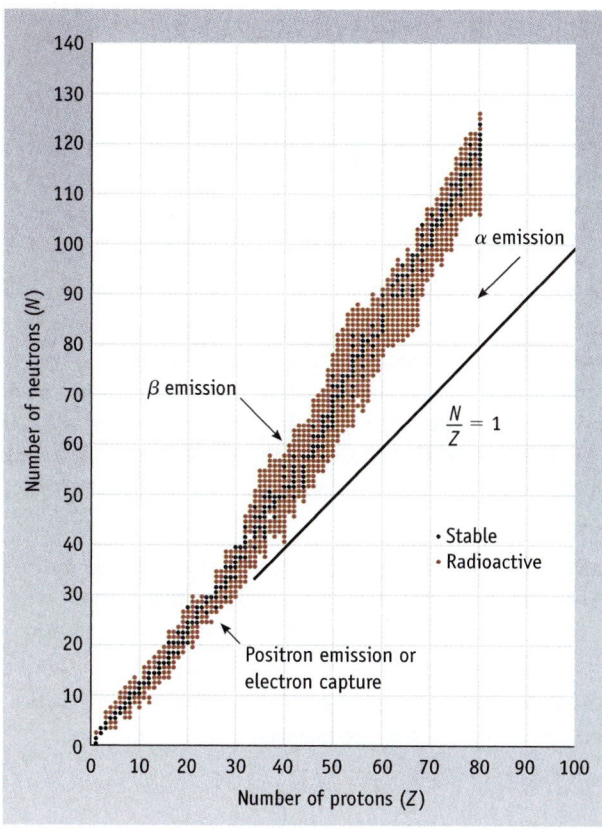

Figure 23.3 **Stable and unstable isotopes.** A graph of the number of neutrons (N) versus the number of protons (Z) for stable (black circles) and radioactive (red circles) isotopes from hydrogen to bismuth. This graph is used to assess criteria for nuclear stability and to predict modes of decay for unstable nuclei.

manner that brings them toward a stable neutron–proton ratio—that is, toward the band of stability.

- All elements beyond bismuth ($Z = 83$) are unstable. To reach the band of stability starting with these elements, a process that decreases the atomic number is needed. Alpha emission is an effective way to lower Z, the atomic number, because each emission decreases the atomic number by 2. For example, americium, the radioactive element used in smoke detectors, decays by α emission:

$$^{243}_{95}\text{Am} \longrightarrow {}^{4}_{2}\alpha + {}^{239}_{93}\text{Np}$$

- Beta emission occurs for isotopes that have a high neutron–proton ratio—that is, isotopes above the band of stability. With β decay, the atomic number increases by 1 and the mass number remains constant, resulting in a lower n/p ratio.

$$^{60}_{27}\text{Co} \longrightarrow {}^{0}_{-1}\beta + {}^{60}_{28}\text{Ni}$$

- Isotopes with a low neutron–proton ratio, below the band of stability, decay by positron emission or by electron capture. Both processes lead to product nuclei with a lower atomic number and the same mass number:

$$^{13}_{7}\text{N} \longrightarrow {}^{0}_{+1}\beta + {}^{13}_{6}\text{C}$$

$$^{41}_{20}\text{Ca} + {}^{0}_{-1}\text{e} \longrightarrow {}^{41}_{19}\text{K}$$

GENERAL
Chemistry ❖ Now™

See the General ChemistryNow CD-ROM or website:
- **Screen 23.3 Modes of Radioactive Decay,** for a tutorial on predicting modes of radioactive decay

---

Example 23.3—Predicting Modes of Radioactive Decay

Problem Identify probable mode(s) of decay for each isotope and write an equation for the decay process.

(a) oxygen-15, $^{15}_{8}\text{O}$ **(b)** uranium-234, $^{234}_{92}\text{U}$ **(c)** fluorine-20, $^{20}_{9}\text{F}$ **(d)** manganese-56, $^{56}_{25}\text{Mn}$

Strategy In parts (a), (c), and (d), compare the mass number with the atomic weight. If the mass number of the isotope is higher than the atomic weight, then there are too many neutrons and β emission is likely. If the mass number is lower than the atomic weight, then there are too few neutrons and positron emission or electron capture is the more likely process. It is not possible to choose between the latter two modes of decay without further information. For part (b), note that isotopes with atomic number greater than 83 are likely to be α emitters.

Solution

(a) Oxygen-15 has 7 neutrons and 8 protons so the n/p ratio is less than 1—too low for ^{15}O to be stable. Nuclei with too few neutrons are expected to decay by either positron emission or electron capture. In this instance, the process is  $^{0}_{+1}\beta$ emission and the equation is $^{15}_{8}\text{O} \longrightarrow {}^{0}_{+1}\beta + {}^{15}_{7}\text{N}$

(b) Alpha emission is a common mode of decay for isotopes of elements with atomic numbers higher than 83. The decay of uranium-234 is one example:

$$^{234}_{92}\text{U} \longrightarrow {}^{230}_{90}\text{Th} + {}^{4}_{2}\alpha$$

(c) Fluorine-20 has 11 neutrons and 9 protons, a high n/p ratio. The ratio is lowered by β emission:

$$^{20}_{9}F \longrightarrow {}^{0}_{-1}\beta + {}^{20}_{10}Ne$$

(d) The atomic weight of manganese is 54.85. The higher mass number, 56, suggests that this radioactive isotope has an excess of neutrons, in which case it would be expected to decay by β emission:

$$^{56}_{25}Mn \longrightarrow {}^{0}_{-1}\beta + {}^{56}_{26}Fe$$

Comment Be aware that predictions made in this manner will be right much of the time, but exceptions will sometimes occur.

Exercise 23.5—Predicting Modes of Radioactive Decay

Write an equation for the probable mode of decay for each of the following unstable isotopes, and write an equation for that nuclear reaction.

(a) silicon-32, $^{32}_{14}Si$

(b) titanium-45, $^{45}_{22}Ti$

(c) plutonium-239, $^{239}_{94}Pu$

(d) potassium-42, $^{42}_{19}K$

Nuclear Binding Energy

An atomic nucleus can contain as many as 83 protons and still be stable. For stability, nuclear binding (attractive) forces must be greater than the electrostatic repulsive forces between the closely packed protons in the nucleus. **Nuclear binding energy,** E_b, is defined as the energy required to separate the nucleus of an atom into protons and neutrons. For example, the nuclear binding energy for deuterium is the energy required to convert one mole of deuterium ($^{2}_{1}H$) nuclei into one mole of protons and one mole of neutrons.

$$^{2}_{1}H \longrightarrow {}^{1}_{1}p + {}^{1}_{0}n \qquad E_b = 2.15 \times 10^8 \text{ kJ/mol}$$

The positive sign for E_b indicates that energy is required for this process. A deuterium nucleus is more stable than an isolated proton and an isolated neutron, just as the H_2 molecule is more stable than two isolated H atoms. Recall, however, that the H—H bond energy is only 436 kJ/mol. The energy holding a proton and a neutron together in a deuterium nucleus, 2.15×10^8 kJ/mol, is about 500,000 times larger than the typical covalent bond energies.

To further understand nuclear binding energy, we turn to an experimental observation and a theory. The experimental observation is that the mass of a nucleus is always less than the sum of the masses of its constituent protons and neutrons. The theory is that the "missing mass," called the **mass defect** [◄ "A Closer Look," page 71], is equated with energy that holds the nuclear particles together.

The mass defect for deuterium is the difference between the mass of a deuterium nucleus and the sum of the masses of a proton and a neutron. Mass spectrometric measurements [◄ page 71] give the accurate masses of these particles to a high level of precision, providing the numbers needed to carry out calculations of mass defects.

Masses of atomic nuclei are not generally listed in reference tables, but masses of atoms are. Calculation of the mass defect can be carried out using masses of atoms instead of masses of nuclei. By using atomic masses, we are including in this calculation the masses of extranuclear electrons in the reactants and the products. Because the same number of extranuclear electrons appears in products and reactants, this does not affect the result. For one mole of deuterium nuclei, the mass defect is found as follows:

$$\ce{^{2}_{1}H} \longrightarrow \ce{^{1}_{1}H} + \ce{^{1}_{0}n}$$

2.01410 g/mol   1.007825 g/mol   1.008665 g/mol

$$\begin{aligned}
\text{Mass defect} = \Delta m &= \text{mass of products} - \text{mass of reactants} \\
&= \left[1.007825 \text{ g/mol} + 1.008665 \text{ g/mol}\right] - 2.01410 \text{ g/mol} \\
&= 0.00239 \text{ g/mol}
\end{aligned}$$

The relationship between mass and energy is contained in Albert Einstein's 1905 theory of special relativity, which holds that mass and energy are different manifestations of the same quantity. Einstein defined the energy–mass relationship: energy is equivalent to mass times the square of the speed of light; that is, $E = mc^2$. In the case of atomic nuclei, it is assumed that the missing mass (the mass defect, Δm) is equated with the binding energy holding the nucleus together.

$$E_b = (\Delta m)c^2 \tag{23.1}$$

If Δm is given in kilograms and the speed of light is given in meters per second, E_b will have units of joules (because $1 \text{ J} = 1 \text{ kg} \cdot \text{m}^2/\text{s}^2$). For the decomposition of one mole of deuterium nuclei to one mole of protons and one mole of neutrons, we have

$$\begin{aligned}
E_b &= (2.39 \times 10^{-6} \text{ kg/mol})(2.998 \times 10^8 \text{ m/s})^2 \\
&= 2.15 \times 10^{11} \text{ J/mol of } \ce{^{2}_{1}H} \text{ nuclei } (= 2.15 \times 10^8 \text{ kJ/mol of } \ce{^{2}_{1}H} \text{ nuclei})
\end{aligned}$$

The nuclear stabilities of different elements are compared using the **binding energy per mole of nucleons.** (**Nucleon** is the general name given to nuclear particles—that is, protons and neutrons.) A deuterium nucleus contains two nucleons, so the binding energy per mole of nucleons, E_b/n, is 2.15×10^8 kJ/mol divided by 2, or 1.08×10^8 kJ/mol nucleon.

$$E_b/n = \left(\frac{2.15 \times 10^8 \text{ kJ}}{\text{mol } \ce{^{2}_{1}H} \text{ nuclei}}\right)\left(\frac{1 \text{ mol } \ce{^{2}_{1}H} \text{ nuclei}}{2 \text{ mol nucleons}}\right)$$

$$E_b/n = 1.08 \times 10^8 \text{ kJ/mol nucleons}$$

The binding energy per nucleon can be calculated for any atom whose mass is known. Then, to compare nuclear stabilities, binding energies per nucleon are plotted as a function of mass number (Figure 23.4). The greater the binding energy per nucleon, the greater the stability of the nucleus. From the graph in Figure 23.4, the point of maximum nuclear stability occurs at a mass of 56 (that is, at iron in the periodic table).

GENERAL
Chemistry·꙰·Now™

See the General ChemistryNow CR-ROM or website:
- **Screen 23.4 Nuclear Stability,** for a simulation on isotope stability
- **Screen 23.5 Binding Energy,** for a tutorial on calculating binding energy

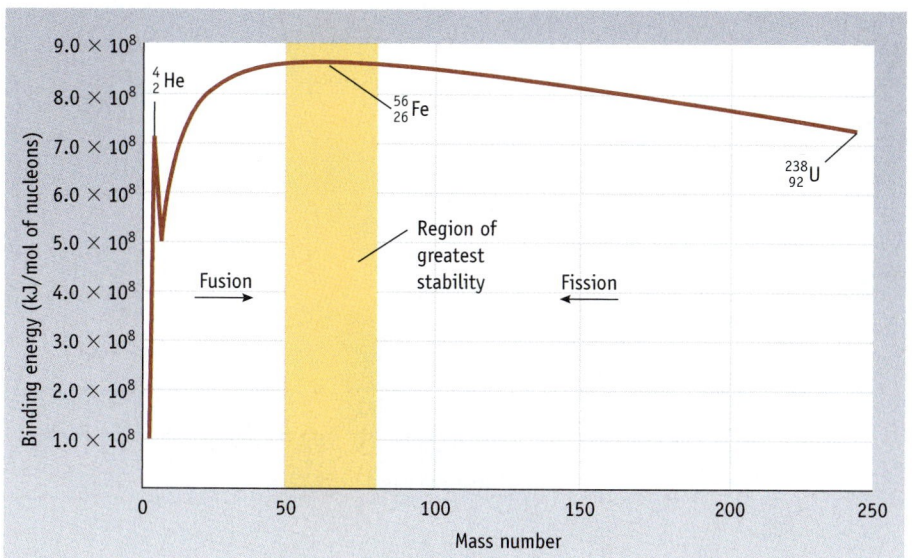

Figure 23.4   Relative stability of nuclei. Binding energy per nucleon for the most stable isotope of elements between hydrogen and uranium is plotted as a function of mass number. (Fission and fusion are discussed on pages 1130–1132.)

Example 23.4—Nuclear Binding Energy

Problem Calculate the binding energy, E_b (in kJ/mol), and the binding energy per nucleon, E_b/n (in kJ/mol nucleon), for carbon-12.

Strategy First determine the mass defect, then use Equation 23.1 to determine the binding energy. There are 12 nuclear particles in carbon-12, so dividing the nuclear binding energy by 12 will give the binding energy per nucleon.

Solution The mass of ^1_1H is 1.007825 g/mol, and the mass of ^1_0n is 1.008665 g/mol. Carbon-12, $^{12}_6C$, is the standard for the atomic masses in the periodic table, and its mass is defined as exactly 12 g/mol (12.000000 g/mol)

$$\Delta m = [(6 \times \text{mass } ^1_1H) + (6 \times \text{mass } ^1_0n)] - \text{mass } ^{12}_6C$$

$$= [(6 \times 1.007825) + (6 \times 1.008665)] - 12.000000$$

$$= 9.8940 \times 10^{-2} \text{ g/mol nuclei}$$

The binding energy is calculated using Equation 23.1. Using the mass in kilograms and the speed of light in meters per second gives the binding energy in joules:

$$E_b = (\Delta m)c^2$$

$$= (9.8940 \times 10^{-5} \text{kg/mol})(2.99792 \times 10^8 \text{m/s})^2$$

$$= 8.89 \times 10^{12} \text{ J/mol nuclei} (= 8.89 \times 10^9 \text{ kJ/mol nuclei})$$

The binding energy per nucleon, E_b/n, is determined by dividing the binding energy by 12 (the number of nucleons)

$$\frac{E_b}{n} = \frac{8.89 \times 10^9 \text{ kJ/mol nuclei}}{12 \text{ mol nucleons/mol nuclei}}$$

$$= 7.41 \times 10^8 \text{ kJ/mol nucleons}$$

Exercise 23.6—Nuclear Binding Energy

Calculate the binding energy per nucleon, in kilojoules per mole, for the formation of lithium-6. The molar mass of ^6_3Li is 6.015125 g/mol.

Figure 23.5   Decay of 20.0 mg of oxygen-15. After each half-life period of 2.0 min, the mass of oxygen-15 decreases by one half.

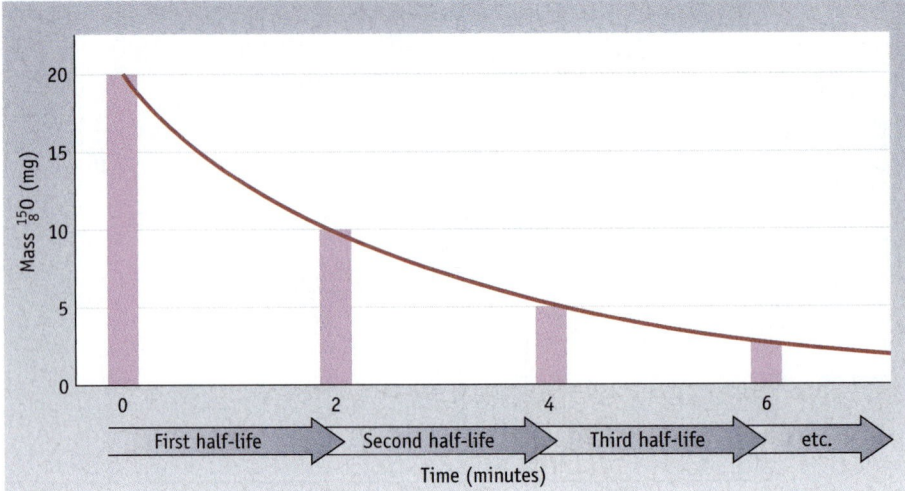

23.4—Rates of Nuclear Decay

Half-Life

When a radioactive isotope is identified, its *half-life* is usually measured. Half-life ($t_{1/2}$) is used in nuclear chemistry in the same way it is used when discussing the kinetics of first-order chemical reactions [◄ Section 15.4]: It is the time required for half of a sample to decay to products (Figure 23.5). Recall that for first-order kinetics the half-life is independent of the amount of sample.

Half-lives for radioactive isotopes cover a wide range of values. Uranium-238 has one of the longer half-lives, 4.47×10^9 years, a length of time close to the age of the earth (estimated at 4.5–4.6×10^9 years). Roughly half of the uranium-238 present when the planet was formed is still around. At the other end of the range of half-lives are isotopes such as element 112, whose 277 isotope has a half-life of 240 microseconds ($1 \, \mu s = 1 \times 10^{-6} \, s$).

Half-life provides an easy way to estimate the time required before a radioactive element is no longer a health hazard. Strontium-90, for example, is a β emitter with a half-life of 29.1 years. Significant quantities of strontium-90 were dispersed into the environment in atmospheric nuclear bomb tests in the 1960s and 1970s; and from the half-life, we know that a little less than half is still around. The health problems associated with strontium-90 arise because calcium and strontium have similar chemical properties. Strontium-90 is taken into the body and deposited in bone, taking the place of calcium. Radiation damage by strontium-90 in bone has been directly linked to bone-related cancers.

■ **Half-Life and Temperature**
Unlike what is observed in chemical kinetics, temperature does not affect the rate of nuclear decay.

Example 23.5—Using Half-Life

Problem Radioactive iodine-131, used to treat hyperthyroidism, has a half-life of 8.04 days.

(a) If you have 8.8 μg (micrograms) of this isotope, what mass remains after 32.2 days?

(b) How long will it take for a sample of iodine-131 to decay to one-eighth of its activity?

(c) Estimate the length of time necessary for the sample to decay to 10% of its original activity.

Strategy This problem asks you to use half-life to qualitatively assess the rate of decay. After one half-life, half of the sample remains. After another half-life, the amount of sample is again decreased by half to one fourth of its original value. (This situation is illustrated in Figure 23.5.) To answer these questions, assess the number of half-lives that have elapsed and use this information to determine the amount of sample remaining.

Solution

(a) The time elapsed, 32.2 days, is 4 half-lives (32.2/8.04 = 4). The amount of iodine-131 has decreased to 1/16 of the original amount [$1/2 \times 1/2 \times 1/2 \times 1/2 = (1/2)^4 = 1/16$]. The amount of iodine remaining is 8.8 μg $\times (1/2)^4$ or 0.55 μg .

(b) After 3 half-lives , the amount of iodine-131 remaining is 1/8 ($= 1/2)^3$ of the original amount. The amount remaining is 8.8 μg $\times (1/2)^3 = 1.1 \mu$g.

(c) After 3 half-lives, 1/8 (12.5%) of the sample remains; after 4 half-lives, 1/16 (6.25%) remains. It will take between 3 and 4 half-lives, between 24.15 and 32.2 days , to decrease the amount of sample to 10% of its original value.

Comment You will find it useful to make approximations as we have done in (c). An exact time can be calculated from the first-order rate law (pages 713 and 1124).

Exercise 23.7—Using Half-Life

Tritium ($^3_1$H), a radioactive isotope of hydrogen has a half-life of 12.3 years.

(a) Starting with 1.5 mg of this isotope, how many milligrams remain after 49.2 years?

(b) How long will it take for a sample of tritium to decay to one eighth of its activity?

(c) Estimate the length of time necessary for the sample to decay to 1% of its original activity.

Kinetics of Nuclear Decay

The rate of nuclear decay is determined from measurements of the **activity** (A) of a sample. Activity refers to the number of disintegrations observed per unit time, a quantity that can be measured readily with devices such as a Geiger–Müller counter (Figure 23.6). *Activity is proportional to the number of radioactive atoms present (N).*

$$A \propto N$$

(23.2)

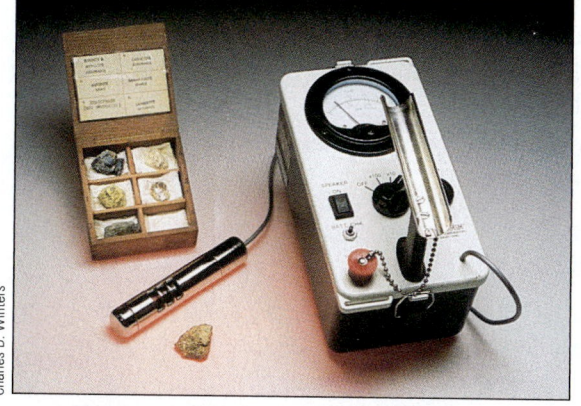
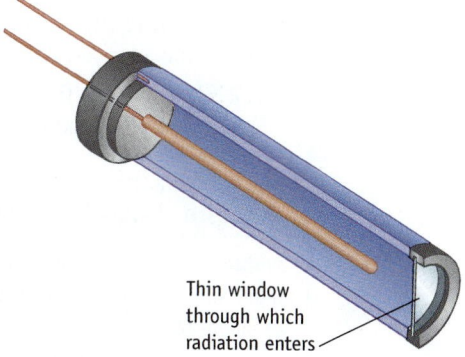

Charles D. Winters

Figure 23.6 A Geiger–Müller counter. A charged particle (an α or β particle) enters the gas-filled tube (diagram at the right) and ionizes the gas. The gaseous ions migrate to electrically charged plates and are recorded as a pulse of electric current. The current is amplified and used to operate a counter. A sample of carnotite, a mineral containing uranium oxide, is also shown in the photograph.

Thin window through which radiation enters

If the number of radioactive nuclei N is reduced by half, the activity of the sample will be half as large. Doubling N will double the activity. This evidence indicates that the rate of decomposition is first order with respect to N. Consequently, the equations describing rates of radioactive decay are the same as those used to describe first-order chemical reactions; the change in the number of radioactive atoms N per unit of time is proportional to N:

$$\frac{\Delta N}{\Delta t} = -kN \qquad (23.3)$$

The integrated rate equation can be written in two ways depending on the data used:

$$\ln\left(\frac{N}{N_0}\right) = -kt \qquad (23.4)$$

or

$$\ln\left(\frac{A}{A_0}\right) = -kt \qquad (23.5)$$

Here, N_0 and A_0 are the number of atoms and the activity of the sample initially, respectively, and N and A are the number of atoms and the activity of the sample after time t, respectively. Thus, N/N_0 is the fraction of atoms remaining after a given time (t), and A/A_0 is the fraction of the activity remaining after the same period. In these equations, k is the rate constant (decay constant) for the isotope in question. The relationship between half-life and the first-order rate constant is the same as seen with chemical kinetics (Equation 15.4, page 719):

$$t_{1/2} = \frac{0.693}{k} \qquad (23.6)$$

Equations 23.3–23.6 are useful in several ways:

- If the activity (A) or the number of radioactive nuclei (N) is measured in the laboratory over some period t, then k can be calculated. The decay constant k can then be used to determine the half-life of the sample.
- If k is known, the fraction of a radioactive sample (N/N_0) still present after some time t has elapsed can be calculated.
- If k is known, the time required for that isotope to decay to a fraction of the original activity (A/A_0) can be calculated.

Example 23.6—Determination of Half-Life

Problem A sample of radon-222 has an initial α particle activity (A_0) of 7.0×10^4 dps (disintegrations per second). After 6.6 days, its activity (A) is 2.1×10^4 dps. What is the half-life of radon-222?

Strategy Values for A, A_0, and t are given. The problem can be solved using Equation 23.5 with k as the unknown. Once k is found, the half-life can be calculated using Equation 23.6.

Solution

$$\ln (2.1 \times 10^4 \text{ dps}/7.0 \times 10^4 \text{ dps}) = -k \text{ (6.6 day)}$$

$$\ln (0.30) = -k(6.6 \text{ day})$$

$$k = 0.18 \text{ days}^{-1}$$

From k we obtain $t_{1/2}$:

$$t_{1/2} = 0.693/0.18 \text{ days}^{-1} = \boxed{3.8 \text{ days}}$$

Comments Notice that the activity decreased to between one half and one fourth of its original value. The 6.6 days of elapsed time represents one full half-life and part of another half-life.

Example 23.7—Time Required for a Radioactive Sample to Partially Decay

Problem Gallium citrate, containing the radioactive isotope gallium-67, is used medically as a tumor-seeking agent. It has a half-life of 78.2 h. How long will it take for a sample of gallium citrate to decay to 10.0% of its original activity?

Strategy Use Equation 23.5 to solve this problem. In this case, the unknown is the time t. The rate constant k is calculated from the half-life using Equation 23.6. Although we do not have specific values of activity, the value of A/A_0 is known. Because A is 10.0% of A_0, the value of A/A_0 is 0.100.

Solution First determine k:

$$k = 0.693/t_{1/2} = 0.693/78.2 \text{ h}$$

$$k = 8.86 \times 10^{-3} \text{ h}^{-1}$$

Then substitute the given values of A/A_0 and k into Equation 23.5:

$$\ln (A/A_0) = -kt$$

$$\ln (0.100) = -(8.86 \times 10^{-3} \text{ h}^{-1})t$$

$$\boxed{t = 2.60 \times 10^2 \text{ h}}$$

Comments The time required is between three half-lives ($3 \times 78.2 \text{ h} = 235 \text{ h}$) and four half-lives ($4 \times 78.2 \text{ h} = 313 \text{ h}$).

Exercise 23.8—Determination of Half-Life

A sample of $Ca_3(PO_4)_2$ containing phosphorus-32 has an activity of 3.35×10^3 dpm. Two days later, the activity is 3.18×10^3 dpm. Calculate the half-life of phosphorus-32.

Exercise 23.9—Time Required for a Radioactive Sample to Partially Decay

A highly radioactive sample of nuclear waste products with a half-life $t_{1/2}$ of 200. years is stored in an underground tank. How long will it take for the activity to diminish from an initial activity of 6.50×10^{12} dpm to a fairly harmless activity of 3.00×10^3 dpm?

Radiocarbon Dating

In certain situations, the age of a material can be determined based on the rate of decay of a radioactive isotope. The best-known example of this procedure is the use of carbon-14 to date historical artifacts.

Carbon is primarily carbon-12 and carbon-13 with isotopic abundances of 98.9% and 1.1%, respectively. In addition, traces of a third isotope, carbon-14, are present to the extent of about 1 in 10^{12} atoms in atmospheric CO_2 and in living materials. Carbon-14 is a β emitter with a half-life of 5730 years. A one-gram sample of carbon from living material will show about 14 disintegrations per minute, not a lot of radioactivity but nevertheless detectable by modern methods.

Carbon-14 is formed in the upper atmosphere by nuclear reactions initiated by neutrons in cosmic radiation:

$$^{14}_{7}N + ^{1}_{0}n \longrightarrow ^{14}_{6}C + ^{1}_{1}H$$

Once formed, carbon-14 is oxidized to $^{14}CO_2$. This product enters the carbon cycle, circulating through the atmosphere, oceans, and biosphere.

The usefulness of carbon-14 for dating comes about in the following way. Plants absorb CO_2 and convert it to organic compounds, thereby incorporating carbon-14 into living tissue. As long as a plant remains alive, this process will continue and the percentage of carbon that is carbon-14 in the plant will equal the percentage in the atmosphere. When the plant dies, carbon-14 will no longer be taken up. Radioactive decay continues, however, with the carbon-14 activity decreasing over time. After 5730 years, the activity will be 7 dpm/g; after 11,460 years, it will be 3.5 dpm/g; and so on. By measuring the activity of a sample, and knowing the half-life of carbon-14, it is possible to calculate when a plant (or an animal that was eating plants) died.

As with all experimental procedures, carbon-14 dating has limitations. The procedure assumes that the amount of carbon-14 in the atmosphere hundreds or thousands of years ago is the same as it is now. We know that this isn't exactly true; the percentage has varied by as much as 10% (Figure 23.7). Furthermore, it is not possible to use carbon-14 to date an object that is less than about 100 years old; the radiation level from carbon-14 will not change enough in this short time period to permit accurate detection of a difference from the initial value. In most instances, the accuracy of the measurement is, in fact, only about ±100 years. Finally, it is not possible to determine ages of objects much older than about 40,000 years. By then, after nearly seven half-lives, the radioactivity will have decreased virtually to zero. But for the span of time between 100 and 40,000 years, this technique has provided important information (Figure 23.8).

See the General ChemistryNow CD-ROM or website:
- **Screen 23.3 Modes of Radioactive Decay,** for an exercise on the Geiger counter
- **Screen 23.6 Half-Life,** for a tutorial on half-life and radiochemical dating

Example 23.8—Radiochemical Dating

Problem To test the concept of carbon-14 dating, J. R. Arnold and W. F. Libby applied this technique to analyze samples of acacia and cyprus wood whose ages were already known. (The acacia wood, which was supplied by the Metropolitan Museum of Art in New York, came from the tomb of Zoser, the first Egyptian pharaoh to be entombed in a pyramid. The cyprus wood was from the tomb of Sneferu.) The average activity based on five determinations was 7.04 dpm per gram of carbon. Assume (as Arnold and Libby did) that the activity of carbon-14, A_0, is 12.6 dpm per gram of carbon. Calculate the approximate age of the sample.

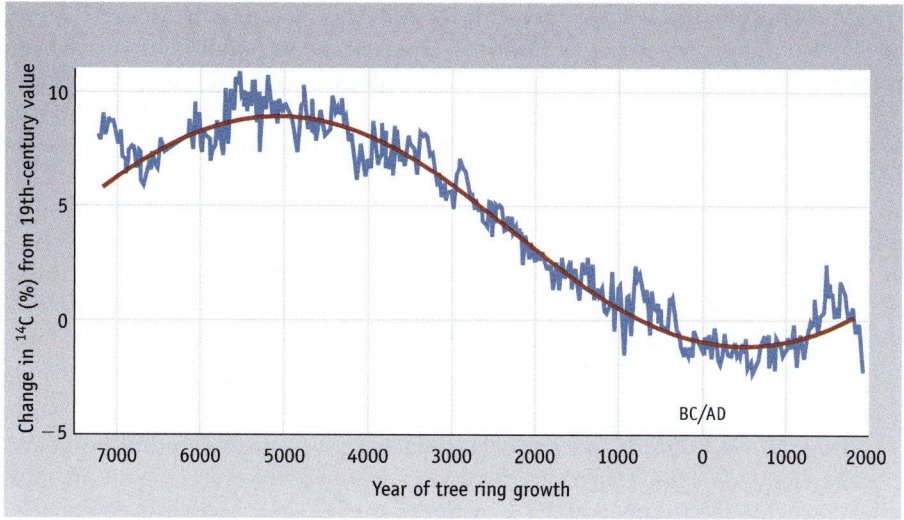

Figure 23.7   Variation of atmospheric carbon-14 activity. The amount of carbon-14 has varied with variation in cosmic ray activity. To obtain the data for the pre-1990 part of the curve shown in this graph, scientists carried out carbon-14 dating of artifacts for which the age was accurately known (often through written records). Similar results can be obtained using carbon-14 dating of tree rings.

Source: Hans E. Suess, La Jolla Radiocarbon Laboratory

Strategy First, determine the rate constant for the decay of carbon-14 from its half-life ($t_{1/2}$ for ^{14}C is 5.73×10^3 years). Then, use Equation 23.5.

Solution

$$k = 0.693/t_{1/2} = 0.693/5730 \text{ yr}$$

$$= 1.21 \times 10^{-4} \text{ yr}^{-1}$$

$$\ln (A/A_0) = -kt$$

$$\ln \left(\frac{7.04 \text{ dpm/g}}{12.6 \text{ dpm/g}} \right) = (-1.21 \times 10^{-4} \text{ yr}^{-1})t$$

$$t = 4.8 \times 10^3 \text{ yr}$$

The wood is about 4800 years old.

Comment This problem uses real data from an early research paper in which the carbon-14 dating method was being tested. The age of the wood was known to be 4750 ± 250 years. (See J. R. Arnold and W. F. Libby: *Science*, Vol. 110, p. 678, 1949.)

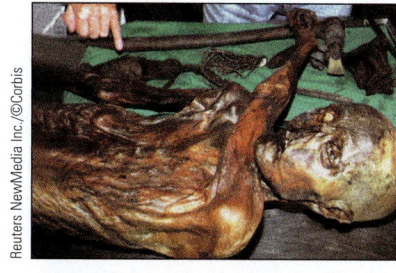

Figure 23.8   The Ice Man. The world's oldest preserved human remains were discovered in the ice of a glacier high in the Alps. Carbon-14 dating techniques allowed scientists to determine that he lived about 5300 years ago.

Exercise 23.10—Radiochemical Dating

A sample of the inner part of a redwood tree felled in 1874 was shown to have ^{14}C activity of 9.32 dpm/g. Calculate the approximate age of the tree when it was cut down. Compare this age with that obtained from tree ring data, which estimated that the tree began to grow in 979 ± 52 BC. Use 13.4 dpm/g for the value of A_0.

23.5—Artificial Nuclear Reactions

How many different isotopes are found on earth? All of the stable isotopes occur naturally. A few unstable (radioactive) isotopes that have long half-lives are found in nature; the best-known examples are uranium-235, uranium-238, and thorium-232. Trace quantities of other radioactive isotopes with short half-lives are present because they are being formed continuously by nuclear reactions. They include isotopes of radium, polonium, and radon, along with other elements produced in various radioactive decay series, and carbon-14, formed in a nuclear reaction initiated by cosmic radiation.

Naturally occurring isotopes account for only a very small fraction of the currently known radioactive isotopes, however. The rest—several thousand—have been synthesized via artificial nuclear reactions, sometimes referred to as transmutation.

The first artificial nuclear reaction was identified by Rutherford about 80 years ago. Recall the classic experiment that led to the nuclear model of the atom [◄ Figure 2.6] in which gold foil was bombarded with α particles. In the years following that experiment, Rutherford and his coworkers bombarded many other elements with α particles. In 1919, one of these experiments led to an unexpected result: when nitrogen atoms were bombarded with α particles, protons were detected among the products. Rutherford correctly concluded that a nuclear reaction had occurred. Nitrogen had undergone a *transmutation* to oxygen:

$$\,^4_2He + \,^{14}_7N \longrightarrow \,^{17}_8O + \,^1_1H$$

During the next decade, other nuclear reactions were discovered by bombarding other elements with α particles. Progress was slow, however, because in most cases α particles are simply scattered by target nuclei. The bombarding particles cannot get close enough to the nucleus to react because of the strong repulsive forces between the positively charged α particle and the positively charged atomic nucleus.

In 1932, two advances were made that greatly extended nuclear reaction chemistry. The first involved the use of particle accelerators to create high-energy particles as projectiles. The second was the use of neutrons as the bombarding particles.

The α particles used in the early studies on nuclear reactions came from naturally radioactive materials such as uranium and had relatively low energies, at least by today's standards. Particles with higher energy were needed, so J. D. Cockcroft (1897–1967) and E. T. S. Walton (1903–1995), working in Rutherford's laboratory in Cambridge, England, turned to protons. Protons are formed when hydrogen atoms ionize in a cathode-ray tube, and it was known that they could be accelerated to higher energy by applying a high voltage. Cockcroft and Walton found that when energetic protons struck a lithium target, the following reaction occurs:

$$\,^7_3Li + \,^1_1p \longrightarrow 2\,^4_2He$$

This was the first example of a reaction initiated by a particle that had been artificially accelerated to high energy. Since this experiment was conducted, the technique has been developed much further, and the use of particle accelerators in nuclear chemistry is now commonplace. Particle accelerators operate on the principle that a charged particle placed between charged plates will be accelerated to a high speed and high energy. Modern examples of this process are seen in the synthesis of the transuranium elements, several of which are described in more detail in "A Closer Look: The Search for New Elements."

Experiments using neutrons as bombarding particles were first carried out in both the United States and Great Britain in 1932. Nitrogen, oxygen, fluorine, and neon were bombarded with energetic neutrons, and α particles were detected among the products. Using neutrons made sense: because neutrons have no charge, it was reasoned that these particles would not be repelled by the positively charged nucleus particles. Thus, neutrons did not need high energies to react.

In 1934, Enrico Fermi (1901–1954) and his coworkers showed that nuclear reactions using neutrons are more favorable if the neutrons have low energy. A low-energy neutron is simply captured by the nucleus, giving a product in which the mass number is increased by one unit. Because of the low energy of the bombarding particle, the product nucleus does not have sufficient energy to fragment in these reactions. The new nucleus is produced in an excited state, however; when the

■ **Discovery of Neutrons**
Neutrons had been predicted to exist for more than a decade before they were identified in 1932 by James Chadwick (1891–1974). Chadwick produced neutrons in a nuclear reaction between α particles and beryllium: $\,^4_2\alpha + \,^9_4Be \longrightarrow \,^{12}_6C + \,^1_0n$.

■ **Glenn T. Seaborg (1912–1999)**
Seaborg figured out that thorium and the elements that followed it fit under the lanthanides in the periodic table. For this insight, he and Edwin McMillan shared the 1951 Nobel Prize in chemistry. Over a 21-year period, Seaborg and his colleagues synthesized 10 new transuranium elements (Pu through Lr). To honor Seaborg's scientific contributions, the name "seaborgium" was assigned to element 106. It marked the first time an element was named for a living person.
Lawrence Berkeley Laboratory

A Closer Look

The Search for New Elements

By 1936, guided first by Mendeleev's predictions and later by atomic theory, chemists had identified all but two of the elements with atomic numbers between 1 and 92. From this point onward, all new elements to be discovered came from artificial nuclear reactions. Two gaps in the periodic table were filled when radioactive technetium and promethium, the last two elements with atomic numbers less than 92, were identified in 1937 and 1942, respectively. The first success in the search for elements with atomic numbers higher than 92 came with the 1940 discovery of neptunium and plutonium.

Since 1950, laboratories in the United States (Lawrence Berkeley National Laboratory), Russia (Joint Institute for Nuclear Research at Dubna, near Moscow), and Europe (Institute for Heavy Ion Research at Darmstadt, Germany) have competed to make new elements. Syntheses of new transuranium elements use a standard methodology. An element of fairly high atomic number is bombarded with a beam of high-energy particles. Initially, neutrons were used; later, helium nuclei and then larger nuclei such as ^{11}B and ^{12}C were employed; and, more recently, highly charged ions of elements such as calcium, chromium, cobalt, and zinc have been chosen. The bombarding particle fuses with the nucleus of the target atom, forming a new nucleus that lasts for a short time before decomposing. New elements are detected by their decomposition products, a signature of particles with specific masses and energies.

By using bigger particles and higher energies, the list of known elements reached 106 by the end of the 1970s. To further extend the search, Russian scientists employed a new idea, matching precisely the energy of the bombarding particle with the energy required to fuse the nuclei. This technique enabled the synthesis of elements 107, 108, and 109 in Darmstadt in the early 1980s, and the synthesis of elements 110, 111, and 112 in the following decade. Lifetimes of these elements were in the millisecond range; the 277 isotope of element 112, for example, had a half-life of 240 μs.

Yet another breakthrough was needed to extend the list further. Scientists have long known that isotopes with specific *magic numbers* of neutrons and protons are more stable. Elements with 2, 8, 20, 50, and 82 protons are members of this category, as are elements with 126 neutrons. The magic numbers correspond to filled shells in the nucleus. Their significance is analogous to the significance of filled shells for electronic structure. Theory had predicted that the next magic numbers would be 114 protons and 184 neutrons. Using this information, researchers discovered element 114 in early 1999. The Dubna group reporting this discovery found that the 289 isotope had an exceptionally long half-life, about 20 s.

At the time this book was written, 116 elements were known. Will research yield further new elements? It would be hard to say no, given past successes in this area of research, but the quest becomes ever more difficult as scientists venture to the very limits of nuclear stability.

Fermilab Visual Media Services, Batavia, IL

Fermilab. The tunnel housing the four-mile-long particle accelerator at Batavia, Illinois

nucleus returns to the ground state, a γ-ray is emitted. Reactions in which a neutron is captured and a γ-ray is emitted are called **(n, γ) reactions.**

The (n, γ) reactions are the source of many of the radioisotopes used in medicine and chemistry. An example is radioactive phosphorus, $^{32}_{15}P$, which is used in chemical studies such as tracing the uptake of phosphorus in the body.

$$^{31}_{15}P + ^{1}_{0}n \longrightarrow ^{32}_{15}P + \gamma$$

Transuranium elements, elements with an atomic number greater than 92, were first made in a nuclear reaction sequence beginning with an (n, γ) reaction. Scientists at the University of California at Berkeley bombarded uranium-238 with

■ **Transuranium Elements in Nature**
Neptunium, plutonium, and americium were unknown prior to their preparation via these nuclear reactions. Later these elements were found to be present in trace quantities in uranium ores.

neutrons. Among the products identified were neptunium-239 and plutonium-239. These new elements were formed when ^{239}U decayed by β radiation.

$$^{238}_{92}\text{U} + ^{1}_{0}\text{n} \longrightarrow ^{239}_{92}\text{U}$$

$$^{239}_{92}\text{U} \longrightarrow ^{239}_{93}\text{Np} + ^{0}_{-1}\beta$$

$$^{239}_{93}\text{Np} \longrightarrow ^{239}_{94}\text{Pu} + ^{0}_{-1}\beta$$

Four years later, a similar reaction sequence was used to make americium-241. Plutonium-239 was found to add two neutrons to form plutonium-241, which decays by β emission to give americium-241.

Example 23.9—Nuclear Reactions

Problem Write equations for the nuclear reactions described below.

(a) Fluorine-19 undergoes an (n, γ) reaction to give a radioactive product that decays by $^{0}_{-1}\beta$ emission. (Write equations for both nuclear reactions.)

(b) A common neutron source is a plutonium–beryllium alloy. Plutonium-239 is an α emitter. When beryllium-9 (the only stable isotope of beryllium) reacts with α particles emitted by plutonium, neutrons are ejected. (Write equations for both reactions.)

Strategy The equations are written so that both mass and charge are balanced.

Solution

(a) $^{19}_{9}\text{F} + ^{1}_{0}\text{n} \longrightarrow ^{20}_{9}\text{F} + \gamma$

$^{20}_{9}\text{F} \longrightarrow ^{20}_{10}\text{Ne} + ^{0}_{-1}\beta$

(b) $^{239}_{94}\text{Pu} \longrightarrow ^{235}_{92}\text{U} + ^{4}_{2}\alpha$

$^{4}_{2}\alpha + ^{9}_{4}\text{Be} \longrightarrow ^{12}_{6}\text{C} + ^{1}_{0}\text{n}$

23.6—Nuclear Fission

In 1938, two chemists, Otto Hahn (1879–1968) and Fritz Strassman (1902–1980), isolated and identified barium in a sample of uranium that had been bombarded with neutrons. How was barium formed? The answer to that question explained one of the most significant scientific discoveries of the 20th century. The uranium nucleus had split into smaller pieces in the process we now call **nuclear fission.**

■ **Fission Reactions**
In the fission of uranium-236, a large number of different fission products (elements) are formed. Barium was the element first identified, and its identification provided the key that led to recognition that fission had occurred.

The details of nuclear fission were unraveled through the work of a number of scientists. They determined that a uranium-235 nucleus initially captured a neutron to form uranium-236. This isotope underwent nuclear fission to produce two new nuclei, one with a mass number around 140 and the other with a mass around 90, along with several neutrons (Figure 23.9). The nuclear reactions that led to formation of barium when a sample of ^{235}U was bombarded with neutrons are

$$^{235}_{92}\text{U} + ^{1}_{0}\text{n} \longrightarrow ^{236}_{92}\text{U}$$

$$^{236}_{92}\text{U} \longrightarrow ^{141}_{56}\text{Ba} + ^{92}_{36}\text{Kr} + 3\,^{1}_{0}\text{n}$$

An important aspect of fission reactions is that they produce more neutrons than are used to initiate the process. Under the right circumstances, these neutrons then serve to continue the reaction. If one or more of these neutrons are captured by another ^{235}U nucleus, then a further reaction can occur, releasing still more neutrons. This sequence repeats over and over. Such a mechanism, in which each step generates a reactant to continue the reaction, is called a **chain reaction.**

A nuclear fission chain reaction has three general steps:

1. *Initiation.* The reaction of a single atom is needed to start the chain. Fission of ^{235}U is initiated by the absorption of a neutron.

2. *Propagation.* This part of the process repeats itself over and over, with each step yielding more product. The fission of ^{236}U releases neutrons that initiate the fission of other uranium atoms.

3. *Termination.* Eventually the chain will end. Termination could occur if the reactant (^{235}U) is used up, or if the neutrons that continue the chain escape from the sample without being captured by ^{235}U.

To harness the energy produced in a nuclear reaction, it is necessary to control the rate at which a fission reaction occurs. This is managed by balancing the propagation and termination steps by limiting the number of neutrons available. In a nuclear reactor, this balance is accomplished by using cadmium rods to absorb neutrons. By withdrawing or inserting the rods, the number of neutrons available to propagate the chain can be changed, and the rate of the fission reaction (and the rate of energy production) can be increased or decreased.

Uranium-235 and plutonium-239 are the fissionable isotopes most commonly used in power reactors. Natural uranium contains only 0.72% of uranium-235; more than 99% of the natural element is uranium-238. The percentage of uranium-235 in natural uranium is too small to sustain a chain reaction, however, so the uranium used for nuclear fuel must be enriched in this isotope. One way to do so is by gaseous diffusion [◀ Section 12.7]. Plutonium, which occurs naturally in only trace quantities, must be made via a nuclear reaction. The raw material for this nuclear synthesis is the more abundant uranium isotope, ^{238}U. Addition of a neutron to ^{238}U gives ^{239}U, which, as noted earlier, undergoes two β emissions to form ^{239}Pu.

Currently, there are 103 operating nuclear power plants in the United States and 435 worldwide. About 20% of this country's electricity (and 17% of the world's energy) comes from nuclear power (Table 23.2). Although one might imagine that nuclear energy would be called upon to meet the ever-increasing needs of society, no new nuclear power plants are under construction in the United States. Among other things, the disasters at Chernobyl (in the former Soviet Union) and Three Mile Island (in Pennsylvania) have sensitized the public to the issue of safety. The cost to construct a nuclear power plant (measured in terms of dollars per kilowatt-hour of power) is considerably more than the cost for a natural gas–powered facility,

■ Lise Meitner (1878–1968)
Meitner's greatest contribution to 20th-century science was her explanation of the process of nuclear fission. She and her nephew, Otto Frisch, also a physicist, published a paper in 1939 that was the first to use the term "nuclear fission." Element number 109 is named meitnerium to honor Meitner's contributions. The leader of the team that discovered this element said that "She should be honored as the most significant woman scientist of [the 20th] century."
AIP-Emilio Segrè Visual Archives, Herzfeld Collection

■ The Atomic Bomb
In an atomic bomb, each nuclear fission step produces 3 neutrons, which leads to about 3 more fissions and 9 more neutrons, which leads to 9 more fission steps and 27 more neutrons, and so on. The rate depends on the number of neutrons, so the nuclear reaction occurs faster and faster as more and more neutrons are formed, leading to an enormous output of energy in a short time span.

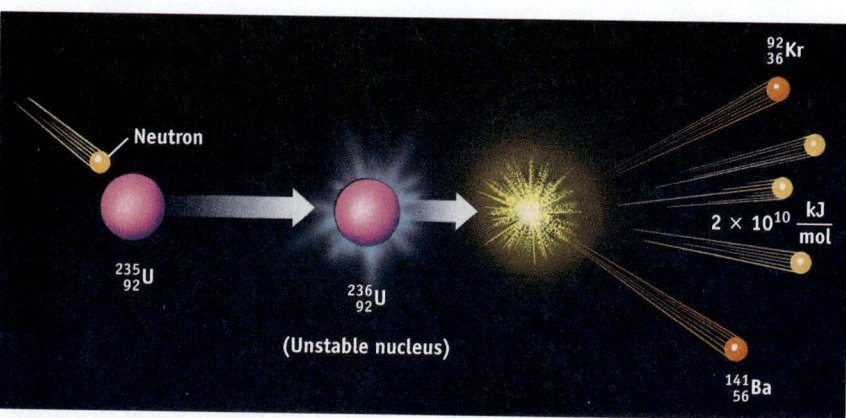

Figure 23.9 Nuclear fission. Neutron capture by $^{235}_{92}U$ produces $^{236}_{92}U$. This isotope undergoes fission, which yields several fragments along with several neutrons. These neutrons initiate further nuclear reactions by adding to other $^{235}_{92}U$ nuclei. The process is highly exothermic, producing about 2×10^{10} kJ/mol.

and the regulatory restrictions for nuclear power are burdensome. Disposal of highly radioactive nuclear waste is another thorny problem, with 20 metric tons of waste being generated per year at each reactor.

In addition to technical problems, nuclear energy production brings with it significant geopolitical security concerns. The process for enriching uranium for use in a reactor is the same process used for generating weapons-grade uranium. Also, some nuclear reactors are designed so that one by-product of their operation is the isotope plutonium-239, which can be removed and used in a nuclear weapon. Despite these problems, nuclear fission is an important part of the energy profile in a number of countries. For example, three-fourths of power production in France and one-third in Japan is nuclear generated.

23.7—Nuclear Fusion

In a **nuclear fusion** reaction, several small nuclei react to form a larger nucleus. Tremendous amounts of energy can be generated by such reactions. An example is the fusion of deuterium and tritium nuclei to form $^4_2$He and a neutron:

$$^2_1H + \,^3_1H \longrightarrow \,^4_2He + \,^1_0n \qquad \Delta E = -1.7 \times 10^9 \text{ kJ/mol}$$

Fusion reactions provide the energy of our sun and other stars. Scientists have long dreamed of being able to harness fusion to provide power. To do so, a temperature of 10^6 to 10^7 K, like that in the interior of the sun, would be required to bring the positively charged nuclei together with enough energy to overcome nuclear repulsions. At the very high temperatures needed for a fusion reaction, matter does not exist as atoms or molecules; instead, matter is in the form of a *plasma* made up of unbound nuclei and electrons.

Three critical requirements must be met before nuclear fusion could represent a viable energy source. First, the temperature must be high enough for fusion to occur. The fusion of deuterium and tritium, for example, requires a temperature of 10^8 K or more. Second, the plasma must be confined long enough to release a net output of energy. Third, the energy must be recovered in some usable form.

Harnessing a nuclear fusion reaction for a peaceful use has not yet been achieved. Nevertheless, many attractive features encourage continuing research in this field. The hydrogen used as "fuel" is cheap and available in almost unlimited amounts. As a further benefit, most radioisotopes produced by fusion have short half-lives, so they remain a radiation hazard for only a short time.

23.8—Radiation Health and Safety

Units for Measuring Radiation

Several units of measurement are used to describe levels and doses of radioactivity. As is the case in everyday life, the units used in the United States are not the same as the SI units of measurement.

In the United States, the degree of radioactivity is often measured in **curies** (Ci). Less commonly used in the United States is the SI unit, the **becquerel** (Bq). Both units measure the number of disintegrations per second; 1 Ci is 3.7×10^{10} dps (disintegrations per second), while 1 Bq represents 1 dps. The curie and the becquerel are used to report the amount of radioactivity when multiple kinds of unstable nuclei are decaying and to report amounts necessary for medical purposes.

By itself, the degree of radioactivity does not provide a good measure of the amount of energy in the radiation or the amount of damage that the radiation can cause to living tissue. Two additional kinds of information are necessary. The first is the amount of energy absorbed; the second is the effectiveness of the particular kind of radiation in causing tissue damage. The amount of energy absorbed by living tissue is measured in **rads.** *Rad* is an acronym for "radiation absorbed dose." One rad represents 0.01 J of energy absorbed per kilogram of tissue. Its SI equivalent is the **gray** (Gy); 1 Gy denotes the absorption of 1 J per kilogram of tissue.

Different forms of radiation cause different amounts of biological damage. The amount of damage depends on how strongly a form of radiation interacts with matter. Alpha particles cannot penetrate the body any farther than the outer layer of skin. If α particles are emitted within the body, however, they will do between 10 and 20 times the amount of damage done by γ-rays, which can go entirely through a human body without being stopped. In determining the amount of biological damage to living tissue, differences in damaging power are accounted for using a "quality factor." This quality factor has been set at 1 for β and γ radiation, 5 for low-energy protons and neutrons, and 20 for α particles or high-energy protons and neutrons.

Biological damage is quantified in a unit called the **rem** (an acronym for "roentgen equivalent man"). A dose of radiation in rem is determined by multiplying the energy absorbed in rads by the quality factor for that kind of radiation. The rad and the rem are very large in comparison to normal exposures to radiation, so it is more common to express exposures in millirems (mrem). The SI equivalent of the rem is the **sievert** (Sv), determined by multiplying the dose in grays by the quality factor.

Radiation: Doses and Effects

Exposure to a small amount of radiation is unavoidable. Earth is constantly being bombarded with radioactive particles from outer space. There is also some exposure to radioactive elements that occur naturally on earth, including ^{14}C, ^{40}K (a radioactive isotope that occurs naturally in 0.0117% abundance), ^{238}U, and ^{232}Th. Radioactive elements in the environment that were created artificially (in the fallout from nuclear bomb tests, for example) also contribute to this exposure. For some people, medical procedures using radioisotopes are a major contributor.

The average dose of background radioactivity in the United States is about 200 mrem per year (Table 23.3). Well over half of that amount comes from natural

■ **The Roentgen**
The roentgen (R) is an older unit of radiation exposure. It is defined as the amount of x-rays or γ radiation that will produce 2.08×10^9 ions in 1 cm^3 of dry air. The roentgen and the rad are similar in size. Also note that roentgenium is the name proposed for element 111.

Table 23.3 Radiation Exposure of an Individual for One Year from Natural and Artificial Sources

	Millirem/Year	Percentage
Natural Sources		
Cosmic radiation	50.0	25.8
The earth	47.0	24.2
Building materials	3.0	1.5
Inhaled from the air	5.0	2.6
Elements found naturally in human tissue	21.0	10.8
Subtotal	**126.0**	**64.9**
Medical Sources		
Diagnostic x-rays	50.0	25.8
Radiotherapy	10.0	5.2
Internal diagnosis	1.0	0.5
Subtotal	**61.0**	**31.5**
Other Artificial Sources		
Nuclear power industry	0.85	0.4
Luminous watch dials, TV tubes	2.0	1.0
Fallout from nuclear tests	4.0	2.1
Subtotal	**6.9**	3.5
Total	**193.9**	99.9

A Closer Look

What Is a Safe Exposure?

Is the exposure to natural background radiation totally without effect? Can you equate the effect of a single dose and the effect of cumulative, smaller doses that are spread out over a long period of time? The assumption generally made is that no "safe maximum dose," or level below which absolutely no damage will occur, exists. However, the accuracy of this assumption has come into question. These issues are not testable with human subjects, and tests based on animal studies are not completely reliable because of the uncertainty of species-to-species variations.

The model used by government regulators to set exposure limits assumes that the relationship between exposure to radiation and incidence of radiation-induced problems, such as cancer, anemia, and immune system problems, is linear. Under this assumption, if a dose of $2x$ rem causes damage in 20% of the population, then a dose of x rem will cause damage in 10% of the population. But is this true? Cells do possess mechanisms for repairing damage. Many scientists believe that this self-repair mechanism renders the human body less susceptible to damage from smaller doses of radiation, because the damage will be repaired as part of the normal course of events. They argue that, at extremely low doses of radiation, the self-repair response results in less damage.

The bottom line is that much still remains to be learned in this area. And the stakes are significant.

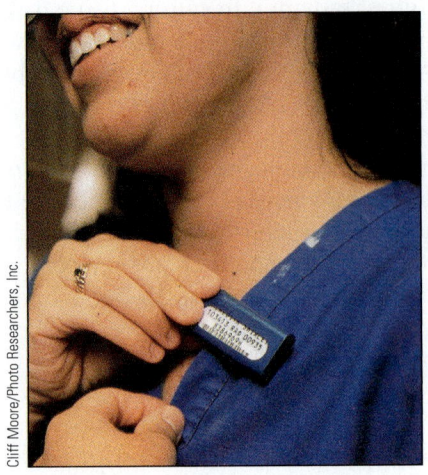

The film badge. These badges, worn by scientists using radioactive materials, are used to monitor cumulative exposure to radiation.

sources over which we have no control. Of the 60–70 mrem per year exposure that comes from artificial sources, nearly 90% is delivered in medical procedures such as x-ray examinations and radiation therapy. Considering the controversy surrounding nuclear power, it is interesting to note that less than 0.5% of the total annual background dose of radiation that the average person receives can be attributed to the nuclear power industry.

Describing the biological effects of a dose of radiation precisely is not a simple matter. The amount of damage done depends not only on the kind of radiation and the amount of energy absorbed, but also on the particular tissues exposed and the rate at which the dose builds up. A great deal has been learned about the effects of radiation on the human body by studying the survivors of the bombs dropped over Japan in World War II and the workers exposed to radiation from the reactor disaster at Chernobyl. From studies of the health of these survivors, we have learned that the effects of radiation are not generally observable below a single dose of 25 rem. At the other extreme, a single dose of >200 rem will be fatal to about half the population (Table 23.4).

Our information is more accurate when dealing with single, large doses than it is for the effects of chronic, smaller doses of radiation. One current issue of debate in the scientific community is how to judge the effects of multiple smaller doses or long-term exposure (see "A Closer Look: What Is A Safe Exposure?").

Table 23.4	Effects of a Single Dose of Radiation
Dose (rem)	**Effect**
0–25	No effect observed
26–50	Small decrease in white blood cell count
51–100	Significant decrease in white blood cell count, lesions
101–200	Loss of hair, nausea
201–500	Hemorrhaging, ulcers, death in 50% of population
500	Death

23.9—Applications of Nuclear Chemistry

We tend to think about nuclear chemistry in terms of power plants and bombs. In truth, radioactive elements are now used in all areas of science and medicine, and they are of ever-increasing importance to our lives. Because describing all of their uses would take several books, we have selected just a few examples to illustrate the diversity of applications of radioactivity.

Nuclear Medicine: Medical Imaging

Diagnostic procedures using nuclear chemistry are essential in medical imaging, which entails the creation of images of specific parts of the body. There are three principal components to constructing a radioisotope-based image:

- A radioactive isotope, administered as the element or incorporated into a compound, that concentrates the radioactive isotope in the tissue to be imaged
- A method of detecting the type of radiation involved
- A computer to assemble the information from the detector into a meaningful image

The choice of a radioisotope and the manner in which it is administered are determined by the tissue in question. A compound containing the isotope must be absorbed more by the diseased tissue than by the rest of the body. Table 23.5 lists radioisotopes that are commonly used in nuclear imaging processes, their half-lives, and the tissues they are used to image. All of the isotopes in Table 23.5 are γ emitters; γ radiation is preferred for imaging because it is less damaging to the body in small doses than either α or β radiation.

Technetium-99m is used in more than 85% of the diagnostic scans done in hospitals each year (see "A Closer Look: Technetium-99m"). The "m" stands for *metastable,* a term used to identify an unstable state that exists for a finite period of time. Recall that atoms in excited electronic states emit visible, infrared, and ultraviolet radiation [◀ Chapter 7]. Similarly, a nucleus in an excited state gives up its excess energy, but in this case a much higher energy is involved and the emission occurs as γ radiation. The γ-rays given off by ^{99m}Tc are detected to produce the image (Figure 23.10).

Another medical imaging technique based on nuclear chemistry is positron emission tomography (PET). In PET, an isotope that decays by positron emission is incorporated into a carrier compound and given to the patient. When emitted, the positron travels no more than a few millimeters before undergoing matter–antimatter annihilation.

$$^{0}_{+1}\beta + {}^{0}_{-1}e \longrightarrow 2\ \gamma$$

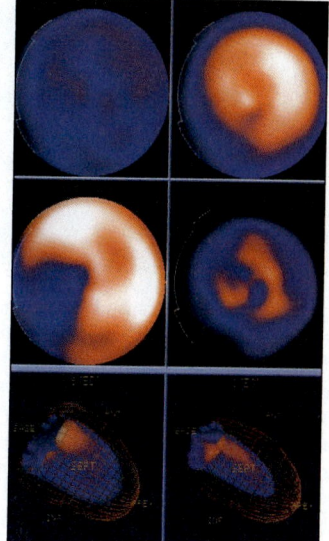

Figure 23.10   Heart imaging with technetium-99m. The radioactive element technetium-99m, a gamma emitter, is injected into a patient's vein in the form of the pertechnetate ion (TcO_4^-) or as a complex ion with an organic ligand. A series of scans of the gamma emissions of the isotope are made while the patient is resting and then again after strenuous exercise. Bright areas in the scans indicate that the isotope is binding to the tissue in that area. The scans in this figure show a normal heart function.

Table 23.5   Radioisotopes Used in Medical Diagnostic Procedures

Radioisotope	Half-Life (h)	Imaging
^{99m}Tc	6.0	Thyroid, brain, kidneys
^{201}Tl	73.0	Heart
^{123}I	13.2	Thyroid
^{67}Ga	78.2	Various tumors and abscesses
^{18}F	1.8	Brain, sites of metabolic activity

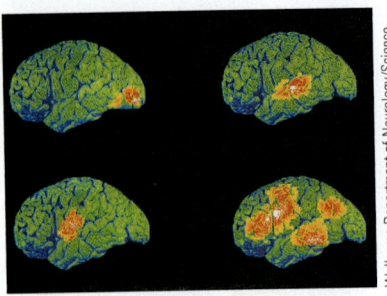

Figure 23.11   PET scans of the brain.
These scans show the left side of the brain; red indicates an area of highest activity. (*upper left*) *Sight* activates the visual area in the occipital cortex at the back of the brain. (*upper right*) *Hearing* activates the auditory area in the superior temporal cortex of the brain. (*lower left*) *Speaking* activates the speech centers in the insula and motor cortex. (*lower right*) *Thinking* about verbs, and speaking them, generates high activity, including in the hearing, speaking, temporal, and parietal areas.

The two emitted γ-rays travel in opposite directions. By determining where high numbers of γ-rays are being emitted, one can construct a map showing where the positron emitter is located in the body.

An isotope often used in PET is ^{15}O. A patient is given gaseous O_2 that contains ^{15}O. This isotope travels throughout the body in the bloodstream, allowing images of the brain and bloodstream (Figure 23.11) to be obtained. Because positron emitters are typically very short-lived, PET facilities must be located near a cyclotron where the radioactive nuclei are prepared and then immediately incorporated into a carrier compound.

Nuclear Medicine: Radiation Therapy

To treat most cancers, it is necessary to use radiation that can penetrate the body to the location of the tumor. Gamma radiation from a cobalt-60 source is commonly used. Unfortunately, the penetrating ability of γ-rays makes it virtually impossible to destroy diseased tissue without also damaging healthy tissue in the process. Nevertheless, this technique is a regularly sanctioned procedure and its successes are well known.

To avoid the side effects associated with more traditional forms of radiation therapy, a new form of treatment has been explored in the last 10 to 15 years, called *boron neutron capture therapy* (BNCT). BNCT is unusual in that boron-10, the isotope of boron used as part of the treatment, is not radioactive. Boron-10 is highly effective in capturing neutrons, however—2500 times better than boron-11, and 8 times better than uranium-235. When the nucleus of a boron-10 atom captures a neutron, the resulting boron-11 nucleus has so much energy that it fragments to form an α particle and a lithium-7 atom. Although the α particles do a great deal of damage, because their penetrating power is so low, the damage remains confined to an area not much larger than one or two cells in diameter.

In a typical BNCT treatment, a solution of a boron compound is injected into the tumor. After a few hours, the tumor is bombarded with neutrons. The α particles are produced only at the site of the tumor, and the production stops when the neutron bombardment ends.

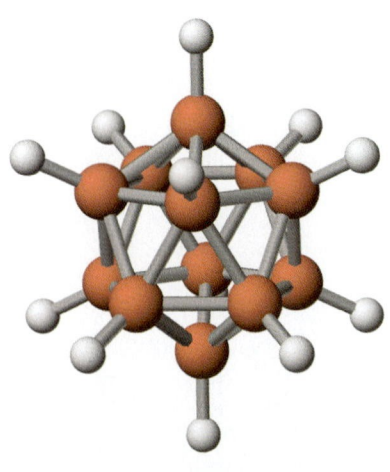

One of the compounds used in BNCT is $Na_2[B_{12}H_{12}]$. The structure of the $B_{12}H_{12}^{2-}$ anion is a regular polyhedron with 20 sides, called an icosahedron.

Analytical Methods: The Use of Radioactive Isotopes as Tracers

Radioactive isotopes can be used to help determine the fate of compounds in the body or in the environment. These studies begin with a compound that contains a radioactive isotope of one of its component elements. In biology, for example, scientists can use radioactive isotopes to measure the uptake of nutrients. Plants take up phosphorus-containing compounds from the soil through their roots. By adding a small amount of radioactive ^{32}P, a β emitter with a half-life of 14.3 days, to fertilizer and then measuring the rate at which the radioactivity appears in the leaves, plant biologists can determine the rate at which phosphorus is taken up. The outcome can assist scientists in identifying hybrid strains of plants that can absorb phosphorus quickly, resulting in faster-maturing crops, better yields per acre, and more food or fiber at less expense.

To measure pesticide levels, a pesticide can be tagged with a radioisotope and then applied to a test field. By counting the disintegrations of the radioactive tracer, information can be obtained about how much pesticide accumulates in the soil, is taken up by the plant, and is carried off in runoff surface water. After these tests are completed, the radioactive isotope decays to harmless levels in a few days or a few weeks because of the short half-lives of the isotopes used.

A Closer Look

Technetium-99m

Technetium was the first new element to be made artificially. One might think that this element would be a chemical rarity but this is not so: its importance in medical imaging has brought a great deal of attention to it. Although all the technetium in the world must be synthesized by nuclear reactions, the element is readily available and even inexpensive; its price in 1999 was $60 per gram, only about four times the price of gold.

Technetium-99m is formed when molybdenum-99 decays by β emission. Technetium-99m decays to its ground state with a half-life of 6.01 h, giving off a 140-KeV γ-ray in the process. (Technetium-99 is itself unstable, decaying to stable ^{99}Ru with a half-life of 2.1×10^5 years.)

Technetium-99m is produced in hospitals using a molybdenum–technetium generator. Sheathed in lead shielding, the generator contains the artificially synthesized isotope ^{99}Mo in the form of molybdate ion, MoO_4^{2-}, adsorbed on a column of alumina, Al_2O_3. In the nuclear reaction MoO_4^{2-} is converted into the pertechnate ion, $^{99m}TcO_4^-$. The $^{99m}TcO_4^-$ is washed from the column using a saline solution. In this process, technetium-99m may be used as the pertechnate ion or converted into other compounds. The pertechnate ion or radiopharmaceuticals made from it are administered intravenously to the patient.

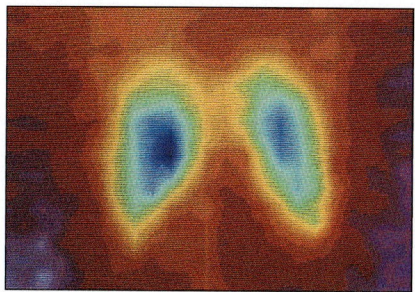

(a) Healthy human thyroid gland.

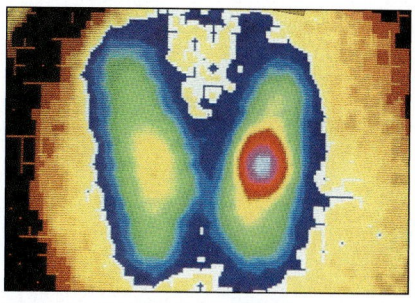

(b) Thyroid gland showing effect of hyperthyroidism.

Thyroid imaging. Technetium-99m concentrates in sites of high activity. Images of this gland, which is located at the base of the neck, were obtained by recording γ-ray emission after the patient was given radioactive technetium-99m. Current technology creates a computer color-enhanced scan.

CNRI/Science Photo Library/Photo Researchers, Inc.

Such small quantities are needed that 1 μg (microgram) of technetium-99m is sufficient for the average hospital's daily imaging needs.

One use of ^{99m}Tc is for imaging the thyroid gland. Because I^-(aq) and TcO_4^-(aq) ions have very similar sizes, the thyroid will (mistakenly) take up TcO_4^-(aq) along with iodide ion. This uptake concentrates ^{99m}Tc in the thyroid and allows a physician to obtain images such as the one shown in the accompanying figure.

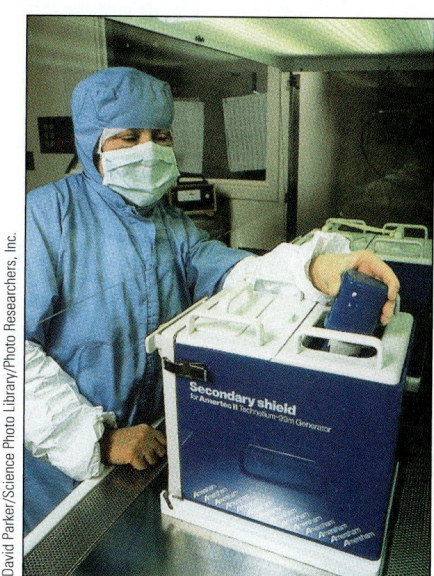

A technetium-99m generator. A technician is loading a sample containing the MoO_4^{2-} ion into the device that will convert molybdate ion to technetium-99m–labeled TcO_4^-.

David Parker/Science Photo Library/Photo Researchers, Inc.

Analytical Methods: Isotope Dilution

Imagine, for the moment, that you wanted to estimate the volume of blood in an animal subject. How might you do this? Obviously, draining the blood and measuring its volume in volumetric glassware is not a desirable option.

One technique uses a method called isotope dilution. In this process, a small amount of radioactive isotope is injected into the bloodstream. After a period of time to allow the isotope to become distributed throughout the body, a blood sample is taken and its radioactivity measured. The calculation used to determine the total blood volume is illustrated in the next example.

Example 23.10—Analysis Using Isotope Dilution

Problem A 1.00-mL solution containing 0.240 μCi of tritium is injected into a dog's bloodstream. After a period of time to allow the isotope to be dispersed, a 1.00-mL sample of blood

Table 23.6   Rare Earth Analysis of Moon Rock Sample 10022 (a fine-grain igneous rock)

Element	Concentration (ppm)
La	26.4
Ce	68
Nd	66
Sm	21.2
Eu	2.04
Gd	25
Tb	4.7
Dy	31.2
Ho	5.5
Er	16
Yb	17.7
Lu	2.55

Source: L. A. Haskin, P. A, Helmke, and R. O. Allen: *Science,* Vol. 167, p. 487, 1970. The concentrations of rare earths in moon rocks were quite similar to the values in terrestrial rocks except that the europium concentration is much depleted. The January 30, 1970, issue of *Science* was devoted to analysis of moon rocks.

is drawn. The radioactivity of this sample is found to be 4.3×10^{-4} μCi/mL. What is the total volume of blood in the dog?

Strategy For any solution, concentration equals the amount of solute divided by volume of solution. In this problem, we relate the activity of the sample (in Ci) to concentration. The total amount of solute is 0.240 μCi, and the concentration (measured on the small sample of blood) is 4.3×10^{-4} μCi/mL. The unknown is the volume.

Solution The blood contains a total of 0.240 μCi of the radioactive material. We can represent its concentration as 0.240 μCi/x, where x is the total blood volume. After dilution in the bloodstream, 1.00 mL of blood is found to have an activity of 4.3×10^{-4} μCi/mL.

$$0.240 \ \mu\text{Ci}/x \ \text{mL} = 4.3 \times 10^{-4} \ \mu\text{Ci}/1.00 \ \text{mL}$$

$$x = 560 \ \text{mL}$$

Exercise 23.11—Analysis by Isotope Dilution

Suppose you hydrolyze a 10.00-g sample of a protein. Next, you add to it a 3.00-mg sample of ^{14}C-labeled threonine, an amino acid with a specific activity of 3000 dpm. After mixing, part of the threonine (60.0 mg) is separated and isolated from the mixture. The activity of the isolated sample is 1200 dpm. How much threonine was present in the original sample?

Space Science: Neutron Activation Analysis and the Moon Rocks

The first manned space mission to the moon brought back a number of samples of soil and rock—a treasure trove for scientists. One of their first tasks was to analyze these samples to determine their identity and composition. Most analytical methods require chemical reactions using at least a small amount of material; however, this was not a desirable option, considering that the moon rocks were at the time the most valuable rocks in the world.

A few lucky scientists got a chance to work on this unique project, and one of the analytical tools they used was **neutron activation analysis.** In this nondestructive process, a sample is irradiated with neutrons. Most isotopes add a neutron to form a new isotope that is one mass unit higher in an excited nuclear state. When the nucleus decays to its ground state, it emits a γ-ray. The energy of the γ-ray identifies the element, and the number of γ-rays can be counted to determine the amount of the element in the sample. Using neutron activation analysis, it is possible to analyze for a number of elements in a single experiment (Table 23.6).

Neutron activation analysis has many other uses. This analytical procedure yields a kind of fingerprint that can be used to identify a substance. For example, this technique has been applied in determining whether an art work is real or fraudulent. Analysis of the pigments in paints on a painting can be carried out without damaging the painting to determine whether the composition resembles modern paints or paints used hundreds of years ago.

Food Science: Food Irradiation

Refrigeration, canning, and chemical additives provide significant protection in terms of food preservation, but in some parts of the world these procedures are unavailable and stored-food spoilage may claim as much as 50% of the food crop. Irradiation with γ-rays from sources such as ^{60}Co and ^{137}Cs is an option for prolonging

Table 23.7	**Approvals of Food Irradiation**
1963	FDA approves irradiation to control insects in wheat and flour.
1964	FDA approves irradiation to inhibit sprouting of white potatoes.
1985	FDA approves irradiation at specific doses to control *Trichinella spiralis* infection of pork.
1986	FDA approves irradiation at specific doses to delay maturation, inhibit growth, and disinfect foods including vegetables and spices.
1992	USDA approves irradiation of poultry to control infection by *Salmonella* and bacteria.
1997	FDA permits use of ionizing radiation to treat refrigerated and frozen meat and other meat products.
2000	USDA approves irradiation of eggs to control *Salmonella* infection.

Source: **www.foodsafety.gov/~fsg/irradiat.html.**

the shelf life of foods. Relatively low levels of radiation may retard the growth of organisms, such as bacteria, molds, and yeasts, that can cause food spoilage. After irradiation, milk in a sealed container has a minimum shelf life of 3 months without refrigeration. Chicken normally has a three-day refrigerated shelf life; after irradiation, it may have a three-week refrigerated shelf life.

Higher levels of radiation, in the 1- to 5-Mrad (1 Mrad = 1×10^6 rad) range, will kill every living organism. Foods irradiated at these levels will keep indefinitely when sealed in plastic or aluminum-foil packages. Ham, beef, turkey, and corned beef sterilized by radiation have been used on many Space Shuttle flights, for example. An astronaut said, "The beautiful thing was that it didn't disturb the taste, which made the meals much better than the freeze-dried and other types of foods we had."

These procedures are not without their opponents, and the public has not fully embraced irradiation of foods yet. An interesting argument favoring this technique is that radiation is less harmful than other methodologies for food preservation. This type of sterilization offers greater safety to food workers because it lessens chances of exposure to harmful chemicals, and it protects the environment by avoiding contamination of water supplies with toxic chemicals.

Food irradiation is commonly used in European countries, Canada, and Mexico. Its use in the United States is currently regulated by the U.S. Food and Drug Administration (FDA) and Department of Agriculture (USDA). In 1997, the FDA approved the irradiation of refrigerated and frozen uncooked meat to control pathogens and extend shelf life, and in 2000 the USDA approved the irradiation of eggs to control *Salmonella* infection (Table 23.7).

Chapter Goals Revisited

Having studied this chapter, you should be able to

Identify radioactive elements and describe natural and artificial reactions

a. Identify α, β, and γ radiation, the three major types of radiation in natural radioactive decay (Section 23.1).

b. Write balanced equations for nuclear reactions (Section 23.2).

c. Predict whether a radioactive isotope will decay by α or β emission, or by positron emission or electron capture (Sections 23.2 and 23.3).

d. Understand the nature and origin of γ radiation (Section 23.2).

GENERAL
Chemistry•Now™

See the General ChemistryNow CD-ROM or website to:

• Check your readiness for an exam by taking the exam-prep quiz and exploring the resources in the personalized Learning Plan it provides

For additional preparation for an examination on this chapter see the **Let's Review** section on pages 998–1011.

Calculate the binding energy and binding energy per nucleon for a particular isotope
  a. Understand how binding energy per nucleon is defined (Section 23.3).
  b. Recognize the significance of a graph of binding energy per nucleon versus mass number (Section 23.3).

Understand the rates of radioactive decay
  a. Understand and use mathematical equations that characterize the radioactive decay process (Section 23.4).
  b. Use the half-life to estimate the time required for an isotope to decay (Section 23.4).

Understand artificial nuclear reactions
  a. Describe procedures used to carry out nuclear reactions (Section 23.5).
  b. Describe nuclear chain reactions, nuclear fission, and nuclear fusion (Sections 23.6 and 23.7).

Understand issues of health and safety with respect to radioactivity
  a. Describe the units used to measure intensity and understand how they pertain to health issues (Section 23.8).

Be aware of some uses of radioactive isotopes in science and medicine (Section 23.9)

Key Equations

Equation 23.1 (page 1120):
The equation relating interconversion of mass (m) and energy (E). This equation is applied in the calculation of binding energy (E_b) for nuclei.

$$E_b = (\Delta m)c^2$$

Equation 23.2 (page 1123):
The activity of a radioactive sample (A) is proportional to the number of radioactive atoms (N).

$$A \propto N$$

Equation 23.4 (page 1124):
The rate law for nuclear decay based on number of radioactive atoms initially present (N_0) and the number N after time t.

$$\ln(N/N_0) = -kt$$

Equation 23.5 (page 1124):
The rate law for nuclear decay based on the measured activity of a sample (A).

$$\ln(A/A_0) = -kt$$

Equation 23.6 (page 1124):
The relationship between the half-life and the rate constant for a nuclear decay process.

$$t_{1/2} = 0.693/k$$

Study Questions

▲ denotes more challenging questions.

■ denotes questions available in the Homework and Goals section of the General ChemistryNow CD-ROM or website.

Blue numbered questions have answers in Appendix O and fully worked solutions in the *Student Solutions Manual*.

Structures of many of the compounds used in these questions are found on the General ChemistryNow CD-ROM or website in the Models folder.

GENERAL
Chemistry⚛Now™ Assess your understanding of this chapter's topics with additional quizzing and conceptual questions at **http://now.brookscole.com/kotz6e**

Reviewing Concepts

1. Some important discoveries in scientific history that contributed to the development of nuclear chemistry are listed below. Briefly, describe each discovery, identify prominent scientists who contributed to it, and comment on the significance of the discovery to the development of this field.
   (a) 1896, the discovery of radioactivity
   (b) 1898, the identification of radium and polonium
   (c) 1918, the first artificial nuclear reaction
   (d) 1932, (n, γ) reactions
   (e) 1939, fission reactions

2. In Chapter 2, the law of conservation of mass was introduced as an important principle in chemistry. The discovery of nuclear reactions forced scientists to modify this law. Explain why, and give an example illustrating that mass is not conserved in a nuclear reaction.

3. A graph of binding energy per nucleon is shown in Figure 23.4. Explain how the data used to construct this graph were obtained.

4. How is Figure 23.3 used to predict the type of decomposition for unstable (radioactive) isotopes?

5. Outline how nuclear reactions are carried out in the laboratory. Describe the artificial nuclear reactions used to make an element with an atomic number greater than 92.

6. What mathematical equations define the rates of decay for radioactive elements?

7. Explain how carbon-14 is used to estimate the ages of archeological artifacts. What are the limitations for use of this technique?

8. Describe how the concept of half-life for nuclear decay is used.

9. What is a radioactive decay series? Explain why radium and polonium are found in uranium ores.

10. The interaction of radiation with matter has both positive and negative consequences. Discuss briefly the hazards of radiation and the way that radiation can be used in medicine.

11. Define the terms *curie*, *rad*, and *rem*.

Practicing Skills

Nuclear Reactions
(See Examples 23.1 and 23.2.)

12. Complete the following nuclear equations. Write the mass number and atomic number for the remaining particle, as well as its symbol.
    (a) $^{54}_{26}\text{Fe} + ^{4}_{2}\text{He} \longrightarrow 2\,^{1}_{0}\text{n} + ?$
    (b) $^{27}_{13}\text{Al} + ^{4}_{2}\text{He} \longrightarrow ^{30}_{15}\text{P} + ?$
    (c) $^{32}_{16}\text{S} + ^{1}_{0}\text{n} \longrightarrow ^{1}_{1}\text{H} + ?$
    (d) $^{96}_{42}\text{Mo} + ^{2}_{1}\text{H} \longrightarrow ^{1}_{0}\text{n} + ?$
    (e) $^{98}_{42}\text{Mo} + ^{1}_{0}\text{n} \longrightarrow ^{99}_{43}\text{Tc} + ?$
    (f) $^{18}_{9}\text{F} \longrightarrow ^{18}_{8}\text{O} + ?$

13. Complete the following nuclear equations. Write the mass number, atomic number, and symbol for the remaining particle.
    (a) $^{9}_{4}\text{Be} + ? \longrightarrow ^{6}_{3}\text{Li} + ^{4}_{2}\text{He}$
    (b) $? + ^{1}_{0}\text{n} \longrightarrow ^{24}_{11}\text{Na} + ^{4}_{2}\text{He}$
    (c) $^{40}_{20}\text{Ca} + ? \longrightarrow ^{40}_{19}\text{K} + ^{1}_{1}\text{H}$
    (d) $^{241}_{95}\text{Am} + ^{4}_{2}\text{He} \longrightarrow ^{243}_{97}\text{Bk} + ?$
    (e) $^{246}_{96}\text{Cm} + ^{12}_{6}\text{C} \longrightarrow 4\,^{1}_{0}\text{n} + ?$
    (f) $^{238}_{92}\text{U} + ? \longrightarrow ^{249}_{100}\text{Fm} + 5\,^{1}_{0}\text{n}$

14. Complete the following nuclear equations. Write the mass number, atomic number, and symbol for the remaining particle.
    (a) $^{111}_{47}\text{Ag} \longrightarrow ^{111}_{48}\text{Cd} + ?$
    (b) $^{87}_{36}\text{Kr} \longrightarrow ^{0}_{-1}\beta + ?$
    (c) $^{231}_{91}\text{Pa} \longrightarrow ^{227}_{89}\text{Ac} + ?$
    (d) $^{230}_{90}\text{Th} \longrightarrow ^{4}_{2}\text{He} + ?$
    (e) $^{82}_{35}\text{Br} \longrightarrow ^{82}_{36}\text{Kr} + ?$
    (f) $? \longrightarrow ^{24}_{12}\text{Mg} + ^{0}_{-1}\beta$

15. Complete the following nuclear equations. Write the mass number, atomic number, and symbol for the remaining particle.
    (a) $^{19}_{10}\text{Ne} \longrightarrow ^{0}_{+1}\beta + ?$
    (b) $^{59}_{26}\text{Fe} \longrightarrow ^{0}_{-1}\beta + ?$
    (c) $^{40}_{19}\text{K} \longrightarrow ^{0}_{-1}\beta + ?$
    (d) $^{37}_{18}\text{Ar} + ^{0}_{-1}\text{e (electron capture)} \longrightarrow ?$
    (e) $^{55}_{26}\text{Fe} + ^{0}_{-1}\text{e (electron capture)} \longrightarrow ?$
    (f) $^{26}_{13}\text{Al} \longrightarrow ^{25}_{12}\text{Mg} + ?$

16. The uranium-235 radioactive decay series, beginning with $^{235}_{92}\text{U}$ and ending with $^{207}_{82}\text{Pb}$, occurs in the following sequence: α, β, α, β, α, α, α, α, β, β, α. Write an equation for each step in this series.

17. The thorium-232 radioactive decay series, beginning with $^{232}_{90}$Th and ending with $^{208}_{82}$Pb, occurs in the following sequence: $\alpha, \beta, \beta, \alpha, \alpha, \alpha, \alpha, \beta, \beta, \alpha$. Write an equation for each step in this series.

Nuclear Stability and Nuclear Decay

(See Examples 23.3 and 23.4.)

18. What particle is emitted in the following nuclear reactions? Write an equation for each reaction.
    (a) Gold-198 decays to mercury-198.
    (b) Radon-222 decays to polonium-218.
    (c) Cesium-137 decays to barium-137.
    (d) Indium-110 decays to cadmium-110.

19. What is the product of the following nuclear decay processes? Write an equation for each process.
    (a) Gallium-67 decays by electron capture.
    (b) Potassium-38 decays with positron emission.
    (c) Technetium-99m decays with γ emission.
    (d) Manganese-56 decays by β emission.

20. Predict the probable mode of decay for each of the following radioactive isotopes, and write an equation to show the products of decay.
    (a) Bromine-80m   (c) Cobalt-61
    (b) Californium-240   (d) Carbon-11

21. Predict the probable mode of decay for each of the following radioactive isotopes, and write an equation to show the products of decay.
    (a) Manganese-54   (c) Silver-110
    (b) Americium-241   (d) Mercury-197m

22. (a) Which of the following nuclei decay by $^{0}_{-1}\beta$ decay?

$$^{3}\text{H} \quad ^{16}\text{O} \quad ^{20}\text{F} \quad ^{13}\text{N}$$

    (b) Which of the following nuclei decays by $^{0}_{+1}\beta$ decay?

$$^{238}\text{U} \quad ^{19}\text{F} \quad ^{22}\text{Na} \quad ^{24}\text{Na}$$

23. (a) Which of the following nuclei decay by $^{0}_{-1}\beta$ decay?

$$^{1}\text{H} \quad ^{23}\text{Mg} \quad ^{32}\text{P} \quad ^{20}\text{Ne}$$

    (b) Which of the following nuclei decay by $^{0}_{+1}\beta$ decay?

$$^{235}\text{U} \quad ^{35}\text{Cl} \quad ^{38}\text{K} \quad ^{24}\text{Na}$$

24. Boron has two stable isotopes, ^{10}B and ^{11}B. Calculate the binding energies per nucleon of these two nuclei. The required masses (in grams per mole) are $^{1}_{1}\text{H} = 1.00783$, $^{1}_{0}\text{n} = 1.00867$, $^{10}_{5}\text{B} = 10.01294$, and $^{11}_{5}\text{B} = 11.00931$.

25. Calculate the binding energy in kilojoules per mole of nucleons of P for the formation of ^{30}P and ^{31}P. The required masses (in grams per mole) are $^{1}_{1}\text{H} = 1.00783$, $^{1}_{0}\text{n} = 1.00867$, $^{30}_{15}\text{P} = 29.97832$, and $^{31}_{15}\text{P} = 30.97376$.

26. Calculate the binding energy per nucleon for calcium-40, and compare your result with the value for calcium-40 in Figure 23.4. Masses needed for this calculation are $^{1}_{1}\text{H} = 1.00783$, $^{1}_{0}\text{n} = 1.00867$, and $^{40}_{20}\text{Ca} = 39.96259$.

27. Calculate the binding energy per nucleon for iron-56. Masses needed for this calculation are $^{1}_{1}\text{H} = 1.00783$, $^{1}_{0}\text{n} = 1.00867$, and $^{56}_{26}\text{Fe} = 55.9349$. Compare the result of your calculation to the value for iron-56 in the graph in Figure 23.4.

28. Calculate the binding energy per mole of nucleons for $^{16}_{8}$O. Masses needed for this calculations are $^{1}_{1}\text{H} = 1.00783$, $^{1}_{0}\text{n} = 1.00867$, and $^{16}_{8}\text{O} = 15.99492$.

29. Calculate the binding energy per nucleon for nitrogen-14. The mass of nitrogen-14 is 14.003074.

Rates of Radioactive Decay

(See Examples 23.5 and 23.6.)

30. Copper acetate containing ^{64}Cu is used to study brain tumors. This isotope has a half-life of 12.7 h. If you begin with 25.0 μg of ^{64}Cu, what mass in micrograms remains after 64 h?

31. Gold-198 is used in the diagnosis of liver problems. The half-life of ^{198}Au is 2.69 days. If you begin with 2.8 μg of this gold isotope, what mass remains after 10.8 days?

32. Iodine-131 is used to treat thyroid cancer.
    (a) The isotope decays by β particle emission. Write a balanced equation for this process.
    (b) Iodine-131 has a half-life of 8.04 days. If you begin with 2.4 μg of radioactive ^{131}I, what mass remains after 40.2 days?

33. Phosphorus-32 is used in the form of Na_2HPO_4 in the treatment of chronic myeloid leukemia, among other things.
    (a) The isotope decays by β particle emission. Write a balanced equation for this process.
    (b) The half-life of ^{32}P is 14.3 days. If you begin with 4.8 μg of radioactive ^{32}P in the form of Na_2HPO_4, what mass remains after 28.6 days (about one month)?

34. Gallium-67 ($t_{1/2} = 78.25$ h) is used in the medical diagnosis of certain kinds of tumors. If you ingest a compound containing 0.015 mg of this isotope, what mass (in milligrams) remains in your body after 13 days? (Assume none is excreted.)

35. Iodine-131 ($t_{1/2} = 8.04$ days), a β emitter, is used to treat thyroid cancer.
    (a) Write an equation for the decomposition of ^{131}I.
    (b) If you ingest a sample of NaI containing ^{131}I, how much time is required for the activity to decrease to 35.0% of its original value?

36. Radon has been the focus of much attention recently because it is often found in homes. Radon-222 emits α particles and has a half-life of 3.82 days.
    (a) Write a balanced equation to show this process.
    (b) How long does it take for a sample of ^{222}Rn to decrease to 20.0% of its original activity?

37. A sample of wood from a Thracian chariot found in an excavation in Bulgaria has a ^{14}C activity of 11.2 dpm/g. Estimate the age of the chariot and the year it was made ($t_{1/2}$ for ^{14}C is 5.73×10^3 years, and the activity of ^{14}C in living material is 14.0 dpm/g).

38. A piece of charred bone found in the ruins of a Native American village has a $^{14}C : {}^{12}C$ ratio that is 72% of the radio found in living organisms. Calculate the age of the bone fragment.

39. Strontium-90 is a hazardous radioactive isotope that resulted from atmospheric nuclear testing. A sample of strontium carbonate containing ^{90}Sr is found to have an activity of 1.0×10^3 dpm. One year later the activity of this sample is 975 dpm.

    (a) Calculate the half-life of strontium-90 from this information.

    (b) How long will it take for the activity of this sample to drop to 1.0% of the initial value?

40. Radioactive cobalt-60 is used extensively in nuclear medicine as a γ-ray source. It is made by a neutron capture reaction from cobalt-59, and it is a β emitter; β emission is accompanied by strong γ radiation. The half-life of cobalt-60 is 5.27 years.

    (a) How long will it take for a cobalt-60 source to decrease to one eighth of its original activity?

    (b) What fraction of the activity of a cobalt-60 source remains after 1.0 year?

41. Scandium occurs in nature as a single isotope, scandium-45. Neutron irradiation produces scandium-46, a β emitter with a half-life of 83.8 days. If the initial activity is 7.0×10^4 dpm, draw a graph showing disintegrations per minute as a function of time during a period of one year.

42. Phosphorus occurs in nature as a single isotope, phosphorus-31. Neutron irradiation of phosphorus-31 produces phosphorus-32, a β emitter with a half-life of 14.28 days. Assume you have a sample containing phosphorus-32 that has a rate of decay of 3.2×10^6 dpm. Draw a graph showing disintegrations per minute as a function of time during a period of one year.

43. Sodium-23 (in a sample of NaCl) is subjected to neutron bombardment in a nuclear reactor to produce ^{24}Na. When removed from the reactor, the sample is radioactive, with β activity of 2.54×10^4 dpm. The decrease in radioactivity over time was studied, producing the following data:

Activity (dpm)	Time (h)
2.54×10^4	0
2.42×10^4	1
2.31×10^4	2
2.00×10^4	5
1.60×10^4	10
1.01×10^4	20

(a) Write equations for the neutron capture reaction and for the reaction in which the product of this reaction decays by β emission.

(b) Determine the half-life of sodium-24.

44. The isotope of polonium that was most likely isolated by Marie Curie in her pioneering studies is polonium-210. A sample of this element was prepared in a nuclear reaction. Initially its activity (α emission) was 7840 dpm. Measuring radioactivity over time produced the following data:

Activity (dpm)	Time (days)
7840	0
7570	7
7300	14
5920	56
5470	72

Determine the half-life of polonium-210.

Nuclear Reactions

(See Example 23.9.)

45. There are two isotopes of americium, both with half-lives sufficiently long to allow the handling of large quantities. Americium-241, with a half-life of 432 years, is an α emitter. It is used in smoke detectors. The isotope is formed from ^{239}Pu by absorption of two neutrons followed by emission of a β particle. Write a balanced equation for this process.

46. Americium-240 is made by bombarding plutonium-239 with α particles. In addition to ^{240}Am, the products are a proton and two neutrons. Write a balanced equation for this process.

47. To synthesize the heavier transuranium elements, a nucleus must be bombarded with a relatively large particle. If you know the products are californium-246 and four neutrons, with what particle would you bombard uranium-238 atoms?

48. Element $^{287}114$ was made by firing a beam of ^{48}Ca ions at ^{242}Pu. Three neutrons were ejected in the reaction. Write a balanced nuclear equation for the synthesis of this superheavy element.

49. Element $^{287}114$ decayed by α emission with a half-life of about 5 s. Write an equation for this process.

50. Deuterium nuclei (^2_1H) are particularly effective as bombarding particles to carry out nuclear reactions. Complete the following equations:

    (a) $^{114}_{48}Cd + {}^2_1H \longrightarrow ? + {}^1_1H$

    (b) $^6_3Li + {}^2_1H \longrightarrow ? + {}^1_0n$

    (c) $^{40}_{20}Ca + {}^2_1H \longrightarrow {}^{38}_{19}K + ?$

    (d) $? + {}^2_1H \longrightarrow {}^{65}_{30}Zn + \gamma$

51. Some of the reactions explored by Rutherford and others are listed below. Identify the unknown species in each reaction.
    (a) $^{14}_{7}N + ^{4}_{2}He \longrightarrow ^{17}_{8}O + ?$
    (b) $^{9}_{4}Be + ^{4}_{2}He \longrightarrow ? + ^{1}_{0}n$
    (c) $? + ^{4}_{2}He \longrightarrow ^{30}_{15}P + ^{1}_{0}n$
    (d) $^{239}_{94}Pu + ^{4}_{2}He \longrightarrow ? + ^{1}_{0}n$

52. Boron is an effective absorber of neutrons. When boron-10 adds a neutron, an α particle is emitted. Write an equation for this nuclear reaction.

53. Tritium, $^{3}_{1}H$, is one of the nuclei used in fusion reactions. This isotope is radioactive, with a half-life of 12.3 years. Like carbon-14, tritium is formed in the upper atmosphere from cosmic radiation, and it is found in trace amounts on earth. To obtain the amounts required for a fusion reaction, however, it must be made via a nuclear reaction. The reaction of $^{6}_{3}Li$ with a neutron produces tritium and an α particle. Write an equation for this nuclear reaction.

General Questions

These questions are not designated as to type or location in the chapter. They may combine several concepts.

54. ▲ A technique to date geological samples uses rubidium-87, a long-lived radioactive isotope of rubidium ($t_{1/2} = 4.8 \times 10^{10}$ years). Rubidium-87 decays by β emission to strontium-87. If the rubidium-87 is part of a rock or mineral, then strontium-87 will remain trapped within the crystalline structure of the rock. The age of the rock dates back to the time when the rock solidified. Chemical analysis of the rock gives the amounts of ^{87}Rb and ^{87}Sr. From these data, the fraction of ^{87}Rb that remains can be calculated.

    Analysis of a stony meteorite determined that 1.8 mmol of ^{87}Rb and 1.6 mmol of ^{87}Sr were present. Estimate the age of the meteorite. (*Hint:* The amount of ^{87}Rb at t_0 is moles ^{87}Rb + moles ^{87}Sr.)

55. The oldest-known fossil found in South Africa has been dated based on the decay of Rb-87.

    $$^{87}Rb \longrightarrow ^{87}Sr + ^{0}_{-1}\beta \qquad t_{1/2} = 4.8 \times 10^{10} \text{ years}$$

    If the ratio of the present quantity of ^{87}Rb to the original quantity is 0.951, calculate the age of the fossil.

56. The age of minerals can sometimes be determined by measuring the amounts of ^{206}Pb and ^{238}U in a sample. This determination assumes that all of the ^{206}Pb in the sample comes from the decay of ^{238}U. The date obtained identifies when the rock solidified. Assume that the ratio of ^{206}Pb to ^{238}U in an igneous rock sample is 0.33. Calculate the age of the rock. ($t_{1/2}$ for ^{238}U is 4.5×10^9 years.)

57. In June 1972, natural fission reactors, which operated billions of years ago, were discovered in Oklo, Gabon. At present, natural uranium contains 0.72% ^{235}U. How many years ago did natural uranium contain 3.0% ^{235}U, the amount needed to sustain a natural reactor? ($t_{1/2}$ for ^{235}U is 7.04×10^8 years.)

58. If a shortage in worldwide supplies of fissionable uranium arose, it would be possible to use other fissionable nuclei. Plutonium, one such fuel, can be made in "breeder" reactors that manufacture more fuel than they consume. The sequence of reactions by which plutonium is made is as follows:
    (a) A ^{238}U nucleus undergoes an (n, γ) to produce ^{239}U.
    (b) ^{239}U decays by β emission ($t_{1/2} = 23.5$ min) to give an isotope of neptunium.
    (c) This neptunium isotope decays by β emission to give a plutonium isotope.
    (d) The plutonium isotope is fissionable. On collision of one of these plutonium isotopes with a neutron, fission occurs, with at least two neutrons and two other nuclei as products.

    Write an equation for each of the nuclear reactions.

59. When a neutron is captured by an atomic nucleus, energy is released as γ radiation. This energy can be calculated based on the change in mass in converting reactants to products. For the nuclear reaction $^{6}_{3}Li + ^{1}_{0}n \longrightarrow ^{7}_{3}Li + \gamma$:
    (a) Calculate the energy evolved in this reaction (per atom). Masses needed for this calculation are $^{6}_{3}Li = 6.01512$, $^{1}_{0}n = 1.00867$, and $^{7}_{3}Li = 7.01600$.
    (b) Use the answer in part (a) to calculate the wavelength of the γ-rays emitted in the reaction.

Summary and Conceptual Questions

60. The average energy output of a good grade of coal is 2.6×10^7 kJ/ton. Fission of 1 mol of ^{235}U releases 2.1×10^{10} kJ. Find the number of tons of coal needed to produce the same energy as 1 lb of ^{235}U. (See Appendix C for conversion factors.)

61. Collision of an electron and a positron results in formation of two γ-rays. In the process, their masses are converted completely into energy.
    (a) Calculate the energy evolved from the annihilation of an electron and a positron, in kilojoules per mole.
    (b) Using Planck's equation (Equation 7.2), determine the frequency of the γ-rays emitted in this process.

62. To measure the volume of the blood system of an animal, the following experiment was done. A 1.0-mL sample of an aqueous solution containing tritium, with an activity of 2.0×10^6 dps, was injected into the animal's bloodstream. After time was allowed for complete circulatory mixing, a 1.0-mL blood sample was withdrawn and found to have an activity of 1.5×10^4 dps. What was the volume of the circulatory system? (The half-life of tritium is 12.3 years, so this experiment assumes that only a negligible amount of tritium has decayed in the time of the experiment.)

63. Suppose that you hydrolyze 4.644 g of a protein to form a mixture of different amino acids. To this is added a 2.80-mg sample of ^{14}C-labeled threonine (one of the amino acids present). The activity of this small sample is 1950 dpm. A chromatographic separation of the amino

acids is carried out, and a small sample of pure threonine is separated. This sample has an activity of 550 dpm. What fraction of the threonine present was separated? What is the total amount of threonine in the sample?

64. The principle underlying the isotope dilution method can be applied to many kinds of problems. Suppose that you, a marine biologist, want to estimate the number of fish in a lake. You release 1000 tagged fish, and after allowing an adequate amount of time for the fish to disperse evenly in the lake, you catch 5250 fish and find that 27 of them have tags. How many fish are in the lake?

65. ▲ Radioactive isotopes are often used as "tracers" to follow an atom through a chemical reaction. The following is an example of this process: acetic acid reacts with methanol, CH_3OH, by eliminating a molecule of H_2O to form methyl acetate, $CH_3CO_2CH_3$. Explain how you would use the radioactive isotope ^{15}O to show whether the oxygen atom in the water product comes from the —OH of the acid or the —OH of the alcohol.

66. Radioactive decay series begin with a very long-lived isotope. For example, the half-life of ^{238}U is 4.5×10^9 years. Each series is identified by the name of the long-lived parent isotope of highest mass.

    (a) The uranium-238 radioactive decay series is sometimes referred to as the $4n + 2$ series because the masses of all 13 members of this series can be expressed by the equation $m = 4n + 2$, where m is the mass number and n is an integer. Explain why the masses are correlated in this way.

    (b) Two other radioactive decay series identified in minerals in the earth's crust are the thorium-232 series and the uranium-235 series. Do the masses of the isotopes in these series conform to a simple mathematical equation? If so, identify the equation.

    (c) Identify the radioactive decay series to which each of the following isotopes belongs: $^{226}_{88}Ra$, $^{215}_{86}At$, $^{228}_{90}Th$, $^{210}_{83}Bi$.

    (d) Evaluation reveals that one series of elements, the $4n + 1$ series, is not present in the earth's crust. Speculate why.

67. ▲ You might wonder how it is possible to determine the half-life of long-lived radioactive isotopes such as ^{238}U. With a half-life of more than 10^9 years, the radioactivity of a sample of uranium will not measurably change in your lifetime. In fact, you can calculate the half-life using the mathematics governing first-order reactions.

    It can be shown that a 1.0-mg sample of ^{238}U decays at the rate of 12 α emissions per second. Set up a mathematical equation for the rate of decay, $\Delta N/\Delta t = -kN$, where N is the number of nuclei in the 1.0-mg sample and $\Delta N/\Delta t$ is 12 dps. Solve this equation for the rate constant for this process, and then relate the rate constant to the half-life of the reaction. Carry out this calculation, and compare your result with the literature value, 4.5×10^9 years.

68. The last unknown element between bismuth and uranium was discovered by Lise Meitner (1878–1968) and Otto Hahn (1879–1968) in 1918. They obtained ^{231}Pa by chemical extraction of pitchblende, in which its concentration is about 1 ppm. This isotope, an α emitter, has a half-life of 3.27×10^4 years.

    (a) Which radioactive decay series (the uranium-235, uranium-238, or thorium-232 series) contains ^{231}Pa as a member?

    (b) Suggest a possible sequence of nuclear reactions starting with the long-lived isotope that eventually forms this isotope.

    (c) What quantity of ore would be required to isolate 1.0 g of ^{231}Pa, assuming 100% yield?

    (d) Write an equation for the radioactive decay process for ^{231}Pa.

Do you need a live tutor for homework problems?
Access vMentor at General ChemistryNow at
http://now.brookscole.com/kotz6e
for one-on-one tutoring from a chemistry expert

List of Appendices

Appendix A

Using Logarithms and the Quadratic Equation

An introductory chemistry course requires basic algebra plus a knowledge of (1) exponential (or scientific) notation, (2) logarithms, and (3) quadratic equations. The use of exponential notation was reviewed in Chapter 1, and this appendix reviews the last two topics.

A.1—Logarithms

Two types of logarithms are used in this text: (1) common logarithms (abbreviated log) whose base is 10 and (2) natural logarithms (abbreviated ln) whose base is e (= 2.71828):

$$\log x = n, \text{ where } x = 10^n$$
$$\ln x = m, \text{ where } x = e^m$$

Most equations in chemistry and physics were developed in natural, or base e, logarithms, and we follow this practice in this text. The relation between log and ln is

$$\ln x = 2.303 \log x$$

Despite the different bases of the two logarithms, they are used in the same manner. What follows is largely a description of the use of common logarithms.

A common logarithm is the power to which you must raise 10 to obtain the number. For example, the log of 100 is 2, since you must raise 10 to the second power to obtain 100. Other examples are

$$\log 1000 = \log (10^3) = 3$$
$$\log 10 = \log (10^1) = 1$$
$$\log 1 = \log (10^0) = 0$$
$$\log 0.1 = \log (10^{-1}) = -1$$
$$\log 0.0001 = \log (10^{-4}) = -4$$

To obtain the common logarithm of a number other than a simple power of 10, you must resort to a log table or an electronic calculator. For example,

$$\log 2.10 = 0.3222, \text{ which means that } 10^{0.3222} = 2.10$$
$$\log 5.16 = 0.7126, \text{ which means that } 10^{0.7126} = 5.16$$
$$\log 3.125 = 0.49485, \text{ which means that } 10^{0.49485} = 3.125$$

To check this on your calculator, enter the number, and then press the "log" key. When using a log table, the logs of the first two numbers can be read directly from the table. The log of the third number (3.125), however, must be interpolated. That is, 3.125 is midway between 3.12 and 3.13, so the log is midway between 0.4942 and 0.4955.

To obtain the natural logarithm ln of the numbers shown here, use a calculator having this function. Enter each number and press "ln:"

$$\ln 2.10 = 0.7419, \text{ which means that } e^{0.7419} = 2.10$$
$$\ln 5.16 = 1.6409, \text{ which means that } e^{1.6409} = 5.16$$

To find the common logarithm of a number greater than 10 or less than 1 with a log table, first express the number in scientific notation. Then find the log of each part of the number and add the logs. For example,

$$\log 241 = \log (2.41 \times 10^2) = \log 2.41 + \log 10^2$$
$$= 0.382 + 2 = 2.382$$
$$\log 0.00573 = \log (5.73 \times 10^{-3}) = \log 5.73 + \log 10^{-3}$$
$$= 0.758 + (-3) = -2.242$$

Significant Figures and Logarithms

Notice that the mantissa has as many significant figures as the number whose log was found. (So that you could more clearly see the result obtained with a calculator or a table, this rule was not strictly followed until the last two examples.)

■ Logarithms and Nomenclature
The number to the left to the decimal in a logarithm is called the **characteristic**, and the number to the right of the decimal is the **mantissa.**

Obtaining Antilogarithms

If you are given the logarithm of a number, and find the number from it, you have obtained the "antilogarithm," or "antilog," of the number. Two common procedures used by electronic calculators to do this are:

Procedure A	**Procedure B**
1. Enter the log or ln.	1. Enter the log or ln.
2. Press 2ndF.	2. Press INV.
3. Press 10^x or e^x.	3. Press log or ln x.

Test one or the other of these procedures with the following examples:

1. Find the number whose log is 5.234:

Recall that log $x = n$, where $x = 10^n$. In this case $n = 5.234$. Enter that number in your calculator, and find the value of 10^n, the antilog. In this case,

$$10^{5.234} = 10^{0.234} \times 10^5 = 1.71 \times 10^5$$

Notice that the characteristic (5) sets the decimal point; it is the power of 10 in the exponential form. The mantissa (0.234) gives the value of the number x. Thus, if you use a log table to find x, you need only look up 0.234 in the table and see that it corresponds to 1.71.

2. Find the number whose log is -3.456:

$$10^{-3.456} = 10^{0.544} \times 10^{-4} = 3.50 \times 10^{-4}$$

Notice here that -3.456 must be expressed as the sum of -4 and $+0.544$.

Mathematical Operations Using Logarithms

Because logarithms are exponents, operations involving them follow the same rules used for exponents. Thus, multiplying two numbers can be done by adding logarithms:

$$\log xy = \log x + \log y$$

For example, we multiply 563 by 125 by adding their logarithms and finding the antilogarithm of the result:

$$\log 563 = 2.751$$
$$\log 125 = \underline{2.097}$$
$$\log xy = 4.848$$
$$xy = 10^{4.848} = 10^4 \times 10^{0.848} = 7.05 \times 10^4$$

One number (x) can be divided by another (y) by subtraction of their logarithms:

$$\log \frac{x}{y} = \log x - \log y$$

For example, to divide 125 by 742,

$$\log 125 = 2.097$$
$$-\log 742 = \underline{2.870}$$
$$\log \frac{x}{y} = -0.773$$

$$\frac{x}{y} = 10^{-0.773} = 10^{0.227} \times 10^{-1} = 1.68 \times 10^{-1}$$

Similarly, powers and roots of numbers can be found using logarithms.

$$\log x^y = y(\log x)$$
$$\log \sqrt[y]{x} = \log x^{1/y} = \frac{1}{y} \log x$$

As an example, find the fourth power of 5.23. We first find the log of 5.23 and then multiply it by 4. The result, 2.874, is the log of the answer. Therefore, we find the antilog of 2.874:

$$(5.23)^4 = ?$$
$$\log (5.23)^4 = 4 \log 5.23 = 4(0.719) = 2.874$$
$$(5.23)^4 = 10^{2.874} = 748$$

As another example, find the fifth root of 1.89×10^{-9}:

$$\sqrt[5]{1.89 \times 10^{-9}} = (1.89 \times 10^{-9})^{1/5} = ?$$
$$\log (1.89 \times 10^{-9})^{1/5} = \frac{1}{5} \log (1.89 \times 10^{-9}) = \frac{1}{5} (-8.724) = -1.745$$

The answer is the antilog of -1.745:

$$(1.89 \times 10^{-9})^{1/5} = 10^{-1.745} = 1.80 \times 10^{-2}$$

A.2—Quadratic Equations

Algebraic equations of the form $ax^2 + bx + c = 0$ are called **quadratic equations.** The coefficients a, b, and c may be either positive or negative. The two roots of the equation may be found using the *quadratic formula*:

$$x = \frac{-b \pm \sqrt{b^2 - 4ac}}{2a}$$

As an example, solve the equation $5x^2 - 3x - 2 = 0$. Here $a = 5$, $b = -3$, and $c = -2$. Therefore,

$$x = \frac{3 \pm \sqrt{(-3)^2 - 4(5)(-2)}}{2(5)}$$

$$= \frac{3 \pm [2(5)/\sqrt{9 - (-40)}]}{10} = \frac{3 \pm \sqrt{49}}{10} = \frac{3 \pm 7}{10}$$

$$= 1 \text{ and } -0.4$$

How do you know which of the two roots is the correct answer? You have to decide in each case which root has physical significance. It is *usually* true in this course, however, that negative values are not significant.

When you have solved a quadratic expression, you should always check your values by substitution into the original equation. In the previous example, we find that $5(1)^2 - 3(1) - 2 = 0$ and that $5(-0.4)^2 - 3(-0.4) - 2 = 0$.

The most likely place you will encounter quadratic equations is in the chapters on chemical equilibria, particularly in Chapters 16 through 18. Here you will often be faced with solving an equation such as

$$1.8 \times 10^{-4} = \frac{x^2}{0.0010 - x}$$

This equation can certainly be solved using the quadratic equation (to give $x = 3.4 \times 10^{-4}$). You may find the *method of successive approximations* to be especially convenient, however. Here we begin by making a reasonable approximation of x. This approximate value is substituted into the original equation, which is then solved to give what is hoped to be a more correct value of x. This process is repeated until the answer converges on a particular value of x—that is, until the value of x derived from two successive approximations is the same.

Step 1: First assume that x is so small that $(0.0010 - x) \approx 0.0010$. This means that

$$x^2 = 1.8 \times 10^{-4}(0.0010)$$
$$x = 4.2 \times 10^{-4} \text{ (to 2 significant figures)}$$

Step 2: Substitute the value of x from Step 1 into the denominator of the original equation, and again solve for x:

$$x^2 = 1.8 \times 10^{-4}(0.0010 - 0.00042)$$
$$x = 3.2 \times 10^{-4}$$

Step 3: Repeat Step 2 using the value of x found in that step:

$$x = \sqrt{1.8 \times 10^{-4}(0.0010 - 0.00032)} = 3.5 \times 10^{-4}$$

Step 4: Continue repeating the calculation, using the value of x found in the previous step:

$$x = \sqrt{1.8 \times 10^{-4}(0.0010 - 0.00035)} = 3.4 \times 10^{-4}$$

Step 5: $\quad x = \sqrt{1.8 \times 10^{-4}(0.0010 - 0.00034)} = 3.4 \times 10^{-4}$

Here we find that iterations after the fourth step give the same value for x, indicating that we have arrived at a valid answer (and the same one obtained from the quadratic formula).

Here are several final thoughts on using the method of successive approximations. First, in some cases the method does not work. Successive steps may give answers that are random or that diverge from the correct value. In Chapters 16 through 18, you confront quadratic equations of the form $K = x^2/(C - x)$. The method of approximations works as long as $K < 4C$ (assuming one begins with $x = 0$ as the first guess, that is, $K \approx x^2/C$). This is always going to be true for weak acids and bases (the topic of Chapters 17 and 18), but it may *not* be the case for problems involving gas-phase equilibria (Chapter 16), where K can be quite large.

Second, values of K in the equation $K = x^2/(C - x)$ are usually known only to two significant figures. We are therefore justified in carrying out successive steps until two answers are the same to two significant figures.

Finally, we highly recommend this method of solving quadratic equations, especially those in Chapters 17 and 18. If your calculator has a memory function, successive approximations can be carried out easily and rapidly.

Appendix B*

Some Important Physical Concepts

B.1—Matter

The tendency to maintain a constant velocity is called inertia. Thus, unless acted on by an unbalanced force, a body at rest remains at rest, and a body in motion remains in motion with uniform velocity. Matter is anything that exhibits inertia; the quantity of matter is its mass.

B.2—Motion

Motion is the change of position or location in space. Objects can have the following classes of motion:

- Translation occurs when the center of mass of an object changes its location. Example: a car moving on the highway.
- Rotation occurs when each point of a moving object moves in a circle about an axis through the center of mass. Examples: a spinning top, a rotating molecule.
- Vibration is a periodic distortion of and then recovery of original shape. Examples: a struck tuning fork, a vibrating molecule.

B.3—Force and Weight

Force is that which changes the velocity of a body; it is defined as

$$\text{Force} = \text{mass} \times \text{acceleration}$$

The SI unit of force is the **newton,** N, whose dimensions are kilograms times meter per second squared ($kg \cdot m/s^2$). A newton is therefore the force needed to change the velocity of a mass of 1 kilogram by 1 meter per second in a time of 1 second.

Because the earth's gravity is not the same everywhere, the weight corresponding to a given mass is not a constant. At any given spot on earth gravity is constant, however, and therefore weight is proportional to mass. When a balance tells us that a given sample (the "unknown") has the same weight as another sample (the "weights," as given by a scale reading or by a total of counterweights), it also tells us that the two masses are equal. The balance is therefore a valid instrument for measuring the mass of an object independently of slight variations in the force of gravity.

---

*Adapted from F. Brescia, J. Arents, H. Meislich, et al.: *General Chemistry,* 5th ed. Philadelphia, Harcourt Brace, 1988.

B.4—Pressure[*]

Pressure is force per unit area. The SI unit, called the pascal, Pa, is

$$1 \text{ pascal} = \frac{1 \text{ newton}}{\text{m}^2} = \frac{1 \text{ kg} \cdot \text{m/s}^2}{\text{m}^2} = \frac{1 \text{ kg}}{\text{m} \cdot \text{s}^2}$$

The International System of Units also recognizes the bar, which is 10^5 Pa and which is close to standard atmospheric pressure (Table 1).

Table 1   Pressure Conversions

From	To	Multiply By
atmosphere	mm Hg	760 mm Hg/atm (exactly)
atmosphere	lb/in^2	14.6960 lb/(in^2 atm)
atmosphere	kPa	101.325 kPa/atm
bar	Pa	10^5 Pa/bar (exactly)
bar	lb/in^2	14.5038 lb/(in^2 bar)
mm Hg	torr	1 torr/mm Hg (exactly)

Chemists also express pressure in terms of the heights of liquid columns, especially water and mercury. This usage is not completely satisfactory, because the pressure exerted by a given column of a given liquid is not a constant but depends on the temperature (which influences the density of the liquid) and the location (which influences gravity). Such units are therefore not part of the SI, and their use is now discouraged. The older units are still used in books and journals, however, and chemists must be familiar with them.

The pressure of a liquid or a gas depends only on the depth (or height) and is exerted equally in all directions. At sea level, the pressure exerted by the earth's atmosphere supports a column of mercury about 0.76 m (76 cm, or 760 mm) high.

One **standard atmosphere** (atm) is the pressure exerted by exactly 76 cm of mercury at 0 °C (density, 13.5951 g/cm^3) and at standard gravity, 9.80665 m/s^2. The **bar** is equivalent to 0.9869 atm. One **torr** is the pressure exerted by exactly 1 mm of mercury at 0 °C and standard gravity.

B.5—Energy and Power

The SI unit of energy is the product of the units of force and distance, or kilograms times meter per second squared (kg $\cdot$ m/s^2) times meters ($\times$ m), which is kg $\cdot$ m^2/s^2; this unit is called the **joule,** J. The joule is thus the work done when a force of 1 newton acts through a distance of 1 meter.

Work may also be done by moving an electric charge in an electric field. When the charge being moved is 1 coulomb (C), and the potential difference between its initial and final positions is 1 volt (V), the work is 1 joule. Thus,

$$1 \text{ joule} = 1 \text{ coulomb volt (CV)}$$

Another unit of electric work that is not part of the International System of Units but is still in use is the **electron volt,** eV, which is the work required to move an electron against a potential difference of 1 volt. (It is also the kinetic energy acquired

[*]See Section 12.1.

by an electron when it is accelerated by a potential difference of 1 volt.) Because the charge on an electron is 1.602×10^{-19} C, we have

$$1 \text{ eV} = 1.602 \times 10^{-19} \text{ CV} \times \frac{1 \text{ J}}{1 \text{ CV}} = 1.602 \times 10^{-19} \text{ J}$$

If this value is multiplied by Avogadro's number, we obtain the energy involved in moving 1 mole of electron charges (1 faraday) in a field produced by a potential difference of 1 volt:

$$1 \frac{\text{eV}}{\text{particle}} = \frac{1.602 \times 10^{-19} \text{ J}}{\text{particle}} \times \frac{6.022 \times 10^{23} \text{ particles}}{\text{mol}} \cdot \frac{1 \text{ kJ}}{1000 \text{ J}} = 96.49 \text{ kJ/mol}$$

Power is the amount of energy delivered per unit time. The SI unit is the watt, W, which is 1 joule per second. One kilowatt, kW, is 1000 W. Watt hours and kilowatt hours are therefore units of energy (Table 2). For example, 1000 watts, or 1 kilowatt, is

$$1.0 \times 10^3 \text{ W} \times \frac{1 \text{ J}}{1 \text{ W} \cdot \text{s}} \cdot \frac{3.6 \times 10^3 \text{ s}}{1 \text{ h}} = 3.6 \times 10^6 \text{ J}$$

Table 2   Energy Conversions

From	To	Multiply By
calorie (cal)	joule	4.184 J/cal (exactly)
kilocalorie (kcal)	cal	10^3 cal/kcal (exactly)
kilocalorie	joule	4.184×10^3 J/kcal (exactly)
liter atmosphere (L·atm)	joule	101.325 J/L·atm
electron volt (eV)	joule	1.60218×10^{-19} J/eV
electron volt per particle	kilojoules per mole	96.485 kJ·particle/eV·mol
coulomb volt (CV)	joule	1 CV/J (exactly)
kilowatt hour (kWh)	kcal	860.4 kcal/kWh
kilowatt hour	joule	3.6×10^6 J/kWh (exactly)
British thermal unit (Btu)	calorie	252 cal/Btu

Appendix C

Abbreviations and Useful Conversion Factors

Table 3 Some Common Abbreviations and Standard Symbols

Term	Abbreviation	Term	Abbreviation
Activation energy	E_a	Face-centered cubic	fcc
Ampere	A	Faraday constant	F
Aqueous solution	aq	Gas constant	R
Atmosphere, unit of pressure	atm	Gibbs free energy	G
Atomic mass unit	u	Standard free energy	$G°$
Avogadro's constant	N_A	Standard free energy of formation	ΔG°_f
Bar, unit of pressure	bar	Free energy change for reaction	ΔG°_{rxn}
Body-centered cubic	bcc	Half-life	$t_{1/2}$
Bohr radius	a_0	Heat	q
Boiling point	bp	Hertz	Hz
Celsius temperature, °C	T	Hour	h
Charge number of an ion	z	Joule	J
Coulomb, electric charge	C	Kelvin	K
Curie, radioactivity	Ci	Kilocalorie	kcal
Cycles per second, hertz	Hz	Liquid	ℓ
Debye, unit of electric dipole	D	Logarithm, base 10	log
Electron	e^-	Logarithm, base e	ln
Electron volt	eV	Minute	min
Electronegativity	χ	Molar	M
Energy	E	Molar mass	M
Enthalpy	H	Mole	mol
Standard enthalpy	$H°$	Osmotic pressure	Π
Standard enthalpy of formation	ΔH°_f	Planck's constant	h
Standard enthalpy of reaction	ΔH°_{rxn}	Pound	lb
Entropy	S	Pressure	
Standard entropy	$S°$	Pascal, unit of pressure	Pa
Entropy change for reaction	ΔS°_{rxn}	In atmospheres	atm
Equilibrium constant	K	In millimeters of mercury	mm Hg
Concentration basis	K_c	Proton number	Z
Pressure basis	K_p	Rate constant	k
Ionization weak acid	K_a	Simple cubic (unit cell)	sc
Ionization weak base	K_b	Standard temperature and pressure	STP
Solubility product	K_{sp}	Volt	V
Formation constant	K_{form}	Watt	W
Ethylenediamine	en	Wavelength	λ

C.1—Fundamental Units of the SI System

The metric system was begun by the French National Assembly in 1790 and has undergone many modifications. The International System of Units or *Système International* (SI), which represents an extension of the metric system, was adopted by the 11th General Conference of Weights and Measures in 1960. It is constructed from seven base units, each of which represents a particular physical quantity (Table 4).

Table 4   SI Fundamental Units

Physical Quantity	Name of Unit	Symbol
Length	meter	m
Mass	kilogram	kg
Time	second	s
Temperature	kelvin	K
Amount of substance	mole	mol
Electric current	ampere	A
Luminous intensity	candela	cd

The first five units listed in Table 4 are particularly useful in general chemistry and are defined as follows:

1. The *meter* was redefined in 1960 to be equal to 1,650,763.73 wavelengths of a certain line in the emission spectrum of krypton-86.
2. The *kilogram* represents the mass of a platinum–iridium block kept at the International Bureau of Weights and Measures at Sèvres, France.
3. The *second* was redefined in 1967 as the duration of 9,192,631,770 periods of a certain line in the microwave spectrum of cesium-133.
4. The *kelvin* is 1/273.15 of the temperature interval between absolute zero and the triple point of water.
5. The *mole* is the amount of substance that contains as many entities as there are atoms in exactly 0.012 kg of carbon-12 (12 g of ^{12}C atoms).

C.2—Prefixes Used with Traditional Metric Units and SI Units

Decimal fractions and multiples of metric and SI units are designated by using the prefixes listed in Table 5. Those most commonly used in general chemistry appear in italics.

C.3—Derived SI Units

In the International System of Units, all physical quantities are represented by appropriate combinations of the base units listed in Table 4. A list of the derived units frequently used in general chemistry is given in Table 6.

Table 5   Traditional Metric and SI Prefixes

Factor	Prefix	Symbol	Factor	Prefix	Symbol
10^{12}	tera	T	10^{-1}	deci	d
10^9	giga	G	10^{-2}	centi	c
10^6	mega	M	10^{-3}	milli	m
10^3	kilo	k	10^{-6}	micro	μ
10^2	hecto	h	10^{-9}	nano	n
10^1	deka	da	10^{-12}	pico	p
			10^{-15}	femto	f
			10^{-18}	atto	a

Table 6   Derived SI Units

Physical Quantity	Name of Unit	Symbol	Definition
Area	square meter	m^2	
Volume	cubic meter	m^3	
Density	kilogram per cubic meter	kg/m^3	
Force	newton	N	$kg \cdot m/s^2$
Pressure	pascal	Pa	N/m^2
Energy	joule	J	$kg \cdot m^2/s^2$
Electric charge	coulomb	C	$A \cdot s$
Electric potential difference	volt	V	$J/(A \cdot s)$

Table 7   Common Units of Mass and Weight

1 Pound = 453.39 Grams

1 kilogram = 1000 grams = 2.205 pounds

1 gram = 1000 milligrams

1 gram = 6.022×10^{23} atomic mass units

1 atomic mass unit = 1.6605×10^{-24} gram

1 short ton = 2000 pounds = 907.2 kilograms

1 long ton = 2240 pounds

1 metric tonne = 1000 kilograms = 2205 pounds

Table 8   Common Units of Length

1 inch = 2.54 centimeters (Exactly)

1 mile = 5280 feet = 1.609 kilometers

1 yard = 36 inches = 0.9144 meter

1 meter = 100 centimeters = 39.37 inches = 3.281 feet = 1.094 yards

1 kilometer = 1000 meters = 1094 yards = 0.6215 mile

1 Ångstrom = 1.0×10^{-8} centimeter = 0.10 nanometer = 100 picometers
$= 1.0 \times 10^{-10}$ meter = 3.937×10^{-9} inch

Table 9   Common Units of Volume

1 quart = 0.9463 liter
1 liter = 1.0567 quarts

1 liter = 1 cubic decimeter = 1000 cubic centimeters = 0.001 cubic meter

1 milliliter = 1 cubic centimeter = 0.001 liter = 1.056×10^{-3} quart

1 cubic foot = 28.316 liters = 29.924 quarts = 7.481 gallons

Appendix D

Physical Constants

Table 10

Quantity	Symbol	Traditional Units	SI Units
Acceleration of gravity	g	980.6 cm/s	9.806 m/s
Atomic mass unit (1/12 the mass of ^{12}C atom)	u	1.6605×10^{-24} g	1.6605×10^{-27} kg
Avogadro's number	N	$6.02214155 \times 10^{23}$ particles/mol	$6.02214155 \times 10^{23}$ particles/mol
Bohr radius	a_0	0.052918 nm 5.2918×10^{-9} cm	5.2918×10^{-11} m
Boltzmann constant	k	1.3807×10^{-16} erg/K	1.3807×10^{-23} J/K
Charge-to-mass ratio of electron	e/m	1.7588×10^{8} C/g	1.7588×10^{11} C/kg
Electronic charge	e	1.6022×10^{-19} C 4.8033×10^{-10} esu	1.6022×10^{-19} C
Electron rest mass	m_e	9.1094×10^{-28} g 0.00054858 amu	9.1094×10^{-31} kg
Faraday constant	F	96,485 C/mol e^- 23.06 kcal/V · mol e^-	96,485 C/mol e^- 96,485 J/V · mol e^-
Gas constant	R	$0.082057 \dfrac{L \cdot atm}{mol \cdot K}$ $1.987 \dfrac{cal}{mol \cdot K}$	$8.3145 \dfrac{Pa \cdot dm^3}{mol \cdot K}$ 8.3145 J/mol · K
Molar volume (STP)	V_m	22.414 L/mol	22.414×10^{-3} m^3/mol 22.414 dm^3/mol
Neutron rest mass	m_n	1.67493×10^{-24} g 1.008665 amu	1.67493×10^{-27} kg
Planck's constant	h	6.6261×10^{-27} erg · s	$6.6260693 \times 10^{-34}$ J · s
Proton rest mass	m_p	1.6726×10^{-24} g 1.007276 amu	1.6726×10^{-27} kg
Rydberg constant	R_α Rhc	3.289×10^{15} cycles/s	1.0974×10^{7} m^{-1} 2.1799×10^{-18} J
Velocity of light (in a vacuum)	c	2.9979×10^{10} cm/s (186,282 miles/s)	2.9979×10^{8} m/s

$\pi = 3.1416$

$e = 2.7183$

$\ln X = 2.303 \log X$

Table 11   Specific Heats and Heat Capacities for Some Common Substances at 25 °C

Substance	Specific Heat (J/g · K)	Molar Heat Capacity (J/mol · K)
Al(s)	0.897	24.2
Ca(s)	0.646	25.9
Cu(s)	0.385	24.5
Fe(s)	0.449	25.1
Hg(ℓ)	0.140	28.0
H_2O(s), ice	2.06	37.1
H_2O(ℓ), water	4.184	75.4
H_2O(g), steam	1.86	33.6
C_6H_6(ℓ), benzene	1.74	136
C_6H_6(g), benzene	1.06	82.4
C_2H_5OH(ℓ), ethanol	2.44	112.3
C_2H_5OH(g), ethanol	1.41	65.4
$(C_2H_5)_2O$(ℓ), diethyl ether	2.33	172.6
$(C_2H_5)_2O$(g), diethyl ether	1.61	119.5

Table 12   Heats of Transformation and Transformation Temperatures of Several Substances

Substance	MP (°C)	Heat of Fusion J/g	Heat of Fusion kJ/mol	BP (°C)	Heat of Vaporization J/g	Heat of Vaporization kJ/mol
Elements*						
Al	660	395	10.7	2518	12083	294
Ca	842	212	8.5	1484	3767	155
Cu	1085	209	13.3	2567	4720	300
Fe	1535	267	13.8	2861	6088	340
Hg	−38.8	11	2.29	357	295	59.1
Compounds						
H_2O	0.00	333	6.09	100.0	2260	40.7
CH_4	−182.5	58.6	0.94	−161.5	511	8.2
C_2H_5OH	−114	109	5.02	78.3	838	38.6
C_6H_6	5.48	127.4	9.95	80.0	393	30.7
$(C_2H_5)_2O$	−116.3	98.1	7.27	34.6	357	26.5

*Data for the elements are taken from J. A. Dean: *Lange's Handbook of Chemistry*, 15th Edition. New York, McGraw Hill Publishers, 1999.

Appendix E

Naming Organic Compounds

It seems a daunting task—to devise a systematic procedure that gives each organic compound a unique name—but that is what has been done. A set of rules was developed to name organic compounds by the International Union of Pure and Applied Chemistry (IUPAC). The IUPAC nomenclature allows chemists to write a name for any compound based on its structure or to identify the formula and structure for a compound from its name. In this book, we have generally used the IUPAC nomenclature scheme when naming compounds.

In addition to the systematic names, many compounds have common names. The common names came into existence before the nomenclature rules were developed, and they have continued in use. For some compounds, these names are so well entrenched that they are used most of the time. One such compound is acetic acid, which is almost always referred to by that name and not by its systematic name, ethanoic acid.

The general procedure for systematic naming of organic compounds begins with the nomenclature for hydrocarbons. Other organic compounds are then named as derivatives of hydrocarbons. Nomenclature rules for simple organic compounds are given in the following section.

E.1—Hydrocarbons

Alkanes

The names of alkanes end in "-ane." When naming a specific alkane, the root of the name identifies the longest carbon chain in a compound. Specific substituent groups attached to this carbon chain are identified by name and position.

Alkanes with chains of from one to ten carbon atoms are given in Table 11.2. After the first four compounds, the names derive from Latin numbers—pentane, hexane, heptane, octane, nonane, decane—and this regular naming continues for higher alkanes. For substituted alkanes, the substituent groups on a hydrocarbon chain must be identified both by a name and by the position of substitution; this information precedes the root of the name. The position is indicated by a number that refers to the carbon atom to which the substituent is attached. (Numbering of the carbon atoms in a chain should begin at the end of the carbon chain that allows the substituent groups to have the lowest numbers.)

Names of hydrocarbon substituents are derived from the name of the hydrocarbon. The group —CH_3, derived by taking a hydrogen from methane, is called the methyl group; the C_2H_5 group is the ethyl group. The nomenclature scheme is easily extended to derivatives of hydrocarbons with other substituent groups such as —Cl (chloro), —NO_2 (nitro), —CN (cyano), —D (deuterio), and so on (Table 13). If two or more of the same substituent groups occur, the prefixes "di-," "tri-," and "tetra-" are added. When different substituent groups are present, they are generally listed in alphabetical order.

Table 13   Names of Common Substituent Groups

Formula	Name	Formula	Name
$-CH_3$	methyl	$-D$	deuterio
$-C_2H_5$	ethyl	$-Cl$	chloro
$-CH_2CH_2CH_3$	1-propyl (*n*-propyl)	$-Br$	bromo
$-CH(CH_3)_2$	2-propyl (isopropyl)	$-F$	fluoro
$-CH=CH_2$	ethenyl (vinyl)	$-CN$	cyano
$-C_6H_5$	phenyl	$-NO_2$	nitro
$-OH$	hydroxo		
$-NH_2$	amino		

Example:

$$\begin{array}{cc} CH_3 & C_2H_5 \\ | & | \\ CH_3CH_2CHCH_2 & CHCH_2CH_3 \end{array}$$

Step	Information to include	Contribution to name
1.	An alkane	name will end in "-ane"
2.	Longest chain is 7 carbons	name as a *heptane*
3.	$-CH_3$ group at carbon 3	*3-methyl*
4.	$-C_2H_5$ group at carbon 5	*5-ethyl*

Name: 5-ethyl-3-methylheptane

Cycloalkanes are named based on the ring size and by adding the prefix "cyclo"; for example, the cycloalkane with a six-member ring of carbons is called cyclohexane.

Alkenes

Alkenes have names ending in "-ene." The name of an alkene must specify the length of the carbon chain and the position of the double bond (and when appropriate, the configuration, either *cis* or *trans*). As with alkanes, both identity and position of substituent groups must be given. The carbon chain is numbered from the end that gives the double bond the lowest number.

Compounds with two double bonds are called dienes and they are named similarly—specifying the positions of the double bonds and the name and position of any substituent groups.

For example, the compound $H_2C=C(CH_3)CH(CH_3)CH_2CH_3$ has a five-carbon chain with a double bond between carbon atoms 1 and 2 and methyl groups on carbon atoms 2 and 3. Its name using IUPAC nomenclature is **2,3-dimethyl-1-pentene.** The compound $CH_3CH=CHCCl_3$ with a *cis* configuration around the double bond is named **1,1,1-trichloro-*cis*-2-butene.** The compound $H_2C=C(Cl)CH=CH_2$ is **2-chloro-1,3-butadiene.**

Alkynes

The naming of alkynes is similar to the naming of alkenes, except that *cis–trans* isomerism isn't a factor. The ending "-yne" on a name identifies a compound as an alkyne.

Benzene Derivatives

The carbon atoms in the six-member ring are numbered 1 through 6, and the name and position of substituent groups are given. The two examples shown here are **1-ethyl-3-methylbenzene** and **1,4-diaminobenzene.**

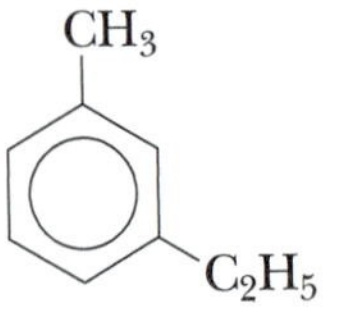

1-ethyl-3-methylbenzene 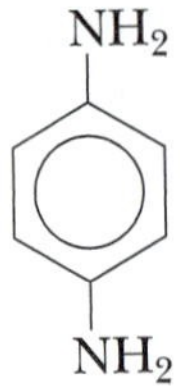   1,4-diaminobenzene

E.2—Derivatives of Hydrocarbons

The names for alcohols, aldehydes, ketones, and acids are based on the name of the hydrocarbon with an appropriate suffix to denote the class of compound, as follows:

- **Alcohols:** Substitute "-ol" for the final "-e" in the name of the hydrocarbon, and designate the position of the —OH group by the number of the carbon atom. For example, $CH_3CH_2CHOHCH_3$ is named as a derivative of the 4-carbon hydrocarbon butane. The —OH group is attached to the second carbon, so the name is 2-butanol.

- **Aldehydes:** Substitute "-al" for the final "-e" in the name of the hydrocarbon. The carbon atom of an aldehyde is, by definition, carbon-1 in the hydrocarbon chain. For example, the compound $CH_3CH(CH_3)CH_2CH_2CHO$ contains a 5-carbon chain with the aldehyde functional group being carbon-1 and the —CH_3 group at position 4; thus the name is **4-methylpentanal.**

- **Ketones:** Substitute "-one" for the final "-e" in the name of the hydrocarbon. The position of the ketone functional group (the carbonyl group) is indicated by the number of the carbon atom. For example, the compound $CH_3COCH_2CH(C_2H_5)CH_2CH_3$ has the carbonyl group at the 2 position and an ethyl group at the 4 position of a 6-carbon chain; its name is **4-ethyl-2-hexanone.**

- **Carboxylic acids (organic acids):** Substitute "-oic" for the final "-e" in the name of the hydrocarbon. The carbon atoms in the longest chain are counted beginning with the carboxylic carbon atom. For example, *trans*-CH_3CH=$CHCH_2CO_2H$ is named as a derivative of *trans*-3-pentene—that is, ***trans*-3-pentenoic acid.**

An **ester** is named as a derivative of the alcohol and acid from which it is made. The name of an ester is obtained by splitting the formula RCO_2R' into two parts, the RCO_2— portion and the —R' portion. The —R' portion comes from the alcohol and is identified by the hydrocarbon group name; derivatives of ethanol, for example, are called *ethyl* esters. The acid part of the compound is named by dropping the "-oic" ending for the acid and replacing it by "-oate." The compound $CH_3CH_2CO_2CH_3$ is named **methyl propanoate.**

Notice that an anion derived from a carboxylic acid by loss of the acidic proton is named the same way. Thus, $CH_3CH_2CO_2^-$ is the **propanoate anion,** and the sodium salt of this anion, $Na(CH_3CH_2CO_2)$, is **sodium propanoate.**

Appendix F

Values for the Ionization Energies and Electron Affinities of the Elements

1A (1)	2A (2)	3B (3)	4B (4)	5B (5)	6B (6)	7B (7)	8B (8,9,10)			1B (11)	2B (12)	3A (13)	4A (14)	5A (15)	6A (16)	7A (17)	8 (18)
H 1312																	He 2371
Li 520	Be 899											B 801	C 1086	N 1402	O 1314	F 1681	Ne 2081
Na 496	Mg 738											Al 578	Si 786	P 1012	S 1000	Cl 1251	Ar 1521
K 419	Ca 599	Sc 631	Ti 658	V 650	Cr 652	Mn 717	Fe 759	Co 758	Ni 757	Cu 745	Zn 906	Ga 579	Ge 762	As 947	Se 941	Br 1140	Kr 1351
Rb 403	Sr 550	Y 617	Zr 661	Nb 664	Mo 685	Tc 702	Ru 711	Rh 720	Pd 804	Ag 731	Cd 868	In 558	Sn 709	Sb 834	Te 869	I 1008	Xe 1170
Cs 377	Ba 503	La 538	Hf 681	Ta 761	W 770	Re 760	Os 840	Ir 880	Pt 870	Au 890	Hg 1007	Tl 589	Pb 715	Bi 703	Po 812	At 890	Rn 1037

Table 14 Electron Affinity Values for Some Elements (kJ/mol)*

H							
−72.77							
Li	Be	B	C	N		O	F
−59.63	0†	−26.7	−121.85	0		−140.98	−328.0
Na	Mg	Al	Si	P		S	Cl
−52.87	0	−42.6	−133.6	−72.07		−200.41	−349.0
K	Ca	Ga	Ge	As		Se	Br
−48.39	0	−30	−120	−78		−194.97	−324.7
Rb	Sr	In	Sn	Sb		Te	I
−46.89	0	−30	−120	−103		−190.16	−295.16
Cs	Ba	Tl	Pb	Bi		Po	At
−45.51	0	−20	−35.1	−91.3		−180	−270

*Data taken from H. Hotop and W. C. Lineberger: *Journal of Physical Chemistry, Reference Data,* Vol. 14, p. 731, 1985. (This paper also includes data for the transition metals.) Some values are known to more than two decimal places.

† Elements with an electron affinity of zero indicate that a stable anion A⁻ of the element does not exist in the gas phase.

Appendix G

Vapor Pressure of Water at Various Temperatures

Table 15 Vapor Pressure of Water at Various Temperatures

Temperature (°C)	Vapor Pressure (torr)	Temperature (°C)	Vapor Pressure (torr)	Temperature (°C)	Vapor Pressure (torr)	Temperature (°C)	Vapor Pressure (torr)
−10	2.1	21	18.7	51	97.2	81	369.7
−9	2.3	22	19.8	52	102.1	82	384.9
−8	2.5	23	21.1	53	107.2	83	400.6
−7	2.7	24	22.4	54	112.5	84	416.8
−6	2.9	25	23.8	55	118.0	85	433.6
−5	3.2	26	25.2	56	123.8	86	450.9
−4	3.4	27	26.7	57	129.8	87	468.7
−3	3.7	28	28.3	58	136.1	88	487.1
−2	4.0	29	30.0	59	142.6	89	506.1
−1	4.3	30	31.8	60	149.4	90	525.8
0	4.6	31	33.7	61	156.4	91	546.1
1	4.9	32	35.7	62	163.8	92	567.0
2	5.3	33	37.7	63	171.4	93	588.6
3	5.7	34	39.9	64	179.3	94	610.9
4	6.1	35	42.2	65	187.5	95	633.9
5	6.5	36	44.6	66	196.1	96	657.6
6	7.0	37	47.1	67	205.0	97	682.1
7	7.5	38	49.7	68	214.2	98	707.3
8	8.0	39	52.4	69	223.7	99	733.2
9	8.6	40	55.3	70	233.7	100	760.0
10	9.2	41	58.3	71	243.9	101	787.6
11	9.8	42	61.5	72	254.6	102	815.9
12	10.5	43	64.8	73	265.7	103	845.1
13	11.2	44	68.3	74	277.2	104	875.1
14	12.0	45	71.9	75	289.1	105	906.1
15	12.8	46	75.7	76	301.4	106	937.9
16	13.6	47	79.6	77	314.1	107	970.6
17	14.5	48	83.7	78	327.3	108	1004.4
18	15.5	49	88.0	79	341.0	109	1038.9
19	16.5	50	92.5	80	355.1	110	1074.6
20	17.5						

Appendix H

Ionization Constants for Weak Acids at 25 °C

Table 16 Ionization Constants for Weak Acids at 25 °C

Acid	Formula and Ionization Equation	K_a
Acetic	$CH_3CO_2H \rightleftharpoons H^+ + CH_3CO_2^-$	1.8×10^{-5}
Arsenic	$H_3AsO_4 \rightleftharpoons H^+ + H_2AsO_4^-$	$K_1 = 2.5 \times 10^{-4}$
	$H_2AsO_4^- \rightleftharpoons H^+ + HAsO_4^{2-}$	$K_2 = 5.6 \times 10^{-8}$
	$HAsO_4^{2-} \rightleftharpoons H^+ + AsO_4^{3-}$	$K_3 = 3.0 \times 10^{-13}$
Arsenous	$H_3AsO_3 \rightleftharpoons H^+ + H_2AsO_3^-$	$K_1 = 6.0 \times 10^{-10}$
	$H_2AsO_3^- \rightleftharpoons H^+ + HAsO_3^{2-}$	$K_2 = 3.0 \times 10^{-14}$
Benzoic	$C_6H_5CO_2H \rightleftharpoons H^+ + C_6H_5CO_2^-$	6.3×10^{-5}
Boric	$H_3BO_3 \rightleftharpoons H^+ + H_2BO_3^-$	$K_1 = 7.3 \times 10^{-10}$
	$H_2BO_3 \rightleftharpoons H^+ + HBO_3^{2-}$	$K_2 = 1.8 \times 10^{-13}$
	$HBO_3^{2-} \rightleftharpoons H^+ + BO_3^{3-}$	$K_3 = 1.6 \times 10^{-14}$
Carbonic	$H_2CO_3 \rightleftharpoons H^+ + HCO_3^-$	$K_1 = 4.2 \times 10^{-7}$
	$HCO_3^- \rightleftharpoons H^+ + CO_3^{2-}$	$K_2 = 4.8 \times 10^{-11}$
Citric	$H_3C_6H_5O_7 \rightleftharpoons H^+ + H_2C_6H_5O_7^-$	$K_1 = 7.4 \times 10^{-3}$
	$H_2C_6H_5O_7^- \rightleftharpoons H^+ + HC_6H_5O_7^{2-}$	$K_2 = 1.7 \times 10^{-5}$
	$HC_6H_5O_7^{2-} \rightleftharpoons H^+ + C_6H_5O_7^{3-}$	$K_3 = 4.0 \times 10^{-7}$
Cyanic	$HOCN \rightleftharpoons H^+ + OCN^-$	3.5×10^{-4}
Formic	$HCO_2H \rightleftharpoons H^+ + HCO_2^-$	1.8×10^{-4}
Hydrazoic	$HN_3 \rightleftharpoons H^+ + N_3^-$	1.9×10^{-5}
Hydrocyanic	$HCN \rightleftharpoons H^+ + CN^-$	4.0×10^{-10}
Hydrofluoric	$HF \rightleftharpoons H^+ + F^-$	7.2×10^{-4}
Hydrogen peroxide	$H_2O_2 \rightleftharpoons H^+ + HO_2^-$	2.4×10^{-12}
Hydrosulfuric	$H_2S \rightleftharpoons H^+ + HS^-$	$K_1 = 1 \times 10^{-7}$
	$HS^- \rightleftharpoons H^+ + S^{2-}$	$K_2 = 1 \times 10^{-19}$
Hypobromous	$HOBr \rightleftharpoons H^+ + OBr^-$	2.5×10^{-9}
Hypochlorous	$HOCl \rightleftharpoons H^+ + OCl^-$	3.5×10^{-8}
Nitrous	$HNO_2 \rightleftharpoons H^+ + NO_2^-$	4.5×10^{-4}
Oxalic	$H_2C_2O_4 \rightleftharpoons H^+ + HC_2O_4^-$	$K_1 = 5.9 \times 10^{-2}$
	$HC_2O_4^- \rightleftharpoons H^+ + C_2O_4^{2-}$	$K_2 = 6.4 \times 10^{-5}$
Phenol	$C_6H_5OH \rightleftharpoons H^+ + C_6H_5O^-$	1.3×10^{-10}

(continued)

Table 16 *(continued)*

Acid	Formula and Ionization Equation	K_a
Phosphoric	$H_3PO_4 \rightleftharpoons H^+ + H_2PO_4^-$	$K_1 = 7.5 \times 10^{-3}$
	$H_2PO_4^- \rightleftharpoons H^+ + HPO_4^{2-}$	$K_2 = 6.2 \times 10^{-8}$
	$HPO_4^{2-} \rightleftharpoons H^+ + PO_4^{3-}$	$K_3 = 3.6 \times 10^{-13}$
Phosphorous	$H_3PO_3 \rightleftharpoons H^+ + H_2PO_3^-$	$K_1 = 1.6 \times 10^{-2}$
	$H_2PO_3^- \rightleftharpoons H^+ + HPO_3^{2-}$	$K_2 = 7.0 \times 10^{-7}$
Selenic	$H_2SeO_4 \rightleftharpoons H^+ + HSeO_4^-$	$K_1 = $ very large
	$HSeO_4^- \rightleftharpoons H^+ + SeO_4^{2-}$	$K_2 = 1.2 \times 10^{-2}$
Selenous	$H_2SeO_3 \rightleftharpoons H^+ + HSeO_3^-$	$K_1 = 2.7 \times 10^{-3}$
	$HSeO_3^- \rightleftharpoons H^+ + SeO_3^{2-}$	$K_2 = 2.5 \times 10^{-7}$
Sulfuric	$H_2SO_4 \rightleftharpoons H^+ + HSO_4^-$	$K_1 = $ very large
	$HSO_4^- \rightleftharpoons H^+ + SO_4^{2-}$	$K_2 = 1.2 \times 10^{-2}$
Sulfurous	$H_2SO_3 \rightleftharpoons H^+ + HSO_3^-$	$K_1 = 1.2 \times 10^{-2}$
	$HSO_3^- \rightleftharpoons H^+ + SO_3^{2-}$	$K_2 = 6.2 \times 10^{-8}$
Tellurous	$H_2TeO_3 \rightleftharpoons H^+ + HTeO_3^-$	$K_1 = 2 \times 10^{-3}$
	$HTeO_3^- \rightleftharpoons H^+ + TeO_3^{2-}$	$K_2 = 1 \times 10^{-8}$

Appendix I

Ionization Constants for Weak Bases at 25 °C

Table 17 Ionization Constants for Weak Bases at 25 °C

Base	Formula and Ionization Equation	K_b
Ammonia	$NH_3 + H_2O \rightleftharpoons NH_4^+ + OH^-$	1.8×10^{-5}
Aniline	$C_6H_5NH_2 + H_2O \rightleftharpoons C_6H_5NH_3^+ + OH^-$	4.0×10^{-10}
Dimethylamine	$(CH_3)_2NH + H_2O \rightleftharpoons (CH_3)_2NH_2^+ + OH^-$	7.4×10^{-4}
Ethylenediamine	$H_2NCH_2CH_2NH_2 + H_2O \rightleftharpoons H_2NCH_2CH_2NH_3^+ + OH^-$	$K_1 = 8.5 \times 10^{-5}$
	$H_2NCH_2CH_2NH_3^+ + H_2O \rightleftharpoons H_3NCH_2CH_2NH_3^{2+} + OH^-$	$K_2 = 2.7 \times 10^{-8}$
Hydrazine	$N_2H_4 + H_2O \rightleftharpoons N_2H_5^+ + OH^-$	$K_1 = 8.5 \times 10^{-7}$
	$N_2H_5^+ + H_2O \rightleftharpoons N_2H_6^{2+} + OH^-$	$K_2 = 8.9 \times 10^{-16}$
Hydroxylamine	$NH_2OH + H_2O \rightleftharpoons NH_3OH^+ + OH^-$	6.6×10^{-9}
Methylamine	$CH_3NH_2 + H_2O \rightleftharpoons CH_3NH_3^+ + OH^-$	5.0×10^{-4}
Pyridine	$C_5H_5N + H_2O \rightleftharpoons C_5H_5NH^+ + OH^-$	1.5×10^{-9}
Trimethylamine	$(CH_3)_3N + H_2O \rightleftharpoons (CH_3)_3NH^+ + OH^-$	7.4×10^{-5}
Ethylamine	$C_2H_5NH_2 + H_2O \rightleftharpoons C_2H_5NH_3^+ + OH^-$	4.3×10^{-4}

Appendix J

Solubility Product Constants for Some Inorganic Compounds at 25 °C

Table 18A Solubility Product Constants (25 °C)

Cation	Compound	K_{sp}	Cation	Compound	K_{sp}
Ba^{2+}	*$BaCrO_4$	1.2×10^{-10}	Mg^{2+}	$MgCO_3$	6.8×10^{-6}
	$BaCO_3$	2.6×10^{-9}		MgF_2	5.2×10^{-11}
	BaF_2	1.8×10^{-7}		$Mg(OH)_2$	5.6×10^{-12}
	*$BaSO_4$	1.1×10^{-10}	Mn^{2+}	$MnCO_3$	2.3×10^{-11}
Ca^{2+}	$CaCO_3$ (calcite)	3.4×10^{-9}		*$Mn(OH)_2$	1.9×10^{-13}
	*CaF_2	5.3×10^{-11}	Hg_2^{2+}	*Hg_2Br_2	6.4×10^{-23}
	*$Ca(OH)_2$	5.5×10^{-5}		Hg_2Cl_2	1.4×10^{-18}
	$CaSO_4$	4.9×10^{-5}		*Hg_2I_2	2.9×10^{-29}
$Cu^{+,2+}$	$CuBr$	6.3×10^{-9}		Hg_2SO_4	6.5×10^{-7}
	CuI	1.3×10^{-12}	Ni^{2+}	$NiCO_3$	1.4×10^{-7}
	$Cu(OH)_2$	2.2×10^{-20}		$Ni(OH)_2$	5.5×10^{-16}
	$CuSCN$	1.8×10^{-13}	Ag^+	*$AgBr$	5.4×10^{-13}
Au^+	$AuCl$	2.0×10^{-13}		*$AgBrO_3$	5.4×10^{-5}
Fe^{2+}	$FeCO_3$	3.1×10^{-11}		$AgCH_3CO_2$	1.9×10^{-3}
	$Fe(OH)_2$	4.9×10^{-17}		$AgCN$	6.0×10^{-17}
Pb^{2+}	$PbBr_2$	6.6×10^{-6}		Ag_2CO_3	8.5×10^{-12}
	$PbCO_3$	7.4×10^{-14}		*$Ag_2C_2O_4$	5.4×10^{-12}
	$PbCl_2$	1.7×10^{-5}		*$AgCl$	1.8×10^{-10}
	$PbCrO_4$	2.8×10^{-13}		Ag_2CrO_4	1.1×10^{-12}
	PbF_2	3.3×10^{-8}		*AgI	8.5×10^{-17}
	PbI_2	9.8×10^{-9}		$AgSCN$	1.0×10^{-12}
	$Pb(OH)_2$	1.4×10^{-15}		*Ag_2SO_4	1.2×10^{-5}
	$PbSO_4$	2.5×10^{-8}			

(continued)

Table 18A *(continued)*

Cation	Compound	K_{sp}	Cation	Compound	K_{sp}
Sr^{2+}	$SrCO_3$	5.6×10^{-10}	Zn^{2+}	$Zn(OH)_2$	3×10^{-17}
	SrF_2	4.3×10^{-9}		$Zn(CN)_2$	8.0×10^{-12}
	$SrSO_4$	3.4×10^{-7}			
Tl^+	$TlBr$	3.7×10^{-6}			
	$TlCl$	1.9×10^{-4}			
	TlI	5.5×10^{-8}			

The values reported in this table were taken from J. A. Dean: *Lange's Handbook of Chemistry,* 15th Edition. New York, McGraw Hill Publishers, 1999. Values have been rounded off to two significant figures.

*Calculated solubility from these K_{sp} values will match experimental solubility for this compound within a factor of 2. Experimental values for solubilities are given in R. W. Clark and J. M. Bonicamp: *Journal of Chemical Education,* Vol. 75, p. 1182, 1998.

Table 18B K_{spa} Values* for Some Metal Sulfides (25 °C)

Substance	K_{spa}
HgS (red)	4×10^{-54}
HgS (black)	2×10^{-53}
Ag_2S	6×10^{-51}
CuS	6×10^{-37}
PbS	3×10^{-28}
CdS	8×10^{-28}
SnS	1×10^{-26}
FeS	6×10^{-19}

*The equilibrium constant value K_{spa} for metal sulfides refers to the equilibrium $MS(s) + H_2O(\ell) \rightleftharpoons M^{2+}(aq) + OH^-(aq) + HS^-(aq)$; see R. J. Myers, *Journal of Chemical Education,* Vol. 63, p. 687, 1986.

Appendix K

Formation Constants for Some Complex Ions in Aqueous Solution

Table 19 Formation Constants for Some Complex Ions in Aqueous Solution

Formation Equilibrium	K
$Ag^+ + 2\ Br^- \rightleftharpoons [AgBr_2]^-$	1.3×10^7
$Ag^+ + 2\ Cl^- \rightleftharpoons [AgCl_2]^-$	2.5×10^5
$Ag^+ + 2\ CN^- \rightleftharpoons [Ag(CN)_2]^-$	5.6×10^{18}
$Ag^+ + 2\ S_2O_3{}^{2-} \rightleftharpoons [Ag(S_2O_3)_2]^{3-}$	2.0×10^{13}
$Ag^+ + 2\ NH_3 \rightleftharpoons [Ag(NH_3)_2]^+$	1.6×10^7
$Al^{3+} + 6\ F^- \rightleftharpoons [AlF_6]^{3-}$	5.0×10^{23}
$Al^{3+} + 4\ OH^- \rightleftharpoons [Al(OH)_4]^-$	7.7×10^{33}
$Au^+ + 2\ CN^- \rightleftharpoons [Au(CN)_2]^-$	2.0×10^{38}
$Cd^{2+} + 4\ CN^- \rightleftharpoons [Cd(CN)_4]^{2-}$	1.3×10^{17}
$Cd^{2+} + 4\ Cl^- \rightleftharpoons [CdCl_4]^{2-}$	1.0×10^4
$Cd^{2+} + 4\ NH_3 \rightleftharpoons [Cd(NH_3)_4]^{2+}$	1.0×10^7
$Co^{2+} + 6\ NH_3 \rightleftharpoons [Co(NH_3)_6]^{2+}$	7.7×10^4
$Cu^+ + 2\ CN^- \rightleftharpoons [Cu(CN)_2]^-$	1.0×10^{16}
$Cu^+ + 2\ Cl^- \rightleftharpoons [CuCl_2]^-$	1.0×10^5
$Cu^{2+} + 4\ NH_3 \rightleftharpoons [Cu(NH_3)_4]^{2+}$	6.8×10^{12}
$Fe^{2+} + 6\ CN^- \rightleftharpoons [Fe(CN)_6]^{4-}$	7.7×10^{36}
$Hg^{2+} + 4\ Cl^- \rightleftharpoons [HgCl_4]^{2-}$	1.2×10^{15}
$Ni^{2+} + 4\ CN^- \rightleftharpoons [Ni(CN)_4]^{2-}$	1.0×10^{31}
$Ni^{2+} + 6\ NH_3 \rightleftharpoons [Ni(NH_3)_6]^{2+}$	5.6×10^8
$Zn^{2+} + 4\ OH^- \rightleftharpoons [Zn(OH)_4]^{2-}$	2.9×10^{15}
$Zn^{2+} + 4\ NH_3 \rightleftharpoons [Zn(NH_3)_4]^{2+}$	2.9×10^9

Appendix L

Selected Thermodynamic Values

Table 20   Selected Thermodynamic Values*

Species	ΔH_f° (298.15 K) (kJ/mol)	S° (298.15 K) (J/K · mol)	ΔG_f° (298.15 K) (kJ/mol)
Aluminum			
Al(s)	0	28.3	0
AlCl$_3$(s)	−705.63	109.29	−630.0
Al$_2$O$_3$(s)	−1675.7	50.92	−1582.3
Barium			
BaCl$_2$(s)	−858.6	123.68	−810.4
BaCO$_3$(s)	−1213	112.1	−1134.41
BaO(s)	−548.1	72.05	−520.38
BaSO$_4$(s)	−1473.2	132.2	−1362.2
Beryllium			
Be(s)	0	9.5	0
Be(OH)$_2$(s)	−902.5	51.9	−815.0
Boron			
BCl$_3$(g)	−402.96	290.17	−387.95
Bromine			
Br(g)	111.884	175.022	82.396
Br$_2$(ℓ)	0	152.2	0
Br$_2$(g)	30.91	245.47	3.12
BrF$_3$(g)	−255.60	292.53	−229.43
HBr(g)	−36.29	198.70	−53.45
Calcium			
Ca(s)	0	41.59	0
Ca(g)	178.2	158.884	144.3
Ca^{2+}(g)	1925.90	—	—
CaC$_2$(s)	−59.8	70.	−64.93
CaCO$_3$(s, calcite)	−1207.6	91.7	−1129.16
CaCl$_2$(s)	−795.8	104.6	−748.1
CaF$_2$(s)	−1219.6	68.87	−1167.3
CaH$_2$(s)	−186.2	42	−147.2
CaO(s)	−635.09	38.2	−603.42
CaS(s)	−482.4	56.5	−477.4
Ca(OH)$_2$(s)	−986.09	83.39	−898.43

(continued)

* Most thermodynamic data are taken from the NIST Webbook at **http://webbook.nist.gov**.

Table 20   (continued)

Species	ΔH_f° (298.15 K) (kJ/mol)	S° (298.15 K) (J/K·mol)	ΔG_f° (298.15 K) (kJ/mol)
Ca(OH)$_2$(aq)	−1002.82	—	−868.07
CaSO$_4$(s)	−1434.52	106.5	−1322.02
Carbon			
C(s, graphite)	0	5.6	0
C(s, diamond)	1.8	2.377	2.900
C(g)	716.67	158.1	671.2
CCl$_4$(ℓ)	−128.4	214.39	−57.63
CCl$_4$(g)	−95.98	309.65	−53.61
CHCl$_3$(ℓ)	−134.47	201.7	−73.66
CHCl$_3$(g)	−103.18	295.61	−70.4
CH$_4$(g, methane)	−74.87	186.26	−50.8
C$_2$H$_2$(g, ethyne)	226.73	200.94	209.20
C$_2$H$_4$(g, ethene)	52.47	219.36	68.35
C$_2$H$_6$(g, ethane)	−83.85	229.2	−31.89
C$_3$H$_8$(g, propane)	−104.7	270.3	−24.4
C$_6$H$_6$(ℓ, benzene)	48.95	173.26	124.21
CH$_3$OH(ℓ, methanol)	−238.4	127.19	−166.14
CH$_3$OH(g, methanol)	−201.0	239.7	−162.5
C$_2$H$_5$OH(ℓ, ethanol)	−277.0	160.7	−174.7
C$_2$H$_5$OH(g, ethanol)	−235.3	282.70	−168.49
CO(g)	−110.525	197.674	−137.168
CO$_2$(g)	−393.509	213.74	−394.359
CS$_2$(ℓ)	89.41	151	65.2
CS$_2$(g)	116.7	237.8	66.61
COCl$_2$(g)	−218.8	283.53	−204.6
Cesium			
Cs(s)	0	85.23	0
Cs$^+$(g)	457.964	—	—
CsCl(s)	−443.04	101.17	−414.53
Chlorine			
Cl(g)	121.3	165.19	105.3
Cl$^-$(g)	−233.13	—	—
Cl$_2$(g)	0	223.08	0
HCl(g)	−92.31	186.2	−95.09
HCl(aq)	−167.159	56.5	−131.26
Chromium			
Cr(s)	0	23.62	0
Cr$_2$O$_3$(s)	−1134.7	80.65	−1052.95
CrCl$_3$(s)	−556.5	123.0	−486.1

(continued)

Table 20   (continued)

Species	ΔH_f° (298.15 K) (kJ/mol)	S° (298.15 K) (J/K · mol)	ΔG_f° (298.15 K) (kJ/mol)
Copper			
Cu(s)	0	33.17	0
CuO(s)	−156.06	42.59	−128.3
CuCl$_2$(s)	−220.1	108.07	−175.7
CuSO$_4$(s)	−769.98	109.05	−660.75
Fluorine			
F$_2$(g)	0	202.8	0
F(g)	78.99	158.754	61.91
F$^-$(g)	−255.39	—	—
F$^-$(aq)	−332.63		−278.79
HF(g)	−273.3	173.779	−273.2
HF(aq)	−332.63	88.7	−278.79
Hydrogen			
H$_2$(g)	0	130.7	0
H(g)	217.965	114.713	203.247
H$^+$(g)	1536.202	—	—
H$_2$O(ℓ)	−285.83	69.95	−237.15
H$_2$O(g)	−241.83	188.84	−228.59
H$_2$O$_2$(ℓ)	−187.78	109.6	−120.35
Iodine			
I$_2$(s)	0	116.135	0
I$_2$(g)	62.438	260.69	19.327
I(g)	106.838	180.791	70.250
I$^-$(g)	−197	—	—
ICl(g)	17.51	247.56	−5.73
Iron			
Fe(s)	0	27.78	0
FeO(s)	−272	—	—
Fe$_2$O$_3$(s, hematite)	−825.5	87.40	−742.2
Fe$_3$O$_4$(s, magnetite)	−1118.4	146.4	−1015.4
FeCl$_2$(s)	−341.79	117.95	−302.30
FeCl$_3$(s)	−399.49	142.3	−344.00
FeS$_2$(s, pyrite)	−178.2	52.93	−166.9
Fe(CO)$_5$(ℓ)	−774.0	338.1	−705.3
Lead			
Pb(s)	0	64.81	0
PbCl$_2$(s)	−359.41	136.0	−314.10
PbO(s, yellow)	−219	66.5	−196
PbO$_2$(s)	−277.4	68.6	−217.39
PbS(s)	−100.4	91.2	−98.7

(continued)

Table 20   *(continued)*

Species	ΔH_f° (298.15 K) (kJ/mol)	S° (298.15 K) (J/K · mol)	ΔG_f° (298.15 K) (kJ/mol)
Lithium			
Li(s)	0	29.12	0
Li^+(g)	685.783	—	—
LiOH(s)	−484.93	42.81	−438.96
LiOH(aq)	−508.48	2.80	−450.58
LiCl(s)	−408.701	59.33	−384.37
Magnesium			
Mg(s)	0	32.67	0
$MgCl_2$(s)	−641.62	89.62	−592.09
$MgCO_3$(s)	−1111.69	65.84	−1028.2
MgO(s)	−601.24	26.85	−568.93
$Mg(OH)_2$(s)	−924.54	63.18	−833.51
MgS(s)	−346.0	50.33	−341.8
Mercury			
Hg(ℓ)	0	76.02	0
$HgCl_2$(s)	−224.3	146.0	−178.6
HgO(s, red)	−90.83	70.29	−58.539
HgS(s, red)	−58.2	82.4	−50.6
Nickel			
Ni(s)	0	29.87	0
NiO(s)	−239.7	37.99	−211.7
$NiCl_2$(s)	−305.332	97.65	−259.032
Nitrogen			
N_2(g)	0	191.56	0
N(g)	472.704	153.298	455.563
NH_3(g)	−45.90	192.77	−16.37
N_2H_4(ℓ)	50.63	121.52	149.45
NH_4Cl(s)	−314.55	94.85	−203.08
NH_4Cl(aq)	−299.66	169.9	−210.57
NH_4NO_3(s)	−365.56	151.08	−183.84
NH_4NO_3(aq)	−339.87	259.8	−190.57
NO(g)	90.29	210.76	86.58
NO_2(g)	33.1	240.04	51.23
N_2O(g)	82.05	219.85	104.20
N_2O_4(g)	9.08	304.38	97.73
NOCl(g)	51.71	261.8	66.08
HNO_3(ℓ)	−174.10	155.60	−80.71
HNO_3(g)	−135.06	266.38	−74.72
HNO_3(aq)	−207.36	146.4	−111.25

(continued)

Table 20   (continued)

Species	ΔH_f° (298.15 K) (kJ/mol)	S° (298.15 K) (J/K · mol)	ΔG_f° (298.15 K) (kJ/mol)
Oxygen			
$O_2(g)$	0	205.07	0
$O(g)$	249.170	161.055	231.731
$O_3(g)$	142.67	238.92	163.2
Phosphorus			
$P_4(s, white)$	0	41.1	0
$P_4(s, red)$	−17.6	22.80	−12.1
$P(g)$	314.64	163.193	278.25
$PH_3(g)$	22.89	210.24	30.91
$PCl_3(g)$	−287.0	311.78	−267.8
$P_4O_{10}(s)$	−2984.0	228.86	−2697.7
$H_3PO_4(\ell)$	−1279.0	110.5	−1119.1
Potassium			
$K(s)$	0	64.63	0
$KCl(s)$	−436.68	82.56	−408.77
$KClO_3(s)$	−397.73	143.1	−296.25
$KI(s)$	−327.90	106.32	−324.892
$KOH(s)$	−424.72	78.9	−378.92
$KOH(aq)$	−482.37	91.6	−440.50
Silicon			
$Si(s)$	0	18.82	0
$SiBr_4(\ell)$	−457.3	277.8	−443.9
$SiC(s)$	−65.3	16.61	−62.8
$SiCl_4(g)$	−662.75	330.86	−622.76
$SiH_4(g)$	34.31	204.65	56.84
$SiF_4(g)$	−1614.94	282.49	−1572.65
$SiO_2(s, quartz)$	−910.86	41.46	−856.97
Silver			
$Ag(s)$	0	42.55	0
$Ag_2O(s)$	−31.1	121.3	−11.32
$AgCl(s)$	−127.01	96.25	−109.76
$AgNO_3(s)$	−124.39	140.92	−33.41
Sodium			
$Na(s)$	0	51.21	0
$Na(g)$	107.3	153.765	76.83
$Na^+(g)$	609.358	—	—
$NaBr(s)$	−361.02	86.82	−348.983
$NaCl(s)$	−411.12	72.11	−384.04
$NaCl(g)$	−181.42	229.79	−201.33
$NaCl(aq)$	−407.27	115.5	−393.133

(continued)

Table 20   *(continued)*

Species	ΔH_f° (298.15 K) (kJ/mol)	S° (298.15 K) (J/K · mol)	ΔG_f° (298.15 K) (kJ/mol)
NaOH(s)	−425.93	64.46	−379.75
NaOH(aq)	−469.15	48.1	−418.09
Na$_2$CO$_3$(s)	−1130.77	134.79	−1048.08
Sulfur			
S(s, rhombic)	0	32.1	0
S(g)	278.98	167.83	236.51
S$_2$Cl$_2$(g)	−18.4	331.5	−31.8
SF$_6$(g)	−1209	291.82	−1105.3
H$_2$S(g)	−20.63	205.79	−33.56
SO$_2$(g)	−296.84	248.21	−300.13
SO$_3$(g)	−395.77	256.77	−371.04
SOCl$_2$(g)	−212.5	309.77	−198.3
H$_2$SO$_4$(ℓ)	−814	156.9	−689.96
H$_2$SO$_4$(aq)	−909.27	20.1	−744.53
Tin			
Sn(s, white)	0	51.08	0
Sn(s, gray)	−2.09	44.14	0.13
SnCl$_4$(ℓ)	−511.3	258.6	−440.15
SnCl$_4$(g)	−471.5	365.8	−432.31
SnO$_2$(s)	−577.63	49.04	−515.88
Titanium			
Ti(s)	0	30.72	0
TiCl$_4$(ℓ)	−804.2	252.34	−737.2
TiCl$_4$(g)	−763.16	354.84	−726.7
TiO$_2$(s)	−939.7	49.92	−884.5
Zinc			
Zn(s)	0	41.63	0
ZnCl$_2$(s)	−415.05	111.46	−369.398
ZnO(s)	−348.28	43.64	−318.30
ZnS(s, sphalerite)	−205.98	57.7	−201.29

Appendix M

Standard Reduction Potentials in Aqueous Solution at 25 °C

Table 21   Standard Reduction Potentials in Aqueous Solution at 25 °C

Acidic Solution	Standard Reduction Potential, $E°$ (volts)
$F_2(g) + 2 e^- \longrightarrow 2 F^-(aq)$	2.87
$Co^{3+}(aq) + e^- \longrightarrow Co^{2+}(aq)$	1.82
$Pb^{4+}(aq) + 2 e^- \longrightarrow Pb^{2+}(aq)$	1.8
$H_2O_2(aq) + 2 H^+(aq) + 2 e^- \longrightarrow 2 H_2O$	1.77
$NiO_2(s) + 4 H^+(aq) + 2 e^- \longrightarrow Ni^{2+}(aq) + 2 H_2O$	1.7
$PbO_2(s) + SO_4^{2-}(aq) + 4 H^+(aq) + 2 e^- \longrightarrow PbSO_4(s) + 2 H_2O$	1.685
$Au^+(aq) + e^- \longrightarrow Au(s)$	1.68
$2 HClO(aq) + 2 H^+(aq) + 2 e^- \longrightarrow Cl_2(g) + 2 H_2O$	1.63
$Ce^{4+}(aq) + e^- \longrightarrow Ce^{3+}(aq)$	1.61
$NaBiO_3(s) + 6 H^+(aq) + 2 e^- \longrightarrow Bi^{3+}(aq) + Na^+(aq) + 3 H_2O$	≈ 1.6
$MnO_4^-(aq) + 8 H^+(aq) + 5 e^- \longrightarrow Mn^{2+}(aq) + 4 H_2O$	1.51
$Au^{3+}(aq) + 3 e^- \longrightarrow Au(s)$	1.50
$ClO_3^-(aq) + 6 H^+(aq) + 5 e^- \longrightarrow \frac{1}{2} Cl_2(g) + 3 H_2O$	1.47
$BrO_3^-(aq) + 6 H^+(aq) + 6 e^- \longrightarrow Br^-(aq) + 3 H_2O$	1.44
$Cl_2(g) + 2 e^- \longrightarrow 2 Cl^-(aq)$	1.36
$Cr_2O_7^{2-}(aq) + 14 H^+(aq) + 6 e^- \longrightarrow 2 Cr^{3+}(aq) + 7 H_2O$	1.33
$N_2H_5^+(aq) + 3 H^+(aq) + 2 e^- \longrightarrow 2 NH_4^+(aq)$	1.24
$MnO_2(s) + 4 H^+(aq) + 2 e^- \longrightarrow Mn^{2+}(aq) + 2 H_2O$	1.23
$O_2(g) + 4 H^+(aq) + 4 e^- \longrightarrow 2 H_2O$	1.229
$Pt^{2+}(aq) + 2 e^- \longrightarrow Pt(s)$	1.2
$IO_3^-(aq) + 6 H^+(aq) + 5 e^- \longrightarrow \frac{1}{2} I_2(aq) + 3 H_2O$	1.195
$ClO_4^-(aq) + 2 H^+(aq) + 2 e^- \longrightarrow ClO_3^-(aq) + H_2O$	1.19
$Br_2(\ell) + 2 e^- \longrightarrow 2 Br^-(aq)$	1.08
$AuCl_4^-(aq) + 3 e^- \longrightarrow Au(s) + 4 Cl^-(aq)$	1.00
$Pd^{2+}(aq) + 2 e^- \longrightarrow Pd(s)$	0.987
$NO_3^-(aq) + 4 H^+(aq) + 3 e^- \longrightarrow NO(g) + 2 H_2O$	0.96
$NO_3^-(aq) + 3 H^+(aq) + 2 e^- \longrightarrow HNO_2(aq) + H_2O$	0.94
$2 Hg^+(aq) + 2 e^- \longrightarrow Hg_2^{2+}(aq)$	0.920
$Hg^{2+}(aq) + 2 e^- \longrightarrow Hg(\ell)$	0.855
$Ag^+(aq) + e^- \longrightarrow Ag(s)$	0.7994
$Hg_2^{2+}(aq) + 2 e^- \longrightarrow 2 Hg(\ell)$	0.789
$Fe^{3+}(aq) + e^- \longrightarrow Fe^{2+}(aq)$	0.771

(continued)

Table 21   *(continued)*

Acidic Solution	Standard Reduction Potential, $E°$ (volts)
$SbCl_6^-(aq) + 2\ e^- \longrightarrow SbCl_4^-(aq) + 2\ Cl^-(aq)$	0.75
$[PtCl_4]^{2+}(aq) + 2\ e^- \longrightarrow Pt(s) + 4\ Cl^-(aq)$	0.73
$O_2(g) + 2\ H^+(aq) + 2\ e^- \longrightarrow H_2O_2(aq)$	0.682
$[PtCl_6]^{2-}(aq) + 2\ e^- \longrightarrow [PtCl_4]^{2-}(aq) + 2\ Cl^-(aq)$	0.68
$H_3AsO_4(aq) + 2\ H^+(aq) + 2\ e^- \longrightarrow H_3AsO_3(aq) + H_2O$	0.58
$I_2(s) + 2\ e^- \longrightarrow 2\ I^-(aq)$	0.535
$TeO_2(s) + 4\ H^+(aq) + 4\ e^- \longrightarrow Te(s) + 2\ H_2O$	0.529
$Cu^+(aq) + e^- \longrightarrow Cu(s)$	0.521
$[RhCl_6]^{3-}(aq) + 3\ e^- \longrightarrow Rh(s) + 6\ Cl^-(aq)$	0.44
$Cu^{2+}(aq) + 2\ e^- \longrightarrow Cu(s)$	0.337
$Hg_2Cl_2(s) + 2\ e^- \longrightarrow 2\ Hg(\ell) + 2\ Cl^-(aq)$	0.27
$AgCl(s) + e^- \longrightarrow Ag(s) + Cl^-(aq)$	0.222
$SO_4^{2-}(aq) + 4\ H^+(aq) + 2\ e^- \longrightarrow SO_2(g) + 2\ H_2O$	0.20
$SO_4^{2-}(aq) + 4\ H^+(aq) + 2\ e^- \longrightarrow H_2SO_3(aq) + H_2O$	0.17
$Cu^{2+}(aq) + e^- \longrightarrow Cu^+(aq)$	0.153
$Sn^{4+}(aq) + 2\ e^- \longrightarrow Sn^{2+}(aq)$	0.15
$S(s) + 2\ H^+ + 2\ e^- \longrightarrow H_2S(aq)$	0.14
$AgBr(s) + e^- \longrightarrow Ag(s) + Br^-(aq)$	0.0713
$2\ H^+(aq) + 2\ e^- \longrightarrow H_2(g)(\text{reference electrode})$	0.0000
$N_2O(g) + 6\ H^+(aq) + H_2O + 4\ e^- \longrightarrow 2\ NH_3OH^+(aq)$	−0.05
$Pb^{2+}(aq) + 2\ e^- \longrightarrow Pb(s)$	−0.126
$Sn^{2+}(aq) + 2\ e^- \longrightarrow Sn(s)$	−0.14
$AgI(s) + e^- \longrightarrow Ag(s) + I^-(aq)$	−0.15
$[SnF_6]^{2-}(aq) + 4\ e^- \longrightarrow Sn(s) + 6\ F^-(aq)$	−0.25
$Ni^{2+}(aq) + 2\ e^- \longrightarrow Ni(s)$	−0.25
$Co^{2+}(aq) + 2\ e^- \longrightarrow Co(s)$	−0.28
$Tl^+(aq) + e^- \longrightarrow Tl(s)$	−0.34
$PbSO_4(s) + 2\ e^- \longrightarrow Pb(s) + SO_4^{2-}(aq)$	−0.356
$Se(s) + 2\ H^+(aq) + 2\ e^- \longrightarrow H_2Se(aq)$	−0.40
$Cd^{2+}(aq) + 2\ e^- \longrightarrow Cd(s)$	−0.403
$Cr^{3+}(aq) + e^- \longrightarrow Cr^{2+}(aq)$	−0.41
$Fe^{2+}(aq) + 2\ e^- \longrightarrow Fe(s)$	−0.44
$2\ CO_2(g) + 2\ H^+(aq) + 2\ e^- \longrightarrow H_2C_2O_4(aq)$	−0.49
$Ga^{3+}(aq) + 3\ e^- \longrightarrow Ga(s)$	−0.53
$HgS(s) + 2\ H^+(aq) + 2\ e^- \longrightarrow Hg(\ell) + H_2S(g)$	−0.72
$Cr^{3+}(aq) + 3\ e^- \longrightarrow Cr(s)$	−0.74
$Zn^{2+}(aq) + 2\ e^- \longrightarrow Zn(s)$	−0.763
$Cr^{2+}(aq) + 2\ e^- \longrightarrow Cr(s)$	−0.91
$FeS(s) + 2\ e^- \longrightarrow Fe(s) + S^{2-}(aq)$	−1.01
$Mn^{2+}(aq) + 2\ e^- \longrightarrow Mn(s)$	−1.18
$V^{2+}(aq) + 2\ e^- \longrightarrow V(s)$	−1.18
$CdS(s) + 2\ e^- \longrightarrow Cd(s) + S^{2-}(aq)$	−1.21
$ZnS(s) + 2\ e^- \longrightarrow Zn(s) + S^{2-}(aq)$	−1.44
$Zr^{4+}(aq) + 4\ e^- \longrightarrow Zr(s)$	−1.53

(continued)

Table 21   *(continued)*

Acidic Solution	Standard Reduction Potential, $E°$ (volts)
$Al^{3+}(aq) + 3\ e^- \longrightarrow Al(s)$	-1.66
$Mg^{2+}(aq) + 2\ e^- \longrightarrow Mg(s)$	-2.37
$Na^+(aq) + e^- \longrightarrow Na(s)$	-2.714
$Ca^{2+}(aq) + 2\ e^- \longrightarrow Ca(s)$	-2.87
$Sr^{2+}(aq) + 2\ e^- \longrightarrow Sr(s)$	-2.89
$Ba^{2+}(aq) + 2\ e^- \longrightarrow Ba(s)$	-2.90
$Rb^+(aq) + e^- \longrightarrow Rb(s)$	-2.925
$K^+(aq) + e^- \longrightarrow K(s)$	-2.925
$Li^+(aq) + e^- \longrightarrow Li(s)$	-3.045

Basic Solution

	Standard Reduction Potential, $E°$ (volts)
$ClO^-(aq) + H_2O + 2\ e^- \longrightarrow Cl^-(aq) + 2\ OH^-(aq)$	0.89
$OOH^-(aq) + H_2O + 2\ e^- \longrightarrow 3\ OH^-(aq)$	0.88
$2\ NH_2OH(aq) + 2\ e^- \longrightarrow N_2H_4(aq) + 2\ OH^-(aq)$	0.74
$ClO_3^-(aq) + 3\ H_2O + 6\ e^- \longrightarrow Cl^-(aq) + 6\ OH^-(aq)$	0.62
$MnO_4^-(aq) + 2\ H_2O + 3\ e^- \longrightarrow MnO_2(s) + 4\ OH^-(aq)$	0.588
$MnO_4^-(aq) + e^- \longrightarrow MnO_4^{2-}(aq)$	0.564
$NiO_2(s) + 2\ H_2O + 2\ e^- \longrightarrow Ni(OH)_2(s) + 2\ OH^-(aq)$	0.49
$Ag_2CrO_4(s) + 2\ e^- \longrightarrow 2\ Ag(s) + CrO_4^{2-}(aq)$	0.446
$O_2(g) + 2\ H_2O + 4\ e^- \longrightarrow 4\ OH^-(aq)$	0.40
$ClO_4^-(aq) + H_2O + 2\ e^- \longrightarrow ClO_3^-(aq) + 2\ OH^-(aq)$	0.36
$Ag_2O(s) + H_2O + 2\ e^- \longrightarrow 2\ Ag(s) + 2\ OH^-(aq)$	0.34
$2\ NO_2^-(aq) + 3\ H_2O + 4\ e^- \longrightarrow N_2O(g) + 6\ OH^-(aq)$	0.15
$N_2H_4(aq) + 2\ H_2O + 2\ e^- \longrightarrow 2\ NH_3(aq) + 2\ OH^-(aq)$	0.10
$[Co(NH_3)_6]^{3+}(aq) + e^- \longrightarrow [Co(NH_3)_6]^{2+}(aq)$	0.10
$HgO(s) + H_2O + 2\ e^- \longrightarrow Hg(\ell) + 2\ OH^-(aq)$	0.0984
$O_2(g) + H_2O + 2\ e^- \longrightarrow OOH^-(aq) + OH^-(aq)$	0.076
$NO_3^-(aq) + H_2O + 2\ e^- \longrightarrow NO_2^-(aq) + 2\ OH^-(aq)$	0.01
$MnO_2(s) + 2\ H_2O + 2\ e^- \longrightarrow Mn(OH)_2(s) + 2\ OH^-(aq)$	-0.05
$CrO_4^{2-}(aq) + 4\ H_2O + 3\ e^- \longrightarrow Cr(OH)_3(s) + 5\ OH^-(aq)$	-0.12
$Cu(OH)_2(s) + 2\ e^- \longrightarrow Cu(s) + 2\ OH^-(aq)$	-0.36
$S(s) + 2\ e^- \longrightarrow S^{2-}(aq)$	-0.48
$Fe(OH)_3(s) + e^- \longrightarrow Fe(OH)_2(s) + OH^-(aq)$	-0.56
$2\ H_2O + 2\ e^- \longrightarrow H_2(g) + 2\ OH^-(aq)$	-0.8277
$2\ NO_3^-(aq) + 2\ H_2O + 2\ e^- \longrightarrow N_2O_4(g) + 4\ OH^-(aq)$	-0.85
$Fe(OH)_2(s) + 2\ e^- \longrightarrow Fe(s) + 2\ OH^-(aq)$	-0.877
$SO_4^{2-}(aq) + H_2O + 2\ e^- \longrightarrow SO_3^{2-}(aq) + 2\ OH^-(aq)$	-0.93
$N_2(g) + 4\ H_2O + 4\ e^- \longrightarrow N_2H_4(aq) + 4\ OH^-(aq)$	-1.15
$[Zn(OH)_4]^{2-}(aq) + 2\ e^- \longrightarrow Zn(s) + 4\ OH^-(aq)$	-1.22
$Zn(OH)_2(s) + 2\ e^- \longrightarrow Zn(s) + 2\ OH^-(aq)$	-1.245
$[Zn(CN)_4]^{2-}(aq) + 2\ e^- \longrightarrow Zn(s) + 4\ CN^-(aq)$	-1.26
$Cr(OH)_3(s) + 3\ e^- \longrightarrow Cr(s) + 3\ OH^-(aq)$	-1.30
$SiO_3^{2-}(aq) + 3\ H_2O + 4\ e^- \longrightarrow Si(s) + 6\ OH^-(aq)$	-1.70

Appendix N

Answers to Exercises

Chapter 1

1.1 Based on appearance, the bottle on the right appears to contain a homogeneous solution. The pile on the left appears to contain a heterogeneous mixture of at least two solid substances.

1.2 (a) Na = sodium; Cl = chlorine; Cr = chromium
(b) Zinc = Zn; nickel = Ni; potassium = K

1.3 (a) Iron: lustrous solid, metallic, good conductor of heat and electricity, malleable, ductile, attracted to a magnet
(b) Water: colorless liquid (at room temperature), melting point is 0 °C and boiling point is 100 °C, density ~ 1 g/cm^2
(c) Table salt: solid, white crystals, soluble in water
(d) Oxygen: colorless gas (at room temperature), low solubility in water

1.4 15.5 g (1 cm^3/1.18 × 10^{-3} g) = 1.31 × 10^4 cm^3

1.5 Density decreases by about 0.025 g/cm^3 for each 10 °C increase in temperature. The density at 30 °C is expected to be about 13.546 g/cm^3 − 0.025 g/cm^3 = 13.521 g/cm^3.

1.6 Chemical changes: the fuel in the campfire burns in air (combustion). Physical changes: water boils. Energy evolved in combustion is transferred to the water, to the water container, and to the surrounding air.

1.7 77 K − 273.15 K (1 °C/K) = −196 °C

1.8 Length: 25.3 cm (1 m/100 cm) = 0.253 m;
25.3 cm (10 mm/1 cm) = 253 mm

Width: 21.6 cm (1 m/100 cm) = 0.216 m;
21.6 cm (10 mm/1 cm) = 216 mm

Area = length × width = (25.3 cm)(21.6 cm) = 546 cm^2

546 cm^2 (1 m/100 cm)^2 = 0.0546 m^2

1.9 Area of sheet = (2.50 cm)^2 = 6.25 cm^2

Volume = 1.656 g (1 cm^3/21.45 g) = 0.07720 cm^3

Thickness = volume/area = (0.07720 cm^3/6.25 cm^2)
= 0.0124 cm

0.0124 cm (10 mm/1 cm) = 0.124 mm

1.10 (a) 750 mL (1 L/1000 mL) = 0.75 L
0.75 L (10 dL/L) = 7.5 dL
(b) 2.0 qt = 0.50 gal
0.50 gal (3.786 L/gal) = 1.9 L
1.9 L (1 dm^3/1 L) = 1.9 dm^3

1.11 (a) Mass in kilograms = 5.59 g (1 kg/1000 g)
= 0.00559 kg

Mass in milligrams = 5.59 g (10^3 mg/g)
= 5.59 × 10^{+3} mg
(b) 0.02 μg/L (1 g/10^6 μg) = 2 × 10^{-8} g/L

1.12 Student A: average = −0.1 °C; average deviation = 0.2 °C; error = −0.1 °C. Student B: average = 273.16 K; average deviation = 0.02 K; error = +0.01 K. Student B's values are more accurate and more precise.

1.13 (a) 2.33 × 10^7 has three significant figures; 50.5 has three significant figures; 200 has one significant figure. (200. would express this number with three significant figures.)
(b) The product of 10.26 and 0.063 is 0.65, a number with two significant figures. (10.26 has four significant figures, whereas 0.063 has two.)
The sum of 10.26 and 0.063 is 10.32. The number 10.26 has only two numbers to the right of the decimal, so the sum must also have two numbers after the decimal.
(c) x = 3.9 × 10^6. The difference between 110.7 and 64 is 47. Dividing 47 by 0.056 and 0.00216 gives an answer with two significant figures.

1.14 (a) 198 cm (1 m/100 cm) = 1.98 m;
198 cm (1 ft/30.48 cm) = 6.50 ft
(b) 2.33 × 10^7 m^2(1 km^2/10^6 m^2) = 23.3 km^2
(c) 19,320 kg/m^3(10^3 g/1 kg)(1 m^3/10^6 cm^3) =
19.32 g/cm^3
(d) 9.0 × 10^3 pc(206,265 AU/1 pc)(1.496 × 10^8 km/
1 AU) = 2.8 × 10^{17} km

1.15 Read from the graph, the mass of 50 beans is about 123 g.

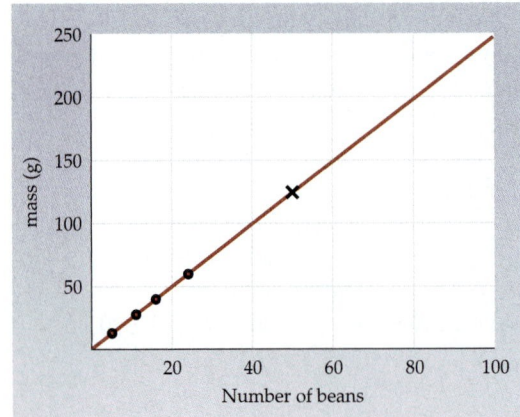

1.16 Change dimensions to centimeters: 7.6 m = 760 cm; 2.74 m = 274 cm; 0.13 mm = 0.013 cm.

Volume of paint = (760 cm)(274 cm)(0.013 cm)
$$= 2.7 \times 10^3 \text{ cm}^3$$

Volume (L) = $(2.7 \times 10^3 \text{ cm}^3)(1 \text{ L}/10^3 \text{ cm}^3) = 2.7 \text{ L}$

Mass = $(2.7 \times 10^3 \text{ cm}^3)(0.914 \text{ g/cm}^3) = 2.5 \times 10^3 \text{ g}$

Chapter 2

2.1 The ratio of the atom radius to the radius of the nucleus is 1×10^5 to 1. If the radius of the atom is 100 m, then the radius of the nucleus is 0.0010 m or 1.0 mm. The head of an ordinary pin has a diameter of about 1.0 mm.

2.2 (a) Mass number with 26 protons and 30 neutrons is 56

   (b) 59.930788 u $(1.661 \times 10^{-24} \text{ g/u}) = 9.955 \times 10^{-23}$ g

   (c) ^{64}Zn has 30 protons, 30 electrons, and $(64 - 30) = 34$ neutrons.

2.3 (a) Argon has an atomic number of 18. ^{36}Ar, ^{38}Ar, ^{40}Ar

   (b) ^{69}Ga: 31 protons, 38 neutrons; ^{71}Ga: 31 protons, 40 neutrons; % abundance of ^{71}Ga = 39.9 %

2.4 $(0.7577)(34.96885 \text{ u}) + (0.2423)(36.96590 \text{ u}) = 35.45$ u

2.5 (a) 1.5 mol Si (28.1 g/mol) = 42 g Si

   (b) 454 g S (1 mol S/32.07 g) = 14.2 mol S

   14.2 mol S $(6.022 \times 10^{23} \text{ atoms/mol}) = 8.53 \times 10^{24}$ atoms S

   (c) (32.07 g S/1 mol S) $(1 \text{ mol S}/6.022 \times 10^{23} \text{ atoms}) = 5.325 \times 10^{-23}$ g/atom

2.6 2.6×10^{24} atoms $(1 \text{ mol}/6.022 \times 10^{23} \text{ atoms})$ (197.0 g Au/1 mol) = 850 g Au

   Volume = 850 g Au $(1 \text{ cm}^3/19.32 \text{ g}) = 44 \text{ cm}^3$

   Volume = 44 cm^3 = (thickness)(area) = (0.10 cm)(area)

   Area = 440 cm^2

   Length = width = $(440 \text{ cm}^2)^{1/2} = 21$ cm

2.7 There are eight elements in the third period. Sodium (Na), magnesium (Mg), and aluminum (Al) are metals. Silicon (Si) is a metalloid. Phosphorus (P), sulfur (S), chlorine (Cl), and argon (Ar) are nonmetals.

Chapter 3

3.1 The molecular formula for styrene is C_8H_8; the condensed formula, $C_6H_5CH{=}CH_2$, contains the same information and is more descriptive of its structure.

3.2 The molecular formula is $C_3H_7NO_2S$. You will often see its formula written as $HSCH_2CH(N^+H_3)CO_2^-$ to emphasize the molecule's structure.

3.3 (a) K^+ is formed if K loses one electron. K^+ has the same number of electrons as Ar.

   (b) Se^{2-} is formed by adding two electrons to an atom of Se. It has the same number of electrons as Kr.

   (c) Ba^{2+} is formed if Ba loses two electrons; Ba^{2+} has the same number of electrons as Xe.

   (d) Cs^+ is formed if Cs loses one electron. It has the same number of electrons as Xe.

3.4 (a) (1) NaF: 1 Na^+ and 1 F^- ion. (2) $Cu(NO_3)_2$: 1 Cu^{2+} and 2 NO_3^- ions. (3) $NaCH_3CO_2$: 1 Na^+ and 1 $CH_3CO_2^-$ ion.

   (b) $FeCl_2$, $FeCl_3$

   (c) Na_2S, Na_3PO_4, BaS, $Ba_3(PO_4)_2$

3.5 (1) (a) NH_4NO_3; (b) $CoSO_4$; (c) $Ni(CN)_2$; (d) V_2O_3; (e) $Ba(CH_3CO_2)_2$; (f) $Ca(ClO)_2$

   (2) (a) magnesium bromide; (b) lithium carbonate; (c) potassium hydrogen sulfite; (d) potassium permanganate; (e) ammonium sulfide; (f) copper(I) chloride and copper(II) chloride

3.6 The force of attraction between ions is proportional to the product of the ion charges (Coulomb's law). The force of attraction between Mg^{2+} and O^{2-} ions in MgO is approximately four times greater than the force of attraction between Na^+ and Cl^- ions in NaCl, so a much higher temperature is required to disrupt the orderly array of ions in crystalline MgO.

3.7 (1) (a) CO_2; (b) PI_3; (c) SCl_2; (d) BF_3; (e) O_2F_2; (f) XeO_3

   (2) (a) dinitrogen tetrafluoride; (b) hydrogen bromide; (c) sulfur tetrafluoride; (d) boron trichloride; (e) tetraphosphorus decaoxide; (f) chlorine trifluoride

3.8 (a) Citric acid: 192.1 g/mol; magnesium carbonate: 84.3 g/mol

   (b) 454 g citric acid (1 mol /192.1 g) = 2.36 mol citric acid

   (c) 0.125 mol $MgCO_3$ (84.3 g/mol) = 10.5 g $MgCO_3$

3.9 (a) 1.00 mol $(NH_4)_2CO_3$ (molar mass 96.09 g/mol) has 28.0 g of N (29.2%), 8.06 g of H (8.39%), 12.0 g of C (12.5%), and 48.0 g of O (50.0%)

   (b) 454 g C_8H_{18} (1 mol C_8H_{18}/114.2 g) (8 mol C/1 mol C_8H_{18})(12.01 g C/1 mol C) = 382 g C

3.10 (a) C_5H_4 (b) $C_2H_4O_2$

3.11 88.17 g C (1 mol C/12.011 g C) = 7.341 mol C

   11.83 g H (1 mol H/1.008 g H) = 11.74 mol H

   11.74 mol H/7.341 mol C = 1.6 mol H/1 mol C = (8/5)(mol H/1 mol C) = 8 mol H/5 mol C

   The empirical formula is C_5H_8. The molar mass, 68.11 g/mol, closely matches this formula, so C_5H_8 is also the molecular formula.

3.12 78.90 g C (1 mol C/12.011 g C) = 6.569 mol C

   10.59 g H (1 mol H/1.008 g H) = 10.51 mol H

   10.51 g O (1 mol O/16.00 g O) = 0.6569 mol O

   10.51 mol H/0.6569 mol O = 16 mol H/1 mol O

   6.569 mol C/0.6569 mol O = 10 mol C/1 mol O

   The empirical formula is $C_{10}H_{16}O$.

3.13 0.586 g K (1 mol K/39.10 g K) = 0.0150 mol K

0.480 g O(1 mol O/16.00 g O) = 0.0300 mol O

The ratio of moles K to moles O atoms is 1 to 2; the empirical formula is KO_2.

3.14 Mass of water lost on heating is 0.235 g − 0.128 g = 0.107 g

0.107 g H_2O (1 mol H_2O/18.016 g H_2O) = 0.00594 mol H_2O

0.128 g $NiCl_2$ (1 mol $NiCl_2$/129.6 g $NiCl_2$) = 0.000988 mol $NiCl_2$

Mole ratio = 0.00594 mol H_2O/0.000988 mol $NiCl_2$ = 6.01

The formula for the hydrate is $NiCl_2 \cdot 6\ H_2O$.

Chapter 4

4.1 (a) Reactants: Al, aluminum, a solid, and Br_2, bromine, a liquid. Product: Al_2Br_6, dialuminum hexabromide, a solid

(b) Stoichiometric coefficients: 2 for Al, 3 for Br_2, and 1 for Al_2Br_6

(c) 8000 atoms of Al requires (3/2)8000 = 12,000 molecules of Br_2

4.2 (a) 2 C_4H_{10}(g) + 13 O_2(g) ⟶ 8 CO_2(g) + 10 H_2O(g)

(b) 2 $Pb(C_2H_5)_4$(ℓ) + 27 O_2 (g) ⟶
      2 PbO(s) + 16 CO_2(g) + 20 H_2O(ℓ)

4.3 454 g C_3H_8 (1 mol C_3H_8/44.10 g C_3H_8) = 10.3 mol C_3H_8

10.3 mol C_3H_8 (5 mol O_2/1 mol C_3H_8) (32.00 g O_2/1 mol O_2) = 1650 g O_2

10.3 mol C_3H_8 (3 mol CO_2/1 mol C_3H_8) (44.01 g CO_2/1 mol CO_2) = 1360 g CO_2

10.3 mol C_3H_8 (4 mol H_2O/1 mol C_3H_8) (18.02 g H_2O/1 mol H_2O) = 742 g H_2O

4.4 Amount C = 125 g C (1 mol C/12.01 g C) = 10.4 mol C

Amount Cl_2 = 125 g Cl_2 (1 mol Cl_2/70.91 g Cl_2)
        = 1.76 mol Cl_2

Mol C/mol Cl_2 = 10.41/1.763 = 5.90

This is more than the 1 : 2 ratio required, so the limiting reactant is Cl_2.

Mass $TiCl_4$ = 1.763 mol Cl_2 (1 mol $TiCl_4$/2 mol Cl_2) (189.7 g $TiCl_4$/1 mol $TiCl_4$) = 167 g $TiCl_4$

4.5 (a) Amount Al = 50.0 g Al (1 mol Al/26.98 g Al)
          = 1.85 mol Al

Amount Fe_2O_3 = 50.0 g Fe_2O_3 (1 mol Fe_2O_3/159.7 g Fe_2O_3)
          = 0.313 mol Fe_2O_3

Mol Al/mol Fe_2O_3 = 1.853/0.3131 = 5.92

This is more than the 2 : 1 ratio required, so the limiting reactant is Fe_2O_3.

Mass Fe = 0.313 mol Fe_2O_3 (2 mol Fe/1 mol Fe_2O_3) (55.85 g Fe/1 mol Fe) = 35.0 g Fe

4.6 Theoretical yield = 125 g CH_3OH (1 mol CH_3OH/32.04 g CH_3OH) (2 mol H_2/1 mol CH_3OH) (2.016 g H_2/1 mol H_2) = 15.7 g H_2

Percent yield = (13.6 g/15.7 g)(100%) = 86.5%

4.7 0.143 g O_2 (1 mol O_2/32.00 g O_2) (3 mol TiO_2/3 mol O_2) (79.88 g TiO_2/1 mol TiO_2) = 0.357 g TiO_2

Percent TiO_2 in sample = (0.357 g/2.367 g)(100%)
        = 15.1%

4.8 1.612 g CO_2 (1 mol CO_2/44.01 g CO_2)(1 mol C/1 mol CO_2) = 0.03663 mol C

0.7425 g H_2O (1 mol H_2O/18.01 g H_2O)(2 mol H/1 mol H_2O) = 0.08243 mol H

0.08243 mol H/0.03663 mol = 2.250 H/1 C = 9 H/4 C

The empirical formula is C_4H_9, which has a molar mass of 57 g/mol. This is one half of the molar mass, so the molecular formula is $(C_4H_9)_2$ or C_8H_{18}.

4.9 0.240 g CO_2 (1 mol CO_2/44.01 g CO_2) (1 mol C/1 mol CO_2)(12.01 g C/1 mol C) = 0.06549 g C

0.0982 g H_2O (1 mol H_2O/18.02 g H_2O)(2 mol H/1 mol H_2O)(1.008 g H/1 mol H) = 0.01099 g H

Mass O (by difference) = 0.1342 g − 0.06549 g − 0.01099 g
        = 0.05772 g

Amount C = 0.06549 g (1 mol C/12.01 g C)
        = 0.00545 mol C

Amount H = 0.01099 g H (1 mol H/1.008 g H)
        = 0.01090 mol H

Amount O = 0.05772 g O (1 mol O/16.00 g O)
        = 0.00361 mol O

To find a whole-number ratio, divide each value by 0.00361; this gives 1.51 mol C : 3.02 mol H : 1 mol O. Multiply each value by 2 and round off to 3 mol C : 6 mol H : 2 mol O. The empirical formula is $C_3H_6O_2$; given the molar mass of 74.1, this is also the molecular formula.

Chapter 5

5.1 Epsom salt is an electrolyte and methanol is a nonelectrolyte.

5.2 (a) $LiNO_3$ is soluble and gives Li^+(aq) and NO_3^-(aq) ions.

(b) $CaCl_2$ is soluble and gives Ca^{2+}(aq) and Cl^-(aq) ions.

(c) CuO is not water-soluble.

(d) $NaCH_3CO_2$ is soluble and gives Na^+(aq) and $CH_3CO_2^-$(aq) ions.

5.3 (a) Na_2CO_3(aq) + $CuCl_2$(aq) ⟶
                2 NaCl(aq) + $CuCO_3$(s)

(b) No reaction; no insoluble compound is produced.

(c) $NiCl_2$(aq) + 2 KOH(aq) ⟶
                $Ni(OH)_2$(s) + 2 KCl(aq)

5.4 (a) $AlCl_3$(aq) + Na_3PO_4(aq) ⟶ $AlPO_4$(s) + 3 NaCl(aq)

Al^{3+}(aq) + PO_4^{3-}(aq) ⟶ $AlPO_4$(s)

(b) $FeCl_3(aq) + 3 KOH(aq) \longrightarrow$
$$Fe(OH)_3(s) + 3 KCl(aq)$$
$Fe^{3+}(aq) + 3 OH^-(aq) \longrightarrow Fe(OH)_3(s)$

(c) $Pb(NO_3)_2(aq) + 2 KCl(aq) \longrightarrow$
$$PbCl_2(s) + 2 KNO_3(aq)$$
$Pb^{2+}(aq) + 2 Cl^-(aq) \longrightarrow PbCl_2(s)$

5.5 (a) $H^+(aq)$ and $NO_3^-(aq)$
(b) $Ba^{2+}(aq)$ and $2 OH^-(aq)$

5.6 Metals form basic oxides; nonmetals form acidic oxides.
(a) SeO_2 is an acidic oxide; (b) MgO is a basic oxide; and (c) P_4O_{10} is an acidic oxide.

5.7 $Mg(OH)_2(s) + 2 HCl(aq) \longrightarrow MgCl_2(aq) + 2 H_2O(\ell)$

Net ionic equation: $Mg(OH)_2(s) + 2 H^+(aq) \longrightarrow$
$$Mg^{2+}(aq) + 2 H_2O(\ell)$$

5.8 (a) $BaCO_3(s) + 2 HNO_3(aq) \longrightarrow$
$$Ba(NO_3)_2(aq) + CO_2(g) + H_2O(\ell)$$
Barium carbonate and nitric acid produce barium nitrate, carbon dioxide, and water.
(b) $(NH_4)_2SO_4(aq) + 2 NaOH(aq) \longrightarrow$
$$2 NH_3(g) + Na_2SO_4(aq) + 2 H_2O(\ell)$$

5.9 (a) Gas-forming reaction: $CuCO_3(s) + H_2SO_4(aq) \longrightarrow$
$$CuSO_4(aq) + H_2O(\ell) + CO_2(g)$$
Net ionic equation: $CuCO_3(s) + 2 H^+(aq) \longrightarrow$
$$Cu^{2+}(aq) + H_2O(\ell) + CO_2(g)$$
(b) Acid–base reaction: $Ba(OH)_2(s) + 2 HNO_3(aq) \longrightarrow$
$$Ba(NO_3)_2(aq) + 2 H_2O(\ell)$$
Net ionic equation: $Ba(OH)_2(s) + 2 H^+(aq) \longrightarrow$
$$Ba^{2+}(aq) + 2 H_2O(\ell)$$
(c) Precipitation reaction: $CuCl_2(aq) + (NH_4)_2S(aq) \longrightarrow$
$$CuS(s) + 2 NH_4Cl(aq)$$
Net ionic equation: $Cu^{2+}(aq) + S^{2-}(aq) \longrightarrow CuS(s)$

5.10 (a) Fe in Fe_2O_3, +3; (b) S in H_2SO_4, +6;
(c) C in CO_3^{2-}, +4; (d) N in NO_2^+, +5

5.11 Dichromate ion is the oxidizing agent and is reduced. (Cr with a +6 oxidation number is reduced to Cr^{3+} with a +3 oxidation number). Ethanol is the reducing agent and is oxidized. (The C atoms in ethanol have an oxidation number of −2, whereas this number is 0 in acetic acid.)

5.12 (a) Acid–base reaction ($H^+ + OH^- \longrightarrow H_2O$).
(b) Oxidation–reduction reaction; Cu is oxidized (oxidation number changes from 0 in Cu to +2 in $CuCl_2$), and Cl_2 is reduced (oxidation number for each Cl changes from 0 to −1). Cu is the reducing agent and Cl_2 is the oxidizing agent.
(c) Gas-forming reaction (gaseous CO_2 is evolved).
(d) Oxidation–reduction reaction; S in $S_2O_3^{2-}$ is oxidized (oxidation number changes from +2 in $S_2O_3^{2-}$ to +2.5 in $S_4O_6^{2-}$), and I_2 is reduced (oxidation number of each I atom changes from 0 to −1). $S_2O_3^{2-}$ is the reducing agent and I_2 is the oxidizing agent.

5.13 26.3 g (1 mol $NaHCO_3$/84.01 g $NaHCO_3$) = 0.313 mol $NaHCO_3$

0.313 mol $NaHCO_3$/0.200 L = 1.57 M

Ion concentrations: $[Na^+] = [HCO_3^-] = 1.57$ M

5.14 First, determine the mass of $AgNO_3$ required.

Amount of $AgNO_3$ required = (0.0200 M)(0.250 L)
$$= 5.00 \times 10^{-3} \text{ mol}$$

Mass of $AgNO_3 = (5.00 \times 10^{-3} \text{ mol})(169.9 \text{ g/mol})$
$$= 0.850 \text{ g } AgNO_3$$

Weigh out 0.850 g $AgNO_3$. Then, dissolve it in a small amount of water in the volumetric flask. After the solid is dissolved, fill the flask to the mark.

5.15 (0.15 M)(0.0060 L) = (0.0100 L)(c_{dilute})
$c_{\text{dilute}} = 0.090$ M

5.16 (2.00 M)(V_{conc}) = (1.00 M)(0.250 L); $V_{\text{conc}} = 0.125$ L

To prepare the solution, measure accurately 125 mL of 2.00 M NaOH into a 250-mL volumetric flask, and add water to give a total volume of 250 mL.

5.17 (a) $pH = -\log(2.6 \times 10^{-2}) = 1.59$
(b) $-\log[H^+] = 3.80$; $[H^+] = 1.5 \times 10^{-4}$ M

5.18 HCl is the limiting reagent.

(0.350 mol HCl/1 L)(0.0750 L)(1 mol CO_2/2 mol HCl) (44.01 g CO_2/1 mol CO_2) = 0.578 g CO_2

5.19 (0.953 mol NaOH/1 L)(0.02833 L NaOH) = 0.0270 mol NaOH

(0.0270 mol NaOH)(1 mol CH_3CO_2H /1 mol NaOH) = 0.0270 mol CH_3CO_2H

(0.0270 mol CH_3CO_2H)(60.05 g/mol) = 1.62 g CH_3CO_2H

0.0270 mol CH_3CO_2H/0.0250 L = 1.08 M

5.20 (0.100 mol HCl/1 L)(0.02967 L) = 0.00297 mol HCl

(0.00297 mol HCl)(1 mol NaOH/1 mol HCl) = 0.00297 mol NaOH

0.00297 mol NaOH/0.0250 L = 0.119 M NaOH

5.21 Mol acid = mol base = (0.323 mol/L)(0.03008 L) = 9.716×10^{-3} mol

Molar mass = 0.856 g acid/9.716×10^{-3} mol acid = 88.1 g/mol

5.22 (0.196 mol $Na_2S_2O_3$/1 L)(0.02030 L) = 0.00398 mol $Na_2S_2O_3$

(0.00398 mol $Na_2S_2O_3$)(1 mol I_2/2 mol $Na_2S_2O_3$) = 0.00199 mol I_2

0.00199 mol I_2 is in excess, and was not used in the reaction with ascorbic acid.

I_2 originally added = (0.0520 mol I_2/1 L)(0.05000 L) = 0.00260 mol I_2

I_2 used in reaction with ascorbic acid = 0.00260 mol − 0.00199 mol = 6.1×10^{-4} mol I_2

(6.1×10^{-4} mol I_2)(1 mol $C_6H_8O_6$ /1 mol I_2)(176.1 g/ 1 mol) = 0.11 g $C_6H_8O_6$

Chapter 6

6.1 The chemical potential energy of a battery can be converted to work (to run a motor), heat (an electric space heater), and light (a light bulb).

6.2 (a) $(3800 \text{ calories})(4.184 \text{ J/calorie}) = 1.6 \times 10^4 \text{ J}$

(b) $(250 \text{ Calories})(1000 \text{ calories/Calorie})(4.184 \text{ J/calorie})(1 \text{ kJ}/1000 \text{ J}) = 1.0 \times 10^3 \text{ kJ}$

6.3 $C = 59.8 \text{ J}/[(25.0 \text{ g})(1.00 \text{ K})] = 2.39 \text{ J/g} \cdot \text{K}$

6.4 $(15.5 \text{ g})(C_{metal})(18.9\ °C - 100.0\ °C) + (55.5 \text{ g})(4.184 \text{ J/g} \cdot \text{K})(18.9\ °C - 16.5\ °C) = 0$

$C_{metal} = 0.44 \text{ J/g} \cdot \text{K}$

6.5 $(400. \text{ g iron})(0.449 \text{ J/g} \cdot \text{K})(32.8\ °C - T_{initial}) + (1000. \text{ g})(4.184 \text{ J/g} \cdot \text{K})(32.8\ °C - 20.0\ °C) = 0$

$T_{initial} = 331\ °C$

6.6 To heat methanol: $(25.0 \text{ g CH}_3\text{OH})(2.53 \text{ J/g} \cdot \text{K})(64.6\ °C - 25.0\ °C) = 2.50 \times 10^3 \text{ J}$

To evaporate methanol:
$(25.0 \text{ g CH}_3\text{OH})(2.00 \times 10^3 \text{ J/g}) = 5.00 \times 10^4 \text{ J}$

Total heat: $0.250 \times 10^4 \text{ J} + 5.00 \times 10^4 \text{ J} = 5.25 \times 10^4 \text{ J}$

6.7 Heat transferred from tea + heat to melt ice = 0

$(250 \text{ g})(4.2 \text{ J/g} \cdot \text{K})(273.2 \text{ K} - 291.4 \text{ K}) + x \text{ g}(333 \text{ J/g}) = 0$

$x = 57.4 \text{ g}$

57 g of ice melts with heat supplied by cooling 250 g of tea from 18.2 °C (291.4 K) to 0 °C (273.2 K)

Mass of ice remaining = mass of ice initially − mass of ice melted

Mass of ice remaining = 75 g − 57 g = 18 g

6.8 (a) $(12.6 \text{ g H}_2\text{O})(1 \text{ mol}/18.02 \text{ g})(285.8 \text{ kJ/mol}) = 2.00 \times 10^2 \text{ kJ}$

(b) $\Delta H = (15.0 \text{ g})(1 \text{ mol}/30.07 \text{ g})(-2857.3 \text{ kJ}/(2 \text{ mol})) = -713 \text{ kJ}$

6.9 Mass of final solution = 400. g; $\Delta T = 27.78\ °C - 25.10\ °C = 2.68\ °C = 2.68 \text{ K}$

Amount of HCl used = amount of NaOH used = $CV = 0.0800 \text{ mol}$

Heat transferred by acid–base reaction + heat absorbed to warm solution = 0

$q_{rxn} + (4.20 \text{ J/g} \cdot \text{K})(400. \text{ g})(2.68 \text{ K}) = 0$

$q_{rxn} = -4.50 \times 10^3 \text{ J}$

This represents the heat evolved in the reaction of 0.0800 mol HCl.

Heat per mole = $\Delta H_{rxn} = -4.50 \text{ kJ}/0.0800 \text{ mol HCl} = -56.3 \text{ kJ/mol of HCl}$

6.10 (a) Heat evolved in reaction + heat absorbed by H$_2$O + heat absorbed by bomb = 0

$q_{rxn} + (1.50 \times 10^3 \text{ g})(4.20 \text{ J/g} \cdot \text{K})(27.32\ °C - 25.00\ °C) + (837 \text{ J/K})(27.32 \text{ K} - 25.00 \text{ K}) = 0$

$q_{rxn} = -16,600 \text{ J}$ (heat evolved in burning 1.0 g sucrose)

(b) Heat per mole = $(-16.6 \text{ kJ/g})(342.2 \text{ g sucrose}/1 \text{ mol sucrose}) = -5650 \text{ kJ/mol}$

6.11 (a) $\text{C(graphite)} + \text{O}_2(g) \longrightarrow \text{CO}_2(g)$ $\Delta H_1 = -393.5 \text{ kJ}$

$\text{CO}_2(g) \longrightarrow \text{C(diamond)} + \text{O}_2(g)$ $\Delta H_2 = +395.4 \text{ kJ}$

Net: $\text{C(graphite)} \longrightarrow \text{C(diamond)}$

$\Delta H_{net} = \Delta H_1 + \Delta H_2 = +1.9 \text{ kJ}$

(b)

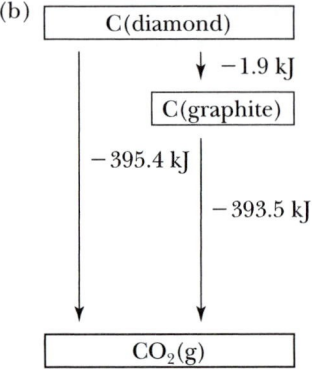

6.12 $\text{C(s)} + \text{O}_2(g) \longrightarrow \text{CO}_2(g)$ $\Delta H_1 = -393.5 \text{ kJ}$

$2 [\text{S(s)} + \text{O}_2(g) \longrightarrow \text{SO}_2(g)]$ $\Delta H_2 = 2(-296.8) = -593.6 \text{ kJ}$

$\text{CO}_2(g) + 2 \text{ SO}_2(g) \longrightarrow \text{CS}_2(g) + 3 \text{ O}_2(g)$

$\Delta H_3 = +1103.9 \text{ kJ}$

Net: $\text{C(s)} + 2 \text{ S(s)} \longrightarrow \text{CS}_2(g)$

$\Delta H_{net} = \Delta H_1 + \Delta H_2 + \Delta H_3 = +116.8 \text{ kJ}$

6.13 Standard states: $\text{Br}_2(\ell)$, $\text{Hg}(\ell)$, $\text{Na}_2\text{SO}_4(s)$, $\text{C}_2\text{H}_5\text{OH}(\ell)$

6.14 $\text{Fe(s)} + \frac{3}{2} \text{Cl}_2(g) \longrightarrow \text{FeCl}_3(s)$

$12 \text{ C(s, graphite)} + 11 \text{ H}_2(g) + \frac{11}{2} \text{O}_2(g) \longrightarrow \text{C}_{12}\text{H}_{22}\text{O}_{11}(s)$

6.15 $\Delta H°_{rxn} = 6 \Delta H°_f[\text{CO}_2(g)] + 3 \Delta H°_f[\text{H}_2\text{O}(\ell)] - \{\Delta H°_f[\text{C}_6\text{H}_6(\ell)] + \frac{15}{2} \Delta H°_f[\text{O}_2(g)]\}$
$= (6 \text{ mol})(-393.5 \text{ kJ/mol}) + (3 \text{ mol})(-285.8 \text{ kJ/mol}) - (1 \text{ mol})(+49.0 \text{ kJ/mol}) - 0$
$= -3267.4 \text{ kJ}$

6.16 (a) $\Delta H°_{rxn} = -2\Delta H°_f[\text{HBr}(g)]$
$= -(2 \text{ mol})(-36.29 \text{ kJ/mol}) = +72.58 \text{ kJ}$

The reaction is reactant-favored.

(b) $\Delta H°_{rxn} = \Delta H°_f[\text{C(graphite)}] - \Delta H°_f[\text{C(diamond)}]$
$= 0 - 1.8 \text{ kJ} = -1.8 \text{ kJ}$

The reaction is product-favored.

Chapter 7

7.1 (a) 10 cm

(b) 5 cm

(c) 4 waves. There are 9 nodes, 7 in the middle and 2 at the ends.

7.2 (a) Highest frequency, violet; lowest frequency, red

(b) The FM radio frequency, 91.7 MHz, is lower than the frequency of a microwave oven, 2.45 GHz.

(c) The wavelength of x-rays is shorter than the wavelength of ultraviolet light.

7.3 Orange light: 6.25×10^2 nm $= 6.25 \times 10^{-7}$ m

$\nu = (2.998 \times 10^8 \text{ m/s})/6.25 \times 10^{-7} \text{ m} = 4.80 \times 10^{14} \text{ s}^{-1}$

$E = (6.626 \times 10^{-34} \text{ J} \cdot \text{s/photon})(4.80 \times 10^{14} \text{ s}^{-1})$
$(6.022 \times 10^{23} \text{ photons/mol})$

$= 1.92 \times 10^5 \text{ J/mol}$

Microwave: $E = (6.626 \times 10^{-34} \text{ J} \cdot \text{s/photon})(2.45 \times 10^9 \text{ s}^{-1})(6.022 \times 10^{23} \text{ photons/mol}) = 0.978 \text{ J/mol}$

E(orange light)$/E$(microwave) $= 1.96 \times 10^5$; orange (625-nm) light is 196,000 times more energetic than 2.45-GHz microwaves.

7.4 (a) E (per atom) $= -Rhc/n^2 = (-2.179 \times 10^{-18})/(3^2)$
$\text{J/atom} = -2.421 \times 10^{-19} \text{ J/atom}$

(b) E (per mol) $= (-2.421 \times 10^{-19} \text{ J/atom})$
$(6.022 \times 10^{23} \text{ atoms/mol})(1 \text{ kJ}/10^3 \text{ J})$
$= -145.8 \text{ kJ/mol}$

7.5 The least energetic line is from the electron transition from $n = 2$ to $n = 1$.

$\Delta E = -Rhc[1/1^2 - 1/2^2]$
$= -(2.179 \times 10^{-18} \text{ J/atom})(3/4)$
$= -1.634 \times 10^{-18} \text{ J/atom}$

$\nu = \Delta E/h = (-1.634 \times 10^{-18} \text{ J/atom})/(6.626 \times 10^{-34} \text{ J} \cdot \text{s})$
$= 2.466 \times 10^{15} \text{ s}^{-1}$

$\lambda = c/\nu = (2.998 \times 10^8 \text{ m/s}^{-1})/(2.466 \times 10^{15} \text{ s}^{-1})$
$= 1.216 \times 10^{-7} \text{ m (or 121.6 nm)}$

7.6 First, calculate the velocity of the neutron:

$v = [2E/m]^{1/2} = [2(6.21 \times 10^{-21} \text{ kg} \cdot \text{m}^2 \text{ s}^{-2})/(1.675 \times 10^{-27} \text{ kg})]^{1/2} = 2720 \text{ m s}^{-1}$

Use this value in the de Broglie equation:

$\lambda = h/mv = (6.626 \times 10^{-34} \text{ kg} \cdot \text{m}^2 \text{ s}^{-2})/(1.675 \times 10^{-31} \text{ kg})(2720 \text{ m s}^{-1}) = 1.45 \times 10^{-6} \text{ m}$

7.7 (a) $\ell = 0$ or 1; (b) $m_\ell = -1$, 0, or $+1$, p subshell;
(c) d subshell; (d) $\ell = 0$ and $m_\ell = 0$; (e) 3 orbitals in the p subshell; (f) 7 values of m_ℓ and 7 orbitals

7.8 (a)

Orbital	n	ℓ
$6s$	6	0
$4p$	4	1
$5d$	5	2
$4f$	4	3

(b) A $4p$ orbital has one nodal plane; a $6d$ orbital has two nodal planes.

Chapter 8

8.1 (a) $4s$ ($n + \ell = 4$) filled before $4p$ ($n + \ell = 5$)

(b) $6s$ ($n + \ell = 6$) filled before $5d$ ($n + \ell = 7$)

(c) $5s$ ($n + \ell = 5$) filled before $4f$ ($n + \ell = 7$)

8.2 (a) chlorine (Cl)

(b) $1s^2 2s^2 2p^6 3s^2 3p^3$

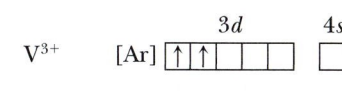

(c) Calcium has two valence electrons in the $4s$ subshell. Quantum numbers for these two electrons are $n = 4$, $\ell = 0$, $m_\ell = 0$, and $m_s = \pm 1/2$

8.3 Obtain the answers from Table 8.3.

8.4

V^{2+} [Ar] $3d$ $4s$

V^{3+} [Ar] $3d$ $4s$

Co^{3+} [Ar] $3d$ $4s$

All three ions are paramagnetic with three, two, and four unpaired electrons, respectively.

8.5 Increasing atomic radius: C $<$ Si $<$ Al

8.6 (a) H—O distance $= 37$ pm $+ 66$ pm $= 103$ pm; H—S distance $= 37$ pm $+ 104$ pm $= 141$ pm

(b) $r_{\text{Br}} = 114$ pm; Br—Cl bond length $= r_{\text{Br}} + r_{\text{Cl}} = 114$ pm $+ 99$ pm $= 213$ pm

8.7 (a) Increasing atomic radius: C $<$ B $<$ Al

(b) Increasing ionization energy: Al $<$ B $<$ C

(c) Carbon is predicted to have the most negative electron affinity.

8.8 Decreasing ionic radius: N^{3-} $>$ O^{2-} $>$ F$^-$. In this series of isoelectronic ions, the size decreases with increased nuclear charge.

8.9 MgCl$_3$, if it existed, would contain one Mg^{3+} ion (and three Cl$^-$ ions). The formation of Mg^{3+} is energetically unfavorable, with a huge input of energy being required to remove the third electron (a core electron).

Chapter 9

9.1 ·Ba· Ba, Group 2A, has 2 valence electrons.

·Äs· As, Group 5A, has 5 valence electrons.

·B̈r: Br, Group 7A, has 7 valence electrons.

9.2 ΔH_f° for Na(g) $= +107.3$ kJ/mol

ΔH_f° for I(g) $= +106.8$ kJ/mol

ΔH° [for Na(g) $\longrightarrow$ Na$^+$(g) $+$ e$^-$] $= +496$ kJ/mol

ΔH° [for I(g) $+$ e$^-$ $\longrightarrow$ I$^-$(g)] $= -295.2$ kJ/mol

ΔH° [for Na$^+$(g) $+$ I$^-$(g)] $= -702$ kJ/mol

The sum of these values = $\Delta H^\circ_f[\text{NaI}(s)] = -287$ kJ/mol; the literature value (from calorimetry) is -287.8 kJ/mol.

9.3

$$\left[\begin{array}{c} H \\ | \\ H-N-H \\ | \\ H \end{array} \right]^+ \qquad :C\equiv O: \qquad [:N\equiv O:]^+ \qquad \left[\begin{array}{c} :\ddot{O}: \\ | \\ :\ddot{O}-S-\ddot{O}: \\ | \\ :\ddot{O}: \end{array} \right]^{2-}$$

9.4

$$H-\underset{\underset{H}{|}}{\overset{\overset{H}{|}}{C}}-\ddot{O}-H \qquad H-\underset{\underset{H}{|}}{N}-\ddot{O}-H$$

methanol   hydroxylamine

9.5

$$\left[H-\ddot{O}-\underset{\underset{:\ddot{O}:}{|}}{\overset{\overset{:\ddot{O}:}{||}}{P}}-\ddot{O}-H \right]^-$$

9.6 (a) The acetylide ion, C_2^{2-}, and the N_2 molecule have the same number of valence electrons (10) and identical electronic structures; that is, they are isoelectronic.

(b) Ozone, O_3, is isoelectronic with NO_2^-; hydroxide ion, OH^-, is isoelectronic with HF.

9.7 Resonance structures for the nitrate ion:

$$\left[\begin{array}{c} :\ddot{O}: \\ \| \\ N \\ \diagup \;\;\; \diagdown \\ :\ddot{O} \qquad \ddot{O}: \end{array} \right]^- \longleftrightarrow \left[\begin{array}{c} :O: \\ \| \\ N \\ \diagup \;\;\; \diagdown \\ :\ddot{O} \qquad \ddot{O}: \end{array} \right]^- \longleftrightarrow \left[\begin{array}{c} :\ddot{O}: \\ | \\ N \\ \diagup \;\;\; \diagdown \\ :\ddot{O} \qquad \ddot{O}: \end{array} \right]^-$$

Lewis electron dot structure for nitric acid, HNO_3:

$$\begin{array}{c} :\ddot{O}-H \\ | \\ N \\ \diagup \;\;\; \diagdown \\ :\ddot{O} \qquad \ddot{O}: \end{array}$$

9.8 $\left[:\ddot{F}-\ddot{Cl}-\ddot{F}: \right]^+$ ClF_2^+, 2 bond pairs and 2 lone pairs.

$\left[:\ddot{F}-\ddot{Cl}-\ddot{F}: \right]^-$ ClF_2^-, 2 bond pairs and 3 lone pairs.

9.9 Tetrahedral geometry around carbon. The Cl—C—Cl bond angle will be close to 109.5°.

9.10 For each species, the electron-pair geometry and the molecular shape are the same. BF_3: trigonal planar; BF_4^-: tetrahedral. Adding F^- to BF_3 adds an electron pair to the central atom and changes the shape.

9.11 The electron-pair geometry around the I atom is trigonal bipyramidal. The molecular geometry of the ion is linear.

$$\left[\begin{array}{c} :\ddot{Cl}: \\ | \\ :I- \\ | \\ :\ddot{Cl}: \end{array} \right]^-$$

9.12 (a) In PO_4^{3-}, there is tetrahedral electron-pair geometry. The molecular geometry is tetrahedral.

$$\left[\begin{array}{c} :\ddot{O}: \\ | \\ :\ddot{O}-P-\ddot{O}: \\ | \\ :\ddot{O}: \end{array} \right]^{3-}$$

(b) In SO_3^{2-}, there is tetrahedral electron-pair geometry. The molecular geometry is trigonal pyramidal.

$$\left[\begin{array}{c} :\ddot{O}-\ddot{S}-\ddot{O}: \\ | \\ :\ddot{O}: \end{array} \right]^{2-}$$

(c) In IF_5, there is octahedral electron-pair geometry. The molecular geometry is square pyramidal.

$$\begin{array}{c} F \\ | \\ F\cdots I \cdots F \\ \diagup \;\; | \;\; \diagdown \\ F \;\;\;\; F \end{array}$$

9.13 (a) CN^-: formal charge on C is -1; formal charge on N is 0.

(b) SO_3: formal charge on S is $+2$; formal charge on each O is $-\frac{2}{3}$.

9.14 (a) The H atom is positive in each case. H—F ($\Delta\chi = 1.8$) is more polar than H—I ($\Delta\chi = 0.5$).

(b) B—F ($\Delta\chi = 2.0$) is more polar than B—C ($\Delta\chi = 0.5$). In B—F, F is the negative pole and B is the positive pole. In B—C, C is the negative pole and B is the positive pole.

(c) C—Si ($\Delta\chi = 0.6$) is more polar than C—S ($\Delta\chi = 0.1$). In C—Si, C is the negative pole and Si is the positive pole. In C—S, S is the negative pole and C the positive pole.

9.15 $\overset{-1}{:\ddot{O}}-\overset{+1}{S}=\overset{0}{\ddot{O}} \longleftrightarrow \overset{0}{\ddot{O}}=\overset{+1}{S}-\overset{-1}{\ddot{O}}:$

The S—O bonds are polar, with the negative end being the O atom. (The O atom is more electronegative than the S atom.) Formal charges show that these bonds are, in fact, polar, with the O atom being the more negative atom.

9.16 (a) $BFCl_2$, polar, negative side is the F atom because F is the most electronegative atom in the molecule.

$$\begin{array}{c} F \\ | \\ B \\ \diagup \;\; \diagdown \\ Cl \;\;\;\; Cl \end{array} \uparrow +$$

(b) NH_2Cl, polar, negative side is the Cl atom.

$$\underset{Cl}{\overset{\delta-}{\;}}\overset{N}{\diagdown}\cdots H^{\delta+} \atop \;\;\;\; \underset{H^{\delta+}}{\diagdown}$$

(c) SCl_2, polar, Cl atoms are on the negative side.

$$\underset{Cl}{\overset{\ddot{S}}{\diagup}}\;\overset{}{\diagdown}\;\underset{Cl}{} \;\uparrow \atop \underset{\delta-}{\;} \;\; \underset{\delta-}{\;} \downarrow$$

9.17 (a) C—N: bond order 1; C=N: bond order 2; C≡N: bond order 3. Bond length: C—N > C=N > C≡N

(b) $\left[:\ddot{O}—\ddot{N}=\ddot{O}: \right]^- \longleftrightarrow \left[\ddot{O}=\ddot{N}—\ddot{O}: \right]^-$

The bond order in NO_2^- is 1.5. Therefore, the NO bond length (124 pm) should be between the length of a N—O single bond (136 pm) and a N=O double bond (115 pm).

9.18 $CH_4(g) + 2\,O_2(g) \longrightarrow CO_2(g) + 2\,H_2O(g)$

Break 4 C—H bonds and 2 O=O bonds:
(4 mol)(413 kJ/mol) + (2 mol)(498 kJ/mol) = 2648 kJ

Make 2 C=O bonds and 4 H—O bonds:
(2 mol)(745 kJ/mol) + (4 mol)(463 kJ/mol) = 3342 kJ

ΔH°_{rxn} = 2648 kJ − 3342 kJ = −694 kJ

Chapter 10

10.1 The oxygen atom in H_3O^+ is sp^3 hybridized. The three O—H bonds are formed by overlap of oxygen sp^3 and hydrogen 1s orbitals. The fourth sp^3 orbital contains a lone pair of electrons.

The carbon and nitrogen atoms in CH_3NH_2 are sp^3 hybridized. The C—H bonds arise from overlap of carbon sp^3 orbitals and hydrogen 1s orbitals. The bond between C and N is formed by overlap of sp^3 orbitals from these atoms. Overlap of nitrogen sp^3 and hydrogen 1s orbitals gives the two N—H bonds, and there is a lone pair in the remaining sp^3 orbital on nitrogen.

10.2 The Lewis structure and the electron-pair and molecular geometries are shown below. The Xe atom is sp^3d^2 hybridized. Lone pairs of electrons reside in two of these orbitals; the four others overlap with 2p orbitals on fluorine, forming sigma bonds.

electron dot structure   molecular geometry

10.3 (a) BH_4^-, tetrahedral electron-pair geometry, sp^3
(b) SF_5^-, octahedral electron-pair geometry, sp^3d^2
(c) SOF_4, trigonal-bipyramidal electron-pair geometry, sp^3d
(d) ClF_3, trigonal-bipyramidal electron-pair geometry, sp^3d
(e) BCl_3, trigonal-planar electron-pair geometry, sp^2
(f) XeO_6^{4-}, octahedral electron-pair geometry, sp^3d^2

10.4 The two CH_3 carbon atoms are sp^3 hybridized and the center carbon atom is sp^2 hybridized. For each of the carbon atoms in the methyl groups, three orbitals are used to form C—H bonds and the fourth is used to bond to the central carbon atom. Overlap of carbon and oxygen sp^2 orbitals gives the sigma bond. The pi bond arises by overlap of p orbitals on these elements.

10.5 A triple bond links the two nitrogen atoms, each of which also has one lone pair. Each nitrogen is sp hybridized. One sp orbital contains the lone pair; the other is used to form the sigma bond between the two atoms. Two pi bonds arise by overlap of p orbitals on the two atoms.

10.6 Bond angles: H—C—H = 109.5°, H—C—C = 109.5°, C—C—N = 180°. Carbon in the CH_3 group is sp^3 hybridized; the central C and the N are sp hybridized. The three C—H bonds form by overlap of an H 1s orbital with one of the sp^3 orbitals of the CH_3 group; the fourth sp^3 orbital overlaps with an sp orbital on the central C to form a sigma bond. The triple bond between C and N is a combination of a sigma bond (the sp orbital on C overlaps with the sp orbital on N) and two pi bonds (overlap of two sets of p orbitals on these elements). The remaining sp orbital on N contains a lone pair.

10.7 H_2^+: $(\sigma 1s)^1$ The ion has a bond order of $\frac{1}{2}$ and is expected to exist. A bond order of $\frac{1}{2}$ is predicted for He_2^+ and H_2^-, both of which are predicted to have electron configurations $(\sigma 1s)^2 (\sigma * 1s)^1$.

10.8 Li_2^- is predicted to have an electron configuration $(\sigma 1s)^2 (\sigma * 1s)^2 (\sigma 2s)^2 (\sigma * 2s)^1$ and a bond order of $\frac{1}{2}$, implying that the ion might exist.

10.9 O_2^+: [core electrons] $(\sigma 2s)^2 (\sigma * 2s)^2 (\pi 2p)^4 (\sigma 2p)^2 (\pi * 2p)^1$. The bond order is 2.5. The ion is paramagnetic with one unpaired electron.

Chapter 11

11.1 (a) Isomers of C_7H_{16}

$CH_3CH_2CH_2CH_2CH_2CH_2CH_3$ heptane

$$CH_3CH_2CH_2CH_2\underset{\underset{CH_3}{|}}{C}HCH_3 \qquad \text{2-methylhexane}$$

$$CH_3CH_2CH_2\underset{\underset{CH_3}{|}}{C}HCH_2CH_3 \qquad \text{3-methylhexane}$$

$$CH_3CH_2\underset{\underset{CH_3}{|}}{C}H\underset{\overset{CH_3}{|}}{C}HCH_3 \qquad \text{2,3-dimethylpentane}$$

$$CH_3CH_2CH_2\underset{\underset{CH_3}{|}}{\overset{\overset{CH_3}{|}}{C}}CH_3 \qquad \text{2,2-dimethylpentane}$$

$$CH_3CH_2\underset{\underset{CH_3}{|}}{\overset{\overset{CH_3}{|}}{C}}CH_2CH_3 \qquad \text{3,3-dimethylpentane}$$

$$CH_3\underset{\underset{CH_3}{|}}{C}HCH_2\underset{\overset{CH_3}{|}}{C}HCH_3 \qquad \text{2,4-dimethylpentane}$$

2-Ethylpentane is pictured on page 484.

$$\underset{\underset{\overset{|}{CH_3}}{\overset{\overset{H_3C\ \ \ CH_3}{|\ \ \ \ \ |}}{CH_3C}}{}}{}\!\!-\!\!CHCH_3 \qquad 2,2,3\text{-trimethylbutane}$$

(b) Two isomers, 3-methylhexane, and 2,3-dimethylpentane, are chiral.

11.2　The names accompany the structures in the answer to Exercise 11.1.

11.3　Isomers of C_6H_{12} in which the longest chain has six C atoms:

$$\overset{H}{\underset{H}{}}\!\!\!C\!\!=\!\!C\!\!\!\underset{CH_2CH_2CH_2CH_3}{\overset{H}{}}$$

$$\overset{H}{\underset{H_3C}{}}\!\!\!C\!\!=\!\!C\!\!\!\underset{CH_2CH_2CH_3}{\overset{H}{}}$$

$$\overset{H}{\underset{H_3C}{}}\!\!\!C\!\!=\!\!C\!\!\!\underset{H}{\overset{CH_2CH_2CH_3}{}}$$

$$\overset{H}{\underset{H_3CCH_2}{}}\!\!\!C\!\!=\!\!C\!\!\!\underset{CH_2CH_3}{\overset{H}{}}$$

$$\overset{H}{\underset{H_3CCH_2}{}}\!\!\!C\!\!=\!\!C\!\!\!\underset{H}{\overset{CH_2CH_3}{}}$$

Names: 1-hexene, *cis*-2-hexene, *trans*-2-hexene, *cis*-3-hexene, *trans*-3-hexene. None of these isomers is chiral.

11.4　(a) $\underset{\underset{H\ \ H}{|\ \ |}}{\overset{\overset{H\ \ H}{|\ \ |}}{H\!-\!C\!-\!C}}\!\!-\!\!Br$　　(b) $\underset{\underset{H\ \ H}{|\ \ |}}{\overset{\overset{Br\ Br}{|\ \ |}}{H_3C\!-\!C\!-\!C}}\!\!-\!\!CH_3$

　　　　bromoethane　　　　　　2,3-dibromobutane

11.5　1,4-diaminobenzene

11.6　$CH_3CH_2CH_2CH_2OH$　1-butanol

$$\underset{\overset{|}{OH}}{CH_3CH_2CHCH_3} \qquad 2\text{-butanol}$$

$$\underset{\overset{|}{CH_3}}{CH_3CHCH_2OH} \qquad 2\text{-methyl-1-propanol}$$

$$\underset{\overset{|}{CH_3}}{\overset{\overset{OH}{|}}{CH_3CCH_3}} \qquad 2\text{-methyl-2-propanol}$$

11.7　(a) $CH_3CH_2CH_2\overset{\overset{O}{\|}}{C}CH_3$　　2-pentanone

$$CH_3CH_2\overset{\overset{O}{\|}}{C}CH_2CH_3 \qquad 3\text{-pentanone}$$

$$CH_3CH_2CH_2CH_2\overset{\overset{O}{\|}}{C}H \qquad \text{pentanal}$$

$$\underset{\overset{|}{CH_3}}{CH_3CHCH_2\overset{\overset{O}{\|}}{C}H} \qquad 3\text{-methylbutanal}$$

(b) $\underset{\overset{|}{OH}}{CH_3CHCH_2CH_2CH_3}$, 2-pentanol

11.8　(a) 1-butanol gives butanal　$CH_3CH_2CH_2\overset{\overset{O}{\|}}{C}H$

(b) 2-butanol gives butanone　$CH_3CH_2\overset{\overset{O}{\|}}{C}CH_3$

(c) 2-methyl-1-propanol gives 2-methylpropanal

$$\underset{\overset{|}{CH_3}}{CH_3\overset{\overset{H}{|}}{C}\!-\!\overset{\overset{O}{\|}}{C}H}$$

11.9　(a) $CH_3CH_2\overset{\overset{O}{\|}}{C}OCH_3$　　methyl propanoate

(b) $CH_3CH_2CH_2\overset{\overset{O}{\|}}{C}OCH_2CH_2CH_2CH_3$

　　　　　　　　　　　butyl butanoate

(c) $CH_3CH_2CH_2CH_2CH_2\overset{\overset{O}{\|}}{C}OCH_2CH_3$

　　　　　　　　　　　ethyl hexanoate

11.10　(a) Propyl acetate is formed from acetic acid and propanol:

$$CH_3\overset{\overset{O}{\|}}{C}OH + CH_3CH_2CH_2OH$$

(b) 3-Methylpentyl benzoate is formed from benzoic acid and 3-methylpentanol:

(c) Ethyl salicylate is formed from salicylic acid and ethanol:

11.11 (a) $CH_3CH_2CH_2OH$: 1-propanol, has an alcohol (—OH) group

CH_3CO_2H: ethanoic acid (acetic acid), has a carboxylic acid (—CO_2H) group

$CH_3CH_2NH_2$: ethylamine, has an amino (—NH_2) group

(b) 1-propyl ethanoate (propyl acetate)

(c) Oxidation of this primary alcohol first gives propanal, CH_3CH_2CHO. Further oxidation gives propanoic acid, $CH_3CH_2CO_2H$.

(d) *N*-ethylacetamide, $CH_3CONHCH_2CH_3$

(e) The amine is protonated by hydrochloric acid, forming ethylammonium chloride, $[CH_3CH_2NH_3]Cl$.

11.12 The polymer is a polyester.

Chapter 12

12.1 0.83 bar (0.82 atm) > 75 kPa (0.74 atm) > 0.63 atm > 250 mm Hg (0.33 atm)

12.2 $P_1 = 55$ mm Hg and $V_1 = 125$ mL; $P_2 = 78$ mm Hg and $V_2 = ?$

$V_2 = V_1(P_1/P_2) = (125 \text{ mL})(55 \text{ mm Hg}/78 \text{ mm Hg})$
$= 88$ mL

12.3 $V_1 = 45$ L and $T_1 = 298$ K; $V_2 = ?$ and $T_2 = 263$ K

$V_2 = V_1(T_2/T_1) = (45 \text{ L})(263 \text{ K}/298 \text{ K}) = 40.$ L

12.4 $V_2 = V_1 (P_1/P_2)(T_2/T_1)$
$= (22 \text{ L})(150 \text{ atm}/0.993 \text{ atm})(295 \text{ K}/304 \text{ K})$
$= 3200$ L

At 5.0 L per balloon, there is sufficient He to fill 640 balloons.

12.5 44.8 L of O_2 is required; 44.8 L of $H_2O(g)$ and 22.4 L $CO_2(g)$ are produced.

12.6 $PV = nRT$

$(750/760 \text{ atm})(V) =$
$(1300 \text{ mol})(0.082057 \text{ L} \cdot \text{atm}/ \text{mol} \cdot \text{K})(296 \text{ K})$

$V = 3.2 \times 10^4$ L

12.7 $d = PM/RT$; $M = dRT/P$

$M = (5.02 \text{ g/L})(0.082057 \text{ L} \cdot \text{atm/mol} \cdot \text{K})(288.2 \text{ K})/$
$(745/760 \text{ atm}) = 121$ g/mol

12.8 $PV = (m/M)RT$; $M = mRT/PV$

$M = (0.105 \text{ g})(0.082057 \text{ L} \cdot \text{atm/mol} \cdot \text{K})(296.2 \text{ K})/$
$[(561/760 \text{ atm})(0.125 \text{ L})] = 27.7$ g/mol

12.9 $n(H_2) = PV/RT = (542/760 \text{ atm})(355 \text{ L})/$
$(0.082057 \text{ L} \cdot \text{atm/mol} \cdot \text{K})(298.2 \text{ K})$

$n(H_2) = 10.3$ mol

$n(NH_3) = (10.3 \text{ mol } H_2)(2 \text{ mol } NH_3/3 \text{ mol } H_2)$
$= 6.87 \text{ mol } NH_3$

$PV = nRT$; $P (125 \text{ L})$
$= (6.87 \text{ mol})(0.082057 \text{ L} \cdot \text{atm}/ \text{mol} \cdot \text{K})(298.2 \text{ K})$

$P (NH_3) = 1.35$ atm

12.10 $P_{halothane} (5.00 \text{ L}) =$
$(0.0760 \text{ mol})(0.082057 \text{ L} \cdot \text{atm/mol} \cdot \text{K}) (298.2 \text{ K})$

$P_{halothane} = 0.372$ atm (or 283 mm Hg)

$P_{oxygen} (5.00 \text{ L}) = (0.734 \text{ mol})(0.082057 \text{ L} \cdot \text{atm/mol} \cdot \text{K})$
(298.2 K)

$P_{oxygen} = 3.59$ atm (or 2730 mm Hg)

$P_{total} = P_{halothane} + P_{oxygen}$
$= 283 \text{ mm Hg} + 2730 \text{ mm Hg} = 3010 \text{ mm Hg}$

12.11 For He: Use Equation 12.9, with $M = 4.00 \times 10^{-3}$ kg/mol, $T = 298$ K, and $R = 8.314$ J/mol $\cdot$ K to calculate the rms speed of 1360 m/s. A similar calculation for N_2, with $M = 28.01 \times 10^{-3}$ kg/mol, gives an rms speed of 515 m/s.

12.12 The molar mass of CH_4 is 16.0 g/mol.

$$\frac{\text{Rate for } CH_4}{\text{Rate for unknown}} = \frac{n \text{ molecules}/1.50 \text{ min}}{n \text{ molecules}/4.73 \text{ min}} = \sqrt{\frac{M_{unknown}}{16.0}}$$

$M_{unknown} = 159$ g/mol

12.13 $P(1.00 \text{ L}) =$
$(10.0 \text{ mol})(0.082057 \text{ L} \cdot \text{atm/mol} \cdot \text{K}) (298 \text{ K})$

$P = 245$ atm (calculated by $PV = nRT$)

$P = 320$ atm (calculated by van der Waals equation)

Chapter 13

13.1 Because F^- is the smaller ion, water molecules can approach most closely and interact more strongly. Thus, F^- should have the more negative heat of hydration.

13.2 Water is a polar solvent, while hexane and CCl_4 are nonpolar. London dispersion forces are the primary forces of attraction between all pairs of dissimilar solvents. For mixtures of water with the other solvents, dipole–induced dipole forces will also be present. When mixed, the three liquids will form two separate layers—the first being water and the second consisting of a mixture of the two nonpolar liquids.

13.3 $H_3C—O$
         with H above and H—O with CH_3 below

Hydrogen bonding in methanol entails the attraction of the hydrogen atom bearing a partial positive charge (δ^+) on one molecule to the oxygen atom bearing a partial negative charge (δ^-) on a second molecule. The strong attractive force of hydrogen bonding will cause the boiling point and the heat of vaporization of methanol to be quite high.

13.4   (a) O_2: induced dipole–induced dipole forces only.
       (b) CH_3OH: strong hydrogen bonding (dipole–dipole forces) as well as induced dipole–induced dipole forces.
       (c) Forces between water molecules: strong hydrogen bonding and induced dipole–induced dipole forces. Between O_2 and H_2O: dipole–induced dipole forces and induced dipole–induced dipole forces.

       Relative strengths: a < forces between O_2 and H_2O in c < b < forces between water molecules in c.

13.5 $(1.00 \times 10^3 \text{ g})(1 \text{ mol}/32.04 \text{ g})(35.2 \text{ kJ/mol}) = 1.10 \times 10^3 \text{ kJ}$

13.6   (a) At 40 °C, the vapor pressure of ethanol is about 120 mm Hg.
       (b) The equilibrium vapor pressure of ethanol at 60 °C is about 320 mm Hg. At 60 °C and 600 mm Hg, ethanol is a liquid. If vapor is present, it will condense to a liquid.

13.7 $PV = nRT$; $P = 0.50 \text{ g } (1 \text{ mol}/18.02 \text{ g})(0.0821 \text{ L} \cdot \text{atm/mol} \cdot \text{K})(333 \text{ K})/5.0 \text{ L}$

       $P = 0.15 \text{ atm (or 120 mm Hg)}$

       The vapor pressure of water at 60 °C is 149.4 mm Hg (Appendix G). The calculated pressure is lower than this, so all the water (0.50 g) evaporates. If 2.0 g of water is used, the calculated pressure, 460 mm Hg, exceeds the vapor pressure. In this case, only part of the water will evaporate.

13.8   Use the Clausius-Clapeyron equation, with $P_1 = 57.0$ mm Hg, $T_1 = 250.4$ K, $P_2 = 534$ mm Hg, and $T_2 = 298.2$ K.
       $\ln [P_2/P_1] =$
       $\Delta H_{vap}/R [1/T_1 - 1/T_2] = \Delta H_{vap}/R[(T_2 - T_1)/T_1 T_2]$
       $\ln [534/57.0] =$
       $\Delta H_{vap}/(0.0083145 \text{ kJ/K} \cdot \text{mol})[47.8/(250.4)(298.2)]$
       $\Delta H_{vap} = 29.1 \text{ kJ/mol}$

13.9   Glycerol is predicted to have a higher viscosity than ethanol. It is a larger molecule than ethanol, and there are higher forces of attraction between molecules because each molecule has three OH groups that hydrogen-bond to other molecules.

13.10 M_2X; In a face-centered cubic unit cell, there are four anions and eight tetrahedral holes in which to place metal ions. All of the tetrahedral holes are inside the unit cell, so the ratio of atoms in the unit cell is 2 : 1.

Chapter 14

14.1   10.0 g sucrose = 0.0292 mol; 250. g H_2O = 13.9 mol
       $X = (0.0292 \text{ mol})/(0.0292 \text{ mol} + 13.9 \text{ mol}) = 0.00210$
       $(0.0292 \text{ mol sucrose})/(0.250 \text{ kg solvent}) = 0.117 \text{ m}$
       % sucrose = (10.0 g sucrose/260. g soln)(× 100%) = 3.85%

14.2 1.08×10^4 ppm ≡ 1.08×10^4 mg NaCl per 1000 g soln
       $(1.08 \times 10^4 \text{ mg Na}/1000 \text{ g soln})(1050 \text{ g soln}/1 \text{ L}) =$
       $1.13 \times 10^4 \text{ mg Na/L} = 11.3 \text{ g Na/L}$
       $(11.3 \text{ g Na/L})(58.44 \text{ g NaCl}/23.0 \text{ g Na}) = 28.7 \text{ g NaCl/L}$

14.3 $\Delta H^\circ_{soln} = \Delta H^\circ_f[\text{NaOH(aq)}] - \Delta H^\circ_f[\text{NaOH(s)}]$
       $= -469.2 \text{ kJ/mol} - (-425.9 \text{ kJ/mol})$
       $= -43.3 \text{ kJ/mol}$

14.4   Solubility (CO_2) = $(4.48 \times 10^{-5} \text{ M/mm Hg})(251 \text{ mm Hg})$
       $= 1.1 \times 10^{-2} \text{ M}$

14.5   The solution contains sucrose [(10.0 g)(1 mol/342.3 g) = 0.0292 mol] in water [(225 g)(1 mol/18.02 g) = 12.5 mol].
       X_{water} = (12.5 mol H_2O)/(12.5 mol + 0.0292 mol) = 0.998
       P_{water} = 0.998(149.4 mm Hg) = 149 mm Hg

14.6 $m = \Delta T_{bp}/K_{bp} = 1.0 \text{ °C}/(0.512 \text{ °C}/m) = 1.95 \text{ m}$
       $(1.95 \text{ mol/kg})(0.125 \text{ kg})(62.02 \text{ g/mol}) = 15 \text{ g glycol}$

14.7   Concentration = (525 g)(1 mol/62.07 g)/(3.00 kg) = 2.82 m
       $\Delta T_{fp} = K_{fp} \times m = (-1.86 \text{ °C}/m)(2.82 \text{ m}) = -5.24 \text{ °C}$
       You will be protected only to about −5 °C and not to −25 °C.

14.8 ΔT_{bp} = 80.23 °C − 80.10 °C = 0.13 °C
       $m = \Delta T_{bp}/K_{bp} = 0.13 \text{ °C}/(2.53 \text{ °C}/m) = 0.051 \text{ m}$
       (0.051 mol/kg)(0.099 kg) = 0.0051 mol
       Molar mass = 0.640 g/0.0051 mol = 130 g/mol
       The formula $C_{10}H_8$ (molar mass = 128.2 g/mol) is the closest match to this value.

14.9   Concentration = (25.0 g NaCl)(1 mol/58.44 g)/(0.525 kg) = 0.815 m
       $\Delta T_{fp} = K_{fp} \times m \times i = (-1.86 \text{ °C}/m)(0.815 \text{ m})(1.85)$
       $= -2.80 \text{ °C}$

14.10 $M = \Pi/RT = [(1.86 \text{ mm Hg})(1 \text{ atm}/760 \text{ mm Hg})]/$
$[(0.08206 \text{ L} \cdot \text{atm/mol} \cdot \text{K})(298 \text{ K})]$
$= 1.00 \times 10^{-4} \text{ M}$

$(1.00 \times 10^{-4} \text{ mol/L})(0.100 \text{ L}) = 1.0 \times 10^{-5} \text{ mol}$

Molar mass $= 1.40 \text{ g}/1.00 \times 10^{-5} \text{ mol} = 1.4 \times 10^{5} \text{ g/mol}$

Assuming the polymer is composed of CH_2 units, the polymer is about 10,000 units long.

Chapter 15

15.1 $-\frac{1}{2}(\Delta[\text{NOCl}]/\Delta t) = \frac{1}{2}(\Delta[\text{NO}]/\Delta t) = \Delta[\text{Cl}_2]/\Delta t$

15.2 For the first two hours:

$-\Delta[\text{sucrose}]/\Delta t = [(0.033 - 0.050) \text{ mol/L}]/(2.0 \text{ h})$
$= 0.0080 \text{ mol/L} \cdot \text{h}$

For the last two hours:

$\Delta[\text{sucrose}]/\Delta t = -[(0.010 - 0.015) \text{ mol/L}]/(2.0 \text{ h})$
$= 0.0025 \text{ mol/L} \cdot \text{h}$

Instantaneous rate at 4 h $= 0.0045 \text{ mol/L} \cdot \text{h}$

15.3 Compare experiments 1 and 2: Doubling $[\text{O}_2]$ causes the rate to double, so the rate is first order in $[\text{O}_2]$. Compare experiments 2 and 4: Doubling $[\text{NO}]$ causes the rate to increase by a factor of 4, so the rate is second order in $[\text{NO}]$. Thus, the rate law is

Rate $= k[\text{NO}]^2[\text{O}_2]$

Using the data in experiment 1 to determine k:

$0.028 \text{ mol/L} \cdot \text{s} = k[0.020 \text{ mol/L}]^2[0.010 \text{ mol/L}]$

$k = 7.0 \times 10^3 \text{ L}^2/\text{mol}^2 \cdot \text{s}$

15.4 Rate $= k[\text{Pt(NH}_3)_2\text{Cl}_2] = (0.090 \text{ h}^{-1})(0.020 \text{ mol/L})$
$= 0.0018 \text{ mol/L} \cdot \text{h}$

The rate of formation of Cl^- is the same value, 0.0018 mol/L $\cdot$ h.

15.5 $\ln([\text{sucrose}]/[\text{sucrose}]_o) = -kt$

$\ln([\text{sucrose}]/[0.010]) = -(0.21 \text{ h}^{-1})(5.0 \text{ h})$

$[\text{sucrose}] = 0.0035 \text{ mol/L}$

15.6 (a) The fraction remaining is $[\text{NO}_2]/[\text{NO}_2]_o$.

$\ln([\text{NO}_2]/[\text{NO}_2]_o) = -(3.6 \times 10^{-3} \text{ s}^{-1})(150 \text{ s})$

$[\text{NO}_2]/[\text{NO}_2]_o = 0.58$

(b) The fraction remaining after the reaction is 99% complete is 0.010.

$\ln(0.010) = -(3.6 \times 10^{-3} \text{ s}^{-1})(t)$

$t = 1300 \text{ s}$

15.7 $1/[\text{HI}] - 1/[\text{HI}]_o = kt$

$1/[\text{HI}] - 1/[0.010 \text{ M}] = (30. \text{ L/mol} \cdot \text{min})(12 \text{ min})$

$[\text{HI}] = 0.0022 \text{ M}$

15.8

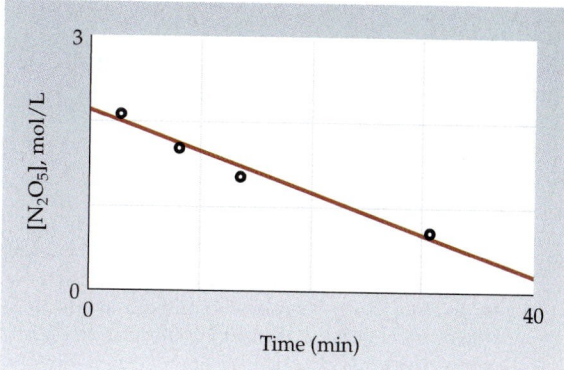

Concentration versus time

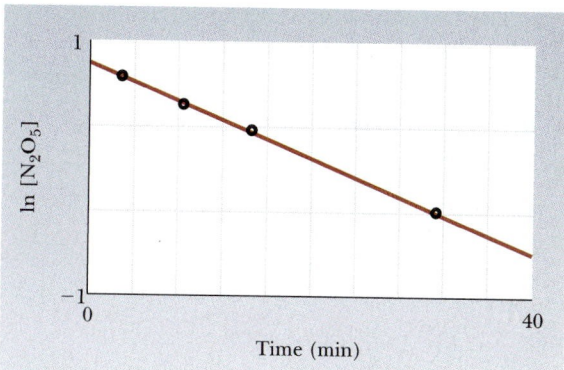

$\ln[\text{N}_2\text{O}_5]$ versus time

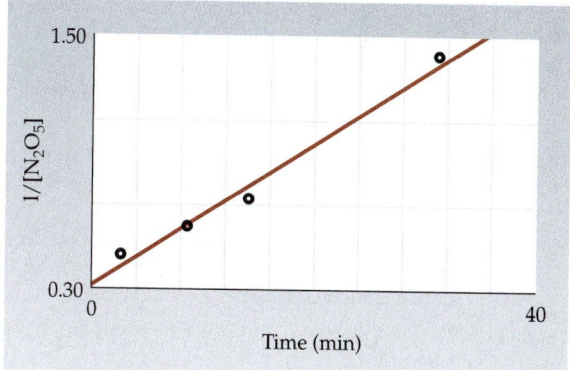

$1/[\text{N}_2\text{O}_5]$ versus time

The plot of $\ln[\text{N}_2\text{O}_5]$ versus time is linear, indicating that this is a first-order reaction. The rate constant is determined from the slope: $k = -\text{slope} = 0.038 \text{ min}^{-1}$.

15.9 (a) For ^{241}Am, $t_{1/2} = 0.693/k = 0.693/(0.0016 \text{ y}^{-1})$
$= 430 \text{ y}$

For ^{125}I, $t_{1/2} = 0.693/(0.011 \text{ d}^{-1}) = 63 \text{ d}$

(b) ^{125}I decays much faster.

(c) $\ln[(n)/(1.6 \times 10^{15} \text{ atoms})] = -(0.011 \text{ d}^{-1})(2.0 \text{ d})$

$n/1.6 \times 10^{15} \text{ atoms} = 0.978; n = 1.57 \times 10^{-15} \text{ atoms}$

Since the answer should have two significant figures, we should round this off to 1.6×10^{15} atoms. The approximately 2% that has decayed is not noticeable within the limits of accuracy of the data presented.

15.10 $\ln(k_2/k_1) = (-E_a/R)(1/T_2 - 1/T_1)$

$\ln[(1.00 \times 10^4)/(4.5 \times 10^3)] = -(E_a/8.315 \times 10^{-3}$ kJ/mol · K)$(1/283\ K - 1/274\ K)$

$E_a = 57$ kJ/mol

15.11 All three steps are bimolecular.

For step 3: Rate = $k[N_2O][H_2]$

When the three equations are added, N_2O_2 (a product in the first step and a reactant in the second step) and N_2O (a product in the second step and a reactant in the third step) cancel, leaving the net equation: $2\ NO(g) + 2\ H_2(g) \longrightarrow N_2(g) + 2\ H_2O(g)$

15.12 (a) $2\ NH_3(aq) + OCl^-(aq) \longrightarrow$
$N_2H_4(aq) + Cl^-(aq) + H_2O(\ell)$

(b) The second step is the rate-determining step.

(c) Rate = $k[NH_2Cl][NH_3]$

(d) NH_2Cl, $N_2H_5^+$, and OH^- are intermediates.

15.13 Overall reaction: $2\ NO_2Cl(g) \longrightarrow 2\ NO_2(g) + Cl_2(g)$

Rate = $k[NO_2Cl]^2/[NO_2]$

The presence of NO_2 causes the reaction rate to decrease.

Chapter 16

16.1 (a) $K = [PCl_3][Cl_2]/[PCl_5]$

(b) $K = [CO]^2/[CO_2]$

(c) $K = [Cu^{2+}][NH_3]^4/[Cu(NH_3)_4^{2+}]$

(d) $K = [H_3O^+][CH_3CO_2^-]/[CH_3CO_2H]$

16.2 (a) Both reactions are reactant-favored ($K \ll 1$).

(b) $[NH_3]$ in the second solution is greater. K for this reaction is larger, so the reactant, $Cd(NH_3)_4^{2+}$, dissociates to a greater extent.

16.3 (a) $Q = [2.18]/[0.97] = 2.3$. The system is not at equilibrium; $Q < K$. To reach equilibrium, [isobutane] will increase and [butane] will decrease.

(b) $Q = [2.60]/[0.75] = 3.5$. The system is not at equilibrium; $Q > K$. To reach equilibrium, [butane] will increase and [isobutane] will decrease.

16.4 $Q = [NO]^2/[N_2][O_2] = [4.2 \times 10^{-3}]^2/[0.50][0.25]$
$= 1.4 \times 10^{-4}$

$Q < K$, so the reaction is not at equilibrium. To reach equilibrium, [NO] will increase and [N_2] and [O_2] will decrease.

16.5 (a)

Equation	$C_6H_{10}I_2$	$\rightleftharpoons$	C_6H_{10}	+	I_2
Initial (M)	0.050		0		0
Change (M)	−0.035		+0.035		+0.035
Equilibrium (M)	0.015		0.035		0.035

(b) $K = (0.035)(0.035)/(0.015) = 0.082$

16.6

Equation	H_2	+	I_2	$\rightleftharpoons$	2 HI
Initial (M)	6.00×10^{-3}		6.00×10^{-3}		0
Change (M)	−x		−x		+2x
Equilibrium (M)	0.00600 − x		0.00600 − x		+2x

$K_c = 33 = \dfrac{(2x)^2}{(0.00600 - x)^2}$

$x = 0.0045$ M, so $[H_2] = [I_2] = 0.0015$ M and $[HI] = 0.0090$ M.

16.7

Equation	C(s) + CO$_2$ (g)	$\rightleftharpoons$	2 CO(g)
Initial (M)	0.012		0
Change (M)	−x		+2x
Equilibrium (M)	0.012 − x		2x

$K_c = 0.021 = \dfrac{(2x)^2}{(0.012 - x)}$

$x = [CO_2] = 0.0057$ M and $2x = [CO] = 0.011$ M

16.8 (a) $K' = K^2 = (2.5 \times 10^{-29})^2 = 6.3 \times 10^{-58}$

(b) $K'' = 1/K^2 = 1/(6.3 \times 10^{-58}) = 1.6 \times 10^{57}$

16.9 Manipulate the equations and equilibrium constants as follows:

$\frac{1}{2}H_2(g) + \frac{1}{2}Br_2(g) \rightleftharpoons HBr(g)$
$\qquad\qquad K_1' = (K_1)^{1/2} = 8.9 \times 10^5$

$H(g) \rightleftharpoons \frac{1}{2}H_2(g) \qquad K_2' = 1/(K_2)^{1/2} = 1.4 \times 10^{20}$

$Br(g) \rightleftharpoons \frac{1}{2}Br_2(g) \qquad K_3' = 1/(K_3)^{1/2} = 2.1 \times 10^7$

Net: $H(g) + Br(g) \rightleftharpoons HBr(g)$
$\qquad\qquad K_{net} = K_1'K_2'K_3' = 2.6 \times 10^{33}$

16.10 (a) [NOCl] decreases with an increase in temperature

(b) [SO_3] decreases with an increase in temperature

16.11

Equation	butane	$\rightleftharpoons$	isobutane
Initial (M)	0.20		0.50
After adding 2.0 M more isobutene	0.20		2.0 + 0.50
Change (M)	+x		−x
Equilibrium (M)	0.20 + x		2.50 − x

$K = \dfrac{[\text{isobutane}]}{[\text{butane}]} = \dfrac{(2.50 - x)}{(0.20 + x)} = 2.50$

Solving for x gives $x = 0.57$ M. Therefore, [isobutene] = 1.93 M and [butane] = 0.77 M.

16.12 (a) Adding H_2 shifts the equilibrium to the right, increasing [NH_3]. Adding NH_3 shifts the equilibrium to the left, increasing [N_2] and [H_2].

(b) An increase in volume shifts the equilibrium to the left.

Chapter 17

17.1 (a) $H_3PO_4(aq) + H_2O(\ell) \rightleftharpoons H_3O^+(aq) + H_2PO_4^-(aq)$

$H_2PO_4^-$ is amphiprotic.

(b) $CN^-(aq) + H_2O(\ell) \rightleftharpoons HCN(aq) + OH^-(aq)$

CN^- is a Brønsted base.

17.2 NO_3^- is the conjugate base of the acid HNO_3; NH_4^+ is the conjugate acid of the base NH_3.

17.3 $[H_3O^+] = 4.0 \times 10^{-3}$ M; $[OH^-] = K_w / [H_3O^+]$
$= 2.5 \times 10^{-12}$

17.4 (a) $pOH = -\log [0.0012] = 2.92$; $pH = 14.00 - pOH = 11.08$

(b) $[H^+] = 4.8 \times 10^{-5}$ M; $[OH^-] = 2.1 \times 10^{-10}$ M

(c) $pOH = 14.00 - 10.46 = 3.54$; $[OH^-] = 2.9 \times 10^{-4}$ M. The solubility of $Sr(OH)_2$ is half of this value, or 1.4×10^{-4} M, because dissolving 1 mol of $Sr(OH)_2$ will give 2 mol of OH^- in solution.

17.5 Answer this question by comparing values of K_a and K_b from Table 17.3.

(a) H_2SO_4 is stronger than H_2SO_3.

(b) $C_6H_5CO_2H$ is a stronger acid than CH_3CO_2H.

(c) The conjugate base of boric acid, $B(OH)_4^-$, is a stronger base than the conjugate base of acetic acid, $CH_3CO_2^-$.

(d) Ammonia is a stronger base than acetate ion.

(e) The conjugate acid of acetate ion, CH_3CO_2H, is a stronger acid than the conjugate acid of ammonia, NH_4^+.

17.6 (a) $pH = 7$

$pH < 7$ (NH_4^+ is an acid)

$pH < 7$ $[Al(H_2O)_6]^{3+}$ is an acid

$pH > 7$ (HPO_4^{2-} is a stronger base than it is an acid)

17.7 (a) $pK_a = -\log [6.3 \times 10^{-5}] = 4.20$

(b) $ClCH_2CO_2H$ is stronger (a pK_a of 2.87 is less than a pK_a of 4.20)

(c) pK_a for NH_4^+, the conjugate acid of NH_3, is $-\log [5.6 \times 10^{-10}] = 9.26$. It is a weaker acid than acetic acid, for which $K_a = 1.8 \times 10^{-5}$.

17.8 K_b for the lactate ion $= K_w / K_a = 7.1 \times 10^{-11}$. It is a slightly stronger base than the formate, nitrite, and fluoride ions, and weaker than the benzoate ion.

17.9 (a) NH_4^+ is a stronger acid than HCO_3^-. CO_3^{2-}, the conjugate base of HCO_3^-, is a stronger base than NH_3, the conjugate base of NH_4^+.

(b) Reactant-favored; the reactants are the weaker acid and base.

(c) Reactant-favored; the reactants are the weaker acid and base.

17.10 $CH_3CO_2H(aq) + HSO_4^-(aq) \rightleftharpoons$
$CH_3CO_2^-(aq) + H_2SO_4(aq)$

The equilibrium favors the weaker acid and base, which in this equation are the reactants.

17.11 (a) The two compounds react and form a solution containing HCN and NaCl. The solution is acidic (HCN is an acid).

(b) $CH_3CO_2H(aq) + SO_3^{2-}(aq) \rightleftharpoons$
$HSO_3^-(aq) + CH_3CO_2^-(aq)$

The solution is acidic, because HSO_3^- is a stronger acid than $CH_3CO_2^-$ is a base.

17.12 From the pH we can calculate $[H_3O^+] = 1.9 \times 10^{-3}$ M. Also, $[butanoate^-] = [H_3O^+] = 1.9 \times 10^{-3}$ M. Use these values along with [butanoic acid] to calculate K_a.

$K_a = [1.9 \times 10^{-3}] [1.9 \times 10^{-3}]/(0.055 - 1.9 \times 10^{-3})$
$= 6.8 \times 10^{-5}$

17.13 $K_a = 1.8 \times 10^{-5} = [x][x]/(0.10 - x)$

$x = [H_3O^+] = [CH_3CO_2^-] = 1.3 \times 10^{-3}$ M; $[CH_3CO_2H] = 0.099$ M; $pH = 2.89$

17.14 $K_a = 7.2 \times 10^{-4} = [x][x]/(0.015 - x)$

The x in the denominator cannot be dropped. This equation must be solved with the quadratic formula or by successive approximations.

$x = [H_3O^+] = [F^-] = 2.9 \times 10^{-3}$ M

$[HF] = 0.015 - 2.9 \times 10^{-3} = 0.012$ M

$pH = 2.54$

17.15 $OCl^-(aq) + H_2O(\ell) = HOCl(aq) + OH^-(aq)$

$K_b = 2.9 \times 10^{-7} = [x][x]/(0.015 - x)$

$x = [OH^-] = [HOCl] = 6.6 \times 10^{-5}$ M

$pOH = 4.18$; $pH = 9.82$

17.16 Equivalent amounts of acid and base were used. The solution will contain $CH_3CO_2^-$ and Na^+. Acetate ion hydrolyzes to a small extent, giving CH_3CO_2H and OH^-. We need to determine $[CH_3CO_2^-]$ and then solve a weak base equilibrium problem to determine $[OH^-]$.

Amount $CH_3CO_2^-$ = moles base = 0.12 M × 0.015 L
$= 1.8 \times 10^{-3}$ mol

Total volume = 0.030 L

so $[CH_3CO_2^-] = (1.8 \times 10^{-3}$ mol$)/0.030$ L = 0.060 M

$CH_3CO_2^-(aq) + H_2O(\ell) \rightleftharpoons CH_3CO_2H(aq) + OH^-(aq)$

$K_b = 5.6 \times 10^{-10} = [x][x]/(0.060 - x)$

$x = [OH^-] = [CH_3CO_2H] = 5.8 \times 10^{-6}$ M

$pOH = 5.24$; $pH = 8.76$

17.17 $H_2C_2O_4(aq) + H_2O(\ell) \rightleftharpoons H_3O^+(aq) + HC_2O_4^-(aq)$

$K_{a_1} = 5.9 \times 10^{-2} = [x][x]/(0.10 - x)$

The x in the denominator cannot be dropped. This equation must be solved with the quadratic formula or by successive approximations.

$x = [H_3O^+] = [HC_2O_4^-] = 5.3 \times 10^{-2}$ M

$pH = 1.28$

$[C_2O_4^{2-}] = K_{a_2} = 6.4 \times 10^{-5}$ M

17.18 (a) Lewis base (electron-pair donor)

(b) Lewis acid (electron-pair acceptor)

(c) Lewis base (electron-pair donor)

(d) Lewis base (electron-pair donor)

17.19 (a) H_2SeO_4

(b) $Fe(H_2O)_6^{3+}$

(c) $HOCl$

(d) Amphetamine is a primary amine and a (weak) base. It is both a Brønsted base and a Lewis base.

Chapter 18

18.1　pH of 0.30 M HCO_2H:

$K_a = [H_3O^+][HCO_2^-]/[HCO_2H]$

$1.8 \times 10^{-4} = [x][x]/[0.30 - x]$

$x = 7.3 \times 10^{-3}$ M; pH = 2.14

pH of 0.30 M formic acid + 0.10 M $NaHCO_2$

$K_a = [H_3O^+][HCO_2^-]/[HCO_2H]$

$1.8 \times 10^{-4} = [x][0.10 + x]/(0.30 - x)$

$x = 5.4 \times 10^{-4}$ M; pH = 3.27

18.2　NaOH: $(0.100 \text{ mol/L})(0.0300 \text{ L}) = 3.00 \times 10^{-3}$ mol

CH_3CO_2H: $(0.100 \text{ mol/L})(0.0450 \text{ L}) = 4.50 \times 10^{-3}$ mol

3.00×10^{-3} mol NaOH reacts with 3.00×10^{-3} mol CH_3CO_2H, forming 3.00×10^{-3} mol $CH_3CO_2^-$; 1.50×10^{-3} mol unreacted CH_3CO_2H remains in solution. The total volume is 75.0 mL. Use these values to calculate $[CH_3CO_2H]$ and $[CH_3CO_2^-]$, and use the concentrations in a weak acid equilibrium calculation to obtain $[H_3O^+]$ and pH.

$[CH_3CO_2H] = 1.5 \times 10^{-3}$ mol/0.075 L = 0.0200 M

$[CH_3CO_2^-] = 3.0 \times 10^{-3}$ mol/0.075 L = 0.0400 M

$K_a = [H_3O^+][CH_3CO_2^-]/[CH_3CO_2H]$

$1.8 \times 10^{-5} = [x][0.0400 + x]/(0.0200 - x)$

$x = [H_3O^+] = 9.0 \times 10^{-6}$ M; pH = 5.05

18.3　pH = pK_a + log {[base]/[acid]}

pH = $-\log (1.8 \times 10^{-4})$ + log {[0.70]/[0.50]}

pH = 3.74 + 0.15 = 3.89

18.4　$(15.0 \text{ g NaHCO}_3)(1 \text{ mol}/84.01 \text{ g}) = 0.179$ mol $NaHCO_3$, and $(18.0 \text{ g Na}_2\text{CO}_3)(1 \text{ mol}/106.0 \text{ g}) = 0.170$ mol Na_2CO_3

pH = pK_a + log {[base]/[acid]}

pH = $-\log (4.8 \times 10^{-11})$ + log {[0.170]/[0.179]}

pH = 10.32 − 0.02 = 10.30

18.5　pH = pK_a + log {[base]/[acid]}

$5.00 = -\log (1.8 \times 10^{-5})$ + log {[base]/[acid]}

$5.00 = 4.74$ + log {[base]/[acid]}

[base]/[acid] = 1.8

To prepare this buffer solution, the ratio [base]/[acid] must equal 1.8. For example, you can dissolve 1.8 mol (148 g) of $NaCH_3CO_2$ and 1.0 mol (60.05 g) of CH_3CO_2H in enough water to make 1.0 L of solution.

18.6　Initial pH (before adding acid):

pH = pK_a + log {[base]/[acid]}

$\quad = -\log (1.8 \times 10^{-4})$ + log {[0.70]/[0.50]}

$\quad = 3.74 + 0.15 = 3.89$

After adding acid, the added HCl will react with the weak base (formate ion) and form more formic acid. The net effect is to change the ratio of [base]/[acid] in the buffer solution.

Initial amount HCO_2H = 0.50 M $\times$ 0.500 L = 0.250 mol

Initial amount HCO_2^- = 0.70 M $\times$ 0.50 L = 0.350 mol

Amount HCl added = 1.0 M $\times$ 0.010 L = 0.010 mol

Amount HCO_2H after HCl addition = 0.250 mol + 0.010 mol = 0.26 mol

Initial amount HCO_2^- after HCl addition = 0.350 mol − 0.010 mol = 0.34 mol

pH = pK_a + log {[base]/[acid]}

pH = $-\log (1.8 \times 10^{-4})$ + log {[0.340]/[0.260]}

pH = 3.74 + 0.12 = 3.86

18.7　After addition of 25.0 mL base, half of the acid has been neutralized.

Initial amount HCl = 0.100 M $\times$ 0.0500 L = 0.00500 mol

Amount NaOH added = 0.100 M $\times$ 0.0250 L = 0.00250 mol

Amount HCl after reaction: 0.00500 − 0.00250 = 0.00250 mol HCl

[HCl] after reaction = 0.00250 mol/0.0750 L = 0.0333 M

This is a strong acid and completely ionized, so $[H_3O^+]$ = 0.0333 M and pH = 1.48.

After 50.50 mL base is added, a small excess of base is present in the 100.5 mL of solution. (Volume of excess base added is 0.50 mL = 5.0×10^{-4} L.)

Amount excess base = 0.100 M $\times$ 5.0×10^{-4} L $= 5.0 \times 10^{-5}$ mol

$[OH^-] = 5.0 \times 10^{-5}$ mol/0.1005 L = 4.9×10^{-4} M

pOH = $-\log (4.9 \times 10^{-4})$ = 3.31; pH = 14.00 − pOH $\quad = 10.69$

18.8　35.0 mL base will partially neutralize the acid.

Initial amount CH_3CO_2H = (0.100 M)(0.1000 L) $= 0.0100$ mol

Amount NaOH added = (0.10 M)(0.035 L) = 0.0035 mol

Amount CH_3CO_2H after reaction = 0.0100 − 0.0035 $= 0.0065$ mol

Amount $CH_3CO_2^-$ after reaction = 0.0035 mol

$[CH_3CO_2H]$ after reaction = 0.0065 mol/0.135 L $= 0.0481$ M

$[CH_3CO_2^-]$ after reaction $= 0.00350 \text{ mol}/0.135 \text{ L}$
$$= 0.0259 \text{ M}$$

$K_a = [H_3O^+][CH_3CO_2^-]/[CH_3CO_2H]$

$1.8 \times 10^{-5} = [x][0.0259 + x]/[0.0481 - x]$

$x = [H_3O^+] = 3.34 \times 10^{-5} \text{ M}; \text{pH} = 4.48$

18.9   75.0 mL acid will partially neutralize the base.
Initial amount $NH_3 = (0.100 \text{ M})(0.1000 \text{ L}) = 0.0100 \text{ mol}$
Amount HCl added $= (0.100 \text{ M})(0.0750 \text{ L}) = 0.00750 \text{ mol}$
Amount NH_3 after reaction $= 0.0100 - 0.00750$
$$= 0.0025 \text{ mol}$$
Amount NH_4^+ after reaction $= 0.00750 \text{ mol}$
Solve using the Henderson-Hasselbach equation; use K_a for the weak acid NH_4^+:
pH $= pK_a + \log\{[\text{base}]/[\text{acid}]\}$
pH $= -\log(5.6 \times 10^{-10}) + \log\{[0.0025]/[0.00750]\}$
pH $= 9.25 - 0.48 = 8.77$

18.10   An indicator that changes color near the pH at the equivalence point is required. Possible indicators include methyl red, bromcresol green, and Eriochrome black T; all change color in the pH range of 5–6.

18.11   (a) $AgI \rightleftharpoons Ag^+ + I^-$
$$K_{sp} = [Ag^+][I^-]; K_{sp} = 8.5 \times 10^{-17}$$
(b) $BaF_2 \rightleftharpoons Ba^{2+} + 2F^-$
$$K_{sp} = [Ba^{2+}][F^-]^2; K_{sp} = 1.8 \times 10^{-7}$$
(c) $Ag_2CO_3 \rightleftharpoons 2Ag^+ + CO_3^{2-}$
$$K_{sp} = [Ag^+]^2[CO_3^{2-}]; K_{sp} = 8.5 \times 10^{-12}$$

18.12 $[Ba^{2+}] = 3.6 \times 10^{-3} \text{ M}; [F^-] = 7.2 \times 10^{-3} \text{ M}$

$K_{sp} = [Ba^{2+}][F^-]^2$

$K_{sp} = [3.6 \times 10^{-3}][7.2 \times 10^{-3}]^2 = 1.9 \times 10^{-7}$

18.13 $Ca(OH)_2 \rightleftharpoons Ca^{2+} + 2OH^-$
$$K_{sp} = [Ca^{2+}][OH^-]^2; K_{sp} = 5.5 \times 10^{-5}$$
$5.5 \times 10^{-5} = [x][2x]^2$ (where $x =$ solubility in mol/L)

$x = 2.4 \times 10^{-2} \text{ mol/L}$

sol. in g/L $= (2.4 \times 10^{-2} \text{ mol/L})(74.1 \text{ g/mol})$
$$= 1.8 \text{ g/L}$$

18.14   (a) AgCl
(b) $Ca(OH)_2$
(c) Because these compounds have different stoichiometries, the most soluble cannot be identified without doing a calculation. The solubility of $Ca(OH)_2$ is 2.4×10^{-2} M (from Exercise 18.13); $Ca(OH)_2$ is more soluble than $CaSO_4$, whose solubility is 7.0×10^{-3} M $\{K_{sp} = [Ca^{2+}][SO_4^{2-}]; 4.9 \times 10^{-5} = [x][x]; x = 7.0 \times 10^{-3} \text{ M}\}$.

18.15   (a) In pure water:
$$K_{sp} = [Ba^{2+}][SO_4^{2-}]; 1.1 \times 10^{-10} = [x][x];$$
$$x = 1.0 \times 10^{-5} \text{ M}$$
(b) In 0.010 M $Ba(NO_3)_2$, which furnishes 0.010 M Ba^{2+} in solution:
$$K_{sp} = [Ba^{2+}][SO_4^{2-}]; 1.1 \times 10^{-10} = [0.010 + x][x];$$
$$x = 1.1 \times 10^{-8} \text{ M}$$

18.16   (a) In pure water:
$$K_{sp} = [Zn^{2+}][CN^-]^2; 8.0 \times 10^{-12} = [x][2x]^2 = 4x^3$$
Solubility $= x = 1.3 \times 10^{-4} \text{ M}$
(b) In 0.10 M $Zn(NO_3)_2$, which furnishes 0.10 M Zn^{2+} in solution:
$$K_{sp} = [Zn^{2+}][CN^-]^2; 8.0 \times 10^{-12} = [0.10 + 2x][2x]^2$$
Solubility $= x = 4.5 \times 10^{-6} \text{ M}$

18.17   When $[Pb^{2+}] = 1.1 \times 10^{-3} \text{ M}$, $[I^-] = 2.2 \times 10^{-3} \text{ M}$.

$Q = [Pb^{2+}][I^-]^2 = [1.1 \times 10^{-3}][2.2 \times 10^{-3}]^2 = 5.3 \times 10^{-9}$

This value is less than K_{sp}, which means that the system has not yet reached equilibrium and more PbI_2 will dissolve.

18.18 $Q = [Sr^{2+}][SO_4^{2-}] = [2.5 \times 10^{-4}][2.5 \times 10^{-4}]$
$$= 6.3 \times 10^{-8}$$

This value is less than K_{sp}, which means that the system has not yet reached equilibrium. Precipitation will not occur.

18.19 $K_{sp} = [Pb^{2+}][I^-]^2$. Let x be the concentration of I^- required at equilibrium. If an amount greater than x is used, precipitation will occur.

$9.8 \times 10^{-9} = [0.050][x]^2$

$x = [I^-] = 4.4 \times 10^{-5} \text{ M}$

Let x be the concentration of Pb^{2+} in solution, in equilibrium with 0.0015 M I^-.

$9.8 \times 10^{-9} = [x][1.5 \times 10^{-3}]^2$

$x = [Pb^{2+}] = 4.4 \times 10^{-3} \text{ M}$

18.20   First determine the concentrations of Ag^+ and Cl^-; then calculate Q and see whether it is greater than or less than K_{sp}. Concentrations are calculated using the final volume, 105 mL, in the equation $C_{dil} \times V_{dil} = C_{conc} \times V_{conc}$.

$[Ag^+](0.105 \text{ L}) = (0.0010 \text{ mol/L})(0.100 \text{ L})$
$$[Ag^+] = 9.5 \times 10^{-4} \text{ M}$$

$[Cl^-](0.105 \text{ L}) = (0.025 \text{ M})(0.005 \text{ L})$
$$[Cl^-] = 1.2 \times 10^{-3} \text{ M}$$

$Q = [Ag^+][Cl^-] = [9.5 \times 10^{-4}][1.2 \times 10^{-3}] = 1.1 \times 10^{-6}$

$Q > K_{sp}$; precipitation occurs.

18.21 $Cu(OH)_2 \rightleftharpoons Cu^{2+} + 2OH^- \qquad K_{sp} = [Cu^{2+}][OH^-]^2$
$Cu^{2+} + 4NH_3 \rightleftharpoons Cu(NH_3)_4^{2+}$
$$K_{form} = [Cu(NH_3)_4^{2+}]/[Cu^{2+}][NH_3]^4$$
Net: $Cu(OH)_2 + 4NH_3 \rightleftharpoons Cu(NH_3)_4^{2+} + 2OH^-$
$K_{net} = K_{sp} \times K_{form} = (2.2 \times 10^{-20})(6.8 \times 10^{12}) = 1.5 \times 10^{-7}$

18.22   (a) Add NaCl(aq). AgCl will precipitate; $CaCl_2$ is soluble.
(b) Add Na_2S(aq) or NaOH(aq) to precipitate FeS or $Fe(OH)_2$; K_2S and KOH are soluble.

Chapter 19

19.1   (a) O_3; larger molecules generally have higher entropies than smaller molecules.

(b) $SnCl_4(g)$; gases have higher entropies than liquids.

19.2 (a) $\Delta S° = \sum S° \text{ (products)} - \sum S° \text{ (reactants)}$

$\Delta S° = S° [NH_4Cl(aq)]) - S° [NH_4Cl(s)]$

$\Delta S° = (1 \text{ mol})(169.9 \text{ J/mol} \cdot \text{K}) - (1 \text{ mol})(94.85 \text{ J/mol} \cdot \text{K}) = 75.1 \text{ J/K}$

A gain in entropy for the formation of a mixture (solution) is expected.

(b) $\Delta S° = 2 S° (NH_3) - [S° (N_2) + 3 S° (H_2)]$

$\Delta S° = (2 \text{ mol})(192.77 \text{ J/mol} \cdot \text{K}) - [(1 \text{ mol})(191.56 \text{ J/mol} \cdot \text{K}) + (3 \text{ mol})(130.7 \text{ J/mol} \cdot \text{K})]$

$\Delta S° = -198.1 \text{ J/K}$

A decrease in entropy is expected because there is a decrease in the number of moles of gases.

19.3 (a) Type 2

(b) Type 3

(c) Type 1

(d) Type 2

19.4 $\Delta S°_{sys} = 2 S° (HCl) - [S° (H_2) + S° (Cl_2)]$

$\Delta S°_{sys} = (2 \text{ mol})(186.2 \text{ J/mol} \cdot \text{K}) - [(1 \text{ mol})(130.7 \text{ J/mol} \cdot \text{K}) + (1 \text{ mol})(223.08 \text{ J/mol} \cdot \text{K})] = 18.6 \text{ J/K}$

$\Delta S°_{surr} = -\Delta H°_{sys}/T = -(-184,620 \text{ J}/298 \text{ K}) = 619.5 \text{ J/K}$

$\Delta S°_{univ} = \Delta S°_{sys} + \Delta S°_{surr} = 18.6 \text{ J/K} + 619.5 \text{ J/K} = 638.1 \text{ J/K}$

19.5 At 298 K, $\Delta S°_{surr} = -(467,900 \text{ J/K})/298 \text{ K} = -1570 \text{ J/K}$

$\Delta S°_{univ} = \Delta S°_{sys} + \Delta S°_{surr} = 560.7 \text{ J/K} - 1570 \text{ J/K} = -1010 \text{ J/K}$

The negative sign indicates that the process is not spontaneous. At higher temperature, the value of $-\Delta H°_{sys}/T$ will be less negative. At a high enough temperature, matter dispersal will outweigh the energy dispersal in the system and the reaction will be spontaneous.

19.6 For the reaction $N_2(g) + 3 H_2(g) \longrightarrow 2 NH_3(g)$:

$\Delta H°_{rxn} = 2 \Delta H°_f$ for $NH_3(g) = (2 \text{ mol})(-45.90 \text{ kJ/mol}) = -91.80 \text{ kJ}$

$\Delta S° = 2 S° (NH_3) - [S° (N_2) + 3 S° (H_2)]$

$\Delta S° = (2 \text{ mol})(192.77 \text{ J/mol} \cdot \text{K}) - [(1 \text{ mol})(191.56 \text{ J/mol} \cdot \text{K}) + (3 \text{ mol})(130.7 \text{ J/mol} \cdot \text{K})]$

$\Delta S° = -198.1 \text{ J/K}$

$\Delta G°_f = \Delta H°_f - T\Delta S° = -91.80 \text{ kJ} - (298 \text{ K})(-0.198 \text{ kJ/K})$

$\Delta G°_f = -32.8 \text{ kJ}$

19.7 $SO_2(g) + \frac{1}{2} O_2(g) \longrightarrow SO_3(g)$

$\Delta G° = \sum \Delta G° \text{ (products)} - \sum \Delta G° \text{ (reactants)}$

$\Delta G° = \Delta G° [SO_3(g)] - \{\Delta G° [SO_2(g)] + \frac{1}{2} \Delta G° [O_2(g)]\}$

$\Delta G° = -371.04 \text{ kJ} - (-300.13 \text{ kJ} + 0) = -70.91 \text{ kJ}$

19.8 $HgO(s) \longrightarrow Hg(\ell) + \frac{1}{2} O_2(g)$; Determine the temperature at which $\Delta G°_f = 0$, in which case $\Delta H° - T\Delta S° = 0$. T is the unknown in this problem.

$\Delta H° = 90.83 \text{ kJ} = [-\Delta H°_f \text{ for } HgO(s)]$

$\Delta S° = S° [Hg(\ell)]) + \frac{1}{2} S° (O_2) - S° (HgO(s))$

$\Delta S° = (1 \text{ mol})(76.02 \text{ J/mol} \cdot \text{K}) + [(0.5 \text{ mol})(205.07 \text{ J/mol} \cdot \text{K}) - (1 \text{ mol})(70.29 \text{ J/mol} \cdot \text{K})] = 108.26 \text{ J/K}$

$\Delta H° - T\Delta S° = 90,830 \text{ J} - T(108.27 \text{ J/K}) = 0$

$T = 839 \text{ K} (566 °C)$

19.9 $C(s) + CO_2(g) \rightleftharpoons 2 CO(g)$

$\Delta G°_{rxn} = 2 \Delta G°_f (CO) - \Delta G°_f (CO_2)$

$\Delta G°_{rxn} = (2 \text{ mol})(-137.17 \text{ kJ/mol}) - (1 \text{ mol})(-394.36 \text{ kJ/mol})$

$\Delta G°_{rxn} = 120.02 \text{ kJ}$

$\Delta G°_{rxn} = -RT \ln K$

$120,020 \text{ J/mol} = -(8.3145 \text{ J/mol} \cdot \text{K})(298 \text{ K})(\ln K)$

$K = 8.94 \times 10^{-22}$

19.10 $\Delta G°_{rxn} = -RT \ln K$

$\Delta G°_{rxn} = -(8.3145 \text{ J/mol} \cdot \text{K})(298 \text{ K}) [\ln (1.6 \times 10^7)]$

$\Delta G°_{rxn} = -41,100 \text{ J/mol} (-41.1 \text{ kJ/mol})$

Chapter 20

20.1 Oxidation half-reaction: $Al(s) \longrightarrow Al^{3+}(aq) + 3 e^-$

Reduction half-reaction: $2 H^+(aq) + 2 e^- \longrightarrow H_2(g)$

Overall reaction: $2 Al(s) + 6 H^+(aq) \longrightarrow 2 Al^{3+}(aq) + 3 H_2(g)$

Al is the reducing agent and is oxidized; $H^+(aq)$ is the oxidizing agent and is reduced.

20.2 $2 VO^{2+}(aq) + Zn(s) + 4 H^+(aq) \longrightarrow Zn^{2+}(aq) + 2 V^{3+}(aq) + 2 H_2O(\ell)$

$2 V^{3+}(aq) + Zn(s) \longrightarrow 2 V^{2+}(aq) + Zn^{2+}(aq)$

20.3 Oxidation half-reaction: $Fe^{2+}(aq) \longrightarrow Fe^{3+}(aq) + e^-$

Reduction half-reaction:

$MnO_4^-(aq) + 8 H^+(aq) + 5 e^- \longrightarrow Mn^{2+}(aq) + 4 H_2O(\ell)$

Overall reaction:

$MnO_4^-(aq) + 8 H^+(aq) + 5 Fe^{2+}(aq) \longrightarrow Mn^{2+}(aq) + 5 Fe^{3+}(aq) + 4 H_2O(\ell)$

20.4 (a) Oxidation half-reaction: $Al(s) + 3 OH^-(aq) \longrightarrow Al(OH)_3(s) + 3 e^-$

Reduction half-reaction: $S(s) + H_2O(\ell) + 2 e^- \longrightarrow HS^-(aq) + OH^-(aq)$

Overall reaction:

$2 Al(s) + 3 S(s) + 3 H_2O(\ell) + 3 OH^-(aq) \longrightarrow 2 Al(OH)_3(s) + 3 HS^-(aq)$

(b) Aluminum is the reducing agent and is oxidized; sulfur is the oxidizing agent and is reduced.

20.5 Construct two half-cells, the first with a silver electrode and a solution containing $Ag^+(aq)$, and the second with a nickel electrode and a solution containing $Ni^{2+}(aq)$. Connect the two half-cells with a salt bridge. When the electrodes are connected through an external circuit, electrons will flow from the anode (the nickel electrode) to the cathode (the silver electrode). The overall cell reaction is $Ni(s) + 2 Ag^+(aq) \longrightarrow Ni^{2+}(aq) + 2 Ag(s)$. To maintain electrical neutrality in the two half-cells,

negative ions will flow from the $Ag|Ag^+$ half-cell to the $Ni|Ni^{2+}$ half-cell, and positive ions will flow in the opposite direction.

20.6 Anode reaction: $Zn(s) \longrightarrow Zn^{2+}(aq) + 2\,e^-$
Cathode reaction: $2\,Ag^+(aq) + 2\,e^- \longrightarrow 2\,Ag(s)$
$E°_{cell} = E°_{cathode} - E°_{anode} = 0.80\,V - (-0.76\,V) = 1.56\,V$

20.7 Mg is easiest to oxidize, and Au is the most difficult. (See Table 20.1.)

20.8 Use the "northwest-southeast rule" or calculate the cell voltage to determine whether a reaction is product-favored. Reactions (a) and (c) are reactant-favored; reactions (b) and (d) are product-favored.

20.9 Overall reaction: $Fe(s) + 2\,H^+(aq) \longrightarrow$
$\qquad\qquad Fe^{2+}(aq) + H_2(g)$ $(E°_{cell} = 0.44\,V, n = 2)$
$E_{cell} = E°_{cell} - (0.0257/n)\,\ln\,\{[Fe^{2+}]P_{H_2}/[H^+]^2\}$
$\qquad = 0.44 - (0.0257/2)\,\ln\,\{[0.024]1.0/[0.056]^2\}$
$\qquad = 0.44\,V - 0.026\,V = 0.41\,V$

20.10 Overall reaction: $2\,Al(s) + 3\,Fe^{2+}(aq) \longrightarrow$
$\qquad\qquad\qquad\qquad 2\,Al^{3+}(aq) + 3\,Fe(s)$
$(E°_{cell} = 1.22\,V, n = 6)$
$E_{cell} = E°_{cell} - (0.0257/n)\,\ln\,\{[Al^{3+}]^2/[Fe^{2+}]^3\}$
$\qquad = 1.22 - (0.0257/6)\,\ln\,\{[0.025]^2/[0.50]^3\}$
$\qquad = 1.22\,V - (-0.023)\,V = 1.24\,V$

20.11 $\Delta G° = -nFE° = -(2\,mol\,e^-)(96,500\,C/mol\,e^-)$
$\qquad\qquad (-0.76\,V)(1\,J/1\,C \cdot V)$
$\qquad = 146,680\,J\ (= 150\,kJ)$

The negative value of $E°$ and the positive value of $\Delta G°$ both indicate a reactant-favored reaction.

20.12 $E°_{cell} = E°_{cathode} - E°_{anode} = 0.80\,V - 0.855\,V = -0.055\,V;$
$n = 2$
$E° = (0.0257/n)\,\ln\,K$
$-0.055 = (0.0257/2)\,\ln\,K$
$K = 0.014$

20.13 Cathode: $Zn^{2+}(aq) + 2\,e^- \longrightarrow Zn(s)$
$\qquad\qquad\qquad\qquad\qquad E°_{cathode} = -0.76\,V$
Anode: $Zn(s) + 4\,CN^-(aq) \longrightarrow [Zn(CN)_4{}^{2-}] + 2\,e^-$
$\qquad\qquad\qquad\qquad\qquad E°_{anode} = -1.26\,V$
Overall: $Zn^{2+}(aq) + 4\,CN^-(aq) \longrightarrow [Zn(CN)_4{}^{2-}]$
$\qquad\qquad\qquad\qquad\qquad E°_{cell} = 0.50\,V$
$E° = (0.0257/n)\,\ln\,K$
$0.50 = (0.0257/2)\,\ln\,K$
$K = 7.9 \times 10^{16}$

20.14 Cathode: $2\,H_2O(\ell) + 2\,e^- \longrightarrow 2\,OH^-(aq) + H_2(g)$
$\qquad\qquad\qquad\qquad\qquad E°_{cathode} = -0.83\,V$
Anode: $4\,OH^-(aq) \longrightarrow O_2(g) + 2\,H_2O(\ell) + 4\,e^-$
$\qquad\qquad\qquad\qquad\qquad E°_{anode} = 0.40\,V$
Overall: $2\,H_2O(\ell) \longrightarrow 2\,H_2(g) + O_2(g)$
$E°_{cell} = E°_{cathode} - E°_{anode} = -0.83\,V - 0.40\,V = -1.23\,V$

This is the minimum voltage needed to cause this reaction to occur.

20.15 O_2 is formed at the anode, by the reaction $2\,H_2O(\ell)$
$\longrightarrow 4\,H^+(aq) + O_2(g) + 4\,e^-$.

$(0.445\,A)(45\,min)(60\,s/min)(1\,C/1\,A \cdot s)$
$(1\,mol\,e^-/96,500\,C)(1\,mol\,O_2/4\,mol\,e^-)(32\,g\,O_2/1\,mol$
$O_2) = 0.10\,g\,O_2$

20.16 The cathode reaction (electrolysis of molten NaCl) is
$Na^+(\ell) + e^- \longrightarrow Na(\ell)$.

$(25 \times 10^3\,A)(60\,min)(60\,s/min)(1\,C/1\,A \cdot s)$
$(1\,mol\,e^-/96,500\,C)(1\,mol\,Na/mol\,e^-)(23\,g\,Na/1\,mol$
$Na) = 21,450\,g\,Na = 21\,kg$

Chapter 21

21.1 (a) $2\,Na(s) + Br_2(\ell) \longrightarrow 2\,NaBr(s)$
(b) $Ca(s) + Se(s) \longrightarrow CaSe(s)$
(c) $4\,K(s) + O_2(g) \longrightarrow 2\,K_2O(s)$

K_2O is one of the possible products of this reaction. The primary product from the reaction of potassium and oxygen is KO_2, potassium superoxide.

(d) $2\,Al(s) + 3\,Cl_2(g) \longrightarrow 2\,AlCl_3(s)$

21.2 (a) H_2Te
(b) Na_3AsO_4
(c) $SeCl_6$
(d) $HBrO_4$

21.3 (a) $NH_4{}^+$ (ammonium ion)
(b) $O_2{}^{2-}$ (peroxide ion)
(c) N_2H_4 (hydrazine)
(d) NF_3 (nitrogen trifluoride)

21.4 (a) ClO is an odd-electron molecule, with Cl having the unlikely oxidation number of +2.
(b) In Na_2Cl, chlorine would have the unlikely charge of 2− (to balance the two positive charges of the two Na^+ ions).
(c) This compound would require either the calcium ion to have the formula Ca^+ or the acetate ion to have the formula $CH_3CO_2{}^{2-}$. In all of its compounds, calcium occurs as the Ca^{2+} ion. The acetate ion, formed from acetic acid by loss of H^+, has a 1− charge.
(d) No octet structure for C_3H_7 can be drawn. This species has an odd number of electrons.

21.5 $CH_4(g) + H_2O(g) \longrightarrow 3\,H_2(g) + CO(g)$

Bonds broken: 4 C—H and 2 O—H (sum = 2578 kJ)

$4\,D(C—H) = 4(413\,kJ) = 1652\,kJ$

$2\,D(O—H) = 4(463\,kJ) = 926\,kJ$

Bonds formed: 3 H—H and 1 C≡O (sum = 2354 kJ)

$3\,D(H—H) = 3(436\,kJ) = 1308\,kJ$

$D(CO) = 1046\,kJ$

Estimated energy of reaction = 2578 kJ − 2354 kJ
$\qquad\qquad\qquad\qquad\qquad = +224\,kJ$

The triple-bond energy is small, relative to the energies of the three single bonds. Six bonds were broken and made, but the incremental energy to form the second and third bonds between C and O is not sufficient to overcome the energy required to break two single bonds.

21.6 Cathode reaction: $Na^+ + e^- \longrightarrow Na(\ell)$; 1 F, or 96,500 C, is required to form 1 mol of Na. There are (24 h) (60 min/h) (60 s/min) = 86,400 s in 1 day. 1000. kg = 1.000×10^6 g.

$(1.000 \times 10^6$ g Na) (1 mol Na/23.00 g Na) (96,500 C/mol Na) (1 A · s/1 C) (1/86,400 s) = 4.855×10^4 A

21.7 Some interesting topics: gemstones of the mineral beryl; uses of Be in the aerospace industry and in nuclear reactors; beryllium–copper alloys; severe health hazards when beryllium or its compounds get into the lungs.

21.8 (a) $Ga(OH)_3(s) + 3 H^+(aq) \longrightarrow Ga^{3+}(aq) + 3 H_2O(\ell)$
$Ga(OH)_3(s) + OH^-(aq) \longrightarrow Ga(OH)_4^-(aq)$
(b) $Ga^{3+}(aq)$ $(K_a = 1.2 \times 10^{-3})$ is stronger acid than $Al^{3+}(aq)$ $(K_a = 7.9 \times 10^{-6})$

21.9

21.10 (a) $:N\equiv N-\ddot{O}: \longleftrightarrow \ddot{N}=N=\ddot{O}:$

The first resonance structure places the negative charge on oxygen; the second places it on the terminal nitrogen. Because oxygen is more electronegative and better able to accommodate the negative charge, the first structure is favored.

(b) $NH_4NO_3(s) \longrightarrow N_2O(g) + 2 H_2O(g)$
$\Delta H°$ (reaction) $= \Delta H_f°$ (products) $- \Delta H_f°$ (reactants)
$\Delta H°$ (reaction) $= \Delta H_f°$ (N_2O) $+ 2 \Delta H_f°$ (H_2O) $- \Delta H_f°$ (NH_4NO_3)
$= 82.05$ kJ $+ 2(-241.83$ kJ)
$- (-365.56$ kJ) $= -36.05$ kJ

The reaction is exothermic.

21.11 First, calculate $\Delta G°$, $\Delta H°$, and $\Delta S°$ for this reaction, using data from Appendix L.
$\Delta G° = \Delta G_f°$ (products) $- \Delta G_f°$ (reactants)
$\Delta G° = 2 \Delta G_f°$ (ZnO) $+ 2 \Delta G_f°$ (SO_2) $- 2 \Delta G_f°$ (ZnS) $- 3 \Delta G_f°(O_2)$
$= 2(-318.30$ kJ) $+ 2(-300.13$ kJ) $- 2(-201.29$ kJ) $- 0 = -834.28$ kJ. The reaction is product-favored at 298 K.
$\Delta H° = 2 \Delta H_f°$ (ZnO) $+ 2 \Delta H_f°$ (SO_2) $- 2 \Delta H_f°$ (ZnS) $- 3 \Delta H_f°(O_2)$
$= 2(-348.28$ kJ) $+ 2(-296.84$ kJ) $- [2(-205.98$ kJ) $+ 0]$
$= -878.28$ kJ
$\Delta S° = 2 S°$ (ZnO) $+ 2 S°$ (SO_2) $- 2 S°$ (ZnS) $- 3 S°$ (O_2)
$= 2(43.64$ J/K) $+ 2(248.21$ J/K) $- [2(57.7$ J/K) $+ 3(205.07$ J/K)] $= -146.9$ J/K

This reaction is enthalpy favored and entropy disfavored. The reaction will become less favored at higher temperatures. See Table 19.2, page 920.

21.12 For the reaction $HX + Ag \longrightarrow AgX + \frac{1}{2} H_2$:
$\Delta G° = \Delta G_f°$ (products) $- \Delta G_f°$ (reactants)
$\Delta G° = \Delta G_f°$ (AgX) $- \Delta G_f°$ (HX)
For HF: $\Delta G° = +79.4$ kJ; reactant-favored
For HCl: $\Delta G° = -14.67$ kJ; product-favored
For HBr: $\Delta G° = -43.45$ kJ; product-favored
For HI: $\Delta G° = -67.75$ kJ; product-favored

Chapter 22

22.1 (a) $Co(NH_3)_3Cl_3$
(b) (i) $K_3[Co(NO_2)_6]$: a complex of cobalt(III) with a coordination number of 6
(ii) $Mn(NH_3)_4Cl_2$: a complex of manganese(II) with a coordination number of 6

22.2 (a) hexaaquanickel(II) sulfate
(b) dicyanobis(ethylenediamine)chromium(III) chloride
(c) potassium amminetrichloroplatinate(II)
(d) potassium dichlorocuprate(I)

22.3 (a) These are geometric isomers (with the NH_3 ligands in *cis* and *trans* positions).
(b) Only a single structure is possible.
(c) Only a single structure is possible.
(d) This compound is chiral; there are two optical isomers.
(e) Only a single structure is possible.
(f) Two structural isomers are possible based on coordination of the NO_2^- ligand through oxygen or nitrogen.

22.4 (a) $[Ru(H_2O)_6]^{2+}$: An octahedral complex of ruthenium(II) (d^6). A low-spin complex has no unpaired electrons and is diamagnetic. A high-spin complex has four unpaired electrons and is paramagnetic.

high-spin Ru^{2+} low-spin Ru^{2+}

(b) $[Ni(NH_3)_6]^{2+}$: An octahedral complex of nickel(II) (d^8). Only one electron configuration is possible; it has two unpaired electrons and is paramagnetic.

Ni^{2+} ion (d^8)

Chapter 23

23.1 (a) $^{222}_{86}Rn \longrightarrow ^{218}_{84}Po + ^{4}_{2}\alpha$

(b) $^{218}_{84}Po \longrightarrow ^{218}_{85}At + ^{0}_{-1}\beta$

23.2 (a) E (per photon) $= h\nu = hc/\lambda$

$E = [(6.626 \times 10^{-34} J \cdot s/photon)(3.00$
$\times 10^{8} m/s)]/(2.0 \times 10^{-12} m)$

$E = 9.94 \times 10^{-14} J/photon$

E (per mole) $= (9.94 \times 10^{-14} J/photon)(6.022$
$\times 10^{23} photons/mol)$

E (per mole) $= 5.99 \times 10^{10} J/mol$

23.3 (a) Emission of six α particles leads to a decrease of 24 in the mass number and a decrease of 12 in the atomic number. Emission of four β particles increases the atomic number by 4, but doesn't affect the mass. The final product of this process has a mass number of $232 - 24 = 208$ and an atomic number of $90 - 12 + 4 = 82$, identifying it as $^{208}_{82}Pb$.

(b) Step 1: $^{232}_{90}Th \longrightarrow ^{228}_{88}Ra + ^{4}_{2}\alpha$

Step 2: $^{228}_{88}Ra \longrightarrow ^{228}_{89}Ac + ^{0}_{-1}\beta$

Step 3: $^{228}_{89}Ac \longrightarrow ^{228}_{90}Th + ^{0}_{-1}\beta$

23.4 (a) $^{0}_{+1}\beta$

(b) $^{41}_{19}K$

(c) $^{0}_{-1}\beta$

(d) $^{22}_{12}Mg$

23.5 (a) $^{32}_{14}Si \longrightarrow ^{32}_{15}P + ^{0}_{-1}\beta$

(b) $^{45}_{22}Ti \longrightarrow ^{45}_{21}Sc + ^{0}_{+1}\beta$ or $^{45}_{22}Ti + ^{0}_{-1}e \longrightarrow ^{45}_{21}Sc$

(c) $^{239}_{94}Pu \longrightarrow ^{235}_{92}U + ^{4}_{2}\alpha$

(d) $^{42}_{19}K \longrightarrow ^{42}_{20}Ca + ^{0}_{-1}\beta$

23.6 $\Delta m = 0.03438$ g/mol

$\Delta E = (3.438 \times 10^{-5} kg/mol)(2.998 \times 10^{8})^{2}$
$= 3.090 \times 10^{12} J/mol (= 3.090 \times 10^{9} kJ/mol)$

$E_{b} = 5.150 \times 10^{8}$ kJ/mol nucleons

23.7 (a) 49.2 years is exactly 4 half-lives; quantity remaining = 1.5 mg$(1/2)^{4} = 0.094$ mg

(b) 3 half-lives, 36.9 years

(c) 1% is between 6 half-lives, 73.8 years (1/64 remains), and 7 half-lives, 86.1 years (1/128 remains)

23.8 $\ln([A]/[A_{o}]) = -kt$

$\ln([3.18 \times 10^{3}]/[3.35 \times 10^{3}]) = -k(2.00$ d)

$k = 0.0260$ d^{-1}

$t_{1/2} = 0.693/k = 0.693/(0.0260$ d$^{-1}) = 26.7$ d

23.9 $k = 0.693/t_{1/2} = 0.693/200.$ y $= 3.47 \times 10^{-3}$ y^{-1}

$\ln([A]/[A_{o}]) = -kt$

$\ln([3.00 \times 10^{3}]/[6.50 \times 10^{12}]) = -(3.47 \times 10^{-3}$ y$^{-1})t$

$\ln(4.62 \times 10^{-10}) = -(3.47 \times 10^{-3}$ y$^{-1})t$

$t = 6190$ y

23.10 $\ln([A]/[A_{o}]) = -kt$

$\ln([9.32]/[13.4]) = -(1.21 \times 10^{-4}$ y$^{-1})t$

$t = 3.00 \times 10^{3}$ y

23.11 3000 dpm$/x = 1200$ dpm$/ 60.0$ mg

$x = 150$ mg

Appendix O

Answers to Selected Study Questions

Chapter 1

1.1 (a) C, carbon (d) P, phosphorus
(b) K, potassium (e) Mg, magnesium
(c) Cl, chlorine (f) Ni, nickel

1.3 (a) Ba, barium (d) Pb, lead
(b) Ti, titanium (e) As, arsenic
(c) Cr, chromium (f) Zn, zinc

1.5 (a) Na (element) and NaCl (compound)
(b) Sugar (compound) and carbon (element)
(c) Gold (element) and gold chloride (compound)

1.7 (a) Physical property (d) Physical property
(b) Chemical property (e) Physical property
(c) Chemical property (f) Physical property

1.9 (a) Physical (colorless) and chemical (burns in air)
(b) Physical (shiny metal, orange liquid) and chemical (reacts with bromine)

1.11 555 g

1.13 2.79 cm^3 or 2.79 mL

1.15 Al, aluminum

1.17 298 K

1.19 (a) 289 K
(b) 97 °C
(c) 310 K

1.21 4.2195×10^4 m; 26.219 miles

1.23 5.3 cm^2; 5.3×10^{-4} m^2

1.25 250 cm^3; 0.25 L, 2.5×10^{-4} m^3; 0.25 dm^3

1.27 2.52×10^3 g

1.29 (a) Method A with all data included: average = 2.4 g/cm^3
Method B with all data included: average = 3.480 g/cm^3
For B the reading of 5.811 can be excluded because it is more than twice as large as all other readings. Using only the first three readings, you find average = 2.703 g/cm^3.
(b) Method A: error = 0.3 g/cm^3 or about 10%
Method B: error = 0.001 g/cm^3 or about 0.04%

(c) Before excluding a data point for B, method A gives more accurate and more precise answer. After excluding data for B, this method gives a more accurate and more precise result.

1.31 (a) Qualitative: blue-green color, solid physical state
Quantitative: density = 2.65 g/cm^3 and mass = 2.5 g
(b) Density, physical state, and color are intensive properties, whereas mass is an extensive property.
(c) Volume = 0.94 cm^3

1.33 (a) Al, Si, O
(b) All are solids. Aluminum metal (Al) and silicon (Si) chips are shiny, whereas the aquamarine crystal is a blue-gray color.

1.35 24.6 K, 27.1 K

1.37 0.197 nm; 197 pm

1.39 (a) 7.5×10^{-6} m; (b) 7.5×10^3 nm; (c) 7.5×10^6 pm

1.41 0.995 g Pt

1.43 50. mg procaine hydrochloride

1.45 The piece of brass has a volume of 18.0 cm^3, so the water volume increases by 18.0 mL. The final water volume is 68.0 mL.

1.47 The experimental density of the necklace is 19 g/cm^3, so the necklace is gold. At $380 per troy ounce it is worth about $820, so $300 is a bargain (and too good to be true).

1.49 The mixture is heterogeneous. Iron is magnetic, so a small magnet will attract the iron chips and remove them from the nonmagnetic sand.

1.51 The density of the plastic is less than that of CCl$_4$, so the plastic will float on the liquid CCl$_4$. Aluminum is more dense than CCl$_4$, so aluminum will sink when placed in CCl$_4$.

1.53 Comparing the two substances on the basis of three properties:

	Sucrose	NaCl
Density (g/cm^3)	1.587	2.164
Melting temperature (°C)	160–186 (decomp)	800
Solubility (g in 100 mL water)	200	36

The melting (or decomposition) temperatures are significantly different, as are their water solubilities.

1.55   Your normal body temperature (about 98.6 °F) is 37 °C. As this is higher than gallium's melting point, the metal will melt in your hand.

1.57   HDPE will float in ethylene glycol, water, acetic acid, and glycerol.

1.59

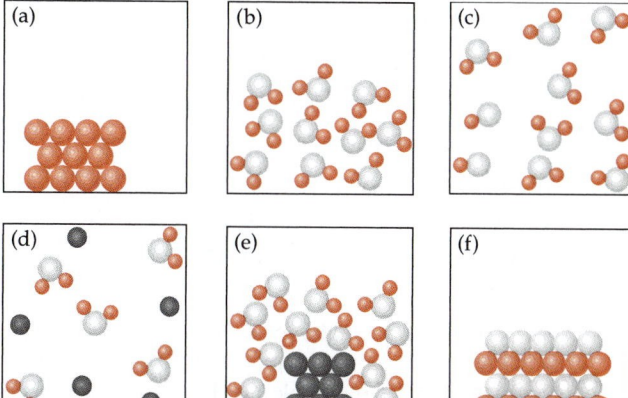

1.61   One could check for an odor, check the boiling or freezing point, or determine the density. If the density is approximately 1 g/cm³ at room temperature, the liquid could be water. If it boils at about 100 °C and freezes about 0 °C, that would be consistent with water. To check for the presence of salt, boil the liquid away. If a substance remains, it could be salt, but further testing would be required.

1.63

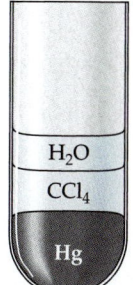

1.65   (a) Solid potassium metal reacts with liquid water to produce gaseous hydrogen and a homogeneous mixture (solution) of potassium hydroxide in liquid water.

(b) The reaction is a chemical change.

(c) The reactants are potassium and water. The products are hydrogen gas and a water (aqueous) solution of potassium hydroxide. Heat and light are also evolved.

(d) Among the qualitative observations are (i) the reaction is violent and (ii) heat and light (a purple flame) are produced.

1.67   (a) The water could be evaporated by heating the solution, leaving the salt behind.

(b) Use a magnet to attract the iron away from lead, which is not magnetic.

(c) Mixing the solids with water will dissolve only the sugar. Filtration would separate the solid sulfur from the sugar solution. Finally, the sugar could be separated from the water by evaporating the water.

1.69   As described on Screen 1.18 of the General ChemistryNow CD-ROM or website, using a reasonably powerful magnet will separate the iron from the cereal.

1.71   (a) The reactants are P_4 and Cl_2 and the product is PCl_3.

(b) The P_4 molecules are tetrahedra (four-sided polyhedra), and the chlorine molecules consist of two atoms. The PCl_3 molecule is a triangular pyramid.

Mathematics: Section 1.8

1.73   (a) 5.4×10^{-2}

(b) 5.462×10^3

(c) 7.92×10^{-4}

1.75   (a) 9.44×10^{-3}

(b) 5.69×10^3

(c) 1.19×10^1 (or 11.9)

1.77   (a) 3   (c) 5

(b) 3   (d) 4

1.79   0.122

1.81   Equation: $y = 248.4x + 0.0022$ (where y is the absorbance and x is the concentration). The slope is 248.4. The concentration is 2.548×10^{-3} when the absorbance is 0.635.

1.83   (a) $x = 0.21$ when $y = 4.0$

(b) $y = 5.6$ when $x = 0.30$

(c) Slope = 18 and intercept = 0.20

(d) When $x = 1.0$, $y = 18$

1.85 $C = 0.0823$

1.87 $T = 295$

1.89   Volume = 0.0854 cm³

1.91   Mass = 22 g

1.93   Correct conversion factor = 0.803 kg/L

Fuel in tank = 7682 L = 6170 kg

Additional fuel needed = 22,300 kg − 6170 kg = 16,130 kg (or 20,100 L)

1.95   Thickness of aluminum foil = 1.8×10^{-2} mm

1.97   Oil layer thickness = 2×10^{-7} cm. This is likely related to the "length" of the oil molecules.

1.99 $(0.546 \text{ g})\left(\dfrac{1 \text{ cm}^3}{8.96 \text{ g}}\right)\left(\dfrac{1 \text{ L}}{1000 \text{ cm}^3}\right) = 6.09 \times 10^{-5} \text{ L}$

1.101   Area $= 9.6 \times 10^3 \text{ m}^2$

1.103   (a) Volume $= 65 \text{ m}^3$ or $6.5 \times 10^4 \text{ L}$

       (b) Mass $= 78$ kg or 170 lb

1.105   (a) Calculated density $= 1.11$ g/mL. The unknown is ethylene glycol.

       (b) If the volume were 3.5 mL, the calculated density would be 1.1 g/mL. Although still indicating ethylene glycol, it is close enough to the density of acetic acid that you would be unsure of the answer. (Fortunately, acetic acid and ethylene glycol have very different odors and could be identified that way.)

1.107 (a,b) Density $= 1.25$ g/L. To three significant figures, three gases (N_2, C_2H_4, and CO) have this density.

       (c) Using the more accurate mass, the density is 1.249 g/L. This eliminates C_2H_4 from consideration, but N_2 and CO are still possible choices.

1.109.   Mass of Hg in the tube $= 0.153$ g

       Volume of Hg in the tube $= 0.0113 \text{ cm}^3$

       Using the formula for the volume of a cylinder, the radius of the cylinder of Hg in the tube is 0.0463 cm, which gives a diameter of 0.0927 cm.

Chapter 2

2.1   Atoms contain the following fundamental particles: protons ($+1$ charge), neutrons (zero charge), and electrons (-1 charge). Protons and neutrons are in the nucleus of an atom. Electrons are the least massive of the three particles.

2.3   The discovery of radioactivity showed that atoms must be divisible; that is, atoms must be composed of even smaller, subatomic particles.

2.5   Exercise 2.1 provides the relative sizes of the nuclear and atomic diameters, with the nuclear radius on the order of 0.001 pm and the atomic radius approximately 100 pm. If the nuclear diameter is 6 cm, then the atomic diameter is 600,000 cm (or 6 km).

2.7   Radon, Rn

2.9   (a) ^{27}Mg, mass number $= 12 + 15 = 27$

       (b) ^{48}Ti, mass number $= 22 + 26 = 48$

       (c) ^{62}Zn, mass number $= 30 + 32 = 62$

2.11   (a) $^{39}_{19}$K

       (b) $^{84}_{36}$Kr

       (c) $^{60}_{27}$Co

2.13

Element	^{24}Mg	^{119}Sn	^{232}Th
Electrons	12	50	90
Protons	12	50	90
Neutrons	12	69	142

2.15 ^{99}Tc has 43 protons, 43 electrons, and 56 neutrons.

2.17 $^{57}_{27}$Co, $^{58}_{27}$Co, $^{60}_{27}$Co

2.19 ^{205}Tl is more abundant (70.5%) than ^{203}Tl (29.5%). Its atomic weight is closer to 205 than to 203.

2.21 $(0.0750)(6.015121) + (0.9250)(7.016003) = 6.94$

2.23   (c), About 50%. Actual percent ^{107}Ag $= 51.839\%$.

2.25 ^{69}Ga, 60.12%; ^{71}Ga, 39.88%

2.27   (a) 68 g Al   (c) 0.60 g Ca

       (b) 0.0698 g Fe   (d) 1.32×10^4 g Ne

2.29   (a) 1.9998 mol Cu   (c) 2.1×10^{-5} mol Am

       (b) 0.0017 mol Li   (d) 0.250 mol Al

2.31   He has the smallest molar mass, and Fe has the largest molar mass. Therefore, 1.0 g of He has the largest number of atoms in these samples, and 1.0 g of Fe has the smallest number of atoms.

2.33 1.0552×10^{-22} g for 1 Cu atom

2.35   Five elements. Nitrogen (N) and phosphorus (P) are nonmetals, arsenic (As) and antimony (Sb) are metalloids, and bismuth (Bi) is a metal.

2.37   8 elements: periods 2 and 3. 18 elements: periods 4 and 5. 32 elements: period 6.

2.39   (a) Nonmetals: C, Cl

       (b) Main group elements: C, Cl, Cs, Ca

       (c) Lanthanides: Ce

       (d) Transition elements: Cr, Co, Cd, Cu, Ce, Cf, and Cm

       (e) Actinides: Cm, Cf

       (f) Gases: Cl

2.41   Metals: Na, Ni, Np.
Metalloids: None in this list
Nonmetals: N, Ne

2.43   Metals: sodium, scandium, strontium, silver, and samarium

       Main group elements: sodium, silicon, sulfur, selenium, strontium

       Transition metals: scandium, silver (some chemists include the lanthanides, such as samarium, in the transition elements)

2.45

Symbol	^{58}Ni	^{33}S	^{20}Ne	^{55}Mn
Protons	28	16	10	25
Neutrons	30	17	10	30
Electrons	28	16	10	25
Name	Nickel	Sulfur	Neon	Manganese

2.47   Potassium has an atomic weight of 39.0983. This mass is close to the mass of the ^{39}K isotope. Therefore, the abundance of ^{41}K is low (6.73%).

2.49   (a) Mg   (c) Si
        (b) H   (d) Fe
        (e) F, Cl, and Br. Chlorine is more abundant.

2.51   (a), (b), and (c) are all possible. (d) is impossible because one atom of S has a mass of 5.325×10^{-23} g. Therefore, one mole of molecules consisting of eight S atoms cannot be less than the mass of one atom.

2.53   (a) Beryllium, magnesium, calcium, strontium, barium, radium
        (b) Sodium, magnesium, aluminum, silicon, phosphorus, sulfur, chlorine, argon
        (c) Carbon   (g) Krypton
        (d) Sulfur   (h) Sulfur
        (e) Iodine   (i) Germanium or arsenic
        (f) Magnesium

2.55   (a) Three elements—Co, Ni, and Cu—have densities of about 9 g/cm^3.
        (b) Boron in the second period and aluminum in the third period have the largest densities. Both are in Group 3A.
        (c) Elements that have very low densities are all gases. These include hydrogen, helium, nitrogen, oxygen, fluorine, neon, chlorine, argon, and krypton.

2.57   (a) 0.5 mol Si
        (b) 0.5 mol Na
        (c) 10 atoms Fe

2.59   Boron; 1.4 mol or 8.4×10^{23} atoms

2.61 9.42×10^{-5} mol Kr; 5.67×10^{19} atoms Kr

2.63   40.2 g H$_2$ (b) < 103 g C (c) < 182 g Al (f) < 210 g Si (d) < 212 g Na (e) < 351 g Fe (a) < 650 g Cl$_2$(g)

2.65   (a) Average charge = 1.59×10^{-19} C; error is about 0.5%

(b) Drop	Number of Electrons
1	1
2	7
3	6
4	10
5	4

2.67   K = 78 weight %; Na = 22 weight %

2.69   The drawing should have two protons and two neutrons in the nucleus and two electrons outside the nucleus. The electrons do not trace a particular path around the nucleus but rather exist as a "cloud" (as in Figure 2.1). The radius of a He atom is 128 pm. If the nucleus is only 1×10^{-5} as large as the atom (see Exercise 2.1), then the nuclear radius is about 0.00128 pm. If you depict the nucleus as a penciled dot on paper (with a radius of about 0.1 mm), then the radius of the atom is 10 m!

2.71   Required data: density of iron, molar mass of iron, Avogadro's number

$$1.0 \text{ cm}^3\left(\frac{7.87 \text{ g}}{1 \text{ cm}^3}\right)\left(\frac{1 \text{ mol}}{55.85 \text{ g}}\right)\left(\frac{6.02 \times 10^{23} \text{ atoms}}{1 \text{ mol}}\right) = 8.5 \times 10^{22} \text{ atoms Fe}$$

2.73   Barium would be more reactive than calcium, so a more vigorous evolution of hydrogen should occur. Reactivity increases on descending the periodic table, at least for Groups 1A and 2A.

2.75   Volume = 3.33×10^3 cm^3. Edge = 14.9 cm.

2.77 4.9×10^{24} atoms Na

2.79 1.0028×10^{23} atoms C. If the accuracy is ±0.0001 g, the maximum mass could be 2.0001 g, which also represents 1.0028×10^{23} atoms C.

2.81   See Screen 2.20 of the General ChemistryNow CD-ROM or website for a possible method.

Chapter 3

3.1   Sulfuric acid, H$_2$SO$_4$. The structure is not flat. Chemists describe the structure as a tetrahedron of O atoms around the S atom.

3.3   Pt(NH$_3$)$_2$Cl$_2$

        structural formula

3.5   (a) Mg^{2+} (c) Ni^{2+}
        (b) Zn^{2+} (d) Ga^{3+}

3.7   (a) Ba^{2+} (e) S^{2-}
        (b) Ti^{4+} (f) ClO$_4^-$
        (c) PO$_4^{3-}$ (g) Co^{2+}
        (d) HCO$_3^-$ (h) SO$_4^{2-}$

3.9   K loses one electron per atom to form a K$^+$ ion. It has the same number of electrons as an Ar atom.

3.11   Ba^{2+} and Br$^-$ ions. Compound formula is BaBr$_2$.

3.13   (a) Two K$^+$ ions and one S^{2-} ion
        (b) One Co^{2+} ion and one SO$_4^{2-}$ ion
        (c) One K$^+$ ion and one MnO$_4^-$ ion
        (d) Three NH$_4^+$ ions and one PO$_4^{3-}$ ion
        (e) One Ca^{2+} ion and two ClO$^-$ ions

3.15   Co^{2+} gives CoO and Co^{3+} gives Co$_2$O$_3$.

3.17   (a) AlCl$_2$ should be AlCl$_3$ (based on one Al^{3+} ion and three Cl$^-$ ions).
        (b) KF$_2$ should be KF (based on one K$^+$ ion and one F$^-$ ion).
        (c) Ga$_2$O$_3$ is correct.
        (d) MgS is correct.

3.19 (a) Potassium sulfide
      (b) Cobalt(II) sulfate
      (c) Ammonium phosphate
      (d) Calcium hypochlorite

3.21 (a) $(NH_4)_2CO_3$ (d) $AlPO_4$
      (b) CaI_2 (e) $AgCH_3CO_2$
      (c) $CuBr_2$

3.23 Compounds with Na^+: Na_2CO_3 and NaI
      Compounds with Ba^{2+}: $BaCO_3$ and BaI_2

3.25 The force of attraction is stronger in NaF than in NaI because the distance between ion centers is smaller in NaF (235 pm) than in NaI (322 pm).

3.27 (a) Nitrogen trifluoride
      (b) Hydrogen iodide
      (c) Boron triiodide
      (d) Phosphorus pentafluoride

3.29 (a) SCl_2 (c) $SiCl_4$
      (b) N_2O_5 (d) B_2O_3

3.31 (a) 159.7 g/mol
      (b) 117.2 g/mol
      (c) 176.1 g/mol

3.33 (a) 290.8 g/mol
      (b) 249.7 g/mol

3.35 (a) 1.53 g
      (b) 4.60 g
      (c) 4.60 g

3.37 60.9 mol CH_3CN

3.39 Amount of SO_3 = 12.5 mol
      Number of molecules = 7.52×10^{24} molecules
      Number of S atoms = 7.52×10^{24} atoms
      Number of O atoms = 2.26×10^{25} atoms

3.41 (a) 86.60% Pb and 13.40% S
      (b) 81.71% C and 18.29% H
      (c) 79.96% C, 9.394% H, and 10.65% O

3.43 86.60% lead. There is 8.66 g of Pb in 10.0 g of PbS.

3.45 66.46% copper in CuS. 15.0 g of CuS is needed to obtain 10.0 g of Cu.

3.47 $C_4H_6O_4$

3.49 (a) CH, 26.0 g/mol; C_2H_2
      (b) CHO, 116.1 g/mol; $C_4H_4O_4$
      (c) CH_2, 112.2 g/mol, C_8H_{16}

3.51 Empirical formula, CH; molecular formula, C_2H_2

3.53 Empirical formula, C_3H_4; molecular formula, C_9H_{12}

3.55 Empirical and molecular formulas, $C_8H_8O_3$

3.57 Formula is $MgSO_4 \cdot 7 H_2O$

3.59 XeF_2

3.61 ZnI_2

3.63 $(NH_4)_2CO_3$, $(NH_4)_2SO_4$, $NiCO_3$, $NiSO_4$

3.65 A strontium atom has 38 electrons, but can lose 2 electrons to form a Sr^{2+} ion. (This characteristic is shared by all Group 2A metals.) The Sr^{2+} ion has 36 electrons, the same number as in Kr.

3.67 All of these compounds have one atom of some element plus three Cl atoms. The highest weight percent of chlorine will occur in the compound having the lightest central element. Here that is B, so BCl_3 should have the highest weight percent of Cl (90.77%).

3.69 All of these compounds have one atom of some element plus one O atom. The highest weight percent of oxygen will occur in the compound having the lightest central element. Here that is C, so CO should have the highest weight percent of O (57.12%).

3.71 Borate anion has the formula BO_3^{3-}.

3.73 3.0×10^{23} molecules represents 0.50 mol adenine. The molar mass of adenine ($C_5H_5N_5$) is 135.13 g/mol, so 0.50 mol adenine has a mass of 68 g. Thus, 0.50 mol has a larger mass than 40.0 g of the compound.

3.75 2×10^{21} molecules of water

3.77 245.8 g/mol. Mass percent: 25.86% Cu, 22.80% N, 5.742% H, 13.05% S, and 32.55% O. In 10.5 g of compound there are 2.72 g Cu and 0.770 g H_2O.

3.79 Empirical formula of malic acid: $C_4H_6O_5$

3.81 FeC_2O_4

3.83 (a) $C_{10}H_{15}NO$, molar mass = 165.23 g/mol
      (b) 72.69% C
      (c) 7.57×10^{-4} mol
      (d) 4.56×10^{20} molecules and 4.56×10^{21} C atoms

3.85 Ionic compounds
      (c) Li_2S, lithium sulfide
      (d) In_2O_3, indium oxide
      (g) CaF_2, calcium fluoride

3.87 (a) NaClO, ionic   (f) $(NH_4)_2SO_3$, ionic
      (b) BI_3 (g) KH_2PO_4, ionic
      (c) $Al(ClO_4)_3$, ionic   (h) S_2Cl_2
      (d) $Ca(CH_3CO_2)_2$, ionic (i) ClF_3
      (e) $KMnO_4$, ionic   (j) PF_3

3.89

Cation	Anion	Name	Formula
Li^+	ClO_4^-	Lithium perchlorate	$LiClO_4$
Al^{3+}	PO_4^{3-}	Aluminum phosphate	$AlPO_4$
Li^+	Br^-	Lithium bromide	$LiBr$
Ba^{2+}	NO_3^-	Barium nitrate	$Ba(NO_3)_2$
Al^{3+}	O^{2-}	Aluminum oxide	Al_2O_3
Fe^{3+}	CO_3^{2-}	Iron(III) carbonate	$Fe_2(CO_3)_3$

3.91 Empirical formula, C_5H_4; molecular formula, $C_{10}H_8$

3.93 Empirical and molecular formulas, $C_5H_{14}N_2$

3.95 $C_9H_7MnO_3$

3.97 1200 kg Cr_2O_3

3.99 Empirical formula, ICl_3; molecular formula, I_2Cl_6

3.101 7.35 kg iron

3.103 (d) Na_2MoO_4

3.105 5.52×10^{-4} mol $C_{21}H_{15}Bi_3O_{12}$; 0.346 g Bi

3.107 The unknown element is carbon.

3.109 $n = 19$

3.111 (a) 0.766 g Ni or 0.0130 mol

(b) NiF_2

(c) Nickel(II) fluoride

3.113 Answer (d) is correct. The other students apparently did not correctly calculate the number of moles of material in 100.0 g or they improperly calculated the ratio of those moles in determining their empirical formula.

3.115 (a) 7.10×10^{-4} mol U was used. The empirical formula is U_3O_8, and 2.37×10^{-4} mol U_3O_8 was obtained.

(b) ^{238}U is more abundant.

(c) The formula of the hydrated compound is $UO_2(NO_3)_2 \cdot 6\,H_2O$.

3.117 First, multiply the volume of the cube (27.0 cm³) by the density of alum (1.757 g/cm³). This gives the mass of alum in the cube. Next, multiply this mass by the weight percent of Al in alum (5.688% Al) to give the mass of Al in the cube. Finally, convert the mass of aluminum to the amount of aluminum (mol) and then use Avogadro's number to calculate the number of Al atoms in the crystal.

$$27.0 \text{ cm}^3 \left(\frac{1.757 \text{ g alum}}{1 \text{ cm}^3}\right)\left(\frac{5.688 \text{ g Al}}{100 \text{ g alum}}\right)$$
$$\times \left(\frac{1 \text{ mol Al}}{26.98 \text{ g Al}}\right)\left(\frac{6.022 \times 10^{23} \text{ atoms Al}}{1 \text{ mole Al}}\right)$$

Chapter 4

4.1 $C_5H_{12}(\ell) + 8\,O_2(g) \longrightarrow 5\,CO_2(g) + 6\,H_2O(g)$

4.3 (a) $4\,Cr(s) + 3\,O_2(g) \longrightarrow 2\,Cr_2O_3(s)$

(b) $Cu_2S(s) + O_2(g) \longrightarrow 2\,Cu(s) + SO_2(g)$

(c) $C_6H_5CH_3(\ell) + 9\,O_2(g) \longrightarrow 4\,H_2O(\ell) + 7\,CO_2(g)$

4.5 (a) $Fe_2O_3(s) + 3\,Mg(s) \longrightarrow 3\,MgO(s) + 2\,Fe(s)$

Reactants = iron(III) oxide, magnesium

Products = magnesium oxide, iron

(b) $AlCl_3(s) + 3\,NaOH(aq) \longrightarrow$
$$Al(OH)_3(s) + 3\,NaCl(aq)$$

Reactants = aluminum chloride, sodium hydroxide

Products = aluminum hydroxide, sodium chloride

(c) $2\,NaNO_3(s) + H_2SO_4(\ell) \longrightarrow$
$$Na_2SO_4(s) + 2\,HNO_3(\ell)$$

Reactants = sodium nitrate, sulfuric acid

Products = sodium sulfate, nitric acid

(d) $NiCO_3(s) + 2\,HNO_3(aq) \longrightarrow$
$$Ni(NO_3)_2(aq) + CO_2(g) + H_2O(\ell)$$

Reactants = nickel(II) carbonate, nitric acid

Products = nickel(II) nitrate, water

4.7 4.5 mol O_2; 310 g Al_2O_3

4.9 22.7 g Br_2; 25.3 g Al_2Br_6

4.11 (a) $4\,Fe(s) + 3\,O_2(g) \longrightarrow 2\,Fe_2O_3(s)$

(b) 3.83 g Fe_2O_3

(c) 1.15 g O_2

4.13 (a) 242 g $CaCO_3$; (b) 329 g $CaSO_4$

4.15

Equation	$2\,PbS(s)$	$+\ 3\,O_2(g)$	$\longrightarrow$	$2\,PbO(s)$	$+\ 2\,SO_2(g)$
Initial (mol)	2.5	0		0	0
Change (mol)	−2.5	(−3/2)(2.5)		(+2/2)(2.5)	(2/2)(2.5)
Final (mol)	0	0		2.5	2.5

The amounts table shows that 2.5 mol PbS requires $3/2(2.5) = 3.75$ mol O_2 and produces 2.5 mol PbO and 2.5 mol SO_2.

4.17 (a) Balanced equation: $4\,Cr(s) + 3\,O_2(g) \longrightarrow$
$$2\,Cr_2O_3(s)$$

(b, c) 0.175 g Cr is equivalent to 0.00337 mol

Equation	$4\,Cr(s)$	$+$	$3\,O_2(g)$	$\longrightarrow$	$2\,Cr_2O_3(s)$
Initial (mol)	0.00337		0		0
Change (mol)	−0.00337		(−3/4)(0.00337)		(2/4)(0.00337)
Final (mol)	0		0		(2/4)(0.00337)

The reaction produces (2/4)(0.00337 mol) Cr_2O_3 or 0.00169 mol. This is equivalent to 0.256 g. Mass of O_2 required = 0.081 g.

4.19   0.11 mol Na_2SO_4 and 0.62 mol C are mixed. Sodium sulfate is the limiting reactant. Therefore, 0.11 mol Na_2S is formed, or 8.2 g.

4.21 F_2 is the limiting reactant.

4.23   (a) CH_4 is the limiting reactant.
        (b) 375 g H_2
        (c) Excess H_2O = 1390 g

4.25   (a) $2 C_6H_{14}(\ell) + 19 O_2(g) \longrightarrow$
        $$12 CO_2(g) + 14 H_2O(g)$$
        (b) O_2 is the limiting reactant. Products are 187 g of CO_2 and 89.2 g of H_2O.
        (c) 154 g of hexane remains

4.27 $(332 \text{ g}/407 \text{ g})(100\%) = 81.6\%$

4.29   (a) 14.3 g $Cu(NH_3)_4SO_4$
        (b) 88.3% yield

4.31   91.9% hydrate

4.33   84.3% $CaCO_3$

4.35   1.467% Tl_2SO_4

4.37   Empirical formula, CH

4.39   Empirical formula, CH_2; molecular formula, C_5H_{10}

4.41   Empirical formula, CH_3O, and molecular formulas, $C_2H_6O_2$

4.43 $Ni(CO)_4$

4.45   (a) $CO_2(g) + 2 NH_3(g) \longrightarrow NH_2CONH_2(s) + H_2O(\ell)$
        (b) $UO_2(s) + 4 HF(aq) \longrightarrow UF_4(s) + 2 H_2O(\ell)$
        $UF_4(s) + F_2(g) \longrightarrow UF_6(s)$
        (c) $TiO_2(s) + 2 Cl_2(g) + 2 C(s) \longrightarrow$
        $$TiCl_4(\ell) + 2 CO(g)$$
        $TiCl_4(\ell) + 2 Mg(s) \longrightarrow Ti(s) + 2 MgCl_2(s)$

4.47   (a) Products = $CO_2(g)$ and $H_2O(g)$
        (b) $2 C_6H_6(\ell) + 15 O_2(g) \longrightarrow 12 CO_2(g) + 6 H_2O(g)$
        (c) 49.28 g O_2
        (d) 65.32 g products (= sum of C_6H_6 mass and O_2 mass)

4.49   71.1 mg

4.51   (a) $2 Fe(s) + 3 Cl_2(g) \longrightarrow 2 FeCl_3(s)$
        (b) 19.0 g Cl_2 required; 29.0 g $FeCl_3$ produced
        (c) 63.7% yield
        (d) 15.3 g $FeCl_3$

4.53   (a) Titanium(IV) chloride, water, titanium(IV) oxide, hydrogen chloride
        (b) 4.60 g H_2O
        (c) 10.2 TiO_2, 18.6 g HCl

4.55   8.33 g NaN_3

4.57   399 g Cu

4.59   The H : B ratio is 1.4 : 1.0. The empirical formula is B_5H_7.

4.61   Empirical formula, $C_{10}H_{20}O$

4.63   (a) $FeCl_2(aq) + Na_2S(aq) \longrightarrow FeS(s) + 2 NaCl(aq)$
        (b) $FeCl_2$
        (c) 28 g FeS produced
        (d) 15 g Na_2S remains
        (e) 40. g Na_2S requires 65 g $FeCl_2$

4.65   The metal is most likely copper, Cu.

4.67 Ti_2O_3 (which could be a mixture of TiO and TiO_2)

4.69   11.48% 2,4-D

4.71   (a) 333 kg $Na_2S_2O_4$
        (b) 369 kg of commercial product

4.73   858 kg H_2SO_4

4.75   1.59 g C_4H_{10} (55.6%) and 1.27 g C_4H_8 (44.4%)

4.77   85.4% $NaHCO_3$

4.79   (a) In the reactions represented by the sloping portion of the graph, Fe is the limiting reactant. At the point at which the yield of product begins to be constant (at 2.0 g Fe), the reactants are present in stoichiometric amounts. That is, 10.6 g of product contains 2.0 g Fe and 8.6 g Br_2.
        (b) 2.0 g Fe = 0.036 mol Fe; 8.6 g Br_2 = 0.054 mol Br_2. The mole ratio is 1.5 mol Br_2 to 1.0 mol Fe.
        (c) The mole ratio is 1.5 mol Br_2/1.0 mol Fe = 3 Br/1 Fe. The empirical formula is $FeBr_3$.
        (d) $2 Fe(s) + 3 Br_2(\ell) \longrightarrow 2 FeBr_3(s)$
        (e) Iron(III) bromide
        (f) Statement (i) is correct.

4.81   (a) 65.02% Pt, 9.34% N, and 23.63% Cl
        (b) 1.31 g NH_3 required and 11.6 g $Pt(NH_3)_2Cl_2$ produced

4.83   See General ChemistryNow CD-ROM or website Screen 4.8 for the calculations.

Chapter 5

5.1   Electrolytes are compounds whose solutions conduct electricity. Substances whose solutions are good electrical conductors are strong electrolytes (such as NaCl), whereas poor electrical conductors are weak electrolytes (such as acetic acid).

5.3   (a) $CuCl_2$
        (b) $AgNO_3$
        (c) All are water-soluble.

5.5   (a) K^+ and OH^- ions   (c) Li^+ and NO_3^- ions
        (b) K^+ and SO_4^{2-} ions   (d) NH_4^+ and SO_4^{2-} ions

5.7   (a) Soluble, Na^+ and CO_3^{2-} ions
        (b) Soluble, Cu^{2+} and SO_4^{2-} ions
        (c) Insoluble
        (d) Soluble, Ba^{2+} and Br^- ions

5.9 $CdCl_2(aq) + 2\,NaOH(aq) \longrightarrow$
$$Cd(OH)_2(s) + 2\,NaCl(aq)$$
$Cd^{2+}(aq) + 2\,OH^-(aq) \longrightarrow Cd(OH)_2(s)$

5.11   (a) $NiCl_2(aq) + (NH_4)_2S(aq) \longrightarrow$
$$NiS(s) + 2\,NH_4Cl(aq)$$
$Ni^{2+}(aq) + S^{2-}(aq) \longrightarrow NiS(s)$

(b) $3\,Mn(NO_3)_2(aq) + 2\,Na_3PO_4(aq) \longrightarrow$
$$Mn_3(PO_4)_2(s) + 6\,NaNO_3(aq)$$
$3\,Mn^{2+}(aq) + 2\,PO_4{}^{3-}(aq) \longrightarrow Mn_3(PO_4)_2(s)$

5.13 $HNO_3(aq) \longrightarrow H^+(aq) + NO_3{}^-(aq)$

5.15 $H_2C_2O_4(aq) \longrightarrow H^+(aq) + HC_2O_4{}^-(aq)$
$HC_2O_4{}^-(aq) \longrightarrow H^+(aq) + C_2O_4{}^{2-}(aq)$

5.17 $MgO(s) + H_2O(\ell) \longrightarrow Mg(OH)_2(s)$

5.19   (a) Acetic acid reacts with magnesium hydroxide to give magnesium acetate and water.
$$2\,CH_3CO_2H(aq) + Mg(OH)_2(s) \longrightarrow$$
$$Mg(CH_3CO_2)_2(aq) + 2\,H_2O(\ell)$$

(b) Perchloric acid reacts with ammonia to give ammonium perchlorate.
$$HClO_4(aq) + NH_3(aq) \longrightarrow NH_4ClO_4(aq)$$

5.21 $Ba(OH)_2(aq) + 2\,HNO_3(aq) \longrightarrow$
$$Ba(NO_3)_2(aq) + 2\,H_2O(\ell)$$

5.23   (a) $(NH_4)_2CO_3(aq) + Cu(NO_3)_2(aq) \longrightarrow$
$$CuCO_3(s) + 2\,NH_4NO_3(aq)$$
$CO_3{}^{2-}(aq) + Cu^{2+}(aq) \longrightarrow CuCO_3(s)$

(b) $Pb(OH)_2(s) + 2\,HCl(aq) \longrightarrow PbCl_2(s) + 2\,H_2O(\ell)$
$Pb(OH)_2(s) + 2\,H^+(aq) + 2\,Cl^-(aq) \longrightarrow$
$$PbCl_2(s) + 2\,H_2O(\ell)$$

(c) $BaCO_3(s) + 2\,HCl(aq) \longrightarrow$
$$BaCl_2(aq) + H_2O(\ell) + CO_2(g)$$
$BaCO_3(s) + 2\,H^+(aq) \longrightarrow$
$$Ba^{2+}(aq) + H_2O(\ell) + CO_2(g)$$

5.25   (a) $AgNO_3(aq) + KI(aq) \longrightarrow AgI(s) + KNO_3(aq)$
$Ag^+(aq) + I^-(aq) \longrightarrow AgI(s)$

(b) $Ba(OH)_2(aq) + 2\,HNO_3(aq) \longrightarrow$
$$Ba(NO_3)_2(aq) + 2\,H_2O(\ell)$$
$OH^-(aq) + H^+(aq) \longrightarrow H_2O(\ell)$

(c) $2\,Na_3PO_4(aq) + 3\,Ni(NO_3)_2(aq) \longrightarrow$
$$Ni_3(PO_4)_2(s) + 6\,NaNO_3(aq)$$
$2\,PO_4{}^{3-}(aq) + 3\,Ni^{2+}(aq) \longrightarrow Ni_3(PO_4)_2(s)$

5.27 $FeCO_3(s) + 2\,HNO_3(aq) \longrightarrow$
$$Fe(NO_3)_2(aq) + CO_2(g) + H_2O(\ell)$$

Iron(II) carbonate reacts with nitric acid to give iron(II) nitrate, carbon dioxide, and water.

5.29   (a) Acid–base
$$Ba(OH)_2(aq) + 2\,HCl(aq) \longrightarrow BaCl_2(aq) + H_2O(\ell)$$
(b) Gas-forming
$$2\,HNO_3(aq) + CoCO_3(s) \longrightarrow$$
$$Co(NO_3)_2(aq) + H_2O(\ell) + CO_2(g)$$

(c) Precipitation
$$2\,Na_3PO_4(aq) + 3\,Cu(NO_3)_2(aq) \longrightarrow$$
$$Cu_3(PO_4)_2(s) + 6\,NaNO_3(aq)$$

5.31   (a) Precipitation
$$MnCl_2(aq) + Na_2S(aq) \longrightarrow MnS(s) + 2\,NaCl(aq)$$
$Mn^{2+}(aq) + S^{2-}(aq) \longrightarrow MnS(s)$

(b) Precipitation
$$K_2CO_3(aq) + ZnCl_2(aq) \longrightarrow ZnCO_3(s) + 2\,KCl(aq)$$
$CO_3{}^{2-}(aq) + Zn^{2+}(aq) \longrightarrow ZnCO_3(s)$

5.33   (a) Precipitation of CuS

(b) Formation of water in an acid–base reaction

5.35   (a) Br $= +5$ and O $= -2$

(b) C $= +3$ each and O $= -2$

(c) F $= -1$

(d) Ca $= +2$ and H $= -1$

(e) H $= +1$, Si $= +4$, and O $= -2$

(f) H $= +1$, S $= +6$, and O $= -2$

5.37   (a) Oxidation–reduction

Zn is oxidized from 0 to $+2$, and N in $NO_3{}^-$ is reduced from $+5$ to $+4$ in NO_2.

(b) Acid–base reaction

(c) Oxidation–reduction

Calcium is oxidized from 0 to $+2$ in $Ca(OH)_2$, and H is reduced from $+1$ in H_2O to 0 in H_2.

5.39   (a) O_2 is the oxidizing agent (as it always is), so C_2H_4 is the reducing agent. As is always the case in a combustion, O_2 oxidizes the other reactant (a C-containing compound), which is reduced.

(b) Si is oxidized from 0 in Si to $+4$ in $SiCl_4$. Cl_2 is reduced from 0 in Cl_2 to -1 in Cl^-.

5.41 $[Na_2CO_3] = 0.254$ M; $[Na^+] = 0.508$ M; $[CO_3{}^{2-}] = 0.254$ M

5.43   0.494 g $KMnO_4$

5.45 5.08×10^3 mL

5.47   (a) 0.50 M $NH_4{}^+$ and 0.25 M $SO_4{}^{2-}$

(b) 0.246 M Na^+ and 0.123 M $CO_3{}^{2-}$

(c) 0.056 M H^+ and 0.056 M $NO_3{}^-$

5.49   A mass of 1.06 g Na_2CO_3 is required. After weighing out this quantity of Na_2CO_3, transfer it to a 500.-mL volumetric flask. Rinse any solid from the neck of the flask while filling the flask with distilled water. Add water until the bottom of the meniscus of the water is at the top of the scribed mark on the neck of the flask.

5.51   0.0750 M

5.53   Method (a) is correct. Method (b) gives an acid concentration of 0.15 M.

5.55 $[H^+] = 10^{-pH} = 4.0 \times 10^{-4}$ M

5.57 HNO_3 is a strong acid, so $[H^+] = 0.0013$ M. pH = 2.89.

5.59

pH	$[H^+]$	Acidic/Basic
(a) 1.00	0.10 M	Acidic
(b) 10.50	3.2×10^{-11} M	Basic
(c) 4.89	1.3×10^{-5} M	Acidic
(d) 7.64	2.3×10^{-8} M	Basic

5.61   268 mL

5.63   210 g NaOH and 190 g Cl_2

5.65   174 mL $Na_2S_2O_3$

5.67   1500 mL $Pb(NO_3)_2$

5.69   44.6 mL

5.71   1.052 M HCl

5.73   104 g/mol

5.75   12.8% Fe

5.77   (a) NaBr, KBr, or other alkali metal bromides; Group 2A bromides; other metal bromides

     (b) $Al(OH)_3$ and transition metal hydroxides

     (c) Alkaline earth carbonates ($CaCO_3$) or transition metal carbonates ($NiCO_3$)

     (d) Metal nitrates are generally water-soluble [e.g., $NaNO_3$, $Ni(NO_3)_2$].

5.79   Water-soluble: $Cu(NO_3)_2$, $CuCl_2$. Water-insoluble: $CuCO_3$, $Cu_3(PO_4)_2$.

5.81   Spectator ion, NO_3^-. Acid–base reaction.

     $2 H^+(aq) + Mg(OH)_2(s) \longrightarrow 2 H_2O(\ell) + Mg^{2+}(aq)$

5.83   (a) Cl_2 is reduced (to Cl^-) and Br^- is oxidized (to Br_2).

     (b) Cl_2 is the oxidizing agent and Br^- is the reducing agent.

     (c) 0.678 g Cl_2

5.85   6 g each of NaCl and Na_2CO_3

5.87   The mass of Na_2CO_3 required is 11 g. Weigh out 11 g Na_2CO_3 and place it in the 500.0-mL flask. Add a small amount of distilled water and mix until the solute dissolves. Add water until the meniscus of the solution rests at the calibrated mark on the neck of the volumetric flask. Cap the flask and swirl to ensure complete mixing.

5.89   In 0.015 M HCl, $[H^+] = 0.015$ M. In a pH 1.2 solution, $[H^+] = 0.06$ M. The pH 1.2 solution has a higher hydrogen ion concentration.

5.91   (a) $MgCO_3(s) + 2 H^+(aq) \longrightarrow$
              $CO_2(g) + Mg^{2+}(aq) + H_2O(\ell)$

     Chloride ion (Cl^-) is the spectator ion.

     (b) Gas-forming reaction

     (c) 0.15 g

5.93   (a) H_2O, NH_3, NH_4^+, and OH^- (and a trace of H^+)

     (b) H_2O, CH_3CO_2H, $CH_3CO_2^-$, and H^+ (and a trace of OH^-)

     (c) H_2O, Na^+, and OH^- (and a trace of H^+)

     (d) H_2O, H^+, and Br^- (and a trace of OH^-)

5.95   15.0 g $NaHCO_3$ requires 1190 mL 0.15 M acetic acid. Therefore, acetic acid is the limiting reactant. (Conversely, 125 mL 0.15 M acetic acid requires only 1.58 g $NaHCO_3$.) 1.54 g $NaCH_3CO_2$ produced.

5.97   3.13 g $Na_2S_2O_3$, 96.8%

5.99   (a) Water-soluble: $Cu(NO_3)_2$ [copper(II) nitrate] or $CuCl_2$ [copper(II) chloride].

     Water-insoluble: CuS [copper(II) sulfide] or $CuCO_3$ [copper(II) carbonate]

     (b) Water-soluble: $BaCl_2$ (barium chloride) or $Ba(NO_3)_2$ (barium nitrate)

     Water-insoluble: $BaSO_4$ (barium sulfate) or barium phosphate [$Ba_3(PO_4)_2$]

5.101   (a) $Pb(NO_3)_2(aq) + 2 KOH(aq) \longrightarrow$
                $Pb(OH)_2(s) + 2 KNO_3(aq)$

     $Pb^{2+}(aq) + 2 OH^-(aq) \longrightarrow Pb(OH)_2(s)$

     (b) $Cu(NO_3)_2(aq) + Na_2CO_3(aq) \longrightarrow$
                $CuCO_3(s) + 2 NaNO_3(aq)$

     $Cu^{2+}(aq) + CO_3^{2-}(aq) \longrightarrow CuCO_3(s)$

5.103   0.029 g $NaHCO_3$

5.105   0.00263 mol HCl reacted; 0.0004 mol HCl remains in 0.500 L; pH = 3.13

5.107   3.7 spoonfuls of $NaHCO_3$ is required.

5.109   1.56 g $CaCO_3$ required; 1.00 g $CaCO_3$ remains; 1.73 g $CaCl_2$ produced

5.111 $x = 6$; $Co(NH_3)_6Cl_3$

5.113   Volume of water in the pool = 7.6×10^4 L

5.115   (a) First reaction: oxidizing agent = Cu^{2+} and reducing agent = I^-

     Second reaction: oxidizing agent = I_3^- and reducing agent = $S_2O_3^{2-}$

     (b) 67.3% copper

5.117   (a) Au, gold, has been oxidized and is the reducing agent.

     O_2, oxygen, has been reduced and is the oxidizing agent.

     (b) 26 L NaCN solution

5.119   If both students base their calculations on the amount of HCl solution pipeted into the flask (20 mL), then the second student's result will be (e) the same as the first student's. However, if the HCl concentration is calculated using the diluted solution volume, student 1 will use a volume of 40 mL and student 2 will use a volume of 80 mL in the calculation. The second student's result will be (c) half that of the first student's.

5.121   100 mL of 0.10 M HCl contains 0.010 mol HCl. This requires 0.0050 mol Zn or 3.17 g for complete reaction. Thus, in flask 2 the reaction just uses all of the Zn and produces 0.0050 mol H_2 gas. In flask 1, containing 7.00 g Zn, some Zn remains after the HCl has been consumed; 0.005 mol H_2 gas is produced. In flask 3, there is insufficient Zn, so less hydrogen is produced.

5.123   (a) Several precipitation reactions are possible:

i. $BaCl_2(aq) + H_2SO_4(aq) \longrightarrow$
      $BaSO_4(s) + 2 HCl(aq)$

ii. $BaCl_2(aq) + Na_2SO_4(aq) \longrightarrow$
      $BaSO_4(s) + 2 NaCl(aq)$

iii. $BaCO_3(aq) + H_2SO_4(aq) \longrightarrow$
      $BaSO_4(s) + CO_2(g) + H_2O(\ell)$

iv. $Ba(OH)_2(aq) + H_2SO_4(aq) \longrightarrow$
      $BaSO_4(s) + 2 H_2O(\ell)$

(b) Gas-forming reaction: reaction (iii) in part (a) is also a gas-forming reaction.

5.125   150 mg/dL. The person is intoxicated.

Chapter 6

6.1   Mechanical energy is used to move the lever, which in turns moves gears. The device produces electrical energy and radiant energy.

6.3 5.0×10^6 J

6.5   170 kcal is equivalent to 710 kJ, considerably greater than 280 kJ.

6.7 0.140 J/g · K

6.9 2.44 kJ

6.11 32.8 °C

6.13 20.7 °C

6.15 47.8 °C

6.17 0.40 J/g · K

6.19 330 kJ

6.21 49.3 kJ

6.23 273 J

6.25 9.97×10^5 J

6.27   Reaction is exothermic because ΔH°_{rxn} is negative. The heat evolved is 2.38 kJ.

6.29 3.3×10^4 kJ

6.31 $\Delta H_{rxn} = -56$ kJ/mol

6.33 0.52 J/g · K

6.35 $\Delta H = +23$ kJ/mol

6.37 297 kJ/mol SO_2

6.39 3.09×10^3 kJ/mol

6.41 0.236 J/g · K

6.43   (a) $\Delta H^\circ_{rxn} = -126$ kJ

(b)

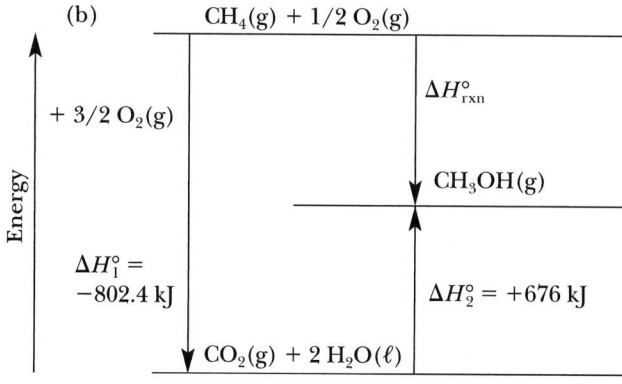

6.45 $\Delta H^\circ_{rxn} = +90.3$ kJ

6.47 $C(s) + 2 H_2(g) + \frac{1}{2} O_2(g) \longrightarrow CH_3OH(\ell)$
       $\Delta H^\circ_f = -238.4$ kJ/mol

6.49   (a) $2 Cr(s) + \frac{3}{2} O_2(g) \longrightarrow Cr_2O_3(s)$
           $\Delta H^\circ_f = -1134.7$ kJ/mol

(b) 2.4 g is equivalent to 0.046 mol Cr. This will produce 26 kJ of heat energy.

6.51   (a) $\Delta H^\circ_{rxn} = -24$ kJ for 1.0 g phosphorus

(b) $\Delta H^\circ_{rxn} = -18$ kJ for 0.2 mol NO

(c) $\Delta H^\circ_{rxn} = -16.9$ kJ for the formation of 2.40 g NaCl(s)

(d) $\Delta H^\circ_{rxn} = -1.8 \times 10^3$ kJ for the oxidation of 250 g iron

6.53   (a) $\Delta H^\circ_{rxn} = -906.2$ kJ (or -226.5 for 1.00 mol NH_3)

(b) The heat evolved is 133 kJ for the oxidation of 10.0 g NH_3.

6.55   (a) $\Delta H^\circ_{rxn} = +80.8$ kJ

(b)

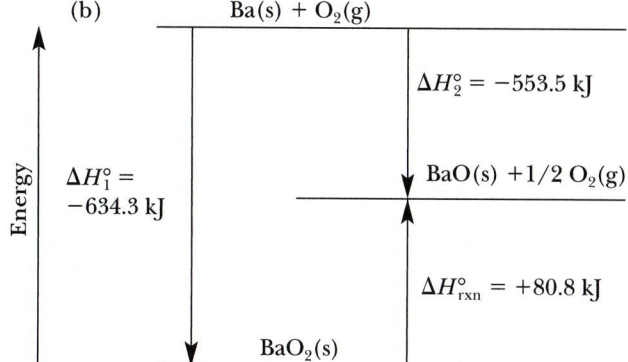

6.57 $\Delta H^\circ_f = +77.7$ kJ/mol for naphthalene

6.59   (a) $\Delta H^\circ_{rxn} = -705.63$ kJ; reaction is expected to be product-favored

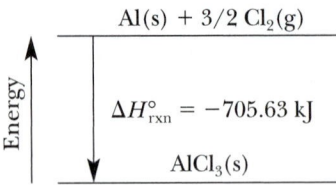

     (b) $\Delta H^\circ_{rxn} = +90.83$ kJ; reaction is expected to be reactant-favored

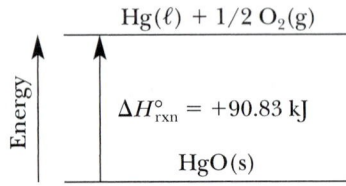

6.61   (a) Exothermic: a process in which heat is transferred from a system to its surroundings. (Heat is evolved in the combustion of methane.)

       Endothermic: a process in which heat is transferred from the surroundings to the system. (Ice melting absorbs heat.)

     (b) System: the object or collection of objects being studied. (A chemical reaction—the system—taking place inside a calorimeter—the surroundings.)

       Surroundings: everything outside the system that can exchange energy with the system. (The calorimeter and everything outside the calorimeter.)

     (c) Specific heat capacity: the quantity of heat required to raise the temperature of 1 gram of a substance by 1 kelvin. (The specific heat capacity of water is 4.184 J/g · K.)

     (d) State function: a quantity that is characterized by changes that do not depend on the path chosen to go from the initial state to the final state. (Enthalpy and internal energy.)

     (e) Standard state: the most stable form of a substance in the physical state that exists at a pressure of 1 bar and at a specified temperature. (The standard state of carbon at 25 °C is graphite.)

     (f) Enthalpy change, ΔH: the difference between the final and initial heat content of a substance at constant pressure. (The enthalpy change for melting ice at 0 °C is 6.00 kJ/mol.)

     (g) Standard enthalpy of formation: the enthalpy change for the formation of one mole of a compound in its standard state directly from the component elements in their standard states. (ΔH°_f for liquid water is -285.83 kJ/mol.)

6.63   (a) System: reaction between methane and oxygen.

       Surroundings: the furnace and the rest of the universe.

       Heat flows from the system to the surroundings.

     (b) System: water drops.

       Surroundings: skin and the rest of the universe.

       Heat flows from the surroundings to the system.

     (c) System: water

       Surroundings: freezer and the rest of the universe.

       Heat flows from the system to the surroundings.

     (d) System: reaction of aluminum and iron (III) oxide.

       Surroundings: flask, laboratory bench, and rest of the universe.

       Heat flows from system to the surroundings.

6.65 $\Delta E = q + w$. ΔE is the change in energy content, q is the heat transferred to or from the system, and w is the work transferred to or from the system.

6.67   Standard state of oxygen is gas, $O_2(g)$.

$O_2(g) \longrightarrow 2\,O(g)$,

$\Delta H^\circ_{rxn} = +498.34$ kJ, endothermic

$\frac{3}{2}\,O_2(g) \longrightarrow O_3(g)$, $\Delta H^\circ_{rxn} = +142.67$ kJ

6.69 $SnBr_2(s) + TiCl_2(s) \longrightarrow SnCl_2(s) + TiBr_2(s)$

                                  $\Delta H^\circ_{rxn} = -4.2$ kJ

     $SnCl_2(s) + Cl_2(g) \longrightarrow SnCl_4(\ell)$ $\Delta H^\circ_{rxn} = -195$ kJ

     $TiCl_4(\ell) \longrightarrow TiCl_2(s) + Cl_2(g)$ $\Delta H^\circ_{rxn} = +273$ kJ

     $SnBr_2(s) + TiCl_4(\ell) \longrightarrow SnCl_4(\ell) + TiBr_2(s)$

                                  $\Delta H^\circ_{net} = +74$ kJ

6.71 $q_{water} = -8400$ kJ and $q_{ethanol} = -9800$ kJ. Ethanol sample gives up more heat than water sample.

6.73 $C_{Ag} = 0.24$ J/g · K

6.75   Mass of ice melted = 75.4 g

6.77   Final temperature = 278 K (4.8 °C)

6.79 $\Delta H_{rxn} = -69$ kJ/mol

6.81   36.0 kJ evolved per mol NH_4NO_3

6.83   (a) When summed, the following equations give the balanced equation for the formation of $B_2H_6(g)$ from the elements.

     $2\,B(s) + \frac{3}{2}\,O_2(g) \longrightarrow B_2O_3(s)$ $\Delta H^\circ_{rxn} = -1271.9$ kJ

     $3\,H_2(g) + \frac{3}{2}\,O_2(g) \longrightarrow 3\,H_2O(g)$ $\Delta H^\circ_{rxn} = -725.4$ kJ

     $B_2O_3(s) + 3\,H_2O(g) \longrightarrow B_2H_6(g) + 3\,O_2(g)$

                                  $\Delta H^\circ_{rxn} = +2032.9$ kJ

     $2\,B(s) + 3\,H_2(g) \longrightarrow B_2H_6(g)$ $\Delta H^\circ_{rxn} = +35.6$ kJ

     (b) The enthalpy of formation of $B_2H_6(g)$ is +35.6 kJ/mol.

     (c)

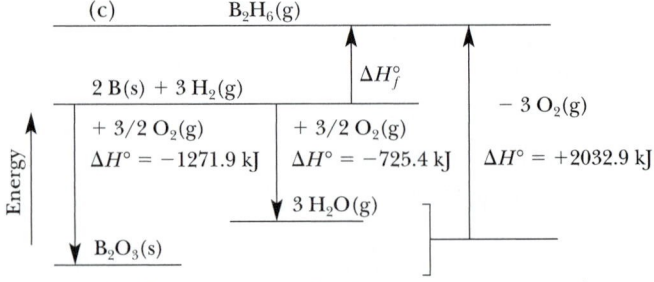

     (d) The formation of $B_2H_6(g)$ is reactant-favored.

6.85   The standard enthalpy change, ΔH°_{rxn}, is -352.88 kJ. The quantity of magnesium needed is 0.43 g.

6.87   (a) $\Delta H^\circ_{rxn} = +131.31$ kJ

(b) Reactant-favored

(c) 1.0932×10^7 kJ

6.89   Assuming $CO_2(g)$ and $H_2O(\ell)$ are the products of combustion:

ΔH°_{rxn} for isooctane is -5461.3 kJ/mol or -47.81 kJ per gram

ΔH°_{rxn} for liquid methanol is -726.77 kJ/mol or -22.682 kJ per gram

6.91   (a) Adding the equations as they are given in the question results in the desired equation for the formation of $SrCO_3(s)$. The calculated $\Delta H^\circ_{rxn} = -1220.$ kJ/mol.

(b) $Sr(s) + 1/2\ O_2(g) + C(graphite) + O_2(g)$

6.93 $\Delta H^\circ_{rxn} = -305.3$ kJ

6.95 3.28×10^4 kJ from 1.00 kg C. 1.00 kg C produces 83.3 mol each of C and H_2 gas. These produce a total heat of 4.37×10^4 kJ. Although the water gas from 1.00 kg of C produces more heat than 1.00 kg C, some carbon must be burned to provide the heat for the water gas reaction in the first place (Question 94).

6.97   Yes, the first law of thermodynamics is a version of the general principle of the conservation of energy applied specifically to a system.

6.99   (a) Product-favored

(b) Reactant-favored

6.101   (a) The temperature of the cooler object increases, and the motions of its particles are faster. The temperature of the warmer object decreases, and its particles move more slowly.

(b) The two objects have the same temperature.

6.103   The enthalpy change for each of the three reactions below is known or can be measured by calorimetry. The three equations sum to give the enthalpy of formation of $CaSO_4(s)$.

$Ca(s) + \frac{1}{2}\ O_2(g) \longrightarrow CaO(s)$
$$\Delta H^\circ_{rxn} = \Delta H^\circ_f = -635.09\ kJ$$

$\frac{1}{8}\ S_8(s) + \frac{3}{2}\ O_2(g) \longrightarrow SO_3(g)$
$$\Delta H^\circ_{rxn} = \Delta H^\circ_f = -395.77\ kJ$$

$CaO(s) + SO_3(g) \longrightarrow CaSO_4(s)$
$$\Delta H^\circ_{rxn} = -402.7\ kJ$$

$Ca(s) + \frac{1}{8}\ S_8(s) + \frac{3}{2}\ O_2(g) \longrightarrow CaSO_4(s)$

$$\Delta H^\circ_{rxn} = \Delta H^\circ_f = -1433.6\ kJ$$

6.105

Metal	Molar Heat Capacity (J/mol · K)
Al	24.2
Fe	25.1
Cu	24.5
Au	25.4

All the metals have a molar heat capacity of 24.8 J/mol · K plus or minus 0.6 J/mol · K. Therefore, assuming the molar heat capacity of Ag is 24.8 J/mol · K, its specific heat capacity is 0.230 J/g · K. This is very close to the experimental value of 0.236 J/g · K.

6.107   120 g CH_4 required

6.109 1.6×10^{11} kJ released to the surroundings. This is equivalent to 3.8×10^4 tons of dynamite.

Chapter 7

7.1   (a) Microwaves

(b) Red light

(c) Infrared

7.3   (a) Green light has a higher frequency than amber light.

(b) $5.04 \times 10^{14}\ s^{-1}$

7.5   Frequency = $6.0 \times 10^{14}\ s^{-1}$; energy per photon = 4.0×10^{-19} J; energy per mol photons = 2.4×10^5 J

7.7   302 kJ/mol photons

7.9   In order of increasing energy: FM station < microwaves < yellow light < x-rays

7.11   Light with a wavelength as long as 600 nm would be sufficient. This is in the visible region.

7.13   (a) The light of shortest wavelength has a wavelength of 253.652 nm.

(b) Frequency = $1.18190 \times 10^{15}\ s^{-1}$. Energy per photon = 7.83139×10^{-19} J/photon.

(c) The lines at 404 and 436 nm are in the visible region of the spectrum.

7.15   The color is violet. $n_{initial} = 6$ and $n_{final} = 2$.

7.17   (a) 10 lines possible

(b) Highest frequency (highest energy), $n = 5$ to $n = 1$

(c) Longest wavelength (lowest energy), $n = 5$ to $n = 4$

7.19   (a) $n = 3$ to $n = 2$

       (b) $n = 4$ to $n = 1$

7.21   Wavelength = 102.6 nm and frequency = $2.923 \times 10^{15} \text{ s}^{-1}$. Light with these properties is in the ultraviolet region.

7.23   Wavelength = 0.29 nm

7.25   The wavelength is 2.2×10^{-25} nm. (Calculated from $\lambda = h / m \cdot v$, where m is the ball's mass in kilograms and v is the velocity.) To have a wavelength of 5.6×10^{-3} nm, the ball would have to travel at 1.2×10^{-21} m/s.

7.27   (a) $n = 4$, $\ell = 0, 1, 2, 3$

       (b) When $\ell = 2$, $m_\ell = -2, -1, 0, 1, 2$

       (c) For a $4s$ orbital, $n = 4$, $\ell = 0$, and $m_\ell = 0$

       (d) For a $4f$ orbital, $n = 4$, $\ell = 3$, and $m_\ell = -3, -2, -1, 0, 1, 2, 3$

7.29   Set 1: $n = 4$, $\ell = 1$, and $m_\ell = -1$

       Set 2: $n = 4$, $\ell = 1$, and $m_\ell = 0$

       Set 3: $n = 4$, $\ell = 1$, and $m_\ell = +1$

7.31   4 subshells. (The number of subshells in a shell is always equal to n.)

7.33   (a) ℓ must have a value no greater than $n - 1$.

       (b) m_ℓ can only equal 0 in this case.

       (c) m_ℓ can only equal 0 in this case.

7.35   (a) None. The quantum number set is not possible. Here m_ℓ can only equal zero.

       (b) 3 orbitals

       (c) 11 orbitals

       (d) 1 orbital

7.37 $2d$ and $3f$ orbitals cannot exist. The $n = 2$ shell consists only of s and p subshells. The $n = 3$ shell consists only of s, p, and d subshells.

7.39   (a) For $2p$: $n = 2$, $\ell = 1$, and $m_\ell = -1, 0$, or $+1$

       (b) For $3d$: $n = 3$, $\ell = 2$, and $m_\ell = -2, -1, 0, +1$, or $+2$

       (c) For $4f$: $n = 4$, $\ell = 3$, and $m_\ell = -3, -2, -1, 0, +1, +2$, or $+3$

7.41 $4d$

7.43   Considering only angular nodes (the planes that pass through the nucleus):

       (a) $2s$ has zero nodal surfaces.

       (b) $5d$ has two nodal surfaces.

       (c) $5f$ has three nodal surfaces.

7.45   (a) Correct

       (b) Incorrect; the intensity of a light beam is independent of frequency and is related to the number of photons of light with a certain energy

       (c) Correct

7.47   Considering only angular nodes (the planes that pass through the nucleus):

s orbital	Zero nodal surface
p orbitals	One nodal surface or plane passing through the nucleus
d orbitals	Two nodal surfaces or planes passing through the nucleus
f orbitals	Three nodal surfaces or planes passing through the nucleus

7.49

ℓ value	Orbital Type
3	f
0	s
1	p
2	d

7.51   Considering only angular nodes (the planes that pass through the nucleus):

Orbital Type	Number of Orbitals in a Given Subshell	Number of Surfaces
s	1	0
p	3	1
d	5	2
f	7	3

7.53   (a) Green light

       (b) Red light has a wavelength of 680 nm, and green light has a wavelength of 500 nm.

       (c) Green light has a higher frequency than red light.

7.55   (a) Wavelength = 0.35 m

       (b) Energy = 0.34 J/mol

       (c) Blue light (with $\lambda = 420$ nm) has an energy of 280 kJ/mol photons.

       (d) Blue light has an energy (per mole of photons) that is 840,000 times greater than a mole of photons from a cell phone.

7.57   The ionization energy for He^+ is 5248 kJ/mol. This is four times the ionization energy for the H atom.

7.59 $1s < 2s = 2p < 3s = 3p = 3d < 4s$

       In the H atom, orbitals in the same shell (e.g., $2s$ and $2p$) have the same energy.

7.61   Frequency = $2.836 \times 10^{20} \text{ s}^{-1}$ and wavelength = 1.057×10^{-12} m

7.63   260 s or 4.3 min

7.65 (a) size

(b) ℓ

(c) more

(d) 7 (when $\ell = 3$ these are f orbitals)

(e) one orbital

(f) (left to right) d, s, and p

(g) $\ell = 0, 1, 2, 3, 4$ (or 5 orbitals)

(h) 16 orbitals (1 s, 3 p, 5 d, and 7 f) ($= n^2$)

7.67 An electron orbiting the nucleus could occupy only certain orbits or energy levels in which it is stable. An electron in an atom will remain in its lowest energy level unless disturbed.

7.69 (c)

7.71 An experiment can be done showing that the electron can behave as a particle, and another experiment can be done showing that it has wave properties. (However, no single experiment shows both properties of the electron.) The modern view of atomic structure is based on the wave properties of the electron.

7.73 (a) and (b)

7.75 Radiation with a wavelength of 93.8 nm is sufficient to raise the electron to the $n = 6$ quantum level (see Figure 7.12). There should be 15 emission lines involving transitions from $n = 6$ to lower energy levels. (There are 5 lines for transitions from $n = 6$ to lower levels, 4 lines for $n = 5$ to lower levels, 3 for $n = 4$ to lower levels, 2 lines for $n = 3$ to lower levels, and 1 line for n = 2 to n = 1.) Wavelengths for many of the lines are given in Figure 7.12. For example, there will be an emission involving an electron moving from $n = 6$ to $n = 2$ with a wavelength of 410.2 nm.

7.77 De Broglie's equation (7.6) states that the wavelength of an object is given by h/mv. Planck's constant, h, has a very small value. A golf ball is a relatively massive object (compared with an electron). Its velocity, while small compared with an electron, is measurable. The quotient h/mv, the wavelength, will be exceedingly small—so small, in fact, that it cannot be measured.

7.79 The pickle glows because it was made by soaking a cucumber in brine, a concentrated solution of NaCl. The sodium atoms in the pickle are excited by the electric current and release energy as yellow light as they return to the ground state. Excited sodium atoms are the source of the yellow light you see in fireworks and in certain kinds of street lighting.

Chapter 8

8.1 (a) Phosphorus: $1s^2 2s^2 2p^6 3s^2 3p^3$

$$\boxed{\uparrow\downarrow}\ \boxed{\uparrow\downarrow}\ \boxed{\uparrow\downarrow}\boxed{\uparrow\downarrow}\boxed{\uparrow\downarrow}\ \boxed{\uparrow\downarrow}\ \boxed{\uparrow}\boxed{\uparrow}\boxed{\uparrow}$$
$$1s \quad 2s \quad 2p \quad\quad 3s \quad 3p$$

The element is in the third period in Group 5A. Therefore, it has five electrons in the third shell.

(b) Chlorine: $1s^2 2s^2 2p^6 3s^2 3p^5$

$$\boxed{\uparrow\downarrow}\ \boxed{\uparrow\downarrow}\ \boxed{\uparrow\downarrow}\boxed{\uparrow\downarrow}\boxed{\uparrow\downarrow}\ \boxed{\uparrow\downarrow}\ \boxed{\uparrow\downarrow}\boxed{\uparrow\downarrow}\boxed{\uparrow}$$
$$1s \quad 2s \quad 2p \quad\quad 3s \quad 3p$$

The element is in the third period and in Group 7A. Therefore, it has seven electrons in the third shell.

8.3 (a) Chromium: $1s^2 2s^2 2p^6 3s^2 3p^6 3d^5 4s^1$

(b) Iron: $1s^2 2s^2 2p^6 3s^2 3p^6 3d^6 4s^2$

8.5 (a) Arsenic: $1s^2 2s^2 2p^6 3s^2 3p^6 3d^{10} 4s^2 4p^3$

$[Ar]3d^{10}4s^2 4p^3$

(b) Krypton: $1s^2 2s^2 2p^6 3s^2 3p^6 3d^{10} 4s^2 4p^6 = [Kr]$

8.7 (a) Tantalum: This is the third element in the transition series in the sixth period. Therefore, it has a core equivalent to Xe plus 2 $6s$ electrons, 14 $4f$ electrons, and 3 $5d$ electrons: $[Xe]4f^{14}5d^3 6s^2$.

(b) Platinum: This is the eighth element in the transition series in the sixth period. Therefore, it has a core equivalent to Xe plus 2 $6s$ electrons, 14 $4f$ electrons, and 8 $5d$ electrons: $[Xe]4f^{14}5d^8 6s^2$. The actual configuration (Table 8.3) is $[Xe]4f^{14}5d^9 6s^1$.

8.9 Americium: $[Rn]5f^7 7s^2$ (see Table 8.3)

8.11 (a) Mg^{2+} ion

$$\boxed{\uparrow\downarrow}\ \boxed{\uparrow\downarrow}\ \boxed{\uparrow\downarrow}\boxed{\uparrow\downarrow}\boxed{\uparrow\downarrow}\ \boxed{}$$
$$1s \quad 2s \quad 2p \quad\quad 3s$$

(b) K^+ ion

$$\boxed{\uparrow\downarrow}\ \boxed{\uparrow\downarrow}\ \boxed{\uparrow\downarrow}\boxed{\uparrow\downarrow}\boxed{\uparrow\downarrow}\ \boxed{\uparrow\downarrow}\ \boxed{\uparrow\downarrow}\boxed{\uparrow\downarrow}\boxed{\uparrow\downarrow}$$
$$1s \quad 2s \quad 2p \quad\quad 3s \quad 3p$$

(c) Cl^- ion (Note that both Cl^- and K^+ have the same configuration; both are equivalent to Ar.)

$$\boxed{\uparrow\downarrow}\ \boxed{\uparrow\downarrow}\ \boxed{\uparrow\downarrow}\boxed{\uparrow\downarrow}\boxed{\uparrow\downarrow}\ \boxed{\uparrow\downarrow}\ \boxed{\uparrow\downarrow}\boxed{\uparrow\downarrow}\boxed{\uparrow\downarrow}$$
$$1s \quad 2s \quad 2p \quad\quad 3s \quad 3p$$

(d) O^{2-} ion

$$\boxed{\uparrow\downarrow}\ \boxed{\uparrow\downarrow}\ \boxed{\uparrow\downarrow}\boxed{\uparrow\downarrow}\boxed{\uparrow\downarrow}$$
$$1s \quad 2s \quad 2p$$

8.13 (a) V (paramagnetic; three unpaired electrons)

$$[Ar]\ \boxed{\uparrow}\boxed{\uparrow}\boxed{\uparrow}\boxed{}\boxed{}\quad \boxed{\uparrow\downarrow}$$
$$\qquad\quad 3d \qquad\qquad 4s$$

(b) V^{2+} ion (paramagnetic, three unpaired electrons)

$$[Ar]\ \boxed{\uparrow}\boxed{\uparrow}\boxed{\uparrow}\boxed{}\boxed{}\quad \boxed{}$$
$$\qquad\quad 3d \qquad\qquad 4s$$

(c) V^{5+} ion. This ion has an electron configuration equivalent to argon. It is diamagnetic, with no unpaired electrons.

8.15   (a) Manganese

[Ar] ↑|↑|↑|↑|↑ |↑↓|
       3d   4s

(b) Manganese(II) ion, Mn^{2+}

[Ar] ↑|↑|↑|↑|↑ |☐|
       3d   4s

(c) The 2+ ion is paramagnetic to the extent of five unpaired electrons.

(d) 5

8.17   (a) The spin quantum number cannot be 0. The set is correct if $m_s = \pm 1/2$.

(b) m_ℓ cannot be larger than ℓ. The set is correct if $m_\ell = -1, 0,$ or $+1$.

(c) ℓ can be no larger than $n - 1$. The set is correct if $\ell = 1$ or 2.

8.19   (a) 14

(b) 2

(c) 0 (because ℓ cannot equal n)

8.21   Magnesium: $1s^2 2s^2 2p^6 3s^2$

|↑↓| |↑↓| |↑↓|↑↓|↑↓| |↑↓|
 1s   2s   2p   3s

Quantum numbers for the two electrons in the 3s orbital:

$n = 3, \ell = 0, m_\ell = 0, m_s = +1/2$

$n = 3, \ell = 0, m_\ell = 0, m_s = -1/2$

8.23   Gallium: $1s^2 2s^2 2p^6 3s^2 3p^6 3d^{10} 4s^2 4p^1$

[Ar] |↑↓|↑↓|↑↓|↑↓|↑↓| |↑↓| |↑| | | |
        3d   4s   4p

Quantum numbers for the 4p electron:

$n = 4, \ell = 1, m_\ell = 1, m_s = +1/2$

8.25   Increasing size: C < B < Al < Na < K

8.27   (a) Cl^-

(b) Al

(c) In

8.29   (c)

8.31   (a) Largest radius, Na

(b) Most negative electron affinity: O

(c) Ionization energy: Na < Mg < P < O

8.33   (a) Increasing ionization energy: S < O < F. S is less than O because the IE decreases down a group. F is greater than O because IE generally increases across a period.

(b) Largest IE: O. See part (a).

(c) Most negative electron affinity: Cl. Electron affinity becomes increasingly more negative across the periodic table and on ascending a group.

(d) Largest I radius, O^{2-}

8.35   (a) Drawing (a) is a ferromagnetic solid, (b) is a diamagnetic solid, and (c) is a paramagnetic solid.

(b) Substance (a) would be most strongly attracted to a magnet, whereas (b) would be least strongly attracted.

8.37   Uranium configuration: $[Rn]5f^3 6d^1 7s^2$

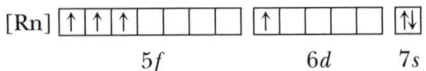

(5f   6d   7s)

Uranium(IV ion, U^{4+}): $[Rn]5f^2$

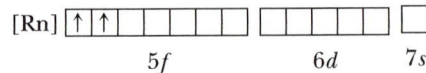

(5f   6d   7s)

Both U and U^{4+} are paramagnetic.

8.39   (a) Atomic number = 20

(b) Total number of s electrons = 8

(c) Total number of p electrons = 12

(d) Total number of d electrons = 0

(e) The element is Ca, calcium, a metal.

8.41   (b) The maximum value of ℓ is $(n - 1)$.

8.43   (a) Neodymium, Nd: $[Xe]4f^4 6s^2$ (Table 8.3)

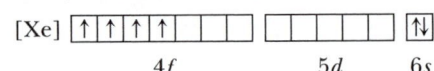

(4f   5d   6s)

Iron, Fe: $[Ar]3d^6 4s^2$

[Ar] |↑↓|↑|↑|↑|↑| |↑↓|
        3d   4s

Boron, B: $1s^2 2s^2 2p^1$

|↑↓| |↑↓| |↑| | |
 1s   2s   2p

(b) All three elements have unpaired electrons and so should be paramagnetic.

(c) Neodymium(III) ion, Nd^{3+}: $[Xe]4f^3$

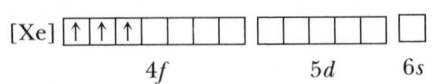

(4f   5d   6s)

Iron(III) ion, Fe^{3+}: $[Ar]3d^5$

[Ar] |↑|↑|↑|↑|↑| |☐|
        3d   4s

Both neodymium(III) and iron(III) have unpaired electrons and are paramagnetic.

8.45   K < Ca < Si < P

8.47   (a) metal   (c) B

(b) B   (d) A

8.49 In^{4+}: Indium has three outer shell electrons, so it is unlikely to form a 4+ ion.

Fe^{6+}: Although iron has eight electrons in its $3d$ and $4s$ orbitals, so ions with a 6+ charge are highly unlikely. The ionization energy is too large.

Sn^{5+}: Tin has four outer shell electrons, so it is unlikely to form a 5+ ion.

8.51 (a) Se   (d) N
(b) Br^- (e) N^{3-}
(c) Na

8.53 (a) Na   (c) $Na < Al < B < C$
(b) C

8.55 (a) Cobalt   (c) 4 unpaired electrons
(b) Paramagnetic

8.57 Li has 3 electrons ($1s^2 2s^1$) and Li^+ has only two electrons ($1s^2$). The ion is smaller than the atom because there are only two electrons to be held by three protons in the ion. Also, an electron in a larger orbital has been removed. Fluorine atoms have 9 electrons and 9 protons ($1s^2 2s^2 2p^5$). The anion, F^-, has one additional electron, which means that 10 electrons must be held by only 9 protons, and the ion is larger than the atom.

8.59
$K\ ([Ar]4s^1) \longrightarrow K^+\ ([Ar])$ IE = 419 kJ/mol
$K^+\ ([Ar]) \longrightarrow K^{2+}\ ([Ne]3s^2 3p^5)$ IE = 3051 kJ/mol

The second electron must be removed from a positive ion and is a core electron, whereas the first electron is removed from a neutral atom.

8.61 (a) In going from one element to the next across the period, the effective nuclear charge increases slightly and the attraction between the nucleus and the electrons increases. (See the General ChemistryNow CD-ROM or website Screen 8.9.)

(b) The size of fourth-period transition elements, for example, is a reflection of the size of the $4s$ orbital. As d electrons are added across the series, protons are added to the nucleus. Adding protons should lead to a decreased atom size, but the effects of the protons are balanced by $3d$ electrons, and the atom size is changed little.

8.63 Arguments for a compound composed of Mg^{2+} and O^{2-}:

(a) Chemical experience suggests that all Group 2A elements form 2+ cations, and that oxygen is typically the O^{2-} ion in its compounds.

(b) Other alkaline earth elements form oxides such as BeO, CaO, and BaO.

A possible experiment is to measure the melting point of the compound. An ionic compound such as NaF (with ions having 1+ and 1− charges) melts at 990 °C, whereas a compound analogous to MgO, CaO, melts at a much higher temperature (2580 °C).

8.65 (a) The effective nuclear charge increases, causing the valence orbital energies to become more negative on moving across the period.

(b) As the valence orbital energies become more negative, it is increasingly difficult to remove an electron from the atom, and the IE increases. Toward the end of the period, the orbital energies have become so negative that removing an electron requires significant energy. Instead, the effective nuclear charge has reached the point that it is energetically more favorable for the atom to gain an electron.

(c) Valence orbital energies:
Li (-530.7 kJ) $<$ Be (-897.3 kJ) $>$ B (-800.8 kJ) $<$ C (-1032 kJ)

It is more difficult to remove an electron from Be than from either Li or B. The energy is more negative for C than for B, so it is more difficult to remove an electron from C than from B.

8.67 The size declines across this series of elements while the mass increases. Thus, the mass per volume—the density—increases.

8.69 (a) Element 113: $[Rn]5f^{14}6d^{10}7s^2 7p^1$
Element 115: $[Rn]5f^{14}6d^{10}7s^2 7p^3$

(b) Element 113 is in Group 3A (with elements such as boron and aluminum), and element 115 is in Group 5A (with elements such as nitrogen and phosphorus).

(c) Americium ($Z = 95$) + argon ($Z = 18$) = element 113

8.71 (a) Sulfur electron configuration

| ↑↓ | ↑↓ | ↑↓ ↑↓ ↑↓ | ↑↓ | ↑↓ ↑ ↑ |
| 1s | 2s | 2p | 3s | 3p |

(b) $n = 3$, $\ell = 1$, $m_\ell = 1$, $m_s = +1/2$

(c) S has the smallest ionization energy and O has the smallest radius.

(d) S is smaller than the S^{2-} ion.

(e) 584 g SCl_2

(f) 10.0 g SCl_2 is the limiting reactant, and 11.6 g $SOCl_2$ can be produced.

(g) $\Delta H_f^\circ\ [SCl_2(g)] = -17.6$ kJ/mol

8.73

Atom Distance	Calculated (pm)	Measured (pm)
B—F	154	130
P—F	6	178
C—H	114	109
C—O	143	150

With the exception of B—F, the agreement is quite good.

Chapter 9

9.1   (a) Group 6A, 6 valence electrons

(b) Group 3A, 3 valence electrons

(c) Group 1A, 1 valence electron

(d) Group 2A, 2 valence electrons

(e) Group 7A, 7 valence electrons

(f) Group 6A, 6 valence electrons

9.3   Group 3A, 3 bonds

Group 4A, 4 bonds

Group 5A, 3 bonds (for a neutral compound)

Group 6A, 2 bonds (for a neutral compound)

Group 7A, 1 bond (for a neutral compound)

9.5   Most negative, MgS; least negative, KI

9.7   Increasing lattice energy: RbI < LiI < LiF < CaO

9.9   As the ion–ion distance decreases, the force of attraction between ions increases. This should make the lattice more stable, and more energy should be required to melt the compound.

9.11   (a) NF_3, 26 valence electrons

:F̈—N̈—F̈:
     |
    :F̈:

(b) ClO_3^-, 26 valence electrons

[:Ö—Cl̈—Ö:]⁻
       |
      :Ö:

(c) HOBr, 14 valence electrons

H—Ö—B̈r:

(d) SO_3^{2-}, 26 valence electrons

[:Ö—S̈—Ö:]²⁻
       |
      :Ö:

9.13   (a) $CHClF_2$, 26 valence electrons

      H
      |
:Cl̈—C—F̈:
      |
     :F̈:

(b) CH_3CO_2H, 24 valence electrons

    H :O:
    | ‖
H—C—C—Ö—H
    |
    H

(c) CH_3CN, 16 valence electrons

    H
    |
H—C—C≡N:
    |
    H

(d) H_2CCCH_2, 16 valence electrons

  H   H
  | |
H—C=C=C—H

9.15   (a) SO_2, 18 valence electrons

:Ö—S̈=Ö ⟷ Ö=S̈—Ö:

(b) NO_2^-, 18 valence electrons

[:Ö—N̈=Ö]⁻ ⟷ [Ö=N̈—Ö:]⁻

(c) SCN^-, 16 valence electrons

[S̈=C=N̈:]⁻ ⟷ [:S̈≡C—N̈:]⁻ ⟷ [:S̈—C≡N̈]⁻

9.17   (a) BrF_3, 28 valence electrons

:F̈:
|
B̈r—F̈:
|
:F̈:

(b) I_3^-, 22 valence electrons

(c) XeO_2F_2, 34 valence electrons

     :F̈:
      |
:Ö—Xe—Ö:
      |
     :F̈:

(d) XeF_3^+, 28 valence electrons

    :F̈:
     |
    Xe—F̈:
     |
    :F̈:

9.19   (a) Electron-pair geometry around N is tetrahedral. Molecular geometry is trigonal pyramidal.

:C̈l—N̈—H
        |
        H

(b) Electron-pair geometry around O is tetrahedral. Molecular geometry is bent.

:C̈l—Ö—C̈l:

(c) Electron-pair geometry around C is linear. Molecular geometry is linear.

[S̈=C=N̈]⁻

(d) Electron-pair geometry around O is tetrahedral. Molecular geometry is bent.

H—Ö—F̈:

9.21   (a) Electron-pair geometry around C is linear. Molecular geometry is linear.

Ö=C=Ö

(b) Electron-pair geometry around N is trigonal planar. Molecular geometry is bent.

$$\left[\ddot{\text{O}}\!-\!\text{N}\!=\!\ddot{\text{O}}\right]^{-}$$

(c) Electron-pair geometry around O is trigonal planar. Molecular geometry is bent.

$$\ddot{\text{O}}\!=\!\ddot{\text{O}}\!-\!\ddot{\text{O}}:$$

(d) Electron-pair geometry around Cl atom is tetrahedral. Molecular geometry is bent.

$$\left[:\ddot{\text{O}}\!-\!\ddot{\text{Cl}}\!-\!\ddot{\text{O}}:\right]^{-}$$

All have two atoms attached to the central atom. As the bond and lone pairs vary, the molecular geometries vary from linear to bent.

9.23 (a) Electron-pair geometry around Cl is trigonal bipyramidal. Molecular geometry is linear.

$$\left[:\ddot{\text{F}}\!-\!\ddot{\text{Cl}}\!-\!\ddot{\text{F}}:\right]^{-}$$

(b) Electron-pair geometry around Cl is trigonal bipyramidal. Molecular geometry is T-shaped.

$$:\ddot{\text{F}}\!-\!\ddot{\text{Cl}}\!-\!\ddot{\text{F}}:$$
$$:\ddot{\text{F}}:$$

(c) Electron-pair geometry around Cl is octahedral. Molecular geometry is square planar.

$$\left[\begin{array}{c}:\ddot{\text{F}}:\\:\ddot{\text{F}}\!-\!\text{Cl}\!-\!\ddot{\text{F}}:\\:\ddot{\text{F}}:\end{array}\right]^{-}$$

(d) Electron-pair geometry around Cl is octahedral. Molecular geometry is square pyramidal.

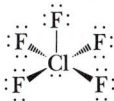

9.25 (a) Ideal O—S—O angle = 120°

(b) 120°

(c) 120°

(d) H—C—H = 109° and C—C—N angle = 180°

9.27 1 = 120°; 2 = 109°; 3 = 120°; 4 = 109°; 5 = 109°

9.29 (a) N = 0; H = 0

(b) P = +1; O = −1

(c) B = −1; H = 0

(d) All are zero.

9.31 (a) N = +1; O = 0.

(b) The central N is 0. The singly bonded O atom is −1, and the doubly bonded O atom is 0.

$$\left[:\ddot{\text{O}}\!-\!\text{N}\!=\!\ddot{\text{O}}\right]^{-}\longleftrightarrow\left[\ddot{\text{O}}\!=\!\text{N}\!-\!\ddot{\text{O}}:\right]^{-}$$

(c) N and F are both 0.

(d) The central N atom is +1, one of the O atoms is −1, and the other two O atoms are both 0.

$$\overset{0}{\text{H}}\!-\!\overset{0}{\ddot{\text{O}}}\!-\!\overset{+1}{\text{N}}\!=\!\overset{0}{\ddot{\text{O}}}$$
$$\underset{-1}{|}$$
$$:\ddot{\text{O}}:$$

9.33 (a) $\overset{\longrightarrow}{\underset{+\delta\ \ -\delta}{\text{C}\!-\!\text{O}}}$ $\overset{\longrightarrow}{\underset{+\delta\ \ -\delta}{\text{C}\!-\!\text{N}}}$ (c) $\overset{\longrightarrow}{\underset{+\delta\ \ -\delta}{\text{B}\!-\!\text{O}}}$ $\overset{\longrightarrow}{\underset{+\delta\ \ -\delta}{\text{B}\!-\!\text{S}}}$

CO is planar   BO is more polar

(b) $\overset{\longrightarrow}{\underset{+\delta\ \ -\delta}{\text{P}\!-\!\text{Cl}}}$ $\overset{\longrightarrow}{\underset{+\delta\ \ -\delta}{\text{P}\!-\!\text{Br}}}$ (d) $\overset{\longrightarrow}{\underset{+\delta\ \ -\delta}{\text{B}\!-\!\text{F}}}$ $\overset{\longrightarrow}{\underset{+\delta\ \ -\delta}{\text{B}\!-\!\text{I}}}$

PCl is more polar   BF is more polar

9.35 (a) CH and CO bonds are polar.

(b) The CO bond is most polar, and O is the most negative atom.

9.37 (a) OH⁻: The formal charge on O is −1 and on H is 0.

(b) BH₄⁻: Even though the formal charge on B is −1 and on H is 0, H is slightly more electronegative than B. The four H atoms are therefore more likely to bear the −1 charge of the ion. The BH bonds are polar, with the H atom being the negative end.

(c) The CH and CO bonds are all polar (but the C—C bond is not). The negative charge in the CO bonds lies on the O atoms.

9.39 Structure C is most reasonable. The charges are as small as possible and the negative charge resides on the more electronegative atom.

$$\overset{-2\ \ +1\ \ +1}{:\ddot{\text{N}}\!-\!\text{N}\!\equiv\!\text{O}:}\longleftrightarrow\overset{-1\ \ +1\ \ 0}{\dot{\text{N}}\!=\!\text{N}\!=\!\ddot{\text{O}}}\longleftrightarrow\overset{0\ \ +1\ \ -1}{:\text{N}\!\equiv\!\text{N}\!-\!\ddot{\text{O}}:}$$
$$\qquad\text{A}\qquad\qquad\qquad\text{B}\qquad\qquad\qquad\text{C}$$

9.41 If an H⁺ ion were to attack NO₂⁻, it would attach to an O atom.

$$\left[\overset{-1\ \ \ 0\ \ \ 0}{:\ddot{\text{O}}\!-\!\text{N}\!=\!\ddot{\text{O}}}\right]^{-}\longleftrightarrow\left[\overset{0\ \ \ 0\ \ \ -1}{\ddot{\text{O}}\!=\!\text{N}\!-\!\ddot{\text{O}}:}\right]^{-}$$

9.43 (i) The most polar bonds are in H₂O (because O and H have the largest difference in electronegativity).

(ii) Not polar: CO₂ and CCl₄

(iii) F

9.45 (a) Not polar; linear molecule

(b) HBF₂, polar trigonal-planar molecule with F atoms at the negative end of the dipole and the H atom at the positive end.

(c) CH₃Cl, polar tetrahedral molecule. The Cl atom is the negative end of the dipole, and the three H atoms are on the positive side of the molecule.

(d) SO₃, nonpolar trigonal-planar molecule

9.47   (a)   C—H bonds, bond order is 1; 1 C=O bond, bond order is 2

      (b)   3 S—O single bonds, bond order is 1

      (c)   2 nitrogen-oxygen double bonds, bond order is 2

      (d)   1 N=O double bond, bond order is 2; 1 N—Cl bond, bond order is 1

9.49   (a) B—Cl   (c) P—O

      (b) C—O   (d) C=O

9.51   NO bond orders: 2 in NO_2^+, 1.5 in NO_2^-; 1.33 in NO_3^-, The NO bond is longest in NO_3^- and shortest in NO_2^+.

9.53   The CO bond in carbon monoxide is a triple bond, so it is both shorter and stronger than the CO double bond in H_2CO.

9.55 $\Delta H^\circ_{rxn} = -126$ kJ

9.57   O—F bond dissociation energy = 192 kJ/mol

9.59

Element	Valence Electrons
Li	1
Ti	4
Zn	2
Si	4
Cl	7

9.61   Ionic: KI and MgS

      Covalent: CS_2 and P_4O_{10}

9.63   Group 2A elements form 2+ ions (such as Ca^{2+}), and Group 7A elements form 1− ions when combined with a metal. Therefore, only $CaCl_2$ is a reasonable formula.

9.65 SeF_4, BrF_4^-, XeF_4

9.67   The C—H bonds in C_2H_2 have a bond order of 1, whereas the carbon–carbon bond has an order of 3. In phosgene the C—Cl bonds are single bonds, whereas the C=O bond is a double bond with an order of 2.

9.69   NO bond order in NO_3^- is 1.33.

9.71   To estimate the enthalpy change we need energies for the following bonds: O=O, H—H, and H—O.

      Energy to break bonds = 498 kJ (for O=O) + 2 × 436 kJ (for H—H) = +1370 kJ.

      Energy evolved when bonds are made = 4 × 463 kJ (for O—H) = −1852 kJ

      Total energy = −482 kJ

9.73   All the molecules in the series have 16 valence electrons and all are linear.

      (a) Ö=C=Ö ⟷ :Ö—C≡O: ⟷ :O≡C—Ö:

      (b)

      [N̈=N=N̈]⁻ ⟷ [:N̈—N≡N:]⁻ ⟷ [:N≡N—N̈:]⁻

9.75   (c)

      [Ö=C=N̈]⁻ ⟷ [:Ö—C≡N:]⁻ ⟷ [:O≡C—N̈:]⁻

9.75   The N—O bonds in NO_2^- have a bond order of 1.5, while in NO_2^+ the bond order is 2. The shorter bonds (110 pm) are the NO bonds with the higher bond order (in NO_2^+), whereas the longer bonds (124 pm) in NO_2^- have a lower bond order.

9.77   The F—Cl—F bond angle in ClF_2^+, which has a tetrahedral electron-pair geometry, is approximately 109°.

      [:F̈—C̈l—F̈:]⁺

      The ClF_2^- ion has a trigonal-bipyramidal electron-pair geometry, with F atoms in the axial positions and the lone pairs in the equatorial positions. Therefore, the F—Cl—F angle is 180°.

      [:F̈—C̈l—F̈:]⁻

9.79   An H^+ ion will attach to an O atom of SO_3^{2-} and not to the S atom. Each O atom has a formal charge of −1, whereas the S atom has a formal charge of +1.

      [:Ö—S̈—Ö: / :Ö:]²⁻

9.81   (a)   Calculation from bond energies: ΔH°_{rxn} = −509 kJ/mol CH_3OH

      (b)   Calculation from thermochemical data: ΔH°_{rxn} = −676 kJ/mol CH_3OH

9.83   (a)

      [C̈=N=Ö]⁻ ⟷ [:C̈—N≡O:]⁻ ⟷ [:C≡N—Ö:]⁻
      −2   1   0   −3   1   0   −1   1   −1

      (b)   The third resonance structure is the most reasonable because the negative formal charge is on the most electronegative atom.

      (c)   Carbon, the least electronegative element in the ion, has a negative formal charge. In addition, all three resonance structures have an unfavorable charge distribution.

9.85

      :—Xe) 120°   F—Cl) 120°
      F   F

      (a) XeF_2 has three lone pairs around the Xe atom. The electron-pair geometry is trigonal bipyramidal. Because lone pairs require more space than bond pairs, it is better to place the lone pairs in the equator of the bipyramid, where the angles between them are 120°.

(b) Like XeF_2, ClF_3 has a trigonal-bipyramidal electron-pair geometry, but only two lone pairs around the Cl. These are better placed in the equatorial plane, where the angle between them is 120°.

9.87   (a) Angle 1 = 109°; angle 2 = 120°; angle 3 = 109°; angle 4 = 109°; angle 5 = 109°

   (b) O—H bonds

9.89 $\Delta H_{rxn} = +146$ kJ $= 2\,(D_{C-N}) + D_{C=O} - [D_{N-N} + D_{C=O}]$

9.91   (a) Two C—H bonds and one O=O bond are broken and two O—C bonds and two H—O bonds are made in the reaction. $\Delta H_{rxn} = -318$ kJ.

   (b) Acetone is polar.

   (c) The O—H hydrogen atoms are the most positive in dihydroxyacetone.

9.93   (a) The C=C bond is stronger than the C—C bond.

   (b) The C—C single bond is longer than the C=C double bond.

   (c) Ethylene is nonpolar, whereas acrolein is polar.

   (d) The reaction is exothermic ($\Delta H_{rxn} = -45$ kJ).

9.95 $\Delta H_{rxn} = -211$ kJ

9.97   (a) Angle 1 = 109°; angle 2 = 120°; angle 3 = 120°; angle 4 = 109°; angle 5 = 109°

   (b) The OH bonds are the most polar bonds in the molecule.

9.99   The molecule can have a pyramidal structure if there are three bond pairs and one lone pair at the corners of a tetrahedron (e.g., NH_3). The bond angles are likely to be slightly less than 109°. The molecule can have a bent structure with two lone pairs and two bond pairs. The bond angle is likely to be less than 109° (e.g., H_2O).

9.101   (a) Odd-electron molecules: BrO (13 electrons) and OH (7 electrons)

   (b) $Br_2(g) \longrightarrow 2\,Br(g)$ $\Delta H_{rxn} = +193$ kJ
     $2\,Br(g) + O_2(g) \longrightarrow 2\,BrO(g)$ $\Delta H_{rxn} = +96$ kJ
     $BrO(g) + H_2O(g) \longrightarrow HOBr(g) + OH(g)$
                                $\Delta H_{rxn} = 0$ kJ

   (c) $\Delta H\,[HOBr(g)] = -101$ kJ/mol

   (d) The reactions in part (b) are endothermic (or thermal-neutral), and the heat of formation in part (c) is exothermic.

9.103   Lattice energy depends directly on ion charges and inversely on the distance between ions. The sizes of the Cl^-, Br^-, and I^- ions fall in a relatively narrow range (181, 196, and 220 pm, respectively), and the ion sizes change by only 15–24 pm from one ion to the next. Therefore, their lattice energies are expected to decrease in a narrow range. The F^- ion (133 pm), is only 74% as large as the Cl^- ion, so the lattice energy of NaF is much more negative.

9.105   (a) BF_3 is not a polar molecule, and replacing one of two F atoms with an H atom (HBF_2 and H_2BF) gives polar molecules.

   (b) $BeCl_2$ is not polar, whereas replacing a Cl atom with a Br atom, gives a polar molecule (BeClBr).

Chapter 10

10.1   The electron-pair and molecular geometries of $CHCl_3$ are both tetrahedral. An sp^3 hybrid orbital on the C atom overlaps a p orbital on a Cl atom to form a sigma bond. A C—H sigma bond is formed by a C atom orbital overlapping an H atom $1s$ orbital.

$$\begin{array}{c} :\ddot{Cl}: \\ | \\ H-C-\ddot{Cl}: \\ | \\ :\ddot{Cl}: \end{array}$$

10.3   (a) sp^2 (c) sp^3

   (b) sp (d) sp^2

10.5   (a) C, sp^3; O, sp^3

   (b) CH_3, sp^3; middle C, sp^2; CH_2, sp^2

   (c) CH_2, sp^3; CO_2H, sp^2; N, sp^3

10.7   (a) Electron-pair geometry is octahedral. Molecular geometry is octahedral. Si: sp^3d^2.

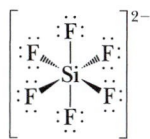

   (b) Electron-pair geometry is trigonal bipyramidal. Molecular geometry is seesaw. Se: sp^3d.

   (c) Electron-pair geometry is trigonal bipyramidal. Molecular geometry is linear. I: sp^3d.

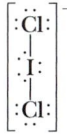

   (d) Electron-pair geometry is octahedral. Molecular geometry is square planar. Xe: sp^3d^2

$$\begin{array}{c} :\ddot{F}\text{...} \quad | \quad \text{...}\ddot{F}: \\ Xe \\ :\ddot{F} \quad | \quad \ddot{F}: \end{array}$$

10.9   There are 32 valence electrons in both HPO_2F_2 and its anion. Both have a tetrahedral molecular geometry, so the P atom in both is sp^3 hybridized.

$$\begin{array}{cc} :\ddot{O}: & \left[\; :\ddot{O}: \;\right]^- \\ \| & \| \\ H-\ddot{O}-P\text{...}\ddot{F}: & :\ddot{O}-P\text{...}\ddot{F}: \\ \diagdown & \diagdown \\ :\ddot{F}: & :\ddot{F}: \end{array}$$

10.11 The C atom is sp^2 hybridized. Two of the sp^2 hybrid orbitals are used to form C—Cl sigma bonds, and the third is used to form the C—O sigma bond. The p orbital not used in the C atom hybrid orbitals is used to form the CO pi bond.

10.13

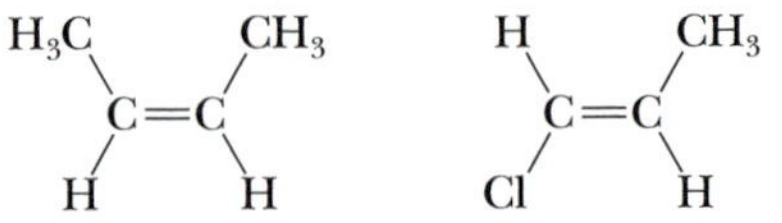

cis isomer   *trans* isomer

10.15 H_2^+ ion: $(\sigma_{1s})^1$. Bond order = 0.5. The bond in H_2^+ is weaker than in H_2 (bond order = 1).

10.17 MO diagram for C_2^{2-} ion:

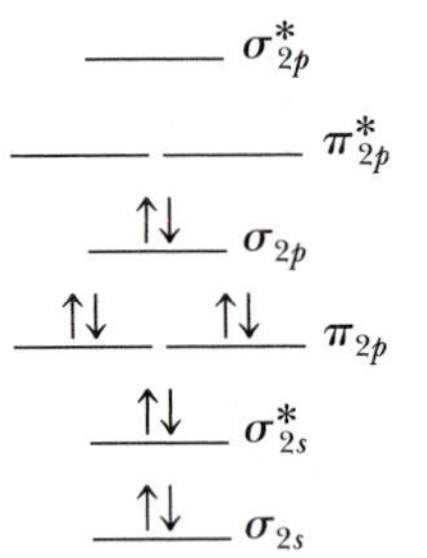

The ion has 10 valence electrons (isoelectronic with N_2). There is one net sigma bond and two net pi bonds, for a bond order of 3. The bond order increases by 1 on going from C_2 to C_2^{2-}. The ion is not paramagnetic.

10.19 (a) CO has 10 valence electrons.

$[core](\sigma_{2s})^2(\sigma_{2s}^*)^2(\pi_{2p})^4(\sigma_{2p})^2$

(b) HOMO, σ_{2p}

(c) Diamagnetic

(d) 1 σ bond and 2 π bonds; bond order is 3

10.21

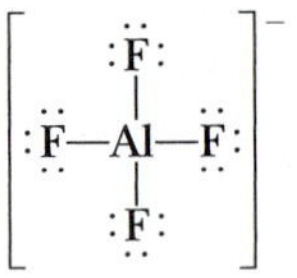

The electron-pair and molecular geometries are both tetrahedral. The Al atom is sp^3 hybridized, and so the Al—F bonds are formed by overlap of an Al sp^3 orbital with a p orbital on the F atom.

10.23

Molecule/Ion	O—S—O Angle	Hybrid Orbitals
SO_2	120°	sp^2
SO_3	120°	sp^2
SO_3^{2-}	109°	sp^3
SO_4^{2-}	109°	sp^3

10.25

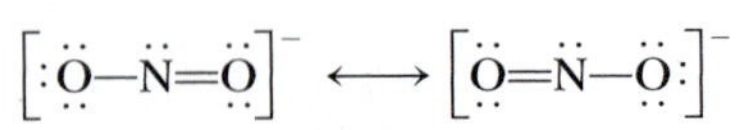

The electron-pair geometry is trigonal planar. The molecular geometry is bent or angular. The O—N—O angle will be about 120°, the average N—O bond order is 3/2, and the N atom is sp^2 hybridized.

10.27 The resonance structures of N_2O, with formal charges, are shown here.

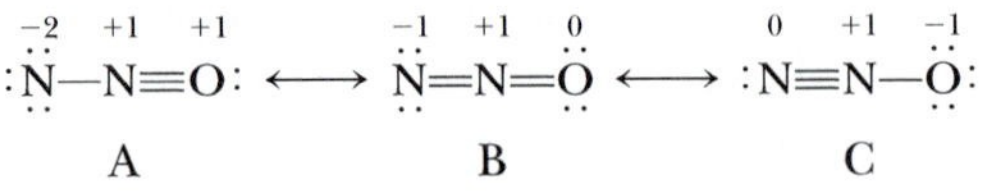

A   B   C

The central N atom is sp hybridized in all structures.

10.29 (a) All three have the formula C_2H_4O. They are usually referred to as structural isomers.

(b) Ethylene oxide: both C atoms are sp^3 hybridized, and the bond angles in the ring are only 60° (which makes this a relatively unstable molecule).

Acetaldehyde: The CH_3 carbon atom has sp^3 hybridization (bond angles of 109°), and the other C atom is sp^2 hybridized (bond angles of 120°).

Vinyl alcohol: Both C atoms are sp^2 hybridized, and the bond angles are 120°.

(c) H—C—H angles in ethylene oxide and acetaldehyde = about 109°. Angle in vinyl alcohol = 120°.

(d) All are polar.

(e) Acetaldehyde has the strongest CO bond, and vinyl alcohol has the strongest C—C bond.

10.31 (a) CH_3 carbon atom: sp^3

C=N carbon atom: sp^2

N atom: sp^2

(b) C—N—O bond angle = 120°

10.33 (a) C(1) = sp^2; O(2) = sp^3; N(3) = sp^3; C(4) = sp^3; P(5) = sp^3

(b) Angle A = 120°; angle B = 109°; C = 109°; angle D = 109°

(c) The P—O and O—H bonds are most polar ($\Delta\chi = 1.4$).

10.35 (a) The geometry about the boron atom is trigonal planar in BF_3, but tetrahedral in H_3N—BF_3.

(b) Boron is sp^2 hybridized in BF_3 but sp^3 hybridized in H_3N—BF_3.

(c) Yes

10.37 (a) Then C=O bond is most polar.

(b) 18 sigma bonds and 5 pi bonds

(c)

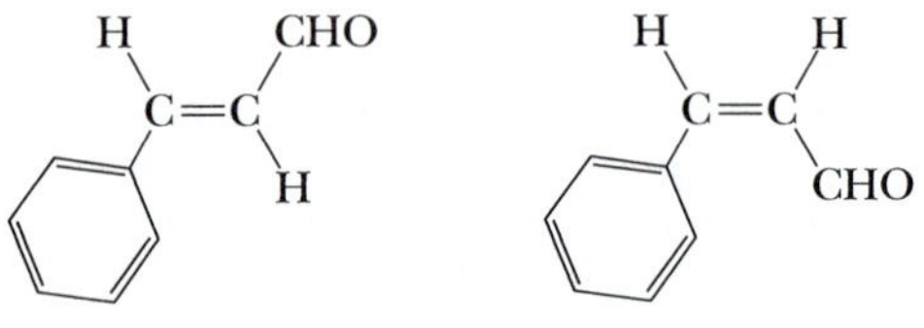

trans isomer   *cis* isomer

(d) All C atoms are sp^2 hybridized.

(e) All bond angles are 120°.

10.39 (a) The Sb in SbF_5 is sp^3d hybridized, whereas it is sp^3d^2 hybridized in SbF_6^-.

(b) The molecular geometry of the H_2F^+ ion is bent or angular, and the F atom is sp^3 hybridized.

$$\left[\begin{array}{c} :\!\overset{|}{\underset{\displaystyle H}{F}}\!\text{---}H \end{array} \right]^-$$

10.41 (a) The peroxide ion has a bond order of 1.

$$\left[:\!\ddot{O}\!\!-\!\!\ddot{O}\!: \right]^{2-}$$

(b) [core electrons] $(\sigma_{2s})^2(\sigma_{2s}^*)^2(\pi_{2p})^4(\sigma_{2p})^2(\pi_{2p}^*)^4$

This configuration also leads to a bond order of 1.

(c) Both theories lead to a diamagnetic ion with a bond order of 1.

10.43 See Table 10.1 on page 463 for the paramagnetism of diatomic molecules.

(a) Paramagnetic diatomic molecules: B_2 and O_2

(b) Bond order of 1: Li_2, B_2, F_2

(c) Bond order of 2: C_2 and O_2

(d) Highest bond order: N_2

10.45 CN has 9 valence electrons.

[core electrons] $(\sigma_{2s})^2(\sigma_{2s}^*)^2(\pi_{2p})^4(\sigma_{2p})^1$

(a) HOMO, σ_{2p}

(b, c) Bond order = 2.5 (0.5 σ bond and 2 π bonds)

(d) Paramagnetic

10.47 (a) All C atoms are sp^3 hybridized.

(b) About 109°

(c) Polar

(d) The six-membered ring cannot be planar owing to the tetrahedral C atoms of the ring. The bond angles are all 109°.

10.49 (a) The keto and enol forms are not resonance structures because both electron pairs and atoms have been rearranged.

(b) In the enol form, the terminal —CH_3 carbon atoms are sp^3 hybridized and the central C atoms are sp^2 hybridized. In the keto form, the terminal —CH_3 carbon atoms and the central C atom are sp^3 hybridized and the two C=O carbon atoms are sp^2 hybridized.

(c) Enol form: The —CH_3 groups have tetrahedral electron-pair and molecular geometries. The other three C atoms all have trigonal-planar electron-pair and molecular geometries.

Keto form: The —CH_3 groups and the central C atom have tetrahedral electron-pair and molecular geometries. The other two C atoms have trigonal-planar electron-pair and molecular geometries.

(d) $H_3C\!-\!\underset{\displaystyle :\!\ddot{O}\!:}{C}\!=\!\underset{\displaystyle :\!\ddot{O}\!:}{C}\!-\!C\!-\!CH_3 \longleftrightarrow H_3C\!-\!\underset{\displaystyle :\!\ddot{O}\!:}{C}\!-\!\underset{\displaystyle :\!\ddot{O}\!:}{C}\!=\!C\!-\!CH_3$

(e) *Cis-trans* isomerism is possible in the enol form.

10.51 A C atom may form, at most, four hybrid orbitals (sp^3). The minimum number is two—for example, the sp hybrid orbitals used by carbon in CO.

10.53 (a) C, sp^2; N, sp^3

(b) The amide or peptide link has two resonance structures (shown here with formal charges on the O and N atoms). Structure B is less favorable owing to the separation of charges.

$$\underset{A}{\underset{\displaystyle R}{\overset{\displaystyle :\ddot{O}:}{\underset{\displaystyle |}{C}}}\!\!-\!\!\underset{\displaystyle R}{\overset{\displaystyle \ddot{N}}{}}\!\!-\!H} \longleftrightarrow \underset{B}{\underset{\displaystyle R}{\overset{\displaystyle :\ddot{O}\overset{\displaystyle -}{:}}{\underset{\displaystyle |}{C}}}\!=\!\underset{\displaystyle R}{\overset{\displaystyle +}{N}}\!\!-\!H}$$

(c) The fact that the amide link is planar indicates that structure B has some importance.

10.55 MO theory is better to use when explaining or understanding the effect of adding energy to molecules. A molecule can absorb energy and an electron can thus be promoted to a higher level. Using MO theory one can see how this can occur. Additionally, MO allows us to understand how a molecule can be paramagnetic.

10.57 (a) The number of hybrid orbitals equals the number of atomic orbitals used in their creation.

(b) No

(c) The hybrid orbitals have an energy that is the weighted average of their constituent atomic orbitals.

(d,e) The hybrid orbital shapes are the same but the hybrids lie along different axes (or in different planes).

10.59 (a) The C atom in the center of the molecule is sp hybridized, so two unhybridized p orbitals remain. These could be the p_x and p_y orbitals, for example. Because these p orbitals lie at 90° angles to each other, they form pi bonds to the end CH_2 groups that are in planes that lie at 90° angles to each other.

(b) The C atoms in benzene are all sp^2 hybridized. These hybrid orbitals all lie in the same plane.

(c) C atom 1 = sp^3; C atoms 2 and 3 = sp^2

Chapter 11

11.1 Heptane

11.3 $C_{14}H_{30}$ is an alkane and C_5H_{10} could be a cycloalkane.

11.5   2,3-dimethylbutane

11.7   (a) 2,3-Dimethylhexane

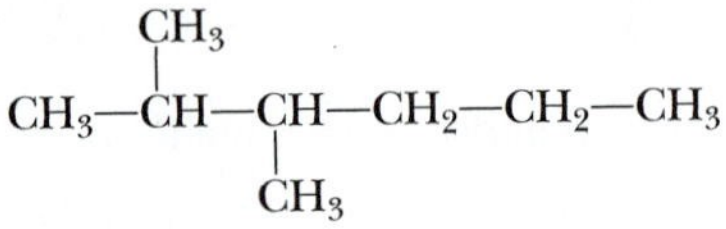

$$CH_3-CH-CH-CH_2-CH_2-CH_3$$
with CH_3 on the second carbon and CH_3 on the third carbon

(b) 2,3-Dimethyloctane

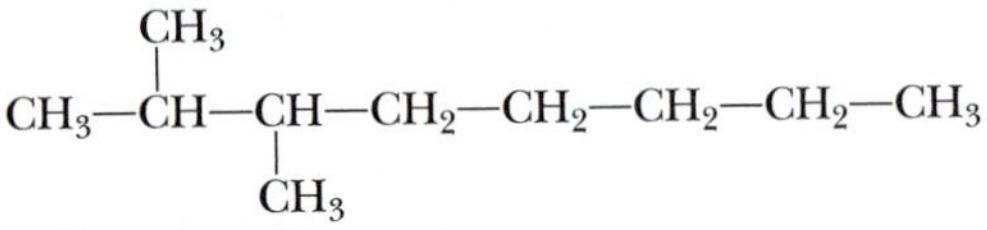

$$CH_3-CH-CH-CH_2-CH_2-CH_2-CH_2-CH_3$$
with CH_3 on the second carbon and CH_3 on the third carbon

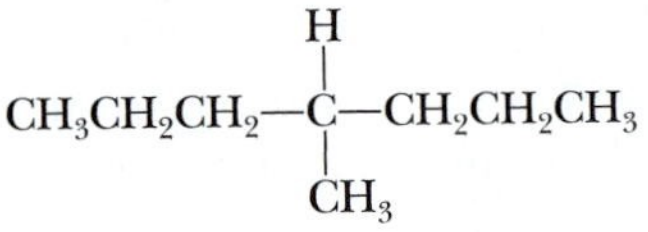

(c) 3-Ethylheptane

$$CH_3-CH_2-CH-CH_2-CH_2-CH_2-CH_3$$
with CH_2CH_3 on the third carbon

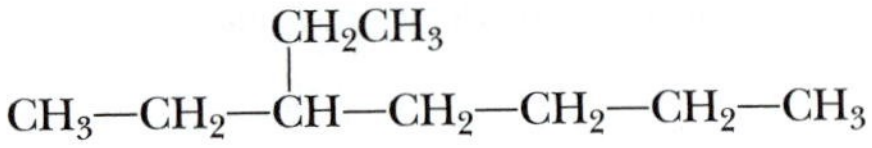

(d) 3-Ethyl-2-methylhexane

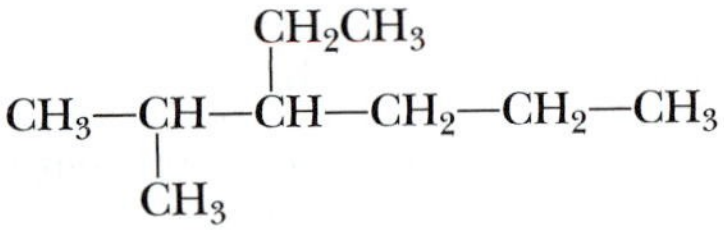

$$CH_3-CH-CH-CH_2-CH_2-CH_3$$
with CH_2CH_3 on the third carbon and CH_3 on the second carbon

11.9

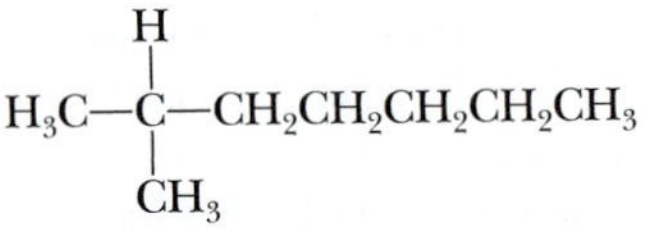

$$H_3C-\overset{H}{\underset{CH_3}{C}}-CH_2CH_2CH_2CH_2CH_3$$ 2-methylheptane

$$CH_3CH_2CH_2-\overset{H}{\underset{CH_3}{C}}-CH_2CH_2CH_3$$ 4-methylheptane

$$CH_3CH_2-\overset{H}{\underset{CH_3}{C^*}}-CH_2CH_2CH_2CH_3$$ 3-methylheptane. The C atom with an asterisk is chiral.

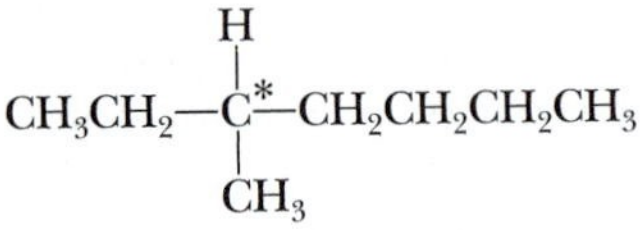

11.11

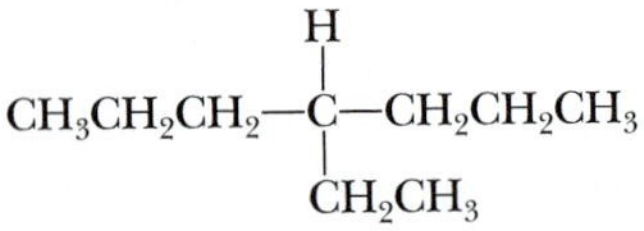

$$CH_3CH_2CH_2-\overset{H}{\underset{CH_2CH_3}{C}}-CH_2CH_2CH_3$$ 4-ethylheptane. The compound is not chiral.

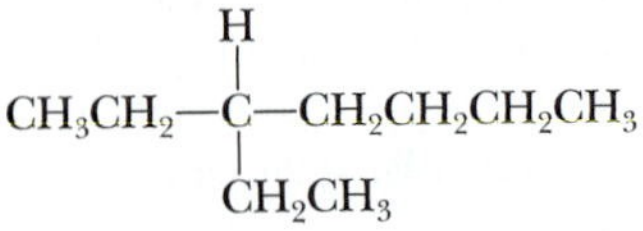

$$CH_3CH_2-\overset{H}{\underset{CH_2CH_3}{C}}-CH_2CH_2CH_2CH_3$$ 3-ethylheptane. Not chiral.

11.13 C_4H_{10}, butane: a low-molecular weight-fuel gas at room temperature and pressure. Slightly soluble in water.

$C_{12}H_{26}$, dodecane: a colorless liquid at room temperature. Expected to be insoluble in water but quite soluble in nonpolar solvents.

11.15

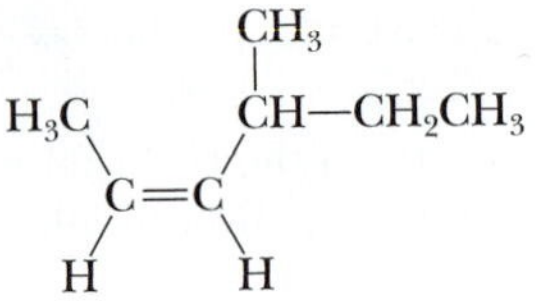

cis-4-methyl-2-hexene

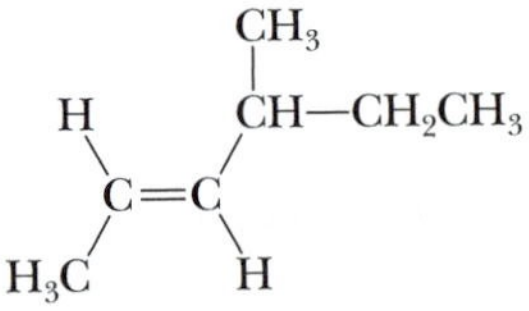

trans-4-methyl-2-hexene

11.17   (a)

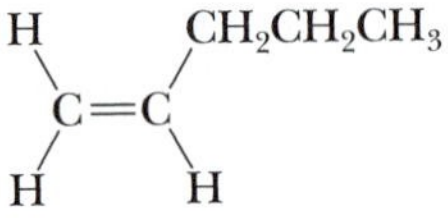

1-pentene

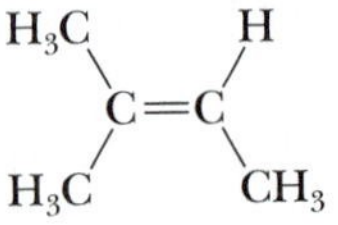

2-methyl-2-butene

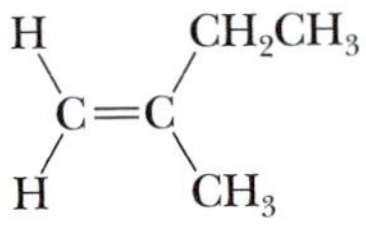

2-methyl-1-butene

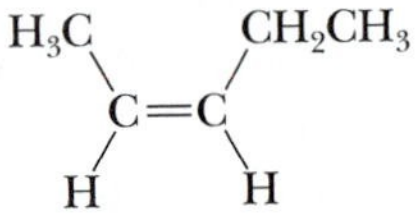

cis-2-pentene

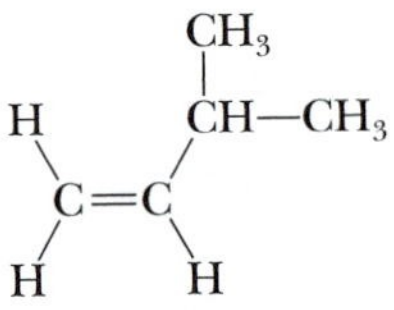

3-methyl-1-butene

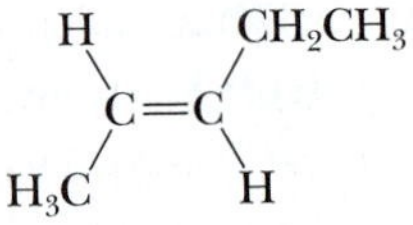

trans-2-pentene

(b)

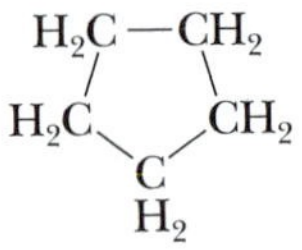

$$H_2C-CH_2$$
$$H_2C \qquad CH_2$$
$$\underset{H_2}{C}$$

cyclopentane

11.19   (a) 1,2-Dibromopropane, $CH_3CHBrCH_2Br$

(b) Pentane, C_5H_{12}

11.21   1-Butene, $CH_3CH_2CH=CH_2$

11.23   Four isomers are possible.

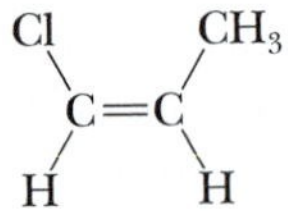

cis-1-chloropropene

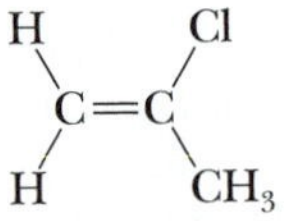

2-chloropropene

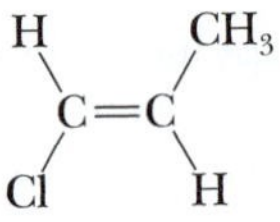

trans-1-chloropropene

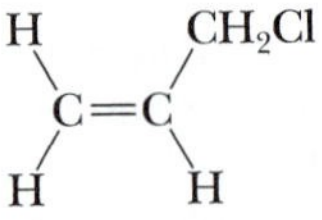

3-chloro-1-propene

11.25

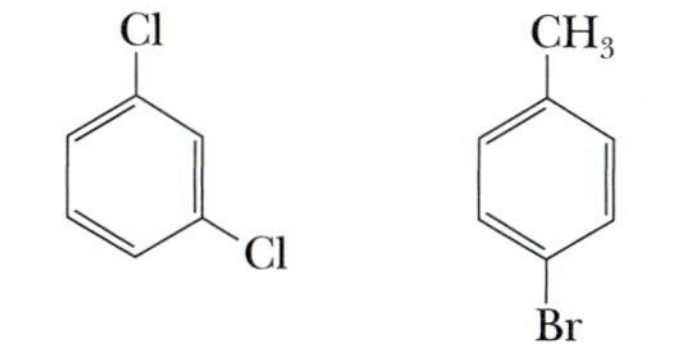

m-dichlorobenzene   *o*-bromotoluene

11.27

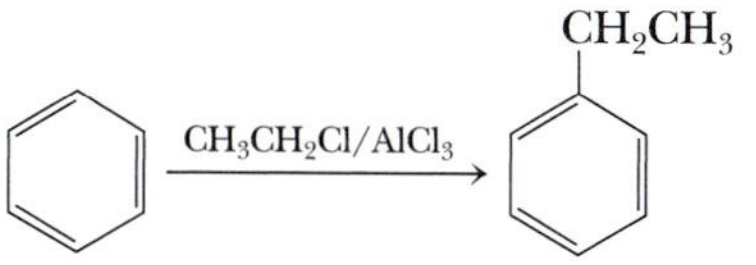

ethylbenzene

11.29

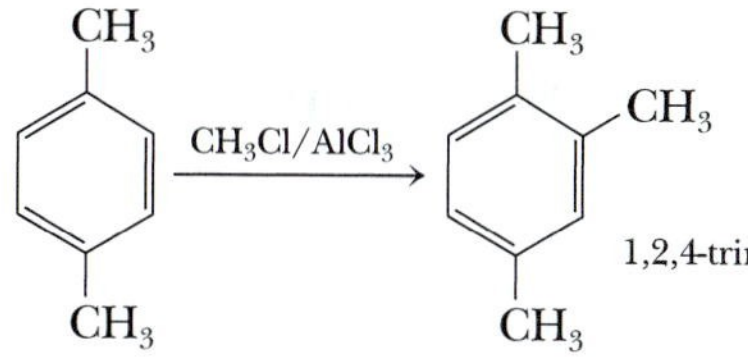

1,2,4-trimethylbenzene

11.31 (a) 1-Propanol, primary
  (b) 1-Butanol, primary
  (c) 2-Methyl-2-propanol, tertiary
  (d) 2-Methyl-2-butanol, tertiary

11.33 (a) Ethylamine, $CH_3CH_2NH_2$
  (b) Dipropylamine, $(CH_3CH_2CH_2)_2NH$

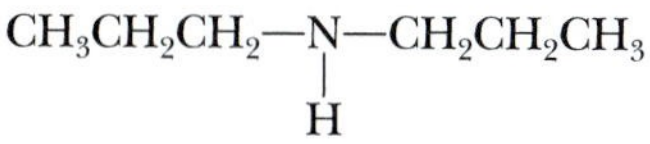

$$CH_3CH_2CH_2-\underset{\underset{H}{|}}{N}-CH_2CH_2CH_3$$

  (c) Butyldimethylamine

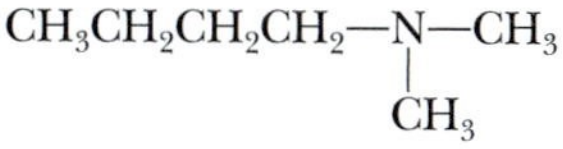

$$CH_3CH_2CH_2CH_2-\underset{\underset{CH_3}{|}}{N}-CH_3$$

  (d) triethylamine

$$CH_3CH_2-\underset{\underset{CH_2CH_3}{|}}{N}-CH_2CH_3$$

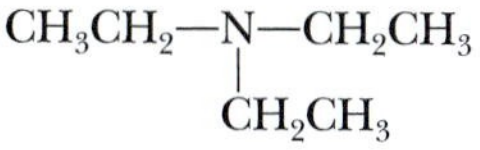

11.35 (a) 1-butanol, $CH_3CH_2CH_2CH_2OH$
  (b) 2-butanol

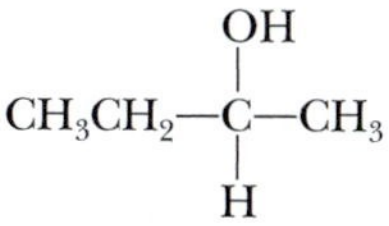

$$CH_3CH_2-\underset{\underset{H}{|}}{\overset{\overset{OH}{|}}{C}}-CH_3$$

  (c) 2-methyl-1-propanol

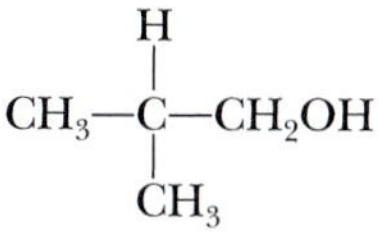

$$CH_3-\underset{\underset{CH_3}{|}}{\overset{\overset{H}{|}}{C}}-CH_2OH$$

  (d) 2-methyl-2-propanol

$$CH_3-\underset{\underset{CH_3}{|}}{\overset{\overset{OH}{|}}{C}}-CH_3$$

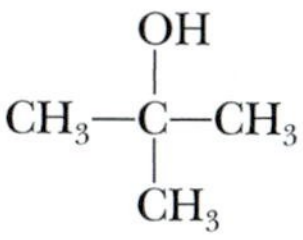

11.37 (a) $C_6H_5NH_2(aq) + HCl(aq) \longrightarrow (C_6H_5NH_3)Cl(aq)$
  (b) $(CH_3)_3N(aq) + H_2SO_4(aq) \longrightarrow$
$$[(CH_3)_3NH]HSO_4(aq)$$

11.39 (a) $CH_3-\overset{\overset{O}{\|}}{C}-CH_2CH_2CH_3$

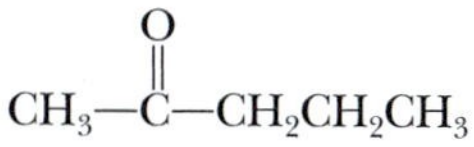

  (b) $H-\overset{\overset{O}{\|}}{C}-CH_2CH_2CH_2CH_2CH_3$

  (c) $CH_3CH_2CH_2CH_2-\overset{\overset{O}{\|}}{C}-OH$

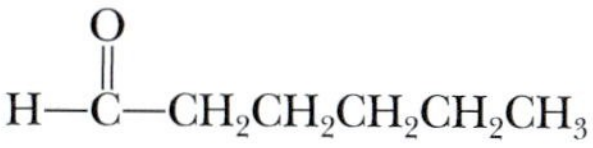

11.41 (a) Acid, 3-methylpentanoic acid
  (b) Ester, methyl propanoate
  (c) Ester, butyl acetate (or butyl ethanoate)
  (d) Acid, *p*-bromobenzoic acid

11.43 (a) Pentanoic acid (see Question 39c)
  (b) 1-Pentanol, $CH_3CH_2CH_2CH_2CH_2OH$
  (c) 2-Octanol

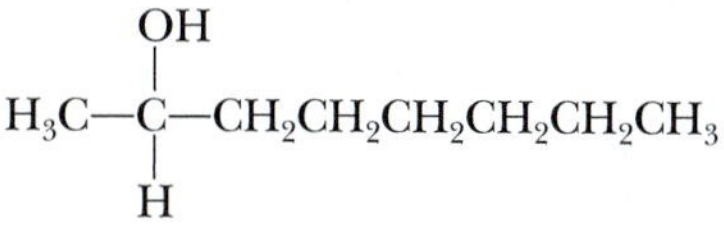

$$H_3C-\underset{\underset{H}{|}}{\overset{\overset{OH}{|}}{C}}-CH_2CH_2CH_2CH_2CH_2CH_3$$

  (d) No reaction. A ketone is not oxidized by $KMnO_4$.

11.45 Step 1: Oxidize 1-propanol to propanoic acid.

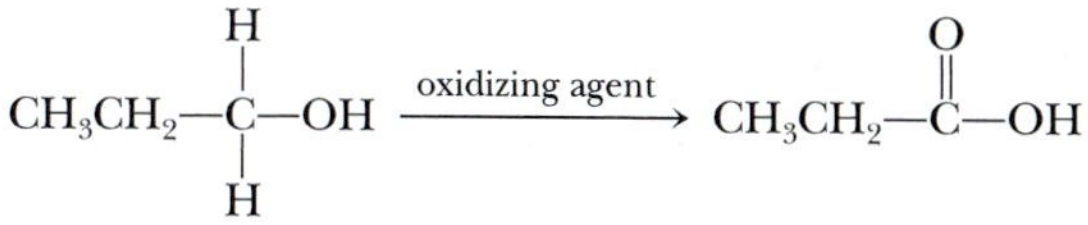

$$CH_3CH_2-\underset{\underset{H}{|}}{\overset{\overset{H}{|}}{C}}-OH \xrightarrow{\text{oxidizing agent}} CH_3CH_2-\overset{\overset{O}{\|}}{C}-OH$$

Step 2: Combine propanoic acid and 1-propanol.

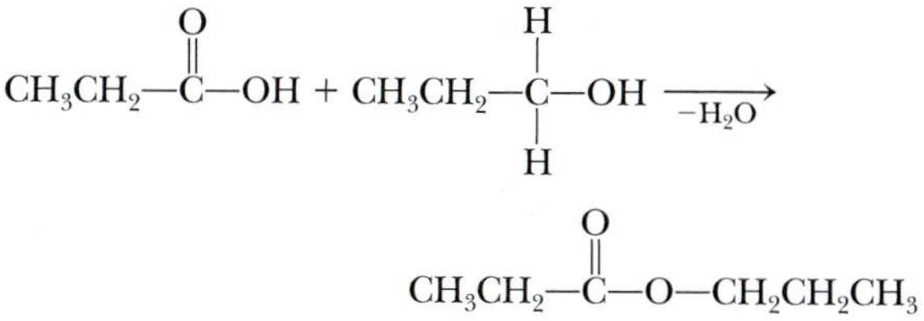

$$CH_3CH_2-\overset{\overset{O}{\|}}{C}-OH + CH_3CH_2-\underset{\underset{H}{|}}{\overset{\overset{H}{|}}{C}}-OH \xrightarrow{-H_2O}$$

$$CH_3CH_2-\overset{\overset{O}{\|}}{C}-O-CH_2CH_2CH_3$$

11.47 Sodium acetate, $NaCH_3CO_2$, and 1-butanol, $CH_3CH_2CH_2CH_2OH$

11.49 (a) Trigonal planar

(b) 120°

(c) The molecule is chiral. There are four different groups around the carbon atom marked 2.

(d) The acidic H atom is the H attached to the CO_2H (carboxyl) group.

11.51 (a) Alcohol (c) Acid

(b) Amide (d) Ester

11.53 (a) Prepare polyvinyl acetate (PVA) from vinylacetate.

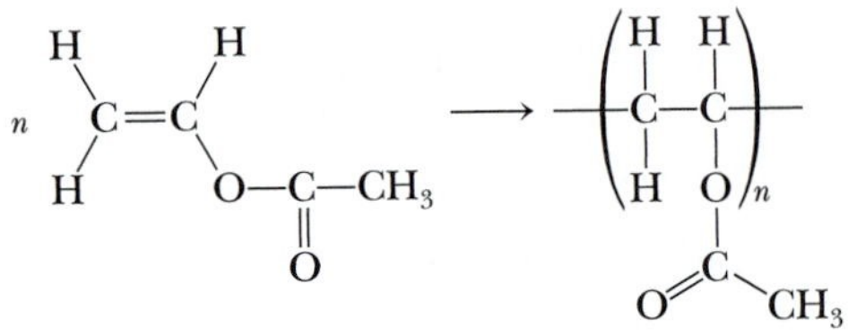

$$n \; \underset{H}{\overset{H}{>}}C=C\underset{O-C-CH_3}{\overset{H}{<}} \longrightarrow \left(\!\!\begin{array}{cc} H & H \\ -C-C- \\ H & O \end{array}\!\!\right)_n$$

(b) The three units of PVA:

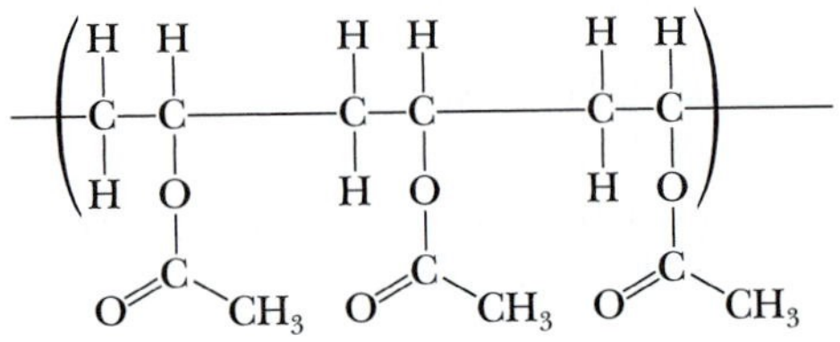

(c) Hydrolysis of polyvinyl alcohol

11.55 Illustrated here is a segment of a copolymer composed of two units of 1,1–dichloroethylene and two units of chloroethylene.

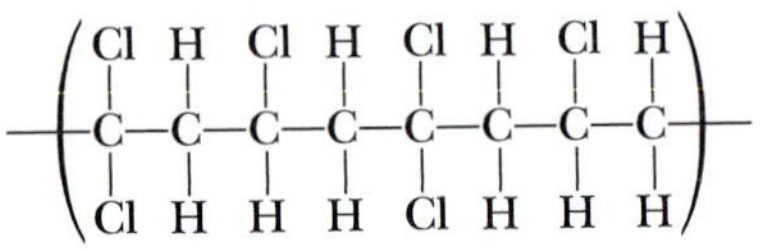

11.57 (a) 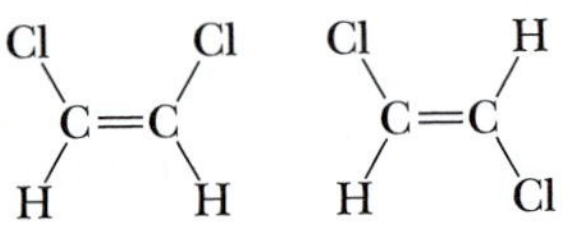 *cis* isomer   *trans* isomer

(b)

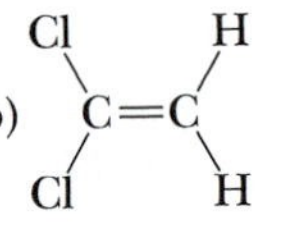

11.59

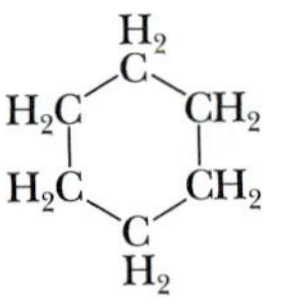

cyclohexane 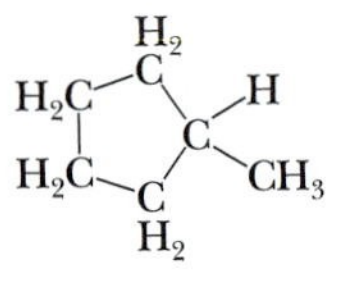   methylcyclopentane

$CH_3CH{=}CHCH_2CH_2CH_3$
2-hexene
Other isomers are possible by moving the double bond and with a branched chain.

11.61

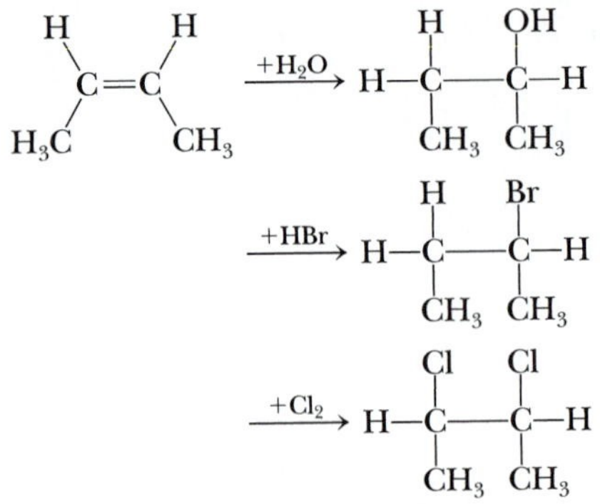

$$+H_2O$$

$$+HBr$$

$$+Cl_2$$

11.63 (a)

$$H_3C-\overset{O}{\overset{\|}{C}}-OH + NaOH \longrightarrow \left[H_3C-\overset{O}{\overset{\|}{C}}-O^-\right]Na^+ + H_2O$$

(b) $H_3C-\overset{H}{\underset{}{N}}-H + HCl \longrightarrow CH_3NH_3{}^+ + Cl^-$

11.65

$$n\, HOCH_2CH_2OH + n\, HO-\overset{O}{\overset{\|}{C}}-\!\!\bigcirc\!\!-\overset{O}{\overset{\|}{C}}-OH \longrightarrow$$

$$\left(\!\!O-\overset{}{C}-\!\!\bigcirc\!\!-\overset{O}{\overset{\|}{C}}-OCH_2CH_2O\!\!\right)_n + n\, H_2O$$

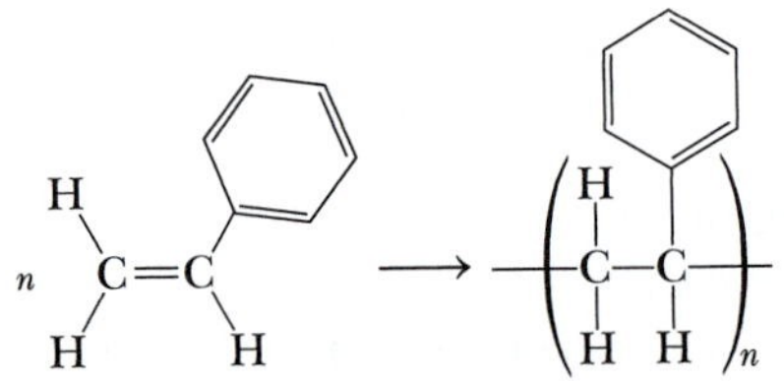

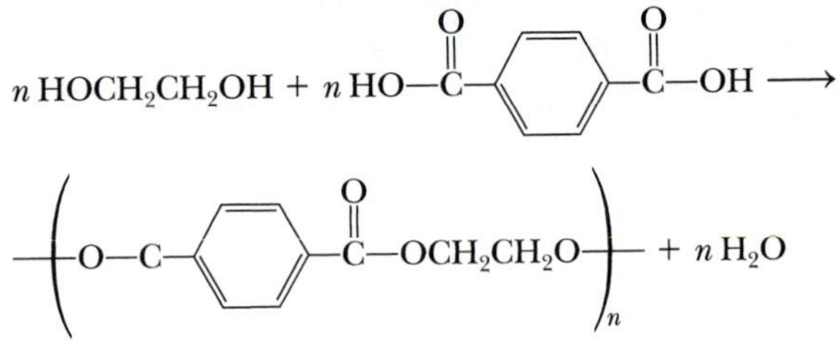

11.67 (a) 2,3-Dimethylpentane

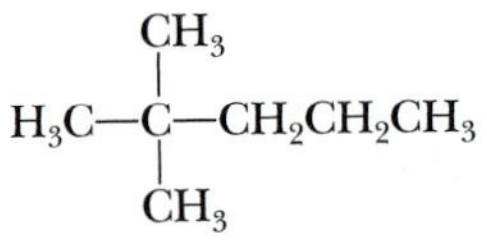

$$H_3C-\overset{CH_3}{\underset{CH_3}{\overset{|}{C}}}-CH_2CH_2CH_3$$

(b) 3,3-Dimethylpentane

$$CH_3CH_2-\overset{CH_2CH_3}{\underset{CH_2CH_3}{\overset{|}{C}}}-CH_2CH_3$$

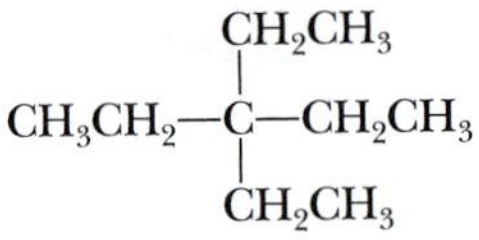

(c) 3-Ethyl-2-methylpentane

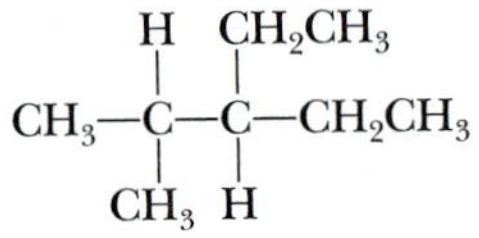

$$CH_3-\overset{H}{\underset{CH_3}{\overset{|}{C}}}-\overset{CH_2CH_3}{\underset{H}{\overset{|}{C}}}-CH_2CH_3$$

(d) 3-Ethylhexane

$$CH_3CH_2-\underset{\underset{H}{|}}{\overset{\overset{CH_2CH_3}{|}}{C}}-CH_2CH_2CH_3$$

11.69

1,1-Dichloropropane $H-\underset{\underset{Cl}{|}}{\overset{\overset{Cl}{|}}{C}}-CH_2CH_3$

1,2-Dichloropropane $H-\underset{\underset{H}{|}}{\overset{\overset{Cl}{|}}{C}}-\underset{\underset{H}{|}}{\overset{\overset{Cl}{|}}{C}}-CH_3$

1,3-Dichloropropane $H-\underset{\underset{H}{|}}{\overset{\overset{Cl}{|}}{C}}-\underset{\underset{H}{|}}{\overset{\overset{H}{|}}{C}}-\underset{\underset{H}{|}}{\overset{\overset{Cl}{|}}{C}}-H$

2,2-Dichloropropane $H-\underset{\underset{H}{|}}{\overset{\overset{H}{|}}{C}}-\underset{\underset{Cl}{|}}{\overset{\overset{Cl}{|}}{C}}-\underset{\underset{H}{|}}{\overset{\overset{H}{|}}{C}}-H$

11.71

1,2,3-trimethylbenzene   1,2,4-trimethylbenzene   1,3,5-trimethylbenzene

11.73 Replace the carboxylic acid group with an H atom

11.75 (a)

$$\underset{H_3C}{\overset{H}{\diagdown}}C=C\underset{\diagdown CH_3}{\overset{H}{\diagup}} \xrightarrow{+H_2} H-\underset{\underset{CH_3}{|}}{\overset{\overset{H}{|}}{C}}-\underset{\underset{CH_3}{|}}{\overset{\overset{H}{|}}{C}}-H$$
butane (not chiral)

(b) $H-\underset{\underset{CH_3}{|}}{\overset{\overset{CH_3}{|}}{C}}-CH_3$

11.77 Compound (b), acetaldehyde, and (c), ethanol, produces acetic acid when oxidized.

11.79

11.81

$$\underset{\substack{\text{glyceryl} \\ \text{trilaurate}}}{\begin{array}{c} H_2C-O-\overset{\overset{O}{\|}}{C}-R \\ HC-O-\overset{\overset{O}{\|}}{C}-R \\ H_2C-O-\overset{\overset{O}{\|}}{C}-R \end{array}} \xrightarrow{\text{NaOH}} \underset{\text{glycerol}}{\begin{array}{c} H_2C-OH \\ HC-OH \\ H_2C-OH \end{array}} + 3\,\underset{\text{sodium laurate}}{R-\overset{\overset{O}{\|}}{C}-O^-\,Na^+}$$

11.83 Cyclohexene, a cyclic alkene, will add Br_2 readily (to give $C_6H_{12}Br_2$). Benzene, however, needs much more stringent conditions to react with bromine; then Br_2 will substitute for H atoms on benzene and not add to the ring.

11.85 (a) The compound is either propanone, a ketone, or propanal, an aldehyde.

$$H-\underset{\underset{H}{|}}{\overset{\overset{H}{|}}{C}}-\overset{\overset{O}{\|}}{C}-\underset{\underset{H}{|}}{\overset{\overset{H}{|}}{C}}-H \qquad H-\underset{\underset{H}{|}}{\overset{\overset{H}{|}}{C}}-\underset{\underset{H}{|}}{\overset{\overset{H}{|}}{C}}-\overset{\overset{O}{\|}}{C}-H$$
propanone (a ketone)   propanal (an aldehyde)

(b) The ketone will not oxidize, but the aldehyde will be oxidized to the acid, $CH_3CH_2CO_2H$. Thus, the unknown is likely propanal.

(c) Propanoic acid

11.87

$$H_2C=C\underset{\underset{H}{|}}{\overset{\overset{CH_3}{|}}{-C}}-\underset{\underset{CH_3}{|}}{\overset{\overset{H}{|}}{C}}-CH_3$$
X = 3,3-dimethyl-2-pentene

$$\downarrow +H_2O$$

$$H_3C-\underset{\underset{H}{|}}{\overset{\overset{OH}{|}}{C}}-\underset{\underset{CH_3}{|}}{\overset{\overset{CH_3}{|}}{C}}-\underset{\underset{H}{|}}{\overset{\overset{H}{|}}{C}}-CH_3 \xrightarrow[\text{agent}]{\text{oxidizing}} H_3C-\overset{\overset{O}{\|}}{C}-\underset{\underset{CH_3}{|}}{\overset{\overset{CH_3}{|}}{C}}-\underset{\underset{H}{|}}{\overset{\overset{H}{|}}{C}}-CH_3$$
Y = 3,3-dimethyl-2-pentanol   3,3-dimethyl-2-pentanone

11.89

methane — four single bonds

formaldehyde — one double bond and two single bonds

allene — two double bonds

acetylene — one single bond and one triple bond

11.93 (a) Cross-linking makes the material very rigid and inflexible.

(b) The OH groups give the polymer a high affinity for water.

(c) Hydrogen bonding allows the chains to form coils and sheets with high tensile strength.

11.95 (a) Ethane heat of combustion = -47.51 kJ/g

Ethanol heat of combustion = -26.82 kJ/g

(b) The heat obtained from the combustion of ethanol is less negative than for ethane, so partially oxidizing ethane to form ethanol decreases the amount of energy per mole available from the combustion of the substance.

11.97 2-Propanol will react with an oxidizing agent such as $KMnO_4$ (to give the ketone), whereas methyl ethyl ether ($CH_3OC_2H_5$) will not react. In addition, the alcohol should be more soluble in water than the ether.

11.99 (a) Empirical formula, CHO

(b) Molecular formula, $C_4H_4O_4$

(c)

(d) All four C atoms are sp^2 hybridized.

(e) $120°$

11.101 (a)

2-bromopropane

(b)

(c)

Adding H_2O to 2-methyl-2-butene gives the same product as in part (b).

11.103 (a) In a substitution reaction, one atom or group of atoms is substituted for another. In an elimination reaction, a small molecule is removed or eliminated from a larger molecule.

(b) The elimination reaction produces an alkene, whereas the hydrogenation reaction has an alkene as a reactant. Both involve a small molecule, either H_2 or H_2O, being added to or eliminated from an organic molecule.

11.105 (a) Double bonds. See page 517.

(b) Termination occurs when the chain reaches 14 atoms. It could have been terminated earlier than this, or it could have continued to grow.

(c) The termination step

(d) Addition

Chapter 12

12.1 (a) 0.58 atm

(b) 0.59 bar

(c) 59 kPa

12.3 (a) 0.754 bar

(b) 650 kPa

(c) 934 kPa

12.5 2.70×10^2 mm Hg

12.7 3.7 L

12.9 250 mm Hg

12.11 3.2×10^2 mm Hg

12.13 9.72 atm

12.15 (a) 75 mL O_2

(b) 150 mL NO_2

12.17 0.919 atm

12.19 $V = 2.9$ L

12.21 1.9×10^6 g He

12.23 3.7×10^{-4} g/L

12.25 34.0 g/mol

12.27 57.5 g/mol

12.29 Molar mass = 74.9 g/mol; B_6H_{10}

12.31 0.039 mol H_2; 0.096 atm; 73 mm Hg

12.33 170 g NaN_3

12.35 1.7 atm O_2

12.37 4.1 atm H_2; 1.6 atm Ar; total pressure = 5.7 atm

12.39 (a) 0.30 mol halothane/1 mol O_2

(b) 3.0×10^2 g halothane

12.41 (a) They all have the same kinetic energy.

(b) The average speed of the H_2 molecules is greater than the average speed of the CO_2 molecules.

(c) The number of CO_2 molecules is greater than the number of H_2 molecules $[n(CO_2) = 1.8n(H_2)]$.

(d) The mass of CO_2 is greater than the mass of H_2.

12.43   Average speed of CO_2 molecule = 3.65×10^4 cm/s

12.45   Average speed (and molar mass) increases in the order $CH_2F_2 < Ar < N_2 < CH_4$.

12.47   (a) F_2 (38 g/mol) effuses faster than CO_2 (44 g/mol)

(b) N_2 (28 g/mol) effuses faster than O_2 (32 g/mol)

(c) C_2H_4 (28.1 g/mol) effuses faster than C_2H_6 (30.1 g/mol)

(d) $CFCl_3$ (137 g/mol) effuses faster than $C_2Cl_2F_4$ (171 g/mol)

12.49   36 g/mol

12.51 P from the van der Waals equation = 29.5 atm

P from the ideal gas law = 49.3 atm

12.53   (a) Standard atmosphere: 1 atm; 760 mm Hg; 101.325 kPa; 1.013 bar.

(b) N_2 partial pressure: 0.780 atm; 593 mm Hg; 79.1 kPa; 0.791 bar.

(c) H_2 pressure: 131 atm; 9.98×10^4 mm Hg; 1.33×10^4 kPa; 133 bar

(d) Air: 0.333 atm; 253 mm Hg; 33.7 kPa; 0.337 bar

12.55 C_2H_7N

12.57 $T = 290.$ K or 17 °C

12.59   (a) There are more molecules of H_2 than atoms of He.

(b) The mass of He is greater than the mass of H_2.

12.61   4 mol

12.63   Ni is the limiting reactant; 1.31 g $Ni(CO)_4$

12.65   (a, b) Sample 4 (He) has the largest number of molecules and sample 3 (H_2 at 27 °C and 760 mm Hg) has the fewest number of molecules.

(c) Sample 2 (Ar)

12.67 $P_{total} = 228$ mm Hg $= P(B_2H_6) + P(O_2)$

Stoichiometry requires that there be three times as many moles of O_2 as B_2H_6, so $P(O_2) = 3\ P(B_2H_6)$. Therefore, 228 mm Hg $= 4\ P(B_2H_6)$ and $P(B_2H_6) = 57$ mm Hg. This means $P(O_2) = 171$ mm Hg. The water vapor pressure is the same as O_2 pressure, or 171 mm Hg.

12.69 S_2F_{10}

12.71   8.54 g $Fe(CO)_5$

12.73   (a) 28.7 g/mol $\cong$ 29 g/mol

(b) X of O_2 = 0.18 and X of N_2 = 0.82

12.75 $P = 5 \times 10^{-8}$ atm

12.77   Molar mass = 86.4 g/mol. The gas is probably ClO_2F.

12.79   Yield = 76.0%

12.81   Weight percent $KClO_3 = 69.1\%$

12.83   (a) $NO_2 < O_2 < NO$

(b) $P(O_2) = 75$ mm Hg

(c) $P(NO_2) = 150$ mm Hg

12.85   The mixture contains 0.22 g CO_2 and 0.77 g CO.

$P(CO_2) = 0.22$ atm; $P(O_2) = 0.12$ atm; $P(CO) = 1.22$ atm

12.87   Formula of the iron compound: $Fe(CO)_5$

12.89 $P(SiH_4) = 40.$ mm Hg and $P(O_2) = 80.$ mm Hg

Pressure after reaction = 80. mm Hg

12.91   (a) $P(B_2H_6) = 0.0160$ atm

(b) $P(H_2) = 0.0320$ atm, so $P_{total} = 0.0480$ atm

12.93   Amount of $Na_2CO_3 = 0.00424$ mol

Amount of $NaHCO_3 = 0.00951$ mol

Amount of CO_2 produced = 0.0138 mol

Volume of CO_2 produced = 0.343 L

12.95   At 20 °C, there is 7.8×10^{-3} g H_2O/L. At 0 °C, there is 4.6×10^{-3} g H_2O/L.

12.97   (a) 10.0 g of O_2 represents more molecules than 10.0 g of CO_2. Therefore, O_2 has the greater partial pressure.

(b) The average speed of the O_2 molecules is greater than the average speed of the CO_2 molecules.

(c) The gases are at the same temperature and so have the same average kinetic energy.

12.99   (a) $P(C_2H_2) > P(CO)$

(b) There are more molecules in the C_2H_2 container than in the CO container.

12.101   (a) Not a gas. A gas would expand to an infinite volume.

(b) Not a gas. A density of 8.2 g/mL is typical of a solid.

(c) Insufficient information

(d) Gas

12.103   (a) 46.0 g NaN_3

(b)

(c) The N_3^- ion is linear.

12.105   The speed of gas molecules is related to the square root of the absolute temperature, so a doubling of the temperature will lead to an increase of about $(2)^{1/2}$ or 1.4.

Chapter 13

13.1   (a) Dipole–dipole interactions (and hydrogen bonds)

(b) Induced dipole–induced dipole forces

(c) Dipole–dipole interactions (and hydrogen bonds)

13.3　(a) Induced dipole–induced dipole forces

(b) Induced dipole–induced dipole forces

(c) Dipole–dipole forces

(d) Dipole–dipole forces (and hydrogen bonding)

13.5　The predicted order of increasing strength is $Ne < CH_4 < CO < CCl_4$. In this case, prediction does not quite agree with reality. The boiling points are Ne (-246 °C) $< CO$ (-192 °C) $< CH_4$ (-162 °C) $< CCl_4$ (77 °C).

13.7　(c) HF; (d) acetic acid; (f) CH_3OH

13.9　(a) LiCl. The Li^+ ion is smaller than Cs^+ (Figure 8.15), which makes the ion-ion forces of attraction stronger in LiCl.

(b) $Mg(NO_3)_2$. The Mg^{2+} ion is smaller than the Na^+ ion (Figure 8.15), and the magnesium ion has a 2+ charge (as opposed to 1+ for sodium). Both of these effects lead to stronger ion–ion forces of attraction in magnesium nitrate.

(c) $NiCl_2$. The nickel(II) ion has a larger charge than Rb^+ and is considerably smaller. Both effects mean that there are stronger ion–ion forces of attraction in nickel(II) chloride.

13.11　$q = +90.1$ kJ

13.13　(a) Water vapor pressure is about 150 mm Hg at 60 °C. (Appendix G gives a value of 149.4 mm Hg at 60 °C.)

(b) 600 mm Hg at about 93 °C

(c) At 70 °C, ethanol has a vapor pressure of about 520 mm Hg, whereas that of water is about 225 mm Hg.

13.15　At 30 °C the vapor pressure of ether is about 590 mm Hg. (This pressure requires 0.23 g of ether in the vapor phase at the given conditions, so there is sufficient ether in the flask.) At 0 °C the vapor pressure is about 160 mm Hg, so some ether condenses when the temperature declines.

13.17　(a) O_2 (-183 °C) (bp of $N_2 = -196$ °C)

(b) SO_2 (-10 °C) (CO_2 sublimes at -78 °C)

(c) HF ($+19.7$ °C) (HI, -35.6 °C)

(d) GeH_4 (-90.0 °C) (SiH_4, -111.8 °C)

13.19　(a) CS_2, about 620 mm Hg; CH_3NO_2, about 80 mm Hg

(b) CS_2, induced dipole–induced dipole forces; CH_3NO_2, dipole–dipole forces

(c) CS_2, about 46 °C; CH_3NO_2, about 100 °C

(d) About 39 °C

(e) About 34 °C

13.21　(a) 80.1 °C

(b) At about 48 °C the liquid has a vapor pressure of 250 mm Hg. The vapor pressure is 650 mm Hg at 75 °C.

(c) 33.5 kJ/mol (from slope of plot)

13.23　Two possible unit cells are illustrated here. The simplest formula is AB_8.

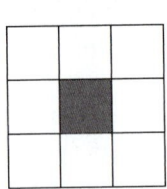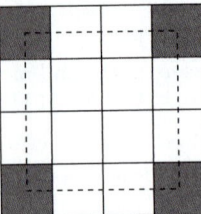

13.25　Ca^{2+} ions at 8 corners = 1 net Ca^{2+} ion

O^{2-} ions in 6 faces = 3 net O^{2-} ions

Ti^{4+} ion in center of unit cell = 1 net Ti^{4+} ion

Formula = $CaTiO_3$

13.27　(a) There are eight O^{2-} ions at the corners and one in the center for a net of two O^{2-} ions per unit cell. There are four Cu ions in the interior in tetrahedral holes. The ratio of ions is Cu_2O.

(b) The oxidation number of copper must be +1.

13.29　(a) 8 C atoms per unit cell. There are 8 corners (= 1 net C atom), 6 faces (= 3 net C atoms), and 4 internal C atoms.

(b) Face-centered cubic (fcc) with C atoms in the tetrahedral holes.

13.31　q (for fusion) = -1.97 kJ; q (for melting) = $+1.97$ kJ

13.33　(a) The density of liquid CO_2 is less than that of solid CO_2.

(b) CO_2 is a gas at 5 atm and 0 °C.

(c) Critical temperature = 31 °C, so CO_2 cannot be liquefied at 45 °C.

13.35　q (to heat the liquid) = 9.4×10^2 kJ

q (to vaporize NH_3) = 1.6×10^4 kJ

q (to heat the vapor) = 8.8×10^2 kJ

q_{total} = 1.8×10^4 kJ

13.37　Yes. The critical temperature (416 K, 143 °C) is well above room temperature.

13.39　$Ar < CO_2 < CH_3OH$

13.41　O_2 phase diagram. (i) Note the slight positive slope of the solid–liquid equilibrium line. It indicates that the density of solid O_2 is greater than that of liquid O_2. (ii) Using the diagram here, the vapor pressure of O_2 at 77 K is between 150 mm Hg and 200 mm Hg.

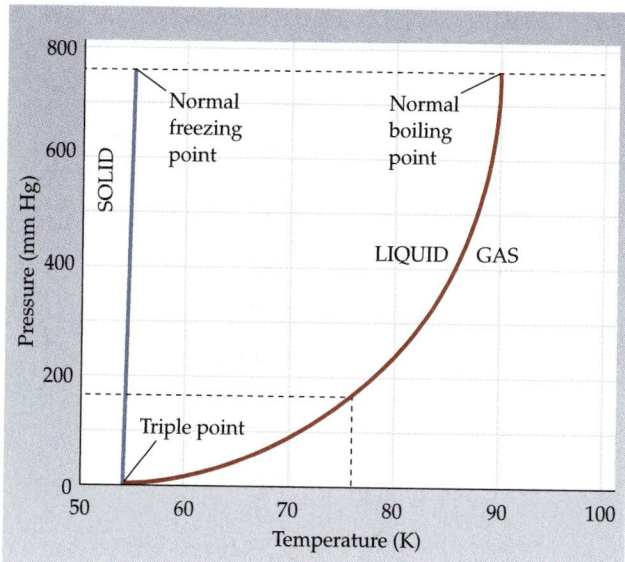

13.43   Less ethanol is available (17 mol) than would be required to completely fill the room with vapor (60. mol), so all of the ethanol will evaporate.

13.45 Li^+ ions are smaller than Cs^+ ions (78 pm and 165 pm, respectively; see Figure 8.15). Thus, there will be a stronger attractive force between Li^+ ion and water molecules than between Cs^+ ions and water molecules.

13.47   (a) 350 mm Hg

(b) Ethanol (lower vapor pressure at every temperature)

(c) 84 °C

(d) CS_2, 46 °C; C_2H_5OH, 78 °C; C_7H_{16}, 99 °C

(e) CS_2, gas; C_2H_5OH, gas; C_7H_{16}, liquid

13.49   Radius of silver = 145 pm

13.51 Ca^{2+}: there are 8 corner Ca^{2+} ions and 1 internal Ca^{2+} ion or a total of 2 Ca^{2+} ions. C atoms: there are 8 C atoms on edges. At 1/4 per atom, there are 2 within the unit cell. There are 2 more C atoms internal to the cell. Thus, there is a total of 4 C atoms per unit cell. The formula is CaC_2.

13.53   Molar enthalpy of vaporization increases with increasing intermolecular forces: C_2H_6 (14.69 kJ/mol; induced dipole) < HCl (16.15 kJ/mol; dipole) < CH_3OH (35.21 kJ/mol, hydrogen bonds) (The molar enthalpies of vaporization here are given at the boiling point of the liquid.)

13.55 1.356×10^{-8} cm (literature value is 1.357×10^{-8} cm)

13.57

Assumed Unit Cell	Calculated Density (g/cm³)
Simple cubic	4.60
Body-centered cubic	5.97
Face-centered cubic	6.52

The calculated density for a body-centered cubic unit cell is closest to the experimental value. In fact, vanadium has a body-centered cubic unit cell.

13.59   Mass of 1 CaF_2 unit calculated from crystal data = 1.2963×10^{-22} g. Divide molar mass of CaF_2 (78.077 g/mol) by mass of 1 CaF_2 to obtain Avogadro's number. Calculated value = 6.0230×10^{23} CaF_2/mol.

13.61   Diagram A leads to a surface coverage of 78.5%. Diagram B leads to 90.7% coverage.

13.63   (a) 70.3 °C

(b)

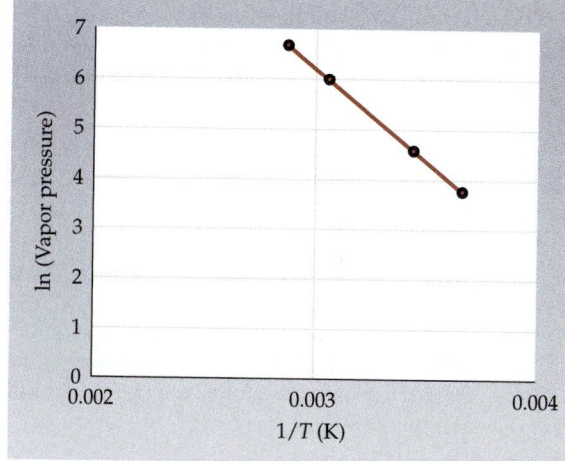

Using the equation for the straight line in the plot

ln P = −3885 (1/T) + 17.949

we calculate that $T = 312.6$ K (39.5 °C) when $P = 250$ mm Hg. When $P = 650$ mm Hg, $T = 338.7$ K (65.5 °C).

(c) Calculated $\Delta H_{vap} = 32.3$ kJ/mol

13.65   Acetone and water can interact by hydrogen bonding.

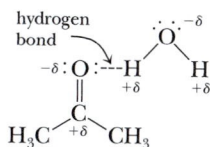

13.67   Glycol's viscosity will be greater than ethanol's owing to the greater hydrogen-bonding capacity of glycol.

13.69   (a) Water has two OH bonds and two lone pairs, whereas the O atom of ethanol has only one OH bond (and two lone pairs). More extensive Hydrogen bonding is likely for water.

(b) Water and ethanol interact extensively through hydrogen bonding, so the volume is expected to be slightly smaller than the sum of the two volumes.

13.71   No. NaCl has a 1:1 ratio of cations and anions in the unit cell, whereas the unit cell of $CaCl_2$ must have a 1:2 ratio of cations to anions.

13.73 Two pieces of evidence for $H_2O(\ell)$ having considerable intermolecular attractive forces:

(a) Based on the boiling points of the Group 6A hydrides (Figure 13.8), the boiling point of water should be approximately −80 °C. The actual boiling point of 100 °C reflects the significant hydrogen bonding that occurs.

(b) Liquid water has a specific heat capacity that is higher than almost any other liquid. This reflects the fact that a relatively larger amount of energy is necessary to overcome intermolecular forces and raise the temperature of the liquid.

13.75 (a) HI, hydrogen iodide

(b) The large iodine atom in HI leads to a significant polarizability for the molecule and thus to a large dispersion force.

(c) The dipole moment of HCl (1.07 D, Table 9.8) is larger than for HI (0.38 D).

(d) HI. See part (b).

13.77 When the can is inverted in cold water, the water vapor pressure in the can, which was approximately 760 mm Hg, drops rapidly—say, to 9 mm Hg at 10 °C. This creates a partial vacuum in the can, and the can is crushed because of the difference in pressure inside the can and the pressure of the atmosphere pressing down on the outside of the can.

13.79 (a) About −27 °C

(b) Pressure is about 6.5 atm.

(c) As the more energetic molecules leave the liquid phase, enter the gas phase, and escape from the tank, only lower-energy molecules remain. These have a lower temperature. The tank is thereby cooled, and water vapor can condense on the tank's surface. As the temperature drops, the vapor pressure of the remaining liquid drops and the flow of gas out of the tank slows.

(d) Cool the tank in dry ice (to −78 °C). The vapor pressure of the liquid is less than one atmosphere, so the tank can be opened safely and the liquid poured out.

13.81   ● = CO_2 molecule

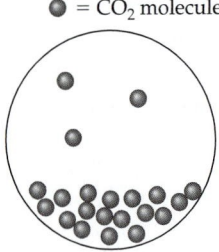

Separate liquid and vapor phases in equilibrium

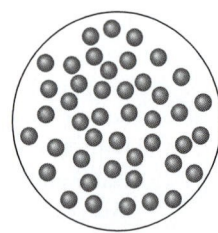

Supercritical CO_2. Distinct liquid and vapor phases not visible. Molecules are closer together than in vapor phase.

13.83 The Zn^{2+} ions are in a face-centered cubic arrangement and the S^{2-} ions fill half of the tetrahedral holes. There are four Zn^{2+} ions and four S^{2-} ions per unit cell, a 1:1 ratio that matches the compound formula.

13.85 (a) The Ca^{2+} ions are in a face-centered arrangement and the F^- ions fill all of the tetrahedral holes.

(b) There are four Ca^{2+} ions and eight F^- ions per unit cell, a 1:2 ratio that matches the compound formula.

(c) The CaF_2 and ZnS structures both have a face-centered cubic arrangement of cations. The ZnS structure has anions in only one-half of the tetrahedral holes, whereas F^- ions fill all of the tetrahedral holes in CaF_2.

13.87 (a) Structure of aspartame:

(b) There are three C=O groups that are highly polar and can interact with H atoms of water. In addition, there are two NH groups and one —OH group that can hydrogen bond.

Chapter 14

14.1 (a) Concentration $(m) = 0.0434\ m$

(b) Mole fraction of acid = 0.000781

(c) Weight percent of acid = 0.509%

14.3 NaI: 0.15 m; 2.2%; $X = 2.7 \times 10^{-3}$

CH_3CH_2OH: 1.1 m; 5.0%; $X = 0.020$

$C_{12}H_{22}O_{11}$: 0.15 m; 4.9%; $X = 2.7 \times 10^{-3}$

14.5 2.65 g Na_2CO_3; $X(Na_2CO_3) = 3.59 \times 10^{-3}$

14.7 220 g glycol; 5.7 m

14.9 16.2 m; 37.1%

14.11 Molality = $2.6 \times 10^{-5}\ m$ (assuming that 1 kg of sea water is equivalent to 1 kg of solvent)

14.13 (b) and (c)

14.15 $\Delta H°_{solution}$ for LiCl = −36.9 kJ/mol. This is an exothermic heat of solution, as compared with the very slightly endothermic value for NaCl.

14.17 Above about 40 °C the solubility increases with temperature; therefore, add more NaCl and raise the temperature.

14.19 See the discussion and data on page 593.

(a) The heat of hydration of LiF is more negative than that for RbF because the Li^+ ion is much smaller than the Rb^+ ion.

(b) The heat of hydration for $Ca(NO_3)_2$ is larger than that for KNO_3 owing to the $+2$ charge on the Ca^{2+} ion (and its smaller size).

(c) The heat of hydration is greater for $CuBr_2$ than for $CsBr$ because Cu^{2+} has a larger charge than Cs^+, and the Cu^{2+} ion is smaller than the Cs^+ ion.

14.21 2×10^{-3} g O_2

14.23   1130 mm Hg or 1.49 atm

14.25   35.0 mm Hg

14.27 $X(H_2O) = 0.869$; 16.7 mol glycol; 1040 g glycol

14.29   Calculated boiling point = 84.2 °C

14.31 $\Delta T_{bp} = 0.808$ °C; solution boiling point = 62.51 °C

14.33   Molality = 0.16 m; 0.0081 mol solute; 1.4 g solute

14.35   Molality = 8.60 m; 28.4%

14.37   Molality = 0.195 m; $\Delta T_{fp} = -0.362$ °C

14.39   Molar mass = 360 g/mol; $C_{20}H_{16}Fe_2$

14.41   Molar mass = 150 g/mol

14.43   Molar mass = 170 g/mol

14.45   Molar mass = 130 g/mol

14.47   Freezing point = −24.6 °C

14.49   0.080 m $CaCl_2$ < 0.10 m NaCl < 0.040 m Na_2SO_4 < 0.10 sugar

14.51   (a) $\Delta T_{fp} = -0.348$ °C; fp = −0.348 °C

(b) $\Delta T_{bp} = +0.0959$ °C; bp = 100.0959 °C

(c) $\Pi = 4.58$ atm

The osmotic pressure is large and can be measured with a small experimental error.

14.53   Molar mass = 6.0×10^3 g/mol

14.55   (a) $BaCl_2(aq) + Na_2SO_4(aq) \longrightarrow BaSO_4(s) + 2\ NaCl(aq)$

(b) Initially the $BaSO_4$ particles form a colloidal suspension.

(c) Over time the particles of $BaSO_4(s)$ grow and precipitate.

14.57 Li_2SO_4 should have a more negative heat of hydration than Cs_2SO_4 because the Li^+ ion is smaller than the Cs^+ ion.

14.59   (a) Increase in vapor pressure of water

0.20 m Na_2SO_4 < 0.50 m sugar < 0.20 m KBr < 0.35 m ethylene glycol

(b) Increase in boiling point

0.35 m ethylene glycol < 0.20 m KBr < 0.50 m sugar < 0.20 m Na_2SO_4

14.61   (a) 0.456 mol DMG and 11.4 mol ethanol; $X(DMG) = 0.0385$

(b) 0.869 m

(c) VP ethanol over the solution at 78.4 °C = 730.7 mm Hg

(d) bp = 79.5 °C

14.63   For ammonia: 23 m; 28%; $X(NH_3) = 0.29$

14.65   0.592 g Na_2SO_4

14.67   (a) 0.20 m KBr   (b) 0.10 m Na_2CO_3

14.69   Freezing point = −11 °C

14.71 4.0×10^2 g/mol

14.73 4.93×10^{-4} mol/L; 1.38×10^{-2} g/L

14.75   (a) Molar mass = 4.9×10^4 g/mol

(b) $\Delta T_{fp} = -3.8 \times 10^{-4}$ °C

14.77   Molar mass in benzene = 1.20×10^2 g/mol; molar mass in water = 62.4 g/mol. The actual molar mass of acetic acid is 60.1 g/mol. In benzene the molecules of acetic acid form "dimers." That is, two molecules form a single unit through hydrogen bonding. See Figure 13.9 on page 601.

14.79 $\Delta H°_{solution}$ $[Li_2SO_4] = -28.0$ kJ/mol

$\Delta H°_{solution}$ $[LiCl] = -36.9$ kJ/mol

$\Delta H°_{solution}$ $[K_2SO_4] = +23.7$ kJ/mol

$\Delta H°_{solution}$ $[KCl] = +17.2$ kJ/mol

Both lithium compounds have exothermic heats of solution, whereas both potassium compounds have endothermic values. Consistent with this is the fact that lithium salts (LiCl) are often more water-soluble than potassium salts (KCl) (see Figure 14.11).

14.81 X(benzene in solution) = 0.67 and X(toluene in solution) = 0.33

$$P_{total} = P_{toluene} + P_{benzene} = 7.3\ \text{mm Hg} + 50.\ \text{mm Hg}$$
$$= 57\ \text{mm Hg}$$

$$X(\text{toluene in vapor}) = \frac{7.3\ \text{mm Hg}}{57\ \text{mm Hg}} = 0.13$$

$$X(\text{benzene in vapor}) = \frac{50.\ \text{mm Hg}}{57\ \text{mm Hg}} = 0.87$$

14.83   The calculated molality at the freezing point of benzene is 0.47 m, whereas it is 0.99 m at the boiling point. A higher molality at the higher temperature indicates more molecules are dissolved. Therefore, assuming benzoic acid forms dimers like acetic acid (Figure 13.9), dimer formation is more prevalent at the lower temperature. In this process two molecules become one entity, lowering the number of separate species in solution and lowering the molality.

14.85 $i = 1.7$. That is, there is 1.7 mol of ions in solution per mole of compound.

14.87 (a) Calculate the number of moles of ions in 10^6 g H_2O: 550. mol Cl^-; 470. mol Na^+; 53.1 mol Mg^{2+}; 9.42 mol SO_4^{2-}; 10.3 mol Ca^{2+}; 9.72 mol K^+; 0.84 mol Br^-. Total moles of ions = 1.103×10^3 per 10^6 g water. This gives ΔT_{fp} of -2.05 °C.

(b) $\Pi = 27.0$ atm. This means that a minimum pressure of 27 atm would have to be used in a reverse osmosis device.

14.89 (a) $i = 2.06$

(b) There are approximately two particles in solution, so $H^+ + HSO_4^{2+}$ best represents H_2SO_4 in aqueous solution.

14.91 (a) Molar mass = 97.6 g/mol; empirical formula, BF_2°, and molecular formula, B_2F_4

(b)

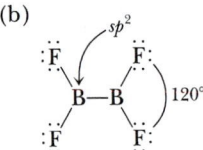

14.93 Colligative properties depend on the number of ions or molecules in solution. Each mole of $CaCl_2$ provides 1.5 times as many ions as each mole of NaCl.

14.95 At 0 °C some solid NaCl remains in the beaker or flask, and Na^+ and Cl^- ions are in solution. As some Na^+ and Cl^- ions are removed from the surface of the solid NaCl, enter the solution, and are hydrated by water, other Na^+ and Cl^- ions move to the solid surface.

14.97 Benzene is a nonpolar solvent. Thus, ionic substances such as $NaNO_3$ and NH_4Cl will certainly not dissolve. However, naphthalene is also nonpolar and resembles benzene in its structure; it should dissolve very well. (A chemical handbook gives a solubility of 33 g naphthalene per 100 g benzene.) Diethyl ether is weakly polar and will also be miscible to some extent with benzene.

14.99 The C—C and C—H bonds in hydrocarbons are nonpolar or weakly polar and tend to make such dispersions hydrophobic (water-hating). The C—O and O—H bonds in starch present opportunities for hydrogen bonding with water. Hence, starch is expected to be more hydrophilic.

14.101 [NaCl] = 1.0 M and [KNO_3] = 0.88 M. The KNO_3 solution has a higher solvent concentration, so solvent will flow from the KNO_3 solution to the NaCl solution.

Chapter 15

15.1 (a) $-\dfrac{1}{2}\dfrac{\Delta[O_3]}{\Delta t} = \dfrac{1}{3}\dfrac{\Delta[O_2]}{\Delta t}$

(b) $-\dfrac{1}{2}\dfrac{\Delta[HOF]}{\Delta t} = \dfrac{1}{2}\dfrac{\Delta[HF]}{\Delta t} = \dfrac{\Delta[O_2]}{\Delta t}$

15.3 $\dfrac{1}{3}\dfrac{\Delta[O_2]}{\Delta t} = -\dfrac{1}{2}\dfrac{\Delta[O_3]}{\Delta t}$ or $\dfrac{\Delta[O_2]}{\Delta t} = -\dfrac{2}{3}\dfrac{\Delta[O_2]}{\Delta t}$

so $\Delta[O_3]/\Delta t = -1.0 \times 10^{-3}$ mol/L · s.

15.5 (a) The graph of [B] (product concentration) versus time shows [B] increasing from zero. The line is curved, indicating the rate changes with time; thus that the rate depends on concentration. Rates for the four 10-s intervals are as follows: 0–10 s, 0.0326 mol/L · s; from 10–20 s, 0.0246 mol/L · s; 20–30 s, 0.0178 mol/L · s; 30–40 s, 0.0140 mol/L · s.

(b) $-\dfrac{\Delta[A]}{\Delta t} = \dfrac{1}{2}\dfrac{\Delta[B]}{\Delta t}$ throughout the reaction

In the interval 10–20 s, $\dfrac{\Delta[A]}{\Delta t} = -0.0123 \dfrac{\text{mol}}{\text{L} \cdot \text{s}}$

(c) Instantaneous rate when [B] = 0.750 mol/L

$= \dfrac{\Delta[B]}{\Delta t} = 0.0163 \dfrac{\text{mol}}{\text{L} \cdot \text{s}}$

15.7 The reaction is second order in A, first order in B, and third order overall.

15.9 (a) Rate = $k[NO_2][O_3]$

(b) If [NO_2] is tripled, the rate triples.

(c) If [O_3] is halved, the rate is halved.

15.11 (a) The reaction is second order in [NO] and first order in [O_2].

(b) $\dfrac{-\Delta[NO]}{\Delta t} = k[NO]^2[O_2]$

(c) $k = 25$ $L^2/mol^2 \cdot s$

(d) $\dfrac{-\Delta[NO]}{\Delta t} = -2.8 \times 10^{-5}$ mol/L · s

(e) When $-\Delta[NO]/\Delta t = 1.0 \times 10^{-4}$ mol/L · s, $\Delta[O_2]/\Delta t = 5.0 \times 10^{-5}$ mol/L · s and $\Delta[NO_2]/\Delta t = 1.0 \times 10^{-4}$ mol/L · s.

15.13 (a) Rate = $k[NO]^2[O_2]$

(b) $k = 50.$ $L^2/mol^2 \cdot h$

(c) Rate = 8.4×10^{-9} mol/L · h

15.15 (a) Rate = $k[CO]^2[O_2]$

(b) Third order overall; 1st order in O_2 and 2nd order in CO.

(c) $k = 5$ $L^2/mol^2 \cdot min$

15.17 $k = 0.0392$ h^{-1}

15.19 5.0×10^2 min

15.21 105 min

15.23 (a) 153 min

(b) 1790 min

15.25 (a) $t_{1/2} = 1400$ s   (b) 4600 s

15.27 4.48×10^{-3} mol (0.260 g) azomethane remains; 0.0300 mol N_2 formed

15.29 Fraction of ^{64}Cu remaining = 0.030

15.31 72 s represents two half-lives, so $t_{1/2} = 36$ s.

15.33 (a) A graph of ln[sucrose] versus time produces a straight line, indicating that the reaction is first order in [sucrose].

(b) $-\Delta[\text{sucrose}]/\Delta t = +k[\text{sucrose}]$; $k = 3.7 \times 10^{-3}$ min^{-1}

(c) At 175 min, [sucrose] = 0.167 M

15.35 The straight line obtained in a graph of $\ln[N_2O]$ versus time indicates a first-order reaction.

$k = (-\text{slope}) = 0.0127$ min^{-1}

The rate when $[N_2O] = 0.035$ mol/L is 4.4×10^{-4} mol/L · min.

15.37 The graph of $1/[NO_2]$ versus time gives a straight line, indicating the reaction is second order with respect to $[NO_2]$ (see Table 15.1 on page 619). The slope of the line is k, so $k = 1.1$ L/mol · s.

15.39 $-\Delta[C_2F_4]/\Delta t = k[C_2F_4]^2 = (0.04$ L/mol · s$)[C_2F_4]^2$

15.41 Activation energy = 102 kJ/mol

15.43 $k = 0.3$ s^{-1}

15.45

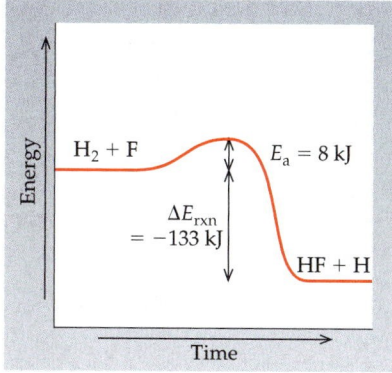

15.47 (a) Rate = $k[NO_3][NO]$

(b) Rate = $k[Cl][H_2]$

(c) Rate = $k[(CH_3)_3CBr]$

15.49 (a) The Second step   (b) Rate = $k[O_3][O]$

15.51 (a) The substances OI$^-$ and HOI cancel out to give the equation for the overall reaction.

(b) Steps 1 and 2 are bimolecular, whereas step 3 is termolecular.

(c) Rate = $k[H_2O_2][I^-]$

(d) OI$^-$ and HOI are intermediates. They are produced and consumed during the reaction and do not appear in the equation for the overall reaction.

15.53 NO_2 is a reactant in the first step and a product in the second step. CO is a reactant in the second step. NO_3 is an intermediate, and CO_2 is a product. NO is a product.

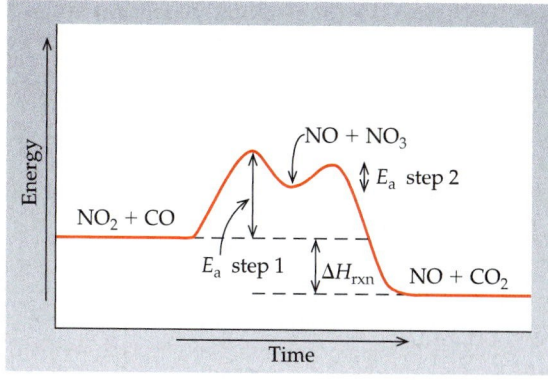

15.55 The reaction rate will double.

15.57 After measuring pH as a function of time, one could then calculate pOH and then [OH$^-$]. Finally, a plot of $1/[OH^-]$ versus time would give a straight line with a slope equal to k.

15.59 (a) Rate = $k[NH_3]$

(b) $k = 0.050$ s^{-1}

(c) Half life = 14 s

15.61 (a) Rate = $k[CO_2]$

(b) $k = 0.028$ s^{-1}

(c) 24 s

15.63 (a) A plot of $1/[C_2F_4]$ versus time indicates the reaction is second order with respect to $[C_2F_4]$. The rate law is Rate = $k[C_2F_4]^2$.

(b) The rate constant (= slope of the line) is about 0.045 L/mol · s. (The graph does not allow a very accurate calculation.)

(c) Using $k = 0.045$ L/mol · s, the concentration after 600 s is 0.03 M (to 1 significant figure).

(d) Time = 2000 s (using k from part a).

15.65 (a) A plot of $1/[NH_4NCO]$ versus time is linear, so the reaction is second order with respect to NH_4NCO.

(b) Slope = $k = 0.0109$ L/mol · min.

(c) $t_{1/2} = 200.$ min

(d) $[NH_4NCO] = 0.0997$ mol/L

15.67 Mechanism 2

15.69 $k = 0.037$ h^{-1} and $t_{1/2} = 19$ h

15.71 (a) After 125 min, 0.251 g remains. After 145, 0.144 g remains.

(b) Time = 43.9 min

(c) Fraction remaining = 0.016

15.73 (a) $2 NO(g) + Br_2(g) \longrightarrow 2 BrNO(g)$

(b) Mechanism 1 is termolecular.

Mechanism 2 has two bimolecular steps.

Mechanism 3 has two bimolecular steps.

(c) Br_2NO is the intermediate in mechanism 2 and N_2O_2 is the intermediate in mechanism 3.

(d) Assuming step 1 in each mechanism is the slow step, the rate equations will all differ. Mechanism 1 would be second order in NO and first order in Br_2. Mechanism 2 would be first order in both NO and Br_2. Mechanism 3 would be second order in NO and zero order in Br_2.

15.75 The rate equation for the slow step is Rate = $k[O_3][O]$. The equilibrium constant, K, for step 1 is $K = [O_2][O]/[O_3]$. Solving this for $[O]$, we have $[O] = K[O_3]/[O_2]$. Substituting the expression for $[O]$ into the rate equation we find

Rate = $k[O_3]\{K[O_3]/[O_2]\} = kK[O_3]^2/[O_2]$

15.77 The slope of the ln k versus $1/T$ plot is -6370. From slope = $-E_a/R$, we derive $E_a = 53.0$ kJ/mol.

15.79 Estimated time at 90 °C = 4.76 min

15.81 After 30 min (one half-life), $P_{HOF} = 50.0$ mm Hg and $P_{total} = 125.0$ mm Hg. After 45 min, $P_{HOF} = 35.4$ mm Hg and $P_{total} = 132$ mm Hg.

15.83 (a) The slow step is unimolecular and the fast step is bimolecular.

(b) Rate = $k[Ni(CO)_4]$. Yes, this agrees with the mechanism proposed.

(c) $[Ni(CO)_3L]$ after 5.0 min = 0.023 M

15.85 The finely divided rhodium metal will have a significantly greater surface area than the small block of metal. This leads to a large increase in the number of reaction sites and vastly increases the reaction rate.

15.87 (a) False. The reaction may occur in a single step but this does not have to be true.

(b) True

(c) False. Raising the temperature increases the value of k.

(d) False. Temperature has no effect on the value of E_a.

(e) False. If the concentrations of both reactants are doubled, the rate will increase by a factor of 4.

(f) True

15.89 (a) True

(b) True

(c) False. As a reaction proceeds, the reactant concentration decreases and the rate decreases.

(d) False. It is possible to have a one-step mechanism for a third-order reaction if the slow, rate-determining step is termolecular.

15.91 (a) Decrease   (d) No change

(b) Increase   (e) No change

(c) No change   (f) No change

15.93 (a) There are three mechanistic steps.

(b) The overall reaction is exothermic.

15.95 (a) The average rate is calculated over a period of time, whereas the instantaneous rate is the rate of reaction at some instant in time.

(b) The reaction rate decreases with time as the dye concentration decreases.

(c) See part (b).

15.97 (a) Molecules must collide with enough energy to overcome the activation energy, and they must be in the correct orientation.

(b) In animation 2 the molecules are moving faster, so they are at a higher temperature.

(c) Less sensitive. The O_3 must collide with NO in the correct orientation for a reaction to occur. The O_3 and N_2 collisions do not depend to the same extent on orientation because N_2 is a symmetrical, diatomic molecule.

15.99 (a) I^- is regenerated during the second step in the mechanism.

(b) The activation energy is smaller for the catalyzed reaction.

Chapter 16

16.1 (a) $K = \dfrac{[H_2O]^2[O_2]}{[H_2O_2]^2}$

(b) $K = \dfrac{[CO_2]}{[CO][O_2]^{1/2}}$

(c) $K = \dfrac{[CO_2]^2}{[CO_2]}$

(d) $K = \dfrac{[CO_2]}{[CO]}$

16.3 $Q = (2.0 \times 10^{-8})^2/(0.020) = 2.0 \times 10^{-14}$

$Q < K$ so the reaction proceeds to the right.

16.5 $Q = 1.0 \times 10^3$, so $Q > K$ and the reaction is not at equilibrium. It proceeds to the left to convert products to reactants.

16.7 $K = 1.2$

16.9 (a) $K = 0.025$

(b) $K = 0.025$

(c) The amount of solid does not affect the equilibrium.

16.11 (a) $[COCl_2] = 0.00308$ M; $[CO] = 0.00712$ M

(b) $K = 144$

16.13 [isobutane] = 0.024 M; [butane] = 0.010 M

16.15 $[I_2] = 6.14 \times 10^{-3}$ M; $[I] = 4.79 \times 10^{-3}$ M

16.17 $[COBr_2] = 0.107$ M; $[CO] = [Br_2] = 0.143$ M

57.1% of the $COBr_2$ has decomposed.

16.19 (b)

16.21 (e), $K_2 = 1/(K_1)^2$

16.23 $K = 13.7$

16.25 (a) Equilibrium shifts to the right
       (b) Equilibrium shifts to the left
       (c) Equilibrium shifts to the right
       (d) Equilibrium shifts to the left

16.27 Equilibrium concentrations are the same under both circumstances: [butane] = 1.1 M and [isobutane] = 2.9 M

16.29 $K = 3.9 \times 10^{-4}$

16.31 For decomposition of $COCl_2$, $K = 1/(K$ for $COCl_2$ formation$) = 1/(6.5 \times 10^{11}) = 1.5 \times 10^{-12}$

16.33 $K = 4$

16.35 Q is less than K, so the system shifts to form more isobutane.

       At equilibrium, [butane] = 0.86 M and [isobutane] = 2.14 M.

16.37 The second equation has been reversed and multiplied by 2.
       (c) $K_2 = 1/K_1^2$

16.39 (a) No change   (d) Shifts right
       (b) Shifts left   (e) Shifts right
       (c) No change

16.41 (a) The equilibrium will shift to the left on adding more Cl_2.
       (b) K is calculated (from the quantities of reactants and products at equilibrium) to be 0.0470. After Cl_2 is added, the concentrations are: $[PCl_5]$ = 0.0199 M, $[PCl_3]$ = 0.0231 M, and $[Cl_2]$ = 0.0403 M.

16.43 $K_p = 0.215$

16.45 $P_{total} = 1.21$ atm

16.47 $[NH_3]$ = 0.67 M; $[N_2]$ = 0.57 M; $[H_2]$ = 1.7 M; P_{total} = 180 atm

16.49 (a) $[NH_3] = [H_2S]$ = 0.013 M
       (b) $[NH_3]$ = 0.027 M and $[H_2S]$ = 0.0067 M

16.51 (a) Fraction dissociated = 0.15
       (b) Fraction dissociated = 0.189. If the pressure decreases, the equilibrium shifts to the right, increasing the fraction of N_2O_4 dissociated.

16.53 (a) The flask containing $(H_3N)B(CH_3)_3$ will have the largest partial pressure of $B(CH_3)_3$.
       (b) $P[B(CH_3)_3] = P(NH_3)$ = 2.1 and $P[(H_3N)B(CH_3)_3]$ = 1.0 atm

       P_{total} = 5.2 atm

       Percent dissociation = 69%

16.55 $P(CO)$ = 0.0010 atm

16.57 1.7×10^{18} O atoms

16.59 Glycerin concentration should be 1.7 M

16.61 (a) $K_p = 0.20$
       (b) When initial $[N_2O_4]$ = 1.00 atm, the equilibrium pressures are $[N_2O_4]$ = 0.80 atm and $[NO_2]$ = 0.40 atm. When initial $[N_2O_4]$ = 0.10 atm, the equilibrium pressures are $[N_2O_4]$ = 0.050 atm and $[NO_2]$ = 0.10 atm. The percent dissociation is now 50.%. This is in accord with Le Chatelier's principle: If the initial pressure of the reactant decreases, the equilibrium shifts to the right, increasing the fraction of the reactant dissociated. See also Question 16.51.

16.63 (a) False. The magnitude of K is always dependent on temperature.
       (b) True
       (c) False. The equilibrium constant for a reaction is the reciprocal of the value of K for its reverse.
       (d) True
       (e) False. Δn = 1 so $K_p = K_c(RT)$

16.65 (a) Product-favored, $K \gg 1$
       (b) Reactant-favored, $K \ll 1$
       (c) Product-favored, $K \gg 1$

16.67 (a) $K_p = K_c$ = 56. Because 2 mol of reactants gives 2 mol of product, Δn does not change and $K_p = K_c$ (see page 264).
       (b,c) Initial $P(H_2) = P(I_2)$ = 2.6 atm and P_{total} = 5.2 atm

       At equilibrium, $P(H_2) = P(I_2)$ = 0.54 atm and $P(HI)$ = 4.1 atm. Therefore, P_{total} = 5.2 atm. The initial total pressure and the equilibrium total pressure are the same owing to the reaction stoichiometry.

16.69 The reaction is endothermic. Adding heat shifts an equilibrium in the endothermic direction.

16.71 An elementary chemical step can occur both in the forward and reverse directions. Solid lead chloride forms when solutions containing lead ions and chloride ions are mixed, and a solution containing lead ions and chloride ions forms when pure lead chloride is placed in water and heated.

Chapter 17

17.1 (a) CN^-, cyanide ion
      (b) SO_4^{2-}, sulfate ion
      (c) F^-, fluoride ion

17.3 (a) $H_3O^+(aq) + NO_3^-(aq)$; $H_3O^+(aq)$ is the conjugate acid of H_2O, and $NO_3^-(aq)$ is the conjugate base of HNO_3.
      (b) $H_3O^+(aq) + SO_4^{2-}(aq)$; $H_3O^+(aq)$ is the conjugate acid of H_2O, and $SO_4^{2-}(aq)$ is the conjugate base of HSO_4^-.
      (c) $H_2O + HF$; H_2O is the conjugate base of H_3O^+, and HF is the conjugate acid of F^-.

17.5   Brønsted acid: $HC_2O_4^-(aq) + H_2O(\ell) \rightleftharpoons$
$H_3O^+(aq) + C_2O_4^{2-}(aq)$

Brønsted base: $HC_2O_4^-(aq) + H_2O(\ell) \rightleftharpoons$
$H_2C_2O_4(aq) + OH^-(aq)$

17.7

Acid (A)	Base (B)	Conjugate Base of A	Conjugate Acid of B
(a) HCO_2H	H_2O	HCO_2^-	H_3O^+
(b) H_2S	NH_3	HS^-	NH_4^+
(c) HSO_4^-	OH^-	SO_4^{2-}	H_2O

17.9 $[H_3O^+] = 1.8 \times 10^{-4}$ M; acidic

17.11   HCl is a strong acid, so $[H_3O^+]$ = concentration of the acid. $[H_3O^+] = 0.0075$ M and $[OH^-] = 1.3 \times 10^{-12}$ M. pH = 2.12.

17.13 $Ba(OH)_2$ is a strong base, so $[OH^-] = 2 \times$ concentration of the base.

$[OH^-] = 3.0 \times 10^{-3}$ M; pOH = 2.52; and pH = 11.48

17.15   (a) The strongest acid is HCO_2H (largest K_a) and the weakest acid is C_6H_5OH (smallest K_a).

(b) The strongest acid (HCO_2H) has the weakest conjugate base.

(c) The weakest acid (C_6H_5OH) has the strongest conjugate base.

17.17   (c) HClO, the weakest acid in this list (Table 17.3), has the strongest conjugate base.

17.19 $CO_3^{2-}(aq) + H_2O(\ell) \longrightarrow HCO_3^-(aq) + OH^-(aq)$

17.21   Highest pH, Na_2S; lowest pH, $AlCl_3$ (which gives the weak acid $[Al(H_2O)_6]^{3+}$ in solution)

17.23 $pK_a = 4.19$

17.25 $K_a = 3.0 \times 10^{-10}$

17.27   2-Chlorobenzoic acid has the smaller pK_a value.

17.29 $K_b = 7.4 \times 10^{-12}$

17.31 $K_b = 6.3 \times 10^{-5}$

17.33 $CH_3CO_2H(aq) + HCO_3^-(aq) \rightleftharpoons$
$CH_3CO_2^-(aq) + H_2CO_3(aq)$

Equilibrium lies predominantly to the right because CH_3CO_2H is a stronger acid than H_2CO_3.

17.35   (a) Left; NH_3 and HBr are the stronger base and acid, respectively.

(b) Left; PO_4^{3-} and CH_3CO_2H are the stronger base and acid, respectively.

(c) Right; $Fe(H_2O)_6^{3+}$ and HCO_3^- are the stronger acid and base, respectively.

17.37   (a) $OH^-(aq) + HPO_4^{2-}(aq) \rightleftharpoons H_2O(\ell) + PO_4^{3-}(aq)$

(b) OH^- is a stronger base than PO_4^{3-}, so the equilibrium will lie to the right. (The predominant species in solution is PO_4^{3-}, so the solution is likely

to be basic because PO_4^{3-} is the conjugate base of a weak acid.)

17.39   (a) $CH_3CO_2H(aq) + HPO_4^{2-}(aq) \rightleftharpoons$
$CH_3CO_2^-(aq) + H_2PO_4^-(aq)$

(b) CH_3CO_2H is a stronger acid than $H_2PO_4^-$, so the equilibrium will lie to the right.

17.41   (a) 2.1×10^{-3} M; (b) $K_a = 3.5 \times 10^{-4}$

17.43 $K_b = 6.6 \times 10^{-9}$

17.45   (a) $[H_3O^+] = 1.6 \times 10^{-4}$ M

(b) Moderately weak; $K_a = 1.1 \times 10^{-5}$

17.47 $[CH_3CO_2^-] = [H_3O^+] = 1.9 \times 10^{-3}$ M and $[CH_3CO_2H] = 0.20$ M

17.49 $[H_3O^+] = [CN^-] = 3.2 \times 10^{-6}$ M; [HCN] = 0.025 M; pH = 5.50

17.51 $[NH_4^+] = [OH^-] = 1.64 \times 10^{-3}$ M; $[NH_3] = 0.15$ M; pH = 11.22

17.53 $[OH^-] = 0.0102$ M; pH = 12.01; pOH = 1.99

17.55   pH = 3.25

17.57 $[H_3O^+] = 1.1 \times 10^{-5}$ M; pH = 4.98

17.59   [HCN] = $[OH^-] = 3.3 \times 10^{-3}$ M; $[H_3O^+] = 3.0 \times 10^{-12}$ M; $[Na^+] = 0.441$ M

17.61 $[H_3O^+] = 1.5 \times 10^{-9}$ M; pH = 8.81

17.63   (a) The reaction produces acetate ion, the conjugate base of acetic acid. The solution is weakly basic. pH is greater than 7.

(b) The reaction produces NH_4^+, the conjugate acid of NH_3. The solution is weakly acidic. pH is less than 7.

(c) The reaction mixes equal molar amounts of strong base and strong acid. The solution will be neutral. pH will be 7.

17.65   (a) pH = 1.17; (b) $[SO_3^{2-}] = 6.2 \times 10^{-8}$ M

17.67   (a) $[OH^-] = [N_2H_5^+] = 9.2 \times 10^{-8}$ M; $[N_2H_6^{2+}] = 8.9 \times 10^{-16}$ M

(b) pH = 9.96

17.69   (a) Lewis base

(b) Lewis acid

(c) Lewis base (owing to lone pair of electrons on the N atom)

17.71   CO is a Lewis base in its reactions with transition metal atoms. It donates a lone pair of electrons on the C atom.

17.73   HOCN should be a stronger acid than HCN because the H atom in HOCN is attached to a highly electronegative O atom. This induces a positive charge on the H atom, making it more readily removed by an interaction with water.

17.75   The S atom is surrounded by four highly electronegative O atoms. The inductive effect of these atoms induces a positive charge on the H atom, making it susceptible to removal by water.

17.77 $pH = 2.671$

17.79 The weaker acid (smaller K_a) will have the higher pH in solution. Thus, the pH of a benzoic acid solution is higher than that of 4-chlorobenzoic acid.

17.81 $H_2S(aq) + CH_3CO_2^-(aq) \rightleftharpoons$
$$CH_3CO_2H(aq) + HS^-(aq)$$
The equilibrium lies to the left and favors the reactants.

17.83 $[X^-] = [H_3O^+] = 3.0 \times 10^{-3}$ M; $[HX] = 0.007$ M; $pH = 2.52$

17.85 $K_a = 1.4 \times 10^{-5}$; $pK_a = 4.86$

17.87 $pH = 5.84$

17.89 (a) Ethylamine is a stronger base than ethanolamine.
(b) For ethylamine, the pH of the solution is 11.82.

17.91 The K_b for pyridine (Appendix I) is 1.5×10^{-9}. Therefore, K_a for the conjugate acid, the pyridinium ion, is $K_a = K_w/K_b = 6.7 \times 10^{-6}$. The pH of the pyridinium hydrochloride solution is 3.39.

17.93 Acidic: $NaHSO_4$, NH_4Br, $FeCl_3$
Neutral: $KClO_4$, $NaNO_3$, $LiBr$
Basic: Na_2CO_3, $(NH_4)_2S$, Na_2HPO_4

17.95 $K_{net} = K_{a_1} \times K_{a_2} = 3.8 \times 10^{-6}$

17.97 For the reaction $HCO_2H(aq) + OH^-(aq) \longrightarrow H_2O(\ell) + HCO_2^-(aq)$, $K_{net} = K_a$ (for HCO_2H) $\times [1/K_w] = 1.8 \times 10^{10}$

17.99 To double the percent ionization, you must dilute 100 mL of solution to 400 mL.

17.101 $pH = 7.97$

17.103 $HCl < NH_4Cl < NaCl < NaCH_3CO_2 < KOH$

17.105 $[H_3O^+] = [(1.12 \times 10^{-3}) \times (3.91 \times 10^{-6}]^{1/2} = 6.62 \times 10^{-5}$ M
$pH = 4.180$

17.107 Water can both accept a proton (a Brønsted base) and donate a lone pair (a Lewis base). Water can also donate a proton (Brønsted acid), but it cannot accept a pair of electrons (and act as a Lewis acid).

17.109 Measure the pH of the 0.1 M solutions of the three bases. The solution containing the strongest base will have the highest pH. The solution having the weakest base will have the lowest pH.

17.111 Species in solution listed in order of decreasing concentration:
$H_2O > HCN$ (0.10 M) $> H_3O^+$ and
CN^- (6.3×10^{-6} M $> OH^-$ (1.6×10^{-9} M)

17.113 Mixing the NaOH and acetic acid solutions gives 60.0 mL of 0.075 M $NaCH_3CO_2$ (sodium acetate). This solution is weakly basic and has water, sodium ion, acetate

ions, acetic acid, and hydrogen and hydroxide ions. In order of decreasing concentration, these are
$H_2O > Na^+$ and $CH_3CO_2^-$ (0.075 M) $> OH^-$ and CH_3CO_2H (6.5×10^{-6} M) $> H_3O^+$ (1.5×10^{-9} M)

17.115 (a) $HClO_4 + H_2SO_4 \rightleftharpoons ClO_4^- + H_3SO_4^+$
(b) The O atoms on sulfuric acid have lone pairs of electrons that can be used to bind to an H^+ ion.

$$H-\overset{\displaystyle :\ddot{O}:}{\underset{\displaystyle :\ddot{O}:}{O-S-O}}-H$$

17.117 The possible cation–anion combinations are NaCl (neutral), NaOH (basic), NH_4Cl (acidic), NH_4OH (basic), HCl (acidic), and H_2O (neutral).
$A = H^+$ solution; $B = NH_4^+$ solution; $C = Na^+$ solution; $Y = Cl^-$ solution; $Z = OH^-$ solution

17.119 (a) Add the three equations.
$$NH_4^+(aq) + H_2O(\ell) \longrightarrow NH_3(aq) + H_3O^+(aq)$$
$$K_1 = K_w/K_b$$
$$CN^-(aq) + H_2O(\ell) \longrightarrow HCN(aq) + OH^-(aq)$$
$$K_2 = K_w/K_a$$
$$H_3O^+(aq) + OH^-(aq) \longrightarrow 2 H_2O(\ell) \quad K_3 = 1/K_w$$
$$\overline{NH_4^+(aq) + CN^-(aq) \longrightarrow NH_3(aq) + HCN(aq)}$$
$K_{net} = K_1K_2K_3 = K_w/K_aK_b$
(b) Substitute expressions for K_w, K_a, and K_b into the equation.
$$[H_3O^+] = \sqrt{\frac{K_wK_a}{K_b}}$$
$$\sqrt{\frac{K_wK_a}{K_b}} = \sqrt{\frac{[H_3O^-][OH^-]\left(\frac{[H_3O^+][NH_3]}{[NH_4^+]}\right)}{\frac{[OH^-][HCN]}{[CN^-]}}}$$
In a solution of NH_4CN, we have $[NH_4^+] = [CN^-]$ and $[NH_3] = [HCN]$. When these and $[OH^-]$, are canceled from the expression, we see it is equal to $[H_3O^+]$.
(c) $pH = 9.33$

Chapter 18

18.1 (a) Decrease pH; (b) increase pH; (c) no change in pH

18.3 $pH = 9.25$

18.5 $pH = 4.38$

18.7 $pH = 9.12$; pH of buffer is lower than the pH of the original solution of NH_3 ($pH = 11.17$).

18.9 4.7 g

18.11 $pH = 4.92$

18.13 (a) $pH = 3.59$; (b) $[HCO_2H]/[HCO_2^-] = 0.45$

18.15   (b), $NH_3 + NH_4Cl$

18.17   The buffer must have a ratio of 0.51 mol NaH_2PO_4 to 1 mol Na_2HPO_4. For example, dissolve 0.51 mol NaH_2PO_4 (61 g) and 1.0 mol Na_2HPO_4 (140 g) in some amount of water.

18.19   (a) pH = 4.95; (b) pH = 5.05

18.21   (a) pH = 9.55; (b) pH = 9.50

18.23   (a) Original pH = 5.62
        (b) $[Na^+]$ = 0.0323 M, $[OH^-]$ = 1.5×10^{-3} M, $[H_3O^+]$ = 6.5×10^{-12} M, and $[C_6H_5O^-]$ = 0.0308 M
        (c) pH = 11.19

18.25   (a) Original NH_3 concentration = 0.0154 M
        (b) At the equivalence point $[H_3O^+]$ = 1.9×10^{-6} M, $[OH^-]$ = 5.3×10^{-9} M, $[NH_4^+]$ = 6.25×10^{-3} M.
        (c) pH at equivalence point = 5.73

18.27   The titration curve begins at pH = 13.00 and drops slowly as HCl is added. Just before the equivalence point (when 30.0 mL of acid has been added), the curve falls steeply. The pH at the equivalence point is exactly 7. Just after the equivalence point, the curve flattens again and begins to approach the final pH of just over 1.0. The total volume at the equivalence point is 60.0 mL.

18.29   (a) Starting pH = 11.12
        (b) pH at equivalence point = 5.28
        (c) pH at midpoint (half-neutralization point) = 9.25
        (d) Methyl red, bromcresol green

        (e)
Acid (mL)	Added pH
5.00	9.85
15.0	9.08
20.0	8.65
22.0	8.39
30.0	2.04

18.31   See Figure 18.10 on page 872.
        (a) Thymol blue or bromphenol blue
        (b) Phenolphthalein
        (c) Methyl red; thymol blue

18.33   (a) Silver chloride, AgCl; lead chloride, $PbCl_2$
        (b) Zinc carbonate, $ZnCO_3$; zinc sulfide, ZnS
        (c) Iron(II) carbonate, $FeCO_3$; iron(II) oxalate, FeC_2O_4

18.35   (a) and (b) are soluble, (c) and (d) are insoluble.

18.37   (a) $AgCN(s) \longrightarrow Ag^+(aq) + CN^-(aq)$, $K_{sp} = [Ag^+][CN^-]$
        (b) $NiCO_3(s) \longrightarrow Ni^{2+}(aq) + CO_3^{2-}(aq)$, $K_{sp} = [Ni^{2+}][CO_3^{2-}]$

18.37   (c) $AuBr_3(s) \longrightarrow Au^{3+}(aq) + 3\ Br^-(aq)$, $K_{sp} = [Au^{3+}][Br^-]^3$

18.39 $K_{sp} = (1.9 \times 10^{-3})^2 = 3.6 \times 10^{-6}$

18.41 $K_{sp} = 4.37 \times 10^{-9}$

18.43 $K_{sp} = 1.4 \times 10^{-15}$

18.45   (a) 9.2×10^{-9} M; (b) 2.2×10^{-6} g/L

18.47   (a) 2.4×10^{-4} M; (b) 0.018 g/L

18.49   Only 2.1×10^{-4} g dissolves.

18.51   (a) $PbCl_2$; (b) FeS; (c) $Fe(OH)_2$

18.53   Solubility in pure water = 1.0×10^{-6} mol/L; solubility in 0.010 M SCN^- = 1.0×10^{-10} mol/L

18.55   (a) Solubility in pure water = 2.2×10^{-6} mg/mL
        (b) Solubility in 0.020 M $AgNO_3$ = 1.0×10^{-12} mg/mL

18.57   (a) PbS
        (b) Ag_2CO_3
        (c) $Al(OH)_3$

18.59 $Q < K_{sp}$, so no precipitate forms.

18.61 $Q > K_{sp}$; $Zn(OH)_2$ will precipitate.

18.63 $[OH^-]$ must exceed 1.0×10^{-5} M.

18.65 $AuCl(s) \rightleftharpoons Au^+(aq) + Cl^-(aq)$
        $Au^+(aq) + 2\ CN^-(aq) \rightleftharpoons Ag(CN)_2^-(aq)$
        Net: $AuCl(s) + 2\ CN^-(aq) \rightleftharpoons$
        $Au(CN)_2^-(aq) + Cl^-(aq)$
        $K_{net} = K_{sp} \times K_f = 4.0 \times 10^{25}$

18.67   (a) Add H_2SO_4, precipitating $BaSO_4$ and leaving $Na^+(aq)$ in solution.
        (b) Add HCl or another source of chloride ion. $PbCl_2$ will precipitate, but $NiCl_2$ is water-soluble.

18.69   (a) $NaBr(aq) + AgNO_3(aq) \longrightarrow$
        $NaNO_3(aq) + AgBr(s)$
        (b) $2\ KCl(aq) + Pb(NO_3)_2(aq) \longrightarrow$
        $2\ KNO_3(aq) + PbCl_2(s)$

18.71 $Q > K_{sp}$, so $BaSO_4$ precipitates.

18.73 $[H_3O^+]$ = 1.9×10^{-10} M; pH = 9.73

18.75 $BaCO_3 < Ag_2CO_3 < Na_2CO_3$

18.77   Original pH = 8.62; dilution will not affect the pH.

18.79   (a) pH = 2.81
        (b) pH at equivalence point = 8.72
        (c) pH at the midpoint = pK_a = 4.62
        (d) Phenolphthalein
        (e) After 10.0 mL, pH = 4.39.
            After 20.0 mL, pH = 5.07.
            After 30.0 mL, pH = 11.84.
        (f) A plot of pH versus volume of NaOH added would begin at a pH of 2.81, rise slightly to the midpoint at pH = 4.62, and then begin to rise more steeply

as the equivalence point is approached (when the volume of NaOH added is 27.0 mL). The pH rises vertically through the equivalence point, and then begins to level off above a pH of about 11.0.

18.81 110 mL NaOH

18.83 The K_b value for ethylamine (4.27×10^{-4}) is found in Appendix I.

(a) pH = 11.89

(b) Midpoint pH = 10.63

(c) pH = 10.15

(d) pH = 5.93 at the equivalence point

(e) pH = 2.13

(g) Alizarin or bromcresol purple (see Figure 18.10)

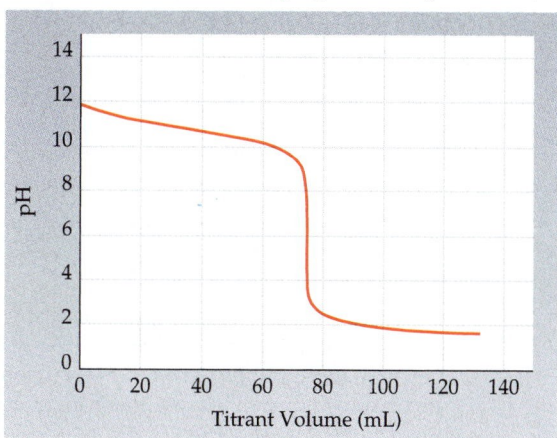

18.85 (a) 0.100 M acetic acid has a pH of 2.87. Adding sodium acetate slowly raises the pH.

(b) Adding $NaNO_3$ to 0.100 M HNO_3 has no effect on the pH.

(c) In part (a), adding the conjugate base of a weak acid creates a buffer solution. In part (b), HNO_3 is a strong acid, and its conjugate base (NO_3^-) is so weak that the base has no effect on the complete ionization of the acid.

18.87 (a) HPO_4^{2-}

(b) 32 g of Na_2HPO_4

(c) Add 10. g of the base, Na_3PO_4, to raise the pH.

18.89 (a) $BaSO_4$ will precipitate first.

(b) $[Ba^{2+}] = 1.8 \times 10^{-7}$ M

18.91 $K = 2.1 \times 10^6$; yes, AgI forms

18.93 (a) $[F^-] = 1.3 \times 10^{-3}$ M; (b) $[Ca^{2+}] = 2.9 \times 10^{-5}$ M

18.95 (a) $PbSO_4$ will precipitate first.

(b) $[Pb^{2+}] = 5.1 \times 10^{-6}$ M

18.97 $Cu(OH)_2$ will dissolve in a non-oxidizing acid such as HCl, whereas CuS will not.

18.99 The strong base (OH^-) is consumed completely in a reaction with the weak acid present in the buffer. In an acetic acid/sodium acetate buffer, the acetic acid reacts with OH^- to produce more of the conjugate base, the acetate ion.

$$OH^-(aq) + CH_3CO_2H(aq) \longrightarrow CH_3CO_2^-(aq) + H_2O(\ell)$$

18.101 When Ag_3PO_4 dissolves slightly, it produces a small concentration of the phosphate ion, PO_4^{3-}. This ion is a strong base and hydrolyzes to HPO_4^{2-}. As this reaction removes the PO_4^{3-} ion from equilibrium with Ag_3PO_4, the equilibrium shifts to the right, producing more PO_4^{3-} and Ag^+ ions. Thus, Ag_3PO_4 dissolves to a greater extent than might be calculated from a K_{sp} value (unless the K_{sp} value was actually determined experimentally).

18.103 (a) Base is added to increase the pH. The added base reacts with acetic acid to form more acetate ions in the mixture. Thus, the fraction of acid declines and the fraction of conjugate base rises (i.e., the ratio $[CH_3CO_2H]/[CH_3CO_2^-]$ decreases) as the pH rises.

(b) At pH = 4, acid predominates (85% acid and 15% acetate ions). At pH = 6, acetate ions predominate (95% acetate ions and 5% acid).

(c) At the point the lines cross, $[CH_3CO_2H] = [CH_3CO_2^-]$. At this point pH = pK_a, so pK_a for acetic acid is 4.75.

18.105 (a) C—C—C angle, 120°; O—C=O, 120°; C—O—H, 109°; C—C—H, 120°

(b) Both the ring C atoms and the C in CO_2H are sp^2 hybridized.

(c) $K_a = 1 \times 10^{-3}$

(d) 10%

(e) pH at half-way point = $pK_a = 3.0$; pH at equivalence point = 7.3

18.107 $PbCl_2(s) \longrightarrow Pb^{2+}(aq) + 2\ Cl^-(aq)$

Adding Cl^- ions to the test tube shifts the equilibrium to the left, forming more $PbCl_2$ and decreasing the concentration of Pb^{2+} ions.

18.109 Decreasing the pH by 1.0 is equivalent to decreasing $[OH^-]$ by a factor of 10. More $Ca(OH)_2$ will dissolve to replace OH^- removed from solution. The hydroxide ion concentration in the equilibrium constant expression for $Ca(OH)_2$ is squared, so decreasing $[OH^-]$ by a factor of 10 results in a solubility increase of 10^2 (i.e., 100).

Chapter 19

19.1 (a) Disorder increases as CO_2 goes from the solid phase to the gas phase.

(b) Liquid water at 50 °C

(c) Ruby

(d) One mole of N_2 at 1 bar

19.3 (a) $CH_3OH(g)$; (b) HBr; (c) $NH_4Cl(aq)$; (d) $HNO_3(g)$

19.5 (a) $\Delta S° = +12.7$ J/K. Disorder increases.

(b) $\Delta S° = -102.55$ J/K. Significant decrease in disorder.

(c) $\Delta S° = +93.2$ J/K. Disorder increases.

(d) $\Delta S° = -129.7$ J/K. The solution is more ordered (with H^+ forming H_3O^+ and hydrogen bonding occurring) than HCl in the gaseous state.

19.7 $\Delta S° = -174.1$ J/K

19.9 (a) $\Delta S° = +9.3$ J/K; (b) $\Delta S° = -293.97$ J/K

19.11 (a) $\Delta S° = -507.3$ J/K; (b) $\Delta S° = +313.25$ J/K

19.13 $\Delta S°_{sys} = -134.18$ J/K; $\Delta H°_{sys} = -662.75$ kJ; $\Delta S°_{surr} = +2222.9$ J/K; $\Delta S°_{univ} = +2088.7$ J/K

19.15 $\Delta S°_{sys} = +163.3$ J/K; $\Delta H°_{sys} = +285.83$ kJ; $\Delta S°_{surr} = -958.68$ J/K; $\Delta S°_{univ} = -795.4$ J/K

The reaction is not spontaneous, because the overall entropy change in the universe is negative. The reaction is disfavored by energy dispersal.

19.17 (a) Type 2. The reaction is enthalpy-favored but entropy-disfavored. It is more favorable at low temperatures.

(b) Type 4. This endothermic reaction is not favored by the enthalpy change nor is it favored by the entropy change. It is not spontaneous under any conditions.

19.19 (a) $\Delta S°_{sys} = +174.75$ J/K; $\Delta H°_{surr} = +116.94$ kJ; $\Delta S°_{surr} = -392.4$ J/K

(b) $\Delta S°_{univ} = -217.67$ J/K. The reaction is not spontaneous at 298 K.

(c) As the temperature increases, $\Delta S°_{surr}$ becomes less important, so $\Delta S°_{univ}$ will become positive at a sufficiently high temperature.

19.21 (a) $\Delta H°_{rxn} = -438$ kJ; $\Delta S°_{rxn} = -201.7$ J/K; $\Delta G°_{rxn} = -378$ kJ

The reaction is product-favored and is enthalpy-driven.

(b) $\Delta H°_{rxn} = -86.61$ kJ; $\Delta S°_{rxn} = -79.4$ J/K; $\Delta G°_{rxn} = -62.9$ kJ

The reaction is product-favored. The enthalpy change favors the reaction.

19.23 (a) $\Delta H°_{rxn} = +116.7$ kJ; $\Delta S°_{rxn} = +168.0$ J/K; $\Delta G°_f = +66.6$ kJ/mol

(b) $\Delta H°_{rxn} = -425.93$ kJ; $\Delta S°_{rxn} = -154.6$ J/K; $\Delta G°_f = -379.82$ kJ/mol

(c) $\Delta H°_{rxn} = +17.51$ kJ; $\Delta S°_{rxn} = +77.95$ J/K; $\Delta G°_f = -5.73$ kJ/mol

19.25 (a) $\Delta G°_{rxn} = -817.54$ kJ; spontaneous

(b) $\Delta G°_{rxn} = +256.6$ kJ; not spontaneous

(c) $\Delta G°_{rxn} = -1101.14$ kJ; spontaneous

19.27 $\Delta G°_f[BaCO_3(s)] = -1134.4$ kJ/mol

19.29 (a) $\Delta H°_{rxn} = +66.2$ kJ; $\Delta S°_{rxn} = -121.62$ J/K; $\Delta G°_{rxn} = +102.5$ kJ

Both the enthalpy and the entropy changes indicate the reaction is not spontaneous. There is no temperature to which it will become spontaneous. This is a case like that in the right panel in Figure 19.12 and is a Type 4 reaction (Table 19.2).

(b) $\Delta H°_{rxn} = -221.05$ kJ; $\Delta S°_{rxn} = +179.1$ J/K; $\Delta G°_{rxn} = -283.99$ kJ

The reaction is favored by both enthalpy and entropy and is product-favored at all temperatures. This is a case like that in the left panel in Figure 19.12 and is a Type 1 reaction.

(c) $\Delta H°_{rxn} = -179.0$ kJ; $\Delta S°_{rxn} = -160.2$ J/K; $\Delta G°_{rxn} = -131.4$ kJ

The reaction is favored by the enthalpy change but disfavored by the entropy change. The reaction becomes less product-favored as the temperature increases; it is a case like the upper line in the middle panel of Figure 19.12.

(d) $\Delta H°_{rxn} = +822.2$ kJ; $\Delta S°_{rxn} = +181.28$ J/K; $\Delta G°_{rxn} = +768.08$ kJ

The reaction is not favored by the enthalpy change but favored by the entropy change. The reaction becomes more product-favored as the temperature increases; it is a case like the lower line in the middle panel of Figure 19.12.

19.31 $\Delta H°_{rxn} = +337.2$ kJ and $\Delta S°_{rxn} = +161.5$ J/K. When $\Delta G°_{rxn} = 0$, $T = 2088$ K.

19.33 $K = 6.8 \times 10^{-16}$. Note that K is very small and that $\Delta G°$ is positive. Both indicate a reactant-favored process.

19.35 $\Delta G°_{rxn} = -100.24$ kJ and $K_p = 3.64 \times 10^{17}$. Both the free energy change and K indicate a product-favored process.

19.37 Reaction 1: $\Delta S°_1 = -80.7$ J/K

Reaction 2: $\Delta S°_2 = -161.60$ J/K

Reaction 3: $\Delta S°_3 = -242.3$ J/K

$\Delta S°_1 + \Delta S°_2 = \Delta S°_3$

19.39 $\Delta H°_{rxn} = -1428.66$ kJ; $\Delta S°_{rxn} = +47.1$ J/K; $\Delta S°_{univ} = +4840$ J/K

Combustion reactions are spontaneous, and this is confirmed by the sign of $\Delta S°_{univ}$.

19.41 The reaction occurs spontaneously and is product-favored. Therefore, $\Delta S°_{univ}$ is positive and $\Delta G°_{rxn}$ is negative. The reaction is likely to be exothermic, so $\Delta H°_{rxn}$ is negative, and $\Delta S°_{surr}$ is positive. $\Delta S°_{sys}$ is expected to be negative because two moles of gas form one mole of solid. The calculated values are as follows:

$\Delta S^\circ_{sys} = -284.2$ J/K

$\Delta H^\circ_{rxn} = -176.34$ kJ

$\Delta S^\circ_{surr} = +591.45$ J/K

$\Delta S^\circ_{univ} = +307.3$ J/K

$\Delta G^\circ_{rxn} = -91.64$ kJ

$K_p = 1.13 \times 10^{16}$

19.43 $K_p = 1.3 \times 10^{29}$ at 298 K ($\Delta G^\circ = -166.1$ kJ). The reaction is already extremely product-favored at 298 K. A higher temperature, however, would make the reaction less product-favored because ΔS°_{rxn} has a negative value (-242.3 J/K).

19.45 At the boiling point, $\Delta G^\circ = 0 = \Delta H^\circ - T\Delta S^\circ$.

Here $\Delta S^\circ = \Delta H^\circ / T = 112$ J/K · mol at 351.15 K.

19.47 For $C_2H_5OH(\ell) \longrightarrow C_2H_5OH(g)$, $\Delta S^\circ = +122.0$ J/K and $\Delta H^\circ = +41.7$ kJ.

$T = \Delta H^\circ / \Delta S^\circ = 341.8$ K or 68.7 °C.

19.49 ΔS°_{rxn} is $+137.2$ J/K. A positive entropy change means that raising the temperature will increase the product favorability of the reaction (because $T\Delta S^\circ$ will become more negative).

19.51 (a) The reaction is endothermic and reactant-favored. Predicted results: $\Delta H^\circ_{rxn} > 0$, $\Delta S^\circ_{surr} < 0$, $\Delta S^\circ_{univ} < 0$, and $\Delta G^\circ_{sys} > 0$. ΔS°_{sys} is > 0 because 1 mol of gas and 2 mol of liquid are produced from 2 mol of solid.

Calculated results: $\Delta H^\circ_{rxn} = +181.66$ kJ, $\Delta S^\circ_{sys} = +216.53$ J/K, $\Delta G^\circ_{sys} = +117$ kJ.

(b) $K_p = 3.1 \times 10^{-21}$ Reaction is reactant-favored.

19.53 The reaction is exothermic, so ΔH°_{rxn} should be negative. Also, a gas and an aqueous solution are formed, so ΔS°_{rxn} should be positive. The calculated values are

$\Delta H^\circ_{rxn} = -183.32$ kJ (with a negative sign as expected)

$\Delta S^\circ_{rxn} = -7.7$ J/K

The entropy change is slightly negative, not positive as predicted. The reason for this is the negative entropy change upon dissolving NaOH. Apparently the OH$^-$ ions in water hydrogen-bond with water molecules and lead to a slight ordering of the system relative to pure water.

19.55 $\Delta H^\circ_{rxn} = +126.03$ kJ; $\Delta S^\circ_{rxn} = +78.2$ J/K; and $\Delta G^\circ_{rxn} = +103$ kJ.

The reaction is not predicted to be spontaneous under standard conditions.

19.57 $\Delta G^\circ_{rxn} = 6.98$ kJ/mol

19.59 $\Delta G^\circ_{rxn} = -98.9$ kJ

The reaction is spontaneous under standard conditions and is enthalpy-driven

19.61 (a) $\Delta G^\circ_{rxn} = +141.82$ kJ, so the reaction is not spontaneous.

(b) $\Delta H^\circ_{rxn} = +197.86$ kJ; $\Delta S^\circ_{rxn} = +187.95$ J/K

$T = \Delta H^\circ_{rxn} / \Delta S^\circ_{rxn} = 1052.7$ K or 779.6 °C

(c) ΔG°_{rxn} at 1500 °C (1773 K) $= -135.4$ kJ

K_p at 1500 °C $= 1 \times 10^4$

19.63 $\Delta S^\circ_{rxn} = -459.0$ J/K; $\Delta H^\circ_{rxn} = -793$ kJ; $\Delta G^\circ_{rxn} = -657$ kJ

The reaction is spontaneous and enthalpy-driven.

19.65 (a) ΔG° at 80.0 °C $= +0.14$ kJ

ΔG° at 110.0 °C $= -0.12$ kJ

Rhombic sulfur is more stable than monoclinic sulfur at 80 °C, but the reverse is true at 110 °C.

(b) $T = 370$ K or about 96 °C

19.67 (a) $\Delta S^\circ_{rxn} = +24.89$ J/K; $\Delta H^\circ_{rxn} = +180.58$ kJ; $\Delta G^\circ_{rxn} = +173.16$ kJ

$K_p = 4.62 \times 10^{-31}$, so the reaction is reactant-favored at 298 K.

(b) At 700 °C (973 K), $\Delta G^\circ_{rxn} = +156.6$ kJ

$K_p = 4 \times 10^{-9}$, so the reaction is still reactant-favored at 700 °C, but less so than at 298 K.

(c) $P(NO) = 6 \times 10^{-5}$; $P(O_2) = P(N_2) = 1$ bar

19.69 $\Delta G^\circ_f [HI(g)] = -10.9$ kJ/mol

19.71 $K_p = P_{Hg(g)}$ at any temperature.

$K_p = 1$ at 620.3 K or 347.2 °C when $P_{Hg(g)} = 1.000$ bar.

T when $P_{Hg(g)} = (1/760)$ bar is 393.3 K or 125.2 °C.

19.73 (a) Reactant-favored (mercury is a liquid under standard conditions)

(b) Product-favored (water vapor will condense to a liquid)

(c) Reactant-favored (a continuous supply of energy is required)

(d) Product-favored (carbon will burn)

(e) Product-favored (salt will dissolve)

(f) Reactant-favored (calcium carbonate is insoluble)

19.75 (a) True

(b) False. Whether an exothermic system is spontaneous also depends on the entropy change for the system.

(c) False. Reactions with $+\Delta H^\circ_{rxn}$ and $+\Delta S^\circ_{rxn}$ are spontaneous at higher temperatures.

(d) True

19.77 Dissolving a solid such as NaCl in water is a spontaneous process. Thus, $\Delta G^\circ < 0$. If $\Delta H^\circ = 0$, then the only way the free energy change can be negative is if ΔS° is positive. Generally the entropy change is the driving force in forming a solution.

19.79 $2\,C_2H_6(g) + 7\,O_2(g) \longrightarrow 4\,CO_2(g) + 6\,H_2O(g)$

(a) Not only is this an exothermic combustion reaction, but there is also an increase in the number of molecules from reactants to products. Therefore, we would predict an increase in $\Delta S°$ for both the system and the surroundings and thus for the universe as well.

(b) The exothermic reaction has $\Delta H°_{rxn} < 0$. Combined with a positive $\Delta S°_{sys}$, the value of $\Delta G°_{rxn}$ is negative.

(c) The value of K_p is likely to be much greater than 1. Further, because $\Delta S°_{sys}$ is positive, the value of K_p will be even larger at a higher temperature. (See the left panel of Figure 19.12.)

19.81 In solid NaCl, the particles are fixed in a solid lattice. When the solid is dissolved, the particles (Na^+ and Cl^- ions) are dispersed throughout the solution.

19.83 Iodine dissolves readily, so the process is spontaneous and $\Delta G°$ must be less than zero. Because $\Delta H° = 0$, the process is entropy-driven.

19.85 (a) $\Delta H°_{rxn} = -352.88$ kJ and $\Delta S°_{rxn} = +21.31$ J/K. Therefore, at 298 K, $\Delta G°_{rxn} = -359.23$ kJ.

(b) 4.84 g of Mg is required.

19.87 (a) $N_2H_4(\ell) + O_2(g) \longrightarrow 2\ H_2O(\ell) + N_2(g)$
O_2 is the oxidizing agent and N_2H_4 is the reducing agent.

(b) $\Delta H°_{rxn} = -622.29$ kJ and $\Delta S°_{rxn} = +4.87$ J/K. Therefore, at 298 K, $\Delta G°_{rxn} = -623.77$ kJ.

(c) 0.0027 K

(d) 7.5 mol O_2

(e) 4.8×10^3 g solution

(f) 7.5 mol $N_2(g)$ occupies 170 L at 273 K and 1.0 atm of pressure.

19.89 (a) $\Delta S°_{sys} = -60.49$ J/K

(b) No

(c) Yes

(d) Yes

(e) Yes. $\Delta S°_{surr}$ always changes with T.

(f) The reaction is spontaneous at 400 K but not at 700 K.

19.91 (a) $\Delta G°$ decreases as temperature increases.

(b) No, the reaction is always spontaneous.

(c) The spontaneity of a reaction is dependent on temperature because $\Delta G°_{rxn}$ is determined in part by the $T\Delta S°$ term.

Chapter 20

20.1 (a) $Cr(s) \longrightarrow Cr^{3+}(aq) + 3\ e^-$

Cr is a reducing agent; this is an oxidation reaction.

(b) $AsH_3(g) \longrightarrow As(s) + 3\ H^+(aq) + 3\ e^-$

AsH_3 is a reducing agent; this is an oxidation reaction.

(c) $VO_3^-(aq) + 6\ H^+(aq) + 3\ e^- \longrightarrow$
$\qquad\qquad\qquad V^{2+}(aq) + 3\ H_2O(\ell)$

$VO_3^-(aq)$ is an oxidizing agent; this is a reduction reaction.

(d) $2\ Ag(s) + 2\ OH^-(aq) \longrightarrow$
$\qquad\qquad\qquad Ag_2O(s) + H_2O(\ell) + 2\ e^-$

Silver is a reducing agent; this is an oxidation reaction.

20.3 (a) $Ag(s) \longrightarrow Ag^+(aq) + e^-$

$\underline{e^- + NO_3^-(aq) + 2\ H^+(aq) \longrightarrow NO_2(g) + H_2O(\ell)}$

$Ag(s) + NO_3^-(aq) + 2\ H^+(aq) \longrightarrow$
$\qquad\qquad\qquad Ag^+(aq) + NO_2(g) + H_2O(\ell)$

(b) $2[MnO_4^-(aq) + 8\ H^+(aq) + 5\ e^- \longrightarrow$
$\qquad\qquad\qquad Mn^{2+}(aq) + 4\ H_2O(\ell)]$

$\underline{5[HSO_3^-(aq) + H_2O(\ell) \longrightarrow}$
$\qquad\qquad\qquad \underline{SO_4^{2-}(aq) + 3\ H^+(aq) + 2\ e^-]}$

$2\ MnO_4^-(aq) + H^+(aq) + 5\ HSO_3^-(aq) \longrightarrow$
$\qquad\qquad 2\ Mn^{2+}(aq) + 3\ H_2O(\ell) + 5\ SO_4^{2-}(aq)$

(c) $4[Zn(s) \longrightarrow Zn^{2+}(aq) + 2\ e^-]$

$\underline{2\ NO_3^-(aq) + 10\ H^+(aq) + 8\ e^- \longrightarrow}$
$\qquad\qquad\qquad \underline{N_2O(g) + 5\ H_2O(\ell)}$

$4\ Zn(s) + 2\ NO_3^-(aq) + 10\ H^+(aq) \longrightarrow$
$\qquad\qquad 4\ Zn^{2+}(aq) + N_2O(g) + 5\ H_2O(\ell)$

(d) $Cr(s) \longrightarrow Cr^{3+}(aq) + 3\ e^-$

$\underline{3\ e^- + NO_3^-(aq) + 4\ H^+(aq) \longrightarrow}$
$\qquad\qquad\qquad \underline{NO(g) + 2\ H_2O(\ell)}$

$Cr(s) + NO_3^-(aq) + 4\ H^+(aq) \longrightarrow$
$\qquad\qquad\qquad Cr^{3+}(aq) + NO(g) + 2\ H_2O(\ell)$

20.5 (a) $2[Al(s) + 4\ OH^-(aq) \longrightarrow Al(OH)_4^-(aq) + 3\ e^-]$

$\underline{3[2\ H_2O(\ell) + 2\ e^- \longrightarrow H_2(g) + 2\ OH^-(aq)]}$

$2\ Al(s) + 2\ OH^-(aq) + 6\ H_2O(\ell) \longrightarrow$
$\qquad\qquad\qquad 2\ Al(OH)_4^-(aq) + 3\ H_2(g)$

(b) $2[CrO_4^{2-}(aq) + 4\ H_2O(\ell) + 3\ e^- \longrightarrow$
$\qquad\qquad\qquad Cr(OH)_3(s) + 5\ OH^-(aq)]$

$\underline{3[SO_3^{2-}(aq) + 2\ OH^-(aq) \longrightarrow}$
$\qquad\qquad\qquad \underline{SO_4^{2-}(aq) + H_2O(\ell) + 2\ e^-]}$

$2\ CrO_4^{2-}(aq) + 3\ SO_3^{2-}(aq) + 5\ H_2O(\ell) \longrightarrow$
$\qquad\qquad 2\ Cr(OH)_3(s) + 3\ SO_4^{2-}(aq) + 4\ OH^-(aq)$

(c) $Zn(s) + 4\ OH^-(aq) \longrightarrow Zn(OH)_4^{2-}(aq) + 2\ e^-$

$\underline{Cu(OH)_2(s) + 2\ e^- \longrightarrow Cu(s) + 2\ OH^-(aq)}$

$Zn(s) + 2\ OH^-(aq) + Cu(OH)_2(s) \longrightarrow$
$\qquad\qquad\qquad Zn(OH)_4^{2-}(aq) + Cu(s)$

(d) $3[HS^-(aq) + OH^-(aq) \longrightarrow S(s) + H_2O(\ell) + 2\ e^-]$

$\underline{ClO_3^-(aq) + 3\ H_2O(\ell) + 6\ e^- \longrightarrow}$
$\qquad\qquad\qquad \underline{Cl^-(aq) + 6\ OH^-(aq)}$

$3\ HS^-(aq) + ClO_3^-(aq) \longrightarrow$
$\qquad\qquad 3\ S(s) + Cl^-(aq) + 3\ OH^-(aq)$

20.7 Electrons flow from the Cr electrode to the Fe electrode. Negative ions move via the salt bridge from the Fe/Fe^{2+} half-cell to the Cr/Cr^{3+} half-cell (and positive ions move in the opposite direction).

Anode (oxidation): $Cr(s) \longrightarrow Cr^{3+}(aq) + 3\ e^-$

Cathode (reduction): $Fe^{2+}(aq) + 2\ e^- \longrightarrow Fe(s)$

20.9 (a) Oxidation: $Fe(s) \longrightarrow Fe^{2+}(aq) + 2\,e^-$

Reduction: $O_2(g) + 4\,H^+(aq) + 4\,e^- \longrightarrow 2\,H_2O(\ell)$

Overall: $2\,Fe(s) + O_2(g) + 4\,H^+(aq) \longrightarrow$
$2\,Fe^{2+}(aq) + 2\,H_2O(\ell)$

(b) Anode, oxidation: $Fe(s) \longrightarrow Fe^{2+}(aq) + 2\,e^-$

Cathode, reduction:
$O_2(g) + 4\,H^+(aq) + 4\,e^- \longrightarrow 2\,H_2O(\ell)$

(c) Electrons flow from the negative anode (Fe) to the positive cathode (site of the O_2 half-reaction). Negative ions move through the salt bridge from the cathode compartment in which the O_2 reduction occurs to the anode compartment in which Fe oxidation occurs (and positive ions move in the opposite direction).

20.11 (a) All are primary batteries, not rechargeable.

(b) Dry cells and alkaline batteries have Zn anodes. Ni-Cd batteries have a cadmium anode.

(c) Dry cells have an acidic environment, whereas the environment is alkaline for alkaline and Ni-Cd cells.

20.13 (a) $E^\circ_{cell} = -1.298$ V; not product-favored

(b) $E^\circ_{cell} = -0.51$ V; not product-favored

(c) $E^\circ_{cell} = -1.023$ V; not product-favored

(d) $E^\circ_{cell} = +0.029$ V; product-favored

20.15 (a) $Sn^{2+}(aq) + 2\,Ag(s) \longrightarrow Sn(s) + 2\,Ag^+(aq)$

$E^\circ_{cell} = -0.94$ V; not product-favored

(b) $3\,Sn^{4+}(aq) + 2\,Al(s) \longrightarrow 3\,Sn^{2+}(aq) + 2\,Al^{3+}(aq)$

$E^\circ_{cell} = +1.81$ V; product-favored

(c) $ClO_3^-(aq) + 6\,Ce^{3+}(aq) + 6\,H^+(aq) \longrightarrow$
$Cl^-(aq) + 6\,Ce^{4+}(aq) + 3\,H_2O(\ell)$

$E^\circ_{cell} = -0.99$ V; not product-favored

(d) $3\,Cu(s) + 2\,NO_3^-(aq) + 8\,H^+(aq) \longrightarrow$
$3\,Cu^{2+}(aq) + 2\,NO(g) + 4\,H_2O(\ell)$

$E^\circ_{cell} = +0.62$ V; product-favored

20.17 (a) Al

(b) Zn and Al

(c) $Fe^{2+}(aq) + Sn(s) \longrightarrow Fe(s) + Sn^{2+}(aq)$; reactant-favored

(d) $Zn^{2+}(aq) + Sn(s) \longrightarrow Zn(s) + Sn^{2+}(aq)$; not product-favored

20.19 Best reducing agent, Zn(s)

20.21 Ag^+

20.23 See Example 20.5, page 971.

(a) F_2, most readily reduced

(b) F_2 and Cl_2

20.25 $E^\circ_{cell} = +0.3923$ V. When $[Zn(OH)_4^{2-}] = [OH^-] = 0.025$ M and $P(H_2) = 1.0$ bar, $E_{cell} = 0.345$ V.

20.27 $E^\circ_{cell} = +1.563$ V and $E_{cell} = +1.58$ V.

20.29 When $E^\circ_{cell} = +1.563$ V, $E_{cell} = 1.48$ V, $n = 2$, and $[Zn^{2+}] = 1.0$ M, the concentration of $Ag^+ = 0.040$ M.

20.31 (a) $\Delta G^\circ = -45.5$ kJ; $K = 9 \times 10^7$

(b) $\Delta G^\circ = +110$ kJ; $K = 4 \times 10^{-19}$

20.33 E°_{cell} for $AgBr(s) \longrightarrow Ag^+(aq) + Br^-(aq)$ is -0.7281.
$K_{sp} = 4.9 \times 10^{-13}$

20.35 $K_{formation} = 2 \times 10^{25}$

20.37 (a) $Fe^{2+}(aq) + 2\,e^- \longrightarrow Fe(s)$
$2[Fe^{2+}(aq) \longrightarrow Fe^{3+}(aq) + e^-]$
$3\,Fe^{2+}(aq) \longrightarrow Fe(s) + 2\,Fe^{3+}(aq)$

(b) $E^\circ_{cell} = -1.21$ V; not product-favored

(c) $K = 1 \times 10^{-41}$

20.39 See Figure 20.18.

20.41 O_2 from the oxidation of water is more likely than F_2. See Example 20.11.

20.43 See Example 20.11.

(a) Cathode: $2\,H_2O(\ell) + 2\,e^- \longrightarrow H_2(g) + 2\,OH^-(aq)$

(b) Anode: $2\,Br^-(aq) \longrightarrow Br_2(\ell) + 2\,e^-$

20.45 Mass of Ni = 0.0334 g

20.47 Time = 2300 s or 38 min

20.49 Time = 250 h

20.51 (a) $UO_2^+(aq) + 4\,H^+(aq) + e^- \longrightarrow$
$U^{4+}(aq) + 2\,H_2O(\ell)$

(b) $ClO_3^-(aq) + 6\,H^+(aq) + 6\,e^- \longrightarrow$
$Cl^-(aq) + 3\,H_2O(\ell)$

(c) $N_2H_4(aq) + 4\,OH^-(aq) \longrightarrow$
$N_2(g) + 4\,H_2O(\ell) + 4\,e^-$

(d) $ClO^-(aq) + H_2O(\ell) + 2\,e^- \longrightarrow$
$Cl^-(aq) + 2\,OH^-(aq)$

20.53 (a,c) The electrode at the right is a magnesium anode. (Magnesium metal supplies electrons and is oxidized to Mg^{2+} ions.) Electrons pass through the wire to the silver cathode, where Ag^+ ions are reduced to silver metal. Nitrate ions move via the salt bridge from the $AgNO_3$ solution to the $Mg(NO_3)_2$ solution (and Na^+ ions move in the opposite direction).

(b) Anode: $Mg(s) \longrightarrow Mg^{2+}(aq) + 2\,e^-$

Cathode: $Ag^+(aq) + e^- \longrightarrow Ag(s)$

Net reaction: $Mg(s) + 2\,Ag^+(aq) \longrightarrow$
$Mg^{2+}(aq) + 2\,Ag(s)$

20.55 (a) For 1.7 V:

Use chromium as the anode to reduce $Ag^+(aq)$ to $Ag(s)$ at the cathode. The cell potential is $+1.71$ V.

(b) For 0.5 V:

Use copper as the anode to reduce silver ions to silver metal at the cathode. The cell potential is $+0.46$ V.

Use silver as the anode to reduce chlorine to chloride ions. The cell potential would be +0.56 V. (In practice, this setup is not likely to work well because the product would be insoluble silver chloride.)

20.57 (a) $Zn^{2+}(aq)$ (c) $Zn(s)$

     (b) $Au^+(aq)$ (d) $Au(s)$

     (e) Yes, $Sn(s)$ will reduce Cu^{2+} (as well as Ag^+ and Au^+).

     (f) No, $Ag(s)$ can only reduce $Au^+(aq)$.

     (g) See part (e).

     (h) $Ag^+(aq)$ can oxidize Cu, Sn, Co, and Zn.

20.59 (a) The cathode is the site of reduction, so the half-reaction must be $2 H^+(aq) + 2 e^- \longrightarrow H_2(g)$. This is the case with the following half-reactions: $Cr^{3+}(aq)|Cr(s)$, $Fe^{2+}(aq)|Fe(s)$, and $Mg^{2+}(aq)|Mg(s)$.

     (b) Choosing from the half-cells in part (a), the reaction of $Mg(s)$ and $H^+(aq)$ would produce the most positive potential (2.37 V), and the reaction of H_2 with Cu^{2+} would produce the least positive potential (+0.337 V).

20.61 8.1×10^5 g Al

20.63 (a) $E^\circ_{anode} = -0.268$ V

     (b) $K_{sp} = 2 \times 10^{-5}$

20.65 $\Delta G^\circ = -562$ kJ

20.67 6700 kWh

20.69 $Ru(NO_3)_2$

20.71 9.5×10^6 g Cl_2 per day

20.73 0.054 g Au

20.75 (a) $2[Ag^+(aq) + e^+ \longrightarrow Ag(s)]$

$$C_6H_5CHO(aq) + H_2O(\ell) \longrightarrow$$
$$C_6H_5CO_2H(aq) + 2 H^+(aq) + 2 e^-$$
$$\overline{}$$
$$2Ag^+(aq) + C_6H_5CHO(aq) + H_2O(\ell) \longrightarrow$$
$$C_6H_5CO_2H(aq) + 2 H^+(aq) + 2 Ag(s)$$

     (b) $3[CH_3CH_2OH(aq) + H_2O(\ell) \longrightarrow$
$$CH_3CO_2H(aq) + 4 H^+(aq) + 4 e^-]$$
$$2[Cr_2O_7{}^{2-}(aq) + 14 H^+(aq) + 6 e^- \longrightarrow$$
$$2 Cr^{3+}(aq) + 7 H_2O(\ell)]$$
$$\overline{}$$
$$3 CH_3CH_2OH(aq) + 2 Cr_2O_7{}^{2-}(aq) + 16 H^+(aq) \longrightarrow$$
$$3 CH_3CO_2H(aq) + 4 Cr^{3+}(aq) + 11 H_2O(\ell)$$

20.77 (a) 0.974 kJ/g

     (b) 0.60 kJ/g

     (c) The silver-zinc battery produces more energy per gram of reactants.

20.79 (a) 92 g HF required; 230 g CF_3SO_2F and 9.3 g H_2 isolated

     (b) H_2 is produced at the cathode.

     (c) 48 kWh

20.81 (a)

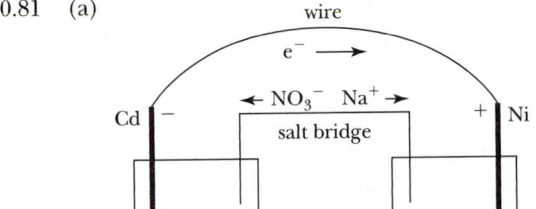

                       Anode   Cathode

     (b) Anode: $Cd(s) \longrightarrow Cd^{2+}(aq) + 2 e^-$

           Cathode: $Ni^{2+}(aq) + 2 e^- \longrightarrow Ni(s)$

           Net: $Cd(s) + Ni^{2+}(aq) \longrightarrow Cd^{2+}(aq) + Ni(s)$

     (c) The anode is negative and the cathode is positive.

     (d) $E^\circ_{cell} = E^\circ_{cathode} - E^\circ_{anode} = (-0.25$ V$) - (-0.40$ V$) = +0.15$ V

     (e) Electrons flow from anode (Cd) to cathode (Ni).

     (f) Na^+ ions move from the anode compartment to the cathode compartment. Anions move in the opposite direction.

     (g) $K = 1 \times 10^5$

     (h) $E_{cell} = 0.21$ V

     (i) 480 h

20.83 I^- is the strongest reducing agent of the three halide ions. Iodide ion reduces Cu^{2+} to Cu^+, forming insoluble $CuI(s)$.

$$2 Cu^{2+}(aq) + 4 I^-(aq) \longrightarrow 2 CuI(s) + I_2(aq)$$

20.85 290 h

20.87 (a) Au^{3+}, Br_2, Hg^{2+}, Ag^+, $Hg_2{}^{2+}$

     (b) Al, Mg, Li

Chapter 21

21.1 $4 Li(s) + O_2(g) \longrightarrow 2 Li_2O(s)$

     $Li_2O(s) + H_2O(\ell) \longrightarrow 2 LiOH(aq)$

     $2 Ca(s) + O_2(g) \longrightarrow 2 CaO(s)$

     $CaO(s) + H_2O(\ell) \longrightarrow Ca(OH)_2(s)$

21.3 These are the elements of Group 3A: boron, B; aluminum, Al; gallium, Ga; indium, In; and thallium, Tl.

21.5 $2 Na(s) + Cl_2(g) \longrightarrow 2 NaCl(s)$

     The reaction is exothermic and the product is ionic. See Figure 1.7.

21.7 The product, NaCl, is a colorless solid and is soluble in water. Other alkali metal chlorides have similar properties.

21.9 Calcium will not exist in the earth's crust because the metal reacts with water.

21.11   Increasing basicity: $CO_2 < SiO_2 < SnO_2$

21.13   (a) $2\,Na(s) + Br_2(\ell) \longrightarrow 2\,NaBr(s)$

(b) $2\,Mg(s) + O_2(g) \longrightarrow 2\,MgO(s)$

(c) $2\,Al(s) + 3\,F_2(g) \longrightarrow 2\,AlF_3(s)$

(d) $C(s) + O_2(g) \longrightarrow CO_2(g)$

21.15 $2\,H_2(g) + O_2(g) \longrightarrow 2\,H_2O(g)$

$H_2(g) + Cl_2(g) \longrightarrow 2\,HCl(g)$

$3\,H_2(g) + N_2(g) \longrightarrow 2\,NH_3(g)$

21.17 $CH_4(g) + H_2O(g) \longrightarrow CO(g) + 3\,H_2(g)$

$\Delta H^{\circ}_{rxn} = +20.62$ kJ; $\Delta S^{\circ}_{sys} = +214.7$ J/K;
$\Delta G^{\circ}_{sys} = -43.4$ kJ (at 298.15 K).

21.19   Step 1: $2\,SO_2(g) + 4\,H_2O(\ell) + 2\,I_2(s) \longrightarrow$
$\qquad\qquad\qquad\qquad\qquad 2\,H_2SO_4(\ell) + 4\,HI(g)$

Step 2: $2\,H_2SO_4(\ell) \longrightarrow 2\,H_2O(\ell) + 2\,SO_2(g) + O_2(g)$

Step 3: $4\,HI(g) \longrightarrow 2\,H_2(g) + 2\,I_2(g)$

Net: $2\,H_2O(\ell) \longrightarrow 2\,H_2(g) + O_2(g)$

21.21 $Na(s) + F_2(g) \longrightarrow 2\,NaF(s)$

$Na(s) + Cl_2(g) \longrightarrow 2\,NaCl(s)$

$Na(s) + Br_2(\ell) \longrightarrow 2\,NaBr(s)$

$Na(s) + I_2(s) \longrightarrow 2\,NaI(s)$

The alkali metal halides are white, crystalline solids. They have high melting and boiling points, and are soluble in water.

21.23 $2\,Cl^-(aq) + 2\,H_2O(\ell) \longrightarrow$
$\qquad\qquad\qquad Cl_2(g) + H_2(g) + 2\,OH^-(aq)$

If this were the only process used to produce chlorine, the mass of Cl_2 reported for industrial production would be 0.88 times the mass of NaOH produced (2 mol NaCl, 117 g, would yield 2 mol NaOH, 80 g, and 1 mol Cl_2, 70 g). The amounts quoted indicate a Cl_2-to-NaOH mass ratio 0.96. Chlorine is presumably also prepared by other routes than this one.

21.25 $2\,Mg(s) + O_2(g) \longrightarrow 2\,MgO(s)$

$3\,Mg(s) + N_2(g) \longrightarrow Mg_3N_2(s)$

21.27 $CaCO_3$ is used in agriculture to neutralize acidic soil, to prepare CaO for use in mortar, and in steel production.

$CaCO_3(s) + H_2O(\ell) + CO_2(g) \longrightarrow$
$\qquad\qquad\qquad\qquad Ca^{2+}(aq) + 2\,HCO_3^-(aq)$

21.29 1.4×10^6 g SO_2

21.31

$B_3O_6^{3-}$ $B_2O_5^{4-}$

21.33   (a) $2\,B_5H_9(g) + 12\,O_2(g) \longrightarrow 5\,B_2O_3(s) + 9\,H_2O(g)$

(b) Heat of combustion of $B_5H_9 = -4341.2$ kJ/mol. This is more than double the heat of combustion of B_2H_6.

(c) Heat of combustion of $C_2H_6(g)$ [to give $CO_2(g)$ and $H_2O(g)$] $= -1428.7$ kJ/mol. C_2H_6 produces 47.5 kJ/g whereas diborane produces much more (73.7 kJ/g).

21.35 $2\,Al(s) + 6\,HCl(aq) \longrightarrow$
$\qquad\qquad\qquad 2\,Al^{3+}(aq) + 6\,Cl^-(aq) + 3\,H_2(g)$

$2\,Al(s) + 3\,Cl_2(g) \longrightarrow 2\,AlCl_3(s)$

$4\,Al(s) + 3\,O_2(g) \longrightarrow 2\,Al_2O_3(s)$

21.37 $2\,Al(s) + 2\,OH^-(aq) + 6\,H_2O(\ell) \longrightarrow$
$\qquad\qquad\qquad\qquad 2\,Al(OH)_4^-(aq) + 3\,H_2(g)$

Volume of H_2 obtained from 13.2 g Al = 18.4 L

21.39 $Al_2O_3(s) + 3\,H_2SO_4(aq) \longrightarrow Al_2(SO_4)_3(s) + 3\,H_2O(\ell)$

Mass of H_2SO_4 required = 860 g and mass of Al_2O_3 required = 298 g

21.41

The ion has tetrahedral geometry. Aluminum is sp^3 hybridized.

21.43 SiO_2 is a network solid, with tetrahedral silicon atoms covalently bonded to four oxygens in an infinite array; CO_2 consists of individual molecules, with oxygen atoms double-bonded to carbon. Melting SiO_2 requires breaking very stable Si—O bonds. Weak intermolecular forces of attraction between CO_2 molecules result in this substance being a gas at ambient conditions.

21.45   (a) $2\,CH_3Cl(g) + Si(s) \longrightarrow (CH_3)_2SiCl_2(\ell)$

(b) 0.823 atm

(c) 12.2 g

21.47   Consider the general decomposition reaction:

$N_xO_y \longrightarrow {}^x/_2\,N_2 + {}^y/_2\,O_2$

The value of ΔG° can be obtained for all N_xO_y molecules because $\Delta G^{\circ}_{rxn} = -\Delta G^{\circ}_f$. These data show that the decomposition reaction is product-favored for all of the nitrogen oxides. All are unstable with respect to decomposition to the elements.

Compound	$-\Delta G_f^\circ$ (kJ/mol)
NO(g)	−86.58
NO_2	−51.23
N_2O	−104.20
N_2O_4	−97.73

21.49 $\Delta H_{rxn}^\circ = -114.4$ kJ; exothermic $\Delta G_{rxn}^\circ = -70.7$ kJ, product-favored

21.51 (a) $N_2H_4(aq) + O_2(g) \longrightarrow N_2(g) + 2\, H_2O(\ell)$
(b) 1.32×10^3 g

21.53 $5\, N_2H_5^+(aq) + 4\, IO_3^-(aq) \longrightarrow$
$\qquad 5\, N_2(g) + 2\, I_2(aq) + H^+(aq) + 12\, H_2O(\ell)$

$E_{net}^\circ = 1.43$ V

21.55 (a) Oxidation number = +3
(b) Diphosphonic acid ($H_4P_2O_5$) should be a diprotic acid (losing the two H atoms attached to O atoms).

$$
\begin{array}{c}
\ddot{\text{O}}: \quad\quad :\ddot{\text{O}}: \\
\| \quad\quad\quad \| \\
\text{H---P---}\ddot{\text{O}}\text{---P---H} \\
| \quad\quad\quad\quad | \\
\text{H---}\ddot{\text{O}}: \quad\quad :\ddot{\text{O}}\text{---H}
\end{array}
$$

21.57 (a) 3.5×10^3 kg SO_2
(b) 4.1×10^3 kg $Ca(OH)_2$

21.59 $\left[:\!\ddot{\text{S}}\!-\!\ddot{\text{S}}\!: \right]^{2-}$

disulfide ion

21.61 $E_{cell}^\circ = E_{cathode}^\circ - E_{anode}^\circ = +1.44$ V $- (+1.51$ V$) = -0.07$ V

The reaction is not product-favored under standard conditions.

21.63 $Cl_2(aq) + 2\, Br^-(aq) \longrightarrow 2\, Cl^-(aq) + Br_2(aq)$

Cl_2 is the oxidizing agent, Br^- is the reducing agent; $E_{cell}^\circ = 0.28$ V.

21.65 The reaction consumes 4.32×10^8 C to produce 8.51×10^4 g F_2.

21.67

Element	Appearance	State
Na, Mg, Al	Silvery metal	Solids
Si	black, shiny metalloid	Solid
P	White, red, and black allotropes; nonmetal	Solid
S	Yellow nonmetal	Solid
Cl	Pale green nonmetal	Gas
Ar	Colorless nonmetal	Gas

21.69 (a) $2\, Na(s) + Cl_2(g) \longrightarrow 2\, NaCl(s)$
$\qquad Mg(s) + Cl_2(g) \longrightarrow MgCl_2(s)$
$\qquad 2\, Al(s) + 3\, Cl_2(g) \longrightarrow 2\, AlCl_3(s)$
$\qquad Si(s) + 2\, Cl_2(g) \longrightarrow SiCl_4(\ell)$
$\qquad P_4(s) + 10\, Cl_2(g) \longrightarrow 4\, PCl_5(s)$ (excess Cl_2)
$\qquad S_8(s) + 8\, Cl_2(g) \longrightarrow 8\, SCl_2(s)$
(b) NaCl and $MgCl_2$ are ionic; the other products are covalent.
(c) $SiCl_4$ is tetrahedral; PCl_5 is trigonal bipyramidal.

$$
\begin{array}{cc}
:\ddot{\text{Cl}}: & :\ddot{\text{Cl}}: \\
| & | \quad\! :\ddot{\text{Cl}}: \\
:\ddot{\text{Cl}}\text{---Si}\cdots\!\ddot{\text{Cl}}: & :\ddot{\text{Cl}}\text{---P}\!\!\diagup^{\,\,\ddots} \\
| & | \quad\! :\ddot{\text{Cl}}: \\
\ddot{\text{Cl}} & \ddot{\text{Cl}}
\end{array}
$$

21.71 (a) $2\, KClO_3(s) \longrightarrow 2\, KCl(s) + 3\, O_2(g)$
(b) $2\, H_2S(g) + 3\, O_2(g) \longrightarrow 2\, H_2O(g) + 2\, SO_2(g)$
(c) $2\, Na(s) + O_2(g) \longrightarrow Na_2O_2(s)$
(d) $P_4(s) + 3\, KOH(aq) + 3\, H_2O(\ell) \longrightarrow$
$\qquad\qquad\qquad\qquad PH_3(g) + 3\, KH_2PO_4(aq)$
(e) $NH_4NO_3(s) \longrightarrow N_2O(g) + 2\, H_2O(g)$
(f) $2\, In(s) + 3\, Br_2(\ell) \longrightarrow 2\, InBr_3(s)$
(g) $SnCl_4(\ell) + 2\, H_2O(\ell) \longrightarrow SnO_2(s) + 4\, HCl(aq)$

21.73 (a)

$$
\begin{array}{c}
:\ddot{\text{Cl}}: :\ddot{\text{Cl}}: \\
| \quad\; | \\
:\ddot{\text{Cl}}\text{---B---B---}\ddot{\text{Cl}}:
\end{array}
$$

(b) Each B atom is surrounded in a trigonal-planar arrangement by another B atom and two Cl atoms. Each B atom is sp^2 hybridized.

21.75 (a) $2\, BaO(s) + O_2(g) \longrightarrow 2\, BaO_2(s)$
(b) $3\, BaO_2(s) + 2\, Fe(s) \longrightarrow 3\, BaO(s) + Fe_2O_3(s)$

21.77 Cathode: $Li^+(\ell) + e^+ \longrightarrow Li(\ell)$

Anode: $2\, H^-(\ell) \longrightarrow H_2(g) + 2\, e^-$

Formation of H_2 at the anode is evidence for the presence of H^-.

21.79 Mg: $\Delta G_{rxn}^\circ = +64.9$ kJ

Ca: $\Delta G_{rxn}^\circ = +131.40$ kJ

Ba: $\Delta G_{rxn}^\circ = +219.4$ kJ

Relative tendency to decompose: $MgCO_3 > CaCO_3 > BaCO_3$

21.81 (a) ΔG_f° should be more negative than $(-95.1$ kJ$) \times n$
(b) Ba, Pb, Ti

21.83 O—F bond energy = 190 kJ/mol

21.85 (a) N_2O_4 is the oxidizing agent (N is reduced from +4 to 0 in N_2), and $H_2NN(CH_3)_2$ is the reducing agent.
(b) 1.3×10^4 kg N_2O_4 is required. Product masses: 5.7×10^3 kg N_2; 4.9×10^3 kg H_2O; 6.0×10^3 kg CO_2.

21.87 (a) The NO bond with a length of 114.2 pm is a double bond. The other two NO bonds (with a length of 121 pm) have a bond order of 1.5 (as there are two resonance structures involving these bonds).

$$:\ddot{O} \quad\quad :\ddot{O}:$$
114.2 pm $\quad$ N—N $\quad$ 121 pm

(b) $K = 1.90$ (c) $\Delta H_f^\circ = 82.9$ kJ/mol

21.89 $\Delta H_{rxn}^\circ = -257.78$ kJ. This reaction is entropy-disfavored, however, with $\Delta S_{rxn}^\circ = -963$ J/K because of the decrease in the number of moles of gases. Combining these values gives $\Delta G_{rxn}^\circ = +29.19$ kJ, indicating that under standard conditions at 298 K the reaction is not spontaneous. (The reaction has a favorable ΔG_{rxn}° at temperatures less than 268 K, indicating that further research on this system might be worthwhile. Note that at that temperature water is a solid.)

21.91 3.5 kWh

21.93 The flask contains a fixed number of moles of gas at the given pressure and temperature. One could burn the mixture because only the H_2 will combust; the argon is untouched. Cooling the gases from combustion would remove water (the combustion product of H_2) and leave only Ar in the gas phase. Measuring its pressure in a calibrated volume at a known temperature would allow one to calculate the amount of Ar that was in the original mixture.

21.95 Generally a sodium fire can be extinguished by smothering it with sand. The worst choice is to use water (which reacts violently with sodium to give H_2 gas and NaOH).

21.97 Nitrogen is a relatively unreactive gas, so it will not participate in any reaction typical of hydrogen or oxygen. The most obvious property of H_2 is that it burns, so attempting to burn a small sample of the gas would immediately confirm or deny the presence of H_2. If O_2 is present, it can be detected by allowing it to react as an oxidizing agent. There are many reactions known with low-valent metals, especially transition metal ions in solution, that can be detected by color changes.

21.99 The reducing ability of the Group 3A metals declines considerably on descending the group, with the largest drop occurring on going from Al to Ga. The reducing ability of gallium and indium are similar, but another large change is observed on going to thallium. In fact, thallium is most stable in the +1 oxidation state. This same tendency for elements to be more stable with lower oxidation numbers is seen in Groups 4A (Ge and Pb) and 5A (Bi).

21.101 A through E, in order: $BaCO_3$, BaO, $CaCO_3$, $BaCl_2$, $BaSO_4$

Chapter 22

22.1 (a) Cr^{3+}: $[Ar]3d^3$, paramagnetic
(b) V^{2+}: $[Ar]3d^3$, paramagnetic
(c) Ni^{2+}: $[Ar]3d^8$, paramagnetic
(d) Cu^+: $[Ar]3d^{10}$, diamagnetic

22.3 (a) Fe^{3+}: $[Ar]3d^5$, isoelectronic with Mn^{2+}
(b) Zn^{2+}: $[Ar]3d^{10}$, isoelectronic with Cu^+
(c) Fe^{2+}: $[Ar]3d^6$, isoelectronic with Co^{3+}
(d) Cr^{3+}: $[Ar]3d^3$, isoelectronic with V^{2+}

22.5 (a) $Cr_2O_3(s) + 2\ Al(s) \longrightarrow Al_2O_3(s) + 2\ Cr(s)$
(b) $TiCl_4(\ell) + 2\ Mg(s) \longrightarrow Ti(s) + 2\ MgCl_2(s)$
(c) $2\ [Ag(CN)_2]^-(aq) + Zn(s) \longrightarrow$
$\quad\quad\quad\quad 2\ Ag(s) + [Zn(CN)_4]^{2-}(aq)$
(d) $3\ Mn_3O_4(s) + 8\ Al(s) \longrightarrow 4\ Al_2O_3(s) + 9\ Mn(s)$

22.7 Monodentate: CH_3NH_2, CH_3CN, N_3^-, Br^-
Bidentate: en, phen (see Figure 22.14)

22.9 (a) Mn^{2+} $\quad\quad$ (c) Co^{3+}
(b) Co^{3+} $\quad\quad$ (d) Cr^{2+}

22.11 $[Ni(en)(NH_3)_3(H_2O)]^{2+}$

22.13 (a) $Ni(en)_2Cl_2$ (en = $H_2NCH_2CH_2NH_2$)
(b) $K_2[PtCl_4]$
(c) $K[Cu(CN)_2]$
(d) $[Fe(NH_3)_4(H_2O)_2]^{2+}$

22.15 (a) Diaquabis(oxalato)nickelate(II) ion
(b) Dibromobis(ethylenediamine)cobalt(III) ion
(c) Amminechlorobis(ethylenediamine)cobalt(III) ion
(d) Diammineoxalatoplatinum(II)

22.17 (a) $[Fe(H_2O)_5OH]^{2+}$
(b) Potassium tetracyanonickelate(II)
(c) Potassium diaquabis(oxalato)chromate(III)
(d) $(NH_4)_2[PtCl_4]$

22.19 (a)

cis $\quad\quad\quad\quad$ trans

(b)

cis $\quad\quad\quad\quad$ trans

(c)

fac $\quad\quad\quad\quad$ mer

(d) 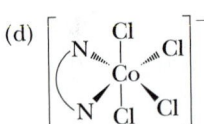 Only one structure possible. (N—N is the bidentate ethylenediamine ligand.)

22.21 For a discussion of chirality, see Chapter 11, pages 477–480).

(a) Fe^{2+} is a chiral center.

(b) Co^{3+} is not a chiral center.

(c) Neither of the two possible isomers is chiral.

(d) No. Square-planar complexes are never chiral.

22.23 (a) $[Mn(CN)_6]^{4-}$: d^5, low-spin complex is paramagnetic.

$\uparrow\downarrow \quad \uparrow\downarrow \quad \uparrow$

(b) $[Co(NH_3)_6]^{3+}$: d^6, low-spin complex is diamagnetic.

$\uparrow\downarrow \quad \uparrow\downarrow \quad \uparrow\downarrow$

(c) $[Fe(H_2O)_6]^{3+}$: d^5, low-spin complex is paramagnetic (1 unpaired electron; same as part a).

(d) $[Cr(en)_3]^{2+}$: d^4, complex is paramagnetic (2 unpaired electrons).

$\uparrow\downarrow \quad \uparrow \quad \uparrow$

22.25 (a) Fe^{2+}, d^6, paramagnetic, 4 unpaired electrons

(b) Co^{2+}, d^7, paramagnetic, 3 unpaired electrons

(c) Mn^{2+}, d^5, paramagnetic, 5 unpaired electrons

(d) Zn^{2+}, d^{10}, diamagnetic, 0 unpaired electrons

22.27 (a) 6; (b) octahedral; (c) +2; (d) 4 unpaired electrons (high spin); (e) paramagnetic

22.29 When $Co_2(SO_4)_3$ dissolves in water, it forms $[Co(H_2O)_6]^{3+}$. Addition of fluoride converts this to $[CoF_6]^{3-}$. The hexaaquo complex is low spin (diamagnetic, no unpaired electrons), and the fluoride complex is high spin (paramagnetic, 4 unpaired electrons.) Notice that fluoride is a weaker field ligand than water.

22.31 The light absorbed is in the blue region of the spectrum (page 1097). Therefore, the light transmitted— which is the color of the solution—is yellow.

22.33 Determine the magnetic properties of the complex. Square-planar Ni^{2+} (d^8) complexes are diamagnetic, whereas tetrahedral complexes are paramagnetic.

22.35 Fe^{2+} has a d^6 configuration. Low-spin octahedral complexes are diamagnetic, whereas high-spin complexes have four unpaired electrons and are paramagnetic.

22.37 Square-planar complexes most often arise from d^8 transition metal ions. Therefore, it is likely that $[Ni(CN)_4]^{2-}$ (Ni^{2+}) and $[Pt(CN)_4]^{2-}$ (Pt^{2+}) are square planar.

22.39 Two geometric isomers are possible.

22.41 Absorbing at 425 nm means the complex is absorbing light in the blue-violet end of the spectrum. Therefore, red and green light are transmitted, and the complex appears yellow (see Figure 22.27).

22.43 (a) Mn^{2+}; (b) 6; (c) octahedral; (d) 5; (e) paramagnetic; (f) cis and trans isomers exist.

22.45 Name: tetraamminedichlorocobalt(III) ion

cis   trans

22.47 $[Co(en)_2(H_2O)Cl]^{2+}$

22.49 (a)

mer   fac

(b)

trans chlorides   cis chlorides

(c)

22.51

H_2O and NH_3 cis, chiral

H_2O cis and NH_3 trans, not chiral

H_2O trans and NH_3 cis, not chiral

22.53 In $[Mn(H_2O)_6]^{2+}$ and $[Mn(CN)_6]^{4-}$, Mn has an oxidation number of $+2$ (Mn is a d^5 ion).

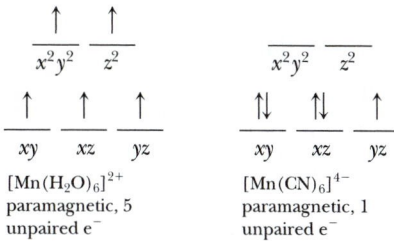

$[Mn(H_2O)_6]^{2+}$
paramagnetic, 5
unpaired e^-

$[Mn(CN)_6]^{4-}$
paramagnetic, 1
unpaired e^-

This shows that Δ_o for CN^- is greater than for H_2O.

22.55 A, dark violet isomer: $[Co(NH_3)_5Br]SO_4$

B, violet-red isomer: $[Co(NH_3)_5(SO_4)]Br$

$[Co(NH_3)_5Br]SO_4(aq) + BaCl_2(aq) \longrightarrow$
$[Co(NH_3)_5Br]Cl_2(aq) + BaSO_4(s)$

22.57 (a) The light absorbed is in the orange region of the spectrum (page 1098). Therefore, the light transmitted (the color of the solution) is blue or cyan.

(b) Using the cobalt(III) complexes in Table 22.3 as a guide, we might place CO_3^{2-} between F^- and the oxalato ion, $C_2O_4^{2-}$.

(c) Δ_o is small, so the complex should be high spin and paramagnetic.

22.59

$\overset{\frown}{N\ O} = H_2N-CH_2-CO_2^-$

enantiometric pair

enantiometric pair

enantiometric pair

22.61 Volume of sulfuric acid required = 1.10 L; mass of TiO_2 obtained = 526 g

22.63 (a) There is 5.41×10^{-4} mol of $UO_2(NO_3)_2$, and this provides 5.41×10^{-4} mol or U^{n+} ions on reduction by Zn. The 5.41×10^{-4} mol U^{n+} requires $2.16 \times$

10^{-4} mol MnO_4^- to reach the equivalence point. This is a ratio of 5 mol of U^{n+} ions to 2 mol MnO_4^- ions. The 2 mol MnO_4^- ions requires 10 mol of e^- (to go to Mn^{2+} ions), so 5 mol of U^{n+} ions provide 10 mol e^- (on going to UO_2^{2+} ions, with a uranium oxidation number of $+6$). This means the U^{n+} ion must be U^{4+}.

(b) $Zn(s) \longrightarrow Zn^{2+}(aq) + 2\ e^-$

$UO_2^{2+}(aq) + 4\ H^+(aq) + 2\ e^- \longrightarrow$
$U^{4+}(aq) + 2\ H_2O(\ell)$

$UO_2^{2+}(aq) + 4\ H^+(aq) + Zn(s) \longrightarrow$
$U^{4+}(aq) + 2\ H_2O(\ell) + Zn^{2+}(aq)$

(c) $5[U^{4+}(aq) + 2\ H_2O(\ell) \longrightarrow$
$UO_2^{2+}(aq) + 4\ H^+(aq) + 2\ e^-]$

$2[MnO_4^-(aq) + 8\ H^+(aq) + 5\ e^- \longrightarrow$
$Mn^{2+}(aq) + 4\ H_2O(\ell)]$

$5\ U^{4+}(aq) + 2\ MnO_4^-(aq) + 2\ H_2O(\ell) \longrightarrow$
$5\ UO_2^{2+}(aq) + 4\ H^+(aq) + 2\ Mn^{2+}(aq)$

22.65

Ion	$K_{formation}$ (ammine complexes)
Co^{2+}	7.7×10^4
Ni^{2+}	5.6×10^8
Cu^{2+}	6.8×10^{12}
Zn^{2+}	2.9×10^9

The data for these hexammine complexes do indeed, verify the Irving-Williams series. In the book *Chemistry of the Elements* (N. N. Greenwood and A. Earnshaw: 2nd edition, p. 908, Oxford, England, Butterworth-Heinemann, 1997), it is stated: "the stabilities of corresponding complexes of the bivalent ions of the first transition series, irrespective of the particular ligand involved, usually vary in the Irving-Williams order, . . . , which is the reverse of the order for the cation radii. These observations are consistent with the view that, at least for metals in oxidation states $+2$ and $+3$, the coordinate bond is largely electrostatic. This was a major factor in the acceptance of the crystal field theory."

Chapter 23

23.12 (a) $^{56}_{28}Ni$; (b) ^1_0n; (c) $^{32}_{15}P$; (d) $^{97}_{43}Tc$; (e) $^0_{-1}\beta$; (f) ^0_1e (positron)

23.14 (a) $^0_{-1}\beta$; (b) $^{87}_{37}Rb$; (c) $^4_2\alpha$; (d) $^{226}_{88}Ra$; (e) $^0_{-1}\beta$; (f) $^{24}_{11}Na$

23.16 $^{235}_{92}U \longrightarrow {}^{231}_{90}Th + {}^4_2\alpha$

$^{231}_{90}Th \longrightarrow {}^{231}_{91}Pa + {}^0_{-1}\beta$

$^{231}_{91}Pa \longrightarrow {}^{227}_{89}Ac + {}^4_2\alpha$

$^{227}_{89}Ac \longrightarrow {}^{227}_{90}Th + {}^0_{-1}\beta$

$^{227}_{90}Th \longrightarrow {}^{223}_{88}Ra + {}^4_2\alpha$

$^{223}_{88}Ra \longrightarrow {}^{219}_{86}Rn + {}^4_2\alpha$

$^{219}_{86}Rn \longrightarrow {}^{215}_{84}Po + {}^4_2\alpha$

$$^{215}_{84}\text{Po} \longrightarrow {}^{211}_{82}\text{Pb} + {}^{4}_{2}\alpha$$

$$^{211}_{82}\text{Pb} \longrightarrow {}^{211}_{83}\text{Bi} + {}^{0}_{-1}\beta$$

$$^{211}_{83}\text{Bi} \longrightarrow {}^{211}_{84}\text{Po} + {}^{0}_{-1}\beta$$

$$^{211}_{84}\text{Po} \longrightarrow {}^{207}_{82}\text{Pb} + {}^{4}_{2}\alpha$$

23.18 (a) $^{198}_{79}\text{Au} \longrightarrow {}^{198}_{80}\text{Hg} + {}^{0}_{-1}\beta$

(b) $^{222}_{86}\text{Rn} \longrightarrow {}^{218}_{84}\text{Po} + {}^{4}_{2}\alpha$

(c) $^{137}_{55}\text{Cs} \longrightarrow {}^{137}_{56}\text{Ba} + {}^{0}_{-1}\beta$

(d) $^{110}_{49}\text{In} \longrightarrow {}^{110}_{48}\text{Cd} + {}^{0}_{1}e$

23.20 (a) $^{80}_{35}\text{Br}$ has a high neutron/proton ratio of 45/35: Beta decay will allow the ratio to decrease: $^{80}_{35}\text{Br} \longrightarrow {}^{80}_{36}\text{Kr} + {}^{0}_{-1}\beta$. Some ^{80m}Br decays by gamma emission.

(b) Alpha decay is likely: $^{240}_{98}\text{Cf} \longrightarrow {}^{236}_{96}\text{Cm} + {}^{4}_{2}\alpha$

(c) Cobalt-61 has a high n:p ratio so beta decay is likely: $^{61}_{27}\text{Co} \longrightarrow {}^{61}_{28}\text{Kr} + {}^{0}_{-1}\beta$

(d) Carbon-11 has only 5 neutrons, so K-capture or positron emission may occur:

$$^{11}_{6}\text{C} + {}^{0}_{-1}e \longrightarrow {}^{11}_{5}\text{B}$$

$$^{11}_{6}\text{C} \longrightarrow {}^{11}_{5}\text{B} + {}^{0}_{1}e$$

23.22 Generally beta decay will occur when the n/p ratio is high, whereas positron emission will occur when the n/p ratio is low.

(a) Beta decay: $^{20}_{9}\text{F} \longrightarrow {}^{20}_{10}\text{Ne} + {}^{0}_{-1}\beta$

$$^{3}_{1}\text{H} \longrightarrow {}^{3}_{2}\text{He} + {}^{0}_{-1}\beta$$

(b) Positron emission:

$$^{22}_{11}\text{Na} \longrightarrow {}^{22}_{10}\text{Ne} + {}^{0}_{1}\beta$$

23.24 Binding energy per nucleon for $^{11}\text{B} = 6.70 \times 10^8$ kJ

Binding energy per nucleon for $^{10}\text{B} = 6.26 \times 10^8$ kJ

23.26 8.256×10^8 kJ/nucleon

23.28 7.700×10^8 kJ/nucleon

23.30 0.781 μg

23.32 (a) $^{131}_{53}\text{I} \longrightarrow {}^{131}_{54}\text{Xe} + {}^{0}_{-1}\beta$

(b) 0.075 μg

23.34 9.5×10^{-4} mg

23.36 (a) $^{222}_{86}\text{Rn} \longrightarrow {}^{218}_{84}\text{Po} + {}^{4}_{2}\alpha$

(b) Time = 8.87 d

23.38 About 2700 years old

23.40 (a) 15.8 yr; (b) 88%

23.42 If $t_{1/2} = 14.28$ d, then $k = 4.854 \times 10^{-2}$ d^{-1}. If the original disintegration rate is 3.2×10^6 dpm, then (from the integrated first-order rate equation), the rate after 365 d is 0.065 dpm. The plot will resemble Figure 23.5.

23.44 Plot ln (activity) versus time. The slope of the plot is $-k$, the rate constant for decay. Here $k = 0.0050$ d^{-1}, so $t_{1/2} = 140$ d.

23.46 $^{239}_{94}\text{Pu} + + {}^{4}_{2}\alpha \longrightarrow {}^{240}_{95}\text{Am} + {}^{1}_{1}\text{H} + 2{}^{1}_{0}n$

23.48 $^{48}_{20}\text{Ca} + {}^{242}_{94}\text{Pu} \longrightarrow {}^{287}_{114}\text{Uuq} + 3{}^{1}_{0}n$

23.50 (a) $^{115}_{48}\text{Cd}$; (b) $^{7}_{4}\text{Be}$; (c) $^{4}_{2}\alpha$; (d) $^{63}_{29}\text{Cu}$

23.52 $^{10}_{5}\text{B} + {}^{1}_{0}n \longrightarrow {}^{7}_{3}\text{Li} + {}^{4}_{2}\alpha$

23.54 Time $= 4.4 \times 10^{10}$ yr

23.56 Time $= 1.9 \times 10^9$ yr

23.58 (a) $^{238}_{92}\text{U} + {}^{1}_{0}n \longrightarrow {}^{239}_{92}\text{U}$

(b) $^{239}_{92}\text{U} \longrightarrow {}^{239}_{93}\text{Np} + {}^{0}_{-1}\beta$

(c) $^{239}_{93}\text{Np} \longrightarrow {}^{239}_{94}\text{Pu} + {}^{0}_{-1}\beta$

(d) $^{239}_{94}\text{Pu} + {}^{1}_{0}n \longrightarrow 2{}^{1}_{0}n + \text{energy} + \text{other nuclei}$

23.60 Energy obtained from 1.000 lb (452.6 g) of $^{235}\text{U} = 4.05 \times 10^{+10}$ kJ

Mass of coal required $= 1.6 \times 10^3$ ton (or about 3 million lb of coal)

23.62 130 mL

23.64 27 tagged fish out of 5250 fish caught represents 0.51% of the fish in the lake. Therefore, 1000 fish put into the lake represents 0.51% of the fish in the lake, or 0.51% of 190,000 fish.

23.66 (a) The mass decreases by 4 units (with an $^{4}_{2}\alpha$ emission), or is unchanged (with a $^{0}_{-1}\beta$ emission) so the only masses possible are 4 units apart.

(b) ^{232}Th series, $m = 4n$; ^{235}U series, $m = 4n + 3$

(c) ^{226}Ra and ^{210}Bi, $4n + 2$ series; ^{215}At, $4n + 3$ series; ^{228}Th, $4n$ series

(d) Each series is headed by a long-lived isotope (on the order of 10^9 years, the age of the earth). The $4n + 1$ series is missing because there is no long-lived isotope in this series. Over geologic time, all of the members of this series have decayed completely.

23.68 (a) The ^{231}Pa isotope belongs to the ^{235}U decay series (see Question 23.66b).

(b) $^{235}_{92}\text{U} \longrightarrow {}^{231}_{90}\text{Th} + {}^{4}_{2}\alpha$

$$^{231}_{90}\text{Th} \longrightarrow {}^{231}_{91}\text{Pa} + {}^{0}_{-1}\beta$$

(c) Pa-231 is present to the extent of 1 part per million. Therefore, 1 million g of pitchblende needs to be used to obtain 1 g of Pa-231.

(d) $^{231}_{91}\text{Pa} \longrightarrow {}^{227}_{89}\text{Ac} + {}^{4}_{2}\alpha$

NOTES

NOTES

NOTES

NOTES

Index/Glossary

Italicized page numbers indicate pages containing illustrations, and those followed by "t" indicate tables. Glossary terms are printed in blue.

Abba, Mohammed Bah, 232
abbreviations, A-10
absolute temperature scale. *See* Kelvin temperature scale.
absolute zero The lowest possible temperature, equivalent to $-273.15\ °C$, used as the zero point of the Kelvin scale, 27, 552
zero entropy at, 912
absorption spectrum A plot of the intensity of light absorbed by a sample as a function of the wavelength of the light, 1100
abundance(s), of elements in Earth's crust, 82t, 87t, 1014
of essential elements in human body, 88t
of isotopes, 69
accuracy The agreement between the measured quantity and the accepted value, 32
acetaldehyde, 504t
structure of, 469
acetaminophen, structure of, 511
acetate ion, buffer solution of, 855t
acetic acid, 505t
as weak electrolyte, 178
buffer solution of, 855t
density of, 52
dimerization of, 793
formation of, 273
hydrogen bonding in, 601
ionization of, 799
equilibrium in, 760
orbital hybridization in, 452
production of, 502
quantitative analysis of, 159
reaction with calcium carbonate, *192*, 193
reaction with ethanol, 771

reaction with sodium bicarbonate, 815
structure of, 186, 477, 502
titration with sodium hydroxide, 864
acetic anhydride, 157
acetone, 504t
hydrogenation of, 423
structure of, *169, 453*, 502
acetonitrile, isomerization of, 750
structure of, 429, 432, *454*
acetylacetonate ion, as ligand, 1082
acetylacetone, enol and keto forms, 472
structure of, 429
acetylcholinesterase, 732t
acetylene, orbital hybridization in, 453
production of, 585
structure of, 477
acetylsalicylic acid. *See* aspirin.
acid(s) A substance that, when dissolved in pure water, increases the concentration of hydrogen ions, 185. *See also* Brønsted acid(s), Lewis acid(s).
bases and, 796–847. *See also* acid–base reaction(s).
Brønsted definition, 799
carboxylic. *See* carboxylic acid(s).
common, 187t
Lewis definition, 828–832
molecular structure of, 832–837
properties of, *186*
reaction with bases, 191–195
strengths of, 807–809
direction of reaction and, 814

strong. *See* strong acid.
weak. *See* weak acid.
acid–base indicator(s), 870–872
acid–base pairs, conjugate, 802, 803t
acid–base reaction(s) An exchange reaction between an acid and a base producing a salt and water, 191–196
characteristics of, 817t
equivalence point of, 217, 862
pH after, 824–826
titration using, 216, 862–872
acid ionization constant (K_a) The equilibrium constant for the ionization of an acid in aqueous solution, 807, 808t
determining, 866
relation to conjugate base ionization constant, 813
values of, A-21t
acidic oxide(s) An oxide of a nonmetal that acts as an acid, 190
acidic solution A solution in which the concentration of hydronium ions is greater than the concentration of hydroxide ion, 804
acidosis, 861
acrolein, formation of, 433
structure of, 430, 470
acrylonitrile, structure of, 100, 428
actinide(s) The series of elements between actinium and rutherfordium in the periodic table, 87, 349
activation energy (E_a) The minimum amount of energy that must be absorbed by a system to cause it to react, 724

experimental, determination, 727–729
reduction by catalyst, 730
activity (A) A measure of the rate of nuclear decay, the number of disintegrations observed in a sample per unit time, 1123
actual yield The measured amount of product obtained from a chemical reaction, 157
addition polymer(s) A synthetic organic polymer formed by directly joining monomer units, 513–517
production from ethylene derivatives, 514t
addition reaction(s), of alkenes and alkynes, 490
adduct, acid–base, 829
adenine, hydrogen bonding to thymine, 603
structure of, 136
adenosine 5′-triphosphate (ATP), 541
adhesive force A force of attraction between molecules of two different substances, 614
adhesives, 654
adipoyl chloride, 518
adrenaline, 833
aerobic fermentation, 497
aerosol, 687t
air, components of, 565t, 577t
density of, 560
fractionation of, 1052
air bags, 548, 556
albite, dissolved by rain water, 175
albumin, precipitation of, 752
alchemy, 1110

I-1

PHYSICAL AND CHEMICAL CONSTANTS

Avogadro's number	$N = 6.02214155 \times 10^{23}/mol$
Electronic charge	$e = 1.60217653 \times 10^{-19}$ C
Faraday's constant	$F = 9.6485338 \times 10^4$ C/mol electrons
Gas constant	$R = 8.314472$ J/K · mol
	$= 0.082057$ L · atm/K · mol

π	$\pi = 3.1415926536$
Planck's constant	$h = 6.6260693 \times 10^{-34}$ J · sec
Speed of light (in a vacuum)	$c = 2.99792458 \times 10^8$ m/sec

USEFUL CONVERSION FACTORS AND RELATIONSHIPS

Length
SI unit: Meter (m)
- 1 kilometer = 1000 meters
  - = 0.62137 mile
- 1 meter = 100 centimeters
- 1 centimeter = 10 millimeters
- 1 nanometer = 1.00×10^{-9} meter
- 1 picometer = 1.00×10^{-12} meter
- 1 inch = 2.54 centimeter (exactly)
- 1 Ångstrom = 1.00×10^{-10} meter

Mass
SI unit: Kilogram (kg)
- 1 kilogram = 1000 grams
- 1 gram = 1000 milligrams
- 1 pound = 453.59237 grams = 16 ounces
- 1 ton = 2000 pounds

Volume
SI unit: Cubic meter (m^3)
- 1 liter (L) = 1.00×10^{-3} m^3
  - = 1000 cm^3
  - = 1.056710 quarts
- 1 gallon = 4.00 quarts

Energy
SI unit: Joule (J)
- 1 joule = 1 kg · m^2/s^2
  - = 0.23901 calorie
  - = 1 C × 1 V
- 1 calorie = 4.184 joules

Pressure
SI unit: Pascal (Pa)
- 1 pascal = 1 N/m^2
  - = 1 kg/m · s^2
- 1 atmosphere = 101.325 kilopascals
  - = 760 mm Hg = 760 torr
  - = 14.70 lb/in^2
  - = 1.01325 bar
- 1 bar = 10^5 Pa (exactly)

Temperature
SI unit: kelvin (K)
- 0 K = −273.15 °C
- K = °C + 273.15°C
- ? °C = (5 °C/9 °F)(°F − 32 °F)
- ? °F = (9 °F/5 °C)(°C) + 32 °F

LOCATION OF USEFUL TABLES AND FIGURES

Atomic and Molecular Properties

Atomic radii	Figures 8.11, 8.12
Bond energies	Table 9.10
Bond lengths	Table 9.9
Electron configurations	Table 8.3
Electronegativity	Figure 9.14
Ionic radii	Figure 8.15
Ionization energies	Figure 8.13, Appendix F

Thermodynamic Properties

Enthalpy, free energy, entropy	Appendix L
Enthalpies of formation (selected values)	Table 6.2
Specific heats	Table 6.1
Lattice energies	Table 9.3

Acids, Bases and Salts

Common acids and bases	Table 5.2
Names and composition of polyatomic ions	Table 3.1
Solubility guidelines	Figure 5.3
Ionization constants for weak acids and bases	Table 17.3

Miscellaneous

Selected alkanes	Table 11.2
Oxidizing and reducing agents	Table 5.4
Charges on common monatomic cations and anions	Figure 3.7
Hybrid orbitals	Figure 10.5
Common polymers	Table 11.12